A book dedicated to the study of the flora of Colorado through education, exploration, documentation, appreciation, and conservation

Aquilegia coerulea (Colorado blue columbine), the Colorado state flower

Flora of Colorado

By Jennifer Ackerfield

Colorado State University Herbarium

Press

2015

First edition 2015
Printed in the United States of America
ISBN 978-1-889878-45-4

Editor: Barney Lipscomb
Botanical Research Institute of Texas Press
1700 University Dr.
Fort Worth, Texas 76107-3400, USA
Production: Jennifer Ackerfield
Cover Design: Elisabeth Owens
Illustrations: Jennifer Ackerfield
Photographs: Jennifer Ackerfield, Lori Brummer, Don Farrar, Ben Legler, Scott Smith, Al Schneider
Map of plant communities (p. 5) courtesy of the USGS Geological Survey, Gap Analysis Program (GAP).
Maps created by Jennifer Ackerfield using ArcMap 10.1.
Photographs of James Cassidy (p. 1) and Charles Spencer Crandall (p. 2) courtesy of the archives at Colorado State University.
Photograph of Jacob Cowen (p. 3) courtesy of Mary Farmer.
Photograph of Harold Harrington (p. 4) courtesy of Tom Jirsa.

Flora of Colorado
By Jennifer Ackerfield

Front cover: photograph *Rudbeckia laciniata* var. *ampla* (by Jennifer Ackerfield)
Inside illustration of *Platanthera tescamnis* courtesy of Carolyn Crawford

Flora of Colorado is published with the generous support of:
Charles Maurer, Mo Ewing, the Colorado Natural Heritage Program, the Colorado Native Plant Society, the San Juan Four Corners Native Plant Society, Steve Popovich, Jan and Charlie Turner, Jack Sosebee, Pat and Heman Adams, Carolyn Sue Savage-Hall, Larry Allison, Gregory Penkowsky, Elizabeth Taylor, Beverley Postmus, Ann Young, Chris Blakeslee, Renee Galeano-Popp, and the many presale orders made possible with the help of the Colorado Native Plant Society.

Distribution of copies by:
Botanical Research Institute of Texas
1700 University Dr.
Fort Worth, Texas 76107-3400, USA
Telephone: 1 817 332 4441
Fax: 1 817 332 4112
Website: http://shop.brit.org
Email: orders@brit.org

PRESS

Botanical Miscellany, No. 41
ISSN: 0833-1475
30 May 2015

DEDICATION

This book is dedicated to all of the botanists in Colorado who came before me, in particular the early curators of the Colorado State University Herbarium (namely Charles Crandall, Jacob Cowen, and Harold Harrington). Without the effort and commitment of these botanists to documenting the flora of Colorado, I could not have completed this project. Their dedication and expertise is evident to all who view their invaluable specimens.

This book is also dedicated to my children, friends, and family for enduring my nonstop plant collecting and consequential absurdly slow hiking over the years. Thank you for waiting!

View into Unaweep Canyon, Colorado

ACKNOWLEDGMENTS

I would like to thank the following people for all their efforts in making the *Flora of Colorado* possible: my Master's advisor Dr. Jun Wen for giving me the opportunity to study with her and begin my work in the herbarium; William F. Jennings for significant taxonomic contributions; my first plant identification teacher Richard (Dick) Walter for instilling a love of collecting and identifying plants; Dr. Ron Hartman at the University of Wyoming for his mentoring and advice; Lori Brummer and Steve Olson for their endless edits and constant vigilance; Ernie Nelson at the University of Wyoming for taxonomic advice; Dr. Mark P. Simmons at Colorado State University for his support; Dr. Neil Snow for his support; Evan Arnold, Elisa Dolan, Chris Miller, John Wendt, and Sean Williams-Perez for compiling distribution maps; Jenna McAleer for assisting with species descriptions; all of the botanists who have worked with past editions of the *Flora* and offered countless critiques and edits including those at the Colorado Natural Heritage Program (David Anderson, Denise Culver, Bernadette Kuhn, Peggy Lyon, Susan Punjabi, and Pam Smith to name a few), Mike Kirkpatrick, Crystal Strouse, Scott Smith, Tim Hogan, Al Schneider, Melissa Islam, and Janet Wingate; Rich Scully for Apiaceae edits; Steve Popovich and Ben Legler for assisting with the key to the Ophioglossaceae; Marika Majack for assisting with aquatic plant distributions and clarifications; Larry Hufford for *Mentzelia* assistance; Mary Farmer for providing information on Jacob Cowen; the Colorado Native Plant Society (CONPS) board members for assistance in gathering funds for publication; every member of the CONPS for their interest and support; Todd Gilligan for InDesign assistance; Elisabeth Owens for designing the beautiful book cover; Brooke Best for edits galore; and last but not least, all of the students in my plant identification class for their abundance of editorial assistance over the years and for making me not forget what it is like to key out a plant for the very first time. This work would not be what it is today without all of your assistance. Thank you!

Funding was generously provided by Charles Maurer, Mo Ewing, the Colorado Natural Heritage Program, the Colorado Native Plant Society, the San Juan Four Corners Native Plant Society, Steve Popovich, Jan and Charlie Turner, Jack Sosebee, Pat and Heman Adams, Carolyn Sue Savage-Hall, Larry Allison, Gregory Penkowsky, Elizabeth Taylor, Beverley Postmus, Ann Young, Chris Blakeslee, Renee Galeano-Popp, and the many presale orders made possible through the assistance of the Colorado Native Plant Society.

TABLE OF CONTENTS

QUICK GUIDE TO THE FAMILIES

NEW COMBINATIONS

New combinations exclusive to the *Flora of Colorado* and not currently in press in other works or journals are made here below.

Asteraceae
Grindelia hirsutula Hook. & Arn. var. ***acutifolia*** (Steyerm.) Ackerfield, comb. nov. (p. 154)
Grindelia acutifolia Steyermark, Ann. Missouri Bot. Gard. 21:498. 1934.
Grindelia hirsutula Hook. & Arn. var. ***decumbens*** (Greene) Ackerfield, comb. nov. (p.154)
Grindelia decumbens Greene, Pittonia 3:102. 1896.
Grindelia hirsutula Hook. & Arn. var. ***revoluta*** (Steyerm.) Ackerfield, comb. nov. (p. 155)
Grindelia revoluta Steyerm., Ann. Missouri Bot. Gard. 21:496. 1934.

Melanthiaceae
Zigadenus paniculatus (Nutt.) S. Wats. var. ***gramineus*** (Rydb.) Ackerfield, comb. nov. (p. 538)
Zigadenus gramineus Rydb., Bull. Torrey Bot. Club 27:535. 1900.

Polemoniaceae
Ipomopsis aggregata (Pursh) V.E. Grant ssp. ***tenuituba*** (Rydb.) Ackerfield, comb. nov. (p. 674)
Gilia tenuituba Rydb., Bull. Torrey Bot. Club 40:472. 1913.

Ranunculaceae
Clematis grosseserrata (Rydb.) Ackerfield, comb. nov. (p. 708)
Atragene grosseserrata Rydb., Bull. Torrey Bot. Club 29:156. 1902.

Adobe hills near Montrose, Colorado

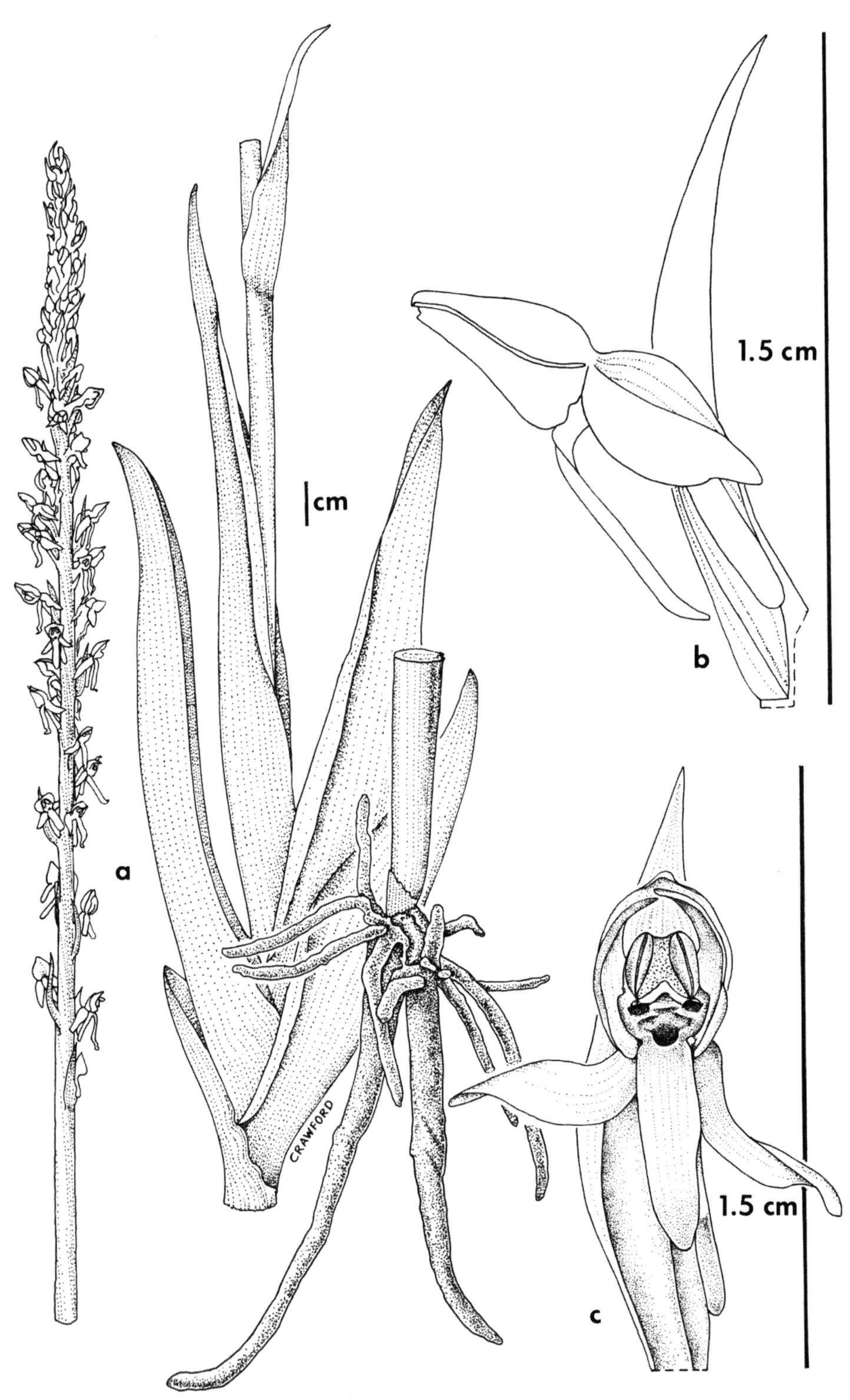

Platanthera tescamnis Sheviak & Jennings
Illustration by Carolyn Crawford

INTRODUCTION

I pretty much always wanted to be a botanist, even before I knew what a botanist was. I just knew that I loved plants. I distinctly remember going on a field trip in kindergarten, our teacher showing us the buds on the trees in the spring, and becoming simply fascinated. I would constantly bring home weeds and flowers from the fields around our house. In high school, I kept an herb garden in my parent's backyard. And when I was an undergraduate in college, I took my first plant identification course. It was like everything I had ever wanted to learn all rolled into one fabulous learning experience – I discovered how to use a dichotomous key, how to make a plant collection, and how to distinguish the major flowering plant families. This is when I first started to see the world of plants as one big puzzle, a puzzle that needed sorting and assembling. I continued to further my botanical education and completed a Master's degree in botany with a concentration in taxonomy and systematics in 2001. Around this time, I also became the collection's manager at the Colorado State University Herbarium, and little did I know that over 15 years later I would still be working here and still loving every minute! I also began teaching plant identification and educating future generations of students in the field of botany, and this is still one of my most passionate pursuits. I love working with students, teaching, and watching their full potential in identifying plants develop and grow and hopefully inspiring them to pursue their own passions in the botanical sciences.

I began writing the *Flora of Colorado* about 12 years ago, as I was attempting to verify the specimens in the CSU Herbarium. I found that for many families or genera (such as *Cirsium* or *Physaria*), misidentifications were quite common. I began asking myself "why?" and "how can misidentifications be eliminated in the future?" I was also utilizing several keys in the process but found that the keys I was working with were oftentimes insufficient, outdated, or difficult to use. Thus, I began writing my own keys based on what I considered to be more useable, major distinguishing morphological characteristics. In particular, I wanted to make keys that students or amateur botanists could use to identify plants with greater ease and efficiency. These keys gradually developed into what is now the *Flora of Colorado*. In addition, supplying the users of the keys with pictures of the major recognition characteristics I felt would enable them to easily see the more difficult characteristics used to delimit species. It is my hope that users of this book will be able to key out plants with confidence and ease (well, with as much ease as possible at least!) and enjoy botany!

THE EARLY CURATORS OF THE COLORADO STATE UNIVERSITY HERBARIUM

The early curators of the Colorado State University Herbarium contributed significantly to our understanding of Colorado botany. In 2003, I decided to find out more about these curators, after seeing their names on so many herbarium specimens. This endeavor ultimately culminated with writing letters to everyone in the town of Hotchkiss, Colorado, with the last name of Hotchkiss in the hopes of finding any information I could about Jacob Hover Cowen (the herbarium's third curator and stepson of Enos Hotchkiss, founder of the town of Hotchkiss). I was delighted to find that Mary Farmer of Hotchkiss had copies of Jacob's diary, numerous photographs, and much family lore. She also made me realize that even today, the memory of Jacob still deeply touches those who know his tragic story. These early curators will forever live on through their work and forever be a part of the herbarium at Colorado State University. Our understanding of the flora of Colorado would not be what it is today without their work and dedication.

The Colorado State University Herbarium (CS), formed in 1883, is the oldest herbarium in the Southern Rocky Mountains and has a rich history. Founded as a military fort in 1864, Fort Collins was incorporated as a town in 1873 (City of Fort Collins Library Archives). In 1879, Colorado Agricultural College or CAC (now known as Colorado State University) was formed. Enrolling freshmen were required to take instruction in botany during their second and third terms and to spend at least two hours of labor each day on the College Farm (Norton, 1981). However, it wasn't until 1883 that CAC officially appointed its first Professor of Botany and Horticulture, James Cassidy.

James Cassidy

"A hard and conscientious worker in the line of duty and was a warm friend, an obliging neighbor, and a gentleman in every sense of the word" (From the obituary of James Cassidy in the November 23, 1889, *Fort Collins Courier*)

James Cassidy was born in England on Aug. 5, 1847. Before becoming Professor of Botany at CAC, Cassidy worked for several years at the Royal Botanic Garden in Regents Park, London. He also worked for Rolliston, one of the oldest nurseries and seedsmen in London at the time. Around 1870, Cassidy came to America, where he lived in New York and worked for Peter Henderson and Company, a seed business. Cassidy was with Henderson and Co. for four years, leaving in 1874 to become a horticulturist at Michigan Agricultural College. Cassidy worked as a florist and horticulturist at Michigan for nine years and married a Michigan native named Marie.

James Cassidy, 1883
Photo courtesy CSU archives

Although Cassidy's interests up to this point seem mainly horticultural, they were in fact very wide-ranging. He came to Colorado in 1883 to teach several botany courses at CAC in addition to horticultural offerings. He also wrote about insects native to Colorado as well as to those injurious to useful vegetation and the means of controlling them (Cassidy, 1888a) and performed experiments on sugar beet and potato farming (Cassidy, 1888b). Cigars were even made from the 1888 crop of tobacco and were a big success at a fair in Denver featuring only Colorado products.

As curator of the herbarium, Cassidy collected plants from the foothills of Larimer County. However, he rarely put any locality information on his herbarium labels. A few of his specimens are at CS, but the majority are housed at the NY Botanical Garden Herbarium (NYBG). While Professor Cassidy was at CAC, the Agricultural Experiment Station was established under the Hatch Act of 1887 (Norton, 1981). Cassidy headed the botany and horticulture research section of the new station. He was especially interested in native grasses and grew 30 varieties at the college greenhouse and farm. In 1889, Cassidy published "Some Colorado Grasses" with chemist David O'Brine (Cassidy & O'Brine, 1889). Two Colorado grasses were named by Cassidy: *Agropyron unilaterale* Cassidy, which is now accepted as *Elymus trachycaulus* (Link) Gould ex Shinners ssp. *subsecundus* (Link) Á. Löve & D. Löve, and *Festuca kingii* Cassidy, which is now accepted as *Leucopoa kingii* (S. Watson) W.A. Weber (Cassidy was unaware that Watson had named this species a few years previous).

James Cassidy died suddenly on November 21, 1889, at the age of 42 after a brief illness. He was buried in Grandview Cemetery with Masonic honors. After Cassidy's sudden death, the college quickly appointed another Michigan graduate, Charles Spencer Crandall, to take over the position of Professor of Botany and Horticulture.

Charles Spencer Crandall

"The department of botany and horticulture is presided over by Professor C.S. Crandall and his well-kept instruments and tools and apparatus generally show that he is adept in his calling." (From "Report of the special examining committee, State Board of Agriculture," Fort Collins Courier, January 26, 1893)

Charles Crandall, 1895
Photo courtesy CSU archives

The son of Dr. Richard Orson Crandall and Maria Louise Cushman, Charles Spencer Crandall was born Oct. 12, 1852, in Waverly, New York. Moving with his parents in 1858 at the young age of six, Crandall spent his boyhood years in LaPorte, Indiana. Crandall entered Michigan Agricultural College, receiving his bachelor's degree in 1873 and moved to Little Traverse, Michigan, shortly thereafter and took up farming. On April 28, 1879, Crandall married Adeline Ocobock, and in 1880 they had a daughter named Lillian. He returned to Michigan Agricultural College in 1885 to get his Master's degree. During this time, he was also the foreman of the gardens in the horticulture department and an instructor in horticulture. Crandall came to Colorado Agricultural College in 1889, just after receiving his Master's degree from Michigan. Crandall's wife Adeline died in 1891 and was buried in LaPorte, Indiana. While at CAC, Crandall met and married Maude Bell, Professor of English at the college. They had two daughters and one son.

As Professor of Botany and Horticulture at CAC, Crandall was responsible for developing a more specialized botany program. In the 1899–1900 College Catalog, Crandall reports that the herbarium contained about 12,000 sheets of North American plants and about 1300 species of the plants of Colorado. It was used freely by students and drawn largely upon for means of illustration. Charles collected numerous specimens while he was curator, either alone or with his assistant Jacob Hover Cowen. His first Colorado collections were made in 1890 in the vicinity of Fort Collins in such places as the College Farm, Dixon Canyon, Horsetooth Mountain, and Poudre Canyon. In 1891, Crandall made a collecting trip in July to Steamboat Springs and reached the Medicine Bow Range of Wyoming in early Aug.

After the establishment of the Experimental Grass Station, Crandall made a multitude of trips in search of promising species of native grasses (Rydberg, 1906). During this time Crandall was also working on a project for the Chicago World's Fair, and numerous collecting trips around the state were made in 1892 in preparation for his project. Crandall's forestry exhibit of wood sections, seeds, plants, and flowers went on to receive a Chicago World's Fair medal in 1893. Crandall never published on any of his collections, but instead they were sent to the NY Botanical Garden where they contributed significantly to the preparation of Per Axel Rydberg's *Flora of Colorado*, published in 1906 by the Colorado Agricultural Experiment Station. At the time of publication, 2912 plant species were included in the flora of Colorado, which was "a number greater than known for any other state except California" (Rydberg, 1906).

Several of Crandall's collections were described by Rydberg as new species, and these types are mostly housed at the NY Botanical Garden. A few species were named for Crandall including *Penstemon crandallii* A. Nelson, *Arabis crandallii* B.L. Rob. [now accepted as *Boechera crandallii* (B.L. Rob.) W.A. Weber], *Saxifraga crandallii* Gand. [now accepted as *Saxifraga flagellaris* Willd. ex Sternb. ssp. *crandallii* (Gand.) Hultén], and *Iliamna crandallii* (Rydb.) Wiggins [which was first described as *Sphaeralcea crandallii* by Rydberg]. Crandall also spent much of his time in the field surveying weeds and collecting timber samples. He wrote the first weed bulletin for the state in 1893 and a book on Russian thistle in 1894. Crandall was also interested in plant diseases, and in 1899 he visited the Arkansas Valley Experiment Station to make a study of cantaloupe rust, already having published a book entitled *Blight and Other Plant Diseases.* Fruit culture was a main focus of his studies, and Crandall pursued this interest more actively at his next post.

Charles left Colorado in 1900 to take up a small fruit plantation near Los Angeles, California. In 1902, he left the farming business for good and accepted a position at the University of Illinois to pursue studies in pomology. Beginning in 1907, Crandall was influential in developing the plant breeding program at Illinois. He retired in 1926 but retained his office space and continued to perform pomology experiments. While at Illinois, Crandall published eight books on fruit culture, the majority of which were on apple breeding and cultivation. Charles Spencer Crandall died at the age of 77 at the home of one of his daughters in Hollywood, California, on July 10, 1929.

Jacob Hover Cowen

"There is only one class of men whom I admire more than our Pilgrim fathers, more than our Revolutionary fathers, and more than those who saved the Union, and they are our American pioneers. I assure that I am not a tenderfoot...I know about pioneers. Why, I arrived in the territory of Colorado twenty-five years ago, and I came straight from heaven."
(From the diary of Jacob Hover Cowen, July 20, 1897)

Jacob Cowen, 1894
Photo courtesy Mary Farmer

After Charles Crandall left CAC, the position of curator was passed to Jacob Hover Cowen. Jacob had been Crandall's assistant from 1895 to 1898 and also received his degree from CAC in 1894. He was born to Jacob Cowen Sr. and Elizabeth McIntyre on Mar. 18, 1872. At the end of the Civil War, Jacob Cowen Sr. and his good friend Enos Throop Hotchkiss took up land from the government in lieu of pay, and the two friends headed west to Colorado. It was in Colorado that Jacob met Elizabeth, who homesteaded with her parents in Conifer, Colorado. In fact, the Meyers Ranch off of Highway 287 and the Jefferson County Open Space around it were all once part of the McIntyre homestead. Jacob and Elizabeth were married on June 25, 1871, and moved to Powderhorn, Colorado, shortly thereafter. Powderhorn was a small town in Gunnison County and close to where Jacob's friend Enos also homesteaded.

All was not to be wedded bliss, however, and soon after their move horrible misfortune befell the newly wedded couple. Jacob Cowen received a fatal blow from a horse and tragically died of blood poisoning on Aug. 1, 1871, just one month after being married. Elizabeth remained in Powderhorn and gave birth to Jacob's son, aptly named Jacob. Jacob Hover Cowen Jr., or "Little Jake" as his family affectionately called him, would go on to be a bright and shining light in the Cowen family. Elizabeth and Little Jake remained in Powderhorn for several more years, and in 1878 Elizabeth married Jacob's old friend Enos Hotchkiss. Enos explored the North Fork Valley northwest of Powderhorn in 1879 and soon founded the town of Hotchkiss. In 1881, he moved his family (which now, in addition to Little Jake, consisted of two more children, Fred and Addie) and a small group of colonists to this new settlement.

In spite of the great distance between them, as soon as he grew old enough to write, Jacob's mother insisted that Little Jake write frequently to his grandmother and other relatives back east. Jacob would send gifts and letters to other Cowen family members, always inscribing them with lovely poems and thoughts. These kind acts would be wonderfully remembered by members of the Cowen family for many years to come, as Jacob became a central, shining member of the family in spite of the many miles of separation.

Jacob was in one of the first classes to graduate from the Delta County school system. Destined for higher learning, Jacob left Hotchkiss around 1890 to begin his education at Colorado Agricultural College. Jacob dutifully wrote home to his mother to tell her of his long journey:

> *"I am here safe and sound. Got into Colorado Springs at 4 P.M. Wednesday...visited Garden of the Gods and Manitou. Saw cog railroad which runs to top of Pike's Peak. Got in Ft. Collins Friday night. The college has rented dormitory quarters ½ mile from college and I and 10 others are numbered to room there. It is quite a walk but more pleasant quarters than the dormitory...I have a very pleasant room-mate. Will get down to hard work – mind work on the morrow."* He ended the letter on a playful note, telling his mother *"Enclosed you'll find a lock of my best girl's hair. If you appear over inquisitive I'll tell you her name next time. Handle with care!"*

Perhaps following in his father's own military footsteps, Jacob was also involved in the Reserve Officer's Training Corps (ROTC) on campus, attaining the rank of Captain and Chief of Artillery by 1895. After graduating from CAC at the top of his class in 1894, Jacob accepted a position as assistant to Charles Spencer Crandall. Jacob assisted Crandall in collecting specimens and general upkeep of the herbarium. He was offered a position of instructor of botany and horticulture at CAC in 1895 and eagerly accepted the post. One of the projects Jacob spearheaded was an effort to prevent the spread of Russian thistle, or *Salsola tragus*, in Colorado by setting potted plants of the weed in window fronts so that people could become familiar with it and thus destroy it on sight.

Jacob's earliest collections in the herbarium date to 1891 from around the Fort Collins area and nearby foothills. Included in these collections is one holotype specimen, *Grindelia serrulata* Rydb., now accepted as *Grindelia squarrosa* (Pursh) Dunal var. *serrulata* (Rydb.) Steyerm. Jacob would often collect plants on his visits home to Hotchkiss in Delta County, collecting extensively near his home on the hills surrounding Hotchkiss as well as along Leroux Creek.

In the fall of 1898, Jacob left Fort Collins for Cornell University to pursue a Master's degree in botany. This trip east also afforded an opportunity to finally meet his Cowen relatives in Pennsylvania. His family back east was likewise thrilled at this long-awaited meeting when they heard the news of his coming. When the womenfolk got wind of his impending arrival, they immediately set forth to make him a befitting gift and decided on *"the nicest quilt that was ever quilted."* Jacob spent the next Christmas with his Cowen relatives, surrounded by grandparents, aunts, uncles, and cousins galore. In fact, he went to not one, but five Christmas dinners!

The turn of the century dawned bright for Jacob as his Master's work at Cornell was nearing its end. He completed his studies in June of 1900, and shortly thereafter CAC offered him the position of head of the department of botany and horticulture. Jacob eagerly accepted this position and was looking forward to beginning his work in his home state of Colorado when tragedy struck. Like father, like son, Jacob's life was cut short too soon. On July 11, 1900, Jacob become critically ill and died the following day at the age of 28 due to acute appendicitis. His death was ironically hastened by none other than blood poisoning, the very thing that had also taken the life of his own father.

The *Rocky Mountain News* paid an inspired tribute to Jacob upon hearing of his death:

"There is sorrow in the death of Professor J.H. Cowen beyond the ordinary. It is only a few days since, in these columns, he was congratulated on his election to the chair of horticulture and botany in the state Agricultural College... The grim destroyer has struck a cruel blow not only at friends and relatives, but at science itself, and has cut short a career that promised to illumine the world the light of as rare a genius as was ever given to mortal man. He was a son of Colorado, and the fires of his brilliant mind were lighted in the halls of Colorado institutions of learning. He perfected himself in his own special branch by most vigorous effort, and attained a personal knowledge of the flora of Colorado and the Rocky Mountains by work in the field which no other man possessed. He won honors in post-graduate work of Eastern universities, and was attracting the attention of the scientists of America when he was unanimously chosen to a professorship in his alma mater. This was the realization of fond ambition. It opened to him a career that promised to reflect honor on himself and the institution. But great indeed are the uncertainties of life. Barely had he grasped the prize and accepted the tribute of his alma mater to his scholarship and genius when he was stricken down by disease, and his life with all of its brilliant promise ended. The State Agricultural College can well afford to enroll his name as among the most illustrious of her graduates, and Colorado can preserve his memory as one of the noblest and brightest of her youth whom death carried to an untimely grave."

Jacob's uncle made the sad trip west, transporting his body back to Colorado from New York for burial, where he was laid to rest in Riverside Cemetery in Hotchkiss.

The specimens collected by Cowen and Crandall were sent to the NY Botanical Garden Herbarium around the turn of the century. These invaluable collections would go on to be the backbone of the first flora for the state, Per Axel Rydberg's *Flora of Colorado* (1906). Approximately 1300 collections made by Cowen are housed at the CSU Herbarium. After Cowen's death, the herbarium went through a long period of inactivity until Ernest Charles Smith became curator in 1928. Smith primarily collected willows, but his young assistant, Harold D. Harrington, would go on to become curator in 1943 and contribute significantly to the flora of Colorado.

Harold David Harrington

H.D. Harrington, 1954
Photo courtesy Tom Jirsa

Harold David (H.D.) Harrington was born in De Motte, Indiana, on Mar. 14, 1903, where he lived until 1909 when his family moved to Mitchell, South Dakota. The family moved yet again in 1911 to Graettinger, Iowa, where Harrington eventually would spend the majority of his childhood with his then seven siblings. Living on a farm, Harrington developed a love of plants. Harrington was destined for greater learning, but by the time he was ready to attend college, the country was in the grips of the Great Depression and money was very tight. To cope with the shortage of funds, Harold and his older brother Elbert would alternate years in school – one would attend school one year while the other worked. While at school, Harrington taught himself to play the violin and even played at dances to earn extra money. He also played the Spanish guitar and the ukulele, which in later years he would bring on collecting trips with students and play around the campfire at night after a long day of botanizing! Harrington also read and wrote poetry and was an avid mystery-story reader.

Harrington completed only two years of college before he was called back to his hometown to teach and coach the football and basketball teams at the high school. He eventually earned his bachelor's degree in 1927 from the University of Northern Iowa. In 1928, he earned his Master's degree and in 1933 his PhD, again from the University of Northern Iowa. In 1933, he married Edith Jirsa in Waterloo, Iowa. Soon after, he was offered a botany position at CSU to teach taxonomy courses. Harrington left Fort Collins a few years later to teach botany and zoology at Chicago Teacher's College but returned in 1943 having been offered the position of curator of the herbarium and professor of botany at CSU. Harrington dutifully served as the curator of the herbarium for the next 25 years.

When Harrington returned to CSU, he was commissioned to produce a modern flora for the state of Colorado. Harrington spent many years piecing together the flora of Colorado and collecting specimens from all over the state. He made over 10,000 collections from every corner of Colorado, discovering many new species to include in the flora and several new species known to science. The CSU Herbarium currently houses approximately 6200 specimens collected by Harrington. Harrington did not actively seek to publish new species of plants but instead would usually send off unusual specimens to experts in those particular families. Harrington collected an unusual *Penstemon* in June of 1955 in Eagle County, sending specimens to William Penland who eventually described it as a new species in 1958. Honoring Harrington, Penland aptly named the species *Penstemon harringtonii* Penland. Another rare species, *Oenothera harringtonii* W.L. Wagner, Stockh. & W.M. Klein was also named for Harrington.

Harrington published the *Manual of the Plants of Colorado* in 1954 after many years of study. This work is still highly regarded as an authority on the flora of Colorado, and at the time of publication approximately 1 out of every 30 species listed in the *Manual* constituted a new record for the state of Colorado. Edith, Harold Harrington's wife, was also a botanist and worked at the Colorado State Seed Laboratory for 15 years. While at CSU together, she aided her husband in what she called "her small way", including hand-typing the entire *Manual*! She also assisted Harrington in plant collecting, photography, and preparing his other publications. The Harringtons wanted to self-publish the book to keep costs down so that botany students would be able to afford the text.

Harold and Edith would go on to visit Europe in 1964, where they visited many botanical gardens and herbaria. They took numerous photographs of wild and native plants, which they shared with the public through educational programs along the Front Range. They were always traveling and eventually made it to every state in the continental United States, even making a trip to Hawaii, New Zealand, Australia, Fiji, Tahiti, and Bora Bora in 1980. In the summer of 1980, the two crossed and re-crossed their beloved Rocky Mountains together for the last time. During his time at Colorado State University, Harrington published 17 books, including *Colorado Ferns and Fern Allies* (1950), *How to Identify Plants* (1957), *Edible Native Plants of the Rocky Mountains* (1967), and *How to Identify Grasses and Grass-like Plants* (1977). In 1955, Harrington also published *The True Aquatic Vascular Plants of Colorado* with Y. Matsumura, who also assisted Harrington in his collections and prepared the illustrations for this book and *Edible Native Plants*. Harrington was known for being a warm-hearted person who deeply cared about his students and botany. He died on January 22, 1981, but like the other curators who came before him, his work lives on through the specimens housed in the CSU Herbarium. Furthermore, the Harold David Harrington Graduate Fellowship was established by Edith to fund students studying plant taxonomy, the field in which Harrington was most passionate and to which he dedicated so many hours of study.

THE MAJOR VEGETATION ZONES AND PLANT COMMUNITIES IN COLORADO

Colorado encompasses nearly 66.6 million acres and supports a variety of habitats and floral diversity, ranging from 3315 ft to 14,431 ft in elevation. Within this boundary, a variety of conditions determines the natural vegetation in an area, including climate, elevation, soil type, and present or past disturbance. Colorado is also divided into the eastern and western slopes by the Continental Divide (delimited by the dark black line), which separates the watersheds that drain into the Pacific Ocean from those that drain into the Atlantic Ocean. A wide diversity of vegetation zones and plant communities are found in Colorado, and the landscape is often divided into the following 6 major vegetation zones: plains/grasslands, foothills, montane, subalpine, alpine, and semi-desert shrublands (Mutel & Emerick, 1992). Within each vegetation zone, some of the major plant communities are discussed (Colorado Natural Heritage Program, 2005).

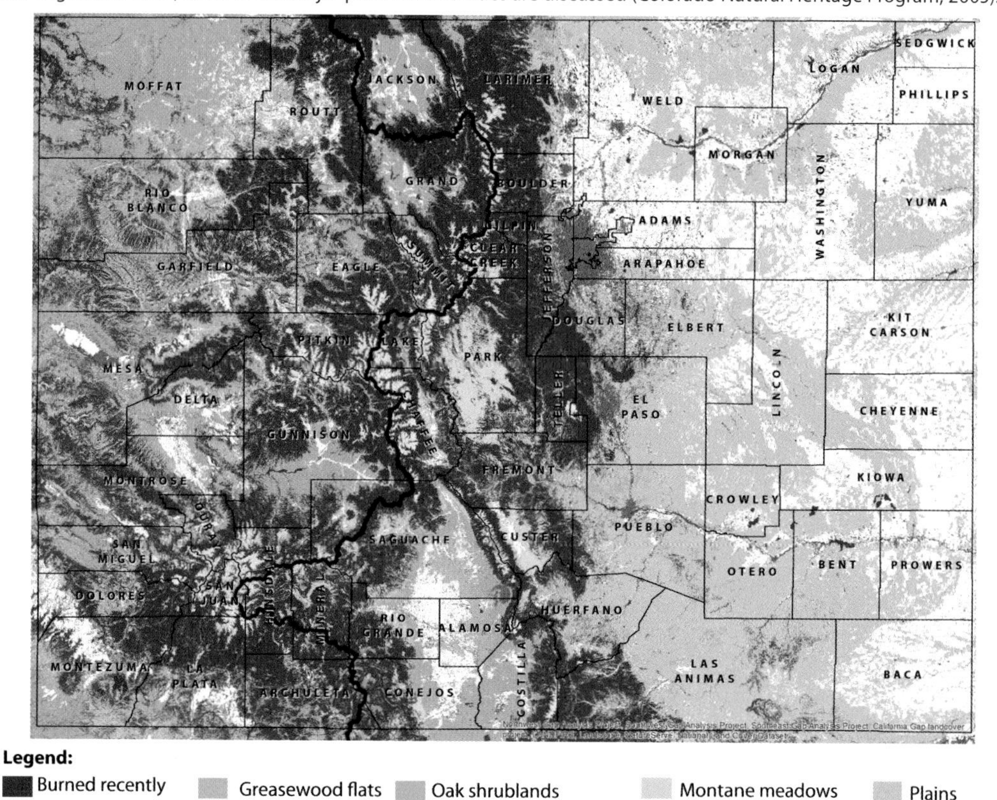

Legend:

Burned recently | Greasewood flats | Oak shrublands | Montane meadows | Plains
Moist meadows | Sand dunes | Montane/foothills forests | Alpine | Urban areas
Water | Cropland | Introduced vegetation | Sagebrush/pinyon-juniper shrublands

Map courtesy of the USGS Geological Survey, Gap Analysis Program (GAP). August 2011. National Land Cover, Version 2.

Plains/grassland

The plains/grassland zone occurs along the eastern third of Colorado and ranges from 3315 ft to approximately 5600 ft in elevation. The plains zone covers a variety of land formations – rolling plains, mesas, sandstone outcroppings, shale barrens, and canyons. While much of the plains/grassland zone has been transformed into farmland and cultivated fields, remnants of the natural plains vegetation remain preserved in areas such as the Pawnee National Grasslands and Comanche National Grasslands.

This ecosystem can be primarily divided into 3 categories: shortgrass/mixed prairie, sandhill prairie, and tallgrass prairie. The shortgrass/mixed prairie vegetation consists primarily of *Bouteloua gracilis* (blue grama) and *Buchloë dactyloides* (buffalograss). Other graminoid species include *Aristida purpurea* (purple three-awn) and *Sporobolus cryptandrus* (sand dropseed), which are often present below an understory of taller grasses such as *Hesperostipa comata* (needle and thread grass), *Pascopyrum smithii* (western wheatgrass), *Schizachyrium scoparium* (little bluestem), and *Andropogon gerardii* (big bluestem). Forbs common to the shortgrass/mixed prairie region include *Sphaeralcea coccinea* (scarlet globemallow), *Machaeranthera tanacetifolia* (tanseyleaf tansy-aster), *Oenothera albicaulis* (whitest evening primrose), *Lygodesmia juncea* (rush skeletonweed), *Opuntia polyacantha* (plains pricklypear), *Liatris punctata* (dotted blazing star), *Ambrosia psilostachya* (western ragweed), *Yucca glauca* (soapweed), *Eriogonum effusum* (spreading buckwheat), *Psoralidium tenuiflorum* (slimflower scurfpea), *Picradeniopsis oppositifolia* (oppositeleaf bahia), and *Helianthus annuus* (sunflower). The most frequent shrubs found on the plains are *Ericameria nauseosa* (rubber rabbitbrush), *Atriplex canescens* (fourwing saltbush), *Krascheninnikovia lanata* (winterfat), and *Gutierrezia sarothrae* (broom snakeweed). *Cylindropuntia imbricata* (tree cholla) is scattered in *Bouteloua gracilis* grasslands in the southeastern counties. Trees of *Populus deltoides* (cottonwood) and *Salix* spp. (willows) are common along streams and in other riparian areas.

The sandhill prairie is characterized by loose sandy soil, often developing dunes, and is primarily dominated by *Artemisia filifolia* (sand sage) (Ramaley, 1939). These areas are scattered across the eastern plains but are particularly prevalent along rivers. *Muhlenbergia pungens* (sandhill muhly), *Muhlenbergia ammophila* (blowout grass), *Achnatherum hymenoides* (Indian ricegrass), and *Psoralidium lanceolatum* (lemon scurfpea) are usually among the first species to inhabit these loose, sandy areas. As the sandy soil becomes more stabilized on these sand hills, species such as *Pascopyrum smithii* (western wheatgrass), *Bouteloua gracilis* (blue grama), *Buchloë dactyloides* (buffalograss), *Panicum virgatum* (switchgrass), *Aristida purpurea* (purple three-awn), *Calamovilfa longifolia* (prairie sandreed), *Sporobolus cryptandrus* (sand dropseed), *Schizachyrium scoparium* (little bluestem), and *Andropogon hallii* (sand bluestem) become the dominant grasses. Other common plant species found in sandhill prairies include *Tradescantia occidentalis* (prairie spiderwort), *Rumex venosus* (wild begonia), *Helianthus petiolaris* (prairie sunflower), *Yucca glauca* (soapweed), *Prunus angustifolia* (sandhill plum), *Eriogonum annuum* (annual wild buckwheat), *Mirabilis linearis* (narrowleaf four o'clock), *Rhus trilobata* (skunkbush sumac), and *Heterotheca villosa* (hairy false goldenaster).

Remnants of tallgrass prairie communities can be found closer to the foothills along the base of the Rocky Mountains and in moist areas along rivers (Bock & Bock, 1998). The dominant species in the tallgrass prairie community include *Andropogon gerardii* (big bluestem), *Schizachyrium scoparium* (little bluestem), *Sorghastrum nutans* (Indian grass), and *Panicum virgatum* (switchgrass), usually in association with shortgrass plants such as *Bouteloua gracilis* (blue grama).

Riparian areas on the eastern plains are usually found (1) in association with alkaline swales, (2) around intermittent ponds or drying pools, and (3) along flood plains of rivers and streams. Common plants found in these areas include *Distichlis spicata* (saltgrass), *Sporobolus airoides* (alkali sacaton), *Hordeum jubatum* (foxtail barley), *Juncus arcticus* (arctic rush), *Carex* spp. (sedges), *Bidens cernua* (nodding beggar-ticks), *Phyla cuneifolia* (wedgeleaf fogfruit), *Spartina gracilis* (alkali cordgrass), *Sagittaria cuneata* (arumleaf arrowhead), *Typha latifolia* (broadleaf cattail), and *Schoenoplectus americanus* (Olney's three-square bulrush).

Foothills

The foothills zone is a transitional zone, separating the plains from the high mountains. It is in this zone that grasslands give way to shrub-type vegetation, pinyon-juniper woodlands, or ponderosa pine forests, and it ranges from approximately 5500–8500 ft in elevation along the eastern slope of the Front Range. Many different landforms make up the foothills – mesas, plateaus, canyons, cliffs, open meadows, and hillsides are all found in the foothills. Shrublands along the eastern edge of the Front Range of the Rocky Mountains consist mainly of *Cercocarpus montanus* (birchleaf mountain mahogany), with *Ericameria nauseosa* (rubber rabbitbrush), *Rhus trilobata* (skunkbush sumac), *Prunus americana* (wild plum), *Prunus virginiana* var. *melanocarpa* (chokecherry), *Ribes cereum* (wax currant), and *Amelanchier* (serviceberry) intermixed. Mixed prairie is usually found in the understory of these shrubs, and common species include *Bouteloua gracilis* (blue grama), *Bouteloua curtipendula* (sideoats grama), *Poa fendleriana* (muttongrass), *Hesperostipa comata* (needle and thread grass), *Heterotheca villosa* (hairy false goldenaster), *Gutierrezia sarothrae* (broom snakeweed), and *Artemisia frigida* (prairie sagewort).

In southern and southwestern Colorado, the shrublands consist primarily of *Quercus gambelii* (Gambel oak), often comingling with sagebrush, pinyon-juniper, or Ponderosa pine. This plant community is also often referred to as montane shrubland. Annual precipitation in oak shrublands is 10–27 inches. In northwestern Colorado, *Quercus gambelii* mixes with *Ceanothus fendleri* (buckbrush) to form large thickets on dry, steep slopes. Shrubs such as *Amelanchier* spp. (serviceberry) are also common in these areas, as well as *Artemisia tridentata* (big sagebrush), *Purshia* spp. (bitterbrush), *Paxistima myrsinites* (Oregon boxleaf), and *Symphoricarpos* spp. (snowberry). Grasses common to oak shrublands include *Bouteloua gracilis* (blue grama), *Bouteloua curtipendula* (sideoats grama), *Aristida* spp. (three-awn), *Muhlenbergia* spp. (muhly), *Koeleria macrantha* (junegrass), *Festuca* spp. (fescue), and *Hesperostipa comata* (needle and thread grass). Numerous forbs occur in oak shrublands, but none have much density of cover.

Also in southern and western Colorado, extensive stands of pinyon-juniper woodlands occur, consisting primarily of *Pinus edulis* (pinyon pine) and either *Juniperus monosperma* (one-seed juniper) or *Juniperus scopulorum* (Rocky Mountain juniper), and are found at elevations ranging from 4900–8000 ft in elevation. Annual precipitation in pinyon-juniper woodlands is usually from 12–22 inches (Springfield, 1976). Other shrubs are also often found in these woodlands, such as *Artemisia tridentata* (big sagebrush), *Artemisia nova* (black sagebrush), *Cercocarpus montanus* (birchleaf mountain mahogany), *Symphoricarpos* spp. (snowberry), *Quercus gambelii* (Gambel oak), *Coleogyne ramosissima* (blackbrush), *Purshia* spp. (bitterbrush), and *Ericameria* spp. (rabbitbrush). In western Colorado, the pinyon-juniper woodlands often give way to dry desert valleys and semi-desert shrublands. Forbs commonly associated with pinyon-juniper woodlands include *Artemisia frigida* (prairie sagewort), *Heterotheca villosa* (hairy false goldenaster), *Eriogonum* spp. (buckwheat), *Hymenoxys richardsonii* (Colorado rubberweed), *Sphaeralcea coccinea* (scarlet globemallow), *Solidago* spp. (goldenrod), *Erigeron* spp. (fleabane), *Draba cuneifolia* (wedgeleaf draba), *Boechera* spp. (rockcress), *Penstemon* spp. (beardtongue), *Phlox* spp. (phlox), *Heliomeris multiflora* (showy goldeneye), and *Oreocarya* spp. (cryptantha). Common grasses include *Festuca arizonica* (Arizona fescue), *Koeleria macrantha* (junegrass), *Bouteloua gracilis* (blue grama), *Poa fendleriana* (muttongrass), *Pseudoroegneria spicata* (bluebunch), *Pascopyrum smithii* (western wheatgrass), *Hilaria jamesii* (galleta), *Achnatherum hymenoides* (Indian ricegrass), *Achnatherum scribneri* (Scribner needlegrass), and *Hesperostipa comata* (needle and thread grass).

Ponderosa pine (*Pinus ponderosa*) forests are distributed primarily along the Front Range, in the southwestern counties, and scattered on the western slope. Douglas-fir (*Pseudotsuga menziesii*) is often found in association with Ponderosa pine, especially on cooler north slopes. Annual precipitation in Ponderosa pine forests is 8–24 inches. The understory often consists of shrubby species such as *Arctostaphylos uva-ursi* (kinnikinnick), *Artemisia tridentata* (big sagebrush), *Cercocarpus montanus* (mountain mahogany), *Prunus virginiana* var. *melanocarpa* (black chokecherry), *Rosa* spp. (rose), *Purshia* spp. (bitterbrush), *Quercus gambelii* (Gambel oak), and *Symphoricarpos* spp. (snowberry). *Festuca arizonica* (Arizona fescue), *Festuca idahoensis* (Idaho fescue), *Muhlenbergia montana* (mountain muhly), *Pseudoroegneria spicata* (bluebunch wheatgrass), *Achnatherum* spp. (needlegrass), *Bouteloua gracilis* (blue grama), *Koeleria macrantha* (junegrass), *Schizachyrium scoparium* (little bluestem), and *Bromus inermis* (smooth brome) are common grasses in ponderosa pine forests. Additional forbs common in ponderosa pine forests include *Carex geyeri* (Geyer's sedge), *Claytonia* spp. (springbeauty), *Anemone patens* (pasqueflower), *Antennaria* spp. (pussytoes), *Collinsia parviflora* (blue-eyed Mary), and *Heterotheca villosa* (hairy false goldenaster).

Montane

The montane zone consists of lodgepole pine forests, spruce-fir forests, aspen forests, and grassland meadows and occurs from approximately 8000–10,500 ft in elevation. Within this zone, precipitation is predominantly in the form of snow, and summers are typically cool. Lodgepole pine (*Pinus contorta*) forests are widespread throughout the northern and central portion of the Rocky Mountains and often have little ground cover below due to a thick layer of needles, especially in dense forests. When the canopy in lodgepole pine forests is not as dense, an understory of species such as *Arctostaphylos uva-ursi* (kinnikinnick), *Juniperus communis* (common juniper), *Ceanothus velutinus* (snowbrush ceanothus), *Linnaea borealis* (twinflower), *Vaccinium scoparium* (dwarf red whortleberry), *Vaccinium cespitosum* (dwarf bilberry), *Vaccinium myrtillus* (whortleberry), *Shepherdia canadensis* (Canadian buffaloberry), *Ribes* spp. (currants), *Carex geyeri* (Geyer's sedge), and *Carex rossii* (Ross' sedge) can be seen.

Spruce-fir forests dominate much of the forest canopy throughout the Rocky Mountains and consist primarily of *Abies bifolia* (Rocky Mountain subalpine fir) and *Picea engelmannii* var. *engelmannii* (Engelmann spruce) alliances. Often, *Pinus contorta* (lodgepole pine) trees are also found mixed in these stands. Understory species may include shrubs such as *Juniperus communis* (juniper), *Berberis repens* (Oregon-grape), *Amelanchier alnifolia* (western serviceberry), *Linnaea borealis* (twinflower), and *Vaccinium scoparium* (dwarf red whortleberry). Grasses including *Achnatherum lettermanii* (Letterman's needlegrass), *Phleum alpinum* (alpine timothy), *Trisetum spicatum* (spike trisetum), and *Calamagrostis canadensis* (bluejoint) are common. Forbs such as *Maianthemum stellatum* (false Solomon's seal), *Cornus canadensis* (bunchberry dogwood), *Erigeron* spp., (fleabane), *Castilleja* spp. (Indian paintbrush), *Ligusticum porteri* (osha), *Hymenoxys hoopesii* (orange sneezeweed), *Carex geyeri* (Geyer's sedge), *Senecio serra* var. *admirabilis* (saw-toothed ragwort), *Saxifraga bronchialis* var. *austromontana* (spotted saxifrage), *Galium boreale* (northern bedstraw), *Pyrola* spp. (wintergreen), *Moneses uniflora* (wood nymph), *Arnica cordifolia* (heartleaf arnica), *Aquilegia elegantula* (western red columbine), *Fragaria virginiana* (mountain strawberry), and *Chimaphila umbellata* (pipsissewa) are common as well.

Often, extensive stands of aspen (*Populus tremuloides*) occur within spruce-fir forests. Aspen forests usually support a wide diversity of forbs and ferns, including *Thalictrum fendleri* (Fendler's meadowrue), *Vicia americana* (American vetch), *Thermopsis rhombifolia* (goldenbanner), *Achillea millefolium* (yarrow), *Delphinium barbeyi* (subalpine larkspur), *Cymopterus lemmonii* (ligusticoid springparsley), *Heracleum maximum* (cowparsnip), *Ligusticum porteri* (osha), *Pedicularis racemosa* var. *alba* (sickletop lousewort), *Aquilegia coerulea* (Colorado blue columbine), *Pteridium aquilinum* var. *pubescens* (hairy bracken fern), *Geranium richardsonii* (Richardson's geranium), *Frasera speciosa* (elkweed), *Lupinus argenteus* (silvery lupine), *Collomia linearis* (tiny trumpet), *Aconitum columbianum* (Columbian monkshood), *Wyethia amplexicaulis* (mule's ears), *Valeriana occidentalis* (western valerian), and *Osmorhiza berteroi* (sweet-cicely). Common grasses in aspen forests include *Bromus carinatus* (California brome), *Calamagrostis rubescens* (pinegrass), *Elymus glaucus* ssp. *glaucus* (blue wildrye), *Festuca thurberi* (Thurber's fescue), and *Elymus trachycaulus* (slender wheatgrass).

Moist meadows and fens are found throughout the Rocky Mountains in montane to subalpine valleys or as narrow strips bordering ponds and streams and support a diverse assortment of plant species. Several plant associations can be found in moist meadows, and they are often dominated by graminoids such as *Juncus arcticus* (arctic rush), *Carex* spp. (sedges), *Calamagrostis stricta* (slimstem reedgrass), *Deschampsia cespitosa* (tufted hairgrass), *Luzula parviflora* (smallflowered woodrush), *Glyceria* spp. (mannagrass), and *Eleocharis* spp. (spikerush). Common forbs and shrubs include *Caltha leptosepala* (marsh marigold), *Trollius laxus* ssp. *albiflorus* (American globeflower), *Ranunculus* spp. (buttercups), *Cardamine cordifolia* (heartleaf bittercress), *Rorippa alpina* (alpine yellow-cress), *Senecio triangularis* (arrowleaf ragwort), *Mimulus guttatus* (yellow monkey-flower), *Dodecatheon pulchellum* (shooting star), *Rhodiola rhodantha* (rose crown), *Salix* spp. (willows), *Betula* spp. (birch), *Pedicularis groenlandica* (elephant's head), *Swertia perennis* (felwort), *Platanthera* spp. (fringed orchid), *Mertensia ciliata* (streamside bluebells), *Bistorta bistortoides* (American bistort), *Micranthes odontoloma* (brook saxifrage), and *Primula parryi* (Parry's primrose).

Subalpine

The subalpine zone is a transitional zone in which the forests of the montane zone give way to true alpine vegetation. It is roughly 10,000–11,500 ft in elevation, and primarily consists of dense forests of Engelmann spruce (*Picea engelmannii*), subalpine fir (*Abies bifolia*), and small stands of bristlecone pine (*Pinus aristata*). Many of the trees in the subalpine zone near treeline are stunted and gnarled due to continual exposure to freezing winds and are often referred to as "krummholz." Flag trees form when the branches on the windward side of the tree are killed or stunted by continual strong, cold winds, giving the tree a flag-like appearance.

Common plants in the subalpine zone include *Primula parryi* (Parry's primrose), *Caltha leptosepala* (marsh marigold), *Trollius laxus* ssp. *albiflorus* (American globeflower), *Ranunculus* spp. (buttercups), *Vaccinium* spp. (whortleberries), and *Carex* spp. (sedges).

Alpine

The alpine zone is the highest zone in Colorado, ranging between 11,500 ft and 14,431 ft in elevation (our highest point in Colorado located on Mt. Elbert). The climate in the alpine is cold and windy, with long winters and very short growing seasons and an annual precipitation between 40–60 inches. In fact, the frost-free season in the alpine is only about 1.5 months long (from July to mid-Aug.). Characteristically, the alpine zone is devoid of trees and is composed mostly of graminoids and herbaceous perennials. In spite of these harsh conditions, the alpine zone supports a wide diversity of plants that have adapted to growing in this environment. Several different plant communities are found in the alpine zone, with scree slopes and exposed ridges, meadows, marshes, and willow-dominated communities the most common. These communities are usually not clear-cut and typically intermingle or overlap within the alpine zone.

Rocky ridgetops, fell fields, talus slopes, and scree slopes (collections of broken rock fragments on mountain slopes) are common formations in the alpine and typically support cushion plants. Cushion plants are well-adapted to the harsh growing conditions of the alpine and grow as compact spreading mats or cushions. They have large, deep taproots and grow very slowly. Cushion plants found in the alpine include *Silene acaulis* (moss campion), *Paronychia pulvinata* (Rocky Mountain nailwort), *Minuartia obtusiloba* (alpine stitchwort), *Eritrichium nanum* (arctic alpine forget-me-not), *Phlox pulvinata* (cushion phlox), and *Trifolium nanum* (dwarf clover). Other plants often found in this alpine community include *Claytonia megarhiza* (alpine spring beauty), *Androsace chamaejasme* ssp. *carinata* (boreal rockjasmine), *Sibbaldia procumbens* (creeping sibbaldia), *Poa alpina* (alpine bluegrass), *Poa glauca* (glaucous bluegrass), *Trisetum spicatum* (spike trisetum), *Luzula spicata* (spiked woodrush), *Podistera eastwoodiae* (Eastwood's podistera), *Potentilla diversifolia* (varileaf cinquefoil), *Potentilla nivea* (snow cinquefoil), *Dryas octopetala* var. *hookeriana* (Hooker's mountain avens), *Chionophila jamesii* (Rocky Mountain snowlover), *Chaenactis douglasii* var. *alpina* (alpine dustymaiden), *Viola adunca* (hook-spurred violet), *Saxifraga rivularis* (weak saxifrage), *Besseya alpina* (alpine kittentails), *Draba* spp., *Sedum lanceolatum* (spearleaf stonecrop), and *Senecio fremontii* var. *blitoides* (dwarf mountain ragwort).

Alpine meadow communities are found on more level areas and are dominated by graminoids such as *Phleum alpinum* (alpine timothy), *Podagrostis humilis* (alpine bentgrass), *Deschampsia cespitosa* (tufted hairgrass), *Trisetum spicatum* (spike trisetum), *Poa alpina* (alpine bluegrass), *Poa arctica* (Arctic bluegrass), *Carex* spp. (sedges), and *Kobresia* spp. (bog sedge). Common forbs in alpine meadows include *Geum rossii* var. *turbinatum* (alpine avens), *Mertensia alpina* (alpine bluebells), *Arnica mollis* (hairy arnica), *Agoseris aurantiaca* (orange agoseris), *Hymenoxys grandiflora* (old man of the mountain), *Achillea millefolium* (yarrow), *Senecio taraxacoides* (dandelion ragwort), *Thalictrum alpinum* (alpine meadowrue), *Bistorta vivipara* (alpine bistort), *Micranthes rhomboidea* (diamondleaf saxifrage), *Castilleja puberula* (shortflower Indian paintbrush), *Castilleja occidentalis* (western Indian paintbrush), *Phacelia sericea* (silky phacelia), *Polemonium viscosum* (sky pilot), *Pedicularis sudetica* ssp. *scopulorum* (sudetic lousewort), *Penstemon whippleanus* (Whipple's penstemon), *Cirsium scopulorum* (mountain thistle), *Artemisia scopulorum* (dwarf sagewort), *Packera werneriifolia* (hoary groundsel), *Erigeron compositus* (cutleaf daisy), *Erigeron melanocephalus* (blackhead daisy), *Erigeron grandiflorus* (Rocky Mountain alpine daisy), *Erysimum capitatum* (sanddune wallflower), *Boechera* spp. (rockcress), *Noccaea fendleri* ssp. *glauca* (alpine pennycress), *Campanula uniflora* (Arctic bellflower), *Cerastium beeringianum* (Bering chickweed), *Vaccinium* spp. (whortleberries), and *Gentiana algida* (Arctic gentian).

Alpine marshes are usually located along the margins of melting snowbanks, in springs, and along streams. Common graminoids include numerous *Carex* (sedge) species, several *Juncus* (rush) species, and *Eriophorum angustifolium* (tall cotton-grass). Common forbs include *Ranunculus adoneus* (alpine buttercup), *Caltha leptosepala* (marsh marigold), *Trollius laxus* ssp. *albiflorus* (American globeflower), *Rhodiola integrifolia* (king's crown), and *Pedicularis groenlandica* (elephant's head).

Dense willow thickets often occupy large expanses in the alpine and typically occur near the lower border of the alpine zone. These communities are the tallest perennial plants in the alpine, and consist primarily of *Salix brachycarpa* var. *brachycarpa* (shortfruit willow), *S. glauca* (grayleaf willow), *S. monticola* (mountain willow), and *S. planifolia* (planeleaf willow). Low-growing, prostrate willows found in the alpine are *Salix calcicola* (lime-loving woolly willow), *S. cascadensis* (cascade willow), *S. petrophila* (alpine willow), and *S. reticulata* var. *nana* (snow willow).

Semi-desert shrublands

On the western slope and in intermountain parks, semi-desert shrublands are common and widespread. Semi-desert shrublands can be divided into three main types: sagebrush-, saltbush-, and greasewood-dominated shrublands.

Sagebrush shrublands occupy much of the intermountain parks (North, Middle, and South Park) in Colorado as well as forming dominant landscapes across the western slope and between mountain ranges. Sagebrush shrublands are found between 7000 and 10,000 ft in elevation, and are typically located on flat or rolling hills with well-drained soils. *Artemisia tridentata* (big sagebrush) or *Artemisia cana* (silver sage) are the dominant sagebrush species in these shrublands. Other shrubs such as *Ericamerica nauseosa* (rabbitbrush), *Chrysothamnus viscidiflorus* (viscid rabbitbrush), *Purshia tridentata* (antelope bitterbrush), *Symphoricarpos rotundifolius* (mountain snowberry), and *Krascheninnikovia lanata* (winterfat) are common in sagebrush shrublands. Common graminoids include *Achnatherum hymenoides* (Indian ricegrass), *Bouteloua gracilis* (blue grama), *Bromus carinatus* (California brome), *Danthonia intermedia* (timber oatgrass), *Danthonia parryi* (Parry's oatgrass), *Elymus elymoides* (squirreltail), *Elymus trachycaulus* (slender wheatgrass), *Festuca idahoensis* (Idaho fescue), *Festuca thurberi* (Thurber's fescue), *Pascopyrum smithii* (western wheatgrass), *Hesperostipa comata* (needle and thread grass), *Hilaria jamesii* (galleta), *Leymus cinereus* (basin wildrye), *Poa secunda* (Sandberg bluegrass), and *Pseudoroegneria spicata* (bluebunch). Herbaceous plants such as *Castilleja flava* (yellow Indian paintbrush), *Castilleja chromosa* (red desert paintbrush), *Tetradymia spinosa* (catclaw horsebrush), *Allium acuminatum* (tapertip onion), *Ipomopsis aggregata* (skyrocket), *Artemisia ludoviciana* (Louisiana sagewort), *Lomatium triternatum* var. *platycarpum* (Great Basin desertparsley), *Agoseris parviflora* (steppe agoseris), *Crepis* spp. (hawk's-beard), *Sphaeralcea coccinea* (scarlet globemallow), *Hymenoxys richardsonii* (rubberweed), *Balsamorhiza sagittata* (arrowleaf balsamroot), *Hymenoxys hoopesii* (orange sneezeweed), *Lupinus* spp. (lupine), *Eriogonum* spp. (buckwheat), *Phlox longifolia* (longleaf phlox), and *Phlox multiflora* (mountain phlox) are scattered throughout sagebrush shrublands.

Saltbush shrublands occur on rolling plains and slopes in western Colorado, particularly in western Moffat County and between Montrose and Grand Junction. These communities are usually associated with shale-derived clay soils and are typically dominated by *Atriplex gardneri* (Gardner's saltbush) and *Atriplex corrugata* (matscale). Other shrubs that occur in saltbush shrublands include *Artemisia pedatifida* (birdfoot sagebrush), *Artemisia spinescens* (budsage), and *Atriplex pleiantha* (four-corners orach). Herbaceous plants such as *Stanleya pinnata* (desert prince's plume), *Oenothera* spp. (evening primrose), *Phlox longifolia* (longleaf phlox), *Eriogonum* spp. (buckwheat), *Lomatium concinnum* (Colorado desert parsley), *Astragalus flavus* (yellow milkvetch), *Astragalus asclepiadoides* (milkweed milkvetch), *Phacelia* spp. (phacelia), *Xylorhiza glabriuscula* (smooth woodyaster), *Eremogone hookeri* (Hooker's sandwort), and *Penstemon retrorsus* (Adobe Hills beardtongue) are scattered within these saltbush shrublands. Common graminoids include *Achnatherum hymenoides* (Indian ricegrass), *Bouteloua gracilis* (blue grama), *Elymus elymoides* (squirreltail), *Pascopyrum smithii* (western wheatgrass), *Elymus lanceolatus* (thickspike wheatgrass), *Pseudoroegneria spicata* (bluebunch), *Bromus tectorum* (cheatgrass), and *Sporobolus airoides* (alkali sacaton).

Greasewood flats occur scattered throughout western Colorado and the San Luis Valley and are dominated by *Sarcobatus vermiculatus* (greasewood). Greasewood flats usually have saline soils and groundwater that is near the soil surface and are typically located on alluvial fans or along streams. They remain dry for the majority of the growing season, with precipitation usually less than 10 inches a year, but flood intermittently. Common shrubs found in greasewood flats include *Atriplex canescens* (fourwing saltbush), *Atriplex confertifolia* (shadscale), *Artemisia spinescens* (budsage), *Ericameria nauseosa* (rabbitbrush), and *Krascheninnikovia lanata* (winterfat). Graminoids common to greasewood flats include *Sporobolus airoides* (alkali sacaton), *Distichlis spicata* (saltgrass), *Bouteloua gracilis* (blue grama), and *Hordeum jubatum* (foxtail barley).

Other communities

Hanging gardens are scattered on the western slope. They are found in alcoves within canyonlands containing seeps (constant water sources) that allow water to continuously drip along cliff faces. Forbs such as *Adiantum capillus-veneris* (common maidenhair), *Mimulus eastwoodiae* (Eastwood's monkey-flower), *Mimulus guttatus* (yellow monkey-flower), *Sullivantia hapemanii* (hanging garden coolwort), *Zigadenus vaginatus* (alcove death camas), *Erigeron kachinensis* (kachina daisy), *Platanthera zothecina* (alcove bog orchid), *Epipactis gigantea* (giant helleborine), *Aquilegia barnebyi* (oil shale columbine), and *Aquilegia micrantha* (Mancos columbine) are found in hanging gardens.

HOW TO USE THIS BOOK

Every project, every analysis, every survey, and every compilation of diversity begins with one thing – the identification of the species around us. This forms a baseline of our knowledge of the floristic diversity in a given area. In order to identify the plants around us, however, you must be able to use a dichotomous key. A dichotomous key is simply a series of paired questions about the plant you are attempting to identify. Each pair of questions (or couplet) gives two mutually exclusive alternatives to choose from. Given the plant at hand, you simply choose the couplet member that best fits and continue working through the key in this manner until you have arrived at the name of the particular species in question. Sounds easy, right? Well, it can be, but it also requires knowing a bit of botanical terminology to help you along the way. Understanding botanical terms forms the basis for plant identification and makes working through a dichotomous key much quicker and easier. I have attempted to make the following dichotomous key as easy to use as possible, keeping botanical terminology to a minimum, but one cannot write a dichotomous key without using some botanical terminology! A fantastic reference for botanical terminology is Harris & Harris (2001) which includes illustrations for every botanical term listed.

An example of a couplet in a dichotomous key:

1a. Fruit an achene enclosed within a fleshy hypanthium (e.g., rose hip); sepals conspicuously constricted above to a linear segment and then expanded at the tip; flowers pink, rose, or yellow...***Rosa***

1b. Fruit an aggregate of drupelets (e.g., raspberry); sepals unlike the above; flowers usually white or sometimes pinkish or rose...***Rubus***

Characteristics in each couplet are separated by semicolons, and the first characteristic listed is typically the one that is the most definitive, useful, or informative in separating the taxa. Make sure you carefully read the entire couplet before making your decision! And of course, go with your gut. Trust your instincts. Don't try and make a characteristic fit onto a plant if it is not there. To view minute characteristics, it is necessary to either have a high-magnification hand lens or a microscope. If a characteristic is not available (e.g., your plant is in fruit and the couplet only lists flowering characteristics), you may need to mark this couplet and take both choices, weighing the outcome in the end. Refer to figures and photographs if these are available, and if you have access to a computer you can easily examine herbarium specimens for each taxon present in Colorado and verify your specimens through the Colorado State University Herbarium website (http://herbarium.biology.colostate.edu/collection/specimens/). You should always verify your specimen first before assigning a name. If you arrive at an unsatisfactory end, simply go back to the place in the key where you had the most uncertainty and begin again from there.

Sometimes characteristics used to delimit taxa overlap. In these cases, we use what is called "a preponderance of evidence" to determine the species, analyzing and weighing the characteristics. This overlap in characteristics can be the result of numerous things including poor species boundaries, hybridization, or morphological variation within species. Additionally, mutant plants can be found in populations from time to time. I distinctly remember a man showing me an *Oenothera* with 3 sepals and 3 petals once! These oddballs of the plant world may not key out easily if at all.

Once you have arrived at the correct identification for the plant you are keying out, the *Flora* gives you the common name, most widely used synonyms (if available), a short description of the taxon, its habitat, range, approximate flowering time, elevation range, and county distribution. The eastern and western slopes of the Continental Divide are denoted with an E and W respectively. Measurements for each taxon are given in metric units. A range of measurements is given for each part, with outliers in parentheses. If not stated, measurement values are for lengths. Attempts were made to provide the user with photographs of either key morphological characteristics (such as the nutlets of *Oreocarya*) or photographs of the more common species one would encounter to aid in identification. Approximately ¼ of the taxa in Colorado are represented with a color photograph. The plant photographs were taken by the author unless otherwise noted, and the photographs of microscopic characteristics were taken at the Colorado State University Herbarium by the author and student interns. The distributions and elevations for each taxon primarily represent our most current records from herbaria in Colorado and the surrounding areas. These distribution maps would not be possible if it were not for the efforts of countless collectors and floristic projects. It is only through these intensive efforts that our true understanding of species distributions can be known. In spite of all these efforts, elevation and range extensions are always possible. Additional records from over 80 herbaria are available for viewing through SEINet (http://swbiodiversity.org/seinet/index.php).

More recently, the multivolume *Flora of North America* series was established, with the goal of documenting every single taxon present in North America, giving dichotomous keys to the species, some illustrations, and lengthy descriptions. However, this series encompasses all of North America north of Mexico; it can be long and cumbersome and quite frankly very difficult to key out plants using this series. A subset of the taxa just for Colorado makes much more sense to someone botanizing here, is easier to transport into the field, and simpler to use overall. The *Flora of North America* series was consulted for species descriptions and taxon delimitations of course, but this work is not meant to be a duplication of this effort. In fact, when making taxon delimitations several factors were taken into consideration, including consulting other regional floras, monographs, original species descriptions, and type specimens. The references cited below each family or genus were the principal ones consulted when writing each treatment. This flora does give short species descriptions for the taxa present, but for more detailed descriptions one should consult the *Flora of North America* or the *Intermountain Flora* series.

Mastering plant identification takes time and patience and of course practice, practice, practice! But in the end, being able to identify the species around us opens up a whole new world to explore, to document, and to embrace. Once you begin keying out plants, you won't be able to stop. And hiking will never be the same again.

NOMENCLATURE

The nomenclature in the *Flora* utilizes scientific names consistent with the International Code of Botanical

Nomenclature, and taxa are organized alphabetically within each family. The task for the taxonomist is to study the relationships of the organisms in question and categorize them accordingly. Scientific names are a concise way of describing species and allow us to categorize plants in a hierarchical, meaningful manner. They are a two-part name consisting of a genus and a specific epithet (or species name) and are referred to as a binomial. For example, the scientific name of a sunflower is *Helianthus annuus* L. *Helianthus* is the genus name for several species of sunflowers, but the specific epithet of *annuus* lets the user know exactly which sunflower is being described. A binomial also includes the authority, or the person who first described that species. In our example, the authority is L., which is an abbreviation for Linnaeus (Carl Linnaeus, who invented the binomial system of nomenclature). This binomial can also be abbreviated to *H. annuus* if we already know that we are referring to the genus *Helianthus*.

Sometimes the scientific name encountered in this flora may differ from the scientific name in another guide or manual. In this case, the scientific name that is not the most currently accepted name is referred to as a synonym. Some species can have a multitude of synonyms while others have none at all. For example, the species *Symphyotrichum porteri* (A. Gray) G.L. Nesom has an older synonym of *Aster porteri* A. Gray. The new name of *Symphyotrichum porteri* reflects our current understanding of the phylogenetic delimitations within the Asteraceae (sunflower) family. The *Flora* does not have a complete list of synonymy for each taxon, but does list the most common synonyms you would encounter for each.

A common name for each species is also included. For example, the common name for *Hymenoxys grandiflora* (Torr. & A. Gray ex A. Gray) Parker (a high alpine member of the Asteraceae family) is old man of the mountain. Common names can be useful for communicating with the general public, and sometimes they do contain informative information. However, there is no formal application process for the generation of common names, and several species can have the same common name. Therefore, for the purposes of plant identification, scientific names are the most widely used.

The *Flora of Colorado* reflects a modern phylogenetic framework with regards to family and generic species delimitations mostly following APG III (Angiosperm Phylogeny Group III system) standards. This is mostly a molecular-based system of plant taxonomy utilizing sequencing of genomic regions of the plants DNA to sort and organize taxa into monophyletic groups (a group containing all and only the known species of a common ancestor). For example, the family Scrophulariaceae is now split into 4 families – the Phrymaceae (*Mimulus*), the Orobanchaceae (including the hemiparasatic members of the old Scrophulariaceae family such as *Castilleja* and *Pedicularis*), the Plantaginaceae (including not only *Plantago* but most of the original members of the Scrophulariaceae family such as *Penstemon* and *Veronica*), and finally the Scrophulariaceae (containing *Limosella*, *Scrophularia*, and *Verbascum*). Such a framework may seem confusing at first but is ultimately necessary as our understanding of the relationships between and within plant families becomes clearer. In an attempt to be consistent with other current floras, this flora also follows the family delimitations used in *The Jepson Manual* (Baldwin et al., 2012), and I chose to be conservative on some family delimitations which are still unclear (such as those between Amaranthaceae and Chenopodiaceae). Families that do not follow current APGIII standards in the flora are: Agavaceae, Alliaceae, Chenopodiaceae, Dipsacaceae, Fumariaceae, Myrsinaceae, Parnassiaceae, Ruscaceae, Themidaceae, Valerianaceae, Viscaceae, and Zannichelliaceae.

HOW TO MAKE A PLANT COLLECTION

Making a plant collection is one of the best ways to become familiar with the flora of an area and to become more confident in your plant identification skills. You should ALWAYS adhere to collecting guidelines and policies of the given area, never collect rare plants, and report rare plant findings to the Colorado Natural Heritage Program (http://www.cnhp.colostate.edu/). Places from which you should NEVER collect specimens without permission include but are not limited to: National and State Parks, National Monuments, private land, and Open Space and other natural areas.

Specimens of plants can be permanent records of a species which occurred at a particular place and time. Such specimens are extremely valuable as a reference for study and comparison and for documentation of the flora for ecological and historical reasons. The value of a specimen depends upon the care taken by the collector in selecting and preparing the specimen as well as providing the necessary data to accompany the specimen.

Supplies you will need:

1. Plant press
2. Straps
3. Sheets of cardboard cut to 12 × 18 inches
4. Newspaper
5. Ziploc bags (large gallon size) or plastic bags
6. Sharpie or marker
7. Digging tool (small spade, etc.)

Collecting plant specimens

When collecting specimens, select the best representative plant in a group. If the plant is very small, you should collect a few specimens to ensure that you get enough material to work with. Specimens should ideally be in flower when collected. After selecting a plant specimen to collect, dig up the entire plant (roots and all) and place the whole plant in a ziploc bag. Collect a few extra flowers for dissecting later. Label the ziploc bag with your collection number (using your trusty Sharpie!) and complete your field notes for that specimen (see below). If the plant has a large taproot, the root can be cut in half and part thrown away. When collecting shrubs and trees, simply clip off a small branch.

Your field notes are an invaluable part of your collection. It is crucial that you fill out field notes containing the following information for each specimen:

COROLLA COLOR:______________________________
HABIT & SIZE:______________________________
SOIL TYPE:______________ SLOPE:______________
SPECIFIC LOCALITY:______________________________
____
ABUNDANCE:______________________________
LATITUDE:______________°N LONGITUDE: ______________°W
ELEVATION:______________ DATE:______________
COLLECTOR:______________ NUMBER:______________

Fill out as much information as possible for each specimen or group of specimens. Take GPS coordinates every ½ mile of collecting or if the habitat changes significantly. You can keep the specimen fresh in a ziploc bag for several hours provided it is not placed directly in the sun. However, it is best if you press your specimens within a few hours of collecting them. Never freeze your specimens!

Pressing plant specimens

The order of materials when you are pressing plants is as follows:

1. Outside press frame
2. Cardboard
3. Newspaper
4. Plant specimen (inside newspaper)
5. Cardboard
6. Newspaper containing second specimen
7. Cardboard (keep doing this until all specimens are pressed)
8. Outside frame

Tighten the press with your two straps placed on either end of the press. Ensure that the press is very tight for the best-looking specimens possible. Succulent or thick specimens may require an extra sheet of newspaper to soak up moisture. Write your collection number (corresponding to your field notes) on the outside of each piece of newspaper. Do not use newspaper that hangs over the edges of your press. Leave specimens in the press for about a week until dry.

Making labels

You will generate your labels using the information from your field notes. Labels are glued in the bottom right corner of your mounted specimens. An example of a good label is:

Flora of Colorado
Larimer County

Asteraceae
Helianthus annuus L.

Flowers yellow; plant erect, 12 in tall; abundant along roadside in disturbed area of no slope in gravelly soil.

Fort Collins: in a vacant field on the northwest side of the road at the intersection of Taft Hill Rd. and Horsetooth Rd.

Latitude: 40.6118° N, Longitude: 105.0753° W.

Date: 1 September 2000 **Elev:** 5000 ft
Collector: Jennifer Ackerfield **No:** 548

COLLECTIONS OF JENNIFER ACKERFIELD

Mounting plant specimens

Specimens are mounted on acid free 11 ½ × 16 ½ inch white rag mounting paper, available from herbarium supply companies and art supply stores. The process for mounting specimens is simple, but achieving the best results takes patience and attention to detail. The trick is to get enough glue that the specimen sticks to the paper, but you never want glue to spill out from under the specimen. The steps are as follows:

1. Pour a dollop of herbarium (or Elmer's white) glue on a glass plate. Using a large brush, spread the glue evenly into a rectangular area. Dilute the glue with some water so that it is not as thick.
2. Determine which side of the plant specimen should face up to display as many features as possible. Place the down side of the specimen on the glue and lightly press it down to get as much of the specimen covered with glue as possible.
3. Gently pull the specimen up from the glue and place it on the sheet of paper. Place the specimen such that space remains on the bottom right for a label. If a label has already been made, mount this with the glue on the bottom right corner of the sheet.
4. Put a spot of glue under any loose pieces of the specimen and press them down.
5. Pencil in the species and the collection number of the specimen in the lower right hand corner, so it can later be matched with its collection data if a label has not already been attached.
6. Lay the sheet upon two pieces of cardboard within a wooden pressing box. Cover with wax paper and two more sheets of cardboard.
7. Repeat the process (2–6) for as many specimens that you have. Add more glue and spread with brush as necessary.
8. On the top of the last piece of cardboard, place additional cardboard and very heavy weights.
9. When finished mounting, wash glass plate with hot water to remove all glue.

FLORISTIC OVERVIEW OF COLORADO

The state of Colorado has a rich floral diversity, encompassing vast expanses of prairie all the way to the alpine, basins, river valleys, sweeping mesas, and adobe hills. In total approximately 3322 taxa (including 645 varieties and subspecies) are currently known from Colorado, with 84% of the flora native to the state.

		Percent of total
Total Natives	2797	84
Total Endemics	108	3
Total Rare	525	16
Total Rare Endemics	90	2.6
Total Invasive	527	16

143 unique families are found in the state of Colorado, with the Asteraceae the largest family represented with a total of 562 taxa. Unless otherwise stated, the term "taxa" includes all species, subspecies, and varieties. The Fabaceae has the most rare species with 60 taxa considered rare or imperiled in the state, and the Brassicaceae has the most endemic species with 24 taxa.

Family	*Total taxa*	*Percent of flora*	*Total native*	*Total endemic*	*Total rare*
Asteraceae	562	16.9	486	14	51
Poaceae	351	10.5	259	1	12
Fabaceae	265	8.0	235	13	60
Brassicaceae	183	5.5	133	24	46
Cyperaceae	163	4.8	155	0	28
Plantaginaceae	104	3.1	91	10	26
Polygonaceae	101	3.0	84	5	18
Rosaceae	101	3.0	80	2	8
Boraginaceae	78	2.3	66	4	17

Caryophyllaceae	72	2.2	47	0	7
Chenopodiaceae	71	2.1	50	2	4
Ranunculaceae	70	2.1	65	3	9
Onagraceae	69	2.1	67	3	11
Polemoniaceae	68	2.0	67	8	15
Apiaceae	65	2.0	59	7	16
Salicaceae	47	1.4	40	1	6
Juncaceae	41	1.2	35	0	5
Lamiaceae	41	1.2	23	0	3

The largest genus in the state of Colorado is the genus *Carex* (Cyperaceae) with 119 taxa, closely followed by the genus *Astragalus* (Fabaceae) with 117 taxa. The genus *Astragalus* also has the most rare taxa with 46 considered rare or imperiled and 12 endemic to the state.

Genus	***Total taxa***	***Total native***	***Total endemic***	***Total rare***
Carex	119	119	0	22
Astragalus	117	115	12	46
Penstemon	65	64	9	22
Eriogonum	51	51	5	17
Erigeron	44	44	0	5
Artemisia	36	32	0	2
Salix	36	30	1	6
Oenothera	35	35	3	9
Potentilla	35	31	1	4
Juncus	34	30	0	4
Muhlenbergia	29	29	0	1
Poa	29	25	0	0
Draba	26	26	7	15
Ranunculus	26	23	0	1
Physaria	25	25	10	11
Mentzelia	24	24	3	7
Oreocarya	24	24	3	13
Chenopodium	22	18	0	1
Phacelia	20	20	4	8
Atriplex	19	16	1	1
Asclepias	19	19	0	6
Cymopterus	18	18	4	5
Cryptantha	15	15	0	0

Taxa formally listed as endangered, threatened, or candidates under the Endangered Species Act:

Family	Taxa	Federal Status	Endemic
Brassicaceae	*Eutrema penlandii* Rollins	T	Yes
Brassicaceae	*Physaria congesta* (Rollins) O'Kane & Al-Shehbaz	T	Yes
Brassicaceae	*Physaria obcordata* Rollins	T	Yes
Cactaceae	*Sclerocactus glaucus* (K. Schum.) L.D. Benson	T	No
Cactaceae	*Sclerocactus mesae-verdae* (Boissev. & C. Davidson) L.D. Benson	T	No
Fabaceae	*Astragalus humillimus* A. Gray	E	No
Fabaceae	*Astragalus microcymbus* Barneby	C	Yes
Fabaceae	*Astragalus osterhoutii* M.E. Jones	E	Yes
Fabaceae	*Astragalus schmollae* Ced. Porter	C	Yes
Fabaceae	*Astragalus tortipes* J.L. Anderson & J.M. Porter	C	Yes
Hydrophyllaceae	*Phacelia formosula* Osterh.	E	Yes
Hydrophyllaceae	*Phacelia scopulina* (A. Nelson) J.T. Howell var. *submutica* (J.T. Howell) R.R. Halse	T	Yes
Onagraceae	*Oenothera coloradensis* (Rydb.) W.L. Wagner & Hoch	T	No
Orchidaceae	*Spiranthes diluvialis* Sheviak	T	No
Plantaginaceae	*Penstemon debilis* O'Kane & J. Anderson	T	Yes
Plantaginaceae	*Penstemon penlandii* W.A. Weber	E	Yes
Polemoniaceae	*Ipomopsis polyantha* (Rydb.) V.E. Grant	E	Yes
Polygonaceae	*Eriogonum pelinophilum* Reveal	E	Yes

KEY TO THE FAMILIES

1a. Plants reproducing by spores, never with true flowers or seeds; ferns and fern allies...**KEY 1** (p. 16)
1b. Plants having true seeds, with flowers or cones; gymnosperms and angiosperms...2

2a. Ovules and seeds borne on the surface of scales, the scales aggregated into a cone which is either fleshy and berry-like or woody; evergreen (rarely deciduous) trees and shrubs with acicular or scale-like leaves; gymnosperms...**KEY 2** (p. 17)
2b. Ovules and seeds enclosed in an ovary; flowering plants, trees, shrubs, vines, and herbs; angiosperms...3

3a. Plants insectivorous, or parasitic (often lacking chlorophyll and not green) or saprophytic on other plants...**KEY 3** (p. 18)
3b. Plants with green leaves (insectivorous or parasitic plants with flowers will key here, too)...4

4a. Plants submerged or floating aquatics...**KEY 4** (p. 18)
4b. Plants terrestrial or emergent aquatics...5

5a. Woody plants (trees, shrubs, or woody vines)...**KEY 5** (p. 20)
5b. Herbaceous plants or only woody at the base (woody plants with flowers will key here, too)...6

6a. Vines (woody or herbaceous) with stems climbing or twining, often with suckers or tendrils (excluding plants merely creeping on the ground)...**KEY 6** (p. 25)
6b. Herbaceous or woody plants, stems not climbing or twining...7

7a. Perianth absent (inflorescence may be subtended by bracts which resemble a calyx, but this subtends many flowers) or of a single whorl, or sometimes reduced to scale-like or bristle-like structures...**KEY 7** (p. 26)
7b. Perianth in 2 or more whorls (generally both the calyx and corolla present, occasionally the sepals intergrading into petals, or both whorls the same in color and texture)...8

8a. Flowers 3-merous (with 3 perianth parts per whorl)...**KEY 8** (p. 30)
8b. Flowers 4- or 5-merous (with 4 or 5 parts per whorl), or in a spiral with the number of parts indefinite, or the flowers with only 2 petals or sepals...9

9a. Petals free to the base (sometimes individually joined to a hypanthium, but free above)...**KEY 9** (p. 32)
9b. Petals united at least at the base...**KEY 10** (p. 38)

KEY 1

FERNS AND FERN ALLIES

1a. Aquatic, floating plants forming mats, with dichotomously branching stems...**AZOLLACEAE** (p. 44)
1b. Plants terrestrial, or if aquatic then unlike the above...2

2a. Stems jointed, with longitudinal grooves, green or brown; papery sheaths surrounding the stem at each node; sporangia in a terminal "cone"...**EQUISETACEAE** (p. 48)
2b. Stems unlike the above; sporangia in a terminal "cone" or not...3

3a. Leaves palmately divided into 4 pinnae (like a four-leaf clover); sori borne in a hard, hairy sporocarp arising from short stalks near the base of the petioles... **MARSILEACEAE** (p. 49)
3b. Leaves not palmately divided into 4 pinnae; sori unlike the above...4

4a. Leaves linear, grass-like; sporangia borne at the widened base of the leaves; plants aquatic or in shallow water...**ISOËTACEAE** (p. 48)
4b. Plants unlike the above in all respects...5

5a. Leaves simple, sessile, lanceolate to linear, usually scale-like and rather rigid, to 8 mm long...6
5b. Leaves unlike the above...7

6a. Sporangia borne in terminal "cones" or solitary in leaf axils; spores all alike; leaves 3–8 mm long, lacking ligules...**LYCOPODIACEAE** (p. 49)
6b. Sporangia borne in 4-sided strobili or rarely in terminal "cones" (these only slightly differentiated from vegetative leaves); spores of 2 kinds: 1 or 4 large meagspores and numerous, tiny microspores in separate sporangia; leaves 1–4 mm long, with minute ligules...**SELAGINELLACEAE** (p. 57)

7a. Leaves linear (grass-like), simple or forked at the tips...**ASPLENIACEAE** (*Asplenium septentrionale*; p. 44)
7b. Leaves unlike the above...8

8a. Lower surface of the leaves covered by a whitish or yellowish powdery coating...**PTERIDACEAE** (*Argyrochosma, Notholaena*; p. 54)
8b. Lower surface of the leaves not covered by a whitish or yellowish powdery coating...9

9a. Leaves densely hairy or scaly below and obscuring the surface; ultimate segments oblong to round, ca. 1–2 mm long...**PTERIDACEAE** (*Cheilanthes*; p. 54)
9b. Leaves not densely hairy or scaly below, or if densely hairy then not obscuring the surface; ultimate segments various...10

10a. Leaves divided into a sterile leaf-like segment (trophophore) and lateral fertile segment (sporophore), with a common stalk; stems and petioles lacking scales...**OPHIOGLOSSACEAE** (p. 50)
10b. Leaves unlike the above, or if dimorphic then lacking a common stalk; stems and petioles often with scales...11

11a. Leaves of 2 kinds, the vegetative leaves triangular and pinnatifid, the fertile leaves shorter, linear, forming bead-like structures; leaf lobes mostly 2.5–5 cm long...**DRYOPTERIDACEAE** (*Onoclea*; p. 45)
11b. Leaves all similar or if dimorphic then unlike the above...12

12a. Leaves merely pinnatifid (not truly compound with distinct leaflets)...**POLYPODIACEAE** (p. 54)
12b. Leaves at least once compound with distinct leaflets (pinnae)...13

13a. Ultimate segments of the pinnae obliquely oblong or fan-shaped; sori discontinuous, borne on the reflexed margins of the lobes on the ultimate segments...**PTERIDACEAE** (*Adiantum*; p. 54)
13b. Leaves and sori unlike the above...14

14a. Petiole and stem lacking scales; plants 3–35 dm tall, often forming dense colonies in forests; sori more or less continuous around an inrolled leaflet margin...**DENNSTAEDTIACEAE** (p. 45)
14b. Petiole or stem with scales, sometimes represented by a tuft-like bunch of "hairs" at the base of the petiole or along the stem; plants various but unlike the above in all respects...15

15a. Leaves of 2 kinds, the fertile longer, of linear ultimate segments with inrolled margins bearing sori, 2–3-pinnate...**PTERIDACEAE** (*Cryptogramma*; p. 54)
15b. Leaves all similar, variously pinnate...16

16a. Sori borne near the leaflet margins, at least partially covered by the inrolled, entire margins; leaves 1–2-pinnate with oblong to oblong-lanceolate leaflets; petioles reddish-brown throughout or nearly so...**PTERIDACEAE** (*Pellaea*; p. 54)
16b. Sori borne on or along veins between the midrib and leaflet or segment margin, the margins usually flat, entire to toothed; leaves and petioles various...17

17a. Sori elongate in outline, usually with a linear indusium; leaves once compound, with entire to crenate or shallowly serrate margins, or rarely 2–3 pinnate...**ASPLENIACEAE** (p. 44)
17b. Sori round to elliptic, or hooked at one end to horseshoe-shaped in outline, indusium present or absent; leaves usually at least twice compound, if once compound then the leaflets sharply toothed or deeply lobed...18

18a. Lower leaf surface hairy with scales along the midrib and rachis; leaves once pinnatifid, the pinnae then deeply lobed with entire-margined segments, or sometimes the uppermost pair with crenate margins...**THELYPTERIDACEAE** (p. 58)
18b. Lower leaf surface and leaves various but unlike the above...**DRYOPTERIDACEAE** (p. 45)

KEY 2

GYMNOSPERMS

1a. Shrubs with grooved, green stems; leaves scale-like, brownish or blackish, triangular, opposite or in whorls of 3 at the joints...**EPHEDRACEAE** (p. 59)
1b. Shrubs or trees, without jointed, grooved stems; leaves needle-like or scale-like and overlapping but not whorled...2

2a. Leaves scale-like or awl-shaped, closely overlapping; female cones grayish-blue, globose and fleshy and berry-like...**CUPRESSACEAE** (p. 59)
2b. Leaves needle-like and linear, spirally arranged or in fascicles; female cones woody...**PINACEAE** (p. 60)

KEY 3

INSECTIVOROUS OR PARASITIC/SAPROPHYTIC PLANTS

1a. Plants insectivorous, capturing insects with sticky glands on the leaves or tiny, underwater bladder traps...2
1b. Plants parasitic or saprophytic, often lacking chlorophyll...3

2a. Aquatic plants capturing insects with underwater bladder traps; leaves finely dissected; flowers zygomorphic, yellow...**LENTIBULARIACEAE** (p. 521)
2b. Terrrestrial plants capturing insects with sticky glandular hairs on the leaves, when triggered causing the slow enclosure of the prey by the folding of the blade; leaves simple, often coiled in bud; flowers actinomorphic, white or pink...**DROSERACEAE** (p. 339)

3a. Plants with a conspicuous host plant, attached to the bark of trees or twining on the aerial stems of herbs...4
3b. Plants attached to the host plant below the surface of the soil, erect herbs...5

4a. Plants green or yellowish, attached to the bark of gymnosperms (*Pinus*, *Pseudotsuga*, *Abies*, and *Juniperus*); stems swollen-jointed and brittle when dry; fruit a berry; ovary inferior...**VISCACEAE** (p. 772)
4b. Plants yellowish or orangish, twining on the aerial stems of herbs; stems not jointed, thin and thread-like; fruit a capsule; ovary superior...**CONVOLVULACEAE** (*Cuscuta*; p. 298)

5a. Flowers actinomorphic...6
5b. Flowers zygomorphic...7

6a. Stems green, glabrous, not thick and stout; petals absent, sepals petaloid and white to pink; stamens 5...**SANTALACEAE** (p. 750)
6b. Stems reddish to purplish-brown, or yellowish to white and drying blackish, glabrous or glandular, often stout and thick; sepals and petals both present; stamens 6–12...**ERICACEAE** (p. 341)

7a. Flowers with the lowest petal enlarged, markedly different from the others (forming the labellum); anthers united with the gynoecium to form a column; petals distinct; ovary inferior...**ORCHIDACEAE** (*Corallorhiza*; p. 560)
7b. Flowers without a labellum; stamens 4, epipetalous but free of one another; petals united below; ovary superior...**OROBANCHACEAE** (p. 565)

KEY 4

AQUATIC PLANTS

1a. Plants free-floating, not rooted in the soil, small and the leaves or plants usually not over 1 cm long, or leaves and plants larger...2
1b. Plants not free-floating, rooted in the soil, the leaves generally larger...5

2a. Leaves over 2 cm long; flowers showy or inconspicuous...3
2b. Leaves smaller; flowers inconspicuous or absent...4

3a. Leaves ovate to round with an abrupt constriction leading to a conspicuously inflated petiole; flowers showy, purple...**PONTEDERIACEAE** (*Eichhornia*; p. 698)
3b. Leaves obovate to spatulate and strongly ribbed, without an inflated petiole; flowers small and inconspicuous, white...**ARACEAE** (p. 94)

4a. Ferns with small leaves (to 2 mm) that overlap on the stem...**AZOLLACEAE** (p. 44)
4b. True leaves absent, plants reduced to small green bodies called fronds, the fronds entire, lanceolate to ovate or round...**ARACEAE** (p. 94)

5a. Leaves trifoliately compound with sheathing bases; flowers white and densely hairy within, in a bracteate raceme...**MENYANTHACEAE** (p. 538)
5b. Leaves simple to dissected or pinnately compound; flowers unlike the above...6

6a. Leaves lobed or finely dissected...7
6b. Leaves simple and undivided, the margins entire or toothed...10

7a. Leaves all submerged, finely dissected, bearing bladders which trap insects; flowers zygomorphic, yellow, and in terminal racemes...**LENTIBULARIACEAE** (p. 521)
7b. Leaves without bladder traps; flowers actinomorphic...8

8a. Flowers conspicuous, perfect, the 5 petals yellow or white; leaves alternate, palmately divided; carpels numerous and distinct...**RANUNCULACEAE** (p. 704)
8b. Flowers inconspicuous, imperfect, with 4 sepals or the perianth absent, greenish; leaves whorled, pinnately or dichotomously dissected; unicarpellate or carpels 4 and united...9

9a. Leaves dichotomously dissected, the margins finely serrulate; perianth absent; ovary unicarpellate...**CERATOPHYLLACEAE** (p. 281)
9b. Leaves pinnately dissected, the margins entire; perianth of 4 sepals; ovary inferior, 4-carpellate...**HALORAGACEAE** (p. 496)

10a. Leaves opposite or whorled...11
10b. Leaves alternate or basal...20

11a. Flowers purple, 4-merous, in axillary or terminal racemes, conspicuous; leaves ovate to ovate-elliptic, toothed; fruit a capsule...**PLANTAGINACEAE** (*Veronica*; p. 598)
11b. Flowers inconspicuous, otherwise unlike the above; leaves and fruit various...12

12a. Leaves both floating and submerged, the submerged leaves very different from the floating leaves in form...13
12b. Leaves all submerged or emergent, but not floating, all leaves about the same in form...14

13a. Floating leaves without parallel veins; flowers imperfect, solitary in leaf axils, consisting of a single pistil and a single stamen; fruit of 4 flattened mericarps often with narrow wings, without a beak...**PLANTAGINACEAE** (*Callitriche*; p. 581)
13b. Floating leaves with distinct parallel veins; flowers perfect, 4-merous, in terminal or axillary spikes; fruit a drupelike achene, not flattened, strongly to weakly beaked...**POTAMOGETONACEAE** (p. 699)

14a. Fruit a capsule; leaves stipulate, opposite although sometimes crowded above and appearing whorled, elliptic to oblong and serrulate, or linear and entire and rather succulent; annuals found in shallow water and muddy shores, not truly submerged; flowers perfect, 3- or 5-merous...**ELATINACEAE** (p. 340)
14b. Fruit an achene, schizocarp, or berrylike; plants unlike the above in all combined characteristics...15

15a. Leaves whorled with 6 or more leaves per whorl, the emergent ones 1–3 cm long, submerged leaves often reduced to short scale-like projections; stems stout and thick, 0.7–2.5 cm wide; flowers clustered in the axils of emerged leaves...**PLANTAGINACEAE (***Hippuris*; p. 582)
15b. Leaves opposite or if whorled then typically with 3 leaves per whorl or rarely up to 5 leaves per whorl; stems thin, relatively weak; unlike the above in all combined characteristics...16

16a. Leaves broader, over 5 mm wide; flowers perfect; fruit a drupelike achene, not flattened, strongly to weakly beaked...**POTAMOGETONACEAE** (p. 699)
16b. Leaves filiform to linear, 0.5–2 mm wide; flowers perfect or imperfect; fruit beaked or not...17

17a. Flowers perfect, 4-merous, in terminal or axillary spikes on long peduncles; lower leaves usually alternate and upper leaves opposite; fruit a drupelike achene, not flattened, strongly to weakly beaked...**POTAMOGETONACEAE** (p. 699)
17b. Flowers imperfect, 3-merous or perianth absent, sessile in leaf axils or pedunculate and with a spathe formed of connate bracts; leaves opposite or whorled; fruit beaked or not...18

18a. Leaves mostly minutely notched at the tip; fruit of flattened mericarps sessile in the leaf axils, with narrow wings, lacking a beak...**PLANTAGINACEAE** (*Callitriche hermaphroditica*; p. 581)
18b. Leaves acute or rounded at the tip; fruit unlike the above...19

19a. Leaves opposite, entire; fruit an achene with a long beak, sharply ridged on the back...**ZANNICHELLIACEAE** (p. 774)
19b. Leaves opposite or whorled with 3 leaves per node, often finely toothed, sometimes remotely so, the teeth small spinulose projections (these are very small, must examine with a hand lens or microscope to see); fruit berrylike or fusiform (with conspicuous areolae in longitudinal rows)...**HYDROCHARITACEAE** (p. 498)

20a. Flowers showy with conspicuous petals or tepals, these white, yellow, pink, red, or purple...21
20b. Flowers inconspicuous, not showy, the perianth sometimes absent...26

21a. Flowers pink or red; stems with a sheathing, scarious and membranous stipule (ocrea) present at the petiole base...**POLYGONACEAE** (*Persicaria*; p. 692)
21b. Flowers yellow, white, or purple, or pinkish; stems without an ocrea...22

22a. Leaves orbicular to ovate with a deeply cordate base, thick and leathery, long-petiolate, large (10–40 cm) and floating; flowers solitary on long pedicels, large (5–10 cm in diam.) with numerous stamens, intergrading from sepals to petals, yellow or white...**NYMPHAEACEAE** (p. 548)
22b. Leaves smaller, linear or lanceolate, ovate, or sagittate or hastate, not thick and leathery; flowers smaller, with distinct sepals and petals or the tepals all similar in appearance...23

23a. Flowers yellow or purple, the tepals all the same in appearance, with a bladeless sheath forming a spathe surrounding the flower; fruit a loculicidal capsule...**PONTEDERIACEAE** (*Heteranthera*; p. 698)
23b. Flowers white, with 2 dissimilar series (separate sepals and petals), without a spathe; fruit a silicle or achene; leaves basal...24

24a. Fruit a silicle or silique; stamens 6, tetradynamous; flowers perfect; leaves pinnately divided with ovate segments, or terete and linear and in a densely packed basal rosette...**BRASSICACEAE** (*Nasturtium*, *Subularia*; p. 219)
24a. Fruit various but not a silicle or silique; stamens 4 or 6–9, not tetradynamous; flowers perfect or imperfect; leaves flat, hastate, sagittate, linear, linear-lanceolate, elliptic, or ovate, not pinnately divided...25

25a. Leaves hastate, sagittate, linear, linear-lanceolate, narrowly elliptic, or ovate; fruit an achene; carpels distinct and numerous; stamens 6–9; flowers 3-merous, not tightly grouped at the base of the plant...**ALISMATACEAE** (p. 66)
25b. Leaves elliptic, mostly 2 cm long; fruit a capsule; carpels united; stamens 4; flowers 5-merous, the flowers small and solitary on short peduncles tightly grouped at the base of the plant...**SCROPHULARIACEAE** (*Limosella*; p. 757)

26a. Flowers imperfect, in dense globose heads, the male flowers above the female flowers; leaves grass-like and linear, 2-ranked, with an open sheath at the base, emergent or floating...**TYPHACEAE** (*Sparganium*; p. 765)
26b. Flowers perfect, in spikes on long peduncles or on short peduncles tightly grouped at the base of the plant; leaves all submerged or all floating, or floating and submerged with the submerged leaves very different from the floating leaves in form; otherwise unlike the above...27

27a. Perianth absent; flowers axillary on long, slender peduncles which coil and elongate at maturity; stipules expanded and tubelike; leaves filiform, all submerged; fruit blackish-gray...**RUPPIACEAE** (p. 739)
27b. Perianth present; flowers not on spiraling peduncles; stipules (if present) not tubelike; leaves filiform or broader, all submerged or floating, or floating and submerged; fruit green or brown...28

28a. Leaves not all basal, if elliptic then usually larger than 2 cm long; fruit an achene; tepals 4; flowers in terminal or axillary spikes, emergent and not grouped at the base of the plant...**POTAMOGETONACEAE** (p. 699)
28b. Leaves all basal, on long petioles, elliptic, to 2 cm long; fruit a capsule; petals 5; flowers small and solitary on short peduncles, tightly grouped at the base of the plant...**SCROPHULARIACEAE** (*Limosella*; p. 757)

KEY 5

WOODY PLANTS

1a. Stems fleshy and succulent, cylindrical or flattened, true leaves absent; spines present at areoles; fruit a berry, often bearing spines; flowers solitary with numerous perianth parts and stamens...**CACTACEAE** (p. 255)
1b. True leaves present, plants with or without spines and thorns; fruit and flowers various...2

2a. Flowers in heads and surrounded by an involucre of bracts (phyllaries); pappus often present surrounding the petals of disk and/or ray flowers; stamens united by the anthers; ovary inferior...**ASTERACEAE** (p. 96)
2b. Flowers not in heads and surrounded by an involucre; pappus not present; stamens not united by the anthers; ovary superior or inferior...3

3a. Leaves in dense basal rosettes, often spine-tipped; flowers with 6 tepals, all similar and white, cream or yellowish...4
3b. Leaves opposite, whorled, or alternate, otherwise unlike the above; flowers various...5

4a. Flowers large, the tepals longer than 5 mm; leaves with a terminal spine, with fibrous margins; seeds flattened...**AGAVACEAE** (p. 64)
4b. Flowers small, the tepals shorter than 5 mm; leaves without a terminal spine, lacking fibrous margins; seeds globose...**RUSCACEAE** (*Nolina*; p. 740)

5a. Young branches and leaves densely pubescent with golden-brown or silver stellate or peltate hairs, especially below; fruit a drupe or berry-like...**ELAEAGNACEAE** (p. 340)
5b. Young branches and leaves without silvery or golden-brown peltate or stellate hairs, or if stellate hairs present these sparse and not densely covering the surface; fruit various...6

6a. Leaves small and scale-like, appressed to the stem; fruit a loculicidal capsule with hairy seeds...**TAMARICACEAE** (p. 764)
6b. Leaves larger and not appressed to the stem, otherwise unlike the above; fruit various...7

7a. Leaves 3–9 cleft from the base, the segments narrowly linear and spinulose; flowers cream to yellowish and often tinged purplish, the corolla salverform and united into a tube below, 12–25 mm long; fruit a loculicidal capsule with 3 valves...**POLEMONIACEAE** (*Linanthus pungens*; p. 677)
7b. Leaves and flowers unlike the above in all respects...8

8a. Leaves all opposite or whorled...9
8b. Leaves alternate (or mostly so)...31

9a. Leaves compound...10
9b. Leaves simple, sometimes lobed...14

10a. Flowers zygomorphic, with an orange-red, thick calyx and orange petals; vine...**BIGNONIACEAE** (p. 203)
10b. Flowers actinomorphic, unlike the above; tree, shrub, or woody vine...11

11a. Perianth in a single series with petaloid sepals (these white, yellow, purple, or bluish); fruit an aggregate of achenes with long, feathery styles; carpels numerous and distinct; stamens numerous; woody vine or shrub...**RANUNCULACEAE** (*Clematis*; p. 707)
11b. Perianth in two series with both sepals and petals present; fruit various but unlike the above; carpels united; stamens various; tree or shrub...12

12a. Flowers imperfect; stamens usually 8; leaves ternately compound or occasionally with 5–7 leaflets; fruit a schizocarp splitting into 2 samaras (maple)...**SAPINDACEAE** (*Acer*; p. 750)
12b. Flowers perfect; stamens 2–6; leaves pinnately compound; fruit a drupe or samara...13

13a. Flowers in a broad, terminal compound cyme, appearing with the leaves, petals white to yellowish; fruit a drupe...**ADOXACEAE** (*Sambucus*; p. 63)
13b. Flowers appearing before the leaves, petals absent; fruit a samara...**OLEACEAE** (*Fraxinus*; p. 548)

14a. Leaf margins entire...15
14b. Leaf margins toothed or lobed...26

15a. Leaves whorled, with 4–6 leaves per node; flowers small, to 4 mm wide, white to ochroleucous, imperfect; ovary inferior...**RUBIACEAE** (*Galium*; p. 737)
15b. Leaves opposite, sometimes fascicled; flowers generally larger, variously colored, perfect; ovary superior or inferior...16

16a. Flowers pink or rose-pink, campanulate with united petals; stamens 10; leaves evergreen, leathery, with a single prominent vein, the margins entire and revolute; low shrub...**ERICACEAE** (*Kalmia*; p. 343)
16b. Flowers white, greenish, yellow, purple, or if pink then unlike the above; stamens 4–6; leaves deciduous or if evergreen then unlike the above...17

17a. Flowers zygomorphic and bilabiate...18
17b. Flowers actinomorphic or if zygomorphic then not bilabiate...20

18a. Tall tree; fruit a long, cylindrical loculicidal capsule; flowers 3–5 cm long, white with 2 yellow stripes and purple spots within; leaves cordate to ovate with a long-acuminate apex, to 15 cm long...**BIGNONIACEAE** (p. 203)
18b. Low shrub; fruit a shorter capsule or schizocarp; flowers smaller, unlike the above; leaves linear to oblong or spatulate to oblanceolate, 0.5–3 cm long...19

19a. Stems 4-angled; leaves linear to oblong, densely white-tomentose; flowers purple to white; calyx with numerous longitudinal veins; stamens 2...**LAMIACEAE** (*Poliomintha*; p. 519)
19b. Stems round; leaves spatulate to oblanceolate, sparsely hairy or glabrous; flowers blue to blue-purple; calyx without longitudinal veins; stamens 4 with a 5th staminode...**PLANTAGINACEAE** (*Penstemon*; p. 583)

20a. Bases of leaf pairs connected by a thin line; leaves linear, crowded and fascicled in the axils of the primary leaves, with revolute margins; flowers white, with 6 stamens...**FRANKENIACEAE** (p. 486)
20b. Bases of leaf pairs not connected by a thin line; leaves various; flowers various, with 2, 4, 5, or many stamens...21

21a. Fruit a samara; flowers appearing before the leaves, petals absent; leaves ovate to oval, usually with rounded tips...**OLEACEAE** (*Fraxinus*; p. 548)
21b. Fruit a follicle, capsule, berry, or drupe; flowers appearing after the leaves, petals present; leaves various...22

22a. Petals united, at least below; flowers white, yellow, purple, or pink...23
22b. Petals distinct; flowers white...25

23a. Ovary inferior; flowers zygomorphic (usually rather weakly so), often paired in the leaf axils; fruit a white, black, red, or orange berry...**CAPRIFOLIACEAE** (*Lonicera*, *Symphoricarpos*; p. 265)
23b. Ovary superior; flowers actinomorphic, not paired in the leaf axils; fruit a follicle, berry, drupe, or loculicidal capsule...24

24a. Plants with milky sap; fruit a follicle with comose seeds; stamens 5...**APOCYNACEAE** (p. 90)
24b. Plants without milky sap; fruit a berry, drupe, or loculicidal capsule; stamens usually 2...**OLEACEAE** (*Forestiera*, *Ligustrum*, *Syringa*; p. 548)

25a. Fruit a drupe; flowers numerous in cymes; leaves mostly over 4 cm long, ovate to elliptic; older stems reddish...**CORNACEAE** (p. 301)
25b. Fruit a capsule; flowers in small cymes or solitary in leaf axils; leaves less than 2 cm long, linear or ovate to elliptic; older stems grayish and new growth often orangish...**HYDRANGEACEAE** (p. 497)

26a. Fruit a schizocarp splitting into 2 samaras; flowers imperfect; leaves palmately 3–5-lobed...**SAPINDACEAE** (*Acer*; p. 750)
26b. Fruit various but unlike the above; flowers perfect or imperfect; leaves undivided or if lobed then with 3 lobes or less...27

27a. Leaves evergreen and leathery, glossy above, with spinulose-serrate to crenate-serrate margins; flowers 1–3 in leaf axils, small and inconspiucous (petals only 1–2 mm long), 4-merous...**CELASTRACEAE** (p. 280)
27b. Leaves deciduous, not leathery or glossy, margins various; flowers larger or the petals absent, 4–5-merous...28

28a. Flowers appearing before the leaves, petals absent; ovary superior; leaves oblanceolate or elliptic, to 5.5 cm long and 2 cm wide; fruit a blue-black, elliptical drupe...**OLEACEAE** (*Forestiera*; p. 548)
28b. Flowers appearing with the leaves, petals present; ovary inferior or half-inferior; leaves various but often larger; fruit a capsule, berry, or drupe...29

29a. Flowers more or less zygomorphic (although weakly so), usually paired in the leaf axils or sometimes in very short spike-like racemes, the petals united (at least below); fruit a white berry...**CAPRIFOLIACEAE** (*Symphoricarpos*; p. 265)
29b. Flowers actinomorphic, in terminal or axillary cymes, the petals united or free; fruit a red or bluish-black drupe or a capsule...30

30a. Fruit a capsule; flowers with distinct petals; leaves densely grayish or whitish-hairy below, simple...**HYDRANGEACEAE** (*Jamesia*; p. 497)
30b. Fruit a red or bluish-black drupe; flowers with united petals; leaves glabrous or densely stellate-hairy below, simple or 3-lobed...**ADOXACEAE** (*Viburnum*; p. 63)

31a. Stems armed with spines, thorns, or prickles...32
31b. Stems unarmed, without spines, thorns, or prickles...43

32a. Leaves compound...33
32b. Leaves simple or merely lobed...34

33a. Leaflet margins entire; flowers zygomorphic or actinomorphic with many showy stamens; fruit a legume...**FABACEAE** (*Caragana*, *Gleditsia*, *Mimosa*, *Prosopis*, *Robinia*; p. 350)
33b. Leaflet margins serrate; flowers actinomorphic; fruit an aggregate of drupelets or achenes...**ROSACEAE** (*Rosa*, *Rubus*; p. 719)

34a. Leaf margins toothed or lobed...35
34b. Leaf margins entire...37

35a. Leaves palmately 5-lobed; fruit a berry with persistent petaloid sepals...**GROSSULARIACEAE** (p. 493)
35b. Leaves toothed or somewhat 3-lobed; fruit a pome, drupe, or a berry without a persistent perianth...36

36a. Flowers yellow, with 6 petals; fruit a berry; spines 1 or 3 per node, 1.5 cm or less in length; leaves with spine-tipped teeth...**BERBERIDACEAE** (p. 201)
36b. Flowers white, with 5 petals; fruit a drupe or a pome; spines 1 per node, 2.5 cm or more in length; leaves without spine-tipped teeth...**ROSACEAE** (*Crataegus*, *Prunus*; p. 719)

37a. Flowers zygomorphic, bright pink-purple with a yellow keel, 8–14 mm long, in racemes...**POLYGALACEAE** (*Polygala subspinosa*; p. 681)
37b. Flowers actinomorphic, variously colored but unlike the above...38

38a. Fruit large (8–13 cm in diam.), with milky juice, yellowish-green with a wrinkled rind, a multiple composed of individual druplets completely enclosed by an enlarged calyx; flowers small and inconspicuous, the petals absent; dioecious tree...**MORACEAE** (*Maclura*; p. 542)
38b. Fruit various but unlike the above, never milky; flowers small or conspicuous and showy; shrubs...39

39a. Flowers yellow...40
39b. Flowers white, purple, or greenish and inconspicuous...41

40a. Flowers with 6 tepals; fruit a red berry; spines 3 per node, 1.5 cm or less in length...**BERBERIDACEAE** (*Berberis fendleri*; p. 201)
40b. Flowers with 4 sepals; fruit an achene; spines unlike the above...**ROSACEAE** (*Coleogyne*; p. 722)

41a. Leaves with 3 veins from the base; flowers white, in corymbose clusters...**RHAMNACEAE** (p. 717)
41b. Leaves unlike the above; flowers purple, green, or yellowish...42

42a. Flowers numerous, small and greenish or yellowish, petals absent; fruit an utricle, sometimes enclosed in a fruiting bract, sometimes winged; plants often with scurfy, mealy hairs...**CHENOPODIACEAE** (*Atriplex*, *Grayia*, *Sarcobatus*; p. 281)
42b. Flowers conspicuous, solitary or in clusters of 2–4, petals present, pink, purple, or greenish to whitish; fruit a red berry; plants without scurfy, mealy hairs...**SOLANACEAE** (*Lycium*; p. 760)

43a. Leaves compound...44
43b. Leaves simple or merely lobed...53

44a. Leaves ternately or palmately compound...45
44b. Leaves pinnately compound...47

45a. Leaves palmately compound, usually with 5 leaflets; woody vine; fruit a berry...**VITACEAE** (*Parthanocissus*; p. 773)
45b. Leaves ternately compound with 3 leaflets; shrub; fruit a drupe or samara...46

46a. Leaflets gland-dotted and malodorous; flowers 4-merous, greenish-white; fruit a round samara...**RUTACEAE** (*Ptelea*; p. 740)
46b. Leaflets without gland-dots, not malodorous; flowers 5-merous, greenish; fruit a drupe (careful, poison ivy in this category!)...**ANACARDIACEAE** (*Rhus*, *Toxicodendron*; p. 71)

47a. Leaflets entire except for 1–5 rounded basal teeth, each tooth with a prominent gland at the tip; plant very malodorous; fruit a samara...**SIMAROUBACEAE** (p. 758)
47b. Leaflets toothed or entire, without gland tips; plants not malodorous; fruit various but not a samara...48

48a. Leaflets with toothed margins...49
48b. Leaflets with entire margins...51

49a. Leaflets with spine-tipped teeth, evergreen and leathery, pinnately compound; flowers yellow, with 6 tepals in 2 whorls; fruit a berry...**BERBERIDACEAE** (p. 201)
49b. Leaflets without spine-tipped teeth, plants otherwise unlike the above in all respects...50

50a. Leaflets very sharply serrate; flowers white, numerous in corymbose cymes, with a short hypanthium; fruit a red to orangish pome, drying purplish...**ROSACEAE** (*Sorbus*; p. 737)
50b. Leaflets serrate but not sharply so; flowers greenish, numerous in a large terminal panicle, without a hypanthium; fruit a red, dry, hairy, and sometimes viscid drupe...**ANACARDIACEAE** (*Rhus*; p. 71)

51a. Flowers zygomorphic, yellow, purple, or rose-purple; fruit a legume or loment...**FABACEAE** (*Amorpha*, *Colutea*, *Dalea formosa*; p. 350)
51b. Flowers actinomorphic, white or yellow; fruit a berry or achene...52

52a. Flowers yellow, the petals 6–14 mm long; leaflets silvery-hairy below and green and sparsely hairy above; fruit an achene...**ROSACEAE** (*Potentilla*; p. 726)
52b. Flowers white, the petals to ca. 3 mm long; leaflets glabrous; fruit a globose berry...**SAPINDACEAE** (*Sapindus*; p. 750)

53a. Woody vines with tendrils...**VITACEAE** (*Vitis*; p. 773)
53b. Shrubs or trees, or woody vines without tendrils...54

54a. Fruit an acorn (a nut with a hard, scaly involucre covering the base); flowers imperfect, staminate flowers in catkins and pistillate flowers solitary or in 3's; leaves pinnately lobed or spine-toothed, rarely entire and with a few leaves having some teeth, generally thick and somewhat leathery...**FAGACEAE** (p. 409)
54b. Fruit various but never an acorn; flowers perfect or imperfect, if in catkins then both the staminate and pistillate flowers in catkins; leaves entire or variously lobed...55

55a. Flowers imperfect, arranged in catkins...56
55b. Flowers perfect or imperfect, variously arranged but not in catkins...57

56a. Fruit a nut or a samara; leaves doubly or simply serrate, ovate to elliptic or nearly orbicular...**BETULACEAE** (p. 202)
56b. Fruit a capsule with comose seeds; leaves entire or serrate, linear, lanceolate, deltoid, or elliptic...**SALICACEAE** (p. 741)

57a. Leaves with 3 main veins arising from the base (this is especially evident on the underside of the leaf), ovate to elliptic; flowers white or cream-colored, in terminal or axillary corymbose cymes; fruit a capsule...**RHAMNACEAE** (*Ceanothus*; p. 717)
57b. Leaves pinnately veined or with only 1 main vein arising from the base, or if with 3 main veins then the leaves lobed; flowers and fruit various...58

58a. Leaves with unequal leaf bases; fruit a drupe or samara; flowers with only sepals, these greenish...59
58b. Leaves with equal leaf bases; fruit various but never a samara; flowers with both petals and sepals, or if only with sepals then these greenish or petaloid...60

59a. Fruit a flat samara; leaves doubly-serrate, cuneate at the base...**ULMACEAE** (p. 766)
59b. Fruit a fleshy drupe; leaves serrate or entire, usually cordate at the base...**CANNABACEAE** (*Celtis;* p. 263)

60a. Leaves toothed or lobed...61
60b. Leaves with entire margins...70

61a. Leaves palmately or pinnately lobed...62
61b. Leaves merely with toothed margins...67

62a. Leaves palmately lobed...63
62b. Leaves pinnately lobed or 3-lobed at the apex...64

63a. Older bark exfoliating; fruit a follicle or aggregate of drupelets (e.g., raspberry); ovary superior; stamens many (over 10)...**ROSACEAE** (*Physocarpus*, *Rubus*; p. 719)
63b. Older bark smooth; fruit a berry with a persistent perianth; ovary inferior; stamens 4 or 5...**GROSSULARIACEAE** (p. 493)

64a. Stamens connivent around the style, the anthers opening from terminal pores; flowers purple; at least some leaves with a pair of rounded lobes at the base; plants climbing or scrambling...**SOLANACEAE** (*Solanum dulcamara*; p. 763)
64b. Stamens not connivent around the style, anthers various; plants otherwise unlike the above in all respects...65

65a. Fruit a red, dry drupe; stamens 5; leaves pinnately lobed, margins crenate…**ANACARDIACEAE** (*Rhus*; p. 71)
65b. Fruit an achene or blackberry-like; stamens 4 or numerous; leaves unlike the above…66

66a. Fruit an achene, often with a persistent style; leaves pinnately lobed or 3-toothed at the apex, with entire margins; stamens numerous…**ROSACEAE** (*Purshia*, *Fallugia*; p. 719)
66b. Fruit blackberry-like; leaves pinnately lobed with toothed margins; stamens 4…**MORACEAE** (*Morus*; p. 542)

67a. Corolla small and brownish, with a disk on which sits the calyx, this deciduous and leaving just the disk behind; fruit a berrylike drupe…**RHAMNACEAE** (*Rhamnus*; p. 717)
67b. Corolla white, cream, or pink, without a disk; fruit various…68

68a. Petals absent, perianth composed of petaloid sepals; fruit blackberry-like; stamens 4…**MORACEAE** (p. 542)
68b. Petals present; fruit not blackberry-like; stamens 8, 10, or numerous…69

69a. Petals united, the corolla urceolate or campanulate; stamens 8 or 10; fruit a berry or capsule enclosed by the fleshy calyx…**ERICACEAE** (*Arctostaphylos*, *Gaultheria*, *Vaccinium*; p. 341)
69b. Petals free; stamens many; fruit a drupe, pome, or achene…**ROSACEAE** (*Amelanchier*, *Cercocarpus*, *Holodiscus*, *Malus*, *Peraphyllum*, *Prunus*; p. 719)

70a. Petals united at least at the base; fruit a septicidal capsule or berry…71
70b. Petals free or absent; fruit an utricle, achene, or follicle…72

71a. Corolla urceolate or campanulate, pink or white; stamens 8 or 10; fruit a capsule or berrylike…**ERICACEAE** (*Arctostaphylos*, *Rhododendron*; p. 341)
71b. Corolla funnelform, pink, purple, greenish, or whitish; stamens 5; fruit a berry…**SOLANACEAE** (p. 759)

72a. Petals absent, the sepals greenish or yellowish (or sometimes absent in pistillate flowers); flowers small and inconspicuous, numerous; stamens usually 5; fruit an utricle, sometimes enclosed in a fruiting bract, sometimes winged; leaves tomentose with stellate hairs, villous, scurfy with mealy hairs, or glabrous…**CHENOPODIACEAE** (*Atriplex*, *Grayia*, *Kochia*, *Krascheninnikovia, Suaeda*; p. 281)
72b. Petals present or if absent then the sepals petaloid; stamens 4–many; fruit a capsule, follicle, or achene; leaves glabrous or hairy, but never scurfy or stellate-hairy…73

73a. Perianth of 6 petaloid tepals (white or yellow), in two whorls of three; flowers 6-numerous from a tubular involucre of fused bracts; fruit a 3-angled or -winged achene…**POLYGONACEAE** (*Eriogonum*; p. 682)
73b. Petals and sepals present; flowers not subtended by an inovlucre; fruit an achene with a long persistent style or a follicle…74

74a. Fruit achenes with long, perisistent styles or 5 follicles; hypanthium present; leaves with revolute margins or if flat then mostly basal and the plant mat-forming…**ROSACEAE** (*Cercocarpus*, *Petrophytum*; p. 719)
74b. Fruit a concavo-convex follicle with numerous longitudinal lines; flowers without a hypanthium, solitary and axillary, small and inconspicuous, the petals to 7 mm long; leaves narrowly elliptic to oblanceolate, ca. 1 cm long, to 4 mm wide, without revolute margins; plants not mat-forming…**CROSSOSOMATACEAE** (p. 302)

KEY 6

VINES

1a. Plants parasitic, twining on the aerial stems of various herbs, yellowish and lacking chlorophyll, the stems thin and thread-like; leaves minute and reduced to scales…**CONVOLVULACEAE** (*Cuscuta*; p. 299)
1b. Plants green, not parasitic; leaves usually conspicuous and present…2

2a. Leaves opposite…3
2b. Leaves alternate…6

3a. Leaves entire, saggitate to hastate; flowers greenish to brown, in umbels…**APOCYNACEAE** (*Sarcostemma*; p. 94)
3b. Leaves palmately lobed or pinnately compound to dissected; flowers unlike the above…4

4a. Leaves palmately 3–5-lobed with toothed margins; flowers imperfect, the pistillate flowers subtended by papery bracts and aggregated into a cone-like structure…**CANNABACEAE** (*Humulus*; p. 263)
4b. Leaves pinnately compound or dissected, the leaflets toothed or entire; flowers perfect or imperfect, not aggregated into cone-like structures…5

5a. Flowers zygomorphic; sepals orange-red and thick; petals orange to orange-red, united, tubular; stamens 5, these markedly unequal; fruit a loculicidal capsule...**BIGNONIACEAE** (*Campsis*; p. 203)
5b. Flowers actinomorphic; sepals purple, bluish, white, or yellow, free; petals absent; stamens numerous; fruit an aggregate of achenes with long, feathery styles...**RANUNCULACEAE** (*Clematis*; p. 707)

6a. Leaves palmately, pinnately, or ternately compound...7
6b. Leaves simple, sometimes lobed (careful, leaves deeply 3-lobed in *Solanum*, but not truly compound)...10

7a. Leaves palmately compound...8
7b. Leaves pinnately or ternately compound...9

8a. Tendrils with adhesive cups; fruit a berry...**VITACEAE** (*Parthenocissus*; p. 773)
8b. Tendrils without adhesive cups; fruit an ovoid pepo with smooth spines...**CUCURBITACEAE** (*Cyclanthera*; p. 303)

9a. Flowers zygomorphic with a banner, wing, and keel, variously colored but not greenish; leaves stipulate, pinnately compound or ternately compound with the terminal leaflet modified into a tendril, the leaflets not shiny; stamens 10, usually diadelphous; fruit a legume or a loment...**FABACEAE** (p. 350)
9b. Flowers actinomorphic, greenish; leaves exstipulate, ternately compound with shiny leaflets; stamens 5, alternate with the petals, equal in size; fruit a white or cream-colored drupe...**ANACARDIACEAE** (*Toxicodendron*; p. 71)

10a. Tendrils present...11
10b. Tendrils absent...13

11a. Flowers 3-merous, in an umbel; leaves simple and entire, palmately parallel-veined with the veins curving inward...**SMILACACEAE** (p. 759)
11b. Flowers 5-merous, the inflorescence solitary to racemose or cymose; leaves simple or palmately lobed with toothed margins, with netted venation...12

12a. Vines woody; ovary superior; fruit a berry...**VITACEAE** (*Vitis*; p. 773)
12b. Vines herbaceous; ovary inferior; fruit a pepo...**CUCURBITACEAE** (p. 303)

13a. Stems with a sheathing, membranous stipule (ocrea) at the petiole base; perianth of tepals, all parts similar, petaloid, greenish or whitish...**POLYGONACEAE** (*Fallopia*, *Polygonum*; p. 681)
13b. Stipules absent; perianth in 2 distinct series (sepals and petals), the petals purple, white, pink, or red...14

14a. Flowers purple; fruit a berry; plants climbing but not twining...**SOLANACEAE** (*Solanum dulcamara*; p. 763)
14b. Flowers white, pink, or red to reddish-orange; fruit a capsule; plants twining and climbing on other plants...**CONVOLVULACEAE** (p. 298)

KEY 7

PERIANTH ABSENT OR OF A SINGLE WHORL

1a. Plants true aquatics, floating or completely submerged in water...see **KEY 4**
1b. Plants terrestrial (sometimes growing in shallow water, but not true aquatics)...2

2a. Plants green or yellowish, parasitic and attached to the bark of gymnosperms (*Pinus*, *Pseudotsuga*, *Abies*, and *Juniperus*); stems swollen-jointed and brittle when dry; fruit a berry; ovary inferior...**VISCACEAE** (p. 772)
2b. Plants not parasitic on gymnosperms...3

3a. Woody trees or shrubs...4
3b. Perennial or annual herbs, sometimes woody at the base but not woody throughout...13

4a. Young branches and leaves with golden-brown or silver stellate or peltate hairs, especially below; fruit a drupe or berry-like...**ELAEAGNACEAE** (p. 339)
4b. Young branches and leaves without silvery or golden-brown peltate hairs; fruit various...5

5a. Leaves whorled, usually with 4 leaves per node; flowers small, to 4 mm wide, white to ochroleucous, imperfect; ovary inferior...**RUBIACEAE** (*Galium*; p. 737)
5b. Leaves alternate or opposite, sometimes fascicled; flowers various...6

6a. Fruit an acorn (a nut with a hard, scaly involucre covering the base); flowers imperfect, staminate flowers in catkins and pistillate flowers solitary or in 3's; leaves pinnately lobed or spine-toothed, rarely entire and with a few leaves having some teeth, generally thick and somewhat leathery...**FAGACEAE** (p. 409)
6b. Fruit various but never an acorn; flowers perfect or imperfect, if in catkins then both the staminate and pistillate flowers in catkins; leaves entire or variously lobed...7

7a. Flowers imperfect, arranged in catkins (these sometimes small)...8
7b. Flowers perfect or imperfect, variously arranged but not in catkins...9

8a. Fruit a nut or a samara; leaves doubly or simply serrate, ovate to elliptic or nearly orbicular...**BETULACEAE** (p. 202)
8b. Fruit a capsule with comose seeds; leaves entire or serrate, linear, lanceolate, deltoid, or elliptic...**SALICACEAE** (p. 741)

9a. Leaves with unequal leaf bases; fruit a drupe or samara...10
9b. Leaves with equal leaf bases; fruit various but never a samara or drupe...11

10a. Fruit a flat samara; leaves doubly-serrate, cuneate at the base...**ULMACEAE** (p. 766)
10b. Fruit a fleshy drupe; leaves serrate or entire, usually cordate at the base...**CANNABACEAE** (*Celtis;* p. 263)

11a. Fruit an achene; stamens numerous...**ROSACEAE** (*Cercocarpus*, *Coleogyne*; p. 719)
11b. Fruit an utricle or multiple of drupelets; stamens 4 or 5...12

12a. Flowers with greenish or yellowish inconspicuous sepals, or the sepals sometimes absent in pistillate flowers; stamens usually 5; fruit an utricle, sometimes enclosed in a fruiting bract, sometimes winged; leaves tomentose with stellate hairs, scurfy, villous, or glabrous...**CHENOPODIACEAE** (*Atriplex*, *Grayia*, *Kochia*, *Krascheninnikovia, Suaeda*; p. 281)
12b. Flowers with petaloid sepals; stamens 4; fruit a multiple of drupelets or large and globose with a wrinkled rind...**MORACEAE** (p. 542)

13a. Leaves scale like and rather inconspicuous, opposite, fused at the base; stems jointed and fleshy, often reddish; flowers with 3 perianth parts, sunken in terminal spikes...**CHENOPODIACEAE** (*Salicornia*; p. 295)
13b. Plants unlike the above in all respects...14

14a. Perianth absent, each flower subtended by a white petaloid bract; flowers in a terminal conic spike, the entire structure subtended by large white or pinkish petal-like bracts (giving the appearance of a single flower); leaves mostly basal, long-petiolate...**SAURURACEAE** (p. 751)
14b. Sepals present, sepaloid or petaloid, sometimes absent and then the ovary 3-carpellate; flowers unlike the above; leaves various...15

15a. Leaves whorled; flowers small, white; plants sometimes prostrate...16
15b. Leaves alternate, opposite, or basal; flowers and plants various...17

16a. Ovary superior; fruit a 3-valved capsule; stems round...**MOLLUGINACEAE** (p. 539)
16b. Ovary inferior; fruit 2 globose mericarps; stems usually 4-angled...**RUBIACEAE** (*Galium*; p. 737)

17a. Plants with stinging hairs...18
17b. Plants without stinging hairs...19

18a. Leaves opposite; ovary unicarpellate...**URTICACEAE** (*Urtica*; p. 767)
18b. Leaves alternate; ovary 3-carpellate...**EUPHORBIACEAE** (*Tragia*; p. 350)

19a. True sepals absent; flowers imperfect, greatly reduced, grouped together in a cyathium which is often subtended by petaloid appendages, the 2–15 staminate flowers consisting of a single stamen, the single pistillate flower consisting of a single stipitate pistil with a 3-carpellate ovary; plants with milky sap...**EUPHORBIACEAE** (*Chamaesyce*, *Euphorbia*; p. 345)
19b. Sepals present (although sometimes small and inconspicuous), these green or petaloid; flowers perfect or imperfect, not in a cyathium; plants with or without milky sap...20

20a. Filaments purple; sepals inconspicuous; corolla absent; flowers in a dense, terminal, villous spike; leaves ovate with crenate to serrate margins, the basal leaves long-petiolate...**PLANTAGINACEAE** (*Besseya wyomingensis*; p. 581)
20b. Plants unlike the above in all respects...21

21a. Flowers several to many, in heads, each flower cluster subtended or surrounded by an involucre of bracts that are separate or fused (often the entire inflorescence giving the appearance of a single flower)...22
21b. Flowers not sessile in heads and subtended by an involucre of bracts...24

22a. Ovary superior (but appearing inferior by the constriction of the perianth tube); flowers white, pinkish, or rose-colored, distinctly separate and not densely packed nor tightly surrounded by the involucre; fruit an anthocarp (achene or utricle enclosed in the base of the calyx)...**NYCTAGINACEAE** (p. 544)
22b. Ovary inferior; flowers various, densely packed and tightly surrounded or subtended by the involucre; fruit various but unlike the above...23

23a. Stamens 4, distinct; corolla 4-lobed; leaves opposite...**DIPSACACEAE** (p. 338)
23b. Stamens 5, united by the anthers; corolla 3- or 5-lobed; leaves alternate, opposite, or basal...**ASTERACEAE** (p. 96)

24a. Plants fernlike with many finely dissected branches, true leaves reduced to scales; fruit a red berry...**ASPARAGACEAE** (p. 96)
24b. Plants not fernlike, otherwise unlike the above in all respects...25

25a. Leaves with parallel venation, linear to linear-lanceolate and grass-like...26
25b. Leaves with pinnate or palmate venation, with just a midvein present, or reduced to scales, not grass-like...32

26a. Inflorescence a cylindrical spadix (numerous flowers on a thickened, fleshy axis), appearing lateral because of the continuing leaf-like spathe; found in moist places (3500–5100 ft)...**ACORACEAE** (p. 63)
26b. Inflorescence not in a spadix; found in moist or dry places (elevation various)...27

27a. Inflorescence a terminal cylindric spike, with the looser staminate flowers above the dense pistillate flowers; plants tall, to ca. 3 m, from stout rhizomes...**TYPHACEAE** (*Typha*; p. 766)
27b. Inflorescence not a terminal cylindric spike, flowers perfect or imperfect; plants unlike the above in all additional characteristics...28

28a. Flowers in dense globose heads, the staminate flowers above the pistillate flowers; fruit a hardened, drupaceous achene with persistent tepals; plants generally found in standing water...**TYPHACEAE** (*Sparganium*; p. 766)
28b. Flowers not in dense globose heads, perfect or imperfect; fruit a caryopsis, capsule, achene, or schizocarp; plants found in water or terrestrial...29

29a. Flowers enclosed in chaffy bracts and aggregated into one- to many-flowered spikelets or spikes; perianth absent or reduced to bristles, scales, or hairs; fruit an achene or caryopsis...30
29b. Flowers not enclosed in chaffy bracts or scales, in narrow spike-like racemes, headlike clusters, or open and cymose; perianth of minute tepals, or the tepals chaffy and scale-like; fruit a capsule or schizocarp...31

30a. Leaves 3-ranked, usually with closed sheaths; stems usually triangular in cross-section; internodes usually solid (a few species hollow with round stems and then the perianth reduced to bristles or scales); flowers subtended by a single bract in staminate flowers, subtended by 2 bracts in pistillate flowers (including the perigynium which encloses the flower)...**CYPERACEAE** (p. 303)
30b. Leaves 2-ranked, usually with open sheaths; stems round or flattened in cross-section; internodes usually hollow; perianth absent; flowers (florets) usually subtended by 2 bracts, the lemma and palea, the florets subtended by 2 glumes, these sometimes reduced...**POACEAE** (p. 600)

31a. Flowers in headlike clusters or open and cymose; perianth scale-like, of white, green, brown, or purplish-black tepals; fruit a loculicidal capsule...**JUNCACEAE** (p. 506)
31b. Flowers in narrow spike-like racemes; perianth of minute, greenish tepals; fruit a schizocarp with 3 or 6 mericarps...**JUNCAGINACEAE** (p. 512)

32a. Perianth petaloid, resembling petals in color and texture (variously colored, sometimes green or yellowish-green, but not leaf-like in texture)...33
32b. Perianth sepaloid, resembling sepals in color and texture (green and leaf-like in texture)...49

33a. Flowers zygomorphic...34
33b. Flowers actinomorphic...35

34a. Carpels united; perianth of 4 petals, in 2 whorls of 2, the outer with a prominent spur or pouch at the base, the inner alike and apically connate over the stigmas and clawed, the 2 sepals small, bractlike and inconspicuous or early-deciduous; fruit a 2-valved capsule...**FUMARIACEAE** (p. 486)
34b. Carpels distinct; perianth of 5 petaloid sepals, the upper one spurred or helmetlike, the 2–4 petals small and enclosed in the petaloid sepals; fruit an aggregate of follicles...**RANUNCULACEAE** (*Aconitum*, *Delphinium*; p. 704)

35a. Ovary half-inferior; flowers greenish and inconspicuous (ca. 3 mm wide); leaves oval to reniform with crenate margins; carpels 2; found along streams and inlets...**SAXIFRAGACEAE** (*Chrysosplenium*; p. 752)
35b. Ovary superior or wholly inferior; plants otherwise unlike the above in all respects...36

36a. Leaves opposite below and alternate above, sessile and succulent; flowers small (3–5 mm long), sessile in the leaf axils, white or pinkish; fruit a subglobose capsule; moist places...**MYRSINACEAE** (*Glaux*; p. 542)
36b. Leaves opposite, alternate, or basal, not succulent; flowers and fruit various; various habitats...37

37a. Leaves opposite...38
37b. Leaves alternate or all basal...42

38a. Flowers solitary in the leaf axils; sepals pink to rose, with a prolonged appendage (hood) at the tip; plants succulent and prostrate; fruit a pyxis...**AIZOACEAE** (p. 66)
38b. Inflorescence various but the flowers not solitary in leaf axils; sepals white, yellow, greenish-yellow, purple, or bluish, without an appendage at the tip; plants not succulent, prostrate or erect; fruit various but not a pyxis...39

39a. Ovary inferior (or appearing inferior by the constriction of the perianth tube)...40
39b. Ovary superior...41

40a. Ovary inferior; carpels 3; flowers in bracteate cymes and corymbs (but usually with no more then 2 bracts subtending the flower clusters); petals white, basally saccate; sepals inrolled and inconspicuous in flower but elongating into a plumose pappus in fruit; fruit an achene; leaves mostly basal with only a few stem leaves present, simple to pinnatifid...**VALERIANACEAE** (p. 767)
40b. Ovary superior but appearing inferior by the constriction of the perianth tube, unicarpellate; flowers white, pinkish, or rose, in clusters subtended by 4–6 involucral bracts; fruit an anthocarp (achene or utricle enclosed in the base of the calyx); leaves various...**NYCTAGINACEAE** (p. 544)

41a. Carpels united; leaves with papery stipules; fruit a capsule; sepals greenish to yellow, awn-tipped; perennial herb...**CARYOPHYLLACEAE** (*Paronychia*; p. 273)
41b. Carpels distinct; leaves exstipulate; fruit an aggregate of achenes with long feathery styles; sepals yellow, purple, white, or bluish, not awn-tipped; perennial herb or vine...**RANUNCULACEAE** (*Clematis*; p. 707)

42a. Ovary inferior...43
42b. Ovary superior...44

43a. Flowers in corymbose clusters, white or pinkish; leaves without a sheathing base, simple and entire; ovary with free-central placentation; fruit a dry drupe with persistent sepals...**SANTALACEAE** (p. 750)
43b. Flowers in compound umbels, variously colored; leaves with a sheathing base, usually compound; ovary with axile placentation; fruit a schizocarp splitting into 2 mericarps, these often suspended on a carpophore...**APIACEAE** (p. 72)

44a. Flowers imperfect (the lower staminate and the upper pistillate), with a hypanthium, in a dense terminal spike; sepals green or pinkish; leaves basal and alternate, pinnately compound, stipulate; hypanthium becoming hardened and woody in fruit...**ROSACEAE** (*Sanguisorba*; p. 737)
44b. Flowers perfect or imperfect, without a hypanthium; sepals variously colored; leaves various but unlike the above in all respects...45

45a. Carpels distinct; stamens many; leaves often with sheathing petioles; fruit an aggregate of follicles or achenes, these often with long styles...**RANUNCULACEAE** (p. 704)
45b. Carpels united; stamens 6–many; leaves without sheathing petioles; fruit a capsule or achene without long styles...46

46a. Leaves linear and almost round in cross-section, somewhat succulent, mostly basal; fruit a 3-valved capsule...**MONTIACEAE** (*Phemeranthus*; p. 541)
46b. Leaves unlike the above, not succulent; fruit a dry drupe, achene, or capsule...47

47a. Stamens 5; flowers in corymbose clusters, white or pinkish; fruit a dry drupe with persistent sepals; plants with deep rhizomes...**SANTALACEAE** (p. 750)
47b. Stamens more than 5; plants otherwise unlike the above in all respects...48

48a. Plants without colored or milky sap; stamens 6–9; stems often with a membranous stipule (ocrea) sheathing at the petiole base; fruit an achene, often enclosed by persistent tepals...**POLYGONACEAE** (p. 681)
48b. Plants usually with milky or colored sap, rarely only with watery sap; stamens 12–many; stems without an ocrea; fruit a capsule...**PAPAVERACEAE** (p. 573)

49a. Leaves palmately compound or palmately lobed; flowers imperfect, the pistillate flowers subtended by a bract; plants glandular-hairy...**CANNABACEAE** (p. 263)
49b. Leaves various but not palmately compound; flowers perfect or imperfect, the pistillate flowers not subtended by a bract; plants usually not glandular-hairy...50

50a. Fruit an elliptic to oval silicle, 2.2–3.5 mm long; stamens 6, tetradynamous; sepals 4...**BRASSICACEAE** (*Lepidium*; p. 241)
50b. Fruit various but not a silicle; stamens not tetradynamous; sepals 4–5...51

51a. Carpels numerous and distinct; stamens numerous, often showy; fruit an aggregate of achenes...**RANUNCULACEAE** (*Myosurus*, *Thalictrum*, *Trautvetteria*; p. 704)
51b. Carpels united or the ovary unicarpellate; fruit various but not an aggregate of achenes...52

52a. Flowers loosely arranged in cymes or sometimes solitary; fruit a capsule; sepals 5; leaves opposite and entire, the nodes at least somewhat swollen...**CARYOPHYLLACEAE** (*Cerastium*, *Stellaria*; p. 266)
52b. Flowers numerous in bracteate clusters or terminal panicles; fruit various but not a capsule; sepals 2–5; leaves alternate or opposite, the nodes usually not conspicuously swollen...53

53a. Ovary unicarpellate; sepals 4; flowers imperfect, in bracteate axillary clusters...**URTICACEAE** (p. 767)
53b. Ovary 2–3-carpellate; sepals 2–5; flowers perfect or imperfect...54

54a. Ovary 3-carpellate; fruit a 3-celled schizocarp or capsule, usually with 3 styles remaining; plants with milky sap, lacking scurfy, mealy hairs...**EUPHORBIACEAE** (p. 345)
54b. Ovary 2-carpellate or rarely 3-carpellate; fruit an achene or utricle; plants lacking milky sap, often with scurfy, mealy hairs...55

55a. Perianth densely white-hairy, lacking hooded appendages; leaves mostly basal with a few stem leaves opposite; plants erect, densely white- to grayish-canescent to tomentose, found on the eastern plains...**AMARANTHACEAE** (*Froelichia*; p. 69)
55b. Perianth glabrous or if densely hairy then with hooked appendages from the back at maturity and the hairs usually tawny; plants otherwise unlike the above, variously distributed...56

56a. Leaves opposite...57
56b. Leaves alternate...58

57a. Flowers not enclosed within a pair of foliaceous bracteoles; prostrate herbs, lacking scurfy (mealy) hairs...**AMARANTHACEAE** (*Guilleminea*, *Tidestromia*; p. 69)
57b. Pistillate flowers lacking a perianth and enclosed by 2 partly or completely united fruiting bracts, these enlarging in fruit and becoming variously thickened and/or appendaged; shrubs or erect herbs, often with scurfy (mealy) hairs...**CHENOPODIACEAE** (*Atriplex*; p. 281)

58a. Sepals dry, papery, and scarious, with a thin, green midvein; leaves ovate to elliptic, with prominent, pinnate venation below, lacking mealy or scurfy hairs...**AMARANTHACEAE** (*Amaranthus*; p. 69)
58b. Sepals often fleshy, otherwise unlike the above; leaves various but unlike the above, often with mealy or scurfy hairs...**CHENOPODIACEAE** (p. 281)

KEY 8

PERIANTH IN 2 WHORLS; FLOWERS 3-MEROUS

1a. Plants aquatic, floating or completely submerged in water...See **KEY 4**
1b. Plants terrestrial...2

2a. Flowers imperfect, greenish-white, in an axillary globose umbel; vine or low herb with tendrils at the nodes; leaves ovate to oval with a cordate or rounded base and acuminate tip; fruit a berry...**SMILACACEAE** (p. 759)
2b. Flowers perfect or rarely imperfect, variously arranged or if in an umbel then subtended by papery bracts; herb or shrub without tendrils; leaves and fruit various...3

3a. Leaves in dense basal rosettes, spine-tipped, the margins with shedding fibers; flowers with 6 tepals, all similar and white, cream, or yellowish; fruit a loculicidal capsule with flat, black seeds...**AGAVACEAE** (*Yucca*; p. 64)
3b. Leaves opposite, whorled, alternate, or basal but not spine-tipped, otherwise unlike the above; flowers and fruit various...4

4a. Flowers strongly zygomorphic...5
4b. Flowers actinomorphic or weakly zygomorphic...6

5a. Ovary inferior; stamens united with the gynoecium to form a column; flowers variously colored...**ORCHIDACEAE** (p. 560)
5b. Ovary superior; stamens 5, not united with the gynoecium; flowers orange with reddish spots...**BALSAMINACEAE** (p. 200)

6a. Ovary inferior...7
6b. Ovary superior...8

7a. Leaves equitant; tepals purple, blue-purple, or rarely white or yellow, in similar or dissimilar whorls, not hairy along their undersides; stamens 3; plants from rhizomes or fibrous roots...**IRIDACEAE** (p. 504)
7b. Leaves not equitant; tepals yellow, all similar in appearance, the outer ones usually hairy along their undersides; stamens 6; plants from small corms...**HYPOXIDACEAE** (p. 504)

8a. Tepals all similar in color...9
8b. Tepals differentiated into 2 series (with 3 petaloid and colorful and 3 sepaloid and green)...19

9a. Shrubs with 1 or 3 spines per node or the leaves pinnately compound with stout spines on the margins; tepals yellow; fruit a berry...**BERBERIDACEAE** (p. 201)
9b. Herbaceous plants, or if shrubs then lacking spines; tepals and fruit various...10

10a. Flowers in an umbel subtended by papery bracts...11
10b. Flowers variously arranged but not in an umbel subtended by papery bracts...12

11a. Tepals distinct; flowers all approximately the same height in the umbel; plants from fibrous-coated or membranous bulbs, onion-scented...**ALLIACEAE** (p. 67)
11b. Tepals connate basally into a funnelform tube; at least a few flowers on elongated pedicels surpassing the other flowers in the umbel; plants from fibrous-coated corms, not onion-scented...**THEMIDACEAE** (p. 764)

12a. Plants 1.5–2 m tall; flowers white, greenish, or cream-colored, numerous in a densely flowered terminal panicle 2–8 dm long; fruit a capsule...**MELANTHIACEAE** (*Veratrum*; p. 537)
12b. Plants unlike the above in all respects...13

13a. Plants from deeply buried, fleshy roots, acaulescent; flowers clustered at ground level; tepals united below and forming a long, slender tube; ovary and fruit subterranean...**AGAVACEAE** (*Leucocrinum*; p. 64)
13b. Plants not from fleshy roots, caulescent or acaulescent; flowers well above-ground level; tepals unlike the above; ovary and fruit above-ground...14

14a. Seeds bearing pits in longitudinal rows; leaves opposite although sometimes crowded above and appearing whorled, elliptic to oblong and serrulate, or linear and entire and succulent; small (to ca. 3 cm tall), annual prostrate plants found in shallow water and on muddy shores...**ELATINACEAE** (p. 340)
14b. Seeds not bearing pits in longitudinal rows; plants otherwise unlike the above in all respects...15

15a. Flowers subtended by a tubular involucre of fused bracts; leaves with pinnate venation or a central midvein...**POLYGONACEAE** (p. 681)
15b. Flowers not subtended by an involucre of fused bracts; leaves with parallel venation or grass-like...16

16a. Flowers solitary or in a terminal cyme; tepals over 5 mm long...**LILIACEAE** (p. 522)
16b. Flowers in a terminal panicle or raceme; tepals 0.5–11 mm long...17

17a. Shrubby plants; leaves all basal, numerous, narrowly linear; known from Las Animas County...**RUSCACEAE** (*Nolina*; p. 740)
17b. Herbaceous; leaves unlike the above; variously distributed...18

18a. Stem leaves evident and similar in size (not greatly reduced upward), lanceolate to oblong-lanceolate (20–40 mm wide); plants from fleshy rhizomes...**RUSCACEAE** (*Maianthemum*; p. 740)
18b. Leaves mostly basal, those of the stem reduced upward, linear (3–18 mm wide); plants from bulbs...**MELANTHIACEAE** (*Zigadenus*; p. 537)

19a. Leaves ternately to pinnately compound; petals white, much smaller than the sepals...**LIMNANTHACEAE** (p. 524)
19b. Leaves not compound; petals various...20

20a. Carpels distinct, in a ring or densely crowded on a globose receptacle; plants generally in moist places or in standing water; petals white...**ALISMATACEAE** (p. 66)
20b. Carpels united; plants usually not in standing water; petals various...21

21a. Petals all purple or blue, or 2 blue and 1 white, less than 2 cm long; inflorescence subtended by a conspicuous spathe or 2 leaf-like bracts...**COMMELINACEAE** (p. 297)
21b. Petals white, pink, or purplish-tinged, over 2 cm long; inflorescence not subtended by a spathe or leaf-like bracts...22

22a. Plants with an elongated scape and 3 leaf-like, ovate, whorled bracts (true leaves absent); petals lacking a bearded or hairy gland at the base...**MELANTHIACEAE** (*Trillium*; p. 537)
22b. Plants unlike the above, with lanceolate to linear, grass-like, true leaves; petals with a prominent, bearded or hairy gland at the base...**LILIACEAE** (*Calochortus*; p. 522)

KEY 9

FLOWERS 4- OR 5-MEROUS; PETALS FREE

1a. Plants capturing insects with sticky glandular hairs on the leaves, when triggered causing the slow enclosure of the prey by the folding of the blade; leaves simple, often coiled in bud; flowers actinomorphic, white or pink; plants of swamps and bogs...**DROSERACEAE** (p. 339)
1b. Plants not insectivorous...2

2a. Plants parasitic or saprophytic, lacking chlorophyll, the stems yellowish to white but drying black...**ERICACEAE** (*Monotropa*; p. 343)
2b. Plants with green stems and/or leaves, not parasitic...3

3a. Stems fleshy and succulent, cylindrical or flattened, true leaves absent; spines present at areoles; fruit a berry, often bearing spines; flowers solitary with numerous perianth parts and stamens...**CACTACEAE** (p. 255)
3b. True leaves present (although sometimes reduced and scale-like); stems generally not fleshy and succulent; plants with or without spines and thorns; fruit and flowers various...4

4a. Aquatic plants, either completely submerged or floating on water...5
4b. Terrestrial plants...8

5a. Leaves lobed or finely dissected...6
5b. Leaves simple and undivided, the margins entire...7

6a. Flowers conspicuous, perfect, yellow or white; leaves alternate, palmately divided or ternately lobed; carpels numerous and distinct...**RANUNCULACEAE** (*Ranunculus*; p. 711)
6b. Flowers inconspicuous, imperfect, greenish; leaves whorled, pinnately dissected; carpels 4 and united...**HALORAGACEAE** (p. 496)

7a. Leaves orbicular to ovate with a deeply cordate base, thick and leathery, long-petiolate, large (10–40 cm) and floating; fruit a leathery capsule; flowers solitary on long pedicels, large (5–10 cm in diam.) with numerous stamens, intergrading from sepals to petals, yellow or white...**NYMPHAEACEAE** (p. 548)
7b. Leaves terete and linear, in a densely packed basal rosette; fruit a silicle; stamens 6, tetradynamous; flowers perfect, white...**BRASSICACEAE** (*Subularia*; p. 254)

8a. Woody trees, shrubs, or vines...9
8b. Perennials or annuals, sometimes woody at the base but not woody throughout...33

9a. Leaves small and scale-like, appressed to the stem; fruit a loculicidal capsule with hairy seeds...**TAMARICACEAE** (p. 764)
9b. Leaves larger and not appressed to the stem, otherwise unlike the above; fruit various...10

10a. Leaves opposite...11
10b. Leaves alternate...16

11a. Leaf margins entire...12
11b. Leaf margins toothed or lobed...14

12a. Bases of leaf pairs connected by a thin line; leaves linear, crowded and fascicled in the axils of the primary leaves, with revolute margins; flowers white, with 6 stamens...**FRANKENIACEAE** (p. 486)
12b. Bases of leaf pairs not connected by a thin line; leaves various; flowers various, with 4–many stamens...13

13a. Fruit a drupe; flowers numerous in cymes; leaves mostly over 4 cm long, ovate to elliptic; older stems reddish...**CORNACEAE** (p. 301)
13b. Fruit a capsule; flowers in small cymes or solitary in leaf axils; leaves less than 2 cm long, linear or ovate to elliptic; older stems grayish and new growth often orangish...**HYDRANGEACEAE** (p. 497)

14a. Fruit a schizocarp splitting into 2 samaras; flowers imperfect; leaves palmately lobed or compound, usually with 5 lobes but occasionally with only 3 lobes...**SAPINDACEAE** (*Acer*; p. 750)
14b. Fruit various but unlike the above; flowers perfect or imperfect; leaves undivided or if lobed then with 3 lobes or less...15

15a. Leaves evergreen and leathery, glossy above, with spinulose-serrate to crenate-serrate margins; flowers 1–3 in leaf axils, small and inconspiucous (petals only 1–2 mm long), 4-merous...**CELASTRACEAE** (p. 280)
15b. Leaves deciduous, not leathery or glossy, densely grayish or whitish-hairy below, the margins toothed; flowers larger, 4–5-merous...**HYDRANGEACEAE** (*Jamesia*; p. 498)

16a. Stems armed with spines, thorns, or prickles...17
16b. Stems unarmed, without spines, thorns, or prickles...21

17a. Leaves compound...18
17b. Leaves simple, the margins entire, toothed, or lobed...19

18a. Leaflet margins entire; flowers with many showy stamens; fruit a legume...**FABACEAE** (p. 350)
18b. Leaflet margins serrate; flowers actinomorphic, with many stamens but also with showy petals; fruit an aggregate of drupelets (e.g., raspberry) or achenes (e.g., rose hip)...**ROSACEAE** (*Rosa*, *Rubus*; p. 719)

19a. Leaf margins toothed or lobed; flowers white, with 5 petals; fruit a drupe or a pome...**ROSACEAE** (*Crataegus*, *Prunus*; p. 719)
19b. Leaf margins entire...20

20a. Flowers zygomorphic, bright pink-purple with a yellow keel, 8–14 mm long, in racemes; leaves without 3 veins arising from the base...**POLYGALACEAE** (*Polygala subspinosa*; p. 681)
20b. Flowers actinomorphic, white, in corymbose clusters; leaves with 3 veins from the base...**RHAMNACEAE** (*Ceanothus fendleri*; p. 717)

21a. Woody vines, often with tendrils; leaves palmately compound or lobed...**VITACEAE** (p. 773)
21b. Trees or shrubs; leaves various...22

22a. Leaves compound...23
22b. Leaves simple or merely lobed...28

23a. Leaves ternately or palmately compound...24
23b. Leaves pinnately compound...25

24a. Leaflets gland-dotted and malodorous; flowers 4-merous, greenish-white; fruit a round samara...**RUTACEAE** (*Ptelea*; p. 740)
24b. Leaflets without gland-dots, not malodorous; flowers 5-merous, greenish; fruit a drupe (careful, poison ivy in this category!)...**ANACARDIACEAE** (*Rhus*, *Toxicodendron*; p. 71)

25a. Leaflets entire except for 1–5 rounded basal teeth, each tooth with a prominent gland at the tip; plant very malodorous; fruit a samara...**SIMAROUBACEAE** (p. 758)
25b. Leaflets toothed or entire, without gland tips; plants not malodorous; fruit various but not a samara...26

26a. Leaflets with entire margins, silvery-hairy below and green and sparsely hairy above; fruit an achene...**ROSACEAE** (*Potentilla fruticosa*; p. 729)
26b. Leaflets with toothed margins, not silvery-hairy below; fruit a drupe or pome...27

27a. Leaflets very sharply serrate; flowers white, numerous in corymbose cymes, with a short hypanthium; fruit a red to orangish pome, drying purplish...**ROSACEAE** (*Sorbus*; p. 737)
27b. Leaflets serrate but not sharply so; flowers greenish, numerous in a large terminal panicle, without a hypanthium; fruit a red, dry, hairy, and sometimes viscid drupe...**ANACARDIACEAE** (*Rhus*; p. 71)

28a. Leaves with 3 main veins arising from the base (this is especially evident on the underside of the leaf), ovate to elliptic; flowers white or cream-colored, in terminal or axillary corymbose cymes; fruit a capsule...**RHAMNACEAE** (*Ceanothus*; p. 717)
28b. Leaves pinnately veined or with only 1 main vein arising from the base, or if with 3 main veins then the leaves lobed; flowers and fruit various...29

29a. Leaves toothed or lobed...30
29b. Leaves with entire margins...32

30a. Leaves palmately or pinnately lobed or 3-lobed at the apex; stamens numerous...**ROSACEAE** (p. 719)
30b. Leaves merely toothed on the margins...31

31a. Corolla small and brownish, with a disk on which sits the calyx, this deciduous and leaving just the disk behind; fruit a berrylike drupe...**RHAMNACEAE** (*Rhamnus*; p. 717)
31b. Corolla white, cream, or pink, without a disk; fruit a drupe, pome, or achene...**ROSACEAE** (p. 719)

32a. Fruit an achene with a long, perisistent style or 5 follicles; flowers with a hypanthium; leaves with revolute margins or if flat then mostly basal and plants mat-forming...**ROSACEAE** (*Cercocarpus*, *Petrophytum*; p. 719)
32b. Fruit a concavo-convex follicle with numerous longitudinal lines; flowers without a hypanthium, solitary and axillary, small and inconspicuous, the petals to 7 mm long; leaves narrowly elliptic to oblanceolate, ca. 1 cm long, to 4 mm wide, without revolute margins; plants not mat-forming...**CROSSOSOMATACEAE** (p. 302)

33a. Petals 6, the upper and lateral ones dissected, the lower 2 entire, greenish-white or greenish-yellow; sepals 4 or 6, linear and subequal; flowers numerous, small, in an elongate, terminal raceme; fruit a capsule opening apically; ovary 3-carpellate...**RESEDACEAE** (p. 717)
33b. Plants unlike the above in all respects...34

34a. Ovary inferior or half-inferior...35
34b. Ovary superior...42

35a. Sepals 2; leaves usually fleshy and succulent, with entire margins, linear and terete or flat and spatulate to obovate; fruit a circumscissile capsule, 1–9 mm in diam...**PORTULACACEAE** (p. 698)
35b. Sepals more than 2; leaves unlike the above in all respects; fruit not a circumscissile capsule...36

36a. Flowers 4-merous or rarely 2-merous...37
36b. Flowers 5-merous...38

37a. Flowers subtended by 4 large, white petaloid bracts; leaves whorled at the top of the stem; fruit a red drupe...**CORNACEAE** (p. 301)
37b. Flowers not subtended by white petaloid bracts; leaves not whorled at the top of the stem, usually opposite; fruit a capsule...**ONAGRACEAE** (p. 549)

38a. Flowers arranged in umbels; fruit a schizocarp or drupe...39
38b. Flowers variously arranged, but not in umbels; fruit a capsule...40

39a. Styles and carpels 2; flowers usually in compound umbels; leaves with sheathing petioles; fruit a schizocarp splitting into 2 mericarps...**APIACEAE** (p. 72)
39b. Styles and carpels 5; flowers in simple umbels, these sometimes arranged in racemes; leaves without sheathing petioles; fruit a drupe...**ARALIACEAE** (p. 96)

40a. Plants caulescent with numerous stem leaves, the leaves not orbicular or reniform; leaf surfaces usually covered in large, multicellular, barbed hairs; outer bark white and often exfoliating; fruit a capsule surmounted by persistent sepals...**LOASACEAE** (p. 526)
40b. Leaves all or mostly basal, often reniform or orbicular; leaf surfaces not covered in large, multicellular, barbed hairs; outer bark not white and exfoliating; fruit unlike the above...41

41a. Staminodes 5 (in branched clusters, each branch tipped with a yellow gland), opposite the petals and alternating with the 5 fertile stamens; carpels united for most their length, the ovary globose with 3–4 stigmas and parietal placentation; flowers solitary on a long peduncle...**PARNASSIACEAE** (p. 576)
41b. Staminodes absent, fertile stamens 5 or 10; carpels united at the base, the ovary with 2–3 stigmas and axile placentation; flowers various...**SAXIFRAGACEAE** (p. 752)

42a. Leaves ternately or palmately compound; sepals and petals 4; fruit a capsule; ovary 2-carpellate, often stipitate; stamens 6–20, equal in length...**CAPPARACEAE** (p. 263)
42b. Plants unlike the above in all respects; if flowers 4-merous with 6 stamens, then the stamens usually tetradynamous...43

43a. Flowers zygomorphic, the flowers distinctly asymmetrical...44
43b. Flowers actinomorphic and symmetrical...50

44a. Carpels distinct; sepals petaloid and showy, purple or bluish, the basal one spurred or the upper helmetlike; petals small and rather inconspicuous; stamens many; fruit an aggregate of follicles...**RANUNCULACEAE** (*Aconitum*, *Delphinium*; p. 704)
44b. Carpels united; flowers unlike the above; fruit a capsule or legume...45

45a. Leaves palmately or pinnately compound...46
45b. Leaves simple, although sometimes deeply dissected but not truly compound...47

46a. Outer petal spurred at the base, the inner petals narrow and apically connate; stamens 6, diadelphous with 3 per set...**FUMARIACEAE** (p. 486)
46b. Petals not spurred, often composed of a banner, 2 wings, and a keel; stamens 10, distinct, monadelphous, or diadelphous with 9 fused and 1 free...**FABACEAE** (p. 350)

47a. Flowers orange-red with reddish spots; sepals 3, the lateral ones small and green and the posterior one petaloid and spurred with a saccate base; ovary 5-carpellate...**BALSAMINACEAE** (p. 200)
47b. Flowers variously colored but unlike the above; sepals 4 or 5, all similar or the inner 2 larger and petaloid; ovary 2–3-carpellate...48

48a. Sepals 4; petals 4, the upper pair ca. ⅓ the length of the lower pair; fruit a silicle...**BRASSICACEAE** (*Iberis*; p. 241)
48b. Sepals 5; petals 3 or 5; fruit not a silicle...49

49a. Sepals 5, the inner 2 large and petaloid; petals 3, not spurred; ovary 2-carpellate; stamens 8, not spurred...**POLYGALACEAE** (p. 681)
49b. Sepals 5, all green; petals 5, the lower one with a basal spur or gibbous; ovary 3-carpellate; stamens 5, often the lower 2 spurred...**VIOLACEAE** (p. 770)

50a. Stamens numerous, more than 10...51
50b. Stamens 10 or fewer...60

51a. Carpels distinct...52
51b. Carpels united or the gynoecium unicarpellate...53

52a. Hypanthium present (although sometimes shallow); leaves often stipulate...**ROSACEAE** (p. 719)
52b. Hypanthium absent; leaves exstipulate...**RANUNCULACEAE** (p. 704)

53a. Leaves opposite...54
53b. Leaves alternate...55

54a. Leaves gland-dotted; flowers yellow; hypanthium absent...**CLUSIACEAE** (p. 296)
54b. Leaves not gland-dotted; flowers purple, rose-purple, white, or pink; hypanthium present...**LYTHRACEAE** (p. 531)

55a. Leaves ternately or pinnately compound, the leaflets with sharply toothed margins; sepals and petals 2–5 mm long, greenish-white to cream-colored; flowers 25 or more in a terminal or axillary raceme; fruit a red or white berry...**RANUNCULACEAE** (*Actaea*; p. 705)
55b. Leaves simple although sometimes lobed, or pinnately to bipinnately compound but the leaflets with entire margins; plants otherwise unlike the above in all respects...56

56a. Leaves pinnately or bipinnately compound; fruit a legume; stamens showy, the filaments often pink or purplish, the sepals and petals small and inconspicuous; gynoecium unicarpellate...**FABACEAE** (p. 350)
56b. Leaves simple, although sometimes pinnately or palmately lobed; fruit various but not a legume; stamens not the showiest part of the flower; gynoecium with 2 or more carpels...57

57a. Stamens monadelphous (united by the filaments and forming a tube around the style); leaves stipulate, often palmately lobed; fruit a loculicidal capsule or schizocarp...**MALVACEAE** (p. 532)
57b. Stamens not as above; leaves exstipulate; fruit a capsule...58

58a. Leaves pinnately or ternately divided or lobed; plants usually with milky or colored sap, rarely only with watery sap; fruit a poricidal or valvate capsule...**PAPAVERACEAE** (p. 573)
58b. Leaves simple and entire; plants without milky or colored sap...59

59a. Plants glabrous; flowers pink, rose, or white; sepals 2 or if more then becoming scarious after flowering; ovary 3–8-carpellate; fruit a circumscissile capsule...**MONTIACEAE** (*Lewisia*; p. 540)
59b. Plants hairy; flowers yellow; sepals 5, the outer 2 smaller than the inner 3; ovary 3-carpellate; fruit a loculicidal or valvate capsule...**CISTACEAE** (p. 296)

60a. Carpels 2 or more, free and distinct...61
60b. Carpels united (at least at the base)...63

61a. Leaves usually compound and stipulate; hypanthium present...**ROSACEAE** (p. 719)
61b. Leaves simple, exstipulate; hypanthium present or not...62

62a. Carpels 4–5; leaves fleshy and succulent; fruit a follicle...**CRASSULACEAE** (p. 301)
62b. Carpels 2–3; leaves not succulent; fruit a capsule...**SAXIFRAGACEAE** (p. 752)

63a. Leaves compound (ternately, palmately, or pinnately)...64
63b. Leaves simple, although sometimes lobed or dissected...67

64a. Leaves ternately compound, leaflets with a rounded notch at the tip...**OXALIDACEAE** (p. 573)
64b. Leaves pinnatifid to pinnately compound, or palmately lobed or compound with more than 3 leaflets...65

65a. Flowers solitary in leaf axils, yellow, white, or orange; leaves pinnately compound or 2-foliate; fruit a schizocarp (often the mericarps with horn-like spines) or capsule...**ZYGOPHYLLACEAE** (p. 774)
65b. Flowers in cymes or racemes, variously colored; leaves various; fruit a schizocarp splitting and coiling up a central beak at maturity or a silicle or silique...66

66a. Sepals and petals 4; fruit a silique or silicle...**BRASSICACEAE** (p. 219)
66b. Sepals and petals 5; fruit a schizocarp splitting and coiling up a central beak...**GERANIACEAE** (p. 492)

67a. Sepals 2, or 6–9 and becoming scarious after flowering; leaves often fleshy...68
67b. Sepals 4–5 (rarely 6 and then unlike the above); leaves fleshy or not...69

68a. Plants prostrate-spreading and much-branched; flowers yellow or reddish; ovary half-inferior; leaves spatulate and flattened or linear and terete...**PORTULACACEAE** (p. 698)
68b. Plants erect or if prostrate-spreading then the basal leaves rhombic or trowel-shaped; flowers white, pink, rose, purplish, or magenta; ovary superior; leaves various...**MONTIACEAE** (p. 539)

69a. Corolla lobes each with 1 or 2 conspicuous, fringed glands on the upper surface at the base; ovary unilocular with parietal placentation on deeply intruded placentae; leaves simple and entire...**GENTIANACEAE** (*Frasera*, *Swertia*; p. 487)
69b. Corolla lobes lacking fringed glands; ovary various but unlike the above; leaves various...70

70a. Sepals and petals 4...71
70b. Sepals and petals 5 (rarely 6)...75

71a. Stamens 6, usually tetradynamous; fruit a silique or silicle...**BRASSICACEAE** (p. 219)
71b. Stamens unlike the above; fruit various but not a silique or silicle...72

72a. Leaves strongly gland-dotted, alternate; fruit a conspicuously gland-dotted capsule; stamens 8...**RUTACEAE** (*Thamnosma*; p. 740)
72b. Leaves not gland-dotted, opposite or alternate; fruit various or if a capsule then not gland-dotted; stamens various...73

73a. Carpels united at the base, free above; plants fleshy and succulent; fruit a follicle...**CRASSULACEAE** (p. 301)
73b. Carpels united for most their length; plants fleshy or not; fruit a capsule...74

74a. Flowers without a hypanthium; seeds bearing pits in longitudinal rows...**ELATINACEAE** (p. 340)
74b. Flowers with a hypanthium; seeds without longitudinal rows of pits...**LYTHRACEAE** (p. 531)

75a. Flowers imperfect, in racemes with the lower 1–3 flowers pistillate and the upper flowers staminate; fruit a 3-celled capsule...**EUPHORBIACEAE** (*Argythamnia*, *Reverchonia*; p. 345)
75b. Flowers perfect, or if imperfect then unlike the above; fruit various...76

76a. Carpels united just at the base, free above; plants fleshy and succulent; fruit a follicle; flowers variously colored but rarely white...**CRASSULACEAE** (p. 301)
76b. Carpels united for most their length; plants usually not fleshy; fruit a schizocarp or capsule; flowers various...77

77a. Leaves palmately lobed or divided; fruit a schizocarp splitting and coiling up a central beak at maturity; sepals distinct...**GERANIACEAE** (p. 492)
77b. Leaves not as above, or if palmately lobed then the sepals united; fruit a capsule...78

78a. Leaves opposite or whorled...79
78b. Leaves alternate or basal...81

79a. Ovary with axile placentation; annual herbs with reddish stems, the entire plant glandular-hairy, the leaf margins glandular-serrate; fruit a globose capsule enclosed in a persistent calyx, the seeds with longitudinal rows of pits; uncommon along drying ponds and mudflats...**ELATINACEAE** (*Bergia*; p. 340)
79b. Ovary with free-central or parietal placentation; plants various but otherwise unlike the above in all respects...80

80a. Ovary with parietal placentation on deeply intruded placentae; flowers either campanulate and blue-purple or rarely white, or rotate and white to pale blue with a darker blue midvein...**GENTIANACEAE** (*Eustoma*, *Lomatogonium*; p. 487)
80b. Ovary with free-central placentation; flowers various but unlike the above in all respects...**CARYOPHYLLACEAE** (p. 266)

81a. Leaves all cauline, lacking a basal rosette...82
81b. Leaves all basal or mostly basal with only a few stem leaves present...84

82a. Flowers white; leaves pinnatifid into linear segments...**NITRARIACEAE** (p. 543)
82b. Flowers yellow, blue, or orange; leaves simple and entire...83

83a. Outer 2 sepals smaller than the inner 3; ovary 3-carpellate; stamens 10–many or 3–8 in cleistogamous flowers; flowers yellow...**CISTACEAE** (p. 296)
83b. Sepals all equal in size or nearly so; ovary 5-carpellate; stamens 5, connate by their filaments basally into a ring; flowers yellow, blue, or orange...**LINACEAE** (p. 525)

84a. Staminodes 5 (in branched clusters, each branch tipped with a yellow gland), opposite the petals and alternating with the 5 fertile stamens; ovary globose with 3–4 stigmas; flowers solitary on long peduncles...**PARNASSIACEAE** (p. 576)
84b. Staminodes absent, fertile stamens 3–10; flowers otherwise unlike the above...85

85a. Ovary 2-carpellate, the carpels united at the base; flowers various but typically not conspicuously downward-pointing; petals often clawed; hypanthium present or absent...**SAXIFRAGACEAE** (p. 752)
85b. Ovary 5-carpellate, the carpels united for most their length; flowers downward-pointing; petals not clawed; hypanthium absent...**ERICACEAE** (p. 341)

KEY 10

FLOWERS 4- OR 5-MEROUS; PETALS UNITED

1a. Plants parasitic or saprophytic, lacking chlorophyll...2
1b. Plants not parasitic or saprophytic, stems and leaves green...4

2a. Plants with a conspicuous above-ground host plant, yellowish or orangish vine twining on the stems of herbs; stems glabrous, thin and thread-like; leaves minute, reduced to scales...**CONVOLVULACEAE** (*Cuscuta*; p. 299)
2b. Plants attached to the host plant below the surface of the soil, erect herbs; stems often glandular viscid hairy, not thin and thread-like; plants otherwise unlike the above...3

3a. Stamens 10; flowers urceolate, in a raceme, nodding on recurved pedicels, the sepals reddish and the petals yellow...**ERICACEAE** (*Pterospora*; p. 343)
3b. Stamens 4; flowers not urceolate, solitary on long pedicels or mostly sessile in a dense spike, yellowish, white, or purplish...**OROBANCHACEAE** (*Conopholis*, *Orobanche*; p. 565)

4a. Plants aquatic...5
4b. Plants terrestrial...7

5a. Leaves simple; flowers pink or white but not densely hairy within, small and solitary on short peduncles, tightly grouped at the base of the plant...**SCROPHULARIACEAE** (*Limosella*; p. 757)
5b. Leaves compound or dissected; flowers yellow or white and densely hairy within, otherwise unlike the above...6

6a. Leaves submerged, finely dissected, bearing tiny bladders which trap insects; flowers zygomorphic, yellow, lowest petal spurred; stamens 2...**LENTIBULARIACEAE** (p. 521)
6b. Leaves emergent, ternately compound with a sheathing base, without bladders; flowers actinomorphic, white, densely hairy within, the petals not spurred; stamens 5...**MENYANTHACEAE** (p. 538)

7a. Stems fleshy and succulent, cylindrical or flattened, true leaves absent; spines present at areoles; fruit a berry, often bearing spines; flowers solitary with numerous perianth parts and stamens; ovary inferior...**CACTACEAE** (p. 255)
7b. True leaves present, plants with or without spines and thorns; fruit and flowers various...8

8a. Flowers subtended by an involucre of fused or distinct bracts, sometimes the entire inflorescence giving the appearance of a single flower...9
8b. Flowers not subtended by an involucre of bracts...15

9a. Flowers zygomorphic, composed of a banner, wings, and keel; stamens 10, diadelphous with 9 fused and 1 free or nearly so; leaves compound with 3 leaflets...**FABACEAE** (*Trifolium*; p. 406)
9b. Flowers actinomorphic or if zygomorphic then not composed of a banner, wings, and keel; plants otherwise unlike the above in all respects...10

10a. Bracts of the involucre extending down from the head forming a sheath around the top of the peduncle; petals and sepals papery and scarious, bright pink (but often drying white), the sepals also plicate; leaves narrowly linear, all basal; rare on alpine...**PLUMBAGINACEAE** (p. 599)
10b. Bracts of the involucre not forming a sheath around the top of the peduncle; plants otherwise unlike the above...11

11a. Ovary superior...12
11b. Ovary inferior...14

12a. Flowers actinomorphic; tepals 6; leaves alternate or rarely the lowermost opposite...**POLYGONACEAE** (*Eriogonum*, *Stenogonum*; p. 681)
12b. Flowers zygomorphic and two-lipped (sometimes obscurely so), with a calyx and corolla; leaves all opposite...13

13a. Plants prostrate or procumbent, often rooting at the nodes; bracts lacking ciliate margins; stems square or not; stamens 4...**VERBENACEAE** (*Phyla*; p. 768)
13b. Plants erect, sometimes shrub-like; bracts with ciliate margins; stems square; stamens 2 or 4...**LAMIACEAE** (*Monarda*, *Monardella*; p. 513)

14a. Stamens 4, distinct; corolla 4-lobed; leaves opposite...**DIPSACACEAE** (p. 338)
14b. Stamens 5, united by the anthers to form a tube; corolla 3- or 5-lobed, rarely 4-lobed; leaves alternate, opposite, or basal...**ASTERACEAE** (p. 96)

15a. Woody vines with tendrils; petals united at the tip; leaves palmately lobed, the margins toothed...**VITACEAE** (p. 773)
15b. Trees, shrubs, or herbaceous planst, or if vines then unlike the above...16

16a. Flowers zygomorphic, the corolla conspicuously asymmetrical...17
16b. Flowers actinomorphic (or nearly so)...36

17a. Fruit a drupaceous capsule with a long, incurved, hooked beak that exceeds the fruit body in length; flowers 8–30 cm long, white with reddish-purple spots and yellow stripes on the inner lower portion of the tube; plants with a fetid odor, densely covered in glandular hairs; leaves suborbicular to reniform, simple, and entire...**PEDALIACEAE** (p. 577)
17b. Fruit various but unlike the above; plants otherwise without the above combination of characteristics...18

18a. Flowers distinctly spurred at the base of the corolla on the lower side, strongly bilabiate, purple, yellow, yellow and orange, or lavender and white...**PLANTAGINACEAE** (*Chaenorrhinum*, *Linaria*; p. 579)
18b. Flowers not spurred but sometimes with a small pouch at the base on the upper side, bilabiate or not, variously colored...19

19a. Leaves opposite or whorled...20
19b. Leaves alternate or all basal...28

20a. Ovary inferior...21
20b. Ovary superior...22

21a. Perennial herbs; stamens 3; sepals inrolled in flower but elongating in fruit; flowers in bracteate cymes or corymbs; fruit achene-like...**VALERIANACEAE** (p. 767)
21b. Shrubs; stamens 5; sepals not inrolled; flowers often paired; fruit a berry...**CAPRIFOLIACEAE** (*Lonicera*, *Symphoricarpos*; p. 265)

22a. Stamens 5, these markedly unequal; woody vine or tree; flowers large, the corolla over 3–8 cm long, orange, or white with purple spots and streaks and with 2 yellow lines on the lowest lobe; fruit a capsule with large, winged seeds; ovary not 4-lobed...**BIGNONIACEAE** (p. 203)
22b. Stamens 5 but not unequal, or fertile and anther-bearing stamens 4; ovary 4-lobed or not; plants various but unlike the above in all respects...23

23a. Ovary 4-lobed, 2- or 4-loculed with 1–2 ovules per locule; stems usually square; fruit of 4 nutlets; stamens 2, or 4 and didynamous...24
23b. Ovary not 4-lobed, 2-loculed with 2–many ovules per locule; stems usually round or square in one species (*Scrophularia*); fruit a capsule; stamens 4 with a 5th sterile, antherless staminode, or 4 and didynamous...25

24a. Corolla tubular and usually bilabiate; inflorescence a determinate cyme or spike, or axillary; style often gynobasic (arising from between the 4 ovary lobes); plants often with a minty odor...**LAMIACEAE** (p. 513)
24b. Corolla salverform and merely irregularly 4–5-lobed or somewhat bilabiate; inflorescence an indeterminate spike or corymb; style terminal on the ovary or nearly so; plants without a minty odor...**VERBENACEAE** (p. 768)

25a. Stems strongly 4-angled; flowers 9–14 mm long, yellowish-green or reddish-brown tinged, the upper lip flat and projecting forward and the lower lip 3-lobed with erect lateral lobes and a deflexed central lobe; leaves triangular-ovate and sharply toothed...**SCROPHULARIACEAE** (*Scrophularia*; p. 757)
25b. Stem more or less round, if somewhat 4-angled then the flowers pink or purple; plants otherwise unlike the above in all respects...26

26a. Calyx lobed to the base or at least to the middle, or if mostly connate then the flowers creamy-white or greenish-white, horizontally flattened at the apex and the leaves mostly basal with reduced stem leaves...**PLANTAGINACEAE** (*Bacopa*, *Chionophila*, *Collinsia*, *Gratiola*, *Penstemon*, *Veronica*, *Lindernia*; p. 579)
26b. Calyx mostly connate or only the upper ⅓-lobed (not lobed to the middle or below); flowers pink, purple, yellow, or red, not horizontally flattened at the apex; leaves not mostly basal...27

27a. Calyx tube 5-angled or pleated; flowers yellow, red, or reddish-violet, often with reddish spots on the inside of the throat; plants usually not blackening when dried...**PHRYMACEAE** (p. 577)
27b. Calyx tube not 5-angled or pleated; flowers yellow with the upper lip forming a hood, or pink to purple or rarely white with 2 yellow lines and reddish spots on the inside of the throat; plants often blackening when dried...**OROBANCHACEAE** (*Agalinis*, *Rhinanthus*; p. 565)

28a. Leaves compound with distinct leaflets...29
28b. Leaves simple, sometimes cleft but not compound with distinct leaflets...30

29a. Stamens 10, diadelphous with 9 fused and 1 free, or monadelphous; petals 5, often the lower, inner two fused to form the keel; fruit a legume or loment; ovary unicarpellate...**FABACEAE** (p. 350)
29b. Stamens 6, diadelphous in 2 sets of 3; petals 4, in 2 whorls of 2, the outer with a prominent spur or pouch at the base, the inner alike and apically connate over the stigmas and clawed; sepals 2, small, bractlike and inconspicuous or early-deciduous; fruit a 2-valved capsule; ovary 2-carpellate...**FUMARIACEAE** (p. 486)

30a. Ovary inferior; flowers red, blue with white stripes, purplish with the lower 3 lobes bright pink with yellow stripes, or rarely white; stamens 5, connate by the anthers and filaments...**CAMPANULACEAE** (*Downingia*, *Lobelia*; p. 260)
30b. Ovary superior; flowers variously colored but unlike the above; stamens 4 or 8, or if 5 then not connate by the anthers and filaments...31

31a. Flowers white or pink-purple with a yellow keel, the upper 2 free and the lower one forming a keel; inflorescence a spike or raceme; stamens 4; fruit an unarmed capsule...**POLYGALACEAE** (p. 681)
31b. Flowers and inflorescence various but none of the petals forming a keel; stamens 8, or if 4 then didynamous or with a 5th staminode; fruit various...32

32a. Stamens 8; flowers reddish-purple, sepals petaloid, petals unequal with the upper 3 long-clawed and the lower 2 broad and thick and sometimes modified into glands, solitary in leaf axils; fruit achene-like, thick-walled, and armed with bristles or spines...**KRAMERIACEAE** (p. 512)
32b. Stamens 2, 4 and didynamous, or 5 and all fertile; flowers unlike the above in all respects; fruit a capsule...33

33a. Stamens 5; corolla nearly actinomoprhic to zygomorphic but not bilabiate...34
33b. Stamens 2 or 4; corolla zygomorphic and bilabiate...35

34a. Flowers greenish-yellow with purple veins; calyx 2-2.5 cm long in fruit, urn-shaped; inflorescence secund; leaves coarsely toothed to pinnatifid, not densely soft-woolly...**SOLANACEAE** (*Hyoscyamus*; p. 760)
34b. Flowers unlike the above; inflorescence not secund; leaves entire to toothed but not pinnatifid, often densely soft-woolly...**SCROPHULARIACEAE** (*Verbascum*; p. 575)

35a. Leaves simple with crenate or toothed margins, oblong, ovate, or elliptic...**PLANTAGINACEAE** (*Besseya*, *Digitalis*; p. 579)
35b. Leaves simple and entire, pinnately lobed with 3–7 linear and entire segments, or pinnatifid to bipinnatifid, if simple with crenate margins then the leaves linear to narrowly lanceolate...**OROBANCHACEAE** (*Castilleja*, *Cordylanthus*, *Orthocarpus*, *Pedicularis*; p. 565)

36a. Ovary inferior or half-inferior...37
36b. Ovary superior...45

37a. Sepals 2; plants succulent; fruit a circumscissle capsule...**PORTULACACEAE** (p. 698)
37b. Sepals more than 2, or if 2 then the plant not succulent; fruit various...38

38a. Leaves opposite or whorled...39
38b. Leaves alternate...42

39a. Leaves whorled or simple and entire, linear, sometimes with membranous, fimbriate stipules at the base of the leaves; flowers 4-merous, white, light blue, or purplish, sometimes densely hairy within...**RUBIACEAE** (p. 737)
39b. Leaves compound, or if simple and entire then not linear or whorled, usually exstipulate; flowers with 2, 3, or 5 parts or if in 4's then unlike the above...40

40a. Stamens 3; sepals inrolled in flower but elongating in fruit; fruit achene-like; perennial herbs; leaves mostly basal, entire to more often pinnately dissected; flowers white, in a bracteate cyme or corymb...**VALERIANACEAE** (p. 767)
40b. Stamens 4 or more; sepals not inrolled in flower; fruit a fleshy or dry drupe, or berry; perennial herbs or shrubs; leaves simple or compound; flowers paired or in terminal cymes...41

41a. Flowers paired, solitary in leaf axils, or in axillary or terminal short spikes, pink or white, usually pointing downward; shrub or subshrub with prostrate stems rooting at the nodes; leaves simple...**CAPRIFOLIACEAE** (*Linnaea*, *Symphoricarpos*; p. 265)
41b. Flowers in terminal cymes, white, yellow, or greenish, erect; perennial herb or shrub; leaves simple or compound...**ADOXACEAE** (p. 63)

42a. Woody shrubs or subshrubs...43
42b. Perennial herbs or vines...44

43a. Leaves simple and entire; stamens 8–12; sepals green, not petaloid, the petals larger than the sepals; flowers without a hypanthium, the corolla campanulate or urceolate...**ERICACEAE** (*Arctostaphylos, Gaultheria, Vaccinium*; p. 341)
43b. Leaves palmately lobed or dissected; stamens 4–5; sepals petaloid, the petals smaller than the sepals; flowers with a hypanthium, the corolla cylindric to narrowly campanulate...**GROSSULARIACEAE** (p. 493)

44a. Erect herbs lacking tendrils; flowers perfect, campanulate; leaves simple, not palmately lobed; fruit a capsule...**CAMPANULACEAE** (*Campanula, Heterocodon, Triodanis*; p. 260)
44b. Prostrate herbs or vines with tendrils; flowers imperfect; leaves palmately lobed or palmately compound; fruit a pepo...**CUCURBITACEAE** (p. 303)

45a. Carpels free or united just at the base; plants succulent...**CRASSULACEAE** (p. 301)
45b. Carpels united for most their length; plants succulent or not...46

46a. Perianth subtended by 3 white, scarious bracts (often mistaken for the sepals); sepals densely woolly; leaves mostly basal, stem leaves opposite and linear, densely or sparsely tomentose; annuals...**AMARANTHACEAE** (*Froelichia*; p. 70)
46b. Perianth unlike the above; plants otherwise unlike the above...47

47a. Carpels 2, united above and free below; plants with milky sap; fruit a follicle with comose seeds; stamens 5, the filaments short and connate into a tube and the anthers usually connivent around the gynoecium or united with the gynoecium and forming a gynostegium...**APOCYNACEAE** (p. 90)
47b. Carpels united at the base for most their length; plants usually without milky sap; fruit various but unlike the above; stamens various but never forming a gynostegium...48

48a. Sepals 2; stamens usually the same number as corolla lobes and opposite them; ovary with free-central or basal placentation; plants often succulent; leaves simple and entire...**PORTULACACEAE** (p. 698)
48b. Sepals more than 2; stamens fewer than the number of corolla lobes, or if the same number then alternate with them; placentation various but not free-central or basal; plants not succulent; leaves various...49

49a. Leaves opposite or whorled...50
49b. Leaves alternate or all basal...59

50a. Ovary 4-lobed; stems usually square; fruit of 4 nutlets...51
50b. Ovary not 4-lobed; stems not square; fruit various but not 4 nutlets...52

51a. Flowers in dense axillary, globose clusters; stamens 2...**LAMIACEAE** (*Lycopus*; p. 517)
51b. Flowers in terminal corymbs or spikes; stamens 4...**VERBENACEAE** (*Glandularia, Verbena*; p. 768)

52a. Small trees or shrubs...53
52b. Herbaceous plants, or only woody at the base...54

53a. Flowers 5-merous, cream to yellowish and often tinged purplish, the corolla salverform and united into a tube below, 12–25 mm long, twisted in bud; leaves 3–9 cleft from the base, the segments narrowly linear and spinulose at the tips; fruit a loculicidal capsule with 3 valves; stamens 5...**POLEMONIACEAE** (*Linanthus*; p. 676)
53b. Flowers 4-merous, purple or white, not twisted in bud; leaves simple; fruit a berry or loculicidal capsule with 2 valves; stamens 2...**OLEACEAE** (*Ligustrum, Syringa*; p. 548)

54a. Stamens 2; flowers yellow, with 5–6- lobes; fruit a circumscissile capsule; leaves lanceolate or narrowly elliptic, 2–5 mm wide, mostly scaberulous...**OLEACEAE** (*Menodora*; p. 549)
54b. Stamens 4 or more, or if 2 then the flowers purple or white; plants otherwise unlike the above in all characteristics...55

55a. Sepals and petals 4-lobed; stamens 2; flowers purple or white, in terminal and/or axillary racemes...**PLANTAGINACEAE** (*Veronica*; p. 598)
55b. Sepals and petals 5-lobed, or if 4-lobed then the stamens 4 or more and the inflorescence solitary, a cymose cluster, or a panicle (flowers not in racemes)...56

56a. Stamens opposite the corolla lobes; ovary 5-carpellate with free-central placentation; flowers yellow or pinkish, borne singly or 2–many in the leaf axils...**MYRSINACEAE** (p. 542)
56b. Stamens alternate with the corolla lobes; ovary 2–3-carpellate, without free-central placentation; flowers various...57

57a. Ovary unilocular with parietal placentation on deeply intruded placentae; leaves opposite or whorled, simple and entire, sessile or the basal petiolate, and often decussate, usually connate at the base or connected by a transverse line on the stem; flowers 4–5-merous, blue, white, pink, or purple, the corolla sometimes plicate...**GENTIANACEAE** (p. 487)
57b. Ovary with 2 or more locules; leaves all opposite but not connate at the base, or alternate below and opposite above; flowers 5-merous, variously colored, plicate or not...58

58a. Ovary 3-carpellate; anthers distinct; flowers twisted in bud; leaves linear or palmately dissected into linear segments, or lanceolate, often spinulose-tipped; fruit a loculicidal capsule...**POLEMONIACEAE** (*Linanthus*, *Microsteris*, *Phlox*; p. 668)
58b. Ovary 2-carpellate; anthers sometimes connivent; flowers not twisted in bud; leaves various but not linear and spinulose-tipped; fruit a berry or capsule...**SOLANACEAE** (p. 759)

59a. Trees; leaves pinnately compound with lanceolate, entire, large leaflets; flowers imperfect; fruit a globose berry...**SAPINDACEAE** (p. 750)
59b. Herbs or shrubs; leaves simple or if compound then unlike the above; flowers perfect; fruit various...60

60a. Corolla bluish-purple, the lobes with 2 orbicular, fringed glands near the base on the upper surface...**GENTIANACEAE** (p. 487)
60b. Corolla variously colored, without 2 fringed glands on the upper surface of the lobes...61

61a. Sepals 4; petals 4-lobed and scarious, white or greenish; leaves all basal or nearly so with parallel veins; fruit a circumscissile capsule...**PLANTAGINACEAE** (*Plantago*; p. 597)
61b. Sepals 5; petals 5-lobed, unlike the above; leaves without parallel veins; fruit various...62

62a. Stamens more than 10, monadelphous (united by their filaments into a tube which surrounds the style); calyx often subtended by an epicalyx; leaves simple, often palmately lobed or dissected, usually with conspicuous stipules; fruit a loculicidal capsule or a schizocarp with 5–15 mericarps...**MALVACEAE** (p. 532)
62b. Stamens not monadelphous, 10 or less; epicalyx absent or rarely present; leaves various, exstipulate; fruit various but not a schizocarp...63

63a. Stamens 4; leaves all basal, on long petioles, elliptic; flowers small, the petals to 4 mm long, white or pinkish, solitary on short peduncles, tightly grouped at the base of the plant; fruit a capsule...**SCROPHULARIACEAE** (*Limosella*; p. 575)
63b. Stamens usually 5, sometimes 4; plants otherwise unlike the above in all respects...64

64a. Stamens 5 and opposite the petals, or 10; ovary 5-carpellate...65
64b. Stamens 4–5 and alternate with the petals; ovary 2–4-carpellate...66

65a. Shrubs or subshrubs (sometimes low and prostrate); leaves sometimes evergreen, not all basal; ovary with axile placentation; flowers urceolate or campanulate...**ERICACEAE** (*Arctostaphylos*, *Kalmia*, *Rhododendron*; p. 341)
65b. Herbs; leaves not evergreen, usually all basal; ovary with free-central placentation; flowers various but never urceolate...**PRIMULACEAE** (p. 702)

66a. Vines, trailing on the ground or climbing on other plants...67
66b. Perennial or annual herbs, or shrubs, not trailing or climbing...68

67a. Flowers purple; fruit a berry; plants not twining; leaves lanceolate to elliptic or some with a pair of rounded basal lobes...**SOLANACEAE** (p. 759)
67b. Flowers white, pink, or red to reddish-orange; fruit a capsule; plants often twining or climbing on other plants; leaves hastate, sagittate, cordate, or linear-lanceolate...**CONVOLVULACEAE** (*Calystegia*, *Convolvulus*, *Ipomoea*; p. 298)

68a. Carpels 3; flowers twisted in bud; fruit a 3-valved capsule; leaves usually linear or dissected into linear segments, or lanceolate, sometimes sharply spine-tipped; calyx often separated by white, hyaline membranes...**POLEMONIACEAE** (p. 668)

68b. Carpels 2 or ovary appearing 4-carpellate; flowers usually not twisted in bud; fruit various but not a 3-valved capsule; leaves various but not spinulose-tipped; calyx unlike the above...69

69a. Ovary 4-lobed; style gynobasic; fruit of 1–4 nutlets usually surrounded by a persistent calyx; hairs often rough, hispid, and pustulose; flowers in cymes, these often coiled...**BORAGINACEAE** (p. 204)

69b. Ovary not 4-lobed; style arising from the top of the ovary; stigma without a broad, disk-like base; fruit a capsule or berry; hairs various, occasionally pustulose; flowers in a coiled cyme or not...70

70a. Flowers in a dense spike or lax raceme, yellow or white, slightly asymmetrical; leaves in a basal rosette and cauline, densely stellate-hairy or if glabrous then the plant glandular-hairy above; filaments often hairy, the hairs yellow or purple...**SCROPHULARIACEAE** (*Verbascum*; p. 757)

70b. Flowers regular; plants otherwise unlike the above in all respects...71

71a. Flowers purple or blue-purple, solitary in leaf axils and present on almost the entire length of the stem, rotate to broadly campanulate, to 12 mm in diam.; herbs to 25 cm tall, densely villous, with several stems from a woody base; fruit an ovoid capsule...**CONVOLVULACEAE** (*Evolvulus*; p. 300)

71b. Plants unlike the above in all respects...72

72a. Flowers funnelform, purplish-pink or pink with a darker throat, large (to 9 cm long); sepals unequal, the outer shorter than the inner; leaves linear to linear-lanceolate; shrubby perennial herb from a large woody root; found in sandy places on the eastern plains...**CONVOLVULACEAE** (*Ipomoea*; p. 300)

72b. Plants unlike the above in all respects...73

73a. Stigma with a broad, disk-like base and often surmounted by a small cone; fruit of 2 or 4 nutlets; style entire; flowers white...**HELIOTROPIACEAE** (p. 497)

73b. Stigma unlike the above; fruit a capsule or berry; style entire or shortly to deeply bifid; flowers various...74

74a. Ovary unilocular with 2 enlarged or intruded parietal placentae, which sometimes meet in the middle but do not join; inflorescence often a coiled cyme; corolla not plicate; style usually deeply divided, rarely almost entire (in *Nama*); fruit a capsule, the calyx not enlarged in fruit; stamens not connivent, often with a pair of scales or appendages at the base of the filament; flowers blue, purple, or white...**HYDROPHYLLACEAE** (p. 499)

74b. Ovary with 2 or more locules and axile placentation, the placentae usually swollen; inflorescence various but not a coiled cyme; corolla often plicate; style undivided; fruit a berry or capsule, the calyx sometimes enlarged in fruit; stamens often connivent, without a pair of scales or appendages at the base; flowers various...**SOLANACEAE** (p. 759)

Foothills near Boulder, Colorado

FERNS AND FERN ALLIES

ASPLENIACEAE Newman – SPLEENWORT FAMILY

Stems erect or nearly so, scaly; *leaves* monomorphic, simple to 4-pinnate, glabrous or often with glandular hairs; *sori* elongate, borne on veins on lower leaf surface; *indusia* present, concealing the sori. (Cronquist et al., 1972; Hickman, 1993; Lellinger, 1985; Tryon & Tryon, 1982)

ASPLENIUM L. – SPLEENWORT

With characteristics of the family. (Moran, 1982)

1a. Leaves linear (grass-like), simple or forked at the tips; sori very elongate, mostly over 1 cm long...***A. septentrionale***
1b. Leaves 1–3-pinnate with ovate, oblong, or rhombic pinnae; sori unlike the above...2

2a. Leaf rachis green throughout or sometimes reddish-brown just at the base, glabrous or sparsely hairy...3
2b. Leaf rachis reddish-brown or blackish throughout, glabrous...4

3a. Leaves at least twice compound (2–3-pinnate); the largest pinnae 15–40 mm long; pinnae margins serrate...***A. adiantum-nigrum***
3b. Leaves once compound or nearly so (sometimes the leaflets with 1 shallow lobe at the base); the largest pinnae 5–8 mm long; pinnae margins crenate...***A. trichomanes-ramosum***

4a. Pinnae ovate, oblong-ovate, or rhombic, the largest 2–8 mm long, with or without a shallow lobe (auricle) at the base...***A. trichomanes* ssp. *trichomanes***
4b. Pinnae oblong, the largest 10–25 mm long, with a shallow lobe (auricle) present at least on the upper margin, at the base of the pinnae...5

5a. Pinnae alternate to subopposite, with crenate to serrate margins...***A. platyneuron***
5b. Pinnae mostly opposite or the pairs slightly offset, with entire to shallowly crenate margins...***A. resiliens***

***Asplenium adiantum-nigrum* L.**, BLACK SPLEENWORT. [*A. andrewsii* A. Nelson]. Leaves 2–3-pinnate, 2.5–10 × 2–6.5 cm, petioles dark reddish-brown proximally; *pinnae* in 4–10 pairs with coarsely incised margins; *sori* 1-numerous per pinna. Rare in rocky crevices and on cliff ledges, known from White Rocks, Boulder Co., 5200–5300 ft. E. (Plate 1)

***Asplenium platyneuron* (L.) Britton, Sterns & Poggenb.**, EBONY SPLEENWORT. Leaves 1-pinnate, 4–50 × 2–5(-7) cm, glabrous, petioles reddish-brown; *pinnae* in 15–45 pairs, the margins crenate to serrulate; *sori* 1–12 pairs per pinna. Rare on sandstone cliffs, 4300–5000 ft. E. (Plate 1)

***Asplenium resiliens* Kunze**, BLACK-STEMMED SPLEENWORT. Leaves 1-pinnate, 9–20 × 1–2.5 cm, glabrous, petioles black; *pinnae* in 20–40 pairs, the margins entire to crenate; *sori* 2–5 pairs per pinna. Rare in rock crevices and on cliffs and caprocks, known from Las Animas and Baca cos., 4400–5000 ft. E.

***Asplenium septentrionale* (L.) Hoffm.**, FORKED SPLEENWORT. Leaves linear, simple, (0.5) 1–4 × 0.1–0.4 cm, petiole dark reddish-brown proximally; *pinnae* with a forked appearance, linear; *sori* mostly 2 or more per pinna. Found in rocky crevices and on cliffs, 4800–10,800 ft. E/W.

Asplenium trichomanes* L. ssp. *trichomanes, MAIDENHAIR SPLEENWORT. Leaves 1-pinnate, 3–22 × 0.5–1.5 cm, glabrous or sparsely hairy, petiole reddish-brown or black; *pinnae* in 15–35 pairs, the margins crenate to serrate or entire; *sori* 2–4 pairs per pinna. Found in rock crevices, 4800–9000 ft. E/W.

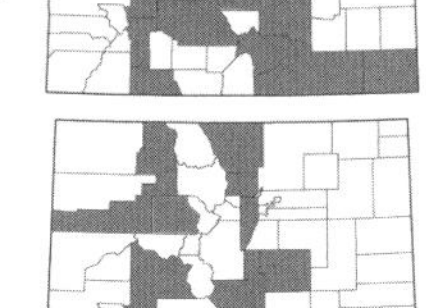

***Asplenium trichomanes-ramosum* L.**, GREEN SLPEENWORT. [*A. viride* L.]. Leaves 1–15 cm long, glabrous or sparsely hairy; petiole reddish-brown below and green above; *pinnae* in 6–21 pairs, the margins crenate; *sori* 2–4 pairs per pinna. Uncommon on limestone cliffs and in rocky crevices, 6000–12,000 ft. E/W.

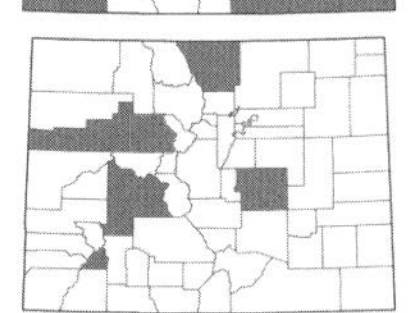

AZOLLACEAE Wettst. – AZOLLA FAMILY

Aquatic, floating plants; *stems* dichotomously branched, forming mats; *leaves* sessile, 2-lobed, the upper lobe emergent, dark red to green and photosynthetic, the lower lobe floating, ovate, slightly larger than the upper lobe, usually translucent.

AZOLLA Lam. – MOSQUITO FERN

With characteristics of the family.

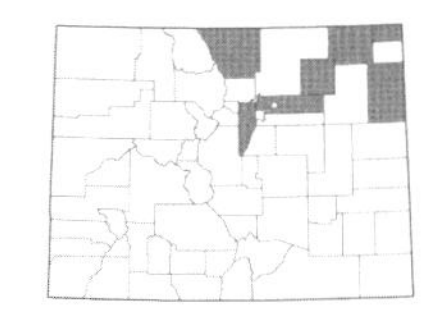

***Azolla mexicana* C. Presl**, MEXICAN MOSQUITO FERN. Annuals, free-floating; *leaves* with the upper lobe 0.7–1.3 mm long, lower lobe larger; *sporocarps* of two kinds, microsporocarps ca. 1.3 mm long with numerous microsporangia, megacarps ca. 0.4 mm long with a single megaspore. Uncommon, found floating on still water, 3500–5500 ft. E.

DENNSTAEDTIACEAE Lotsy – BRACKEN FERN FAMILY

Stems hairy; *leaves* monomorphic, 2–4-pinnate; *false indusium* covering the sporangia at the leaflet margins; *true indusium* present but obscure; *sporangia* more or less continuous around the underside of the leaflet margins.

PTERIDIUM Gleditsch ex Scop. – BRACKEN FERN

With characteristics of the family.

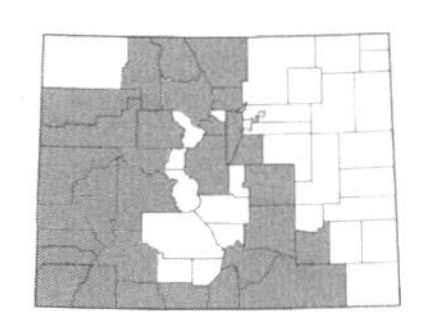

***Pteridium aquilinum* (L.) Kuhn var. *pubescens* Underw.**, HAIRY BRACKEN FERN. Leaves broadly triangular, 2–8 dm long, the ultimate segments oblong, hairy below between the midvein and margin; *sori* marginal. Fairly common along streams and in aspen and Douglas-fir forests, 6000–11,000 ft. E/W. (Plate 1)

DRYOPTERIDACEAE Herter – WOOD FERN FAMILY

Stems creeping or erect; *leaves* monomorphic or dimorphic, simple to 1–5-pinnate or more; *sori* borne on lower leaf surface on veins or at vein tips (usually not marginal); *indusia* present or absent. (Harrington, 1950; Lellinger, 1985; Wherry, 1938)

1a. Leaves once compound or nearly so, or rarely the lower pinnae divided at the base into 1–3 pairs of lobes...2
1b. Leaves at least twice compound...3

2a. Pinnae entire or sinuate; leaves of 2 kinds, the vegetative leaves triangular and pinnatifid (not divided into distinct leaflets), the fertile leaves shorter, linear, forming bead-like structures; petioles lacking scales...***Onoclea***
2b. Pinnae with sharply toothed margins; leaves once compound with distinct leaflets, not of 2 kinds; petioles densely scaly...***Polystichum***

3a. Fronds broadly triangular (widest at the base)...4
3b. Fronds elliptic or oblong (widest at the middle or the same width throughout)...6

4a. Indusium present; fronds mostly over 30 cm long...***Dryopteris*** (*expansa*)
4b. Indusium absent or inconspicuous; fronds mostly 5–15 cm long...5

5a. Frond clearly ternately divided; ultimate divisions with entire or crenate margins...***Gymnocarpium***
5b. Frond triangular or inconspicuously ternately divided; ultimate divisions with serrate or pinnatifid margins...***Cystopteris*** (*montana*)

6a. Indusium round, almost completely covering the sori; petiole scales of two different kinds, broad and scale-like and narrow and hair-like; fronds 3–12 dm long...***Dryopteris*** (*filix-mas*)
6b. Indusium absent, of filiform segments, or attached laterally and hood-like; petiole scales all broad and scale-like or absent; fronds various...7

7a. Indusium divided into narrow, filiform segments, or of relatively broad segments arranged around the sori; numerous old, persistent leaf bases present at the base of leaf clusters; fronds often densely glandular...***Woodsia***
7b. Indusium absent or hood-like and covering the sori; fronds sometimes sparsely glandular...8

8a. Lower petiole densely covered with scales over 5 mm in length; indusium absent, or if present then elongate or crescent-shaped...***Athyrium***
8b. Lower petiole not densely covered with scales, or if some scales present these less than 5 mm long; indusium present but often inconspicuous on mature leaves, laterally attached with an ovate, hood-like covering over the sori (and bend back at maturity)...***Cystopteris***

ATHYRIUM Roth – LADY-FERN

Stems short-creeping or erect; *leaves* monomorphic, 1–4-pinnate, usually glabrous; *sori* round to elongate, in a single row between the midrib and margin; *indusia* absent or present, shaped like the sori, entire or fimbriate.

1a. Indusium absent; sori round; fronds about the same width throughout (mostly less than 5 cm wide), the pinnae crowded...***A. alpestre* var. *americanum***
1b. Indusium present; sori crescent-shaped or elongate; fronds widest at about the middle (mostly over 8 cm wide), the pinnae not crowded on the rachis...***A. filix-femina* var. *californicum***

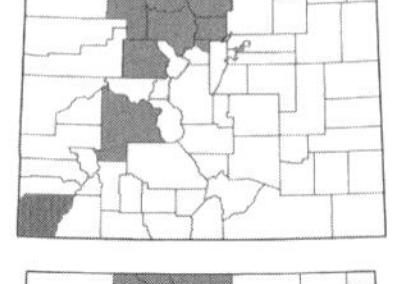

***Athyrium alpestre* (Hoppe) Chairv. var. *americanum* Butters**, ALPINE LADY-FERN. [*A. distentifolium* Tausch ex Opiz ssp. *americanum* (Butters) Hultén]. Leaves 2–3-pinnate, 15–55 × 3–25 cm; *sori* round; *indusium* absent; *rachis* with scales. Rare on rocky slopes, in meadows, and spruce forests, 8800–11,800 ft. E/W. (Plate 1)

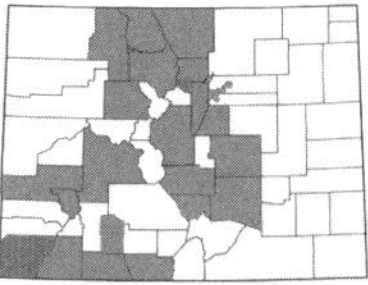

***Athyrium filix-femina* (L.) Roth ex Mert. var. *californicum* Butters**, COMMON LADY-FERN. Leaves 2-pinnate, 18–30 × 5–50 cm; *sori* horseshoe-shaped; *indusium* present, dentate or ciliate; *rachis* glabrous, glandular, or hairy. Found on moist forest slopes, along streams, and in meadows, 6500–10,500 ft. E/W. (Plate 1)

CYSTOPTERIS Bernh. – BLADDER FERN

Stems short to long-creeping; *leaves* monomorphic, 1–3-pinnate-pinnatifid; *sori* in a single row between the midrib and margin, round; *indusia* hood-like, attached to the receptacle base. (Blasdell, 1963)

1a. Leaves triangular (widest at the base), about as wide as long...***C. montana***
1b. Leaves elliptic (widest in the middle) or rarely triangular, 2–3 times as long as wide...2

2a. Rachis of leaves glandular and often with misshapen bulblets...***C. utahensis***
2b. Rachis of leaves lacking glandular hairs and bulblets...***C. fragilis***

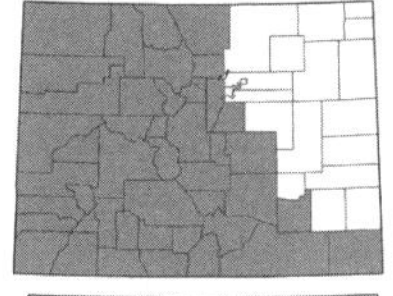

***Cystopteris fragilis* (L.) Bernh.**, BRITTLE BLADDER FERN. [*C. reevesiana* Lellinger; *C. tenuis* (Michx.) Desv.]. Leaves 1-pinnate-pinnatifid, to 40 cm, widest at or just below the middle; *pinnae* with serrate margins; *indusium* ovate to lanceolate, lacking glandular hairs. Common in rocky crevices, cliffs, and at the base of rocks, 4500–14,000 ft. E/W. (Plate 1)

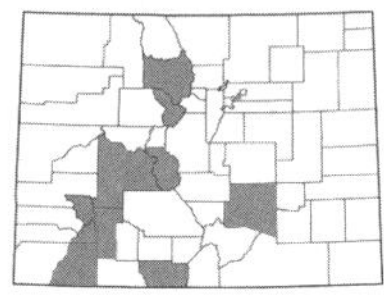

***Cystopteris montana* (Lam.) Bernh. ex Desv.**, MOUNTAIN BLADDER FERN. Leaves 3-pinnate-pinnatifid, to 45 cm, widest toward the base; *pinnae* ascending, with serrate margins; *indusium* cup-shaped, with glandular hairs along the margin. Uncommon in shady, moist spruce-fir forests and along streams, 9000–11,500 ft. E/W.

***Cystopteris utahensis* Windham & Haufler**, UTAH BLADDER FERN. Leaves 2-pinnate-pinnatifid, to 45 cm; *pinnae* perpendicular to the apex, with serrate margins; *indusium* cup-shaped, with glandular hairs. Rare in crevices of sandstone ledges, known from Moffat Co., 5500–6000 ft. W.

DRYOPTERIS Adans. – WOOD FERN

Stems short-creeping or erect, scaly; *leaves* monomorphic, 1–3-pinnate-pinnatifid, leathery, glabrous; *sori* in a single row between the margin and midrib, round; *indusia* round to reniform, glabrous.

1a. Leaves widest at the base (triangular), three times compound; petiole scales all broad and scale-like...***D. expansa***
1b. Leaves widest near the middle (elliptic), twice compound or sometimes more at the base; petiole scales of two different kinds, broad and scale-like and narrow and hair-like...***D. filix-mas***

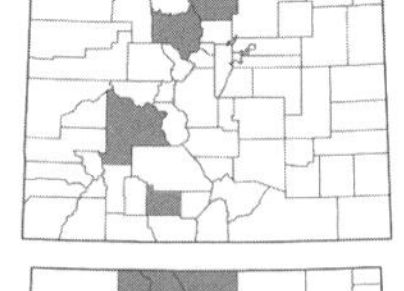

***Dryopteris expansa* (C. Presl) Frazer-Jenkins & Jermy**, SPREADING WOOD FERN. Leaves 3-pinnate-pinnatifid, to 90 cm long, deltate-ovate, usually lacking glandular hairs; *pinnule* margins serrate; *indusium* lacking glands or sparsely glandular. Rare in moist, dense spruce-fir forests and at cliff bases, 9000–10,500 ft. E/W.

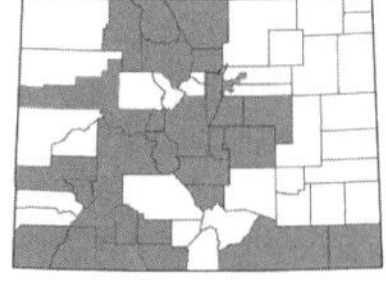

***Dryopteris filix-mas* (L.) Schott**, MALE FERN. Leaves pinnate-pinnatifid or 2-pinnate at the base, 15–100 cm long, lacking glandular hairs; *pinnule* margins serrate to lobed; *indusium* lacking glands. Common in dense forests, in rocky crevices, and on talus slopes, 4400–10,700 ft. E/W. (Plate 1)

GYMNOCARPIUM Newman – OAK FERN

Stems long-creeping; *leaves* monomorphic, 2–3-pinnate-pinnatifid, glabrous; *sori* in a single row between the midrib and margin, round; *indusia* absent.

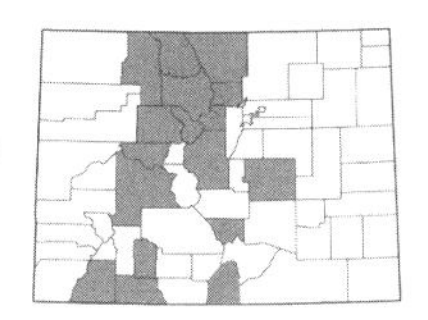

***Gymnocarpium dryopteris* (L.) Newman**, WESTERN OAK FERN. Perennials with creeping rhizomes; *leaves* 5–17 cm long, 6–20 cm wide, deltoid, the ultimate segments to 1.5 cm long, with crenate margins; *petioles* to 25 cm long, scaly near the base. Found in moist, shady forests, often growing along streams, 7500–10,500 ft. E/W. (Plate 1)

ONOCLEA L. – SENSITIVE FERN

Stems creeping; *leaves* strongly dimorphic, glabrous, sterile leaves pinnate, fertile leaves persistent, 2-pinnate, shorter and becoming dark brown or black at maturity; *sori* enclosed by the strongly revolute margins of fertile leaves; *indusia* obscure.

***Onoclea sensibilis* L.**, SENSITIVE FERN. Perennials with long-creeping rhizomes; *leaves* dimorphic; *sterile leaves* broadly triangular, 5–35 cm long, pinnate with a broadly winged rachis, green; *fertile leaves* 2–18 cm long, lance-oblong, the pinnules rolled into spherical segments, dark brown or black. Rare in moist, shady places, known from a single 1948 collection in Douglas Co., 7000ft. E.

POLYSTICHUM Roth – HOLLY FERN

Stems short-creeping or erect, scaly; *leaves* monomorphic (ours), 1–3-pinnate; *sori* in a single row between the midrib and margins, round; *indusia* round, peltate.

1a. Pinnae undivided (although usually oblique with one side larger at the base), rarely overlapping...***P. lonchitis***
1b. At least some pinnae divided at the base into 1–3 pairs of lobes, usually overlapping...***P. scopulinum***

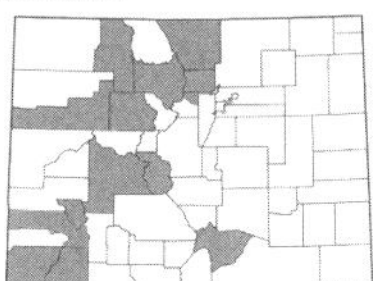

***Polystichum lonchitis* (L.) Roth**, NORTHERN HOLLY FERN. Leaves 1-pinnate, 5–60 cm long; *pinnae* oblong to lanceolate, entire, rarely overlapping, with serrulate-spiny margins; *indusium* entire or minutely erose. Found on rocky cliffs, in rock crevices, and on talus slopes, 7000–11,500 ft. E/W. (Plate 1)

***Polystichum scopulinum* (D.C. Eaton) Maxon**, MOUNTAIN HOLLY FERN. Leaves 1-pinnate-pinnatifid, 15–40 cm long; *pinnae* oblong-lanceolate, overlapping, lobed, with serrulate margins with the teeth curved inward; *indusium* entire or ciliate. Rare in rock crevices and on rocky slopes, known from Moffat Co., 6000–6500 ft. W.

WOODSIA R. Br. – CLIFF FERN

Stems compact to short-creeping, scaly; *leaves* monomorphic, 1–2-pinnate-pinnatifid; *sori* in a single row between the midrib and margin, round; *indusia* basal, thin, opening at the top and becoming dissected into spreading segments.

1a. Pinnae with multicellular, flattened nonglandular hairs (these concentrated along the midrib) in addition to glandular hairs...***W. scopulina***
1b. Pinnae with only glandular hairs or nearly glabrous...2

2a. Indusium of relatively broad segments...***W. plummerae***
2b. Indusium of narrow, filamentous segments...3

3a. Margins of the pinnae divisions with translucent projections on the teeth; fronds glabrescent or sparsely glandular; petiole light brown to straw-colored at maturity...***W. neomexicana***
3b. Margins of the pinnae divisions usually lacking translucent projections on the teeth; fronds usually moderately to densely glandular; petiole usually reddish-brown to dark purple towards the base at maturity...***W. oregana***

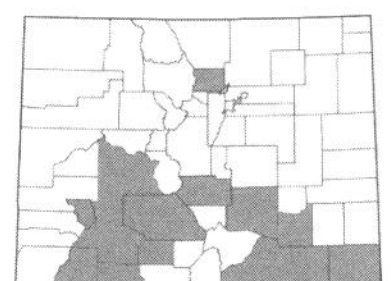

***Woodsia neomexicana* Windham**, NEW MEXICO CLIFF FERN. Leaves pinnate-pinnatifid, glabrous to sparsely glandular, 4–30 × 1.5–6 cm, petiole brown to straw-colored; *pinnules* dentate, sometimes shallowly lobed; *indusium* of narrow, filamentous segments. Uncommon in rock crevices and on cliffs and ledges, 5500–8500 ft. E/W. (Plate 1)

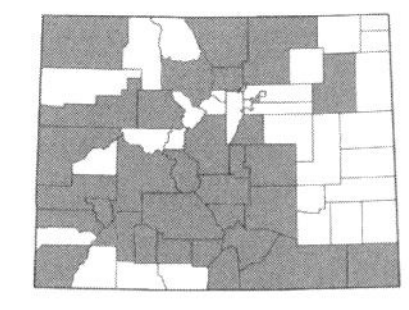

***Woodsia oregana* D.C. Eaton ssp. *cathcartiana* (B.L. Rob.) Windham**, OREGON CLIFF FERN. Leaves pinnate-pinnatifid or 2-pinnate proximally, 4–25 × 1–4 cm, sparsely to moderately glandular, petiole reddish-brown to dark purple below at maturity; *pinnules* dentate, often shallowly lobed; *indusium* of narrow segments. Common at the base of rocks, in rocky crevices, and on cliffs, 4000–10,600 ft. E/W.

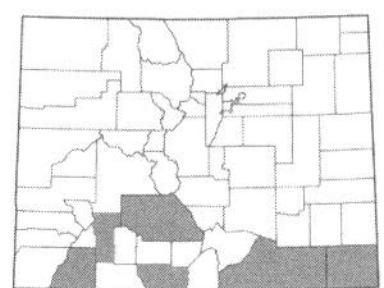

***Woodsia plummerae* Lemmon**, PLUMMER'S CLIFF FERN. Leaves usually 2-pinnate below, 5–25 × 1.5–6 cm, densely glandular, petiole reddish-brown to dark purple; *pinnules* dentate, often shallowly lobed; *indusium* of broader segments. Uncommon in rock crevices and on cliffs, 4500–9000 ft. E/W.

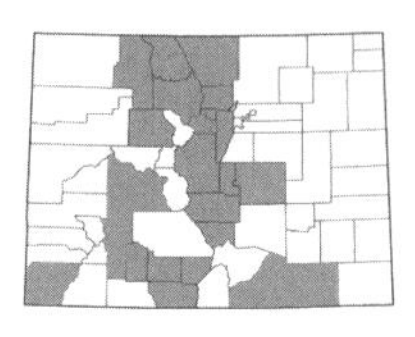

***Woodsia scopulina* D.C. Eaton**, Rocky Mountain woodsia. Leaves 2-pinnate below, 9–35 × 1–8 cm, glandular, petiole reddish-brown to dark purple below; *pinnules* dentate, often shallowly lobed; *indusium* of narrow segments. Found in crevices and on cliffs, 5500–11,000 ft. E/W.

EQUISETACEAE Michx. ex DC.– HORSETAIL FAMILY

Perennial plants with jointed, longitudinally ribbed, green stems; *leaves* small, whorled, reduced and scale-like, not photosynthetic; *branches* (if present) whorled; *sporangia* borne in terminal "cones," or sometimes on reproductive, brown stems; *spores* uniform, minute. (Harrington, 1950; Hauke, 1993)

EQUISETUM L. – HORSETAIL; SCOURING RUSH

With characteristics of the family.

1a. Stems with whorls of branches at the nodes, or the stems brown and unbranched...***E. arvense***
1b. Stems unbranched and green...2

2a. Leaf sheaths with a dark band at the base...***E. hyemale* ssp. *affine***
2b. Leaf sheaths lacking a dark band at the base...3

3a. Leaf sheaths usually lacking persistent teeth at the tips (these shed early, sometimes present at lower nodes); stems usually more robust, the sheaths 7–15 mm wide...***E. laevigatum***
3b. Leaf sheaths usually with persistent teeth at the tips; stems usually not as robust, the sheaths mostly 1–6 mm wide...***E. variegatum* ssp. *variegatum***

***Equisetum arvense* L.**, field horsetail. Stems dimorphic; *vegetative stems* green, branched in whorls; *fertile stems* brown, unbranched; *strobili* 5–40 mm long. Common in forests, along streams, on floodplains, and along pond margins, 5000–11,500 ft. E/W. (Plate 1)

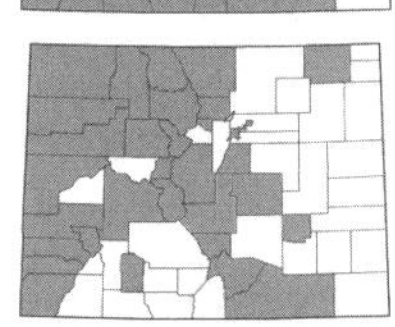

***Equisetum hyemale* L. ssp. *affine* (Engelm.) Calder & Roy L. Taylor**, scouring-rush horsetail. Stems monomorphic, green, unbranched; *sheaths* with a black lower band and black teeth; *strobili* 5–25 mm long. Common in moist forests and along streams and lake shores, 3500–10,000 ft. E/W.

Equisetum ×*ferrissii* Clute is a hybrid between *E. hyemale* and *E. laevigatum*. It is characterized by having white, misshapen spores instead of green, spherical spores, and a dark band present towards the middle of the sheath.

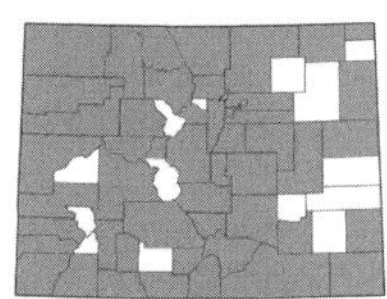

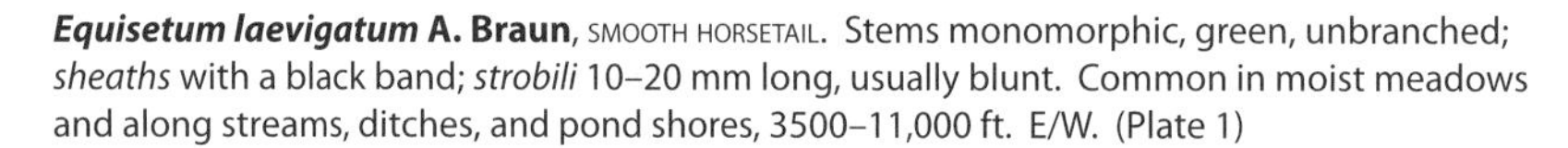

***Equisetum laevigatum* A. Braun**, smooth horsetail. Stems monomorphic, green, unbranched; *sheaths* with a black band; *strobili* 10–20 mm long, usually blunt. Common in moist meadows and along streams, ditches, and pond shores, 3500–11,000 ft. E/W. (Plate 1)

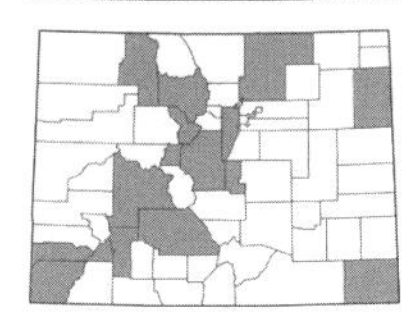

Equisetum variegatum* Schleich. ex F. Weber & D. Mohr ssp. *variegatum, variegated scouring rush. Stems monomorphic, green, branched at the base; *sheaths* with a terminal black band, teeth white with a black center; *strobili* 5–10 mm long, sharp-pointed at the tip. Uncommon in moist meadows, along streams, on floodplains, and in other moist places, 3900–11,500 ft. E/W.

ISOËTACEAE Dumort. – QUILLWORT FAMILY

Grass-like perennial herbs; *leaves* linear, simple, broadened at the base, containing 4 transversely septate, longitudinal air channels around a central vascular strand; *sporangia* solitary, adaxial on the leaf base, with a velum partly to completely covering the adaxial surface of the sporangium; *spores* dissimilar. (Lellinger, 1985; Taylor et al., 1993)

ISOËTES L. – QUILLWORT

With characteristics of the family. All three *Isoëtes* species can hybridize, and proper identifications are all but impossible if megaspores are not present. (Dorn, 1972)

1a. Megaspores with thin, sharp spines (echinate)...***I. echinospora***
1b. Megaspores wrinkled to tuberculate or with some of the tubercles becoming confluent forming low to high ridges...2

2a. Megaspores mostly 0.3- 0.5 mm wide, wrinkled to tuberculate, some of the tubercles may become confluent to form short, low ridges...***I. bolanderi***
2b. Megaspores mostly 0.5–0.7 mm wide, with high ridges or jagged crests...***I. occidentalis***

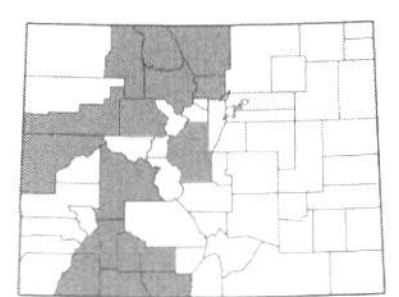

***Isoëtes bolanderi* Engelm.**, BOLANDER QUILLWORT. Leaves to 20 cm long, bright green; *megaspores* white, rugulate to tuberculate. Found in lakes and ponds, our most common *Isoëtes*, 8500–11,700 ft. E/W.

***Isoëtes echinospora* Durieu**, SPINY-SPORED QUILLWORT. [*I. setacea* Lam. ssp. *muricata* (Durieu) Holub]. Leaves to 25 (-40) cm long, bright green to reddish-green; *megaspores* white, echinate with thin, sharp spines. Uncommon in mountain lakes and ponds, 9000–11,500 ft. E. (no map)

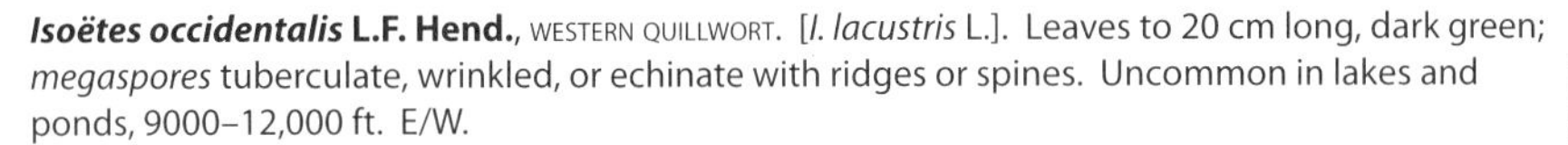

***Isoëtes occidentalis* L.F. Hend.**, WESTERN QUILLWORT. [*I. lacustris* L.]. Leaves to 20 cm long, dark green; *megaspores* tuberculate, wrinkled, or echinate with ridges or spines. Uncommon in lakes and ponds, 9000–12,000 ft. E/W.

LYCOPODIACEAE P. Beauv. ex Mirb. – CLUBMOSS FAMILY

Terrestrial plants; *rhizomes* widely creeping, branched; *aerial stems* scattered, usually conspicuously leafy, simple or branched; *leaves* with 1 central unbranched vein; *sporophylls* unmodified and green, or reduced and aggregated into cones; *sporangia* solitary, reniform to globose; *spores* uniform. (Harrington, 1950; Wagner & Beitel, 1993)

1a. Sporangia borne in the axils of unmodified leaves which are similar to those of the stem; horizontal stems absent, the branches in a cluster; leaves mostly in 8 longitudinal rows...***Huperzia***

1b. Sporangia borne in the axils of yellowish, ovate, scale-like leaves (terminal cones) which are very different from the leaves of the sterile stems; horizontal stems present; leaves in 4–8 longitudinal rows...2

2a. Leaves of branches imbricate, in 4 longitudinal rows...***Diphasiastrum***

2b. Leaves of branches not imbricate, mostly in 5–8 longitudinal rows...***Lycopodium***

DIPHASIASTRUM Holub – FALSE DIPHASIUM

Horizontal stems present; *leaves* mostly imbricate, linear to lanceolate, scale-like; *sporangia* in the axils of highly modified, reduced sporophylls aggregated into cones.

***Dlphaslastrum alplnum* (L.) Holub**, ALPINE CLUBMOSS. [*Lycopodlum alplnum* L.]. Leaves In 4 longltudlnal rows, 1.5–4 × 0.5–1.5 mm, upperside leaves appressed and lateral leaves strongly divergent; *sporophylls* 2–3.5 × 1.5–3 mm, with tapering tips. Rare in the alpine and spruce forests, reported for Colorado but no specimens have been seen.

HUPERZIA Bernh. – FIRMOSS

Horizontal stems absent; *leaves* not imbricate; *sporangia* in the axils of unmodified leaves.

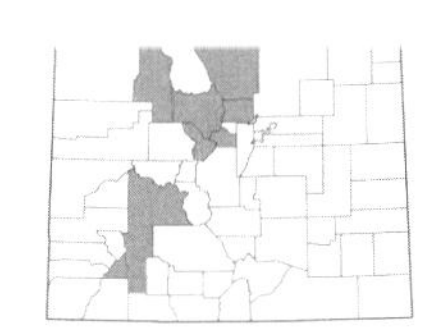

***Huperzia haleakalea* (Brack.) Holub**, ALPINE FIRMOSS. Leaves mostly in 8 longitudinal rows, lanceolate to oblanceolate, 3–6 (7) mm long, yellow-green to greener at the stem tip, appressed to spreading-ascending. Uncommon in moist tundra and spruce forests, 10,000–12,000 ft. E/W.

LYCOPODIUM L. – CLUBMOSS

Horizontal stems present; *leaves* not imbricate, linear to linear-lanceolate; *sporangia* in the axils of highly modified, reduced sporophylls aggregated into cones.

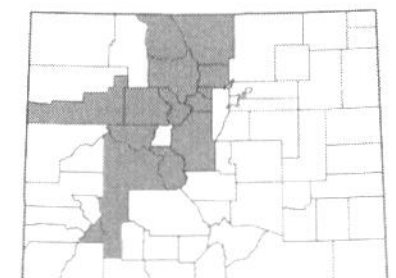

***Lycopodium annotinum* L.**, STIFF CLUBMOSS. [*L. dubium* Zoëga]. Leaves mostly in 5–8 longitudinal rows, (2.5) 5–8 × 0.5–1.2 mm, the margins minutely toothed in the distal ½; *sporophylls* (1.5) 3.5 × 0.7 (2) mm, abruptly narrowed to a pointed tip. Locally common in moist spruce-fir forests, 8000–13,000 ft. E/W. (Plate 2)

MARSILEACEAE Mirb. – WATER-CLOVER FAMILY

Aquatic plants; *leaves* alternate, long-petioled, palmately divided into 4 pinnae (ours); *sori* borne in a hard, pubescent sporocarp arising from short stalks near the base of the petioles. (Lellinger, 1985)

MARSILEA L. – WATER-CLOVER

With characteristics of the family.

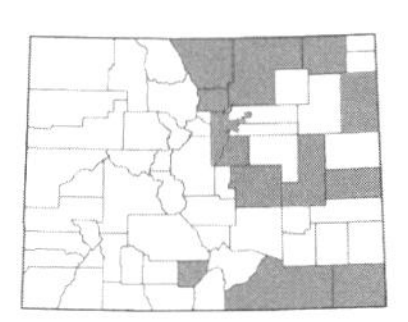

***Marsilea vestita* Hook. & Grev.**, HAIRY WATERCLOVER. Perennials; *leaves* often folded; *pinnae* 5–25 mm long, glabrous to hairy, with entire margins; *sporocarps* solitary on short peduncles, 5–8 mm long, densely hairy to glabrate, splitting into 2 valves at maturity and releasing 10–20 sori. Found in temporary pools, low swales, ditches, shallow water, and in fields, 4000–7500 ft. E/W.

OPHIOGLOSSACEAE Martinov – ADDER'S-TONGUE FAMILY

Terrestrial (ours), perennial plants; *stems* simple, unbranched, erect; *blades* usually 1 or sometimes 2, divided into a sterile leaf-like segment (trophophore) and lateral fertile segment (sporophore), with a common stalk; *trophophore* blades compound or simple; *sporophore* simple or pinnately branched, with the sporangia arranged in a terminal spike or panicle; *sporangia* thick-walled, lacking an indusium and annulus; *spores* uniform. (Clausen, 1938; Farrar, 2002; Farrar & Popovich *in* Weber & Wittmann, 2012; Stensvold, 2007; Wagner & Wagner, 1981, 1983, 1986)

1a. Trophophore 1–2 pinnately divided (the segments often fan-shaped) and to 4.5 cm wide, rarely simple (in a reduced *B. simplex*)...***Botrychium***
1b. Trophophore usually 3–4-pinnately divided (sometimes 2-pinnately divided in small plants), often over 6 cm wide...2

Botrychium (once-pinnate, fan-leaflet form):

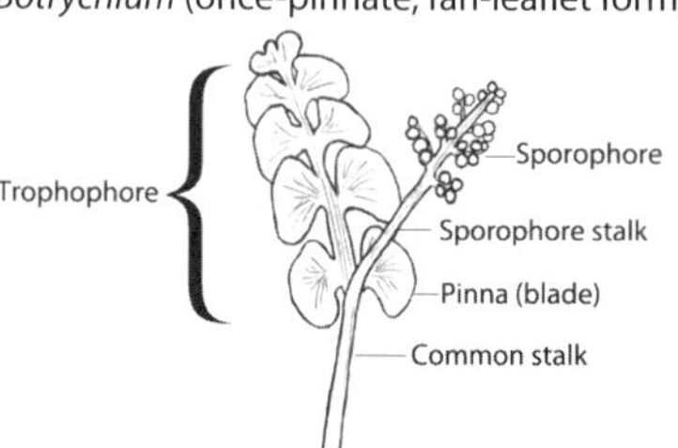

2a. Pinnules deeply lobed into linear, serrate segments with acute tips; trophophore sessile, thin; sporophore and trophophore joined well above ground-level, with a long common stalk...***Botrypus***
2b. Pinnules obliquely ovate, usually with entire margins or sometimes shallowly crenulate, rounded at the tips; trophophore stalked, thick and leathery; sporophore and trophophore joined at or near ground-level...***Sceptridium***

BOTRYCHIUM Sw. – MOONWORT

Trophophore sessile or stalked; *blades* 1–2-pinnate; *pinnae* spreading to ascending, linear to fan-shaped, the margins entire to toothed or lobed, the venation pinnate with a central midrib or arranged like the ribs of a fan (dichotomous); *sporophores* 1–2-pinnate.

Our *Botrychium* tend to prefer subalpine habitats with a history of past disturbance (now stabilized), such as ski slopes, in meadows near old cabins, and roadcuts. *Botrychium* are quite complex and morphologically variable, and thus difficult to accommodate in a key. Identification of *Botrychium* species is a difficult process and oftentimes requires genotyping to be certain of the proper identification. There are three basic forms in *Botrychium* plants: 1) a once-pinnate, fan-leaflet form (example: *B. neolunaria*); 2) a triangular, bipinnate form (example: *B. lanceolatum* ssp. *lanceolatum*); and 3) an intermediate, pinnate-pinnatifid form derived from the ancestral hybridization between *B. lanceolatum* and species with the fan-leaflet form (example: *B. echo*). The bipinnate and intermediate forms are often referred to as the midribbed species because of the presence of a central vein in the pinnae. The fan-leaflet species lack a central midrib and have multiple parallel veins of equal size or nearly so.

1a. Leaf lacking a trophophore, divided into 2 more or less equal sporophore segments instead...***B. paradoxum***
1b. Leaf with a distinct trophophore and sporophore...2

2a. Basal pinnae usually enlarged (much larger than the adjacent pinnae) and more or less distinctly stalked, pinnately divided into distinct fan-shaped segments, or the trophophore simple with a single blade; common stalk usually shorter than the trophophore...***B. simplex***
2b. Basal pinnae unlike the above; trophophores never simple, at least 1-pinnately divided; common stalk shorter or more often as long or longer than the trophophore...3

3a. Basal pinnae pinnately divided or lobed, with pinnate venation and a distinct central midrib (the pinnae never fan-shaped)...4
3b. Basal pinnae linear to fan-shaped or broadly rounded and entire, forked, or sometimes fan-shaped and palmately lobed, with venation like the ribs of a fan lacking a central midrib...7

4a. Trophophore triangular or pentangular in outline, with 3 main segments, each pinnately lobed and about equal in length (the basal pinnae nearly as large as the remainder of the trophophore)...***B. lanceolatum* ssp. *lanceolatum***
4b. Trophophore ovate to oblong in outline, otherwise unlike the above...5

5a. Plants dull or glaucous blue-green; sporophore usually ternately divided into three main branches (at least in larger plants); common stalk usually with a red stripe...***B. hesperium***
5b. Plants usually lustrous and shiny green; sporophore pinnately branched; common stalk usually lacking a red stripe...6

6a. Lobes of the basal pinnae mostly linear-spatulate; basal pinnae widest near the middle (elliptic)...***B. echo***
6b. Lobes of the basal pinnae mostly obliquely ovate or fan-shaped; basal pinnae widest towards the base...***B. pinnatum***

7a. Pinnae mostly linear to linear-spatulate or forked into 2 to several linear segments; sporophore stalk usually less than ⅓ the total length of the trophophore; trophophore sessile...***B. campestre* ssp. *lineare***
7b. Pinnae mostly fan-shaped (or if linear and stubby then the sporophore stalk ½ or more of the total length of the trophophore); trophophore sessile to stalked...8

8a. Outer pinnae margins finely crenulate to regularly sharply toothed with more or less evenly spaced teeth (sometimes also cleft)...9
8b. Outer pinnae margins entire to undulate or lobed with rounded sinuses, or irregularly and very coarsely toothed, but not finely crenulate or regularly sharply toothed...11

9a. Trophophore sessile or short-stalked; sporophore stalk to ½ the total length of the trophophore...***B. ascendens***
9b. Trophophore evidently stalked; sporophore stalk usually ½ or more of the total length of the trophophore...10

10a. Pinnae ascending, wedge-shaped to narrowly fan-shaped, the outer margins irregularly toothed or scalloped; plants yellow-green to whitish-green, usually pale and glaucous when fresh...***B. furculatum***
10b. Pinnae widely spreading, broadly fan-shaped, the outer margins regularly finely crenulate to dentate with numerous short, sharp teeth; plants green to yellowish-green, usually neither pale nor glaucous when fresh...***B. lunaria* var. *crenulatum***

11a. Basal pinnae broadly fan-shaped or round...12
11b. Basal pinnae mostly narrowly fan-shaped to spatulate or strongly cuneate (gradually tapering to the base), or ovate with concave sides...13

12a. Pinnae fan-shaped, with definite outer corners; plants duller, grass-green; sporophore branches usually ascending, with the stalks usually longer than the height of the trophophore; common stalk to 3.5 cm; trophophore sessile or with a short stalk to 1 mm long...***B. neolunaria***
12b. Pinnae broadly rounded, with nearly no outer corners; plants glossy green; sporophore branches usually spreading outwards, with the stalks shorter than or equal to the height of the trophophore; common stalk 0–3 cm (average of 1.5) cm long; trophophore sessile or with a stalk to 10 mm...***B. tunux***

13a. Sporophore stalk less than ½ the length of the total trophophore; trophophore sessile or nearly so; blades ovate to deltate in outline (widest at the base); pinnae spatulate with rounded outer margins...***B. spathulatum***
13b. Sporophore stalk ½ of the total length of the trophophore or more; trophophore usually on a stalk over 3 mm long (seldom sessile); blades ovate to oblong in outline; pinnae wedge-shaped to fan-shaped, usually with definite corners along the outer margins...14

14a. Basal pinnae with narrow stalks (about ¼ the overall width of the pinnae), entire to shallowly lobed with the middle lobe usually the largest; plants dull green but neither pale nor glaucous when fresh...***B. minganense***
14b. Basal pinnae with wide stalks (about ½ the overall width of the pinnae), entire to irregularly toothed or scalloped, or lobed with the upper lobes usually the largest; plants yellowish-green to whitish-green, usually pale and glaucous when fresh...***B. furculatum***

***Botrychium ascendens* W.H. Wagner**, UPSWEPT MOONWORT. Trophophore stalk 3–10 mm; *blades* oblong to oblong-lanceolate, 1-pinnate, to 6 cm, yellow-green; *pinnae* to 5 pairs, strongly ascending, well-separated, the basal pinna pair about equal in size and shape to the adjacent pair, obliquely narrowly cuneate, undivided to the tip, the margins sharply toothed and sometimes shallowly incised, the apex rounded; *venation* like ribs of a fan with no midrib; *sporophores* 2-pinnate, the stalk to ½ the total length of the tropophore, 1.3–2 times the length of the trophophore in total length (including the stalk). Uncommon on open, subalpine slopes, 10,500–11,600 ft. E/W.

***Botrychium campestre* W.H. Wagner & Farrar ssp. *lineare* (W.H. Wagner) Farrar ined.**, PRAIRIE MOONWORT. [*B. lineare* W.H. Wagner]. Trophophore stalk usually absent; *blades* oblong, glaucous, longitudinally folded, 1-pinnate, to 4 cm; *pinnae* to 5(9) pairs, spreading, usually remote, the largest strongly asymmetrical and typically bifid, mostly linear to linear-spatulate, the basal pair about equal in size and shape to the adjacent pair, undivided to the tip, the margins mostly shallowly crenate or dentate; *venation* like ribs of a fan; *sporophores* 1-pinnate, the stalk usually less than ⅓ the total length of the trophophore, 1–1.5 times the length of the trophophore (including the stalk). Found on open, subalpine slopes along the Continental Divide and in grasslands near Bonny Reservoir on the eastern plains, 3800–11,700 ft. E/W.

Populations of *B.* "*furcatum*" (undescribed and previously thought to be a distinct species) are included here.

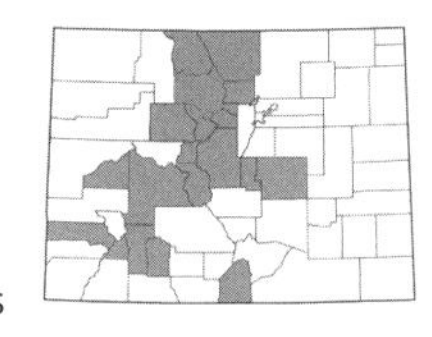

***Botrychium echo* W.H. Wagner**, ECHO MOONWORT. Trophophore stalk 0–4 mm; *blades* oblong to oblong-deltate, 1-2-pinnate, to 4 cm long, shiny green and firm; *pinnae* to 4+ pairs, spreading to somewhat ascending, well-separated, the basal pinna pair cleft into a smaller lobe and larger lobe, oblanceolate to linear-spatulate, with more or less parallel sides, divided to the tip, usually shallowly lobed; *venation* pinnate; *sporophores* 1–2-pinnate, 1–2 times the length of the trophophore (including the stalk). Common, easily overlooked in subalpine and alpine meadows and on rocky slopes, 9500–11,700 ft. E/W.

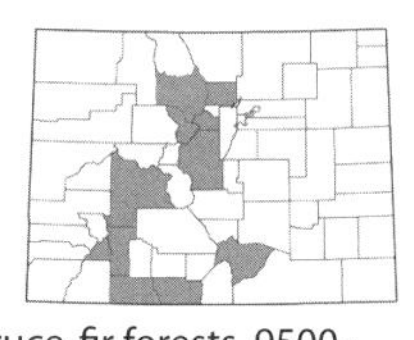

***Botrychium furculatum* Popovich & Farrar ined.**, FURCULATE MOONWORT. Trophophore stalk 3–8 mm; *blades* oblong, 1-pinnate, whitish-green to yellow-green, to 4 cm long; *pinnae* to 5 pairs, ascending-spreading, the basal pinna pair about equal in size and shape to the adjacent pair, narrowly fan-shaped to spatulate or sometimes short and stubby, the outer margins entire to irregularly toothed or scalloped; *venation* like the ribs of a fan; *sporophores* 1–2.3 times the length of the trophophore (including the stalk), the stalk about ½ the total length of the trophophore or more. Common but easily overlooked in subalpine meadows and openings in spruce-fir forests, 9500–11,000 ft. E/W.

Colorado plants originally identified as *B. pallidum* W.H. Wagner have been shown to be a distinct and separate species, now referred to as *B. furculatum*.

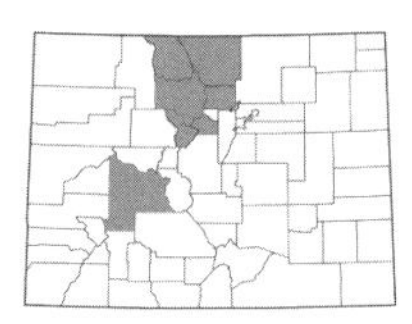

***Botrychium hesperium* (Maxon & R.T. Clausen) W.H. Wagner & Lellinger**, WESTERN MOONWORT. Trophopore stalk 0–3 (10) mm; *blades* oblong-linear to deltate, 1–2-pinnate, to 6 cm long, gray-green and dull; *pinnae* to 6 pairs, ascending, the basal pinna pair usually much larger and more divided than the adjacent pair, lobed to the tip, the basal pair oblong with lobed margins and the others broadly spatulate with entire margins or shallowly lobed, apex rounded; *venation* pinnate; *sporophores* 1–3-pinnate, 2–3 times the length of the trophophore (including the stalk). Common but easily overlooked in subalpine meadows and openings in spruce-fir forests, 8500–11,500 ft. E/W.

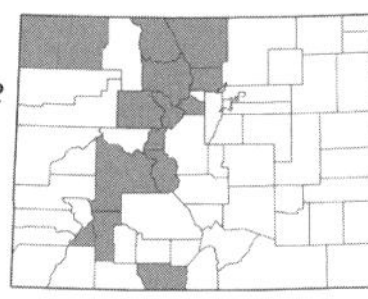

Botrychium lanceolatum* (S.G. Gmel.) Angstr. ssp. *lanceolatum, TRIANGLE MOONWORT. Trophophore stalk 0–1 mm; *blades* deltate, 1–2-pinnate, dull to shiny green or yellow-green, to 6 cm long; *pinnae* to 5 pairs, ascending, linear to broadly lanceolate, entire to divided to the tip, the margins with lobes or linear segments; *venation* pinnate; *sporophores* 1–3-pinnate, 1–2.5 times the length of the trophophore (including the stalk), divided into several equally long branches. Common but easily overlooked in subalpine meadows, openings in forests, and on rocky slopes, 9000–12,200 ft. E/W.

An all-green phenotype accommodated here may prove to be a distinct species, *B.* "*viride*." Colorado specimens generally have a red phenotype with a maroon or maroon-striped common stalk.

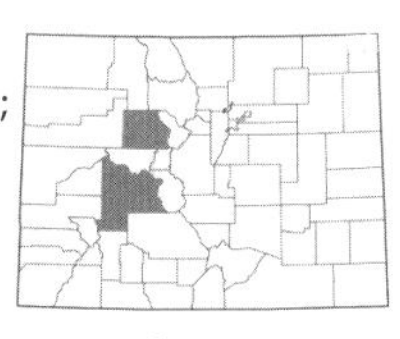

***Botrychium lunaria* (L.) Sw. var. *crenulatum* (W.H. Wagner) Stensvold ined.**, CRENULATE MOONWORT. [*B. crenulatum* W.H. Wagner]. Trophophore stalk 0.5–7 mm; *blades* oblong, 1-pinnate, yellow-green; *pinnae* to 5 pairs, well-separated and spreading, the basal pinna pair about equal in size and shape to the adjacent pair, broadly fan-shaped, undivided to the tip, the margins mostly crenulate to dentate, the apex rounded, the apical lobe linear to linear-cuneate and well-separated from adjacent lobes; *venation* like ribs of a fan with no midrib; *sporophores* 1–2-pinnate, 1.3–3 times the length of the trophophore (including the stalk). Rare on moist subalpine and alpine slopes, 10,500–12,000 ft. W.

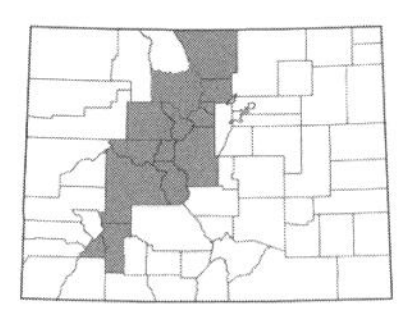

***Botrychium minganense* Vict.**, MINGAN MOONWORT. Trophophore stalk 0–2 cm; *blades* oblong to linear, 1-pinnate, to 10 cm, dull green; *pinnae* to 10 pairs, horizontal or slightly spreading, the basal pinna pair about equal in size and shape to the adjacent pair or occasionally the basal pinnae elongate, lobed to the tip, narrowly fan-shaped to ovate with concave sides, the margins nearly entire to shallowly crenate or sometimes pinnately lobed or divided, the apex rounded; *venation* like ribs of a fan; *sporophores* 1-pinnate (2-pinnate in robust plants), the stalk about ½ the total length of the trophophore or more, 1.5–2.5 times the length of the trophophore (including the stalk). Common, easily overlooked in subalpine meadows, forests, and on rocky slopes, 9500–12,000 ft. E/W.

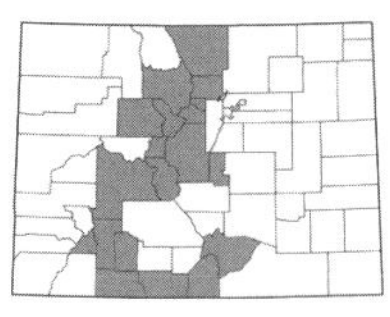

***Botrychium neolunaria* Stensvold & Farrar ined.**, COMMON MOONWORT. Trophophore stalk 0–1 mm; *blades* oblong, 1-pinnate, dark green, thick; *pinnae* to 9 pairs, mostly overlapping, spreading, the basal pinna pair about equal in size and shape to the adjacent pair, broadly fan-shaped, undivided to the tip, the margins usually entire or undulate, apical lobe usually cuneate to spatulate, notched; *venation* like ribs of a fan; *sporophores* 1–2-pinnate, 0.8–2 times the length of the trophophore (including the stalk). Common but easily overlooked in subalpine meadows, forest openings, and on rocky slopes, 9000–12,500 ft. E/W.

This is the species we have long referred to as *B. lunaria* (L.) Swartz, which is actually restricted in North America to Canada, Alaska, and Greenland. However, a specimen of *B. lunaria* was recently confirmed for north-central Wyoming.

***Botrychium paradoxum* W.H. Wagner**, PECULIAR MOONWORT. Trophophore replaced by a second sporophore; *sporophores* double, 1-pinnate, 0.5–4 cm long. Rare on subalpine slopes, currently known from one collection near Crested Butte, Gunnison Co., 9000–9500 ft. W.

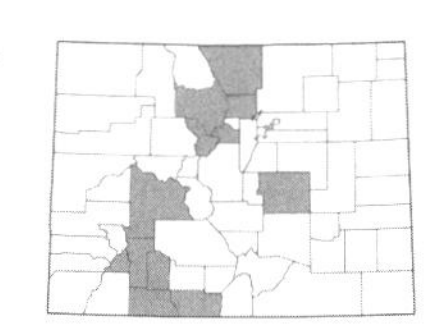

***Botrychium pinnatum* H. St. John**, NORTHERN MOONWORT. Trophopore stalk 0–2 mm; *blades* oblong-deltate, 1–2-pinnate, to 8 cm long, bright shiny green, thin; *pinnae* to 7 pairs, slightly ascending, the basal pinna pair about equal in size and shape to the adjacent pair, obliquely ovate to lanceolate-oblong or spatulate, deeply and regularly pinnatifid, lobed to the tip, the margins entire to shallowly crenate, the apex truncate to acute; *venation* pinnate; *sporophores* 2-pinnate, 1–2 times the length of the trophophore (including the stalk). Uncommon in forest openings and on moist subalpine slopes, 9500–11,000 ft. E/W.

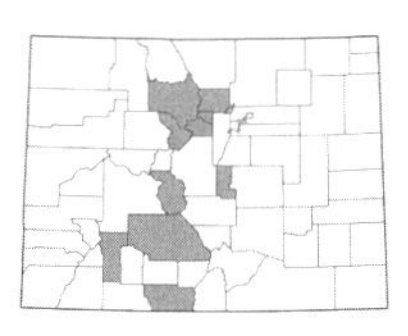

***Botrychium simplex* E. Hitchc.**, LEAST MOONWORT. Trophophore stalk 0–3 cm, 0–1.5 times the length of the trophophore rachis; *blades* linear to ovate-oblong to oblong overall or triangular with pinnae arranged ternately, to 7 cm long, simple to 2-pinnate, dull to bright green or whitish-green; *pinnae* to 7 pairs, spreading to ascending, cuneate to fan-shaped, often strongly asymmetric, the basal pinna pair often much larger and more complex than the adjacent pair, undivided to divided to the tip, the margins entire or shallowly sinuate, the apex rounded and undivided to strongly divided and plane; *venation* pinnate; *sporophores* mostly 1-pinnate, 1–8 times the length of trophophores. Uncommon in moist subalpine meadows and forest openings and along streams, 7500–11,500 ft. E/W.

There are two varieties of *B. simplex* in Colorado:

1a. Pinnae usually more rounded, often confluent with broad attachments to the rachis; trophophore simple or 1-pinnate with the basal pinnae much larger than the adjacent pinnae...**var. *simplex***

1b. Pinnae usually more fan-shaped, not often confluent to the rachis, with narrower attachments; trophophore usually 2-pinnate with the basal pair of pinnae pinnately divided into distinct fan-shaped segments...**var. *compositum* (Lash) Milde**. Less common than the previous, currently known only from near Wolf Creek Pass (Mineral Co.).

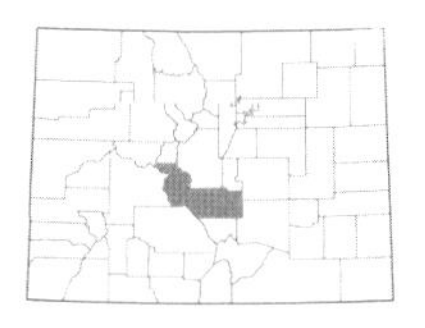

***Botrychium spathulatum* W.H. Wagner**, SPATULATE MOONWORT. Trophophore stalk 0–1 mm; *blades* narrowly deltate, 1-pinnate, to 8 cm, yellow green, thick and leathery; *pinnae* to 8 pairs, ascending, remote, the basal pinna pair about equal in size and shape or often larger to the adjacent pair, mostly spatulate and rounded or 2-cleft, lobed to unlobed to the tip, the margins mostly entire or sometimes shallowly incised, the apex rounded-notched; *venation* like ribs of a fan; *sporophore* 1–2-pinnate, the stalk less than ½ the total length of the trophophore, 1.2–2 times the length of the trophophore (including the stalk). Uncommon in subalpine meadows and on rocky slopes, 11,000–11,700 ft. E/W.

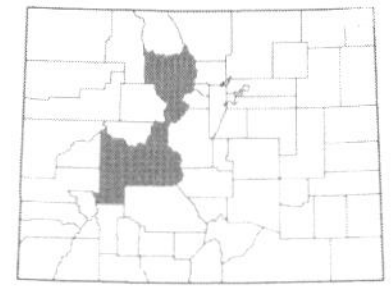

***Botrychium tunux* Stensvold & Farrar**, MOOSEWORT. Trophophore stalk 0–1 cm; *blades* ovate, 1-pinnate, to 7 cm long, yellow-green to dark glossy green and leathery; *pinnae* mostly 3–4 pairs, separated or slightly overlapping, the basal pinna pair broadly rounded and often asymmetrical, the margins entire or sometimes incised or lobed with rounded sinuses; *venation* like ribs of a fan; *sporophores* 1–2-pinnate, shorter than or equal to the trophophore in length (including the stalk). Uncommon on subalpine slopes, 10,000–11,700 ft. E/W.

Often found in Colorado with *B. campestre* ssp. *lineare*, and hybrids between the two have been reported.

BOTRYPUS Rich. – RATTLESNAKE-FERN

Trophophore sessile; *blades* 3–4-pinnate; *pinnae* to 12 pairs, lanceolate; *pinnules* lanceolate, deeply lobed into linear, serrate ultimate segments with pointed tips, midrib present; *sporophores* 2-pinnate, 0.5–1.5 (2) times the length of the trophophore (including the stalk).

***Botrypus virginianus* (L.) Michx.**, VIRGINIA RATTLESNAKE-FERN. [*Botrychium virginianum* (L.) Sw.]. With characteristics of the genus. Rare in cool, moist ravines and canyons, 6000–9500 ft. E/W.

SCEPTRIDIUM Lyon – GRAPE-FERN

Trophophore stalk 2–15 cm, 0.3–1.2 times the length of the trophophore rachis; *blades* 2–3-pinnate, leathery; *pinnae* to 10 pairs; *pinnules* obliquely ovate, rounded at the apex, usually with entire margins or sometimes shallowly crenulate, with pinnate venation; *sporophores* 2–3-pinnate, 1–1.2 times the length of the trophophore (including the stalk).

***Sceptridium multifidum* (S.G. Gmel.) M. Nishida ex Tagawa**, LEATHERY GRAPE-FERN. [*Botrychium multifidum* (S.G. Gmel.) Trevis.]. With characteristics of the genus. Rare in fens, along streams, and in moist upper montane to subalpine meadows, 6700–9500 ft. E/W.

POLYPODIACEAE J. Presl & C. Presl – TRUE FERN FAMILY

Perennials from creeping, scaly rhizomes; *leaves* monomorphic, pinnatifid, glabrous to puberulent below; *sori* round or elliptic, in 1–3 rows on either side of the midrib; *indusia* absent. (Lellinger, 1985; Harrington, 1950)

POLYPODIUM L. – POLYPODY

With characteristics of the family.

1a. Fronds triangular-ovate (widest near the base), mostly with 6–7 pairs of pinnae...***P. hesperium***
1b. Fronds oblong and about the same width throughout, mostly with 8–12 pairs of pinnae...***P. saximontanum***

***Polypodium hesperium* Maxon**, WESTERN POLYPODY. Leaves to 35 cm, glabrous, with entire to crenulate margins, widest near the base; *petiole* 1.5–8 cm long; *sori* oval. Rare on rock ledges and cliffs, known from the southwestern counties, 8000–8700 ft. W. (Plate 2)

***Polypodium saximontanum* Windham**, ROCKY MOUNTAIN POLYPODY. Leaves to 25 cm, glabrous, widest near the middle; *sori* circular. Found on granite cliff faces and in rocky crevices, 5800–9800 ft. E/W. (Plate 2)

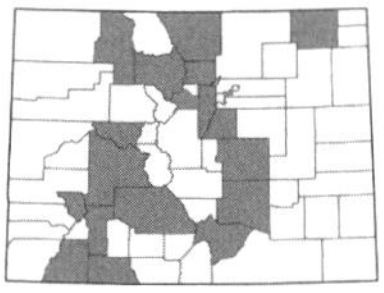

PTERIDACEAE E.D.M. Kirchn. – MAIDENHAIR FERN FAMILY

Terrestrial, usually perennial herbs; *stems* with hairs and/or scales, creeping or compact, branched or unbranched; *leaves* dimorphic or monomorphic, 1–6-pinnate; *sori* borne abaxially on veins, sometimes with a false indusium formed from a reflexed or revolute leaf margin; *spores* uniform. (Lellinger, 1985; Harrington, 1950)

1a. Lower surface of the leaves covered by a whitish or yellowish powdery coating...2
1b. Lower surface of the leaves not covered by a whitish or yellowish powdery coating...3

2a. Lower leaf surface whitish; ultimate segments of pinnae ovate to oval, 2–4 mm long...***Argyrochosma***
2b. Lower leaf surface yellow; ultimate segments of pinnae elliptic or linear-oblong, mostly over 10 mm long...***Notholaena***

3a. Leaves densely hairy or scaly below...4
3b. Leaves not densely hairy or scaly below...5

4a. Leaves 1-pinnate to pinnate-pinnatifid throughout...***Astrolepis***
4b. Leaves 3–4-pinnate at least at the base...***Cheilanthes***

5a. Ultimate divisions of the pinnae obliquely oblong and lobed on the upper side and entire on the lower side, or fan-shaped; sori discontinuous and borne on the reflexed margins of the upper lobes of the ultimate segments...***Adiantum***
5b. Ultimate divisions of the pinnae unlike the above; sori continuous, borne along the entire margin of fertile ultimate segments...6

6a. Leaves of 2 kinds, the fertile longer and of linear ultimate segments with inrolled margins bearing sori, 2–3-pinnate; rachis green or straw-colored throughout or green above and dark brown below...***Cryptogramma***
6b. Leaves all similar, 1–2-pinnate; rachis usually reddish-brown throughout, sometimes green above...***Pellaea***

ADIANTUM L. – MAIDENHAIR FERN

Leaves 2–3-pinnate, usually monomorphic, glabrous, the ultimate segments rhombic, fan-shaped, round, or oblong; *false indusium* marginal, concealing the sporangia; *sporangia* borne along the abaxial surface of the false indusium.

1a. Ultimate leaf segments obliquely oblong, lobed on the upper side and entire on the lower side...***A. aleuticum***
1b. Ultimate leaf segments fan-shaped or irregularly rhombic, unlobed...***A. capillus-veneris***

***Adiantum aleuticum* (Rupr.) C.A. Paris**, ALEUTIAN MAIDENHAIR FERN. Leaves monomorphic, 2–7-pinnate below, 5–45 × 5–45 cm, the blade fan-shaped; *ultimate segments* oblong or reniform, the margin lobed; *false indusium* oblong to crescent-shaped, 0.2–3.5 mm long, glabrous. Rare in shady spruce forests, known from Ouray and San Juan cos., 8500–9600 ft. W. (Plate 2)

***Adiantum capillus-veneris* L.**, COMMON MAIDENHAIR FERN. Leaves monomorphic, 3–4-pinnate below, 10–45 × 4–15 cm, the blade lanceolate; *ultimate segments* cuneate or fan-shaped; *false indusium* oblong or crescent-shaped, 1–3 (7) mm long, glabrous. Rare on moist cliffs and near springs and seepages, 4200–7800 ft. E/W.

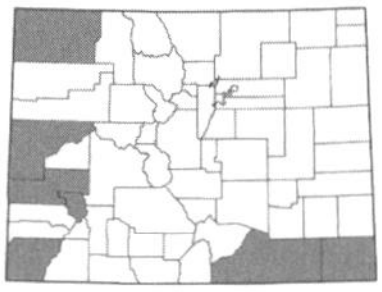

ARGYROCHOSMA (J. Sm.) Windham – FALSE CLOAK FERN

Leaves monomorphic, 2–6-pinnate, abaxially covered by a whitish powdery coating (ours); *false indusium* narrow, marginal, concealing sporangia; *sporangia* scattered along veins of lower leaf surface, often submarginal.

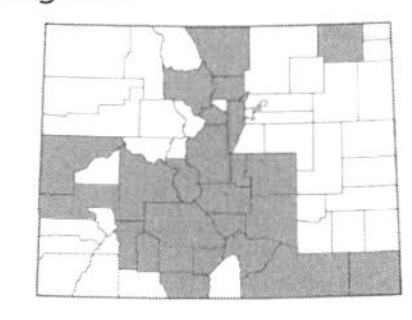

***Argyrochosma fendleri* (Kunze) Windham**, FENDLER'S FALSE CLOAK FERN. [*Notholaena fendleri* Kunze]. Leaves 5–25 cm long; *petioles* dark brown. Found in rocky crevices and on cliffs, 5300–9600 ft. E/W. (Plate 2)

ASTROLEPIS D.M. Benham & Windham – CLOAKFERN

Leaves monomorphic, linear to linear-oblong, 1-pinnate to pinnatifid, leathery, the upper surface with stellate or coarsely ciliate scales, or glabrescent when mature; *false indusium* absent; *sporangia* scattered along veins near margins.

1a. Largest pinnae mostly 4–7 mm long; abaxial scales ovate, mostly 0.5–1 mm long; adaxial scales attached in the middle (peltate)...***A. cochisensis* ssp. *cochisensis***

1b. Largest pinnae mostly 7–15 mm long; abaxial scales lanceolate, mostly 1–1.5 mm long; adaxial scales attached at the base...***A. integerrima***

Astrolepis cochisensis* (Goodd.) D.M. Benham & Windham ssp. *cochisensis, COCHISE SCALY CLOAKERN. [*Notholaena cochisensis* Goodd.]. Leaves 7–40 cm long, 1-pinnate to pinnate-pinnatifid, with 20–50 pinna pairs; *pinnae* oblong, the largest usually 4–7 mm, entire or lobed, with abaxial scales concealing the surface. Rare in rocky crevices, 5000–5200 ft. E.

***Astrolepis integerrima* (Hook.) D.M. Benham & Windham**, HYBRID CLOAKERN. [*Notholaena integerrima* (Hook.) Hevly]. Leaves 1-pinnate to pinnate-pinnatifid, 8–45 cm long, with 20–45 pinna pairs; *pinnae* oblong to ovate, the largest usually 7–15 mm long, entire or lobed, with abaxial scales concealing the surface. Occurs close to the border of New Mexico and Colorado, but is not yet reported for the state.

CHEILANTHES Sw. – LIP FERN

Leaves monomorphic, pinnate-pinnatifid to 4-pinnate, usually hairy and/or scaly below, the ultimate segments bead-like; *false indusium* marginal, formed by the reflexed edge of the ultimate leaf segment; *sporangia* marginal, separate or narrowly confluent.

1a. Leaves densely hairy below (with or without scales)...2

1b. Leaves scaly below but lacking hairs, or sometimes sparsely hairy...3

2a. Ultimate leaf segments glabrous above...***C. feei***

2b. Ultimate leaf segments densely to sparsely tomentose above...***C. eatonii***

3a. Scales on the leaf rachis not ciliate...***C. fendleri***

3b. Scales on the leaf rachis ciliate, at least toward the base...***C. wootonii***

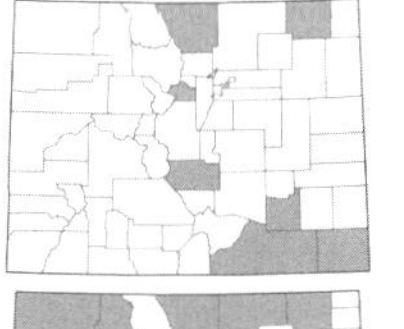

***Cheilanthes eatonii* Baker**, EATON'S LIP FERN. Leaves 6–35 cm long, 3–4-pinnate below, petiole dark brown, with scales and hairs below; *ultimate segments* oval, 1–3 mm long, densely tomentose below. Found in rock crevices, at the base of boulders, and on cliffs, 3500–6500 ft. E. (Plate 2)

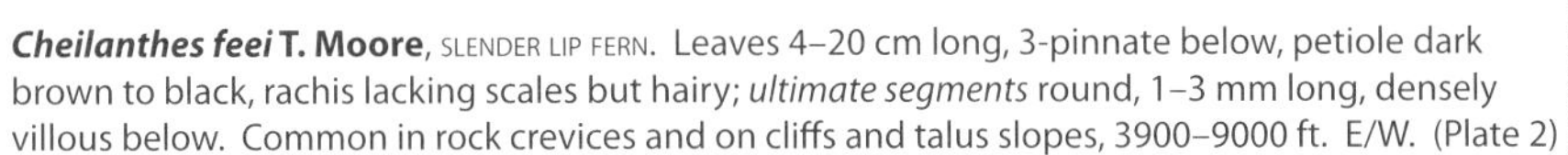

***Cheilanthes feei* T. Moore**, SLENDER LIP FERN. Leaves 4–20 cm long, 3-pinnate below, petiole dark brown to black, rachis lacking scales but hairy; *ultimate segments* round, 1–3 mm long, densely villous below. Common in rock crevices and on cliffs and talus slopes, 3900–9000 ft. E/W. (Plate 2)

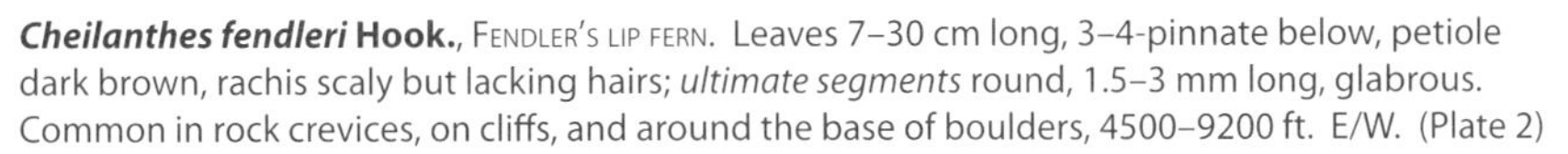

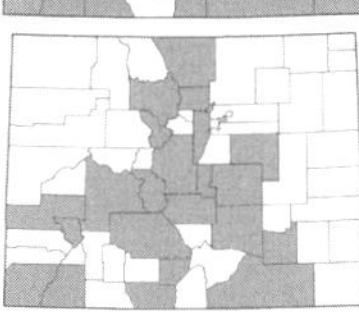

***Cheilanthes fendleri* Hook.**, FENDLER'S LIP FERN. Leaves 7–30 cm long, 3–4-pinnate below, petiole dark brown, rachis scaly but lacking hairs; *ultimate segments* round, 1.5–3 mm long, glabrous. Common in rock crevices, on cliffs, and around the base of boulders, 4500–9200 ft. E/W. (Plate 2)

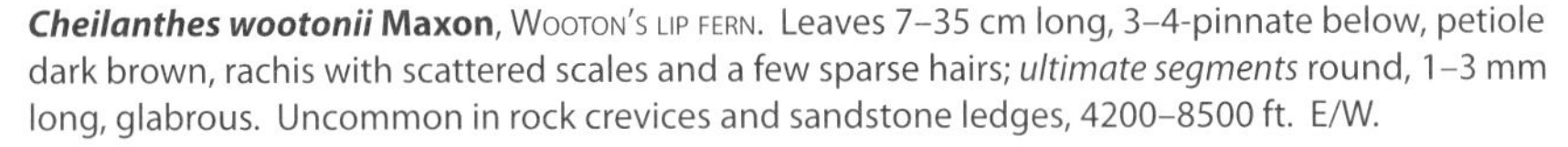

***Cheilanthes wootonii* Maxon**, WOOTON'S LIP FERN. Leaves 7–35 cm long, 3–4-pinnate below, petiole dark brown, rachis with scattered scales and a few sparse hairs; *ultimate segments* round, 1–3 mm long, glabrous. Uncommon in rock crevices and sandstone ledges, 4200–8500 ft. E/W.

CRYPTOGRAMMA R. Br. – PARSELY FERN; ROCK BRAKE

Leaves 2–4-pinnate, dimorphic; *sterile leaves* shorter than fertile leaves, ultimate segments of sterile leaves ovate, elliptic, or fan-shaped; *fertile leaves* lanceolate to linear with reflexed margins forming a false indusium extending over the entire length of the segments; *sporangia* usually hidden by false indusium, scattered along veins on abaxial leaf surface.

1a. Leaves densely clustered, firm...***C. acrostichoides***

1b. Leaves scattered along the stem, thin and delicate...***C. stelleri***

***Cryptogramma acrostichoides* R. Br.**, AMERICAN PARSELY FERN. Sterile leaves 3–20 cm long, 2–3-pinnate, the ultimate segments oblong to ovate-lanceolate; *fertile leaves* 5–30 cm long, 2-pinnate, the ultimate segments linear, with revolute margins covering sporangia. Common in rock crevices, on cliffs, and on talus slopes, 6200–12,500 ft. E/W. (Plate 2)

***Cryptogramma stelleri* (S.G. Gmel.) Prantl**, SLENDER ROCK BRAKE. Sterile leaves 3–15 cm long, mostly 2-pinnate, the ultimate segments ovate-lanceolate to fan-shaped, shallowly lobed above; *fertile leaves* 5–20 cm long, mostly 2-pinnate, the ultimate segments lanceolate to linear, the margins reflexed and covering the sporangia. Rare on cliffs and rock ledges, 8500–11,000 ft. W.

NOTHOLAENA R. Br. – CLOAK FERN

Perennials; *leaves* monomorphic, 4-pinnate, leathery, covered abaxially by a whitish or yellowish powdery coating; *false indusium* narrow, marginal; *sporangia* on submarginal vein tips.

***Notholaena standleyi* Maxon**, STAR CLOAK FERN. Leaves 5–30 cm long, broadly pentagonal, with a yellowish powdery coating below; *petiole* glabrous, with a few scales at the base, equal to or longer than the blade. Uncommon in rocky crevices and on cliffs, 4000–5500 ft. E. (Plate 2)

PELLAEA Link – CLIFF-BRAKE

Leaves monomorphic to somewhat dimorphic, 1–4-pinnate below, usually leathery; *false indusium* narrow, marginal, at least partially concealing the sporangia; *sporangia* scattered along veins near segment margins. (Tryon & Britton, 1958)

1a. Ultimate leaf segments with a short, bristle-tip; stem scales bicolored with black centers and brown margins; leaves twice compound (at least below)...2
1b. Ultimate leaf segments lacking a short bristle-tip; stem scales uniformly reddish-brown; leaves once compound or twice compound below...3

2a. Larger pinnae divided into 10 or more ultimate segments...***P. truncata***
2b. Larger pinnae usually divided into 3–5 segments...***P. wrightiana***

3a. One side of the leaf rachis densely pubescent with curved, upturned hairs...***P. atropurpurea***
3b. Leaf rachis glabrous or sparsely pubescent with long hairs...4

4a. Lower pinnae usually with an enlarged, upper lobe (giving the pinnae a mitten-shape); upper ½ to ¼ of the rachis green; sporangia sessile or nearly so...***P. breweri***
4b. Lower pinnae entire or clearly divided into 2–3 distinct ultimate divisions; rachis reddish-brown or brown throughout; at least some sporangia long-stalked...5

5a. Rachis and petiole sparsely hairy with long, divergent hairs; ultimate segments of the pinnae sparsely hairy below along the midrib...***P. gastonyi***
5b. Rachis and petiole glabrous or nearly so; ultimate segments of the pinnae glabrous or occasionally with sparse, hair-like scales below along the midrib...***P. glabella***

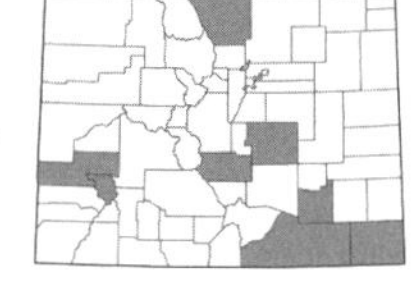

***Pellaea atropurpurea* (L.) Link**, PURPLE CLIFF-BRAKE. Leaves somewhat dimorphic, sterile leaves shorter and not as divided as fertile leaves, 5–50 cm long, usually 2-pinnate below, rachis densely hairy above; *ultimate segments* linear-oblong, 10–75 mm long, the margins weakly recurved, borders whitish, crenulate. Uncommon in rock crevices and on limestone cliffs, 4000–8900 ft. E/W.

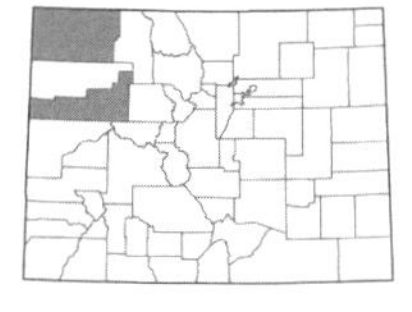

***Pellaea breweri* D.C. Eaton**, BREWER'S CLIFF-BRAKE. Leaves monomorphic, 2.5–20 cm long, pinnate-pinnatifid below, rachis glabrous to sparsely hairy; *ultimate segments* lanceolate-deltate, 5–25 mm long, margins recurved, borders whitish, erose-toothed. Rare on rimrock and cliffs, usually on limestone, 6000–11,200 ft. W.

***Pellaea gastonyi* Windham**, GASTONY'S CLIFF-BRAKE. Leaves somewhat dimorphic, sterile leaves shorter than fertile leaves, 8–25 cm long, 2-pinnate below, rachis sparsely hairy; *ultimate segments* oblong-lanceolate, 7–30 mm long, margins recurved on fertile segments, borders whitish, crenulate. Rare on limestone cliffs and ledges, known from Larimer Co., 5800–6200 ft. E.

Pellaea glabella **Mett. ex Kuhn**, SMOOTH CLIFF-BRAKE. Leaves monomorphic, 2–40 cm long, the rachis glabrous or nearly so; *ultimate segments* oblong-lanceolate, 5–20 mm long, margins recurved, borders whitish, erose-toothed. (Plate 2)

There are two subspecies of *P. glabella* in Colorado:

1a. Sporangia containing 32 spores; ultimate segments glabrous or nearly so...**ssp. *simplex* (Butters) Á. Löve & D. Löve**. Uncommon on limestone cliffs and outcrops, 5300–7800 ft. E/W.

1b. Sporangia containing 64 spores...**ssp. *occidentalis* (E.E. Nelson) Windham**. Rare on limestone cliffs and ledges, 5500–6200 ft. E (but could be present on the western slope).

Pellaea truncata **Goodd.**, SPINY CLIFF-BRAKE. Leaves somewhat dimorphic, sterile leaves shorter than fertile leaves, 8–40 cm long, 2-pinnate below, rachis glabrous; *ultimate segments* narrowly oblong, 4–10 mm long, margins recurved, borders whitish, entire or nearly so. Rare on cliffs and rocky outcrops, known from Fremont Co., 5200–7000 ft. E. (Plate 2)

Pellaea wrightiana **Hook.**, WRIGHT'S CLIFF-BRAKE. Leaves monomorphic, 6–40 cm long, 2-pinnate below, rachis glabrous; *ultimate segments* narrowly oblong, 5–20 mm long, margins recurved, borders whitish, crenulate. Rare on cliffs, canyons, and rocky slopes, 4700–6500 ft. E.

SELAGINELLACEAE Willk. – SPIKE-MOSS FAMILY

Terrestrial, usually evergreen herbs; *stems* prostrate or erect, leafy, branching dichotomously or monopodially; *leaves* simple, single-veined, imbricate, arranged in several ranks; *sporophylls* only slightly differentiated from vegetative leaves, arranged in terminal cones; *sporangia* solitary in the axils of sporophylls; *spores* of 2 types. (Lellinger, 1985; Harrington, 1950; Valdespino, 1993)

SELAGINELLA P. Beauv. – SPIKE-MOSS

With characteristics of the family.

1a. Terminal strobili (clusters of overlapping fertile leaves) cylindric; fertile leaves spreading; leaves with soft, spiny projections on the margins, lacking a bristle tip and dorsal groove...***S. selaginoides***

1b. Terminal strobili 4-sided; fertile leaves usually appressed; leaves ciliate-margined but not spiny, with a dorsal groove and the apex often bristle-tipped...2

2a. Leaves lacking a bristle tip (or if present then less than 0.06 mm long), tightly appressed and 1–2 mm long...***S. mutica***

2b. Leaves with a white, yellowish, or transparent bristle tip (0.3–1 mm long), otherwise various...3

3a. Aerial stems erect, the plants forming erect, dense tufts, not rooting along a creeping stem; leaves of rhizomatous (underground) stem loosely appressed and often incurved, leaves of aerial stem tightly appressed...***S. weatherbiana***

3b. Aerial stems prostrate or creeping (sometimes the short lateral branches strongly ascending but the main stem creeping), rooting along the stem; leaves unlike the above...4

4a. Plants densely matted with short, compact branches, the lateral branches strongly curved-ascending; leaves on the lower side of the stem somewhat longer than the others...***S. densa***

4b. Plants loosely matted with long, spreading branches; leaves not obviously longer on one side of the stem than the other...***S. underwoodii***

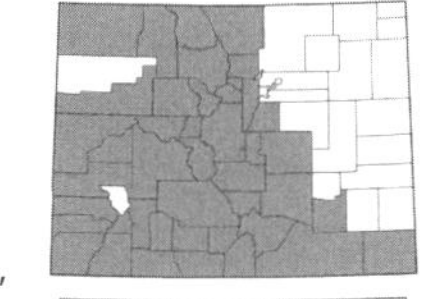

Selaginella densa **Rydb.**, ROCKY MOUNTAIN SPIKE-MOSS. [*S. scopulorum* Maxon; *S. standleyi* Maxon]. Plants forming cushions or loose mats; *leaves* in whorls of 4–6, linear-lanceolate, 2–5 mm long; *strobili* 1–3 cm long. Common in dry grasslands, rocky forests, and on rocky slopes and outcrops, cliffs, and alpine ridges and boulder fields, 4000–13,000 ft. E/W. (Plate 4)

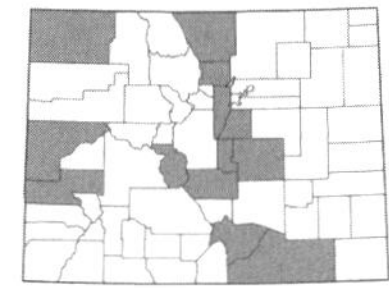

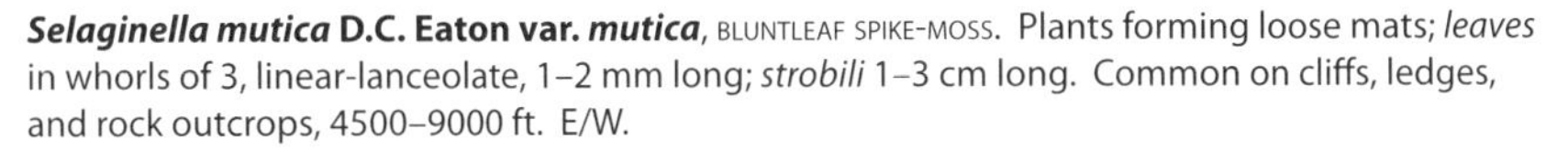

Selaginella mutica **D.C. Eaton var. *mutica***, BLUNTLEAF SPIKE-MOSS. Plants forming loose mats; *leaves* in whorls of 3, linear-lanceolate, 1–2 mm long; *strobili* 1–3 cm long. Common on cliffs, ledges, and rock outcrops, 4500–9000 ft. E/W.

Selaginella selaginoides **(L.) Schrank & Mart.**, NORTHERN SPIKE-MOSS. Plants forming loose to dense mats; *leaves* spirally arranged, lanceolate, 3–4.5 mm long; *strobili* 1–3 (5) cm long. Rare on mossy stream banks and in bogs, reported for Jackson Co., but no specimens have been seen. E.

***Selaginella underwoodii* Hieron.**, Underwood's spike-moss. Plants forming loose mats; *leaves* in whorls of 3–4, linear-lanceolate, 2–3.5 mm long; *strobili* 0.5–3.5 mm long. Found on rocky slopes, cliffs, rock outcrops, and ledges, 4500–10,500 ft. E/W.

***Selaginella weatherbiana* R. Tryon**, Weatherby's spike-moss. Plants forming clumps; *leaves* dimorphic, rhizomatous stem leaves loosely appressed and scale-like, aerial stem leaves tightly appressed and linear-lanceolate, 1.7–2.5 mm long; *strobili* 1–3 cm long. Found on rocky outcrops and cliffs, 5300–11,300 ft. E.

THELYPTERIDACEAE Ching ex Pic. Serm. – marsh fern family

Stems long-creeping; *leaves* monomorphic, 2–3-pinnatifid, pubescent below; *indusia* absent (ours); *sori* round to oblong, not marginal.

PHEGOPTERIS (C. Presl) Fée – beech fern

With characteristics of the family.

***Phegopteris connectilis* (Michx.) Watt**, long beechfern. Leaves 15–60 cm long; *pinnae* deeply pinnatifid; *ultimate segments* entire to crenate; *petiole* (8) 15–35 cm long, with brown scales at the base. Rare on rock ledges, known from a single collection in Gunnison Co., 8500–9000 ft. W.

Pinyon pine overlooking Great Sand Dunes National Park, Colorado

GYMNOSPERMS

CUPRESSACEAE Gray – CYPRESS FAMILY

Evergreen shrubs or trees; *leaves* scale-like or subulate and linear, opposite or in whorls of 3; *staminate cones* catkin-like, globose to ovoid; *pistillate cones* of 2–6 united, fleshy scales, bluish, glaucous, berry-like; *seeds* wingless.

JUNIPERUS L. – JUNIPER

With characteristics of the family. (Adams, 2014)

1a. Leaves all needle-like or awl-shaped (no scale-like leaves present), in whorls of 3; cones axillary; prostrate or low shrubs mostly to 3 m tall…***J. communis* var. *depressa***
1b. Leaves mostly scale-like and opposite (sometimes awl-shaped leaves present as well); cones terminal; erect shrubs or trees to 7 m tall…2

2a. Leaf margins entire…***J. scopulorum***
2b. Leaf margins denticulate or minutely ciliate (at 20×)…3

3a. Scalelike leaves with a conspicuous resin gland below; pistillate cones with lots of resin and juicy when crushed…***J. monosperma***
3b. Scalelike leaves lacking a conspicuous resin gland below; pistillate cones dry, mealy, and fibrous at maturity…***J. osteosperma***

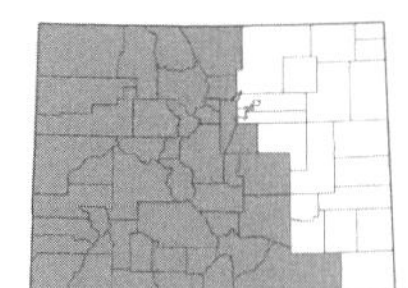

***Juniperus communis* L. var. *depressa* Pursh**, COMMON JUNIPER. Shrubs, low or spreading, to 3 m; *leaves* 5–10 (15) mm long, green, with a white band on the upper surface, mostly in whorls of 3, awl-shaped; *cones* 6–13 mm diam., bluish-black at maturity. Common in aspen or spruce-fir forests, 5500–12,700 ft. E/W.

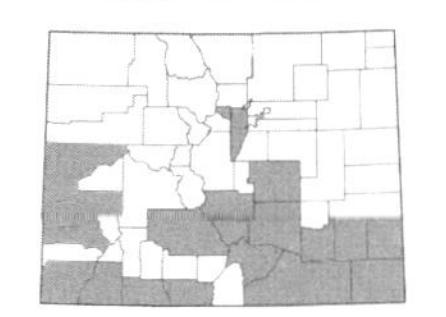

***Juniperus monosperma* (Engelm.) Sarg.**, ONE-SEED JUNIPER. [*Sabina monosperma* (Engelm.) Rydb.]. Trees or shrubs, 2–7 m; leaves opposite or sometimes in 3's, scale-like 1–3 mm long, denticulate, awl-shaped to 5 mm long; *cones* 4–7 mm diam., dark blue to blue-purple at maturity. Found on dry, rocky slopes, known from the southern counties, 4000–6700 ft. E/W.

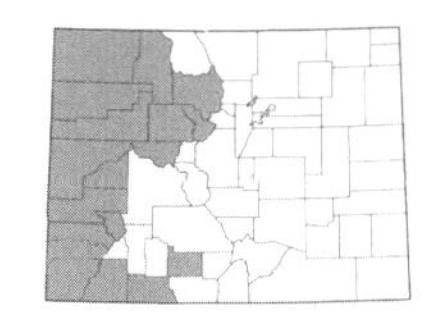

***Juniperus osteosperma* (Torr.) Little**, UTAH JUNIPER. [*Sabina osteosperma* (Torr.) Antoine]. Trees or shrubs, 2–4 m; *leaves* opposite or sometimes in 3's, scale-like 1–3 mm long, denticulate on the margins, awl-shaped 2–8 mm long, *cones* 6–12 mm diam., blue-purple at maturity. Found on dry, rocky slopes, sometimes with pinyon pine or sagebrush, common on the western slope, 4300–7700 ft. W.

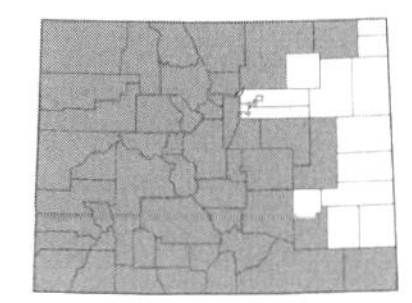

***Juniperus scopulorum* Sarg.**, ROCKY MOUNTAIN JUNIPER. [*Sabina scopulorum* (Sarg.) Rydb.]. Trees, 3–6 m; *leaves* opposite or sometimes in 3's, green to blue-green, scale-like 0.5–4 mm long, awl-shaped 3–8 mm long, entire; *cones* 4–6 mm diam., bluish at maturity. Common on dry slopes, often with sagebrush, pinyon pine, ponderosa pine, or oak communities, 4000–10,000 ft. E/W. (Plate 4)

EPHEDRACEAE Dumort. – EPHEDRA FAMILY

Dioecious shrubs with green, striate stems; *leaves* scarious and scale-like, opposite or in whorls of 3; *staminate cones* compound, whorled at the nodes or terminal, with 2–8 microsporophylls; *pistillate cones* solitary or whorled, subtended by bracts; *seeds* 1–3.

EPHEDRA L. – MORMON TEA; JOINTFIR

With characteristics of the family. (Cutler, 1939)

1a. Leaves and bracts mostly in whorls of 3 per node; branches whorled; pistillate cone scales scarious throughout…***E. torreyana***
1b. Leaves and bracts 2 per node; branches opposite, sometimes becoming whorled above; pistillate cone scales scarious on the margins only…***E. viridis***

***Ephedra torreyana* S. Watson**, TORREY'S EPHEDRA. Branches solitary or mostly whorled at the nodes; *leaves* whorled, 2–5 mm long, connate for ⅔ their length; *male cones* solitary or whorled, 6–8 mm long; *female cones* solitary to several per node, 9–13 mm long, sessile, the bracts scarious. Found on dry, rocky or sandy slopes, 4500–6500 ft. W.

***Ephedra viridis* Coville**, GREEN EPHEDRA; MORMON TEA. Branches opposite or sometimes whorled above; *leaves* opposite, 1.5–4 mm long; *male cones* 2 or more, sessile, 5–7 mm long; *female cones* sessile or pedunculate, 6–10 mm long, the bracts scarious on the margins. Found on dry, rocky or sandy slopes and on mesa tops, 4500–7500 ft. W. (Plate 4)

There are two varieties of *E. viridis* in Colorado:

1a. Pistillate cones on peduncles 5–25 mm long; young stems usually viscid (with grains of sand adhering); plants rhizomatous...**var. *viscida* (Cutler) L.D. Benson**, CUTLER'S EPHEDRA. [*E. cutleri* Peebles]. Less common than var. *viridis*, 5000–7500 ft. W.

1b. Pistillate cones sessile or if on peduncles these up to 4 mm long; young stems not viscid; plants not rhizomatous...**var. *viridis***. Common in the western counties, 4500–7500 ft. W.

PINACEAE Spreng. ex F. Rudolphi – PINE FAMILY

Evergreen (ours) trees; *leaves* linear or acicular, often fascicled; *staminate cones* small, not woody, with spirally arranged microsporophylls; *pistillate cones* woody, with spirally arranged scales, each subtended by a bract and with 2 ovules on the adaxial side; *seeds* winged or not. (Critchfield, 1966; Thieret, 1993)

1a. Leaves present...**Key 1**

1b. Only cones present...**Key 2**

Key 1
(*Leaves present*)

1a. Leaves fascicled in clusters of 2–5...***Pinus***

1b. Leaves borne single on the stem...2

2a. Leaves somewhat 4-angled, with sharp tips; twigs and branchlets rough where the leaves have fallen off (the leaves deciduous above the persistent base); cones pendulous, the scales thin and usually erose at the tips...***Picea***

2b. Leaves flat, with blunt tips; twigs and branchlets smooth where leaves have fallen off; cones unlike the above...3

3a. Leaves about the same width throughout; leaf scars circular or nearly so; terminal buds resinous; cones erect, deciduous at maturity...***Abies***

3b. Leaves narrowed to a very short petiole at the base; leaf scars elliptic; terminal buds not resinous; cones pendulous, the scales subtended and surpassed by a conspicuous, 3-lobed bract...***Pseudotsuga***

Key 2
(*Only cones present*)

1a. Cone scales subtended and surpassed by a conspicuous, 3-lobed bract...***Pseudotsuga***

1b. Cone scales not subtended and surpassed by a conspicuous, 3-lobed bract...2

2a. Cones erect, the scales deciduous at maturity (rarely seen)...***Abies***

2b. Cones pendulous or spreading, the scales persistent...3

3a. Scales thin, usually erose at the tips, never with a slender bristle tip...***Picea***

3b. Scales thick, woody, not erose at the tips, sometimes with a slender bristle tip...***Pinus***

ABIES Mill. – FIR

Leaves not fascicled, spirally arranged, with a nearly circular leaf scar; *pistillate cones* erect, the woody scales deciduous at maturity.

1a. Leaves (those on lower branches) 3–6 cm long; cross-section of a leaf showing resin ducts located near the margins of the leaves near the lower epidermis; scales of pistillate cones 30–35 mm wide, grayish or yellowish-green...***A. concolor***

1b. Leaves (those on lower branches) 2–3 cm long; cross-section of a leaf showing resin ducts located about halfway between the lower and upper epidermis and midway between the leaf margin and midvein; scales of pistillate cones 12–25 mm wide, brownish-purple to purple...***A. bifolia***

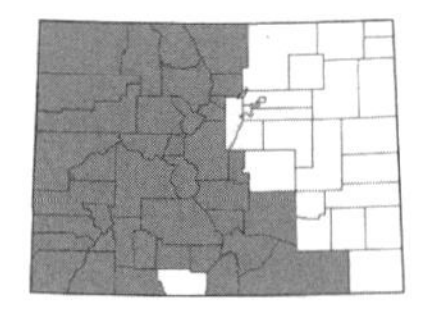

***Abies bifolia* A. Murray**, ROCKY MOUNTAIN SUBALPINE FIR. Trees to 35 m; *bark* gray, smooth; *leaves* solitary, 1.1–2.5 cm long; *pistillate cones* 5–10 cm long, sessile, dark purplish-blue. Found in the mountains and subalpine forests, often in association with *Picea engelmannii*, 7800–12,500 ft. E/W. (Plate 4)

This is the species that we have long called *A. lasiocarpa* (Hook.) Nutt., a species of the Pacific Northwest.

Abies concolor **(Gord. & Glend.) Hildebr.**, WHITE FIR. Trees to 80 m; *bark* gray, smooth; *leaves* solitary, 1.5–6 cm long; *pistillate cones* 7–12 cm long, reddish, purple, or greenish. Found in the southcentral and southwestern counties, 7500–11,000 ft. E/W.

PICEA A. Dietr. – SPRUCE

Leaves not fascicled, spirally arranged, square in cross-section or flattened, soon deciduous from cut branches, leaving short pegs on the twig after falling; *pistillate cones* pendulous, with thin, woody, persistent scales. (Beidleman, 1960; Mitton & Andalora, 1981)

1a. Pistillate cones 3–7 (8) cm long, the scale extending 3–8 mm beyond the seed-wing impression; young twigs finely pubescent with brownish-yellow hairs, rarely glabrous...***P. engelmannii*** **var.** ***engelmannii***
1b. Pistillate cones 5–11 (12) cm long, the scale extending 8–10 mm beyond the seed-wing impression; young twigs usually glabrous...***P. pungens***

Picea engelmannii **Parry ex Engelm. var.** ***engelmannii***, ENGELMANN SPRUCE. Trees to 45 m; *bark* gray or reddish-brown, scaly; *leaves* bluish-green, 1.3–3 cm long; *pistillate cones* purplish-brown to brown, 3–7 (8) cm long. Common in the mountains, usually found with *Abies bifolia* in subalpine forests, 7300–12,000 ft. E/W.

Picea pungens **Engelm.**, COLORADO BLUE SPRUCE. Trees to 50 m; *bark* gray-brown or brownish, scaly; *leaves* bluish-green, 1.2–3 cm long; *pistillate cones* brown, 5–11 (12) cm long. Common in the mountains, often found with *Pseudotsuga menziesii* or *Abies concolor*, 7000–9500 (11,000) ft. E/W. (Plate 4)

This is the state tree of Colorado and is often planted as an ornamental.

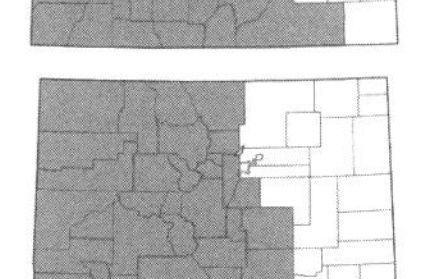

PINUS L. – PINE

Leaves acicular, in fascicles of 2–5 (ours); *pistillate cones* pendulous or spreading, with woody, persistent scales. (Andresen & Steinhoff, 1971; Haller, 1965)

1a. Leaves 2–3 per fascicle...2
1b. Leaves 5 per fascicle...4

2a. Leaves mostly 9–20 (30) cm long, usually 3 per fascicle but sometimes 2; bark thick and deeply furrowed, reddish-brown...***P. ponderosa*** **var.** ***scopulorum***
2b. Leaves 1.5–9 cm long, usually 2 per fascicle but sometimes 3; bark unlike the above...3

3a. Trees to 15 m tall, often with a twisted or crooked trunk; leaves 2–4 (5) cm long; scales of pistillate cones not prickly near the tip; seeds large (10–16 mm long) and wingless...***P. edulis***
3b. Trees with a long, bare, slender trunk, to 30 m tall; leaves (2.5) 3–7 (8.5) cm long; scales of pistillate cones prickly near the tip; seeds 4–5 mm long with a prominent wing...***P. contorta*** **var.** ***latifolia***

4a. Leaves 3–4 cm long, covered with drops of resin; scales of pistillate cones with a slender bristle tip 4–10 mm long...***P. aristata***
4b. Leaves 2–9 cm long, not covered with drops of resin; scales of pistillate cones lacking a bristle tip...5

5a. Cones 6–15 cm long; terminal, exposed portion of cone scales not recurved...***P. flexilis***
5b. Cones 15–25 cm long; terminal, exposed portion of cone scales recurved...***P. strobiformis***

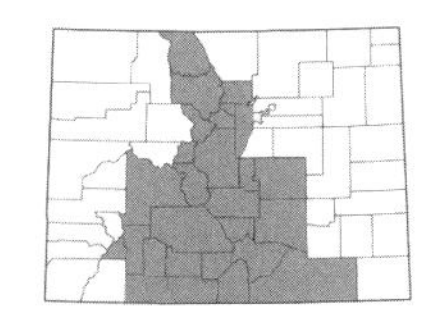

Pinus aristata **Engelm.**, BRISTLECONE PINE. Trees to 15 m; *bark* gray to reddish-brown, ridged; *leaves* 5 per fascicle, 3–4 cm long, covered with drops and scales of resin; *pistillate cones* 6–11 cm long, purplish to brown, umbo central, extended into slender prickle 4–10 mm long. Scattered in subalpine forests, 8300–13,000 ft. E/W. (Plate 4)

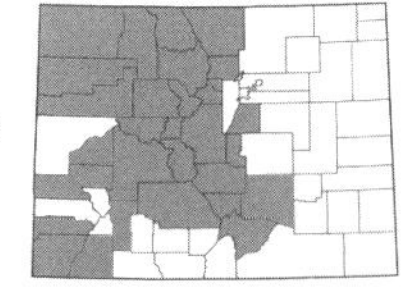

Pinus contorta **Douglas ex Loud. var.** ***latifolia*** **Engelm.**, LODGEPOLE PINE. Trees to 50 m; *bark* grayish to red-brown, scaly, not ridged; *leaves* 2 per fascicle, (2.5) 3–7 (8.5) cm long; *pistillate cones* 3–6 cm long, umbo central, prickle short and barely elongate. Dominant tree in montane forests, 7500–11,600 ft. E/W. (Plate 4)

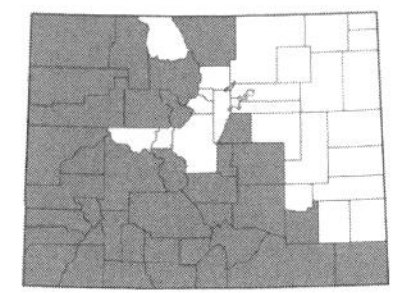

Pinus edulis **Engelm.**, PINYON PINE. Shrubs or trees to 20 m; *bark* reddish-brown, furrowed; *leaves* usually 2 per fascicle, 2–4 (5) cm long; *pistillate cones* depressed-ovoid to nearly globose, 3–5 cm long, seeds wingless. Common on dry slopes, mesas, and in canyons, often with junipers, 4000–9500 ft. E/W. (Plate 4)

***Pinus flexilis* E. James**, LIMBER PINE. Trees to 25 m; *bark* gray, nearly smooth to furrowed; *leaves* 5 per fascicle, 3–7 cm long; *pistillate cones* 6–15 cm long, umbo terminal, unarmed, seeds winged. Scattered in montane to subalpine forests, often growing in rocky crevices, 5200–12,000 ft. E/W. (Plate 4)

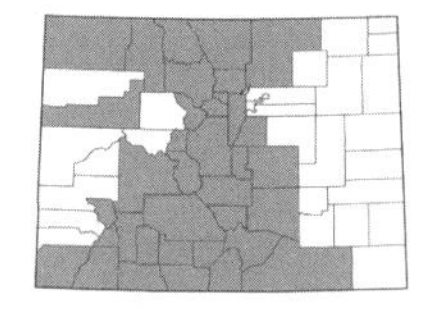

***Pinus ponderosa* Douglas ex Laws. var. *scopulorum* Engelm.**, PONDEROSA PINE. Trees to 25 m; *bark* reddish-brown, deeply furrowed; *leaves* 2–3 per fascicle, 9–20 (30) cm long; *pistillate cones* 5–10 cm long, umbo narrowing to a stout prickle. Common in the foothills and canyons, sometimes with Douglas fir, 4600–9600 ft. E/W. (Plate 4)

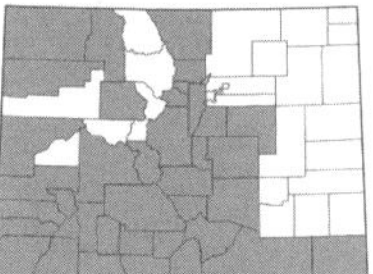

***Pinus strobiformis* Engelm.**, SOUTHWESTERN WHITE PINE. Trees to 30 m; *bark* gray, furrowed; *leaves* 5 per fascicle, 4–9 cm long; *pistillate cones* 15–25 cm long, umbo terminal, unarmed. Found in montane and subalpine forests in the southwestern and southcentral counties, 6000–11,000 ft. E/W.

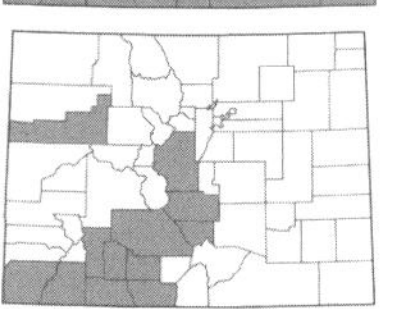

PSEUDOTSUGA Carrière – DOUGLAS-FIR

Leaves not fascicled, spirally arranged, flat; *pistillate cones* pendulous, the persistent scales surpassed by a conspicuous, subtending 3-lobed bract.

***Pseudotsuga menziesii* (Mirb.) Franco var. *glauca* (Beissn.) Franco**, DOUGLAS-FIR. Trees to 50 m; *leaves* 1.5–3 (4) cm × 1–1.5 mm, green to bluish-green; *pistillate cones* 4–7 cm long with spreading bracts. Common in the mountains, forming dominant forests with *Abies concolor* (fir) or *Picea* (spruce), 5500–10,500 ft. E/W. (Plate 4)

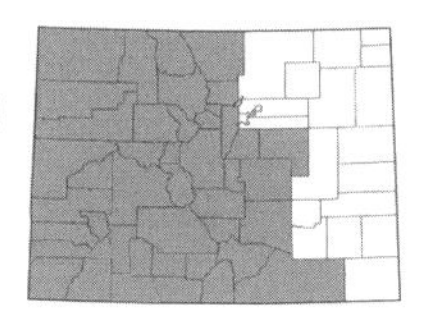

ANGIOSPERMS

ACORACEAE Martinov – SWEETFLAG FAMILY

Erect, perennial, herbaceous plants; *leaves* alternate; *inflorescences* in spadices, subtended by a leaf-like spathe, spadix cylindrical, apex obtuse; *flowers* perfect; *fruit* a berry with a leathery pericarp.

ACORUS L. – SWEETFLAG

With characteristics of the family.

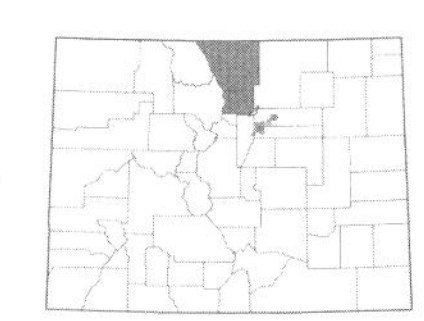

***Acorus calamus* L.**, SWEETFLAG. Leaves 2-ranked, folding along prominent midrib, linear to ensiform, sheathing at base, sessile, 2–7.5 × 7–20 dm; *infloresence* 5–10 cm long, green; *flowers* greenish-yellow; *fruits* not produced in North America. Rare in wet meadows and ditches on the plains and along the northern Front Range. 3500–5100 ft. June–July. E.

ADOXACEAE E. Mey. – MOSCHATEL FAMILY

Herbs, shrubs, or small trees; *leaves* opposite, simple to compound, *flowers* perfect, actinomorphic; *fruit* a fleshy drupe with 1, 3, or 5 stones or small pits. (Donoghue et al., 1992; Judd et al., 1994; Olmstead et al., 1993)

1a. Small (5–15 cm tall), herbaceous plants of rocky scree slopes and streambanks in the higher mountains; leaves once or twice ternately compound or parted...***Adoxa***
1b. Shrubs or small trees (usually 1 m or more in height); leaves pinnately compound or simple...2

2a. Leaves pinnately compound...***Sambucus***
2b. Leaves simple...***Viburnum***

ADOXA L. – ADOXA

Perennial herbs from rhizomes; *leaves* 3-parted or ternately compound; *flowers* in a terminal, head-like cyme; *sepals* 2 or 3; *petals* small and inconspicuous, yellowish-green, sympetalous.

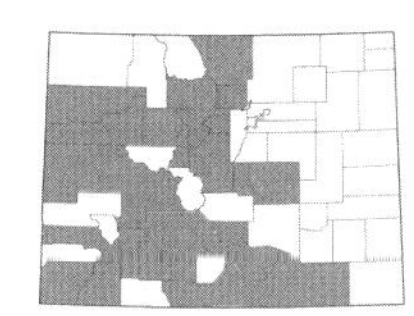

***Adoxa moschatellina* L.**, MUSKROOT. Plants 5–15 cm; *cauline leaves* in one pair, generally above midpoint of stem; *basal leaves* variable in size averaging 8 cm long; *leaflets* round-toothed and mucronate; *calyx* 2- to 4-lobed, persistent in fruit; *corolla* 4- to 6-lobed, lobes 1.5–3 mm long. Infrequent on rocky scree slopes and along forested streambanks in the middle to higher mountains, 7200–13,000 ft. June–Aug. E/W.

SAMBUCUS L. – ELDERBERRY

Shrubs or small trees; *leaves* pinnately compound; *flowers* in broad, terminal compound cymes; *sepals* 5, generally small and inconspicuous; *petals* white to yellowish; *fruit* with 3–5 stones.

1a. Inflorescence a pyramidal cyme, as long or longer than wide; fruit black, yellow, red, or orange, not glaucous...***S. racemosa***
1b. Inflorescence a flat-topped or broadly rounded cyme, wider than long; fruit blue , purplish-black, or bluish-black, glaucous or not...2

2a. Plants not stoloniferous, usually in one small clump; fruit dark blue, glaucous, usually over 5 mm wide; leaflets usually glabrous below...***S. caerulea***
2b. Plants stoloniferous and forming large clumps; fruit purplish-black, not glaucous, mostly 4–5 mm wide; leaflets usually hairy below... ***S. canadensis***

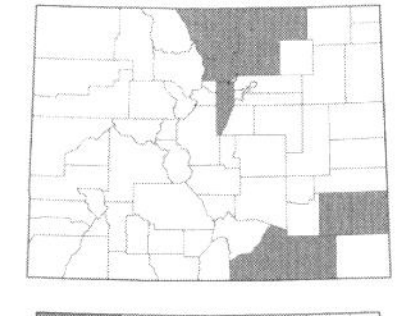

***Sambucus canadensis* L.**, COMMON ELDERBERRY. [*S. nigra* L. ssp. *canadensis* (L.) R. Bolli]. Shrubs to 5 dm; *leaflets* 5–9, lanceolate to elliptic, serrate, usually hairy below; *cymes* corymbose; *fruit* purplish-black, 4–5 mm diam., not glaucous. Locally common plants escaping from cultivation, found along the Front Range and in southeastern Colorado, 3500–5500 ft. May–July. E. Introduced.

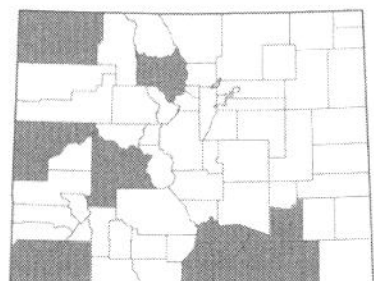

***Sambucus caerulea* Raf.**, BLUE ELDERBERRY. [*S. nigra* L. ssp. *cerulea* (Raf.) R. Bolli]. Shrubs, 2–4 dm; *leaflets* 5–9, 3–15 cm long, lanceolate to elliptic, serrate, usually glabrous below; *cyme* corymbose, 7–30 cm wide; *fruit* 4–6 mm diam., bluish-black or purplish-black, glaucous. Found along creeks, on open slopes, 4500–9000 ft. June–Aug. E/W.

***Sambucus racemosa* L.**, RED ELDERBERRY. Shrubs, 0.5–2 m; *leaflets* 5–7, lanceolate to elliptic, serrate; *cyme* paniculate, 3–10 cm long; *fruit* black or red. Common along streams, on moist slopes, and in aspen forests, 6500–12,000 ft. May–Aug. E/W. (Plate 5)

There are 2 varieties of *S. racemosa* in Colorado that are difficult to tell apart unless in fruit:

1a. Fruit purple or black...**var. *melanocarpa* (A. Gray) McMinn**, BLACK ELDERBERRY. [*S. melanocarpa* A. Gray].
1b. Fruit red or yellowish...**var. *microbotrys* (Rydb.) Kearn. & Peeb.**, RED ELDERBERRY.

VIBURNUM L. – VIBURNUM

Shrubs or small trees; *leaves* simple; *flowers* in terminal, flat-topped cymes; *sepals* 5; *petals* white; *fruit* with one stone.

1a. At least some of the leaves 3-lobed; fruit red or orange...***V. edule***
1b. Leaves all simple; fruit bluish-black or red...2

2a. Leaves densely stellate-hairy, especially below; fruit red...***V. lantana***
2b. Leaves essentially glabrous; fruit bluish-black, glaucous...***V. lentago***

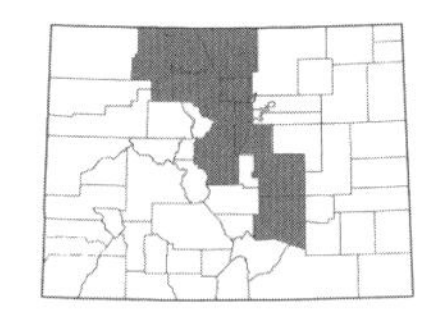

***Viburnum edule* (Michx.) Raf.**, SQUASHBERRY. [*V. pauciflorum* Pylaie ex Torr. & A. Gray]. Shrubs to 2 m; *leaves* broadly ovate, serrate, mostly 3-lobed, 3–10 cm long; *flowers* 2–3 mm long; *fruit* red or orange, 7–10 mm diam. Locally common along streams and in moist, shaded places, especially common along the Front Range, 6500–9800 ft. May–Aug. E/W.

***Viburnum lantana* L.**, WAYFARING TREE. Shrubs to 3 m; *leaves* ovate to obovate, serrate, unlobed, stellate-hairy below, 5–10 cm long; *flowers* 4 mm long; *fruit* red, 8–10 mm diam. Escaped from cultivation in gulches near Boulder, 5500–6600 ft. May–July. E. Introduced.

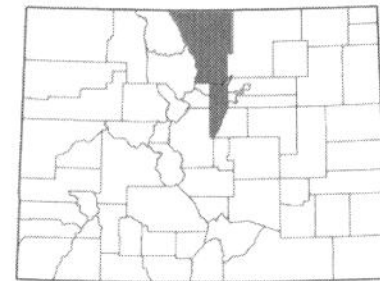

***Viburnum lentago* L.**, NANNY-BERRY. Shrubs to 4 m; *leaves* ovate to elliptic, serrate, unlobed, 3–9 cm long, glabrous; *flowers* 2–4 mm long; *fruit* bluish-black, glaucous, 5–10 mm diam. Escaped from cultivation in canyons along the Front Range, 5000–6500 ft. May–July. Introduced.

AGAVACEAE Dumort. – AGAVE FAMILY

Perennial herbs; *leaves* simple, in basal rosettes (at least ours); *flowers* actinomorphic, 6-merous, perfect or imperfect; *ovary* inferior or superior, 3-carpellate; *fruit* a berry or capsule.

1a. Plants shrubby; leaves in basal rosettes, thick, with a terminal spine and fibrous margins; flowers in terminal racemes or panicles; seeds flattened...***Yucca***
1b. Low perennial herbs; leaves thin, lacking a terminal spine and fibrous margins; flowers clustered at ground level; seeds angled...***Leucocrinum***

LEUCOCRINUM Nutt. ex A. Gray – SAND LILY; STAR LILY

Perennial herbs from deeply buried, fleshy roots; *leaves* linear, thin and flexible; *flowers* clustered at ground level; *ovary* superior; *fruit* a 3-angled loculicidal capsule; *seeds* angled.

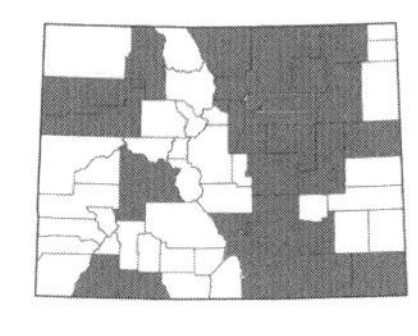

***Leucocrinum montanum* Nutt. ex A. Gray**, COMMON SAND LILY. Plants 5–10 cm; *leaves* 10–20 cm long, spreading, basal, sheathed at base by membranous bracts; *perianth* 5–10 cm long, more or less salverform; *tepals* white, 2–2.5 cm long; *fruits* 5–8 mm long; *seeds* 3–4 mm long, angled. Common on grassy slopes, in prairies and grasslands, and in open forests, 3500–8000 ft. April–June. E/W. (Plate 5)

YUCCA L. – SPANISH BAYONET

Perennial shrubs; *leaves* usually thick and fleshy, narrow with an expanded base, with a terminal spinose tip, margins often fibrous; *flowers* in terminal racemes or panicles; *ovary* superior; *fruit* fleshy and baccate or capsular; *seeds* flattened. (Hess & Robbins, 2002; Reveal, 1977; Wooton & Standley, 1913)

Yucca identification from herbarium specimens is difficult. There are several characters which are necessary to have in order to correctly identify a *Yucca*, and these are rarely all on a single specimen sheet. Proper identification of yuccas requires observation of the style color, whether the inflorescence is included within the leaves or not, flower color, and the degree of constriction (if present) in the fruit. Notes should be made on *Yucca* specimen labels regarding any of the above characteristics.

1a. Fruit fleshy, indehiscent; flowers long-campanulate; leaves 2–6 cm wide, thick and stiff with coarsely filiferous margins and terminating in a very stiff, sharp terminal spine...***Y. baccata* var. *baccata***
1b. Fruit a dry capsule, dehiscent; flowers spherically campanulate or globose; leaves mostly narrow (0.5–2 cm wide), or if wider then not thick and somewhat flexible, finely filiferous on the margins...2

2a. Style dark or bright green (this usually shows on herbarium specimens as darker than the ovary)...***Y. glauca***
2b. Style white or pale greenish-white (this usually shows on herbarium specimens as paler than the ovary)...3

3a. Fruit not constricted or slightly so; inflorescence included or barely exceeding the leaves, 3–5 dm long...***Y. baileyi* var. *baileyi***
3b. Fruit conspicuously constricted in the middle or distally; inflorescence well above the leaves, 3.5–18 dm long...4

4a. Leaves long and narrow (typically 25–50 cm long), mostly linear (the same width throughout the leaf); flowers white to greenish-white; inflorescence 8–20 dm long...***Y. angustissima* var. *angustissima***
4b. Leaves wide and short, usually lanceolate (widest near the middle, sometimes only slightly so) or sometimes linear (typically 15–40 cm long); flowers yellow or greenish-yellow to white; inflorescence 3.5–7 dm long...5

5a. Flowers white to greenish-white; leaves 0.2–2 cm wide; plants of the eastern slope...***Y. neomexicana***
5b. Flowers yellow or greenish-yellow; leaves 1.5–4.5 cm wide; plants of the western slope...***Y. harrimaniae***

Yucca angustissima* Engelm. ex Trel. var. *angustissima, NARROWLEAF YUCCA Leaves 1.5–5 dm × 4–20 mm; *inflorescence* 8–20 dm long; *flowers* white to greenish-white, the perianth segments 3–6 cm long; *style* white to pale green; *fruit* dehiscent, moderately to deeply constricted, 3.5–6 cm long. Found in dry, open places, 4500–6500 ft. May–July. W.
(See discussion under *Y. baileyi*)

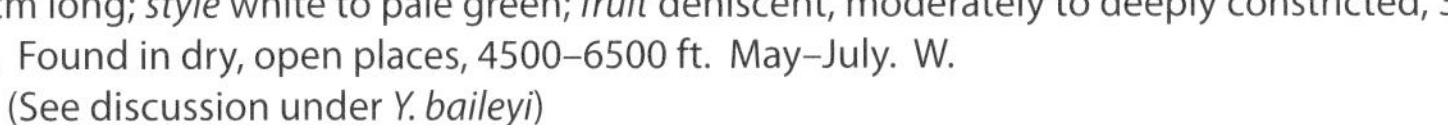

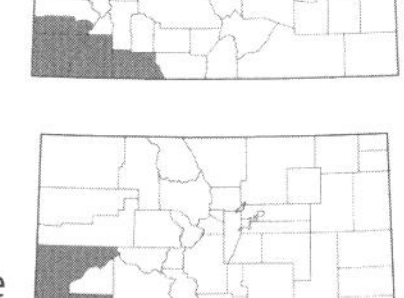

Yucca baccata* Torr. var. *baccata, BANANA YUCCA Leaves 3–10 dm × 2–6 cm, bluish-green, margins usually with coarse fibers; *inflorescence* 3.5–7 dm long; *flowers* white to cream, perianth segments 4–9 cm long; *fruit* fleshy, indehiscent, not or only slightly constricted, 8–20 cm long. Found on the western slope (mostly in the southwestern counties) on dry slopes and in pinyon-juniper and oak communities, 5400–7500 ft. May–June. E/W. (Plate 5)

Yucca baileyi* Wooton & Standl. var. *baileyi, NAVAJO YUCCA Leaves 2.5–5 dm × 2–8 cm; *inflorescence* 3–5 dm long; *flowers* whitish-green, the segments 4–6 cm long; *style* white to pale green; *fruit* dehiscent, not or only slightly constricted, 4–7 cm long. Found in dry woodlands or grasslands, 5500–7000 ft. May–July. W.

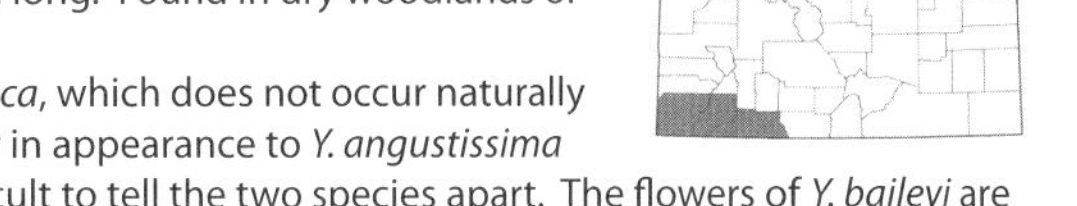

Weber and Wittmann (2001) group *Y. baileyi* with *Y. glauca*, which does not occur naturally on the western slope (see below). *Yucca baileyi* is very similar in appearance to *Y. angustissima* but without constricted fruit, and without fruit, it is very difficult to tell the two species apart. The flowers of *Y. baileyi* are usually longer, 5–6.5 cm long as opposed to 3–5.5 cm long in *Y. angustissima.*

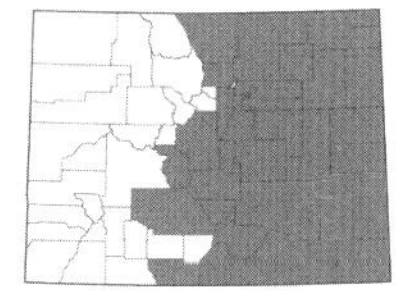

***Yucca glauca* Nutt.**, GREAT PLAINS YUCCA; SOAPWEED. Leaves 4–6 dm long; *inflorescence* 5–10 dm long; *flowers* white to greenish-white, the segments 4.5–6 cm long; *style* dark green; *fruit* rarely constricted, 5–9 cm long. Common in sandy places and dry slopes on the plains, foothills, and San Luis Valley, often found in disturbed areas, 3800–9000 ft. May–July (Oct.). E. Possibly introduced on the western slope as a cultivated plant but not naturally occurring there. (Plate 5)

The fruit of *Y. glauca* is not constricted or only slightly so, and the only other *Yucca* to share this characteristic is *Y. baileyi*. *Yucca baileyi* has been erroneously reported as occurring along the Front Range by Hess and Robbins (2002) but only occurs in southwestern Colorado (see above). The only other species of *Yucca* to occur on the eastern slope is *Y. neomexicana*. It can be readily distinguished from *Y. glauca* by the leaves, which are 40–70 cm long in *Y. glauca* and only 25–30 cm long in *Y. neomexicana*.

***Yucca harrimaniae* Trel.**, SPANISH BAYONET. Leaves 2–5 dm × 1.5–4.5 cm; *inflorescence* 3.5–7 dm long; *flowers* yellow or greenish-yellow, the segments 4–5.5 cm long; *style* pale green; *fruit* deeply constricted, 3–6 cm long. Found on dry desert slopes, foothills, and plateaus, 4500–8000 ft. May–June. W.

***Yucca neomexicana* Wooton & Standl.**, NEW MEXICO SPANISH BAYONET. [*Y. harrimaniae* Trel. var. *neomexicana* (Wooton & Standl.) Reveal]. Leaves 1.5–4.5 dm × 0.2–2 cm; *inflorescence* 4–7 dm long; *flowers* white to greenish-white, the segments 3–5 cm long; *style* pale green; *fruit* deeply constricted in the middle, 3–4.5 cm long. Found on dry slopes, 4500–5000 ft. E.

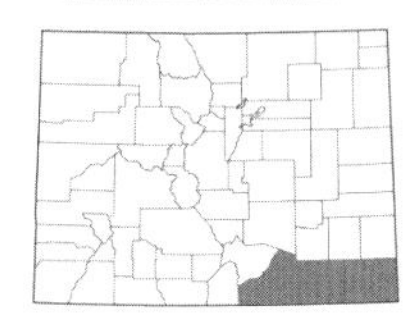

AIZOACEAE Martinov – CARPETWEED; FIG-MARIGOLD FAMILY

Herbs, generally succulent; *leaves* opposite or alternate, simple, fleshy; *flowers* perfect, actinomorphic, solitary in the leaf axils or axillary in a cymose or spike-like cluster; *petals* absent; *fruit* generally a capsule with numerous seeds.

1a. Stipules absent; ovary 3–5-loculed; stamens numerous…***Sesuvium***
1b. Stipules present; ovary 1–2-loculed; stamens 5–10…***Trianthema***

SESUVIUM L. – SEA-PURSLANE

Leaves opposite, exstipulate; *flowers* solitary in the leaf axils; *sepals* 5-lobed, usually horned on the back near the tip; *stamens* numerous; *carpels* 3–5; *ovary* superior; *fruit* an ovoid circumscissile capsule.

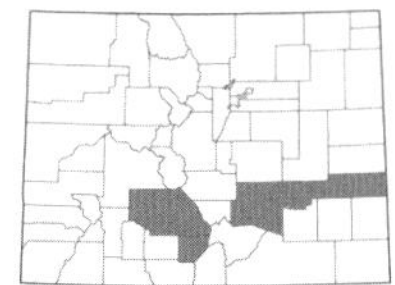

***Sesuvium verrucosum* Raf.**, WESTERN SEA-PURSLANE. Perennials, prostrate, glabrous, generally papillate; *leaves* oblanceolate to oblong ovate or linear-oblong, clasping at base, 0.5–4 cm long, less than 1 cm wide; *sepals* 4–10 mm long; *fruits* 4–5 mm long, 3 mm in diam. Uncommon in alkaline wetlands and on dry flats in the San Luis and Arkansas River Valleys, 3900–7600 ft. May–Aug. E.

TRIANTHEMA L. – HORSE-PURSLANE

Annuals, prostrate herbs; *leaves* opposite, stipulate; *flowers* solitary in the leaf axils; *sepals* 5; *stamens* 5–10; *carpels* 1–2; *ovary* superior; *fruit* a cylindric or turbinate circumscissile capsule.

***Trianthema portulacastrum* L.**, DESERT HORSE-PURSLANE. Plants diffusely branched, to 10 dm; *leaves* in unequal pairs, elliptic to orbiculate, 1–4 cm long, 3 cm wide, narrowed to a long slender petiole; *sepals* 3–5 mm long; *petals* 2.5 mm long, purple adaxially; *fruit* 4–5 mm long. Found in disturbed places, reported for the southeastern counties, 3500–4500 ft. May–Sept. E/W. Introduced.

ALISMATACEAE Vent. – WATER-PLANTAIN FAMILY

Perennial aquatic or semiaquatic herbs; *leaves* submerged to emergent, alternate, long-petiolate, with sheathing bases, the blades simple and entire, usually linear or sagittate; *flowers* perfect or occasionally imperfect, actinomorphic, bracteate, often in whorls of 3; *fruit* usually an achene. (Haynes & Hellquist, 2000; Hellquist & Crow, 1981)

1a. Flowers in diffusely branched panicles; leaves ovate to narrowly elliptic; achenes arranged in a single ring…***Alisma***
1b. Flowers in whorls of 3 along a central axis; leaves sagittate, hastate, or rarely linear; achenes densely crowded in a globose head…***Sagittaria***

ALISMA L. – WATER-PLANTAIN

Leaves basal; *flowers* perfect, in compound, terminal, whorled panicles; *sepals* green, persistent; *petals* white (usually), deciduous; *stamens* 6–9; *pistils* 6–28, the styles lateral, persistent; *fruit* a strongly compressed achene, ribbed on the back.

1a. Leaves linear to narrowly elliptic, less than 3 cm wide; achenes with 2 grooves near the tip…***A. gramineum***
1b. Leaves broader, ovate to elliptic, 2–20 cm wide; achenes with one central groove at the tip…2

2a. Petals 3.5–6 mm long, longer than the sepals; fruiting heads larger, 4–7 mm in diam.…***A. triviale***
2b. Petals 1–3 mm long, equaling the sepals; fruiting heads smaller, 2–4 mm in diam.…***A. subcordatum***

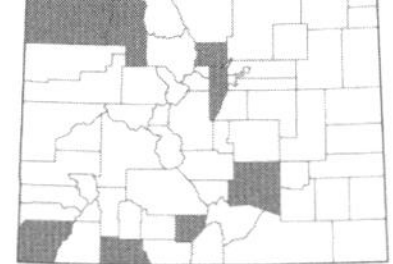

***Alisma gramineum* Lej.**, NARROWLEAF WATER-PLANTAIN. Leaves linear to narrowly elliptic, 3–30 × 0.2–2.5 cm; *sepals* 2–3 mm long; *petals* white, 2–4 mm long; *fruiting heads* 3–6 mm diam.; *achenes* 2–2.7 mm long. Infrequent in wet places such as pond borders, marshes, and on mud flats, at the water's edge or almost completely submerged, 5400–8000 ft. July–Sept. E/W.

***Alisma subcordatum* Raf.**, SOUTHERN WATER-PLANTAIN. Leaves ovate to elliptic, to 10 cm wide; *sepals* 1.5–3 mm long; *petals* white, 1–3 mm long; *fruiting heads* 2–4 mm diam.; *achenes* 1.5–2.2 mm long. Found in wet places such as shallow ponds, ditches, and marshes, reported for Colorado from the northeastern counties, 3500–4500 ft. June–Sept. E.

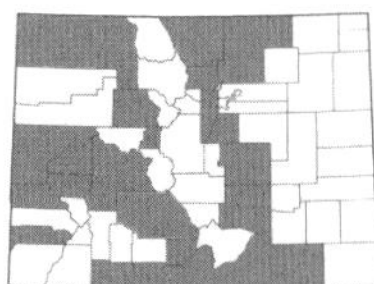

***Alisma triviale* Pursh**, NORTHERN WATER-PLANTAIN. [*A. plantago-aquatica* L. var. *americanum* Schult.]. Leaves elliptic to ovate, 4–20 cm long; *sepals* 3–6 mm long; *petals* white, 3.5–6 mm long; *fruiting heads* 4–7 mm diam.; *achenes* 2–3 mm long. Common along pond shores, in ditches and marshes, and on mud flats, rarely in deep water, 5000–10,000 ft. July–Sept. E/W.

SAGITTARIA L. – ARROWHEAD

Leaves basal, floating or submerged or emergent; *flowers* sometimes perfect but mostly imperfect; *sepals* green, usually reflexed, persistent although often inconspicuous; *petals* white, deciduous; *stamens* many; *pistils* many, crowded and spirally arranged on a globose receptacle; *fruits* bilaterally compressed, beaked achenes, often winged on the margins. (Bogin, 1955; Smith, 1895)

1a. Sepals erect and enclosing the flower in pistillate plants; fruiting pedicels recurved...***S. calycina* var. *calycina***
1b. Sepals recurved or spreading, not enclosing the flower in pistillate plants; fruiting pedicels spreading or ascending...2

2a. Leaves linear to linear-oblanceolate, not lobed; filaments hairy, shorter than the anthers...***S. graminea* var. *graminea***
2b. Leaves cordate, sagittate, or hastate; filaments glabrous, longer than the anthers...3

3a. Beak of the achenes horizontal...***S. latifolia***
3b. Beak of the achenes erect or recurved...4

4a. Achene beaks recurved, prominent (0.4–1.7 mm long)...***S. brevirostra***
4b. Achene beaks erect and straight, minute (0.1–0.4 mm long)...***S. cuneata***

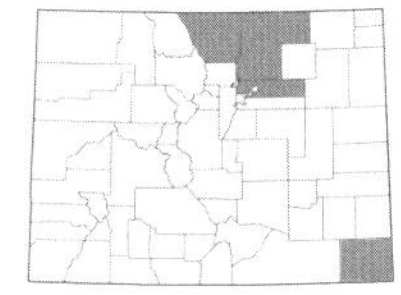

***Sagittaria brevirostra* Mack. & Bush**, SHORTBEAK ARROWHEAD. Leaves sagittate, 5–20 × 2–8 cm; *inflorescence* of 5–12 whorls; *flowers* to 3.5 cm diam.; *fruiting heads* 1.2–2.5 cm diam.; *achenes* 2–3.1 mm long, with a recurved beak 0.4–1.7 mm long. Found along muddy shorelines and streamsides, ca. 4500–5500 ft. June–Sept. E/W.

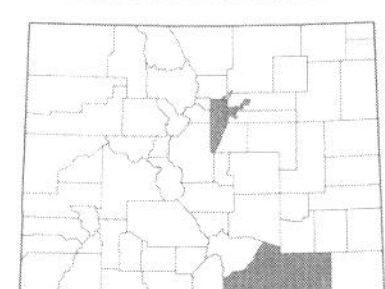

Sagittaria calycina* Engelm. var. *calycina, HOODED ARROWHEAD. [*S. montevidensis* Cham. & Schltdl. ssp. *calycina* (Engelm.) Bogin]. Leaves hastate to sagittate, 2.5–18 × 0.5–22 cm; *inflorescence* of 1–15 whorls; *flowers* 2–5 cm diam.; *fruiting heads* 1.2–2.3 cm diam.; *achenes* 2–4.3 mm long, beaked. Locally common in wet places such as along pond shores, 5000–6000 ft. May–Sept. E.

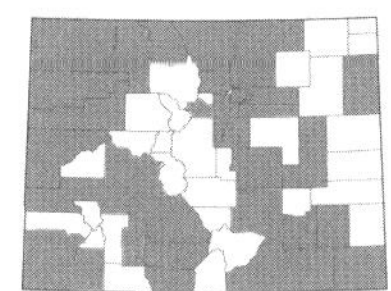

***Sagittaria cuneata* Sheldon**, ARUMLEAF ARROWHEAD. Leaves cordate to sagittate, 7.5–9 × 3.5–4 cm; *inflorescence* of 2–10 whorls; *flowers* to 2.5 cm diam.; *fruiting heads* 0.8–1.5 cm diam.; *achenes* 1.8–2.6 mm long, with a straight, erect beak 0.1–0.4 mm long. Common along shorelines and slow-moving streams and in swampy places, 3500–10,000 ft. June–Aug. E/W.

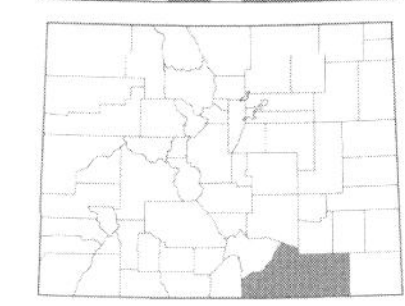

Sagittaria graminea* Michx. var. *graminea, GRASSY ARROWHEAD. Leaves linear to linear-oblanceolate, 2.5–17.5 × 0.2–4 cm; *inflorescence* of 1–12 whorls; *flowers* to 2.5 cm diam.; *fruiting heads* 0.6–1.5 cm diam.; *achenes* 1.5–2.8 mm long, with a lateral, erect beak. Infrequent along pond shores, 5700–6000 ft. June–Sept. E.

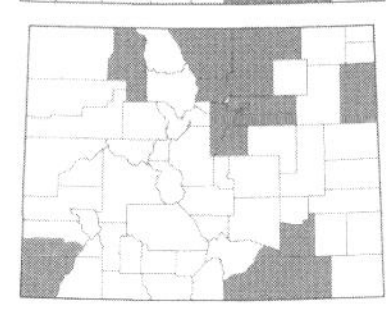

***Sagittaria latifolia* Willd.**, BROADLEAF ARROWHEAD. Leaves sagittate or rarely hastate, 1.5–30 × 2–17 cm; *inflorescence* of 3–9 whorls; *flowers* to 4 cm diam.; *fruiting heads* 1–1.7 cm diam.; *achenes* 2.5–3.5 mm long, with a lateral, horizontal beak. Common along pond shores, in muddy ditches, and in swampy areas on the plains and in the foothills, 3500–6000 ft. July–Sept. E/W.

ALLIACEAE Borkh. – ONION FAMILY

Perennial herbs from bulbs; *leaves* basal, terete or flattened; *inflorescence* umbellate; *flowers* perfect, all perianth parts alike; *tepals* 6; *stamens* epipetalous, the filaments usually broad at the base and fused into a ring; *fruit* a loculicidal capsule. (Cronquist & Ownbey, 1977; McNeal, Jr. & Jacobsen, 2002)

ALLIUM L. – ONION

With characteristics of the family.

1a. Outer bulb coats persisting as coarse fibers...2
1b. Outer bulb coats membranous and not coarsely fibrous...5

2a. Involucre bracts of the scape mostly 1-nerved...3
2b. Involucre bracts of the scape 3–7-nerved...4

3a. Leaves usually 3 per bulb; tepals usually pink, rarely white; cell of each seed coat with a minute, central papilla...***A. geyeri***
3b. Leaves usually 2 per bulb; tepals white, rarely pink; cell of each seed coat without a central papilla...***A. textile***

4a. Leaves round and hollow, 2–7 mm wide; scape 20–50 cm; umbels with mostly 30–50 flowers; tepals purplish, although often drying pink, or sometimes white...***A. schoenoprasum***
4b. Leaves semi-round or concavo-convex, 1–3 mm wide, solid; scape 5–20 cm; umbels with mostly 10–20 flowers; tepals pink to nearly white except for the darker midrib...***A. macropetalum***

5a. Plants 5–10 dm; flowers mostly or completely replaced by bulbils, if present the tepals whitish-green; leaves 5–20 mm wide, flat...***A. sativum***
5b. Plants usually shorter, although sometimes to 6 dm; flowers usually not replaced by bulbils, tepals white, pink, or rose-purple to lilac; leaves 1–8 mm wide, round to concavo-convex or nearly flat...6

6a. Umbels nodding; stamens exserted...***A. cernuum***
6b. Umbels erect; stamens exserted or not...7

7a. One leaf per scape, round or nearly so, the tip usually curled; outer bulb coats (at least some) with contorted reticulations...***A. nevadense***
7b. Leaves 2 or more per scape, flat to concavo-convex or round; outer bulb coats with elongate or square cells...8

8a. Bulb elongate, terminating in a short, iris-like, stout rhizome; outer bulb coats with elongate cells that are several times longer than wide in regular, vertical rows; leaves flat; involucre bracts connate at the base and often along one side, ovate...***A. brevistylum***
8b. Bulb ovoid to subglobose, if rhizomes present then these slender and secondary; outer bulb coat sometimes with elongate cells, but these not several times longer than wide in regular rows; leaves concavo-convex to round or sometimes flat; involucre bracts separate or nearly so, lanceolate to ovate...9

9a. Leaves round and hollow, 2–7 mm wide; tepals purple, although often drying pink, or rarely white...***A. schoenoprasum***
9b. Leaves flat to concavo-convex, channeled, not hollow, mostly 1–3 mm wide; tepals pink to rose-purple or white...10

10a. Leaves longer than the scape; tepals white with a conspicuous green to red midrib...***A. brandegeei***
10b. Leaves shorter than the scape; tepals pink, rose-purple (rarely white)...***A. acuminatum***

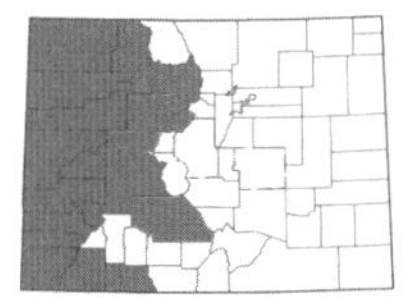

***Allium acuminatum* Hook.**, TAPERTIP ONION. Bulb outer coats not fibrous-shredded; *leaves* 2–4; *bracts* 2, 10–25 mm long, 3–7-veined; *tepals* pink, rose-purple, or sometimes white, with recurved tips. Common in dry, open places, often with pinyon-juniper or sagebrush, not yet reported for the eastern slope but to be expected in North Park, 5200–8500 ft. May–July. W. (Plate 5)

***Allium brandegeei* S. Watson**, BRANDEGEE'S ONION. Bulb outer coats membranous with hexagonal reticulations; *leaves* 2; *bracts* 2, 7–10-veined; *tepals* white with a purple or green midvein. Locally common in dry, sandy or rocky soil, not yet reported for the eastern slope but to be expected in North Park, 6500–10,000 ft. May–July. W.

***Allium brevistylum* S. Watson**, SHORTSTYLE ONION. Bulb outer coats membranous with elongate cells in vertical rows; *leaves* 2–5; *bracts* 2, 3–5-veined; *tepals* rose pink, the tips recurved. Common along streams and in wet meadows in the northcentral counties, 7000–9500 ft. June–Aug. E/W.

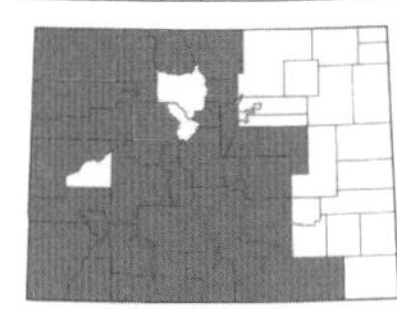

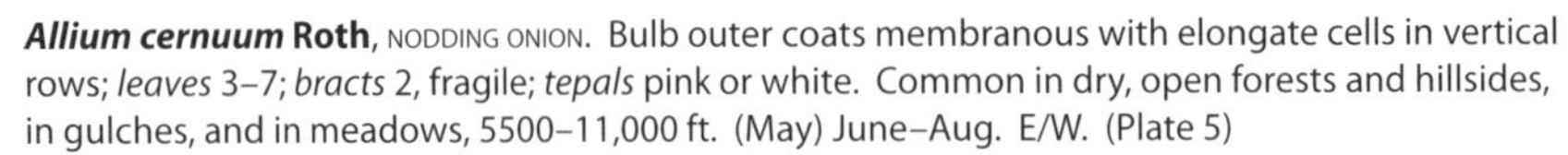

***Allium cernuum* Roth**, NODDING ONION. Bulb outer coats membranous with elongate cells in vertical rows; *leaves* 3–7; *bracts* 2, fragile; *tepals* pink or white. Common in dry, open forests and hillsides, in gulches, and in meadows, 5500–11,000 ft. (May) June–Aug. E/W. (Plate 5)

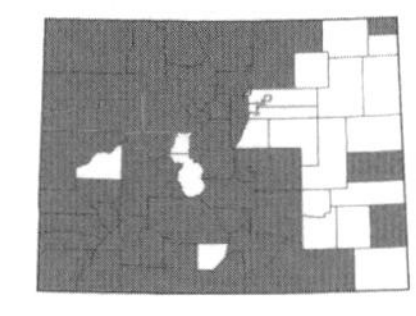

***Allium geyeri* S. Watson**, GEYER'S ONION. Bulb outer coats fibrous; *leaves* 2–3; *bracts* 2–3, 1-veined; *tepals* pink or sometimes white. Common on moist slopes, along streams, in meadows, and alpine, 4500–13,500 ft. May–Aug. E/W.

There are two varieties of *A. geyeri* in Colorado:

1a. Flowers usually 10–25 per umbel, not replaced by bulbils...**var. *geyeri***
1b. Flowers usually fewer than 10 per umbel, the others replaced by bulbils...**var. *tenerum* M.E. Jones**, [*A. rubrum* Osterh.].

***Allium macropetalum* Rydb.**, SAN JUAN ONION. Bulb outer coats fibrous; *leaves* 2, 1-3 mm wide; *bracts* 2–3, 3–5-veined; *tepals* pink. Locally common on dry hills, usually in clay or shale, known from the southwestern counties, 4900–6900 ft. April–June. W.

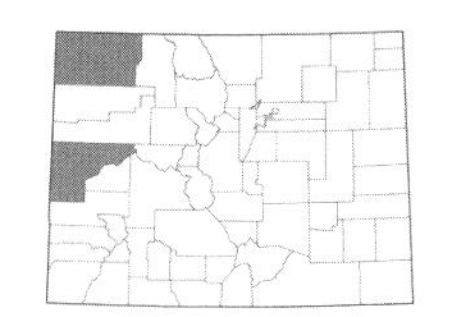

Allium nevadense **S. Watson**, Nevada onion. Bulb outer coats membranous; *leaf* solitary; *bracts* 2–3, 3–7-veined; *tepals* pink, rose, or white. Locally common in sandy or clay soil of dry, open places, 4500–7600 ft. April–June. W.

Allium sativum **L.**, garlic. Bulb outer coats membranous; *leaves* flat, 5–20 mm wide, the stems leafy to or above the middle; *umbel* producing ovoid bulbils and few flowers, deciduous in one piece. Uncommon along roadsides near Boulder, 5500–6000 ft. June–Aug. E. Introduced.

Allium schoenoprasum **L.**, wild chives. Bulb outer coats fibrous; *leaves* 2, 2-7 mm wide; *bracts* 2, 3–7-veined; *tepals* purple (drying pink) or white. Uncommon along streams and in wet meadows in mountain parks of Jackson County, 7500–8500 ft. June–July. E.

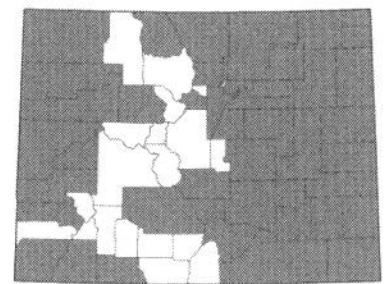

Allium textile **A. Nelson & J.F. Macbr.**, textile onion. Bulb outer coats fibrous; *leaves* 2; *bracts* 3, 1-veined; *tepals* white or pale pink. Common on dry slopes and the eastern plains, 3500–8500 ft. May–July. E/W.

AMARANTHACEAE Juss. – pigweed family

Annual herbs, dioecious or monoecious; *leaves* alternate or opposite, simple, exstipulate, usually entire; *flowers* perfect or imperfect, inconspicuous, actinomorphic; *sepals* usually 5, often persistent, usually scarious; *petals* absent; *stamens* as many as the sepals; *ovary* superior, unicarpellate; *fruit* an utricle. (Sauer, 1955, 1972; Sauer & Davidson, 1961)

1a. Plants densely stellate-hairy; leaves opposite, ovate to obovate; plants prostrate; flowers perfect...***Tidestromia***
1b. Plants not stellate-hairy; plants otherwise various...2

2a. Perianth segments densely white-hairy; leaves mostly basal, linear to linear-lanceolate or narrowly elliptic, canescent to tomentose; flowers in a terminal spicate-raceme...***Froelichia***
2b. Perianth segments glabrous or nearly so; leaves various but unlike the above in all respects; flowers variously arranged...3

3a. Flowers perfect; sepals and bracts white and scarious throughout; leaves opposite, the pair of unequal size, lacking prominent, pinnate venation below; plants prostrate, mat-forming...***Guilleminea***
3b. Flowers imperfect; sepals and bracts not white and scarious throughout; leaves alternate, with prominent, pinnate venation below (the pale veins markedly contrasting with the green leaf blade); plants prostrate or erect...***Amaranthus***

AMARANTHUS L. – pigweed

Plants dioecious or monoecious; *leaves* alternate, simple; *inflorescences* of dense terminal or axillary spikes or clusters; *flowers* subtended by a bract and 2 bracteoles, these usually thin, dry, and transparent, often concealing the perianth.

1a. Inflorescence of axillary clusters only; plants diffusely branched, prostrate or low bushy herbs...2
1b. Inflorescence mostly terminal (although axillary clusters may be present); plants erect with a dominant main stem...3

2a. Plants prostrate; bracts with a shortly excurrent midrib, about equal in length or scarcely exceeding the sepals...***A. blitoides***
2b. Plants low, bushy herbs; bracts with a long-excurrent midrib, more than twice as long as the sepals...***A. albus***

3a. Plants monoecious with staminate and pistillate flowers intermingled or in nearly separate inflorescences...4
3b. Plants dioecious with staminate and pistillate flowers in separate plants...7

4a. Sepals of pistillate flowers mostly obtuse to truncate, notched at the apex...5
4b. Sepals of pistillate flowers acute or acuminate...6

5a. Plant villous at least in the inflorescence; flowers in dense, robust panicles of spikes...***A. retroflexus***
5b. Plant glabrous or nearly so; flowers in slender, more or less interrupted, leafy spikes...***A. wrightii***

6a. Bracts only slightly exceeding the sepals (3–4 mm long), with a long-excurrent midrib; inflorescence lax...***A. hybridus***
6b. Bracts far exceeding the sepals (5 mm long), with a moderately excurrent midrib; inflorescence stiff...***A. powellii***

7a. Plants pistillate...8
7b. Plants staminate...9

8a. Sepals obtuse; bract midvein scarcely excurrent; bracts 1.5–2.5 mm long, equal in length to the sepals...***A. arenicola***
8b. Outer sepals acute; bract midvein excurrent into a rigid spine; bracts 4–6 mm long and longer than the sepals...***A. palmeri***

9a. Bract midvein scarcely excurrent...***A. arenicola***
9b. Bract midvein conspicuously excurrent...***A. palmeri***

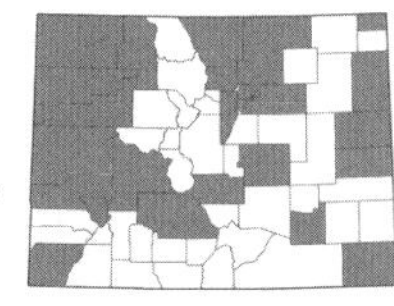

***Amaranthus albus* L.**, TUMBLE PIGWEED. [*A. pubescens* (Uline & Bray) Rydb.]. Monoecious; *stems* erect, 2–10 dm, bush-branched; *leaves* 1.5–6 cm long, obovate to elliptic or lanceolate; *inflorescence* of short axillary clusters, bracts 2–4 mm long; *sepals* 3, 1–2 mm long. Found on roadsides, waste places, and in fields, 4000–6000 ft. July–Sept. E/W. Introduced.

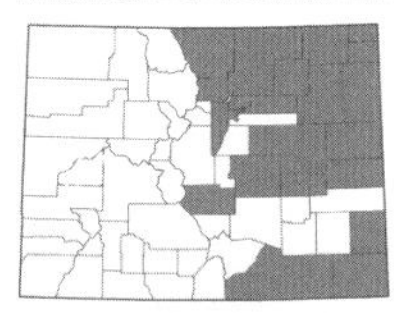

***Amaranthus arenicola* I.M. Johnst.**, SANDHILL AMARANTH. [*A. torreyi* (A. Gray) Benth.]. Dioecious; *stems* erect, 5–25 dm; *leaves* 1.5–8 cm long, oval-oblong to oblanceolate; *inflorescence* a terminal, dense spike or thyrse, bracts 1.5–2.5 mm long; *sepals* 5, 1.5–2.5 mm long. Common in waste places, along roadsides, in fields, and on sand dunes, 3400–5500 ft. July–Sept. E.

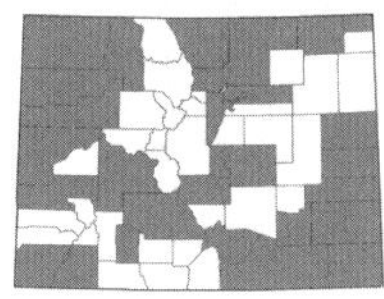

***Amaranthus blitoides* S. Watson**, MAT AMARANTH. [*A. graecizans* L.]. Monoecious; *stems* prostrate, forming mats; *leaves* oval to obovate, 0.8–4 cm long; *inflorescence* of dense axillary clusters, bracts 2.5–3 mm long; *sepals* (4) 5, 2.5–3 mm long. Locally common in waste places, dry prairies, fields, and roadsides, 3500–8700 ft. July–Sept. E/W.

***Amaranthus hybridus* L.**, SLENDER PIGWEED. Monoecious; *stems* erect, 5–20 dm; *leaves* lanceolate to rhombic-ovate, 1–15 cm long; *inflorescence* a terminal panicle, bracts 3–4 mm long; *sepals* 5, 1.5–2 mm long. Found in waste places, along roadsides, and in fields and pastures, reported for the eastern plains. July–Sept. E.

***Amaranthus palmeri* S. Watson**, CARELESSWEED. Dioecious; *stems* erect, 10–25 dm; *leaves* rhombic-ovate to lanceolate, 1–8 cm long; *inflorescence* a terminal spike or also with lower axillary clusters, bracts 4–6 mm long; *sepals* 5, 2–5 mm long. Infrequent to locally common in fields, pastures, along roadsides, and in waste places, 3900–6000 ft. July–Sept. E.

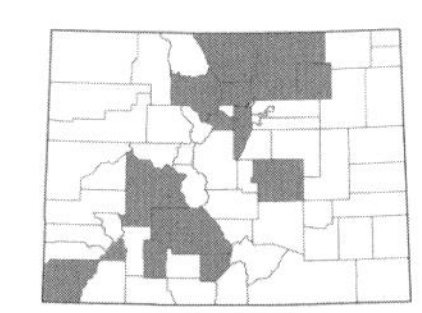

***Amaranthus powellii* S. Watson**, POWELL'S AMARANTH. Monoecious; *stems* erect, 6–20 dm; *leaves* rhombic-ovate to lanceolate, 2–12 cm long; *inflorescence* of terminal and axillary panicles, bracts 2.5–5 mm long; *sepals* 3–5, 1.2–3 mm long. Found in fields and pastures, along streams, roadsides, and in waste places, 4900–8800 ft. July–Sept. E/W.

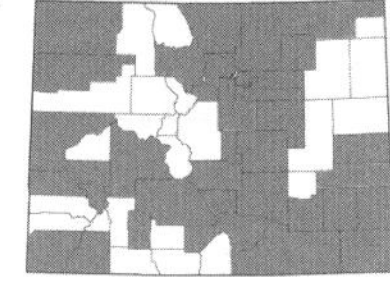

***Amaranthus retroflexus* L.**, REDROOT AMARANTH. Monoecious; *stems* erect, 3–30 dm; *leaves* lanceolate to ovate, 2–10 cm long; *inflorescence* of terminal or axillary paniculate spikes, bracts 3.5–5 mm long; *sepals* 5, 2.5–3.5 mm long. Common in fields and pastures, waste places, and along roadsides, 4500–8000 ft. July–Sept. E/W. Introduced. (Plate 5)

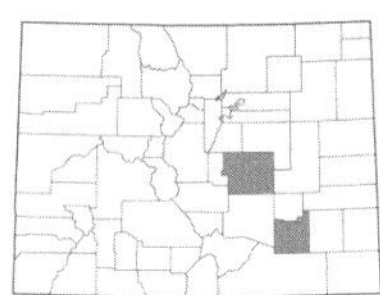

***Amaranthus wrightii* S. Watson**, WRIGHT'S AMARANTH. Monoecious; *stems* erect, 2–10 dm; *leaves* rhombic-ovate to lanceolate, 1.5–6 cm long; *inflorescence* of terminal and axillary spikes, bracts to 4 mm; *sepals* 5, 1.5–2 mm long. Uncommon in fields and pastures and along roadsides, 4000–5500 ft. July–Sept. E.

FROELICHIA Moench – COTTONSNAKE

Annual herbaceous plants with woolly stems; *leaves* narrow, opposite; *flowers* perfect, in spikes, each flower subtended by a scarious bract and 2 bractlets; *sepals* densely woolly.

1a. Lateral spikes sessile; sepals with 2 lateral rows of spine-like projections...***F. gracilis***
1b. Some of the lateral spikes usually peduncled; sepals with dentate or erose wings...***F. floridana***

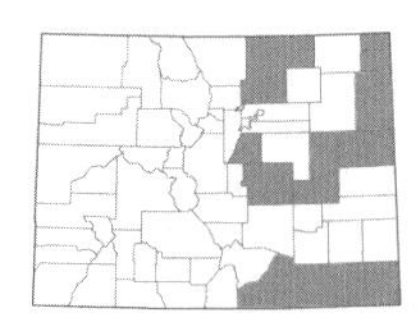

***Froelichia floridana* (Nutt.) Moq.**, PLAINS COTTONSNAKE. Leaves linear to oblanceolate, 3–12 cm × 5–4 mm, tomentose; *inflorescence* of dense spikes, lateral spikes usually peduncled; *sepals* 4–6 mm long, with 2 lateral dentate crests or wings. Not as common as *F. gracilis*, found on the eastern plains on sand dunes, in sandy prairies, and in stream valleys, 3500–6000 ft. July–Sept. E.

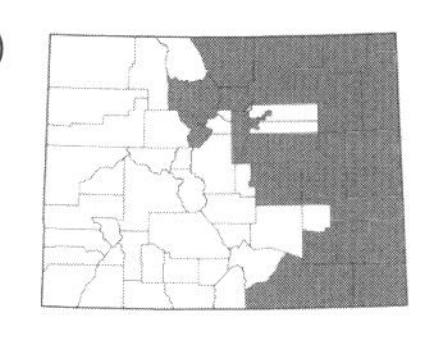

Froelichia gracilis **(Hook.) Moq.**, SLENDER SNAKECOTTON. Leaves linear to elliptic-lanceolate, 3–10 (12) cm × 2–10 mm, tomentose; *inflorescence* of sparsely branched spikes, lateral spikes sessile; *sepals* 2.5–5 mm long, with 2 lateral rows of distinct spines. Locally abundant in the eastern plains on sand dunes, in sandy prairies, in stream valleys, or occasionally present in rocky open woodlands, 3500–9500 ft. July–Sept. E. (Plate 5)

GUILLEMINEA Kunth – COTTONFLOWER

Prostrate or erect perennial herbaceous plants; *basal leaves* in rosettes; *cauline leaves* opposite; *flowers* in an axillary head or spike, these often densely aggregated at the nodes; *flowers* perfect, with 2 bracts and a bractlet subtending each.

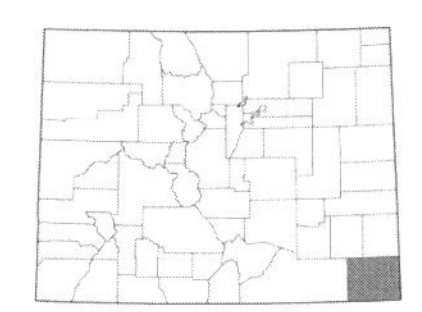

Guilleminea densa **(Willd.) Moq.**, DENSE COTTONFLOWER. Plants highly branched, usually lanate, with stems 0.5–3(6) dm long; *basal leaves* lanceolate to spatulate, acute,10–25 mm × 5–15 mm, early-deciduous; *cauline leaves* lanceolate to narrowly ovate, 3–30 mm × 1–12 mm; *flowers* 1.2–1.5 mm long; *fruit* glabrous; *seeds* 0.5–0.7 mm long, brown. Found in dry, sandy soil or rocky, flat prairies, 4200–4500 ft. June–Aug. E.

TIDESTROMIA Standl. – HONEYSWEET

Prostrate annuals; *leaves* opposite, densely stellate-hairy, becoming glabrate in age; *flowers* in small axillary clusters; *flowers* perfect, subtended by 3 hyaline and hairy bracts.

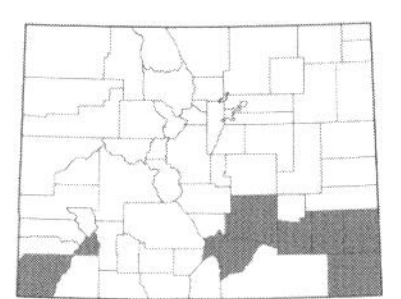

Tidestromia lanuginosa **(Nutt.) Standl.**, WOOLLY TIDESTROMIA. Plants to 5 dm; *leaves* 6–32 mm × 9–30 mm, petiole to 2.5 cm long, blade gray-green; *flowers* 1.5–3 mm long, tepals yellow; *fruit* 1.3–1.6 mm long, seeds brown-red. Locally common in sandy or gravelly places, especially common on sand dunes in the Arkansas River Valley, 3600–5500 ft. July–Sept. E/W.

ANACARDIACEAE R. Br. – CASHEW OR SUMAC FAMILY

Woody shrubs or vines; *leaves* alternate, pinnately or ternately compound; *inflorescence* a terminal or axillary thyrse or panicle, eventually cymose; *flowers* small, actinomorphic, perfect or imperfect and then with reduced parts of the opposite sex; *fruit* a drupe, often with a waxy or oily mesocarp.

1a. Leaves ternately compound, shiny, terminal leaflet petiolate; fruit white...***Toxicodendron***
1b. Leaves simple, ternately compound, or pinnately compound, dull, terminal leaflet sessile; fruit red...***Rhus***

RHUS L. – SUMAC

Shrubs or small trees; *leaves* deciduous or evergreen, alternate, odd-pinnately or ternately compound; *inflorescence* a dense terminal or axillary thyrse or panicle; *flowers* yellow, appearing before or after the leaves, each subtended by a bract; *sepals* 5, fused at the base; *petals* 5, distinct; *stamens* 5; *ovary* 3-carpellate; *fruit* a reddish drupe with glandular hairs. (Barkley, 1937)

1a. Leaves ternately compound with 3 leaflets or simple or nearly so; inflorescence small, axillary on branches; flowers appearing before the leaves...***R. trilobata***
1b. Leaves pinnately compound with 7 or more leaflets; inflorescence large, terminal; flowers appearing after the leaves...2

2a. Twigs and petioles glabrous; fruit with short hairs; native shrub of foothills...***R. glabra***
2b. Twigs and petioles densely hairy; fruit with long hairs; cultivated and escaping...***R. typhina***

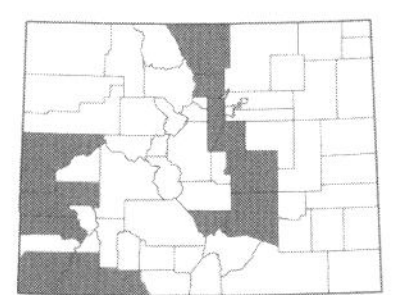

Rhus glabra **L.**, SMOOTH SUMAC. Leaves odd-pinnately compound, 10–30 cm long; *leaflets* 7–21, lanceolate, 1.5–8 cm long, serrate, glabrate; *inflorescence* a dense pyramidal panicle, 7–20 cm long; *drupes* red, viscid-hairy. Found on rocky slopes of the foothills, 5000–7500 ft. May–July. E/W.

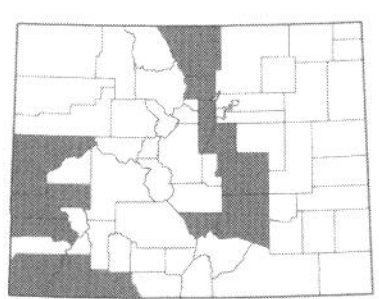

Rhus trilobata **Nutt.**, SKUNKBUSH SUMAC. [*R. aromatica* Aiton]. Leaves simple or ternately compound; *leaflets* 0.8–9.5 cm long, glabrous or minutely hairy; *inflorescence* a globular raceme; *drupes* reddish-orange, 5–8 mm long. Common in canyons and on rocky slopes, 3500–9000 ft. April–June. E/W. (Plate 5)

There are three varieties of *R. trilobata* in Colorado:

1a. Lateral leaflets small, no more than ½ as long as the terminal one and more commonly wholly suppressed such that the leaves appear simple...**var. *simplicifolia* (Greene) Barkl.**, SKUNKBUSH SUMAC. Found in canyons and along streams on the western slope, 4500–9000 ft.
1b. Lateral leaflets large, usually ½ – ¾ as long as the terminal one, distinctly ternately compound...2

2a. Leaves and young twigs densely pilose with long, soft hairs...**var. *pilosissima* Engelm.**, SKUNKBUSH SUMAC. Common in the southern counties, 4500–6000 ft.
2b. Leaves and young twigs glabrous or hairy, but not pilose...**var. *trilobata*.**, SKUNKBUSH SUMAC. Common in canyons and on rocky slopes of the foothills and plains, 3500–9000 ft. The most common *R. trilobata* variety.

***Rhus typhina* L.**, STAGHORN SUMAC. [*R. hirta* (L.) Sudworth]. Leaves odd-pinnately compound, 15–35 cm long; *leaflets* 11–31, 5–10 cm long, serrate or laciniate; *inflorescence* a dense terminal cluster 7–20 cm long; *drupes* red, densely hairy. Introduced ornamental occasionally escaping from cultivation, 5000–6000 ft. May–July. E/W. Introduced.

TOXICODENDRON Mill. – POISON IVY; POISON-OAK

Dioecious shrubs or vines, poisonous; *leaves* ternately compound, shiny; *inflorescence* appearing raceme-like; *sepals* 5, fused at the base; *petals* 5, distinct, cream-colored; *stamens* 5; *ovary* 3-carpellate; *fruit* a round drupe, white to yellow.

***Toxicodendron rydbergii* (Small ex Rydb.) Greene**, WESTERN POISON IVY. Plants 3–20 dm, often thicket-forming; *terminal leaflets* broadly ovate, rhombic, or suborbicular, the margins dentate, undulate or notched; *lateral leaflets* inequilateral; *fruits* (4) 5–6 (7) mm diam. Found in shaded canyons or along streams or on open, rocky slopes, along roadsides, and in open woods of the plains and foothills, 4500–8500 ft. May–June. E/W. (Plate 5)

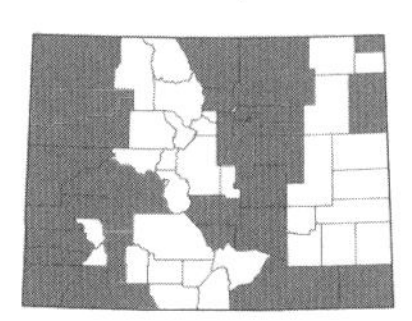

APIACEAE Lindl. – CARROT FAMILY

Herbs; *stems* commonly hollow; *leaves* alternate or basal (sometimes on a pseudoscape), usually compound; *flowers* usually in compound or simple umbels, rarely in heads, usually perfect, actinomorphic; *sepals* small or obsolete, commonly represented by small teeth around the top of the ovary; *petals* 5, distinct, usually white or yellow; *fruit* a schizocarp composed of 2 mericarps, united by their faces or commissure, each mericarp usually with 3 dorsal and 2 marginal ribs, the marginal ones often winged, the dorsal ones less often so, or sometimes the ribs obscure or absent. (Coulter & Rose, 1900; Cronquist, 1997; Mathias & Constance, 1944–45)

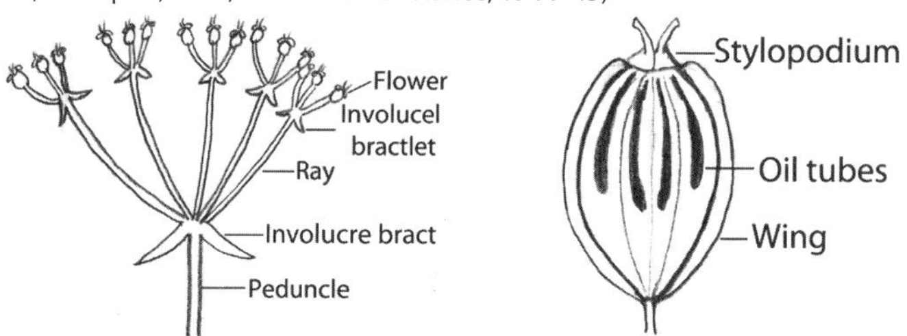

1a. Leaves or leaflets spiny-toothed; flowers sessile in dense heads subtended by toothed involucel bractlets...***Eryngium***
1b. Leaves not spiny-toothed; flowers in compound umbels...2

2a. Leaves palmately cleft with 5–9 toothed segments...***Sanicula***
2b. Leaves variously compound, but never palmately compound...3

3a. Basal leaves simple...4
3b. Basal leaves variously compound, never simple...5

4a. Basal leaves linear or lanceolate, entire; stem leaves similar...***Bupleurum***
4b. Basal leaves cordate, toothed to crenate; stem leaves mostly 3-foliate...***Zizia***

5a. Fruit and ovary covered with hooked bristles; involucre bracts pinnatifid into filiform segments...***Daucus***
5b. Fruit and ovary glabrous, variously hairy, or bristly-hispid; involucre bracts not pinnatifid into filiform segments or lacking...6

6a. Stylopodium present; plants tall (usually over 3 dm tall), caulescent with at least 3 stem leaves or rarely subacaulescent with fewer stem leaves; plants of wet places such as moist meadows, irrigation ditches, streamsides, or sometimes in drier middle elevations in the mountains...7
6b. Stylopodium absent (except in the alpine genus *Podistera*, and then plants acaulescent); plants usually short (under 3 dm tall) but in a few species up to 6 dm tall, usually acaulescent or short-caulescent, or in a few species caulescent with a few stem leaves; plants of alpine meadows, sagebrush, pinyon-juniper, or oak communities, growing on clay soil, or in rocky or sandy places of the plains and foothills, a few species found growing in middle elevations in the mountains...20

7a. Principal leaves with more or less well-defined leaflets, not dissected into small or very narrow fern-like segments, mostly once or twice compound; leaflet margins toothed...8
7b. Principal leaves dissected into small or very narrow fern-like segments, without well-defined leaflets; leaflet margins entire, not toothed...16

8a. Fruits 4–5 times longer than wide, glabrous or with bristly hairs (especially at the base); fruits and roots with a distinct anise odor...***Osmorhiza***
8b. Fruits not significantly longer than wide, glabrous or rarely slightly hairy; roots lacking an anise odor...9

9a. Flowers yellow; 4 dark linear lines from oil tubes visible through the fruit wall, these not tear-drop-shaped at the base...***Pastinaca***
9b. Flowers white; oil tubes visible through the fruit wall but ending in a tear-drop shape, or not visible through the fruit wall...10

10a. Leaflets usually 3, these large (often over 7 cm wide); leaves thinly tomentose or villous at least on the lower surfaces; oil tubes tear-drop-shaped, reaching only partway from the apex toward the base of the fruit, readily visible through the fruit wall...***Heracleum***
10b. Leaflets more than 3, these less than 7 cm wide (except rarely the terminal one); leaves glabrous or rarely scabrous; oil tubes not as above...11

11a. Plants stoloniferous, found growing in water or very moist areas; ribs of fruit inconspicuous, obscured in a thick corky fruit wall; leaflets commonly with a large center leaflet and two much smaller leaflets or lobes on either side; involucre bracts well-developed...***Berula***
11b. Plants not stoloniferous, found in various growing conditions, in wet or dry areas; ribs of fruit filiform, winged, or corky, but the fruit wall not thick or corky; leaflets and involucre bracts various...12

12a. Ultimate leaf divisions narrow, 1–4 mm; plants of dry mountain meadows and slopes; root crowned with numerous stringy, brown fibers...***Ligusticum*** (*porteri*)
12b. Ultimate leaf divisions broader; plants of usually wet places; root crowned with stringy, brown fibers or not...13

13a. Fruit ribbed, subterete, somewhat compressed laterally (at right angles to the commissure); ultimate leaf divisions lance-linear to lanceolate...14
13b. Fruit with winged lateral ribs (sometimes narrowly so), compressed dorsally (parallel to the commissure); ultimate leaf divisions usually broader, ovate to ovate-lanceolate...15

14a. Leaves once pinnately compound; involucre bracts well-developed; roots fibrous...***Sium***
14b. Leaves 1–3-ternate-pinnately compound; involucre bracts lacking or of inconspicuous, narrow bracts; roots generally tuberous-thickened...***Cicuta***

15a. Plants without a taproot or well-developed caudex; leaflet margins usually crenate (at least the leaflets of the more basal leaves)...***Oxypolis***
15b. Plants with a taproot or well-developed caudex; leaflet margins serrate...***Angelica***

16a. Leaves dissected with mostly narrow and elongate ultimate segments...***Perideridia***
16b. Leaves variously dissected, but never into long linear segments...17

17a. Stems with purple spots; robust weed, 0.5–3 m tall; carpophore undivided (this can be seen by removing each mericarp of the fruit)...***Conium***
17b. Stems without purple spots; native plants or if introduced then less than 1 m tall; carpophore bifid...18

18a. Root crowned with numerous stringy brown fibers; native plants commonly found in dry meadows, rarely in wet meadows, 7000–12,500 ft...***Ligusticum***
18b. Root not crowned with numerous stringy brown fibers; native or weedy plants of various habitats...19

19a. Weedy biennials found in disturbed places, along roadsides, and in dry meadows; involucel bractlets lacking or inconspicuous; fruit somewhat compressed laterally...***Carum***
19b. Native perennials of streamsides and wet meadows; involucel bractlets 3–6, linear or filiform, 2–8 mm long; fruit compressed dorsally...***Conioselinum***

20a. Stylopodium present; leaflet groups distinctly fan-shaped; flowers bright yellow; plants of subalpine and alpine meadows…***Podistera***
20b. Stylopodium lacking; leaflets not as above; flowers various; plants of various habitats…21

21a. Plants arising from a globose root that is generally positioned approximately 2–7 cm below ground; leaves ternate with linear leaflets; umbels longer than wide…***Orogenia***
21b. Plants arising from a woody taproot often surmounted by a branching caudex; leaves various; umbels wider than long…22

If you have fruit, continue to the following key. If not, use Key 2.

Key 1

(*Fruit present*)

22a. Fruit dorsally compressed, lateral ribs more or less winged (these are sometimes narrow); flowers yellow or white…23
22b. Fruit subterete or somewhat laterally compressed, lateral ribs not winged but sometimes corky-thickened; flowers yellow…24

23a. Dorsal ribs of the fruit more or less winged (rarely seen as only minute projections)…***Cymopterus***
23b. Dorsal ribs of the fruit wingless…***Lomatium***

24a. Fruit ovate; plants caulescent with stem leaves; leaf divisions narrowly linear, stiff and broom-like…***Harbouria***
24b. Fruit oblong to oblong-ovate; plants acaulescent, short-caulescent, or caulescent; leaf divisions various but not stiff and broom-like…25

25a. Ultimate leaf divisions broadly oblong, ovate-oblong, obovate, or orbicular, green, not glaucous; fruit with 1 oil tube in the intervals between the ribs…***Aletes***
25b. Ultimate leaf divisions linear or if oblong then usually glaucous or bluish in color; fruit with 3 or more oil tubes in the intervals between the ribs…26

26a. Rays of the umbel not reflexed or widely spreading in the fruiting stage; surface below the umbels hirtellous-scabrous or glabrous; ultimate leaf divisions oblong or linear; oil tubes 3–4 in the intervals between the ribs of the fruit, up to 6 on the commissure; found on the eastern slope on dry hills or sandy plains…***Musineon***
26b. Rays of the umbel reflexed or widely spreading in the fruiting stage; surface below the umbels glabrous; ultimate leaf divisions linear; oil tubes numerous and scattered in the fruit; endemic to southcentral Colorado where it is usually found on cliffs and ledges in pinyon-juniper woodlands…***Neoparrya***

Key 2

(*Mature fruit not present*)

This key is long, but includes all the members of the following genera: *Aletes*, *Cymopterus* (including former members of *Aletes*, *Oreoxis*, *Pseudocymopterus*, and *Pteryxia*), *Harbouria*, *Lomatium*, *Musineon*, and *Neoparrya*. Locality and habitat information is given here to assist in identification and ease of use, consult the species descriptions under each genus for more detailed information and synonymy.

1a. Flowers white (sometimes pinkish- or purplish-tinged)…2
1b. Flowers yellow or purple…10

2a. Involucel bractlets white…3
2b. Involucel bractlets green (can be scarious-margined, but not completely white) or purple…5

3a. Mature peduncles shorter than or just equal to the height of the leaves; vein of bractlet green and thick…***Cymopterus montanus* Nutt. ex Torr. & A. Gray**, MOUNTAIN SPRINGPARSLEY. Common in sandy places on the eastern plains and foothills, 3800–6200 ft. April–May. E. (Plate 6)
3b. Mature peduncles equal to or more commonly exceeding the height of the leaves; bractlet vein(s) narrow, green or purple…4

4a. Bractlets with mostly 1 vein arising from the base, occasionally with 2 additional veins, but these shorter and not equal to the central vein…***Cymopterus bulbosus* A. Nelson,** BULBOUS SPRINGPARSLEY. Typically found on heavy clay soil, often associated with sagebrush or pinyon-juniper communities, 4500–8500 ft. Mar.–June. W.
4b. Bractlets with mostly 3 but up to 5 veins arising from the base, these more or less equal, parallel, and flaring distally…***Cymopterus constancei* R.L. Hartm.**, CONSTANCE'S SPRINGPARSLEY. Typically found in grasslands or sagebrush-grasslands on loose, sandy or loamy soil, 5500–8700 ft. April–May. E/W.

5a. Leaves glabrous, although sometimes roughened on the margins and veins...6
5b. Leaves minutely hairy, tomentose, or villous...8

6a. Taproot with a simple or relatively few-branched, surficial or subterranean herbaceous crown; plants with a conspicuous pseudoscape, the leaves and peduncles all beginning at a point elevated above the taproot on a stem-like structure (the pseudoscape is sometimes deeply buried, with the leaves spreading out and lying at ground level); peduncles usually shorter than the leaves...***Cymopterus acaulis*** **(Pursh) Raf.**, PLAINS SPRINGPARSLEY. Common in gravelly and sandy soil of the plains and foothills, 3700–7500 ft. April–June. E.
6b. Taproot surmounted by a freely branching, surficial woody caudex; plants lacking a pseudoscape; peduncles longer than the leaves...7

7a. Plants of the western slope; ultimate leaf divisions broadly linear to subcuneate, 1–5 mm long...***Cymopterus terebinthus*** **(Hook.) Torr. & A. Gray var. *albiflorus* (Torr. & A. Gray) M.E. Jones**, AROMATIC SPRINGPARSLEY. Uncommon on the western slope (Moffat and Routt cos.) in dry, open, rocky or sandy places, 5500–7500 ft. May–June. W.
These usually have yellow flowers, but they can fade to white.
7b. Plants of the eastern slope; ultimate leaf divisions narrowly linear, 2–30 mm long...***Musineon tenuifolium*** **Nutt.**, SLENDER WILDPARSLEY. Locally common in sandy or sometimes rocky soil on the eastern plains to foothills, especially common at Pawnee Buttes (Weld co.), 4500–6200 ft. May–June. E.
These usually have yellow flowers, but rarely white-flowered plants have been observed.

8a. Involucel bracts tomentose or villous...***Lomatium macrocarpum*** **(Nutt. ex Torr. & A. Gray) J.M. Coult. & Rose**, BIGSEED LOMATIUM. Found on open, rocky, dry slopes, scattered on the western slope, 5500–8500 ft. May–June. W.
8b. Involucel bracts glabrous or minutely hairy...9

9a. Anthers red; bractlets of the involucel obovate to linear-lanceolate, glabrous; common on the eastern slope, with scattered occurrences on the western slope...***Lomatium orientale*** **J.M. Coult. & Rose**, SALT-AND-PEPPER. Common on open slopes from the plains to the foothills on the eastern slope and with scattered occurrences on the western slope, blooming in early spring, 4800–9000 ft. April–June. E/W. (Plate 6)
9b. Anthers usually yellow; bractlets of the involucel filiform or linear, minutely hairy; plants of the western slope only...***Lomatium juniperinum*** **(M.E. Jones) J.M. Coult. & Rose**, JUNIPER-LOMATIUM. Found on dry, rocky slopes and rocky openings in aspen forests, usually associated with juniper-sagebrush communities, known from the northwestern counties, 7000–8600 ft. May–June. W.

10a. Leaves biternately compound, the ultimate leaf divisions long and linear, mostly 30–60 times as long as wide; plants densely minutely hairy throughout...***Lomatium triternatum*** **(Pursh) J.M. Coult. & Rose var. *platycarpum* (Torr.) Boivin**, GREAT BASIN DESERTPARSLEY. Common in sagebrush meadows and on open gravelly slopes, 5500–10,000 ft. May–July. W. (Plate 7)
10b. Leaves variously pinnately or ternately compound, but the ultimate leaf divisions less than 15 times as long as wide; plants glabrous to minutely hairy or villous throughout...11

11a. Rays reflexed...***Neoparrya lithophila*** **Mathias**, BILL'S NEOPARRYA. Locally common in southcentral Colorado in the upper San Luis Valley where it is found growing in rock crevices and on ledges, often associated with pinyon-juniper woodlands, 7000–10,000 ft. May–July. E. Endemic.
11b. Rays ascending-spreading...12

12a. Plants with a conspicuous pseudoscape, the leaves and peduncles all beginning at a point elevated above the taproot on a stem-like structure (*careful*—care must be taken when collecting as the pseudoscape can be deeply buried, with the leaves spreading out and lying at ground level)...13
12b. Plants acaulescent with the leaves and peduncles arising from the top of the taproot or caudex, or caulescent with 1–2 stem leaves, usually lacking a pseudoscape or the pseudoscape very short and inconspicuous, essentially absent...19

13a. Involucel bractlets ovate and conspicuous, sometimes 3-toothed at the apex, as long or longer than the flowers...14
13b. Involucel bractlets filiform and inconspicuous, longer or shorter than the flowers...15

14a. Rays mostly longer than 10 mm, the middle umbels not sessile; involucre bracts usually 5 mm or less in length, 1.5 mm or less in width...***Lomatium concinnum* (Osterh.) Mathias**, COLORADO DESERT-PARSLEY. Associated with sagebrush communities on adobe hills and in rocky soils derived from Mancos shale, known from Delta, Montrose, and Ouray cos., 5500–7000 ft. May. W. Endemic. (Plate 6)
14b. Rays mostly shorter than 10 mm, the middle umbels sessile; involucre bracts usually 5 mm or longer, 2 mm or more in width...***Cymopterus fendleri* A. Gray**, FENDLER'S SPRINGPARSLEY. Common on the western slope in clay soil or gravelly soil often in association with juniper-sagebrush or pinyon-juniper communities, 4500–7000 ft. April–June. W. (Plate 6)

15a. Plants of the eastern slope; surface below the umbels usually hirtellous-scabrous...***Musineon divaricatum* (Pursh) Raf.**, LEAFY WILDPARSLEY. Common in sandy soil on the eastern plains and foothills, 4500–6000 ft. April–June. E. (Plate 7)
15b. Plants of the western slope; surface below the umbels glabrous...16

16a. Leaves very finely dissected, the ultimate leaf divisions very numerous, filiform, about 0.5 mm wide or less, mostly 1–4 mm long...***Lomatium bicolor* (S. Watson) J.M. Coult. & Rose**, LITTLE-HEADS LOMATIUM. Uncommon on dry slopes and in meadows, 6500–8000 ft. May–July. W.
16b. Ultimate leaf divisions not very finely dissected, mostly over 0.5 mm wide, linear to ovate...17

17a. Ultimate leaf divisions at least 6 times longer than wide...***Lomatium leptocarpum* (Torr. & A. Gray) J.M. Coult. & Rose**, GUMBO-LOMATIUM. Found on open slopes and meadows, especially common in heavy clay soils, 6200–8000 ft. May–June. W.
17b. Ultimate leaf divisions about 2–3 times as long as wide...18

18a. Ultimate leaf divisions narrowly ovate or linear, spreading and not overlapping with the central rachis, usually 1 mm or less in width...***Cymopterus planosus* (Osterh.) Mathias**, ROCKY MOUNTAIN SPRINGPARSLEY. Found on open gravelly slopes and in juniper-sagebrush communities, 6500–9500 ft. May–June. W. Endemic.
18b. Ultimate leaf divisions ovate, erect and overlapping with the central rachis, usually 1.5–4 mm wide...***Cymopterus longipes* S. Watson**, LONGSTALK SPRINGPARSLEY. Found in open places, often in association with pinyon-juniper or sagebrush communities. April–May.
This species is not reported from Colorado yet but occurs right up to the border especially along Moffat County.

19a. Rays conspicuously unequal in length, with some outer rays at least twice as long as the inner rays, or the inner umbels completely sessile...20
19b. Rays equal or nearly so, or if unequal then the longer rays less than twice as long as the shorter rays...33

20a. Ovary and fruit villous; plants villous below the umbel and usually densely villous throughout...***Lomatium foeniculaceum* (Nutt.) J.M. Coult. & Rose**, DESERT BISCUITROOT. Found in dry, open places, 5400–6200 ft. May. E/W.
20b. Ovary and fruit usually glabrous or sometimes minutely hairy; plants glabrous, or shortly hairy just below the umbel or throughout but scabrous or minutely hairy, not villous...21

21a. Ultimate leaf divisions broad, usually 5 mm or more in width, 3-toothed at apex, appearing as acute, mucronate lobes of a broad leaflet...***Cymopterus duchesnensis* M.E. Jones**, DUCHESNE SPRINGPARSLEY. Found in desert shrub, sagebrush, and juniper communities on sandy clay and clay soils derived from shales, known from Moffat and Rio Blanco cos., 4700–6800 ft. April–May. W.
21b. Ultimate leaf divisions narrower, usually 4 mm or less in width, filiform, linear, ovate, or triangular in shape, not 3-toothed at the apex...22

22a. Leaves densely minutely hairy or shortly hairy, especially on the margins...23
22b. Leaves glabrous or sometimes with some sparse short hairs, but not densely minutely hairy...24

23a. Peduncles glabrous; ultimate leaf divisions ovate to oblong, crowded, as a whole often appearing fan-shaped, densely scaberulent...***Lomatium eastwoodiae* (J.M. Coult. & Rose) J.F. Macbr.**, EASTWOOD'S BISCUITROOT. Found on the Uncompahgre Plateau in pinyon-juniper woodlands and in sandy soil, 4600–7000 ft. Mar.–April. W. Endemic.
23b. Peduncles minutely hairy; ultimate leaf divisions linear, not crowded and appearing fan-shaped as a whole, mostly minutely hairy along the margins...***Lomatium juniperinum* (M.E. Jones) J.M. Coult. & Rose**, JUNIPER-LOMATIUM. Found on dry, rocky slopes and rocky openings in aspen forests but most often associated with juniper-sagebrush communities, 7000–8600 ft. May–June. W.

24a. Ultimate leaf divisions extremely narrow and small, no greater than 0.5 mm in width and up to 7 mm long, usually very numerous with several hundred to a thousand or more per leaf…25
24b. Ultimate leaf divisions usually greater than 0.5 mm in width and longer than 5 mm, the larger ones generally 1 mm wide or more, and less numerous…26

25a. Pedicels 7–15 mm long at maturity; plants from a long taproot usually surmounted by a branching caudex…***Lomatium grayi*** **(J.M. Coult. & Rose) J.M. Coult. & Rose var. *grayi***, MILFOIL LOMATIUM. Found on dry, open, rocky slopes, 5000–9700 ft. April–May. W. (Plate 6)
25b. Pedicels 1–2.5 mm long at maturity; plants from a short, round root and mostly subterranean crown or short pseudoscape…***Lomatium bicolor*** **(S. Watson) J.M. Coult. & Rose**, LITTLE-HEADS LOMATIUM. Uncommon on dry slopes and in meadows, 6500–8000 ft. May–July. W.

26a. Ultimate leaf divisions triangular in shape, flaring outward; petioles purple or reddish…***Cymopterus purpureus*** **S. Watson**, COLORADO PLATEAU SPRINGPARSLEY. Common in clay or gravelly soil, often in association with juniper-sagebrush or pinyon-juniper communities, 4800–7500 ft. May–June. W.
26b. Ultimate leaf divisions unlike the above; petioles usually greenish or brown…27

27a. Leaves appearing very open and skeleton-like, leaf segments not closely crowded on the rachis, 1–3.5 mm long; involucel bractlets inconspicuous and linear, much shorter than the flowers…***Cymopterus petraeus*** **M.E. Jones**, ROCK SPRINGPARSLEY. [*Aletes petraea* (M.E. Jones) W.A. Weber]. Uncommon (Moffat and San Juan cos.) in open, rocky places, 7000–7500 ft. May–June. W.
27b. Leaves not appearing skeleton-like, leaf segments more or less crowded on the rachis, 3–50 mm long; involucel bractlets linear to linear-lanceolate, about equaling or longer than the flowers…28

28a. Longer rays less than 10 mm long; ultimate leaf divisions linear, mostly over 10 times longer than wide…***Cymopterus sessiliflorus*** **(W.L. Theob. & C.C. Tseng) R.L. Hartm.**, SESSILEFLOWER SPRINGPARSLEY. [*Aletes sessiliflorus* W.L. Theob. & C.C. Tseng; *C. macdougalii* (J.M. Coult. & Rose) Tidestr. ssp. *breviradiatus* W.L. Theob. & C.C. Tseng]. Found on sandstone ledges and in clay soil, known from Montezuma and La Plata cos., 6000–7000 ft. April–Aug. W.
28b. Longer rays usually over 15 mm long; ultimate leaf divisions linear or narrowly ovate, mostly less than 10 times longer than wide…29

29a. Leaf surface scabrous or minutely hairy, especially along the margins and veins, rarely glabrous; plants caulescent, lacking a pseudoscape, often over 5 dm tall…***Lomatium dissectum*** **(Nutt.) Mathias & Constance var. *eatonii* (J.M. Coult. & Rose) Cronquist**, GIANT LOMATIUM. Found on rocky slopes and in dry meadows, 6000–9500 ft. May–June. E/W.
29b. Leaf surface glabrous, rarely scaberulent; plants acaulescent or sometimes shortly caulescent and then usually with a pseudoscape, usually less than 5 dm tall…30

30a. Taproot with a simple or relatively few-branched, surficial or subterranean herbaceous crown, sometimes with a short pseudoscape; plants not appearing densely tufted…***Lomatium leptocarpum*** **(Torr. & A. Gray) J.M. Coult. & Rose**, GUMBO-LOMATIUM. [*L. bicolor* (S. Watson) J.M. Coult. & Rose var. *leptocarpum* (Torr. & A. Gray) SchlesJ. Sm.]. Found on open slopes and meadows, especially common in heavy clay soils, 6200–8000 ft. May–June. W.
30b. Taproot surmounted by a freely branching, surficial woody caudex; plants appearing densely tufted…31

31a. Leaves broadly ovate to ovate-oblong in general outline; leaflets generally without a sharp-pointed tip, grayish in color; involucel bractlets about equaling the flowers, 2–6 mm long; known from Moffat County…***Cymopterus terebinthus*** **(Hook.) Torr. & A. Gray var. *albiflorus* (Torr. & A. Gray) M.E. Jones**, AROMATIC SPRINGPARSLEY. Uncommon in dry, open, rocky or sandy places, 5500–7500 ft. May–June. W.
31b. Leaves narrowly oblong in general outline; leaflets generally with a sharp-pointed tip, green in color; involucel bractlets mostly longer than the flowers, 3–15 mm long; not confined to Moffat County…32

32a. Leaf segments more erect than spreading…***Cymopterus anisatus*** **A. Gray**, ANISE SPRINGPARSLEY. [*Aletes anisatus* (A. Gray) W.L. Theob. & C.C. Tseng]. Found in gravelly or sandy soil, 6200–10000 ft. May–July. E/W. Endemic.
32b. Leaf segments widely spreading…***Cymopterus longilobus*** **(Rydb.) W.A. Weber**, MOUNTAIN SPRINGPARSLEY. Found on rocky soil, primarily known from the Piceance Basin (Rio Blanco County), 6300–11,000 ft. May–July. E/W.

33a. Involucel bractlets purple or purple-tinted at least at the apex, obovate and 3-toothed at the tip…***Cymopterus bakeri*** **(J.M. Coult. & Rose) M.E. Jones**, BAKER'S ALPINEPARSLEY. Plants of alpine meadows in the southern mountains, 12,000–14,000 ft. June–Aug. E/W. Endemic.
33b. Involucel bractlets green, sometimes purple-tinted but then not 3-toothed at the tip…34

34a. Involucel bractlets conspicuous and longer than the flowers, 5–10 mm in length; leaves once pinnately compound with lanceolate leaflets 5–40 mm long...***Lomatium latilobum* (Rydb.) Mathias**, CANYONLANDS LOMATIUM. Found in sandy soils of pinyon-juniper and desert shrub communities, Mesa County, 5000–7000 ft. April–May. W.
34b. Involucel bractlets inconspicuous, usually 5 mm or less in length, usually shorter than the flowers but sometimes longer; leaves unlike the above in all respects...35

35a. Ultimate leaf divisions ovate, usually over 2 mm wide, 3-toothed at the tips and often fan-shaped in outline...36
35b. Ultimate leaf divisions unlike the above...37

36a. Mature peduncles shorter than the height of the leaves; umbels few-flowered (mostly 2–5 per umbel), mostly 0.5 cm wide...***Aletes humilis* J.M. Coult. & Rose**, COLORADO ALETES. Uncommon, usually found growing in cracks of rocks, Larimer and Boulder cos., 6500–8700 ft. April–June. E. Endemic.
36b. Mature peduncles longer than the height of the leaves; umbels with several flowers, mostly over 1 cm wide...***Aletes acaulis* (Torr.) J.M. Coult. & Rose**, STEMLESS INDIAN PARSLEY. Common, usually found growing in cracks of rocks in canyons of the foothills, 7000–9000 ft. May–July. E. (Plate 5)

37a. Surface below the umbel hirtellous-scabrous or minutely hairy with granular projections, the remaining peduncle glabrous...38
37b. Surface below the umbel glabrous or if hirtellous-scabrous or minutely hairy then the remaining peduncle also hairy...39

38a. Leaf divisions linear, stiff and broom-like; roots usually crowned with numerous stringy brown fibers...***Harbouria trachypleura* (A. Gray) J.M. Coult. & Rose**, WHISKBROOM PARSLEY. Common on open rocky places, 4400–8600 ft. April–June. E.
38b. Leaf divisions usually lanceolate to broadly lanceolate or sometimes linear, not stiff and broom-like; roots not crowned with numerous stringy brown fibers, although base of plants may have brown sheaths (but these not stringy)...***Cymopterus lemmonii* (J.M. Coult. & Rose) Dorn**, LIGUSTICOID SPRINGPARSLEY. Common in dry, wooded slopes from the mountains to alpine meadows, 6000–14,000 ft. May–July. E/W. (Plate 6)

39a. Plants mat-forming, of alpine and subalpine meadows, above 9500 ft; umbel about 1 cm wide in flower...40
39b. Plants not mat-forming, variously distributed, but not of alpine or subalpine meadows, to 9500 ft in elevation; umbel various...41

40a. Ovary (and fruit) scabrous-puberulent; oil tubes usually solitary in the intervals between the fruit ribs...***Cymopterus alpinus* A. Gray**, ALPINE SPRINGPARSLEY. Common in alpine and subalpine meadows in the mountains, 9400–14,000 ft. June–Aug. E/W. (Plate 6)
40b. Ovary (and fruit) glabrous; oil tubes more than 2 in the intervals between the fruit ribs...***Cymopterus humilis* (Raf.) Tidestr. & Kittell**, PIKE'S PEAK ALPINE PARSELY. Rare, found only on Pikes Peak, 11,900–14,000 ft. June–Aug. E. Endemic.

41a. Leaf surface scabrous or minutely hairy, especially along the margins and veins...42
41b. Leaf surface glabrous...43

42a. Umbels usually not over 2 cm in diam. in flower or fruit (usually 1 cm wide or less at anthesis); plants acaulescent, less than 5 dm tall, of Montezuma County...***Cymopterus sessiliflorus* (W.L. Theob. & C.C. Tseng) R.L. Hartm.**, SESSILEFLOWER SPRINGPARSLEY. [*Aletes sessiliflorus* W.L. Theob. & C.C. Tseng; *C. macdougalii* (J.M. Coult. & Rose) Tidestr. ssp. *breviradiatus* W.L. Theob. & C.C. Tseng]. Found on sandstone ledges and in clay soil, known from Montezuma and La Plata cos., 6000–7000 ft. April–Aug. W.
42b. Umbels over 2 cm in diam. in flower or fruit; plants usually caulescent, often over 5 dm tall, not restricted to Montezuma County...***Lomatium dissectum* (Nutt.) Mathias & Constance var. *eatonii* (J.M. Coult. & Rose) Cronquist**, GIANT LOMATIUM. Found on rocky slopes and in dry meadows, common on the western slope, known on the eastern slope from North Park, 6000–9500 ft. May–June. E/W.

43a. Ovary and surface below the umbels glabrous; rays usually 15 mm or more in length...***Lomatium nuttallii* (A. Gray) J.F. Macbr.**, NUTTALL'S BISCUITROOT. Known on the eastern slope so far only from the chalk bluffs, northern Weld County, 5600 ft; known on the western slope in sagebrush shrublands from Middle Park near Kremmling, 7500–8000 ft. May–June. E/W.
43b. Ovary and surface below the umbels minutely scabrous or granular with short projections; rays usually less than 15 mm in length...***Musineon tenuifolium* (Nutt. ex Torr. & A. Gray) J.M. Coult. & Rose**, SLENDER WILDPARSLEY. Locally common in sandy or rocky soil on the eastern plains to foothills, 4500–6200 ft. May–June. E. (Plate 7)

ALETES J.M. Coult. & Rose – Indian parsley

Acaulescent, perennial herbs; *leaves* pinnately to bipinnately compound; *involucre* usually lacking; *involucel* of narrow bractlets; *flowers* yellow; *stylopodium* lacking; *fruit* slightly flattened laterally, ribs reduced and obscure or corky-winged.

1a. Mature peduncles shorter than the height of the leaves; umbels few-flowered (mostly 2–5 per umbel), mostly 0.5 cm wide...***A. humilis***

1b. Mature peduncles longer than the height of the leaves; umbels with several flowers, over 1 cm wide...***A. acaulis***

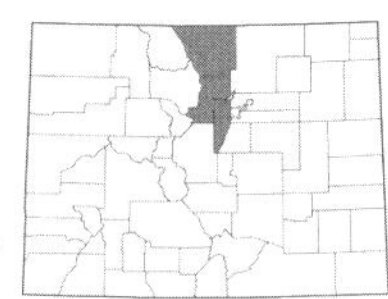

***Aletes acaulis* (Torr.) J.M. Coult. & Rose**, stemless Indian parsley. Plants 0.5–3.5 dm, minutely hairy; *leaflets* 4–15 mm long, lanceolate to orbicular, pinnately lobed and dentate; *involucel bractlets* 2–3 mm long, lanceolate or linear; *rays* 8–15, subequal, spreading to reflexed; *fruit* 4–7 mm long, with corky wings. Locally common, usually found growing in cracks of rocks in canyons of the foothills, 7000–9000 ft. May–July. E. (Plate 5)

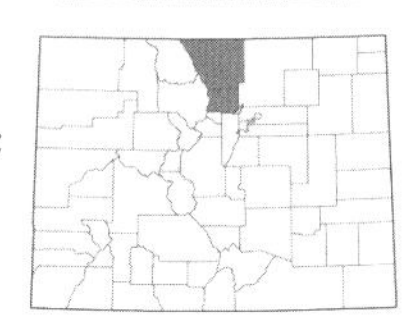

***Aletes humilis* J.M. Coult. & Rose**, Colorado aletes. Plants 0.2–1 dm, glabrous to minutely hairy; *leaflets* linear to ovate or ovate-oblong; *involucel bractlets* 2–4 mm long, linear; *rays* 4–6, subequal; *fruit* 3–4 mm long, with inconspicuous ribs. Rare, usually found growing in cracks of rocks, 6500–8700 ft. April–June. E. Endemic.

ANGELICA L. – angelica

Perennial herbs; *leaves* pinnately to ternately 1–3 times compound; *involucre* and involucel of narrow and scarious bracts or bractlets, or absent; *flowers* white, pinkish, or purplish-brown; *stylopodium* broadly conic; *fruit* strongly compressed dorsally, with a broad commissure, dorsal ribs inconspicuous to narrowly winged, lateral ribs thin-winged or corky-winged.

1a. Plants usually over 10 dm; umbels globose in shape; flowers white; fruit with numerous oil tubes...***A. ampla***

1b. Plants under 10 dm; umbels rather flat-topped; flowers white, pinkish, or purplish-brown; fruit with oil tubes solitary (or rarely in pairs) in the intervals...2

2a. Bractlets of involucels conspicuous, linear-lanceolate to lanceolate, usually over 1 mm wide; leaflets 1–5 cm long; flowers purplish-brown...***A. grayi***

2b. Bractlets of involucels absent; leaflets 3–9 cm long; flowers white or pinkish...***A. pinnata***

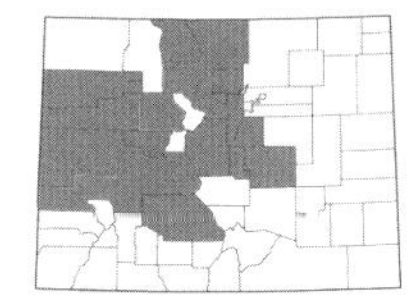

***Angelica ampla* A. Nelson**, giant angelica. Plants usually over 10 dm; *leaves* ternate, then bipinnate; *leaflets* 3–20 cm long, ovate, serrate; *involucel bractlets* few, filiform; *rays* 30–45; *umbel* globular; *flowers* white; *fruit* 7–8 mm long, oblong-oval, the ribs narrowly winged. Common in moist places, along streams, and in the mountains, 6000–9000 ft. July–Aug. E/W.

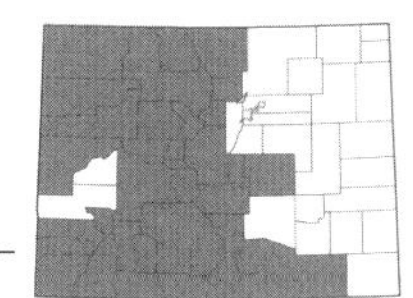

***Angelica grayi* (J.M. Coult. & Rose) J.M. Coult. & Rose**, Gray's angelica. Plants 2–6 dm; *leaves* pinnate to bipinnate or ternate-pinnate, scabrous; *leaflets* 1–5 cm long, lanceolate to ovate, serrate; *involucel bractlets* 5–18 mm long, linear-lanceolate; *rays* numerous; *umbel* flat-topped; *flowers* purplish-brown; *fruit* 4–5 mm long, oval, the dorsal ribs narrowly winged and lateral ribs with broader wings. Common in mountain meadows and on alpine slopes, 9800–13,500 ft. June–Aug. E/W. (Plate 5)

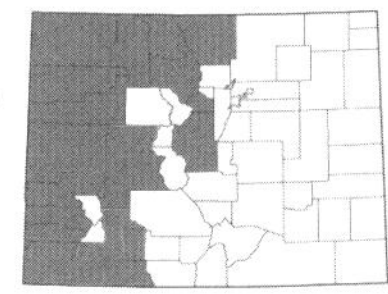

***Angelica pinnata* S. Watson**, small-leaf angelica. Plants 2.5–9 dm, glabrous; *leaves* pinnately compound or incompletely bipinnate; *leaflets* 3–9 cm long, lanceolate to ovate-lanceolate, serrate or rarely entire; *involucel bractlets* absent; *rays* 6–25, unequal, 1–6 cm long; *umbel* flat-topped; *flowers* white or pinkish; *fruit* 3–6 mm long, nearly orbicular, the dorsal ribs narrowly winged and lateral ribs broadly winged. Locally common along streamsides and in aspen groves, 5000–9500 ft. June–Aug. E/W.

BERULA W.D.J. Koch – water parsnip

Stoloniferous, caulescent, perennial herbs; *leaves* pinnately compound, leaflets subentire to toothed or cleft, the submerged leaves often filiform-dissected; *involucre* of conspicuous, narrow, entire or toothed bracts; *flowers* white; *stylopodium* present; *fruit* elliptic to orbicular, somewhat compressed laterally, ribs obscure in the thick corky pericarp.

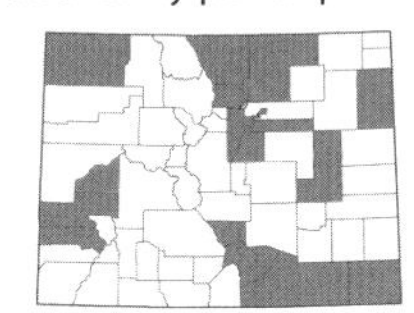

***Berula erecta* (Huds.) Coville**, cutleaf waterparsnip. Plants to 15 dm, glabrous; *leaves* petioled, petiole 4–12 cm long, blade 1.5 dm long; *leaflets* oblong to ovate, in 7–12 pairs, 1–8 cm long; *sepals* minute, persistent; *petals* white; *fruit* 1.5–2 mm long, oil tubes embedded deeply in fruit wall. Localized in wet places or in shallow water in the valleys and plains, 3500–6300 ft. July–Sept. E/W. Introduced. (Plate 6)

BUPLEURUM L. – THOROW WAX

Caulescent, perennial herbs; *leaves* simple and entire; *involucre* and involucel conspicuous; *flowers* yellow or sometimes purple; *stylopodium* present; *fruit* glabrous, compressed laterally, ribbed laterally.

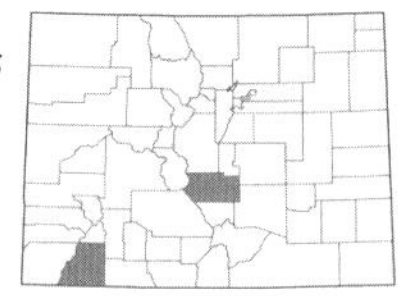

***Bupleurum americanum* J.M. Coult. & Rose**, AMERICAN THOROW WAX. Plants 5–50 cm, glabrous; *leaves* linear to narrowly lanceolate, 2–12 cm long; *rays* 1–5 cm long; *involucre* of lanceolate bracts, 5–15 mm long; *involucel* of ovate bractlets, 2–5 mm long; *flowers* up to 1 mm long; *fruit* 2–3 mm long, glaucous, oil tubes numerous. Uncommon in rocky soil of the alpine, 12,500–13,000 ft. July–Aug. W.

CARUM L. – CARAWAY

Perennial herbs; *leaves* pinnately dissected; *rays* few, spreading-ascending, unequal; *flowers* white or pinkish; *stylopodium* present; *fruit* oblong to broadly elliptic-oblong, compressed laterally, glabrous, evidently ribbed.

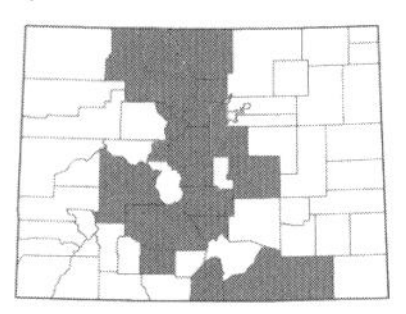

***Carum carvi* L.**, CARAWAY. Plants 3–6 dm, glabrous, taprooted; *leaves* 3 to 4 times pinnately divided, divisions filiform to narrowly linear, 8–15 cm long; *rays* 1–4 cm long; *involucre* and involucel minute or absent; *petals* ca. 1 mm long; *fruit* 3–4 mm long. Locally common in disturbed places in the foothills and mountains, 5000–8500 ft. June–July. E/W. Introduced.

CICUTA L. – WATER HEMLOCK

Caulescent, perennial herbs; *leaves* once-pinnately to thrice-pinnately compound; *involucre* absent, or of a few inconspicuous narrow bracts; *involucel* of several narrow bractlets or absent; *flowers* white or greenish; *stylopodium* present; *fruit* ovate to orbicular, compressed laterally, the ribs rather prominent and corky.

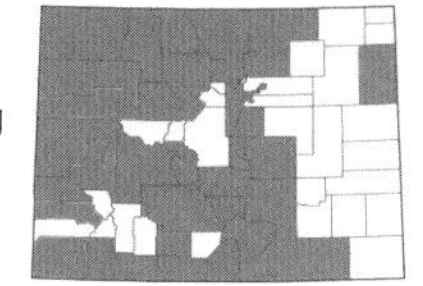

***Cicuta maculata* L.**, SPOTTED WATER HEMLOCK. Plants 5–25 dm, often glaucous, with sap drying red-brown; *leaflets* linear to narrowly ovate, usually 3–10 cm long and 6–30 cm wide; *rays* 2–6 cm long at maturity; *petals* wide with narrowing tips; *fruit* 2–4 mm long. Locally common in moist places such as marshes and ditches, 3400–8000 ft. June–Aug. E/W. (Plate 6)

One of our most poisonous plants in Colorado.

CONIOSELINUM Hoffm. – HEMLOCKPARSLEY

Caulescent, perennial herbs; *leaves* pinnately or ternate-pinnately decompound; *involucre* of a few narrow or leafy bracts or absent; *flowers* white; *stylopodium* present; *fruit* glabrous, compressed dorsally with a broad commissure, the lateral ribs evident and broadly thin-winged, the dorsal ribs low and corky.

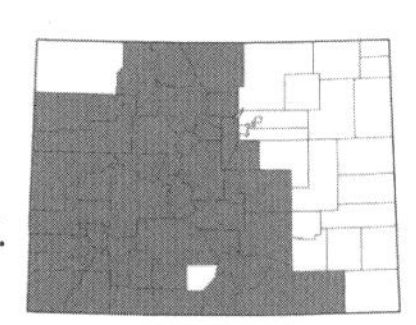

***Conioselinum scopulorum* (A. Gray) J.M. Coult. & Rose**, ROCKY MOUNTAIN HEMLOCKPARSLEY. Plants 3–12 dm, stems arising from 2 to many tuberous roots, simple or sparingly branched; *leaves* with blades 5–20 cm long, secondary leaflets 1.5–4 cm long; *rays* 2–4 cm long; *fruit* 3.5–5.5 mm long. Common in the mountains along streams and in moist meadows, 7000–12,500 ft. July–Sept. E/W.

CONIUM L. – POISON HEMLOCK

Caulescent, biennial herbs; *stems* purple-spotted; *leaves* large, pinnately or ternate-pinnately dissected, with small fern-like ultimate segments; *involucre* and involucel of several small bracts or bractlets or absent; *flowers* white; *stylopodium* present; *fruit* glabrous, flattened laterally, the ribs prominent, raised, often wavy and somewhat crenate.

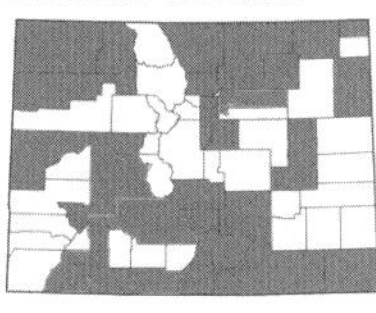

***Conium maculatum* L.**, POISON HEMLOCK. Plants 5–30 dm; *leaf blade* 1.5–3 dm long; *rays* 1.5–5 cm long, numbering 10–20; *fruit* 2–2.5 mm long. Common, tall weed of roadside ditches and moist, disturbed sites, 3600–8700 ft. June–Aug. E/W. Introduced.

CYMOPTERUS Raf. – SPRINGPARSLEY

Perennial herbs; *leaves* pinnately or ternately cleft or compound or dissected; *involucre* in most species lacking or inconspicuous, the involucel in most species inconspicuous or strongly oblique; *flowers* usually yellow or white; *stylopodium* lacking; *fruit* compressed dorsally, oblong to elliptic or orbicular, dorsal and lateral ribs winged.

1a. Bractlets of the involucel large, showy, white to purple in color, scarious, with 1–3 prominent veins, tending to form a cup around the umbellet; flowers never yellow…2

1b. Bractlets of the involucel small, inconspicuous, or if larger then not scarious and white but herbaceous or coriaceous and green or purplish; flowers mostly yellow, but can be white or purplish-colored…4

2a. Mature peduncles shorter than or just equal to the height of the leaves, wings of the fruit conspicuously enlarged at the base in cross-section; vein of bractlet green and thick…***C. montanus***

2b. Mature peduncles equal to or more commonly exceeding the height of the leaves, wings of the fruit not conspicuously enlarged at the base in cross-section; bractlet vein(s) narrow, green or purple…3

3a. Bractlets with mostly 1 vein arising from the base, occasionally with 2 additional veins, but these shorter and not equal to the central vein…***C. bulbosus***
3b. Bractlets with mostly 3 but up to 5 veins arising from the base, these more or less equal, parallel, and flaring distally…***C. constancei***

4a. Stems conspicuously scabrous-hirtellous just below the umbel and often just below the nodes, the rest of the stem glabrous or nearly so; plants caulescent and leafy-stemmed when well-developed…***C. lemmonii***
4b. Stems glabrous or if short-hairy below the umbel then the rest of the peduncle short-hairy as well…5

5a. Plants with a conspicuous pseudoscape, the leaves and peduncles all beginning at a point elevated above the taproot on a stem-like structure (the pseudoscape is sometimes deeply buried, with the leaves spreading out and lying at ground level)…6
5b. Plants acaulescent (lacking stem leaves, the leaves and peduncles arising from the top of the taproot or caudex), or caulescent with 1–2 stem leaves, usually lacking a pseudoscape or the pseudoscape very short and nearly inconspicuous…9

6a. Involucel bractlets ovate and conspicuous, often 3-toothed at the apex, as long or longer than the flowers…7
6b. Involucel bractlets filiform and inconspicuous, longer or shorter than the flowers…8

7a. Flowers white; mature peduncles shorter than or just equal to the height of the leaves; plants of the eastern slope…***C. acaulis***
7b. Flowers yellow; mature peduncles exceeding the height of the leaves; plants of the western slope…***C. fendleri***

8a. Ultimate leaf divisions narrowly ovate or linear, spreading and not overlapping with the central rachis, usually 1 mm or less in width…***C. planosus***
8b. Ultimate leaf divisions ovate, erect and overlapping with the central rachis, usually 1.5–4 mm in width…***C. longipes***

9a. Involucel bractlets purple or purple-tinted at least at the apex, obovate and 3-toothed at the tip; alpine…***C. bakeri***
9b. Involucel bractlets green, sometimes purple-tinted but then not 3-toothed at the tip; variously distributed…10

10a. Ultimate leaf divisions broad, usually 5 mm or more in width, 3-toothed at apex, appearing as acute, mucronate lobes of a broad leaflet…***C. duchesnensis***
10b. Ultimate leaf divisions narrower, usually 4 mm or less in width, filiform, linear, ovate, or triangular in shape, not 3-toothed at the apex…11

11a. Ultimate leaf divisions triangular, flaring outward, bright green; petioles purple or reddish…***C. purpureus***
11b. Ultimate leaf divisions unlike the above; petioles usually greenish or brown…12

12a. Plants mat-forming; found in alpine and subalpine meadows above 9500 ft; umbels usually 2 cm or less wide, the rays equal or nearly so in length…13
12b. Plants not mat-forming; usually found below 10,500 ft; umbels and rays various, but usually over 2 cm wide…14

13a. Ovary (and fruit) scabrous-puberulent; oil tubes usually solitary in the intervals between the fruit ribs; plants widely distributed throughout the subalpine and alpine…***C. alpinus***
13b. Ovary (and fruit) glabrous; oil tubes more than 2 in the intervals between the fruit ribs; plants only known from the alpine tundra on Pikes Peak (El Paso Co.)…***C. humilis***

14a. Leaves appearing very open and skeleton-like, leaf segments not closely crowded on the rachis, 1–3.5 mm long; involucel bractlets inconspicuous and linear, much shorter than the flowers…***C. petraeus***
14b. Leaves not appearing skeleton-like, leaf segments more or less crowded on the rachis, 3–50 mm long; involucel bractlets linear to linear-lanceolate, about equaling or longer than the flowers…15

15a. Longer rays less than 10 mm long; known from Montezuma and La Plata counties…***C. sessiliflorus***
15b. Longer rays mostly over 15 mm long; variously distributed but not known from Montezuma or La Plata counties…16

16a. Leaves broadly ovate to ovate-oblong in general outline; leaflets generally without a sharp-pointed tip, grayish in color; involucel bractlets about equaling the flowers, 2–6 mm long; known from Moffat and Routt counties...***C. terebinthus***
16b. Leaves narrowly oblong in general outline; leaflets generally with a sharp-pointed tip, green in color; involucel bractlets mostly longer than the flowers, 3–15 mm long; variously distributed...17

17a. Leaf segments more erect than spreading...***C. anisatus***
17b. Leaf segments widely spreading...***C. longilobus***

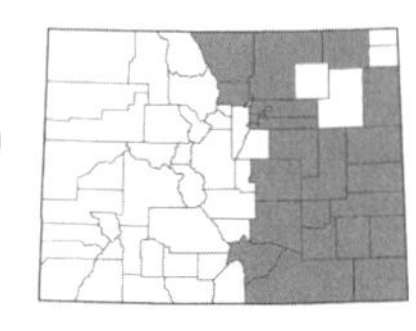

Cymopterus acaulis **(Pursh) Raf.**, PLAINS SPRINGPARSLEY. Plants 3–30 cm; *stems* with pseudoscape, this 0.5–5.5 cm long; *leaves* basal, 2-3-pinnate, ultimate divisions linear, to 15 mm long and 2 mm wide, glabrous; *peduncles* 2–14 cm long; *rays* 6–9, 1–10 mm long; *involucel bractlets* 3–8 (11) mm long, entire or with 2–3 teeth; *petals* white; *fruit* 5–10 mm long. Common in gravelly and sandy soil of the plains and foothills, 3500–7500 ft. April–June. E.

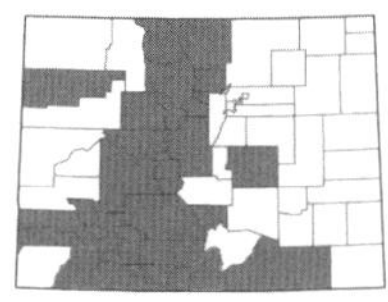

Cymopterus alpinus **A. Gray**, ALPINE SPRINGPARSLEY. [*Oreoxis alpina* (A. Gray) J.M. Coult. & Rose]. Plants 2.5–12 cm; *leaves* basal, mostly bipinnate, the ultimate divisions linear to narrowly elliptic, 1–6 mm × 0.4–1.5 mm, scabrous; *peduncles* 2–10.5 cm long; *involucel bractlets* 5–9, 1–4 mm long; *petals* yellow, fading to white in age; *fruit* 4–5 mm long, corky-winged. Common in alpine and subalpine meadows in the mountains, 9400–14,000 ft. June–Aug. E/W. (Plate 6)

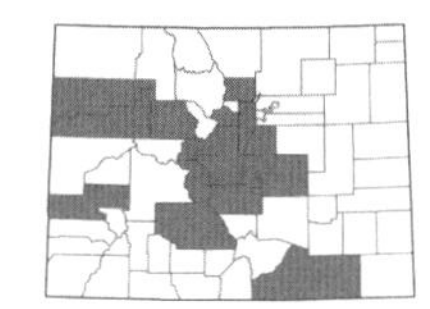

Cymopterus anisatus **A. Gray**, ANISE SPRINGPARSLEY. [*Aletes anisatus* (A. Gray) W.L. Theob. & C.C. Tseng; *Pteryxia anisata* (A. Gray) Mathias & Constance]. Plants 10–35 cm; *leaves* basal, bipinnate, the ultimate divisions 1–6 mm long, rigid, glabrous; *involucel bractlets* 0.3–1.5 cm long, linear-lanceolate; *petals* yellow; *fruit* 4–6 mm long, lateral wings narrower than the body, the dorsal wings conspicuously narrower than the lateral. Found in gravelly or sandy soil in the mountains, 6200–10,000 ft. May–July. E/W. Endemic.

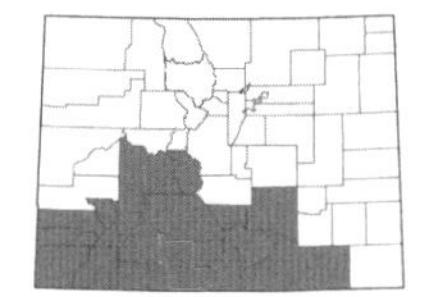

Cymopterus bakeri **(J.M. Coult. & Rose) M.E. Jones**, BAKER'S ALPINEPARSLEY. [*Oreoxis bakeri* J.M. Coult. & Rose]. Plants 1–12 cm, usually minutely hairy at the base of umbels and rays; *leaves* usually bipinnate, glabrous, the ultimate divisions linear; *involucel bractlets* 3–5 mm long, purple, 3-toothed at the apex; *petals* yellow; *fruit* 2–4 mm long, the wings linear-oblong. Plants of alpine meadows in the southern mountains, 12,000–14,000 ft. June–Aug. E/W. Endemic.

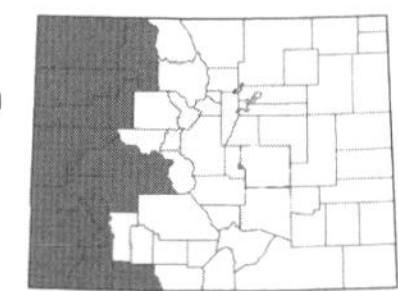

Cymopterus bulbosus **A. Nelson**, BULBOUS SPRINGPARSLEY. Plants 8–27 cm, glabrous and glaucous; *pseudoscape* obsolete or to 6.5 cm long; *leaves* basal, 2-3-pinnate, the ultimate divisions 1–10 mm long; *involucel bractlets* white, 1-2-nerved, more or less united; *petals* white to purplish-pink; *fruit* 6–11 mm long, the wings 1.7–3 mm wide. Typically found on heavy clay soil, often associated with sagebrush or pinyon-juniper communities, 4500–8500 ft. Mar.–June. W.

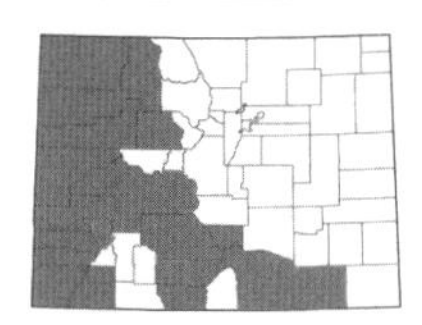

Cymopterus constancei **R.L. Hartm.**, CONSTANCE'S SPRINGPARSLEY. Plants 3–18 cm, glabrous; *pseudoscape* conspicuous; *leaves* 2-3-pinnate, the ultimate divisions oblong to oblanceolate with rounded tips; *involucel bractlets* white, mostly 3-veined; *petals* white to purple; *fruit* dorsally winged. Typically found in grasslands or sagebrush on loose, sandy or loamy soil, 5500–8800 ft. April–May. E/W.

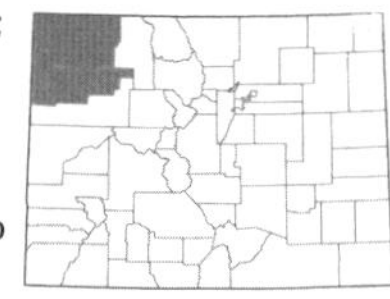

Cymopterus duchesnensis **M.E. Jones**, DUCHESNE SPRINGPARSLEY. Plants 7–30 cm; *pseudoscape* absent; *leaves* basal or a few cauline, pinnate, the ultimate divisions appearing as acute, mucronate lobes of a broad leaflet; *involucel bractlets* absent or 1–5 mm long, linear; *petals* yellow; *fruit* 5–9 mm long, the wings 2–2.5 mm wide, papery. Found in desert shrub, sagebrush, and juniper communities and sandy clay and clay soils derived from shales, known from Moffat and Rio Blanco cos., 4600–6800 ft. April–May. W.

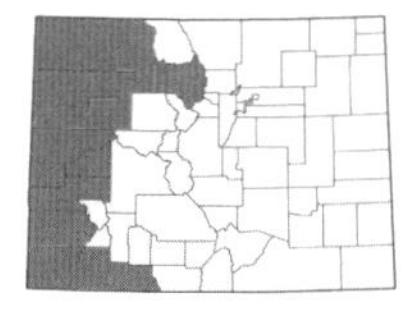

Cymopterus fendleri **A. Gray**, FENDLER'S SPRINGPARSLEY. Plants 4–30 cm; *pseudoscape* present; *leaves* 2-4-pinnate, the ultimate divisions linear, 0.5–5 mm long; *involucel bractlets* linear to ovate-oblong, 3-toothed at the apex; *petals* yellow; *fruit* 5–10 mm long, the wings slightly corky, to 2 mm wide. Common in clay or gravelly soil often with juniper-sagebrush or pinyon-juniper communities, 4500–7000 ft. April–June. W. (Plate 6)

Cymopterus humilis **(Raf.) Tidestr. & Kittell**, Pike's Peak alpine parsley. [*Oreoxis humilis* Raf.]. Plants 2–10 cm, minutely hairy in the inflorescence; *leaves* basal, glabrous, the ultimate divisions linear to narrowly elliptic, 1–6 mm long; *involucel bractlets* 1–4 mm long, linear; *petals* yellow; *fruit* 2–4 mm long, corky-winged. Rare, found only on Pikes Peak, 11,900–14,000 ft. June–Aug. E. Endemic.

Cymopterus lemmonii **(J.M. Coult. & Rose) Dorn**, ligusticoid springparsley. [*Pseudocymopterus montanus* (A. Gray) J.M. Coult. & Rose]. Plants 8–50 cm, glabrous except minutely hairy below the peduncle; *leaves* basal and often 1–2 cauline, mostly bipinnate, the ultimate divisions linear to narrowly elliptic, 2–20 mm long; *involucel bractlets* absent or of 1–2 small bractlets; *petals* yellow or rarely reddish; *fruit* 3–6 mm long, the wings to 1.5 mm wide. Common on wooded slopes and in meadows, 6000–14,000 ft. May–July. E/W. (Plate 6)

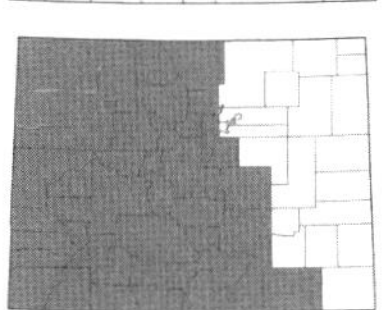

Cymopterus longilobus **(Rydb.) W.A. Weber**, mountain springparsley. [*Pteryxia hendersonii* (J.M. Coult. & Rose) Mathias & Constance]. Plants 5–40 cm, glabrous; *leaves* bipinnate, the ultimate divisions linear, 1–15 mm long; *involucel bractlets* linear, 2–12 mm long; *petals* yellow; *fruit* 3–6 mm long, wings 0.5–1.5 mm wide. Found on rocky soil, 6000–11,000 ft. May–July. E/W.

Cymopterus longipes **S. Watson**, longstalk springparsley. Plants 7–30 (50) cm; *pseudoscape* 4–24 cm long; *leaves* 1–3 times pinnate, the ultimate lobes 1–5 mm long; *involucel bractlets* to 7 mm long, linear; *petals* yellow or white; *fruit* 4–6 mm long, wings 1–2 mm wide. Not reported from Colorado yet but could occur in Moffat Co., often in association with pinyon-juniper or sagebrush communities. April–May.

Cymopterus montanus **Nutt. ex Torr. & A. Gray**, mountain springparsley. Plants 5–30 cm; *pseudoscape* present; *leaves* 2–3-pinnate, the ultimate divisions obtuse, 1–2 mm long; *involucel bractlets* white, ovate-oblong, 1-nerved; *petals* white to purple; *fruit* 5–12 mm long, with wings about twice as wide as the body. Common in sandy places on the eastern plains and foothills, 3800–6200 ft. April–May. E. (Plate 6)

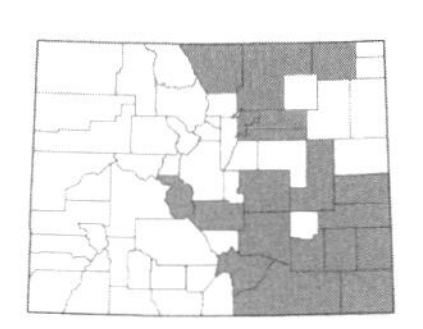

Cymopterus petraeus **M.E. Jones**, rock springparsley. [*Aletes petraea* (M.E. Jones) W.A. Weber; *Pteryxia petraea* (M.E. Jones) J.M. Coult. & Rose]. Plants 15–35 cm, glabrous; *leaves* caulescent, bipinnate, the ultimate divisions linear, 1–3.5 mm long; *involucel bractlets* inconspicuous, linear; *petals* yellow; *fruit* dorsally winged. Uncommon (Moffat and San Juan cos.) in open, rocky places, 7000–7500 ft. May–June. W.

Cymopterus planosus **(Osterh.) Mathias**, Rocky Mountain springparsley. Plants 10–30 cm, glabrous; *pseudoscape* present; *leaves* 3-pinnate, the ultimate divisions acute, incurved; *involucel bractlets* linear; *petals* red-purple or yellow; *fruit* 5–7 mm long, with narrow wings. Found on open gravelly slopes and in juniper-sagebrush communities, 6500–9500 ft. May–June. W. Endemic.

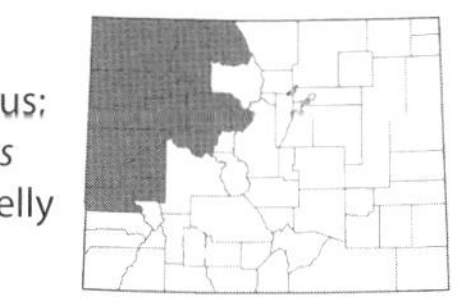

Cymopterus purpureus **S. Watson**, Colorado Plateau springparsley. Plants 5–25 cm, glabrous; *pseudoscape* absent; *leaves* 3–4-pinnate, the ultimate divisions triangular, acute, 1–5 mm long; *involucel bractlets* linear, 2–4 mm long; *petals* yellow, drying purple; *fruit* 4–8 mm long with wings 1.5–4 mm wide. Common in clay or gravelly soil, often in association with sagebrush or pinyon-juniper communities, 4800–7500 ft. May–June. W.

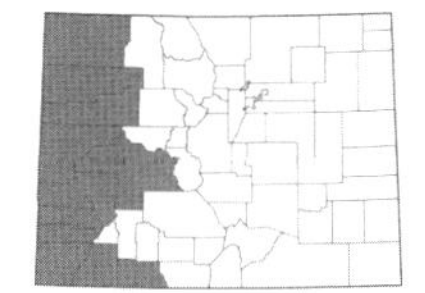

Cymopterus sessiliflorus **(W.L. Theob. & C.C. Tseng) R.L. Hartm.**, sessileflower springparsley. [*Aletes sessiliflorus* W.L. Theob. & C.C. Tseng; *C. macdougalii* (J.M. Coult. & Rose) Tidestr. ssp. *breviradiatus* W.L. Theob. & C.C. Tseng]. Plants 6–20 cm, glabrous; *pseudoscape* absent; *leaves* pinnate to bipinnate, the ultimate divisions linear, 3–25 mm long; *petals* yellow; *fruit* with narrow wings. Found on sandstone ledges and in clay soil, 6000–7000 ft. April–Aug. W.

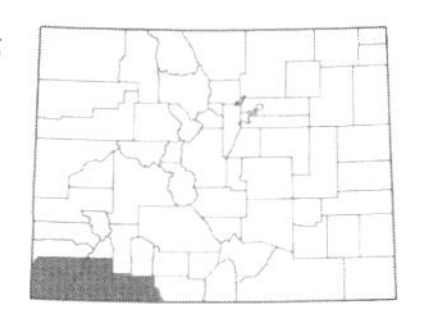

Cymopterus terebinthus **(Hook.) Torr. & A. Gray var. *albiflorus* (Torr. & A. Gray) M.E. Jones**, aromatic springparsley. [*Pteryxia terebinthina* (Hook.) J.M. Coult. & Rose var. *albiflora* (Torr. & A. Gray) Mathias]. Plants 12–40 cm, glabrous; *pseudoscape* absent; *leaves* 2–4-pinnate, the ultimate divisions 1–5 mm long; *petals* yellow, fading white; *fruit* 5–8 mm long, wings 0.5–2 mm wide, dorsal ones often reduced. Uncommon in dry, open, rocky or sandy places, 5500–7500 ft. May–June. W.

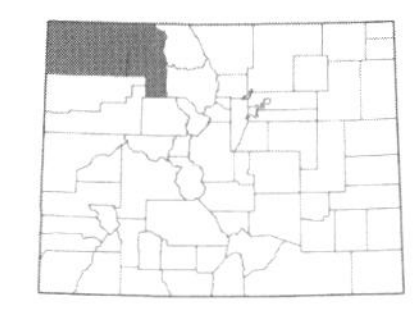

DAUCUS L. – carrot

Caulescent herbs; *leaves* with small, fern-like ultimate segments; *involucre* of conspicuous, foliaceous, pinnately compound bracts; *involucel* of numberous toothed or entire bractlets, or lacking; *petals* white; *stylopodium* present; *fruit* oblong to ovoid, flattened dorsally, ribbed, glochidiate or barbed prickles present along alternate ribs.

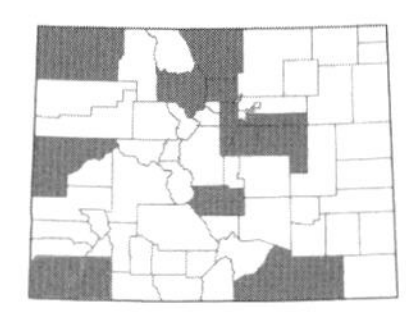

***Daucus carota* L.**, QUEENE ANNE'S LACE. Biennials, 4–15 dm; *leaves* broadly lanceolate, 5–20 cm long, ultimate divisions 2–12 mm long; *involucre* of deflexed bracts, 4–40 mm long, *rays* numerous, unequal, 3–7.5 cm long; *fruit* 3–4 mm long. Uncommon weed found along roadsides and in irrigation ditches, 5000–7500 ft. July–Sept. E/W. Introduced.

ERYNGIUM L. – ERYNGO

Perennial herbs; *leaves* entire to deeply cleft, usually spinose-toothed; *inflorescence* of bracteolate heads; *involucre* of one or more series of entire to cleft bracts subtending the head; *stylopodium* absent; *fruit* globose to obovoid, flattened laterally, variously covered with scales or tubercles.

***Eryngium planum* L.**, PLAINS ERYNGO. Plants caulescent, robust; *basal leaves* leathery, widely oblong, blades 10–15 cm long; *petals* blue. Uncommon weed, found along roadside, in ditches, and in moist, disturbed sites, known from one old collection in Denver, 5000–6000 ft. July–Sept. E. Introduced.

HARBOURIA J.M. Coult. & Rose – WHISKBROOM PARSLEY

Caulescent, perennial herbs; *leaves* mostly basal, pinnately dissected with the ultimate divisions linear and distinct, stiff; *involucre* lacking or of inconspicuous bracts; *involucel* of several linear bractlets; *flowers* yellow; *stylopodium* absent; *fruit* ovoid, laterally compressed, granular-roughened, with prominent, corky ribs.

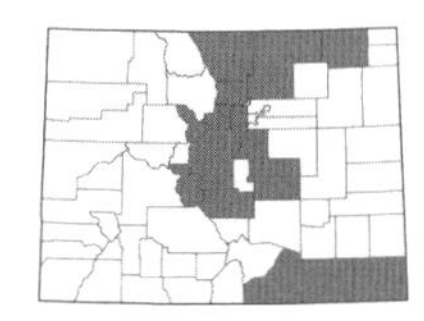

***Harbouria trachypleura* (A. Gray) J.M. Coult. & Rose**, WHISKBROOM PARSLEY. Plants 0.8–5.5 dm, hirsute below and often in umbel, otherwise glabrous; *leaves* widely oblong, ultimate divisions 2–30 mm long; *rays* spreading; *fruit* 3–6 mm long. Common on open rocky places, 4400–8600 ft. April–June (July). E. (Plate 6)

HERACLEUM L. – COWPARSNIP

Caulescent, biennial or perennial herbs; *leaves* large, petiolate, ternately or pinnately compound; *involucre* lacking or of a few narrow bracts; *rays* unequal; *flowers* white; *stylopodium* present; *fruit* orbicular to elliptic, usually hairy, strongly flattened dorsally, dorsal ribs narrow, lateral ribs broadly winged, the oil tubes readily visible as brown bands on the surface.

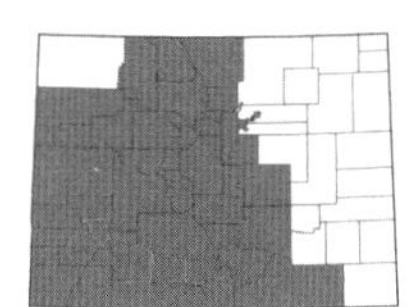

***Heracleum maximum* Bartr.**, COMMON COWPARSNIP. [*H. sphondylium* L. ssp. *montanum* (Schleich. ex Gaudin) Briq.]. Perennials, tomentose, stems grooved, to 1–3 m; *leaves* ternate, orbicular to reniform, blades 2–5 dm long, 2–5 dm wide; *leaflets* ovate to orbicular, 15–40 × 10–35 cm; *involucral bracts* 5–20 mm long if present; *petals* (2) 4–8.5 mm long, some deeply bilobed; *fruit* 8–12 mm long. Common in moist areas such as along streams or in swampy places, 5000–10,500 ft. June–Aug. E/W. (Plate 6)

LIGUSTICUM L. – LOVAGE

Caulescent or acaulescent perennial herbs; *leaves* ternately or ternate-pinnately compound to dissected; *involucre* and involucel lacking, or of a few inconspicuous narrow bracts or bractlets; *rays* spreading-ascending; *flowers* white; *stylopodium* present, low-conic; *fruit* oblong, somewhat compressed laterally, glabrous, with evident ribs, these often narrowly winged.

1a. Ultimate leaf divisions broad, usually over 3 mm wide; leaves less dissected than next...***L. porteri***
(*Note: Lomatium dissectum* may key to here if you were unsure of whether a stylopodium was present, but it can be distinguished from *L. porteri* by having a single midvein in the ultimate leaf divisions while *L. porteri* has pinnate venation in the ultimate leaf divisions.)
1b. Ultimate leaf divisions narrow, mostly 1–3 mm wide; leaves dissected into numerous segments...2

2a. Plants generally 5–13 dm tall, relatively robust, caulescent with 1 or more well-developed stem leaves; fruit 5–8 mm long...***L. filicinum***
2b. Plants generally small and slender, 1–7 dm tall, scapose or subscapose with 1 stem leaf that is much reduced or lacking all together; fruit 2–5 mm long...***L. tenuifolium***

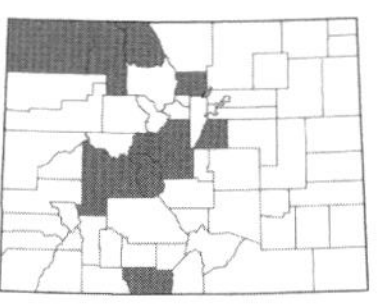

***Ligusticum filicinum* S. Watson**, UTAH LIGUSTICUM. Plants 5–13 dm, root crowned with fibrous, persisting petiole bases (brown fibers), glabrous; *leaves* tripinnate, 12–30 cm long, ultimate leaf divisions 1–18 mm long, linear to deltoid; *rays* 2.5–8 cm long; *petals* white; *fruit* 5–8 mm long. Found in open meadows or (more often dry) wooded slopes and ridges, 7000–10,000 ft. July–Sept. E/W.

Ligusticum porteri **J.M. Coult. & Rose**, OSHA. Plants 5–13 dm, root crowned with numerous stringy, brown fibers, glabrous; *leaves* tripinnate, 15–30 cm long, the ultimate divisions ovate; *rays* 2–8 cm long; *petals* white; *fruit* 5–8 mm long. Common in meadows, on open gravelly slopes, or in aspen or coniferous woods, 7200–12,000 ft. June–Aug. E/W. (Plate 6)

Ligusticum tenuifolium **S. Watson**, SLENDER LIGUSTICUM. Plants 1–7 dm, glabrous; *leaves* bipinnate to tripinnate, 5–20 cm long, the ultimate divisions linear; *rays* 5–30 mm long; *petals* white; *fruit* 2–5 mm long. Uncommon in meadows, along streambanks, and on moist slopes, 8000–12,500 ft. July–Sept. E/W.

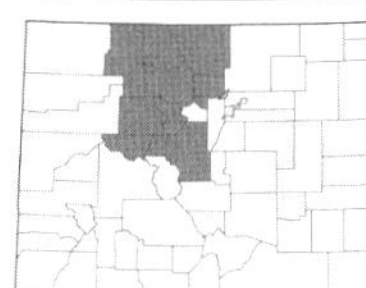

LOMATIUM Raf. – BISCUITROOT

Perennial herbs; *leaves* variously compound; *involucre* lacking or inconspicuous; *involucel* of often conspicuous bractlets, sometimes absent; *flowers* mostly yellow, sometimes white, rarely purple; *stylopodium* lacking; *fruit* flattened dorsally, lateral ribs with thin or corky wings, dorsal ribs evident to obsolete, but only very narrowly (if at all) winged. (Dorn & Hartman, 1988)

1a. Leaves biternately compound; ultimate leaf divisions long and linear, mostly 30–60 times as long as wide...***L. triternatum* var. *platycarpum***
1b. Leaves not biternately compound; ultimate leaf divisions short, usually less than 8 times as long as wide...2

2a. Plants caulescent, usually robust and tall, well over 2 dm in height; usually with more than 1 elongate internode; wings of the fruit narrow and corky-thickened; oil tubes in fruit obscure; flowers yellow or purplish; leaves often granular-scabrous, sometimes glabrous...***L. dissectum***
2b. Plants usually low and slender, occasionally to 4 dm in height, leaves chiefly basal or with one short internode; wings of the fruit narrow or broad, thin and membranous (but not corky-thickened); oil tubes in fruit conspicuous; flowers yellow or white; leaves various...3

3a. Ultimate leaf divisions broad, 2–7 (10) mm wide; leaves mostly only once pinnately compound with 2–6 pairs of lateral leaflets, these entire or some of them again bifid; plants glabrous or obscurely granular-scabrous; flowers yellow...***L. latilobum***
3b. Ultimate leaf divisions narrow, less than 2 mmwide and more commonly averaging 1 mm wide; leaves variously compound but rarely only pinnately compound; plants glabrous or hairy; flowers yellow or white...4

4a. Ultimate leaf divisions extremely narrow and small, no greater than 0.5 mm in width and to 7 mm long, usually very numerous with several hundred to a thousand or more per leaf; flowers yellow...5
4b. Ultimate leaf divisions usually greater than 0.5 mm in width, the larger ones generally 1 mm wide or more, and less numerous; flowers yellow or white...7

5a. Plants densely hairy...***L. foeniculaceum***
5b. Plants glabrous...6

6a. Pedicels 7–15 mm long at maturity; plants from a long taproot usually surmounted by a branching caudex...***L. grayi***
6b. Pedicels 1–2.5 mm long at maturity; plants from a short, round root and mostly subterranean crown or short pseudoscape...***L. bicolor***

7a. Leaves densely villous, minutely hairy, or granular-scabrous; flowers yellow or white...8
7b. Leaves glabrous, sometimes obscurely granular-scabrous; flowers yellow...12

8a. Involucel bractlets glabrous or sparsely short granular-scabrous...9
8b. Involucel bractlets densely hairy or minutely hairy, at least on the outer side...10

9a. Flowers white; anthers red; ultimate leaf divisions oblong to linear, not appearing fan-shaped as a whole, densely short-hairy to shortly tomentose...***L. orientale***
9b. Flowers yellow; anthers yellow; ultimate leaf divisions ovate to oblong, crowded, as a whole often appearing fan-shaped, densely scaberulent...***L. eastwoodiae***

10a. Involucel bractlets and leaves puberulent; fruit 6–12 mm long, glabrous or sparsely scaberulous...***L. juniperinum***
10b. Involucel bractlets and leaves villous or tomentose; fruit 5–25 mm long, glabrous or hairy...11

11a. Flowers yellow; fruit elliptic to suborbicular, 5–11 mm long, short-hairy...***L. foeniculaceum***
11b. Flowers white or purplish-white; fruit usually narrowly elliptic or oblong, (7) 10–25 mm long, glabrous or minutely hairy...***L. macrocarpum***

12a. Involucel bractlets conspicuous, ovate or narrowly elliptic, often fused at the base; leaves with more broad than linear ultimate leaf divisions; fruit ovate to orbicular...***L. concinnum***
12b. Involucel bractlets inconspicuous, linear or filiform, not fused at the base; leaves with narrowly linear ultimate leaf divisions; fruit narrowly oblong...13

13a. Rays long (4–10 cm), conspicuously unequal in length; fruit 8–12 mm long and narrow (2–3 mm wide); oil tubes usually solitary in the fruit intervals...***L. leptocarpum***
13b. Rays short (1.5–4.5 cm), more or less equal in length; fruit 5–8 mm long (not long and narrow); oil tubes 3–5 in the fruit intervals...***L. nuttallii***

***Lomatium bicolor* (S. Watson) J.M. Coult. & Rose**, LITTLE-HEADS LOMATIUM. Plants 10–50 cm, glabrous, caulescent or acaulescent; *leaves* triternate or 3-pinnate, the ultimate divisions very numerous (more than 300), linear, 1–4 mm long; *rays* 1–10 cm long, very unequal; *involucel bractlets* absent or linear; *petals* yellow; *fruit* 8–11 mm long, lateral wings 0.4–0.6 mm wide. Uncommon on dry slopes and in meadows, 6500–8000 ft. May–July. W.

***Lomatium concinnum* (Osterh.) Mathias**, COLORADO DESERT-PARSLEY. Plants 12–25 cm, glabrous; *leaves* mostly bipinnate, the ultimate divisions narrowly elliptic or linear-lanceolate, 2–11 mm long; *rays* subequal, 2.5–4 cm long; *involucel bractlets* ovate to narrowly elliptic, often fused below; *petals* yellow; *fruit* with lateral wings. Associated with sagebrush communities on adobe hills and in rocky soils derived from Mancos Formation shale, 5500–7000 ft. May. W. Endemic. (Plate 6)

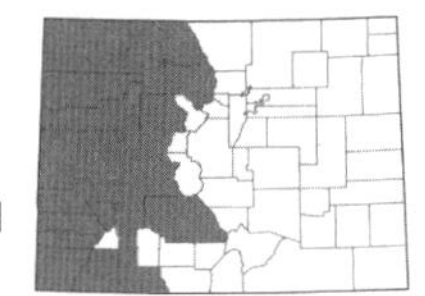

***Lomatium dissectum* (Nutt.) Mathias & Constance var. *eatonii* (J.M. Coult. & Rose) Cronquist**, GIANT LOMATIUM. Plants 3–13 dm, caulescent, minutely hairy or glabrous; *leaves* ternate-pinnate, the ultimate divisions 1–12 mm long, linear; *rays* 4–10 cm long, equal or subequal; *involucel bractlets* 3–15 mm long; *petals* white; *fruit* 7–13 mm long, lateral wings 1–2 mm wide. On rocky slopes and in dry meadows, 6000–9500 ft. May–June. E/W.

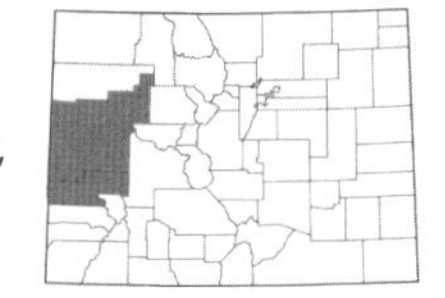

***Lomatium eastwoodiae* (J.M. Coult. & Rose) J.F. Macbr.**, EASTWOOD'S BISCUITROOT. [*Aletes eastwoodiae* (J.M. Coult. & Rose) W.A. Weber]. Plants 10–15 cm, scabrous; *leaves* 3–7 cm long, 1-2-pinnate, with 5–7 remote pairs of segments; *rays* 1–3 cm long, unqual; *involucel bractlets* few, linear; *petals* yellow; *fruit* 8–10 mm long. Found on the Uncompahgre Plateau in pinyon-juniper woodlands and in sandy soil; 4600–7000 ft. Mar.–April. W. Endemic.

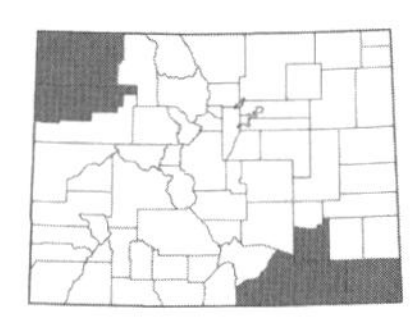

***Lomatium foeniculaceum* (Nutt.) J.M. Coult. & Rose**, DESERT BISCUITROOT. Plants 5–30 cm, densely hairy; *leaves* ternate-pinnately dissected, 2–13 cm long, the ultimate divisions numerous (usually over 500), 1–4 mm long; *rays* 0.2–7 cm long; *involucel bractlets* linear, 2–6 mm long; *petals* yellow; *fruit* 5–11 mm long, lateral wings 1–2 mm wide. Found in dry, open places, 5400–6200 ft. May. E/W.

There are two varieties of *L. foeniculaceum* in Colorado:

1a. Petioles 1-2.5 cm long...**var. *macdougalii* (J.M. Coult. & Rose) W.L. Theob.**. Plants of the western slope (Moffat, Rio Blanco cos), found on clay-shale soil.
1b. Petioles 3-8 cm long...**var. *foeniculaceum***. Plants of the eastern slope.

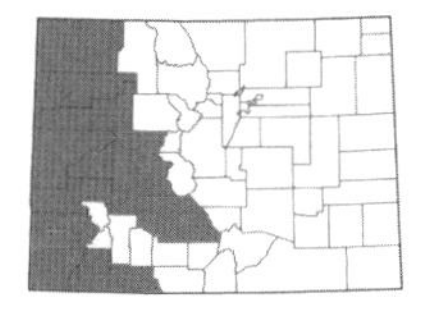

Lomatium grayi* (J.M. Coult. & Rose) J.M. Coult. & Rose var. *grayi, MILFOIL LOMATIUM. Plants 8–40 (80) cm, glabrous; *leaves* ternate-pinnately dissected, 7–20 cm long, the ultimate divisions very numerous, 1–3 mm long; *rays* 1.5–8 cm long; *involucel bractlets* filiform, 3–5 mm long; *petals* yellow, fading whitish; *fruit* 6–12 mm long, lateral wings 2 mm wide. Found on dry, open, rocky slopes, 5000–9700 ft. April–May. W. (Plate 6)

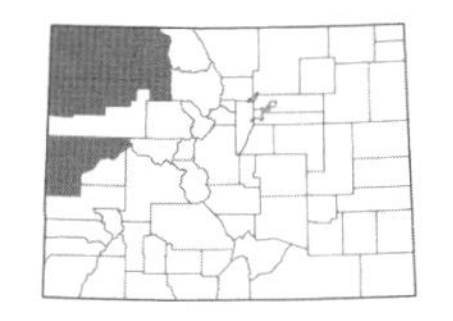

***Lomatium juniperinum* (M.E. Jones) J.M. Coult. & Rose**, JUNIPER-LOMATIUM. Plants 8–30 cm, minutely hairy to glabrous; *leaves* ternate-pinnately dissected, 2.5–10 cm long, the ultimate divisions 1–7 mm long, linear; *rays* 1–8 cm long; *involucel bractlets* 1–4.5 mm long, linear; *petals* white or yellow; *fruit* 6–12 mm long, with lateral wings 0.5–1.5 mm wide. Found on dry, rocky slopes and rocky openings in aspen forests, usually associated with juniper-sagebrush communities, 7000–8600 ft. May–June. W.

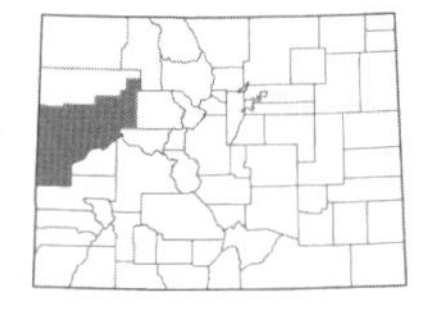

***Lomatium latilobum* (Rydb.) Mathias**, CANYONLANDS LOMATIUM. [*Aletes latilobus* (Rydb.) W.A. Weber]. Plants 6–30 cm, glabrous; *leaves* 1–10 cm long, pinnate with mostly 3–4 pairs of lateral lanceolate leaflets 5–40 mm long; *rays* 0.5–2 cm long; *involucel bractlets* 5–10 mm long, linear; *petals* yellow; *fruit* 8–12 mm long, lateral wings 1 mm wide. Found in sandy soils of pinyon-juniper and desert shrub communities, 5000–7000 ft. April–May. W.

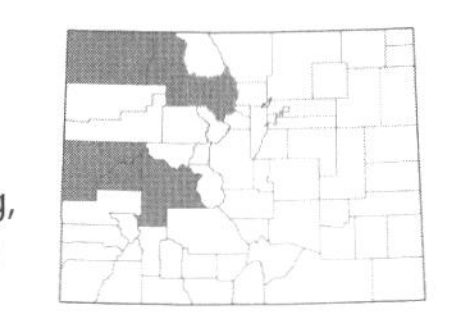

Lomatium leptocarpum **(Torr. & A. Gray) J.M. Coult. & Rose**, GUMBO-LOMATIUM. [*L. bicolor* (S. Watson) J.M. Coult. & Rose var. *leptocarpum* (Torr. & A. Gray) SchlesJ. Sm.]. Plants 10–50 cm, glabrous to scaberulent; *leaves* ternate-pinnately dissected, the ultimate divisions linear, 0.5–45 mm long; *rays* 4–10 cm long, unequal; *involucel bractlets* linear; *petals* yellow; *fruit* 8–12 mm long, with lateral wings ca. 0.5 mm wide. Found on open slopes and meadows, especially common in heavy clay soils, 6200–8000 ft. May–June. W.

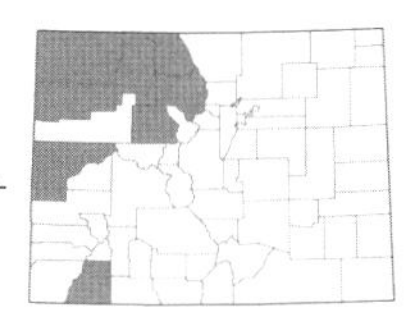

Lomatium macrocarpum **(Nutt. ex Torr. & A. Gray) J.M. Coult. & Rose**, BIGSEED LOMATIUM. Plants 10–30 cm, tomentose-villous to glabrate; *leaves* pinnately to ternate-pinnately dissected, 3–6 cm long, ultimate divisions 1.5–6 mm long, linear; *rays* 1–6 cm long, unequal; *involucel bractlets* linear-lanceolate, 2–10 mm long; *petals* white or purplish-white; *fruit* (7) 10–25 mm long, lateral wings 1–2 mm wide. Found on open, rocky, dry slopes, 5500–8500 ft. May–June. W.

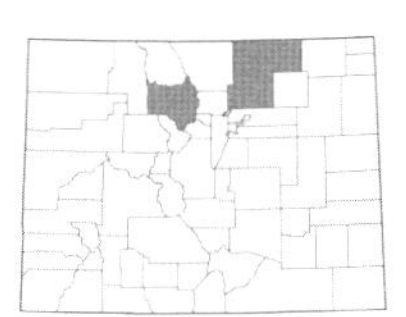

Lomatium nuttallii **(A. Gray) J.F. Macbr.**, NUTTALL'S BISCUITROOT. [*Aletes nuttallii* (A. Gray) W.A. Weber]. Plants 15–42 cm, glabrous; *leaves* 1–2-pinnate or ternate-pinnately dissected, the ultimate divisions linear, 10–50 mm long; *rays* 1.5–4.5 cm long, unequal or subequal; *involucel bractlets* linear; *petals* yellow; *fruit* 5–8 mm long, lateral wings 0.5–1 mm wide. Uncommon on dry slopes and chalk bluffs, 5500–7500 ft. May–June. E/W.

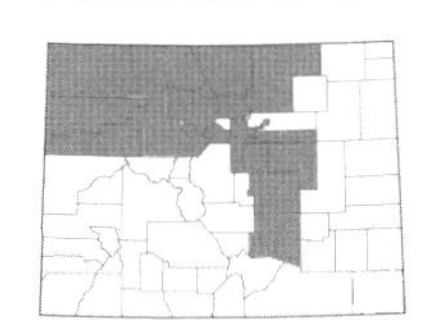

Lomatium orientale **J.M. Coult. & Rose**, SALT-AND-PEPPER. Plants 10–40 cm, soft-puberulent; *leaves* tripinnate, 3–8 cm long, the ultimate divisions linear, 2–5 mm long; *rays* 1.2–5.5 cm long, subequal; *involucel bractlets* linear-lanceolate, 2–4 mm long; *petals* white, anthers red; *fruit* 6–9 mm long, lateral wings 0.5–1 mm wide. Common on open slopes from the plains to the foothills on the eastern slope and with scattered occurrences on the western slope, blooming in early spring, 4800–9000 ft. April–June. E/W. (Plate 6)

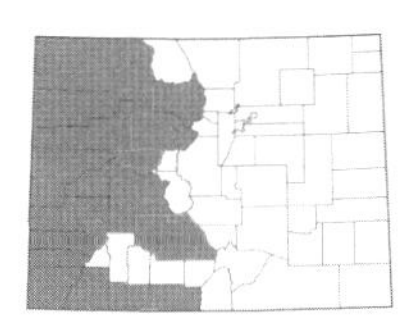

Lomatium triternatum **(Pursh) J.M. Coult. & Rose var. *platycarpum* (Torr.) Boivin**, GREAT BASIN DESERTPARSLEY. [*L. simplex* (Nutt.) J.F. Macbr.]. Plants 1.5–6 dm, minutely hairy; *leaves* biternate, with 9–20 leaflets, the ultimate divisions 1–13 cm long; *rays* 2–10 cm long; *involucel bractlets* 1–10 mm long, linear; *petals* yellow; *fruit* 8–15 mm long, lateral wings 1–2 mm wide. Common in sagebrush meadows and on open gravelly slopes, 5500–10,000 ft. May–July. W. (Plate 7)

MUSINEON Raf. – WILDPARSLEY

Low, caulescent or acaulescent, perennial herbs, sometimes with a pseudoscape; *leaves* pinnately or ternately compound; *involucre* usually lacking or sometimes conspicuous; *rays* spreading, subequal; *flowers* yellow; *stylopodium* absent; *fruit* ovoid to narrowly oblong, flattened laterally, glandular-scabrous, the ribs evident but not winged.

1a. Ultimate leaf divisions narrowly linear, green; taproot surmounted by a freely branching, surficial woody caudex; plants acaulescent without a pseudoscape...***M. tenuifolium***

1b. Ultimate leaf divisions oblong, green but often somewhat bluish or glaucous; taproot with a simple or relatively few-branched, surficial or subterranean herbaceous crown; plants caulescent with a deeply buried pseudoscape...***M. divaricatum***

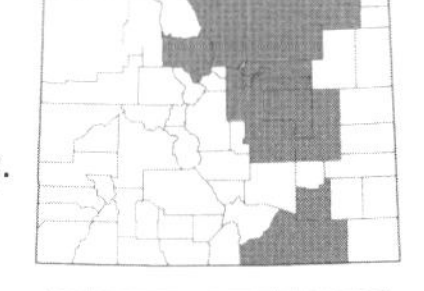

Musineon divaricatum **(Pursh) Raf.**, LEAFY WILDPARSLEY. Plants with a short pseudoscape, glabrous or somewhat scabrous; *leaves* 1–2-pinnate or ternate-pinnate, 2–12 cm long, the ultimate divisions 3–15 mm long, oblong; *involucel bractlets* 1–7 mm long; *petals* yellow; *fruit* 3–6 mm long. Common in sandy soil on the eastern plains and foothills, 4500–6000 ft. April–June. E. (Plate 7)

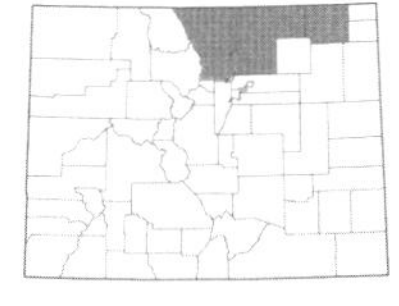

Musineon tenuifolium **(Nutt. ex Torr. & A. Gray) J.M. Coult. & Rose**, SLENDER WILDPARSLEY. [*Aletes tenuifolius* (Nutt.) W.A. Weber]. Plants lacking a pseudoscape, hirtellous below the umbel, otherwise glabrous; *leaves* 1–3-pinnate, the ultimate divisions linear, 2–30 mm long; *involucel bractlets* 1–3 mm long or absent; *petals* yellow; *fruit* 2–5 mm long. Locally common in sandy or rocky soil, 4500–6200 ft. May–June. E. (Plate 7)

NEOPARRYA Mathias – NEOPARRYA

Acaulescent perennials; *leaves* pinnately compound with linear leaflets; *involucre* lacking; *involucel* of inconspicuous, narrow bractlets; *rays* few, reflexed; *flowers* yellow; *stylopodium* lacking; *fruit* compressed laterally, ribs unwinged.

Neoparrya lithophila **Mathias**, BILL'S NEOPARRYA. [*Aletes lithophilus* (Mathias) W.A. Weber]. Plants ca. 1.5 dm; *leaves* 8–10 cm long, glabrous; *leaflets* 5–20 mm long; *rays* 5–15 mm long; *involucel bractlets* narrowly lanceolate, ca. 3 mm long; *fruit* 3–5 mm long. Locally in rock crevices and on ledges, 7000–10,000 ft. May–July. E. Endemic.

OROGENIA S. Watson – Indian potato

Acaulescent or short-caulescent, perennial herbs from globose tubers; *leaves* mostly once- or thrice-pinnately compound; *involucre* lacking; *involucel* of conspicuous bractlets or lacking; *rays* spreading, unequal; *flowers* white; *stylopodium* lacking; *fruit* glabrous, dorsal ribs filiform and evident or obscure, lateral ribs corky winged.

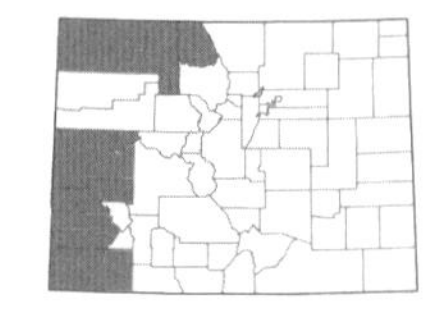

***Orogenia linearifolia* S. Watson**, Great Basin Indian potato. Tuber about 1.5 cm thick, usually located 2–7 cm below the surface; *scape* rising to slightly above the soil surface; *leaves* numbering 2 or 3, ultimate segments usually 1–4.5 cm long and 0.5–4 mm wide; *pedicels* up to 2 mm long; *fruit* 3–4 mm long. Uncommon or easily overlooked, found growing in sagebrush meadows, 7500–9500 ft. Mar.–June. E/W.

OSMORHIZA Raf. – sweet-cicely

Perennial herbs; *leaves* ternately or ternate-pinnately compound; *involucre* lacking or of 1 or more narrow, foliaceous bracts; *involucel* of many narrow, foliaceous, reflexed bractlets or lacking; *flowers* white, yellow, purple, or pinkish; *stylopodium* conic to depressed; *fruit* narrow, linear-oblong to clavate, obtuse to short-beaked at the apex, ribs narrow. (Constance & Shan, 1948; Lowry & Jones, 1979, 1984)

1a. Fruit glabrous, the base obtuse; flowers yellow to greenish-yellow; plants robust, 6–13 dm tall; stems densely clustered…***O. occidentalis***
1b. Fruit hispid with retrorse bristles, especially at the base, attenuate at the base; flowers white to greenish-white; plants slender to rather stout, 1–8 dm tall; stems not densely clustered…2

2a. Involucel bractlets conspicuous, 5–15 mm long; style (including the stylopodium) 2–3.6 mm long…***O. longistylis***
2b. Involucel bractlets inconspicuous or absent, or of a solitary bract to 12 mm long; style (including the stylopodium) 0.2–0.7 mm long…3

3a. Fruit linear-oblong, beaked at the apex; rays and pedicels spreading-ascending…***O. berteroi***
3b. Fruit club-shaped, obtuse at the apex; rays and pedicels strongly divaricate (widely spread apart, away from the main axis) to nearly reflexed…***O. depauperata***

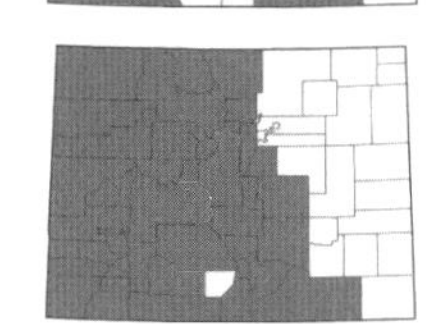

***Osmorhiza berteroi* DC.**, sweet-cicely. [*O. chilensis* Hook. & Arn.]. Plants 2–8 dm; *leaves* biternate, the leaflets 2–8 cm long, lobed and toothed; *involucel bractlets* absent; *rays* 3–7, 2.5–10 cm long, spreading-ascending; *petals* white; *fruit* 16–25 mm long, linear-clavate, hispid, with a beak 1–2 mm long. Locally common along streams and in shaded woods, 6300–11,000 ft. June–Aug. E/W. (Plate 7)

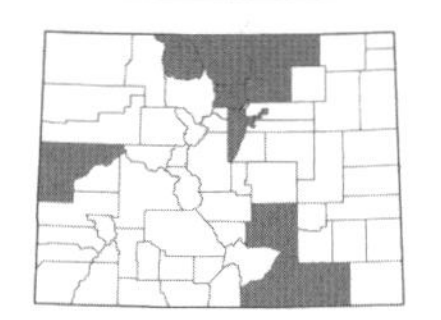

***Osmorhiza depauperata* Phil.**, blunt sweet-cicely. Plants 1–7 dm; *leaves* biternate, the leaflets 1–5 cm long, lobed to toothed; *involucel bractlets* absent or of a solitary bract to 12 mm long; *rays* 3–5, 1.5–9 cm long, widely spreading to nearly reflexed; *petals* white; *fruit* 11–18 mm long, linear-clavate (club-shaped), hispid, rounded at the tip. Common in woods and along streams, 6900–12,500 ft. May–Aug. E/W. (Plate 7)

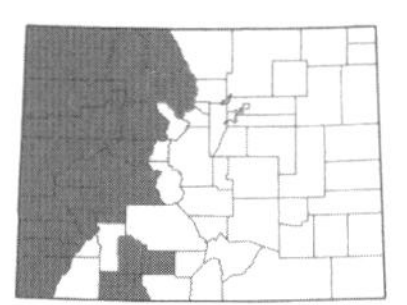

***Osmorhiza longistylis* (Torr.) DC.**, longstyle sweetroot. Plants 2–6 dm; *leaves* biternate, the leaflets 2–9 cm long, lobed to toothed; *involucel bractlets* conspicuous, 5–15 mm long; *rays* 1–5 cm long, spreading-ascending; *petals* white; *fruit* linear-oblong, 10–15 mm long, the style 2–3.6 mm long. Rare in cool canyons and ravines in the foothills, 5000–8300 ft. May–June. E.

***Osmorhiza occidentalis* (Nutt. ex Torr. & A. Gray) Torr.**, licorice-flavored sweet-cicely. Plants 6–13 dm; *leaves* biternate to bipinnate, the leaflets lobed to toothed, 2–9 cm long; *involucel bractlets* absent; *rays* 7–13, 2–7 cm long; *petals* greenish-yellow; *fruit* linear, glabrous, 15–20 mm long. Common in open meadows, woods, and moist slopes, 6000–11,000 ft. May–Aug. E/W.

OXYPOLIS Raf. – cowbane

Caulescent perennials; *cauline leaves* pinnately compound; *involucre* lacking or rarely a single bract; *involucel* of a few slender bractlets or lacking; *rays* ascending; *flowers* white to purplish; *stylopodium* conic; *fruit* strongly flattened dorsally, dorsal ribs filiform, lateral ribs broadly thin-winged with a nerve at the inner margin on the dorsal side so that the mericarp appears to have 5 dorsal ribs (instead of 3) as well as the 2 marginal wings.

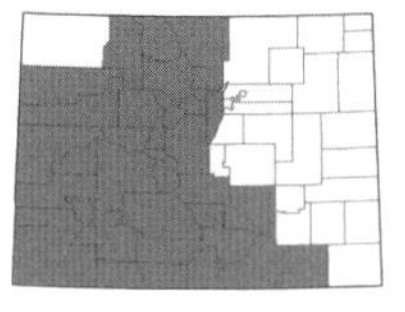

***Oxypolis fendleri* (A. Gray) Heller**, Fendler's cowbane. Plants 3–10 dm, glabrous; *leaflets* of lower leaves ovate to lanceolate, 2–7 cm long, of middle and upper leaves reduced; *rays* 5–20, unequal, to 3 cm long in anthesis, 3–7 cm long in fruit; *fruit* 3.5–5 mm long. Common in wet places such as along streambanks, found in the middle mountains, 7800–12,500 ft. June–Aug. E/W. (Plate 7)

PASTINACA L. – PARSNIP

Caulescent, biennial or perennial herbs; *leaves* pinnately compound, large, with broad, toothed to lobed leaflets; *rays* and pedicels spreading-ascending; *flowers* yellow; *stylopodium* depressed-conic; *fruit* ovoid to obovate, glabrous, strongly compressed dorsally, dorsal ribs filiform and inconspicuous, lateral ribs narrowly thin-winged.

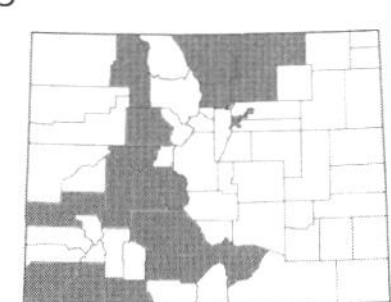

***Pastinaca sativa* L.**, WILD PARSNIP. Plants aromatic, biennials, tap-rooted, 3–10 dm; *basal leaves* up to 5 dm long; *leaflets* up to 13 cm long, up to 10 cm wide; *rays* mostly 15–25, unequal, 2–10 cm long; *fruit* 5–6 mm long. Weedy plants found growing in disturbed sites, along roadsides, in ditch banks, and in neglected agricultural land, 5000–8000 ft. June–Aug. E/W. Introduced.

PERIDERIDIA Rchb. – YAMPA

Caulescent, perennial herbs; *leaves* pinnately or ternate-pinnately compound or dissected; *involucre* of few to several small, narrow, more or less scarious bracts; *rays* spreading-ascending; *flowers* white or pink; *stylopodium* low-conic; *fruit* laterally compressed, glabrous, with filiform ribs.

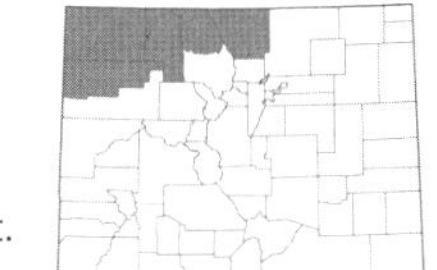

***Perideridia gairdneri* (Hook. & Arn.) Mathias ssp. *borealis* T.I. Chuang & Constance**, GARDNER'S YAMPA. Plants 3–12 dm; *basal leaves* oblong to ovate in outline, generally once pinnate, blade 20–35 cm long; *basal leaflets* 2–12 cm long; *rays* unequal 1.5–7 cm long; *petals* 5- or 7-veined; *fruit* 2.5–3.5 mm long. Found in wet meadows of the Yampa Valley and North Park, 6500–8500 ft. July–Aug. E/W.

PODISTERA S. Watson – PODISTERA

Low, perennial herbs; *leaves* basal, pinnatifid, ultimate leaf divisions deeply 2–3-lobed; *involucre* lacking or of numerous slender bracts; *rays* few, short, spreading; *flowers* yellow, greenish-yellow, or purplish; *stylopodium* conic; *fruit* slightly compressed laterally, glabrous, ribs evident but not winged.

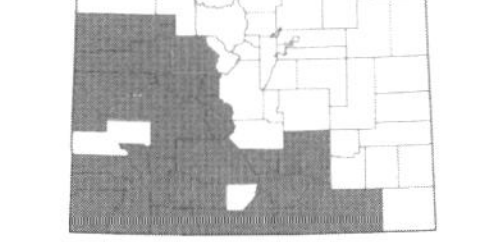

***Podistera eastwoodiae* (J.M. Coult. & Rose) Mathias & Constance**, EASTWOOD'S PODISTERA. Plants glabrous; *leaves* mostly 3–12 cm long, leaflets arranged in 3–7 pairs; *rays* unequal, to 5 mm at maturity; *involucel bractlets* mostly 4–7 mm long; *fruit* 3–4 mm long. Locally common in subalpine and alpine meadows, 10,000–13,000 ft. July–Aug. E/W. (Plate 7)

SANICULA L. – SNAKEROOT

Caulescent, biennial or perennial herbs; *leaves* palmately 3–5-cleft or appearing to be 7-parted; *involucre* foliaceous, often appearing as opposite sessile leaves; *flowers* white or greenish-white to yellow; *stylopodium* flattened and disk-like or lacking; *fruit* globose to ovoid-oblong, somewhat compressed laterally, covered with hooked prickles or tubercles, ribs lacking.

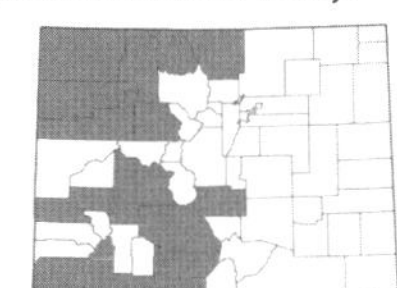

***Sanicula marilandica* L.**, MARYLAND SANICULA. Perennials, 2–10 dm, glabrous; *basal leaves* 2–10 cm long; *fruit* 4–6 mm long. Uncommon in cool shaded ravines and canyons and along streams, 5200–7500 ft. June–July. E/W. (Plate 7)

SIUM L. – WATERPARSNIP

Caulescent, perennial herbs; *leaves* pinnately compound, the leaflets narrow, serrate or incised; *involucre* of conspicuous, linear, entire or incised, unequal, often reflexed bracts; *rays* few, subequal, spreading-ascending; *flowers* white; *stylopodium* depressed or rarely conic; *fruit* glabrous, somewhat compressed laterally, ribs prominent and corky.

***Sium suave* Walter**, HEMLOCK WATERPARSNIP. Plants glabrous, 5–12 dm; *leaflets* linear to lanceolate, 1–4 cm long; *involucre bracts* 6–10, 3–15 mm long; *involucel bractlets* 4–8; *rays* 10–20, 1.5–3 cm long; *fruit* 2–3 mm long. Locally common in swampy places and in shallow water, 6000–8000 ft. July–Sept. E/W.

ZIZIA W.D.J. Koch – GOLDEN ALEXANDERS

Perennial herbs; *leaves* petiolate, dimorphic; *involucre* lacking; *rays* rather few, spreading; *flowers* bright yellow; *stylopodium* lacking; *fruit* glabrous, somewhat laterally compressed, the ribs conspicuous and filiform but not much raised.

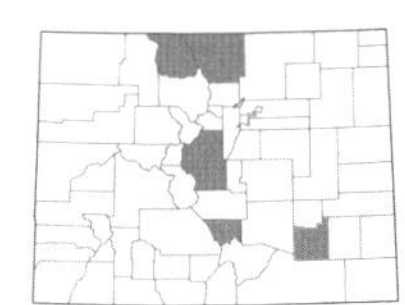

***Zizia aptera* (A. Gray) Fernald**, MEADOW ZIZIA. Plants glabrous, 3–7 dm; *basal leaves* long-petiolate, 2.5–10 cm long; *cauline leaves* ternately divided into lanceolate divisions, coarsely serrate; *rays* 12–18, unequal, 1–3 cm long; *fruit* 2–4 mm long. Locally common in moist or wet meadows and along streambanks, 4300–8000 ft. June–July. E.

APOCYNACEAE Juss. – DOGBANE FAMILY

Trees, shrubs, lianas, or perennial herbaceous plants, usually lactiferous; *leaves* usually opposite, simple; *inflorescence* usually a panicle or cyme, or solitary; *flowers* perfect, actinomorphic; *calyx* 5-lobed; *corolla* 5-lobed, often with a corona; *stamens* 5, epipetalous, anthers usually connivent around the gynoecium or united with the gynoecium and forming a gynostegium; *fruit* a pair of follicles; *seeds* usually comose.

Included here are the former members of the Asclepiadaceae or milkweed family.

1a. Corolla funnelform or salverform in shape; corona without well-developed filament appendages (i.e., the hood and horn, not a true corona); filaments distinct...2
1b. Corolla rotate or tubular with reflexed lobes or rarely salverform in shape; corona usually with well-developed filament appendages (i.e., the hood and horn); filaments usually connate into a tube...3

2a. Flowers blue; anthers free from the stigma; leaves alternate; seeds glabrous...***Amsonia***
2b. Flowers pink, white, or greenish-white; anthers adnate to and connivent around the stigmatic cap; leaves opposite or whorled; seeds comose...***Apocynum***

3a. Stems twining, vines; leaves linear, hastate or sagittate at the base...***Sarcostemma***
3b. Stems prostrate or erect, but not twining; leaves various but never hastate or sagittate at the base...***Asclepias***

AMSONIA Walter – BLUESTAR

Perennial herbs; *leaves* alternate, entire; *flowers* salverform; *calyx* deeply 5-parted, fused in the lower portion; *corolla* blue, without a corona; *stamens* epipetalous, the anthers free from the stigmatic cap, unappendaged; *pistils* 2, united by their styles, the stigmatic cap cylindric; *fruit* a pair of follicles.

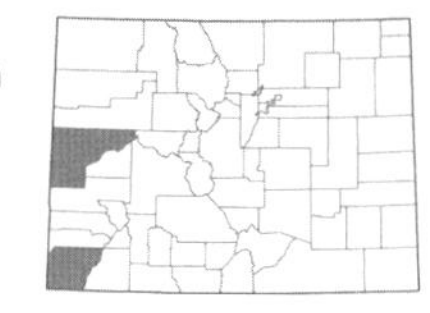

***Amsonia jonesii* Woodson**, JONES' BLUESTAR. Plants 2–5 dm; *leaves* ovate to broadly ovate, 3–6.5 cm long, glabrous, leathery; *calyx* lobes 1–3 mm long; *corolla* lobes 4.5–8 mm long; *follicles* 7–9 cm long. Infrequent in sandy, gravelly, or sometimes clay soil, often associated with sagebrush or pinyon-juniper communities (Mesa and Montezuma cos.), 4900–6000 ft. April–May. W.

APOCYNUM L. – DOGBANE, INDIAN HEMP

Perennial herbs; *leaves* opposite; *calyx* adnate in the lower half; *corolla* white, greenish-white, or pink, the corona present as triangular appendages; *stamens* epipetalous, the anthers sagittate and adnate to and connivent around the stigmatic cap; *pistils* 2, united by their styles to form a clavuncle; *fruit* a pair of usually pendulous follicles.

1a. Corolla more than 2 times as long as the calyx, mostly 5–12 mm long (including the lobes), lobes spreading or reflexed; leaves drooping to spreading to slightly ascending...2
1b. Corolla less than 2 times as long as the calyx, mostly 2.5–6 mm long, lobes erect to spreading; leaves ascending to spreading...3

2a. Leaves drooping at maturity; corolla lobes reflexed...***A. androsaemifolium***
2b. Leaves spreading to slightly ascending at maturity; corolla lobes spreading...***A. ×floribundum***

3a. Leaves usually spreading, although rarely ascending; corolla lobes spreading... ***A. ×floribundum***
3b. Leaves ascending; corolla lobes erect...***A. cannabinum***

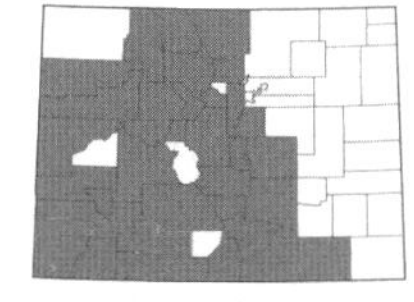

***Apocynum androsaemifolium* L.**, SPREADING DOGBANE. Plants 1.5–5 (7.5) dm; *leaves* ovate to lanceolate-ovate, 2–10 × 1–6 cm, drooping; *corolla* white with pink veins, 5–12 mm long, the lobes usually reflexed; *follicles* 6–15 cm long. Common along moist roadsides, in ditches, on wooded slopes, and in open woods, 5000–10,000 ft. June–Aug. E/W.

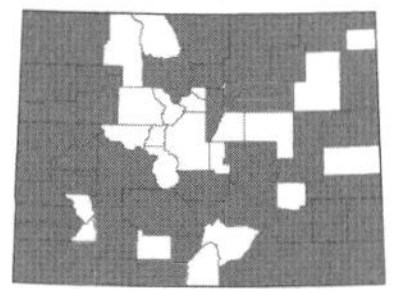

***Apocynum cannabinum* L.**, INDIAN HEMP. [including *A. sibericum* Jacq.]. Plants 3–12 dm; *leaves* lanceolate to ovate, 2–14 × 1–7 cm, ascending; *corolla* white or greenish-white, 3–6 mm long, the lobes usually erect; *follicles* 12–20 cm long. Found on disturbed sand or gravel bars of rivers and ditchbanks, 3400–7500 ft. June–Aug. E/W. (Plate 7)

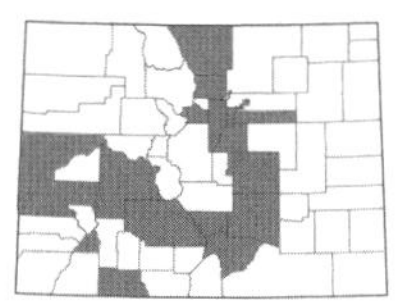

***Apocynum ×floribundum* Greene**, DOGBANE. Plants 2–5 dm; *leaves* ovate to oblong-lanceolate, 6–10 × 1.5–2 cm, usually spreading; *corolla* white or with pink veins, 3–6 mm long, the lobes spreading; *follicles* 7–15 cm long. Found in open woods and on gravelly wooded slopes, 5000–9000 ft. June–Aug. E/W.

A hybrid between *A. androsaemifolium* and *A. cannabinum*.

ASCLEPIAS L. – MILKWEED

Perennial herbs; *leaves* usually decussate; *calyx* reflexed, connate at the base; *corolla* usually reflexed; *corona* of 5 hoods attached at the base to the staminal column, each often bearing a horn; *stamens* coherent; *pistils* 2, united by a common stigmatic head, the stigmatic head adnate to the anthers and forming the gynostegium; *fruit* a pair of follicles. (Sundell, 1990, 1993; Woodson, 1941, 1954)

Asclepias flower:

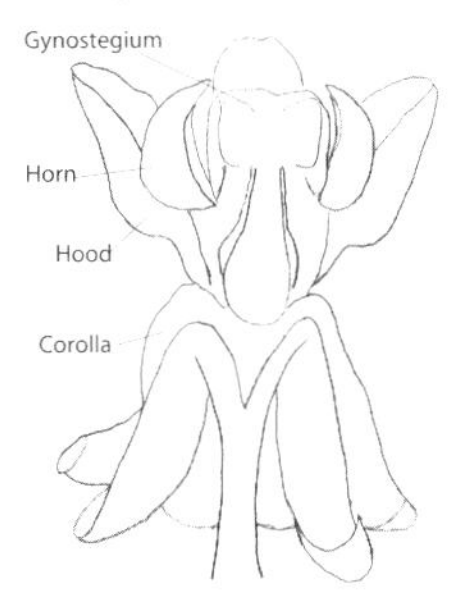

1a. Corolla lobes orange-red to orange-yellow in color; stems conspicuously spreading-hirsute...***A. tuberosa* ssp. *interior***
1b. Corolla lobes greenish, white, yellowish, or purplish, but never orange or red; stems glabrous, minutely hairy, or tomentose...2

2a. Hood reddish-violet to purple and corolla pale yellowish-green or pale yellow...3
2b. Hood and corolla variously colored, but lacking the above combination...4

3a. Leaves broadly ovate to suborbicular; corolla lobes conspicuously reflexed; pedicels glabrous...***A. cryptoceras***
3b. Leaves linear-lanceolate to lanceolate; corolla lobes spreading to erect, not reflexed; pedicels hairy...***A. asperula***

4a. Leaves all filiform or linear, mostly 1–4 mm wide, opposite or whorled...5
4b. At least the lower leaves lanceolate, ovate, broadly oblong, elliptic, obovate, or subquadrate (the upper leaves sometimes linear), at least the larger ones over 5 mm wide, never whorled...10

5a. Leaves strictly opposite; hood purple below and yellowish-white on the sides; inflorescences few-flowered with 1–5 flowers per umbel...***A. macrotis***
5b. Leaves alternate or appearing whorled; hoods yellowish-green, green, or white, not purplish below; umbels 4–many-flowered...6

6a. Leaves densely crowded, 2–5 cm long; plants low, 0.5–3 dm; stems 1–several...***A. pumila***
6b. Leaves not densely crowded, 5–25 cm long; plants usually taller, mostly 3–12 dm; stems solitary or few...7

7a. Hoods of the corona with an obvious horn; hoods and corolla usually white, or sometimes the corolla yellowish or cream...***A. subverticillata***
7b. Hoods of the corona lacking a horn or with a very short, small crest included within the hood; hoods greenish to white, corolla greenish-white to yellowish-green, sometimes purplish below...8

8a. Hoods about as long as the anther head, deeply three-lobed at the apex; stems minutely hairy to glabrate; corolla greenish-white or yellowish-green, not purplish-tinged below...***A. stenophylla***
8b. Hoods shorter than the anther head, broadly emarginated (with a notch at the apex); stems glabrous, rarely sparsely pilose; corolla greenish or yellowish-green and usually purplish-tinged below...9

9a. Gynostegium rounded or truncate, not apically depressed; hoods with a minute horn or crest toward the base...***A. rusbyi***
9b. Gynostegium apically depressed; hoods without a horn or crest...***A. engelmanniana***

10a. Hoods of the corona lacking horns, greenish-yellow; corolla greenish...***A. viridiflora***
10b. Hoods of the corona with a horn, variously colored; corolla variously colored...11

11a. Leaves broadly obovate to broadly elliptic or subquadrate, cordate at the tip with a small mucronate apex, 40–120 mm wide; corolla and hoods greenish to greenish-yellow; pedicels densely white-tomentose but sometimes becoming glabrate in age...12
11b. Leaves variously shaped with an acute to rounded tip, 5–100 mm wide; corolla and hoods variously colored; pedicels white-tomentose or not...13

12a. Leaves subsessile, if petioles present, these usually less than 5 mm in length; hoods entire; plants tomentulose when young, soon becoming glabrate...***A. latifolia***
12b. Leaves petiolate, petioles 5–13 mm in length; hoods broadly 2-lobed at the tip; plants usually persistently minutely hairy or tomentulose...***A. arenaria***

13a. Pedicels densely white-tomentose; hoods 10–13 mm long; corolla 8–15 mm long; leaves mostly ovate, (3) 4–10 cm wide...***A. speciosa***
13b. Pedicels glabrous to hairy but not densely white-tomentose; hoods 1.2–10 mm long; corolla 3–14 mm long; leaves various...14

14a. Hoods 7–10 mm long, about twice the length of the column, very narrowly oblong in the lower half, prominently flaring above; corolla and hoods greenish-white...***A. oenotheroides***
14b. Hoods 1.2–6 mm long, not twice the length of the column, broad at the base and abruptly narrowed into a linear tip or ovate; corolla greenish to purplish or pink; hoods white, pink, or yellowish...15

15a. Stems hairy in lines decurrent from the nodes and on the petioles; plants 4–15 dm tall, from a shallow, fibrous root system; follicles erect on erect pedicels; corolla light to dark pink, rarely white; hoods white or dark pink on older flowers...***A. incarnata***
15b. Stems uniformly hairy or glabrate; plants 0.5–5 (6) dm tall, usually from woody rootstock or a thickened base; follicles erect on deflexed or spreading pedicels; corolla greenish, purple, or pink; hoods white to yellowish or pale pink...16

16a. Hoods 5–6 mm long; peduncles 1–4 (6) cm long; leaves petiolate 4–10 (15) mm, 6–14 cm long, uniformly hairy to glabrate, not densely hairy along the margins; plants 2–5 (6) dm tall...***A. hallii***
16b. Hoods 1.2–4 mm long; peduncles absent or to 0.5 cm long, the umbels usually sessile and directly subtended by leaves; leaves petiolate 1–5 mm, 1–6 cm long, densely hairy along the margins; plants 0.5–5 dm tall...17

17a. Hoods 1.2–2.2 mm long, yellowish to pinkish, truncate or broadly rounded at the apex; corolla purple to purplish-brown, 3–4 mm long...***A. uncialis* ssp. *uncialis***
17b. Hoods 2.5–4 mm long, white to yellow, acute or somewhat rounded at the apex; corolla yellow, green, pinkish, or purple, 4–6 mm long...18

18a. Plants sparsely to moderately short-hairy, more densely so on the leaf margins; corolla pale green to pinkish; seeds 6–10 mm long; plants of the southeastern plains...***A. involucrata***
18b. Plants densely tomentose; corolla pale yellow to yellowish-green; seeds 10–14 mm long; plants of the extreme southwestern counties...***A. macrosperma***

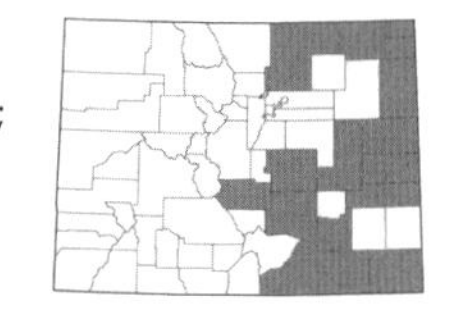

***Asclepias arenaria* Torr.**, SAND MILKWEED. Plants 2–6 dm, tomentose to glabrate; *leaves* 5–10 × 4–8 cm, obovate to oblong-oval with a broadly rounded, retuse tip; *umbel* densely white-tomentose; *calyx* 4–7 mm long; *corolla* 7–10 mm long, greenish; *column* 2 × 3.5 mm; *hood* 3–5 mm long, greenish, with horns; *follicles* 5–10 cm long. Common in sandy soil of the plains, 3800–5000 ft. June–Aug. E.

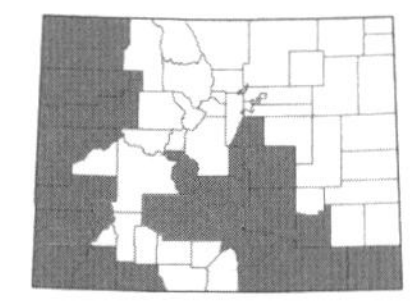

Asclepias asperula* (Decne.) Woodson ssp. *asperula, ANTELOPEHORNS; SPIDER MILKWEED. Plants (2) 3–7 (9) dm, scabrous to glabrous; *leaves* 8–20 × 0.8–2.5 cm, lanceolate to linear-lanceolate with an acute tip; *umbel* glandular-hairy; *calyx* 3.5–5.5 mm long; *corolla* 9–12 mm long, yellowish-green; *hood* 8–10 mm long, reddish-violet, lacking horns; *follicles* 6–10 cm long. Found in sandy or gravelly soil on the prairie, in pinyon-juniper communities, or in open ponderosa pine forests of the plains and foothills, 4000–7600 ft. May–July. E/W.

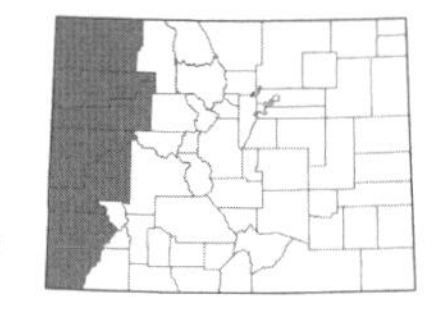

Asclepias cryptoceras* S. Watson ssp. *cryptoceras, PALLID MILKWEED; COW CABBAGE. Plants 0.3–3.5 dm, glabrous and glaucous; *leaves* 2–8 × 3–7 cm, broadly ovate to suborbicular; *umbel* glabrous; *calyx* 4.5–8 mm long; *corolla* 9–15 mm long, pale yellow or greenish-yellow; *column* 2.2–4 mm high; *hood* 3.5–9 mm long, reddish-violet, horns absent or included in the hood; *follicles* 5.5–7 cm long. Locally common on dry, open places in sandy, clay, and serpentine soil, in sagebrush or shadscale, or pinyon-juniper and aspen communities, 4700–7500 ft. May–July. W. (Plate 7)

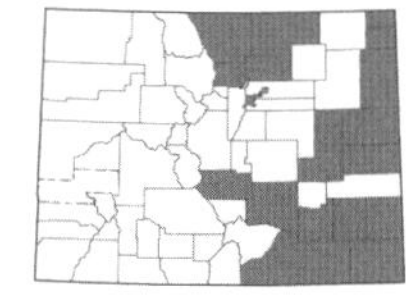

***Asclepias engelmanniana* Woodson**, ENGELMANN'S MILKWEED. Plants 6–15 dm; *leaves* 5–25 × 0.1–0.8 cm, linear, sessile, glabrous; *umbel* spreading-hairy; *calyx* 3.5–5 mm long; *corolla* 5–7 mm long, greenish or with purplish tinges; *hood* 2–3 mm long, yellowish-green, horns absent; *follicles* 7–12 cm long. Common in sandy, rocky, or calcareous soil of the plains and foothills on the eastern slope, 3700–6500. June–Aug. E. (Plate 7)

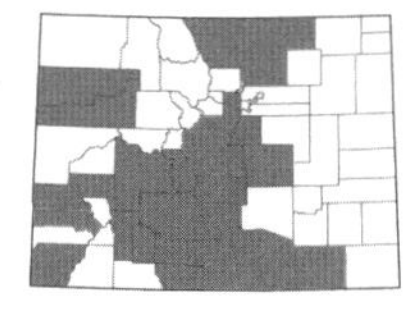

***Asclepias hallii* A. Gray**, HALL'S MILKWEED. Plants 2–5 (6) dm; *leaves* 6–14 × 1.5–4 cm, lanceolate to ovate, tomentulose to glabrate; *umbel* sparsely villose; *calyx* 2–4 mm long; *corolla* 5.5–8 mm long, pink, reddish-cream, or purple; *column* 0.4–1 mm high; *hood* 5–6 mm long, pale rose to cream or white, horns present; *follicles* 8–12 cm long. Common in sandy and gravelly soil of roadsides and sagebrush, pinyon-juniper, or cottonwood communities, 7400–10,000 ft. June–Aug. E/W.

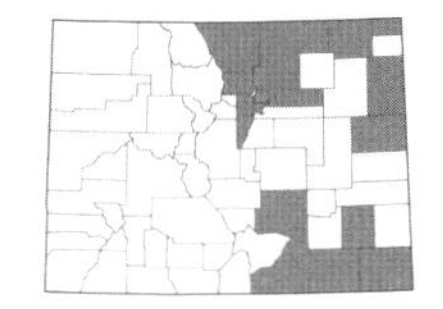

***Asclepias incarnata* L.**, SWAMP MILKWEED. Plants 4–15 dm; *leaves* (5) 10–17 × 1–3 (6) cm, lanceolate; *umbel* sparsely pilose; *calyx* 1.5–2.5 mm long; *corolla* 3–5.5 mm long, pink or rarely white; *column* 1–1.5 mm high; *hood* 1.8–2.5 mm long, white or dark pink, horns present; *follicles* 5–9 cm long. Locally common along ditches, streams, in marshes, and in other wet areas of the plains and foothills, 3400–5500 ft. July–Sept. E. (Plate 7)

***Asclepias involucrata* Engelm. ex Torr.**, DWARF MILKWEED. Plants 1–2 dm, short-hairy; *leaves* lanceolate to narrowly ovate, 1–6 × 0.5–2.5 cm; *calyx* 2.5–3 mm long; *corolla* 4–6 mm long, greenish to pink or purple; *column* 1–1.5 mm tall; *hoods* 2.2–4.5 mm long, white to yellowish, horns present; *follicles* 3.5–6 cm long. Uncommon in sandy or rocky soil, 4000–5000 ft. April–June. E.

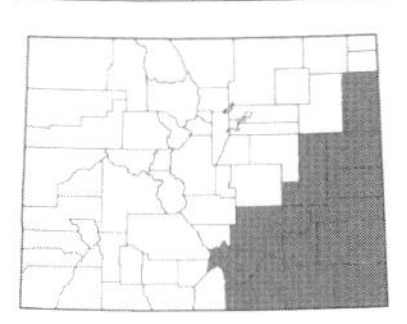

***Asclepias latifolia* (Torr.) Raf.**, BROADLEAF MILKWEED. Plants 2–6 (8) dm; *leaves* ovate to elliptic or broadly obovate, with a mucronate tip, 6–15 (20) × 5–14 cm; *umbels* white-tomentose; *calyx* 3–5 mm long; *corolla* 7.5–10 (12) mm long, pale green or sometimes purplish-tinged; *column* 0.8–1.8 mm high; *hoods* 2.5–4 mm long, greenish-yellow, horns present; *follicles* 6–9 cm long. Found in sandy, clay, or calcareous soils in ditches, along roadsides, and in open prairie, 3500–4500 ft. June–Aug. E. (Plate 8)

***Asclepias macrosperma* Eastw.**, LARGE-SEEDED MILKWEED. Plants 0.8–5 dm, densely tomentose; *leaves* 2–6.5 × 1–3 cm, ovate to lanceolate upward; *calyx* 3 mm long; *corolla* 5–6 mm long, pale green; *column* 0.7–1 mm long; *hoods* 2.5–3 mm long; *follicles* 4–7 cm long. Uncommon in dry, sandy places, 4500–5000 ft. April–June. W.

Although Sundell (1990) groups *A. macrosperma* with *A. involucrata*, the two are very different morphologically and have not been shown to intergrade. Furthermore, their distributions are quite different and do not overlap.

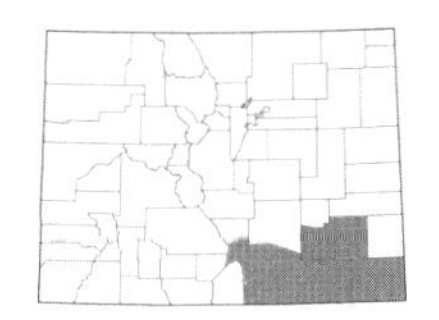

***Asclepias macrotis* Torr.**, LONGHOOD MILKWEED. Plants 1.5–3.5 dm; *leaves* narrowly linear, 2–9 cm × 0.5–1.5 mm, sessile; *calyx* 2–3 mm long; *corolla* 4–5 mm long, green to purple-tipped; *column* ca. 0.5 mm tall; *hoods* 4–6 mm long, purple below and yellowish white along the sides, horns present; *follicles* 4–8 cm long. Infrequent in dry, sandy or gravelly soil of the southeastern plains, 4500–5500 ft. June–Aug. E.

***Asclepias oenotheroides* Cham. & Schltdl.**, ZIZOTES MILKWEED; SIDECLUSTER MILKWEED. Plants 0.4–5 dm; *leaves* oblong-lanceolate to ovate, 4–12 × 1–6 cm; *calyx* 3–4 mm long; *corolla* 8–14 mm long, greenish-white; *column* 1.5 mm high; *hoods* 7–10 mm long; *follicles* 7–9 cm long. Infrequent in sandy or rocky soil, 5000–5500 ft. June–July. E.

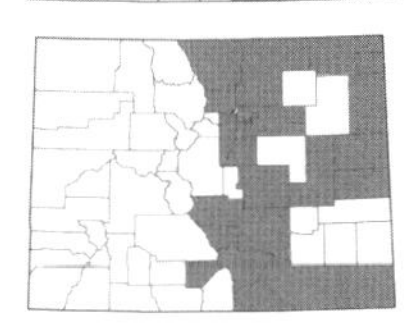

***Asclepias pumila* (A. Gray) Vail**, PLAINS MILKWEED. Plants 0.5–3 dm; *leaves* irregularly whorled, linear-filiform, 2–5 cm × 1 mm, sessile; *calyx* ca. 2 mm long; *corolla* 3–5 mm long, greenish-white; *column* 1–1.5 mm high; *hoods* 1–1.5 mm long, whitish, horns present; *follicles* 3–8 cm long. Common in sandy, clay, or rocky calcareous soil of the plains and foothills, 3500–6000 ft. July–Sept. E.

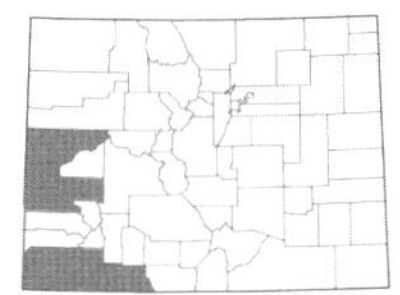

***Asclepias rusbyi* (Vail) Woodson**, RUSBY'S MILKWEED. Plants 3–10 dm, glabrous; *leaves* narrowly linear or filiform, 7–14 (19) cm × 1–5 (8) mm; *calyx* 3–5 mm long; *corolla* 4.5–6.5 mm long, yellow-orange or pale green; *column* 0.5–1.3 mm high; *hoods* 2–3 mm long, white to yellow, horns present as a small crest within the hood; *follicles* 7–10 cm long. Infrequent in rocky soil, often on adobe flats, 7000–7500 ft. July–Aug. W.

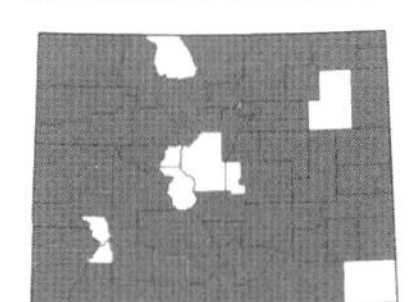

***Asclepias speciosa* Torr.**, SHOWY MILKWEED; COMMON MILKWEED. Plants 4–12 dm, tomentulose; *leaves* (6) 10–20 × (3) 4–10 cm, usually ovate; *umbel* densely white-lanate; *calyx* 4–6 mm long; *corolla* 8–15 mm long, purple to pink or rarely white; *column* ca. 0.5 mm high; *hoods* 10–13 mm long, pinkish-cream, horns present; *follicles* 6–12 cm long. Common in sandy, loamy, or gravelly soil along roadsides, in ditches, in fields, and along streams, 3400–8600 ft. (May) June–Aug. E/W. (Plate 8)

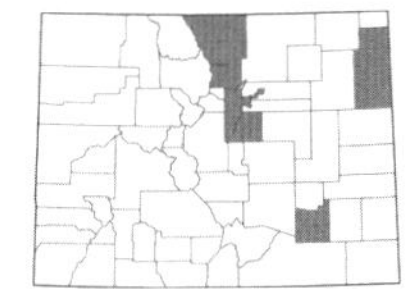

***Asclepias stenophylla* A. Gray**, SLIMLEAF MILKWEED. Plants 2–8 dm, glabrate or minutely hairy; *leaves* narrowly linear, 5–14 cm × 2–4 mm, sessile; *calyx* 2–3 mm long; *corolla* 4–6 mm long, greenish-white to yellowish-green; *hoods* 3–4 mm long, greenish, horns absent; *follicles* 7–12 cm long. Infrequent in sandy, gravelly, or often calcareous soil of the plains and foothills, 4500–6500 ft. June–Aug. E.

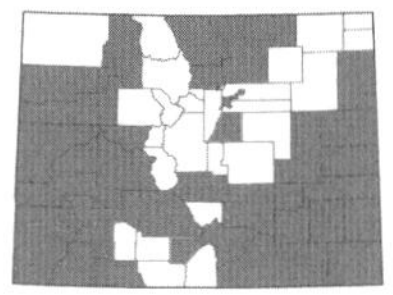

***Asclepias subverticillata* (A. Gray) Vail**, HORSETAIL MILKWEED. Plants 3–7 (12) dm; *leaves* 4–13 cm × 1–4 mm, linear, mostly whorled; *calyx* 1.5–2.5 mm long; *corolla* 3–5 mm long, white to cream; *column* 0.6–1.1 mm high; *hoods* 1.2–2 mm long, white, horns present; *follicles* 5–9 cm long. Common in sandy or rocky soil, along ditches and streams, and in open ponderosa pine or sagebrush communities, 3400–8600 ft. June–Aug. E/W.

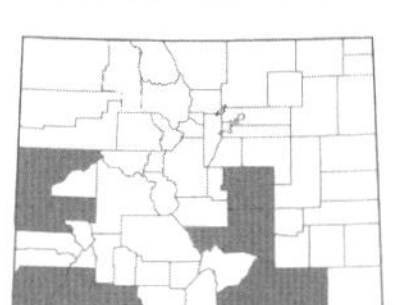

***Asclepias tuberosa* L. ssp. *interior* Woodson**, BUTTERFLY MILKWEED. Plants 2–5 (9) dm, spreading-hispid; *leaves* 4–9 cm × 4–15 mm, lanceolate; *calyx* 2–3 mm long; *corolla* 6–8 mm long, orange to yellow-orange; *column* 0.6–2 mm high; *hoods* 4–6 mm long, yellow to orange, horns present; *follicles* 7–12 cm long. Found in sandy, gravelly, or calcareous soil of the plains, in open woodlands, and in pinyon-juniper communities, 5000–7000 ft. June–Aug. E/W. (Plate 8)

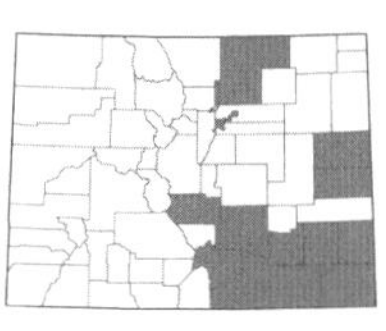

Asclepias uncialis* Greene ssp. *uncialis, WHEEL MILKWEED. Plants 0.5–1 dm, with short, curly hairs; *leaves* lanceolate, 1–5 cm × 1–25 mm, sessile; *calyx* 2–4 mm long; *corolla* 3–4 mm long, purple to purplish-brown; *column* 0.4–0.6 mm high; *hoods* 1.2–2.2 mm long, yellowish to pinkish, horns present; *follicles* 4–6 cm long. Uncommon in sandy or gravelly soil of the plains and outer foothills, usually inconspicuous, 4000–8000 ft. April–June. E.

Asclepias uncialis ssp. *ruthiae* (Maguire) Kartesz & Gandhi is known from the Four Corners region and has been reported from Utah, Arizona, and New Mexico. It could also occur in Montezuma Co. and should be looked for in desert scrub and pinyon-juniper communities. Subspecies *ruthiae* differs by having larger flowers (corolla lobes 4–6 mm long vs. 3–4 mm long in ssp. *uncialis*) and mostly ovate leaves as opposed to lanceolate or linear leaves.

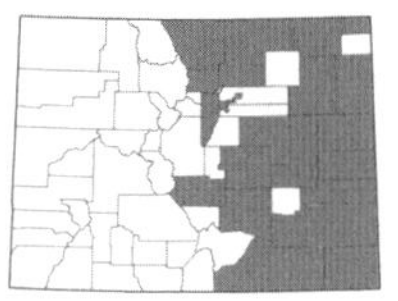

***Asclepias viridiflora* Raf.**, GREEN COMET MILKWEED. Plants 1–6 dm, glabrate to tomentose; *leaves* 3–14 cm × 2–17 mm, lanceolate to obovate or ovate; *calyx* 2–4 mm long; *corolla* 4–7 mm long, greenish; *hoods* 2.5–4 mm long, greenish-yellow, horns absent; *follicles* 5–15 cm long. Common in sandy, rocky, or calcareous soil of the plains and foothills, 3500–7000 ft. June–Aug. E. (Plate 8)

SARCOSTEMMA R. Br. – TWINEVINE

Prostrate vines; *leaves* opposite; *calyx* deeply 5-parted, adnate to the petals, lobes widely spreading; *corona* of two parts, a ring at the base of the petals surrounding the gynostegium and 5 distinct, inflated vesicles adnate to the petals or staminal column; *stamens* petaloid.

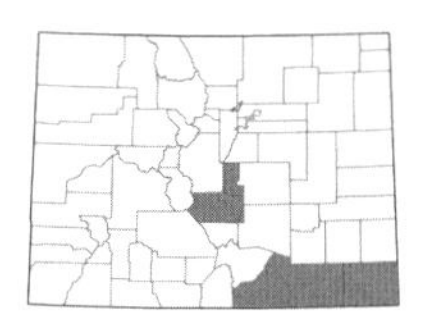

***Sarcostemma crispum* Benth.**, WAVYLEAF TWINEVINE. [*Funastrum crispum* (Benth.) Schltr.]. Perennials, moderately to densely minutely hairy or glabrous; *leaves* lanceolate to linear, 2.5–9 cm long, minutely hairy, margins crisped; *calyx* lobes 3–5 mm long, green to purple; *corolla* 3–5 mm long, purple; *follicles* 8.5–12.5 cm long. Uncommon in rocky soil of open wooded slopes in the foothills or in sandy soil of the plains, 4000–5700 ft. June–Aug. E.

ARACEAE Juss. – ARUM FAMILY

Perennial herbs, aquatic, ours often reduced to small green bodies called fronds that are rootless or with 1–21 roots borne on the lower surface, reproducing vegetatively; *leaves* absent or alternate, petiolate or sessile, simple or compound; *flowers* rarely produced or in terminal spadices; *tepals* absent or 4–8 (12), usually distinct; *stamens* 1–12; *pistil* 1; *ovary* superior, 1–3-carpellate; *fruit* an utricle, berry, drupe, or capsule.

1a. Leaves over 2 cm long, obovate to spatulate and strongly ribbed...***Pistia***
1b. True leaves absent, plants reduced to small green bodies called fronds, the fronds entire, lanceolate to ovate or round...2

2a. Fronds with 1–21 roots and at least 1 vein...3
2b. Roots and veins absent on fronds...4

3a. Roots 1 per frond; fronds with 1–3 (5) veins...***Lemna***
3b. Roots 2 or more per frond; fronds with 7–16 veins...***Spirodela***

4a. Fronds 3–15 mm long, the base narrowed to a stalk...***Lemna*** (*trisulca*)
4b. Fronds 0.5–1.5 mm long, the base not narrowed to a stalk...***Wolffia***

LEMNA L. – DUCKWEED

Fronds 1–20, floating or submerged, flattened, lanceolate-ovate to orbicular, with 1–5 veins; *roots* 1 per frond; *flowers* borne in a pouch, with 2 staminate and 1 pistillate flower per pouch. (Landolt, 2000)

1a. Fronds usually submerged, narrowly ovate, the base narrowed to a stalk; roots sometimes not developed...***L. trisulca***
1b. Fronds floating, ovate to obovate, the base not narrowed to a stalk; roots present...2

2a. Fronds with a single vein, this sometimes indistinct; anthocyanin absent...***L. minuta***
2b. Fronds with 3 veins; lower and upper surfaces often reddish...***L. minor***

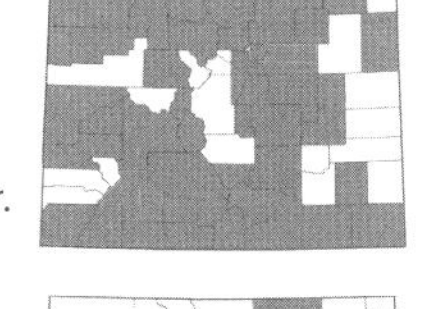

***Lemna minor* L.**, COMMON DUCKWEED. [*L. turionifera* Landolt]. Fronds floating, 1 or 2–5+, ovate, 1–8 mm, 1.3–2 times as long as wide; *veins* 3 (5). Common and scattered across the state, 3500–9800 ft. E/W.

Lemna gibba L. has been reported for the state, but these collections are most likely *L. minor*. *Lemna gibba* differs in having 4–5 veins on the fronds which are often gibbous.

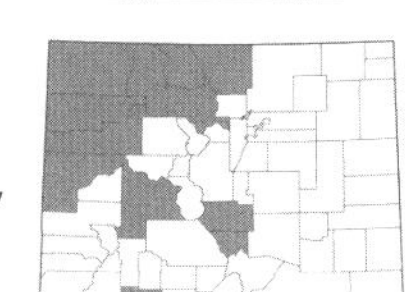

***Lemna minuta* Kunth**, LEAST DUCKWEED. [*L. minima* Phil. ex Hegelm.; *L. minuscula* Herter]. Fronds floating, 1–2, obovate, 0.8–4 mm long, 1–2 times as long as wide; *veins* 1, sometimes indistinct. Our records scattered across the state, 3800–9500 ft. E/W.

Lemna valdiviana Phil. has been reported for the state, but these collections are most likely *L. minuta*. *Lemna valdiviana* differs from *L. minuta* in having a prominent, longer vein.

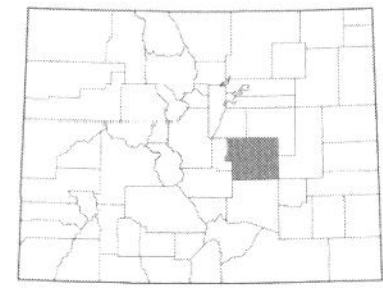

***Lemna trisulca* L.**, STAR DUCKWEED. Fronds usually submerged, 3–50, narrowly ovate, 3–15 mm long, 2–3.5 times as long as wide, abruptly narrowed at the base into a green stalk; *veins* (1) 3. Uncommon, our records scattered across the state but apparently absent from the eastern plains, 6100–10,000 ft. E/W.

PISTIA L. – WATER LETTUCE

Aquatic, floating herbs; *leaves* sessile, in a dense rosette, obovate to spatulate and strongly ribbed; *flowers* imperfect.

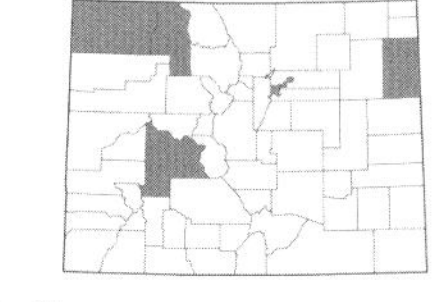

***Pistia stratiotes* L.**, WATER LETTUCE. Plants stoloniferous, *leaves* up to 20 cm long, less than 7 cm wide, velvety-hairy; *inflorescence* a spadix, spathe pale to white green, spathe exceeding spadix; *staminate flowers* in single whorl around central stalk; *pistillate flowers* solitary. Found recently in a lake in El Paso Co., probably not persisting, 7500 ft. July–Sept. E. Introduced.

SPIRODELA Schleid. – DUCKMEAT

Fronds 2–10, floating, flattened, obovate to orbicular, with 5–15 veins; *roots* 2–21 per frond; *flowers* surrounded by a sac-like pouch, with 2 or 3 staminate and 1 pistillate flower per pouch.

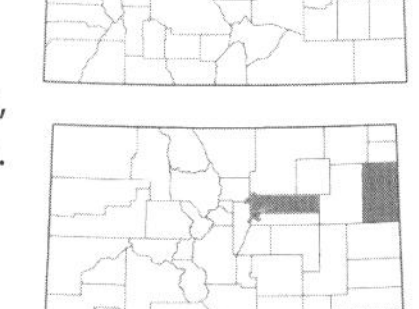

***Spirodela polyrrhiza* (L.) Schleid.**, COMMON DUCKMEAT. Plants producing buds in autumn; *fronds* round to obovate, 2–10 mm wide; *fruits* 1–1.5 mm long, winged. Uncommon on ponds and shallow pools, with occurrences scattered across the state, 4400–9300 ft. E/W.

WOLFFIA Horkel ex Schleid. – WATERMEAL

Fronds 1 or 2, floating or submerged, globular; *roots* absent; *flowers* 2, 1 staminate and 1 pistillate.

1a. Fronds boat-shaped, 1.3–2 times as long as wide, 0.7–1 times as deep as wide, upper surface intensely green, with 50–100 stomates...***W. borealis***
1b. Fronds nearly globular, 1–1.3 times as long as wide, 1–1.3 times as deep as wide, upper surface transparently green, with 1–10 (30) stomates...***W. columbiana***

***Wolffia borealis* (Engelm.) Landolt & Wildi ex Gandhi, Wiersema & Brouillet**, NORTHERN WATERMEAL. Fronds boat-shaped, 0.7–1.5 mm long, 1.3–2 times as long as wide, 0.7–1 times as deep as wide, intensely green; *stomates* 50–100. Uncommon in slow-moving water, 3600–5200 ft. E.

***Wolffia columbiana* H. Karst.**, COLUMBIAN WATERMEAL. Fronds globular or nearly so, 0.5–1.4 mm long, 1–1.3 times as long as wide, 1–1.3 times as deep as wide, transparently green; *stomates* 1–10 (30). Uncommon in ditches and ponds , 3600–5200 ft. June–Aug. E.

ARALIACEAE Juss. – GINSENG FAMILY

Perennial herbs or shrubs; *leaves* alternate; *flowers* small, actinomorphic, perfect or imperfect, usually in umbels; *sepals* 5; *petals* 5, usually distinct; *stamens* usually 5, alternate with the petals; *pistil* 1; *ovary* inferior with 2–5 locules with 1 ovule per locule; *styles* connate or distinct, usually swollen at the base to form a stylopodium; *fruit* most commonly a drupe or sometimes a berry.

ARALIA L. – SPIKENARD

Unarmed or prickly perennial herbs or shrubs; *leaves* pinnately or ternately compound or decompound, the leaflets toothed; *flowers* greenish-white; *fruit* a drupe, purple or black.

1a. Umbels in a globose corymb; leaves usually with 5 leaflets, basal and long-petioled...***A. nudicaulis***
1b. Umbels in a large compound panicle; leaves usually with 7 leaflets, present on the stem...***A. racemosa***

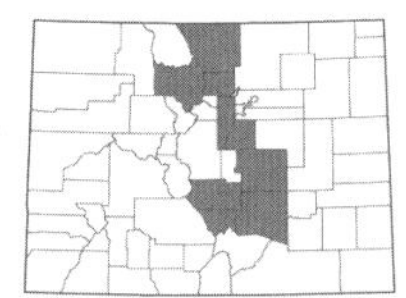

***Aralia nudicaulis* L.**, WILD SARSPARILLA. Perennials, appearing acaulescent; *leaves* solitary, petiole 15–40 cm long, blade with 3 digitate primary divisions, each division pinnately 3–5-foliate, leaflets usually 5, 5–15 cm long, with finely serrate margins; *inflorescence* of 2–6 umbels, rays 2–7 cm long; *flowers* greenish-white, petals 1–2 mm long; *drupes* dark purple, 4–8 mm diam. Uncommon in woodlands and cool ravines, 5600–9000 ft. June–Aug. E/W.

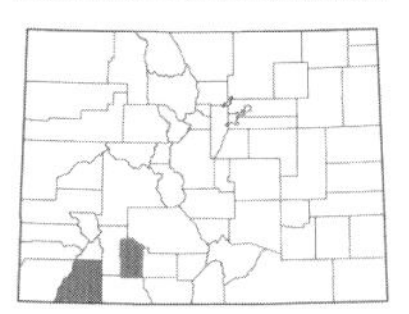

***Aralia racemosa* L. ssp. *bicrenata* (Wooton & Standl.) S.L. Welsh & N.D. Atwood**, AMERICAN SPIKENARD. Perennials, caulescent; *leaves* with 3 primary divisions, these pinnately compound, with serrate margins; *inflorescence* a panicle of numerous umbels; *flowers* greenish-white; *drupe* purplish-black, 4–6 mm diam. Rare in moist woodlands and ravines, 7300–8000 ft. June–Aug. W.

ASPARAGACEAE Juss. – ASPARAGUS FAMILY

Perennial herbs from rhizomes, the stems green and photosynthetic; *leaves* alternate, scale-like; *flowers* perfect or imperfect, actinomorphic, small, solitary or paired along the stem or in racemes or umbels; *tepals* 6, all alike and petaloid, greenish-yellow; *stamens* 6; *ovary* superior, 3-carpellate; *fruit* a red to purplish-black berry.

ASPARAGUS L. – ASPARAGUS

With characteristics of the family.

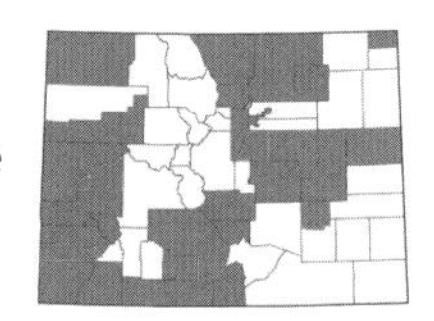

***Asparagus officinalis* L.**, GARDEN ASPARAGUS. Plants 1.5–2.5 m, with finely dissected branches; *leaves* 3–4 mm, base hardened; *infloresence* of axillary racemes, 1–3 flowered; *tepals* 3–8 mm long; *berries* red, 6–10 mm. Locally common in grasslands, meadows, and along ditches, often in waste places, 3700–7800 ft. May–June. E/W.

ASTERACEAE Bercht. & J. Presl – SUNFLOWER FAMILY

Annual, biennial, or perennial herbs or shrubs; *leaves* simple to compound or dissected; *inflorescence* a head surrounded by one or more series of involucre bracts (phyllaries) and composed of only disk flowers (discoid heads), ray flowers and disk flowers (radiate heads), or only ligulate ray flowers (ligulate heads); *flowers* perfect or imperfect, sessile on a receptacle; *receptacle* sometimes with chaff, or naked; *disk flowers* perfect or functionally staminate, actinomorphic, corolla tubular with 5 (rarely 4) teeth; *ray flowers of radiate heads* pistillate or sterile, zygomorphic, corolla often with 2–3 terminal teeth, located along the margin of a radiate head with disk flowers in the center; *ray flowers of ligulate heads* perfect, zygomorphic, corolla with 5 terminal teeth (plants with this type of ray flower have milky sap); *sepals* reduced to a pappus, this of simple or plumose hairs or bristles, scales, stout awns, a projecting ring or crown, or absent; *stamens* 5, anthers usually united into a tube surrounding the style; *ovary* inferior; *fruit* an achene or cypsela.

Asteraceae head:

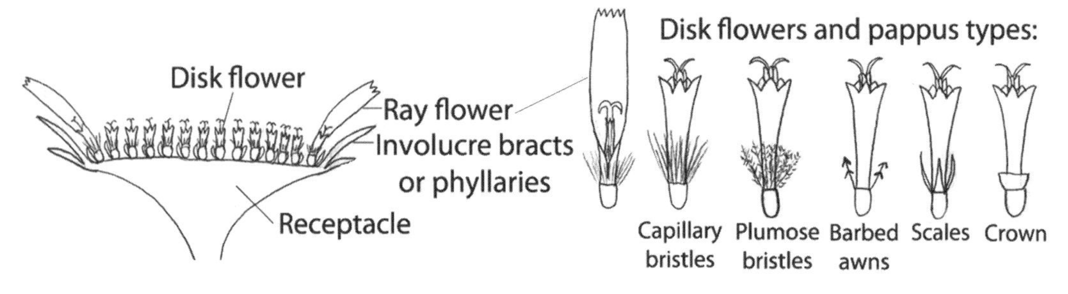

1a. Heads composed entirely of ray flowers (ligulate heads), these usually perfect and with 5 terminal teeth; plants with milky or colored sap...**Key 1**
1b. Heads with disk and ray flowers or only disk flowers, the ray flowers usually with 2–3 terminal teeth if present, or sometimes disciform heads present with disk flowers and small, inconspicuous ray flowers; plants with watery sap...2

2a. Heads with only disk flowers, lacking ray flowers (species with very small and inconspicuous rays shorter than or only slightly surpassing the disk flowers will key here also)...3
2b. Heads with both ray and disk flowers...4

3a. Pappus of capillary or plumose bristles...**Key 2**
3b. Pappus of scales, awns, a short crown, very short and rigid scale-like bristles, or absent...**Key 3**

4a. Ray flowers white, blue, purple, or pink...**Key 4**
4b. Ray flowers yellow or orange...5

5a. Pappus of capillary or barbellate bristles, or of an outer series of short scales and inner series of capillary bristles; receptacle naked or fimbrillate...**Key 5**
5b. Pappus of awns, scales, a short crown, or absent; receptacle naked, chaffy, or bristly...**Key 6**

Key 1

(Heads ligulate; sap milky or colored)

1a. Pappus of simple capillary bristles, intermixed with attenuate scales in one species...2
1b. Pappus of feathery plumose bristles, scales, scales and plumose bristles, or absent...14

2a. Achenes strongly flattened; leaves often prickly-margined, alternate and evidently cauline...3
2b. Achenes not strongly flattened, terete or angular; leaves never prickly-margined, mostly basal or alternate and cauline...4

3a. Achenes beakless; pappus not inserted on a prominent disk at the summit of the achene; heads with 85–250 ray flowers...***Sonchus***
3b. Achenes beaked or if beakless then the ray flowers blue or purple; pappus inserted on an expanded disk at the summit of the achenes; heads with 11–50 ray flowers...***Lactuca***

4a. Ray flowers yellow to orange (sometimes drying to pinkish in *Agoseris*); leaves mostly basal or alternate with the stem leaves reduced...5
4b. Ray flowers pink, purple, or white; leaves opposite below and alternate above or alternate and cauline (these reduced or not)...10

5a. Heads solitary; leaves all basal...6
5b. Heads more than one and paniculately arranged; leaves mostly basal and usually with a few, reduced stem leaves...8

6a. Achenes tapering upward but not beaked; pappus of capillary bristles intermixed with short, attenuate scales; leaves with wavy, white-ciliate margins...***Nothocalais***
6b. Achenes with a long, prominent beak; pappus of capillary bristles; leaves unlike the above...7

7a. Achenes prominently 10-nerved, without tubercles or sharp projections; ray flowers yellow or orange, or sometimes drying pinkish...***Agoseris***
7b. Achenes 4–5-angled or -nerved, with tubercles or rough, small sharp projections toward the top; ray flowers yellow...***Taraxacum***

8a. Achenes with a toothed, white-cartilaginous ring at the summit; leaves pinnatifid with the lobes softly spinulose-toothed; pappus bristles tending to be united at the base and falling as a single unit; plants of the western slope...***Malacothrix***
8b. Achenes without a toothed, white-cartilaginous ring at the summit; leaves entire to pinnatifid, but without spinulose-toothed lobes; pappus bristles falling separately, not united at the base; plants of the eastern or western slope...9

9a. Perennials from a short to elongate rhizome, caudex, or crown with fibrous roots; leaves entire or nearly so, all basal or with a few reduced stem leaves; ray flowers yellow or orange; pappus tawny or brownish or occasionally white…***Hieracium***
9b. Annuals, biennials, or perennials from a taproot or a taproot giving rise to several slender, rhizome-like adventitious stems; leaves pinnatifid, toothed, or rarely entire; ray flowers yellow; pappus white or nearly so…***Crepis***

10a. Leaves over 1 cm wide…11
10b. Leaves small and scale-like or reduced to bracts or linear and less than 1 cm wide…12

11a. Stem leaves well-developed; ray flowers pink to purple…***Prenanthes***
11b. Leaves mostly basal, 1–2 reduced stem leaves sometimes present; ray flowers white…***Hieracium***

12a. Perennials…***Lygodesmia***
12b. Annuals…13

13a. Leaves opposite below and long-linear, alternate above and reduced to bracts; heads larger, the involucre 12–18 mm high…***Shinnersoseris***
13b. Leaves all alternate and reduced to small bracts; heads smaller, the involucre 3.5–5 mm high…***Prenanthella***

14a. Pappus of one or two series of short scales or absent…15
14b. Pappus of plumose bristles, or both bristles and scales…16

15a. Pappus absent; ray flowers yellow; annuals…***Lapsana***
15b. Pappus of one or two series of short scales; ray flowers blue or rarely white or pink; perennials…***Cichorium***

16a. Pappus only of plumose bristles…17
16b. Pappus of bristles and scales, or of narrow scales each bearing a long, white plumose terminal bristle…20

17a. Ray flowers pink, purple, or white; leaves deeply pinnatifid, or often withered and absent at anthesis; achene beakless; involucre bracts in 2 series, the larger principal bracts in a single series and the outer bracts much shorter and calyculate…***Stephanomeria***
17b. Ray flowers yellow; leaves various but usually present at anthesis; achene beaked or if beakless then the involucre bracts imbricate in 3 or more series…18

18a. Leaves mostly or all basal; receptacle chaffy…***Hypochaeris***
18b. Stem leaves well-developed; receptacle naked…19

19a. Leaves all entire, linear and somewhat grass-like; achenes beaked…***Tragopogon***
19b. Basal and lower leaves pinnatifid, upper leaves often entire and linear; achenes beakless…***Scorzonera***

20a. Basal leaves oblanceolate to elliptic-oblong, to 6 cm wide; pappus of 20–40 unequal bristles plus an outer whorl of 10 short, hyaline scales; achenes obscurely ribbed…***Krigia***
20b. Basal leaves long and narrow, to 2 cm wide; pappus of 15–20 narrow, bifid scales, each bearing a long, white plumose terminal bristle; achenes usually prominently 8–10-ribbed…***Microseris***

Key 2

(Heads discoid or rays small and inconspicuous; pappus of capillary or plumose bristles)

1a. Plants with spiny-margined leaves and/or spine-tipped or fringed involucre bracts, often thistle-like; receptacle densely bristly or naked and alveolate…2
1b. Plants not spiny and thistle-like; receptacle naked or chaffy in one species…5

2a. Leaves lacking spiny margins, deeply pinnatifid with narrow lobes; heads usually smaller, the involucre 8–15 mm in height; involucre bracts with a dark tip and pectinate-fringed or spinose-ciliate margins, sometimes also with a short spine tip at the apex…***Centaurea*** (*diffusa, stoebe*)
2b. Leaves with spiny margins, variously shaped; heads various but usually larger; involucre bracts unlike the above…3

3a. Pappus of plumose (feather-like) bristles; receptacle densely bristly…***Cirsium***
3b. Pappus of barbellate bristles; receptacle densely bristly or naked…4

4a. Receptacles alveolate, naked or with very short bristles…***Onopordum***
4b. Receptacles not alveolate, densely bristly…***Carduus***

5a. Involucre bracts dotted with brownish translucent oil glands; ill-scented annuals; leaves pinnately dissected...***Dyssodia***
5b. Involucre bracts not dotted with brownish translucent oil glands; plants otherwise unlike the above...6

6a. Involucre bracts essentially uniseriate and equal, sometimes subtended by a few reduced calyculate bracts at the base of the involucre...7
6b. Involucre bracts imbricate or subequal in 2 or more series...10

7a. Spinescent or unarmed, low shrubs...***Tetradymia***
7b. Perennials...8

8a. Leaves triangular, cordate-sagittate at the base, densely white-tomentose below; stem leaves reduced and scale-like; disk flowers white to pale yellow...***Petasites***
8b. Leaves lanceolate, oblanceolate, obovate, or elliptic, not densely white-tomentose below; stem leaves not scale-like; disk flowers orange, orange-yellow, or yellow...9

9a. Most leaves basal, elliptic to ovate with entire to crenate-dentate margins, the stem leaves deeply pinnately lobed...***Packera*** (*debilis*)
9b. Leaves unlike the above...***Senecio***

10a. Involucre bracts longitudinally striate with 3 or more essentially parallel lines (this sometimes faint in *Ageratina*), usually rather dry...11
10b. Involucre bracts without longitudinally striate lines, dry or herbaceous...13

11a. Achenes 10-ribbed...***Brickellia***
11b. Achenes mostly 5-angled or -nerved...12

12a. Leaves whorled (mostly in 3's or 4's); involucre 6.5–9 mm high, often purplish; disk flowers purple...***Eutrochium***
12b. Leaves opposite or alternate; involucre 2.5–4 mm high, greenish; disk flowers white or whitish...***Ageratina***

13a. Involucre bracts sticky and gummy, firm, the tips reflexed...***Grindelia***
13b. Involucre bracts not sticky and gummy, the tips usually not reflexed...14

14a. Disk corollas purple to pink or rarely white; involucre bracts often anthocyanic...15
14b. Disk corollas yellow, ochroleucous, or brownish; involucre bracts usually not anthocyanic...17

15a. Pappus in a single series of plumose or barbellate bristles; leaves elongate, linear to linear-ovate, conspicuously punctate with resinous dots...***Liatris***
15b. Pappus double, the outer of short rigid bristles or scales and the inner of long, slender capillary or plumose bristles; leaves various but not punctate with resinous dots...16

16a. Receptacles chaffy with long, narrow bracts; inner pappus of plumose bristles; plants low, less than 2.2 dm; found on the alpine above 10,000 ft...***Saussurea***
16b. Receptacles naked; inner pappus of capillary bristles; plants 3–15 dm; found on the eastern plains between 3500–7000 ft...***Vernonia***

17a. Stems and leaves tomentose and white-woolly; involucre bracts white or pink and scarious (at least at the tip), often woolly at the base...18
17b. Both stems and leaves not white-woolly; involucre bracts herbaceous or herbaceous at the tip and chartaceous at the base, but not white or pink and scarious, or woolly at the base...21

18a. Annuals; heads embedded in dense woolly hairs...***Gnaphalium***
18b. Perennials; heads not embedded in dense woolly hairs...19

19a. Heads basically all alike with numerous, pistillate outer flowers and few, perfect inner flowers; from taproots...***Pseudognaphalium***
19b. Heads unlike, plants dioecious or polygamo-dioecious; fibrous-rooted, often with rhizomes or stolons, and often forming mats...20

20a. Basal leaves withering and early-deciduous, stem leaves well-developed and bicolored with a darker adaxial surface; plants polygamo-dioecious, some heads with a few central staminate flowers surrounded by several rows of pistillate flowers...***Anaphalis***
20b. Basal leaves tufted and persistent, stem leaves reduced and usually not bicolored; plants dioecious, the heads only with staminate or pistillate flowers...***Antennaria***

21a. Leaves opposite...***Arnica*** (*parryi*)
21b. Leaves alternate...22

22a. Shrubs, subshrubs, or perennial herbs (mostly with woody bases)...23
22b. Annual herbs...29

23a. Leaves dentate to serrate with spinulose teeth, the teeth or lobes often bristle-tipped...***Xanthisma*** (*grindelioides*)
23b. Leaves entire or sometimes toothed but without spinulose teeth...24

24a. Plants dioecious, the heads of one plant either staminate or pistillate...***Baccharis***
24b. Plants monoecious, the heads bisexual...25

25a. Stems with white-tomentose hairs (these closely appressed in some species but can be detected by scratching the stem to remove the tomentum)...***Ericameria***
25b. Stems glabrous or variously hairy but not closely appressed-white-tomentose...26

26a. Involucre bracts with a herbaceous center and scarious margins, subequal or slightly imbricate; plants 0.2–3 dm tall, with mostly basal leaves, these linear or pinnately cleft into linear segments...***Erigeron*** (*concinnus, mancus*)
26b. Involucre bracts chartaceous with a greenish or brownish tip, or large and foliaceous; plants otherwise unlike the above in all respects...27

27a. Disk corollas with a long, slender tube and abruptly dilated short throat; involucre turbinate, the bracts closely imbricate in several series and chartaceous with a greenish or brownish acute tip; plants 2.5–8 dm; plants of the western slope...***Isocoma***
27b. Disk corollas gradually expanded from the base, without a strongly differentiated tube and throat; involucre campanulate or broadly hemispheric, the bracts imbricate and chartaceous with a greenish acute to acuminate or cuspidate tip, or not much imbricate and then large and foliaceous; plants usually 5 dm or less tall; plants of the eastern or western slope...28

28a. Involucre bracts usually strongly vertically arranged (the bracts from each lower row consecutively stacked with the upper rows) with obtuse to acute or acuminate tips, the midrib usually keeled; disk flowers few, 3–8; plants of the eastern or western slope...***Chrysothamnus***
28b. Involucre bracts not strongly vertically arranged with sharply cuspidate tips, without a keeled midrib; disk flowers 15–20; plants of the eastern slope...***Oönopsis***

29a. Involucres 4–12 mm high, the bracts mostly green and herbaceous or with a small yellowish, chartaceous base...30
29b. Involucres 3–5 mm high, the bracts brown or green but scarcely herbaceous...31

30a. Achenes several-nerved; both basal and stem leaves linear...***Symphyotrichum*** (*ciliatum*)
30b. Achenes 2-nerved; basal leaves oblanceolate to spatulate...***Erigeron*** (*acris* and *lonchophyllus*)

31a. Leaves entire or some of the lower ones irregularly few-toothed, tapering to a short petiole, subglabrous to spreading-hirsute...***Conyza***
31b. Leaves evidently toothed or sublobate, sessile and clasping, glandular or viscid and/or loosely arachnoid-hairy...***Laennecia***

Key 3

(Heads discoid or rays small and inconspicuous; pappus of scales, awns, a short crown, scale-like bristles, or absent)

1a. Heads unisexual, one kind pistillate with a fused nut-like or bur-like involucre often armed with prickles or spines, the other staminate and generally above the pistillate with a poorly developed distinct or connate involucre...2
1b. Heads bisexual, the involucre not fused and nut-like or bur-like, although the bracts sometimes armed with a hooked bristle or spine...3

2a. Involucre of pistillate heads forming a bur with numerous hooked prickles; involucre of staminate heads distinct or absent...***Xanthium***
2b. Involucre of pistillate heads with one or more series of tubercules or straight spines; involucre of staminate heads connate...***Ambrosia***

3a. At least the innermost involucre bracts with tipped with hooked bristles, spines, or conspicuously erose or fringed at the tips or along the margins...4
3b. Involucre bracts without a bristle or spine tip, not conspicuously erose or fringed along the margins...7

4a. Involucre bracts with an inward pointing hooked bristle; basal leaves large, 3–6 dm long and 3–4 dm wide, the bases mostly cordate, green and glabrate above and pale and arachnoid-tomentose below, with long, stout hollow petioles...***Arctium***
4b. Involucre bracts lacking inward pointing hooked bristles; plants otherwise various but unlike the above...5

5a. Leaves without spiny margins; involucre bracts spine-tipped or with erose or fringed margins...***Centaurea***
5b. Leaves with spiny margins; receptacle chaffy...6

6a. Heads 1-flowered, these grouped in tight, spheric clusters; disk corollas blue or sometimes white...***Echinops***
6b. Heads with many flowers, these not grouped in spheric secondary heads; disk corollas yellow or orange...***Carthamus***

7a. Receptacles enlarged, hemispheric to conic or columnar...8
7b. Receptacles flat to convex, not evidently enlarged...10

8a. Disk flowers dark purple to black; receptacle columnar...***Rudbeckia*** (*montana*)
8b. Disk flowers yellow, blue, or white; receptacle hemispheric to conic...9

9a. Leaves cauline, pinnatifid to pinnately dissected; plants aromatic; pappus a short crown or obscure...***Matricaria***
9b. Leaves mostly basal, simple and entire; plants not aromatic; pappus of aristate scales...***Tetraneuris*** (*acaulis*)

10a. Heads in densely white-woolly glomerules; involucre bracts absent, the apparent bracts actually the chaff of the receptacle...***Evax***
10b. Heads not in densely white-woolly glomerules; involucre bracts present...11

11a. Involucre bracts sticky and gummy, firm, the tips reflexed...***Grindelia***
11b. Involucre bracts not sticky and gummy, the tips usually not reflexed...12

12a. Receptacle densely bristly; the outer involucre bracts with longitudinally striate lines and a broadly rounded, scarious tip, the inner bracts narrower and tapering to an acuminate, plumose tip; disk flowers pink or purple...***Acroptilon***
12b. Receptacle chaffy or naked, sometimes hairy, or rarely somewhat bristly but not densely so; involucre bracts and disk flowers unlike the above in all respects...13

13a. Very low, mat-forming, essentially stemless perennial herbs; heads more or less sessile and hidden among the leaf bases...***Parthenium***
13b. Annual or perennial herbs or shrubs unlike the above...14

14a. Leaves cordate-hastate to triangular, opposite or nearly so; disk corollas 4-toothed; involucre bracts fused for half their length or more with only the attenuate apices separate...***Pericome***
14b. Leaves various but not cordate-hastate or triangular, opposite or alternate; disk corollas usually 5-toothed; involucre bracts various...15

15a. Involucre bracts in 2 series, the inner ones enlarging and ultimately ovate to orbicular in fruit; plants gland-dotted and canescent throughout; achenes winged...***Dicoria***
15b. Involucre unlike the above; plants gland-dotted or not; achenes not winged or if so then some achenes also wingless...16

16a. Leaves with entire margins...17
16b. Leaves with toothed to pinnatifid margins or the leaves compound...22

17a. Plants essentially with all basal leaves...18
17b. Plants with stem leaves...19

18a. Leaves broadly ovate, 3–4 cm long; pappus absent or rarely of 2 bristle-like awns; heads mostly 3–4 cm wide...***Encelia***
18b. Leaves ovate to suborbicular, 0.4–1.5 cm long; pappus of hyaline scales; heads mostly 1–1.5 cm wide...***Chamaechaenactis***

19a. Disk flowers purple or pinkish; heads on long, densely glandular peduncles...***Palafoxia*** (*rosea*)
19b. Disk flowers yellow or white; heads unlike the above...20

20a. Annual herbs from slender taproots; involucre bracts mostly 4, equal and uniseriate, herbaceous, clasping the achenes such that the involucre appears deeply furrowed; plants densely glandular and malodorous...***Madia***
20b. Shrubs or perennial herbs; involucre bracts unlike the above; plants sometimes glandular but not malodorous...21

21a. Involucre bracts fused below to form a cup with a toothed or lobed apex; receptacle chaffy...***Iva***
21b. Involucre bracts distinct, imbricate in 2–several series, at least the inner ones scarious or with scarious margins; receptacle naked or hairy...***Artemisia***

22a. Involucre bracts dotted with brownish translucent oil glands; ill-scented annuals; leaves pinnately dissected...***Dyssodia***
22b. Involucre bracts not dotted with brownish translucent oil glands; plants otherwise unlike the above...23

23a. Pappus of scales...24
23b. Pappus absent, or of awns or a short crown...26

24a. Leaves opposite below and alternate above; low annuals; leaves linear or dissected into linear segments...***Schkuhria***
24b. Leaves alternate or mostly basal; annual, biennial, or perennial herbs...25

25a. Involucre bracts with broad, scarious tips...***Hymenopappus***
25b. Involucre bracts green and narrow, without scarious tips...***Chaenactis***

26a. Leaves opposite or sometimes just the upper ones alternate...27
26b. Leaves alternate or crowded at the base...31

27a. Leaves pinnately dissected into linear or filiform segments...28
27b. Leaves not pinnately dissected into linear or filiform segments...29

28a. Inner involucre bracts basally connate for about ⅓ of their length; perennials; disk flowers yellow with reddish-brown veins; achenes all similar...***Thelesperma*** (*megapotamicum*)
28b. Inner involucre bracts distinct; annuals; disk flowers yellow; achenes of two different kinds, the outer oval, incurved, with winged margins and devoid of pappus, and the inner slender, attenuate, not winged, with a pappus of 2 or 3 widely divergent awns...***Heterosperma***

29a. Heads small (the involucre 1.5–3 mm tall) and numerous in a panicle-like array; leaves ovate with toothed margins...***Cyclachaena***
29b. Heads larger or not numerous in a panicle-like array; leaves various but unlike the above...30

30a. Pappus absent; receptacle naked; heads with one small (ca. 2 mm long), yellow ray flower and 5–6 (8) disk flowers; leaves lanceolate, 3-nerved...***Flaveria***
30b. Pappus of 2–4 retrorsely barbed awns (at least on the inner flowers); receptacle chaffy; heads unlike the above; leaves compound or if lanceolate then not 3-nerved...***Bidens***

31a. Heads in spikes, panicles, or racemes...32
31b. Heads in flat-topped clusters or in a dense, globose cluster...33

32a. Receptacles naked or with long hairs; disk flowers pale yellow...***Artemisia***
32b. Receptacles chaffy; disk flowers whitish...***Oxytenia***

33a. Plants taller, 3–15 dm high; leaves not mostly basal with well-developed stem leaves present, entire or with a few reduced basal pinnae or pinnatifid…***Tanacetum***
33b. Plants shorter, 0.5–2 dm high; leaves 3-toothed or -lobed or deeply 3-lobed with narrow divisions…***Sphaeromeria***

Key 4

(Heads radiate; rays white, blue, purple, or pink)

1a. Very low, mat-forming, essentially stemless perennial herbs; heads small and more or less sessile and hidden among the leaf bases; rays small and inconspicuous, not surpassing the disk flowers…***Parthenium***
1b. Annual or perennial herbs or shrubs without the above set of characteristics…2

2a. Receptacle densely bristly…3
2b. Receptacle chaffy or naked, but not densely bristly…4

3a. Marginal flowers with a falsely radiate corolla; involucre bracts usually with fringed, erose, or spinose margins and tips…***Centaurea***
3b. Marginal flowers truly radiate; involucre bracts lacking fringed, erose, or spinose margins and tips…***Gaillardia***

4a. Pappus of the disk flowers of capillary or barbellate bristles, sometimes also double with a series of short hairs or that of the ray flowers sometimes reduced to a crown; receptacle naked…5
4b. Pappus of scales, awns, a short crown, or absent; receptacle naked or chaffy…29

5a. Leaves mostly basal, triangular, cordate-sagittate at the base, densely white-tomentose on the abaxial side; stem leaves reduced and scale-like…***Petasites***
5b. Stem leaves present and usually not scale-like, basal leaves various but unlike the above…6

6a. Rays short and inconspicuous, barely surpassing the disk flowers; annuals or short-lived perennials…7
6b. Rays well-developed and surpassing the disk flowers; annuals or perennials…10

7a. Involucre 4–12 mm high, the bracts mostly herbaceous with a small chartaceous base…8
7b. Involucre 3–5 mm high, the bracts brown or green and scarcely herbaceous…9

8a. Achenes several-nerved; both basal and stem leaves linear…***Symphyotrichum*** (*ciliatum*)
8b. Achenes 2-nerved; basal leaves oblanceolate to spatulate…***Erigeron*** (*acris*, *lonchophyllus*)

9a. Leaves entire or some of the lower ones irregularly few-toothed, tapering to a short petiole, subglabrous to spreading-hirsute…***Conyza***
9b. Leaves evidently toothed or sublobate, sessile and clasping, glandular or viscid and/or loosely arachnoid-hairy…***Laennecia***

10a. Leaves scale-like, mostly less than 1 cm long, entire with a ciliate margin; involucre bracts green but dry with a thin, hyaline margin; low plants of desert steppes and the eastern plains…***Chaetopappa***
10b. Leaves not scale-like, usually more than 1 cm long, or if small then pinnatifid or toothed; involucre bracts various but not green and dry with a hyaline margin; low or tall plants of various habitats…11

11a. Involucre bracts nearly equal or subequal (the bracts more or less appearing to be in a single series), green and herbaceous throughout and lacking a hard, yellowish or whitish chartaceous base…12
11b. Involucre bracts distinctly imbricate (clearly in 2 or more series), or with a hard, yellowish or whitish chartaceous base…15

12a. Leaves spinulose-toothed; heads solitary, 10–25 mm wide; ray flowers rose-purple to purple…***Xanthisma***
12b. Leaves entire or sometimes toothed, but not spinulose-toothed; heads and ray flowers various…13

13a. Involucre bracts, peduncles, and upper stems and leaves densely covered in glandular hairs; stem leaves linear to linear-lanceolate with revolute margins, sessile; achenes 7–10-nerved; found in moist, alkaline places…***Almutaster***
13b. Plants unlike the above in all respects, variously distributed…14

14a. Corolla lobes and style-branch appendages of disk flowers lanceolate; ray flowers 1.5–3 mm wide; involucre bracts hairy but not glandular-stipitate; pappus a single series of capillary bristles; plants uncommon in subalpine and alpine meadows…***Aster***
14b. Corolla lobes and style-branch appendages of disk flowers triangular; ray flowers usually 1 mm wide or if 1.5–3 mm wide then the involucre bracts densely glandular-stipitate; pappus a single series of capillary bristles or also with shorter outer setae or scales; plants of various habitats…***Erigeron***

15a. Leaves mostly toothed to pinnatifid or pinnately dissected; involucre bracts (at least the lower) often reflexed or spreading and glandular-hairy...16
15b. Leaf margins entire or sometimes just a few of the lower shallowly toothed; involucre bracts usually ascending or sometimes spreading to reflexed, sometimes glandular-hairy...21

16a. Leaf teeth lacking spinulose tips on the margins; involucre bracts with ascending tips...***Symphyotrichum***
16b. Leaf teeth with spinulose tips; involucre bracts often reflexed...17

17a. Leaves deeply pinnatifid to bipinnatifid; heads solitary, large (the disk 6–20 mm in diam. when pressed); ray flowers blue to purple; plants densely glandular-stipitate throughout...***Machaeranthera***
17b. Leaves simple and toothed, or if deeply pinnatifid then plants unlike the above in all respects...18

18a. Leaves rigid, thick, sessile, oblong to oblong-ovate; involucre bracts with ascending tips; known only from Raton Pass in Las Animas County...***Herrickia*** (*horrida*)
18b. Leaves unlike the above in all respects, sessile and narrower and lanceolate or pinnatifid, or petiolate and spatulate to oblanceolate; at least the lower involucre bracts often with reflexed tips; plants variously distributed...19

19a. Heads solitary, large (the disk usually 10–25 mm in diam. when pressed); involucre bracts hairy but not glandular; rays purple or rose-purple...***Xanthisma*** (*coloradense*)
19b. Heads usually smaller (the disk mostly 6–10 mm wide), usually not solitary; involucre bracts glandular or sometimes hairy; rays blue to purple...20

20a. Pappus essentially absent on ray flowers or about half as long as that of the disk flowers; at least the lower leaves usually deeply pinnatifid...***Arida***
20b. Pappus of numerous capillary bristles on both ray and disk flowers; lower leaves not deeply pinnatifid...***Dieteria***

21a. Involucre bracts with distinct scarious, ciliate-fringed margins; plants usually low, rarely exceeding 2 dm tall, often acaulescent; pappus of disk flowers of barbellate bristles or short bristle-like scales, that of ray flowers similar but usually shorter or sometimes reduced to a crown...***Townsendia***
21b. Involucre bracts with less conspicuous scarious margins, these not ciliate-fringed; plants low to tall, caulescent or acaulescent; pappus of numerous capillary bristles or sometimes also with short outer setae or scales...22

22a. Heads large, the disk 12–35 mm wide and involucre 7–20 mm high, solitary on naked or subnaked peduncles; plants of the western slope...***Xylorhiza***
22b. Heads smaller, few to numerous in open clusters or rarely solitary on leafy peduncles; plants of the eastern or western slope...23

23a. Upper stems, leaves, and/or involucre bracts glandular-hairy (sometimes minutely so)...24
23b. Stems, leaves, and involucre bracts glabrous or variously hairy but not glandular...26

24a. Inner involucre bracts purple-tipped, margined, or purple throughout, with a strong purple-green midvein; achenes flattened, laterally 1–2-ribbed, sometimes with 1–2 nerves on each face...***Eucephalus***
24b. Inner involucre bracts with a green and herbaceous tip or purple-tinged but without a strong central midvein; achenes somewhat compressed or round, the margins not ribbed, usually 3–10-nerved on each face...25

25a. Involucre bracts, peduncles, and upper stems and leaves densely covered in glandular hairs; stem leaves linear to linear-lanceolate with revolute margins, sessile; achenes 7–10-nerved; usually found in moist, alkaline places...***Almutaster***
25b. Plants unlike the above in all respects, variously distributed...***Symphyotrichum***

26a. Inner involucre bracts dry and chartaceous, the tips not herbaceous but with a prominent greenish-brown midrib...27
26b. Inner involucre bracts with a green, herbaceous tip or nearly herbaceous throughout...28

27a. Leaves glaucous and glabrous, giving them a grayish-green color; outer involucre bracts obtuse or broadly acutish...***Herrickia*** (*glauca*)
27b. Leaves not glaucous, green; outer involucre bracts acute...***Eucephalus***

28a. Inflorescence flat-topped; larger leaves 3-nerved, linear to linear-lanceolate; at least some pappus bristles subclavate and thickened toward the apex; achenes usually with 8–10 nerves...***Solidago*** (*ptarmicoides*)
28b. Inflorescence various but not flat-topped; leaves variously shaped; pappus bristles not thickened toward the apex; achenes with 2–5 (10) nerves...***Symphyotrichum***

29a. Leaves simple with entire margins…30
29b. At least some leaves compound, pinnatifid, dissected, or simple with toothed or lobed margins…32

30a. Upper stem and involucre bracts glandular-hairy; rays 3–5, with deeply trifid tips; plants of the eastern plains…***Palafoxia***
30b. Upper stem and involucre bracts lacking glandular hairs; rays usually more than 5, lacking deeply trifid tips; plants variously distributed…31

31a. Receptacle chaffy, enlarged and conical or hemispheric in shape; plants usually densely hispid, rarely glabrous, perennial herbs 3 to 12 dm tall…***Echinacea***
31b. Receptacle naked, not enlarged and conical or hemispherical; plants not densely hispid, low and often acaulescent, annual or perennial herbs mostly less than 2 dm tall…***Townsendia***

32a. Leaves mostly basal (with 1–2 reduced stem leaves present)…***Hymenopappus*** (*newberryi*)
32b. Stem leaves well-developed, with more than 2 present…33

33a. Leaves opposite…34
33b. Leaves alternate…36

34a. Leaves pinnately dissected into narrow, filiform segments; achenes evidently beaked; pappus of 2–4 retrorsely hispid awns; ray flowers white, pink, rose, or purple…***Cosmos***
34b. Leaves linear-oblong, simple to pinnately lobed but not dissected into filiform segments; achenes not beaked; pappus absent or of scales; ray flowers white…35

35a. Leaves linear-oblong, entire to pinnately lobed; involucre bracts in 2 series, the outer bracts ovate and fused for about ½ their length, the inner bracts each enclosing a ray achene and expanded upward and hood-like; ray flowers 8–13, the ligule 7–13 mm long; plants of the southeastern plains…***Melampodium***
35b. Leaves ovate, serrate; involucre bracts unlike the above; ray flowers 5–6, the ligule to 2.5 mm long; weedy plants found in gardens and fields along the Front Range…***Galinsoga***

36a. Heads numerous in a flat-topped, umbel-like cyme, small (the involucre 3–5 mm high); ligule mostly less than 4 mm long; leaves finely 1–2-pinnately dissected into fern-like segments…***Achillea***
36b. Heads not in a flat-topped cyme, usually larger (the involucre over 5 mm high), unlike the above; leaves various…37

37a. Ray flowers purple, dark red, or pink; receptacle chaffy, globular, columnar, or conic in shape; pappus of 1–2 toothlike projections or a short crown…38
37b. Ray flowers white; receptacle naked or chaffy, flat, convex, or conical in shape; pappus absent…39

38a. Leaves ovate to lanceolate, with serrate or dentate margins; rays 3–8 cm long; plants usually densely hispid…***Echinacea*** (*purpurea*)
38b. Leaves pinnately dissected into narrow segments; rays 0.3–1 cm long; plants strigose and gland-dotted…***Ratibida*** (*tagetes*)

39a. Leaves simple and toothed or pinnately dissected into wider segments (not linear or filiform); achenes usually 10-ribbed; rhizomatous perennials…***Leucanthemum***
39b. Leaves bipinnatifid into linear or filiform segments; achenes with 2 marginal and 1 ventral, strongly thickened, almost wing-like ribs or 9–10 ribs with the furrows often gland-dotted; annuals or sometimes biennials or short-lived perennials, but not from rhizomes…40

40a. Receptacle naked; achenes with 2 marginal and 1 ventral, strongly thickened, almost wing-like ribs…***Tripleurospermum***
40b. Receptacle chaffy (sometimes only in the center of the head); achenes with 9–10 ribs with the furrows often gland-dotted…***Anthemis***

Key 5

(Heads radiate; rays yellow or orange; pappus of capillary or barbellate bristles, or of short scales and capillary bristles)

1a. Involucre bracts dotted with brownish translucent oil glands; ill-scented annuals; leaves pinnately dissected into narrow segments; rays inconspicuous…***Dyssodia***
1b. Plants unlike the above in all respects…2

2a. Leaves opposite or occasionally the uppermost alternate...***Arnica***
2b. Leaves alternate or all basal...3

3a. Involucre bracts in one row and essentially equal, sometimes with some smaller, calyculate bracteoles at the base of the involucre...4
3b. Involucre bracts imbricate or subequal in 2 or more rows...5

4a. Plants from taprooted or rhizomatous caudices with branching fibrous lateral roots; stem leaves progressively reduced distally; heads erect; involucre bracts rarely black-tipped, and if so then middle stem leaves not clasping...***Packera***
4b. Plants from button-like or lateral rhizomes with unbranched and fleshy fibrous roots; stem leaves basically equally distributed along the stem or if reduced distally then heads nodding and/or involucre bracts with black tips...***Senecio***

5a. Heads numerous and small (the involucre bracts 2–12 mm high) in paniculate or corymbiform inflorescences; involucre bracts chartaceous and sometimes with a greenish or brownish tip...6
5b. Heads few and larger (the involucre bracts over 10 mm high) in various inflorescences; involucre bracts green and herbaceous or with a chartaceous base and green herbaceous tip...8

6a. Leaves variously hairy or glabrous, but not resinous or glandular-punctate, variously shaped, sometimes linear or narrowly lanceolate throughout the stem...***Solidago***
6b. Leaves resinous or glandular-punctate, linear or narrowly lanceolate throughout the stem...7

7a. Suffrutescent herbs from a woody, branched caudex; leaves resin-dotted; achenes glabrous...***Petradoria***
7b. Perennial herbs from rhizomes; leaves glandular-punctate with small, dark spots; achenes hairy...***Euthamia***

8a. Involucre bracts sticky and gummy, firm, the tips reflexed...***Grindelia***
8b. Involucre bracts not sticky and gummy, the tips reflexed or erect...9

9a. Leaves pinnatifid or pinnately dissected or toothed with the teeth bristle-tipped...10
9b. Leaves simple with entire margins or if toothed then without bristle-tips...12

10a. Leaves and/or stems strigose-puberulent to tomentose, narrow (1–10 mm wide), pinnatifid or toothed...***Xanthisma***
10b. Leaves and stems glabrous or sparsely to densely glandular, usually wider (5–20+ mm wide), toothed...11

11a. Leaves glabrous; achenes glabrous; plants known from Baca County...***Prionopsis***
11b. Leaves sparsely to densely glandular; achenes hairy; plants found on the eastern plains...***Rayjacksonia***

12a. Stem leaves well-developed and not greatly reduced in size from the lower leaves...13
12b. Leaves chiefly basal, stem leaves distinctly and greatly reduced in size from the lower leaves...17

13a. Pappus of disk flowers double with the outer of short scales or bristles and inner of capillary bristles, often the pappus of ray flowers absent; leaves hispid-strigose or strigose-canescent with or without glandular hairs...***Heterotheca***
13b. Pappus of disk flowers of a single series of capillary bristles, pappus of ray flowers not absent; leaves glabrous, sparsely hairy, or with ciliate or glandular margins...14

14a. Heads 3 or more in a corymbiform array; involucre bracts imbricate, the outer herbaceous and foliaceous and the inner with a chartaceous base and short foliaceous tip, often at least some of the tips reflexed; leaves ovate-lanceolate to elliptic, usually clasping, glabrous or inconspicuously hairy; plants common in the mountains and in open meadows above 7500 ft...***Oreochrysum***
14b. Heads solitary or rarely 2 per flowering stem, or if more then the heads arranged in a paniculiform or spiciform array, if heads numerous and in a corymbiform array then the leaves linear; involucre bracts various but unlike the above in all respects, the tips erect; plants variously distributed...15

15a. Pappus white; leaves with 3–5 nerves; stems often mat-forming; leaves with ciliate or glandular margins; low plants (0.3–1.5 dm tall) of the alpine...***Tonestus***
15b. Pappus tawny or brown; leaves usually 1-nerved, sometimes 3-nerved in *Oönopsis foliosa*; stems not mat-forming; leaves glabrous to hairy; plants usually of lower elevations...16

16a. Involucre bracts acuminate to cuspidate; leaves linear to narrowly oblong or lanceolate, mostly 4–8 cm long and 5–15 mm wide; plants of the eastern slope...***Oönopsis***
16b. Involucre bracts broad with a bluntly rounded tip, rarely lanceolate with an acute tip; leaves broadly lanceolate to oblanceolate or narrowly elliptic, mostly 10–35 cm long and 10–16 mm wide; plants mostly found on the western slope...***Pyrrocoma*** (*crocea*)

17a. Stems curved or decumbent at the base, usually with 5 or more reduced stem leaves; pappus usually tawny or brown...***Pyrrocoma***
17b. Stems erect, leafless or with fewer than 5 stem leaves; pappus white...***Stenotus***

Key 6

(Heads radiate; rays yellow or orange; pappus of awns, scales, a short crown, or absent)

1a. Involucre bracts sticky and gummy, firm, the tips reflexed; pappus of 2–several separate, firm, deciduous awns...***Grindelia***
1b. Involucre bracts not sticky, gummy, and firm, the tips not reflexed or sometimes reflexed-spreading in age; pappus various...2

2a. Involucre bracts and leaves with conspicuous yellow-brownish oil glands; annuals, often aromatic and ill-scented plants...3
2b. Involucre bracts and leaves without conspicuous yellow-brownish oil glands, although sometimes finely glandular-punctate; annuals or perenniasl, aromatic or not...5

3a. Involucre bracts uniseriate, narrowly linear and strongly keeled to the tip; leaves nearly linear...***Pectis***
3b. Involucre bracts in 2 or more series, not narrowly linear and strongly keeled; leaves pinnately dissected...4

4a. Ray flowers conspicuous and showy, yellow, 4–6 mm long; involucre bracts fused for more than ½ their length; plants not ill-scented...***Thymophylla***
4b. Ray flowers inconspicuous (1.5–2.5 mm long); plants usually ill-scented (fetid marigold)...***Dyssodia***

5a. Involucre bracts broad and hemispheric, herbaceous throughout, imbricate in a few series; ray flowers yellow with red longitudinal lines on the abaxial side...***Berlandiera***
5b. Involucre bracts not broad, herbaceous, and hemispheric; ray flowers with or without red longitudinal lines...6

6a. Leaves alternate, deeply pinnatifid to laciniate, the lowermost leaves up to 4 dm long; heads several in an elongated raceme; involucre bracts all more or less the same and herbaceous; plants 4–30 dm tall, known from Yuma and El Paso counties...***Silphium***
6b. Plants unlike the above in all respects, variously distributed...7

7a. Involucre bracts in two distinct, dissimilar series (typically with an outer series of linear, foliaceous bracts and an inner series of oval, often membranous, striate bracts); achenes of disk flowers flattened parallel to the involucre bracts (at right angles to the radius of the head); leaves toothed to dissected or laciniate, not 3-nerved...8
7b. Involucre bracts subequal or imbricate in two or more series and not conspicuously dimorphic, or if appearing dimorphic then the leaves 3-nerved (look on the abaxial side at the base) and entire to slightly toothed on the margins; achenes of disk flowers either not much flattened or flattened at right angles to the involucre bracts (parallel to the radius of the head)...12

8a. Leaves alternate, pinnatifid or laciniate; plants of the eastern plains...***Engelmannia***
8b. Leaves opposite, simple or pinnatifid to compound; plants variously distributed...9

9a. Achenes of two different kinds, the outer oval, incurved, with winged margins and devoid of pappus, and the inner slender, attenuate, not winged, with a pappus of 2 or 3 widely divergent awns; ray flowers few, usually about 2 or 3; known from Las Animas County...***Heterosperma***
9b. Achenes all similar; ray flowers more than 3; variously distributed...10

10a. Inner involucre bracts basally connate for about ⅓ of their length...***Thelesperma***
10b. Inner involucre bracts distinct...11

11a. Pappus of 2–4 retrorsely barbed awns...***Bidens***
11b. Pappus of short teeth or absent...***Coreopsis***

12a. Ray flowers small (surpassing the disk flowers but only 1–3 mm long) or inconspicuous (shorter than or barely surpassing the disk flowers)...13
12b. Ray flowers longer than 3 mm, conspicuous and surpassing the disk flowers...16

13a. Leaves pinnately dissected; receptacle chaffy; heads arranged in a flat-topped or dome-shaped corymbiform cyme...***Achillea***
13b. Leaves entire or merely toothed, but not dissected; receptacle naked or with a single series of chaff between the ray and disk flowers; heads in a flat-topped corymbiform cyme or not...14

14a. Shrubs or subshrubs; leaves all alternate, linear or linear-filiform; heads small, the involucre 2–5 mm high...***Gutierrezia***
14b. Annual or perennial herbs; leaves opposite at least below, linear to lanceolate; heads usually larger, the involucre 5–9 mm high...15

15a. Leaves decussately opposite, lanceolate with serrulate margins, 3-nerved; involucre bracts usually 3, not clasping the achenes; ray flowers one per head...***Flaveria***
15b. Leaves opposite below and alternate above, narrowly lanceolate with entire margins, 1-nerved; involucre bracts mostly 4, clasping the achenes such that the heads appear deeply furrowed; ray flowers 1–3 per head...***Madia***

16a. Receptacle alveolate with numerous long, stiff bristles that do not individually subtend the flowers; pappus of scales, usually each with a long, prominent awn...***Gaillardia***
16b. Receptacle naked, with a few scattered bristles, or chaffy; pappus various...17

17a. Receptacle naked, with a few scattered bristles, or with a single series of chaff between the ray and disk flowers near the edge of the head...18
17b. Receptacle chaffy, at least in the center...25

18a. Leaves opposite, ternately dissected into linear segements which are entire or dissected again or toothed, canescent-puberulent...***Picradeniopsis***
18b. Leaves alternate or mostly basal, and unlike the above...19

19a. Pappus absent; leaves pinnatifid...20
19b. Pappus of scales, sometimes these with a prominent awn tip; leaves entire or pinnatifid...21

20a. Stem glandular-hairy, especially above; involucre bracts herbaceous, elliptic to obovate with an abruptly acuminate apex, glandular-hairy or somewhat viscid...***Bahia***
20b. Stem glabrous; involucre bracts mostly scarious with a green midrib, ovate to oblong without an abruptly acuminate apex, glabrous...***Chrysanthemum***

21a. Rays persistent on the achenes and becoming papery; pappus of broad, round-tipped scales; receptacle flat; plants usually white villous-tomentose throughout, not resinous-punctate...***Psilostrophe***
21b. Rays deciduous and not persistent on the achenes or becoming papery; pappus of scales that are either aristate or awn-tipped or with a thickened midrib; receptacle convex to subglobose or hemispheric; plants variously hairy or glabrous, often resinous-punctate with impressed glands...22

22a. Involucre bracts and ray flowers reflexed at maturity; stem leaves well-developed...***Helenium***
22b. Involucre bracts erect at maturity and ray flowers erect to spreading at maturity; leaves all basal or the stem leaves well-developed...23

23a. Peduncle conspicuously glandular; receptacle flat or slightly convex; leaves simple, lance-elliptic to broadly ovate or subrotund, mostly 2–10 cm long and 0.5–3.5 cm wide, glabrous to shortly strigose below; plants of the western slope...***Platyschkuhria***
23b. Peduncle glabrous to hairy but not glandular; receptacle hemispheric to conic; leaves simple and usually narrower or if wider then unlike the above, or pinnately dissected, variously hairy or glabrous; plants of the eastern or western slope...24

24a. Leaves simple, all basal or basal and cauline in one species, linear to narrowly oblanceolate or spatulate; plants usually with a dense tuft of brownish or white hairs at the base of the leaves...***Tetraneuris***
24b. Leaves pinnatifid or divided into 2–5 linear lobes or 3–7 segments, or if simple then oblanceolate and 10–30 cm long, basal and cauline; plants sometimes densely hairy in old leaf bases...***Hymenoxys***

25a. Plants scapose or subscapose with the leaves essentially all basal…26
25b. Plants leafy-stemmed, although the basal ones may be larger than the stem leaves…27

26a. Leaves simple, entire, round or rotund-ovate; ray flowers neutral and sterile; plants known from Mesa County…***Enceliopsis***
26b. Leaves pinnatifid or simple but triangular-hastate; ray flowers pistillate and fertile; plants variously distributed but not restricted to Mesa County…***Balsamorhiza***

27a. Leaves pinnately dissected or trilobed…28
27b. Leaves simple and not dissected, although the margins may be toothed…30

28a. Leaves fern-like and pinnatifid into narrow segments (mostly 1–3 mm wide), these segments again deeply toothed or cleft; receptacle hemispheric; introduced plants…***Anthemis***
28b. Leaves pinnatifid into wider segments, these sometimes toothed or cleft, or if narrower then the receptacle long and columnar and the segments not toothed; native plants…29

29a. Leaves pinnatifid into linear, narrow segments; receptacle columnar (cone flower), 1.5–4.5 cm long and 2–4.5 times as long as wide; both the disk and ray flowers subtended by chaffy bracts…***Ratibida***
29b. Leaves pinnatifid into lanceolate or elliptic, wider segments, or trilobed; receptacle hemispheric, up to 4 cm long in fruit but usually less than 2–4.5 times as long as wide (not columnar); only the disk flowers subtended by chaffy bracts…***Rudbeckia***

30a. Involucre bracts imbricate in several series, broad, pale and scarious with a greenish band near the tip; chaffy bracts enclosing the achenes; ray flowers yellow, becoming papery, and persistent on the achenes; disk flowers red or green; leaves opposite, linear, entire, and 3-nerved; plants of the southeastern plains…***Zinnia***
30b. Involucre bracts subequal or imbricate, mostly green and herbaceous; otherwise plants unlike the above in all respects, variously distributed…31

31a. Leaves softly strigose-canescent and whitish below, deltoid-ovate, the margins toothed or subentire; achenes conspicuously wing-margined…***Verbesina***
31b. Leaves not strigose-canescent and whitish below, variously shaped; achenes not wing-margined or only thinly or slightly wing-margined…32

32a. Receptacle conic and elongating at maturity to about 2 cm long and 2.5 cm wide; ray flowers persistent on the achenes; leaves opposite…***Heliopsis***
32b. Receptacle flat, convex, or hemispheric, less than 2 cm long; ray flowers not persistent on the achenes; leaves alternate or opposite below and alternate above…33

33a. Pappus of 2 awns or scales, occasionally with additional smaller scales between the awns, or a crown of scales often prolonged into awns…34
33b. Pappus absent…36

34a. Achenes strongly compressed at right angles to the involucre bracts, thin-edged or slightly wing-margined; pappus of 2 persistent slender awns or scales…***Helianthella***
34b. Achenes thick and compressed-quadrangular or moderately compressed at right angles to the involucre bracts, not wing-margined or thin-edged; pappus of 2 early-deciduous awns with smaller scales between them or a crown of scales often prolonged into awns…35

35a. Plants from deep-seated taproots surmounted by a simple or much-branched woody caudex; ray flowers pistillate and fertile with well-developed styles; pappus a crown of scales often prolonged into awns…***Wyethia***
35b. Plants from taproots, rhizomes with or without tubers, or a cluster of fibrous roots; ray flowers neutral and sterile without functional styles; pappus of 2 early-deciduous awns with some smaller scales between…***Helianthus***

36a. Plants strigose to minutely hairy; leaves opposite below and alternate above, usually with smaller leaves in the leaf axils; stem usually red…***Heliomeris***
36b. Plants hispid and rough with spreading, stiff hairs; leaves alternate, without smaller leaves in the leaf axils; stem usually green…***Rudbeckia*** (*hirta*)

ACHILLEA L. – YARROW

Perennial herbs; *leaves* alternate, pinnately dissected or entire; *heads* radiate; *involucre bracts* imbricate in several series, dry, with scarious or hyaline margins and a green midrib; *receptacle* chaffy; *ray flowers* pistillate and fertile, white or occasionally pink or yellow; *pappus* none; *achenes* flattened parallel to the involucral bracts. (Mulligan & Bassett, 1959; Tryl, 1975)

1a. Leaves linear to narrowly lanceolate, unlobed and not finely pinnately dissected...***A. ptarmica***
1b. Leaves lobed or dissected...2

2a. Ray flowers yellow...***A. filipendulina***
2b. Ray flowers white to pinkish...***A. millefolium***

Achillea filipendulina **Lam.**, FERNLEAF YARROW. An introduced plant escaped from cultivation in Boulder and along I-70 especially in Clear Creek County near Idaho Springs, 5500–7500 ft. June–Aug. E. Introduced.

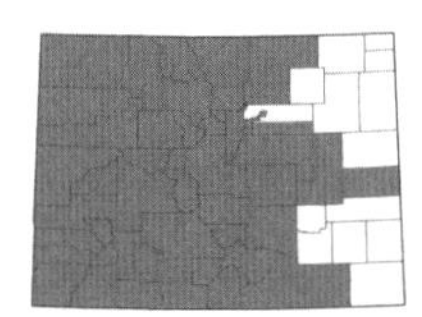

Achillea millefolium **L.**, COMMON YARROW. Plants 0.6–6.5 dm, densely tomentose to glabrate; *leaves* lanceolate to oblong, 3.5–35 × 0.5–3.5 cm, finely 1-2-pinnately dissected; *heads* 10–100, in a corymbiform array; *involucre bracts* in 3 series; *rays* white or light pink, 1.5–3 mm long. Common in gravelly soil along roadsides, in prairies, on open slopes, and in mountain meadows. A highly variable species occurring from the high plains to the alpine, 4800–13,200 ft. June–Oct. E/W. (Plate 8)

Achillea ptarmica **L.**, SNEEZEWEED. Plants 3–6 dm, glabrate to villous or tomentose; *leaves* linear to narrowly lanceolate, 3–10 × 0.3–0.5 cm, serrulate-crenate; *heads* 3–20, in corymbiform arrays; *involucre bracts* in 3 series; *rays* white, 4–6 mm long. Uncommon weed found escaped from cultivation along roadsides, 5500–8000 ft. June–Sept. E. Introduced.

ACROPTILON Cass. – HARDHEADS

Perennials; *leaves* alternate, basal; *heads* discoid; *involucre bracts* in 6–8 series, the outer with scarious tips, the inner with ciliate or plumose tips; *receptacle* densely bristly; *disk flowers* blue, pink, or white; *ray flowers* absent; *pappus* of unequal flattened bristles, barbed at the tips; *achenes* slightly compressed, glabrous, smooth.

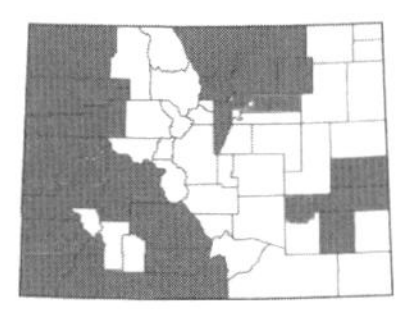

Acroptilon repens **(L.) DC.**, RUSSIAN KNAPWEED. [*Centaurea repens* L.]. Plants 2–10 dm, from rhizomes, finely arachnoid-tomentose; *basal and lower cauline leaves* oblong, 4–15 cm long; *upper cauline leaves* narrowly lanceolate or oblong, 1–7 cm long; *involucre bracts* 9–17 mm long; *corollas* 11–14 mm long, white to pink or purple; *achenes* 3–4 mm long. Found in disturbed places and along roadsides, 3900–7500 ft. June–Sept. E/W. Introduced.

AGERATINA SPACH – SNAKEROOT

Herbs or shrubs; *leaves* mostly opposite; *heads* discoid; *involucre bracts* imbricate or subequal, green to chartaceous or weakly coriaceous; *receptacle* naked, flat or conic; *disk flowers* blue, pink, purple, or white; *ray flowers* absent; *pappus* of capillary bristles; *achenes* prismatic, usually 5-angled.

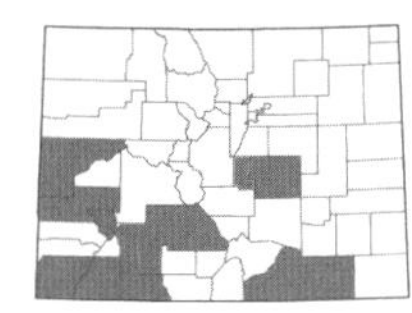

Ageratina herbacea **(A. Gray) R.M. King & H. Rob.**, FRAGRANT SNAKEROOT. [*Eupatorium herbaceum* (A. Gray) Green]. Plants (2) 3–6 (8) dm, from a woody caudex; *leaves* opposite, deltoid to widely lanceolate or ovate, 2–5 (7) cm long, dentate, abaxially hispidulous to glabrate; *involucre bracts* 4–5 mm long; *disk flowers* white; *achenes* 2–3 mm long. Found on rocky hillsides, in forest openings, and in meadows, 5200–9000 ft. Aug.–Oct. E/W.

AGOSERIS Raf. – FALSE DANDELION; MOUNTAIN DANDELION

Lactiferous herbs; *leaves* mostly basal, entire to pinnatifid; *heads* solitary; *involucre bracts* subequal or imbricate, subherbaceous or partly scarious; *receptacle* naked or chaffy; *disk flowers* absent; *ray flowers* yellow, orange, or sometimes pinkish (especially when dry); *pappus* of capillary bristles; *achenes* prominently nerved (usually 10), usually beaked. (G.I. Baird, 2006)

1a. Annuals with slender taproots; hairs on the involucre bracts multicellular with purple cross-walls...***A. heterophylla* var. *heterophylla***
1b. Perennials; hairs on the involucre bracts white without purple cross-walls...2

2a. Ray flowers orange (although often drying purplish); achene beak slender, (2) 5–10 mm long...***A. aurantiaca***
2b. Ray flowers yellow (often drying pinkish); achene beak usually stout, 1–4 mm long, or if beak longer than the body then leaves laciniate (deeply cleft or toothed along the margin)...3

3a. Leaves laciniate (deeply cleft or toothed along the margin); achene beak 3–10 mm long...***A. parviflora***
3b. Leaves not laciniate, the margins usually entire but sometimes weakly toothed; achene beak 1–4 mm long...***A. glauca***

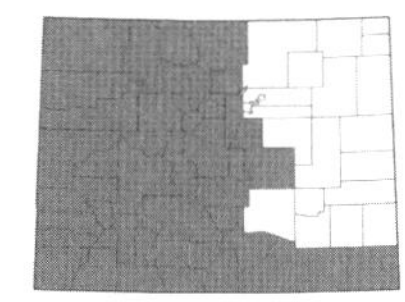

***Agoseris aurantiaca* (Hook.) Greene**, ORANGE AGOSERIS. Perennials; *leaves* 7–38 cm long, lanceolate to oblanceolate, the margins entire to laciniately pinnatifid, glabrous to sparsely villous; *involucre bracts* often with purplish-black splotches; *rays* usually orange, 4–12 mm long; *achenes* 8–18 mm long, abruptly or gradually tapered to slender beaks (2) 5–10 mm long. Common in forest openings, mountain meadows, and on subalpine slopes, 5400–12,500 ft. June–Aug. E/W.

There are two varieties of *A. aurantiaca* in Colorado:
1a. Involucre bracts conspicuously imbricate, broader (most 3–6 mm wide), usually abruptly narrowed or rounded at the tip...**var. *purpurea* (A. Gray) Cronquist**
1b. Involucre bracts inconspicuously imbricate or not imbricate, narrower (most under 3 mm wide), usually gradually tapering to the tip...**var. *aurantiaca***

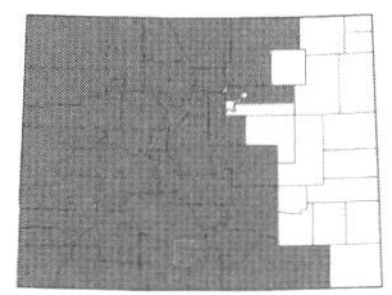

***Agoseris glauca* (Pursh) Raf.**, PALE AGOSERIS. Perennials; *leaves* 2–46 cm long, lanceolate to oblanceolate, entire or weakly toothed; *involucre bracts* often with purplish-black spots; *rays* yellow, 6–24 mm long; *achenes* 7–15 mm long, tapered to stout beaks 1–4 mm long. Common in open, often rocky places and along streams in the mountains, in sagebrush, pinyon-juniper, and oak communities, or in open meadows, 5500–13,800 ft. June–Aug. E/W. (Plate 8)

There are two basically well-defined varieties of *A. glauca* in Colorado. *Agoseris parviflora* is sometimes referred to as a variety of *A. glauca* (var. *laciniata*); however it differs in several morphological characteristics. Hybrids can occur between the two species, but *A. parviflora* can also form hybrids with *A. aurantiaca* as well:
1a. Leaves glabrous or sometimes sparsely ciliate along the margins, glabrous just below the involucre...**var. *glauca***, PALE AGOSERIS. In open, often rocky places and along streams in the mountains, 7000–12,500 ft. June–Aug. E/W.
1b. Leaves and the surface just below the involucre hairy...**var. *dasycephala* (Torr. & A. Gray) Jeps.**, PALE AGOSERIS. [including *A. glauca* (Pursh) Raf. var. *agrestis* (Osterh). Q. Jones ex Cronquist]. Common in open meadows, 5500–12,500 ft. June–Aug. E/W.

Agoseris heterophylla* (Nutt.) Greene var. *heterophylla, ANNUAL AGOSERIS. Annuals; *leaves* oblanceolate to spatulate, 1–25 cm long, entire or lobed, glabrous to densely hairy; *involucre bracts* glandular-stipitate or sometimes glabrous; *rays* yellow, 2–15 mm long; *achenes* 7–16 mm long tapering to a beak 5–11 mm long. Dry, open slopes and in sagebrush communities; known only from Moffat Co. in northwestern Colorado, 7500–7600 ft. May–June. W.

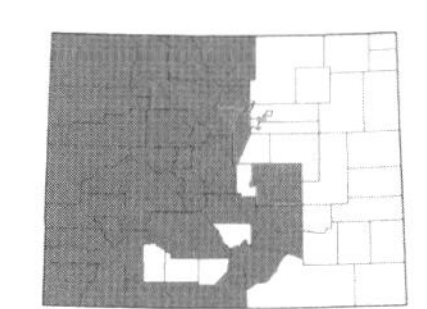

***Agoseris parviflora* (Nutt.) D. Dietr.**, STEPPE AGOSERIS. [*A. glauca* (Pursh) Raf. var. *laciniata* (DC. Eaton) Kuntze]. Perennials; *leaves* linear-lanceolate to oblanceolate, 5–25 cm long, usually lobed or sometimes entire; *involucre bracts* eglandular; *rays* yellow, 10–20 mm long; *achenes* 9–18 mm long, narrowed to a beak 3–10 mm long. Common in sagebrush, pinyon-juniper, and oak communities, often in clay soil, 5500–12,500 ft. June–Aug. E/W.

ALMUTASTER Á. Löve & D. Löve – ALKALI MARSH ASTER

Perennial herbs; *leaves* alternate, simple; *heads* radiate; *involucre bracts* in 3–4 series, subequal, densely short glandular-stipitate, herbaceous with a scarious margin; *receptacle* naked; *disk flowers* yellow; *ray flowers* white to purple; *pappus* of a single series of barbellate bristles; *achenes* 7–10-nerved.

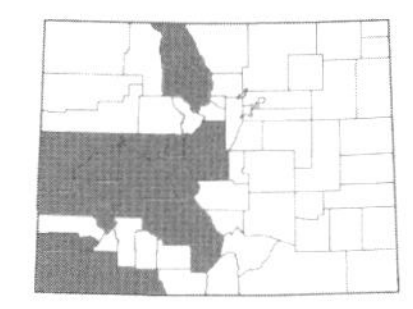

***Almutaster pauciflorus* (Nutt.) Á. Löve & D. Löve**, FEW-FLOWERED ASTER. [*Aster pauciflorus* Nutt.]. Plants 3–12 dm, with lengthy rhizomes; *leaves* linear, 6–10 cm long, glabrous; *involucre* 4–7 mm high; *corollas* of the disk flowers 4–6 mm long; *achenes* hairy. Found in meadows or moist places especially where the soil is alkaline, 4700–10,700 ft. July–Sept. E/W.

AMBROSIA L. – RAGWEED

Caulescent herbs or subshrubs; *leaves* opposite or alternate, entire to lobed or pinnatifid; *heads* unisexual, small, discoid; *involucre bracts* of staminate heads in a single series, more or less connate and unarmed, bracts of pistillate heads fused into a hard, bur-like structure; *receptacle* of staminate heads chaffy; *disk flowers* numerous in staminate heads, 1 or 2 in pistillate heads; *ray flowers* absent; *pappus* absent; *achenes* wholly enclosed in the involucre body. (Payne, 1963, 1964)
1a. Leaves large (up to 25 cm in length), 3–5 cleft or unlobed, opposite; margins serrate and leaves green...***A. trifida***
1b. Leaves smaller, variously cleft or pinnatifid usually into numerous segments, alternate or opposite only from the base to the middle of the plant; margins entire or if toothed then the leaves gray...2

2a. Leaves with silvery-gray or whitish hairs (at least below)...3
2b. Leaves green on both the upper and lower surface...6

3a. Leaves with broad lobes, margins serrate...***A. grayi***
3b. Leaves with narrow and often linear lobes, margins entire to revolute...4

4a. Leaf margins entire, both the upper and lower leaf surfaces grayish...***A. confertifolia***
4b. Leaf margins often revolute, upper leaf surface green and lower surface white-woolly hairy...5

5a. Leaves sessile, 1.2–3.5 cm long; pistillate heads normally 1-flowered...***A. linearis***
5b. Leaves petiolate, 5–10 (18) cm long; pistillate heads normally 2-flowered...***A. tomentosa***

6a. Staminate involucre with dark ridges on some or all of the lobes...***A. acanthicarpa***
6b. Staminate involucre without dark ridges on the lobes...7

7a. Lobes of the staminate flowers with dark-lined margins; taprooted annuals...***A. artemisiifolia***
7b. Lobes of the staminate flowers without dark-lined margins; perennials from creeping rootstock...8

8a. Pistillate involucre usually with 2–several series of spines; ultimate leaf divisions very narrow...***A. confertifolia***
8b. Pistillate involucre with 1 series of short tubercles or these reduced and obsolete; ultimate leaf divisions wider...***A. psilostachya***

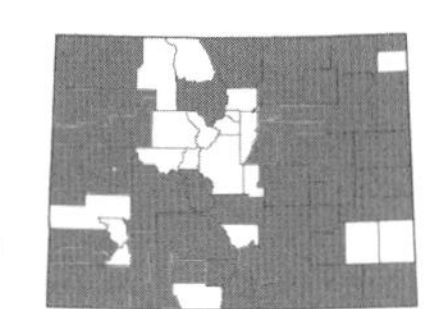

***Ambrosia acanthicarpa* Hook.**, ANNUAL BURSAGE. Annuals, 1–8 dm; *leaves* deltate, opposite below and alternate above, 2–8 cm long, 1–2-pinnately lobed, hispid to sericeous below, hispid and gland-dotted above; *pistillate heads* with 1 floret, spines in several series; *staminate heads* with thickened dark strips directed into the upper lobes. Common in open places, often in sandy soil, 4000–6500 ft. July–Sept. E/W. (Plate 8)

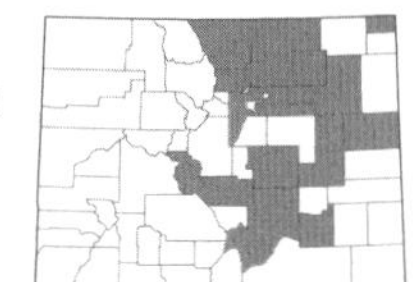

***Ambrosia artemisiifolia* L.**, COMMON RAGWEED. [*A. elatior* L.]. Annuals, 1–6 (15) dm; *leaves* opposite below and alternate above, 2–6 cm long, deltate to elliptic, 1–2-pinnately lobed, coarse-hairy and gland-dotted; *pistillate heads* with 1 floret, and a single series of erect spines on the involucre; *staminate heads* usually without black strips. Common in disturbed areas and open fields of the plains and foothills, 4000–7000 ft. July–Sept. E.

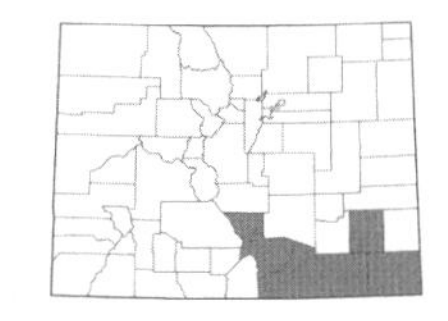

***Ambrosia confertifolia* DC.**, WEAKLEAF BURR RAGWEED. Perennials, 2–8 (15) dm; *leaves* mostly alternate, lanceolate to ovate, 2–4 pinnately lobed, 4–9 (15) cm long, coarse-hairy to sericeous and gland-dotted; *pistillate heads* with 1 floret, and spines in several series on the involucre; *staminate heads* with resinous-dotted or short-hairy involucre bracts. Locally common in the southeastern counties in disturbed open sites such as fallow fields, 3800–5000 ft. Aug.–Oct. E.

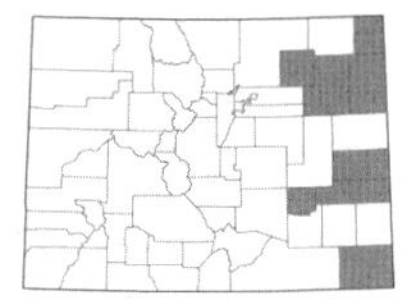

***Ambrosia grayi* (A. Nelson) Shinners**, WOOLLYLEAF BURR RAGWEED. Perennials, 1–4 dm; *leaves* mostly alternate, elliptic to ovate, 4–6 (10) cm long, 1–2-pinnately lobed, coarse-hairy to sericeous; *pistillate heads* with 2 florets, spines in several series on the involucre; *staminate heads* with tomentose, 5–9-lobed involucre bracts. Locally common on the eastern plains in open fields and along roadsides, 3500–4500 ft. July–Sept. E.

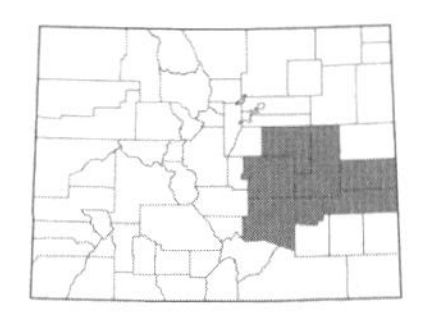

***Ambrosia linearis* (Rydb.) W.W. Payne**, STREAKED BURR RAGWEED. Perennials, 2–5 dm; *leaves* mostly alternate, linear or some 1-pinnate, 1.2–3.5 cm long, with entire margins, strigillose; *pistillate heads* with 1 floret, spines in several series on the involucre; *staminate heads* with short-hairy, 5–9-lobed involucre bracts. Rare in sandy or sandy clay soil on the eastern plains, often in moist places such as along streams or pond margins, 4300–6700 ft. June–Aug. E. Endemic.

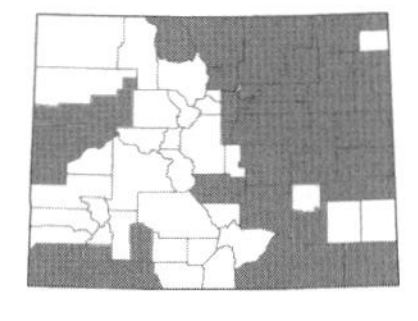

***Ambrosia psilostachya* DC.**, WESTERN RAGWEED. [*A. coronopifolia* Torr. & A. Gray]. Perennials, 1–7 (10) dm; *leaves* opposite below, alternate above, lanceolate to deltate, 2–7 (14) cm long, pinnatifid to 1-pinnately lobed; *pistillate heads* with 1 floret, spines in a single series of short tubercles on the involucre or obsolete; *pistillate heads* with a single floret, proximal to staminate heads; *staminate heads* with hispid involucre bracts, 5–15(30+) florets. Common in disturbed sites and open fields, often in moist soil, 4000–6200 ft. July–Sept. E/W.

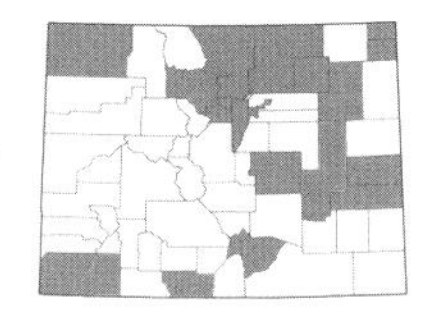

***Ambrosia tomentosa* Nutt.**, PERENNIAL BURSAGE; SKELETON-LEAF BURSAGE. Perennials, 1–4 dm; *leaves* mostly alternate, elliptic, 5–10 (18) cm long, 1-3-pinnately lobed, silvery-gray; *pistillate heads* with 2 florets, spines in several series on the involucre; *staminate heads* with short-hairy involucre bracts. Common along roadsides and streams and in disturbed places from the plains to upper foothills, 4000–8600 ft. June–Sept. E/W.

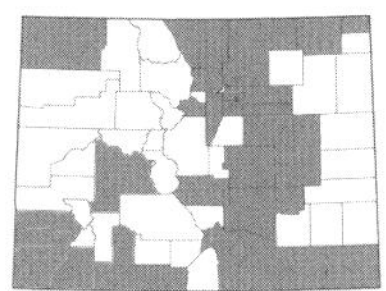

Ambrosia trifida* L. var. *trifida, GREAT RAGWEED. Annuals, 3–15 (40) dm; *leaves* mostly opposite, palmately 3-5-lobed, 4–20 (25) cm long; *pistillate heads* with 1 floret, each ridge of the involucre ending in a small spine; *staminate heads* with involucre bracts with 3 dark strips from the center to the margin. Common in disturbed areas such as along ditches, stream banks, and roadsides, especially in moist soil, 4000–8300 ft. July–Sept. E/W. (Plate 8)

ANAPHALIS DC. – PEARLY EVERLASTING

Caulescent, dioecious or polygamo-dioecious, perennial herbs; *leaves* alternate, simple, entire; *heads* discoid; *involucre bracts* imbricate in several series, the outer series ovate, white and scarious, the inner series less scarious; *receptacle* naked; *disk flowers* numerous; *ray flowers* absent; *pappus* of capillary bristles; *achenes* glabrous.

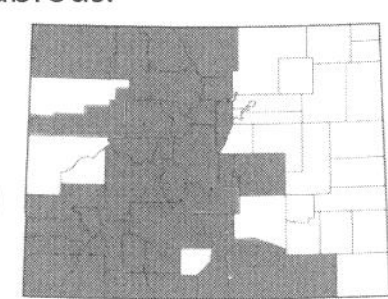

***Anaphalis margaritacea* (L.) Benth. & Hook.**, PEARLY EVERLASTING. Plants from slender rhizomes, 2–8 (12) dm; *leaves* linear to widely lanceolate, 3–10 (15) cm long, tomentose or glabrate below, green and glabrate above; *involucre* 5–7 mm high; *achenes* 0.5–1 mm long. Common in mountain meadows and forest openings from the montane to subalpine, 7800–11,500 ft. July–Sept. E/W. (Plate 8)

ANTENNARIA Gaertn. – PUSSY-TOES; EVERLASTING

Dioecious, white-woolly, perennial herbs; *leaves* simple, mostly entire; *heads* unisexual, discoid; *involucre bracts* imbricate in several series, scarious and dry at least at the tip, white or sometimes colored; *receptacle* naked; *disk flowers* numerous; *ray flowers* absent; *pappus* of capillary bristles; *achenes* round or compressed. (Bayer, 1982, 1987)

1a. Heads solitary and terminal; plants low, mostly less than 10 cm tall...2
1b. Heads few to many in capitate or corymbose clusters; plants usually well over 10 cm in height...4

2a. Plants lacking stolons; involucre bracts blackish-green or brownish and often with a colorless margin and apex...***A. dimorpha***
2b. Plants with stolons (*careful*—these short in *A. rosulata*!); involucre bracts (at least the inner ones) white or occasionally pink...3

3a. Leaves 5–10 mm long; stems often 1 cm long or less; the heads scarcely elevated above the basal leaves...***A. rosulata***
3b. Leaves 10–35 mm long; stems typically over 1 cm in length; the heads usually elevated well above the basal leaves...***A. parvifolia***

4a. Plants lacking stolons, often with several stems arising from a branched, subterranean rhizome or caudex...5
4b. Plants stoloniferous, usually mat-forming...7

5a. Involucre bracts glabrous or nearly so, the lower portion of the bracts scarious, not white (the bracts usually light brown or greenish-brown below and white or pink at the tips) ...***A. luzuloides* ssp. *luzuloides***
5b. Involucre bracts hairy, especially at the base, the lower portion of the bracts white (sometimes with a dark spot at the base) or brownish...6

6a. Involucre white, although sometimes with a small brown spot at the base, 6–8 mm high...***A. anaphaloides***
6b. Involucre darker, brown or black in color, although sometimes with a small whitish tip, 5–12 mm high...***A. pulcherrima***

7a. Leaves glabrous and green above, or distinctly less hairy than the lower side...8
7b. Leaves white-tomentose on both sides, with more or less uniform hairs...9

8a. Involucre bracts widest above the middle, the tips abruptly acute; generally smaller, usually about 15 cm or less in height...***A. marginata***
8b. Involucre bracts narrow and pointed; generally more robust and taller, 10–40 cm in height...***A. neglecta***

9a. Plants small and low, less than 3 cm tall, the heads scarcely elevated above the basal leaves...***A. rosulata***
9b. Plants taller, the heads well above the basal leaves...10

10a. Involucre tips of the scarious bracts white or pink (sometimes with a dark spot at the base); plants generally found from the foothills to montane, only seldom alpine...11
10b. Involucre tips of the scarious bracts (at least the outer and middle ones) brown (sometimes streaked with pink) to blackish, rarely the tips all white or yellowish; plants generally of the subalpine and alpine...14

11a. Involucre bracts with a conspicuous dark spot at the base of the scarious portion; basal leaves narrowly oblanceolate...***A. corymbosa***
11b. Involucre bracts lacking a conspicuous dark spot at the base of the scarious portion; basal leaves usually broader, cuneate-oblanceolate or spatulate...12

12a. Involucre 7–11 mm in height (heads larger); plants 2–15 cm tall...***A. parvifolia***
12b. Involucre usually less than 8 mm in height (heads smaller); plants 10–45 cm tall...13

13a. Involucre bracts rose- or pinkish-colored at least near the apex...***A. rosea***
13b. Involucre bracts whitish or yellowish-green at the apex...***A. microphylla***

14a. Outer involucre bracts blackish or blackish-green throughout, usually sharp-pointed at the apex...***A. media***
14b. Outer involucre bracts brown, or rarely white or yellowish, usually blunt at the apex...***A. umbrinella***

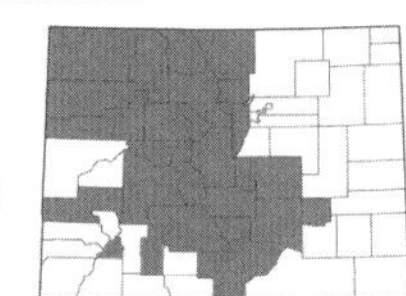

***Antennaria anaphaloides* Rydb.**, PEARLY PUSSY-TOES. [*A. pulcherrima* (Hook.) Greene ssp. *anaphaloides* (Rydb.) W.A. Weber]. Plants 20–50 cm, lacking stolons; *leaves* lance-elliptic, 5–20 × 0.5–2 cm, loosely tomentose; *heads* several in a corymb; *involucre* 6–8 mm high, the bracts white with a brown spot at the base. Common in mountain meadows, dry woods, aspen forests, and sagebrush communities, 6000–12,000 ft. June–Aug. E/W.

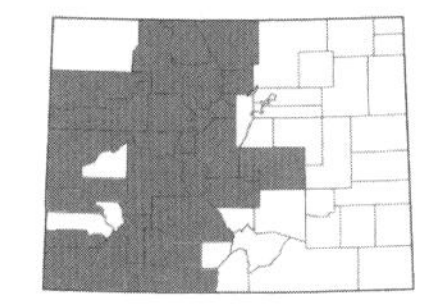

***Antennaria corymbosa* E.E. Nelson**, FLAT-TOP PUSSY-TOES. Plants 15–30 cm, with stolons; *leaves* 3–3.5 cm long, densely white-tomentose; *heads* few in a small corymb; *involucre* 4–5 mm high, the bracts with a brown spot at the base with white tips. Common along streambanks, in moist meadows, and in open woods, 8000–11,200 ft. June–Aug. E/W.

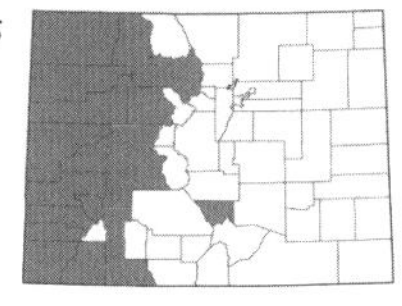

***Antennaria dimorpha* (Nutt.) Torr. & A. Gray**, LOW PUSSY-TOES. Plants 1–3 cm, lacking stolons; *leaves* linear to oblanceolate, 1–3 cm long, white-tomentose; *heads* solitary; *involucre* 5–7 mm high, the bracts brown or blackish-green distally. Dry, rocky, open places in the lower elevations, often in sagebrush communities, 6000–8000 ft. May–June. W.

Antennaria luzuloides* Torr. & A. Gray ssp. *luzuloides, RUSH PUSSY-TOES. Plants 15–70 cm, lacking stolons; *leaves* linear-oblanceolate to oblanceolate, 2–8 cm long, tomentose; *heads* numerous; *involucre* 4–5 mm high, the bracts greenish or brown below and white above. Found in open meadows, known from Grand and Routt cos., 8000–9300 ft. June–Aug. E/W.

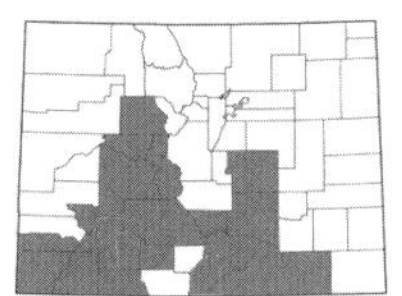

***Antennaria marginata* Greene**, WHITEMARGIN PUSSY-TOES. [*A. fendleri* Greene]. Plants 2–15 (25) cm, with stolons; *leaves* oblanceolate to spatulate, 1–3 cm long, green and glabrous above and white-tomentose below; *heads* numerous in a compact cluster; *involucre* 6–9 mm high, the bracts brownish at the base and white above. Found in dry places, often with ponderosa or pinyon pine, 5100–8800 ft. June–Aug. E/W.

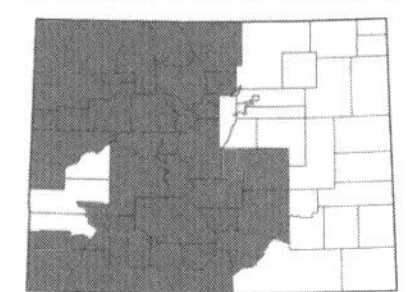

***Antennaria media* Greene**, ROCKY MOUNTAIN PUSSY-TOES. Plants 5–15 cm, with stolons; *leaves* 0.6–2.5 cm long, oblanceolate to spatulate, tomentose; *heads* numerous in a subcapitate cyme; *involucre* 4–7 mm high, the bracts blackish-green with white-scarious tips. Common on rocky, alpine slopes, 10,000–14,000 ft. July–Aug. E/W.

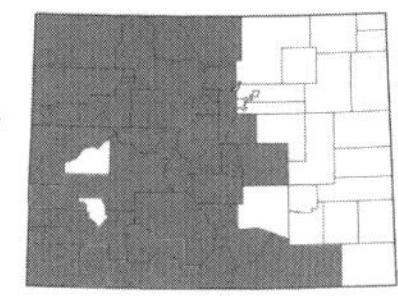

***Antennaria microphylla* Rydb.**, LITTLELEAF PUSSY-TOES. Plants 10–40 cm, with stolons; *leaves* spatulate, 0.5–1.5 mm long, tomentose; *heads* 3–10, in a corymbose array; *involucre* 4–8 mm high, the bracts brown or brownish-green at the base and white above. Common on open slopes and in forest openings, along streambanks, and in sagebrush meadows, 6000–11,500 ft. May–Aug. E/W.

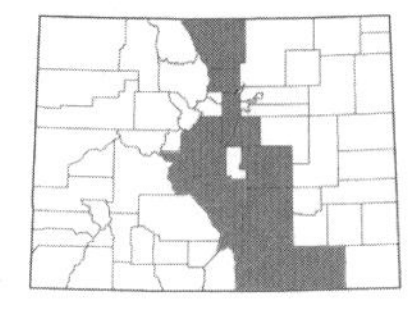

***Antennaria neglecta* Greene**, FIELD PUSSY-TOES. [including *A. howellii* Greene ssp. *neodioica* (Greene) R.J. Bayer]. Plants 10–40 cm, with stolons; *leaves* oblanceolate to spatulate, 1.5–6.5 cm long, becoming glabrous above and white-tomentose below; *heads* several in a dense cyme; *involucre* 5–9 mm high, the bracts white above. Found in open meadows and ponderosa pine forests along the Front Range, 5500–9000 ft. May–June. E.

***Antennaria parvifolia* Nutt.**, SMALL-LEAF PUSSY-TOES. Plants 2–15 cm, with stolons; *leaves* oblanceolate to spatulate, 1–3.5 cm long, white-tomentose; *heads* 3–8 or rarely solitary, in a corymbiform array; *involucre* 7–11 mm high, the bracts white or occasionally pinkish. Common in open meadows and on rocky slopes, 4600–9500 ft. May–July. E/W.

***Antennaria pulcherrima* (Hook.) Greene**, SHOWY PUSSY-TOES. Plants 20–65 cm, lacking stolons; *leaves* oblanceolate to lanceolate, 5–20 cm long, gray-tomentose; *heads* 3–30 in a corymbiform array; *involucre* 5–12 mm high, the bracts black or brown below. Found in wet mountain meadows, along streambanks, and in willow-thickets, 7500–9500 ft. June–Aug. E/W.

***Antennaria rosea* Greene**, ROSY PUSSY-TOES. Plants 10–45 cm, with stolons; *leaves* oblanceolate to spatulate, 15–20 cm long, tomentose; *heads* 3–20 in corymbiform arrays; *involucre* 4–7 (10) mm high, the bracts pink or reddish below and white above. Common in open meadows, on rocky slopes, and in wooded forests, 5000–12,000 ft. May–Aug. E/W.

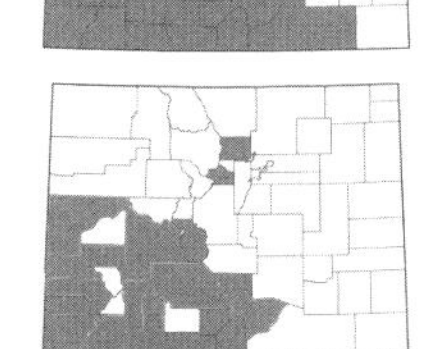

***Antennaria rosulata* Rydb.**, KAIBAB PUSSY-TOES. Plants 1–3 cm, with short stolons; *leaves* spatulate, 0.5–1 cm long, white-tomentose; *heads* solitary or 2–3; *involucre* 5–9 mm high, the outer bracts green and inner white-scarious. Infrequent on open rocky slopes and in dry meadows, often associated with sagebrush communities, 7000–9600 ft. May–June. E/W.

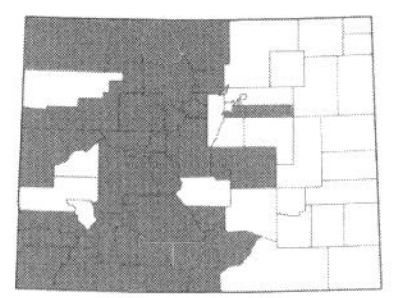

***Antennaria umbrinella* Rydb.**, UMBER PUSSY-TOES. Plants 7–20 cm, with stolons; *leaves* spatulate to narrowly oblanceolate, 1–2 cm long, white-tomentose; *heads* 3–8, in a dense cyme; *involucre* 5–6 mm high, the outer bracts brown or greenish and the inner with white tips. Common on rocky slopes, dry meadows, and in open forests, 7500–13,000 ft. June–Aug. E/W. (Plate 8)

ANTHEMIS L. – CHAMOMILE, DOG FENNEL

Herbs; *leaves* alternate, pinnately dissected into linear or filiform ultimate segments; *heads* radiate (ours); *involucre bracts* numerous in two to several series, imbricate, scarious-margined; *receptacle* chaffy throughout or only near the center; *disk flowers* yellow; *ray flowers* white or yellow; *pappus* absent or a short crown; *achenes* round or 4- or 5-angled.

1a. Ray flowers yellow, usually more than 20; heads large, the disk mostly 10–20 mm broad...***A. tinctoria***
1b. Ray flowers white, fewer than 20; heads smaller, the disk 5–13 mm broad...2

2a. Receptacle chaffy throughout, subtending each of the disk flowers; stems light-pubescent to villous-pubescent; ray flowers pistillate and fertile...***A. arvensis***
2b. Receptacle chaffy only in the center; stems usually glabrous or nearly so, or somewhat hairy above; ray flowers sterile...***A. cotula***

***Anthemis arvensis* L.**, CORN CHAMOMILE. Annuals, lightly hairy to villous; *leaves* 1–2-pinnately lobed, 15–35 × 8–16 mm; *involucre* 6–13 mm in diam., villous; *disk flowers* 2–4 mm; *ray flowers* 5–15 mm, white, fertile; *pappus* absent or a short crown; *achenes* 1.7–2 mm, the ribs smooth or weakly tuberculate. Found in waste places and along roadsides, known from Gunnison Co., 7500–7800 ft. May–July. W. Introduced.

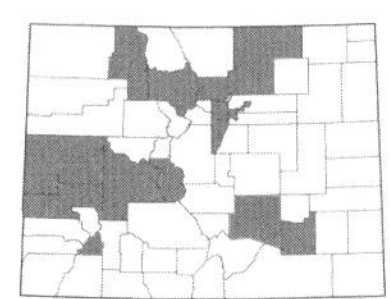

***Anthemis cotula* L.**, DOG FENNEL. Annuals, often ill-sented, glabrous or somewhat hairy above; *leaves* 1–2-pinnately lobed, 25–55 × 15–30 mm; *involucre* 5–9 mm in diam., villous; *disk flowers* 2–3 mm, sparsely gland-dotted; *ray flowers* 5–15 mm long, white, sterile; *pappus* absent; *achenes* 1.3–2 mm long, the ribs usually tuberculate. Locally common in disturbed places such as along roadsides and in fields, 5000–7000 ft. June–Sept. E/W. Introduced.

This species is often mistaken for *Tripleurospermum perforatum* (Mérat) M. Lainz, which has no receptacle chaff, a short crown of pappus, and is odorless or nearly so as opposed to *A. cotula* which has a distinctly fetid scent.

***Anthemis tinctoria* L.**, YELLOW CHAMOMILE. [*Cota tinctoria* (L.) J. Gay ex Gussone]. Short-lived perennials, villous to sericeous; *leaves* 1–2-pinnately lobed, 10–70 mm long; *involucre* (5) 10–20 mm in diam., villous; *disk flowers* 3.5–4 mm long; *ray flowers* 6–15 mm long, yellow, fertile or sterile; *pappus* a short crown; *achenes* 1.8–2.2 mm long. Found along roadsides and in disturbed areas, known from near Aspen in Pitkin Co., 7800–8000 ft. July–Aug. W. Introduced.

ARCTIUM L. – BURDOCK

Biennial herbs; *leaves* large, alternate; *heads* discoid; *involucre* subglobose to globose, the bracts imbricate in several series, each tip with an inward pointing hooked bristle; *receptacle* densely bristly; *disk flowers* pink or purplish or rarely white; *ray flowers* absent; *pappus* of deciduous, short, rigid scale-like bristles; *achenes* oblong, slightly compressed, with many nerves.

1a. Heads sessile or peduncles short (up to 4 cm long); inflorescence racemose; petioles hollow...***A. minus***
1b. Peduncles 3–10 cm long; inflorescence corymbose; petioles solid...2

2a. Involucre bracts glabrous or subglabrous...***A. lappa***
2b. Involucre bracts densely tomentose...***A. tomentosum***

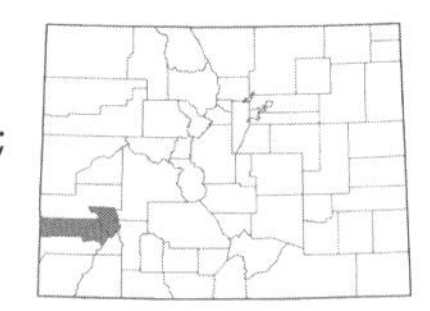

***Arctium lappa* L.**, GREAT BURDOCK. Plants 8–30 dm; *basal leaves* long-petiolate, blades 2–8 dm long, cordate-ovate, thinly gray-tomentose below and green above, with dentate to subentire margins; *heads* in corymbiform clusters, long-pedunculate, 2.5–4.5 cm wide; *involucre bracts* glabrous to loosely cobwebby; *pappus* 2–5 mm long. Uncommon in waste places and along roadsides, 7800–9000 ft. July–Oct. W. Introduced.

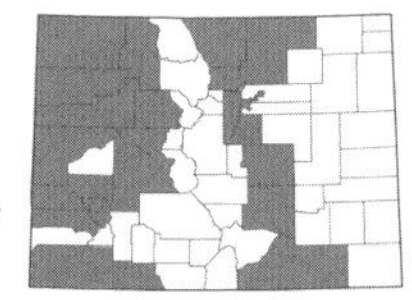

***Arctium minus* Bernh.**, COMMON BURDOCK. Plants 5–30 dm; *basal leaves* long-petiolate, petioles hollow, blades 3–6 dm long, thinly gray-tomentose below and green above, the margins coarsely dentate; *heads* in racemose clusters, sessile or on short peduncles to 4 cm long, 1.5–4 cm wide; *involucre bracts* minutely hairy, often glandular; *pappus* 1–3.5 mm long. Common in waste places, along roadsides, and in fields, often in moist places, 4500–7500 ft. July–Sept. E/W. Introduced.

***Arctium tomentosum* Mill.**, WOOLLY BURDOCK. Plants to 25 dm; *basal leaves* long-petiolate, blades 3–4 dm long, white-tomentose below and green above, the margins coarsely dentate to subentire; *heads* in corymbiform clusters, long-pedunculate, 1.5–2.5 cm wide; *involucre bracts* densely tomentose; *pappus* 1–3 mm long. Uncommon in waste places and other disturbed areas, known from two old collections near Denver, 4500–5200 ft. July–Sept. E. Introduced.

ARIDA (R.L. Hartm.) D.R. Morgan & R.L. Hartm. – DESERT TANSY-ASTER

Herbs; *leaves* alternate, simple; *heads* radiate or discoid; *involucre bracts* imbricate, the base hard and tannish or purplish and the apex herbaceous; *receptacle* naked or with scales; *disk flowers* yellow; *ray flowers* white, pink, purple, or blue; *pappus* of bristles in 2–3 series; *achenes* narrowly oblong, with 8–13 ribs on each face. (Morgan & Hartman, 2003)

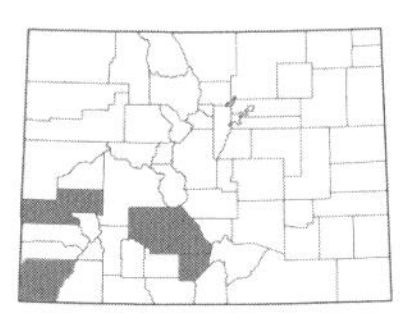

***Arida parviflora* (A. Gray) D.R. Morgan & R.L. Hartm.**, SMALLFLOWER TANSY-ASTER. [*Machaeranthera parviflora* A. Gray]. Annuals or short-lived perennials, 1–5 dm; *leaves* lanceolate to oblong, 1–3 cm long, glabrous or sparsely stipitate-glandular; *disk flowers* 4–5 mm long; *ray florets* 6–8 mm long, blue to purple; *achenes* 1.5–2 mm long. Rare in open places in alkaline soil, often in association with greasewood, 5400–7500 ft. July–Oct. E/W.

ARNICA L. – ARNICA

Perennial herbs; *leaves* mostly, simple; *heads* radiate or rarely discoid; *involucre bracts* herbaceous, more or less in two series, subequal; *receptacle* fimbrillate or hirsute; *disk flowers* yellow to orange; *ray flowers* yellow to orange; *pappus* of numerous white to tawny, barbellate to subplumose, capillary bristles; *achenes* cylindric to fusiform, 5–10-nerved. (Maguire, 1943; Wolf & Denford, 1984)

1a. Heads lacking ray flowers, or rarely with some very short ray flowers...***A. parryi***
1b. Heads with conspicuous ray flowers (1–3 cm long)...2

2a. Stem leaves of mostly 5–12 pairs, gradually reduced upwards; heads generally 5–20 (rarely 3) per stem...3
2b. Stem leaves of mostly 2–4 (rarely 5) pairs (not counting the basal rosette if present), often strongly reduced upwards; heads less than 5 per stem (except in *A. gracilis*)...4

3a. Involucre bract tips obtuse or acuminate, conspicuously pilose with an apical or subapical tuft of long hairs; involucre bracts with tangled, white hairs at the base in addition to glandular hairs throughout; the lowermost leaves usually petiolate...***A. chamissonis***
3b. Involucre bract tips sharply acute, not pilose within; involucre bracts with glandular hairs and little or no additional hairs; the lowermost leaves sessile...***A. longifolia***

4a. Pappus subplumose, tawny...***A. mollis***
4b. Pappus merely barbellate, white or nearly so...5

5a. Leaf blades broad, mostly 1–2.5 times as long as wide, cordate, elliptic, or ovate...6
5b. Leaf blades narrow, mostly 3–10 times as long as wide, lanceolate, oblanceolate, or narrowly elliptic...8

6a. Leaves cordate, the lower leaves long-petiolate; heads generally solitary but sometimes up to 3 per stem; peduncles below the head generally densely white pilose...***A. cordifolia***
6b. Leaves ovate to elliptic, sessile or sometimes the lower short-petiolate; heads generally 3–15 per stem or occasionally solitary; peduncles below the head sparingly short-hairy to glandular-hairy...7

7a. Leaves seldom over 2.5 cm wide, the lower stem ones often petiolate; heads more numerous, usually 5–15 per stem; heads generally smaller, involucre bracts 7–13 mm in height; disk flowers 10–25 per head; achenes black...***A. gracilis***
7b. Leaves 1.5–8 cm wide, usually sessile; heads usually 1–5 per stem; heads generally larger, involucre bracts 10–18 mm in height; disk flowers 20–90 per head; achenes brown...***A. latifolia***

8a. Leaves mostly basal with tufts of long brown woolly hairs in the axils of the basal leaves (very soft to the touch, even if not clearly visible); plants generally taller than the next, 2–6 dm in height; plants of lower elevations, from the foothills to montane...***A. fulgens***
8b. Leaves mostly cauline and the basal leaves never with tufts of brown hairs in the axils; plants generally smaller than the previous, 0.5–3 dm in height; plants of higher elevations, from the subalpine to alpine...***A. rydbergii***

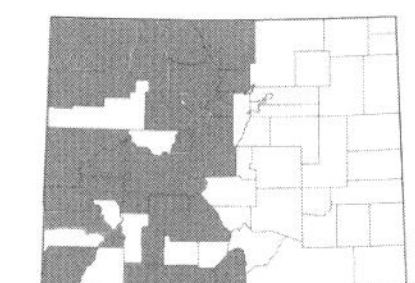

***Arnica chamissonis* Less.**, Chamisso arnica Plants 2–10 dm, hairy or subtomentose; *leaves* all cauline, with 5–10 pairs per stem, lanceolate to oblanceolate, with entire to denticulate margins, 5–30 × 1–4 cm, the upper sessile and the lowermost petiolate; *heads* 5–15; *involucre* 8–13 mm high, the bracts with a subapical tuft of long white hairs internally; *rays* 1.5–2 cm long; *pappus* barbellate, tawny. Common in mountain meadows and moist places, 7500–10,500 ft. July–Sept. E/W.

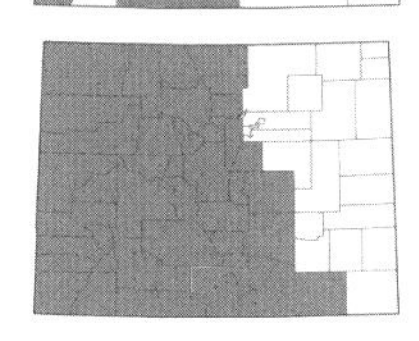

***Arnica cordifolia* Hook.**, heartleaf arnica Plants 1–6 dm, glandular-hairy to loosely white-hairy; *leaves* cauline, with 2–4 pairs per stem, cordate, with dentate margins, 4–12 × 3–9 cm, long-petiolate; *heads* 1–3 (7); *involucre* 13–20 mm high, with long, spreading white hairs; *rays* 1.5–3 cm long; *pappus* barbellate, white or stramineous. Common in moist, shaded forests and along streams from the foothills to subalpine, 6000–12,000 ft. (May) June–Sept. E/W. (Plate 8)

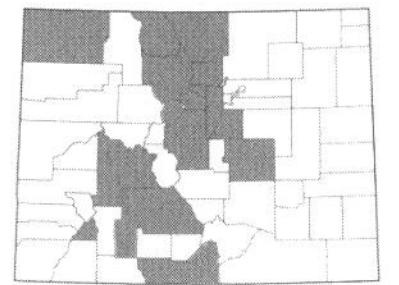

***Arnica fulgens* Pursh**, foothill arnica Plants 2–6 dm, glandular-stipitate and often also hairy; *leaves* cauline and basal, with 2–4 pairs per stem, oblanceolate to narrowly elliptic, entire or nearly so, 3–12 × 1–4 cm, sessile above; *heads* 1–3; *involucre* 10–15 mm high, glandular and hairy; *rays* 1.5–2.5 cm long; *pappus* barbellate, white or stramineous. Common in open meadows and on rocky slopes, especially along the Front Range, 5300–9500 ft. May–Aug. E/W.

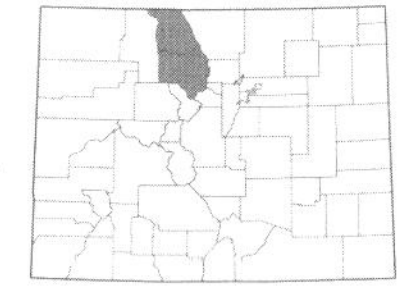

***Arnica gracilis* Rydb.**, smallhead arnica [*A. latifolia* Bong. var. *gracilis* (Rydb.) Cronquist]. Plants 1–3 dm, glandular and sometimes hairy; *leaves* cauline and basal, 2–4 pairs per stem, elliptic, entire or sometimes toothed; *heads* (1) 3–9; *involucre* 7–13 mm high, glandular; *rays* 1–2.5 cm long; *pappus* barbellate, white. Infrequent on rocky slopes and ridges in the mountains, 8500–11,000 ft. July–Aug. E/W.

There are probably more accounts of this species in Colorado, but it has been grouped with *A. latifolia* in many cases. *Arnica gracilis* arose as a hybrid between *A. cordifolia* and *A. latifolia*, but is morphologically and ecologically distinct from both its parents and is therefore appropriate to recognize at the specific level.

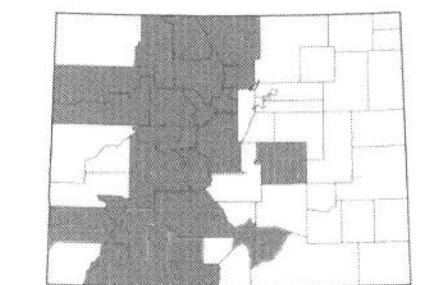

***Arnica latifolia* Bong.**, broadleaf arnica Plants 1–6 dm, glandular; *leaves* cauline, 2–4 pairs per stem, elliptic, usually toothed, 2–14 × 1.5–8 cm, sessile or nearly so; *heads* 1–3; *involucre* 10–18 mm high, glandular; *rays* 1–2.5 cm long; *pappus* barbellate, white. Locally common in moist forests, meadows, and open places in the mountains, 8800–13,000 ft. July–Aug. E/W.

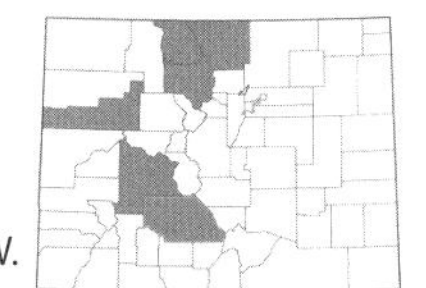

***Arnica longifolia* D.C. Eaton**, spearleaf arnica Plants 3–6 dm, sparsely glandular to puberulous above; *leaves* cauline, with 5–7 pairs per stem, lanceolate to lance-elliptic, entire or slightly toothed, 5–12 × 1–2.5 cm, sessile or shortly petiolate; *heads* 7–20 (35); *involucre* 7–10 mm high, glandular; *rays* 1–2 cm long; *pappus* barbellate to shortly subplumose, tawny. Infrequent in well-drained soil around seeps, springs, and along cliffs and river banks, 9000–11,500 ft. July–Oct. E/W.

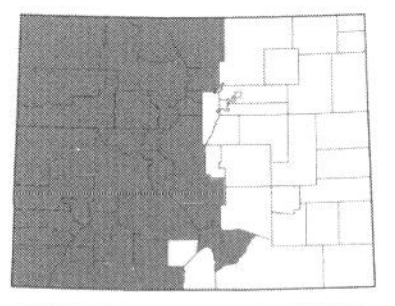

***Arnica mollis* Hook.**, hairy arnica [*A. ovata* Greene]. Plants 2–6 dm, minutely hairy to glandular; *leaves* cauline, 3–4 (5) pairs per stem, ovate to lanceolate or oblanceolate, irregularly toothed to entire, to 25 cm long, sessile or the lower shortly petiolate; *heads* 1–3; *involucre* 10–16 mm high, long-hairy below and glandular above; *rays* 1.5–2.5 cm long; *pappus* subplumose, tawny. Common in subalpine meadows, on rocky slopes, and in moist places in the mountains, 7500–13,300 ft. (June) July–Sept. E/W.

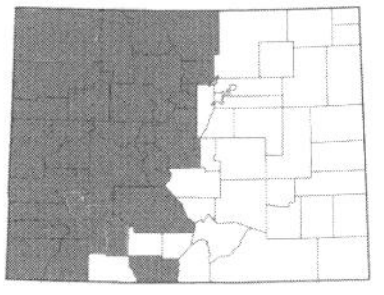

***Arnica parryi* A. Gray**, Parry's arnica. Plants 2–6 dm, glandular above; *leaves* cauline and basal, 2–4 pairs per stem, lanceolate, entire to toothed, 5–20 × 1.5–6 cm, the lowermost petiolate; *heads* 3–9; *involucre* 10–14 mm high, minutely hairy and glandular; *rays* absent; *pappus* barbellate, tawny. Common in open forests and meadows, 8000–12,500 ft. July–Sept. E/W.

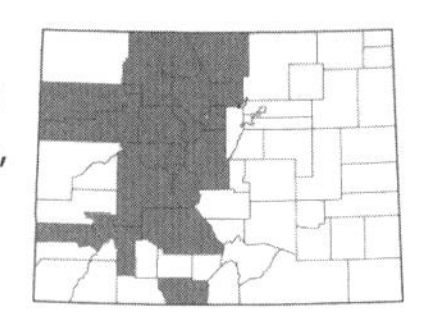

Arnica rydbergii **Greene**, Rydberg's arnica. Plants 0.5–3 dm, glandular and short-hairy or subglabrous; *leaves* cauline and basal, 3–4 pairs per stem, oblanceolate to lanceolate, entire, 3–10 × 0.5–2.5 cm, sessile or the lower shortly winged-petiolate; *heads* 1–3 (5); *involucre* 9–13 mm high, glandular and sparsely long-hairy to subglabrous; *rays* 1–2 (3) cm long; *pappus* barbellate, white. Common in alpine meadows and on rocky slopes, 10,000–13,500 ft. July–Aug. E/W.

ARTEMISIA L. – wormwood, mugwort, sagebrush

Shrubs or herbs; *leaves* alternate, entire to lobed or dissected; *heads* discoid or disciform; *involucre bracts* imbricate in 2–several series, dry, at least the inner ones scarious or with scarious or hyaline margins; *receptacle* naked or with long hairs; *disk flowers* with a short, tubular corolla; *ray flowers* usually absent; *pappus* usually absent; *achenes* usually glabrous. (Beetle, 1960; Estes, 1969; Hall & Clements, 1923; Keck, 1946; Komkven et al., 1998; Ward, 1953)

1a. Shrubs or subshrubs, always woody at least at the base...2
1b. Annual or perennial herbs, sometimes the stems thickened at the base but not distinctly woody...16

2a. Receptacles with long woolly hairs between the flowers (artificially resembling a pappus, although none is present)...3
2b. Receptacles naked, no long woolly hairs present between the flowers...4

3a. Leaves 0.5–1.5 cm long, dense at the base as well as cauline, 2–3-ternately (or subpinnately) divided, the ultimate divisions narrowly linear to filiform, less than 1 mm wide; plants mat-forming...***A. frigida***
3b. Leaves 3–8 cm long, all cauline and not dense at the base, 2–3-pinnately divided, the ultimate divisions lanceolate, oblanceolate, or narrowly oblong, 1.5–4 mm wide; plants not mat-forming...***A. absinthium***

4a. Most leaves shallowly 3–5-toothed to deeply 3-cleft at the apex (the upper stem leaves usually reduced and entire)...5
4b. Leaves all pinnatifid, ternately divided, 3–5-palmately cleft, or entire, the ultimate leaf divisions often filiform or narrowly linear...11

5a. Leaves mainly all entire, only a few 3-toothed or shallowly cleft at the apex...***A. cana***
5b. Leaves generally all 3–5-toothed at the apex, or the upper stem leaves sometimes reduced and entire...6

6a. Leaves deeply 3-cleft at the apex, the axis of the leaf blade slender throughout and not expanded beneath the cleft portion...***A. tripartita* var. *tripartita***
6b. Leaves shallowly cleft to toothed at the apex, or if more deeply cleft then the axis of the leaf blade expanded beneath the cleft portion...7

7a. Outer involucre bracts closely tomentose, inner bracts glabrous and resinous, yellowish-green; leaves with small, green, glandular swellings that show through the tomentum as scattered green dots, usually less than 1–1.5 cm in length...***A. nova***
7b. Both outer and inner involucre bracts tomentose; leaves not gland-dotted or scarcely so, of various lengths but usually greater than 1 cm...8

8a. Inflorescence a spike; heads generally 5–30 per stem...9
8b. Inflorescence broader, a panicle (although sometimes narrow), or the panicle with short but spreading or ascending glomerulate branches; heads very numerous per stem...10

9a. Leaves smaller, 0.4–2 cm long; heads smaller with 4–8 flowers; generally shorter than the next, 2–5 dm...***A. arbuscula***
9b. Leaves larger, 2–5 cm long; heads larger with 8–27 flowers; generally taller, 3–8 dm...***A. spiciformis***

10a. Heads on short but spreading or ascending glomerulate branches, erect to deflexed; each head usually with 2 perfect flowers and 1 marginal pistillate flower, or occasionally the 3 flowers all perfect; generally shorter than the next, 1–3 or occasionally up to 10 dm...***A. bigelovii***
10b. Heads not in glomerulate branches, erect; each head with (3) 4–8 (11) flowers, these all perfect and potentially fertile; generally taller, 4–30 dm...***A. tridentata***

11a. Leaves pinnately divided...12
11b. Leaves entire or variously ternately or palmately divided...13

12a. Plants short, 0.5–2 dm; leaves 0.3–1.2 cm long, resinous-glandular...***A. pygmaea***
12b. Plants 5–20 dm; leaves 3–8 cm long, not resinous-glandular...***A. abrotanum***

13a. Heads crowded on short racemes or spikes, these barely if at all surpassing the height of the plant; plants appearing thorny from the presence of old lateral branches...***A. spinescens***
13b. Inflorescence branches extending well above the leaves; plants not appearing thorny ...14

14a. Inflorescence broad, the heads arranged in a panicle; leaves divided into long, narrow filiform segments; heads nodding...***A. filifolia***
14b. Inflorescence narrow, the heads arranged in a spike or narrow panicle; leaves various but not divided into long, narrow filiform segments; heads erect...15

15a. Plants taller, 2.5–12 (20) dm; leaves mostly all entire or a few with 3–5-toothed apices, 2–5.5 (7) cm long; plants variously distributed...***A. cana***
15b. Plants short, 0.5–2 dm; basal leaves ternately divided, mostly less than 1 cm in length; plants known from Moffat County...***A. pedatifida***

16a. Involucre bracts with distinct dark brown or black margins; heads few, arranged in a spike or narrow raceme; plants of the subalpine and alpine...17
16b. Involucre bracts without distinct dark margins; heads and inflorescence various; plants variously distributed...19

17a. Receptacles naked, no long woolly hairs present between the flowers; corollas long-hairy...***A. arctica***
17b. Receptacles with long woolly hairs between the flowers (artificially resembling a pappus, although none is present); corollas glabrous...18

18a. Basal leaves once-pinnately parted or cleft at the apex; stem leaves entire; heads relatively few, usually 1–5, these larger (6–11 mm broad); disk flowers 30–130 (average 85) per head; corollas usually glabrous...***A. pattersonii***
18b. Basal leaves bipinnately compound or ternate-pinnatifid; stem leaves pinnatifid to entire; heads 5–12, these smaller (3–6 mm broad); disk flowers 10–30 (average 18) per head; corollas usually hairy...***A. scopulorum***

19a. Receptacles with long woolly hairs between the flowers (artificially resembling a pappus, although none is present)...20
19b. Receptacles naked, no long woolly hairs present between the flowers...21

20a. Leaves 0.5–1.5 cm long, dense at the base as well as cauline, 2–3-ternately (or subpinnately) divided, the ultimate divisions narrowly linear to filiform, less than 1 mm wide; plants mat-forming...***A. frigida***
20b. Leaves 3–8 cm long, all cauline and not dense at the base, 2–3-pinnately divided, the ultimate divisions lanceolate, oblanceolate, or narrowly oblong, 1.5–4 mm wide; plants not mat-forming...***A. absinthium***

21a. Leaves glabrous, not at all tomentose, or occasionally sparingly and loosely villous...22
21b. Leaves tomentose (at least below), conspicuously villous, or resinous-glandular...25

22a. Leaves entire to rarely a few cleft; disk flowers functionally staminate and sterile, only the marginal flowers pistillate and fertile...***A. dracunculus***
22b. Leaves variously pinnatifid; disk flowers all potentially fertile, the marginal ones pistillate and the inner ones perfect...23

23a. Plants small, 0.8–3 dm; perennials from woody caudices or rootstock with well-developed basal leaves; heads larger, 4–7 mm broad...***A. parryi***
23b. Plants tall, 3–30 dm; annuals or biennials from more or less well-developed taproots, without well-developed basal leaves; heads smaller, 1–2 mm broad...24

24a. Inflorescence broad and open, panicle-like; heads pedunculate; stem usually green; known from one collection in Boulder County...***A. annua***
24b. Inflorescence dense on spike-like or spiciform branches; heads sessile or obscurely pedunculate; stem reddish; plants variously distributed...***A. biennis* var. *biennis***

25a. Stem leaves and a basal cluster of leaves present (these sometimes withered or lacking in the second year of growth in *A. campestris* var. *caudata*), pinnatifid (at least at the base); stems often reddish-tinged...26
25b. Leaves all cauline with no basal cluster, entire to variously pinnatifid; stems green to gray or rarely reddish...27

26a. Heads larger, mostly 4–5 mm broad with 45–90 disk flowers per head; leaves generally bicolored (green and glabrous above, gray-tomentose below)...***A. franserioides***
26b. Heads smaller, mostly 1–3 mm broad with 5–30 disk flowers per head; leaves not bicolored, green and villous to pilose on both sides...***A. campestris***

27a. Leaves 2–3 times pinnatifid...28
27b. Leaves entire, toothed, or only once pinnatifid or incised...30

28a. Inflorescence broad, paniculate...***A. abrotanum***
28b. Inflorescence narrow, spicate...29

29a. The ultimate leaf divisions obtuse at the apex; heads with 45–90 disk flowers; involucres 2.5–3 mm high...***A. franserioides***
29b. The ultimate leaf divisions acute at the apex; heads with 6–20 disk flowers; involucres 3.4–4.5 mm high...***A. michauxiana***

30a. Leaves mostly under 1.5 cm in length, dissected to the midrib into slender (0.5–1 mm wide) segments...***A. carruthii***
30b. Leaves mostly over 2 cm in length (but if shorter, then the inflorescence broad and 3 cm or more wide, not narrow), entire or lobed to dissected in wider (1.5–4 mm at the narrowest) segments ...31

31a. Leaves entire, linear to linear-lanceolate; plants without rhizomes...***A. longifolia***
31b. Leaves lanceolate or elliptic (if entire), lobed or toothed on the upper portion of the leaf, or the entire leaf deeply parted or divided; plants arising from rhizomes...***A. ludoviciana***

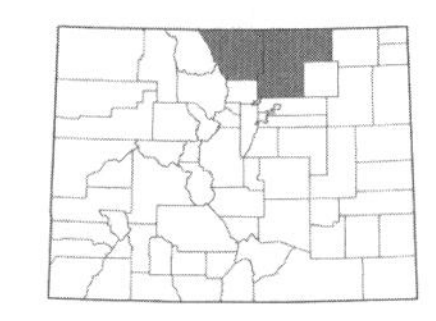

***Artemisia abrotanum* L.**, GARDEN SAGEBRUSH; SOUTHERNWOOD. Shrubs, 5–20 dm; *leaves* pinnatifid, the ultimate segments linear or filiform and 0.5–1.5 mm wide, tomentulose below and green and glabrous above; *heads* numerous, arranged in a panicle; *involucre* 1.5–3.5 mm high; *receptacle* naked; *disk flowers* 10–20+. A common Old World plant occasionally found escaped from cultivation or in disturbed areas, 4500–5000 ft. July–Sept. E. Introduced.

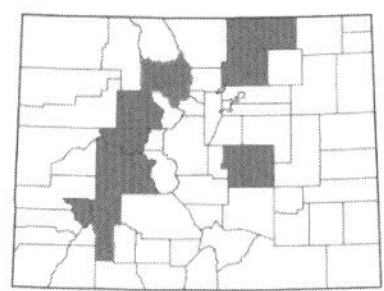

***Artemisia absinthium* L.**, ABSINTHE WORMWOOD. Perennials, 4–12 dm; *leaves* 2–3 times pinnatifid into oblong, obtuse segments 1.5–4 mm wide, silvery-sericeous on both sides; *heads* numerous, arranged in a panicle; *involucre* 2–3 mm high, sericeous; *receptacle* long-hairy; *disk flowers* 9–50. A very fragrant sagewort found in disturbed sites such as along roadsides and in ditch banks, 5000–8500 ft. July–Oct. E/W. Introduced.

Artemisia absinthium is the primary ingredient in the mind-altering elixir absinth, and the toxic chemical thujone present in the plant gives the drink its hallucinogenic qualities. The green color of absinth comes from the presence of chlorophyll in the drink.

***Artemisia annua* L.**, SWEET WORMWOOD; ANNUAL SAGEWORT. Annuals, 3–30 dm, glabrous; *leaves* 2–3 times pinnatifid into linear or lanceolate, sharply toothed segments; *heads* numerous, arranged in a broad and open panicle; *involucre* 1–2 mm high; *receptacle* naked; *disk flowers* 5–30. Found in disturbed places, known from one collection in Boulder, 5000–6000 ft. Aug.–Oct. E. Introduced.

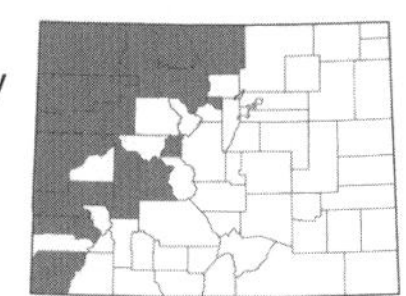

***Artemisia arbuscula* Nutt.**, LITTLE SAGEBRUSH. [*Seriphidium arbusculum* (Nutt.) W.A. Weber]. Shrubs, 2–5 dm, appressed-tomentose to villous-tomentose; *leaves* shallowly 3-lobed to deeply terminally lobed, 0.4–2 cm long; *heads* 10–20 per stem, erect, arranged in a spike; *involucre* 3–6 mm high; *receptacle* naked; *disk flowers* 4–8. Found in dry soil, commonly in less favorable conditions than *A. tridentata*, 5300–9700 ft. July–Sept. E/W.

There are two subspecies of *A. arbuscula* in Colorado, each ecologically, if not morphologically, distinct:

1a. Leaves not deeply cleft, the lobes less than ½ the length of the blade; involucres 2–4.5 mm in diam....**ssp. *arbuscula***. Found on the western slope in rocky soil. W.
1b. Leaves deeply cleft; involucres 1.5–2.5 mm in diam....**ssp. *longiloba* (Osterh.) L.M. Shultz**. Found on the eastern slope in alkaline clay soil and may occur in Routt County, and especially common in Jackson County near Coalmont. E.

***Artemisia arctica* Less.**, BOREAL SAGEBRUSH. [*A. norvegica* Fr. var. *saxatilis* (Besser) Jeps.]. Perennials, 2–6 dm, loosely villous to nearly glabrous; *leaves* pinnately dissected into linear segments; *heads* 6–30, nodding, arranged in a narrow panicle or spike; *involucre* 4–7 mm high, the bracts prominently dark-margined; *receptacle* naked; *disk flowers* 30–75. Common in alpine meadows and on rocky slopes, 10,000–13,000 ft. July–Sept. E/W.

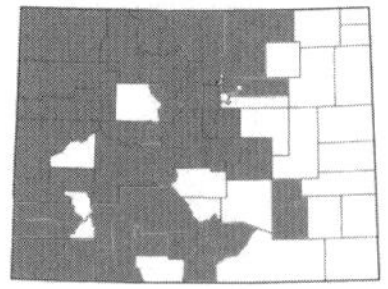

Artemisia biennis* Willd. var. *biennis, BIENNIAL SAGEWORT. Annuals or biennials, 3–30 dm, glabrous; *leaves* pinnately divided into oblong to oblanceolate, sharply toothed segments; *heads* numerous, erect, arranged in a dense spike-like panicle; *involucre* 2–3 mm high; *receptacle* naked; *disk flowers* 6–40. Found in disturbed places, 5000–11,500 ft. Aug.–Oct. E/W. Introduced.

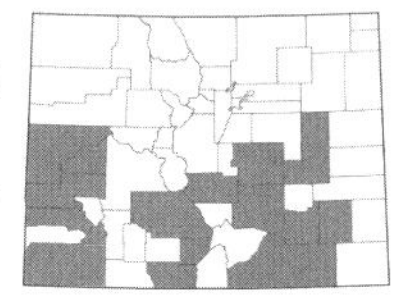

***Artemisia bigelovii* A. Gray**, BIGELOW'S SAGEBRUSH; FLAT SAGEBRUSH. Shrubs, 1–3 (10) dm, sericeous-tomentose; *leaves* mostly 3-toothed, 0.3–2.3 cm long; *heads* 10–20, sometimes deflexed, arranged in a narrow panicle; *involucre* 2–3.5 mm high; *receptacle* naked; *disk flowers* 3–4, usually with 2 perfect flowers and 1 marginal pistillate flower. Found on rocky soil, in canyons, and in dry places, 4000–6000 ft. July–Sept. E/W.

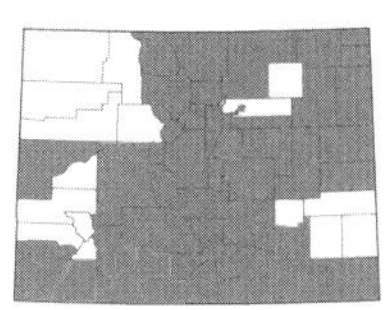

***Artemisia campestris* L.**, FIELD SAGEWORT. Perennials,1.5–7 dm, stems often reddish; *leaves* 2–3-pinnatifid or partly ternate into linear segments, villous to glabrous on both sides; *heads* numerous, arranged in an elongate panicle; *involucre* 2–4 mm high; *receptacle* naked; *disk flowers* 5–30. 4000–13,000 ft. July–Oct. E/W.

Within *A. campestris* there are three rather distinct varieties, the most unique of which is the alpine var. *purshii* (Hook.) Cronquist. The other two varieties, var. *caudata* (Michx.) Palmer & Steyerm. and var. *pacifica* (Nutt.) M. Peck, are more difficult to tell apart, and conclusive identifications sometimes cannot be determined:

1a. Heads fewer and larger (involucre bracts 3–4 mm high and 3.5–5 mm wide) in a narrow, condensed, spike-like panicle; perennials from a branching caudices... **var. *purshii* (Hook.) Cronquist,** BOREAL SAGEWORT [*A. borealis* Pall., *A. spithamaea* Pursh, *Oligosporus groenlandicus* (Hornem.) Á. Löve & D. Löve]. Common in the subalpine and alpine on rocky slopes and in meadows, 9500–13,000 ft. July–Aug. E/W.

1b. Heads numerous and smaller (involucre bracts 2–3 mm high and 2–3.5 mm wide), in a broad to narrow, open, elongate panicle; biennials or perennials from taproots or branching caudices...2

2a. Plants biennial with mostly 1 stem; basal leaves withered at the time of flowering... **var. *caudata* (Michx.) Palmer & Steyerm.**, FIELD SAGEWORT. [*A. caudata* Michx., *Oligosporus caudatus* (Michx.) Poljakov]. Common in sandy places, in fields, and along roadsides on the eastern plains, 4000–5700 ft. July–Sept. E.

2b. Plants perennial with several stems; basal leaves present and not withered at the time of flowering... **var. *pacifica* (Nutt.) M. Peck**, FIELD SAGEWORT. [*A. pacifica* Nutt., *Oligosporus pacificus* (Nutt.) Poljakov]. Common in dry, sandy soil, grasslands, and in open meadows, often found at higher elevations than var. *caudata*, 4500–10,500 ft. July–Oct. E/W.

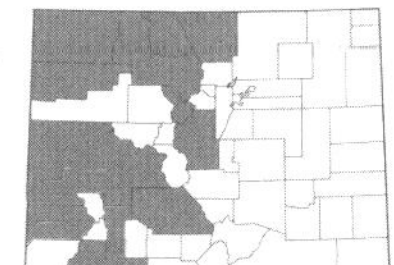

***Artemisia cana* Pursh**, HOARY SAGEBRUSH; SILVER SAGE. [*Seriphidium canum* (Pursh) W.A. Weber]. Shrubs, 2.5–12 (20) dm, appressed-tomentose; *leaves* entire or some 3-lobed at the tip, 2–5.5 (7) cm long; *heads* numerous, erect, arranged in a narrow spike; *involucre* 3–6 mm high; *receptacle* naked; *disk flowers* 8–16 (20). Found in dry or more often moist meadows and on stream terraces, 6000–10,500 ft. July–Sept. E/W.

Artemisia cana forms intermediates with *A. tridentata* var. *vaseyana* and *A. spiciformis* where their habitats overlap. There are two subspecies of *A. cana* in Colorado:

1a. Leaves 2–8 cm long, entire; plants 10–15 dm; plants primarily found on the eastern slope...**ssp. *cana***

1b. Leaves (1.5) 2–3 cm long, often irregularly lobed; plants 5–7 (9) dm; primarily distributed on the western slope...**ssp. *viscidula* (Osterh.) Beetle**

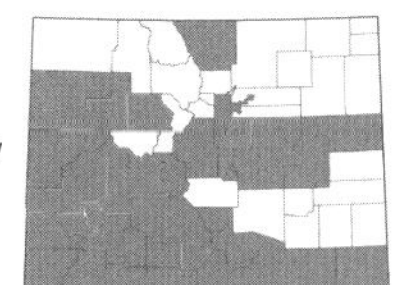

***Artemisia carruthii* Alph. Wood ex J.H. Carruth**, CARRUTH'S WORMWOOD. [*A. wrightii* A. Gray]. Perennials, 2–7 dm; *leaves* pinnately few-cleft or trifid into narrow, entire segments, closely tomentose or green and subglabrous above; *heads* numerous, erect, arranged in a spike or narrow panicle; *involucre* 2.3–3.5 mm high; *receptacle* naked; *disk flowers* 6–30. Found on dry slopes and in ponderosa forests, 5000–10,000 ft. July–Sept. E/W.

Artemisia carruthii appears to intergrade with *A. ludoviciana*, and some specimens are intermediate between the two.

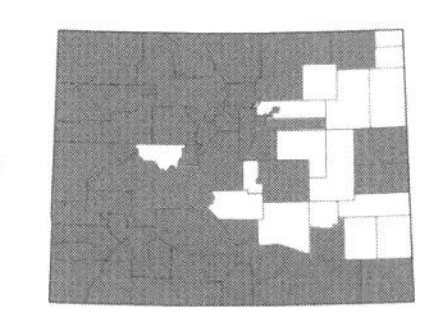

***Artemisia dracunculus* L.**, TARRAGON; DRAGON SAGEWORT. Perennials, (2) 5–15 dm, glabrous; *leaves* usually entire, linear, (1.2) 3–8 cm long; *heads* numerous, erect or nodding, arranged in a panicle; *involucre* 2–4 mm high; *receptacle* naked; *disk flowers* 10–30. Common from the plains to higher elevations in the mountains, 4000–11,500 ft. July–Oct. E/W.

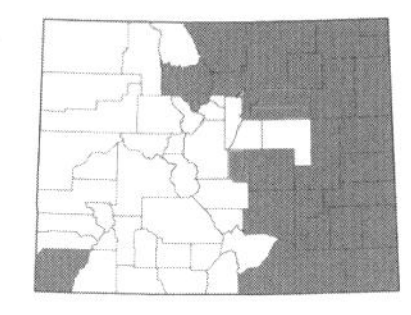

***Artemisia filifolia* Torr.**, SAND SAGE; OLD-MAN SAGEBRUSH. [*Oligosporus filifolius* (Torr.) Poljakov]. Shrubs, 5–15 dm; *leaves* ternately divided into filiform segments, closely tomentose or becoming green and glabrate; *heads* numerous, nodding, arranged in a panicle; *involucre* 1.5–2.5 mm high; *receptacle* naked; *disk flowers* 1–6. Common in sandy places and on open, dry slopes, 3500–5500 ft. July–Sept. E/W.

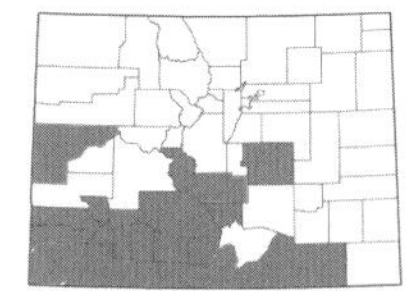

***Artemisia franserioides* Greene**, FOREST SAGEWORT. Perennials, 3–10 dm; *leaves* twice pinnately divided into lanceolate, obtuse segments, glabrous or sparsely hairy above, tomentose below; *heads* 10-numerous, nodding, arranged in a narrow spike; *involucre* 2.5–3 mm high; *receptacle* naked; *disk flowers* 45–90. Common in shady aspen or spruce forest openings in the southern mountains, 7500–11,500 ft. July–Sept. E/W.

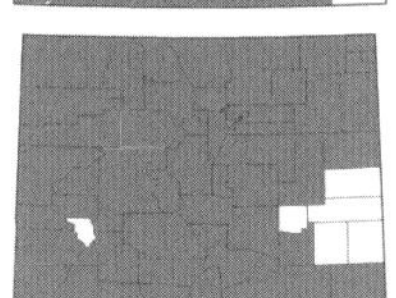

***Artemisia frigida* Willd.**, PRAIRIE SAGEWORT, FRINGED SAGEBRUSH. Shrubs or subshrubs, 1–5 dm, sericeous-tomentose; *leaves* 2–3-ternately divided into linear segments; *heads* numerous, nodding, arranged in a panicle; *involucre* 2–3.5 mm high; *receptacle* long-hairy; *disk flowers* 10–50. Common and abundant in open, dry, sometimes rocky places; one of the most characteristic fall plants of the plains, foothills, and mountains, 4000–10,000 (12,000) ft. July–Oct. E/W. (Plate 8)

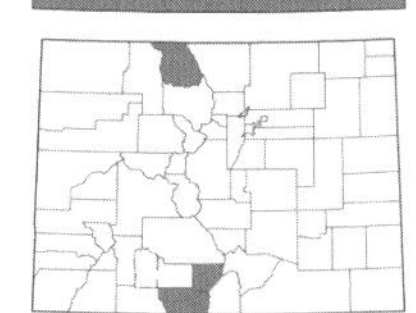

***Artemisia longifolia* Nutt.**, LONGLEAF WORMWOOD. Perennials, 1–8 dm; *leaves* usually entire, linear or narrowly lanceolate, often with strongly revolute margins, 3–12 cm long, tomentose below and glabrescent above; *heads* numerous, erect, arranged in a narrow panicle; *involucre* 4–5 mm high; *receptacle* naked; *heads* 3–26. Found in dry, open places, often in heavy clay soil, known from North Park and the San Luis Valley, 7500–8600 ft. July–Sept. E.

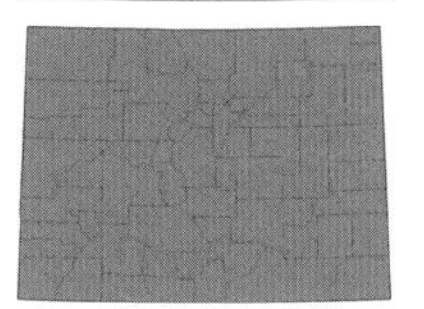

***Artemisia ludoviciana* Nutt.**, LOUISIANA SAGEWORT. Perennials, 2–10 dm; *leaves* entire, lobed, or pinnately incised into lanceolate or linear segments, white-tomentose or sometimes green and glabrous above; *heads* numerous, arranged in a spike or open panicle; *involucre* 2.5–3.5 mm high; *receptacle* naked; *disk flowers* 8–13. Common and widespread in dry, open places, 4000–9500 (11,200) ft. July–Oct. E/W. (Plate 9)

Artemisia ludoviciana is a highly complex species, and several varieties and races have been described over the years. In Colorado, there are four, more or less distinct, varieties:

1a. Inflorescence a broad, open panicle…2
1b. Inflorescence a narrow, congested, spicate panicle…3

2a. Leaves less than 2.5 cm long, entire or more often with 2 to several slender lateral segments…**var. *albula* (Wooton) Shinners**, WHITE SAGEBRUSH. 4500–8600 ft. July–Sept. E/W.
2b. Leaves greater than 2 cm in length, entire or more often with a few elongate, narrow lateral segments…**var. *mexicana* (Willd. ex Spreng.) A. Gray**, MEXICAN SAGEBRUSH. 4800–8500 ft. July–Sept. E/W.

3a. Leaves entire or a few toothed or shallowly lobed on the upper portion of the leaf, if divided further then the divisions not to the midrib…**var. *ludoviciana***, LOUISIANA WORMWOOD. 4000–9500 ft. July–Oct. E/W.
3b. Leaves deeply parted or divided, the divisions usually to the midrib…**var. *incompta* (Nutt.) Cronquist**, WHITE SAGEBRUSH. 4400–9500 (11,200) ft. July–Sept. E/W.

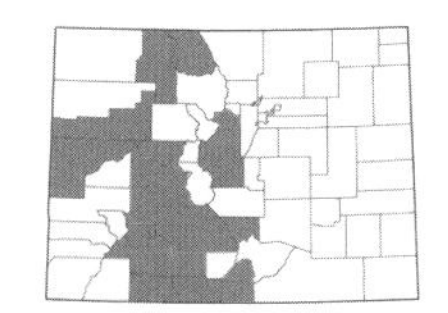

***Artemisia michauxiana* Besser**, MICHAUX'S WORMWOOD. Perennials, 1.5–5 (6) dm; *leaves* bipinnatifid into linear segments, tomentose below and green above, sometimes resinous-glandular; *heads* numerous, arranged in a spike; *involucre* 3.4–4.5 mm high; *receptacle* naked; *disk flowers* 6–20. Found on rocky alpine slopes and in spruce-lodgepole pine forest openings, 9000–12,000 ft. July–Oct. E/W.

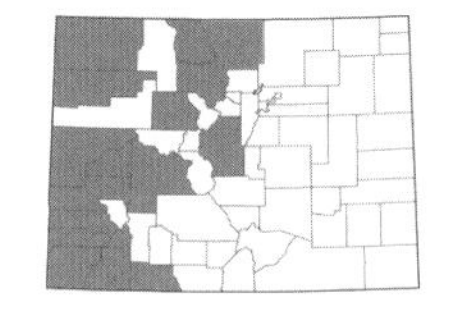

***Artemisia nova* A. Nelson**, BLACK SAGEBRUSH. [*Seriphidium novum* (A. Nelson) W.A. Weber]. Shrubs, 1–3 dm; *leaves* 3-lobed at the apex, closely tomentose with small, green glandular swellings (dots), 0.5–2 cm long; *heads* numerous, erect, arranged in a spike; *involucre* 3.5–5 mm high, some resinous and glabrous; *receptacle* naked; *disk flowers* (2) 3–5 (6). Found on rocky ridges and typically in poorer soil than *A. tridentata*, more common on the western slope, found on the eastern slope in North Park and South Park, 5500–9600 ft. July–Sept. E/W. (Plate 9)

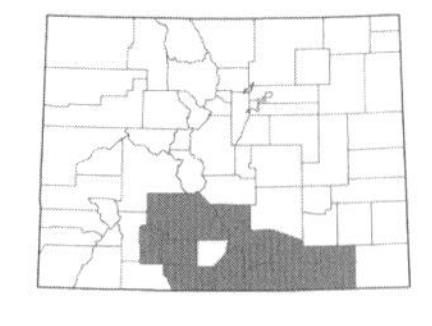

***Artemisia parryi* A. Gray**, PARRY'S WORMWOOD. [*A. laciniata* Willd. ssp. *parryi* (A. Gray) W.A. Weber]. Perennials, 0.8–3 dm; *leaves* bipinnatifid into oblong or lance-oblong segments, 2–7 cm long, loosely villous to glabrous; *heads* numerous, nodding, arranged in a raceme-like or spike-like inflorescence; *involucre* 2.5–4 mm high; *receptacle* naked; *disk flowers* 20–50. Found on subalpine and alpine slopes in the southern mountains, 9300–12,000 ft.

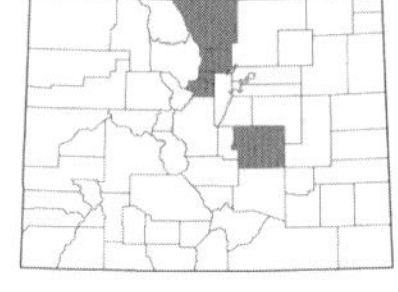

***Artemisia pattersonii* A. Gray**, ALPINE SAGEWORT. Perennials, 0.8–2 dm, silky canescent; *lower leaves* once-pinnately divided at the tip, upper leaves entire, linear, 2–4 cm long; *heads* 1–5, nodding or horizontal, arranged in a raceme-like or spike-like inflorescence; *involucre* to 5 mm high, villous with dark-lined margins; *receptacle* long-hairy; *disk flowers* 32–100. Found in open, rocky places in the alpine, 10,500–14,000 ft. July–Aug. E/W.

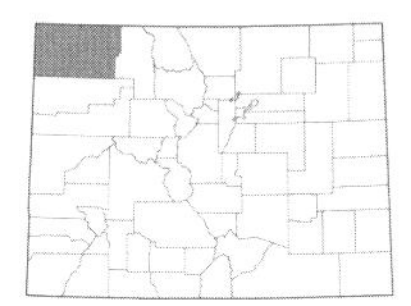

Artemisia pedatifida **Nutt.**, BIRDFOOT SAGEBRUSH. [*Oligosporus pedatifidus* (Nutt.) Poljakov]. Subshrubs, 0.5–1.5 dm, appressed-tomentose; *leaves* 1–2 times ternately divided into linear segments, 1–2 cm long; *heads* few, erect, arraned in a spike or narrow raceme; *involucre* 3–4 mm high; *receptacle* naked; *disk flowers* 5–9. Found on dry plateaus and ridges and often in clay soil, 6000–7000 ft. May–July. W.

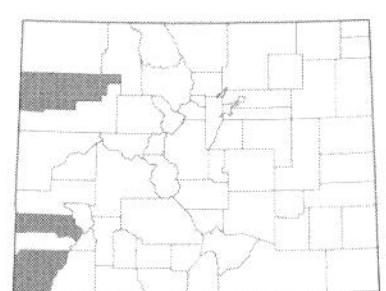

Artemisia pygmaea **A. Gray**, PYGMY SAGEBRUSH. [*Seriphidium pygmaeum* (A. Gray) W.A. Weber)]. Subshrubs, 0.5–2 dm; *leaves* deeply pinnatifid with 2–5 pairs of lateral segments to 4 mm long, or some 3-lobed or the upper entire, 0.3–1.2 cm long; *heads* 4–15, erect, arranged in a spike or narrow raceme; *involucre* 4–6.5 mm high, often resinous-glandular, firm; *receptacle* naked; *disk flowers* 3–5. Uncommon on shale slopes and dry mesas, limited to calcareous, shale soils, 4900–6500 ft. Aug.–Sept. W.

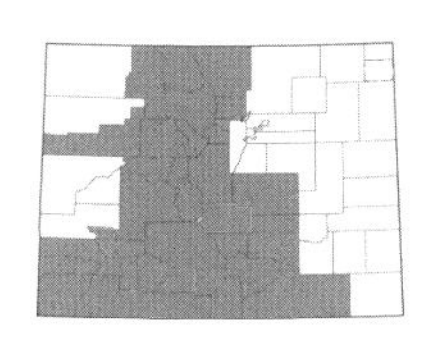

Artemisia scopulorum **A. Gray**, DWARF SAGEWORT. Perennials, 0.5–3.5 dm; *leaves* bipinnatifid or ternate-pinnatifid into linear segments, 0.5–3.5 cm long, sericeous-tomentose; *heads* 5–12, nodding, arranged in a spike; *involucre* 4–6 mm high, the bracts with dark-lined margins; *receptacle* long-hairy; *disk flowers* 15–30. Common in open, rocky places and meadows in the alpine, 10,000–14,000 ft. July–Sept. E/W. (Plate 9)

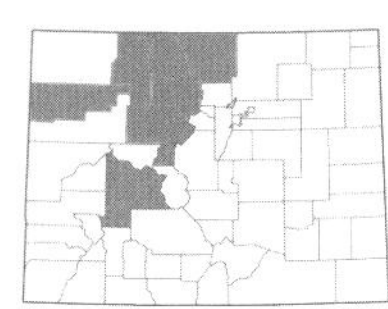

Artemisia spiciformis **Osterh.**, SNOWFIELD SAGEBRUSH. [*A. tridentata* Nutt. ssp. *spiciformis* (Osterh.) Kartesz & Gandhi]. Shrubs, 3–8 dm; *leaves* shallowly to deeply 3-lobed or 5-toothed at the tips, long-cuneate, 2–5 cm long, sericeous-tomentose; *heads* 5–25, erect, arranged in a narrow panicle; *involucre* 5–6.5 mm high; *receptacle* naked; *disk flowers* 8–20 (27). Found in dry to moist meadows and forest openings, often where drifting snow accumulates, at typically higher elevations in the mountains, 7500–10,000 ft. July–Sept. E/W.

Our plants were long called *A. rothrockii* A. Gray, which is a species from California. Hall and Clements (1923) and Weber (1996) group *A. spiciformis* with *A. tridentata* var. *vaseyana* [*Seriphidium vaseyanum* (Rydb.) W.A. Weber]. *Artemisia spiciformis* can be delimited from *A. tridentata* by the presence of a much narrower, spicate inflorescence and fewer, larger heads per stem.

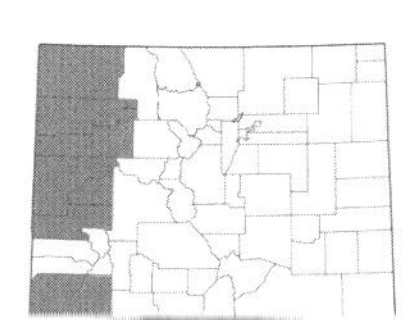

Artemisia spinescens **DC. Eaton**, BUDSAGE. [*Picrothamnus desertorum* Nutt.]. Shrubs, 0.5–5 dm, villous; *leaves* palmately 3–5-parted, the divisions cleft into narrow segments, to 2 cm long; *heads* numerous, nodding, arranged in a raceme or spike, barely surpassing the leaves; *involucre* 2–3.5 mm high, the bracts obscurely imbricate and few; *receptacle* naked; *disk flowers* 5–20. A common low shrub found on dry plains and hills, 5000–8000 ft. April–June. W. (Plate 9)

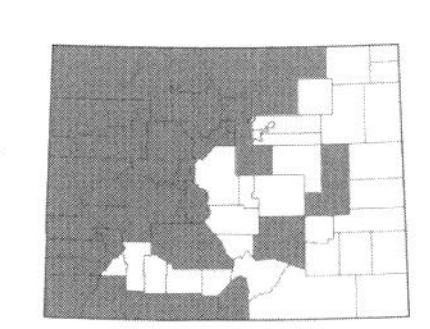

Artemisia tridentata **Nutt.**, BIG OR COMMON SAGEBRUSH. [*Seriphidium tridentatum* (Nutt.) W.A. Weber]. Shrubs, 4–35+ dm, sericeous-tomentose; *leaves* 3-lobed at the tip with a cuneate base, 0.5–5 cm long; *heads* numerous, erect, arranged in a panicle; *involucre* 3–5 mm high; *receptacle* naked; *disk flowers* usually 3–9 (11). Common and widespread in dry, sometimes rocky places of the plains, foothills, and intermountain parks and in open places in the mountains where its range extends to timberline, 4700–10,000 (12,000) ft. July–Oct. E/W.

There are three ecotypic varieties of *A. tridentata* in Colorado. However these are quite difficult to determine from herbarium specimens unless careful notes to plant height and inflorescence characteristics are made in the field. Intermediates are known to form where the varieties meet:

1a. Plants taller, 10–30 dm; inflorescence relatively large and broad...**var. *tridentata***, BIG SAGEBRUSH.
1b. Plants shorter, generally less than 10 dm; inflorescence relatively narrow...2

2a. Inflorescence mostly arising to about the same level, the plant appearing flat-topped when fresh; leaves with a sweet, camphor-like odor when crushed... **var. *vaseyana* (Rydb.) B. Boivin**, VASEY'S SAGEBRUSH. [*Seriphidium vaseyanum* (Rydb.) W.A. Weber; including var. *pauciflora* Winward & Goodrich].
2b. Inflorescence arising to different heights; leaves with a spicy odor when crushed... **var. *wyomingensis* (Beetle & A.L. Young) S.L. Welsh**, WYOMING SAGEBRUSH.

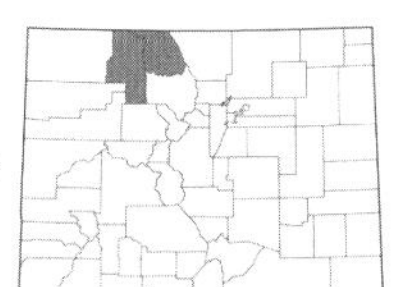

Artemisia tripartita **Rydb. var. *tripartita***, THREETIP SAGEBRUSH. [*Artemisia argillosa* Beetle; *Seriphidium tripartitum* (Rydb.) W.A. Weber]. Shrubs, 2–10 dm, closely tomentose; *leaves* deeply 3-lobed into linear segments, 0.6–2 cm long; *heads* numerous, erect, arranged in a narrow panicle; *involucre* 3–4 mm high; *receptacle* naked; *disk flowers* 4–8. Found near Coalmont in Jackson Co. and in the Yampa River Valley south of Oak Creek in Routt Co., 8000–9000 ft. July–Sept. E/W.

Specimens of this species were originally called *A. argillosa*, which is a synonym of *A. tripartita* var. *tripartita* and not *A. cana* var. *viscidula*.

ASTER L. – ASTER

Perennial herbs; *leaves* alternate, simple; *heads* radiate; *involucre bracts* subequal in 2–4 series, herbaceous and also often with purple tips; *receptacle* naked; *disk flowers* yellow; *ray flowers* white, blue, purple, or pink; *pappus* of capillary bristles; *achenes* usually several-nerved.

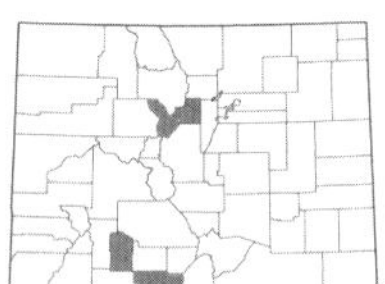

***Aster alpinus* L. var. *vierhapperi* (Onno) Cronquist**, ALPINE ASTER. Plants 2.5–32 (40) cm; *leaves* densely villous, glandular or eglandular, 1–11 cm long; *heads* solitary; *involucre bracts* 6–11 mm high; *pappus* 5–6 mm long; *achenes* (2) 2.5–3.2 mm long. Uncommon in rocky subalpine and alpine meadows, 8500–12,500 ft. June–Aug. E/W.

BACCHARIS L. – FALSE WILLOW; GROUNDSEL-TREE

Dioecious shrubs or subshrubs; *leaves* simple, alternate; *heads* discoid; *involucre bracts* subequal to imbricate in several series, chartaceous to subherbaceous, usually with scarious margins; *receptacle* naked; *disk flowers* white to yellowish-brown or light brown; *ray flowers* absent; *pappus* of capillary bristles; *achenes* 4- to 10-ribbed. (Mahler & Waterfall, 1964)

1a. Leaves numerous, the upper not reduced to bracts, finely gland-dotted; heads numerous and crowded in leafy-bracteate terminal inflorescences, smaller (involucre 4–9 mm high); pappus whitish; plants distinctly woody throughout...***B. salicina***
1b. Leaves not crowded, the upper reduced to linear bracts, not resinous; heads terminating the individual branches, not closely crowded, larger (involucre 5–14 mm high); pappus reddish to tawny; plants only woody at the base...***B. wrightii***

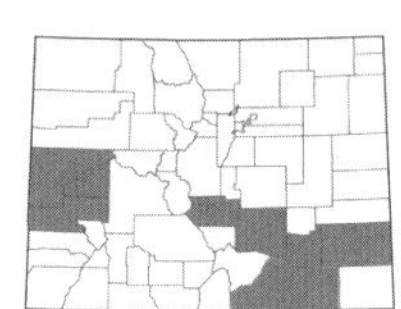

***Baccharis salicina* Torr. & A. Gray**, GREAT PLAINS FALSE WILLOW. Shrubs, 10–30 dm; *leaves* oblong to oblanceolate, 2.5–7 cm long, the margins usually irregularly toothed above, finely gland-dotted; *heads* 100–200+, in a paniculate array; *involucre* 4–9 mm high, the bracts gland-dotted, resinous, otherwise glabrous; *pappus* 8–12 mm long, whitish. Found in sandy or saline soil along streams and in meadows on the plains, 3400–5500 ft. May–Sept. E/W.

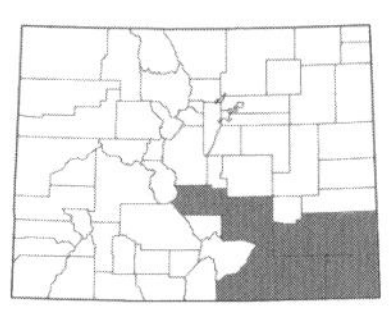

***Baccharis wrightii* A. Gray**, WRIGHT'S FALSE WILLOW. Shrubs or subshrubs, 1–8 dm; *leaves* oblanceolate to narrowly oblong, 0.5–1 (2.5) cm long, with entire or finely toothed margins, not resinous; *heads* solitary and terminal on each branch; *involucre* 5–14 mm high, glabrous, eglandular; *pappus* 15–20 mm long, tawny or reddish. Found in sandy, dry soil on the plains and rocky places of canyons and mesas, 4000–5900 ft. May–June. E.

BAHIA Lag. – BAHIA

Biennials or short-lived perennials herbs; *leaves* once or twice pinnatifid, alternate; *heads* radiate; *involucre bracts* subequal in one to two series, herbaceous and green; *receptacle* naked, flat; *ray flowers* yellow; *pappus* absent; *achenes* 4-angled, slender, usually glabrous.

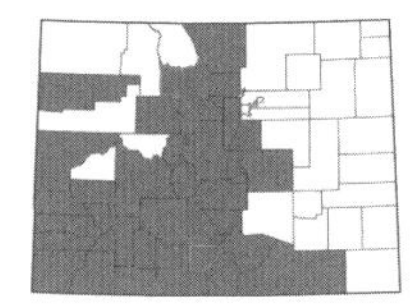

***Bahia dissecta* (A. Gray) Britton**, CUTLEAF. [*Amauriopsis dissecta* (A. Gray) Rydb.]. Plants 2–8 dm, with glandular stems; *leaves* distally reduced, minutely hairy to subglabrous, the larger leaves 3–15 cm long, with ultimate segments up to 2 cm long and 4 mm wide; *involucre* 4–8 mm high, glandular-hairy; *achenes* 3–4.5 mm long, black. Common along roadsides and in open places in the mountains in sandy or gravelly soil, 6000–9800 ft. July–Oct. E/W. (Plate 9)

BALSAMORHIZA Hook. ex Nutt. – BALSAMROOT

Perennial scapose or subscapose herbs; *heads* usually solitary, radiate; *involucre bracts* imbricate to subequal in several series, herbaceous and green; *receptacle* chaffy, this enclosing the achenes, broadly convex; *disk flowers* numerous, yellow; *ray flowers* usually yellow; *pappus* absent; *achenes* 4-angled, radially compressed. (Ownbey & Weber, 1943; Sharp, 1935)

1a. Leaves sagittate, entire, with silvery, felt-like tomentum (at least when young)...***B. sagittata***
1b. Leaves pinnatifid, with hirsute and glandular hairs...***B. hispidula***

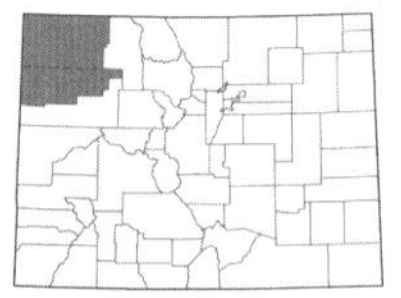

***Balsamorhiza hispidula* Sharp**, HAIRY BALSAMROOT. [*B. hookeri* Nutt. var. *hispidula* (Sharp) Cronquist]. Plants 1–3 dm; *leaves* 1–2-pinnatifid, 0.5–10 cm long, hirsute and gland-dotted; *involucre* 15–30 mm diam.; *ray flowers* 1.5–3 (4.5) cm long. Found in rocky, sandy soil, 6000–8800 ft. May–June. W.

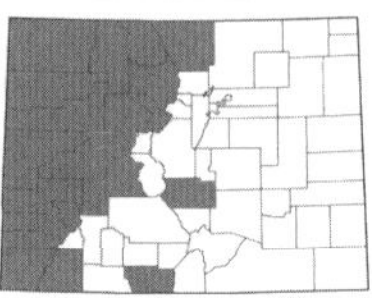

***Balsamorhiza sagittata* (Pursh) Nutt.**, ARROWLEAF BALSAMROOT. Plants (1.5) 2–5 (6.5) dm; *leaves* sagittate, sericeous and usually gland-dotted, 5–25 cm long; *involucre* 12–25 mm diam.; *ray flowers* 2–4 cm long. Common in sagebrush meadows, 5500–9300 ft. May–July. E/W. (Plate 9)

BERLANDIERA DC. – GREEN EYES

Perennials; *leaves* alternate, simple; *heads* radiate; *involucre bracts* broad and hemispheric, herbaceous, imbricate in a few series; *receptacle* chaffy; *disk flowers* yellow or red to maroon; *ray flowers* yellow to orange-yellow, with 9–12 prominent abaxial veins; *pappus* absent or an inconspicuous crown; *achenes* obcompressed, adhering to the two adjacent chaffy bracts and associated disk flowers plus the subtending involucre bract, these all falling together.

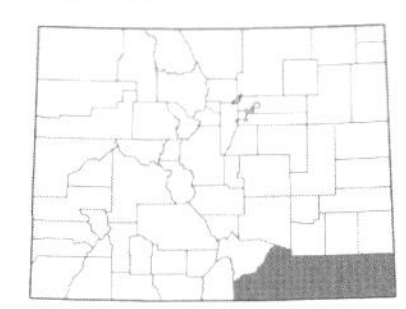

Berlandiera lyrata **Benth.**, CHOCOLATE FLOWER. Plants to 12 dm; *leaves* oblanceolate or obovate to spatulate, velutinous, those of midstem lyrate-pinnatifid; *peduncles* scabrous; *achenes* 4.5–6 mm long. Common on dry, rocky limestone soil of the southeastern plains, 3500–5200 ft. May–Aug. E.

BIDENS L. – BEGGAR-TICKS

Herbs; *leaves* opposite; *heads* discoid or radiate; *involucre bracts* in two distinct series, the outer herbaceous and the inner usually striate; *receptacle* chaffy; *disk flowers* yellow or orange; *ray flowers* (if present) neutral or pistillate, yellow to white or pink; *pappus* of 2–4 awns or teeth, these usually retrorsely barbed; *achenes* flattened or sometimes 4-angled, hairy. (Sherff, 1937)

1a. Leaves simple...2
1b. Leaves pinnately dissected, or pinnately or trifoliately compound...3

2a. Leaves sessile or basally connate; heads nodding in age; corolla of disk flowers 5-lobed...***B. cernua***
2b. At least some stem leaves petiolate; heads erect; corolla of disk flowers 4-lobed or rarely 5-lobed...***B. tripartita***

3a. Leaves once-pinnately or trifoliately compound with 3–5 distinct leaflets...4
3b. Leaves 2–3-times pinnately compound or dissected...5

4a. Outer involucre bracts mostly 8 (ranging from 5–10); heads smaller, the disk about 10 mm across; disk flowers orange-yellow...***B. frondosa***
4b. Outer involucre bracts mostly 13 (ranging from 10–16 or more); heads larger, the disk 15–25 mm across; disk flowers yellow...***B. vulgata***

5a. Inner involucre bracts hispid-hirsute, the outer glabrous or nearly so; ultimate leaf segments very narrow, usually less than 2 mm wide, linear...***B. tenuisecta***
5b. Involucre bracts with ciliate margins, hairy only at the base near the peduncle; ultimate leaf segments broader, usually greater than 2.5 mm wide, oblong with a cuneate base...***B. bigelovii***

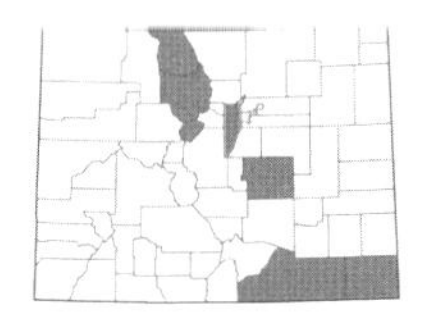

Bidens bigelovii **A. Gray**, BIGELOW'S BEGGAR-TICKS. Annuals, 2–8 dm, glabrous or nearly so; *leaves* 2.5–9 cm long, 2–3-pinnatifid; *involucre* 2.5–7 mm high, the outer bracts linear and the inner lanceolate; *rays* 1–3 (7) mm long and whitish or absent; *pappus* of 2 retrorsely barbed awns 2–4 mm long; *achenes* narrowly linear in the middle and the outer ones linear-cuneate. Found in wet soil along streams and pond edges and in drier soil of canyons and hillsides, 4000–9000 ft. July–Sept. E.

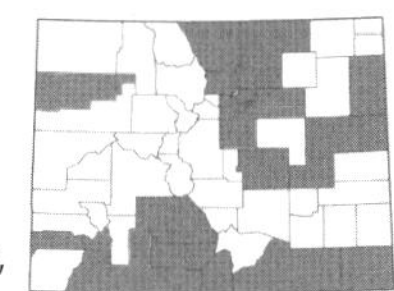

Bidens cernua **L.**, NODDING BEGGAR-TICKS. Annuals, 2–10 (40) dm, glabrous or the stem scabrous-hispid; *leaves* lanceolate to lance-ovate, sessile, the margins usually toothed, 4–20 cm long; *involucre* of lance-linear to ovate bracts 2–10 mm long; *rays* 2–15 mm long and yellow or sometimes absent; *pappus* of (2) 4 retrorsely barbed awns 2–4 mm long; *achenes* 4-angled, narrowly cuneate. Common along streams, ditches, or in other low, wet places or disturbed areas, 4400–8900 ft. July–Oct. E. Introduced.

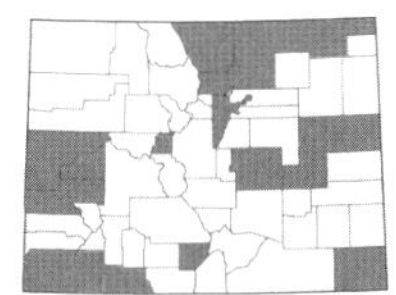

Bidens frondosa **L.**, DEVIL'S BEGGAR-TICKS. Annuals, 2–12 (18) dm, glabrous or nearly so; *leaves* pinnately compound into 3–5 lanceolate or lance-ovate segments, petiolate, 3–20 cm long; *involucre* of linear-spatulate outer bracts and ovate to lanceolate inner bracts; *rays* 2–3.5 mm long and yellow or absent; *pappus* of 2 antrorsely or retrorsely barbed awns 2–5 mm long; *achenes* narrowly cuneate, flat, with 1-nerve on each face. Common in disturbed areas, along ditches, especially in wet soil, 3900–7000 ft. June–Sept. E/W. Introduced in Colorado.

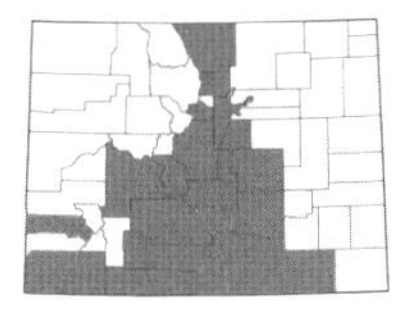

Bidens tenuisecta **A. Gray**, SLIMLOBE BEGGAR-TICKS. Annuals, 1–4 dm, glabrous or nearly so; *leaves* pinnately 2–3 times dissected into small, linear segments, 2–5 cm long, petiolate; *involucre* of lance-linear bracts, hispid-hirsute; *rays* 4–6 mm long and yellow; *pappus* of 2 retrorsely barbed awns 1–3 mm long; *achenes* weakly 4-angled, coarsely ribbed. Found in typically wet soil along streams and ditches, along roadsides, and in disturbed places, 5000–9500 ft. July–Sept. E/W. (Plate 9)

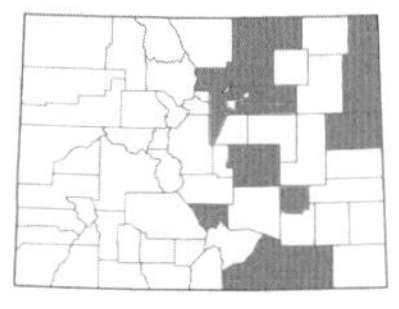

Bidens tripartita **L.**, STRAW-STEM BEGGAR-TICKS. [*B. comosa* (A. Gray) Wieg.]. Annuals, 2–15 dm, glabrous or nearly so; *leaves* elliptic-lanceolate, 3–15 cm long, sessile or petiolate; *involucre* of ovate bracts; *rays* 4–8 mm long and yellow or absent; *pappus* absent or of 2–3 retrorsely barbed awns 2–3 (6) mm long; *achenes* very flat, usually smooth. Found along streams, ditches, and in moist disturbed areas, 4000–6500 ft. July–Oct. E.

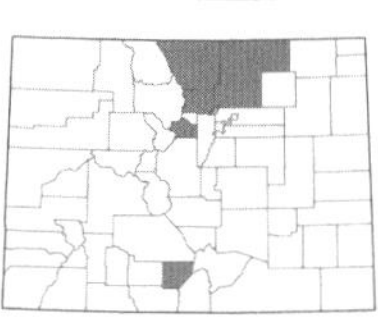

Bidens vulgata **Greene**, TALL BEGGAR-TICKS. Annuals, 3–15 dm, glabrous to minutely hairy; *leaves* pinnatifid into 3–5 lanceolate segments, 5–15 cm long, petiolate, the margins toothed; *involucre* of lanceolate to ovate bracts; *rays* 2.5–3.5 mm long and yellow or absent; *pappus* of 2 retrorsely barbed awns 3–4 (7) mm long; *achenes* flat. Found along ditches and streams, and in moist disturbed areas, 5000–7500 ft. June–Sept. E.

BRICKELLIA Elliott – BRICKELLBUSH; FALSE BONESET

Perennial herbs or shrubs; *leaves* alternate or opposite, simple; *heads* discoid; *involucre bracts* longitudinally striate, imbricate in several series or subequal; *receptacle* naked; *disk flowers* white or ochroleucous or pale green; *ray flowers* absent; *pappus* of numerous capillary, plumose, or barbellate bristles; *achenes* 10-ribbed. (Robinson, 1917)

1a. Leaves triangular to deltoid-ovate, mostly 2 cm or more in length, distinctly petiolate...2
1b. Leaves lanceolate, or if ovate then less than 2 cm in length, sessile or shortly petiolate...3

2a. Shrubs or subshrubs; outer involucre bracts acute, without a slender appendage at the tip; heads with 8–12 flowers...***B. californica***
2b. Perennial herbs; outer involucre bracts with a short, slender appendage at the tip; heads with mostly 20–40 flowers...***B. grandiflora***

3a. Leaves ovate to oblong-ovate, small (mostly less than 2 cm in length); outer and middle involucre bracts with a short, green, mostly reflexed tip...***B. microphylla***
3b. Leaves lanceolate to narrowly ovate or oblong, 1–11 cm in length; outer and middle involucre bracts without a reflexed green tip...4

4a. Heads with mostly 4–6 flowers; involucre 5–7 mm high; leaves shiny green and appearing varnished, glabrous or nearly so; stems white...***B. longifolia***
4b. Heads with 8 or more flowers; involucre 8 mm or more high; leaves not shiny green, minutely hairy to scabrous; stems green to brown...5

5a. Pappus barbellate; heads large with involucre bracts 10–20 mm high and 40–50 flowers per head...***B. oblongifolia* var. *linifolia***
5b. Pappus plumose to subplumose; heads smaller with involucre bracts 7–14 mm high and mostly 30 flowers or less per head...6

6a. Outer and middle involucre bracts with a bristle tip or abruptly acuminate; cilia on pappus shorter, about 0.2–0.3 mm long; 8–10 flowers per head...***B. brachyphylla***
6b. Outer and middle involucre bracts without a bristle tip, slender and acute; cilia on pappus longer, about 0.4–0.5 mm long; 15–34 flowers per head...***B. eupatorioides***

Brickellia brachyphylla **A. Gray**, PLUMED BRICKELLBUSH. Perennials or subshrubs, 2.5–8 dm; *leaves* lanceolate or narrowly ovate, minutely hairy and gland-dotted, 1–4 cm long, with serrate to nearly entire margins; *involucre* 7–9 mm high, the bract tips abruptly acuminate to nearly aristate into a short bristle; *disk flowers* 8–10 per head. Uncommon in rocky soil and on sandstone ledges, 4000–7500 ft. July–Sept. E/W.

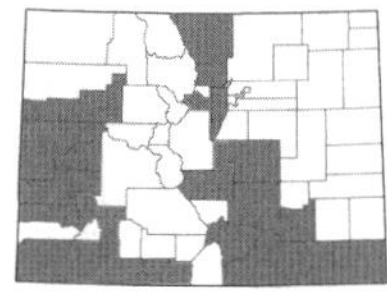

Brickellia californica **(Torr. & A. Gray) A. Gray**, CALIFORNIA BRICKELLBUSH. Shrubs, 4–10 dm; *leaves* triangular, usually with a cordate or truncate base, scabrous-puberulent, 1–8 cm long, with toothed or shallowly lobed margins; *involucre* 5–8 mm high, the bract tips rounded to acutish; *disk flowers* 8–12 per head. Found in open, dry, rocky places, 4500–9200 ft. July–Sept. E/W.

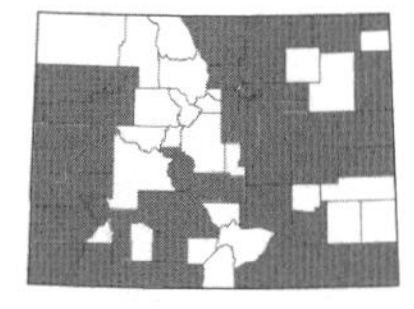

Brickellia eupatorioides **(L.) Shinners**, FALSE BONESET. [*Kuhnia eupatorioides* L.]. Perennials, 3–10 dm; *leaves* lanceolate, scabrous above, glandular-punctate, 2–10 cm long, entire to irregularly toothed; *involucre* 8–14 mm high, the bract tips acute and slender; *disk flowers* 15–34. Common in dry, open places, 3500–9000 ft. July–Sept. E/W. (Plate 9)

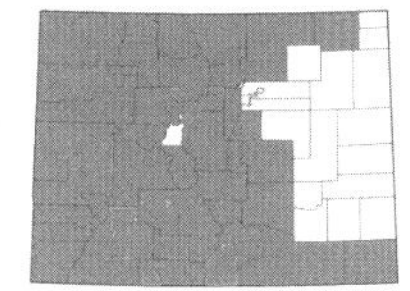

***Brickellia grandiflora* (Hook.) Nutt.**, TASSELFLOWER BRICKELLBUSH. Perennials, 2.5–7 dm; *leaves* triangular to subhastate, minutely hairy to glabrous, 2–7 cm long, with crenate-dentate margins; *involucre* 7–12 mm high, the outer bracts with a slender appendage at the tip; *disk flowers* usually 20–40. Common in rocky soil and forest openings, 4800–12,000 ft. July–Sept. E/W.

***Brickellia longifolia* S. Watson**, LONGLEAF BRICKELLBUSH. Shrubs, 2–25 dm; *leaves* lanceolate, glabrous or nearly so, usually 2.5–11 cm long, entire; *involucre* 5–7 mm high, the outer bracts with acute tips and the inner with broad blunt tips; *disk flowers* (3) 4–6. Found in rocky or sandy soil, known from Mesa, Delta, and Montrose cos., 4800–5500 ft. Aug.–Oct. W.

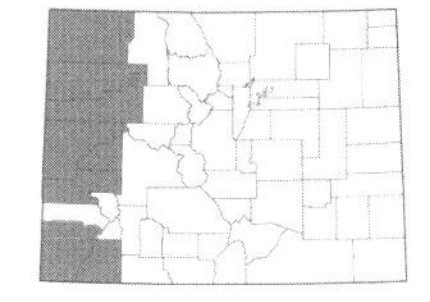

***Brickellia microphylla* (Nutt.) A. Gray var. *scabra* A. Gray**, LITTLELEAF BRICKELLBUSH. Shrubs or subshrubs, 3–8 dm; *leaves* ovate to oblong-ovate, scabrous, 0.5–2 cm long, entire or toothed; *involucre* 7–12 mm high, the outer and middle bracts with a short, green, spreading tip; *disk flowers* 10–26. Found in dry, open, rocky or sandy soil, often with pinyon-juniper, 4300–7000 ft. July–Sept. W.

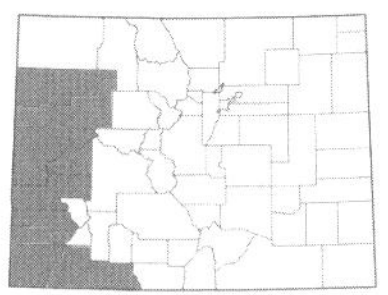

***Brickellia oblongifolia* Nutt. var. *linifolia* (D.C. Eaton) B.L. Rob.**, MOJAVE BRICKELLBUSH. Perennials or subshrubs, 1–6 dm; *leaves* lance-linear to elliptic, minutely hairy, 1.4–4 cm long, entire or nearly so; *involucre* 10–20 mm high, the bracts acute; *disk flowers* 40–50. Found on sandstone ledges, shale, or rocky hillsides, often with pinyon-juniper, 4500–9700 ft. May–July. W.

CARDUUS L. – PLUMELESS THISTLE

Annual or biennial spiny herbs; *leaves* alternate, bases strongly decurrent; *heads* discoid; *involucre bracts* imbricate in several series, the tips spreading to reflexed, the midrib excurrent as a spine; *receptacle* bristly; *disk flowers* purplish or reddish or rarely white; *ray flowers* absent; *pappus* of barbellate bristles; *achenes* with longitudinal lines, glabrous, basifixed.

1a. Heads smaller, 1–2.5 cm in diam.; outer involucre bracts narrowly lanceolate, 1–2 mm wide at the base, without a constriction in the middle...***C. acanthoides***

1b. Heads larger, 3–7 cm in diam.; outer involucre bracts ovate-lanceolate to lanceolate, 2–10 mm wide, with a shallow constriction in the middle...***C. nutans***

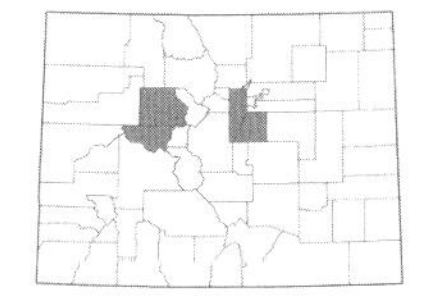

***Carduus acanthoides* L.**, PLUMELESS THISTLE. Biennials, 3–15 dm, with strongly spiny-winged stems; *leaves* 10–30 cm long, deeply 1–2-pinnately lobed or spiny-toothed, sparsely hairy; *involucre* 1–2.5 cm diam.; *disk flowers* 13–20 mm long, purple; *pappus* 11–13 mm long; *achenes* 2.5–3 mm long. Found in disturbed places, open fields, and along roadsides, 6900–8300 ft. July–Oct. E/W. Introduced.

The first specimens of *C. acanthoides* documented for Colorado were from Jefferson County in 1957. Although not as widespread as the next, *C. acanthoides* is another relatively recently introduced plant native to Europe and is becoming a troublesome weed.

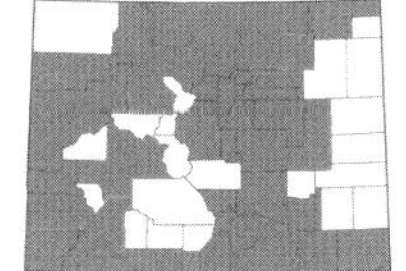

***Carduus nutans* L.**, MUSK THISTLE; NODDING THISTLE. [*Carduus nutans* L. ssp. *macrolepis* (Peterm.) Kazmi]. Annuals or biennials, 4–20+ dm, with spiny-winged stems; *leaves* 10–40 cm long, 1–2-pinnately lobed; *involucre* 3–7 cm diam.; *disk flowers* 15–28 mm long, purple; *pappus* 13–25 mm long; *achenes* 4–5 mm long. Common in disturbed places, open fields and meadows, and along roadsides, 3500–8500 ft. June–Oct. E/W. Introduced. (Plate 9)

Although *C. nutans* is a common and widespread plant today, it is a relatively recently introduced plant. When Harrington (1954) published his *Manual of the Plants of Colorado*, he stated that "[*C. nutans*] has been reported in northeastern Arizona and may well be in Colorado." Although the first specimens of *C. nutans* appeared in 1953 in Boulder and Jefferson cos., *C. nutans* is now incredibly widespread. *Carduus nutans* has been reported for over 50% of the counties in Colorado and is widespread on both the eastern and western slopes.

CARTHAMUS L. – SAFFLOWER

Annual herbs; *leaves* simple, alternate to subopposite; *heads* discoid; *involucre bracts* imbricate in several series, rather foliaceous, the inner with a spine tip; *receptacle* chaffy; *disk flowers* with the corollas deeply 5-lobed and bright orange to yellow; *ray flowers* absent; *pappus* usually lacking; *achenes* 4-angled.

***Carthamus tinctorius* L.**, SAFFLOWER. Plants 3–10 dm, with striate stems; *leaves* sessile, clasping, broadly elliptic to ovate or oblong, spine-tipped; *involucre* 2–4 cm in diam.; *disk flowers* 2–3 cm long; *achenes* white, 6–9 mm long. This species does not appear to escape from cultivation in Colorado but is included here because many people see the bright flowers along the roadsides. 4000–5000 ft. July–Aug. E. Introduced.

Cultivated for the achenes which are harvested for oil and birdseed and a dye which is extracted from the flowers.

CENTAUREA L. – STAR-THISTLE, KNAPWEED

Herbs; *leaves* alternate, usually pinnatifid or toothed; *heads* discoid; *involucre bracts* imbricate in several series, the margins and tips often fringed or spinose; *receptacle* densely bristly; *disk flowers* all perfect or sometimes the marginal ones sterile with an enlarged, irregular, falsely radiate corolla, the corollas purple, blue, yellow, or white; *ray flowers* absent; *pappus* lacking or of one to several series of graduated bristles or narrow scales; *achenes* 4-angled or compressed.

1a. Flowers yellow...2
1b. Flowers white, lavender, or blue...3

2a. Involucre bracts spine-tipped, the spine 11–22 mm long; leaves gray-tomentose and scabrous to short-bristly, stem leaves long-decurrent on the stem...***C. solstitialis***
2b. Involucre bracts not spine-tipped, the margins lacerate fringed or less often tipped by weak spines that are less than 2 mm long; leaves thinly arachnoid-hairy, with glandular dots, the bases sessile or shortly decurrent...***C. macrocephala***

3a. Involucre bracts spine-tipped, the central spine 1–5 mm long, and also with marginal fringed spines; leaves more or less pinnatifid with narrow lobes...4
3b. Involucre bracts not spine-tipped, usually with erose or fringed tips and margins, sometimes with marginal fringed spines but without a terminal spine; leaves entire to lobed or pinnatifid...5

4a. Flowers white or less often lavender, 12–13 mm long; involucre 10–13 mm high...***C. diffusa***
4b. Flowers pink or pale lavender, 7–9 mm long; involucre 7–8 mm high...***C. virgata***

5a. Upper stem leaves pinnatifid with narrow lobes...***C. stoebe* ssp. *micranthos***
5b. Upper stem leaves simple and entire or sometimes irregularly toothed or shallowly lobed, but not regularly pinnatifid...6

6a. Involucre bracts completely brown, the outer ones with upward curving, narrowly ciliate margins; flowers purple or rarely white...***C. jacea***
6b. Involucre bracts green to pale yellow with brown erose or fringed margins; flowers usually blue but sometimes purple, pink, or white...7

7a. Leaves not decurrent on the stem; lower leaves often lobed, upper stem leaves narrow, about 0.3–1 cm wide; annuals or winter-annuals...***C. cyanus***
7b. Leaves decurrent on the stem; lower leaves usually not lobed, upper stem leaves wider, mostly 2–4 cm wide; perennials...***C. montana***

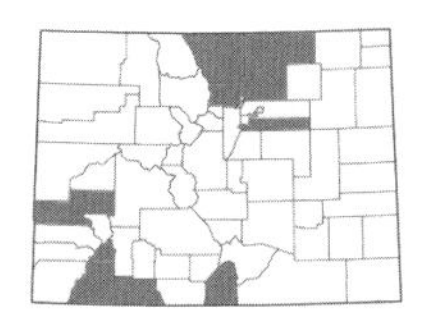

***Centaurea cyanus* L.**, BACHELOR'S BUTTON, CORNFLOWER. Annuals, 2–10 dm, loosely tomentose; *leaves* entire or with remote linear lobes, linear to lanceolate, 3–10 cm long; *involucre* 10–16 mm high, the bracts green with white to dark brown or black scarious, fringed margins; *disk flowers* blue or sometimes white or purple, the outer enlarged and ray-like; *pappus* 1–4 mm long. A commonly cultivated plant occasionally escaping into adjacent roadsides and meadows but usually not persisting, 5000–6000 ft. May–Sept. E/W. Introduced.

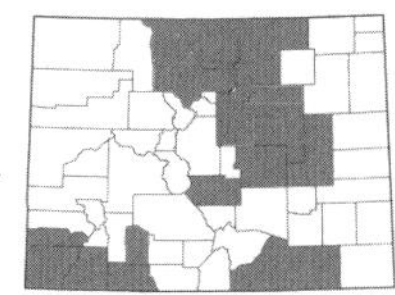

***Centaurea diffusa* Lam.**, DIFFUSE KNAPWEED. [*Acosta diffusa* (Lam.) Sojak]. Biennials or perennials, 2–8 dm; *leaves* 2-pinnately divided into narrow segments, 10–20 cm long, resin-dotted and hispidulous; *involucre* 10–13 mm high, the bracts green with prominent parallel veins, the margins fringed with slender spines, the tip with a spine 1–3 mm long; *disk flowers* 12–13, creamy-white or rarely pink or pale purple; *pappus* absent or of small scales. Found in disturbed areas, along roadsides, and along streams and ditches, especially common in the Denver-Boulder area, 5000–8200 ft. July–Sept. E/W. Introduced. (Plate 9)

This species can hybridize with *C. stoebe* ssp. *micranthos*, and the resulting hybrid is referred to as *C.* ×*psammogena* G. Gayer. The hybrid is extremely variable in form and may be intermediate between the two species, or more closely resemble one parent than the other. *Centaurea stoebe* ssp. *micranthos* has a pappus of 1–2 series of white, unequal stiff bristles while *C. diffusa* lacks a pappus or it is less than 0.5 mm long. Hybrids seem to always have a pappus present and usually have dark-margined phyllaries like *C. stoebe* ssp. *micranthos*, but the spines can be absent or present.

***Centaurea jacea* L.**, BROWN KNAPWEED. [*C. debauxii* Gren. & Godr. ssp. *thuillieri* Dostál; *C. pratensis* Thuill., non Salisb.]. Perennials, 3–15 dm; *leaves* oblanceolate to elliptic, 5–25 cm long, the margins entire or shallowly toothed; *involucre* 15–18 mm high, the bracts usually hidden by expanded appendages, the appendages brown, entire to dentate or dissected into filiform lobes; *disk flowers* purple; *pappus* absent or of scales 0.5–1 mm long. Known from one location in Colorado near Steamboat Springs, 7300–7500 ft. July–Sept. W. Introduced.

***Centaurea macrocephala* Puschk. ex Willd.**, BIGHEAD KNAPWEED. [*Grossheimia macrocephala* (Mussin-Puschkin) Sosnowsky & Takhtajan]. Perennials, 5–17 dm; *leaves* oblanceolate to narrowly ovate, 10–30 cm long, with entire or shallowly toothed margins, sessile or shortly decurrent; *involucre* 25–35 mm high, the bracts lacerate-fringed and sometimes with a weak spine 1–2 mm long at the tip, glabrous; *disk flowers* numerous, yellow; *pappus* of scales, 5–8 mm long. Escaped from cultivation near Boulder and Nederland, 5500–8500 ft. July–Sept. E. Introduced.

***Centaurea montana* L.**, MOUNTAIN CORNFLOWER. Perennials, 2.5–8 dm; *leaves* ovate to oblong, 10–30 cm long, entire or remotely toothed or lobed, decurrent; *involucre* 20–25 mm high, the bracts with dark pectinate-fringed appendages; *disk flowers* 25–60, blue to purple; *pappus* of bristles 0.5–1.5 mm long. Uncommon in disturbed places and along roadsides where it has escaped from cultivation, known from one collection made near Glenwood Canyon in Garfield Co., 6000–6500 ft. July–Sept. W. Introduced.

***Centaurea solstitialis* L.**, YELLOW STAR-THISTLE. [*Leucantha solstitialis* (L.) Á. Löve & D. Löve]. Annuals, 1–10 dm, gray-tomentose; *leaves* pinnately lobed below and entire above, long-decurrent, 1–15 cm long; *involucre* 13–17 mm high, the bracts with spiny margins and tipped with a central spine 1–2.5 cm long; *disk flowers* numerous, yellow; *pappus* of bristles 2–4 mm long. Found in waste places and along roadsides, with the last collections made in 1951 near Boulder, 5000–6000 ft. July–Sept. E/W. Introduced.

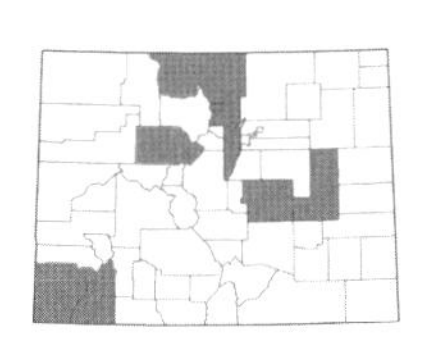

***Centaurea stoebe* L. ssp. *micranthos* (S.G. Gmelin ex Gugler) Hayek**, SPOTTED KNAPWEED. [*C. biebersteinii* DC.; *C. maculosa* Lam.; *Acosta maculosa* (Lam.) Holub]. Biennials or perennials, 3–10 dm; *leaves* deeply 1–2-lobed or the upper entire to lobed, shortly hispid and resin-dotted; *involucre* 10–13 mm high, the bracts parallel-veined, margins dark and fringed with slender teeth or lobes; *disk flowers* pink or purple; *pappus* of bristles 1–2 mm long. Found in waste places and along roadsides, 5000–8000 ft. July–Oct. E/W. Introduced.

Can hybridize with *C. diffusa*, and the resulting hybrid is referred to as *C.* ×*psammogena* G. Gayer. The hybrid is extremely variable in form and may be intermediate between the two species or more closely resemble one parent than the other. *Centaurea stoebe* ssp. *micranthos* has a pappus of 1–2 series of white, unequal stiff bristles while *C. diffusa* lacks a pappus or it is less than 0.5 mm long. Hybrids seem to always have a pappus present and usually have dark-margined phyllaries like *C. stoebe* ssp. *micranthos*, but the spines can be absent or present.

***Centaurea virgata* Lam.**, SQUARROSE KNAPWEED. Perennials, 2–5 dm; *leaves* deeply 1–2-lobed or the upper entire to lobed, resin-dotted, 10–15 cm long; *involucre* 7–8 mm high, the bracts fringed with slender spines and a spine tip 1–3 mm long; *disk flowers* few, pink or purple; *pappus* of bristles 2–2.5 mm long. Reported for Colorado, but no specimens have been seen. June–Sept. Introduced.

CHAENACTIS DC. – FALSE YARROW; PINCUSHION

Herbs; *leaves* alternate; *heads* discoid; *involucre bracts* equal or occasionally imbricate, herbaceous or subherbaceous; *receptacle* naked (ours); *disk flowers* white or pale cream or yellow, sometimes the marginal ones with a slightly enlarged, irregular, shortly subradiate corolla; *ray flowers* absent; *pappus* of scales, rarely absent; *achenes* terete, somewhat compressed. (Mooring, 1980; Stockwell, 1940)

1a. Pappus of 4 scales; marginal corollas commonly enlarged and shortly subradiate; annuals...***C. stevioides***
1b. Pappus of 8–16 scales; marginal corollas not enlarged and shortly subradiate; biennials or perennials...***C. douglasii***

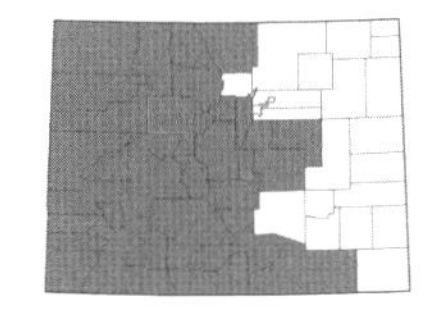

***Chaenactis douglasii* (Hook.) Hook. & Arn.**, DUSTYMAIDEN. Biennials or perennials, 5–60 cm, arachnoid-tomentose; *leaves* basal or basal and cauline, usually 2-pinnately lobed, the ultimate segments often involute or twisted; *involucre* obconic to hemispheric, the bracts 9–17 mm long, outer usually glandular-stipitate; *disk flowers* 5–8 mm long; *pappus* scales 3–6 mm long; *achenes* 5–8 mm long. Found in sandy or rocky soil, from pinyon-juniper and sagebrush communities to scree slopes in the alpine. 5300–14,000 ft. May–Aug. E/W. (Plate 9)

A very variable, highly complex species often separated into numerous varieties. Most of the varieties are ill-defined, however, and are included here within *C. douglasii* var. *douglasii*. The following varieties do intergrade at lower elevations:

1a. Leaves chiefly basal; peduncles unbranched or slightly branched just at the apex, terminating in 1–3 (4) heads; plants short, to 1 dm; true perennials with numerous slender branching caudices each terminating in a rosette of leaves...**var. *alpina* A. Gray**, ALPINE DUSTYMAIDEN; PINCHUSION. [*C. alpina* (A. Gray) M.E. Jones; *C. leucopsis* Greene]. Found in the alpine on scree slopes or in rocky places, occasionally found in the subalpine, 9500–13,500 ft. June–Aug. E/W.
1b. Basal and stem leaves present; peduncles branching with 2–4 heads per branch; plants taller, mostly 1–5 dm; biennials or short-lived perennials usually without a branching caudex...**var. *douglasii***, FALSE YARROW. [including var. *achilleaefolia* (Hook. & Arn.) A. Nelson and var. *montana* M.E. Jones]. Found in sandy or rocky places, often in pinyon-juniper or sagebrush, 5300–9000 (10,000) ft. May–Aug. (Sept.). E/W.

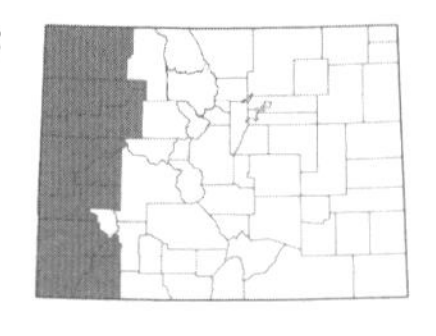

***Chaenactis stevioides* Hook. & Arn.**, Steve's dustymaiden. Annuals, 4–40 cm, arachnoid-tomentose; *leaves* basal and cauline, 1–2-pinnately lobed, the ultimate segments involute and twisted; *involucre* obconic to hemispheric, the bracts 5.5–10 mm long, glandular-stipitate; *disk flowers* 4.5–6.5 mm long; *pappus* scales 1.5–6 mm long; *achenes* 4–6.5 mm long. Found in sandy or clay soil, often in association with pinyon-juniper communities, 4500–6500 ft. May–July. W.

CHAETOPAPPA DC. – leastdaisy

Herbs; *leaves* alternate, simple, entire; *heads* radiate; *involucre bracts* imbricate in 2–6 series, green, rather dry, the margins scarious; *receptacle* naked; *disk flowers* yellow; *ray flowers* white to pinkish or drying purple; *pappus* of a hyaline crown or scales, capillary bristles, a few awn-like bristles, or a combination of alternating scales and bristles, or reduced to an inconspicuous crown and appearing absent; *achenes* terete to strongly compressed, 2–10-ribbed. (Nesom, 1988)

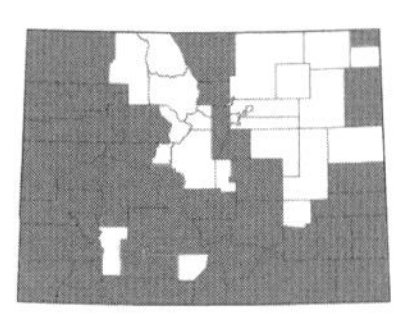

***Chaetopappa ericoides* (Torr.) G.L. Nesom**, rose heath. [*Leucelene ericoides* (Torr. & A. Gray) Greene]. Perennials, 6–12 cm, from rhizomes; *leaves* narrowly oblanceolate to lanceolate, 4–12 mm long, usually glandular-stipitate, coriaceous; *involucre* 6–7.5 mm high; *pappus* of one series of barbellate bristles; *achenes* 2.5–3.5 mm long. Found in dry, sandy, open places, especially common on the desert steppes and the grasslands of the eastern plains, 3500–8500 ft. May–July. E/W. (Plate 9)

CHAMAECHAENACTIS Rydb. – chamaechaenactis

Perennial herbs; *leaves* simple, entire or nearly so, petiolate, all basal; *heads* solitary at the end of scapose peduncles, discoid; *involucre bracts* subequal or biseriate, the inner series with scarious margins and anthocyanic tips; *receptacle* naked; *disk flowers* white or cream-colored; *ray flowers* absent; *pappus* of 7–10 hyaline scales; *achenes* multiribbed, densely hairy. (Preece & Turner, 1953)

***Chamaechaenactis scaposa* Rydb.**, fullstem. Plants 2–7 (9) cm, from a branching caudex; *leaves* elliptic, ovate, or rounded, 4–15 mm long, adaxial side sometimes glabrescent, canescent beneath; *involucre* 11–17 mm high; *disk flowers* 10–30+. Found on clay shale slopes and rocky soil, often in association with pinyon-juniper, 4500–7200 ft. May–June. W.

CHRYSANTHEMUM L. – chrysanthemum; daisy

Annual or perennial herbs; *leaves* alternate; *heads* radiate; *involucre bracts* imbricate in 2–4 series, the margins scarious, green throughout or only with a green midrib; *receptacle* naked; *disk flowers* yellow and 5-lobed; *ray flowers* white, yellow, pink, or reddish; *pappus* a short crown or lacking; *achenes* usually 10-ribbed.

***Chrysanthemum coronarium* L.**, crown daisy. [*C. roseum* Adams; *Glebionis coronaria* (L.) Cassini ex Spach]. Annuals; *leaves* oblong to obovate, most 20–55 mm long, two to three times pinnatifid, glabrous; *ray flowers* yellow, the corolla tips sometimes white, 15–25 mm; *achenes* 2.5–3 mm long. A cultivated plant, rarely escaping along roadsides or near old homesteads, known from one collection made near Morrison in 1905, 6000 ft. June–Aug. E. Introduced.

CHRYSOTHAMNUS Nutt. – rabbitbrush

Shrubs or half-shrubs; *leaves* alternate, simple, entire or scabrous-serrulate; *heads* discoid; *involucre bracts* imbricate in several series, usually keeled, often chartaceous, usually with a green herbaceous tip; *receptacle* naked; *disk flowers* yellow; *ray flowers* absent; *pappus* of numerous capillary bristles; *achenes* somewhat angled, narrow. (Anderson, 1978; 1980; Hall & Clements, 1923)

1a. Heads larger, involucre bracts 20–30, 9–15 mm in length; disk flowers 7–14 mm long; achenes glabrous or minutely hairy towards the tip...2

1b. Heads smaller, involucre bracts 15–20, 4–8 mm in length; disk flowers 4–7 mm long; achenes hairy or glabrous...3

2a. Leaves and involucre bracts finely minutely hairy; disk flowers 7–11 mm long, not surpassed by the pappus; plants generally smaller, 1–3 (5) dm in height...***C. depressus***

2b. Leaves and involucre bracts glabrous or with ciliate-scabrous margins; disk flowers 9–14 mm long, surpassed by the pappus; plants generally taller, 2.5–10 dm in height...***C. baileyi***

3a. Achenes glabrous or somewhat glandular or hairy just near the tip, longitudinally striate; leaves with numerous resinous dots on both the adaxial and abaxial sides...***C. vaseyi***

3b. Achenes hairy, not longitudinally striate; leaves without resinous dots or with resinous dots only numerous on the abaxial side of the leaf...4

4a. Leaves flat and not twisted, lanceolate and 3–8 mm wide; plants tall shrubs, 7–35 dm in height; the upper stem glabrous or nearly so...***C. linifolius***
4b. Leaves usually twisted, filiform to linear or if lanceolate then plants shorter, 1–15 dm in height and the upper stem densely minutely hairy to hispidulous...5

5a. Involucre bracts acuminate, usually with a short setulose tip; leaves filiform to linear, 0.5–1 mm wide, with 1 nerve...***C. greenei***
5b. Involucre bracts acute or obtuse, without a setulose tip; leaves linear to oblong or lanceolate, 1–10 mm wide, with 1–5 nerves...***C. viscidiflorus***

***Chrysothamnus baileyi* Wooton & Standl.**, BAILEY'S RABBITBRUSH. [*C. pulchellus* (A. Gray) Greene var. *baileyi* (Wooton & Standl.) Hall; *Lorandersonia baileyi* (Wooton & Standl.) Urbatsch]. Shrubs, 2.5–10 dm, glabrous or minutely hairy; *leaves* linear, flat, 1-nerved, 0.4–3.5 cm long; *involucre* 10–15 mm high, the bracts in 5 series, keeled; *disk flowers* 5, 9–14 mm long; *pappus* 9–12 mm long; *achenes* 5–7 mm long, usually sparsely hairy. Found in sandy soil on the southeastern plains, 3900–4200 ft. July–Sept. E.

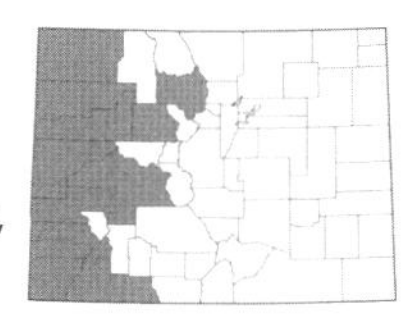

***Chrysothamnus depressus* Nutt.**, DWARF RABBITBRUSH. Shrubs or subshrubs, 1–3 (5) dm, short-hairy; *leaves* sessile, linear to oblanceolate, 1-veined, flat, 1–3 cm; *involucre* 9–15 mm high, the bracts imbricate in 5–6 ranks, strongly keeled; *disk flowers* 5–6, 7–11 mm long; *pappus* 5.5–7.5 mm long; *achenes* 5–6.5 mm long. Found in dry, often sandy soil, usually in association with sagebrush, oak, or pinyon-juniper communities, 5500–8700 ft. July–Sept. E/W.

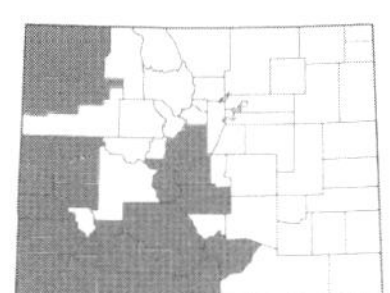

***Chrysothamnus greenei* (A. Gray) Greene**, GREENE'S RABBITBRUSH. Shrubs, 1–5 dm, glabrous; *leaves* sessile, linear, 1–4 cm long, flat, veins inconspicuous; *involucre* 5–8 mm high, the bracts weakly keeled, imbricate; *disk flowers* 4–5, 3.7–5.5 mm long; *pappus* 3.7–5 mm long; *achenes* 3–4 mm long, densely hairy. Found in dry, open places in sandy or rocky soil, 5000–8500 ft. July–Sept. E/W. (Plate 10)

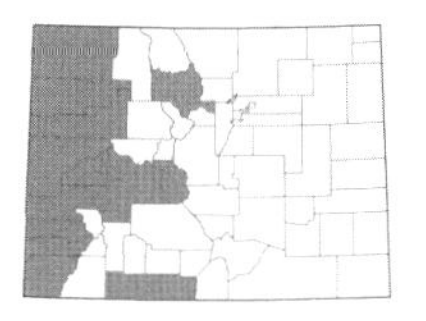

***Chrysothamnus linifolius* Greene**, SPREADING RABBITBRUSH. [*C. viscidiflorus* (Hook.) Nutt. ssp. *linifolius* (Greene) Hall & Clements; *Lorandersonia linifolia* (Greene) Urbatsch]. Shrubs, 7–35 dm, glabrous to scabrous; *leaves* linear to lanceolate, 2–7.5 cm long, flat, with 3 evident veins; *involucre* 4.5–7 mm high, the bracts in 3–4 series, keeled; *disk flowers* 4–6, 4–5.5 mm long; *pappus* 4.5–7 mm long; *achenes* 2.5–3.5 mm long, densely hairy. Found in sandy soil of floodplains and dry washes or along the margins of streams and rivers, 4600–9000 ft. July–Sept. W.

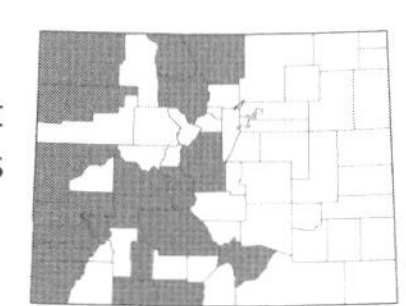

***Chrysothamnus vaseyi* (A. Gray) Greene**, VASEY'S RABBITBRUSH. Shrubs, 1–3 dm, glabrous to minutely hairy, usually resin-dotted; *leaves* sessile, linear to oblanceolate, 1–4 cm long, usually not twisted; *involucre* 5–8 mm high, the bracts weakly imbricate in 3–4 series, weakly keeled, glabrous or gland-dotted; *disk flowers* 5–7, 4–6.5 mm long; *pappus* 3.5–5 mm long; *achenes* 4–5 mm long, glabrous. Found in dry, open places in sandy or rocky soil, often in association with sagebrush or pinyon-juniper communities, 6000–8500 ft. July–Sept. E/W.

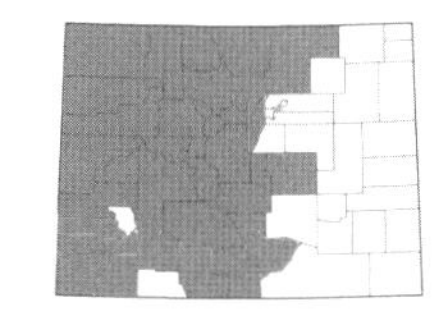

***Chrysothamnus viscidiflorus* (Hook.) Nutt.**, VISCID RABBITBRUSH. Shrubs, 1–15 dm, glabrous to minutely hairy, sometimes resin-dotted; *leaves* filiform to lanceolate, 1–5-veined, flat or twisted, 1–7.5 cm long; *involucre* 4–7 mm high, the bracts in 3–5 series, keeled, chartaceous; *disk flowers* 4–5 (14), 3.5–6.5 mm long; *pappus* 3.5–6 mm long; *achenes* 2.5–4.2 mm long, hairy. Common in open places in sandy or rocky soil, often in association with sagebrush or pinyon-juniper communities but sometimes with aspen, 4600–12,000 ft. July–Sept. E/W.

There are two well-defined varieties of *C. viscidiflorus* in Colorado. Occasionally the two varieties intergrade, resulting in a plant with sparse puberulence on the upper stem. I typically place the sparsely hairy specimens under var. *viscidiflorus*, as var. *lanceolatus* is very densely minutely hairy. The var. *puberulus* (D.C. Eaton) Hall & Clements was erroneously reported for Colorado by Hall & Clements (1923). The eastern limit of this variety is central Utah:

1a. Stems and leaves glabrous or with scabrous-ciliate margins, sometimes sparsely minutely hairy... **var. *viscidiflorus***, VISCID RABBITBRUSH. [including *C. viscidiflorus* (Hook.) Nutt. ssp. *axillaris* (Keck) L.C. Anderson; and *C. viscidiflorus* (Hook.) Nutt. ssp. *stenophyllus* (A. Gray) Hall & Clements]. Widespread throughout the western slope and northcentral counties, apparently absent from the San Luis Valley, 4600–9500 (11,500) ft. E/W.
1b. Upper stems and leaves densely minutely hairy to hispidulous...**var. *lanceolatus* (Nutt.) Greene**, LANCELEAF RABBITBRUSH. Widespread throughout the central and western counties, including the San Luis Valley, 5500–10,500 ft. E/W.

CICHORIUM L. – CHICORY

Lactiferous herbs; *leaves* alternate, the cauline ones usually reduced and bractlike; *heads* ligulate; *involucre bracts* in two series, green and herbaceous or the outer with a tan base; *receptacle* naked; *disk flowers* absent; *ray flowers* blue or sometimes white to pinkish; *pappus* of short scales; *achenes* striate-nerved.

***Cichorium intybus* L.**, COMMON CHICORY. Perennials, 4–20 dm, glabrous to bristly; *basal leaves* oblong to elliptic, 5–35 cm long; *involucre bracts* of the outer series 4–7 mm long, those of the inner series 6–12 mm long; *achenes* 1.5–2.5 mm long. A common weed found along roadsides, in disturbed places, and in fields, 4200–7000 ft. June–Oct. E/W. Introduced. (Plate 10)

CIRSIUM Mill. – THISTLE

Spiny herbs; *leaves* alternate, spiny-margined; *heads* discoid; *involucre bracts* imbricated in several series or subequal, usually the midrib strong and excurrent as a spine; *receptacle* densely bristly; *disk flowers* perfect and fertile (except in *C. arvense* which is dioecious), purplish, pink, reddish, yellowish, or white; *ray flowers* absent; *pappus* of plumose bristles; *achenes* glabrous, nerveless, usually flattened. (Cronquist, 1994; Gardner, 1974; Gray, 1874; Howell, 1959; Keil, 2004; Kelch & Baldwin, 2003; Rydberg, 1922)

1a. Plants dioecious; heads unisexual and small, the pappus of pistillate heads longer than the corollas and that of the staminate heads shorter than the corollas; involucre 10–20 mm high and 5–10 mm wide…***C. arvense***
1b. Plants monoecious; heads perfect and usually larger…2

2a. Upper leaf surface hispid-scabrous, not at all tomentose or arachnoid-pubescent; leaves with strongly decurrent bases, the base extending from one node to the next…***C. vulgare***
2b. Upper leaf surface glabrous, glabrate, to tomentose (sometimes thinly so), but never scabrous; leaves with a shortly decurrent base, this usually not extending from node to node, or the leaves sessile and simply clasping the stem…3

3a. Plants acaulescent or nearly so…***C. scariosum***
3b. Plants caulescent with a conspicuous leafy stem…4

4a. Outer and middle involucre bracts with spine-tips reflexed to spreading, the body with a dark, well-developed, glutinous dorsal ridge (except in *C. neomexicanum*); heads usually indented around the top of the peduncle…5
4b. Outer and middle involucre bract spine-tips ascending to somewhat spreading, the body with or without a dark, well-developed, glutinous dorsal ridge; heads rarely indented around the top of the peduncle…10

5a. Involucre bracts without a dark, well-developed, glutinous dorsal ridge; the outer involucre bracts strongly reflexed with the reflexed tip including not only the spine but a significant portion of the green bract as well…***C. neomexicanum***
5b. Involucre bracts with a dark, well-developed, glutinous dorsal ridge; the reflexed spine tip not including a significant portion of the green bract…6

6a. Middle and upper leaves decurrent on the stem for more than 1 cm…7
6b. Middle and upper leaves sessile and clasping or decurrent on the stem for less than 1 cm…8

7a. Corolla lobes 5.5–7 mm long; disk flowers ochroleucous or rarely pale lavender or pink; outer involucre bracts lanceolate with spines 2–4 mm long…***C. canescens***
7b. Corolla lobes 9–14 mm long; disk flowers reddish-purple, purple, pink, or rarely white; outer involucre bracts ovate with spines 4–12 mm long…***C. ochrocentrum***

8a. Plants from horizontal creeping roots; achenes 3–5 mm long, 1.5–2 mm wide, yellow apical collar of achenes conspicuous, 0.3–0.7 mm long; spines on the tips of involucre bracts 2–4 mm long, weak and slender; usually in wetter places than the next…***C. flodmanii***
8b. Plants from deep stout taproots; achenes 4–8 mm long, 2–3 mm wide, yellow apical collar of achenes inconspicuous, 0.2 mm long or less or lacking; spines on the tips of involucre bracts 3–7 mm long, robust and stout; in drier places than the previous…9

9a. Heads smaller, involucre 18–30 mm high; involucre bracts mostly ovate, in 3–4 (5) rows…***C. tracyi***
9b. Heads larger, involucre 25–50 mm high; involucre bracts lanceolate to narrowly ovate, in 5–7 rows…***C. undulatum***

10a. Leaves glabrous on both sides or only sparingly tomentose along the abaxial midrib, tripinnately divided into numerous lanceolate, narrow segments (mostly 2–3 mm wide), the midstrip narrow (about 2 mm)...***C. ownbeyi***
10b. Leaves tomentose below (usually densely so) and glabrate or tomentose above, pinnately divided or lobed into ovate, elliptic or triangular, wider (greater than 3 mm wide) segments, the midstrip wider (4–10+ mm); if glabrate above and below then the leaves shallowly lobed...11

11a. Leaves densely white-tomentose on both sides, the lobes sometimes relatively closely spaced without deep sinuses between them; corolla 18–24 mm long and style tip 3.5–5 mm long...***C. barnebyi***
11b. Leaves white-tomentose or arachnoid-pubescent below and green or gray and glabrate above, if densely white-tomentose on both sides then the corolla 25–33 mm long and the style tip 2–3 mm long (*C. arizonicum*); leaves shallowly or deeply lobed...12

12a. Corolla usually purplish red, the lobes (8) 10–17 mm long, at least twice as long as the corolla throat; style tip short, 1.2–3 (4.5) mm long; involucres often cylindric...***C. arizonicum* var. *bipinnatum***
12b. Corolla white, greenish-yellow, pink, or lavender, the lobes shorter, 4–9 mm long; style tip mostly longer, 3–6 mm long; involucres usually campanulate...13

13a. Leaves very shallowly lobed, almost entire, glabrous above and arachnoid-pilose below (but not evenly white-tomentose), sessile and clasping on the stem; corolla greenish-yellow; involucre bracts densely hairy...***C. parryi***
13b. Leaves deeply pinnatifid, or if shallowly lobed then the corolla not greenish-yellow, glabrous or sparsely pubescent above, glabrous or tomentose below (at least thinly so), usually decurrent; involucre bracts densely hairy or not...14

14a. Involucre bracts densely arachnoid- or woolly-pubescent, or with septate, multicellular hairs along the margins that horizontally connect adjacent phyllaries...15
14b. Involucre bracts glabrous or thinly arachnoid-pubescent on the margins, but without septate, multicellular hairs connecting adjacent phyllaries...17

15a. Heads numerous, sessile or short-pedunculate in a dense, massive, nodding terminal spiciform or racemiform inflorescence; involucre bracts densely arachnoid woolly pubescent, the hairs obscuring the bracts; flowers yellow, white, or less often pinkish...***C. scopulorum***
15b. Heads in erect cymose or racemiform inflorescences or terminal clusters, usually fewer (6 or less); involucre bracts arachnoid-pubescent, but this usually not so dense that it obscures the bracts; flowers pink, purple, or white...16

16a. Heads in open, spike-like inflorescences, usually not closely subtended by a whorl of leaves which overtop the heads; leaves usually flat, less spiny with more widely spaced lobes, and widely and evenly spaced on the stem...***C. osterhoutii***
16b. Heads in dense subcapitate or dense terminal spike-like clusters, closely subtended by a whorl of leaves which overtop the heads; leaves undulate, densely spiny, with numerous closely spaced lobes, and tightly grouped on the stem...***C. eatonii***

17a. Corollas rose or reddish-purple; middle and inner bracts with erose or lacerate margins at the apex; spines on outer involucre bracts spreading to reflexed; leaves sessile and clasping, not decurrent...***C. perplexans***
17b. Middle and inner bracts without erose or fringed margins at the apex, or if with erose apices then the corollas white to ochroleucous or pale pink; outer bracts lacking spines or spine-tipped but the tips spreading to erect; leaves usually decurrent...18

18a. Heads in dense terminal clusters or racemiform arrays, closely subtended by leafy bracts which overtop the heads (these usually in a whorl below the heads); biennials; stems usually fleshy and thick; involucre bracts without a dark, narrow glutinous ridge, the inner bracts often with erose or fringed tips...***C. scariosum***
18b. Heads in loose clusters or racemiform arrays but not closely subtended by leafy bracts which overtop the heads; perennials; stems not fleshy and thick; involucre bracts usually with a dark, narrow glutinous ridge, the inner bracts with or without erose or fringed tips...19

19a. Middle stem leaves sessile and not decurrent on the stem; involucre bracts relatively wide and more ovate, mostly 2.5–4 mm wide at the middle and mostly carrying this width well towards the apex...***C. wheeleri***
19b. Middle stem leaves decurrent to 2 cm on the stem; involucre bracts narrower, lanceolate, mostly 1.5–2.5 mm wide at the middle and tapering to the apex...20

20a. Inner involucre bracts often with erose or fringed apices; achenes without a distinct apical collar; corollas creamy-white to pale pink; leaves with 5–9 pairs of lobes, usually with a wider midrib...***C. clavatum***
20b. Inner involucre bracts entire; achenes usually with a narrow, yellow apical collar; corollas usually pink or purple, occasionally creamy-white; leaves usually with 8–14 pairs of lobes, usually deeply lobed with a narrow midrib (3–6 mm wide)...***C. pulcherrimum* var. *pulcherrimum***

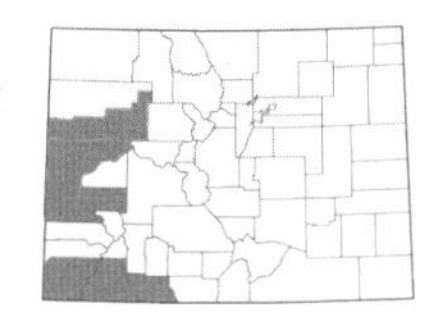

***Cirsium arizonicum* (A. Gray) Petr. var. *bipinnatum* (Eastw.) D.J. Keil**, Cainville thistle. [*C. calcareum* (M.E. Jones) Wooton & Standl.; *C. pulchellum* Wooton & Standl.]. Plants 3–12 dm; *leaves* to 40 (60) cm long, glabrate above and tomentose below, deeply pinnatifid, decurrent 1.5–2 cm on the stem; *involucre* 20–30 mm high, not glutinous, often tomentose on the margins, spines 1.5–6 mm long; *disk flowers* reddish-purple to purple, rarely nearly white, 25–33 mm long with lobes (8) 10–17 mm long; *pappus* 15–25 mm long; *achenes* ca. 6 mm long. Uncommon in dry, open places and canyon bottoms, occasionally in hanging gardens, typically in sandy soil, 5000–8400 ft. June–Sept. W.

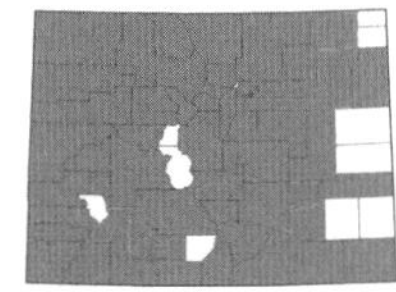

***Cirsium arvense* (L.) Scop.**, Canada thistle. [*Breea arvensis* (L.) Less.]. Plants dioecious, 3–20 dm; *leaves* 5–18 cm long, shallowly to deeply pinnately lobed or occasionally unlobed, white-tomentose below and green and glabrate above; *involucre* 10–20 mm high, glabrous to arachnoid-hairy, spines ca. 1 mm long; *disk flowers* pink-purple or rarely white, 12–24 mm long; *pappus* 15 mm long; *achenes* 2.5–4 mm long. Common in disturbed places, along roadsides, and in fields and meadows, 4000–9800 ft. June–Sept. E/W. Introduced. (Plate 10)

***Cirsium barnebyi* S.L. Welsh & Neese**, Barneby's thistle. Plants 2–6 dm; *leaves* 10–30 cm long, deeply pinnatifid, spiny-toothed, white-tomentose on both sides, decurrent up to 3 cm on the stem; *involucre* (15) 18–27 mm high, with or without a glutinous dorsal ridge, spines to 6 mm long; *disk flowers* lavender-purple, 18–24 mm long; *pappus* ca. 15 mm long. Restricted to shale slopes, usually in pinyon-juniper, 5900–8000 ft. June–Aug. W.

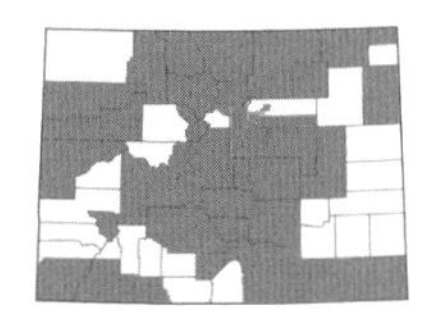

***Cirsium canescens* Nutt.**, prairie thistle. [*C. plattense* (Rydb.) Cockerell ex Daniels]. Plants 4–8 dm; *leaves* deeply pinnatifid, 12–30 cm long, green and arachnoid above and densely white-tomentose below, decurrent; *involucre* 30–40 mm high, with a dark dorsal ridge, spines 2–4 mm long, the outer and middle reflexed; *disk flowers* ochroleucous, 24–30 mm long; *pappus* 18–30 mm long; *achenes* 5–7 mm long. Found in gravelly soil of dry places, especially common on dry, open grasslands of the eastern plains, but also found along roadsides and in open meadows in the mountains, 3500–9500 ft. June–Sept. E/W. (Plate 10)

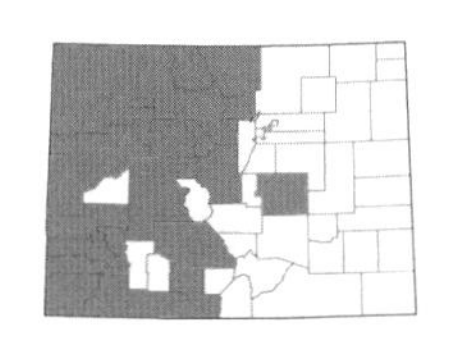

***Cirsium clavatum* (M.E. Jones) Petr.**, Fish Lake thistle. [*C. griseum* (Rydb.) K. Schum.; *C. modestum* (Osterh.) Cockerell; *C. oreophilum* (Rydb.) K. Schum.; *C. scapanolepis* Petr.; *C. spathulifolium* Rydb.]. Plants 2–10 dm; *leaves* pinnatifid into broad segments, glabrous or thinly tomentose below, 10–30 cm long, decurrent to 2 cm; *involucre* 15–25 mm high, lacking a glutinous dorsal ridge, spines not reflexed, the margins sometimes erose or fringed; *disk flowers* lavender-purple or white, 16–25 mm long; *pappus* 10–18 mm; *achenes* 4.5–7 mm long. Common in rocky soil of mountain meadows and forest clearings, 6200–10,800 ft. June–Sept. E/W. (Plate 10)

Specimens of *C. clavatum* from Colorado have traditionally been placed in *C. eatonii*. However they are significantly different in leaf spininess and orientation as well as involucre bract spine length. *Cirsium clavatum* is closely related to *C. pulcherrimum*, and the two species might form hybrids in the northern counties, but in general *C. pulcherrimum* is located to the west of *C. clavatum*. The erose or fringed bracts typically characteristic of *C. centaureae* (var. *americanum*) simply fall within the variation of *C. clavatum* as a whole, as some specimens have all bracts erose while others only have a few inner bracts with erose tips. Otherwise, var. *americanum* and var. *clavatum* overlap in almost every other vegetative feature. In addition, the range of *C. centaureae* completely overlaps that of *C. clavatum*:

1a. Involucre bracts with erose or fringed apices...**var. *americanum* (A. Gray) D.J. Keil**, Rocky Mountain fringed thistle. [*C. centaureae* (Rydb.) K. Schum.]. Common in mountain meadows, along streams, and in forest openings, (6200) 7000–10,800 ft. July–Sept. E/W.
1b. Involucre bracts with entire apices...**var. *clavatum***, Fish Lake thistle. Common in rocky soil of mountain meadows and forest clearings, 6800–10,000 ft. June–Sept. E/W.

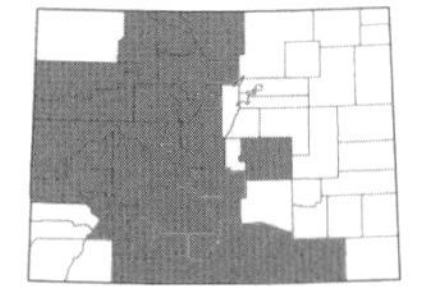

***Cirsium eatonii* (A. Gray) B.L. Rob.**, Eaton's thistle. [*C. tweedyi* (Rydb.) Petr.]. Plants 2–7 (10) dm; *leaves* pinnatifid, the segments cleft and spiny-toothed, 10–35 cm long, glabrous or sparsely hairy above, white-tomentose below, decurrent 1–3 cm on the stem; *involucre* 20–40 mm high, with multicellular hairs on the margins of the bracts, spines 5–20 mm long, not reflexed; *disk flowers* lavender-purple, 18–25 mm long; *pappus* 10–15 mm long; *achenes* 4–5 mm long. Common on rocky open slopes at high elevations, frequent along the Continental Divide, 9500–13,000 ft. July–Sept. E/W. (Plate 10)

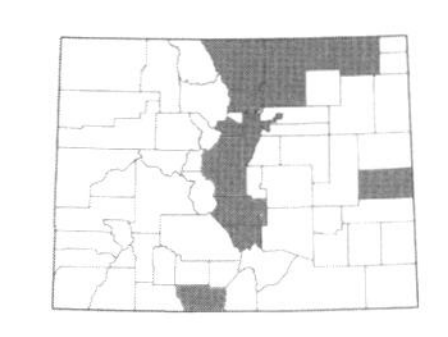

Cirsium flodmanii **(Rydb.) Arthur**, Flodman's thistle. Plants 3–10 dm; *leaves* pinnatifid or sometimes lobed to entire, 12–25 cm long, gray-tomentose below and green and arachnoid above, sessile and clasping; *involucre* 20–30 mm high, with a dark dorsal ridge, spines 2–4 mm long, reflexed to spreading; *disk flowers* purple or pink, rarely white, 21–36 mm long; *pappus* 20–30 mm long; *achenes* 3–5 mm long. Typically found in moist habitats such as along streams, near springs, and in wet meadows, 4800–8500 ft. July–Sept. E.

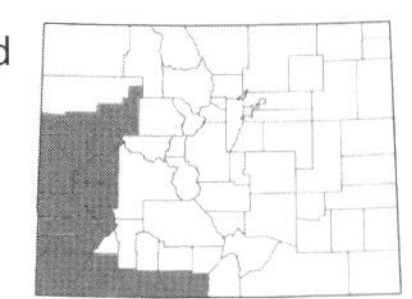

Cirsium neomexicanum **A. Gray**, New Mexico thistle. Plants 5–20 dm; *leaves* pinnatifid with a broad central midstrip, 10–35 cm long, thinly to densely tomentose on both sides, decurrent 1–3 cm; *involucre* 20–35 mm high, tomentose, the outer reflexed from the middle, weakly spine-tipped; *disk flowers* white or pale purple, 22–35 mm long; *pappus* 20–25 mm long; *achenes* 5–6 mm long. Found in southwestern Colorado on dry hillsides or along roadsides, often in clay or shale soil, 4500–7000 ft. May–July. W.

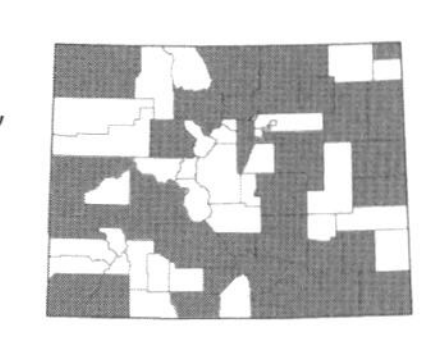

Cirsium ochrocentrum **A. Gray**, yellowspine thistle. Plants 2–10 dm; *leaves* pinnatifid with a broad central midstrip, densely white-tomentose below and green and arachnoid above, 8–22 cm long, decurrent; *involucre* 25–45 mm long, with a well-developed dark glutinous dorsal ridge, spines 4–12 mm long, the outer reflexed; *disk flowers* 30–45 mm long, purple, pink, or rarely white; *pappus* 17–30 mm long; *achenes* 7–8 mm long. Common in dry soil of the eastern plains, along roadsides, and on dry, open slopes, 3500–9500 ft. June–Sept. E/W.

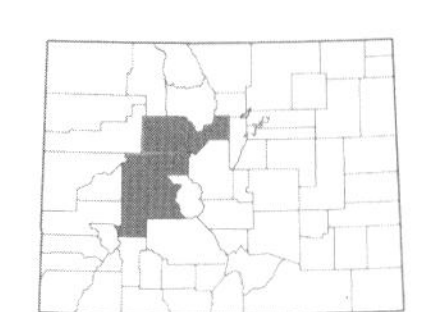

Cirsium osterhoutii **(Rydb.) Petr.**, Osterhout's thistle. [*C. araneans* Rydb.]. Plants to 10 dm; *leaves* pinnatifid, the segments cleft and spiny-toothed, to 35 cm long, glabrous or sparsely hairy above, white-tomentose below, decurrent to 3 cm on the stem; *involucre* 20–30 mm high, with multicellular hairs on the margins of the bracts, with spines 2–5 mm long, not reflexed; *disk flowers* white to pinkish; *pappus* 14–20 mm long; *achenes* 5–6 mm long. Found on gravelly and rocky open slopes, in open forests, and along streams, known from high elevations in the central mountains along the Continental Divide, where it is often found growing near *C. eatoni*, 9500–13,000 ft. July–Sept. E/W. Endemic. (Plate 10)

Cirsium ownbeyi **S.L. Welsh**, Ownbey's thistle. Plants 3–7 dm, glabrous; *leaves* deeply pinnatifid into numerous narrow segments, 10–30 cm long, decurrent 1–2 cm; *involucre* 18–25 mm long, lacking a glutinous dorsal ridge, spines 3–8 mm long, not reflexed; *disk flowers* lavender or white, 16–20 mm long; *pappus* 13–17 mm long; *achenes* 4–5 mm long. Found in open places in rocky, sandy, or clay soil, known only from Moffat County, 5500–6200 ft. June–Aug. W.

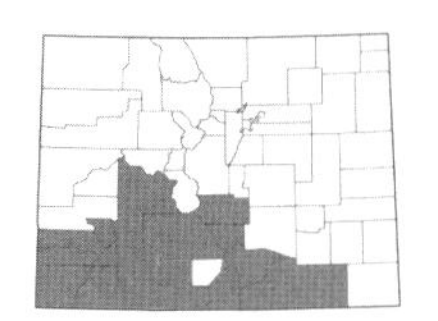

Cirsium parryi **(A. Gray) Petr.**, Parry's thistle. [*C. pallidum* (Wooton & Standl.) Wooton & Standl.]. Plants 3–20 dm; *leaves* pinnatifid with a broad central midstrip, glabrate above and arachnoid below, 10–30 cm long, not decurrent; *involucre* 20–30 mm high, densely arachnoid-pubescent, lacking a glutinous dorsal ridge, the margins often erose, spines 4–7 mm long; *disk flowers* greenish-yellow, 11–17 mm long; *pappus* 9–15 mm long; *achenes* 4–6 mm long. Found in open subalpine forests and alpine meadows, 8500–13,500 ft. June–Aug. E/W.

This species can apparently hybridize with *C. canescens* Nutt. in southern Colorado.

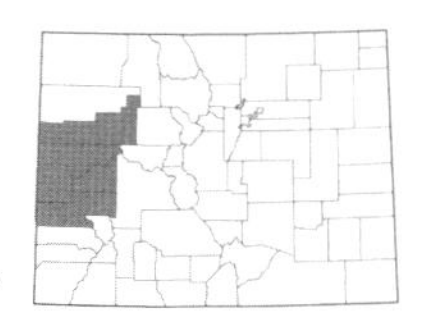

Cirsium perplexans **(Rydb.) Petr.**, Rocky Mountain thistle. [*C. vernale* (Osterh.) Cockerell]. Plants 2–6 dm; *leaves* dentate or sparingly lobed, glabrate above and thinly tomentose below, 15–30 cm long, sessile; *involucre* 25–35 mm high, glabrate, with a glutinous dorsal ridge, the margins conspicuously erose, spines 1–3 mm long; *disk flowers* purple or reddish-purple, 16–22 mm long; *pappus* 15–17 mm long; *achenes* 4–5 mm long. Found on barren, dry hillsides in shale or clay soil, often with pinyon-juniper or sagebrush, 5000–8000 ft. June–July. W. Endemic.

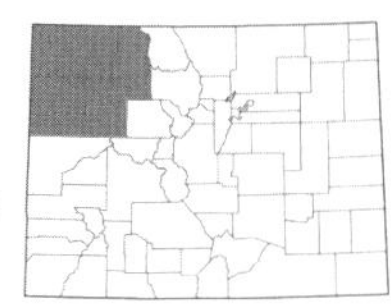

Cirsium pulcherrimum **(Rydb.) K. Schum. var. *pulcherrimum***, Wyoming thistle. Plants 3–8 dm; *leaves* deeply pinnatifid, to 40 cm long, white-tomentose below and glabrate above, sometimes shortly decurrent; *involucre* 15–25 mm high, lacking an evident glutinous dorsal ridge, spines 3–12 mm long, not reflexed; *disk flowers* pink-purple or sometimes white, 19–25 mm long; *pappus* 15–20 mm long; *achenes* 4–5 mm long. Common on shale outcroppings and rocky ridgetops, 5600–8600 ft. May–July. W.

See discussion under *C. clavatum*. *Cirsium pulcherrimum* (at least in Colorado) is very similar to *C. barnebyi* except that the leaves are green on one side. The two species apparently do not grow together.

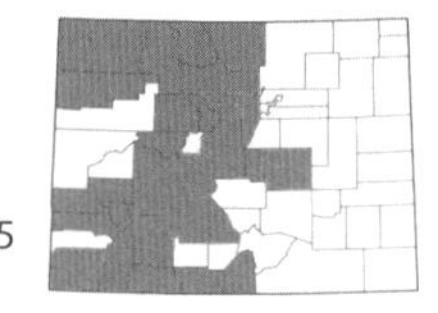

***Cirsium scariosum* Nutt.**, ELK THISTLE; MEADOW THISTLE. [*C. acaulescens* K. Schum.; *C. coloradense* Cockerell ex Daniels; *C. drummondii* misapplied by authors, not Torr. & A. Gray; *C. erosum* (Rydb.) K. Schum.; *C. tioganum* (Congd.) Petr.]. Plants acaulescent or 3–10 dm; *leaves* pinnatifid with a broad central midstrip, 10–40 cm long, tomentose below and glabrate and green above, sessile; *involucre* 20–35 mm high, lacking a glutinous dorsal ridge, often erose on the margins, spines 2–5 mm long, not reflexed; *disk flowers* white to lavender or pinkish-purple, 20–35 mm long; *pappus* 15–30 mm long; *achenes* 4–6 mm long. Common in forest clearings, along streams and roadsides, and in wet meadows, 7000–12,500 ft. June–Sept. E/W.

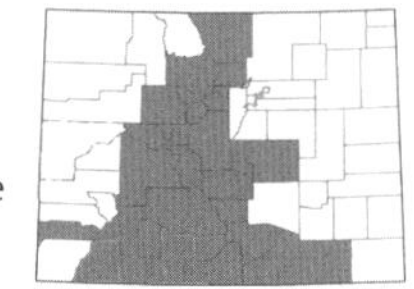

***Cirsium scopulorum* (Greene) Cockerell ex Daniels**, MOUNTAIN THISTLE. Plants 2–8 dm; *leaves* pinnatifid, 3–35 cm long, tomentose below and glabrate above, sometimes decurrent to 4 cm; *involucre* 30–35 mm high, densely arachnoid-hairy, spines 10–18 mm long; *disk flowers* white to light pink, 18–25 mm long; *pappus* 10–15 mm long; *achenes* 4–6 mm long. Common in subalpine and alpine meadows, frequent along the Continental Divide, (9500) 10,500–13,500 ft. July–Sept. E/W. (Plate 10)

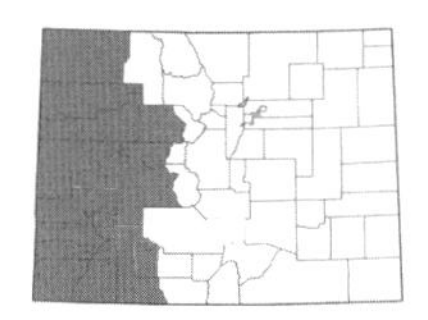

***Cirsium tracyi* (Rydb.) Petr.**, TRACY'S THISTLE. [*C. undulatum* (Pursh) Spreng. var. *tracyi* (Rydb.) S.L. Welsh]. Plants 3–15 dm; *leaves* pinnately lobed, sometimes shallowly so, 10–40 cm long, white-tomentose below and glabrate above, sessile or very shortly decurrent (less than 1 cm); *involucre* 18–30 mm high, with a well-developed glutinous dorsal ridge, spines 3–7 mm long, reflexed to spreading; *disk flowers* pink-purple or sometimes white, 22–30 mm long; *pappus* 18–20 mm long; *achenes* 5–8 mm long. Common on dry, open places such as on hillsides, in canyons, and along roads and railroad tracks, 4500–8500 (10,500) ft. June–Aug. W.

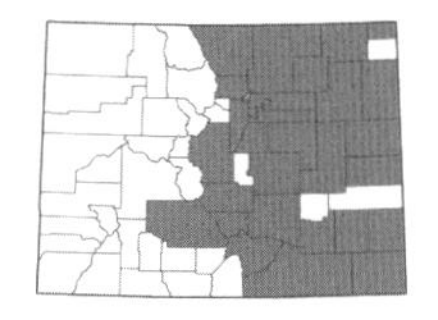

***Cirsium undulatum* (Nutt.) Spreng.**, WAVYLEAF THISTLE. Plants 3–15 dm; *leaves* coarsely dentate to pinnatifid, 10–45 cm long, white-tomentose below and glabrate above, sessile or shortly decurrent; *involucre* 25–50 mm high, with a well-developed glutinous dorsal ridge, spines 3–5 mm long, outer reflexed to spreading; *disk flowers* 25–36 mm long, purple, pink, or sometimes white; *pappus* 22–40 mm long; *achenes* 4–7 mm long. Common in sandy or gravelly soil of the plains and open hillsides, 3500–9000 ft. June–Aug. E. (Plate 10)

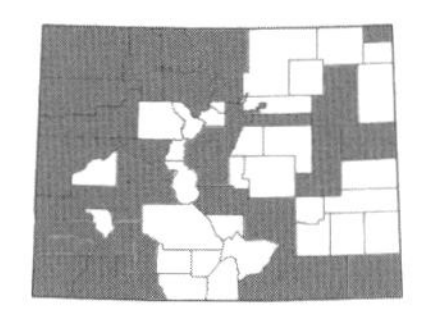

***Cirsium vulgare* (Savi) Ten.**, BULL THISTLE. Plants 5–15 dm; *leaves* pinnatifid, white-tomentose below and scabrous-hispid above, 12–30 (50) cm long, decurrent; *involucre* 25–40 mm long, lacking a glutinous dorsal ridge, spines 2–5 mm long; *disk flowers* dark purple, 25–35 mm long; *pappus* 20–30 mm long; *achenes* 3–5 mm long. Found in disturbed places, along roadsides, and in moist fields, 4000–9500 ft. E/W. Introduced.

***Cirsium wheeleri* (A. Gray) Petr.**, WHEELER'S THISTLE. [*C. olivescens* (Rydb.) Petr.]. Plants 3–10 dm; *leaves* shallowly to deeply pinnatifid, 10–25 cm long, white-tomentose below and green and glabrate above, sessile; *involucre* 17–30 mm high, lacking a glutinous dorsal ridge, spines 3–7 mm long; *disk flowers* purple, pink, or rarely white, 21–30 mm long; *pappus* 15–25 mm long; *achenes* 5–6 mm long. Reported for southwestern Colorado. July–Sept. W.

CONYZA Less. – HORSEWEED

Annual herbs; *leaves* alternate, petiolate; *heads* disciform; *involucre bracts* imbricate, scarcely herbaceous; *receptacle* flat, naked; *disk flowers* yellow; *pistillate flowers* numerous, rayless or with a short, inconspicuous, white or pinkish ray; *pappus* of capillary bristles; *achenes* 2-nerved, scarcely contracted at the top.

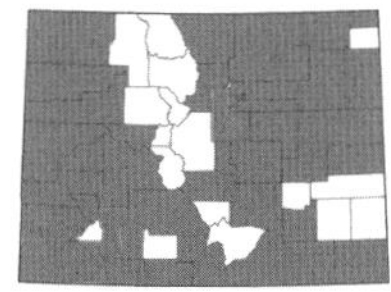

***Conyza canadensis* (L.) Cronquist**, HORSEWEED. Plants 2–200 (350) dm, strigose, bristly, or glabrous; *leaves* oblanceolate to linear, 2–5 (10) cm long, toothed to entire; *involucres* 3–4 mm high; *pappus* 2–3 mm long; *achenes* 1–1.5 mm long. Common weed found in disturbed places, along roadsides, and in fields throughout the state, 3500–8000 ft. July–Sept. E/W. Introduced.

COREOPSIS L. – TICKSEED

Annual or perennial herbs; *leaves* opposite; *heads* radiate; *involucre bracts* in two series and dimorphic, fused at the base, the outer bracts shorter, narrower and more herbaceous than the striate inner bracts; *receptacle* chaffy; *disk flowers* 4- or 5-lobed or toothed; *ray flowers* usually yellow; *pappus* absent or of 2 short awns or teeth or a minute crown; *achenes* flattened. (Sherff, 1936; Smith, 1976)

1a. Ray flowers with a red spot at the base; disk corollas 4-lobed...***C. tinctoria***
1b. Ray flowers without a red spot at the base; disk corollas 5-lobed...2

2a. Lower leaves well-developed, simple and entire or with 1 or 2 pairs of lateral lobes, the stem leaves reduced; disk flowers 6–7.5 mm long...***C. lanceolata***
2b. Leaves well-developed along the stem, pinnately divided into narrow, linear or lance-linear segments; disk flowers 3–5 mm long...***C. grandiflora***

***Coreopsis grandiflora* Hogg ex Sweet**, BIGFLOWER COREOPSIS. Perennials, 4–6+ dm; *leaves* usually pinnately lobed into linear or lance-linear segments, 1.5–5 (9) cm long; *disk flowers* 3–5 mm long, 5-lobed; *ray flowers* 1.2–2.5+ cm long, yellow; *achenes* 2–4 mm long, winged. Seen only in cultivation but could escape onto nearby roadsides. May–July. Introduced.

***Coreopsis lanceolata* L.**, LANCELEAF TICKSEED. Perennials, 1–3 (6) dm; *leaves* simple and entire or with 1–2 lateral lobes, 5–12 cm long; *disk flowers* 6–7.5 mm long, 5-lobed; *ray flowers* 15–35 mm long, yellow; *achenes* 2.5–4 mm long, winged. Occasionally escaping from cultivation. June–Aug. Introduced

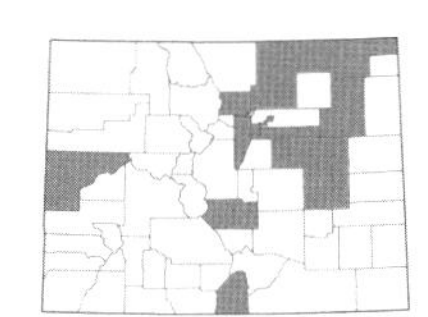

***Coreopsis tinctoria* Nutt.**, PLAINS COREOPSIS. Annuals, (1) 3–8 (15) dm; *leaves* 1-2-pinnate, rarely simple, the ultimate segments linear, 1–5 cm long; *disk flowers* 2–4 mm long, 4-lobed; *ray flowers* yellow with a red spot at the base, 1.2–2 cm long; *achenes* 1.5–4 mm long, wings absent or not. Common on the plains and along the Front Range, especially along roadsides, occasionally escaping from cultivation at higher elevations, 3700–6000 (7800) ft. July–Oct. E/W. (Plate 10)

Western slope records are plants escaped from cultivation.

COSMOS Cav. – COSMOS

Annual herbs; *leaves* opposite, pinnately dissected into narrow segments; *heads* radiate; *involucre bracts* in two series and dimorphic, the outer bracts narrower than the inner, the inner with scarious margins; *receptacle* chaffy; *disk flowers* yellow; *ray flowers* pink, rose, or white; *pappus* absent or of 2–4 retrorsely hispid awns; *achenes* beaked.
1a. Rays 1.5–5 cm long; involucre over 10 mm high...***C. bipinnatus***
1b. Rays to 1.2 cm; involucre less than 10 mm high...***C. parviflorus***

***Cosmos bipinnatus* Cav.**, GARDEN COSMOS. Plants 3–20 dm, glabrous or sparsely hairy; *leaves* 6–11 cm long; *involucre* 7–15 mm diam.; *disk flowers* 5–7 mm long; *ray flowers* 1.5–5 cm long, white, pink, or purple; *pappus* absent or of awns 1–3 mm long; *achenes* 7–16 mm long. A common garden plant occasionally escaping from cultivation, especially in the Denver-Boulder-Ft. Collins region, 5000–6000 ft. June–Oct. E. Introduced.

***Cosmos parviflorus* Pers.**, SOUTHWESTERN COSMOS. Plants 3–9 dm, glabrous or sparsely hairy; *leaves* 2.5–6.5 cm long; *disk flowers* 4–5 mm long; *ray flowers* 5–12 mm long, white, pink, or purple; *pappus* of awns 2–3 mm long; *achenes* 9–16 mm long. Found along roadsides and in dry open places, distributed in the southern counties, 6000–7800 ft. June–Sept. E/W.

CREPIS L. – HAWK'S-BEARD

Lactiferous herbs; *leaves* alternate or basal, entire to pinnatifid; *heads* ligulate; *involucre bracts* in one or two series, if in two series then the outer bracts reduced; *receptacle* naked or rarely chaffy; *disk flowers* absent; *ray flowers* ligulate, yellow (ours); *pappus* of numerous capillary bristles; *achenes* terete or nearly so, 10- to 20-ribbed. (Babcock, 1947)
1a. Annuals; stem leaves entire, auriculate-clasping; involucre 5–8 mm high; achenes 1.5–2.5 mm long...***C. capillaris***
1b. Perennials; stem leaves sessile, not auriculate-clasping, pinnatifid or narrow and inconspicuous; involucre 8–21 mm high; achenes 4–12 mm long...2

2a. Plants low, no more than 2 dm tall, the taproot giving rise to numerous slender, rhizome-like or stolon-like apical or lateral stems; leaves entire, orbicular or spatulate, to 8.5 mm long (excluding the long petiole), generally clustered at the ground level on the stem tips; heads on a short scape usually no higher than the leaves...***C. nana***
2b. Plants mostly taller, to 8 dm, without numerous slender rhizome-like stems; leaves pinnatifid or toothed, longer than 10 mm, not clustered at the stem tips; heads on scapes well above the basal leaves...3

3a. Stems and leaves glabrous or hispid, but not tomentose; stem leaves generally all reduced to narrow, inconspicuous leaves...***C. runcinata***
3b. Stems and leaves grayish-tomentose or minutely hairy, especially when young, or sometimes glandular-hispid or bristly; the 1–3 stem leaves well-developed and at least the lower resembling the basal leaves...4

4a. Involucre bracts glabrous or the outer slightly hairy at the tip or along the margins; heads generally smaller with flowers mostly 5–10 per head and 5–7 inner involucre bracts...***C. acuminata***
4b. Involucre bracts tomentose, sparsely hairy, bristly, or glandular; heads generally larger with 10 or more flowers per head and 8–15 inner involucre bracts...5

5a. Lower part of the stem with numerous stout, spreading, glandless bristles interspersed in the tomentum...***C. modocensis* ssp. *modocensis***
5b. Lower part of the stem without spreading bristles or the bristles present interspersed in the tomentum with gland-tips...6

6a. Basal leaves deeply pinnatifid, with linear or lance-linear, mostly entire segments, the segments conspicuously curved and usually pointing upwards...***C. atribarba***
6b. Basal leaves pinnatifid with lanceolate to deltoid, mostly toothed segments, or the leaves merely runcinately toothed, the segments usually straight or somewhat curved...7

7a. Heads narrowly cylindric (3–5.5 mm wide) with 7–14 (mostly 8–10) flowers and 7 or 8 inner involucre bracts; inner involucre bracts with black bristles mostly at the tips; plants taller, 3–7 dm in height...***C. intermedia***
7b. Heads broadly cylindric (5–10 mm wide) with 10–40 (mostly 12–25) flowers and 8–13 inner involucre bracts; inner involucre bracts with black bristles evenly dispersed throughout; plants shorter, 0.5–4 dm in height...***C. occidentalis***

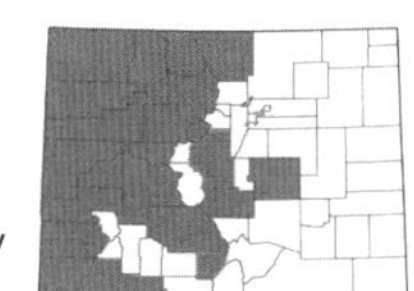

***Crepis acuminata* Nutt.**, TAPERTIP HAWK'S-BEARD. [*Psilochenia acuminata* (Nutt.) W.A. Weber]. Perennials, 2–7 dm; *leaves* pinnately lobed, the segments entire or sometimes toothed, gray-tomentose to glabrate, 10–40 cm long; *heads* (12) 20–100+; *involucre* 8–16 mm high, glabrous; *ray flowers* 10–18 mm long; *achenes* 6–9 mm long. Common in open, dry places in rocky or sandy soil, often with sagebrush, (4800) 6500–9500 ft. June–Aug. E/W.

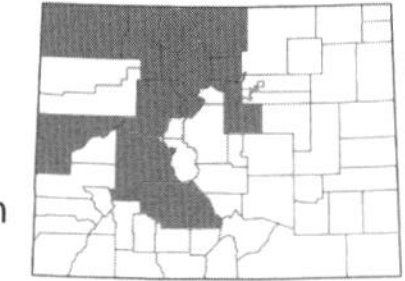

***Crepis atribarba* A. Heller**, SLENDER HAWK'S-BEARD. [*Psilochenia atribarba* (A. Heller) W.A. Weber]. Perennials, 1.5–8 dm; *leaves* deeply pinnatifid with linear, mostly entire ultimate segments, gray-tomentose when young turning glabrate in age, 10–35 cm long; *heads* 3–30 (40); *involucre* 8–15 mm high, gray-tomentose, often with black bristles; *ray flowers* 10–18 mm long; *achenes* 6–10 mm long. Found in dry, open places, often with sagebrush, 5500–10,000 ft. May–July. E/W. (Plate 10)

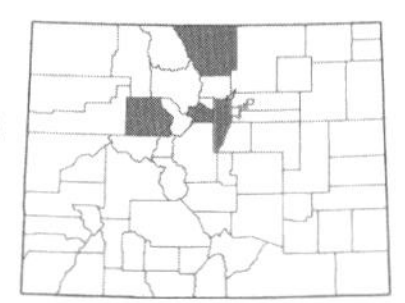

***Crepis capillaris* (L.) Wallr.**, SMOOTH HAWK'S-BEARD. Annuals or biennials, 1–9 dm; *leaves* runcinate-pinnatifid or bipinnately divided, the cauline leaves usually auriculate, hispid or glabrous, 3–10 cm long; *heads* numerous; *involucre* 5–8 mm high, tomentose-puberulent and often with black hairs; *ray flowers* 8–12 mm long; *achenes* 1.5–2.5 mm long, brown. Uncommon in disturbed places, lawns, and meadows, usually in moist soil, 5000–6600 (10,500) ft. May–Sept. E. Introduced.

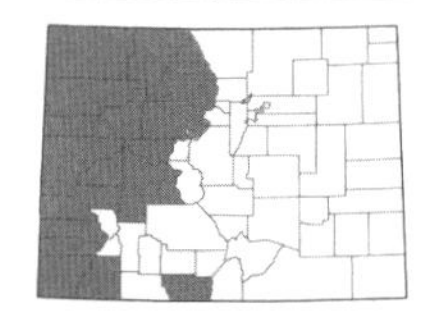

***Crepis intermedia* A. Gray**, LIMESTONE HAWK'S-BEARD. [*Psilochenia intermedia* (A. Gray) W.A. Weber]. Perennials, 3–7 dm; *leaves* pinnatifid, the ultimate segments entire to toothed, gray-tomentose, 10–40 cm long; *heads* usually 10–60; *involucre* 10–16 mm high, lightly tomentose; *ray flowers* 14–30 mm long; *achenes* 5.5–9 mm long. Found in dry, open places in rocky soil, often with sagebrush, not as common as the next, 6000–10,000 ft. May–July. W.

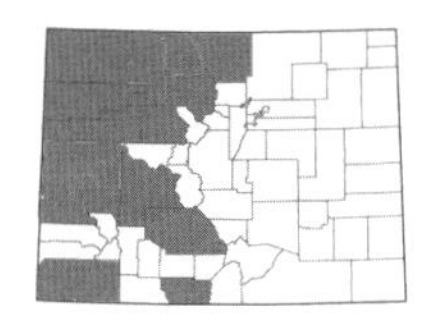

Crepis modocensis* Greene ssp. *modocensis, MODOC HAWK'S-BEARD. [*Psilochenia modocensis* (Greene) W.A. Weber]. Perennials, 1–4 dm tall, the basal part of the stem conspicuously setose; *leaves* deeply pinnatifid, the ultimate segments toothed and lanceolate, tomentulose when young, 5–15 (20) cm long; *heads* (1) 2–20; *involucre* 11–16 mm high, gray-tomentose, often with black bristles; *ray flowers* 13–22 mm long; *achenes* 7–12 mm long. Found in dry, open, rocky places, often with sagebrush, 5300–9800 ft. May–July. E/W.

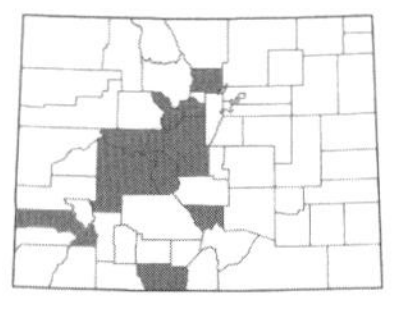

***Crepis nana* Richardson**, DWARF ALPINE HAWK'S-BEARD. [*Askellia nana* (Richardson) W.A. Weber]. Perennials to 1(2) dm, glabrous and glaucous; *leaves* spatulate to orbicular, to 8.5 cm long; *heads* numerous, usually not surpassing the leaves; *involucre* 8–13 mm high, glabrous; *ray flowers* 7–9 mm long; *achenes* 7–12 mm long. Uncommon on scree and talus slopes in the alpine, 10,000–14,000 ft. July–Sept. E/W.

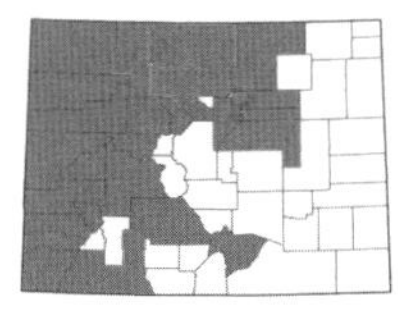

***Crepis occidentalis* Nutt.**, LARGEFLOWER HAWK'S-BEARD. [*Psilochenia occidentalis* (Nutt.) Nutt.]. Perennials, 0.5–4 dm; *leaves* toothed to deeply pinnatifid with toothed ultimate segments, densely gray-tomentose, glabrate in age, often glandular-hirsute above, 10–30 cm long; *heads* 2–25; *involucre* 11–20 mm high, often with black setae; *ray flowers* to 20 mm long; *achenes* 6–10 mm long. Common in dry, open rocky places, often with sagebrush, 4600–9000 ft. May–July. E/W.

There are three intergrading subspecies of *C. occidentalis* in Colorado:
1a. Involucre bracts without glandular hairs...**ssp. *conjuncta* Babcock & Stebbins**
1b. Involucre bracts glandular-stipitate...2

2a. Involucre bracts glandular without black setae; florets 18–30; phyllaries 7–8 or 10–13...**ssp. *occidentalis***
2b. Involucre bracts with dark or black, glandular setae; florets 10–14; phyllaries 8...**ssp. *costata* (A. Gray) Babcock & Stebbins**

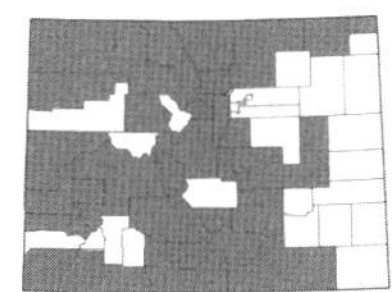

***Crepis runcinata* (E. James) Torr. & A. Gray**, FIDDLELEAF HAWK'S-BEARD. [*Psilochenia runcinata* (E. James) Á. Löve & D. Löve]. Perennials, mostly 2–7 dm; *leaves* pinnatifid to entire, glabrous or somewhat hispid, 3–15 (30) cm long; *involucre* 7–21 mm high, glabrous or glandular; *ray flowers* 10–20 mm long; *achenes* 4–7 mm long. Common in open places, meadows, usually found in moist soil, 4500–12,000 ft. May–Aug. E/W.

There are three subspecies of *C. runcinata* in Colorado:
1a. Involucre bracts glabrous or sparsely tomentose-puberulent (especially on the margins)...**ssp. *glauca* (Nutt.) Babcock & Stebbins**
1b. Involucre bracts with glandular hairs or bristles...2

2a. Leaves wider, the larger ones 3–8 cm across; heads usually more numerous, about 10–25 or more...**ssp. *hispidulosa* (Howell ex Rydb.) Babcock & Stebbins**
2b. Leaves seldom wider than 3.5 cm; heads few, rarely up to 10...**ssp. *runcinata***

CYCLACHAENA Fresen. – SUMPWEED

Annual herbs; *leaves* mostly opposite; *heads* discoid; *involucre bracts* subequal or imbricate in 2–3 series; *receptacle* chaffy; *disk flowers* pistillate or staminate, the pistillate flowers 5 and peripheral, the staminate flowers 5–10 and central, 5-lobed, with abortive pistils; *ray flowers* absent; *pappus* absent; *achenes* compressed parallel to the involucre bracts.

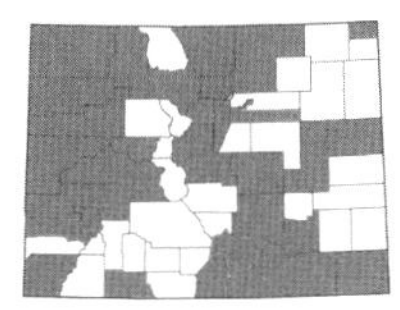

***Cyclachaena xanthifolia* (Nutt.) Fresen.**, GIANT SUMPWEED. [*Iva xanthifolia* Nutt.]. Plants 4–20 dm; *leaves* long-petiolate, ovate, 6–12 (20+) cm long, scabrous above, sericeous below; *infloresence* a large paniculate cluster; *involucre* 1.5–3 mm high; *staminate flowers* 8–20; *pistillate flowers* typically 5, the corollas greatly reduced; *achenes* 2–2.5 mm long. Common in sandy or rocky soil of moist places on the plains and in valleys, 4000–8500 ft. Aug.–Oct. E/W.

DICORIA Torr. ex A. Gray – TWINBUGS

Annuals, diffusely branched herbs; *leaves* simple; *heads* disciform with 1–4 marginal pistillate flowers and 5–20 central staminate flowers, or discoid; *involucre bracts* 5, subherbaceous, distinct; *receptacle* chaffy; *disk flowers* with a corolla; *pistillate flowers* lacking a corolla; *pappus* lacking; *achenes* compressed slightly parallel to the involucre bracts, narrowly wing-margined.

***Dicoria canescens* A. Gray**, DESERT TWINBUGS. Plants 2–8 dm, canescent; *leaves* petiolate, 1–3 (12) cm long; *involucres* 2–3+ mm high, the outer bracts lanceolate to ovate, the inner bracts obovate to orbiculate with each typically enclosing a fruit; *staminate flowers* 2.5–3 mm long; *achenes* 3–8+ mm long. Found in sandy soil in the desert and on sand dunes.

The type specimens were collected by Brandegee near the Colorado and Utah line in the San Juan Valley and on the sands of the San Juan River near the Utah line. However, no specimens of this species have been collected in Colorado since, and the locality is probably in Utah and not Colorado. If the locality was from Colorado, then this species is most likely to be present in Montezuma Co. near Four Corners, 4700–5000 ft. Aug.–Oct. W.

DIETERIA Nutt. – BIENNIAL TANSY-ASTER

Herbs or subshrubs; *leaves* alternate, simple; *heads* radiate or discoid; *involucre bracts* imbricate in 3–12 series, the base hard and tannish and the apex herbaceous, often bristle-tipped; *receptacle* naked; *disk flowers* yellow; *ray flowers* white, pink, purple, or blue, or absent; *pappus* of barbellate bristles in 1–3 series, persistent; *achenes* flattened, smooth or with 8–12 ribs. (Morgan & Hartman, 2003)
1a. Involucre bracts and peduncles both with conspicuous, well-developed glandular hairs; involucre bracts with long-acuminate tips; middle leaves lanceolate to oblanceolate, 5–15 mm wide...***D. bigelovii* var. *bigelovii***
1b. Involucre bracts canescent or with some glandular hairs but rarely with conspicuous glandular hairs on both the bracts and peduncles; involucre bracts with acute tips; middle leaves linear to linear-lanceolate or linear-oblanceolate, generally narrower and mostly 1.5–5 mm wide...***D. canescens***

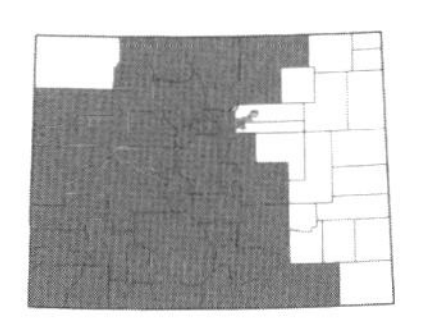

Dieteria bigelovii* (A. Gray) D.R. Morgan & R.L. Hartm. var. *bigelovii, BIGELOW'S TANSY-ASTER. [*Machaeranthera bigelovii* (A. Gray) Greene var. *bigelovii*; *Machaeranthera pattersonii* (A. Gray) Greene]. Biennials or perennials, 1–10 dm; *leaves* lanceolate to oblanceolate, denticulate to nearly entire, 4–20 cm long, glabrous to sparsely hairy and often glandular-stipitate; *involucre* 8–15 mm high, glandular, mostly with reflexed tips; *ray flowers* 10–25 mm long, lavender-blue to purple. Common along roadsides, on open slopes, and in meadows and forest clearings, 5000–11,200 ft. June–Oct. E/W. (Plate 10)

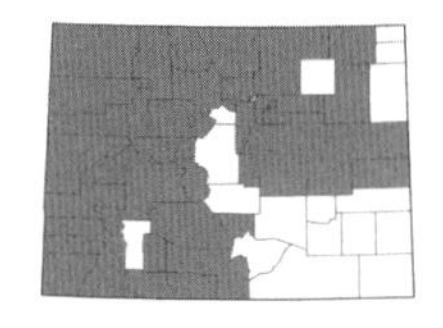

***Dieteria canescens* (Pursh) Nutt.**, HOARY TANSY-ASTER. [*Machaeranthera canescens* (Pursh) A. Gray]. Biennials or perennials, 1–5 dm; *leaves* linear-oblanceolate to spatulate, toothed to nearly entire, finely canescent-puberulent, sometimes also glandular-stipitate to nearly glabrous, to 10 cm long; *involucre* 5.5–12 mm high, canescent-puberulent and/or glandular, the bracts appressed to reflexed; *ray flowers* blue-purple to pink, 5–15 mm long. Common along roadsides, on open hillsides, and in meadows, 4400–12,000 ft. July–Oct. E/W.

DYSSODIA Cav. – DYSSODIA

Annuals or perennials; *leaves* usually pinnately dissected; *heads* radiate; *involucre bracts* in 2 series but equal; *receptacle* naked; *disk flowers* yellow to orange; *ray flowers* yellow to orange; *pappus* of blunt or awned scales; *achenes* striate-angled.

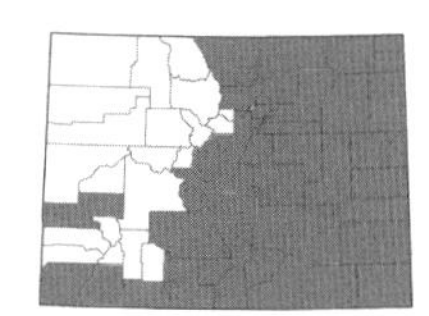

***Dyssodia papposa* (Vent.) Hitchc.**, FETID MARIGOLD. [*Tagetes papposa* Vent.]. Plants 10–30 (70+) cm, malodorous; *leaves* (1) 2–3 times pinnate, 1.5–5 cm long, glabrous or minutely hairy, dotted with oil-glands, the uppermost often alternate; *involucre* 6–10 mm high, the bracts with 1–7 oil-glands; *disk flowers* about 3 mm long; *ray flowers* 1.5–2.5 mm long; *achenes* 3–3.5 mm long. Common in disturbed places, along roadsides, and in fields and gardens, 3500–7900 ft. June–Oct. E/W. (Plate 11)

Once you smell this plant, you will understand how it got its name "fetid marigold"!

ECHINACEA Moench – PURPLE CONEFLOWER

Perennial herbs; *leaves* alternate; *heads* radiate; *involucre bracts* in 3 or 4 series, imbricate, narrow, herbaceous and intergrading into the chaffy bracts; *receptacle* conic or hemispheric, chaffy; *disk flowers* yellowish or purplish; *ray flowers* pink, purple, or white; *pappus* a crown; *achenes* 4-angled.

1a. Leaves ovate to ovate-lanceolate, mostly 2–3 times longer than wide, the margins coarsely to sharply serrate; ray flowers reddish-purple or rarely pink; plants 6–18 dm tall...***E. purpurea***

1b. Leaves narrowly elliptic-lanceolate to oblong, mostly more than 5 times longer than wide, the margins entire or nearly so; ray flowers light pink to light purple; plants 1–7 dm tall...***E. angustifolia***

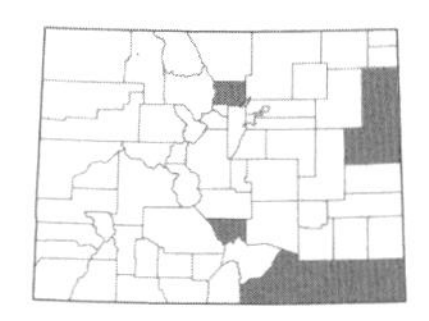

***Echinacea angustifolia* DC.**, BLACKSAMSON ECHINACEA. Plants to 7 dm, hispid; *leaves* narrowly elliptic-lanceolate to oblong, 7–30 cm long, entire; *disk flowers* 5–7 mm long, usually purple; *ray flowers* 1.5–4 cm long, pink to purple, the ligule reflexed; *pappus* to 1 mm long; *achenes* 4–5 mm long, usually glabrous. Found in rocky places on the plains, occasionally planted in gardens and escaping, 3800–5300 (8200) ft. June–Aug. E.

***Echinacea purpurea* (L.) Moench**, PURPLE CONEFLOWER. Plants 6–18 dm, hispid or sometimes glabrous; *leaves* ovate to ovate-lanceolate, 5–30 cm long, usually serrate; *disk flowers* 4.5–6 mm long, greenish to pink or purple; *ray flowers* 3–8 cm long, pink to purple, spreading or reflexed; *pappus* ca. 1.2 mm long; *achenes* 3.5–5 mm long, usually glabrous. Commonly cultivated and occasionally escaping, especially in the Denver-Boulder area, 5000–6000 ft. June–Aug. E. Introduced.

ECHINOPS L. – GLOBETHISTLE

Perennials; *leaves* alternate, pinnatifid or less often dentate; *heads* discoid, crowded into a globose cluster resembling a single head; *receptacle* chaffy; *involucre bracts* with outer ones bristle-like and inner ones fimbriate; *disk flowers* white or blue; *ray flowers* absent; *pappus* cup-shaped, the edges toothed or scale-like; *achenes* hairy.

***Echinops sphaerocephalus* L.**, GREAT GLOBETHISTLE. Plants 1–2 m, spiny; *leaves* oblong-elliptic to narrowly obovate, lobes lanceolate to deltoid, arachnoid-tomentose below and green and glandular above, tipped with slender spines; *involucres* 15–25 mm high, the bracts spine-tipped; *corollas* 12–14 mm long; *pappus* of ciliate scales, 1–1.5 mm long; *achenes* 7–8 mm long. A common garden plant sometimes escaping cultivation, 5000–6700 ft. July–Sept. E/W. Introduced.

ENCELIA Adans. – NODDINGHEAD

Perennial herbs or subshrubs; *leaves* simple; *heads* radiate or discoid; *involucre bracts* subequal in 2 or 3 series, herbaceous; *receptacle* chaffy; *disk flowers* perfect and fertile, numerous, yellow; *ray flowers* (when present) sterile, large and yellow; *pappus* of 2 short awns with or without small scales between, or lacking; *achenes* strongly flattened.

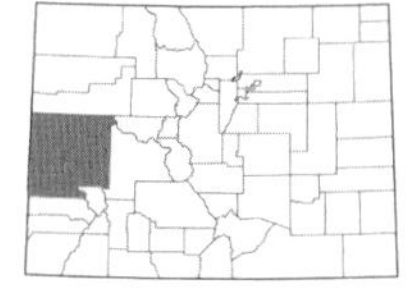

***Encelia nutans* Eastw.**, NODDINGHEAD. [*Enceliopsis nutans* (Eastw.) A. Nelson]. Plants 1–2.5 dm; *leaves* basal, broadly ovate, 3–4 cm long, glabrous or strigose; *heads* borne singly, nodding at maturity; *involucre* 12–22 mm high; *disk flowers* 6–8 mm long; *ray flowers* lacking; *pappus* usually lacking, sometimes of 2 bristle-like awns. Found in dry, open places in adobe, shale, sandy, or gravelly soil, 4600–5500 ft. April–June. W. (Plate 11)

ENCELIOPSIS A. Nelson – SUNRAY

Scapose or subscapose perennial herbs; *leaves* simple, entire, often silvery-hairy; *heads* radiate; *involucre bracts* subequal in 2 or 3 series, herbaceous; *receptacle* chaffy; *disk flowers* yellow; *ray flowers* (when present) sterile, large and yellow; *pappus* of 2 short awns with or without small scales between, or lacking; *achenes* strongly flattened, not winged.

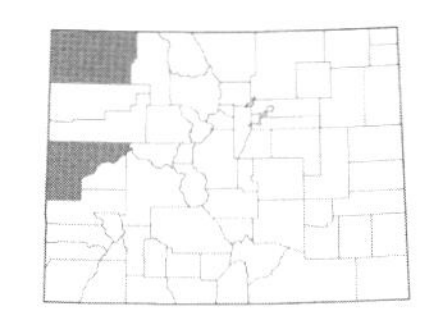

***Enceliopsis nudicaulis* A. Nelson**, SILVERLEAF SUNRAY. Plants 1–4 dm; *leaves* suborbiculate to ovate, 2–6 cm long; *peduncles* 15–45 cm long; *involucre* 10–20 mm high, the bracts in 3–5 series, narrowly lanceolate; *ray flowers* 20–40 mm long; *pappus* 1–1.5 mm long; *achenes* ca. 9 mm long, silky-hairy. Uncommon on open clay hills, 5500–5700 ft. May–June. W.

ENGELMANNIA A. Gray ex Nutt. – ENGELMANN'S DAISY

Perennials; *leaves* alternate; *heads* radiate; *involucre bracts* imbricate in 2 or 3 series, the outer bracts linear, the middle suborbicular with linear tips, the inner ovate or obovate; *receptacle* flat, chaffy; *disk flowers* perfect but sterile, partly enclosed by the receptacle chaff; *ray flowers* 8–10, fertile, yellow; *pappus* an irregular crown; *achenes* flat, not winged, with 1 nerve on each side.

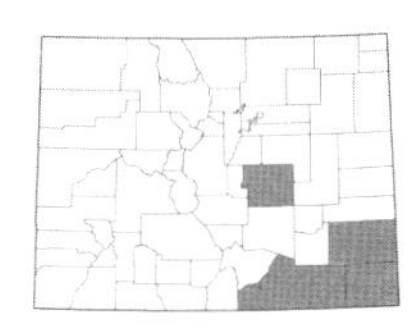

***Engelmannia pinnatifida* A. Gray ex Nutt.**, ENGELMANN'S DAISY. [*E. peristenia* (Raf.) Goodman & C.A. Lawson]. Plants 2–5 dm, densely hispid; *basal leaves* (1) 2–3 dm long, 1-pinnate; *cauline leaves* 8–30 cm long, pinnately lobed; *involucres* 6–10 mm high; *disk flowers* 3.5–5 mm long; *ray flowers* ca. 8, 10–16 mm long, each subtended by an inner involucral bract; *achenes* 3–4 mm long. Found in sandy or rocky soil of the plains and along roadsides, 3500–6200 ft. May–Aug. E.

ERICAMERIA Nutt. – GOLDENBUSH

Shrubs or half-shrubs with densely tomentose stems (ours); *leaves* alternate, simple, sessile or nearly so; *heads* discoid (ours); *involucre bracts* imbricate in several series, keeled or not, often chartaceous, usually with a green herbaceous tip; *receptacle* naked; *disk flowers* yellow; *ray flowers* absent (ours); *pappus* of capillary bristles; *achenes* somewhat angled, slender. (Nesom, 1990; Nesom & Baird, 1993; 1995)

1a. Leaves with glandular hairs; outer involucre bracts large and herbaceous, surpassing the brown inner bracts; heads solitary to few on each branch...***E. discoidea* var. *discoidea***
1b. Leaves tomentose or subglabrous; outer and inner involucre bracts the same, the outer usually not surpassing the inner bracts; heads in racemes, cymes, or corymbiform clusters, numerous on each branch...2

2a. Involucre bracts obtuse to acute, without an attenuate tip, or if an attenuate tip is present then the bracts chartaceous and strongly keeled; heads with 4–6 flowers; inflorescence cymose...***E. nauseosa***
2b. Involucre bracts with an attenuate tip, more or less membranous and weakly keeled; heads with 4–20 flowers; inflorescence more or less racemose...***E. parryi***

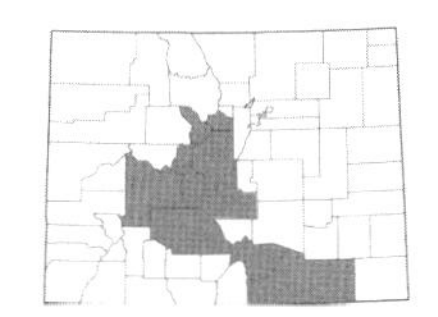

Ericameria discoidea* (Nutt.) G.L. Nesom var. *discoidea, COBWEBBY GOLDENBUSH; WHITESTEM GOLDENBUSH. [*Macronema discoidea* Nutt.; *Haplopappus macronema* A. Gray]. Plants 1.5–4 dm; *stems* with felt-like, white tomentum; *leaves* oblong to oblanceolate or elliptic, 1–4 cm long, stipiate-glandular, margins usually wavy; *involucre* 8–15 mm high, glandular, not well-imbricate; *disk flowers* 10–25, 8–11 mm long, hairy. Found on rocky alpine and subalpine slopes and in dry meadows, 8700–14,000 ft. July–Aug. E/W.

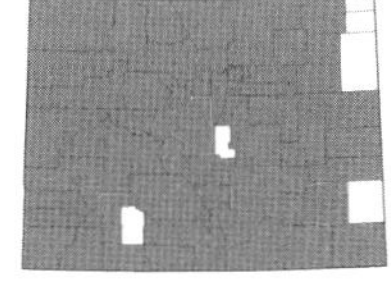

***Ericameria nauseosa* (Pall. ex Pursh) G.L. Nesom & G.I. Baird**, RUBBER RABBITBRUSH. [*Chrysothamnus nauseosus* (Pall. ex Pursh) Britton]. Plants 2–20 dm; *stems* with close tomentum, white to green; *leaves* filiform to linear, tomentose to subglabrous, often gland-dotted, 1–7 cm long; *involucre* 6–16 mm high, the bracts well-imbricate and keeled; *disk flowers* 4–6, 7–12 mm long, hairy or glabrous. Common and widespread in dry, sandy or rocky, open places, 3900–9500 ft. July–Oct. E/W. (Plate 11)

Ericameria nauseosa is a highly variable species that can be more or less separated into the following varieties:

1a. Achenes glabrous; stems nearly leafless at flowering...2
1b. Achenes hairy; stems leafless or not at flowering...3

2a. Corolla 5–8.5 mm long; stems yellowish-green; leaves glabrate...**var. *leiosperma* (A. Gray) G.L. Nesom & G.I. Baird**. [*Chrysothamnus nauseosus* (Pall. ex Pursh) Britton var. *leiospermus* (A. Gray) Hall]. Occurring at the eastern range of its distribution in Moffat and Montezuma cos., 6000–7000 ft. W.
2b. Corolla 9–11 mm long; stems whitish; leaves tomentulose...**var. *bigelovii* (A. Gray) G.L. Nesom & G.I. Baird**. [*Chrysothamnus nauseosus* (Pall. ex Pursh) Britton ssp. *bigelovii* (A. Gray) Hall & Clements]. Scattered in the San Luis Valley, 6000–7800 ft. E.

3a. Corolla lobes hairy, although sometimes only sparsely so; leaves few and reduced, mostly less than 1 mm wide and 2.5 cm long, stems often leafless at flowering; corolla 7–8.5 mm long...**var. *juncea* (Greene) G.L. Nesom & G.I. Baird**. [*Chrysothamnus nauseosus* (Pall. ex Pursh) Britton var. *junceus* (Greene) Hall & Clements]. Found in southwestern Colorado, 5000–7000 ft. W.
3b. Corolla lobes glabrous; leaves well-developed and more numerous, 0.5–3 mm wide and mostly 3 cm or more long, stems leafy at flowering; corolla various...4

4a. Involucre bracts glabrous or with ciliate margins, sometimes just the outer sparsely pubescent...5
4b. Involucre bracts pubescent to tomentose (sometimes only thinly so), or occasionally only the very short outermost ones glabrous...7

5a. Corolla 11–13 mm long, the tube sparsely pubescent with a few hairs 1–3 mm long...**var. *ammophila* L.C. Anderson**. Found in the San Luis Valley (Alamosa and Saguache cos.), 7500–8100 ft. E.
Rarely found growing with var. *oreophila*, which has yellowish-green leaves and smaller flowers and typically grows in more disturbed or saline sites where the two overlap.
5b. Corolla 6–9 mm long, the tube glabrous or minutely hairy with shorter hairs...6

6a. Leaves broadly linear, 1–3 mm wide and 1–5-nerved; corolla lobes 0.6–1.5 mm long; style appendages longer than the stigmatic portions...**var. *graveolens* (Nutt.) Reveal & Schuyler**. [*Chrysothamnus nauseosus* (Pall. ex Pursh) Britton ssp. *graveolens* (Nutt.) Piper; *C. nauseosus* (Pall. ex Pursh) Britton var. *glabratus* (A. Gray) Cronquist]. Common and widespread throughout the state, 3900–9500 ft. E/W.
6b. Leaves narrowly linear, 0.5–1 mm wide and 1-nerved; corolla lobes 1.5–2.5 mm long; style appendages shorter than or equaling the stigmatic portions...**var. *oreophila* (A. Nelson) G.L. Nesom & G.I. Baird**. [*Chrysothamnus nauseosus* (Pall. ex Pursh) Britton ssp. *consimilis* (Greene) Hall & Clements; C. *nauseosus* (Pall. ex Pursh) Britton var. *oreophilus* (A. Nelson) Hall & Clements]. Found in the central counties and western slope, especially common in the San Luis Valley, 6000–8500 ft. E/W.

7a. Corolla 6–8.5 mm long; involucre bracts 6.5–9 mm high; plants 2–10 dm tall...**var. *nauseosa***. [*Chrysothamnus nauseosus* (Pall. ex Pursh) Britton var. *nauseosus*]. Widespread throughout the central mountains and eastern plains, scattered on the western slope, 4500–8500 ft. E/W.
7b. Corolla mostly 9–11 mm long; involucre bracts 8–11 mm high; plants 6–20 dm tall...**var. *speciosa* (Nutt.) G.L. Nesom & G.I. Baird**. [*Chrysothamnus nauseosus* (Pall. ex Pursh) Britton var. *albicaulis* (Nutt.) Rydb.; C. *nauseosus* (Pall. ex Pursh) Britton ssp. *speciosus* (Nutt.) Hall & Clements]. Found in northwestern Colorado (with one additional record from Mesa County), 5000–8000 ft. W.

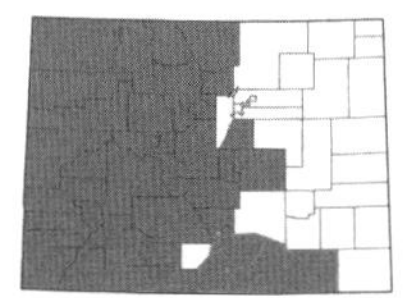

***Ericameria parryi* (A. Gray) G.L. Nesom & G.I. Baird**, Parry's rabbitbrush. [*Chrysothamnus parryi* (A. Gray) Greene]. Plants 2–8 dm; *stems* with close tomentum, gray-white to gray-green; *leaves* linear, gray-tomentose to glabrous, 1–8 cm long; *involucre* 9–18 mm high, the bracts weakly imbricate, slightly keeled; *disk flowers* 4–20, 8–11 mm long, hairy. Common in dry, open places in sandy or rocky soil, widespread but absent from the eastern plains, 5100–12,000 ft. July–Sept. E/W.

There are 3 varieties of *E. parryi* in Colorado:

1a. Flowers 8–20 per head; leaves 3–8 cm long and 2–3 mm wide...**var. *parryi***. [*Chrysothamnus parryi* (A. Gray) Greene ssp. *parryi*]. The most common variety, found throughout the central and western counties, 7000–11,000 ft. E/W.
1b. Flowers 5–7 per head; leaves 2–4 cm long and 1 mm wide...2

2a. Upper leaves overtopping the inflorescence; corolla pale yellow, 8–10.7 mm long, the tubes hairy...**var. *howardii* (Parry ex A. Gray) G.L. Nesom & G.I. Baird**. [*Chrysothamnus parryi* (A. Gray) Greene ssp. *howardii* (Parry ex A. Gray) Hall & Clements]. Scattered across the state, especially on the western slope, 5100–9600 ft. E/W.
2b. Upper leaves shorter than the inflorescence; corolla clear yellow, 10–11 mm long, the tubes glabrous...**var. *attenuata* (M.E. Jones) G.L. Nesom & G.I. Baird**. [*Chrysothamnus parryi* (A. Gray) Greene ssp. *attenuatus* (M.E. Jones) Hall & Clements; *C. parryi* (A. Gray) Greene ssp. *nevadensis* (A. Gray) Kittell]. Known only from Montezuma Co., 5500–5700 ft. W.

ERIGERON L. – fleabane; daisy

Herbs; *leaves* alternate, simple; *heads* usually radiate; *involucre bracts* narrow, herbaceous and subequal; *receptacle* naked; *disk flowers* yellow; *ray flowers* numerous, with narrow ligules, white, pink, purple, or blue, rarely absent; *pappus* of capillary bristles, commonly with some short outer setae; *achenes* usually 2-nerved but up to 14-nerved, commonly flattened. (Cronquist, 1947; Nesom, 2004; Nesom & Murray, 2004)

1a. Ray flowers very numerous with an erect, very narrow (mostly less than 0.5 mm wide), short (2–4.5 mm long) ligule or sometimes the inner pistillate flowers rayless…2
1b. Ray flowers few to numerous but without an erect ligule, the ray usually longer (4–15 mm) and wider (0.5–4 mm)…3

2a. Stem leaves usually linear; pistillate ray flowers not in two series; inflorescence racemiform; plants usually found in moist places…***E. lonchophyllus***
2b. Stem leaves usually broader than linear; pistillate ray flowers in two series with the outer shortly ligulate and the inner essentially eligulate; inflorescence more or less corymbiform; plants usually found in dry places…***E. acris***

3a. Ray flowers absent or with an inconspicuous, white or yellowish ligule shorter than the disk…4
3b. Ray flowers present…5

4a. Leaves linear, entire, densely spreading-hirsute…***E. aphanactis***
4b. Leaves pinnately cleft into linear segments, mostly glabrous with ciliate margins…***E. mancus***

5a. Leaves all pinnatifid, ternately dissected, or 3-lobed at the tips (sometimes a few, reduced, entire linear cauline leaves present), mostly basal…6
5b. Leaves entire or coarsely toothed, or sometimes with a few leaves 3-lobed to pinnatifid, mostly basal or not…9

6a. Leaves pinnatifid, nearly glabrous except for the ciliate margins or sometimes with a few appressed hairs or glands; ray flowers purple or blue…***E. pinnatisectus***
6b. Leaves 1–4 times ternately or palmately dissected or 3-lobed or 3-toothed at the apex, pubescent with glandular, hirsute, or pilose hairs; ray flowers usually white or sometimes pink or blue…7

7a. Leaves 1–4 times ternately or palmately dissected into linear segments; plants with a stout, branching caudex…***E. compositus***
7b. Leaves 3-lobed or 3-toothed into obtuse or lanceolate segments; plants with a slender, branching caudex, this sometimes divided into several rhizome-like branches…8

8a. Involucre 9–13 mm high, the bracts usually densely woolly at the base of the head or throughout; leaves entire to 3-toothed at the apex, the tip of the leaves and/or lobes acute…***E. lanatus***
8b. Involucre 5–7 mm high, the bracts glandular and hirsute or villous-hirsute but not woolly; leaves 3-lobed at the apex, the tips of the lobes rounded…***E. vagus***

9a. Pappus of the ray and disk flowers unlike, that of the disk flowers of long bristles and short outer setae, that of the ray flowers only of short setae and lacking long bristles; heads small (the involucre 2–5 mm high and 5–12 mm wide), several to numerous in an open, wide-branching corymbiform cluster; rays white…***E. strigosus* var. *strigosus***
9b. Pappus of the ray and disk flowers alike; heads and rays various, usually not in an open, wide-branching corymbiform cluster…10

10a. At least the lower leaves toothed; involucre more or less villous-hirsute covered with multicellular hairs with black cross-walls; rays white…***E. coulteri***
10b. Leaves entire; involucre sometimes with multicellular hairs with black cross-walls; rays various…11

11a. Stem leaves lanceolate to elliptic or ovate, usually ample and well-developed (mostly over 1 cm wide), little if any reduced above, lacking a basal cluster of leaves; plants tall (mostly over 2 dm high)…12
11b. Leaves mostly linear or oblanceolate, either mostly basal or with a basal cluster and with stem leaves not well-developed or reduced, or with well-developed, linear to narrowly oblanceolate stem leaves; plants various…21

12a. Involucre bracts not at all glandular…13
12b. Involucre bracts glandular (at least at the base), often with longer hairs, too…15

13a. Involucre bracts woolly or long-villous with multicellular hairs; stems glandular above; leaves entire, villous-hirsute on both sides, the upper ones occasionally glandular; perennials…***E. elatior***
13b. Involucre bracts not woolly or long-villous, with short hairs or nearly glabrous; stems not glandular above; leaves entire or toothed, nearly glabrous to hirsute, the upper ones not glandular; biennials or perennials…14

14a. Disk corollas 4–6 mm long; heads usually larger, the disk 5–15 mm wide; rays 1 mm wide; pappus double, the outer setulose and sometimes very scanty...***E. glabellus***
14b. Disk corollas 2.5–3.5 mm long; heads usually smaller, the disk 4–6 mm wide; rays very narrow, 0.2–0.6 mm wide; pappus simple, of 20–30 bristles that are often shorter than the disk corollas...***E. philadelphicus* var. *philadelphicus***

15a. Leaves mostly basal, the middle and upper stem leaves smaller, the basal and lower leaves oblanceolate and the middle and upper leaves lanceolate or oblong; stems usually curved at the base...***E. formosissimus***
15b. Leaves unlike the above in all respects; stems usually erect, not curved at the base...16

16a. Rays 2–4 mm wide (*careful*—can dry narrower); pappus of 20–30 bristles or rarely with a few outer setae (not double); stem densely appressed white-villous under heads; involucre surface usually with reddish-purple-tipped glandular hairs...***E. glacialis***
16b. Rays usually narrower, to 2 mm wide; pappus double with 15–30 inner bristles and shorter outer setae; stem usually unlike the above; involucre glands reddish-purple-tipped or not...17

17a. Leaves without long-ciliate margins, glabrous or the middle and upper glandular; rays 1–2 mm wide...***E. eximius***
17b. Leaves with long-ciliate margins, hairy or glabrous to glandular; rays 1 mm wide or less...18

18a. Stems glabrous or sometimes slightly hairy or glandular; leaves glabrous except for the ciliate margins or sometimes sparsely hairy or the uppermost occasionally glandular...***E. speciosus***
18b. Stems pubescent with spreading hairs, often glandular above; leaves hairy throughout or only densely hairy on the veins, at least the uppermost with glandular hairs as well...19

19a. Leaves densely hairy on the veins or hairy throughout, only the uppermost rarely a little glandular...***E. subtrinervis***
19b. Leaves glandular and hirsute...20

20a. Disk corollas 4.5–5 mm long; stems usually not densely white spreading-villous below the heads...***E. uintahensis***
20b. Disk corollas 3–4 mm long; stems usually densely white spreading-villous below the heads...***E. vreelandii***

21a. Stems glabrous, slender and lax; basal leaves broadly oblanceolate or ovate, glabrous...***E. kachinensis***
21b. Stems variously hairy or glandular (at least under the head), usually not slender and lax; leaves variously shaped, usually not glabrous...22

22a. Involucre bracts and upper stem with multicellular hairs with black or dark purplish-black cross-walls...23
22b. Involucre bracts and/or upper stem lacking multicellular hairs, or if present then these with clear, purplish-red, or reddish cross-walls...24

23a. Rays very short and narrow, 3.5–6 mm long and 0.5–1 mm wide and usually barely exceeding the disk corollas; disk corollas 3–5 mm long; stems with mostly spreading hairs...***E. humilis***
23b. Rays longer and wider, 7–11 mm long and 1.4–2 mm wide and conspicuously exceeding the disk corollas; disk corollas 2.4–3.5 mm long; stems with mostly ascending hairs...***E. melanocephalus***

24a. Stems with spreading or tangled (pointing in all directions) hairs or mostly glandular...25
24b. Stems with appressed or ascending hairs (at above the middle), not glandular...34

25a. Leaves more or less densely glandular, or occasionally obscurely so and the leaves stiffly ciliate, mostly or all basal with a few linear, reduced stem leaves present; stems glandular and also hirsute with spreading hairs (at least above); rays usually purple or blue...***E. vetensis***
25b. Leaves variously hairy to glabrate, if glandular then only the upper stem leaves glandular; stems variously hairy to glabrate, or sometimes glandular or viscid under the heads; rays various...26

26a. Leaves and stems densely pubescent with spreading, hirsute hairs mostly 1 mm long, especially on the petioles; leaves mostly less than 3 mm wide; heads more or less hemispheric; rays usually white...27
26b. Leaves and stems unlike the above in all respects; heads and rays various...28

27a. Outer pappus of inconspicuous bristles...***E. pumilus***
27b. Outer pappus of narrow to broad scales...***E. concinnus***

28a. Annuals, biennials, or short-lived perennials with short, simple caudices and taproots; involucres 3–5 mm high, lacking multicellular or glandular hairs; stems leafy with linear to narrowly oblanceolate leaves mostly less than 3 mm wide...29
28b. Perennials usually with stout, freely branching but sometimes with a simple to slightly branching caudices, or from deep rhizomes, or if biennials or short-lived perennials then from fibrous roots; involucres 5–13 mm high, often with multicellular or glandular hairs; stems and leaves various...30

29a. Branches from the lower half of the main stems often developing into long, trailing leafy stoloniferous branches; basal leaves present at flowering time...***E. tracyi***
29b. Stems usually erect and without leafy stoloniferous branches, often branched above the middle; basal leaves usually withered and not present at flowering time...***E. divergens***

30a. Stems and leaves loosely woolly-villous with long tangled hairs, usually densely so under the head; involucre 9–13 mm high, the bracts moderately to densely lanate-woolly; leaves entire to 3-toothed at the apex, the tip of the leaves and/or lobes acute; rays usually white...***E. lanatus***
30b. Stems and leaves not loosely woolly-villous with tangled hairs; involucre 5–10 mm high, the bracts often with multicellular hairs or glandular but not lanate-woolly; leaves oblanceolate to spatulate and rounded at the apex, or acute and entire to irregularly toothed on the margins but not 3-toothed at the apex; rays various...31

31a. Involucre surface with hirsute or strigose hairs, not at all glandular or multicellular; stem not glandular-hairy above; basal leaves usually with an acute tip, entire or sometimes irregularly toothed on the margins...***E. glabellus***
31b. Involucre surface glandular (at least at the base of the head) or with multicellular hairs; stem (at least above) with glandular or viscid hairs, usually also with some spreading hairs; basal leaves oblanceolate to spatulate with a rounded tip, entire...32

32a. Leaves densely pubescent with short spreading hairs, often grayish in appearance; known from the northern counties (Moffat, Routt, Jackson, and Larimer)...***E. caespitosus***
32b. Leaves glabrous with ciliate margins or sparsely to moderately hirsute-pilose, not grayish, or the uppermost ones sometimes glandular...33

33a. Stems usually erect or slightly curved at the base; involucre glandular and moderately to densely woolly-villous with multicellular hairs (especially at the base), these often with reddish or purplish cross-walls...***E. grandiflorus***
33b. Stems distinctly curved at the base; involucre glandular and/or hirsute with few multicellular hairs, these lacking reddish or purplish cross-walls...***E. formosissimus***

34a. Involucre bracts finely glandular or viscid, lacking eglandular hairs; leaves glabrous or nearly so, sometimes with a few short strigose hairs...***E. leiomerus***
34b. Involucre bracts hirsute to strigose with eglandular hairs, sometimes also glandular or viscid; leaves usually variously hirsute or strigose, occasionally glabrate...35

35a. Stems leafy throughout, lacking a basal cluster of leaves at flowering time; plants annuals or biennials from taproots; pappus simple, of 15 very fragile bristles; heads numerous, the involucre 5–8 mm wide; plants of the eastern slope...***E. bellidiastrum***
35b. Plants short- to long-lived perennials, if biennials then unlike the above in all respects, usually with a basal cluster of leaves; pappus double (with inner longer bristles and outer shorter setae) or if simple then plants perennials; heads various; variously distributed...36

36a. At least some branches from the lower half of the main stems developing into long, trailing leafy stoloniferous branches; plants biennials or short-lived perennials from a taproot...***E. flagellaris***
36b. Trailing, leafy stoloniferous branches absent; plants perennials or if biennials or short-lived perennials then from fibrous roots...37

37a. Basal leaves broadly lanceolate to oblanceolate, 3–20 mm wide, usually some of the margins irregularly toothed; involucre bracts hirsute or strigose, without glandular or viscid hairs; plants generally taller, 1.5–7 dm high, from a simple or slightly branched caudex...***E. glabellus***
37b. Basal leaves linear or linear-oblanceolate, if oblanceolate and wider than at least some with 3 nerves, the leaves entire; involucre bracts often hirsute as well as glandular or viscid; plants generally shorter, 0.5–3.5 (5) dm high, usually from a stout, branched caudex...38

38a. Involucre bracts with at least some multicellular hairs with reddish-purple cross-walls; basal leaves linear to narrowly oblanceolate, to 2.5 mm wide, glabrous above and sparsely strigose below...***E. radicatus***
38b. Involucre bracts lacking multicellular hairs with purplish cross-walls; basal leaves various...39

39a. Stems (scapes) naked or with 2 or 3 very reduced leaves; leaves 0.4–2 cm long, narrowly linear and less than 1 mm wide; rays usually white or sometimes pink; achenes hairy on the nerves but otherwise glabrous...***E. consimilis***
39b. Stems usually with at least 2 more or less well-developed leaves; leaves usually over 2 cm long, variously shaped, usually over 1 mm wide; rays various; achenes hairy throughout or glabrous...40

40a. Basal leaves usually distinctly 3-nerved, sparsely to moderately strigose or hirsute; rays usually white; pappus simple with a few inconspicuous outer setae...***E. eatonii* var. *eatonii***
40b. Basal leaves only 1-nerved or obscurely 3-nerved, variously hairy; rays and pappus various...41

41a. Ovary and achenes glabrous; leaves and stems densely grayish-strigose...***E. canus***
41b. Ovary and achenes hairy, at least on the nerves; leaves and stems various...42

42a. Caudex branches bearing numerous fibrous roots, lacking a well-defined taproot; rays purple or blue; plants found in meadows above 8000 ft in elevation...***E. ursinus***
42b. Caudex branches not bearing numerous fibrous roots, with a taproot; rays various; plants usually found below 8000 ft in elevation or not found in subalpine and alpine meadows...43

43a. Pappus simple, of 15–25 bristles and sometimes a few intermingled short setae; leaves glabrous above and sparsely strigose below, linear to narrowly oblanceolate (to 3 mm wide); rays white or sometimes pink...***E. nematophyllus***
43b. Pappus usually double (sometimes the outer setae inconspicuous); leaves hairy above and below, otherwise various; rays various...44

44a. Basal leaves with ciliate margins, usually green (not grayish or silvery) in appearance; rays usually white or sometimes pink...***E. engelmannii***
44b. Basal leaves lacking ciliate margins, usually grayish or silvery in appearance; rays purple, blue, pink, or sometimes white...45

45a. Involucre bracts hirsute-villous with crinkled, somewhat flattened multicellular hairs and obscurely viscid; achenes 2–5-nerved...***E. pulcherrimus***
45b. Involucre bracts strigose, sometimes only sparsely so, and often finely glandular; achenes 4–8-nerved...46

46a. Basal leaves silvery-strigose, sometimes becoming greener in age, densely tufted and persistent at flowering time; lower stem not densely strigose and white at the base; achenes 6–8-nerved; heads generally larger with the disk 10–18 mm wide and involucre bracts 5.5–9 mm high; ray flowers pale lavender to blue-lavender; known from Montezuma County...***E. argentatus***
46b. Basal leaves gray-green, not silvery, usually withered by flowering time and not forming a conspicuous dense tuft; lower stem often densely strigose and white at the base; achenes 4–6-nerved; heads generally smaller with the disk 5–15 mm wide and involucre bracts 3–7 mm high; ray flowers blue, pink, or white; plants not restricted to Montezuma County...47

47a. Ray flowers 10–14 (20), the ligule 4–8 mm long and 1–1.8 mm wide; disk corollas viscid-puberulent with blunt-tipped hairs...***E. sparsifolius***
47b. Ray flowers 28–40, the ligule 4–18 mm long and 1.4–2.7 mm wide; disk corollas sparsely strigose-villous with needle-like hairs...***E. utahensis***

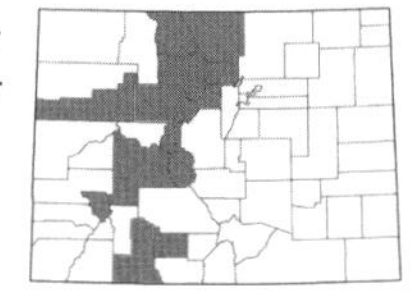

***Erigeron acris* L.**, BITTER FLEABANE. [*Trimorpha acris* (L.) A. Gray; including *E. nivalis* Nutt.]. Biennials or short-lived perennials, 0.5–8 dm; *leaves* oblanceolate to spatulate or narrowly lanceolate, entire or remotely toothed, to 10 cm long; *involucre* 5–12 mm high, finely glandular, hirsute, or both; *disk flowers* 4–6.2 mm long; *ray flowers* numerous, pink, purple, or white, the outer 2.5–4.5 mm long and erect; *pappus* of 25–35 bristles. Found in open spruce forests, dry meadows, and on rocky slopes, 8000–12,500 ft. E/W.

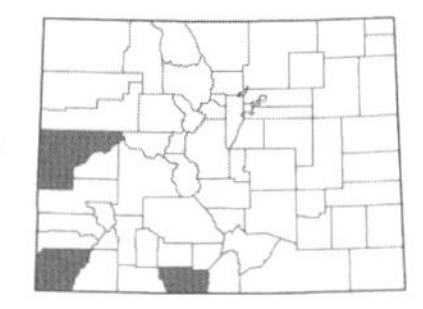

***Erigeron aphanactis* (A. Gray) Greene**, RAYLESS SHAGGY DAISY. [*E. concinnus* (Hook. & Arn.) Torr. & A. Gray var. *aphanactis* A. Gray]. Perennials, 0.2–3 dm; *leaves* linear-oblanceolate to spatulate, entire, to 8 cm long, spreading-hirsute; *involucre* 4–6 mm high, spreading-hirsute and often finely glandular; *disk flowers* 2.8–5 mm long, yellow; *ray flowers* absent; *pappus* double. Uncommon in open, rocky places, 4600–8000 ft. May–July. E/W.

***Erigeron argentatus* A. Gray**, SILVER DAISY. Perennials, 0.6–4 dm; *leaves* oblanceolate to linear, silvery-strigose, entire, to 7 cm long; *involucre* 5.5–9 mm high, silvery-strigose and often finely glandular; *disk flowers* 3.8–5.6 mm long; *ray flowers* 9–15 mm long, blue or sometimes purple or pink, rarely white; *pappus* double. Uncommon in dry, open places, known from Montezuma Co., 6000–6500 ft. May–June. W.

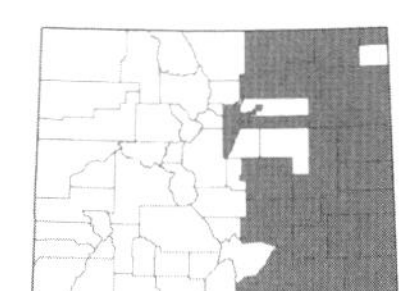

***Erigeron bellidiastrum* Nutt.**, PRETTY DAISY. Annuals or biennials, 1–5 dm; *leaves* linear to oblanceolate, entire or occasionally 3-toothed, to 4 cm long, finely hirsute with incurved hairs; *involucre* 3–5 mm high, hirsute with curved hairs; *disk flowers* 2.5–3 mm long; *ray flowers* 4–6 mm long, white or pink; *pappus* of 15 very fragile bristles. Common in dry grasslands and sandy soil, 3400–8800 ft. May–Sept. E.

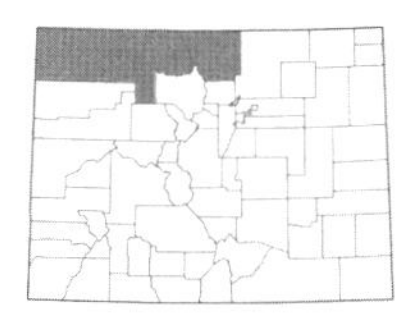

***Erigeron caespitosus* Nutt.**, TUFTED DAISY. Perennials, 0.5–3 dm; *leaves* oblanceolate to oblong-ovate or linear, entire, densely hirtellous or canescent, to 12 cm long; *involucre* 4–7 mm high, glandular and canescent; *disk flowers* 3–4.5 mm long; *ray flowers* 5–15 mm long, blue, white, or pink; *pappus* double. Locally common on rocky hillsides and in dry, open meadows, 6000–9700 ft. July–Sept. E/W. (Plate 11)

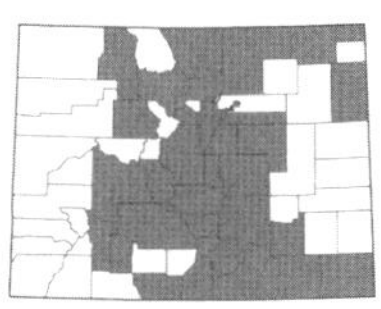

***Erigeron canus* A. Gray**, HOARY DAISY. Perennials, 0.5–3.5 dm; *leaves* oblanceolate to linear, entire, densely strigose, grayish-white, to 10 cm long; *involucre* 5–7 mm high, canescent with short, spreading hairs; *disk flowers* 3.5–5.6 mm long; *ray flowers* 7–12 mm long, white or blue; *pappus* double. Found in sandy or rocky soil of dry, open places of the eastern plains and outer foothills, 3500–9500 ft. May–July. E/W.

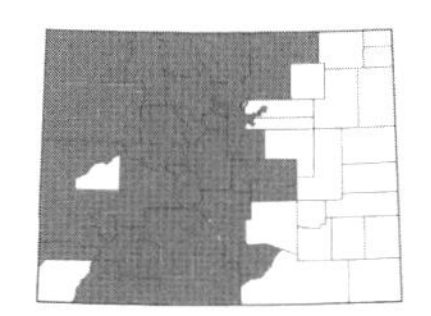

***Erigeron compositus* Pursh**, CUTLEAF DAISY. Perennials to 2.5 dm; *leaves* ternately compound, minutely glandular and hispid, 0.5–5 (7) cm long; *involucre* 5–10 mm high, hirsute and minutely glandular; *disk flowers* 3–5 mm long; *ray flowers* to 12 mm long, white, pink, or blue; *pappus* of 12–20 simple bristles. Common on rocky slopes and in meadows and other open places, 5300–14,000 ft. (April) June–Aug. (Sept.). E/W. (Plate 11)

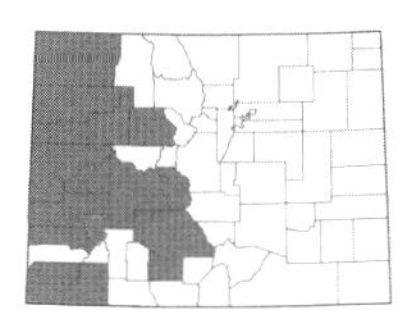

***Erigeron concinnus* (Hook. & Arn.) Torr. & A. Gray**, NAVAJO DAISY. [*E. pumilus* Nutt. var. *concinnus* (Hook. & Arn.) Dorn]. Perennials, 0.5–5 dm; *leaves* oblanceolate, mostly basal, hirsute, entire, to 8 cm long; *involucre* 4–7 mm high, hirsute and finely glandular; *disk flowers* 3–5 mm long; *ray flowers* 6–15 mm long, white, pink, or deep blue; *pappus* double. Common in sandy or rocky soil of dry, open places, often in association with sagebrush, 4300–8900 ft. May–June (Aug.). E/W.

***Erigeron consimilis* Cronquist**, CUSHION-DAISY. [*E. compactus* S.F. Blake]. Perennials, 0.3–1.2 dm; *leaves* linear, entire, basal, densely strigose, 0.4–2 cm long; *involucre* 6–8.5 mm high, hispid with short, spreading hairs; *disk flowers* 4–6 mm long; *ray flowers* 7–11 mm long, white or pink; *pappus* double. Found in gravelly and sandy soil of dry, open places, and on gypsum hills, known from Moffat County, 5700–6800 ft. May–July. W.

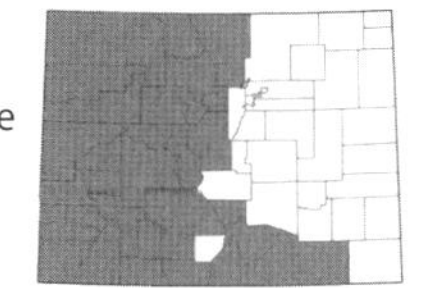

***Erigeron coulteri* Porter**, COULTER'S DAISY. Perennials, 1–6 dm; *leaves* evidently cauline, oblanceolate to elliptic, entire or the lower toothed, to 15 cm, hirsute; *involucre* 7–10 mm high, glandular and hirsute, the hairs with black cross-walls; *disk flowers* 3–4.5 mm long; *ray flowers* 9–24 mm long, white; *pappus* of 20–25 bristles. Common in open mountain meadows, along streams, and in spruce or aspen forests, 7500–14,000 ft. (June) July–Sept. E/W. (Plate 11)

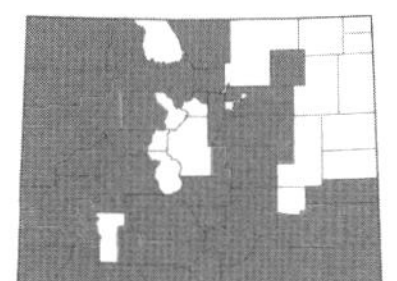

***Erigeron divergens* Torr. & A. Gray**, SPREADING DAISY. Annuals, biennials, or short-lived perennials, 1–7 dm; *leaves* evidently cauline, linear to oblanceolate, entire or coarsely toothed, hirsute, to 4 cm long; *involucre* 4–5 mm high, densely hirsute; *disk flowers* 2–3 mm long; *ray flowers* 5–10 mm long, white, pink, or blue; *pappus* double. Common in dry to somewhat moist open meadows and slopes or in sparsely wooded forests, 4000–8500 (10,300) ft. May–Sept. E/W.

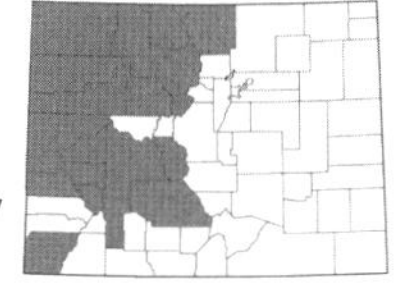

Erigeron eatonii* A. Gray var. *eatonii, EATON'S DAISY. Perennials, 0.5–3 dm; *leaves* mostly basal, linear, 3-nerved, strigose, to 15 cm long; *involucre* 5–8 mm high, minutely to conspicuously glandular and hirsute; *disk flowers* 3.5–5 mm long; *ray flowers* 5–10 mm long, white or occasionally pink; *pappus* of 15–20 fragile bristles. Common in gravelly or sandy soil of open places, often in sagebrush or pinyon-juniper communities, 6000–10,000 ft. May–July. E/W. (Plate 11)

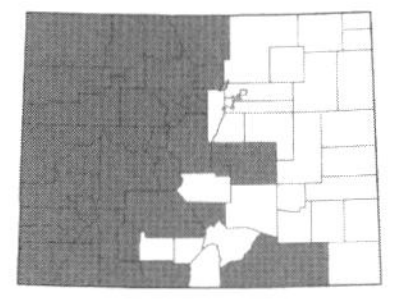

***Erigeron elatior* (A. Gray) Greene**, TALL DAISY. Perennials, 2–6 dm; *leaves* cauline, oblanceolate to narrowly elliptic, entire, villous-hirsute, to 10 cm long; *involucre* 7–12 mm high, woolly or long-villous; *disk flowers* 4–5.5 mm long; *ray flowers* 10–20 mm long, pink to purple; *pappus* double. Common in meadows and in aspen and spruce forests, 6500–13,500 ft. June–Aug. E/W.

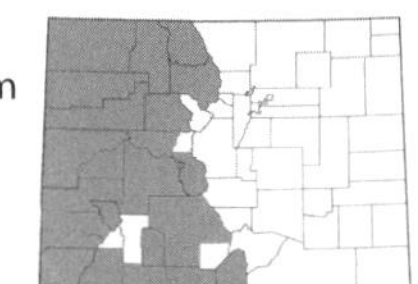

***Erigeron engelmannii* A. Nelson**, ENGELMANN'S DAISY. Perennials, 0.3–3 dm, with short caudex branches; *leaves* mostly basal, linear-oblanceolate, entire, hirsute, to 10 cm long; *involucre* 4–8 mm high, hirsute and finely glandular; *disk flowers* 2.5–4.5 mm long; *ray flowers* 5–14 mm long, white or sometimes pink; *pappus* double. Common in sandy or rocky soil of open places, 4800–9000 ft. May–July (Aug.). E/W.

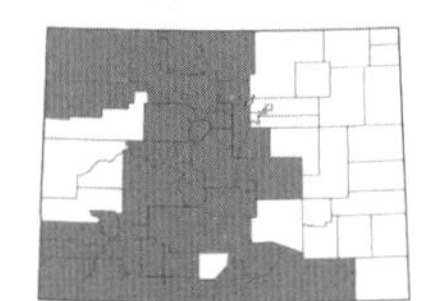

***Erigeron eximius* Greene**, SPLENDID DAISY. Perennials, 1.5–6 dm; *leaves* oblanceolate to ovate or oblong, 3-nerved, entire, lower glabrous and middle glandular, 3–15 cm long; *involucre* 7–9 mm high, glandular and sometimes hairy; *disk flowers* 4.5–6 mm long; *ray flowers* 12–20 mm long, blue or rose-purple, rarely white; *pappus* double. Common in mountain forests and meadows, 6500–12,000 ft. July–Sept. E/W.

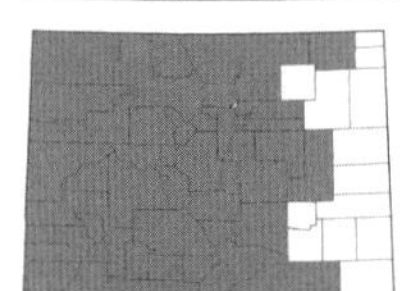

***Erigeron flagellaris* A. Gray**, TRAILING DAISY. Biennials or short-lived perennials, 0.5–4 dm; *leaves* linear to linear-oblanceolate, some stems becoming long, trailing leafy and stoloniferous, entire to pinnately lobed, hirsute, to 5 cm long; *involucre* 3.5–5 mm high, finely glandular and hirsute; *disk flowers* 2.5–3.5 mm long; *ray flowers* 5–10 mm long, white or occasionally pink or blue; *pappus* double. Common in open meadows and dry slopes, 5000–11,800 ft. May–July. E/W.

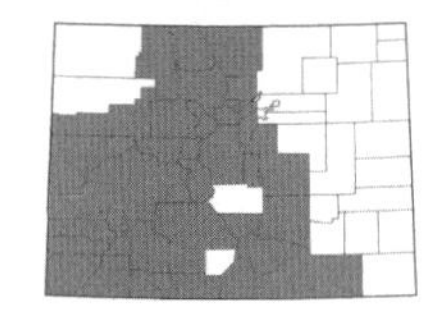

***Erigeron formosissimus* Greene**, BEAUTIFUL DAISY. Perennials, 1–5.5 dm; *leaves* evidently cauline, oblanceolate to spatulate or oblong, entire, uppermost glandular, 2–15 cm long; *involucre* 5–8 mm high, glandular and often hirsute; *disk flowers* 3.5–4.5 mm long; *ray flowers* 8–15 mm long, blue, violet, or rarely white; *pappus* double. Common in mountain meadows and forests, (5000) 7000–12,000 ft. June–Sept. E/W.

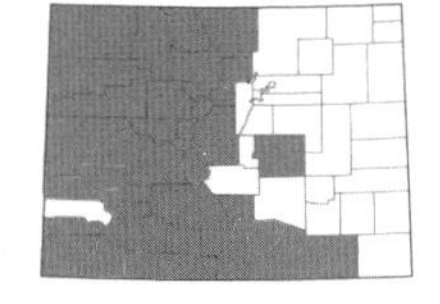

***Erigeron glabellus* Nutt.**, SMOOTH DAISY. Biennials or perennials, 1.5–7 dm; *leaves* evidently cauline, oblanceolate to lanceolate, hirsute or strigose, 4–15 cm long; *involucre* 5–9 mm high, hirsute or strigose; *disk flowers* 4–6 mm long; *ray flowers* 8–15 mm long, violet, blue, pink, or rarely white; *pappus* double. Common in wet or dry meadows, along streams, and in forest openings, 5500–12,500 ft. June–Sept. E/W.

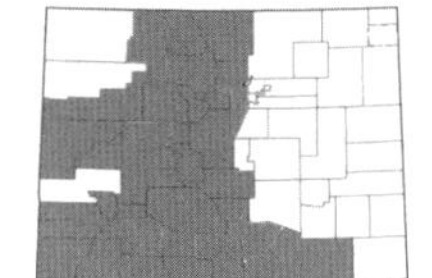

***Erigeron glacialis* (Nutt.) A. Nelson.**, GLACIAL DAISY. Perennials, 0.5–7 dm; *leaves* evidently cauline, oblanceolate to spatulate, glabrous or long-ciliate on the margins, to 20 cm long; *involucre* 7–11 mm high, glandular; *disk flowers* 4–6 mm long; *ray flowers* 8–25 mm long, 2–4 mm wide, rose-purple or pink; *pappus* simple. Common in mountain meadows, along streams, and in forest openings, 8000–14,300 ft. (June) July–Sept. E/W. (Plate 11)

This is the *Erigeron* that was formerly referred to as *E. peregrinus* (Banks ex Pursh) Greene, which is restricted to the northwest states and Canada.

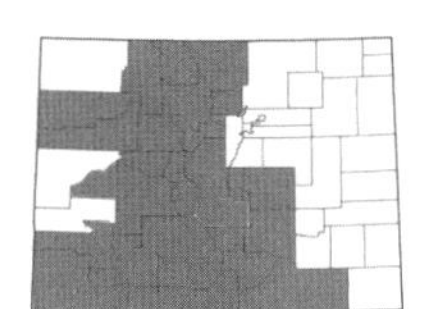

***Erigeron grandiflorus* Hook.**, ROCKY MOUNTAIN ALPINE DAISY. [*E. simplex* Greene]. Perennials, 0.2–2 dm; *leaves* oblanceolate to spatulate, glabrous or somewhat hirsute, with ciliate margins, entire, mostly basal, to 8 cm long; *involucre* 5–8 mm high, viscid and white-villous with multicellular hairs, the cross-walls clear or reddish; *disk flowers* 3–4 mm long; *ray flowers* 7–11 mm long, purple, pink, or rarely white; *pappus* double. Common in alpine and subalpine meadows, 9500–14,000 ft. July–Sept. E/W. (Plate 11)

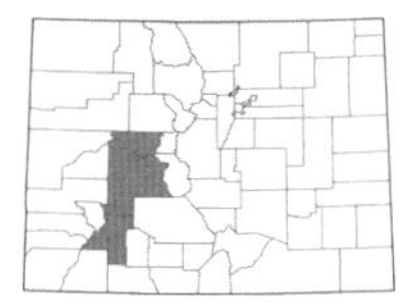

***Erigeron humilis* Graham**, ARCTIC ALPINE DAISY. Perennials, 0.2–2.5 dm; *leaves* mostly basal, oblanceolate to spatulate, entire, villous, to 8 cm; *involucre* 6–9 mm high, densely woolly-villous, the hairs with dark purple or black cross-walls; *disk flowers* 3–5 mm long; *ray flowers* 3.5–6 mm long, white to purple; *pappus* of 20–30 bristles. Uncommon on mossy turf between rocks in the alpine, 11,500–14,000 ft. June–Aug. W.

***Erigeron kachinensis* S.L. Welsh & Moore**, KACHINA DAISY. Perennials to 2 dm; *leaves* mostly basal, oblanceolate to ovate, entire, glabrous, to 6 cm long; *involucre* 3–5 mm high, glandular; *disk flowers* 2.5–3.5 mm long; *ray flowers* 4–7 mm long, usually white or sometimes purple; *pappus* of 15 bristles. Rare in saline soils in alcoves and seeps in canyon walls, 4800–6600 ft. May–July. W.

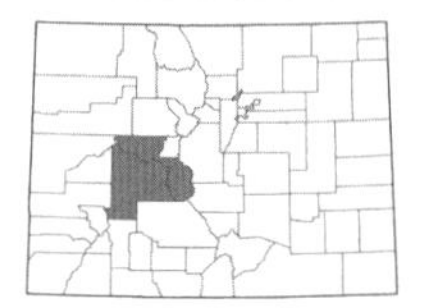

***Erigeron lanatus* Hook.**, WOOLLY DAISY. Perennials to 0.5 dm; *leaves* basal, oblanceolate to narrowly elliptic, pilose or woolly-villous, often 3-toothed at the apex, to 3 cm long; *involucre* 9–13 mm high, lanate-woolly; *disk flowers* 5–6.5 mm long; *ray flowers* 8–11 mm long, white or sometimes pink; *pappus* double. Rare on scree slopes in the alpine, 12,500–13,500 ft. July–Aug. E/W.

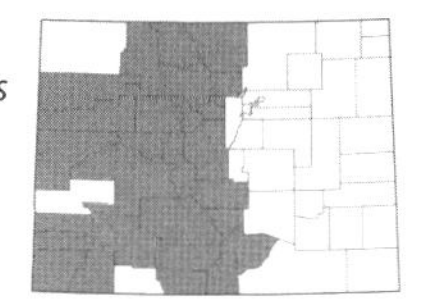

***Erigeron leiomerus* A. Gray**, ROCKSLIDE DAISY. Perennials to 1 dm; *leaves* mostly basal, spatulate to oblanceolate, entire, glabrous or nearly so, to 7 cm; *involucre* 4–6 mm, finely glandular; *disk flowers* 3–4.5 mm long; *ray flowers* 6–11 mm long, deep blue-violet or white; *pappus* double. Found in rocky alpine and subalpine meadows and on scree slopes and rock ledges, (7200) 9000–13,500 ft. June–Aug. E/W. (Plate 11)

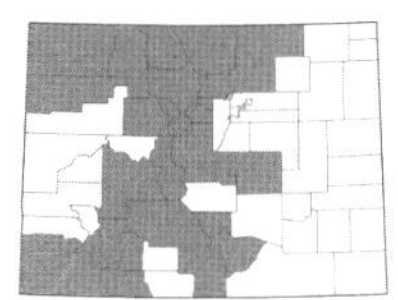

***Erigeron lonchophyllus* Hook.**, SHORTRAY FLEABANE; LONGLEAF DAISY. [*Trimorpha lonchophylla* (Hook.) G.L. Nesom]. Biennials or short-lived perennials, 0.2–6 dm; *leaves* oblanceolate to spatulate, entire, hirsute, to 15 cm; *involucre* 4–9 mm, hirsute; *disk flowers* 3.5–5 mm long; *ray flowers* 2–3 mm long, erect, white or sometimes pinkish; *pappus* of 20–30 bristles. Found in moist meadows, along the edges of ponds, and around springs, 5200–11,500 ft. July–Aug. E/W.

***Erigeron mancus* Rydb.**, LA SAL DAISY. Perennials to 0.7 dm; *leaves* mostly basal, pinnatifid with linear lobes, glabrous or minutely glandular, 1.2–4 cm long; *involucre* 5–6.5 mm high, hirsute and minutely glandular; *disk flowers* 3.5–4.5 mm long; *ray flowers* absent; *pappus* double. This is not reported for Colorado but occurs in Utah next to the Colorado border at higher elevations in rocky slopes and fell fields in the alpine. July–Aug.

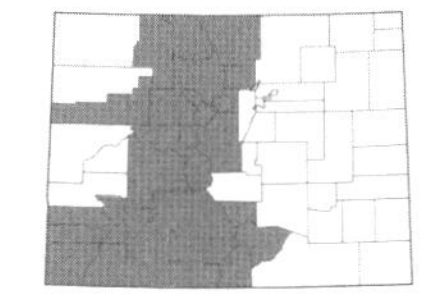

***Erigeron melanocephalus* (A. Nelson) A. Nelson**, BLACKHEAD DAISY. Perennials, 0.5–1.5 dm; *leaves* oblanceolate to spatulate, mostly basal, glabrous or somewhat hirsute, to 6 cm; *involucre* 6–9 mm high, densely villous with multicellular hairs, these with black or dark purple cross-walls; *disk flowers* 2.4–3.5 mm long; *ray flowers* 7–11 mm long, white or pink; *pappus* of 20–25 bristles. Common in alpine and subalpine meadows, 9300–14,000 ft. July–Sept. E/W. (Plate 11)

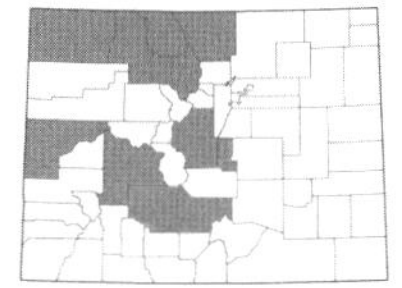

***Erigeron nematophyllus* Rydb.**, NEEDLELEAF DAISY. [*E. wilkenii* O'Kane]. Perennials, 0.5–1.5 dm; *leaves* mostly basal, linear or linear-oblanceolate, entire, sparsely hairy to glabrate, margins ciliate toward base, 2–8 cm long; *involucre* 4–6.5 mm high, strigose; *disk flowers* 3.5–4.5 mm long; *ray flowers* 4–8 mm long, white or pink; *pappus* of 15–25 bristles. Found in sandy or rocky soil of open places, often with sagebrush, 5200–10,000 ft. May–July. E/W.

Erigeron philadelphicus* L. var. *philadelphicus, PHILADELPHIA FLEABANE. Biennials or short-lived perennials, 2–7 dm; *leaves* evidently cauline, oblanceolate to obovate, sometimes toothed, villous or sometimes glabrate, to 15 cm long; *involucre* 4–6 mm high, hirsute to glabrate; *disk flowers* 2.5–3.5 mm long; *ray flowers* 5–10 mm long, pink to rose-purple or white; *pappus* of 20–30 bristles. Uncommon in dry to moist meadows, 6400–6600 ft. May–July. E/W.

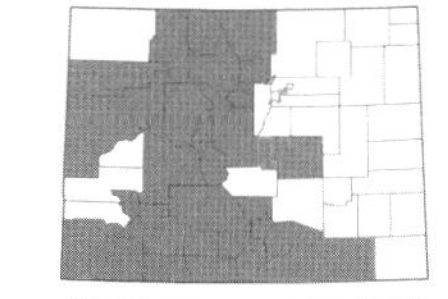

***Erigeron pinnatisectus* (A. Gray) A. Nelson**, FEATHERLEAF DAISY. Perennials, 0.5–1 dm, with woody caudex branches; *leaves* mostly basal, pinnatifid, nearly glabrous with ciliate petioles; *involucre* 5.5–8 mm high, glandular and usually hirsute; *disk flowers* 3.5–4.5 mm long; *ray flowers* 7–12 mm long, purplish to blue; *pappus* of 25–30 bristles. Common in rocky subalpine and alpine meadows, 8900–13,500 ft. June–Aug. E/W. (Plate 11)

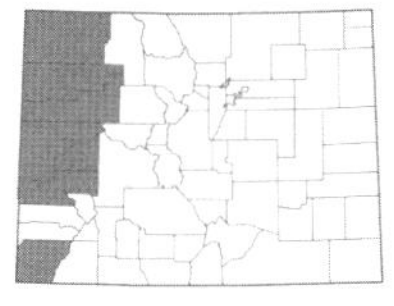

***Erigeron pulcherrimus* Heller**, BASIN DAISY. Perennials, 0.5–3.5 dm, with short caudex branches; *leaves* mostly basal, linear to narrowly oblanceolate, grayish-strigose, to 7 cm; *involucre* 6–9 mm high, hirsute-villous with crinkled multicellular hairs; *disk flowers* 4–6 mm long; *ray flowers* 8–15 mm long, pink, white, or sometimes violet; *pappus* of 30–50 bristles. Uncommon on clay slopes and sandy soil, 4400–6700 ft. May–July. W.

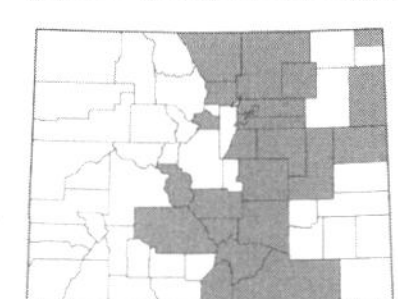

***Erigeron pumilus* Nutt.**, SHAGGY DAISY. Perennials, 0.5–3 dm; *leaves* evidently cauline, oblanceolate, hirsute, to 8 cm; *involucre* 7–15 mm high, spreading-hirsute and finely glandular; *disk flowers* 3–5 mm long; *ray flowers* 6–15 mm long, white; *pappus* double. Common in dry grasslands and sandy soil, 3400–8700 ft. May–July (Aug.). E. (Plate 12)

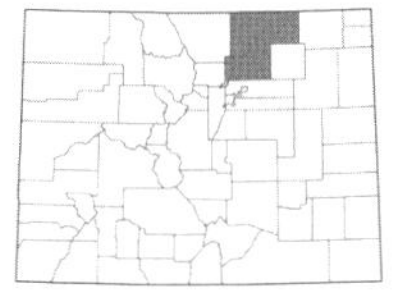

***Erigeron radicatus* Hook.**, TAPROOT DAISY. [Originally identified as *E. ochroleucus* Nutt.]. Perennials, 0.2–1.2 dm; *leaves* mostly basal, linear to narrowly oblanceolate, sparsely strigose to glabrate,1–6 cm long; *involucre* 4–7 mm high, villous, the hairs usually with purplish cross walls; *disk flowers* 2–4 mm long; *ray flowers* 4–8 mm long, usually white; *pappus* double. Uncommon on the plains, 5500–6000 ft. May–Aug. E.

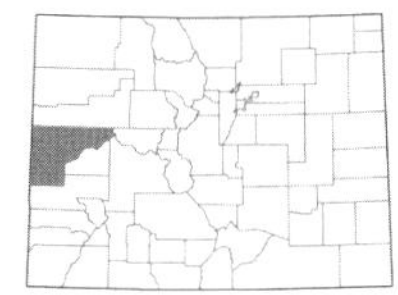

***Erigeron sparsifolius* Eastw.**, UTAH FLEABANE. [*E. utahensis* A. Gray var. *sparsifolius* (Eastw.) Cronquist]. Perennials, 1–5.5 dm; *leaves* oblanceolate to spatulate, 2–5 cm long, closely strigose; *involucre* 3–5 mm high, sparsely strigose to glabrous, densely minutely glandular; *disk flowers* 2–3.5 mm long; *ray flowers* 4–8 mm long, white to blue; *pappus* of 20–25 bristles. Uncommon in sandy or rocky soil of dry, open places, and sandstone outcrops, often with pinyon-juniper, 4800–6900 ft. June–Sept. W.

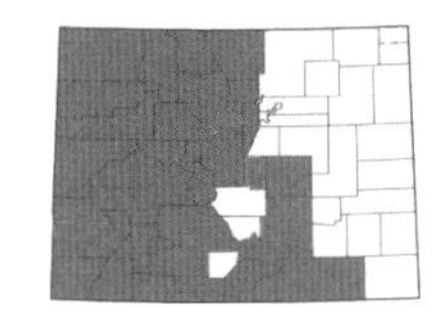

Erigeron speciosus **(Lindl.) DC.**, ASPEN DAISY. Perennials, 1.5–8 dm; *leaves* evidently cauline, oblanceolate to oblong or broadly ovate, glabrous with ciliate margins, 5–15 cm long; *involucre* 6–9 mm high, glandular and sparsely hairy; *disk flowers* 4–5 mm long; *ray flowers* 9–18 mm long, blue-violet to violet; *pappus* double. Common in mountain meadows and aspen forests, 5500–12,000 ft. June–Sept. E/W. (Plate 12)

Erigeron strigosus **Muhl. ex Willd. var. *strigosus***, PRAIRIE FLEABANE. Annuals or biennials, 3–7 dm; *leaves* evidently cauline, oblanceolate to elliptic or lanceolate, entire or sometimes toothed, strigose to hirsute, to 15 cm long; *involucre* 2–5 mm high, finely glandular and hairy; *disk flowers* 1.5–3 mm long; *ray flowers* to 6 mm long, white or sometimes pink or bluish; *pappus* double. Uncommon along the Front Range and in Las Animas Co. in rocky soil and in open meadows, often in disturbed sites, 5000–7700 ft. July–Sept. E.

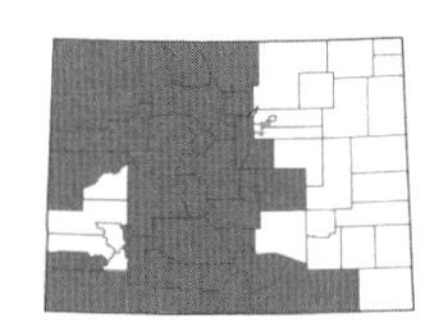

Erigeron subtrinervis **Rydb. ex Porter & Britton**, THREENERVE DAISY. Perennials, 1.5–9 dm; *leaves* evidently cauline, oblanceolate to oblong or ovate, entire, densely hairy, 4–13 cm long; *involucre* 6–9 mm high, glandular and hirsute; *disk flowers* 4–5 mm long; *ray flowers* 7–18 mm long, blue-violet or rose-purple; *pappus* double. Common in mountain meadows and forests, typically in drier places than *E. speciosus*, 6000–12,500 ft. June–Sept. E/W.

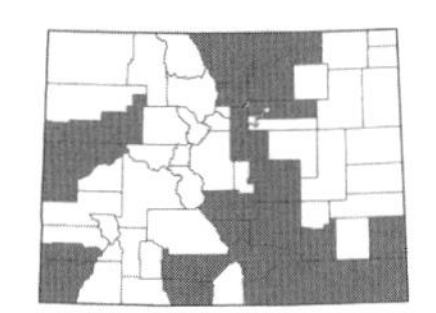

Erigeron tracyi **Greene**, RUNNING DAISY. [*E. colomexicanus* A. Nelson]. Annuals, biennials, or short-lived perennials, 1–4 dm; *leaves* oblanceolate to linear, entire or few-toothed, the stems forming stoloniferous branches, 1–3 (6) cm long; *involucre* 3–5 mm high, hirsute with curved hairs; *disk flowers* 2–2.5 mm long; *ray flowers* 4–6 mm long, white or pink; *pappus* double. Common on the eastern slope in open meadows and on dry slopes, with scattered occurrences on the western slope, 4000–9000 ft. April–July. E/W.

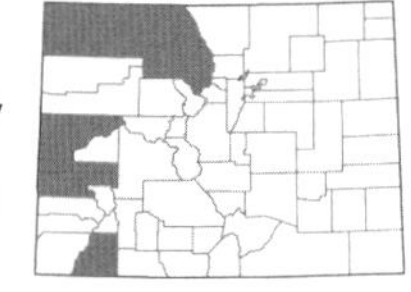

Erigeron uintahensis **Cronquist**, UINTA DAISY. Perennials, 2–5 dm, usually with thick, woody, branched caudices; *leaves* oblanceolate to ovate or oblong, glandular and hirsute, margins ciliate, 3–10 cm long; *involucre* 5.5–7 mm × 12–18 mm, densely minutely glandular and with a few villous hairs; *disk flowers* 4.5–5 mm long; *ray flowers* 9–15 mm long, blue or rose-lavender; *pappus* double. Uncommon on forested plateaus, 6400–8800 ft. July–Sept. W.

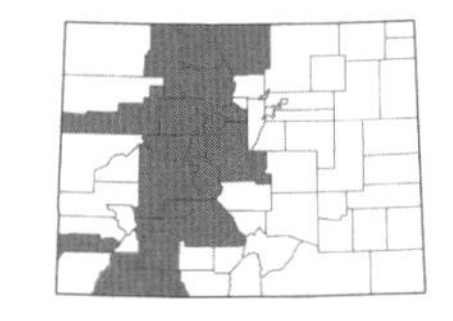

Erigeron ursinus **D.C. Eaton**, BEAR RIVER DAISY. Perennials, 0.5–2.5 dm, the caudices diffusely branched with an extensive system of long, slender rhizomelike branches; *leaves* oblanceolate to lanceolate, glabrous or sometimes sparsely strigose, to 12 cm long; *involucre* 5–7 mm high, glandular or viscid and hirsute; *disk flowers* 3–5 mm long; *ray flowers* 7–15 mm long, blue, pink, or purple; *pappus* double. Uncommon in alpine and subalpine meadows and on grassy slopes, 8000–13,500 ft. June–Sept. E/W.

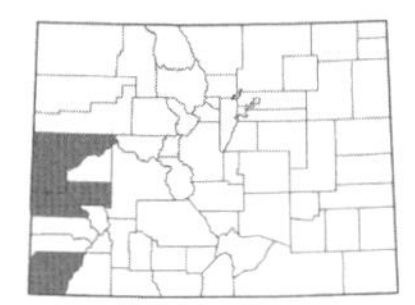

Erigeron utahensis **A. Gray**, UTAH DAISY. Perennials, 1–5 dm, with branched caudices; *leaves* oblanceolate to linear, strigose, gray-green, to 10 cm long; *involucre* 3–6 mm high, strigose and often finely glandular; *disk flowers* 3–4.6 mm long; *ray flowers* 4–18 mm long, blue, pink, or white; *pappus* double. Uncommon in sandy or rocky soil of dry, open places, ledges, and crevices and sandstone outcrops, 4800–6000 ft. May–July. W.

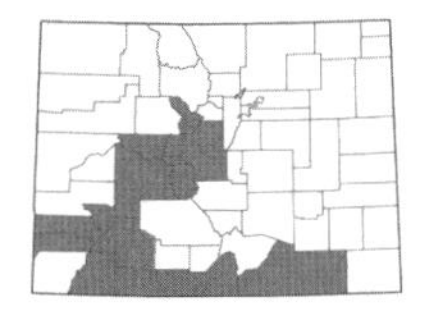

Erigeron vagus **Payson**, RAMBLING DAISY. Perennials to 0.5 dm, the caudices diffusely branched with an extensive system of long, slender rhizomelike branches; *leaves* 3-lobed, obtuse, spreading-hirsute and usually glandular, to 3 cm long; *involucre* 5–7 mm high, glandular and hirsute; *disk flowers* 3–4 mm long; *ray flowers* 4–7 mm long, white or pink; *pappus* simple. Uncommon on scree slopes in the alpine, 11,000–13,500 ft. July–Aug. E/W.

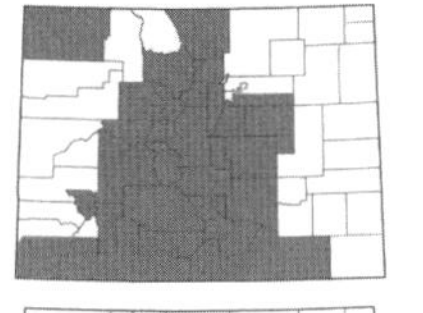

Erigeron vetensis **Rydb.**, EARLY BLUETOP DAISY. Perennials, 0.5–2.5 dm; *leaves* oblanceolate to linear, usually densely glandular, margins ciliate, to 15 cm long; *involucre* 4–8 mm high, glandular and usually sparsely hirsute; *disk flowers* 3–5.5 mm long; *ray flowers* 6–16 mm long, blue, purplish, or pink-purple; *pappus* double. Common in rocky and sandy soil on rocky outcroppings and slopes and in dry meadows, 5700–11,800 ft. May–July. E/W. (Plate 12)

Erigeron vreelandii **Greene**, VREELAND'S ERIGERON. [*E. platyphyllus* Greene]. Perennials, 3–8 dm; *leaves* oblanceolate to oblong or ovate, densely glandular, margins ciliate, to 15 cm long; *involucre* 6–9 mm high, glandular; *disk flowers* 3–4 mm long; *ray flowers* 9–20 mm long, blue or rarely rose; *pappus* double. Uncommon in mountain meadows and forests of the southern mountains, 9000–13,000 ft. July–Aug. E.

EUCEPHALUS Nutt. – ASTER

Perennial herbs; *leaves* alternate, sessile, entire; *heads* radiate (ours); *involucre bracts* usually imbricate or sometimes equal, base chartaceous; *receptacle* naked; *disk flowers* yellow; *ray flowers* white, blue, purple, or pink; *pappus* of 2 (3) series of capillary or barbellate bristles; *achenes* usually several-nerved.

1a. Ray flowers white or occasionally pink; leaves broader, mostly 1.5–4.5 cm wide (although rarely only 1 cm wide)...***E. engelmannii***

1b. Ray flowers blue, violet, or purple; leaves narrower, less than 2 cm wide...***E. elegans***

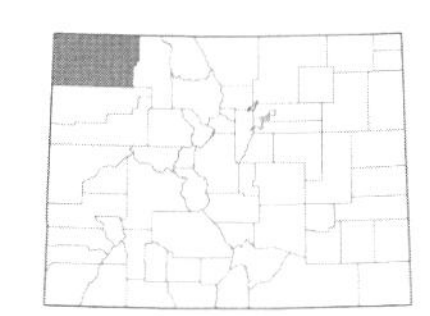

***Eucephalus elegans* Nutt.**, ELEGANT ASTER. [*Aster perelegans* A. Nelson & J.F. Macbr.]. Perennials, 3–7 dm; *leaves* linear-oblong to lanceolate, 2–6 cm long, scabrous and short-stipitate glandular; *involucre* 6–9 mm high, the bracts glandular and minutely hairy, in 3–5 series; *ray flowers* purple. Known only from Moffat Co. where it is found on dry, open slopes, 8500–8600 ft. July–Sept. W.

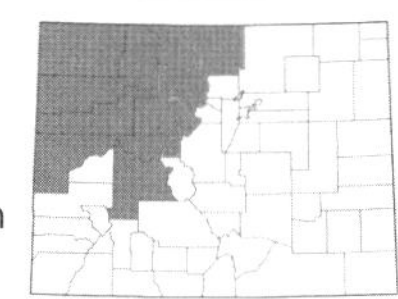

***Eucephalus engelmannii* (D.C. Eaton) Greene**, ENGELMANN'S ASTER. [*Aster engelmannii* (D.C. Eaton) A. Gray]. Perennials, 5–15 dm; *leaves* elliptic to lance-ovate, glabrous or abaxially villous and/or glandular, to 10 cm long; *involucre* 7–10 mm high, the bracts hairy to glandular, appearing ciliate toward the tips, in 4–6 series, usually reddish apically; *ray flowers* white to pink. Common in meadows and spruce-fir and aspen forests in the mountains, 7500–11,500 ft. July–Sept. E/W.

EUTHAMIA (Nutt.) Cass. – GOLDENTOP

Perennial herbs or subshrubs; *leaves* alternate, entire, glandular-punctate; *heads* radiate; *involucre bracts* imbricate in several series, chartaceous, yellowish- or greenish-tipped, more or less glutinous; *receptacle* more or less fimbrillate; *ray flowers* usually more numerous than the disk flowers, yellow; *pappus* of capillary bristles; *achenes* with several nerves, nearly terete. (Taylor & Taylor, 1983)

1a. Inflorescence elongate or rounded and interrupted, with lateral corymbiform clusters arising from the axils of well-developed leafy bracts; plants often more than 1 m tall...***E. occidentalis***

1b. Inflorescence broad and flat-topped, without lateral corymbiform clusters; plants usually less than 1 m tall...2

2a. Heads relatively small with fewer than 20 florets; leaves prominently 1-nerved or obscurely 3-nerved, but without additional parallel lateral nerves; herbage usually conspicuously resinous-glutinous...***E. gymnospermoides***

2b. Heads larger with 20–40 or more florets; leaves prominently 3-nerved, or the larger ones with 1 or 2 additional pairs of less-prominent lateral nerves; herbage smooth and not glutinous although glandular-punctate...***E. graminifolia***

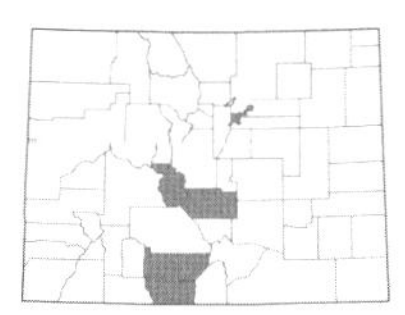

***Euthamia graminifolia* (L.) Nutt.**, FLAT-TOP GOLDENROD. [*Solidago graminifolia* (L.) Salisb.]. Perennials, 3–15 dm; *leaves* 3–5-nerved, linear to lanceolate, 3.5–13 cm long, with scabrous margins, little or obscurely gland-dotted; *heads* in a flat-topped array; *involucre* 3–5.5 mm high; *disk flowers* 2.5–3.5 mm. Usually found along streams and ditches in sandy soil, although sometimes found in drier sites, known from the San Luis Valley and near Buena Vista, 6800–7700 ft. July–Sept. E.

***Euthamia gymnospermoides* Greene**, VISCID EUTHAMIA. [*Solidago gymnospermoides* (Greene) Fernald]. Perennials or subshrubs, 4–15 dm; *leaves* 3–5-nerved, linear to lanceolate, 4–12 cm long, with scabrous margins, gland-dotted; *heads* in a flat-topped array; *involucre* 4–6.2 mm high, the bracts green-tipped, linear; *disk flowers* 3–4.8 mm long. Uncommon in rocky or sandy soil of dry places, known from one record near Julesburg (Sedgwick Co.), 3400–3500 ft. Aug.–Oct. E.

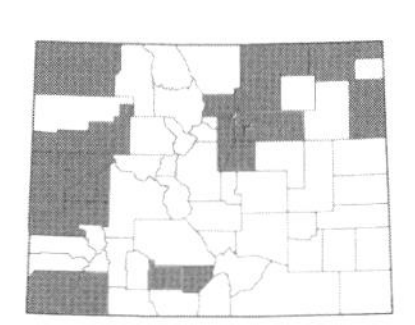

***Euthamia occidentalis* Nutt.**, WESTERN GOLDENTOP. [*Solidago occidentalis* (Nutt.) Torr. & A. Gray]. Perennials or subshrubs, 4–20 dm; *leaves* 3–5-nerved, linear, 8–10 cm long, with scabrous margins, gland-dotted and often sparsely hairy; *heads* in narrow, elongate arrays; *involucre* 3.5–5 mm high, the bracts yellowish-brown, lanceolate to linear; *disk flowers* 3–4.2 mm long. Found in moist soil and along rivers and ditches, 3400–7000 ft. Aug.–Oct. E/W.

EUTROCHIUM Raf. – JOEPYEWEED

Perennials; *leaves* toothed; *heads* discoid; *involucre bracts* imbricate or subequal, green to chartaceous or weakly coriaceous; *receptacle* naked, flat or convex; *disk flowers* blue, pink, purple, or rarely white; *ray flowers* absent; *pappus* of capillary bristles; *achenes* usually 5-angled.

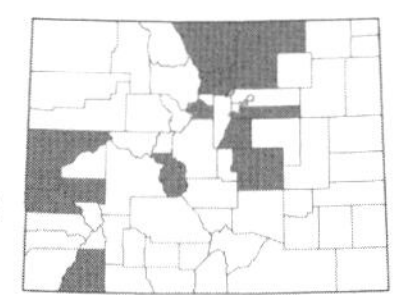

***Eutrochium maculatum* (L.) E.E. Lamont,** SPOTTED JOEPYEWEED. [*Eupatorium maculatum* L. var. *bruneri* (A. Gray) Breit]. Plants 6–20 dm, the stems purple-spotted to purple throughout; *leaves* in whorls, usually in 4's or 5's, 6–15 cm long, narrowly to broadly lanceolate, margins serrate, gland-dotted below; *involucre bracts* purplish; *disk flowers* purple; *achenes* 3–5 mm long. Found in moist places such as ditches, wet meadows, along streams, and in swamps, 5000–7500 ft. July–Sept. E/W.

EVAX Gaertn. – PYGMY CUDWEED

Annuals, low woolly herbs; *leaves* alternate, entire; *heads* in dense woolly glomerules, these with internal leaves whose tips protrude from between the heads, discoid; *involucre bracts* absent; *receptacle* chaffy; *outer disk flowers* with a minute tubular corolla; *central disk flowers* few; *ray flowers* absent; *pappus* absent; *achenes* with sharp edges.

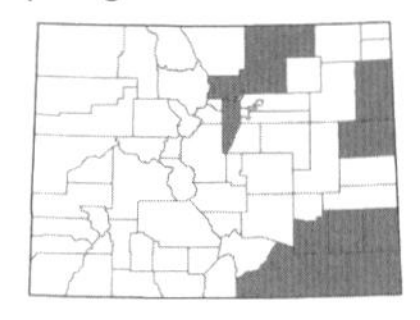

***Evax prolifera* Nutt. ex DC.**, RABBIT-TOBACCO. Plants 3–15 cm, sericeous; *leaves* narrowly oblanceolate to spatulate, 3–15 mm long; *achenes* compressed, 0.6–0.9 mm long. Infrequent or easily overlooked on dry places of the plains, often in disturbed or overgrazed places, 3400–5500 ft. May–July. E.

FLAVERIA Juss. – YELLOWTOPS

Herbs (ours); *leaves* opposite, sessile, simple; *heads* small, numerous, crowded into small glomerules which are aggregated into a terminal cluster, radiate (ours); *involucre bracts* subequal; *receptacle* naked, small; *disk flowers* yellow; *ray flowers* usually 1 per head, yellowish, inconspicuous; *pappus* absent or of a few scales; *achenes* small, black, 8–10-ribbed.

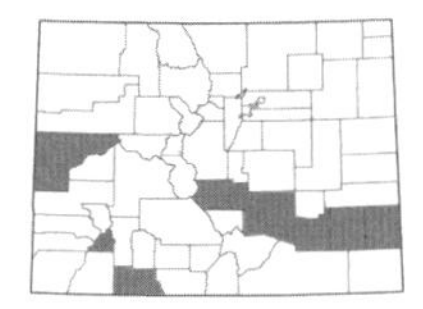

***Flaveria campestris* J.R. Johnst.**, ALKALI YELLOWTOPS. Annuals, 3–9 dm, mostly glabrous; *leaves* lanceolate to lance-linear, 3–9 cm long, margins weakly serrate to or subentire; *involucre bracts* 3, 5–7 mm long; *disk flowers* 5–6 (8); *pappus* absent; *achenes* 2.8–3.6 mm long. Uncommon along ditches and along the margins of streams and ponds, usually in alkaline soil, 3500–5000 ft. July–Sept. E/W.

GAILLARDIA Foug. – BLANKETFLOWER

Herbs; *leaves* alternate or basal; *heads* radiate; *involucre bracts* in 2 or 3 series, with chartaceous bases and herbaceous tips; *receptacle* alveolate with long stiff setae, or the setae absent; *disk flowers* woolly-villous; *ray flowers* yellow to partly or wholly red or purple, 3-toothed; *pappus* of scales, usually with an awn; *achenes* partly or wholly covered with long hairs. (Biddulph, 1944; Stoutamire, 1977; Turner & Whalen, 1975)

1a. Leaves deeply pinnatifid into narrow segments; awn of pappus no more than ½ as long as the body of the scale...***G. pinnatifida***
1b. Leaves entire, toothed, or shallowly lobed to pinnatifid into broad segments; awn of pappus usually equaling or exceeding the length of the scale, sometimes shorter than the scale but more than ½ as long...2

2a. Corollas of the disk flowers yellow; awn of the pappus scale shorter than to about as long as the body...***G. spathulata***
2b. Corollas of the disk flowers purple or brownish-purple to dark red, or if yellow then the awn of the pappus scale twice as long as the body...3

3a. Ray flowers yellow or with some purple or red at the base; awn of the pappus scale twice as long as the body...***G. aristata***
3b. Ray flowers reddish-purple or with a yellow tip, rarely all yellow; awn of the pappus scale about equaling the length of the body...***G. pulchella***

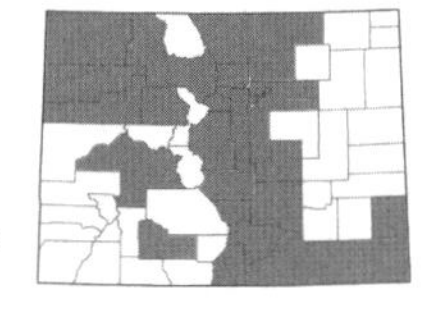

***Gaillardia aristata* Pursh**, BLANKETFLOWER. Perennials, 3–6 dm; *leaves* linear-oblong to lance-ovate, entire to toothed or somewhat pinnatifid, 5–20 cm long, hirsute and resinous-glandular; *disk flowers* 4.5–5.5 mm long, brownish-purple or rarely yellow; *ray flowers* 1–3.5 cm long, yellow or with purple at the base; *pappus* scales 5–6 mm long, the awn twice as long as the body. Very common in dry to moderately moist, open places such as in meadows and along roadsides, often found in disturbed habitats, 4900–9500 ft. May–Sept. (Oct.). E/W. (Plate 12)

Gaillardia ×*grandiflora* Van Houtte is a hybrid between *G. aristata* and *G. pulchella* that is commonly planted in gardens and along roadsides. The ray flowers of the hybrid vary between all yellow and all reddish-purple but are usually yellow just at the tip and reddish-purple the rest of the length. The hybrid resembles *G. aristata* in habit but is more vigorous. *Gaillardia pulchella* is an annual or short-lived perennial, and *G. aristata* and *G.* ×*grandiflora* are perennials.

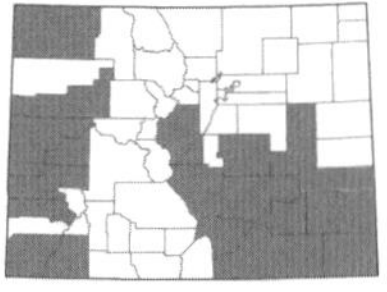

***Gaillardia pinnatifida* Torr.**, RED DOME BLANKETFLOWER. Perennials, 1–5 dm; *leaves* deeply pinnatifid, to 12 cm long, villous and resinous-glandular; *disk flowers* reddish to brownish-purple; *ray flowers* 1–2 cm long, yellow; *pappus* scales 3–7 mm long, the awn no more than ½ as long as the body. Common in dry, rocky, sandy, or clay soil of open places, 4300–7200 ft. May–Aug. (Sept.). E/W.

***Gaillardia pulchella* Foug.**, INDIAN BLANKETFLOWER. Annuals or short-lived perennials, 1–6 dm; *leaves* lanceolate or oblong, entire to toothed or somewhat pinnatifid, 2–8 cm long, glandular-villous; *disk flowers* brownish-purple; *ray flowers* 1–2 cm long, usually reddish-purple with a yellow tip; *pappus* scales 5–6 mm long, the awn about equal in length with the body. Scattered on dry, open prairie, known from the southeastern counties, 3300–4600 ft. May–Aug. E. (Plate 12)

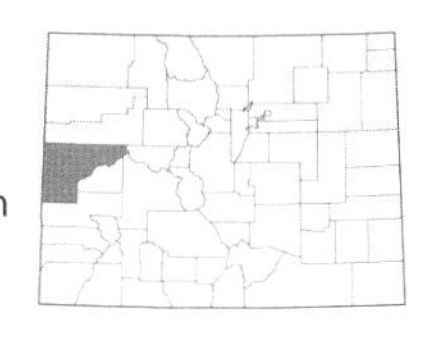

***Gaillardia spathulata* A. Gray**, WESTERN BLANKETFLOWER. Perennials, 1–4 dm; *leaves* spatulate to elliptic, entire to toothed or somewhat pinnatifid, 2–6 cm long, sparsely hairy and evidently resinous-glandular; *disk flowers* yellow; *ray flowers* 1–2 cm long, yellow; *pappus* scales 6–7 mm long, the awn shorter than to about equal to the body in length. Infrequent in dry, open places in sandy, rocky, or clay soil, 6500–7000 ft. May–July (Sept.). W.

GALINSOGA Ruiz & Pav. – QUICKWEED; GALLANT-SOLDIER

Annual herbs; *leaves* opposite; *heads* radiate; *involucre bracts* few in 1–3 poorly defined series, several-nerved; *receptacle* chaffy; *disk flowers* yellow to greenish-yellow; *ray flowers* white to purple, usually subtended by an involucre bract which is often fused to 2 adjacent receptacle bracts; *pappus* of ray flowers of scales that are fimbriate or awn-tipped, that of the disk flowers reduced or absent; *achenes* 4-angled, black. (Canne, 1977)

1a. Outer involucre bracts 1 or 2, with herbaceous margins; inner chaffy involucre bracts entire or weakly 2–3-parted, early-deciduous; ray ligule 1–2.8 mm long…***G. quadriradiata***

1b. Outer involucre bracts 2–4, with scarious margins; inner chaffy involucre bracts deeply 3-parted and late-deciduous; ray ligule 0.5–1.8 (2) mm long…***G. parviflora* var. *parviflora***

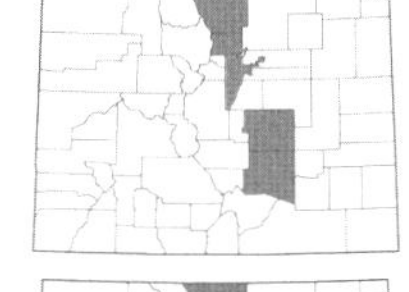

Galinsoga parviflora* Cav. var. *parviflora, QUICKWEED. Plants 0.4–6 dm; *leaves* 0.7–11 cm long, lanceolate to ovate; *involucre* 2.5–5 mm diam., the outer 1 or 2 bracts with herbaceous margins, the inner bracts entire or weakly 2–3-parted and falling early; *ray flowers* mostly 5, white, 0.5–1.8 (2) mm long. Uncommon weed in gardens and fields, 4700–6100 ft. June–Sept. E. Introduced.

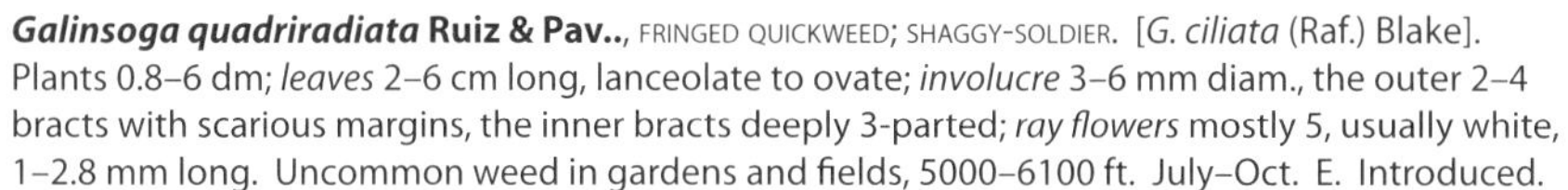

***Galinsoga quadriradiata* Ruiz & Pav..**, FRINGED QUICKWEED; SHAGGY-SOLDIER. [*G. ciliata* (Raf.) Blake]. Plants 0.8–6 dm; *leaves* 2–6 cm long, lanceolate to ovate; *involucre* 3–6 mm diam., the outer 2–4 bracts with scarious margins, the inner bracts deeply 3-parted; *ray flowers* mostly 5, usually white, 1–2.8 mm long. Uncommon weed in gardens and fields, 5000–6100 ft. July–Oct. E. Introduced.

GNAPHALIUM L. – CUDWEED

Annual herbs, generally white-woolly; *leaves* alternate, entire, sessile; *heads* in dense glomerules at the ends of branches and in leaf axils, disciform; *involucre bracts* slightly imbricate, usually white at the tip; *receptacle* naked; *disk flowers* yellowish or whitish; *ray flowers* absent; *pappus* of capillary bristles; *achenes* small, nerveless, round or flat.

1a. Leaves lanceolate to oblanceolate or oblong, generally wider, 2–10 mm in width; leaves subtending the heads not much longer than the inflorescence…***G. palustre***

1b. Leaves linear or linear-oblanceolate, narrow, 0.5–3 (5) mm wide; leaves subtending the heads much longer than the inflorescence…2

2a. Heads in axillary, spiciform glomerules; leaves linear to linear-oblanceolate…***G. exilifolium***

2b. Heads in capitate glomerules terminating the branches or sometimes with axillary glomerules; leaves oblanceolate…***G. uliginosum***

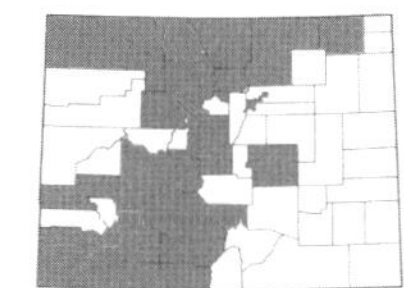

***Gnaphalium exilifolium* A. Nelson**, SLENDER CUDWEED. Annuals, 3–25 cm; *leaves* linear to linear-oblanceolate, 0.4–5 cm long, loosely tomentose; *involucre* 2.5–3.5 mm high, the bracts brownish with woolly bases, the inner with white, acute tips. Found along streams and pond margins, 4500–10,800 ft. July–Oct. E/W.

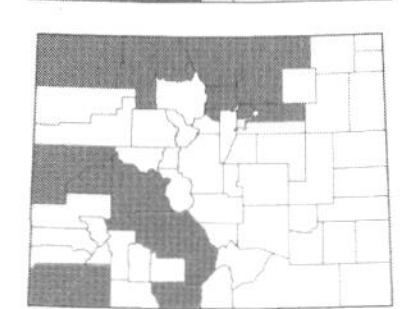

***Gnaphalium palustre* Nutt.**, DIFFUSE CUDWEED. Annuals, 1–30 cm; *leaves* oblanceolate to lanceolate or oblong, 1–3.5 mm long, woolly-tomentose; *involucre* 2.5–4 mm high, the bracts brownish with woolly bases, the inner with white, blunt tips. Found in sandy soil of moist places along streams and pond margins, sometimes in alkali soil, 5000–9000 ft. June–Oct. E/W.

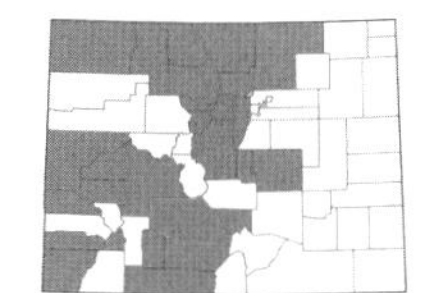

***Gnaphalium uliginosum* L.**, MARSH CUDWEED. Annuals, 3–25 cm; *leaves* oblanceolate, 1–5 cm long, loosely tomentose; *involucre* 2–4 mm high, the bracts brownish with woolly bases, the inner bracts with white, acute tips. Found along streams and pond margins and in swampy places, sometimes in disturbed areas, 5000–10,000 ft. July–Oct. E/W. Introduced.

Gnaphalium exilifolium and *G. uliginosum* are very close morphologically, but *G. exilifolium* is considered a New World native while *G. uliginosum* is a native of Europe (although it is unclear whether some of the North American plants are native).

GRINDELIA Willd. – GUMWEED

Herbs; *leaves* alternate, sessile and often clasping, the margins entire to sharply toothed; *heads* radiate or discoid; *involucre bracts* imbricate, sticky and gummy, firm, with a herbaceous, reflexed tip; *receptacle* naked; *disk flowers* yellow; *ray flowers* yellow to orange; *pappus* of 2–several firm awns; *achenes* smooth, compressed to quadrangular. (Dunford, 1986; Steyermark, 1937; Strother & Wetter, 2006)

1a. Ray flowers present...2
1b. Ray flowers absent, heads only with disk flowers...5

2a. Middle and upper involucre bracts erect and appressed...***G. arizonica***
2b. Middle and upper involucre bracts reflexed...3

3a. Stem leaves closely and evenly crenulate-serrate or closely serrulate to only remotely closely serrulate, the teeth mostly 1–2 (2.5) mm apart, if entire then the involucre surface conspicuously and abundantly resinous...***G. squarrosa***
3b. Stem leaves sharply and coarsely and more remotely dentate to serrate or the upper and middle sometimes entire or serrate near the apex only, the teeth mostly 2.5–5 mm apart, if entire then the involucre surface only moderately resinous...4

4a. Pappus awns 4–8, closely serrulate; plants mostly of the northern and central east-slope counties...***G. subalpina***
4b. Pappus awns mostly 2–3 (4), entire or rather remotely serrulate; plants mostly of the southern eastern slope counties (south of Colorado Springs)...***G. hirsutula***

5a. Middle and outer involucre bracts with a loosely or moderately reflexed tip, not as tightly curled as the next; heads depressed-hemispheric or broadly bowl-shaped, usually much broader than high; achenes 2–3 mm long, smooth to striated on the angles or deeply furrowed or ribbed, the ridges sometimes wrinkled...***G. nuda***
5b. Middle and outer involucre bracts with a strongly revolute, tightly curled tip; heads deeply bowl-shaped, usually as high or higher than broad; achenes 3–5.5 mm long, slightly nerved or smooth...6

6a. Stigmas linear-lanceolate; leaves conspicuously resinous-punctate...***G. fastigiata***
6b. Stigmas oblong-lanceolate or oblong; leaves less conspicuously resinous-punctate...***G. inornata***

***Grindelia arizonica* A. Gray**, ARIZONA GUMWEED. [including *G. laciniata* Rydb.]. Plants 4–8 dm; *leaves* oblong to oblanceolate or ovate-lanceolate, with entire to serrate or denticulate margins, 2–7 cm long; *involucre* 7–10 mm high, the bracts erect and appressed; *ray flowers* 12–20; *achenes* 2.3–3.5 mm long. Found on dry hills and mesas, sometimes along rivers, 6200–7600 ft. June–Sept. W.

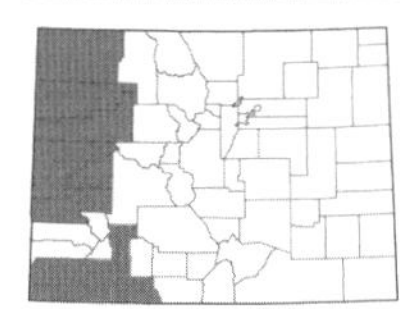

***Grindelia fastigiata* Greene**, POINTED GUMWEED. Plants 3–15 dm; *leaves* oblanceolate to oblong-spatulate, 1.5–15 cm long, the margins entire to dentate or closely serrate; *involucre* 8–11 mm high, the upper tips closely and strongly revolute; *ray flowers* absent; *achenes* 3.5–5 mm long. Found in dry, open places, often in sandy soil, 4500–8500 ft. July–Sept. W.

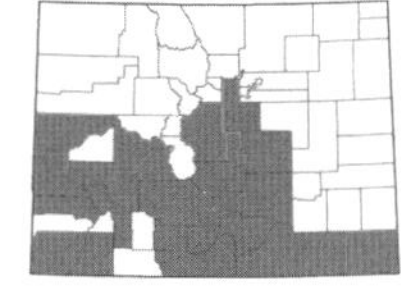

***Grindelia hirsutula* Hook. & Arn.**, HIRSUTE GUMWEED. [including *G. acutifolia* Steyerm.; *G. decumbens* Greene; *G. revoluta* Steyerm.]. Plants 1.5–8 dm; *leaves* oblong to oblong-lanceolate or ovate-lanceolate, 2–8.5 cm long, the margins entire to dentate; *involucre* 8–15 mm high, the upper bracts strongly reflexed or revolute; *ray flowers* 12–37; *achenes* 3–5 mm long. Found on dry, open hillsides and canyon slopes, 4500–9600 ft. July–Sept. E/W.

Grindelia acutifolia and *G. revoluta* are included within *G. hirsutula* by Strother and Wetter (2006). I have also included *G. decumbens* within the broadly defined *G. hirsutula* because all of the characters for *G. decumbens* and *G. revoluta* overlap. The main delimiting character supporting maintaining these 3 as separate species seems to be geographic locality. The rayless *G. inornata* and *G. fastigiata* are also included within the *G. hirsutula* complex by Strother and Wetter. I have chosen to use a varietal ranking to categorize the morphological variation and geographical differences between the following forms of *G. hirsutula* and choose to keep the rayless forms as separate species:

1a. Lower leaves with small glandular hairs on the lower sides along the main vein, the margins mostly conspicuously scabridulous...**var. *acutifolia* (Steyerm.) Ackerfield, comb. nov.** (see p. ix), RATON GUMWEED. [*G. acutifolia* Steyerm.]. Uncommon in dry, open places, found near Trinidad in the Raton Mesa region, 6500–9000 ft. July–Sept. E.
1b. Lower leaves without glandular hairs, the margins mostly scarcely scabridulous...2

2a. Pappus awns mostly 2 on central florets and 3–4 on outer disk and ray florets; involucre 8–11 mm high, the bracts slightly to moderately resinous; disk width 6–15 mm; ray flowers 12–24...**var. *decumbens* (Greene) Ackerfield, comb. nov.** (see p. ix), RECLINED GUMWEED. [*G. decumbens* Greene]. Found on dry, open hillsides and canyon slopes, in the San Luis Valley and Gunnison watershed, 5800–9600 ft. July–Sept. E/W.
2b. Pappus awns 2 per floret; involucre 10–13 mm high, the bracts conspicuously and abundantly resinous; disk width 13–25 mm; ray flowers 21–37...**var. *revoluta* (Steyerm.) Ackerfield, comb. nov.** (see p. ix), ROLLED GUMWEED. [*G. revoluta* Steyerm.]. Found in open, dry places from El Paso County south to northern Huerfano County, 5400–6500 ft. July–Sept. E.

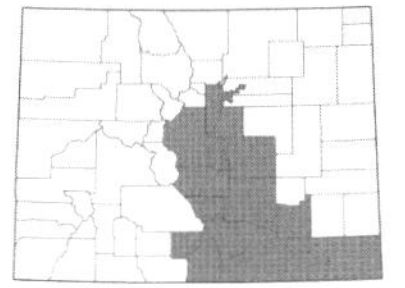

***Grindelia inornata* Greene**, COLORADO GUMWEED. Plants 2.5–8 dm; *leaves* obovate-oblong, 2–5.5 cm long, the margins dentate; *involucre* 8–12 mm high, the upper bracts strongly reflexed; *ray flowers* absent; *achenes* 3–5 mm long. Locally common on dry hillsides and slopes, 4500–8000 ft. July–Sept. E. Endemic.

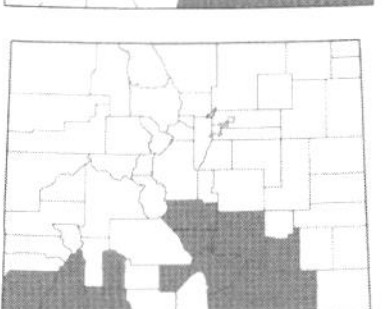

***Grindelia nuda* Wood**, CURLYCUP GUMWEED. Plants 1.5–6 dm; *leaves* oval to broadly oblong, closely and evenly crenate, 1.5–4.5 cm long; *involucre* 8–15 mm high, with loosely or moderately reflexed tips; *ray flowers* absent; *achenes* 2–3 mm long. Found in dry, open places, 4300–8000 ft. July–Sept. E/W.

Strother and Wetter (2006) group *G. nuda* and *G. aphanactis* within a broader defined *G. squarrosa*. However, I choose to keep the discoid plants as separate varieties based on the presence of ray flowers as well as several additional morphological characteristics:

1a. Stem leaves oval to ovate or broadly oblong, 1.5–3 times longer than wide, closely and evenly crenulate-serrate or closely serrulate to only remotely closely serrulate, the teeth mostly 1 mm apart, rarely the teeth wider; achenes smooth to striated on the angles; pappus awns usually entire or remotely serrulate; stems often greenish...**var. *nuda***, CURLYTOP GUMWEED. [*G. pinnatifida* Wooton & Standley]. Uncommon in dry, open places on the plains, known from Pueblo County but suspected in the other southeastern counties, 4300–5000 ft. July–Sept. E.

1b. Stem leaves oblanceolate to oblong, mostly 5–10 times longer than wide, entire or finely to coarsely dentate or the lower sometimes pinnatifid, the teeth mostly 2 mm or more apart; achenes deeply furrowed or ribbed, the ridges sometimes wrinkled; pappus awns closely serrulate to setulose-serrulate; stems usually reddish...**var. *aphanactis* (Rydb.) G.L. Nesom**, CURLYTOP GUMWEED. Found in moist to dry fields, meadows, along roadsides, or sandy places along streams, scattered in the southern counties, 4700–8000 ft. July–Sept. E/W.

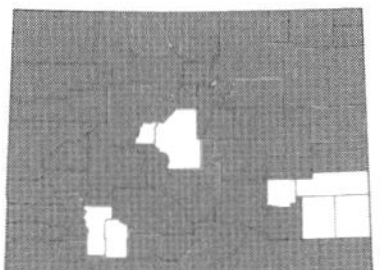

***Grindelia squarrosa* (Pursh) Dunal**, CURLYCUP GUMWEED. Plants 1–10 dm; *leaves* oblong to oblanceolate, 3–7 cm long, the margins closely and evenly crenulate-serrate, entire, or remotely serrulate; *involucre* 6–11 mm high, the upper strongly reflexed; *ray flowers* 22–36; *achenes* 2–3.5 mm long. Very common in dry, open places, 3500–8500 ft. June–Oct. E/W. (Plate 12)

Strother and Wetter (2006) combine some of the rayless *Grindelias* (*G. nuda* and *G. aphanactis*) with *G. squarrosa* and completely eliminate the following varieties. However, even though the following varieties intergrade, there are significant morphological differences between them. I choose to keep a narrower circumscription of the group as completed by Steyermark (1937):

1a. Leaves mostly entire to remotely serrulate, the lower often irregularly toothed or somewhat pinnatifid...**var. *quasiperennis* Lunell**

1b. Leaves mostly closely crenulate-serrulate with short obtuse teeth...2

2a. Leaves relatively narrow, the middle and upper ones mostly 5–8 times as long as wide, linear-oblong to oblanceolate...**var. *serrulata* (Rydb.) Steyerm.**

2b. Leaves relatively broad, the middle and upper ones mostly 2–4 times as long as wide, ovate or oblong...**var. *squarrosa***

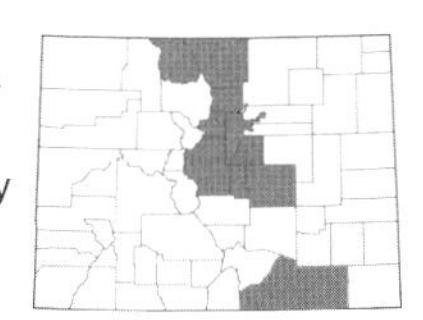

***Grindelia subalpina* Greene**, SUBALPINE GUMWEED. Plants 1.5–4 dm; *leaves* oblanceolate-oblong, 1.5–6 cm long, the margins coarsely serrate or dentate; *involucre* 8–11 mm high, the tips moderately reflexed; *ray flowers* 18–27; *achenes* 2.5–3.5 mm long. Common in dry, open places, often in sandy or rocky soil, 5500–9500 ft. June–Oct. E. (Plate 12)

GUTIERREZIA Lag. – SNAKEWEED

Herbs, shrubs, or subshrubs; *leaves* simple, alternate, entire; *heads* radiate; *involucre bracts* imbricate in 2–4 series, pale and chartaceous with a greenish tip; *receptacle* small, naked, somewhat alveolate, flat, convex or hemispheric; *disk* yellow; *ray flowers* few and usually short, yellow or white; *pappus* of scales in 1 or 2 series; *achenes* obovoid, oblong, or cylindrical. (Lane, 1985; Schneider et al., 2008; Solbrig, 1960; Suh & Simpson, 1990)

1a. Heads shortly pedunculate, the peduncles 1–4 cm long; plants known from Mancos shale slopes in Dolores County...***G. elegans***

1b. Heads sessile or subsessile; plants variously distributed...2

2a. Ray flowers usually 1 (or 2 but very rarely if disk flowers 2); disk flowers usually 1 (or 2 but only very rarely if ray flowers 2); involucres cylindric with 4–6 phyllaries...***G. microcephala***

2b. Ray flowers (2) 3–5 (8); disk flowers (2) 3–5 (9); involucres narrowly obconic with 8 or more phyllaries...***G. sarothrae***

***Gutierrezia elegans* A. Schneider & P. Lyon**, LONE MESA SNAKEWEED. Subshrubs, 0.7–1.5 dm; *leaves* linear-lanceolate, to 1.6 cm long, hirtellous; *heads* shortly pedunculate, 2–8 in flat-topped arrays or sometimes solitary; *involucre* 2.5–3 mm diam., the bracts gland-dotted; *disk flowers* 6–9; *ray flowers* 6–8, 3–5 mm long; *achenes* 1–2 mm long. A newly described species found on barren Mancos shale, currently known only from Lone Mesa State Park and the surrounding area in Dolores Co., 7500–8000 ft. July–Sept. W. Endemic.

***Gutierrezia microcephala* (DC.) A. Gray**, THREADLEAF SNAKEWEED. Subshrubs, 2–14 dm; *leaves* linear to lanceolate, glabrous or somewhat hispid; *heads* sessile or subsessile, 2–6 in compact, flat-topped arrays; *involucre* 1–1.5 mm diam.; *disk flowers* 1 or rarely 2; *ray flowers* 1 or rarely 2; *achenes* 1–2.5 mm long. Uncommon in dry, open places, often in sandy soil, 4500–6000 ft. July–Oct. W.

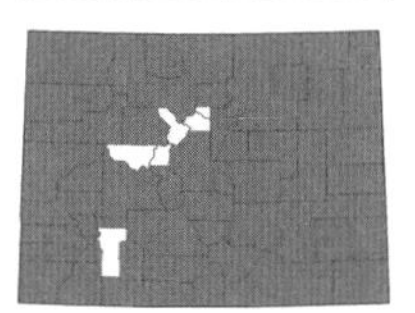

***Gutierrezia sarothrae* (Pursh) Britton & Rusby**, BROOM SNAKEWEED. Subshrubs, 1–6 (10) dm; *leaves* linear to lanceolate, glabrous to hispidulous; *heads* sessile or subsessile in dense, flat-topped arrays; *involucre* 1.5–2 (3) mm diam.; *disk flowers* (2) 3–5 (9); *ray flowers* (2) 3–5 (8), 3–5.5 mm long; *achenes* 0.8–2.2 mm long. Very common in dry, open places, sometimes with sagebrush or rabbitbrush, 3500–9500 ft. July–Oct. E/W. (Plate 12)

This species is sometimes confused with *Chrysothamnus viscidiflorus*, which lacks ray flowers and often has twisted leaves.

HELENIUM L. – SNEEZEWEED

Herbs; *leaves* simple, alternate; *heads* usually radiate (ours); *involucre bracts* subequal in 2–4 series, more or less herbaceous, the outer reflexed; *receptacle* naked or with a few scattered bristles; *disk flowers* perfect and fertile; *ray flowers* 3-lobed at the apex, or absent; *pappus* of 5–10 scarious or hyaline, awn-tipped thin scales; *achenes* obpyramidal, 4–5-angled. (Bierner, 1972; Rock, 1957)

1a. Disk flowers yellow; perennial with fibrous roots or rhizomes...***H. autumnale***
1b. Disk flowers reddish-brown; annual from a taproot...***H. microcephalum* var. *microcephalum***

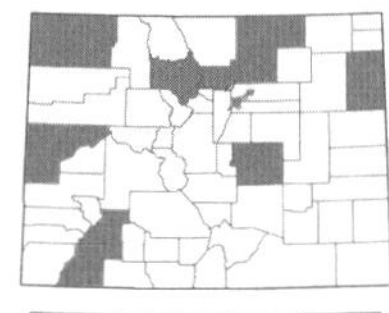

***Helenium autumnale* L.**, MOUNTAIN SNEEZEWEED. Perennials, 5–13 dm; *leaves* lanceolate to obovate, entire to dentate or weakly lobed, hairy; *involucre* 8–20 mm high, the bracts hairy; *disk flowers* yellow, 2.4–4 mm long; *ray flowers* 10–23 mm long; *pappus* of 4–7 aristate scales 0.5–1.5 mm long; *achenes* 1–2 mm, hairy. Locally common in moist places, 3500–8600 ft. July–Sept. E/W.

Helenium microcephalum* DC. var. *microcephalum, SMALLHEAD SNEEZEWEED. Annuals, 2–12 dm; *leaves* narrowly elliptic to oblong-elliptic, serrate to deeply toothed or lobed, glabrous to hairy; *involucre* 4–8 mm high, the bracts hairy; *disk flowers* mostly reddish-brown, 1.2–2.5 mm long; *ray flowers* 2.6–9 mm long; *pappus* of 6 scales 0.3–0.7 mm long; *achenes* 0.7–1.5 mm long, hairy. Uncommon in moist places in disturbed sites, 5000–5500 ft. July–Sept. E.

HELIANTHELLA Torr. & A. Gray – LITTLE SUNFLOWER

Herbs; *leaves* simple, entire; *heads* radiate; *involucre bracts* subequal or imbricate; *receptacle* chaffy, the scarious bracts persistent and clasping the achenes; *disk flowers* yellow, purple, or brownish-purple; *ray flowers* yellow; *pappus* of 2 persistent slender awns or scales, or absent; *achenes* strongly compressed at right angles to the involucre bracts. (Weber, 1952)

1a. Heads numerous in a corymbose inflorescence; rays 6–12 mm long and 2–4 mm wide, little exceeding the disk flowers; disk flowers purple...***H. microcephala***
1b. Heads solitary or few; rays 15–45 mm long and 5–12 mm wide, conspicuously and greatly exceeding the disk flowers; disk flowers yellow...2

2a. Heads erect at anthesis; receptacle chaff firm and chartaceous...***H. uniflora***
2b. Heads nodding or turned to the side at anthesis; receptacle chaff soft and scarious...3

3a. Plants with a basal cluster of leaves; leaves lanceolate to lanceolate-ovate, mostly 0.8–2.3 cm wide; heads smaller, the disk 15–20 mm wide; ray flowers mostly 8–10...***H. parryi***
3b. Plants usually without a basal cluster of leaves; leaves ovate-lanceolate to elliptic-lanceolate, 1.5–10 cm wide; heads larger, the disk 25–40 mm wide; ray flowers mostly 13–21...***H. quinquenervis***

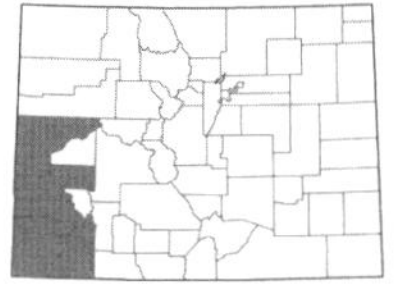

***Helianthella microcephala* (A. Gray) A. Gray**, PURPLE-DISK LITTLE SUNFLOWER. Plants 2–8 dm; *leaves* mostly 3-nerved, linear-oblanceolate to oblong, 6–25 cm, hispid to glabrate; *heads* in corymbiform arrays; *disk flowers* purple to brown; *ray flowers* 6–12 × 2–4 mm, little exceeding the disk flowers. Uncommon in dry, open places, often with pinyon-juniper, 5400–6500 ft. July–Sept. W.

***Helianthella parryi* A. Gray**, Parry's little sunflower. Plants 2–5 dm; *leaves* mostly 5-nerved, oblanceolate, hispid; *heads* usually solitary, nodding; *disk flowers* yellow; *ray flowers* 25–30 mm long. Locally common in aspen and pine forests from the montane to timberline, 7000–12,000 ft. June–Sept. E/W.

***Helianthella quinquenervis* (Hook.) A. Gray**, fivenerve little sunflower. Plants 3–15 dm; *leaves* 3–5-nerved, ovate-lanceolate to elliptic, 10–50 cm long, hirsute to glabrate; *heads* usually solitary, nodding; *disk flowers* yellow; *ray flowers* 25–40 mm long. Common in mountain meadows and aspen forests, sometimes on open sagebrush slopes, 7200–13,500 ft. June–Sept. E/W. (Plate 12)

***Helianthella uniflora* (Nutt.) Torr. & A. Gray**, common little sunflower. Plants 4–12 dm; *leaves* mostly 3-nerved, lanceolate to elliptic, hirsute to glabrate, 12–25 cm long; *heads* usually solitary, erect; *disk flowers* yellow; *ray flowers* 15–45 mm long. Found in open meadows, woods, and shrubby slopes, often in sandy soil, 6500–9500 ft. June–Aug. E/W.

HELIANTHUS L. – sunflower

Herbs; *leaves* simple, usually coarse; *heads* radiate; *involucre bracts* subequal to well-imbricate, usually herbaceous; *receptacle* chaffy, the chaff clasping the achenes; *disk flowers* yellowish to reddish or brown; *ray flowers* yellow; *pappus* of 2 early-deciduous awns with additional smaller scales occasionally between the awns; *achenes* usually glabrous. (Heiser, 1969)

1a. Taprooted annuals...2
1b. Perennials with rhizomes, erect branching crowns, or tuberous-thickened roots...3

2a. Involucre bracts ovate or ovate-lanceolate, hispid and usually ciliate on the margins, the tip abruptly attenuate...***H. annuus***
2b. Involucre bracts lanceolate to ovate-lanceolate, appressed short-hairy, usually glabrous on the margins, the tip usually tapering but sometimes attenuate...***H. petiolaris***

3a. Heads in a spiciform or racemiform inflorescence; upper and sometimes middle leaves folded lengthwise...***H. maximiliani***
3b. Heads few or solitary at the ends of branches or several in a corymbiform inflorescence; leaves not folded...4

4a. Leaves broadly lanceolate to ovate, 7–15 cm wide, conspicuously petiolate (petiole 2–8 cm long)...***H. tuberosus***
4b. Leaves lanceolate to ovate, usually less than 5 cm wide, sessile or short-petiolate (petiolate about 1 cm long)...5

5a. Leaf margins usually revolute or ruffled, the margins usually ciliate; leaves glabrous or with scattered stout hairs, usually blue-green glaucous...***H. ciliaris***
5b. Leaf margins not revolute, green and not glacous, the margins sometimes ciliate...6

6a. Involucre bracts conspicuously imbricate in several series, the tips more or less appressed; disk flowers usually red, rarely all yellow...***H. rigidus* ssp. *subrhomboideus***
6b. Involucre bracts all about the same length and not conspicuously imbricate, the tips loosely spreading; disk flowers yellow...7

7a. Stems arising from a tough, erect, taprooted crown; middle and lower stems scabrous-hispid to strigose; leaves mostly ovate but sometimes lanceolate; disk 10–14 mm wide...***H. pumilus***
7b. Stems arising from a thick, short, fascicled tuber-like rhizome; middle and lower stems glabrous to sparsely hispid; leaves lanceolate or lance-linear; disk 12–25 mm wide...***H. nuttallii***

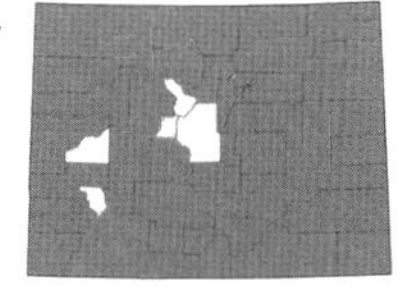

***Helianthus annuus* L.**, common sunflower. Annuals, 10–30 dm; *leaves* ovate to ovate-oblong, hispid, 10–40 × 5–35 cm, the margins serrate; *involucre* 15–40 (200) mm diam., the bracts 13–25 mm long, the margins usually ciliate, usually hispid; *disk flowers* 5–8 mm long, reddish-purple or rarely yellow; *ray flowers* 25–50 mm long; *pappus* of scales 2–3.5 mm long. Very common in open, dry to moderately moist soil, along roadsides, especially common in disturbed places, 3300–9000 ft. June–Oct. E/W. (Plate 12)

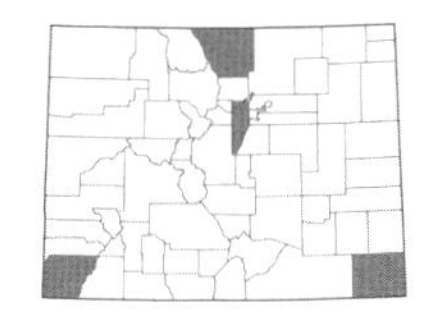

***Helianthus ciliaris* DC.**, Texas blueweed. Perennials, 3–7 dm; *leaves* linear-oblong to lanceolate-ovate, glabrous to hispid, 3–8 × 0.5–2.2 cm, the margins entire to serrate; *involucre* 12–25 mm diam., the bracts 3–8 mm long, the margins ciliate, faces glabrate to hispid; *disk flowers* 4–6 mm long, reddish; *ray flowers* 8–9 mm long; *pappus* of scales 1.2–1.5 mm long. Uncommon in dry to moist open places, usually in disturbed sites, 4400–6000 ft. June–Sept. E/W. Introduced.

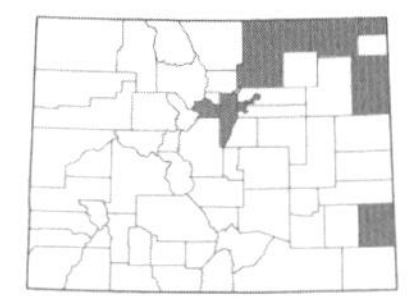

Helianthus maximiliani **Schrad.**, MAXIMILIAN SUNFLOWER. Perennials, 5–30 dm; *leaves* lanceolate, scabrous to hispid, the margins entire or sometimes serrate, 10–30 × 2–6 cm; *involucre* 13–30 mm diam., the bracts 14–20 mm long, pilose; *disk flowers* 5–7 mm long, yellow; *ray flowers* 15–40 mm long; *pappus* of scales 3–4 mm long. Native to the open, moist prairie of the eastern plains but commonly planted in gardens where it can escape cultivation, 3300–6000 ft. Aug.–Oct. E.

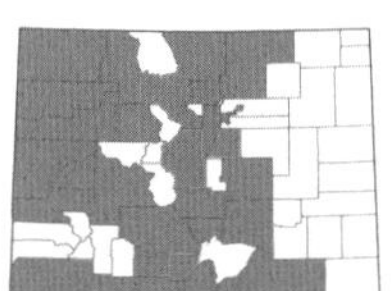

Helianthus nuttallii **Torr. & A. Gray**, NUTTALL'S SUNFLOWER. Perennials, 10–40 dm; *leaves* lanceolate to ovate, hispid to villous, the margins entire or serrate, 4–20 × 0.8–4 cm; *involucre* 10–20 mm diam., the bracts 8–15 mm long, margins ciliate, faces usually strigose; *disk flowers* 5–7 mm long, yellow; *ray flowers* 10–25 mm long; *pappus* of scales 2–4.5 mm long. Common in wet places such as ditches, moist meadows, and along streams or pond borders, 4500–8500 ft. June–Oct. E/W.

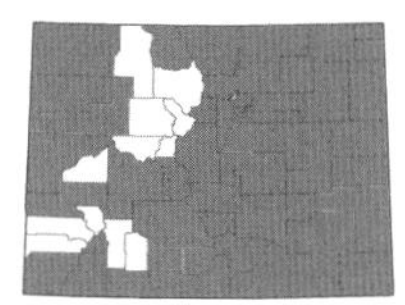

Helianthus petiolaris **Nutt.**, PRAIRIE SUNFLOWER. Annuals, 4–20 dm; *leaves* lanceolate to ovate, hispid, 4–15 × 1–8 cm; *involucre* 10–25 mm diam., the bracts 10–14 mm long, appressed short-hairy, the margins glabrous; *disk flowers* reddish-purple, 4.5–6 mm long; *ray flowers* 15–20 mm long; *pappus* of scales 1.5–3 mm long. Common in open, dry places, sometimes present in disturbed places, 3300–9000 ft. June–Oct. E/W.

There are two varieties of *H. petiolaris* in Colorado:

1a. Involucre bracts 14–30, oblong-lanceolate to lanceolate, 12–20 mm long and 2–3.5 mm wide; ray flowers 10–20; peduncles usually with a leafy bract subtending the head; lower leaves usually more than twice as long as wide...**ssp. *fallax* Heiser**

1b. Involucre bracts 18–45, ovate-lanceolate or oblong-lanceolate, 10–14 mm long and 3–5 mm wide; ray flowers 15–30; peduncles usually naked; lower leaves usually about twice as long as wide...**ssp. *petiolaris***

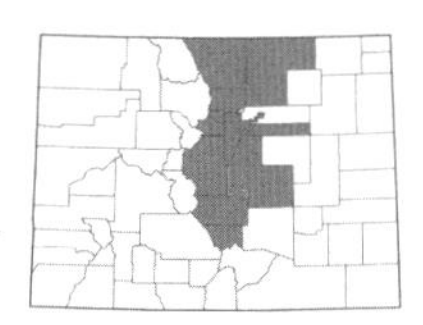

Helianthus pumilus **Nutt.**, LITTLE SUNFLOWER. Perennials, 3–10 dm; *leaves* lanceolate to ovate, hispid, 4–15 × 1–5 cm; *involucre* 7–15 mm diam., the bracts 3.5–8 mm long, hispid; *disk flowers* 5–6 mm long, yellow; *ray flowers* 15–20 mm long; *pappus* of scales 4–5 mm long. Common in dry, open meadows and grasslands, often in rocky soil, 5000–9500 ft. June–Oct. E. (Plate 12)

Specimens of *H. annuus* or *H. petiolaris* may key to here if it is difficult to distinguish whether the plant is annual or perennial. However, the annual species have reddish-purple disk flowers while *H. pumilus* has yellow disk flowers.

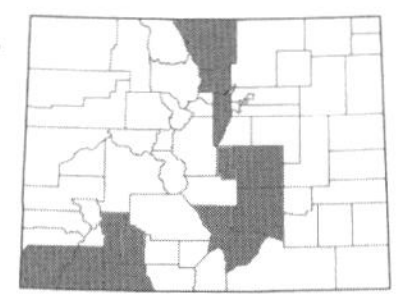

Helianthus rigidus **Desf. ssp. *subrhomboideus* (Rydb.) Heiser**, STIFF SUNFLOWER. [*H. pauciflorus* Nutt. ssp. *subrhomboideus* (Rydb.) O. Spring & E. Schilling]. Perennials, 5–12 dm; *leaves* linear-lanceolate to rhombic-ovate, sparsely hispid, 5–12 × 2–4 cm; *involucre* 15–25 mm diam., the bracts 6–10 mm long, hispid to glabrate, with ciliate margins; *disk flowers* 6.5–7 mm long, red to reddish-purple; *ray flowers* 20–35 mm long; *pappus* of scales 4–5 mm long. Locally common in dry grasslands and meadows, sometimes in open ponderosa pine forests, 5000–7200 ft. July–Sept. E/W.

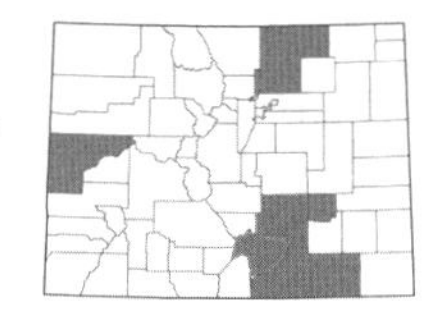

Helianthus tuberosus **L.**, JERUSALEM ARTICHOKE. Perennials, 10–30 dm; *leaves* broadly lanceolate to ovate, hirsute to scabrous, 10–25 × 7–15 cm; *involucre* 8–12 mm diam., the bracts 8–15 mm long, hispid, margins ciliate; *disk flowers* 6–7 mm long, yellow; *ray flowers* 25–40 mm long; *pappus* of scales 2–3 mm long. Uncommon escapee from cultivation, usually found in moist soil of disturbed places, 4300–8500 ft. Aug.–Oct. E/W. Introduced.

HELIOMERIS Nutt. – GOLDENEYE

Herbs; *leaves* simple, opposite below and alternate above, sessile or subsessile; *heads* radiate; *involucre bracts* subequal to evidently imbricate in 2–3 series, more or less herbaceous; *receptacle* chaffy, the chaffy bracts clasping the achenes; *disk flowers* yellow; *ray flowers* yellow, pubescent dorsally; *pappus* absent; *achenes* weakly 4-angled, glabrous.

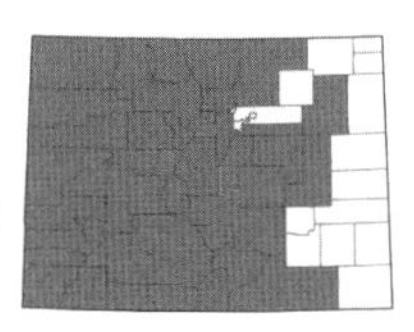

Heliomeris multiflora **Nutt.**, SHOWY GOLDENEYE. [*Viguiera multiflora* (Nutt.) Blake]. Perennials, 2–12 dm; *leaves* linear, lanceolate-linear, elliptic, or ovate, 1–9 × 2–28 (30) mm, with ciliate margins, the faces shortly hairy, sometimes gland-dotted below; *involucre* 6–14 mm diam.; *disk flowers* 50+, 3–4 mm long; *ray flowers* 5–14, 7–20 mm long, gland-dotted below; *achenes* 1.2–3 mm long, black or gray. Common in meadows and desert-steppes where it is often found with pinyon-juniper or sagebrush, 4500–11,000 ft. May–Sept. E/W. (Plate 12)

There are two well-marked but often intergrading varieties of *H. multiflora*:

1a. Leaves broader, the principal ones 8–25 (30) mm wide and (2) 3–8 times as long as wide, the margins usually not conspicuously revolute...**var. *multiflora***. Usually found at higher elevations or in moister habitats. E/W.

1b. Leaves narrower, the principal ones 2–8 mm wide and (5) 7–20 times as long as wide, the margins usually conspicuously revolute...**var. *nevadensis* (A. Nelson) W.F. Yates**, NEVADA GOLDENEYE. Usually found in drier places and often with pinyon-juniper or sagebrush. W.

HELIOPSIS Pers. – OX-EYE

Perennial herbs; *leaves* opposite, serrate; *heads* radiate; *involucre bracts* in 2–3 poorly defined series; *receptacle* convex or conic and elongating at maturity, chaffy, the chaffy bracts clasping, subtending the disk and ray flowers; *disk flowers* perfect and fertile; *ray flowers* yellow; *pappus* a short crown or absent; *achenes* 3–4-angled.

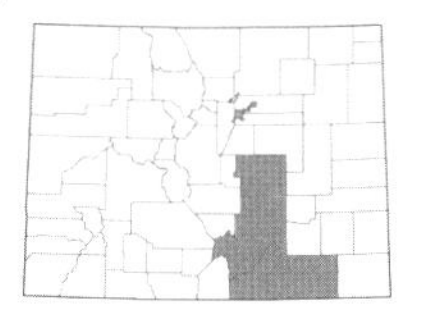

Heliopsis helianthoides **(L.) Sweet var. *scabra*** **(Dunal) Fernald**, SMOOTH OX-EYE. Plants 3–15 dm; *leaves* ovate to broadly lanceolate, 5–15 cm long, scabrous or glabrous; *heads* solitary; *involucre bracts* subequal; *disk flowers* yellow or brown to purple; *achenes* brown. Uncommon on gravelly slopes and at the edges of fields and in waste places, often cultivated, 6000–8600 ft. July–Oct. E.

HERRICKIA Wooton & Standl. – HERRICKA

Perennial herbs or subshrubs; *leaves* alternate; *heads* radiate; *involucre bracts* subequal to imbricate in 3–6 series, conspicuously keeled, the tip herbaceous and green and the base chartaceous; *receptacle* naked; *disk flowers* yellow but sometimes drying purplish; *ray flowers* white to purple; *pappus* simple, strongly barbellate bristles; *achenes* compressed, striate.

1a. Leaves with entire margins; involucre bracts appressed...***H. glauca*** **var. *glauca***
1b. Leaves with sharply spinulose-serrate margins; involucre bracts spreading...***H. horrida***

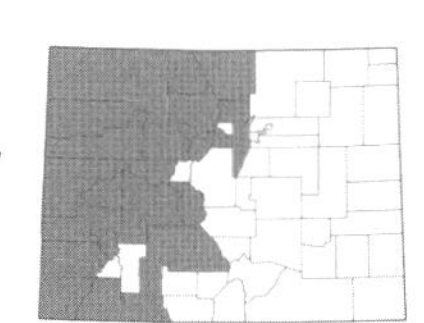

Herrickia glauca **(Nutt.) Brouillet var. *glauca***, GRAY ASTER. [*Aster glaucodes* S.F. Blake; *Eucephalus glaucus* Nutt.; *Eurybia glauca* (Nutt.) G.L. Nesom]. Perennials, 2–7 dm; *leaves* oblong to lanceolate, 4–12 cm long, glabrous, margins entire; *involucre* 6–9 mm diam.; *disk flowers* 6.5–7.5 mm long, yellow; *ray flowers* 8–18 mm long, purple; *pappus* of bristles 6–7 mm long. Found along roadsides and in open, rocky places in the mountains, often in association with pinyon-juniper communities, 5000–10,000 ft. July–Sept. E/W.

Herrickia horrida **Wooton & Standl.**, HORRID HERRICKIA. [*Aster horridus* (Wooton & Standl.) Blake; *Eurybia horrida* (Wooton & Standl.) G.L. Nesom]. Perennials or subshrubs, 3–6 dm; *leaves* oblong to nearly orbiculate, 1–5 cm long, thick and rigid, the margins sharply spinose-serrate, glandular-stipitate and scabrous; *involucre* 8–12 mm diam.; *disk flowers* 8–9 mm long, yellow; *ray flowers* 15–25 mm long, purple; *pappus* of bristles 7–8 mm long. Uncommon on rocky hillsides and on the sides of canyons, often with oak, found near the Raton Pass area, 5500–9000 ft. July–Oct. E.

HETEROSPERMA Cav. – HETEROSPERMA

Annual herbs; *leaves* opposite; *heads* radiate; *involucre bracts* in 2 series, the outer 3–5 bracts linear and herbaceous, the inner series oval, striate, and membranous; *receptacle* chaffy; *disk flowers* yellow; *ray flowers* yellow; *pappus* of 2 or 3 deciduous awns or absent; *achenes* of 2 different kinds, the outer with corky wings and the inner tapering to barbellate beaks.

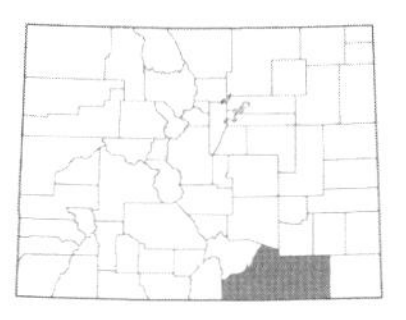

Heterosperma pinnatum **Cav.**, WINGPETAL. Plants 1–4 (7) dm; *leaves* usually 1–4 cm long, the ultimate divisions 0.5–1 (3) mm wide; *disk flowers* ca. 2.5 mm long; *ray flowers* 1–3 (8), ca. 1–2 mm long; *pappus* 0.5–1 (3) mm long; *achenes* 5–18 mm long, glabrous. Uncommon in open places on rocky hills and dry grasslands, known only from the Mesa de Maya region, 5700–5800 ft. Aug.–Sept. E.

HETEROTHECA Cass. – FALSE GOLDENASTER

Herbs; *leaves* alternate, simple; *heads* radiate (ours); *involucre bracts* imbricate in several series, greenish but scarcely herbaceous; *receptacle* naked; *disk flowers* yellow; *ray flowers* yellow, absent in one species; *pappus* of ray flowers absent or double, of disk flowers double; *achenes* of ray flowers often 3-angled, glabrous or nearly so, of disk flowers villous-hirsute. (Harms, 1965; Semple, 1996; Wagenknecht, 1960)

1a. Leaves ovate to elliptic ovate, with usually toothed or sometimes subentire margins; annuals or short-lived perennials; pappus absent on ray flowers and present on disk flowers...***H. subaxillaris*** **ssp. *latifolia***
1b. Leaves linear, oblanceolate (sometimes narrowly so) to oblong, with entire margins or rarely with 1–2 apical teeth; perennials; pappus present on both ray and disk flowers...2

2a. Leaves linear to narrowly oblanceolate, mostly 1.5–3 mm wide, densely compacted on the upper half or third of the stem, the lower half of the stem usually lacking leaves; stems with glandular hairs, especially above...***H. stenophylla***
2b. Leaves oblong to oblanceolate (sometimes narrowly so), leaves generally well-distributed along the stem or absent from just the lower portion but not absent from half or more of the lower stem; stems lacking glandular hairs or if present then the leaves oblong or over 3 mm wide...3

3a. Stems and leaves silvery to silvery-gray pubescent with closely appressed, usually nonglandular hairs; heads on long peduncles, with a few sparse, small leaves but not closely subtended by leaves or bracts...***H. zionensis***
3b. Stems and leaves not densely silvery pubescent, often with glandular hairs; heads terminating leafy stems or sometimes on long peduncles...4

4a. Heads usually solitary...***H. pumila***
4b. Heads mostly two or more per stem...5

5a. Heads subtended by one or more leafy bracts that conspicuously exceed the involucre; plants usually glandular...***H. foliosa***
5b. Heads not subtended by leafy bracts that conspicuously exceed the involucre; plants glandular or not...***H. villosa***

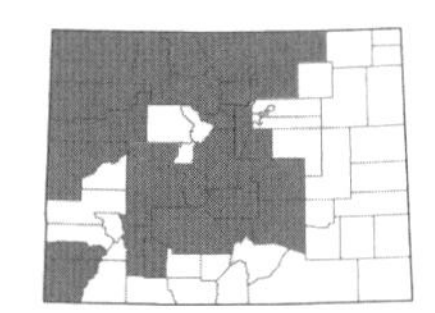

***Heterotheca foliosa* (Nutt.) Shinners**, FOLIOSE FALSE GOLDENASTER. [*H. fulcrata* (Greene) Shinners, misapplied]. Perennials, 1.5–6.5 dm; *leaves* oblanceolate to ovate, hispid-strigose and usually glandular-stipitate, 1.5–5.5 cm long; *heads* 1–15 (43), subtended by modified bracts which overtop the involucre; *involucre* 6–20 mm high, strigose and eglandular to sparsely glandular-stipitate; *ray flowers* 7.5–16 mm long. Common in open meadows and forests, 4500–10,500 ft. July–Oct. E/W.

Heterotheca foliosa and *H. villosa* (see below) form intermediates with heads subtended by small, narrowly lanceolate bracts that just reach the top of the head or barely exceed it. Intergrades are very common, and the two species may not be distinct. *Heterotheca pumila* is sometimes recognized at the species level or as a variety of *H. villosa*. In my opinion, the overtopping bracts subtending the heads clearly place it as more closely related to *H. foliosa*. It can be very difficult to determine high elevation specimens of *H. foliosa* from *H. pumila*.

***Heterotheca pumila* (Greene) Semple**, ALPINE FALSE GOLDENASTER. Perennials, 0.7–3.5 dm; *leaves* narrowly oblanceolate, 2.5–5.6 cm long, hispid-strigose and sparsely glandular-stipitate; *heads* usually solitary, sometimes to 9, usually subtended by modified bracts which overtop the involucre; *involucre* 6.5–12 mm high, strigose, eglandular or sparsely glandular-stipitate; *ray flowers* 10–15 mm long. Common in the mountains, (9000) 10,000–13,000 ft. July–Aug. (Sept.). E/W.

***Heterotheca stenophylla* (A. Gray) Shinners**, STIFFLEAF FALSE GOLDENASTER. Perennials, 2–6 dm; *leaves* linear to narrowly oblanceolate, hispid-strigose, glandular-stipitate, 1–3 cm long; *heads* (1) 2–16 (30); *involucre* 4.5–10 mm high, sparsely to densely glandular and strigose; *ray flowers* 5.5–15 mm long. Found in sandy soil or rocky prairies and hills, 3500–3600 ft. June–Sept. E.

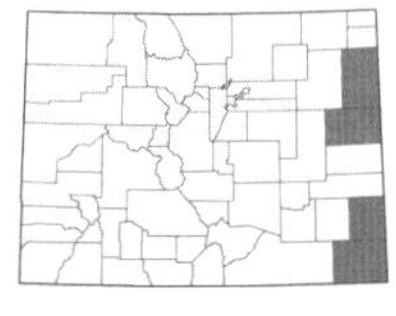

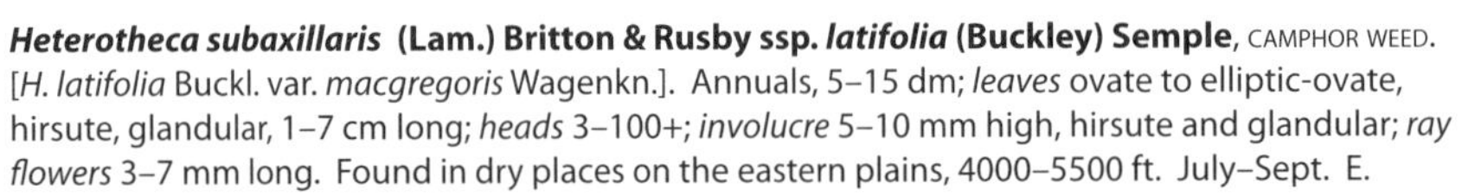

***Heterotheca subaxillaris* (Lam.) Britton & Rusby ssp. *latifolia* (Buckley) Semple**, CAMPHOR WEED. [*H. latifolia* Buckl. var. *macgregoris* Wagenkn.]. Annuals, 5–15 dm; *leaves* ovate to elliptic-ovate, hirsute, glandular, 1–7 cm long; *heads* 3–100+; *involucre* 5–10 mm high, hirsute and glandular; *ray flowers* 3–7 mm long. Found in dry places on the eastern plains, 4000–5500 ft. July–Sept. E.

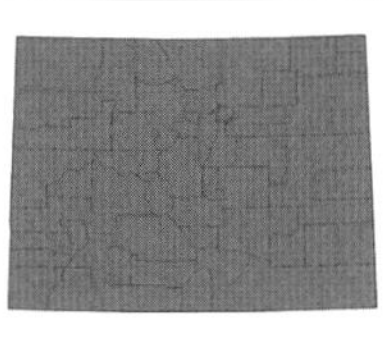

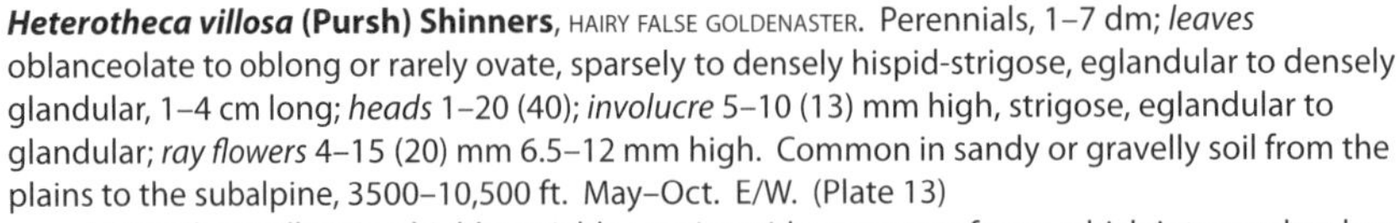

***Heterotheca villosa* (Pursh) Shinners**, HAIRY FALSE GOLDENASTER. Perennials, 1–7 dm; *leaves* oblanceolate to oblong or rarely ovate, sparsely to densely hispid-strigose, eglandular to densely glandular, 1–4 cm long; *heads* 1–20 (40); *involucre* 5–10 (13) mm high, strigose, eglandular to glandular; *ray flowers* 4–15 (20) mm 6.5–12 mm high. Common in sandy or gravelly soil from the plains to the subalpine, 3500–10,500 ft. May–Oct. E/W. (Plate 13)

Heterotheca villosa is a highly variable species with numerous forms which intergrade where their boundaries meet. It can also form hybrids with *H. foliosa* (see above):

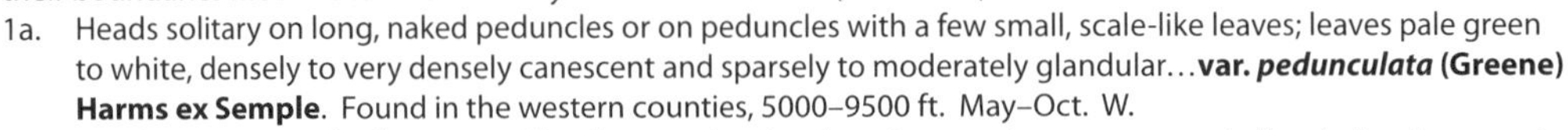

1a. Heads solitary on long, naked peduncles or on peduncles with a few small, scale-like leaves; leaves pale green to white, densely to very densely canescent and sparsely to moderately glandular...**var. *pedunculata* (Greene) Harms ex Semple**. Found in the western counties, 5000–9500 ft. May–Oct. W.
1b. Heads terminating leafy stems, or if on long peduncles then the stem leaves contorted after drying; leaves not densely canescent...2

2a. Leaves mostly oblong to oblong-lanceolate, the margins usually contorted after drying; heads sometimes on naked peduncles...**var. *nana* (A. Gray) Semple**. [*Heterotheca horrida* (Rydb.) Harms]. Common throughout the state, especially on the eastern plains and along the Front Range, 4300–9500 ft. May–Oct. E/W.
2b. Leaves narrowly to broadly oblanceolate or lanceolate to deltoid-lanceolate, the margins usually not contorted after drying; heads terminating leafy stems...3

3a. Leaves narrowly linear-oblanceolate, to 6 mm wide; plants with a densely leafy appearance due to the presence of fascicles of short leaves arising in the axes of the many mid- and upper stem leaves; stems lacking glandular hairs...**var. *angustifolia* (Rydb.) Harms**. [*H. stenophylla* (A. Gray) Shinners var. *angustifolia* (Rydb.) Semple]. Uncommon on the eastern plains, 3500–4500 ft. May–Oct. E.
3b. Leaves oblanceolate to lanceolate or deltoid-lanceolate, if narrow and linear-oblanceolate then the stems with some glandular hairs; plants usually without fascicles of short leaves arising from the leaf axes...4

4a. Leaves and involucre bracts eglandular...**var. *villosa***. Common on the eastern plains and along the base of the Front Range, 3500–7500 (9500) ft. May–Oct. E.
4b. Leaves and involucre bracts sparsely to densely glandular...5

5a. Upper stem leaves lanceolate to deltoid-lanceolate with an acute tip; involucre bracts very sparsely strigose...**var. *scabra* (Eastw.) Semple**. [*H. horrida* (Rydb.) Harms ssp. *cinerascens* (Blake) Semple; *H. polothrix* G.L. Nesom]. Uncommon in Colorado, 4700–6000 ft. May–Oct. W.
5b. Upper stem leaves narrowly to broadly oblanceolate; involucre bracts usually moderately strigose...**var. *minor* (Hook.) Semple**. [*H. villosa* (Pursh) Shinners var. *hispida* (Hook.) Harms]. Common throughout the state, 6000–10,000 ft. May–Oct. E/W

***Heterotheca zionensis* Semple**, ZION FALSE GOLDENASTER. Perennials, 1.7–7 dm; *leaves* oblanceolate to ovate or oblong, hispid-strigose, sparsely to moderately glandular, 1–3 cm long; *heads* 7–30 (45); *involucre* 4.5–8 mm high, sparsely to densely glandular and strigose; *ray flowers* 6.5–12 mm 6.5–12 mm high. Uncommon in sandy or rocky soil of open places, 5600–6400 ft. June–Aug. (Sept.). W.

HIERACIUM L. – HAWKWEED

Perennial herbs; *leaves* alternate or basal, usually with some stellate hairs; *heads* ligulate; *involucre bracts* in 1–2 series, obscurely to conspicuously imbricate, often with stellate hairs; *receptacle* naked; *disk flowers* absent; *ray flowers* yellow, orange, or white; *pappus* simple, of numerous capillary bristles; *achenes* weakly to strongly ribbed. (Beaman, 1990; Fernald, 1943)

1a. Leaves glabrous or nearly so; flowers yellow; involucre bracts black-hirsute and glandular...***H. triste***
1b. Leaves hairy; flowers and involucre bracts various...2

2a. Flowers white; stems usually loosely long-hairy at least at the base, becoming glabrous upward...***H. albiflorum***
2b. Flowers orange or yellow; stems hairy throughout...3

3a. Involucre bracts black-hirsute and glandular; flowers orange or yellow-orange (often drying reddish or purple); mature achenes 1.5–2 mm long...***H. aurantiacum***
3b. Involucre bracts with matted, tomentose hairs and longer, golden bristles; flowers yellow; mature achenes 5–7 mm long...***H. fendleri***

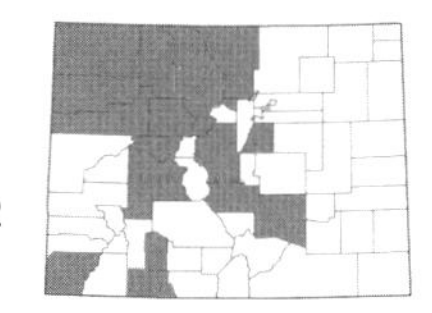

***Hieracium albiflorum* Hook.**, WHITE HAWKWEED. [*Chlorocrepis albiflora* (Hook.) W.A. Weber]. Plants 3–8 dm; *leaves* oblong to oblanceolate, sparsely to moderately long-hairy or the upper glabrate, 4–17 cm long, the margins entire to few-toothed; *involucre* 6–11 mm high, blackish-green, sparsely glandular or pubescent to glabrate; *ray flowers* 13–34, white; *achenes* 2.5–3.5 mm 6.5–12 mm high. Common in open woods and meadows and on moist to dry slopes, 6000–11,500 ft. July–Sept. E/W.

***Hieracium aurantiacum* L.**, ORANGE HAWKWEED. Plants 2–7 dm; *leaves* oblanceolate to narrowly elliptic, densely soft-hairy, 5–20 cm long, the margins entire or few-toothed; *involucre* 5–9 mm high, glandular and black-hairy; *ray flowers* 30–45, orange; *achenes* 1.5–2 mm 6.5–12 mm high. Uncommon in mountain meadows and forests, often in more disturbed places, 7000–10,000 ft. June–Aug. E/W. Introduced.

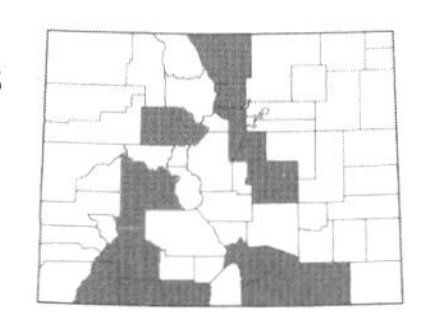

***Hieracium fendleri* Sch. Bip.**, YELLOW HAWKWEED. [*Chlorocrepis fendleri* (Sch. Bip.) W.A. Weber]. Plants 1–5 dm; *leaves* oblanceolate to elliptic, glaucous with sparsely long, spreading bristles, 3–15 cm long, the margins entire or few-toothed; *involucre* 9–15 mm high, sparsely setose and thinly stellate; *ray flowers* yellow; *achenes* 5–7 mm 6.5–12 mm high. Locally common in dry or moist forests, 6800–9500 ft. June–Aug. E/W.

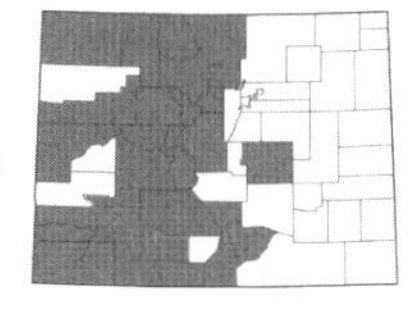

***Hieracium triste* Willd. ex Spreng.**, SLENDER HAWKWEED. [*Chlorocrepis tristis* (Willd. ex Spreng.) Á. Löve & D. Löve; *H. gracile* Hook.]. Plants 1–3 dm; *leaves* obovate to oblong, glabrous or minutely hairy, 2–8 cm long, the margins entire or nearly so; *involucre* 7–10 mm high, densely black-hirsute at least near the base, glandular; *ray flowers* 20–35, yellow; *achenes* 2–3 mm 6.5–12 mm high. Common in subalpine and alpine forests and meadows, (8000) 9500–13,000 ft. June–Sept. E/W.

HYMENOPAPPUS L'Hér. – HYMENOPAPPUS

Biennial or perennial herbs; *leaves* alternate or basal; *heads* discoid or radiate; *involucre bracts* in 2–3 subequal series, somewhat herbaceous below, scarious or hyaline toward the broad or rounded tip; *receptacle* naked or chaffy; *disk flowers* yellow or white; *ray flowers* white or absent; *pappus* of scales or absent; *achenes* 4-angled. (Turner, 1956)

1a. Ray flowers present, white and conspicuous; receptacle chaffy; achenes glabrous...***H. newberryi***
1b. Ray flowers absent; receptacle usually naked or rarely chaffy with 6–10 well-developed scales; achenes hairy (at least on the angles)...2

2a. Perennials from woody, branched taproots bearing several crowns; stem leaves typically less (absent or to 9); achenes densely long-hairy throughout...***H. filifolius***
2b. Biennials from unbranched taproots; stem leaves usually more, 8–40; achenes long-hairy principally on the angles...3

3a. Disk flowers bright yellow; ultimate leaf segments broader, 2–6 mm wide; involucre bracts 4–5 (6) mm high...***H. flavescens***
3b. Disk flowers white to cream-colored; ultimate leaf segments linear to filiform, 0.5–1.5 mm wide; involucre bracts 5–8 mm high...***H. tenuifolius***

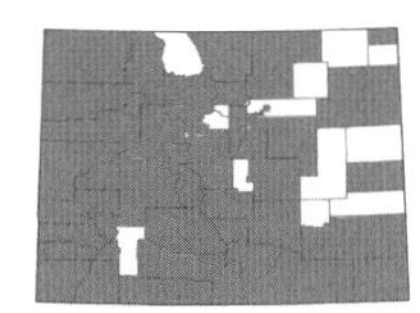

Hymenopappus filifolius **Hook.**, FINELEAF HYMENOPAPPUS. Perennials, 0.5–10 dm; *leaves* bipinnately dissected, the ultimate segments linear to filiform, tomentose to glabrate, 3–20 cm long; *heads* 1–30 (60) per stem; *involucre* 3–14 mm high, white or yellowish-membranous; *disk flowers* 2–7 mm, densely glandular to glabrate; *ray flowers* absent; *achenes* densely hairy, 3–7 mm. Common in dry, sandy or gravelly soil, 3500–9200 ft. May–Sept. E/W. (Plate 13)

There are 13 varieties of *H. filifolius*, 6 of which are present in Colorado:

1a. Leaf bases inconspicuously tomentose to glabrous; stem leaves 0–2...**var. *parvulus* (Greene) B.L. Turner**, FINELEAF HYMENOPAPPUS. Found on bare, sandy or gravelly slopes, 5600–8800 ft. June–Aug. E/W. Endemic.
1b. Leaf bases conspicuously and densely tomentose; stem leaves 0–9...2

2a. Ultimate leaf divisions very short, only 1–5 mm long; heads relatively small and few-flowered with 10–30 (average of 20) florets per head; involucre bracts 3–7 mm high; disk corollas 2–3 (3.5) mm high; plants found on the western slope...3
2b. Ultimate leaf divisions usually longer, (3) 4–30 mm long; heads mostly larger with 20–60 florets per head; involucre bracts (5) 6–14 mm high; disk corollas 3–7 mm high; plants of the eastern or western slope...4

3a. Stem leaves 3–9; plants mostly 2–4 dm tall...**var. *pauciflorus* (I.M. Johnst.) B.L. Turner**, FINELEAF HYMENOPAPPUS. Found in dry, rocky and sandy soil in the southwestern corner of the state (La Plata and Montezuma cos.), 5700–6800 ft. May–Sept. W.
3b. Stem leaves 0–4; plants mostly 1–2.5 dm tall...**var. *luteus* (Nutt.) B.L. Turner**, FINELEAF HYMENOPAPPUS. Found in sandy soil of dry slopes, distributed mostly in Moffat County but scattered throughout the western slope, 5000–7000 ft. June–July. W.

4a. Heads larger, the involucre (8) 9–14 mm high; anthers 3–4 mm long...**var. *megacephalus* B.L. Turner**, LARGE-HEADED HYMENOPAPPUS. Found in dry, sandy or gravelly soil, scattered on the western slope, 5000–7000 ft. May–Aug. W.
4b. Heads smaller, the involucre 5–8 (9) mm high; anthers 2–3 mm long...5

5a. Heads 1–6 per stem; stem leaves (0) 2–4; leaf divisions mostly 2–8 mm long...**var. *cinereus* (Rydb.) I.M. Johnst.**, FINELEAF HYMENOPAPPUS. Found on exposed, mostly vegetated hillsides in clay or sandy soil, widespread throughout the state, 3600–9200 ft. May–Sept. E/W.
5b. Heads 5–50 per stem; stem leaves 3–8; leaf divisions mostly (5) 10–30 mm long...**var. *polycephalus* (Osterh.) B.L. Turner**, MANYHEAD HYMENOPAPPUS. Found on sandy plains and grasslands of the eastern plains, 3800–5000 ft. June–Aug. E.

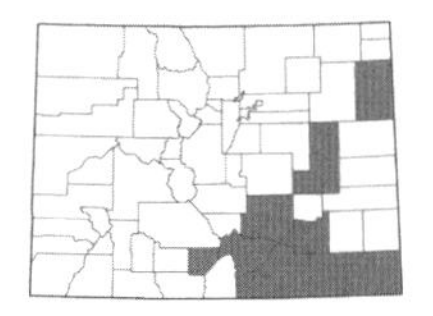

Hymenopappus flavescens **A. Gray var. *flavescens***, COLLEGE-FLOWER. Biennials, 4–9 dm; *leaves* bipinnately dissected with broad ultimate segments, sparsely tomentose to glabrate, 6–15 cm long; *heads* 30–100 per stem; *involucre* 4–5 (6) mm high, yellow-membranous at tips; *disk flowers* 2.5–3.5 mm long, glandular; *ray flowers* absent; *achenes* hairy mostly on the edges, 3–4 mm long. Found on dry, sandy hills and rocky slopes, 3800–5600 ft. May–Aug. E.

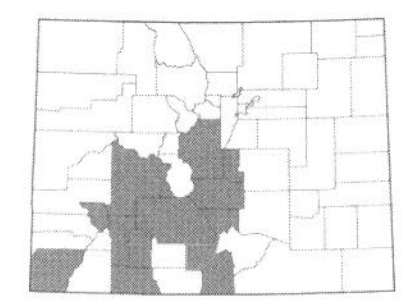

***Hymenopappus newberryi* (A. Gray) I.M. Johnst.**, NEWBERRY'S HYMENOPAPPUS. Perennials, 2–6 dm; *leaves* bipinnately dissected with linear ultimate segments, tomentose to glabrate, 12–25 cm long; *heads* 3–8 per stem; *involucre* 8–10 mm high, yellow-membranous at tips; *disk flowers* 3–4 mm long, sparsely glandular to glabrate; *ray flowers* present, white; *achenes* black, glabrous, 3–4 mm long. Locally common on dry hills in sandy or rocky soil of mountainous areas, often with pinyon-juniper or spruce/aspen forests, 6500–10,500 ft. June–Sept. E/W.

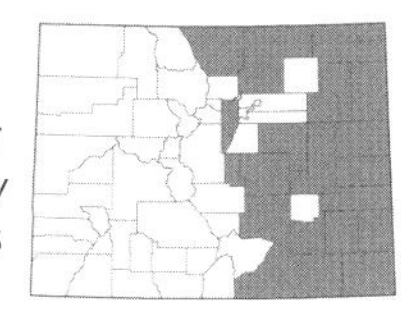

***Hymenopappus tenuifolius* Pursh**, CHALK HILL HYMENOPAPPUS. Biennials, 4–15 dm; *leaves* bipinnately dissected into linear or filiform segments, tomentose to glabrate, 8–15 cm long; *heads* 20–200 per stem; *involucre* 5–8 mm high, yellow-membranous at tips; *disk flowers* 2–3 mm long, glandular; *ray flowers* absent; *achenes* hairy mostly on edges, 3.5–4.5 mm long. Common in sandy or sometimes rocky soil on the eastern plains, 3400–6000 ft. May–Aug. E.

HYMENOXYS Cass. – RUBBERWEED

Annual, biennial, or perennial herbs; *leaves* alternate; *heads* radiate (ours); *involucre bracts* subequal in 2–4 series or unequal in 2 series, herbaceous; *receptacle* naked; *disk flowers* yellow; *ray flowers* yellow or yellow-orange or orange; *pappus* of awned or aristate scales; *achenes* mostly 5-angled. (Bierner, 1994; 2001; 2004; Bierner & Jansen, 1998)

1a. Annuals from slender taproots...***H. odorata***
1b. Perennials from branched, woody caudices or short rhizomes...2

2a. Leaves simple, 10–50 mm wide; flowers orange to orange-yellow or sometimes yellow, the rays numbering mostly 17–25...***H. hoopesii***
2b. At least some leaves lobed, divided, or pinnatifid, 5.5 mm or less wide; flowers yellow or sometimes yellow-orange (and then the number of ray flowers 10–16)...3

3a. Heads solitary or few per plant, larger (mostly 17–35 mm wide); the surface under the heads densely tomentose...4
3b. Heads numerous to several per plant, smaller (mostly 5–20 mm wide); the surface under the heads not densely tomentose...5

4a. Stems glabrate at the base and woolly toward the apex; leaves glabrate to sparsely villous, simple or with 2–3 linear lobes...***H. brandegeei***
4b. Stems arachnoid to woolly-pubescent throughout; leaves arachnoid to woolly-pubescent, most once or twice pinnatifid...***H. grandiflora***

5a. Leaf segments 2–5.5 mm wide, and often some leaves undivided; flowers yellow or yellow-orange...***H. helenioides***
5b. Leaf segments to 2 mm wide, the leaves all divided; flowers yellow...***H. richardsonii***

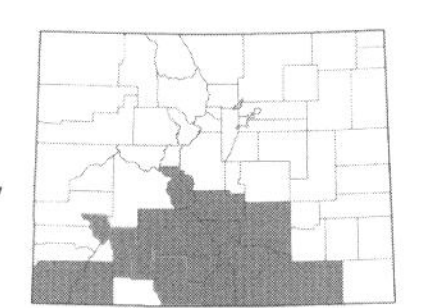

***Hymenoxys brandegeei* Porter ex A. Gray**, BRANDEGEE'S RUBBERWEED. [*Tetraneuris brandegeei* (Porter ex A. Gray) Parker]. Perennials to 2 dm; *leaves* simple or usually with 2–3 linear lobes, glabrate to villous; *heads* solitary, the peduncle densely tomentose under the heads; *involucre* 7–13 mm high, tomentose; *disk flowers* 4–5.5 mm long; *ray flowers* 10–20 mm long, yellow; *achenes* 2.5–3 mm long. Uncommon in the alpine of the southern mountains, 11,000–13,000 ft. June–Aug. E/W.

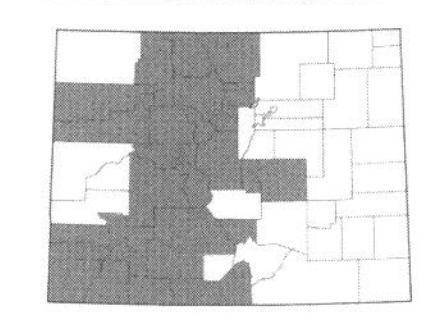

***Hymenoxys grandiflora* (Torr. & A. Gray ex A. Gray) Parker**, OLD MAN OF THE MOUNTAIN. [*Tetraneuris grandiflora* (Torr. & A. Gray ex A. Gray) Parker]. Perennials, 1–3 dm; *leaves* usually 1–2-pinnatifid with linear segments, arachnoid-woolly; *heads* solitary or few; *involucre* 9–14 mm high; *disk flowers* 5–6 mm long; *ray flowers* 15–30 m long m, yellow; *achenes* 3–4 mm long. Common and widespread in the alpine, (9000) 10,000–14,000 ft. June–Aug. (Sept.). E/W. (Plate 13)

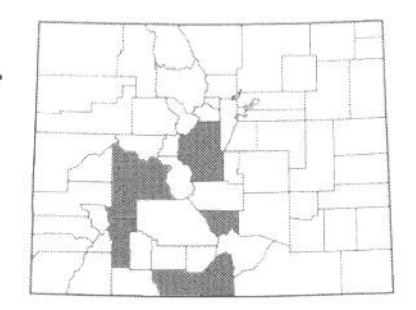

***Hymenoxys helenioides* (Rydb.) Cockerell**, INTERMOUNTAIN RUBBERWEED. [*Picradenia helenioides* Rydb.]. Perennials, 1–5 (10) dm; *leaves* simple or 3-lobed, glabrous to sparsely hairy, 7–20 cm long; *heads* 5–50+; *involucre* 5–7 mm high, sparsely to moderately hairy and gland-dotted; *disk flowers* 3.5–5.5 mm long; *ray flowers* 17–30 mm long, yellow to yellow-orange; *achenes* 2.5–3.5 mm long. Uncommon in open meadows and along roadsides, 5600–10,700 ft. June–Aug. E/W.

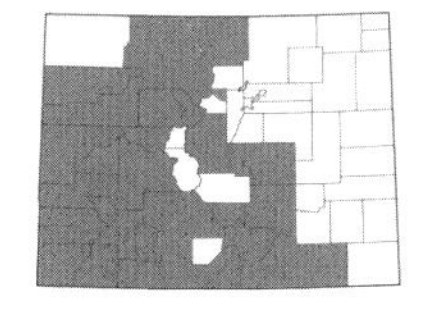

***Hymenoxys hoopesii* (A. Gray) Bierner**, ORANGE SNEEZEWEED; OWL'S CLAWS. [*Dugaldia hoopesii* Rydb.; *Helenium hoopesii* A. Gray]. Perennials, 2–10 dm; *leaves* simple, oblanceolate, loosely villous-tomentose to glabrate, 10–30 cm long; *heads* solitary to 12; *involucre* 6–10 mm high; *disk flowers* 4–5.5 mm long; *ray flowers* 15–30 mm long, orange or orange-yellow; *achenes* 3.5–4.5 mm long. Common in moist places along streams, in meadows, and on open slopes, 7000–11,500 ft. June–Aug. E/W.

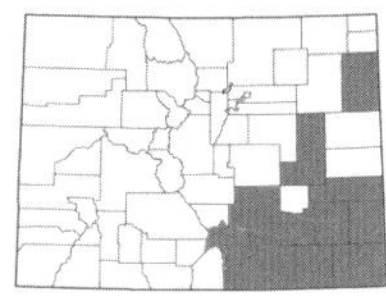

***Hymenoxys odorata* DC.**, BITTER RUBBERWEED. [*Picradenia odorata* (DC.) Britton]. Annuals, 1–8 dm; *leaves* deeply divided into 3–5 linear lobes, hairy to minutely hairy, mostly 2–5 cm long; *heads* 15–300+; *involucre* 3–5 mm high, gland-dotted; *disk flowers* 2.5–4 mm long; *ray flowers* 5–10 mm long, yellow; *achenes* 1.5–2.5 mm long. Found in dry, open places on the eastern plains, 3700–5000 ft. April–June. E.

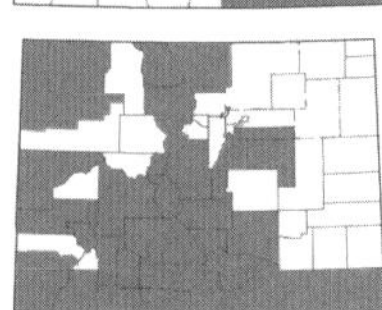

***Hymenoxys richardsonii* (Hook.) Cockerell**, COLORADO RUBBERWEED. Perennials, 0.7–3.5 dm; *leaves* cleft into 3–5 long, linear or linear-filiform segments, sparsely hairy to glabrate, usually punctate; *heads* 5–300+; *involucre* 4–8 mm high, often glandular punctate; *disk flowers* 3–5 mm long; *ray flowers* 7–15 mm long, yellow; *achenes* 2–3 mm long. Common in dry, open or sparsely wooded places, often in rocky or sandy soil, 5400–10,500 ft. June–Sept. E/W. (Plate 13)

There are two varieties of *H. richardsonii* in Colorado, each occupying their own geographic locality for the most part but wholly confluent where their ranges overlap in Park and Summit cos.:

1a. Heads relatively few and larger, usually 1–4 per stem and the involucre 5–8 mm high; rays 8–14; plants 0.7–2.4 dm tall...**var. *richardsonii***. Found in Moffat, Jackson, Grand, and Larimer cos., 7800–10,500 ft. June–July. E/W.

1b. Heads relatively more numerous and smaller, usually more than 4 per stem and the involucre 4–5 mm high; plants 1.9–3.4 dm tall...**var. *floribunda* (A. Gray) Parker**. Found in the central and southwestern and southcentral counties, 5700–10,500 ft. June–Sept. E/W.

HYPOCHAERIS L. – CAT'S-EAR

Annual or perennial, lactiferous herbs; *leaves* primarily basal; *heads* ligulate; *involucre bracts* imbricate, greenish-black, the inner with hyaline margins; *receptacle* chaffy; *disk flowers* absent; *ray flowers* usually yellow; *pappus* of one or two rows of plumose bristles; *achenes* several-nerved, subterete, long-beaked.

***Hypochaeris radicata* L.**, SPOTTED CAT'S-EAR. Perennials, 1–6 dm; *leaves* all basal, oblanceolate, 5–35 cm long, toothed or pinnatifid, spreading-hispid; *heads* (1) 2–7; *involucre bracts* 3–20 mm long, narrowly lanceolate, glabrous or hispid medially; *ray flowers* 10–15 mm long; *pappus* to 10–12 mm long; *achenes* 6–10 mm long, with 10–12 barbellate ribs. Uncommon weed found in lawns and pavement cracks, currently known only from Boulder, 5400–5500 ft. May–Sept. E. Introduced.

ISOCOMA Nutt. – GOLDENBUSH

Perennial subshrubs, often gultinous; *leaves* alternate, entire or with spinulose-tipped teeth or lobes; *heads* discoid; *involucre bracts* imbricate, scarcely herbaceous; *receptacle* naked; *disk flowers* yellow; *ray flowers* absent; *pappus* of numerous barbellate bristles; *achenes* mostly 4–10-ribbed, sometimes the ribs very thick and resinous, broadly to narrowly turbinate.

***Isocoma rusbyi* Greene**, RUSBY'S GOLDENBUSH. [*Haplopappus rusbyi* (Greene) Cronquist]. Plants 2.5–8 dm; *leaves* linear-oblong or narrowly oblong-oblanceolate, 3–8 cm long, slightly glandular, usually entire; *involucre* 6–9 mm high, glandular-gutinous; *disk flowers* 5–6.5 mm long; *achenes* with long, ascending hairs. Uncommon in rocky or sandy soil of dry, open places, 5100–6500 ft. Aug.–Oct. W.

Isocoma rusbyi is often included in *I. drummondii*, a species with somewhat smaller heads and larger, shorter corollas, and more densely hairy achenes with thicker, resinous ribs. *Isocoma drummondii* is also widely separated by geographical range as it is only found in southeastern Texas. *Isocoma pluriflora* is sometimes reported for Colorado based on an old type specimen from the Long Expedition that was most likely collected in northeastern New Mexico and not Colorado.

IVA L. – MARSH ELDER

Herbs or shrubs; *leaves* opposite or alternate; *heads* discoid; *involucre bracts* subequal or imbricate in 1–3 series; *receptacle* chaffy; *disk flowers* pistillate or staminate, the pistillate flowers 1–4 and peripheral, the staminate flowers 5–22 and central; *ray flowers* absent; *pappus* absent; *achenes* compressed parallel to the involucre bracts, usually gland-dotted.

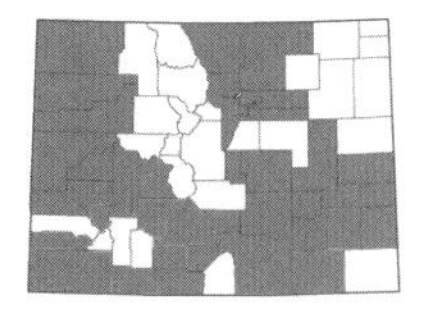

***Iva axillaris* Pursh**, POVERTY WEED. Perennials, 1–4 (6) dm, from rhizomes; *leaves* elliptic to obovate or spatulate, 15-25 (45) mm long, usually strigose to minutely scabrous, gland-dotted, *involucres* 2–3 mm high, the outer more or less connate; *achenes* 2.5–3 mm long. Common in dry, open, sometimes alkaline places on the plains and in valleys, often with pinyon-juniper, 3500–7500 ft. May–Sept. E/W.

KRIGIA Schreb. – DWARF DANDELION

Annuals or perennials; *leaves* alternate or basal; *heads* ligulate; *involucre bracts* in 1–2 rows, nearly equal, herbaceous; *receptacle* naked; *disk flowers* absent; *ray flowers* yellow to orange; *pappus* absent or of a few scales, or of an inner ring of bristles and an outer ring of short scales; *achenes* ovoid-columnar to prismatic, the apex not beaked.

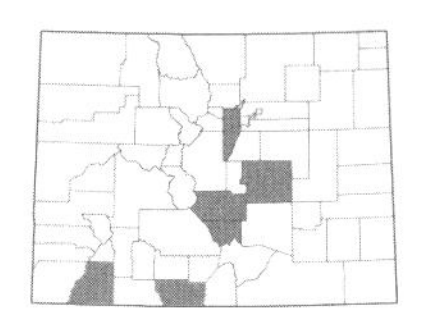

Krigia biflora **(Walter) S.F. Blake**, TWO-FLOWER DWARF DANDELION. Perennials, 2–6 (8) dm, from fibrous roots, glabrous and somewhat glaucous; *basal leaves* oblanceolate to elliptic-oblong, to 25 cm long and 6 cm wide, glabrous, usually glaucous, the margins entire or toothed; *cauline leaves* clasping; *involucres* 7–14 mm high; *ray flowers* 15–25 mm long; *pappus* of ca. 10 outer scales and 20–40 barbellate, unequal bristles; *achenes* 2–3 mm long. Uncommon in moist meadows, 7000–8500 ft. June–Aug. E/W.

LACTUCA L. – LETTUCE

Lactiferous herbs; *leaves* alternate, entire to variously pinnatifid or lobed; *heads* ligulate; *involucre bracts* imbricate in 3–4 series, more or less herbaceous; *receptacle* naked; *disk flowers* absent; *ray flowers* yellow, blue, violet, or rarely white; *pappus* of numerous white or brown capillary bristles; *achenes* strongly flattened, winged or ribbed, beaked or beakless. (Prince & Carter, 1977; Stebbins, 1939)

1a. Leaf margins and often the abaxial midrib with numerous, small, sharp spines or prickles...2
1b. Leaf margins without numerous, small spines or prickles (sometimes the lobes sharp-pointed, but not prickly), the abaxial midrib glabrous or sometimes hirsute in lower leaves...3

2a. Achenes with 1 conspicuous nerve on each face, occasionally with an additional, less prominent pair; stem glabrous; flowers usually yellow but sometimes blue...***L. ludoviciana***
2b. Achenes with 5–7 conspicuous nerves on each face; usually with stiff, sharp bristles on the lower ⅓ of the stem; flowers yellow with a dark blue stripe on the abaxial side...***L. serriola***

3a. Achenes with 1 (3) conspicuous nerve on each face, occasionally with an additional, less prominent pair...4
3b. Achenes with 4–7 conspicuous nerves on each face...5

4a. Leaves mostly pinnatifid with lanceolate segments; involucre 8–10 mm high; flowers yellow or sometimes drying purplish...***L. canadensis***
4b. Leaves mostly entire or with a few recurving linear-lanceolate segments; involucre 12–20 mm high; flowers blue or purple, sometimes yellow, or rarely white...***L. graminifolia***

5a. Flowers yellow, sometimes with a dark blue stripe on the abaxial side...***L. saligna***
5b. Flowers blue or purple...6

6a. Leaves pinnatifid to deeply lobed with a broadly triangular apex, the lobes sharp-pointed; plants 15–20 dm; involucre 7–12 mm high; pappus usually brown but rarely white...***L. biennis***
6b. Leaves entire or to pinnately lobed, but lacking a broadly triangular apex, the lobes not sharp-pointed; plants 1.5–10 dm; involucre 12–20 mm high; pappus white...***L. tatarica* var. *pulchella***

Lactuca biennis **(Moench) Fernald**, TALL BLUE LETTUCE. [*L. spicata* (Lam.) A.S. Hitchc.]. Annuals or biennials, 15–20 dm; *leaves* pinnatifid to deeply lobed with a broadly triangular apex, the lobes sharp-pointed, often hispid; *involucre* 7–12 mm high; *ray flowers* blue with a purple tip, the ligule 4 mm long; *pappus* brown or rarely white; *achenes* with 4–5 nerves on both sides. Uncommon in forest clearings and along streams, 6500–8500 ft. June–Aug. E/W.

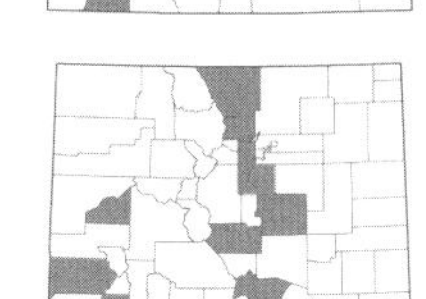

Lactuca canadensis **L.**, CANADA LETTUCE. Biennials, 5–30 dm; *leaves* weakly dentate to pinnatifid or deeply lobed, not prickly-margined, sometimes sparsely pilose; *involucre* 8–12 mm high; *ray flowers* yellow, the ligule 3–4 mm long; *pappus* white; *achenes* with 1 nerve on each side. Uncommon in open forests and meadows, sometimes in ravines or disturbed places, 5500–7500 ft. (June) July–Sept. E/W.

Lactuca graminifolia **Michx.**, GRASSLEAF LETTUCE. Biennials, 3–15 dm; *leaves* linear to linear-lanceolate, entire or with recurving lanceolate lobes, not prickly-margined; *involucre* 12–20 mm high; *ray flowers* blue or purple, sometimes yellow, rarely white; *pappus* white; *achenes* with 1 nerve on each side. Uncommon on open slopes and in meadows, 7500–8000 ft. July–Sept. W.

I am not convinced that this species occurs in Colorado (although I have not confirmed the identification of the specimen). It is only known from near Ouray, but the northernmost specimen in New Mexico occurs east of the Continental Divide.

Lactuca ludoviciana **(Nutt.) Riddell**, BIANNUAL LETTUCE. Biennials, 1.5–20 dm; *leaves* dentate or pinnatifid to deeply lobed, prickly-margined; *involucre* 12–17 mm high; *ray flowers* yellow or sometimes bluish; *pappus* white; *achenes* with 1 conspicuous nerve on each side. Uncommon on the plains and in the northern foothills, 3500–6500 ft. June–Aug. E.

***Lactuca saligna* L.**, WILLOWLEAF LETTUCE. Annuals or biennials, 1.5–10 dm; *leaves* entire to pinnatifid, with 1–2 pairs of linear lobes near the base, not prickly-margined; *involucre* 6–10 mm high; *ray flowers* yellow with a dark blue stripe on the abaxial side; *pappus* white; *achenes* with 5–7 nerves on each side. This is not yet reported for Colorado but is becoming a troublesome weed elsewhere and could eventually occur in the state. Introduced.

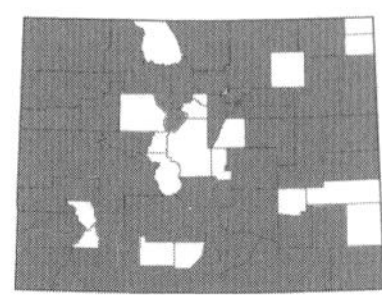

***Lactuca serriola* L.**, PRICKLY LETTUCE. Annuals or biennials, 1.5–15 dm; *leaves* entire to dentate or deeply pinnate-lobed, prickly-margined; *involucre* 8–12 mm high; *ray flowers* yellow with a dark blue stripe on the lower side; *pappus* white; *achenes* with 5–7 nerves on each side. Common in disturbed sites, open grasslands, and forest clearings, 3500–8500 ft. July–Oct. E/W. Introduced.

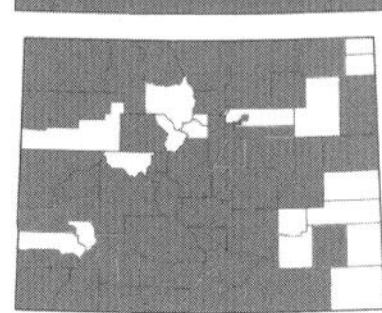

***Lactuca tatarica* (L.) C.A. Mey. var. *pulchella* (Pursh) Breitung**, BLUE LETTUCE. [*L. oblongifolia* Nutt.; *L. pulchella* (Pursh) DC.; *Mulgedium pulchellum* (Pursh) G. Don; *M. pulchellum* (Pursh) G. Don in R. Sweet]. Perennials, 2–10 dm; *leaves* entire or pinnately lobed, not prickly-margined; *involucre* 12–20 mm high; *ray flowers* blue or purple; *pappus* white; *achenes* with 4–6 nerves on each side. Common in open meadows, along roadsides, and in moist places along streams and ditches, 3500–10,000 ft. July–Sept. E/W. (Plate 13)

LAENNECIA Cass. – LAENNECIA

Annual herbs; *leaves* alternate, sessile and clasping, glandular; *heads* disciform or inconspicuously radiate; *involucre bracts* not much imbricate, scarcely herbaceous; *receptacle* flat, naked; *disk flowers* few, yellow; *pistillate flowers* numerous, corolla slender, rayless or with a short, inconspicuous, white or pinkish ray; *pappus* of capillary bristles; *achenes* 2-nerved. (Nesom, 1990)

1a. Pistillate flowers with a short ligule (about 0.5–1 mm long); leaves loosely arachnoid-tomentose; heads usually in an elongate, narrow, thyrsoid inflorescence...***L. schiedeana***

1b. Pistillate flowers without a ligule; leaves not arachnoid-tomentose; heads in an elongate, wider, thyrsoid-paniculiform inflorescence...***L. coulteri***

***Laennecia coulteri* (A. Gray) G.L. Nesom**, CONYZA. [*Conyza coulteri* A. Gray]. Plants 1–15 dm; *leaves* spatulate to oblong, 2–5 cm long, villous and glandular-viscid, the margins lobed to coarsely toothed or entire; *involucre* 2.5–3.5 mm high; *disk flowers* 5–20; *pistillate flowers* 60–100, lacking a ligule; *achenes* 0.5–1 mm long. Found along ditches, dry stream beds, and in pine forests, reported for Colorado along the Front Range. July–Sept. E.

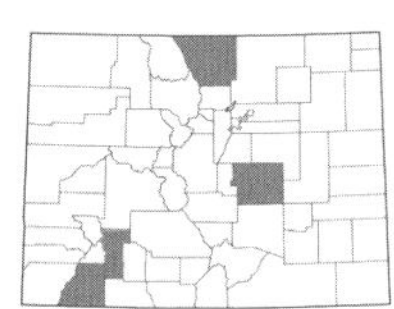

***Laennecia schiedeana* (Less.) G.L. Nesom**, PINELAND MARSHTAIL. [*Conyza schiedeana* (Less.) Cronquist]. Plants 1–5 (10) dm; *leaves* spatulate to oblong or linear, 2–5 cm long, tomentose and glandular, the margins usually entire or sometimes lobed distally; *involucre* 4–6 mm high; *disk flowers* 5–10; *pistillate flowers* 60–100+, the ligule 0.5–1 mm long; *achenes* 1–1.5 mm long. Uncommon in open pine forests, with occurrences scattered across the state, 8000–9500 ft. July–Sept. E/W.

LAPSANA L. – NIPPLEWORT

Annual or perennial, lactiferous herbs; *leaves* alternate, entire to variously dentate or pinnatifid, petiolate; *heads* small, ligulate; *involucre bracts* in 2 series, the outer minute and few, the inner subequal and keeled; *receptacle* naked; *disk flowers* absent; *ray flowers* yellow; *pappus* absent; *achenes* narrow, glabrous, with many nerves.

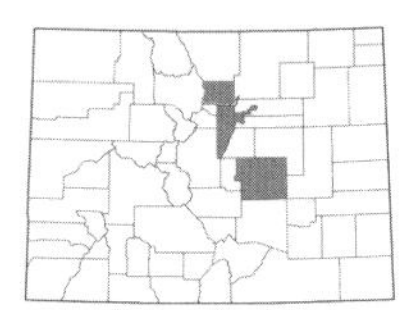

***Lapsana communis* L.**, COMMON NIPPLEWORT. Annuals, 15–100 (150) cm, hirsute to nearly glabrous; *leaves* ovate or rounded-obtuse, lowermost lyrate, 1–15 (30) cm long; *involucre* 5–8 mm high, the bracts 3–9 mm long; *ray flowers* 8–15, 1–10 mm long; *achenes* 3–5 mm long. Uncommon weed in disturbed forests, 5000–6300 ft. June–Aug. E. Introduced.

LEUCANTHEMUM Mill. – DAISY

Perennial herbs; *leaves* alternate, simple; *heads* radiate; *involucre bracts* imbricate in 2–4 series, the margins and tips scarious; *receptacle* convex, naked; *disk flowers* yellow; *ray flowers* white; *pappus* lacking; *achenes* 10-ribbed.

1a. Leaves smaller, 4–15 cm long and 0.4–1 cm wide, the upper leaves auriculate-clasping at the base; heads smaller, the disk mostly 10–20 mm wide...***L. vulgare***

1b. Leaves larger, 10–16 cm long and 2–3 cm wide, the upper leaves sessile but not auriculate-clasping; heads larger, the disk mostly 20–30 mm wide...***L. maximum***

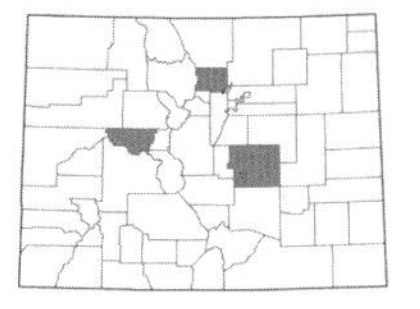

***Leucanthemum maximum* DC.**, MAX CHRYSANTHEMUM. [*Chrysanthemum maximum* Ramond]. Plants 2–8 dm; *leaves* obovate to spatulate or lanceolate to linear, 10–16 cm long, the margins usually toothed; *involucre* 20–30 mm diam.; *ray flowers* 20–40 mm long; *achenes* 2–4 mm long. Uncommon along roadsides or near homes where it escapes from cultivation, 5000–6500 ft. June–Aug. E. Introduced.

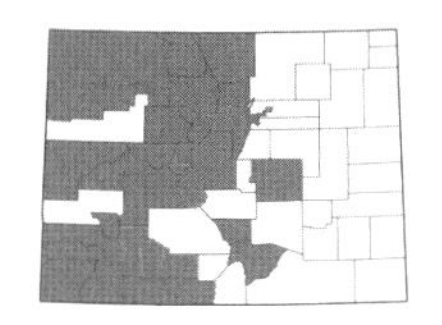

***Leucanthemum vulgare* Lam.**, OXEYE DAISY. [*Chrysanthemum leucanthemum* L.]. Plants 1–10 dm; *leaves* obovate to spatulate or oblanceolate, 4–15 cm long, the upper leaves auriculate-clasping, margins usually pinnately lobed or toothed; *involucre* 10–20 mm diam.; *ray flowers* 12–30 mm long; *achenes* 1.5–2.5 mm long. Widespread, found along roadsides, in disturbed areas, old mining camps, and in mountain meadows, 4800–10,500 ft. May–Sept. E/W. Introduced.

LIATRIS Gaertn. ex Schreb. – BLAZING STAR; GAY-FEATHER

Perennial herbs; *leaves* alternate, simple, entire, generally conspicuously punctate; *heads* discoid; *involucre bracts* imbricate, herbaceous, the margins scarious; *receptacle* naked; *disk flowers* purple, pink, or rarely white; *pappus* of plumose or barbellate bristles, in one or more series; *achenes* about 10-ribbed, somewhat cylindrical but pointed at the base. (Gaiser, 1946; Nesom, 2005)

1a. Involucre bracts rounded at the tip, this scarious and usually broadly lacerate or fringed; heads broadly campanulate to hemispheric; stem reddish, glabrous below with white appressed hairs above...***L. ligulistylis***
1b. Involucre bract tips acute to acuminate, ciliate-margined or narrowly scarious-margined; heads cylindric to weakly campanulate; stem green to yellowish-brown or the upper sometimes reddish, glabrous or nearly so...2

2a. Pappus barbellate; corolla about 6 mm long, glabrous inside; leaves lax, the lower 5-veined...***L. lancifolia***
2b. Pappus plumose; corolla 9–12 mm long, pilose or pubescent inside; leaves rigid, often coarsely ciliate-margined, 1–3-veined (very rarely a couple of leaves 5-veined)...3

3a. Involucre bracts erect; 4–8 flowers per head...***L. punctata***
3b. Involucre bracts spreading to reflexed; (10) 25–40 flowers per head...***L. squarrosa* var. *glabrata***

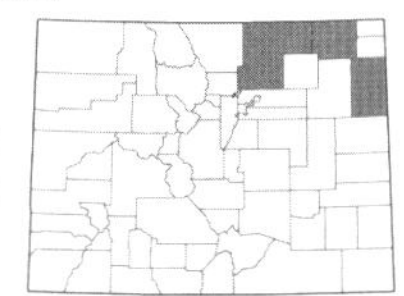

***Liatris lancifolia* (Greene) Kittell**, LANCELEAF BLAZING STAR. Plants (2) 4–8 dm; *leaves* broadly linear, the lower 5-veined, glabrous, 10–30 cm long; *involucre* 7–11 mm high, the bracts erect, acute; *disk flowers* 12–15, ca. 6 mm long, purple, glabrous; *pappus* barbellate, 5 mm long. Uncommon on the eastern plains, 3500–4500 ft. July–Sept. E.

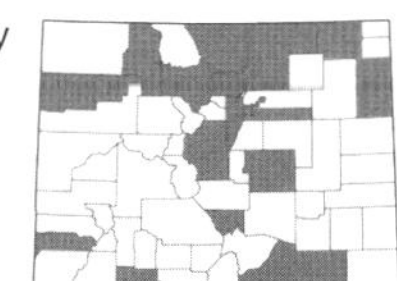

***Liatris ligulistylis* (A. Nelson) K. Schum.**, ROCKY MOUNTAIN BLAZING STAR. Plants 1–6 dm; *leaves* broadly linear, glabrous to hispid on the midvein to densely hairy on both sides, 8–25 cm long; *involucre* 13–20 mm high, the bracts scarious and sometimes broadly lacerate, purple; *disk flowers* 30–70, 9–11 mm long, purple, glabrous; *pappus* barbellate, 8–10 mm long. Uncommon or locally common in wet meadows and along streams, 4000–8000 ft. July–Sept. E/W.

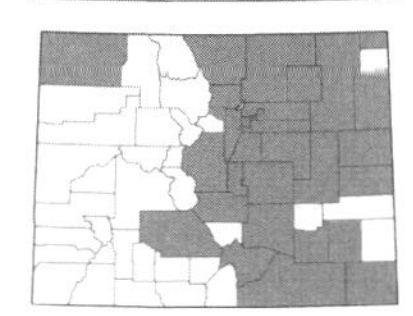

***Liatris punctata* Hook.**, DOTTED BLAZING STAR. Plants 1.5–8 dm; *leaves* linear, 1–3-veined, glabrous with ciliate margins, 8–15 cm long; *involucre* 15–20 mm high, the bracts ciliate or membranous-margined, rounded; *disk flowers* 4–8, 9–12 mm long, purple or rarely white, pilose inside; *pappus* plumose, 9–11 mm long. Common and widespread in open grasslands and meadows, especially common on the eastern plains, 3500–8000 ft. (July) Aug.–Oct. E/W. (Plate 13)

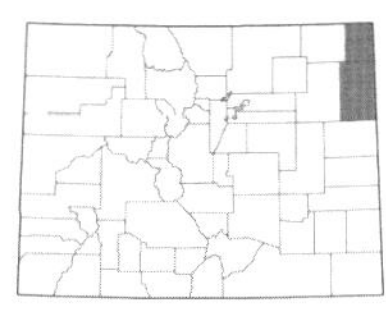

***Liatris squarrosa* (L.) Michx. var. *glabrata* (Rydb.) Gaiser**, SCALY BLAZING STAR. Plants 3–6 dm; *leaves* linear, 1–3-veined, rigid and glabrous, 10–25 cm long; *involucre* 12–30 mm high, the bracts ciliate or membranous-margined; *disk flowers* (10) 25–40, 9–11 mm long, purple, pilose inside; *pappus* plumose, 7–8 mm long. Uncommon on sandy hills of the northeastern plains, 3500–4000 ft. July–Sept. E.

LYGODESMIA D. Don – SKELETONWEED; RUSH-PINK

Perennial, lactiferous herbs; *leaves* simple, alternate; *heads* ligulate; *involucre bracts* in 2 conspicuous series, the outer much shorter than the inner; *receptacle* naked, scabrous; *disk flowers* absent; *ray flowers* pink, lavender, or white; *pappus* of capillary bristles; *achenes* columnar to subcylindrical, ridged or smooth, not beaked. (Tomb, 1980)

1a. Ray flowers less conspicuous, 1–1.5 cm long and about 4 mm wide; involucre bracts 1.3–1.6 cm high, the apex unappendaged; lower leaves about 4 cm long or less; sap milky yellow; often with small, spherical galls on the stem...***L. juncea***
1b. Ray flowers large and showy, 1.6–2.5 cm long and 6–10 mm wide; involucre bracts 1.8–2.5 cm high, the apex appendaged or not; lower leaves 2.5–16 cm long; sap milky white; usually without galls...***L. grandiflora***

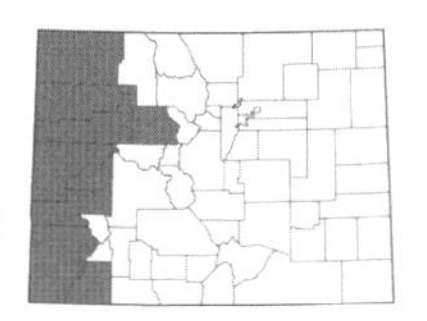

***Lygodesmia grandiflora* (Nutt.) Torr. & A. Gray**, LARGEFLOWER SKELETONWEED. Plants 0.5–5 dm; *leaves* linear, the upper often reduced to scales, 0.5–15 cm long, margins entire; *involucre* 18–25 mm high, the bracts appendaged or not at the tips; *ray flowers* 1.6–2.5 cm long, white, pink, or lavender; *pappus* 10–14 mm long; *achenes* 10–28 mm long. Common in dry, open places, often in sandy or shale soil, 4300–8000 ft. May–July. W. (Plate 13)

There are 3 varieties of *L. grandiflora* in Colorado:

1a. Principal involucre bracts 8–9; heads with 6–12 flowers...**var. *grandiflora***, LARGEFLOWER SKELETONWEED. Common in dry, open places throughout the western counties, often in sandy or shale soil, 4500–8000 ft. May–July. W.
1b. Principal involucre bracts 5; heads with 5 (rarely 7) flowers...2

2a. Involucre bract tip usually unappendaged; leaves 3–6 mm wide...**var. *arizonica* (Tomb) S.L. Welsh**, ARIZONA SKELETONWEED. [*L. arizonica* Tomb]. Infrequent in dry, open places, known from Montezuma Co., 4700–5700 ft. April–June. W.
2b. Involucre bract tip with a small keel-shaped appendage; leaves 1.5–3.5 mm wide...**var. *doloresensis* (Tomb) S.L. Welsh**, DOLORES RIVER SKELETONWEED. [*L. doloresensis* Tomb]. Infrequent in dry, open places in sandy soil, known from Mesa Co., 4300–4900 ft. May–June. W. Endemic.

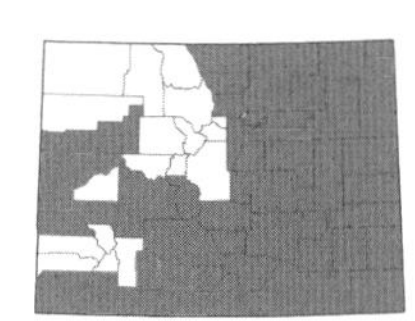

***Lygodesmia juncea* (Pursh) D. Don ex Hook.**, RUSH SKELETONWEED. Plants 1–7 dm; *leaves* linear, 5–30 (60) mm long, margins entire, glabrous, reduced to scales above; *involucre* 13–26 mm high, the bracts not appendaged at the tip; *ray flowers* 1–1.5 cm long, lavender to light pink or rarely white; *pappus* 6–9 mm long; *achenes* 6–10 mm long. Common in dry, open places often in sandy soil, especially common on the eastern plains and outer foothills, 3500–9000 ft. June–Sept. E/W. (Plate 13)

MACHAERANTHERA Nees – TANSY-ASTER

Herbs; *leaves* alternate, simple, deeply pinnatifid or lobed, glandular-stipitate; *heads* radiate; *involucre bracts* imbricate, spreading or reflexed, often bristle-tipped; *receptacle* naked or with a few scales; *disk flowers* yellow; *ray flowers* white, pink, purple, or blue; *pappus* of filiform bristles; *achenes* usually obovate, smooth or with ribs.

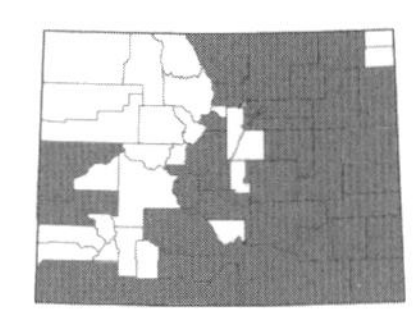

***Machaeranthera tanacetifolia* (Kunth) Nees**, TANSEYLEAF TANSY-ASTER. Taprooted annuals or biennials, 5–100 cm; *leaves* 3–12 cm long, 1–2 pinnate; *involucre bracts* in 3–6 series, 4–11 mm long, tips spreading to reflexed (appressed); *ray flowers* blue or purple; *pappus* 2–8 mm long; *achenes* 2–3.5 (4) mm long. Common in sandy or rocky soil on the plains and dry, open places in valleys, 3400–7000 ft. May–Sept. E/W. (Plate 13)

MADIA Molina – TARWEED

Herbs, more or less glandular and strong-scented; *leaves* simple; *heads* radiate or rarely discoid; *involucre bracts* equal, uniseriate, subherbaceous; *receptacle* with a single series of chaff between the ray and disk flowers; *disk flowers* yellow; *ray flowers* yellow or purplish or absent; *pappus* usually absent; *achenes* usually radially compressed.

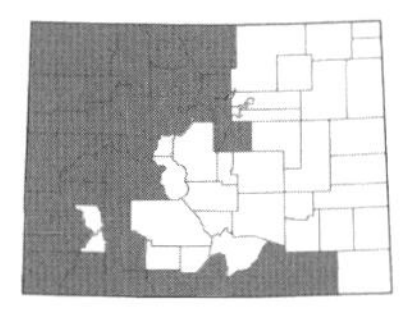

***Madia glomerata* Hook.**, MOUNTAIN TARWEED. Plants 5–120 cm; *leaves* linear to narrowly linear, 2–10 cm long; *heads* usually in tightly grouped glomerules; *involucre bracts* pilose and glandular-hairy; *disk flowers* 3–4.5 mm long, hairy; *ray flowers* 1–3 mm long; *achenes* black. Common in dry, open places, 5500–9600 ft. July–Sept. E/W. Introduced in Colorado.

MALACOTHRIX DC. – DESERT DANDELION

Lactiferous herbs; *leaves* mostly basal, the margins usually toothed to lobed; *heads* ligulate; *involucre bracts* imbricate, narrowly to broadly scarious-margined; *receptacle* naked, glabrous, pubescent, or with setose bristles; *disk flowers* absent; *ray flowers* yellow, white, or pink; *pappus* of capillary bristles, eventually falling as a single unit; *achenes* 10–16 ribbed with 5 of the ribs often more prominent, truncate or tapering toward the base. (Williams, 1957)

1a. Pappus of 15–25 teeth forming a short crown; achenes mostly 1.7–3 mm long with a minutely toothed, white-cartilaginous ring at the summit; receptacle with 3–5 mm long persistent bristles...***M. sonchoides***
1b. Pappus of 12–15 teeth forming a short crown and 0–6 longer bristles; achenes mostly (2.5) 3–4 mm long, with a more conspicuous and deeply toothed cartilaginous ring at the summit; receptacle without bristles or with a few, short and deciduous bristles...***M. torreyi***

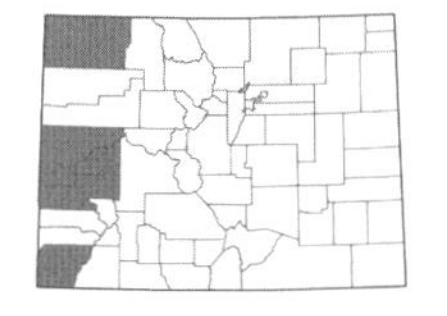

***Malacothrix sonchoides* (Nutt.) Torr. & A. Gray**, SOWTHISTLE DESERT DANDELION. Annuals, 0.5–2.5 (5) dm; *leaves* pinnately lobed, the ultimate segments triangular or oblong with dentate margins, glabrous or arachnoid when young; *involucre* 7–13 mm high; *receptacle* bristly; *ray flowers* 10–16 mm long, yellow; *pappus* of 15–25 teeth; *achenes* 1.7–3 mm long. Found in sandy soil in open, dry places, often in or about sand dunes, 4500–5700 ft. April–June. W.

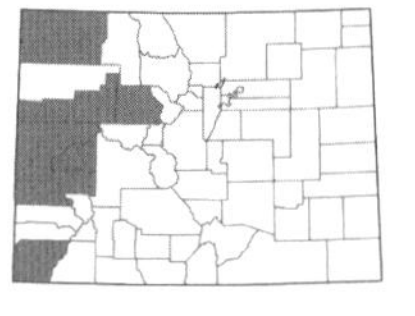

***Malacothrix torreyi* A. Gray**, TORREY'S DESERT DANDELION. [*M. sonchoides* (Nutt.) Torr. & A. Gray var. *torreyi* (A. Gray) E. Williams]. Annuals, 0.5–2.5 (4) dm; *leaves* pinnately lobed, the ultimate segments triangular to linear with toothed margins, glabrous or sparsely arachnoid; *involucre* 8–14 mm high; *receptacle* sparsely bristly or naked; *ray flowers* 14–20 mm long, yellow; *pappus* of 12–15 teeth and 0–6 bristles; *achenes* (2.5) 3–4 mm long, with a prominent ring at the summit. Found in open, dry places, 4800–7000 ft. April–June. W. (Plate 13)

MATRICARIA L. – CHAMOMILE

Annual herbs; *leaves* alternate, pinnatifid; *heads* radiate or discoid; *involucre bracts* subequal in (2) 3–4 (5) series, usually with hyaline-scarious margins; *receptacle* naked; *disk flowers* yellow; *pappus* a short crown or absent; *achenes* laterally compressed to subcylindric, usually nerved on the margins and ventrally and nerveless dorsally.

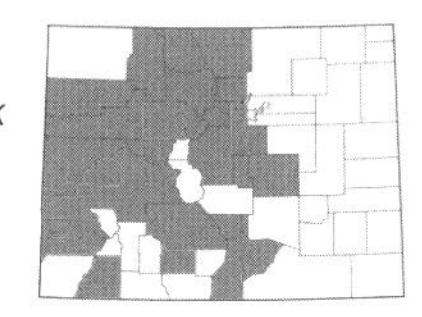

***Matricaria discoidea* DC.**, PINEAPPLE WEED. Plants (1) 4–40 (50) cm, sweet-scented, branching from base; *leaves* up to 6.5 (8.5) cm long; *heads* discoid, usually solitary; *involucre* 2.5–3.8 mm high; *disk flowers* 1–2 mm long, green-yellow, *ray flowers* absent; *achenes* light brown, nerves white, with narrow brown glands extending to or at least near the bottom of the achene. Common along roadsides and in disturbed places, 6000–10,500 ft. May–Sept. E/W. Introduced.

MELAMPODIUM L. – BLACKFOOT

Herbs; *leaves* opposite; *heads* solitary, radiate; *involucre bracts* in 2 series, the outer series of 2–5 more or less equal bracts that are distinct or connate, the inner series each enclosing a ray achene and often expanded upward into a hood; *receptacle* chaffy; *disk flowers* yellow; *ray flowers* yellow or white; *pappus* absent; *achenes* laterally compressed.

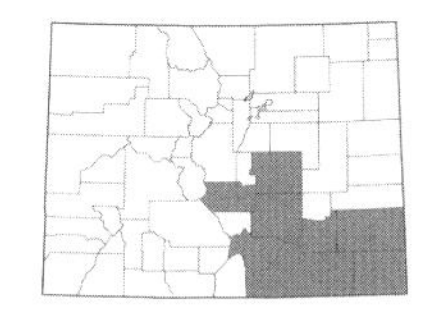

***Melampodium leucanthum* Torr. & A. Gray**, BLACK-FOOT DAISY. [*M. strigosum* Stuessy]. Perennials or subshrubs, 1.2–4 (6) dm; *leaves* linear-oblong, 20–35 (45) mm long, pinnately lobed to entire; *outer involucre bracts* 5–7 mm long, connate; *ray flowers* white, 7–13 mm long; *achenes* 1.5–2.6 mm long. Common in sandy or rocky soil of grasslands and mesas, 3800–7200 ft. May–Aug. E. (Plate 13)

MICROSERIS D. Don – SILVERPUFFS

Annual or perennial herbs; *leaves* alternate; *heads* ligulate; *involucre bracts* herbaceous, subequal to imbricate; *receptacle* naked; *disk flowers* absent; *ray flowers* yellow; *pappus* of 2–many parts (usually 5), usually with a slender, bristle-like awn, or of very narrow bristle-like scales; *achenes* not beaked, prominently 8–10-ribbed or -nerved.

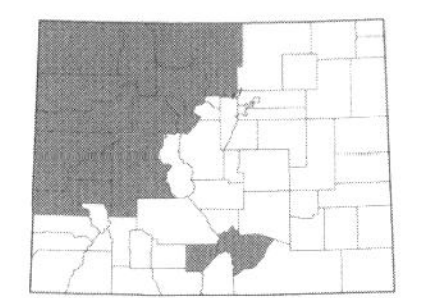

***Microseris nutans* (Hook.) Sch. Bip.**, NODDING SILVERPUFFS. [*Ptilocalais nutans* (Hook.) Greene; *Scorzonella nutans* Hook.]. Perennials, 1–7 dm; *leaves* linear to oblanceolate, 5–30 cm long, glabrous or minutely hairy; *involucre bracts* linear to deltate, usually black-hairy; *ray flowers* 5+ mm beyond phyllaries; *pappus* of plumose bristles intermixed with slender scales, the bristles 4–7 mm long; *achenes* 3.5–8 mm long. Found in mostly open places such as on hillsides and in mountain meadows, 5500–10,000 ft. May–July. E/W.

NOTHOCALAIS (A. Gray) Greene – PRAIRIE-DANDELION; FALSE-AGOSERIS

Perennials, lactiferous plants; *leaves* basal; *heads* solitary, ligulate; *involucre bracts* subequal in 2–3 series; *receptacle* naked; *disk flowers* absent; *ray flowers* yellow; *pappus* of capillary bristles intermixed with attenuate scales; *achenes* cylindric, mostly 8–10-ribbed, beakless.

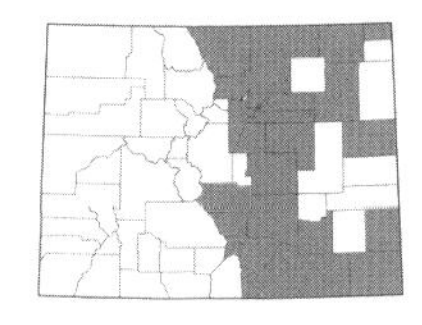

***Nothocalais cuspidata* (Pursh) Greene**, SHARPPOINT PRAIRIE-DANDELION. Plants 7–35 cm; *leaves* narrowly lanceolate, 7–30 cm, the margins entire, weakly crispate and ciliolate; *involucre* 17–27 mm high, the bracts often red-striped; *ray flowers* 15–25 mm long; *achenes* 7–10 mm long. Common in open places of the plains and outer foothills, 3800–7000 ft. April–June. E. (Plate 13)

ONOPORDUM L. – SCOTCH THISTLE; COTTON THISTLE

Biennials, spiny herbs; *leaves* lobed, winged-decurrent on the stem; *heads* discoid; *involucre bracts* imbricate, spine-tipped; *disk flowers* purple, pink, or white; *ray flowers* absent; *pappus* of barbellate bristles; *achenes* 4–5-ribbed, 4-angled.

1a. Leaves tomentose-woolly; involucre bracts tomentose...***O. acanthium***

1b. Leaves glandular-puberulent but not at all tomentose; involucre bracts glabrous...***O. tauricum***

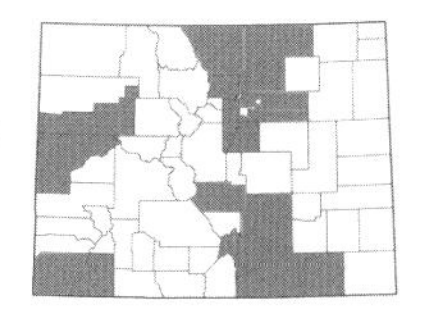

***Onopordum acanthium* L.**, SCOTCH COTTONTHISTLE. Plants to 30 dm; *leaves* 1–5 dm, dentate to shallowly lobed, the ultimate segments triangular, with spines 5–20 mm long, tomentose-woolly; *involucre* tomentose, with spines 2–5 mm long; *disk flowers* 20–25 mm long, purple or pinkish-white; *pappus* 6–9 mm long; *achenes* 4–5 mm long. Locally common in disturbed places and along roadsides, 5200–7000 ft. June–Aug. E/W. Introduced.

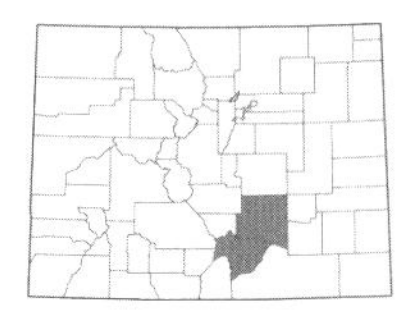

***Onopordum tauricum* Willd.**, BULL THISTLE. Plants to 30 dm; *leaves* 1–2.5 dm, pinnately lobed, the ultimate segments triangular, spiny, glandular-puberulent; *involucre* glabrous, with spines 4 mm long; *disk flowers* 25–30 mm long, purple-pink; *pappus* 8–10 mm long; *achenes* 5–6 mm long. Found in waste places and other disturbed sites and along roadsides, 5500–5800 ft. June–Aug. E/W. Introduced.

OÖNOPSIS Nutt. – FALSE GOLDENWEED

Perennial herbs; *leaves* alternate, simple; *heads* radiate or discoid; *involucre bracts* imbricate in 3–6 series, leafy and herbaceous or mostly chartaceous with a greenish or brownish spot below the tip; *receptacle* naked; *disk flowers* yellow; *ray flowers* yellow, or sometimes absent; *pappus* of rigid capillary bristles; *achenes* prismatic or narrowly turbinate, 5–8-ribbed. (Brown & Nesom, 2006; Hall, 1928)

1a. Ray flowers present…2
1b. Ray flowers absent…3

2a. Heads 2–6; plants 1–4 dm; involucre 17–25 mm high…***O. foliosa***
2b. Heads 1 (rarely 2) per stem; plants 1–1.6 dm; involucre 15–17 mm high…***O. puebloensis***

3a. Leaves wide (10 mm); involucre 17–20 mm high and 18–20 mm wide; plants typically with 1 head per stem (or less often 2), this subtended by large, foliaceous bracts…***O. monocephala***
3b. Leaves narrower (1–7 mm wide); involucre 8–16 mm high and 8–15 mm wide; plants typically with more than 3 heads per stem not subtended by large, foliaceous bracts…4

4a. Leaves very narrow, 1–3 mm wide; disk flowers 7–28; involucre 8–10 mm high and 6–8 mm wide…***O. engelmannii***
4b. Leaves 4–7 mm wide; disk flowers 10–50; involucre12–16 mm high and 10–15 mm wide…***O. wardii***

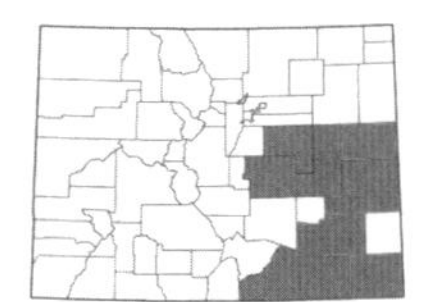

***Oönopsis engelmannii* (A. Gray) Greene**, ENGELMANN'S FALSE GOLDENWEED. Plants 0.6–3 dm; *leaves* linear to linear-oblanceolate, mostly 1–9 cm long; *heads* few to numerous, not subtended by foliaceous bracts; *involucre* 8–10 mm high; *disk flowers* 7–28; *ray flowers* absent; *achenes* 5–6 mm long. Found in open places on the eastern plains, 3700–6200 ft. Aug.–Oct. E.

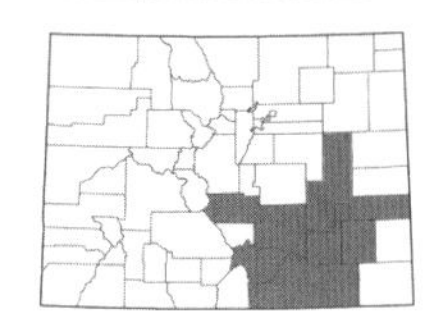

***Oönopsis foliosa* (A. Gray) Greene**, LEAFY FALSE GOLDENWEED. Plants 1–4 dm; *leaves* oblanceolate to oblong-oblanceolate, 2.5–10 cm long; *heads* 2–6, subtended by large foliaceous bracts; *involucre* 17–25 mm high; *disk flowers* 53–150; *ray flowers* 15–30; *achenes* 5–7 mm long. Found on shale and limestone outcroppings, 4000–6000 ft. June–Aug. E.

***Oönopsis monocephala* A. Nelson**, RAVEN RIDGE FALSE GOLDENWEED. [*O. foliosa* (A. Gray) Greene var. *monocephala* (A. Nelson) Kartesz & Gandhi]. Plants 1–4 dm; *leaves* oblanceolate to narrowly obovate, 3–10 cm long; *heads* usually solitary, subtended by large, foliaceous bracts; *involucre* 17–20 mm high; *disk flowers* 50–150; *ray flowers* absent; *achenes* 5–7 mm long. Found on open shale slopes, 6000–6500 ft. June–Aug. E. Endemic.

Oönopsis monocephala and *O. foliosa* can hybridize, creating plants with a gradual transition from discoid to radiate heads. They are referred to as varieties of *O. foliosa* by Brown & Nesom (2006).

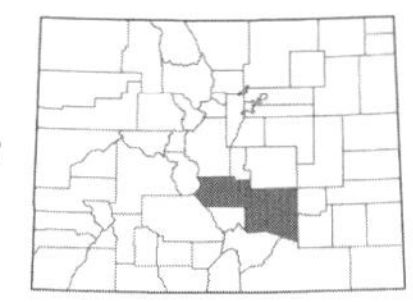

***Oönopsis puebloensis* Brown & Evans**, PUEBLO FALSE GOLDENWEED. Plants 1–1.6 dm; *leaves* linear-lanceolate to linear-oblanceolate; *heads* usually solitary, subtended by foliaceous bracts; *involucre* 15–17 mm high; *ray flowers* 12–30. Uncommon on shale outcroppings, 5000–5500 ft. June–Aug. E. Endemic.

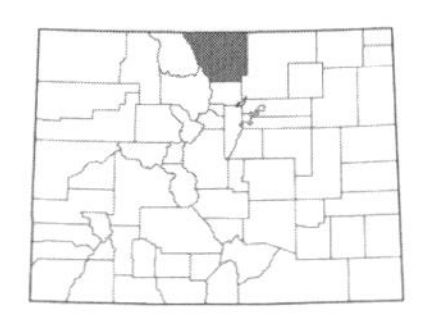

***Oönopsis wardii* (A. Gray) Greene**, WARD'S FALSE GOLDENWEED. Plants 0.5–3 dm; *leaves* oblanceolate to narrowly oblong, 2–12 cm long; *heads* 3–12, not subtended by foliaceous bracts; *involucre* 12–16 mm high; *disk flowers* 10–50; *ray flowers* absent; *achenes* 4.5–7 mm long. Found on open shale or clay slopes, recently discovered in Colorado in Larimer Co. by Holnholz Lakes, 7800–7900 ft. July–Aug. E.

OREOCHRYSUM Rydb. – GOLDENROD

Perennial herbs; *leaves* simple, alternate, entire; *heads* radiate, in a flat-topped terminal cluster; *involucre bracts* not much imbricate, the outer and often middle green and foliaceous, the inner with a chartaceous base and foliaceous tip; *receptacle* naked, alveolate; *disk flowers* yellow; *ray flowers* yellow; *pappus* of capillary bristles; *achenes* 5–7-nerved, glabrous.

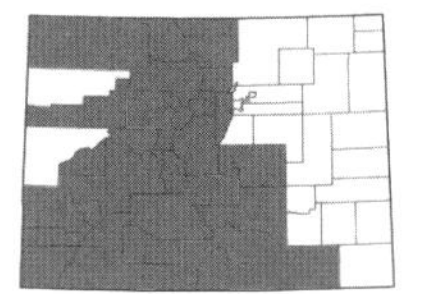

***Oreochrysum parryi* (A. Gray) Rydb.**, PARRY'S GOLDENROD. [*Haplopappus parryi* A. Gray; *Solidago parryi* (A. Gray) Greene]. Plants 1.5–6 (10) dm, from long, slender rhizomes; *lower leaves* mostly 4–15 cm long, oblanceolate to spatulate, 1-nerved; *disk flowers* 25–37; *ray flowers* 12–20; *pappus* persistent, the bristles equal. Common in the mountains on open or wooded slopes and in open meadows, 7800–12,500 ft. July–Sept. E/W.

OXYTENIA Nutt. – COPPER WEED

Subshrubs or shrubs; *leaves* alternate, gland-dotted; *heads* small, discoid; *involucre bracts* imbricate in 2–3 series; *receptacle* chaffy; *disk flowers* pistillate or staminate, the pistillate flowers 5 and peripheral, the staminate flowers 10–25 and central; *ray flowers* absent; *pappus* absent; *achenes* obovoid, obcompressed or weakly 3–4-angled, smooth.

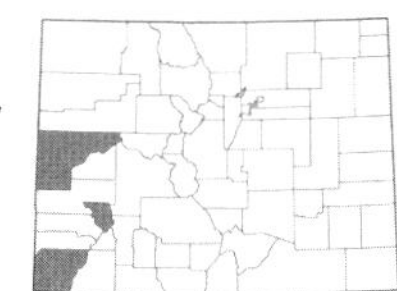

***Oxytenia acerosa* Nutt.**, COPPERWEED. [*Iva acerosa* (Nutt.) R.C. Jackson]. Plants 5–20 dm; *leaves* pinnately divided into 3–7+ lobes or entire distally, the ultimate divisions linear to filiform, usually sericeous to minutely strigose, sometimes glabrous; *disk flowers* whitish, 2.5–3 mm long; *achenes* 1.5–2.5 mm long. Locally common in moist, alkaline soil such as in seeps, along creeks, and in the bottom of dry washes, 4500–6500 ft. (July) Aug.–Oct. W.

PACKERA Á. Löve & D. Löve – GROUNDSEL; RAGWORT

Perennials, biennial, or winter annuals; *leaves* alternate; *heads* radiate or discoid; *involucre bracts* uniseriate, often with some smaller bracteoles at the base of the involucre, without black-tipped apices or rarely so; *receptacle* naked; *disk flowers* yellow to orange or reddish; *ray flowers* yellow to orange, or absent; *pappus* of capillary bristles; *achenes* 5- to 10-nerved.

1a. Stems and involucre bracts anthocyanic (purplish-red); stems with one terminal head; leaves reniform or subreniform with crenate or rarely wavy margins, typically anthocyanic below especially on the veins and margin…***P. porteri***

1b. Stems and involucre bracts usually not anthocyanic, occasionally anthocyanic at just the tips of the involucre bracts; usually more than one head per stem; leaves various but unlike the above…2

2a. Basal and stem leaves nearly all deeply and evenly pinnately dissected or runcinate-pinnatifid…3

2b. Basal leaves with entire, toothed, or crenate margins (in *P. plattensis* and *P. tridenticulata* a few basal leaves may have pinnatisect margins, but the majority of the leaves will have entire or merely toothed margins), stem leaves entire to pinnatisect or sublyrate…4

3a. Leaves and stem loose-tomentose (at least on the lower leaf side), the dissected lobes rounded at the tip and entire…***P. fendleri***

3b. Leaves and stem more or less glabrous, the dissected lobes either acute and entire or toothed…***P. multilobata***

4a. Plants essentially scapose, the stem leaves reduced to small, linear bracts; low (0.2–2 dm tall) plants…***P. werneriifolia***

4b. Plants with at least a few well-developed stem leaves; plants 1 dm or more tall…5

5a. Basal and stem leaves more or less densely and evenly whitish-tomentose (at least below), often subglabrescent on the upper surface, the margins entire or sometimes the upper weakly dentate…***P. cana***

5b. Leaves thinly tomentose or densely tomentose along the midvein, at least some of the leaves with pinnatifid, toothed, crenate, or lyrate margins…6

6a. Ray flowers absent…***P. debilis***

6b. Ray flowers present…7

7a. Basal leaves lanceolate or narrowly oblanceolate (about 3 times longer than wide), usually 3-toothed at the apex, tapering at the base…8

7b. Basal leaves ovate to orbicular or oblong, usually not 3-toothed at the apex, cordate, truncate, or tapering at the base…10

8a. Leaves mostly permanently but loosely tomentose, especially below, also with tomentum in the leaf axils…***P. neomexicana* var. *mutabilis***

8b. Leaves glabrous, or with some tomentum in the leaf axils…9

9a. Smaller bracteoles at the base of the involucre conspicuous and numerous but small; stem leaves dentate to pinnatifid; plants typically of shortgrass prairies or sagebrush scrub habitats to 8500 ft in elevation…***P. tridenticulata***

9b. Smaller bracteoles at the base of the involucre few and inconspicuous; stem leaves more or less all pinnatifid; plants typically of wet mountain meadows and streambanks, and open woodlands from 8000–10,700 ft in elevation…***P. paupercula***

10a. Basal leaves with mostly entire or wavy margins, or just a few shallowly crenate; plants glabrous throughout or occasionally with small tufts of tomentum in the leaf axils; rays yellow to orange; plants of mountain meadows and alpine...***P. crocata***
10b. Basal leaves with mostly crenate, dentate, or even pinnatifid margins; plants glabrous or often tomentose on the stem and in the leaf axils, often the leaves tomentose as well; rays yellow; plants of various habitats...11

11a. Leaves densely floccose-tomentose below along the midvein and petiole; at least some basal leaves usually deeply pinnatifid to sublyrate; pappus 6–7.5 mm long; plants of the eastern plains and foothills (to 7500 ft in elevation)...***P. plattensis***
11b. Leaves unlike the above; pappus 3–6 mm long; plants variously distributed, but absent from the eastern plains...12

12a. Leaves relatively thick and turgid, often hairy or sometimes glabrous; stem leaves usually lobed just at the base, otherwise entire and toothed at the apex; basal leaves tapering or subcordate bases...***P. streptanthifolia***
12b. Leaves not thick and turgid, glabrous; stem leaves usually pinnatifid; basal leaves with truncate, subcordate, or cordate bases...***P. pseudaurea***

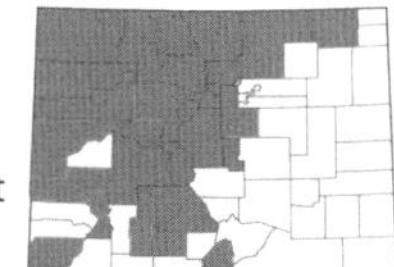

Packera cana **(Hook.) W.A. Weber & Á. Löve**, WOOLLY GROUNDSEL. [*Senecio canus* Hook.]. Plants 1–4 dm, white-tomentose; *leaves* ovate to elliptic or lanceolate, 2.5–5 × 0.4–3 cm, white-tomentose, margins entire or the upper weakly dentate; *heads* 6–15 per stem; *involucre bracts* 4–8 mm long; *ray flowers* 6–13 mm long, yellow; *achenes* glabrous. Common on rocky subalpine slopes, in forest openings, and in open places on the high plains and in sagebrush meadows, 3900–12,600 ft. June–Aug. E/W. (Plate 14)

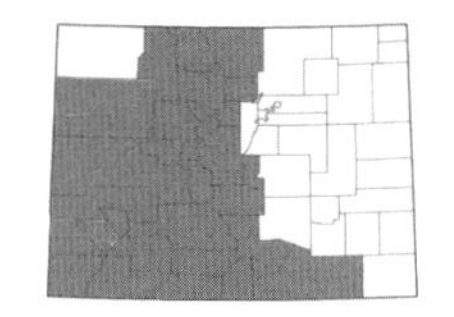

Packera crocata **(Rydb.) W.A. Weber & Á. Löve**, SAFFRON RAGWORT. [*P. dimorphophylla* (Greene) W.A. Weber & Á. Löve; *Senecio crocatus* Rydb.]. Plants 1–9 dm, glabrous or sometimes with tufts of tomentum in leaf axils; *basal leaves* ovate to oblanceolate with tapering bases, 2–9 × 1–4 cm; *cauline leaves* pinnate-lobed or lyrate; *heads* 1–15 per stem; *involucre bracts* 5–8 mm long; *ray flowers* 5–8 mm long, yellow to orange; *achenes* glabrous. Common in moist meadows and on rocky subalpine slopes, 8500–13,000 ft. June–Aug. E/W. (Plate 14)

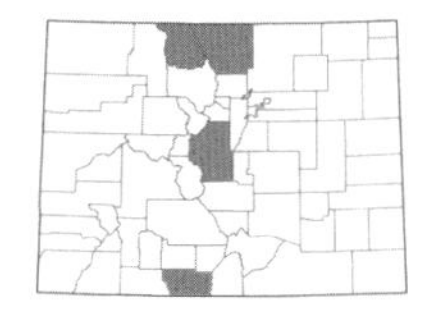

Packera debilis **(Nutt.) W.A. Weber & Á. Löve**, WEAK GROUNDSEL. [*Senecio debilis* Nutt.]. Plants 1–5 dm, glabrous or lightly floccose-tomentose; *basal leaves* elliptic to ovate, 2–5 × 1–3 cm; *cauline leaves* pinnatifid; *heads* 6–25 per stem; *involucre bracts* 6–8 mm long; *ray flowers* absent; *achenes* glabrous. Infrequent in moist meadows and fens and along creeks, 7500–9500 ft. June–Aug. E.

Often misidentified as *P. pauciflora*, which does not occur in Colorado.

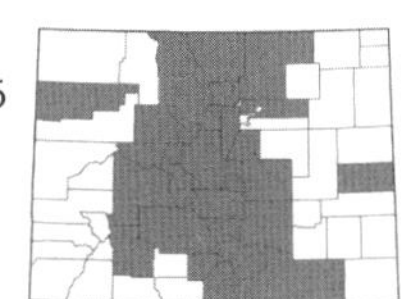

Packera fendleri **(A. Gray) W.A. Weber & Á. Löve**, FENDLER'S RAGWORT. [*Senecio fendleri* A. Gray]. Plants 1–4 dm, floccose-tomentose; *leaves* lanceolate to oblanceolate with pinnatifid margins, 3–6 × 1–3 cm; *heads* 6–25 per stem; *involucre bracts* 5–7 mm long; *ray flowers* 5–7 mm long, yellow; *achenes* glabrous. Common in rocky soil of the foothills and in forest openings and meadows, occasionally found on the eastern plains, 5000–11,500 ft. May–Sept. E/W. (Plate 14)

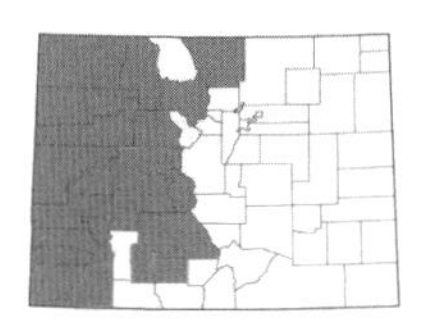

Packera multilobata **(Torr. & A. Gray ex A. Gray) W.A. Weber & Á. Löve**, LOBELEAF GROUNDSEL. [*Senecio multilobatus* Torr. & A. Gray ex A. Gray]. Plants 1–5 (9) dm, thinly tomentose; *leaves* obovate to oblanceolate, deeply pinnatifid or lyrate, 4–8 × 1–3 cm; *heads* 10–30 per stem; *involucre bracts* 5–10 mm long; *ray flowers* 7–10 mm long, yellow; *achenes* glabrous or hirtellous on the angles. Common in mostly dry soil of sagebrush meadows, pinyon-juniper woodlands, 5000–9700 ft. May–July. W.

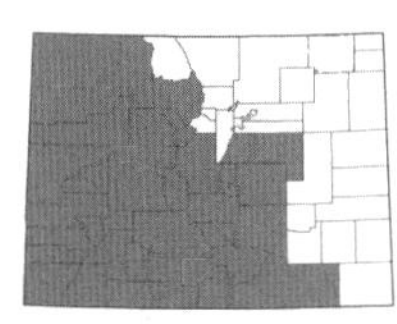

Packera neomexicana **(A. Gray) W.A. Weber & Á. Löve var. *mutabilis* (Greene) W.A. Weber & Á. Löve**, NEW MEXICO GROUNDSEL. [*Senecio neomexicanus* A. Gray var. *mutabilis* (Greene) T.M. Barkl.]. Plants 2–6 dm, tomentose to glabrate; *basal leaves* lanceolate to oblanceolate, 2–6 × 1–3 cm, loosely tomentose, dentate to crenate; *cauline leaves* entire to shallowly toothed; *heads* 3–20 per stem; *involucre bracts* 4–7 mm long; *ray flowers* 4–10 mm long, yellow; *achenes* usually hirtellous on the angles. Common in meadows and open forests, 6000–10,500 ft. May–Aug. E/W.

Hybridizes with *P. tridenticulata* in southcentral Colorado near Alamosa. Specimens of *P. tridenticulata* from the eastern plains can resemble *P. neomexicana* var. *mutabilis* but are completely glabrous.

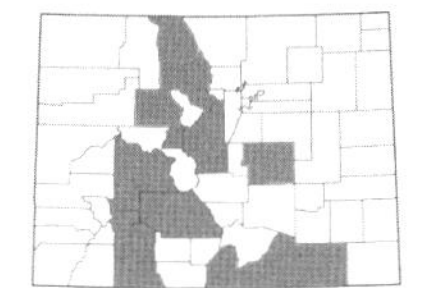

Packera paupercula **(Michx.) Á. Löve & D. Löve**, BALSAM GROUNDSEL. [*Senecio pauperculus* Michx.]. Plants 2–5 dm, lightly tomentose when young; *basal leaves* lanceolate to oblanceolate, 3–6 × 1–2 cm, glabrate; *cauline leaves* usually pinnatifid; *heads* 2–12 per stem; *involucre bracts* 5–9 mm long; *ray flowers* 5–10 mm long, yellow; *achenes* glabrous or occasionally hirtellous on the angles. Infrequent in moist meadows and along streams and in open forests, 7500–10,700 ft. July–Aug. E/W.

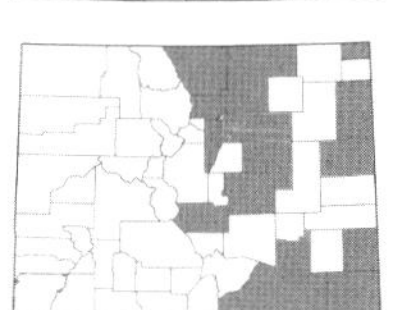

Packera plattensis **(Nutt.) W.A. Weber & Á. Löve**, PRAIRIE GROUNDSEL. [*Senecio plattensis* Nutt.]. Plants 2–6 dm, floccose-tomentose; *basal leaves* ovate to elliptic, subentire to crenate, dentate, or sublyrate, 2–7 × 1–3 cm; *cauline leaves* pinnatifid to lyrate; *heads* 6–20 per stem; *involucre bracts* 5–7 mm long; *ray flowers* 9–10 mm long, yellow; *achenes* hirtellous on the angles or occasionally glabrous. Common on the eastern plains and foothills, 3400–7500 ft. May–June. E.

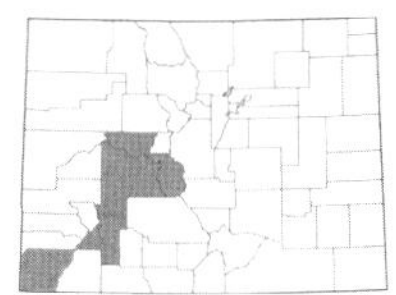

Packera porteri **(Greene) C. Jeffrey**, PORTER'S GROUNDSEL. [*Senecio porteri* Greene]. Plants 0.3–1 dm, glabrous and often anthocyanic; *leaves* reniform, crenate to wavy, 0.5–1.5 × 0.5–2.5 cm; *heads* solitary; *involucre bracts* 5–7 mm long, usually anthocyanic; *ray flowers* 8–10 mm long, yellow; *achenes* glabrous. Infrequent on rocky alpine slopes and ridges, 11,000–13,500 ft. July–Aug. E/W. Endemic.

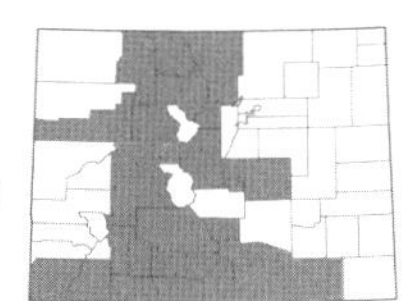

Packera pseudaurea **(Rydb.) W.A. Weber & Á. Löve var. *flavula*** **(Greene) W.A. Weber & Á. Löve**, FALSEGOLD GROUNDSEL. [*Senecio pseudaureus* Rydb. var. *flavulus* (Greene) Greenm.]. Plants 3–7 dm, glabrous or sometimes with tomentum in the leaf axils; *basal leaves* ovate to elliptic, serrate or dentate, 2–10 × 2–6 cm; *cauline leaves* pinnatifid; *heads* 5–30 per stem; *involucre bracts* 3–8 mm long; *ray flowers* 6–10 mm long, yellow; *achenes* glabrous. Common in moist meadows, along streams, and in forest openings, 5600–10,500 ft. May–July. E/W.

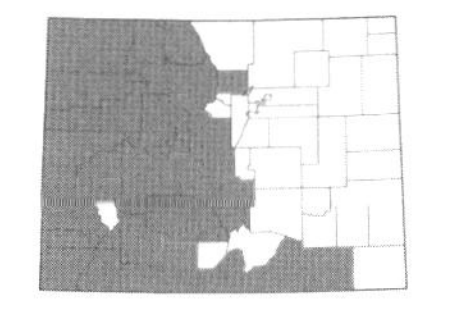

Packera streptanthifolia **(Greene) W.A. Weber & Á. Löve**, ROCKY MOUNTAIN GROUNDSEL. [*Senecio streptanthifolius* Greene]. Plants 1–7 dm, glabrous or lightly tomentose; *basal leaves* oblong to ovate or orbicular, with dentate to crenate margins, 1–6 × 1–5 cm; *cauline leaves* pinnatifid at the base; *heads* 6–20 per stem; *involucre bracts* 4–8 mm long; *ray flowers* 5–10 mm long, yellow; *achenes* glabrous. Common in a wide range of habitats from wooded slopes and mountain meadows to rocky, dry hillsides, 6500–11,500 ft. June–Aug. E/W.

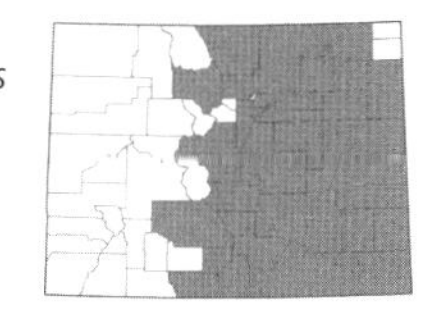

Packera tridenticulata **(Rydb.) W.A. Weber & Á. Löve**, THREETOOTH RAGWORT. [*Senecio tridenticulatus* Rydb.]. Plants 1–3 dm, glabrous or with tufts of tomentum in the leaf axils; *leaves* lanceolate to oblanceolate, entire or dentate at the apex, 2–4 × 0.5–1.5 cm; *heads* 4–15; *involucre bracts* 6–10 mm long; *ray flowers* 5–8 mm long, yellow; *achenes* glabrous or lightly hirtellous on the angles. Common in dry gravelly or sandy soils, 3500–8500 ft. May–July. E/W.

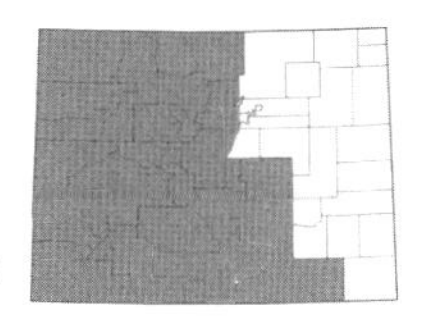

Packera werneriifolia **(A. Gray) W.A. Weber & Á. Löve**, HOARY GROUNDSEL. [*P. mancosana* Yeatts, B. Schneid., & Al Schneid.; *Senecio werneriifolius* (A. Gray) A. Gray]. Plants 0.2–2 dm, glabrous or thinly tomentose; *basal leaves* lanceolate, elliptic, or suborbicular, 1.5–4 × 0.5–2.5 cm; *cauline leaves* mostly reduced to linear bracts; *heads* 1–5; *involucre bracts* 6–10 mm long; *ray flowers* 5–10 mm long, yellow to orange; *achenes* glabrous. Common in subalpine and alpine meadows, forest openings, and on dry, rocky slopes, 6500–14,100 ft. June–Aug. E/W. (Plate 14)

PALAFOXIA Lag. – PALAFOX

Annual or perennial herbs; *leaves* alternate, firm-membranous, the margins entire; *heads* discoid or radiate in 3 species; *involucre bracts* subequal in 2–3 series, thickish to membranous, often with purplish margins; *receptacle* naked; *disk flowers* white to purple or violet; *ray flowers* pink or pale to dark violet, or absent; *pappus* of 4–12 scarious scales; *achenes* 4-angled. (Turner & Morris, 1976)

1a. Ray flowers present; pappus of ray flowers distinctly shorter than that of disk flowers (that of ray flowers less than 1 mm long and that of disk flowers 7–9 mm long)...***P. sphacelata***

1b. Ray flowers absent; pappus of 8 scarious scales, all the same height...***P. rosea***

Palafoxia rosea **(Bush) Cory**, ROSY PALAFOX. Annuals, 1–5 dm; *leaves* linear-lanceolate, 3–6 cm long; *involucre* usually shortly hairy and glandular-stipitate; *disk flowers* 7–10 mm long; *ray flowers* absent; *pappus* of 8 scarious scales 3–8 mm long. Found in sandy soil of the southeastern plains, 4000–5500 ft. June–Sept. E.

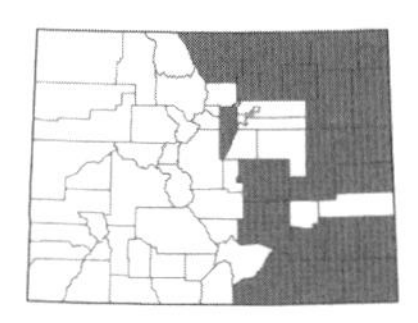

***Palafoxia sphacelata* (Nutt. ex Torr.) Cory**, OTHAKE. Annuals, 1–9 dm; *leaves* lanceolate, 3–9 cm long; *involucre* shortly hairy and/or glandular-stipitate; *disk flowers* 10–14 mm long; *ray flowers* 12–25 mm long; *pappus* of ray flowers shorter than that of disk flowers, 1 mm (ray flowers) or 7–9 mm (disk flowers) long. Common in sandy soil of the eastern plains, 3500–5700 ft. July–Sept. E.

PARTHENIUM L. – FEVERFEW

Herbs or shrubs; *leaves* alternate; *heads* small, inconspicuously radiate; *involucre bracts* in 2 series of 5 each, scarious with a greenish tip; *receptacle* chaffy; *disk flowers* staminate; *ray flowers* white; *pappus* of awns or scales, or absent; *achenes* of ray flowers flattened at right angles to the radius of the head, the achene and attached 2 sterile florets and bracts falling together. (Rollins, 1950)

1a. Short ray flowers present, 1–2 mm long...***P. ligulatum***
1b. Ray flowers absent...***P. alpinum***

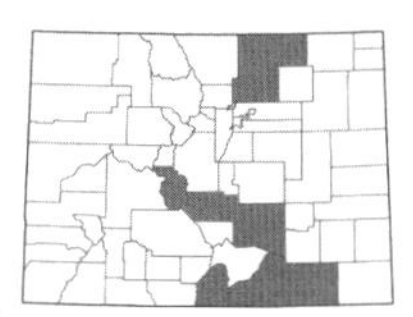

***Parthenium alpinum* (Nutt.) Torr. & A. Gray**, ALPINE FEVERFEW. [*Bolophyta alpina* Nutt.]. Perennials, 1–2 cm, forming mats; *leaves* spatulate to oblanceolate, 0.6–2 (3.5) cm long, closely sericeous and inconspicuously gland-dotted; *heads* solitary; *disk flowers* 18–30; *pistillate flowers* 5–8, lacking a ligule. Rare or easily overlooked on ridges and open shale bluffs, usually with other cushion plants, 4800–5800 ft. April–May. E. (Plate 14)

There are two very morphologically similar but geographically distinct varieties of *P. alpinum* in Colorado:

1a. Heads shortly pedunculate to sessile, the peduncles less than 1 cm in length...**var. *alpinum***, ALPINE FEVERFEW. Known from Weld Co., 5400–5800 ft.
1b. Heads pedunculate, the peduncles 1–3 cm long...**var. *tetraneuris* (Barneby) Rollins**, ARKANSAS RIVER FEVERFEW. [*Bolophyta tetraneuris* (Barneby) W.A. Weber; *Parthenium tetraneuris* Barneby]. Known from Chaffee, Costilla, Fremont, Las Animas, and Pueblo cos., 4800–5600 ft. Endemic.

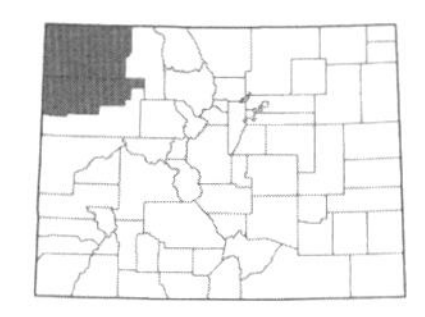

***Parthenium ligulatum* (M.E. Jones) Barneby**, COLORADO FEVERFEW. [*Bolophyta ligulata* (M.E. Jones) W.A. Weber]. Perennials, 1–2 cm, forming mats; *leaves* spatulate to oblanceolate, 0.5–2 cm long, closely sericeous and obscurely gland-dotted; *heads* solitary; *disk flowers* 15–25; *pistillate flowers* 5, with a short ligule 1–2 mm long. Rare or easily overlooked on barren shale knolls, 5400–6500 ft. April–May. W.

PECTIS L. – CINCHWEED

Herbs; *leaves* opposite, simple and entire, punctate; *heads* radiate; *involucre bracts* uniseriate; *receptacle* naked; *disk flowers* yellow; *ray flowers* yellow; *pappus* of scales, awns, or bristles, or reduced to an irregular crown, that of ray flowers often differing from that of the disk flowers; *achenes* cylindrical, several-ribbed, black.

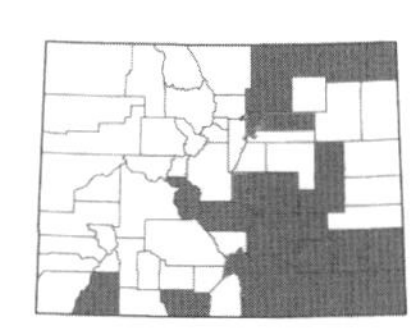

Pectis angustifolia* Torr. var. *angustifolia, LEMONSCENT. Annuals, 1–20 cm, with yellowish oil glands, lemon-scented; *leaves* 10–45 mm long, linear, crowded toward the tips of the branches; *involucre* 4–5 mm high, the bracts with multiple oil glands; *disk flowers* 2.5–3.5 mm long; *ray flowers* 8, 3–5 mm long; *achenes* 2.5–4 mm long. Found in dry, open places, usually in sandy soil, 3500–7800 ft. June–Sept. E/W.

PERICOME A. Gray – PERICOME

Perennial herbs or subshrubs; *leaves* opposite, simple; *heads* discoid; *involucre bracts* essentially uniseriate, subherbaceous, fused for half their length or more; *receptacle* naked; *disk flowers* yellow; *ray flowers* absent; *pappus* a crown of hyaline scales, sometimes with 1–2 bristles; *achenes* black, radially compressed, ciliate on the margins, nerveless.

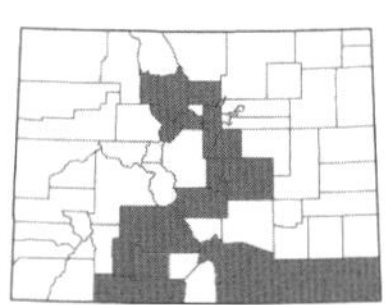

***Pericome caudata* A. Gray**, MOUNTAIN TAIL-LEAF. Plants mostly 6–15 dm; *leaves* 3–12 cm long, mostly deltate-hastate, scabrous-puberulent and/or glandular to suglabrous, usually conspicuously gland-dotted, entire or irregularly toothed; *involucre bracts* 4.5–7 mm long, hairy at tips; *disk flowers* 3–5 mm long; *achenes* 3.5–5 mm long. Found on open, rocky hillsides, 4400–10,000 ft. (July) Aug.–Oct. E/W.

PETASITES Mill. – SWEET COLTSFOOT

Imperfectly dioecious, perennial herbs; *leaves* mostly basal, long-petioled, cauline leaves scale-like; *heads* discoid or weakly radiate; *involucre bracts* uniseriate, equal; *receptacle* naked; *staminate flowers* discoid, many in staminate heads, 0-few in the center of pistillate heads; *pistillate flowers* fertile, corollas either discoid or sometimes the marginal ones ligulate, white to pale yellow or pale pink; *pappus* of capillary bristles; *achenes* linear, 5- to 10-ribbed.

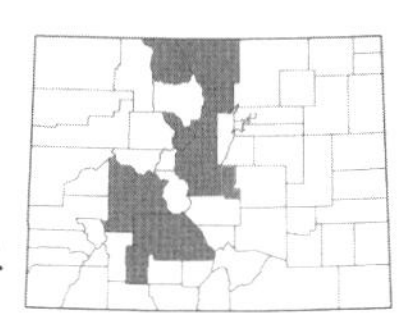

***Petasites frigidus* (L.) Fries var. *sagittatus* (Banks ex Pursh) Cherniawsky**, ARROWLEAF SWEET COLTSFOOT. [*P. sagittatus* (Banks ex Pursh) A. Gray]. Plants from rhizomes; *leaves* 2–34 cm long, sagittate or deltoid to oblong, toothed, usually densely tomentose below; *staminate flowers* 1.1–7.7 mm long; *pistillate flowers* 0.6–5.4 mm long; *pappus* to 17.3 mm long. Uncommon in wet marshes and meadows, along creeks and streams, and in ditches, 7500–10,600 ft. May–June. E/W.

The flowers usually appear before the emergence of the basal leaves.

PETRADORIA Greene – ROCK GOLDENROD

Suffrutescent herbs; *leaves* alternate, entire, the margins finely scabrous; *heads* radiate (ours); *involucre bracts* imbricate in several series and aligned in vertical ranks, keeled, firm; *receptacle* small, naked; *disk flowers* functionally staminate, (1) 2–4 (5), yellow; *ray flowers* yellow; *pappus* of capillary bristles; *achenes* inconspicuously 6–9-nerved.

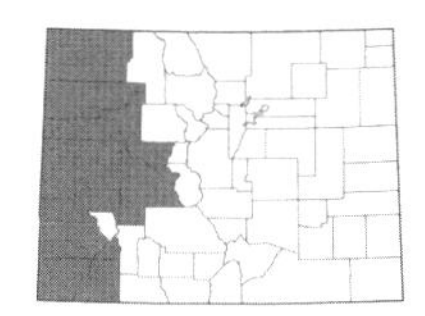

Petradoria pumila* (Nutt.) Greene ssp. *pumila, GRASSY ROCK GOLDENROD. [*Solidago petradoria* S.F. Blake]. Plants to 3 dm; *leaves* broadly linear to oblanceolate, 2–12 mm long, usually 3–5-veined, the margins scabrous; *heads* dense, in flat-topped clusters; *involucres* 6–9.5 mm high; *disk flowers* 4.5–6.2 mm long; *ray flowers* (1) 2–3, 2.5–4.5 mm long; *achenes* 4–5 mm long. Found in open, dry places, often with pinyon-juniper, 5700–8300 ft. June–Aug. W.

PICRADENIOPSIS Rydb. – BAHIA

Perennial herbs; *leaves* opposite, palmately to pinnately dissected with the lobes often further divided; *heads* radiate; *involucre bracts* subequal in 2 more or less distinct series, the outer keeled; *receptacle* naked; *disk flowers* yellow; *ray flowers* yellow; *pappus* a crown of about 8 scales; *achenes* narrowly obpyramidal. (Ellison, 1964; Steussy et al., 1973)

1a. Achenes conspicuously gland-dotted; pappus scales more or less ovate with a short midrib becoming obscured upward and rarely exserted as a bristle...***P. oppositifolia***

1b. Achenes hairy, not gland-dotted; pappus scales lanceolate with a well-developed midrib that often extends into a prominent bristle at the tip...***P. woodhousei***

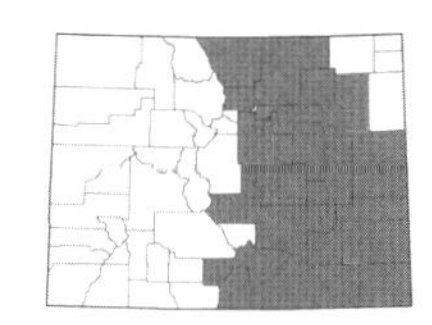

***Picradeniopsis oppositifolia* (Nutt.) Rydb.**, OPPOSITELEAF BAHIA. [*Bahia oppositifolia* (Nutt.) DC.]. Plants 0.3–2 dm; *leaves* linear to lanceolate, 1–3 cm long, hairy and gland-dotted; *involucres* 5–7 mm; *disk flowers* 3.5–5 mm long; *ray flowers* 3–5 mm long; *pappus* ovate with a short midrib becoming obscured upward and rarely exserted as a bristle, 0.5–1.5 mm long; *achenes* 3–5 mm long, gland-dotted. Common in sandy or gravelly soil of open places on the plains and outer foothills, 3500–7000 ft. June–Sept. (Oct.). E.

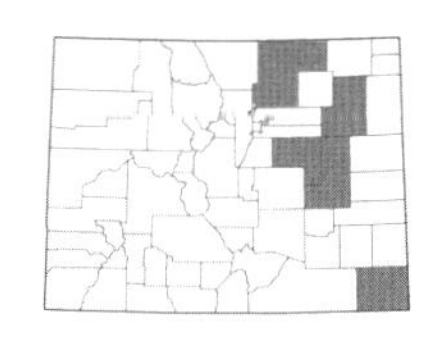

***Picradeniopsis woodhousei* (A. Gray) Rydb.**, WOODHOUSE'S BAHIA. Plants 0.3–2 dm; *leaves* linear to lanceolate, 0.8–2.5 cm long, hairy and gland-dotted; *involucres* 5–7 mm long; *disk flowers* 3–3.5 mm; *ray flowers* 2–5 mm long; *pappus* lanceolate or linear, with a well-developed midrib that is often extended into a prominent bristle; *achenes* 3–4 mm long, usually hairy and not gland-dotted. Uncommon in sandy soil of the plains, 4000–5200 ft. June–Sept. E.

PLATYSCHKUHRIA Rydb. – BASIN DAISY

Perennial herbs; *leaves* alternate or sometimes opposite above; *heads* radiate; *involucre bracts* equal or subequal in 2 series, more or less herbaceous; *receptacle* naked; *disk flowers* yellow; *ray flowers* yellow; *pappus* of 8–16 scales with lacerate margins and midribs sometimes excurrent as a short awn; *achenes* narrowly obpyramidal.

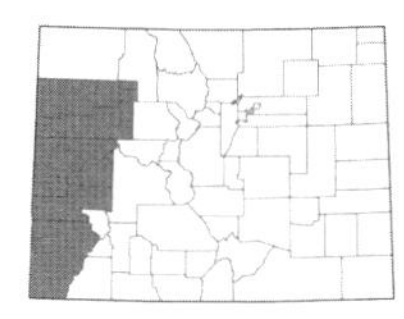

***Platyschkuhria integrifolia* (A. Gray) Rydb.**, BASIN DAISY. [*Bahia nudicaulis* A. Gray]. Plants to 5 dm; *leaves* lanceolate or sometimes ovate, 2–10 cm × 5–35 mm, the margins entire, gland-dotted and sparsely to densely scabrous; *involucres* 9–12 × 12–30 mm; *disk flowers* 3–7 mm long; *ray flowers* 6–15 mm long; *pappus* 0.5–4 mm long; *achenes* (1) 3–5 (8) mm long. Found on shale slopes, in clay soil, and on sandy mesas, 4200–6000 ft. May–June. W.

There are two varieties of *P. integrifolia* in Colorado:

1a. Leaves mostly basal, or extending up the stem but never to the tip; disk flowers (31) 39–82...**var. *desertorum* (M.E. Jones) W. Ellison**

1b. Leaves extending up the stem to the tip (rarely only midway); disk flowers 25–58...**var. *oblongifolia* (A. Gray) W. Ellison**

PRENANTHELLA Rydb. – PRENANTHELLA

Annual herbs; *leaves* alternate, sessile; *heads* ligulate; *involucre bracts* in 2 series, the 1–2 outer bracts minute and few, the 3–4 inner bracts subequal and linear; *receptacle* naked; *disk flowers* absent; *ray flowers* light pink; *pappus* of capillary bristles, connate at the base and falling as a single unit; *achenes* columnar, truncate at both ends, 5-ridged.

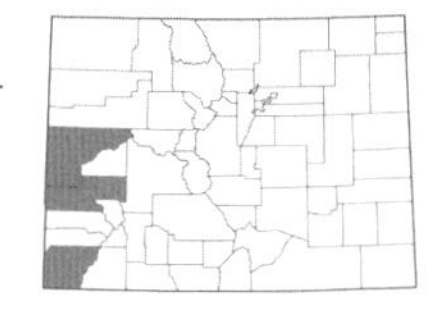

***Prenanthella exigua* (A. Gray) Rydb.**, BRIGHT-WHITE. [*Lygodesmia exigua* (A. Gray) A. Gray]. Plants 5–30 cm, with milky sap; *leaves* spoon-shaped to oblanceolate, 1–3 cm long, the tips often spine-tipped, reduced above; *involucre bracts* in 2 series, the outer 1 mm long, the inner 3–5 mm long; *ray flowers* 1.5–2 mm long; *achenes* 2.5–3.5 mm long. Uncommon in sandy soil of washes and canyons, often found with sagebrush and pinyon-juniper, 4600–5500 ft. April–June. W.

PRENANTHES L. – RATTLESNAKE-ROOT; WHITE LETTUCE

Perennial, lactiferous herbs; *leaves* alternate, entire to dentate; *heads* ligulate; *involucre bracts* in 1–2 series and nearly all equal; *receptacle* naked; *disk flowers* absent; *ray flowers* pink to lavender or cream; *pappus* of capillary bristles; *achenes* cylindric or oblong, weakly angled, the apex not beaked, indistinctly ribbed.

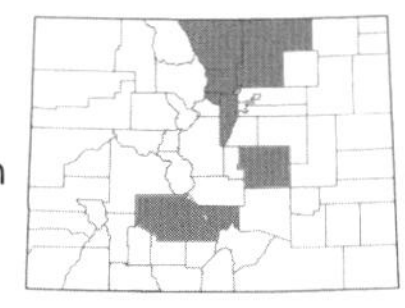

***Prenanthes racemosa* Michx.**, WHITE LETTUCE. Plants 3–17.5 dm, from tuberous, fusiform taproots; *leaves* oblanceolate to spatulate, 4–25 cm long, coriaceous, glabrous, the petioles winged; *involucre bracts* 7–14, green to purple, 10–12 mm long; *ray flowers* 7–13 mm long; *pappus* 6–7 mm long; *achenes* 5–6 mm long. Uncommon in moist places such as bogs, fens, and along streams, 5500–10,000 ft. Aug.–Sept. E.

PRIONOPSIS Nutt. – SPANISH GOLD

Annual herbs; *leaves* sessile, spinulose-dentate to serrate; *heads* radiate; *involucre bracts* imbricate in 4 or 5 series, the outermost squarrose-spreading; *receptacle* naked; *disk flowers* yellow; *ray flowers* yellow; *pappus* of numerous rigid bristles, yellowish to reddish-brown; *achenes* laterally compressed, ellipsoid to oblong.

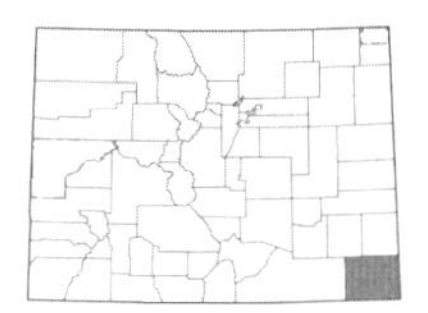

***Prionopsis ciliata* (Nutt.) Nutt.**, GOLDENWEED. [*Grindelia papposa* G.L. Nesom & Suh; *Haplopappus ciliatus* (Nutt.) DC.]. Plants 3–8 (10) dm, glabrous; *cauline leaves* oblong to narrowly obovate, 3–5 (8) cm long, dentate, glabrous, obscurely gland-dotted or not; *disk flowers* ca. 7 mm long; *ray flowers* 12–15+ mm long; *pappus* ca. 7 mm long; *achenes* 2–4 mm long. Found in open sandy or rocky places along the Cimarron River, 3800–4300 ft. Aug.–Oct. E.

Similar in appearance to *Grindelia* but without resinous-punctate leaves and phyllaries.

PSEUDOGNAPHALIUM Kirp. – CUDWEED

Annual to perennial (ours) herbs, generally white-woolly throughout; *leaves* alternate, entire, sessile; *heads* disciform; *involucre bracts* imbricate, usually white or scarious throughout or at the tip; *receptacle* naked; *disk flowers* yellowish, whitish, or rarely pink, the outer florets pistillate and fertile, the central florets few and perfect; *ray flowers* absent; *pappus* of capillary bristles which are sometimes thickened at the apex; *achenes* nerveless.

1a. Upper surface of the leaves glandular-hairy…***P. macounii***
1b. Stems and leaf surfaces tomentose but not glandular…2

2a. Involucre bracts yellow; disk flowers very numerous (150–200 or more) per head; pappus tending to cohere in small groups by means of tiny, interlocking basal hairs…***P. stramineum***
2b. Involucre bracts white; disk flowers fewer (20–60) per head; pappus distinct, falling separately…***P. canescens***

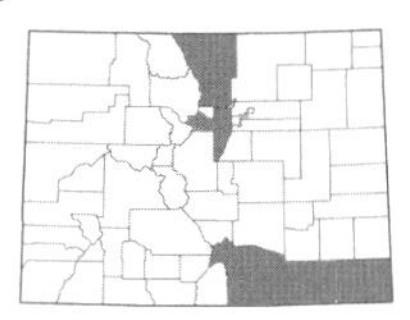

***Pseudognaphalium canescens* (DC.) W.A. Weber**, WRIGHT'S CUDWEED. [*Gnaphalium wrightii* A. Gray]. Plants 1–10 dm; *leaves* oblanceolate, 2–5 cm long, tomentose; *involucre* 4–5 mm high, the bracts white, glabrous; *disk flowers* 20–60; *pappus* distinct. Usually found in dry, open places in sandy soil or rock crevices, uncommon, 4700–6500 ft. Aug.–Oct. E.

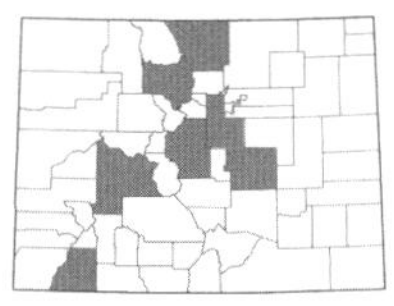

***Pseudognaphalium macounii* (Greene) Kartesz**, MACOUN'S CUDWEED. [*Gnaphalium macounii* Greene]. Plants 4–9 dm; *leaves* oblanceolate to lanceolate, stipitate glandular above and tomentose below, 3–10 cm long; *involucre* 4.5–5.5 mm high, the bracts creamy or brownish, glabrous; *disk flowers* 45–100+. Found in open places in ponderosa pone or lodgepole pine forests, sometimes near bogs, 5800–10,000 ft. July–Sept. E/W.

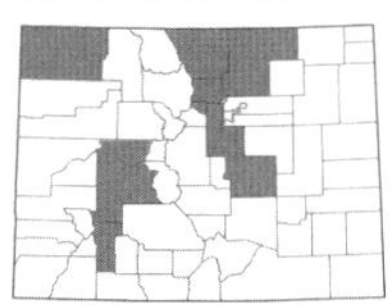

***Pseudognaphalium stramineum* (Kunth) W.A. Weber**, WINGED CUDWEED. [*Gnaphalium chilense* Spreng.; *Gnaphalium stramineum* Kunth]. Plants 3–8 dm; *leaves* oblanceolate to oblong, 2–9 cm long, tomentose; *involucre* 4–6 mm high, the bracts whitish to yellowish, glabrous; *disk flowers* 150–200+; *pappus* loosely coherent basally, falling in clusters. Usually found in moist places along streams and pond margins, often in disturbed places, 5000–9000 ft. June–Sept. E/W.

PSILOSTROPHE DC. – PAPER FLOWER

Herbs or shrubs; *leaves* alternate, entire to pinnately lobed; *heads* radiate; *involucre bracts* in 2 series, the outer subherbaceous, linear-lanceolate, the inner smaller and more scarious, or sometimes the inner undifferentiated from the outer; *receptacle* naked; *disk flowers* 5–25; *ray flowers* 3–7, yellow; *pappus* of hyaline scales; *achenes* linear. (Heiser, 1944)

1a. Heads larger, the involucre bracts 7–10 mm high, pappus shorter, less than half as long as the disk corollas...***P. bakeri***
1b. Heads smaller, the involucre bracts 4–7 mm high, pappus longer, half to as long as the disk corollas...***P. tagetina***

Psilostrophe bakeri **Greene**, COLORADO PAPER FLOWER. Plants 0.5–2 dm, arachnoid-villous; *involucre* 7–10 mm high; *disk flowers* (10) 12–20; *ray flowers* 4–5 (8), 8–15 mm long; *pappus* of 4–5 scales 1.5–2 mm long. Found on dry, open hillsides, 4500–6100 ft. May–June. W. (Plate 14)

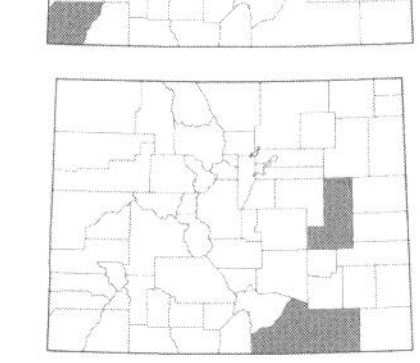

Psilostrophe tagetina **(Nutt.) Greene**, MARIGOLD PAPER FLOWER. Plants 1–3 (6) dm, arachnoid-villous; *involucre* 4–7 mm high; *disk flowers* 6–10 (12); *ray flowers* 3–5, (3) 7–15 mm long; *pappus* of 4–8 scales 2–3 mm long. Uncommon in sandy soil on the eastern plains, apparently not persisting, 4500–4900 ft. June–Aug. E.

PYRROCOMA Hook. – GOLDENWEED

Perennial herbs; *leaves* chiefly basal, simple; *heads* radiate; *involucre bracts* nearly equal or subequal in 2 to several series, herbaceous or green only at the base with scarious or coriaceous margins; *receptacle* naked; *disk flowers* yellow; *ray flowers* yellow; *pappus* of capillary bristles, generally rigid, unequal; *achenes* 3–4-angled.

1a. Stems generally erect, sometimes somewhat curved at the base; involucre bracts large and herbaceous, 2.5–8 mm wide; achenes glabrous, 5–8 mm long; rays 9–35 mm long...***P. crocea* var. *crocea***
1b. Stems usually curved or decumbent at the base; involucre bracts wholly herbaceous but narrower (mostly about 2 mm wide) or herbaceous at the tip with a chartaceous base; achenes sparsely to evidently strigose or rarely glabrous, 2–6 mm long; rays 6–18 mm long...2

2a. Involucre bracts evidently imbricate in several series, the outer not more than half as long as the inner, all with a herbaceous, green tip and chartaceous base; heads usually 5–20 (50), only 1–4 per stem in smaller plants...***P. lanceolata* var. *lanceolata***
2b. Involucre bracts slightly to not at all imbricate, the outer more than half as long as the inner, the outer often green and herbaceous throughout, or the involucre bracts wholly herbaceous and somewhat imbricate in 2–4 series; heads usually solitary or seldom 2–4 per stem...3

3a. Involucre bracts obovate to ovate, about 2 mm wide, generally wholly green and herbaceous, somewhat imbricate in 2–4 series; achenes quadrangular and 4-nerved, 4–6 mm long; basal leaves usually with entire and often villous-ciliate margins, without tufts of hairs in the axils...***P. clementis* var. *clementis***
3b. Involucre bracts narrowly linear, the outer often green and herbaceous throughout but the inner with a chartaceous base, slightly to not at all imbricate; achenes inconspicuously multinerved and sometimes few-angled, 2–4 mm long; basal leaves usually with toothed margins that are scarcely or not at all ciliate, often with tufts of hair in the axils...***P. uniflora* var. *uniflora***

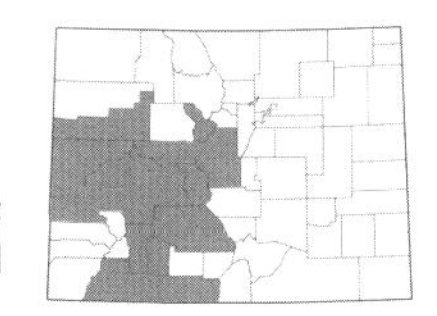

Pyrrocoma clementis **Rydb. var. *clementis***, TRANQUIL GOLDENWEED. [*Haplopappus clementis* (Rydb.) Blake]. Plants 2–3 (4) dm; *leaves* lanceolate to oblanceolate, 5–14 cm long, entire or sometimes toothed, ciliate, glabrous or sometimes sparsely hairy; *heads* usualy solitary; *involucre* 8–15 mm hgih, the bracts obovate to ovate, green and herbaceous, in 2–4 series; *disk flowers* 6–8 mm long; *ray flowers* 10–18 mm long; *pappus* tawny, 6–8 mm long; *achenes* 4–6 mm long, 4-angled. Found in meadows and the alpine, 9200–13,000 ft. July–Aug. E/W.

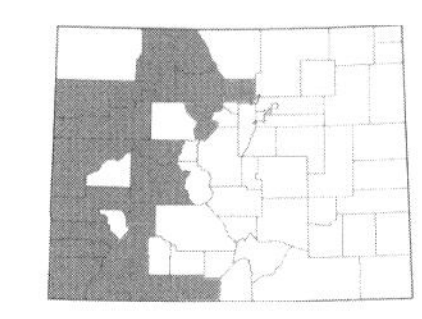

Pyrrocoma crocea **(A. Gray) Greene var. *crocea***, CURLYHEAD GOLDENWEED. [*Haplopappus croceus* A. Gray]. Plants 1–8 dm; *leaves* oblanceolate to narrowly elliptic, 8–45 cm long, entire, lacking cilia, glabrous or rarely sparsely hairy; *heads* usualy solitary; *involucre* 10–20 mm high, the bracts large and herbaceous; *disk flowers* 7–13 mm long; *ray flowers* 9–35 mm long; *pappus* tawny, 6–12 mm long; *achenes* compressed, 5–8 mm long. Found in meadows, forest openings, and along roadsides, 7000–11,800 ft. (June) July–Sept. E/W.

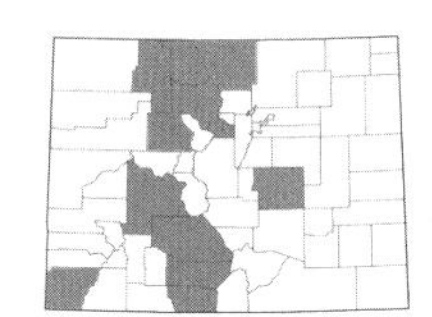

Pyrrocoma lanceolata **(Hook.) Greene var. *lanceolata***, LANCELEAF GOLDENWEED. [*Haplopappus lanceolatus* (Hook.) Torr. & A. Gray]. Plants 2–5 dm; *leaves* lanceolate to narrowly elliptic, sharply toothed, lacking cilia, sparsely tomentose to glabrous, sometimes glandular; *heads* (1) 5–20+; *involucre* 7–10 mm high, the bracts with chartaceous bases and green tips, in 3–4 series; *disk flowers* 5–7 mm long; *ray flowers* 6–11 mm long; *pappus* tawny, 5–7 mm long; *achenes* 4-angled, 3.5–4.5 mm long. Found in alkaline meadows and moist places, 7500–8800 ft. July–Sept. E/W.

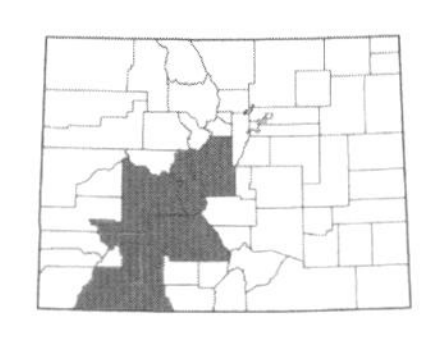

Pyrrocoma uniflora* (Hook.) Greene var. *uniflora, PLANTAIN GOLDENWEED. [*Haplopappus uniflorus* (Hook.) Torr. & A. Gray]. Plants 0.7–4 dm; *leaves* linear to elliptic, 4–15 cm long, sharply toothed to laciniate, glabrous to tomentose; *heads* usualy solitary, sometimes 2–4; *involucre* 6–13 mm high, the bracts green throughout or the inner with a chartaceous base, slightly imbricate; *disk flowers* 5–8 mm long; *ray flowers* 7–11 mm long; *pappus* tawny, 5–7 mm long; *achenes* obscurely angled, 2–4 mm long. Found in wet or dry meadows, 8000–13,800 ft. July–Aug. E/W.

RATIBIDA Raf. – PRAIRIE CONEFLOWER

Perennials; *leaves* alternate, entire to pinnatifid; *heads* radiate; *involucre bracts* in 2 series, the outer short, ovate and the inner linear-lanceolate; *receptacle* globular or columnar, chaffy; *disk flowers* yellow to purplish; *ray flowers* yellow, purplish-yellow, or purple; *pappus* a crown, teeth, or absent; *achenes* radially compressed. (Richards, 1968)

1a. Receptacle long-columnar, to 4.5 cm; rays 7–35 mm long; pappus of 1 or 2 teeth or rarely absent; heads solitary or several at the ends of long peduncles...***R. columnifera***
1b. Receptacle globular, 0.8–1.5 cm; rays 4–8 mm long; pappus a thick crown; heads on short peduncles and often closely clustered...***R. tagetes***

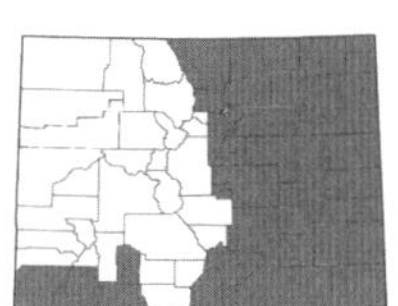

***Ratibida columnifera* (Nutt.) Wooton & Standl.**, PRAIRIE CONEFLOWER. Plants 3–10 dm; *leaves* 2–15 cm long, 1–2-pinnatifid, the ultimate segments linear to narrowly ovate, hirsute and gland-dotted; *disk flowers* greenish-yellow and often purplish distally, 1–2.5 mm long; *ray flowers* yellow, purplish-yellow, or reddish-maroon, sometimes bicolored, 7–35 mm long; *pappus* of 1–2 teeth or absent; *achenes* 1.2–3 mm long. Common on the plains, in fields, along roadsides, and in open places in the foothills, 3500–9000 ft. June–Oct. E/W. (Plate 14)

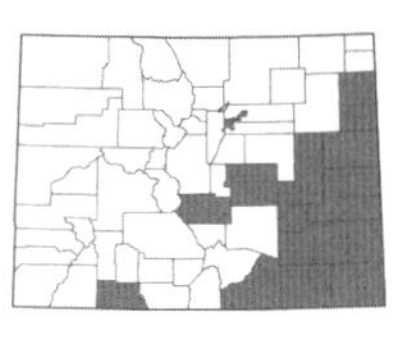

***Ratibida tagetes* (E. James) Barnhart**, SHORT-RAY PRAIRIE CONEFLOWER. Plants to 6 dm; *leaves* 0.5–9 cm long, 1–2-pinnatifid, the ultimate segments linear to narrowly obovate, hirsute and gland-dotted; *disk flowers* greenish-yellow, sometimes purplish distally, 1.2–2.5 mm long; *ray flowers* yellow or purplish, 3–8 mm long; *pappus* a thick crown; *achenes* 1.9–3 mm long. Common in sandy soil and grasslands of the eastern plains, 3500–6000 ft. June–Oct. E/W.

RAYJACKSONIA R.L. Hartm. & M.A.Lane – CAMPHOR-DAISY

Herbs; *leaves* alternate, usually glandular-stipitate; *heads* radiate; *involucre bracts* imbricate in 4–5 series, the base chartaceous and the tip green, the apex erect to recurved and often bristle-tipped; *receptacle* naked; *disk flowers* yellow; *ray flowers* yellow; *pappus* of barbellate bristles; *achenes* often strongly ribbed. (Lane & Hartman, 1996)

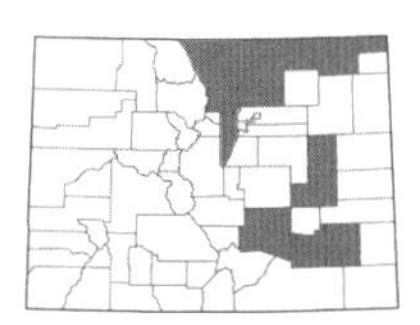

***Rayjacksonia annua* (Rydb.) R.L. Hartm. & M.A. Lane**, VISCID CAMPHOR-DAISY. [*Haplopappus annuus* (Rydb.) Cory; *Haplopappus phyllocephalus* DC. ssp. *annuus* (Rydb.) Hall; *Machaeranthera annua* (Rydb.) Shinners]. Annuals, 2–10 dm; *leaves* oblanceolate to oblong, 2.5–5+ cm long, spine-toothed; *disk flowers* 3–5 mm long; *ray flowers* (13) 20–36, 6–11 (14) mm long. Found in sandy soil of the eastern plains, often in weedy areas or along roadsides, 3500–5300 ft. Aug.–Oct. E.

RUDBECKIA L. – CONEFLOWER

Annual or perennial herbs; *leaves* alternate; *heads* radiate or discoid; *involucre bracts* subequal or of irregular lengths, green, more or less herbaceous; *receptacle* hemispheric to columnar, chaffy; *disk flowers* yellow to dark purple, narrowed to a distinct tube at the base; *ray flowers* neutral, yellow to orange; *pappus* a short crown or absent; *achenes* 4-angled, glabrous.

1a. Ray flowers absent; disk flowers maroon below and greenish above...***R. montana***
1b. Ray flowers present; disk flowers yellowish-green throughout or sometimes purplish-brown above...2

2a. Pappus absent; leaves not pinnatifid or lobed, hirsute...***R. hirta* var. *pulcherrima***
2b. Pappus a short crown; leaves deeply 3-lobed or parted or pinnatifid, glabrous to variously hirsute...3

3a. Rays 3–6 cm long; receptacle bracts (chaff) canescent near the tip...***R. laciniata* var. *ampla***
3b. Rays mostly 1–2 cm long; receptacle bracts glabrous and abruptly contracted into a short, distinct awn tip...***R. triloba* var. *triloba***

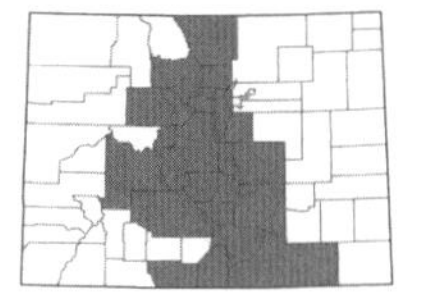

***Rudbeckia hirta* L. var. *pulcherrima* Farw.**, BLACK-EYED SUSAN. Plants 3–10 dm; *leaves* lanceolate to elliptic, 8–15 cm long, hirsute, the margins usually entire; *involucre bracts* to 3 cm long, hirsute; *receptacle* hemispheric to ovoid; *disk flowers* 3–4.5 mm long, yellowish-green below and purplish-brown above; *ray flowers* 1.5–4.5 cm long, yellow or sometimes maroon; *pappus* absent; *achenes* 1.5–3 mm long. Common in meadows and forests, 5200–10,000 ft. June–Sept. E/W. (Plate 14)

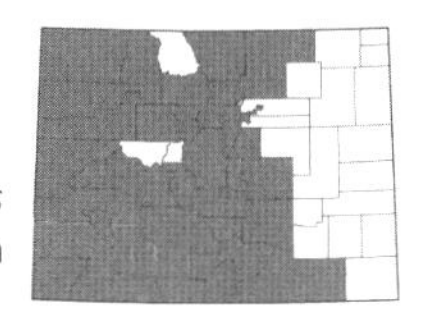

***Rudbeckia laciniata* L. var. *ampla* (A. Nelson) Cronquist**, CUTLEAF CONEFLOWER. Plants 5–30 dm; *leaves* 15–50 cm long, mostly trilobed or deeply pinnatifid or pinnately compound, with 3–11 lobes or leaflets, glabrous or somewhat hairy below; *involucre bracts* to 2 cm long, glabrous or hairy; *receptacle* ovoid, elongating in fruit; *disk flowers* 3.5–5 mm long, yellowish-green; *ray flowers* 3–6 cm long, yellow; *pappus* a short crown; *achenes* 3–5 mm long. Common along streams and in moist meadows, 5000–9000 ft. June–Sept. E/W.

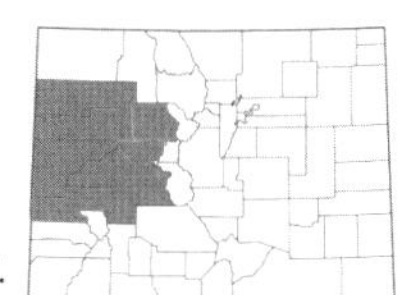

***Rudbeckia montana* A. Gray**, MONTANE CONEFLOWER. [*R. occidentalis* Nutt. var. *montana* (A. Gray) Perdue]. Plants 5–20 dm; *leaves* 8–25 cm long, glabrous or short-hairy, ovate, usually pinnatifid to pinnately lobed; *involucre bracts* to 4 cm long; *receptacle* columnar to ovoid; *disk flowers* 4–5 mm long, maroon below and greenish above; *ray flowers* absent; *pappus* a short crown; *achenes* 5–7 mm long. Found along streams and in aspen forests and meadows, 6400–10,500 ft. July–Sept. W.

Rudbeckia triloba* L. var. *triloba, BROWN-EYED SUSAN. Plants to 15 dm; *leaves* 2–30 cm long, usually 3–5-lobed, toothed, hirsute; *involucre bracts* to 1.5 cm, hirsute; *receptacle* conic to hemispheric; *disk flowers* 3–4 mm long, yellowish-green below and purplish-brown above; *ray flowers* 0.8–3 cm long, yellow and often with maroon spots below, ca. 1–2 cm long; *pappus* a minute crown; *achenes* 2–3 mm long. Escaped from cultivation near Boulder, 5000–5500 ft. July–Sept. E. Introduced.

SAUSSUREA DC. – SAW-WORT

Perennial herbs; *leaves* alternate; *heads* discoid; *involucre bracts* imbricate; *receptacle* chaffy with long, narrow bracts, rarely naked; *disk flowers* blue or purple, with a slender tube and long, narrow lobes; *ray flowers* absent; *pappus* double, of short, rigid outer bristles and plumose inner bristles; *achenes* oblong.

***Saussurea weberi* Hultén**, WEBER'S SAW-WORT. Plants 5–22 cm; *leaves* elliptic or lanceolate to ovate, 2–8 cm long; *heads* 1–15 in subcapitate clusters; *involucre bracts* in 3–4 series, unequal, pilose; *disk flowers* purple, 10–12 mm long; *pappus bristles* 1–11 mm long; *achenes* 3–5 mm long. Uncommon in gravelly soil of the alpine and on scree slopes, 10,500–14,300 ft. July–Aug. E/W.

SCHKUHRIA Roth – FALSE THREADLEAF

Annual or perennial herbs; *leaves* opposite below, alternate above, usually gland-dotted; *heads* discoid (ours); *involucre bracts* about equal, more or less herbaceous with scarious margins; *receptacle* naked; *disk flowers* yellow; *ray flowers* absent (ours); *pappus* of usually 8 scales; *achenes* usually 4-angled, hairy on the angles.

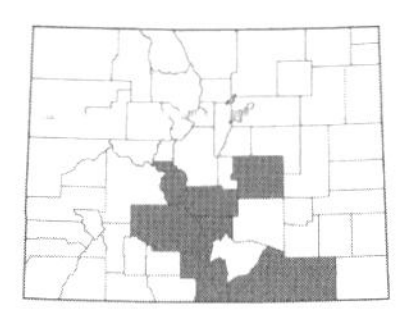

***Schkuhria multiflora* Hook. & Arn.**, MANYFLOWER FLASE THREADLEAF. [*Bahia neomexicana* (A. Gray) A. Gray]. Plants 3–12 (25+) cm; *leaves* linear or lobed, 1–3 cm long; *involucre bracts* hirtellous and gland-dotted; *disk flowers* 1–2 mm long; *pappus* 1–2 mm long; *achenes* 3 mm long, black or brown. Locally common in dry, sandy soil, 5700–8500 ft. Aug.–Sept. E.

SCORZONERA L. – FALSE SALSIFY

Lactiferous herbs; *leaves* alternate, entire to pinnately lobed or dissected; *heads* ligulate; *involucre bracts* imbricate in several series, green and herbaceous; *receptacle* naked; *disk flowers* absent; *ray flowers* readily withering, yellow, purple, or white; *pappus* of plumose bristles; *achenes* cylindric, not beaked, glabrous.

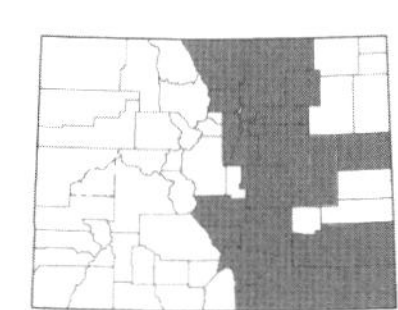

***Scorzonera laciniata* L.**, CUTLEAF VIPERGRASS. [*Podospermum laciniatum* (L.) DC.]. Plants glabrous to sparsely floccose; *leaves* pinnatisect into linear lobes, 7–20 cm long; *involucre* 7–20 mm high; *ray flowers* yellow; *achenes* 8–17 mm long. Common in disturbed sites, open fields, and on grassy plains, 3700–7500 ft. May–July. E. Introduced.

SENECIO L. – RAGWORT

Herbs; *leaves* alternate; *heads* radiate; *involucre bracts* uniseriate, with black-tipped apices, often with smaller bracteoles at the base; *receptacle* naked; *disk flowers* yellow to orange or reddish; *ray flowers* yellow to orange; *pappus* of capillary bristles; *achenes* subterete, 5- to 10-nerved, glabrous or hirtellous on the angles. (Barkley, 1960; Trock, 2003)

1a. Ray flowers absent…2
1b. Ray flowers present…5

2a. Annuals from distinct taproots…***S. vulgaris***
2b. Perennials from woody caudices or crowns…3

3a. Heads large, the disk 15–25 mm wide (when pressed), the mostly 21 (sometimes 13) involucre bracts 7–14 mm long…***S. bigelovii* var. *hallii***
3b. Heads smaller, the disk less than 15 mm wide, the 5, 8, or 13 involucre bracts 3–9 mm long…4

4a. Heads fewer (usually about 8–16) and larger (involucre bracts 5–9 mm long), nodding; leaves entire to shallowly dentate...***S. pudicus***
4b. Heads more numerous (about 25–60) and smaller (involucre bracts 3–5 mm long), erect; leaves sharply dentate to incised-dentate...***S. rapifolius***

5a. Leaves filiform or narrowly linear, or pinnately divided into long linear-filiform segments, sometimes a few of the lower leaves lanceolate but the majority of the leaves linear-filiform; achenes hairy...6
5b. Leaves wider and variously shaped but not narrowly linear or filiform, if pinnatifid then not divided into long linear-filiform segments; achenes glabrous or minutely hirtellous, especially on the angles...9

6a. Plants tomentose...***S. flaccidus* var. *flaccidus***
6b. Plants glabrous...7

7a. Leaves entire or rarely with one or two pairs of linear-filiform lobes...***S. spartioides***
7b. Leaves pinnately divided into linear-filiform segments...8

8a. Heads larger, the disk 7–12 mm wide; involucre bracts usually 13...***S. riddellii***
8b. Heads smaller, the disk 3–6 mm across; involucre bracts usually 8...***S. multicapitatus***

9a. Plants floccose-tomentose (cobwebby hairy) to silvery-canescent, sometimes becoming glabrescent in age but retaining the floccose-tomentose pubescence in the leaf axils and on the peduncles in the heads...10
9b. Plants glabrous, or sometimes somewhat minutely hairy toward the base or with a few short hairs irregularly disposed...13

10a. Heads few (1–3), large (10–20 mm wide), and nodding...11
10b. Heads more numerous, smaller (less than 10 mm wide), erect in a flat-topped cyme...12

11a. Leaves floccose-tomentose above but not below; ray flowers (10) 15–25 mm long...***S. amplectens***
11b. Leaves floccose-tomentose above and below; ray flowers 8–12 mm long...***S. taraxacoides***

12a. Leaves dentate with small, dark cartilaginous denticles on the tips of the teeth; involucre bracts 5 or 8; ray flowers 3 or 5, 5–8 mm long...***S. atratus***
12b. Leaves entire or dentate but without cartilaginous denticles; involucre bracts 13 or 21; ray flowers 8 or 13, 6–15 mm long...***S. integerrimus***

13a. Heads nodding (at least in bud), solitary or few (mostly 1–3) per stem, large (10–20 mm wide); ray flowers about 13 per head...14
13b. Heads erect, solitary to numerous per stem, 4–15 mm wide; ray flowers 5 or 8 per head...15

14a. Leaves strongly dentate to denticulate on the margins, ovate to obovate or broadly lanceolate; only the petioles sometimes purplish-tinged; rays (10) 15–25 mm long, about 1.5–2 times as long as the involucre bracts...***S. amplectens***
14b. Leaves weakly dentate to entire on the margins, ovate-rotund; both leaves and petioles usually purplish-tinged; rays usually 10–14 mm long, about the same length as the involucre bracts...***S. soldanella***

15a. Leaves mostly all basal, broadly ovate or obovate to lanceolate, the stem leaves few with the very lowermost similar to the basal leaves and the upper ones reduced to 2–4 sessile bracts; leaf margins wavy to subentire or dentate, sometimes with small, dark cartilaginous denticles...***S. wootonii***
15b. Leaves not mostly all basal or reduced up the stem, variously shaped, if the stem leaves are strongly reduced then they are not reduced to sessile bracts; leaf margins entire to dentate or pinnatifid, with or without dark, cartilaginous denticles...16

16a. Short-lived perennials from short taproots; leaves irregularly incised or laciniate to pinnatifid, the segments also often with some small teeth...***S. eremophilus* var. *kingii***
16b. Perennials with fibrous-rooted caudices; leaves entire to dentate or sharply toothed but not laciniate or pinnatifid...17

17a. Leaves narrowly or broadly triangular to triangular-hastate or triangular-cordate with a serrate margin, if merely lance-elliptic then the teeth more widely spaced (mostly 3–4 mm from tip to tip)...***S. triangularis***
17b. Leaves variously shaped but not triangular, or if lance-elliptic then the teeth more closely spaced (about 1 mm from tip to tip of each tooth)...18

18a. Stems usually several, clustered from a woody caudex which ultimately surmounts a taproot; leaves sharply toothed and thick; involucre bracts 9–12 mm long, usually not black-tipped; plants 0.5–3 dm tall...***S. fremontii* var. *blitoides***
18b. Stems few to several from a fibrous-rooted woody caudex; leaves entire to toothed but not as sharply so as above, sometimes thick or even glaucous; involucre bracts 5–10 mm long, often black-tipped; plants 2–20 dm tall...19

19a. Stem leaves short-petiolate to petiolate, lanceolate to lance-elliptic with a finely serrate margin (the teeth mostly 1 mm from tip to tip), not strongly reduced on the stem...***S. serra* var. *admirabilis***
19b. Stem leaves sessile, sometimes clasping or with a short auriculate base, elliptic to broadly oblanceolate with an entire to dentate margin, often strongly reduced on the stem...20

20a. Heads fewer in an open inflorescence, usually 4–12 per stem; leaves not glaucous, 2.5–12 cm long, usually sharply dentate; ray flowers 6–12 mm long...***S. crassulus***
20b. Heads numerous and crowded, usually 20–40 or more per stem; leaves thick and glaucous with a paler midrib, 5–20 cm long, usually entire; ray flowers 3–8 mm long...***S. hydrophilus***

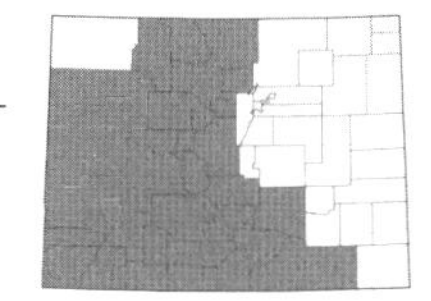

***Senecio amplectens* A. Gray**, SHOWY ALPINE RAGWORT. [*Ligularia amplectens* (A. Gray) W.A. Weber]. Perennials, 1–6 dm, glabrous; *leaves* ovate to obovate, 3.5–7 × 2.5–5 cm, petiolate, often purplish-tinged, with dentate margins; *heads* 1–5 (10), nodding; *involucre bracts* 10–15 mm long; *ray flowers* ca. 13, (10) 15–25 mm long; *achenes* glabrous. Common in open forests, mountain meadows, and the alpine, 8500–14,400 ft. July–Aug. E/W. (Plate 14)

There are two more or less well-defined varieties of *S. amplectens* in Colorado:

1a. Plants taller, mostly 4–6 dm high; principal leaves mostly lower and middle cauline; involucre bracts frequently dark with a few black hairs on the abaxial surface...**var. *amplectens***
1b. Plants shorter, up to 2 dm high; principal leaves basal; involucre bracts glabrous and sometimes purplish-tinged...**var. *holmii* (Greene) Harrington**

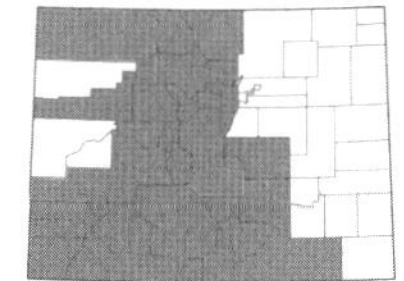

***Senecio atratus* Greene**, TALL BLACKTIP RAGWORT. Perennials, (2) 3.5–8 dm, floccose-tomentose to canescent; *leaves* oblong-ovate to oblanceolate, (5) 10–30 × 1.5–5 cm, with shallowly dentate margins, the teeth dark and cartilaginous; *heads* 2–60, erect; *involucre bracts* 6–8 mm long; *ray flowers* 3–5, 5–8 mm long; *achenes* glabrous. Common in rocky soil along roadsides and on open slopes, 7000–13,200 ft. July–Sept. E/W. (Plate 14)

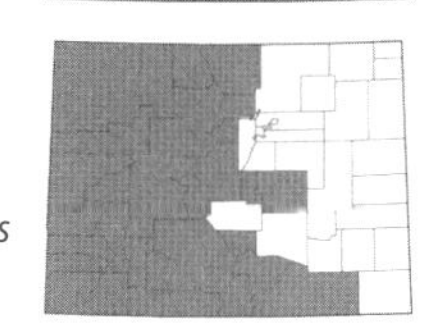

***Senecio bigelovii* A. Gray var. *hallii* A. Gray**, HALL'S RAGWORT. [*Ligularia bigelovii* (A. Gray) W.A. Weber var. *hallii* (A. Gray) W.A. Weber]. Perennials, (2) 4–10 (12) dm, floccose-tomentose; *leaves* lanceolate to obovate, 7–15 × 2–5 cm, subentire to dentate; *heads* 3–20, nodding; *involucre bracts* 7–14 mm long; *ray flowers* absent; *achenes* glabrous. Common in dry, open places such as open forests, hillsides, and meadows, 7800–12,000 ft. July–Sept. E/W.

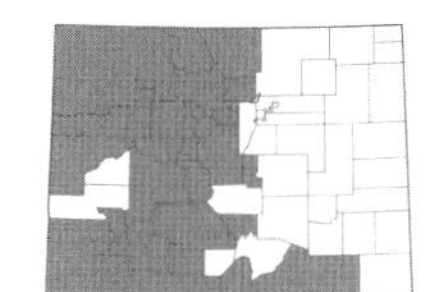

***Senecio crassulus* A. Gray**, MOUNTAIN MEADOW RAGWORT. Perennials, 2–7 dm, glabrous; *leaves* elliptic to broadly oblanceolate, 2.5–12 × 1–5 cm, entire to sharply dentate; *heads* 4–12, erect; *involucre bracts* 5–9 mm long; *ray flowers* 8 or 13, 6–12 mm long; *achenes* glabrous. Common in mountain meadows, on subalpine and alpine slopes, and in open forests, 8500–14,000 ft. June–Sept. E/W.

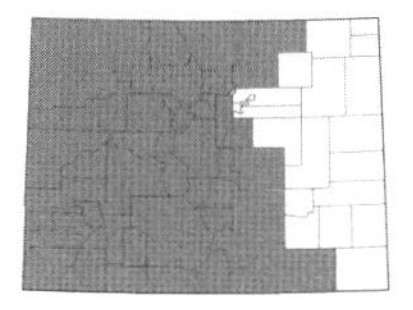

***Senecio eremophilus* Richardson var. *kingii* (Rydb.) Greenm.**, CUT-LEAVED GROUNDSEL. Perennials, 3–10 dm, glabrous; *leaves* irregularly laciniate to pinnatifid, 4–15 × 1–5 cm, the ultimate segments often with a few small teeth; *heads* numerous, erect; *involucre bracts* 4–6 mm long; *ray flowers* 8 or 13, 6–10 mm long; *achenes* usually glabrous. Common along roadsides, in meadows, along streams, and in open forests, 6000–11,500 ft. June–Sept. E/W. (Plate 14)

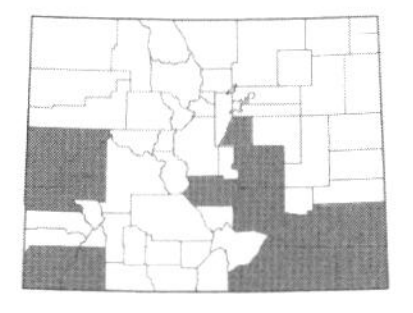

Senecio flaccidus* Less. var. *flaccidus, THREADLEAF RAGWORT. Perennials, 3–10 dm, tomentose, shrubby; *leaves* narrowly linear, simple or deeply pinnatifid into long, linear segments; *heads* 3–20, erect; *involucre bracts* 5–10 mm long; *ray flowers* 8 or 13, 10–15 (20) mm long; *achenes* hairy. Common in dry, open places in sandy or rocky soil, found in the southern counties, 3500–6500 ft. May–Oct. E/W.

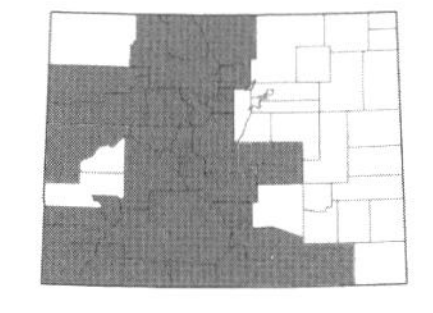

***Senecio fremontii* Torr. & A. Gray var. *blitoides* (Greene) Cronquist**, DWARF MOUNTAIN RAGWORT. Perennials, 0.5–3 dm, glabrous; *leaves* obovate to spatulate, sharply toothed, 1–6 × 1–3 cm; *heads* 1–5, erect; *involucre bracts* 9–12 mm long; *ray flowers* 8, 6–12 mm long; *achenes* glabrous or strigose on the angles. Common in rocky soil of alpine and subalpine slopes, 9500–13,500 ft. July–Sept. E/W. (Plate 14)

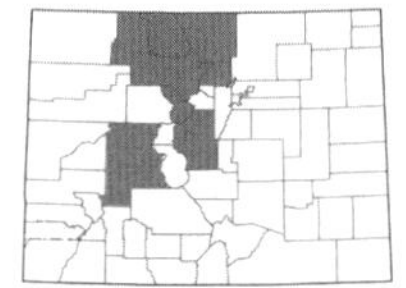

***Senecio hydrophilus* Nutt.**, ALKALI-MARSH RAGWORT. Perennials, 4–15 (20) dm, glabrous, glaucous, with hollow stems; *leaves* thick, elliptic to oblanceolate, 5–20 × 2–10 cm, with entire or denticulate margins; *heads* 20–40 (80), erect, in corymbiform arrays; *involucre bracts* 5–8 mm long; *ray flowers* ca. 5, 3–8 mm long; *achenes* glabrous. Found in wet meadows and marshes, 6300–9200 ft. July–Aug. E/W.

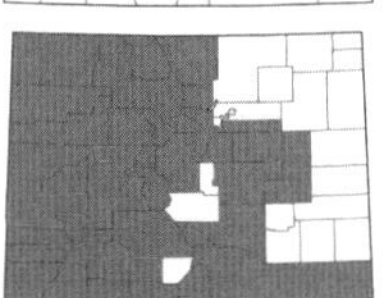

***Senecio integerrimus* Nutt.**, LAMBS-TONGUE RAGWORT; GAUGE-PLANT. Perennials, 2–7 dm, floccose-tomentose, glabrate in age; *leaves* lanceolate, elliptic, to suborbiculate, 6–25 × 1–6 cm, with entire or irregularly dentate margins; *heads* 6–20 (40), erect, in corymbiform arrays; *involucre bracts* 5–10 mm long; *ray flowers* usually 8 or sometimes 13, 6–15 mm long; *achenes* glabrous or hairy on the angles. Common in open places from the foothills to alpine, 5500–13,000 ft. May–July. E/W.

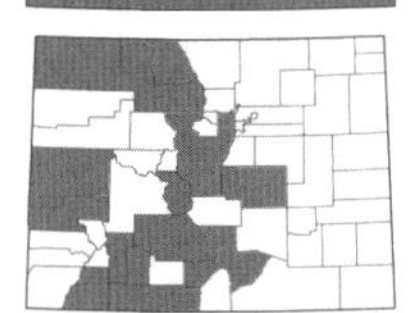

***Senecio multicapitatus* Greene**, BROOM-LIKE RAGWORT. [*Senecio spartioides* Torr. & A. Gray var. *multicapitatus* (Greenm. ex Rydb.) S.L. Welsh]. Perennials or subshrubs, 3–13 dm, glabrous; *leaves* pinnately divided into long, linear segments, to 7 cm long; *heads* 10–20 (60), erect, in compound corymbiform arrays; *involucre bracts* 5–8 mm long; *ray flowers* ca. 5 (13), 8–12 mm long; *achenes* usually hairy or sometimes glabrous. Common in sandy soil of open places, 4500–8800 ft. July–Oct. E/W.

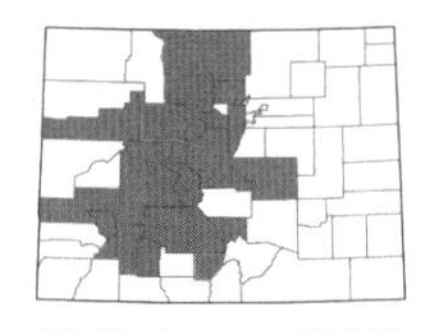

***Senecio pudicus* Greene**, BASHFUL RAGWORT. [*Ligularia pudica* (Greene) W.A. Weber]. Perennials, 1.5–4 (6) dm, glabrous; *leaves* narrowly lanceolate to oblanceolate, 8–15 (20) cm × 1–4 cm, entire or shallowly dentate; *heads* (3) 8–16 (40), nodding, in racemiform or paniculate arrays; *involucre bracts* 5–9 mm long; *ray flowers* absent; *achenes* glabrous. Found on rocky slopes in the mountains, (6500) 8000–12,000 ft. July–Sept. E/W.

***Senecio rapifolius* Nutt.**, OPEN-WOODS RAGWORT. Perennials, 2–6 dm, glabrous, often purplish-tinged; *leaves* obovate to elliptic, 4–9 cm × 2–5 cm, sharply dentate with some teeth calloused; *heads* 25–60, erect, in cymose clusters of 3–12; *involucre bracts* (5) 8, 3–5 mm long; *ray flowers* absent; *achenes* glabrous. Infrequent in rocky soil of open woods and along canyon roadsides, 6000–8500 ft. July–Sept. E.

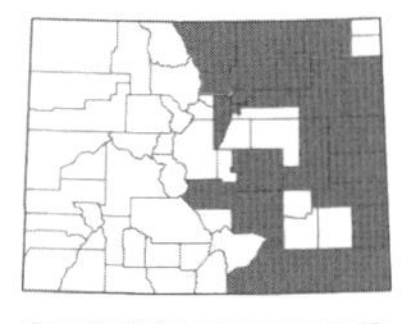

***Senecio riddellii* Torr. & A. Gray**, RIDDELL'S RAGWORT. [*Senecio spartioides* Torr. & A. Gray var. *fremontii* (Torr. & A. Gray) Greenm. ex L.O. Williams]. Subshrubs, 3–10 dm, glabrous; *leaves* irregularly pinnatifid into linear-filiform segments, 4–9 cm × 0.1–0.5 cm; *heads* 5–25, erect, in corymbiform arrays; *involucre bracts* ca. 13, 7–12 mm long; *ray flowers* ca. 8, 8–10 mm long; *achenes* hairy. Common in dry, open places on the plains and in the foothills, 3500–6500 ft. Aug.–Oct. E.

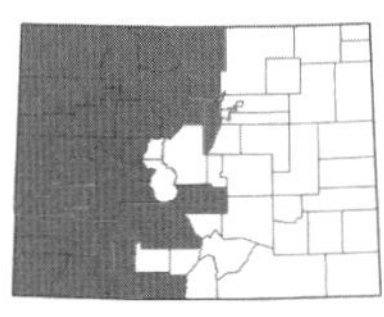

***Senecio serra* Hook. var. *admirabilis* (Greene) A. Nelson**, SAW-TOOTHED RAGWORT; TALL RAGWORT. Perennials, 5–20 dm, glabrous or somewhat sparsely hairy toward the base; *leaves* oblanceolate to lanceolate or elliptic, 7–15 × 1–4 cm, usually sharply serrate; *heads* (12) 30–50, erect; *involucre bracts* 7–9 mm long; *ray flowers* 7–10 mm long; *achenes* glabrous. Found in meadows, along streams, and in open aspen forests, 6200–10,500 ft. July–Sept. E/W.

Very narrowly triangular-leaved specimens of *S. triangularis* may key out to here, but the leaf margins of *S. triangularis* are more deeply and strongly serrate and the teeth wider spaced than those of *S. serra* var. *admirabilis*.

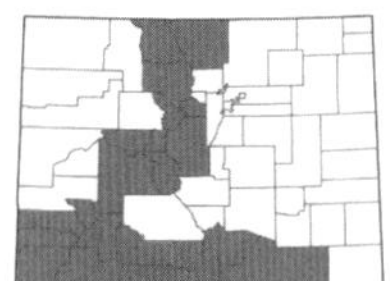

***Senecio soldanella* A. Gray**, COLORADO RAGWORT. [*Ligularia soldanella* (A. Gray) W.A. Weber]. Perennials, 0.5–2 dm, glabrous or with a few short hairs above; *leaves* obovate to ovate, 1.5–5 × 1.5–5 cm, purplish-tinged, weakly dentate to subentire; *heads* solitary, nodding; *involucre bracts* (8) 10–16 mm long; *ray flowers* ca. 13, 10–14 mm long; *achenes* glabrous. Found in rocky alpine and on scree slopes, 11,200–14,000 ft. July–Aug. E/W.

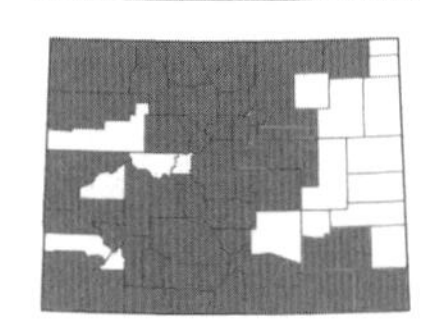

***Senecio spartioides* Torr. & A. Gray**, NARROW-LEAVED BUTTERWEED. Perennials or subshrubs, 2–8 dm, glabrous; *leaves* narrowly linear, entire or rarely with one or two pairs of filiform lobes, 5–10 × 0.2 cm; *heads* 10–20 (60), erect, in compound corymbiform arrays; *involucre bracts* 6–10 mm long; *ray flowers* 5 (13) 8–12 mm long; *achenes* usually hairy or sometimes glabrous. Common in dry, open places, widespread across the state, 4000–9500 ft. July–Oct. E/W. (Plate 15)

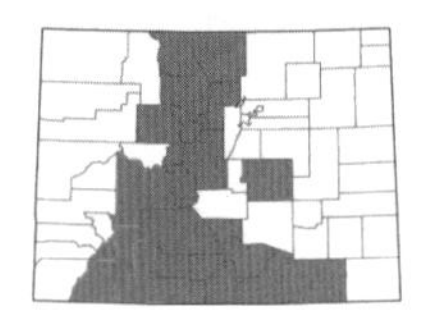

***Senecio taraxacoides* (A. Gray) Greene**, DANDELION RAGWORT. [*Ligularia taraxacoides* (A. Gray) W.A. Weber]. Perennials, 0.5–1.5 dm, floccose-tomentose, becoming glabrate in age; *leaves* oblanceolate to obovate, 1.5–6 × 1–2.5 cm, sharply and deeply dentate; *heads* 1–3 (5), nodding; *involulcre bracts* 7–10 mm long; *ray flowers* ca. 13, 8–12 mm long; *achenes* glabrous. Uncommon in rocky soil and on scree slopes in the alpine, 10,500–14,000 ft. July–Aug. E/W. (Plate 15)

***Senecio triangularis* Hook.**, ARROWLEAF RAGWORT. Perennials, 3–15 (20) dm, usually glabrous; *leaves* triangular, 4–20 × 2–10 cm, sharply serrate, short-petiolate; *heads* 10–30 (60), erect, in corymbiform arrays; *involucre bracts* ca. 13, 7–10 mm long; *ray flowers* ca. 8, 7–13 mm long; *achenes* glabrous. Common along streams and in moist meadows in the mountains, 6500–13,500 ft. July–Sept. E/W. (Plate 15)

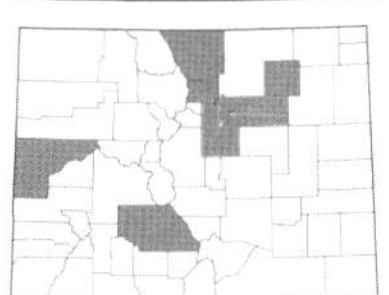

***Senecio vulgaris* L.**, OLD MAN IN THE SPRING; GROUNDSEL. Perennials, 1–6 dm, glabrous or lightly tomentose when young; *leaves* obovate to oblanceolate, deeply and irregularly dentate to pinnatisect, 2–10 × 0.5–2 (4) cm; *heads* 8–20, erect; *involucre bracts* 4–6 mm long; *ray flowers* absent; *achenes* sparsely hairy. Found in disturbed places, especially common in yards and gardens, 4600–7500 ft. Mar.–Sept. E/W. Introduced.

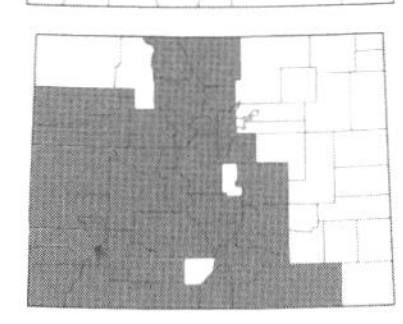

***Senecio wootonii* Greene**, WOOTON'S RAGWORT. Perennials, 1.5–5 dm, glabrous; *leaves* lanceolate to ovate or obovate, (3) 4–10 (12) × 1.5–3 (4) cm, subentire to dentate, often with dark cartilaginous denticles on the teeth; *heads* (3) 8–24, erect; *involucre bracts* 6–9 mm long, *ray flowers* 7–10 mm long; *achenes* glabrous. Found in open forests and meadows, 8000–11,500 ft. June–Aug. E/W.

SHINNERSOSERIS Tomb – BEAKED SKELETONWEED

Annual herbs; *leaves* opposite below, alternate above, simple, entire; *heads* ligulate; *involucre bracts* more or less herbaceous, in 3 series; *receptacle* naked; *disk flowers* absent; *ray flowers* lavender with white or yellow tips; *pappus* of capillary bristles; *achenes* abruptly narrowed just below the apex.

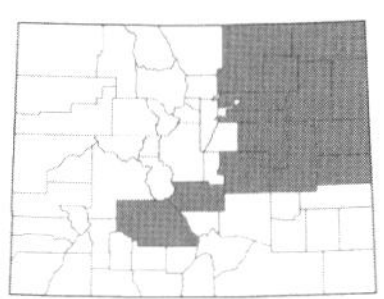

***Shinnersoseris rostrata* (A. Gray) Tomb**, BEAKED SKELETONWEED. [*Lygodesmia rostrata* (A. Gray) A. Gray]. Plants 1–4 (8) dm, lactiferous; *leaves* 6–15 cm long, uppermost reduced to scales, *involucre bracts* 12–18 mm long; *ray flowers* 8–11, 6 mm long; *pappus* 6–8 mm long; *achenes* 8–10 mm long. Uncommon in sandy soil, 3500–5700 (8000) ft. July–Sept. E.

SILPHIUM L. – ROSINWEED

Perennials, coarse herbs; *leaves* alternate or opposite; *heads* radiate; *involucre bracts* subequal or imbricate, more or less herbaceous; *receptacle* chaffy; *disk flowers* yellow; *ray flowers* yellow; *pappus* of 2 awns extending from the wings of the achene or absent; *achenes* strongly flattened at right angles, the margin winged. (Perry, 1937)

1a. Leaves mostly opposite, with an entire or toothed margin, not deeply lobed; heads in an open, short, terminal cymose inflorescence...***S. integrifolium* var. *laeve***
1b. Leaves alternate, deeply pinnately lobed; heads in an elongate, racemiform inflorescence...***S. laciniatum***

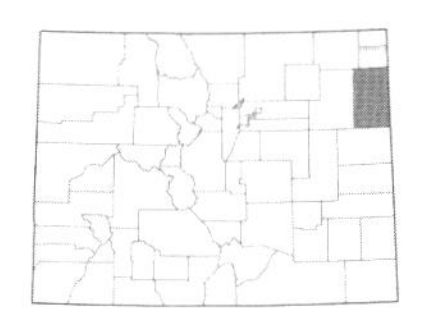

***Silphium integrifolium* Michx. var. *laeve* Torr. & A. Gray**, WHOLELEAF ROSINWEED. Plants 4–20 dm; *leaves* mostly opposite, lanceolate to ovate, 2–20 cm long, sessile and sometimes clasping, the margins entire to finely toothed; *involucre* 1–2 cm high, the bracts in 2–3 series, the outer appressed, glabrous; *ray flowers* 2–5 cm long; *achenes* 9–15 mm long. Rare on the eastern plains, known from one collection near Wray in Yuma Co., 3500–4000 ft. June–Sept. E.

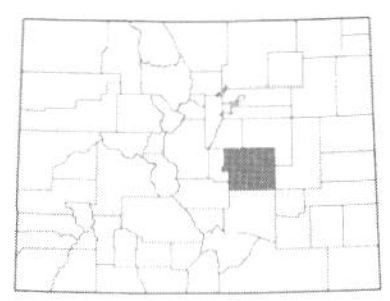

***Silphium laciniatum* L.**, COMPASS PLANT. Plants 10–30 dm; *leaves* alternate, 20–400 cm long, deeply pinnately lobed, hirsute; *involucre* 2–4 cm high, hispid, the bracts in 2–3 series, the outer reflexed or appressed; *ray flowers* 2–5 cm long; *achenes* 10–20 mm long. Rare in open meadows and along roadsides, typically preferring areas of mild disturbance, known from El Paso County near Palmer Lake, 7000–7500 ft. June–Sept. E.

SOLIDAGO L. – GOLDENROD

Perennial herbs; *leaves* alternate, simple; *heads* radiate; *involucre bracts* imbricate in several series or not much imbricate, more or less chartaceous at the base with green tips; *receptacle* naked; *disk flowers* yellow; *ray flowers* white or yellow; *pappus* of capillary bristles; *achenes* subterete or angled, short-pubescent to glabrous. (Croat, 1972; Nesom, 1993, 1999; Taylor & Taylor 1983; 1984)

1a. Ray flowers white, 7–8 mm long; heads "aster-like"...***S. ptarmicoides***
1b. Ray flowers yellow, usually less than 5 mm long; heads smaller, not "aster-like"...2

2a. Involucre bracts with 3–5 parallel veins; heads in a flat-topped corymbiform inflorescence; plants densely pubescent with short, spreading hairs; leaves uninerved, elliptic-oblong to broadly ovate...***S. rigida* var. *humilis***
2b. Involucre bracts without parallel veins; heads in a paniculate or thyrsiform inflorescence; plants variously pubescent but usually not densely so with short, spreading hairs; leaves often trinerved, variously shaped...3

3a. Leaves glabrous or with ciliate leaf margins; stems glabrous or occasionally minutely hairy above or in the inflorescence...4
3b. Leaves puberulent to hirsute (at least below), the margins ciliate or not; stems evenly short-hairy to hirsute, sometimes glabrate toward the base...8

4a. Basal leaves strongly ciliate-margined on the petiolar base; involucre bracts thin and narrow, long-pointed and not much imbricate with the outer bracts almost as long as the inner...***S. multiradiata* var. *scopulorum***
4b. Basal leaves lacking or without ciliate margins or the margins sometimes very weakly ciliate on the petiolar base; involucre bracts well-imbricate with the outer bracts much shorter than the inner, rounded to subacute or blunt at the apex...5

5a. Leaves mostly 3-nerved (the middle nerve the strongest with 2 fainter lateral nerves from the leaf base, sometimes just the upper uninerved), narrowly elliptic to lanceolate or broadly linear...6
5b. Leaves uninerved, at least the lower oblanceolate to obovate or ovate to broadly lanceolate...7

6a. Leaf margins sharply serrate for at least ½ the length of the leaf; plants (5) 10–15 dm or more tall; achenes short-hairy...***S. gigantea***
6b. Leaf margins entire or (especially those of the lower leaves) remotely toothed (on the upper ⅓ of the leaf or less), rarely toothed for ½ of the margin or more; plants generally shorter, 1.5–9 dm; achenes glabrous or occasionally sparsely hairy...***S. missouriensis***

7a. Lower and basal leaves oblanceolate to spatulate or obovate; disk flowers 8–17 (25); achenes hairy...***S. simplex* var. *simplex***
7b. Lower and basal leaves ovate to broadly lanceolate or oblong; disk flowers usually 4 or 5 but rarely up to 11; achenes glabrous...***S. speciosa* var. *pallida***

8a. Basal leaves well-developed and present at flowering time, the stem leaves smaller than the basal and progressively reduced in size upward (note that specimens collected late in the season may have withered basal leaves)...9
8b. Leaves mainly all cauline without well-developed basal leaves present at flowering time, these not greatly reduced in size upward...10

9a. Disk flowers 8–20, greater than the number of ray flowers...***S. nana***
9b. Disk flowers 3–6 (9), fewer than or equal to the number of ray flowers...***S. nemoralis* var. *decemflora***

10a. Leaves 1-nerved or rarely obscurely 3-nerved, the margins mostly entire...11
10b. Leaves 3-nerved, rarely obscurely so, the margins serrate to irregularly toothed (at least remotely so)...12

11a. Involucre bracts slightly viscid with granular or minutely stipitate glands (sometimes this is only present on the phyllary margins)...***S. wrightii* var. *adenophora***
11b. Involucre bracts not viscid with granular or minutely stipitate glands...***S. nemoralis* var. *decemflora***

12a. Most leaves sharply serrate for ½ of the margin or more, linear-lanceolate to lanceolate; ray flowers mostly 13, ranging from (5) 10–18...***S. canadensis***
12b. Leaf margins mostly entire or serrate or irregularly toothed only on the upper ½ to ⅓, lanceolate, elliptic, or oblanceolate; ray flowers mostly 8, ranging from 6–10...13

13a. Leaves grouped tightly on the stem (such that the internodes are usually hidden by leaves when pressed), generally elliptic to broadly ovate or obovate; plants usually found at lower elevations (3400–7000 ft)...***S. mollis***
13b. Leaves more sparsely distributed and not grouped tightly on the stem (such that the internodes are usually quite visible when pressed), generally lanceolate to ovate or narrowly oblanceolate; plants found from 5500–10,500 ft in elevation...***S. velutina* ssp. *sparsiflora***

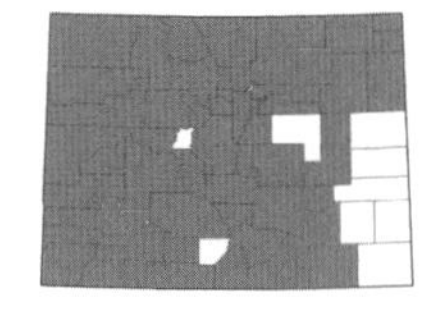

***Solidago canadensis* L.**, Canada goldenrod. [*S. altissima* L.]. Plants 2.5–20 dm; *leaves* linear-lanceolate to lance-elliptic, 3–15 cm long, 3-nerved, usually sharply serrate, densely short-hairy below and glabrous or shortly hairy above; *involucre* 2.5–5 mm high, the bracts imbricate; *disk flowers* 2–3 mm long; *ray flowers* (5) 10–18, 1.5–2.5 mm long; *pappus* 1.5–2.2 mm long; *achenes* 1–1.5 mm long. Common and widespread in open places from the plains and valleys to the mountains, 3500–9800 ft. July–Oct. E/W. (Plate 15)

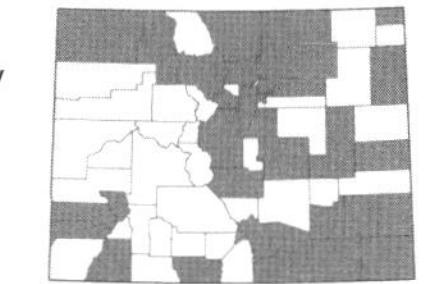

***Solidago gigantea* Ait.**, GIANT GOLDENROD. [*S. serotina* Ait.]. Plants (5) 10–15+ dm; *leaves* narrowly elliptic to lanceolate, 6–17 cm long, glabrous to somewhat hairy on the nerves, 3-nerved, sharply serrate; *involucre* 2.5–4 mm high, the bracts in 3–4 series; *disk flowers* 2.5–4 mm long; *ray flowers* (8) 10–18, 1–3 mm long; *pappus* 2–2.5 mm long; *achenes* 1.3–1.5 mm long. Common in moist places, especially on the eastern plains, 3400–8800 ft. July–Oct. E/W.

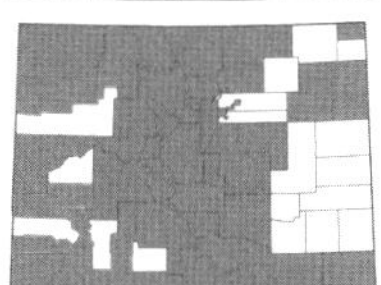

***Solidago missouriensis* Nutt.**, MISSOURI GOLDENROD. Plants 1.5–9 dm; *leaves* linear-lanceolate to narrowly elliptic, 4–25 cm long, glabrous, entire to serrate, 3-nerved (often weakly so); *involucre* 3–5 mm high; *disk flowers* 2–4 mm long; *ray flowers* 5–11, 2–3 mm long; *pappus* 2.5–3 mm long; *achenes* 1–2 mm long. Common and widespread in open meadows and forests and on the plains, 3700–10,000 ft. July–Oct. E/W. (Plate 15)

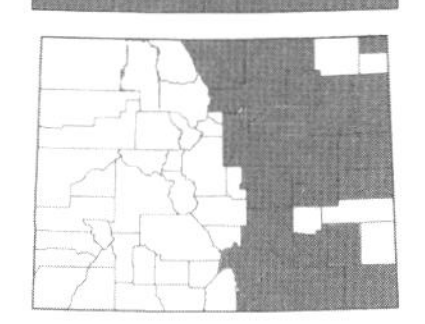

***Solidago mollis* Bartl.**, SOFT GOLDENROD. Plants 1–5 (7) dm; *leaves* elliptic to ovate, 3–8 cm long, 3-nerved, hirsute, toothed on the upper third of the leaf; *involucre* 3.5–6 mm high; *disk flowers* 2.5–4 mm long; *ray flowers* 6–10, 1–2 mm long; *pappus* 2–3 mm long; *achenes* 1.5–2 mm long. Common in sandy or rocky soil of the plains and outer foothills, 3400–7000 ft. July–Oct. E.

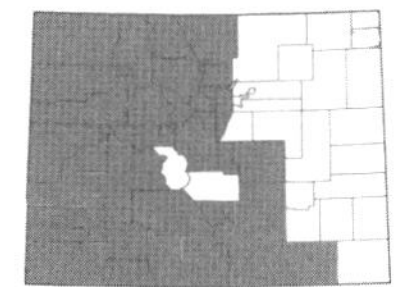

***Solidago multiradiata* Ait. var. *scopulorum* A. Gray**, ROCKY MOUNTAIN GOLDENROD. Plants 0.5–5 dm; *leaves* oblanceolate to spatulate, 2–10 cm long, petiole margins strongly ciliate, entire to remotely toothed, glabrous; *involucre* 4–7 mm high; *disk flowers* 3–5 mm long; *ray flowers* 12–18, 3–4 mm long; *pappus* 3–4 mm long; *achenes* 1.5–4 mm long. Common in rocky soil of mountain meadows, forests, and the alpine, 6500–13,000 ft. July–Sept. E/W.

Can hybridize with *S. simplex* at high elevations.

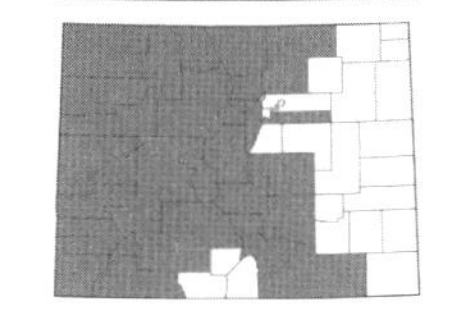

***Solidago nana* Nutt.**, BABY GOLDENROD. Plants 1–5 dm; *leaves* oblanceolate to spatulate, 2–10 cm long, weakly or scarcely 3-nerved, short-hairy, entire to remotely toothed; *involucre* 4–6 mm high; *disk flowers* 4–5 mm long; *ray flowers* (5) 6–10, ca. 3 mm long; *pappus* 3.5–4 mm long; *achenes* 2–3 mm long. Common in meadows and on open or slightly forested slopes, 5000–9800 ft. July–Sept. E/W.

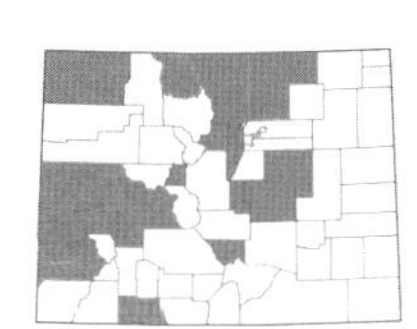

***Solidago nemoralis* Ait. var. *decemflora* (DC.) Brammall ex Semple**, GRAY GOLDENROD. [*S. nemoralis* Ait. var. *longipetiolata* (Mack. & Bush) Palmer & Steyerm.]. Plants 1–10 (13) dm; *leaves* oblanceolate to narrowly ovate, 5–20 cm long, 1-nerved or obscurely 3-nerved, densely short-hairy, margins subentire to crenate-dentate; *involucre* 4–6 mm high; *disk flowers* 2.5–4.5 mm long; *ray flowers* 5–11, 2.8–5.5 mm long; *pappus* 2–4 mm long; *achenes* 0.5–2 mm long. Locally common on open slopes and in meadows, 5000–7500 ft. July–Sept. E/W.

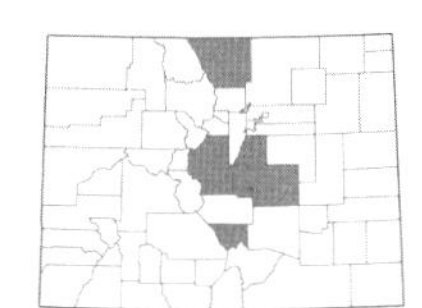

***Solidago ptarmicoides* (Nees) Boivin**, PRAIRIE GOLDENROD. [*Aster ptarmicoides* (Nees) Torr. & A. Gray; *Oligoneuron album* (Nutt.) G.L. Nesom; *Unamia alba* (Nutt.) Rydb.]. Plants 1–7 dm; *leaves* linear-lanceolate, 3–20 cm long, 3-nerved, entire or nearly so, glabrous or somewhat short-hairy; *involucre* 5–7 mm high, imbricate in 3–4 series; *disk flowers* 3.8–4 mm long; *ray flowers* 10–20, 7–8 mm long, white; *pappus* 3–4 mm long, clavate at the tips; *achenes* 1–1.5 mm long. Uncommon in forest clearings and open meadows, 6500–9000 ft. July–Sept. E.

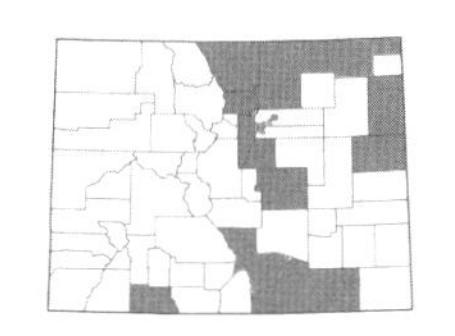

***Solidago rigida* L. var. *humilis* Porter**, STIFF GOLDENROD. [*Oligoneuron rigidum* (L.) Small var. *humile* (Porter) G.L. Nesom; *Solidago canescens* (Rydb.) Friesner]. Plants 2–16 dm; *leaves* elliptic-oblong to lanceolate, 5–15 (25) cm long, 1-nerved, densely hairy, entire or obscurely toothed; *involucre* 5–9 mm high, the bracts conspicuously striate; *disk flowers* 4–6 mm long; *ray flowers* 6–13, 1.4–5.5 mm long; *pappus* 3–4 mm long; *achenes* 0.8–2 mm long. Common along roadsides and in open meadows, 3500–8800 ft. Aug.–Oct. E/W. (Plate 15)

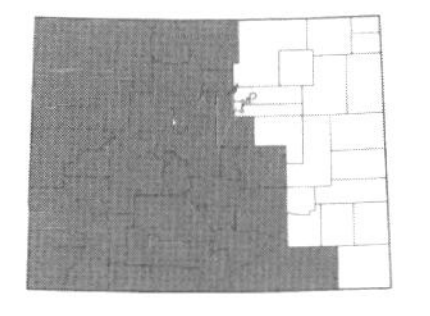

Solidago simplex* Kunth var. *simplex, MT. ALBERT GOLDENROD. [*S. spathulata* DC.]. Plants 0.5–6 dm; *leaves* oblanceolate to spatulate, to 15 cm long, glabrous, toothed to crenate (mostly in the upper half to third); *involucre* 4–7 mm high; *disk flowers* 4–5 mm long; *ray flowers* (6) 8–12, mostly 3–4 mm long; *pappus* 2–5 mm long; *achenes* 2–3.2 mm long. Common in mountain meadows, forests, and the alpine, 6000–13,000 ft. July–Sept. E/W. (Plate 15)

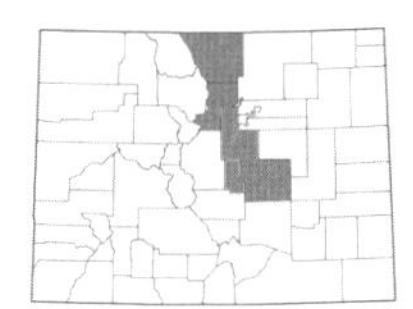

***Solidago speciosa* Nutt. var. *pallida* Porter**, SHOWY GOLDENROD. Plants 2–15 (20) dm; *leaves* ovate to lanceolate or oblong, 5–15 (30) cm long, usually entire, 1-nerved, glabrous; *involucre* 3–6 mm high; *disk flowers* 2.5–4 mm long; *ray flowers* 4–8 (11), 3–4 mm long; *pappus* 3–4.5 mm long; *achenes* 1.5–2.5 mm long. Uncommon or locally common in open places of the foothills in sandy or rocky soil, 5400–9000 ft. Aug.–Oct. E.

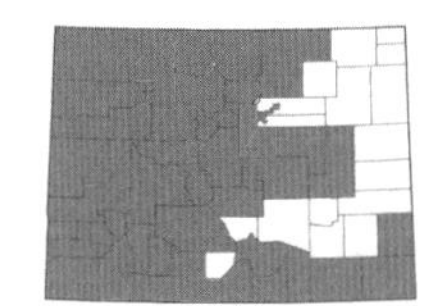

***Solidago velutina* DC. ssp. *sparsiflora* (A. Gray) Semple**, THREE-NERVE GOLDENROD. [*S. sparsiflora* A. Gray]. Plants 1.5–8 (15) dm; *leaves* elliptic to spatulate or oblanceolate, 5–12 cm long, usually entire, short-hairy, 3-nerved; *involucre* 3–4.5 mm high; *disk flowers* 3.5–6 mm long; *ray flowers* 6–10, 3–6 mm long; *pappus* 2.5–4.7 mm long; *achenes* 0.7–3 mm long. Common on open slopes and in mountain meadows and forest clearings, 5500–10,500 ft. July–Oct. E/W.

***Solidago wrightii* A. Gray var. *adenophora* Blake**, WRIGHT'S GOLDENROD. Plants 2–6 dm; *leaves* elliptic to oblong, to 8 cm long, short-hairy, mostly 1-nerved, margins entire; *involucre* 5–6 mm high; *disk flowers* 3–4 mm long; *ray flowers* 6–10, 3–5 mm long; *pappus* 3–4 mm long; *achenes* 1.5–2.5 mm long. Uncommon on bluffs and sandstone rocks of the plains, 4500–5700 ft. E.

SONCHUS L. – SOW-THISTLE

Lactiferous herbs; *leaves* alternate, usually prickly-margined; *heads* ligulate; *involucre bracts* imbricate in several series; *receptacle* naked; *disk flowers* absent; *ray flowers* yellow; *pappus* of capillary bristles that tend to fall as a single unit; *achenes* flattened, prominently 6–20-ribbed, narrowed at the tip but not beaked, glabrous but often wrinkled.

1a. Perennials with deep, extensive creeping root systems; heads larger, 3–5 cm wide...2
1b. Annuals from taproots; heads smaller, mostly 1.5–2.5 cm wide...3

2a. Involucre bracts with glandular hairs and sometimes irregularly tomentose, 14–22 mm high...***S. arvensis***
2b. Involucre bracts without glandular hairs, 10–15 mm high...***S. uliginosus***

3a. Achenes evidently wrinkled; leaf auricles acute at the base...***S. oleraceus***
3b. Achenes not wrinkled and the margins may have minute, recurved spinules; leaf auricles rounded at the base...***S. asper***

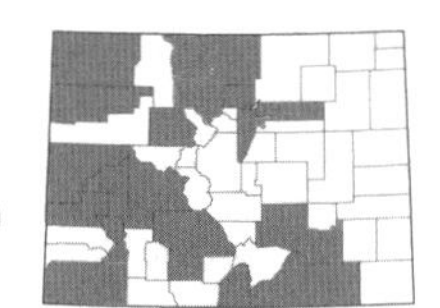

***Sonchus arvensis* L.**, FIELD SOW-THISTLE. Perennials, 4–20 dm; *leaves* pinnately lobed or pinnatifid, glabrous and prickly-margined, 6–40 cm long; *involucre* 14–22 mm high, glandular and sometimes irregularly tomentose; *pappus* 8–14 mm long; *achenes* 2.5–3.5 mm long, with 5+ prominent ribs on each face, strongly wrinkled. Found in wet places such as along ditches and in wet meadows, 4500–9000 ft. July–Sept. E/W. Introduced.

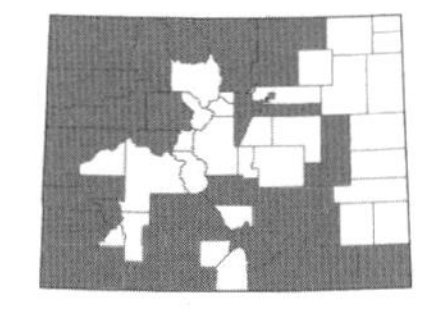

***Sonchus asper* (L.) Hill**, SPINY SOW-THISTLE. Annuals, 1–10 (15) dm; *leaves* pinnatifid or pinnately lobed, auriculate-clasping with round auricles, to 2.5 cm long, glabrous and prickly-margined; *involucre* to 15 mm high, usually glandular; *pappus* 6–9 mm long; *achenes* 2–3 mm long, with 3 (5) ribs on each face, not wrinkled. A weedy plant found in disturbed places and in cultivated plantings, sometimes found in moist places along ditches and in wet meadows, 4000–8700 ft. June–Sept. E/W. Introduced.

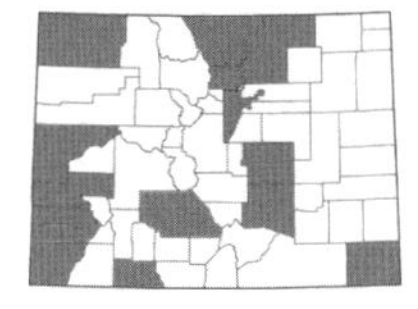

***Sonchus oleraceus* L.**, COMMON SOW-THISTLE. Annuals, 1–20 dm; *leaves* pinnatifid or pinnately lobed, auriculate-clasping with acute auricles, 6–35 cm long, glabrous and prickly-margined; *involucre* 9–15 mm high, glabrous or tomentose and/or glandular; *pappus* 5–8 mm long; *achenes* 2.5–3 mm long, 3–5-ribbed on each face, transverse-rugose. Found in disturbed places and as a weed in gardens, 4500–7500 ft. May–Sept. E/W. Introduced.

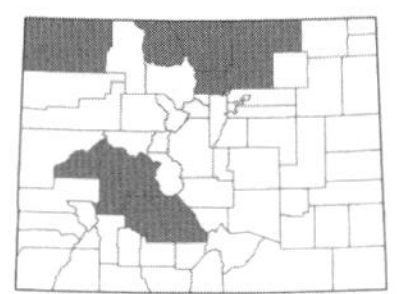

***Sonchus uliginosus* M. Bieb.**, MOIST SOW-THISTLE. Perennials, 4–20 dm; *leaves* pinnately lobed or pinnatifid, 6–40 cm long, glabrous and prickly-margined; *involucre* 10–15 mm high, glabrous or nearly so; *pappus* 8–14 mm long; *achenes* 2.5–3.5 mm long, with 5+ prominent ribs on each face, wrinkled. Found in moist places such as along ditches and streams and around lake margins, 4500–8500 ft. July–Sept. E/W. Introduced.

SPHAEROMERIA Nutt. – CHICKEN SAGE

Perennial herbs; *leaves* alternate, usually crowded at the base; *heads* discoid; *involucre bracts* imbricate to nearly equal in 2–3 series, with scarious margins and greenish center; *receptacle* naked; *disk flowers* perfect and fertile, the marginal flowers pistillate and fertile, yellow, 5-lobed; *ray flowers* absent; *pappus* usually absent; *achenes* subcylindric, 3–5- or 10-ribbed. (Holmgren et al., 1976)

1a. Heads sessile in a dense, globose capitate cluster; leaves with linear segments...***S. capitata***
1b. Heads pedunculate or subsessile in a subcapitate to loosely corymbose cluster; leaves with wider, cuneate segments...***S. argentea***

***Sphaeromeria argentea* Nutt.**, SILVER CHICKEN SAGE. Plants 0.5–2 dm; *leaves* 3-lobed with cuneate segments, 7–15 mm long, silvery-canescent; *involucre* 3–5 (7) mm high; *achenes* 2–3 mm long, 3–5-ribbed, glabrous or sparsely gland-dotted. Found in open, dry places in sandy to clay soil, known from Moffat Co., 6500–7000 ft. June–July. W.

Sphaeromeria capitata **Nutt.**, ROCK TANSY. Plants 0.5–2 dm; *leaves* lobed with linear segments, 8–20 mm long, tomentose; *involucre* 2–4 mm high; *achenes* 1–2 mm long, 3-5-ribbed, glabrous or sparsely gland-dotted. Found in sandy soil of open, dry places, 7500–7900 ft. June–July. W.

STENOTUS Nutt. – MOCK GOLDENWEED

Perennial herbs or subshrubs; *leaves* chiefly basal, crowded at branch tips, alternate, simple, entire; *heads* radiate; *involucre bracts* subequal in 2–3 series, with a greenish center or tip and scarious margin; *receptacle* naked; *disk flowers* yellow; *ray flowers* yellow; *pappus* of white capillary bristles; *achenes* compressed, densely hairy.

1a. Involucre bracts acute or acuminate, only slightly or moderately imbricate, the center and not just the tips green; leaf margins stiffly ciliate...***S. acaulis***
1b. Involucre bracts rounded to obtuse, strongly imbricate, with conspicuous green tips; leaf margins sometimes ciliate...***S. armerioides* var. *armerioides***

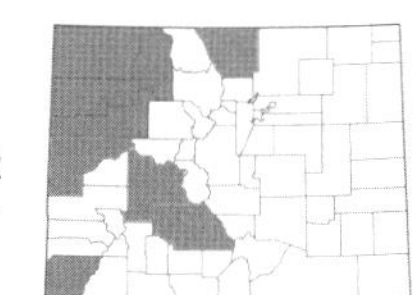

Stenotus acaulis **(Nutt.) Nutt.**, STEMLESS MOCK GOLDENWEED. Perennials, 0.3–2 dm; *leaves* linear to oblanceolate, 0.5–8.5 cm long, glabrous or scabrous, eglandular or glandular-stipitate, margins stiffly ciliate; *heads* 1–2 (4); *involucre* 9–20 mm high; *disk flowers* 6–10 mm long; *ray flowers* 5.5–12 (16) mm long; *pappus* 4–8 mm long; *achenes* 2–5 mm long. Found in dry, open places in shale or sandy soil, 5200–8500 ft. May–June. E/W.

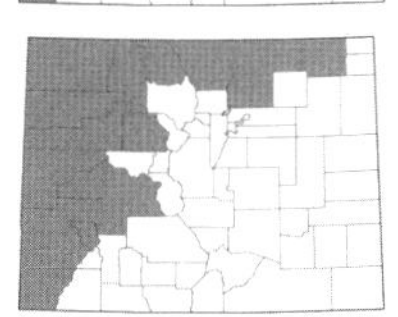

Stenotus armerioides **Nutt. var. *armerioides***, THRIFT MOCK GOLDENWEED. Perennials, 1–3 dm; *leaves* narrowly oblanceolate to spatulate, 1.7–9 cm long, margins sometimes ciliate, gland-dotted; *heads* 1–2 (4); *disk flowers* 4–9 mm long; *ray flowers* 5.5–19 mm long; *pappus* 3.5–7.5 mm long; *achenes* 2–6 mm long. Found in dry, open places in sandy or rocky soil, 4500–8000 ft. May–June. E/W. (Plate 15)

STEPHANOMERIA Nutt. – WIRE LETTUCE; SKELETONWEED

Annual or perennial, lactiferous herbs; *leaves* alternate, entire to toothed or pinnatifid, the upper bractlike; *heads* ligulate; *involucre bracts* in 2 series, the larger principal bracts 3–8 and in a single series, the outer bracts much shorter and calyculate; *receptacle* naked; *disk flowers* absent; *ray flowers* pink to lavender or white; *pappus* of plumose bristles; *achenes* columnar, beakless, 5-ribbed. (Gottlieb, 1972)

1a. Leaves linear and filiform, entire or with a few scattered, slender spreading lobes or teeth...2
1b. Leaves evidently runcinate-pinnatifid or regularly toothed along the margins...3

2a. Pappus bright white, plumose to the base or nearly so...***S. tenuifolia***
2b. Pappus tawny or golden, naked for 1–2 mm at the base and plumose above...***S. pauciflora***

3a. Perennials 0.5–2 (2.5) dm; pappus bristles plumose to the base or nearly so...***S. runcinata***
3b. Annuals 1–8 dm; pappus bristles plumose only on the upper ⅔ to ⅖...***S. exigua* ssp. *exigua***

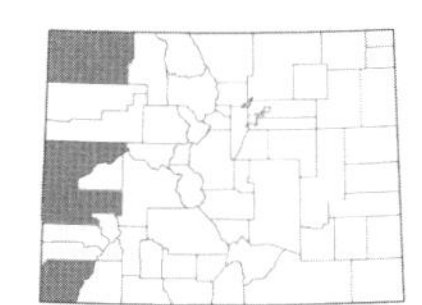

Stephanomeria exigua **Nutt. ssp. *exigua***, SMALL WIRE LETTUCE. Annuals, 1–8 dm; *leaves* toothed to pinnatifid or bipinnatifid, soon deciduous, to 6 cm long; *involucre* 5–10 mm high; *ray flowers* 3–8, pink-lavender; *pappus* white or nearly so, usually thickened at the base and connate into 5 groups; *achenes* 3–4 mm long, rugose-tuberculate, with 5 broad ribs. Uncommon in sandy soil of dry, open places, 4800–7500 ft. May–Sept. W.

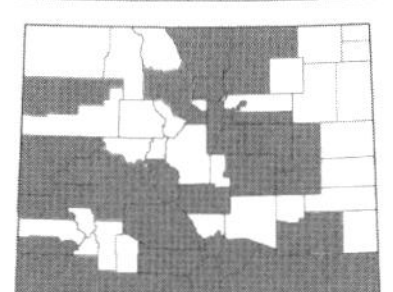

Stephanomeria pauciflora **(Torr.) A. Nelson**, BROWNPLUME WIRE LETTUCE. Perennials, 2–7 dm; *leaves* runcinate and pinnately lobed, to 5 cm long; *involucre* 7–10 mm high; *ray flowers* (4) 5–6, pink or lavender; *pappus* tawny or golden; *achenes* 3.5–4.5 mm long, with 5 broad ribs, scabrous, roughened, or cross-wrinkled. Common in sandy soil of dry, open places, 3500–8500 ft. May–Sept. E/W.

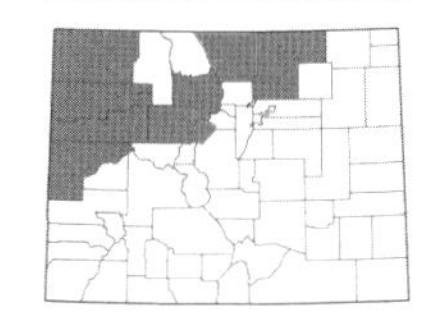

Stephanomeria runcinata **Nutt.**, DESERT WIRE LETTUCE. Perennials, 0.5–2 (2.5) dm; *leaves* runcinate-pinnatifid, 1.5–7 cm long; *involucre* 9–12 mm high; *ray flowers* 5–6, pink; *pappus* bright white; *achenes* 3–5 mm long, with 5 broad ribs, smooth or rugose-tuberculate and pitted. Uncommon in sandy or rocky soil of dry, open places, 5000–8500 ft. June–Sept. E/W.

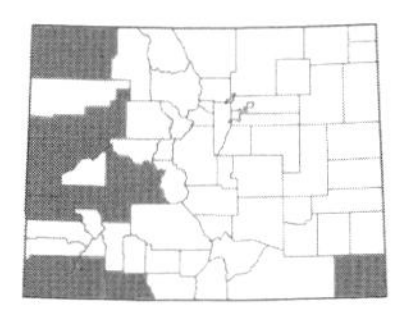

Stephanomeria tenuifolia **(Raf.) Hall**, NARROWLEAF WIRE LETTUCE. [*S. minor* (Hook.) Nutt.]. Perennials, 1–7 dm; *leaves* entire or sparsely toothed, linear to filiform, to 8 cm long; *involucre* 5–11 mm high; *ray flowers* 3–5 (6), pink, lavender, or white; *pappus* bright white; *achenes* 4–6 mm long, longitudinally ribbed and grooved. Uncommon in dry, sandy or rocky soil, 4400–8500 ft. July–Oct. E/W.

SYMPHYOTRICHUM Nees – ASTER

Herbs; *leaves* alternate, simple; *heads* usually radiate; *involucre bracts* in two or more series, imbricate, green at the tip and chartaceous below; *receptacle* naked; *disk flowers* yellow, purple, pink or occasionally white; *ray flowers* white, blue, purple, or pink, sometimes the ligule very short and inconspicuous and the heads appearing discoid; *pappus* of capillary bristles or double with an inner series of capillary bristles and an outer series of short hairs; *achenes* usually several-nerved. (Allen, 1985; Allen et al., 1983; Cronquist, 1943; Dean & Chambers, 1983; Jones, 1975; 1978; 1980; Nesom, 1997; Semple & Brouillet, 1980; Semple & Chmielewski, 1987; Wetmore & Delisle, 1939)

This is a very complex genus with many species exhibiting weak reproductive barriers. Some species are allopolyploids formed from hybridization between two other species and can backcross to one or more parent resulting in a plant with traits from one or more species. Other species hybridize where their ranges overlap, again resulting in a hybrid with traits of one or more species. In such cases, determinations can be quite difficult and specimen annotation labels often read "with influence of other species"!

1a. Annuals from taproots; rays inconspicuous (less than 2 mm long) and barely surpassing the disk flowers or essentially absent...2
1b. Perennials with woody caudices or creeping rhizomes; rays conspicuous and well-developed, far surpassing the disk flowers...3

2a. Rays 1.5–2 mm long, pinkish; involucre bracts oblong to narrowly oblanceolate...***S. frondosum***
2b. Rays essentially absent; involucre bracts acute to acuminate...***S. ciliatum***

3a. Stems and/or involucre bracts with glandular hairs...4
3b. Stems and involucre bracts without glandular hairs, instead glabrous to variously hairy ...8

4a. Leaves linear or narrowly oblong, more than 7 times as long as wide, usually less than 4 mm wide...5
4b. Leaves lanceolate, oblong, or elliptic, less than 5 times as long as wide and the principal leaves over 4 mm wide...7

5a. Leaf margins with coarse, bristly-ciliate hairs, these spaced approximately 1 mm apart...***S. fendleri***
5b. Leaf margins without ciliate hairs, or if present then these not long and coarse and very closely spaced together...6

6a. Involucre bracts hairy, not glandular...***S. ×amethystinum***
6b. Involucre bracts glandular...***S. campestre***

7a. Leaf bases not auriculate but sometimes slightly clasping; involucre bracts oblanceolate, obtuse, or rarely acute, the tips green, not at all purple or purple-margined; stem not reddish...***S. oblongifolium***
7b. At least some of the leaf bases auriculate or cordate-clasping; involucre bracts acute or acuminate, usually with purple margins or tips; upper stem often reddish or purplish...***S. novae-angliae***

8a. Involucre bracts and most leaves with a white spinulose tip; ray flowers usually white or sometimes violet or pink...9
8b. Involucre bracts and leaves without a white spinulose tip; ray flowers various...12

9a. Stems and leaves glabrous, although often ciliate-margined; basal leaf cluster usually present at the time of flowering; stem leaves usually a mixture of long leaves and short leaves...***S. porteri***
9b. Stems and leaves hairy; basal leaf cluster not present at the time of flowering or rarely so; stem leaves all generally the same length...10

10a. Ray flowers violet, blue, pink, or very seldom white; involucre bracts with a small spinulose tip, the green portion ovate nearly orbicular; heads usually larger (involucre 6–15 mm wide)...***S. ascendens***
10b. Ray flowers white; involucre bracts with a conspicuous spinulose tip, the green portion lanceolate; heads usually smaller (involucre 4.5–9.0 mm wide)...11

11a. Heads secund (situated on one side of the stem; this is easiest to see on unpressed specimens) and relatively small with an involucre height of 2.5–5 (average 4) mm; disk flowers 5–25 per head...***S. ericoides***
11b. Heads usually not secund and larger with an involucre height of 4.5–8 (average 5.5) mm; disk flowers 13–36 per head...***S. falcatum***

12a. Outer involucre bracts equaling or surpassing the inner involucre bracts in length, mostly foliaceous...13
12b. Outer involucre bracts not equaling or surpassing the inner involucre bracts in length, foliaceous or not...15

13a. Inflorescence a long panicle with numerous leaves and many heads; ray flowers pink or white (sometimes drying purple)...***S. eatonii***
13b. Inflorescence with few heads or if many then the inflorescence a cymose-panicle with reduced leaves; ray flowers pink, purple, blue, or violet...14

14a. Middle stem leaves narrowly lanceolate, mostly less than 1 cm wide; involucre bracts narrow, never leafy, sometimes purple-tipped but rarely purple-margined, found below 10,500 ft in elevation; stems erect...***S. spathulatum***
14b. Middle stem leaves lanceolate to ovate, mostly over 1 cm wide; involucre bracts wider and leafy (if bracts narrow and linear and leaves less than 1 cm wide then the bracts usually purple-tipped and -margined, and plants with decumbent to erect stems found in the alpine typically above 10,000 ft)...***S. foliaceum***

15a. Outer involucre bracts obtuse (the green portion ovate to nearly orbicular)...***S. ascendens***
15b. Outer involucre bracts acute...16

16a. Stems glabrous, or sometimes inconspicuously hairy in short lines in the inflorescence...17
16b. Stems hairy, at least in lines decurrent from the leaf base, and uniformly so in the inflorescence...18

17a. Leaves lanceolate to ovate, or sometimes the uppermost linear-lanceolate (less than 5 times as long as wide), 5–45 mm wide; ray flowers usually blue or purple, seldom white; achenes glabrous; plants usually from thick, woody caudex and short rhizomes...***S. laeve* var. *geyeri***
17b. Leaves linear to linear-lanceolate (usually 6 or more times longer than wide), 1–12 mm wide; ray flowers white or sometimes purple; achenes sparsely hairy; plants from slender, long rhizomes...***S. boreale***

18a. Hairs on the stem in lines decurrent from the leaf bases and not uniform under the heads; heads usually many in much-branched paniculiform arrays and smaller...***S. lanceolatum* ssp. *hesperium***
18b. Hairs on the stem uniform or if in lines, then uniform under the heads; heads usually fewer (3–10) on few branches and larger...***S. spathulatum***

Symphyotrichum ×amethystinum **(Nutt.) G.L. Nesom**. [*Aster ×amethystinus* Nutt.]. A hybrid between *S. novae-angliae* and *S. ericoides*, occurring where their ranges overlap, 5000–6000 ft. Aug.–Oct. E.

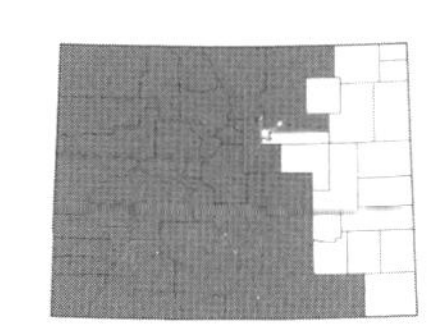

Symphyotrichum ascendens **(Lindl.) G.L. Nesom**, WESTERN ASTER. [*Aster ascendens* Lindl.; *Virgulaster ascendens* (Lindl.) Semple]. Perennials, 1–12 dm; *leaves* linear or narrowly lanceolate, 2–10 cm long, entire, glabrous to hairy; *involucre* (4) 5–7 mm high, the tips of bracts rounded or sometimes minutely mucronate; *ray flowers* 25–65 (80), 6–10 mm long, pink or violet to blue, occasionally white. Common in a wide variety of habitats including along roadsides, in meadows, along streams, and in sagebrush, 5000–10,500 ft. July–Sept. (Oct.). E/W.

Symphyotrichum ascendens is an allopolyploid derived from the hybridization of *S. spathulatum* and *S. falcatum* (Allen, 1985). It is a variable species, probably from the result of backcrosses to the original parents or even to another closely related species. It tends to resemble *S. spathulatum* morphologically more than *S. falcatum*.

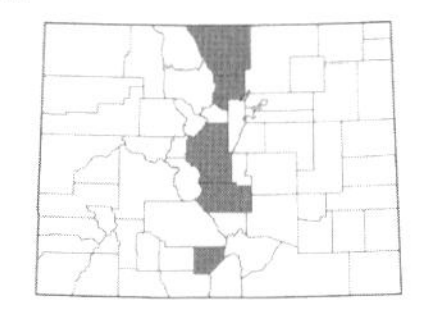

Symphyotrichum boreale **(Torr. & A. Gray) Á. Löve & D. Löve**, SLENDER WHITE BOG ASTER. [*Aster borealis* (Torr. & A. Gray) Prov.]. Perennials, 1.5–7 dm; *leaves* linear, 2–10 cm long, glabrous or the margins scabrous; *involucre* 5–7 mm high, acute or the outer obtuse, glabrous to sparsely hairy; *ray flowers* 25–35 (50), 7–11 mm long, white, pink, or lavender. Uncommon in fens, marshes, and along streams, 5000–9500 ft. July–Sept. E/W.

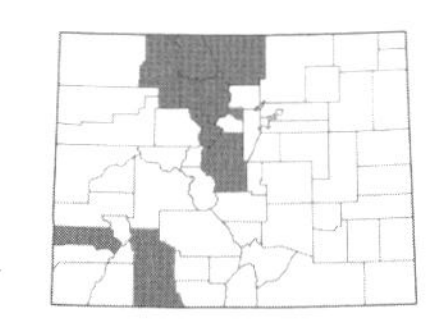

Symphyotrichum campestre **(Nutt.) G.L. Nesom,** WESTERN MEADOW ASTER. [*Aster campestris* Nutt.; *Virgulus campestris* (Nutt.) Reveal & Keener]. Perennials, 1–4 dm; *leaves* linear-oblanceolate, 1–8 cm long, glabrous to sparsely scabrous to glandular-stipitate, mucronate or white spinulose; *involucre* 5.5–8 mm high, sparsely to densely glandular-stipitate; *ray flowers* 15–30, 5–15 mm long, purple. Infrequent in dry meadows and open places in the mountains, 7000–9500 ft. July–Sept. E/W.

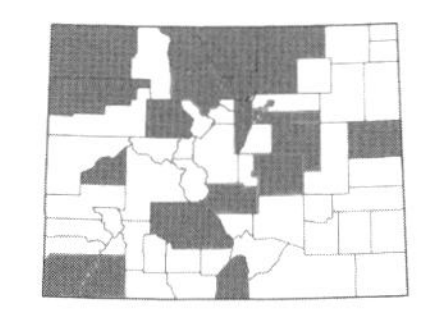

Symphyotrichum ciliatum **(Ledeb.) G.L. Nesom**, RAYLESS ALKALI ASTER. [*Aster brachyactis* S.F. Blake; *Brachyactis ciliata* (Ledeb.) Ledeb. ssp. *angusta* (Lindl.) A.G. Jones]. Annuals, 0.7–7 dm; *leaves* linear-oblanceolate to spatulate, 3–8 (15) cm long, glabrous, margins ciliate to scabrous; *involucre* 5–7 (11) mm high, the bracts acute, glabrous; *ray flowers* absent. Usually found along the borders of lakes or streams in moist, saline soil, 4500–8500 ft. Aug.–Oct. E/W.

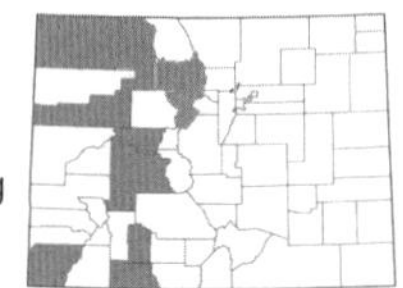

***Symphyotrichum eatonii* (A. Gray) G.L. Nesom**, EATON'S ASTER. [*Aster bracteolatus* Nutt.; *Aster eatonii* (A. Gray) Howell]. Perennials, (2.7) 6–15 dm; *leaves* linear to narrowly elliptic or oblanceolate, 0.8–15 cm long, glabrous to minutely hairy, the margins ciliate; *involucre* 4.5–8 (10) mm high, the bracts mucronate; *ray flowers* 20–40, 5–12 mm long, pink or white. Infrequent along streams, moist meadows, and on open slopes in the mountains, 6300–10,000 ft. July–Sept. W.

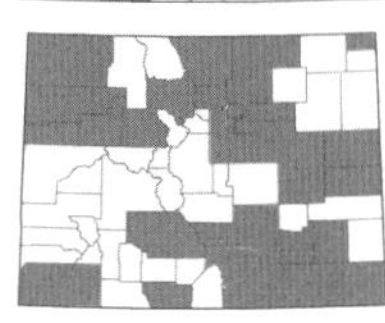

***Symphyotrichum ericoides* (L.) G.L. Nesom**, WHITE ASTER. [*Aster ericoides* L.; *Virgulus ericoides* (L.) Reveal & Keener]. Perennials, 3–10 dm; *leaves* linear to lanceolate, 5–7 cm long, hirsute, white spinulose at the tips; *involucre* 2.5–5 mm high, the bracts white spinulose-tipped; *ray flowers* 10–18, 3–8 mm long, white. Common in fields, along roadsides, and in open plains, 3500–8000 ft. July–Oct. E/W. (Plate 15)

Stem hair type has been used as criteria for separating *S. ericoides* (hairs appressed or ascending) from *S. falcatum* (hairs spreading). However, Jones (1978) notes that in populations of *S. ericoides*, colonies of plants with both pubescence types were found occurring side by side and that the plants were otherwise indistinguishable from each other. In addition, the stem pubescence can vary between both types on a single specimen! Although *S. ericoides* and *S. falcatum* can cross and produce hybrid plants, the occurrences of these are uncommon, indicating a strong reproductive barrier between the two species.

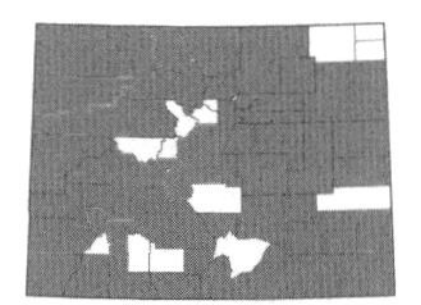

***Symphyotrichum falcatum* (Lindl.) G.L. Nesom**, WHITE PRAIRIE ASTER. [*Aster falcatus* Lindl.; *Virgulus falcatus* (Lindl.) Reveal & Keener]. Perennials, 2–8 dm; *leaves* linear to narrowly lanceolate, 2–8 cm long, white spinulose-tipped; *involucre* 4.5–8 mm high, the bracts white spinulose-tipped; *ray flowers* 17–35, 6–8 mm long, white. Common in fields, along roadsides, and in open plains and meadows, 3500–10,000 ft. July–Oct. E/W. (Plate 15)

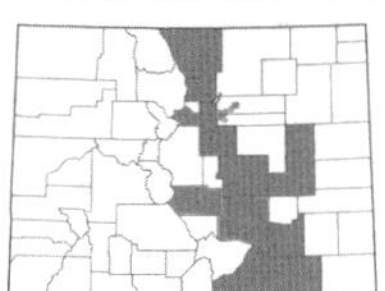

***Symphyotrichum fendleri* (A. Gray) G.L. Nesom**, FENDLER'S ASTER. [*Aster fendleri* A. Gray]. Perennials, 0.5–3 dm; *leaves* linear-oblanceolate to linear-lanceolate, 1–4 cm long, white spinulose-tipped, sometimes glandular-stipitate; *involucre* 4–7 mm high, the bracts usually glandular-stipitate; *ray flowers* purple, 5–10 mm long. Found on dry, rocky slopes and plains, 5000–6500 ft. Aug.–Oct. E.

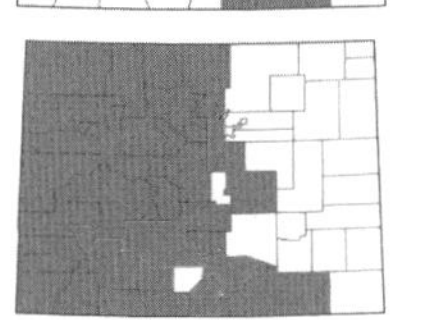

***Symphyotrichum foliaceum* (DC.) G.L. Nesom**, LEAFY ASTER. [*Aster foliaceus* Lindl.]. Perennials, 0.5–10 dm; *leaves* oblanceolate, lanceolate, or ovate, 1.8–16 cm long; *involucre* 6–12 mm high, the bracts usually large and leafy, acute to rounded at the tips; *ray flowers* 15–60, 9–20 mm long, pink, purple, or blue. Common in moist meadows, along streams, and in wooded places in the mountains, 5700–14,000 ft. July–Sept. E/W.

There are three varieties of *S. foliaceum* in Colorado. These can intergrade, but for the most part they are distinct:

1a. Plants shorter, less than 2 dm; involucre bracts with purple tips and margins; middle stem leaves mostly less than 1 cm in width...**var. *apricum* (A. Gray) G.L. Nesom**, ALPINE LEAFY ASTER. Found in alpine and subalpine slopes and meadows and along streams, (9200) 10,000–13,800 ft. July–Sept. E/W.

1b. Plants taller, greater than 2 dm; involucre bracts typically without purple tips and margins; middle stem leaves mostly greater than 1 cm in width...2

2a. Involucre bracts oblong to ovate with an obtuse to acutish tip, the outer foliaceous ones broadly lanceolate to ovate with a rounded apex...**var. *canbyi* A. Gray (G.L. Nesom)**, CANBY'S ASTER. Found in typically drier habitats such as in woods and on open slopes, usually at lower elevations than var. *apricum*, 5700–9800 ft. July–Sept. E/W.

2b. Involucre bracts linear with an acute to acuminate tip, the outer foliaceous ones linear with a very acute apex...**var. *parryi* (D.C. Eaton) G.L. Nesom**, PARRY'S ASTER. Found in moist habitats such as along streams and in wet meadows, usually at lower elevations than var. *apricum*, 7000–10,500 ft. July–Sept. E/W.

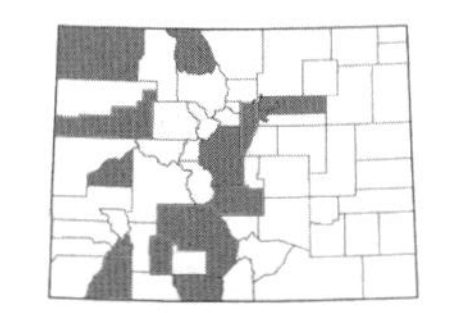

***Symphyotrichum frondosum* (Nutt.) G.L. Nesom**, SHORT-RAYED ALKALI ASTER. [*Aster frondosus* (Nutt.) Torr. & A. Gray; *Brachyactis frondosa* A. Gray]. Annuals, 0.5–14 dm; *leaves* oblanceolate to linear, 1–11 cm long, glabrous; *involucre* 5–9 mm high, the bracts glabrous; *ray flowers* 90–110, 1.5–2 mm long, pink to pinkish-white. Usually found along the borders of lakes or ponds in moist, saline soil, 5000–9000 ft. Aug.–Oct. E/W.

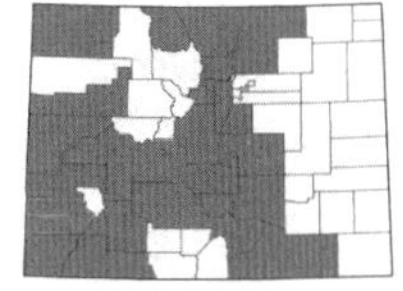

***Symphyotrichum laeve* (L.) Á. Löve & D. Löve var. *geyeri* (A. Gray) G.L. Nesom**, SMOOTH BLUE ASTER. [*Aster laevis* L. var. *geyeri* A. Gray]. Perennials, 4–12 dm; *leaves* linear-oblanceolate to lanceolate or elliptic, 0.8–14 cm long, glabrous; *involucre* 5–8 mm high, the bracts glabrous; *ray flowers* 15–30, 6–9 mm long, purple or blue. Common in open meadows, along streams, and in forest openings, 5000–10,500 ft. July–Sept. E/W. (Plate 15)

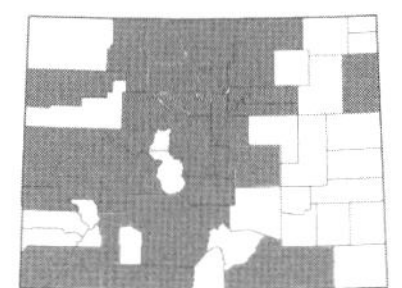

Symphyotrichum lanceolatum **(Willd.) G.L. Nesom ssp.** ***hesperium*** **(A. Gray) G.L. Nesom**, WESTERN LINED ASTER. [*Aster hesperius* A. Gray]. Perennials, 3–15 (20) dm; *leaves* linear, lanceolate-ovate, oblanceolate, or obovate, 4–15 cm long, glabrous; *involucre* 3–8 mm high, the outer bracts sometimes foliaceous; *ray flowers* 16–50, white to pinkish or pale purple. Common along streams and ditches and in moist meadows, 3500–10,000 ft. July–Sept. (Oct.). E/W.

Symphyotrichum foliaceum (see above) specimens with influence of other *Symphyotrichum* species may key out to here because the outer involucre bracts are not equal to or longer than the inner involucre bracts in these hybrids. However, this species can be separated from *S. lanceolatum* by the presence of fewer, larger, and showier heads (involucre width 10–20 mm as opposed to an involucre width of 7–10 (12) mm in *S. lanceolatum*).

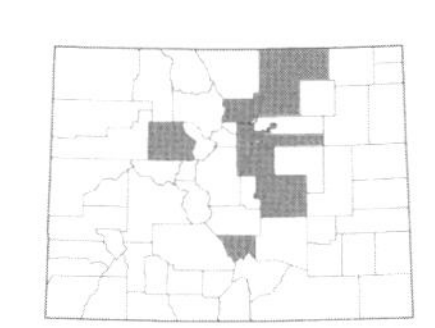

Symphyotrichum novae-angliae **(L.) G.L. Nesom**, NEW ENGLAND ASTER. [*Aster novae-angliae* L.; *Virgulus novae-angliae* (L.) Reveal & Keener]. Perennials, 3–12 dm; *leaves* spatulate to oblanceolate, oblong or lanceolate, 2–10 cm long, hairy and glandular-stipitate; *involucre* 7–10 (15) mm high, the bracts glandular-stipitate; *ray flowers* 40–75 (100), 9–15 mm long, dark pink to purple, rarely white. Found along roadsides and in open meadows, more or less found in the Denver-Boulder area but can also escape from cultivation and is sometimes intentionally introduced, 5000–7800 ft. Aug.–Oct. E.

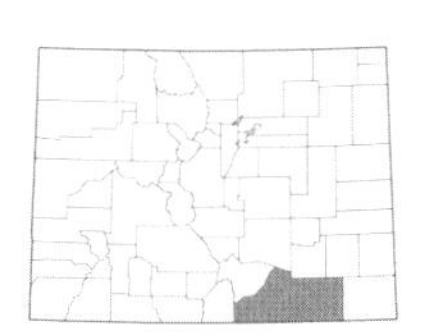

Symphyotrichum oblongifolium **(Nutt.) G.L. Nesom**, AROMATIC ASTER. [*Aster oblongifolius* Nutt.; *Virgulus oblongifolius* (Nutt.) Reveal & Keener]. Perennials, 1–10 dm; *leaves* linear-lanceolate to oblong or oblanceolate, 2–10 cm long, glandular-stipitate; *involucre* 5–9 mm high, the bracts hairy and glandular-stipitate; *ray flowers* 25–35, 9–15 mm long, purple or rose-purple. Found in dry, rocky open sites and mesas, known only from southern Las Animas Co. very close to the New Mexico border, 5800–7800 ft. Aug.–Oct. E.

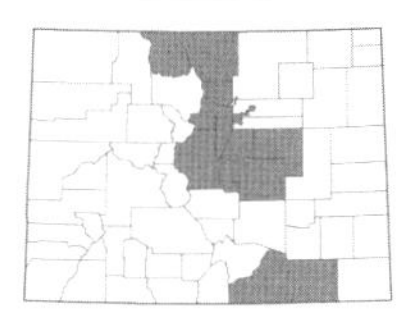

Symphyotrichum porteri **(A. Gray) G.L. Nesom**, SMOOTH WHITE ASTER. [*Aster porteri* A. Gray]. Perennials, 2–4 dm; *leaves* linear to linear-oblanceolate, 5–10 cm long, white spinulose-tipped; *involucre* 4–5 mm high, the bracts white spinulose-tipped; *ray flowers* 4–8 mm long, white. Common in open fields, meadows, along roadsides, and in ponderosa pine forests, 4900–9500 ft. July–Oct. E.

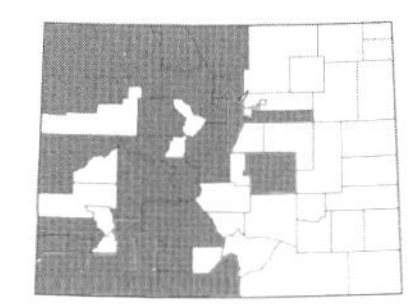

Symphyotrichum spathulatum **(Lindl.) G.L. Nesom**, WESTERN MOUNTAIN ASTER. [*Aster occidentalis* (Nutt.) Torr. & A. Gray; *Aster spathulatus* Lindl. var. *spathulatus*]. Perennials, 1–8.5 dm; *leaves* oblanceolate to lanceolate or elliptic, 1–15 cm long, glabrous; *involucre* 5–8 mm high; *ray flowers* 15–50, 6–15 mm long, purple to blue. Infrequent along streams and in moist meadows in the mountains, 6000–10,500 ft. July–Oct. E/W.

Symphyotrichum spathulatum freely hybridizes with *S. foliaceum*, especially var. *parryi* and sometimes var. *apricum* where their habitats overlap.

TANACETUM L. – TANSY

Annual or perennial herbs; *leaves* alternate; *heads* in a corymbiform inflorescence, discoid or disciform; *involucre bracts* imbricate in 2–3 series, the inner at least with scarious margins and tips; *receptacle* naked; *disk flowers* tubular, yellow, the corolla 5-lobed; *ray flowers* usually lacking or less than 1 cm long; *pappus* a short crown or lacking; *achenes* 3- to 10-ribbed.

1a. Leaves simple, occasionally the stem leaves with a few basal pinnae, the margins crenate-dentate...***T. balsamita***
1b. Leaves pinnatifid to bipinnatifid, the pinnae lobed or dissected...2

2a. Heads with well-developed ray flowers...***T. parthenium***
2b. Heads with only disk flowers present...***T. vulgare***

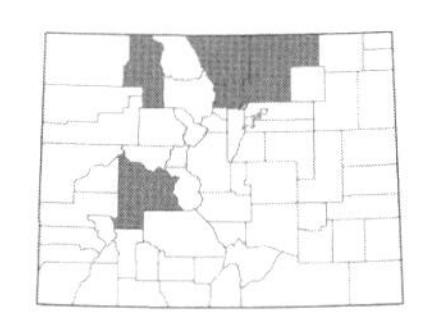

Tanacetum balsamita **L.**, COSTMARY. [*Balsamita major* Desf.; *Chrysanthemum balsamita* L.]. Perennials, 3–10 dm; *leaves* simple or occasionally with 1–4 lobes near the base, the margins crenate-dentate, 10–20 cm long; *disk flowers* 2 mm long; *ray flowers* usually absent; *pappus* a short crown. Cultivated in gardens and rarely escaping where it is found in fields and along roadsides and ditches, 5000–7700 ft. Aug.–Oct. E. Introduced.

Tanacetum parthenium **(L.) Sch. Bip.**, FEVERFEW. [*Chrysanthemum parthenium* Bernh.]. Perennials, 2–6 (8) dm; *leaves* 1–2-pinnately lobed, the margins toothed to pinnatifid, gland-dotted and short-hairy, 4–10 cm long; *disk flowers* 2 mm long; *ray flowers* 2–8 (12) mm long; *pappus* absent or a short crown. Occasionally escaping from cultivation, 5000–6000 ft. June–Sept. E. Introduced.

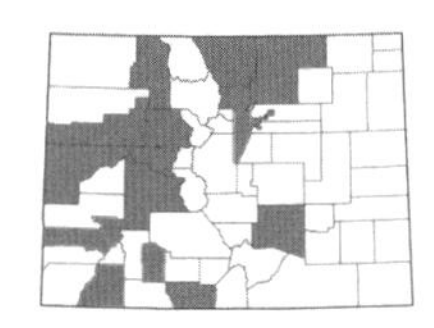

***Tanacetum vulgare* L.**, TANSY. Perennials, 4–15 dm; *leaves* oblong to elliptic or oval, 4–20 cm long, pinnatifid, the margins toothed, gland-dotted, glabrous to sparsely hairy; *disk flowers* 2–3 mm long; *ray flowers* absent; *pappus* a short crown. Commonly cultivated in gardens and occasionally escaping where it is found along roadsides and ditches and in meadows, 5000–8500 ft. July–Sept. E/W. Introduced.

TARAXACUM F.H. Wigg. – DANDELION

Perennials, scapose, lactiferous herbs; *leaves* basal; *heads* solitary, ligulate; *involucre bracts* biseriate, the outer shorter and generally reflexed, the inner bracts of some with a hooded appendage near the apex; *receptacle* naked; *disk flowers* absent; *ray flowers* yellow (ours); *pappus* of capillary bristles; *achenes* columnar, 4–5-angled or -ribbed, with a slender beak. (Brouillet, 2006; Kirschner & Stepanek, 1987; Richards, 1985; Sherff, 1920)

1a. Achenes tawny to brown or black at maturity, sometimes with a reddish tip...2
1b. Achenes reddish or reddish-purple at maturity...4

2a. Leaves mostly lobed more than halfway to the midrib, without a broadly winged base; outer involucre bracts usually reflexed; introduced weedy plants...***T. officinale***
2b. Leaves mostly lobed less than halfway to the midrib or entire and unlobed, usually with a broadly winged base; outer involucre bracts erect to spreading; native plants of higher elevations...3

3a. Heads smaller, the involucre usually 6–11 (13) mm high; inner involucre bracts not or only scarcely corniculate with a hooded appendage; achenes black or brown at maturity...***T. scopulorum***
3b. Heads usually larger, the involucre (8) 11–20 mm high; involucre bracts usually corniculate with a hooded appendage; achenes light brown to straw-colored at maturity...***T. ceratophorum***

4a. Leaves mostly lobed more than halfway to the midrib; outer involucre bracts usually reflexed; inner involucre bracts often corniculate with a hooded appendage; introduced weedy plants...***T. laevigatum***
4b. Leaves mostly lobed less than halfway to the midrib or more commonly entire and unlobed; outer involucre bracts erect to spreading; inner involucre bracts rarely corniculate; native plants of the mountains...***T. eriophorum***

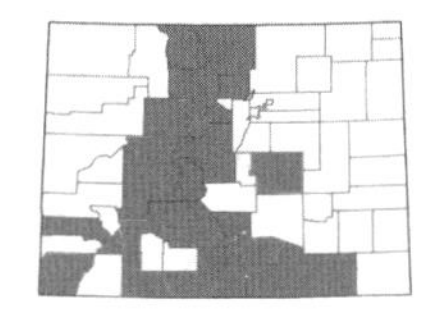

***Taraxacum ceratophorum* (Ledeb.) DC.**, NORTHERN DANDELION. [*T. ovinum* Rydb.]. Plants 0.5–5 dm; *leaves* narrowly oblanceolate to linear-oblong, 4–30 cm long, the margins sinuate-dentate to pinnatifid; *involucre* (8) 11–20 mm high, the outer usually corniculate; *ray flowers* 10–22 mm long; *achenes* 4.5–5 mm long, straw-colored to light brown. Common in subalpine and alpine meadows and on rocky ridges, 8700–13,000 ft. June–Aug. E/W.

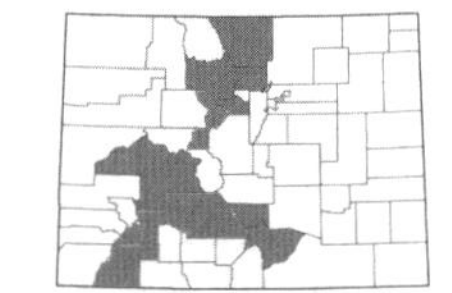

***Taraxacum eriophorum* Rydb.**, WOOLBEARING DANDELION. Plants 0.5–5 dm; *leaves* oblanceolate to obovate or runcinate, 5–25 cm long, the margins mostly lobed less than halfway to the midrib; *involucre* 10–25 mm high, the outer bracts erect to spreading, the inner rarely corniculate; *ray flowers* 12–16 mm long; *achenes* 3–5 mm long, reddish or reddish-purple. Infrequent on alpine slopes, 10,000–14,000 ft. June–Aug. E/W.

Included within a very broad *T. ceratophorum* by Brouillet (2006).

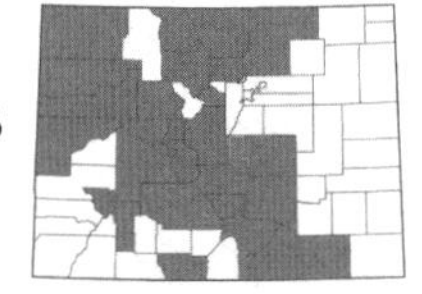

***Taraxacum laevigatum* (Willd.) DC.**, ROCK DANDELION. Plants 0.5–5 dm; *leaves* oblanceolate to obovate or runcinate, 5–25 cm long, the margins mostly lobed more than halfway to the midrib; *involucre* 10–25 mm high, the outer usually reflexed, the inner often corniculate; *ray flowers* 12–16 mm long; *achenes* 2.2–4 mm long, with ca. 15 ribs, reddish or reddish-purple. Found in disturbed places such as lawns, gardens, and fields, 5000–10,500 ft. April–Oct. E/W. Introduced.

Very similar in appearance and distribution to *T. officinale* but with reddish achenes.

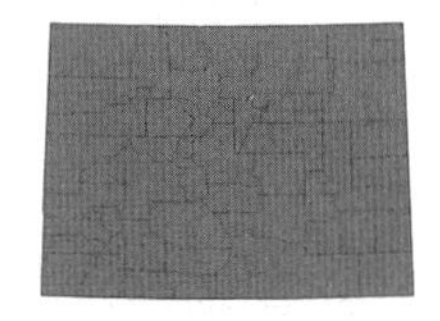

***Taraxacum officinale* F.H. Wigg.**, COMMON DANDELION. Plants 0.5–5 (6) dm; *leaves* oblanceolate to obovate, 4–45 cm long, the margins mostly lobed more than halfway to the midrib; *involucre* 14–25 mm high, the outer bracts usually reflexed; *ray flowers* 15–22 mm long; *achenes* 2–3 (4) mm long, with a slender beak 7–9 mm long and 4–12 ribs, brown. Common and widespread in a variety of habitats, although usually found in disturbed or waste places or lawns and cultivated plantings, 3500–10,500 ft. Mar.–Oct. E/W. Introduced. (Plate 15)

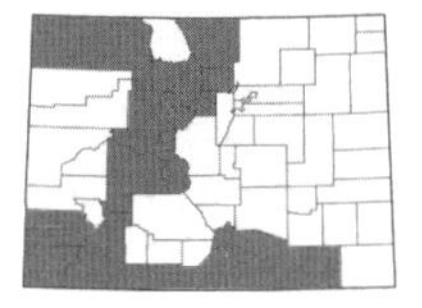

***Taraxacum scopulorum* (A. Gray) Rydb.**, ALPINE DANDELION. [*T. lyratum* (Ledeb.) DC.]. Plants 0.1–0.5 dm; *leaves* narrowly oblanceolate to runcinate, 1–4 cm long, the margins entire to dentate or shallowly pinnatifid; *involucre* 6–11 (13) mm high, the outer bracts erect to spreading; *ray flowers* 7.5–9 mm long; *achenes* 2.8–3.5 mm long, with a stout beak 2.8–4.5 mm long and 7–14 ribs, blackish or brown. Common in subalpine and alpine meadows and rocky ridges, 9500–14,000 ft. June–Aug. E/W.

TETRADYMIA DC. – HORSEBRUSH

Shrubs; *leaves* alternate, entire, often forming spines; *heads* discoid; *involucre bracts* 4–5 or rarely 6, equal to subequal; *receptacle* naked; *disk flowers* yellow, each corolla lobe often bearing a linear nerve (oil duct); *ray flowers* absent; *pappus* of barbellate or capillary bristles, slender squamellae, or absent; *achenes* prismatic to fusiform, slender, obscurely 5-nerved. (Strother, 1974)

1a. Unarmed shrubs without the primary leaves forming rigid spines...***T. canescens***
1b. Spinescent shrubs with the primary leaves forming rigid spines...2

2a. Achenes with hairs not obscuring the well-developed pappus of 75–100 capillary bristles; young twigs with alternating longitudinal strips of densely and thinly tomentose hairs...***T. nuttallii***
2b. Achenes with long hairs that obscure and hide the pappus of 25 capillary bristles; twigs densely and uniformly tomentose...***T. spinosa***

Tetradymia canescens **DC.**, COMMON HORSEBRUSH. Shrubs, 1–8 dm, unarmed; *leaves* lanceolate to spatulate, 0.5–4 cm long, tomentose; *involucre* 6–12 mm high, with 4 bracts; *disk flowers* 4, 7–15 mm long; *achenes* 3–5 mm long, glabrous or hairy. Common in dry, open places, usually with sagebrush, pinyon-juniper, or oak shrubs, 4500–9500 ft. July–Oct. E/W. (Plate 16)

Tetradymia nuttallii **Torr. & A. Gray**, NUTTALL'S HORSEBRUSH. Shrubs, 1–12 dm, spiny; *leaves* spatulate, 1–2 cm long, glabrous to tomentose, the primary leaves forming spines; *involucre* 6–9 mm high, with 4 bracts; *disk flowers* 8–10 mm long; *achenes* 4–6 mm long, densely hairy. Infrequent in dry, open places, usually on gypsum hills and shale soil, 5300–6500 ft. May–June. W.

Tetradymia spinosa **Hook. & Arn.**, CATCLAW HORSEBRUSH. Shrubs, 1–10 dm, spiny; *leaves* linear to spatulate, 0.3–2.5 cm long, the primary leaves forming spines; *involucre* 8–12 mm high, with 4–6 bracts; *disk flowers* 6–10 mm long; *achenes* 6–8 mm long, with long hairs. Found in sandy soil of dry, open places, usually in association with *Atriplex*, sagebrush, or rarely pinyon-juniper woodlands, 4500–7000 ft. May–June. W.

TETRANEURIS – FOUR-NERVE DAISY

Annual or perennial herbs; *leaves* all basal or basal and cauline, alternate, simple; *heads* radiate or discoid; *involucre bracts* in 3 series, subequal, herbaceous; *receptacle* naked; *disk flowers* yellow to purple-red, 5-lobed; *ray flowers* pistillate and fertile or absent, yellow, usually 3-lobed; *pappus* usually of aristate scales; *achenes* mostly 5-angled, hairy. (Bierner & Jansen, 1998; Bierner & Turner, 2003)

1a. Both basal and stem leaves present...***T. ivesiana***
1b. Leaves all basal...2

2a. Caudices not thickened distally, relatively narrow and the same width throughout...***T. scaposa* var. *scaposa***
2b. Caudices thickened distally...3

3a. Outer involucre bract margins usually conspicuously scarious; leaves densely gland-dotted...***T. torreyana***
3b. Outer involucre bract margins not scarious or sometimes inconspicuously and very narrowly so; leaves eglandular or sparsely to densely gland-dotted...***T. acaulis***

Tetraneuris acaulis **(Pursh) Greene**, STEMLESS FOUR-NERVE DAISY. [*Hymenoxys acaulis* (Pursh) Parker]. Plants 0.2–3.5 dm; *leaves* spatulate to linear-oblanceolate, basal, sparsely to densely hairy, sparsely to moderately gland-dotted or eglandular, 0.5–10 cm long; *involucre* 5–10 mm high, sparsely to densely hairy, sparsely to densely gland-dotted or eglandular, the outer margins not scarious or very narrowly so; *disk flowers* 2.7–4.3 mm long; *ray flowers* 9–17 mm long. Common in dry, open places from the plains to alpine, 3500–14,000 ft. May–Aug. E/W. (Plate 16)

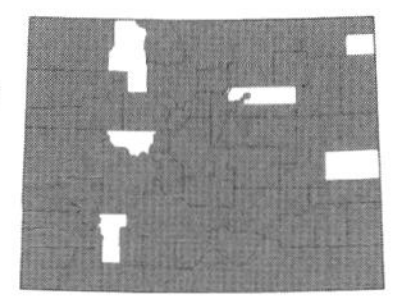

Occasionally, discoid plants of *T. acaulis* are found. *Tetraneuris acaulis* is a widespread and variable species, separable into the following varieties:

1a. Leaves eglandular or inconspicuously gland-dotted, glabrous or sparsely to moderately hairy, not strigose-canescent or sericeous...**var. *epunctata* (A. Nelson) Kartesz**, STEMLESS FOUR-NERVE DAISY. Locally common in dry, open places in sandy or rocky soil, 4900–9000 (10,000) ft. May–July. W.
Often forms intermediates with var. *arizonica* and *T. ivesiana*.
1b. Leaves densely gland-dotted, or if inconspicuously or sparsely gland-dotted, then also strigose-canescent and sometimes sericeous and usually densely hairy...2

2a. Leaves densely gland-dotted, not strigose-canescent or sericeous; plants exclusively of the western slope...**var. *arizonica* (Greene) Parker**, ARIZONA FOUR-NERVE DAISY. Uncommon in dry, open places, often with sagebrush, 5700–8500 ft. May–July. W.

2b. Leaves sparsely to densely gland-dotted, usually densely hairy and strigose-canescent or sericeous; plants mostly of the eastern slope...3

3a. Peduncles densely lanate below the heads, mostly 0.5–8 cm long...**var. *caespitosa* A. Nelson**, CAESPITOSE FOUR-NERVE DAISY. Common in the alpine, (8300) 10,000–14,000 ft. June–Aug. E/W.
3b. Peduncles not densely lanate below the heads, mostly 8–20 cm long...**var. *acaulis***, STEMLESS FOUR-NERVE DAISY. Common on the plains, in the foothills, and in the San Luis Valley, 3500–9500 ft. May–July. E/W.

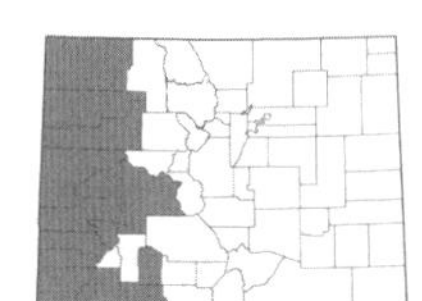

***Tetraneuris ivesiana* Greene**, IVES' FOUR-NERVE DAISY. [*Hymenoxys ivesiana* (Greene) Parker]. Plants 1–3 dm; *leaves* linear to linear-oblanceolate, glabrous to moderately hairy, strongly gland-dotted, basal and cauline, 5–19 cm long; *involucre* 8–12 mm high, the bracts sparsely to densely hairy and gland-dotted; *disk flowers* 3–4.5 mm long; *ray flowers* 10–20 mm long. Common in dry, open places, often with sagebrush or pinyon-juniper, 4900–9500 ft. May–Aug. W.

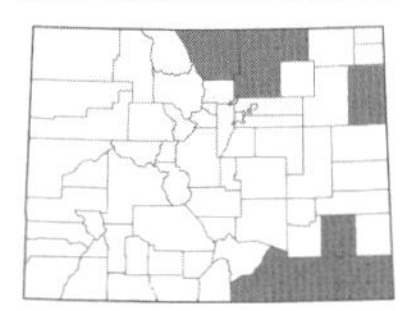

Tetraneuris scaposa* (DC.) Greene var. *scaposa, STEMMY FOUR-NERVE DAISY. [*Hymenoxys scaposa* (DC.) Parker var. *scaposa*]. Plants 1.4–4.5 dm; *leaves* spatulate to linear-oblanceolate, basal, sparsely to densely hairy and densely gland-dotted, 12–40 cm long; *involucre* 5–10 mm high, the bracts eglandular or sometimes gland-dotted; *disk flowers* 2.5–3.5 mm long; *ray flowers* 9.5–22 mm long. Found in open, rocky or sandy places on the plains, 3500–5500 ft. April–June (July). E.

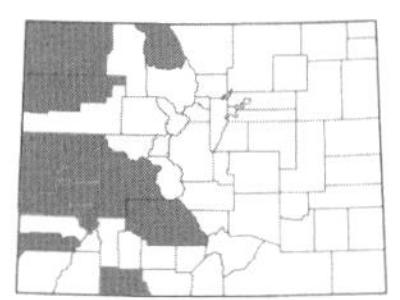

***Tetraneuris torreyana* (Nutt.) Greene**, TORREY'S FOUR-NERVE DAISY. [*Hymenoxys torreyana* (Nutt.) Parker]. Plants 0.2–2 dm; *leaves* spatulate to oblanceolate, glabrous or hairy, densely gland-dotted, basal; *involucre* 10–15 mm high, hairy, eglandular or sparsely gland-dotted; *disk flowers* 4–5 mm long; *ray flowers* 11–17 mm long. Uncommon or locally common in open, dry places in rocky or sandy soil, 6200–9000 ft. May–June. E/W.

THELESPERMA Less. – GREENTHREAD

Annual or perennial herbs; *leaves* opposite or sometimes upper alternate, 1 (3)-pinnately lobed; *heads* discoid or radiate; *involucre bracts* biseriate and dimorphic; *receptacle* chaffy; *disk flowers* yellow to reddish-brown; *ray flowers* yellow or absent; *pappus* of 2 or sometimes 3 retrorsely barbed awns, small teeth, or absent; *achenes* more or less compressed parallel to the involucre bracts.

1a. Ray flowers absent, or rarely present but very small and inconspicuous; leaves well-distributed along the stem...***T. megapotamicum***
1b. Ray flowers present and conspicuous; leaves mostly basal or well-distributed along the stem...2

2a. Leaves mostly basal; ultimate leaf segments broadly linear to lanceolate and wider (1–5 mm)...***T. subnudum***
2b. Leaves evenly distributed along the stem; ultimate leaf segments filiform and narrower (1 mm or less wide)...***T. filifolium***

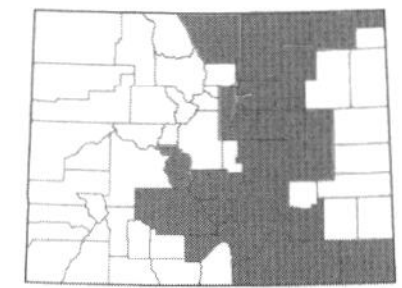

***Thelesperma filifolium* (Hook.) A. Gray**, STIFF GREENTHREAD. Annuals, 1–4 (7) dm; *leaves* with filiform ultimate lobes, mostly 0.5–1 mm wide; *disk flowers* reddish-brown to yellow with reddish-brown veins; *ray flowers* 8, yellow, 12–20 mm long; *achenes* 3.5–4.5 mm long. Common in sandy or rocky soil of open places, often in disturbed sites, 3500–7500 ft. June–Sept. E.

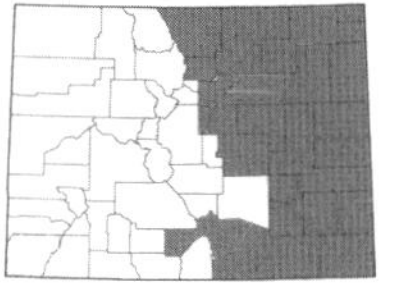

***Thelesperma megapotamicum* (Spreng.) Kuntze**, HOPI TEA GREENTHREAD. Perennials or subshrubs, 2–8 dm; *leaves* with linear to filiform ultimate lobes, 0.5–1 (2.5) mm wide; *disk flowers* yellow, usually with reddish-brown veins; *ray flowers* absent; *achenes* 5–8 mm long. Common in open places of the eastern plains and outer foothills, 3500–7500 ft. June–Sept. E. (Plate 16)

***Thelesperma subnudum* A. Gray**, NAVAJO TEA. Perennials, 1–4 dm; *leaves* with ultimate lobes lanceolate to oblanceolate, mostly 1–5 mm wide; *disk flowers* yellow, sometimes with reddish-brown veins; *ray flowers* absent or usually 8, yellow, (6) 12–20 mm; *achenes* 5–7 mm long. Infrequent in dry, open places, 4300–8000 ft. May–July. E/W.

Rarely, discoid *T. subnudum* specimens are found. These can be distinguished from *T. megapotamicum* by the mostly basal leaves which are pinnatifid to bipinnatifid. In *T. megapotamicum*, the leaves are well-distributed along the stem and are usually trifid or simple.

THYMOPHYLLA Lag. – PRICKLYLEAF

Annual or perennial herbs or subshrubs; *leaves* gland-dotted; *heads* radiate (ours); *involucre bracts* not much imbricate in 2–3 series, fused for about ½ their length, gland-dotted; *receptacle* naked; *disk flowers* yellow; *ray flowers* few or absent, yellow or white; *pappus* of scales, each tipped with an awn or dissected into bristles; *achenes* obpyramidal to cylindric.

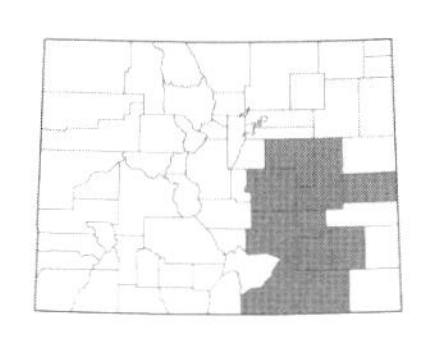

Thymophylla aurea **(A. Gray) Greene ex Britton var. *aurea*,** MANY-AWN PRICKLYLEAF. [*Dyssodia aurea* (A. Gray) A. Nelson]. Annuals to 20 cm, branching from bases; *leaves* alternate, lobed into 5–13 linear segments; *peduncles* 1–3 cm long; *ray flowers* yellow, 4–6 mm long; *pappus* 0.3–0.6 mm long; *achenes* ca. 3 mm long. Infrequent in open places on the eastern plains, 4300–5400 ft. June–Oct. E.

TONESTUS A. Nelson – SERPENTWEED

Perennials; *leaves* alternate and basal; *heads* radiate (ours); *involucre bracts* nearly equal in 3–4 series, the outer usually foliaceous and intergrading into the upper leaves; *receptacle* naked; *disk flowers* yellow; *ray flowers* yellow or absent; *pappus* of barbellate bristles; *achenes* narrowly oblong to subcylindric, somewhat compressed, usually ribbed. (G.L. Nesom & Morgan, 1990)

1a. Outer involucre bracts linear-lanceolate with acute tips; stems and leaves densely glandular; achenes glabrous or sparsely hairy distally...***T. lyallii***
1b. Outer involucre bracts ovate to elliptic with rounded or shortly mucronate tips; stems and leaves not glandular or sparsely glandular, although hairy along the margins and on the stem; achenes villous...***T. pygmaeus***

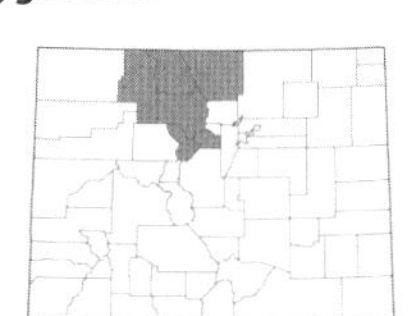

Tonestus lyallii **(A. Gray) A. Nelson,** LYALL'S GOLDENWEED. [*Haplopappus lyallii* A. Gray]. Plants 4–9 (15) cm; *leaves* linear to spatulate or oblong, 1.2–8.5 cm long, glandular-stipitate, entire; *involucre* 11–22 mm high, the outer bracts linear-lanceolate with acute tips, glandular-stipitate distally; *disk flowers* 5.5–8.5 mm long; *ray flowers* 6.5–8 mm long; *achenes* 2.5–5.5 mm long, glabrous or sparsely hairy. Uncommon in rocky soil of the alpine, 11,000–13,000 ft. July–Sept. E/W.

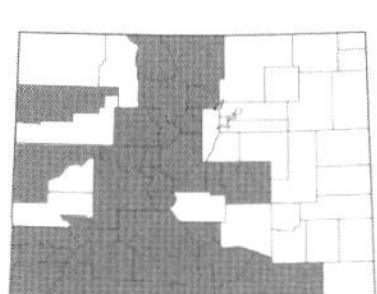

Tonestus pygmaeus **(Torr. & A. Gray) A. Nelson,** PYGMY GOLDENWEED. [*Haplopappus pygmaeus* (Torr. & A. Gray) A. Gray]. Plants 1–9 cm; *leaves* linear to spatulate or oblong, 1–5 (9.5) cm long, eglandular or sparsely glandular-stipitate, entire; *involucre* 8–20 mm high, the outer bracts ovate to elliptic with rounded or shortly mucronate tips; *disk flowers* 4.5–7.5 mm long; *ray flowers* 6.5–8.5 mm long; *achenes* 2–5 mm long, villous. Locally common in rocky soil of the alpine, 10,000–14,000 ft. July–Aug. E/W. (Plate 16)

TOWNSENDIA Hook. – EASTER DAISY; TOWNSEND DAISY

Herbs; *leaves* alternate; *heads* radiate; *involucre bracts* imbricate in 2–7 series, usually scarious-ciliate-margined; *receptacle* naked; *disk flowers* yellow or purplish; *ray flowers* white to pink; *pappus* of disk flowers of barbellate bristles, that of ray flowers sometimes reduced to a crown or set of short bristles; *achenes* compressed, 2–3 ribbed. (Beaman, 1957)

1a. Involucre bracts usually reddish-purple and anthocyanic throughout, elliptical, ovate, obovate, or broadly lanceolate; ray flowers blue or white; heads usually sessile or nearly so, or sometimes on short peduncles...***T. rothrockii***
1b. Involucre bracts sometimes anthocyanic along the margins or at the tips but not throughout, variously shaped; ray flowers white, pink, or blue; heads sessile or on evident peduncles...2

2a. Heads sessile or nearly so on very short, inconspicuous peduncles, often prominently surpassed by the basal leaves...3
2b. Heads evidently pedunculate, this naked or leafy, and not surpassed by the basal leaves...6

3a. Pappus of ray flowers shorter than that of disk flowers; leaves narrowly spatulate to oblanceolate; ray flowers 8–22...***T. incana***
3b. Pappus of ray flowers equal to or longer than that of disk flowers; leaves various; ray flowers 15–40...4

4a. Heads large, the disk 15–30 mm wide and involucre bracts 10–22 mm high; ray flowers 12–22 mm long; disk corollas 6–11 mm long...***T. exscapa***
4b. Heads smaller, the disk 10–15 mm wide and involucre bracts 7–12 mm high; ray flowers 8–14 mm long; disk corollas 3–7 mm long...5

5a. Involucre bracts mostly acuminate, terminated by a tuft of tangled cilia...***T. hookeri***
5b. Involucre bracts mostly acute, without a terminal tuft of cilia...***T. leptotes***

6a. Heads solitary at the ends of long, naked peduncles; plants glabrous or slightly strigose...***T. glabella***
6b. Heads solitary or corymbose on leafy peduncles or stems; plants strigose to villous or canescent...7

7a. Stems densely, conspicuously, and permanently white-canescent; disk pappus as long or longer than the disk corollas...***T. incana***
7b. Stems strigose to villous, rarely canescent and if so then the disk pappus shorter than the disk corollas...8

8a. Disk pappus of short squamellae or bristles and mostly 2–4 (8) coarse, stiff barbellate bristles to 4 mm longer than the squamellae; ray flowers purple to blue...***T. eximia***
8b. Disk pappus of 12 or more barbellate bristles; ray flowers white, pink, or light lavender...9

9a. Involucre bracts with long, attenuate-acuminate tips; heads usually larger, the disk (13) 15–30 mm wide and involucre bracts 10–18 mm high...***T. grandiflora***
9b. Involucre bract tips acute or rarely slightly acuminate; heads smaller, the disk 6–15 mm wide and involucre bracts 5–10 mm high...10

10a. Perennials; leaves usually nearly linear or sometimes narrowly oblanceolate; plants of the eastern slope...***T. fendleri***
10b. Annuals or biennials; leaves oblanceolate or spatulate; plants of the western slope...11

11a. Annuals; involucre bracts not anthocyanic on the margins and tips; disk pappus shorter than the disk corollas...***T. annua***
11b. Biennials; involucre bracts usually anthocyanic on the margins and tips; disk pappus as long or longer than the disk corollas...***T. strigosa***

Townsendia annua **Beaman**, ANNUAL EASTER DAISY. Annuals, 0.2–2.5 dm tall, caulescent; *leaves* oblanceolate to spatulate, lightly strigose, 1–3 cm long; *heads* pedunculate on inconspicuous peduncles or terminating leafy stems; *involucre* 5–7.5 mm high, the bracts obtuse to acute or rarely slightly acuminate; *disk flowers* 2.3–3.7 mm long; *ray flowers* 5–9 mm long, white to pink or light lavender; *pappus* of rays 1 mm, of disks mostly 2–3 mm long; *achenes* 2–3 mm long. Found in open, dry places on clay hills and in alkaline soil, 5000–8200 ft. May–Aug. W.

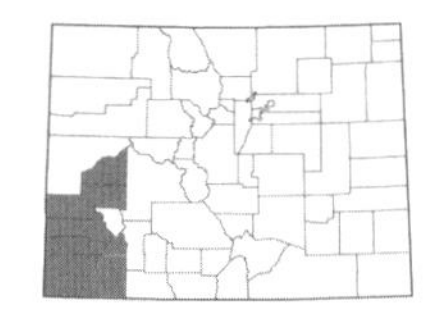

Townsendia eximia **A. Gray**, TALL EASTER DAISY. Biennials or short-lived perennials, to 5 dm tall, caulescent; *leaves* oblanceolate, lightly strigose to nearly glabrous, to 13 cm long; *heads* pedunculate on leafy stems; *involucre* bracts acuminate, most apiculate; *disk flowers* 3.5–5 mm long; *ray flowers* 8–20 mm long, blue; *pappus* of rays a short corona, of disks lanceolate scales 0.5–1 mm long and a few scales 1–4 mm long; *achenes* 3–4.5 mm long. Found in sandy soil of dry, open places, 7300–11,000 ft. June–Aug. E.

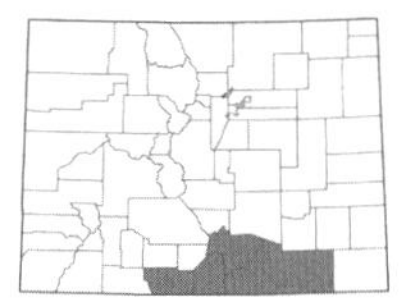

Townsendia exscapa **(Richardson) Porter**, STEMLESS EASTER DAISY. Perennials, acaulescent; *leaves* oblanceolate, strigose to subglabrate, to 8 cm long; *heads* sessile or very shortly pedunculate; *involucre* 10–22 mm high, the bracts linear to lanceolate; *disk flowers* 6–11 mm long; *ray flowers* 12–22 mm long, white or pink; *pappus* 6–12 mm long; *achenes* 3.5–6.5 mm long. Common in dry, open places and openings in ponderosa forests, 3500–9200 ft. April–June. E/W.

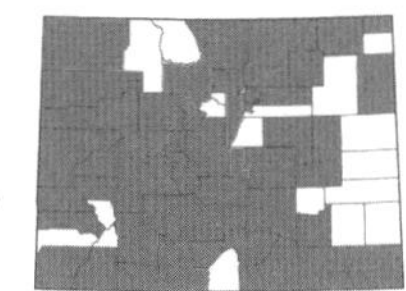

Townsendia fendleri **A. Gray**, FENDLER'S EASTER DAISY. Perennials to 3 dm, caulescent; *leaves* oblanceolate to nearly linear, strigose, to 3.5 cm long; *heads* pedunculate; *involucre* 5–8.5 mm high, the bracts lanceolate, acute; *disk flowers* 2–3.5 mm long; *ray flowers* 4.5–10 mm long, white to pinkish; *pappus* of rays 0.2–1 mm long, of disks 2.5–3 mm long; *achenes* 2–3.2 mm long. Uncommon in dry, open places, often with pinyon-juniper, 7300–7700. July–Sept. E.

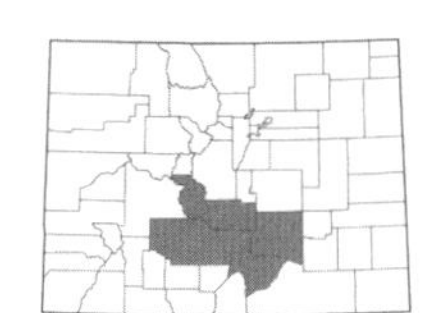

Townsendia glabella **A. Gray**, SMOOTH EASTER DAISY. Perennials, acaulescent; *leaves* oblanceolate, glabrous or slightly strigose, to 6.5 cm long; *heads* solitary, long-pedunculate; *involucre* 7.5–12.5 mm high, the bracts lanceolate, acute; *disk flowers* 3.5–5.3 mm long; *ray flowers* 8–14 mm long, white to pinkish or light purple; *pappus* of rays 0.5–1.8 mm long, of disks 5–7 mm long; *achenes* 2–4 mm long. Found in dry places and on shale slopes, 6300–8700 ft. May–July. W.

Townsendia grandiflora **Nutt.**, LARGEFLOWER EASTER DAISY. Biennials, 1.5–3 dm tall, caulescent; *leaves* spatulate, strigose-pilose, to 5 cm long; *heads* pedunculate; *involucre* 10–18 mm high, the bracts lanceolate with bristly-acuminate tips; *disk flowers* 4–6 mm long; *ray flowers* 3–4.5 mm long, white; *pappus* of rays a corona 0.1–0.5 mm long or scales 1–2.5 mm long, of disks 4–6 mm long; *achenes* 3.5–4 mm long. Common in open forests and on dry slopes, 5000–11,000 ft. May–Aug. E. (Plate 16)

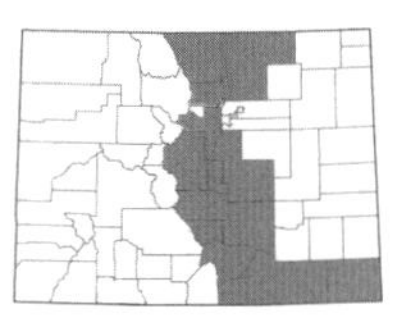

Townsendia hookeri **Beaman**, HOOKER'S EASTER DAISY. Perennials, acaulescent; *leaves* linear to narrowly oblanceolate, strigose, to 4.5 cm; *heads* sessile; *involucre* 7–12 mm high, the bracts linear, acute to acuminate, terminated by a tuft of tangled cilia; *disk flowers* 4–6.5 mm long; *ray flowers* 8.5–14 mm long, white to pinkish; *pappus* of rays (1) 5–7 mm long, of disks 5.5–7.5 mm; *achenes* 3.5–4.5 mm long. Common in dry, open places, 5000–9000 ft. Mar.–May. E/W. (Plate 16)

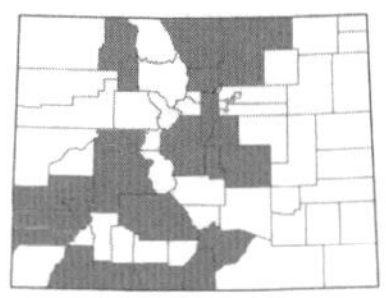

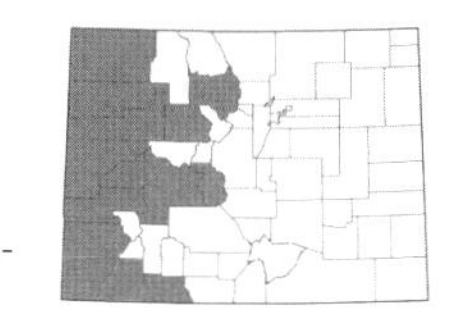

***Townsendia incana* Nutt.**, HOARY EASTER DAISY. Perennials to 1.2 dm, caulescent, the stem densely white-canescent; *leaves* spatulate to oblanceolate, strigose, 1–4 cm long; *heads* on short peduncles; *involucre* 6–12 mm high, the bracts acute; *disk flowers* 3.5–6.5 mm long; *ray flowers* 7–13 mm long, white to pinkish; *pappus* of rays 0.3–0.6 or 4–6 mm long, of disks 4–6 mm long; *achenes* 3.5–4.5 mm long. Common in sandy or shale soil of dry, open places, often with pinyon-juniper and sagebrush, 4700–8200 ft. May–July. W. (Plate 16)

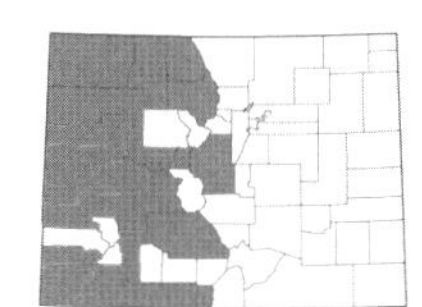

***Townsendia leptotes* (A. Gray) Osterh.**, COMMON EASTER DAISY. Perennials, acaulescent; *leaves* linear to oblanceolate, strigose to subglabrous, to 6 cm long; *heads* sessile or shortly pedunculate; *involucre* 8–11 mm high, the bracts acute; *disk flowers* 3–7 mm long; *ray flowers* 8–14 mm long, white, pink, or blue; *pappus* of rays 0–1 or 5–8 mm long, of disks 5–8 mm long; *achenes* 3–4 mm long. Common in dry, open places in rocky soil, on sagebrush plateaus or alpine, 7000–13,500 ft. April–June. E/W.

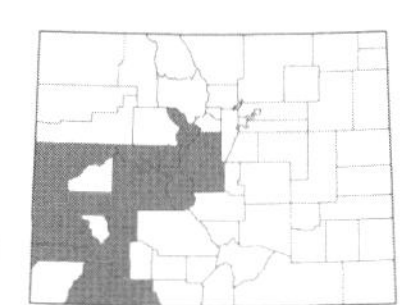

***Townsendia rothrockii* A. Gray ex Rothrock**, ROTHROCK'S EASTER DAISY. Perennials, acaulescent; *leaves* spatulate, thickened, glabrous or lightly strigose, 1–3.5 cm long; *heads* sessile or very shortly pedunculate; *involucre* 8–12 mm high, the bracts ovate to obovate, obtuse to acute, anthocyanic; *disk flowers* 3.3–5 mm long; *ray flowers* 8–16 mm long, blue to purplish; *pappus* of rays 0.5–1.5 mm long, of disks 3–6 mm long; *achenes* 3.5–4.5 mm long. Found in dry, open places in rocky soil, especially in alpine fell fields, 8000–13,500 ft. June–Aug. E/W. Endemic. (Plate 16)

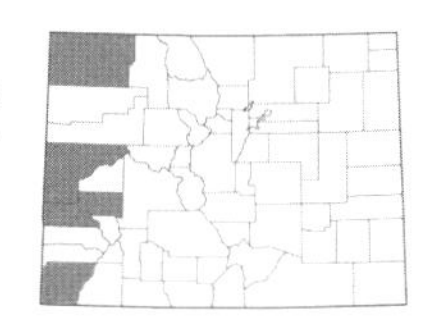

***Townsendia strigosa* Nutt.**, HAIRY EASTER DAISY. Biennials to 2 dm, caulescent; *leaves* oblanceolate to spatulate, strigose, to 5 cm long; *heads* pedunculate; *involucre* 6–10 mm high, the bracts acute to slightly acuminate, usually anthocyanic on the margins and tips; *disk flowers* 3–5 mm long; *ray flowers* 5–15 mm long, white to pink; *pappus* of rays 0.5–1.5 mm long, of disks 3.5–5.5 mm long; *achenes* 3–4 mm long. Uncommon in sandy or clay soil of dry, open places, 5000–6700 ft. May–June. W.

TRAGOPOGON L. – GOAT'S BEARD; SALSIFY

Annual, biennial, or perennial herbs; *leaves* alternate, entire, linear to linear lanceolate, grass like; *heads* solitary, ligulate; *involucre bracts* uniseriate or subequal; *receptacle* naked; *disk flowers* absent; *ray flowers* yellow or purple; *pappus* of a single series of plumose bristles united at the base, the plume-branches interwebbed; *achenes* 5- or 10-nerved, slender-beaked or the outer obscurely beaked.

1a. Flowers purple...***T. porrifolius***
1b. Flowers yellow...2

2a. Peduncles not enlarged at flowering time; outer ligules equal to or longer than the involucre bracts...***T. pratensis***
2b. Peduncles enlarged and hollow at flowering time; outer ligules shorter than the involucre bracts...***T. dubius***

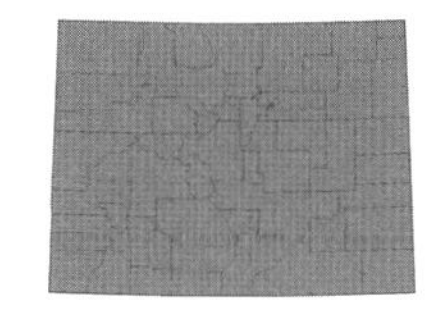

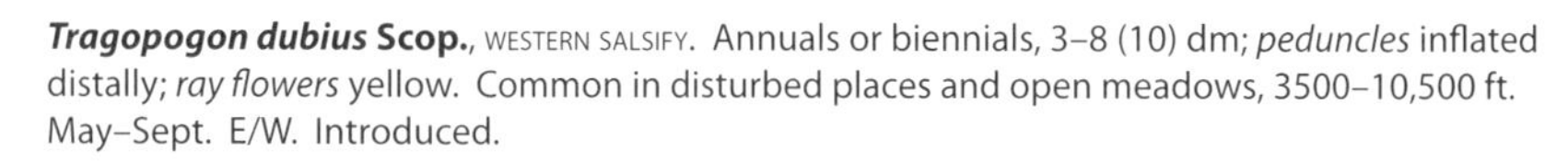

***Tragopogon dubius* Scop.**, WESTERN SALSIFY. Annuals or biennials, 3–8 (10) dm; *peduncles* inflated distally; *ray flowers* yellow. Common in disturbed places and open meadows, 3500–10,500 ft. May–Sept. E/W. Introduced.

***Tragopogon porrifolius* L.**, SALSIFY; VEGETABLE OYSTER. Biennials, 4–10 (15) dm; *peduncles* inflated distally; *ray flowers* purple. Found in disturbed places, along roadsides, and near the edges of fields, 4000–9000 ft. May–July. E/W. Introduced.

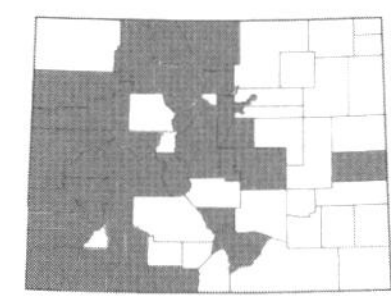

***Tragopogon pratensis* L.**, MEADOW SALSIFY. Biennials, (1.5) 4–10 dm; *peduncles* little or not inflated distally (may be inflated in fruit); *ray flowers* yellow. Locally common in disturbed places and along roadsides and fields, 4200–9000 ft. May–Sept. E/W. Introduced.

TRIPLEUROSPERMUM Sch. Bip. – MAYWEED

Annual, biennial, or perennial herbs; *leaves* alternate; *heads* radiate; *involucre bracts* subequal in 2–5 series, with hyaline-scarious margins; *receptacle* naked; *disk flowers* yellow or greenish, the corollas 5-lobed; *ray flowers* white; *pappus* absent; *achenes* laterally compressed to subcylindric.

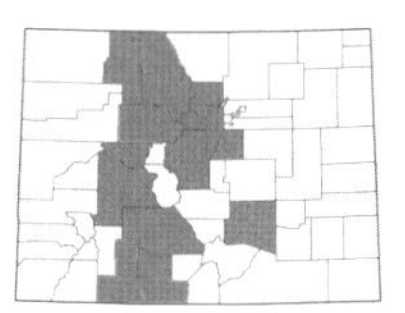

***Tripleurospermum inodorum* (L.) Sch. Bip.**, WILD CHAMOMILE. [*Matricaria maritima* L.; *Matricaria perforata* Merat; *T. perforata* (Merat) M. Lainz]. Usually annuals, (5) 30–60 (80) cm; *leaves* 2–8 cm long, the ultimate lobes filiform; *involucre bracts* oblong, the margins colorless to light brown; *disk flowers* 1–2.5 mm long; *ray flowers* 10–25, (4) 10–13 (20) mm long; *achenes* light brown. Common in disturbed places, 4600–10,500 ft. July–Sept. E/W. Introduced.

VERBESINA L. – CROWNBEARD

Herbs or shrubs; *leaves* opposite or alternate, simple; *heads* radiate; *involucre bracts* subequal or slightly imbricate, somewhat herbaceous; *receptacle* chaffy; *disk flowers* yellow; *ray flowers* yellow or white; *pappus* of 2 awns and occasionaly also with a few short scales; *achenes* flattened radially, often winged.

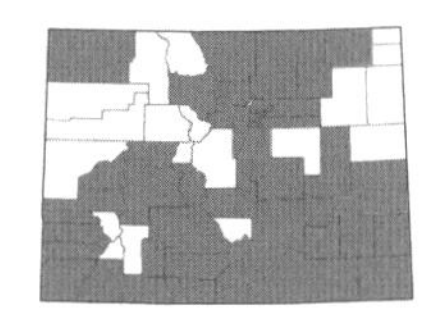

***Verbesina encelioides* (Cav.) A. Gray ssp. *exauriculata* (B.L. Rob. & Greenm.) J.R. Coleman**, GOLDEN CROWNBEARD. Plants 1.5–13 dm, malodorous; *leaves* mostly alternate, lanceolate to deltate-ovate, 3–8 (12+) cm long, scabrellate to sericeous, coarsely dentate to subentire; *involucre bracts* 8–10 mm long, narrowly ovate to linear; *ray flowers* 8–10 (20+) mm long; *achenes* 3.3–5+ mm long. Common in sandy or rocky soil, along roadsides, in disturbed places, and near fields, 3400–9000 ft. Aug.–Oct. E/W. Introduced. (Plate 16)

VERNONIA Schreb. – IRONWEED

Perennial herbs (ours); *leaves* alternate, simple; *heads* discoid; *involucre bracts* imbricate, the inner bracts longer and variously tipped; *receptacle* naked; *disk flowers* purple or rarely white; *ray flowers* absent; *pappus* double, the inner of numerous slender bristles and the outer of short scales or bristles, usually brownish or purplish; *achenes* ribbed or smooth. (Jones & Faust, 1978)

1a. Leaves not conspicuously pitted and punctate below…***V. baldwinii***
1b. Leaves conspicuously pitted and punctate below…2

2a. Leaves lanceolate, mostly 0.4–4.5 cm wide, the margins serrate; inner involucre bracts tips subacute to rounded…***V. fasciculata***
2b. Leaves linear to broadly linear, mostly 0.3–1 cm wide, the margins entire or with scattered short, stout teeth; inner involucre bracts tips acute to acuminate, the tips 0.5–1 mm long…***V. marginata***

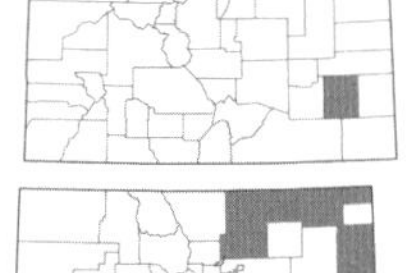

***Vernonia baldwinii* Torr.**, WESTERN IRONWEED. Plants 6–15 dm; *leaves* elliptic to ovate-lanceolate, 5–18 cm long, short-hairy to tomentose, serrate; *heads* in corymbiform arrays; *peduncles* 1–25 mm long; *involucre* 4–6 (8) mm high; *achenes* 2.5–3 mm long. Uncommon on the eastern plains, usually found along roadsides, in ditches, and in disturbed areas, 3500–5300 ft. July–Sept. E.

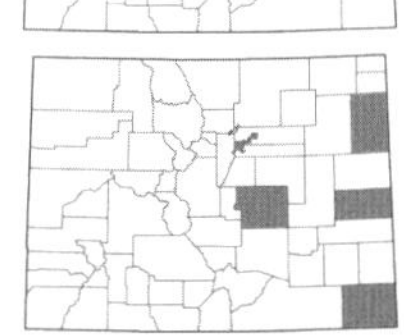

***Vernonia fasciculata* Michx.**, PRAIRIE IRONWEED. Plants 3–15 dm; *leaves* lanceolate, 5–12 (20) cm long, conspicuously pitted below (with awl-shaped hairs in pits), serrate; *heads* in corymbiform arrays; *peduncles* 1–8 (12) mm long; *involucre* 5–8 mm high; *achenes* 3–4 mm long. Uncommon on open prairies of the eastern plains, 3500–5000 ft. July–Sept. E.

***Vernonia marginata* (Torr.) Raf.**, PLAINS IRONWEED. Plants 3–7 dm; *leaves* linear to broadly linear, 5–18 cm long, conspicuously pitted below (with awl-shaped hairs in pits), entire or with scattered short, stout teeth on the margins; *heads* in corymbiform arrays; *peduncles* (3) 10–35 mm long; *involucre* 7–11 mm high; *achenes* 4–5 mm long. Uncommon in open places on the eastern plains, 3500–7000 ft. July–Sept. E.

WYETHIA Nutt. – MULE'S EARS

Perennial herbs; *leaves* alternate, simple; *heads* radiate (ours); *involucre bracts* subequal in several series, herbaceous or coriaceous, the outer often long and foliaceous; *receptacle* chaffy; *disk flowers* yellow; *ray flowers* yellow (ours); *pappus* a crown of unequal, laciniate scales, these often prolonged into awns; *achenes* of disk flowers radially compressed-quadrangular. (Weber, 1946)

1a. Leaves narrowly oblong to lanceolate or nearly linear, narrower (0.5–2 cm wide), the lowermost reduced; older stems whitish…***W. scabra***
1b. Leaves elliptic to lance-ovate, wider (3–16 cm wide), the basal ones the largest; older stems not whitish…2

2a. Leaves, stem, and involucre bracts villous or hirsute to glabrate but not resin-varnished; stem leaves short-petiolate, rarely sessile and clasping…***W. arizonica***
2b. Leaves, stem, and involucre bracts glabrous and resin-varnished; stem leaves sessile and clasping, rarely short-petiolate…***W. amplexicaulis***

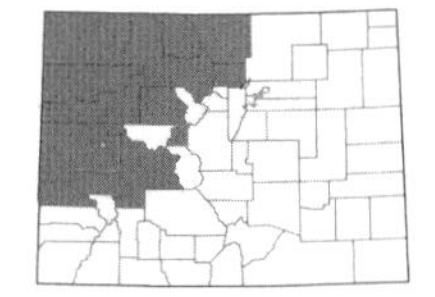

***Wyethia amplexicaulis* (Nutt.) Nutt.**, MULE'S EARS. Plants 2.5–5 (10) dm; *leaves* lanceolate-elliptic to oblong-lanceolate, 15–40 cm long, resin-varnished and glabrous, cauline leaves sessile and clasping or rarely short-petiolate; *involucre* 15–30 mm diam.; *ray flowers* 25–60 mm long; *achenes* 8–9 mm long. Common in wet to dry, open meadows and on open hills and slopes, often with sagebrush, 6500–10,500 ft. June–Aug. E/W. (Plate 16)

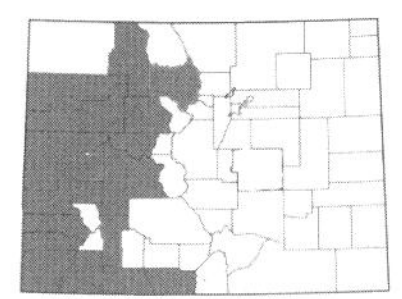

***Wyethia arizonica* A. Gray**, ARIZONA MULE'S EARS. Plants 2–3 (10) dm; *leaves* lanceolate to elliptic, 12–30 cm long, scabrous to hirsute, sometimes glabrate, the cauline leaves usually short-petiolate; *involucre* 18–30 mm diam.; *ray flowers* (25) 35–50 mm long; *achenes* 9–10 mm long. Found on dry hills and slopes, often with sagebrush, 5400–9000 ft. May–July. W.

Wyethia arizonica and *W. amplexicaulis* commonly hybridize where their ranges meet.

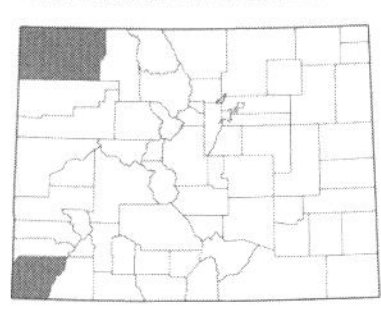

***Wyethia scabra* Hook.**, WHITESTEM SUNFLOWER. [*Scabrethia scabra* (Hook.) W.A. Weber]. Plants 2–6 dm; *leaves* narrowly oblong to lanceolate or nearly linear, hispid to scabrous or sometimes glabrescent; *involucre* (15) 20–35 mm diam.; *ray flowers* 15–50 mm long; *achenes* 7–9 mm long. Uncommon on dry, open slopes in sandy soil, 5000–6500 ft. June–July. W.

XANTHISMA DC. – SLEEPY-DAISY

Herbs or subshrubs; *leaves* alternate, simple; *heads* radiate or discoid; *involucre bracts* imbricate or unequal, the base hard and tannish or purple-tinged and the apex herbaceous, spreading, or reflexed; *receptacle* naked; *disk flowers* yellow; *ray flowers* yellow, white, pink, purple, or blue, or absent; *pappus* of bristles; *achenes* narrowly obovoid to oblong. (Turner & Hartman, 1976)

1a. Ray flowers absent...***X. grindelioides***
1b. Ray flowers present...2

2a. Ray flowers purple or rose-purple; heads larger, the disks 10–25 mm wide; leaves simple with toothed margins...***X. coloradoense***
2b. Ray flowers yellow; heads smaller, the disks 6–12 mm wide; leaves simple and toothed or pinnately dissected...3

3a. Annual or winter annual with simple, toothed leaves...***X. gracile***
3b. Perennial with pinnately dissected leaves...***X. spinulosum***

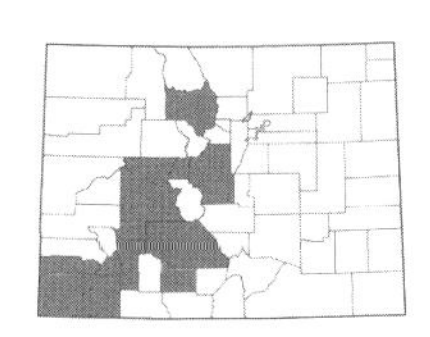

***Xanthisma coloradoense* (A. Gray) D.R. Morgan & R.L. Hartm.**, COLORADO SLEEPY-DAISY. [*Machaeranthera coloradoensis* (A. Gray) Osterh.]. Perennials, 0.3–1.4 dm; *leaves* oblanceolate to spatulate, 1–8 cm long, the margins toothed, each tooth bristle-tipped; *involucre* 5–8 mm high; *disk flowers* 4.5–6.5 mm long; *ray flowers* 9–15 mm long, pink to purple; *achenes* 1.5–3 mm long. Uncommon in gravelly soil of open places such as on slopes and rocky outcrops and in alpine meadows, 8500–12,700 ft. July–Aug. E/W.

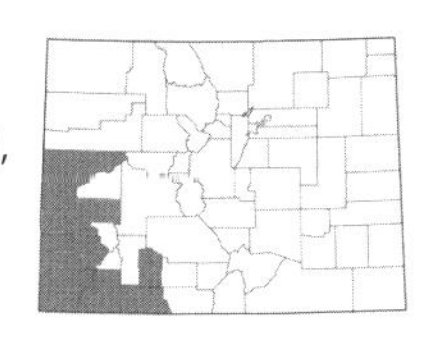

***Xanthisma gracile* (Nutt.) D.R. Morgan & R.L. Hartm.**, SLENDER GOLDENWEED. [*Haplopappus gracilis* (Nutt.) A. Gray; *Machaeranthera gracilis* (Nutt.) Shinners]. Annuals, 0.5–5 dm; *leaves* 1–2-pinnatifid, 2–6 cm long, the margins toothed, each tooth bristle-tipped, hairy; *involucre* 6–8 mm high; *disk flowers* 4–5 mm long; *ray flowers* 6–10 mm long, yellow; *achenes* 1.5–3 mm long. Uncommon in sandy soil of open places, often in disturbed areas, 5000–7500 ft. July–Sept. W.

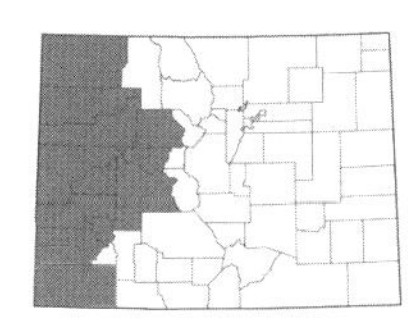

Xanthisma grindelioides* (Nutt.) D.R. Morgan & R.L. Hartm. var. *grindelioides, GOLDENWEED. [*Haplopappus nuttallii* Torr. & A. Gray; *Machaeranthera grindelioides* (Nutt.) Shinners var. *grindelioides*]. Subshrubs, 0.3–3.5 dm; *leaves* lanceolate to oblong or narrowly ovate, 7–60 cm long, the margins coarsely toothed, each tooth bristle-tipped, hairy and often glandular-stipitate; *involucre* 0.5–1 cm high; *disk flowers* 5–8.5 mm long; *ray flowers* absent; *achenes* 2–3.5 mm long. Found in open, dry places, in sandy, rocky, or clay soil, often in association with pinyon-juniper, 4300–9000 ft. May–Sept. W.

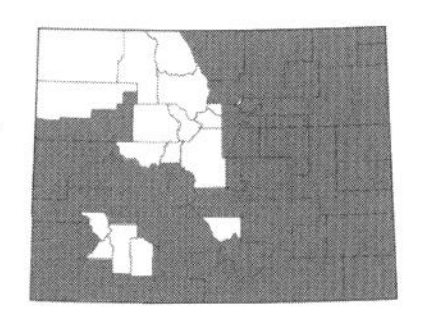

***Xanthisma spinulosum* (Pursh) D.R. Morgan & R.L. Hartm.**, SPINY GOLDENWEED. [*Aster pinnatifidus* (Hook.) Kuntze; *Haplopappus spinulosus* (Pursh) DC.; *Machaeranthera pinnatifida* (Hook.) Shinners]. Perennials or subshrubs, 1–10 dm; *leaves* 1–2-pinnatifid, 0.2–8 cm long, the margins lobed to toothed, each tooth bristle-tipped, glabrous to hairy, sometimes glandular-stipitate; *involucre* 6–10 mm high; *disk flowers* 4–5 mm long; *ray flowers* 5–12 mm long, yellow; *achenes* 1.8–2.5 mm long. Common in open, dry places, sandy soil, often in disturbed areas, 3400–8000 ft. May–Oct. E/W. (Plate 16)

A highly complex, variable species separated into numerous varieties, 3 of which occur in Colorado:

1a. Heads mostly solitary on elongate peduncles, large (the disk 15–25 mm wide)...**var. *paradoxum* (B.L. Turner & R.L. Hartm.) D.R. Morgan & R.L. Hartm.**
1b. Heads mostly 2–many per stem on relatively short peduncles, smaller (the disk 8–15 mm wide)...2

2a. Stems stiffly erect, branched in the upper ⅓; leaves usually glabrous or sparsely tomentose, without glandular hairs, ascending, pinnatifid…**var. *glaberrimum* (Rydb.) D.R. Morgan & R.L. Hartm.**
2b. Stems spreading to sprawling, rarely stiffly erect, typically much-branched in the lower half; leaves hairy or rarely glabrous and usually glandular-stipitate, spreading to loosely ascending, pinnatifid or double pinnatifid…**var. *spinulosum***

XANTHIUM L. – COCKLEBUR

Annual herbs; *leaves* alternate; *heads* unisexual, discoid; *involucre bracts* few and distinct and subherbacous or absent in staminate heads, fused into a 2-chambered prickly bur in pistillate heads; *receptacle* chaffy; *disk flowers* staminate or pistillate, pistillate flowers lacking a corolla; *ray flowers* absent; *pappus* absent.

1a. Leaves densely silvery-sericeous below, lanceolate and tapering to the base; nodes with a 3-forked axillary spine…***X. spinosum***
1b. Leaves not silvery below, ovate to suborbicular and cordate to truncate at the base; nodes unarmed…***X. strumarium***

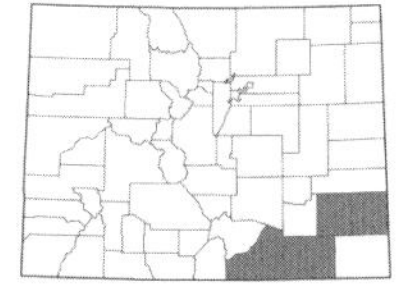

***Xanthium spinosum* L.**, SPINY COCKLEBUR. [*Acanthoxanthium spinosum* (L.) Fourr.]. Plants 1–6 (15) dm; *stems* with a 3-forked axillary spine at each node; *leaves* lanceolate, 4–12 cm long, densely silvery-sericeous below. Uncommon in disturbed places, fields, and along roadsides, 3500–5500 ft. July–Sept. E. Introduced.

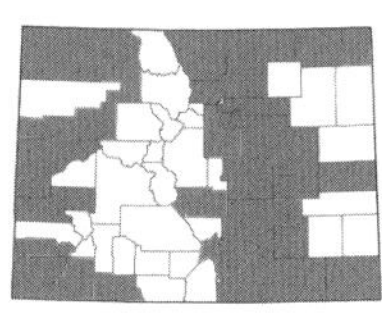

***Xanthium strumarium* L.**, COMMON COCKLEBUR. Plants 1–10 (20) dm; *stems* unarmed; *leaves* deltate to ovate or suborbiculate or palmately lobed, 4–15 (20) cm long, not silvery-sericeous below. Common in disturbed places, in fields, and along roadsides, 3500–8500 ft. July–Oct. E/W. Introduced.

XYLORHIZA Nutt. – WOODYASTER

Perennial herbs to small shrubs; *leaves* simple, alternate, sessile, with conspicuous white midribs; *heads* radiate; *involucre bracts* unequal or subequal, green with scarious margins and a yellowish chartaceous base; *receptacle* naked; *disk flowers* yellow; *ray flowers* light blue or white; *pappus* of bristles; *achenes* usually laterally compressed, woolly-pubescent. (Nesom, 2002; Watson, 1977)

1a. Stems leafy usually for ¾ or more of their length; peduncles mostly 5 cm or less in length (2–6 cm); involucre 7–14 mm high, 12–25 mm wide…***X. glabriuscula***
1b. Stems leafy usually only on the lower half; peduncles greater than 5 cm in length (6–20 cm); involucre 12–20 mm high, 20–35 mm wide…***X. venusta***

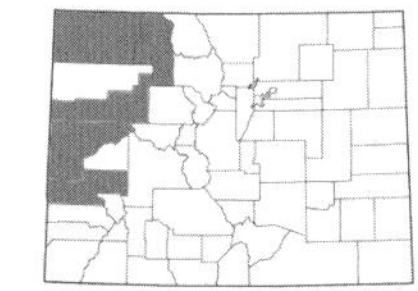

***Xylorhiza glabriuscula* Nutt.**, SMOOTH WOODYASTER. [*Aster glabriuscula* (Nutt.) Torr. & A. Gray]. Subshrubs, 1–2.5 dm; *leaves* oblanceolate, glabrous to short-hairy; *peduncles* 2–6 cm long; *involucre* 7–14 mm high; *ray flowers* white or light blue. Found on seleniferous clay soil, often with sagebrush or saltbush, 5400–6200 ft. May–June. W.

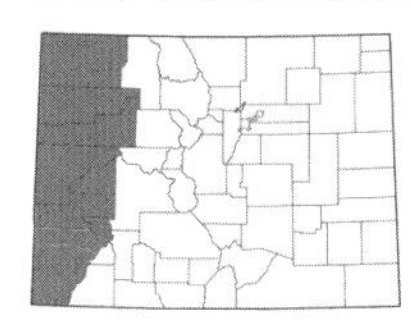

***Xylorhiza venusta* (M.E. Jones) Heller**, CHARMING WOODYASTER. [*Aster venustus* M.E. Jones]. Perennials or subshrubs, 1–4 dm; *leaves* oblanceolate to spatulate, glabrate to densely hairy; *peduncles* 6–20 cm long; *involucre* 12–20 mm high; *ray flowers* white to pale purple. Common in seleniferous clay soil, often with saltbush, 4800–7000 ft. April–June. W.

These two species can form hybrids where their ranges overlap.

ZINNIA L. – ZINNIA

Herbs or low shrubs; *leaves* opposite, simple, entire; *heads* radiate (ours); *involucre bracts* imbricate in several series, broad, scarious with a greenish band near the tip; *receptacle* chaffy; *disk flowers* yellow to red; *ray flowers* yellow, white, orange, red, or purple; *pappus* of 1–4 equal or unequal awns or absent; *achenes* radially compressed or angular.

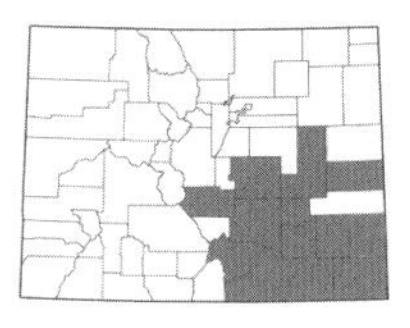

***Zinnia grandiflora* Nutt.**, ROCKY MOUNTAIN ZINNIA. Subshrubs, 8–22 cm; *leaves* linear, 1–3 cm long, strigose to scabrous; *ray flowers* yellow, to 18 mm long; *achenes* 4–5 mm long. Common in dry, open places of the southeastern plains and valleys, 3500–7000 ft. June–Oct. E. (Plate 16)

BALSAMINACEAE A. Rich. – JEWELWEED FAMILY

Herbaceous plants; *leaves* usually alternate, simple; *flowers* perfect, zygomorphic; *sepals* 3 or rarely 5, the lateral ones small and green, the posterior one somewhat petaloid and spurred and saccate; *petals* 5, but often seeming 3 by the fusion of the lateral pairs to form two bilobed petals; *stamens* 5, the filaments connate at least above, anthers connate or connivent; *ovary* superior; *fruit* a loculicidal, explosive capsule dehiscent into 5 spirally coiled valves (*Impatiens*).

IMPATIENS L. – TOUCH-ME-NOT

With characteristics of the family.

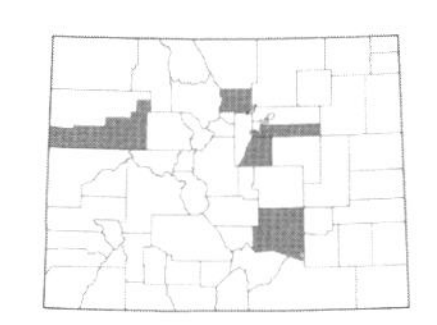

***Impatiens capensis* Meerb.**, JEWELWEED; ORANGE TOUCH-ME-NOT. Plants hollow-stemmed, annuals, 5–15 dm; *leaves* ovate to elliptic, 3–10 cm long, upper surface usually green, lower surfaces pale or glaucous, crenate; *flowers* orange to red, usually crimson-spotted, with a spur 6–9 mm long; *capsules* 4–5 mm long, with corky longitudinal ridges. Found in shaded, moist places such as along irrigation ditches and streams, 5000–6500 ft. May–Sept. E/W. Introduced.

BERBERIDACEAE Juss. – BARBERRY FAMILY

Perennials or shrubs; *leaves* alternate, simple or pinnately or ternately compound, the margins entire or dentate and usually prickly; *flowers* perfect, actinomorphic; *sepals* 6, distinct, 2-whorled, often petaloid; *petals* 6–12, distinct, 2–4-whorled, yellow or orange; *stamens* 6, 2-whorled; *pistil* 1; *ovary* superior; *fruit* a follicle, berry, or utricle.

BERBERIS L. – BARBERRY; OREGON-GRAPE

Evergreen or deciduous shrubs or subshrubs; *inflorescence* terminal, usually in racemes or rarely solitary or umbels; *flowers* yellow, 3-merous; *sepals* early-deciduous, 6; *petals* 6; *fruit* a berry. (Ahrendt, 1961; Loconte & Estes, 1989; Meacham, 1980; Whittemore, 1997)

1a. Leaves simple, those of long-shoots modified into spines (stems with branched spines at the base of the leaf clusters), margins with relatively weak spines…2
1b. Leaves pinnately compound, the margins with stout spines; stems without spines…3

2a. Racemes 10–20-flowered; leaves obovate to elliptic, the margins spinulose-serrate; bark of second year stems gray…***B. vulgaris***
2b. Racemes 4–10 (15)-flowered; leaves narrowly oblanceolate to spatulate, the margins entire to spinulose-serrate; bark of second year stems purple or reddish…***B. fendleri***

3a. Plants low, 0.2–2 dm high; racemes dense, 25–50-flowered; berries blue…***B. repens***
3b. Plants tall, shrubs or even small trees, 10–45 dm high; racemes loose, 3–7-flowered; berries yellow, red, purple, or brownish…4

4a. Berries yellow or red to brown, dry and inflated, 12–18 mm in diam.; filaments with a distal pair of recurved lateral teeth…***B. fremontii***
4b. Berries purple or red, solid and juicy, 5–8 mm in diam.; filaments without a distal pair of recurved lateral teeth…***B. haematocarpa***

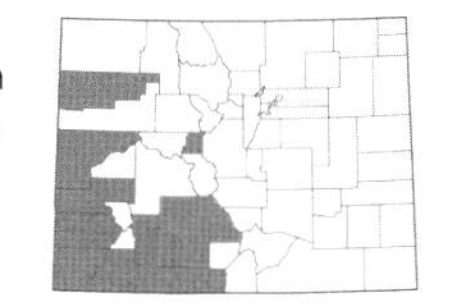

***Berberis fendleri* A. Gray**, FENDLER'S BARBERRY. Deciduous shrubs, 1–2 m; *stems* of second years with purple or reddish bark; *spines* present; *leaves* simple, narrowly oblanceolate to spatulate, 1.7–5 × 0.6–1.7 cm, the margins entire or spinulose-serrate; *inflorescence* 4–10 (15)-flowered; *berries* red, 6–8 mm diam.. Found along rivers, on moist hillsides, and on pinyon-juniper slopes, 5000–9000 ft. May–July. E/W.

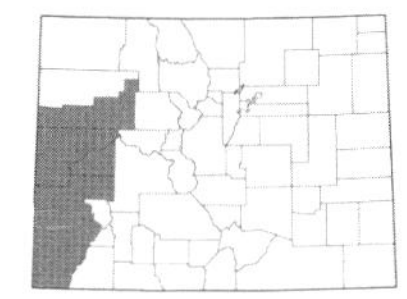

***Berberis fremontii* Torr.**, FREMONT'S BARBERRY. [*Mahonia fremontii* (Torr.) Fedde]. Evergreen shrubs, 1–4.5 m; *stems* of second years with light brown to grayish-purple bark; *spines* absent; *leaves* pinnately compound with 5–9 (11) leaflets, these elliptic to ovate, the margins spinulose-serrate; *inflorescence* 3–6-flowered; *berries* yellow, red, or brown, 12–18 mm diam. Found on rocky hillsides and pinyon-juniper slopes, 4500–7000 ft. April–June. W.

***Berberis haematocarpa* Wooton**, RED BARBERRY. [*Mahonia haematocarpa* (Wooton) Fedde]. Evergreen shrubs, 1–4 m; *stems* of second years with grayish-purple bark; *spines* absent; *leaves* pinnately compound with 3–9 leaflets, these ovate to oblong, the margins spinulose-serrate; *inflorescence* 3–7-flowered; *berries* red or purple, 5–8 mm diam. Generally found on chaparral and oak scrub slopes, known only from one old 1902 collection near Trinidad, 6000 ft. Mar.–June. E.

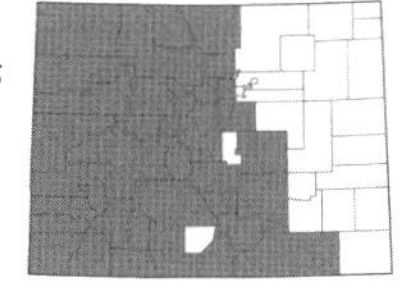

***Berberis repens* Lindl.**, OREGON-GRAPE; CREEPING BARBERRY. [*Mahonia repens* (Lindl.) G. Don]. Evergreen shrubs, 0.2–2 dm; *stems* of second years with grayish or purplish-brown bark; *spines* absent; *leaves* pinnately compound with (3) 5–7 leaflets, these ovate to elliptic, spinulose-serrate; *inflorescence* dense, 25–50-flowered; *berries* blue, glaucous, 6–10 mm diam. Common and widespread throughout the foothills and mountains, 5500–10,500 ft. April–June. E/W. (Plate 17)

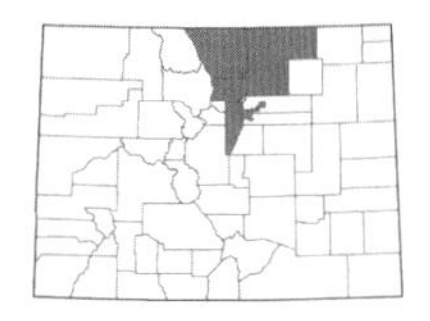

***Berberis vulgaris* L.**, COMMON BARBERRY. Deciduous shrubs 1–3 m; *stems* of second years with gray bark; *spines* present; *leaves* simple, obovate to elliptic, 2–8 × 1–3 cm, the margins spinulose-serrate; *inflorescence* 10–20-flowered; *berries* red or purple, 10–11 mm diam. Locally common near Boulder, older specimens prior to 1925 were collected from the Poudre River near Fort Collins and near Denver, but the population near Boulder on the Enchanted Mesa is apparently the only place in Colorado where it persists, 5000–6000 ft. May–June. E. Introduced.

BETULACEAE Gray – BIRCH FAMILY

Monoecious trees and shrubs; *leaves* alternate, petioled, simple; *flowers* imperfect, crowded in unisexual catkins; *staminate flowers* subtended by scale bracts, the perianth lacking, in catkins; *pistillate flowers* subtended by nearly foliaceous or woody bracts, consisting of 1 pistil and a minute or absent perianth, grouped in 2–3-flowered clusters in catkins; *ovary* inferior; *fruit* nuts, nutlets, or winged samaras, often subtended or enclosed by a foliaceous involucre developed from 2–3 bracts. (Chenz et al., 1999; Fernald, 1945; Furlow, 1997)

1a. Fruit a nut enclosed by a conspicuously foliaceous involucre with coarsely toothed or fringed lobes; pistillate flowers enclosed in bud scales except for 2 protruding styles; pistillate catkins distal to staminate catkins, small and ovoid, consisting of small clusters of flowers and bracts; prominent lenticels absent...***Corylus***
1b. Fruit a winged samara; pistillate flowers not enclosed in bud scales; pistillate catkins proximal to staminate catkins, conelike with woody, persistent bracts or with firm but deciduous scales; bark usually with prominent lenticels...2

2a. Pistillate catkins conelike with persistent, woody bracts; pistillate flowers usually 2 per bract scale...***Alnus***
2b. Pistillate catkins with firm but not woody, deciduous scales; pistillate flowers 3 per bract scale...***Betula***

ALNUS Mill. – ALDER

Leaves with serrate, doubly serrate, serrulate, or nearly entire margins; *staminate flowers* 3 per bract scale, stamens 2–4; *staminate catkins* slender, pendulous, lateral, in racemose clusters or solitary; *pistillate flowers* usually 2 per bract scale; *pistillate catkins* erect to pendulous, proximal to staminate catkins, with persistent, woody bracts; *fruit* a thin-winged samara (nutlet).

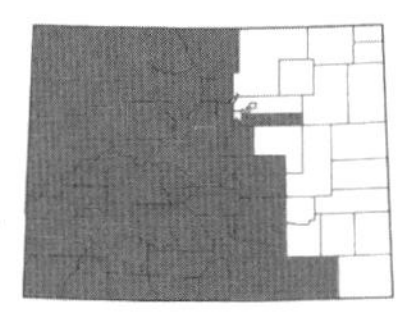

***Alnus incana* (L.) Moench ssp. *tenuifolia* (Nutt.) Brietung**, THINLEAF ALDER. [*Alnus tenuifolia* Nutt.]. Plants to 12 m; *leaves* orbiculate to elliptic , 4–10 cm long, 2.5–8 cm wide, glabrous abaxially to sparsely hairy, margins doubly serrate to nearly lobulate or crenate, *staminate catkins* 5–20 cm long; *pistillate catkins* 5–20 mm long. Common along streams, bordering lakes and wet meadows, and in moist gulches, 5000–10,000 ft. April–July. E/W. (Plate 17)

BETULA L. – BIRCH

Leaves with serrate, doubly serrate, or crenate margins; *staminate flowers* in clusters of 3 per scale bract, stamens 2–3; *staminate catkins* slender, pendulous or spreading; *pistillate flowers* in clusters of 3 per bract scale; *pistillate catkins* mostly solitary, erect, proximal to staminate catkins, with firm, deciduous scales; *fruit* a 2-winged samara (nutlet). (Dugal, 1966)

1a. Leaf apex obtuse to rounded; leaves obovate to orbiculate, 0.5–3 cm long; leaf margins dentate-crenate with rounded teeth; shrubs to 3 m; samaras with wings narrower than the body...***B. glandulosa***
1b. Leaf apex acute to acuminate; leaves ovate to rhombic, 2–10 cm long; leaf margins serrate or doubly serrate; shrubs or small trees to 10 m, or trees to 30 m; samaras with wings as broad or broader than the body...2

2a. Shrubs or small trees to 10 m; bark smooth, not exfoliating, dark reddish-brown to bronze; leaves with 2–6 pairs of lateral veins...***B. occidentalis***
2b. Trees to 30 m; bark exfoliating or flaking off in layers, mature bark white or pale brown; leaves with 5–18 pairs of lateral veins...3

3a. Leaves broadly ovate to rhombic, 3–7 cm long; bark exfoliating or flaking off as long strands; samaras with wings much broader than the body...***B. pendula***
3b. Leaves ovate, 5–9 (12) cm long; bark exfoliating in paper-thin sheets; samaras with wings as broad as or slightly broader than the body...***B. papyrifera* var. *papyrifera***

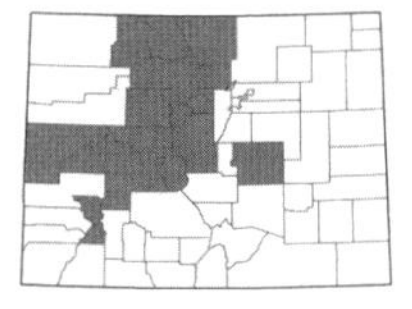

***Betula glandulosa* Michx.**, DWARF BIRCH. Shrubs to 3 m; *bark* dark brown, smooth; *twigs* glabrous to sparsely hairy, usually with large, warty, resinous glands; *leaves* obovate to orbiculate, 0.5–3 cm long, the margins dentate-crenate; *infructescence* 1–2.5 × 0.5–1.2 cm; *samaras* with wings narrower than body. Common along streams and in wet fens and meadows, in willow thickets, and in subalpine to alpine meadows, 8000–12,500 ft. April–June. E/W.

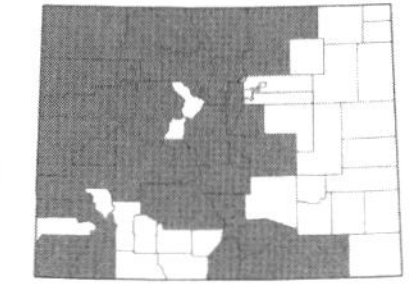

Betula occidentalis **Hook.**, WATER BIRCH. [*B. fontinalis* Sarg.]. Shrubs or small trees to 10 m; *bark* dark reddish-brown to bronze, smooth; *twigs* glabrous to sparsely hairy, with reddish, resinous glands; *leaves* ovate to rhombic-ovate, 2–6 cm long, the margins sharply and coarsely toothed; *infructescence* 2–4 × 0.8–1.5 cm; *samaras* with wings broader than body. Common along streams and in wet gulches, 5500–9500 ft. April–June. E/W.

Betula glandulosa and *B. occidentalis* can hybridize, producing plants with larger leaves than typical of *B. glandulosa* with more numerous, irregularly serrate teeth along the margins. This hybrid is called *Betula ×eastwoodiae* Sarg.

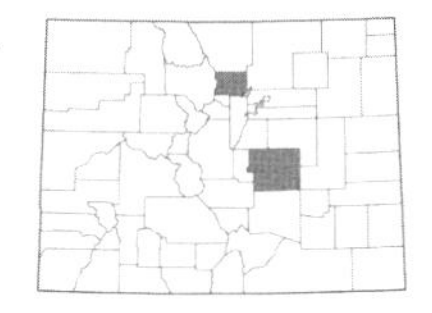

Betula papyrifera **Marshall var. *papyrifera***, PAPER BIRCH. Trees to 30 m; *bark* of mature trees white, exfoliating in paper-thin sheets; *twigs* slightly hairy, sometimes with small, resinous glands; *leaves* ovate, 5–9 (12) cm long, with sharply toothed margins; *infructescence* 2.5–5 × 0.6–1.2 cm; *samaras* with wings as broad as or slightly broader than body. Uncommon on the slopes of Green Mountain in Boulder County and along streams in Engelmann Canyon in El Paso County, 6800–7300 ft. May–July. E.

Betula pendula **Roth**, EUROPEAN WHITE BIRCH. Trees to 25 m; *bark* white, smooth, exfoliating as long strands; *twigs* glabrous, with small resinous glands; *leaves* ovate to rhombic, 3–7 cm long, with doubly serrate margins; *infructescence* 2–3.5 × 0.6–1 cm; *samaras* with wings much broader than body. A cultivated ornamental sometimes escaping from cultivation, known to persist in Jefferson County, 5000–5500 ft. May–July. E. Introduced.

CORYLUS L. – HAZELNUT

Small trees or shrubs; *leaves* with doubly serrate margins; *staminate flowers* 3 per bract scale, stamens 4; *staminate catkins* on short shoots lateral on branchlets; *pistillate flowers* 2 per bract scale, enclosed by bud scales except for the 2 elongate, protruding styles; *pistillate catkins* small, ovoid, distal to staminate catkins; *fruit* a nut, enclosed by a foliaceous involucre.

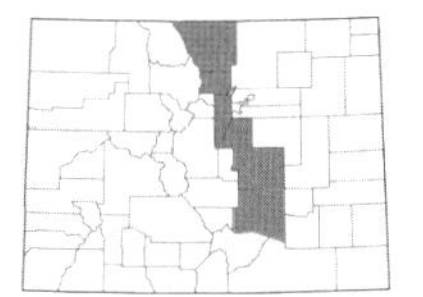

Corylus cornuta **Marshall var. *cornuta***, BEAKED HAZELNUT. Plants to 6 m; *bark* smooth, light brown; *twigs* glabrous to slightly hairy, eglandular; *leaves* narrowly elliptic to ovate or obovate, 5–12 cm long, the apices usually acuminate; *nuts* clustered in groups of 2–6, the involucral beak tubular, narrow, and bristly. Uncommon in cool ravines and gulches, along creeks, and as an understory in Douglas-fir woodlands, 5500–7500 ft. June–Aug. E. (Plate 17)

BIGNONIACEAE Juss. – BIGNONIA FAMILY

Trees, shrubs, or lianas; *leaves* usually compound or sometimes simple, exstipulate; *flowers* perfect, zygomoprhic; *calyx* 5-lobed, the lobes unequal but not bilabiate, or bilabiate; *corolla* 5-lobed, bilabiate or actinomorphic, imbricate; *stamens* 5, markedly unequal, distinct; *ovary* superior, 2-carpellate; *fruit* a loculicidal or septicidal capsule.

1a. Woody vines; leaves compound with toothed leaflets...***Campsis***
1b. Trees; leaves simple and entire with a long-acuminate apex...***Catalpa***

CAMPSIS Lour. – TRUMPET CREEPER

Lianas; *leaves* opposite, pinnately compound with toothed leaflets; *flowers* orange; *calyx* orange-red, coriaceous; *corolla* zygomorphic; *fruit* a loculicidal capsule with numerous winged seeds.

Campsis radicans **(L.) Seem.**, TRUMPET VINE. Stems ca. 10 m; *leaves* to 30+ cm long; *leaflets* 5–13, widely lanceolate, serrate, finely pubescent along abaxial veins; *calyx* up to 2.5 cm long, tubular; *corolla* 4× longer than the calyx, tubular-funnelform; *capsules* 1–2 dm long. Cultivated as an ornamental but could possibly escape along roadsides. June–Sept. Introduced.

CATALPA Scop. – INDIAN CIGAR TREE

Trees; *leaves* opposite or sometimes whorled, simple and entire or coarsely lobed; *flowers* white with purple spots, pink, or yellowish, in cymes; *calyx* bilabiate; *corolla* bilabiate with 2 smaller upper lobes and 3 larger lower lobes, gibbous on the lower side; *fruit* a long, cylindrical loculicidal capsule.

Catalpa speciosa **Warder**, NORTHERN CATALPA; CATAWBA-TREE. Plants to 30 m; *leaves* opposite, ovate, 15–30+ cm long, green and glabrous adaxially, densely hairy abaxially; *calyx* of two distinct lobes, ca. 1 cm long; *corolla* (1) 5 (7) cm long, 4 cm wide, lobes undulate; *capsules* 25–45 (60) cm long, brown; *seeds* 3–4 cm long, winged, flattened. Cultivated in towns, could possibly escape near old homesteads or on roadsides, to be expected along the Front Range at low elevations. May–July. Introduced.

BORAGINACEAE Juss. – BORAGE FAMILY

Annuals, biennial, and perennial herbs (ours), usually rough-hairy; *leaves* simple or pinnately compound or dissected, usually alternate, or sometimes the lower ones opposite, or rarely whorled, exstipulate; *flowers* perfect, actinomorphic or rarely slightly zygomorphic, solitary or in cymose inflorescences, the cymes often helicoid; *calyx* 5-lobed, usually persistent in fruit; *corolla* 5-lobed, usually funnelform to salverform, the lobes usually imbricate or convolute, sometimes crested or hairy-appendaged in the throat; *stamens* 5, epipetalous; *ovary* superior, 2-carpellate, deeply 4-lobed; *style* gynobasic; *fruit* 1–4 nutlets.

Species delimitations in this family are based primarily on fruit characteristics. In many instances, if mature nutlets are not present, precise identification to species may not be possible. However, thankfully fruit can usually be found at the base of the inflorescences even on very early flowering plants. *Echium vulgare* L. has long been reported for Colorado based on a single old collection from near Rocky Ford in Otero Co. (RM). However, examination of this specimen shows that it is an *Anchusa*.

1a. Weak prostrate or climbing-scrambling annuals with retrorsely prickly-hispid stems; calyx enlarged in fruit and becoming firmly chartaceous, folded and flattened, and prominently veiny; flowers axillary in the leaves or forks of the branches...***Asperugo***
1b. Perennials, biennial, or annual plants lacking retrorsely prickly-hispid stems; calyx not enlarged in fruit and otherwise unlike the above; flowers in terminal cymes or racemes, solitary, or rarely axillary...2

2a. Cushion or mat-like alpine plants; leaves 5–10 mm long, loosely soft and long-hairy (the hairs more numerous at the tips); flowers blue (rarely white) with yellow crests in the center; nutlets with a lacerate-toothed to entire upturned margin on the dorsal side...***Eritrichium***
2b. Plants not cushion or mat-like, or if so then the flowers white or yellow and plants not of the alpine; leaves various but not soft, long-hairy; nutlets (if present) unlike the above...3

3a. Leaf petioles decurrent on the stem, leaving the stem winged; base of the nutlet with an enlarged, swollen and thickened rim, leaving a distinct pit on the gynobase; tall, coarse weedy perennials; flowers white, purple, rose, or yellowish...***Symphytum***
3b. Stem not winged from decurrent leaf petioles; nutlets, plants, and flowers various...4

4a. Styles bifid; prostrate, dichotomously branched annuals; leaves ovate to suborbicular, clustered in small rosettes; flowers minute, the corolla barely longer than the calyx, pink or lavender or sometimes nearly white...***Tiquilia***
4b. Styles entire; plants otherwise unlike the above...5

5a. Corolla whitish-green to yellowish, tubular, with 5 erect, hairy lobes constricted around the long-exserted style; leaves strongly and prominently veined, especially below; nutlets ovoid, smooth and white; plants coarsely hispid-hairy...***Onosmodium***
5b. Corolla various but the lobes not hairy or constricted around a long-exserted style; leaves without numerous prominent veins; nutlets variously pitted or rugose, or if smooth and white then flowers yellow and salverform; plants variously hairy or glabrous...6

6a. Flowers reddish-purple or rarely white; nutlets separating and spreading at maturity, the entire surface covered with barbed prickles, depressed ovoid or orbicular; plants softly pilose or canescent (lacking coarsely hirsute or pustulate hairs), erect, annual or biennial herbs...***Cynoglossum***
6b. Flowers blue, purple, pink, yellow, orange, or white; nutlet surface not entirely covered with barbed prickles, if present these restricted to the tips or margins; plants various...7

7a. Nutlets with hooked or barbed bristles or prickles...8
7b. Nutlets without hooked or barbed bristles or prickles, or merely with a lacerate margin...10

8a. Prostrate or weakly ascending annuals; nutlets spreading in pairs, with prominent, upturned winged margins with hooked bristles near the tip...***Pectocarya***
8b. Erect annuals, biennials, or perennials; nutlets not as above...9

9a. Pedicels erect or nearly so in fruit; racemes bracteate; annuals...***Lappula***
9b. Pedicels recurved in fruit; racemes naked or nearly so; perennials or biennials...***Hackelia***

10a. Flowers yellow, orange, greenish-white, or white (occasionally with a blue or purple eye)...11
10b. Flowers blue, purple, or pink...16

11a. Low annuals (to 2 dm tall) with several, mostly prostrate stems from the base; nutlets with an elevated, club-shaped scar near the base and a well-developed ventral keel extending from the tip to the middle or base, sometimes with distally branched bristles, rugose-tuberculate or rarely smooth and shiny; flowers small and white, only about 1–2 mm wide, in an irregularly bracteate cyme...***Plagiobothrys***
11b. Erect annuals or erect to spreading perennials; nutlets and flowers unlike the above...12

12a. Nutlets whitish-gray, attached by their base to a flat gynobase, leaving a large scar usually surrounded by an evident sharp rim at the base of the nutlet, also with an evident ventral keel; flowers yellow or orange, white in one weedy annual species...***Lithospermum***
12b. Nutlets variously attached above the base to a broadly to narrowly pyramidal gynobase, without an evident rim at the base of the nutlet; flowers usually white, sometimes yellow or orange...13

13a. Plants glabrous to strigose; corolla usually blue but occasionally white...***Mertensia***
13b. Plants coarsely pubescent with hispid or pustulate hairs and bristles or densely strigose; corolla white, yellow, or orange...14

14a. Flowers yellow or orange, without fornices and with an open throat; cotyledons deeply 2-cleft and appearing as 4; uncommon, weedy annuals...***Amsinckia***
14b. Flowers white or yellow, with conspicuous fornices closing off the throat; cotyledons entire; widespread native annual, biennial, or perennial plants...15

15a. Annuals from a slender taproot; plants without a conspicuous tuft of basal leaves; leaves usually linear or sometimes narrowly oblanceolate; corolla minute, the limb 0.5–2 mm wide; cymes in condensed head-like inflorescences or solitary or paired in the axils of the upper branches and then usually elongate and open...***Cryptantha***
15b. Perennials or biennials from a woody taproot and often with a branching caudex; plants more or less with a conspicuous tuft of basal leaves; leaves oblanceolate or sometimes linear-oblanceolate; corolla conspicuous, the limb 4–16 mm wide; flowers aggregated into a dense, terminal, cylindric to capitate thyrse...***Oreocarya***

16a. Corolla tubular to campanulate with a well-defined tube or throat and erect lobes sometimes with just the tips recurved, blue, purple, pink, or rarely white, without a yellow center; glabrous to strigose perennial, native plants...***Mertensia***
16b. Corolla salverform or broadly funnelform with spreading lobes and an ill-defined tube or throat, blue to purple, often with a yellow or white center; plants often with pustulate hairs, or sometimes glabrous to strigose...17

17a. Base of the nutlet with an enlarged, swollen and thickened rim, leaving a distinct pit on the gynobase; tall, weedy perennials mostly 3–8 dm tall, with coarsely hirsute or pustulate hairs; corolla 6–20 mm wide...***Anchusa***
17b. Base of the nutlet not enlarged with a thickened rim, and not leaving a pit on the gynobase; glabrous, strigose to hirsute perennials or annuals, native or weedy plants to 5 dm; corolla 2–10 mm wide...***Myosotis***

AMSINCKIA Lehm. – FIDDLENECK

Annuals from taproots, bristly-hairy; *leaves* alternate, entire; *calyx* cleft to the base or nearly so, persistent; *corolla* yellow or orange, funnelform, the throat marked red; *stamens* included; *nutlets* 4, with a small scar near the base of the ventral side, on a short-pyramidal gynobase, with a well-developed ventral keel extending from near the tip to near or below the middle. (Macbride, 1917; Ray & Chisaki, 1957)

1a. Corolla limb 5–10 mm wide; fornices hairy, large and well-developed; corolla yellow to yellow-orange; stamens attached below the middle of the corolla tube; plants to 3 dm tall...***A. lycopsoides***
1b. Corolla limb 1–3 mm wide; fornices absent; corolla pale yellow; stamens attached above the middle of the corolla tube; plants 1.5–7 dm tall...***A. menziesii* var. *menziesii***

***Amsinckia lycopsoides* Lehm.**, TARWEED FIDDLENECK. Plants to 3 dm; *leaves* linear to oblong-lanceolate, to 10 cm long; *calyx* 6–10 mm long; *corolla* yellow to yellow-orange, with well-developed, large fornices, the limb 5–10 mm wide; *stamens* attached below the middle of the corolla tube; *nutlets* 2.5–3 mm long, sharp-tubercled. Uncommon in disturbed areas, our 2 records from Clear Creek Co. around Silver Plume, 9000 ft. June–Aug. E. Introduced.

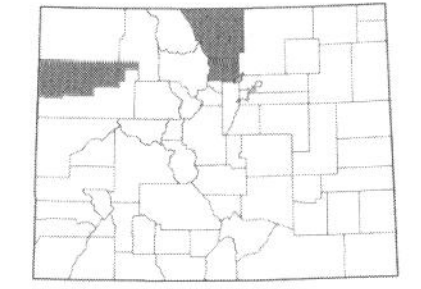

Amsinckia menziesii* (Lehm.) A. Nelson & J.F. Macbr. var. *menziesii, MENZIES' FIDDLENECK. Plants 1.5–7 dm; *leaves* lanceolate to oblong, to 12 cm long; *calyx* 5–10 mm long; *corolla* pale yellow, lacking fornices, the limb 1–3 mm wide; *stamens* attached above the middle of the corolla tube; *nutlets* 2–3.5 mm long, sharp-tubercled. Uncommon in disturbed places, 7500–8500 ft. June–Aug. E/W. Introduced.

ANCHUSA L. – ALKANET; BUGLOSS

Annual, biennial, or perennial caulescent herbs; *leaves* entire; *calyx* shallowly to deeply cleft; *corolla* blue or purple, funnelform or salverform, the throat often poorly defined; *stamens* included; *nutlets* 4, their attachment surrounded by a prominent, thickened annular ring leaving a pit on the low gynobase.

1a. Calyx lobes linear, much longer than the corolla tube; corolla about 10–15 mm long, the limb 12–20 mm wide...***A. azurea***
1b. Calyx lobes lanceolate to narrowly triangular, about as long as the corolla tube; corolla about 6–11 mm long, the limb 6–11 mm wide...***A. officinalis***

***Anchusa azurea* Mill.**, COMMON ALKANET. Plants 5–8 dm; *leaves* lance-linear, 5–15 mm wide; *calyx* 6–9 mm long, 8–12 mm long in fruit; *corolla* 10–15 mm long, the limb 12–20 mm wide, blue with white appendages; *nutlets* 2–4 mm wide at the base. Found as a weed along roadsides and vacant lots and in other disturbed areas, 4200–8000 ft. May–July. E. Introduced.

***Anchusa officinalis* L.**, COMMON BUGLOSS. Plants 4–7 dm; *leaves* lance-linear to oblanceolate, 6–20 cm long and 10–25 mm wide; *calyx* 5–7 mm long, 8–12 mm long in fruit; *corolla* 6–11 mm long, the limb 6–11 mm wide, blue-violet to dark blue with white appendages; *nutlets* 2–3.5 mm wide at the base. Found as a weed in disturbed places, 5000–6000 ft. May–July. E. Introduced. (Plate 17)

ASPERUGO L. – MADWORT

Annual plants with retrorsely prickly-hispid stems; *leaves* alternate or the upper often opposite; *flowers* in or near the axils of the leaves or bracts; *calyx* 5-lobed, much enlarged and reticulate in fruit; *corolla* blue; *nutlets* 4, obliquely compressed, granular-tuberculate.

***Asperugo procumbens* L.**, CATCHWEED. Stems to ca. 6 dm long; *leaves* ovate to elliptic or oblanceolate, 2–7 cm long, strigose-hispid; *calyx* ca. 3 mm long in anthesis, ca. 15 mm long in fruit; *corolla* ca. 3 mm wide, funnelform to campanulate; *nutlets* brown, enclosed within the enlarged calyx. Found in gardens, vacant lots, and disturbed places, often in moist areas such as along streams and ditches, 5000–7500 ft. May–July. E/W. Introduced.

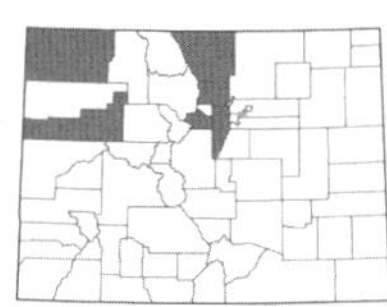

CRYPTANTHA Lehm. ex G. Don – CRYPTANTHA

Annual herbs, usually bristly-hairy; *leaves* alternate, entire; *flowers* in cymes; *calyx* enlarging in fruit; *corolla* white, usually with yellow fornices, usually salverform with a spreading limb but sometimes funnelform with erect limbs, minute; *stamens* 5, included or the tips just reaching the orifice of the corolla-tube; *nutlets* 1–4, smooth or variously roughened. (Cronquist et al., 1984; Johnston, 1925)

1a. Flowers aggregated into dense, terminal, head-like inflorescences; plants forming low cushions mostly 0.2–0.6 dm; calyx circumscissle just below the middle...***C. circumscissa***
1b. Flowers in solitary or paired cymes in the axils of the upper branches, these usually elongate and open; plants not cushion-forming, 0.5–3 dm; calyx not circumscissle just below the middle...2

2a. At least some of the nutlets with broad wings on the margins...***C. pterocarya***
2b. All nutlets without broad wings on the margins (although sometimes the margins are sharply raised)...3

3a. All of the nutlets smooth and shiny, sometimes with a sharply raised margin (at 10×)...4
3b. Nutlets all tuberculate, papillate, finely granular, or roughened, or at least some of them so, without sharply raised margins (at 10×)...8

4a. Usually only 1 nutlet maturing, or occasionally 2–3 in some flowers; calyx densely white-pubescent with long, soft hairs and few intermingled bristles; cymes not much elongate, only about 0.5–2 cm long at maturity with the flowers mostly aggregated at the tips...***C. gracilis***
4b. All 4 nutlets maturing; calyx hirsute-strigose with coarser hairs intermingled with numerous bristles; cymes elongate or compact...5

5a. Nutlets with the scar on the ventral side lying near one margin and not in the middle (excentric)...***C. affinis***
5b. Nutlets with the scar on the ventral side in the middle of the nutlet...6

6a. Nutlets with a conspicuous, sharply angled raised margin; cymes smaller, elongating only 1–3 cm at maturity...***C. watsonii***
6b. Nutlets with rounded margins; cymes usually elongating 3–15 cm at maturity...7

7a. Nutlets lanceolate with a narrow body (mostly 0.5–0.8 mm wide) and narrow, tapering tip...***C. fendleri***
7b. Nutlets ovate with a wider body (mostly 0.8–1.2 mm wide)...***C. torreyana***

8a. Calyx bent and recurved at maturity; only 1 nutlet maturing per flower, this inwardly curved in alignment with the calyx; style much shorter than the nutlets...***C. recurvata***
8b. Calyx ascending or slightly spreading at maturity; usually all 4 nutlets maturing or seldom only 2 or 3, these not curved; style about equaling the nutlets...9

9a. Nutlets of 2 kinds, either dissimilar in size or surface texture(either similar nutlets tuberculate and the odd nutlet slightly larger and finely granulate or papillate, or the odd nutlet larger and tuberculate but otherwise similar to the other nutlets); plants of the eastern or western slope...10
9b. Nutlets all alike, either tuberculate or papillate with upcurved papillae; plants of the western slope (Moffat or Rio Blanco cos.)...13

10a. Odd nutlet tuberculate, similar to the 3 other nutlets in texture, but larger...***C. rudis***
10b. Odd nutlet smooth or papillate-granulate, distinctly different in texture than the 3 other nutlets...11

11a. Nutlets lanceolate-ovate, the 3 similar nutlets 1.8–2.5 mm long, with a shallowly impressed scar on the ventral side abruptly opening into a small triangular basal areola...***C. kelseyana***
11b. Nutlets mostly ovate, the 3 similar nutlets 1.2–1.5 mm long, deeply excavated on the ventral side forming a little pit...12

12a. Odd nutlet closely papillate-granulate; flowering inflorescences bracteate throughout (look like small leaves subtending most of the flowers in the spike)...***C. minima***
12b. Odd nutlet spinular-muricate with small projections present in addition to the underlying finely granulate surface; flowering inflorescences naked or bracteate at the base only...***C. crassisepala* var. *elachantha***

13a. Nutlets broadly ovate, the surface finely granulate and coarsely tuberculate...***C. ambigua***
13b. Nutlets lanceolate, the surface dotted with upcurved papillae especially towards the tip...***C. scoparia***

***Cryptantha affinis* (A. Gray) Greene**, QUILL CRYPTANTHA. Plants 0.5–3 dm; *leaves* linear to narrowly oblanceolate, 1–3.5 (5) cm long, strigose and often pustulate; *calyx* 2.5–4 mm long at maturity; *corolla* white, 1–2 mm wide; *nutlets* 4, 1.6–2 mm long, with an excentric scar, lying off-center, not opening into a basal areola. Recently discovered in Grand Co., usually in sagebrush, 8400–9300 ft. June–July. W. (Plate 18)

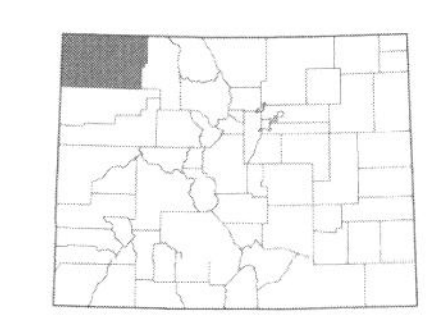

***Cryptantha ambigua* (A. Gray) Greene**, BASIN CRYPTANTHA. Plants 0.5–3 dm; *leaves* linear to narrowly lanceolate, 1–4 cm long, hirsute and strigose; *calyx* 4–6 mm long; *corolla* white, 1–2 mm wide; *nutlets* 4, 1.2–2 mm long, finely granulate and tuberculate or rarely smooth, the scar closed or nearly so. Uncommon in sandy soil of open, dry places, often with pinyon-juniper communities, 5500–9000 ft. May–July. W.

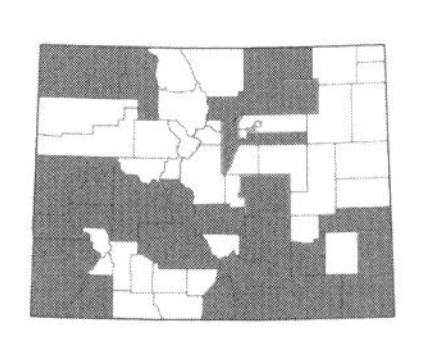

***Cryptantha circumscissa* (Hook. & Arn.) I.M. Johnst.**, CUSHION CRYPTANTHA. Plants 0.2–0.6 (1) dm; *leaves* linear, strigose to hirsute; *calyx* 2–3 mm long, circumscissile just below the middle; *corolla* 0.5–1.5 mm wide; *nutlets* 4, narrowly triangular-ovate, 1–1.3 mm long, smooth or finely roughened, the scar closed. Uncommon in dry, open, often sandy places, 5900–6500 ft. April–July. W.

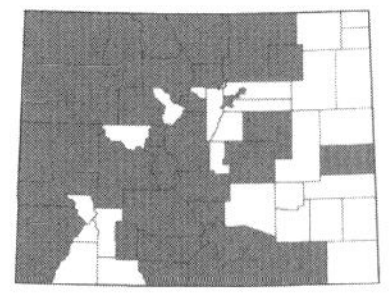

***Cryptantha crassisepala* (Torr. & A. Gray) Greene var. *elachantha* I.M. Johnst.**, THICKSEPAL CRYPTANTHA. Plants 0.5–1.5 (3) dm; *leaves* linear to narrowly oblanceolate, 1–4 cm long, hirsute and pustulate; *calyx* 4–6.5 mm long, with a strongly thickened midrib; *corolla* 1.5–2.5 mm wide, white; *nutlets* usually 4, with 3 ovate, 1.2–1.5 mm long, and tuberculate with a deeply excavated pit on the ventral side, and 1 lance-ovate, 2–2.5 mm long, and minutely speculate-papillate. Locally common in dry, sandy places, sometimes in disturbed sites, 4000–7500 ft. April–June. E/W. (Plate 18)

***Cryptantha fendleri* (A. Gray) Greene**, SAND-DUNE CRYPTANTHA. [*C. pattersonii* (A. Gray) Greene]. Plants 1–5 dm; *leaves* narrowly oblanceolate, 2–5 cm long, strigose and hirsute; *calyx* 4–6 mm long; *corolla* white, 1 mm wide; *nutlets* 4, lanceolate, smooth and shiny, the scar closed except for a basal areola. Found in sandy soil of open, dry places, 4900–9000 ft. May–July. E/W. (Plate 18)

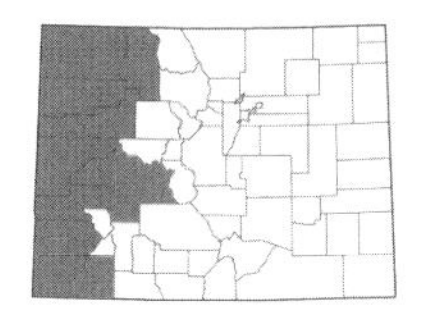

***Cryptantha gracilis* Osterh.**, SLENDER CRYPTANTHA. Plants 1–3 dm; *leaves* linear, 1–3.5 cm long, stiffly hirsute; *calyx* 2–3 mm long, densely white-pubescent with rather long, soft hairs; *corolla* white, 1–2 mm wide; *nutlets* usually 1, lanceolate, smooth and shiny, the scar closed or narrowly open below. Found on open, dry slopes, often with pinyon-juniper or sagebrush communities, 5000–7300 ft. April–July. W.

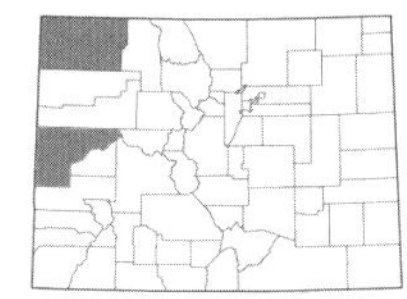

***Cryptantha kelseyana* Greene**, Kelsey's cryptantha. Plants 0.5–2.5 dm; *leaves* linear to narrowly oblanceolate, 1–4 cm long, strigose and hirsute; *calyx* 5–7 mm long; *corolla* white, 1–2 mm wide; *nutlets* 4, lance-ovate, 1.8–2.5 mm long (with 1 nutlet longer), finely and obscurely granular or tuberculate, the scar narrow or nearly closed. Found on dry, open, sandy places, 4500–7800 ft. June–July. W. (Plate 18)

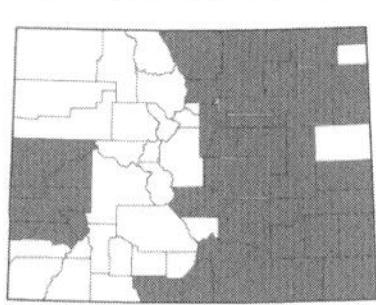

***Cryptantha minima* Rydb.**, little cryptantha. Plants 1–2 dm; *leaves* oblanceolate, 1–3 cm long, strigose and hirsute, pustulate; *calyx* 5–7 (9) mm long, the midrib thickened and indurate; *corolla* white, 1–1.5 mm wide; *nutlets* 4, ovate, heteromorphous, 3 nutlets strongly tuberculate with an excavated groove on the ventral side, 1.2–1.5 mm long, and one nutlet 2–3 mm long, and finely and closely papillate-granulate. Common in dry, sandy places, along roadsides, or sometimes in disturbed areas, 3500–8000 ft. April–June. E/W. (Plate 18)

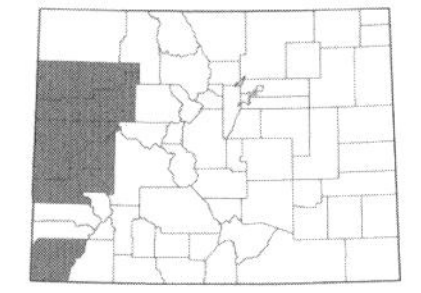

***Cryptantha pterocarya* (Torr.) Greene**, wing-nut cryptantha. Plants 1–4 dm; *leaves* linear-lanceolate, 1–5 cm long, strigose or hirsute; *calyx* 4–8 mm long; *corolla* white, 0.5–2 mm wide; *nutlets* 4, oblong-lanceolate, winged or sometimes the axial one wingless, 2.2–3.2 mm long. Uncommon in pinyon-juniper on dry slopes, 4600–6500 ft. May–June. W. (Plate 18)

There are two varieties of *C. pterocarya* in Colorado:

1a. All of the nutlets winged...**var. *cycloptera* (Greene) J.F. Macbr.**
1b. One of the 4 nutlets wingless...**var. *pterocarya***

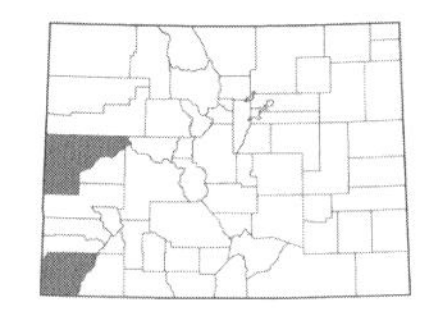

***Cryptantha recurvata* Cov.**, bent-nut cryptantha. Plants 0.5–4 dm; *leaves* linear to lanceolate-oblong, 1–3 cm long, strigose; *calyx* 2.5–3.5 mm long, bent and recurved at maturity; *corolla* white, 1–2 mm wide; *nutlets* 1, oblong-lanceolate, inwardly curved, finely granular-tuberculate, 1.6–2 mm long. Uncommon in open, dry places, 4500–5200 ft. April–June. W.

***Cryptantha rudis* A. Nelson ex Brand**, tuberculate cryptantha. Plants 0.5–2.5 dm; *leaves* linear to narrowly oblanceolate, 1–5 cm long, hirsute; *calyx* 4–7 mm long; *corolla* white, 1–2 mm wide; *nutlets* 4, lanceolate, tuberculate, heteromorphous, 3 nutlets 1.2–1.8 mm long and 1 nutlet 2–3 mm long. Found in dry, open sandy places, 5500–7500 ft. June–July. W. (Plate 18)

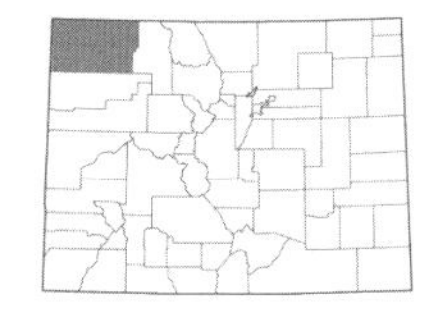

***Cryptantha scoparia* A. Nelson**, pinyon desert cryptantha. Plants 0.5–3 dm; *leaves* linear to lanceolate, 2–4 cm long, strigose and hispid; *calyx* 4–6 mm long; *corolla* white; *nutlets* usually 4, lanceolate, 1.6–2 mm long, with forwardly directed papillae especially towards the tip. Uncommon in dry, open places in sandy soil, usually with pinyon-juniper or sagebrush communities, 5500–6000 ft. May–June. W. (Plate 18)

***Cryptantha torreyana* (A. Gray) Greene**, Torrey's cryptantha. Plants 1–4 dm; *leaves* linear to narrowly oblong, 1–5 cm long, hirsute; *calyx* 4–8 mm long; *corolla* white, 1–2 mm wide; *nutlets* 4, ovate, 1.3–2.3 mm long, smooth and shiny. Probably present in Grand Co. (see note below) and elsewhere in the northwestern counties in dry places, often with sagebrush, 8500–9000 ft. May–Aug. W.

Possibly found in Colorado in Grand Co. near Granby. However the fruit on the specimen (*Erin Foley 197*/RM) isn't mature enough to conclusively identify, and the specimen could be *C. affinis*.

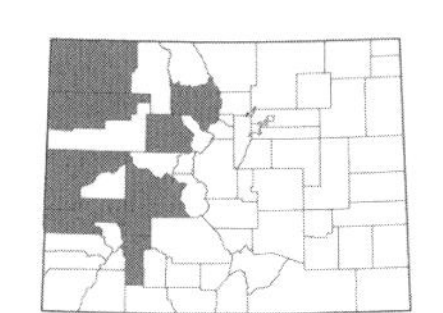

***Cryptantha watsonii* (A. Gray) Greene**, Watson's cryptantha. Plants 1–3 dm; *leaves* linear to oblanceolate, 1–5 cm long, hirsute; *calyx* 2–4 mm long; *corolla* white, 1–2 mm wide; *nutlets* 4, lanceolate, 1.3–2 mm long, smooth and sometimes shiny, with evident, sharply angled raised margins, with a small triangular scar at the base on the ventral side. Found in open, dry places, often with pinyon-juniper or sagebrush communities, 6300–8000 ft. June–Aug. W.

Similar to *C. kelseyana* in having a small triangular scar at the base of the nutlets instead of deeply excavated pit, but all the nutlets are tuberculate.

CYNOGLOSSUM L. – hound's tongue

Biennial or perennial herbs; *leaves* entire, alternate; *calyx* deeply cleft to below the middle; *corolla* violet to reddish-purple, funnelform or salverform; *stamens* included or to the corolla throat; *nutlets* 4, covered with short, glochidiate prickles, depressed ovoid with a flat or convex dorsal side, with a broad scar not extending below the middle of the ventral surface.

***Cynoglossum officinale* L.**, gypsyflower. Biennials, to 12 dm; *leaves* elliptic to oblanceolate, 8–20 cm long; *calyx* 4–6 mm long in anthesis, to 1 cm long in fruit, hirsute; *corolla* 3–5 mm long; *nutlets* 1–4, 5–7 mm long, brown. Common along roadsides, in vacant lots, and in other disturbed places, with occurrences scattered across the state, 5000–9000 (11,000) ft. May–Aug. E/W. Introduced.

ERITRICHIUM Schrad. ex Gaudin – ALPINE FORGET-ME-NOT

Caespitose perennials; *leaves* entire, small, densely crowded on short stems; *calyx* cleft essentially to the base; *corolla* blue or rarely white, salverform; *stamens* included; *nutlets* 1–4, smooth, entire or toothed, or with a lacerate-toothed flange on the back of the nutlet above the base.

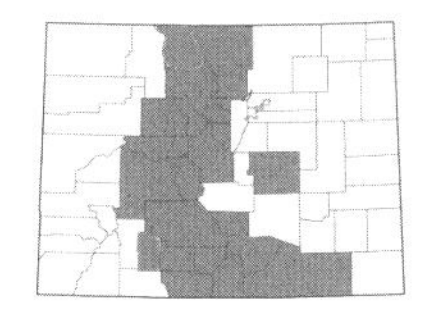

***Eritrichium nanum* (Vill.) Schrad. ex Gaudin var. *elongatum* (Rydb.) Cronquist**, ARCTIC ALPINE FORGET-ME-NOT. [*E. nanum* Vill. Schrad. ex Gaudin var. *argenteum* (Wight) I.M. Johnst..]. Plants mat-forming, to 1 dm; *leaves* oblong to ovate, to 1 cm long, hairy, tufting at the tips; *corolla limb* 4–8 mm wide. Common in the alpine, 10,000–14,000 ft. June–Aug. E/W. (Plate 17)

HACKELIA Opiz – WILD FORGET-ME-NOT; STICKSEED

Biennials, perennial, or rarely annual herbs; *leaves* alternate, entire; *calyx* cleft to the base; *corolla* blue or white or sometimes ochroleucous; *stamens* included; *nutlets* 4, the margin with glochidiate prickles, these sometimes connate below forming a cupulate border, attached ventrally to the gynobase by a broad scar. (Gentry, 1974; Gentry & Carr, 1976)

1a. Leaves with evident pustulate-based hairs; slender biennials from short taproots...2
1b. Leaves without pustulate-based hairs or sometimes with a few short pustulate-based hairs; biennial or perennial plants, usually more robust, from taproots and often with a branching caudex...3

2a. Flowers larger, the limb 6–8 mm wide and the tube 1.6–2 mm long; stem surface hispid; nutlets with intramarginal prickles that are shorter than the marginal ones, rarely absent...***H. gracilenta***
2b. Flowers smaller, the limb 1.5–2.5 mm wide and the tube 0.8–0.9 mm long; stem surface reflexed or spreading-hairy, sometimes canescent throughout; nutlets without intramarginal prickles but the marginal prickles alternating long and short on each side...***H. besseyi***

3a. Marginal prickles connate at the base forming a wing along the edge of the nutlet; usually only one or seldom up to 3 intramarginal prickles present, or these absent; biennial or short-lived perennial with the basal leaves withering early...***H. floribunda***
3b. Marginal prickles distinct or only slightly connate at the base; intramarginal prickles 4–9, seldom fewer; perennial with the basal leaves usually persistent...***H. micrantha***

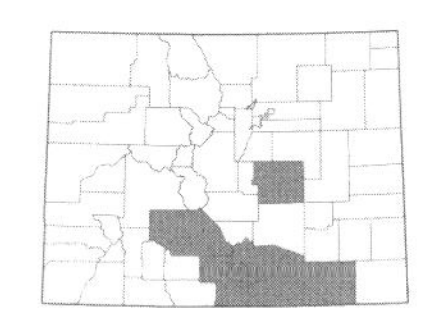

***Hackelia besseyi* (Rydb.) J.L. Gentry**, BESSEY'S STICKSEED. Biennials, 3–7.5 (11) dm; *leaves* oblanceolate to lanceolate, 2–9 (13) cm long, hispid-hirsute, pustulate; *calyx* lobes 1 mm long; *corolla* blue, 1.5–2.5 mm wide; *nutlets* 2–2.5 mm long, ovate to ovate-lanceolate, hispidulous dorsally, with 8–13 marginal prickles on each side, alternating long and short, distinct or slightly connate at the base, 0.5–1.5 mm long. Found on dry hillsides, often in pinyon-juniper woodlands, 6000–8000 ft. June–Aug. E.

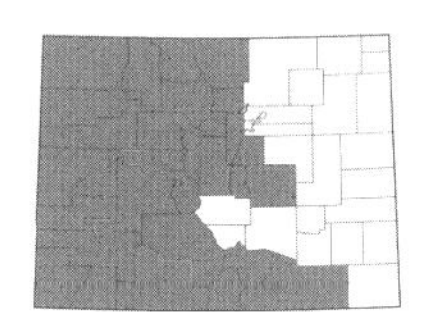

***Hackelia floribunda* (Lehm.) I.M. Johnst.**, MANYFLOWER STICKSEED. Biennials or short-lived perennials, 3–14 dm; *leaves* oblanceolate to narrowly elliptic, 3–30 cm long, hirsute, not pustulate; *calyx* lobes 1.5–2.5 mm long; *corolla* blue, rarely white, (2.5) 4–7.5 mm wide; *nutlets* ovate to ovate-lanceolate, hispidulous, with 5–9 marginal prickles on each side 1.5–3.5 mm long, connate ⅓–½ their length, 2–4 mm long. Common in moist places along streams, in aspen groves, open meadows and woodlands, and on dry hillsides, 5600–10,500 ft. May–Aug. E/W. (Plate 18)

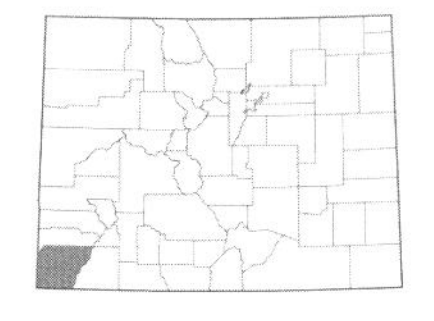

***Hackelia gracilenta* (Eastw.) I.M. Johnst.**, MESA VERDE STICKSEED. Biennials, (1.5) 3–9 dm; *leaves* oblanceolate to narrowly oblong, 2.5–12 (17) cm long, pustulate; *calyx* lobes 2–3 mm long; *corolla* blue with a white center, 6–8 mm wide; *nutlets* lance-ovate, 2–3 mm long, with intramarginal prickles shorter than the marginal ones, marginal prickles 5–6 on each side, 1–3 mm long. Uncommon on dry, rocky or oak brush canyon slopes or sometimes in pinyon-juniper woodlands, 6500–8000 ft. May–June. W. Endemic.

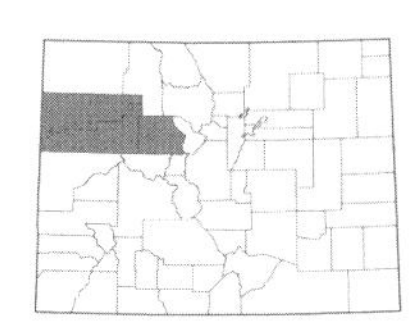

***Hackelia micrantha* (Eastw.) J.L. Gentry**, JESSICA STICKTIGHT. Perennials, 3–11 dm; *leaves* oblanceolate to narrowly elliptic, 4–30 cm long, hirsute, not pustulate; *calyx* lobes 1–3 mm long; *corolla* blue or rarely white, (3) 5–10 mm wide; *nutlets* ovate, 3–4.5 mm long, with intramarginal prickles, marginal prickles 4–9 on each side, 2–4.5 mm long and distinct or slightly connate at their bases. Uncommon in dry, open forests or on open hillsides, or sometimes along streams, 9000–10,600 ft. June–Aug. W.

LAPPULA Moench – STICKSEED

Annuals; *leaves* alternate, entire; *flowers* in bracteate, helicoid cymes; *corolla* blue or white, funnelform, with fornices; *calyx* cleft to the base; *stamens* included; *nutlets* 4, with 1–2 rows of glochidiate prickles on the dorsal margin, often connate and forming a cupulate border, attached to the gynobase along the length of the ventral keel.

1a. Prickles on the margin of nutlets in a single row, sometimes fused at the base to form a cupulate border; flowers blue or white, the limb 1.5–2.5 mm wide...***L. occidentalis***
1b. Prickles on the margin of nutlets in at least 2 rows, not fused at the base; flowers usually blue, the limb 2–4 mm wide...***L. squarrosa***

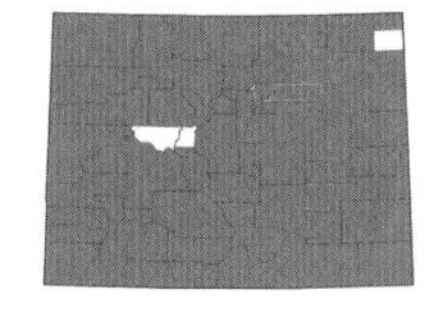

***Lappula occidentalis* (S. Watson) Greene**, WESTERN STICKSEED. [*L. redowskii* (Hornem.) Greene]. Plants 0.5–5 (7) dm; *leaves* linear to oblanceolate, to 6 cm long, strigose; *corolla* blue or white, 1.5–2.5 mm wide; *nutlets* with marginal prickles in a single row, sometimes fused and forming a cupulate border. Common and widespread in disturbed areas, along roadsides, and in open prairie and meadows, 3500–10,000 ft. April–Aug. E/W. Introduced. (Plate 18)

There are two varieties of *L. occidentalis* in Colorado, both of which are common:
1a. Marginal prickles distinct, not fused at the base...**var. *occidentalis***
1b. Marginal prickles on some of the nutlets fused at the base to form a cupulate, sometimes swollen, border...**var. *cupulata* (A. Gray) Higgins**, FLATSPINE STICKSEED. [*L. marginata* (M. Bieb.) Gürke; *L. texana* (Scheele) Britton].

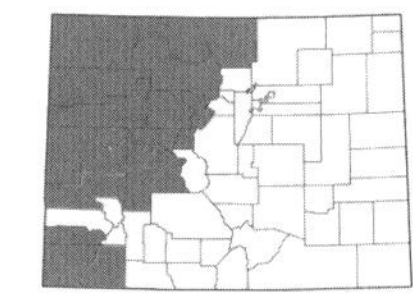

***Lappula squarrosa* (Retz.) Dumort.**, EUROPEAN STICKSEED. [*L. echinata* Gilib.; *L. fremontii* (Torr.) Greene]. Plants 1–7 dm; *leaves* linear to oblanceolate, to 8 cm long; *corolla* blue, 2–4 mm wide; *nutlets* with marginal prickles in at least 2 rows, not fused at the base. Found in disturbed places, along roadsides, and in open places, 5000–9000 ft. May–Aug. E/W. Introduced. (Plate 18)

LITHOSPERMUM L. – STONESEED; PUCCOON

Annual or more often perennial herbs; *leaves* alternate, entire; *calyx* deeply cleft; *corolla* yellow, orange, white, or violet, funnelform or salverform, with or without fornices; *stamens* included or partly exserted; *nutlets* 4, smooth or wrinkled, basally attached, the gynobase flat or low-pyramidal, usually with a prominent ventral keel. (Johnston, 1952)
1a. Annuals; corolla white, small and barely (if at all) exceeding the calyx, the limb 2–4 mm wide; nutlets gray-brown, wrinkled, pitted, and often tuberculate...***L. arvense***
1b. Perennials; corolla yellow or orange, conspicuous and much exceeding the calyx (*L. incisum* can have small cleistogamous flowers later in the season), the limb 5–20 mm wide; nutlets gray to white, smooth or sparsely pitted...2

2a. Corolla lobes evidently erose to fimbriate, the tube 15–35 (40) mm long; small but evident fornices present in the corolla-throat...***L. incisum***
2b. Corolla lobes entire or nearly so, the tube 4–15 mm long; fornices absent...3

3a. Flowers bright yellow-orange, crowded in terminal cymes at the end of branch tips; leaves harshly and densely strigose; flowers salverform, with the limb of the corolla mostly 10–20 mm wide; plants often drying black...***L. caroliniense* var. *croceum***
3b. Flowers yellow, in the axils of reduced upper leaves or in leafy-bracteate clusters in the upper axils; leaves strigose to spreading-hirsute; flowers tubular-funnelform, salverform, or funnelform-salverform, the limb of the corolla mostly 5–13 mm wide; plants usually not drying black...4

4a. Corolla tube 4–7 mm long; flowers salverform or funnelform-salverform with more spreading lobes; inflorescence with numerous, long, crowded leaves; nutlets 3.5–6 (8) mm long...***L. ruderale***
4b. Corolla tube 8–15 mm long; flowers tubular-funnelform; leaves in the inflorescence greatly reduced; nutlets about 3 mm long...***L. multiflorum***

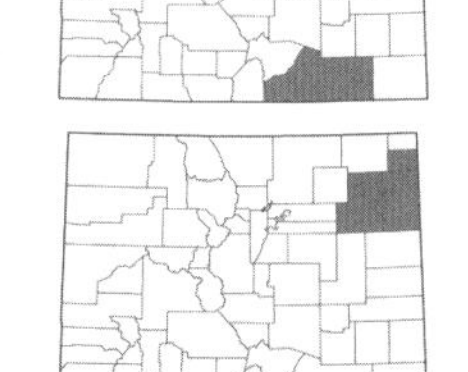

***Lithospermum arvense* L.**, GROMWELL. [*Buglossoides arvensis* (L.) I.M. Johnst.]. Annuals, 1–7 dm; *leaves* oblanceolate to linear-oblong, 1.5–6 cm long, strigose; *corolla* white or bluish-white, 2–4 mm wide; *nutlets* 3 mm long, wrinkled, pitted, sometimes tuberculate. Uncommon in disturbed areas and along roadsides, 4500–6000 ft. April–June. E. Introduced.

***Lithospermum caroliniense* (J.F. Gmel.) MacMill. var. *croceum* (Fernald) Cronquist**, CAROLINA PUCCOON. Perennials, 2–5 dm; *leaves* linear to lanceolate, 2–6 cm long, strigose; *corolla* bright yellow-orange, 10–20 mm wide; *nutlets* smooth, shiny, white. Infrequent on sandy soil of the plains, 3500–4000 ft. May–July. E.

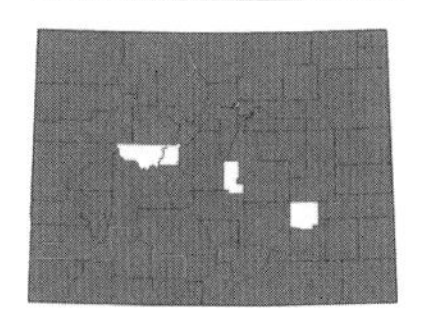

***Lithospermum incisum* Lehm.**, PLAINS STONESEED. Perennials, 0.5–3 dm; *leaves* linear-oblong to lanceolate, 1.5–6 cm long, strigose; *corolla* bright yellow, the lobes erose, 10–15 (20) mm wide; *nutlets* 3–3.5 mm long, sparsely pitted. Common on the plains and on open slopes and meadows, 3500–9500 ft. April–July. E/W. (Plate 17, 18)

Lithospermum incisum produces small, cleistogamous flowers later in the season. The corolla of these flowers is only 2–6 mm long or is absent. However, it can easily be distinguished from *L. arvense* by its perennial habit.

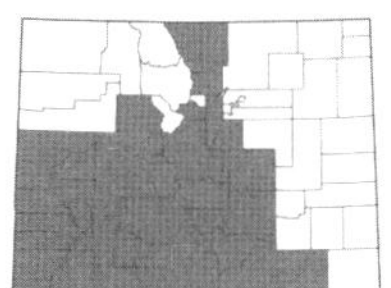

Lithospermum multiflorum **Torr. ex A. Gray**, SOUTHWESTERN STONESEED. Perennials, 1.5–7 dm; *leaves* linear to lanceolate or oblong, 2–6 cm long, strigose to spreading-hirsute; *corolla* bright yellow, 5–11 mm wide; *nutlets* ca. 3 mm long, sparsely pitted or smooth. Common in open woodlands and meadows, 5500–11,000 ft. May–Aug. E/W.

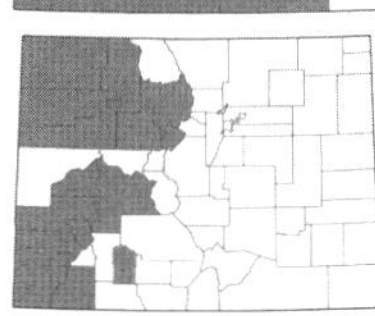

Lithospermum ruderale **Douglas ex Lehm.**, WESTERN STONESEED. Perennials, 2–6 dm; *leaves* linear to lanceolate, strigose to spreading-hirsute; *corolla* light yellow, 8–13 mm wide; *nutlets* 3.5–6 (8) mm long, smooth and shining. Common in open places, often with sagebrush or pinyon-juniper, 6000–9700 ft. May–Aug. W.

MERTENSIA Roth – BLUEBELLS

Perennial herbs; *leaves* entire, alternate; *calyx* cleft to the base or nearly so; *corolla* tubular to campanulate, usually blue; *stamens* included; *nutlets* 4, rugose, attached laterally to the gynobase near or below the middle. (Williams, 1937)

A recent paper by Nazaire and Hufford (2014) on the phylogenetic systematics of the genus *Mertensia* recommends a narrow circumscription of the genus. The author recognizes that this will approximately double the number of recognized species in this genus for Colorado, but at the time of press did not have adequate time or resources to fully develop a key incorporating these findings.

1a. Plants relatively tall (4–15 dm) and robust; leaves broadly elliptic to ovate with evident lateral veins in the stem leaves; plants typically found in moist or shaded habitats...2

1b. Plants relatively short (mostly 4 dm or less in height); leaves usually lance-ovate or elliptic without evident lateral veins in the stem leaves (lateral veins are sometimes present in the lower leaves of *M. oblongifolia*, but the plants are to 4 dm tall and of dry habitats); plants usually found in dry habitats...4

2a. Leaves strigose above, the hairs often somewhat bulbous-thickened at the base or sometimes evidently pustulate; calyx glabrous to more often loosely strigose on the back as well as ciliate on the margins...***M. franciscana***

2b. Leaves glabrous or pustulate (but the pustules not supporting hairs) above; calyx glabrous on the back with ciliate margins...3

3a. Calyx 1.5–4 mm long, less than ¼ the length of the corolla tube; anthers usually 2 mm or less in length; inflorescence usually tightly packed with numerous flowers...***M. ciliata***

3b. Calyx 3–8 mm long, ⅓–⅔ the length of the corolla tube; anthers 2–3.5 mm long; inflorescence usually looser with fewer flowers...***M. arizonica***

4a. Filaments much shorter than the anthers (often so short that the anthers appear almost sessile), attached in the corolla tube with the anther tips reaching just to the level of the fornices...5

4b. Filaments as long or longer than the anthers, attached near the throat of the corolla tube at or just below the level of the fornices with the anthers elevated above the fornices and not contained wholly in the corolla tube...7

5a. Leaves glabrous or pustulate (but the pustules not supporting hairs) above and below...***M. humilis***

5b. Leaves short-strigose above, glabrous below...6

6a. Calyx sparsely strigose on the back (especially towards the base) in addition to ciliate on the margins; plants of the western slope, at or below 10,000 ft in elevation...***M. brevistyla***

6b. Calyx glabrous on the back and ciliate on the margins; plants of the eastern slope at or above 11,000 ft in elevation...***M. alpina***

7a. Corolla tube glabrous within; lower leaves often with some lateral veins...***M. oblongifolia***

7b. Corolla tube with a ring of hairs in the middle or near the base, or with scattered hairs over much of the inner surface of the tube; lower leaves without lateral veins...8

8a. Upper surface of the leaves with strigose hairs directed toward the margins of the leaves...***M. fusiformis***

8b. Upper surface of the leaves glabrous, pustulate, or if strigose then the hairs directed forward toward the apex of the leaves...***M. lanceolata***

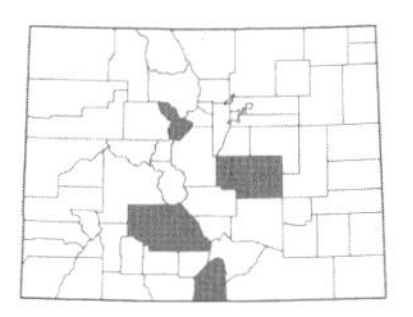

Mertensia alpina **(Torr.) G. Don**, ALPINE BLUEBELLS. Plants 0.2–3 dm; *leaves* linear-lanceolate to oblong, 1–7 × 0.5–3 cm, short-strigose above, glabrous below, without evident lateral veins; *calyx* 2–5 mm long; *corolla* tube 3–6 mm long; *stamens* with filaments shorter than anthers, the anther tips coming to just below the level of the fornices. Locally common in the alpine, especially common on Pike's Peak, 11,000–14,000 ft. June–Aug. E.

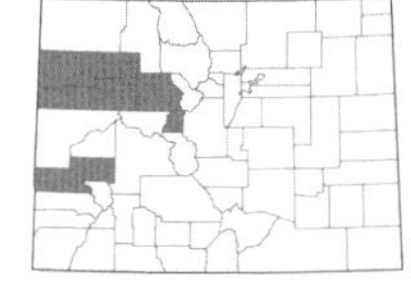

***Mertensia arizonica* Greene**, ASPEN BLUEBELLS. Plants 4–10 (15) dm; *leaves* lanceolate to elliptic-ovate, 5–15 × 1.5–5 (7) cm, glabrous and often pustulate on upper surface, with evident lateral veins; *calyx* 3–8 mm long; *corolla* tube 6–9 mm long; *stamens* with filaments attached at the level of the fornices, about as long as the anthers. Locally common in moist places along streams and creeks, 6800–8500 ft. June–Aug. W.

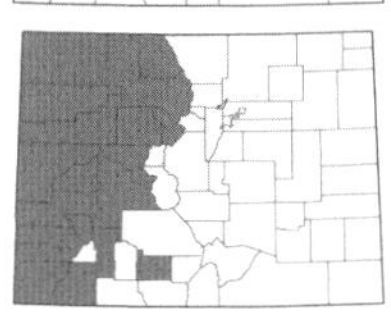

***Mertensia brevistyla* S. Watson**, SHORTSTYLE BLUEBELLS. Plants 1–4 dm; *leaves* narrowly oblong to elliptic or spatulate, 2–9 × 0.5–2.5 cm, glabrous below and strigose above, without evident lateral veins; *calyx* 2.5–4 mm long; *corolla* tube 2.5–4 mm long; *stamens* with filaments very short, the anthers virtually sessile, usually with just the tips reaching the level of the fornices. Found in meadows and on dry slopes, often found with oak or sagebrush, 6300–10,000 ft. April–June (July). W.

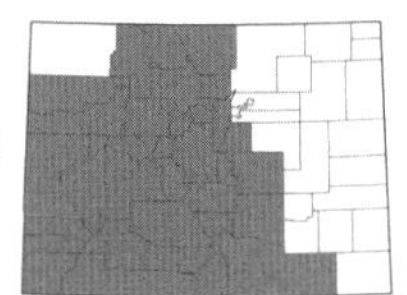

***Mertensia ciliata* (James ex Torr.) G. Don**, STREAMSIDE BLUEBELLS. Plants 4–10 (15) dm; *leaves* ovate to lance-ovate, 3–15 × 1–5 cm, glabrous and often pustulate on upper surface, with evident lateral veins, with ciliate margins; *calyx* 1.5–4 mm long; *corolla* tube 6–8 mm long; *stamens* with filaments attached at or shortly below the fornices, mostly longer than the anthers. Common in moist places along streams and creeks, 6000–12,500 ft. June–Aug. E/W. (Plate 17)

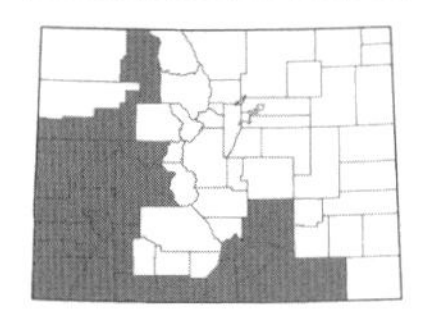

***Mertensia franciscana* Heller**, FRANCISCAN BLUEBELLS. Plants 4–10 (15) dm; *leaves* ovate to elliptic or lance-ovate, 5–10 (20) × 2–6 (9) cm, strigose above, occasionally pustulate, glabrous or loosely strigose below, with evident lateral veins; *calyx* 2.5–4 mm long; *corolla* tube 5–9 mm long; *stamens* with filaments attached at the level of the fornices, equaling or slightly shorter than the anthers. Common along streams and in moist, shaded places, 6500–10,800 ft. June–Aug. E/W.

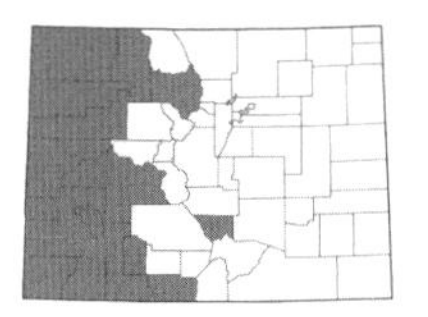

***Mertensia fusiformis* Greene**, SPINDLE-ROOTED BLUEBELLS. Plants 1–3 dm; *leaves* oblong to oblanceolate, 2–9 × 0.5–2 cm, the upper surface strigose with parallel hairs directed toward the margins, without evident lateral veins; *calyx* 3–6 mm long; *corolla* tube 4–7 mm long; *stamens* with filaments attached about at the level of the sinuses, about equal to or longer than the anthers. Common on open slopes, 5600–11,000 ft. June–Aug. E/W.

It can be difficult to tell this and *M. lanceolata* apart, and it might be better to place *M. fusiformis* as a variety of *M. lanceolata*.

***Mertensia humilis* Rydb.**, ROCKY MOUNTAIN BLUEBELLS. Plants 0.4–2 dm; *leaves* linear-oblong to lanceolate-oblong, 1.5–6 × 0.5–1.5 cm, glabrous or pustulate, without evident lateral veins; *calyx* 2.5–5 mm long; *corolla* tube 3–6 mm long; *stamens* with filaments shorter than the anthers, the anther tips coming to just below the level of the fornices. Uncommon in sagebrush meadows, 8000–8500 ft. May–June. E.

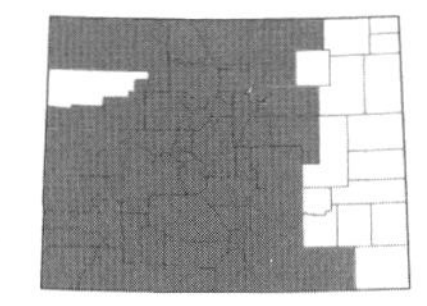

***Mertensia lanceolata* DC.**, PRAIRIE BLUEBELLS. Plants 1–4.5 dm; *leaves* lanceolate to oblong-elliptic, 1.5–10 × 0.2–3 cm, strigose, pustulate or glabrous above, without evident lateral veins; *calyx* 2–9 mm long; *corolla* tube 3–6.5 mm long; *stamens* with filaments 1.5–4 mm long. Common and widespread throughout the state, usually in open meadows and on dry slopes, 5000–14,000 ft. April–Aug. E/W. (Plate 17)

This is a highly variable species, with leaf pubescence varying from completely glabrous to hairy on both surfaces. Several varieties of *M. lanceolata* have been proposed based primarily on this characteristic. However, this characteristic is highly variable, and thus I choose to be conservative and separate *M. lanceolata* only into the following varieties:

1a. Calyx divided to the base; filaments usually shorter than the anthers; typically above 9500 ft in elevation…**var. *nivalis* (S. Watson) L.C. Higgins**, SNOWY BLUEBELLS. [*M. viridis* (A. Nelson) A. Nelson]. Common in alpine, occasionally found in lower elevations, 8500–14,000 ft. May–Aug. E/W.

1b. Calyx not quite divided to the base, often divided less than halfway; filaments usually longer than the anthers; typically below 9500 ft in elevation…**var. *lanceolata***. [*M. bakeri* Greene; *M. viridis* (A. Nelson) A. Nelson var. *cynoglossoides* (Greene) J.F. Macbr.]. Common from the high plains to mountains, 5000–10,000 ft. April–Aug. E/W.

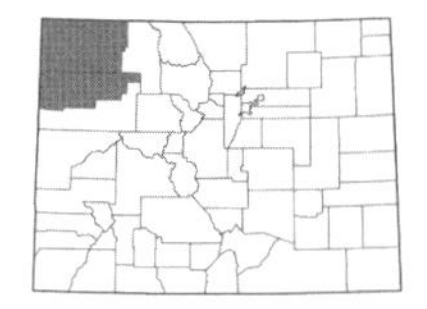

***Mertensia oblongifolia* (Nutt.) G. Don**, SAGEBRUSH BLUEBELLS. Plants 1–4 dm; *leaves* elliptic to oblanceolate, 2–15 × 0.7–6 cm, glabrous and usually pustulate above, without evident lateral veins; *calyx* 2.5–6 mm long; *corolla* tube 5–12 mm long; *stamens* with filaments 1.5–4 mm long, attached at or just below the level of the fornices, usually longer than the anthers. Found on dry slopes, often with sagebrush, 5600–8700 ft. May–June. W.

MYOSOTIS L. – FORGET-ME-NOT

Annual or perennial herbs usually with slender stems; *leaves* entire, alternate; *calyx* cleft to the middle or below, with a definite tube; *corolla* salverform or broadly funnelform, usually blue, or white or rarely yellow; *stamens* included or excluded, adnate to the corolla tube; *nutlets* 4, smooth, with a sharp, raised margin, the attachment scar flat.

1a. Calyx tube with closely strigose hairs, these neither spreading nor hooked at the tip; lower leaves oblanceolate, 7–20 mm wide; stems often creeping at the base and stoloniferous...***M. scorpioides***
1b. Calyx tube with some loose, spreading hairs that are hooked at the tip; basal leaves linear-oblanceolate to spatulate, to 10 mm wide; stems erect, not creeping at the base or stoloniferous...2

2a. Perennials, native plants found at high elevations in the mountains; corolla limb 4–8 mm wide...***M. asiatica***
2b. Biennials or annuals, weedy plants found at low elevations along the Front Range; corolla limb 2–4 mm wide...***M. arvensis***

Myosotis arvensis **(L.) Hill**, FIELD FORGET-ME-NOT. Annuals or biennials, 1–4 dm; *leaves* oblanceolate to oblong, 1–6 × 0.3–1.6 cm, strigose; *corolla* blue or sometimes white, the limb 2–4 mm wide; *nutlets* brown or black, surpassing the style. Uncommon in disturbed places, known from a collection near Fort Collins, 4900–5500 ft. June–Aug. E. Introduced.

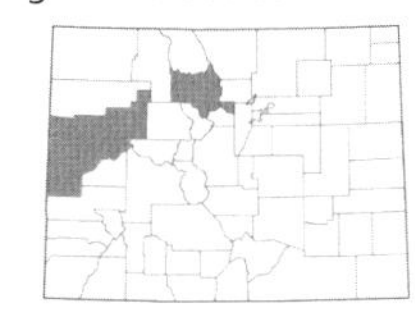

Myosotis asiatica **(Vestergr.) Schischk. & Serg.**, ALPINE FORGET-ME-NOT. [*M. alpestre* F.W. Schmidt]. Perennials, 0.5–4 dm; *leaves* oblanceolate to elliptic or oblong, to 13 × 1.3 cm, hirsute to hirsute-strigose; *corolla* blue or rarely white, the limb mostly 4–8 mm wide; *nutlets* black, surpassing the style. Uncommon on alpine ridges, 10,000–11,500 ft. June–Aug. W.

Myosotis scorpioides **L.**, TRUE FORGET-ME-NOT. Perennials, 2–6 dm, often stoloniferous; *leaves* oblanceolate, lance-oblong to elliptic, 2.5–8 × 0.7–2 cm, strigose; *corolla* blue, the limb 5–10 mm wide; *nutlets* blackish, usually shorter than the style. Uncommon along creeks, 5300–8000 ft. June–Aug. E. Introduced.

ONOSMODIUM Michx. – MARBLESEED

Coarse perennial herbs; *leaves* alternate, entire; *flowers* in bracteate helicoid cymes; *calyx* 5-lobed, cleft nearly to the base, persistent in fruit; *corolla* whitish-green to yellowish, tubular; *stamens* included; *style* long-exserted; *nutlets* 1–4, ovoid, smooth or rarely pitted, white, attached by their bases to a flat gynobase, with a small, flat scar.

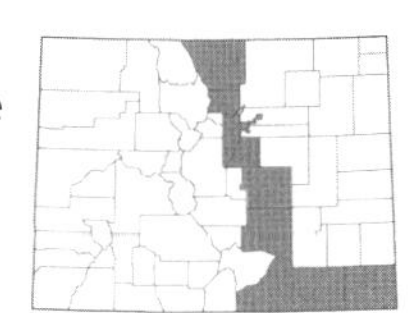

Onosmodium bejariense **DC. ex A. DC. var. *occidentale* (Mack.) B.L. Turner**, WESTERN MARBLESEED. [*Onosmodium molle* Michx. var. *occidentale* (Mack.) I.M. Johnst.]. Plants 5–12 dm; *leaves* lanceolate to elliptic or ovate, 5–12 cm long, prominently 5- to 7-nerved; *calyx* 8–9 mm long; *corolla* 11–15 mm long; *style* 10–18 mm long; *nutlets* slightly or not constricted at the base. Common in dry, open places on the plains and lower foothills, 4000–6100 ft. May–July. E. (Plate 17)

OREOCARYA Greene – CRYPTANTHA

Biennial or perennial herbs, usually bristly-hairy; *leaves* alternate, entire, alternate or opposite below; *calyx* lobed to below the middle or almost to the base; *corolla* white to pale yellow or yellow, usually with yellow fornices; *stamens* 5; *nutlets* 1–4, variously roughened, without hooked prickles, affixed laterally to an erect gynobase. (Eastwood, 1903; Higgins, 1971; Payson, 1927; Reveal & Broome, 2006)

1a. Corolla tubes 4–14 mm long, distincly longer than the calyx lobes in flower...2
1b. Corolla tubes to 7 mm long, usually shorter than or just equal to the calyx lobes or sometimes barely surpassing them...9

2a. Corollas bright yellow; nutlets smooth and glossy; leaves linear-oblanceolate and densely appressed-strigose...***O. flava***
2b. Corollas white with yellow fornices or occasionally pale yellow (*careful*—sometimes plants with white flowers can dry yellow); nutlets rugose to tuberculate; leaves various...3

3a. Usually only 1 or 2 nutlets maturing, the surface uniformly and densely muricate-tuberculate on both sides; nutlet margins on the ventral side not prominently raised; leaves densely appressed-strigose and silvery-pubescent and usually without pustulate bristles on the abaxial side or these inconspicuous within the appressed-strigose pubescence...***O. nitida***
3b. All 4 nutlets usually maturing, the surface rugose or rugose-tuberculate; nutlet margins raised or not; leaves coarsely strigose to sericeous-strigose, not silvery-pubescent, and with conspicuous pustulate bristles at least on the abaxial side...4

4a. Leaves with numerous, conspicuous pustulate bristles on both sides...5
4b. Leaves with conspicuous pustulate bristles on the abaxial side only (sometimes with a few pustulate bristles above, but these inconspicuous)...6

5a. Corolla limb flaring and never opening out flat at anthesis; nutlets slightly or obscurely roughened on both sides or smooth ventrally; the scar closed without raised margins; several leaves (5–10 nodes) present on the flowering stem before flowers begin...***O. rollinsii***
5b. Corolla limb spreading out flat at anthesis; nutlets moderately to strongly rugose-tuberculate on both sides; the scar closed with slightly elevated margins; flowers numerous, usually beginning just above the basal leaves (often overlapping with the tips of the leaves in pressed specimens)...***O. longiflora***

6a. Corolla tubes 4–6 mm long; nutlet scars closed or very nearly so with conspicuously ornate-raised margins; style only surpassing the nutlets by (0.5) 1–2.2 mm; biennials or perennials from taproots with 1–several stems from the base, usually with 1 stem distinctly longer than the others...***O. bakeri***
6b. Corolla tubes 7–12 mm long; nutlet scars open with raised-ornate margins; style surpassing the nutlets by 2.5–9 mm; perennials from taproots and with branching caudices, the stems all about equal in height...7

7a. Plants 0.5–3 dm tall, not low and densely tufted; leaves larger, 3–10 cm long and 3–15 mm wide...***O. flavoculata***
7b. Plants low and densely tufted, 1 dm or less high; leaves small, to 3.5 cm long and 3 mm wide...8

8a. Leaves strigose or sericeous-strigose on both surfaces, with pustulate hairs also present below...***O. paradoxa***
8b. Leaves with pustulate hairs on the lower surfaces and glabrous on the upper surfaces...***O. revealii***

9a. Nutlets smooth and glossy on both surfaces (at 10×); plants with several slender stems from the base; thyrse of several widely spreading helicoid cymes that elongate and straighten at maturity giving the inflorescence an open appearance; basal leaves linear-oblanceolate with appressed strigose, scarcely hispid hairs, and stem leaves well-developed and often as large as the basal leaves...10
9b. Nutlets variously roughened, muricate, or tuberculate on at least the dorsal surface (back), or if smooth then the thyrse long and cylindric with conspicuous foliar bracts that much exceed the helicoid cymes; plants and leaves various but the thyrse usually more congested...11

10a. Leaves glabrous on the upper surfaces, the lower surfaces dense with conspicuous pustulate hairs...***O. pustulosa***
10b. Upper leaf surfaces hairy, the lower surface swithout numerous conspicuous pustulate hairs...***O. suffruticosa***

11a. Perennials from a woody taproot and branched caudex, with 1–several stems usually all about the same height; plants often low, 1 dm or less in height...12
11b. Biennials or perennials from a taproot, with 1–several stems and often with one stem larger than the others; plants typically over 1 dm in height...19

12a. Leaves with conspicuous pustulate bristles on the abaxial side (pustulate bristles appear better on older leaves with less hairs on the abaxial side), green to grayish and variously hairy; 1–4 nutlets maturing...13
12b. Leaves without pustulate bristles (or these few and inconspicuous), silvery-gray with appressed hairs and short bristles; only 1 or 2 nutlets maturing...17

13a. Style surpassing the nutlets by 1.5–2 mm; leaves with spreading bristles, and pustulate bristles on both the adaxial and abaxial sides; plants (0.5) 1–2.5 dm tall...14
13b. Style about equaling the nutlets or surpassing them by less than 1 mm; leaves with more or less appressed bristles, and pustulate bristles only on the abaxial side; plants low perennials, mostly 0.1–1 dm tall...16

14a. Leaves narrowly oblanceolate to nearly linear; thyrse cylindric and densely compact; plants of southcentral Colorado...***O. weberi***
14b. Leaves oblanceolate to spatulate; thyrse cylindric to subcapitate but looser; plants of Garfield and Mesa counties...15

15a. Nutelt scars straight and closed, extending almost to the apex, without elevated margins; nutlets 2–2.6 mm long; crests conspicuous at the base within the corolla tube...***O. aperta***
15b. Nutlet scars open and conspicuous with prominently raised-ornate margins; nutlets 2.5–3.5 mm long; crests inconspicuous at the base within the corolla tube...***O. mensana***

16a. Caudex with numerous, slender, weak or curved-ascending stems; nutlets with raised-ornate margins along the scar; calyx sparsely setose with few bristles along the margins and veins and with short, loose hairs underneath...***O. osterhoutii***
16b. Caudex with stout, erect stems; nutlets without elevated margins along the scar; calyx densely setose with numerous bristles...***O. humilis***

17a. Tall perennials, 1–3.5 dm; corolla 9–13 mm wide; leaves broadly oblanceolate or spatulate...***O. breviflora***
17b. Low perennials, 0.2–1 dm; corolla 4–6 mm wide; leaves narrowly oblanceolate...18

18a. Nutlets densely muricate dorsally and similarly so ventrally but with the papillae less elevated; plants of the eastern slope...***O. cana***
18b. Nutlets strongly raised-rugose and somewhat tuberculate dorsally, and tuberculate ventrally; plants of the western slope...***O. caespitosa***

19a. Corollas funnelform, the limbs flaring and never opening out flat at anthesis; style surpassing the nutlets by 4–5 mm; nutlets narrowly triangular-ovate; pustulate bristles present on both leaf surfaces...***O. rollinsii***
19b. Corollas salverform with the limbs spreading out flat at anthesis; style only surpassing the nutlets by as much as 2.5 (3) mm; nutlets ovate or lance-ovate; pustulate bristles on both leaf surfaces or only below...20

20a. Ventral surfaces of nutlets smooth or sometimes with just a few tubercles; basal leaves mostly narrowly linear-oblanceolate or occasionally oblanceolate in *O. stricta*; thyrse densely cylindric with tight, compact helicoid cymes or tending to open out a bit at maturity...21
20b. Ventral surfaces of nutlets rugose or tuberculate or sometimes almost smooth in *O. thyrsiflora* (and then the thyrse broad and rounded in outline); basal leaves oblanceolate; thyrse usually open at maturity...23

21a. Thyrse long-cylindric (to 9 dm long) with conspicuous linear-oblanceolate foliar bracts that much exceed the helicoid cymes; stems usually solitary and unbranched...***O. virgata***
21b. Thyrse shorter, without conspicuous linear-oblanceolate foliar bracts that much exceed the helicoid cymes; stems 1–several per plant...22

22a. Nutlets 2–2.5 mm long, the scars open and triangular; plants caespitose with numerous stems; plants 1–2 dm tall...***O. weberi***
22b. Nutlets 3.5–4 mm long, the scars closed or narrowly open; plants solitary or with 2–3 stems; plants typically taller, (1) 1.5–4 dm tall...***O. stricta***

23a. Plants of the eastern slope...24
23b. Plants of the western slope...25

24a. Nutlet scars open and triangular at the base; nutlets rather glossy and slightly bent and forming a small opening at the middle of their contact surface while attached to the style; ventral nutlet surfaces low-rugose and tuberculate to almost smooth; inflorescence very broad and rounded in outline at maturity...***O. thyrsiflora***
24b. Nutlet scars closed; nutlets dull or somewhat glossy, straight and not forming a small opening along their contact surface while attached to the style; ventral nutlet surfaces rugose and tuberculate but never smooth; inflorescence narrower...***O. celosioides***

25a. Nutlet scars with prominently ornate-raised margins; nutlets conspicuously rugose and prominently sculptured with high ridges on both sides (resembling the annulus of a fern sporangium in outline from the side); style surpassing the nutlets by (0.5) 1–2.5 mm...26
25b. Nutlet scars without raised margins; nutlets rugose or roughened, but not as densely so or the sculptures as high as above, the ventral side with more subdued roughening; style surpassing the nutlets by 0.2–1 mm...27

26a. Nutlet scars conspicuously open; corolla tubes 3–4 mm long; fornices 0.5 mm high, obviously wider than high; basal leaves rather loosely tufted...***O. mensana***
26b. Nutlet scars closed or nearly so; corolla tubes 4–6 mm long; fornices 1–1.5 mm high, longer than wide; basal leaves rather tightly tufted...***O. bakeri***

27a. Upper leaf surfaces without pustulate bristles or pustulate bristles few; dorsal nutlet surfaces irregularly roughened with many low tubercles and a few higher ridges (surface more smooth than bumpy); plants with 1–several stems from the base...***O. sericea***
27b. Upper leaf surfaces with pustulate bristles; dorsal nutlet surfaces closely rugose-retuculate (surface more bumpy than smooth); stems usually solitary...***O. elata***

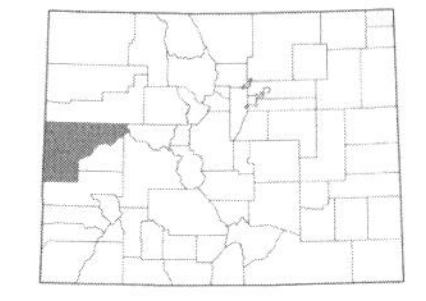

Oreocarya aperta **Eastw.**, Grand Junction cryptantha. [*Cryptantha aperta* (Eastw.) Payson]. Perennials, 1–2 dm; *leaves* oblanceolate to spatulate, strigose and pustulate-setose, to 3 cm long; *calyx* 8–10 mm long in fruit; *corolla* white, 6 mm wide; *nutlets* 4, ovate, 2–2.6 × 1.4–1.6 mm, tuberculate or rugose dorsally, irregularly roughened ventrally, the scar straight, closed, with no elevated margin. Possibly extinct as it hasn't been seen since the type collections were made in 1892 near Grand Junction. Most likely grows on clay soil, about 4600 ft. May–June. W. Endemic.

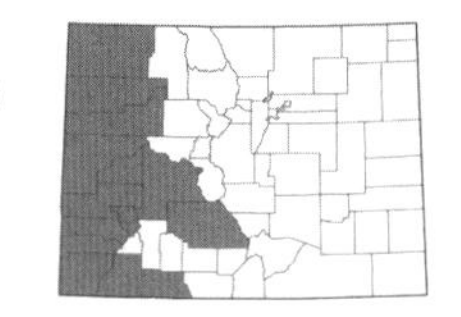

***Oreocarya bakeri* Greene**, BAKER'S CRYPTANTHA. [*Crypthanta bakeri* (Greene) Payson]. Biennials or perennials, 1–4 dm; *leaves* oblanceolate or spatulate, 4–9 × 0.4–2 cm, densely sericeous-strigose and pustulate below; *calyx* 3.5–5 mm long in flower, 5–8 mm long in fruit; *corolla* white with yellow fornices, 7–9 mm wide, tube 4–6 mm long; *nutlets* 4, ovate, 2.4–3.5 mm long, rugose on both sides, the scar closed with conspicuously ornate-raised margins. Found on dry hills and slopes, often with pinyon-juniper or sagebrush, 5400–8600 ft. May–July. W. (Plate 19)

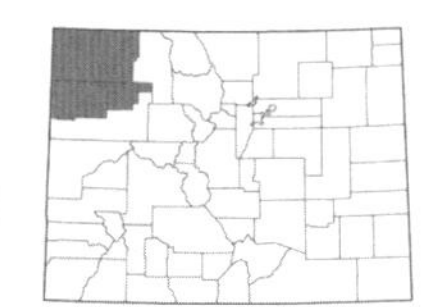

***Oreocarya breviflora* Osterh.**, UINTA BASIN CRYPTANTHA. [*Cryptantha breviflora* (Osterh.) Payson]. Perennials, 1–3.5 dm; *leaves* oblanceolate, 2–10 × 0.3–1.2 cm, sericeous; *calyx* 4–6 mm long in flowers, 6–9 mm long in fruit; *corolla* white with yellow fornices, 9–13 mm wide; *nutlets* 1 or 2, ovate, 3–4.5 mm long, conspicuously muricate on both sides, the scar narrow. Uncommon in red sandy soil of open, dry places, often with sagebrush communities, 5500–8300 ft. May–July. W.

***Oreocarya caespitosa* A. Nelson**, TUFTED CRYPTANTHA. [*Cryptantha caespitosa* (A. Nelson) Payson]. Perennials, 0.5–1 dm; *leaves* broadly oblanceolate, 2–5 cm long, softly gray-hairy; *calyx* 3–4 mm long in flower, 4–5 mm long in fruit; *corolla* white with yellow fornices, 4–6 mm wide; *nutlets* usually 2, truncate-ovate, 2.5–3.5 mm long, strongly rugose dorsally and ventrally, with a narrowly triangular scar. Uncommon on gypsum shale outcroppings, often with pinyon-juniper or sagebrush communities, 6000–8100 ft. April–June. W. (Plate 19)

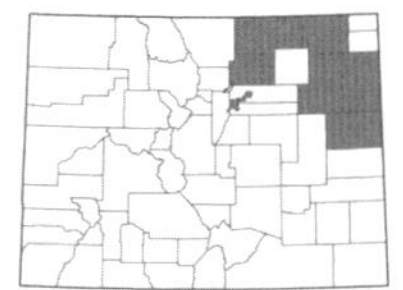

***Oreocarya cana* A. Nelson**, SILVER-MOUNDED CRYPTANTHA. [*Cryptantha cana* (A. Nelson) Payson]. Perennials, 0.2–0.5 dm; *leaves* narrowly oblanceolate, 2–6 × 0.3–1 cm, silky strigose; *calyx* ca. 3 mm long in flower, 5.5–6 mm long in fruit; *corolla* white, 6 mm wide; *nutlets* usually 1, lanceolate, 3–3.5 mm long, densely muricate with elongated papillae dorsally and ventrally, the scar narrowly triangular. Locally common on open sandstone outcroppings and sandy bluffs, 4000–5700 ft. May–June. E. (Plate 17, 19)

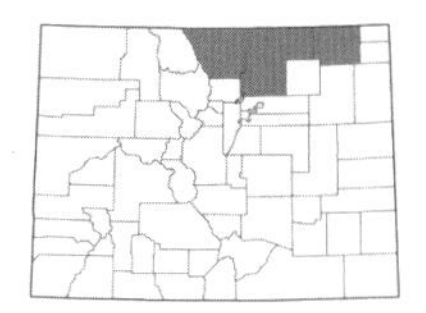

***Oreocarya celosioides* Eastw.**, BUTTECANDLE. [*Cryptantha celosioides* (Eastw.) Payson]. Biennials or perennials, 1–5 dm; *leaves* spatulate to oblanceolate, 2–8 × 0.4–1.5 cm, sericeous-strigose, pustulate; *calyx* ca. 5 mm long in flower, to 12 mm long in fruit; *corolla* white with yellow fornices, 7–12 mm wide; *nutlets* 2–4, ovate, 3–5 mm long, tuberculate and rugose dorsally and ventrally, the scar closed. Uncommon in rocky or sandy soil of the northeastern plains, 3800–5300 ft. May–July. E. (Plate 19)

***Oreocarya elata* Eastw.**, TALL CRYPTANTHA. [*Cryptantha elata* (Eastw.) Payson]. Perennials, 3–5 dm; *leaves* oblanceolate to linear-oblong, 2–5 × 0.4–1.3 cm, sericeous and pustulate; *calyx* 2.5–4 mm long in flower, 6–8 mm long in fruit; *corolla* white with yellow fornices, 6–10 mm wide; *nutlets* usually 4, ovate, 3.5–4.5 mm long, rugose-reticulate, with a sharp margin, the scar closed or nearly so. Uncommon on clay soil of Mancos shale with other salt desert shrub plants, 4500–5200 ft. April–June. W. (Plate 19)

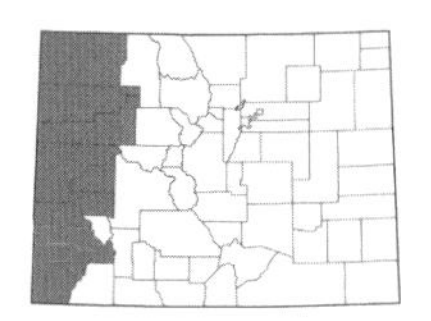

***Oreocarya flava* A. Nelson**, PLATEAU YELLOW CRYPTANTHA. [*Cryptantha flava* (A. Nelson) Payson]. Perennials, 1.5–4 dm; *leaves* linear-oblanceolate, 2–9 × 0.2–1 cm, strigose; *calyx* 6–10 mm long in flower, 9–12 mm long in fruit; *corolla* bright yellow, 7–11 mm wide; *nutlets* usually 1, lanceolate, narrowly winged, 3.5–4.5 mm long, smooth and glossy. Found in sandy soil, often with pinyon-juniper or sagebrush communities, 4300–7500 ft. May–July. W. (Plate 19)

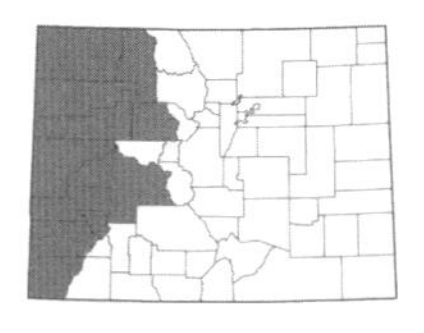

***Oreocarya flavoculata* A. Nelson**, ROUGHSEED CRYPTANTHA. [*Cryptantha flavoculata* (A. Nelson) Payson]. Perennials, 0.5–3 dm; *leaves* oblanceolate to spatulate, 3–10 × 0.3–1.5 cm, strigose to sericeous, pustulate below; *calyx* 5–7 mm long in flower, 8–10 mm long in fruit; *corolla* white or pale yellow, 7–12 mm wide; *nutlets* 4, lance-ovate, 2.5–4 mm long, rugose, the scar open and conspicuous with prominently raised-ornate margins. Common in open, dry places on a variety of soil types, often with pinyon-juniper or sagebrush communities, 5000–8500 ft. May–June. W.

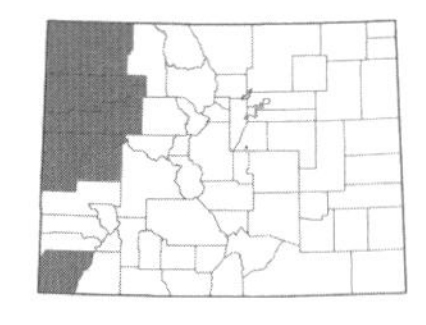

***Oreocarya humilis* (A. Gray) Greene**, ROUNDSPIKE CRYPTANTHA. [*Cryptantha humilis* (A. Gray) Payson]. Perennials, 0.1–0.3 dm; *leaves* oblanceolate, 1.5–6 × 0.2–1 cm, silvery-sericeous; *calyx* 3–5 mm long in flower, 5–9 mm long in fruit; *corolla* white with yellow fornices, 6–10 mm wide; *nutlets* 1–4, lance-ovate, 2.5–4.5 mm long, finely tuberculate, the scar open and elongate without raised margins. Found on white or red sandy soil of open, dry places, often with sagebrush or pinyon-juniper communities, 4600–8000 ft. April–June. W.

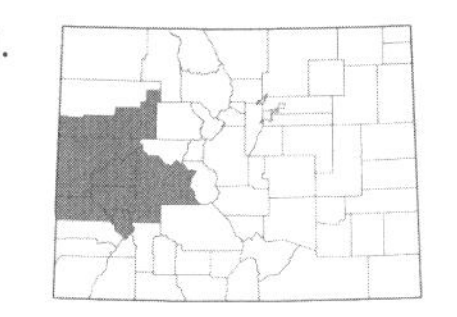

***Oreocarya longiflora* A. Nelson**, LONGFLOWER CRYPTANTHA. [*Cryptantha longiflora* (A. Nelson) Payson]. Perennials, 1–4 dm; *leaves* oblanceolate to spatulate, to 7 cm long, silky-sericeous and strongly pustulate; *calyx* 7–13 mm long in flower, 10–16 mm long in fruit; *corolla* white with yellow fornices, 7–11 mm wide; *nutlets* 4, lance-ovate, 3–4 mm long, rugose-tuberculate, the closed or narrowly open. Found on dry, open sandy, clay, or shale slopes, often with pinyon-juniper communities, 4600–6500 ft. April–June. W.

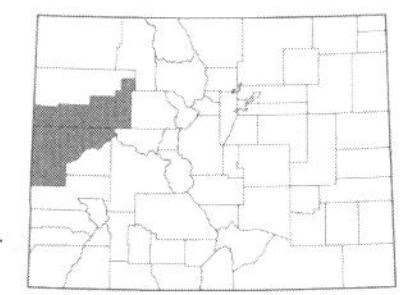

***Oreocarya mensana* M.E. Jones**, SOUTHWESTERN CRYPTANTHA. [*Cryptantha mensana* (M.E. Jones) Payson]. Biennials or perennials, 0.5–2 dm; *leaves* oblanceolate, 2–7 × 0.2–1 cm, strigose and pustulate; *calyx* 3–5 mm long in flower, 6–8 mm long in fruit; *corolla* white with yellow fornices, 7–10 mm wide, tube 3–4 mm long; *nutlets* 1–4, ovate, 2.5–3.5 mm long, tuberculate and rugose, the scar open with prominently raised-ornate margins. Rare on dry, open clay slopes, and pinyon-juniper forests, 6000–6500 ft. May–June. W.

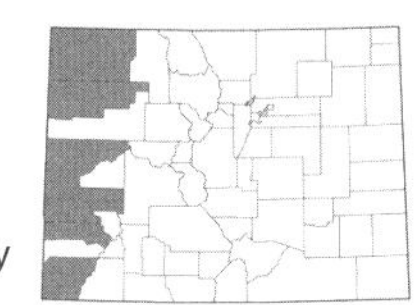

***Oreocarya nitida* Greene**, TAWNY CRYPTANTHA. [*Cryptantha fulvocanescens* (S. Watson) Payson var. *nitida* (Greene) R.C. Sivinski]. Perennials, 1–3 dm; *leaves* oblancolate, 2–6 × 0.4–1 cm, silvery-pubescent, not pustulate; *calyx* 5–8 mm long in flower, 8–12 mm long in fruit; *corolla* white with yellow fornices, 7–10 mm wide; *nutlets* usually 1 or 2, ovate, 3–4 mm long, densely muricate-tuberculate, the scar narrow or nearly closed. Found on dry, sandstone outcrops and open, sandy slopes, 4500–8000 ft. May–June. W.

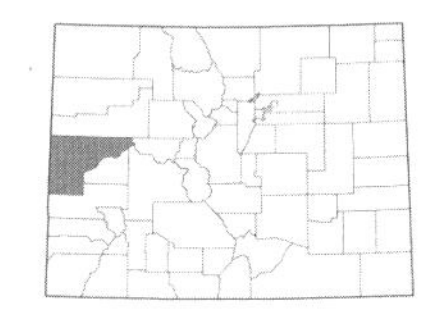

***Oreocarya osterhoutii* Payson**, OSTERHOUT'S CRYPTANTHA. [*Cryptantha osterhoutii* (Payson) Payson]. Low perennials, to 1 dm; *leaves* oblanceolate, 1–2.5 × 0.2–0.6 cm, strigose; *calyx* 3–4 mm long in flower, 5–6 mm long in fruit; *corolla* white with yellow fornices, 6–9 mm wide; *nutlets* 1–4, ovate, ca. 3 mm long, rugose-tuberculate, the scar open with elevated-ornate margins. Uncommon in dry, open places in reddish sandy soil, often with juniper or sagebrush communities, 4500–6100 ft. April–June. W. (Plate 17)

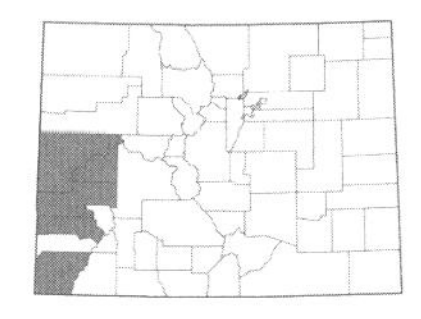

***Oreocarya paradoxa* A. Nelson**, PARADOX VALLEY CRYPTANTHA. [*Cryptantha paradoxa* (A. Nelson) Payson]. Low, densely tufted perennial, to 1 dm; *leaves* oblanceolate to spatulate, to 3 cm long, sericeous-strigose above and pustulate below; *calyx* 4–6 mm long in flower, 6–8 mm long in fruit; *corolla* white with yellow fornices, 9–12 mm wide; *nutlets* 4, ovate, 2–3 mm long, rugose, the scar open with somewhat elevated margins. Uncommon on shale and sandstone slopes, 4800–6000 ft. April–June. W.

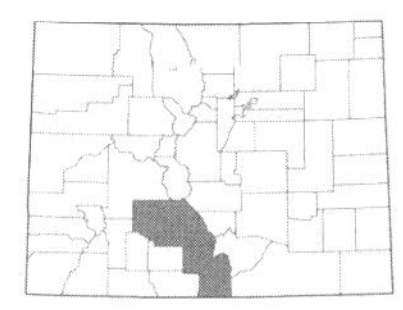

***Oreocarya pustulosa* Rydb.**, SAN JUAN CRYPTANTHA. [*Cryptantha cinerea* (Greene) Cronquist var. *pustulosa* (Rydb.) Higgins]. Perennials, 1–6 dm; *leaves* oblanceolate to lance-linear, 3–10 × 0.3–1 cm, glabrous above and pustulate below; *calyx* 3–4 mm long in flower, 5–7 mm long in fruit; *corolla* white with yellow fornices, 5–8 mm wide; *nutlets* 1–4, ovate, 1.8–2.5 mm long, smooth and glossy. Found in southcentral Colorado on dry slopes, 6000–8500 ft. May–Aug. E/W.

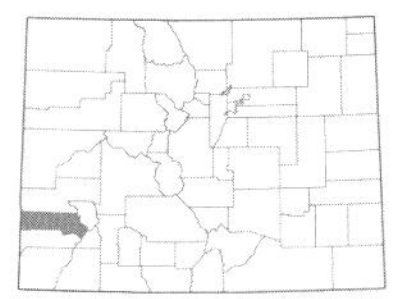

***Oreocarya revealii* W.A. Weber & R.C. Wittmann**, GYPSUM VALLEY CRYPTANTHA. [*Cryptantha gypsophila* Reveal & C.R. Broome]. Perennials, 0.3–2.5 dm; *leaves* oblanceolate to narrowly spatulate, 0.1–3 cm long, glabrous above and pustulate below; *calyx* 4–6 mm long in flower and 6–9 mm long in fruit; *corolla* white with yellow fornices, 8–12 mm wide; *nutlets* usually 4, ovate, 2.2–3 mm long, rugose-tuberculate, the scar open with somewhat elevated margins. Locally common on gypsum outcrops, 5000–6500 ft. April–June. W. Endemic.

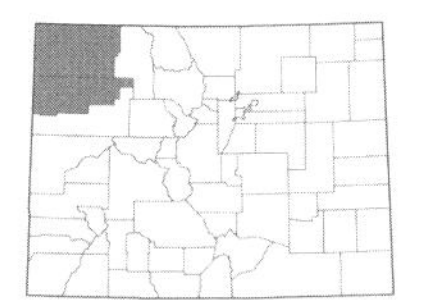

***Oreocarya rollinsii* (I.M. Johnst.) W.A. Weber**, ROLLINS' CRYPTANTHA. [*Cryptantha rollinsii* I.M. Johnst.]. Biennials or short-lived perennials, 1–3.5 dm; *leaves* oblanceolate, setose and pustulate; *calyx* 6–8 mm long in flower, 7–10 mm long in fruit; *corolla* white, 6–8 mm wide; *nutlets* usually 4, narrowly triangular-ovate, 2–4 mm long, slightly roughened or smooth ventrally, with a sharp edge, the scar closed or nearly so except at the base. Uncommon on dry, open white shale slopes, 5300–6000 ft. May–June. W. (Plate 19)

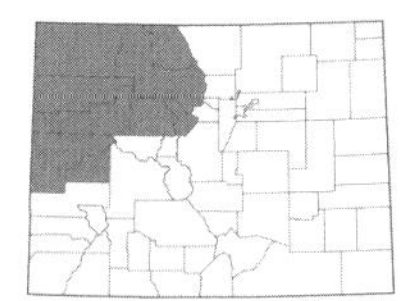

***Oreocarya sericea* (A. Gray) Greene**, SILKY CRYPTANTHA. [*Cryptantha sericea* (A. Gray) Payson]. Perennials, 1.5–5 dm; *leaves* oblanceolate, 2–10 × 0.4–2 cm, strigose and pustulate below; *calyx* 2.5–4 mm long in flower, 6–9 mm long in fruit; *corolla* white with yellow fornices, 7–10 mm wide; *nutlets* usually 4, ovate, 2.5–3.5 mm long, irregularly roughened and tuberculate, the scar closed or nearly so. Common in sandy or clay soil, 5000–8700 ft. W. (Plate 19)

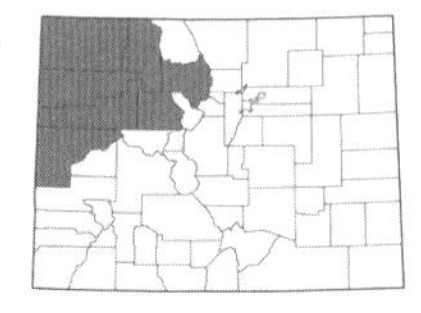

***Oreocarya stricta* Osterh.**, Yampa River cryptantha. [*Cryptantha stricta* (Osterh.) Payson]. Perennials, (1) 1.5–4 dm; *leaves* oblanceolate or linear-oblanceolate, 4.5–7 × 0.3–1 cm, strigose; *calyx* 4–5 mm long in flower, 6–9 mm long in fruit; *corolla* white or ochroleucous, 8–11 mm wide; *nutlets* 4, ovate, 3.5–4 mm long, tuberculate or ridged dorsally, ventrally smooth with a closed scar. Found in sandy or clay soil of dry, open places and on shale outcroppings, often with pinyon-juniper communities, 5500–9200 ft. May–July. W. (Plate 19)

Occasional specimens of *O. sericea* with very subdued muriculate roughening on the ventral side of the nutlets will key to *O. stricta*. However, the two species can easily be distinguished by the hairs on the leaves – *O. sericea* has densely silky-strigose hairs and few or no pustulate bristles on the upper leaf surface while *O. stricta* has rougher, setose hairs with numerous spreading, pustulate bristles about equally numerous on both sides.

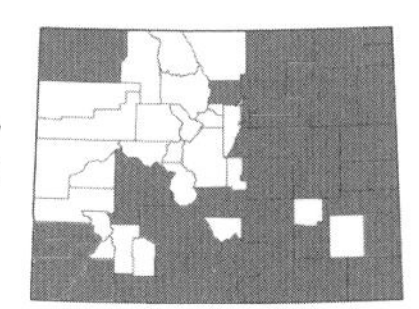

***Oreocarya suffruticosa* (Torr.) Greene**, James' cryptantha. [*Cryptantha cinerea* (Greene) Cronquist var. *jamesii* (Torr.) Cronquist]. Perennials, 1–6 dm; *leaves* oblanceolate to linear, 2–15 × 0.2–1.5 cm, hirsute and appressed-hairy; *calyx* 3–4 mm long in flower, 5–7 mm long in fruit; *corolla* white with yellow fornices, 5–8 mm wide; *nutlets* 1–4, ovate, 1.8–2.5 mm long, smooth and glossy. Common in dry, sandy places, 3500–8000 ft. April–Aug. E/W. (Plate 19)

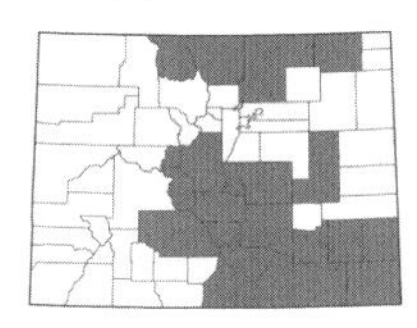

***Oreocarya thyrsiflora* Greene**, calcareous cryptantha. [*Cryptantha thyrsiflora* (Greene) Payson]. Perennials, 2–4 dm; *leaves* oblanceolate, setose and strigose, 4–10 × 0.5–1.5 cm; *calyx* 3–4 mm long in flower, 6–9 mm long in fruit; *corolla* white with yellow fornices, 5–8 mm wide; *nutlets* 2–4, ovate, 2.5–3.5 mm long, rugose and tuberculate, the scar open, straight, and narrow. Common in rocky soil and on bluffs and outcroppings, abundant in dry pastures and along roadsides, 3800–9600 ft. May–July (Sept.). E. (Plate 17)

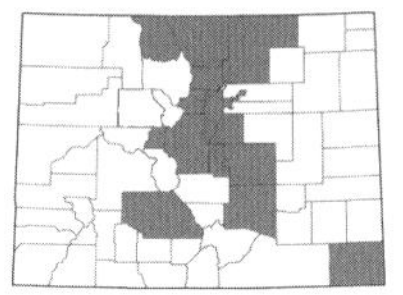

***Oreocarya virgata* (Porter) Greene**, miner's candle. [*Cryptantha virgata* (Porter) Payson]. Biennials, 2.5–10 dm; *leaves* narrowly oblanceolate, 3–12 × 0.4–1.5 cm, hirsute-hispid; *calyx* 3.5–4 mm long in flower, 10–12 mm long in fruit; *corolla* white with yellow fornices, 8–11 mm wide; *nutlets* usually 4, ovate, 2.5–3.5 mm long, with conspicuous low ridges dorsally or sometimes nearly smooth, ventrally smooth or with a few tubercles, the scar straight and nearly closed. Common in sandy soil of the foothills and mountains along the Front Range, 4500–9300 ft. May–July. E. (Plate 19)

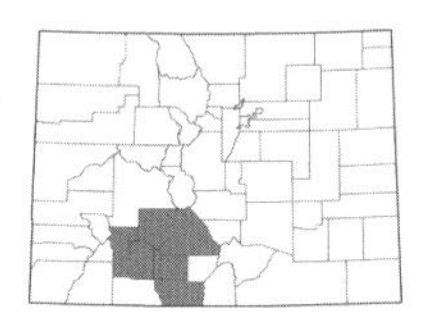

***Oreocarya weberi* (I.M. Johnst.) W.A. Weber**, Weber's cryptantha. [*Cryptantha weberi* I.M. Johnst.]. Perennials, 1–2 dm; *leaves* narrowly oblanceolate to nearly linear, 3–8 × 0.3–0.7 cm, hispid-villous; *calyx* 3–4 mm long in flower, 5–7 mm long in fruit; *corolla* white with yellow fornices; *nutlets* 4, ovate, 2–2.5 mm long, tuberculate and with short transverse ridges dorsally, nearly smooth ventrally, the scar open and triangular. Found in rocky soil, often with sagebrush, 7700–9500 ft. June–Aug. E/W. Endemic. (Plate 19)

PECTOCARYA DC. ex Meisn. – combseed

Annual herbs; *leaves* opposite below, alternate above; *calyx* deeply cleft, spreading or reflexed in fruit; *corolla* white, the tube shorter than the calyx, with prominent fornices; *stamens* included; *nutlets* 4, thin and flattened, commonly forming 2 pairs, or widely divergent, spreading from the broad, low gynobase at maturity, the margins with uncinate bristles.

***Pectocarya penicillata* (Hook. & Arn.) A. DC.**, sleeping combseed. Plants prostrate or decumbent; *stems* 1–25 cm long; *leaves* linear, to 2 (4) cm long; *calyx* slightly irregular in fruit; *corolla* up to 1 mm wide; *nutlets* 1.5–2 mm long, glabrous to minutely hairy. Not reported for Colorado, but occurs very close to the border in Wyoming and could possibly occur in dry, sandy places in northwestern Moffat County. April–June.

PLAGIOBOTHRYS Fisch. & C.A. Mey. – popcorn flower

Herbs; *leaves* opposite below, alternate above; *flowers* in racemes or spikes; *calyx* cleft to the middle or below; *corolla* white, with well-developed fornices, salverform with a short tube; *stamens* included; *nutlets* 1–4, rugose. (Johnston, 1932)

1a. Calyx lobes becoming elongate and thickened, curved toward the same side of the fruit; scar of nutlet very nearly basal...***P. leptocladus***
1b. Calyx with symmetrical lobes, these not thickened or much elongate in fruit; scar of nutlet lateral or basilateral (more lateral than basal)...***P. scouleri* var. *hispidulus***

***Plagiobothrys leptocladus* (Greene) I.M. Johnst.**, fine-branched popcorn flower. Plants with stems 0.5–3 dm long, prostrate; *leaves* linear to narrowly oblanceolate, to 6 cm long; *calyx* 4–7 mm long in fruit, curving; *corolla* 1–2 mm wide; *nutlets* lanceolate, 1.5–2.5 mm long, tuberulcate-rugose and usually bristly. Not reported for Colorado, but found in southern Wyoming and may occur in northern Moffat or Routt cos. near drying pools or on moist clay soil. May–Aug.

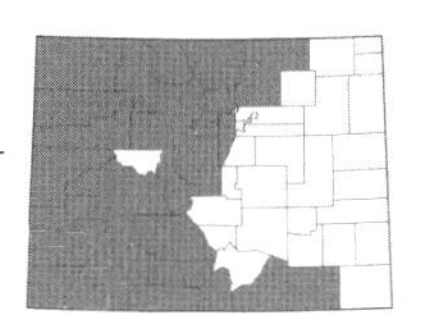

***Plagiobothrys scouleri* (Hook. & Arn.) I.M. Johnst. var. *hispidulus* (Greene) Dorn**, SLEEPING POPCORN FLOWER. Plants with stems to 2 dm long, prostrate; *leaves* linear, to 6.5 cm long; *calyx* 2–4 mm long in fruit; *corolla* 1–2 mm wide; *nutlets* ovate to lance-ovate, 1.5–2.2 mm long, tuberculate-rugose and usually bristly. Common along drying pond margins and mudholes or in moist, muddy soil in open meadows, 5000–10,700 ft. May–Sept. E/W.

SYMPHYTUM L. – COMFREY

Perennials; *leaves* alternate, entire; *corolla* white to yellowish or pink or blue, tubular-campanulate, with a well-defined throat and short lobes; *stamens* inserted at the level of the fornices within the tube; *nutlets* 1–4, smooth to finely rugose or tuberculate, the scar at the base flat with a thick, toothed rim.

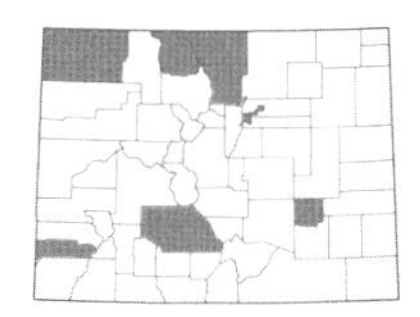

***Symphytum officinale* L.**, COMMON COMFREY. Plants 5–10 dm, coarse; *leaves* lanceolate to ovate, 5–30 cm long, decurrent; *calyx* 3–6 mm long in flower, up to 9 mm long in fruit; *corolla* 15–25 mm long, usually lavender-blue, tubular; *nutlets* 4–5 mm long. Uncommon weed in disturbed sites, old gardens, and along roadsides, very difficult to eradicate, 5000–7200 ft. May–Aug. E/W. Introduced.

TIQUILIA Pers. – CRINKLEMAT

Annual or perennial herbs, often prostrate; *leaves* alternate, mostly entire; *flowers* solitary or in axillary clusters, sessile; *calyx* deeply 5-lobed; *corolla* yellow, white, or purplish, fornices absent; *stamens* included; *style* terminal from an entire ovary, terminal but arising from between a 4-grooved ovary, or gynobasic; *nutlets* 1–4.

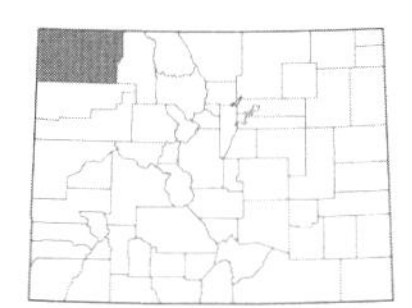

***Tiquilia nuttallii* (Hook.) A.T. Richardson**, ROSETTE TIQUILIA. Annuals, prostrate to 3 to 4 dm across; *stems* branching dichotomously; *leaves* broadly elliptic to ovate, 3–9 mm long; *calyx* 2–3 mm long in flower, to 3–4.5 mm long in fruit; *corolla* 3–4 mm long, white to pink or lavender; *nutlets* ca. 1 mm long, smooth, ventrally connate over halfway to the top. Uncommon in dry, sandy soil of dunes, 5800–6500 ft. May–Aug. W.

BRASSICACEAE Burnett – MUSTARD FAMILY

Annual, biennial, or perennial herbs, or rarely subshrubs; *leaves* usually alternate or sometimes opposite, simple to dissected, or rarely compound, exstipulate; *flowers* perfect, usually actinomorphic; *sepals* 4; *petals* 4, rarely rudimentary or absent, often clawed; *stamens* usually 6, tetradynamous; *ovary* superior, 2-carpellate, 2-locular with a false septum connecting 2 placentae; *fruit* a silicle or silique with a papery replum. (Al-Shehbaz, 2010; Rollins, 1993)

Typical Brassicaceae flower:

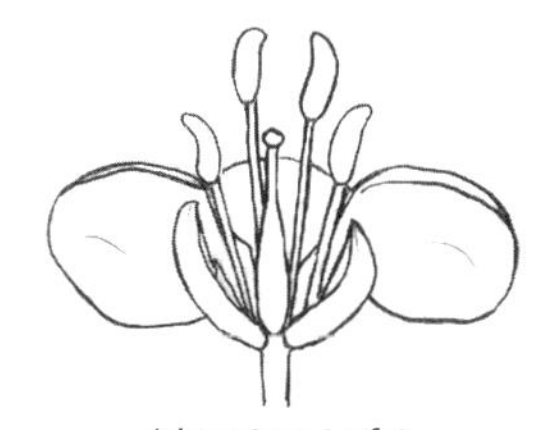

(showing 4 of 6 tetradynamous stamens)

1a. Plants with at least some branched, Y-shaped, forked, stellate, or dolabriform hairs at the base of the stem or on the lower leaves (sometimes mixed with simple or glandular hairs)...**Key 1**
1b. Plants glabrous or with simple, unbranched or glandular hairs...2

2a. Stem leaves sessile with a sagittate base, or amplexicaul to auriculate and clasping the stem...**Key 2**
2b. Stem leaves not sagittate, amplexicaul, or auriculate and clasping the stem, sessile or petiolate, or sometimes absent...**Key 3**

Key 1
(Plants with branched, forked, stellate, or dolabriform hairs)

1a. Plants scapose with all basal leaves or with a single stem leaf...2
1b. Plants with at least two stem leaves...3

2a. Fruit somewhat round in cross-section; flowers white or purplish; plants with a mixture of simple and 2–3-forked hairs...***Braya*** (*glabella*)
2b. Fruit conspicuously flattened (or inflated just at the base); flowers yellow or white; hairs various...***Draba***

3a. Leaves pinnately or bipinnately compound, or pinnately cleft with divisions all the way to the midrib...4
3b. Leaves simple, toothed, or pinnatifid but not all the way to the midrib...5

4a. Perennials from a branched caudex; flowers white; plants never glandular...***Smelowskia***
4b. Annuals or biennials from a taproot; flowers yellow or whitish; plants often glandular...***Descurainia***

5a. Stem leaves sessile with a sagittate base, or amplexicaul or auriculate and clasping the stem...6
5b. Stem leaves not sagittate, amplexicaul, or auriculate and clasping the stem, sessile or petiolate...11

6a. Fruit a triangular-obcordate silicle; basal leaves pinnately lobed or toothed; plants with sessile, 3–5-rayed, stellate hairs, or these mixed with simple ones...***Capsella***
6b. Fruit a linear silique or ovoid silicle; plants otherwise unlike the above in all respects...7

7a. Fruit an ovoid silicle; ovary and fruit hairy (sometimes becoming glabrous with age); flowers yellow...***Camelina***
7b. Fruit a linear silique; ovary and fruit glabrous or sometimes hairy; flowers white, pink, purple, or yellow...8

8a. Flowers yellow or creamy-white; fruit 4-angled in cross-section, erect and closely appressed to the rachis; basal leaves with a mixture of simple and Y-shaped hairs; stem leaves glabrous; plants mostly 4–12 dm tall...***Turritis***
8b. Flowers white, pink, or purple; fruit flattened or round in cross-section, if erect and closely appressed to the rachis then the basal leaves with dolabriform, sessile, 2-rayed hairs; plants of various heights...9

9a. Fruit round in cross-section; hairs Y-shaped, sometimes mixed with simple ones...***Transberingia***
9b. Fruit flattened; hairs various...10

10a. Stem leaves numerous, with toothed margins; basal leaves mostly withered at flowering time; lower stems usually hirsute...***Arabis***
10b. Stem leaves with mostly entire margins; basal leaves present, forming a rosette at flowering time; lower stems usually not hirsute...***Boechera***

11a. Stems and leaves mostly with sessile, 2-rayed, dolabriform hairs (these sometimes mixed with sessile 3–4-rayed hairs)...12
11b. Stems and leaves with stellate hairs (with 5 or more arms), or forked, Y-shaped, or branched hairs (these sometimes mixed with simple hairs)...14

12a. Fruit an orbicular or obovate silicle; flowers white, pink, or purple; plants usually cultivated, sometimes escaping, found below 7000 ft in elevation...***Lobularia***
12b. Fruit a linear or elliptic silique; flowers yellow, orange, or if pink-purple then found in the subalpine and alpine zones...13

13a. Fruit flattened, elliptic or lanceolate, 0.5–1.3 cm long; stem leaves ovate to lanceolate, usually less than 2 cm long; petals 4–6.5 mm long; ovules 12–24 per ovary...***Draba*** (*malpighiacea, spectabilis*)
13b. Fruit 4-angled, linear, 1–13 cm long; stem leaves linear to lanceolate or oblanceolate, 1.5–15.5 cm long; petals 3–25 mm long; ovules 20–120 per ovary...***Erysimum***

14a. Leaves with sessile, stellate or stellate-scale-like hairs with 5 or more arms (these sometimes mixed with simple or forked hairs); fruit a silicle; flowers yellow or white...15
14b. Leaves with forked, Y-shaped, or branched hairs (these sometimes mixed with simple hairs); fruit and flowers various...18

15a. Fruit usually inflated or somewhat flattened (mostly at the apex), mostly round in cross-section; flowers usually yellow or sometimes white to pale yellow; plants usually with a dense rosette of basal leaves...***Physaria***
15b. Fruit flattened parallel to the septum; flowers white or yellow; plants usually lacking a dense basal rosette...16

16a. Fruit about 3–4 times longer than wide, sometimes twisted, glabrous or pubescent with simple or 2–4-rayed hairs; ovules 16–60 per ovary; stem leaves usually few (usually 8 or less)...***Draba***
16b. Fruit 1–2 times as long as wide, never twisted, stellate-hairy; ovules 1–16 per ovary; stem leaves usually numerous (usually more than 8)...17

17a. Fruit orbicular or broadly elliptic; petals yellow or white, with entire tips; ovules 1–2 per ovary...***Alyssum***
17b. Fruit oblong or elliptic; petals white, the tips strongly bifid; ovules 4–16 per ovary...***Berteroa***

18a. Fruit an ovoid, nut-like silicle with the style curved away from the rachis; petals white, 1–1.5 mm long; plants with a mixture of stalked, 2-forked and simple hairs; annuals...***Euclidium***
18b. Fruit unlike the above, the style not curved away from the rachis; petals white, pink, purple, or yellow, 1.5–20 mm long; plants with various hairs; annuals, biennials, or perennials...19

19a. Fruit of 2 ovoid-oblong or suborbicular parts, breaking into two 1-seeded units at maturity; at least some basal leaves pinnately lobed, others with coarsely dentate margins; fruiting pedicels horizontally spreading...***Dimorphocarpa***
19b. Fruit variously shaped but not breaking into two 1-seeded units at maturity; leaves and fruiting pedicels various...20

20a. Stem leaves ovate to oblong, the larger ones over 9 cm long; lateral pair of sepals strongly saccate at the base; petals purple, the blade (13) 15–20 (22) mm long and the claw 6–12 mm long; plants 4–8 (11) dm tall...***Hesperis***
20b. Stem leaves unlike the above; lateral pair of sepals usually not strongly saccate at the base; petals white, pink, or purple, 1.5–12 mm long; plants 0.4–15 dm tall...21

21a. Fruit strongly flattened parallel to the septum, flattened and twisted, or sometimes flattened distally but inflated basally; plants usually with a basal rosette of leaves...22
21b. Fruit 4-angled or round in cross-section; basal leaves usually withering with age, or present in *Braya*...23

22a. Fruit elliptic, oblong, or ovoid, less than 12 mm long; flowers usually yellow or sometimes white...***Draba***
22b. Fruit linear or narrowly oblong, usually more than 12 mm long; flowers white, pink, or purple...***Boechera***

23a. Plants with a basal rosette of leaves at anthesis; fruit 1–3 cm long; plants 0.5–2 dm tall, found above 11,000 ft in elevation...***Braya*** (*humilis*)
23b. Basal leaves withering with age; fruit 2–10 cm long; plants 2–15 dm tall, found below 9000 ft in elevation...24

24a. Flowers white with purple tips; ovules 90–250 per ovary; fruit usually glabrous or sparsely hairy; perennials...***Pennellia***
24b. Flowers purple throughout; ovules 40–80 per ovary; fruit usually densely hairy, rarely glabrous; annuals...***Strigosella***

Key 2

(*Plants glabrous or with simple hairs; leaves with a sagittate or auriculate base*)

1a. Leaves all basal, linear to narrowly lanceolate with entire margins, glabrous; plants aquatic; flowers white; fruit a silicle...***Subularia***
1b. Plants unlike the above in all respects...2

2a. At least some leaves pinnately lobed; flowers pale to bright yellow or rarely white (or fading to white)...3
2b. Leaves entire or toothed; flowers usually white or purple, or occasionally yellow...6

3a. Leaves strongly dimorphic, the lower leaves pinnatifid with linear or filiform segments and the upper leaves entire with deeply cordate-amplexicaul bases (nearly perfoliate)...***Lepidium*** (*perfoliatum*)
3b. Leaves unlike the above...4

4a. Fruit an ovoid to elliptic silicle or a silique less than 12 mm long at maturity; plants erect, decumbent, or prostrate; petals 0.5–5.5 mm long...***Rorippa***
4b. Fruit a linear silique mostly over 15 mm long at maturity; plants erect; petals 2.5–25 mm long...5

5a. Petals 5–10 mm long, the claw not conspicuously differentiated from the limb; terminal lobe of basal leaves usually less than 5 cm long; mature fruit 0.7–4.5 cm long...***Barbarea***
5b. Petals 6–25 mm long, the claw conspicuously differentiated from the limb; terminal lobe of basal leaves usually 5 cm or longer; mature fruit (2.5) 3–11 cm long...***Brassica***

6a. Stem leaves with entire margins; flowers white, purple, brownish, or yellow...7
6b. Stem leaves with toothed margins; flowers white...14

7a. Ovules solitary in the ovaries; fruit 1-seeded, clavate, winged apically; flowers yellow...***Isatis***
7b. Ovules 2 to many per ovary; fruit 2–many-seeded, winged or not apically; flowers white, purple, brownish, or yellow...8

8a. Fruit an obovate or obcordate silicle, usually winged apically; flowers usually white or sometimes pinkish-purple, the petals 3.4–8.5 mm long; perennials...***Noccaea***
8b. Fruit a linear silique, not winged; flowers and plants various...9

9a. Stem leaves mostly oblong with a rounded or obtuse apex and deeply auriculate or amplexicaul bases...10
9b. Stem leaves various with an acute or acuminate apex...11

10a. Petals white or yellow, the blade wider than the claw; mature fruit 2–2.5 mm wide...***Conringia***
10b. Petals purple to brownish, the claw wider than the blade; mature fruit 2.5–6 mm wide...***Streptanthus***

11a. Siliques sessile...***Boechera*** (*grahamii*)
11b. Siliques stipitate 0.3–25 mm...12

12a. Fruit stipitate (6) 11–25 mm (and ovary on a distinct gynophore), horizontally spreading or curved downward; stamens nearly equal...***Stanleya*** (*viridiflora*)
12b. Fruit stipitate 0.3–8 mm, spreading, divaricate-ascending, or erect; stamens tetradynamous...13

13a. Stigma entire; stamens oblong; fruit 1–5.5 (7) cm long; flowers white or purple...***Thelypodium*** (*paniculatum, sagittatum*)
13b. Stigma 2-lobed; stamens linear; fruit 4–9 cm long; flowers white, purple, or yellow...***Thelypodiopsis***

14a. Fruit a linear silique; base of the stem densely hirsute, becoming glabrous above...***Arabis***
14b. Fruit an elliptic, ovate, or obovate silicle; base of the stem glabrous or the plants uniformly hairy throughout...15

15a. Silicles with broadly winged margins, (5) 7–20 mm wide; ovules 6–16 per ovary...***Thlaspi***
15b. Silicles apically winged or not, but not winged on the margins, 3–6.5 (7) mm wide; ovules usually 1 or 2 per ovary...***Lepidium***

Key 3

(Plants glabrous or with simple hairs; leaves lacking a sagittate or auriculate base)

1a. Basal leaves large, (10) 20–45 (60) cm long, broadly oblong or ovate, with coarsely toothed margins, on petioles to 60 cm long...***Armoracia***
1b. Basal leaves smaller, or if larger then pinnatifid to pinnately lobed...2

2a. Plants with glandular hairs and papillose; weedy, annuals with a rank smell; flowers purple; fruit a silique with a long, beaklike style...***Chorispora***
2b. Plants lacking glandular hairs and papillae; otherwise unlike the above in all respects...3

3a. Ovary and fruit long-stipitate on a gynophore 5–20 mm long; stamens equal or nearly so; flowers yellow to whitish...***Stanleya***
3b. Ovary and fruit sessile or shortly stipitate to 4 (5) mm; stamens usually tetradynamous; flowers various...4

4a. Fruit reflexed, pendant, or curved downward, a linear silique...5
4b. Fruit ascending or horizontally spreading but not reflexed or curved downward, or reflexed but not a linear silique...7

5a. Stem leaves usually petiolate; silique stipitate on a short gynophore 0.2–2 (5) mm long...***Thelypodium***
5b. Stem leaves sessile; silique sessile...6

6a. Basal leaves absent at flowering time; flowers yellow or white (often with purplish veins); fruiting pedicels strongly reflexed...***Streptanthella***
6b. Rosette of basal leaves present at flowering time; flowers lavender or white; fruiting pedicels horizontally spreading or curved downward...***Boechera***

7a. At least some leaves conspicuously toothed, pinnatifid, or pinnately lobed or compound...8
7b. Leaves all simple with entire margins (or rarely a few leaves with 1 or 2 teeth on the margin)...31

8a. Leaf margins all toothed less than halfway to the midrib...9
8b. At least some leaf margins toothed over halfway to the midrib, pinnatifid, or pinnately lobed or compound...16

9a. Most leaves (especially those towards the base) cordate or reniform...10
9b. Leaves variously shaped but not cordate or reniform...12

10a. Flowers purple; fruit a large, suborbicular silicle 3–5 cm long and 2–3.5 cm wide...***Lunaria***
10b. Flowers white; fruit a linear silique...11

11a. Tips of the lobes or teeth on the leaf margins with a small, slender point...***Cardamine*** (*cordifolia*)
11b. Tips of the lobes or teeth on the leaf margins lacking a small, slender point...***Alliaria***

12a. Fruit a linear silique, stipitate on a gynophore 0.2–4 (5) mm long...13
12b. Fruit an ovate, elliptic, or orbicular silicle, sessile...14

13a. Petals white or rarely lavender; stem leaves lanceolate or linear-lanceolate; silique lacking a terminal, beaklike segment…***Thelypodium*** (*laxiflorum*, *wrightii*)
13b. Petals yellow to orange-yellow; stem leaves ovate to oblong or just the very upper ones lanceolate; silique with a terminal, beaklike segment 0.5–3 mm long…***Brassica*** (*elongata*)

14a. Flowers zygomorphic, the lower, longer pair of petals 5–8 mm long; silicle 4–6 (7) mm long with a style 0.8–2 mm long…***Iberis***
14b. Flowers actinomorphic, the petals 0.3–4 mm long; silicle 2–10 mm long with a style 0.05–0.6 mm long or obsolete…15

15a. Silicles ovate to orbicular; plants with more than 2 stem leaves…***Lepidium***
15b. Silicles lanceolate to ovate-lanceolate or elliptic-lanceolate; plants usually scapose or rarely with 2 stem leaves…***Draba***

16a. Stems hollow and strongly inflated, to 3 cm in diam. at its widest point; basal leaves strongly dimorphic from the stem leaves, at least some leaves pinnatifid or pinnately lobed and some leaves entire; stem leaves linear, simple, with an entire margin; flowers purple or brownish; sepals densely hairy…***Caulanthus***
16b. Plants unlike the above…17

17a. Fruit ovate, ovoid, obovate, globose, oblong, elliptic (or ovary with this shape), or if linear then less than 12 mm long…18
17b. Fruit a linear to lanceolate silique (or ovary linear), more than 12 mm long…20

18a. Ovules 1–2 per ovary; silicle apically winged, globose to elliptic; flowers usually white or sometimes yellow…***Lepidium***
18b. Ovules 10–80 per ovary; silicle not apically winged, clavate to oblong, ovoid, or sometimes globose; flowers usually yellow or sometimes white or pink…19

19a. Fruit flattened; flowers white…***Hornungia***
19b. Fruit round in cross-section, flowers usually yellow or occasionally white or pink…***Rorippa***

20a. Plants scapose or nearly so, strongly scented; flowers yellow or rarely purple…***Diplotaxis***
20b. Plants with stem leaves, or if scapose then the leaves pinnately compound, usually not strongly scented; flowers various…21

21a. Stem leaves pinnately compound; plants aquatic, growing in shallow water and rooting at the lower nodes; flowers white…***Nasturtium***
21b. Stem leaves pinnatifid or pinnately lobed, or if pinnately compound then the plants not rooting at the lower nodes, usually not aquatic but sometimes growing in moist places; flowers various…22

22a. Fruit with a conspicuous terminal beak 10–50 mm long; basal leaves lyrate or pinnately lobed with 1–4 lobes on each side, and the terminal lobe much larger than the lateral…23
22b. Fruit lacking a terminal beak or if a terminal beak present then less than 10 mm long; basal leaves various…25

23a. Fruit valves each with a single prominent vein; terminal segment of silique (beak) (4) 5–10 (15) mm long; stems glabrous; fruit glabrous; flowers yellow…***Brassica*** (*juncea*)
23b. Fruit valves each with 2–5 prominent veins; terminal segment of silique (7) 10–50 mm long; stems usually hairy (at least below), or sometimes glabrous; fruit glabrous or hairy; flowers various…24

24a. Fruit indehiscent, 3–15 mm wide; flowers purple, pink, white, or yellow, usually with veins darker than the rest of the petal; roots often thick and fleshy…***Raphanus***
24b. Fruit dehiscent, 1–5.5 (6.5) mm wide; flowers yellow, without darker veins in the petals; roots not thick and fleshy…***Sinapis***

25a. Flowers white or rarely lavender…26
25b. Flowers yellow…27

26a. Siliques sessile; leaves sometimes pinnately compound or if pinnately dissected then the few segments linear to narrowly elliptic and mostly 2–8 mm wide…***Cardamine***
26b. Siliques stipitate on a short gynophore 0.2–4 (5) mm long; leaves unlike the above…***Thelypodium*** (*laxiflorum*, *wrightii*)

27a. Stem leaves mostly linear and entire, only a few pinnatifid or toothed; fruit narrowly linear, 0.9–1.2 mm wide; perennials…***Sisymbrium*** (*linifolium*)
27b. Stem leaves unlike the above; fruit various; annuals or perennials…28

28a. Lower stems and/or petioles of lower leaves sparsely to densely hispid; fruit narrowly linear, 15–90 mm long at maturity; petals 2.5–8 mm long…***Sisymbrium***
28b. Lower stems and petioles of lower leaves glabrous or sparsely hirsute, or if hispid then the fruit oblong and 10 mm or less in length and petals 1.5–3 mm long…29

29a. Ovary and immature fruit 2-segmented (terminal segment 1.5–15 mm long, seedless, tapering to a slender style); fruit linear; basal leaves 6–30 (80) cm long with 1–3 lobes on each side…***Brassica*** (*juncea, nigra*)
29b. Ovary and immature fruit unsegmented; fruit linear, oblong, or elliptic; basal leaves various…30

30a. Stigma 2-lobed; stems erect; fruit usually linear…***Sisymbrium***
30b. Stigma entire; stems often prostrate or decumbent or the branches ascending, sometimes erect; fruit linear, oblong, or elliptic…***Rorippa***

31a. Fruit a linear to lanceolate silique (or ovary linear), more than 12 mm long; stem leaves linear to lanceolate, 2.5–15 cm long; petals 4.5–18 mm long…32
31b. Fruit ovate, ovoid, obovate, globose, oblong, or elliptic (or ovary with this shape), or if linear then less than 12 mm long; stem leaves various or absent; petals 0.5–3 mm long…34

32a. Flowers yellow; usually with at least one or two basal leaves with toothed margins; perennials from deep-seated rhizomes…***Sisymbrium*** (*linifolium*)
32b. Flowers white or purple; leaves with entire margins; biennials from taproots or perennials lacking a rhizome…33

33a. Stigmas entire; petals (4.5) 6–9 (13) mm long; ovules 14–40 per ovary; fruit usually somewhat constricted between the seeds…***Thelypodium*** (*integrifolium*)
33b. Stigmas 2-lobed; petals (10) 12–16 (18) mm long; ovules 80–110 per ovary; fruit smooth…***Hesperidanthus***

34a. Plants scapose or with 1 or 2 stem leaves, with a dense rosette of basal leaves; flowers yellow or white; fruit flattened parallel to the septum…***Draba***
34b. Plants with stem leaves, lacking a dense rosette of basal leaves; flowers white; fruit flattened at a right angle to the septum…35

35a. Fruit linear to narrowly oblong, 7–25 mm long; plants either with long-petioled basal leaves or the stem leaves with sagittate-amplexicaul clasping bases; known from Park County…***Eutrema***
35b. Fruit oblong, elliptic, or ovoid, 1.6–5 mm long; plants unlike the above; not restricted to Park County…36

36a. Ovules 1–2 per ovary; perennials 0.7–15 dm tall…***Lepidium*** (*alyssoides, latifolium*)
36b. Ovules 10–24 per ovary; annuals (0.2) 0.5–2.2 (3) dm tall…***Hornungia***

ALLIARIA Heist. ex Fabr. – ALLIARIA

Biennial herb, glabrous or with a few simple hairs; *leaves* alternate, basal and cauline, petiolate, simple with toothed margins; *flowers* white; *fruit* a linear, 4-angled, widely divergent silique.

Alliaria petiolata **(M. Bieb.) Cavara & Grande**, GARLIC MUSTARD. Plants to 1 m; *basal leaves* reniform or cordate, the upper sometimes deltate, (6) 15–88 (118) mm wide; *sepals* (2) 2.5–3.5 (4.5) mm long; *petals* (2.5) 4–8 (9) mm long; *fruit* (2) 3–7 (8) cm long. Uncommon weed found along creeks in shady, disturbed areas, 5200–6500 ft. April–June. E. Introduced.

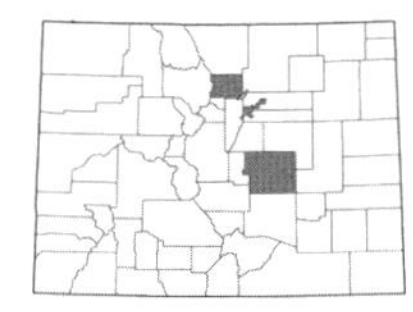

ALYSSUM L. – MADWORT

Annual, biennial, or perennial herbs, with stellate hairs; *leaves* alternate, basal and cauline, sessile or petiolate, with entire margins; *flowers* white, cream, or yellowish; *fruit* an elliptic to orbicular silicle, flattened parallel to the septum, sessile, glabrous or hairy.

1a. Perennials; ovules 1 per ovary…***A. murale***
1b. Annuals; ovules 2 per ovary…2

2a. Fruit glabrous…***A. desertorum***
2b. Fruit stellate-hairy…3

3a. Sepals persistent in fruit; filaments not winged...***A. alyssoides***
3b. Sepals deciduous; at least some filaments broadly winged...***A. simplex***

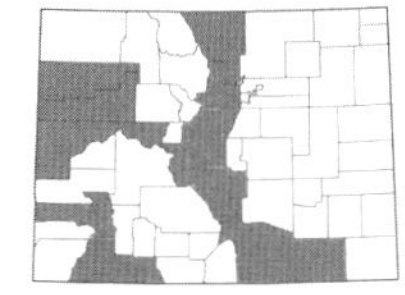

Alyssum alyssoides **(L.) L.**, PALE MADWORT. Annuals, 0.5–5 dm; *leaves* linear to oblanceolate, 3–4.5 cm long; *sepals* 1.5–3 mm long, persistent; *petals* 2–4 mm long, white or pale yellow; *fruit* orbicular, 2–5 mm diam., stellate-hairy. Found along roadsides, in vacant lots, disturbed areas, and on dry slopes, 5200–8500 ft. May–July. E/W. Introduced. (Plate 20)

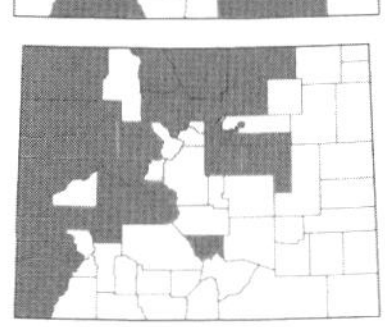

Alyssum desertorum **Stapf**, DESERT MADWORT. Annuals, 0.5–2.5 dm; *leaves* linear to narrowly oblanceolate, 0.3–3 cm long; *sepals* 1.4–2 mm long; *petals* 2–2.5 mm long, pale yellow; *fruit* orbicular, 2–4.5 mm diam., glabrous. Found in fields, meadows, sagebrush flats, along roadsides, vacant lots, and in disturbed areas, 4800–9000 ft. April–July. E/W. Introduced. (Plate 20)

Alyssum murale **Waldst. & Kit.**, YELLOWTUFT. Perennials, 2.5–7 dm; *leaves* linear to narrowly oblanceolate, 0.5–3 cm long; *sepals* 1.2–2 mm long; *petals* 2.5–3.5 mm long, yellow; *fruit* orbicular to broadly elliptic, 3.5–5 × 2.5–5 mm, stellate-hairy. Uncommon along roadsides and in disturbed areas, known from Boulder Co., 5800–6500 ft. May–July. E. Introduced.

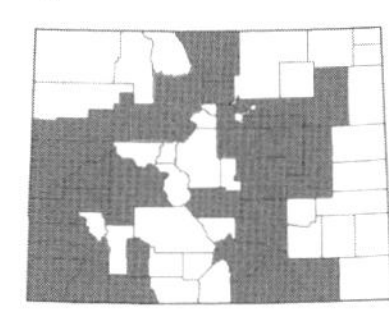

Alyssum simplex **Rudolphi**, ALYSSUM. [*A. minus* (L.) Rothm.]. Annuals, 0.5–3.5 dm; *leaves* oblanceolate to elliptic-lanceolate, 0.5–3 cm long; *sepals* 1.7–2.5 mm long; *petals* 1.8–3 mm long, pale yellow; *fruit* orbicular, 3.5–7 mm diam., stellate-hairy. Common along roadsides, in fields, disturbed areas, and on dry hillsides, 4400–9500 ft. April–June. E/W. Introduced. (Plate 20)

ARABIS L. – ROCKCRESS

Herbs, usually densely hirsute (ours) with a mixture of simple and stalked or sessile, forked hairs; *leaves* basal and cauline, the margins entire or dentate; *flowers* white; *fruit* a silique, usually sessile, torulose (ours), glabrous.

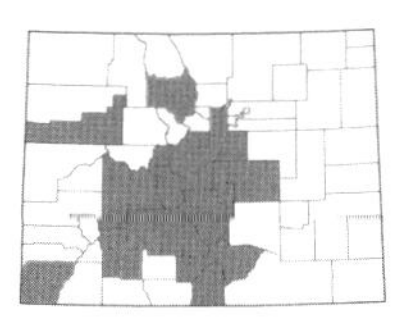

Arabis pycnocarpa **M. Hopkins var. *pycnocarpa***, HAIRY ROCKCRESS. [*A. hirsuta* (L.) Scop. var. *pycnocarpa* (M. Hopkins) Rollins]. Biennials or perennials, 1–8 dm; *basal leaves* oblong to obovate or spatulate, 2–10 cm long, sometimes undulate; *cauline leaves* ovate to oblong or lanceolate (linear), (1) 1.5–6 (8) cm long; *sepals* 2–4 mm long; *petals* 3.5–5 (5.5) mm long; *fruit* (3.5) 4–5.8 (6) mm long. Common on cliffs and ledges and in forests, meadows, and grasslands, 4800–10,500 ft. May–July. E/W.

ARMORACIA G. Gaertn., B. Mey. & Scherb. – ARMORACIA

Perennial herb, glabrous; *leaves* basal and cauline, the margins crenate, serrate, or pinnatifid; *flowers* white; *fruit* a silicle, sessile, subglobose to elliptic, style obsolete or to 0.5 mm.

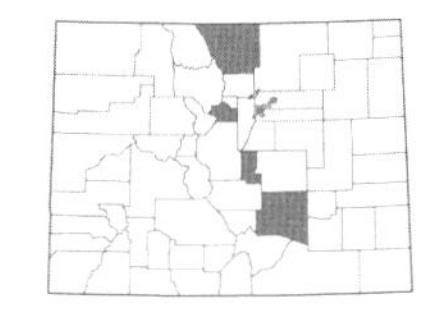

Armoracia rusticana **G. Gaertn., B. Mey., & Scherb.**, HORSERADISH. Plants 5–12 (20) dm, deeply rooted; *basal leaves* oblong to oblong-ovate, (10) 20–45 (60) cm long; *cauline leaves* lanceolate, smaller than basal; *fruiting pedicels* ascending, 8–20 mm long; *sepals* 2–4 mm long; *petals* 5–7 (8) mm long; *fruits* up to 6 mm in diam. Uncommon near old homesteads and cabins and along ditches, 5500–10,500 ft. May–July. E. Introduced.

BARBAREA W.T. Aiton – WINTERCRESS

Biennial or perennial herbs, glabrous or sparsely pubescent with a few simple hairs; *leaves* basal and cauline, petiolate or sessile, the margins entire, crenate, or pinnately lobed; *flowers* yellow; *fruit* a silique, linear, sessile or shortly stipitate, smooth or torulose, terete or somewhat 4-angled, usually glabrous.

1a. Mature fruit (2.5) 3.1–4.5 cm long with a stout style 0.5–1.2 (2) mm long; auricles of stem leaves usually hairy or rarely glabrous...***B. orthoceras***
1b. Mature fruit (0.7) 1.5–3 cm long with a slender style (1) 1.5–3.5 mm long; auricles of stem leaves glabrous...***B. vulgaris***

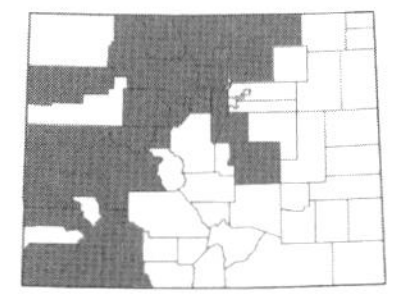

Barbarea orthoceras **Ledeb.**, AMERICAN YELLOW-ROCKET. Plants 2–6 (10) dm; *leaves* lyrate-pinnatifid with 1–4 lateral lobes, cuneate at the base, 1.5–6 (11) cm long; *sepals* 2.5–3.5 mm long, the lateral pair slightly saccate at the base; *petals* 5–8 mm long; *fruit* (2.5) 3.1–4.5 cm long with a stout style 0.5–1.2 (2) mm long, erect, torulose. Found along streams and creeks, ditches, roadsides, in meadows, and in disturbed areas, 5000–10,000 ft. May–July. E/W. Introduced. (Plate 20)

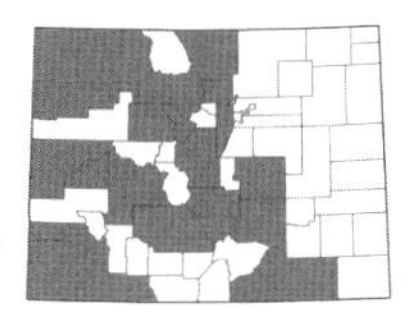

Barbarea vulgaris **W.T. Aiton**, YELLOW-ROCKET. Plants 1.5–10 dm; *leaves* lyrate-pinnatifid, with 1–5 lobes on each side, cuneate at the base, 1–10 cm long; *sepals* 3–5 mm long; *petals* 5–10 mm long; *fruit* (0.7) 1.5–3 cm long with a style (1) 1.5–3.5 mm, erect, torulose. Found along ditches, streams, in meadows, and disturbed areas, 5000–9500 ft. May–July. E/W. Introduced.

BERTEROA DC. – HOARY ALYSSUM

Annual or biennial herbs, with a mixture of stellate and simple hairs; *leaves* basal and cauline, petiolate or sessile, the margins entire; *flowers* white; *fruit* a silicle, somewhat flattened parallel to the septum, stellate-hairy.

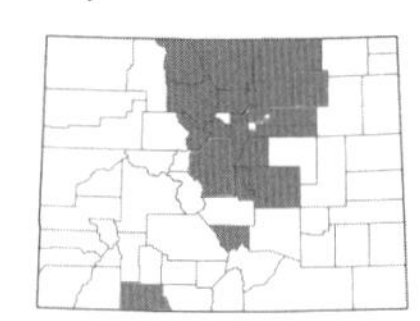

Berteroa incana **(L.) DC.**, HOARY ALYSSUM. Plants (2) 3–8 (11) dm, densely pubescent with stellate and simple hairs; *basal leaves* (2.5) 3.5–8 (10) cm long, oblanceolate; *sepals* 2–2.5 mm long; *petals* (4) 5–6.5 (8) mm long, obcordate; *fruit* (4) 5–8.5 (10) mm long, 2–4 mm wide, the style 1–4 mm long. Weedy plants found in disturbed areas, 5000–9500 ft. June–Oct. E/W. Introduced. (Plate 20)

BOECHERA Á. Löve & D. Löve – ROCKCRESS

Perennial or rarely biennial herbs, glabrous or pubescent with simple, dolabriform, or forked hairs; *leaves* basal and cauline, sessile or petiolate, the margins entire or dentate; *flowers* white, pink, lavender, or purple; *fruit* a silique, usually sessile, usually linear, smooth or torulose, usually glabrous or rarely hairy. (Rollins, 1941, 1982)

1a. Stem leaves lacking auriculate bases…2
1b. Stem leaves with auriculate bases…6

2a. Fruit (and ovary) densely hairy (especially when young), biseriate, divaricate-descending to reflexed but usually not appressed to the rachis…***B. formosa***
2b. Fruit (and ovary) glabrous, or if sparsely hairy then the fruit uniseriate and strongly reflexed and usually appressed to the rachis…3

3a. Fruit strongly reflexed and usually appressed to the rachis; stem leaves 15–40, usually concealing the lower stem; inflorescences arising from the center of the basal rosette, usually 1 per rosette…***B. retrofracta***
3b. Fruit divaricate-ascending or pendant, not appressed to the rachis; stem leaves 2–12, usually not concealing the lower stem; inflorescences various…4

4a. Fruit secund (one-sided), divaricate-ascending to slightly descending; flowers deep purple or lavender; lower stem glabrous or sparsely pubescent with 2–6-rayed hairs; inflorescences arising from the center of the basal rosette, usually 1 per rosette…***B. lemmonii***
4b. Fruit not secund or rarely weakly secund, pendant; flowers white or pale lavender; lower stem glabrous or with at least some simple hairs; inflorescences arising laterally from the basal rosette, usually 2–5 per rosette…5

5a. Basal leaves with simple and stalked, 2-rayed hairs; lower stem with simple hairs…***B. pendulina***
5b. Basal leaves glabrous or sparsely pubescent with stalked, 2–3-rayed hairs; lower stem with simple and stalked, 2-rayed hairs…***B. oxylobula***

6a. Fruit (and ovary) sparsely hairy, at least in the upper half (look on younger fruit)…7
6b. Fruit (and ovary) glabrous throughout…8

7a. Fruit strongly reflexed and usually appressed to the rachis; stem leaves 15–40, usually concealing the lower stem…***B. retrofracta***
7b. Fruit horizontal, widely spreading; stem leaves 3–8, usually not concealing the lower stem…***B. duchesnensis***

8a. Fruit erect and appressed to the rachis; basal leaves with sessile, 2-rayed (dolabriform) hairs, rarely glabrous…***B. stricta***
8b. Fruit various but if divaricate-ascending then not appressed to the rachis; basal leaves with stalked hairs, or if the hairs sessile then at least some more than 2-rayed, or the surface glabrous…9

9a. Basal leaves glabrous or with simple and short-stalked 2–3-rayed hairs; lower stem surfaces with simple and 2-rayed hairs…10
9b. Basal leaves pubescent with sessile or stalked 2–8-rayed hairs, lacking simple hairs (the margins can be ciliate with simple hairs however); lower stem surfaces with 2–8-rayed hairs or glabrous, with or without simple hairs…13

10a. Fruit horizontal or divaricate-ascending; seeds uniseriate; known from Moffat County…***B. glareosa***
10b. Fruit pendant; seeds biseriate; not restricted to Moffat County…11

11a. Lower stem surfaces with simple hairs only; basal leaves hairy; ovules 40–70 (90) per ovary; stem leaf auricles to 0.7 mm long…***B. pendulina***
11b. Lower stem surfaces with simple and stalked, 2-rayed hairs; basal leaves glabrous or hairy; ovules 90–128 per ovary; stem leaf auricles 0.5–3 mm long…12

12a. Basal leaves 5–15 (20) mm wide, with toothed margins; petals usually lavender, rarely white, 5–9 mm long...***B. fendleri***
12b. Basal leaves 1.5–4 mm wide, with entire margins; petals usually white, rarely pale lavender, 3–4 mm long...***B. spatifolia***

13a. Fruit one-sided (secund), usually horizontal or ascending; plants of the alpine, above 12,000 ft in elevation; flowers deep purple to lavender...14
13b. Fruit not one-sided, or if secund then the fruit strongly reflexed; plants to 10,500 ft in elevation; flowers white or lavender...15

14a. Basal leaves with subsessile, 2–6-rayed hairs; lower stem glabrous or with sessile or subsessile, 2–3-rayed hairs; stem leaves with auricles 0.7–2.5 mm long; ovules 44–100 per ovary; petals 6–8 mm long...***B. drepanoloba***
14b. Basal leaves with short-stalked, 3–9-rayed hairs; lower stem glabrous or with short-stalked, 2–6-rayed hairs; stem leaves with auricles 0.1–0.5 mm long; ovules 28–40 (-44) per ovary; petals 3.5–6 mm long...***B. lemmonii***

15a. Lower stems with sessile or short-stalked, branched hairs and simple hairs, or glabrous...16
15b. Lower stems with short-stalked, branched hairs only (2–8-rayed), simple hairs absent...18

16a. Inflorescences arising from the center of the basal rosette, usually 1 per rosette, 16–88-flowered; stem leaves (10)13–52; hairs on leaves and lower stem sessile or subsessile...***B. grahamii***
16b. Inflorescences arising laterally from the basal rosette, usually 2–5 per rosette, 7–20-flowered; stem leaves 3–9; hairs on leaves and lower stem short-stalked...17

17a. Fruit usually widely pendant (rarely horizontal); fruiting pedicels (7) 10–18 mm long; seeds irregularly biseriate; ovules 60–96 per ovary...***B. gracilenta***
17b. Fruit horizontal to divaricate-descending; fruiting pedicels 4–7 mm long; seeds uniseriate; ovules 36–54 per ovary...***B. gunnisoniana***

18a. Fruit reflexed or pendant...19
18b. Fruit ascending or divaricate-ascending...21

19a. Inflorescences arising laterally from the basal rosette, usually 2–5 per rosette, 6–15 (25)-flowered; fruit widely pendant; ovules 48–74 per ovary...***B. lignifera***
19b. Inflorescences arising from the center of the basal rosette, usually 1 per rosette, 15–80-flowered; fruit reflexed or pendant; ovules 60–128 per ovary...20

20a. Fruit biseriate or irregularly biseriate, reflexed or pendant but only rarely appressed to the rachis; fruiting pedicels reflexed or descending but not abruptly recurved at the base (arched, not bent)...***B. consanguinea***
20b. Fruit uniseriate, strongly reflexed and usually appressed to the rachis; fruiting pedicels abruptly recurved at the base (bent at the base)...***B. retrofracta***

21a. Inflorescences arising from the center of the basal rosette, usually 1 per rosette...***B. fernaldiana* ssp. *vivariensis***
21b. Inflorescences arising laterally from the basal rosette, usually 2–5 per rosette...22

22a. Fruit 0.6–0.9 mm wide; lower stem with 5–8-rayed hairs; flowers usually white; stem leaf auricles 0.1–0.5 mm long; long-lived perennials with a somewhat woody caudex...***B. crandallii***
22b. Fruit 1–1.1 mm wide; lower stem with 2–6-rayed hairs; flowers usually lavender; stem leaf auricles (0.5) 1–2 mm long; short-lived perennials, mostly lacking woody caudices...***B. pallidifolia***

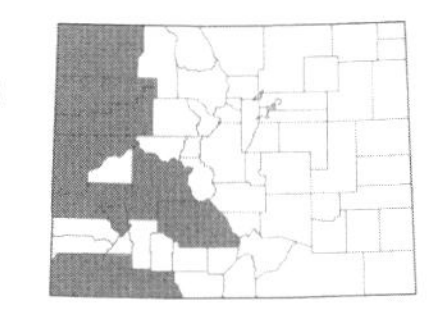

***Boechera consanguinea* (Greene) Windham & Al-Shehbaz**, SECUND ROCKCRESS. [*Arabis consanguinea* Greene]. Plants 1.5–5 dm; *stems* usually 1 per caudex branch, with 2–6-rayed hairs below; *basal leaves* oblanceolate, 2–10 mm wide, with short-stalked, 4–8-rayed hairs; *cauline leaves* with auricles 1–3 mm long; *petals* pale lavender, 5–8.5 mm long; *fruit* reflexed to pendant, 4–6 cm long, glabrous, the seeds biseriate to sub-biseriate. Found on rocky slopes and dry hillsides, often in pinyon-juniper and sagebrush, 6000–9000 ft. May–June. E/W. (Plate 20)

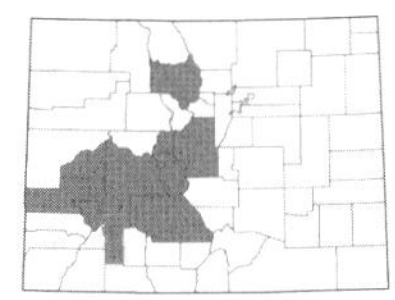

***Boechera crandallii* (B.L. Robinson) W.A. Weber**, CRANDALL'S ROCKCRESS. [*Arabis crandallii* B.L. Robinson]. Plants 1–4 dm; *stems* usually 2–5 per caudex branch, with 5–8-rayed hairs below; *basal leaves* narrowly oblanceolate, 1.5–3.5 (5) mm wide, with simple and 5–8-rayed hairs; *cauline leaves* with auricles 0.1–0.5 mm long; *petals* white, 5–7 mm long; *fruit* ascending, 3–5.5 cm long, glabrous, the seeds uniseriate. Found on rocky slopes and canyonsides, 6500–9000 ft. May–June. E/W. Endemic.

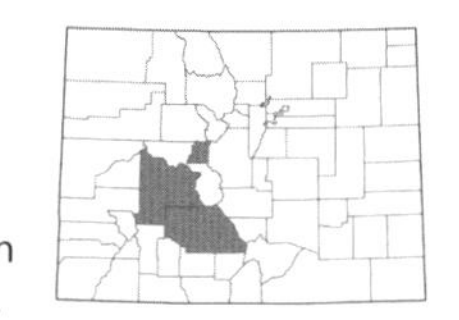

Boechera drepanoloba **(Greene) Windham & Al-Shehbaz**, SOLDIER ROCKCRESS. [*Arabis lemmonii* S. Watson var. *drepanoloba* (Greene) Rollins]. Plants 1–4 dm; *stems* usually 1 per caudex branch, glabrous or sparsely hairy below with 2–3-rayed hairs; *basal leaves* oblanceolate, 2–6 mm wide, with 2–6-rayed hairs; *cauline leaves* with auricles 0.7–2.5 mm long; *petals* purple, 6–8 mm long; *fruit* ascending to horizontal, secund, 3–6 cm long, glabrous, the seeds uniseriate. Uncommon in the alpine and on talus slopes, often near melting snowbanks, 12,000–14,000 ft. July–Aug. E/W.

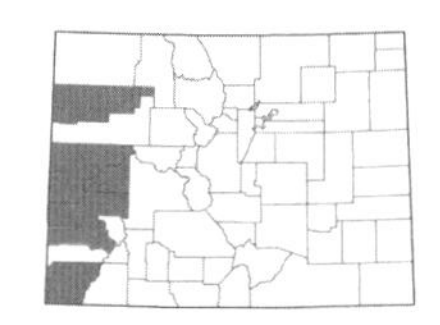

Boechera duchesnensis **(Rollins) Windham, Al-Shehbaz, & Allphin**, DUCHESNE ROCKCRESS. [*Arabis pulchra* M.E. Jones ex S. Watson var. *duchesnensis* Rollins]. Plants 1.5–4.5 dm; *stems* usually 1 per caudex branch, with 2–7-rayed hairs below; *basal leaves* linear-oblanceolate, 2–5 mm wide with simple and short-stalked, 4–8-rayed hairs; *cauline leaves* with auricles 0.7–2 mm long; *petals* whitish to pale purple, 7–10 mm long; *fruit* horizontal, 3.5–5 cm long, hairy distally, with sub-biseriate seeds. Found on dry slopes, often with sagebrush, 4900–6100 ft. April–May. W.

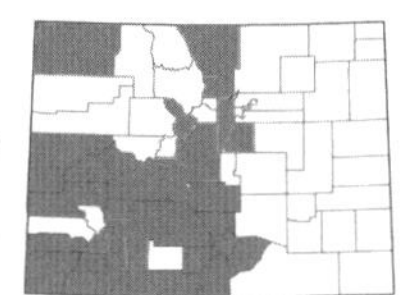

Boechera fendleri **(S. Watson) W.A. Weber**, FENDLER'S ROCKCRESS. [*Arabis fendleri* (S. Watson) Greene]. Plants 1.5–6 (8) dm; *stems* 1–7 per caudex branch, with simple and 2-rayed hairs below; *basal leaves* oblanceolate, 5–15 (20) mm wide, with simple and long-stalked, 2-rayed hairs; *cauline leaves* with auricles 0.8–3 mm long; *petals* usually lavender, 5–9 mm long; *fruit* widely pendant, 3–6 cm long, glabrous, the seeds biseriate. Found on dry, rocky slopes, often with pinyon-juniper or scrub oak, 5500–8000 ft. April–July. E/W.

Boechera fernaldiana **(Rollins) W.A. Weber ssp. *vivariensis* (S.L. Welsh) Windham & Al-Shehbaz**, PARK ROCKCRESS. [*Arabis fernaldiana* Rollins]. Plants 1–4 dm; *stems* usually 1 per caudex branch, with 4–8-rayed hairs sometimes mixed with 2-rayed hairs below; *basal leaves* oblanceolate, 1–4 mm wide, with simple and short-stalked, 4–8-rayed hairs; *cauline leaves* with auricles 0.3–2 mm long; *petals* white, 7–12 mm long; *fruit* divaricate-ascending, 3.5–7 cm long, glabrous, the seeds uniseriate. Found on dry, sandy slopes, 5500–7500 ft. May–June. W.

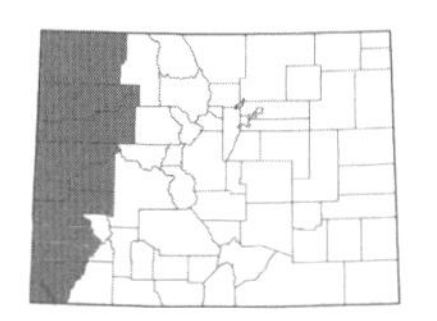

Boechera formosa **(Greene) Windham & Al-Shehbaz**, BEAUTIFUL ROCKCRESS. [*Arabis pulchra* M.E. Jones ex S. Watson var. *pallens* M.E. Jones]. Plants 2–5.5 dm; *stems* usually 1 per caudex branch, with 4–7-rayed hairs below; *basal leaves* linear to linear-oblanceolate, 2–4 mm wide, with simple and 4–8-rayed hairs; *cauline leaves* lacking auricles; *petals* white to pale lavender, 8–18 mm long; *fruit* reflexed or descending, 4–7 cm long, hairy, the seeds biseriate. Found on rocky or dry slopes, often with sagebrush or pinyon-juniper, 4900–6200 ft. April–May. W.

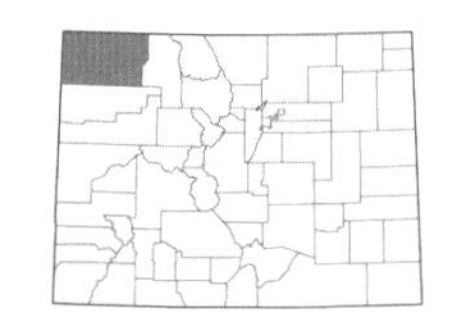

Boechera glareosa **Dorn**, ROCKCRESS. Plants 0.8–4 dm; *stems* usually 2–6 per caudex branch, with simple and 2-rayed hairs below; *basal leaves* narrowly oblanceolate, 1–5 mm wide, with simple and short-stalked, 2–3-rayed hairs; *cauline leaves* with auricles 0.3–0.6 mm long; *petals* lavender, 3.5–6 mm long; *fruit* divaricate-ascending, 2.5–4 cm long, glabrous, the seeds uniseriate. Found on gravelly conglomerate and limestone outcrops, 6700–8000 ft. May–June. W.

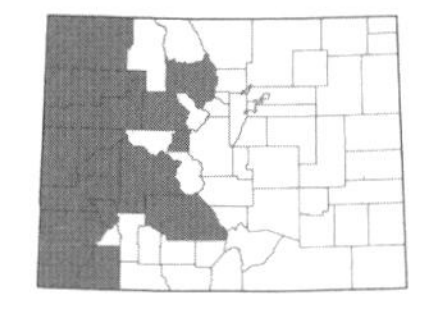

Boechera gracilenta **(Greene) Windham & Al-Shehbaz**, PERENNIAL ROCKCRESS. [*Arabis selbyi* Rydb.]. Plants 1.5–5.5 dm; *stems* 1–9 per caudex branch, with 2–4-rayed and simple hairs below; *basal leaves* oblanceolate, 2–9 mm wide, with simple and 4–6-rayed hairs; *cauline leaves* with auricles 1–3 mm long; *petals* lavender, 6–7 mm long; *fruit* widely pendant, 4–7 cm long, glabrous, the seeds sub-biseriate. Found on rocky slopes and in dry, sandy soil, often in pinyon-juniper communities, 4700–8500 ft. April–May. W. (Plate 20)

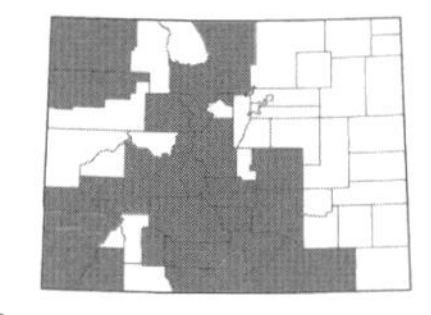

Boechera grahamii **(Lehm.) Windham & Al-Shehbaz**, GRAHAM'S ROCKCRESS. [*Arabis drummondii* A. Gray var. *brachycarpa* (Torr. & A. Gray) A. Gray]. Plants 1.5–12 dm; *stems* usually 1 per caudex branch, with simple and 2–3 (6)-rayed hairs below; *basal leaves* oblanceolate, 1.5–10 (20) mm wide, with simple and 2–4 (7)-rayed hairs; *cauline leaves* with auricles 1–5 mm long; *petals* white, 5.5–8 mm long; *fruit* divaricate-ascending to pendant, 3.5–9 cm long, glabrous, the seeds uniseriate to sub-biseriate. Found on dry slopes, in grasslands, and forests, 5000–10,500 ft. May–Aug. E/W. (Plate 20)

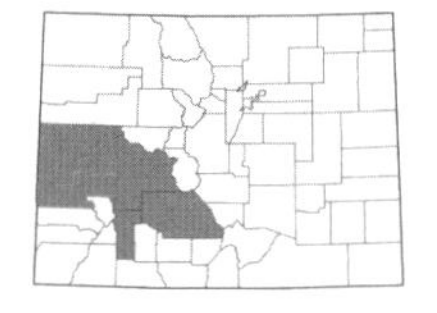

Boechera gunnisoniana **(Rollins) W.A. Weber**, GUNNISON'S ROCKCRESS. [*Arabis gunnisoniana* Rollins]. Plants 0.8–2.5 dm; *stems* usually 2–6 per caudex branch, with 2–4-rayed and few simple hairs below; *basal leaves* linear-oblanceolate, 1–4 mm wide, with 3–6-rayed hairs; *cauline leaves* with auricles 0.2–1 mm long; *petals* white or lavender, 4–6 mm long; *fruit* horizontal to divaricate-descending, 2.5–4 cm, glabrous, the seeds uniseriate. Found on rocky slopes and ridgetops, usually with sagebrush, 6900–8800 ft. May–June. W. Endemic.

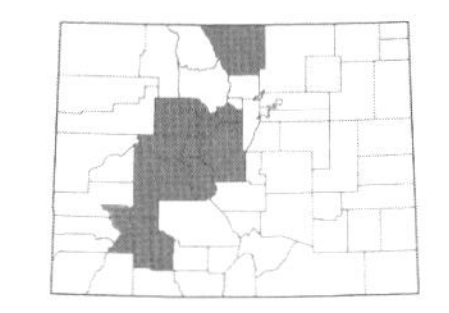

***Boechera lemmonii* (S. Watson) W.A. Weber**, Lemmon's rockcress. [*Arabis lemmonii* S. Watson]. Plants 0.5–2.5 dm; *stems* usually 1 per caudex branch, with 2–6-rayed hairs below or glabrous; *basal leaves* oblanceolate, 1.5–5 mm wide, with 3–9-rayed hairs; *cauline leaves* with auricles absent or 0.1–0.5 mm long; *petals* lavender, 3.5–6 mm long; *fruit* divaricate-ascending, secund, 1.6–4 cm long, glabrous, the seeds uniseriate. Found in alpine, on scree slopes and fell fields, 12,000–14,000 ft. June–Aug. E/W.

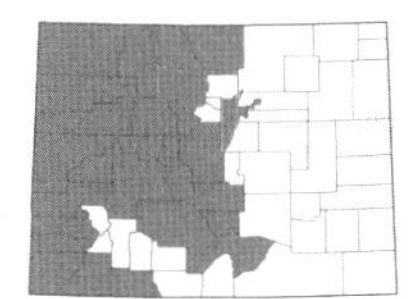

***Boechera lignifera* (A. Nelson) W.A. Weber**, desert rockcress. [*Arabis lignifera* A. Nelson]. Plants 1.2–5 dm; *stems* usually 2–5 per caudex branch, with 4–7-rayed hairs below; *basal leaves* narrowly oblanceolate, 2–5 (8) mm wide, with 3–7-rayed hairs; *cauline leaves* with auricles 0.5–2 mm long; *petals* whitish, 5–7 mm long; *fruit* widely pendant, 2.5–5.5 cm long, glabrous, the seeds uniseriate. Common on rocky or dry, sandy slopes, usually in sagebrush, pinyon-juniper, or oak shrublands, 5000–8600 ft. May–June. E/W.

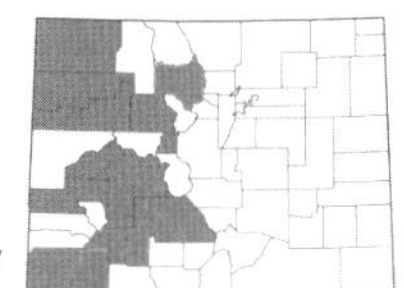

***Boechera oxylobula* (Greene) W.A. Weber**, Glenwood Springs rockcress. [*Arabis demissa* Greene; *A. oxylobula* Greene]. Plants 0.4–2.5 dm; *stems* usually 3–7 per caudex branch, glabrous or with simple and 2-rayed hairs below; *basal leaves* linear to linear-oblanceolate, 1–2.5 mm wide, with 2–3-rayed hairs; *cauline leaves* lacking auricles; *petals* white to pale lavender, 4–5 mm long; *fruit* pendant, glabrous, 1.5–3.5 cm long, the seeds uniseriate. Found on rocky slopes, cliffs, and ridges, 6800–11,800 ft. May–July. E/W. Endemic.

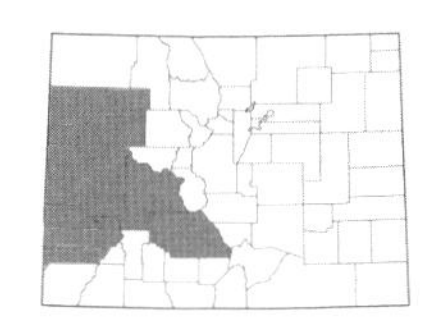

***Boechera pallidifolia* (Rollins) W.A. Weber**, Gunnison County rockcress. [*Arabis pallidifolia* Rollins]. Plants 1–4 dm; *stems* usually 2–5 per caudex branch, with 2–6-rayed hairs below; *basal leaves* oblanceolate, 5–13 mm wide, with simple and 4–8-rayed hairs; *cauline leaves* with auricles 1–2 mm long; *petals* lavender, 5–9 mm long; *fruit* usually ascending, 2.5–6 cm long, glabrous, the seeds uniseriate. Found on rocky slopes and in sandy soil, usually with sagebrush or pinyon-juniper, 6000–8500 ft. May–June. W.

Commonly hybridizes with *B. crandallii*, *B. fendleri*, *B. formosa*, and *B. pendulina*.

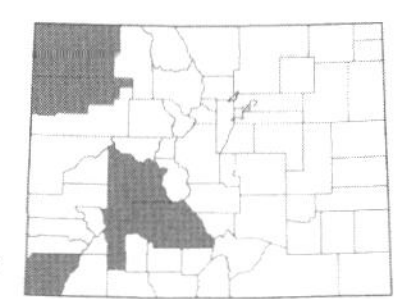

***Boechera pendulina* (Greene) W.A. Weber**, rabbit ear rockcress. [*Arabis pendulina* Greene]. Plants 0.5–3.5 dm; *stems* usually 2–6 per caudex branch, with simple hairs below; *basal leaves* oblanceolate, 1.5–6 mm wide, with simple and 2-rayed hairs; *cauline leaves* lacking auricles or rarely to 0.7 mm long; *petals* whitish to pale lavender, 4–6 mm long; *fruit* widely pendant, 2–4 cm long, glabrous, the seeds biseriate. Found on rocky slopes and dry hillsides, often with sagebrush, pinyon-juniper, or shrub communities, 5500–10,000 ft. April–June. E/W. (Plate 20)

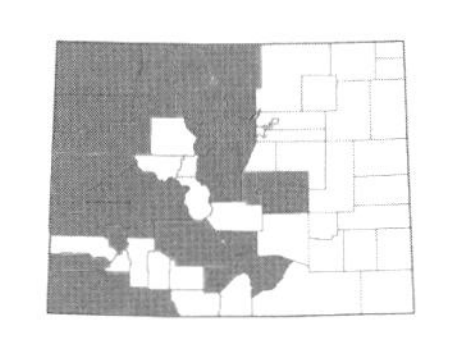

***Boechera retrofracta* (Graham) Á. Löve & D. Löve**, reflexed rockcress. [This is the species that we have called *Arabis holboellii* Hornem., which is actually restricted to Greenland]. Plants 1.5–10 dm; *stems* usually 1 per caudex branch, with 2–8-rayed hairs below; *basal leaves* oblanceolate, 2–7 mm wide, with 5–10-rayed hairs; *cauline leaves* with auricles 0.5–2.5 mm long; *petals* white to lavender, 4–8 mm long; *fruit* strongly reflexed, 3.5–9 mm long, glabrous or sparsely hairy, the seeds uniseriate. Common in forests and on dry hillsides, often in sagebrush or pinyon-juniper, 6000–10,500 ft. May–July. E/W.

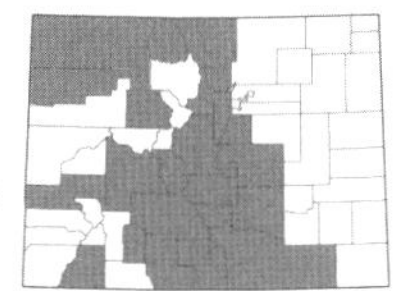

***Boechera spatifolia* (Rydb.) Windham & Al-Shehbaz**, spoonleaf rockcress. [*Arabis fendleri* var. *spatifolia* (Rydb.) Rollins]. Plants 1.5–5 dm; *stems* 1–5 per caudex branch, with simple and 2-rayed hairs below; *basal leaves* narrowly oblanceolate, 1.5–4 mm wide, with simple and 2-rayed hairs; *cauline leaves* with auricles 0.5–1.5 mm long; *petals* usually white, 3–4 mm long; *fruit* pendant, 3–6 cm long, glabrous, the seeds biseriate. Found on rocky slopes, dry hillsides, and in open forests, often with sagebrush or pinyon-juniper, 5500–10,500 ft. May–July. E/W. (Plate 20)

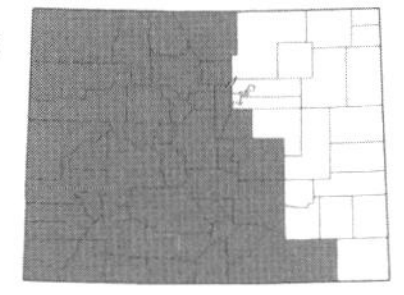

***Boechera stricta* (Graham) Al-Shehbaz**, Drummond's rockcress. [*Arabis drummondii* A. Gray]. Plants 1.5–10 dm; *stems* usually 1–4 per caudex branch, usually with 2-rayed hairs below or glabrous; *basal leaves* oblanceolate, 1.5–8 (14) mm wide, with 2-rayed, dolabriform hairs; *cauline leaves* with auricles 0.5–3 mm long; *petals* white to pale lavender, 5–11 mm long; *fruit* erect, appressed to the rachis, 4–10 cm long, glabrous, the seeds biseriate. Common on rocky slopes, in open forests, meadows, and alpine, 7000–13,000 ft. May–Aug. E/W.

BRASSICA L. – mustard

Annual, biennial, or perennial herbs, glabrous or pubescent with simple hairs; *leaves* basal and cauline, petiolate or sessile, with entire, dentate, or lyrate-pinnatifid margins; *flowers* yellow or rarely white; *fruit* a silique, sessile or stipitate, torulose or smooth, terete or 4-angled, terminated by a conspicuous beak.

1a. Stem leaves with auriculate or amplexicaul and clasping bases, sessile...2
1b. Stem leaves not auriculate or clasping at the base, sessile or petiolate...3

2a. Petals 10–16 mm long; flower buds overtopping or equal to open flowers; plants glaucous...***B. napus***
2b. Petals 6–10 (13) mm long; flower buds shorter than or equal to open flowers; plants green or slightly glaucous...***B. rapa***

3a. Fruit and ovary stipitate on a gynophore 1.5–4 (5) mm long (in fruit); basal leaves toothed but not lobed; biennials or short-lived perennials, often woody at the base...***B. elongata***
3b. Fruit and ovary sessile; basal leaves pinnatifid to pinnately lobed; annuals...4

4a. Fruiting pedicels erect and ascending, often appressed to the rachis; plants sparsely to densely hairy below; terminal segment of fruit (1) 2–6 mm long...***B. nigra***
4b. Fruiting pedicels spreading or somewhat ascending, but not appressed to the rachis; plants glabrous below; terminal segment of fruit (4) 5–10 (15) mm long...***B. juncea***

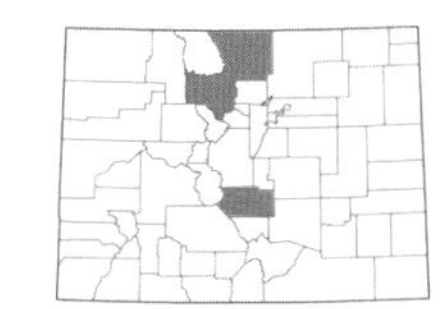

Brassica elongata **Ehrh.**, ELONGATED MUSTARD. Biennials or perennials, 5–10 dm; *basal leaves* unlobed, obovate to elliptic, toothed, to 30 cm long; *cauline leaves* sessile, oblong to lanceolate, to 10 cm long; *petals* (5) 7–10 mm long; *fruit* (1.5) 2–4.5 cm long, stipitate on a gynophore 1.5–4 (5) mm long, terminal segment 0.5–3 mm long, spreading to ascending, terete. Uncommon along roadsides and in fields, 5000–8000 ft. June–Aug. E/W. Introduced.

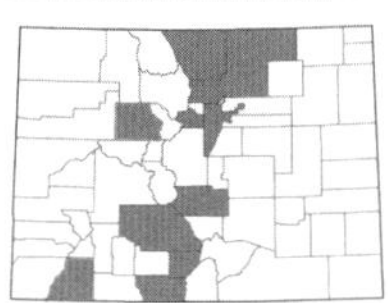

Brassica juncea **(L.) Czern.**, INDIAN MUSTARD. Annuals, 2–10 dm; *basal leaves* pinnatifid or pinnately lobed, mostly to 35 cm long; *cauline leaves* sessile, oblong to lanceolate, with dentate to pinnately lobed margins; *petals* (7) 9–13 mm long; *fruit* sessile, (2) 3–6 cm long, terminal segment (4) 5–10 (15) mm long, spreading to ascending. Uncommon along roadsides, in old fields and gardens, and in disturbed areas, often found in moist places, 5000–7500 ft. May–Aug. E/W. Introduced.

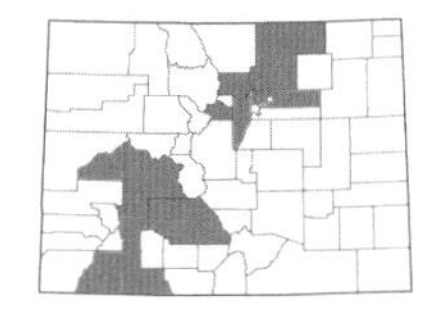

Brassica napus **L.**, RAPESEED. Annual or biennial, 3–13 dm; *basal leaves* pinnately lobed, to 40 cm long; *cauline leaves* auriculate or amplexicaul, with entire margins; *petals* 10–16 mm long; *fruit* (3.5) 5–11 cm long, with a terminal segment (5) 9–16 mm long, spreading to ascending, sessile. Uncommon along roadsides, in disturbed areas, abandoned fields, and old gardens, escaping cultivation, 5000–9000 ft. May–Aug. E/W. Introduced.

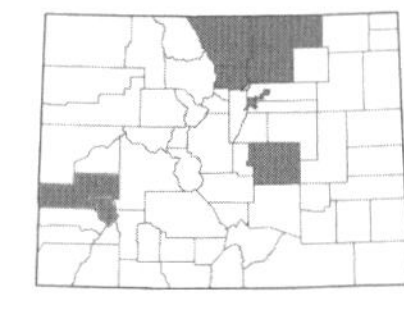

Brassica nigra **(L.) W.D.J. Koch**, BLACK MUSTARD. Annuals, 3–20 dm; *basal leaves* lyrate-pinnatifid, to 30 cm long; *cauline leaves* sessile, lanceolate to ovate, with entire to toothed margins; *petals* 7–13 mm long; *fruit* 1–3 cm long, the terminal segment (1) 2–6 mm long, erect-ascending, usually appressed to the rachis, 4-angled. Uncommon in fields, gardens, along roadsides, and in disturbed areas, 5000–8000 ft. May–Sept. E/W. Introduced.

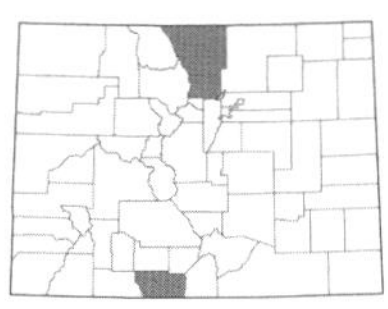

Brassica rapa **L.**, FIELD MUSTARD; CANOLA. Annual or biennial, 3–10 dm; *basal leaves* lyrate-pinnatifid, to 50 cm long, usually setose; *cauline leaves* auriculate to amplexicaul; *petals* 6–10 (13) mm long; *fruit* (2) 3–11 cm long with a terminal segment 8–22 mm long, ascending to spreading, terete. Uncommon along roadsides, in old gardens, and fields, cultivated and sometimes escaping, 5000–9000 ft. April–Aug. E/W. Introduced.

BRAYA Sternb. & Hoppe – NORTHERN ROCKCRESS

Perennial herbs, sometimes glabrous but usually pubescent with a mixture of simple and dolabriform or forked hairs; *leaves* basal and sometimes with a few cauline, petiolate or sessile, with entire or dentate margins; *flowers* white, pink, or purple; *fruit* a silique or silicle, sessile, glabrous or hairy, terete.
1a. Leaves all basal or stems with a single stem leaf...***B. glabella* ssp. *glabella***
1b. Plants with 2–4 stem leaves...***B. humilis* ssp. *humilis***

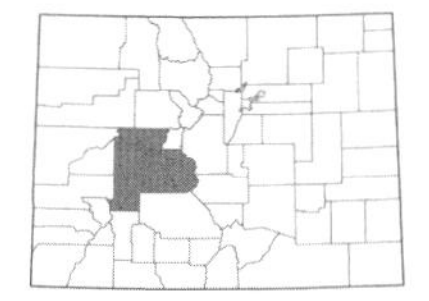

Braya glabella **Richardson ssp. *glabella***, SMOOTH NORTHERN ROCKCRESS. Plants 0.5–2 dm; *basal leaves* spatulate to linear-oblanceolate, to 8 cm long, the margins usually entire; *cauline leaves* absent or 1; *petals* white or purplish, 2–5 mm long; *fruit* oblong, 0.3–1.5 cm long, (0.8) 1–3.5 mm wide, glabrous to hairy with simple or 2-rayed hairs. Uncommon on rocky soil of the alpine, often in disturbed old mining areas, 12,000–13,000 ft. July–Aug. E/W.

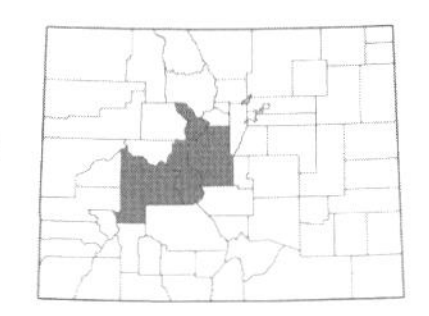

Braya humilis* (C.A. Mey.) B.L. Rob. ssp. *humilis, LOW NORTHERN ROCKCRESS. Plants 0.5–3 dm; *basal leaves* spatulate to obovate or oblong, 0.3–3.5 cm long, the margins entire to pinnatifid, hairy or rarely glabrous; *cauline leaves* 3 or more; *petals* white, pink, or purple, 2.5–7 mm long, 0.6–1.3 mm wide; *fruit* linear, 1–3 cm long, with 2-rayed hairs or rarely glabrescent. Uncommon on rocky soil of alpine, often in disturbed old mining areas, 11,400–13,000 ft. E/W.

CAMELINA Crantz – FALSE FLAX

Annual or biennial herbs, glabrous or pubescent with longer simple or short-stalked and shorter forked to substellate hairs; *leaves* basal and cauline, petiolate or subsessile, with entire or toothed to rarely lobed margins; *flowers* yellow or rarely white; *fruit* a silicle, obovoid or pyriform, shortly stipitate, hairy.

1a. Leaves primarily with simple hairs; lower stem with mostly simple hairs or these mixed with branched ones; mature fruit 3.5–7 mm long, each valve with an obscure midvein...2
1b. Leaves primarily with forked hairs or glabrescent; lower stem glabrous or hairs mostly small and branched, rarely mixed with simple hairs; mature fruit 7–11 (13) mm long, each valve with a prominent midvein...3

2a. Petals pale yellow, 2.5–4 (6) mm long...***C. microcarpa***
2b. Petals white or creamy-white, 5–9 mm long...***C. rumelica***

3a. Fruit wider than long (depressed globose), 5.5–9 mm wide; stem leaf margins lobed or coarsely dentate...***C. alyssum***
3b. Fruit longer than wide (broadly obovoid to pyriform), 4–6 mm wide; stem leaf margins entire or rarely remotely denticulate...***C. sativa***

***Camelina alyssum* (Mill.) Thell.**, FALSE FLAX. Plants 2–7 (10) dm, glabrous or with minute, branched hairs below; *cauline leaves* lanceolate to oblong, lobed or coarsely dentate, (1.5) 2.5–7 (10) cm long, sagittate or auriculate, with forked hairs; *petals* 3.5–6.5 mm long; *fruit* depressed globose, 7–11 × 5.5–9 mm. Reported for Colorado as found in fields, along roadsides, and in disturbed areas, but no specimens have been seen. May–June. Introduced.

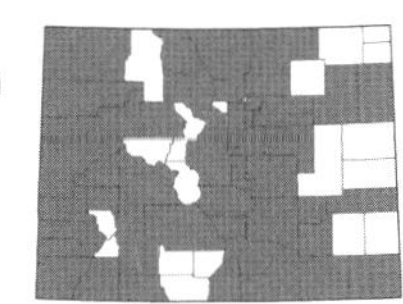

***Camelina microcarpa* Andrz. ex DC.**, LITTLEPOD FALSE FLAX. Plants (1) 2–8 (10) dm, densely hairy with simple hairs often mixing with branched ones below; *cauline leaves* lanceolate to oblong, with a sagittate base, usually entire, 1–5.5 (7) cm long, with simple hairs; *petals* pale yellow, 2.5–4 (6) mm long; *fruit* 3.5–5 (7) × 2–4 (5) mm. Common in meadows and grasslands, open forests, along roadsides, and in disturbed areas, 3800–8500 ft. April–June. E/W. Introduced.

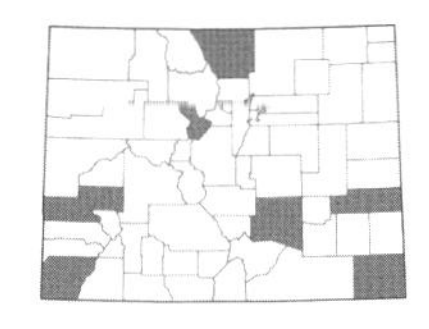

***Camelina rumelica* Velen.**, GRACEFUL FALSE FLAX. Plants 1.5–5 (6) dm, densely hirsute below with simple hairs and a few branched ones; *cauline leaves* lanceolate to oblong, with a sagittate base, usually entire, (1) 2–6 (9) cm long, with simple hairs; *petals* white or creamy-white, 5–9 mm long; *fruit* 5–7 × 3.5–5 mm. Locally common in disturbed areas, along roadsides, and in fields, 4000–8500 ft. May–June. E/W. Introduced.

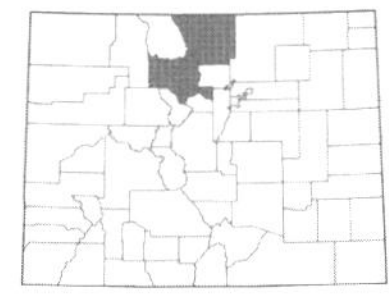

***Camelina sativa* (L.) Crantz**, GOLD-OF-PLEASURE. Plants 1.5–10 dm, glabrous or sparsely hairy with branched hairs below; *cauline leaves* lanceolate to oblong, 1–9 cm long, the base sagittate, margins usually entire, glabrescent or with sparse, forked hairs; *petals* yellow, 3.5–6 mm long; *fruit* broadly obovoid to pyriform, 7–9 (13) × 4–6 mm. Uncommon along roadsides, in fields, and in disturbed areas, 6000–8900 ft. May–June. E/W. Introduced.

CAPSELLA Medik. – CAPSELLA

Annual or biennial herbs; *leaves* with pinnately lobed margins or the upper leaves sometimes entire or denticulate; *flowers* white; *fruit* a silicle, triangular-obcordate, truncate or broadly notched at the apex, glabrous.

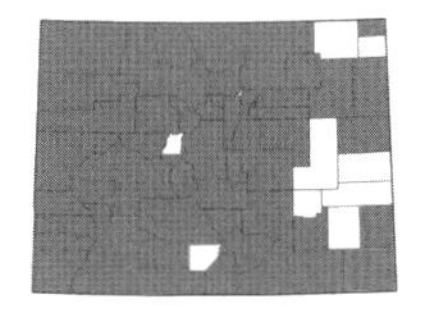

***Capsella bursa-pastoris* (L.) Medik.**, SHEPHERD'S PURSE. Plants (0.2) 1–5 (7) dm, the stellate hairs 3–5-rayed, simple hairs generally occurring at base; *basal leaves* oblong or oblanceolate, (0.5) 1.5–10 (15) cm long; *cauline leaves* oblong, lanceolate, or linear, 1–5.5 (8) cm long; *sepals* 1.5–2 mm long; *petals* (1.5) 2–4 (5) mm long; *fruit* (3) 4–9 (10) mm long. Weedy plants found in lawns, gardens, along roadsides, and disturbed areas, 4000–10,500 ft. April–June. E/W. Introduced.

CARDAMINE L. – BITTERCRESS

Annual, biennial, or perennial herbs, glabrous or pubescent with simple hairs; *leaves* basal or cauline, with entire to variously lobed or toothed margins; *flowers* white or purple; *fruit* a silique, linear, sessile, glabrous.

1a. Leaves simple and cordate with toothed or sinuate margins...***C. cordifolia***
1b. Leaves pinnately or ternately compound...2

2a. Perennials from slender rhizomes; fruiting pedicels (7) 10–20 mm long; stems glabrous or rarely sparsely hairy below...***C. breweri***
2b. Annuals or biennials lacking rhizomes; fruiting pedicels 2–10 (13) mm long; stems usually sparsely hairy below, rarely glabrous...3

3a. Plants with a basal rosette of leaves present at anthesis...***C. oligosperma***
3b. Plants lacking a basal rosette of leaves, these few and often withered by anthesis...***C. pensylvanica***

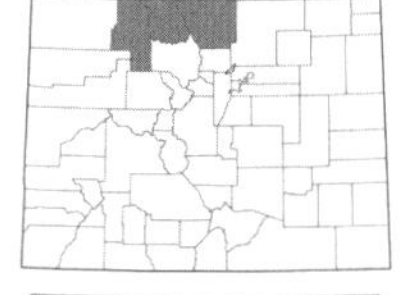

***Cardamine breweri* S. Watson**, BREWER'S BITTERCRESS. Perennials from slender rhizomes, 1–6 (7) dm, glabrous or rarely sparsely hairy below; *cauline leaves* pinnatifid; *fruiting pedicels* (7) 10–20 mm long; *petals* 3.5–6 (7) mm long; *fruit* 1.5–3.5 cm × 1–1.5 mm. Uncommon along streams and in moist forests, 6500–9500 ft. May–July. E/W.

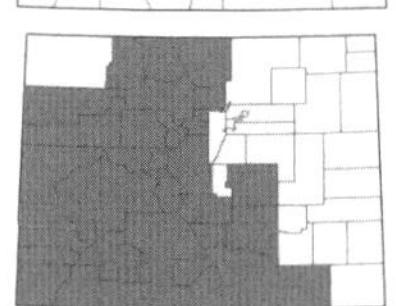

***Cardamine cordifolia* A. Gray**, HEARTLEAF BITTERCRESS. Perennials from rhizomes, 2–10 dm, glabrous to densely short-hairy below; *cauline leaves* simple, cordate to reniform, 2–9 cm long; *fruiting pedicels* (7) 10–20 mm; *petals* 7–12 mm long; *fruit* 2–4 cm × 1–2 mm. Common along streams, lake margins, and in moist meadows, 6000–13,500 ft. June–Aug. E/W. (Plate 20)

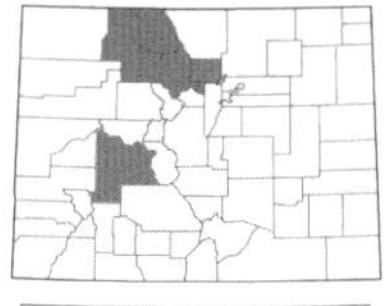

***Cardamine oligosperma* Nutt.**, LITTLE WESTERN BITTERCRESS. Annuals or biennials, lacking rhizomes, 0.5–4 dm, usually hairy throughout; *cauline leaves* pinnately compound; *fruiting pedicels* (2) 3–9 (12) mm; *petals* 2.5–3.5 mm long; *fruit* 1.3–3 cm × 1–1.7 mm. Uncommon in moist meadows and along streams, 7000–10,500 ft. June–Aug. E/W.

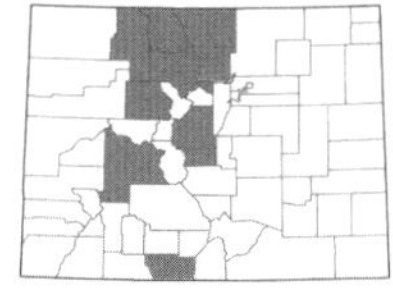

***Cardamine pensylvanica* Muhl. ex Willd.**, PENNSYLVANIA BITTERCRESS. Annuals or biennials, lacking rhizomes, 0.5–6 dm, sparsely hairy below; *cauline leaves* pinnately compound; *fruiting pedicels* (3) 4–13 mm; *petals* 2–4 mm long; *fruit* 1.5–3 cm × 0.8–1 mm. Uncommon in moist meadows and forests, and along streams and lake margins, 7500–10,500 ft. June–Aug. E/W.

CAULANTHUS S. Watson – WILD CABBAGE

Perennial (ours) herbs, glabrous or sparsely pubescent with simple hairs; *leaves* basal and cauline; *flowers* purple (ours) with white, purplish, or greenish sepals; *fruit* a silique, sessile, linear, glabrous or sparsely hairy.

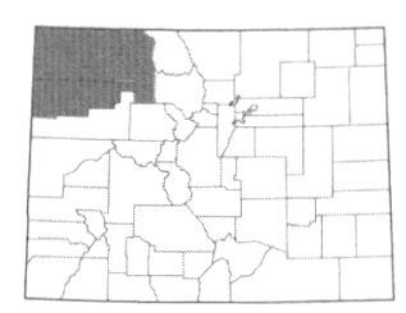

***Caulanthus crassicaulis* (Torr.) S. Watson**, THICK-STEM WILD CABBAGE. Plants 2–10 dm; *basal leaves* arranged in a rosette, obovate to oblanceolate, 1–12 cm long, entire to dentate or variously pinnately lobed; *cauline leaves* linear to narrowly oblanceolate, entire; *sepals* 7.5–14 mm long; *petals* 10–15 mm long; *fruit* erect to ascending, 4.5–14 cm long. Uncommon on clay or sandy soil, usually with pinyon-juniper or sagebrush, 5500–8000 ft. May–June. W.

CHORISPORA R. Br. ex DC. – CHORISPORA

Annual herbs; *leaves* basal, sometimes cauline, petiolate, with undulate-dentate or rarely entire margins; *flowers* usually purple; *fruit* a silique, linear, terete, splitting into 1-seeded segments at maturity, usually glandular.

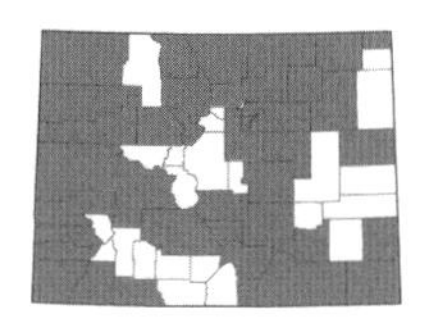

***Chorispora tenella* (Pall.) DC.**, BLUE MUSTARD. Plants (0.5) 1–4 (5.6) dm, papillose, occasionally pubescent; *basal leaves* oblanceolate to oblong, (1.5) 2.5–8 (13) cm long, glandular; *sepals* (3) 4–5 (6) mm long; *petals* 8–10 (12) cm long, long-clawed; *fruit* (1.4) 1.8–2.5 (3) cm long, spreading to ascending. Common weedy plants found on open slopes, along roadsides, in fields, vacant lots, and in other disturbed areas, 4000–9500 ft. April–June. E/W. Introduced. (Plate 20)

CONRINGIA Heist. ex Fabr. – HARE'S-EAR MUSTARD

Annual herbs, glabrous and usually glaucous; *leaves* basal and cauline, the margins usually entire; *flowers* white; *fruit* a silique, dehiscent, sessile, linear, 4-angled to cylindrical.

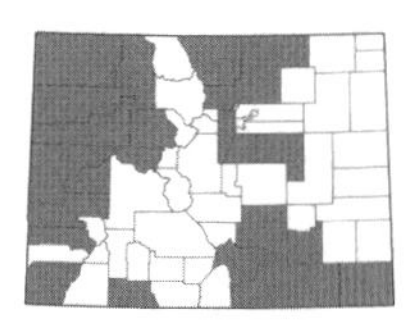

***Conringia orientalis* (L.) Dumort.**, HARE'S-EAR MUSTARD. Plants (1) 3–7 dm; *basal leaves* oblanceolate to obovate, 5–9 cm long, pale green; *cauline leaves* oblong to elliptic or lanceolate, (1) 3–10 (15) cm long; *sepals* 6–8 mm long; *petals* 7–12 mm long; *fruit* (5) 8–14 cm. Weedy plants found in disturbed areas, grasslands, and along roadsides, 4000–7000 ft. May–July. E/W. Introduced.

DESCURAINIA Webb & Bethel. – TANSY MUSTARD

Annual, biennial, or perennial herbs, glabrous or pubescent with branched hairs intermixed with simple hairs and/or short-stalked glands; *leaves* basal and cauline, petiolate or sessile, with entire or toothed margins; *flowers* yellow or whitish; *fruit* a silique (ours), linear or clavate, sessile, terete or nearly so, usually glabrous. (Detling, 1939)

1a. Racemes not elongating in fruit; short-lived perennials 1–15 cm tall, usually unbranched or rarely branched above; ovules 4–8 per ovary...***D. kenheilii***
1b. Racemes considerably elongated in fruit; annuals or biennials, (8) 13–62 (85) cm tall, usually branched above; ovules 4–48 per ovary...2

2a. Leaves (at least the lower) 2–3 times pinnately compound; fruit linear, (12) 15–30 mm long, torulose, the septum with a broad central longitudinal band appearing as 2 or 3 veins; plants lacking glandular hairs...***D. sophia***
2b. Leaves 1-pinnately compound, or sometimes 2 times pinnately compound (and then the plants usually glandular); fruit 3–20 mm long, the septum not veined or sometimes with a distinct midvein, otherwise various...3

3a. Fruiting pedicels erect to erect-ascending, straight, the fruit (especially the upper ones) appressed close to the raceme axis; septa usually with a distinct midvein...***D. incana***
3b. Fruiting pedicels spreading or ascending, the fruit not appressed close to the raceme axis; septa not veined...4

4a. Fruit with seeds biseriate (at least in the middle), usually clavate (widest towards the apex) and 1.2–2.2 mm wide; leaves usually densely hairy; plants usually glandular...***D. pinnata* ssp. *brachycarpa***
4b. Fruit with seeds uniseriate throughout, linear or fusiform (widest in the middle), 0.8–1.3 mm wide; leaves glabrous or hairy; plants glandular or not...5

5a. Fruit (2) 3–5 (6) mm long, fusiform (widest in the middle); ovules 4–12 per ovary; plants eglandular...***D. californica***
5b. Fruit 8–20 mm long, linear; ovules 14–32 per ovary; plants glandular or not...6

6a. Upper stem leaves with oblong to lanceolate lateral lobes and dentate to denticulate margins; plants glandular or not, but not canescent; fruiting pedicels 3–10 (12) mm long...***D. incisa* ssp. *incisa***
6b. Upper stem leaves with linear or filiform lateral lobes and entire margins; plants usually eglandular, often canescent; fruiting pedicels (8) 10–25 (30) mm long...7

7a. Fruit strongly curved inward, slightly torulose; plants usually canescent...***D. incisa* ssp. *paysonii***
7b. Fruit straight or slightly curved inward, not torulose; plants not canescent...***D. longepedicellata***

***Descurainia californica* (A. Gray) O.E. Schulz**, Sierra tansy mustard. Annuals or biennials, 1.5–10.5 dm; *stems* eglandular, usually hairy; *basal leaves* pinnate, 1.5–6 cm long, with entire to toothed or lobed margins; *fruiting pedicels* ascending to divaricate, 3–10 mm long; *petals* 1–2 mm long; *fruit* fusiform, (2) 3–5 (6) × 1–1.3 mm, not torulose. Found in open forests, on dry slopes, often with sagebrush or pinyon-juniper, 5500–10,000 ft. June–Aug. E/W.

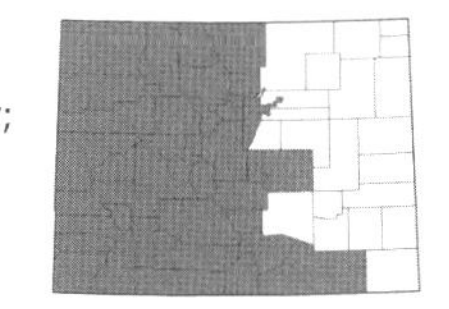

***Descurainia incana* (Bernh. ex Fisch. & C.A. Mey.) Dorn**, mountain tansy mustard. [*D. richardsonii* O.E. Schulz ssp. *procera* (Green) Detling]. Biennials, (1.5) 2–12 dm; *stems* usually eglandular, hairy; *basal leaves* pinnatifid, 1.5–10 (13) cm long; *fruiting pedicels* erect, 2–8 (11) mm long; *petals* 1–2 mm long; *fruit* erect, often appressed to the rachis, (4) 5–10 (15) × 0.7–1.5 mm, slightly torulose. Common in forests, meadows, and along streams, 5300–11,500 ft. May–Aug. E/W.

***Descurainia incisa* (Engelm. ex A. Gray) Britton**, mountain tansy mustard. Annuals, 1.5–8.5 (10) dm; *stems* glandular or eglandular, sparsely to densely hairy; *basal leaves* pinnate, 1.5–10 cm long; *fruiting pedicels* ascending to horizontal, (3) 5–25 mm long; *petals* 1.7–3 mm long; *fruit* erect to ascending, 8–20 × 0.9–1.3 mm, slightly torulose.

There are two subspecies of *D. incisa* in Colorado:

1a. Upper stem leaves with oblong to lanceolate lateral lobes and dentate to denticulate margins; plants glandular or not, but not canescent...**ssp. *incisa***. [*D. incana* ssp. *incisa* (Engelm.) Kartesz & Gandhi]. Common in meadows, along streams, in forests, on dry slopes, and in pinyon-juniper or sagebrush, 5500–10,000 ft. June–Aug. E/W.
1b. Upper stem leaves with linear or filiform lateral lobes and entire margins; plants usually eglandular and canescent...**ssp. *paysonii* (Detling) Rollins**. [*D. pinnata* ssp. *paysonii* Detling]. Found on dry slopes, in disturbed areas, and with pinyon-juniper, 5500–8000 ft. May–June. E/W.

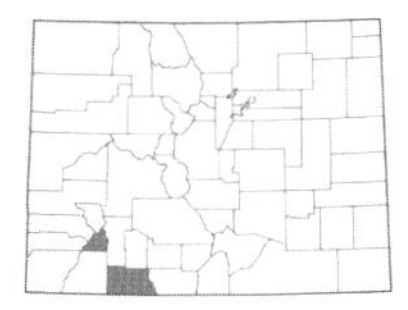

***Descurainia kenheilii* Al-Shehbaz**, Heil's tansy mustard. Perennials, 0.1–0.15 dm; *stems* eglandular, sparsely hairy; *basal leaves* pinnate, 1–2.5 cm long; *fruiting pedicels* erect to ascending, 1–1.5 mm long; *petals* 1–1.5 mm long; *fruit* erect, 6–10 × 1–1.3 mm, torulose. Rare on talus slopes and in the alpine, 11,500–12,500 ft. Aug-Sept. W. Endemic.

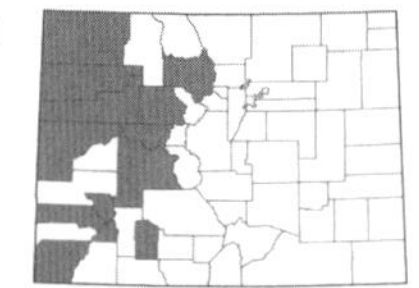

Descurainia longepedicellata **(Fournier) O.E. Schulz**, WESTERN TANSY MUSTARD. [*D. pinnata* ssp. *filipes* (A. Gray) Detling]. Annuals, (1.5) 3–6 (8.5) dm; *stems* usually eglandular, sparsely to moderately hairy; *basal leaves* pinnate, 1.5–7 cm long; *fruiting pedicels* divaricate to horizontal, (7) 10–15 (20) mm long; *petals* 1.7–2.5 cm long; *fruit* erect, (9) 12–17 × 0.8–1.1 mm long, not torulose. Found on dry slopes, often with pinyon-juniper or sagebrush, 5500–8000 ft. May–July. W.

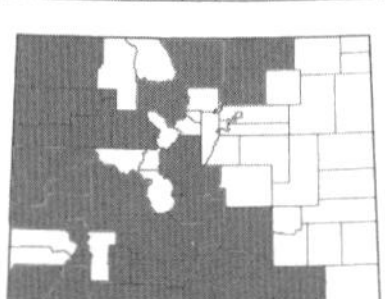

Descurainia pinnata **(Walter) Britton ssp. *brachycarpa* (Richardson) Detling**, PINNATE TANSY MUSTARD. [*D. ramosissima* Rollins]. Annuals, (0.8) 1–6 (9) dm; *stems* usually glandular; *basal leaves* 1–2-pinnate, 1–15 cm long; *fruiting pedicels* usually ascending or divaricate to horizontal, 4–18 (23) mm; *petals* 1–3 mm long; *fruit* erect to ascending, usually clavate, 4–15 (17) × 1.2–2.2 mm, not torulose. Found in grasslands, on dry slopes, in sagebrush or pinyon-juniper, or in disturbed areas, 3500–7000 ft. Mar.–July. E/W.

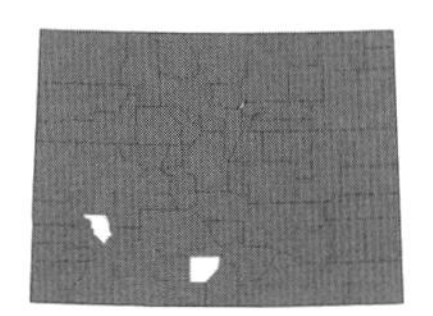

Descurainia sophia **(L.) Webb ex Prantl**, FLIXWEED. Annuals, (1) 2–8 (10) dm; *stems* eglandular, sparsely to densely hairy; *basal leaves* 2–3-pinnate, to 15 cm long; *fruiting pedicels* ascending to divaricate, (5) 8–15 (20) mm long; *petals* 2–3 mm long; *fruit* erect to divaricate-ascending, (12) 15–30 × 0.5–1 mm, torulose. Common in disturbed areas, fields, along roadsides and railroads, in grasslands and shrublands, 3500–10,500 ft. Mar.–June. E/W. Introduced.

DIMORPHOCARPA Rollins – SPECTACLE POD

Annual, biennial, or perennial herbs, with branched, stellate hairs mixed with minutely stalked, dentritic ones; *leaves* basal and cauline; *flowers* white or lavender; *fruit* a silicle, densely hairy.

Dimorphocarpa wislizeni **(Engelm.) Rollins**, TOURIST PLANT. Plants 1–4 (6) dm, gray-tomentose; *basal leaves* lanceolate to narrowly so, (2) 3–7 (10) cm long, pinnately lobed to coarsely dentate; *cauline leaves* linear to narrowly lanceolate; *sepals* (2.5) 3–4 mm long; *petals* 4–7 (9) mm long; *fruit* broader than long, 11–18 mm wide. Uncommon in sandstone canyons, 5500–6200 ft. Mar.–May. W.

DIPLOTAXIS DC. – WALL ROCKET

Annual, biennial, or perennial (ours) herbs; *leaves* mostly basal, with dentate to pinnatifid or lyrately lobed margins; *flowers* yellow to lavender; *fruit* a silique, linear, sessile, glabrous, the terminal segment beaklike.

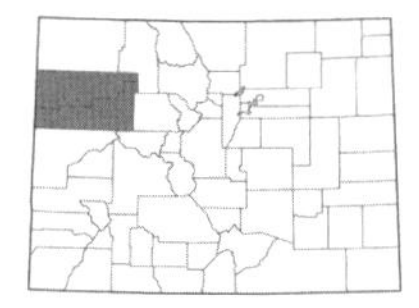

Diplotaxis muralis **(L.) DC.**, ANNUAL WALL ROCKET. Plants (0.5) 2–5 (6) dm long, generally scapose, glabrous or hirsute-hispid; *basal leaves* elliptic to obovate, 2–9 cm long; *sepals* 3–5.5 mm long; *petals* 5–8 (10) mm long, yellow; *fruit* (1.5) 2.5–4.5 cm long, torulose. Uncommon in disturbed areas, along roadsides, and on shale slopes, perhaps not persisting, 5500–7000 ft. May–Aug. W. Introduced.

DRABA L. – WHITLOW GRASS

Herbs, glabrous or pubescent with simple, forked, stellate, or dolabriform hairs, or more than one kind present; *leaves* usually basal and cauline, with entire or toothed margins; *flowers* yellow or white; *fruit* a silicle or silique. (Hitchcock, 1941; Mulligan, 1975; Payson, 1917; Price, 1980; Price & Rollins, 1991)

1a. Leaves glabrous above and below, or sometimes glabrous with ciliate margins; fruit glabrous...2
1b. Leaves hairy at least below (look towards the tips of the leaves for hairs), the margins ciliate or not; fruit glabrous or hairy...6

2a. Rachis hairy; stems usually with 2–9 leaves, hairy at least above; sepals 3–6 mm long; flowers yellow...3
2b. Rachis glabrous; stems usually lacking leaves or rarely 1 or 2 present, usually glabrous throughout; sepals 1.5–4 mm long; flowers white or yellow...4

3a. Basal leaves 2.5–8 (10) mm wide, oblanceolate or narrowly elliptic; racemes lacking bracts; style (0.4) 0.7–1.2 (1.5) mm long...***D. crassa***
3b. Basal leaves 0.3–2 (3) mm wide, linear to linear-oblanceolate; racemes bracteate; style 0.2–0.6 mm long...***D. graminea***

4a. Fruit ovate, 2.5–4 mm wide; basal leaves sessile, with stiffly ciliate margins; plants strongly perennial, caespitose, from a branched caudex...***D. globosa***
4b. Fruit elliptic, lanceolate, or oblong-ovate, 1.5–2.5 mm wide; basal leaves narrowed at the base into a petiole (can be obscure), with ciliate margins or not; plants various, with or without a branched caudex...5

5a. Flowers yellow, but can fade white; annuals or biennials with a simple caudex, rarely perennials with a few-branched caudex...***D. crassifolia***
5b. Flowers white; perennials with a branching caudex or occasionally approaching an annual or biennial habit with a simple caudex...***D. fladnizensis***

6a. Annuals from slender taproots (lacking a caudex); style 0.01–0.3 mm long; basal leaves pubescent below with 2–4-rayed and/or simple hairs...7
6b. Biennials or perennials with simple or branching caudices; style 0.01–2.5 mm long; basal leaves pubescent below with simple, dolabriform, pectinate, stellate, or 2–12-rayed hairs...12

7a. Rachis and pedicels conspicuously pubescent with simple and 2–4-rayed hairs...8
7b. Rachis and pedicels glabrous (or rarely sparsely hairy)...9

8a. Flowers white; rachis not conspicuously elongated in fruit, the fruit usually crowded in a terminal cluster; stem leaves absent or up to 6 but primarily grouped on the lower ⅓ of the stem...***D. cuneifolia***
8b. Flowers yellow, sometimes fading to white on dried specimens; rachis conspicuously elongated in fruit with evenly distributed fruit; stem leaves 3–10 (17), evenly distributed along the stem...***D. rectifructa***

9a. Leaves glabrous or sparsely pubescent with simple and 2-rayed hairs; stem leaves usually absent or rarely 1; plants 0.1–1.5 dm tall...***D. crassifolia***
9b. Leaves sparsely to densely pubescent with 2–4-rayed hairs on the lower surface (rarely with simple hairs along the midvein), and with simple and/or 2–3-rayed hairs above; stem leaves absent or not; plants 0.1–6 dm tall...10

10a. Flowers white; rachis not elongated in fruit, the fruit clustered in the upper ⅓ of the rachis; stem leaves usually absent or rarely 1–3; basal leaves entire...***D. reptans***
10b. Flowers yellow; rachis elongated in fruit with the fruit evenly distributed along the rachis; stem leaves 1–12 or rarely absent; basal leaves entire or toothed...11

11a. Stem leaves 4–13; fruiting pedicels 1.5–7 times longer than the fruit; basal leaves rosulate or not; sepals 0.7–1.5 mm long...***D. nemorosa***
11b. Stem leaves 1–3 (5) or rarely absent; fruiting pedicels equal to or shorter than the fruit; basal leaves rosulate; sepals 1.4–2 mm long...***D. albertina***

12a. Lower leaf surface of basal leaves with at least some simple hairs, these sometimes mixed with stalked or sessile 2- or 3-rayed hairs, the petiole margin ciliate (usually strongly so)...13
12b. Lower leaf surface of basal leaves lacking simple hairs, the hairs present 2–12-rayed, pectinate, dolabriform, or stellate, the margin of the petiole ciliate or not...19

13a. Lower leaf surface with simple, appressed, and sessile 2–4-rayed hairs; style 0.2–0.3 mm long; stem leaves 1–3; plants known from Summit County...***D. weberi***
13b. Lower leaf surface with simple and stalked 2–4-rayed hairs; style 0.01–2.5 mm long; stem leaves absent or 1–15; plants variously distributed...14

14a. Flowers white...***D. fladnizensis***
14b. Flowers yellow, sometimes fading to white on dried specimens...15

15a. Stems and rachis moderately to densely hairy; fruit hairy or glabrous...16
15b. Stems and rachis glabrous (or rarely sparsely hairy); fruit glabrous...17

16a. Petals 4–7.5 mm long; sepals 2–3 mm long; at least some fruit usually twisted (to 3 turns); stem leaves 2–15; plants 2–25 cm tall...***D. streptocarpa***
16b. Petals 3–5 mm long; sepals 1.5–2.5 mm long; fruit flat or slightly twisted; stem leaves 1–4; plants 0.8–6 cm tall...***D. grayana***

17a. Style 0.3–1 mm long; sepals pubescent with simple hairs; perennials with branched caudices...***D. exunguiculata***
17b. Style 0.01–0.12 mm long; sepals glabrous or sparsely pubescent with simple hairs; biennials or short-lived perennials with simple caudices, rarely perennials with few-branched caudices...18

18a. Leaves glabrous or sparsely pubescent with simple and 2-rayed hairs on the lower surface; stem leaves usually absent or rarely 1; plants 0.1–1.5 dm tall...***D. crassifolia***
18b. Leaves sparsely to densely pubescent with 2–4-rayed hairs on the lower surface (rarely with simple hairs along the midvein); stem leaves usually 1–3, rarely absent; plants 0.3–3 (4) dm tall...***D. albertina***

19a. Lower leaf surfaces or stems with at least some dolabriform hairs (2-rayed, sessile, attached in the middle, and appearing to lie flat on the surface) or with sessile, 4-rayed hairs in which the longest rays parallel the midvein; hairs on the sepals dolabriform or absent; flowers yellow; plants with 4–15 stem leaves...20
19b. Hairs on the leaves and stem various but not dolabriform or if sessile and 4-rayed then unlike the above; sepals without dolabriform hairs; flowers yellow or white; plants scapose or not...21

20a. Leaves exclusively with dolabriform hairs; style 0.8–1.2 mm long...***D. malpighiacea***
20b. Leaves with sessile, 4-rayed hairs or a mixture of 4-rayed hairs and dolabriform hairs (the lateral rays sometimes small and reduced to tiny spurs); style (0.5) 1–2.7 mm long...***D. spectabilis***

21a. Leaves pubescent below with at least some hairs pectinate or doubly stellate (with a central axis and several smaller rays coming off perpendicular to the central axis); plants scapose or with one stem leaf...22
21b. Leaves variously pubescent but lacking pectinate or doubly stellate hairs; plants scapose or not...24

22a. Leaves pubescent with sessile, pectinate or doubly pectinate hairs; fruit inflated at the base, ovate...***D. oligosperma***
22b. Leaves pubescent below with stalked (sometimes minutely so), pectinate hairs and sometimes also with 4–6-rayed hairs; fruit not inflated at the base, linear, lanceolate, or ovate...23

23a. Petals yellow, but sometimes fading to white, 4–6 mm long; style 0.3–0.9 mm long; fruit ovate to lanceolate...***D. incerta***
23b. Petals white, 2–3.5 mm long; style 0.1–0.3 mm long; fruit linear to narrowly lanceolate or oblong...***D. lonchocarpa***

24a. Leaves with 2 different kinds of hairs, that of the lower surface of stellate or 2–4-rayed hairs and that of the upper surface glabrous or simple and sometimes also with 2-rayed hairs; style 0.01–0.3 mm long; fruit glabrous or sparsely short-hairy with simple and 2-rayed hairs, usually not twisted; stem leaves absent or 1–3; flowers yellow...25
24b. Leaves with the same kind of hairs above and below; style 0.1–3.5 mm long; fruit usually pubescent with 2–7-rayed and sometimes simple hairs, or occasionally glabrous, twisted or flat; stem leaves 2–30, or absent in *D. ventosa*; flowers yellow or white...26

25a. Style 0.01–0.12 mm long (nearly obsolete); plants short-lived perennials from a simple caudex, 0.3–3 (4) dm tall; plants variously distributed...***D. albertina***
25b. Style 0.2–0.3 mm long; plants long-lived perennials from a well-developed caudex, 0.2–0.6 (1) dm; plants known from Summit County...***D. weberi***

26a. Fruit suborbicular to broadly ovate (nearly as wide as long), inflated at the base, flat; plants scapose; flowers yellow...***D. ventosa***
26b. Fruit flattened throughout, flat or twisted; plants with (1) 2–30 stem leaves; flowers yellow or white...27

27a. Flowers white...28
27b. Flowers yellow...30

28a. Style (0.7) 1–2.3 mm long; fruit conspicuously twisted with at least two turns, with simple and 2–4-rayed hairs; basal leaves with entire margins...***D. smithii***
28b. Style 0.1–0.6 (0.8) mm long; fruit flat or slightly twisted (with a single turn), with simple and 2–4-rayed hairs or the hairs all 3–7-rayed; basal leaves with entire or dentate margins...29

29a. Hairs on the fruit simple and 2–4-rayed; petals 4–6 mm long...***D. borealis***
29b. Hairs on the fruit all 3–7-rayed; petals 2.3–4 mm long...***D. cana***

30a. Stem leaves 1–5; perennials with much-branched caudices; plants 0.1–1.3 dm tall...***D. streptobrachia***
30b. Stem leaves 5–30; biennials or perennials with simple or few-branched caudices; plants (0.1) 1–5 dm tall...31

31a. Style 0.5–1.5 mm long; petals 3.5–5 mm long; stem leaves usually entire or sometimes with dentate margins; widespread...***D. aurea***
31b. Style 1.3–3.5 mm long; petals 5–7 mm long; stem leaves usually with dentate margins; found in the mountains southwest of Pueblo...***D. helleriana***

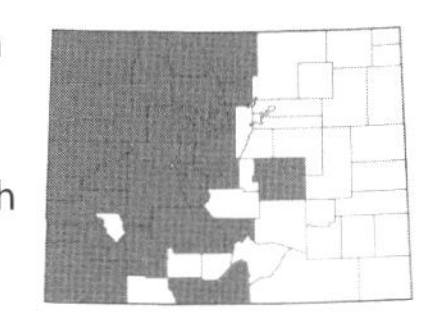

Draba albertina **Greene**, SLENDER DRABA. Annuals, biennials, or perennials, 0.3–3 (4) dm; *stems* with simple or few 2-rayed hairs below; *basal leaves* obovate to oblanceolate, (0.3) 1–3 cm × (1) 2–6 (9) mm, abaxially with 2–4-rayed hairs, adaxially with simple hairs and some 2-rayed hairs, with ciliate, entire to toothed margins; *cauline leaves* (0) 1–3 (5); *sepals* 1.4–2 mm long, glabrous or with sparse, simple hairs; *petals* 2–3.2 mm long, yellow; *fruit* (4) 6–12 (15) × 1–2 mm, glabrous; *style* 0.01–0.12 mm long. Common in meadows, grasslands, forests, and along streams, 7500–12,500 ft. May–July. E/W.

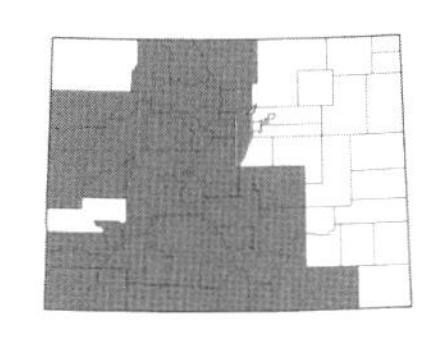

Draba aurea **Vahl ex Hornem.**, GOLDEN DRABA. Perennials, 0.5–3.5 (5) dm; *stems* hairy with simple and 3–6-rayed hairs throughout; *basal leaves* oblanceolate to obovate, (0.5) 1–4 (5) cm × (1) 2–7 (10) mm, with (2) 4–7-rayed hairs, the margins entire or toothed; *cauline leaves* 5–25; *sepals* 2–3 mm long, with simple and branched hairs; *petals* 3.5–5 mm long, yellow; *fruit* (6) 9–15 × 2–3.5 mm, hairy with simple and 2–4-rayed hairs; *style* 0.5–1.5 mm long. Common in the alpine, meadows, and forests, 8200–13,500 ft. June–Aug. E/W. (Plate 21)

Draba borealis **DC.**, BOREAL DRABA. Perennials, 0.5–3.5 (5.5) dm; *stems* hairy with simple and 2–8-rayed hairs; *basal leaves* ovate to oblanceolate, (0.5) 1–5 (6) cm × 3–10 (25) mm, with (2) 4–6-rayed hairs, sometimes appearing to 10-rayed, the margins toothed; *cauline leaves* (2) 3–7 (12); *sepals* 2–3 mm long, with simple hairs; *petals* 4–6 mm long, white; *fruit* (5) 7–12 × 2.5–4.5 mm, hairy with simple and 2–4-rayed hairs; *style* 0.2–0.8 mm long.

Draba borealis has been reported for Colorado but does not occur in the state.

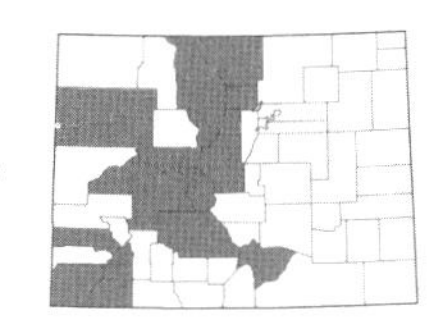

Draba cana **Rydb.**, CUSHION DRABA. [*Draba breweri* S. Watson var. *cana* (Rydb.) Rollins]. Perennials, (0.5) 1–3.5 dm; *stems* hairy with simple and 4–10-rayed hairs; *basal leaves* linear to oblanceolate, (0.5) 0.8–2 (3.5) cm × 1.5–4 (11) mm, with 4–12-rayed hairs, the margins entire to toothed, ciliate; *cauline leaves* 3–10 (17); *sepals* 1.5–2 mm long, with simple and few-rayed hairs; *petals* 2.3–4 mm long, white; *fruit* (5-) 6–11 × 1.5–2 (2.5) mm, with 3–7-rayed hairs; *style* 0.1–0.6 mm long. Common in the alpine and occasionally in subalpine forests, 10,000–14,000 ft. June–Aug. E/W.

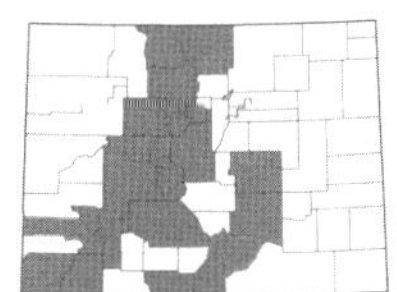

Draba crassa **Rydb.**, THICKLEAF DRABA. Perennials, 0.5–1.5 dm; *stems* glabrous below and hairy above with simple and 2-rayed hairs; *basal leaves* oblanceolate, 2–6 (7) cm × 2.5–8 (10) mm, glabrous, with entire, ciliate margins; *cauline leaves* 2–4 (6); racemes (4) 8–25-flowered, with simple and 2-rayed hairs; *sepals* 2–3.3 mm long, with simple and 2-rayed hairs; *petals* 3.5–6 mm long, yellow; *fruit* (7) 8–14 × 3–5 mm, glabrous; *style* (0.4) 0.7–1.2 (1.5) mm long. Common in the alpine and scree slopes throughout the mountains, 10,500–14,000 ft. July–Aug. E/W.

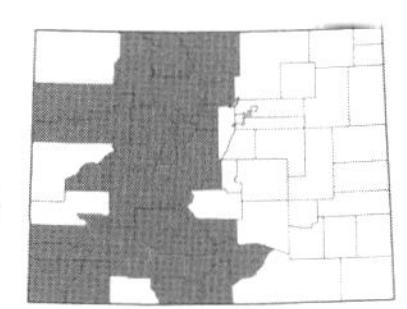

Draba crassifolia **Graham**, SNOWBED DRABA. Annuals or perennials, 0.1–1.5 dm; *stems* usually glabrous, rarely with simple hairs below; *basal leaves* oblanceolate, (0.2) 0.5–2.5 (3) cm × (1) 2–4 (6) mm, glabrous or with simple and 2-rayed hairs, the margins usually entire, sometimes ciliate; *cauline leaves* usually absent, rarely 1; *sepals* 1–2 mm long, glabrous; *petals* 1.5–3 mm long, yellow, fading white; *fruit* (3) 5–10 × 1.5–2.5 mm, glabrous; *style* 0.02–0.1 mm long. Common in the alpine and in subalpine spruce-fir forests and meadows, 9000–13,500 ft. June–Aug. E/W.

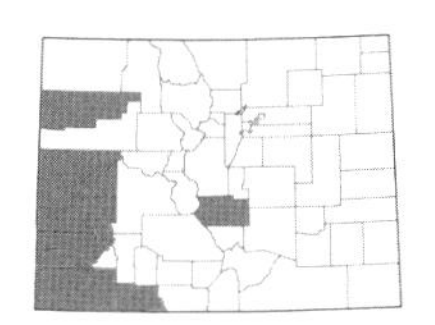

Draba cuneifolia **Nutt. ex Torr. & A. Gray**, WEDGELEAF DRABA. Annuals, 0.2–3.5 dm; *stems* with 3–4-rayed hairs, sometimes with simple ones; *basal leaves* oblanceolate, (0.4) 1–3.5 (5) cm × (2) 6–20 (28) mm, with 2–4-rayed hairs abaxially, and 2–4-rayed and simple hairs adaxially, margins dentate; *cauline leaves* 0–6; *sepals* 1.5–2.5 mm long, glabrous or with simple hairs; *petals* 2–5 mm long, white; *fruit* (3) 6–12 (16) × 1.7–2.7 (3) mm, glabrous or with simple or 2-rayed hairs; *style* 0.01–0.4 mm long. Found on dry, rocky slopes in pinyon-juniper, 4700–7000 ft. Mar.–May. E/W.

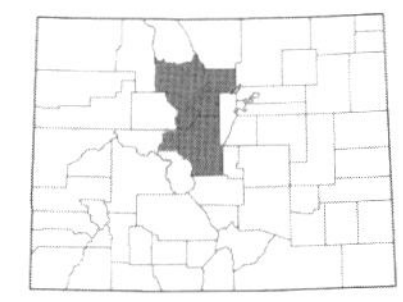

Draba exunguiculata **(O.E. Schulz) C.L. Hitchc.**, CLAWLESS DRABA. Perennials, 0.15–0.7 dm; *stems* glabrous or with simple hairs; *basal leaves* (0.5-)0.8–2(-2.5) cm × 1–3 mm, with simple and 2–3-rayed hairs below and glabrous or with simple hairs above, the margins entire, ciliate; *cauline leaves* 1–4; *sepals* 2–2.5 mm long; *petals* 2–3 mm long, yellow; *fruit* 5–13 × 1.5–3 mm, glabrous; *style* 0.3–1 mm long. Uncommon in the alpine, 11,500–13,000 ft. July–Aug. E/W. Endemic. (Plate 21)

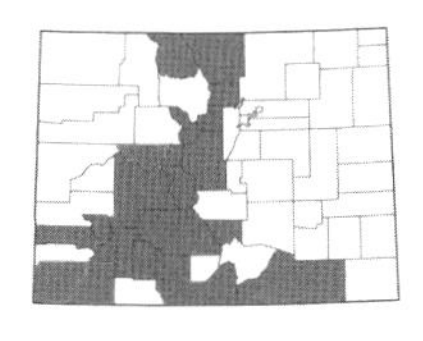

Draba fladnizensis **Wulfen**, AUSTRIAN DRABA. Perennials, 0.2–1.3 dm; *stems* glabrous; *basal leaves* (0.3) 0.4–1.2 (1.6) cm × 1–3 (4) mm, glabrous or with simple and/or 2-rayed hairs below, often glabrous above, the margins usually entire; *cauline leaves* absent or 1–2; *sepals* 1.2–2.2 mm long; *petals* 2–2.5 mm long, white; *fruit* 3–8 (9) × 1.5–2 mm, glabrous; *style* 0.05–0.3 mm long. Locally common in the alpine, on scree slopes, and in rocky crevices, 11,000–14,000 ft. June–Aug. E/W.

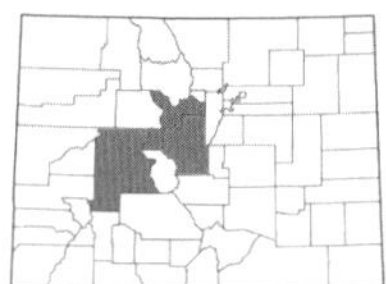

***Draba globosa* Payson**, beavertip draba. Perennials, 0.1–0.5 dm; *stems* glabrous; *basal leaves* oblanceolate to linear, (0.2) 0.3–0.8 cm × 0.5–1.6 (2) mm, glabrous, margins ciliate; *cauline leaves* absent; *sepals* 2–3 mm long, glabrous; *petals* 2.5–4 mm long, white or pale yellow; *fruit* ovate, 4.5–8 × 2.5–4 mm, glabrous; *style* 0.2–0.6 mm long. Rare in the alpine, 11,500–13,000 ft. July–Aug. E/W.

***Draba graminea* Greene**, Rocky Mountain draba. Perennials, 0.1–0.8 dm; *stems* usually with simple or 2-rayed hairs, rarely glabrous; *basal leaves* (0.5) 1–4 cm × 0.3–2 (3) mm, glabrous with a ciliate petiole; *cauline leaves* (1) 3–10 (12); *sepals* 1.5–2.5 mm long, glabrous or with simple and 2-rayed hairs; *petals* 3–5 mm long, yellow; *fruit* plane or slightly twisted, 5–11 × 2.5–5 mm, glabrous; *style* 0.2–0.7 mm long. Locally common in the alpine, 11,500–13,500 ft. July–Aug. E/W. Endemic.

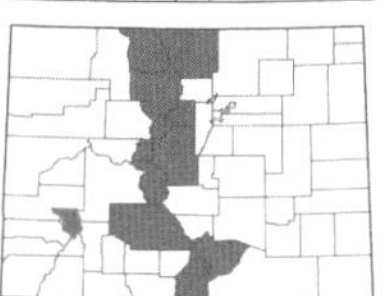

***Draba grayana* (Rydb.) C.L. Hitchc.**, A. Gray's draba. Perennials, 0.08–0.6 dm; *stems* with simple and 2–3-rayed hairs; *basal leaves* (0.4) 0.6–1.5 (2) cm × 1–2 mm, with simple and 2–3-rayed hairs below, glabrate or with simple hairs above; *cauline leaves* 1–4; *sepals* 1.5–2 mm long, with simple and 2–3-rayed hairs; *petals* 3–4.5 mm long, yellow; *fruit* 4–12 × 1.5–3 mm, glabrous or with simple hairs; *style* 0.4–1.2 mm long. Found in the alpine, 11,500–14,100 ft. July–Aug. E/W. Endemic.

***Draba helleriana* Greene**, Heller's draba. Biennials or perennials, (0.5) 1.5–5 dm; *stems* with simple and 3–5-rayed hairs; *basal leaves* oblanceolate, 0.9–4.1 (5.2) cm × 2–7 (10) mm, with cruciform and 3–5-rayed hairs below and cruciform and simple hairs above; *cauline leaves* (8) 12–30 (40); *sepals* 2.5–4 mm long, with simple and 2-rayed hairs; *petals* 5–7 mm long, yellow; *fruit* twisted or plane, 5–15 × 2–3.5 mm, with simple and 2-rayed hairs; *style* 1.3–3.5 mm long. Locally common in shrublands, spruce forests, and on alpine slopes, 7400–12,500 ft. June–Aug. E/W.

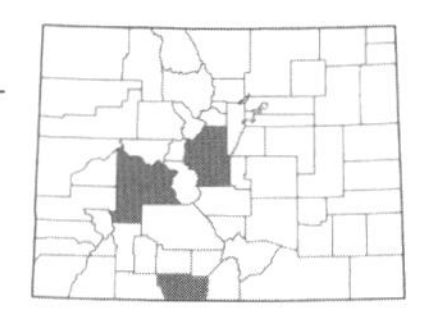

***Draba incerta* Payson**, Yellowstone draba. Perennials, 0.2–1.5 (2) dm; *stems* usually with simple and 2–5-rayed hairs, sometimes glabrous above; *basal leaves* linear to narrowly oblanceolate, 0.5–2 cm × 1.5–4 (5) mm, usually with pectinate hairs and sometimes with 4–6-rayed ones; *cauline leaves* usually absent, rarely 1; *sepals* 2.5–4 mm long, with simple and 2–3-rayed hairs; *petals* 4–6 mm long, yellow, fading white; *fruit* 5–10 mm, glabrous or with simple and 2-rayed hairs; *style* 0.3–0.9 mm long. Rare in the alpine on limestone, 12,000–13,000 ft. July–Aug. E/W. (Plate 21)

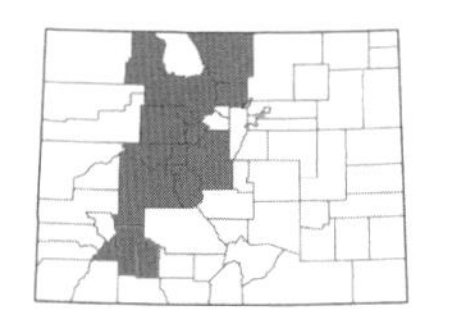

***Draba lonchocarpa* Rydb.**, lancepod draba. Perennials, 0.1–1 dm; *stems* glabrous or with 8–12-rayed hairs; *basal leaves* oblanceolate, 0.2–1.5 cm × 1–3 (5) mm, with stellate, 8–12-rayed hairs below and glabrous or with simple and branched hairs above; *cauline leaves* absent or sometimes 1 (4); *sepals* 1.5–2 mm long, with simple and 2–5-rayed hairs; *petals* 2–3.5 mm long, white; *fruit* 6–15 (18) × 1–2 (3) mm, glabrous or with simple and 2-rayed hairs; *style* 0.1–0.3 mm long. Uncommon in the alpine, often on limestone, 10,500–14,000 ft. July–Aug. E/W.

***Draba malpighiacea* Windham & Al-Shehbaz**, malpighiaceous draba. Perennials, 0.5–1.5 dm; *stems* with malpighiaceous hairs; *basal leaves* oblanceolate, 0.5–1.8 cm × 2–4 mm, with malpighiaceous hairs; *cauline leaves* 4–13; *sepals* 2.5–3 mm long, with simple and malpighiaceous hairs; *petals* 4–5 mm long, yellow; *fruit* 5–10 × 1.5–2.2 mm, glabrous; *style* 0.8–1.2 mm long. Uncommon in the alpine, forests, and on rocky outcrops, 9000–11,500 ft. June–Aug. W. Endemic.

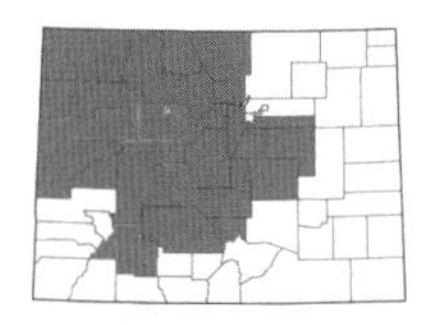

***Draba nemorosa* L.**, woodland draba. Annuals, 0.5–5 dm; *stems* with simple and 2–4-rayed hairs below; *basal leaves* oblanceolate, (0.4) 1–3.5 (5) cm × (2) 5–15 (20) mm, with 2–3-rayed and cruciform hairs; *cauline leaves* 4–13; *sepals* 0.7–1.5 mm long; *petals* 1.7–2.5 mm long, yellow; *fruit* oblong, (3) 5–8 (10) × 1.5–2.5 mm, glabrous or with simple hairs; *style* 0.01–0.1 mm long. Common in meadows, on rocky outcrops, and along streams, 5400–10,800 ft. April–June. E/W.

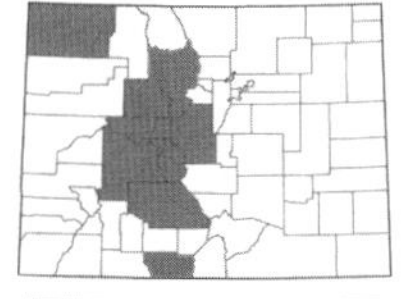

***Draba oligosperma* Hook.**, fewseed draba. [*D. juniperina* Dorn]. Perennials, 0.1–0.6 (1) dm; *stems* glabrous or with pectinate hairs; *basal leaves* 0.2–1.5 cm × 0.4–1.5 mm, with sessile, pectinate hairs; *cauline leaves* absent; *sepals* 1.5–3 mm long; *petals* 2.5–4 mm long, usually yellow; *fruit* ovoid, 3–7 × 2–4 mm, usually with simple and sometimes 2-rayed hairs; *style* 0.1–1 mm long. Found in the alpine, on dry slopes, and rocky outcrops, 6000–14,000 ft. May–July. E/W. (Plate 21)

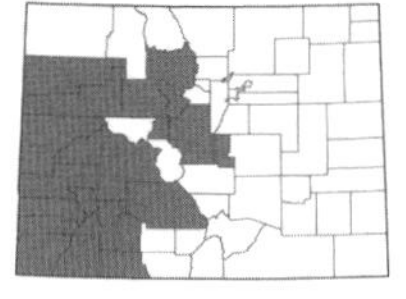

***Draba rectifructa* C.L. Hitchc.**, mountain draba. Annuals, 0.5–3.5 dm; *stems* with simple and 2–4-rayed hairs; *basal leaves* oblanceolate, 1–3 × 2–10 mm, with simple and 2–4-rayed hairs below and simple and 2-rayed hairs above; *cauline leaves* 3–15; *sepals* 1.3–2 mm long, with simple and 2-rayed hairs; *petals* 1.5–3 mm long, yellow; *fruit* lanceolate, 6–10 × 1.3–2 mm, with simple hairs; *style* 0.01–0.1 mm long. Found in forests, meadows, and on open slopes, 7300–9500 ft. June–July. E/W.

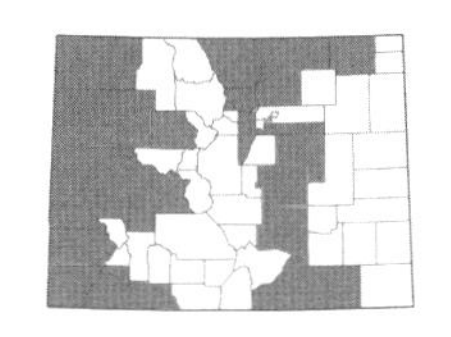

***Draba reptans* (Lam.) Fernald**, Carolina draba. Annuals, 0.1–1.5 dm; *stems* with 2 (3)-rayed hairs and sometimes simple hairs below; *basal leaves* spatulate to suborbicular, 0.5–2.5 cm × 1.5–10 mm, with 2–4-rayed hairs below and simple and 2-rayed hairs above; *cauline leaves* usually absent, rarely 1–3, similar to basal; *sepals* 1.5–2.3 mm long, with simple hairs; *petals* 2–4.5 mm long, white; *fruit* linear to linear-oblong, 7–18 × 1.2–2.3 mm, glabrous or with simple and few 2-rayed hairs; *style* 0.02–0.1 mm long. Found on dry slopes, rocky outcrops, and in grasslands, 3900–8000 ft. April–May. E/W. (Plate 21)

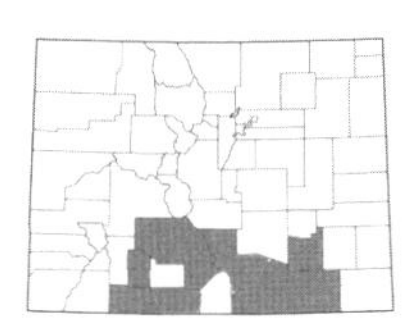

***Draba smithii* Gilg ex O.E. Schulz**, Smith's draba. Perennials, 0.5–2.5 dm; *stems* usually unbranched, with 4–12-rayed hairs throughout; *basal leaves* obovate, 0.5–1.5 (2.5) cm × (1) 2–5 (7) mm, with short-stalked, 5–12-rayed hairs; *cauline leaves* 3–8, with 2–5-rayed hairs; *sepals* 2–2.5 mm long; *petals* 4–6 mm long, white; *fruit* 5–9 × 2–3 mm, twisted, flattened, with simple and 2–4-rayed hairs; *style* (0.7) 1–2.3 mm long. Rare on talus and scree slopes, and in rocky crevices in the southern counties, 7700–12,000 ft. June–July. E/W.

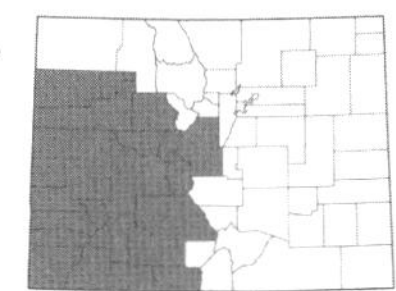

***Draba spectabilis* Greene**, showy draba. Perennials, 0.7–4 (5.3) dm; *stems* with simple, dolabriform, or sessile and 4-rayed hairs; *basal leaves* oblanceolate, 1–4.5 (6) cm × (3) 5–15 mm, with sessile, 4-rayed hairs, the lateral rays sometimes reduced to tiny spurs; *cauline leaves* 4–15; *sepals* 2.2–4 mm long, glabrous or with simple and malpighiaceous hairs; *petals* 4–6.5 mm long, yellow; *fruit* 6–13 × 2–3.5 mm, glabrous or with simple and 2-rayed hairs; *style* (0.5) 1–2.7 mm long. Common in forests, meadows, and the alpine, 8000–13,000 ft. June–Aug. E/W. (Plate 21)

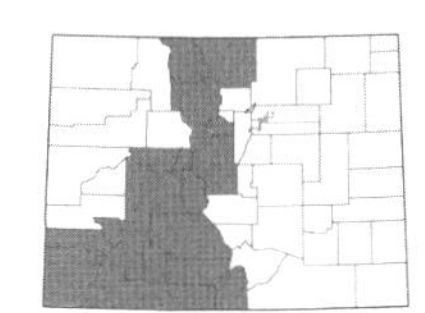

***Draba streptobrachia* R.A. Price**, alpine draba. Perennials, 0.1–1.3 dm; *stems* with 3–5-rayed, stellate, subsessile hairs; *basal leaves* oblanceolate, 0.4–4 cm × 1–5 mm, with short-stalked, 3–8-rayed hairs; *cauline leaves* 1–5; *sepals* 2–3 mm long, with simple and 2–4-rayed hairs; *petals* 3–5 mm long, yellow; *fruit* (3) 5–10 × 2–4 mm, glabrous or with simple and 2–4-rayed hairs; *style* 0.3–0.8 (1.3) mm long. Locally common in alpine meadows and on scree slopes and in rocky crevices, 10,800–13,000 ft. July–Aug. E/W.

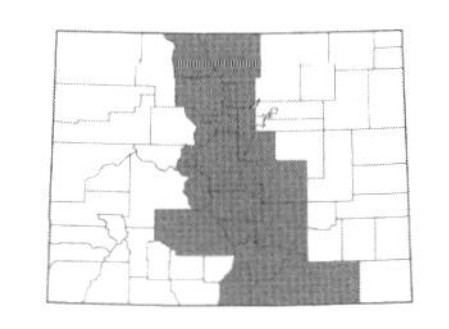

***Draba streptocarpa* A. Gray**, pretty draba. Perennials, 0.2–2.5 dm; *stems* glabrous or with simple and 2-rayed hairs; *basal leaves* oblanceolate, 0.5–3.8 cm × 1.5–6 mm, with simple and 2–4-rayed hairs below and simple hairs above; *cauline leaves* 2–15; *sepals* 2.5–3.5 mm long, with simple and 2-rayed hairs; *petals* 4–7.5 mm long, yellow; *fruit* usually strongly twisted, 5–16 × 2–3 mm, with simple hairs along the margin; *style* 0.8–2.5 mm long. Common in the alpine and forests, 7500–14,000 ft. June–Aug. E/W.

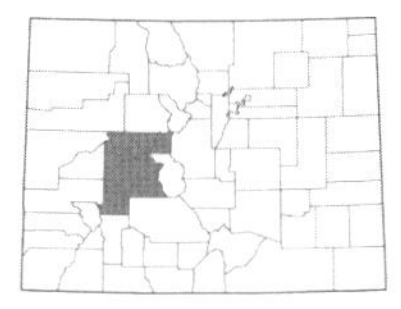

***Draba ventosa* A. Gray**, Wind River draba. Perennials, 0.1–0.5 dm; *stems* with 2–6-rayed hairs throughout; *basal leaves* oblanceolate to obovate, 0.4–1 cm × 1.5–4.5 mm, petiole base and margin not ciliate, with 2–6-rayed hairs; *cauline leaves* absent; *sepals* 2–2.5 mm long, with 2–6-rayed hairs; *petals* 3.5–5.5 mm long, yellow; *fruit* suborbicular to ovate, 4–8 × 3.5–5 mm, inflated basally, densely pubescent with 2–6-rayed hairs; *style* 0.5–1.5 mm long. Rare in the alpine and on scree slopes, 12,000–14,000 ft. July–Aug. E/W.

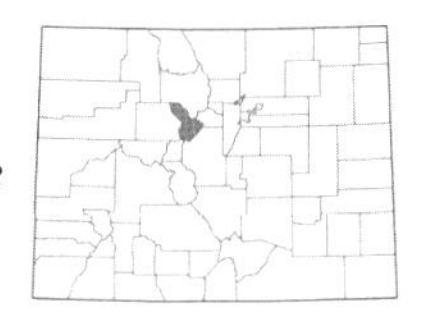

***Draba weberi* R.A. Price & Rollins**, Weber's draba. Perennials, 0.2–0.6 (1) dm; *stems* glabrous or with simple or 2–3-rayed hairs; *basal leaves* linear-oblanceolate, 0.5–1.5 × 0.8–1.7 mm, with simple and sessile or nearly so 2–4-rayed hairs below, glabrous or with simple hairs above; *cauline leaves* 1–3; *sepals* 1.5–2 mm long, with simple and 2-rayed hairs; *petals* 3–4 mm long, yellow; *fruit* ovate, 4–8 × 2–3 mm, glabrous; *style* 0.2–0.4 mm long. Rare in rocky crevices along streams, 11,000–12,000 ft. June–July. E/W. Endemic.

ERYSIMUM L. – wallflower

Annual, biennial, or perennial herbs, with sessile, appressed, 2-rayed or stellate hairs; *leaves* with entire or toothed margins; *flowers* yellow, orange, or pink-purple; *fruit* a silique, usually sessile, linear, more or less 4-angled, hairy.

1a. Petals 12–30 mm long; sepals 7–15 mm long, the lateral pair saccate basally; anthers 2.5–4 mm long, linear…2
1b. Petals 3–10 mm long; sepals 1.8–6 mm long, the lateral pair not or slightly saccate basally; anthers 0.5–2 mm long, oblong to linear…3

2a. Fruit widely spreading horizontally or divaricate; flowers yellow…***E. asperum***
2b. Fruit ascending-divaricate or ascending to erect; flowers yellow, yellow-orange, or purplish…***E. capitatum***

3a. Fruit densely hairy inside; sepals 1.8–3.3 mm long; petals 3–5.5 mm long…***E. cheiranthoides***
3b. Fruit glabrous or sparsely hairy inside; sepals 4–6 mm long; petals 6–10 mm long…4

4a. Fruiting pedicels slightly narrower than the fruit; fruit surface with 2–4-rayed hairs; biennials or short-lived perennials...***E. inconspicuum***
4b. Fruiting pedicels as wide as the fruit; fruit surface with mostly 2-rayed hairs and fewer 3-rayed hairs; annuals...***E. repandum***

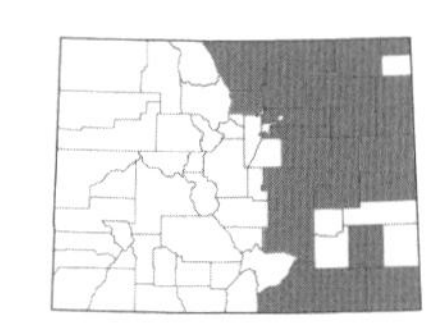

***Erysimum asperum* (Nutt.) DC.**, WESTERN WALLFLOWER. Biennials, 1–7 dm; *leaves* 2–10 cm × 0.5–2 mm, with 2–3-rayed hairs; *fruiting pedicels* horizontal, 5–20 mm; *sepals* 8–12 mm long; *petals* yellow, 13–23 mm long; *fruit* widely spreading, 5–14 cm × 1.2–3 mm, with 2-rayed hairs outside, glabrous inside; *style* 1–4 mm long. Found in sandy soil and grasslands of the eastern plains, 3500–5500 ft. May–June. E. (Plate 21)

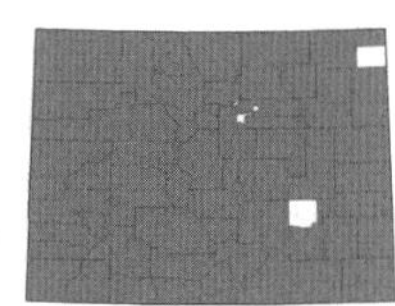

***Erysimum capitatum* (Douglas ex Hook.) Greene**, SAND DUNE WALLFLOWER. Biennials or perennials, 1–12 dm; *leaves* 2–20 cm × 3–15 (30) mm, with 2–4 (7)-rayed hairs; *fruiting pedicels* divaricate to ascending, 4–17 (25) mm; *sepals* 7–15 mm long; *petals* yellow, orange, or purplish-pink, 12–30 mm long; *fruit* divaricate or ascending, 3.5–15 cm × 1.3–3 mm, with 2–5-rayed hairs outside, glabrous inside; *style* 0.2–3 mm long. Common on the plains, in the foothills, mountain meadows, and the alpine; and in sand dunes and on mesas and dry hillsides on the western slope, 3500–13,500 ft. May–Aug. E/W. (Plate 21)

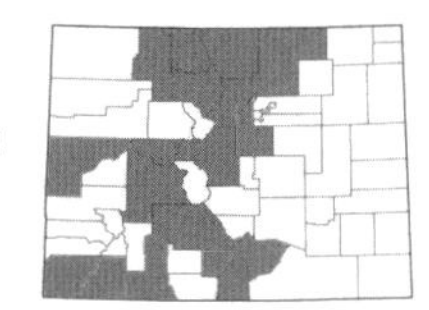

***Erysimum cheiranthoides* L.**, WORMSEED WALLFLOWER. Annuals, 0.8–10 (15) dm; *leaves* 2–7 (11) cm × (2) 5–12 (20) mm, mostly with 3–4-rayed hairs sometimes mixed with 5-rayed ones; *fruiting pedicels* divaricate or ascending, 5–16 mm; *sepals* 1.8–3.3 mm long; *petals* yellow, 3–5.5 mm long; *fruit* divaricate-ascending, 1–2.5 (4) cm × 1–1.3 mm, with 3–5-rayed hairs outside, densely hairy inside; *style* 0.5–1.5 mm long. Uncommon along roadsides, in meadows, and in moist forests, 5000–9000 ft. June–Aug. E/W. Introduced.

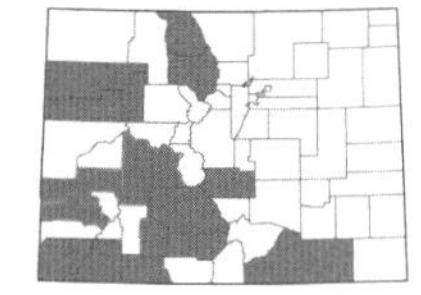

***Erysimum inconspicuum* (S. Watson) MacMill.**, SHY WALLFLOWER. Biennials or perennials, 1.5–7 dm; *leaves* 1.5–8 cm × 2–8 mm, with 2–3-rayed hairs; *fruiting pedicels* divaricate-ascending, 4–9 (15) mm; *sepals* 4–6 mm long; *petals* 6–10 mm long, yellow; *fruit* ascending, 3–6 cm × 1–1.8 mm, with 2–4-rayed hairs outside, glabrous inside; *style* 0.7–3 mm long. Uncommon in dry grasslands and sandy soil, 6000–8500 ft. June–Aug. E/W.

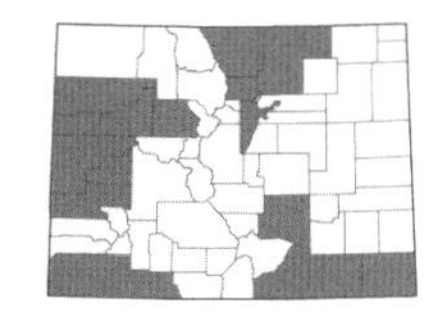

***Erysimum repandum* L.**, SPREADING WALLFLOWER. Annuals, (0.5) 1.5–5 (7) dm; *leaves* 1–10 cm × (2) 5–15 mm, with 2-rayed hairs mixed with fewer 3-rayed ones; *fruiting pedicels* divaricate, 2–6 mm; *sepals* 4–6 mm long; *petals* yellow, 6–8 mm long; *fruit* widely spreading to divaricate-ascending, 2–10 cm, with mostly 2-rayed hairs, sometimes glabrous, sometimes hairy inside; *style* 1–4 mm long. Found along roadsides, in moist disturbed areas, grasslands, and on dry hillsides, 4800–7500 ft. April–June. E/W. Introduced.

EUCLIDIUM W.T. Aiton – MUSTARD

Annual herbs, with stalked, 2-forked hairs mixed with fewer, simple or rarely 3-forked hairs; *leaves* with entire, toothed, or rarely pinnatifid margins; *flowers* white; *fruit* a silicle, indehiscent, sessile, ovoid, scabrous.

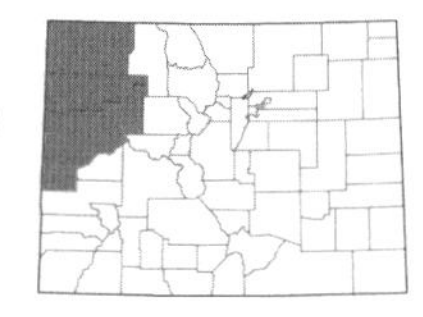

***Euclidium syriacum* (L.) R. Br.**, SYRIAN MUSTARD. Plants (0.4) 1–4 (4.5) dm; *leaves* primarily cauline, oblong to lance-oblong or elliptic, (1) 1.5–7 (9) cm long (distally smaller), *sepals* 0.6–0.9 mm long; *petals* 0.9–1.3 mm long; *fruit* 2–2.5 mm long, 1.5–2 mm wide. Uncommon weedy plants found along roadsides and in disturbed areas, 6000–8200 ft. May–June. W. Introduced.

EUTREMA R. Br. – EUTREMA

Annual or perennial herbs, glabrous and often glaucous; *leaves* with entire or palmately lobed margins, rarely pinnatifid; *flowers* white; *fruit* a silique, sessile or shortly stipitate, linear or oblong, terete or somewhat 4-angled, glabrous.

1a. Perennials; stem leaves shortly petiolate or sessile with acute bases...***E. penlandii***
1b. Annuals; stem leaves with auriculate or sessile, sagittate bases...***E. salsugineum***

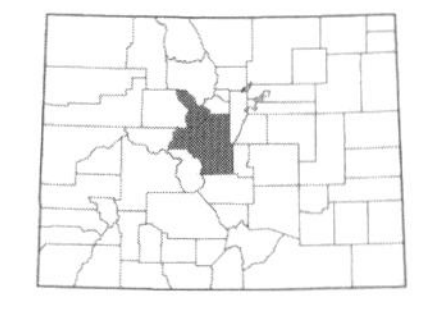

***Eutrema penlandii* Rollins**, PENLAND'S EUTREMA. Perennials, 0.5–3 (4.5) dm; *leaves* linear to ovate, (0.7) 1–4 cm × (1) 3–10 mm, margins entire, sessile or shortly petiolate; *sepals* 1.5–3 mm long; *petals* 3–5 mm long; *fruit* stipitate, 0.8–2.5 cm × 2–3 mm. Uncommon in the alpine, 11,800–14,000 ft. July–Aug. E/W. Endemic.

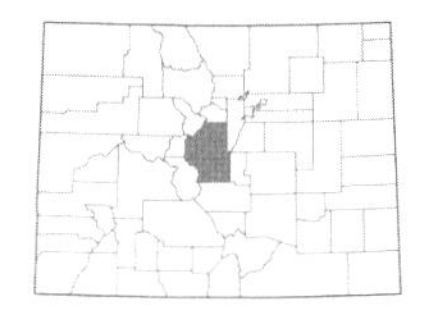

***Eutrema salsugineum* (Pall.) Al-Shehbaz & Warwick**, SALTWATER CRESS. [*Arabidopsis salsuginea* (Pall.) N. Busch; *Thellungiella salsuginea* (Pall.) O.E. Schulz]. Annuals, 0.6–4 dm; *leaves* obovate to oblong, 0.5–1.5 (2.5) cm × 2–5 mm, margins usually entire, auriculate or sagittate at the base; *sepals* 1–1.5 mm long; *petals* 2–3 mm long; *fruit* sessile, 0.7–2 cm × 0.7–1 mm. Rare on alkaline ground and salty marshes, 8400–9500 ft. May–Aug. E.

HESPERIDANTHUS (B.L. Rob.) Rydb. – PLAINSMUSTARD

Perennial herbs, glabrous; *leaves* linear to linear-lanceolate (ours), usually entire; *flowers* purple, claw differentiated from the blade; *fruit* a silique, linear, terete, glabrous.

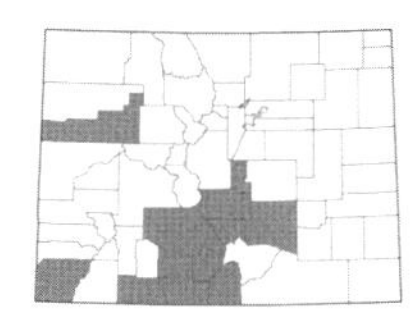

***Hesperidanthus linearifolius* (A. Gray) Rydb.**, SLIMLEAF PLAINSMUSTARD. [*Schoenocrambe linearifolia* (A. Gray) Rollins]. Plants (2.5) 3.5–10 (15) dm; *cauline leaves* (2.5) 3.5–12 (15) cm long, distally reduced; *sepals* (4) 4.5–7 mm long, purplish; *petals* (10) 12–16 (18) mm long, blade and claw distinctly differentiated; *fruit* (3.5) 4–9 (11) cm long. Found on dry hillsides and mesas, shale barrens, sometimes with sagebrush, 5500–9500 ft. June–Sept. E/W.

HESPERIS L. – DAME'S ROCKET

Biennial or perennial herbs, glabrous or pubescent with simple and branched hairs; *leaves* with entire or dentate margins; *flowers* purple or bluish-purple, rarely white; *fruit* a silique, sessile, linear, torulose, glabrous.

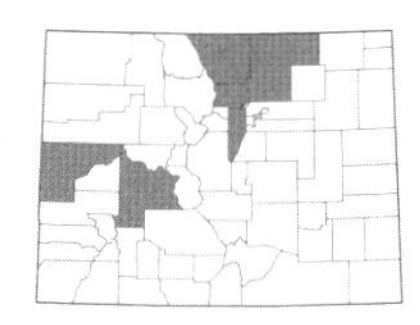

***Hesperis matronalis* L.**, DAME'S ROCKET. Plants 4–8 (11) dm, pubescent at least proximally; *leaves* narrowly oblong to lanceolate or broadly ovate, (2) 4–15 (20) cm long; *sepals* 5–8 mm long; *petals* (13) 15–20 (22) mm long; *fruit* (4) 6–10 (14) cm long. Weedy plants found along roadsides, in meadows, vacant lots, gardens, and disturbed areas, 4000–8500 ft. May–July. E/W. Introduced.

HORNUNGIA Rchb. – HUTCHINSIA

Annual herbs, glabrous or puberulent with minute forked hairs mixed with simple ones; *leaves* with entire, toothed, or pinnatifid margins; *flowers* white; *fruit* a silicle, sessile, oblong to elliptic or obovoid, keeled, glabrous.

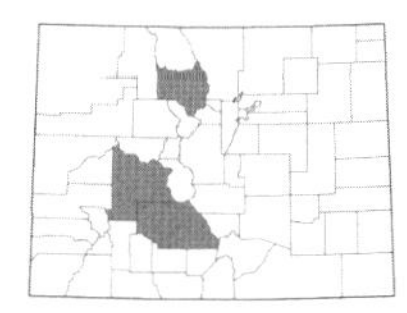

***Hornungia procumbens* (L.) Hayek**, PROSTRATE HUTCHINSIA. [*Hutchinsia procumbens* (L.) Desv.]. Plants (2) 5–22 (30) cm; *basal leaves* obovate to oblanceolate, (0.2) 1–1.2.5 (4) cm long; *cauline leaves* smaller; *sepals* 0.6–1.1 mm long; *petals* 0.6–1.2 mm long; *fruits* (0.2) 0.3–0.4 (0.5) cm long. Uncommon in wetlands, meadows, and on open slopes, 7000–8500 ft. April–June. E/W.

IBERIS L. – CANDYTUFT

Annual (ours) herbs, glabrous or hairy; *leaves* with entire to toothed or crenate margins; *flowers* white, pink, or purple; *fruit* a silicle, ovate to nearly orbicular, keeled, glabrous.

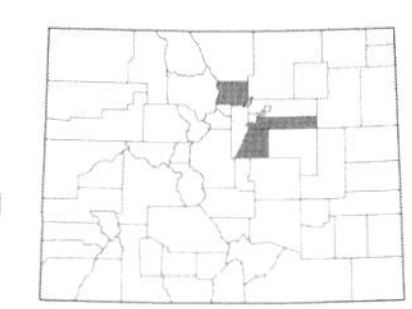

***Iberis amara* L.**, ANNUAL CANDYTUFT. Plants 1–4 dm; *leaves* cauline, spatulate or oblanceolate to oblong, 2–6 cm long; *sepals* 1.5–3 mm long; *abaxial petals* 5–8 mm long; *adaxial petals* 2–5 mm long; *fruit* 4–6 (7) mm long, apically notched. Common garden plants sometimes found escaping along roadsides or near homes, 5000–6500 ft. May–Sept. E. Introduced.

ISATIS L. – WOAD

Biennials, glabrous and often glaucous or hairy; *leaves* with entire or toothed margins; *flowers* yellow; *fruit* a silique or silicle, black or dark brown, with a conspicuous wing all around the margin or distally, glabrous or hairy.

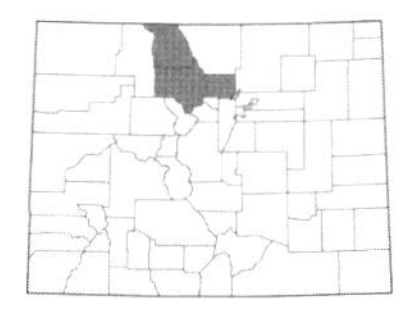

***Isatis tinctoria* L.**, DYER'S WOAD. Plants (3) 4–10 (15) dm; *basal leaves* oblanceolate or oblong, (2.5) 5–20 (25) cm long, entire to repand or dentate; *cauline leaves* lanceolate or oblong, entire; *sepals* 1.5–2.8 mm long; *petals* 2.5–4 mm long; *fruit* (0.9) 1.1–2 (2.7) cm long, 3–6 (10) mm wide. Weedy plants found along roadsides and in disturbed areas, 5500–8500 ft. April–July. E/W. Introduced.

LEPIDIUM L. – PEPPERGRASS

Annual, biennial, or perennial herbs, glabrous or pubescent with simple hairs; *leaves* with entire, toothed, or pinnately divided margins; *flowers* white, yellow, or greenish; *fruit* a silicle, flattened contrary to the septum or inflated and terete. (Hitchcock, 1936)

1a. Leaves of two kinds, the lower leaves pinnatifid into linear segments and the upper leaves deeply cordate-clasping the stem (nearly perfoliate); annuals...***L. perfoliatum***
1b. Leaves unlike the above; annuals, biennials, or perennials...2

2a. Stem leaves with an auriculate, clasping, or sagittate base...3
2b. Stem leaves lacking an auriculate, clasping, or sagittate base...7

3a. Plants not rhizomatous (annual or perennial); fruit apically broadly winged, with an apical notch 0.2–0.6 mm deep; inflorescence much-elongated in fruit with a hirsute rachis...4
3b. Plants rhizomatous; fruit not apically winged, lacking an apical notch; inflorescence usually rather flat-topped, the rachis glabrous to hairy...5

4a. Annuals from a taproot; style 0.2–0.5 (0.7) mm long, slightly exserted from or included in the apical notch...***L. campestre***
4b. Perennials with a branched caudex; style (0.6) 1–1.5 mm long, well-exserted beyond the apical notch...***L. heterophyllum***

5a. Fruit (and ovary) densely hairy...***L. appelianum***
5b. Fruit (and ovary) glabrous...6

6a. Mature fruit cordate at the base...***L. draba***
6b. Mature fruit nearly globose or obovoid, not cordate at the base...***L. chalepensis***

7a. Subshrubs or perennials (or rarely biennials from a taproot); stamens 6; petals 1.8–4 mm long; style 0.1–0.8 mm long, exserted beyond the apical notch (or sometimes the apical notch absent)...8
7b. Annuals from taproots, rarely perennials with woody caudices; stamens 2; petals usually absent or rudimentary, if present then 0.2–2 mm long; style obsolete or to 0.2 mm long and included in the apical notch...13

8a. Basal leaves not pinnately lobed or pinnatifid...9
8b. Basal leaves pinnately lobed or pinnatifid...10

9a. Mature fruit with an apical notch 0.1–0.2 mm deep; style 0.2–0.6 mm long; basal leaves oblanceolate to spatulate with a crenate or serrate-crenate margin, or 3-toothed at the apex, with mostly rounded tips, rosulate on sterile shoots...***L. crenatum***
9b. Mature fruit usually lacking an apical notch, or the notch to 0.1 mm deep; style 0.05–0.15 mm long; basal leaves elliptic to ovate or oblong, acute or somewhat rounded at the tips, not rosulate on sterile shoots...***L. latifolium***

10a. Pedicels minutely hairy throughout...***L. huberi***
10b. Pedicels minutely hairy adaxially or occasionally glabrous...11

11a. Stem leaves mostly pinnatifid or pinnately lobed...***L. montanum***
11b. Stem leaves entire or toothed, or rarely a few of the lower ones pinnately lobed...12

12a. Stem leaves linear, mostly 1–2 mm wide; plants 0.7–5 (6) dm tall...***L. alyssoides***
12b. Stem leaves lanceolate or oblanceolate, mostly 4–10 mm wide; plants (3.5) 4.5–16 dm tall...***L. eastwoodiae***

13a. Fruit hirsute to hispid (sometimes only on the margin); plants hirsute or hispid throughout...***L. lasiocarpum* ssp. *lasiocarpum***
13b. Fruit glabrous or minutely hairy along the margin; plants glabrous to minutely hairy throughout...14

14a. Fruit elliptic (longer than wide), the valves acute at the apex; perennial or biennial...15
14b. Fruit nearly orbicular (about the same length and width), the valves rounded at the apex; annual or biennial...16

15a. Stems several from the base...***L. paysonii***
15b. Stems simple from the base and branched above...***L. ramosissimum***

16a. Petals absent or rudimentary, less than 1 mm long...***L. densiflorum***
16b. Petals (when present) 1–2.5 mm long...***L. virginicum***

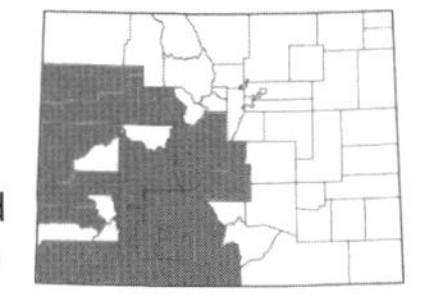

Lepidium alyssoides **A. Gray**, MESA PEPPERWORT. Perennials or subshrubs, 0.7–5 (6) dm; *basal leaves* pinnately lobed, 1.5–10 cm long; *cauline leaves* linear, entire, sessile; *sepals* 1–2 mm long; *petals* 2–3 mm long, white; *fruit* broadly ovate, 2–4 × 1.5–3 mm, apically winged, apical notch 0.1–0.4 mm, glabrous; *style* 0.2–0.6 mm long, exserted beyond apical notch. Found in pinyon-juniper and sagebrush, in meadows, on dry slopes, and along roadsides, 5500–9600 ft. May–Aug. E/W. (Plate 21)

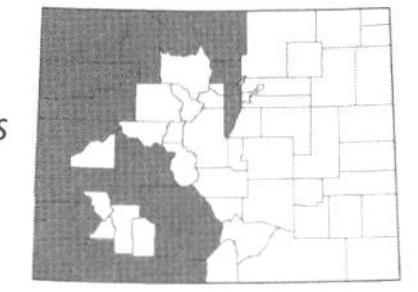

Lepidium appelianum **Al-Shehbaz**, HAIRY WHITETOP. [*Cardaria pubescens* (Meyer) Jarmol.]. Perennials, 1–3.5 (5) dm; *basal leaves* obovate to oblanceolate, 2–7 cm long, margins dentate; *cauline leaves* lanceolate to oblong, margins dentate or nearly entire, sagittate or auriculate; *sepals* 1.4–2 mm long; *petals* 2.5–4 mm long, white; *fruit* globose, (2) 3–4.5 (5) mm diam., not apically winged, apical notch absent, densely hairy; *style* 0.5–1.5 mm long. Found along roadsides, in fields, and on open flats, 4800–8500 ft. May–July. E/W. Introduced. (Plate 21)

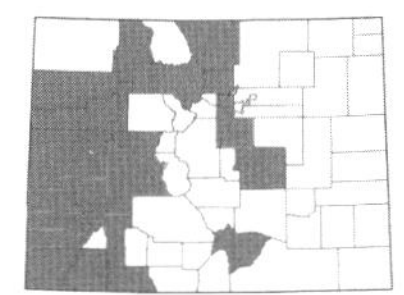

Lepidium campestre **(L.) W.T. Aiton**, FIELD PEPPERWEED. Annuals, 0.8–6 dm; *basal leaves* oblanceolate to oblong, 1–8 cm long, margins entire to pinnatifid; *cauline leaves* lanceolate to oblong, sagitttate or auriculate at the base, margins dentate or nearly entire; *sepals* 1–1.8 mm long; *petals* 1.5–3 mm long, white; *fruit* oblong to ovate, 4–6.5 × 3–5 mm, broadly winged apically, with an apical notch 0.2–0.6 mm deep, papillate; *style* 0.2–0.5 (0.7) mm long, exserted or included in the apical notch. Found in disturbed areas, meadows, grasslands, along roadsides, and on open slopes, 5200–9000 ft. May–July. E/W. Introduced. (Plate 21)

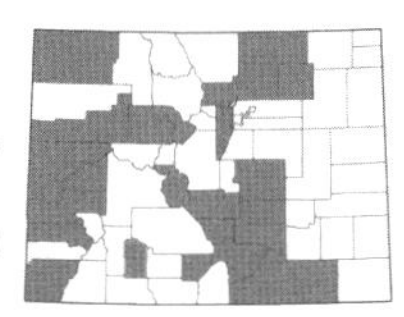

Lepidium chalepensis **L.**, LENSPOD WHITETOP. [*Cardaria chalapensis* (L.) Hand.-Mazz.]. Perennials, 1–7 dm; *basal leaves* obovate to ovate, 2–9 cm long, dentate; *cauline leaves* lanceolate to oblanceolate, 2–10 cm long, sagittate or auriculate; *sepals* 1.7–3 mm long; *petals* 3–5 mm long, white; *fruit* subglobose, 3.5–6 × 3.5–6 mm, not winged apically, apical notch absent, glabrous; *style* 1–2.3 mm long. Found in disturbed areas, roadsides, and on open slopes, 4300–10,000 ft. May–June. E/W. Introduced.

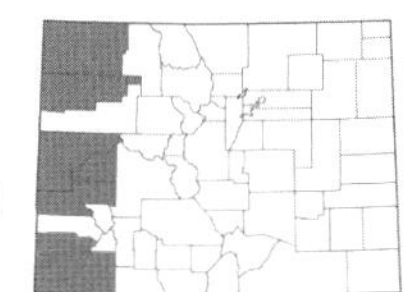

Lepidium crenatum **(Greene) Rydb.**, ALKALI PEPPERWEED. Perennials or subshrubs, 2–10 dm; *basal leaves* 2–10 cm long, oblanceolate to spatulate, crenate; *cauline leaves* 1–3.5 cm long, cuneate, entire; *sepals* 1–2 mm long; *petals* 2–3 mm long, white; *fruit* ovate, 2.5–4 × 2–3 mm, winged apically, with an apical notch 0.1–0.2 mm deep, glabrous; *style* 0.2–0.6 mm long, exserted beyond apical notch. Found in pinyon-juniper, sagebrush meadows, and on mesas, 5400–7500 ft. May–Sept. W. (Plate 21)

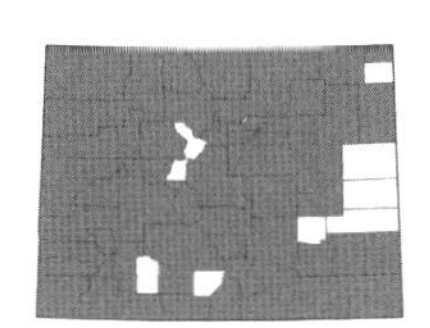

Lepidium densiflorum **Schrad.**, COMMON PEPPERWEED. Annuals or biennials, 2–6 dm; *basal leaves* oblanceolate to oblong, 2–10 cm long, serrate or pinnatifid; *cauline leaves* linear to oblanceolate, 1–7 cm long, cuneate, entire to irregularly toothed; *sepals* 0.5–1 mm long; *petals* 0.3–1 mm long or absent, white; *fruit* obovate, 2–3.5 × 1.5–3 mm, apically winged, with an apical notch 0.2–0.4 mm deep, glabrous or sparsely hairy; *style* 0.1–0.2 mm long, included in the apical notch. Common in disturbed areas, fields, grasslands, sagebrush, open slopes, and along roadsides, 3400–10,000 ft. May–Aug. E/W. Introduced. (Plate 22)

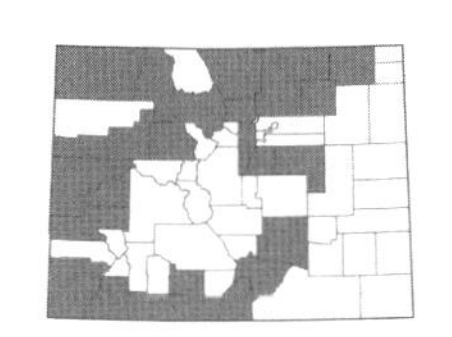

Lepidium draba **L.**, WHITETOP. [*Cardaria draba* (L.) Desv.]. Perennials, 1–8 dm; *basal leaves* obovate to spatulate, (1.5) 3–12 cm long, entire to toothed or sinuate; *cauline leaves* lanceolate to oblong or obovate, (1) 3–10 (15) cm long, auriculate or sagittate, entire to toothed; *sepals* 1.5–2.5 mm long; *petals* 2.5–4.5 mm long, white; *fruit* cordate, 2–4 × 3.5–5.5 mm, not winged apically, apical notch absent, glabrous; *style* (0.6) 1–2 mm long. Common along roadsides, in disturbed areas, fields, and on open slopes, 3700–8700 ft. May–June. E/W. Introduced. (Plate 22)

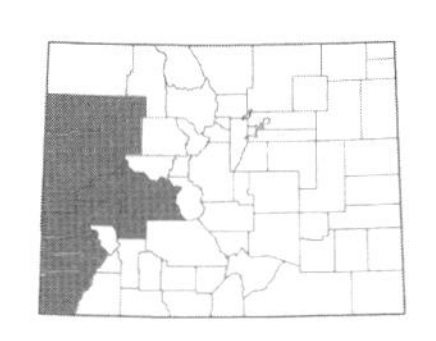

Lepidium eastwoodiae **Wooton**, EASTWOOD'S PEPPERWORT. Perennials or subshrubs, (3.5) 4.5–16 dm; *basal leaves* (2) 3–7 (9) cm long, pinnatifid; *cauline leaves* linear to oblanceolate, 3–7 cm long, usually entire, sessile; *sepals* 0.8–1.5 mm long; *petals* 2- 3.5 (4) mm long, white; *fruit* ovate, 2–4 × 1.8–3 mm, winged apically, with an apical notch 0.1–0.2 mm deep, glabrous; *style* 0.2–0.7 mm long, exserted beyond the apical notch. Uncommon in pinyon-juniper, sagebrush, or in canyons, 4800–8000 ft. July–Sept. W.

Lepidium heterophyllum **Benth.**, PURPLE-ANTHER FIELD PEPPERWEED. Perennials, 1–5 dm; *basal leaves* oblanceolate to oblong, 1–4.5 cm long, entire to toothed; *cauline leaves* oblong to lanceolate, 1–3.5 cm long, sagittate or auriculate, toothed; *sepals* 1.5–2.2 mm long; *petals* 1.8–2.8 mm long, white; *fruit* oblong to ovate, 4–5.5 × 3.5–4 mm, broadly winged apically, with an apical notch 0.2–0.3 mm deep; *style* (0.6) 1–1.5 mm long, exserted beyond apical notch. Reported for Colorado, but no specimens have been seen. May–June.

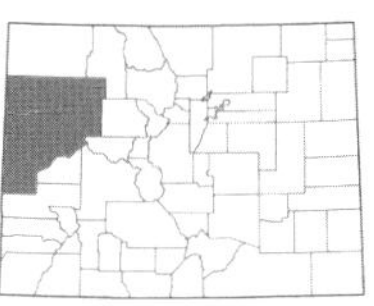

Lepidium huberi **S.L. Welsh & Goodrich**, HUBER'S PEPPERWEED. Perennials or subshrubs, 2–7 dm; *basal leaves* lanceolate to ovate, pinnatifid, 1–4.5 cm long; *cauline leaves* lanceolate, 1.8–3.5 cm long, sessile; *sepals* 1.2–2 mm long; *petals* 2–3.2 mm long, white; *fruit* obovate, 2–3.3 × 1.8–2.5 mm, apically winged, with an apical notch 0.1–0.2 mm deep, glabrous; *style* 0.2–0.8 mm long, exserted beyond the apical notch. Found in sagebrush and pinyon-juniper, 5800–7500 ft. June–Aug. W.

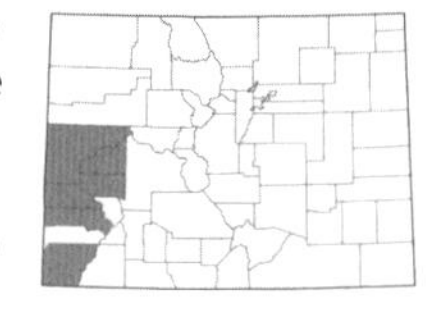

Lepidium lasiocarpum* Nutt. ssp. *lasiocarpum, SHAGGYFRUIT PEPPERWEED. Annuals, 0.2–3.5 dm, hairy; *basal leaves* spatulate to oblanceolate, pinnatifid, (0.5) 1–3.5 (5) cm long; *cauline leaves* lanceolate or oblanceolate, sessile, 1–3.5 (5) cm long; *sepals* 1–1.5 mm long; *petals* 0.3–2 mm long, white, or sometimes absent; *fruit* ovate, 2.5–4.5 mm long, apically winged, with an apical notch 0.2–0.7 mm deep, hispid; *style* obsolete or to 0.1 mm. Uncommon in pinyon-juniper, sagebrush, and dry, open places, 4800–5800 ft. April–June. W. (Plate 22)

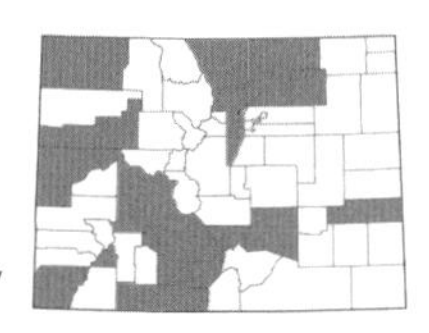

***Lepidium latifolium* L.**, BROAD-LEAVED PEPPERWEED. Perennials, (2) 3.5–15 dm; *basal leaves* elliptic to oblong, (2) 3.5–15 (25) cm long, margins entire or toothed; *cauline leaves* oblong to lanceolate, (1) 2–10 (12) cm long, sessile; *sepals* 1–1.5 mm long; *petals* 1.8–2.5 mm long, white; *fruit* elliptic to ovate, 1.6–2.7 × 1.3–1.8 mm, not winged apically, with an apical notch 0.1 mm deep or absent, glabrous or sparsely hairy; *style* 0.05–0.15 mm long, exserted if present. Found in disturbed areas, along ditches and roadsides, and in grasslands, 3900–7600 ft. June–Oct. E/W. Introduced.

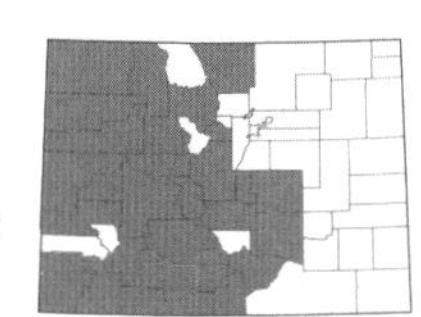

***Lepidium montanum* Nutt.**, MOUNTAIN PEPPERWEED. Annuals, biennials, or perennials, 0.5–5 (7) dm; *basal leaves* 1–2-pinnatifid, (0.8) 1.5–5 (6) cm long; *cauline leaves* similar to basal or undivided and linear, sessile; *sepals* 1.2–2 mm long; *petals* 2.2–4 mm long, white; *fruit* ovate to suborbicular, 2–4.5 (5) × (1.5) 1.8–4 mm, apically winged, with an apical notch 0.1–0.3 mm deep, glabrous; *style* 0.2–0.9 mm long, usually exserted beyond the apical notch. Common in pinyon-juniper and sagebrush, and on dry slopes, 4500–9800 ft. May–Sept. E/W. (Plate 22)

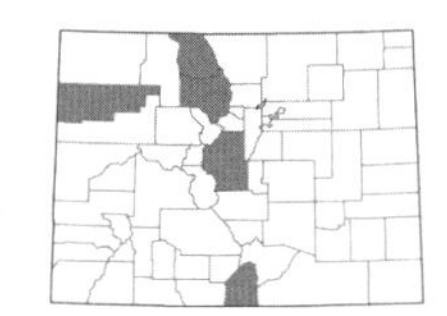

***Lepidium paysonii* Rollins**, PAYSON'S PEPPERWEED. Perennials, 0.5–2.5 dm, densely minutely hairy; *basal leaves* oblanceolate, 1–3.5 cm long, with toothed margins or rarely with 1–2 lateral lobes; *cauline leaves* linear to narrowly oblanceolate, sessile, entire or toothed distally; *sepals* 0.6–1 mm long; *petals* absent or rudimentary and 0.3–0.6 mm long; *fruit* elliptic, 2.4–3 × 1.6–2 mm, apically winged, with an apical notch 0.2–0.3 mm deep, minutely hairy; *style* obsolete or to 0.1 mm long. Found in sagebrush, pine, and dry, open places, 8000–10,000 ft. June–Aug. E/W.

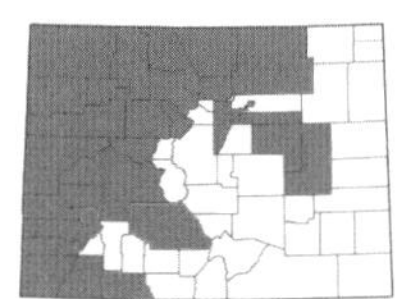

***Lepidium perfoliatum* L.**, CLASPING PEPPERWEED. Annuals or biennials, 1–4.5 (5.6) dm; *basal leaves* 2–3-pinnatifid, (1) 3–10 (15) cm long; *cauline leaves* ovate to suborbicular, deeply amplexicaul and nearly perfoliate at the base, margins entire; *sepals* 0.8–1.3 mm long; *petals* 1–2 mm long, pale yellow; *fruit* orbicular to obovate, 3–5 × 3–4 mm, apically winged, with an apical notch 0.1–0.3 mm deep; *style* 0.1–0.4 mm long, equal to or slightly exserted beyond the apical notch. Common along streams, and in disturbed areas, and meadows, 4500–9000 ft. April–June. E/W. Introduced. (Plate 22)

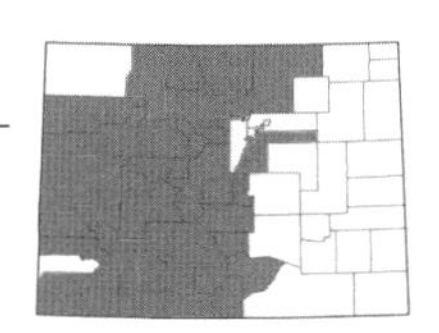

***Lepidium ramosissimum* A. Nelson**, MANY-BRANCHED PEPPERWEED. Annuals or biennials, 0.8–5.5 (7.5) dm; *basal leaves* oblanceolate to pinnatifid, 2–5 cm; *cauline leaves* linear to oblanceolate, (0.6) 1.2–4.5 (6) cm, sessile, margins toothed to entire; *sepals* 0.6–1 mm long; *petals* absent or rudimentary and 0.2–0.8 mm long, white; *fruit* elliptic, 2.2–3.2 × 1.7–2 mm, apically winged, with an apical notch 0.1–0.4 mm deep; *style* usually obsolete. Common in disturbed areas, open fields, pine forests, and along roadsides, 6500–10,500 ft. June–Aug. E/W.

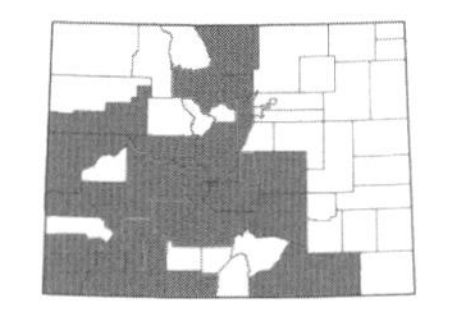

***Lepidium virginicum* L.**, VIRGINIA PEPPERWEED. Annuals, (0.6) 1.5–5.5 (7) dm; *basal leaves* spatulate to obovate, (1) 2.5–10 (15) cm long, margins pinnatifid to dentate; *cauline leaves* linear to oblanceolate, 1–6 cm long, sessile, with entire or toothed margins; *sepals* 0.5–1 mm long; *petals* 1–2.5 mm long, rarely absent; *fruit* orbicular, 2.5–4 mm diam., apically winged, with an apical notch 0.2–0.5 mm deep; *style* 0.1–0.2 mm long, included in the apical notch. Found in disturbed areas, forests, along roadsides, and on dry slopes, 5000–9500 ft. May–Aug. E/W. Introduced.

LOBULARIA Desv. – LOBULARIA

Herbs with appressed, dolabriform hairs; *leaves* cauline with entire margins; *flowers* white, pink, or purple; *fruit* a silicle, sessile or shortly stipitate, orbicular, obovate, or elliptic, hairy.

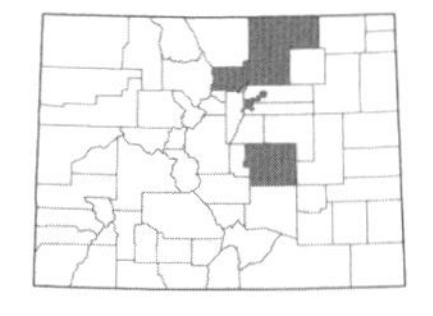

***Lobularia maritima* (L.) Desv.**, SWEET ALYSSUM. [*Koniga maritima* (L.) R. Br.]. Annuals or perennials, 0.5–4 dm; *leaves* linear to narrowly lanceolate, 1–2.5 (4.2) cm long; *sepals* 1.4–1.7 (2.4) mm long, often purplish; *petals* (1.9) 2.3–3.1 mm long; *fruit* (1.9) 2.3–2.7 (4.2) mm long. Garden plants sometimes escaping and found along roadsides, 5000–6600 ft. July–Sept. E. Introduced.

LUNARIA L. – HONESTY

Annual, biennial, or perennial herbs, glabrous or nearly so; *leaves* with coarsely toothed margins; *flowers* purple; *fruit* a silicle, long-stipitate, suborbicular, glabrous.

***Lunaria annua* L.**, MONEY PLANT. Annuals or biennials, (3) 4–10 (12) dm; *basal leaves* broadly cordate to cordate ovate, (1.5) 3–12 (18) cm long, *cauline leaves* similar, distally reduced and sessile; *sepals* (5) 6–9 (10) mm long; *petals* (15) 17–25 (30) mm long; *fruit* 3–5 × 2–3 (3.5) cm. Uncommon along South Boulder Creek in Boulder, often cultivated but rarely escaping and persisting, 5500–6000 ft. May–June. E. Introduced.

NASTURTIUM R. Br. – NASTURTIUM

Perennial herbs, often aquatic, usually glabrous; *leaves* cauline, with simple and entire margins or pinnately compound; *flowers* white or rarely pink; *fruit* a silique, sessile, usually linear, terete, glabrous.

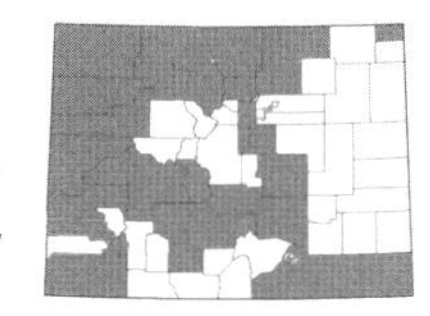

***Nasturtium officinale* W.T. Aiton**, WATERCRESS. [*Rorippa nasturtium-aquaticum* (L.) Hayek]. Plants glabrous or sparsely pubescent; *stems* 1–11 (20) dm long; *leaves* pinnately compound, 2–15 (20) cm long; *leaflets* 3–9 (13), (0.4) 1–4 (5) cm long; *sepals* 2–3.5 mm long; *petals* 2.8–4.5 (6) mm long; *fruit* (0.6) 1–1.8 (2.5) cm long. Common in slow-moving streams, ditches, and along lake margins, 3800–9500 ft. May–Aug. E/W. (Plate 22)

NOCCAEA Moench – NOCCAEA

Biennial or perennial herbs, glabrous and often glaucous; *leaves* with entire to toothed margins; *flowers* white or sometimes pinkish-purple; *fruit* a silicle, winged or rarely not, obovate to obcordate.

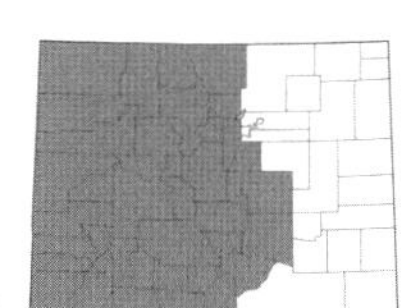

***Noccaea fendleri* (A. Gray) Holub ssp. *glauca* (A. Nelson) Al-Shehbaz & M. Koch**, ALPINE PENNYCRESS. [*Thlaspi montanum* L. var. *fendleri* (A. Gray) P.K. Holmgren]. Plants (0.1) 0.5–3.2 (4.5) dm; *basal leaves* ovate to oblong, 4–9 (15) mm wide; *cauline leaves* (4) 7–16 (21); *petals* (3.4) 4–7 (8.5) mm long; *fruit* (2.5) 5–8 (12) mm long. Common on rocky slopes, in the alpine, and in forest openings, 5200–14,300 ft. April–Aug. E/W. (Plate 22)

PENNELLIA Nieuwl. – MOCK THELYPODY

Perennial herbs, glabrous or pubescent with simple, stalked, forked, or branched hairs; *leaves* with entire or toothed margins; *flowers* white or purplish; *fruit* a silique, sessile or shortly stipitate, linear, glabrous or hairy.

1a. Racemes one-sided; fruiting pedicels reflexed; fruit 5–10 cm long; petals (4) 5–7 (12) mm long...***P. longifolia***
1b. Racemes not one-sided; fruiting pedicels straight and erect or ascending; fruit (1.5) 2.2–5.8 cm long; petals 1.5–4 (5) mm long...***P. micrantha***

***Pennellia longifolia* (Benth.) Rollins**, LONGLEAF MOCK THELYPODY. Plants 4–15 dm, with simple and 2-rayed hairs mixed with dendritic hairs; *cauline leaves* linear, 3–9 cm × 0.8–5 mm, glabrous; *sepals* 4–9 mm long; *petals* (4) 5–7 (12) mm long, purplish at the tips; *fruit* pendant, straight, 5–10 cm × 1–1.3 mm. Rare in forests, 8000–8500 ft. July–Sept. W.

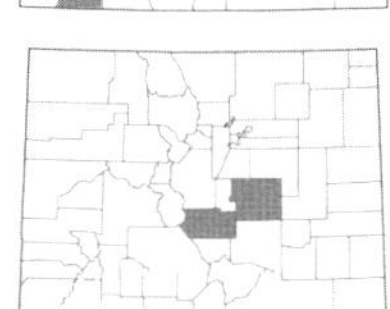

***Pennellia micrantha* (A. Gray) Nieuwl.**, MOUNTAIN MOCK THELYPODY. Plants 2–12 dm, the hairs mostly dendritic, mixed with simple and 2-rayed hairs; *cauline leaves* oblanceolate, 3–10 cm × 1–20 mm, glabrous; *sepals* 3–4 mm long; *petals* 1.5–4 (5) mm long; *fruit* erect to ascending, straight, (1.5) 2.2–5.8 cm long. Uncommon in forests and meadows, 6000–10,000 ft. July–Sept. E.

PHYSARIA Rchb. – BLADDERPOD

Annual, biennial, or perennial herbs, usually pubescent with sessile, stellate hairs, rarely with simple hairs; *leaves* usually with entire margins, sometimes pinnatifid or dentate; *flowers* usually yellow, rarely white or purple; *fruit* a silicle, sessile, inflated or not, often terete, hairy or glabrous. (Al-Shehbaz & O'Kane, 2002; Grady & O'Kane, 2007; Payson, 1921; Rollins, 1939, 1981)

1a. Fruit (and ovary) glabrous throughout...2
1b. Fruit (and ovary) pubescent with stellate hairs...4

2a. Basal leaves linear to narrowly elliptic (to 4 mm wide)...***P. fendleri***
2b. Basal leaves suborbicular, elliptic, rhombic, or ovate (over 4 mm wide)...3

3a. Fruiting pedicels sigmoid or slightly curved; inflorescence elongate...***P. pruinosa***
3b. Fruiting pedicels mostly straight; inflorescence compact, subumbellate or densely corymbiform...***P. ovalifolia***

4a. Plants strictly alpine, found in tundra above 11,000 ft in elevation...5
4b. Plants not present in the alpine, found below 11,000 ft in elevation...6

5a. Fruit 4–11 mm long, the apex conspicuously notched; fruiting pedicels 7–11 mm long...***P. alpina***
5b. Fruit 3–5 mm long, the apex rounded or only slightly notched; fruiting pedicels 1.8–3.5 mm long...***P. scrotiformis***

6a. Inflorescence not exceeding the leaves; plants to 4 cm tall, forming small, dense, button-like mounds; plants from Rio Blanco Co. and the northeastern plains; leaves all linear-oblanceolate; fruit apex entire...7
6b. Inflorescence usually obviously exceeding the leaves, or if included then the plants otherwise unlike the above; plants sometimes mound-forming, but the mounds usually not small and button-like...8

7a. Hairs with the rays free nearly to the base; ovules 8–12 per ovary; plants of the eastern slope...***P. reediana***
7b. Hairs with the rays fused in the lower ⅓; ovules 4 per ovary; plants of the western slope (Rio Blanco County)...***P. congesta***

8a. Mature fruiting pedicels mostly recurved (such that the mature fruit is pendant)...9
8b. Mature fruiting pedicels S-curved (sigmoid), ascending, or sometimes horizontal...12

9a. Basal leaves linear to narrowly lanceolate, mostly 2 mm or less wide and usually with involute margins; simple hairs often present just at the very base of the outer leaves...***P. ludoviciana***
9b. Basal leaves spatulate, ovate, or elliptic, mostly over 4 mm wide, flat or folded but not involute; simple hairs absent at the base of leaves...10

10a. Mature fruit with a deeply notched apex (didymous); basal leaves usually with a dentate margin (with 2–6 teeth); petals 9–11 mm long...***P. floribunda***
10b. Mature fruit with an entire apex; basal leaves usually with entire margins or sometimes shallowly dentate with 1–2 teeth; petals 4–9 mm long...11

11a. Inner stems erect...***P. arenosa***
11b. Inner stems usually prostrate or decumbent...***P. parviflora***

12a. Plants of the eastern slope...13
12b. Plants of the western slope...19

13a. Fruit entire at the apex, not or only slightly inflated; ovules 4–20 per ovary...14
13b. Fruit notched at the apex (didymous), usually inflated; ovules 4 per ovary...16

14a. Basal leaves suborbicular, obovate, or elliptic, mostly over 4 mm wide (rarely narrowly elliptic and 2–3 mm wide), usually with a well-defined petiole; stem leaves usually secund; ovules (8) 12–24 per ovary...***P. montana***
14b. Basal leaves linear or narrowly spatulate to narrowly elliptic, 1–4 mm wide, usually with an ill-defined petiole; stem leaves usually not secund; ovules 4–8 per ovary...15

15a. Basal leaves 2–4 mm wide, to 7 (10) cm long, flat or folded but not involute; outer stems often trailing and decumbent...***P. calcicola***
15b. Basal leaves less than 2 mm wide, all less than 3 cm long, with involute margins; stems all erect...***P. parvula***

16a. Plants of the northeastern plains; fruit broader at the apex and appearing flared...***P. brassicoides***
16b. Plants of the foothills, absent from the northeastern plains; fruit not conspicuously broader at the apex or appearing flared from the base...17

17a. At least some basal leaf margins deeply and broadly incised, rarely almost entire...***P. vitulifera***
17b. Basal leaf margins shallowly dentate or entire...18

18a. Basal leaf bases gradually tapering to an ill-defined petiole; fruit 2–8 mm wide...***P. bellii***
18b. Basal leaf bases abruptly narrowed with a well-defined petiole; fruit (4) 8–20 mm wide... ***P. acutifolia***

19a. Plants with several stems from the base and a much-branched caudex, forming dense mats 0.5–3 dm across; ovules 2 per ovary; known from San Miguel County...***P. pulvinata***
19b. Plants unlike the above, not forming dense mats; ovules 4–20 per ovary; variously distributed...20

20a. Fruit entire at the apex, not or only slightly inflated; ovules 4–20 per ovary...21
20b. Fruit notched at the apex (didymous), usually inflated; ovules 4 per ovary...25

21a. Basal leaves linear to narrowly spatulate, mostly 2 mm or less wide with involute margins...***P. parvula***
21b. Basal leaves rhombic, obovate, linear-oblanceolate, narrowly elliptic, or nearly orbicular, mostly over 2 mm wide, flat or folded but not involute...22

22a. Petals creamy-white (sometimes drying yellow); fruit with spreading hairs; Montrose and Ouray counties...***P. vicina***
22b. Petals yellow; fruit usually with appressed hairs; not restricted to Montrose and Ouray counties...23

23a. Fruit flattened at the apex; ovules 4–6 per ovary; plants dense and forming mounds, usually with a branching caudex...***P. subumbellata***
23b. Fruit not flattened at the apex; ovules (8) 12–20 per ovary; plants not forming mounds, lacking a branching caudex...24

24a. Fruit densely hairy, the hairs usually completely covering the surface; basal leaves suborbicular, obovate, or elliptic...***P. montana***
24b. Fruit sparsely hairy, the hairs not completely covering the surface; basal leaves oblanceolate to elliptic...***P. rectipes***

25a. Hairs with the rays fused nearly to the tip; fruit obcordate...***P. obcordata***
25b. Hairs with the rays distinct or fused at the base; fruit not obcordate...26

26a. Basal leaves nearly orbicular to obovate with a rounded apex, 2–8 cm long; fruit (4) 6–15 mm long and (4) 8–20 mm wide, the wall thin and papery...***P. acutifolia***
26b. Basal leaves oblanceolate, narrowly elliptic, or triangular, 2–4 cm long with an acute apex; fruit 2–5 (8) mm long and 4–8 (10) mm wide, the wall rather thick and leathery... ***P. rollinsii***

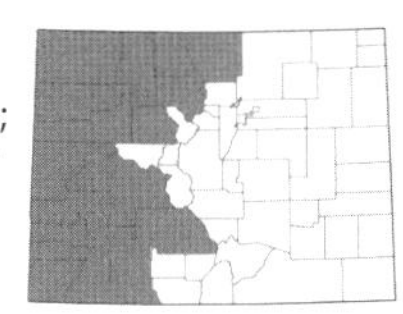

Physaria acutifolia **Rydb.**, Rydberg's twinpod. Plants 0.4–2 dm; *basal leaves* obovate to orbicular, 2–9 cm long, the margins usually entire; *cauline leaves* spatulate, 1–3 cm long, with entire margins; *fruiting pedicels* 6–12 mm long, divaricate, sigmoid to nearly straight; *sepals* 4–7.5 mm long; *petals* 6–11 mm long; *fruit* didymous, (4) 6–15 × (4) 8–20 mm, with deep apical and basal sinuses, with appressed hairs; *ovules* (2) 4 per ovary. Common on shale or clay slopes, dry hillsides, and in sagebrush and pinyon-juniper, 4500–10,000 ft. May–June. E/W. (Plate 22)

Immature *P. floribunda* can key to *P. acutifolia*. However, the basal leaves of *P. floribunda* are usually toothed to pinnatifid while those of *P. acutifolia* are usually entire.

Physaria alpina **Rollins**, alpine twinpod. Plants 0.3–0.8 dm; *basal leaves* obovate to ovate, 1.5–3.5 cm long, the margins entire or obscurely toothed; *cauline leaves* oblanceolate to spatulate, entire; *fruiting pedicels* 7–11 mm long, spreading to ascending, straight or slightly curved; *sepals* 7–9 mm long; *petals* 10–15 mm long; *fruit* didymous, 4–11 × 10–13 mm, densely hairy; *ovules* 4 per ovary. Found in rocky alpine, 11,400–13,500 ft. June–July. E/W. Endemic.

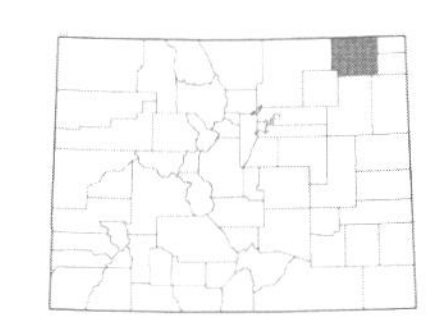

Physaria arenosa **(Richardson) O'Kane & Al-Shehbaz**, Great Plains bladderpod. [*Lesquerella arenosa* (Richardson) Rydb.]. Plants 0.5–2 (3) dm; *basal leaves* oblanceolate, 1.5–5 (7) cm long, with entire or shallowly toothed margins; *cauline leaves* linear to elliptic, 0.5–3 cm long, usually entire; *fruiting pedicels* usually sharply recurved, 5–15 (20) mm long; *sepals* 4–7 mm long; *petals* 6–9 mm long; *fruit* subglobose to ellipsoid, 3.5–6.5 mm long, densely hairy; *ovules* (4) 8–10 per ovary. Uncommon on clay ridges and in open places in grasslands, 3800–4000 ft. May–June. E.

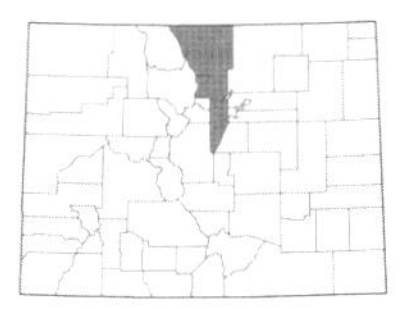

Physaria bellii **G.A. Mulligan**, Front Range twinpod. Plants 0.5–1.3 dm; *basal leaves* broadly obovate, 1.5–7.5 cm long, with shallowly toothed margins; *cauline leaves* oblanceolate to obovate, 1–2.5 cm long, with entire margins; *fruiting pedicels* spreading to ascending, 7–12 mm long; *sepals* 4–8 mm long; *petals* 9–13 mm long; *fruit* didymous, 4–9 × 2–8 mm, hairy; *ovules* 4 per ovary. Locally common on shale slopes or limestone soil, 5100–5800 ft. Mar.–May. E. Endemic. (Plate 22)

Physaria brassicoides **Rydb.**, mustard twinpod. Plants 0.2–1.5 dm; *basal leaves* orbicular to obovate, 2–6 cm, with repand or rarely entire margins; *cauline leaves* oblanceolate to spatulate, 1–2 cm long, entire; *fruiting pedicels* spreading, 5–12 mm long; *sepals* 6–8 mm long; *petals* 9–12 mm long; *fruit* didymous, (6) 10–20 × 10–23 mm, hairy; *ovules* 4 per ovary. Uncommon on sandy hillsides and bluffs, 5500–5700 ft. May–June. E.

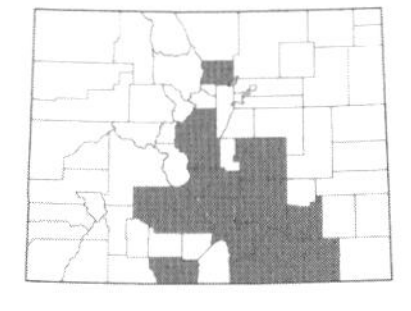

Physaria calcicola **(Rollins) O'Kane & Al-Shehbaz**, Rocky Mountain bladderpod. [*Lesquerella calcicola* Rollins]. Plants 1–3 dm; *basal leaves* linear, 2–7 (10) cm long, the margins entire, repand, or shallowly toothed; *cauline leaves* linear to spatulate, 1–3 (4.5) cm long, sometimes involute; *fruiting pedicels* spreading, sigmoid, 8–15 mm long; *sepals* 4.5–7 mm long; *petals* 7–10 mm long; *fruit* oblong to ovate, 5–9 mm long, sparsely hairy; *ovules* 4–8 per ovary. Found on shale, limestone, or gypsum slopes, often in pinyon-juniper woodlands, 4800–6900 ft. May–June. E.

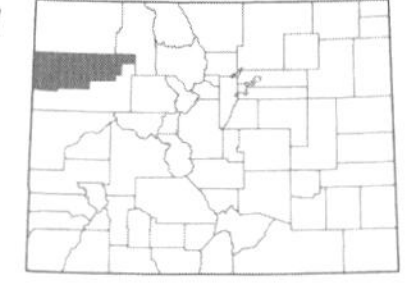

***Physaria congesta* (Rollins) O'Kane & Al-Shehbaz**, Dudley Bluffs bladderpod. [*Lesquerella congesta* Rollins]. Plants to 0.15 dm; *basal leaves* linear-oblanceolate, 0.6–1.5 cm long, entire; *cauline leaves* similar to basal; *fruiting pedicels* erect, straight or slightly curved, 3–6 mm long; *sepals* 3–4 mm long; *petals* 5–6 mm long; *fruit* ovate, 4–5 mm long, densely hairy; *ovules* 4 per ovary. Found on barren white shale slopes, with pinyon-juniper, 6000–6700 ft. April–May. W. Endemic.

This species is listed as threatened under the Endangered Species Act.

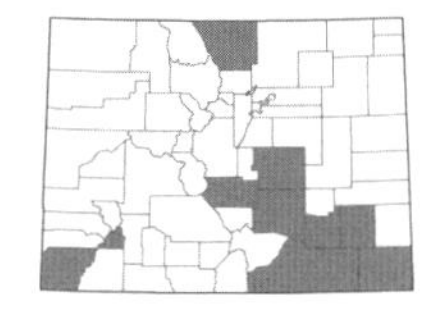

***Physaria fendleri* (A. Gray) O'Kane & Al-Shehbaz**, Fendler's bladderpod. [*Lesquerella fendleri* (A. Gray) S. Watson]. Plants 0.3–3 (4) dm; *basal leaves* linear to ovate, 1–4 (8) cm, the margins entire to coarsely toothed; *cauline leaves* linear to oblanceolate, 0.5–2.5 cm long; *fruiting pedicels* divaricate to erect, 8–20 (40) mm long; *sepals* 5–8 mm long; *petals* 8–12 mm long; *fruit* globose to ovoid, 5–8 mm long, glabrous; *ovules* (12) 20–35 (40) per ovary. Found in grasslands and on rocky outcrops, dry slopes, and barrens, 4000–6500 ft. April–June. E/W.

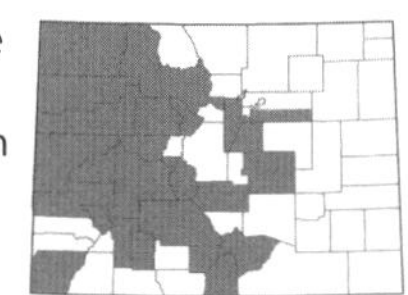

***Physaria floribunda* Rydb.**, point-tip twinpod. Plants 1–2 dm; *basal leaves* oblanceolate, 3–8 cm, the margins usually toothed or pinnatifid, rarely subentire; *cauline leaves* linear-oblanceolate, 1–3 cm long, usually entire; *fruiting pedicels* recurved, 6–15 mm long; *sepals* 5–7 mm long; *petals* 9–11 mm long; *fruit* oblong to ovate, didymous, hairy; *ovules* 4 per ovary. Common on dry hillsides, shale slopes, and in sagebrush, pinyon-juniper, or oak, 5300–10,000 ft. May–June. E/W. (Plate 22)

Immature *P. floribunda* can key to *P. acutifolia*. However, the basal leaves of *P. floribunda* are usually dentate to pinnatifid while those of *P. acutifolia* are usually entire.

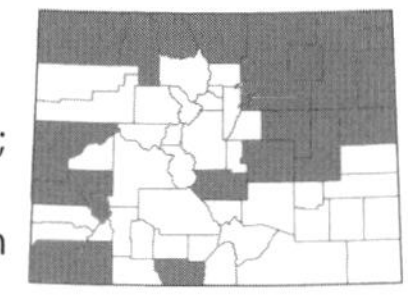

***Physaria ludoviciana* (Nutt.) O'Kane & Al-Shehbaz**, foothill bladderpod. [*Lesquerella ludoviciana* (Nutt.) S. Watson]. Plants 1–3.5 (5) dm; *basal leaves* linear to narrowly lanceolate, (1) 2–6 (9) cm long, the margins usually entire; *cauline leaves* linear to narrowly oblanceolate, (1) 2–4 (8) cm long; *fruiting pedicels* usually recurved, (5) 10–25 mm long; *sepals* 4–8 mm long; *petals* (5) 6.5–10 mm long; *fruit* subglobose to obovoid, 3–6 mm long, hairy; *ovules* (4) 8–12 (16) per ovary. Common on dry hillsides, rocky outcrops, sand dunes, and in grasslands, 3500–7500 ft. May–July. E/W.

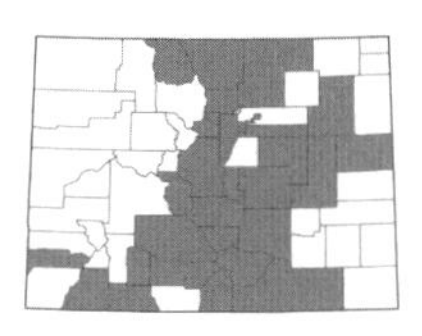

***Physaria montana* (A. Gray) Greene**, mountain bladderpod. [*Lesquerella montana* (A. Gray) S. Watson]. Plants 0.5–2 (3.5) dm; *basal leaves* elliptic to obovate or suborbicular, (1) 2–7 cm, the margins entire to shallowly toothed; *cauline leaves* linear to rhombic, 1–2.5 (4) cm; *fruiting pedicels* usually sigmoid, 5–15 (20) mm; *sepals* 5–8.5 mm; *petals* (6) 7.5–12 mm; *fruit* ovoid to elliptic, 6–12 mm, densely hairy; *ovules* (8) 12–24 per ovary. Common on dry slopes, rocky outcrops, and in grasslands, sagebrush, and forests, 4500–10,500 ft. April–July. E/W. (Plate 22)

***Physaria obcordata* Rollins**, Piceance twinpod. Plants 1.2–2 dm; *basal leaves* oblanceolate, 4–8 cm long, the margins entire to shallowly sinuate-dentate; *cauline leaves* narrowly lanceolate, the margins entire; *fruiting pedicels* widely spreading to recurved, 1–1.5 cm long; *sepals* 5–7 mm long; *petals* 7–10 mm long; *fruit* obcordate, 4–7 × 3–6 mm, pendant, didymous, hairy; *ovules* usually 4 per ovary. Rare on white shale outcrops, 5900–7500 ft. May–June. W. Endemic. (Plate 22)

This species is listed as threatened under the Endangered Species Act.

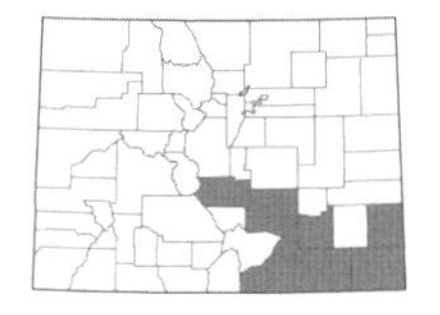

***Physaria ovalifolia* (Rydb.) O'Kane & Al-Shehbaz**, roundleaf bladderpod. [*Lesquerella ovalifolia* Rydb.]. Plants 0.5–2.5 dm; *basal leaves* elliptic to deltate or suborbicular, 0.5–2 (6.5) cm long, the margins entire to shallowly toothed; *cauline leaves* narrowly elliptic to obovate, 0.5–2.5 (4) cm long; *fruiting pedicels* usually spreading, 5–18 mm long; *sepals* 4.5–7 (8.5) mm long; *petals* 6.5–15 mm long; *fruit* subglobose to ellipsoid, 4–9 mm, glabrous; *ovules* 8–16 per ovary. Found on limestone breaks and sandstone outcrops, in grasslands and juniper, 3700–6000 ft. April–May. E.

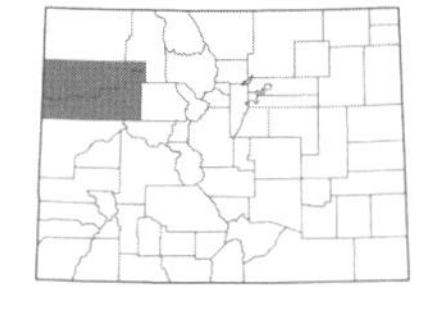

***Physaria parviflora* (Rollins) O'Kane & Al-Shehbaz**, Piceance bladderpod. [*Lesquerella parviflora* Rollins]. Plants 1–3 dm; *basal leaves* broadly obovate, 1–2 cm long, margins entire or with 1–2 teeth; *cauline leaves* oblong to oblanceolate, similar to basal; *fruiting pedicels* recurved, 6–10 (12) mm long; *sepals* 2–4 mm long; *petals* 4–7 mm long; *fruit* subglobose to elliptic, usually pendant, 3–4 mm long, densely hairy; *ovules* 4 per ovary. Found on barren shale slopes, 6800–8900 ft. June–July. W. (Plate 23)

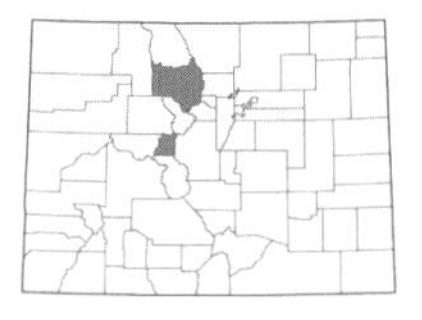

***Physaria parvula* (Greene) O'Kane & Al-Shehbaz**, pygmy bladderpod. [*Lesquerella parvula* Greene]. Plants 0.3–1.5 (3) dm; *basal leaves* linear to narrowly spatulate, 1–4 cm long, the margins involute; *cauline leaves* similar to basal; *fruiting pedicels* ascending, 2–10 mm long; *sepals* 3.5–7 mm long; *petals* 5–6 mm long; *fruit* ovoid, 4–5 mm long, hairy; *ovules* 4–8 per ovary. Uncommon on dry slopes, 7000–9800 ft. May–July. E/W.

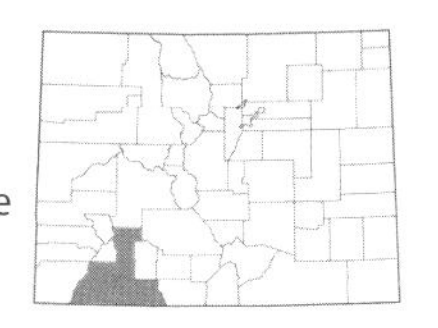

***Physaria pruinosa* (Greene) O'Kane & Al-Shehbaz**, PAGOSA BLADDERPOD. [*Lesquerella pruinosa* Greene]. Plants to 2 dm; *basal leaves* suborbicular, obovate, or rhombic, 4–8 cm long, margins entire to shallowly toothed; *cauline leaves* obovate to rhombic, 0.8–2.5 cm long; *fruiting pedicels* ascending to spreading, 8–11 mm long; *sepals* 5–6 mm long; *petals* 8–9 mm long; *fruit* subglobose to elliptic, 6–9 mm long, glabrous; *ovules* 4–9 (12) per ovary. Rare on Mancos shale hillsides, 6500–8300 ft. May–June. W.

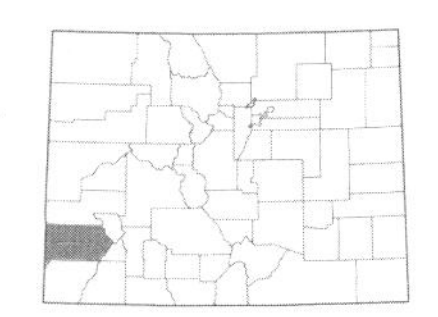

***Physaria pulvinata* O'Kane & Reveal**, CUSHION BLADDERPOD. Plants to 7 dm, spreading; *basal leaves* linear-oblanceolate to narrowly elliptic, 0.8–1.5 cm long, the margins entire; *cauline leaves* similar to basal; *fruiting pedicels* strongly sigmoid, 5–10 mm long; *sepals* 2.5–4 mm long; *petals* 4–7 mm long; *fruit* elliptic, 4–6 mm long, densely hairy; *ovules* 2 per ovary. Found on shale outcrops, with sagebrush and junipers, 7500–8500 ft. May–June. W. Endemic.

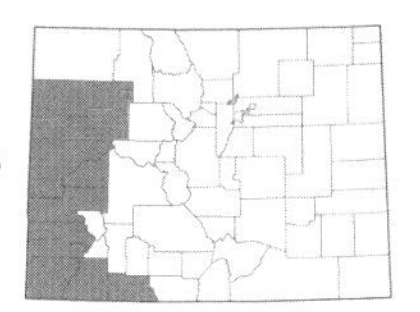

***Physaria rectipes* (Wooton & Standl.) O'Kane & Al-Shehbaz**, STRAIGHT BLADDERPOD. [*Lesquerella rectipes* Wooton & Standl.]. Plants 0.5–3 (6) dm; *basal leaves* narrowly oblanceolate to elliptic, 1–7 (12) cm long, the margins entire or shallowly toothed; *cauline leaves* usually secund, spatulate, 1–3 (4.5) cm long; *fruiting pedicels* spreading, sometimes slightly recurved, 5–15 mm long; *sepals* 4–8 (9) mm long; *petals* 7–10 (16) mm long; *fruit* ovoid to subglobose, 4–7 (9) mm long, sparsely hairy; *ovules* (8) 12–16 (20) per ovary. Common on dry hillsides and rocky slopes, often in sagebrush or pinyon-juniper, 5000–8500 ft. May–June. W.

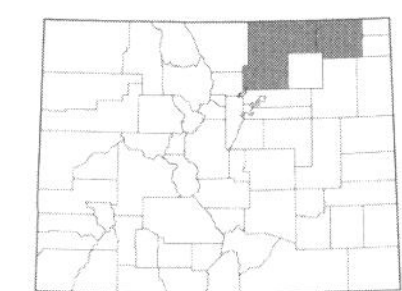

***Physaria reediana* O'Kane & Al-Shehbaz**, REED'S BLADDERPOD. [*Lesquerella alpina* (Torr. & A. Gray) S. Watson]. Plants 0.2–0.4 dm; *basal leaves* linear-oblanceolate, 1.2–3 cm long, the margins entire; *cauline leaves* similar to basal; *fruiting pedicels* ascending, 3–5.5 mm long; *sepals* 4–5 mm long; *petals* 6–9 mm long; *fruit* lanceolate, 4–5 mm long, hairy; *ovules* 8–12 per ovary. Found on chalk bluffs, barren outcrops, and gravelly slopes, 4500–5800 ft. April–June. E. (Plate 23)

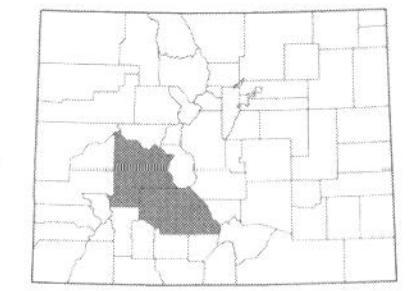

***Physaria rollinsii* G.A. Mulligan**, ROLLINS' TWINPOD. Plants 0.5–1 dm; *basal leaves* oblanceolate, narrowly elliptic, or triangular, 2–3.5 cm long, the margins entire or few-toothed; *cauline leaves* oblanceolate, 1–1.5 cm long; *fruiting pedicels* spreading, 5–8 mm long; *sepals* 5–7 mm long; *petals* 8–10 mm long; *fruit* 2–5 (8) × 4–8 (10) mm, hairy; *ovules* 4 per ovary. Found on dry hillsides and rocky ridges, in sagebrush, 7500–8700 ft. May–June. W. Endemic.

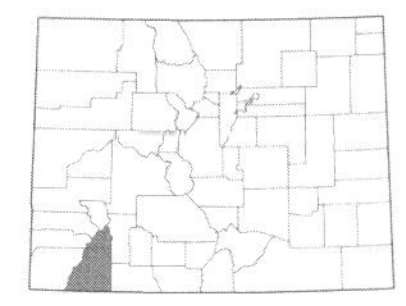

***Physaria scrotiformis* O'Kane**, WEST SILVER BLADDERPOD. Plants 0.08–0.3 dm; *basal leaves* oblanceolate to rhombic, 0.6–2.7 cm long, the margins entire; *cauline leaves* oblanceolate to elliptic, 0.3–0.5 cm long, margins entire; *fruiting pedicels* ascending, 1.8–3.5 mm long; *sepals* 3.5–5 mm long; *petals* 4.5–9 mm long; *fruit* ovoid, 3–5 mm long, didymous, hairy; *ovules* 4–6 (8) per ovary. Uncommon in the alpine on limestone, 11,400–12,500 ft. June–July. W. Endemic.

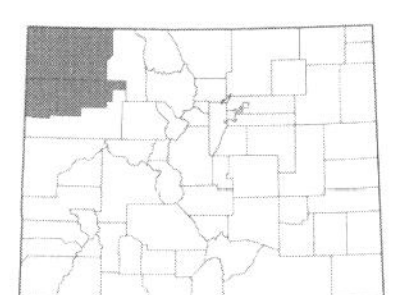

***Physaria subumbellata* (Rollins) O'Kane & Al-Shehbaz**, PARASOL BLADDERPOD. [*Lesquerella subumbellata* Rollins]. Plants 0.1–0.6 dm; *basal leaves* obovate to rhombic, 2–4 cm long, the margins entire; *cauline leaves* linear-oblanceolate, margins entire; *fruiting pedicels* ascending, 3–5 mm long; *sepals* 3.5–7 mm long; *petals* 4–7 mm long; *fruit* ovate to suborbicular, 3–4 mm long, hairy; *ovules* 4–6 per ovary. Found on clay hillsides, rocky ridges, and sandy slopes, usually in pinyon-juniper, 6000–7500 ft. April–June. W.

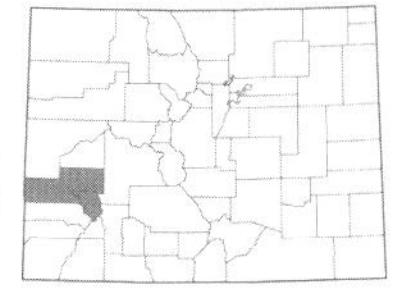

***Physaria vicina* (J. Anderson, Reveal & Rollins) O'Kane & Al-Shehbaz**, UNCOMPAGHRE BLADDERPOD. [*Lesquerella vicina* J. Anderson, Reveal & Rollins]. Plants 1–2.5 dm; *basal leaves* ovate to rhombic or nearly orbicular, 2–7 cm long, the margins usually entire; *cauline leaves* elliptic, 0.7–2.5 cm long, margins entire; *fruiting pedicels* ascending, (4) 6–12 mm long; *sepals* 4–6 mm long; *petals* 6–10 mm long; *fruit* ovoid to subglobose, 5–7 mm long, densely hairy; *ovules* 8–12 per ovary. Uncommon on barren clay or Mancos shale slopes, 5900–7500 ft. May–June. W. Endemic.

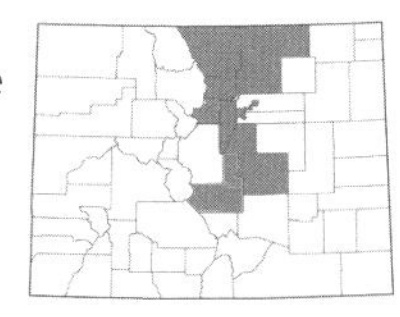

***Physaria vitulifera* Rydb.**, FIDDLELEAF TWINPOD. Plants 1–2 dm; *basal leaves* obovate, 3–6 cm long, the margins deeply and broadly incised; *cauline leaves* similar to basal, the margins sometimes entire; *fruiting pedicels* 6–10 mm long, sigmoid; *sepals* 6–8 mm long; *petals* to 10 mm long; *fruit* elliptic to suborbicular, didymous, 5–7 mm long, hairy; *ovules* 4 per ovary. Found on rocky slopes and dry hillsides, 5000–9800 ft. April–June. E. (Plate 23)

RAPHANUS L. – RADISH

Annual or biennial herbs, glabrous or with simple hairs; *leaves* lyrate or pinnatifid; *flowers* white, yellow, pink, or purple; *fruit* a silique, indehiscent, sessile, breaking into 1-seeded segments, with a weakly distinct beak.

1a. Flowers yellow or sometimes creamy-white; fruit conspicuously constricted between the seeds; roots not fleshy...***R. raphanistrum***
1b. Flowers purple or pink, or sometimes white; fruit smooth or slightly constricted between the seeds; roots usually fleshy and red, reddish-purple, or white (radish)...***R. sativus***

***Raphanus raphanistrum* L.**, WILD RADISH. Plants 2–8 dm, lacking fleshy roots; *basal leaves* oblong to obovate, the margins lyrate to pinnatifid with 1–4 lobes on each side, or sometimes undivided, 3–20 cm × 10–50 mm; *cauline leaves* undivided; *sepals* 7–11 mm long; *petals* yellow or creamy-white, 15–25 mm long; *fruit* cylindric, strongly constricted between the seeds, the valvate segment 1–1.5 mm long, the terminal segment 1.5–11 (14) cm long. Uncommon in old gardens and disturbed places, 7000–8000 ft. May–July. W. Introduced.

***Raphanus sativus* L.**, CULTIVATED RADISH. Plants (1) 4–13 dm, with fleshy roots; *basal leaves* oblong to obovate or spatulate, the margins lyrate or pinnatifid with 1–12 lobes on each side or sometimes undivided, 2–60 × 1–20 cm; *cauline leaves* usually undivided; *sepals* 5.5–10 mm long; *petals* purple or pink, sometimes white, 15–25 mm long; *fruit* lanceolate to fusiform, slightly constricted between seeds or smooth, the valvate segment 1–3.5 mm long, the terminal segment (1) 3–15 (25) cm long. Common garden plants occasionally escaping along roadsides or persisting in old gardens, 5000–6000 ft. May–July. E. Introduced.

RORIPPA Scop. – YELLOW-CRESS

Annual, biennial, or perennial herbs; *leaves* with entire, dentate, lyrate, or pinnatifid margins; *flowers* yellow or sometimes white or pink; *fruit* a silique or silicle, usually sessile, linear, ovoid, elliptic, to globose, glabrous or hairy.

Rorippa coloradensis Stuckey was collected in 1875 by T.S. Brandegee, but this specimen is actually a misidentification of the weed *Sinapis alba* L.

1a. Plants sparsely to moderately pubescent with oval, white, and inflated (mealy) hairs (these primarily found at the base of the stem and midvein on the underside of the leaves), perennials from deep-seated rhizomes or with creeping roots...***R. sinuata***
1b. Plants glabrous or hirsute, lacking white, inflated hairs, annuals or perennials, rarely with creeping roots...2

2a. Fruit linear, 10–20 (25) mm long; petals (2.2) 2.8–6 mm long; perennials...***R. sylvestris***
2b. Fruit lanceolate, oblong, ovoid, or globose, 2–10 mm long; petals 0.5–3 mm long; annuals or perennials...3

3a. Fruit globose or nearly so, 1.2–3 mm in diam....***R. sphaerocarpa***
3b. Fruit linear, oblong, lanceolate, or ovoid, longer than wide...4

4a. Plants with erect, usually solitary stems (0.5) 1–10 (14) dm; stem leaves with auriculate bases; stems often densely hirsute...***R. palustris***
4b. Plants with prostrate or decumbent stems, or prostrate with ascending tips, few to many from the base, rarely erect and solitary; stem leaves auriculate or not; stems usually glabrous...5

5a. Fruit surface minutely papillate; petals 0.5–0.8 mm long; stem leaves lacking auriculate bases...***R. tenerrima***
5b. Fruit glabrous; petals 0.5–2 mm long; stem leaves often with auriculate bases...6

6a. Annuals or biennials; stem leaves with an auriculate or amplexicaul and clasping base; siliques constricted near the middle...***R. curvipes***
6b. Perennials; stem leaves lacking an auriculate or amplexicaul base; siliques not constricted near the middle...***R. alpina***

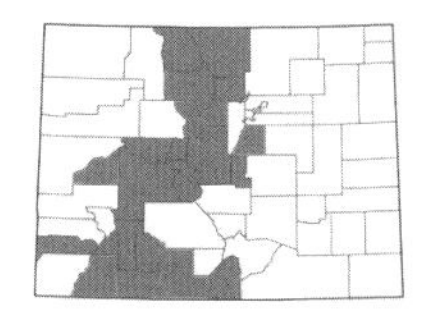

***Rorippa alpina* (S. Watson) Rydb.**, ALPINE YELLOW-CRESS. Perennials, usually prostrate or decumbent, 0.3–2.5 dm, glabrous; *basal leaves* (0.6) 1–4 cm × 2–8 (15) mm, the margins pinnatifid or toothed; *cauline leaves* not auriculate, the margin entire, toothed, or repand; *sepals* 1–2 mm long; *petals* 1.3–2 mm long; *fruit* oblong to lanceolate, 3–8 × 1.5–2.5 mm, glabrous. Common along the shores of lakes and ponds, streams, seepages, melting snowbanks, and in moist meadows, 7800–14,000 ft. July–Sept. E/W. (Plate 23)

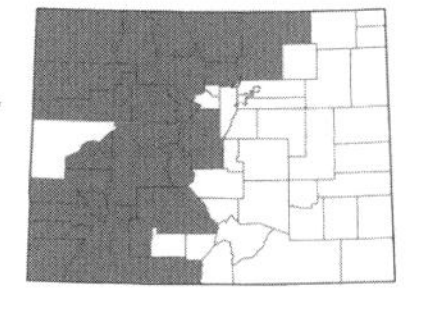

***Rorippa curvipes* Greene**, BLUNTLEAF YELLOW-CRESS. Annuals or rarely perennials, prostrate to ascending, 1–4.5 dm, glabrous or hirsute; *basal leaves* pinnatifid; *cauline leaves* oblong to obovate, (2) 3.5–12 cm × (5) 10–35 mm, usually auriculate, the margins pinnatifid to toothed or sometimes entire; *sepals* 0.8–1.8 mm long; *petals* 0.5–1.8 mm long; *fruit* ovoid, 2–8.5 × (0.5) 1–2.5 mm, glabrous. Common along the margins of lakes and ponds, streams, ditches, fields, and in moist depressions, 4800–10,500 ft. May–Sept. E/W.

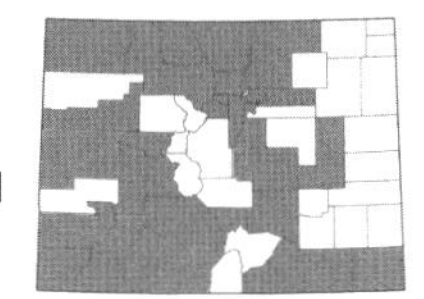

Rorippa palustris **(L.) Besser**, BOG YELLOW-CRESS. Annuals or rarely perennials, erect, (0.5) 1–10 (14) dm, glabrous or hirsute; *basal leaves* (4) 6–20 (30) × 1–5 (8) cm, the margins lyrate-pinnatifid; *cauline leaves* lyrate-pinnatifid, auriculate or amplexicaul; *sepals* 1.5–2.5 mm long; *petals*1.5–3 mm long; *fruit* oblong to elliptic, (2.5) 4–10 × 1.5–3.5 mm. Found along the margins of ponds and lakes, ditches, streams, and other moist areas, 5000–10,500 ft. June–Aug. E/W.

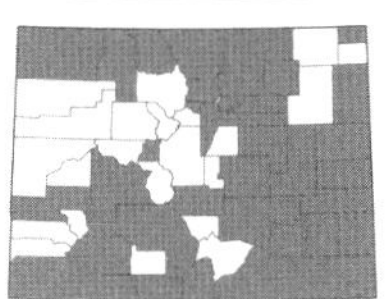

Rorippa sinuata **(Hook.) Hitchc.**, SPREADING YELLOW-CRESS. Perennials, prostrate or decumbent, 1–4.5 dm, with oval, white, inflated hairs; *basal leaves* pinnatifid; *cauline leaves* (1.5) 2.5–6.5 (9) × 0.5–2 cm, usually auriculate, pinnatifid to dentate or sometimes entire; *sepals* 2.2–4 mm long; *petals* 3–5.5 mm long; *fruit* oblong to linear, 4–15 × 1–2.5 mm, glabrous or hairy. Common along the margins of lakes, streams, in fields, ditches, and moist depressions, 3500–8500 ft. May–July. E/W.

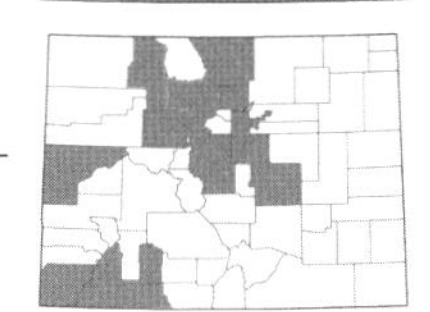

Rorippa sphaerocarpa **(A. Gray) Britton**, ROUNDFRUIT YELLOW-CRESS. Annuals or rarely biennials, 1–5 dm, erect or decumbent, glabrous or hirsute; *basal leaves* pinnatifid; *cauline leaves* 4–9 (12) × 1–3 cm, sometimes auriculate, the margins pinnatifid to subentire; *sepals* 0.7–1.3 mm long; *petals* 0.6–1.2 mm long; *fruit* globose or nearly so, 1.2–3 mm diam., glabrous. Uncommon along the shores of ponds and lakes, streams, ditches, and moist places, 5000–9700 ft. June–Aug. E/W.

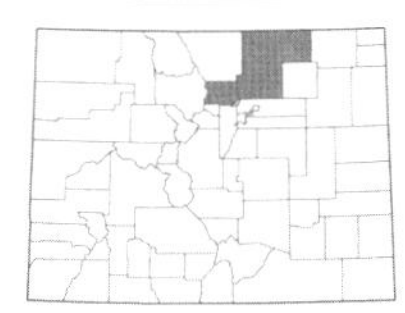

Rorippa sylvestris **(L.) Besser**, CREEPING YELLOW-CRESS. Perennials, prostrate to suberect, (0.5) 1–10 dm, glabrous or sparsely hairy; *basal leaves* pinnatifid; *cauline leaves* deeply pinnatifid, 3.5–15 (20) × 1–5 (6) cm, usually not auriculate; *sepals* 1.8–3.5 mm long; *petals* (2.2) 2.8–6 mm long; *fruit* linear, 10–20 (25) × 1–1.5 mm, glabrous. Uncommon weed in gardens and plantings, 4800–6000 ft. June–Aug. E. Introduced.

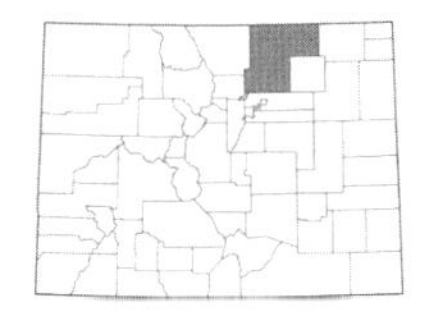

Rorippa tenerrima **Greene**, MODAC YELLOW-CRESS. Annuals, prostrate or decumbent, 0.7–4 dm, glabrous; *basal leaves* pinnatifid; *cauline leaves* 2–10 × 0.7–2 (3) cm, not auriculate, pinnatifid, dentate, or sometimes entire; *sepals* 0.7–1.3 mm long; *petals* 0.5–0.8 mm long; *fruit* lanceolate to oblong, 3–8 × 1–1.7 mm, often constricted in the middle, papillate. Uncommon in moist places, 4500–5500 ft. June–Aug. E.

SINAPIS L. – MUSTARD

Annual herbs, glabrous or with simple, hispid hairs; *leaves* usually dentate, lyrate, or pinnatifid; *flowers* yellow; *fruit* a silique, dehiscent, sessile, linear to lanceolate, terete or nearly so, usually with a distinct beak, glabrous or hairy.

1a. Fruit with 2 types of hairs present, (2) 3–6.5 mm wide; fruiting pedicels spreading…***S. alba***
1b. Fruit with only 1 type of hair present, 1.5–3.5 (4) mm wide; fruiting pedicels ascending…***S. arvensis***

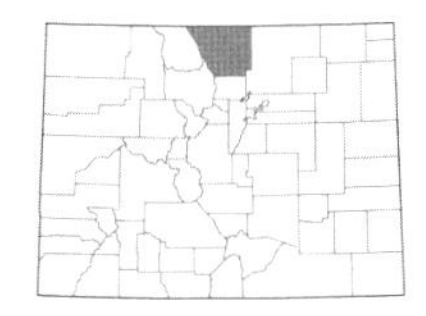

Sinapis alba **L.**, WHITE MUSTARD. Plants 0.2–1 (2.2) dm, usually hispid; *basal leaves* lanceolate to oblong, (3.5) 5–15 (20) cm long, with lyrate or pinnatifid margins; *cauline leaves* 2–4.5 cm long, usually with toothed margins; *sepals* 4–8 mm long; *petals* pale yellow, 7–14 mm long; *fruit* lanceolate, 1.5–5 cm × (2) 3–6.5 mm, with hairs of 2 types present. Cultivated and occasionally escaping along roadsides, 5000–5500 ft. May–Aug. E. Introduced.

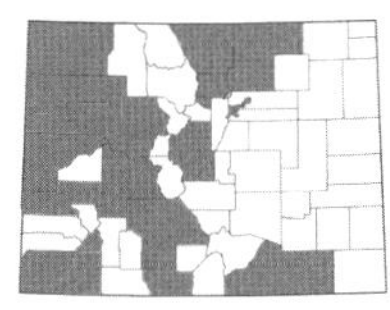

Sinapis arvensis **L.**, WILD MUSTARD. [*Brassica arvensis* (L.) Rabenh.]. Plants (0.5) 2–10 (20) dm, hispid or glabrous; *basal leaves* lanceolate to oblong or obovate, (3) 4–20 (25) cm long, with pinnatifid, lyrate, or sometimes undivided margins; *cauline leaves* with pinnatifid to entire margins; *sepals* 4.5–7 mm long; *petals* bright yellow, 8–12 (17) mm long; *fruit* linear, (1.5) 2–5 (6) cm × 1.5–3.5 (4) mm, with one type of hair present. Found along roadsides, in fields, and in disturbed areas, 4500–8500 ft. June–Sept. E/W. Introduced.

SISYMBRIUM L. – HEDGEMUSTARD

Annual or biennial herbs, glabrous or pubescent with simple hairs; *leaves* with dentate, sinuate, lyrate, or pinnately lobed margins; *flowers* yellow; *fruit* a silique, usually sessile, linear or rarely lanceolate or subulate, usually glabrous. (Payson, 1922)

1a. Perennials from rhizomes; stem leaves linear to filiform, usually with entire margins or rarely toothed or pinnately lobed; stems usually glabrous at the base…***S. linifolium***
1b. Annuals from taproots; stem leaves pinnatisect, pinnatifid, or runcinate; stems often hairy at the base…2

2a. Upper stem leaves with narrowly linear or filiform lobes; sepals with a small, hood-shaped appendage at the tip; ovules 90–120 per ovary…***S. altissimum***
2b. Stem leaves unlike the above; sepals lacking a hood-shaped appendage at the tip; ovules 10–90 per ovary…3

3a. Fruit oblong-linear, (7) 10–15 (18) mm long; fruiting pedicels erect and appressed to the rachis; ovules 10–20 per ovary; style (0.8) 1–2 mm long...***S. officinale***
3b. Fruit narrowly linear, 20–40 (50) mm long; fruiting pedicels spreading or ascending but not appressed to the rachis; ovules 40–90 per ovary; style 0.2–0.7 mm long...4

4a. Stems usually densely hispid at the base; petals 6–8 mm long; young fruits not overtopping flowers...***S. loeselii***
4b. Stems usually glabrous or sometimes sparsely at the base; petals 2.5–4 mm long; young fruits overtopping flowers...***S. irio***

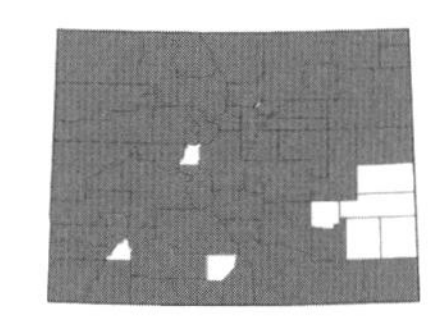

***Sisymbrium altissimum* L.**, TALL TUMBLEMUSTARD. Annuals, (2) 4–15 dm, hirsute below and glabrous above; *basal leaves* (2) 5–20 (35) × (1) 2–8 (10) cm, with pinnatifid or runcinate margins; *cauline leaves* similar to basal, with linear to filiform lobes; *sepals* cucullate, 4–6 mm long; *petals* (5) 6–10 mm long; *fruit* linear, (4.5) 6–10 (12) cm × 1–2 mm, glabrous. Common in disturbed areas, fields, grasslands, and along roadsides, 3500–9000 ft. May–Oct. E/W. Introduced.

***Sisymbrium irio* L.**, LONDON ROCKET. Annuals, (1) 2–7 dm, glabrous or sparsely hairy below; *basal leaves* (1.5) 3–15 × (0.5) 1–6 (9) cm, with pinnatifid to runcinate margins; *cauline leaves* similar to basal, the margins 1–3-lobed or entire; *sepals* 2–2.5 mm long; *petals* 2.5–4 mm long; *fruit* narrowly linear, (2.5) 3–4 (5) cm × 0.9–1.2 mm, glabrous. Not currently known from Colorado but could occur in the western counties. Introduced.

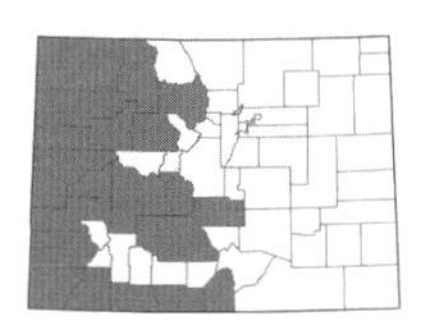

***Sisymbrium linifolium* (Nutt.) Nutt. ex Torr. & A. Gray**, FLAXLEAF HEDGE MUSTARD. [*Schoenocrambe linifolia* (Nutt.) Greene]. Perennials, (1.5) 3–10 dm, glabrous or rarely sparsely hairy; *basal leaves* 1.5–6 cm, with entire or pinnatifid margins; *cauline leaves* usually linear or filiform; *sepals* 3–7 mm long; *petals* (6) 8–12 mm long; *fruit* linear, (2.5) 3.5–6.5 cm × 0.9–1.2 mm, glabrous. Common on dry hills, cliffs, often in pinyon-juniper or sagebrush, 4500–9200 ft. May–July. E/W.

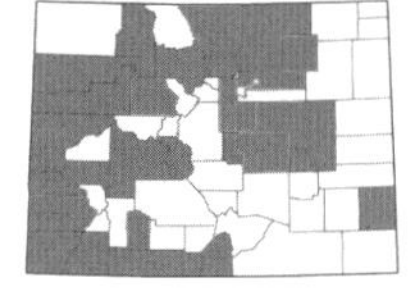

***Sisymbrium loeselii* L.**, SMALL TUMBLEWEED MUSTARD. Annuals, (2) 3.5–15 (17.5) dm, densely hispid below, usually glabrous above; *basal leaves* (1.5) 2.5–8 (12) × (1) 2–5 (7) cm, with pinnatifid, lyrate, or runcinate margins; *cauline leaves* similar to basal, with toothed or entire margins; *sepals* 3–4 mm long; *petals* 6–8 mm long; *fruit* narrowly linear, 2–3.5 (5) cm × 0.9–1.2 mm, usually glabrous. Found along roadsides, in disturbed areas, and along streams, 3800–8000 ft. May–Sept. E/W. Introduced.

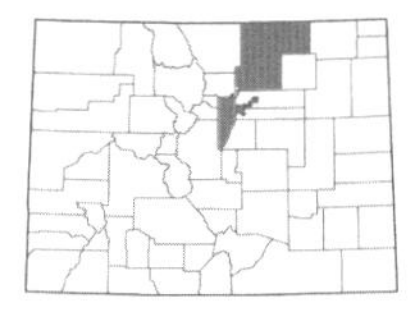

***Sisymbrium officinale* (L.) Scop.**, HEDGEMUSTARD. Annuals, 2.5–8 (11) dm, usually hirsute; *basal leaves* (2) 3–10 (15) cm long, with pinnatifid, lyrate, or runcinate margins; *cauline leaves* similar to basal, with toothed or entire margins; *sepals* 2–2.5 mm long; *petals* 2.5–4 mm long; *fruit* oblong-linear, (0.7) 1–1.5 (1.8) cm × 1–1.5 mm, glabrous or hairy. Uncommon in gardens, fields, and along roadsides, 5000–5500 ft. May–Sept. E. Introduced.

SMELOWSKIA C.A. Mey. – FALSE CANDYTUFT
Perennial herbs, usually pubescent with simple hairs; *leaves* 1- or 2-pinnatifid; *flowers* usually white; *fruit* a silique, linear, oblong, spatulate, or oblanceolate, 4-angled, usually glabrous.

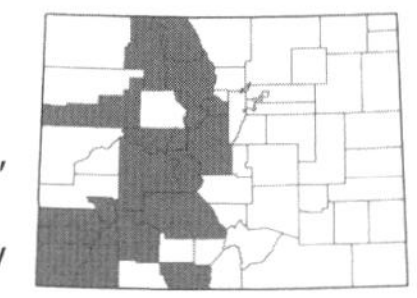

***Smelowskia americana* Rydb.**, AMERICAN FALSE CANDYTUFT. [*S. calycina* (Steph. ex Willd.) C.A. Mey. var. *americana* (Regel & Herd.) Drury & Rollins]. Plants (4) 6–20 (27) dm; *basal leaves* pinnately divided, oblong to oval, (0.5) 1–3.5 (5.2) cm long, pubescence variously dense; *cauline leaves* similar to basal, reduced distally; *sepals* 2–3.5 mm long; *petals* 3.5–6.5 mm long; *fruit* 5–13 mm long. Locally common in the alpine and on subalpine slopes, 10,200–14,000 ft. June–Aug. E/W. (Plate 23)

STANLEYA Nutt. – PRINCE'S PLUME
Annual, biennial, or perennial herbs, shrubs, or subshrubs, glabrous or pubescent with simple hairs; *leaves* with entire, lyrately lobed, or pinnatifid margins; *flowers* yellow or white; *fruit* a silique, long-stipitate, linear, terete or nearly so, glabrous.
1a. Stem leaves sessile with sagittate or auriculate bases, simple...***S. viridiflora***
1b. Stem leaves petiolate, often at least some lyrately lobed...2

2a. Petals pale yellow to white, broadly ovate to orbicular, (2.5) 3–6 mm wide; fruit ascending or somewhat erect; plants glaucous...***S. albescens***
2b. Petals bright yellow or yellow-orange, oblong to oblanceolate, 0.8–3 mm wide; fruit spreading horizontally or rarely ascending; plants glaucous or not...***S. pinnata***

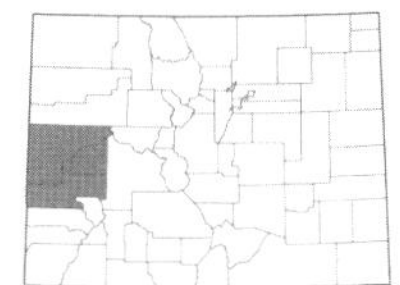

***Stanleya albescens* M.E. Jones**, WHITE PRINCE'S PLUME. Biennials, 2–8 (10) dm, glabrous and glaucous; *basal leaves* (1.5) 3–20 × (1) 2–8 cm, the margins pinnatifid, lyrate, or runcinate; *cauline leaves* similar to basal, the margins lyrate-pinnatifid or entire, petiolate; *sepals* (9) 11–18 mm long; *petals* pale yellow to white, 12–18 mm long; *gynophore* 10–25 mm long; *fruit* 2.3–6 cm × 1–2 mm. Uncommon on barren clay and shale hillsides, 4500–7500 ft. April–June. W.

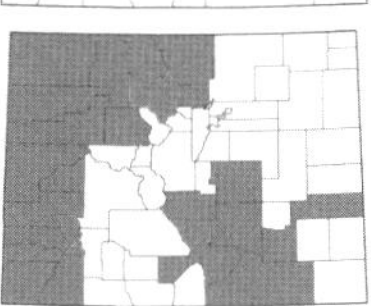

***Stanleya pinnata* (Pursh) Britton**, DESERT PRINCE'S PLUME. Perennials, subshrubs, or shrubs, (1.2) 3–15 dm; *basal leaves* withered by flowering; *cauline leaves* 3–15 cm, entire, dentate, or pinnately lobed; *sepals* 8–16 mm long; *petals* yellow, 8–20 mm long; *gynophore* 7–28 mm long; *fruit* 3–9 cm × 1.5–3 mm. Found on dry hillsides, barren slopes, and in pinyon-juniper or sagebrush in selenium-rich soil, 4000–8300 ft. April–Sept. E/W. (Plate 23)

There are 3 varieties of *S. pinnata* in Colorado:

1a. Leaves usually entire or the uppermost dentate...**var. *integrifolia* (James ex Torr.) Rollins**. W.
1b. Leaves mostly pinnately lobed or deeply cleft...2

2a. Lower leaves without a pair of lobes at the base; mature fruit 4–8 cm long, stipitate 7–25 mm...**var. *pinnata***. E/W. Our most common variety, widespread.
2b. Leaflets or lobes of the lower leaves mostly with a pair of lobes at the base; mature fruit 2.5–5 cm long, stipitate 4–11 mm...**var. *bipinnata* (Greene) Rollins**. Rare, known from northern Larimer Co. but potentially present in Fremont and Moffat cos., 7000–8500 ft. June–July. E.

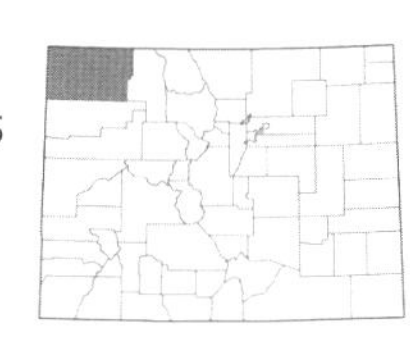

***Stanleya viridiflora* Nutt.**, GREEN PRINCE'S PLUME. Perennials, (2.5) 4–14 dm; *basal leaves* (2.2) 5–20 × 1–4 (6) cm, the margins entire, dentate, or rarely lyrate-pinnatifid; *cauline leaves* sessile, (2) 3.5–8.5 (11) cm, with entire margins; *sepals* 12–18 mm long; *petals* white to pale yellow, 13–20 mm long; *gynophore* (6) 11–25 mm long; *fruit* 3–7 cm × 1.2–2 mm. Rare on barren shale, clay, or gypsum hillsides, 5500–8300 ft. June–Aug. W.

STREPTANTHELLA Rydb. – STREPTANTHELLA

Annual herbs, glabrous; *leaves* with entire, toothed, or pinnatifid margins; *flowers* white or yellow; *fruit* a silique, sessile, linear, glabrous.

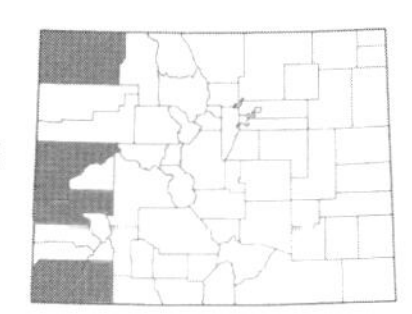

***Streptanthella longirostris* (S. Watson) Rydb.**, LONGBEAK STREPTANTHELLA. Plants 1.5–5 dm; *leaves* (2) 3–6 cm long, the lower narrowly oblanceolate and sinuate-dentate to entire, the upper linear and usually entire; *sepals* 2–3 mm long; *petals* white to yellowish with purple veins, 3.5–6 (7) mm long; *fruiting pedicels* 1.5–3 mm long, becoming reflexed; *fruit* 3.5–4.5 cm × ca. 1.5 mm, with a beaklike, indehiscent tip ca. 3.5 mm long, reflexed. Uncommon on shale slopes, sandstone caprocks, and in sandy soil, 4700–6800 ft. April–May. W.

STREPTANTHUS Nutt. – TWISTFLOWER

Perennial herbs (ours), usually glabrous; *flowers* purple to brownish (ours); *fruit* a silique, subsessile or shortly stipitate, linear, glabrous.

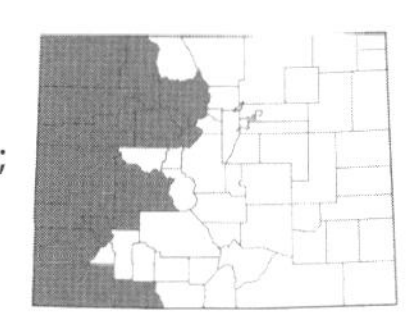

Streptanthus cordatus* Nutt. var. *cordatus, HEARTLEAF TWISTFLOWER. Plants 3–8 dm, glaucous; *basal leaves* spatulate-obovate, 2.5–7 cm long, petiolate, dentate; *cauline leaves* oblong to suborbicular or rarely lanceolate, 2–6 cm long, with an auriculate and sagittate base, entire or sparsely toothed; *sepals* 5–10 (12) mm long; *petals* 10–16 mm long; *fruit* 5–8 cm × 3–4 mm, broadly flattened. Common in pinyon-juniper, sagebrush, found on sandy or clay slopes and shale barrens, 4700–8500 ft. May–July. W.

STRIGOSELLA Boiss. – STRIGOSELLA

Annual herbs, pubescent with simple and short-stalked, forked or branched hairs; *leaves* usually with entire or toothed margins; *flowers* pink or purple or rarely white; *fruit* a silique, subsessile, linear, 4-angled.

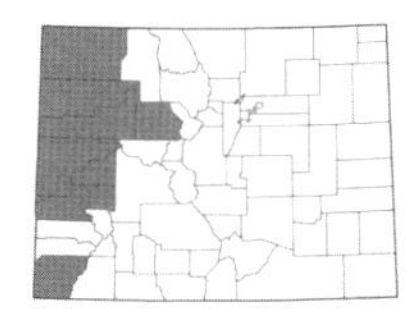

***Strigosella africana* (L.) Botsch.**, AFRICAN MUSTARD. [*Malcolmia africana* (L.) W.T. Aiton]. Plants (0.4) 1.5–3 (5) dm; *cauline leaves* oblanceolate to oblong or elliptic, (0.5) 1.5–6 (10) cm × (3) 10–25 (35) mm, with short-stalked, forked or subdendritic hairs, sometimes with simple ones; *sepals* 3.5–5 mm long; *petals* (6.5) 8–10 (12) mm long; *fruit* (2.5) 3.5–5.5 (7) cm × 1–1.3 mm, usually with forked hairs. Uncommon weedy plants found on shale slopes, sandstone ledges, and dry hillsides, 4700–6500 ft. April–June. W.

SUBULARIA L. – AWLWORT

Annuals, aquatic herbs, glabrous; *leaves* all basal, with entire margins; *flowers* white; *fruit* a silicle, obovoid to elliptic, terete or slightly inflated, glabrous.

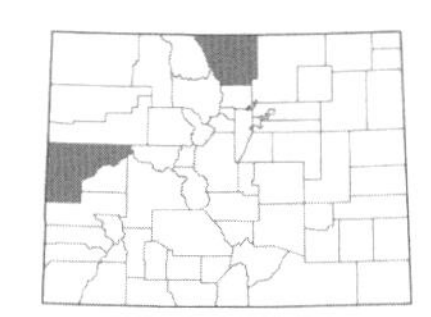

***Subularia aquatica* L. var. *americana* (Mulligan & Calder) Bovin**, AMERICAN AWLWORT. Plants scapose; *leaves* narrowly subulate, 1–5 (7) cm long, entire; *scapes* 2–12 (18) cm long; *sepals* 0.5–1 mm long; *petals* 1.2–1.5 mm long; *fruit* 0.2–0.4 (0.5) cm × 1.2–2.5 mm. Rare aquatic plants found in subalpine lakes and shallow ponds, 10,000–11,000 ft. July–Sept. E/W.

THELYPODIOPSIS Rydb. – TUMBLE MUSTARD

Annuals, biennial, or rarely perennial herbs, glabrous or pubescent with simple hairs; *leaves* with entire or toothed margins; *flowers* white, purple, or rarely yellow; *fruit* a silique, stipitate, linear, terete, usually glabrous.

1a. Sepals and petals yellow; plants glabrous or sparsely hairy at the base…***T. aurea***
1b. Sepals and petals purple, or rarely the petals whitish; plants pilose throughout or at least at the base…2

2a. Fruit stipitate on a stout gynophore 0.3–1.5 mm long…***T. elegans***
2b. Fruit stipitate on a slender gynophore 3–6 mm long…***T. juniperorum***

***Thelypodiopsis aurea* (Eastw.) Rydb.**, DURANGO TUMBLE MUSTARD. Annuals or perennials, (1.5) 2–6 dm; *basal leaves* 2–7 × 0.5–2.2 cm, oblanceolate, with irregularly toothed margins; *cauline leaves* auriculate, lanceolate to oblong, with entire margins; *sepals* 5–8 mm long; *petals* yellow, 7–13 mm long; *gynophore* 2–6 (8) mm long; *fruit* 5–8 (9) cm × 1.2–1.7 mm. Uncommon on dry, sagebrush or clay slopes, 4900–6800 ft. April–May. W.

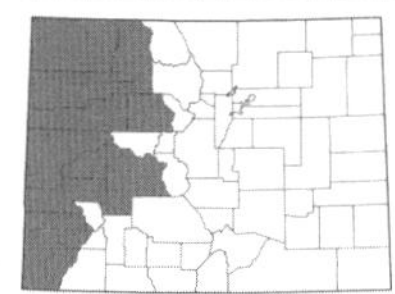

***Thelypodiopsis elegans* (M.E. Jones) Rydb.**, WESTWATER TUMBLE MUSTARD. Annuals or biennials, 1.5–9 (12) dm, hairy at least basally; *basal leaves* withering soon, oblanceolate, 1–5.8 × 0.5–2.2 cm, the margins entire to toothed; *cauline leaves* auriculate, ovate to oblong, with usually entire margins; *sepals* 4–7 mm long; *petals* purple to white, 7.5–14 mm long; *gynophore* 0.3–1.5 mm long; *fruit* 4–9 cm × 1.2–1.5 mm. Common on barren, clay, shale, or gypsum slopes, sometimes with pinyon-juniper, 4600–8700 ft. April–June. W.

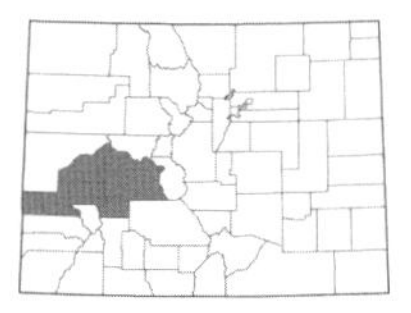

***Thelypodiopsis juniperorum* (Payson) Rydb.**, JUNIPER TUMBLE MUSTARD. Annuals, 1.5–10 dm, hairy basally; *basal leaves* withering soon, 5–15 × 1–2 cm, with entire to toothed margins; *cauline leaves* auriculate, oblong, with entire margins; *sepals* 5–7 mm long; *petals* purple, 14–17 mm long; *gynophore* 3–6 mm long; *fruit* 5–9 cm × 1–1.2 mm. Rare on dry slopes with sagebrush or pinyon-juniper, 6400–8500 ft. May–June. W. Endemic.

THELYPODIUM Endl. – THELYPODY

Herbs, glabrous or pubescent with simple hairs; *leaves* with entire, toothed, lyrate, or pinnatifid margins; *flowers* white or purple; *fruit* a silique, stipitate, linear, terete, somewhat 4-angled, or flattened, glabrous. (Payson, 1922)

1a. Stem leaves sessile with a sagittate or amplexicaul and clasping base; fruit erect…2
1b. Stem leaves petiolate or sessile but lacking a sagittate or amplexicaul base; fruit various…3

2a. Seeds plump, not flattened; mature silique (0.8) 1.2–2.3 mm wide; petals lavender or purple, 2.5–5 (6) mm wide, the margin not crisped…***T. paniculatum***
2b. Seeds flattened; mature silique (0.5) 0.8–1 (1.2) mm wide; petals white or lavender to purple, (0.5) 1–3 (4) mm wide, the margins slightly crisped between the blade and claw…***T. sagittatum* ssp. *sagittatum***

3a. Stem leaves sessile…***T. integrifolium***
3b. Stem leaves petiolate…4

4a. Stamens with all filaments equal in length and the anthers circinately coiled; stems glabrous below; plants of the eastern slope…***T. wrightii***
4b. Stamens tetradynamous, the filaments of unequal lengths, the anthers not circinately coiled; stems hairy or glabrous below; plants of the western slope…***T. laxiflorum***

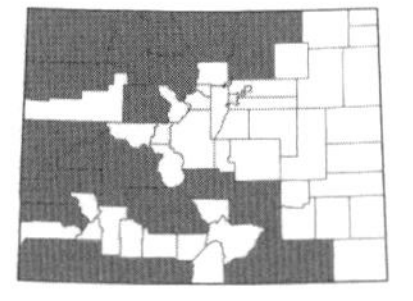

***Thelypodium integrifolium* (Nutt.) Endl. ex Walp.**, ENTIRE-LEAVED THELYPODY. Biennials, (2) 4–17 (28) dm, glabrous; *basal leaves* lanceolate to oblong or spatulate, (3.7) 5–30 (50) × (1.2) 1.6–8 (14) cm, margins usually entire; *cauline leaves* linear to narrowly lanceolate, not auriculate, margins entire; *sepals* 3–5.5 (7.5) mm long; *petals* white or purple, (4.5) 6–9 (13) mm long; *gynophore* 0.5–2 (5.5) mm long; *fruit* (0.8) 1.4–6.5 (8) cm × 1–1.5 (2) mm. Common on dry hillsides, in pinyon-juniper and sagebrush, and near streams and seeps, 4500–8500 ft. June–Sept. E/W.

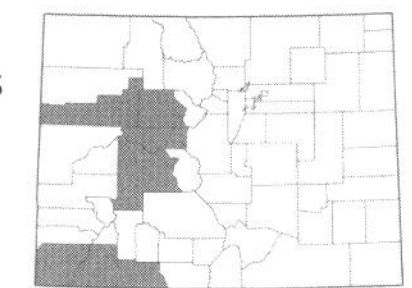

***Thelypodium laxiflorum* Al-Shehbaz**, DROOPFLOWER THELYPODY. Biennials, (1.5) 3–15 (23) dm, glabrous or hirsute below; *basal leaves* oblanceolate, (4) 7–20 (30) × (0.6) 1–4 (10) cm, the margins pinnately lobed; *cauline leaves* linear to lanceolate, the margins usually entire; *sepals* 2.5–5.5 mm long; *petals* 5–7.5 (9.5) mm long, usually white; *gynophore* 0.5–0.8 (1.5) mm long; *fruit* often reflexed, (2) 3–7 cm × 0.7–1.5 mm. Found on rocky hillsides, cliffs and shale outcroppings, in canyons, and along roadsides, 5500–9500 ft. May–Sept. W.

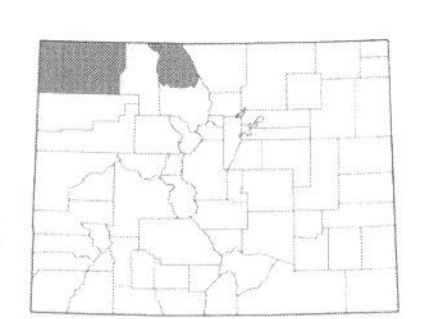

***Thelypodium paniculatum* A. Nelson**, NORTHWESTERN THELYPODY. Biennials or perennials, (1.4) 2–7 dm, glabrous or sparsely hairy; *basal leaves* oblanceolate to oblong, (3) 6–20 × (0.6) 1–2.5 (4) cm, the margins entire; *cauline leaves* lanceolate, sagittate, the margins entire; *sepals* 3–6 mm long; *petals* (6.5) 8–12 (14) mm long, purple; *gynophore* 0.5–0.8 (1.2) mm long; *fruit* erect, 1–3.5 (5) cm × (0.8) 1.5–2.3 mm. Rare in moist meadows and along streams, 7000–8000 ft. June–July. E/W.

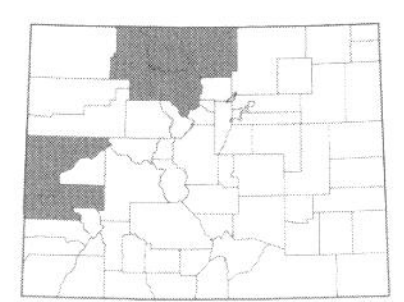

Thelypodium sagittatum* (Nutt. ex Torr. & A. Gray) Endl. ex Walp. ssp. *sagittatum, ARROW THELYPODY. Biennials or perennials, (2) 3–8 (12) dm, glabrous to hairy; *basal leaves* oblanceolate to ovate, (2) 6.5–20 (30) × 1–4 (5) cm, the margins entire; *cauline leaves* lanceolate, with sagittate or amplexicaul bases, the margins entire; *sepals* 2.5–6 (10) mm long; *petals* (5) 7–15 (19) mm long, white to purple; *gynophore* 0.5–1 (2) mm long; *fruit* erect, 1–5.5 (7) cm × (0.5) 0.8–1 (1.2) mm. Rare along streams, in moist meadows, and on alkaline flats, 5000–8700 ft. May–July. E/W.

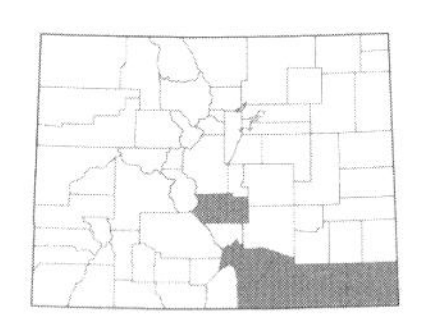

***Thelypodium wrightii* A. Gray**, WRIGHT'S THELYPODY. Biennials, (6) 9.5–23 (28) dm, glabrous; *basal leaves* lanceolate, (6.5) 9–23 (28) × 2.8–5.5 (7.5) cm, the margins pinnately lobed; *cauline leaves* lanceolate, the margins usually entire or toothed; *sepals* (3) 4–7 mm long; *petals* 4–8 (9) mm long, usually white; *gynophore* 0.2–2 (5) mm long; *fruit* reflexed or spreading, (2.5) 4–7.5 (9) cm × 1–1.5 mm. Found on rocky outcrops and hillsides, 4400–7500 ft. June–Sept. E.

THLASPI L. – PENNYCRESS

Annual herbs, glabrous (ours); *leaves* with entire or coarsely toothed margins; *flowers* white; *fruit* a silicle, sessile, obovate to suborbicular with a deeply notched apex, the margins broadly winged, glabrous.

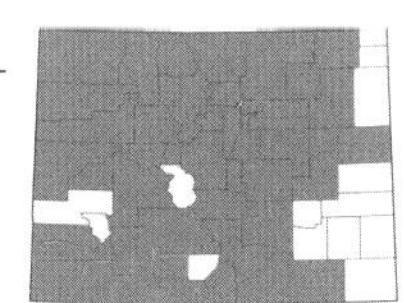

***Thlaspi arvense* L.**, FIELD PENNYCRESS. Plants 1–5 dm; *basal leaves* 2–6 cm long, oblanceolate, sinuate-dentate to lyrate, short-petiolate; *cauline leaves* similar to basal, reduced distally, sessile and auriculate; *sepals* 1.5–2.5 mm long; *petals* 3–4 mm long; *fruit* strongly obcompressed, 10–17 mm long with a 1.5–2.5 mm deep sinus. Common weedy plants found along roadsides, in meadows, fields, and disturbed areas, 3700–10,500 ft. May–Aug. E/W. Introduced.

TRANSBERINGIA Al-Shehbaz & O'Kane – TRANSBERINGIA

Perennial herbs; *flowers* white; *fruit* a silique, sessile or nearly so, linear, terete, glabrous.

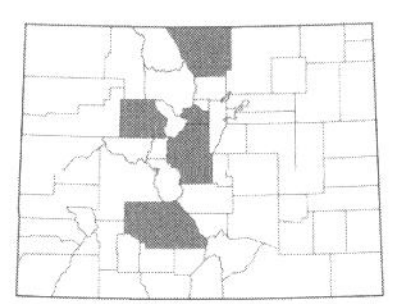

***Transberingia bursifolia* (DC.) Al-Shehbaz & O'Kane ssp. *virgata* (Nutt.) Al-Shehbaz & O'Kane**, ROD TRANSBERINGIA. [*Halimolobos virgata* (Nutt.) O.E. Schulz]. Plants 1–4 dm; *basal leaves* narrowly oblanceolate to obovate, 2–6 cm × 5–10 mm, sinuate-dentate to subentire, densely covered with forked hairs; *cauline leaves* 1.5–4 cm × 2–7 mm, sessile and auriculate; *sepals* 2–3 mm long; *petals* 3–4 mm long; *fruit* erect, 2–4 cm long. Uncommon in grasslands, fens, and meadows, 8000–11,000 ft. June–Aug. E/W.

TURRITIS L. – TOWER-MUSTARD

Biennial or rarely perennial herbs, glabrous or pubescent with simple and/or stalked or forked hairs; *flowers* yellow or creamy-white, rarely pinkish; *fruit* a silique, sessile, linear, glabrous.

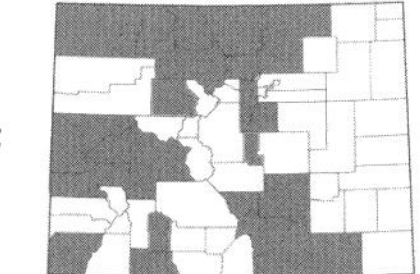

***Turritis glabra* L.**, TOWER ROCKCRESS. [*Arabis glabra* (L.) Bernh.]. Plants 4–12 dm; *cauline leaves* lanceolate to ovate, 4–15 cm long, sessile and auriculate to sagittate, entire or the lower toothed; *sepals* (2.5) 3–5 mm long; *petals* 5–8.5 mm long; *fruit* (3) 4–10 (12.5) cm × 0.7–1.5 mm, erect. Common in meadows, along streams, and in forests, 5600–10,000 ft. May–July. E/W.

CACTACEAE Juss. – CACTUS FAMILY

Trees, shrubs, or perennial herbs, often growing in xeric habitats; *stems* succulent, cylindric, columnar, or flattened, often jointed, and ribbed, smooth, or tuberculate; *leaves* generally reduced to spines; *spines* flexible to rigid, borne in areoles, sometimes glochids (barbed bristles) present in the areole as well; *flowers* perfect, solitary, actinomorphic, with a hypanthium; *perianth* of numerous intergrading sepal-like outer tepals to petal-like inner tepals; *stamens* numerous; *carpels* 3 to many; *ovary* inferior; *fruit* a berry. (Anderson, 2001; Boissev. & Davidson, 1940; Parfitt & Gibson, 2003)

1a. Areoles with glochids and spines; stems segmented and generally branched, these round and cylindric, or flat and ovoid...2
1b. Areoles only with spines; stems unsegmented and typically unbranched, cylindric to spheric...3

2a. Stem segments round and long-cylindric, with prominent tubercles present; spines with a deciduous sheath...***Cylindropuntia***
2b. Stem segments subcylindric to more often flattened and ovoid, without prominent tubercles; spines without a sheath...***Opuntia***

3a. Stems merely ribbed (tubercles, if present, confluent into ribs and not distinct); flowers and fruits bearing sharp, stiff spines from all or most areoles, or spineless...4
3b. Stems tuberculate, the tubercles distinct and not confluent into ribs; flowers and fruits spineless or with soft trichomes or hair-like spines...5

4a. Flowers and fruit with spines from all or most areoles; flowers red, white, pink, orange, yellow, or brown; widespread...***Echinocereus***
4b. Flowers and fruit without spines; flowers yellow to cream or rarely pink; known only from Montezuma County...***Sclerocactus*** (*mesae-verdae*)

5a. Flowers and fruit axillary at the base of the tubercles; tubercles well-developed and prominent, 5–25 mm high, with a groove extending from the areole on the adaxial side of the tubercle; plants often deep-seated in the soil, with ½ of the plant or less showing; spines straight...***Coryphantha***
5b. Flowers and fruit arising from the tubercle apex or on the upper side of tubercles adjacent to the areole; otherwise unlike the above in all respects...6

6a. Spines straight; stems ovoid to globose; plants of the eastern or western slope...***Pediocactus***
6b. Spines hooked or if straight then the stems cylindric; plants of the western slope...***Sclerocactus***

CORYPHANTHA (Engelm.) Lem. – BEEHIVE CACTUS

Stems unsegmented, the tubercles distinct, each tubercle apex bearing an areole and a cluster of spines; *areoles* circular, often woolly, bearing spines, a groove extending from the areole on the adaxial side of the tubercle; *spines* 3–95 per areole, stiff, rigid and needle-like, central spines present or absent; *perianth* with pink, purple, rose, or yellow inner tepals; *fruit* fleshy. (Zimmerman, 1985)

1a. Inner tepals usually pink to reddish-pink or magenta, or occasionally white or pale greenish with magenta tips; fruit green but slowly turning brownish-red, 12–28 mm in length; stems usually more than ½ above the soil...***C. vivipara***
1b. Inner tepals pale green or yellow-green with midstripes of green, pink, or brown; fruit orange-red to scarlet, 6.5–10 mm in length; stems very deep-seated in the soil, becoming almost flat-topped and completely subterranean in the winter...***C. missouriensis***

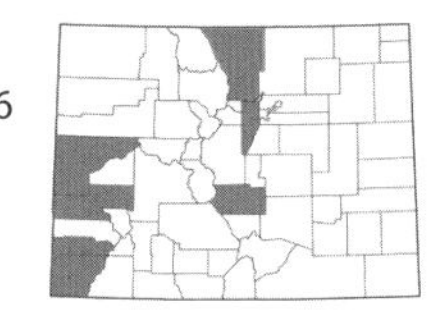

***Coryphantha missouriensis* (Sweet) Britton & Rose**, MISSOURI FOXTAIL CACTUS. [*Escobaria missouriensis* (Sweet) D.R. Hunt]. Plants 2–8 × (1) 1.8–8 (10) cm; *tubercles* pronounced, 5–12 × 3–6 mm; *spines* 6–20 per areole, the radial 4–16 mm long, inner central spine 8–20 mm long; *flowers* greenish-yellow with midstripes of green, or rose-pink, 18–50 mm long; *fruit* 6.5–10 × 5–9 mm, orange-red. Found in dry, open areas of the foothills, 5200–7000 ft. May–June. E/W.

Coryphantha missouriensis blooms in May, but the fruit does not mature until the following spring.

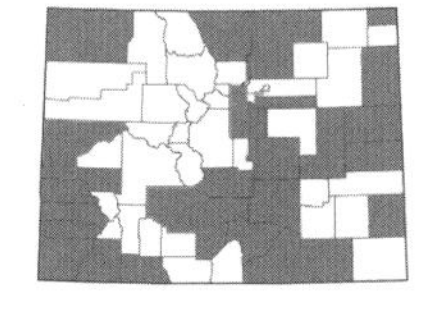

***Coryphantha vivipara* (Nutt.) Britton & Rose**, PINCUSHION CACTUS. [*Escobaria vivipara* (Nutt.) Buxbaum]. Plants 2.5–20 × 3–11 cm; *tubercles* pronounced, 8–25 × 3–8 mm; *spines* 11–55 per areole, the radial ones 7–22 mm long, inner central one 9–25 mm long; *flowers* pink to reddish-pink, sometimes pale greenish but with magenta tips, 20–57 mm long; *fruit* 12–28 × 7–20 mm, green or reddish-brown. Common in open, dry places, 4500–8300 ft. May–July. E/W. (Plate 23)

CYLINDROPUNTIA (Engelm.) F.M. Knuth – CHOLLA

Stems segmented, cylindric, glabrous, tuberculate; *areoles* circular to elliptic, or obovate to rhombic, bearing spines or sometimes nearly spineless, and glochids, often woolly; *spines* round, rigid and needle-like; *flowers* borne in areoles of previous years growth; *perianth* green, yellow-green, red, or magenta; *fruit* fleshy or dry. (Pinkava, 1999)

1a. Flowers dark pink to magenta or red-magenta; spines (5) 8–15 (30) per areole or nearly spineless, typically longer, 8–30 (40) mm; stem segments usually larger, 10–40 × 1.5–4 cm...***C. imbricata* var. *imbricata***
1b. Flowers yellow to green-yellow or pale lime green; spines (1) 3–8 (10) per areole, typically shorter, 2–8 mm; stem segments usually smaller, 3–9 (15) × 0.5–1.5 (2) cm...***C. whipplei***

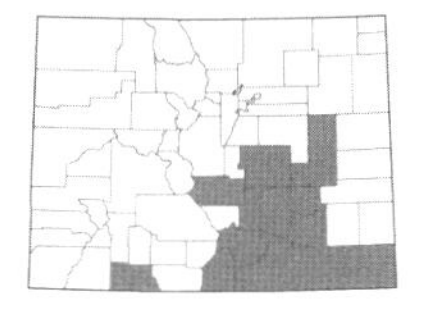

Cylindropuntia imbricata **(Haw.) F.M. Knuth var. *imbricata***, TREE CHOLLA. [*Opuntia imbricata* (Haw.) DC. var. *imbricata*]. Plants (1) 1.5–5 dm; *tubercles* 1.5–5 cm; *spines* (5) 8–15 (30) per areole or nearly spineless, 8–30 (40) mm long; *glochids* 0.5–3 mm long, in a dense tuft; *flowers* dark pink to red-magenta, 15–35 mm long; *filaments* green basally to pink or magenta distally; *fruit* 24–45 mm long, yellow. Found in dry, open places, native to the southeastern counties, 4400–6500 ft. May–July. E/W. (Plate 23)

This species is often cultivated in gardens, and plants near Boulder at the American Legion park have persisted untended for many years, with one collection from this locality dating back to 1948. Also see discussion below.

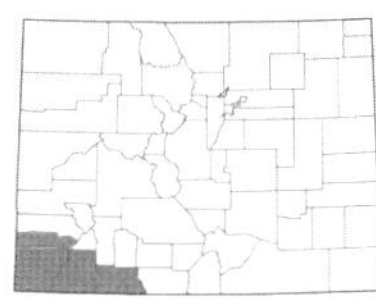

Cylindropuntia whipplei **(Engelm. & J.M. Bigelow) F.M. Knuth**, WHIPPLE'S CHOLLA. [*Opuntia whipplei* Engelm. & J.M. Bigelow]. Plants 3–7 (15) dm; *tubercles* 0.5–1 cm; *spines* (1) 3–8 (10) per areole, 2–8 mm long; *glochids* 1–3 mm long, in a dense tuft; *flowers* yellow to yellow-green, 15–30 mm long; *filaments* yellow or yellow-green; *fruit* 18–25 mm long, yellow to yellow-green. Found in dry, open places in the southwestern counties, 5500–6800 ft. May–June. W.

Cylindropuntia imbricata and *C. whipplei* can hybridize, and these hybrids are found in northwestern New Mexico and 20 miles west of Pagosa Springs on US 160 in Archuleta County. The hybrids are characterized by having pale orange inner tepals and a bushy habit. For the most part, however, these two species remain geographically separated in CO.

ECHINOCEREUS Engelm. – STRAWBERRY CACTUS; HEDGEHOG CACTUS

Stems unsegmented, tuberculate or ribbed; *areoles* usually with spines; *spines* (0) 4–55, straight or curved, radial spines (0) 4–38 (45) per areole, central spines 0–17 per areole; *flowers* borne laterally below the apex of the stem above the areole; *perianth* with pink, red, orange, yellow, greenish, or brown inner tepals; *stigma* lobes 5–22; *fruit* indehiscent or dehiscent. (Arp, 1973; Taylor, 1985)

1a. Flowers yellow to greenish-yellow or brown, 2–3.5 cm long; fruit small, 6–17 mm in length...***E. viridiflorus***
1b. Flowers crimson, orange-red, magenta, pink, white, or green, 1.4–10 cm long; fruit generally larger, 15–40 (72) mm in length...2

2a. Flowers red, crimson, or orange-red, the tips thick and rigid; fruit green to yellowish with white pulp...***E. triglochidiatus***
2b. Flowers pink, white, magenta, purplish, or green, rarely red (and then the areoles mostly 1–6 mm apart), the tips relatively thin and delicate; fruit red to purplish with red pulp or green with white pulp...3

3a. Fruit green with white pulp; central spines shorter than the longest radial spines or absent; radial spines 12–26 and central spines absent or up to 7; plants of the eastern slope...***E. reichenbachii***
3b. Fruit bright red to purplish with red pulp; central spines mostly longer than the longest radial spines; radial spines (2) 4–10 (12) and central spines absent or 1–3; plants of the western slope...***E. fendleri***

Echinocereus fendleri **Small**, STRAWBERRY CACTUS. Stems ovoid to cylindric, 0.8–1.7 (3) dm; *ribs* (5) 6–14; *spines* (2) 4–12 (16) per areole, (8) 11–40 mm long; *flowers* magenta or sometimes white, 3–7 cm long; *fruit* bright red or purplish with red pulp, 20–30 (50) mm long. Uncommon in dry, open places, often with sagebrush, 5200–6700 ft. May–June. W.

Echinocereus reichenbachii **(Terscheck ex Walpers) Britton & Rose**, LACE HEDGEHOG CACTUS. Stems cylindric, 1–3 (4) dm; *ribs* 10–19; *spines* 12–26, 2–10 mm long; *flowers* pink to magenta, white, red, or greenish, 4.5–8 (12) cm long; *fruit* green with white pulp, 15–28 mm long. Uncommon on open limestone or shale bluffs and slopes, 4200–5500 ft. April–May. E.

Echinocereus triglochidiatus **Engelm.**, KING'S CROWN CACTUS. Stems cylindric to spheric, 0.5–7 dm; *ribs* 5–14; *spines* (1) 3–16 (20) per areole, (3) 5–8 cm long; *flowers* red, crimson, or orange-red, 5–10 cm long; *fruit* green or yellowish with white pulp, (15) 20–40 mm long. Found in dry, open places, often with pinyon-juniper or sagebrush, 4500–8800 ft. May–June. E/W. (Plate 23)

Echinocereus triglochidiatus is sometimes split into 2 separate species, *E. triglochidiatus* and *E. coccineus*. These differ in chromosome number (*E. coccineus* is tetraploid while *E. triglochidiatus* is diploid) but otherwise overlap in every morphological characteristic. The best character to separate the two taxa is whether the spines are angled (*triglochidiatus*) or round (*coccineus*). Several specimens, however, have spines angled at the base and round towards the tips and have an intermediate number of ribs. Furthermore, the type specimens for each taxon were collected on the same day in the same locality! Despite the polyploid nature of this complex, it seems more logical to place *E. coccineus* as a variety under *E. triglochidiatus*:

1a. Spines usually round or rarely flat, but usually not angled; ribs usually 6–14...**var. *melanacanthus* (Engelm.) L.D. Benson**, CLARET-CUP CACTUS. [*E. coccineus* Engelm.]. Found in dry, open places, often with pinyon-juniper or sagebrush, known from the western counties, southcentral counties, and San Luis Valley, 5000–7000 ft. May–June. E/W.
1b. Spines 3–6-angled; ribs usually 5–8...**var. *triglochidiatus***, KING'S CROWN CACTUS; SCARLET CUP CACTUS. Found in dry, open places, often with pinyon-juniper or sagebrush, widespread across the western counties, 4500–8800 ft. May–June. W.

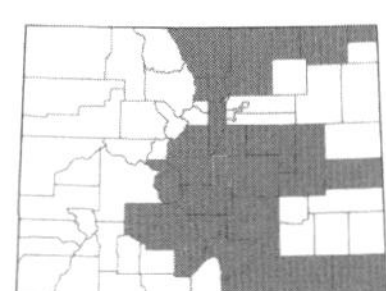

***Echinocereus viridiflorus* Engelm.**, NYLON HEDGEHOG CACTUS. Stems spheric to short cylindric, (0.3) 1–3.5 dm; *ribs* 10–20 with prominent crests; *spines* 12–40 per areole, 3–40 mm long; *flowers* yellow, 20–35 mm long; *fruit* yellowish-green to dark purple or reddish with white pulp, 6–17 mm long. Common in dry, open places, 3500–8500 ft. May–June. E. (Plate 23)

OPUNTIA (L.) Mill. – PRICKLYPEAR

Stems segmented, usually flattened; *areoles* bearing spines and glochids; *spines* mostly 0–15 per areole, not sheathed, round or angled to flattened, usually rigid and needle-like; *flowers* borne in areoles of previous year's growth; *perianth* with yellow to orange, pink, red, or magenta tepals, these sometimes bicolored; *fruit* fleshy or dry. (Arp, 1973; Grant & Grant, 1979)

1a. Plants low, 2–10 cm tall, and mat-forming; areoles 0.3–0.6 cm apart, 3–5 per diagonal row across the midstem segment; stem segments easily detached, mostly terete or slightly flattened, typically small, 2–6 × 1.5–3 cm; fruit dry and tan with distal areoles bearing 1–6 short spines...***O. fragilis***
1b. Plants 10–100 cm; areoles 5–11 per diagonal row across the midstem segment, close set or widely separated; stem segments flattened, not easily detached; fruit fleshy, spineless, and green, yellow, or reddish, or if dry and tan then the areoles bearing 4–15 spines and the stem segments mostly 8–12 × 5–11 cm...2

2a. Spines usually numerous, 6–12 per areole, sometimes so dense they obscure the stem segments (pads), variously distributed on the segments; areoles rather closely set, mostly 1.3 cm or less apart (up to 1.7 mm apart in var. *hystricina*); fruit dry and tan at maturity, cylindric, usually bearing spines (these can be seen on the ovary if the fruit is not developed)...***O. polyacantha***
2b. Spines usually less numerous, 1–6 on most areoles, usually only in the uppermost areoles; areoles spaced 1.2–4 cm apart; fruit fleshy, magenta to purple-red, green, or yellow at maturity, subspheric to ovoid or obovoid, spineless...3

3a. Segments (pads) often larger, 10–25 × 7–20 cm, glacous and bluish-green or sometimes reddish-purple in winter, not cross-wrinkled; areoles spaced mostly 2–4 cm apart; plants to 30 cm or more high; fruit not narrowed at the base...***O. phaeacantha***
3b. Segments (pads) generally smaller, 4–10 × 3.5–10 cm, bluish-green to pale green or deep green, usually cross-wrinkled; areoles spaced mostly 1.2–2 cm apart; plants 7.5–15 cm high; fruit narrowed at the base...***O. macrorhiza***

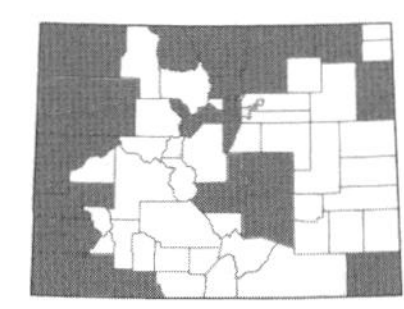

***Opuntia fragilis* (Nutt.) Haw.**, BRITTLE PRICKLYPEAR. Plants 0.2–1 dm, forming mats, the segments 2–5.5 × 1.5–3 cm; *areoles* 0.3–0.6 cm apart, with 3–5 per diagonal row across the midstem segment; *spines* 3–8 per areole, 8–24 mm long; *glochids* in a crescent at the adaxial areole margin, to 3 mm; *flowers* yellow, 20–25 mm long; *filaments* white or red; *fruit* tan, dry, 10–30 mm long. Found in dry, open places, 3900–7000 ft. June–Aug. E/W.

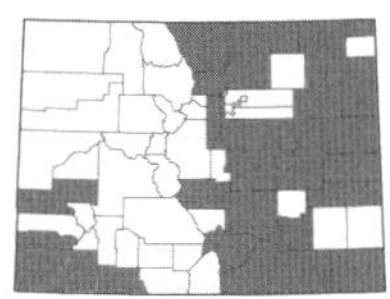

***Opuntia macrorhiza* Engelm.**, WESTERN PRICKLYPEAR. [*O. tortispina* Engelm. & J.M. Bigelow]. Plants 0.8–3 dm, the segments 5–11 × 3.5–7.5 cm; *areoles* 1–2 cm apart, with 5–7 areoles per diagonal row across the midstem segment; *spines* (0) 1–4 per areole, to 60 mm long; *glochids* in a dense tuft, to 5 mm; *flowers* yellow, 25–45 mm long; *filaments* yellow; *fruit* fleshy, green to yellow or reddish, 25–40 mm long. Found in open, dry places, 3600–7000 ft. May–Aug. E/W.

Some consider *O. tortispina* to be a separate species from *O. macrorhiza*. *Opuntia tortispina* is a highly variable taxon that has risen from hybridization of *O. macrorhiza* with *O. polyacantha*. *Opuntia tortispina* supposedly has greenish to yellow-green stigma lobes and white to pale green styles while *O. macrorhiza* has yellowish stigma lobes and white styles and fruit that are narrrowed at the base. I find it extremely difficult to tell these two apart, as they are usually lacking fruit. The stigma color is variable within *O. macrorhiza*, too, and thus I am including *O. tortispina* within *O. macrorhiza*.

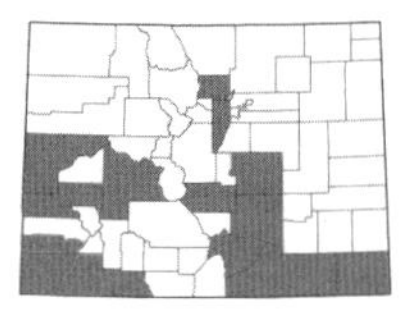

***Opuntia phaeacantha* Engelm.**, TULIP PRICKLYPEAR. Plants 3–10 dm, the segments 10–25 × 7–20 cm; *areoles* 2–3 cm apart, with 5–7 areoles per diagonal row across the midstem segment; *spines* (0) 2–8 per areole, 30–80 mm long; *glochids* in dense tufts, to 5 mm; *flowers* yellow, 30–45 mm long; *filaments* green to yellow; *fruit* fleshy, reddish or purple, 30–40 mm long. Found in dry, open places, 4600–7500 ft. June–July. E/W.

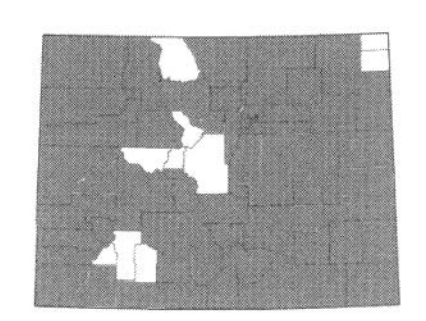

***Opuntia polyacantha* Haw.**, PLAINS PRICKLYPEAR. [*O. erinacea* (Engelm. & J.M. Bigelow) Parfitt, a name widely misapplied to other varieties of *O. polyacantha* and now accepted as a variety of *O. polyacantha* (var. *erinacea*)]. Plants 1–5 dm, the segments (4) 6.5–15 × 4–11 cm; *areoles* close set, 0.6–1.3 cm apart, with 6–11 areoles per diagonal row across the midstem; *spines* 2–6 per areole, 20–95 mm long; *glochids* in a narrow crescent at the adaxial edge of the areole, to 10 mm; *flowers* yellow to orange or magenta, 25–40 mm long; *filaments* white, yellow, or reddish; *fruit* dry, tan, 15–45 mm long. Common in dry, open places, 3500–9200 ft. June–Aug. E/W. (Plate 23)

Opuntia polyacantha is a highly variable species that has historically been split into many different varieties or species. Many of these "new" species were based on single specimens with unique individual characteristics. However, these varieties and species are now recognized as variations of form within the variable *O. polyacantha* complex:

1a. Major central 1–3 spines gradating in length from the other shorter spines, 4–9.5 cm long and pointing out instead of down; the spines on lower areoles spreading, nearly straight, and not flexuous or threadlike; areoles 1–2 cm apart...**var. *hystricina* (Engelm. & J.M. Bigelow) B.D. Parfitt**, PORCUPINE PRICKLYPEAR. [*O. erinacea* Engelm. & J.M. Bigelow var. *hystricina* (Engelm. & J.M. Bigelow) L.D. Benson; *O. polyacantha* var. *rufispina* (Engelm. & J.M. Bigelow ex Engelm.) L.D. Benson; *O. rhodantha* K. Schum.]. Found in the western counties and San Luis Valley, 4500–7300 ft. June–Aug. E/W.

1b. Areoles with distinct long, central spines and shorter spines; spines on lower areoles straight and rigid or deflexed and flexuous; areoles 1 cm or less apart...2

2a. Major spines long (3.8–7.5 cm), flexible and threadlike, curving on the lower areoles, especially dense and tangled at the stem joints, often the spines so dense that they obscure the stem; areoles about 0.5 cm apart...**var. *trichophora* (Engelm. & J.M. Bigelow) J.M. Coult.**, GRIZZLY BEAR CACTUS. Found in Las Animas Co. at the Pinyon Canyon Maneuver site and in the southwestern counties; plants intermediate with var. *polyacantha* with long spines that are not as densely covering the pad are found in the Wet Mountain Valley in Custer Co., 5000–7000 ft. June–Aug. E/W.

2b. Major spines shorter, 2.5–3.8 (5) cm in length, stiff, rigid, and deflexed or ascending at the tips of stems; areoles 1 cm or less apart...**var. *polyacantha***, STARVATION PRICKLYPEAR. [*O. haecockiae* G. Arp; *O. polyacantha* var. *juniperina* (Britton & Rose) L.D. Benson; *O. rutila* Nutt.; *O. schweriniana* K. Schum.]. Common throughout the state, on the eastern plains, along the Front Range, and in the San Luis Valley, 3600–9200 ft. June–Aug. E/W.

PEDIOCACTUS Britton & Rose – HEDGEHOG CACTUS

Stems unsegmented, cylindric to spheric, with spirally arranged tubercles; *areoles* with spines; *spines* 3–45, generally rigid, round to flat, spreading, erect, straight, or recurved; *flowers* borne at the tubercle apex on one side of the areole; *perianth* with yellow, pink, magenta, cream, or white inner tepals and greenish outer tepals; *fruit* dehiscent vertically or circumscissally. (Benson, 1961; 1962; Heil & Schleser, 1981)

1a. Spines 3–13 mm long; 4–11 dark, reddish-brown central spines present in each areole in addition to 15–35 whitish radial spines; plants 2.5–15 (25) cm tall and 2.5–15 cm in diam.; fruit 6–11 mm long...***P. simpsonii***

1b. Spines 1–1.5 mm long; the 18–26 whitish spines present all radial; plants small, 0.7–5.5 cm tall and 1–3 cm in diam.; fruit about 4 mm long...***P. knowltonii***

***Pediocactus knowltonii* L.D. Benson**, KNOWLTON'S MINIATURE CACTUS. Stems spherical to short cylindric, 0.7–5.5 cm; *spines* 18–26 per areole, 1–1.5 mm long; *flowers* pink, 10–35 mm long; *fruit* green, drying reddish-brown, 4 mm long. Known in New Mexico within 50 yards of the Colorado border and could potentially be present in Colorado in La Plata Co. near La Boca. April–May.

One collection of *P. knowltonii* exists from "Colorado" (*Peterson & Knight, 83-7*, CS), but this collection was mostly likely made in New Mexico as Benson (1962) notes *Pediocactus knowltonii* "was found in [New Mexico] one place within 50 yards of the state line. However, a subsequent visit...did not reveal *Pediocactus knowltonii* to occur in a habitat in Colorado similar to the one in New Mexico a stone's throw away." One of the rarest cacti in the U.S.

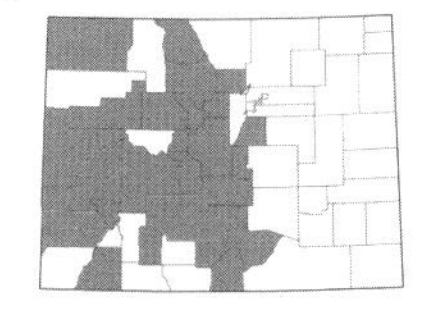

***Pediocactus simpsonii* (Engelm.) Britton & Rose**, MOUNTAIN CACTUS. Stems spherical to ovoid, 2.5–15 (25) cm; *spines* 4–35 per areole, 3–13 mm long; *flowers* usually pink, 10–25 mm long; *fruit* green, drying reddish-brown, 6–11 mm long. Common in dry, open places from the plains to mountains, often found with pinyon-juniper or sagebrush, 5200–10,500 ft. April–June. E/W. (Plate 23)

Small, immature plants of *P. simpsonii* can resemble *P. knowltonii* but soon develop the dark central spines that are characteristic of *P. simpsonii*.

SCLEROCACTUS Britton & Rose – FISHHOOK CACTUS

Stems unsegmented, subglobose, ovoid, spheric, or cylindric, usually with tubercles coalescent into ribs; *spines* 2–17 (29), radial spines 2–11 (18) per areole, central spines lacking or 1–6 (11) per areole; *perianth* with white, yellowish, pink, or purple inner tepals and greenish or brownish outer tepals; *fruit* dehiscent along vertical slits or a basal pore. (Heil & Porter, 1994)

1a. Usually only spreading radial spines present, these relatively short (6–13 mm long) and straw-colored; flowers yellow to cream or rarely pink; tubercles inconspicuous on the ribs of older plants...***S. mesae-verdae***
1b. 1–9 central spines present in addition to the radial spines, these straight or hooked, generally longer (6–72 mm long), straw-colored to brown, reddish, or black; flowers purple, pink, rose, yellow, or rarely white; tubercles conspicuous and evident on the ribs...2

2a. Central 1–3 spines straight, without a hooked tip; flowers pink...***S. glaucus***
2b. Central spines mostly 4–6, with a hooked tip; flowers purple, rose, pink, yellow, or rarely white...***S. parviflorus***

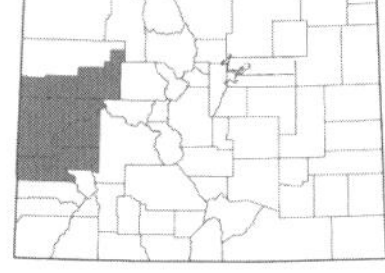

Sclerocactus glaucus **(K. Schum.) L.D. Benson**, Uinta Basin hookless cactus. Stems cylindric, 0.3–3 dm; *ribs* (8) 12–13 (15); *spines* 2–8 (12) per areole, to 26 mm long; *flowers* pink, (30) 40–50 mm long; *filaments* white or green; *fruit* 9–25 (30) mm long. Rare in dry places, often with pinyon-juniper or sagebrush, 4500–6000 ft. April–May. W. Endemic.

Occasionally a couple of areoles will have a central spine with a hook. However, these are not common and are often formed as an artifact of pressing. Listed as threatened under the Endangered Species Act.

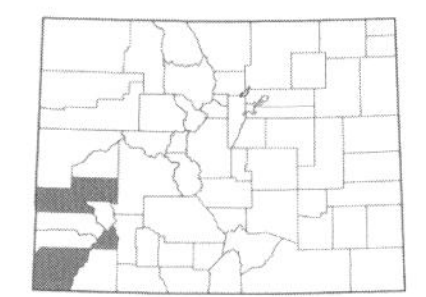

Sclerocactus mesae-verdae **(Boissev. & C. Davidson) L.D. Benson**, Mesa Verde fishhook cactus. [*Coloradoa mesae-verdae* Boissev. & C. Davidson ex Marshall & Bock]. Stems spheric to ovoid, 0.3–2 dm; *ribs* 13–17; *spines* 7–14 per areole, usually only radial spines present, 6–13 mm long; *flowers* yellow to cream or rarely pink, 10–25 mm long; *filaments* yellow or white; *fruit* (4) 8–10 mm long. Rare in open, barren, dry places on shale or clay soil, 4000–5000 ft. April–May. W.

Rarely specimens will have 1 central spine that is straight or hooked. Listed as threatened under the Endangered Species Act.

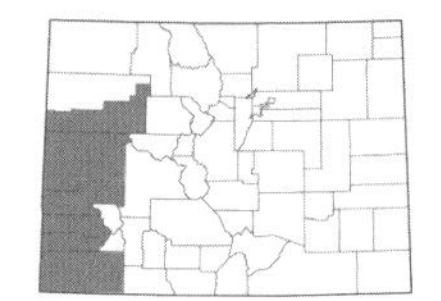

Sclerocactus parviflorus **Clover & Jotter**, smallflower fishhook cactus. [*S. whipplei* (Engelm. & J.M. Bigelow) Britton & Rose]. Stems cylindric or spheric, 0.4–4.5 dm; *ribs* (10) 13–15 (16); *spines* 8–20 per areole, with 4–6 central spines, 6–72 mm long; *flowers* purple, rose, pink, yellow, or rarely white, (20) 30–60 mm long; *filaments* purple, yellow, or green; *fruit* 10–30 mm long. Found in open, dry places, on clay, shale, or adobe hillsides, and often with pinyon-juniper, 4500–6500 ft. April–June (Sept.). W.

CAMPANULACEAE Juss. – bellflower family

Campanula flower:

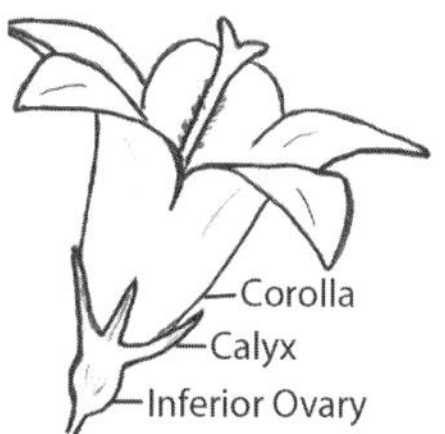

Herbs or shrubs; *leaves* usually alternate or sometimes opposite or whorled, petiolate, usually simple or rarely pinnately compound, exstipulate; *flowers* perfect, actinomorphic or zygomorphic; *calyx* 5-lobed; *corolla* 5-lobed; *stamens* 5, alternating with the petals; *pistil* 1; *stigmas* 2, 3, or 5; *ovary* inferior; *fruit* a capsule or a berry.
1a. Flowers zygomorphic, the corolla two-lipped; stamens connate by the anthers and filaments into a tube...2
1b. Flowers actinomorphic; stamens distinct...3

2a. Corolla 3–5 (7) mm long, white, pale blue, pink, or purple, the lower lip white or pink with 2 yellow spots and 3 small purple spots on the throat, sometimes the purple spots forming a dark band; flowers sessile but appearing pedicellate from an elongate hypanthium; annual to 1.5 dm tall...***Downingia***
2b. Corolla 15–45 mm long, red or blue, rarely white; flowers pedicellate; perennial to 15 dm tall...***Lobelia***

3a. Perennials; flowers pedicellate and terminal and solitary or in a raceme, or sessile in a dense terminal cluster, generally larger, the corolla 0.5–3 cm long, the lowermost flowers not cleistogamous...***Campanula***
3b. Annuals; flowers sessile in leaf axils, the corolla 0.4–1 cm long, the lowermost flowers cleistogamous...4

4a. Corolla shallowly lobed; flowers in false spikes, small, 3–5 mm long; leaves mostly 4–7 mm long, oval to orbicular with serrate margins...***Heterocodon***
4b. Corolla lobed to below the middle; flowers solitary to few in leaf axils, usually larger, 5–10 mm long; leaves mostly 5–20 mm long, linear or ovate with crenate to serrate margins...***Triodanis***

CAMPANULA L. – bellflower; harebell

Perennial herbs (ours); *leaves* alternate, petiolate, simple; *flowers* actinomorphic, campanulate, funnelform, or rotate; *corolla* blue to purple; *stamens* distinct; *ovary* 3–5-carpellate; *fruit* a poricidal capsule.
1a. Flowers numerous in long racemes or dense globose clusters; leaves with serrate margins, only the very upper leaves entire and bracteate; corollas generally larger, 15–30 mm long...2
1b. Flowers solitary or few in loose racemes or panicles; leaves entire or a few remotely serrate or obscurely toothed, or the basal dentate and stem leaves entire; corollas generally smaller, 3–20 mm long...3

2a. Flowers nodding in racemes; calyx deflexed at anthesis, especially on lower flowers in the raceme; plants 4–15 dm tall...***C. rapunculoides***
2b. Flowers erect in dense terminal and axillary globose clusters; calyx not deflexed; plants 2.5–5 dm tall...***C. glomerata***

3a. Calyx 1–2.3 mm long; corolla pale blue, almost white, 3–8 mm long...***C. aparinoides***
3b. Calyx 2.5–15 mm long; corolla blue to blue-purple, rarely white, 5–22 mm long...4

4a. Corolla lobes ⅓–¼ the length of the corolla, not extending to the midpoint of the corolla; capsule nodding, opening by pores near the base; flowers in a loose raceme or panicle, solitary at high elevations; basal leaves oblanceolate to nearly orbicular and usually dentate (these withering early and rarely collected) while the stem leaves are linear to lanceolate and entire...***C. rotundifolia***
4b. Corolla lobes ½ the length of the corolla; capsule erect, opening by pores near the apex; flowers solitary; basal leaves lanceolate or narrowly elliptic, entire to toothed, upper leaves similar but reduced...5

5a. Flowers campanulate with spreading lobes, generally larger, the corolla 17–23 mm long and the sepals mostly 8–15 mm long, rarely shorter; leaves with stiff cilia on the lower portions of the margins of the basal and middle leaves...***C. parryi* var. *parryi***
5b. Flowers narrowly campanulate, generally smaller, the corolla 5–9 (12) mm long and the sepals 2.5–6 mm long, rarely larger; leaves glabrous...***C. uniflora***

***Campanula aparinoides* Pursh**, MARSH BELLFLOWER. Plants 1.5–5 dm; *leaves* linear to oblanceolate, usually greatly reduced above, 1–4.5 × 0.1–0.5 cm, glabrous or hispid on the midvein below, the margins entire to remotely toothed; *inflorescence* solitary on long pedicels; *calyx* 1–2.3 mm long; *corolla* 3–8 mm long, pale blue to nearly white, with lobes about ½ the corolla length. Rare in moist meadows and along streams, known from 2 very old collections in Colorado along the South Platte, 5000 ft. Possibly extirpated from Colorado (see note below). July–Aug. E.

The first collection of this species in Colorado was made by Hall and Harbour in 1862, labeled as collected in South Park but probably collected along the Platte as South Park would be out of range and habitat for this species. The other collection was made by Alice Eastwood sometime prior to 1893 for her *Popular Flora of Denver*. Eastwood notes that this was found several years ago near Smith's Bridge along the Platte in Denver but hasn't been seen since. These collections may represent depauperate forms of *C. rotundifolia*, which rarely can have pale blue or white flowers. If these collections are of *C. aparinoides*, then this species is apparently extirpated from Colorado.

***Campanula glomerata* L.**, DANE'S BLOOD. Plants 2.5–5 dm; *leaves* lanceolate to lance-ovate, 5–10 × 2–3 cm, sessile and decurrent above, scabrous, with toothed margins; *inflorescence* of dense axillary, globose clusters; *calyx* 7–11 mm long; *corolla* 18–30 mm long, blue-purple or rarely white, the lobes about ½ to nearly as long as the tube. Currently known from vacant lots in Marble, Gunnison Co., and near Shadow Mt. Lake, Grand Co., but expected elsewhere, 7800–8500 ft. June–Aug. W. Introduced.

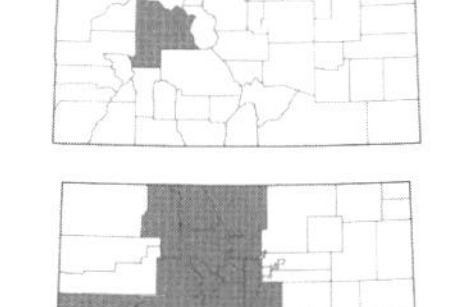

Campanula parryi* A. Gray var. *parryi, ROCKY MOUNTAIN BELLFLOWER. Plants 0.7–2.5 dm; *leaves* lanceolate to narrowly elliptic, 1–3.5 × 0.1–1.2 cm; *inflorescence* solitary; *calyx* (5.5) 8–15 (19) mm long; *corolla* 17–23 mm long, blue to blue-purple, the lobes about ½ the corolla length. Found in moist meadows and along creeks and streams, 6500–13,500 ft. June–Aug. E/W. (Plate 24)

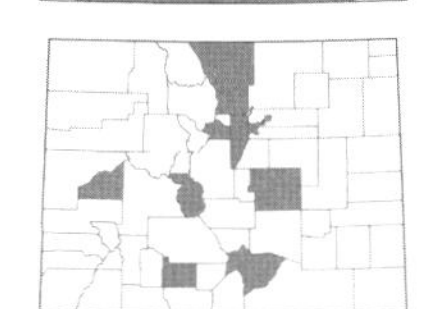

***Campanula rapunculoides* L.**, ROVER BELLFLOWER. Plants 4–15 dm; *leaves* lanceolate to ovate, 3–15 × 1–5 cm, glabrous above and usually hispid below, the margins irregularly double-serrate; *inflorescence* a raceme; *calyx* 5–8 mm long; *corolla* 15–30 mm long, blue or blue-purple, the lobes about ⅓–½ the corolla length. A garden plant escaping and found along roadsides and trails, 5000–7500 ft. June–Aug. E/W. Introduced. (Plate 24)

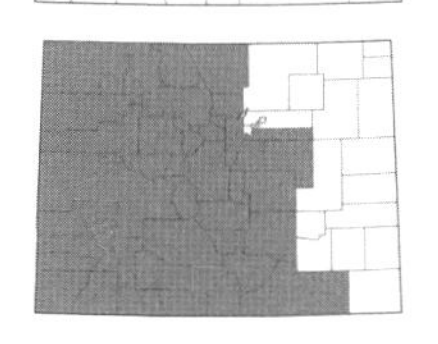

***Campanula rotundifolia* L.**, BLUEBELLS OF SCOTLAND; HAREBELL. [*C. gieseckiana* Vest ex Schult.; *C. groenlandica* Berlin]. Plants (0.6) 1–5 (7.5) dm; *leaves* oblanceolate to orbicular below and linear to lanceolate above, 0.7–3 × 0.4–1.2 cm, the margins of basal leaves dentate and of cauline leaves entire; *inflorescence* solitary or a loose raceme; *calyx* 4–8 mm long; *corolla* 10–20 mm long, blue or blue-purple, the lobes about ¼–⅓ the length of the corolla. Common in dry, rocky soil of slopes and meadows, 5000–13,500 ft. June–Aug. E/W. (Plate 24)

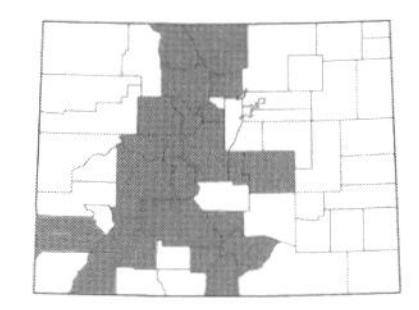

***Campanula uniflora* L.**, ARCTIC BELLFLOWER. Plants 0.4–1 dm; *leaves* elliptic to oblanceolate, 1–3.5 × 0.2–0.8 cm, glabrous, the margins shallowly lobed or obscurely toothed; *inflorescence* solitary; *calyx* 2.5–6 mm long; *corolla* 5–9 (12) mm long, blue to blue-purple, the lobes about ½ the corolla length. Found in dry alpine meadows and scree slopes, 11,500–14,000 ft. July–Aug. E/W.

DOWNINGIA Torr. – CALICOFLOWER

Annuals; *leaves* alternate, sessile; *flowers* zygomorphic, bilabiate, sessile in bract axils but appearing pedicellate from an elongate hypanthium; *stamens* connate by the anthers and filaments, distinct from petals; *ovary* 2-carpellate; *fruit* a capsule splitting by 3–5 longitudinal slits.

Downingia laeta **(Greene) Greene**, GREAT BASIN CALICOFLOWER. Plants 3–15 cm; *leaves* linear-lanceolate to narrowly oblong, 3–20 × 0.5–2 mm; *corolla* blue to purplish, the lower lip marked with yellow and rose-purple spots; *capsules* 20–40 mm long. Not reported for Colorado but occurs very close to the border in southern Wyoming. To be sought for in Larimer, Jackson, Routt, and Moffat cos. in moist places such as mud flats, meadows, along streams and ditches, and along the shores of vernal pools. June–Aug.

HETEROCODON Nutt. – HETEROCODON

Annual herbs; *leaves* alternate, sessile, simple; *flowers* actinomorphic, solitary, short-campanulate, the lowermost cleistogamous; *corolla* blue; *stamens* distinct; *ovary* 3-carpellate; *fruit* a poricidal capsule.

Heterocodon rariflorum **Nutt.**, RAREFLOWER HETEROCODON. Plants to 3 dm; *leaves* oval to orbicular, 2–10 mm long, serrate, sessile; *corolla* 3–5 mm, shallowly lobed, cylindric. Uncommon in moist places along streams, known from a single collection by Alice Eastwood from near Steamboat Springs in July 1891, 6500–6800 ft. Should be sought for in the Elkhead Mountains and around Fish Creek Falls near Steamboat Springs. June–Aug. W.

LOBELIA L. – LOBELIA

Herbs or shrubs; *leaves* alternate, simple; *flowers* zygomorphic, bilabiate; *corolla* blue, red, yellow, or white; *stamens* connate by the anthers and filaments, distinct from the petals; *ovary* 2-carpellate; *fruit* a capsule splitting by 2 valves near the tip. (McVaugh, 1936)

1a. Flowers deep red; filament tube red, (15) 18–30 mm long…***L. cardinalis***
1b. Flowers blue with white stripes or rarely white; filament tube blue, 12–15 mm long…***L. siphilitica*** **var. *ludoviciana***

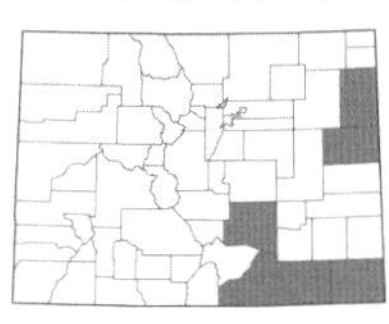

Lobelia cardinalis **L.**, CARDINAL FLOWER. Plants to 1.5 m; *leaves* 2.5–20 × 0.5–5 cm, lanceolate to ovate-lanceolate, the margins toothed, usually glabrous; *calyx* 9–15 mm long; *corolla* red, 2–3.5 cm long; *filament tube* red, (15) 18–30 mm long; *capsules* 5–9 mm long. Uncommon in moist places such as along creeks and on floodplains, 3500–4900 ft. Aug.–Sept. E.

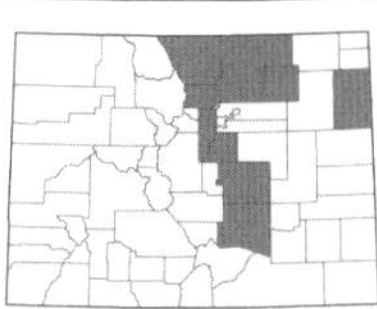

Lobelia siphilitica **L. var. *ludoviciana*** **A. DC.**, GREAT BLUE LOBELIA. Plants to 1.5 m; *leaves* 2–15 × 0.6–5 cm, lanceolate to oblanceolate, the margins toothed to subentire, usually glabrous; *calyx* 8–20 mm long; *corolla* blue with white stripes or rarely white, 1.5–3.5 cm long; *filament tube* blue, 12–15 mm long; *capsules* 6–9 mm long. Uncommon in moist places such as wet meadows and along creeks and ditches, 3500–6800 ft. Aug.–Sept. E.

TRIODANIS Raf. – VENUS' LOOKING-GLASS

Annual herbs; *leaves* alternate, the upper sessile and clasping, simple; *flowers* actinomorphic, the lowermost cleistogamous; *corolla* blue to purple; *stamens* distinct; *ovary* usually 3-carpellate, rarely 2-carpellate; *fruit* a poricidal capsule. (McVaugh, 1945)

1a. Leaves linear to lanceolate or narrowly elliptic, with entire to crenate margins, not clasping the stem…***T. leptocarpa***
1b. Leaves cordate to broadly ovate with crenate to serrate margins, clasping the stem or not…2

2a. Middle leaves not clasping; capsules elliptic to oblong or narrowly obovoid, 5–9 mm long, with linear pores in the middle or higher on the capsule…***T. holzingeri***
2b. Middle leaves clasping; capsules ovoid, 3.5–6 (8) mm long, with elliptic pores below the middle of the capsule…***T. perfoliata***

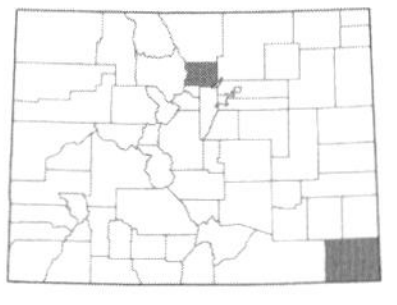

Triodanis holzingeri **McVaugh**, HOLZINGER'S VENUS' LOOKING-GLASS. Plants 1–10 dm; *leaves* cordate to broadly ovate, 0.5–3 × 0.8–3 cm, sessile, the middle leaves not clasping; *corolla* usually purple, the tube to 2.5 mm long and the lobes 4–7 mm long; *capsules* 5–9 mm long, with linear pores at the middle or higher, these 0.2–0.4 mm wide. Uncommon in sandy or rocky soil of open places on the plains, often in disturbed areas, 4000–5200 ft. May–July. E.

Triodanis leptocarpa **Nieuwl.**, SLIMPOD VENUS' LOOKING-GLASS. Plants 1–7 dm; *leaves* linear to lanceolate or narrowly elliptic, 1–3 × 0.2–1 cm, sessile; *corolla* purple or sometimes white-streaked, the tube 0.5–1.5 mm long and the lobes 3–7 mm long; *capsules* 8–25 mm long, dehiscing by longitudinal slits or a single pore near the tip. Uncommon in sandy or rocky soil of open places, usually in disturbed sites, 5000–6000 ft. May–July. E.

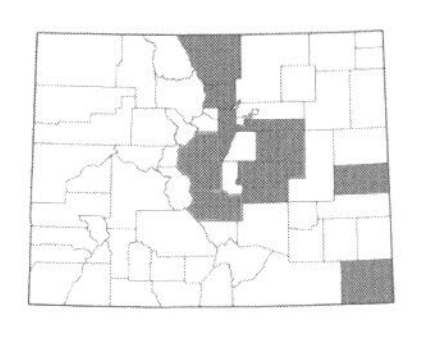

Triodanis perfoliata **(L.) Nieuwl.**, CLASPING VENUS' LOOKING-GLASS. Plants 1–10 dm; *leaves* cordate to broadly ovate, 0.5–3 × 0.8–2.5 cm, sessile and clasping, the margins crenate to serrate; *corolla* purple or sometimes white, the tube 1–2.5 mm long and lobes 4.5–7 mm long; *capsules* 3.5–6 (8) mm long, with pores at the middle or below, these 0.5–1.5 mm wide. Uncommon in sandy or rocky soil of open places, often in disturbed or burned areas, 5000–7800 ft. May–Aug. E.

CANNABACEAE Martinov – HEMP FAMILY

Annual or perennial herbs or trees; *leaves* opposite or alternate, simple to palmately lobed or compound, stipulate; *flowers* imperfect; *staminate flowers* with 5 sepals and 5 stamens opposite the sepals, green or whitish; *pistillate flowers* solitary or in few-flowered clusters, sometimes paired, often covered or subtended by a bract, perianth inconspicuous, with a 2-branched stigma; *pistil* 1; *ovary* superior, 2-carpellate; *fruit* an achene or fleshy drupe.

1a. Trees; leaves simple, usually asymmetrically cordate at the base; fruit a fleshy drupe...***Celtis***
1b. Annual herbs or vines; leaves unlike the above; fruit an achene enclosed within a persistent bract...2

2a. Stems erect; leaves palmately compound...***Cannabis***
2b. Stems twining; leaves simple, usually palmately lobed...***Humulus***

CANNABIS L. – HEMP; MARIJUANA

Dioecious annual herbs from taproots; *leaves* palmately compound; *pistillate flowers* subtended and enclosed by a bract; *fruit* a lenticular achene enclosed within a persistent bract. (Small & Cronquist, 1976)

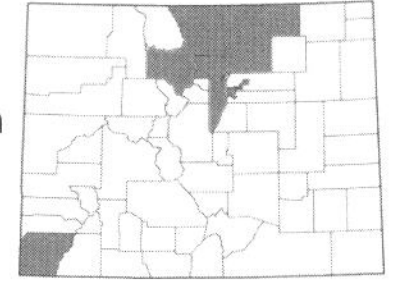

Cannabis sativa **L. ssp. *sativa***, MARIJUANA. Plants to 4 m; *leaflets* linear-lanceolate or linear, mostly 4–15 cm long, serrate, sparsely to densely hairy; *staminate flowers* on pedicels 0.5–2 mm long, with sepals 2.5–4 mm long; *pistillate flowers* sessile, enclosed by glandular-stipitate bracts; *achenes* 2–4 mm long, smooth. Cultivated, sometimes escaping, 5000–8600 ft. June–Aug. E/W. Introduced.

CELTIS L. – HACKBERRY

Monoecious trees; *leaves* entire to serrate, with pinnate venation; *staminate flowers* in cymose clusters or fascicles; *pistillate flowers* solitary or in few-flowered clusters; *fruit* a fleshy drupe.

Celtis reticulata **Torr.**, NETLEAF HACKBERRY. [*C. laevigata* Willd. var. *reticulata* (Torr.) L.D. Benson]. Trees to 15 m; *leaves* ovate, 3–5 (7) × 1.5–4 cm, cordate at the base, with entire or somewhat toothed margins above the middle, green and scabrous above, yellow-green below; *drupes* red, ca. 1 cm diam. Found on rocky slopes and in canyons, 3500–6600 ft. April–Sept. E/W.

The leaves and twigs are often infested by insect galls.

HUMULUS L. – HOP

Twining vines; *leaves* simple, palmately lobed; *pistillate flowers* in pairs, each pair subtended by a large, foliaceous bract, these imbricate in short spikes or drooping racemes; *fruit* an achene enclosed within a persistent bract.

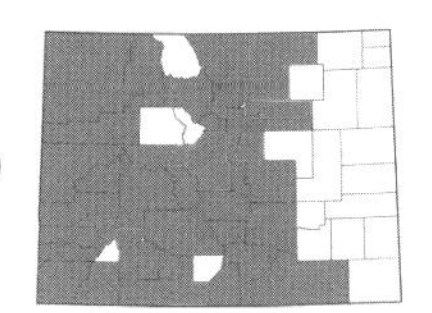

Humulus neomexicanus **Rydb.**, NEW MEXICAN HOP. Vines with scabrous stems; *leaves* 5–15 cm long and nearly as wide, hirsute, with serrate margins; *staminate flowers* on pedicels 0.5–3.5 mm long, with sepals 1.5–3 mm long; *pistillate flowers* subtended by ovate, papery (at maturity) bracts 7–20 mm long; *achenes* 2–3 mm long, smooth. Found in cool, moist, shaded canyons, sometimes on rocky outcrops, 4800–8500 ft. June–Aug. E/W. (Plate 24)

CAPPARACEAE Juss. – CAPER FAMILY

Herbs (ours); *leaves* alternate, petiolate, palmately compound or rarely simple above, stipulate or exstipulate; *flowers* perfect, actinomorphic or zygomorphic, solitary in axils or in terminal racemes; *sepals* 4, distinct; *petals* 4, distinct, often clawed; *stamens* 6–many, distinct; *pistil* 1; *ovary* superior, 2-carpellate, often stipitate; *fruit* a 2-valved capsule.

1a. Flowers white, ochroleucous, or pale pink, usually notched at the tip; plants densely viscid-hairy; stamens 8–20; fruit erect at maturity, elongate and broadly linear...***Polanisia***
1b. Flowers pink, pink-purple, or yellow, rounded or pointed at the tip; plants glabrous; stamens 6; fruit elongate and broadly linear and pendulous at maturity or rhomboidal and divaricately ascending...2

2a. Fruit elongate (longer than wide) and pendulous at maturity; flowers pink or pink-purple or if yellow then the leaves mostly 5-foliate...***Cleome***
2b. Fruit wider than long, rhomboidal and often horned distally, divaricately ascending at maturity; flowers yellow...***Cleomella***

CLEOME L. – SPIDERFLOWER

Annual herbs; *leaves* palmately 3–7-foliate, exstipulate; *flowers* in terminal bracteate racemes; *petals* yellow, pink, purplish, or rarely white; *stamens* usually 6; *fruit* borne on a slender gynophore.

1a. Flowers yellow; leaves mostly 5-foliate, occasionally 3–4-foliate...***C. lutea* var. *lutea***
1b. Flowers pink or pink-purple; leaves 3-foliate...2

2a. Leaflets linear to linear-lanceolate, less than 1.5 mm wide and often folded along the midrib; petals 4–7 mm long; stamens equal to the petals in length; fruit 0.9–2 cm long...***C. multicaulis***
2b. Leaflets lanceolate to elliptic, 5–10 mm wide; petals 8–12 mm long; stamens conspicuously longer than the petals; fruit 2–8 cm long...***C. serrulata***

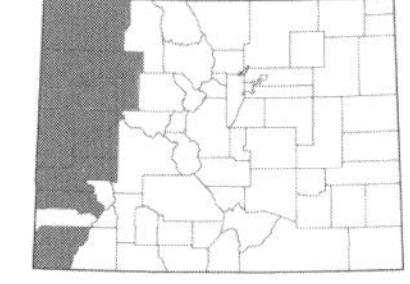

Cleome lutea* Hook. var. *lutea, YELLOW SPIDERFLOWER. Plants 5–10 dm; *leaflets* 5 or occasionaly 3 in upper leaves, 3–5 cm long; *petals* yellow, 5–8 mm long; *capsules* linear, 15–35 mm long, on a gynophore 10–15 mm long. Found in dry, open places such as canyon bottoms and adobe flats, sometimes in disturbed areas, 4500–7200 ft. May–July. W.

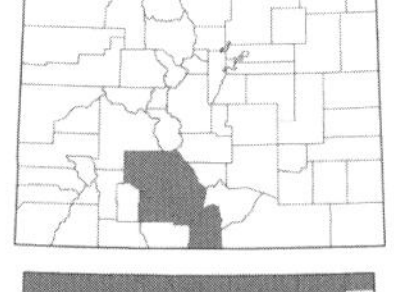

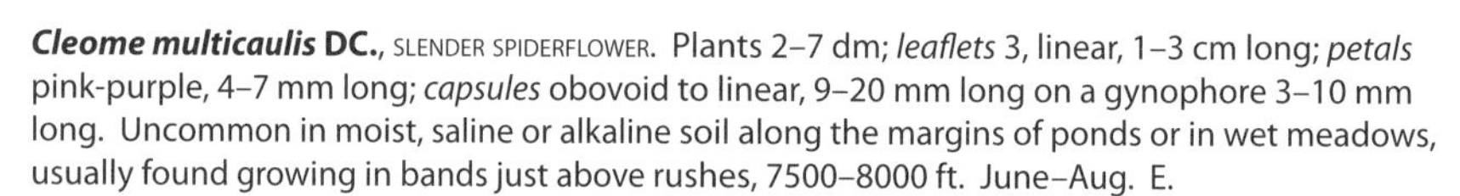

***Cleome multicaulis* DC.**, SLENDER SPIDERFLOWER. Plants 2–7 dm; *leaflets* 3, linear, 1–3 cm long; *petals* pink-purple, 4–7 mm long; *capsules* obovoid to linear, 9–20 mm long on a gynophore 3–10 mm long. Uncommon in moist, saline or alkaline soil along the margins of ponds or in wet meadows, usually found growing in bands just above rushes, 7500–8000 ft. June–Aug. E.

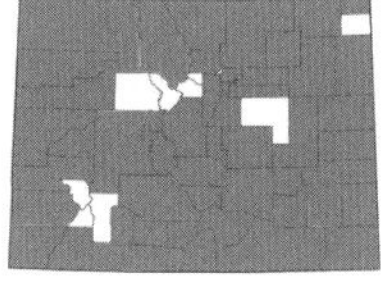

***Cleome serrulata* Pursh**, ROCKY MOUNTAIN BEEPLANT. Plants 2–15 dm; *leaflets* 3, 2–6 cm long; *petals* pink to purplish or rarely white, 8–12 mm long; *capsules* linear to fusiform, 2–8 cm long, on a gynophore 11–23 mm long. Common in sandy soil, often found along roadsides, 3500–9000 ft. June–Sept. E/W. (Plate 24)

CLEOMELLA DC. – STINKWEED

Annual herbs; *leaves* palmately 3-foliate; *flowers* in terminal bracteate racemes; *petals* yellow, closed in bud; *stamens* 6; *fruit* borne on a slender gynophore.

1a. Tall plants (6–26 dm); leaflets 2.5–6 cm long, linear-elliptic...***C. angustifolia***
1b. Smaller plants (1–4 dm); leaflets 0.8–2.6 cm long, elliptic, ovate, or ovate-oblong...***C. palmeriana***

***Cleomella angustifolia* Torr.**, NARROWLEAF RHOMBOPOD. Plants 6–26 dm; *leaflets* 2.5–6 cm long, linear-elliptic; *petals* 4–6 mm long; *capsules* rhomboid-obdeltoid, 5–10 mm long, on a gynophore 4–7 mm long. Found in sandy places, known from 2 old collections near Julesburg along the Platte River, 3500 ft. June–Aug. E.

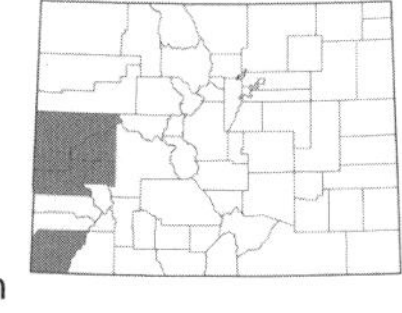

***Cleomella palmeriana* M.E. Jones**, ROCKY MOUNTAIN STINKWEED. [*C. cornuta* Rydb.; *C. montrosae* Payson]. Plants 1–4 dm; *leaflets* 0.8–2.6 cm long, elliptic, ovate, or ovate-oblong; *petals* 2–4 mm long; *capsules* ovoid to globose, 3–5 mm long, on a gynophore about 2 mm long. Found on clay slopes and adobe hills, 4800–6500 ft. May–July. W.

The type of *Cleomella hillmanii* A. Nelson var. *goodrichii* (S.L. Welsh) P.K. Holmgren was taken about 7 miles west of the Colorado state line at 5300 ft in elevation, and it could very well occur in Moffat Co. It differs from *C. palmeriana* in having well-defined bracts extending nearly to the top of the inflorescence, longer petioles (2–6 cm vs. 0.4–2.5 cm in *C. palmeriana*), and wider fruit (6–10.5 mm wide vs. 3–5.5 mm wide in *C. palmeriana*) with conspicuous horns.

POLANISIA Raf. – CLAMMYWEED

Annual herbs with a rank odor, viscid-pubescent; *leaves* palmately 3-foliate; *flowers* in bracteate terminal racemes; *petals* white, ochroleucous, or purplish-tinged, open in bud and not covering the stamens; *stamens* 6–20, of unequal length, all adaxial; *fruit* sessile or short-stipitate. (Iltis, 1958)

1a. Leaflets linear to linear-lanceolate, 1–4 mm wide; petals 2–4 mm long and laciniate at the tip...***P. jamesii***
1b. Leaflets oblanceolate to lanceolate-elliptic, 4–20 mm wide; petals 8–16 mm long and usually emarginate at the tip...***P. dodecandra* ssp. *trachysperma***

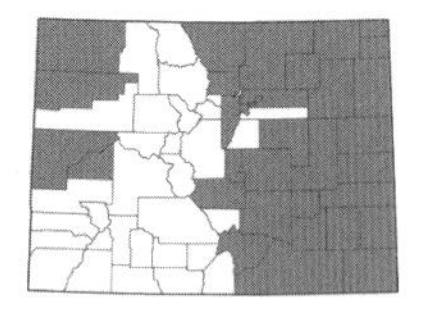

***Polanisia dodecandra* (L.) DC. ssp. *trachysperma* (Torr. & A. Gray) Iltis**, RED-WHISKER CLAMMYWEED. Plants 2–8 dm; *leaflets* 2–4 cm long, oblanceolate; *petals* white or sometimes purplish-tinged, 8–16 mm long; *capsules* 2–7 cm long. Found in sandy soil, often in disturbed areas, 3500–6800 ft. July–Oct. E/W. (Plate 24)

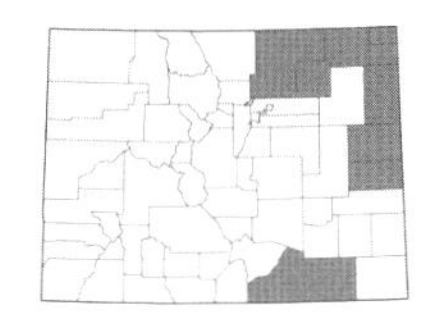

Polanisia jamesii **(Torr. & A. Gray) Iltis**, JAMES' CLAMMYWEED. Plants 1–4 dm; *leaflets* linear to narrowly lanceolate, 2–4 cm long; *petals* white or ochroleucous, 2–4 mm long; *capsules* linear-cylindric, 2–3 cm long. Found in sandy places on the eastern plains, 3500–4700 ft. June–Sept. E.

CAPRIFOLIACEAE Juss. – HONEYSUCKLE FAMILY

Trees, shrubs or subshrubs, or woody vines; *leaves* opposite, simple; *flowers* perfect or imperfect, usually borne in pairs; *calyx* usually small and 5-lobed; *corolla* usually 5-lobed, sometimes bilabiate; *ovary* inferior, 2–8-carpellate; *fruit* a capsule, berry, or dry drupe.

1a. Creeping subshrubs with prostrate stems, rooting at the nodes; stamens 4; flowers in terminal pairs on long peduncles...***Linnaea***
1b. Erect shrubs; stamens 5; flowers in axillary pairs or terminal clusters...2

2a. Axillary pairs of flowers subtended by a pair of bracts; flowers irregular, yellow, yellowish-red, pink, or white; berries black, orange, or red, with many seeds...***Lonicera***
2b. Pairs of flowers not subtended by a pair of bracts; flowers pink or white; berries white, with 2 seeds...***Symphoricarpos***

LINNAEA L. – TWINFLOWER

Evergreen subshrubs; *leaves* opposite, exstipulate, the margins entire or toothed; *flowers* actinomorphic or nearly so, pink or pinkish, borne in pairs; *calyx* 5-lobed; *corolla* 5-lobed; *stamens* 4, with 2 long and 2 short; *ovary* 3-carpellate; *fruit* a dry drupe.

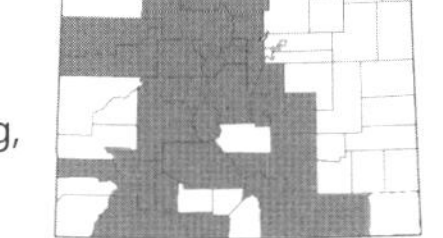

Linnaea borealis **L. var. *longiflora*** **Torr.**, TWINFLOWER. Stems trailing, rooting at the nodes; *leaves* elliptic to obovate or nearly orbicular, 0.5–2 cm × 3–15 mm, entire or few-toothed, glabrous to hairy; *peduncles* glandular-stipitate and often hairy; *corolla* 6–12 mm long; *drupes* 1.5–3 mm long, hairy. Locally common in moist forests and on alpine slopes, 7500–13,000 ft. June–Sept. E/W. (Plate 24)

LONICERA L. – HONEYSUCKLE

Shrubs or woody vines; *leaves* opposite, the margins usually entire; *flowers* actinomorphic or nearly so to zygomorphic and bilabiate, borne in axillary pairs or in terminal clusters; *calyx* 5-lobed; *corolla* 5-lobed; *stamens* 5; *ovary* 2–3-carpellate; *fruit* a berry.

1a. Flowers yellow or yellowish-red, hairy externally, each pair subtended by large (over 5 mm wide), green and foliaceous, glandular bracts; fruit red, becoming black at maturity...***L. involucrata***
1b. Flowers pink, white, ochroleucous, or light yellow, glabrous, each pair subtended by narrow (to 1 mm wide), eglandular bracts; fruit red or orange...2

2a. Flowers pale yellow to ochroleucous; leaves rounded to obtuse at the tips...***L. utahensis***
2b. Flowers rose-pink or white; leaves acute or acuminate at the tips...3

3a. Plants glabrous; flowers rose-pink...***L. tatarica***
3b. Plants hairy, especially on the underside of the leaves; flowers white or tinged with light pink...***L. morrowii***

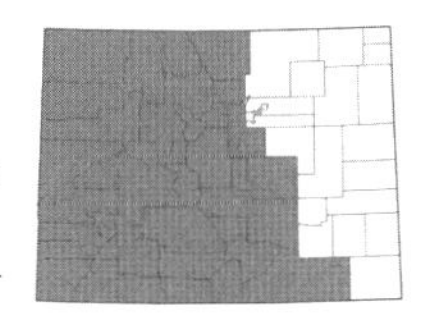

Lonicera involucrata **(Richardson) Banks ex Spreng.**, BLACK TWINBERRY. Shrubs 0.5–2 (3) m; *leaves* elliptic to elliptic-ovate, 5–14 × 2–8 cm, glabrous above and usually hirsute below; *bracts* (8) 10–15 (20) mm long, foliaceous or anthocyanic, these enclosing another pair of bracts, both pairs spreading in fruit; *corolla* yellow or yellowish-red, hairy, ca. 1–1.5 cm long; *berries* red, becoming black at maturity. Common along streams and creeks and in meadows, often with willows, 6000–11,800 ft. June–Aug. E/W.

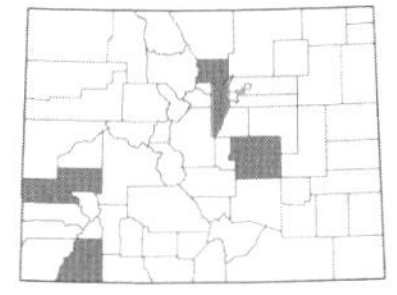

Lonicera morrowii **A. Gray**, MORROW'S HONEYSUCKLE. Shrubs to 3 m; *leaves* elliptic to oblong, hirsute below; *bracts* minute; *corolla* white to light pink, 1.5–2 cm long; *berries* red or reddish-orange. Uncommon in canyons of the foothills and disturbed areas, 5500–7000 ft. May–Aug. E/W. Introduced.

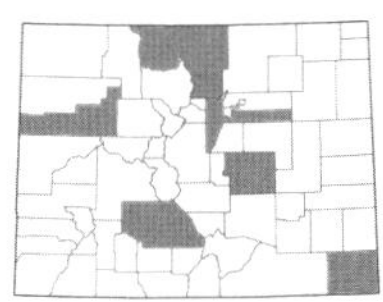

Lonicera tatarica **L.**, TATARIAN HONEYSUCKLE. Shrubs to 3 m; *leaves* ovate to elliptic, 2.5–5 (6) cm long, glabrous; *bracts* minute; *corolla* rose-pink, 11–20 mm long; *berries* orange or reddish-orange. Uncommon in canyons of the foothills and disturbed areas, 4300–8500 ft. April–June. E/W. Introduced.

Lonicera utahensis **S. Watson**, UTAH HONEYSUCKLE. Shrubs to 2 m; *leaves* elliptic to oblong, ca. 2–8 cm long, glabrous above and glabrous or hirsute below; *bracts* minute, 1–3 mm long; *corolla* pale yellow to nearly white, 1–2 cm long; *berries* red. Not reported for Colorado but to be expected near the border of Utah. June–July.

SYMPHORICARPOS Duham. – SNOWBERRY

Shrubs; *leaves* opposite, the margins entire or toothed to lobed; *flowers* actinomorphic, in a terminal cluster or axillary pairs, funnelform; *calyx* (4-) 5-lobed; *corolla* (4-) 5-lobed; *stamens* (4) 5; *ovary* 4-carpellate; *fruit* a berry. (Jones, 1940)

1a. Corolla densely hairy within at the level of the insertion of the filaments (near the top of the corolla tube), campanulate with lobes nearly as long or longer than the tube...2
1b. Corolla glabrous within or densely hairy within below the level of insertion of the filaments (near the base of the corolla tube), campanulate to salverform with lobes shorter than the tube...3

2a. Stamens and style exserted from the corolla; flowers several, 6–15 in dense, spicate clusters in the leaf axils; corolla lobes evidently spreading; leaves mostly 2.5–8 cm long...***S. occidentalis***
2b. Stamens and style included in the corolla; flowers generally fewer, 2–3 (5) in the leaf axils; corolla lobes scarcely spreading; leaves 1–3 (5) cm long...***S. albus***

3a. Anthers sessile or nearly so; style 4–7 mm long, stiffly hairy above the middle; corolla salverform, the lobes abruptly spreading; leaves 0.6–1.5 (2) cm long, oblanceolate or narrowly elliptic although sometimes broadly elliptic and up to 9 mm wide...***S. longiflorus***
3b. Anthers on short but evident filaments; style 2–4 mm long, glabrous; corolla elongate-campanulate or subsalverform, the lobes slightly spreading; leaves 0.7–3.5 (5) cm long, elliptic to ovate, 3–25 mm wide...***S. rotundifolius***

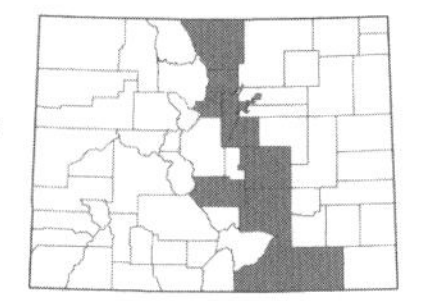

Symphoricarpos albus **(L.) S.F. Blake**, WHITE SNOWBERRY. Shrubs to 1 m; *leaves* ovate to elliptic, 1–3 (5) cm long, entire or with few irregular teeth; *flowers* 2–3 (5) in the leaf axils; *corolla* 5–8 mm long, the lobes shallow and scarcely spreading, densely hairy within; *berries* 7–9 mm long. Locally common on open hillsides and in forests and meadows along the Front Range, 5500–9800 ft. June–July. E. (Plate 24)

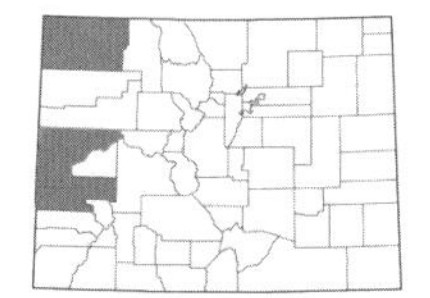

Symphoricarpos longiflorus **A. Gray**, DESERT SNOWBERRY. Shrubs to 1.5 (2) m; *leaves* oblanceolate to elliptic, 0.6–1.5 (2) cm long, entire or with 1–2 teeth; *flowers* in a few-flowered, terminal raceme; *corolla* tube 7–14 mm long with spreading lobes 2–4 mm long, glabrous within; *berries* 8–10 mm long. Uncommon on dry slopes and sandy ledges, 5000–7000 ft. May–July. W.

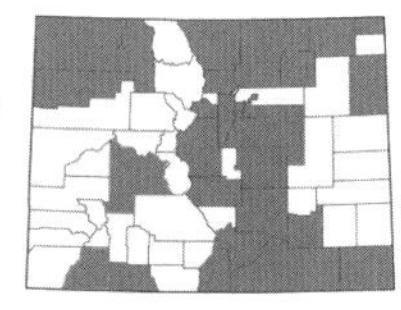

Symphoricarpos occidentalis **Hook.**, WOLFBERRY. Shrubs 0.3–1 m; *leaves* ovate to elliptic, mostly 2.5–8 cm long, entire or with a few irregular teeth; *flowers* 6–15 in dense, spicate clusters in the leaf axils; *corolla* 5–8 mm long, the lobes as long or longer than the tube, spreading, densely hairy within; *berries* 6–9 mm long. Common near streams and lakes and in meadows, 3500–8500 ft. June–Aug. E/W. (Plate 24)

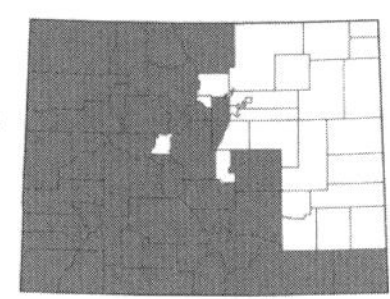

Symphoricarpos rotundifolius **A. Gray**, MOUNTAIN SNOWBERRY. [*S. oreophilus* A. Gray; *S. palmeri* G.N. Jones]. Shrubs to 1.5 (2) m; *leaves* elliptic to ovate, 0.7–3.5 (5) cm long, entire or occasionally few-toothed or lobed; *flowers* in few-flowered, terminal racemes; *corolla* 6–13 mm long, the lobes ¼–½ as long as the tube, glabrous or often hairy below the level of insertion of the filaments; *berries* mostly 7–10 mm long. Common on dry slopes, in meadows, canyons, and along roadsides and occasionally along creeks, 5500–11,00 ft. June–Aug. E/W. (Plate 24)

CARYOPHYLLACEAE Juss. – PINK FAMILY

Annuals, biennials, or perennials; *leaves* opposite, whorled, or rarely alternate, simple and entire, stipulate or exstipulate; *flowers* usually perfect, actinomorphic, usually in terminal or axillary cymes; *pistil* 1; *ovary* superior, with free-central or basal placentation; *fruit* a capsule opening by valves, or a one-seeded utricle. (Rabeler & Hartman, 2005)

1a. Papery stipules present...2
1b. Stipules absent...6

2a. Leaves all whorled, linear; styles 5; petals white, the apex entire...***Spergula***
2b. Leaves mostly opposite, sometimes with axillary fascicles of leaves present, variously shaped; styles 1–3; petals absent or present and various...3

3a. Petals absent, the sepals greenish or yellowish; styles 1–2; fruit an indehiscent utricle…4
3b. Petals present (although sometimes shorter than the calyx), white or pink to rose; styles (2) 3; fruit a capsule opening by 3 valves…5

4a. Sepals usually awn-tipped or hooded; flowers in terminal cymes or solitary at the end of shoots; native perennials…***Paronychia***
4b. Sepals not awn-tipped or hooded; flowers axillary, opposite a leaf; weedy annuals or perennials…***Herniaria***

5a. Sepal tips hooded; petals white, the apex bifid; stamens 5…***Drymaria***
5b. Sepals lacking hooded tips; petals white or pink to rose, the apex entire; stamens 1–10 but usually not 5…***Spergularia***

6a. Sepals united at the base to form a campanulate or cylindric cup or tube, with distinct teeth or lobes at the top; flowers white, red, purple, or pink…7
6b. Sepals all distinct or nearly so; flowers white to yellowish or rarely pink…12

7a. Calyx 1–5 mm long; inflorescence a much-branched, open cyme with numerous flowers…***Gypsophila***
7b. Calyx usually well over 5 mm long, or if 5 mm long then the flowers solitary and terminal or in a few-flowered axillary cyme…8

8a. Flowers closely subtended by 2 bracts; petals with toothed tips, sometimes the margins of the petals toothed as well…***Dianthus***
8b. Flowers not closely subtended by 2 bracts; petal tips entire or shortly bifid, but not toothed…9

9a. Calyx lobes 12–45 mm long, linear to linear-lanceolate, green and foliaceous, longer than the petals; plants with long spreading, pilose hairs, often giving the plants a grayish or silvery appearance; flowers reddish-purple or purple and paler at the base, rarely all white, each petal with 3 darker purplish veins; styles (4) 5…***Agrostemma***
9b. Calyx lobes shorter than 12 mm, otherwise unlike the above in all respects; hairs unlike the above or plants glabrous; flowers various but unlike the above; styles 2–5…10

10a. Calyx tube with 5 prominent, green, winged angles; plants glabrous and glaucous; middle stem leaves clasping at the stem; stamens adnate to the petals; flowers pink to purple, usually numerous in an open cyme…***Vaccaria***
10b. Calyx tube lacking green wings on the angles; plants variously pubescent; middle stem leaves petiolate or sessile but not clasping; stamens distinct; flowers white, pink, reddish, or purple; inflorescence various…11

11a. Petal limb usually shorter than 8 mm, if longer then the tip bifid over half its length; calyx often with 10–20 prominent, green or purple veins, sometimes inflated; leaves usually with 1 vein or if 3-veined then also with lateral veins present as well; styles 3–5…***Silene***
11b. Petal limb 8–15 mm long, the tips entire or shortly notched; calyx lacking prominent veins, not inflated; leaves usually 3-veined or sometimes with 1 vein; styles 2 or rarely 3…***Saponaria***

12a. Petals erose but not cleft at the tips; stamens 3–5; flowers in a terminal, umbellate cyme with all pedicels arising from the same point or nearly so, the pedicels strongly reflexed after flowering but ultimately erect in fruit…***Holosteum***
12b. Petals entire or notched to deeply bifid, but not erose, or absent; stamens 4–10; flowers variously arranged or sometimes in a subumbellate cyme but then the petals absent; pedicels various but generally unlike the above…13

13a. Inflorescence a densely compact, round, head-like cyme, the pedicels mostly less than 4 mm long and often hidden…***Eremogone*** (*congesta*, *hookeri*)
13b. Inflorescence an open cyme or the flowers solitary…14

14a. Stems with small, downward-pointing hairs throughout or in lines, sometimes glandular-hairy as well, or spreading puberulent with small, peglike hairs (look under 20× magnification); styles 3…15
14b. Stems glabrous, papillate-scabrid on the stem angles, pubescent with pilose, spreading hairs, or glandular-hairy but lacking retrorse hairs; styles 3–5…16

15a. Pedicels glabrous or spreading puberulent with small, peglike hairs; seeds 1–2.2 mm long, with a white and spongy appendage present; sepals tips rounded or acute; perennials from slender rhizomes…***Moehringia***
15b. Pedicels with small, downward-pointing hairs; seeds 0.4–0.8 mm long, lacking an appendage; sepal tips acute; perennials from slender rhizomes or annuals from slender taproots…***Arenaria***

16a. Sepals glandular-hairy; perennials from rhizomes with spherical or elongate, tuberous thickenings; leaves linear-lanceolate to lanceolate, spreading at right angles, sessile, 2–10 cm long (usually over 3 cm long); petals bifid on the upper ⅓, about twice as long as the sepals; anthers purple...***Pseudostellaria***
16b. Sepals glabrous or if glandular-hairy then the leaves unlike the above in all respects; annuals or perennials, rhizomes lacking or if present then without tuberous thickenings; petals entire or bifid, shorter than to twice as long as the sepals, or absent; anthers yellow or purple...17

17a. Leaves linear, filiform, or needle-like, mostly 1.5 (2) mm wide; perennials, often caespitose or cushion- or mat-forming...18
17b. Leaves lanceolate to elliptic or ovate-oblong, usually at least some leaves more than 2 mm wide; annuals or perennials, mat-forming or not...23

18a. Petals distinctly bifid at the tips, sometimes nearly to the base, rarely absent and then the styles 3; leaves not mostly basal or greatly reduced on the stem, plants not densely caespitose or cushion-forming, sometimes mat-forming; stems 4-angled (sometimes difficult to see on pressed specimens) or round...19
18b. Petals entire or slightly notched at the tips, or if absent then the styles 4 or 5; leaves mostly basal and often reduced on the stem, or plants densely caespitose or cushion- to mat-forming; stems round...20

19a. Small tufts of leaves present in the axils of lower leaves; styles 5; petals about twice as long as the sepals, bifid to near the middle; stems round...***Cerastium*** (*arvense*)
19b. Small tufts of leaves absent in the axils of lower leaves; styles 3; petals equal to twice as long as the sepals, deeply bifid nearly to the base or rarely absent; stems 4-angled (sometimes difficult to see on pressed specimens)...***Stellaria***

20a. Styles 4 or 5; capsules opening by 4 or 5 valves; stamens 4, 5, 8, or 10; petals absent or 4–5, 1–3 mm long...***Sagina***
20b. Styles 3; capsules opening by 3 valves or 6 teeth; stamens 10; 5 petals present, 1.5–9 mm long...21

21a. Leaves lacking a spinose tip; plants small, mostly to 3 cm tall, found in the alpine and spruce-fir forests above 9000 ft in elevation; flowers solitary or in 2–7-flowered cymes; leaves 1–10 mm long; seeds 0.4–1 mm long...***Minuartia***
21b. Leaves with a small, spinose tip; plants generally larger, mostly over 3 cm tall, variously distributed at 4400–13,000 ft in elevation; flowers in 3–35-flowered cymes, rarely solitary; leaves usually longer, 5–30 mm in length; seeds 1.2–3 mm long...22

22a. Leaves glabrous to sparsely hairy but not glandular-hairy, mostly basal or with a basal tuft, often reduced upwards on the stem; plants caespitose or densely matted, the stems erect; capsules opening by 6 teeth...***Eremogone*** (*eastwoodiae*, *fendleri*)
22b. Leaves glandular-hairy, mostly basal or not; plants in mats with trailing stems; capsules opening by 3 valves...***Minuartia***

23a. Styles 5; petals bifid to near the middle; stems round, subglabrous or more often hairy or glandular, at least distally...***Cerastium***
23b. Styles 3; petals bifid to near the base or absent; stems 4-angled, usually glabrous or minutely papillate-scabrid on the angles, sometimes softly hairy or the hairs in lines and sometimes glandular...***Stellaria***

Stellaria flower:

AGROSTEMMA L. – CORNCOCKLE
Annual herbs; *leaves* exstipulate; *flowers* usually perfect; *calyx* 5-lobed, connate at the base; *petals* 5, reddish-purple or rarely white, clawed; *stamens* 10, 5 epipetalous; *styles* (4) 5; *stigmas* (4) 5; *fruit* an ovoid capsule opening by (4) 5 teeth.

***Agrostemma githago* L.**, COMMON CORNCOCKLE. Plants 3–10 dm; *leaves* linear to lanceolate, 4–15 cm long; *inflorescence* solitary and axillary; *calyx* 2.5–6 cm long, the tube 1.2–1.7 cm, spreading-hairy; *petals* 1.5–4 cm long; *capsules* 1.5–2.5 cm long. Not yet reported for Colorado, but a weed in the surrounding states, found in disturbed places, along roadsides, and in fields, and to be expected. Introduced.

ARENARIA L. – SANDWORT
Annual or perennial herbs; *leaves* usually sessile, exstipulate; *flowers* perfect, in terminal or axillary open cymes, or solitary; *sepals* 5, distinct; *petals* 5, white, not clawed, the apex entire; *stamens* 10, staminodes absent; *styles* 3; *stigmas* 3; *fruit* an ovoid to cylindric capsule opening by 6 teeth.
1a. Annuals; leaves elliptic to ovate, ...***A. serpyllifolia***
1b. Perennials; leaves linear-lanceolate to narrowly elliptic...***A. lanuginosa* var. *saxosa***

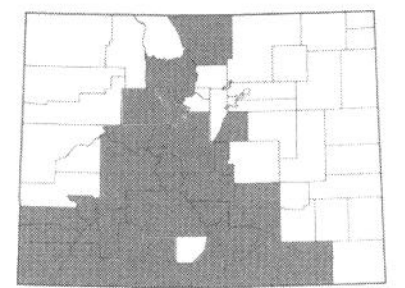

***Arenaria lanuginosa* (Michx.) Rohrb. var. *saxosa* (A. Gray) Zarucchi, R.L. Hartm., & Rabeler,** SPREADING SANDWORT. Perennials, 0.5–6 dm; *leaves* linear-lanceolate to narrowly elliptic, 3–35 × 2–14 mm, ciliate on margins; *flowers* in axillary cymes; *sepals* 2–5 mm long; *petals* 1.5–6 mm long; *capsules* ovoid, 3–6 mm long. Found in meadows, forests, along streams, and on open slopes, 7800–12,500 ft. June–Sept. E/W.

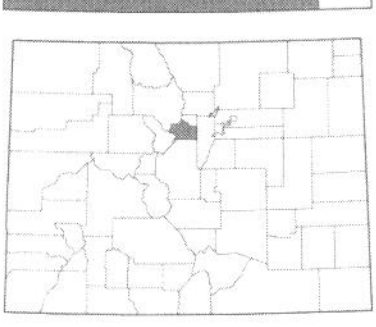

***Arenaria serpyllifolia* L.,** THYMELEAF SANDWORT. Annuals, 0.3–4 dm; *leaves* elliptic to ovate, 2–7 × 1–4 mm, ciliate on the margins; *flowers* in terminal cymes; *sepals* 2–3 (4) mm long; *petals* 0.6–3 mm long; *capsules* 3–3.5 mm long. Uncommon, known from a single collection made in Clear Creek Co., found in disturbed places, 7500–8000 ft. April–June. E. Introduced.

CERASTIUM L. – MOUSE-EAR; CHICKWEED

Annual or perennial herbs; *leaves* exstipulate; *flowers* usually perfect; *sepals* (4) 5, distinct; *petals* (4) 5 or rarely absent, white, clawed, 2-cleft or notched at the apex; *stamens* usually 10, occasionally 5 or 8 or fewer; *styles* (3) 5 (6); *stigmas* (3) 5 (6); *fruit* an often curved capsule opening at the apex by twice as many valves as the style number.

1a. Middle and lower leaves with tufts of leaves in the axils; petals twice as long as the sepals; capsules usually shorter than the sepals, rarely slightly longer...***C. arvense* ssp. *strictum***
1b. Middle and lower leaves lacking tufts of leaves in the axils; petals usually shorter to slightly longer than the sepals, rarely twice as long; capsules at least twice as long as the sepals...2

2a. Sepals with only eglandular, multicellar hairs, or a mixture of glandular hairs and eglandular, multicellular hairs; stems often decumbent and rooting at the nodes, usually with widely spaced internodes...***C. fontanum* ssp. *vulgare***
2b. Sepals with only glandular hairs; stems usually erect, the internodes widely spaced or not...3

3a. Petals 3–4 mm long, equal to or shorter than the sepals; pedicels often becoming deflexed at the base, 3–10 mm long; annuals...***C. brachypodum***
3b. Petals 6–12 mm long, equal to or longer than the sepals; pedicels erect, 8–55 mm long; perennials...***C. beeringianum***

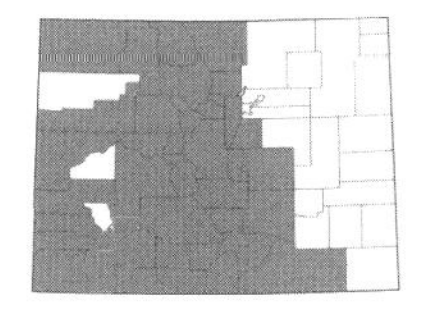

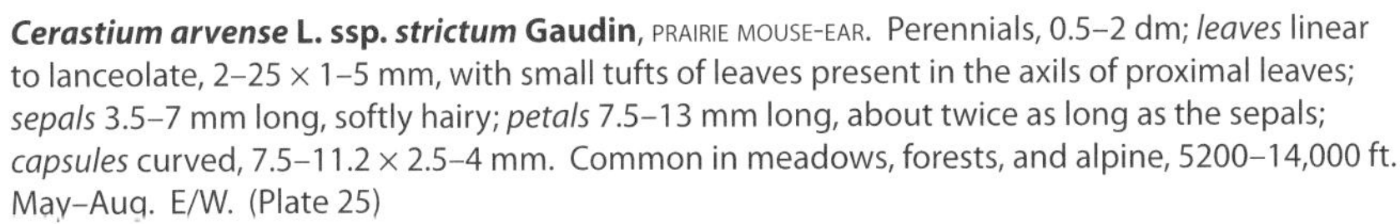

***Cerastium arvense* L. ssp. *strictum* Gaudin,** PRAIRIE MOUSE-EAR. Perennials, 0.5–2 dm; *leaves* linear to lanceolate, 2–25 × 1–5 mm, with small tufts of leaves present in the axils of proximal leaves; *sepals* 3.5–7 mm long, softly hairy; *petals* 7.5–13 mm long, about twice as long as the sepals; *capsules* curved, 7.5–11.2 × 2.5–4 mm. Common in meadows, forests, and alpine, 5200–14,000 ft. May–Aug. E/W. (Plate 25)

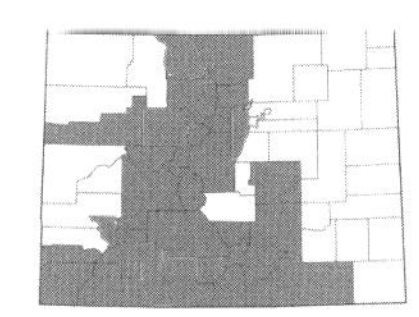

***Cerastium beeringianum* Cham. & Schltdl.,** BERING CHICKWEED. Perennials, 0.5–2.5 dm; *leaves* lanceolate to oblanceolate, 5–20 × 2–5 mm, small axillary tufts of leaves absent; *sepals* 3–7 mm long, densely glandular; *petals* 6–12 mm long, about equaling the sepals in length or rarely to twice as long; *capsules* curved, 8–12 mm long. Common in subalpine spruce-fir forests and alpine, 9500–14,100 ft. June–Aug. E/W.

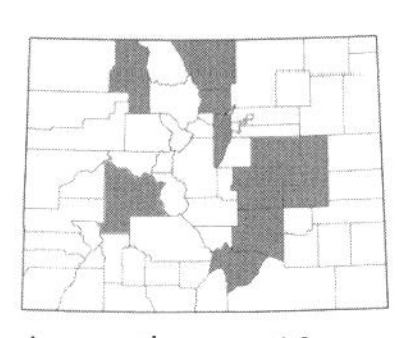

***Cerastium brachypodum* (Engelm. ex A. Gray) B.L. Rob.,** SHORTSTALK CHICKWEED. Annuals, 0.5–2.5 dm; *leaves* lanceolate to oblanceolate or narrowly elliptic, 5–30 × 2–8 mm, small axillary tufts of leaves absent; *sepals* 3–5 mm long, glandular; *petals* 3–4 mm long, shorter than or equal to the sepals; *capsules* curved, 6–12 mm long. Uncommon on grassy slopes, in disturbed meadows, and moist places, 5000–8500 ft. May–June. E/W.

Cerastium nutans Raf. could occur in Colorado. It differs from *C. brachypodum* in having pedicels mostly over 10 mm long, conspicuously longer than the capsules, and usually becoming sharply deflexed towards the tip (near the capsule) in fruit. *Cerastium brachypodum* has pedicels mostly 3–10 mm long, shorter than or equaling the capsules in length, and usually becoming deflexed at the base in fruit.

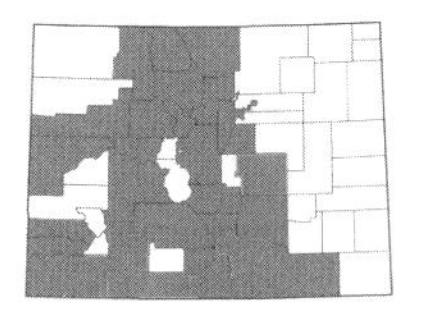

***Cerastium fontanum* Baumg. ssp. *vulgare* (Hartm.) Greuter & Burdet,** COMMON MOUSE-EAR. Perennials or annuals, 1–4.5 dm; *leaves* elliptic to oblong, 10–25 (40) × 3–8 (12) mm, small axillary tufts of leaves absent; *sepals* 5–7 mm long, with eglandular hairs; *petals* 1–1.5 times as long as the sepals; *capsules* curved, 9–17 mm long. Common in disturbed places, along roadsides, along streams, and in moist meadows, 5400–10,500 ft. June–Aug. E/W. Introduced.

DIANTHUS L. – PINK

Annual, biennial, or usually perennial herbs; *leaves* exstipulate; *flowers* perfect, in a terminal, headlike cyme; *calyx* 5-lobed, connate basally into a tube; *petals* 5, pink, red, white, or purple, sometimes with a darker center, clawed, the apex dentate (ours) or fimbriate; *stamens* 10; *styles* 2; *stigmas* 2; *fruit* an ovoid to cylindric capsule opening by 4 teeth.

1a. Petal limbs narrowly elliptic, 4–5 mm long, tapering to the tip (about the same width at the tip as at the base of the limb), toothed at the tip and along the margin; bracteoles subtending each flower linear to lanceolate with an acute tip; sepals conspicuously hairy; stems usually hairy; flowers usually 3–6 per cyme, rarely solitary, pink or purplish-pink with white dots (rarely all white)...***D. armeria***
1b. Petal limbs flaring outward, 5–10 mm long, wider at the tip than at the base, toothed only at the tip; bracteoles subtending each flower ovate with a long-aristate tip; sepals glabrous or minutely hairy; stems glabrous or sparsely hairy; flowers usually solitary or sometimes 2–4 per cyme, variously colored...***D. deltoides* ssp. *deltoides***

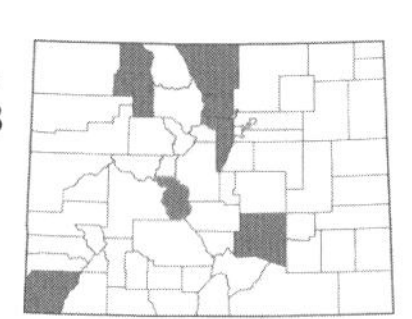

***Dianthus armeria* L.**, Deptford pink. Annuals or biennials, 2–7 dm, dichotomously branched above, usually hairy; *leaves* linear, 3–10 cm long; *flowers* 3–6 per cyme or rarely solitary, subtended by 1–3 pairs of linear to lanceolate-attenuate bracteoles; *petals* pink, purplish-pink with white dots, or rarely white, the limb narrowly elliptic, 4–5 mm long, tapering to the tip, dentate. Uncommon in disturbed places, ditches, and meadows, 5500–9500 ft. June–Aug. E/W. Introduced. (Plate 25)

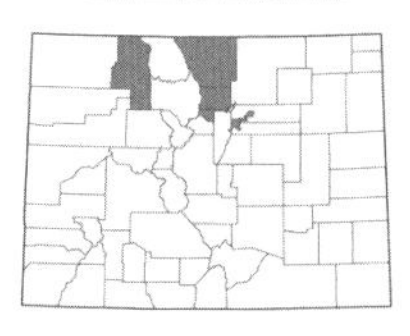

Dianthus deltoides* L. ssp. *deltoides, maiden pink. Perennials, 1–5 dm, glabrous or minutely hairy; *leaves* oblanceolate to narrowly lanceolate, 1.5–4 cm long; *flowers* solitary or 2–4 per cyme, subtended by an ovate bracteole with a long-aristate tip; *petals* white, red, purple, or red-purple, 5–10 mm long, the tips toothed. Uncommon in meadows and along roadsides, often escaping from cultivation, 5000–8500 ft. June–Aug. E/W. Introduced.

DRYMARIA Willd. ex Roem. & Schult. – drymary

Annual or perennial herbs; *leaves* stipulate; *flowers* perfect, in terminal or axillary bracteate cymes or umbellate clusters, or solitary; *sepals* 5, distinct, the apex hooded (ours); *petals* (3) 5 or sometimes absent, white, the apex 2–4-lobed; *stamens* 5; *styles* 3; *stigmas* 3 or sometimes 2; *fruit* a round or elliptical capsule. (Duke, 1961)

1a. Plants 0.5–5 cm in height; sepal tips obtuse or rounded; stem leaves oblong, the basal leaves orbicular or spatulate, 1–4 mm wide; style shallowly cleft; capsules about equaling the inner sepals in length...***D. depressa***
1b. Plants 8–25 cm in height; sepal tips acute or acuminate; leaves all linear to narrowly oblong, to 1 mm wide; style deeply cleft; capsules shorter than the inner sepals...***D. leptophylla* var. *leptophylla***

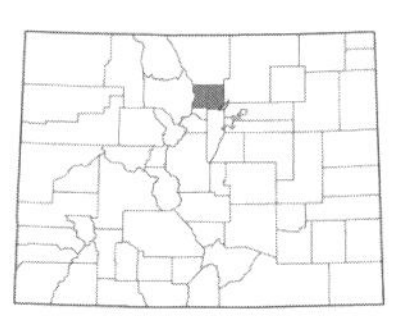

***Drymaria depressa* Greene**, pinewoods drymary. [*D. effusa* A. Gray var. *depressa* (Greene) J.A. Duke]. Annuals, 0.5–5 cm; *leaves* orbiculate to spatulate or oblong, 0.3–1 cm × 0.2–3 mm; *inflorescence* a 3–25+-flowered cyme; *sepals* 1.8–2.3 mm long, with 3 prominent, parallel veins, the tips obtuse or rounded; *petals* 1.5–3 mm long, bifid. Uncommon on drying mudflats and in sandy soil, known from along Sandbeach Lake trail in Rocky Mountain National Park, 9500–9700 ft. Aug.–Sept. E.

Drymaria leptophylla* (Cham. & Schltdl.) Fenzl ex Rohrb. var. *leptophylla, canyon drymary. Annuals, 8–25 cm; *leaves* linear to narrowly oblong, often involute, to 2.5 cm long; *inflorescence* a many-flowered terminal cyme; *sepals* 1.5–3.5 mm long, with 3 prominent, parallel veins, the tips acute or acuminate; *petals* to 2.5 mm long, bifid. Reported for Colorado, but no specimens have been seen.

EREMOGONE Fenzl – sandwort

Perennials; *leaves* needle-like or filiform, exstipulate; *flowers* perfect; *sepals* 5, distinct or nearly so; *petals* 5, white to yellowish or pink to brownish, sometimes clawed, apex entire or erose; *stamens* 10; *styles* 3; *stigmas* 3; *fruit* opening by 6 teeth. (Maguire, 1947)

1a. Plants forming dense, rounded cushions; stems scabrid-puberulent; leaves 0.3–4 cm long...***E. hookeri***
1b. Plants tufted or sometimes matted, but not forming dense, rounded cushions; stems glabrous or glandular-stipitate; leaves 1–11 cm long...2

2a. Inflorescence a compact, capitate cyme, the pedicels mostly less than 4 mm long and often hidden; petals 1.5–2 times as long as the sepals in length...***E. congesta***
2b. Inflorescence an open cyme, the pedicels mostly over 5 mm long and conspicuous; petals 0.9–1.3 times as long as the sepals in length...3

3a. Sepals glabrous or sparsely glandular-stipitate; nectaries longitudinally rectangular and cleft or notched at the tip, over 0.5 mm long; leaves 1–3.5 cm long...***E. eastwoodiae***
3b. Sepals densely glandular-stipitate; nectaries rounded at the base of the filaments, less than 0.5 mm long; leaves 1–11 cm long...***E. fendleri***

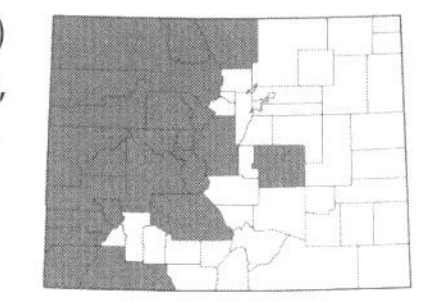

Eremogone congesta **(Nutt.) Ikonn.**, BALLHEAD SANDWORT. [*Arenaria congesta* Nutt.]. Plants 0.3–4 (5) dm; *leaves* 1–14 cm × 0.4–2 mm, glabrous; *inflorescence* a congested cyme; *sepals* 3–6.5 mm long, ovate to lanceolate with a shortly acute to rounded tip, glabrous; *petals* white, 5–8 (10) mm long, the tip entire to slightly emarginated; *capsules* 3.5–6 mm long. Found in meadows and open slopes in sandy or rocky soil, often with sagebrush, 6800–12,600 ft. June–Aug. E/W.

There are two varieties of *E. congesta* in Colorado:

1a. Pedicels evident, 1–4 mm long…**var. *lithophila* (Rydb.) Dorn**
1b. Flowers sessile or nearly so, the pedicels not evident…**var. *congesta***

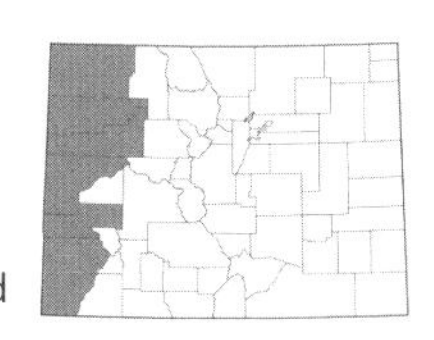

Eremogone eastwoodiae **(Rydb.) Ikonn.**, EASTWOOD'S SANDWORT. [*Arenaria eastwoodiae* Rydb.]. Plants 0.8–2.5 dm; *leaves* 1–3.5 cm × 0.5–0.7 mm, glabrous to minutely hairy; *inflorescence* a 3–17-flowered, open cyme; *sepals* 3.5–6.5 mm long, acute to acuminate, glabrous or sparsely glandular-stipitate; *petals* yellowish-white or brownish to pink, 4–6.5 mm long, cleft or emarginated at the tips; *capsules* 4–6 mm long. Found on dry, open slopes, barren ridgetops, and bluffs, in sandy or rocky soil, 4400–9200 ft. May–July. W.

There are two varieties of *E. eastwoodiae* in Colorado. *Eremogone kingii* (S. Watson) Ikonn. [*Arenaria kingii*] has long been reported for Colorado but does not occur in the state. Most specimens identified as such are actually misidentifications of *E. eastwoodiae*:

1a. Stems and pedicels glabrous…**var. *eastwoodiae***
1b. Stems and pedicels glandular-stipitate…**var. *adenophora* (Kearney & Peebles) R.L. Hartm. & Rabeler**

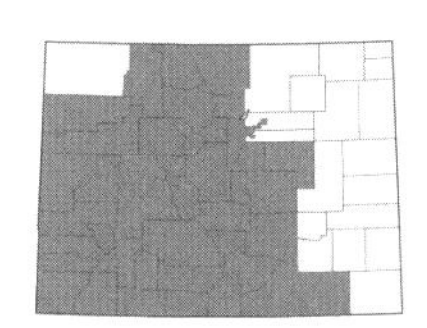

Eremogone fendleri **(A. Gray) Ikonn.**, FENDLER'S SANDWORT. [*Arenaria fendleri* A. Gray]. Plants (0.2) 1–4 dm; *leaves* 1–11 cm × 0.2–0.4 mm, glabrous to minutely hairy; *inflorescence* an open cyme; *sepals* 4–7.5 mm long, acuminate, glandular-stipitate; *petals* white, 4–8 mm long, entire to erose at the tips; *capsules* 5–7 mm long. Common in sandy or rocky soil of open slopes, meadows, forests, and the alpine, 5400–13,000 ft. June–Aug. E/W. (Plate 25)

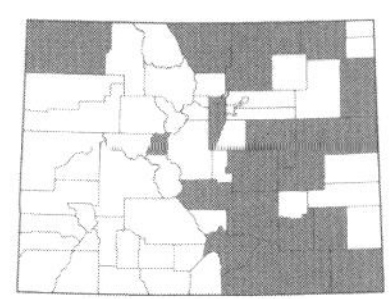

Eremogone hookeri **(Nutt.) W.A. Weber**, HOOKER'S SANDWORT. [*Arenaria hookeri* Nutt.]. Plants forming dense cushions, 0.1–2 dm; *leaves* 0.3–4 cm × 0.5–1.5 mm, glabrous; *inflorescence* a congested cyme; *sepals* 5–10 mm long, acute to acuminate, glabrous or minutely hairy; *petals* white, 4.5–8.5 mm long, with a rounded to obtuse apex; *capsules* to 4 mm long. Found on sandstone outcroppings, sandy bluffs, and limestone breaks, 3500–7800 ft. May–July. E/W. (Plate 25)

There are two distinct varieties of *E. hookeri* in Colorado:

1a. Capitate cymes mostly 1.5 cm or less wide; sepals mostly 0.5–0.6 cm long; basal leaves 0.3–1.5 cm long…**var. *hookeri***. [*Arenaria hookeri* ssp. *desertorum* (Maguire) W.A. Weber]. 3500–7800 ft. E/W.
1b. Capitate cymes mostly 3–4 cm wide; sepals mostly 0.9–1 cm long; basal leaves 2–4 cm long…**var. *pinetorum* (A. Nelson) Dorn**. 3500–7200 ft. E.

GYPSOPHILA L. – BABY'S BREATH

Annual or perennial herbs; *leaves* exstipulate; *flowers* perfect, in diffuse cymes or thyrses; *calyx* 5-lobed, connate basally into a cup; *petals* 5, white, pink, or rose-purple, with a poorly defined claw, the apex entire to bifid; *stamens* 10; *styles* 2 (3); *stigmas* 2 (3); *fruit* a globose or ellipsoid capsule usually opening by 4 valves. (Barkoudah, 1962)

1a. Inflorescence with glandular hairs on the pedicels; leaf bases clasping and surrounding the stem; petals 4–6 mm long…***G. scorzonerifolia***
1b. Inflorescence glabrous on the pedicels; leaf bases not clasping or only the lower leaves with clasping bases; petals 1–4 or 6–15 mm long…2

2a. Petals 6–15 mm long; leaves at the base of the plant with clasping bases that surround the stem, the upper leaves usually only with rounded bases; annuals, 0.4–6 dm tall…***G. elegans***
2b. Petals 1–4 mm long; leaf bases not clasping; perennials, 4–10 dm tall…***G. paniculata***

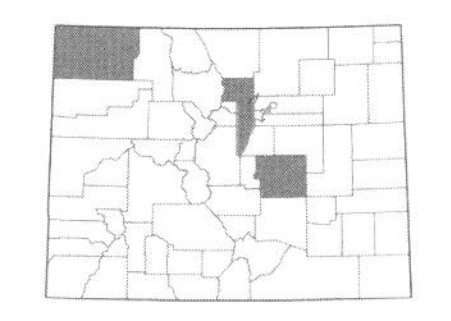

Gypsophila elegans **Bieb.**, SHOWY BABY'S BREATH. Annuals, 0.4–6 dm, glabrous; *leaves* linear-lanceolate to oblong, 1.5–7 cm × (1) 3–16 mm, the lower leaves with clasping bases; *calyx* 2.5–5 mm long; *petals* white or sometimes with pinkish veins, rarely pink, 6–15 mm long. Uncommon in disturbed places and along roadsides, apparently introduced by reseeding projects, 5500–7500 ft. June–Sept. E/W. Introduced.

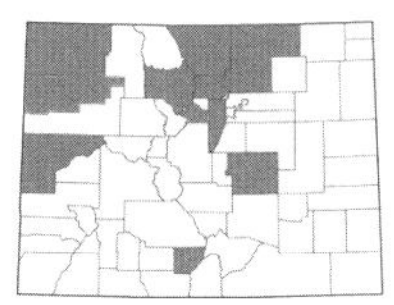

Gypsophila paniculata **L.**, TALL BABY'S BREATH. Perennials, 4–10 dm, glabrous or glandular to minutely hairy below; *leaves* linear-lanceolate to narrowly oblong, 2–9 cm × 2–10 mm, the bases not clasping; *calyx* 1–3 mm long; *petals* white or rarely light pink, 1–4 mm long. Uncommon along roadsides and in other disturbed places, 4500–7500 ft. June–Aug. E/W. Introduced.

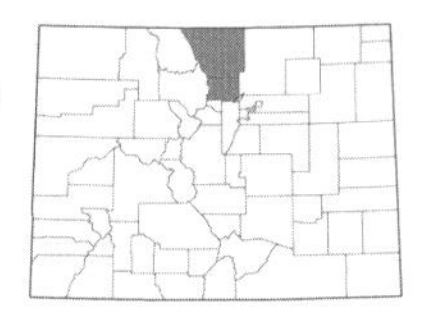

***Gypsophila scorzonerifolia* Ser.**, GLANDULAR BABY'S BREATH. Perennials, 5–20 dm, glabrous below and sparsely glandular above; *leaves* narrowly oblong to narrowly ovate, 2–15 cm × 7–22 (35) mm, the bases clasping; *calyx* 2.5–4 mm long, glandular; *petals* white with pink tinges to light pinkish, 4–6 mm long. Uncommon in disturbed places, 5000–5500 ft. July–Sept. E. Introduced.

HERNIARIA L. – RUPTUREWORT

Herbs; *leaves* stipulate; *flowers* perfect, in densely clustered cymes on short lateral branches opposite a leaf; *sepals* 5, greenish, distinct; *petals* absent; *stamens* 4–5; *styles* 2; *stigmas* 2; *fruit* an utricle.

***Herniaria glabra* L.**, GREEN CARPET RUPTUREWORT. Plants prostrate, 5–35 cm; *leaves* obovate-elliptic to orbiculate, 3–7 (10) mm long, usually glabrous; *sepals* 0.5–0.6 mm long, glabrous; *stamens* 5, alternating with 5 petaloid staminodes 0.5 mm long. Uncommon weed, sometimes planted and escaping, recently found in Boulder Co., 5500–6000 ft. April–June. E. Introduced.

HOLOSTEUM L. – JAGGED CHICKWEED

Annual herbs; *leaves* exstipulate; *flowers* perfect or sometimes imperfect, in terminal, umbellate cymes; *sepals* 5, distinct; *petals* 5, white to pink, clawed, the apex erose but not cleft; *stamens* 3–5, staminodes absent; *styles* 3 (5); *stigmas* 3 (5); *fruit* an ovoid to cylindric capsule, usually opening by 6 teeth.

Holosteum umbellatum* L. ssp. *umbellatum, COMMON JAGGED CHICKWEED. Plants 1–3 dm, glandular-stipitate or sometimes only in the middle, otherwise glabrous; *leaves* oblong to oblanceolate, 1–2.5 cm long; *sepals* ca. 3 mm long; *petals* just longer than the sepals. Uncommon in disturbed places, along ditches and paths, 5000–6000 ft. April–June. E/W. Introduced.

MINUARTIA L. – STITCHWORT

Annual or perennial herbs; *leaves* usually connate basally, exstipulate; *flowers* perfect; *sepals* 5, distinct; *petals* usually 5, usually white or rarely pink or purple, clawed or not, apex entire to notched; *stamens* 10; *styles* 3; *stigmas* 3; *fruit* an ovoid to elliptical capsule opening by 3 valves.

1a. Plants glabrous...2
1b. Plants glandular-hairy...3

2a. Sepals 3.5–5.5 mm long; seeds tuberculate, 0.7–1 mm long; capsule 3–4 mm long...***M. macrantha***
2b. Sepals (1.5) 2.5–3.2 mm long (to 4 mm long in fruit though); seeds smooth or obscurely tuberculate, 0.4–0.6 mm long; capsule 2.5–3.2 mm long...***M. stricta***

3a. Sepals with a hooded tip; petals 1.2–2 times as long as the sepals; seeds obscurely tuberculate; flowers usually solitary or sometimes in a 2–3-flowered cyme...***M. obtusiloba***
3b. Sepals tips acute to acuminate, not hooded; petals 0.5–1.3 times as long as the sepals; seeds tuberculate; flowers usually in a 3–30-flowered cyme, rarely solitary...4

4a. Leaves 1-veined abaxially, often recurved; sepals 3–7 mm long, usually 1-veined; inflorescence a (3) 6–30-flowered cyme; seeds 1.5–2.7 mm long, oblong-elliptic...***M. nuttallii* var. *nuttallii***
4b. Leaves 3-veined abaxially, straight; sepals 2.5–3.2 mm long, 3-veined; inflorescence a 3–7-flowered cyme or rarely solitary; seeds 0.4–0.5 mm long, suborbiculate...***M. rubella***

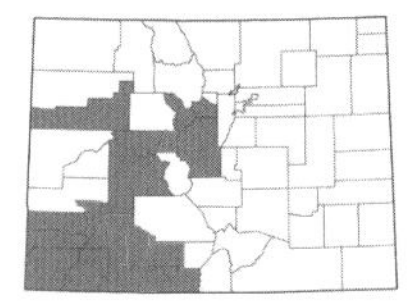

***Minuartia macrantha* (Rydb.) House**, HOUSE'S STITCHWORT. [*Alsinanthe macrantha* (Rydb.) W.A. Weber]. Perennials, caespitose or mat-forming, 2–15 cm, glabrous; *leaves* linear to subulate, 5–10 × 0.5–1.2 mm; *sepals* 3.5–5.5 mm long, acute to acuminate; *petals* 0.7–1.8 times as long as the sepals, entire; *capsules* 3–4 mm long, shorter than the sepals. Found in the alpine and spruce-fir forests, 8900–14,500 ft. July–Sept. E/W.

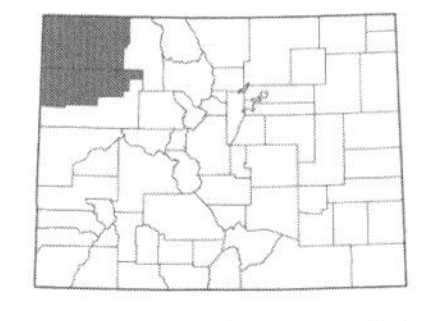

Minuartia nuttallii* (Pax) Briq. var. *nuttallii, NUTTALL'S SANDWORT. [*Minuopsis nuttallii* (Pax) W.A. Weber]. Perennials, mat-forming, 2–20 cm, densely glandular-hairy; *leaves* linear to lanceolate, 5–20 × 0.5–1.5 mm; *sepals* 3–7 mm long, acute to acuminate; *petals* 0.5–1.6 times as long as the sepals; *capsules* 5 mm long, usually shorter than the sepals. Uncommon on sandy and gravelly slopes, 6000–7600 ft. June–July. W.

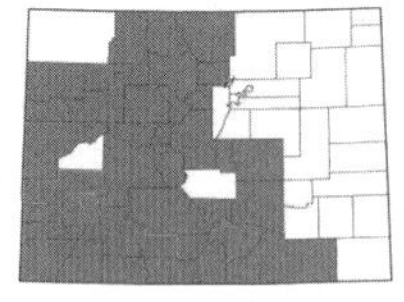

***Minuartia obtusiloba* (Rydb.) House**, ALPINE STITCHWORT. [*Arenaria obtusiloba* (Rydb.) Fernald; *Lidia obtusiloba* (Rydb.) Á. Löve & D. Löve]. Perennials, caespitose to mat-forming, 1–12 cm, glandular-stipitate; *leaves* acicular to linear, 1–8 × 0.4–1 mm; *sepals* 3–6.5 mm long, tips hooded; *petals* 1.2–2 times as long as the sepals; *capsules* 3.5–6 mm long, equal to the sepals. Common in the alpine and occasionally in subalpine spruce-fir forests, 10,000–14,000 ft. June–Sept. E/W. (Plate 25)

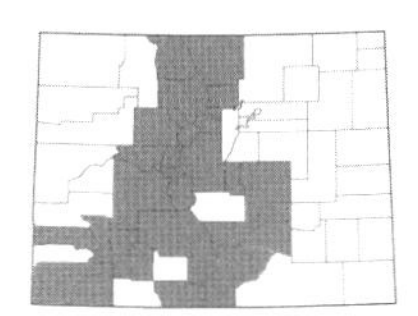

Minuartia rubella **(Wahlenb.) Hiern**, REDDISH SANDWORT. [*Tryphane rubella* (Wahlenb.) Rchb.]. Perennials, caespitose or mat-forming, erect, 2–8 (18) cm, glandular-stipitate; *leaves* subulate, 1.5–10 × 0.3–1.3 mm; *sepals* 2.5–3.2 mm long, acute to acuminate; *petals* 0.8–1.3 times as long as the sepals, entire; *capsules* 4.5–5 mm long, longer than the sepals. Found in the alpine, on rocky ridges, and in spruce-fir forests, 9500–14,000 ft. June–Sept. E/W.

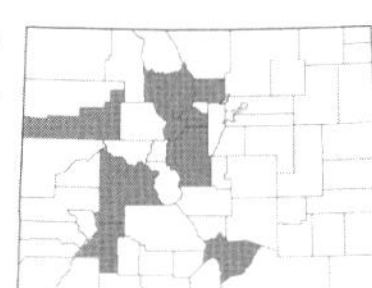

Minuartia stricta **(Sw.) Hiern**, ROCK SANDWORT. [*Alsinanthe stricta* (Sw.) Rchb.]. Perennials, caespitose or mat-forming, erect or somewhat procumbent, (0.8) 3–12 cm, glabrous; *leaves* linear to subulate, 2–10 × 0.3–1.5 mm; *sepals* (1.5) 2.5–3.2 mm long in flower, acute to acuminate; *petals* 0.6–1 times as long as the sepals, entire, or rudimentary or absent; *capsules* 2.5–3.2 mm long, shorter than or equal to the sepals. Uncommon in the alpine, 10,500–14,000 ft. July–Aug. E/W.

MOEHRINGIA L. – SANDWORT

Perennial (ours) herbs; *leaves* exstipulate; *flowers* perfect; *sepals* (4) 5, distinct; *petals* (4) 5, white, not clawed, the apex entire; *stamens* (8) 10, staminodes absent; *styles* 3; *stigmas* 3; *fruit* a capsule opening by 6 teeth.

1a. Small hairs on the stem spreading; sepals tips acute to acuminate; leaves with acute tips; petals shorter than or to 1.5 times the length of the sepals...***M. macrophylla***

1b. Small hairs on the stem pointed downward; sepals tips usually rounded or obtuse; leaves with rounded or obtuse tips; petals mostly twice as long as the sepals...***M. lateriflora***

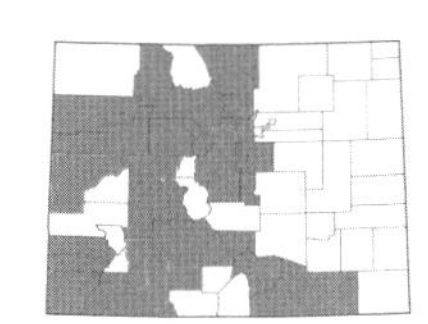

Moehringia lateriflora **(L.) Fenzl**, BLUNTLEAF SANDWORT. Plants 5–30 cm, retrorsely hairy; *leaves* elliptic to oblanceolate, 6–35 × (2) 5–10 (17) mm, the margins granular or minutely ciliate, rounded to obtuse at the tips; *sepals* 1.7–3 mm long; *petals* 3–6 mm long, ca. 2 times as long as the sepals; *capsules* 3–5 mm long. Locally common along streams and in shady forests, 6800–9700 ft. June–Aug. E/W.

Closely resembling *Silene menziesii* Hook. except that the sepals are distinct.

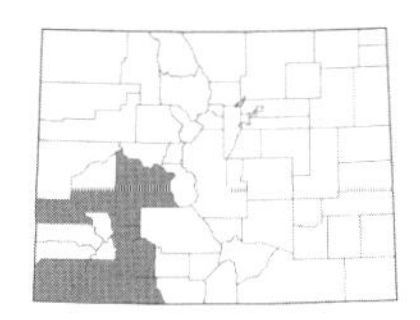

Moehringia macrophylla **(Hook.) Fenzl**, LARGELEAF SANDWORT. Plants 2–18 cm, with peglike hairs; *leaves* lanceolate to elliptic, (8) 15–50 (70) × 2–9 mm, the margins smooth to granular or ciliate, acute at the tips; *sepals* 2.8–6 mm long; *petals* 2–6 mm long; *capsules* ca. 5 mm long, equal to the sepals. Found on rocky ridges and in forests, usually in dry places, 6500–10,500 ft. June–July. W.

PARONYCHIA Mill. – NAILWORT

Annual, biennial, or perennial herbs; *leaves* crowded, stipulate; *flowers* perfect or rarely imperfect; *calyx* (3) 5-lobed or sometimes absent, connate at the base, the apex usually with an adaxial hood and prominent awn; *petals* absent; *stamens* 5; *styles* 1–2 (3), partly connate; *stigmas* 2 (3); *fruit* an ovoid to globose or rarely 4-angled utricle. (Chaudhri, 1968; Core, 1941; Hartman, 1974)

1a. Plants erect or the branches ascending, not caespitose or mat-forming; inflorescence a 20–70-flowered cyme; leaves linear, mostly over 10 mm long...***P. jamesii***

1b. Plants prostrate or sprawling, caespitose or mat-forming, sometimes cushion-forming; inflorescence solitary or a 3–7-flowered cyme; leaves linear to elliptic, mostly less than 10 mm long...2

2a. Leaves narrowly elliptic to oblong, glabrous; stipules ovate; flowers usually solitary; sepal awn tips erect; plants usually above 10,000 ft in elevation...***P. pulvinata***

2b. Leaves linear, sparsely hairy to glabrous; stipules lanceolate; inflorescence a 3–7-flowered cyme or sometimes solitary; sepal awn tips erect or spreading; plants usually below 9000 ft in elevation...3

3a. Leaves 4–7.5 mm long, with an acute or shortly cuspidate tip, tightly overlapping; flowers often solitary or in pairs, sometimes the cymes 3–6-flowered; sepal awn tips erect or somewhat spreading; fruit densely hairy in the upper half...***P. sessiliflora***

3b. Leaves 8–20 mm long, with a shortly cuspidate tip, not as tightly overlapping; cymes 3–7-flowered; sepal awn tips spreading; fruit glabrous...***P. depressa***

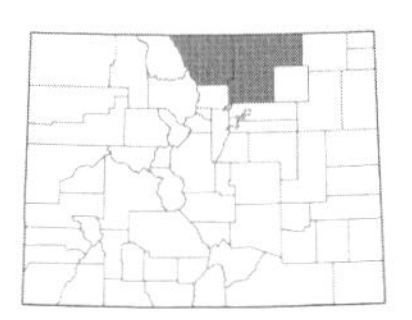

Paronychia depressa **Nutt. ex Torr. & A. Gray**, SPREADING NAILWORT. Perennials, often matted, prostrate, much-branched, 8–15 cm; *leaves* linear, 8–20 × 0.5–1 mm, minutely hairy; *inflorescence* a 3–7-flowered cyme; *calyx* 1.7–2 mm long; *utricles* 0.8–1 mm long, glabrous. Found on dry hills, rocky or sandstone outcroppings, and in dry grasslands, 5000–7500 ft. May–Aug. E.

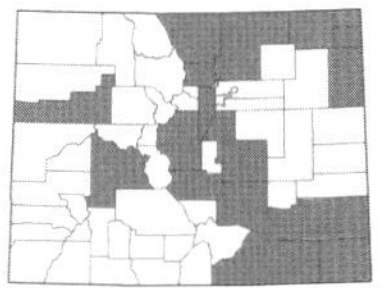

Paronychia jamesii **Torr. & A. Gray**, JAMES' NAILWORT. Perennials, not mat-forming, much-branched, 10–35 cm; *leaves* linear, 7–30 × 0.5–1 mm, minutely hairy; *inflorescence* a 20–70-flowered cyme; *calyx* 1.3–2 mm long; *utricles* 0.8–1 mm long, glabrous. Common on rocky outcroppings, sandstone, rocky plains and slopes, and in dry grasslands, 3500–8600 ft. May–July. E/W. (Plate 25)

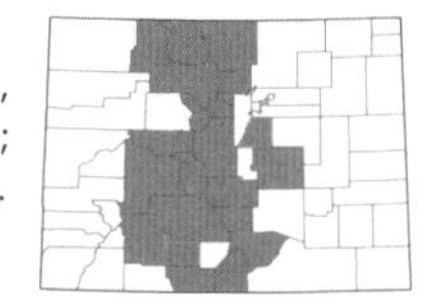

***Paronychia pulvinata* A. Gray**, ROCKY MOUNTAIN NAILWORT. Perennials, densely caespitose and cushion-forming, prostrate, 5–10 cm; *leaves* narrowly elliptic to narrowly oblong, 2–5 × 0.2–2 mm, fleshy, glabrous; *inflorescence* usually solitary, almost hidden by the leaves; *calyx* 1.5–1.8 mm long; *utricles* 1.3–1.5 mm long, glabrous. Found on alpine and rocky slopes, 9800–14,000 ft. June–Aug. E/W. (Plate 25)

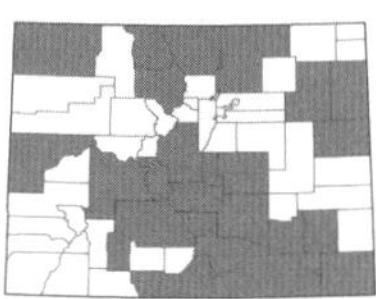

***Paronychia sessiliflora* Nutt.**, CREEPING NAILWORT. Perennials, densely caespitose and mat-forming, 5–25 cm; *leaves* linear to subulate, 4–7.5 × 0.5–0.8 mm, minutely hairy to glabrous; *inflorescence* a 3–6-flowered cyme; *calyx* 1.5–2 mm long; *utricles* 1.3–1.5 mm long, densely hairy in the upper half. Found on dry, rocky or sandy slopes, sandstone or shale outcroppings, and bluffs, 4000–9000 ft. May–Aug. E/W.

PSEUDOSTELLARIA Pax – STICKY STARWORT

Perennials from rhizomes or running rootstocks; *leaves* exstipulate; *flowers* perfect, in terminal, open cymes; *sepals* 5, distinct, herbaceous with white, scarious margins; *petals* 5, white, not clawed, the apex 2-cleft and often deeply so; *stamens* 5 or 10; *styles* 3; *stigmas* 3; *fruit* an ovoid capsule opening by 6 recoiled valves. (Weber & Hartman, 1979)

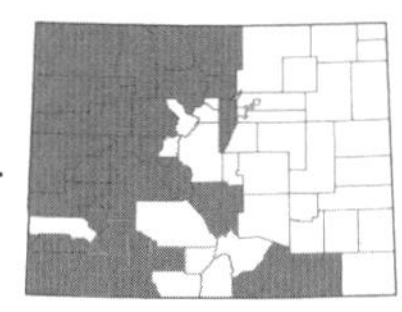

***Pseudostellaria jamesiana* (Torr.) W.A. Weber & R.L. Hartm.**, TUBER STARWORT. [*Stellaria jamesiana* Torr.]. Plants 1.2–6 dm, glabrous or glandular-stipitate; *leaves* linear to lanceolate, 1.5–10 (15) × 0.2–2 cm; *sepals* 3–6 (7) mm long, glandular-stipitate; *petals* 7–10 mm long; *capsules* 4–5 mm long. Common in meadows and forests, and along streams, 5400–10,000 ft. May–July. E/W. (Plate 25)

SAGINA L. – PEARLWORT

Annual or perennial herbs; *leaves* exstipulate; *flowers* perfect; *sepals* 4 or 5, distinct; *petals* 4 or 5 or sometimes absent, white, clawed or not, the apex entire; *stamens* 4, 5, 8, or 10, staminodes absent; *styles* 4 or 5; *stigmas* 4 or 5; *fruit* a globose to ovoid capsule opening by 4 or 5 valves. (Crow, 1978)

1a. Petals 2.5–3 mm long, longer than or occasionally just equaling the sepals; plants forming cushions; hyaline sepal margins purple, at least at the tips…***S. caespitosa***

1b. Petals 0.8–2 mm long, shorter than or equaling the sepals; plants not forming cushions; hyaline sepal margins white, rarely purple in some alpine plants…2

2a. Most of the flowers 4-merous; petals 0.8–1 (1.5) mm long or sometimes absent; capsules only slightly longer than the sepals; tufts of leaves often present in the axils of leaves; stems usually procumbent and rooting at the nodes, usually mat-forming, the procumbent stems often giving rise to secondary tufts at the rooting nodes…***S. procumbens***

2b. Flowers 5-merous or rarely 4-merous; petals (1) 1.5–2 mm long; capsules 1.5–2 times as long as the sepals; tufts of leaves absent from the axils of leaves; stems usually ascending, tufted or sometimes caespitose in the alpine…***S. saginoides***

***Sagina caespitosa* (J. Vahl) Lange**, TUFTED PEARLWORT. Perennials, mat- or cushion-forming; *leaves* linear to subulate, 2–13 mm long, glabrous; *flowers* 5-merous or sometimes 4- and 5-merous; *sepals* 2–2.5 mm long, glabrous or glandular at the base; *petals* 2.5–3 mm long, longer than the sepals or rarely equal to; *capsules* 3–3.5 mm long, longer than the sepals. Uncommon or inconspicuous and overlooked in the alpine, 11,000–13,500 ft. July–Aug. E.

Resembling a small *Minuartia obtusiloba* except that there are 4 styles present instead of 3. *Sagina nivalis* (Lindblom) Fries could also be present in Colorado. It is similar to *S. caespitosa* in forming cushions but differs in having petals equal to or shorter than the sepals, and the majority of the flowers are 4-merous.

***Sagina procumbens* L.**, MATTED PEARLWORT. Perennials, usually mat-forming, procumbent or sometimes ascending, glabrous; *leaves* linear, 4–17 mm long; *flowers* 4-merous or sometimes with a few 5-merous; *sepals* 1.5–2.5 mm long, glabrous; *petals* 0.8–1 (1.5) mm long, shorter than or equal to the sepals; *capsules* 1.5–3 mm long, just longer than the sepals. Uncommon in disturbed places, sidewalk cracks, known from Clear Creek Co., 5200–5500 ft. June–Aug. E. Introduced.

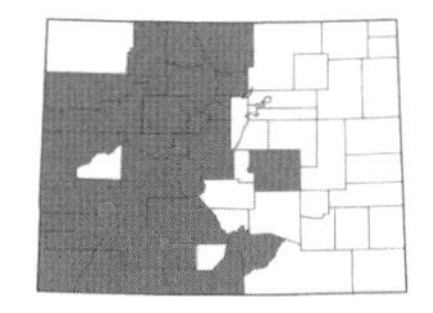

***Sagina saginoides* (L.) H. Karst.**, ARCTIC PEARLWORT. Perennials, caespitose or tufted, procumbent to ascending, glabrous; *leaves* linear, 4–20 mm long; *flowers* 5-merous or rarely some 4-merous; *sepals* 2–2.5 mm long, glabrous; *petals* (1) 1.5–2 mm long, shorter than or equal to the sepals; *capsules* 2.5–3.5 mm long, 1.5–2 times the sepals in length. Common but inconspicuous along streams, in moist areas, and in alpine, (6000) 8500–13,500 ft. June–Aug. E/W.

SAPONARIA L. – SOAPWORT

Annual or perennial herbs; *leaves* exstipulate; *flowers* perfect; *calyx* 5-lobed, connate basally; *petals* 5, pink to white, clawed, apex entire or emarginate; *stamens* 10; *styles* 2 (3); *stigmas* 2 (3); *fruit* a capsule usually opening by 4 apical teeth.

1a. Leaves 1-veined; stems trailing or ascending, much-branched; calyx 7–12 mm long, usually purple, glandular-hairy...***S. ocymoides***
1b. Leaves with 3 prominent veins; stems erect, simple or branched above; calyx 15–25 mm long, usually green or sometimes purplish, glabrous or with some scattered hairs...***S. officinalis***

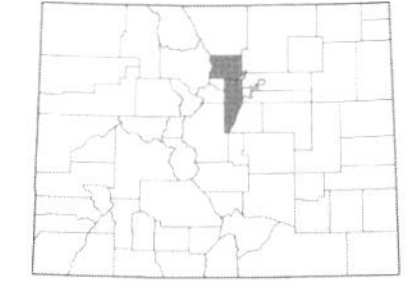

Saponaria ocymoides **L.**, ROCK SOAPWORT. Perennials, the stems trailing or ascending, much-branched, 0.5–2.5 dm; *leaves* spatulate to lanceolate-ovate, 0.6–2.5 × 0.3–1.5 cm, 1-veined; *calyx* 7–12 mm long, usually purple, glandular-hairy; *petals* red, pink, or white, the limb 8–15 mm long; *capsules* 6–8 mm long. A commonly cultivated rock garden plant, found in disturbed areas along roadsides and in gardens, 5500–7000 ft. May–July. E. Introduced.

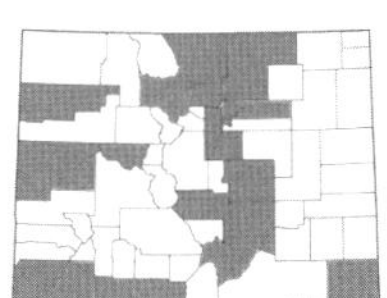

Saponaria officinalis **L.**, BOUNCING BET. Perennials, erect, branched above, 3–9 dm; *leaves* elliptic to oblanceolate or ovate, 3–15 × 1.5–5 cm, 3-veined; *calyx* 15–25 mm long, glabrous or with scattered hairs; *petals* white or pink, the limb 8–15 mm long; *capsules* 15–20 mm long. Common in disturbed places, along roadsides, and in floodplains and gravel bars of streams, 4500–8500 ft. June–Sept. E/W. Introduced. (Plate 25)

SILENE L. – CATCHFLY

Herbs; *leaves* exstipulate; *flowers* usually perfect; *calyx* 5-lobed, connate into a tube; *petals* 5, white, pink, red, or purple, clawed, the apex usually 2-lobed or fimbriate; *stamens* usually 10; *styles* 3 or 5, rarely 4; *stigmas* 3 or 5, rarely 4; *fruit* an ovoid to globose capsule longitudinally dehiscent at the apex with 3–5 valves splitting into 6–10 equal teeth. (Hitchcock & Maguire, 1947; McNeill, 1978; Williams, 1896)

Silene scaposa B.L. Rob. is excluded from this key, although shown present in Colorado in the *Flora of North America*. This species is present in the Pacific Northwest (Oregon, Idaho, Nevada), and no specimens have been reported from Colorado.

1a. Calyx and inflorescence glabrous; stems glabrous or sometimes hairy towards the base, or sometimes with a dark sticky band on the stem below the upper pair of leaves...2
1b. Calyx, inflorescence, and stems hairy or glandular-hairy...5

2a. Plants densely caespitose and cushion- or mat-forming, strongly perennial, 3–6 (15) cm; flowers solitary, pink or rarely white; seeds light brown...***S. acaulis***
2b. Plants erect, not cushion- or mat-forming, simple or branched, annual or short-lived perennials, to 80 cm; flowers 3–many, in open cymes, white, sometimes tinged with red, or rarely completely red; seeds grayish to black...3

3a. Petals white and usually tinged with pink, purple or red, rarely completely purple; calyx prominently 10-veined with dark green veins, 5–9 mm long; stems often with a dark sticky band on the stem below the upper leaf pair...***S. antirrhina***
3b. Petals white; calyx with obscure veins, 9–13 mm long; stems lacking a dark sticky band on the stem...4

4a. Filaments purple; calyx indented at the base around the peduncle; stems simple and branched below the inflorescence; seeds grayish-brown, 0.6–1 mm, with concentric rings of papillae...***S. csereii***
4b. Filaments white; calyx not indented around the peduncle; stems rarely simple, usually several to many; seeds black, 1–1.5 mm, finely tuberculate...***S. vulgaris***

5a. Stem, leaves, and calyx coarsely hispid; stamens long-exserted; inflorescence many-flowered with one sessile (or nearly so) flower per node; seeds rugose with concave faces...***S. dichotoma* ssp. *dichotoma***
5b. Stems, leaves, and calyx variously hairy or glandular, but not hispid; stamens included to slightly exserted; inflorescence and seeds various but unlike the above...6

6a. Calyx with multicellular, glandular hairs with purple cross-walls, with 10 prominent, purple veins; flowers usually solitary, or sometimes to 4 (8) per stem; perennials, often found above 10,000 ft in elevation...7
6b. Calyx with hairs lacking purple cross-walls (or if present then the flowers in whorls of 2–5), with 10–20 prominent, green or purplish veins, or the veins obscure; inflorescence and plants various, but the flowers usually not solitary; annuals or perennials, at various elevations...10

7a. Pedicels nodding in flower; seeds with a broad wing, as wide as the body of the seed or nearly so, 1.5–2 mm in diam. (including the wing); calyx conspicuously inflated, papery, 6–10 mm wide in flower, enlarging to 16 mm wide in fruit; flowers usually solitary, purplish-pink...***S. uralensis* ssp. *uralensis***
7b. Pedicels erect in flower, sometimes spreading in fruit; seeds with a narrow wing or the wing lacking, 0.5–1.3 mm in diam.; calyx not conspicuously inflated, thicker in texture, 5–8 mm wide and only slightly enlarging in fruit; flowers solitary or not, white, pink, purple, or reddish-purple...8

8a. Seeds with a narrow wing; calyx 7–10 (12) mm long; leaves glabrous with ciliate margins, or with multicellular hairs; flowers white or pink...***S. hitchguirei***
8b. Seeds lacking a wing; calyx 12–18 mm long; leaves pubescent on both sides, the hairs short, retrorse, not multicellular; flowers pink, purple, or reddish-purple...9

9a. Plants 1.5–6 dm; flowers 1–4 (8) per stem...***S. drummondii* ssp. *striata***
9b. Plants 0.7–2 dm; flowers usually solitary, the inflorescence rarely with up to 3-flowers...***S. kingii***

10a. Calyx 5–8 mm long, usually obscurely 10-veined; leaves elliptic-lanceolate to elliptic, broadest at or above the middle; flowers few in an axillary cyme or solitary and terminal...***S. menziesii***
10b. Calyx 10–25 mm long, prominently 10–20-veined; leaves and flowers various...11

11a. Calyx with long (mostly 1–1.5 mm), multicellular hairs in addition to shorter, glandular hairs; lower stems with spreading, long, multicellular hairs; basal leaves often withering by flowering time, the cauline leaves lanceolate or more often elliptic; annuals or short-lived perennials...12
11b. Czlyx with short, glandular hairs, lacking long, multicellular hairs; lower stems usually with shorter, retrorsely puberulent hairs; basal leaves often present at flowering time, the cauline leaves linear to narrowly oblanceolate; perennials...13

12a. Flowers perfect; styles 3; calyx 10-veined, mostly 3 mm wide in flower and to 10 mm wide in fruit, with linear to linear-lanceolate teeth 5–10 mm long...***S. noctiflora***
12b. Flowers imperfect; styles 5 or rarely 4 in pistillate flowers; calyx 10-veined in staminate flowers, 20-veined in pistillate flowers, mostly 3–8 mm wide in flower and 8–15 mm wide in fruit, with deltoid teeth to 6 mm long...***S. latifolia***

13a. Styles 3–4; flowers arranged in 3–7 flowering nodes, each node composed of a usually sessile, 2–10-flowered cyme subtended by a pair of narrowly lanceolate bracts, the flowers in each node with stout pedicels mostly shorter than or equaling the calyx in length; base of the petals and filaments usually ciliate; capsule with 3 or 6 teeth...***S. scouleri***
13b. Styles usually 5, rarely 4; flowers 1–several in a narrow, elongate cyme or occasionally in flowering nodes, usually at least some of the flowers on pedicels over twice as long as the calyx; base of the petals and filaments glabrous; capsule usually with 5 teeth...***S. drummondii***

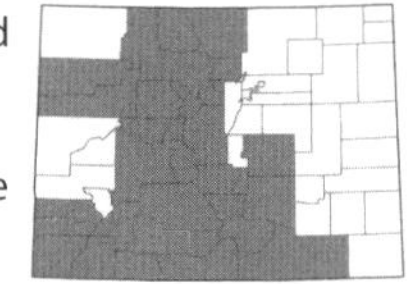

Silene acaulis **(L.) Jacq.**, MOSS CAMPION. Perennials, mat- or cushion-forming, with a much-branched caudex, 3–6 (15) cm; *leaves* mostly basal, linear to lanceolate, 0.4–1.5 cm × 0.8–1.5 mm, glabrous to scabrous; *inflorescence* solitary; *calyx* 10-veined, (5) 7–10 mm long, glabrous; *petals* bright pink or rarely white, the limb entire to shallowly bifid, 2.5–3.5 mm long; *styles* 3; *capsules* equal to twice as long as the calyx; *seeds* light brown, 0.8–1.2 mm diam. Common in the alpine, 10,500–14,000 ft. June–Aug. E/W. (Plate 25)

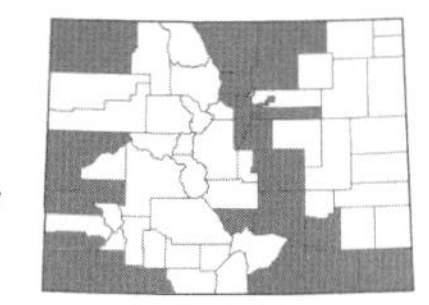

Silene antirrhina **L.**, SLEEPY CATCHFLY. Annuals to 8 dm; *leaves* oblanceolate to linear, 1–9 cm × 2–15 mm, scabrous or rarely glabrous; *inflorescence* an open, many-flowered cyme; *calyx* 10-veined, 5–9 mm long, glabrous; *petals* white or sometimes with red tinges, usually bifid, the limb ca. 2.5 mm long; *styles* 3; *capsules* equal to the calyx; *seeds* gray-black, 0.5–0.8 mm diam. Found in rocky soil, dry grasslands, open forests, 4000–8000 ft. June–Aug. E/W.

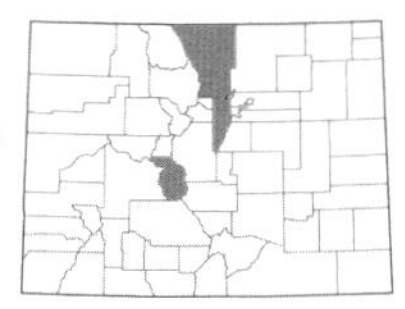

Silene csereii **Baumg.**, BIENNIAL CAMPION. Annuals or biennials to 6.5 dm; *leaves* spatulate to ovate-lanceolate, 3–7 cm × 7–30 mm, glabrous; *inflorescence* a many-flowered, open cyme; *calyx* ca. 20-veined, abruptly contracted at the base, 7–13 mm long, glabrous; *petals* white, the limb deeply bifid into 2 lobes ca. 5 mm long; *styles* 3; *capsules* equal to the calyx; *seeds* grayish-brown, 0.6–1 mm diam., with concentric rings of papillae. Uncommon in disturbed areas and meadows, 6000–7800 ft. June–Aug. E. Introduced.

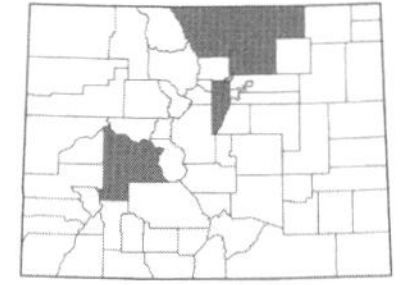

Silene dichotoma **Ehrh. ssp. *dichotoma***, FORKED CATCHFLY. Annuals, (2) 5–10 dm; *leaves* lanceolate to spatulate, 1.5–10 cm × 3–30 mm, hispid; *inflorescence* a many-flowered, open cyme, sometimes glandular; *calyx* 10-veined, not inflated, (7) 10–15 mm long, setose; *petals* white or rarely pink, the limb deeply bifid, 5–9 mm long; *styles* 3; *capsules* shorter than the calyx; *seeds* grayish-brown to black, ca. 1 mm diam. Found in disturbed areas, meadows, and along roadsides, 5000–10,800 ft. June–Aug. E/W. Introduced.

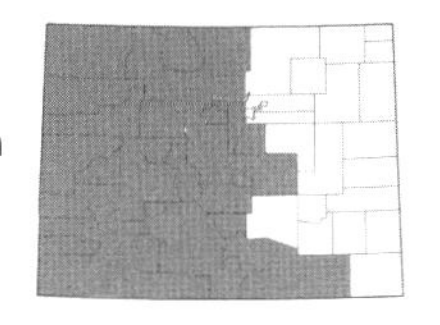

***Silene drummondii* Hook.**, DRUMMOND'S CATCHFLY. [*Gastrolychnis drummondii* (Hook.) Á. Löve & D. Löve]. Perennials; *leaves* linear, lanceolate to elliptic, 3–10 cm × 2–12 mm, hairy; *inflorescence* a 1–20-flowered cyme, strongly glandular or densely hairy; *calyx* 10-veined, not inflated, 12–18 mm long in fruit; *petals* white to pink or reddish-purple, equal to 1.5 times the length of the calyx, the limb not well-differentiated; *styles* (4) 5; *capsules* 12–15 mm long; *seeds* dark brown, 0.7–1 mm diam. Common in meadows, forests, and the alpine, 6500–13,000 ft. June–Aug. E/W. (Plate 25)

There are two subspecies of *S. drummondii* in Colorado:

1a. Calyx narrow, cylindrical; claw of the petal not exceeding the calyx in length; inflorescences (1) 3–10-flowered...**ssp. *drummondii***. Common in meadows, forests, and on open slopes, 7500–11,000 ft. June–Aug. E/W.

1b. Calyx with an enlarged base; claw of the petal longer than the calyx in length; inflorescences 1–4 (8)-flowered...**ssp. *striata* (Rydb.) J.K. Morton**. [*Lychnis striata* Rydb.]. Common in alpine meadows, forests, and rocky soil, 6500–13,000 ft. July–Aug. E/W.

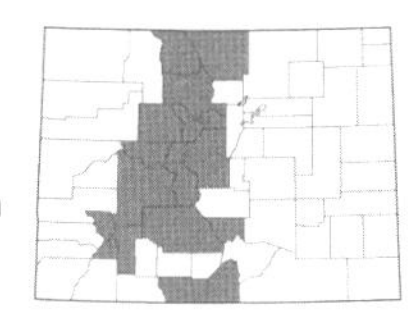

***Silene hitchguirei* Bocq.**, MOUNTAIN CAMPION. [*Gastrolychnis apetala* (L.) Tolm. & Kozhanch. ssp. *uralensis* (Rupr.) Á. Löve & D. Löve; *S. uralensis* (Ruprecht) Bocq. ssp. *montana* (S. Watson) McNeill]. Perennials, 0.2–1 (1.2) dm; *leaves* oblanceolate to spatulate, to 2.5 cm × 4 mm, glabrous with ciliate margins; *inflorescence* usually solitary; *calyx* prominently veined, not inflated, 7–10 (12) mm long, densely hairy, the hairs with purple cross-walls; *petals* white to pink, slightly longer than the calyx, the limb not well-differentiated from the claw; *styles* 5; *capsules* equal to the calyx; *seeds* brown, 0.5–1.3 mm diam. Locally common in the alpine, 11,000–14,000 ft. July–Aug. E/W.

***Silene kingii* (S. Watson) Bocquet**, KING'S CATCHFLY. [*Gastrolychnis kingii* (S. Watson) W.A. Weber]. Perennials, 0.7–2 dm; *leaves* linear to lanceolate or narrowly oblanceolate, 1–4 cm × 1.5–4 mm; *inflorescence* usually solitary; *calyx* 10-veined, somewhat inflated, 12–14 mm long in flower, glandular and hairy, the hairs with purple cross-walls; *petals* pink to purple, just longer than the calyx; *styles* (4) 5; *capsules* equal to the calyx; *seeds* dark brown, 0.7–1 mm diam. To be expected in the alpine.

The specimen of *S. kingii* reported from Colorado is actually a specimen of *S. hitchguirei*.

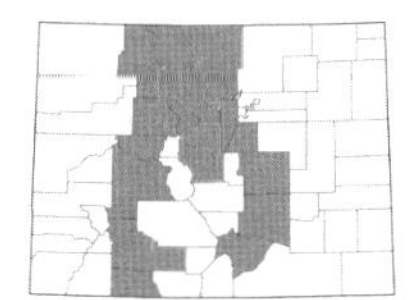

***Silene latifolia* Poir.**, WHITE CAMPION. [*Lychnis alba* Mill.; *Melandrium dioicum* (L.) Cosson & Germain]. Annuals or short-lived perennials to 10 dm; *leaves* lanceolate to elliptic, 3–12 cm × 6–30 mm, hirsute; *inflorescence* a many-flowered cyme; *calyx* 10-veined in staminate flowers, 20-veined in pistillate flowers, 10–25 mm long in fruit, hirsute and glandular; *petals* white, twice as long as the calyx; *styles* (4) 5; *capsules* equal to the calyx; *seeds* gray-brown, ca. 1.5 mm diam. Common in disturbed areas, meadows, floodplains, and along roadsides, 4900–10,500 ft. June–Sept. E/W. Introduced.

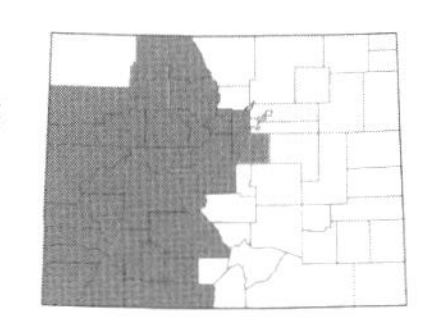

***Silene menziesii* Hook.**, MENZIES' CATCHFLY. Perennials, 0.5–3 (7) dm; *leaves* elliptic-lanceolate to elliptic, 2–6 (10) cm × 3–20 (35) mm, puberulent to hairy; *inflorescence* a cyme, or flowers solitary; *calyx* obscurely 10-veined, 5–8 mm long, hairy and glandular; *petals* white, equal to 1.5 times the length of the calyx; *styles* 3 (4); *capsules* just longer than the calyx; *seeds* black, 0.5–1 mm diam. Found in forests and along shady streams and creeks, 6500–11,000 ft. June–July. E/W. (Plate 25)

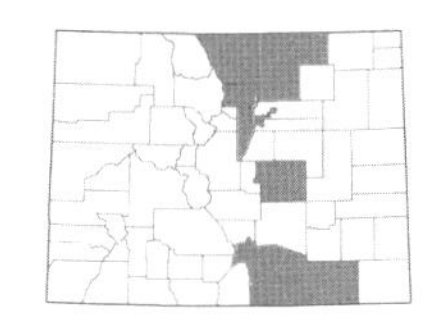

***Silene noctiflora* L.**, NIGHT-FLOWERING CATCHFLY. [*Melandrium noctiflorum* (L.) Fries]. Annuals to 8 dm; *leaves* lanceolate to elliptic, 1–11 cm × 3–40 mm, densely hairy; *inflorescence* a 3–15-flowered cyme; *calyx* 10-veined, 15–25 (40) mm long in flower, hairy and glandular; *petals* white, the limb deeply bifid; *styles* 3; *capsules* equal to or just longer than the calyx; *seeds* dark brown to black, 0.8–1 mm diam. Found in disturbed areas and along roadsides, 5000–9000 ft. June–Sept. E. Introduced.

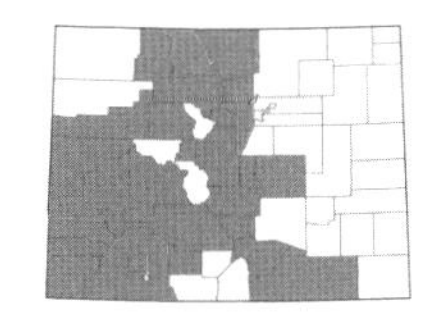

***Silene scouleri* Hook.**, SIMPLE CAMPION. Perennials, 1–8 dm; *leaves* lanceolate to oblanceolate or rarely linear, 6–25 cm × 4–30 mm, retorsely puberulent; *inflorescence* an open cyme, the flowers paired or in many-flowered whorls; *calyx* 10-veined, 8–20 mm long, sometimes the hairs with purple cross-walls; *petals* white to pink, the limb deeply 2–4-lobed, 2.5–8 mm long; *styles* 3–4; *capsules* equal to or just longer than the calyx; *seeds* brown to gray-brown, 1–1.5 mm diam. Found in meadows, forests, and the alpine, 8000–12,500 ft. July–Aug. E/W.

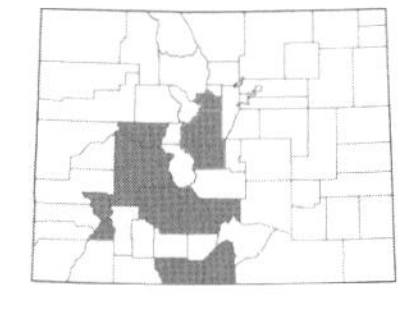

Silene uralensis* (Ruprecht) Bocq. ssp. *uralensis, NODDING CAMPION. [*Gastrolychnis uralensis* (Rupr.) Bocq.]. Perennials to 3 dm; *leaves* linear to spatulate or lanceolate, 0.5–5 cm × 1–5 mm; *inflorescence* usually solitary; *calyx* 10-veined, inflated, 11–20 mm long, hairs with purple cross-walls, *petals* pink to purple, the limb not well-differentiated from the claw; *styles* 5; capsules equal to just longer than the calyx; *seeds* brown, broadly winged, 1.5–2.5 mm diam. (including wing). Uncommon in the alpine, 11,000–13,500 ft. July–Aug. E/W.

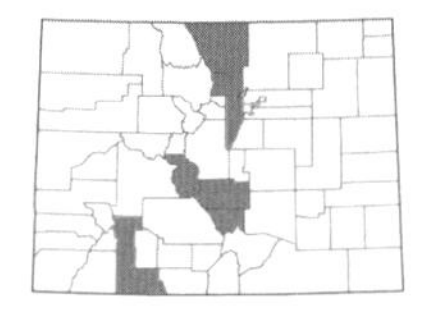

***Silene vulgaris* (Moench) Garcke**, MAIDEN'S TEARS. Perennials, 2–8 dm, glaucous and glabrous; *leaves* lanceolate to oblong, 2–8 cm × 5–30 mm; *inflorescence* a many-flowered, open cyme; *calyx* with obscure venation, inflated, 9–12 mm long in flower, to 18 mm long in fruit; *petals* white, about twice as long as the calyx; *styles* 3; *capsules* equal to the calyx; *seeds* black, 1–1.5 mm diam. Uncommon in disturbed areas and along roadsides, 7000–9000 ft. July–Sept. E/W. Introduced.

SPERGULA L. – SPURRY

Annual herbs; *leaves* opposite but appearing whorled, stipulate; *flowers* usually perfect; *sepals* 5, distinct, silvery; *petals* 5, white, the apex entire; *stamens* 5 or 10; *styles* 5; *stigmas* 5; *fruit* a 5-valved capsule.

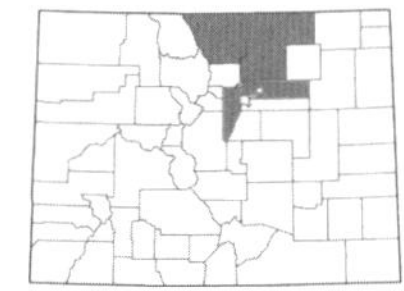

***Spergula arvensis* L.**, STICKWORT; CORN SPURRY. Plants 1–5 dm; *leaves* linear, 1–5 cm long, with a strongly inrolled margin; *sepals* 2.5–5 mm long, glandular; *petals* 2.5–4 mm long; *capsules* ovoid. Uncommon in disturbed places, 5000–7600 ft. June–Aug. E. Introduced.

SPERGULARIA J. Presl & C. Presl – SANDSPURRY

Annual or perennial herbs; *leaves* stipulate; *sepals* 5, green, connate on the lower ⅕; *petals* 5, white to pink, the apex entire; *stamens* 1–10; *styles* 3; *stigmas* 3; *fruit* a 3-valved capsule opening to the base or nearly so.

1a. Entire plant densely glandular-hairy; stamens (1) 2–3 (5)...***S. salina***
1b. Leaves glabrous, plants glandular-hairy in inflorescence only; stamens 6–10...2

2a. Leaves mostly 2–3 times as long as the stipules; flowers pink; seeds papillate, lacking a winged margin; capsules 3.5–5 mm long...***S. rubra***
2b. Leaves over 6 times as long as the stipules; flowers white or pink at the tips; seeds smooth, usually with a conspicuous winged margin; capsules (4.5) 5.5–7 mm long...***S. media***

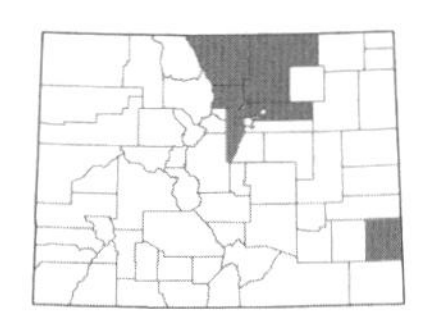

***Spergularia media* (L.) C. Presl**, GREATER SEA-SPURREY. [*S. maritima* (All.) Choiv.]. Annuals or short-lived perennials, 0.7–3 dm; *leaves* linear, 0.5–3.5 cm long, apiculate to spine-tipped; *sepals* 2.5–5 mm long; *petals* white to pink at the tips, shorter than to equal to the sepals; *stamens* 9–10; *capsules* (4.5) 5.5–7 mm long; *seeds* reddish-brown, not papillate, lacking a submarginal groove, 0.8–1.1 mm, usually with a wing 0.3–0.5 mm wide. Uncommon in disturbed pastures, on dry salt flats, and in alkaline soil of moist, low places, 4300–6000 ft. June–Aug. E. Introduced.

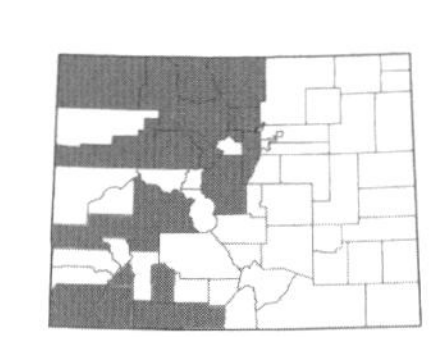

***Spergularia rubra* J. Presl & C. Presl**, RED SANDSPURRY. Annuals or short-lived perennials, 0.4–2.5 dm, erect or prostrate; *leaves* filiform to linear, 0.4–1.5 cm long, apiculate to spine-tipped; *sepals* 2–4 mm long; *petals* pink, about equal to the sepals; *stamens* 6–10; *capsules* 3.5–5 mm long; *seeds* reddish-brown to dark brown, papillate, with a submarginal groove, 0.4–0.6 mm, lacking a winged margin. Found in disturbed places, along roadsides, in forest clear-cuts, and in parking lots, 6000–11,500 ft. June–Aug. E/W. Introduced.

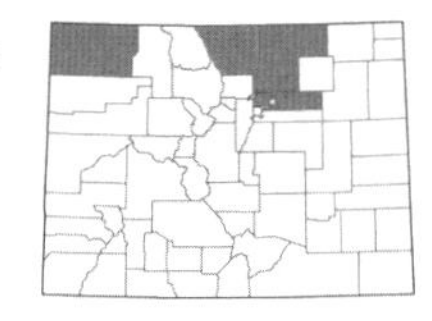

***Spergularia salina* J. Presl & C. Presl**, SALT-MARSH SANDSPURRY. [*S. marina* (L.) Griseb.]. Annuals, erect to prostrate, 0.8–3 dm; *leaves* linear, 0.8–4 cm long, apiculate; *sepals* 2.5–5 mm long; *petals* white to pink, shorter to equal to the sepals; *stamens* (1) 2–5; *capsules* 2.5–6.5 mm long; *seeds* brown to reddish-brown, with a submarginal groove, 0.5–0.8 mm, usually lacking a wing. Uncommon in disturbed pastures, sandy soil along rivers, and on alkaline soil, 5000–6000 ft. July–Oct. (Nov.). E/W. Introduced.

STELLARIA L. – STARWORT

Annual or perennial herbs; *leaves* exstipulate; *flowers* usually perfect; *sepals* (4) 5, distinct; *petals* (1) 5 or absent, white, not clawed, the apex deeply 2-cleft; *stamens* (1) 5 or 10, rarely absent; *styles* (2) 3 (5); *stigmas* (2) 3 (5); *fruit* a globose to oblong capsule, opening nearly to the base by twice as many valves as there are styles. (Boivin, 1956; Chinnappa & Morton, 1991; Morton & Rabeler, 1989)

1a. Stems with a line of hairs extending from one internode to the next; sepals pubescent with multicellular hairs; leaves ovate to elliptic, the lower and middle leaves distinctly petiolate, the upper leaves sessile and sometimes clasping; stamens 1–5 (8)...2
1b. Stems glabrous or the angles minutely papillate-scabrid, rarely uniformly pilose; sepals glabrous or sometimes the margins ciliate; leaves linear-lanceolate to elliptic or sometimes ovate, usually sessile or rarely shortly petiolate; stamens 5–10...3

2a. Sepals 4.5–6 mm long, green; stamens 3–5 (8); plants usually green; petals usually present but sometimes absent; seeds reddish-brown...***S. media***
2b. Sepals 3–4 mm long, sometimes reddish at the tips and base; stamens 1–3; plants usually yellowish-green; petals usually absent; seeds pale yellowish-brown...***S. pallida***

3a. Leaves and angles (especially at the nodes) on the stem papillate-scabrid with warty projections (look under 20× magnification); inflorescence a 2–many-flowered, axillary cyme subtended by wholly scarious bracts...***S. longifolia***
3b. Leaves and angles on the stem glabrous or seldom pilose, or the leaves with ciliate margins but the stems not papillate-scabrid; inflorescence various...4

4a. Petals absent or inconspicuous and much shorter than the sepals...5
4b. Petals equaling or longer than the sepals, conspicuous...9

5a. Leaves broadly ovate to suborbicular, 0.2–1.2 cm long; sepals usually 4, ovate, slightly scarious-margined or scarious margins absent; flowers solitary in upper leaf axils; bracts absent; seeds blackish...***S. obtusa***
5b. Leaves linear-lanceolate to elliptic (if ovate, then the flowers in 2–5-flowered cymes or solitary and terminal), 0.1–9 cm long; sepals usually 5, lanceolate to ovate, usually with a distinct scarious margin; bracts present or sometimes absent; seeds brown...6

6a. Plants purplish-tinged, especially in the leaves and sepals; internodes usually shorter than the leaves; sepals prominently 3-veined; flowers solitary in upper leaf axils; plants 2–10 cm tall...***S. irrigua***
6b. Plants green; internodes longer than the leaves; sepals 1–3-veined; flowers in cymes, or solitary and terminal; plants 5–50 cm tall...7

7a. Flowers in a 2–21-flowered, subumbellate cyme, rarely solitary, the pedicels sharply reflexed at the base at maturity; capsules conic, opening by 6 valves; bracts subtending the inflorescence usually completely scarious...***S. umbellata***
7b. Flowers solitary and terminal or in a 2–5-flowered cyme that is not subumbellate, the pedicels ascending; capsules ovoid to globose, opening by 3 valves; bracts subtending the inflorescence green and foliaceous or green with scarious margins...8

8a. Mature capsules green or tan; sepals 2–2.5 mm long in flower, to 3.5 mm long in fruit, prominently 1–3-veined; styles usually 0.9–1.6 mm long; leaves usually lanceolate-elliptic or linear-lanceolate...***S. borealis* ssp. *borealis***
8b. Mature capsules dark purple; sepals 1.2–1.5 mm long in flower and to 2.5 mm long in fruit, usually obscurely 1-veined; styles usually 0.5–0.9 mm long; leaves elliptic to ovate...***S. calycantha***

9a. Leaves with an obscure midrib; bracts subtending the inflorescence green and foliaceous; flowers usually solitary, terminal and in leaf axils or in few-flowered cymes...***S. crassifolia***
9b. Leaves with a prominent midrib; bracts subtending the inflorescence completely scarious or green with scarious margins; flowers in 2–30-flowered cymes or solitary...10

10a. Sepals 3-veined, the veins prominent and forming ridges, the margins usually ciliate; flowers in a 5–many-flowered cyme; seeds prominently tuberculate; weedy plants found below 6000 ft in elevation...***S. graminea***
10b. Sepals 1–3-veined, the veins not forming ridges, the margins usually glabrous but occasionally ciliate; flowers solitary or in a 3–many-flowered cyme; seeds smooth or shallowly tuberculate; native plants found above 7000 ft in elevation...***S. longipes* ssp. *longipes***

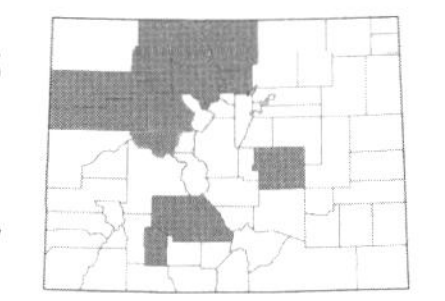

***Stellaria borealis* J.M. Bigelow ssp. *borealis*,** BOREAL STARWORT. Perennials, rhizomatous, (0.5) 2.5–5 dm; *leaves* linear-lanceolate to lanceolate-elliptic, 2–3 cm long; *sepals* 2–3.5 mm long, glabrous; *petals* 5, rarely absent, 1–3 mm long; *capsules* 3–7 mm long, longer than the sepals, dark brown; *seeds* brown, 0.7–1 mm diam. Uncommon along streams, on rocky slopes, and in willow thickets, 7000–11,000 ft. June–Aug. E/W.

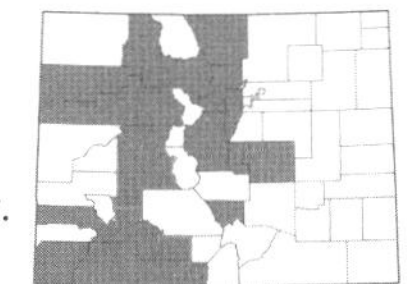

***Stellaria calycantha* (Ledeb.) Bong.,** NORTHERN STARWORT. Perennials, with slender rhizomes, the stems weak, to 3.5 dm; *leaves* ovate to elliptic, 0.5–2.5 cm long; *sepals* 1.2–2.5 mm long, glabrous; *petals* absent or 1–1.5 mm long; *capsules* green, 3–5 mm long; *seeds* brown, 0.5–1 mm diam. Uncommon in fens, willow thickets, along lake shores, and on shady, moist slopes, 7500–12,000 ft. June–Aug. E/W.

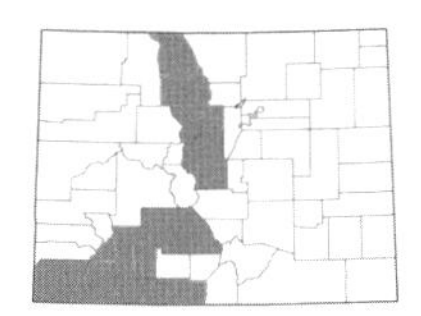

***Stellaria crassifolia* Ehrh.,** THICK-LEAVED STARWORT. Perennials, 0.3–3 dm, weak-stemmed; *leaves* elliptic-lanceolate to linear-lanceolate, 0.2–1 (1.5) cm long; *sepals* 3–4 mm long, usually glabrous; *petals* 5, 2.5–5 mm long; *capsules* tan or yellowish, 4–5 mm long; *seeds* reddish-brown, 0.7–1 mm diam. Uncommon in meadows, along streams, and in moist places, 8800–11,500 ft. June–Aug. E/W.

***Stellaria graminea* L.**, GRASS-LIKE STARWORT. Perennials, with slender rhizomes, weak-stemmed, 2–9 dm; *leaves* linear-lanceolate, 1.5–4 cm × 1–6 mm; *sepals* 3–7 mm long, glabrous; *petals* 5, 3–7 mm long; *capsules* green or yellowish, 5–7 mm long; *seeds* reddish-brown, ca. 1 mm diam., wrinkled. Uncommon weed in lawns, ditches, and wet meadows, known from Boulder Co., 5200–5400 ft. June–July. E. Introduced.

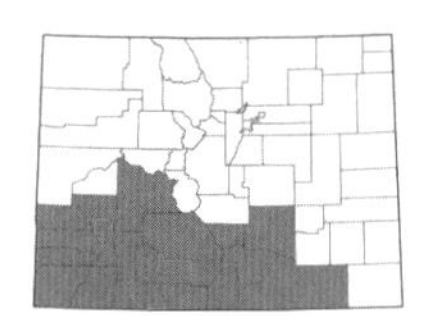

***Stellaria irrigua* Bunge**, ALTAI STARWORT. Perennials, rhizomatous, 0.2–1 dm; *leaves* lanceolate to oblanceolate or elliptic, 0.1–1 cm × 0.5–4 mm, purplish-tinged; *sepals* 3–4 mm long, glabrous, purplish-tinged; *petals* 5, 2 mm long, shorter than the sepals; *capsules* green to yellowish, ca. 3 mm long; *seeds* brown, 1–1.2 mm diam., the margins thickened with shallow, longitudinal ridges. Uncommon in rocky alpine and on scree slopes, 10,000–14,000 ft. June–Aug. E/W.

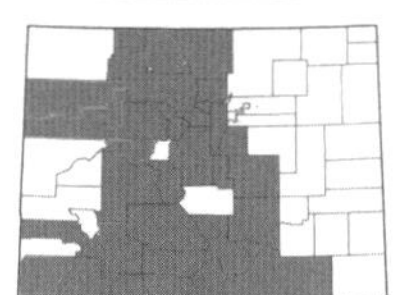

***Stellaria longifolia* Muhl. ex Willd.**, LONG-LEAVED STARWORT. Perennials, rhizomatous, 1–3.5 dm, the angles on the stem minutely papillate-scabrid; *leaves* linear to elliptic-lanceolate, 0.8–4 cm × 1–3 mm; *sepals* 2–4 mm long, glabrous; *petals* 5, 2–3.5 mm long; *capsules* yellowish to dark purple, 3–6 mm long; *seeds* brown, 0.7–0.8 mm diam., slightly rugose. Common in moist meadows, along streams and lakes, and in marshes, 6400–10,500 ft. June–Aug. E/W.

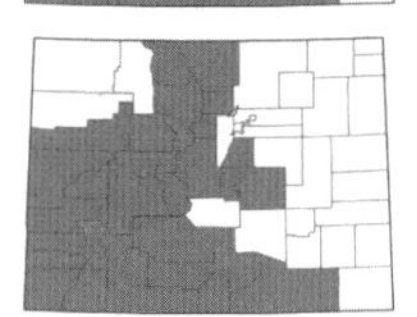

Stellaria longipes* Goldie ssp. *longipes, LONG-STALKED STARWORT. Perennials, rhizomatous, 0.3–3.2 dm; *leaves* linear-lanceolate to lanceolate, 0.4–4 cm × 1–4 mm; *sepals* 3.5–5 mm long, glabrous or hairy; *petals* 5, 3–8 mm long; *capsules* black to purple or yellowish, 4–6 mm long; *seeds* brown, 0.6–1 mm diam. Common in meadows, forests, along streams, and in the alpine, 7500–13,000 ft. June–Aug. E/W.

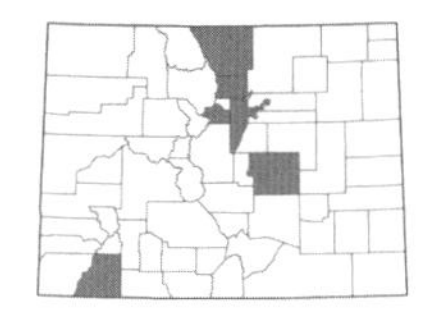

***Stellaria media* (L.) Vill.**, COMMON CHICKWEED. Annuals or short-lived perennials, decumbent or ascending, 0.5–4 dm, the stem with a line of hairs from each internode; *leaves* ovate to elliptic, 0.5–4 cm × 2–20 mm; *sepals* 4.5–6 mm long, usually glandular; *petals* absent or 5, 1–4 mm long, shorter than or equal to the sepals; *capsules* yellowish to green, 3–5 mm long; *seeds* reddish-brown, 1–1.3 mm diam., tuberculate. Found in disturbed places, lawns, and along streams, 5000–8000 ft. April–Sept. E/W. Introduced.

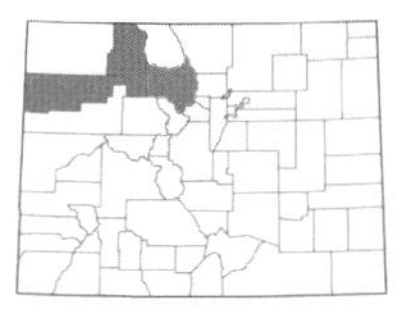

***Stellaria obtusa* Engelm.**, ROCKY MOUNTAIN STARWORT. Perennials, rhizomatous, prostrate, 0.3–2.5 dm; *leaves* elliptic to broadly ovate, 0.2–1.2 cm × 1–7 mm; *sepals* 1.5–3.5 mm long, glabrous; *petals* absent; *capsules* yellowish to green, 2.3–3.5 mm long, about twice as long as the sepals; *seeds* blackish, 0.5–0.7 mm diam. Uncommon in spruce-fir and aspen forests and along streams, 8500–10,500 ft. June–Aug. W.

***Stellaria pallida* (Dumort.) Crépin**, LESSER CHICKWEED. Annuals, prostrate, 1–2 (4) dm, the stem with a line of hairs from each internode; *leaves* ovate to elliptic, 0.3–2 cm × 1–7 mm; *sepals* 3–4 mm long, hairy; *petals* absent; *capsules* pale yellowish, 2–5 mm long, equal to slightly longer than the sepals; *seeds* yellowish-brown, 0.5–1 mm diam., tuberculate. Uncommon in disturbed places, along roadsides, reported for Colorado but no specimens have been seen. May–June. Introduced.

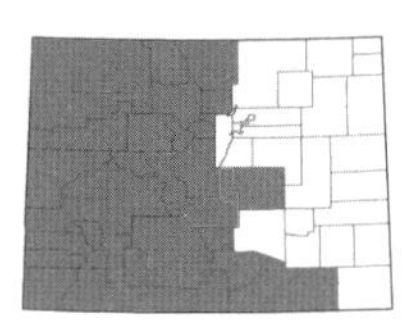

***Stellaria umbellata* Turcz.**, UMBELLATE STARWORT. Perennials, rhizomatous, 0.5–2 (4) dm; *leaves* lanceolate to elliptic, 3–9 cm × 1–3 mm; *inflorescences* subumbellate; *sepals* 2.5–3 mm long; *petals* absent; *capsules* 3–5 mm long, longer than the sepals; *seeds* brown, 0.5–0.7 mm diam., rugose. Common in meadows, along streams, and in the alpine, 7000–14,000 ft. June–Aug. E/W.

VACCARIA Wolf – SOAPWORT

Annual herbs; *leaves* exstipulate; *flowers* perfect; *calyx* 5-lobed; *petals* 5, pink to purple, clawed, the apex entire to shortly bifid; *stamens* 10; *styles* 2 (3); *stigmas* 2 (3); *fruit* a capsule opening by 4 teeth.

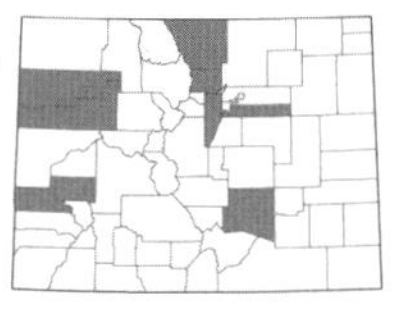

***Vaccaria hispanica* (Mill.) Rauschert**, COW SOAPWORT. [*V. pyramidata* Medic.]. Plants 2–10 dm; *leaves* lanceolate to ovate-lanceolate, 2–10 cm long; *calyx* 9–18 mm long, with winged angles or ridges; *petals* with a limb 3–8 mm long; *seeds* 2–2.5 mm wide, black, papillose. Found in disturbed places, along roadsides, and near old homesteads, 4800–7000 ft. June–Sept. E/W. Introduced.

CELASTRACEAE R. Br. – STAFFTREE FAMILY

Trees, shrubs, and lianas; *leaves* alternate or opposite, often leathery, simple and entire; *flowers* perfect, actinomorphic, aggregated in cymes or fascicles; *sepals* 4–5, distinct; *petals* 4–5, distinct; *stamens* 3–5, sometimes 3–5 staminodes present; *pistil* 1; *style* 1; *stigma* 2–5-lobed; *ovary* superior, 2–5-carpellate; *fruit* a loculicidal capsule, berry, drupe, samara, or achene-like.

PAXISTIMA Raf. – MOUNTAIN LOVER

Shrubs; *leaves* opposite, evergreen, leathery, stipulate; *sepals* 4; *petals* 4; *stigma* 2-lobed; *ovary* 2-carpellate; *fruit* a loculicidal capsule.

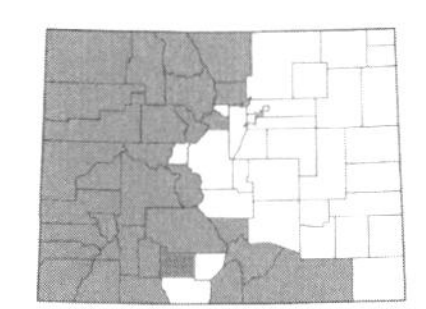

***Paxistima myrsinites* (Pursh) Raf.**, OREGON BOXLEAF; MOUNTAIN LOVER. [*Pachystima myrsinites* (Pursh) Raf.]. Low shrubs, spreading to prostrate, with stems mostly 3–8 dm long; *leaves* oblanceolate to elliptic or ovate, 0.8–4 cm × 2–15 mm, sharply spinulose-serrate to crenate-serrate; *petals* 1–2 mm long, reddish-brown; *capsules* 4–5 mm long. Found in spruce-fir forests, 5800–11,000 ft. June–Aug. E/W.

CERATOPHYLLACEAE Gray – HORNWORT FAMILY

Aquatic, perennial herbs; *leaves* whorled, sessile, exstipulate, dichotomously dissected, submerged, the margins serrulate; *flowers* imperfect, sessile and solitary in leaf axils, subtended by an involucre of 8–15 linear bracts; *perianth* absent; *stamens* 3–50; *pistil* 1; *ovary* superior, unicarpellate; *fruit* an achene.

CERATOPHYLLUM L. – HORNWORT

Submerged perennials; *leaves* 3–11 per whorl; *achenes* ellipsoid, smooth or tuberculate, usually with spines.

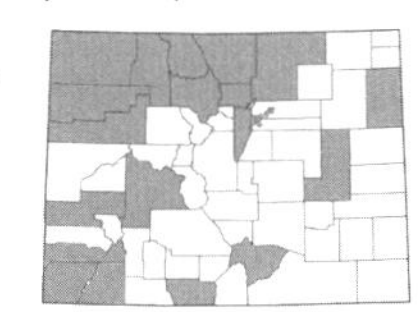

***Ceratophyllum demersum* L.**, COON'S TAIL. Stems to 10+ dm; *leaves* 5–12 per node, dichotomously 1–2× divided into linear segments, antrorsely serrate; *achenes* 4–5 mm long, with 2 basal spines. Found submerged in lakes, ponds, and slow-moving streams, scattered across the state, 3500–9500 ft. June–Sept. E/W.

Usually found without fruit since its primary means of reproduction is asexual.

CHENOPODIACEAE Vent. – GOOSEFOOT FAMILY

Herbs, subshrubs, or shrubs, sometimes succulent, often with scurfy hairs, plants monoecious or dioecious; *leaves* simple, alternate or opposite, exstipulate; *flowers* perfect or imperfect, inconspicuous; *sepals* 1–5, persistent, enclosing the fruit at maturity; *petals* absent; *stamens* as many or fewer than the sepals, opposite the sepals; *pistil* 1; *ovary* superior, 1–3-carpellate; *fruit* an utricle.

1a. Shrubs or subshrubs, or perennials, woody at least at the base…2
1b. Annuals from a taproot (rarely short-lived perennials, but these not woody at the base)…9

2a. Plants densely covered with white to tawny tomentose, stellate hairs; staminate flowers 4-merous with pinkish stamens; pistillate flowers enclosed in densely hairy bracts; leaves linear to narrowly lanceolate with revolute margins…***Krascheninnikovia***
2b. Plants glabrous, scurfy, minutely hairy, or villous; staminate flowers usually 5-merous, stamens variously colored; pistillate flowers not enclosed in densely hairy bracts; leaves various…3

3a. Leaves linear, usually succulent and fleshy, terete to flat…4
3b. At least some leaves wider, fleshy or not, flat…7

4a. Shrubs with thorns extending at right angles from the main stem…***Sarcobatus***
4b. Shrubs lacking thorns or herbs…5

5a. Flowers imperfect, lacking a perianth at least in pistillate flowers; fruiting bracts completely enclosing the pistillate flowers, usually samaralike with a winged, entire margin, the faces smooth and lacking appendages…***Zuckia***
5b. Flowers mostly perfect, with a perianth; fruiting bracts unlike the above…6

6a. Fruiting perianth hairy or rarely glabrate, forming a 1.5–3 mm long, horizontal, scarious wing at maturity; leaves villous (sometimes becoming glabrate); leaves not white-margined…***Kochia*** (*americana*)
6b. Fruiting perianth glabrous or sometimes hairy, lacking a wing; leaves glabrous or rarely villous; leaves often with a thin, white margin…***Suaeda*** (*nigra*)

7a. Main stems with whitish, exfoliating strips of bark; branches usually spine-tipped; fruiting bracts flattened, orbicular to elliptic, samaralike with a wing completely surrounding the pistillate flowers, smooth and lacking appendages on the faces; leaves fleshy or leathery…***Grayia***
7b. Main stems lacking whitish, exfoliating strips of bark; branches usually not spine-tipped (or if spine-tipped then unlike the above in all other respects); fruiting bracts various; leaves fleshy or not…8

8a. Fruiting bracts completely enclosing the pistillate flowers, usually samaralike with a winged, entire margin, the faces smooth and lacking appendages; staminate flowers usually with 4 perianth lobes; leaves often succulent and fleshy, usually with small, round buds in the axils…***Zuckia***
8b. Fruiting bracts not completely enclosing the pistillate flowers, free at least at the tip, not winged around the margin or if winged then the margins toothed at least towards the apex, the faces smooth or with appendages; staminate flowers usually with 5 perianth lobes; leaves not fleshy, usually lacking small, round buds in the axils…***Atriplex***

9a. Leaves opposite, fused at the base, scale-like and rather inconspicuous; stems jointed and fleshy, usually reddish; flowers with 3 perianth parts, sunken in terminal spikes…***Salicornia***
9b. Leaves not both opposite and scale-like, conspicuous and well-developed; stems not jointed, reddish or not; flowers unlike the above..10

10a. Plants stellate-hairy…11
10b. Plants glabrous or variously pubescent, but lacking stellate hairs…12

11a. Leaves petiolate, narrowly ovate; fruit a dark red or brownish, oval to obovate utricle topped with a whitish or light brown, 2-lobed appendage (must remove persistent perianth to see); flowers imperfect, in glomerules at the ends of branches, with 3–5 perianth segments…***Axyris***
11b. Leaves sessile, linear to narrowly lanceolate; fruit a strongly flattened achene; flowers perfect, solitary in the axils of bracts, lacking a perianth…***Corispermum***

12a. Leaves with a short spine or bristle at the tip, filiform to linear or linear-lanceolate with entire margins…13
12b. Leaves lacking a spine or bristle tip, variously shaped…15

13a. Plants glaucous and glabrous except for tufts of white hairs in the axils of bracts and leaves; bracts of the inflorescence similar to the leaves, terete, succulent and linear; flowers in groups of 3 in axillary glomerules, with a 5-parted, membranous but firm, fan-shaped perianth usually with a pinkish or reddish center; plants with an erect, main stem and several additional stems spreading from the base (sometimes forming an X pattern with 4 lateral stems spreading)…***Halogeton***
13b. Plants glabrous to pubescent but lacking tufts of white hairs in the axils of bracts and leaves, or the hairs branched; bracts of the inflorescence well-differentiated from the leaves, ovate-lanceolate with a whitish, scarious margin at least at the base; leaves various; flowers solitary in the bract axils, lacking a perianth or the perianth unlike the above; plants usually not spreading like the above…14

14a. Bracts and leaves with sharp, stiff spine tips; fruit not flat, lacking a wing or prominently thin-winged; plants often profusely branched from or near the base (forming tumbleweeds)…***Salsola***
14b. Bracts and leaves with flexuous or weak spine or bristle tips; fruit relatively flat, ovate to elliptic, narrowly winged or not winged; plants not forming tumbleweeds…***Corispermum***

15a. Perianth with an erose, continuous, membranous wing encircling the flower at maturity, becoming reddish in age; plants cobwebby pubescent, not scurfy, becoming glabrate in age; most leaves sharply toothed, the teeth usually ending in a spinulose point, lanceolate to oblong-ovate; inflorescence a diffusely branched panicle with numerous flowers…***Cycloloma***
15b. Perianth unlike the above or absent; plants not cobwebby pubescent, often scurfy; leaves various but usually unlike the above; inflorescence various…16

16a. Leaves succulent and fleshy, linear to linear-lanceolate, often with a narrow, whitish margin; perianth segments usually fleshy and succulent as well, usually with at least 1 segment larger than the others…***Suaeda***
16b. Leaves variously shaped or if linear to linear-lanceolate then not succulent and fleshy; perianth not fleshy and all segments equal in size, or absent…17

17a. Flowers subtended by an ovate, acuminate, scarious-margined bract, solitary in the axils of bracts, perfect, lacking a perianth; fruit 2–5 mm long, relatively flat, ovate to elliptic, narrowly winged or not winged; leaves linear to linear-lanceolate…***Corispermum***
17b. Flowers not subtended by bracts like the above; fruit various but unlike the above; leaves various…18

18a. Perianth densely hairy, the segments with hooked appendages from the back at maturity, the hairs usually tawny with age; leaves villous on both sides…***Bassia***
18b. Perianth glabrous or nearly so; leaves various but not villous on both sides…19

19a. Inflorescence bracts and leaves long-ciliate on the margins (the hairs mostly 1–3 mm long), usually turning tawny in age; rachis of the inflorescence densely long-hairy; leaves linear to linear-lanceolate or narrowly obovate...***Kochia*** (*scoparia*)
19b. Inflorescence bracts and leaves not long-ciliate on the margins; leaves various...20

20a. Pistillate flowers lacking a perianth and enclosed by 2 partly or completely united fruiting bracts; flowers imperfect...21
20b. Flowers perfect or imperfect, with a perianth or subtended by a single bract, not enclosed by fruiting bracts, the seeds brown to black with a free or attached pericarp...22

21a. Fruiting bracts vertically keeled, united almost to the apex, ovate, with winged margins, the faces lacking appendages; leaves suborbicular to ovate with a cuneate base, the margins mostly toothed...***Suckleya***
21b. Fruiting bracts not vertically keeled, otherwise unlike the above, the faces smooth or with appendages; leaves various...***Atriplex***

22a. Flowers imperfect, with 2–3 inconspicuous perianth segments, solitary in axils or rarely in 2–4-flowered clusters; leaves thick and succulent, oblong, entire; plants with a slender stem and branches, delicate, dichotomously branched, often turning reddish in age...***Micromonolepis***
22b. Flowers perfect, with a conspicuous 3–5-lobed perianth or lacking a perianth and subtended by a single leaf-like bract, in glomerules or clusters or panicles, but not solitary in the leaf axils; leaves not succulent, variously shaped; plants not dichotomously branched, otherwise various...23

23a. Flowers lacking a perianth and subtended by a single leaf-like bract; leaves hastately lobed at the base, lanceolate to narrowly elliptic; inflorescence bracteate throughout...***Monolepis***
23b. Flowers with a 3–5-lobed perianth, sometimes the segments fleshy and reddish; leaves variously shaped; inflorescence bracteate throughout or not...24

24a. Plants glandular-hairy or the lower leaves and sepal lobes with yellow resinous dots, usually strongly aromatic; leaves mostly pinnately lobed or lyrate-sinuate...***Dysphania***
24b. Plants glabrous or with mealy (scurfy) hairs, aromatic or not; leaves various...***Chenopodium***

ATRIPLEX L. – SALTBUSH

Herbs or shrubs, monoecious or dioecious; *leaves* alternate or opposite; *flowers* imperfect, usually clustered in glomerules or spikes; *staminate flowers* ebracteate, with a 3–5-parted calyx and 3–5 stamens; *pistillate flowers* subtended by 2 accrescent bracts that enclose the fruit, the calyx absent or rarely 3–5-parted; *fruit* an utricle, the pericarp free from the seed. (Bassett et al., 1983; Brown, 1956; Hall & Clements, 1923; Wagner & Aldon, 1978; Weber, 1950)

1a. Perennials, dioecious or seldom monoecious shrubs or subshrubs...2
1b. Annuals, monoecious or seldom dioecious herbs...6

2a. Shrubs with branches and twigs becoming spiny, leafless and sharp-tipped; fruiting bracts orbicular or broadly elliptic, united to the middle, with a terminal and somewhat triangular foliaceous tip, smooth and lacking wings or tubercles, 4–12 mm long; leaves orbiculate to elliptic or ovate...***A. confertifolia***
2b. Shrubs or subshrubs lacking spiny branches and twigs; fruiting bracts united beyond the middle or if united to the middle then 4–5 mm long and sharply toothed (*A. obovata*), otherwise unlike the above in all respects; leaves various...3

3a. Shrubs low-spreading and mat-forming, prostrate, mostly to 1.5 dm tall, often rooting at the nodes; leaves usually smaller, linear-oblanceolate or narrowly oblong, 3–18 mm long and 2–3 (6) mm wide, mostly opposite or just the upper alternate; fruiting bracts 3–5 mm long, with numerous small, wart-like tubercles to 2 mm long (rarely smooth); staminate flowers in a nearly naked terminal spike 1–8 cm long; pistillate flowers in terminal spikes...***A. corrugata***
3b. Shrubs or subshrubs 2–20 dm tall, erect, or if to 1 dm tall, prostrate and mat-forming then the leaves elliptic, oblong, or spatulate and usually over 5 mm wide; leaves usually mostly alternate; fruiting bracts 4–25 mm long, with wings or tubercles, or smooth; staminate flowers in terminal spikes or panicles 2–30 cm long; pistillate flowers in terminal spikes or panicles...4

4a. Fruiting bracts united to the middle, wider than long, 4–5 mm long, the apex and margins sharply toothed; stem and leaves densely silvery-scurfy, obovate to broadly elliptic to orbiculate; stems strictly erect from a much-branched, decumbent base, forming subshrubs 2–8 dm tall...***A. obovata***
4b. Fruiting bracts united throughout or beyond the middle, usually longer than wide, 2–25 mm long, the apex various; stems and leaves not densely silvery-scurfy, variously shaped; stems spreading or erect, forming shrubs or subshrubs 1–20 dm tall...5

5a. Fruiting bracts conspicuously 4-winged with the wings extending the entire length of the fruiting body, the margins of the wings toothed or entire, the apex toothed, smooth, the body (5) 8–25 mm long; shrubs usually woody throughout and mostly 8–20 dm tall (rarely 2–5 dm tall); leaves 3–8 mm wide...***A. canescens***
5b. Fruiting bracts lacking wings, smooth or tuberculate, 2–9 mm long; subshrubs usually woody just at the base and 1–5 dm; leaves 2–25 mm wide...***A. gardneri***

6a. Leaves usually green and glabrous above and below, occasionally somewhat white and scurfy below...7
6b. Leaves finely to densely scurfy above and below, white to grayish...13

7a. Perianth of staminate flowers cup-shaped, usually pink or reddish, each lobe with a fleshy crest on the back; leaves sessile, lanceolate to elliptic, glabrous, thick and succulent; staminate flowers usually in clusters arranged in a terminal, simple spike; fruiting bracts small and difficult to locate, about 2 mm long, united to the tip, smooth...***A. suckleyi***
7b. Perianth of staminate flowers not cup-shaped, usually green or yellowish, the lobes lacking a fleshy crest on the back; leaves petiolate or shortly petiolate below and sessile above, variously shaped; staminate flowers in various inflorescences; fruiting bracts various...8

8a. Leaves entire, triangular-ovate to orbicular, 5–20 mm long; fruiting bracts triangular-ovate to orbicular, smooth; plants 0.5–3 dm tall, of the west slope...9
8b. Usually at least a few leaves irregularly toothed or hastately lobed, lanceolate, ovate-lanceolate, triangular, or hastate, 15–125 mm long; fruiting bracts various; plants 1–20 dm tall, of the east or west slope...10

9a. Fruiting bract orbicular (or nearly so) with a yellowish-brown center and green wing completely surrounding the bract, united to the tip, samaralike, enclosing a single flower; staminate flowers above the fruiting bracts; pistillate flowers lacking a perianth; leaves often drying bright green...***A. graciliflora***
9b. Fruiting bract triangular-ovate, leaf-like, united to the middle, enclosing 2–5 flowers; staminate flowers intermixed with the fruiting bracts; pistillate flowers with a thin, membranous 5-lobed perianth; leaves often drying dark green...***A. pleiantha***

10a. Fruiting bracts orbicular (or nearly so) to ovate, with a wing completely surrounding the central body (samaralike), smooth and lacking appendages, 5–18 mm long or with very small bracts (2 mm) in addition to the larger ones...11
10b. Fruiting bracts triangular to ovate or rhombic, lacking a wing and not samaralike, often toothed and tuberculate on the faces, mostly 3–10 mm long...12

11a. Mature fruiting bracts of two conspicuously different sizes, the larger ones 5–6 mm long and the smaller ones 2 mm long; seeds all vertical; leaves mostly all triangular...***A. heterosperma***
11b. Mature fruiting bracts not of two conspicuously different sizes, 5–18 mm long, with most 10 mm or longer; seeds horizontal or vertical; leaves a mixture of lanceolate to lance-ovate or the larger ones triangular...***A. hortensis***

12a. Fruiting bracts united only at the base, triangular to ovate or hastate to slightly rhombic with a truncate or broadly rounded base, usually toothed, sometimes slightly thick and spongy at the base, the lateral angles obscure; leaves sometimes scurfy below, thick or thin...***A. subspicata***
12b. Fruiting bracts united just to below the middle, rhombic with a cuneate base, usually with entire margins or obscurely toothed, not thick or spongy at the base, the lateral angles sharp and well-defined; leaves green above and below, thin...***A. patula***

13a. Leaves linear to narrowly oblong, 1–3 mm wide, entire, sessile and alternate; staminate and pistillate flowers intermixed in axillary clusters of leaves...***A. wolfii***
13b. Leaves variously shaped but not linear, over 3 mm wide, margins entire or not, sessile or petiolate, alternate or opposite; staminate flowers various...14

14a. Most leaves sinuate-dentate, sessile or shortly petiolate, prominently 3-veined from the base; fruiting bracts ovate to rhombic, toothed on the margins, sharply tuberculate to smooth on the surfaces, united to the middle...***A. rosea***
14b. Most leaves entire or some few-toothed or lobed, or if the majority sinuate-dentate then the leaves conspicuously petiolate, prominently 3-veined or not; fruiting bracts various...15

15a. Flowers in naked spikes; seeds dimorphic, the larger ones brown and 1.5–3 mm wide, the smaller ones black and shiny and 1–2 mm wide; leaves sometimes thick and prominently 3-veined from the base, linear-lanceolate, ovate-lanceolate to triangular-hastate...***A. subspicata***
15b. Flowers in leafy spikes or in leaf axils; seeds all similar; leaves thick or not, variously shaped...16

16a. Fruiting bracts of two kinds, the larger ones globose, 4–6 mm long, on stipes mostly 4–8 mm long and densely covered with prominent horn-like appendages, the smaller ones 3–4 mm long, urn-shaped, sessile with a truncate apex, smooth or notched to toothed only at the apex; leaves cordate-ovate to triangular-ovate...***A. saccaria***
16b. Fruiting bracts all similar, sessile or nearly so; leaves various...17

17a. Leaves prominently 3-veined from the base (especially abaxially), ovate to rhombic, usually cuneate or rounded at the base; fruiting bracts fiddle-shaped with 2 lateral, rounded, smooth lobes adjacent to an apex with a short, central tooth, the faces below usually with densely covered with appendages or sometimes smooth; plants often dioecious...***A. powellii***
17b. Leaves not prominently 3-veined or sometimes 3-veined but the veins spreading to adjacent lobes, triangular or ovate to triangular-ovate and usually truncate or cordate at the base; fruiting bracts various but unlike the above in all respects; plants monoecious...18

18a. Fruiting bracts 2–3 mm long, broadly cuneate with a truncate apex, the apex at least 3-toothed, the faces usually smooth and unappendaged...***A. truncata***
18b. Fruiting bracts (2.5) 4–11.5 mm long, orbicular or nearly so, usually with many teeth at the apex and along the margins, appendaged or smooth...***A. argentea***

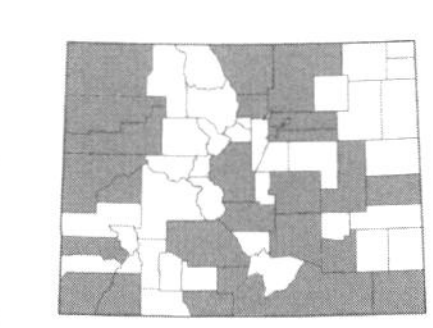

Atriplex argentea **Nutt.**, SILVERSCALE SALTBUSH. [*A. pachypoda* Stutz & G.L. Chu]. Annuals, 0.5–6 dm, monoecious; *leaves* elliptic, lanceolate, or deltoid, 0.5–7.5 × 0.4–5 cm, the base subhastate to acute, the margins entire or nearly so, scurfy; *staminate flowers* 5-merous, borne in distal axils or in short dense spikes or panicles; *fruiting bracts* cuneate-orbicular, (2.5) 4–11.5 × 2–9 (14) mm, the face smooth or tuberculate, the margins dentate to laciniate; *seeds* 1.5–2 mm diam., brown. Common on dry slopes, shortgrass prairie, alkalai flats, 4000–8500 ft. June–Oct. E/W. (Plate 26)

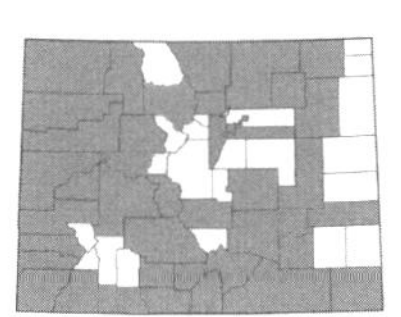

Atriplex canescens **(Pursh) Nutt.**, FOURWING SALTBUSH. [*A. gardneri* (Moq.) D. Dietr. var. *aptera* (A. Nelson) S.L. Welsh & Crompton; *A. nuttallii* S. Watson var. *nuttallii*]. Shrubs, mostly 8–20 dm, usually dioecious; *leaves* alternate, sessile, scurfy, linear, oblanceolate, oblong, or obovate, 1–4 × 0.3–0.8 cm, the margins entire; *staminate flowers* borne in panicles or axillary spikes; *fruiting bracts* (5) 8–25 mm long, on stipes 1–8 mm long, with 4 conspicuous wings, the wings entire to toothed; *seeds* 1.5–2.5 mm diam. Common on dry slopes and the plains, 3800–8800 ft. May–Aug. E/W. (Plate 26)

Atriplex garrettii Rydb. var. *garrettii* could possibly occur in Colorado and shares the 4-winged fruiting bracts with *A. canescens*. However, the leaves of *A. garrettii* are usually orbiculate or obovate to ovate and 15–30 mm wide, and the shrubs are mainly 2–6 dm tall. Can form hybrids with *A. gardneri* and *A. confertifolia*. It is difficult to tell *A. canescens* from *A. gardneri* when only staminate plants are available.

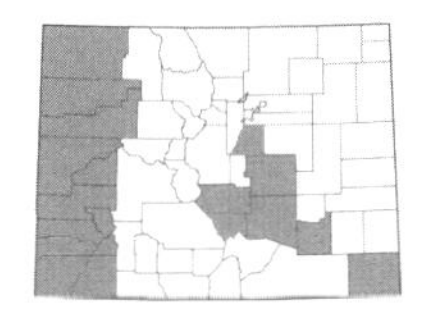

Atriplex confertifolia **(Torr. & Frém.) S. Watson**, SHADSCALE. Shrubs, 3–8 dm, spinescent, dioecious; *leaves* alternate, petiolate, orbicular to ovate or elliptic, 0.9–3 (4.5) × 0.4–2.5 cm, the margins entire; *staminate flowers* in axillary clusters or panicles 3–15 cm long; *fruiting bracts* sessile or nearly so, 4–12 mm long, the faces smooth, the margins entire or toothed; *seeds* 1.5–2 mm diam. Common on rocky, clay, or shale slopes, often with pinyon-juniper or sagebrush, 4500–7200 ft. May–July. E/W. (Plate 26)

Can form hybrids with *A. canescens*, *A. garrettii*, *A. corrugata*, and *A. gardneri*.

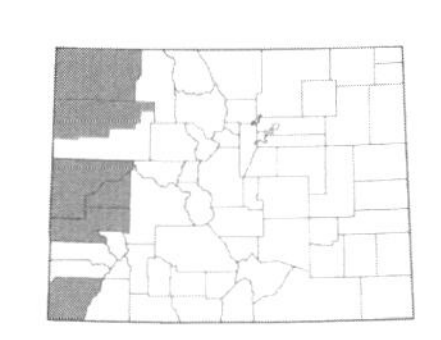

Atriplex corrugata **S. Watson**, MATSCALE. Low, spreading shrubs, 0.3–1.5 dm, dioecious; *leaves* opposite below and alternate above, sessile, linear-oblanceolate or narrowly oblong, 0.3–1.8 × 0.2–0.3 (0.6) cm, the margins entire; *staminate flowers* borne in spikes; *fruiting bracts* sessile or nearly so, 3–5 mm long, united to above the middle, usually densely tuberculate, rarely smooth, the margins entire or undulate; *seeds* 1.5 mm diam. Found on dry shale, clay, or gypsum slopes, 4700–7000 ft. May–July. W. (Plate 26)

Can form hybrids with *A. confertifolia* and *A. gardneri*.

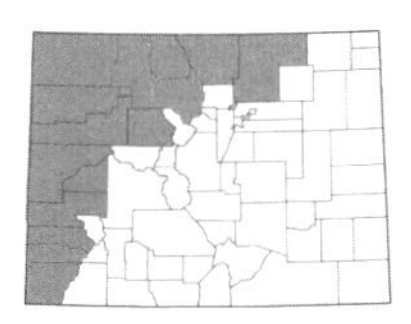

Atriplex gardneri **(Moq.) D. Dietr.**, GARDNER'S SALTBUSH. [*A. cuneata* A. Nelson; *A. nuttallii* S. Watson var. *gardneri* (Moq.-Tandon) R.J. Davis]. Shrubs or subshrubs, prostrate to ascending, 1–5 dm; *leaves* alternate or opposite, sessile or petiolate, linear to oblanceolate, obovate, or orbiculate, 0.5–5.5 × 0.2–2.5 cm, the margins usually entire; *staminate flowers* borne in spikes or panicles 2–30 cm long; *fruiting bracts* 2–9 × 2–9 mm, the faces usually with tubercles or wings aligned in 4 rows, the apex toothed; *seeds* 1.5–2.5 mm diam. Common on dry slopes, often with sagebrush, 4500–9500 ft. May–Aug. E/W. (Plate 26)

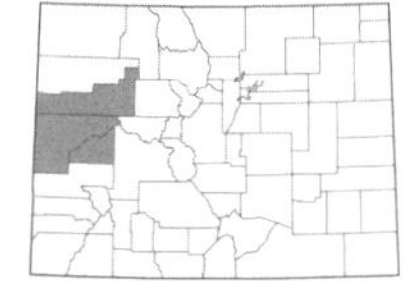

***Atriplex graciliflora* M.E. Jones**, Blue Valley orach. Annuals, 1–5 dm, monoecious; *leaves* mostly alternate, petiolate, cordate-ovate to orbicular or deltoid, 0.8–2.5 × 0.5–2 cm, truncate to cordate at the base, the margins entire; *staminate flowers* in loose terminal panicles; *fruiting bracts* on stipes 2–6 mm long, samaralike, suborbicular to cordate, winged, 6–16 mm wide, the faces smooth, the wings entire to undulate; *seeds* ca. 3 mm diam., white. Locally common on shale hills and clay slopes, 4600–5700 ft. May–June. W. (Plate 26)

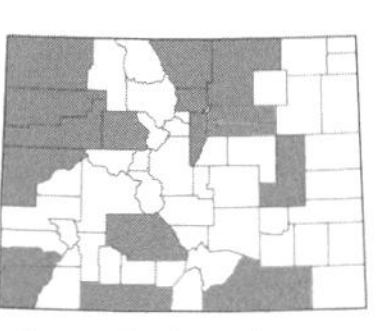

***Atriplex heterosperma* Bunge**, twoscale saltbush. [*A. micrantha* Ledeb.]. Annuals, 5–15 dm, monoecious; *leaves* mostly alternate, petiolate, triangular, hastately lobed at the base, the margins subentire or irregularly toothed, farinose when young, becoming glabrous; *staminate flowers* 5-merous; *fruiting bracts* orbiculate-ovate, dimorphic, the larger ones 5–6 mm long and the smaller ones ca. 2 mm long, the faces smooth, margins entire; *seeds* of larger bracts yellowish to reddish-brown, 2–3 mm long, of smaller bracts black and ca. 1.5 mm diam. A weedy plant found along ditches, in gardens, in floodplains, and disturbed places, 4500–7600 ft. June–Aug. E/W. Introduced. (Plate 26)

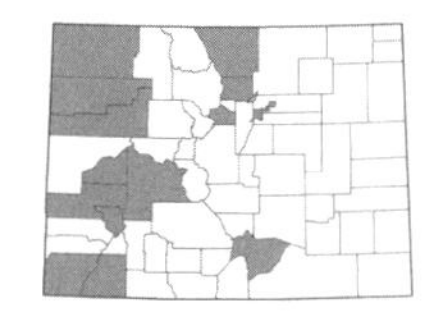

***Atriplex hortensis* L.**, garden orach. Annuals, 5–20 dm, monoecious; *leaves* opposite or alternate, petiolate, ovate to cordate, sometimes cordate-hastate to acute at the base, 1.5–18 × 0.8–13 cm; *staminate flowers* 5-merous; *fruiting bracts* orbicular to ovate, samaralike, 5–18 mm long, united at the base only, the faces smooth, margins entire; *seeds* dimorphic, 3–4.5 mm diam. or 1–2 mm diam. A weedy plant found along railroads, in gardens, fields, and vacant lots, and in other disturbed places, 5000–7500 ft. June–Aug. E/W. Introduced.

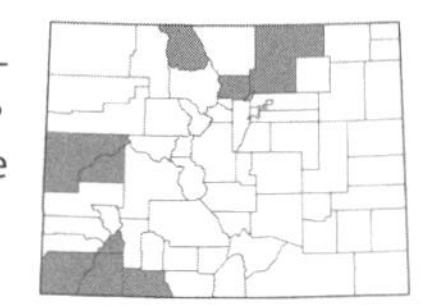

***Atriplex obovata* Moq.**, broadscale. Shrubs, 2–8 dm, dioecious; *leaves* alternate, petiolate, oblong-ovate to elliptic or orbicular, 0.8–3.5 × 0.6–2 cm, the margins usually entire, gray-green; *staminate flowers* borne in panicles 6–30 cm long; *fruiting bracts* sessile or nearly so, 4–5 × 5–9 mm, the base cuneate, the apex toothed with 2–6 teeth subtending, margin sharply toothed, the faces usually smooth; *seeds* 2.5–3 mm diam. Uncommon in sandy soil, 4700–6500 ft. May–Aug. W. (Plate 26)

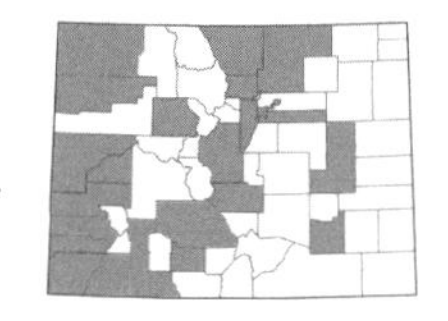

***Atriplex patula* L.**, spear saltbush. Annuals, 1.5–10 dm, monoecious; *leaves* alternate or opposite, petiolate, ovate, deltoid, or lanceolate, 2.5–12 × 0.3–4 (7.5) cm, cordate, hastate, truncate, or acute basally, the margins entire to toothed, glabrous or scurfy; *staminate flowers* with 4–5 sepals; *fruiting bracts* deltoid to rhombic, 2–12 × 3–9 mm, the margins entire or sparsely toothed, faces tuberculate; *seeds* dimorphic, 2.5–3.5 mm diam. or 1–2 mm diam. A weedy plant found in gardens, fields, and other disturbed places, 4900–7000. June–Aug. E/W. Introduced.

***Atriplex pleiantha* W.A. Weber**, four-corners orach. [*Proatriplex pleiantha* (W.A. Weber) Stutz & G.L. Chu]. Annuals, 0.5–1.5 dm; *leaves* alternate or subopposite, petiolate, ovate to suborbicular, 0.5–2 cm long and about as wide, the margins entire; *staminate flowers* in short terminal spikes; *fruiting bracts* short-stipitate, ovate, 3–7 mm long, entire; *seeds* ca. 1.5 mm diam. Locally common on Mancos shale slopes, 4800–5000 ft. May–June. W. (Plate 26)

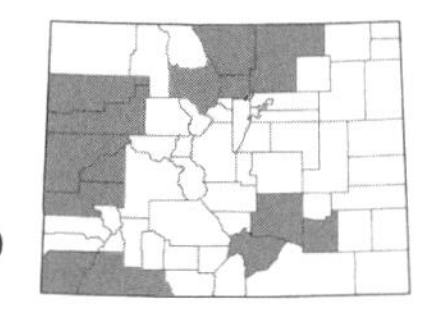

***Atriplex powellii* S. Watson**, Powell's saltweed. Annuals, 1–5 (7) dm, usually dioecious; *leaves* alternate, petiolate or sessile upward, ovate to rhombic or orbicular, 0.4–5 × 0.2–3 cm, the margins entire; *staminate flowers* with a 4–5-lobed calyx; *fruiting bracts* sessile, ovate to oblong, 1.5–5.5 × 1.5–5 mm, united to the apex, the apex tridentate to cuspidate, faces usually tuberculate; *seeds* 1–2 mm diam. Found on alkali flats, shale slopes, and gypsum hills, 4700–7000 ft. May–Aug. E/W. (Plate 26)

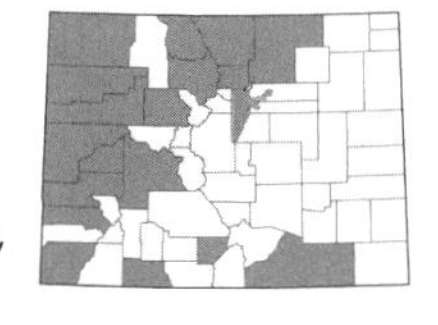

***Atriplex rosea* L.**, tumbling saltweed. Annuals, 2–10 (20) dm, monoecious; *leaves* alternate, lanceolate to ovate, mostly 1.2–8 × 0.6–5 cm, the margin irregularly toothed and often subhastately lobed; *staminate flowers* with 4–5 sepals; *fruiting bracts* sessile or shortly stipitate, ovate to rhombic, united to the middle, the margins dentate, the faces tuberculate to nearly smooth; *seeds* dimorphic, 2–2.5 mm diam. or 1–2 mm diam. Found on floodplains and alkali flats, along roadsides, and in disturbed areas, 5000–9000 ft. July–Sept. E/W. Introduced. (Plate 26)

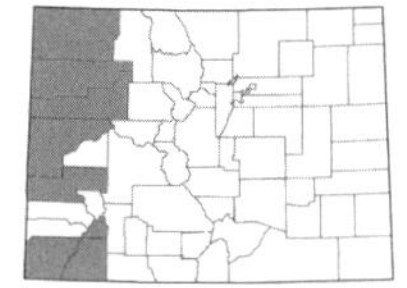

***Atriplex saccaria* S. Watson**, stalked orach. Annuals, 0.4–5 dm, monoecious; *leaves* alternate, cordate-ovate to subreniform or oval, 0.6–4 × 0.4–3 cm, truncate to subcordate or cuneate at the base, the margins entire or sometimes undulate-dentate; *staminate flowers* in distal axils or borne on short, terminal panicles, 5-merous; *fruiting bracts* dimorphic, united to the base, the larger ones 4–6 mm long, irregularly toothed and densely covered with flat hornlike appendages, the smaller ones 3–4 mm long, dentate only at the tips, otherwise the faces smooth; *seeds* 1.5–2.3 mm diam. Uncommon on barren slopes and dry hills, 4800–6200 ft. May–Aug. W. (Plate 26)

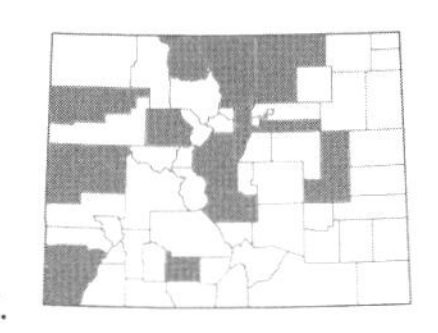

***Atriplex subspicata* (Nutt.) Rydb.**, SALINE SALTBUSH. [*A. dioica* Raf.]. Annuals, 5–15 dm, monoecious; *leaves* opposite proximally, lanceolate to triangular-ovate, usually with a hastate base, 3–12 (19) cm long, the margins entire to toothed, scurfy but glabrous in age; *fruiting bracts* sessile, triangular to ovate with a truncate base, 3–10 mm long, the margins united at the base and entire or with short teeth, smooth or with 2 tubercles; *seeds* larger ones 1.5–3 mm diam., smaller ones 1–2 mm diam. Found on alkaline flats and drying shores of ponds, 5000–8500 ft. July–Sept. E/W. (Plate 27)

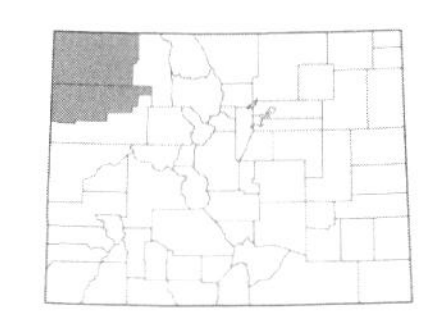

***Atriplex suckleyi* (Torr.) Rydb.**, SUCKLEY'S ORACH. [*Endelopis suckleyi* Torr.]. Herbs, 0.3–4 dm; *leaves* alternate, sessile, lanceolate to elliptic, 0.7–3.5 × 0.4–1.1 cm, succulent, the margins entire; *staminate flowers* in glomerules in distal axils or in short terminal spikes; *fruiting bracts* sessile, ovate, ca. 2 mm long, united to the apex, smooth; *seeds* ca. 1.5 mm diam. Uncommon on clay or shale slopes, 5500–6500 ft. June–Aug. W.

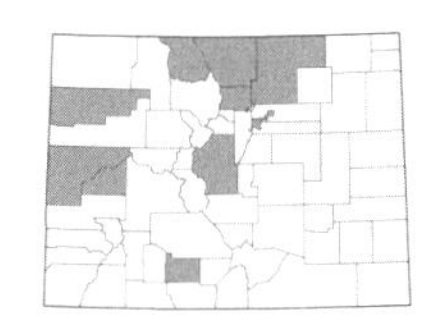

***Atriplex truncata* (Torr. ex S. Watson) A. Gray**, WEDGE ORACH. Annuals, 3–8 dm, monoecious; *leaves* alternate, petiolate below and sessile above, ovate to deltate, 0.4–3 × 0.3–3 cm, truncate or subhastate to rounded at the base, the margins entire or toothed; *staminate flowers* in glomerules in distal axils, sepals 3–5; *fruiting bracts* 2–3 mm long, truncate at the apex with at least 3 teeth at the summit, the faces usually smooth; *seeds* 1–2 mm diam. Uncommon on alkali flats and saline soil, 4700–8500 ft. July–Sept. E/W. (Plate 27)

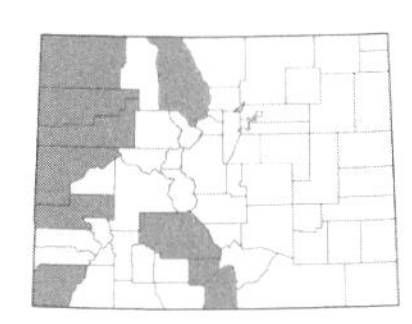

***Atriplex wolfii* S. Watson**, SLENDER ORACH. Annuals, 0.5–3.5 dm, monoecious; *leaves* alternate, sessile, linear to narrowly lanceolate, 0.4–2.5 × 0.1–0.3 cm; *staminate flowers* in axillary clusters, sepals 5; *fruiting bracts* sessile, ovate to cuneate, 1.5–3 × 1–2 mm, the apex truncate to attenuate, faces tuberculate or smooth; *seeds* 1–2 mm diam. Uncommon on alkaline flats, marshy areas, and in sandy soil along roadsides, 4500–8500 ft. June–Aug. E/W. (Plate 27)

There are two rather distinct varieties of *A. wolfii* in Colorado:

1a. Uppermost lateral lobes of fruiting bracts much larger than the middle lobe...**var. *wolfii***. Found mostly in the southcentral counties. E/W.

1b. Uppermost lateral lobes of the fruiting bracts not larger than the middle lobe...**var. *tenuissima* (A. Nelson) S.L. Welsh**. [*A. tenuissima* A. Nelson] Found in the northwestern counties. W.

AXYRIS L. – RUSSIAN PIGWEED

Annuals, monoecious herbs with tawny hairs; *leaves* alternate, entire; *flowers* imperfect; *staminate flowers* with 3–5 perianth segments and 3–5 stamens, in glomerules at the ends of branches; *pistillate flowers* with 3–4 perianth segments and 2 stigmas, each flower subtended by 2 bracteoles, found in the axils of bracts below the staminate flowers; *fruit* an utricle enclosed by persistent sepals, winged at the apex.

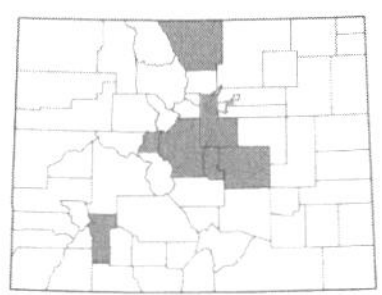

***Axyris amaranthoides* L.**, RUSSIAN PIGWEED. Plants (0.5) 3–9 dm; *leaves* petiolate, narrowly lanceolate to narrowly ovate,1.5–3 × 5–10 cm; *utricles* reddish or dark brown, 2.5–3 mm long, with a 2-lobed apical wing. Uncommon weedy plants found in disturbed places, fields, and along roadsides, 7500–9800 ft. July–Sept. E/W. Introduced. (Plate 27)

BASSIA All. – SMOOTHERWEED

Annual herbs with hirsute or villous hairs or becoming glabrate; *leaves* alternate, entire; *flowers* perfect or pistillate, in glomerules in spikes; *sepals* 5; *fruit* a flattened achene, enclosed by the membranous calyx.

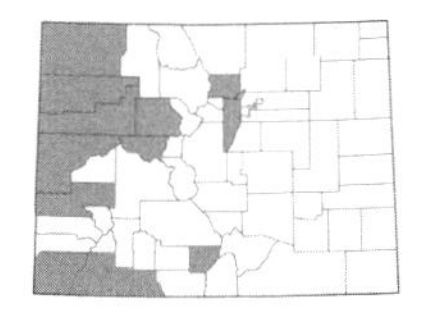

***Bassia hyssopifolia* (Pall.) Kuntze**, FIVEHORN SMOOTHERWEED. [*Kochia hyssopifolia* (Pall.) Schrad.]. Plants 0.5–10 dm; *leaves* sessile, linear to lanceolate or lance-elliptic, 0.6–5 × 0.1–0.4 cm; *sepals* densely woolly with curved, hooked spines ca. 1 mm long; *fruit* 1–1.5 mm diam. Weedy plants found in disturbed places, irrigation ditches, along the borders of drying alkaline ponds, and along roadsides, 4500–6500 ft. Aug.–Oct. E/W. Introduced. (Plate 27)

CHENOPODIUM L. – GOOSEFOOT

Annual or perennial herbs, often with mealy hairs (small, white inflated hairs) or glabrous; *leaves* alternate; *flowers* perfect or rarely imperfect, actinomorphic, in spicate glomerules; *calyx* (3) 5-lobed, sometimes becoming fleshy; *petals* absent; *fruit* an utricle or achene, usually enclosed by the sepals, the pericarp free or attached to the seed. (Aellen & Just, 1943; Bassett & Crompton, 1982; Crawford, 1975; 1977; Crawford & Wilson, 1986; Mosyakin & Clements, 1996; Wahl, 1954)

Writing a key to *Chenopodium* is quite a challenge. The best characters to use to separate species are based on pericarp characteristics. However, plants are usually collected in flower and not fruit! There are some species pairs which really can't be determined without fruit – namely *C. album* and *C. berlandieri*, *C. leptophyllum* and *C. hians*, and *C. incanum* and *C. watsonii*.

1a. Leaves and sepals glabrous; leaves usually triangular or deltate with dentate margins...2
1b. Leaves and/or sepals sparsely to densely farinose, at least abaxially (if leaves and sepals appearing glabrous, then leaves linear); leaves linear, ovate, deltate, or triangular with entire to dentate margins...7

2a. Sepals red and fleshy at maturity...3
2b. Sepals green or sometimes somewhat reddish at maturity, but not fleshy...4

3a. Bracts subtending glomerules throughout the inflorescence; stamens usually 1...***C. foliosum***
3b. Bracts absent in the distal ½ of the inflorescence; stamens 3...***C. capitatum***

4a. Sepals 5; stamens 5; fruit horizontal; flowers in terminal and lateral panicles; plants 3–15 dm tall...***C. simplex***
4b. Sepals 3 or 4; stamens 1–3; fruit vertical; flowers in dense glomerules, these sessile on terminal or lateral spikes; plants 0.1–10 dm tall...5

5a. Glomerules sessile on unbranched terminal spikes, occasionally also with a few axillary spikes...***C. overi***
5b. Glomerules sessile on numerous axillary spikes or panicles as well as a terminal spike...6

6a. Sepals free nearly to the base...***C. rubrum***
6b. Sepals connate nearly to the tip...***C. chenopodioides***

7a. Leaves linear to linear-lanceolate or ovate-lanceolate, or oblong (2–4 times longer than broad), with entire margins or sometimes also with a pair of lobes near the base on some leaves...8
7b. Leaves ovate, oblong, deltate, rhombic, triangular (to 2 times longer than broad), the margins entire with a pair of basal lobes or toothed throughout...16

8a. Leaves 3-veined from the base (the lateral veins can be faint; look on the largest leaves), linear, linear-lanceolate, oblong, or ovate-lanceolate, the margins entire or occasionally with a pair of lobes at the base...9
8b. Leaves 1-veined, linear, the margins entire...13

9a. Pericarp firmly attached to the seed (flaking off in small pieces or remaining attached when pried with a dissecting needle)...10
9b. Pericarp free from the seed (entire pericarp separating easily when pried with a dissecting needle)...11

10a. Inflorescence of crowded glomerules, leafy to near the tip; plants usually strictly upright or if branched from the base the branches strongly upright; leaves mostly narrowly lanceolate...***C. hians***
10b. Inflorescence of loosely spaced glomerules; plants open with spreading branches; leaves mostly oblong or the larger ones deltoid-rhombic...***C. atrovirens***

11a. Sepals enclosing the fruit at maturity; stems much-branched from the base, usually spreading; leaves entire and unlobed, moderately to densely farinose (whitish) above; glomerules usually densely packed and present in the top ½–¾ of the plant...***C. desiccatum***
11b. Sepals spreading to expose the fruit at maturity; stems solitary, branched above, or branched from the base (and then the glomerules loosely spaced in the inflorescence and not densely packed), erect to ascending; leaves entire or often at least some with a pair of lobes near the base, sparsely farinose to almost glabrous above (green above and whitish below)...12

12a. Leaves 1.5–3 times longer than wide, ovate, oblong, oval, or triangular; stems ascending to erect, usually branched at the base; glomerules loosely spaced in the inflorescence and generally not crowded...***C. atrovirens***
12b. Leaves 3–5 times longer than wide, linear to narrowly lanceolate or oblong-elliptic; stems strictly erect, simple or branching above; glomerules densely packed and present on the top ⅓–½ of the plant...***C. pratericola***

13a. Leaves sparsely farinose abaxially and glabrous adaxially (green); sepals connate into a 0.3–0.5 mm tube (to about the widest part of the seed); fruit 1.2–1.6 mm in diam.; glomerules widely spaced; uncommon plants of the eastern slope...14
13b. Leaves densely farinose abaxially, sparsely to moderately farinose adaxially (whitish); sepals distinct; fruit 0.8–1.1 mm in diam.; glomerules widely spaced to densely packed; found on either slope...15

14a. Pericarp firmly attached to the seed (coming off in small pieces or remaining attached when pried with a dissecting needle); sepals with wavy upper margins, spreading and not covering the fruit at maturity, not keeled...***C. cycloides***
14b. Pericarp free from the seed (entire pericarp separating easily when pried with a dissecting needle); sepals without wavy margins, largely covering the fruit at maturity, keeled...***C. subglabrum***

15a. Pericarp firmly attached to the seed (coming off in small pieces or remaining attached when pried with a dissecting needle); leaves linear, 1–3 mm wide; stems solitary or branched from the base and erect...***C. leptophyllum***
15b. Pericarp free from the seed (entire pericarp separating easily when pried with a dissecting needle); leaves linear, narrowly lanceolate, oblong-elliptic, or ovate-lanceolate, 4–6 mm wide; stems usually branched from the base and spreading...***C. desiccatum***

16a. Sepals 3–4, glabrous, not keeled; leaves densely farinose below, green and glabrous above, the margins entire to more often undulate-dentate...***C. glaucum***
16b. Sepals 5, farinose, keeled or not; leaves various but unlike the above in all respects...17

17a. At least some leaves toothed above the basal lobes...18
17b. Leaves entire and usually with a pair of basal lobes or single lobe at the base...20

18a. Leaves thin, broadly triangular with a broad, rounded apex, glabrous above and sparsely farinose below...***C. fremontii***
18b. Leaves unlike the above in all respects...19

19a. Fruit surface honeycombed, with a large, distinct yellow area at the base of the style; seeds honeycomb-pitted; leaves generally thicker...***C. berlandieri* var. *zschackii***
19b. Fruit surface smooth to papillate or roughened, without a large yellow area at the base of the style; seeds smooth to granulate; leaves generally thinner...***C. album***

20a. Leaves (at least the majority) 1.5–4 times longer than wide, ovate, oblong, or sometimes just the larger ones triangular, the margins entire and often at least some with a pair of basal lobes; sepals exposing the fruit at maturity...21
20b. Leaves as wide or nearly as wide as long (measuring the width across the widest point, usually at the basal lobes), broadly triangular, the margins entire and usually with a pair of basal lobes; sepals exposing or completely covering the fruit at maturity...22

21a. Inflorescence of loosely spaced glomerules; plants open with spreading branches; leaves mostly oblong or the larger ones triangular; pericarp usually free from the seed (entire pericarp separating easily when pried with a dissecting needle) or sometimes attached...***C. atrovirens***
21b. Inflorescence of crowded glomerules; plants usually strictly upright or if branched from the base then the branches strongly upright; leaves mostly narrowly lanceolate or narrowly oblong; pericarp firmly attached to the seed (coming off in small pieces or remaining attached when pried with a dissecting needle)...***C. hians***

22a. Leaves thin, glabrous above and sparsely farinose below; stems solitary or branched above, seldom few-branched from the base; pericarp free from the seed (entire pericarp separating easily when pried with a dissecting needle)...***C. fremontii***
22b. Leaves thick, densely farinose below and usually above as well; stems usually densely branched from the base; pericarp free or firmly attached...23

23a. Fruit surface honeycombed, often whitened; seeds coarsely honeycombed, often whitened; pericarp firmly attached to the seed (coming off in small pieces or remaining attached when pried with a dissecting needle); plants very strong-smelling (smelling like rotten fish)...***C. watsonii***
23b. Fruit surface smooth, black to brown; seeds black and smooth; pericarp free from the seed (entire pericarp separating easily when pried with a dissecting needle); plants not conspicuously strong-smelling...***C. incanum* var. *incanum***

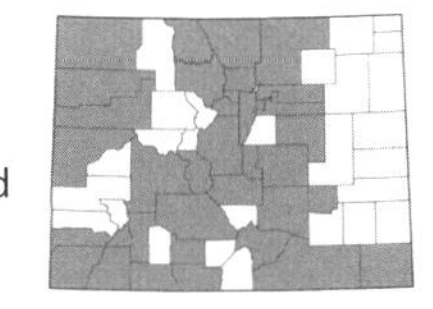

***Chenopodium album* L.**, LAMBSQUARTERS. Plants 1–30 dm; *leaves* rhombic to ovate-lanceolate or broadly oblong, 1–5.5 (12) × 0.5–4 (8) cm, the margins toothed, farinose below; *glomerules* 3–4 mm diam.; *sepals* 5, ca. 1 mm long, farinose; *stamens* 5; *fruit* usually with a free pericarp, smooth to papillate; *seeds* 0.9–1.5 mm diam., smooth. Common in disturbed places, along roadsides, and open places, 4800–9000 ft. June–Oct. E/W. Introduced. (Plate 28)

There are 3 varieties of *C. album* in Colorado:

1a. Pericarp firmly attached to the seed; leaf apices obtuse; sepals largely covering fruit at maturity...**var. *striatum* Krašan**, LATE-FLOWERING GOOSEFOOT. [*C. strictum* Roth]. Uncommon in the northeastern counties, 5000 ft. Aug.–Sept. E.
1b. Pericarp usually free from the seed; leaf apices acute to subobtuse; sepals reflexed and exposing fruit at maturity...2

2a. Inflorescence highly branched, of flexuous, delicate and spreading panicles, the glomerules loosely spaced...**var. *missouriense* (Aellen) I.J. Bassett & C.W. Crompton**, MISSOURI LAMBSQUARTERS. [*C. missouriense* Aellen]. Uncommon in disturbed places, known from Boulder and Montezuma cos., 5000–7000 ft. Sept. E/W.

2b. Inflorescence sparsely branched, erect, the glomerules densely packed...**var. *album***, COMMON LAMBSQUARTERS. Common, scattered across the state, 4800–9000 ft. June–Oct. E/W.

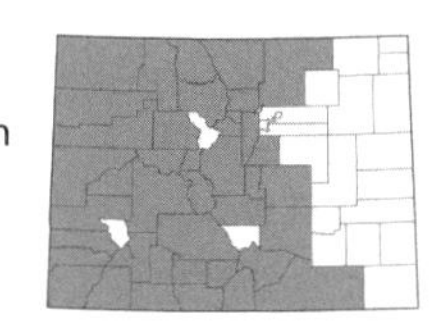

***Chenopodium atrovirens* Rydb.**, PINYON GOOSEFOOT. Plants 0.7–6.5 dm; *leaves* ovate, oblong, oval, or sometimes triangular, 1–3 × 0.4–2.2 cm, 1.5–3 times as long as wide, the margins entire or with a basal lobe, farinose below; *glomerules* 2–8 × 1–1.5 cm; *sepals* 5, 0.8–0.9 × 0.6–0.8 mm, farinose; *stamens* 5; *fruit* with a free or adherent pericarp, smooth; *seeds* 0.9–1.3 mm diam., wrinkled. Found in open places, usually in sandy soil, 6000–10,800 ft. July–Sept. E/W. (Plate 28)

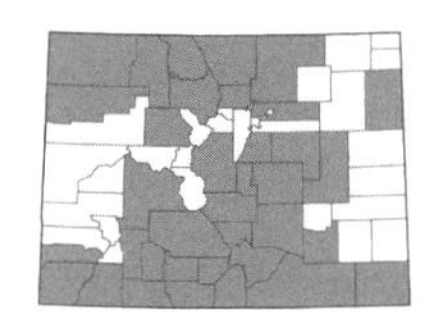

***Chenopodium berlandieri* Moq. var. *zschackii* (J. Murray) J. Murray ex Aschers.**, ZSCHACK'S GOOSEFOOT. Plants 1–10.5 dm; *leaves* deltate or rhombic, 3-lobed, 1.7–4 cm long, the margins toothed with basal lobes, farinose; *glomerules* 4–7 mm diam.; *sepals* 5, 0.7–1.5 × 0.7–1.3 mm, farinose; *stamens* 5; *fruit* with a free or adherent pericarp, alveolate-rugose; *seeds* 1.2–1.5 mm diam., honeycomb-pitted. Common in disturbed places, along creeks, and open places, 4000–10,800 ft. June–Sept. E/W. (Plate 28)

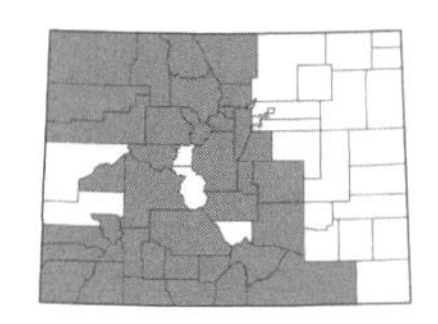

***Chenopodium capitatum* (L.) Aschers.**, BLITE GOOSEFOOT. [*Blitum capitatum* L.]. Plants 1.5–10 dm; *leaves* lanceolate, ovate, triangular, or triangular-hastate, 2.5–10 × 1–9 cm, the margins sharply dentate or entire, glabrous; *glomerules* 3–10 mm diam.; *sepals* 3, 0.6–0.9 × 0.4–0.5 mm, glabrous, red and fleshy; *stamens* 3; *fruit* smooth; *seeds* 0.7–1.2 mm diam., reticulate-punctate. Locally common in disturbed sites in the mountains, 6000–11,000 ft. June–Sept. E/W. Introduced.

***Chenopodium chenopodioides* (L.) Aellen**, LOW GOOSEFOOT. [*Oxybasis chenopodioides* (L.) S. Fuentes, Uotila & Borsch]. Plants 0.1–3.5 dm, prostrate to erect; *leaves* deltate, 0.8–6 × 0.2–3.5 cm, the margins entire or broadly dentate, glabrous; *glomerules* 3–4 mm diam.; *sepals* 3, 0.1–0.5 × 0.3–0.4 mm, glabrous; *stamens* 1; *fruit* with a free pericarp, reticulate-punctate; *seeds* 0.6–0.9 mm diam., smooth. Uncommon in moist or disturbed places such as along the borders of lakes and ponds, 5000–5500 ft. July–Sept. E.

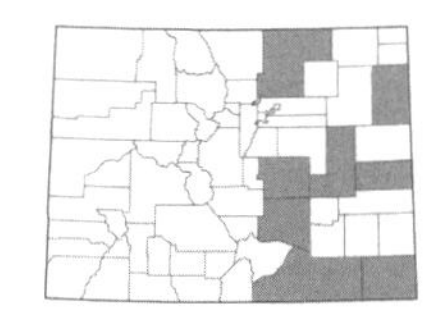

***Chenopodium cycloides* A. Nelson**, SANDHILL GOOSEFOOT. Plants 3–8 dm, erect; *leaves* linear, 1–3 × 0.1–0.2 cm, the margins entire, glabrous above and sparsely farinose below; *glomerules* widely spaced; *sepals* (4) 5, 0.7–0.8 × 0.7–1 mm, farinose; *stamens* 5; *fruit* with an adherent pericarp, minutely tuberculate; *seeds* 1.3–1.5 mm diam., wrinkled. Rare in sandy soil of the eastern plains, 4000–5500 ft. June–Sept. E. (Plate 28)

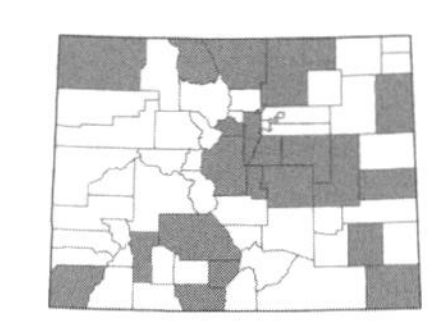

***Chenopodium desiccatum* A. Nelson**, ARIDLAND GOOSEFOOT. Plants 1–1.4 (6) dm; *leaves* linear, narrowly lanceolate, oblong-elliptic, or ovate-lanceolate, 1.5–2.5 × 0.4–0.6 cm, at least 3 times as long as wide, the margins entire and unlobed, sparsely farinose above, densely-white-farinose below; *glomerules* densely packed; *sepals* 5, 0.8–1 mm long, densely farinose; *stamens* 5; *fruit* with a free pericarp, smooth; *seeds* 0.8–1.1 mm diam., warty. Found in dry, open places, along roadsides, sometimes in disturbed places, 4500–8000 ft. June–Sept. E/W.

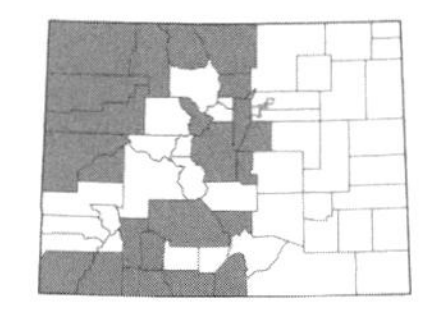

***Chenopodium foliosum* (Moench) Aschers.**, LEAFY GOOSEFOOT. [*Blitum virgatum* L.]. Plants 1.4–6 (8) dm; *leaves* narrowly triangular, oblong-triangular, or deltate, 1.7–7.5 (9) × 0.8–3.5 cm, the margins laciniate- or sinuate-dentate, glabrous; *glomerules* 3–8 mm diam.; *sepals* 3 (4), 0.5–0.7 × 0.3–0.7 mm, glabrous, red and fleshy; *stamens* 1; *fruit* smooth; *seeds* 1–1.2 mm diam., minutely reticulate-punctate. Uncommon in disturbed sites in the mountains, 6700–8500 ft. July–Sept. E/W. Introduced. (Plate 28)

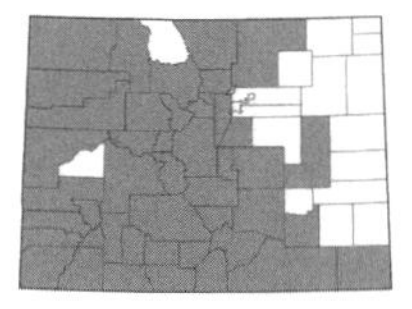

***Chenopodium fremontii* S. Watson**, FREMONT'S GOOSEFOOT. Plants 1–8 dm; *leaves* broadly triangular or occasionally ovate to elliptic, 0.7–6 cm, the margins entire or with a pair of basal lobes or teeth, farinose below; *glomerules* 2–5 mm diam.; *sepals* 5, 0.7–1 × 0.5–0.9 mm, farinose to subglabrous; *stamens* 5; *fruit* with a free pericarp, warty-smooth; *seeds* 1–1.3 mm diam., smooth. Found in shaded areas under pines and other shrubs, sometimes in moist soil near streams, and along roadsides, 5000–10,000 ft. July–Sept. E/W. (Plate 28)

Seldomly, farinose forms of *C. fremontii* approaching *C. incanum* will be found. In these instances, *C. fremontii* can be distinguished from *C. incanum* by the presence of sepals which expose the fruits at maturity and are less strongly keeled. *Chenopodium incanum* has sepals which cover the fruits at maturity and are strongly keeled.

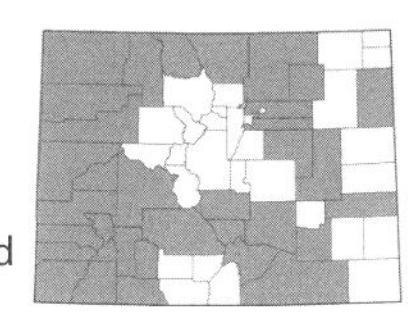

***Chenopodium glaucum* L.**, OAKLEAF GOOSEFOOT. [*Oxybasis glauca* (L.) S. Fuentes, Uotila & Borsch]. Plants 0.5–2.5 (4) dm, often prostrate; *leaves* ovate or lanceolate to oblong, 0.5–4 × 0.3–1.5 cm, the margins undulate-dentate, densely farinose, glaucous below; *glomerules* 1.8–2.5 mm diam.; *sepals* 3–4, 0.5–0.7 × 0.4–0.5 mm; *stamens* 1; *fruit* with a free pericarp, smooth; *seeds* 0.6–1.1 mm diam., rugose-punctate. Common in disturbed, moist places such as along the shores of lakes and ponds, 4000–8000 ft. June–Oct. E/W. Introduced. (Plate 28)

There are two varieties of *C. glaucum* in Colorado:

1a. Bracts absent, at least on the terminal ½ of the inflorescence; leaves usually with obtuse apices...**var. *glaucum***, OAKLEAF GOOSEFOOT. Found in the northeastern counties and in Gunnison Co. on the western slope (specimens from Gunnison Co. are intermediate between var. *glaucum* and var. *salinum*; this distribution only from CS specimens), 4300–8000 ft. E/W.

1b. Leaflike bracts present throughout the inflorescence; leaves usually with acute apices...**var. *salinum* (Standl.) Boivin**, ROCKY MOUNTAIN GOOSEFOOT. [*C. salinum* Standl.]. Mostly found on the western slope, with scattered occurrences on the eastern slope (Larimer and Otero cos.), 4200–7600 ft. E/W.

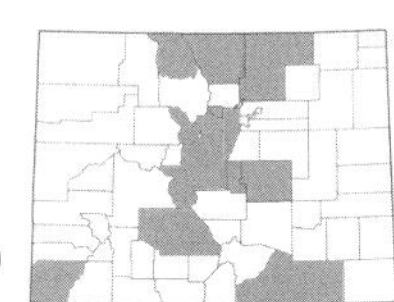

***Chenopodium hians* Standl.**, HIANS GOOSEFOOT. [*C. incognitum* H.A. Wahl (in part)]. Plants 1–4.5 (8) dm; *leaves* elliptic-oblong or narrowly lanceolate, 1–2.5 (3) × 0.3–0.6 (0.8) cm, the margins entire, densely farinose below; *glomerules* densely spaced; *sepals* 5, 0.8–1 × 0.6–0.8 mm, densely farinose; *stamens* 5; *fruit* with an adherent pericarp, minutely tuberculate; *seeds* 1–1.4 mm diam., wrinkled. Found in open meadows, along roadsides, sometimes in disturbed places, 6500–10,800 ft. June–Sept. E/W. (Plate 28)

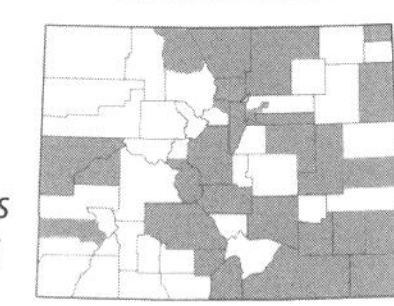

Chenopodium incanum* (S. Watson) Heller var. *incanum, MEALY GOOSEFOOT. Plants 0.6–1.5 (3) dm; *leaves* triangular or broadly to narrowly ovate, 1–1.5 (2.3) × 0.5–1.6 cm, with acute basal teeth or lobes, farinose below; *glomerules* 3.5–12 × 1–7 cm; *sepals* 5, 0.8–1.1 × 0.7–1 mm, farinose; *stamens* 5; *fruit* with a free pericarp, smooth; *seeds* 0.9–1.25 mm diam., wrinkled. Common in sandy soil of open places, sometimes in disturbed areas, 3500–8500 ft. June–Sept. E/W.

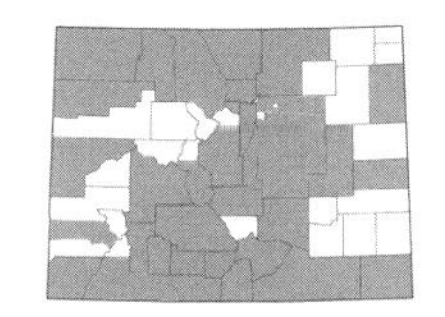

***Chenopodium leptophyllum* (Moq.) Nutt. ex S. Watson**, NARROWLEAF GOOSEFOOT. Plants 1–4 dm; *leaves* linear, 0.7–2.6 (3) × 0.1–0.3 cm, the margins entire, densely farinose below; *glomerules* widely spaced; *sepals* 5, 0.8–1 × 0.5–0.6 mm, densely farinose; *stamens* 5; *fruit* with an adherent pericarp, smooth; *seeds* 0.9–1.1 mm diam., finely wrinkled. Found in open places, often in sandy soil, sometimes in disturbed areas or fields, 5000–9000 ft. July–Sept. E/W.

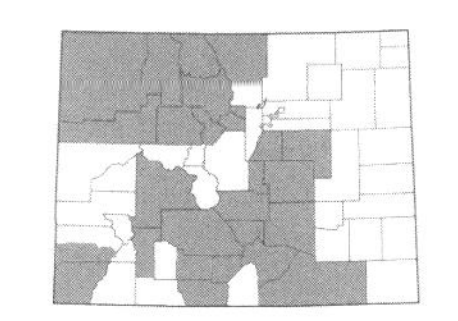

***Chenopodium overi* Aellen**, OVER'S GOOSEFOOT. Plants 1.5–10 dm; *leaves* lanceolate to ovate or triangular, 2.5–10 × 1–9 cm, cuneate to truncate at the base, with entire to toothed margins, glabrous; *glomerules* membranous, sometimes red, 3–5 mm diam.; *sepals* 3, 0.6–1 mm, glabrous; *stamens* 3; *fruit* smooth; *seeds* 0.7–1.2 mm diam. Found on open slopes, along roadsides, and in open meadows, sometimes in disturbed places, 7000–10,500 ft. June–Aug. E/W. (Plate 28)

Resembling *Monolepis nuttalliana*, which has bracts throughout the inflorescence while *C. overi* has bracts only in the lower ½–¾ of the inflorescence.

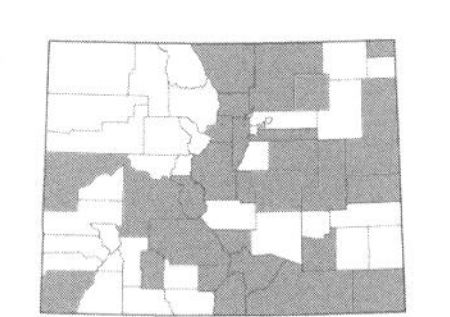

***Chenopodium pratericola* Rydb.**, DESERT GOOSEFOOT. [*C. desiccatum* A. Nelson var. *leptophylloides* (J. Murr) H.A. Wahl]. Plants 2–8 dm; *leaves* linear to narrowly lanceolate or oblong-elliptic, 1.5–4.2 (6) × 0.4–1 (1.4) cm, the margins entire or with a pair of lobes near the base, densely to sparsely farinose below; *sepals* (4) 5, 0.8–1 × 0.5–0.7 mm, densely farinose; *stamens* (4) 5; *fruit* with a free pericarp, smooth; *seeds* 0.9–1.3 mm diam., wrinkled. Found in open places, often in sandy or alkaline soil, 3500–8600 ft. June–Sept. E/W.

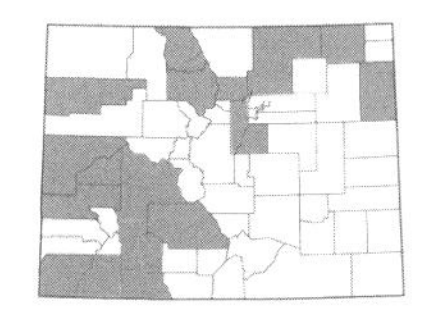

***Chenopodium rubrum* L.**, RED GOOSEFOOT. [*C. humile* Hook.; *Oxybasis rubra* (L.) S. Fuentes, Uotila & Borsch]. Plants 0.1–6 (8) dm; *leaves* triangular to rhombic, 1–9 × 1–6 cm, the margins dentate or entire, glabrous; *glomerules* 2–5 mm diam.; *sepals* 3 or 4, 0.8–1 × 0.4–0.8 mm, glabrous; *stamens* 2–3; *fruit* with a free pericarp, reticulate-punctate; *seeds* 0.6–1 (1.2) mm diam., smooth. Uncommon in moist or disturbed places, usually in alkaline or saline soil, 4900–7500 ft. July–Sept. E/W. (Plate 28)

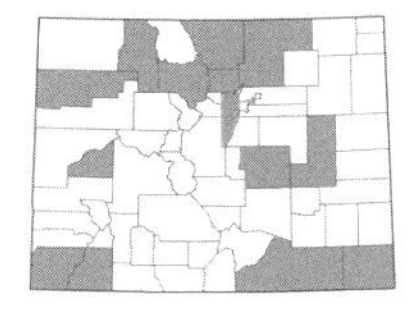

***Chenopodium simplex* (Torr.) Raf.**, MAPLE-LEAF GOOSEFOOT. [*Chenopodiastrum simplex* (Torr.) S. Fuentes, Uotila & Borsch; *C. gigantospermum* Aellen]. Plants 3–15 dm; *leaves* ovate to triangular, 3.5–15 × 2–9 cm, margins sinuate-dentate to entire, glabrous; *glomerules* 0.5–2 mm diam.; *sepals* 5, 0.7–1 × 0.4–0.6 mm, glabrous; *stamens* 5; *fruit* with an adherent pericarp, smooth; *seeds* 1.3–1.9 mm diam., honeycombed to smooth. Common along ditches, pond borders, in shady forests, and in cool ravines, 3600–7000 ft. July–Sept. E/W. (Plate28).

Chenopodium subglabrum **(S. Watson) A. Nelson**, SMOOTH GOOSEFOOT. Plants 1–5.5 dm; *leaves* linear, 1–3 (5) × 0.1–0.2 (0.4) cm, the margins entire, glabrous; *glomerules* widely spaced; *sepals* 5, 1–1.4 mm long, sparsely farinose; *stamens* 5; *fruit* with a free pericarp, smooth; *seeds* 1.2–1.6 mm diam., smooth and shiny. Rare in sandy soil of the eastern plains, 3800–5700 ft. July–Sept. E.

Chenopodium watsonii **A. Nelson**, WATSON'S GOOSEFOOT. Plants 1–4.5 dm, smelling of rotting fish; *leaves* rounded-triangular to rounded-rhombic or ovate, 1–2.6 × 0.5–2.9 cm, the margins entire or with 1–2 pairs of teeth at base, densely farinose above and below; *sepals* 5, 1–1.4 × 0.9–1.2 mm, farinose; *stamens* 5; *fruit* with an adherent pericarp, honeycombed; *seeds* 0.9–1.3 mm diam., coarsely honeycombed. Found in sandy soil of open places, 4500–5500 ft. June–Aug. E/W. (Plate 28)

CORISPERMUM L. – BUGSEED

Annuals, branched herbs, glabrous or pubescent with nearly stellate hairs; *leaves* alternate, sessile, the margins entire; *flowers* perfect, solitary in the axils of leaf-like bracts; *sepals* absent or 1 (3); *fruit* a strongly flattened achene, vertical, sessile, the margins with or without wings, glabrous.

Note: When using the following key, measure the wing from the abaxial side (the convex side) of the fruit. The adaxial side (the flat side) often has a pale margin that could be confused for a true wing.

1a. Fruit (4) 4.5–5.2 mm long, with a narrow, thick, non-translucent wing (or sometimes translucent only at the margin) 0.1–0.2 (0.3) mm wide; inflorescences dense and compact, ovoid to oblong-obovate...***C. navicula***
1b. Fruit 1.8–4.5 mm long (if longer than 2.5 mm, then with a distinct wing), wingless or with a narrow margin about 0.1 mm wide, or with a thin, translucent wing 0.2–0.6 mm wide; inflorescences various...2

2a. Fruit 1.8–3.5 mm long, wingless or nearly so, or with a narrow wing about 0.1 mm wide...3
2b. Fruit 2.5–4.5 mm long, with a distinct, thin, whitish or pale yellowish wing 0.2–0.6 mm wide...4

3a. Fruit usually lacking reddish-brown spots and whitish warts, shiny; inflorescences linear or indistinctly oblong-linear, usually not strongly condensed at the apex, usually not interrupted in the lower half...***C. hyssopifolium***
3b. Fruit usually with reddish-brown spots and occasionally with whitish warts, dull; inflorescences usually widening toward the apex, usually condensed in the upper half and sometimes interrupted in the lower half...***C. villosum***

4a. Inflorescences narrowly linear or linear, usually loose and not dense, interrupted at the base such that the lowermost bracts are little or not overlapping; bracts usually narrower than the fruit...***C. americanum***
4b. Inflorescences ovoid to oblong-obovate, usually compact and dense, rarely loose but then condensed at least at the apex, the lower most bracts overlapping and usually as wide as the fruit...***C. welshii***

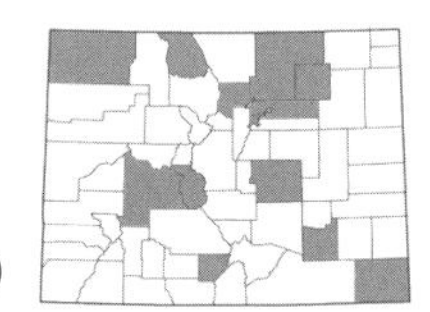

Corispermum americanum **(Nutt.) Nutt.**, AMERICAN BUGSEED. Plants 1–3.5 (5) dm, with sparsely stellate or dendroid hairs, becoming glabrous; *leaves* linear, 1.5–4 × 0.1–0.3 cm; *inflorescences* linear to narrowly linear, interrupted at the base; *bracts* 0.5–2 (3.5) × 0.2–0.7 cm; *sepals* 1; *fruit* obovate, 2.5–4.5 × 2–3.5 mm, wing 0.2–0.4 mm wide. Found on sandy soil of dunes and hills or along sandy roadsides, our most common *Corispermum*, 4900–8600 ft. Aug.–Oct. E/W. (Plate 27)

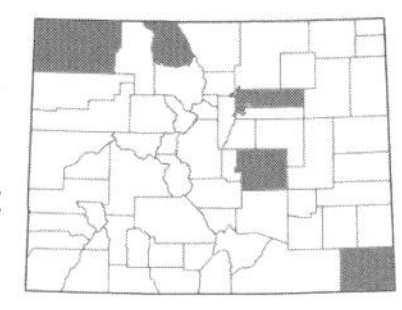

Corispermum hyssopifolium **L.**, HYSSOP-LEAVED BUGSEED. Plants 1–3.5 (5.5) dm, branched from the base, sparsely to densely covered with stellate or dendroid hairs, becoming glabrous; *leaves* linear to narrowly lanceolate, 1.5–4 × 0.2–0.5 cm; *inflorescences* compact, oblong-linear to linear, usually not strongly condensed distally; *bracts* 0.5–2 × 0.3–0.7 cm; *sepals* 1; *fruit* elliptic to obovate-elliptic or orbiculate, 2.2–3.5 × 1.7–2.8 mm, the wing 0.1 mm wide (when present). Uncommon in sandy soil, sandbars, and along sandy roadsides, 4800–8000 ft. Aug.–Oct. E/W. Introduced. (Plate 27)

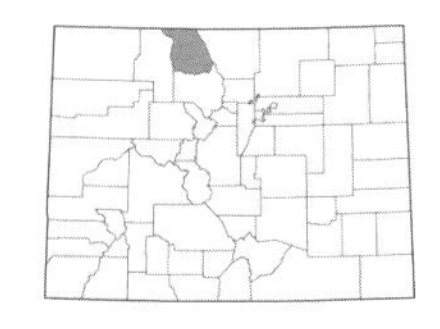

Corispermum navicula **Mosyakin**, BOAT-SHAPED BUGSEED. Plants 0.5–2 dm, sparsely covered with stellate or dendroid hairs or nearly glabrous; *leaves* linear to narrowly lanceolate, 1.5–4 × 0.1–0.5 cm; *inflorescences* compact, ovoid to oblong-obovate; *bracts* 0.5–2 × 0.2–0.6 cm; *sepals* 1; *fruit* (4) 4.5–5.2 × 2.5–3 mm, with a thick, non-transulcent (or sometimes translucent only at the margin) wing 0.1–0.2 (0.3) mm wide. Found on sand hills and dunes, known only from the sand dunes east of Cowdrey, Jackson Co., 8200–8600 ft. Aug.–Oct. E. Endemic. (Plate 27)

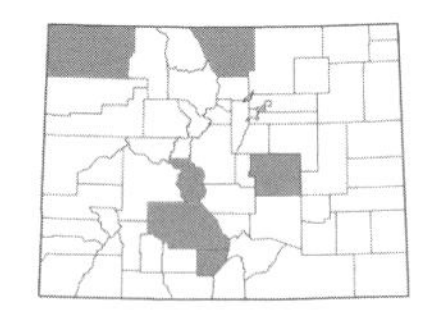

Corispermum villosum **Rydb.**, HAIRY BUGSEED. Plants (0.5) 1–3.5 dm, sparsely to densely covered with stellate or dendroid hairs, becoming glabrous; *leaves* linear to linear-oblanceolate, 1–3.5 × 0.1–0.3 cm; *inflorescences* usually clavate, dense, condensed in the upper half; *bracts* 0.5–1.5 (2.5) × 0.3–1 cm; *sepals* 1; *fruit* elliptic to obovate-elliptic, 1.8–3.2 × 1.5–2 mm, the wing absent or to 0.1 mm wide. Found on sand dunes and hills, 5500–8000 ft. Aug.–Oct. E. (Plate 27)

***Corispermum welshii* Mosyakin**, S.L. WELSH'S BUGSEED. Plants 1–3.5 dm, branched from the base, sparsely to densely covered with stellate or dendroid hairs; *leaves* linear to narrowly lanceolate, 1–6 × 0.2–0.5 cm; *inflorescences* usually dense, ovoid to oblong-obovate, rarely loose; *bracts* narrowly ovate to ovate, 1–3 × 0.3–0.8 cm; *sepals* 1; *fruit* obovate, 3.5–4.6 × 3–3.6 mm, with a thin wing 0.3–0.6 mm wide. Probably occurs in Colorado on sand dunes or hills, but yet not reported for the state.

CYCLOLOMA Moq. – CYCLOLOMA

Annual herbs, not fleshy, with much-branched, striate stems; *leaves* alternate, sessile or petiolate; *flowers* perfect or a few pistillate, in interrupted, linear spikes that appear paniculate; *calyx* 5-lobed, connate to above the middle, winged; *fruit* enclosed in the persistent perianth, depressed-globose.

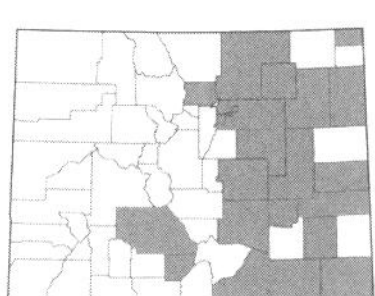

***Cycloloma atriplicifolium* (Spreng.) J.M. Coult.**, WINGED PIGWEED. Plants 0.5–8 dm; *leaves* oblong to oblong-ovate or lanceolate, 2–8 × 0.5–2 cm, the margins sinuately dentate, cuneate; *sepals* 2–4.5 mm diam., tomentose or becoming glabrous and reddish; *fruit* ca. 2 mm diam.; *seeds* 1.5–2 mm diam., tomentose, black. Found in sandy soil and along roadsides on the eastern plains, 4000–5500 ft. July–Sept. E. (Plate 27)

DYSPHANIA R. Br. – DYSPHANIA

Annual herbs or short-lived perennials, usually glandular-hairy, aromatic; *leaves* alternate, often pinnately lobed or lyrate-sinuate; *flowers* perfect or rarely imperfect, arranged in glomerules; *calyx* 1-5-lobed, connate at the base or distinct, or fused to form a sac surrounding the fruit; *fruit* an achene usually enclosed in a persistent perianth.

1a. Sepals glabrous or sparsely glandular; stigmas 3; stamens 4–5; flowers sessile in small glomerules, these arranged in spikes; leaves mostly over 4 cm long...***D. ambrosioides***
1b. Sepals densely glandular-hairy or with golden sessile glands; stigmas 2; stamens 1–3 (5); flowers on short pedicels, solitary or in few-flowered glomerules, arranged in terminal or axillary cymes or thyrses; leaves to 4.5 cm long...2

2a. Sepals with golden, sessile glands and with a single horn-like appendage on the back; stems sparsely pubescent with short hairs, these occasionally with glandular heads...***D. graveolens***
2b. Sepals with stalked glandular hairs, lacking a horn-like appendage on the back; stems densely glandular-hairy...***D. botrys***

***Dysphania ambrosioides* (L.) Mosyakin & Clemants**, MEXICAN TEA. [*Chenopodium ambrosioides* L.; *Teloxys ambrosioides* (L.) W.A. Weber]. Plants 3–15 dm, glandular; *leaves* lanceolate to ovate, (2) 4–8 (12) × 0.5–5 cm, the margins entire to toothed or laciniate, gland-dotted; *inflorescence* of lateral spikes, glomerules 1.5–2.3 mm diam.; *bracts* leaf-like, 0.3–2.5 cm long; *sepals* 4–5, connate to the middle, glabrous or glandular; *stigmas* 3; *seeds* 0.6–1 × 0.4–0.5 mm. Not yet reported for Colorado, but found as a weed in the surrounding states. Introduced.

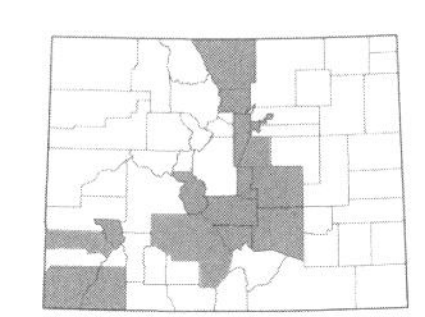

***Dysphania botrys* (L.) Mosyakin & Clemants**, JERUSALEM OAK. [*Chenopodium botrys* L.; *Teloxys botrys* (L.) W.A. Weber]. Plants 1–6 (10) dm, glandular; *leaves* pinnatifid, lyrate-sinuate, or rarely entire, 1.3–4 × 0.6–2.7 cm, glandular-hairy; *inflorescence* of axillary cymes usually arranged in terminal thyrses, 12–24 cm long; *bracts* absent; *sepals* 5, distinct to nearly the base, with stalked glandular hairs; *stigmas* 2; *seeds* 0.5–0.8 × 0.5–0.7 mm. Found in disturbed places, on dry slopes, and mud flats, 4500–7500 ft. July–Sept. E/W. Introduced. (Plate 28)

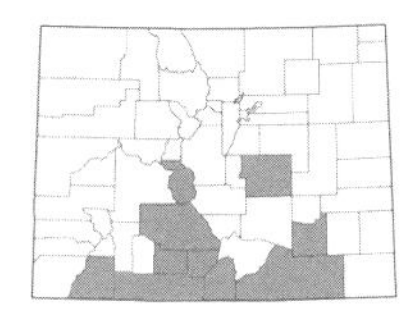

***Dysphania graveolens* (Willd.) Mosyakin & Clemants**, FETID GOOSEFOOT. [*Chenopodium graveolens* Willd.; *Teloxys graveolens* (Willd.) W.A. Weber]. Plants 2.3–5.2 dm, sparsely hairy, rarely glandular; *leaves* pinnatifid or the distal entire, 1.7–4.5 × 0.7–2.6 cm, with sessile glands above; *inflorescence* of terminal compound cymes, 8.5–22 cm long; *bracts* leaf-like, 2–10 × 0.1–0.6 mm, usually absent in fruit; *sepals* 5, distinct to nearly the base, with a single horn-like appendage on the back, with sessile golden glands; *stigmas* 2; *seeds* 0.6–1 × 0.5–0.7 mm. Found in the shade of pines, junipers, and cottonwoods, 5000–9200 ft. July–Sept. E/W. Introduced.

GRAYIA Hook. & Arn. – HOPSAGE

Shrubs, with thorns; *leaves* alternate, entire; *flowers* imperfect; *staminate flowers* with a 4–5-parted perianth and 4–5 stamens, in 2–5-flowered axillary clusters; *pistillate flowers* with 2 stigmas, enclosed by 2 fused bracteoles, in 1–several-flowered clusters per bract; *fruit* enclosed by the 2 fused bracts, becoming flattened and samaralike.

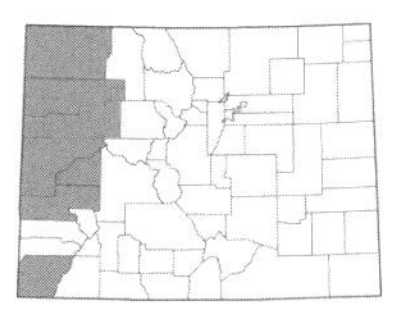

***Grayia spinosa* (Hook.) Moq.**, SPINY HOPSAGE. [*Atriplex grayi* Collotzi ex W.A. Weber]. Plants 3–10 (15) dm, scurfy when young, becoming glabrate; *leaves* oblanceolate, 0.5–2.5 (4) cm long; *staminate flowers* with perianth segments 1.5–2 mm long; *fruiting bracts* connate, 7.5–14 × 6–12 mm, winged, margins entire. Found on dry slopes in pinyon-juniper and sagebrush, 4000–7000 ft. May–June. W. (Plate 27)

HALOGETON C.A. Mey. – SALTLOVER

Annual herbs, glabrous except for tufts of white hairs in the axils; *leaves* alternate, fleshy, terete, with a slender bristle tip; *flowers* perfect and imperfect, in axillary glomerules; *sepals* 5, distinct nearly to the base, the segments winged in fruit; *fruit* an utricle enclosed in a persistent perianth.

***Halogeton glomeratus* (M. Bieb.) C.A. Mey.**, SALTLOVER. Plants 1–4 dm, spreading to decumbent; *leaves* linear, 4–17 mm long; *sepals* with a 2–3 mm long claw and 2–4 mm wide blade; *utricles* vertical, dimorphic, the central one 1–2 mm long and brown, lateral ones 0.5–1 mm long and black. Found on drying soil, barren buttes, alkaline flats, and in disturbed places, 4500–7000 ft. June–Aug. E/W. Introduced. (Plate 27)

Halogeton glomeratus is highly toxic to livestock and is able to withstand very high levels of salinity.

KOCHIA Roth – KOCHIA

Annual or perennial herbs and subshrubs, glabrous to densely tomentose; *leaves* alternate, with entire margins; *flowers* perfect or pistillate; *sepals* 5, with an adaxial, horizontal, membranous wing or the wing reduced to minutely winged tubercles; *fruit* an utricle enclosed in the persistent perianth.

1a. Annual herb from a taproot; leaves usually lanceolate to narrowly obovate or sometimes linear, flat, usually not tightly overlapping on the stem, usually 3–5-veined; stems branched throughout; perianth glabrous or nearly so...***K. scoparia***

1b. Perennial subshrub with a woody base; leaves linear, usually terete, tightly overlapping on the stem, with a single vein or the vein obscure; stems simple or only branched at the base; perianth often tomentose or sometimes nearly glabrous...***K. americana***

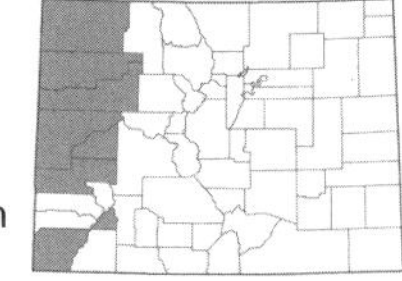

***Kochia americana* S. Watson**, GREEN MOLLY. [*Bassia americana* (S. Watson) Scott; *Neokochia americana* (S. Watson) G.L. Chu & S.C. Sand.]. Subshrubs, branched near the base, 0.5–3.5 (5) dm; *leaves* linear, usually terete, 4–25 × 0.5–2 mm, sericeous to glabrous; *bracts* 4–15 × 0.5–1 mm; *sepals* white-tomentose to nearly glabrous, with fan-shaped wings; *fruit* ca. 4 mm diam. Found on dry hills and knolls, sometimes on alkaline soil, 4500–6500 ft. May–Aug. W. (Plate 29)

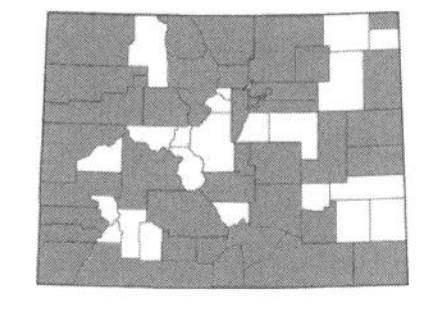

***Kochia scoparia* (L.) Schrad.**, BURNING BUSH; MEXICAN FIREWEED. [*Bassia sieversiana* (Pall.) W.A. Weber]. Annuals, simple to branched throughout, 2–12 dm; *leaves* 8–50 × 1–6 mm, flat, appressed-hairy; *sepals* glabrous or sparsely hairy, perfect flowers with wings or tubercles. Common in disturbed places and along roadsides, often on alkaline soil, 3400–9700 ft. June–Oct. E/W. Introduced.

KRASCHENINNIKOVIA Gueldenst. – WINTERFAT

Subshrubs, monoecious or dioecious, densely tomentose with stellate hairs; *leaves* alternate, entire, not fleshy; *flowers* imperfect; *staminate flowers* with a 4 (5) parted perianth and 4–5 stamens; *pistillate flowers* lacking a perianth and with 2 styles, enclosed in 2 tomentose, proximally connate bracts; *fruit* a flat utricle.

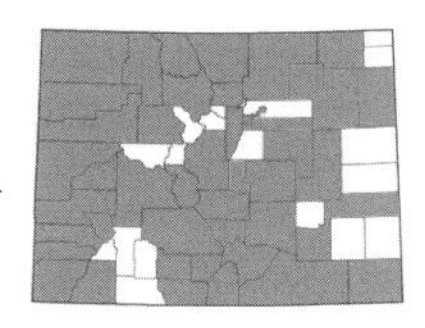

***Krascheninnikovia lanata* (Pursh) A. Meeuse & A. Smit**, WINTERFAT. [*Ceratoides lanata* (Pursh) J.T. Howell; *Eurotia lanata* (Pursh) Moq.]. Plants 1.5–5 dm; *leaves* 1–4 × 1.5–3.5 (5) cm, the margins usually strongly revolute; *fruiting bracts* 4–7.5 mm long; *utricles* 2.5–3.5 mm long, densely hairy. Common on the plains and foothills, on open slopes, and in pinyon-juniper and sagebrush, 4000–9500 ft. May–July. E/W. (Plate 29)

MICROMONOLEPIS Ulbr. – SMALL POVERTYWEED

Annuals, monoecious herbs; *leaves* alternate, succulent; *flowers* imperfect, usually solitary in leaf axils; *staminate flowers* with (1) 2–3 sepals and 1–2 stamens; *pistillate flowers* with (1) 2–3 sepals and 2 stigmas; *fruit* an utricle.

***Micromonolepis pusilla* (Torr. ex S. Watson) Ulbr.**, SMALL POVERTYWEED. [*Monolepis pusilla* Torr. ex S. Watson]. Plants 0.4–1.5 (2) dm, the stems dichotomously branched; *leaves* oblanceolate, 2.5–8.5 × 0.2–2.5 mm; *utricles* 0.6–0.8 mm long. Uncommon on alkaline flats, known from an old collection near Grand Junction, 4500–5000 ft. May–July. W. (Plate 29)

MONOLEPIS Schrad. – POVERTYWEED

Annual herbs, sparsely farinose to glabrate; *leaves* alternate, with a pair of spreading lobes below or unlobed; *flowers* perfect or pistillate, clustered in dense, sessile, axillary glomerules; *sepal* 1, persistent, bractlike and green; *stamens* 1 or absent in pistillate flowers; *fruit* a more or less flattened utricle.

***Monolepis nuttalliana* (Schult.) Greene**, NUTTALL'S POVERTYWEED. Plants 0.5–2 (5) dm, prostrate to ascending; *leaves* linear to narrowly elliptic or narrowly triangular, 1–4 cm × 2–20 (25) mm; *utricles* 1–1.5 mm; *seeds* dark brown to black. Common on open slopes and hills, plains, sometimes with sagebrush or pinyon-juniper, often in disturbed places, 3800–11,000 ft. May–July. E/W. (Plate 29)

SALICORNIA L. – PICKLEWEED

Annuals, glabrous; *leaves* simple, reduced and scale-like, sessile, opposite; *flowers* perfect, in terminal spikes, sessile in 3-flowered, opposite cymes; *sepals* usually 3, fused except for the very tips; *fruit* enclosed in the perianth.

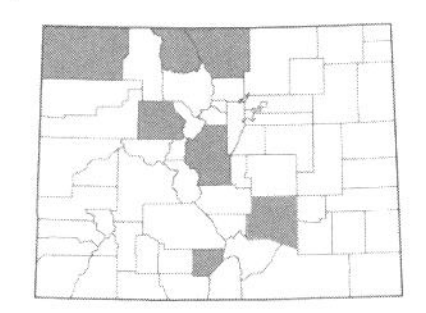

***Salicornia rubra* A. Nelson**, RED SWAMPFIRE. Plants 0.5–2.5 dm, the stems often turning completely red at maturity, erect; *inflorescence* 5–35 (50) mm long, 1.8–3.5 mm wide, with 4–10 (19) fertile nodes; *seeds* 1–2 mm long, with hooked, curved hairs. Found on the margins of drying alkaline ponds and in alkaline soil of wet meadows, 5500–8700 ft. July–Sept. E/W. (Plate 29)

SALSOLA L. – RUSSIAN THISTLE

Annual herbs; *leaves* alternate, the apex often spinose; *flowers* perfect, solitary in axils of spinescent bracts; *sepals* 5; *fruit* an utricle enclosed in the persistent perianth, each segment with a transverse, dorsal wing or wingless.

1a. Inflorescence dense, not interrupted, the bracts strongly imbricate and ascending or just the tips spreading at maturity (but not strongly reflexed); fruit lacking wings, hard and indurate, or with a narrow, erose wing at maturity...***S. collina***

1b. Inflorescence interrupted at least towards the base, the bracts not imbricate and strongly reflexed at maturity; fruit prominently thin-winged...2

2a. Tips of the perianth segments long-acuminate and spine-tipped, forming a narrow columnar beak at maturity; fruiting perianth usually 7–12 mm in diam.; leaves sometimes fleshy...***S. paulsenii***

2b. Tips of the perianth segments rounded or weakly acuminate, not forming a narrow columnar beak at maturity; fruiting perianth usually 4–10 mm in diam.; leaves not fleshy...***S. tragus***

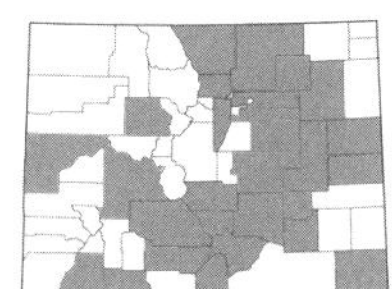

***Salsola collina* Pall.**, PALL.' TUMBLEWEED. [*Kali collina* (Pall.) Akhani & E.H. Roalson]. Plants 1–10 dm, branched above the base; *leaves* filiform or narrowly linear, ca. 1 mm wide; *inflorescence* dense, not interrupted, the bracts strongly imbricate; *perianth segments* wingless or with a narrow, erose wing; *fruiting perianth* ca. 2–5 mm diam. Found in disturbed places, along roadsides, in fields, 3500–8500 ft. July–Sept. E/W. Introduced.

***Salsola paulsenii* Litv.**, BARBWIRE RUSSIAN THISTLE. [*Kali paulsenii* (Litv.) Akhani & E.H. Roalson]. Plants 1–8 (10) dm, profusely branched from the base; *leaves* filiform to narrowly linear, ca. 1 mm wide; *inflorescence* interrupted, the bracts not imbricate, strongly reflexed at maturity; *perianth segments* prominently winged; *fruiting perianth* 7–12 mm diam. Uncommon in disturbed areas and along roadsides, 4500–6000 ft. July–Sept. W. Introduced.

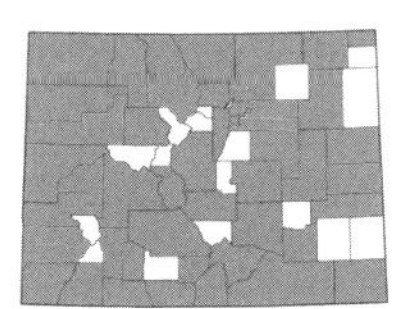

***Salsola tragus* L.**, RUSSIAN THISTLE; TUMBLEWEED. [*Kali tragus* (L.) Scop.; *S. australis* R. Br.; *S. iberica* (Senn. & Pau) Botschantzev ex Czerepanov]. Plants (0.5) 1–10 dm, profusely branched from the base; *leaves* filiform to narrowly linear, ca. to 1 mm wide; *inflorescence* interrupted, the bracts not imbricate, strongly reflexed; *perianth segments* with prominent, membranous wings; *fruiting perianth* ca. 4–10 mm diam. Common in disturbed areas, along roadsides, in fields, forming huge masses along fencerows, 3500–9000 ft. July–Sept. E/W. Introduced. (Plate 29)

SARCOBATUS Nees – GREASEWOOD

Monoecious shrubs with axillary thorns; *leaves* usually alternate, linear, fleshy; *flowers* imperfect; *staminate flowers* lacking sepals, with (1) 2–4 (5) stamens each arising from near a stalked, peltate, scarious scale, in a terminal spike; *pistillate flowers* with 2 stigmas and a cuplike, shallowly lobed or entire perianth that develops into membranous wings in fruit, solitary or paired in the axils of leaves; *fruit* an achene, winged above the middle.

***Sarcobatus vermiculatus* (Hook.) Torr.**, GREASEWOOD. Plants 1–2 (5) m; *leaves* (0.3) 1.5–5 cm long, usually glabrous; *fruit* 3–5 mm long with a wing 0.7–1.4 cm diam.; *seeds* 1.8–2 mm diam. Common in dry places such as on alkaline flats, open slopes, and along roadsides, often with pinyon-juniper or sagebrush, 4500–8500 ft. May–Aug. E/W. (Plate 29)

SUAEDA Forssk. ex J.F. Gmel. – SEEPWEED

Annual herbs, subshrubs, or shrubs, usually fleshy; *leaves* alternate or opposite; *flowers* perfect or imperfect, sessile; *sepals* 5, distinct to nearly completely connate, sometimes irregular with 1 or 3 segments larger; *fruit* a compressed utricle, enclosed by the perianth.

1a. Perennials from a woody caudex or subshrubs with a woody base, occasionally annuals; perianth actinomorphic, all segments equal, not keeled on the back and lacking wings at the base, merely rounded at the tips; bracts usually narrowed at the base...***S. nigra***

1b. Annuals; perianth irregular with 1–3 segments larger, the segments keeled on the back, thin-winged at the base, and horned or hooded at the tips; bracts usually widened at the base...***S. calceoliformis***

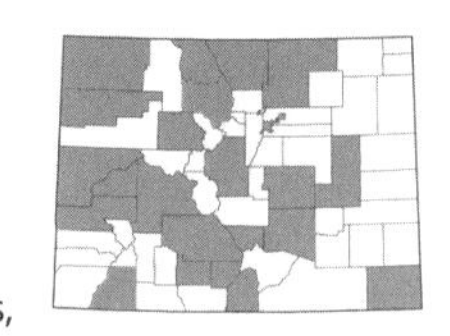

***Suaeda calceoliformis* (Hook.) Moq.**, PURSH'S SEEPWEED. [*S. depressa* (Pursh) S. Watson; *S. occidentalis* (S. Watson) S. Watson]. Annuals, 0.5–8 (10) dm, prostrate to erect, glaucous; *leaves* linear-lanceolate, (5) 10–40 × 0.2–15 mm; *flowers* perfect, the perianth irregular or zygomorphic with 1–3 segments larger, 1–4 mm diam., each segment transversely winged at the base and horned or hooded above; *seeds* dimorphic, 0.8–1.7 mm diam., black and shiny or 1–1.5 mm diam. and brown. Found on alkaline or saline flats or along the margins of lakes or drying ponds, 4500–9000 ft. June–Sept. E/W. (Plate 29)

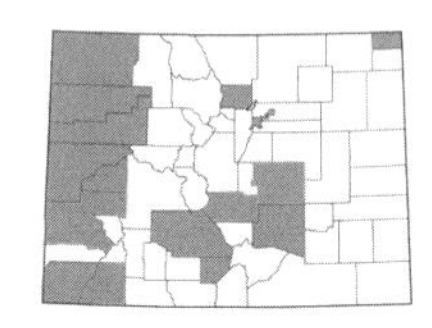

***Suaeda nigra* J.F. Macbr.**, BUSH SEEPWEED. [*S. moquinii* (Torr.) Greene; *S. torreyana* S. Watson]. Perennials or subshrubs, occasionally annuals, 2–15 dm, erect, glabrous or loosely hairy; *leaves* linear, (5) 10–30 × 1–2 mm; *flowers* mostly perfect, the perianth actinomorphic, 0.7–2 mm diam., rounded at the tips; *seeds* 0.5–2 mm diam., brown or black. Found on alkaline or saline flats and dry hillsides, 4000–6000 ft. June–Sept. E/W. (Plate 29)

SUCKLEYA A. Gray – SUCKLEYA

Annuals, monoecious herbs; *leaves* alternate, petiolate; *flowers* imperfect, in axillary glomerules; *staminate flowers* usually with 4 perianth lobes, with 2 segments longer than the others, and 4 stamens; *pistillate flowers* with 4 perianth lobes becoming marginally fused, and 2 stigmas; *fruit* an utricle, enclosed in an enlarged, compressed perianth.

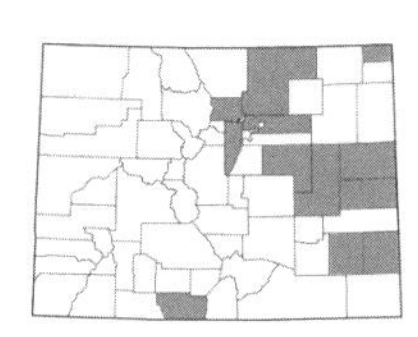

***Suckleya suckleyana* Rydb.**, POISON SUCKLEYA. Plants 0.5–3 dm, prostrate or ascending, stems succulent; *leaves* rhombic-ovate to suborbicular, 1–3 × 0.5–2 cm, toothed, glabrous or sparsely scurfy; *fruiting bracts* winged, 5–6 mm long, with toothed margins. Found along the margins of lakes and ponds, in dried lake bottoms and dry beds of seasonal pools, and in fields, 3500–8500 ft. June–Sept. E. (Plate 29)

ZUCKIA Standl. – ZUCKIA

Shrubs or subshrubs, dioecious or monecious, more or less scurfy, thorns absent; *leaves* alternate; *flowers* imperfect; *staminate flowers* with a 4–5-lobed perianth and 4–5 stamens, 2–5 in axillary clusters, with a linear bract subtending each spike but not individually bracteate; *pistillate flowers* with the perianth absent and 2 stigmas, each flower enclosed by 2 bracteoles, these laterally flattened, sometimes unequally 6-ridged, with 2 slightly enlarged wings; *fruit* vertical or horizontal.

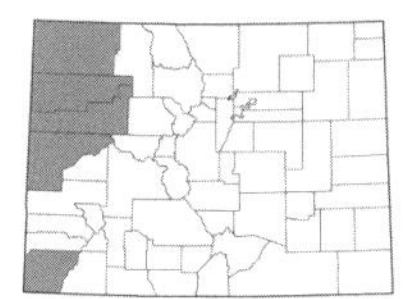

***Zuckia brandegeei* (A. Gray) S.L. Welsh & Stutz**, SILTBUSH. Shrubs, 0.5–5 dm; *leaves* linear to elliptic, ovate, or obovate,1.3–8 × 1.5–4 cm, the margins usually entire; *fruiting bracts* 3.4–9 mm diam., 2-winged; *achenes* 1–2.2 mm long. Found in dry places, canyons, and on open slopes, 4800–7000 ft. May–June. W. (Plate 29)

CISTACEAE Juss. – ROCKROSE FAMILY

Shrubs and herbs; *leaves* simple, entire; *flowers* perfect, actinomorphic; *sepals* 3 or 5, distinct; *petals* 3 or 5 or absent in cleistogamous flowers, distinct, contorted, white, yellow, or red; *stamens* 10–many, with distinct filaments; *pistil* 1; *ovary* superior, unilocular, 3–10-carpellate; *fruit* a loculicidal or valvate capsule.

HELIANTHEMUM Mill. – FROSTWEED

Herbs or shrubs; *leaves* alternate (ours), exstipulate (ours); *sepals* 5, the outer 2 smaller than the inner 3; *petals* 5, yellow; *stamens* 10–many (3–8 in cleistogamous flowers); *ovary* 3-carpellate.

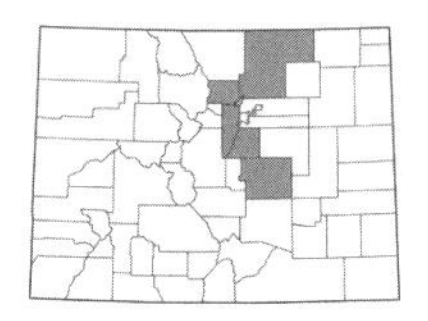

***Helianthemum bicknellii* Fernald**, HOARY FROSTWEED. [*Crocanthemum bicknellii* (Fernald) Barnhart]. Plants mostly 2–6 dm; *leaves* linear-oblong to oblanceolate, mostly 2–3 cm long; *flowers* of two kinds, the first ones large, yellow, with numerous stamens, 1.5–2.5 cm wide, the later ones cleistogamous, smaller with 3–10 stamens; *capsules* of chasmogamous flowers 4–5 mm diam. Infrequent in dry pine forests and open meadows, 5200–7500 ft. June–Aug. E.

CLUSIACEAE Lindl. – ST. JOHN'S-WORT FAMILY

Trees, shrubs, herbs, and lianas; *leaves* simple, entire, opposite or whorled, often gland-dotted, exstipulate; *flowers* perfect or imperfect, actinomorphic; *sepals* 2–10, distinct; *petals* 4–12, distinct, yellow, pink, or white; *stamens* 4–many, the filaments distinct or connate into 3–5 groups; *pistil* 1; *ovary* superior, 3–5-carpellate; *fruit* a septicidal capsule.

HYPERICUM L. – ST. JOHN'S-WORT

Perennial herbs; *leaves* opposite, sessile, gland-dotted; *flowers* perfect; *sepals* 5; *petals* 5, yellow; *stamens* 4–many.

1a. Petals 4–7 mm long, about the same length as the sepals, not black gland-dotted along the margins; stamens 15–35 with distinct filaments; cymes not leafy-bracteate...***H. majus***
1b. Petals 6–15 mm long, about twice as long as the sepals, black gland-dotted along the margins; stamens 75–100, connate at the base into 3–5 groups; cymes leafy-bracteate...2

2a. Leaves oval or elliptic; plants with simple stems or occasionally branched above; cymes few-flowered...***H. scouleri* ssp. *nortoniae***
2b. Leaves linear to lanceolate or narrowly oblanceolate; plants branched above and often at the base as well; cymes many-flowered...***H. perforatum***

***Hypericum majus* (A. Gray) Britton**, GREATER ST. JOHN'S-WORT. Plants to 5 (7) dm; *leaves* linear to lanceolate, to 4 cm long; *cymes* not leafy-bracteate; *petals* 4–7 mm long, not black gland-dotted along the margins; *stamens* 15–35 with distinct filaments; *capsules* 3–7 mm long, purplish, ovoid. Uncommon along the margins of ponds or on floodplains, 5000–7600 ft. July–Sept. E.

***Hypericum perforatum* L.**, COMMON ST. JOHN'S-WORT. Plants to 7 (15) dm; *leaves* linear to lanceolate or narrowly oblanceolate, to 1.5 cm long; *cymes* leafy-bracteate; *petals* 8–12 mm long, with black dots along the margins; *stamens* 75–100, connate at the base into 3 fascicles. Found in dry, rocky meadows, and along roadsides and streams, 5200–9800 ft. June–Aug. E/W. Introduced. (Plate 30)

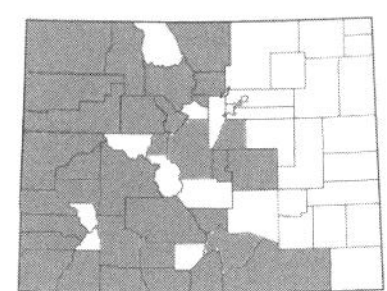

***Hypericum scouleri* Hook. ssp. *nortoniae* (M.E. Jones) J. Gillett**, NORTON'S ST. JOHN'S-WORT. [*H. formosum* Kunth]. Plants 2–7 dm; *leaves* elliptic to oval, 1–3.5 cm long; *cymes* leafy-bracteate; *petals* 6–15 mm long, with black dots along the margins; *stamens* 75–100, connate at the base into 3 fascicles. Found in moist meadows, ditches, and along pond margins and streams, 5500–10,500 ft. July–Sept. E/W.

COMMELINACEAE Mirb. – SPIDERWORT FAMILY

Annual or perennial herbs; *leaves* alternate, basal or cauline, with entire margins; *flowers* perfect, perfect and staminate, or rarely perfect and pistillate; *sepals* 3, occasionally petaloid but usually green, distinct or basally connate; *petals* 3, distinct; *stamens* 6, sometimes staminodes present, rarely absent; *pistil* 1; *ovary* superior, 2–3-loculed; *fruit* a loculicidal capsule. (Brashier, 1966)

1a. Flowers blue; inflorescence subtended by a conspicuous spathe enclosing the flower buds; fertile stamens 3...***Commelina***
1b. Flowers purple to purplish-blue; inflorescence subtended by 2 leaf-like bracts; fertile stamens 6...***Tradescantia***

COMMELINA L. – DAYFLOWER

Perennial or annual herbs; *leaves* 2-ranked or spirally arranged; *flowers* perfect and staminate, zygomorphic, the inflorescences enclosed in spathes when in bud; *sepals* unequal; *petals* blue or occasionally lavender, yellow, or white, the lowermost petal often a different color than the upper 2, unequal with the lowermost petal smaller; *stamens* 6, with 3 fertile and 3 staminodial and infertile; *fruit* a 2–3-valved or 2–3-locular capsule, with 1–2 seeds per locule or valve.

1a. Spathe with a long, narrow tip (1.5 times the length of the body of the spathe), the margins distinct to the base; petals dark blue; capsules 5–6 mm long...***C. dianthifolia***
1b. Spathe without a long, narrow tip, the margins connate basally; 2 distal petals blue and the smaller, proximal petal white; capsules 3.5–4.5 mm long...***C. erecta***

***Commelina dianthifolia* Redouté**, BIRDBILL DAYFLOWER. Perennials to 10 dm; *leaves* linear to narrowly lanceolate, 4–15 × 0.4–1 cm; *spathe* 2.5–8 cm long, with an acuminate tip, the margins distinct; *petals* dark blue; *capsules* 5–6 mm long. Uncommon on rocky ledges and canyon bottoms, 6000–8500 ft. June–Sept. E/W.

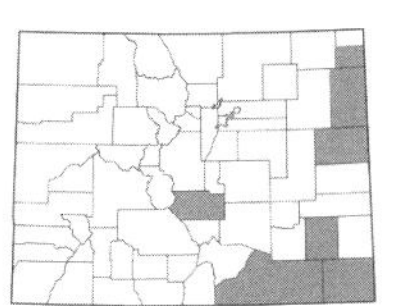

***Commelina erecta* L.**, WHITEMOUTH DAYFLOWER. Perennials to 10 dm; *leaves* linear to lanceolate, 5–15 × 0.3–4 cm; *spathe* 1–2.5 (4) cm long, acute or sometimes acuminate at the tip, the margins connate basally; *petals* bicolored, the distal ones blue and the proximal petal white; *capsules* 3.5–4.5 mm long. Found along roadsides and on open prairie, 3500–6500 ft. June–Aug. E.

TRADESCANTIA L. – SPIDERWORT; WANDERING-JEW

Perennial herbs; *leaves* cauline, sessile, 2-ranked or spirally arranged; *flowers* perfect, actinomorphic, the inflorescences subtended by a spathe-like bract; *sepals* usually distinct, subequal; *petals* white, pink, blue, or violet, equal, rarely clawed; *stamens* 6, all fertile, equal; *ovary* 3-loculed; *fruit* a 3-valved capsule, with 2 seeds per valve.

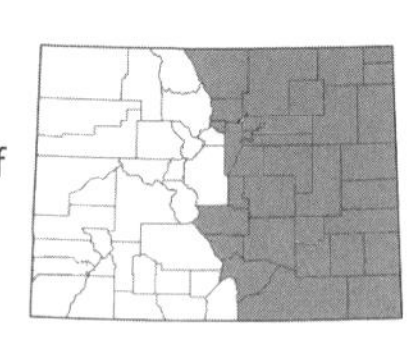

Tradescantia occidentalis **(Britton) Smyth**, PRAIRIE SPIDERWORT. Plants to 5 dm; *leaves* linear-lanceolate, 5–50 × 0.2–3 cm; *bracts* foliaceous, 6–20 cm long; *sepals* 8–13 mm long, glandular; *petals* purple to blue or magenta, 6–16 mm long; *capsules* 4–7 mm long. Common in sandy soil of the eastern plains and lower foothills, 3500–7800 ft. May–Aug. E. (Plate 30)

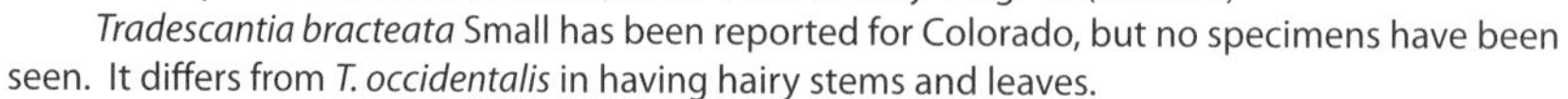

Tradescantia bracteata Small has been reported for Colorado, but no specimens have been seen. It differs from *T. occidentalis* in having hairy stems and leaves.

CONVOLVULACEAE Juss. – MORNING-GLORY FAMILY

Herbs, vines, shrubs, or rarely trees; *leaves* alternate, exstipulate; *flowers* perfect, actinomorphic; *sepals* 5, distinct or united at the base; *corolla* 5-lobed, plicate; *stamens* 5, distinct, adnate to the base of the corolla; *pistil* 1; *stigmas* 2; *ovary* superior, 2–3-carpellate; *fruit* a capsule.

1a. Plants parasitic, twining on the aerial stems of various herbs, yellowish and lacking chlorophyll, the stems thin and thread-like; leaves minute and reduced to scales...***Cuscuta***
1b. Plants green, not parasitic, otherwise unlike the above in all respects...2

2a. Flowers purple or blue-purple, rotate to broadly campanulate; erect herbs to 25 cm tall, with several stems arising from a woody base...***Evolvulus***
2b. Flowers white, pink, purplish-pink, or red, funnelform, salverform, or campanulate; trailing or climbing vines or prostrate plants, or somewhat shrubby and taller than 25 cm...3

3a. Flowers red, purplish-pink, or pink, the throat darker than the limb; leaves linear to linear-lanceolate or 3-lobed with a cordate base...***Ipomoea***
3b. Flowers white or pink with white stripes; leaves hastate, sagittate, or sometimes merely with a cordate base but not 3-lobed...4

4a. Flowers larger, 4–6 cm long; calyx enclosed by 2 foliaceous bracts...***Calystegia***
4b. Flowers smaller, 1.2–3 cm long; calyx not enclosed by bracts...***Convolvulus***

CALYSTEGIA R. Br. – FALSE BINDWEED

Herbs, vines; *leaves* simple, ovate, sagittate, hastate, or reniform; *flowers* solitary in axils, subtended by 2 leaf-like bracts covering most of the calyx; *corolla* campanulate to funnelform, white or pink with white stripes; *fruit* a capsule, surrounded partly by the enlarged sepals and bracts.

1a. Plants glabrous or with a few hairs on the petioles; leaves 2–3-dentate at the base (angled); peduncles 3–13 cm long...***C. sepium***
1b. Plants finely hairy; leaves rounded at the base or rarely angled; peduncles 3–5 cm long...***C. macounii***

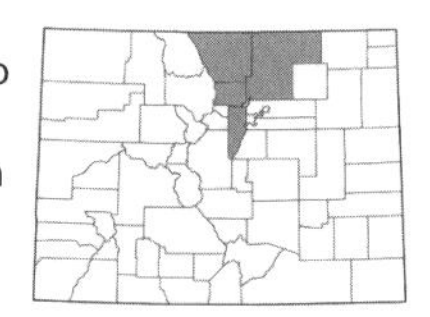

Calystegia macounii **(Greene) Brummitt**, MACOUN'S FALSE BINDWEED. Leaves 2–6 × 1–5 cm, cordate to subsagittate, the basal lobes usually rounded; *flowers* solitary, on peduncles 3–5 cm long; *bracts* 2–2.5 × 1–1.5 cm, mostly obtuse; *calyx* 1–1.2 cm long; *corollas* funnelform, 4–5 cm long. Found in moist places, along ditches, and in disturbed places, 5000–6000 ft. May–July. E.

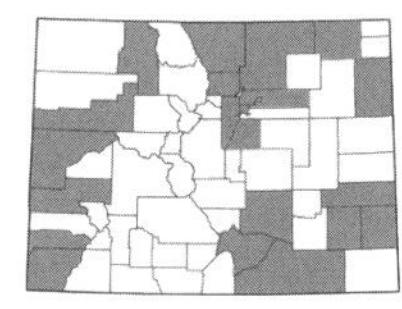

Calystegia sepium **(L.) R. Br.**, HEDGE FALSE BINDWEED. Leaves 2–15 × 1–9 cm, cordate-sagittate to hastate, the basal lobes angled; *flowers* solitary, on peduncles 3–13 cm long; *bracts* 1.4–2.5 × 1–1.8 cm, mostly acute; *calyx* 1–1.5 cm long; *corollas* funnelform, white or tinged with pink, 4.5–5.8 cm long. Found along roadsides, along moist creeks or ditches, and in disturbed areas, 3600–7700 ft. May–July. E/W. (Plate 30)

CONVOLVULUS L. – BINDWEED

Herbs, usually vines, or shrubs; *leaves* simple to lobed, sagittate, hastate, or ovate; *flowers* solitary in axils or in loose cymes; *corolla* campanulate to funnelform, white, pink, or pink with white stripes; *fruit* a 2–4-seeded capsule.

1a. Calyx 3–5 mm long, glabrous or inconspicuously minutely hairy; basal lobes of leaves entire or rarely a few with some teeth...***C. arvensis***
1b. Calyx 6–12 mm long, densely hairy; basal lobes of at least some leaves with teeth or lobes...***C. equitans***

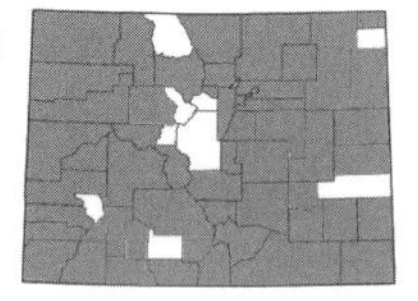

Convolvulus arvensis **L.**, FIELD BINDWEED. Leaves ovate to elliptic, cordate, hastate, or sagittate at the base, 1–10 × 0.3–6 cm, the basal lobes usually entire; *bracts* 2–3 (9) mm long; *calyx* 3–5 × 2–3 mm, glabrous or inconspicuously hairy; *corollas* campanulate, white or with pink stripes, 1.2–2.5 cm long; *capsules* 5–7 mm diam. Common along roadsides, in fields, and other disturbed places, one of our most noxious weeds. 3500–8500 ft. April–Oct. E/W. Introduced.

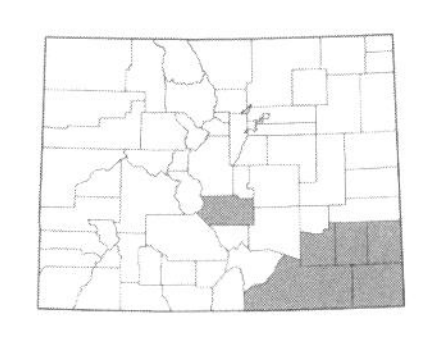

Convolvulus equitans **Benth.**, TEXAS BINDWEED. Leaves ovate to elliptic or narrowly oblong, hastate or sagittate at the base, 1–7 × 0.2–4 cm, the basal lobes toothed or lobed; *bracts* scale-like; *calyx* 6–12 mm long, densely hairy; *corollas* campanulate, white to pinkish, 1.5–3 cm long; *capsules* 7–8 mm diam. Found on dry hills and plains of the southeastern counties, 4300–5600 ft. May–Sept. E.

CUSCUTA L. – DODDER

Parasitic, chlorophyll-less herbs with twining stems, attached to hosts by haustoria; *leaves* minute and reduced to scales, alternate, sessile, simple; *calyx* (3-) 5-lobed; *corolla* (3-) 5-lobed, white or pink. (Yuncker, 1932)

The distribution of each species of *Cuscuta* is based on herbarium specimens, but the ranges are likely greater and more collections of this genus need to be made.

1a. Scales on the inner surface of the petals lacking...***C. occidentalis***
1b. Scales on the inner surface of the petals present, usually fringed or sometimes denticulate...2

2a. Stigma linear, about as long as the style...***C. approximata***
2b. Stigma capitate, appearing as a small ball on the tip of the style...3

3a. Flowers pedicellate, subtended by 1–3 large ovate to orbicular bracts which resemble the sepals; sepals distinct to the base or nearly so; style 1.5–3 mm long...***C. cuspidata***
3b. Flowers sessile or pedicellate, ebracteate, with small, scale-like bracts, or if larger then all parts denticulate; sepals fused at the base to form a short tube; style 0.3–1.5 mm long...4

4a. Petals erose-denticulate; scales erose-denticulate; capsules conical and longer than wide, surrounded by the withered corolla, 1-seeded...***C. denticulata***
4b. Petals entire; scales evidently fringed; capsules depressed-ovoid and wider than long, or if ovoid and longer than wide then only loosely cupped below by the withered calyx, 2–4-seeded...5

5a. Capsules large, 3–6 mm long, longer than wide, ovoid and often beaked, only loosely cupped below by the withered calyx; corolla lobes obtuse or rounded...***C. megalocarpa***
5b. Capsules generally smaller, depressed globose and not beaked, wider than long, surrounded by the withered corolla or loosely cupped below; corolla lobes acute...6

6a. Capsules protruding from the corolla and only loosely cupped below, not conspicuously thickened at the top; flowers sessile or short-pedicellate; corolla glabrous and smooth...***C. pentagona***
6b. Capsules more or less hidden by the corolla, thickened at the top; flowers pedicellate; corolla papillate, granulate, or glabrous and smooth...7

7a. Corolla tube extending beyond the calyx; corolla lobes erect to spreading, shorter than the tube; corolla 3–4 mm long, usually granulate to papillate or sometimes glabrous; seeds mostly 1.5–1.7 mm long...***C. indecora***
7b. Corolla tube about as long as the calyx and enclosed within; corolla lobes reflexed, as long or longer than the tube; corolla 2–3 mm long, usually glabrous or rarely slightly minutely hairy; seeds mostly 1 mm long...***C. umbellata***

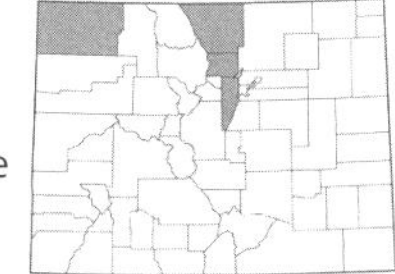

Cuscuta approximata **Bab.**, ALFALFA DODDER. [*C. epithymum* (L.) L. ssp. *approximata* (Bab.) Rouy]. Inflorescence of compact, few-several-flowered glomerules with sessile flowers; *calyx* 2–2.5 mm long, enclosing the corolla, the lobes wider than long, fleshy, abruptly narrowed to a recurved apex, reticulate-veiny upon drying; *corolla* lobes 1.5–2 mm long, shorter than the tube, with entire margins; *style* 0.3–0.5 mm long; *stigma* elongate and slender; *capsules* globose-depressed. Found especially on legumes (alfalfa) but found on other hosts as well, 5000–5500 ft. July–Sept. E/W.

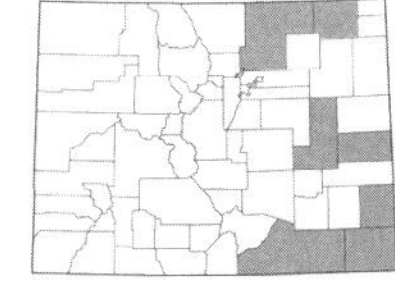

Cuscuta cuspidata **Engelm.**, CUSP DODDER. [*Grammica cuspidata* (Engelm.) Hadac & Chrtek]. Inflorescence bracteate, loose, of paniculate glomerules of subsessile or pedicellate flowers; *calyx* 1.5–2 mm long, the margin erose to serrulate; *corolla* lobes ca. 1.5 mm, spreading to reflexed, entire or sometimes serrate apically; *style* 1.5–3 mm long; *stigma* capitate; *capsules* globose; *seeds* ca. 1.4 mm long. Found mostly on Asteraceae hosts (*Helianthus*) but sometimes found on other hosts as well, 4000–5500 ft. July–Sept. E. (Plate 30)

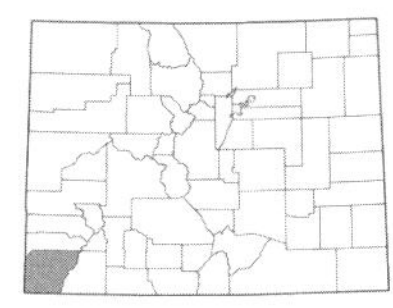

Cuscuta denticulata **Engelm.**, DESERT DODDER. [*Grammica denticulata* (Engelm.) W.A. Weber]. Inflorescence bracteate, the flowers single and scattered or congested into few-flowered glomerules; *calyx* 1.2–1.7 mm long, finely erose-denticulate; *corolla* lobes ca. 1.5 mm, the margins erose-denticulate; *style* 0.3–0.5 mm long; *stigma* capitate; *capsules* conical. Found on various desert shrubs, especially *Artemisia*, *Chrysothamnus*, *Eriogonum*, and *Gutierrezia*, 4600–5000 ft. June–Aug. W.

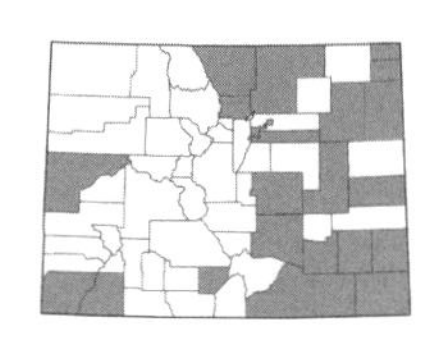

***Cuscuta indecora* Choisy**, BIGSEED ALFALFA DODDER. [*Grammica indecora* (Choisy) W.A. Weber]. Inflorescence loose to compact, of paniculately cymose glomerules, flowers pedicellate; *calyx* 1.5–3 mm long, the lobes triangular; *corolla* 3–4 mm long with lobes 1–1.5 mm long, triangular-ovate, erect to spreading; *style* 0.5–1.5 mm long; *stigma* capitate; *capsules* depressed-globose, thickened at the top; *seeds* ca. 1.7 mm long. Found on a wide range of hosts, our most common dodder, scattered across the state but especially common on the eastern plains, 3400–7500 ft. June–Sept. E/W. (Plate 30)

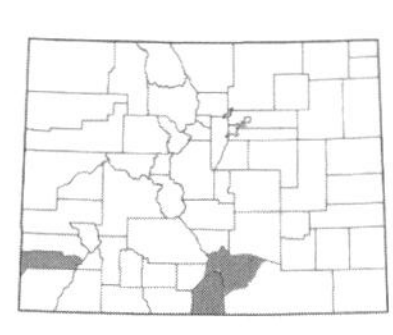

***Cuscuta megalocarpa* Rydb.**, BIGFRUIT DODDER. [*C. curta* (Engelm.) Rydb.; *C. gronovii* Willd. ex Schult. var. *curta* Engelm.]. Inflorescence compact to dense, of few-flowered, cymose-paniculate glomerules of pedicellate flowers; *calyx* 1–1.5 mm long, entire or slightly erose; *corolla* lobes 0.7–1 mm long, spreading to reflexed, entire, obtuse to rounded; *style* 0.3–0.7 mm long; *stigma* capitate; *capsules* ovoid and often beaked, 3–6 mm long. Found on a wide range of hosts, 7000–7500 ft. July–Aug. E/W.

Cuscuta gronovii Willd. ex Schult. could be in southeastern Colorado. It keys here to *C. megalocarpa* but differs in having smaller capsules and seeds (1.5 mm long vs. 2–3 mm long in *C. megalocarpa*).

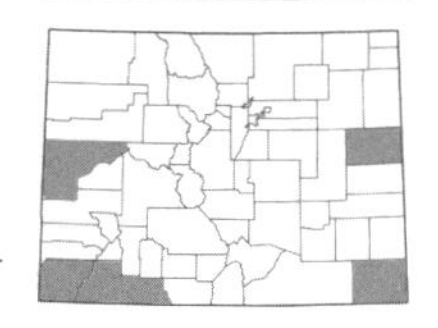

***Cuscuta occidentalis* Millsp.**, DODDER. [*C. californica* Hook. & Arn. var. *breviflora* Engelm.]. Inflorescence of small, compact glomerules with sessile (or nearly so) flowers; *calyx* 1.5–2 mm long, acuminate, entire; *corolla* lobes 1.5–2 mm long, erect or spreading, acuminate; *style* 1–1.5 mm; *stigma* capitate; *capsules* globose. Found on a wide range of hosts, 5000–5500 ft. May–Aug. W.

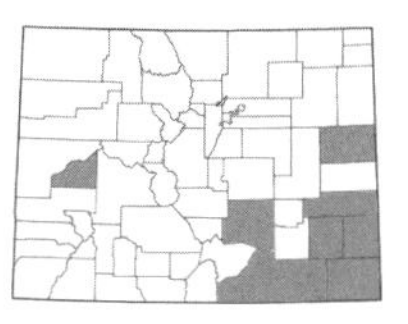

***Cuscuta pentagona* Engelm.**, FIVE-ANGLED DODDER. [*C. campestris* (Yuncker) Hadac & Chrtek; *Grammica pentagona* (Engelm.) W.A. Weber]. Inflorescence of loose, cymose clusters of pedicellate flowers; *calyx* lobes broadly ovate, obtuse, angled in appearance; *corolla* lobes lanceolate, spreading, acute; *style* equal to or shorter than the ovary; *stigma* capitate; *capsules* globose or depressed-globose; *seeds* ca. 1 mm long. Uncommon on a wide range of hosts, 4000–5000 ft. July–Oct. E/W.

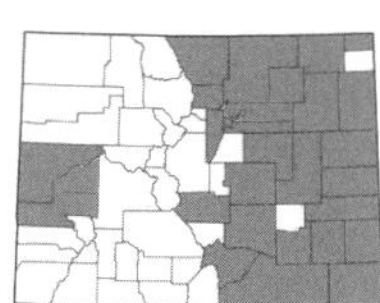

***Cuscuta umbellata* Kunth**, FLATGLOBE DODDER. [*Grammica umbellata* (Kunth) Hadac & Chrtek]. Inflorescence of pedicellate flowers in dense, compound cymes; *calyx* equal to or longer than the corolla, triangular, acute; *corolla* 2–3 mm long, the lobes lanceolate, reflexed, acute; *style* longer than the ovary; *stigma* capitate; *capsules* depressed-globose; seeds ca. 1 mm long, angled. Found on halophytic hosts (*Kochia*, *Salsola*), 3500–5000 ft. July–Sept. E/W. (Plate 30)

EVOLVULUS L. – DWARF MORNING-GLORY

Perennial herbs; *leaves* sessile or subsessile, simple, entire; *flowers* solitary in leaf axils or few in cymes; *sepals* 5; *corolla* rotate to shortly funnelform, purple to blue or white; *fruit* an ovoid capsule with 1–4 seeds.

***Evolvulus nuttallianus* Schult.**, SHAGGY DWARF MORNING-GLORY. Plants 1–1.5 dm, densely villous; *leaves* linear-oblong to narrowly lanceolate, 0.8–2 cm long; *sepals* 4–5 mm long, villous; *corolla* rotate to campanulate, 8–12 mm diam., purple or blue-purple. Common in sandy places on the plains and lower foothills, 3500–6000 ft. June–Aug. E/W. (Plate 30)

IPOMOEA L. – MORNING-GLORY

Annual or perennial herbs or vines; *leaves* petiolate or sessile, simple to palmately compound; *flowers* solitary in axils or terminal; *sepals* 5, equal or unequal; *corolla* funnelform to salverform, rarely campanulate, purple, red, orange, or sometimes white; *fruit* a 1–4-seeded capsule. (House, 1908)

1a. Shrub-like perennials; leaves linear to linear-lanceolate, entire; corollas bright pink to purplish-red, 5–10 cm long...***I. leptophylla***
1b. Annuals with twining stems, viney; leaves usually 3–5-lobed, rarely unlobed, with cordate bases; corollas red, reddish-orange, blue, white, or purple, 1.8–5 cm long...2

2a. Flowers red to reddish-orange, 1.8–3.5 cm long, with salverform corollas; sepals to 5.5 mm long, glabrous...***I. cristulata***
2b. Flowers blue, white, or purple, 2.5–5 cm long, with funnelform corollas; sepals 8–17 mm long, hispid...***I. purpurea***

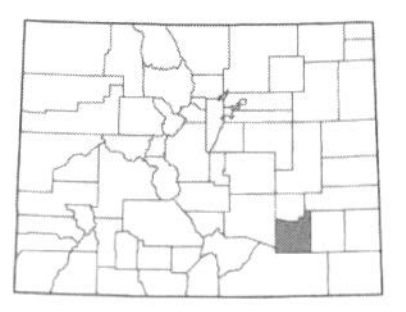

***Ipomoea cristulata* Hallier f.**, TRANSPECOS MORNING-GLORY. Annuals with twining stems; *leaves* ovate, 3–5-lobed, the margins irregularly toothed, 1.5–10 × 1–7 cm, with a cordate base; *sepals* unequal, the outer 3–3.5 mm long and the inner 4–5.5 mm long, glabrous; *corollas* salverform, 1.8–3.5 cm long, red to reddish-orange; *capsules* 7–8 mm diam. Recently found growing on trees in a cultivated orchard near Rocky Ford, 4200–4500 ft. July–Sept. E.

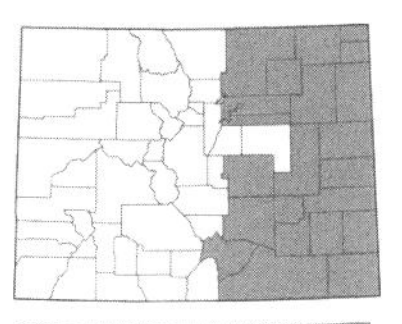

Ipomoea leptophylla **Torr.**, BUSH MORNING-GLORY. Shrub-like perennials, 0.3–1.2 m; *leaves* linear to linear-lanceolate, 3–15 × 0.2–0.8 cm, entire, acute; *sepals* unequal, 5–10 mm long; *corollas* funnelform, 5–9 cm long, bright pink to purple-red; *capsules* 10–15 mm long. Common in sandy places on the eastern plains, 3400–5200 ft. June–Aug. E. (Plate 30)

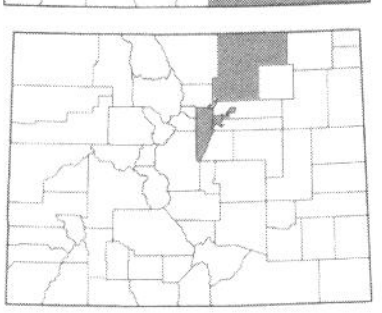

Ipomoea purpurea **(L.) Roth**, COMMON MORNING-GLORY. Annuals with twining stems; *leaves* ovate, 3–5-lobed or unlobed, to 11 cm long, cordate basally; *sepals* subequal, 8–17 mm long, hispid; *corollas* funnelform, white, blue, or purple, 2.5–5 cm long; *capsules* ca. 10 mm diam. Uncommon weed found in disturbed areas, parks, and in fields, 5000–5500 ft. June–Sept. E. Introduced.

CORNACEAE Bercht. & J. Presl – DOGWOOD FAMILY

Trees, shrubs, or perennial herbs; *leaves* simple and usually entire, exstipulate; *flowers* perfect or imperfect, actinomorphic; *sepals* 4–5, distinct, usually no more than small teeth; *petals* 4–5, distinct; *stamens* 4–5, distinct, alternating with the petals; *pistil* 1; *styles* 1 or 2–4; *ovary* inferior, 2–3-carpellate; *fruit* a drupe or berry.

CORNUS L. – DOGWOOD

Perennial herbs, shrubs, or small trees; *leaves* opposite or rarely alternate; *flowers* perfect, in terminal corymbose cymes or solitary; *sepals* 4, small; *petals* 4, white; *stamens* 4; *fruit* a white drupe.

1a. Low-growing herbaceous plants or subshrubs woody only at the base, usually less than 2 dm high; leaves whorled at the top of the stem; flowers subtended by 4 large white bracts; drupes red...***C. canadensis***
1b. Shrubs usually well over 2 dm high; leaves not whorled at the top of the stem; flowers numerous in a terminal corymbose cyme; drupes white...***C. sericea* ssp. *sericea***

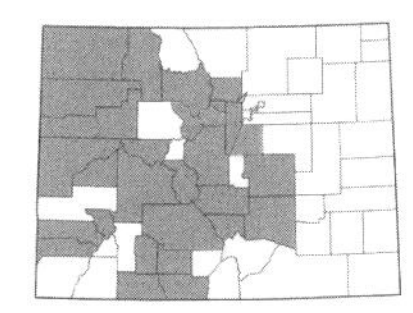

Cornus canadensis **L.**, BUNCHBERRY DOGWOOD. [*Chamaepericlymenum canadense* (L.) Asch. & Graebner]. Perennials or subshrubs, 0.5–2 dm; *leaves* elliptic to ovate, in a terminal whorl, 2–8 cm long; *flowers* in a semi-capitate cyme, subtended by 4 large white bracts 1–2.5 cm long; *petals* 1–1.5 mm long, greenish-white; *drupes* red, 6–8 mm diam. Uncommon in moist, shaded spruce-fir forests or along creeks, 6900–10,000 ft. June–Aug. E/W. (Plate 30)

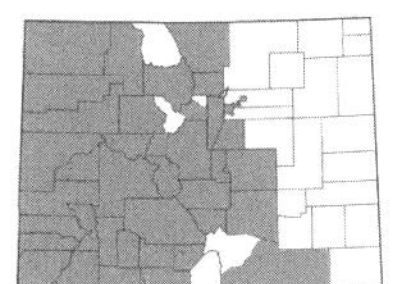

Cornus sericea **L. ssp. *sericea***, REDOSIER DOGWOOD. [*Cornus stolonifera* Michx.; *Swida sericea* (L.) Holub]. Shrubs mostly 1.5–4 m, at least the younger branches red to yellowish; *leaves* lanceolate to elliptic, 1–12 cm long, sparsely hairy; *flowers* in a terminal corymbose cyme; *sepals* ca. 0.5 mm long; *petals* white, 2–3.5 mm long; *drupes* white, 7–9 mm diam. Locally common in moist gulches and cool ravines, sometimes planted, 5000–9800 ft. May–Aug. E/W.

CRASSULACEAE J. St.-Hil. – STONECROP FAMILIY

Herbs or shrubs, often succulent; *leaves* alternate, opposite, or whorled, simple, exstipulate; *flowers* perfect, actinomorphic; *sepals* 4 or 5, distinct; *petals* 4 or 5, distinct, white, yellow, pink, reddish, or purple; *stamens* 4, 5, 8, or 10, epipetalous or not; *pistils* 4 or 5, with a nectariferous appendage near the base; *ovary* superior; *fruit* a follicle.

1a. Small annuals; stems usually prostrate; flowers minute and inconspicuous, solitary in the axils of the leaves...***Crassula***
1b. Perennials; stems erect; flowers larger and conspicuous, aggregated in cymes...2

2a. Flowers yellow, or if pink, white, or red then the leaves opposite...***Sedum***
2b. Flowers pink, white, or red; leaves alternate...***Rhodiola***

Crassulaceae flower:

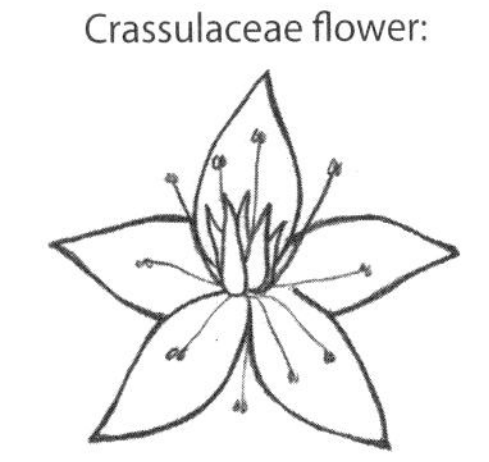

CRASSULA L. – PYGMYWEED

Annual herbs (ours); *leaves* opposite, entire to toothed, fleshy; *flowers* perfect, minute, solitary in leaf axils; *sepals* 4; *petals* 4, white; *stamens* 4; *pistils* 4, distinct.

Crassula aquatica **(L.) Schoenl.**, WATER PYGMYWEED. [*Tillaea aquatica* L.]. Plants usually prostrate, 1–8 cm long; *leaves* linear to narrowly oblanceolate, 1–6 × 0.2–1 mm; *sepals* 0.5–1 mm long, connate to near the middle; *petals* 1–2 mm long, whitish. Rare in shallow water and moist soil around vernal pools, we have one old record from Twin Lakes (Rydberg, 1906) and reported by T.S. Brandegee (1875) from probably near San Luis Lakes. This species has recently been discovered in Boulder County but was previously thought extirpated in the state. 5000–6800 ft. July–Aug. E.

RHODIOLA L. – STONECROP

Perennials; *leaves* alternate, entire to toothed, fleshy; *flowers* perfect or rarely imperfect, in terminal or axillary cymes; *sepals* 4 or usually 5; *petals* 4 or usually 5; *stamens* twice the number of sepals in number; *pistils* 4 or usually 5.

1a. Petals deep red to purple, 2.4–4 mm long; sepals 1.6–3 mm long...***R. integrifolia***
1b. Petals pink to rose or white, 7–12 mm long; sepals 4–7 mm long...***R. rhodantha***

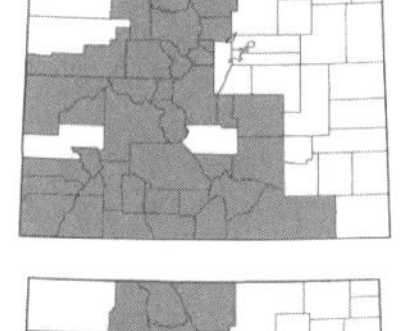

***Rhodiola integrifolia* Raf.**, KING'S CROWN. [*Sedum integrifolium* (Raf.) A. Nelson; *Tolmachevia integrifolia* (Raf.) Á. Löve & D. Löve]. Plants 0.3–3 (4.5) dm; *leaves* flat, oblanceolate to ovate or elliptic, scale-like below, 4–40 × 2–15 mm, entire to toothed; *sepals* 1.6–3 mm long, dark purple; *petals* 2.4–4 mm long, dark purple to red. Common along streams and in wet meadows and in alpine rock crevices, 9000–14,000 ft. June–Aug. E/W. (Plate 30)

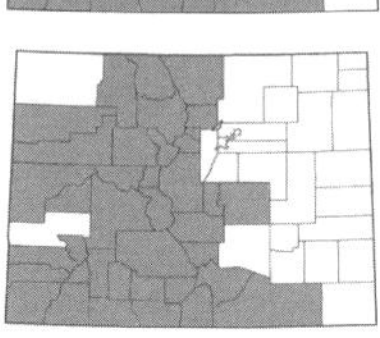

***Rhodiola rhodantha* (A. Gray) Jacobsen**, ROSE CROWN. [*Clementsia rhodantha* (A. Gray) Rose; *Sedum rhodanthum* A. Gray]. Plants 0.5–4 dm; *leaves* flat, oblanceolate to ovate, 6–30 × 2–6 mm, usually entire; *sepals* 4–7 mm long, usually pinkish; *petals* 7–12 mm long, pink to rose or white. Common along streams and in wet meadows, 8500–13,500 ft. June–Aug. E/W.

SEDUM L. – STONECROP

Perennials; *leaves* opposite or alternate, entire to toothed, fleshy; *flowers* perfect or rarely imperfect, in terminal or axillary cymes; *sepals* 4 or usually 5; *petals* 4 or usually 5; *stamens* 8 or 10; *pistils* 4 or usually 5. (Clausen, 1975)

1a. Flowers pink, white, red, or purple; leaves opposite, 10–30 mm long, flat and broad, orbicular, obovate, or oblong, with a sharply cuneate base and coarsely toothed apex, with glandular margins...***S. spurium***
1b. Flowers yellow; leaves alternate, generally smaller, to 10 mm long, subterete to terete, linear to lanceolate or ovate, otherwise unlike the above...2

2a. Stem leaves linear to lanceolate, falling early, not imbricate, 3–20 mm long...***S. lanceolatum***
2b. Stem leaves ovate, persistent, densely imbricate, 2–5 mm long...***S. acre***

***Sedum acre* L.**, GOLDMOSS STONECROP. Evergreen perennial, mat-forming, 0.5–1 dm; *leaves* alternate, ovate, terete, densely imbricate, 2–5 mm long; *petals* yellow, 6–8 mm long. Cultivated plants sometimes escaping, 7500–8000 ft. June–Aug. E/W. Introduced.

***Sedum lanceolatum* Torr.**, SPEARLEAF STONECROP. [*S. stenopetalum* Pursh]. Plants mostly 0.4–2.5 dm; *leaves* alternate, linear to lanceolate, subterete, 3–20 mm long, not densely imbricate; *petals* yellow, 5–8 mm long. Common in rocky soil and on dry slopes, 5300–14,000 ft. May–Sept. E/W. (Plate 30)

***Sedum spurium* Bieb.**, TWO-ROW STONECROP. Plants 1–2 dm; *leaves* opposite, orbicular to obovate or oblong, 10–30 mm long, sharply cuneate at the base, flat, the margins glandular and coarsely toothed above; *petals* pink, white, red, or purple, 8–11 mm long. Cultivated garden plants sometimes escaping along roadsides, 3600–6500 ft. June–Aug. E/W. Introduced.

CROSSOSOMATACEAE Engl. – CROSSOSOMA FAMILY

Shrubs or small trees; *leaves* usually alternate, simple, coriaceous; *flowers* perfect or some imperfect, actinomorphic, solitary, with a well-developed hypanthium; *sepals* (3) 4–6, distinct; *petals* (3) 4–6, distinct, sometimes clawed; *stamens* 4–50, distinct, 1-3-whorled; *pistils* 1–5 (9), distinct; *ovary* superior; *fruit* a follicle.

FORSELLESIA Greene – GREASEBUSH

Much-branched, xerophytic shrubs, often spinescent; *leaves* entire, minutely stipulate; *flowers* perfect or some imperfect; *sepals* 4–6; *petals* 4–6; *stamens* 4–10; *pistils* 1 (2), sessile. (Ensign, 1942)

1a. New growth yellowish-green; petals 1–2 mm wide, not constricted below the apex...***F. planitierum***
1b. New growth grayish-green; petals usually narrower, about 1 mm wide, somewhat constricted below the apex...***F. meionandra***

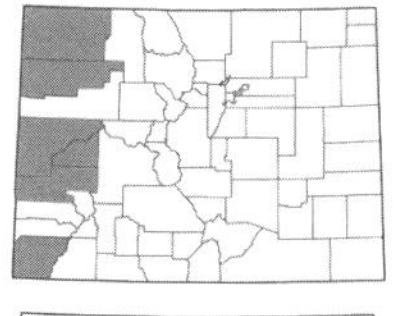

***Forsellesia meionandra* A. Heller**, SPINY GREASEBUSH. [*Glossopetalon meionandrum* Koehne]. Spinescent shrubs, mostly 1–8 (10) dm tall, the new grown grayish-green; *leaves* oblanceolate to elliptic, mostly 7–17 × 1.5–6 mm; *sepals* 1–3 mm long; *petals* white, 4–7 mm long; *follicles* 3.5–5 mm long. Found on sandstone ledges and shale slopes, 4900–7100 ft. May–June. W.

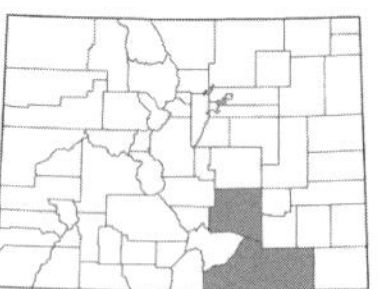

***Forsellesia planitierum* Ensign**, PLAINS GREASEBUSH. [*Glossopetalon planitierum* (Ensign) St. John]. Spinescent shrubs, 4–12 dm tall, the new growth yellowish-green; *leaves* oblanceolate to narrowly elliptic, (4) 6–14 × 2–4.5 mm; *sepals* 1.6–2.2 mm long; *petals* white, 4–6 mm long; *follicles* 5–7 mm long. Found on limestone outcrops and sandstone ledges, 5200–5500 ft. May–June. E.

CUCURBITACEAE Juss. – CUCUMBER FAMILY

Herbs or rarely shrubs, monoecious or dioecious, usually climbing and trailing, usually with tendrils; *leaves* alternate, petiolate, simple or compound, often palmately lobed, exstipulate; *flowers* imperfect, usually actinomorphic; *sepals* 5, distinct; *petals* 5, distinct or more often connate, green, white, yellow, or orange; *stamens* 3 or 5, adnate to the hypanthium, filaments distinct or connate, anthers usually connate or connivent; *pistil* 1; *stigmas* 2–3 (5); *ovary* inferior; *fruit* a pepo.

1a. Leaves simple and triangular-ovate; flowers large (over 5 cm long), solitary, yellow to yellow-orange; fruit smooth and globose, green with paler stripes; perennial, prostrate herb with a fetid odor...***Cucurbita***
1b. Leaves simple and palmately lobed or compound; flowers smaller (to 1 cm long), white to greenish, the pistillate flower solitary and the staminate flowers in racemes or panicles, with both arising from the same axil; fruit ovoid, with spines or prickles; annual, usually climbing herb, lacking a fetid odor...2

2a. Leaves palmately compound with 3–7 leaflets or deeply palmately lobed almost to the base; fruit narrowly ovoid, 2–3 cm long, covered with long, smooth spines; seeds 6–8 mm long...***Cyclanthera***
2b. Leaves simple and palmately lobed, the lobes shallower and only about ⅓ the length of the leaf; fruit ovoid, 3–5 cm long, covered with weak prickles; seeds 12–20 mm long...***Echinocystis***

CUCURBITA L. – GOURD

Annual or perennial monoecious herbs, trailing or climbing, with tendrils; *leaves* simple, usually palmately lobed; *flowers* solitary, yellow to orange; *stamens* 3, with distinct filaments and connate anthers; *stigmas* 3–5.

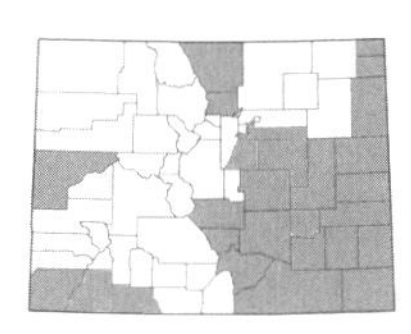

***Cucurbita foetidissima* Kunth**, BUFFALO GOURD; STINKING GOURD. Trailing perennials, to several meters long; *leaves* triangular-ovate, cordate or rounded at the base, acute, 1–2.5 (3) dm long, irregularly toothed or lobed along the margin, grayish-green, foul-smelling; *corolla* 5-lobed, yellow, to 1 dm long; *fruit* globose, 5–10 cm diam., smooth; *seeds* 6–9 mm long. Found along roadsides, in fields, and disturbed prairies, 3500–5500 ft. June–Aug. E/W. (Plate 30)

CYCLANTHERA Schrad. – CYCLANTHERA

Annual herbs, climbing or trailing, with tendrils; *leaves* compound; *flowers* solitary (pistillate) or in racemes or panicles (staminate), both in the same axils, white; *stamens* united into a central column; *stigmas* 5.

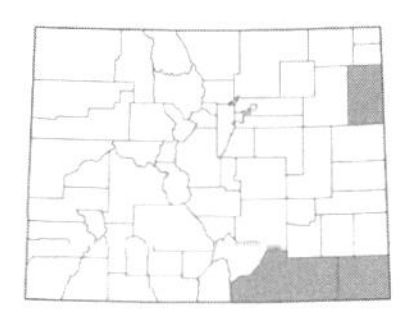

***Cyclanthera dissecta* (Torr. & A. Gray) Arn.**, CUTLEAF CYCLANTHERA. Leaves palmately compound with 3–7 leaflets or deeply palmately lobed, the leaflets to 6 cm long, elliptic-lanceolate; *corolla* 5-lobed; *fruit* 2–3 cm long, narrowly ovoid, covered with long, smooth spines; *seeds* 6–8 mm long. Found climbing over bushes and small trees, usually in drier parts of riparian areas, 4000–5100 ft. July–Aug. E.

ECHINOCYSTIS Torr. & A. Gray – WILD CUCUMBER

Annual herbs, with tendrils; *leaves* simple, palmately lobed; *flowers* solitary (pistillate) or in racemes or panicles (staminate), both in the same axils; *stamens* 3, filaments united into a column, anthers connivent.

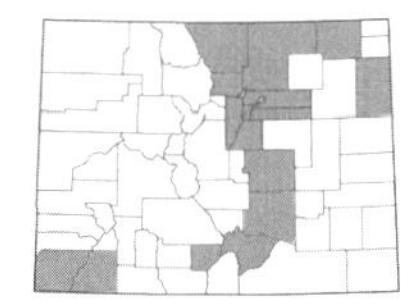

***Echinocystis lobata* (Michx.) Torr. & A. Gray**, WILD CUCUMBER. [*Micrampelis lobata* (Michx.) Greene]. Leaves palmately 3–7-lobed; *corolla* rotate, 0.8–1 cm wide, greenish or white; *fruit* ovoid, 3–5 cm long, covered with weak prickles; *seeds* 12–20 mm long. Found climbing on fences or on bushes, usually in moist areas along creeks or ditches, 3500–7500 ft. July–Sept. E/W.

CYPERACEAE Juss. – SEDGE FAMILY

Annual, biennial, or perennial herbs; *stems* triangular or terete, usually solid; *leaves* linear, basal and/or cauline, alternate, usually 3-ranked, the sheath closed; *flowers* perfect or imperfect, in spikelets, subtended by a scale (bract); *spikelets* 1–several-flowered; *perianth* absent or reduced to bristles or scales; *stamens* usually 3; *ovary* 2–3-carpellate; *fruit* an achene. (Porter, 1964; Smith & Durrell, 1944)

1a. Achene and ovary enclosed in a perigynium (sac) with a tiny opening at the top through which the stigmas emerge, or the perigynium split on one side nearly to the base; perigynium subtended by a scale...2
1b. Achene and ovary not enclosed in a perigynium...3

2a. Perigynium closed except for the tiny opening at the top...***Carex***
2b. Perigynium open, wrapped around the achene or split on one side nearly to the base...***Kobresia***

3a. Perianth bristles numerous (more than 10), long and hair-like, the entire inflorescence resembling a ball of cotton at maturity...***Eriophorum***
3b. Perianth bristles usually less than 10 (rarely to 12), not long and hair-like...4

4a. Spikelets flattened, with the scales in 2 ranks; perianth absent...***Cyperus***
4b. Spikelets not flattened, the scales spirally arranged; perianth of scales or bristles usually present...5

5a. Spikelets solitary and terminal at the tips of each stem; leaves usually reduced to sheaths at the base...6
5b. Spikelets not solitary, or if solitary then subtended by involucral bracts that are at least twice as long as the spikelet, or the spikelets appearing lateral along the stem; leaves reduced or not...7

6a. Achenes with a conical or flattened cap at the tip; leaves reduced to sheaths (*careful*—in *Eleocharis parvula* the spikelets are sometimes not present on the tips of all the stems, giving the appearance of basal leaves); perianth absent or of up to 7 bristles...***Eleocharis***
6b. Achenes lacking a conical or flattened cap at the tip; sheaths with short leaves; perianth absent...***Trichophorum***

7a. Achenes tipped with a narrowly triangular tubercle 0.5–1.2 mm long; spikelets nearly white or sometimes pale brown; perianth bristles 10–12...***Rhynchospora***
7b. Achenes not tipped with a narrowly triangular tubercle; spikelets brown or green; perianth bristles absent or less than 10...8

8a. Small, delicate annual 3–8 cm tall with narrow, filiform leaves and stems; achenes smooth, subtended by a clear scale in addition to a conspicuous outer scale; inflorescence of a solitary, lateral spikelet 2–6 mm long (or sometimes 2–3 spikelets present); perianth bristles absent...***Lipocarpha***
8b. Plants unlike the above, or if annual then the achenes with prominent horizontal ridges; achenes not subtended by a clear scale; inflorescence terminal or if lateral then usually of more than one spikelet; perianth bristles present or absent...9

9a. Inflorescence subtended by one bract, this usually erect, appearing to be a continuation of the stem (giving the inflorescence a lateral appearance)...10
9b. Inflorescence subtended by 2 or more bracts, these usually very unequal and spreading (generally giving the inflorescence a terminal appearance)...13

10a. Tufted annuals; stems slender, mostly less than 1.5 mm wide; achenes with prominent horizontal ridges, sharply 3-sided...***Schoenoplectus*** (*saximontanus*)
10b. Perennials, usually with conspicuous rhizomes; stems stout, mostly over 2 mm wide; achenes smooth or reticulate but lacking horizontal ridges, plano-convex or weakly 3-sided...11

11a. Stems sharply triangular...***Schoenoplectus*** (*americanus*, *pungens*)
11b. Stems round...12

12a. Spikelets sessile; leaves narrow, 0.5–2 mm wide; perianth bristles 1–3...***Amphiscirpus***
12b. Inflorescence 2–4-times branched; leaves over 2 mm wide; perianth bristles 6...***Schoenoplectus***

13a. Achenes reticulate with numerous longitudinal ribs and finer horizontal ridges; stems round; leaves narrowly linear or filiform, 1–3 mm wide...***Fimbristylis***
13b. Achenes smooth or minutely reticulate but lacking longitudinal ribs and horizontal ridges; stems triangular, rarely obscurely so; leaves mostly 6–20 mm wide...14

14a. Spikelets 10–25 mm long, usually few (mostly 3–40), most or all sessile...***Bolboschoenus***
14b. Spikelets 3–10 mm long, 3–40, usually numerous (more than 100), arranged in a branching inflorescence...***Scirpus***

AMPHISCIRPUS Oteng-Yeb. – BULRUSH

Perennial herbs; *stems* terete; *leaves* all basal, ligules ciliate; *flowers* perfect; *perianth* of 1–6 bristles.

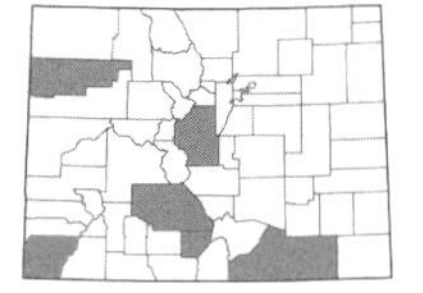

***Amphiscirpus nevadensis* (S. Watson) Oteng-Yeb.**, NEVADA BULRUSH. [*Scirpus nevadensis* S. Watson]. Plants 1–7 dm, from rhizomes; *leaves* mostly 3–30 cm × 0.5–2 mm, sharply acute; *inflorescences* with a foliose involucre bract 1–15 cm long; *spikelets* 5–20 × 3–5 mm; *scales* reddish-brown with paler midribs, the proximal scales to 15 mm long, other scales ca. 4 × 3 mm; *perianth* bristles brown; *achenes* 2–2.3 × 1.5–1.7 mm. Found on alkaline salt flats, 7500–8500 ft. May–Sept. E/W. (Plate 39)

BOLBOSCHOENUS (Asch.) Palla in Hallier & Brand – TUBEROUS BULRUSH

Perennial herbs; *stems* triangular; *leaves* basal and cauline, ligules absent; *flowers* perfect; *perianth* of 3–6 bristles.

1a. Spikelets on evident, elongated branches (such that the inflorescence is umbellate in appearance); perianth bristles equal to the length of the achene...***B. fluviatilis***

1b. Spikelets all sessile; perianth bristles to ½ the length of the achene...***B. maritimus* ssp. *paludosus***

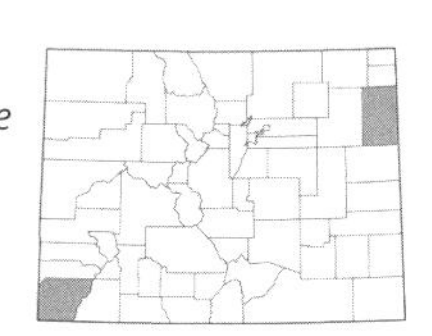

Bolboschoenus fluviatilis **(Torr.) Soják**, RIVER BULRUSH. [*Scirpus fluviatilis* (Torr.) A. Gray]. Plants 10–20 dm; *leaves* to 22 mm wide; *involucral bracts* 3–6, longer than the inflorescence; *inflorescence* umbellate; *spikelets* 10–25 mm long; *scales* orange-brown to tan, 7–10 mm long, bifid at the apex with awns 2–3 mm long; *perianth* bristles equal to the achene; *achenes* 3.8–5.5 × 2–3 mm, trigonous, obovoid, with a beak 0.2–0.8 mm. Rare in floodplains, marshes, and wetlands, 3900–5500 ft. June–Aug. E/W.

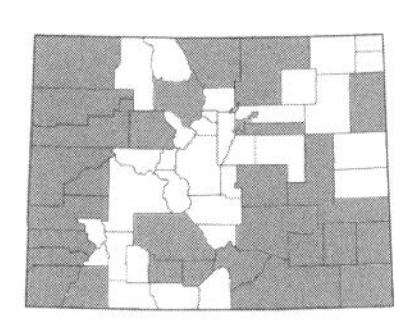

Bolboschoenus maritimus **(L.) Palla ssp. *paludosus*** **(A. Nelson) Á.Löve & D.Löve**, SALT-MARSH BULRUSH. [*Scirpus maritimus* L. var. *paludosus* (A. Nelson) Kükenthal]. Plants 5–15 dm; leaves to 12 mm wide; *inflorescence* of sessile spikelets; *spikelets* 7–40 mm long, ovoid to lance-ovoid; *scales* orange-brown to tan, 5–8 mm long, the apex bifid with an awn 1–3 mm long; *perianth* bristles to ½ the length of the achene; *achenes* 2.3–4 mm long with a beak 0.1–0.4 mm long, lenticular. Common in wet depressions, ditches, marshes, and along pond shores, 3500–8000 ft. May–Aug. E/W. (Plate 39)

CAREX L. – SEDGE

Perennial herbs; *stems* usually triangular; *leaves* basal and cauline or all basal, ligules present; *flowers* imperfect, pistillate flowers enclosed in a perigynium; *perianth* absent. (Hermann, 1959, 1970; Hurd et al., 1998; Johnson, 1964; Murray, 1969)

1a. Lower pistillate scales large and leaf-like, mostly over 2 cm long (2–3 times the length of the spike); spike terminal and solitary, with 2–3 inconspicuous male flowers and 2–6 female flowers; leaves flat, 2–6 mm wide; known from Larimer, Boulder, Jefferson, and Douglas counties...2

1b. Pistillate scales not large and leaf-like; plants otherwise unlike the above in all respects; variously distributed...3

2a. Beak of perigynia 1.9–3 mm long, smooth; perigynia 4.8–6.6 mm long, gradually tapering at the tip...***C. backii***

2b. Beak of perigynia 0.6–1.2 mm long, scabrous; perigynia 3.2–5 mm long, abruptly tapered at the tip...***C. saximontana***

Carex perigynium:

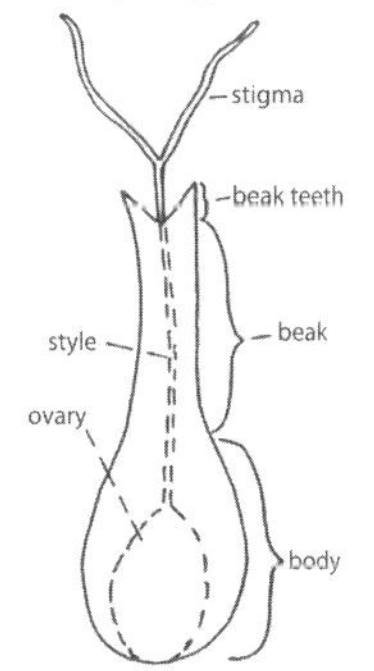

3a. Staminate flowers in a slender, cylindrical spike separated from the few (1–3) lower pistillate flowers by a short but conspicuous exposed rachis; perigynia 5–7 mm long, beakless; stigmas 3; leaves flat, 1–4 mm wide; common in woodlands...***C. geyeri***

3b. Plants unlike the above in all respects...4

4a. Spikes solitary and terminal on each stem (never in a dense globose head)...**Key 1**

4b. Spikes more than one per stem, sometimes crowded into a dense globose or ovoid head and appearing solitary...5

5a. Perigynium surface sparsely to densely hairy (at least in the upper third)...**Key 2**

5b. Perigynium surface glabrous, although the margins sometimes ciliate or serrulate...6

6a. Bract subtending the lowest spike with a well-developed sheath usually at least 5 mm long (species included here key elsewhere too)...**Key 3**

6b. Bract subtending the lowest spike absent or lacking a sheath, or with a short sheath less than 5 mm long...7

7a. Staminate and pistillate flowers on completely separate spikes, the terminal 1–3 spikes staminate (occasionally with just a few perigynia at the base or tip) and the lower lateral spikes pistillate (sometimes with a few upper staminate flowers); spikes often cylindric and at least the lowest pistillate spikes evidently pedunculate...**Key 4**

7b. Staminate and pistillate flowers in the same spikes or the spikes all staminate or pistillate (plants dioecious); spikes variously shaped, mostly sessile or nearly so...8

8a. Spikes all staminate or pistillate (plants dioecious)...**Key 5**

8b. Spikes with both staminate and pistillate flowers...9

9a. Stigmas 3 (check numerous perigynia, these can easily break); achenes mostly 3-sided; pistillate scales usually black, purplish-black, or dark brown-black...**Key 6**
9b. Stigmas 2; achenes mostly lenticular; pistillate scales various...10

10a. Perigynia with winged margins at least above the middle and on the beak, usually serrulate-margined above the middle; spikes gynaecandrous...**Key 7**
10b. Perigynia lacking a distinctly winged margin (sometimes the margins thin-edged, raised, or sharp-edged), otherwise various, or appearing to be wing-margined but the spikes androgynous...11

11a. Spikes androgynous...**Key 8**
11b. Spikes gynaecandrous...**Key 9**

Key 1
(Spikes solitary)

1a. Spikes entirely staminate...2
1b. Spikes with both male and female flowers or entirely pistillate...3

2a. Leaves filiform, 0.4–1 mm wide...***C. gynocrates***
2b. Leaves flat, 1–4 mm wide...***C. scirpoidea***

3a. Perigynia short-hairy to hirsute (at least above the middle or near the beak); spike linear-cylindric...4
3b. Perigynia glabrous throughout or serrulate just on the margins above; spike various...6

4a. Leaves flat, 1–4 mm wide; spike with only female flowers; stems arising singly or loosely caespitose from creeping rhizomes...***C. scirpoidea***
4b. Leaves filiform or folded above, mostly 0.3–1 mm wide (but up to 2 mm wide near the base); spikes with both male and female flowers present; plants densely caespitose and lacking creeping rhizomes...5

5a. Leaves filiform throughout; lowest pistillate scale usually awnless or short-awned...***C. filifolia***
5b. Leaves folded, 1–2 mm wide at the base; lowest pistillate scale usually awned (to 2 mm)...***C. oreocharis***

6a. Rachilla bristle-like and exserted 0.5–2.8 mm beyond the opening of the perigynium beak; perigynia strongly reflexed, soon deciduous, linear-lanceolate; leaves less than 1 mm wide...***C. microglochin***
6b. Rachilla lacking or included in the perigynia; perigynia ascending or widely spreading (sometimes just a couple of perigynia at the base pointing down, but otherwise spreading or ascending); plants otherwise unlike the above in all respects...7

7a. Perigynia beakless with a rounded apex, oval-elliptic, pale green to yellowish-green throughout; spike with 2–9 female flowers and 2–7 male flowers...***C. leptalea***
7b. Perigynia with a short to long but definite beak, otherwise various...8

8a. Perigynia margins serrulate on upper ⅓ to ½ (sometimes obscurely so; check several perigynia); leaves stiff and wiry, 0.25–0.5 mm wide; plants densely caespitose without creeping rhizomes...***C. nardina***
8b. Perigynia margins entire throughout; leaves various; plants with or without rhizomes...9

9a. Stigmas 2; achenes lenticular...10
9b. Stigmas 3 (*careful*—look at several perigynia as stigmas can break off); achenes 3-sided...11

10a. Perigynia ovate to orbicular, broadly rounded at the base, flattened, spreading or ascending but the lowest few not recurved, nerveless; spikes androgynous; rachilla well-developed...***C. capitata***
10b. Perigynia elliptic to lanceolate, narrowed and tapering at the base, biconvex at maturity, plump and glossy, soon widely spreading and the lowest one or two often recurved, obscurely to often evidently nerved; spikes sometimes entirely pistillate; rachilla obsolete...***C. gynocrates***

11a. Perigynia with an entire, obscurely hyaline-tipped beak at the tip; leaves flat, 1–3 mm wide, the tips soon becoming brown and dry and usually coiled downward or flexuous; achenes dark brown or black; rachilla absent...***C. rupestris***
11b. Perigynia with a bidentate, obliquely cleft, or obliquely cut beak at the tip, the beak hyaline or not; leaves flat or involute, 0.25–4 mm wide, the tips usually not coiled downward; achenes various; rachilla absent or well-developed...12

12a. Perigynia prominently and finely several-nerved on both surfaces, plump and coriaceous, glossy, dark brown or blackish-brown throughout; achenes tan to yellowish...***C. obtusata***
12b. Perigynia nerveless or obscurely nerved, otherwise unlike the above in all respects...13

13a. Leaves flat, 1–4 mm wide; perigynia dark brown or black above, paler and greenish or stramineous at the base, with an evident, slender stipe...14
13b. Leaves involute, 0.2–0.6 mm wide; perigynia various...15

14a. Perigynia ascending; leaves mostly to 1.5 mm wide; plants densely caespitose, lacking evident rhizomes...***C. micropoda***
14b. Perigynia becoming widely spreading and deflexed at maturity; leaves (1.5) 2–4 mm wide; plants with evident rhizomes, loosely or not caespitose...***C. nigricans***

15a. Rachilla obsolete; perigynia lanceolate, tapering to a narrow apex, dark brown or black above and green to stramineous below...***C. micropoda***
15b. Rachilla well-developed, at least half as long as the achene (look for the impression of the rachilla through the wall of the perigynia); perigynia unlike the above...16

16a. Plants densely caespitose without creeping rhizomes; perigynia 2.5–4.5 mm long, with a cylindric, truncately cut hyaline beak...***C. elynoides***
16b. Plants with evident rhizomes, the stems arising singly or loosely caespitose; perigynia 4–7.5 mm long, with a short, obliquely cleft beak...***C. engelmannii***

Key 2
(Perigynia hairy)

1a. Perigynia hairy only near the very tip; perigynium beak very short (to 0.2 mm long) or absent; spikes cylindric, mostly 1–3 cm long...***C. parryana***
1b. Perigynia hairy throughout or above the middle; plants otherwise various...2

2a. Pistillate spikes cylindric, 1.5–6 cm long, clearly separated from the staminate spikes by about 2 cm; plants mostly 30–110 cm tall... ***C. pellita***
2b. Pistillate spikes ovoid to short cylindric, less than 1.5 cm long, often not separated from the staminate spikes by more than 0.5 cm; plants 5–50 cm tall...3

3a. Bracts subtending the spikes lacking a blade, reduced to a sheath; perigynium with a short beak 0.2–0.4 mm long; bases of plants not fibrous, lacking the remnants of old leaves...***C. concinna***
3b. Bracts subtending the spikes with a definite blade or long awn, sometimes this surpassing the staminate spike; perigynium with a beak 0.4–1.7 mm long; bases of plants often fibrous (from the remnants of old leaves present)...4

4a. Plants with culms of two types, some short, mostly hidden in the basal leaves and bearing only pistillate spikes, and others elevated above the leaves with both staminate and pistillate spikes (bisexual)...5
4b. Plants with fertile culms all alike, bearing both staminate and pistillate spikes...8

5a. Lowest pistillate bract on a bisexual culm (the one subtending the lowest pistillate spike) much shorter than the entire inflorescence...***C. geophila***
5b. Lowest pistillate bract on a bisexual culm equal to or longer than the inflorescence...6

6a. Pistillate spikes of bisexual culms with 1–3 flowers; perigynium beak inconspicuously or not ciliate-serrulate along the margins...***C. pityophila***
6b. Pistillate spikes of bisexual culms usually with more than 3 flowers; perigynium beak with ciliate-serrulate margins...7

7a. Perigynium beak 0.25–0.75 mm long; perigynia 2.5–3.5 mm long...***C. deflexa* var. *boottii***
7b. Perigynium beak 1–1.7 mm long; perigynia 3.5–4.5 mm long...***C. rossii***

8a. Scales longer than the perigynium body...***C. inops* ssp. *heliophila***
8b. Scales shorter than the perigynium body...***C. peckii***

Key 3

(*Bract subtending the lowest spike with a well-developed sheath at least 5 mm long*)

1a. Perigynium golden-yellow or orange at maturity, or covered with a whitish powdery coating, lacking a beak, somewhat inflated and fleshy; lowest bract subtending the inflorescence usually sheathing at the base and longer than the inflorescence...***C. aurea***
1b. Perigynium not golden-yellow or orange, or covered with a whitish powdery coating, with or without a beak, not inflated or fleshy; plants otherwise various...2

2a. Perigynium beak serrulate along the margins; terminal spike gynecandrous or rarely staminate; scales black or brown...***C. fuliginosa***
2b. Perigynium beak absent or not serrulate along the margins; terminal spike usually staminate; scales pale brown to reddish-brown...3

3a. Lateral pistillate spikes on nodding, slender peduncles; perigynia nerveless except for 2 marginal nerves...***C. capillaris***
3b. Lateral spikes erect or ascending, not nodding on slender peduncles; perigynia with evident nerves or nerveless and densely papillose...4

4a. Perigynia mostly spreading (or the lower even somewhat reflexed), the beak 0.6–1.5 mm long; lateral spikes globose to short-cylindric; lowest bract usually at least 3 times longer than the inflorescence...***C. viridula* ssp. *viridula***
4b. Perigynia ascending, the beak absent or to 0.4 mm long; lateral spikes cylindric; lowest bract shorter to longer than the inflorescence...5

5a. Perigynia nerveless or the nerves obscure, densely papillose; pistillate spikes with 3–20 flowers...***C. livida***
5b. Perigynia distinctly nerved; pistillate spikes with 10–50 (85) flowers...6

6a. Plants densely caespitose; veins on the perigynia slightly impressed; lowest bract blade slightly shorter to much longer than the terminal spike; achenes 1.8–2.5 mm long...***C. conoidea***
6b. Plants arising singly or a few together from a well-developed, creeping rhizome; veins on the perigynia raised; lowest bract blade shorter than the terminal spike; achenes 1.3–2 mm long...***C. crawei***

Key 4

(*Staminate and pistillate flowers on separate spikes*)

1a. Perigynium with an evident beak 1.2–6.5 mm long, shallowly to deeply bidentate; scales never black or purplish-brown; stigmas 3...2
1b. Perigynium beak absent or inconspicuous, or shorter (0.1–1) mm long, entire or shallowly bidentate; scales green, brown, purplish-brown, or black; stigmas 2 or 3...9

2a. Perigynium beak equal to or longer than the body; perigynium 2-ribbed, otherwise nerveless...***C. sprengelii***
2b. Perigynium beak shorter than the body; perigynium strongly 6–20-veined...3

3a. Perigynium beak deeply bidentate with spreading or divergent tips 1.2–3 mm long; leaf sheaths usually hairy or scabrous at least at the tips, or occasionally glabrous...***C. atherodes***
3b. Perigynium beak bidentate but the teeth straight and 0.2–1.3 mm long; leaf sheaths glabrous...4

4a. Pistillate scales with with a long, serrulate awn tip (the awn much longer than the small, often inconspicuous scale body); lower pistillate spikes on nodding, slender peduncles...***C. hystericina***
4b. Pistillate scales awnless or acuminate; lower pistillate spikes usually erect...5

5a. Bract subtending the lowest pistillate spike 3–9 times longer than the inflorescence; with a single terminal staminate spike...6
5b. Bract subtending the lowest pistillate spike to 2 times longer than the inflorescence; staminate spikes 2–5...7

6a. Perigynium 7–10 mm long, the beak 2–4.5 mm long; pistillate spikes 1.5–8 cm long...***C. retrorsa***
6b. Perigynium 2–4.5 mm long, the beak 0.6–1.5 mm long; pistillate spikes 0.4–1.8 cm long...***C. viridula* ssp. *viridula***

7a. Leaves 4.5–12 mm wide, the ligules as long as wide, the basal sheaths spongy-thickened; perigynia spreading; plants from elongate rhizomes...***C. utriculata***
7b. Leaves 1–7 mm wide, the ligules longer than wide, the basal sheaths not spongy-thickened; perigynia ascending to spreading-ascending; plants caespitose, from short rhizomes...8

8a. Perigynia 7.5–10 mm long, leathery; beak indistinct...***C. exsiccata***
8b. Perigynia (3.5) 4–11 mm long, papery; beak distinct...***C. vesicaria***

9a. Perigynium golden-yellow or orange at maturity, or covered with a whitish powdery coating, somewhat inflated and fleshy; lowest bract subtending the inflorescence usually sheathing at the base and longer than the inflorescence...***C. aurea***
9b. Perigynium not golden-yellow or orange, or covered with a whitish powdery coating, not inflated or fleshy; plants otherwise various...10

10a. Leaf blades and/or sheaths hairy; spikes mostly less than 1.5 cm long...***C. torreyi***
10b. Leaf blades and sheaths glabrous; spikes various...11

11a. Lateral pistillate spikes on long, slender nodding peduncles...12
11b. Lateral pistillate spikes sessile or nearly so, not on long, slender nodding peduncles...14

12a. Perigynium with a beak about 1 mm long; roots glabrous...***C. capillaris***
12b. Perigynium essentially beakless or the beak to 0.2 mm long; roots densely covered with yellowish-brown felty hairs (carefully wash away mud on the roots to see)...13

13a. Pistillate scales obtuse or with an acute tip...***C. limosa***
13b. Pistillate scales usually awn-tipped...***C. magellanica* ssp. *irrigua***

14a. Stigmas usually 2 (rarely some 3); achenes usually lenticular...15
14b. Stigmas 3; achenes 3-sided...20

15a. Perigynia nerveless or with obscure, indistinct nerves...16
15b. Perigynia evidently 3–9-nerved on both surfaces...18

16a. Style tough, hard and bony, continuous with the achene, persistent; perigynia oblong to elliptic-ovate, 3–5 mm long, yellowish below and reddish-black above; pistillate spikes oblong to short-cylindric...***C. saxatilis***
16b. Style delicate, jointed with the achene, usually falling; perigynia orbicular to obovoid or ovate, 1.8–3.6 (4) mm long, green to tan and often speckled with purplish spots or reddish-black above; pistillate spikes long-cylindric to oblong...17

17a. Perigynia orbicular to broadly obovoid, rounded at the base and abruptly beaked, green to tan and often tinged with dark purple above; pistillate scale usually black throughout; primary bract subtending the inflorescence with a purplish-black band at the base or occasionally purplish-black throughout, 1–7 cm long, shorter than the terminal spike; spikes usually crowded or sometimes just the lowest one separated...***C. scopulorum***
17b. Perigynia elliptic to ovate, tapering at the base and tip, green or tan and often with purplish-red speckles; pistillate scale usually with a pale midrib; primary bract subtending the inflorescence usually green throughout, 3–15 cm long, usually longer than the terminal spike; spikes well-separated...***C. aquatilis***

18a. Leaves blue-green, glaucous; scales often with the midrib extending to a serrulate awn; perigynia persistent...***C. nebrascensis***
18b. Leaves green, not glaucous; scales awnless; perigynia early-deciduous...19

19a. Lowest bract subtending the inflorescence equal in length to the inflorescence; perigynia 3–5-nerved on each surface; pistillate scales equal to the perigynia...***C. emoryi***
19b. Lowest bract subtending the inflorescence longer than the inflorescence; perigynia 5–7-nerved on each surface; pistillate scales mostly shorter than the perigynia...***C. lenticularis* var. *lipocarpa***

20a. Pistillate scales black to dark reddish-brown, without a white hyaline margin...21
20b. Pistillate scales green to brown, or if dark reddish-brown then with a white hyaline margin...22

21a. Perigynia spreading (at least the lower ones) to spreading-ascending, appearing plump and conspicuous in the spike; achenes 1.8–2.5 mm long...***C. raynoldsii***
21b. Perigynia ascending, flattened or somewhat inflated but not appearing plump; achenes 1.2–1.7 mm long...***C. paysonis***

22a. Perigynia mostly spreading (or the lower even somewhat reflexed), the beak 0.6–1.5 mm long; lateral spikes globose to short-cylindric (0.5–1.4 cm long); lowest bract usually at least 3 times longer than the inflorescence...***C. viridula* ssp. *viridula***
22b. Perigynia ascending, the beak absent or to 0.4 mm long; lateral spikes cylindric (most over 1 cm long); lowest bract shorter to longer than the inflorescence...23

23a. Perigynia nerveless or the nerves obscure...24
23b. Perigynia distinctly nerved...25

24a. Perigynia green or yellowish-green, 2.5–5 mm long, narrowly elliptic; pistillate scales with a broad greenish-yellow center and narrow brown margins...***C. livida***
24b. Perigynia reddish-tinged and granular-roughened at the apex, 1.5–3 mm long, broadly obovoid; pistillate scales reddish-brown with a narrow green midrib and white margins...***C. parryana***

25a. Plants densely caespitose; veins on the perigynia slightly impressed; lowest bract blade slightly shorter to much longer than the terminal spike; achenes 1.8–2.5 mm long...***C. conoidea***
25b. Plants arising singly or a few together from a well-developed, creeping rhizome; veins on the perigynia raised; lowest bract blade shorter than the terminal spike; achenes 1.3–2 mm long...***C. crawei***

Key 5

(Spikes unisexual and plants dioecious)

1a. Perigynia small, 1.7–2.4 (2.8) mm long, with a small beak 0.2–0.6 mm long and slightly serrulate, raised and narrowly winged greenish margins above; pistillate scales dark reddish-brown to dark brown, completely concealing the perigynia...***C. simulata***
1b. Perigynia 2.5–7 mm long, with a beak 0.5–2 mm long, with or without serrulate or narrowly winged margins; pistillate scales various...2

2a. Basal leaf sheaths and rhizome dark black to black-brown or reddish-brown; spikes arranged in an elongate head; plants 20–80+ cm; anther tips bristly on staminate plants...***C. praegracilis***
2b. Basal leaf sheaths and rhizome grayish, pale brown, or brownish (but not reddish-brown or black-brown); spikes arranged in an ovoid-fusiform or oblong-cylindric head; plants (6) 10–40 cm; anther tips not bristly...3

3a. Perigynia 3.5–7 mm long with a prominent beak 1–1.8 mm long; spikes aggregated into a head 1.5–4.5 cm long; style 1.8–3.5 mm long, exserted; leaves 1–3 mm wide...***C. douglasii***
3b. Perigynia 2.4–4 mm long with a short beak 0.3–1 mm long; spikes aggregated into a head 0.8–1.7 cm long; style to 1.5 mm long, included; leaves 0.5–2 mm wide...***C. duriuscula***

Key 6

(Stigmas 3; pistillate scales usually black or purplish-black)

1a. Pistillate scales often longer than the perigynia, with a prominent awn tip 0.5–3 mm long; plants long-rhizomatous; perigynia densely papillate...***C. buxbaumii***
1b. Pistillate scales acute or with a short-mucronate tip less than 0.5 mm long; plants lacking long rhizomes; perigynia various...2

2a. Spikes all sessile, forming a dense terminal ovoid or oblong cluster (often in clusters of 3)...3
2b. Spikes (at least the lowest pistillate spike) on short to long stalks, not forming a dense cluster (although sometimes forming a dense cluster with the lowest spike approximate and separate)...6

3a. Perigynia oblong-obovoid, not strongly flattened (at least somewhat inflated), 1.5–1.75 mm wide, the beak 0.8–1 mm long...***C. nelsonii***
3b. Perigynia broadly ovate, obovate, elliptic, to oblong, strongly flattened, 2–4 mm wide, the beak 0.3–0.8 mm long...4

4a. Perigynium beak 0.5–0.8 mm long, deeply bidentate; heads usually nodding...***C. pelocarpa***
4b. Perigynium beak 0.3–0.5 mm long, entire to shallowly or obscurely bidentate; heads nodding or erect...5

5a. Heads nodding or pendant; perigynium margins and beak smooth; pistillate scales usually longer than the perigynia...***C. chalciolepis***
5b. Heads strictly erect; perigynium margins and beak granular-roughened or serrulate; pistillate scales usually shorter than to as long as the perigynia...***C. nova***

6a. Bract subtending the lowest spike with a well-developed sheath 0.7–2 cm long; perigynium with a poorly defined, tapering beak with serrulate margins...***C. fuliginosa***
6b. Bract subtending the lowest spike absent or lacking a sheath, or with a short sheath less than 0.5 cm long; perigynium various...7

7a. Lateral spikes linear to oblong-linear, not crowded, drooping on long, slender peduncles 1.5–4 cm long, the terminal spikes short-pedunculate...***C. bella***
7b. Lateral spikes not drooping on long slender peduncles, sometimes the whole inflorescence nodding, otherwise unlike the above...8

8a. Perigynia 1.5–3 mm long, rough-granular or ciliate-serrulate towards the tip; pistillate scales light to dark brown with a broad, white-hyaline margin; spikes linear to linear-oblong, 2–4 mm wide...***C. parryana***
8b. Perigynia 2–4 mm long, the margins smooth throughout; pistillate scales usually black to purplish-black or sometimes blackish-brown, with or without a white-hyaline margin; spikes various but usually not linear, or wider than 4 mm...9

9a. Lower spikes nodding or spreading (on thin, weak peduncles); peduncle of lowest spike usually as long or longer than the spike...10
9b. Lower spikes erect, not nodding or spreading; peduncle of lowest spike usually shorter than the spike...12

10a. Perigynia ovate; pistillate scales usually longer than the perigynia (giving a shaggy appearance to the spikes)...***C. chalciolepis***
10b. Perigynia obovate or broadly elliptic; pistillate scales shorter or sometimes longer than the perigynia...11

11a. Perigynia light green or pale yellow; terminal spike with ⅓–⅔ staminate flowers at the base...***C. heteroneura***
11b. Perigynia light to dark brown; terminal spike with fewer than ¼ staminate flowers at the base...***C. epapillosa***

12a. Pistillate scales black or dark purplish-brown to the margins, or sometimes just the upper margins very slightly white-hyaline; perigynia narrowly oblong, 3.5–5 mm long...***C. atrosquama***
12b. Pistillate scales with conspicuous white-hyaline margins; perigynia ovate to obovate or elliptic, 2–3.5 mm long...13

13a. Perigynia 2.5–3.5 mm long and 1.5–2.5 mm wide, reddish-black; pistillate scales equal or nearly so to the perigynia in length...***C. albonigra***
13b. Perigynia 2–3 mm long and 1–1.3 mm wide, usually green or becoming brown to reddish-black in age; pistillate scales shorter than to equaling the perigynia...14

14a. Perigynia strongly papillose in the upper half; perigynium beak spikes usually crowded towards the tip of the stem forming a dense terminal cluster...***C. norvegica***
14b. Perigynia smooth to sparsely papillose in the upper half; spikes not crowded at the tip of the stem, clearly separate...***C. stevenii***

Key 7

(Stigmas 2; perigynia with winged margins; spikes gynaecandrous)

1a. At least the lowest bract subtending the spikes much longer than the inflorescence, more or less leaf-like, 1–20 cm long...2
1b. All or most bracts subtending the spikes shorter than the inflorescence, or absent...3

2a. Perigynia 2.8–4 (4.8) mm long, the distance from beak tip to achene 1.9–2.5 mm...***C. athrostachya***
2b. Perigynia (4.5) 5.5–7.3 mm long, the distance from beak tip to achene 3–5 mm...***C. sychnocephala***

3a. Spikes densely aggregated into an ovoid to triangular-ovoid head (sometimes just the lowest one more distant), individually distinct or not...4
3b. Spikes all well-separated or loosely aggregated into an elongate, cylindric or oblong head, individually distinct and ovoid to ovoid-cylindric in outline...15

4a. Perigynia broadly ovate, 2.7–3.8 mm wide and (5.5) 6–7.2 mm long (1.8–2.3 times as long as wide); beak flat...***C. egglestonii***
4b. Perigynia lanceolate to lance-ovate or broadly ovate, 1–2.7 mm wide and 2.8–7 mm long; beak cylindric for at least 0.2 mm from the tip...5

5a. Pistillate scales with a broad white hyaline margin (to 0.3 mm wide), often similar in size to and nearly concealing the perigynia; perigynium beak abaxial suture with a white margin...6
5b. Pistillate scales lacking a white hyaline margin, with a very narrow white hyaline strip along the edge, or with a golden-hyaline margin, usually shorter and narrower than the perigynia; perigynium beak abaxial suture inconspicuous or with a white margin...11

6a. Perigynia boat-shaped with the wings adaxially incurved, lanceolate or occasionally narrowly ovate, 1–1.2 mm wide, with a wing 0.05–0.2 mm wide...***C. leporinella***
6b. Perigynia not boat-shaped with the wings adaxially incurved (at least at maturity), narrowly to broadly ovate or obovate, (1.1) 1.2–2.7 mm wide, with a wing 0.2–0.6 mm wide...7

7a. Proximal inflorescence internode (the lowest internode present between spikes) 4–10 mm long...8
7b. Proximal inflorescence internode 0.5–3.7 (4.8) mm long...9

8a. Pistillate scales 3.7–5.1 mm long; perigynia lacking a glossy metallic sheen, the abaxial suture on the beak with a conspicuous white margin...***C. phaeocephala***
8b. Pistillate scales 2.2–3.7 (4.2) mm long; perigynia with a glossy metallic sheen, the abaxial suture on the beak usually inconspicuous...***C. pachystachya***

9a. Pistillate scale usually concealing the perigynia; perigynia flat except over the achene, dull; heads with several layers of papery staminate bracts at the base of the spike...***C. arapahoensis***
9b. Pistillate scale usually shorter and narrower than the perigynia; perigynia plano-convex to biconvex or rarely somewhat flat around the achene, with a glossy metallic sheen; heads lacking several layers of papery staminate bracts...10

10a. Perigynia with dark wings conspicuously contrasting with the lighter body, the abaxial suture on the beak with a conspicuous white margin; plants (0.9) 1.5–6 dm tall...***C. macloviana***
10b. Perigynia with wings green or similar in color to the body, the abaxial suture on the beak usually inconspicuous; plants 1.5–10 dm tall...***C. pachystachya***

11a. Perigynia plano-convex (the achene appearing to fill most of the body)...12
11b. Perigynia flat except over the achene...14

12a. Perigynia 5.3–7.1 mm long; distance from beak tip to achene 2.9–4 mm...***C. ebenea***
12b. Perigynia 2.8–5.1 mm long; distance from beak tip to achene 1.5–3.1 mm...13

13a. Perigynia narrowly lanceolate to lanceolate, 1–1.3 (1.5) mm wide...***C. stenoptila***
13b. Perigynia ovate to elliptic-ovate, 1.1–2.3 mm wide...***C. pachystachya***

14a. Perigynia (2.8) 3.4–4.5 (5.2) mm long; distance from beak tip to achene 1.5–2.5 (2.8) mm; plants 2–10 dm tall...***C. microptera***
14b. Perigynia 4.2–6.5 mm long; distance from beak tip to achene (2.3) 2.6–3.8 mm; plants 1–3 (4) dm tall...***C. haydeniana***

15a. Perigynium beak flat, strongly margined and serrulate to the tip...16
15b. Perigynium beak round (cylindric) for at least 0.4 mm from the tip, entire for at least 0.3 mm...25

16a. Perigynia (5.5) 6–8 mm long; distance from the beak tip to the achene over 3 mm...17
16b. Perigynia less than 6 mm long; distance from the beak tip to the achene less than 3 mm...19

17a. Perigynia plano-convex or concavo-convex, 6–8 mm long, with numerous (ca. 10–20) conspicuous nerves abaxially; the terminal spike often strongly clavate (club-shaped)...***C. petasata***
17b. Perigynia usually flat except over the achene, to 7.2 mm long, veinless or with 5–6 veins abaxially; the terminal spike not club-shaped...18

18a. Perigynia (5.5) 6–7.2 mm long, broadly ovate, 1.8–2.3 times as long as wide, usually coppery-tinged on the margins; pistillate scales 4.7–6.7 mm long; achenes 1.9–2.3 mm long...***C. egglestonii***
18b. Perigynia 4.2–6.8 mm long, lanceolate, at least 3 times as long as wide, not coppery-tinged on the margins; pistillate scales 3.4–4 mm long; achenes 1.3–1.7 mm long...***C. scoparia* var. *scoparia***

19a. Perigynia (3) 3.2–5.5 (5.7) mm long and 2–3.5 mm wide, 1–1.5 times as long as wide (broadly elliptic to ovate or nearly orbicular); individual spikes globose to elliptic…20
19b. Perigynia at least twice as long as wide, or narrower and only 1–2 mm wide (lanceolate to ovate or elliptic); spikes various…21

20a. Perigynia veinless or obscurely 1–5-veined abaxially; spikes mostly 5–7 on larger stems; achenes 1.3–2 mm wide…***C. brevior***
20b. Perigynia conspicuously at least 5-veined abaxially; spikes mostly 2–4; achenes 1–1.3 mm wide…***C. molesta***

21a. Perigynia 4.2–6.8 × 1.2–2 mm, lanceolate and at least 3 times as long as wide, flat except over the achene, the margin 0.2–0.6 mm wide…***C. scoparia* var. *scoparia***
21b. Perigynia 2.5–4.8 mm long, usually less than 3 times as long as wide, usually plano-convex, or flat except over the achene in *C. crawfordii* and then the margin 0.1–0.2 mm wide and perigynia 3.4–4 (4.7) × 0.9–1.3 mm…22

22a. Perigynia usually flat except over the achene or occasionally plano-convex, 3.4–4 (4.7) × 0.9–1.3 mm, 0.15–0.35 mm thick…***C. crawfordii***
22b. Perigynia plano-convex or concavo-convex, 2.5–4.5 × (1) 1.2–2.2 mm, 0.35–0.6 mm thick…23

23a. Pistillate scales longer and as wide as and concealing the perigynia, 4.4–6 mm long; perigynia 3.8–4.8 mm long…***C. xerantica***
23b. Pistillate scales shorter and/or narrower than the perigynia, 2.3–3.5 mm long…24

24a. Proximal inflorescence internode (the lowest one) 1–4 mm, the second internode 1.4–3.5 mm…***C. bebbii***
24b. Proximal inflorescence internode (4) 7–20 mm long, the second internode (3) 6–10 mm long…***C. tenera* var. *tenera***

25a. Perigynia 6–8 mm long, the distance from the beak tip to the achene (2.8) 3.2–4.6 mm; pistillate scales 5.8–7.6 mm long; the terminal spike often strongly clavate (club-shaped and widest at the tip)…***C. petasata***
25b. Perigynia 2.8–6 mm long, the distance from the beak tip to the achene 1.5–3 mm; pistillate scales 2.2–5.8 mm long; the terminal spike not clavate…26

26a. Perigynia boat-shaped with the wings adaxially incurved, lanceolate or occasionally narrowly ovate, 1–1.2 mm wide, with a wing 0.05–0.2 mm wide…***C. leporinella***
26b. Perigynia not boat-shaped with the wings adaxially incurved (at least at maturity), usually narrowly to broadly ovate, occasionally lanceolate, (1.1) 1.2–2.7 mm wide, with a wing 0.2–0.6 mm wide…27

27a. Perigynia conspicuously 3–8-veined adaxially (ventrally), with at least 3 veins longer than the achene; spikes tapered to attenuate at the base…***C. tahoensis***
27b. Perigynia lacking veins adaxially or with 1–4 (7) conspicuous veins, but the veins shorter than the achene; spikes truncate, rounded, acute, or sometimes attenuate at the base…28

28a. Pistillate scales 2.2–3.7 (4.2) mm long; perigynia rounded at the base, with a glossy metallic sheen, the abaxial suture on the beak usually inconspicuous…***C. pachystachya***
28b. Pistillate scales (3.7) 3.8–6 mm long; perigynia usually tapering toward a blunt base, lacking a glossy metallic sheen, the abaxial suture on the beak conspicuous or not…29

29a. Inflorescence usually dense, not constricted at regular intervals and giving a beaded appearance, the second inflorescence internode 2–5 mm…***C. phaeocephala***
29b. Inflorescence usually open, usually flexuous or nodding, often cylindrical and constricted at regular intervals (giving a beaded appearance), the second inflorescence internode (2.5) 4–10 mm…***C. praticola***

Key 8

(*Stigmas 2; perigynia lacking winged margins; spikes androgynous*)

1a. Pistillate scales black or purplish-black; primary bract subtending the inflorescence with a purplish-black band at the base or occasionally purplish-black throughout, 1–7 cm long, shorter than the terminal spike…***C. scopulorum***
1b. Pistillate scales green to brown or brownish-black; primary bract lacking a purplish-black band at the base…2

2a. Spikes 2–15 in a dense terminal globose to ovoid or short-oblong cluster, sometimes so closely aggregated that they appear indistinct…3
2b. Spikes well-separated or in an elongated cluster…10

3a. Perigynia serrulate along the margins to the middle or below, brown with green margins...***C. hoodii***
3b. Perigynia smooth along the margins or if serrulate then only along the beak, variously colored...4

4a. Leaves 4–5 per stem, borne on the lower part of the stem but not clustered at the base, usually with a ventrally cross-corrugated sheath near the mouth of the leaf sheath (on the side of the stem opposite the leaf blade); perigynium beak tapered, about half the length of the body...***C. neurophora***
4b. Leaves clustered near the base of the stem, lacking a ventrally cross-corrugated sheath; perigynium beak various...5

5a. Stems arising singly or few together from slender, long-creeping, well-developed rhizomes...6
5b. Stems loosely to densely caespitose from short rhizomes...8

6a. Pistillate scales larger than and concealing the perigynia; perigynia 1.7–2.4 (2.8) mm long, yellowish-brown, with green margins when young...***C. simulata***
6b. Pistillate scales shorter than and not concealing the perigynia; perigynia 2.4–4 mm long, variously colored, sometimes dark reddish-brown...7

7a. Perigynia 1.5–2 mm wide, broadly ovate to nearly orbicular, usually dark reddish-brown at maturity; spikes in an ovoid or oblong-cylindric head (usually longer than wide)...***C. duriuscula***
7b. Perigynia 1–1.5 mm wide, narrowly elliptic, usually pale yellowish-brown with darker margins; spikes in an ovoid to nearly orbicular head (usually about as wide as long)...***C. incurviformis***

8a. Perigynia broadly elliptic to ovate-elliptic, 1.9–2.6 mm wide, with an ill-defined beak and a stipe at the base to 0.2 mm long...***C. perglobosa***
8b. Perigynia lanceolate to narrowly ovate or obovate, 1–1.8 mm wide (if over 1.5 mm wide then with a well-defined stipe at the base 0.2–0.6 mm long), with a tapering or well-defined beak...9

9a. Perigynia with several strong and prominent nerves nearly the entire length, conspicuously swollen at the base with spongy pith...***C. jonesii***
9b. Perigynia inconspicuously nerved, or the nerves only prominent towards the base of the perigynia, not conspicuously swollen at the base...***C. vernacula***

10a. Perigynia 1–3 per spike, 2–3 mm long; inflorescence narrowly linear (mostly 3–4 mm wide) with small individual spikes spaced well apart from each other...***C. disperma***
10b. Perigynia usually at least 3 or more in most spikes; inflorescence usually wider than 4 mm or the spikes not well-separated...11

11a. Pistillate scales with a conspicuous awn 1–5 mm long; spikes usually with short bracts subtending each spike and with a long lowest bract to 5 cm long; leaf sheaths usually conspicuously cross-rugose on the ventral side (the side opposite the blade)...***C. vulpinoidea***
11b. Pistillate scales awnless or with a short awn less than 1 mm long; spikes unlike the above; leaf sheaths sometimes cross-rugose on the ventral side...12

12a. Perigynia conspicuously widely spreading to reflexed at maturity, 3.6–6 mm long, giving the spikes a prickly appearance...13
12b. Perigynia ascending or spreading-ascending, of various lengths, the spikes lacking a prickly appearance...14

13a. Perigynium beak about ½ the total length of the perigynium...***C. stipata* var. *stipata***
13b. Perigynium beak distinctly less than ½ the total length of the perigynium...***C. gravida***

14a. Perigynium dorsally bulged such that the marginal vein is actually on the ventral face (the marginal nerve displaced onto the ventral side), otherwise nerveless, 2.5–4 mm long, glossy at maturity...***C. vallicola***
14b. Perigynium with both marginal nerve on the sides of the perigynia, not with one displaced, otherwise various...15

15a. Plants arising singly or a few together from long-creeping rhizomes; perigynia often concealed or nearly so by the pistillate scales...16
15b. Plants densely clustered from short rhizomes; perigynia usually exposed and longer and/or wider than the pistillate scales...18

16a. Basal leaf sheaths dark purplish-brown to nearly black; perigynium not thin-margined above...***C. praegracilis***
16b. Basal leaf sheaths brown; perigynium thin-margined above...17

17a. Leaf sheaths with a veined, green inner band, the ligules 2.2–8 mm long; perigynia 2.5–4.5 mm long...***C. sartwellii***
17b. Leaf sheaths with a white-hyaline inner band, the ligules 0.7–2.5 mm long; perigynia 3–6 mm long...***C. siccata***

18a. Ventral surface of the leaf sheaths conspicuously red dotted; perigynia (2) 2.3–2.5 (2.9) mm long, the beak usually with a conspicuous white hyaline dorsal suture flap ...***C. diandra***
18b. Ventral surface of the leaf sheaths not or rarely red dotted; perigynia 2.5–5.5 mm long, the beak lacking a white hyaline dorsal suture flap, sometimes with a brown dorsal suture flap or stripe ...19

19a. Perigynia nerveless on each face; widest leaves 1.5–2.5 (3) mm wide; plants with short, black rhizomes...***C. occidentalis***
19b. Perigynia strongly few-nerved on the dorsal surface; widest leaves (3) 4–8 mm wide; plants with short, brown rhizomes ...***C. gravida***

Key 9

(*Stigmas 2; perigynia lacking winged margins; spikes gynaecandrous*)

1a. Pistillate scales black or purplish-black; primary bract subtending the inflorescence with a purplish-black band at the base or occasionally purplish-black throughout, 1–7 cm long, shorter than the terminal spike...***C. scopulorum***
1b. Pistillate scales green to brown or brownish-black; primary bract lacking a purplish-black band at the base...2

2a. Perigynia 2.5–3.5 mm long, lacking a beak or nearly so; spikes 2–3 (4) per stem; pistillate scales with a narrow green center and wide, white-hyaline margin...***C. tenuiflora***
2b. Perigynia 1.5–5 mm long, with at least a short beak 0.5 mm long; spikes usually more than 3 per stem; pistillate scales various...3

3a. Spikes 3–6 in a dense terminal spherical to ovoid cluster, usually individually indistinct; pistillate scales dark brown to brownish-black...***C. illota***
3b. Spikes clustered into a head but individually distinct, or distinct and in an elongated cluster; pistillate scales various...4

4a. Perigynia conspicuously widely spreading to reflexed, giving the spikes a prickly appearance...5
4b. Perigynia ascending or spreading-ascending, the spikes lacking a prickly appearance...7

5a. Perigynia 2.9–3.6 (4) mm long, the beak about half the total length of the perigynium or more...***C. echinata* ssp. *echinata***
5b. Perigynia 1.9–3.7 mm long, the beak less than half the total length of the perigynium...6

6a. Perigynia margins sparsely serrulate with a few teeth or more often entire above; usually 3–8-veined ventrally...***C. laeviculmis***
6b. Perigynia margins conspicuously serrulate with many teeth above; usually veinless ventrally or rarely with up to 6 faint veins...***C. interior***

7a. Perigynia 3.3–5 mm long, with beaks 1–2.1 mm long (about ⅓ to ½ the length of the perigynia); lowest bract leaf-like and setaceous-prolonged, but shorter than the inflorescence...8
7b. Perigynia 1.5–3.7 mm long, with beaks 0.3–1.1 (1.3) mm long (to ⅓ the length of the perigynia); lowest bract various...10

8a. Perigynia margins sparsely serrulate with a few teeth or more often entire above, the beak 0.4–1.1 (1.3) mm long; leaf blades 1–2.5 mm wide...***C. laeviculmis***
8b. Perigynia margins conspicuously serrulate above, the beak 1–2.1 mm long; widest leaf blades 2.4–6 mm...9

9a. Perigynia beak 1.1–2.7 mm long; ligule of distal leaf 0.9–2.2 mm long, as long as wide...***C. deweyana***
9b. Perigynia beak 1–1.5 (1.7) mm long; ligule of distal leaf (2.5) 3.4–7 mm long, 1.5–3 times as long as wide...***C. leptopoda***

10a. Spikes clustered into an oblong or oval head but distinct, reddish-brown; pistillate scales largely concealing the perigynia...***C. lachenalii***
10b. Spikes well-separated, variously colored; pistillate scales usually shorter than or subequal to the perigynia but not concealing them...11

11a. Perigynium beak conspicuously serrulate, 0.5–0.7 mm long, with a prominent, white-hyaline dorsal suture flap; spikes with 5–10 perigynia…***C. brunnescens***
11b. Perigynium beak smooth and entire or sparsely serrulate, 0.2–1.1 (1.3) mm long, the dorsal suture flap obsolete or various but usually unlike the above; spikes with (5) 8–20 (30) perigynia…12

12a. Perigynia 2.3–3.7 mm long with a beak 0.4–1.1 (1.3) mm long (1.8–3 times as long as wide), sometimes with a spongy-thickened base but not thickened to ½ the length of the perigynium body, usually not papillate…***C. laeviculmis***
12b. Perigynia 1.5–3 mm long with a very short beak 0.2–0.5 mm long, spongy-thickened at least in the lower ½ of the body, papillate at least above …13

13a. Pistillate scales white-hyaline with a green midvein, or sometimes pale brown in age; spikes gray-green or light green (giving an overall silvery appearance to the inflorescence)…***C. canescens* ssp. *canescens***
13b. Pistillate scales brown with white-hyaline margins and a light center; spikes green when young but golden brown at maturity (giving an overall golden appearance to the inflorescence at maturity)…***C. praeceptorum***

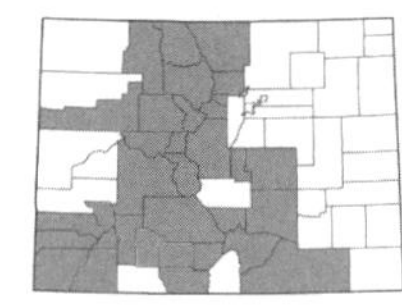

***Carex albonigra* Mack.**, BLACK-AND-WHITE SCALED SEDGE. Plants 1–3 dm, with short rhizomes; *leaves* flat, 2–7 mm wide; *bracts* with the lowest leaf-like, equal to or shorter than the inflorescence; *spikes* 2–4, closely aggregated into a headlike inflorescence but individually distinguishable, the terminal gynaecandrous and the lateral pistillate; *pistillate scales* ovate, equal to or shorter than the perigynia, reddish-black or blackish-purple with the upper margins white-hyaline; *perigynia* elliptic-obovate, 2.5–3.5 × 1.5–2.5 mm, reddish-black, beaks 0.1–0.5 mm long; *stigmas* 3; *achenes* 1.3–2 × 0.7–1.3 mm. Common in meadows and the alpine, 10,000–13,800 ft. July–Aug. E/W. (Plate 31)

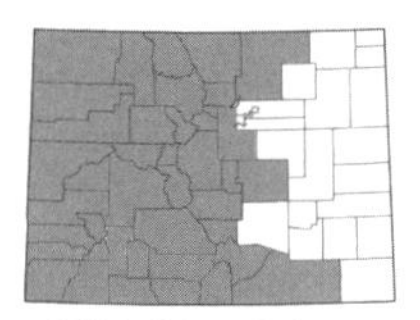

***Carex aquatilis* Wahlenb.**, WATER SEDGE. Plants to 15 dm, rhizomatous, phyllopodic; *leaves* flat, green to glaucous-green, 2–7 mm wide; *bracts* sheathless, the lowest leaf-like, usually equal to or longer than the inflorescence; *spikes* 3–7, the terminal 1–3 staminate, the lateral pistillate; *pistillate scales* ovate, with an obtuse to acuminate tip, reddish-brown to purplish-black with a paler midrib; *perigynia* ovate, 2–3.6 × 1.2–2.2 mm, greenish-yellow, only marginal nerves present, the beak 0.1–0.3 mm long, entire; *stigmas* 2; *achenes* obovate, 1–2 × 0.7–1.6 mm. Common along streams and lake margins, in marshes and bogs, and in the moist alpine, 7000–13,000 ft. June–Sept. E/W. (Plate 31)

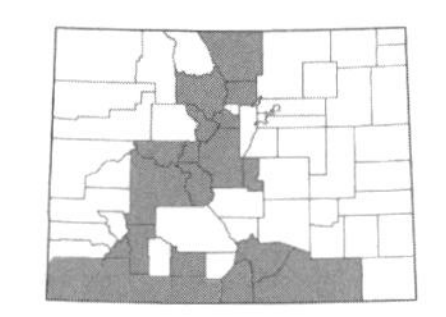

***Carex arapahoensis* Clokey**, ARAPAHO SEDGE. Plants 1.5–4 dm, tufted; *leaves* flat, 2.3–4 mm wide; *bracts* inconspicuous, the lowest excurrent as a serrulate awn, shorter than the inflorescence; *spikes* 3–6, gynaecandrous, sessile, aggregated into a suborbicular head; *pistillate scales* ovate, 4.3–5.6 mm long, nearly concealing the perigynia, brown with white-hyaline margins; *perigynia* ovate, 4–5.5 × 2–2.7 mm, winged, serrulate above the middle, flat except over the achene, with numerous fine nerves, the beak ca. 1 mm long, terete; *stigmas* 2; *achenes* 1.6–2.1 × 1–1.5 mm. Common in meadows and on alpine slopes, 10,300–14,000 ft. July–Aug. E/W. (Plate 31)

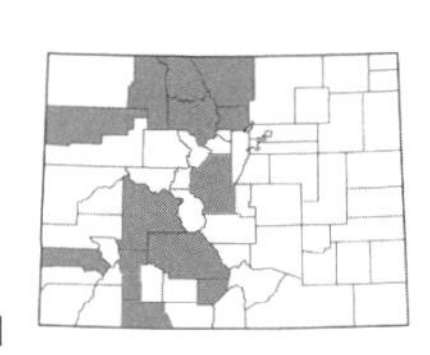

***Carex atherodes* Spreng.**, AWNED SEDGE. Plants 3–15 dm, rhizomatous, aphyllopodic; *leaves* flat, 3–12 mm wide; *bracts* sometimes leaf-like, the lower longer than the inflorescence; *spikes* several, sessile or nearly so, the upper staminate and the lower pistillate; *pistillate scales* lanceolate, with an aristate tip; *perigynia* lanceolate, 7–10 × 1.7–2.5 mm, pale green to tan, with conspicuous nerves, the beak 1.2–3 mm long with slender, divergent teeth; *stigmas* 3; *achenes* obovoid, 2–3.2 × 1.2–1.5 mm, continuous with the persistent, bony style. Found along the margins of ponds and streams, and in moist meadows, 5500–8500 ft. July–Sept. E/W. (Plate 31)

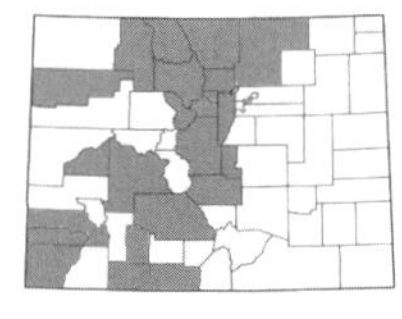

***Carex athrostachya* Olney**, SLENDERBEAK SEDGE. Plants 0.5–10 dm, caespitose; *leaves* flat, 1.5–4 mm wide; *bracts* with the lowest equal to or longer than the inflorescence, setaceous-prolonged, the upper reduced; *spikes* 4–20, gynaecandrous, sessile, densely crowded in an ovoid to globose head; *pistillate scales* oblong-ovate, acute, slightly shorter than the perigynia, brown with a pale center and white-hyaline margins; *perigynia* lanceolate, 2.8–4 (4.8) × 0.9–1.8 mm, wing-margined, serrulate distally, faintly nerved dorsally, beak 0.5–0.8 mm long; *stigmas* 2; *achenes* 0.8–1.2 × 0.7–1.1 mm. Found in moist meadows, forests, and along streams, 7500–10,600 ft. June–Sept. E/W. (Plate 31)

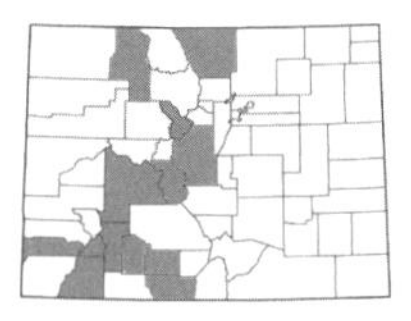

***Carex atrosquama* Mack.**, DARK-SCALED SEDGE. Plants 1.5–5 dm, with short rhizomes; *leaves* flat, 1.5–3.5 mm wide; *bracts* mostly inconspicuous; *spikes* 3–4, the terminal gynaecandrous and the lateral pistillate; *pistillate scales* ovate-oblong, shorter than the perigynia, reddish-brown to black; *perigynia* oblong, 3.5–5 × 1.5–1.8 mm, reddish-brown with greenish margins, nerveless, the beak 0.5 mm long, reddish-black, terete; *stigmas* 3; *achenes* 1.5–2 × ca. 1 mm. Uncommon in meadows, forests, and along streams, 10,000–13,000 ft. July–Aug. E/W.

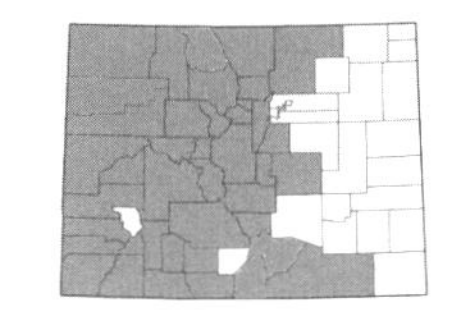

***Carex aurea* Nutt.**, GOLDEN SEDGE. Plants to 5.5 dm, rhizomatous, phyllopodic or somewhat aphyllopodic; *leaves* flat, 1–5 mm wide; *bracts* leaf-like, usually longer than the inflorescence; *spikes* 4–6, the upper staminate and the lower pistillate; *pistillate scales* ovate, shorter than the perigynia, reddish-brown with a pale center; *perigynia* orbicular, 1.7–3 × 1.5–2 mm, golden-yellow at maturity, obscurely to conspicuously nerved, strongly whitish-papillate, beaks absent; *stigmas* 2; *achenes* 1.3–2 × 1–1.6 mm, brown. Common along streams and in moist meadows, 5000–12,000 ft. June–Aug. E/W. (Plate 31)

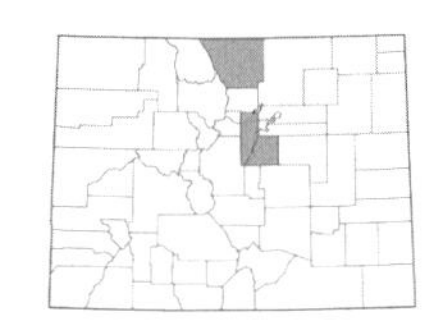

***Carex backii* Boott**, BACK'S SEDGE. Plants to 2.5 dm, tufted with very short rhizomes, phyllopodic; *leaves* flat, 2–6 mm wide; *bracts* absent; *spikes* 1–3, androgynous, few-flowered; *pistillate scales* leaf-like, the lower scales surpassing the inflorescence and concealing the perigynia; *perigynia* oblong-ovoid, 4.8–6.6 × 1.9–3.2 mm, greenish, smooth, with obscure nerves, the beaks 1.9–3.2 mm long; *stigmas* 3; *achenes* ca. 3 mm long, greenish to brown-black. Rare on rocky slopes, and in dry canyons and woodlands, 6000–7000 ft. May–July. E.

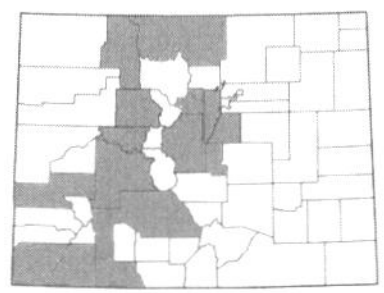

***Carex bebbii* (L.H. Bailey) Olney ex Fernald**, BEBB'S SEDGE. Plants 2–9 dm, caespitose, aphyllopodic; *leaves* flat, 2–4.5 mm wide; *bracts* with the lowest setaceous-prolonged and shorter than the inflorescence, upper reduced; *spikes* 3–12, gynaecandrous, sessile, aggregated into a linear-oblong head; *pistillate scales* oblong-lanceolate, 2.5–3.5 mm long, narrower than the perigynia, mostly hyaline-scarious throughout with a brown center and brown margins; *perigynia* ovate, 2.5–3.8 × (1) 1.2–2 mm, wing-margined, serrulate, plano-convex, distinctly nerved dorsally, the beak 0.75–1 mm long, tapering, flat; *stigmas* 2; *achenes* 1–1.5 × 0.6–0.8 mm. Found in moist meadows and along streams and ponds, 5600–10,500 ft. July–Sept. E/W. (Plate 31)

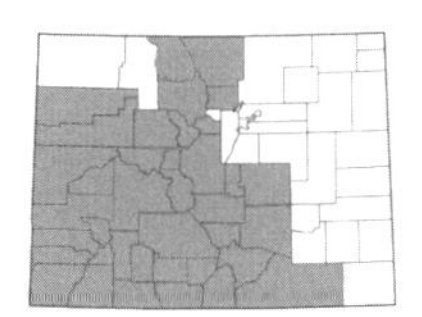

***Carex bella* L.H. Bailey**, BEAUTIFUL SEDGE. Plants to 9 dm tall, rhizomatous, phyllopodic; *leaves* flat, 3–6 mm wide; *bracts* with the lowest leaf-like, equal to or longer than the inflorescence; *spikes* 3–4, drooping on slender peduncles, gynaecandrous; *pistillate scales* ovate, reddish-brown with a pale midrib and white-hyaline margins; *perigynia* oval to obovoid, 2.5–4 × 1.7–2 mm, yellowish-green, with 2 marginal nerves, beaks 0.2–0.5 mm long, purplish to brown; *stigmas* 3; *achenes* ovoid, 1.8–2.5 × 1.7–1.8 mm. Common in moist meadows, open spruce forests, and alpine, 8300–12,500 ft. June–Aug. E/W. (Plate 31)

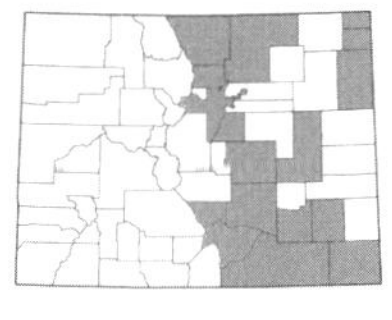

***Carex brevior* (Dewey) Mack.**, SHORT-BEAKED SEDGE. Plants 2–10 dm, caespitose, aphyllopodic; *leaves* flat to canaliculated, 1.5–4 mm wide; *bracts* with the lowest setaceous-prolonged and shorter than the inflorescence; *spikes* 3–10, gynaecandrous, sessile, distinguishable; *pistillate scales* ovate-lanceolate, 2.6–4.3 mm long, acuminate, shorter than the perigynia, brownish with narrow hyaline margins; *perigynia* ovate to suborbicular, 3.2–5.5 × 2.3–3.5 mm, wing-margined, serrulate, plano-convex, greenish-white, prominently nerved dorsally, the beak 0.8–1.5 mm long; *stigmas* 2; *achenes* 1.6–2.2 × 1.3–2 mm. Common in moist meadows, grasslands, on floodplains, and along streams, 3500–9000 ft. May–July. E. (Plate 31)

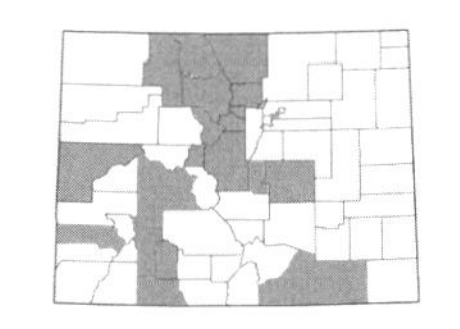

***Carex brunnescens* (Pers.) Poir.**, BROWNISH SEDGE. Plants 2–6 dm, caespitose with short rhizomes, slightly aphyllopodic; *leaves* flat, 1–2.5 mm wide; *bracts* short and inconspicuous or the lowest setaceous and shorter than to longer than the inflorescence; *spikes* 4–9, gynaecandrous, sessile, distinct; *pistillate scales* ovate, acute, shorter than the perigynia, brownish with white-hyaline margins; *perigynia* ovate, 1.7–2.7 × 1–1.5 mm, with raised margins, with several nerves dorsally, the beak 0.5–0.7 mm long, serrulate, with a prominent dorsal suture; *stigmas* 2; *achenes* 1.2–1.5 × 1 mm. Uncommon in fens, along streams, and in moist meadows, 8000–10,500 ft. July–Aug. E/W. (Plate 31)

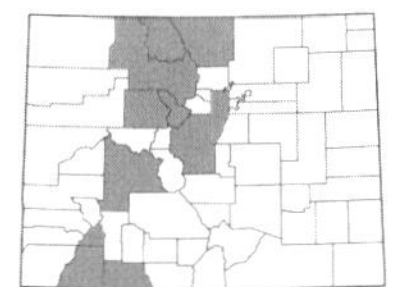

***Carex buxbaumii* Wahlenb.**, BUXBAUM'S SEDGE. Plants 2.5–10 dm, rhizomatous, aphyllopodic; *leaves* flat, 1.5–4 mm wide; *bracts* dark-auricled, the lowest shorter to longer than the inflorescence; *spikes* 2–5, erect, the terminal gynaecandrous and the lateral pistillate; *pistillate scales* lanceolate, awn-tipped, reddish-black with a pale midrib, narrower than the perigynia; *perigynia* elliptic to obovoid, 2.5–4.3 × 1.4–2.1 mm, green, inconspicuously nerved, densely papillate, the beak 0.2 mm long, reddish-black; *stigmas* 3; *achenes* 1.5–2.2 × 1–1.5 mm, blackish-brown. Found in moist meadows, along pond margins, and in peat bogs, 8500–10,500 ft. July–Aug. E/W. (Plate 31)

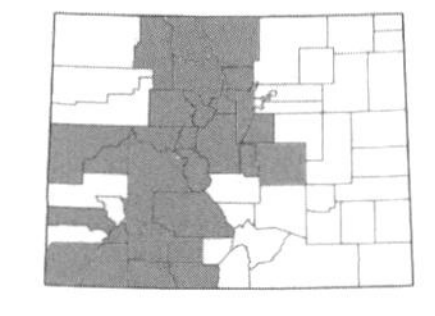

***Carex canescens* L. ssp. *canescens*,** GRAY SEDGE. Plants 1–8 dm, shortly rhizomatous, slightly aphyllopodic; *leaves* flat, 1.5–4 mm wide; *bracts* inconspicuous or the lowest ones setaceous-tipped and sometimes longer the inflorescence; *spikes* 4–8, gynaecandrous, sessile, distinguishable but aggregated; *pistillate scales* ovate, shorter than the perigynia, white-hyaline with a green center; *perigynia* ovoid-oblong, 1.8–3 × 0.9–1.8 mm, with raised margins, minutely serrulate distally, golden-yellow to brownish, with dark nerves, the beaks 0.2–0.5 mm long; *stigmas* 2; *achenes* 1.2–1.5 × 0.8–1 mm. Common in moist meadows, marshes, and along lake margins, 8000–12,000 ft. June–Aug. E/W. (Plate 31)

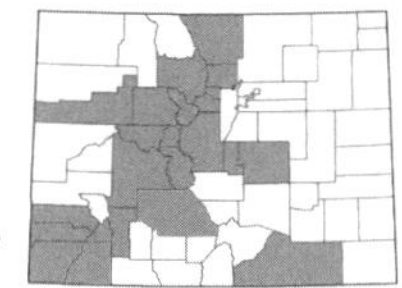

***Carex capillaris* L.,** HAIR SEDGE. Plants to 6 dm, lacking rhizomes, phyllopodic; *leaves* flat, 0.5–4 mm wide; *bracts* with the lowest leaf-like and shorter to longer than the inflorescence; *spikes* 2–4, nodding or loosely spreading, the terminal staminate and the lateral pistillate; *pistillate scales* ovate, shorter than the perigynia, brown or greenish with a white-hyaline margin distally; *perigynia* ovate-lanceolate, 2–4 × 0.75–1.2 mm, brown to greenish-brown, obliquely 2-nerved, the beaks 1 mm long, entire; *stigmas* 3; *achenes* 1.2–1.5 × 0.7 mm, jointed to the style. Common on alpine slopes, near snowmelt, and along streams, 8500–13,500 ft. June–Aug. E/W. (Plate 32)

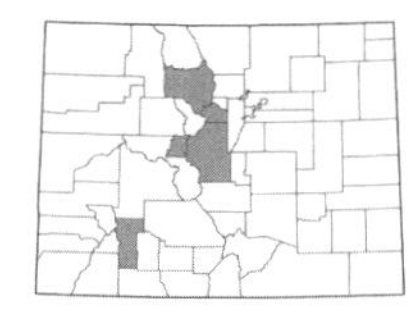

***Carex capitata* L.,** CAPITATE SEDGE. Plants 1–3.5 dm, rhizomatous, aphyllopodic; *leaves* involute, 1 mm wide; *bracts* absent; *spikes* solitary, androgynous; *pistillate scales* orbicular to ovate, shorter than the perigynia, dark brownish-red with white-hyaline margins; *perigynia* ovate to orbicular, 2–3.5 × 1.3–2 mm, yellowish with a reddish-brown tip, the nerves absent or a few dorsally, beaks 0.6–1 mm long; *stigmas* 2; *achenes* 1–2 × 0.5–1.2 mm, the rachilla well-developed. Uncommon in bogs and moist tundra, 11,000–12,500 ft. July–Aug. E/W. (Plate 32)

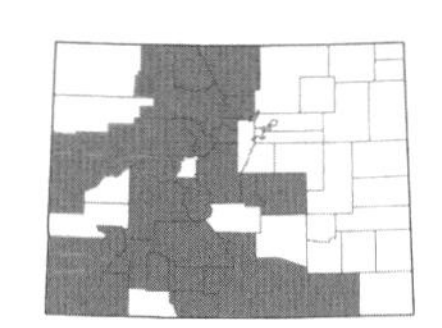

***Carex chalciolepis* Holm,** HOLM'S SEDGE. [*C. atrata* L. var. *chalciolepis* (Holm) Kuek.]. Plants 2–5 dm, caespitose, aphyllopodic; *leaves* flat, 2–5 mm wide; *bracts* with the lowest leaf-like, shorter to longer than the inflorescence; *spikes* 3–7, gynaecandrous, nodding; *pistillate scales* lanceolate, acute to acuminate, reddish-brown to blackish with a paler midrib; *perigynia* broadly ovate, 3–4 × 2.2–3.5 mm, brownish-purple with greenish margins, nerveless, the beak 0.3–0.5 mm long; *stigmas* 3; *achenes* 1.2–2.3 × 0.7–1 mm. Found along streams and pond margins, in moist meadows, and in the alpine, 9000–12,500 ft. July–Sept. E/W. (Plate 32)

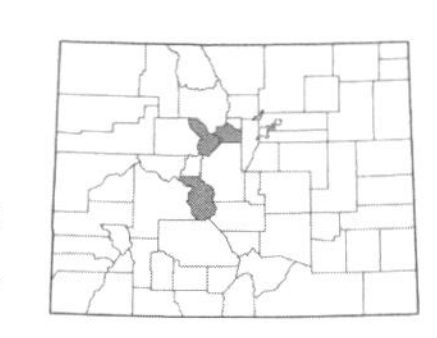

***Carex concinna* R. Br.,** LOW NORTHERN SEDGE. Plants 0.5–2 dm, shortly rhizomatous; *leaves* flat, 1–3 mm wide; *bracts* inconspicuous; *spikes* few-flowered, the terminal staminate and the lateral pistillate; *pistillate scales* obtuse, minutely ciliate, reddish-brown with white-hyaline margins; *perigynia* elliptic, 2.2–3.3 × 1.1–1.4 mm, hairy, several-nerved, the beak 0.2–0.4 mm long; *stigmas* 3; *achenes* elliptic, 1.7–2 × 1–1.3 mm. Rare in moist spruce-fir forests, 8800–10,500 ft. June–Aug. E/W. (Plate 32)

***Carex conoidea* Willd.,** OPEN-FIELD SEDGE. Plants 0.2–7.5 dm, caespitose; *leaves* flat, 2.5–4 (5.6) mm wide; *bracts* with the lowest shorter to longer than the inflorescence; *spikes* (2) 3–6, the terminal spike staminate, lateral spikes pistillate; *pistillate scales* reddish-brown, awned or rarely awnless, with white-hyaline margins; *perigynia* oblong-ovoid to orbicular, 2.5–4 × 1.2–1.8 mm, green, conspicuously nerved, the beak absent or to 0.2 mm long; *stigmas* 3; *achenes* 1.8–2.5 × 1–1.4 mm. Rare in moist meadows and along streams, 7500–8000 ft. June–Aug. E. (Plate 32)

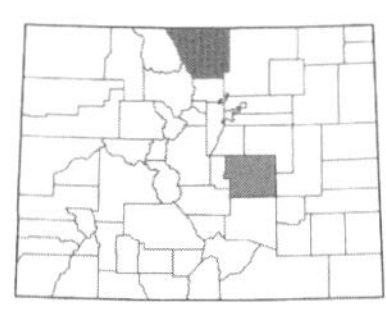

***Carex crawei* Dewey,** CRAWE'S SEDGE. Plants to 3(4) dm, rhizomatous, slightly phyllopodic; *leaves* flat, 1.5–3 mm wide; *bracts* with the lowest leaf-like and shorter than the inflorescence; *spikes* 3–5, the terminal staminate and the lateral pistillate or androgynous; *pistillate scales* ovate with the midrib often a short awn, equal to or shorter than the perigynia, reddish-brown with a pale center and hyaline margins; *perigynia* ovoid, 2–3.5 × 1.2–2 mm, reddish-brown to yellowish-green, with numerous nerves, the beak 0.4 mm long; *stigmas* 3; *achenes* 1.3–2 × 1.2–1.3 mm. Rare in moist meadows and along streams, 5500–7000 ft. June–Aug. E. (Plate 32)

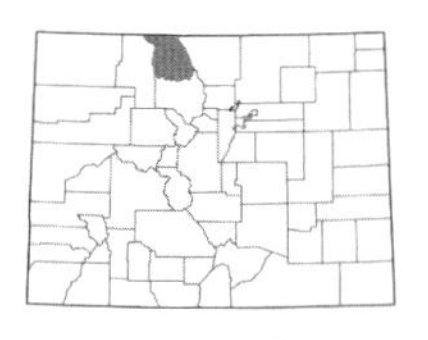

***Carex crawfordii* Fernald,** CRAWFORD'S SEDGE. Plants 1–8 dm, caespitose, aphyllopodic; *leaves* flat or channeled, 1–4 mm wide; *bracts* with the lowest setaceous-prolonged, shorter to equal to the inflorescence; *spikes* 3–15, gynaecandrous, sessile, closely aggregated into an oblong head; *pistillate scales* ovate-lanceolate, 3–3.8 mm long, shorter to equal but narrower than the perigynia, brown with a hyaline center; *perigynia* lanceolate-subulate, 3–4(4.7) × 0.9–1.3 mm, brown to pale green, narrowly winged, serrulate above the middle, flat except over the achene or plano-convex, faintly nerved, the beak flat and serrulate; *stigmas* 2; *achenes* 0.8–1.5 × 0.5–0.8 mm. Rare on dry, forest slopes and in moist places, 8700–9000 ft. June–Aug. E.

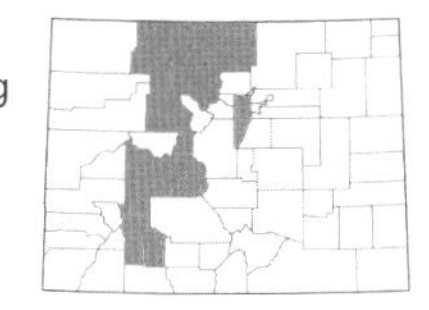

Carex deflexa **Hornem. var. *boottii*** **L.H. Bailey**, BOOTT'S SEDGE. [*C. brevipes* W. Boott]. Plants to 2 dm, with short rhizomes; *leaves* flat above, channeled below, 1.5–2.5 mm wide; *bracts* subtending pistillate spikes on bisexual culms longer than the inflorescence; *staminate spikes* terminal, 4–12 mm long; *pistillate spikes* below the staminate on bisexual culms, with 6–20 perigynia; *pistillate scales* reddish-brown, shorter than the perigynia; *perigynia* ovate, 2.5–3.5 × 1.2–1.5 mm, green, hairy, with a ciliate-serrulate beak 0.25–0.75 mm long; *stigmas* 3; *achenes* 1.5–2 mm long. Found on dry slopes and in alpine grasslands and open spruce forests, 7400–12,000 ft. May–July. E/W. (Plate 32)

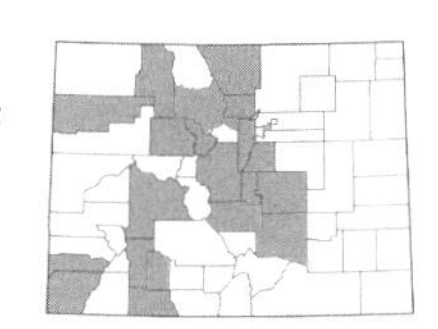

Carex deweyana **Schwein.**, DEWEY'S SEDGE. Plants 2–12 dm, with short rhizomes, slightly aphyllopodic; *leaves* flat, 2–5 mm wide; *bracts* inconspicuous or the lowest setaceous-prolonged; *spikes* 2–10, gynaecandrous, sessile, loosely aggregated; *pistillate scales* ovate, sometimes awn-tipped, shorter than the perigynia, brown with a greenish center; *perigynia* oblong-lanceolate, 2.8–4.8 × 1–1.5 mm, greenish to tan, nerved dorsally, the beak prominent, 1.1–2.7 mm long, serrulate; *stigmas* 2; *achenes* 1.5–2.2 × 1.2–1.8 mm. Uncommon along streams and in moist forests, 6400–9000 ft. June–Aug. E/W. (Plate 32)

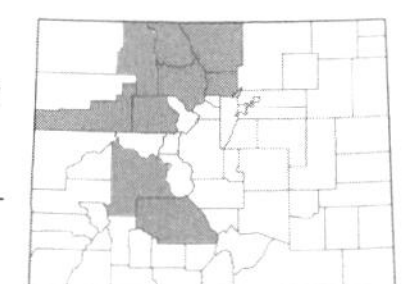

Carex diandra **Schrank**, LESSER PANICLED SEDGE. Plants 3–10 dm, with short rhizomes, aphyllopodic; *leaves* 1–3 mm wide; *bracts* shorter than the inflorescence; *spikes* numerous, androgynous, closely aggregated into a cylindric head; *pistillate scales* ovate, shorter than or equal to the perigynia, brown with a pale midrib and white-hyaline margins; *perigynia* lance-ovate, (2) 2.3–2.5 (2.9) × 0.9–1.4 mm, dark brown, with a prominent dorsal suture flap, the beak serrulate; *stigmas* 2; *achenes* obovate, 1–1.4 × 0.8–1 mm, jointed to the style. Uncommon along the edges of ponds and in fens, often forming floating mats in the center of lakes, 8000–10,000 ft. July–Aug. E/W. (Plate 32)

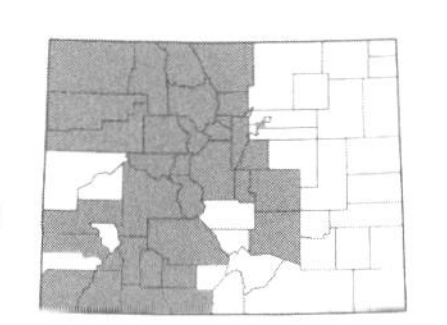

Carex disperma **Dewey**, SOFT-LEAVED SEDGE. Plants 1–6 dm, rhizomatous, slightly aphyllopodic; *leaves* flat, 0.7–2 mm wide; *bracts* inconspicuous or the lower ones filiform-foliaceous and to 2 cm long; *spikes* androgynous, sessile, 2–6-flowered with 1–3 perigynia; *pistillate scales* ovate, acute, shorter than to equal to the perigynia, mostly scarious; *perigynia* elliptic-ovate, plump, 2–3 × 1–1.6 mm, greenish to golden brown, with numerous nerves, the beaks very short or absent; *stigmas* 2; *achenes* 1–2 × 0.9–1.3 mm. Common along streams and in fens, moist meadows, and shady forests, 6500–11,000 ft. June–Aug. E/W. (Plate 32)

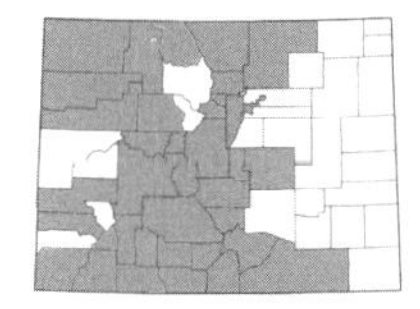

Carex douglasii **Boott**, DOUGLAS' SEDGE. Plants to 3 dm, rhizomatous, phyllopodic, dioecious; *leaves* involute or flattened at the base, 1–3 mm wide; *bracts* shorter than the inflorescence; *spikes* several, sessile, closely aggregated; *pistillate scales* lanceolate with an acuminate tip, concealing the perigynia, stramineous with a pale midrib; *perigynia* ovate-lanceolate, 3.5–7 × 1.3–2.5 mm, finely serrulate distally, obscurely nerved, the beaks tapered, serrulate, to 1.8 mm long; *stigmas* 2; *achenes* 1.4–2 × 1–1.5 mm, brown. Common along streams and in meadows, 4800–10,000 ft. May–Sept. E/W. (Plate 32)

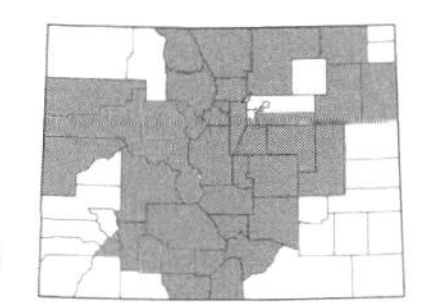

Carex duriuscula **C.A. Mey.**, NEEDLELEAF SEDGE. [*C. stenophylla* Wahlenb.]. Plants (0.6) 1–3.5 dm, rhizomatous; *leaves* involute, 0.5–2 mm wide; *spikes* 3–8, androgynous, aggregated into an ovoid head; *pistillate scales* ovate, acute to acuminate, dark reddish-brown with hyaline margins; *perigynia* ovate to nearly orbicular, 2.4–4 × 1.5–2 mm, dark reddish-brown to straw-colored, inconspicuously nerved, the beak 0.3–1 mm long; *stigmas* 2; *achenes* 1.5–1.7 × 1.5 mm. Common in grasslands and forests, on dry hillsides, 3800–10,000 ft. April–Aug. E/W. (Plate 32)

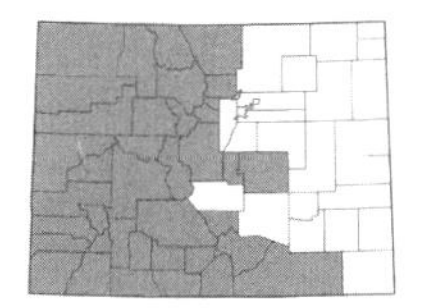

Carex ebenea **Rydb.**, EBONY SEDGE. Plants 1–5.5 dm, caespitose, slightly aphyllopodic; *leaves* flat, 2–4 mm wide; *bracts* shorter than the inflorescence; *spikes* 5–10, gynaecandrous, sessile, densely aggregated into an ovoid head; *pistillate scales* ovate-lanceolate, 3.6–4.7 mm, acute to obtuse, shorter than the perigynia, brownish-black to black; *perigynia* lanceolate, 5.3–7.1 × 1–1.5 mm, coppery brown with green margins, narrowly wing-margined or thin-edged, plano-convex, finely nerved, the beak 0.9–1.5 mm long, cylindric; *stigmas* 2; *achenes* 1.5–2 × 0.8–1.1 mm. Common along streams, in meadows, and in alpine, 9000–13,500 ft. July–Sept. E/W. (Plate 33)

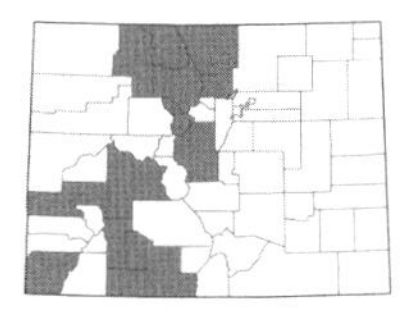

Carex echinata **Murray ssp. *echinata***, STAR SEDGE. [*C. angustior* Mack.]. Plants 1–6 dm, caespitose; *leaves* flat, 1–3 mm wide; *bracts* inconspicuous to setaceous-prolonged; *spikes* 3–6, remote or clustered, gynaecandrous or sometimes the lateral pistillate, perigynia widely spreading to deflexed; *pistillate scales* ovate, acute, shorter than the perigynia, scarious with a greenish midrib; *perigynia* lance-triangular, 2.9–3.6 (4) × 0.8–2.1 mm, green or stramineous, serrulate distally, few to many-nerved, the beak 1.1–1.6 mm long; *stigmas* 2; *achenes* 1–1.6 × 0.8–1.3 mm. Uncommon in moist meadows and forests, fens, and along pond margins, 8000–11,000 ft. July–Aug. E/W. (Plate 33)

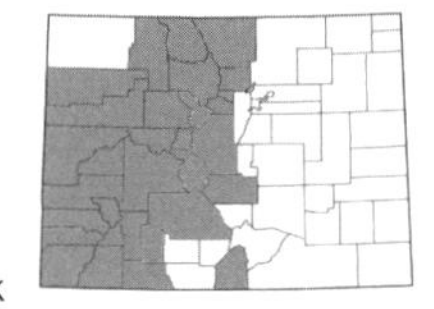

Carex egglestonii **Mack.**, EGGLESTON'S SEDGE. Plants 1.5–6 dm, with short rhizomes, slightly aphyllopodic; *leaves* flat, 2–6 mm wide; *bracts* with the lowest setaceous-prolonged, shorter than the inflorescence, upper reduced; *spikes* 3–6, gynaecandrous, sessile, distinguishable but aggregated into an ovoid head; *pistillate scales* ovate-lanceolate, 4.7–6.7 mm long, acute, shorter than the perigynia, brown with white-hyaline margins; *perigynia* ovate, (5.5) 6–7.2 × 2.7–3.8 mm, winged, serrulate, usually flat except over the achene, coppery-green, obscurely nerved, the beak 1.5 mm long, flat; *stigmas* 2; *achenes* 1.9–2.3 × 1–1.6 mm. Common in meadows, spruce-fir forests, along streams, and in the alpine, 8500–12,000 ft. July–Sept. E/W. (Plate 33)

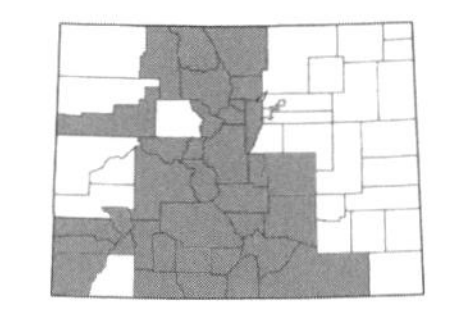

Carex elynoides **Holm**, BLACKROOT SEDGE. Plants to 1.5 dm, lacking rhizomes, phyllopodic; *leaves* involute, stiff, to 0.5 mm wide; *bracts* absent; *spikes* solitary, androgynous, linear, to 15 mm long; *pistillate scales* wider and equal to or longer than the perigynia, light to dark brown with a paler midvein and white-hyaline margins; *perigynia* elliptic to obpyramidal, 2.5–4.5 × 1.8–2.2 mm, obscurely 2-ribbed, with a slender, cylindrical beak 0.4–1 mm long; *stigmas* 3; *achenes* 1.5–2 × 1.2–2 mm, oblong-obovoid. Common in the alpine and meadows, 9000–14,000 ft. June–Aug. E/W. (Plate 33)

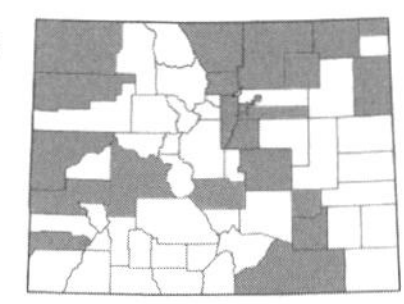

Carex emoryi **Dewey**, EMORY'S SEDGE. Plants 3–12 dm, rhizomatous; *leaves* flat, 2–6 mm wide; *bracts* with the lowest equal to the inflorescence; *spikes* 3–7, the terminal spike staminate and lateral spikes pistillate; *pistillate scales* reddish-brown, equal to the perigynia; *perigynia* elliptic to ovate, 1.7–3.2 × 1–2 mm, green, with 3–5 nerves on each face, the beak 0.2 mm long; *stigmas* 2; *achenes* ca. 1.5 × 1 mm. Found along ditches, streams, and pond margins and in floodplains and moist meadows, 3500–6500 ft. May–July. E/W. (Plate 33)

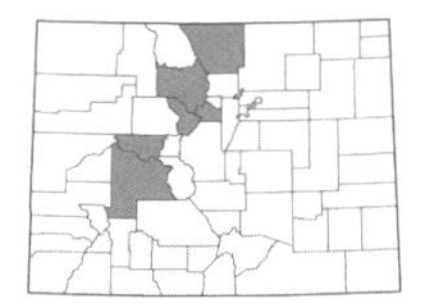

Carex engelmannii **L.H. Bailey**, ENGELMANN'S SEDGE. [*C. breweri* Boott var. *paddoensis* (Suksd.) Cronquist]. Plants 0.5–3 dm, rhizomatous, slightly phyllopodic; *leaves* involute or channeled, ca. 1 mm wide; *bracts* absent; *spikes* solitary, androgynous; *pistillate scales* ovate-oblong, acute, equal to or shorter than the perigynia, brown with hyaline margins; *perigynia* ovate to elliptic, 4–7.5 × 2–4.8 mm, dark brown with a pale base, obscurely nerved, the beak short; *stigmas* 3; *achenes* 1–2 × 0.8–1.2 mm, dark brown, with a well-developed rachilla. Uncommon in the alpine, near snowmelt areas, and along streams, 10,500–13,000 ft. July–Sept. E/W. (Plate 33)

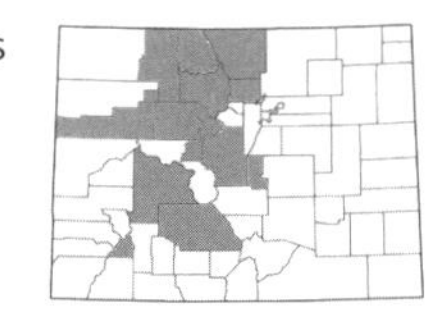

Carex epapillosa **Mack.**, DIFFERENT-NERVE SEDGE. [*C. atrata* L. var. *epapillosa* (Mack.) F.J. Herm.]. Plants 2.5–6 dm, caespitose; *leaves* flat, 2–6 mm wide; *bracts* with the lowest leaf-like, shorter to longer than the inflorescence; *spikes* 3–6, the terminal spike gynaecandrous, lateral spikes pistillate, usually nodding; *pistillate scales* lanceolate, dark brown with white hyaline margins; *perigynia* obovate to orbicular, 3.5–4 × 2–3 mm, brown, nerveless, the beak 0.3–0.5 mm long, bidentate; *stigmas* 3; *achenes* 1.2–2.3 × 0.7–1 mm. Uncommon along streams and pond margins, in meadows, and alpine, 9000–12,500 ft. July–Sept. E/W.

Carex exsiccata **L.H. Bailey**, WESTERN INFLATED SEDGE. Plants 3–10 dm, caespitose; *leaves* flat, 2.5–6 mm wide; *bracts* with the lowest longer than the inflorescence; *spikes* 4–7, the terminal 2–3 spikes staminate and the lateral spikes pistillate, cylindric; *pistillate scales* lanceolate to ovate, shorter than the perigynia, reddish-brown; *perigynia* lanceolate, 7.5–10 × 1.5–2.5 mm, strongly nerved, the beak 1.5–3 mm long, bidentate; *stigmas* 3; *achenes* trigonous. Rare in marshes, and along streams and pond margins, 7300–9000 ft. June–Aug. W. (Plate 33)

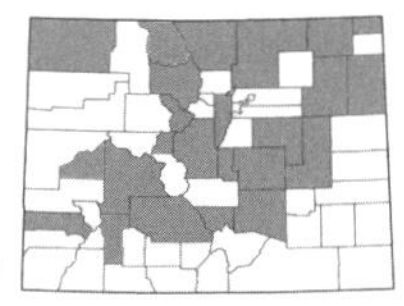

Carex filifolia **Nutt.**, THREADLEAF SEDGE. Plants to 3 dm, lacking rhizomes, phyllopodic; *leaves* involute, stiff, to 0.4 mm wide; *bracts* absent; *spikes* solitary, androgynous; *pistillate scales* broadly obovate, concealing the perigynia, brown with wide white-hyaline margins; *perigynia* 3–4.5 × 2–2.5 mm, minutely hairy at least above the middle, obscurely 2-ribbed, with a stout, beak 0.1–0.8 mm long; *stigmas* 3; *achenes* 2.2–3 mm, obovoid. Found in grasslands, pinyon-juniper woodlands, and on dry slopes, 3500–9500 ft. April–June. E/W. (Plate 33)

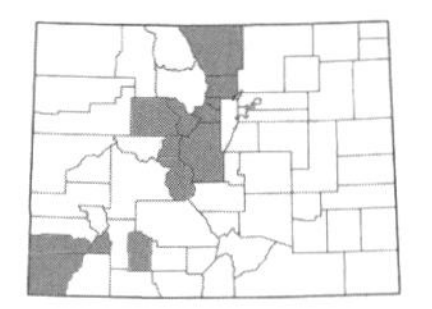

Carex fuliginosa **Schkuhr**, SHORT-LEAVED SEDGE. [*C. misandra* R. Br.]. Plants to 3 dm, lacking creeping rhizomes, phyllopodic; *leaves* flat, 1–4 mm wide; *bracts* with the lowest leaf-like, shorter than the inflorescence; *spikes* 3–4, nodding, the terminal gynaecandrous or pistillate, the lateral pistillate; *pistillate scales* ovate, shorter than the perigynia, brownish to black with a white-hyaline margin; *perigynia* lanceolate, 3.3–5.5 × 1–1.3 mm, ciliate-serrulate above the middle, with few obscure nerves, the beak tapered, poorly defined; *stigmas* 3; *achenes* 1.4–2 × 0.75–1 mm. Found in the alpine, 11,000–13,000 ft. July–Aug. E/W. (Plate 33)

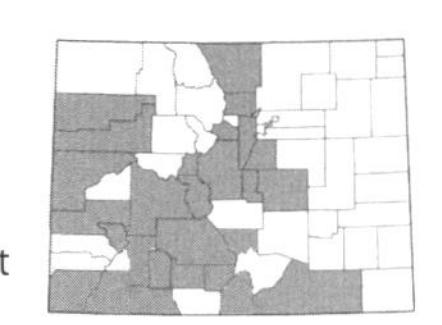

***Carex geophila* Mack.**, GROUND-LOVING SEDGE. Plants 0.2–1 dm, rhizomatous, phyllopodic; *leaves* flat, 1.5–2.5 mm wide, numerous; *bracts* subtending pistillate spikes on bisexual culms shorter than the inflorescence; *staminate spikes* terminal, 5–15 mm long; *pistillate spikes* 2–5, mostly 5–15-flowered; *pistillate scales* ovate with an acute tip, equal to or shorter than the perigynia; *perigynia* subglobose, with a stipitate base, 3.2–4 × 1.7–1.8 mm, minutely hairy, with 2 prominent lateral nerves, with a serrulate beak 0.4–1.2 mm long; *stigmas* 3; *achenes* 1.6–3 × 1.7–1.8 mm. Found in forests and on dry slopes, 5500–10,500 ft. May–Aug. E/W. (Plate 33)

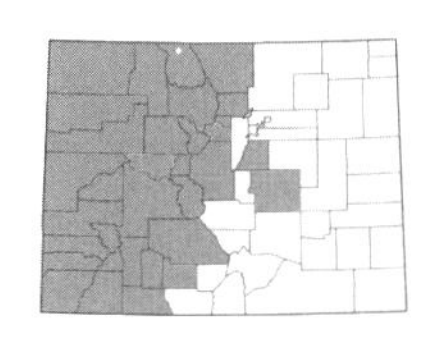

***Carex geyeri* W. Boott**, GEYER'S SEDGE. Plants 1–5 dm, rhizomatous, aphyllopodic; *leaves* flat, 1.5–4 mm wide; *bracts* absent; *spikes* solitary, androgynous, the staminate portion slender, mostly 1–3 cm long; *pistillate scales* narrowly oblong, hyaline-margined, the lower short-awned and longer than the perigynia; *perigynia* 1–3, at the base of the spike, 5–7 mm long, ellipsoid or obovoid, tapering at the base, greenish to brownish, 2-ribbed (otherwise nerveless), the beak obsolete or very short; *stigmas* 3; *achenes* 4–6.2 × 1–3 mm, filling the perigynium. Common in aspen, oak, pine, or spruce forests, 6500–12,700 ft. May–Aug. E/W. (Plate 33)

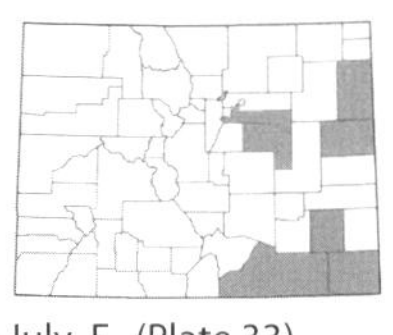

***Carex gravida* L.H. Bailey**, HEAVY SEDGE. Plants 3–6 dm, caespitose; *leaves* flat, (3) 4–8 mm wide, the sheath green-and-white-mottled; *bracts* inconspicuous; *spikes* densely aggregated into a cylindric head, andryogynous; *pistillate scales* ovate, brown with a green center, shorter than to slightly longer than the perigynium; *perigynia* ovate-lanceolate, 3–5 × 2–3 mm, nerveless or 3–7-nerved abaxially, the beak 0.6–1.6 mm long, serrulate; *stigmas* 2; *achenes* 1.8–2 × 1.6–2 mm. Found along pond margins and river bottoms, in moist canyons, and on sandstone rimrock, 3500–5500 ft. May–July E. (Plate 33)

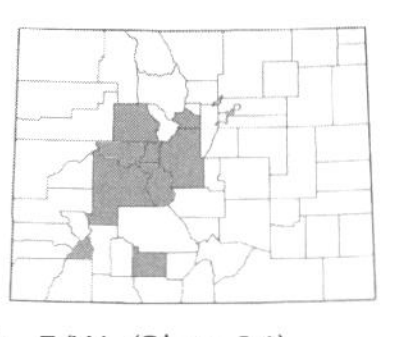

***Carex gynocrates* Wormsk.**, NORTHERN BOG SEDGE. [*C. dioica* L. ssp. *gynocrates* (Wormsk.) Hultén]. Plants to 3 dm, rhizomatous, phyllopodic; *leaves* involute, 3–15 cm × 0.4–1 mm; *bracts* absent; *spikes* solitary, androgynous or almost wholly staminate or pistillate; *pistillate scales* broadly ovate, with a brown center and lighter margins; *perigynia* elliptic to lanceolate, 2.5–4 × 1.5–2 mm, widely spreading, short-stipitate, glossy, plump, with numerous conspicuous nerves or the nerves obscure, with an abruptly contracted beak 0.5 mm long; *stigmas* 2; *achenes* ovate to obovate, yellowish-brown, 1.5–2 × 1.2–1.5 mm. Uncommon in bogs, fens, and along streams, 8500–12,000 ft. E/W. (Plate 34)

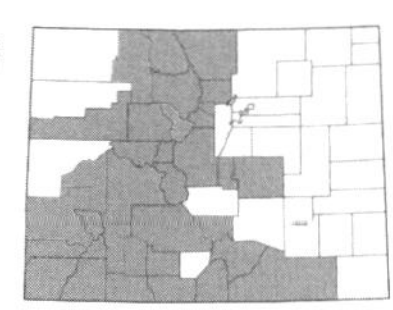

***Carex haydeniana* Olney**, HAYDEN'S SEDGE. Plants 1–3 (4) dm, caespitose; *leaves* flat, 1.5–4 mm wide; *bracts* reduced or setaceous but shorter than the inflorescence; *spikes* 4–7 (9), gynaecandrous, sessile, densely aggregated into an ovoid head; *pistillate scales* ovate, 3–4.8 mm, acute, shorter than the perigynia, brownish-black; *perigynia* ovate-lanceolate to ovate, 4.2–6.5 × 1.5–2.6 mm, flat except over the achene, brownish-black in the center and stramineous along the margins, prominently to inconspicuously nerved, the beaks half the length of the body, cylindric; *stigmas* 2; *achenes* 1.3–1.8 × 0.75–1.1 mm. Common in the alpine and in meadows, 9000–14,000 ft. July–Sept. E/W. (Plate 34)

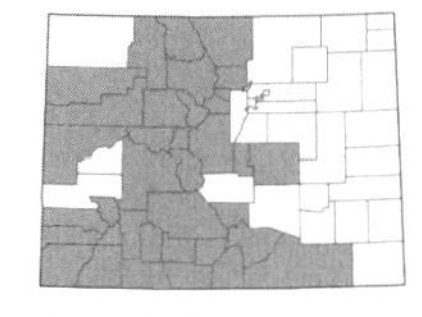

***Carex heteroneura* W. Boott**, DIFFERENT-NERVE SEDGE. [*C. atrata* L. var. *erecta* W. Boott.]. Plants 2.5–6 dm, caespitose; *leaves* flat, 2–5 mm wide; *bracts* with the lowest shorter to longer than the inflorescence; *spikes* usually erect or the proximal ones sometimes pendant, separate but aggregated, gynaecandrous; *pistillate scales* lanceolate, dark brown to black, usually shorter than the perigynia; *perigynia* ovate to orbicular, 2.5–3.5 × 1.7–2.5 mm, green or pale yellow, nerveless or obscurely nerved, the beak 0.3–0.5 mm long; *stigmas* 3; *achenes* filling less than ½ of the perigynia. Found along streams and pond margins, in meadows, and alpine, 9000–13,000 ft. July–Sept. E/W.

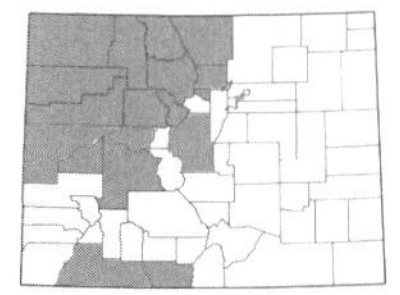

***Carex hoodii* W. Boott**, HOOD'S SEDGE. Plants 2–8 dm, from short, black roots; *leaves* flat, 1.5–3.5 mm wide; *bracts* shorter than the inflorescence; *spikes* 4–8, androgynous, sessile, loosely to closely aggregated into an ovoid or cylindric head; *pistillate scales* narrowly ovate, shorter than the perigynia, stramineous or brown with hyaline margins; *perigynia* elliptic, 3.5–5 × 1.5–2.5 mm, serrulate above the middle, brown with green sharp-edged margins, nerves inconspicuous, the beak 0.7–1.5 mm long, serrulate; *stigmas* 2; *achenes* 1.7–2 × 1.3–1.7 mm. Common in meadows, along streams and pond margins, and in aspen forests, 6000–11,500 ft. June–Sept. E/W. (Plate 34)

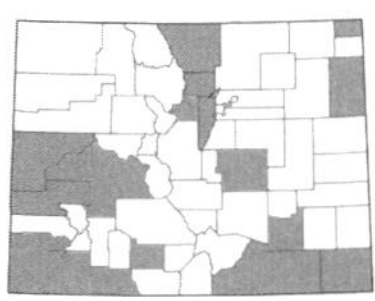

***Carex hystericina* Muhl. ex Willd.**, BOTTLEBRUSH SEDGE. Plants 1.5–10 dm, with short rhizomes, somewhat aphyllopodic; *leaves* flat, 2–10 mm wide; *bracts* leaf-like, the lowermost longer than the inflorescence; *spikes* several, the upper staminate and the lower pistillate; *pistillate scales* ovate, 1–6 mm long, with a long awn 2–6 mm long, ciliate-serrulate distally; *perigynia* lanceolate to lance-ovoid, 5–7 × 1.5–2 mm, conspicuously nerved, the beaks 2–2.5 mm long; *stigmas* 3; *achenes* obovoid-triangular, 1.2–2 × 0.9–1.3 mm, brown, continuous with the bony style. Uncommon in ditches, along the margins of lakes and streams, and in meadows, 3500–7000 ft. June–Aug. E/W. (Plate 34)

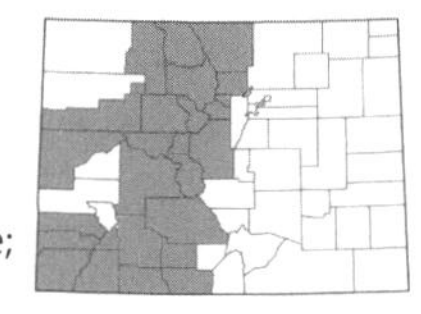

Carex illota **L.H. Bailey**, SHEEP SEDGE. Plants 1–3 (4) dm, shortly rhizomatous, slightly aphyllopodic; *leaves* flat or channeled, 1–3 mm wide; *bracts* inconspicuous; *spikes* 3–6, gynaecandrous, sessile, closely aggregated into a dense suborbicular head; *pistillate scales* ovate, obtuse, shorter than the perigynia, dark brown to greenish-black; *perigynia* lanceolate to narrowly ovate, 2.5–3.2 × 0.9–1.3 mm, greenish to brown, faintly nerved, the beak 0.8–2 mm long, dark on the dorsal suture; *stigmas* 2; *achenes* 1.2–1.5 × 0.8–1 mm. Common along streams, in moist meadows, and in moist alpine, 9500–13,000 ft. July–Sept. E/W. (Plate 34)

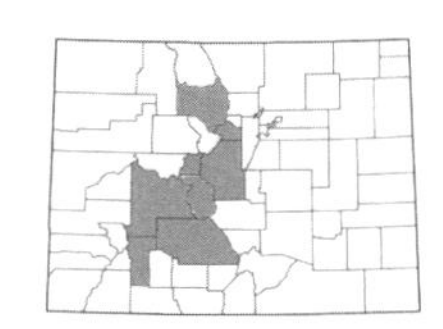

Carex incurviformis **Mack.**, INCURVED SEDGE. Plants 0.2–1.2 dm, rhizomatous; *leaves* involute, 0.5–1.5 mm wide; *bracts* inconspicuous; *spikes* densely aggregated into a hemispheric head, androgynous; *pistillate scales* ovate, brown with white-hyaline margins, shorter than the perigynia; *perigynia* narrowly elliptic, 3–4 × 1–1.5 mm, strongly nerved, the beak 0.4–0.9 mm long, slightly scabrous-margined; *stigmas* 2; *achenes* 1.4–1.7 × 0.8–1.3 mm. Found in the alpine, 11,500–14,000 ft. July–Sept. E/W. (Plate 34)

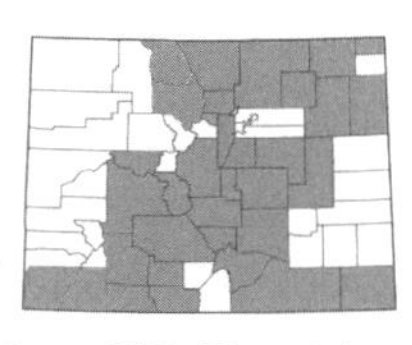

Carex inops **L.H. Bailey ssp. *heliophila* (Mack.) Crins**, SUN SEDGE. Plants 1.3–5 dm, rhizomatous; *leaves* flat to folded, 0.7–4.5 mm wide; *bracts* with the lower leaf-like and shorter than the inflorescence or inconspicuous; *spikes* aggregate, the terminal staminate or sometimes gynaecandrous, the lateral pistillate; *pistillate scales* ovate to elliptic, reddish-brown with white-hyaline margins, equal to the perigynia; *perigynia* obovoid, 2.8–4.5 × 1.5–2.2 mm, hairy, yellowish-green to brown, nerveless, the beak 0.4–1.3 mm long, serrulate; *stigmas* 3; *achenes* 1.4–2.5 × 1.5–2.2 mm. Common on dry slopes, in grasslands, and oak or pine forests, 3800–10,500 ft. April–Sept. E/W. (Plate 34)

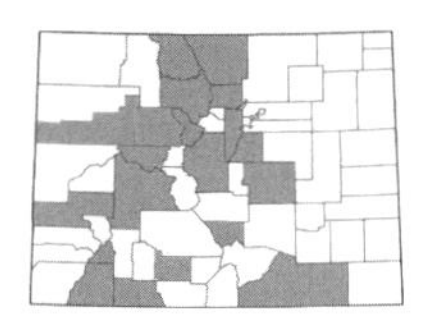

Carex interior **L.H. Bailey**, INLAND SEDGE. Plants 1.5–5 dm, from short, dark rhizomes, slightly aphyllopodic; *leaves* flat or somewhat channeled, 1–3 mm wide; *bracts* inconspicuous; *spikes* 2–6, gynaecandrous, sessile, remote, with few, widely spreading perigynia; *pistillate scales* ovoid, obtuse, shorter than the perigynia, brown with broad white-hyaline margins; *perigynia* ovate to lance-triangular, 1.9–3.5 × 1.5–2 mm, plump, brown at maturity, the beak 0.5–1 mm long, serrulate; *stigmas* 2; *achenes* 1.2–1.8 × 0.9–1.5 mm, jointed to the style. Common in moist meadows, fens, and along streams, 5200–10,000 ft. May–Aug. E/W. (Plate 34)

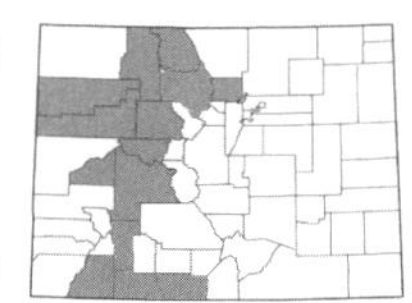

Carex jonesii **L.H. Bailey**, JONES' SEDGE. Plants 1.5–6 dm, shortly rhizomatous, slightly aphyllopodic; *leaves* flat, 1.5–3 mm wide; *bracts* short and inconspicuous; *spikes* 4–8, androgynous, sessile, densely aggregated into an orbicular or oblong head; *pistillate scales* ovate, acute, shorter than or equal to the perigynia, brownish-black with a pale midrib; *perigynia* ovate-lanceolate, 2.5–5 × 1–1.4 mm, brown, with conspicuous nerves, the beak tapering, sharp-edged, smooth to serrulate, 0.5–2 mm long; *stigmas* 2; *achenes* 1–2 × 0.5–1 mm. Uncommon along streams and lakes and in meadows and the alpine, 8000–13,000 ft. July–Sept. E/W. (Plate 34)

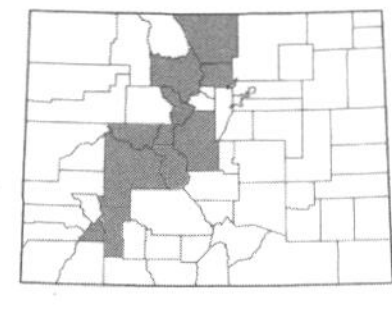

Carex lachenalii **Schkuhr**, TWO-PARTED SEDGE. [*C. bipartita* All.]. Plants 0.5–3 dm, shortly rhizomatous, slightly aphyllopodic; *leaves* flat, revolute, 1–2.5 mm wide; *bracts* inconspicuous, shorter than the inflorescence; *spikes* 1–4, gynaecandrous, sessile, aggregated into an ovoid head, distinct; *pistillate scales* oblong-ovate, obtuse, largely concealing the perigynia, brown with white-hyaline margins; *perigynia* narrowly ovate, 2–3 × 1–1.5 mm, brownish, finely nerved, the beak 0.5 mm long, with a conspicuous dorsal suture; *stigmas* 2; *achenes* 1.2–1.5 × 0.8–0.9 mm. Found along melting snowbanks, lake margins, and in alpine wetlands, 10,500–13,000 ft. July–Aug. E/W. (Plate 34)

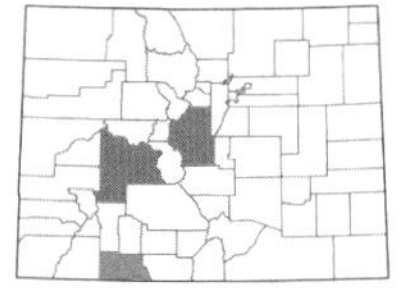

Carex laeviculmis **Meinsh.**, SMOOTHSTEM SEDGE. Plants 1.4–6.6 dm, caespitose; *leaves* 1–2.5 mm wide; *bracts* with the lowest 3.8–20 (35) mm long, usually shorter than the inflorescence; *spikes* 4–6 (7), usually gynaecandrous or sometimes the lateral pistillate; *pistillate scales* brownish, equal to or shorter than the perigynia; *perigynia* ovate, 2.3–3.7 × 1.1–1.5 mm, green to brownish, the beak 0.4–1.1 (1.3) mm long, serrulate or entire; *stigmas* 2; *achenes* 1.3–2 × 1–1.2 mm. Rare in moist meadows, spruce-fir forests, and along streams, 9500–10,500 ft. June–Aug. E/W.

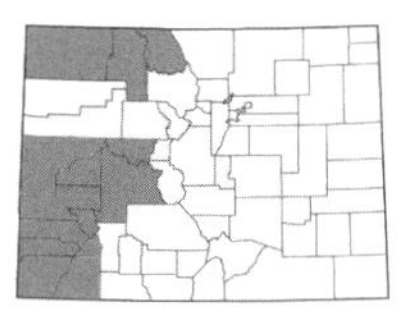

Carex lenticularis **Michx. var. *lipocarpa* (Holm) L.A. Standl.**, LENS SEDGE. Plants 1–8 dm, caespitose, usually phyllopodic; *leaves* flat, 1–4 mm wide; *bracts* leaf-like, equal to or longer than the inflorescence; *spikes* 3–6, the terminal staminate and the lateral pistillate, cylindric; *pistillate scales* ovate, narrower than the perigynia, reddish-brown to blackish with a pale center and hyaline margins; *perigynia* ovate to elliptic, 1.8–3.5 × 1–2 mm, green, few-nerved, the beak 0.1–0.3 mm long, entire; *stigmas* 2; *achene* oblong-triangular, 1–1.5 × 0.8–1.6 mm, brown. Found along streams and lake margins and in moist meadows, 5500–10,500 ft. June–Aug. E/W. (Plate 34)

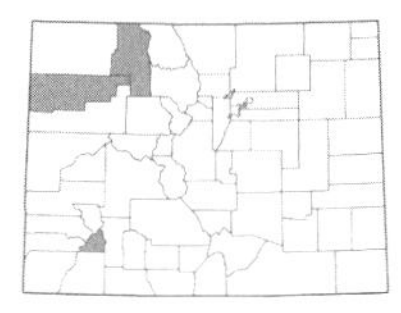

Carex leporinella **Mack.**, SIERRA-HARE SEDGE. Plants 1–3 dm, caespitose, aphyllopodic; *leaves* involute or folded, 0.7–1.5 mm wide; *bracts* reduced with the lower prolonged and shorter than to equal to the inflorescence; *spikes* 3–10, gynaecandrous, sessile, distinct but aggregated into an oblong head; *pistillate scales* oblong-ovate, (3) 3.5–5 mm long, usually concealing the perigynia, reddish-brown with a paler midrib and white-hyaline margins; *perigynia* elliptic-ovate, 3.5–4.2 × 1–1.2 mm, narrowly winged, serrulate above, plano-convex, stramineous, finely nerved, the beak ca. 1 mm long, serrulate, cylindric; *stigmas* 2; *achenes* 1.4–2 × 0.7–1 mm. Rare in fens and moist meadows, 9500–11,500 ft. July–Sept. W.

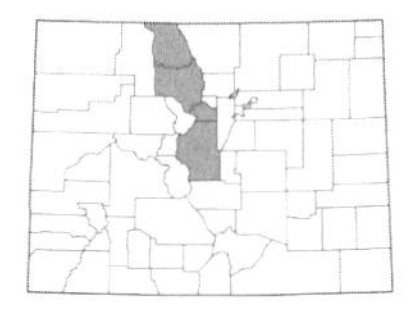

Carex leptalea **Wahlenb.**, BRISTLESTALKED SEDGE. Plants to 6 dm, rhizomatous, aphyllopodic; *leaves* flat, to 1.3 mm wide; *bracts* absent; *spikes* solitary, androgynous, to 16 mm long; *pistillate scales* ovate, shorter than the perigynia, brown or greenish, deciduous; *perigynia* oval-elliptic, 2.5–5 × 1–1.5 mm, green or stramineous, with numerous nerves, lacking a beak; *stigmas* 3; *achenes* 1.3–2 × 0.8–1 mm, brown or yellowish, nearly filling the perigynium. Rare in fens, along pond margins, and near seepages, 8500–10,500 ft. June–Aug. E/W. (Plate 34)

Carex leptopoda **Mack.**, TAPERFRUIT SHORTSCALE SEDGE. Plants 2–7 dm, caespitose with short rhizomes, slightly aphyllopodic; *leaves* flat, 2.4–6 mm wide; *bracts* with the lowest setaceous-prolonged, shorter than the inflorescence; *spikes* 5–7, gynaecandrous, loosely aggregated; *pistillate scales* with a green midvein and white-hyaline margins; *perigynia* ovate, 3.3–4 × 1.1–1.5 mm, greenish, weakly to prominently nerved adaxially, the beak 1–1.5 (1.7) mm long, serrulate; *stigmas* 2; *achenes* 1.5–2 × 0.9–1.4 mm. Uncommon along streams and in moist meadows, 7000–9000 ft. June–Aug. E/W.

Carex leptopoda is often considered to be synonymous with or a variety of *C. deweyana.*

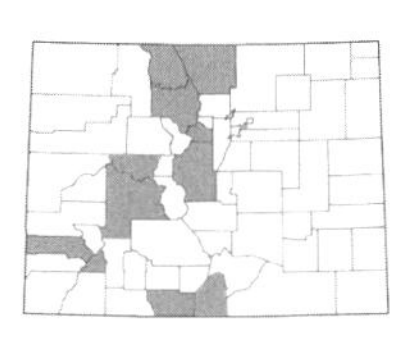

Carex limosa **L.**, MUD SEDGE. Plants to 6 dm, rhizomatous, roots covered with yellowish-brown felty tomentum, usually aphyllopodic; *leaves* channeled, 1–3 mm wide; *bracts* with the lowest leaf-like, 2–10 cm long; *spikes* 2–4, on nodding peduncles, the terminal staminate and the lateral pistillate; *pistillate scales* often ovate, shorter than or equal to the perigynia, reddish-brown; *perigynia* ovoid, 2.3–4.2 × 2 mm, yellowish-green, with conspicuous nerves, the beaks absent or to 0.2 mm long; *stigmas* 3; *achenes* oblong-ovoid, 1.5–2.7 × 1–2 mm. Found in fens and moist meadows, along streams and ponds margins, and forming mats in the middle of lakes, 9000–11,700 ft. July–Sept. E/W. (Plate 35)

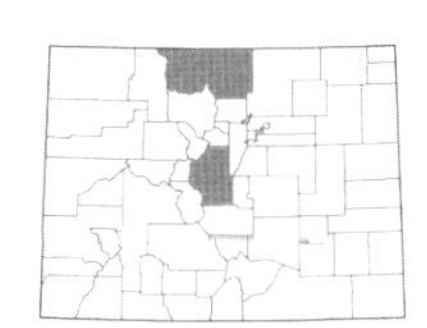

Carex livida **(Wahlenb.) Willd.**, LIVID SEDGE. Plants 0.5–6 dm, rhizomatous, phyllopodic; *leaves* sometimes channeled, 0.5–3.5 mm wide; *bracts* with the lowest leaf-like, equal to or longer than the inflorescence; *spikes* 2–4, the terminal staminate and the lateral pistillate, loosely flowered; *pistillate scales* ovate, equal to or shorter than the perigynia, coppery brown or greenish with hyaline margins; *perigynia* narrowly elliptic, 2.5–5 × 1.2–2.4 mm, green, the nerves absent or inconspicuous, the beak absent or to 0.2 mm long; *stigmas* 3; *achenes* 2–3.5 × 1–1.8 mm, jointed to the style. Rare in fens and peatlands, 9000–10,000 ft. June–Aug. E/W. (Plate 35)

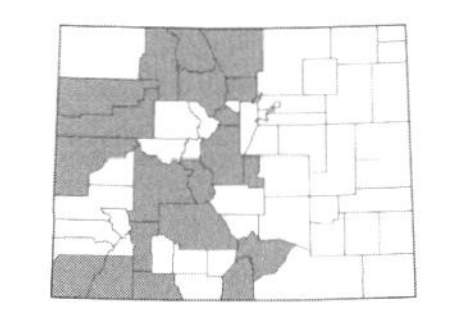

Carex macloviana **d'Urv.**, BROWN SEDGE. [*C. subfusca* W. Boott]. Plants (0.9) 1.5–6 dm, caespitose with short rhizomes, aphyllopodic; *leaves* flat, 2–4 mm wide; *bracts* inconspicuous; *spikes* (3) 5–9, gynaecandrous, densely aggregated into an ovoid head; *pistillate scales* ovate, 2.7–3.5 (4) mm long, shorter than the perigynia, greenish-brown with a paler center and margins; *perigynia* ovate, 3.5–4.5 × 1.1–2 mm, plano-convex or flat, reddish-brown with green margins, wing-margined, finely nerved, the beak 0.5–1 mm long, cylindric; *stigmas* 2; *achenes* 1.2–2 × 0.8–1.1 mm. Found in meadows, along ponds, and in alpine, 8000–12,000 ft. June–Sept. E/W.

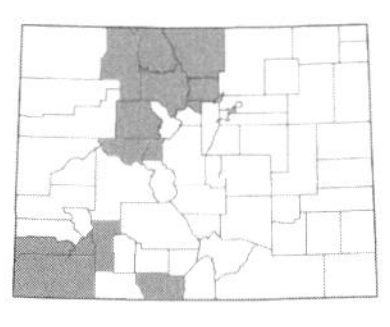

Carex magellanica **Lam. ssp. *irrigua* (Wahlenb.) Hultén**, LITTLE SEDGE. [*C. paupercula* Michx.]. Plants 1.5–7 dm, rhizomatous, roots covered with yellowish-brown felty tomentum, phyllopodic or aphyllopodic; *leaves* flat, 2–4 mm wide; *spikes* 1–5, on nodding peduncles, the terminal staminate and the lateral pistillate; *pistillate scales* lanceolate to ovate-lanceolate, longer than the perigynia, brown; *perigynia* elliptic to ovate, 2.2–3.8 × 1.7–2.2 mm, yellowish-green and usually darker at the tip, with obscure or prominent nerves, beaks absent or to 0.1 mm long; *stigmas* 3; *achenes* ca. 2 × 1.2 mm. Found along creeks and lake shores and in bogs and moist meadows, 8300–11,500 ft. July–Sept. E/W. (Plate 35)

Carex microglochin **Wahlenb.**, SMALL-TIPPED SEDGE. Plants 0.2–2.5 dm, rhizomatous; *leaves* involute, 0.3–0.6 mm wide; *bracts* absent; *spikes* solitary, androgynous, the perigynia soon deflexed; *pistillate scales* ovate-triangular, shorter than the perigynia, brown, early-deciduous; *perigynia* linear-lanceolate, 3–5 × 0.8–1 mm, brownish to green, conspicuously nerved, the beak tapering; *stigmas* 3; *achenes* 2.5 × 0.5–0.8 mm, continuous with the style, the rachilla longer than the achene. Found in fens and bogs, along lake margins, and in moist tundra, 10,000–12,000 ft. July–Sept. E/W. (Plate 35)

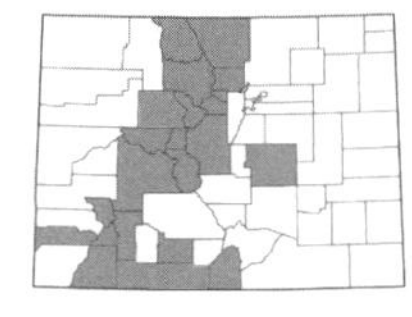

***Carex micropoda* C.A. Mey.**, Pyrenean sedge. [*C. pyrenaica* Wahlenb.]. Plants to 3 dm, lacking rhizomes, aphyllopodic; *leaves* flat, 2–10 cm × 0.25–1.5 mm; *bracts* absent; *spikes* solitary, androgynous, to 20 mm long; *pistillate scales* oblong to ovate with acute to obtuse tips, brown to blackish-brown with pale midveins and narrow hyaline margins, shorter than the perigynia; *perigynia* elliptic to lanceolate, stipitate, dark brown distally and green below, 3–4.5 × 0.7–1.5 mm, lacking nerves, with an obscure beak 0.5 mm long; *stigmas* 3; *achenes* 1.2–1.8 × 0.7–1 mm. Common in moist meadows, along streams, and in the alpine, 10,000–13,500 ft. July–Sept. E/W. (Plate 35)

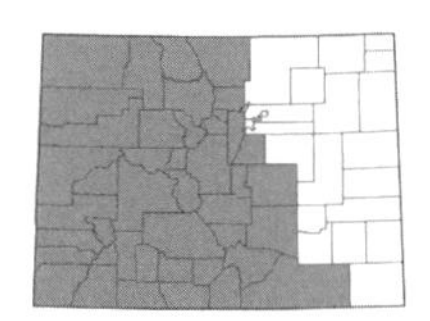

***Carex microptera* Mack.**, small-winged sedge. Plants 2–10 dm, caespitose; *leaves* flat, 2–4.5 mm wide; *spikes* 5–10, gynaecandrous, densely aggregated into an ovoid or suborbicular head; *pistillate scales* ovate-lanceolate, 2.4–3.5 mm long, acute, shorter than the perigynia, brown or greenish-black; *perigynia* lanceolate to ovate-lanceolate, (2.8) 3.4–4.5 (5.2) × 1.1–2.4 mm, brown with green margins, wing-margined, serrulate above the middle, flat except over the achene or plano-convex, several-nerved, the beak cylindric; *stigmas* 2; *achenes* 1–1.6 × 0.7–1.5 mm. Common along streams, in meadows, and in forests, 5500–12,500 ft. June–Sept. E/W. (Plate 35)

***Carex molesta* Mack. ex Bright**, troublesome sedge. Plants 3.5–10 dm, caespitose; *leaves* 2–3 mm wide, the sheaths green-and-white-mottled dorsally; *bracts* aristate, shorter than the inflorescence; *spikes* 2–4 (5), gynaecandrous, aggregate; *pistillate scales* ovate, 2.9–3.5 mm long, shorter than the perigynia, brown with a green or pale midvein; *perigynia* elliptic, (3) 3.3–5 (5.7) × 1.8–3 mm, pale brown to green, strongly wing-margined, plano-convex, nerveless to strongly nerved ventrally, the beak flat, ca. ½ the length of the body, serrulate; *stigmas* 2; *achenes* 1.3–1.7 × 1–1.3 mm. Reported for Colorado for Baca and Las Animas cos., 4000–4500 ft. May–July. E.

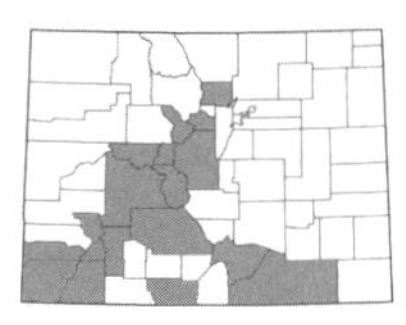

***Carex nardina* Fries**, spikenard sedge. [*C. hepburnii* Boott.]. Plants 1–1.5 dm, caespitose; *leaves* involute, stiff, 1–10 cm × 0.25–0.5 mm; *bracts* absent; *spikes* solitary, androgynous, to 1.5 cm long; *pistillate scales* ovate to nearly orbicular, tips obtuse to acute, brown with a paler midvein and hyaline margins, equaling the perigynia; *perigynia* obovate to lanceolate, with thin, sharply angled margins, obscurely serrulate or glabrous distally, 3–4.5 × 1.2–2 mm, nerves obscure to conspicuous, the beak serrulate, to 0.8 mm long; *stigmas* 2 or 3; *achenes* 1.6–2 × 1–1.3 mm, rachilla well-developed. Found on alpine ridges and scree slopes, 11,000–14,000 ft. July–Aug. E/W. (Plate 35)

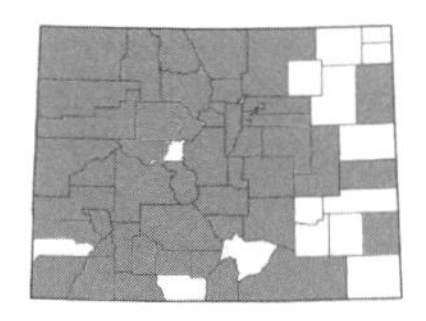

***Carex nebrascensis* Dewey**, Nebraska sedge. Plants 2–12 dm, rhizomatous; *leaves* flat, 3–12 mm wide, usually glaucous and bluish-green; *bracts* leaf-like, the lowest usually longer than the inflorescence; *spikes* 3–6, the terminal 1 or 2 staminate and the lateral pistillate, erect; *pistillate scales* lanceolate with an acuminate tip, narrower than the perigynia, reddish-brown or blackish with a pale midrib and often with hyaline margins; *perigynia* oblong-ovate, 2.7–4 × 1.5–2.5 mm, yellowish-brown to brown, with 5–10 nerves on each surface, the beak 0.2–1 mm long; *stigmas* 2; *achene* 1.2–2.5 × 0.9–1.8 mm. Common along streams, ditches, and lake margins and in moist meadows and fields, marshes, and floodplains, 3500–10,500 ft. June–Sept. E/W. (Plate 35)

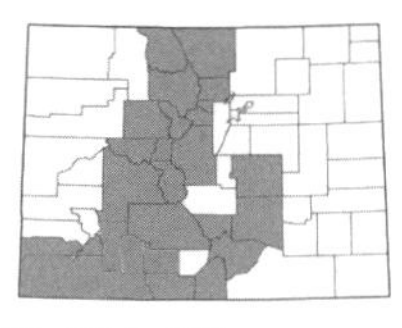

***Carex nelsonii* Mack.**, Nelson's sedge. Plants 1.5–3 dm, with short rhizomes, slightly aphyllopodic; *leaves* flat, 3–4 mm wide; *bracts* inconspicuous or the lowest slightly prolonged; *spikes* 2–4, the terminal gynaecandrous and the lateral pistillate, closely aggregated into a dense obovoid head; *pistillate scales* ovate, shorter than the perigynia, black; *perigynia* oblong-obovoid, ca. 4 × 1.5 mm, dark reddish-black with yellowish-green margins, nerves absent on the faces, the beak ca. 1 mm long, terete, sparsely serrulate; *stigmas* 3; *achenes* ca. 1.5 × 0.75 mm. Common along lake margins and melting snowbanks and in moist meadows and alpine, 8500–13,500 ft. June–Aug. E/W. (Plate 35)

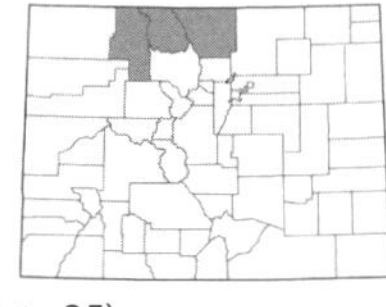

***Carex neurophora* Mack.**, alpine nerved sedge. Plants 2–8 dm, caespitose, aphyllopodic; *leaves* flat, 1.5–3.5 mm wide, cross-corrugate ventrally; *bracts* inconspicuous; spikes 5–10, androgynous, sessile, densely aggregated into an ovoid head; *pistillate scales* ovate, shorter than the perigynia, brown with a pale midrib and hyaline margins; *perigynia* lance-triangular, 2.9–4 × 0.8–1.5 mm, brown, sharp-edged, minutely serrulate above, with numerous nerves, the beaks about half the length of the body, minutely serrulate; *stigmas* 2; *achenes* 1.1–1.6 × 0.8–1 mm, jointed to the style. Rare along streams and in swamps and moist meadows, 8500–11,000 ft. July–Aug. E/W. (Plate 35)

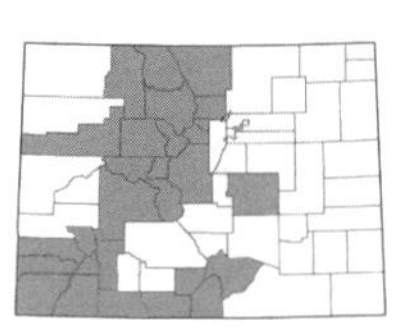

***Carex nigricans* C.A. Mey.**, black alpine sedge. Plants to 3 dm, rhizomatous; *leaves* flat, 4–13 cm × (1.5) 2–4 mm; *bracts* absent; *spikes* solitary, androgynous, to 20 mm; *pistillate scales* ovate, blackish-brown, sometimes with a pale midrib; *perigynia* lanceolate, stipitate, blackish-brown distally and greenish below, 3–4.5 × 0.8–1.5 mm, beaks ca. 0.5 mm long, terete; *stigmas* 3 or 2; *achenes* 1–2 × 0.6–1 mm, the rachilla obsolete. Common in the alpine, moist meadows, near seepages, and along lake margins, 10,000–13,500 ft. July–Sept. E/W. (Plate 35)

Carex norvegica **Retz.**, Norway sedge. Plants to 8 dm, rhizomatous, aphyllopodic; *leaves* flat, 1.5–3 mm wide; *bracts* usually leaf-like, shorter or longer than the inflorescence; *spikes* 2–5, the terminal gynaecandrous and the lateral pistillate; *pistillate scales* ovate, equal to or shorter than the perigynia, black with white-hyaline margins; *perigynia* obovate to elliptic, 2–3 × 1–1.3 mm, yellowish-green, with few nerves or these absent, beaks 0.3–0.5 mm long, black; *stigmas* 3; *achenes* 1.2–1.7 × 0.7–1 mm. Found in the alpine, 11,500–13,500 ft. June–Aug. E/W.

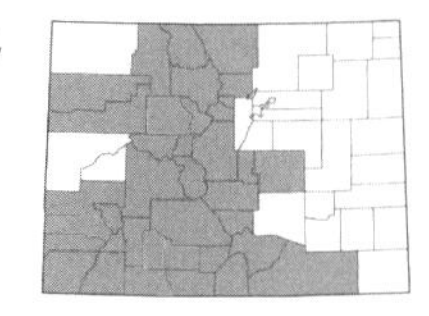

Carex nova **L.H. Bailey**, black sedge. Plants to 6 dm, rhizomatous, slightly aphyllopodic; *leaves* flat, 2–5 mm wide; *bracts* shorter than the inflorescence; *spikes* 3–5, closely aggregated in a dense capitate inflorescence, the terminal gynaecandrous and the lateral pistillate; *pistillate scales* ovate-oblong, equal to or shorter than the perigynia, dark brownish-black to purplish-black throughout; *perigynia* elliptic to oblong, strongly flattened, 2.8–4.5 × 2–3.5 mm, green to dark purple, marginal nerves present, the beaks 0.3–0.6 mm long, black; *stigmas* 3; *achenes* 1.5–2 × 1 mm, much smaller than the perigynia. Common in meadows, along streams, in fens, and in spruce-fir forests, 9500–12,500 ft. July–Sept. E/W. (Plate 35)

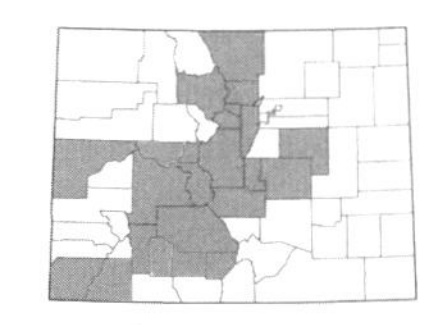

Carex obtusata **Lilj.**, obtuse sedge. Plants to 2 dm, rhizomatous, aphyllopodic; *leaves* flat, 0.5–1.5 mm wide; *bracts* absent; *spikes* solitary, androgynous, to 15 mm long; *pistillate scales* ovate with a sharply acute tip, shorter or longer than the perigynia, brown with white-hyaline margins; *perigynia* obovoid, glossy, dark brown, 3–4 × 1.5–2 mm, with numerous, prominent nerves, the beak 0.5–1 mm; *stigmas* 3; *achenes* to 1.75 mm long, ca. 1 mm wide, yellowish to tan. Found in grasslands, meadows, dry forests, on rocky ridges, and in alpine, 7500–13,000 ft. June–Aug. E/W. (Plate 36)

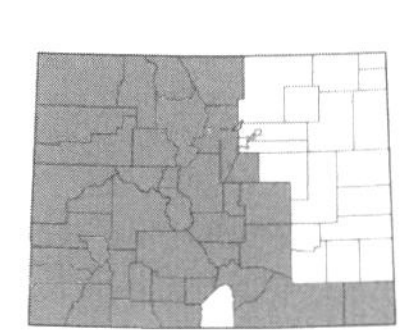

Carex occidentalis **L.H. Bailey**, western sedge. Plants 1.5–8 dm, from black rhizomes, aphyllopodic; *leaves* flat, 1.5–2.5 (3) mm wide; *bracts* absent or shorter than the inflorescence; *spikes* androgynous, the upper aggregated and indistinguishable, the lower separate; *pistillate scales* ovate with an acuminate tip, largely concealing the perigynia; *perigynia* oblong-elliptic, 2.5–4.5 × 1.5–1.9 mm, serrulate above the middle, green to stramineous, nerves inconspicuous, the beaks serrulate, flat, 0.2–1.2 mm long; *stigmas* 2; *achenes* 1.3–2.4 × 1–1.5 mm. Common in dry forests, canyons, grasslands, and along streams, 5000–10,500 ft. June–Aug. E/W. (Plate 36)

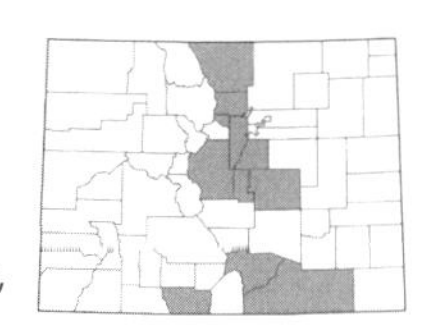

Carex oreocharis **Holm**, grassyslope sedge. Plants 1–5 dm, caespitose with short rhizomes; *leaves* channeled or folded, 1–2 mm wide; *bracts* absent; *spikes* solitary, androgynous; *pistillate scales* orbiculate, shorter than the perigynia, brown with white-hyaline margins; *perigynia* obovate to ovate, 3–4.5 × 1.4–2.2 mm, whitish-green to golden-brown, hirsute, the beak 0.4–1 mm long; *stigmas* 2; *achenes* 2.2–3 × 1.2–2 mm. Uncommon on dry slopes and in grasslands and meadows, 5900–10,500 ft. May–Aug. E. (Plate 36)

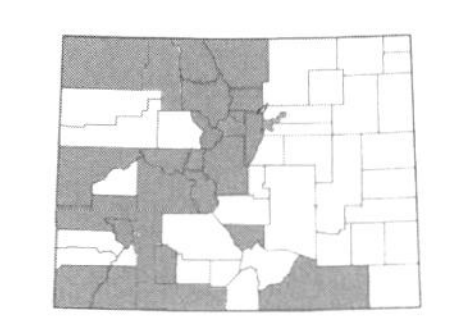

Carex pachystachya **Cham. ex Steud.**, chamisso sedge. Plants 1.5–10 dm, with short, black rhizomes, aphyllopodic; *leaves* flat, (1.5) 2–6 mm wide; *bracts* inconspicuous; *spikes* 4–12, gynaecandrous, sessile, closely aggregated into an orbicular or oblong head; *pistillate scales* ovate, 2.2–3.7 (4.2) mm long, acute, shorter than the perigynia, brown or blackish; *perigynia* ovate, 2.8–5 × 1.1–2.3 mm, wing-margined, serrulate distally, plano-convex, tan with green margins, finely nerved dorsally, the beak less than half the length of the body, cylindric, dark brown; *stigmas* 2; *achenes* 1.2–2 × 0.8–1.5 mm. Common along streams and in meadows, 5500–11,000 ft. June–Aug. E/W. (Plate 36)

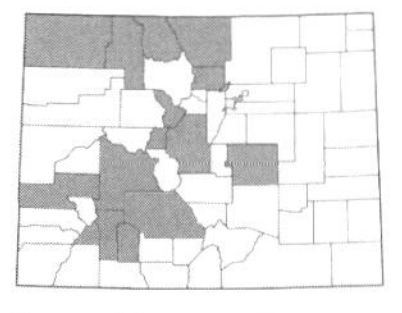

Carex parryana **Dewey**, Parry's sedge. Plants to 6 dm, rhizomatous, phyllopodic; *leaves* flat, 2–4 mm wide; *bracts* usually shorter than the inflorescence; *spikes* 1–5, the terminal often staminate, sometimes gynaecandrous, the lateral pistillate; *pistillate scales* suborbicular, equal to the perigynia and concealing them, dark reddish-brown with white-hyaline margins; *perigynia* obovoid, 1.5–3 × 1–1.5 mm, rough-granular or minutely serrulate towards the tip, yellowish and dark reddish-brown above the middle, with conspicuous marginal nerves, the beak 0.1–0.2 mm long or obsolete, often strongly ciliate; *stigmas* 3; *achenes* 1.4–1.8 × 1–1.3 mm, jointed to the style. Found in meadows, on moist slopes, and around lakes and streams, 5200–10,000 ft. June–Aug. E/W. (Plate 36)

There are two varieties of *C. parryana* in Colorado:

1a. Lateral spikes 2 or more, at least one spike as large as the terminal spike...**var. *parryana***

1b. Lateral spikes 1 or more, all smaller than the terminal spike...**var. *unica*** **L.H. Bailey**, [*C. hallii* Olney].

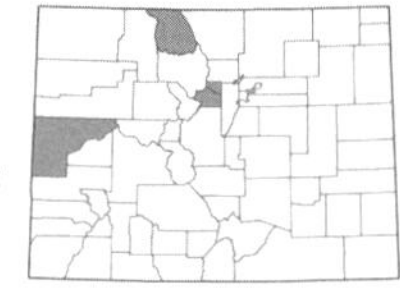

***Carex paysonis* Clokey**, Payson's sedge. Plants to 5 dm, rhizomatous, aphyllopodic; *leaves* flat, 2–8 mm wide; *bracts* with the lower 1 or 2 leaf-like, dark-auricled, equal to or shorter than the inflorescence; *spikes* 2–6, the terminal staminate or sometimes androgynous, the lateral pistillate or androgynous; *pistillate scales* ovate, equal to or shorter than the perigynia, dark reddish-brown; *perigynia* elliptic, 2–4.3 × 1.5–2.2 mm, yellowish below and reddish-brown above, the nerves few, prominent to obscure, beaks 0.2–0.5 mm long; *stigmas* 3; *achenes* 1.2–1.7 × 1 mm, much shorter than the perigynia. Rare in moist meadows and on rocky slopes, 10,000–13,000 ft. June–Aug. E/W. (Plate 36)

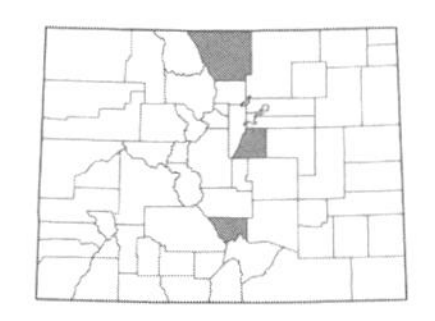

***Carex peckii* Howe**, Peck's sedge. Plants 2–5 dm, loosely caespitose or shortly rhizomatous; *leaves* 1–3.3 mm wide; *bracts* with the lowest leaf-like, shorter than the inflorescence; *spikes* 2–3, the terminal staminate and the lateral pistillate or androgynous; *pistillate scales* ovate, reddish-brown with white-hyaline margins, shorter than the perigynia; *perigynia* elliptic to ovoid, 3.2–4.2 × 1–1.3 mm, greenish, hirsute, nerveless, the beak 0.7–1 mm long, serrulate; *stigmas* 3; *achenes* 2–2.5 × 1–1.3 mm. Rare along streams and in moist forests, 6500–7500 ft. June–Aug. E. (Plate 36)

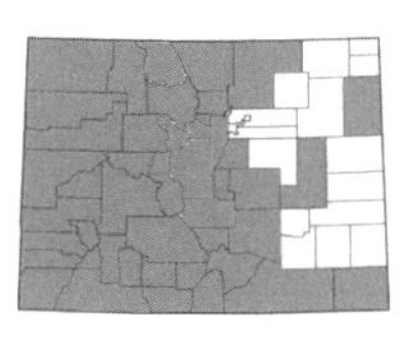

***Carex pellita* Willd.**, woolly sedge. [*C. lanuginosa* Michx.]. Plants to 11 dm, rhizomatous, aphyllopodic; *leaves* flat, 2–5 mm wide; *bracts* leaf-like, equal to or slightly longer than the inflorescence; *spikes* several, the terminal staminate and the lateral pistillate; *pistillate scales* lanceolate, long-acuminate with a densely hairy awn, reddish-brown with a paler center; *perigynia* ovoid, 2.5–5 × 1.5–2 mm, densely short-hairy, brown to reddish-brown, often paler below, the nerves numerous but hidden by hairs, beaks 0.5–1.5 mm long; *stigmas* 3; *achenes* 1.5–2 × 1–1.5 mm. Common along streams, lake margins, and in fens and meadows, 3500–10,500 ft. May–Aug. E/W. (Plate 36)

Carex lasiocarpa Ehrh. is similar to *C. pellita* but does not occur in Colorado.

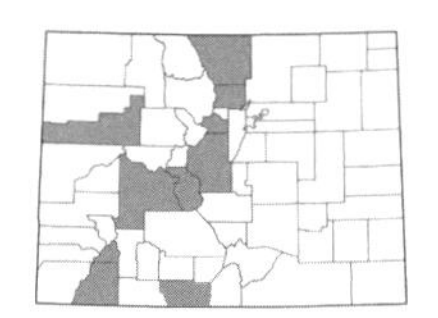

***Carex pelocarpa* F.J. Herm.**, duskyseed sedge. Plants 1–4 dm, caespitose, aphyllopodic; *leaves* flat, 3–4 mm wide; *bracts* with the lowest shorter to longer than the inflorescence, leaf-like; *spikes* gynaecandrous, in a dense terminal ovoid cluster or sometimes the proximal spikes separate and distinct; *pistillate scales* lanceolate, shorter than the perigynia, blackish-brown; *perigynia* obovate to orbicular, 3.5–4.5 × 2.5–3 mm, yellowish with greenish margins, nerveless, the beak 0.5–0.8 mm long, dark; *stigmas* 3; *achenes* filling less than ½ of the perigynia. Found along streams, in meadows, and on talus slopes, 9000–12,000 ft. July–Sept. E/W.

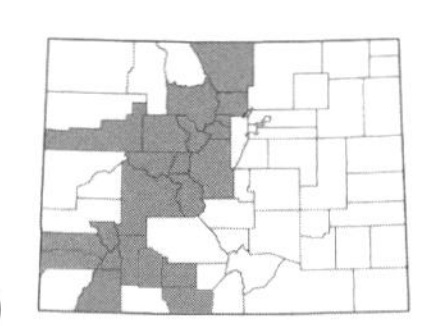

***Carex perglobosa* Mack.**, globe sedge. Plants 0.3–2 dm, rhizomatous, phyllopodic; *leaves* flat, 0.75–3 mm wide; *bracts* absent or inconspicuous to almost equaling the inflorescence; *spikes* androgynous, sessile, densely compacted into a globose head 10–18 mm long; *pistillate scales* ovate, equal to the perigynia, brown with a pale midrib and hyaline margins; *perigynia* elliptic, 3.6–4.7 × 1.9–2.6 mm, thin margined, nerves faint on the faces, the beak ill-defined; *stigmas* 2; *achenes* 1.4–1.7 × 1–1.2 mm. Common in the alpine, 11,000–13,500 ft. July–Sept. E/W. (Plate 36)

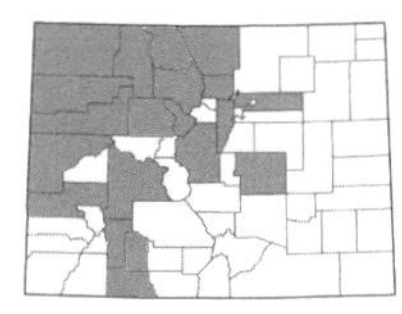

***Carex petasata* Dewey**, Liddon sedge. Plants 3–9 dm, usually caespitose; *leaves* flat, 2–5 mm wide; *bracts* short and inconspicuous; *spikes* 3–6, gynaecandrous, sessile, loosely aggregated; *pistillate scales* ovate, 5.8–7.6 mm long, largely concealing the perigynia, coppery-brown with white-hyaline margins; *perigynia* oblong-lanceolate, 6–8 × 1.7–2.4 mm, serrulate, wing-margined, plano-convex or concavo-convex, stramineous to pale green with green margins, with numerous nerves on both sides, the beak cylindric; *stigmas* 2; *achenes* 2.2–3 × 1–1.8 mm. Common on open slopes, in grasslands, forests, and oak shrublands, 5500–8500 ft. June–Aug. E/W. (Plate 36)

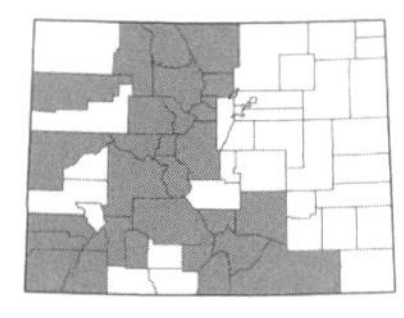

***Carex phaeocephala* Piper**, dunhead sedge. Plants 0.5–3 (4) dm, caespitose, aphyllopodic; *leaves* flat to folded, 0.5–3 mm wide; *bracts* inconspicuous; *spikes* 3–7, gynaecandrous, sessile, closely aggregated but distinguishable in an ovoid head; *pistillate scales* ovate, 3.7–5.1 mm long, acute, largely concealing the perigynia, reddish-brown with hyaline margins; *perigynia* oblong to ovate, 3.8–5.2 × 1.5–2.3 mm, winged, serrulate, flat or plano-convex, reddish-brown with green margins, conspicuously nerved dorsally, the beak tapering, cylindric; *stigmas* 2; *achenes* 1.5–2.3 × 0.8–1.3 mm. Common in the alpine, meadows, and on open slopes, 8300–13,000 ft. June–Sept. E/W. (Plate 36)

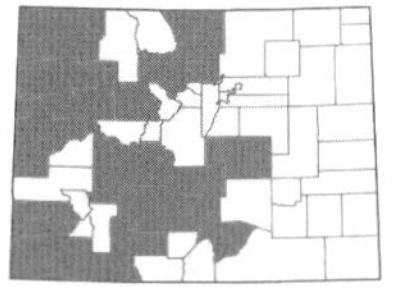

***Carex pityophila* Mack.**, loving sedge. Plants 3–15 cm, with short rhizomes; *leaves* flat above and channeled toward the base, 0.7–1.5 mm wide; *bracts* subtending pistillate spikes on bisexual culms longer than the inflorescence; *staminate spikes* terminal, 4–8 mm long; *pistillate spikes* 2–5, 2–5-flowered on pistillate culms and 1–3-flowered on bisexual culms; *pistillate scales* ovate, with a green center and hyaline margins, nearly equal to the perigynia; *perigynia* ovoid to elliptic, 3.3–4.5 × 1.4–1.8 mm, hairy, green, conspicuously stipitate, with a beak 0.75–1 mm long; *stigmas* 3; *achenes* 1.9–3 × 1.5–1.8 mm. Found on dry slopes and in pinyon pine forests, 6700–10,500 ft. May–Aug. E/W. (Plate 36)

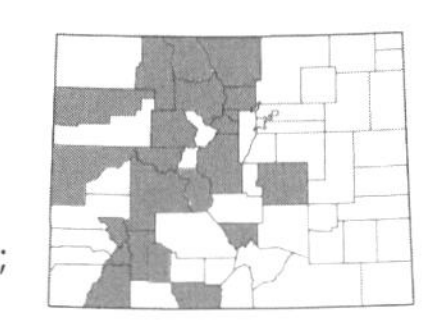

***Carex praeceptorum* Mack.**, EARLY SEDGE. Plants 1–3 dm, rhizomatous, aphyllopodic; *leaves* canaliculate, 1.2–2.5 mm wide; *bracts* inconspicuous; *spikes* 4–6, gynaecandrous, sessile, distinguishable but closely aggregated; *pistillate scales* ovate, shorter than the perigynia, brown with a paler center and hyaline margins; *perigynia* ovate, 1.5–2.5 × 1–1.2 mm, golden brown, white-punctate or rough, with prominent nerves, the beak 0.25–0.5 mm long, minutely serrulate; *stigmas* 2; *achenes* 1.2–1.5 × 0.8–1 mm. Rather uncommon in bogs, moist meadows, and along streams and lake margins, 9000–12,500 ft. July–Aug. E/W. (Plate 37)

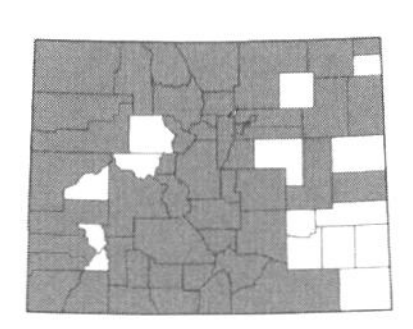

***Carex praegracilis* W. Boott**, CLUSTERED FIELD SEDGE. Plants 2–8 dm, from blackish creeping rhizomes, somewhat aphyllopodic; *leaves* flat or channeled, 1–3 mm wide; *bracts* inconspicuous, shorter than the inflorescence; *spikes* 5–15, androgynous or pistillate, aggregated into a cylindric head; *pistillate scales* ovate with an acuminate tip, concealing the perigynia, brown with a pale midrib and hyaline margins; *perigynia* ovate-lanceolate, 2.8–4 × 1.3–1.6 mm, serrulate distally, brown with green margins, the nerves faint, the beak prominent, 0.6–1.3 mm long; *stigmas* 2; *achenes* 1.2–2 × 1–1.4 mm, brown. Common along streams and in meadows and forests, 3500–10,500 ft. May–Sept. E/W. (Plate 37)

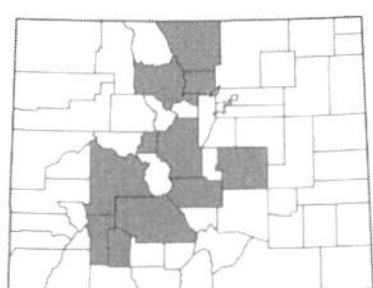

***Carex praticola* Rydb.**, MEADOW SEDGE. Plants 2–10 dm, caespitose, aphyllopodic; *leaves* flat, 1–4 mm wide; *bracts* with the lowest setaceous-prolonged, shorter than the inflorescence; *spikes* 2–7, gynaecandrous, sessile, approximate; *pistillate scales* narrowly ovate, (3.4) 4.2–5.8 mm long, largely concealing the perigynia, brown with white-hyaline margins; *perigynia* ovate-lanceolate, (3.7) 4.5–6 × 1.2–2 mm, wing-margined, serrulate, flat or plano-convex, green to stramineous, with evident nerves dorsally, the beak cylindric, ⅓–½ the length of the body; *stigmas* 2; *achenes* 1.5–2.7 × 0.9–2 mm, jointed to the style. Found in meadows, aspen forests, and along streams and ponds, 6500–11,000 ft. June–Aug. E/W. (Plate 37)

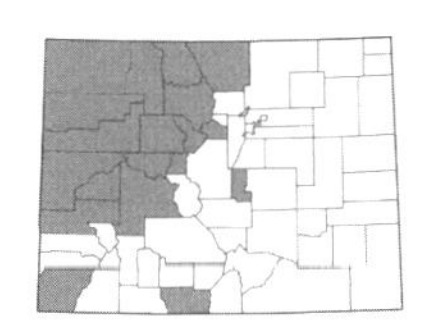

***Carex raynoldsii* Dewey**, RAYNOLD'S SEDGE. Plants to 8 dm, rhizomatous, phyllopodic or slightly aphyllopodic; *leaves* flat, 2–8 mm wide; *bracts* sometimes leaf-like, the lowest equal to or shorter than the inflorescence; *spikes* several, the terminal staminate and the lateral pistillate; *pistillate scales* ovate, shorter than the perigynia, dark reddish-black; *perigynia* oval, 3–4.5 × 1.5–2 mm, yellowish green and brownish red above the middle, nerves several or obscure, beaks 0.1–0.5 mm long; *stigmas* 3; *achenes* 1.8–2.5 × 1–1.7 mm. Found along streams and lake shores and in bogs, meadows, and open spruce forests, 7500–11,200 ft. June–Aug. E/W. (Plate 37)

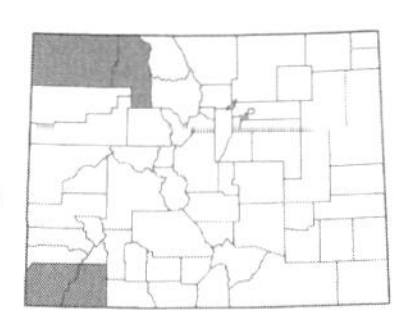

***Carex retrorsa* Schwein.**, KNOTSHEATH SEDGE. Plants 2–10 dm, with very short rhizomes, aphyllopodic; *leaves* flat, 3–10 mm wide; *bracts* leaf-like, several times longer than the inflorescence; *spikes* several, the upper staminate and the lower pistillate; *pistillate scales* lanceolate, brownish with a paler center; *perigynia* lance-ovoid, 7–10 mm long, yellowish-green or brownish, conspicuously nerved, the beak 2–4.5 mm long, bidentate; *stigmas* 3; *achenes* oblong-obovoid, 1.9–2.5 × 1.2–1.3 mm, brownish, continuous with the persistent, bony style. Rare along streams and pond margins, 6000–8200 ft. July–Sept. W. (Plate 37)

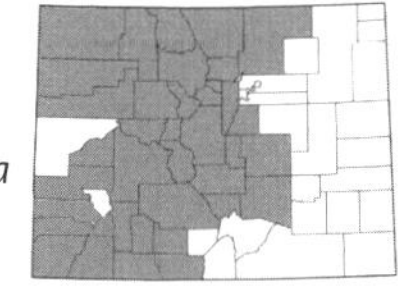

***Carex rossii* F. Boott**, ROSS' SEDGE. [*C. brevipes* W. Boott]. Plants 0.5–4 dm, with very short rhizomes, aphyllopodic; *leaves* flat, 1–4 mm wide; *bracts* leaf-like, the lower longer than the inflorescence; *spikes* 2–6, the upper terminal one staminate and the lower few-flowered and pistillate; *pistillate scales* ovate with an acute or awned tip, with a green center and reddish-brown margins; *perigynia* subglobose with a stipelike base, 3.5–4.5 × 1–2.5 mm, finely hairy, usually greenish, facial nerves absent, the beak 1–1.7 mm long, ciliate-serrulate; *stigmas* 3; *achenes* ovoid, 1.3–2.5 × 1–1.7 mm. Found on dry slopes, in pine or spruce-fir forests, and the alpine, 5400–12,500 ft. May–Aug. E/W. (Plate 37)

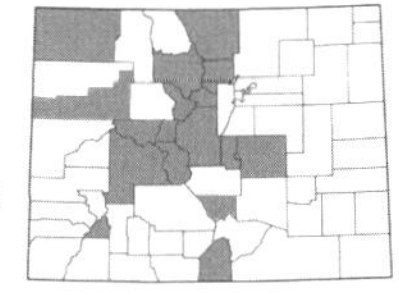

***Carex rupestris* All.**, CURLY SEDGE. Plants 0.4–1.5 dm, rhizomatous, slightly phyllopodic; *leaves* flat, 1–3 mm wide; *bracts* absent; *spikes* solitary, androgynous, to 2 cm long; *pistillate scales* ovate, largely concealing the perigynia, brownish with hyaline margins; *perigynia* oblong-obovoid, 3–4 × 1.7–1.8 mm, brownish above and green below, with conspicuous nerves, the beak terete, ca. 0.2 mm long; *stigmas* 3; *achenes* oblong, dark brown or black, ca. 2.5 mm long. Found on dry slopes and in the alpine, 8400–13,000 ft. July–Aug. E/W. (Plate 37)

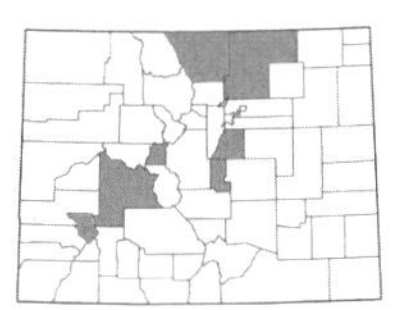

***Carex sartwellii* Dewey**, SARTWELL'S SEDGE. Plants (3) 4–8 dm, rhizomatous; *leaves* flat, 2–4 mm wide; *bracts* inconspicuous; *spikes* androgynous, aggregated into a cylindric head; *pistillate bracts* ovate, brown with hyaline margins, nearly equal to the perigynia; *perigynia* 2.5–4.5 × 1.3–2 mm, brown or tan, prominently nerved, the beak 0.4–1.2 mm long, serrulate; *stigmas* 2; *achenes* ca. 1.5 × 1 mm. Rare along creeks and pond margins and in moist meadows, 7000–10,500 ft. June–Sept. E/W. (Plate 37)

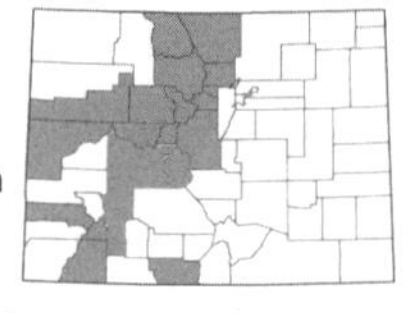

***Carex saxatilis* L.**, ROCK SEDGE. Plants 2–8 dm, rhizomatous, phyllopodic; *leaves* flat, 1.5–5 mm wide; *bracts* leaf-like, 3–15 cm long, sometimes longer than the inflorescence; *spikes* sessile or pedunculate, the upper staminate and the lower pistillate; *pistillate scales* ovate-lanceolate, dark reddish-black with a lighter midrib; *perigynia* ovoid to elliptic, 3–5 × 2 mm, brownish-yellow with a darker upper half, with obscure or no nerves, the beak 0.5 mm; *stigmas* 2 or sometimes 3; *achenes* ovoid, 1.7–2 × 1.5–1.7 mm, yellowish, continuous with the persistent, bony style. Common along streams and lakes, in fens, near melting snowbanks, and in alpine meadows, 8500–12,500 ft. July–Sept. E/W. (Plate 37)

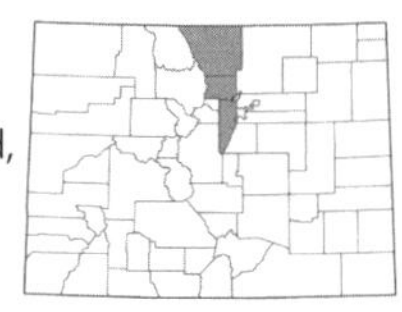

***Carex saximontana* Mack.**, ROCKY MOUNTAIN SEDGE. Plants to 3.5 dm, caespitose; *leaves* flat, 2–5 mm wide; *bracts* leaf-like, longer than the inflorescence; *spikes* solitary, androgynous, few-flowered; *pistillate scales* leaf-like and essentially concealing the perigynia, 7–35 mm long; *perigynia* obovoid, 3.2–5 × 1.6–2.5 mm, green, faintly nerved, the beak 0.6–1.2 mm long, serrulate; *stigmas* 3; *achenes* 2.5–3 × 1.5–2.4 mm. Rare in woodlands and dry canyons, 6000–7500 ft. May–July. E. (Plate 37)

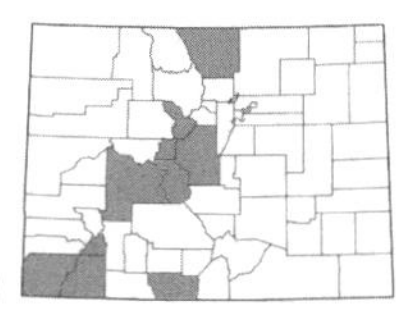

***Carex scirpoidea* Michx.**, NORTHERN SINGLE-SPIKE SEDGE. Plants to 4 (8) dm, rhizomatous, dioecious; *leaves* flat, 1–4 mm wide; *bracts* absent; *spikes* usually solitary, wholly pistillate or staminate; *pistillate scales* ovate, brownish-black with a pale midrib and white-hyaline margins; *perigynia* ovoid, 2–3.5 × 1–1.7 mm, short-hairy, reddish-purple-brown above the middle, obscurely nerved, with a slender beak 0.25 mm long; *achenes* ovate to obovoid, 1.5–2 × 0.75–1 mm. Found in fens, along streams, on rocky ridges, and in the moist alpine, 9000–14,000 ft. July–Sept. E/W. (Plate 37)

There are two subspecies of *C. scirpoidea* in Colorado:

1a. Leaf sheaths and bases from the previous year's leaves persistent; plants not caespitose...**ssp. *pseudoscirpoidea* (Rydb.) Dunlop**

1b. Leaf sheaths and bases from previous year's leaves absent; plants caespitose...**ssp. *scirpoidea***

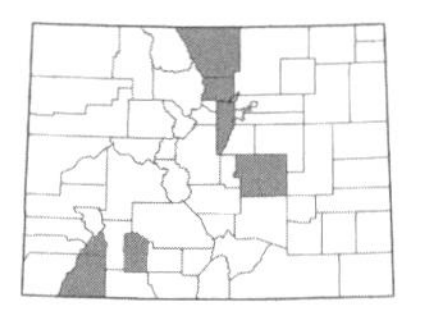

Carex scoparia* Schkuhr ex Willd. var. *scoparia, BROOM SEDGE. Plants 1.5–10 dm, caespitose from short rhizomes; *leaves* 1–3 mm wide; *bracts* inconspicuous or the lower sometimes setaceous-prolonged; *spikes* 3–12, gynaecandrous, sessile, distinguishable; *pistillate scales* lanceolate to ovate, 3.4–4 mm long, narrower than the perigynia, white hyaline with a brownish center; *perigynia* ovate-lanceolate, 4.2–6.8 × 1.2–2 mm, wing-margined, serrulate, flat except over the achene, stramineous to pale green, several-nerved, the beak tapering, flattish; *stigmas* 2; *achenes* 1.3–1.7 × 0.7–0.9 mm. Found along streams and lakes and in moist meadows, 5800–8700 ft. June–Aug. E. (Plate 37)

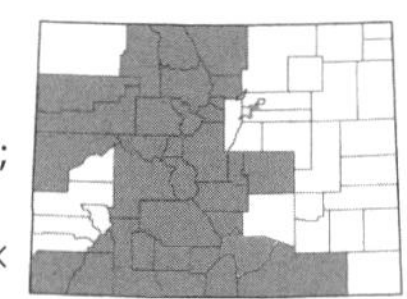

***Carex scopulorum* Holm**, CLIFF SEDGE. Plants 1–4 (6) dm, rhizomatous; *leaves* flat, 2–7 mm wide; *bracts* black-auriculate, the lowest leaf-like, shorter than the inflorescence; *spikes* 3–6, terminal staminate or androgynous, rarely gynaecandrous, the lateral pistillate or sometimes androgynous; *pistillate scales* obovate, black to purplish-black; *perigynia* orbicular to obovoid, 1.8–4 × 1.5–2.5 mm, green to coppery-tan, nerves absent, beak 0.1–0.4 mm long; *stigmas* 2 or 3; *achenes* 1.2–1.8 × 1–1.5 mm. Common in the alpine and meadows, 9000–13,500 ft. June–Aug. E/W. (Plate 38)

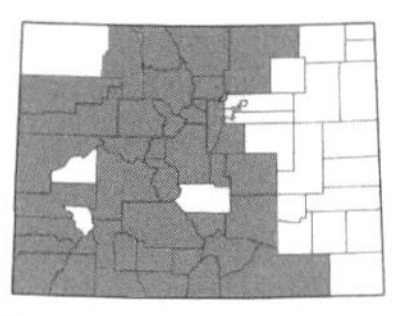

***Carex siccata* Dewey**, DRY SEDGE. [*C. foenea* Willd.]. Plants 1.5–6 (9) dm, rhizomatous, aphyllopodic; *leaves* flat, 1–3 mm wide; *bracts* inconspicuous to evident, shorter than the inflorescence; *spikes* androgynous, aggregated into a cylindric head; *pistillate scales* ovate, concealing the perigynia, brown with a paler midrib, usually hyaline-margined; *perigynia* lanceolate-ovate, 3–6 × 1.5–2 mm, serrulate most the length, greenish-brown, with several nerves, the beaks flattened, serrulate, about ½ as long as the body; *stigmas* 2; *achenes* 1.7–2.2 mm long. Common along streams, on grassy slopes, in forests and meadows, and in the alpine, 5000–13,000 ft. June–Aug. E/W. (Plate 38)

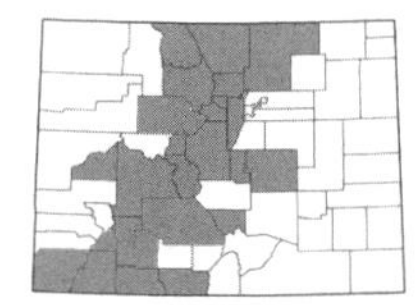

***Carex simulata* Mack.**, ANALOGUE SEDGE. Plants 1–9 dm, rhizomatous; *leaves* 1–4 mm wide; *bracts* inconspicuous, shorter than the inflorescence; *spikes* androgynous or unisexual, aggregated into an oblong head; *pistillate scales* ovate, covering the perigynia, green to brown with a lighter midrib and hyaline margins; *perigynia* ovate, 1.7–2.4 (2.8) × 1.3–1.6 mm, with raised margins, narrowly winged, serrulate at the junction of the body and beak, yellowish-brown, nerves absent or few, beak 0.2–0.6 mm long; *stigmas* 2; *achenes* 1–2.3 × 0.7–1 mm. Found along streams and in moist meadows, 5000–10,000 ft. May–Sept. E/W. (Plate 38)

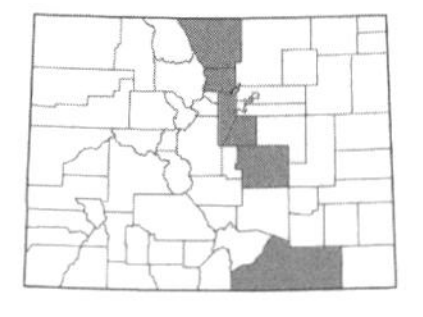

***Carex sprengelii* Dewey ex Spreng.**, SPRENGEL'S SEDGE. Plants 3–9 dm, rhizomatous; *leaves* flat, 2.5–4 mm wide; *spikes* 4–5, the terminal spike staminate and the lateral spikes pistillate, nodding or drooping at maturity; *pistillate scales* ovate-oblong, shorter than the perigynia, with a green midvein and hyaline margins; *perigynia* ovoid-ellipsoid, 4.5–6.5 × 1.5–2 mm, tan to greenish, 2-ribbed, the beak nearly as long as the body, serrulate; *stigmas* 3; *achenes* 2–2.5 × 1.7–1.8 mm. Uncommon along streams and lakes and in forests, 5000–8000 ft. May–July. E. (Plate 38)

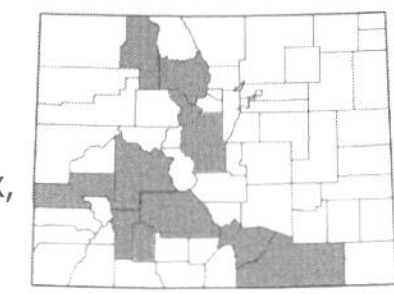

***Carex stenoptila* F.J. Herm.**, RIVERBANK SEDGE. Plants 1.7–7 dm, caespitose; *leaves* flat, 2–3.8 mm wide; *bracts* inconspicuous or bristle-like, shorter than the inflorescence; *spikes* 5–10, densely aggregated into an ovoid head; *pistillate scales* lanceolate to ovate, 3.1–3.8 mm long, shorter than the perigynia, brown or golden; *perigynia* lanceolate, 4–5.1 × 1–1.5 mm, wing-margined, biconvex, green or tan, conspicuously nerved, the beak cylindric, dark; *stigmas* 2; *achenes* 1.4–1.8 × 0.6–1 mm. Uncommon along streams and in meadows, 8000–12,000 ft. June–Aug. E/W. (Plate 38)

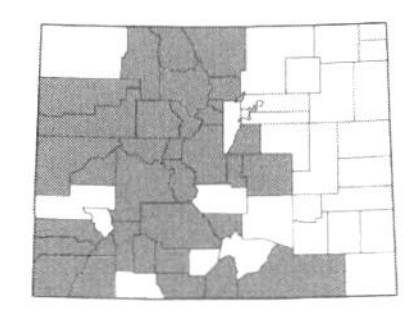

***Carex stevenii* (Holm) Kalela**, NORWAY SEDGE. [*C. norvegica* Retz. ssp. *stevenii* (Holm) E. Murray]. Plants 2–8 dm, rhizomatous, aphyllopodic; *leaves* flat, 1.5–3 mm wide; *bracts* usually leaf-like, shorter to longer than the inflorescence; *spikes* 2–5, the terminal gynaecandrous and the lateral pistillate; *pistillate scales* ovate, obtuse, equal to or shorter than the perigynia, purplish-black with white-hyaline margins; *perigynia* obovate to elliptic, 2–3 × 1–1.3 mm, coppery brown to pale green, the nerves inconspicuous, the beak 0.3–0.5 mm long, black; *stigmas* 3; *achenes* 1.2–1.7 × 0.7–1 mm. Common along streams and in meadows and aspen forests, 8000–12,000 ft. June–Aug. E/W. (Plate 38)

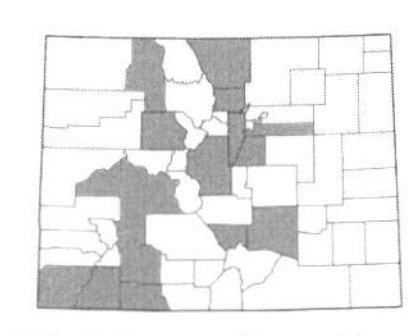

Carex stipata* Muhl. ex Willd. var. *stipata, AWLFRUIT SEDGE. Plants 3.5–10 dm, from short rhizomes, slightly aphyllopodic; *leaves* flat, 4–11 mm wide, cross-wrinkled ventrally; *bracts* short and inconspicuous or sometimes longer than the inflorescence; *spikes* numerous, androgynous, sessile, aggregated into an oblong-linear to ovoid head; *pistillate scales* ovate, acuminate, shorter than the perigynia, with broad hyaline margins; *perigynia* lance-triangular, 3.6–6 × 1.5–1.8 mm, serrulate above, sharp-edged, stramineous to greenish-yellow or brown, with prominent nerves, the beak tapered, ill-defined, serrulate, about the length of the body; *stigmas* 2; *achenes* 1.3–2.3 × 1.3–1.8 mm. Found in moist meadows and along streams and ditches, 5200–8000 ft. May–Aug. E/W. (Plate 38)

***Carex sychnocephala* Carey**, MANYHEAD SEDGE. Plants 0.8–4 dm, caespitose; *leaves* 1.5–4 mm wide; *bracts* leaf-like, longer than the inflorescence; *spikes* 3–8, densely clustered, gynaecandrous; *pistillate scales* lanceolate to ovate, shorter to longer than the perigynia, with a green midvein and white-hyaline margins; *perigynia* lanceolate, (4.5) 5.5–7.3 × 0.7–1.2 mm, greenish to light brown, wing-margined, conspicuously nerved abaxially, the beak 3–4.5 mm long, serrulate; *stigmas* 2; *achenes* 1–1.8 × 0.6–0.8 mm. Rare along the margins of drying ponds, 9000–10,000 ft. June–Sept. E.

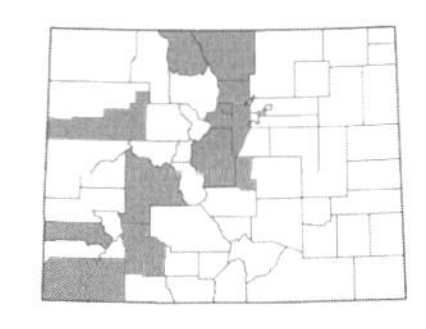

***Carex tahoensis* Smiley**, TAHOE SEDGE. Plants 1.5–4.5 dm, caespitose; *leaves* 1.5–3 mm wide; *bracts* scale-like or bristle-like, shorter than the inflorescence; *spikes* 4–6, distinct, gynaecandrous; *pistillate scales* ovate, 4–5 mm long, equal to the perigynia, reddish-brown with white-hyaline margins; *perigynia* lanceolate to ovate, (3.7) 4.5–6 × 1.5–2.6 mm, wing-margined, plano-convex, conspicuously nerved, the beak serrulate; *stigmas* 2; *achenes* 1.9–2.5 × 1.2–1.6 mm. Uncommon in meadows, grasslands, alpine, and on open slopes, 7000–12,000 ft. June–Aug. E/W.

Carex tenera* Dewey var. *tenera, QUILL SEDGE. Plants 2–9 dm, densely caespitose; *leaves* 1.3–3 mm wide; *bracts* scale-like, shorter than the inflorescence; spikes 3–8, loosely aggregated, ovoid, gynaecandrous; *pistillate scales* ovate, hyaline with a green or brown center, shorter than the perigynia; *perigynia* ovate, plano-convex, 2.8–4.5 × 1.4–2 mm (2–2.3 times as long as wide), brown, wing-margined, serrulate above the middle, plano-convex, with a flat, serrulate beak; *stigmas* 2; *achenes* 1.3–1.7 × 0.8–1.1 mm. Rare in moist meadows, known from a single collection near Wolf Creek Pass (Mineral Co.), 7700–8000 ft. June–Aug. W.

***Carex tenuiflora* Wahlenb.**, SPARSEFLOWER SEDGE. Plants 1–5 dm, loosely caespitose; *leaves* flat or channeled, 0.5–2 mm wide; *bracts* inconspicuous to bristle-like, shorter than the inflorescence; *spikes* 2–4, approximate to densely aggregated in an ovoid head, gynaecandrous; *pistillate scales* ovate, with a green, 3-nerved center and white-hyaline margins; *perigynia* obovate-elliptic, 2.5–3.5 × 1.5–1.8 mm, grayish-green, obscurely few-nerved, the beak absent or nearly so; *stigmas* 2; *achenes* 1.5–2 × 1.2–1.5 mm. Rare in fens, 9000–10,000 ft. July–Aug. E. (Plate 38)

***Carex torreyi* Tuck.**, TORREY'S SEDGE. Plants 1.5–5 dm, caespitose; *leaves* flat, 1.5–3 mm wide, the sheaths pilose; *bracts* inconspicuous or the lower leaf-like and shorter than the inflorescence; *spikes* 2–4, the terminal spike staminate and the lateral spikes gynaecandrous or pistillate; *pistillate scales* ovate, shorter than the perigynia; *perigynia* ovate to obovate, 2.2–3.2 × 1.5–2.2 mm, yellowish-green, strongly nerved, the beak 0.2–0.5 mm long; *stigmas* 3; *achenes* 2–2.5 × 1.5–1.9 mm. Rare on shrubby slopes and in fir and pine forests, 5500–7500 ft. June–Aug. E. (Plate 38)

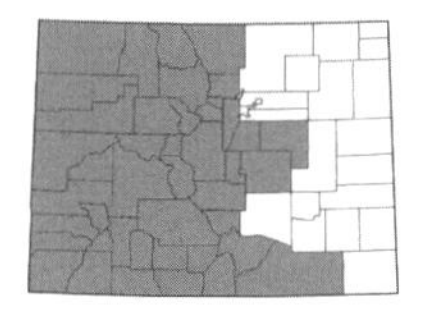

Carex utriculata **Boott**, BEAKED SEDGE. Plants 3–12 dm, rhizomatous, slightly aphyllopodic; *leaves* flat, 4.5–12 mm wide, thick; *bracts* leaf-like, longer than the inflorescence; *spikes* sessile to short-peduncled, the terminal 2–4 staminate and the lateral pistillate or androgynous, cylindric; *pistillate scales* lanceolate, acute to awn-tipped, reddish-brown with a pale center; *perigynia* ovoid, inflated, 3.5–8 × 1.3–3.5 mm, yellowish-green to reddish-brown, prominently nerved, the beak 1–2 mm long, bidentate; *stigmas* 3; *achenes* 1–2 × 1–1.3 mm, the style persistent. Common along streams and lake margins, in bogs and marshes, and in meadows, 5800–11,000 ft. June–Sept. E/W. (Plate 38)

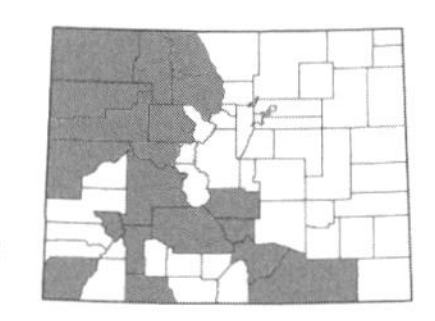

Carex vallicola **Dewey**, VALLEY SEDGE. Plants 1.5–6 dm, rhizomatous, somewhat aphyllopodic; *leaves* flat or canaliculate, 0.5–2 mm wide; *bracts* shorter than the inflorescence; *spikes* androgynous, loosely aggregated into a cylindric head; *pistillate scales* ovate with an acute tip, shorter than the perigynia, brown with a broad white-hyaline margin and pale center; *perigynia* oblong-elliptic, 2.5–4 × 1.5–2.5 mm, greenish or coppery-tinged, the nerves ventro-marginal, the beaks 0.6–1 mm long, minutely serrulate; *stigmas* 2; *achenes* 1.6–2.5 × 1.4–2 mm. Common on dry slopes and along springs, 5900–9600 ft. May–Aug. E/W. (Plate 38)

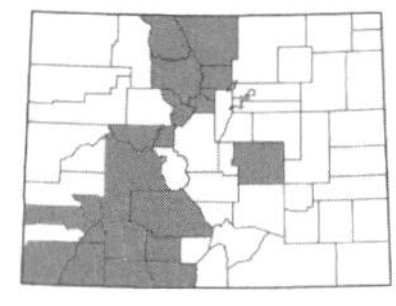

Carex vernacula **L.H. Bailey**, NATIVE SEDGE. [*C. foetida* All. var. *vernacula* (Bailery) Kük.]. Plants to 3 dm, caespitose or shortly rhizomatous, phyllopodic; *leaves* flat, 2–4 mm wide; *bracts* short and inconspicuous or absent; *spikes* androgynous, sessile, densely compacted into a globose or ovoid head 8–16 mm long; *pistillate scales* ovate, usually concealing the perigynia, brown with a pale midrib; *perigynia* lanceolate to obovate, flattened, 3.2–4.5 × 1–1.8 mm, brown with pale green margins, the nerves absent or conspicuous towards the base on each face, the beak prominent, 0.8–1.7 mm long, with a stipe 0.2–0.6 mm long at the base; *stigmas* 2; *achenes* ovate, 1–1.5 × 0.7–1.1 mm. Found in moist meadows and along melting snowbanks and alpine streams, 9000–13,000 ft. July–Aug. E/W. (Plate 38)

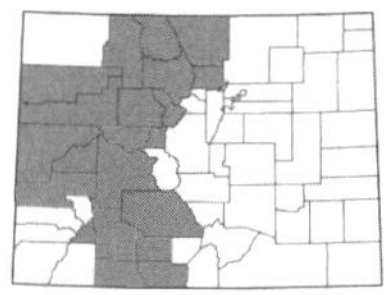

Carex vesicaria **L.**, BLISTER SEDGE. Plants 3–10 dm, with short rhizomes, aphyllopodic; *leaves* flat, 1–7 mm wide; *bracts* leaf-like, longer than the inflorescence; *spikes* several, the upper staminate and the lower pistillate; *pistillate scales* lanceolate to ovate, the tip acuminate to shortly awned, reddish-brown with a paler center and hyaline margins; *perigynia* lanceolate to ovate-lanceolate, (3.5) 4–11 × 1.7–3 mm, reddish-brown to yellowish-green, conspicuously nerved, the beak 1.8–2.5 mm long; *stigmas* 3; *achenes* obovoid, 1.7–3 × 1.1–1.8 mm, continuous with the persistent, bony style. Found along streams and lake shores, in moist forests, and in meadows, 7500–11,000 ft. July–Sept. E/W. (Plate 39)

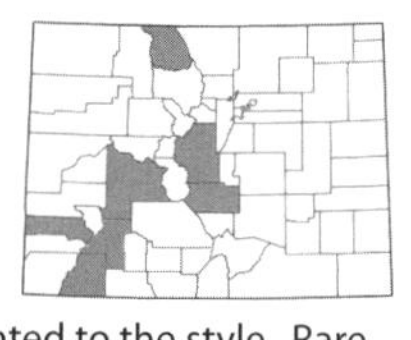

Carex viridula **Michx. ssp. *viridula***, GREEN SEDGE. [*C. oederi* Retz.]. Plants to 4 dm, caespitose, phyllopodic; *leaves* flat, 1–3 mm wide; *bracts* leaf-like, longer than the inflorescence; *spikes* 3–6, sessile or nearly so, the terminal staminate and the lateral pistillate, remote or aggregated; *pistillate scales* obovate, shorter than the perigynia, brown with a paler midrib and narrow, white-hyaline margins; *perigynia* obovoid, 2–4.5 × 0.9–1.5 mm, greenish to yellow, with conspicuous nerves, the beaks 0.6–1.5 mm long; *stigmas* 3; *achenes* truncate at the apex, 1–2 × 0.8–1.1 mm, jointed to the style. Rare in fens, bogs, and moist meadows, 7500–10,700 ft. July–Sept. E/W. (Plate 39)

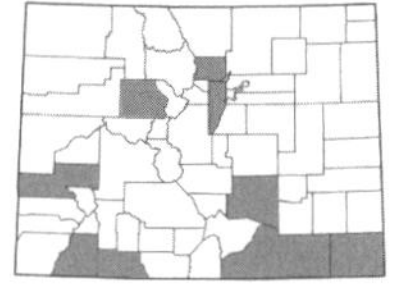

Carex vulpinoidea **Michx.**, FOX SEDGE. Plants 2–10 dm, with stout rhizomes, slightly aphyllopodic; *leaves* flat, 2–5 mm wide; *bracts* prolonged, usually shorter than the inflorescence, conspicuous; *spikes* numerous, androgynous, sessile, densely aggregated into a cylindric head; *pistillate scales* ovate, awned, shorter than or equal to the perigynia, brown with a pale midrib and hyaline margins; *perigynia* ovate, 1.7–3 × 1–1.2 mm, sharp-edged to narrowly winged above, serrulate distally, brown and greenish distally, obscurely nerved, the beak 0.7–1.8 mm long, serrulate, flattened; *stigmas* 2; *achenes* 1.2–1.6 × 0.8–1 mm, jointed to the style. Found in wetlands and along springs and streams, 4000–6000 ft. E/W. (Plate 39)

Carex xerantica **L.H. Bailey**, DRYLAND SEDGE. Plants 3–7 dm, tufted, aphyllopodic; *leaves* flat, 1–3 mm wide; *bracts* inconspicuous or the lowest setaceous-prolonged and shorter than the inflorescence; *spikes* 3–5, gynaecandrous, sessile, approximate but distinguishable; *pistillate scales* ovate, shorter than the perigynia or sometimes concealing them; *perigynia* ovate, 3.8–4.8 × 1.4–2.2 mm, winged, serrulate, plano-convex or concavo-convex, brown or greenish, prominently nerved dorsally, the beak 2.2–3.6 mm long, flattened; *stigmas* 2; *achenes* 1.7–2 × 1–1.5 mm. Possibly present on the eastern plains, but reports of this are primarily based on misidentifications of *C. petasata* and *C. tahoensis*. June–Aug. (Plate 39)

CYPERUS L. – FLATSEDGE

Herbs; *stems* terete or triangular; *leaves* usually basal, ligules absent; *flowers* perfect; *spikelets* flattened, with the scales in 2 ranks; *perianth* absent.

1a. Stigmas 2 (rarely 3); spikelets 3–5 (8) per capitate cluster; stamens 2 (3); achenes lenticular, black or dark green; annuals...***C. bipartitus***

1b. Stigmas 3; spikelets (1) 5–60 (100) per cluster; stamens 1 or 2; achenes usually 3-sided, variously colored; annuals or perennials...2

2a. Floral scales 3-nerved (with 1 midrib and 2 lateral nerves), the tips mostly recurved-acuminate; spikelets in a hemispheric cluster lacking a conspicuous rachis; annuals...***C. acuminatus***

2b. Floral scales with more than 3 nerves, the tips appressed and ascending, spreading, or recurved; spikelets and plants various...3

3a. Floral scales with a recurved tip terminating in a slender, short sharp point or awn 0.5–1 mm long; small annual to 1.6 dm tall...***C. squarrosus***

3b. Floral scales appressed and ascending or spreading, sometimes with an excurrent sharp point to 1 mm long, or if the tips recurved with a sharp point then the plants perennial with rhizomes and 2–7 dm; plants otherwise various...4

4a. Floral scales small, 1.2–1.6 mm long; achenes 0.5–1 mm long; annuals...***C. erythrorhizos***

4b. Floral scales 1.8–5 (6) mm long; achenes 1–2.5 mm long; annuals or perennials...5

5a. Rachilla of spikelets winged (the wing 0.3–0.5 mm wide, giving the rachilla a 4-angled appearance); usually found in moist places such as along drying pond margins, in sand bars, and on muddy shores...6

5b. Rachilla of spikelets wingless; usually found in dry places such as forests, grasslands, on sand dunes, and rocky slopes...8

6a. Perennials with slender rhizomes terminating in small tubers (but these delicate and easily breaking off); floral scales 1.8–3 (3.5) mm long; anthers 1–1.5 (2.1) mm long; achenes 1.3–2 mm long...***C. esculentus***

6b. Annuals or short-lived perennials, lacking tubers; floral scales (2) 2.2–5 (6) mm long; anthers 0.2–0.7 mm long;...7

7a. Floral scales 1–2.8 (3.2) mm long; stems not swollen and cormlike at the base; spikelets easily breaking into sections at the base of each scale, with each section consisting of a scale, the next lower internode, the attached wings, and achene...***C. odoratus***

7b. Floral scales 3.2–5 (6) mm long; stems usually swollen and cormlike at the base; spikelets unlike the above...***C. strigosus***

8a. Floral scales with a stout, excurved mucro (short, sharp point) 0.3–0.6 mm long; spikelet clusters sessile, lacking rays; rhizomes not irregularly thickened...***C. fendlerianus***

8b. Floral scales entire, or the mucro straight or ascending and 0.3–1 mm long; usually at least some spikelet clusters on rays (often the central one sessile, but the others on conspicuous rays), sometimes the rays all absent and the clusters all sessile; rhizomes usually irregularly thickened (appearing knotted or beaded)...9

9a. Floral scales with an entire apex or the tip with a short, sharp point 0.05–0.2 mm long; stems glabrous throughout; spikelets 15–60 per cluster...***C. lupulinus***

9b. Uppermost floral scales with a sharp point 0.3–1 mm (the lower scales with a shorter mucro 0.1–0.4 mm long); stems usually scabridulous at least in the upper half, rarely glabrous throughout; spikelets 5–20 per cluster...***C. schweinitzii***

Cyperus acuminatus **Torr. & Hook.**, TAPERTIP FLATSEDGE. Annuals, 0.2–3.5 dm; *leaves* 8–12 (16) cm × 1–2 (4) mm; *involucral bracts* 2–4, longer than the inflorescence; *inflorescence* of dense, hemispheric clusters; *spikelets* ovoid, 4–7 × 2–3 mm, 10–24-flowered; *scales* 1–2 mm long, strongly 3-nerved, the tip recurved-acuminate; *achenes* stipitate, elliptic, 0.8–1.1 × 0.3–0.4 mm, papillose. Locally common along the margins of ponds, 4500–5700 ft. July–Oct. E. (Plate 39)

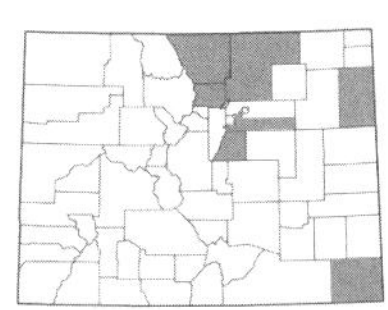

Cyperus bipartitus **Torr.**, SLENDER FLATSEDGE. [*C. rivularis* Kunth]. Annuals, 0.5–2 (3) dm; *leaves* 1–8 cm × 1–2 mm; *involucral bracts* 1–2 (3), leaf-like; *inflorescence* of capitate clusters, each with 3–5 (8) spikelets; *spikelets* oblong, 8–18 × 2–3 mm, 10–30-flowered; *scales* 1.9–2.7 mm long, obtuse, anthocyanic with a pale midrib; *achenes* sessile, lenticular, obovoid to ovoid, 1–1.5 × 0.6–0.8 mm, black or dark green. Uncommon along creeks, in moist meadows, and along drying margins, 3500–5500 ft. July–Sept. E.

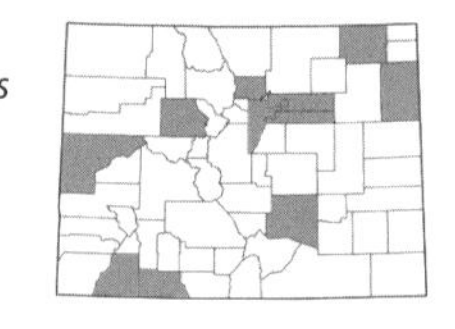

***Cyperus erythrorhizos* Muhl.**, REDROOT FLATSEDGE. Annuals, 0.5–2.5 (10) dm; *leaves* 5–25 (90) cm × 2–6 (11) mm; *involucral bracts* usually 4–10, unequal, leaf-like; *inflorescence* umbelliform; *spikelets* linear, quadrangular, 3–10 × 1–1.5 mm, 8–30-flowered; *scales* 1.2–1.6 mm long, reddish-brown with a green midrib; *achenes* sessile, trigonous, ovoid, 0.5–1 × 0.4–0.6 mm, gray to brown. Uncommon along the drying margins of ponds and lakes, 3500–6700 ft. June–Oct. E/W. Introduced. (Plate 39)

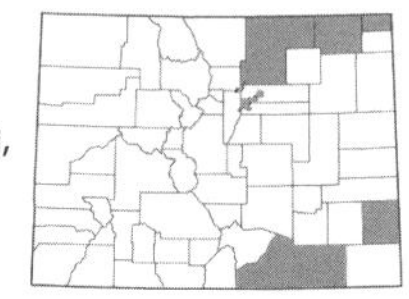

***Cyperus esculentus* L.**, YELLOW NUTSEDGE. Perennials, 3–7 dm; *leaves* (6) 20–40 (80) cm × 2–5 (6.5) mm; *involucral bracts* 3–6, unequal, shorter to longer than the inflorescence; *inflorescence* umbelliform; *spikelets* linear, (5) 10–20 (55) × 1.2–2 (3) mm, 8–20-flowered; *scales* 1.8–3.4 mm long, golden brown to reddish; *rachilla* winged; *achenes* sessile, ellipsoid, trigonous, 1.3–2 mm long, brown. Uncommon along drying pond margins and in sand bars, 3500–5000 ft. June–Aug. E. Introduced. (Plate 39)

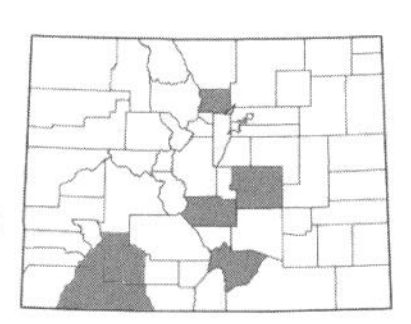

***Cyperus fendlerianus* Boeckeler**, FENDLER'S FLATSEDGE. [*Mariscus fendlerianus* (Boeckeler) T. Koyama]. Perennials, rhizomatous, (0.7) 2–8 dm; *leaves* (12) 20–40 (55) cm × 2–5 (7) mm; *involucral bracts* (2) 3–6 (10), mostly 6–20 cm long; *inflorescence* umbelliform; *spikelets* oblong-lanceolate, quadrangular, 5–8 (10) × 2–4 mm; *scales* 2.4–3 mm long, with a stout, excurved mucro 0.3–0.6 mm long; *achenes* obovoid, 1.6–2 mm long, brown to reddish-brown. Uncommon in grasslands, pine forests, and on rocky slopes and ledges, 5500–8500 ft. June–Aug. E/W. (Plate 39)

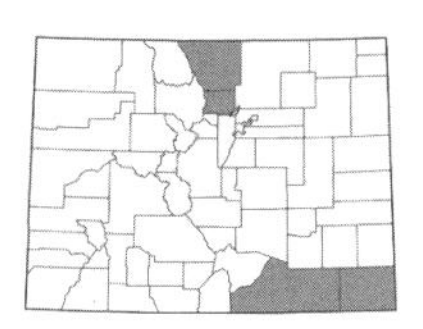

***Cyperus lupulinus* (Spreng.) Marcks**, GREAT PLAINS FLATSEDGE. [*Mariscus filiculmis* (Vahl) T. Koyama]. Perennials, with knotted rhizomes, 1–5 dm; *leaves* 5–40 cm × 1–3.5 mm; *involucral bracts* 2–4, 6–25 cm long; *inflorescence* a single, sessile, capitate cluster; *spikelets* oblong-lanceolate, (3) 6–22 × 2.5–4 mm, 25+-flowered; *scales* ovate-elliptic, 2.5–4 mm long, entire to mucronate; *achenes* sessile, ellipsoid, trigonous, 1.7–2.2 mm long, black or dark brown. Uncommon in grasslands, canyons, and on rocky slopes, 4000–6000 ft. June–Aug. E. (Plate 39)

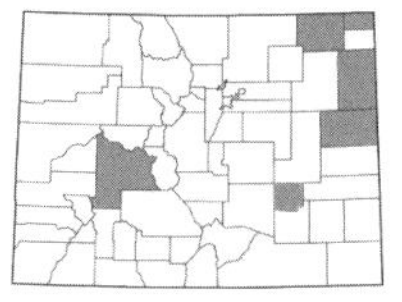

***Cyperus odoratus* L.**, FRAGRANT FLATSEDGE. Annuals or short-lived perennials, to 1 m; *leaves* 5–30 (60) cm × 4–12 mm; *involucral bracts* 5–8, mostly 10–30 cm long, spreading to ascending; *inflorescence* of a dense, capitate cluster; *spikelets* linear, (5) 8–15 (30) × 0.8–1.5 mm, easily breaking into sections at the base of each scale; *scales* ovate to elliptic, 1–2.8 (3.2) mm long; *achenes* stipitate, trigonous, oblong to ellipsoid, 1.2–2 mm long, black or brown. Locally common in muddy shores, sandy riverbanks, and floodplains, 3500–5000 ft. June–Oct. E/W. Introduced. (Plate 39)

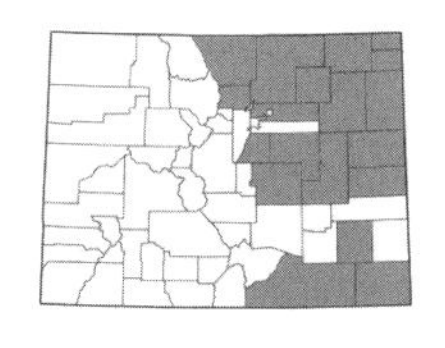

***Cyperus schweinitzii* Torr.**, SCHWEINITZ'S FLATSEDGE. [*Mariscus schweinitzii* (Torr.) T. Koyama]. Perennials, rhizomatous, (1) 2–5 dm; *leaves* 20–35 cm × 2–6 mm; *involucral bracts* 3–7, 8–25 cm long; *inflorescence* umbellate; *spikelets* oblong to narrowly oblong, 7–10 (18) × 2.8–4.5 mm, to 18-flowered; *scales* ovate, 2.3–3.2 mm long, terminated in a sharp point 0.3–1 mm long; *achenes* trigonous, ellipsoid, 2–2.4 mm long, brown to black. Found in dry ravines, sandy floodplains and dunes, on dry slopes, and ridgetops, 3500–6000 ft. June–Sept. E. (Plate 40)

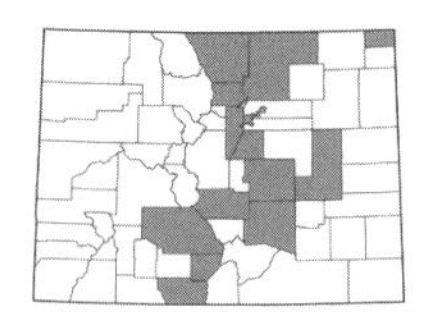

***Cyperus squarrosus* L.**, BEARDED FLATSEDGE. [*C. aristatus* Rottb.]. Annuals, 0.2–1.6 dm; *leaves* 5–15 × 0.5–2.5 mm; *involucral bracts* 2–4, 1–15 cm long; *inflorescence* of capitate clusters, the terminal cluster sessile and others (if present) pedunculate; *spikelets* oblong to narrowly ovoid, 2.5–10 (20) × 1.3–2.2 mm (excluding awns); *scales* with a slender, recurved tip terminating in a slender, sharp point or awn 0.5–1 mm long; *achenes* stipitate, trigonous, 0.7–1 mm long, brown to black. Found along streams, pond shores, in floodplains, and other moist places, 3500–7600 ft. July–Oct. E/W. (Plate 40)

***Cyperus strigosus* L.**, STRAW-COLORED FLATSEDGE. Short-lived perennials, 2–6 (9) dm; *leaves* (10) 20–40 cm × 1–4 (8) mm; *involucral bracts* 3–7, leaf-like, 10–40 cm long; *inflorescence* of spike-like or capitate clusters, some clusters sessile; *spikelets* narrowly lanceolate, 5–30 × 0.6–1 mm, 10–16-flowered; *scales* 3.2–5 (6) mm long, oblong-obovate; *achenes* trigonous, narrowly oblong, 1.5–2.5 mm long, purplish-brown. Reported for Colorado, but no specimens have been seen.

ELEOCHARIS R. Br. – SPIKERUSH

Annual or perennial herbs; *stems* round or 3–5-angled; *leaves* basal, 2 per stem, ligules absent, blades absent or reduced to a short tooth at the tip of the sheath; *flowers* perfect; *spikelets* spirally arranged, terminal; *perianth* usually of 3–6 bristles or absent; *achenes* with a conical or flattened cap at the tip. (Beetle, 1938; Svenson, 1932, 1934, 1937, 1939, 1947)

When measuring the achenes, exclude the beak or tubercle at the apex in the measurement.

1a. Achenes with distinct longitudinal ridges and numerous cross-ridges, 0.7–1.1 mm high with a tubercle forming a distinct apical cap 0.1–0.2 mm high; lowermost floral scale not markedly different from the other scales; perianth bristles usually absent, rarely 2–4; stigmas 3...2
1b. Achenes lacking longitudinal ridges and numerous cross-ridges, smooth or finely wrinkled; lowermost floral scales often markedly different from the other scales and clasping the stem; plants otherwise various...3

2a. Floral scales 1.5–2.5 (3.5) mm; culms usually not compressed, filiform to capillary, the margins not minutely serrulate; plants 1–50 cm tall...***E. acicularis***
2b. Floral scales (2.2) 2.7–3.5 mm; culms distinctly compressed, usually inrolled when dry (rectangular in cross-section), the margins and ridges often minutely serrulate at 20–30×; plants 10–40 cm tall...***E. wolfii***

3a. Stems filiform, 0.1–0.3 mm wide; small, perennial plants 2–6 (10) cm tall with slender rhizomes; floral scales 1.3–2 (2.5) mm long; perianth bristles 6; spikelets 2–4 mm long, these sometimes lacking in deeper water; stigmas 3; achenes 3-sided, 0.9–1.2 mm long...***E. parvula***
3b. Stems not filiform, mostly more than 0.3 mm wide, usually ridged (especially when dry) or angled; plants mostly over 5 cm tall (or if smaller, then tufted annual plants from fibrous roots); floral scales 2–6 mm long; plants otherwise various...4

4a. Tufted annuals with fibrous roots; achenes 0.3–1.5 mm long, with a small tubercle 0.1–0.2 mm long or the tubercle flat and strongly compressed; spikelets usually with numerous flowers, averaging 50–100 per spikelet...5
4b. Perennials with stolons or rhizomes, often mat-forming; achenes various; spikelets with 4–100 flowers per spikelet...7

5a. Achenes black and shiny, 0.3–0.5mm long, with a minute, conic tubercle at the apex; floral scales 0.6–1.3 mm long; perianth bristles usually 4 or sometimes 6, or absent, translucent or white, to about half the length of the achene when present...***E. atropurpurea***
5b. Achenes yellow or brown, 0.9–1.5 mm long, with a flat, strongly compressed, deltoid tubercle at the apex; floral scales 1.5–2.5 mm long; perianth bristles 5–8 or absent, brown, mostly as long as or exceeding the length of the achene when present...6

6a. Tubercles at the apex of achenes ½–⅔ the width of the achene, mostly to ¼ as high as the achene (0.1–0.4 mm); floral scales mostly keeled ...***E. engelmannii***
6b. Tubercles at the apex of achenes more than ⅔ the width of the achene, ⅓–½ as high as the achene (0.35–0.5 mm); floral scales mostly not keeled...***E. obtusa***

7a. Stigmas mostly 2 (rarely some with 3); achenes biconvex to lenticular...***E. palustris***
7b. Stigmas mostly 3; achenes 3-sided or terete...8

8a. Achenes narrowed at the tips into a thick, beaklike region, the tubercle at the apex of the achenes similar to and merging with the achene apex in color and texture (not forming a cap), or rarely the tubercle absent; floral scales with an entire apex...9
8b. Achenes not narrowed at the tips into a thick beaklike region, the tubercle at the apex of the forming a distinct apical cap clearly differentiated from the achene apex; floral scales with an entire, notched, or bifid apex...11

9a. Stems more or less distinctly flattened (at least distally), at least some of them usually arching and rooting at the tips (stoloniferous); floral scales (and flowers) 10–40 per spikelet...***E. rostellata***
9b. Stems terete or 3-angled (sometimes slightly compressed), not stoloniferous; floral scales (and flowers) 2–10 (12) per spikelet...10

10a. Achenes 0.7–1.5 (3) mm long; lowermost floral scale usually subtending a flower; perianth bristles 3–6 or rarely absent, often unequal in length...***E. pauciflora***
10b. Achenes 2–3 mm long; lowermost floral scale usually empty; perianth bristles 6, equal in length...***E. suksdorfiana***

11a. Floral scales all with an entire, acute apex; stems subterete, usually with 6 prominent ridges (when dry); tubercle often 3-lobed if viewed from the top; perianth bristles 3–6...***E. bolanderi***
11b. At least some floral scales notched or bifid (at least the lowermost scales); stems subterete to greatly compressed and flattened, with 2–12 ridges; tubercle not 3-lobed as viewed from the top, often rudimentary; perianth bristles absent or to 5...***E. compressa***

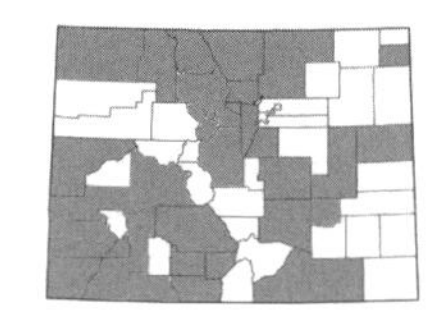

Eleocharis acicularis **(L.) Roem. & Schult.**, NEEDLE SPIKERUSH. Perennials, rhizomatous, 0.1–5 dm; *spikelets* 2–9 mm long; *scales* 4–25, reddish to purplish-brown or stramineous with a green midrib, 1.5–2.5 (3.5) mm long; *perianth bristles* usually absent; *stigmas* 3; *achenes* white to yellowish, with distinct longitudinal ridges and numerous cross-ridges, 0.7–1.1 mm long, with a tubercle forming a distinct apical cap 0.1–0.2 mm high. Common along streams, margins of ponds and pools, and in creek bottoms, 3500–10,000 ft. July–Oct. E/W. (Plate 40)

Eleocharis atropurpurea **(Retz.) Kuntz**, PURPLE SPIKERUSH. Annuals, 0.2–2 (4) dm; *spikelets* ovoid to elliptic, 2–6 (8) mm long; *scales* to 100, reddish-brown to stramineous, 0.6–1.3 mm long, with a rounded to acute tip; *perianth bristles* usually 4, vestigial to ½ as long as the achene; *stigmas* 3; *achenes* black, smooth, obovoid, 0.3–0.5 mm long, with a minute, conic tubercle at the apex. Reported for the eastern plains of Colorado, but no specimens have been seen. June–Sept. E.

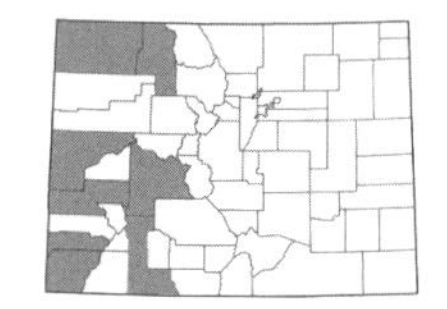

Eleocharis bolanderi **A. Gray**, BOLANDER'S SPIKERUSH. Perennials, with short rhizomes, 1–3 dm; *spikelets* 3–8 mm long; *scales* 8–30, black to dark brown with a greenish or pale midrib, 2–3 mm long; *perianth bristles* 3–6, vestigial to ½ as long as the achene; *stigmas* 3; *achenes* golden-yellow or rarely dark brown, minutely roughened, the tubercle pyramidal, lower than wide, often 3-lobed if viewed from the top. Uncommon in moist meadows, along streams, and near springs, 7000–9000 ft. May–July. W. (Plate 40)

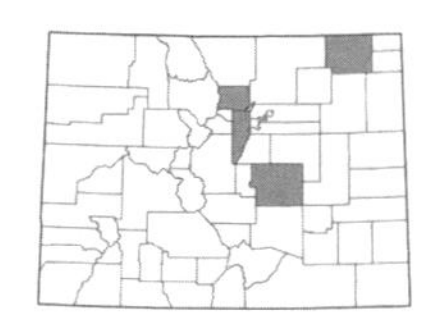

Eleocharis compressa **Sull.**, FLATSTEM SPIKERUSH. Perennials, rhizomatous, 0.8–4.5 dm; *spikelets* 4–8 mm long; *scales* 20–60, brown or sometimes colorless, with a paler midrib, usually bifid or notched at the apex, 2–3 (4) mm long; *perianth bristles* 0–5, equal to the achene in length; *stigmas* 3; *achenes* golden-brown to brown, reticulate, 0.8–1.5 mm long, the tubercle usually depressed and often rudimentary. Found in moist meadows, grasslands, and on sandy floodplains, 4000–7600 ft. May–July. E. (Plate 40)

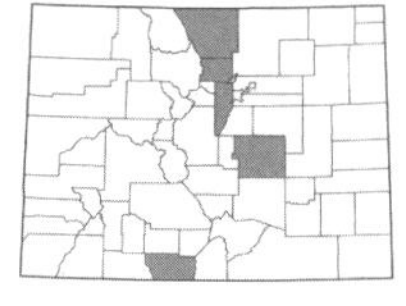

Eleocharis engelmannii **Steud.**, ENGELMANN'S SPIKERUSH. Annuals, tufted, to 4 dm; *spikelets* 5–10 (20) mm long; *scales* 25–100, brown to golden-brown, 2–3 mm long; *perianth bristles* 5–8 or absent, vestigial to longer than the tubercle; *stigmas* 2–3; *achenes* 0.9–1.5 mm long, with a flat, depressed tubercle 0.1–0.3 mm high. Uncommon along the drying margins of ponds and lakes, 5000–7500 ft. June–Oct. E. (Plate 40)

Eleocharis obtusa **(Willd.) Schult.**, BLUNT SPIKERUSH. Annuals, 0.3–5 (9) dm; *spikelets* (2) 5–13 mm long; *scales* 15–150+, golden-brown, 1.5–2.5 mm long; *perianth bristles* 5–7 or rarely absent, usually longer than the tubercle; *stigmas* 2–3; *achenes* yellow to brown, 0.7–1.2 mm long, with a flat, deltoid tubercle 0.35–0.5 mm. Found along the margins of ponds and lakes, 5000–7500 ft. June–Oct. E. (Plate 40)

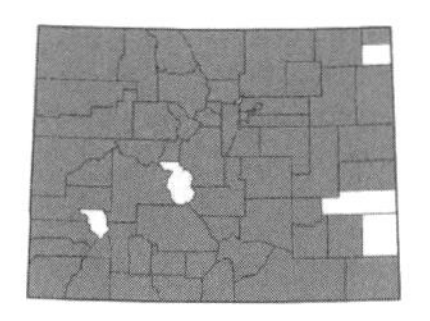

Eleocharis palustris **(L.) Roem. & Schult.**, COMMON SPIKERUSH. [*E. erythropoda* Steudel; *E. macrostachya* Britton ex Small; *E. xyridiformis* Fernald & Brackett]. Perennials, with long rhizomes, 3–12 dm; *spikelets* ovoid, 5–25 mm long; *scales* 30–100+, brown with green midribs, 3–5 mm long, with acute tips; *perianth bristles* usually 4, sometimes absent, shorter or longer than the achene; *stigmas* 2; *achenes* brown, smooth, 1–2 mm long, with a pyramidal tubercle 0.3–0.7 mm high. Common along ditches, streams, pond margins, and in moist meadows, 3500–10,500 ft. May–Aug. E/W. (Plate 40)

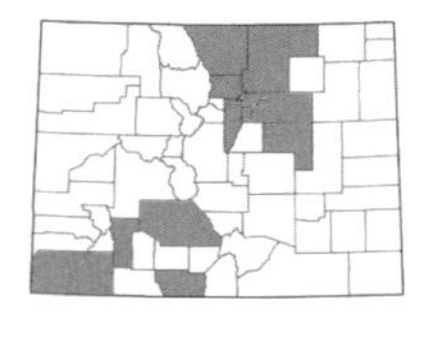

Eleocharis parvula **(Roem. & Schult.) Link ex Bluff, Nees & Schauer**, SMALL SPIKERUSH. [*E. coloradoensis* (Britton) Gilly]. Perennials from slender rhizomes, to 2–6 (10) cm; *spikelets* 2–4 mm long; *scales* 6–10, brown to stramineous, 1.3–2 (2.5) mm long; *perianth bristles* 6, usually equal to or longer than the achene; *stigmas* 3; *achenes* brown to stramineous, smooth, 0.9–1.2 mm long, with a confluent tubercle 0.1–0.2 mm long. Found on muddy shores of drying pools and ponds, 3900–10,500 ft. July–Sept. E/W.

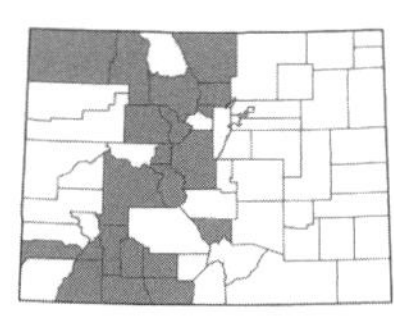

Eleocharis pauciflora **(Lightf.) Link**, FEWFLOWER SPIKERUSH. [*E. quinqueflora* (Hartmann) O. Schwarz]. Perennials from slender rhizomes, 0.5–3.5 dm; *spikelets* 3–8 mm long, the lowermost floral scale usually subtending a flower; *scales* 3–10, brown with pale margins, 2.5–6 mm long; *perianth bristles* 3–6 or rarely absent, usually equal to the achene; *stigmas* 3; *achenes* brown to gray-brown, 0.7–1.5 (3) mm long, with a confluent tubercle 0.3–0.4 mm high. Common along lake and pond margins, streams, in moist meadows, 6000–12,500 ft. June–Sept. E/W. (Plate 40)

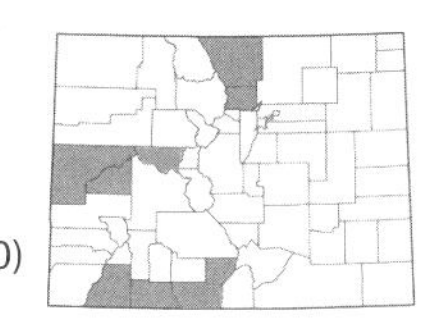

Eleocharis rostellata **Torr.**, BEAKED SPIKERUSH. Perennials from short rhizomes, 2–10 dm; *spikelets* 5–17 mm long; *scales* 10–40, brown to stramineous with a pale midrib, 3.5–6 mm long; *perianth bristles* usually equal to the achene and tubercle; *stigmas* 3; *achenes* brown to greenish-brown, 1.5–3 mm long with a confluent tubercle about ⅓ the length of the achene. Uncommon in swamps and bogs, near springs, and in moist meadows, 5500–9000 ft. June–Sept. E/W. (Plate 40)

Eleocharis suksdorfiana **P. Beauv.**, SUKSDORF'S SPIKERUSH. Perennials, rhizomatous, (0.5) 1–4 dm; *spikelets* 5–10 mm long, the lowermost floral scale usually empty; *scales* 8–12, 3.5–5 mm long; *perianth bristles* 6, equal to or longer than the achene; *stigmas* 3; *achenes* brown to gray-brown, 2–3 mm long, with a confluent tubercle 0.4–0.5 mm high. Reported for Colorado, but no specimens have been seen.

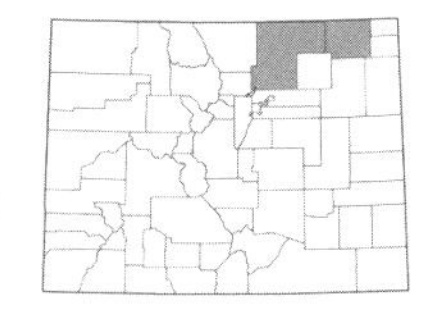

Eleocharis wolfii **(A. Gray) A. Gray**, WOLF'S SPIKERUSH. Perennials, rhizomatous, 1–4 dm; *spikelets* 3–10 mm long; *scales* 15–30, golden-brown or stramineous to colorless, (2.2) 2.7–3.5 mm long; *perianth bristles* absent; *stigmas* 3; *achenes* white to pale brown, with distinct longitudinal ridges and numerous cross-ridges, 0.7–1.1 mm long, with a pyramidal, depressed tubercle 0.1–0.15 mm high. Rare along the margins of ponds and in low swales, 4000–4500 ft. June–Sept. E.

ERIOPHORUM L. – COTTON-GRASS

Perennial herbs; *stems* round or triangular; *leaves* basal and cauline, ligules present; *flowers* perfect; *perianth* of 10–25 smooth bristles.

1a. Spikelets solitary at the apex of the stem...2
1b. Spikelets several at the apex of the stem, usually at least some of them on pedicels...3

2a. Perianth bristles usually reddish-brown (at least partly) or sometimes white; anthers (0.6) 1.5–3 mm long; hyaline margins of fertile scales 1 mm or more wide...***E. chamissonis***
2b. Perianth bristles bright white; anthers 0.5–1.5 mm long; hyaline margins of fertile scales to 1 mm wide...***E. scheuchzeri***

3a. Leaf blades 3-sided throughout, 1–2 mm wide; distal leaf blade 1–4 cm long, shorter than the sheath; perianth bristles 10–15 mm long; scales 3–4 mm long...***E. gracile***
3b. Leaf blades flat but 3-sided toward the tips, 1.5–6 mm wide; distal leaf blade much longer than the sheath; perianth bristles 15–30 mm long; scales 4–10 mm long...4

4a. Scale midrib fading toward the tip, not reaching the end of the scale; anthers 2–5 mm long; scales 5–10 mm long...***E. angustifolium***
4b. Scale midrib prominent and enlarged at the tip, reaching the end of the scale or nearly so; anthers 0.8–2 mm long; scales 4–6 mm long...***E. viridicarinatum***

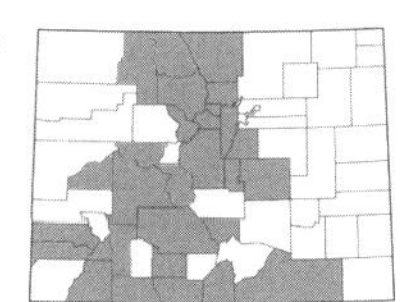

Eriophorum angustifolium **Honck.**, TALL COTTON-GRASS. Plants to 10 dm; *leaves* flat with a trigonous-channeled tip, ca. 1.5–6 mm wide; *inflorescence* with 1–3 blade-bearing involucral bracts, the longest 1–12 cm; *spikelets* several, 10–20 mm long in flower, 20–50 mm long in fruit; *scales* 5–10 mm long, the midrib fading toward the tip; *perianth bristles* 10+, white to tawny, 15–30 mm long; *achenes* 2–5 mm long, black. Common in bogs, fens, moist meadows, and along lake margins, 7000–12,500 ft. June–Sept. E/W. (Plate 40)

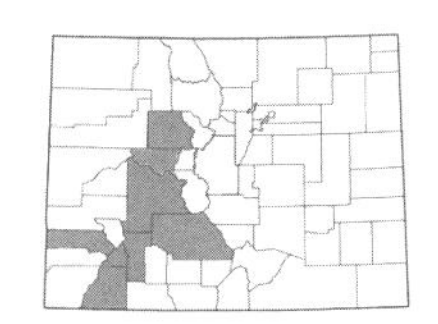

Eriophorum chamissonis **C.A. Mey.**, CHAMISSO'S COTTON-GRASS. Plants 2–8 dm; *leaves* trigonous-channeled in cross-section,1–2 mm wide; *inflorescence* lacking involucral bracts; *spikelets* solitary, 15–20 (40) mm; *scales* 4–20 mm long, black to purplish-brown with hyaline margins ca. 1+ mm wide; *perianth bristles* 10+, usually reddish-brown at least in part or sometimes white, 20–40 mm long; *achenes* 2–3 mm long. Found in fens and on lake margins, 10,000–12,000 ft. June–Aug. W.

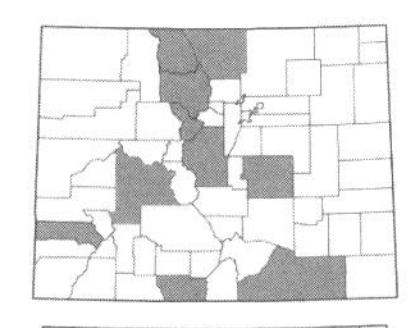

Eriophorum gracile **W.D.J. Koch in Roth**, SLENDER COTTON-GRASS. Plants 2–6 dm; *leaves* trigonous-channeled in cross-section,1–2 mm wide; *inflorescence* with a solitary blade-bearing involucral bract 0.6–2 cm long; *spikelets* 2–5, 7–10 mm long in flower, 15–25 mm long in fruit; *scales* 3–4 mm long, black; *perianth bristles* 10+, white, 10–15 mm long; *achenes* 1.5–3 mm long. Uncommon in fens and moist meadows and along lake margins, 8000–12,500 ft. June–Aug. E/W.

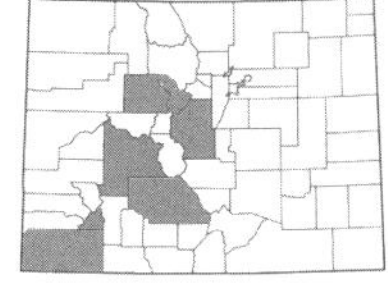

Eriophorum scheuchzeri **Hoppe**, WHITE COTTON-GRASS. Plants 0.5–3.5 (7) dm; *leaves* involute or channeled, ca. 1 mm wide; *inflorescence* lacking involucral bracts; *spikelets* solitary, 8–12 (40) mm long; *scales* 4–10 mm long, gray to blackish with hyaline margins to 1 mm wide, the midrib not reaching the tip; *perianth bristles* 10+, bright white, 15–30 mm long; *achenes* 0.4–2.5 mm long. Found in moist meadows and the alpine and fens, 10,200–12,700 ft. June–Aug. E/W.

***Eriophorum viridicarinatum* (Engelm.) Fernald**, THINLEAF COTTON-GRASS. Plants 2–9 dm; *leaves* flat or trigonous distally, 2–6 mm wide; *inflorescence* with 2–4 blade-bearing involucral bracts, to 7 cm long; *spikelets* several, 6–10 mm long in flower, 15–30 mm long in fruit; *scales* 4–6 mm long, greenish-black to gray sometimes with scarious margins to 0.1 mm wide, the midrib prominent and enlarged at the tip; *perianth bristles* 10+, white to tawny, 15–25 mm long; *achenes* 2.5–3.5 mm long. Reported for Colorado, but no specimens have been seen.

FIMBRISTYLIS Vahl – FIMBRY

Perennial herbs; *stems* terete; *leaves* basal, ligule absent; *flowers* perfect; *perianth* absent.

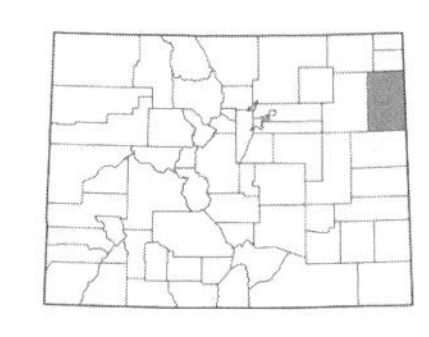

***Fimbristylis puberula* Vahl var. *interior* (Britton) Kral**, HAIRY FIMBRY. Plants to 10 dm; *leaves* 1–2 mm wide, usually involute; *inflorescence* of several spikelets in terminal umbellate cymes, subtended by several involucral bracts; *spikelets* 5–10 mm long; *scales* 2.5–3.5 mm long; *achenes* brown or yellowish, ca. 1 mm long, with several longitudinal ribs and finer horizontal lines. Rare on floodplains and in grassy meadows, 3500–4000 ft. June–Aug. E.

KOBRESIA Willd. – BOG SEDGE

Perennial herbs; *stems* triangular; *leaves* basal and cauline, filiform to involute, ligules present; *flowers* imperfect, perigynium present and enclosing the pistillate flower; *perianth* absent.

1a. Inflorescences usually compound, the terminal upper spikelets staminate, the lower spikelets 1-flowered and pistillate or sometimes 2-flowered and androgynous; basal leaf sheaths usually with the remains of the blades attached...***K. simpliciuscula***

1b. Inflorescences simple, the upper spikelets staminate, the lower spikelets usually androgynous or rarely 1-flowered and pistillate; basal leaf sheaths usually bladeless...2

2a. Perigynium 2–3.5 mm long; scales 2–3.5 mm long with the midvein visibile nearly to the tip; inflorescence 2–3 mm wide...***K. myosuroides***

2b. Perigynium 3.5–5.5 mm long; scales 3.5–5.5 mm long with the midvein visible only near the base; inflorescence 4–8 mm wide...***K. sibirica***

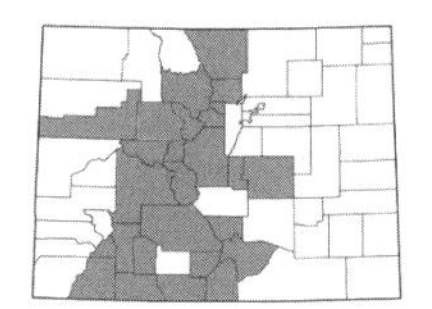

***Kobresia myosuroides* (Vill.) Fiori**, BELLARDI'S BOG SEDGE. [*K. bellardii* (All.) Degland]. Plants 0.5–2 (3.5) dm; *leaves* filiform, 2–20 cm × 0.2–0.5 mm; *inflorescence* simple, 2–3 mm wide, the upper spikelets staminate and the lower spikelets usually androgynous or rarely 1-flowered and pistillate; *scales* 2–3.5 mm long, brown with hyaline margins, the midvein visible nearly to the tip; *perigynia* 2–3.5 mm long, the margins free to the base; *achenes* 2–3 mm long. Common in alpine and meadows, 9200–13,000 ft. June–Sept. E/W. (Plate 40)

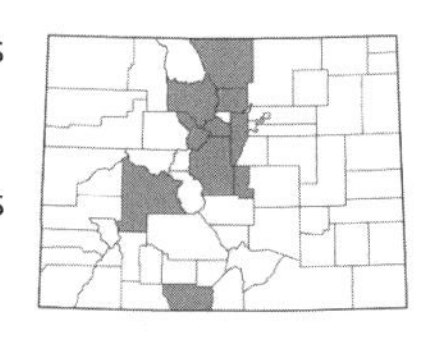

***Kobresia sibirica* Turcz. ex Boeckeler**, SIBERIAN BOG SEDGE. [*K. schoenoides* (C.A. Mey.) Steud.]. Plants 0.5–4 dm; *leaves* filiform, 2–15 cm × 0.25–0.5 mm; *inflorescence* simple, 4–8 mm wide, the upper spikelets staminate and the lower spikelets androgynous or 1-flowered and pistillate; *scales* 3.5–5.5 mm long, the midvein distinct only near the base; *perigynia* 3.5–5.5 mm long, the margins connate near the base; *achenes* 2.5–4 mm long. Uncommon in moist alpine, 11,500–13,500 ft. July–Sept. E/W.

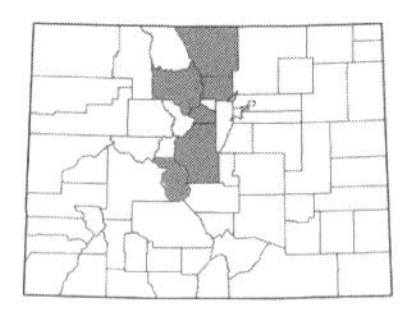

***Kobresia simpliciuscula* Mack.**, SIMPLE BOG SEDGE. Plants 0.5–3.5 dm; *leaves* 2–20 cm × 0.2–1.5 mm; *inflorescence* usually compound, the terminal upper spikelets staminate and the lower spikelets 1-flowered and pistillate or sometimes 2-flowered and androgynous; *scales* 2–3 mm long; *perigynia* 2.5–3.2 mm long, the margins free to the base; *achenes* 2–3 mm long. Uncommon in moist alpine and forming tussocks in fens, 9200–13,000 ft. July–Sept. E/W.

LIPOCARPHA R. Br. in Tuckey – HALFCHAFF SEDGE

Annual herbs; *stems* round; *leaves* basal, ligules absent; *flowers* perfect; *perianth* of 1–2 scales.

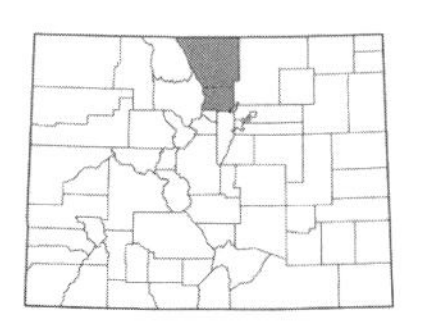

***Lipocarpha aristulata* (Coville) G.C. Tucker**, AWNED HALFCHAFF SEDGE. [*Hemicarpha micrantha* (Vahl) Pax var. *aristulata* Coville]. Plants 2–15 cm; *leaves* filiform, 1–6 cm × 0.4–0.6 mm; *inflorescence* of 1–2 ovoid clusters 1–5 (8) mm long, subtended by 1–2 bracts, the longest erect and 1–3.5 cm long; *spikelets* arranged in a tight spiral; *scales* solitary per spikelet, 1–1.6 mm long, long-acuminate and curving outward; *stigmas* 2; *achenes* 0.5–0.8 × 0.25–0.4 mm. Uncommon along drying pond margins and in wetlands, 5000–6000 ft. July–Oct. E. (Plate 41)

RHYNCHOSPORA Vahl – BEAK-RUSH

Perennial herbs (ours); *stems* round or obscurely triangular; *leaves* basal and cauline, ligules present or absent; *flowers* perfect; *perianth* of 2–12 bristles.

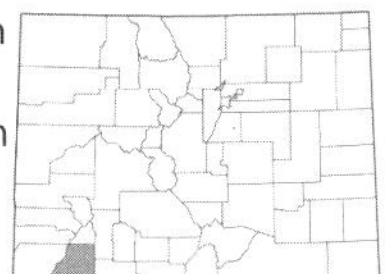

***Rhynchospora alba* (L.) Vahl**, WHITE BEAK-RUSH. Plants 0.6–7.5 dm; *leaves* filiform to linear, 0.5–1.5 mm wide; *inflorescence* of 1–3 clusters, subtended by leafy bracts; *spikelets* 3.5–5.5 mm long, white to pale brown; *scales* 3–4 mm long; *perianth bristles* 10–12, longer than the achene; *achenes* 2.3–3 mm long, tuberculate, brown. Rare in mossy fens, 8500–8600 ft. July–Oct. W. (Plate 41)

SCHOENOPLECTUS (Rchb.) Palla – NAKED-STEMMED BULRUSH

Annual or perennial herbs; *stems* triangular or terete; *leaves* basal, rarely cauline, ligules membranous; *flowers* perfect; *perianth* of 0–6 bristles.

1a. Tufted annual (rarely perennial) with very short, inconspicuous rhizomes; stems slender, mostly less than 1.5 mm wide; achenes with prominent horizontal ridges, sharply 3-sided; spikelet scale tips entire, acute to acuminate...***S. saximontanus***

1b. Perennials, usually with conspicuous rhizomes; stems stout, mostly over 2 mm wide; achenes smooth, plano-convex or weakly 3-sided; spikelet scale tips notched and awned...2

2a. Stems round; inflorescence 2–4-times branched...3

2b. Stems triangular; inflorescences capitate, sessile, usually not branched...4

3a. Spikelet scales 3.5–4 mm long, with mostly strongly contorted awns 0.5–2 mm long (these often broken off); spikelets never all solitary...***S. acutus***

3b. Spikelet scales 2–3.5 mm long, with straight or bent awns 0.2–0.8 mm long; spikelets often all solitary...***S. tabernaemontani***

4a. Bract subtending the inflorescence 1–6 cm long, secondary involucral bracts lacking blades; spikelet scales 2.5–4 mm long with an apical notch 0.1–0.4 mm deep; achenes 1.5–3 mm long...***S. americanus***

4b. Bract subtending the inflorescence 3–20 cm long, secondary involucral bracts with blades (resembling large scales); spikelet scales 3.5–6 mm long with an apical notch 0.5–1 mm deep; achenes (2) 2.5–3.5 mm long...***S. pungens***

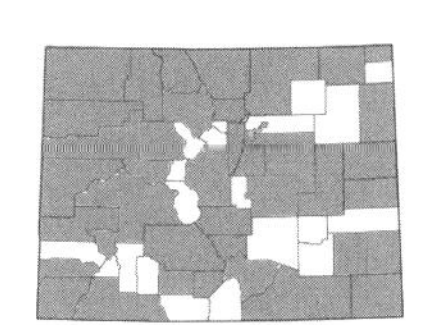

***Schoenoplectus acutus* (Muhl. ex J.M. Bigelow) Á. Löve & D. Löve**, HARDSTEM BULRUSH. [*Scirpus acutus* Muhl. ex J.M. Bigelow]. Perennials to 4 m; *stems* terete; *involucral bract* usually solitary, appearing as a continuation of the stem; *inflorescence* 2–3 times branched, *spikelets* in clusters of 2–8 or solitary, 10–15 mm long; *scales* 3.5–4 mm long, usually with strongly contorted awns 0.5–2 mm long; *bristles* 6, shorter than to equal to the achene; *achenes* obovoid, 1.5–3 × 1.2–1.7 mm, beak 0.2–0.4 mm long. Common along lake margins and streams, 3500–8000 ft. May–Sept. E/W. (Plate 41)

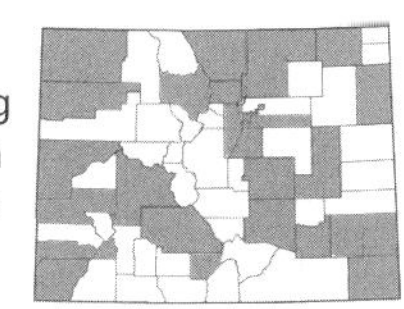

***Schoenoplectus americanus* (Pers.) Volkart ex Schinz & R. Keller**, OLNEY'S THREE-SQUARE BULRUSH. [*Scirpus americanus* Pers.]. Perennials to 2.5 m; *stems* triangular; *involucral bract* solitary, appearing as a continuation of the stem, 1–6 cm long; *inflorescence* capitate; *spikelets* 2–20, sessile, to 15 mm long; *scales* 2.5–4 mm long with an apical notch 0.1–0.4 mm deep; *bristles* 4–6, shorter to equal to the achene; *achenes* obovoid, 1.8–2.8 × 1.3–2 mm with a beak 0.1–0.3 mm long. Found in moist places along streams, ditches, and pond margins, 4500–8000 ft. May–Sept. E/W.

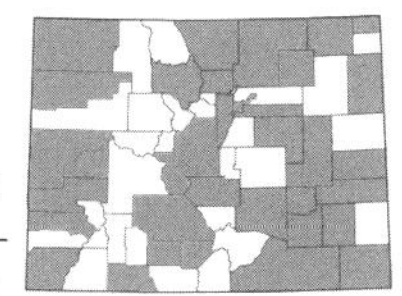

***Schoenoplectus pungens* (Vahl) Palla**, COMMON THREE-SQUARE BULRUSH. [*Scirpus pungens* Vahl]. Perennials to 2 m; *stems* triangular; *involucral bract* 3–20 cm long, appearing as a continuation of the stem, with up to 2 smaller bracts usually present and resembling large scales; *inflorescence* capitate; *spikelets* 1–5 (10), sessile, 5–23 mm long; *scales* 3.5–6 mm long with an apical notch 0.5–1 mm deep; *bristles* 4–8, equal to the achene; *achenes* obovoid, (2) 2.5–3.5 × 1.3–2.3 with a beak 0.1–0.5 mm long. Common along lake and pond margins and streams, 3500–8000 ft. May–Sept. E/W.

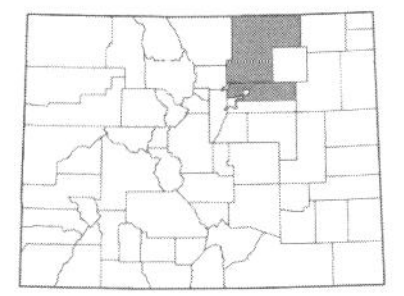

***Schoenoplectus saximontanus* (Fernald) J. Raynal**, ROCKY MOUNTAIN BULRUSH. [*Scirpus saximontanus* Fernald]. Annuals to 4 dm; *stems* terete, slender, mostly less than 1.5 mm wide; *involucral bract* appearing as a continuation of the stem, with additional smaller bracts below the inflorescences; *inflorescence* capitate; *spikelets* 1–20, 6–20 mm long; *scales* 2.2–3.5 mm long; *bristles* often absent; *achenes* sharply 3-sided, 2.2–3 mm long (including the beak), usually with prominent horizontal ridges. Rare along the shore of reservoirs and ponds, 4500–5000 ft. July–Oct. E.

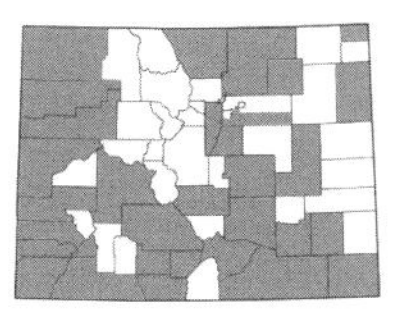

***Schoenoplectus tabernaemontani* (K.C. Gmel.) Palla**, SOFTSTEM BULRUSH. [*Scirpus validus* Vahl]. Perennials to 3 m; *stems* terete; *involucral bract* 1–8 cm, appearing as a continuation of the stem; *inflorescence* 2–4-times branched; *spikelets* often all solitary or in clusters of 2–4 (7), 3–17 mm long; *scales* 2–3.5 mm long, with a straight or bent awn 0.2–0.8 mm long; *bristles* 6, equal to the achene; *achenes* obovoid, 1.5–2.8 × 1.2–1.7 mm with a beak 0.2–0.4 mm. Common along the margins of lakes and ponds, in streams and ditches, and other riparian areas, 3500–7000 ft. May–Sept. E/W.

SCIRPUS L. – BULRUSH

Perennial herbs; *stems* usually triangular; *leaves* basal and cauline or all cauline, ligules present; *flowers* perfect; *perianth* of 3–6 bristles. The filaments often persist after the anthers have fallen and can be mistaken for perianth bristles. (Beetle, 1940; 1941; 1942; 1943; 1944)

1a. Spikelets solitary on long pedicels (drooping), except the central spikelet of each cyme which is sessile; perianth bristles about twice as long as the achene, strongly contorted, smooth...***S. pendulus***
1b. Spikelets sessile, 3–130, in dense clusters (the largest clusters with 6 or more spikelets); perianth bristles shorter than to 1.5 times the length of the achene, straight or curved but not contorted, toothed at least in the upper ⅓...2

2a. Stigmas usually 2; achenes usually lenticular; inflorescence scales usually with an ill-defined midstrip that is sometimes shortly excurrent as a very small point; basal leaf sheaths usually reddish-purple-tinged...***S. microcarpus***
2b. Stigmas usually 3; achenes usually 3-sided; inflorescence scales with a conspicuous, thickened midrib that is exserted as a short awn mostly about 0.5 mm long; basal leaf sheaths green or pale, usually not reddish-purple-tinged...***S. pallidus***

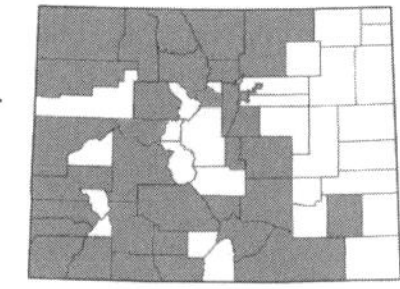

***Scirpus microcarpus* J. Presl & C. Presl**, PANICLED BULRUSH. Plants to 15 dm; *involucral bracts* usually 3, foliaceous, usually longer than the inflorescence; *inflorescence* of dense clusters of sessile spikelets; *spikelets* 2–8 mm long; *scales* 1–3.5 mm long, black or greenish; *bristles* usually 4, shorter to longer than the achene; *stigmas* usually 2; *achenes* obovate, 0.7–1.6 × 0.8–1 mm, white. Found in moist places, along creeks, in ditches, 3800–9000 ft. June–Aug. E/W.

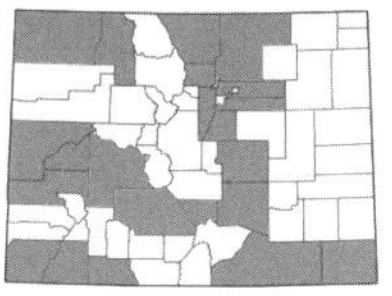

***Scirpus pallidus* (Britton) Fernald**, CLOAKED BULRUSH. Plants to 15 dm; *involucral bracts* numerous and foliaceous; *inflorescence* a terminal umbel, the spikelets sessile and aggregated into dense clusters; *spikelets* 4–5 mm long; *scales* 1.6–2.8 mm long, black or brownish; *bristles* 6, equal to the achene; *stigmas* usually 3; *achenes* usually 3-sided, 0.8–1.2 × 0.4–0.6 mm, white or pale brown. Found in moist places, 4500–6500 ft. June–Aug. E/W. (Plate 41)

***Scirpus pendulus* Muhl.**, RUFOUS BULRUSH. Inflorescence a terminal, sessile cluster of spikelets and sometimes also with 1–2 lateral inflorescences; *spikelets* 5–10 mm long; *scales* ca. 2 mm long, brown to reddish-brown; *bristles* 6, strongly contorted, about twice as long as the achene; *stigmas* 3; *achenes* biconvex or slightly 3-sided, 1–1.2 × 0.6–0.8 mm, brown. Uncommon in moist places and fields, known from Boulder Co. where it has escaped cultivation, 5200–5500 ft. June–Aug. E. Introduced.

TRICHOPHORUM Pers. – BULRUSH

Perennial herbs; *stems* round (ours); *leaves* basal or nearly so, ligules present; *flowers* perfect; *perianth* absent (ours).

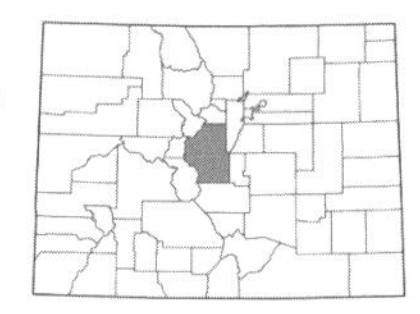

***Trichophorum pumilum* Schinz & Thell.**, ROLLAND'S BULRUSH. [*Scirpus pumilus* Vahl]. Plants caespitose with long, slender rhizomes; *leaves* 2–8 × 0.4–0.5 mm, shorter than stems at flowering; *involucral bract* 1, scale-like; *inflorescence* of a single terminal spikelet; *spikelets* 3–6-flowered, 3–4.6 mm long; *scales* brown, obtuse; *bristles* absent; *stigmas* 3; *achenes* 3-sided to plano-convex, 1.2–2 × 0.8–1.2 mm. Rare on peat hummocks in rich fens, 9300–11,000 ft. June–Sept. E.

DIPSACACEAE Juss. – TEASEL FAMILY

Herbs; *leaves* usually opposite; *flowers* perfect, zygomorphic, aggregated in dense, involucrate heads; *sepals* 4 or 5 or represented by pappus-like bristles; *corolla* 4- or 5-lobed; *stamens* 4, epipetalous; *pistil* 1; *ovary* inferior, 2-carpellate but 1 carpel obsolete so that the ovary is unilocular; *fruit* an achene enclosed in the involucel and crowned by a persistent calyx.

1a. Stems and leaves prickly; heads elongate, ovoid to cylindric; involucre bracts linear to lanceolate, upcurved, spine-tipped, surpassing the head or not; receptacle chaffy...***Dipsacus***
1b. Stems and leaves not prickly; heads convex or subglobose; involucre bracts ovate to narrowly lanceolate, ascending to reflexed, not spine-tipped, about equaling the head; receptacle hairy...***Knautia***

DIPSACUS L. – TEASEL

Herbs with prickly stems; *leaves* sessile, entire to pinnately dissected; *involucre bracts* linear, spine-tipped, subtending dense heads; *receptacle* chaffy, with rigid bracts that are acuminate into an awn which surpasses and subtends each flower; *calyx* 4-lobed, short, cupulate; *corolla* 4-lobed, white to purple.

1a. Stem leaves entire or with crenate-serrate margins, lanceolate to oblanceolate, often basally connate but not forming a cup; corolla purple to lilac or rarely white; involucre bracts linear, surpassing or equaling the head...***D. fullonum***
1b. Stem leaves pinnatifid to bipinnatifid dissected, the bases of opposite leaves joined together forming a cup; corolla white or rarely lilac; involucre bracts lanceolate to linear-lanceolate, usually shorter than or sometimes just barely surpassing the head...***D. laciniatus***

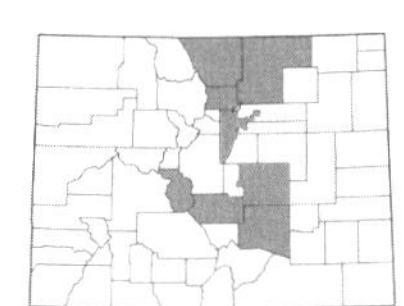

***Dipsacus fullonum* L.**, COMMON TEASEL. [*D. sylvestris* Huds.]. Biennials, 5–20 (30) dm; *leaves* lanceolate to oblanceolate, to 3 (4) dm × 4–10 cm, usually with prickly margins and nerves, entire or crenate-serrate; *heads* 3–10 cm high, 3–5 cm wide; *involucre bracts* linear, upcurved, as long as or longer than the head; *achenes* 3–8 mm long. Found in moist, disturbed places such as along roadsides and in wet ditches, 4600–7000 ft. June–Sept. E/W. Introduced.

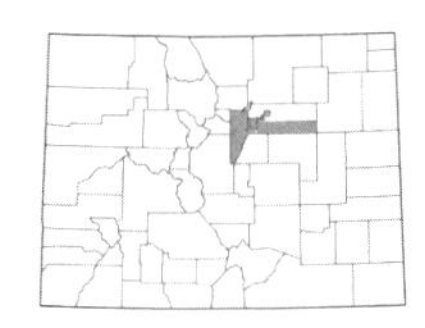

***Dipsacus laciniatus* L.**, CUTLEAF TEASEL. Biennials, to 25 dm; *leaves* pinnatifid to bipinnatifid, prickly, the bases of paired leaves joined together forming a cup; *heads* 3–10 cm high; *involucre bracts* linear to linear-lanceolate, 2–11 cm long, usually shorter than or sometimes just surpassing the head; *achenes* ca. 3 mm long. Found in moist, disturbed places, 5000–5500 ft. July–Sept. E. Introduced.

KNAUTIA L. – SCABIOSA

Herbs without prickly stems; *leaves* simple or pinnatifid; *involucre bracts* herbaceous, ovate to lanceolate, about equaling the heads; *receptacle* densely hairy, without bracts; *calyx* represented by pappus-like bristles that are shortly connate below; *corolla* 4-lobed, funnelform to narrowly campanulate, purple.

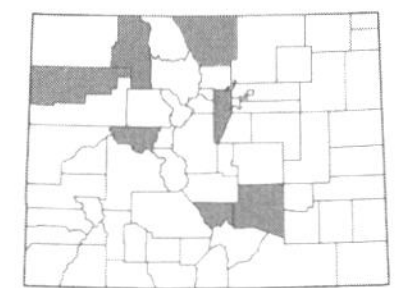

***Knautia arvensis* (L.) J.M. Coult.**, BLUEBUTTONS; FIELD SCABIOSA. Perennials, 4–10 dm, hirsute; *leaves* simple to lyrate-pinnatifid or the upper deeply pinnatifid; *heads* 1.5–4 cm wide; *involucre bracts* lanceolate to ovate, 8–16 mm long; *corolla* purple to lilac, 8–18 mm long; *achenes* 5–6 mm long. Cultivated, sometimes escaping and becoming established along roadsides, 6800–8800 ft. June–Aug. E/W. Introduced.

DROSERACEAE Salisb. – SUNDEW FAMILY

Herbs or subshrubs, carnivorous; *leaves* simple, often coiled in bud, in a basal rosette, the blade modified into an insect trap, each trap covered with sticky glandular hairs and sensitive bristles (ours), when triggered causing the enclosure of the prey by folding of the blade; *flowers* perfect, actinomorphic; *calyx* 4- or 5-lobed; *petals* 4–5, distinct; *stamens* usually 5, distinct or basally connate; *ovary* superior; *styles* 1 or 3 (5); *fruit* a capsule.

DROSERA L. – SUNDEW

Herbs; *flowers* 5-merous.
1a. Leaves 15–50 mm long, oblanceolate, ascending to erect; flowers white, 8–12 mm long...***D. anglica***
1b. Leaves 3–15 mm long, more or less orbicular, spreading; flowers white to pink, 4–6 mm long...***D. rotundifolia***

***Drosera anglica* Huds.**, ENGLISH SUNDEW. [*Drosera longifolia* L.]. Perennials; *leaves* oblanceolate, 1.5–5 cm × 2–7 mm, ascending to erect; *calyx* 4–6 mm long; *petals* 8–12 mm long, white. Uncommon in acidic fens, known only from La Plata Co., where it was first collected in Colorado in 2006, 8500–8600 ft. July–Aug. W.

***Drosera rotundifolia* L.**, ROUNDLEAF SUNDEW. Perennials; *leaves* more or less orbicular, 0.3–1.5 cm × 4–20 mm, spreading; *calyx* 4–6 mm long; *petals* 4–6 mm long, white to pink. Uncommon in acidic fens and ponds and on floating peat mats, 9000–9800 ft. July–Aug. E/W.

ELAEAGNACEAE Juss. – OLEASTER FAMILY

Shrubs or trees with silver or golden-brown stellate or peltate hairs; *leaves* usually alternate, petiolate, simple and entire, exstipulate; *flowers* perfect or imperfect, actinomorphic; *sepals* 4, united into a hypanthium, petaloid or green; *petals* absent; *stamens* 4 or 8, inserted on the hypanthium within the calyx tube; *pistil* 1; *ovary* superior though often appearing inferior, unicarpellate, with basal placentation; *fruit* an achene enclosed in the persistent hypanthium and drupelike.
1a. Leaves and branches alternate; flowers perfect or only a few of them staminate; fruit drupaceous with a hard inner layer...***Elaeagnus***
1b. Leaves and branches opposite or sometimes subopposite; flowers imperfect, plants dioecious; fruit berry-like without a well-developed stone pit...***Shepherdia***

ELAEAGNUS L. – OLEASTER

Shrubs or trees with silvery or golden-brown stellate or peltate hairs on young branches; *leaves* alternate; *sepals* spreading, yellow inside, the hypanthium constricted into a tube above the ovary; *stamens* 4; *fruit* covered with silvery-gray hairs.

1a. Leaves lance-linear or narrowly elliptic, mostly 3–8 times as long as wide, the lower side of young leaves lacking brown gland dots; young twigs densely silvery-scaly hairy; nectary disk well-developed forming a short tube around the style; trees mostly 5–12 m tall...***E. angustifolia***

1b. Leaves ovate-oblong to broadly elliptic, mostly 1.5–3 times as long as wide, the lower side of young leaves covered with brown gland dots; young twigs densely brownish-scaly hairy; nectary disk not well-developed; shrubs mostly 1–5 m tall...***E. commutata***

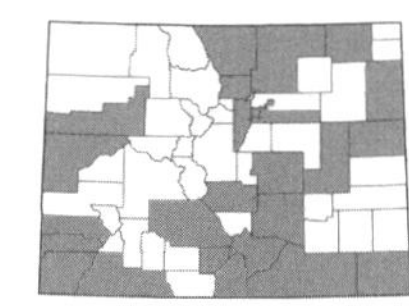

Elaeagnus angustifolia **L.**, RUSSIAN OLIVE. Trees, mostly 5–12 m; *leaves* lance-linear to narrowly elliptic, 2–9 cm × 5–38 mm, silvery with scale-stellate hairs above and below; *flowers* 8–12 mm long; *nectary disk* forming a short tube around the style; *fruit* 10–20 mm diam. Often cultivated and escaping where it is found along roadsides, streams, or floodplains, 4400–7500 ft. May–July. E/W. Introduced.

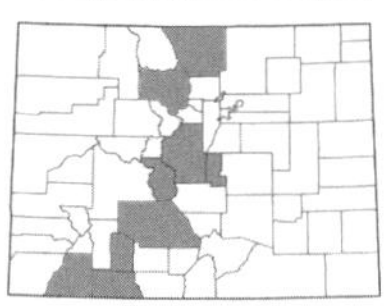

Elaeagnus commutata **Bernh. ex Rydb.**, SILVERBERRY. Shrubs, mostly 1–5 m; *leaves* ovate-oblong to broadly elliptic, 1.3–7 cm × 5–30 mm, silvery above and below and with brown gland dots below; *flowers* 10–15 mm long; *nectary disk* not well-developed; *fruit* 6–10 mm diam. Found on open slopes, 7500–9700 ft. June–July. E/W.

SHEPHERDIA Nutt. – BUFFALOBERRY

Dioecious shrubs or small trees with brownish or silvery peltate or stellate hairs, older branches scaly; *leaves* opposite; *flowers* imperfect; *sepals* with glandular thickenings at the base of the lobes, forming an urn-shaped hypanthium and covering the ovary in pistillate flowers, open in staminate flowers; *stamens* 8; *fruit* a fleshy drupelike achene.

1a. Leaves silvery above and below with closely spaced scale-stellate hairs, usually less hairy above; leaf bases more or less acute...***S. argentea***

1b. Leaves green and sparsely stellate above and whitish below with scattered brownish scale-stellate hairs; leaf bases rounded to obtuse...***S. canadensis***

Shepherdia argentea **(Pursh) Nutt.**, SILVER BUFFALOBERRY. Shrubs, 2–4 m; *leaves* lanceolate to oblong or elliptic, 5–60 × 3–15 mm, silvery above and below with closely spaced scale-stellate hairs, acute at the base; *flowers* 2.5–4 mm long; *fruit* 4–7 mm, red. Common in moist places along rivers and in canyon bottoms, 4500–7500 ft. April–May. E/W.

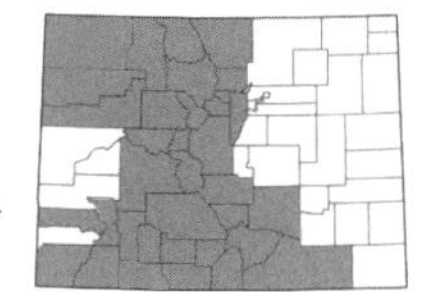

Shepherdia canadensis **Nutt.**, CANADIAN BUFFALOBERRY. Shrubs, 0.5–2 m; *leaves* lanceolate to ovate, 5–80 × 3–30 (50) mm, green and sparsely stellate above and whitish below with scattered brownish scale-stellate hairs, rounded to obtuse at the base; *flowers* 2–3 mm long; *fruit* 4–7 mm, red. Common in pine forests, often in shaded places, scattered throughout the mountains, 7500–11,600 ft. May–July. E/W. (Plate 41)

ELATINACEAE Dum. – WATERWORT FAMILY

Annual herbs, aquatic or found in moist places; *leaves* opposite and decussate, simple, stipulate; *flowers* perfect, actinomorphic, axillary; *sepals* 2–5, distinct; *petals* 2–5 (the same number as sepals), distinct; *stamens* 2–10 (the same number as or twice the number of sepals), distinct; *pistil* 1; *ovary* superior, 2–5-carpellate; *fruit* a capsule.

1a. Leaves elliptic to oblong, the margins serrulate; plants erect or ascending, glandular-hairy; flowers with 5 sepals and petals, the sepals white-scarious-margined...***Bergia***

1b. Leaves linear to spatulate, the margins entire; plants prostrate and mat-forming, glabrous; flowers with 3 sepals and petals, the sepals not scarious-margined...***Elatine***

BERGIA L. – BERGIA

Annual herbs; *leaves* opposite and sometimes crowded above and then apparently whorled; *sepals* 5; *petals* 5, white, not exceeding the sepals; *stamens* 5 or 10; *fruit* a globose capsule included in the calyx.

Bergia texana **Seub. ex Walp.**, TEXAS BERGIA. Plants 1–3 (4) dm, branching from the base; *leaves* oblong to elliptic, to 3 cm long, serrulate, glandular; *sepals* 3–4 mm long; *capsules* ca. 3 mm diam. Uncommon along the margins of drying ponds and mudflats, 5000–5600 ft. July–Oct. E.

ELATINE L. – WATERWORT

Aquatic or terrestrial herbs found in moist places; *leaves* opposite, sessile or petiolate; *sepals* 3–4; *petals* 3–4; *stamens* 3 or 8; *ovary* 3-4-carpellate; *fruit* a septicidal capsule.

1a. Leaves rounded at the apex; seeds mostly with 9–15 pits per longitudinal row...***E. brachysperma***
1b. Leaves notched at the apex; seeds mostly with 15–25 pits per longitudinal row...***E. rubella***

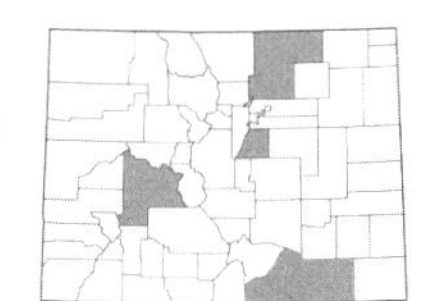

Elatine brachysperma **A. Gray**, SHORTSEED WATERWORT. Plants prostrate or sometimes erect, rooting at the nodes, glabrous; *leaves* linear to spatulate or narrowly oblong, entire, rounded at the apex; *sepals* ca. 2 mm long; *seeds* mostly with 9–15 pits per longitudinal row. Uncommon in moist soil along the margins of ponds and mudflats, 5000–6000 ft. June–Aug. E/W.

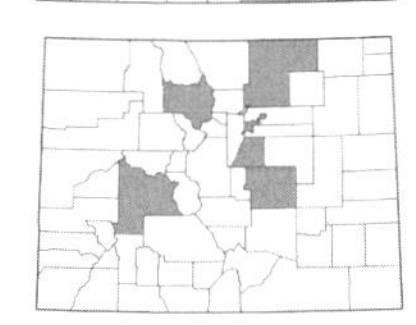

Elatine rubella **Rydb.**, SOUTHWESTERN WATERWORT. [*E. triandra* Schkuhr]. Plants prostrate or sometimes erect, the stems to 2 dm long, rooting at the nodes, glabrous; *leaves* linear to spatulate, 3–6 (12) mm long, punctate above, with entire margins, notched at the apex; *sepals* ca. 2 mm long; *seeds* mostly with 15–25 pits per longitudinal row. Rare in moist soil along the margins of ponds and mudflats, 5000–6000 ft. June–Aug. E/W.

ERICACEAE Juss. – HEATH FAMILY

Herbs, shrubs, or subshrubs; *leaves* alternate or opposite, or mostly basal, often evergreen and leathery, exstipulate; *flowers* perfect, actinomorphic, the corolla urceolate, rotate, or campanulate; *sepals* (4) 5, distinct or shortly connate; *petals* (4) 5, connate or distinct; *stamens* 8–10, distinct; *ovary* superior or inferior, 4–5-carpellate; *fruit* a capsule, berry, or a drupe. (Wallace, 1975)

1a. Plants lacking chlorophyll, the stems either reddish to purplish-brown, or white to pinkish and drying black...2
1b. Plants green, with chlorophyll...3

2a. Petals distinct; stems glabrous, white or pinkish and often turning black in drying...***Monotropa***
2b. Petals united, the corolla urn-shaped; stems stout and glandular-hairy, reddish-brown and not turning black in drying...***Pterospora***

3a. Petals distinct; flowers downward-pointing; plants herbaceous...4
3b. Petals united, the corolla urn-shaped or campanulate, downward-pointing or not; plants subshrubs or shrubs...7

4a. Flowers 4–10 in a terminal umbelliform corymb; stem leaves evident, oblanceolate or oblong-oblanceolate, broadest above the middle, with sharply serrate margins; style straight and very short, nearly immersed in the depressed top of the ovary; stamens with dilated filaments that are ciliate on the margins and maroon anthers...***Chimaphila***
4b. Flowers solitary or 4–25 in a terminal raceme; leaves basal or clustered at the base, elliptic to orbicular and broadest at the middle or below, the margins entire to serrate; style straight but over 2 mm long or sigmoidally curved with an upturned end; stamens without dilated filaments...5

5a. Flowers solitary, waxy-white; stigma peltate with 5 prominent erect lobes; stamens paired and opposite the petals; valves of the capsule not connected by fibers at maturity...***Moneses***
5b. Flowers 4–25 in a terminal raceme, white, greenish-white, or pink; stigma capitate or peltate with 5 spreading lobes; stamens alternate and opposite the petals, not paired; valves of the capsule connected by cobwebby fibers at maturity...6

6a. Flowers in a strongly one-sided raceme; style straight, 3–5.5 mm long; petals with 2 basal tubercles on the inner surface...***Orthilia***
6b. Flowers not in a one-sided raceme; style sigmoidally curved with an upturned end or straight but 0.8–1.5 mm long; petals without tubercles...***Pyrola***

7a. Flowers white, open-campanulate, the petals 10–15 mm long; leaves and sepals hirsute and with glandular-ciliate margins; tall shrub to 2 m; fruit a septicidal capsule...***Rhododendron***
7b. Flowers white or pink, urn-shaped, rotate, or campanulate, the petals to 10 mm long; leaves and sepals glabrous or with ciliate margins, rarely with a few short hairs; low shrubs, often prostrate, to 3 dm, or sometimes taller erect shrubs to 2 m; fruit a septicidal capsule or a berry, berry-like, or a drupe...8

8a. Leaves opposite, the margins entire and often revolute; flowers on long, slender pedicels (10–30 mm long), deep pink, rotate; fruit a septicidal capsule...***Kalmia***
8b. Leave alternate, the margins toothed or entire but not revolute; flowers not on long pedicels, white to pink, urn-shaped to campanulate; fruit a berry, berry-like, or a drupe...9

9a. Leaves with entire margins, evergreen and leathery; flowers in terminal racemes or panicles...***Arctostaphylos***
9b. Leaves with toothed or ciliate margins, sometimes the teeth ending in a slender bristle or the bristle arising from the inconspicuously toothed margin, deciduous or evergreen but not leathery; flowers solitary in the leaf axils...10

10a. Ovary inferior; corolla urn-shaped; leaves obovate to ovate; calyx pale green, small and inconspicuous, not enlarging in fruit; fruit red or blue-black; plants erect-ascending, not creeping...***Vaccinium***
10b. Ovary superior; corolla campanulate; leaves orbicular or broadly elliptic; calyx often reddish (at least at the tips), almost as long as the corolla, enlarging and becoming fleshy in fruit; fruit red; plants usually low and creeping...***Gaultheria***

ARCTOSTAPHYLOS Adans. – BEARBERRY; MANZANITA

Low shrubs; *leaves* alternate, evergreen and leathery, the margins entire (ours); *flowers* urceolate; *sepals* (4) 5, persistent, distinct; *petals* (4) 5, connate; *stamens* (8) 10; *ovary* superior; *fruit* a fleshy or mealy berry-like drupe. (Adams, 1940; Eastwood, 1934; Packer & Denford, 1974; Rosatti, 1987)

1a. Leaves elliptic-ovate to subrotund, 1–4 cm wide, obtuse or truncate at the base, bright green on both sides; bark reddish-brown; erect shrub to 2 m tall; fruit dull orange...***A. patula***
1b. Leaves oblanceolate to obovate, 0.3–1 cm wide, with a cuneate base, dark green above and lighter green below; bark brown; prostrate shrub to 0.2 m tall; fruit bright or dull red...***A. uva-ursi***

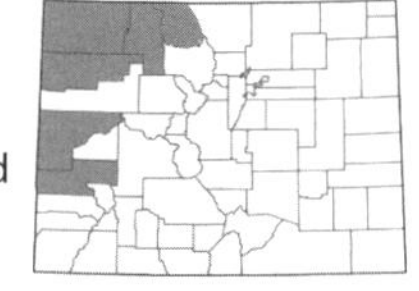

Arctostaphylos patula **Greene**, GREENLEAF MANZANITA. [*A. pinetorum* Rollins]. Erect shrubs to 2 m; *leaves* elliptic-ovate to subrotund, 1–4 cm wide, with an obtuse or truncate base, bright green above and below; *corolla* 5–8 mm, white or pink; *fruit* dull orange. Found in open pine forests and on sandstone outcroppings, 7000–9500 ft. May–Aug. E/W.

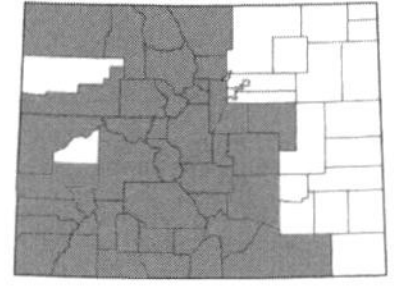

Arctostaphylos uva-ursi **(L.) Spreng.**, KINNIKINNICK. Prostrate shrubs, forming mats 1–2 m across; *leaves* oblanceolate to obovate, 0.3–1 cm wide, the base cuneate, darker green above; *corolla* 4–5.5 mm long, pink to white; *fruit* red. Common ground cover in forests, 6100–13,000 ft. May–Aug. E/W. (Plate 41)

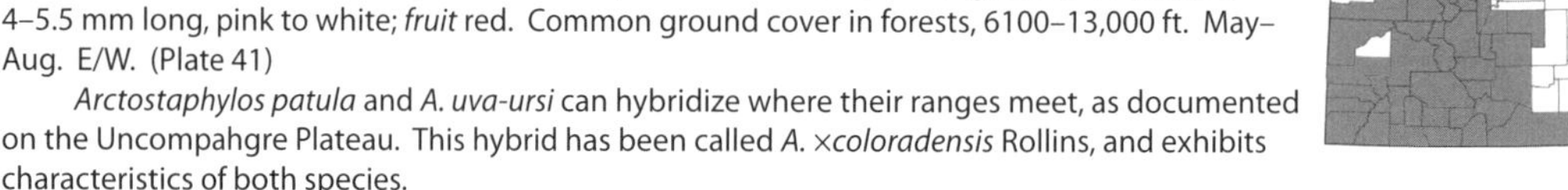

Arctostaphylos patula and *A. uva-ursi* can hybridize where their ranges meet, as documented on the Uncompahgre Plateau. This hybrid has been called *A.* ×*coloradensis* Rollins, and exhibits characteristics of both species.

CHIMAPHILA Pursh – PIPSISSEWA

Leaves alternate to subopposite or whorled; *flowers* actinomorphic, in terminal corymbs or umbels; *sepals* persistent, distinct; *petals* white, pink, or rose-purple; *stamens* 10, opposite and alternate with the petals; *ovary* 5-carpellate; *style* very short and inconspicuous, nearly immersed in the depressed top of the ovary; *stigma* peltate, broad and flat, without evident lobes.

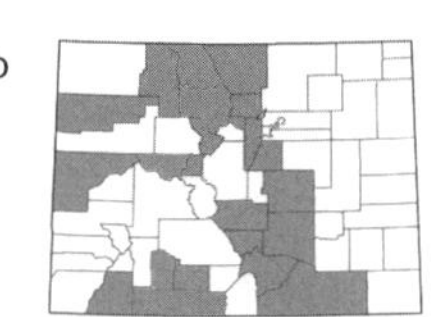

Chimaphila umbellata **(L.) W.P.C. Barton**, PIPSISSEWA. Plants to 3 dm; *leaves* whorled, lanceolate to oblanceolate, 3–7 cm long, leathery, sharply toothed; *petals* pink or reddish-pink, 5–7 mm long. Common understory in coniferous forests, usually in moist places, 7300–11,000 ft. June–Sept. E/W.

GAULTHERIA L. – SNOWBERRY

Low shrubs; *leaves* alternate, evergreen, the margins usually serrate or crenate; *flowers* urceolate or campanulate; *stamens* 10, included; *ovary* superior (ours); *fruit* a capsule closely surrounded by the persistent, fleshy calyx forming a dry or mealy berry-like fruit.

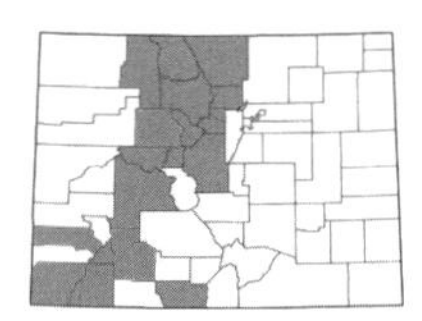

Gaultheria humifusa **(Graham) Rydb.**, ALPINE SPICY-WINTERGREEN. Prostrate plants with stems to 20 cm long; *leaves* orbicular to broadly elliptic, 6–17 mm long; *corolla* campanulate, 3–4 mm long, pink or sometimes white; *fruit* red. Found along moist streambanks, in wet meadows, and in moist spruce woods, 9000–12,500 ft. July–Aug. E/W.

KALMIA L. – LAUREL

Low shrubs (ours); *leaves* alternate or opposite, the margins entire and often revolute; *flowers* rotate; *sepals* 5, distinct, persistent; *petals* 5, connate; *stamens* 10, included; *ovary* superior; *fruit* a septicidal capsule.

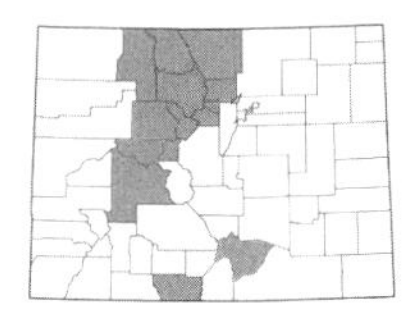

***Kalmia microphylla* (Hook.) A. Heller**, ALPINE LAUREL. Low shrubs 5–17 cm; *leaves* opposite, elliptic, 6–20 (30) mm long, the margins revolute; *inflorescence* a 2–6-flowered corymb with pedicels 10–30 mm long; *corolla* campanulate, pink, 11–14 mm wide. Found on streambanks and lake margins, and in moist meadows at higher elevations, 9500–12,500 ft. July–Aug. E/W. (Plate 41)

MONESES Salisb. ex Gray – WOOD NYMPH

Perennials; *leaves* basal; *flowers* solitary, nodding; *sepals* persistent, distinct; *petals* white or sometimes tinged with pink; *stamens* (8) 10, paired and opposite the petals; *ovary* superior, (4) 5-carpellate, subglobose and concave at the top; *style* straight and well-developed; *stigma* peltate with 5 prominent erect lobes.

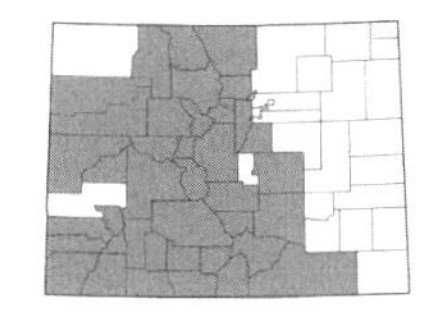

***Moneses uniflora* (L.) A. Gray**, WOOD NYMPH. [*Pyrola uniflora* L.]. Plants 0.5–1 dm; *leaves* ovate to obovate, 0.8–4 cm long, serrate or crenate; *sepals* 1.5–2.5 cm long, ciliate-fringed; *petals* waxy-white, 7–10 mm long. Common in moist, shaded forests, in bogs, and along shaded streams, 8000–14,000 ft. July–Sept. E/W. (Plate 41)

MONOTROPA L. – INDIANPIPE

Perennials, lacking chlorophyll, usually white or pinkish, drying black; *leaves* scale-like; *sepals* 2–5, distinct; *petals* 3–6, distinct; *stamens* 6–12, the anthers awnless; *ovary* superior; *fruit* a capsule.

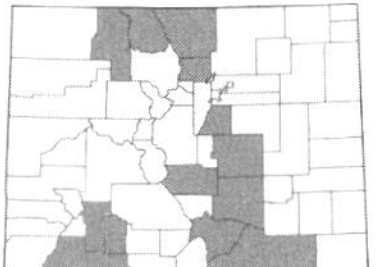

***Monotropa hypopithys* L.**, PINESAP. [*Hypopitys monotropa* Crantz]. Plants glabrous; *sepals* usually 5, 8–10 mm long; *petals* usually 5, 10–12 mm long; *stamens* 10. Uncommon in pine and spruce forests, 6400–9600 ft. July–Sept. E/W.

ORTHILIA Raf. – ORTHILIA

Perennials; *leaves* basal; *flowers* in a strongly one-sided terminal raceme; *sepals* persistent, distinct; *petals* white or greenish; *stamens* 10, alternate and opposite with the petals; *ovary* superior, 5-carpellate; *style* straight and well-developed, exserted; *stigma* peltate and crateriform with 5 prominent spreading lobes.

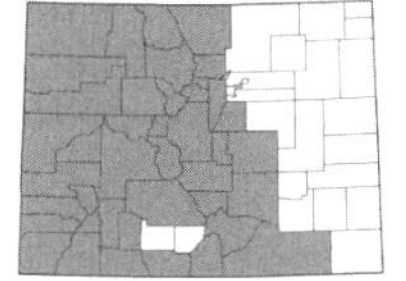

***Orthilia secunda* (L.) House**, SIDEBELLS. [*Pyrola secunda* L.]. Plants 0.5–2 dm; *leaves* ovate to orbicular or elliptic, 1–3 cm long, crenate-serrate; *sepals* 0.5–1.5 mm long; *petals* 4–6 mm long. Common in moist coniferous forests, 7000–13,000 ft. June–Aug. E/W. (Plate 41)

PTEROSPORA Nutt. – PINEDROPS

Perennials, lacking chlorophyll, usually reddish to purplish-brown and glandular-hairy; *leaves* scale-like; *calyx* 5-lobed; *corolla* 5-lobed, urceolate; *stamens* 10, the anthers with deflexed awns; *ovary* superior; *fruit* a capsule.

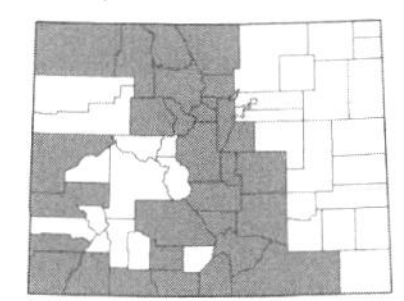

***Pterospora andromedea* Nutt.**, PINEDROPS. Plants 2–9 dm; *inflorescence* a raceme with nodding flowers; *corolla* yellow, 4–6 mm long; *capsules* 5-lobed. Common in dry pine forests, 6100–11,000 ft. July–Aug. E/W. (Plate 41)

PYROLA L. – WINTERGREEN

Leaves basal; *flowers* actinomorphic or zygomorphic, in terminal racemes; *sepals* persistent, distinct; *petals* white, cream, or greenish; *stamens* 10, alternate and opposite with the petals; *ovary* superior, 5-carpellate; *stigma* peltate or capitate.

1a. Style short, (0.5) 0.8–1.5 (1.8) mm long, straight and erect, without a collar beneath the stigma; stigma peltate, with 5 short lobes; anthers without the terminal pores on tubules...***P. minor***

1b. Style elongate, 4–10 mm long, sigmoidally curved with an upturned end, with an evident or inconspicuous collar just beneath the stigma; stigma capitate, minutely 5-lobed; anthers with the terminal pores on tubules...2

2a. Leaves dark green and prominently white- or gray-mottled along the midrib and main veins on the upper surface, the lower surface usually purplish-red; flowers white to greenish-white or sometimes pink-tinged...***P. picta***

2b. Leaves green, the veins not white-mottled, the lower surface sometimes purplish-red but usually paler green; flowers white to pale green or pink to rosy...3

3a. Flowers white to cream or pale green, mostly 2–10 in a raceme; terminal tubules 0.5–0.9 mm long and prominent on the anthers; leaves usually smaller, mostly 1–3 × 0.5–3 cm...***P. chlorantha***

3b. Flowers pink to rosy, mostly 10–25 in a raceme; terminal tubules 0.2–0.4 mm long; leaves usually larger, mostly 3–7 × 3–6 cm...***P. asarifolia***

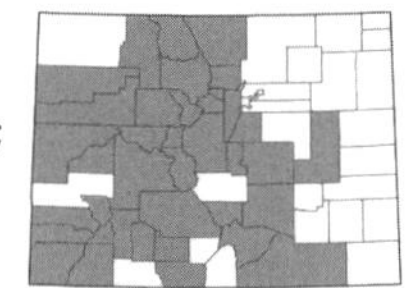

Pyrola asarifolia **Michx.**, PINK WINTERGREEN. Plants 1.3–4 dm; *leaves* orbicular to ovate or broaldy elliptic, mostly 3–7 × 3–6 cm, rounded at the tip, entire or serrulate; *racemes* mostly 10–25-flowered; *petals* pink to rose, 5–7 mm long; *anthers* with terminal tubules 0.2–0.4 mm long; *stigma* capitate, 0.7–1.5 mm wide; *style* exserted, 7–10 mm long. Common along streambanks and in moist, shaded coniferous forests, 7000–11,000 ft. June–Aug. E/W.

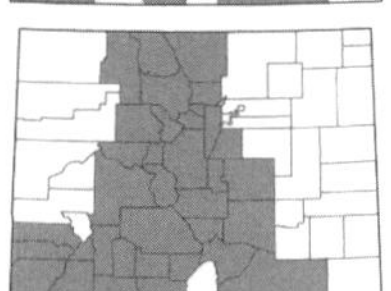

Pyrola chlorantha **Sw.**, GREEN-FLOWERED WINTERGREEN. Plants 1–2.5 dm; *leaves* oval to obovate, rounded, mostly 1–3 × 0.5–3 cm, entire to crenate-serrate; *racemes* mostly 2–10-flowered; *petals* white to cream or pale green, 5–7 mm long; *anthers* with terminal tubules 0.5–0.9 mm long and prominent on the anthers; *stigma* capitate, 1–1.5 mm wide; *style* exserted, (4) 5–7 mm long. Common in usually moist, shaded coniferous forests, 7200–11,000 ft. June–Aug. E/W. (Plate 41)

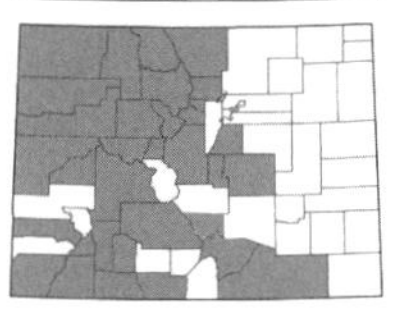

Pyrola minor **L.**, LESSER WINTERGREEN. Plants 0.8–2.5 dm; *leaves* oval to elliptic, rounded, 1–3 cm long, entire to crenate; *racemes* mostly 5–13-flowered; *petals* pale pink or cream, 3.5–4.5 mm long; *anthers* lacking terminal pores on the tubules; *stigma* peltate, with 5 short, spreading lobes; *style* included, straight, (0.5) 0.8–1.5 (1.8) mm long. Locally common in moist coniferous forests and bogs, 8000–12,000 ft. June–Aug. E/W.

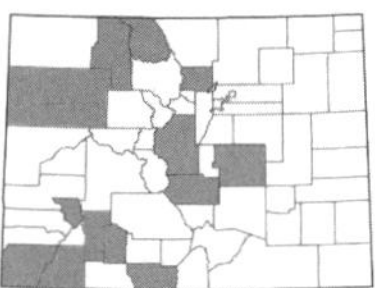

Pyrola picta **Sm.**, WHITE-VEINED WINTERGREEN. Plants 1–2.5 dm; *leaves* oval to elliptic, dark green with white or gray mottling along the veins on the upper surface, 2–7 cm long; *racemes* 2–7-flowered; *petals* greenish-white to cream, 7–8 mm long; *anthers* with terminal tubules; *stigma* capitate, 1–1.5 mm wide, with erect lobes; *style* exserted, 4–9 mm long. Rare in moist, shaded coniferous forests, 6000–9800 ft. June–Aug. E/W.

RHODODENDRON L. – RHODODENDRON; AZALEA

Shrubs; *leaves* alternate, deciduous (ours) or evergreen, the margins entire; *flowers* campanulate; *calyx* 5-lobed, persistent; *corolla* 5-lobed; *stamens* 10; *ovary* superior; *fruit* a septicidal capsule.

Rhododendron albiflorum **Hook.**, WHITE AZALEA. [*Azaleastrum albiflorum* (Hook.) Rydb.]. Shrubs to 2 m; *leaves* elliptic to ovate or obovate, 10–20 mm long, hirsute and glandular-ciliate; *calyx* glandular; *corolla* campanulate, white or pink-tinged at the tips, 10–15 mm long. Uncommon in spruce-fir forests, 9000–12,000 ft. June–Aug. E/W.

VACCINIUM L. – BLUEBERRY

Shrubs (ours); *leaves* alternate, deciduous (ours); *flowers* urceolate to campanulate; *calyx* (4-)-5-lobed, the lobes sometimes small or inconspicuous; *corolla* (4-)-5-lobed; *stamens* 8 or 10, the anther sacs often with a pair of long dorsal spurs; *ovary* inferior or partly inferior; *fruit* a berry with persistent calyx lobes (although infrequently seen with fruit in Colorado). (Camp, 1945; VanderKloet & Dickinson, 1999)

1a. Twigs of the current season round and not angled or inconspicuously angled, reddish-green, yellow-green, or brown, but not bright green; mature berries blue to black, usually glaucous...***V. cespitosum***

1b. Twigs of the current season sharply angled with the angles appearing as raised ridges or narrow wings on the twig, bright green; mature berries red or orange-red or blue-black but usually not glaucous...2

2a. Leaves larger, the larger ones 1.5–4 cm long and 7–16 mm wide; mature berries blue or blue-black; flowers larger, the corolla 4–5 mm long and 5–7 mm wide; plants not broom-like in aspect...***V. myrtillus***

2b. Leaves smaller, mostly 0.5–1.5 cm long and 3–9 mm wide; mature berries red or orange-red, sometimes drying bluish; flowers smaller, the corolla 2–4 mm long and 3–4 mm wide; plants usually broom-like in aspect with numerous narrow, rigid branches...***V. scoparium***

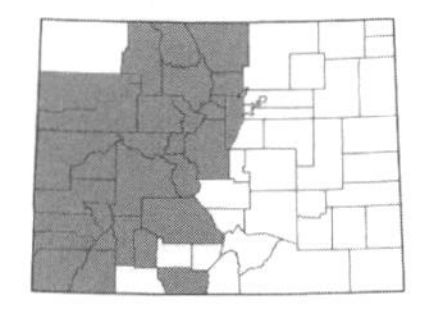

Vaccinium cespitosum **Michx.**, DWARF BLUEBERRY. Plants 1–3 dm; *twigs* of the current season round, reddish-green, yellow-green, or brown; *leaves* oblanceolate to obovate, 7–40 × 3–20 mm, the margins serrulate above the middle; *corolla* urceolate, 5–6 mm long, pink to whitish; *berries* blue to black, usually glaucous, 5–8 mm diam. Found in moist forests and meadows and on rocky ridges, 8000–13,000 ft. June–Aug. E/W.

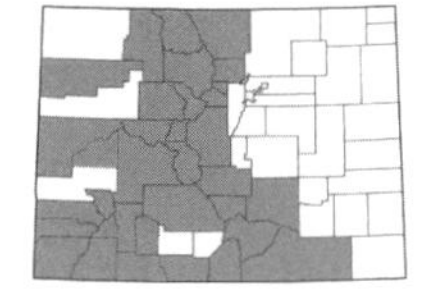

Vaccinium myrtillus **L.**, WHORTLEBERRY. Plants 0.5–3 dm; *twigs* of the current season sharply angled, bright green; *leaves* elliptic to ovate, the larger ones 1.5–4 cm × 7–16 mm, the margins sharply serrulate; *calyx* ca. 0.5 mm; *corolla* urceolate, 4–5 cm long, pink; *berries* blue-black or purplish-black, not glaucous, 5–8 mm diam. Common on forest floors and on rocky alpine ridges, 7500–12,500 ft. June–Aug. E/W.

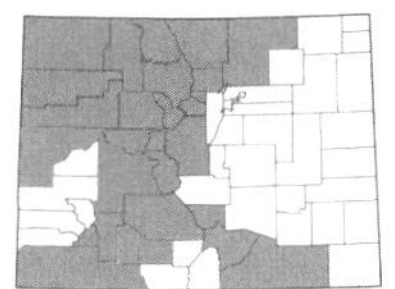
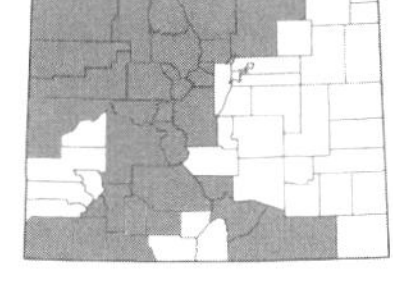

***Vaccinium scoparium* Leiberg ex Coville**, DWARF RED WHORTLEBERRY. Plants 1–3 dm; *twigs* of the current season sharply angled, bright green; *leaves* elliptic to ovate, mostly 0.5–1.5 cm × 3–9 mm, serrulate; *corolla* urceolate, 2–4 cm long, pink; *berries* red or orange-red, sometimes drying bluish, not glaucous, 4–6 mm diam. Common on forest floors and meadows, on alpine ridges, and along the margins of high elevation lakes, 8000–12,700 ft. June–Aug. E/W. (Plate 41)

EUPHORBIACEAE Juss. – SPURGE FAMILY

Herbs (ours), usually with milky or watery latex, monoecious or dioecious; *leaves* alternate or opposite, usually stipulate; *flowers* imperfect, sometimes arranged in a cyathium; *sepals* 4–6 or absent; *petals* 4–6 or absent, distinct; *stamens* 1–10; *pistil* 1; *ovary* superior, 3-carpellate; *fruit* a capsule.

1a. Leaves and stems covered in stellate hairs; dioecious annuals...***Croton***
1b. Leaves and stems lacking stellate hairs; monoecious annuals or perennials...2

2a. Plants pubescent with stinging hairs; leaves sharply serrate...***Tragia***
2b. Plants lacking stinging hairs; leaves various...3

3a. Flowers arranged in a cyathium (resembling a single flower, with a single, central pistillate flower surrounded by several staminate flowers consisting of a single stamen, sometimes with petaloid appendages on the marginal glands); plants with milky juice; variously distributed...4
3b. Flowers not arranged in a cyathium; plants usually with watery juice; plants of the eastern plains...5

Euphorbia cyathium:

Petaloid appendage
Nectar gland
Male flower
Female flower

4a. Leaves opposite, less than 2.5 cm long; stems prostrate or erect...***Chamaesyce***
4b. Leaves alternate or if opposite then usually over 2.5 cm long; stems erect...***Euphorbia***

5a. Plants pubescent, usually some hairs dolabriform; leaves usually 3-veined from the base...***Argythamnia***
5b. Plants glabrous; leaves not 3-veined from the base...6

6a. Leaves linear to narrowly oblong, entire; staminate flowers with a 4-lobed calyx; flowers axillary to leaves; annuals...***Reverchonia***
6b. Leaves elliptic to lanceolate, toothed or crenate with a small gland in each notch; staminate flowers with a cuplike calyx; flowers in a terminal spike-like thyrse; perennials...***Stillingia***

ARGYTHAMNIA P. Br. – SILVERBUSH; WILD MERCURY

Perennials, monoecious herbs, usually pubescent; *leaves* alternate, simple; *inflorescence* of axillary bracteate racemes, the lower 1–3 flowers pistillate and the upper flowers staminate; *staminate flowers* with 5 sepals, 5 petals, 5 glands, and 7–10 stamens; *pistillate flowers* with 5 sepals, 5 petals (or these lacking), and 5 glands.

1a. Stems usually trailing or spreading, freely branching; inflorescence shorter than the leaves; leaves 5–15 mm wide...***A. humilis***
1b. Stems erect, unbranched; inflorescence longer than or equal to the leaves; leaves 10–40 mm wide...***A. mercurialina***

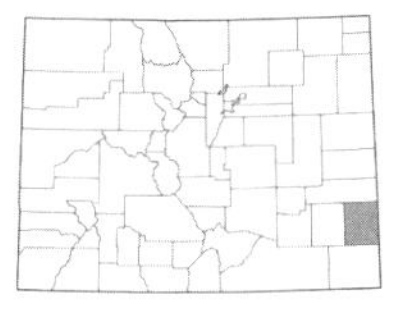

***Argythamnia humilis* (Engelm. & A. Gray) Müll. Arg.**, LOW SILVERBUSH. Plants to 4 dm, usually trailing or spreading; *leaves* oblanceolate to elliptic, 1–5.5 cm × 5–15 mm, usually hairy, entire or nearly so; *pistillate flowers* with sepals ca. 3 mm long, the petals absent or 1–5; *capsules* 5–7 mm diam.; *seeds* ca. 2.5 mm long, roughened to reticulately ridged. Rare on shortgrass prairie, known from a single old collection (1879) from Granada in Prowers Co., 3500–3600 ft. June–Aug. E.

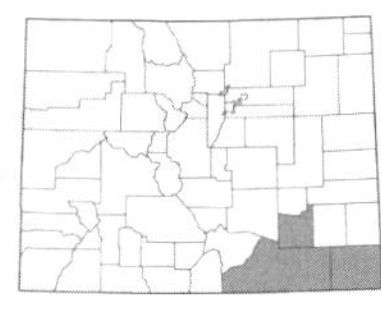

***Argythamnia mercurialina* (Nutt.) Müll. Arg.**, TALL SILVERBUSH. Plants 3–7 dm, erect; *leaves* lanceolate to elliptic, 3–8 cm × 10–40 mm, hairy, entire; *pistillate flowers* with sepals ca. 4.5 mm long, the petals usually absent or 1–5; *fruit* ca. 10 mm diam; *seeds* 5 mm long, irregularly reticulate. Uncommon on sandy or rocky soil of the shortgrass prairie, 4000–4500 ft. May–July. E.

CHAMAESYCE A. Gray – SANDMAT

Annual or perennial, prostrate, monoecious herbs with milky latex; *leaves* opposite, simple; *flowers* in a cyathium, often with petaloid appendages on the marginal glands, lacking sepals and petals; *staminate flowers* of a single stamen; *pistillate flowers* solitary in the center of the cyathium. (Wheeler, 1941)

1a. Leaves hairy (at least below, near the base) or stems uniformly hairy...2
1b. Leaves and stems glabrous (or rarely the stem with lines of short hairs)...6

2a. Leaves entire; capsule 2–2.5 mm long; perennials...***C. lata***
2b. Leaves serrulate, especially toward the tip; capsule 1–2.3 mm long; annuals from taproots...3

3a. Capsules (and ovaries) glabrous; plants erect, 3–8 dm tall...***C. nutans***
3b. Capsules (and ovaries) hairy; plants prostrate with spreading branches...4

4a. Styles entire or slightly notched at the tip; glands 5 with yellowish-white to reddish petaloid appendages; capsules 1.4–2.5 mm long; seeds pitted and mottled...***C. stictospora***
4b. Styles divided about ⅓ its length or deeply divided to near the base (forming 6 stigmas); glands 4 with a small, inconspicuous white or pinkish-white appendage or the appendage lacking; capsules 1–1.5 mm long; seeds not pitted or mottled...5

5a. Seeds smooth or obscurely cross-wrinkled between the angles; capsules mostly uniformly hairy; styles divided about ⅓ their length...***C. maculata***
5b. Seeds prominently cross-wrinkled between the angles; capsules mostly hairy on the angles; styles bifid to the base or nearly so...***C. prostrata***

6a. Leaves all linear to linear-oblong, 5–15 times as long as wide...7
6b. The larger leaves mostly ovate, elliptic, oblong, or nearly orbicular, 1.5–3 times as long as wide...9

7a. Petaloid appendages of glands conspicuous, white to pinkish, 0.5–2.5 mm long; leaves 1–5 mm wide...***C. missurica***
7b. Petaloid appendages of glands lacking or yellowish and to 0.5 mm long; leaves 1–1.5 mm wide...8

8a. Capsules 2–2.5 mm long; cyathium involucre 1–2 mm; seeds 1.4–2 mm long with rounded angles...***C. parryi***
8b. Capsules 1.2–1.7 mm long; cyathium involucre 0.5–0.7 mm; seeds 1–1.3 mm long, strongly angled...***C. revoluta***

9a. Stipules forming a prominent whitish or reddish-white hyaline scale about 1 mm long...10
9b. Stipules deeply divided or dissected into linear segments...11

10a. Petaloid appendages conspicuous, 0.5–1.3 mm long (mostly 1–3 times longer than the gland is wide); capsules 1.5–2.3 mm long; perennials from woody taproots...***C. albomarginata***
10b. Petaloid appendages small, to about 0.5 mm long (mostly as long as the glands are wide) or rarely lacking; capsules 1–1.5 mm long; annuals from slender taproots...***C. serpens***

11a. Perennials from deeply buried caudices, erect or prostrate; leaves suborbicular, elliptic, or ovate, with usually symmetrical bases...***C. fendleri***
11b. Annuals, prostrate herbs from slender taproots; leaves oblong to ovate-oblong or elliptic, the bases asymmetrical to nearly symmetrical...12

12a. Leaf bases nearly symmetrical; leaf margins entire; seeds 1.3–1.6 mm long, smooth...***C. geyeri***
12b. Leaf bases conspicuously asymmetrical; leaf margins entire or often minutely serrulate at least at the tips; seeds 0.9–1.3 mm long, smooth or roughened...13

13a. Seeds distinctly cross-ridged; staminate flowers 1–5 (usually 4) per cyathium...***C. glyptosperma***
13b. Seeds smooth or irregularly roughened or pitted; staminate flowers 5–18 per cyathium...***C. serpyllifolia***

Chamaesyce albomarginata **(Torr. & A. Gray) Small**, WHITE-MARGINED SANDMAT. Perennials, prostrate; *leaves* glabrous, broadly elliptic to suborbicular, 2–8 mm long, entire with a narrowly cartilaginous margin; *inflorescence* of solitary cyathia; *glands* 4, greenish to maroon, with a white appendage ca. 0.5–1.3 mm long; *capsules* 1.5–2.3 mm long; *seeds* 1.2–1.7 mm long, prismatic. Doubtfully present in Colorado, its presence based on a misidentification. Found to the south and west of Colorado.

Chamaesyce fendleri **(Torr. & A. Gray) Small**, FENDLER'S SANDMAT. Perennials, erect or prostrate; *leaves* glabrous, ovate, elliptic or suborbicular, (3) 5–11 mm long, entire; *inflorescence* of solitary cyathia; *glands* 4, yellowish to maroon, with a white appendage 0.5–1 mm long (rarely absent); *capsules* 2–2.5 mm long; *seeds* 1.5–2 mm long, ovoid-prismatic. Common on rocky or sandstone outcrops and ledges, grasslands, and pinyon-juniper, 4000–7500 ft. May–Oct. E/W. (Plate 42)

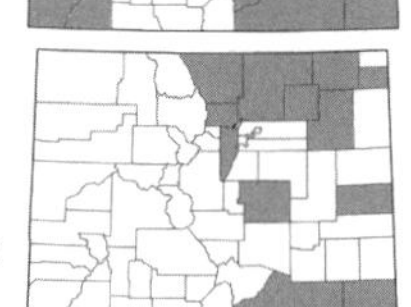

Chamaesyce geyeri **(Engelm.) Small**, GEYER'S SANDMAT. Annuals, prostrate; *leaves* glabrous, oblong to ovate-oblong, 4–12 mm long, entire; *inflorescence* of solitary cyathia in the upper forks; *glands* 4, 0.2–0.6 mm long, with white to reddish appendages; *capsules* 1.5–2 mm long; *seeds* 1.3–1.6 mm long, smooth. Found on sand hills and dunes, 3800–5500 ft. July–Sept. E.

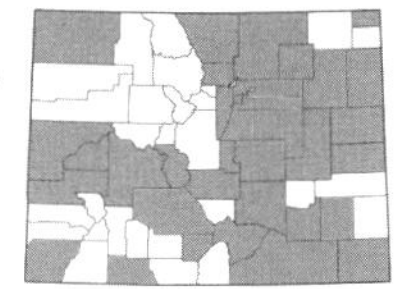

Chamaesyce glyptosperma **(Engelm.) Small**, RIBSEED SANDMAT. Annuals, prostrate; *leaves* glabrous, oblong to obovate-oblong, 2–8 mm long, entire or serrulate; *inflorescence* of cyathia crowded into short, leafy-bracteate axillary clusters, with several cyathia per node; *glands* 4, minute, with a very short appendage; *capsules* 1.4–1.7 mm long, glabrous; *seeds* 1–1.3 mm long, quadrangular and cross-rugose. Common in sandy soil, fields, and disturbed areas, 4000–8000 ft. July–Sept. E/W.

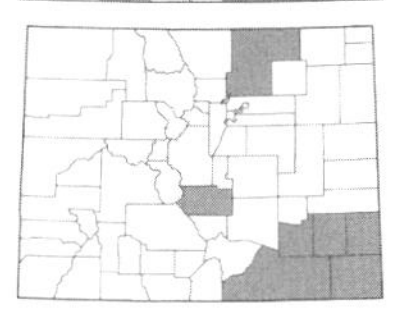

Chamaesyce lata **(Engelm.) Small**, HOARY SANDMAT. Perennials, prostrate; *leaves* hairy, deltoid and usually falcate, 4–12 × 3–9 mm, entire; *inflorescence* of solitary cyathia in the upper forks; *glands* 4–5, 0.5–0.8 mm long, with well-developed appendages; *capsules* 2–2.5 mm long, hairy; *seeds* 1.7–2 mm long, somewhat quadrangular, smooth. Found on sandy prairies and sandstone and limestone outcrops, 3700–5000 ft. May–July. E.

Chamaesyce maculata **(L.) Small**, SPOTTED SANDMAT. [*C. supina* (Raf.) Moldenke]. Annuals, prostrate; *leaves* hairy below, oblong to ovate-oblong, 5–15 mm long, inconspicuously serrulate; *inflorescence* of cyathia crowded into short, leafy-bracteate axillary clusters, with several cyathia per node; *glands* 4, with an inconspicuous appendage; *capsules* ca. 1.5 mm long, hairy; *seeds* ca. 1 mm long, prominently angled. Found as a weed in gardens, lawns, and sidewalk cracks, 4500–5500 ft. May–Oct. E/W. Introduced.

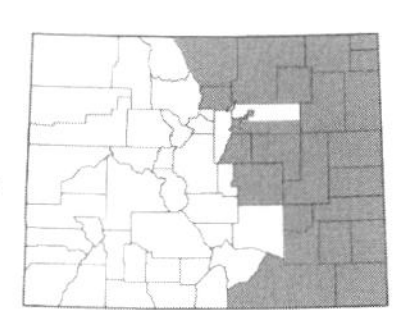

Chamaesyce missurica **(Raf.) Shinners**, PRAIRIE SANDMAT. Annuals, prostrate; *leaves* glabrous, linear to oblong, 10–30 × 1–5 mm, entire, often revolute; *inflorescence* of solitary cyathia in the upper forks; *glands* 4, with white or pinkish appendages 0.5–2.5 mm long; *capsules* 2–2.5 mm long; *seeds* 1.5–2.2 mm, smooth to rugose. Found in sandy soil and grasslands, 3500–6500 ft. May–Oct. E.

Chamaesyce nutans **(Lag.) Small**, EYEBANE. Annuals, erect, 3–8 dm; *leaves* hairy below at least near the base, oblong to oblong-lanceolate, often with a red splotch above, 8–40 mm long, serrate; *inflorescence* of cyathia solitary in the upper forks; *glands* 4, 0.3–0.5 mm diam., with minute to conspicuous appendages; *capsules* 1.5–2.3 mm long, glabrous; *seeds* 1–1.5 mm long, irregularly wrinkled. May be in Colorado, but no specimens have been reported, a weedy plant. May–Oct. Introduced.

Chamaesyce parryi **(Engelm.) Rydb.**, DUNE SPURGE. Annuals, prostrate; *leaves* glabrous, linear, 1–3 cm long, entire; *inflorescence* of solitary cyathia in the upper forks; *glands* 4, prominent, with a yellowish petaloid appendage to 0.5 mm long; *capsules* 2–2.5 mm long; *seeds* 1.4–2 mm long, smooth, with rounded angles. Found in sandy soil, 6500–7500 ft. May–Sept. W.

Chamaesyce prostrata **(Aiton) Small**, PROSTRATE SANDMAT. Annuals, prostrate; *leaves* hairy, oblong to ovate-oblong, (3) 5–11 mm long, serrulate; *inflorescence* of cyathia crowded into short, leafy-bracteate axillary clusters; *glands* 4, lacking appendages or the appendage a pinkish-white rim; *capsules* 1–1.5 mm long, hairy; *seeds* cross-wrinkled with high and narrow ridges. Found in disturbed areas, prairie, and along roadsides, 4500–5500 ft. May–Oct. E/W. Introduced.

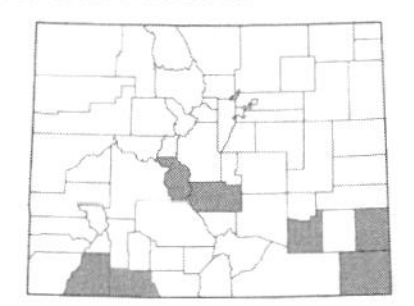

Chamaesyce revoluta **(Engelm.) Small**, THREADSTEM SANDMAT. Annuals, spreading; *leaves* glabrous, linear, 7–25 mm long, entire, revolute; *inflorescence* of solitary, terminal cyathia in the upper forks; *glands* 4, appendage lacking or minute; *capsules* 1.2–1.7 mm long; *seeds* 1–1.3 mm long, strongly angled, transversely ridged to nearly smooth. Found in grasslands, 4500–7000 ft. July–Sept. E/W.

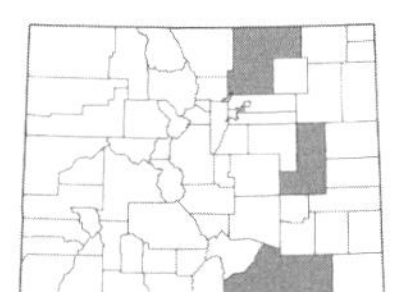

Chamaesyce serpens **(Kunth) Small**, MATTED SANDMAT. Annual, prostrate; *leaves* glabrous, ovate to oblong or orbicular, 2–8 mm long, entire; *inflorescence* of cyathia solitary in the upper forks and nodes; *glands* 4, ca. 0.5 mm long, with white to pinkish appendages, or rarely these absent; *capsules* 1–1.5 mm long; *seeds* 0.9–1 mm long, quadrangular, smooth. Uncommon in sandy soil, fields, and shortgrass prairie, 4900–5500 ft. July–Oct. E. Introduced.

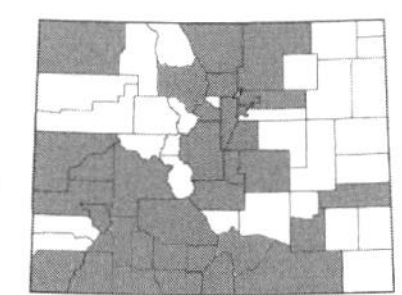

Chamaesyce serpyllifolia **(Pers.) Small**, THYMELEAF SANDMAT. Annuals, prostrate; *leaves* glabrous, oblong to elliptic, 3–15 mm long, minutely serrulate towards the apex or sometimes entire; *inflorescence* of cyathia crowded into short, leafy-bracteate axillary clusters, with several cyathia per node; *glands* 4, yellow-green to reddish-purple, with a short, white appendage; *capsules* 1.3–2 mm long; *seeds* 0.9–1.3 mm long, smooth or irregularly roughened. Common in grasslands and prairies, drying depressions, disturbed areas, and sidewalks, 3800–9500 ft. June–Sept. E/W.

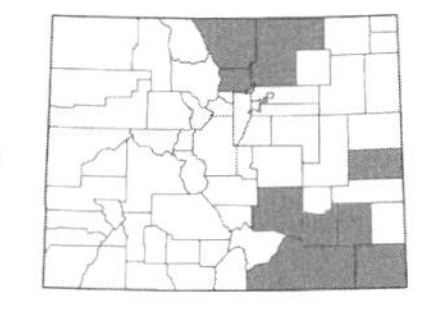

Chamaesyce stictospora **(Engelm.) Small**, SLIMSEED SANDMAT. Annuals, prostrate; *leaves* villous, oblong to orbicular, 3–10 mm long, serrulate; *inflorescence* of solitary cyathia at the nodes or congested leafy lateral branches; *glands* 5, with yellowish-white or reddish appendages; *capsules* 1.4–2.5 mm long, hairy; *seeds* 1–1.5 mm long, quadrangular, with shallow pits or low ridges. Found in fields, grasslands, along roadsides, and in disturbed areas, 4000–6000 ft. June–Oct. E.

CROTON L. – CROTON

Annuals, dioecious (ours) herbs; *leaves* alternate, simple; *inflorescence* of axillary or terminal spikes or racemes; *flowers* not in a cyathium; *staminate flowers* with 5 sepals and petals, and 5–10 stamens; *pistillate flowers* with 5–9 sepals, lacking petals.

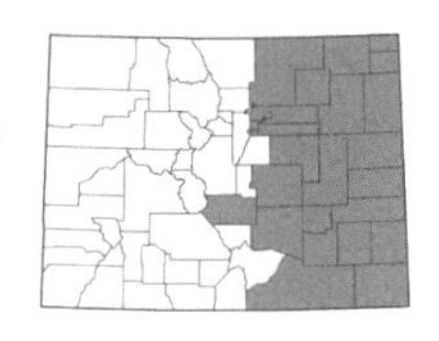

***Croton texensis* (Kl.) Müll. Arg.**, TEXAS CROTON. Plants 2–8 dm, densely stellate-hairy; *leaves* linear-oblong to ovate-oblong, 1–8 cm long, entire; *staminate flowers* with a calyx 2–4 mm wide and no petals; *pistillate flowers* with a calyx 2.5–4 mm wide; *capsules* 4–6 mm. Common in sandy soil on the eastern plains, could be present in Montezuma Co., 3500–6000 ft. June–Sept. E. (Plate 42)

EUPHORBIA L. – SPURGE

Annual or perennial, erect, monoecious herbs with milky latex; *leaves* usually alternate or sometimes opposite, simple; *flowers* in a cyathium, often with petaloid appendages on the marginal glands, lacking sepals and petals; *staminate flowers* of 1 stamen; *pistillate flowers* solitary in the center of the cyathium. (Croizat, 1945)

1a. Most leaves toothed or lobed, at least on the upper half; annuals...2
1b. Leaves all entire; annuals or perennials...4

2a. Leaves alternate, oblong to spatulate (widest above the middle); stems glabrous; capsules strongly warty-tuberculate...***E. spathulata***
2b. Leaves opposite, ovate, lanceolate, or linear; stems with sparsely hairy or with short hairs and scattered multicellular hairs; capsules smooth or hairy...3

3a. Glands with white, petaloid appendages; stems sparsely hairy...***E. exstipulata***
3b. Glands lacking petaloid appendages; stems with short hairs and scattered multicellular hairs...***E. dentata***

4a. Bracteate leaves in the inflorescence with white margins or entirely white; cyathia with 5 white appendages (resembling petals); leaves oblong to ovate, 1.5–10 cm long and 1–3 cm wide...***E. marginata***
4b. Bracteate leaves (if present) green throughout; cyathia and leaves various...5

5a. Leaves mostly linear to narrowly elliptic, usually over 5 times as long as wide...6
5b. Leaves orbicular to spatulate, obovate, or elliptic (never linear), less than 5 times as long as wide...8

6a. Leaves all opposite; cyathia solitary in the upper bract axils and forks of the inflorescence; annual from a slender taproot...***E. hexagona***
6b. Lower leaves alternate (bracts subtending the inflorescence branches opposite); cyathia in umbels with evident rays, ultimately dichotomous and subtended by a pair of yellowish-green bracts; perennial herbs...7

7a. Main leaves 3–10 cm long and 3–10 mm wide, not especially crowded on the stem; plants 3–9 dm...***E. esula***
7b. Main leaves 1–3 cm long and 0.5–3 mm wide, crowded on the stem; plants 1–4 dm...***E. cyparissias***

8a. Leaves densely crowded on the stem (such that the internodes are not visible), obovate with a prominently mucronate tip, rather succulent; capsule 5–7 mm long...***E. myrsinites***
8b. Leaves unlike the above; capsule 2–5 mm long...9

9a. Annual from a slender taproot; capsule with longitudinal raised crests...***E. peplus***
9b. Perennial herbs; capsules lacking raised crests...10

10a. Glands irregularly toothed along the entire margin, lacking marginal horns...***E. incisa***
10b. Glands entire with 2 marginal horns on either side...***E. brachycera***

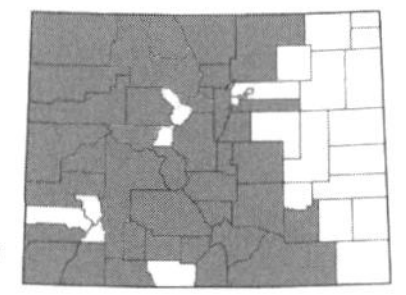

***Euphorbia brachycera* Engelm.**, HORNED SPURGE. [*Tithymalus robustus* (Engelm.) Small]. Perennials, 1–4.5 dm; *leaves* alternate, orbicular to spatulate, 0.7–2.8 cm × 3–25 mm, the margins entire; *inflorescence* of dichotomous branches with foliose, opposite bracts subtending each dichotomy; *glands* 4, yellowish, 1–2 mm wide, with 2 marginal horns on either side; *capsules* 3–5 mm long, glabrous or hairy; *seeds* 2.5–3.5 mm long, pitted to nearly smooth, whitish. Common in sandy soil, dry grasslands, and on rocky slopes, 4700–9500 ft. April–Aug. E/W. (Plate 42)

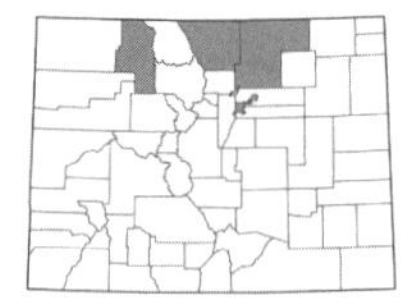

***Euphorbia cyparissias* L.**, CYPRESS SPURGE. [*Tithymalus cyparissias* (L.) Lam.]. Perennials, 1–4 dm; *leaves* alternate, linear to linear-spatulate, crowded, 1–3 cm × 0.5–3 mm, entire; *inflorescence* umbellate; *glands* 4, yellowish, 1–1.5 mm wide, horned on each side; *capsules* ca. 3 mm long, glandular-roughened; *seeds* 1.5–2 mm long, smooth. Cultivated and sometimes escaping in disturbed areas, fields, and along roadsides, 5000–6500 ft. May–Aug. E/W. Introduced.

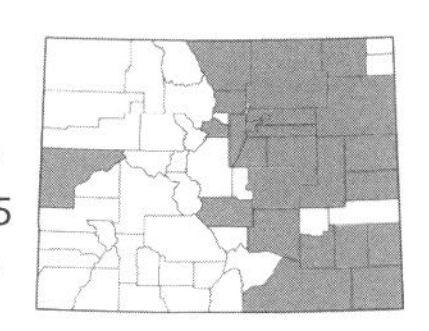

Euphorbia dentata **Michx.**, TOOTHED SPURGE. [*Poinsettia dentata* (Michx.) Klotzsch & Garcke]. Annuals, 1–6 dm; *leaves* opposite, linear to ovate, 1.5–6 cm long, with dentate margins, on petioles 5–25 mm long; *inflorescence* congested at the summit of the stems, the involucres nearly sessile; *glands* (1) 2, lacking appendages, or the central involucre rarely with 5 glands; *capsules* 3–5 mm long, glabrous or sparsely hairy; *seeds* 2.5–3 mm long, sharply angled, tuberculate. Common in grasslands, disturbed areas, gardens, fields, and along roadsides, 3500–6400 ft. June–Sept. E/W. (Plate 42)

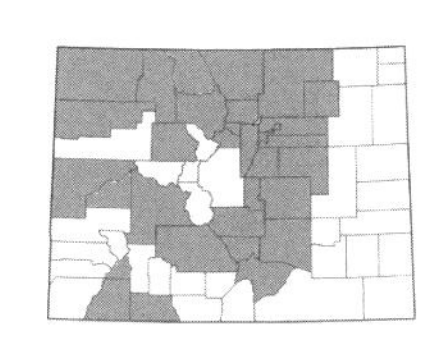

Euphorbia esula **L.**, LEAFY SPURGE. [*Tithymalus esula* (L.) Scop.]. Perennials from stout, forking rhizomes, 3–9 dm; *leaves* alternate, oblanceolate to narrowly oblong or linear, 3–10 cm × 3–10 mm, entire; *inflorescence* umbellate; *glands* 4, yellowish, prominently horned on both sides; *capsules* 3–3.5 mm long; *seeds* 2–2.5 mm long, smooth. Common in disturbed areas, fields, grasslands, on floodplains, and along roadsides and streams, 3500–9000 ft. May–Sept. E/W. Introduced.

Euphorbia exstipulata **Engelm.**, SQUARE-SEED SPURGE. [*Zygophyllidium exstipulatum* (Engelm.) Wooton & Standl.]. Annuals; *leaves* opposite, dimorphic, the basal linear and the upper oblanceolate, 1–5 cm × 2–8 (10) mm, serrulate towards the apex; *inflorescence* of solitary cyathia in the stem forks or leaf axils; *glands* 5, greenish, 0.8–1 mm wide, with white appendages ca. 0.2–0.4 mm long; *capsules* 2.3–3 mm long, minutely hairy; *seeds* 2–2.2 mm long, 4-angled and transversely ridged. Reported for Colorado, but no specimens have been seen. Found on rocky limestone ridges.

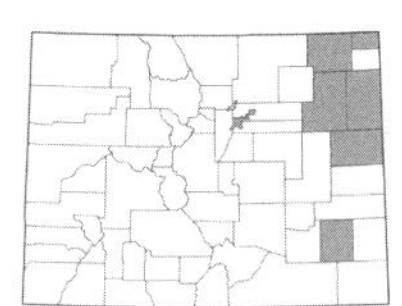

Euphorbia hexagona **Nutt. ex Spreng.**, SIX-ANGLED SPURGE. [*Zygophyllidium hexagonum* (Nutt.) Small]. Annuals, 2–5 (10) dm; *leaves* opposite, linear to lanceolate or oblong, 1–7 cm long, entire; *inflorescence* of solitary cyathia in the upper leaf axils and stem forks; *glands* 5, with white or greenish-white appendages; *capsules* 3–5 mm long; *seeds* 2.5–3.3 mm long, tuberculate or roughened. Found on sandy prairie, dunes, and along roadsides, 3400–5000 ft. June–Aug. E.

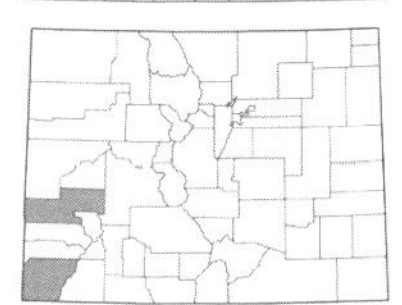

Euphorbia incisa **Engelm.**, MOJAVE SPURGE. [*Tithymalus incisa* (Engelm.) J.B.S. Norton]. Perennials, 1–4 dm; *leaves* alternate, ovate to elliptic or obovate, 0.6–2 cm long, entire; *inflorescence* of dichotomous branches with foliose, opposite bracts subtending each dichotomy; *glands* 4, yellowish, 1–2 mm long, fan-shaped with scalloped or toothed tips; *capsules* 4–5 mm long; *seeds* 2–3 mm long, white to gray. Uncommon in sandy soil and on shale slopes, 6500–7500 ft. May–July. W.

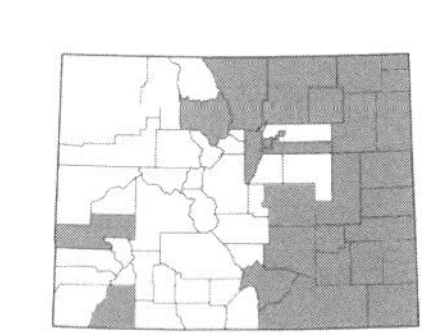

Euphorbia marginata **Pursh**, SNOW-ON-THE-MOUNTAIN. [*Agaloma marginata* (Pursh) Á. Löve & D. Löve]. Annuals, 2–10 dm; *leaves* alternate, ovate to elliptic or oblong, 1.5–10 cm × 10–30 mm, entire; *inflorescence* cymose, with white-margined or entirely white bracteate leaves; *glands* 5, with white appendages 2–3 mm long; *capsules* 5–8 mm long, usually hairy; *seeds* ca. 4 mm long. Common in shortgrass prairie, disturbed areas, fields, floodplains, and along roadsides, 3500–7000 ft. June–Oct. E/W. (Plate 42)

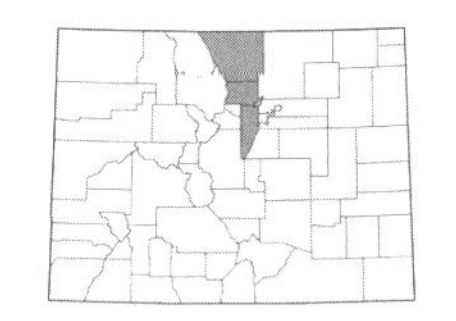

Euphorbia myrsinites **L.**, MYRTLE SPURGE. [*Tithymalus myrsinites* (L.) Hill]. Perennials, sprawling, the stems mostly 0.6–4.5 dm long; *leaves* alternate, obovate to obovate-oblong, thick and fleshy, densely crowded, 0.8–3 cm × 6–25 mm, entire; *inflorescence* umbellate; *glands* 4, often distally expanded, with marginal horns; *capsules* 5–7 mm long, glabrous; *seeds* 3–5 mm long, sculptured to nearly smooth. Cultivated and escaping onto nearby roadsides, mesas, fields, and disturbed areas, 5000–6000 ft. April–May. E. Introduced.

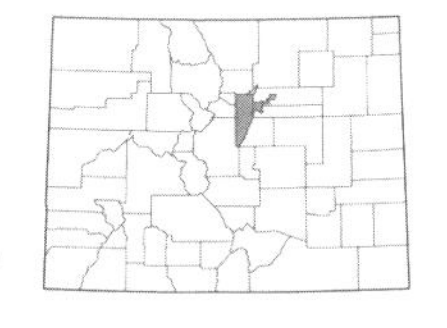

Euphorbia peplus **L.**, PETTY SPURGE. [*Tithymalus peplus* (L.) Hill]. Annuals, 1–3 dm; *leaves* elliptic to obovate, mostly 0.3–2.5 cm × 3–15 mm, entire; *inflorescence* umbellate with dichotomously branched rays; *glands* 4, each with a prominent horn at each end, the outer margins distinctly concave; *capsules* ca. 2 mm long, each lobe with a pair of raised ridge-crests running down the dorsal side; *seeds* ca. 1.5 mm long, the outer faces with 3–4 large pits, the inner faces with 2 deep longitudinal grooves. Uncommon in gardens, known from the Denver area, 5000–5500 ft. May–Sept. E. Introduced.

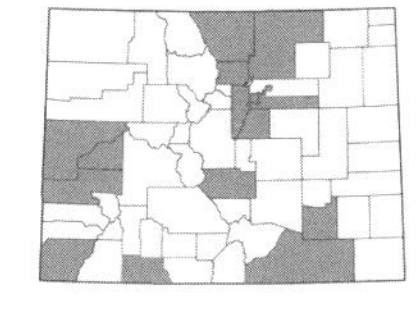

Euphorbia spathulata **Lam.**, WARTY SPURGE. [*Tithymalus spatulatus* (Lam.) W.A. Weber]. Annuals, 1–6 dm; *leaves* alternate, oblong to spatulate, 0.6–4 cm × 4–20 mm, serrulate above the middle; *inflorescence* umbellate with dichotomously branched rays, each dichotomy subtended by a pair of opposite bracts; *glands* 4, yellowish-green, entire; *capsules* 2–2.5 mm long, warty-tuberculate; *seeds* 1.3–2 mm long, brown, reticulate. Found in dry grasslands, and on shale ridgetops and rocky slopes, 4400–6600 ft. May–June. E/W.

REVERCHONIA A. Gray – REVERCHONIA

Annuals, monoecious, glabrous herbs; *leaves* alternate, simple and entire; *flowers* in cymes, with each cymule consisting of a central pistillate flower and 4–6 lateral staminate flowers, reddish, lacking petals; *staminate flowers* with a 4-lobed calyx and 2 stamens; *pistillate flowers* with a 6-lobed calyx.

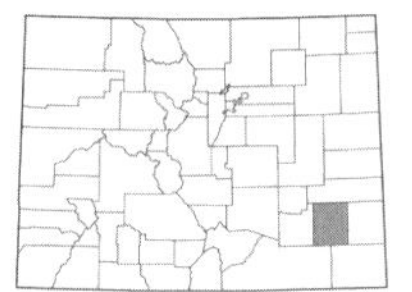

***Reverchonia arenaria* A. Gray**, SAND REVERCHONIA. Plants 2–5 dm, with a glaucous-white, smooth stem; *leaves* linear to elliptic, 1.5–3.5 (4.5) cm long, glabrous, entire; *flowers* reddish, in cymules wih 1 central pistillate flower and 4–6 lateral staminate flowers; *capsules* 7–9 mm diam.; *seeds* 4.5–6.6 mm diam., reddish-brown. Found on sand dunes, known from Bent Co. on the south shore of John Martin Reservoir, 3800–4000 ft. May–Sept. E.

STILLINGIA Garden ex L. – TOOTHLEAF

Perennials, monoecious, glabrous herbs; *leaves* alternate, simple, toothed or crenulate with a small gland in each notch; *flowers* in terminal spike-like thyrses, lacking petals; *staminate flowers* with the sepals fused into a cup and 2 stamens; *pistillate flowers* with a 3-lobed calyx.

***Stillingia sylvatica* L.**, QUEEN'S DELIGHT. Plants 3–6 (8) dm; *leaves* lanceolate to narrowly elliptic, 3.5–8 (12) cm long; *capsules* ca. 12 mm long; *seeds* ca. 8 mm diam., white, smooth. Uncommon on sand dunes, known from Baca Co. along the Cimarron River, 3500–3700 ft. May–Aug. E.

TRAGIA L. – NOSEBURN

Perennials, monoecious herbs with stinging hairs; *leaves* alternate, simple, toothed; *flowers* in bracteate racemes, lacking petals; *staminate flowers* above the pistillate, with a 3–5-lobed calyx; *pistillate flowers* with a 3–8-lobed calyx.

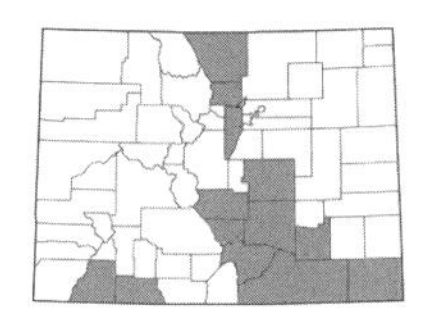

***Tragia ramosa* Torr.**, NOSEBURN. Plants 1–3 (5) dm; *leaves* ovate to lance-ovate, truncate, 1–3 cm × 2–15 mm; *capsules* 3–4 mm long; *seeds* 2.5–3.5 mm diam. Found on dry slopes, in rock crevices, dry grasslands, and pinyon-juniper woodlands, 4000–7000 ft. May–Sept. E/W. (Plate 42)

The leaves, when touched, produce a stinging sensation in the hand very similar to that of stinging nettle.

FABACEAE Lindl. – PEA FAMILY

Herbs, shrubs, or trees; *leaves* usually opposite, simple or compound, stipulate; *flowers* perfect, actinomorphic or zygomorphic; *calyx* 5-lobed; *petals* 5 (an upper petal called the banner, 2 lateral petals called the wings, and 2 lower petals united to form the keel), or 5 segments and distinct, or united basally, rarely 1 or absent; *stamens* 5, 10, or many, diadelphous (9 united by their filaments plus 1 free or nearly so), monadelphous, or distinct; *ovary* unicarpellate, superior; *fruit* a legume.

Fabaceae flower with one wing petal removed:

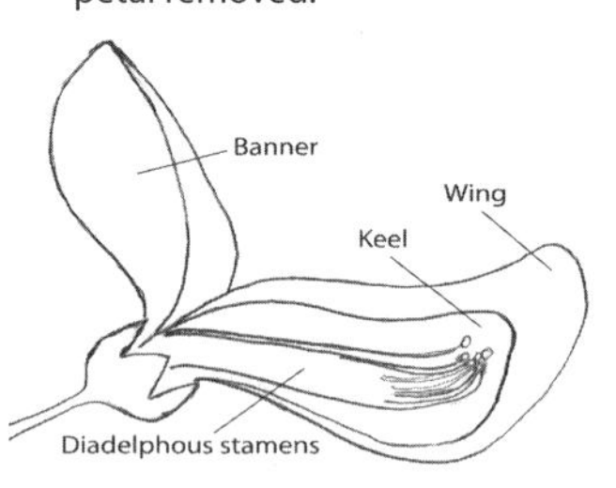

1a. Trees or shrubs with woody stems...2
1b. Perennial or annual herbs, sometimes shrublike and woody below but the stems not woody throughout...12

2a. Corolla lacking; leaflets filiform, the margins involute or folded; fruit 5–6 mm long, strongly glandular-punctate...***Parryella***
2b. Corolla present; plants otherwise unlike the above in all respects...3

3a. Leaves simple, subtending a spine; flowers red-purple...***Alhagi***
3b. Leaves pinnately or palmately compound; stems with spines or without; flowers various but usually not red-purple...4

4a. Leaves odd-pinnately compound (with a terminal leaflet)...5
4b. Leaves even-pinnately compound (without a terminal leaflet) or palmately compound...9

5a. Young stems densely covered with convex, orange glands, pedicels also with a pair of orange glands; stems armed with divaricate thorns; pod with a beak approximately 1 mm long; flowers purple or purplish-pink, in terminal racemes...***Psorothamnus***
5b. Plants lacking orange glands, glandular-punctate with black dots or without; stems armed or unarmed; pods and flowers various...6

6a. Leaflets very small, 1–5 mm long, glandular punctate with black dots; calyx with dense tawny hairs, the teeth 4.5–8.5 mm long; petals bicolored when young, the banner yellowish but soon turning rose-purple like the wings and keel...***Dalea*** *(formosa)*
6b. Leaflets, usually larger, 5–35 mm long (mostly over 10 mm long), glandular punctate or not; plants otherwise unlike the above in all respects...7

7a. Flowers yellow, the corolla 1.5–2 cm long; fruit bladdery-inflated, 5–7 cm × 2–3 cm (when pressed)...***Colutea***
7b. Flowers various but not yellow; fruit not bladdery-inflated...8

8a. Corolla small, 4.5–6 mm long, consisting only of the banner, purple to rose-purple; stamens 10, monadelphous only at the very base; unarmed shrubs; leaves often glandular punctate...***Amorpha***
8b. Corolla large, 15–25 mm long, consisting of a banner, wing, and keel, white or pink-purple; stamens 10, diadelphous; trees or shrubs, often armed with spines; leaves not glandular punctate...***Robinia***

9a. Leaves dimorphic, those fasciculate from spurs merely pinnately compound while those on long shoots bipinnately compound; fruit 2–4 dm long, hairy at first but becoming shiny brown and glabrous, often twisted or contorted; flowers imperfect, yellowish-green; tree, usually armed with spines...***Gleditsia***
9b. Leaves not dimorphic, either all bipinnately compound or all pinnately compound; fruit to 2 (3) dm long, otherwise unlike the above; flowers perfect, yellow, yellowish-green, or pinkish; tree or shrub, armed or unarmed...10

10a. Leaves pinnately or palmately compound; flowers zygomorphic, yellow; stamens included in the petals...***Caragana***
10b. Leaves bipinnately compound; flowers actinomorphic; stamens showy and well-exserted beyond the petals...11

11a. Leaflets 2–4 (5) mm long; inflorescence of globose, pinkish heads; stem armed with broad-based, straight or recurved prickles; fruit 3–5 cm long...***Mimosa*** *(borealis)*
11b. Leaflets 10–35 (60) mm long; inflorescence of yellowish-green spikes; stem armed with long, straight spines that are not broad-based; fruit 10–20 (30) cm long...***Prosopis***

12a. Leaves ternately compound with 3 leaflets, palmately compound with 3 or more leaflets, or simple, lacking tendrils...13
12b. Leaves pinnately or bipinnately compound with 4 or more leaflets, or occasionally with only 2 leaflets and the terminal leaflet replaced by a tendril...25

13a. Leaflets with toothed margins, at least distally...14
13b. Leaflets or leaves with entire margins...16

14a. Leaflets all attached at a single point; flowers in heads or occasionally solitary, sometimes subtended by an involucre; plants acaulescent or caulescent...***Trifolium***
14b. Middle leaflet not attached at the same point as the lateral leaflets, prolonged beyond the lateral leaflets on a petiolule; flowers in axillary racemes or clustered in heads but never subtended by an involucre; plants caulescent...15

15a. Fruit spirally coiled; flowers in a compact axillary raceme or head, purple, yellow, or rarely white...***Medicago***
15b. Fruit straight, ovoid, 3–4 mm long; flowers loose in an elongate axillary raceme (usually over 5 cm long), yellow or white...***Melilotus***

16a. Flowers yellow, sometimes fading reddish or brownish when dried...17
16b. Flowers purple, pinkish, rose, or white...19

17a. Stamens all distinct; large and foliaceous stipules present; flowers yellow, 18–25 mm long; leaflets usually larger, 20–80 mm long...***Thermopsis***
17b. Stamens united by the filaments into a group of 9 with 1 free; stipules absent, gland-like, or small and filiform; flowers usually smaller or if larger then drying reddish; leaflets usually smaller, 3–24 mm long...18

18a. Flowers in a dense terminal spike 1.5–6 cm long; calyx densely pubescent with tawny hairs, the calyx teeth 5–19 mm long; leaflets silky-pilose...***Dalea*** *(jamesii)*
18b. Flowers in a 1–5-flowered umbel; calyx glabrous or closely appressed-hairy with white or grayish hairs, the calyx teeth 1.5–3 mm long; leaflets glabrous or strigulose...***Lotus***

19a. Prostrate or climbing vine; leaflets 3, mostly lanceolate-oblong, the middle one elongated on a petiolule and subtended by a pair of small, stipule-like bracts (stipels), the lateral leaflets also subtended by a pair of stipule-like bracts; stipules with striate nerves; flowers purple to rose...***Strophostyles***
19b. Erect herb; leaflets 3 or more, not subtended by a pair of stipule-like bracts (stipels) or very rarely with stipels, or simple; stipules without striate nerves; flowers various...20

20a. Fruit a loment with 1–2 (3) segments, the margins and surface minutely hairy with hairs hooked at the tips; petioles, inflorescence, and stem usually densely minutely hairy with hairs hooked at the tips; leaflets 3, the middle leaflet prolonged on a petiolule, subtended by stipule-like bracts (stipels)...***Desmodium***
20b. Fruit a legume unlike the above; plants without hairs hooked at the tips; leaflets 3 or more, without stipule-like bracts (stipels)...21

21a. Stamens monadelphous, alternating long and short; leaves palmately compound, usually with 5–12 leaflets but rarely with as few as 3 leaflets; inflorescence a terminal raceme; plants lacking glandular-punctate dots...***Lupinus***
21b. Stamens diadelphous, with 9 fused by their filaments and 1 free; leaves simple, or palmately or ternately compound with 3–7 leaflets; inflorescence various; plants sometimes glandular-punctate with black dots...22

22a. Leaves with 3–7 leaflets (often with 3 leaflets above and 5 leaflets below), usually glandular-punctate with black dots; plants usually caulescent, not mat-forming; ovules and seeds 1; fruit globose or ovoid, usually glandular-punctate...23
22b. Leaves usually with 3 leaflets or simple, lacking glandular-punctate, black dots; plants usually acaulescent, often mat-forming; ovules 2 or more, seeds 1 or more; fruit various, not glandular-punctate...24

23a. Fruit enclosed in the enlarged calyx except for the beak, the body papery and brittle; calyx usually 3 mm or more in length; flowers usually larger, 7–21 mm in length; plants often from a taproot with a fusiform-thickened portion below the surface of the soil...***Pediomelum***
23b. Fruit not enclosed in an enlarged calyx, the body rather thick and coriaceous; calyx small, usually less than 3 mm in length; flowers generally small, 4.5–8 mm in length; plants usually from rhizomes, without fusiform-thickened portions below the soil...***Psoralidium***

24a. Flowers usually 6 or more in a head, rarely fewer or solitary; leaflets 3...***Trifolium***
24b. Flowers 1-numerous, solitary or in a raceme; leaflets 3 or leaves simple...***Astragalus***

25a. Leaves bipinnately compound; flowers actinomorphic and arranged in heads or irregular to zygomorphic but not differentiated into a banner, wing, and keel; stamens distinct...26
25b. Leaves pinnately compound; flowers zygomorphic, differentiated into a banner, wing, and keel or occasionally consisting only of the banner; stamens diadelphous, monadelphous, or sometimes distinct...29

26a. Flowers actinomorphic, arranged in heads; corolla sympetalous, very small and inconspicuous; stamens 5–10 and exserted well beyond the corolla, showy...27
26b. Flowers irregular to zygomorphic, in racemes; corolla yellow (often drying orange), conspicuous, the petals distinct and showy; stamens 10, subequal to the petals...28

27a. Filaments white with yellow anthers; plants lacking recurved prickles...***Desmanthus***
27b. Filaments rose-purple or pink with yellow anthers; plants armed with numerous recurved prickles...***Mimosa***

28a. Plant densely glandular-punctate with black dots...***Pomaria***
28b. Plant lacking glandular-punctate, black dots...***Hoffmannseggia***

29a. Terminal leaflet modified into a tendril or short bristle; plants often with trailing or climbing stems...30
29b. Terminal leaflet not modified into a tendril or bristle; plants only occasionally trailing or climbing...31

30a. Style round, with a ring of hairs below the stigma; flowers generally smaller, 0.45–2.5 cm long, the banner usually reflexed below (the last ⅓ of the petal) or occasionally at the middle; stems not winged...***Vicia***
30b. Style flattened, hairy along one side only; flowers generally larger, (1) 1.5–3 cm long, the banner usually reflexed at or above the middle of the petal; stems sometimes winged...***Lathyrus***

31a. Inflorescence a head-like umbel, rarely paired (and then bright yellow)...32
31b. Inflorescence various, but not a head-like umbel...33

32a. Flowers bright yellow; inflorescence with (2) 4–8 flowers per umbel; leaflets 5, the lowest pair resembling stipules giving the leaves a ternately compound appearance...***Lotus***
32b. Flowers white to pink or purplish; inflorescence with 10–15 flowers per umbel; leaflets 9–19...***Securigera***

33a. Vine, glabrous to densely hairy; leaflets ovate to lance-ovate, 4–10 cm long; flowers purplish-red or brownish-red...***Apios***
33b. Herb; plants otherwise unlike the above...34

34a. Corolla consisting only of the banner, blue-purple; plants somewhat shrubby; inflorescence a dense terminal, spike-like raceme, these often several in the axils of the upper leaves and forming a compound cluster...***Amorpha***
34b. Corolla consisting of more than a solitary banner, variously colored; plants usually not somewhat shrubby; inflorescence various...35

35a. Pod densely covered with stout, hooked prickles; inflorescence a dense axillary spike; flowers greenish-white or ochroleucous; leaves conspicuously glandular-punctate under magnification...***Glycyrrhiza***
35b. Pod not covered with hooked bristles; plants otherwise unlike the above in all respects...36

36a. Stamens 5 or 8–10, monadelphous; plants conspicuously glandular-punctate (at least under magnification); keel blades sometimes distinct and exposing the stamens...***Dalea***
36b. Stamens 10, diadelphous, distinct, or monadelphous; plants not glandular-punctate; keel blades united, generally enclosing the stamens...37

37a. Stipules sagittately lobed and irregularly toothed at the base; stamens monadelphous; flowers pinkish-purple to blue-purple or white...***Galega***
37b. Stipules entire, not lobed at the base; stamens distinct, diadelphous, or monadelphous with the upper stamen free at the base and united above; flowers various...38

38a. Stamens distinct; plants rhizomatous, silky-canescent; flowers white or ochroleucous; fruit constricted between the seeds, the segments often separated by sterile sections, villous at first but becoming glabrate in maturity; keel beaked, not rounded...***Sophora***
38b. Stamens diadelphous or monadelphous; plants otherwise generally unlike the above in all respects; keel beaked or rounded...39

39a. Keel petals abruptly narrowed at the tip to an erect or recurving beak; plants acaulescent or shortly caulescent...***Oxytropis***
39b. Keel petals not beaked; plants caulescent or sometimes acaulescent or shortly caulescent (in some *Astragalus* species)...40

40a. Style barbellate below the stigma; pod bladdery-inflated, spherical or ovoid, stipitate, with numerous seeds; plants rhizomatous, 4–10 dm high; flowers brick red or orange-red, drying lavender-brown...***Sphaerophysa***
40b. Style glabrous; plants otherwise unlike the above in all respects...41

41a. Stamens monadelphous with the upper stamen free at the base and united above; pod 1-seeded, thick and coarsely reticulate with short prickles on the margin, 6–7 mm long; keel significantly longer than the small wings (these barely ⅓ as long as the keel); flowers pink to pink-purple with reddish veins...***Onobrychis***
41b. Stamens diadelphous; pod with more than 1 seed; otherwise unlike the above...42

42a. Fruit a loment, constricted between the seeds into circular or broadly elliptic segments; keel broadly truncate and conspicuously longer than the wings...***Hedysarum***
42b. Fruit a legume; keel rounded or occasionally notched, usually shorter than or equal to the wings in length, if longer then usually just the tip protruding from below the wings...***Astragalus***

ALHAGI Gagnebin – ALHAGI

Much-branched, thorny shrubs, glabrous; *leaves* alternate, simple; *flowers* in racemes; *calyx* short-toothed; *petals* 5, composed of a banner, wing, and keel, reddish-purple; *stamens* 10, diadelphous; *legumes* indehiscent, torulose, constricted between the seeds.

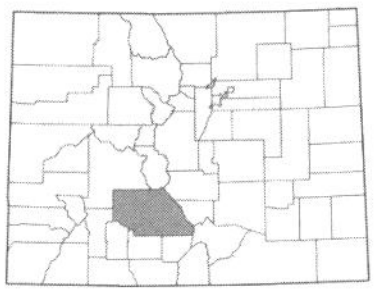

***Alhagi maurorum* Medik.**, CAMEL THORN. Shrubs 4–15 dm; *leaves* elliptic to narrowly obovate, 0.7–2 cm long; *calyx* 2–3 mm long; *flowers* 8–9 mm long; *fruit* shortly stipitate, 10–30 mm long, glabrous. Rare along ditches and roadsides, known from an old collection from the San Luis Valley near Center (Saguache Co.) and evidently now extirpated, 7000–8000 ft. May–Aug. E. Introduced.

AMORPHA L. – FALSE INDIGO

Shrubs or subshrubs; *leaves* odd-pinnately compound, gland-dotted; *flowers* in terminal spike-like racemes; *petals* 1, only the banner present, wrapping around the stamen and style, purple; *stamens* 10, monadelphous at the base; *legumes* gland-dotted. (Palmer, 1931)

1a. Leaves and calyx tube densely soft gray-hairy; leaflets mostly 15–22 pairs plus the terminal 1, 0.7–2 cm long; racemes usually several in the axils of the upper leaves and forming a compound cluster; legume 3–5 mm long...***A. canescens***

1b. Leaves and calyx tube glabrous or minutely hairy; leaflets mostly 4–13 pairs plus the terminal 1, 0.6–5 cm long; racemes solitary or forming a compound cluster; legume 4.5–7 mm long...2

2a. Plants usually less than 1 m tall (mostly 3–8 dm tall); leaflets conspicuously gland-dotted, mostly 0.6–1.5 cm long; racemes usually solitary at the tips of branches...***A. nana***

2b. Plants usually over 1 m tall (1–3.5 m); leaflets eglandular or rarely conspicuously gland-dotted, mostly 2–5 cm long; racemes usually several in the axils of upper leaves, forming a compound cluster, sometimes solitary...***A. fruticosa***

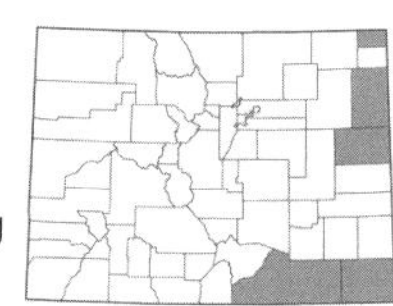

***Amorpha canescens* Pursh**, LEAD PLANT. Subshrubs, 3–5 (10) dm; *leaflets* (9) 15-numerous, elliptic-oblong to elliptic-obovate, 0.7–2 cm long, gray-canescent, glandular-hairy; *inflorescence* of racemes clustered in upper leaf axils; *calyx* 3–5 mm long; *flowers* violet, the banner 4.5–6 mm long; *legumes* half-ovoid, 3–5 mm long, villous. Found on rocky slopes, in open prairies, and along roadsides, scattered on the eastern plains, 3500–6000 (7600) ft. June–Aug. E. (Plate 42)

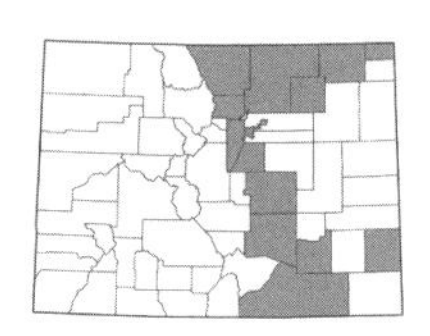

***Amorpha fruticosa* L.**, FALSE INDIGO. Shrubs to small trees, 10–30 (40) dm; *leaflets* 9–23 (31), elliptic, oblong, or nearly lanceolate, 2–5 cm long, eglandular or rarely conspicuously gland-dotted, the midrib often extended as a short mucro; *inflorescence* solitary to several racemes clustered in upper leaf axils; *calyx* 2.5–4 (5) mm long; *flowers* violet to red-purple, the banner 5–6 mm long; *legumes* curved or erect, 5–7 mm long, glabrous or occasionally hairy. Found along streams, in open meadows and prairies, and along roadsides, 3800–6000 ft. May–July. E.

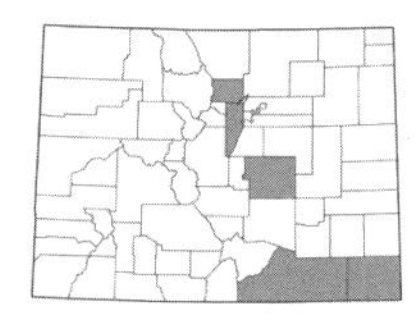

***Amorpha nana* Nutt.**, DWARF WILD INDIGO. Subshrubs, 3–8 dm; *leaflets* 9–25 (31), elliptic to elliptic-oblong, 0.6–1.5 cm long, conspicuously gland-dotted, the midvein extended as a slender mucro; *inflorescence* a densely flowered raceme; *calyx* 3.5–4 mm long, lobes 1.8–2 mm long; *flowers* purple, the banner 4.5–6 mm long; *legumes* half-ovate, 4.5–7 mm, conspicuously glandular-punctate. Found on rocky slopes and in open prairies, 5200–7200 ft. May–July. E.

APIOS Fabr. – GROUNDNUT

Sprawling or twining perennials; *leaves* alternate, pinnately compound with entire leaflets; *flowers* zygomorphic, in axillary racemes; *calyx* 5-lobed, bilabiate, the upper 4 lobes very short or absent and the lower lobe longer; *petals* 5, composed of a banner, wing, and keel, the wings connate to the keel; *stamens* 10, diadelphous; *legumes* sessile, flat.

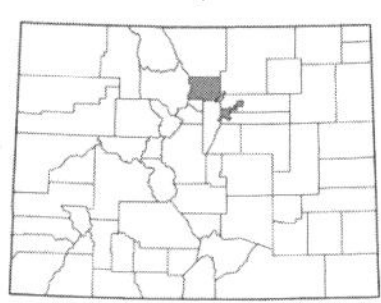

***Apios americana* Medik.**, GROUNDNUT. Vines with stems 1–3 m long; *leaflets* 5–7 (9), lanceolate to obovate, 4–9 cm long; *calyx* tube 2.5–3 mm long; *corolla* reddish-purple, 0.9–1.3 cm long; *legumes* linear, 5–12 cm long. Uncommon in moist, shady thickets and along streams, 5100–5400 ft. July–Sept. E.

ASTRAGALUS L. – MILK VETCH

Annual or perennial herbs, caulescent or acaulescent; *leaves* alternate, usually odd-pinnately compound, sometimes ternately compound, rarely simple; *flowers* zygomorphic; *petals* 5, composed of a banner, wing, and keel, the keel usually obtuse or subacute at the apex; *stamens* 10, diadelphous; *legumes* sessile to stipitate, unilocular or bilocular through the intrusion of the dorsal suture, the most common forms being (1) laterally compressed and often flat, (2) triquetrous-compressed and triangular in cross-section, or (3) subterete or obcompressed with the dorsal and ventral faces often sulcate, usually ellipsoid, subterete, cordate, or narrowly oblong in cross-section, sometimes bladdery-inflated with papery walls. (Barneby, 1964)

The following key is quite long, but you should be able to key out most *Astragalus* species in fruit or flower (there are a few species which do require mature fruit and a few which require flowers to accurately identify). For almost each couplet, there are a few species which are intermediate and can be keyed either way – if questionable, follow either choice.

1a. Leaflets 3–9, tipped with a yellowish or pale spine or bristle, usually thick and rigid with a prominent midrib below, continuous with the rachis; flowers 1–3, subsessile in the leaf axils, 4–9 mm long...***A. kentrophyta***

1b. Leaflets not spine-tipped, or if spine-tipped then the leaves filiform and simple and mostly basal; plants otherwise unlike the above...2

2a. Plants acaulescent to shortly caulescent, often tufted, the stem shorter than the leaves and the internodes tightly packed together and often concealed by stipules; peduncles arising from near the base of the plant or the flowers subsessile in the leaf axils...**Key 1**
2b. Plants evidently caulescent, with a conspicuous stem bearing leaves and elongated internodes; peduncles arising higher on the stem of the plant...3

3a. Some or all the leaves simple, or reduced to filiform phyllodes (often appearing juncaceous or rush-like)...**Key 2**
3b. Leaves all compound...4

4a. Plants with mostly dolabriform hairs (hairs attached near the middle, look under magnification and gently move a hair at one end to see where it is attached, in general dolabriform hairs appear very narrowly rhombic)...**Key 3**
4b. Plants with basifixed hairs (hairs attached at the base), rarely with a few obscure and inconspicuous dolabriform hairs...5

5a. Plants with mature fruit (*note*: there are a few species which are intermediate for each couplet of the following fruit keys – if questionable, follow either choice)...6
5b. Plants without mature fruit...12

6a. Calyx inflated, papery, enclosing the small (5–7 mm long) fruit; inflorescence a spicate raceme of 30-numerous, densely crowded, spreading flowers, on a robust peduncle 9–17 cm long; leaflets linear to linear-oblong, on pressed specimens usually all ascending from the rachis; flowers ochroleucous or pale yellow...***A. oocalycis***
6b. Calyx not inflated and enclosing the fruit; plants otherwise various...7

7a. Ventral face of the fruit strongly 2-grooved (grooved lengthwise on the fruit along either side of the prominent, raised suture); fruit deflexed, stipitate 1.5–5 mm; plants erect in clumps or spreading with ascending tips, the stems 1.5–5 dm long, usually strongly selenium-scented; racemes with numerous spreading and soon declined flowers...***A. bisulcatus***
7b. Ventral face of the fruit not 2-grooved; fruit and plants various...8

8a. Plants annual from a slender taproot, usually basally branched; fruit bilocular, linear and gently incurved ¼–½ circle, not bladdery-inflated, 12–20 mm long, sessile or nearly so, glabrous or minutely strigulose...***A. nuttallianus* var. *micranthiformis***
8b. Plants perennial or if annual from a slender taproot then the fruit unlike the above in all respects...9

9a. Fruit laterally compressed (the sutures outlining the fruit), usually flat or nearly so (sometimes becoming slightly turgid when the seeds are fully developed), rarely laterally compressed and conspicuously turgid and then the fruit erect and stipitate...**Key 4**
9b. Fruit obcompressed or dorsiventrally compressed (with sutures usually evident on the dorsal and ventral faces of the fruit), sometimes laterally compressed just at the beak but not throughout, at least somewhat inflated or turgid, obcordate to subterete or 3-angled in cross-section...10

10a. Fruit bladdery-inflated, the walls thinly or thickly papery, in cross-section usually subterete...**Key 5**
10b. Fruit not bladdery-inflated, ellipsoid, cordate, or triangular in cross-section, or if inflated and subterete in cross-section then the walls thick and coriaceous, woody, or fleshy but not papery...11

11a. Fruit uniformly hairy (strigulose to villous or hirsute with white or black hairs, look under magnification of 10×)...**Key 6**
11b. Fruit glabrous or nearly so (sometimes minutely hairy just at the apex, but not uniformly hairy)...**Key 7**

12a. Calyx tube 4.9–5.8 mm long with short (0.2–0.7 mm), broadly low-deltoid teeth, the teeth with ciliate margins but otherwise the calyx nearly glabrous; leaflets ovate to elliptic, 7–15 mm wide; flowers ochroleucous...***A. americanus***
12b. Calyx with longer teeth, or if the teeth shorter than 1 mm then the tube shorter than 5 mm or the leaflets linear to linear-oblong or narrowly elliptic; flowers various...13

13a. Leaflets 15–27, densely silvery-cinereous, 2–4 (9) mm long, obovate or elliptic, and densely crowded on a leaf rachis 1–3 (4) cm long; plants 2–6 cm high, usually mat-forming from a subterranean caudex; fruit stipitate 1–4 mm...***A. lutosus***
13b. Plants unlike the above in all respects...14

14a. Wing petals cleft at the apex into 2 unequal lobes; flowers whitish with a purple-tipped keel, mostly 7.5–12.5 mm long; leaflets mostly acute; leaves mostly sessile or subsessile; plants usually villous or silky-strigose but occasionally glabrate; fruit laterally compressed and flat, stipitate 2.5–8 mm...***A. australis***
14b. Wing petals obtuse, truncate, or minutely notched at the apex; plants otherwise various...15

15a. Lower stipules connate into a sheath opposite the petiole...16
15b. Lower stipules free, sometimes amplexicaul but not connate...17

16a. Flowers white, ochroleucous, or yellow, sometimes just with a purple-tipped keel or purple veins in the banner...**Key 8**
16b. Flowers purple, pink-purple, red-purple, or bicolored with a purplish banner and white wings or with a white banner and purple wings, with or without a purple-tipped keel...**Key 9**

17a. Flowers 5–12.5 mm long; calyx tube 1.5–5 mm long...**Key 10**
17b. Flowers 13–30 mm long; calyx tube 4–12 mm long...**Key 11**

Key 1

(*Plants acaulescent or shortly caulescent*)

1a. Leaves all simple or only a few pinnately compound; hairs dolabriform; flowers 7–9 mm long; fruit laterally compressed although sometimes turgid...2
1b. Leaves mostly ternately or pinnately compound, rarely a few of the outer leaves simple; hairs dolabriform or basifixed; flowers and fruit various...3

2a. Inflorescence an elongated raceme 15–25 cm long with 7–23 loose, well-spaced flowers; leaves 5–10 cm long, filiform with inrolled margins; calyx teeth spreading...***A. chloödes***
2b. Inflorescence shortly racemose, mostly to 6 cm long; leaves 0.5–5 cm long, spatulate or linear, flat or involute; calyx teeth erect...***A. spatulatus***

3a. Leaves ternately compound with 3 leaflets (or mostly with 3 leaflets, sometimes a few up to 5 leaflets); plants densely silvery-hairy; fruit enclosed within or partly exserted from the persistent calyx; inflorescence of 1–4 flowers subsessile and usually clustered in the leaf axils; stipules large, membranous, connate, imbricate and usually concealing the internodes...4
3b. Usually some leaves with at least 5 leaflets; inflorescence a subumbellate or subcapitate cluster of 2–8 flowers; if plants silvery-hairy then the fruit well-exserted from the calyx or the stipules free...8

4a. Flowers white with a purple-tipped keel or occasionally blue-purple (pale yellow when dry), 12.5–28 mm long, subsessile in the leaf axils; calyx tube 5.8–16 mm long; fruit (6) 6.5–10 mm long...5
4b. Flowers pink or purplish (but often drying yellowish), 5–11.5 mm long, on a short but evident peduncle, included or exserted above the leaves; calyx tube 2–4 mm long; fruit 4–6.5 mm long...6

5a. Petals glabrous; banner 16–28 mm long; calyx tube 6.5–15 mm long; blooming April–May...***A. gilviflorus***
5b. Petals villous dorsally; banner 12.5–17 mm long; calyx tube 6–6.5 mm long; blooming June–July...***A. hyalinus***

6a. Stipules glabrous but ciliate on the margins at the apex; plants of the western slope...***A. aretioides***
6b. At least the outer stipules hairy dorsally; plants of the eastern slope...7

7a. Mature fruit early-deciduous with the pedicels, immature fruit usually evident; flowers generally smaller, 5–7 mm long, usually exserted above the leaves...***A. sericoleucus***
7b. Fruit inconspicuous, hidden by the stipules, often persistent past maturity; flowers generally larger, 7–11.5 mm long, usually included in the leaves...***A. tridactylicus***

8a. Plants with mostly dolabriform hairs (attached near the middle, look under magnification and gently move a hair at one end to see where it is attached, in general dolabriform hairs appear very narrowly rhombic)...9
8b. Plants with basifixed hairs (attached at the base), rarely with a few obscure dolabriform hairs...18

9a. Leaflets very small, 0.8–2 mm long, elliptic but usually folded and less than 1 mm wide; plants 0.5–2 cm tall, with persistent old leaf rachises giving the plant a spinescent appearance; racemes 1–3-flowered, the flowers included in the leaves, purplish but drying pink; fruit 4–5 mm long...***A. humillimus***
9b. Leaflets over 2 mm long, variously shaped but usually 1 mm or more wide; plants 1–25 cm tall, usually lacking persistent old leaf rachises; racemes with 3-numerous flowers, included or exserted from the leaves, variously colored; fruit 7–40 mm long...10

10a. Leaflets linear to narrowly elliptic, mostly involute or folded; lower stipules connate opposite the petiole...11
10b. Leaflets obovate, oblanceolate, or elliptic, flat; lower stipules free (connate in one hybrid)...12

11a. Flowers 2–8 in a subumbellate raceme, bright pink-purple; fruit 15–30 (35) mm long, oblong, usually green- or purple-mottled; leaves with 3–5 (7) leaflets...***A. detritalis***
11b. Flowers 10–many in an elongated raceme, yellowish or cream, usually with a purple-tipped keel; fruit 7–12 (14) mm long, ovoid to elliptic-oblong, brownish, not green- or purple-mottled; leaves with (7) 8–17 leaflets...***A. flavus***

12a. Wing tips notched at the apex; flowers whitish (drying ochroleucous), pink, or bright purple, the wing tips always white or pale and the keel purple-tipped; fruit bilocular or nearly so, oblong, straight or falcate...***A. calycosus***
12b. Wing tips entire at the apex; flowers various; fruit unilocular or if bilocular then obovoid to spherical...13

13a. Calyx campanulate, the tube 3–4.5 mm long; flowers sometimes of two kinds, the open flowers 9–14 mm long and cleistogamous flowers 4–7 mm long (these are usually hidden by the basal leaves), usually whitish to ochroleucous or sometimes purplish; fruit villous, not purple-mottled; plants of the eastern slope...***A. lotiflorus***
13b. Calyx cylindric, the tube 5–10 mm long; flowers never of two kinds, 13–30 mm long, pink-purple or ochroleucous with a purple-tipped keel; fruit glabrate or strigose, sometimes purple-mottled; plants of the eastern or western slope...14
(note: mature fruit is needed to ensure proper identification of the following species*)*

14a. Fruit bilocular, obovoid to almost globose, 10–15 mm long × 10–12 mm wide, fleshy and becoming spongy or somewhat thick at maturity, not mottled...***A. anisus***
14b. Fruit unilocular, ovoid-ellipsoid to lanceolate, straight or incurved, 12–45 mm long, if fleshy and becoming spongy then purple-mottled...15

15a. Fruit purple-mottled, lunately incurved, at least somewhat fleshy when young; flowers ochroleucous to creamy-white and often with a purple-tipped keel, pale lilac, or bright pink-purple; plants of the western slope...16
15b. Fruit not purple-mottled, straight or lunately incurved, not fleshy; flowers bright pink-purple; plants of the eastern or western slope...17

16a. Pod thickly succulent and fleshy when young, becoming spongy and pithy in age, the walls of the pod 1–3 mm thick at maturity; flowers ochroleucous to creamy-white and often with a purple-tipped keel or bright pink-purple...***A. chamaeleuce* var. *chamaeleuce***
16b. Pod somewhat fleshy when young, but not spongy or pithy at maturity, the walls about 0.5 mm thick; flowers pale lilac...***A. piscator***

17a. Fruit persistent (the pod remaining attached to the receptacle even after dehiscence, sometimes seen remaining attached to the previous year's peduncle), obcompressed or quadrangular to laterally compressed in the beak, rarely dorsally flattened in the middle, straight or sometimes lunately incurved but then not laterally compressed at the base; keel 11–18.5 mm long; strongly perennial with a more or less woody caudex, occasionally appearing as a short-lived perennial...***A. missouriensis***
17b. Fruit deciduous (the pod detaching from the receptacle and dehiscent on the ground), laterally compressed at both ends and dorsally flattened in the middle, usually lunately incurved; keel (14) 19–23 mm long; short-lived perennials, often appearing annual, or rarely distinctly perennial with a more or less woody caudex...***A. amphioxys* var. *vespertinus***

18a. Leaves mostly with 3–5 leaflets (rarely a few with 7)...19
18b. Leaves mostly with 7-numerous leaflets (rarely only a few leaves with 5 leaflets)...20

19a. Fruit thinly hirsute or villous, the surface visible, the hairs 0.5–1.5 mm long; leaflets usually larger, the smaller ones mostly 10–20 mm long...***A. musiniensis***
19b. Fruit densely villous-hirsute, the hairs usually concealing the surface, the hairs to 2.5–4.5 mm long; leaflets usually smaller, the smaller ones mostly 5–10 (20) mm long...***A. newberryi* var. *newberryi***

20a. Lower stipules connate into a sheath opposite the petiole...21
20b. Lower stipules free or amplexicaul but not connate...25

21a. Fruit bladdery-inflated, papery, purplish-mottled; plants from a woody, branching caudex and stout taproot, thatched in appearance due to the presence of persistent leaf rachises; terminal leaflet continuous with the rachis; flowers 5–7 mm long; calyx tube 1.5–2 mm long...***A. jejunus***
21b. Fruit not bladdery-inflated, papery to coriaceous, purplish-mottled or not; plants from a slender taproot and/or loosely spreading rhizome-like caudex, usually not thatched; terminal leaflet petiolulate and not confluent with the rachis; flowers 6–14 mm long; calyx tube 2–4 mm long...22

22a. Leaflets with obtuse or retuse tips, mostly ovate or elliptic; stems arising from a slender taproot and subterranean caudex; flowers purplish or bicolored with white wings...23
22b. Leaflets with acute tips, linear to narrowly elliptic; stems arising from a root-crown located at the surface of the soil and woody taproot; flowers whitish or pale purple with a purple-tipped keel...24

23a. Fruit ascending, sessile or nearly so; keel shorter than the wings; flowers pink-purple or pale lilac with the banner purple-veined; racemes usually 2–4 (6)-flowered...***A. molybdenus***
23b. Fruit pendulous, stipitate 1.4–3.5 mm; keel as long or longer than the wings in length; flowers with a purple banner and keel tip and white wings; racemes (5) 7–20-flowered...***A. alpinus***

24a. Fruit laterally compressed, not mottled; keel usually longer than the wings (sometimes barely so), the purple tip conspicuous; terminal leaflet usually longer or remote from the uppermost pair; plants usually larger, 3–40 cm in height; racemes usually elongated, 3–20-flowered; plants common and variously distributed...***A. miser* var. *oblongifolius***
24b. Fruit obcompressed, purplish-mottled; keel shorter than the wings; terminal leaflet not remote from the uppermost pair; plants usually very small, to 5 cm in height; racemes subcapitate, 2–5-flowered; plants known from Montezuma County...***A. deterior***

25a. Calyx tube 1.5–2.5 mm long; flowers small, 4–7 mm long, white with a purple-tipped keel; peduncles elongate, filiform, to 15 cm long; leaflets linear-filiform to linear-elliptic with an acute apex, the terminal leaflet longer than the subtending pair; fruit plumply obovoid, 10–20 mm long...***A. brandegeei***
25b. Calyx tube 3–15 mm long; flowers 7–23 mm long, variously colored; peduncles not elongate and filiform; leaflets elliptic to obovate or oblong, the terminal leaflet not conspicuously longer; fruit various, 6–40 mm long...26

26a. Fruit bladdery-inflated, usually purple-mottled; plants glabrate or sparsely strigulose; leaflets broadly obovate or elliptic...***A. megacarpus***
26b. Fruit not bladdery-inflated; plants tomentose, villous, or strigulose but evidently pubescent; leaflets various...27

27a. Plants with spreading hairs, the leaflets usually glabrate above (at least in the middle), not silvery; calyx tube campanulate or rarely shortly cylindric, 4–6 (7) mm long and the lobes mostly 4–5 mm long; flowers ochroleucous and usually with a purple-tipped keel...***A. parryi***
27b. Plants with appressed or appressed-ascending hairs, sometimes silvery pubescent; calyx tube various; flowers pink-purple or ochroleucous with a purple-tipped keel...28

28a. Fruit narrow, 2.5–6 mm wide, narrowly oblong, linear-oblong to oblanceolate or narrowly ellipsoid, usually red- or purple-mottled; plants strigulose-villous; leaflets usually folded and thus appearing linear-elliptic; flowers 7–16 mm long; plants of southwestern Colorado (southern Garfield to Montezuma cos.)...29
28b. Fruit usually wider, 5–13 mm wide, ovoid-acuminate to broadly ellipsoid or ovoid, tapering to a laterally compressed, incurved beak, not mottled; plants tomentose or sericeous; leaflets elliptic to obovate or broadly ovate, flat; flowers 9–28 mm long; plants variously distributed but not limited to southwestern Colorado...31

29a. Fruit loosely villous with spreading hairs, declined or spreading, lunately incurved, 6–16 mm long; inflorescence usually well-exserted above the leaves...***A. desperatus***
29b. Fruit strigulose with appressed hairs, ascending or spreading, straight or incurved, 12–30 mm long; inflorescence usually included or slightly exserted above the leaves...30

30a. Fruit 3-angled in cross-section, the dorsal face narrowly but deeply sulcate, semibilocular or almost bilocular, usually lunately incurved, 15–30 mm long; flowers pink-purple...***A. monumentalis* var. *cottamii***
30b. Fruit obcompressed to weakly 3-angled in cross-section, the dorsal face flat and smooth or shallowly sulcate, unilocular or nearly so, usually straight or somewhat incurved just at the tip, 12–20 mm long; flowers bicolored with a white banner and purple wings and keel...***A. naturitensis***

31a. Calyx tube 3–4.5 mm long, campanulate; flowers 9–14 mm long, usually ochroleucous or whitish, sometimes purple; leaflets usually elliptic with an acute tip; plants usually with some dolabriform hairs...***A. lotiflorus***
31b. Calyx tube 6–15 mm long, cylindric or if campanulate then over 9 mm long; flowers 12–28 mm long, variously colored; leaflets with obtuse or acute tips; plants only with basifixed hairs...32

32a. Fruit glabrous or sometimes minutely hairy just at the tip; inflorescence with 10-numerous flowers...33
32b. Fruit strigulose to tomentose (look under 10× magnification) and uniformly hairy throughout; inflorescence with 3-numerous flowers...34

33a. Leaflets densely silky-tomentose on both sides, the hairs turning tawny in age; calyx densely villous; fruit bilocular with a unilocular beak...***A. mollissimus***
33b. Leaflets greenish-cinereous or silky-canescent, glabrate above, not turning tawny; calyx thinly villous; fruit unilocular or nearly bilocular...***A. iodopetalus***

34a. Leaflets mostly acute; fruit densely tomentose with hairs 2.5–4 mm long, usually obscuring the surface, unilocular; flowers ochroleucous or whitish with a purple-tipped keel; calyx whitish and usually with some black hairs...***A. purshii***
34b. Leaflets mostly obtuse; fruit with hairs not obscuring the surface or if hairs densely tomentose and obscuring the surface then the fruit bilocular and the hairs usually shorter; flowers pink purple or ochroleucous with a purple-tipped keel; calyx various...35

35a. Fruit bilocular, densely tomentose with spreading hairs up to 1.5–2 mm long, the hairs usually obscuring the fruit surface; flowers pink-purple; inflorescence (7) 12–25-flowered...***A. thompsoniae***
35b. Fruit unilocular, apressed hairy or loosely villous with spreading hairs, the surface visible; flowers pink-purple (and then plants mostly of the eastern slope) or ochroleucous with a purple-tipped keel; inflorescence 3–15-flowered...36

36a. Flowers whitish and purple-tinged with a purple-tipped keel; fruit 15–30 mm long; leaflets narrowly obovate, 4–14 mm long...***A. argophyllus* var. *martinii***
36b. Flowers pink-purple; fruit 25–40 mm long; leaflets broadly obovate to elliptic-ovate, 5–25 mm long...***A. shortianus***

Key 2

(Some or all leaves simple or reduced to filiform phyllodia)

1a. Leaves broadly ovate or orbicular, 1–3 cm wide, glabrous, glaucous and leathery; flowers greenish-yellow with a purple-tipped keel or purplish with whitish wing-tips, 18–22 mm long; fruit ellipsoid, 2.4–3.5 cm long, evidently stipitate, the stipe 1–2 cm long...***A. asclepiadoides***
1b. Leaves filiform, linear, or narrowly elliptic, less than 1 cm wide, glabrous or variously hairy but not glaucous and leathery; flowers various, 5.5–20 mm long; fruit various, generally sessile or if stipitate then the stipe to 1.2 cm long...2

2a. Fruit bladdery-inflated, elliptic to broadly ovoid, glabrous, purple-mottled; inflorescence included; plants 1–2 (4) dm tall, from a deep rhizomatous system and slender underground caudex; flowers 6–10 mm long...***A. ceramicus***
2b. Fruit not bladdery-inflated, linear-oblong to oblanceolate or ellipsoid-oblong, glabrous or hairy, sometimes mottled; inflorescence usually exserted or occasionally included; plants (1) 1.5–9 dm tall, rhizomatous or not; flowers 5.6–20 mm long...3

3a. Flowers strongly upswept, the petals all abruptly upturned 90 degrees, 7–12 mm long; fruit strigulose, pendulous...***A. convallarius***
3b. Flowers not strongly upswept, the wing and keel not abruptly upturned 90 degrees, 5.6–23 mm long; fruit glabrous or rarely inconspicuously strigulose, erect or pendulous...4

4a. Leaves mostly simple and phyllodial, only a few lower ones sometimes pinnately compound (juncaceous in appearance); fruit not mottled...5
4b. Leaves mostly pinnately compound, with only a few of the uppermost leaves simple and phyllodial; fruit sometimes mottled...8

5a. Fruit ascending-erect; flowers white with a purple-tipped keel; calyx lobes 1.5–2 mm long...***A. linifolius***
5b. Fruit pendulous (at least some); flowers usually pink-purple with white wing tips or if white with a purple-tipped keel then the calyx lobes only 0.8–1 mm long...6

6a. Fruit stipitate 4–9 mm, glabrous; calyx tube thinly to densely black-strigulose; flowers pink-purple, often with paler wing tips…***A. coltonii* var. *coltonii***
6b. Fruit sessile, glabrous or strigulose; calyx tube white-strigulose but sometimes mixed with some black hairs, or glabrous with ciliate teeth; flowers various…7

7a. Stems glabrous or nearly so, 4–5 dm long; fruit initially fleshy, becoming thickly coriaceous or woody, glabrous, longitudinally wrinkled, 12–20 (25) mm long; calyx lobes 0.8–1 mm long; flowers white with a purple-tipped keel or less often pink-purple with paler tips…***A. rafaelensis***
7b. Stems initially strigulose, becoming glabrate with age, the growing tips usually cinereous, 1.5–3 dm long; fruit initially fleshy, becoming stiffly papery, finely strigulose, reticulate but not wrinkled, 20–35 mm long; calyx lobes 1–3 mm long; flowers pink-purple with paler tips or less often white with a purple-tipped keel…***A. saurinus***

8a. Flowers 5.6–8 mm long, purplish with white wings; fruit (0.6) 0.9–1.5 cm long, sessile or substipitate, the stipe only to 1.7 mm long, purple-mottled when young…***A. wingatanus***
8b. Flowers 13–23 mm long, whitish or ochroleucous or bicolored with a pink-purple banner and white wings; fruit 2–4 cm long, stipitate or sessile, mottled or not…9

9a. Flowers whitish or ochroleucous; fruit obcompressed, stipitate 4–12 mm, glabrous, sometimes mottled…***A. lonchocarpus***
9b. Flowers bicolored with a pink-purple banner and white wings; fruit laterally compressed, sessile or substipitate to 1 mm, inconspicuously strigose, not mottled…***A. saurinus***

Key 3

(Plants with dolabriform hairs)

1a. Terminal leaflet continuous with the rachis, leaves linear-filiform or narrowly elliptic; fruit bladdery-inflated, elliptic to broadly ovoid, glabrous, purple-mottled; plants 1–2 (4) dm tall, from an deep rhizomatous system and slender underground caudex; flowers 6–10 mm long…***A. ceramicus* var. *ceramicus***
1b. Terminal leaflet of all leaves jointed to the rachis, leaves elliptic, lanceolate, or lanceolate-oblong, if linear or narrowly elliptic then usually folded; fruit not bladdery-inflated, variously shaped, glabrous or strigose, not purple-mottled; plants 1–15 dm; flowers 9–17.5 mm long…2

2a. Flowers at first ascending but soon spreading to ultimately deflexed at full anthesis (but fruit is erect or reflexed), ochroleucous or greenish-yellow to greenish-white, in a dense raceme; plants stout and robust, 1.5–15 dm tall, arising singly or few together from woody rhizomes or clump-forming from a root crown; leaflets 1–5 cm long; fruit bilocular…3
2b. Flowers ascending or spreading at full anthesis, purplish, yellowish, or white, in a dense or loose raceme; plants occasionally stout and robust, 0.5–6 dm tall, from a root crown or branching caudex; leaflets 0.2–2.5 cm long; fruit unilocular or if bilocular then both sharply 3-angled and erect…4

3a. Calyx tube 4.5–10.5 mm long, the teeth 1.2–4.5 mm long; fruit erect, terete, straight or nearly so; sometimes all stipules connate…***A. canadensis***
3b. Calyx tube 3–3.5 mm long, the teeth 0.5–1 mm long; fruit reflexed, incurved, three-angled; only the lower stipules connate…***A. falcatus***

4a. Flowers subsessile in a dense subcapitate or cylindric raceme, purplish, white or ochroleucous; calyx tube 5–7 mm long; fruit incompletely bilocular, sharply three-angled at the middle; leaflets green or occasionally silvery canescent; plants of the eastern or western slope…***A. laxmannii* var. *robustior***
4b. Flowers on a short but evident pedicel, in an elongated, usually loose raceme, yellowish, light purple, or whitish; calyx tube 1.5–5 mm long; fruit unilocular, not sharply 3-angled at the middle, sometimes slightly 3-angled just below the beak; leaflets usually grayish or silvery cinereous or canescent; plants of the western slope…5

5a. Flowers 6–8 mm long, 1–3 in a loose raceme with the axis to 1 cm long, light pink-purple with a purple-tipped keel; rachis commonly excurrent as a short bristle; fruit papery at maturity, often mottled; plants usually prostrate and mat-forming; plants known from Montrose County…***A. sesquiflorus***
5b. Flowers 9–18 mm long, 5–many in a subcapitate cluster or an elongate, loose raceme with the axis usually longer than 1 cm, yellowish, whitish, or ochroleucous; rachis not excurrent as a short bristle; fruit coriaceous or stiffly papery at maturity, mottled or not; plants erect or prostrate and mat-forming; plants variously distributed…6

6a. Leaflets mostly folded; flowers yellow to cream, the keel usually purple-tipped; fruit erect, straight, 7–12 (14) mm long, brownish, lacking green or purple dots; plants erect or decumbent with ascending tips…***A. flavus***
6b. Leaflets flat; flowers whitish or ochroleucous, often drying purple; fruit spreading-ascending, lunately incurved, 15–18 mm long, green or purple dotted; plants prostrate or weakly ascending…***A. humistratus* var. *humistratus***

Key 4

(Fruit laterally compressed, flat or slightly turgid, rarely very turgid and then erect and stipitate)

1a. Lower stipules connate into a sheath opposite the petiole, black or sometimes dark brown with a black band at the base of the sheath; calyx tube 2–2.6 (3) mm long; flowers 6–9 (11) mm long, white; fruit 7–17 mm long, greenish and often with reddish or brown mottling when young, becoming dark brown or black at maturity...***A. tenellus***
1b. Lower stipules free (but sometimes amplexicaul), or if connate then not black; flowers various; fruit usually not purplish-mottled, or if mottled then not becoming dark brown or black at maturity...2

2a. Fruit stipitate 2–20 mm (the stipe either exserted above the calyx or included, therefore removal of the calyx is needed to see if a stipe is present)...3
2b. Fruit sessile or substipitate to 1 mm...7

3a. Fruit erect, laterally compressed but quite turgid when mature, the pedicels twisted, stipitate 12–16 mm; stems cinereous, especially below; leaflets linear, rather sparsely positioned on the rachis...***A. tortipes***
3b. Fruit deflexed or pendulous, sometimes spreading but never erect, sessile or the stipe shorter; plants and leaflets various...4

4a. Lower stipules conspicuously several-nerved, amplexicaul or shortly connate; leaflets usually acute and subsessile or inconspicuously petiolate; flowers 7.5–12.6 (14.5) mm long, the wing petals cleft at the apex into 2 unequal lobes...***A. australis***
4b. Lower stipules free or connate but not conspicuously several-nerved; leaflets petiolate; flowers (12) 13–22 mm long, the wing petal obtuse, truncate, or slightly notched at the apex; plants generally of lower elevations...5

5a. Pedicels 1–2.5 mm long in fruit, 0.5–1.5 mm long in flower; plants usually densely strigulose, cinereous or greenish-cinereous, sometimes less hairy in age; flowers purple or pink-purple; calyx tube 4–6.7 mm long...***A. coltonii* var. *moabensis***
5b. Pedicels 3–6 mm long in fruit, 2.3–5 mm long in flower; plants glabrous or thinly strigulose throughout; flowers white or yellow; calyx tube 5–10 mm long...6

6a. Fruit stipitate 2–6 mm, glabrous; plants glabrous, lightly hairy in the inflorescence and on the lower stem; lower stipules connate; calyx tube 7–10 mm long; flowers white...***A. osterhoutii***
6b. Fruit stipitate 8–15 mm, thinly strigulose; plants finely but thinly strigulose throughout; stipules free; calyx tube 5–6.5 mm long; flowers pale yellow...***A. ripleyi***

7a. Fruit glabrous, greenish- or purplish-mottled, (6) 9–15 mm long; calyx tube (1.5) 1.6–2.6 mm long...***A. wingatanus***
7b. Fruit thinly to densely strigulose, rarely mottled, 5–35 mm long; calyx tube 2.5–7 mm long...8

8a. Leaflets mostly linear-involute or thread-like, a few lower leaflets sometimes linear-elliptic, 1–10 cm long; plants of the western slope...9
8b. Leaflets linear-oblong to linear-elliptic or oblong-oblanceolate, never thread-like, 0.3–3 cm long; plants of the eastern or western slope...10

9a. Fruit 2–3 (5) mm wide; flowers strongly upswept, the petals all strongly upturned 90 degrees, 7–12 mm long, whitish, ochroleucous, or purplish; calyx tube 3–5 mm long...***A. convallarius***
9b. Fruit 4.5–6 mm wide; flowers not strongly upswept, 18–23 mm long, pink-purple with white wings; calyx tube 5–7 mm long...***A. saurinus***

10a. Fruit ovoid-ellipsoid, (4) 5–13 mm long, densely pubescent with black or black and white hairs, rarely with all white hairs; keel 4–5.7 mm long...***A. eucosmus***
10b. Fruit oblanceolate to linear-oblanceolate or narrowly ovoid-ellipsoid, (8) 11–25 mm long, thinly to densely pubescent with white hairs; keel 5–8.5 mm long...11

11a. Plants usually subacaulescent, the stems not flexuous or zigzag distally; terminal leaflet usually longer or remote from the uppermost pair; purple-tipped keel usually longer than the wings (sometimes barely so)...***A. miser* var. *oblongifolius***
11b. Plants evidently caulescent, the stems usually flexuous or zigzag distally, sparsely foliose with well-separated internodes; terminal leaflet not considerably larger than and directly subtended by the uppermost pair of leaflets; keel shorter than the wings...***A. flexuosus***

Key 5

(Fruit bladdery-inflated, papery)

1a. Fruit conspicuously and usually densely villous or hirsute with hairs 1–2 mm long, lunately incurved or ovoid to ellipsoid, moderately inflated with thick papery walls...2
1b. Fruit glabrous, minutely hairy, or strigulose with hairs to 0.5 mm long, straight or incurved, conspicuously inflated with thin papery walls or moderately inflated with thick walls...3

2a. Fruit bilocular and persistent, densely crowded into an oblong or subcapitate head, ascending or spreading (when very crowded), brownish or greenish with black hairs when young but black with white hairs at maturity, pustulate-hirsute, ovoid to ellipsoid...***A. cicer***
2b. Fruit unilocular and deciduous, not crowded into a subcapitate head, spreading or deflexed, with white hairs, ellipsoid-lanceolate and usually lunately incurved...***A. pubentissimus***

3a. Fruit surface purplish-mottled, sometimes less so on older fruit, but remaining purplish-mottled near the apex or along the dorsal suture...4
3b. Fruit surface not purplish-mottled (sometimes purple-speckled, but not mottled with purple splotches)...8

4a. Fruit stipitate 1.5–10 mm (the stipe either exserted above the calyx or included, therefore removal of the calyx is needed to see if a stipe is present)...5
4b. Fruit sessile or nearly so, if substipitate then the short stipe much less than 1 mm long...6

5a. Leaflets linear-filiform to narrowly elliptic-oblong, the terminal leaflet continuous with the rachis; calyx tube 2–4 mm long; flowers 6–10 mm long; plants from an deep rhizomatous system and slender underground caudex; lowermost stipules connate...***A. ceramicus***
5b. Leaflets broadly obovate, elliptic, or suborbicular, the terminal leaflet petiolulate; calyx tube 4–9 mm long; flowers 11–19 mm long; plants from branched caudices; lowermost stipules free but amplexicaul...***A. oophorus* var. *caulescens***

6a. Fruit bilocular, deciduous, contracted at the tip into a beak (3) 4–15 mm long; lower stipules free; flowers 8–20 mm long; stems generally not sparsely foliose; leaflets usually flat...***A. lentiginosus***
6b. Fruit unilocular, persistent, contracted at the tip into a beak 0.8–2.5 mm long; lower stipules connate into a sheath opposite the petiole; flowers 6.5–11 mm long; stems generally sparsely foliose with long internodes, often flexuous and zigzag above; leaflets often folded...7

7a. Fruit 0.5–0.9 cm in diam., turgid to inflated; stems and calyx greenish-cinereous; flowers 9–11 mm long...***A. flexuous* var. *greenei***
7b. Fruit 1.2–2 cm in diam., distinctly inflated; stems and calyx silvery-cinereous with appressed hairs; flowers 6.5–8 mm long...***A. fucatus***

8a. Leaflets densely silvery-cinereous, 2–4 (9) mm long, obovate or elliptic, (11) 15–27 and densely crowded on a leaf rachis 1–3 (4) cm long; plants only 2–6 cm high, usually mat-forming from a subterranean caudex; fruit stipitate 1–4 mm...***A. lutosus***
8b. Leaflets glabrous or slightly hairy, occasionally cinereous, usually larger, variously shaped, the leaf rachis generally longer; plants various; fruit sessile or stipitate...9

9a. Calyx tube 4.9–5.8 mm long with short (0.2–0.7 mm), broadly low-deltoid teeth, the teeth margins ciliate but otherwise the calyx nearly glabrous; fruit stipitate 4.5–7 mm...***A. americanus***
9b. Calyx teeth 1 mm or more in length; fruit sessile or stipitate...10

10a. Lower stipules connate into a sheath opposite the petiole, the upper connate at the base only or nearly free; caulescent plants with stems 1–6 dm long, generally sparsely foliose, often flexuous and zigzag at least above, mat-forming or not from a subterranean root-crown...11
10b. Stipules all free, the lowermost sometimes amplexicaul; plants with stems 0.5–3 (4) dm long, usually basally branched and tufted or mat-forming...12

11a. Flowers 9–11 mm long; calyx tube 2.8–4.5 mm long; fruit sessile or nearly so, the stipe to 0.5 mm in length if present...***A. flexuous* var. *greenei***
11b. Flowers 13–18 mm long; calyx tube 4.5–6.5 (7) mm long; fruit stipitate on a short stipe 1–2 mm long (hidden by the calyx)...***A. hallii* var. *hallii***

12a. Fruit sessile...13
12b. Fruit stipitate 1–7 mm (peel back the calyx to see if a stipe is present)...16

13a. Fruit bilocular (sometimes unilocular just at the beak), ascending or spreading, contracted at the tip into an incurved beak (3) 4–15 mm long…***A. lentiginosus***
13b. Fruit unilocular, deflexed to spreading, the beak to 5 mm long…14

14a. Fruit half-ovoid in profile, slightly or conspicuously curved; plants of western Colorado north of Montrose County…***A. geyeri***
14b. Fruit ovoid to ellipsoid, straight; plants of southcentral and southwestern Colorado…15

15a. Fruit hairy, 1–1.5 (2) cm long and 0.6–1.2 (1.4) cm in diam.; plants shortly villous throughout with spreading or ascending hairs; keel petal 3.5–4.3 mm long…***A. cerussatus***
15b. Fruit inconspicuously hairy or glabrous, 1.5–4 cm long and (0.8) 1.2–2.4 cm in diam.; plants sometimes densely pubescent when young with appressed hairs, but usually green and glabrate in age; keel petal 4.4–6.5 mm long…***A. allochrous* var. *playanus***

16a. Plants strigulose (especially on new growth); fruit stipitate 1–2.5 mm; flowers 7–10 mm long, white; calyx tube 3–3.5 mm long…***A. wetherillii***
16b. Plants glabrate; fruit stipitate 1.5–7 mm; flowers 15–22 mm long, white to purple or pink-purple; calyx tube 2.3–9 mm long…17

17a. Flowers white or yellowish-white; calyx tube 5–6 mm long…***A. debequaeus***
17b. Flowers pink-purple or purple; calyx tube 6–10 mm long…18

18a. Plants from a branching caudex and taproot, with spreading-ascending or mat-forming stems; fruit generally humistrate…***A. eastwoodiae***
18b. Plants lacking a branching caudex, with erect stems; fruit ascending or spreading…***A. preussii***

Key 6

(Fruit uniformly hairy, not laterally compressed or bladdery-inflated)

1a. Fruit tightly clustered in a dense oblong or subcapitate raceme, the axis of the inflorescence hidden by the fruit, ascending or sometimes spreading when very crowded, bilocular, persistent; lower stipules connate…2
1b. Fruit in a loose or few-flowered raceme, pendulous, humistrate (on the ground) or sometimes ascending, or if in a more densely flowered, compact raceme then pendulous to deflexed, bilocular or unilocular; lower stipules connate or free…3

2a. Fruit 10–14 mm long and 7–9 mm wide, brownish or greenish with black hairs when young but black with white hairs at maturity, pustulate-hirsute; flowers ochroleucous; calyx teeth 1.5–2 mm long; plants ascending or decumbent, the stems 3–8 dm long…***A. cicer***
2b. Fruit 7–10 mm long and 3–4 mm wide, black at maturity and densely villous with white hairs 1–2 mm long; flowers pink-purple; calyx teeth 2.5–4 (5) mm long; plants erect, somewhat tufted or in patches, the stems 0.5–3 (4) dm long…***A. agrestis***

3a. Fruit nearly globose with a sharp beak, 1–3 cm in length, inflated but woody at full maturity (fleshy and reddish when immature), bilocular and persistent…***A. plattensis***
3b. Fruit unlike the above in all respects…4

4a. Fruit surface with at least some black hairs (often densely black-hairy); plants of alpine or mountain meadows, 7500–13,500 ft in elevation; lower stipules connate opposite the petiole…5
4b. Fruit surface with all white or tawny hairs; plants usually of lower elevations, variously distributed; lower stipules free or connate…10

5a. Fruit ascending or spreading to humistrate, sessile or stipitate to 0.8 mm (but usually shorter)…6
5b. Fruit pendulous or occasionally spreading and then stipitate 1.5–5 mm…7

6a. Plants low, 1–6 (10) cm high, from subterranean caudices; fruit not inflated; racemes usually 2–4 (5)-flowered…***A. molybdenus***
6b. Plants generally taller, 10–40 cm high, from a superficial crown; fruit slightly inflated; racemes usually 4–15-flowered…***A. bodinii***

7a. Racemes usually 2–3-flowered, sometimes up to 5-flowered; flowers white with a purple-tipped keel; fruit subsessile or stipitate to 1.5 mm; plants weak-stemmed, delicate, rhizomatous from a subterranean caudex…***A. leptaleus***
7b. Racemes usually 7–25-flowered, rarely some with as few as 5 flowers; flowers purplish or bicolored with a purple banner and white wings, with a purple-tipped keel or not; if plants weak-stemmed from a slender taproot and subterranean caudex then the fruit stipitate 1.5–3.5 mm…8

8a. Fruit sessile or nearly so, or occasionally with a small stipe present up to 0.4 mm in length, ovoid-ellipsoid, (4) 5–13 mm long; flowers (4.1) 5.5–7.6 mm long…***A. eucosmus***
8b. Fruit shortly stipitate 1–5.5 mm (remove the calyx to see stipe), ellipsoid or oblong-ellipsoid, 7–25 mm long; flowers 7–14 mm long…9

9a. Stems arising singly or a few together from a slender taproot and subterranean caudex; fruit evidently sulcate dorsally, 7–14 mm long and 2.5–4 mm wide; keel as long or longer than the wings in length; leaflets (11) 15-25 (27)…***A. alpinus***
9b. Stems several from a root-crown located at the surface of the soil; fruit inconspicuously sulcate dorsally, 8–25 mm long and 3.5–6 mm wide; keel much shorter than the wings; leaflets (5) 7–17…***A. robbinsii* var. *minor***

10a. Stems and leaflets with spreading hairs, the leaflets usually glabrate above (at least in the middle), not silvery; flowers ochroleucous and usually with a purple-tipped keel…***A. parryi***
10b. Plants with appressed or appressed-ascending hairs, sometimes silvery-hairy; flowers various…11

11a. Fruit narrowly ellipsoid and lunately incurved, with spreading, villous hairs 1–2 mm long…***A. pubentissimus***
11b. Fruit with appressed, strigulose to villous hairs, rarely with loosely villous, spreading hairs and then the fruit ovoid-acuminate and straight or abruptly curved at the tip…12

12a. Lower stipules several-nerved, ovate-obtuse, clasping (less obvious on fruiting specimens); inflorescence a subcapitate raceme; flowers 14–18 mm long, yellowish-white or purplish; stems spreading-decumbent; leaflets obovate to broadly oblanceolate, the tips mostly obtuse…***A. cibarius***
12b. Lower stipules unlike the above; inflorescence usually an elongated raceme; flowers 4–22 mm long (but longer-flowered species fall out of the range of *A. cibarius*), variously colored; stems and leaflets various…13

13a. Lower stipules free, although sometimes strongly amplexicaul and clasping the stem…14
13b. Lower stipules connate into a sheath opposite the petiole, becoming less united upward…23

14a. Plants usually prostrate and shortly caulescent, cinereous to sericeous, usually silvery but becoming green in age; fruit ovoid-acuminate, densely villous…15
14b. Plants usually erect, caulescent or sometimes shortly caulescent, glabrous or strigulose, or occasionally cinereous; fruit variously shaped, strigulose…16

15a. Calyx tube campanulate, 3–4.5 mm long; flowers 4–14 mm long; plants of the eastern slope…***A. lotiflorus***
15b. Calyx tube cylindric, 6.5–12 mm long; flowers 18–22 mm long; plants of the western slope…***A. argophyllus* var. *martinii***

16a. Leaves elliptic to broadly or narrowly obovate, or oblong-elliptic; fruit spreading or ascending…17
16b. Leaves linear to linear-elliptic or narrowly oblong; fruit deflexed or humistrate…20

17a. Plants of the eastern slope; fruit triangular in cross-section, lunately curved, usually mottled, sessile or nearly so, 5–25 mm long and 2.5–5 mm wide; flowers 5–8 mm long, mostly 2–9 per inflorescence; calyx tube 1.5–3 mm long, uniformly strigose with black and white hairs; plants prostrate to decumbent…***A. sparsiflorus***
17b. Plants of the western slope; fruit subterete in cross-section, straight or lunately curved, sessile or stipitate 1–5 mm, 15–35 mm long and 4–15 mm wide; flowers 7–20 mm long, often numerous; calyx tube 3–7 mm long, glabrate to white- or black-strigulose; plants usually erect…18

18a. Fruit bilocular, usually lunately curved, often purplish-mottled; flowers purple…***A. lentiginosus* var. *palans***
18b. Fruit unilocular, straight, sometimes purple-speckled but not purplish-mottled; flowers whitish or ochroleucous with a purple-tipped keel…19

19a. Calyx tube 3–3.5 mm long; flowers 4–9 per inflorescence, 7–10 mm long, whitish; fruit often purple-speckled; leaves broadly obovate; fruit 15–20 mm long…***A. wetherillii***
19b. Calyx tube 4–6.5 mm long; flowers 10–30 per inflorescence, 15–22 mm long, ochroleucous and usually with a purple-tipped keel; leaves oblong-elliptic or cuneate-obovate; fruit 20–35 mm long…***A. praelongus***

20a. Fruit stipitate 5–10 mm, linear-oblanceolate, slightly curved, 2.5–3.5 (4) cm long; flowers ochroleucous; leaves linear to linear-elliptic or oblong, usually cinereous; plants known from Montezuma County...***A. schmollae***
20b. Fruit sessile or nearly so; flowers and leaves various; variously distributed...21

21a. Calyx tube 1.5–2.5 mm long with lobes 1–2.5 mm long; flowers small, 4–7 mm long, white with a purple-tipped keel; later peduncles elongate, filiform, to 15 cm long; fruit plumply obovoid, 10–20 mm long; leaflets linear-filiform to linear-elliptic with an acute apex, the terminal leaflet longer than the subtending pair; plants green, not cinereous...***A. brandegeei***
21b. Calyx tube 3–5 mm long with lobes 0.5–1 mm long; flowers 8–12 mm long, reddish-purple or reddish-pink, sometimes with white wings; peduncles not elongate and filiform; fruit oblong to oblong-oblanceolate, 20–35 mm long; leaflets linear to narrowly oblong, usually with a rounded apex, the terminal leaflet not conspicuously larger; plants often cinereous...22

22a. Terminal leaflet of most leaves confluent with the rachis; flowers not strongly upswept, bright reddish-purple with white wing tips...***A. duchesnensis***
22b. Terminal leaflet petiolulate and not confluent with the rachis; flowers strongly upswept, the petals all abruptly upturned 90 degrees, red-purple or reddish...***A. cronquistii***

23a. Leaflets acute at the apex, narrowly elliptic to linear; flowers white with a purple-tipped keel; inflorescence 1–5 (10)-flowered; stems not sparsely foliose...24
23b. Leaflets rounded or obcordate at the apex, various; flowers various; inflorescence usually 7-numerous-flowered; stems generally sparsely foliose, with well-spaced internodes...25

24a. Calyx tube 1.5–2.5 mm long; flowers small, 4–7 mm long; peduncles elongate, filiform, to 15 cm long; fruit bilocular or nearly so...***A. brandegeei***
24b. Calyx tube 2.7–3.4 mm long; flowers 8.5–11.8 mm long; peduncles not elongate and filiform; fruit unilocular...***A. leptaleus***

25a. Fruit 4–9 mm long, ovate to ellipsoid; calyx tube 1.4–2.5 mm long; flowers 5.3–8.5 mm long...26
25b. Fruit 10–25 mm long (rarely as short as 8 mm long, and then linear-oblong to linear-oblanceolate); calyx tube 2.5–8 mm long; flowers 7–19 mm long...27

26a. Fruit obtuse at the base; leaflets linear, linear-oblong, or narrowly oblanceolate, flat or folded, 0.5–2.5 cm long, loosely spaced on a rachis 2.5–7 cm long; peduncles 3–9 cm long; racemes mostly 12–45-flowered; flowers purplish...***A. gracilis***
26b. Fruit acute at the base; leaflets oblong-obovate or narrowly elliptic, folded, 0.6–0.9 cm long, rather crowded on a rachis 2–4 cm long; peduncles 1–3.5 cm long; racemes mostly 7–14-flowered; flowers lilac-tinged white with a purple-tipped keel...***A. microcymbus***

27a. Flowers 7.2–11 mm long; calyx tube 2.4–4.2 mm long; fruit usually linear-oblong to linear-oblanceolate and 2.7–5 mm wide, less often ovate-elliptic and 5–9 mm wide...***A. flexuosus***
27b. Flowers 13–19 mm long; calyx tube 4.5–8 mm long; fruit ovoid, oblong, or elliptic-lanceolate, 5–12 mm wide...28

28a. Plants silky-canescent or cinereous throughout; fruit sessile or nearly so, becoming coriaceous...***A. puniceus* var. *puniceus***
28b. Plants sparsely strigulose to glabrate throughout; fruit stipitate 1–2 mm, usually becoming papery but sometimes coriaceous...***A. hallii* var. *hallii***

Key 7

(Fruit glabrous or slightly minutely hairy at the apex, not laterally compressed or bladdery-inflated)

1a. Fruit subglobose with a short, filiform beak, thick and succulent, becoming coriaceous or woody in age, usually reddish on the upper side, humistrate, bilocular; plants decumbent or spreading-ascending, the stems 1–5 dm long...***A. crassicarpus***
1b. Fruit variously shaped but not subglobose; plants various...2

2a. Stems and lower side of leaves densely spreading-hairy; fruit stipitate (3) 5–12 mm, pendulous, linear-oblong to linear-oblanceolate; inflorescence an elongated raceme of numerous (15–30) flowers; plants tall, 3–7 dm, usually in clumps; flowers white with a purple-tipped keel...***A. drummondii***
2b. Stems and leaves glabrous or with appressed or ascending hairs; fruit and flowers various...3

3a. Inflorescence a subcapitate raceme with 5–12 (15) flowers; lower stipules several-nerved, ovate-obtuse, clasping (less obvious on fruiting specimens but quite conspicuous on flowering ones); calyx densely black-hairy; stems spreading-decumbent; leaflets obovate to broadly oblanceolate, obtuse...***A. cibarius***
3b. Inflorescence an elongated raceme; lower stipules usually unlike the above; calyx glabrous or white or black pubescent; stems and leaflets various...4

4a. Fruit triangular in cross-section (3-angled), unilocular, stipitate 3–5 mm, the dorsal face usually deeply sulcate; at least the largest leaflets (1.8) 2–3 cm long; calyx tube gibbous at the base, with filiform lobes 2–6 (10) mm long; racemes densely and numerously flowered; flowers ochroleucous with a purple-tipped keel; plants robust and clump-forming, to 5 (6) dm tall, seleniferous...***A. racemosus***
4b. Fruit subterete to obcompressed in cross-section, or if triangular then the fruit bilocular and stipitate 4–9 mm, and the leaflets 0.4–1.8 cm long (*A. scopulorum*); plants various...5

5a. Calyx tube 4.9–5.8 mm long with short (0.2–0.7 mm), broadly low-deltoid teeth, the teeth with ciliate margins but otherwise the calyx nearly glabrous; leaflets ovate to elliptic, 7–15 mm wide; fruit stipitate 4.5–7 mm, inflated and subterete in cross-section...***A. americanus***
5b. Calyx with longer teeth, or if the teeth shorter than 1 mm then the tube shorter than 5 mm or the leaflets linear to linear-oblong; fruit sessile or stipitate...6

6a. Lowermost stipules free, although sometimes amplexicaul...7
6b. Lowermost stipules connate into a sheath opposite the petiole...14

7a. Fruit sessile or nearly so...8
7b. Fruit stipitate 1.5–12 mm...10

8a. Fruit bilocular, usually lunately curved and purplish-mottled; flowers purple...***A. lentiginosus* var. *palans***
8b. Fruit unilocular or nearly so, straight, not mottled; flowers ochroleucous with a purple-tipped keel or whitish...9

9a. Calyx teeth divergent or recurved in age, usually narrowly triangular with a bristle-like tip, the tube 6.5–9 mm long, sparsely to densely white- to black-strigulose; flowers white, aging ochroleucous when dried; leaflets strigulose below...***A. pattersonii***
Note: This species and the next are very difficult to tell apart when only in mature fruit. The fruit of both species is quite variable.
9b. Calyx teeth pointing forward and straight, usually deltoid and without a bristle-like tip, the tube 4–6.5 mm long, glabrous or sparsely strigulose with black hairs, lacking white hairs; flowers ochroleucous or yellowish-green; leaflets glabrous or slightly strigulose below...***A. praelongus* var. *praelongus***

10a. Leaflets filiform-involute or linear to linear-oblanceolate, 1–3 mm wide; terminal leaflet continuous with the rachis; plants densely strigulose, usually cinereous or silvery-canescent; fruit pendulous, narrowly oblong to narrowly oblanceolate, obcompressed and flattened dorsally, stipitate 4–12 mm...***A. lonchocarpus***
10b. Leaflets elliptic-ovate to oblong or narrowly elliptic; terminal leaflet petiolulate; plants glabrous or nearly so; fruit ascending, spreading, or humistrate, ovoid to oblong-ellipsoid to narrowly oblong, subterete or slightly obcompressed, stipitate 1.5–7 (8) mm...11

11a. Calyx tube green to yellowish, usually glabrate; flowers ochroleucous or yellowish-green, usually with a purple-tipped keel; plants robust, to 10 dm tall...***A. praelongus* var. *lonchopus***
11b. Calyx tube green to purple, usually black-strigulose; flowers various; plants erect or spreading, to 3 (4) dm tall...12

12a. Flowers white or yellowish-white; calyx tube 5–6 mm long...***A. debequaeus***
12b. Flowers pink-purple or purple; calyx tube 6–10 mm long...13

13a. Plants from a branching caudex and taproot, with spreading-ascending or mat-forming stems; fruit generally humistrate...***A. eastwoodiae***
13b. Plants lacking a branching caudex, with erect stems; fruit ascending or spreading...***A. preussii***

14a. Calyx tube 1.8–2.5 mm long; flowers 6–7 mm long, whitish with a purple banner and purple-tipped keel; fruit 10–15 mm long, linear-ellipsoid and cuneate at both ends, stipitate 1.2–2 mm; leaflets 7–11, linear to linear-oblanceolate (at least above), 0.6–2.2 cm long...***A. proximus***
14b. Calyx tube 4–12 mm long, if as short as 2.5 mm then the fruit sessile or shortly stipitate to 1.3 mm and the leaflets usually more numerous (*A. flexuosus*); flowers 7–28 mm long, variously colored; fruit and leaflets various...15

15a. Leaflets filiform, linear or narrowly oblong-lanceolate, mostly 2–5 cm long, often curved, the terminal leaflet confluent with the rachis; fruit woody at maturity, sessile…***A. pectinatus***
15b. Leaflets narrowly oblong, linear-elliptic, or oval-oblong, 0.3–1.8 cm long, not curved, the terminal leaflet petiolulate; fruit papery or coriaceous at maturity, sessile or stipitate to 2 mm…16

16a. Fruit stipitate 4–9 mm, (1.8) 2.5–3.5 cm long, bilocular, triangular in cross-section (although this difficult to see on herbarium specimens); leaflets mostly acute; calyx tube 6.5–8.5 mm long…***A. scopulorum***
16b. Fruit sessile or stipitate to 2 mm, (0.8) 1–2.5 cm long, unilocular, subterete to obcompressed; leaflets obtuse or retuse; calyx tube 2.5–6.5 (7) mm long…17

17a. Flowers 7.2–11 mm long; calyx tube 2.4–4.2 mm long; fruit linear-oblong to linear-oblanceolate and 2.7–5 mm wide…***A. flexuosus***
17b. Flowers 13–18 mm long; calyx tube 4.5–6.5 (7) mm long; fruit ovoid, oblong, or elliptic-lanceolate, (4) 6–12 mm wide…***A. hallii* var. *hallii***

Key 8

(Lower stipules connate into a sheath opposite the petiole; flowers white or yellowish)

1a. Calyx soon inflated and papery, enclosing the small (5–7 mm long) fruit; inflorescence a spicate raceme of 30-numerous, densely crowded, spreading flowers, on a robust peduncle 9–17 cm long; leaflets linear to linear-oblong, on pressed specimens usually all ascending from the rachis; known from La Plata and Archuleta counties…***A. oocalycis***
1b. Calyx not inflated; plants otherwise unlike the above, variously distributed…2

2a. Flowers tightly clustered in a dense oblong or subcapitate raceme, the axis of the inflorescence hidden, ascending or sometimes spreading when very crowded; fruit black with white hairs at maturity, pustulate-hirsute…***A. cicer***
2b. Flowers in a loose or few-flowered raceme, spreading to deflexed, or ascending but soon deflexed…3

3a. Flowers less than 13 mm in length; calyx tube 1.4–4.5 mm long…4
3b. Flowers 13 mm or more in length; calyx tube mostly over 4.5 mm long…13

4a. Lower stipules black or sometimes dark brown with a black band at the base of the sheath; keel 4.3–6 mm long; calyx tube 2–2.6 (3) mm long; fruit laterally compressed, 7–17 mm long, substipitate or stipitate 1.5–5 mm, greenish and often with brown or red mottling when young, becoming dark brown or black at maturity…***A. tenellus***
4b. Lower stipules not black; keel 3.5–8.5 mm long; calyx tube 1.5–5.7 mm long; if fruit laterally compressed then unlike the above…5

5a. Keel 3.5–4.5 mm long; calyx tube 1.5–2.5 mm long; flowers 4–7 mm long…6
5b. Keel 5.5–9 mm long; calyx tube 2.5–5.5 mm long; flowers 6–12 mm long…8

6a. Peduncles elongate and filiform, to 15 cm long; flowers 1–5 per raceme, 4–7 mm long with a keel 3.5–4.5 mm long; leaflets linear-filiform to linear-elliptic with an acute tip, the terminal leaflet longer than and remote from the subtending pair; plants usually shortly caulescent…***A. brandegeei***
6b. Peduncles not both elongate and filiform; flowers usually 7 or more per raceme; leaflets oblong to elliptic, the terminal leaflet not longer and remote from the subtending pair; plants shortly or evidently caulescent with a well-developed stem…7

7a. Calyx tube 1.4–1.9 mm long with short teeth 0.5–0.7 mm long; stems prostrate to weakly ascending, purplish, several and diffusely branching, with well-spaced internodes; plants conspicuously strigulose to subvillous; plants of dry places in Gunnison Co…***A. microcymbus***
7b. Calyx tube 2.5 mm or longer with teeth 0.7–2 mm long; stems erect or sometimes weakly ascending, sometimes purplish at the base, not diffusely branching with well-spaced internodes; plants thinly strigulose; plants of moist places in the mountains…***A. eucosmus***

8a. Flowers strongly upswept, the petals all abruptly upturned 90 degrees, 7–12 mm long; terminal leaflet absent or confluent with the rachis, the leaflets filiform or linear-involute…***A. convallarius***
8b. Flowers not strongly upswept; terminal leaflet not confluent with the rachis or if confluent with the rachis then the leaflets not filiform or linear-involute…9

9a. Inflorescence with numerous (25–80), usually densely packed flowers; stems several, erect or decumbent-ascending, 1.5–8 dm long, from a thick, woody taproot; ventral face of the fruit strongly 2-grooved (grooved lengthwise along either side of the prominent, raised suture)...***A. bisulcatus***
9b. Inflorescence with fewer flowers (1–25), usually loosely packed; stems erect or decumbent, 0.2–4.5 dm long, rhizomatous from a subterranean caudex or from a root crown located at the surface of the soil; ventral face of the fruit not 2-grooved...10

10a. Plants rhizomatous from a subterranean caudex; racemes usually 2–3-flowered, but sometimes up to 5-flowered; plants weak-stemmed, delicate...***A. leptaleus***
10b. Plants from a root crown located at the surface of the soil; racemes 3–25-flowered; plants weak-stemmed or not...11

11a. Purple-tipped keel usually longer than the wings (sometimes slightly so); leaflets linear to linear-elliptic, the terminal leaflet usually longer and remote from the uppermost pair; calyx tube 2.2–3 mm long; plants generally shortly caulescent; fruit laterally compressed and generally flat, linear to linear-oblong...***A. miser* var. *oblongifolius***
11b. Keel shorter than and hidden by the wings, purple-tipped or not; leaflets ovate, oblong, lance-oblong, or oblanceolate, the terminal leaflet not longer than and remote from the uppermost pair; calyx tube 2.5–4.5 mm long; plants generally evidently caulescent with a well-developed stem; fruit obcompressed (but sometimes appearing laterally compressed on herbarium specimens)...12

12a. Flowers (4.1) 5.5–7.6 mm long, the wings 4.5–6.3 mm long; stem not purplish except at the very base; calyx usually thinly black-hairy; fruit sessile or stipitate 0.4 mm or less...***A. eucosmus***
12b. Flowers 7–12 mm long, the wings 6–9.3 mm long; stem often purplish, lighter brown at the base and darker above; calyx usually densely black-hairy; fruit stipitate 0.5–5.5 mm...***A. robbinsii* var. *minor***

13a. Stems and lower side of leaves densely spreading-hairy; fruit stipitate (3) 5–12 mm, pendulous, linear-oblong to linear-oblanceolate; inflorescence an elongated raceme of numerous (15–30) flowers; plants tall, 3–7 dm, usually in clumps...***A. drummondii***
13b. Stems and leaves glabrous or with appressed or ascending hairs; fruit and plants various...14

14a. Stems densely cinereous and white, especially below; fruit laterally compressed but quite turgid when mature, stipitate 12–16 mm, the pedicels becoming twisted so that the fruit is erect; flowers yellow; leaflets linear, rather sparsely positioned on the rachis...***A. tortipes***
14b. Stems not densely cinereous and white; fruit pendulous or spreading, not laterally compressed, obcompressed, or trigonous or subterete in cross-section, sessile or stipitate to 9 mm; flowers white, ochroleuous, or dull yellow; leaflets various...15

15a. Leaflets filiform, linear or narrowly oblong-lanceolate, mostly 2–5 cm long, often curved, the terminal leaflet confluent with the rachis...16
15b. Leaflets elliptic to oblong or oval-oblong, sometimes linear-elliptic, mostly 0.5–2.5 cm long, not curved, the terminal leaflet petiolulate and not confluent with the rachis...17

16a. Leaflets usually shorter, mostly about 1 (4) cm long, usually straight, not involute although the margins elevated; fruit laterally compressed and flat, stipitate 2–6 mm...***A. osterhoutii***
16b. Leaflets usually longer, mostly over 2 cm long, often curved, involute or the margins simply elevated; fruit obcompressed and subterete, sessile...***A. pectinatus***

17a. Racemes (5) 10–22-flowered; calyx tube usually with black hairs, rarely with all white hairs; leaflets usually shorter, (0.2) 0.4–1.8 cm long; fruit bilocular, triangular in cross-section (although this difficult to see on herbarium specimens); plants from a subterranean root-crown, the stems diffuse and decumbent-ascending, rather slender, usually not seleniferous...***A. scopulorum***
17b. Racemes 20–80-flowered; calyx tube usually with white hairs, occasionally with some black hairs or wholly with black hairs or glabrate; leaflets usually longer, mostly 1.5–3 cm long; fruit unilocular, triangular or not in cross-section; plants from a thick, multicipital root-crown and taproot, robust and clump-forming, the stems stout, erect or decumbent-ascending, seleniferous...18

18a. Calyx tube 4–5.5 mm long, pallid or reddish-purple; ventral face of the fruit strongly 2-grooved (grooved lengthwise along either side of the prominent, raised suture)...***A. bisulcatus***
18b. Calyx tube 5–9 mm long, pallid or pinkish; fruit triangular in cross-section (3-angled)...***A. racemosus***

Key 9

(Lower stipules connate into a sheath opposite the petiole; flowers purplish or bicolored)

1a. Flowers strictly ascending, densely packed into a shortly ovoid or globose cluster; fruit densely white-villous with hairs 1–2 mm long…***A. agrestis***
1b. Flowers spreading to deflexed, not packed into ovoid or globose clusters; fruit glabrous or strigulose, if densely villous then the hairs usually black and/or shorter than 1 mm in length…2

2a. Keel as long or usually longer than the wings in length; flowers with a purple banner and keel tip and white wings; racemes (5) 7–20-flowered; fruit usually densely black-hairy; plants arising singly or a few together from a slender taproot and subterranean caudex…***A. alpinus***
2b. Keel conspicuously shorter than the wings; flowers, fruit, and plants various…3

3a. Inflorescence of numerous (mostly 30–80) flowers, usually densely packed; calyx tube usually gibbous dorsally behind the pedicel, often purplish, the teeth bristle-like and 1.5–5 mm long; plants seleniferous, robust, clustered and clump-forming, erect or decumbent-ascending, 1.5–8 dm tall, the stems not flexuous or zigzag; ventral face of the fruit strongly 2-grooved (grooved lengthwise along either side of the prominent, raised suture)…***A. bisulcatus* var. *bisulcatus***
3b. Inflorescence of fewer flowers (or if more, then loosely packed and the stems flexuous and zigzag distally), usually loosely packed; calyx tube usually not gibbous (sometimes gibbous in *A. hallii* and then the teeth broadly triangular and 0.7–2 mm long), variously colored; plants not seleniferous, otherwise various; fruit unlike the above…4

4a. Flowers 5–12 mm long; calyx tube 1.5–4.5 mm long…5
4b. Flowers 13–23 mm long; calyx tube 4.5–8 mm long…13

5a. Plants low, 1–6 (10) cm high, from subterranean caudices, loosely tufted or matted; racemes usually 2–4 (5)-flowered…***A. molybdenus***
5b. Plants mostly 10–80 cm; racemes 5–30-flowered…6

6a. Stems and calyx densely silvery-cinereous with appressed hairs; flowers reddish-purple; fruit bladdery-inflated, mottled; leaflets oblanceolate, often involute or folded; reported for Montezuma County…***A. fucatus***
6b. Stems and calyx not densely silvery-cinereous; flowers pinkish-purple, purple, or bicolored; fruit not bladdery-inflated; leaflets various; variously distributed…7

7a. Stems usually flexuous or zigzag distally, often only sparsely leafy; leaflets linear to linear-oblong or narrowly elliptic; plants of dry places or occasionally of dry mountain meadows; calyx tube strigulose with white hairs or sometimes with all black hairs; fruit with white hairs or glabrous…8
7b. Stems not flexuous or zigzag distally, usually not sparsely leafy; leaflets ovate to oblong-ovate, oblanceolate, or elliptic; plants of moist meadows, aspen groves, and alpine ridges in the mountains (above 7500 ft); calyx tube usually strigulose with all black hairs or rarely with all white hairs; fruit usually with at least some black hairs, rarely with all white hairs…11

8a. Ovules 12–25; wings mostly 7–10 mm long; keel 5–8.2 mm long; calyx tube 2.4–4.2 mm long; fruit linear-oblong to oblanceolate or linear-ellipsoid, rarely plumply ellipsoid…***A. flexuosus***
8b. Ovules 4–10; wings 4–7 mm long; keel 3.5–6 mm long; calyx tube 1.5–2.7 mm long; fruit ovate, ellipsoid, or linear-ellipsoid…9

9a. Ovary and fruit hairy; fruit ovate or ovate-elliptic, 4–8.5 mm long, obtuse at the base, sessile; plants of the eastern slope…***A. gracilis***
9b. Ovary and fruit glabrous; fruit linear-ellipsoid or oblong-ellipsoid, mostly 9–15 mm long, cuneate at the base, subsessile or stipitate to 2 mm; plants of the western slope…10

10a. Fruit laterally compressed, greenish- or purplish-mottled; flowers purplish with paler, whitish wings…***A. wingatanus***
10b. Fruit obcompressed, not greenish- or purplish-mottled; flowers whitish with a purple banner and purple-tipped keel…***A. proximus***

11a. Stem purplish or dark brown above, lighter brown at the base; fruit stipitate 0.5–5.5 mm; flowers pale purple with whitish claws; plants thinly to densely hairy…***A. robbinsii* var. *minor***
11b. Stem light brown above and darker brown or purplish at the base; fruit sessile or stipitate less than 0.5 mm; flowers usually darker purple or pinkish; plants glabrate or thinly hairy…12

12a. Flowers (4.1) 5.5–7.6 mm long; fruit pendulous...***A. eucosmus***
12b. Flowers 8–11 mm long; fruit ascending or often humistrate...***A. bodinii***

13a. Leaflets linear or filiform-involute, the terminal leaflet confluent with the rachis; flowers 18–23 mm long, bicolored with a pinkish-purple banner and keel and white wings; fruit laterally compressed...***A. saurinus***
13b. Leaflets obovate, elliptic or oblanceolate, the terminal leaflet petiolulate; flowers 13–19 mm long, variously colored; fruit obcompressed or subterete...14

14a. Plants sparsely strigulose to glabrate throughout; fruit sparsely hairy; plants found above 7000 ft in elevation...***A. hallii* var. *hallii***
14b. Plants silky-canescent, cinereous, or villous with hairs 1 mm long throughout (except glabrate on the upper leaf surface); fruit villous; plants of the eastern plains, found below 7000 ft in elevation...15

15a. Plants villous with hairs about 1–1.5 mm long; inflorescence of 6–12 flowers; fruit nearly globose with a sharp beak, 1–3 cm in length, bilocular and persistent...***A. plattensis***
15b. Plants sericeous-canescent; inflorescence of (5) 10–20 flowers; fruit oblong to narrowly elliptic, unilocular...***A. puniceus* var. *puniceus***

Key 10

(Lower stipules free; flowers 5–12.5 mm long)

1a. Plants shortly caulescent, to 1–10 (30) cm; peduncles elongate, filiform, to 15 cm long; leaflets linear-elliptic with an acute apex, the terminal leaflet longer than the subtending pair; flowers 4–7 mm long, whitish with a purple-tipped keel...***A. brandegeei***
1b. Plants caulescent with an evident stem, 2–50 cm; peduncles not elongate and filiform; leaflets various but usually unlike the above; flowers various...2

2a. Leaflets narrowly oblong with a truncate apex; flowers strongly upswept, the petals abruptly upturned 90 degrees, reddish-purple; calyx tube 3–4 mm long with small lobes 0.5–1 mm long...***A. cronquistii***
2b. Leaflets unlike the above; flowers not strongly upswept, variously colored; calyx various...3

3a. Leaflets linear or linear-filiform, some of the lower leaflets sometimes narrowly oblanceolate, the terminal leaflet of most leaves confluent with the rachis; flowers 10–12 mm long, reddish-purple with white wings...***A. duchesnensis***
3b. Leaflets oblong-lanceolate to elliptic or obovate, the terminal leaflet petiolulate and not confluent with the rachis; flowers unlike the above...4

4a. Lower stipules broadly ovate and mostly obtuse or nearly so, several-nerved; plants of the central and southwestern mountains above 8500 ft in elevation; fruit with black hairs; perennials...5
4b. Lower stipules triangular to lance-acuminate, not several-nerved; plants variously distributed but usually not in the mountains, to 9000 ft in elevation; fruit without black hairs; annuals, weak perennials, or occasionally perennials...6

5a. Flowers (4) 5.5–7.6 mm long, the wings 4.5–6.3 mm long; stem not purplish except at the very base; calyx usually thinly black-hairy; fruit sessile or stipitate 0.4 mm or less...***A. eucosmus***
5b. Flowers 7–12 mm long, the wings 6–9.3 mm long; stem often purplish, lighter brown at the base and darker above; calyx usually densely black-hairy; fruit stipitate 0.5–5.5 mm...***A. robbinsii* var. *minor***

6a. Plants cinereous or canescent, at least the younger leaflets silvery below, usually at least some hairs dolabriform; flowers sometimes of two kinds, the open flowers 9–14 mm long and cleistogamous flowers 4–7 mm long (these are usually hidden by the basal leaves); plants of the eastern plains... ***A. lotiflorus***
6b. Plants glabrous to strigulose or villous, or if canescent and silvery then the flowers 5–7 mm long, the hairs all basifixed; plants variously distributed but not of the eastern plains...7

7a. Flowers 9–12 mm long, pink-purple; fruit densely spreading-villous with hairs 1–2 mm long; plants of the northwestern counties (Moffat and Rio Blanco)...***A. pubentissimus***
7b. Flowers whitish and sometimes with a purple-tipped keel or pale lilac, or if pink-purple then 4–8 mm long; fruit not densely spreading-villous; plants usually not of the northwestern counties...8

8a. Flowers bright purple with a large pale spot in the banner center, 4–7.5 mm long; fruit bilocular, linear and gently incurved, not bladdery-inflated, glabrous or minutely strigulose...***A. nuttallianus* var. *micranthiformis***
8b. Flowers white or merely tinged with pink or purple, or pale lilac, sometimes also with a purple-tipped keel, 5–10 mm long; fruit unlike the above...9

9a. Plants shortly villous throughout with spreading or ascending hairs; calyx villous with white hairs; flowers pale purple or whitish with purplish tips, 5–6 mm long; leaflets narrowly oblanceolate; fruit bladdery-inflated, ovoid, 10–15 (20) mm long...***A. cerussatus***
9b. Plants glabrate or strigulose with appressed hairs; calyx strigulose with white or mixed black and white hairs; flowers whitish or faintly pink-purple, 5–10 mm long; leaflets various; fruit bladdery-inflated or not, 5–40 mm long...10

10a. Plants of the eastern slope along the Front Range; fruit 2.5–5 mm wide, not bladdery-inflated, usually purplish-mottled; leaflets obovate to obcordate or elliptic...***A. sparsiflorus***
10b. Plants of the western slope or if on the eastern slope then of the southcentral counties; fruit 7–20 mm wide, bladdery-inflated or inflated but thickly papery, not mottled; leaflets various...11

11a. Keel 6.5–8 mm long; calyx tube 3–3.5 mm long; leaflets obovate, flat; fruit stipitate 1–2 mm, inflated but not strongly bladdery and the walls thickly papery...***A. wetherillii***
11b. Keel 3.5–6 mm long; calyx tube 1.5–3 mm long; leaflets linear-oblong, oblanceolate, or elliptic, usually folded; fruit sessile, bladdery-inflated with papery walls...12

12a. Calyx lobes 0.5–1.5 mm long; keel 3.5–5 mm long; fruit strigulose, half-ovoid in profile, 7–10 mm wide; plants 0.2–2 dm tall, uncommon on the western slope (known from Delta and Moffat cos.)...***A. geyeri* var. *geyeri***
12b. Calyx lobes 2–3.5 mm long; keel 4.5–6.5 mm long; fruit inconspicuously hairy, ovoid in profile, (8) 12–24 mm wide; plants 1–5 dm tall, known from the southern counties (La Plata, Archuleta, and Conejos)...***A. allochrous* var. *playanus***

Key 11

(Lower stipules free; flowers 13–30 mm long)

1a. Stems and lower side of leaves densely spreading-hairy; flowers white or yellowish; keel purple-tipped...2
1b. Stems and leaves glabrous or with appressed or ascending hairs; flowers various...3

2a. Plants tall, 3–7 dm, usually in clumps; inflorescence an elongated raceme of numerous (15–30) flowers; fruit glabrous, stipitate (3) 5–12 mm, pendulous, linear-oblong to linear-oblanceolate...***A. drummondii***
2b. Plants usually prostrate, 0.1–2 dm tall, usually shortly caulescent; inflorescence of 4–8 flowers; fruit villous, sessile, ascending or humistrate, lanceolate and incurved...***A. parryi***

3a. Plants of the eastern slope, mostly of the plains and lower foothills; leaflets elliptic to oblong...4
3b. Plants of the western slope or southcentral counties (Conejos, Costilla), or if elsewhere on the eastern slope then the leaflets linear to filiform...5

4a. Calyx tube subcylindric to cylindric, 5–9 mm long; flowers never of two kinds, 16–25 mm long; plants glabrate to strigulose, but not silvery-cinereous; fruit subglobose with a short, filiform beak, glabrous, thick and succulent, usually reddish on the upper side, becoming coriaceous or woody in age...***A. crassicarpus***
4b. Calyx tube campanulate, 3–4.5 mm long; flowers sometimes of two kinds, the open flowers 9–14 mm long and cleistogamous flowers 4–7 mm long (usually hidden by the basal leaves); plants usually silvery-cinereous, at least on younger leaflets; fruit ovoid-acuminate or ellipsoid, pubescent, otherwise unlike the above... ***A. lotiflorus***

5a. Flowers white, ochroleucous, or yellow throughout, although sometimes also with purple veins in the banner or a purple-tipped keel...6
5b. Flowers purple, pink-purple, whitish with rose or purplish tips, or bicolored with a purple banner and white wings, the keel tip purple or not...13

6a. Plants glabrous or thinly strigulose; leaflets elliptic, broadly obovate, or elliptic-oblong, sometimes narrowly oblanceolate and then mostly 4–7 mm long; calyx tube glabrous or sparsely strigulose with a few black or white hairs, sometimes uniformly strigulose; fruit spreading, ascending, humistrate, or rarely pendulous, sessile or stipitate 2–10 mm...7
6b. Plants densely strigulose to villous, cinereous or greenish; leaflets filiform, linear, or linear-oblong, rarely oblong-lanceolate, mostly over 8 mm long; calyx tube uniformly strigulose with white or black hairs; fruit pendulous, stipitate 4–19 mm...11

7a. Plants from a slender, branching caudex, clump-forming, mostly 1.4–3 dm; leaflets narrowly elliptic to oblanceolate, mostly 4–7 mm long; calyx tube 5–6 mm long with teeth 1.3–2 mm long; fruit stipitate 2–2.5 mm; plants endemic to the Colorado River Valley…***A. debequaeus***
7b. Plants from a superficial caudex or woody crown and taproot, clump-forming or not, 1–10 dm; leaflets elliptic, broadly obovate, suborbicular, or oblong-elliptic, mostly 7–21 mm long; calyx tube 4–9 mm long with teeth 1.3–6 mm long; fruit sessile or stipitate 3–10 mm; plants variously distributed…8

8a. Inflorescence of numerous (10–30), densely packed flowers; calyx tube dorsally gibbous at the base; plants seleniferous, erect, to 10 dm tall…9
8b. Inflorescence of 5–15 flowers but loosely packed; calyx tube not gibbous dorsally; plants usually not seleniferous, decumbent to decumbent-ascending, 1–3 dm tall…10

9a. Calyx teeth divergent or recurved in age, usually narrowly triangular with a bristle-like tip, the tube 6.5–9 mm long, sparsely to densely white- or black-strigulose; flowers white, but drying ochroleucous; leaflets sparsely hairy below…***A. pattersonii***
9b. Calyx teeth pointing forward and straight, usually deltoid and without a bristle-like tip, the tube 4–6.5 mm long, glabrous or sparsely strigulose with black hairs, lacking white hairs; flowers ochroleucous or yellowish-green; leaflets glabrous or sparsely hairy below…***A. praelongus***

10a. Calyx tube glabrous or sparsely strigulose with a few black hairs dorsally; wing petals 16.5–19 mm long; peduncles 5–12 cm long; fruit bladdery-inflated, papery, stipitate 4–10 mm, usually purple- or reddish-mottled…***A. oophorus* var. *caulescens***
10b. Calyx tube usually uniformly pubescent with black or white hairs; wing petals 12–17.5 mm long; peduncles 1–5 cm long; fruit little to strongly inflated, leathery or stiffly papery, sessile, purple-speckled or rarely purple-mottled…***A. lentiginosus* var. *chartaceus***

11a. Leaflets 13–17, glabrous above, the terminal leaflet continuous with the rachis; fruit laterally compressed…***A. ripleyi***
11b. Leaflets pubescent above, if glabrous then also minutely red-dotted, if the terminal leaflet continuous with the rachis then the leaflets only 3–9; fruit obcompressed…12

12a. Leaflets 3–9, the terminal leaflet continuous with the rachis; calyx strigulose with white hairs, rarely with a few black hairs; fruit glabrous…***A. lonchocarpus***
12b. Leaflets 11–15 (17), the terminal leaflet petiolulate and jointed; calyx strigulose with black hairs; fruit uniformly hairy…***A. schmollae***

13a. Plants usually densely strigulose, cinereous or greenish-cinereous, sometimes less pubescent in age…14
13b. Plants glabrous or thinly strigulose…17

14a. Leaflets linear to narrowly elliptic, folded or involute; plants erect; calyx tube 4–6.7 mm long with lobes 0.6–1.8 (2.3) mm long; fruit narrowly oblanceolate, pendulous, stipitate 4–8 mm…***A. coltonii* var. *moabensis***
14b. Leaflets obovate to elliptic or oblanceolate, sometimes narrowly elliptic, flat; plants decumbent to decumbent-ascending; calyx tube over 7 mm long or if shorter then the lobes 1.5–3.5 mm long; fruit ovoid-acuminate or ellipsoid, ascending or humistrate, sessile or stipitate 1–1.5 mm…15

15a. Calyx tube 4–7 mm long, densely strigulose with black hairs; flowers 14–18 mm long…***A. cibarius***
15b. Calyx tube 7–11 mm long, usually with white hairs but sometimes with black hairs; flowers 18–22 mm long…16

16a. Flowers whitish and purple-tinged with a purple-tipped keel; keel 16–19 mm long; fruit hairy…***A. argophyllus* var. *martinii***
16b. Flowers purple or reddish-purple; keel 12–16 mm long; fruit glabrous…***A. iodopetalus***

17a. Plants from a branching caudex and taproot, with spreading-ascending or mat-forming stems; leaflets narrowly elliptic to oblanceolate, mostly less than 2 mm wide…***A. eastwoodiae***
17b. Plants lacking a branching caudex, with decumbent, ascending, or erect stems; leaflets obovate to broadly ovate or oblanceolate, mostly over 3 mm wide…18

18a. Fruit sessile, bilocular with a unilocular beak; keel mostly 10–15 mm long…***A. lentiginosus***
18b. Fruit stipitate 2–7 mm, unilocular throughout; keel mostly over 17 mm long…***A. preussii***

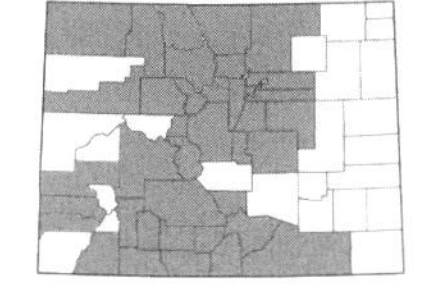

***Astragalus agrestis* Douglas ex G. Don**, PURPLE MILKVETCH. Caulescent perennials, 0.5–3 (4) dm; *leaves* pinnately compound with (9) 13–19 leaflets; *leaflets* elliptic to oblong, 5–20 mm long, strigulose; *inflorescence* a dense ovoid or globose head, (3) 5–15-flowered; *calyx* tube 4–8 mm long, lobes 2.5–4 (5) mm long, villous with white or black hairs; *flowers* pink-purple, blue-lavender, or bicolored, the banner 14–20 mm long; *legumes* ovoid to elliptic, obcompressed, 7–10 × 3–4 mm, ultimately black and densely white-villous. Common in meadows, grasslands, along streams, found from the high plains to upper montane meadows and forests, 5000–10,500 ft. May–July. E/W. (Plate 42)

Similar in appearance and distribution to *A. adsurgens* var. *robustior*, which has dolabriform hairs and flowers usually in an elongated head while *A. agrestis* is sparsely pubescent with basifixed hairs and flowers usually in a tightly clustered, more spherical head.

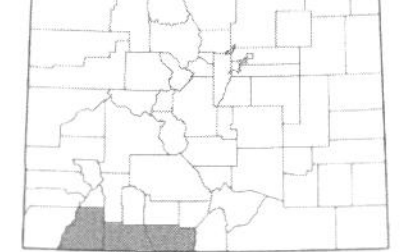

***Astragalus allochrous* A. Gray var. *playanus* Isely**, WOOTON'S MILKVETCH. [*A. wootonii* Sheldon]. Annuals or short-lived perennials, 1–5 dm; *leaves* pinnately compound with (7) 11–23 leaflets; *leaflets* linear-oblanceolate to oblong-obovate, usually folded, 5–20 mm long; *inflorescence* a loosely 2–10 (15)-flowered raceme; *calyx* tube 2–3 mm, lobes 2–3.5 mm long, with white or sometimes black hairs; *flowers* whitish, sometimes tinged with pink or purple, the banner 5–9 mm long; *legumes* bladdery-inflated, ovoid, inconspicuously hairy, 1.5–4 × (0.8) 1.2–2.4 cm. Uncommon in rocky soil and dry grasslands, 7000–9000 ft. May–June. E/W.

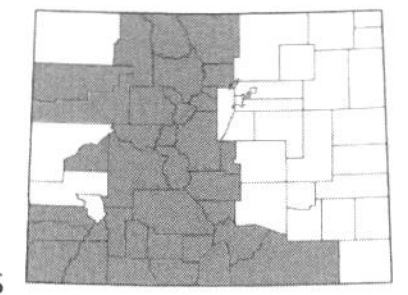

***Astragalus alpinus* L.**, ALPINE MILKVETCH. Caulescent or apparently acaulescent perennials, 0.5–2.5 dm; *leaves* pinnately compound with (11) 15–25 (27) leaflets; *leaflets* oval-ovate to suborbicular, 0.4–2 cm long, often glabrate to sparsely hairy above, more densely hairy below; *inflorescence* a (5) 7–17 (23)-flowered raceme; *calyx* tube 2–4.1 mm long, lobes 0.9–2.8 mm long, densely to sparsely strigulose with black and white hairs; *flowers* bicolored, banner and keel tip purple, wings whitish, the banner 7–14 mm; *legumes* ellipsoid, 3-angled, 7–14 × 2.5–4 mm, strigulose or villous, with black, mixed black and white, or rarely all white hairs. Common in alpine, aspen groves, meadows, along streams, and in spruce-pine forests, 7500–12,700 ft. June–Aug. E/W. (Plate 42)

***Astragalus americanus* (Hook.) M.E. Jones**, AMERICAN MILKVETCH. Caulescent perennials, 3–10 dm; *leaves* pinnately compound with (7) 9–17 leaflets; *leaflets* ovate to oblong or elliptic-oblanceolate, 2–5 × 0.7–1.5 cm, glabrous or sparsely strigulose below ; *inflorescence* a loosely 10–25-flowered raceme; *calyx* tube 4.9–5.8 mm long, lobes 0.2–0.7 mm long, glabrous to glabrate with white or partly black-ciliate teeth; *flowers* white or ochroleucous, the banner 11–14 mm long; *legumes* obliquely ellipsoid, 2–2.8 mm long, 6.5–9.5 mm diam., stipitate 4.5–7 mm, glabrous or minutely black-hairy. Known from a very old collection by Hall and Harbour on the headwaters of the South Platte and not seen in Colorado since 1860. June–Aug.

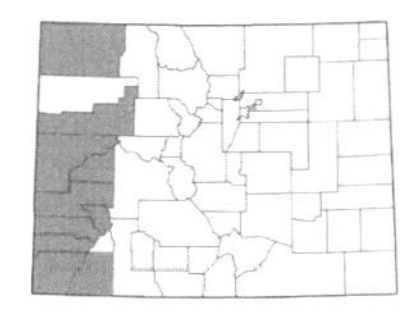

***Astragalus amphioxys* A. Gray var. *vespertinus* (Sheldon) M.E. Jones**, CRESCENT MILKVETCH. Subacaulescent or prostrate-caulescent perennials, 1–5 (15) cm; *leaves* pinnately compound with (5) 11–17 (21) leaflets; *leaflets* elliptic or obovate to oblanceolate, 4–15 (20) mm long, dolabriform pubescent, strigose, cinereous; *inflorescence* included or exserted, 5–12-flowered; *calyx* tube 5–10 mm long, lobes 1–3 mm long; *flowers* pink-purple, the banner 13–28 mm; *legumes* lanceolate-attenuate, obcompressed in the middle and laterally compressed at ends, or obcompressed throughout, (1.5) 2–4 (5) cm × 5–9 mm, occasionally mottled, thinly pubescent or glabrate. Common in rocky soil and on dry hillsides, often with sagebrush, 4400–8500 ft. April–June. W.

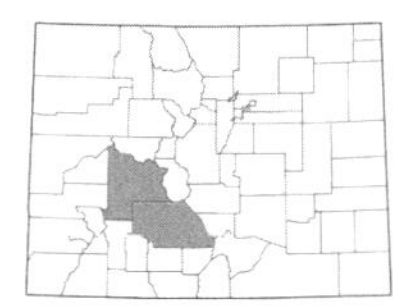

***Astragalus anisus* M.E. Jones**, GUNNISON MILKVETCH. Subacaulescent or shortly prostrate-caulescent perennials; *leaves* pinnately compound with 11–15 leaflets; *leaflets* obovate, 4–10 mm long, dolabriform pubescent, cinereous or silvery; *inflorescence* included or exserted, 3–8-flowered; *calyx* tube 9–10 mm long, lobes 1.5–3 mm long; *flowers* pink-purple, the banner 15–19 mm long; *legumes* obovoid to subspheroid, slightly obcompressed, 1–1.5 × 1–1.2 cm, strigose. Locally common on sagebrush slopes along the Gunnison River, 7400–9500 ft. May–June. W. Endemic.

Plants without fruit can be mistaken for *A. missouriensis*, which usually has a dark, purple-tinged calyx and does not occur in the same area as *A. anisus*.

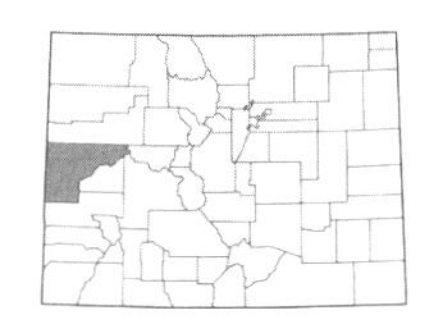

***Astragalus aretioides* (M.E. Jones) Barneby**, CUSHION MILKVETCH. [*Orophaca aretioides* (M.E. Jones) Rydb.]. Mat-forming perennials; *leaves* ternately compound; *leaflets* 4–7 mm long, silvery pubescent; *inflorescence* 2–3-flowered; *calyx* tube 2–2.3 mm long, lobes 1.2–2 mm long; *flowers* pink-purple to violet, the banner 6.5–8 mm long; *legumes* ovoid-ellipsoid, 4–4.5 mm long. Uncommon on barren, gypsum hillsides, 6500–7000 ft. May–June. W.

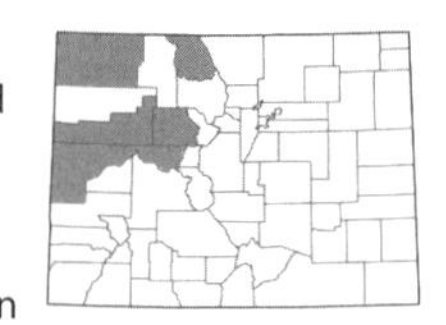

Astragalus argophyllus **Nutt. var.** ***martinii*** **M.E. Jones**, SILVERLEAF MILKVETCH. Caulescent or subacaulescent perennials, sometimes mat-forming, 0.5–10 (20) cm; *leaves* pinnately compound with (9) 11–19 leaflets; *leaflets* obovate to narrowly elliptic, 4–14 mm long; *inflorescence* subumbellate, 3-8-flowered; *calyx* tube 6.5–12 mm long, lobes 1.5–3 (5) mm long; *flowers* ochroleucous to purplish-tinged, maculate, the banner 18–22 mm long; *legumes* ovoid-acuminate, 15–30 × 7–12 mm, strigulose or villous. Locally common on sandy or rocky soil, often with sagebrush or pinyon-juniper, 6000–8500 ft. April–May. E/W.

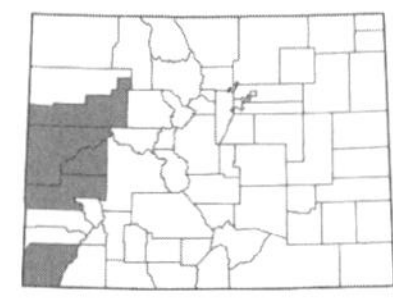

Astragalus asclepiadoides **M.E. Jones**, MILKWEED MILKVETCH. Caulescent perennials, 1–6 dm; *leaves* simple, broadly ovate or cordate-orbicular, to 3 cm wide, glabrous, glaucous; *calyx* tube 7–11 mm long, lobes 1.5–3 mm long, glabrous; *flowers* purple or yellowish-green with pale wing tips, the banner 18–22 mm; *legumes* ovoid to ellipsoid, somewhat inflated, 2.4–3.5 × 1–1.5 cm, stipitate, often purple-speckled, glabrous. Found on shale or clay slopes, 4500–6000 ft. May–June. W. (Plate 42)

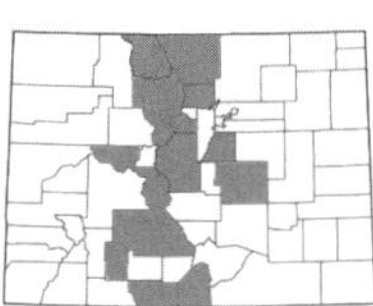

Astragalus australis **(L.) Lam.**, INDIAN MILKVETCH. [*A. aboriginum* Richardson]. Caulescent perennials, 0.2–2.6 dm; *leaves* pinnately compound with 7–15 leaflets; *leaflets* oblong or linear-elliptic, occasionally linear, 0.3–2.7 (3.5) × 0.1–0.7 cm, hairy or glabrous above, hairy below; *inflorescence* a (2) 6–30-flowered raceme; *calyx* tube (2.1) 2.7–3.9 (5) mm long, lobes 1–4 mm long, loosely strigulose to densely villous with black, fuscous, and white hairs; *flowers* whitish or cream, occasionally with a distally purple-tinged banner and maculate keel tip, the banner 7.5–12.6 (14.5) × 4–8 mm; *legumes* obliquely narrow-elliptic, 10–26 (31) × 3–7 mm, greenish with red or purple dots glabrous or sometimes strigulose. Uncommon in alpine meadows and on rocky slopes, or in ponderosa pine forests, 6500–13,200 ft. June–Aug. E/W.

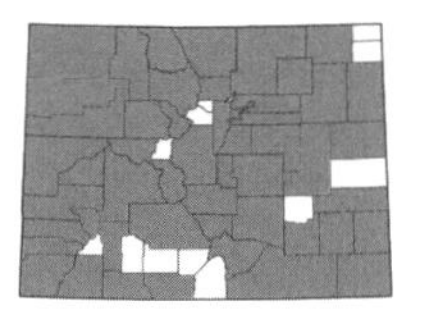

Astragalus bisulcatus **(Hook.) A. Gray**, TWO-GROOVED MILKVETCH. Erect or ascending perennials, 1.5–8 dm; *leaves* pinnately compound with (13) 15-numerous leaflets; *leaflets* elliptic to linear, 5–25 (30) mm long, glabrate above; *inflorescence* exserted, numerous-flowered; *calyx* tube 3–5.5 mm long, lobes (1.5) 2–4 (5) mm long, often red; *flowers* purple, white, or white with a maculate keel, the banner 7–16 (18) mm long; *legumes* oblong or shortly ellipsoid, obcompressed, the ventral face strongly 2-grooved, 6–16 (20) × 2–4.5 mm, strigulose to glabrate. Common on dry slopes and grasslands, shale, usually on seleniferous soil, 3500–8800 ft. May–July. E/W. (Plate 42)

There are two varieties of *A. bisulcatus* in Colorado:

1a. Fruit linear to narrowly oblong, (8) 10–18 mm long, glabrous or strigose; flowers 10–17.5 mm long, purple or white…**var.** ***bisulcatus***, TWO-GROOVED MILKVETCH. Common on grasslands, dry slopes, and shale outcroppings, 3500–8800 ft. May–July. E/W.

1b. Fruit elliptic to shortly oblong, 6–9.5 mm long, strigose; flowers 7–11 mm long, white or ochroleucous…**var.** ***haydenianus*** **(A. Gray) Barneby**, HAYDEN'S MILKVETCH. [*A. haydenianus* A. Gray]. Common on dry hillsides, shale slopes, and sagebrush flats, 6000–8500 ft. May–July. W.

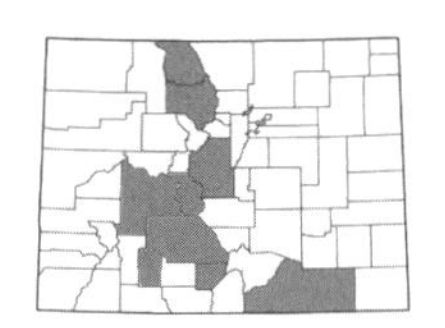

Astragalus bodinii **Sheldon**, BODIN'S MILKVETCH. Sprawling perennials, to 4 dm; *leaves* pinnately compound with 11–17 leaflets; *leaflets* oblanceolate or broadly ovate to rhombic, 5–15 mm long, glabrate above; *inflorescence* 4–15-flowered; *calyx* tube 2.5–3.5 mm long, lobes 1–1.5 mm long; *flowers* purple, the banner 8–11 mm long; *legumes* ovoid to shortly lanceolate-ellipsoid, slightly inflated, subterete to obscurely 3-angled, (5) 6–10 × 3–5 mm, strigulose. Uncommon in moist meadows, along streams, and in aspen groves, 7500–9800 ft. June–Aug. E/W.

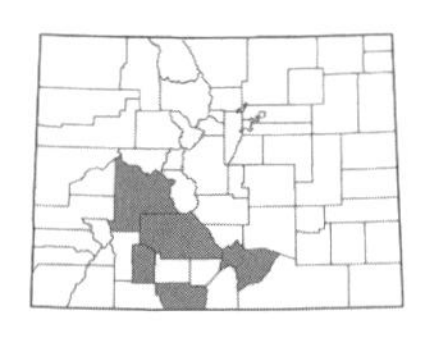

Astragalus brandegeei **Porter**, BRANDEGEE'S MILKVETCH. Subacaulescent or shortly decumbent-caulescent perennials, 1–10 (30) cm; *leaves* pinnately compound with 5–13 leaflets; *leaflets* oblong to linear, 5–20 mm long, strigulose with basifixed hairs; *inflorescence* scapiform and included or terminal from foliose stems, 1-5-flowered; *calyx* tube 1.5–2.5 mm long, lobes 1–2.5 mm long, white or partly black-strigulose; *flowers* white with lavender striations and maculate tips, the banner 4–7 mm long; *legumes* obovoid to short-cylindric, subterete or obcompressed, 1–2 cm × 3.5–5 mm, strigulose. Uncommon in rocky meadows and gravelly flats, usually with pinyon-juniper or sometimes in oak, 5400–8800 ft. April–Sept. E/W.

Can be mistaken for *A. miser* which has larger flowers (6–10 mm long vs. 4–7 mm long), a longer keel (6–8.5 mm long vs. 3.5–4.5 mm long in *A. brandegeei*), and lanceolate or oblanceolate fruit.

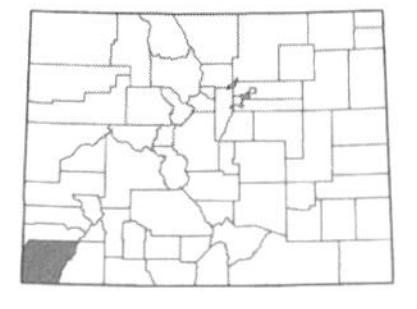

Astragalus calycosus **Torr. ex S. Watson**, TORREY'S MILKVETCH. Subacaulescent, matted perennials; *leaves* ternately or pinnately compound with (1) 3–13 leaflets; *leaflets* cuneate-obovate to oblanceolate, 2–15 (20) mm long, cinereous-silvery with dolabriform hairs; *inflorescence* exserted, 2–15-flowered; *calyx* tube 3–6 mm long, lobes 0.5–3 (4) mm long, densely strigulose with white and usually some black hairs; *flowers* white, pink, or bright purple, the wing tips white or pallid, the keel tip maculate, the banner 9.5–16.5 (20) mm long; *legumes* oblong, laterally or triquetrously compressed, 12–20 × (2.5) 3–4 mm, strigulose. Uncommon in rocky or shale soil of pinyon-juniper woodlands, 5400–6800 ft. April–May. W.

There are two varieties of *A. calycosus* possibly present in Colorado:

1a. Leaflets 3–7, often palmately trifoliate or with 1–2 pairs of lateral leaflets on a rachis to 1 cm long; raceme 1–3-flowered...**var. *calycosus***. Possibly present in Dolores or San Miguel cos. in southwestern Colorado.

1b. Leaflets (5) 7–13, pinnately compound, the rachis 0.5–4 cm long; raceme 5–15-flowered...**var. *scaposus* (A. Gray) M.E. Jones**. Found in rocky soil of pinyon-juniper woodlands, known from Montezuma Co., 5400–6800 ft. April–May. W.

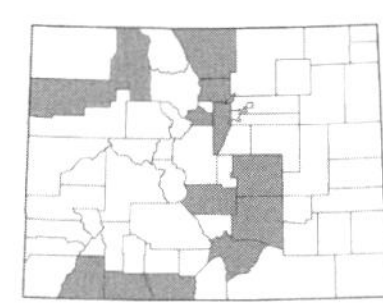

***Astragalus canadensis* L.**, Canadian milkvetch. Caulescent perennials, 1.5–12 dm; *leaves* pinnately compound with (7) 13–35 leaflets; *leaflets* lanceolate, lance-oblong, or elliptic, 10–52 mm long, strigose or glabrous above, strigose below; *inflorescence* numerous-flowered racemes; *calyx* tube 4.5–10.5 mm long, lobes 1.2–4.5 mm long, strigose; *flowers* ochroleucous, the banner (11.5) 13.2–17.5 mm long; *legumes* cylindric, 10–20 mm long, strigose or glabrous. Uncommon in moist meadows, along creeks, and in forests, 5100–8300 ft. June–Sept. E/W.

There are two varieties of *A. canadensis* in Colorado:

1a. Pod terete, not grooved dorsally, glabrous; stems 4–12 dm tall...**var. *canadensis***. Found throughout the range of the species in Colorado, 5100–8000 ft.

1b. Pod deeply grooved dorsally, strigulose but sometimes becoming glabrate in age; stems 1–5.5 dm tall...**var. *brevidens* (Gand.) Barneby**. At its southern limit in northcentral Colorado, 7000–8300 ft. Could possibly be in Moffat Co. as well.

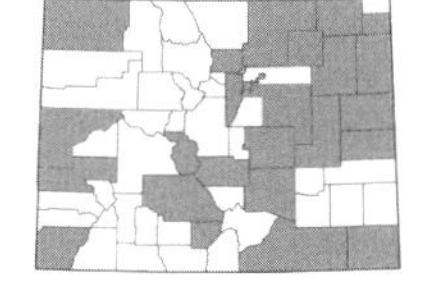

***Astragalus ceramicus* Sheldon**, painted milkvetch. Caulescent perennials, 1–2 (4) dm, the stems zigzag to flexuous; *leaves* phyllodial without leaflets or pinnately compound with up to 11 leaflets; *leaflets* linear-filiform or elliptic-oblong, 0.5–3 cm long, with dolabriform or basifixed hairs; *inflorescence* included, 3–9 (15)-flowered; *calyx* tube 2–4 mm long, lobes 1–2 (3) mm long; *flowers* ochroleucous to pinkish or lavender-striate, the keel usually purple-tipped, the banner 6–10 mm long; *legumes* ellipsoid to broadly ovoid, bladdery-inflated, subterete, (1.5) 2–3 × (0.6) 0.8–2.3 cm, purple-mottled, glabrous. Found in sandy soil, dunes, and on sandstone outcroppings, 3600–8000 ft. May–Aug. (Sept.). E/W. (Plate 42)

There are two varieties of *A. ceramicus* in Colorado:

1a. Pubescence composed mostly of dolabriform hairs; leaves mostly pinnately compound or only the upper reduced to a leafstalk; body of the pod to 3 cm long...**var. *ceramicus***. Found in the western counties (Moffat and Montezuma), 5000–6200 ft. May–June (Sept.). W.

1b. Pubescence composed mostly of basifixed hairs; leaves mostly simple and phyllodial except for a few of the lowermost; body of the pod (2) 3–5 cm long...**var. *filifolius* (A. Gray) F.J. Herm.** Found on the eastern plains and in the San Luis Valley, 3600–8000 ft. May–Aug. E.

***Astragalus cerussatus* Sheldon**, powdery milkvetch. Spreading to ascending short-lived perennials or annuals, 0.5–3 dm; *leaves* pinnately compound with 13–19 leaflets; *leaflets* oblong-oblanceolate, 4–14 mm long, with basifixed hairs; *inflorescence* included or exserted, 2–6-flowered; *calyx* tube 1.5–2 mm long, lobes 1.5–2.5 mm long; *flowers* pale lilac or whitish with lilac tips, the banner 5–6 mlong; *legumes* ovoid to ellipsoid, bladdery-inflated, subterete to slightly obcompressed, 1–1.5 (2) cm × 6–12 (14) mm, stramineous, minutely hairy. Locally common in sandy or rocky soil, grasslands, or sometimes with pinyon-juniper or sagebrush, 7500–9000 ft. May–June. E/W. Endemic.

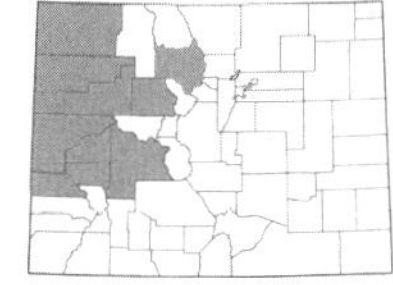

Astragalus chamaeleuce* A. Gray var. *chamaeleuce, cicada milkvetch. Acaulescent or subacaulescent perennials; *leaves* pinnately compound with 5–15 leaflets; *leaflets* obovate to elliptic, 4–5 mm long, pubescent with dolabriform hairs; *inflorescence* scapose, subumbellate, or racemose, 4–9-flowered; *calyx* tube 6.5–10 mm long, lobes 1.5–3 mm long; *flowers* pink-purple or ochroleucous with a maculate keel, the banner 17–22 mm long; *legumes* half-ellipsoid, incurved, or obliquely arcuate-lanceolate, subterete to slightly obcompressed, 1.5–3.5 (4) cm × 7–15 mm, mottled and reticulate, strigulose. Locally common on shale or rocky slopes and barren hillsides, or with pinyon-juniper, 4600–7500 ft. May–June. W.

Without fruit, *A. chamaeleuce* can be easily misidentified as *A. amphioxys*. However, *A. chamaeleuce* often has ochroleucous flowers with a purple-tipped keel and usually has fewer leaflets (5–15) while *A. amphioxys* usually has 11–17 leaflets and purple or rose-purple flowers.

***Astragalus chloödes* Barneby**, grass milkvetch. Subacaulescent perennials, 1.5–2.5 dm; *leaves* simple, filiform, involute, 5–10 cm long, with dolabriform hairs; *inflorescence* of 15–25-flowered racemes; *calyx* tube ca. 3 mm long, lobes 2–3 (4) mm long; *flowers* purple, the banner 8–9 mm long; *legumes* obliquely half-elliptic or shortly oblong-lanceolate, 8–11 × 2–3 mm, glabrous or strigose. Not reported for Colorado but occurring in Utah very close to the border of Moffat Co.

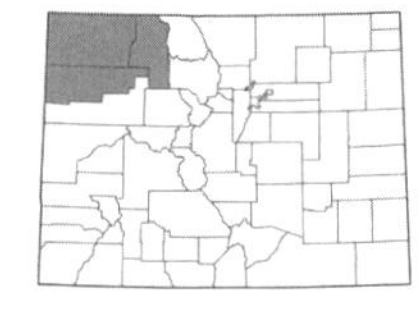

***Astragalus cibarius* Sheldon**, BROWSE MILKVETCH. Perennials, 0.5–2 dm; *leaves* pinnately compound with 13–17 (21) leaflets; *leaflets* obovate to oblanceolate, glabrate above; *inflorescence* included or exserted, subcapitate to shortly racemose, 5–12 (15)-flowered; *calyx* tube 4–7 mm long, lobes 1.5–3 mm long; *flowers* yellowish-white with a lavender tinge, the banner 14–18 mm long; *legumes* oblong-arcuate or nearly straight, (1.5) 2–3 (3.5) cm × 7–9 mm, glabrate to strigose. Locally common on sagebrush slopes, 5900–7100 ft. May–June. W.

Similar to *A. lentiginosus* var. *palans* in habit and in having an almost bilocular fruit. However the fruit of that species is often mottled and the inflorescence is an elongate raceme.

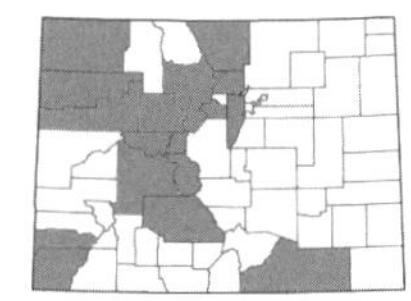

***Astragalus cicer* L.**, CHICKPEA MILKVETCH. Caulescent perennials, 3–8 dm; *leaves* pinnately compound with 17-numerous leaflets; *leaflets* elliptic to oblong, 0.5–3 cm long, often glabrate above; *inflorescence* included to shortly exserted, compactly ovoid to short-spicate, 10-numerous-flowered; *calyx* tube 4–5 mm long, lobes 1.5–2 mm long; *flowers* pink-purple or ochroleucous with a maculate keel, the banner 12–15 mm long; *legumes* ovoid to ellipsoid, inflated and sub-bladdery, subterete, 10–14 × 7–9 mm, straw-colored to black, pustulate-hirsute. Found in disturbed areas, along roadsides, and near trailheads and campgrounds, 5600–11,000 ft. June–Aug. E/W. Introduced.

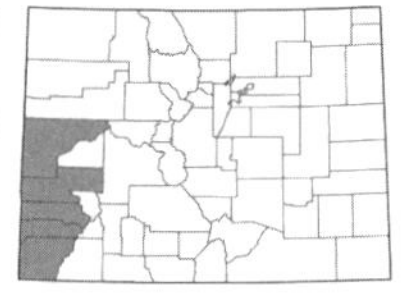

***Astragalus coltonii* M.E. Jones,** COLTON'S MILKVETCH. Caulescent perennials, 1–4 dm; *leaves* pinnately compound (at least below) with 1–5 leaflets; *leaflets* linear to linear-oblong, involute, 4–10 (14) mm long, glabrous or hairy above, hairy below; *inflorescence* exserted, (2) 5–20-flowered racemes; *calyx* tube 4–6.7 mm long, lobes 0.6–1.8 (2.3) mm long, densely to thinly black-strigulose; *flowers* pink-purple with paler wing tips, drying bluish, the banner 12–18.5 mm long; *legumes* oblong or oblong-oblanceolate, laterally compressed, 2.5–3.2 cm × 3.5–5.2 mm, stipitate 4–9 mm, glabrous.

There are two varieties of *A. coltonii* in Colorado:

1a. Leaves distant, linear to filiform, the terminal leaflet usually continuous with the rachis; plants sparsely foliose or juncaceous…**var. *coltonii***. Uncommon on sandy soil of mesas and badlands, 5800–6000 ft. April–June. W.

1b. Leaves oblong, the terminal leaflet articulate; plants conspicuously foliose…**var. *moabensis* M.E. Jones**. Found in sagebrush meadows or on dry slopes with pinyon-juniper and along roadsides, 4700–8500 ft. April–June. W.

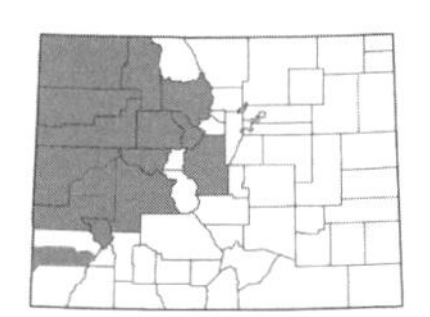

***Astragalus convallarius* Greene**, LESSER RUSHY MILKVETCH. Caulescent perennials, rhizomatous, (1) 2–5 (7) dm; *leaves* simple, or pinnately compound with up to 13 leaflets; *leaflets* filiform or linear-involute, 2–10 cm long; *inflorescence* included or exserted, 5-many-flowered; *calyx* tube 3–5 mm long, lobes 0.5–1 (1.5) mm long; *flowers* white, ochroleucous, purple-veined, or purplish, the banner 7–12 mm long; *legumes* narrowly oblong or oblanceolate, 1.3–3.5 (4) cm × 2–3 (5) mm, strigulose. Common on open, dry hillsides, shale slopes, occasionally to aspen woodland, but usually with pinyon-juniper, oak, or sagebrush, 4700–9200 ft. May–June. W. (Plate 43)

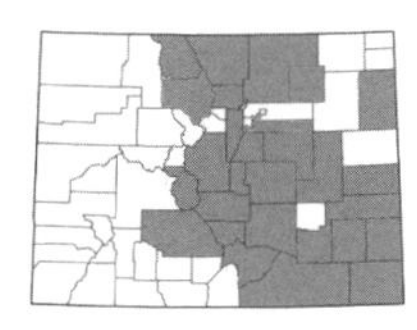

***Astragalus crassicarpus* Nutt.**, GROUNDPLUM. Decumbent or ascending perennials, 1–5 dm; *leaves* pinnately compound with 13-numerous leaflets; *leaflets* broadly elliptic, oblong, or oblanceolate, 0.5–1.5 cm long, glabrate above; *inflorescence* included or shortly exserted, (5) 8–20-flowered; *calyx* tube 5–9 mm long, lobes 2–3 (4) mm long, strigulose to tomentose; *flowers* pink-purple, bicolored with a maculate keel, or greenish-white, the banner 16–25 mm long; *legumes* subglobose, (1) 1.5–2.5 cm long, often reddish on upper side, glabrous. Common on the shortgrass prairie and grassy slopes of the foothills and north and south parks, 3500–9500 ft. April–June. E/W. (Plate 43)

***Astragalus cronquistii* Barneby**, CRONQUIST'S MILKVETCH. Decumbent perennials, rhizomatous, 2–4 dm; *leaves* pinnately compound with 7–13 leaflets; *leaflets* elliptic to narrowly oblong, 6–20 mm long, glabrate above; *inflorescence* exserted, 5–20-flowered; *calyx* tube 3–4 mm long, lobes 0.5–1 mm long; *flowers* red-purple or reddish, the banner 8–9 mm long; *legumes* oblong to oblong-oblanceolate, 3-angled, 2–3 cm × 3.5–5 mm, strigulose. Uncommon in sandy and gravelly soil, often on shale, 4800–5900 ft. April–May. W.

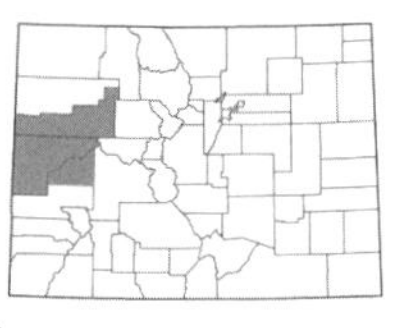

***Astragalus debequaeus* S.L. Welsh**, DEBEQUE MILKVETCH. Clump-forming perennials, 14–30 cm; *leaves* pinnately compound with 13–21 leaflets; *leaflets* elliptic to oblanceolate, glabrous; *inflorescence* racemes; *calyx* tube 5–6 mm long, lobes 1.3–2 mm long, sparsely black strigose; *flowers* white to yellowish-white, the banner 17–21 mm long; *legumes* oblong to lance-ellipsoid, inflated, 15–23 × 6–11 mm, stipitate 2–2.5 mm, straw-colored, scabrid-pubescent to glabrous. Found on dry slopes, often in clay soil, usually with pinyon-juniper or sagebrush, endemic to the Colorado River Valley near Debeque (Garfield and Mesa cos.), 5100–6400 ft. April–May. W. Endemic.

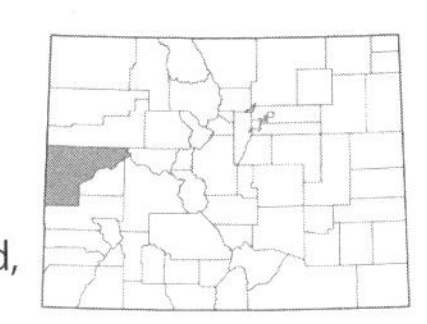

***Astragalus desperatus* M.E. Jones**, RIMROCK MILKVETCH. Subacaulescent or shortly caulescent perennials; *leaves* pinnately compound with 7–13 (15) leaflets; *leaflets* elliptic to oblanceolate, often folded, 2–12 mm long; *inflorescence* included or exserted, 5–12-flowered racemes; *calyx* tube 3–7 mm long, lobes 1–2.5 mm long; *flowers* purple or bicolored, the banner 7–14 mm long; *legumes* ellipsoid-lunate to hamate-incurved, proximally obcompressed, 6–16 × 4–6 mm, mottled, wrinkled, loosely villous. Found in gravelly or sandy soil of rock ledges, washes, canyons, and at the bases of cliffs, 4500–7000 ft. April–May. W.

There are two varieties of *A. desperatus* in Colorado:

1a. Corolla 7–10 mm long...**var. *desperatus***.
1b. Corolla 13–16 mm long...**var. *neeseae* Barneby**. [*A. equisolensis* Neese & S.L. Welsh].

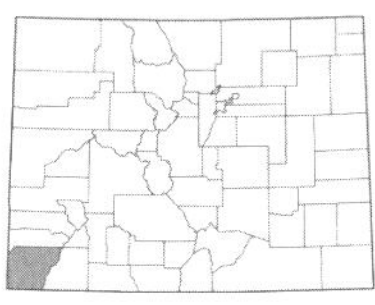

***Astragalus deterior* (Barneby) Barneby**, CLIFF PALACE MILKVETCH. Subacaulescent to shortly caulescent perennials, 0.5–5 cm; *leaves* pinnately compound with 11–15 leaflets; *leaflets* elliptic to narrowly elliptic, usually folded, 3–9 mm long; *inflorescence* subscapose, 2–5-flowered racemes; *calyx* tube 3–4 mm long, lobes 1–2 mm long; *flowers* light lilac or ochroleucous with a maculate keel, the banner 10–12 mm long; *legumes* oblong to slightly oblanceolate, obcompressed, obcordate, or 3-angled, 1.3–1.8 cm × 3.5–5 mm, mottled, slightly reticulate, strigose. Uncommon on sandstone rimrock ledges of mesas, known from Mesa Verde National Park (Montezuma Co.), 6400–7000 ft. April–May. W. Endemic.

Astragalus humillimus resembles *A. deterior* but has persistent old leaf rachises giving the plant a spinescent appearance and has dolabriform hairs.

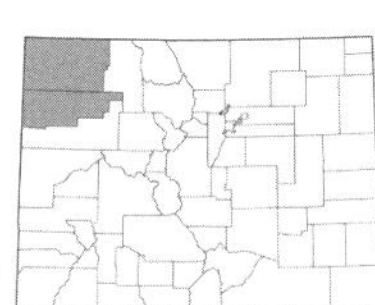

***Astragalus detritalis* M.E. Jones**, DEBRIS MILKVETCH. Subacaulescent perennials, caespitose-matted or pulvinate; *leaves* phyllodial or pinnately compound with 3–5 (7) leaflets; *leaflets* oblong to linear, flat to involute, pubescent with dolabriform hairs; *inflorescence* 2–8-flowered racemes; *calyx* tube 3–5 mm long, lobes 2–4 mm long; *flowers* bright pink-purple, the banner 13–19 mm long; *legumes* oblong to linear, laterally compressed, 1.5–3 (3.5) cm × 2–3.5 mm, often mottled, strigose. Locally common on white shale bluffs and slopes and open, barren places, 5400–8700 ft. April–June. W.

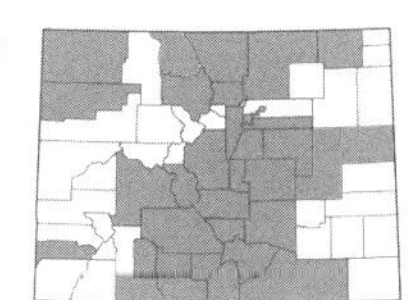

***Astragalus drummondii* Douglas ex Hook.**, DRUMMOND'S MILKVETCH. Caulescent perennials, 3–7 dm; *leaves* pinnately compound with 17- many leaflets; *leaflets* elliptic-oblong or oblanceolate, 5–25 mm long, often glabrate above except at edges; *inflorescence* exserted, numerous-flowered; *calyx* tube 4.5–7 mm long, lobes 1.5–3 (4) mm long, white or black-hairy; *flowers* white with a maculate keel, ochroleucous when dried, the banner 15–25 mm long; *legumes* oblong to oblanceolate, obcordate in cross-section, (1) 1.5–3 cm × 3–5 mm, glabrous. Common on the shortgrass prairie, grassy slopes, and sagebrush flats to pinyon-juniper, 4200–9500 ft. May–Aug. E/W. (Plate 43)

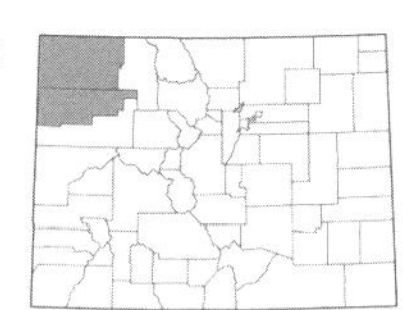

***Astragalus duchesnensis* M.E. Jones**, DUCHESNE MILKVETCH. Shortly rhizomatous perennials, to 3 dm; *leaves* pinnately compound with ca. 7–11 leaflets; *leaflets* oblong-lanceolate or linear, 5–25 mm long, often glabrate above; *inflorescence* exserted or included, loosely 6–20-flowered; *calyx* tube 3–5 mm long, lobes 0.5–1 mm long; *flowers* reddish-purple with white wings, the banner 10–12 mm long; *legumes* oblong or oblong-oblanceolate, obcompressed or 3-angled at base, laterally compressed apically, 2.5–3.5 cm × 3–5 mm, strigulose. Uncommon on sandy bluffs, often with pinyon-juniper, 4600–6400 ft. April–June. W.

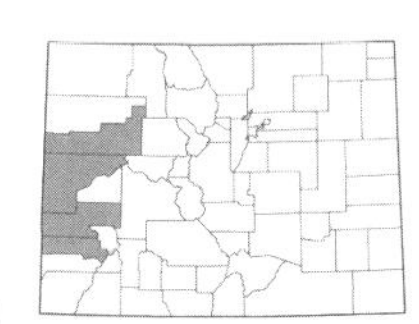

***Astragalus eastwoodiae* M.E. Jones**, EASTWOOD'S MILKVETCH. Caulescent perennials, 3–15 (30) cm; *leaves* pinnately compound with 15-numerous leaflets; *leaflets* elliptic to lanceolate or oblanceolate, 3–12 mm long; *inflorescence* included or shortly exserted, 4–8-flowered; *calyx* tube (6) 8–9 mm long, lobes 1.5–2.5 mm long, slightly black-hairy; *flowers* pink-purple, the banner 18–21 mm long; *legumes* ovoid to oblong-ellipsoid, subterete or slightly obcompressed, 1.5–3.5 cm × 8–12 (15) mm, glabrate or minutely hairy. Found on dry slopes and in dry creek beds, often on clay soil, usually with pinyon-juniper, 4700–6700 ft. April–May. W.

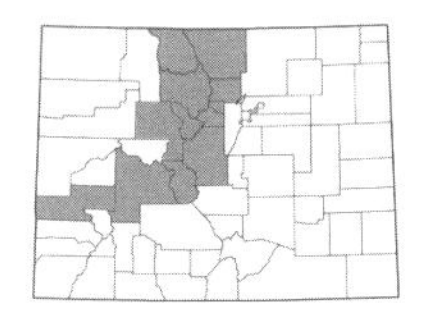

***Astragalus eucosmus* B.L. Rob.**, ELEGANT MILKVETCH. Caulescent perennials, (0.6) 1–4 dm; *leaves* pinnately compound with 9–15 (17) leaflets; *leaflets* oblong, lance-oblong, or oblanceolate, (0.4) 0.6–2 (3) cm long, glabrous or strigulose above, strigulose below; *inflorescence* a compact to lax (5) 7–25-flowered raceme; *calyx* tube 2.5–3.5 mm long, lobes 0.7–2 mm long, strigulose or pilose with black, gray, or white hairs; *flowers* purplish, purple-tipped, pinkish, or whitish, the banner (4.1) 5.5–7.6 × 2.8–3.7 mm; *legumes* ovoid-ellipsoid, (4) 5–13 mm long, (2.3) 2.5–5.5 mm diam., densely strigose-pilose with black, gray, white, or black and white hairs. Uncommon in moist meadows, along streams, willow thickets, in aspen groves, and along rivers, 8500–10,500 ft. June–July. E/W.

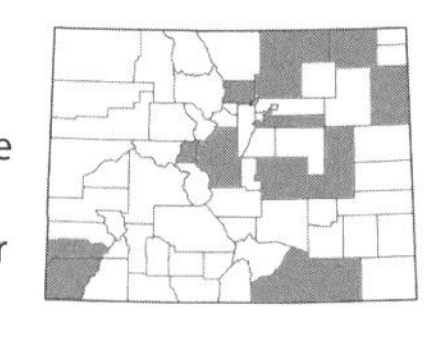

Astragalus falcatus **Lam.**, Russian milkvetch. Caulescent perennials, 4–7 dm; *leaves* pinnately compound with numerous leaflets; *leaflets* elliptic-oblong, 1–3 cm long; *inflorescence* exserted, numerous-flowered; *calyx* tube 3–3.5 mm long, lobes 0.5–1 mm long; *flowers* greenish-yellow, the banner 9–11 mm long; *legumes* oblong or lanceolate-falcate, triquetrous-compressed, 1.5–3 cm × 3–4 mm, strigulose. Uncommon along roadsides and in old gardens, apparently introduced for revegetation and as a forage crop, 3500–8000 ft. May–Aug. E/W. Introduced.

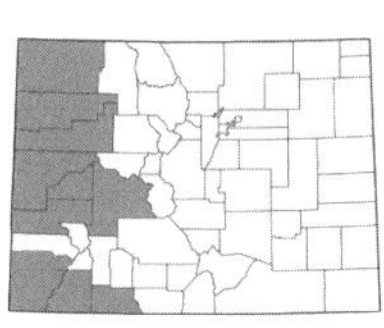

Astragalus flavus **Nutt.**, yellow milkvetch. [*A. confertiflorus* A. Gray]. Subacaulescent to caulescent perennials, to 2.5 dm; *leaves* pinnately compound with (7) 8–17 leaflets; *leaflets* narrowly elliptic to linear-oblong, often involute or folded, glabrate or hairy above; *inflorescence* an exserted, 10-numerous-flowered spicate raceme; *calyx* tube 3–5 mm long, lobes 1–4 (5) mm long; *flowers* yellow-white or pink-purple, the keel usually purple, the banner 9–15 (18) mm long; *legumes* ovoid to ellipsoid-oblong, obcompressed, 7–12 (14) × 4–5 mm strigulose or subvillous. Common on clay soil or shale, often with sagebrush or pinyon-juniper, 4700–7500 ft. April–June. W. (Plate 43)

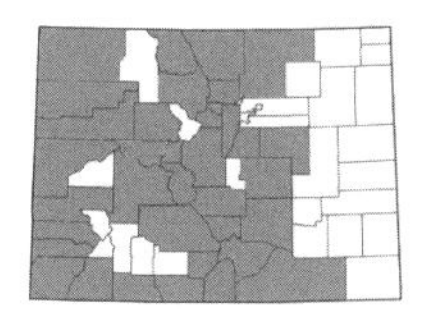

Astragalus flexuosus **Douglas ex G. Don**, flexible milkvetch. Caulescent perennials, (1) 1.5–6 dm; *leaves* pinnately compound with (9) 11–25 leaflets; *leaflets* linear, oblong, or obcordate, (0.2) 0.3–1.9 cm long, glabrous or hairy above, densely hairy below; *inflorescence* a loosely (7) 10–30-flowered raceme; *calyx* tube 2.4–4.2 mm long, lobes 0.5–2 mm long, strigulose-pilose with white, black, or mixed black and white hairs; *flowers* whitish to purple, the banner 7.2–11 × 4.4–7.8 mm; *legumes* linear-oblong, ovate-elliptic, or linear-oblanceolate to elliptic, (8) 11–24 mm long, 2.7–9 mm diam., sparsely to densely strigulose, villous, or glabrous. Common in meadows, grasslands, forest openings, and on sagebrush and pinyon-juniper slopes, 4500–10,300 ft. May–Aug. E/W. (Plate 43)

There are three varieties of *A. flexuosus* in Colorado:

1a. Fruit somewhat inflated or turgid, ovoid to ellipsoid, mostly 5–9 mm in diam., hairy; keel 7–8.2 mm...**var. *greenei* (A. Gray) Barneby**. Uncommon in dry places, reported for the northwestern counties. May–June. W.
1b. Fruit not inflated, oblong to oblanceolate, 2.5–4.8 mm in diam., glabrous or hairy; keel 5–7.5 mm long...2

2a. Fruit 10–15 mm long; keel 5–5.5 mm long; calyx tube 2.4–2.7 mm long...**var. *diehlii* (M.E. Jones) Barneby**. Found on open pinyon-juniper or sagebrush slopes and in rocky or shale soil, known from the westernmost counties (Mesa south to Montezuma), 5400–8000 ft. May–July. W.
2b. Fruit (10) 12–25 mm long; keel 5.3–7.5 mm long; calyx tube 2.7–4.5 mm long...**var. *flexuosus***. Common in meadows, grasslands, forest openings, and on dry slopes, 4500–10,300 ft. May–Aug. E/W.

Astragalus fucatus **Barneby**, Hopi milkvetch. Caulescent perennials, to 4 dm; *leaves* pinnately compound with 9–15 leaflets; *leaflets* linear to oblanceolate, (4) 6–12 (15) mm long, glabrate above; *inflorescence* 10–15-flowered; *calyx* tube 2–3 mm long, lobes 1–1.5 mm long, white-hairy; *flowers* red-purple to rose or bicolored, the banner 6.5–8 mm long; *legumes* ovoid, bladdery-inflated, slightly obcompressed, 2–3 × 1.2–2 cm, mottled, strigose. Uncommon in sandy, open places and washes, reported for Montezuma Co. near the Four Corners area, 4500–5500 ft. May–July. W.

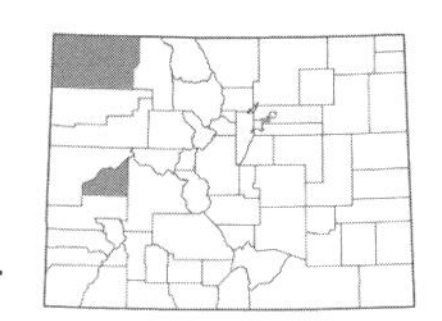

Astragalus geyeri **A. Gray var. *geyeri***, Geyer's milkvetch. Annuals, 0.2–2 dm; *leaves* pinnately compound with 5–9 (13) leaflets; *leaflets* elliptic, oblanceolate, or oblong, usually folded, 5–15 mm long, glabrate above; *inflorescence* 3–8-flowered; *calyx* tube 1.5–2.5 mm long, lobes 0.5–1.5 mm long; *flowers* white to whitish-purple, the keel slightly maculate, the banner 5–7.5 mm long; *legumes* half-ovoid or lunate-incurved, bladdery-inflated, 3-angled, 15–25 × 7–10 mm, strigulose. Uncommon in sandy soil, often with sagebrush or juniper, 5000–6000 ft. May–June. W.

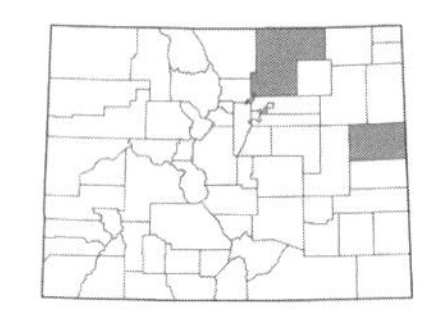

Astragalus gilviflorus **Sheldon**, plains milkvetch. [*A. triphyllus* Pursh; *Orophaca triphylla* (Eaton & Wright) Britton]. Acaulescent or subacaulescent perennials, caespitose to mounded; *leaves* ternately compound; *leaflets* spatulate or obovate-elliptic to oblong-lanceolate, 0.5–3 (3.5) cm long, silvery pubescent; *inflorescence* 1–3-flowered axillary clusters; *calyx* tube 6.5–15 mm long, lobes shorter than the tube; *flowers* white or blue-purple with a purplish-tipped keel, the banner 16–28 mm; *legumes* ovoid-ellipsoid, (6) 6.5–10 × 2–6.5 mm, densely strigose-hirsute. Locally common on barren hillsides and sandstone outcroppings, 4200–5800 ft. April–May. E.

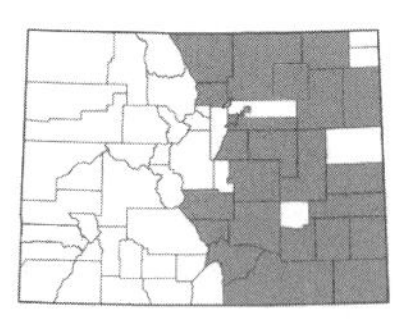

Astragalus gracilis **Nutt.**, slender milkvetch. Caulescent perennials, 1.5–4 dm; *leaves* pinnately compound with 9–17 leaflets; *leaflets* linear, linear-oblong, or oblanceolate, (0.3) 0.5–2 (2.5) cm long, glabrous to glabrate above; *inflorescence* a loosely (6) 12–40 (55)-flowered raceme; *calyx* tube 1.5–2.7 mm long, lobes 0.4–0.9 mm long, strigulose with white or a few black hairs; *flowers* pale lilac or purplish, the banner 5.3–8.4 × 4–5.8 mm; *legumes* ovate, obliquely suborbicular, or ovate-elliptic, (4) 4.3–8 (9) mm long, 2.2–3.6 mm diam., densely strigulose or minutely villous. Common on the shortgrass prairie, rocky slopes, and found occasionally in pinyon-juniper, 3500–6500 ft. May–July. E.

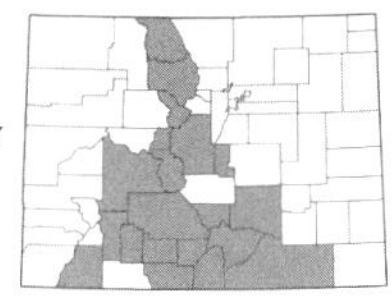

Astragalus hallii* A. Gray var. *hallii, HALL'S MILKVETCH. Ascending or decumbent-spreading perennials, rhizomatous; *leaves* pinnately compound with 11-numerous leaflets; *leaflets* elliptic-oblong or oblanceolate, 4–14 mm long, glabrate above; *inflorescence* 10-numerous-flowered; *calyx* tube 4.5–6.5 (7) mm long, lobes 1–2 (3.5) mm long; *flowers* purple to lilac or whitish, the banner 13–18 mm long; *legumes* oblong or ovoid to ellipsoid-oblanceolate, subterete to moderately obcompressed, 15–25 × (4) 6–12 mm, stipitate 1–2 mm, strigulose. Common in meadows and on dry slopes, often with sagebrush, 7200–10,800 ft. June–Aug. E/W.

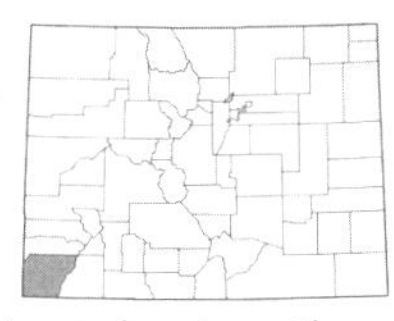

***Astragalus humillimus* A. Gray**, MANCOS MILKVETCH. Subacaulescent perennials, 0.5–2 cm; *leaves* pinnately compound with 7–11 leaflets; *leaflets* obovate to elliptic-oblong, folded, 0.8–2 mm long, cinereous with dolabriform hairs; *inflorescence* included, 1–3-flowered; *calyx* tube 2.2 mm long, lobes 0.8 mm long; *flowers* purplish, drying pink; *legumes* oblong-ellipsoid, laterally compressed, 4–5 × 2 mm, strigulose. Rare on sandstone ledges of mesas, usually with pinyon-juniper, 5500–6100 ft. April–June. W.

Although the type specimen of *A. humillimus* was collected by Brandegee in 1875, it was not located again until 1981. Listed as endangered under the Endangered Species Act.

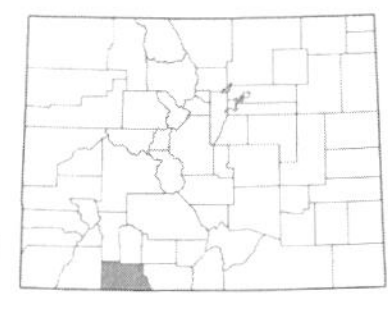

Astragalus humistratus* A. Gray var. *humistratus, GROUNDCOVER MILKVETCH. Mat-forming perennials, (1) 2–6 dm; *leaves* pinnately compound with (9) 11–17 leaflets; *leaflets* elliptic to oblong-lanceolate, 3–15 mm long, with dolabriform hairs, glabrate to glabrous above; *inflorescence* slightly or considerably exserted, (3) 6–many-flowered; *calyx* tube 2–4 mm long, lobes greater than or nearly equal to tube length; *flowers* greenish-white or ochroleucous, purple-striate, drying purple, the banner 9.5–12 mm long; *legumes* oblong-lanceolate, proximally 3-angled, dorsally obcompressed, 15–18 × 4–6 mm, often mottled, villous. Uncommon in rocky soil, reaching its northern limit near Pagosa Springs (Archuleta Co.), 7000–7500 ft. June–Aug. W.

It is probable that a hybrid between *A. humistratus* and *A. missouriensis* is present in southwestern Colorado. Plants in this area have characteristics of both species – the lower stipules are connate and the plant is strongly caulescent but prostrate as in *A. humistratus*. However the straight, usually glabrous or sparsely pubescent pod is more similar to *A. missouriensis*. The hybrid is here referred to as *A. missouriensis* Nutt. var. *humistratus* Isely.

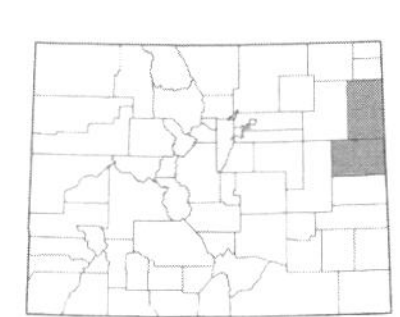

***Astragalus hyalinus* M.E. Jones**, SUMMER MILKVETCH. [*Orophaca hyalina* (M.E. Jones) Isely]. Perennials, shortly caulescent, mound-forming, 2–7 cm diam.; *leaves* ternately compound; *leaflets* oblanceolate, 0.3–1.2 cm long, strigose and silvery; *inflorescence* of 1–2 axillary flowers; *calyx* tube 6–6.5 mm long; *flowers* ochroleucous or bicolored, 12–17 mm long; *legumes* enclosed in the calyx, ovoid-ellipsoid, pilose. Uncommon on limestone bluffs, barren hillsides, and sandstone outcroppings, 3500–5500 ft. June–July. E.

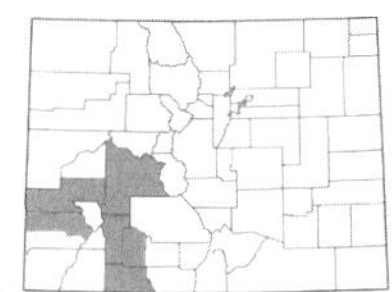

***Astragalus iodopetalus* (Rydb.) Barneby**, VIOLET MILKVETCH. Prostrate perennials; *leaves* pinnately compound with 17-numerous leaflets; *leaflets* obovate or elliptic to oblanceolate, 4–15 (20) mm long, glabrate above; *inflorescence* shortly racemose with 10-numerous flowers; *calyx* tube 7–10 mm long, lobes 3–5 mm long; *flowers* purple, the banner 18–21 mm long; *legumes* ovoid-acuminate to ellipsoid-lanceolate, proximally obcompressed, laterally compressed distally, (1.5) 2–3 cm × 8–10 mm, wrinkled, glabrous. Uncommon on dry hillsides, often with oak or sagebrush, 6000–8000 ft. May–June. W.

Similar in appearance to *A. shortianus* but with glabrous fruit and more numerous leaflets (17-numerous vs. 9–17 in *A. shortianus*).

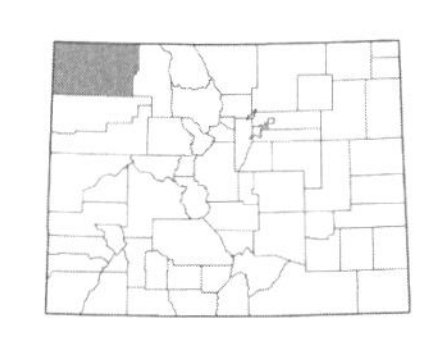

***Astragalus jejunus* S. Watson**, STARVELING MILKVETCH. Subacaulescent or shortly caulescent perennials, rarely exceeding 1 dm; *leaves* pinnately with 9–17 leaflets; *leaflets* linear or ovate to elliptic-lanceolate, 1–5 mm long, often folded; *inflorescence* exserted, 3–6-flowered; *calyx* tube 1.5–2 mm long, lobes 0.5–1 mm long; *flowers* pink, lavender, or bicolored, the banner 5–7 mm long; *legumes* subglobose to obovoid, bladdery-inflated, (0.8) 1–1.5 (1.8) × 0.8–1.2 cm, mottled, inconspicuously strigulose. Uncommon in sandy or clay soil, often with sagebrush or juniper, 6500–7000 ft. May–June. W.

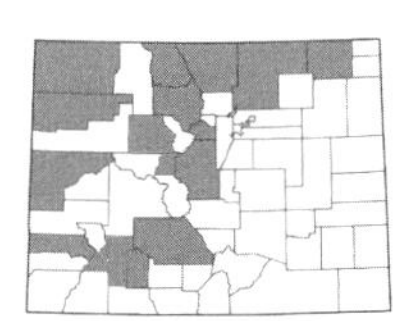

***Astragalus kentrophyta* A. Gray**, SPINY MILKVETCH. Perennials, caespitose or mat-forming, to 4 dm diam.; *leaves* pinnately to subpalmately compound, with 3–7 (9) leaflets, with dolabriform or basifixed hairs; *leaflets* elliptic-lanceolate or oblanceolate, flat to involute, 0.3–2 cm long, stiffly spine-tipped; inflorescence short, of 1–3 flowers; *calyx tube* 1.5–3 mm long, lobes 0.5–3.5 (5) mm long; *flowers* white or pink-purple, 4–9 mm long; *legumes* elliptic to lenticular-ovoid, laterally compressed or subterete, 4–9 × 1.5–4 mm. Locally common in sandy or rocky soil, on sandstone outcroppings, or sometimes with pinyon-juniper, 4000–10,800 ft. May–Aug. (Sept.). E/W.

Astraglus kentrophyta is quite unique with its spinose, confluent leaflets and few, small flowers. There are 3 varieties of *A. kentrophyta* in Colorado:

1a. Plants not mat-forming, with evident internodes (2–4 cm); hairs dolabriform; flowers whitish or rarely purplish-tinged…**var. *elatus* S. Watson**, TALL SPINY MILKVETCH. Found in sandy soil, often with pinyon-juniper, known from the western counties (San Miguel, Mesa, Rio Blanco, Moffat), 4500–7000 ft. May–Aug. (Sept.). W.

1b. Plants mat-forming or prostrate, with short internodes (0.5–1.5 cm); hairs basifixed or obscurely dolabriform below; flowers white or purplish-pink…2

2a. Flowers white, 4–5.5 mm long; leaflets with basifixed hairs; ovules 2–3…**var. *kentrophyta***, SPINY MILKVETCH. Locally common in sandy soil and on sandstone outcroppings, known from the high plains in the northeastern counties (Logan, Weld), 4000–5300 ft. May–June. E.

2b. Flowers purplish-pink or sometimes ochroleucous to whitish with a purplish keel tip, (4) 5–9 mm long; leaflets with some obscurely dolabriform hairs below; ovules 5–8…**var. *tegetarius* (S. Watson) Dorn**, MAT MILKVETCH. [var. *implexus* (Canby ex Porter & J.M. Coult.) Barneby]. Found in sandy or clay soil and on rocky slopes, known from the central counties (Clear Creek, Grand, Gunnison, Jackson, Larimer, Mineral, Park, Saguache), 8000–10,800 ft. June–Aug. E/W.

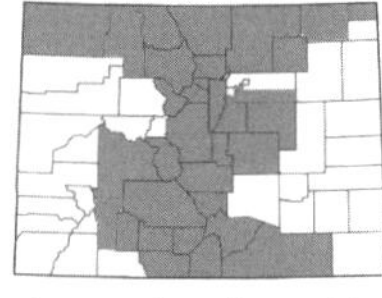

***Astragalus laxmannii* Jacq. var. *robustior* (Hook.) Barneby & S.L. Welsh**, PRAIRIE MILKVETCH. [*A. adsurgens* Pall. var. *robustior* Hook.; *A. striatus* Nutt.]. Caulescent perennials, 1–3 (4) dm; *leaves* pinnately compound with (9) 11–21 (25) leaflets; *leaflets* elliptic-lanceolate to oblong, 8–25 mm long, glabrate or hairy above; *inflorescence* subcapitate to densely spicate with numerous flowers; *calyx* tube 5–7 mm long, lobes 1.5–3 (4) mm long; *flowers* blue-lavender to purple, white, or ochroleucous, the banner 12–15 mm long; *legumes* ellipsoid to ellipsoid-lanceolate, 7–12 × 2.5–4 mm, closely strigulose. Common in sandy soil and on sandy outcroppings on the shortgrass prairie, in sagebrush, and in open forests and meadows in the mountains, 4000–10,500 ft. June–Aug. E/W. (Plate 43)

Astragalus canadensis in early anthesis can be easily mistaken for *A. laxmannii* as the flowers have not yet reflexed. However, *A. canadensis* stems arise singly or few together from woody rhizomes, while *A. laxmannii* stems are several from a woody taproot and shortly branching caudex. Both have bilocular, erect fruit, but the fruit of *A. canadensis* is terete while that of *A. laxmanii* is three-angled at the middle.

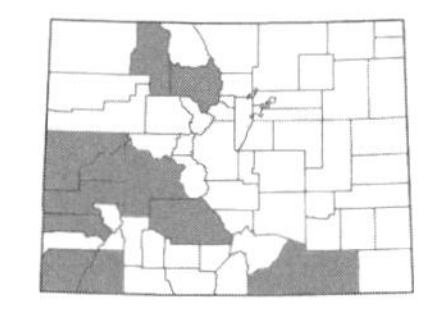

***Astragalus lentiginosus* Douglas ex Hook.**, FRECKLED MILKVETCH. Decumbent to erect perennials, 1–5 (10) dm; *leaves* pinnately compound with (3) 9-numerous leaflets; *leaflets* broadly ovate to narrowly elliptic-lanceolate, often folded; *inflorescence* included or exserted, subcapitate with 4-numerous flowers; *calyx* tube 2.5–7.5 mm long, lobes 1–4 mm long; *flowers* white, bicolored, or pink-purple, the banner 8–20 mm long; *legumes* subglobose, oblong, or ovoid to lanceolate, subterete to obcompressed, often mottled, glabrous to canescent. Found in rocky or sandy soil of dry places, 4300–8000 ft. April–June. E/W. (Plate 43)

Astragalus lentiginosus forms a very large species complex with numerous varieties. Isely (1998) even states that "no two people will identify [intermediates] in quite the same way." There are 3 varieties known from Colorado, which intergrade where their ranges overlap:

1a. Fruit turgid or merely scarcely inflated, linear-lanceolate to linear-oblong in profile, 4–7.5 mm in diam.; racemes sometimes greatly elongating in fruit, the axis (1) 2–14 cm long in fruit; flowers purplish…**var. *palans* (M.E. Jones) M.E. Jones**. The most common variety, found in rocky or sandy soil, canyons, in washes, and along roadsides, known from the westernmost counties, 4700–8000 ft. May–June. W.

1b. Fruit strongly inflated, plumply ovoid to subglobose or lance-acuminate in profile, 7–18 mm in diam.; racemes not or little elongating in fruit, the axis 1–4 (5) cm long in fruit; flowers whitish or purple…2

2a. Flowers purple; fruit usually purplish-mottled or red-tinged…**var. *diphysus* (A. Gray) M.E. Jones**. [*A. lentiginosus* var. *albiflorus* Schoener]. Found on dry slopes, barely entering Colorado near Mesa Verde (Montezuma Co.) and in Las Animas Co., 4300–7000 ft. April–June. E/W.

2b. Flowers whitish, the keel and wing tips sometimes tinged pink or purplish; fruit rarely mottled…**var. *chartaceus* M.E. Jones**. [*A. lentiginosus* var. *platyphyllidius* (Rydb.) M.E. Peck]. Found in sandy, dry places, 6500–7500 ft. April–May. W.

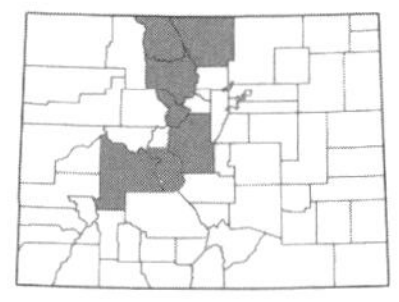

***Astragalus leptaleus* A. Gray**, PARK MILKVETCH. Caulescent perennials, 0.5–2 (3) dm; *leaves* pinnately compound with (9) 15–23 (27) leaflets; *leaflets* narrowly elliptic, lanceolate, or ovate, 0.3–1.5 cm long, glabrous above; *inflorescence* a loosely 2–5-flowered raceme; *calyx* tube 2.7–3.4 mm long, lobes 1.1–2.5 mm long, densely to thinly black or white-strigulose; *flowers* white, the keel tip dull bluish-purple, the banner 8.5–11.8 × 4.8–7.2 mm; *legumes* oblong-elliptic to lance-elliptic, obcompressed and bluntly 3-angled, 8–14 mm long, 2.5–4 mm diam., green to brown, sparsely black or white-strigulose. Locally common but easily overlooked in moist meadows, 7800–9800 ft. June–Aug. E/W.

Astragalus linifolius **Osterh.**, GRAND JUNCTION MILKVETCH. Caulescent perennials; *leaves* reduced to a filiform leafstalk, the lowermost with 1–2 confluent leaflets; *leaflets* linear-involute, slightly strigulose; *inflorescence* exserted, loosely 7–12-flowered; *calyx* tube ca. 5 mm long, lobes 1.5–2 mm long; *flowers* white with a purple-tipped keel, the banner ca. 15 mm long; *legumes* ellipsoid-oblong, laterally compressed, 1.2–1.5 cm × 4.5–6 mm, glabrous. Found in rocky soil and on dry hillsides, usually with pinyon-juniper, 5400–6500 ft. May–June. W. Endemic. (Plate 43)

Very close morphologically to *A. rafaelensis*, basically only differing in the erect fruit.

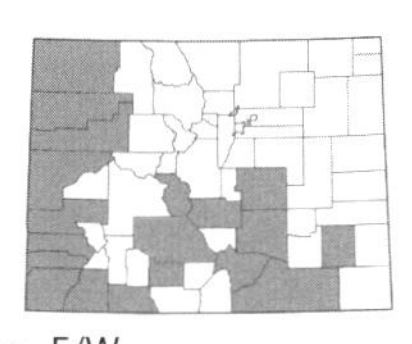

Astragalus lonchocarpus **Torr.**, RUSHY MILKVETCH. Caulescent perennials, rhizomatous, 3–9 dm; *leaves* pinnately compound with 3–9 leaflets, or phyllodial; *leaflets* filiform-involute or linear and folded, 0.5–2.5 cm long, strigulose; *inflorescence* densely flowered, often secund; *calyx* tube 6–8 mm long, lobes 1–2 (2.5) mm long, loosely strigose; *flowers* white but aging ochroleucous or dull orange, the banner 13–20 (23) mm; *legumes* narrowly oblong to slightly oblanceolate, obcompressed, 2.5–4 cm × 3.5–6 mm, stipitate 4–12 mm, sometimes mottled, glabrous. Found on shale slopes and barren, dry hillsides, often with oak or pinyon-juniper, 4300–8700 ft. May–June. E/W.

Astragalus hamiltonii Ced. Porter is closely related to *A. lonchocarpus* and occurs close to the Colorado border near Dinosaur National Monument in Moffat Co. It differs from *A. lonchocarpus* in having leaflets oblong-oblanceolate, flat, 4–8 (9) mm wide that are densely silvery pubescent above and greenish below.

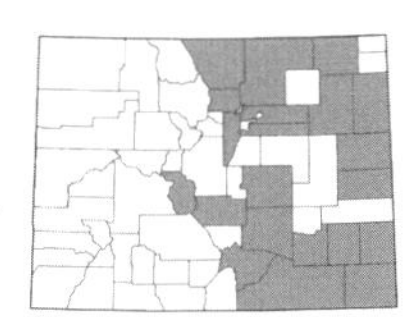

Astragalus lotiflorus **Hook.**, LOTUS MILKVETCH. Subacaulescent or caulescent perennials, 0.5–8 (12) cm; *leaves* pinnately compound with (5) 7–15 leaflets; *leaflets* elliptic to oblong, 0.5–2 cm long, cinereous with dolabriform hairs; *inflorescence* 3–15 (20)-flowered subcapitate racemes; *calyx* tube 3–4.5 mm long, lobes 2–3.5 (5) mm long; *flowers* ochroleucous, whitish and lavender-tinged, bicolored, or purple, the banner 4–14 mm long; *legumes* ovoid-acuminate, ellipsoid, or oblong-lanceolate, obcompressed or subterete, (1.2) 1.5–3 cm × (5) 6–8 mm, villous. Found in sandy or rocky soil, on sandstone bluffs, and on barren slopes, 3600–5800 (7900) ft. April–June. E.

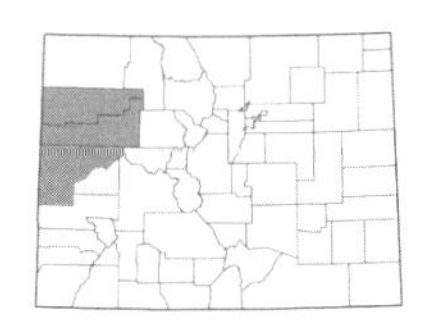

Astragalus lutosus **M.E. Jones**, DRAGON MILKVETCH. Prostrate perennials, shortly rhizomatous, 2–6 cm; *leaves* pinnately compound with (11) 15–27 leaflets; *leaflets* obovate or elliptic, reduced distally, 2–4 (9) mm long, cinereous; *inflorescence* included or exserted ,1–4-flowered; *calyx* tube 4–7 mm long, lobes 1–3 mm long, white or black-hairy; *flowers* white or pinkish with a maculate keel, the banner 8–16 mm long; *legumes* ellipsoid, bladdery-inflated, 1.5–3.5 cm × 8–15 mm, stipitate 1–4 mm, strigulose. Found on barren, shale slopes, 5800–8600 ft. May–Aug. W.

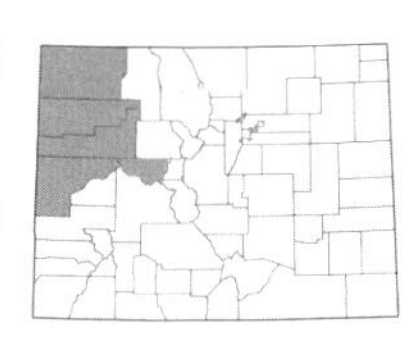

Astragalus megacarpus **(Nutt.) A. Gray**, GREAT BLADDERY MILKVETCH. Subacaulescent perennials, 0.5–5 cm; *leaves* pinnately compound with 9–17 leaflets; *leaflets* broadly obovate or elliptic, 5–15 mm long; *inflorescence* included, 2–5 (7)-flowered; *calyx* tube 6–8 (10) mm long, lobes 1.5–3 (5) mm long, strigulose; *flowers* white, bicolored, or pink-purple, drying violet, the banner 17–22 cm long; *legumes* ovoid-ellipsoid or ovoid-lanceolate, bladdery-inflated, subterete, 4–8 × 1.5–4 cm, often mottled, strigulose to glabrate. Locally common in dry washes or rocky slopes, usually in clay or shale, often with pinyon-juniper, 5800–7600 ft. April–May. W.

Resembling *A. oophorus* var. *caulescens* which has peduncles arising from the middle and upper leaf axils, a glabrous or thinly strigulose calyx, and bladdery-inflated fruit 2–5 cm long that is stipitate 4–10 mm. *Astragalus megacarpus* has a densely strigulose calyx and fruit 4–8 cm long that is stipitate 1.5–4 mm.

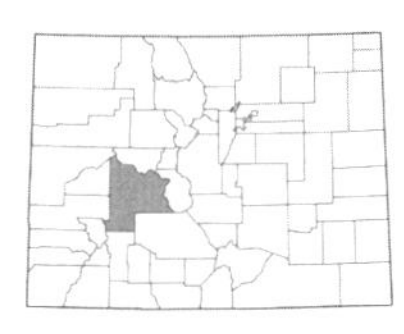

Astragalus microcymbus **Barneby**, SKIFF MILKVETCH. Caulescent perennials, 2.5–6 dm; *leaves* pinnately compound with 9–15 leaflets; *leaflets* oblong-obovate or obovate-cuneate, folded, 0.3–0.9 cm long, thinly hairy above, more densely hairy below; *inflorescence* a loosely (3) 7–14-flowered raceme; *calyx* tube 1.4–1.9 mm long, lobes 0.5–0.7 mm long, loosely strigulose with white and some grayish hairs; *flowers* lilac-tinged white, the keel dull purple-tipped, the banner 5.6–5.8 mm long; *legumes* ellipsoid or lance-ellipsoid, obcompressed, 6–9 mm long, (2.5) 3–3.3 mm diam., green with purple speckles, white-villous. Rare along roadsides and on dry, sagebrush slopes, 7500–8500 ft. May–July. W. Endemic.

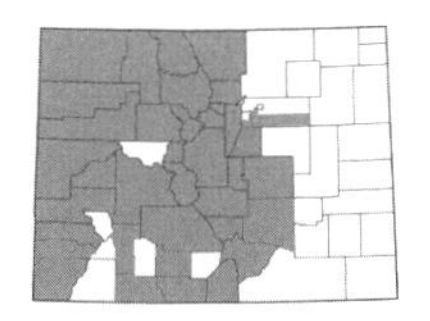

Astragalus miser **Douglas ex Hook. var.** ***oblongifolius*** **(Rydb.) Cronquist**, TIMBER MILKVETCH. Subacaulescent or caulescent perennials, 0.3–0.4 dm; *leaves* pinnately compound with (9) 11–19 (21) leaflets; *leaflets* linear to linear-elliptic, (3) 5–30 (42) mm long, glabrous above; *inflorescence* 3–15 (25)-flowered; *calyx* tube 2.2–3 mm long, lobes (0.8) 1–2.5 mm long; *flowers* whitish, occasionally with purple veins, the banner (5.9) 6.5–10 mm long; *legumes* oblanceolate, strigulose, sometimes purple-dotted or -tinged, but not mottled, (1.2) 1.5–2.5 cm × (1.9) 2.3–4 mm. Common in mountain meadows, dry or moist forests, alpine ridges, or on dry slopes with sagebrush or pinyon-juniper, 5200–11,300 ft. May–Aug. E/W.

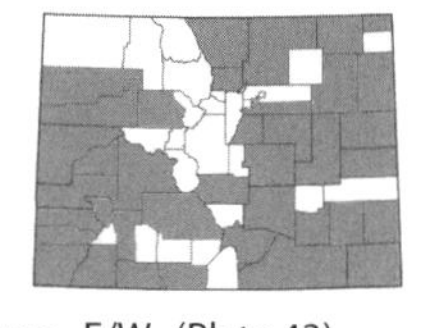

***Astragalus missouriensis* Nutt.**, Missouri milkvetch. Subacaulescent to caulescent perennials, 0.1–1 (2) dm; *leaves* pinnately compound with (7) 11–19 leaflets; *leaflets* obovate or elliptic, 5–15 mm, cinereous with dolabriform hairs; *inflorescence* included or exserted, subumbellate, 3–10 (13)-flowered; *calyx* tube (3.5) 6–9 (10) mm long, lobes (1) 1.5–4 mm long, purplish; *flowers* pink-purple, the banner (7.5) 14–25 (30) mm long; *legumes* ovoid-ellipsoid to oblong-lanceolate, subterete or obcompressed, or laterally compressed to almost quadrangular, (10) 12–25 (28) × (4) 6–9 mm, brown to black, hairy or glabrate. Common in sandy or rocky soil, 3500–8500 ft. April–June. E/W. (Plate 43)

There are three varieties of *A. missouriensis* in Colorado. *Astragalus missouriensis* var. *amphibolus* may hybridize with *A. amphioxys*, or perhaps the variety *amphibolus* is the result of past hybridization between *A. missouriensis* and *A. amphioxys*. The variety has the habit of *A. missouriensis* but the characteristic incurved pods of *A. amphioxys* except that they are not laterally compressed at the base.

1a. Lower stipules connate; plants prostrate and distinctly caulescent...**var. *humistratus* Isely**. Known from Archuleta, Hinsdale, and La Plata cos., 6200–8500 ft. May–June. W.

1b. Lower stipules free; plants erect or the longer stems somewhat decumbent, shortly caulescent...2

2a. Pods straight or nearly so...**var. *missouriensis***. Common in sandy or rocky soil of the eastern plains and foothills, scattered in the southwestern counties where it is largely replaced by var. *amphibolus*, 3500–7300 ft. April–June. E/W.

2b. Pods lunately incurved...**var. *amphibolus* Barneby**. Found in rocky soil and on dry slopes in the southwestern counties, sporadically occurring northward in Garfield Co., often with pinyon-juniper or sagebrush, 5000–8500 ft. May–June. W.

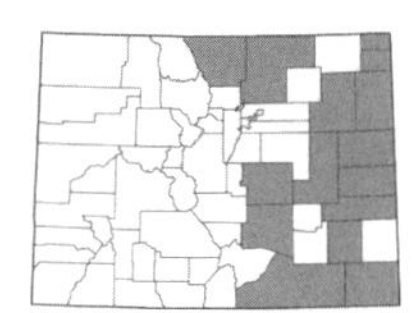

***Astragalus mollissimus* Torr.**, woolly locoweed. Subacaulescent to shortly caulescent perennials, 1.5–15 cm; *leaves* pinnately compound with 15-numerous leaflets; *leaflets* obovate, ovate, elliptic, or rhombic, tomentose; *inflorescence* included or exserted,15-numerous-flowered; *calyx* tube 5–10 mm long, lobes 2–5 mm long; *flowers* purple, bicolored (white-tipped), or ochroleucous, the banner 12–20 mm long; *legumes* ellipsoid, oblong-ellipsoid, or lanceolate, obcompressed to subterete, 10–25 × 4–8 (9) mm, glabrous or puberulent-tipped. Found on the shortgrass prairie, 3500–5700 ft. May–June. E.

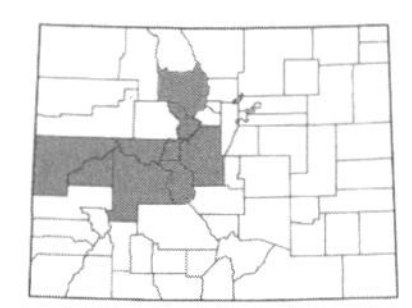

***Astragalus molybdenus* Barneby**, Leadville milkvetch. Loosely tufted or matted perennials, 1–6 (10) cm; *leaves* pinnately compound with (9) 17–25 leaflets; *leaflets* ovate, ovate-oblong, or elliptic, folded or involute, 0.2–1 cm long, thinly to densely strigulose-pilose with appressed and ascending hairs; *inflorescence* a loosely 2–4 (5)-flowered raceme; *calyx* tube 3–4.2 mm long, lobes 2–3 mm long, densely black-strigulose; *flowers* pink-purple, dull lilac, or whitish with lilac-tinged banner, the banner 10.7–12.5 × 5.2–7.2 mm; *legumes* obliquely ovoid or ovoid-ellipsoid, 7–11 mm long, 3.5 mm diam., densely black or rarely partly white-strigulose. Locally common on rocky alpine slopes and ridges, 11,000–13,500 ft. June–Aug. E/W.

***Astragalus monumentalis* Barneby var. *cottamii* (S.L. Welsh) Isely**, Cottam's milkvetch. [*A. cottamii* S.L. Welsh]. Subacaulescent or caulescent perennials, short-creeping and tufted; *leaves* pinnately compound with 9–15 (19) leaflets; *leaflets* obovate to oblong, 2–10 mm long; *inflorescence* exserted, subcapitate to racemose, 3–12-flowered; *calyx* tube 4.5–5.5 mm; *flowers* pink-purple, striate, the banner 8–16 mm; *legumes* oblong to lanceolate-attenuate, 3-angled-compressed, 1.5–3 cm × 2.5–4.5 mm, mottled, strigulose. Uncommon on mesas, 5400–6500 ft. April–May. W.

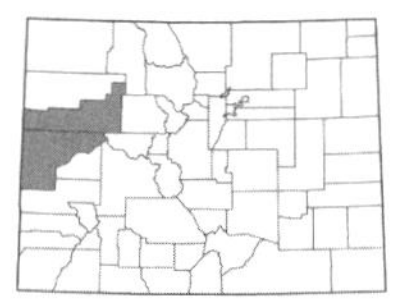

***Astragalus musiniensis* M.E. Jones**, Ferron's milkvetch. Acaulescent perennials, to 1 dm; *leaves* ternately or pinnately compound with (1) 3–5 leaflets; *leaflets* elliptic, rhombic, or oblanceolate, 1–2 (3) cm long; *inflorescence* included or exserted, subumbellate, 2–4 (6)-flowered; *calyx* tube 9–12 mm long, lobes 2–3 mm long, dark hairy; *flowers* lilac to purple, the banner 20–25 mm long; *legumes* broadly ovoid-lanceolate or ellipsoid, obcompressed, 2–3.5 × 1–1.5 cm, thinly pilose. Uncommon on dry hillsides, bluffs, usually on shale or sandstone, often with pinyon-juniper or desert shrubs, 4600–7000 ft. April–June. W.

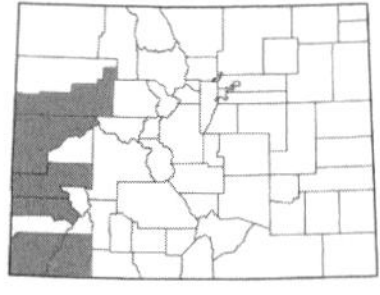

***Astragalus naturitensis* Payson**, Naturita milkvetch. Subacaulescent perennials, tufted; *leaves* pinnately compound with 9–15 leaflets; *leaflets* obovate to elliptic, mostly folded, 2–7 mm long; *inflorescence* subcapitate or racemose, 4–9-flowered; *calyx* tube 4–6 mm long, lobes 1–1.5 mm long, mixed white and black hairs; *flowers* bicolored, the banner whitish-striate, the keel and wings purple or purple-tipped, the banner 12–15 mm long; *legumes* narrowly ellipsoid to oblong-lanceolate, obcompressed to obcompressed-3-angled, 1.2–2 cm × (3) 4–6 mm, mottled, strigulose. Uncommon on sandstone ledges and rimrock, occasionally on sandy slopes with pinyon-juniper, 5000–7000 ft. April–June. W.

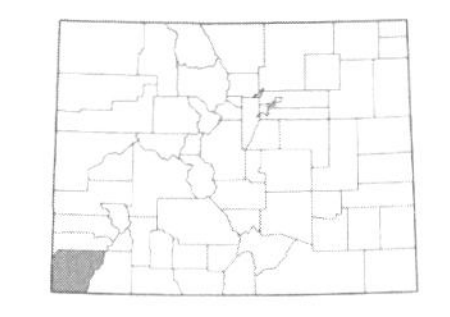

Astragalus newberryi **A. Gray var.** ***newberryi***, Newberry's milkvetch. Perennials; *leaves* ternately or pinnately compound with 3–5, rarely 1–3, leaflets; *leaflets* ovate or elliptic-obovate, 5–20 mm; long *inflorescence* included or exserted, subcapitate or shortly racemose, 2–6 (8)-flowered; *calyx* tube 6–13 mm long, lobes 1.5–5 (6) mm long, black to light hairs; *flowers* pink-purple to bicolored, the banner 15–30 mm long; *legumes* obliquely semi-ovoid to oblong, obcompressed, laterally compressed apically, (1) 1.5–2.5 (3) cm × 7–10 mm, villous and tomentose with white to tawny hairs. Uncommon on shale slopes, bluffs, and badlands, 4900–6100 ft. April–June. W.

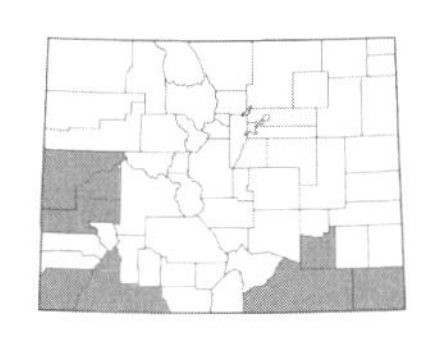

Astragalus nuttallianus **DC. var.** ***micranthiformis*** **Barneby**, turkeypeas. Ascending to prostrate annuals, 0.3–4 dm; *leaves* pinnately compound with (7) 9–19 leaflets; *leaflets* obovate-cuneate, elliptic, or oblong, 2–10 (15) mm long; *inflorescence* subcapitate or shortly racemose, 1–8-flowered; *calyx* tube 2–3 mm long, lobes 1.2–2.5 mm long, loosely strigulose with black, white, or mixed black and white hairs; *flowers* bright purple with a pale-eyed banner, the banner 4–7.5 mm long; *legumes* oblong to linear-oblanceolate, ultimately subterete, 1–2.5 cm × 2–3.5 mm, greenish to dark brown or black, glabrous or strigulose. Uncommon on shale and gravelly slopes, usually with pinyon-juniper, 4300–6700 ft. April–May. E/W.

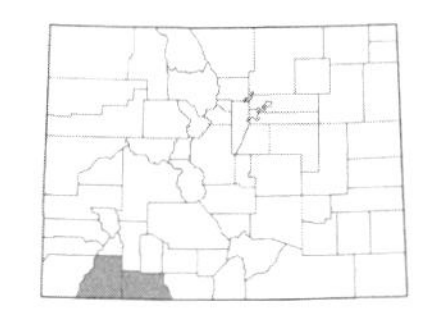

Astragalus oocalycis **M.E. Jones**, arboles milkvetch. Caulescent perennials, 4–6 (8) dm; *leaves* pinnately compound with numerous leaflets; *leaflets* oblong to linear, 1.5–3.5 cm long; *inflorescence* exserted, spicate, 20-numerous-flowered; *calyx* tube 8–9 mm long, lobes ca. 2 mm long, villous in flower; *flowers* ochroleucous, the banner 14–16 mm long; *legumes* half-ellipsoid or boat-shaped, obcompressed, 5–7 × 3.5 mm, glabrous. Uncommon on shale slopes and dry knolls, 4600–7000 ft. June–July. W.

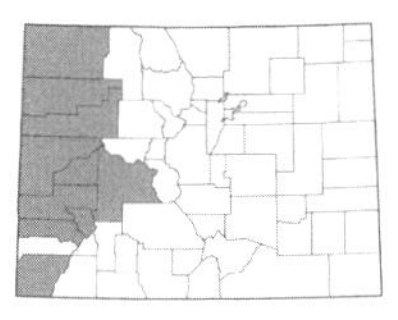

Astragalus oophorus **S. Watson var.** ***caulescens*** **(M.E. Jones) M.E. Jones**, egg milkvetch. Caulescent perennials, 1–3 dm; *leaves* pinnately compound with (7) 9–19 leaflets; *leaflets* suborbicular, elliptic, or elliptic-oblong, 4–21 mm long, margins with a few appressed hairs; *inflorescence* a 5–10-flowered; *calyx* tube 4–9 mm long, lobes 2–3 (4) mm long, glabrous or with few black hairs; *flowers* ochroleucous to lemon-yellow or purple and white-tipped, the banner 11–19 mm long; *legumes* ovoid-ellipsoid, bladdery-inflated, 2–5 × 1–2 (2.5) cm, stipitate 4–10 mm, mottled, glabrous. Common on dry slopes, often with sagebrush or pinyon-juniper, 5200–8500 ft. May–June. W.

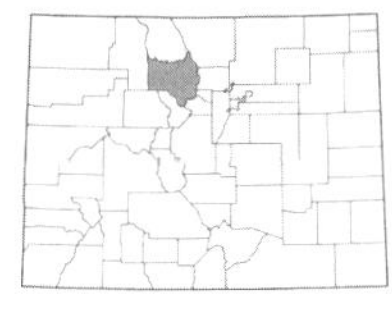

Astragalus osterhoutii **M.E. Jones**, Osterhout's milkvetch. Erect or ascending perennials, 3–4 (5) dm; *leaves* pinnately compound with 7–13 (15) leaflets; *leaflets* narrowly oblanceolate to linear, with elevated margins, 1 (4) cm long; *inflorescence* loosely 12–25-flowered; *calyx* tube 7–10 mm long, lobes 0.8–2 mm long, strigulose with black or mixed black and white hairs; *flowers* white, the banner 17–22 mm long; *legumes* oblong-lanceolate, laterally compressed and flat, 2–4 cm × 4 mm, stipitate 2–6 mm, glabrous. Uncommon on clay or shale, seleniferous slopes, known from the Muddy and Troublesome Creek drainages in Grand Co., 7400–7900 ft. June–Aug. W. Endemic.

Listed as endangered under the Endangered Species Act.

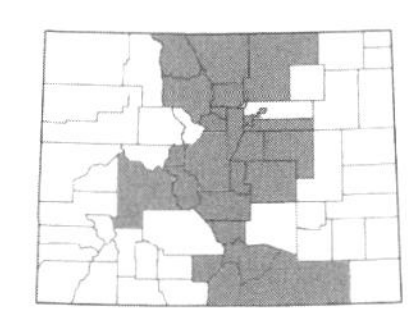

Astragalus parryi **A. Gray**, Parry's milkvetch. Acaulescent or subacaulescent perennials, 1–20 cm; *leaves* pinnately compound with 4–12 (14) leaflets; *leaflets* elliptic to broadly ovate, 4–15 mm long; *inflorescence* 4–8-flowered; *calyx* tube 4–6 (7) mm long, lobes (2.5) 4–5 mm long; *flowers* white or ochroleucous, or lavender with a maculate keel, the banner 15–20 mm long; *legumes* lunate-lanceolate, obcompressed, (2) 2.5–3 cm × 8–10 mm, villous. Common in grasslands, rocky or sandy slopes, dry meadows, and forest openings, 5000–10,000 ft. May–July. E/W. (Plate 43)

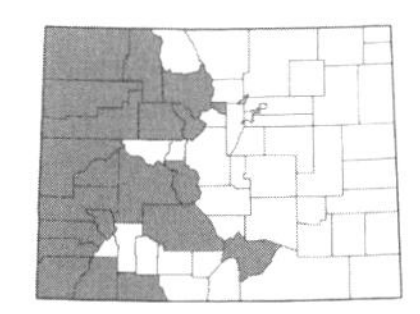

Astragalus pattersonii **A. Gray**, Patterson's milkvetch. Erect perennials, clump-forming, to 7 dm; *leaves* pinnately compound with 13-numerous leaflets; *leaflets* elliptic to oblong, 0.5–3 cm long, glabrous above; *inflorescence* exserted, numerous-flowered; *calyx* tube 6.5–9 mm long, lobes 2–4 (5) mm long, white or mixed white- and black-hairy; *flowers* white, aging yellowish, the banner 15–19 mm long; *legumes* ellipsoid to ellipsoid-lanceolate, subterete or slightly obcompressed, 1.5–2.3 cm × 6–8 (12) mm, glabrate. Locally common on clay or shale slopes and grassy hillsides, 5000–8500 ft. May–July. E/W. (Plate 43)

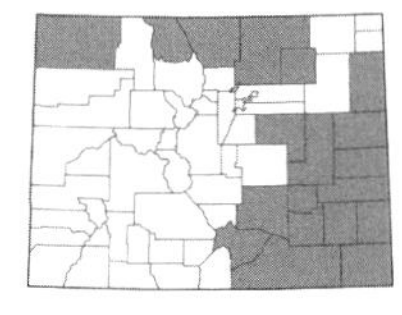

Astragalus pectinatus **(Douglas ex Hook.) Douglas ex G. Don**, narrowleaf milkvetch. Decumbent to ascending perennials, to 6 dm; *leaves* pinnately compound with 11–17 leaflets; *leaflets* linear or filiform, often involute, 1.5–5 cm long; *inflorescence* 10-numerous-flowered; *calyx* tube 6–9 mm long, lobes 1–3 mm long, often black-hairy; *flowers* ochroleucous, the banner 18–22 (30) mm long; *legumes* ovoid-ellipsoid to oblong, subterete or slightly obcompressed, 12–18 (24) × 5–7 mm, glabrous. Common in grasslands and dry places, 3500–8200 ft. May–June. E/W. (Plate 43)

There are two varieties of *A. pectinatus* in Colorado:

1a. Leaflets usually filiform to linear, usually involute, rarely wider than 2 mm; flowers 18–22 (24) mm long...**var. *pectinatus***. Common in sandy or clay soil of grasslands, common on the eastern plains and also found in northern Jackson Co. at 8200 ft, 3500–5300 (8200) ft. May–June. E.

1b. Leaflets narrowly oblong-lanceolate, not involute; flowers 24–30 mm long...**var. *platyphyllus* M.E. Jones**. [*A. nelsonianus* Barneby]. Uncommon in sandy soil, known from northern Moffat Co., 6500–7000 ft. May–June. W.

***Astragalus piscator* Barneby & S.L. Welsh,** FISHER MILKVETCH. Acaulescent or subacaulescent perennials; *leaves* pinnately compound with 5–11 (13) leaflets; *leaflets* elliptic to lanceolate, 6–14 (20) mm long, with dolabriform hairs; *inflorescence* included or exserted, 4–10-flowered; *calyx* tube 8–10 mm long, lobes 2–3.5 (4) mm long, strigose with black or mixed black and white hairs; *flowers* lavender or lilac, the banner 16–24 mm long; *legumes* lanceolate, laterally compressed, 2.5–3.5 cm × 8–12 mm, mottled at maturity, strigose. Uncommon in rocky or clay soil, known from western Mesa Co. near Gateway, 4400–5500 ft. April–June. W.

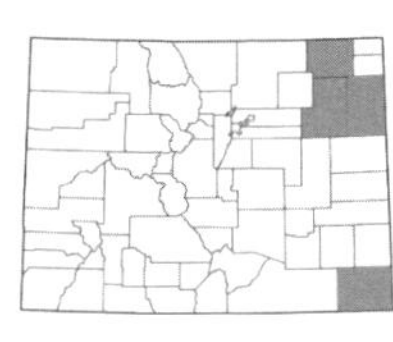

***Astragalus plattensis* Nutt.,** PLATTE RIVER MILKVETCH. Decumbent or ascending perennials, 1–2 (3) dm; *leaves* pinnately compound with 13-numerous leaflets; *leaflets* obovate to oblong, 4–15 mm long, glabrous above; *inflorescence* slightly exserted, subcapitate to shortly racemose, 6–12-flowered; *calyx* tube 6–8 mm long, lobes 2.5–3.5 (5) mm long, with white or mixed black and white hairs; *flowers* light pink-purple or lavender and purple-tipped, pale when dried, the banner 14–19 mm long; *legumes* ovoid or ellipsoid, inflated, subterete or obcompressed, (1) 1.2–3 × 1–1.3 cm, reddish on upper sides, villous. Uncommon in sandy soil of the eastern plains, 3500–4500 ft. April–June. E.

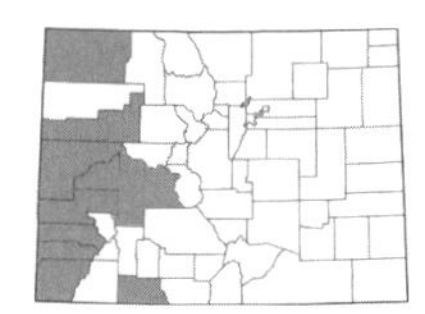

***Astragalus praelongus* Sheldon,** STINKING MILKVETCH. Clump-forming perennials, to 1 m; *leaves* pinnately compound with 13-numerous leaflets; *leaflets* cuneate-obovate, oblong-elliptic, or narrowly oblong, 1–3.5 cm long; *inflorescence* well-exserted with 10–30 flowers; *calyx* tube 4–6.5 mm long, lobes 1.3–5 (6) mm long, glabrate; *flowers* ochroleucous to yellowish, the keel sometimes maculate, the banner 15–22 mm long; *legumes* broadly ovoid-ellipsoid to narrowly oblong or oblanceolate, subterete, 2–3.5 × 0.6–1.5 cm, hairy to glabrous. Locally common on shale or clay slopes, dry hillsides, badlands, and along roadsides, sometimes with pinyon-juniper, 4500–6900 ft. April–June. W.

There are two varieties of *A. praelongus* in Colorado. The previously recognized var. *ellisae* (Rydb.) Barneby is included within var. *praelongus*:

1a. Fruit sessile or shortly stipitate to 3 mm...**var. *praelongus***

1b. Fruit stipitate 4–8 mm...**var. *lonchopus* Barneby**. Not reported for Colorado yet, but probably occurring near the Four Corners area.

***Astragalus preussii* A. Gray,** PREUSS' MILKVETCH. Perennials, (0.5) 1–3 (4) dm, glabrate; *leaves* pinnately compound with (9) 11-numerous leaflets; *leaflets* narrowly obovate to oblong-oblanceolate, 0.5–2.5 cm long; *inflorescence* exserted with 6-numerous ascending flowers; *calyx* tube 6–10 mm long, dark purple to pinkish, usually black-strigulose, lobes 1.5–2.5 mm long; *flowers* pink-purple to purple, 15–22 mm long; *legumes* stipitate 2–7 mm, inflated, stiffly papery or coriaceous, ovoid to ellipsoid, 1.5–3 cm × 8–13 mm, glabrous or minutely hairy. Uncommon on rocky, dry slopes and in desert valleys, 4500–4600 ft. April–May. W.

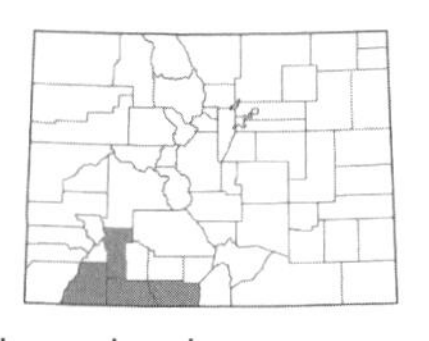

***Astragalus proximus* (Rydb.) Wooton & Standl.,** AZTEC MILKVETCH. Caulescent perennials, 1.5–4.5 dm; *leaves* pinnately compound with 7–11 leaflets; *leaflets* narrowly linear, linear-oblanceolate, or filiform, 0.6–2.2 cm long, glabrous above, thinly strigulose below; *inflorescence* very loosely (7) 12–40-flowered; *calyx* tube 1.8–2.5 mm long, lobes 0.6–1.1 mm long, strigulose with black, mixed black and white, or rarely all white hairs; *flowers* whitish, the banner and keel tip purplish, the banner 6–7 × 3.5–4.8 mm; *legumes* linear-ellipsoid, 10–15 mm long, 2.3–3.2 mm diam., stipitate 1.2–2 mm, greenish to greenish-stramineous, glabrous. Uncommon on dry slopes, sometimes with sagebrush or junipers, 5400–8700 ft. April–July. W.

Resembling a small-flowered *A. flexuosus*, but the leaflets are usually fewer, the fruit with a longer stipe, and the plants more erect than those of *A. flexuosus*.

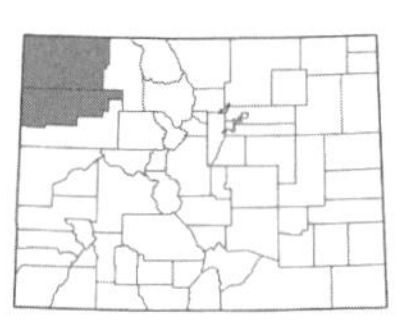

***Astragalus pubentissimus* Torr. & A. Gray,** GREEN RIVER MILKVETCH. Annuals or short-lived perennials, 5–20 cm; *leaves* pinnately compound with 7–15 leaflets; *leaflets* obovate-elliptic to oblong, often folded, 3–12 (15) mm long; *inflorescence* exserted or included, 4–10 (12)-flowered; *calyx* tube 3–4 mm long, lobes 2–3 mm long; *flowers* pink-purple, or rarely white with a maculate keel, the banner 9–12 mm long; *legumes* narrowly ellipsoid and lunately incurved, subterete, 1–2 cm × 4–8 mm, loosely villous. Uncommon on sandstone ridges and in washes, 5200–6000 ft. May–June. W.

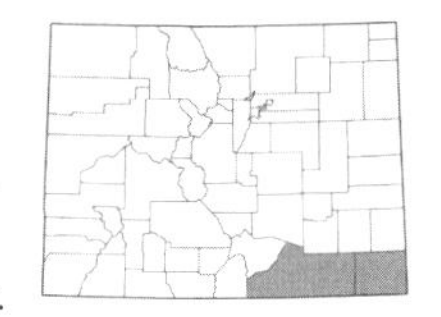

Astragalus puniceus* Osterh. var. *puniceus, Trinidad milkvetch. Decumbent or assurgent perennials, 1–5 dm; *leaves* pinnately compound with 11-numerous leaflets; *leaflets* obovate, elliptic, or elliptic-oblanceolate, 5–15 mm long; *inflorescence* included or shortly exserted, (5) 10–20-flowered; *calyx* tube 5–8 mm long, lobes (1) 2–3 mm long; *flowers* pink-purple, the banner 13–19 mm long; *legumes* oblong to ellipsoid-lanceolate, terete or obcompressed, 1.5–2.5 cm × 5–9 mm, often mottled, villous. Uncommon on dry, rocky slopes and in grasslands, 5000–6500 ft. May–July. E.

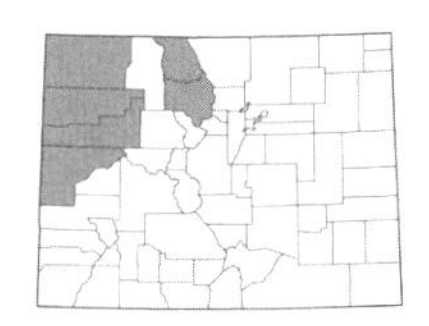

Astragalus purshii* Douglas ex Hook. var. *purshii, woollypod milkvetch. Acaulescent to shortly caulescent perennials, 0.1–1 (2) dm; *leaves* pinnately compound with (5) 7–19 leaflets; *leaflets* broadly obovate to elliptic-oblong or oblanceolate, 1–10 (15) mm long; *inflorescence* included or exserted, subumbellate to shortly racemose, 3–10-flowered; *calyx* tube 9–13 mm long, lobes 1–6 mm long, purple, white, or mixed white and dark hairs; *flowers* ochroleucous or whitish with a maculate keel, the banner 19–28 mm long; *legumes* obliquely ovoid-acuminate, ovoid-lunate, or reniform, 1.3–2.5 cm, densely white to tawny-tomentose or densely villous. Found on badlands and dry hillsides, often with sagebrush or pinyon-juniper, 5500–8000 ft. April–May. E/W.

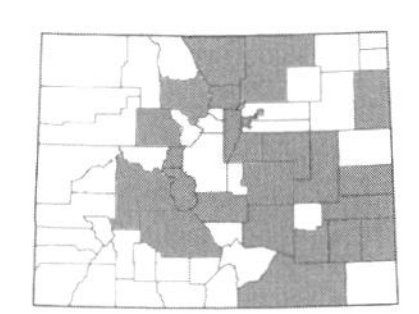

***Astragalus racemosus* Pursh**, cream milkvetch. Clump-forming perennials, to 5 (6) dm; *leaves* pinnately compound with 11-numerous leaflets; *leaflets* broadly elliptic to oblong or linear, 1–3 cm long, glabrate above, glabrate or strigose below; *inflorescence* exserted, numerous-flowered; *calyx* tube 5–9 mm long, lobes 2–6 (10) mm long; *flowers* yellowish-ochroleucous to white, occasionally pink-tipped or with a maculate keel, the banner 13–20 mm long; *legumes* oblong, 3-angled, (10) 15–30 × 3–7 mm, stipitate 3–5 mm, glabrous to strigulose. Common on the shortgrass prairie and grassy slopes, 3500–7900 ft. May–June. E/W.

There are possibly two varieties of *A. racemosus* in Colorado:

1a. Dorsal face deeply sulcate; calyx sparsely to conspicuously pubescent with white hairs; plants of the eastern slope...**var. *racemosus***. Range and habit as given above.

1b. Dorsal face shallowly sulcate or flat; calyx glabrous or sparsely pubescent with black or white hairs; plants of the western slope...**var. *treleasei* Ced. Porter**, Trelease's milkvetch. Not reported for Colorado but occurs close to the border of Moffat Co. in adjacent Utah.

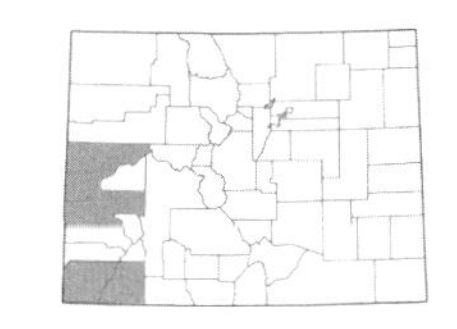

***Astragalus rafaelensis* M.E. Jones**, San Rafael milkvetch. Erect or ultimately drooping perennials, 4–5 dm; *leaves* reduced and simple, or pinnately compound with 2–4 leaflets; *leaflets* linear and involute; *inflorescence* (5) 6–12-flowered; *calyx* tube 5 mm long, lobes 0.8–1 mm long, glabrate to strigose, ciliate lobes; *flowers* bicolored, usually pink-purple and white-tipped or white with a maculate keel, the banner 17–20 mm long; *legumes* ellipsoid to oblong, laterally compressed, 12–20 (25) × 5–7 mm, glabrous. Found on rocky slopes, 4500–6000 ft. April–May. W.

***Astragalus ripleyi* Barneby**, Ripley's milkvetch. Erect perennials, 4–10 dm; *leaves* pinnately compound with 13–17 leaflets; *leaflets* linear-oblong, 8–30 mm long; *inflorescence* exserted, numerous-flowered; *calyx* tube 5–6.5 mm long, lobes 0.5–1 mm long, mixed dark and light strigulose; *flowers* yellow, the banner 14–17 mm long; *legumes* oblong-lanceolate, laterally compressed and flat, 2–3 cm × 4–6 mm, stipitate 8–15 mm, sparsely hairy. Uncommon in meadows and in grasslands along the edges of pine woodlands, 8200–9300 ft. June–July. E.

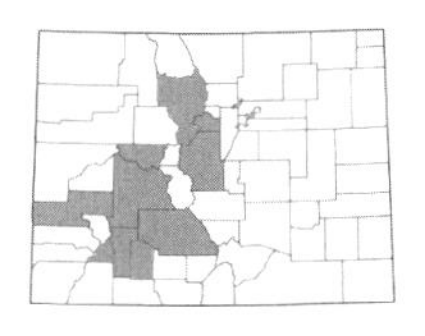

***Astragalus robbinsii* (Oakes) A. Gray var. *minor* (Hook.) Barneby**, Robbins' milkvetch. Caulescent perennials, (0.7) 1–4.5 (6) dm; *leaves* pinnately compound with (5) 7–17 leaflets; *leaflets* lance-oblong, oblong-elliptic, or ovate, (0.3) 0.5–3.2 cm long, glabrous to glabrate above, strigulose to pilose-villous below; *inflorescence* a (3) 6–21 (25)-flowered raceme; *calyx* tube (2.4) 2.6–4.5 mm long, lobes 0.8–3.4 mm long, thinly to densely strigose with black, mixed black and white, or rarely all white hairs; *flowers* pale purple with whitish claws, rarely whitish with a purple keel tip, the banner 7–12 × 3.8–7 mm; *legumes* ellipsoid to oblong-ellipsoid, 8–25 mm long, sparsely to densely pilose with black or fuscous hairs, occasionally mixed with white hairs. Uncommon on alpine ridges and scree slopes, mountain meadows, and along streams, 8500–13,000 ft. July–Aug. E/W.

***Astragalus saurinus* Barneby**, Dinosaur milkvetch. Caulescent perennials, 1.5–3 dm; *leaves* simple and phyllodial or pinnately compound with up to 9 leaflets; *leaflets* linear to filiform-involute, 1–2 cm long, cinereous to glabrate; *inflorescence* an exserted, 5–15-flowered raceme; *calyx* tube 5–7 mm long, lobes 1–3 mm long, villous to substrigose; *flowers* bicolored with pink-purple banner and keel and white wings, the banner 18–23 mm long; *legumes* oblong to oblanceolate, laterally compressed, 2–3.5 cm × 4.5–6 mm, sessile or substipitate to 1 mm, finely strigulose. Not reported for Colorado but occurs close to the border in Dinosaur National Monument in Uintah County, to be sought after in the White River Valley of western Rio Blanco Co. in sandy clay soil of badlands and canyons.

***Astragalus schmollae* Ced. Porter**, Schmoll's milkvetch. Erect perennials, to 5 dm; *leaves* pinnately compound with 11–15 (17) leaflets; *leaflets* oblong to linear, 6–20 mm long, cinereous, strigulose with basifixed hairs; *inflorescence* exserted, 10-numerous-flowered; *calyx* tube 4.5–6 mm long, lobes 1–1.5 mm long, black strigulose; *flowers* ochroleucous, the banner 15–17 mm long; *legumes* oblong, obcompressed, 2.5–3.5 (4) cm × 4–5 mm, stipitate 5–10 mm, strigulose. Rare on sandy or rocky soil, often with pinyon-juniper, known from Mesa Verde National Park, 6800–7000 ft. May–June. W. Endemic.

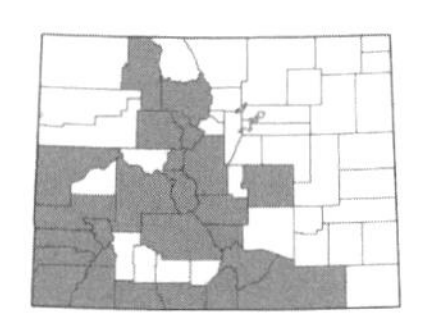

***Astragalus scopulorum* Porter**, Rocky Mountain milkvetch. Caulescent perennials, 1.5–4.5 dm; *leaves* pinnately compound with (7) 15–29 (35) leaflets; *leaflets* oval-oblong to linear-elliptic or oblanceolate, (0.2) 0.4–1.8 cm long, glabrous above, thinly strigulose below; *inflorescence* a (5) 10–22-flowered raceme, secund; *calyx* tube 6.5–8.5 mm long, lobes 1.5–3.5 (6) mm long, strigulose with black or occasionally with some or all white hairs; *flowers* ochroleucous, sometimes the keel tip slightly maculate, the banner 18.3–23.5 × 6–8.5 mm; *legumes* linear-oblong to linear-oblanceolate, (1.8) 2.5–3.5 cm long, 3–6.5 mm diam., stipitate 4–9 mm, green, purple-speckled, greenish-stramineous, or brown, glabrous. Common in sagebrush, pinyon-juniper, or occasionally in aspen forests, 6300–11,000 ft. May–July. E/W.

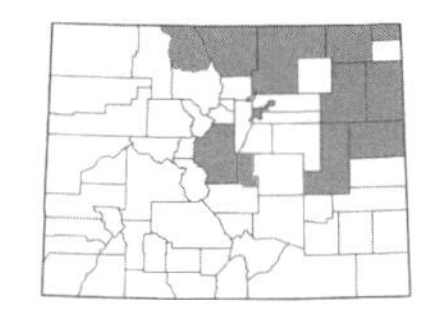

***Astragalus sericoleucus* A. Gray**, silky milkvetch. [*Orophaca sericea* (Nutt.) Britton]. Caulescent perennials, prostrate; *leaves* ternately compound; *leaflets* 2–6 (10) mm long, silvery pubescent; *inflorescence* exserted, 2–4 (6)-flowered; *calyx* tube 2–2.5 mm long, lobes shorter than the tube; *flowers* light or bright pink-purple, the banner 5–7 mm long; *legumes* ovoid-acuminate to lanceolate, laterally compressed, 4–6.5 mm long, chartaceous, villous. Common on sandstone outcroppings, bluffs, and rocky openings in the prairie, 3600–5500 ft. April–June. E. (Plate 44)

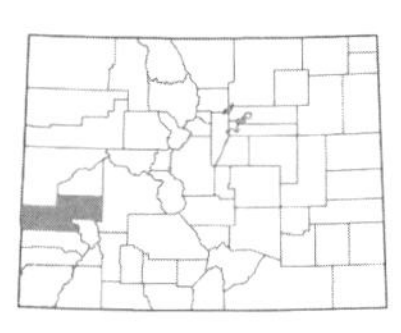

***Astragalus sesquiflorus* S. Watson**, sandstone milkvetch. Subacaulescent perennials, 3–10 cm; *leaves* pinnately compound with 7–11 leaflets; *leaflets* obovate to narrowly elliptic, 2–6 (10) mm long, villous with dolabriform hairs; *inflorescence* included or slightly exserted, 1–3-flowered; *calyx* tube 1.5–2.5 mm long, lobes 2–3 mm long; *flowers* light purple, the keel maculate, the banner 6–8 mm long; *legumes* half-ovoid to oblanceolate-incurved, slightly inflated, laterally compressed dorsally, 6–10 × 3–4 mm, often mottled, strigulose-puberulent. Uncommon in rocky soil and rock crevices, usually with pinyon-juniper, barely entering Colorado in Montrose Co. near Hwy 141 along the Dolores River, 4800–5500 ft. May–July. W.

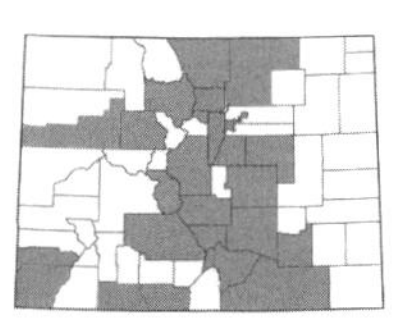

***Astragalus shortianus* Nutt.**, Short's milkvetch. Acaulescent to shortly caulescent perennials; *leaves* pinnately compound with 9–17 leaflets; *leaflets* elliptic, obovate, or rhombic-obovate, 0.5–2.5 cm long; *inflorescence* shortly racemose, 6–15-flowered; *calyx* tube 9–15 mm long, lobes 2–4 mm long, villous; *flowers* pink-purple, the banner 18–23 mm long; *legumes* ovoid-acuminate or lanceolate, obcompressed, 2.5–4 (4.5) × 1–1.3 cm, minutely hairy. Common in grasslands, rocky slopes, meadows, and forest openings, 4500–9800 ft. April–June. E/W. (Plate 44)

Sometimes misidentified as *A. missouriensis*, which has dolabriform hairs, usually straight, shorter fruit (12–25 mm long), and a usually dark, purple-tinged calyx with some black hairs.

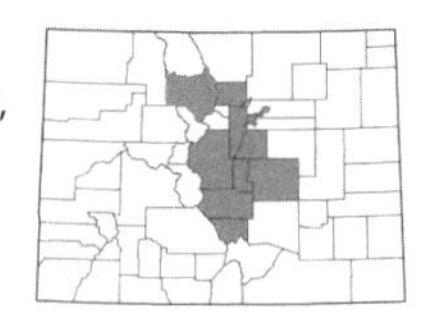

***Astragalus sparsiflorus* A. Gray**, Front Range milkvetch. Prostrate to decumbent perennials, 0.5–3.5 dm; *leaves* pinnately compound with 9–17 leaflets; *leaflets* broadly obovate or obcordate to elliptic, 3–15 mm long, glabrate above; *inflorescence* 2–9-flowered; *calyx* tube 1.5–3 mm long, lobes 1–2 mm long, with mixed black and white hairs; *flowers* whitish with a maculate keel, the banner 5–8 mm long; *legumes* falcate, oblong, or lunate, 3-angled, 5–25 × 2.5–5 mm, mottled, strigulose. Uncommon in meadows, open slopes, and rocky hillsides, 5000–8200 ft. May–Aug. E/W.

There are two well-marked varieties of *A. sparsiflorus*, sometimes found growing at the same locality:

1a. Leaflets of upper leaves oblong-oblanceolate or oblong-obovate, 4–15 mm long; calyx tube 2–3 mm long, the lobes 1.2–1.7 mm long; fruit 10–26 mm long...**var. *majusculus* A. Gray**

1b. Leaflets of upper leaves broadly elliptic to suborbicular, 2–7 mm long; calyx tube 1.5–2 mm long, the lobes 1.5–2 mm long; fruit (5) 6–8 mm long...**var. *sparsiflorus***

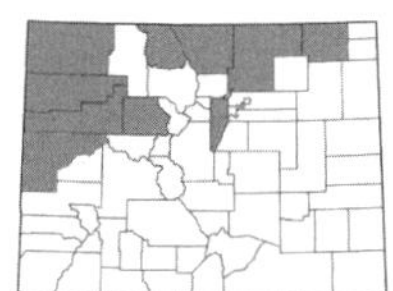

***Astragalus spatulatus* Sheldon**, tufted milkvetch. Subacaulescent perennials, caespitose-tufted, mat-forming , or pulvinate; *leaves* simple or sometimes a few compound, 1.5–4 (6) cm long; *leaflets* oblanceolate to linear or narrowly spatulate, cinereous-silvery; *inflorescence* included or exserted, 3–10-flowered; *calyx* tube 2–3 mm long, lobes 2–2.5 mm long, with white to black hairs; *flowers* pink-purple, the banner 7–9 mm long; *legumes* ovoid-acuminate, laterally compressed, 5–12 (15) × 1.5–3 mm, strigose. Found in gravelly or sandy soil of open places, outcroppings, and bluffs, 4400–9000 ft. May–June. E/W. (Plate 44)

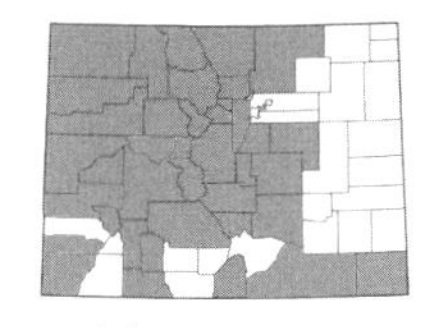

***Astragalus tenellus* Pursh**, LOOSEFLOWER MILKVETCH. Perennials, 2–4 dm; *leaves* pinnately compound with 13-numerous leaflets; *leaflets* elliptic to linear, 6–20 mm long, glabrate above; *inflorescence* included or exserted, (5) numerous-flowered; *calyx* tube 2–2.6 (3) mm long, lobes 0.7–1.5 (2.5) mm long, black-hairy; *flowers* whitish or ochroleucous, often with a maculate keel, the banner 6–9 (11) mm long; *legumes* elliptic to elliptic-oblong, laterally compressed, 7–17 × 2.5–4.5 mm, greenish and mottled to black, glabrate or strigose. Common in forest openings, sagebrush meadows, grassy slopes, bluffs, and dry grasslands of the high plains, 4900–11,000 ft. May–Aug. E/W. (Plate 44)

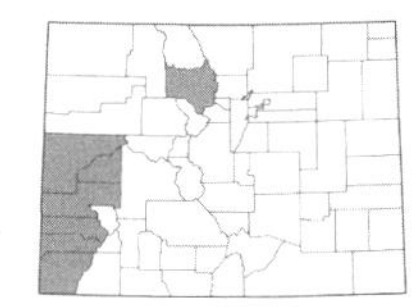

***Astragalus thompsoniae* S. Watson**, THOMPSON'S MILKVETCH. [*A. mollissimus* Torr. var. *thompsoniae* (S. Watson) Barneby]. Perennials, 1–3 dm; *leaves* pinnately compound with numerous leaflets; *leaflets* broadly obovate to elliptic, 5–12 (15) mm long, tomentose; *inflorescence* included or exserted, subscapose, (7) 12–25-flowered; *calyx* tube (8) 9–11 mm long, lobes 1.5–3 (4) mm long; *flowers* pink-purple, the banner 18–22 mm long; *legumes* half-ellipsoid, obliquely ovoid, or ovoid-lanceolate, obcompressed, 12–20 × 6–10 mm, thinly to densely white-tomentose. Found on dry slopes, usually with pinyon-juniper, 4500–7800 ft. April–June. W. (Plate 44)

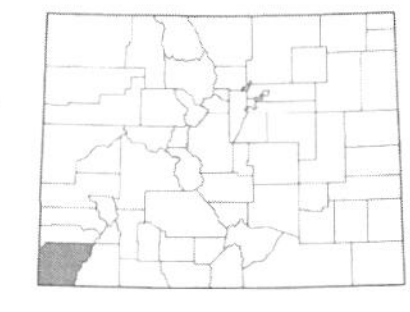

***Astragalus tortipes* J.L. Anderson & J.M. Porter**, SLEEPING UTE MILKVETCH. Ascending to erect perennials, (3) 6–8 dm; *leaves* pinnately compound with 7–15 leaflets; *leaflets* linear; *inflorescence* loosely 1–25-flowered; *calyx* tube campanulate, lobes shorter than the tube; *flowers* yellow, the banner 14–18 mm long; *legumes* laterally compressed, 2–3 cm × 6–9 mm, stipitate 12–16 mm. Rare on open, dry ridges and slopes, 5400–5700 ft. April–May. W. Endemic.

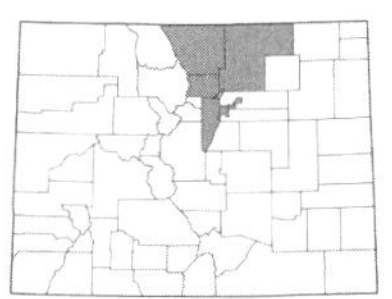

***Astragalus tridactylicus* A. Gray**, FOOTHILL MILKVETCH. [*Orophaca tridactylica* (A. Gray) Rydb.]. Acaulescent or shortly caulescent perennials; *leaves* ternately compound; *leaflets* oblanceolate, 10–20 mm long, silvery pubescent; *inflorescence* included, 2–4-flowered; *calyx* tube 2.5–3.8 (4) mm long, lobes 1.8–3.5 mm long; *flowers* pink-purple, the banner 7–11.5 mm long; *legumes* ovoid-acuminate to lanceolate, laterally compressed, 4–6.5 mm long, chartaceous, villous. Common on rocky outcroppings, bluffs, and open hillsides, 4900–7300 ft. April–June. E.

Leaflet length is not a good character to separate *A. sericoleucus* from *A. tridactylicus* because the leaflets of *A. tridactylicus* are dimorphic, and leaves maturing with and after the flowers have longer leaflets than the leaves before them. These two species can be quite difficult to separate and may even hybridize. Isely (1998) even states that these species (including *A. aretioides*) are "so similar that they perhaps should be treated as geographic phases of one species."

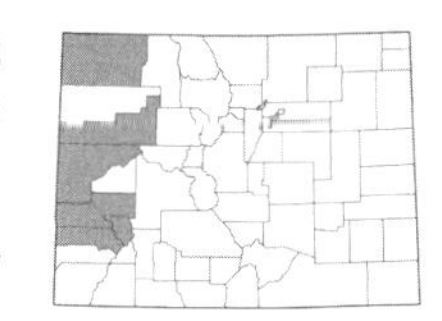

***Astragalus wetherillii* M.E. Jones**, WETHERILL'S MILKVETCH. Annuals or short-lived perennials, 0.5–2.5 dm; *leaves* pinnately compound with 9–15 leaflets; *leaflets* obovate, 5–12 (15) mm long, glabrate above; *inflorescence* included, 4-9-flowered; *calyx* tube 3–3.5 mm long, lobes 2 mm long; *flowers* whitish, the banner 7–10 mm long; *legumes* ovoid-lanceolate or half-ovoid, inflated but scarcely bladdery, slightly obcompressed, 1.5–2 × 0.8–1 cm, stipitate 1–2.5 mm, strigulose. Found on dry slopes, often in clay soil, usually with pinyon-juniper or sagebrush, 5200–7400 ft. May–June. W.

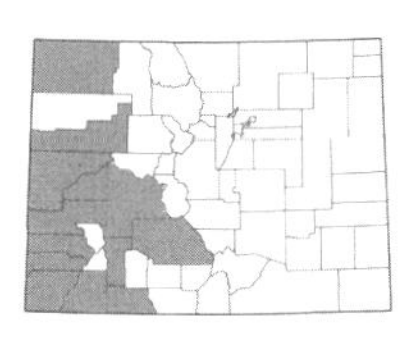

***Astragalus wingatanus* S. Watson**, FORT WINGATE MILKVETCH. Caulescent perennials, 1.5–4 (6.5) dm; *leaves* pinnately compound with 3–15 (17) leaflets; *leaflets* linear, narrowly elliptic, or filiform, (0.25) 0.4–1.8 cm long, glabrous above, thinly strigulose below; *inflorescence* a (5) 10–35-flowered raceme; *calyx* tube (1.5) 1.6–2.6 mm long, lobes 1.2–1.8 (2) mm long, strigulose with black, white, or mixed hairs; *flowers* whitish with purplish banner and keel tip or bicolored with purple petals and white wing tips, the banner 5.6–8 × 4–6 mm long; *legumes* ellipsoid to oblong-ellipsoid, (0.6) 0.9–1.5 cm long, (2.5) 3–4.5 mm diam., sessile or substipitate, greenish, purple-speckled or mottled, or stramineous, glabrous. Found on dry hillsides and clay or shale slopes, often with pinyon-juniper or sagebrush, 4700–7500 ft. April–June. W. (Plate 44)

Resembling *A. flexuosus* var. *diehlii*, which can have somewhat laterally compressed pods. However, *A. flexuosus* var. *diehlii* has a longer keel (5–5.5 mm vs. 3.5–5.4 mm) and strigulose, sessile fruit, while *A. wingatanus* has glabrous, subsessile or shortly stipitate fruit.

CARAGANA Fabr. – PEASHRUB

Shrubs or small trees; *leaves* alternate, even-pinnately compound, fascicled, the rachis extended as a spine or bristle, not gland-dotted, the stipules spiny or small and deciduous; *flowers* solitary or few and clustered on short branches; *calyx* obscurely toothed; *petals* 5, composed of a banner, wing, and keel, yellow; *stamens* 10, diadelphous; *legumes* sessile.

1a. Leaves even-pinnately compound; leaflets 8–12, obovate to elliptic, (0.5) 1–3 cm long; nodes sometimes with a pair of spines... ***C. arborescens***

1b. Leaves palmately compound; leaflets 3–5, narrowly lanceolate to oblanceolate, 0.5–1.5 cm long; nodes usually with a single spine... ***C. aurantiaca***

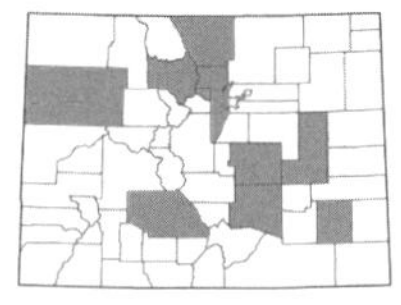

***Caragana arborescens* Lam.**, SIBERIAN PEASHRUB. Shrubs to 4 m, the nodes sometimes with a pair of spines; *leaves* even-pinnately compound; *leaflets* 8–12, obovate to elliptic, (0.5) 1–3 cm long; *calyx* 5–6 cm long; *flowers* 1.5–2 cm long; *legumes* 4–5.5 cm long, glabrous. Uncommon along roadsides, near old buildings, and in hedge rows, 3800–8200 ft. May–June. E/W. Introduced.

***Caragana aurantiaca* Koehne**, DWARF PEASHRUB. Shrubs, 0.5–1 m, the nodes usually with a single spine; *leaves* palmately compound; *leaflets* lanceolate to oblanceolate, 0.5–1.5 cm long; *calyx* 5–6.5 cm long; *flowers* 1.5–2.5 cm long; *legumes* 2.5–4 cm long, glabrous. Uncommon along roadsides and near old buildings, 6500–8500 ft. June–July. E. Introduced.

CASSIA L. – SENNA

Annual or perennial herbs; *leaves* spiral or distichous, pinnately compound with 2 to numerous leaflets, the stipules early-deciduous or persistent; *flowers* in axillary racemes or terminal panicles; *petals* 5, subequal or 1 larger; *stamens* 10; *legumes* erect or pendant.

***Cassia fasciculata* (Michx.) Greene**, PARTRIDGE PEA. [*Chamaecrista fasciculata* (Michx.) Greene]. Annuals, 1–10 dm; *leaflets* 5–18, oblong, 5–15 mm long, glabrous or sparsely hairy; *calyx* 9–13 mm; *flowers* 2.5–4 cm diam., yellow; *fruit* 3–6 (8) cm long, flat, glabrous to hairy. Not reported for Colorado but likely present in Prowers Co. along the Arkansas River in sandy soil. June–Oct.

COLUTEA L. – COLUTEA

Shrubs; *leaves* alternate, odd-pinnately compound, the leaflets subopposite, the stipules small and free; *flowers* in axillary racemes; *petals* 5, composed of a banner, wing, and keel, yellow, orange-brown, or red; *stamens* 10, diadelphous; *legumes* bladdery-inflated, obliquely ellipsoid.

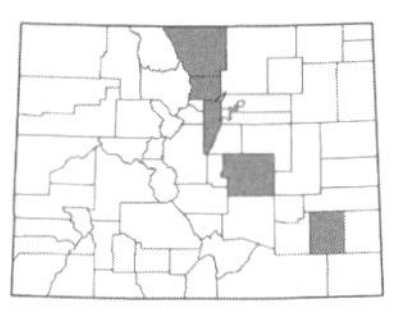

***Colutea arborescens* L.**, BLADDER-SENNA. Shrubs, 1–2 m; *leaflets* 9–13, ovate to elliptic, 1.4–2 cm long, strigose below; *calyx* 6–7 mm long with lobes 1–2 mm; *flowers* 1.5–2 cm long, yellow; *legumes* 5–7 × 2–3 cm, stipitate, glabrous or slightly hairy. Cultivated plants sometimes escaping to nearby roadsides or established around old buildings, 3900–6800 ft. May–July. E. Introduced.

DALEA L. – PRAIRIE CLOVER

Annual or perennial herbs, sometimes low shrubs; *leaves* alternate, odd-pinnately compound or ternately compound, with entire leaflets, gland-dotted; *flowers* zygomorphic, in dense terminal spikes or racemes; *petals* 5, composed of a banner, wing, and keel, clawed, the banner inserted on the rim of the floral cup, the other petals with claws adnate to the tube of the filaments; *stamens* 5, monadelphous; *legumes* often included in the persistent calyx. (Barneby, 1977; Wiggins, 1940)

1a. Plants silky-canescent, tomentose, or densely villous throughout, or densely pilose at least on the lower portion of the stem…2
1b. Plants glabrous or nearly so to the inflorescence…8

2a. Flowers yellow, often fading brownish or purplish-brown when dried…3
2b. Flowers purple, reddish-purple, pink, or white…5

3a. Leaves 3-foliate; calyx teeth 5–19 mm long; plants 0.1–1.2 dm tall…***D. jamesii***
3b. Leaves 5–7-foliate, rarely a few leaves 3-foliate; calyx teeth 2–5 mm long; plants 1–7.5 dm tall…4

4a. Stems erect, 2–7.5 dm; spikes 1.3–1.6 cm in diam.; banner 6–8.5 mm long, the petals remaining yellow after drying…***D. aurea***
4b. Stems diffuse and spreading, 1–2 (3) dm; spikes 0.5–0.9 mm in diam.; banner 4–5.5 mm long, the petals often fading from yellow to brownish after drying…***D. nana* var. *nana***

5a. Leaflets linear to linear-oblanceolate, the margins usually involute; roots brown…6
5b. Leaflets obovate to oblanceolate or elliptic, without involute margins; roots yellowish to reddish-orange…7

6a. Stems pilose throughout; spikes remaining dense, often conelike, the axis hidden in pressed plants; legumes 2.1–2.5 mm long…***D. purpurea***
6b. Stems densely pilose below and glabrous or nearly so above; spikes becoming loose, the axis visible in pressed plants; legumes 2.8–3.5 mm long…***D. tenuifolia***

7a. Stems prostrate; calyx with white hairs, the teeth 1.5–2.3 mm long; banner 3–4.3 mm long, the petals purple to reddish-violet; spikes relatively loose...***D. lanata***
7b. Stems erect or diffuse; calyx with usually tawny or sometimes white hairs, the teeth to 1.4 mm long; banner 4.5–5.5 mm long, the petals rose-purple to lavender or pinkish, sometimes nearly white; spikes dense...***D. villosa* var. *villosa***

8a. Flowers purple or rose-purple...9
8b. Flowers white or sometimes tinged with blue or fading to light yellow...10

9a. Shrubs; leaflets 1–5 mm long, obovate to oblanceolate; calyx with tawny hairs, the teeth 4.5–8.5 mm long; petals bicolored when young, the banner yellowish but aging rose-purple like the wings and keel...***D. formosa***
9b. Perennials; leaflets larger, 10–24 mm long, linear to linear-oblanceolate; calyx with white or tawny hairs, the teeth 1–2 mm long; petals all rose-purple or purple...***D. purpurea***

10a. Spikes remotely flowered, appearing 2-ranked, the axis clearly visible; bract subtending the calyx prominent, obovate to broadly ovate, 3–4 mm long, blackish-green with a white or pale margin; calyx teeth 3.3–4.6 mm long...***D. enneandra***
10b. Spikes densely flowered, conelike or cylindric, the axis not clearly visible; bract subtending the calyx mostly inconspicuous, narrowly lanceolate or triangular, without a white or pale margin; calyx teeth 0.6–2.5 mm long...11

11a. Annuals from slender taproots; calyx with black nerves and white hairs; stamens 9 or 10; leaves with 8–24 pairs of leaflets...***D. leporina***
11b. Perennials from thick taproots and short caudices; calyx without black nerves, glabrous or the hairs white or tawny; stamens 5; leaves with 3–13 pairs of leaflets...12

12a. Spikes subglobose, 0.4–1.5 cm long; calyx glabrous...***D. multiflora***
12b. Spikes cylindric, 1–20 cm long; calyx glabrous below and hairy above or hairy throughout...13

13a. Calyx densely pilose throughout with tawny hairs 1.6–2.4 mm long; axis of the inflorescence densely pilose; leaflets with acute tips; legumes pilose, not exserted; spikes 3–20 cm long...***D. cylindriceps***
13b. Calyx glabrous below and pubescent above with white hairs to 0.3 mm long; leaflets with obtuse or mucronate tips; legumes glabrous or with short hairs above the middle, usually exserted; spikes 1–6.5 cm long...***D. candida* var. *olgophylla***

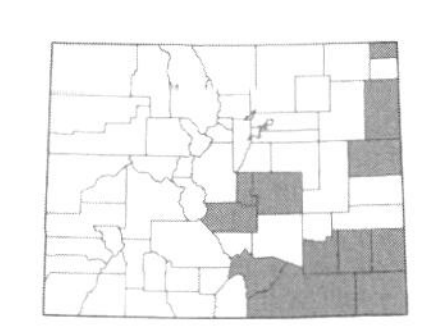

***Dalea aurea* Nutt. ex Pursh**, GOLDEN PRAIRIE CLOVER. Erect perennials, 2–7.5 dm; *leaves* pinnately compound with (3) 5–7 leaflets; *leaflets* obovate to oblanceolate, 2.5–5.5 mm long, pilose, often finely glandular; *inflorescence* of dense spikes, 1.5–5 cm long, 1.3–1.6 cm diam.; *calyx* tube 2–3 mm long, pilose; *flowers* yellow, the banner 6–8.5 mm long; *legumes* obovate or half-ovate, with 2 rows of short bristles on the apical margin. Uncommon to locally common in sandy soil of the eastern plains, 3500–6200 ft. June–Aug. E.

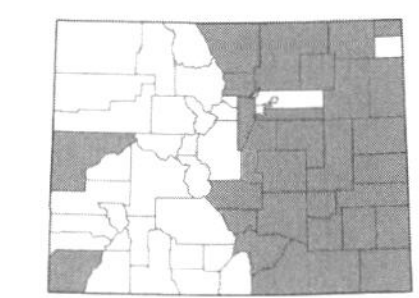

***Dalea candida* Michx. ex Willd. var. *olgophylla* (Torr.) Shinners**, WHITE PRAIRIE CLOVER. Erect perennials, 3–9 dm; *leaves* pinnately compound with 5–9 leaflets; *leaflets* elliptic to oblanceolate with obtuse or mucronate apices, 0.8–2.5 cm long, glabrous; *inflorescence* of ovoid to cylindrical compact spikes, 1–6.5 cm long, 0.7–0.8 cm diam.; *calyx* 2–2.6 mm long, glabrous below, white-hairy above; *flowers* white, ca. 6–7 mm long; *legumes* obovate or half-ovate, glabrous or with short hairs distally. Common in sandy or rocky soil and along roadsides throughout the eastern plains to the lower foothills, 3400–7200 ft. June–Sept. E/W. (Plate 44)

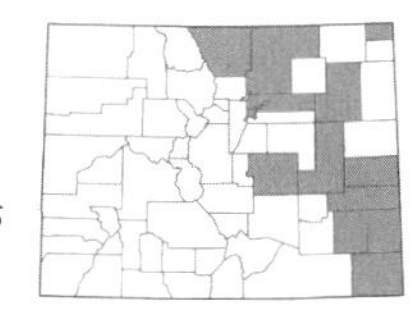

***Dalea cylindriceps* Barneby**, ANDEAN PRAIRIE CLOVER. Erect perennials, 2–7 dm; *leaves* pinnately compound with 7–9 leaflets; *leaflets* elliptic-oblong to elliptic-lanceolate with acute tips, 1.2–2.5 cm long, glabrous; *inflorescence* of cylindric spikes, 3–20 cm long, 0.9–1.2 cm diam.; *calyx* 2–2.3 mm long, densely pilose with tawny hairs; *flowers* whitish to whitish-pink, ca. 6 mm long; *legumes* obovate or half-ovate, pilose. Locally common in sandy soil of the eastern plains, 3600–5300 ft. June–Sept. E.

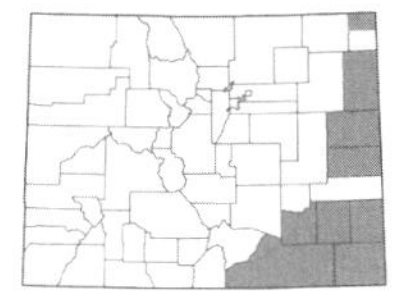

***Dalea enneandra* Nutt.**, NINE-ANTHER PRAIRIE CLOVER. Erect perennials, 5–10 (15) dm; *leaves* pinnately compound with 5–11 (13) leaflets; *leaflets* elliptic to oblong-oblanceolate, glabrous; *inflorescence* of slender spikes with remotely spaced flowers, 2–15 cm long; *calyx* 3–3.5 mm long, pilose to plumose; *flowers* white, the banner with a proximal greenish spot, 7–9 mm long; *legumes* obovate or half-ovate. Uncommon to locally common in sandy soil and limestone bluffs, 3400–5600 ft. June–Aug. E.

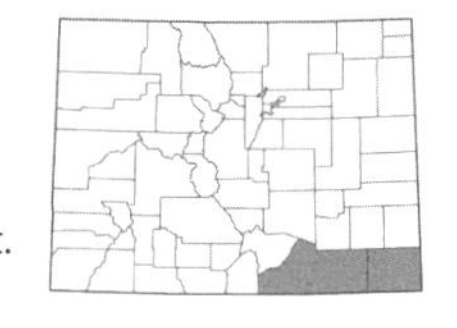

***Dalea formosa* Torr.**, FEATHERPLUME. Subshrubs to shrubs, 2–10 dm; *leaves* pinnately compound with 5–13 leaflets; *leaflets* obovate to oblanceolate, 1–5 mm long, glabrous; *inflorescence* of short, subcapitate spikes with 2–8 (12) flowers, 1–1.5 cm diam.; *calyx* tube 3–5 mm long, pilose to plumose with tawny hairs, glandular; *flowers* rose-purple or bicolored with a yellow banner; *legumes* obovate or half-ovate. Uncommon on rocky soil and barren outcroppings, 4000–5200 ft. May–July. E.

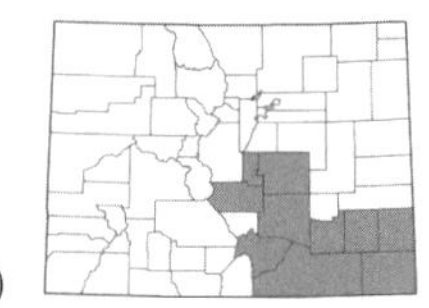

***Dalea jamesii* (Torr.) Torr. & A. Gray**, JAMES' PRAIRIE CLOVER. Ascending or decumbent perennials, 0.1–1.2 dm; *leaves* ternately compound; *leaflets* obovate to oblanceolate, 5–15 mm long, silky-canescent; *inflorescence* of ovoid to cylindric spikes, 1–8 cm long, 15–20 mm diam.; *calyx* 2.5–3.5 mm long, pilose; *flowers* yellow but aging orange to brown or purplish-brown, 11–14 mm long; *legumes* obovate or half-ovate, with 2 rows of short bristles on the apical margin. Uncommon to locally common on rocky soil and sandstone outcroppings, 4000–6600 ft. May–July. E. (Plate 44)

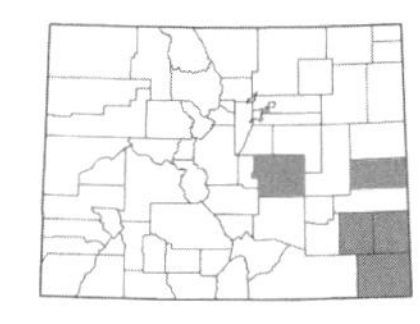

***Dalea lanata* Spreng.**, WOOLLY PRAIRIE CLOVER. Prostrate, occasionally mat-forming perennials, stems 2–6 dm long; *leaves* pinnately compound with 5–15 leaflets; *leaflets* obovate or oblanceolate or elliptic, 3–12 mm long; *inflorescence* compact to elongate-cylindric, 2–7 cm long; *calyx* 2–2.5 mm long, white-villous; *flowers* purple to reddish-violet, 6 mm, the banner 3–4.3 mm long; *legumes* obovate or half-ovate. Uncommon in sandy soil and on dunes, 3400–6000 ft. June–Aug. E.

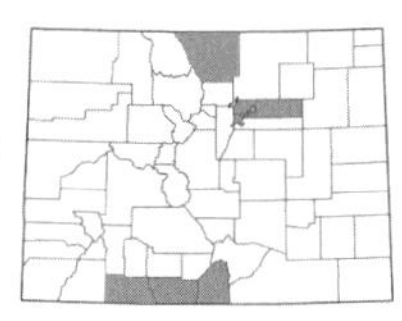

***Dalea leporina* (Ait.) Bullock**, FOXTAIL PRAIRIE CLOVER. Annuals, 1.5–12 dm; *leaves* pinnately compound with 8–24 pairs of leaflets; *leaflets* elliptic-obovate to oblong-oblanceolate, (3) 5–12 mm long, glabrous; *inflorescence* of densely flowered ovoid to cylindric spikes, 2–12 cm long; *calyx* 2–2.5 mm long, white-hairy with black nerves, glandular; *flowers* white to blue-lavender, 5–7 mm long; *legumes* obovate or half-ovate. Uncommon in fields and along roadsides, often in disturbed places and having a tendency to be somewhat weedy, 5000–7600 ft. July–Sept. E/W.

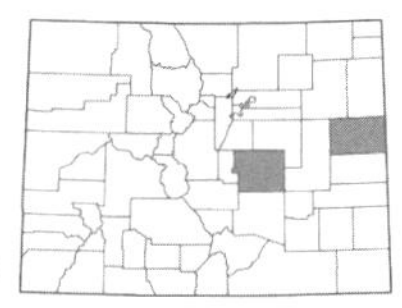

***Dalea multiflora* (Nutt.) Shinners**, ROUNDHEAD PRAIRIE CLOVER. Erect to spreading perennials, 3–8 dm; *leaves* pinnately compound with (5) 7–11 leaflets; *leaflets* oblong-oblanceolate or elliptic-oblong, 6–15 mm long, glabrous; *inflorescence* of subglobose spikes, 0.4–1.5 cm long; *calyx* 2.2–2.5 mm long, glabrous; *flowers* white, ca. 8 mm long; *legumes* obovate or half-ovate. Uncommon on rocky soil and along roadsides, known from an old collection in El Paso Co. and from a collection along I-70 in Kit Carson Co. that was likely the result of introduction from a seeding program, 4100–6000 ft. June–Sept. E.

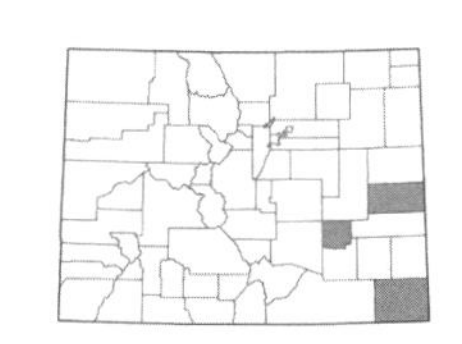

Dalea nana* Torr. ex A. Gray var. *nana, DWARF PRAIRIE CLOVER. Decumbent to ascending perennials, 1–2 (3) dm; *leaves* pinnately compound with (3) 5–7 leaflets; *leaflets* oblanceolate to obovate-elliptic, 5–10 (15) mm long, strigose to pilose; *inflorescence* of compact conical to short cylindric spikes, 1–3 cm long; *calyx* tube 2–2.5 (2.8) mm long, pilose, glandular; *flowers* yellow but aging reddish to brownish, the banner 4–5.5 mm long; *legumes* obovate or half-ovoid, keeled proximally, with 2 rows of short hairs on prow margins. Uncommon on sandy soil of the southeastern plains, 3600–4500 ft. May–Sept. E.

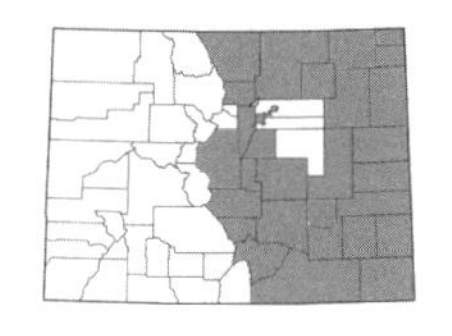

***Dalea purpurea* Vent.**, PURPLE PRAIRIE CLOVER. Perennials, 2–8 dm; *leaves* pinnately compound with 3–5 leaflets; *leaflets* linear to linear-oblanceolate, involute or folded, 1–2.4 cm long; *inflorescence* of dense, conic to cylindric spikes, 1–7 cm long; *calyx* tube 2–3.5 mm long, hairs white to tawny; *flowers* rose-purple to purple, 6–7 mm long; *legumes* obovate or half-ovoid, 2.1–2.5 mm long. Common in dry grasslands in sandy or rocky soil, 3500–7700 ft. May–Aug. E. (Plate 44)

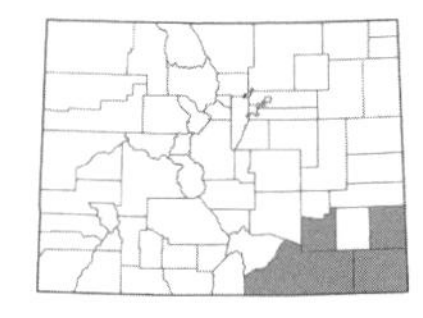

***Dalea tenuifolia* (A. Gray) Shinners**, SLIMLEAF PRAIRIE CLOVER. Erect or prostrate perennials, 2–6 (9) dm; *leaves* pinnately compound with 3–5 leaflets; *leaflets* linear to linear-oblanceolate, with involute margins, 0.7–2 cm long; *inflorescence* of loosely flowered oblong spikes, 3–10 cm long; *calyx* tube (1.5) 2–2.5 mm long, pilose; *flowers* lavender to violet-purple, ca. 6 mm long; *legumes* obovate or half-ovoid, 2.8–3.5 mm long. Uncommon on limestone outcroppings or rocky bluffs, 4000–5300 ft. May–July. E.

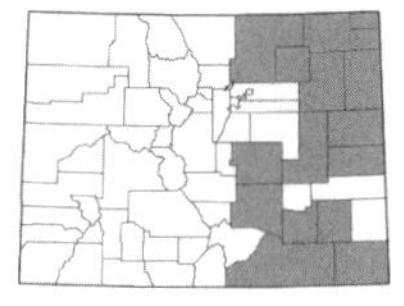

Dalea villosa* (Nutt.) Spreng. var. *villosa, SILKY PRAIRIE CLOVER. Erect to ascending perennials, 2–7 dm; *leaves* pinnately compound with 11–17 (19) leaflets; *leaflets* oblanceolate to elliptic-oblong, 5–15 mm long, cinereous; *inflorescence* of dense, cylindric spikes; *calyx* tube 2–2.7 mm long, pilose with tawny to white hairs; *flowers* rose-purple to lavender or pinkish, occasionally whitish, 5–6 mm long, the banner 4.5–5.5 mm long; *legumes* obovate to half-ovoid. Locally common in sandy soil and on dunes, 3400–5400 ft. July–Sept. E.

DESMANTHUS Willd. – BUNDLEFLOWER

Perennial herbs, unarmed; *leaves* bipinnately compound, the leaflets numerous and entire, stipules small; *flowers* actinomorphic, in axillary, pedunculate heads; *calyx* 5-lobed, small; *petals* 5, distinct or connate near the base; *stamens* 5 or 10, distinct, long-exserted; *legumes* glabrous, straight or often curved, in subglobose clusters.

1a. Legumes falcate, 3–5 times longer than wide; stipules 6–10 mm long, filiform; leaflets 7–12 (15) pairs per pinna; stamens 5...***D. illinoensis***

1b. Legumes straight or nearly so, 7 times or more longer than wide; stipules absent or to 2 mm long; leaflets (2) 3–6 pairs per pinna; stamens 10...***D. cooleyi***

Desmanthus cooleyi **(Eaton) Branner & Coville**, COOLEY'S BUNDLEFLOWER. Ascending to prostrate perennials; *leaves* bipinnately compound with (2) 3–6 pairs of pinnae, each pinna with 12–30 leaflets; *leaflets* elliptic-oblong, 2–4 mm long, glabrous or margins ciliate; *inflorescence* of axillary heads with 10–20 flowers, 1.5–2 cm diam.; *flowers* whitish; *legumes* straight, 7 times or more longer than wide, glabrous. Uncommon in sandy soil and along roadsides, 4000–4500 ft. June–Sept. E.

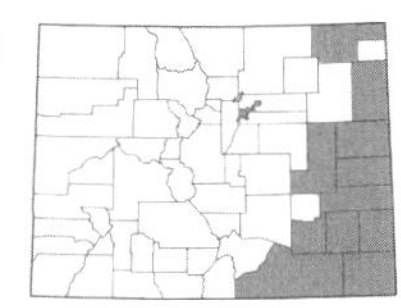

Desmanthus illinoensis **(Michx.) MacMill. ex B.L. Rob. & Fernald**, ILLINOIS BUNDLEFLOWER. Ascending to erect perennials, to 13 dm; *leaves* bipinnately compound with 7–12 (15) pairs of pinnae, each pinna with 15–25 (30) leaflets; *leaflets* oblong, 2–3 mm long; *inflorescence* of axillary heads with numerous flowers, 1 cm diam.; *flowers* whitish, ca. 2 mm long; *legumes* falcate, 3–5 times longer than wide, black at maturity, glabrous. Found in open prairie, along roadsides and streams, or sometimes in disturbed places, 3400–5000 ft. June–Sept. E.

DESMODIUM Desv. – TICKCLOVER; TICK-TREFOIL

Herbaceous perennials (ours) with trailing or prostrate stems; *leaves* alternate, pinnately trifoliate; *flowers* in racemes or panicles; *petals* 5, composed of a banner, wing, and keel; *stamens* 10, monadelphous or diadelphous; *legumes* stipitate, constricted between the seeds, indehiscent.

Desmodium obtusum **(Muhl. ex Willd.) DC.**, STIFF TICK-TREFOIL. [*Desmodium rigidum* (Elliot) DC.]. Ascending to erect or prostrate perennials, 5–15 dm; *leaves* pinnately trifoliate; *leaflets* lanceolate to elliptic-ovate, (2.5) 4–6 (7.5) cm long, glabrous or minutely hairy above, glabrate to appressed hairy below; *inflorescence* of terminal panicles; *flowers* pinkish-purple to pinkish-white; *legumes* dorsally sinuate and proximally notched. Very rare in rocky, open prairie, known from a single, old (1871) collection made in Denver and presumably extinct in the state, 5200 ft. Aug.–Sept. E.

GALEGA L. – PROFESSOR-WEED

Perennials, caulescent herbs; *leaves* alternate, odd-pinnately compound, with sagittate stipules; *flowers* zygomorphic; *petals* 5, composed of a banner, wing, and keel, blue-purple; *stamens* monadelphous; *legumes* sessile.

Galega officinalis **L.**, GOATSRUE; PROFESSOR-WEED. Erect perennials, 3–8 (12) dm; *leaflets* 11–15 (19), elliptic to oblong, 3–6 cm long; *inflorescence* of axillary racemes with numerous flowers; *calyx* 4–5 mm long, setaceous; *flowers* blue-purple to pink-lavender, 8–12 cm long; *legumes* linear, subterete, 3–5 cm × 3–4 mm, longitudinally striate. Cultivated ornamental, occasionally escaping and reported for Colorado although no specimens have been seen. June–Aug. Introduced.

GLEDITSIA L. – HONEY LOCUST

Trees; *leaves* alternate, pinnate to bipinnately compound, stipules lacking; *petals* 3(5); *stamens* (4)5–7; *legumes* stipitate.

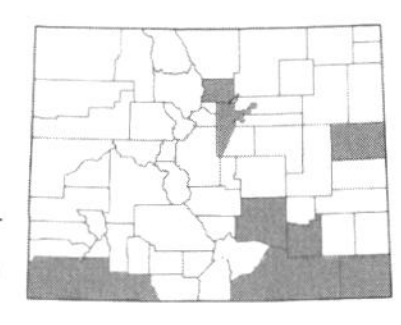

Gleditsia triacanthos **L.**, HONEY LOCUST. Armed or unarmed, thorns to 2 dm; *leaves* 1–2 pinnate, pinnate leaves with 10–14 pairs of leaflets, bipinnate leaves with 3–6 (8) pairs of pinnae, each pinna with (2) 5–8 pairs leaflets; *leaflets* elliptic-oblong, 1.3–2.5 cm long, glabrate; *staminate inflorescence* of crowded racemes with numerous flowers; *calyx* yellowish-green; *flowers* yellowish-green; *legumes* oblong or ovate. Cultivated tree sometimes escaping along roadsides, 4300–5700 ft. April–June. E/W. Introduced.

GLYCYRRHIZA L. – LICORICE

Perennial herbs; *leaves* alternate, odd-pinnately compound, with gland-dotted leaflets; *flowers* zygomorphic; *petals* 5, composed of a banner, wing, and keel, white to cream; *stamens* 10, diadelphous; *legumes* sessile, indehiscent.

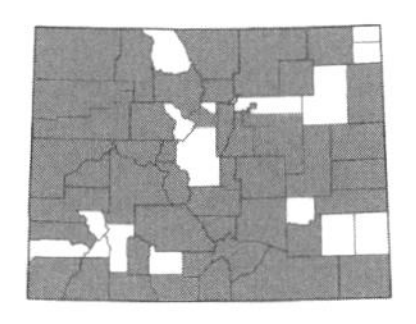

Glycyrrhiza lepidota **Pursh**, WILD LICORICE. Plants 0.5–1 m; *leaflets* 9–19, ovate-lanceolate to lanceolate, (2) 2.5–4 cm long; *inflorescence* of racemes with 10-numerous crowded, ascending-spreading flowers; *calyx* 8–10 mm long, villous with glandular hairs; *flowers* ochroleucous to greenish-white; *legumes* ovate or oblong, densely covered with hooked prickles. Common along streams and ditches, roadsides, and in disturbed areas, 3800–8600 ft. June–Aug. E/W. (Plate 44)

HEDYSARUM L. – SWEETVETCH

Perennials, caulescent herbs; *leaves* alternate, odd-pinnately compound, the leaflets nearly sessile; *flowers* zygomorphic, in axillary racemes; *petals* 5, composed of a banner, wing, and keel, the banner and wings shorter than the keel, purple-red to pink or rarely white; *stamens* 10, diadelphous; *legumes* constricted between the seeds, indehiscent.

1a. Secondary veins prominent above and below on the leaves; fruit narrowly wing-margined, the segments elliptic-obovate…***H. occidentale***

1b. Secondary veins faint or lacking, only the midrib prominent below on the leaves; fruit not wing-margined, the segments orbicular…***H. boreale***

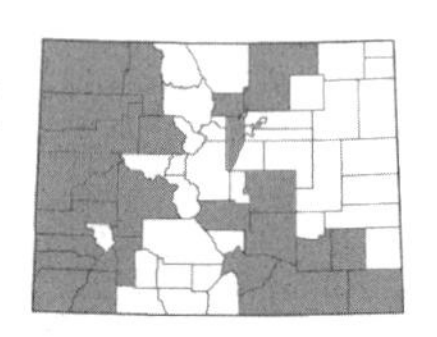

Hedysarum boreale **Nutt.**, UTAH SWEETVETCH. Erect or ascending perennials, 1.5–6 dm; *leaflets* 7–15, elliptic or oblong, 1–3.5 cm long, with faint venation; *inflorescence* with 5-numerous ascending or spreading flowers; *calyx* tube 2.5–3.5 mm long, conspicuously hairy; *flowers* purple-red to pink, or occasionally white, 1.2–2 (2.5) cm long; *legumes* with 2–7 segments, each segment elliptic or orbicular, transversely striate and reticulate. Common in rocky or sandy soil, 4000–9000 ft. May–July. E/W. (Plate 44)

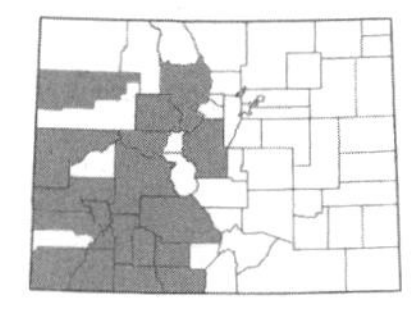

Hedysarum occidentale **Greene**, WESTERN SWEETVETCH. Erect-ascending perennials, 3–8 dm; *leaflets* 11–17, ovate to elliptic-lanceolate, 1–4 cm long, with evident secondary venation; *inflorescence* with reflexed flowers; *calyx* tube 2.5–4 mm long, strigose or glabrate; *flowers* red-purple to pink, sometimes white, 15–22 mm long; *legumes* with (1) 2–5 segments, each segment suborbicular to elliptic, reticulate. Found in open forests and on rocky slopes, 7300–12,600 ft. June–Aug. E/W.

HOFFMANNSEGGIA Cav. – RUSH-PEA

Perennial herbs; *leaves* alternate, odd-bipinnately compound, lacking glandular dots; *flowers* nearly actinomorphic, in a terminal raceme; *corolla* 5-lobed, the uppermost dissimilar, yellow; *stamens* 10, distinct, the filaments nearly equal to the petals; *legumes* flat, straight or falcate.

1a. Upper stem and inflorescence with stipitate glands; petal claws with stipitate glands…***H. glauca***

1b. Upper stem and inflorescence pilose or finely short-hairy; petal claws glabrous…2

2a. Fruit 13–17 mm wide, arched downward or crumpled at maturity; leaflets pilose below and less so above; longest petal 8–11 mm long; plants from creeping rhizomes…***H. repens***

2b. Fruit 4.5–7.5 mm wide, slightly to strongly falcate; leaflets pilose below and glabrous above; longest petal 6–9 mm long; plants from a woody caudex…***H. drepanocarpa***

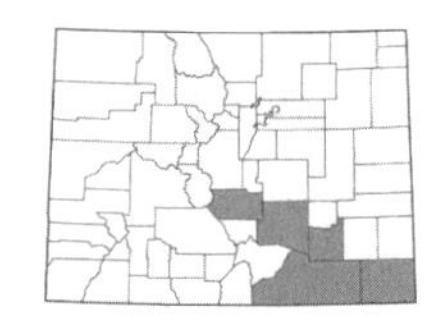

Hoffmannseggia drepanocarpa **A. Gray**, SICKLEPOD RUSH-PEA. [*Caesalpinia drepanocarpa* (A. Gray) Fisher]. Caulescent to subscapose perennials, 6–20 cm; *leaves* mostly basal, bipinnately compound, each pinna with 8–16 leaflets; *leaflets* obovate to elliptic-oblong, 1.5–6 mm long, glabrous above, pilose below; *inflorescence* of short terminal racemes with 4–10 flowers; *calyx* lobes 3–5 mm long; *flowers* pale to orange-yellow; *legumes* oblong, weakly to strongly falcate, 2–4 cm × 5–8 mm, reddish-brown, reticulate. Uncommon in dry grasslands and sandy soil, 4200–5300 ft. May–June. E.

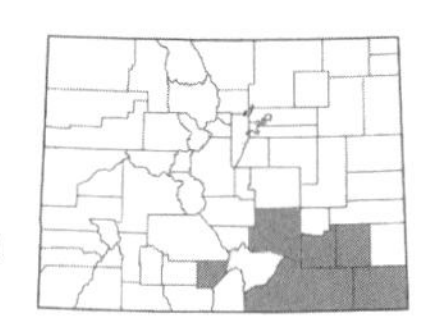

Hoffmannseggia glauca **(Ort.) Eifert**, INDIAN RUSH-PEA; PIG NUT. [*Hoffmannseggia densiflora* Benth.]. Subscapose to caulescent perennials, 0.5–2 (5) dm; *leaves* mostly basal, bipinnately compound, each pinna with 12–22 leaflets; *leaflets* elliptic-oblong, 2.5–6 mm long, strigulose to glabrous; *inflorescence* of terminal racemes with 5–15 flowers, glandular; *calyx* 6–7 mm long; *flowers* orange-yellow with a red-marked banner, glandular; *legumes* oblong to falcate, flat, 2–4 cm × 6–8 mm, often glandular. Uncommon on rocky soil and in fields, 4200–4800 ft. June–Sept. E.

Hoffmannseggia repens **(Eastw.) Cockerell**, CREEPING RUSH-PEA. [*Caesalpinia repens* Eastw.]. Subacaulescent perennials, 5–10 cm; *leaves* mostly basal, bipinnately compound, each pinna with 4–14 crowded leaflets; *leaflets* obovate-elliptic to shortly oblong, 4–9 mm long, weakly punctate; *inflorescence* of terminal, subcapitate racemes with numerous, crowded flowers; *calyx* lobes 6–8 mm long; *flowers* golden-yellow with a red-spotted banner; *legumes* ovate to oblong, flat, 2–5 × 1–1.5 cm, canescent to glabrate, reticulate. Uncommon in sandy soil, known from a single Colorado collection made near the Utah state line along I-70, 4400–4500 ft. April–June. W.

LATHYRUS L. – PEA

Annual or perennial herbs, usually with trailing or climbing stems; *leaves* alternate, even-pinnately compound, the rachis extending into a tendril or short bristle; *flowers* zygomorphic, in axillary racemes; *petals* 5, composed of a banner, wing, and keel; *stamens* 10, diadelphous; *legumes* usually flattened. (Hitchcock, 1952)

1a. Leaflets 2 (a single pair); stems winged…2

1b. Leaflets 4 or more (rarely 2); stems not winged…3

2a. Banner 20–25 mm long; leaflets narrowly to broadly lance-elliptic; stipules 3–10 mm wide; petioles broadly winged; fruit hairy...***L. latifolius***
2b. Banner 12–18 mm long; leaflets linear-lanceolate to lanceolate, sometimes lance-elliptic; stipules 1–2 (4) mm wide; petioles narrowly winged; fruit glabrous...***L. sylvestris***

3a. Petals white, sometimes the banner with purplish or pinkish veins...***L. lanszwertii***
3b. Petals purple, pink-purple, or bluish-purple, rarely white (and then the stipules large and herbaceous, 5 mm or more wide)...4

4a. Tendrils reduced to a simple, straight or curved (but not coiled) bristle to 6 mm long; banner 20–27 mm long; plants usually densely hirsute-pilose to villous or sometimes glabrous...***L. polymorphus***
4b. Tendrils usually coiling at the tip, simple or forked, more than 6 mm long; banner 9–30 mm long; plants glabrous or villous throughout with fine hairs...5

5a. Stipules large and foliaceous, 5–20 mm wide; keel conspicuously shorter than the wings; plants glabrous throughout except for occasionally ciliate on the margins of the calyx teeth...***L. pauciflorus* var. *utahensis***
5b. Stipules herbaceous but not large and foliaceous, to 4 mm wide; keel equal or subequal to the wings; plants glabrous to hairy, the calyx glabrous with ciliate margins or sparsely hairy...6

6a. Banner 8–15 mm long; calyx 5–10 mm long, glabrous or with some ciliate hairs on the margins of the teeth; petals pale lavender to pink-purple...***L. lanszwertii* var. *lanszwertii***
6b. Banner (14) 16–29 mm long; calyx 5–15 mm long, usually sparsely hairy, rarely just with ciliate hairs on the margins of the teeth; petals bright pink or blue-purple...7

7a. Calyx 5–8 (8.5) mm long, the tube 3.5–5.5 mm long; style 4–7.5 mm long; fruit attenuate at the base but not truly stipitate...***L. brachycalyx* var. *zionis***
7b. Calyx (8.5) 9–12 mm long, the tube (5) 5.5–7 mm long; style 7–9 mm long; fruit stipitate, the stipe 4–6 mm long...***L. eucosmus***

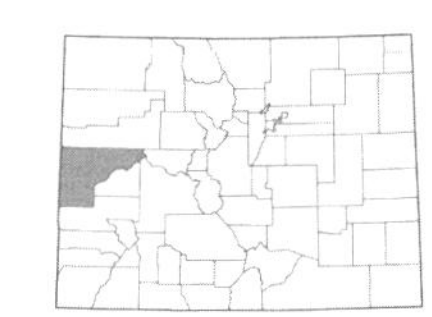

***Lathyrus brachycalyx* Rydb. var. *zionis* (C.L. Hitchc.) S.L. Welsh**, Zion pea. Sprawling or erect perennials, stems 1–5 dm long; *leaflets* 6–10 (12), elliptic to oblong or occasionally linear, 1–5 (8) cm long; *tendrils* typically short and simple; *inflorescence* with 2–5 flowers; *calyx* 5–8 (8.5) mm long, usually minutely hairy; *flowers* various shades of pink to pink-purple or sometimes white, the banner (14) 16–29 mm long; *legumes* proximally tapered, appearing stipitate, 3–5 cm × 5–7 mm, glabrous. Uncommon in sandy or gravelly soil, 4600–7000 ft. April–June. W.

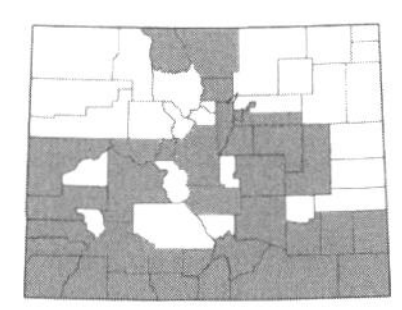

***Lathyrus eucosmus* Butters & H. St. John**, seemly pea. Erect perennials, 1.5–5 dm; *leaflets* 6–10, narrowly elliptic to oblong-lanceolate, 2.5–7.5 cm long, coriaceous, with apparent parallel venation; *tendrils* poorly developed or slightly prehensile; *inflorescence* with 2–5 distant flowers; *calyx* (8.5) 9–12 mm long, usually minutely hairy; *flowers* lavender to purple or bicolored, the banner (14) 16–29 mm long; *legumes* oblong, 3–4 cm × 8–10 mm, laterally compressed. Common in the shortgrass prairie and canyons, in open woods and meadows, and oak or pinyon-juniper, 4500–8500 ft. May–July. E/W.

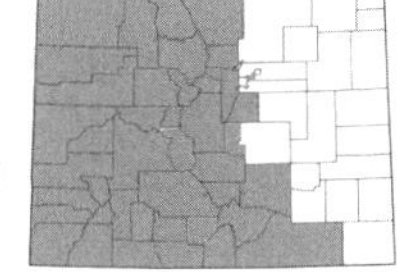

***Lathyrus lanszwertii* Kellogg**, Lanszwert's pea. Climbing, sprawling or erect perennials, 2–12 dm; *leaflets* 4–8 (10), linear to elliptic, occasionally ovate, 2–8 cm long; *tendrils* reduced to a bristle or prehensile and coiled; *inflorescence* with 2–10 flowers; *calyx* 5–10 mm long, glabrous or villous; *flowers* white, lavender, light pink to pink-purple, or cream-yellow, 7–20 mm long; *legumes* oblong or oblong-lanceolate, laterally compressed, 2–4 cm × 7–9 mm, glabrous. Common in meadows, aspen and spruce-fir forests, and on open slopes, 6500–10,500 ft. May–July. E/W. (Plate 44)

Lathyrus lanszwertii is a highly variable species. There are three varieties present in Colorado:

1a. Flowers pink-purple...**var. *lanszwertii***, Lanszwert's pea. Uncommon in aspen or Douglas fir forests and mountain meadows, currently known only from western Jackson Co., 8000–9500 ft. June–Aug. E/W?
1b. Flowers white, sometimes the banner with pinkish or purplish veins...2

2a. Tendril reduced to a simple, straight or slightly curved (but not coiled) bristle to 6 mm long...**var. *leucanthus* (Rydb.) Dorn**, whiteflower pea. [*L. arizonicus* Britton; *L. leucanthus* Rydb.]. Found in meadows, spruce-fir forests, and on open slopes, known from the southcentral and southwestern counties, 7000–10,500 ft. May–July. E/W.
2b. Tendril often coiling at the tip, simple or forked, more than 6 mm long...**var. *laetivirens* (Greene ex Rydb.) S.L. Welsh**, aspen pea. [*L. laetivirens* Greene ex Rydb.; *L. lanszwertii* var. *pallescens* Barneby]. Common in meadows and aspen and spruce-fir forests, widespread in the mountains, 6500–10,500 ft. May–July. E/W.

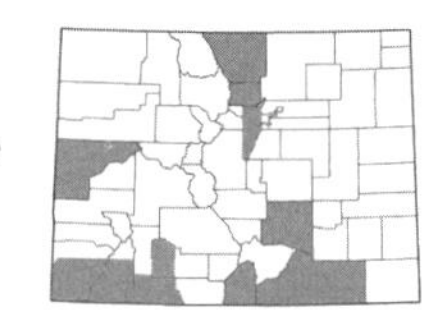

***Lathyrus latifolius* L.**, EVERLASTING PEA. Viney or sprawling perennials, to 2 m, stems conspicuously winged; *leaflets* 2, narrowly to broadly lance-elliptic, 5–12 (15) cm long; *tendrils* prehensile; *inflorescence* with (2) 4–15 crowded or distant flowers; *calyx* 10–12 mm long; *flowers* deep purple to white or pink, the banner 20–25 mm long; *legumes* oblong or oblong-lanceolate, laterally compressed, 6–9 cm. Often cultivated and escaping to nearby roadsides and fences, also found in disturbed areas and meadows, 4900–6500 ft. May–Sept. E/W. Introduced.

***Lathyrus pauciflorus* Fernald var. *utahensis* (M.E. Jones) Piper ex M. Peck**, UTAH PEA. Scandent, trailing or erect perennials, 3–9 dm; *leaflets* 6–10 (12), oval to elliptic or linear-filiform; *tendrils* usually coiled, occasionally reduced to a curved bristle; *inflorescence* with 2–10 well-spaced flowers, usually with long, arcuate peduncles; *calyx* glabrous, usually shiny; *flowers* various shades of lavender to purple, the keel shorter than the wings; *legumes* oblong or oblong-lanceolate, laterally compressed, 3–6 cm × 6–10 mm, glabrous. Uncommon in oak and pinyon-juniper forests, 7000–8500 ft. May–July. W.

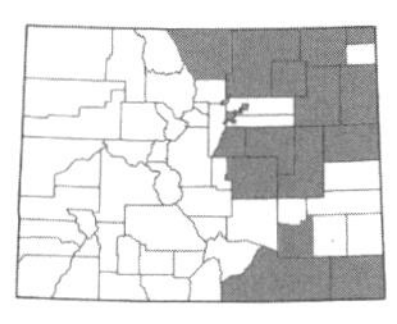

***Lathyrus polymorphus* Nutt.**, MANYSTEM PEA. Erect perennials, 1–3 (4) dm; *leaflets* (4) 6–10, narrowly oblong-linear, (1.5) 2–3 (4) cm long, glabrous or hairy; *tendrils* absent or reduced to short bristles; *inflorescence* with 2–5 (8) flowers; *calyx* 7–11 mm long; *flowers* purple or bicolored, occasionally white, 2–2.8 cm long; *legumes* oblong or oblong-lanceolate, laterally compressed, 3–6 cm × 7–10 (12) mm, glabrous. Common in sandy soil on the prairie, 3500–6500 ft. May–July. E. (Plate 45)

There are two varieties of *L. polymorphus* in Colorado, sometimes found growing right next to each other:

1a. Plants glabrous...**var. *polymorphus***, MANYSTEM PEA. Uncommon, 3500–6000 ft. May–July. E.
1b. Plants densely hairy...**var. *incanus* (J.G. Sm. & Rydb.) C.L. Hitchc.**, HOARY PEA. Common, 3500–6500 ft. May–July. E.

***Lathyrus sylvestris* L.**, FLAT PEA. Viney perennials, 5–20 dm, stems narrowly winged; *leaflets* 2, oblong-lanceolate to narrowly lanceolate, 4–10 (15) cm long; *tendrils* prehensile; *inflorescence* with 4–12 long-pedunculate flowers; *flowers* various shades of pink or pink-purple, occasionally white, the banner 12–18 mm long; *legumes* oblong or oblong-lanceolate, laterally compressed, 5–7 cm × 6–8 mm, glabrous. Cultivated plants occasionally escaping to nearby roadsides and fences, known from near Boulder, 5500–6000 ft. May–Sept. E. Introduced.

LOTUS L. – TREFOIL

Annual or perennial herbs; *leaves* alternate, ternately or palmately compound; *flowers* zygomorphic, in axillary umbellate clusters or solitary, yellow, white, or orange-red; *petals* 5, composed of a banner, wing, and keel; *stamens* 10, diadelphous; *legumes* terete, dehiscent.

1a. Stems and leaves densely strigose with appressed hairs; leaves palmately 3–5-foliate; flowers solitary or sometimes paired in leaf axils, yellow with the banner often reddish, fading orange...***L. wrightii***
1b. Stems and leaves glabrous or sparsely hairy; leaves pinnately compound with 5 leaflets, sessile on the stem such that the lowest pair of leaflets resemble foliaceous stipules; flowers paired or more often in umbels, yellow throughout, not fading to orange...2

2a. Leaflets obovate, mostly 4–6 mm wide; umbels usually 5–7-flowered...***L. corniculatus***
2b. Leaflets narrowly oblanceolate, mostly 2–3 mm wide; umbels usually 2–5-flowered...***L. tenuis***

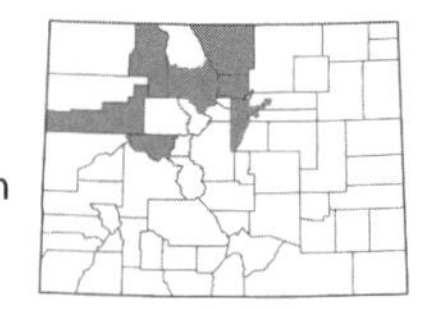

***Lotus corniculatus* L.**, BIRD'S FOOT TREFOIL. Ascending or procumbent perennials, 0.5–4 dm; *leaflets* obovate to narrowly oblong, 5–15 cm × 4–6 mm, glabrous to sparsely hairy; *inflorescence* with (2) 4–8 flowers; *calyx* tube 2.5–3.5 mm long; *flowers* bright yellow, 10–14 mm long; *legumes* narrowly oblong, terete, 1.5–3.5 cm long, corniculate. Found in meadows, along roadsides, and in disturbed areas, 5000–9500 ft. May–Aug. E/W. Introduced.

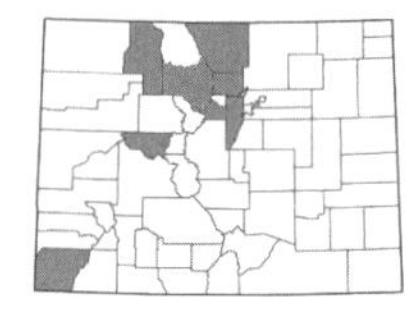

***Lotus tenuis* Waldst. & Kit. ex Willd.**, NARROWLEAF BIRD'S FOOT TREFOIL. [*L. glaber* Mill.]. Ascending or procumbent perennials, 0.5–4 dm; *leaflets* narrowly oblanceolate, 5–15 cm × 2–3 mm, glabrous to sparsely hairy; *inflorescence* with 2–5 (8) flowers; *calyx* tube 2.5–3.5 mm long; *flowers* bright yellow, 10–14 mm long; *legumes* narrowly oblong, terete, 1.5–3.5 cm long, corniculate. Uncommon in meadows and disturbed places, 5200–9000 ft. May–Aug. E/W. Introduced.

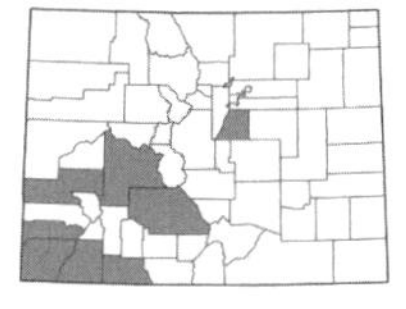

***Lotus wrightii* (A. Gray) Greene**, SCRUB LOTUS. Erect or sprawling perennials, 2–4 (8) dm; *leaflets* oblanceolate or linear to filiform, 8–20 mm, densely strigose with appressed hairs; *inflorescence* with 1 (2) flowers; *calyx* tube 4–5 mm long; *flowers* yellow but aging reddish, 1–1.5 (1.8) cm long; *legumes* oblong, flat, 2–3.5 cm × 1.5–3 mm, strigose or glabrate. Found in rocky soil of open pine forests and along roadsides, 5000–8500 ft. June–Aug. E/W.

LUPINUS L. – LUPINE

Annual or perennial herbs; *leaves* alternate, palmately compound, petiolate; *flowers* zygomorphic, in terminal racemes, usually blue-purple or blue, sometimes white; *calyx* bilabiate; *petals* 5, composed of a banner, wing, and keel; *stamens* 10, monadelphous, alternating long and short; *legumes* laterally compressed, somewhat flattened, hairy. (Detling, 1951; Dunn, 1959; Phillips, 1955)

1a. Annuals to 2 dm tall, softly pilose except for glabrous on the upper surface of the leaflets; cotyledons usually present at flowering time...2
1b. Perennials; cotyledons not present at flowering time...4

2a. Upper lip of the calyx 2.5–4.5 mm long, more than half as long as the lower lip and nearly equal to it; racemes ovoid to shortly cylindric, mostly 0.3–2 cm long...***L. kingii***
2b. Upper lip of the calyx 0.2–2 mm long, less than half as long as the lower lip; racemes subcapitate to shortly cylindric or elongate, to 11 cm long...3

3a. Plants acaulescent or nearly so, 2.5–8 cm high; pedicels 0.2–1.3 mm long; racemes subcapitate to shortly cylindric, the axis becoming 0.3–2.5 (3) cm long; fruit 7–13 mm long...***L. brevicaulis***
3b. Plants usually caulescent, 5–20 cm high; pedicels 1.3–3 mm long; racemes elongate, the axis becoming (1) 1.5–11 cm long; fruit 15–20 mm long...***L. pusillus***

4a. Plants acaulescent or shortly caulescent, the stems and inflorescence included within the leaves; banner not reflexed such that the flower appears closed, narrow, 2.5–3 mm wide; usually a few bracts persistent at anthesis...***L. lepidus* var. *utahensis***
4b. Plants strongly caulescent with an evident stem, stems (1) 1.5–12 dm tall and not shorter than the longer leaves, or if shortly caulescent then the inflorescence well above the leaves; banner various; bracts usually absent at anthesis...5

5a. Banner recurved from a point 3 mm or less below the apex (tip of the flower), giving the flower a closed or only shallowly gaping appearance in profile...6
5b. Banner recurved at or below the midpoint of the flower, 3–6 mm below the apex, giving the flower a widely gaping appearance in profile...8

6a. Plants densely villous or velvety pubescent throughout, often the hairs becoming tawny after drying; racemes densely flowered, somewhat looser in age...***L. leucophyllus***
6b. Plants not densely villous or velvety pubescent throughout, if hairs present then these not drying tawny; racemes various...7

7a. Calyx tube shortly spurred on the upper side at the base (0.5–3 mm long), at least on open flowers (flowers in bud may not have a developed spur yet); banner usually hairy dorsally near the middle (this is easiest to see on unopened flower buds)...***L. caudatus***
7b. Calyx tube symmetrical, or sometimes gibbous at the base on the upper side but the sac less than 1 mm deep; banner usually glabrous or rarely hairy dorsally near the middle...***L. argenteus***

8a. Banner hairy on the dorsal side, at least in the middle (this is easiest to see on unopened flower buds)...***L. sericeus***
8b. Banner glabrous on the dorsal side...9

9a. Basal leaves absent at anthesis; flowers bicolored, the banner purplish with a conspicuous dark blue-purple spot in the middle, in age flowers turn brown; petioles usually glabrous or appressed-hairy, lacking spreading, yellowish hairs...***L. plattensis***
9b. Long-petioled basal leaves present at anthesis, forming a basal leaf tuft; flowers blue with a pale white or yellowish eyespot, blue to blue-purple, or whitish throughout; petioles of basal leaves and lower stem often with yellowish, spreading hairs...10

10a. Stems and petioles with appressed hairs; flowers white, cream, or light yellow, rarely purple-tinged; known from Montrose County...***L. crassus***
10b. Stems and petioles with spreading hairs; flowers blue to purple or rarely white; plants variously distributed...11

11a. Plants from rhizomes; leaflets 5–8...***L. ammophilus***
11b. Plants from a superficial woody caudex; leaflets 8–15...***L. polyphyllus***

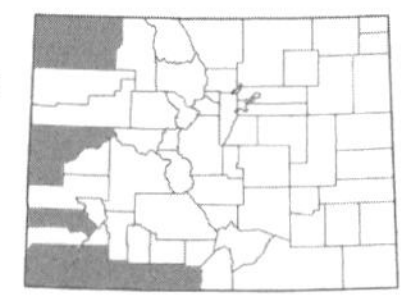

***Lupinus ammophilus* Greene**, SAND LUPINE. Erect perennials, 2–5 dm; *leaflets* 5–8, broadly oblong-oblanceolate, 3–5 cm long, glabrous to glabrate above; *inflorescence* with numerous flowers; *calyx* 4–6 mm long; *flowers* blue-violet, 10–15 mm long, the banner orbicular; *legumes* 2–3 × 5–6 mm, hairy. Found in the southwestern counties in dry areas with sagebrush, oak, or pinyon-juniper, 5000–7500 ft. May–June. W.

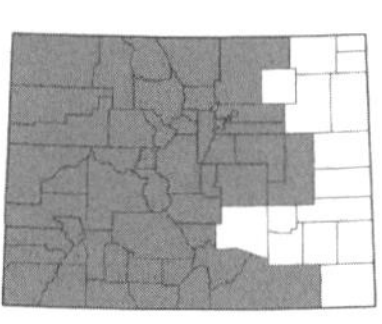

***Lupinus argenteus* Pursh**, SILVERY LUPINE. Erect or ascending perennials, 2–6 dm; *leaflets* 5–9, oblanceolate or oblong-oblanceolate, folded or flat, hairy or glabrate above, hairy below; *calyx* 4–10 mm long, sometimes gibbous; *flowers* various shades of blue-violet or lavender, occasionally white, 5–14 mm long, the banner glabrous or inconspicuously hairy near the middle; *legumes* 2–2.5 cm × 5–9 mm, pilose. Common in dry, open places or moist mountain meadows and forest openings, 4800–10,500 ft. May–Sept. E/W. (Plate 45)

Lupinus argenteus is a large complex with numerous intergrading varieties and can also hybridize with *L. caudatus* (see discussion under *L. caudatus*):

1a. Banner pale blue with a dark brown eyespot and dark-spotted keel; flowers 5–7 mm in length...**var. *fulvomaculatus* (Payson) Barneby**, LODGEPOLE LUPINE. [*L. fulvomaculatus* Payson]. Usually found in moist places in the mountains, scattered in the central mountains and on the western slope, 7000–9000 ft. June–Aug. E/W.
1b. Banner with a pale eyespot or without an eyespot, variously colored; flowers various...2

2a. Banner hairy on the back (this is often difficult to see); inflorescence generally loose...**var. *holosericeus* (Nutt.) Barneby**, HOLO LUPINE. [*L. holosericeus* Nutt.]. Found on sagebrush plains, known from Middle Park and North Park, 7000–8000 ft. May–Aug. W.
2b. Banner glabrous on the back; inflorescence various...3

3a. Plants of dry places from the foothills to mountains; leaflets usually folded...**var. *argenteus***, SILVERY LUPINE. [*L. argenteus* var. *moabensis* (D. Dunn & Harmon) S.L. Welsh; *L. argenteus* var. *myrianthus* (Greene) Isely]. Common in dry, open places, along roadsides, and occasionally in open prairie, scattered throughout the state (except mostly absent from the eastern plains), usually replaced at higher elevations by var. *rubricaulis*, 4800–10,500 ft. May–Sept. E/W.
Varieties *moabensis* and *myrianthus* are often separated from var. *argenteus* by the size of the flowers, but as this is the only difference between the them they are here included within var. *argenteus*.
3b. Plants of moist, cool mountain meadows, along streambanks or other wet places, and aspen and lodgepole pine forests to timberline; leaflets usually broad and flat...4

4a. Inflorescence terminal and solitary; wings 7.5–11 mm long; stems simple...**var. *rubricaulis* (Greene) S.L. Welsh**, SILVERY LUPINE. [*L. depressus* Rydb.; *L. rubricaulis* Greene]. Common in mountain meadows and forest openings, found in the central mountains and scattered on the western slope, 6300–11,000 ft. May–Sept. E/W.
4b. Inflorescence branching, of many lateral racemes maturing with or soon after the central terminal raceme; wings 5–7.5 mm long; stems usually branching distally ...**var. *parviflorus* (Nutt. ex Hook. & Arn.) C.L. Hitchc.**, LODGEPOLE LUPINE. [*L. parviflorus* Nutt. ex Hook. & Arn. ssp. *parviflorus*]. Usually found in cool meadows or forests, occasionally in sagebrush flats, scattered in the western and central counties, less common than var. *rubricaulis*, 6500–10,000 ft. June–Sept. E/W.

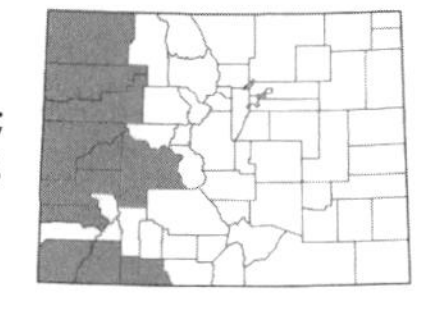

***Lupinus brevicaulis* S. Watson**, SHORTSTEM LUPINE. Acaulescent to shortly caulescent annuals, to ca. 2.5–8 cm; *leaflets* 5–7, cuneate-obovate to narrowly oblanceolate, 1–1.5 cm long, villous or pilose; *inflorescence* with 3–15 flowers; *calyx* 3–5 mm long; *flowers* purple-blue or occasionally white, 5–8 mm long; *legumes* ovate to elliptic, 7–13 × 4–6 mm, pilose. Found in dry, open places in sandy or rocky soil, 4700–6600 ft. April–June. W.

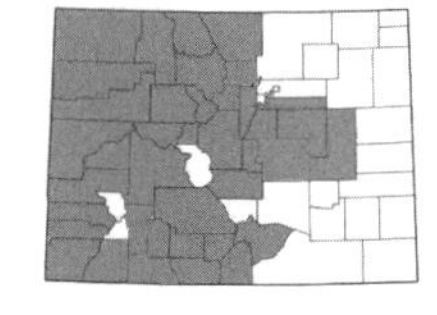

***Lupinus caudatus* Kellogg**, TAILCUP LUPINE. Caulescent, ascending perennials, 2–8 dm; *leaflets* 5–9, elliptic to narrowly lanceolate, flat or folded, 5–7 times as long as wide, hairy; *inflorescence* with 10-numerous flowers; *calyx* shortly spurred; *flowers* purple or sometimes white, 8–14 mm long, the banner usually hairy dorsally near the middle; *legumes* 2–3 cm × 8–9 mm. Common on dry, open hillsides and along roadsides, occasionally found in moist meadows, 5300–10,500 ft. May–Sept. E/W.

Lupinus caudatus is often included within the large *L. argenteus* complex. However, the short spur and usually pubescent banner are good characters to separate it from *L. argenteus*. *Lupinus sericeus* approaches *L. caudatus* (especially var. *utahensis*) in having a pubescent banner, but the flowers are widely gaping in profile and lack a spur. In addition, *L. sericeus* does not have long-petioled basal leaves present at anthesis like *L. caudatus* var. *utahensis*. Hybrids between *L. caudatus* and *L. argenteus* can form, and the calyx is gibbous to slightly spurred at the base, but the banner is usually glabrous. There are 2 well-marked+ but ultimately intergrading varieties of *L. caudatus* in Colorado:

1a. Long-petioled basal leaves lacking; petioles of the lowest stem leaves only 1.5–3 cm long; banner slightly hairy dorsally near the middle...**var. *argophyllus* (A. Gray) L. Phillips**, KELLOGG'S SPURRED LUPINE. [*L. aduncus* Greene; *L. argenteus* Pursh var. *argophyllus* (A. Gray) S. Watson]. Common on dry hillsides, occasionally in moist meadows, scattered throughout the central mountains and western counties, 5300–10,500 ft. May–Sept. E/W.

1b. Long-petioled basal leaves present when just beginning to flower, sometimes lacking at maturity; petioles of the lowest stem or basal leaves 3–12 mm long; banner densely hairy dorsally near or below the middle...**var. *utahensis* (S. Watson) S.L. Welsh**, UTAH SPURRED LUPINE. [*L. argenteus* Pursh var. *utahensis* (S. Watson) Barneby]. Common on open, dry hillsides, found mostly in the westcentral and northwestern counties, 5800–10,000 ft. May–Aug. W.

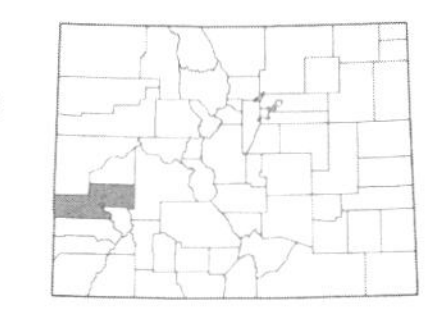

***Lupinus crassus* Payson**, PARADOX LUPINE. [*L. ammophilus* Greene var. *crassus* (Payson) Isely]. Erect perennials, 2–5 dm; *leaflets* 5–8, broadly oblong-oblanceolate, 3–5 cm long, glabrous or glabrate above; *inflorescence* with numerous flowers; *calyx* 4–6 mm long; *flowers* white to light yellow, rarely purple-tinged, 10–15 mm long; *legumes* laterally compressed, 2–3 × 5–6 mm, pilose. Found on adobe hills and in sandy soil of dry hillsides, often with sagebrush or pinyon-juniper, known from Paradox Valley and vicinity, 5000–6000 ft. May–June. W. Endemic.

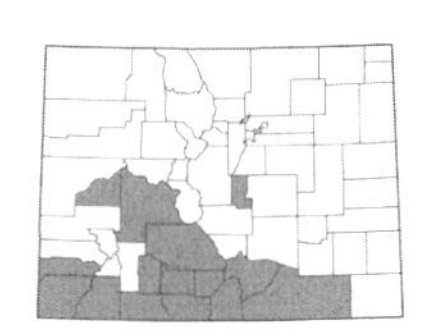

***Lupinus kingii* S. Watson**, KING'S LUPINE. Caulescent or subacaulescent annuals, 1–2 dm; *leaflets* 4–7, elliptic to oblanceolate, 1–2 cm long, slightly glaucous, glabrous above; *inflorescence* with 3–10 (15) flowers, 4–20 mm long; *calyx* 5–7 mm long; *flowers* blue or violet but aging white, 6–8 mm long, the banner barely reflexed; *legumes* ovate-acuminate, 10–13 × 4–5.5 mm, pilose. Locally common in dry, open places, 6500–10,000 ft. June–Aug. E/W.

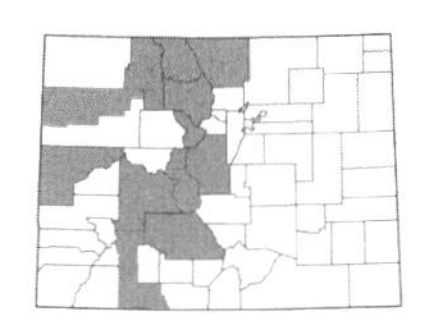

***Lupinus lepidus* Douglas ex Lindl. var. *utahensis* (S. Watson) C.L. Hitchc.**, UTAH LUPINE. [*L. caespitosus* Nutt. var. *utahensis* (S. Watson) Cox]. Acaulescent perennials, 0.5–3 (8) dm; *leaflets* 5–9, narrowly to broadly oblanceolate, 1–3 cm long, glabrate to sericeous or strigose above, sericeous or strigose below; *inflorescence* of sessile, terminal racemes; *calyx* 4.5–7 mm long; *flowers* blue to lavender or occasionally white, the banner with a large pale spot; *legumes* elliptic-oblong, 1–1.5 cm × 3–5 mm, short-hairy. Found in sandy or rocky soil of dry, open places, often with sagebrush, 6000–10,000 ft. May–Aug. E/W.

***Lupinus leucophyllus* Douglas ex Lindl.**, VELVET LUPINE. Erect perennials, 3–10 dm; *leaflets* 6–11, oblanceolate, 3–6 cm long, densely hairy above and below; *inflorescence* with numerous flowers, 1–2 dm × 1–3 cm; *calyx* 5–8 mm; *flowers* lavender, light blue or whitish, 6–10 mm long, the banner only somewhat reflexed; *legumes* shortly oblong, 1.5–2.5 cm × 5–8 mm. Uncommon on dry, open slopes, our one record from Utah 0.4 mi west of the Colorado-Utah state line in Moffat Co., and it is possible that it occurs in Colorado as well. July–Sept.

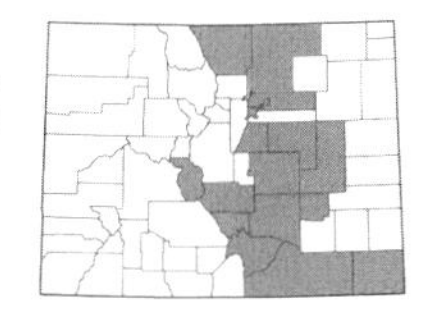

***Lupinus plattensis* S. Watson**, NEBRASKA LUPINE. Erect to ascending perennials, 1–5 dm; *leaflets* 7–9, oblanceolate to spatulate, 2–4 cm long, glabrate above, pubescent below; *inflorescence* 0.5–2 dm long; *calyx* 8–9 mm long, sericeous; *flowers* violet or bicolored, with a dark-eyed banner, aging brown; *legumes* oblong, 2–4 cm × 8–9 mm, villous with appressed hairs. Locally common on the high plains in sandy or rocky soil, 4200–8800 ft. May–July (Aug.). E. (Plate 45)

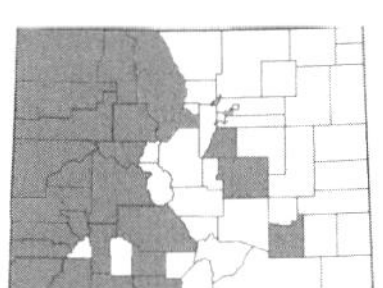

***Lupinus polyphyllus* Lindl.**, BIGLEAF LUPINE. Perennials, 2–10 (15) dm; *leaflets* 8–12, oblanceolate, glabrate above; *inflorescence* with numerous flowers, 0.5–2 dm long; *calyx* 4–7 mm long, villous, glandular; *flowers* blue to violet, bicolored or pale, with a white-eyed banner; *legumes* 2.5–4 × 0.7 cm, pilose. Common in dry places, often with sagebrush, 6000–10,000 ft. May–July. E/W.

There are 2 varieties of *L. polyphyllus* in Colorado:

1a. Leaves hairy on both sides, often folded and appearing linear-oblanceolate...**var. *humicola* (A. Nelson) Barneby**, BIGLEAF LUPINE. [*L. wyethii* S. Watson]. Uncommon in the western counties in dry places, often with sagebrush, 7000–9500 ft. May–July. E/W.

1b. Leaves glabrous above or only thinly hairy along the margins, usually flat and broadly to narrowly oblanceolate...**var. *prunophilus* (M.E. Jones) L. Phillips**, HAIRY BIGLEAF LUPINE. [*L. greenei* A. Nels; *L. prunophilus* M.E. Jones]. Widespread, especially in the northwestern counties (usually Gunnison Co. north), found in dry places along roadsides, often in sagebrush and grasslands, 6000–10,000 ft. May–July. E/W.

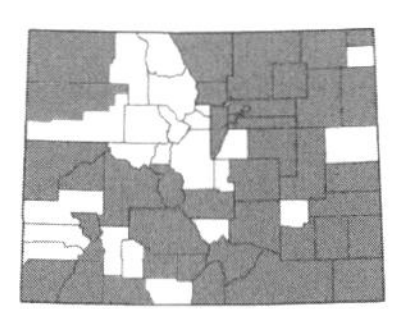

***Lupinus pusillus* Pursh**, RUSTY LUPINE. Acaulescent to subacaulescent annuals, 5–20 cm; *leaflets* 5–7, obovate-elliptic to oblanceolate, 1–4 cm long, glabrous above; *inflorescence* with numerous flowers, to 4 cm long; *calyx* 4–5 mm long, glabrous or pilose; *flowers* various shades of blue-purple or bicolored, 5–10 (12) mm long; *legumes* ovate-acuminate to shortly oblong, 15–20 × 6–7 mm, hirsute. Common throughout the eastern plains, San Luis Valley, and on rocky, dry slopes, 3500–8500 ft. April–July. E/W. (Plate 45)

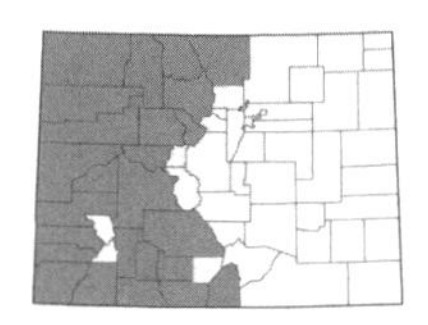

***Lupinus sericeus* Pursh var. *sericeus*,** SILKY LUPINE. [*L. bakeri* Greene]. Erect perennials, 4–15 dm; *leaflets* 6–9, oblanceolate, 3–7 cm long, villous or strigulose above and below, or thinly villous or strigulose above; *inflorescence* with numerous, well-spaced flowers; *calyx* 6–10 cm long; *flowers* purple-violet with a pale-eyed banner that darkens with age, 10–13 mm; *legumes* oblong, 2–3 cm × 6–8 mm, hairy. Common in dry, open places, and in meadows, 6300–10,000 ft. May–Aug. E/W.

MEDICAGO L. – ALFALFA

Annual or perennial, caulescent herbs with erect or prostrate stems; *leaves* alternate, pinnately trifoliate, the leaflets serrate distally; *flowers* zygomorphic, in axillary racemes or heads; *petals* 5, composed of a banner, wing, and keel; *stamens* 10, diadelphous; *fruit* curved to spirally coiled, indehiscent.

1a. Flowers purple or rarely white...***M. sativa***
1b. Flowers yellow...2

2a. Flowers 2–3 mm long; sepals 1–1.5 mm long; leaflets mostly 0.8–1.5 cm long...***M. lupulina***
2b. Flowers 7–11 mm long; sepals mostly 4–6 mm long; leaflets mostly 1–4 cm long ...***M. falcata***

***Medicago falcata* L.,** YELLOW ALFALFA. Prostrate or erect perennials, 3–8 dm; *leaflets* oblanceolate to elliptic, 1–4 cm long; *inflorescence* of axillary racemes with (5) 10–20 flowers; *calyx* 4–6 mm long; *flowers* yellow, 7–11 mm long; *legumes* oblong, falcate or hamate, laterally compressed, 5–10 × 2–4 mm, loosely strigulose. Infrequent along roadsides and in disturbed areas, 6000–8000 ft. May–Oct. W. Introduced.

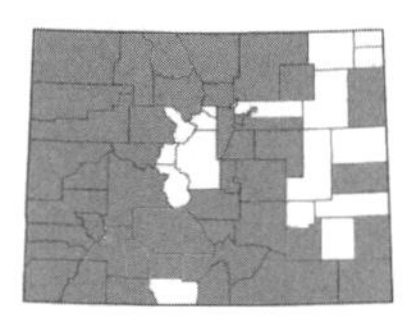

***Medicago lupulina* L.,** BLACK MEDICK. Prostrate or ascending annuals, (0.5) 1–4 dm; *leaflets* rhombic-obovate, oblanceolate or obovate, 0.8–1.5 mm long; *inflorescence* with 10-numerous tiny flowers; *calyx* 1–1.5 mm long; *flowers* yellow, 2–3 mm long; *legumes* reniform-incurved, 2–2.5 × 1.5–2 mm, black at maturity, glabrous to occasionally hairy. Common along roadsides, in fields, forest openings, and disturbed areas, 3900–9000 ft. April–Oct. E/W. Introduced.

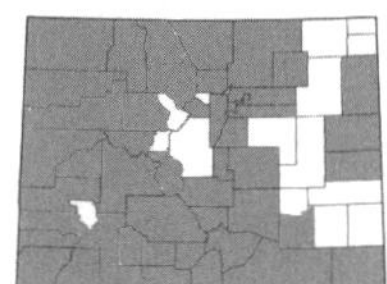

***Medicago sativa* L.,** ALFALFA. Erect or sometimes decumbent perennials, 2–8 (10) dm; *leaflets* obovate to oblong-oblanceolate or narrowly lanceolate, 1–2.5 cm long; *inflorescence* of axillary racemes with 8-numerous crowded flowers; *calyx* 4–5.5 mm long; *flowers* purple or rarely white, 8–11 mm long; *legumes* coiled with (1) 2–3 turns or occasionally falcate, 4–6 (8) mm diam., glabrous to hairy. Common along roadsides, in fields, and disturbed areas, 4000–9500 ft. May–Oct. E/W. Introduced. (Plate 45)

MELILOTUS L. – SWEET CLOVER

Annuals, biennial, or short-lived perennial herbs; *leaves* alternate, pinnately trifoliate, the leaflets toothed distally; *flowers* zygomorphic, in axillary racemes; *calyx* 5-lobed, *petals* 5, composed of a banner, wing, and keel, white or yellow; *stamens* 10, diadelphous; *legumes* straight, indehiscent, 1-2-seeded.

1a. Flowers white, the banner conspicuously longer than the wings; seeds never mottled...***M. albus***
1b. Flowers yellow, the banner nearly equal to the wings; seeds sometimes purple-mottled...***M. officinalis***

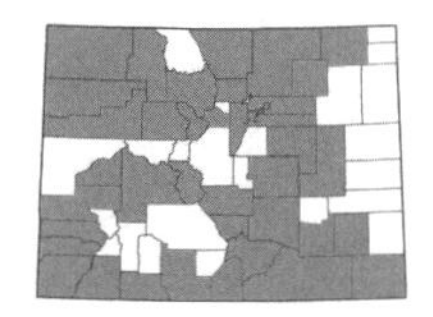

***Melilotus albus* Medik.,** WHITE SWEET CLOVER. Biennials or annuals; *leaflets* obovate to elliptic or elliptic-oblong, 1.2–5 cm long; *inflorescence* of axillary racemes, 4–12 cm long; *calyx* ca. 2 mm long; *flowers* white, (3.5) 4–5 mm long; *legumes* ovoid, compressed, 3–4 × 1.5–2 mm, reticulate. Common along roadsides, in fields and pastures, and in weedy or disturbed areas, 3400–8500 ft. May–Oct. E/W. Introduced.

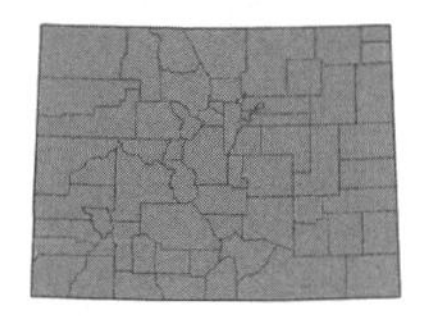

***Melilotus officinalis* (L.) Pall.,** YELLOW SWEET CLOVER. Biennials, 0.5–2 dm; *leaflets* obovate, ovate or elliptic, 1–2.5 cm long; *inflorescence* of racemes, 4–12 cm long; *calyx* 2–2.5 mm long; *flowers* yellow, 5–7 mm long; *legumes* ovoid, compressed, 3–4 × 1.5–2 mm, cross-striate or cross-rugose. Common along roadsides, in fields, and in disturbed and weedy areas, 3500–9000 ft. May–Oct. E/W. Introduced.

MIMOSA L. – CATCLAW; SENSITIVE BRIER

Shrubs or perennial herbs, armed with stout, recurved prickles; *leaves* alternate, bipinnately compound, the leaflets entire and often touch sensitive; *flowers* actinomorphic, in axillary, globose heads; *calyx* 5-lobed; *petals* 5, connate or distinct; *stamens* 8–10, distinct or united below, exserted; *legumes* flattened or linear, often prickly.

1a. Much-branched shrub to 2 m tall; stems with alternating, stout recurved prickles; legumes constricted between the seeds, the margins sometimes with prickles but otherwise the fruit glabrous...***M. borealis***
1b. Perennial herb, usually prostrate or sprawling; stems with numerous weak, recurved prickles; legumes not constricted between the seeds, covered with numerous recurved prickles...2

2a. Leaflets with only the midrib evident; stems usually minutely hairy; seeds obovate, mostly 6–7 mm long...***M. rupertiana***
2b. Leaflets with an evident midrib and lateral veins, at least in the upper half; stems glabrous; seeds quadrate-rhombic, mostly 4 mm long...***M. nuttallii***

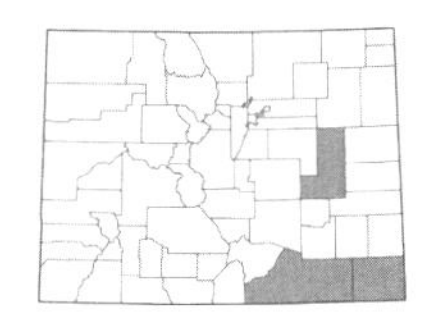

***Mimosa borealis* A. Gray**, PINK MIMOSA. Shrubs to 2 m; *leaves* with 6–16 leaflets per pinna; *leaflets* elliptic to elliptic-oblong, 2–4 (5) mm long, glabrous to occasionally minutely hairy; *inflorescence* of globose heads, 9–15 mm diam.; *calyx* 0.4–0.8 mm long; *flowers* rose to pinkish, 2–2.5 (3) mm long; *legumes* oblong, curved and contorted, 3–5 cm × 6–8 mm, yellow-brown, prickly or not. Uncommon in rocky soil and in canyons, 4000–5200 ft. May–July. E.

***Mimosa nuttallii* (DC.) B.L. Turner**, NUTTALL'S CATCLAW. [*Schrankia nuttallii* (DC. ex Britton & Rose) Standl.; *S. uncinata* Willd.]. Sprawling perennials; *leaves* with 10–18 pinnae, each pinna with 20–50 leaflets; *leaflets* with evident midrib and lateral veins; *inflorescence* of 1–3 heads clustered in leaf axils; *calyx* campanulate, reduced; *flowers* pink; *legumes* oblong to linear, usually quadrangular, densely prickled. Uncommon on sandy soil, reported for Colorado from Baca Co., 3600–4000 ft. May–Aug. E. (Plate 45)

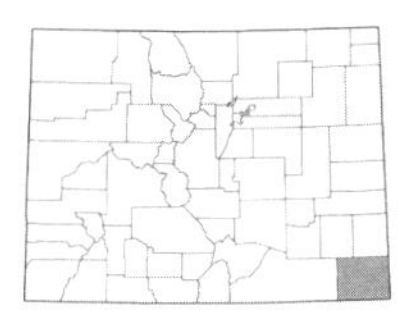

***Mimosa rupertiana* B.L. Turner**, EASTERN CATCLAW. [*M. quadrivalvis* L. var. *occidentalis* (Wooton & Standl.) Barneby; *Schrankia occidentalis* (Wooton & Standl.) Standl.]. Sprawling perennials; *leaves* with 2–4 pinnae, each pinna with 20–50 leaflets; *leaflets* with only an evident midrib; *inflorescence* of 1–3 heads clustered in leaf axils; *calyx* campanulate, reduced; *flowers* pink; *legumes* oblong to linear, usually quadrangular, 5–12 cm long, densely prickled. Uncommon in sandy soil, 3600–4500 ft. May–July. E.

ONOBRYCHIS Mill. – SAINFOIN

Perennials, caulescent herbs; *leaves* alternate, odd-pinnately compound; *flowers* zygomorphic, in axillary racemes, each subtended by a bract; *calyx* 5-lobed; *petals* 5, composed of a banner, wing, and keel, purple to pink, the keel longer than the wings; *stamens* 10, diadelphous; *fruit* a loment reduced to 1 segment, armed with prickles.

***Onobrychis viciifolia* Scop.**, SAINFOIN. Plants 3–8 dm; *leaflets* 15–21, obovate to narrowly elliptic, 1–2.5 cm, red-dotted above; *calyx* 5–7 mm; *flowers* dull pink to pink-purple, 8–15 mm; *legumes* obovate, 6–7 mm, rugose-reticulate, shortly prickled. Uncommon in fields and along roadsides where it is used for revegetation and soil improvement, scattered throughout the state, 5000–9000 ft. May–Aug. E/W. Introduced.

OXYTROPIS DC. – LOCOWEED

Perennials, acaulescent or sometimes caulescent herbs; *leaves* alternate, odd-pinnately compound; *flowers* zygomorphic, in racemes or spikes; *calyx* 5-lobed; *petals* 5, composed of a banner, wing, and keel, the keel with a prominent ascending beak, white, ochroleucous, pink, or purplish; *stamens* 10, diadelphous; *legumes* sessile or stipitate, sometimes inflated. (Barneby, 1952; Porter, 1951)

1a. Calyx teeth, undersides of the leaflets and bracts, pod, and sometimes the whole plant with wart-like hairs, giving the plant a viscid appearance, also with white and sometimes black, glandless hairs...***O. viscida***
1b. Plants without wart-like hairs...2

2a. Flowers white to ochroleucous or pale yellow, sometimes with a purple keel tip...3
2b. Flowers purple, blue-purple, pink-purple, or pale purple...5

3a. Flowers pendulous at late anthesis (spreading at early anthesis; if picked early, look at the lowest flowers in the raceme to see if any are beginning to nod); flowers smaller, the banner 6–11 mm and the keel 7–8.5 mm in length; stipules shortly adnate to the base of the petiole for up to 3 mm; fruit pendulous, stipitate on a short stipe to 2 mm long...***O. deflexa* var. *sericea***
3b. Flowers ascending throughout anthesis; flowers larger, the banner 12–26 mm and the keel 10–19 mm in length; stipules adnate to the base of the petiole for half their length or more, or for at least 4 mm; fruit erect, sessile or nearly so...4

4a. Keel tip usually with a purple spot; flowers generally larger, the banner (15) 18–22 (25) mm long; wing petals conspicuously obliquely dilated upward and 5–8 mm wide near the apex; keel 13–19 mm long; fruit rigid, sometimes almost woody, and stiffly leathery or chartaceous at maturity...***O. sericea* var. *sericea***
4b. Keel tip lacking a purple spot; flowers generally smaller, the banner (10) 14–18 (20) mm long; wing petals not conspicuously dilated upward, mostly 3–4.5 mm wide near the apex; keel 10–14.5 mm long; fruit not rigid, with thin, fleshy valves becoming papery at maturity...***O. campestris***

5a. Flowers pendulous at late anthesis (spreading at early anthesis; if picked early, look at the lowest flowers in the raceme to see if any are beginning to nod); pods pendulous; flowers usually smaller, the banner 6–11 mm and the keel 4–8.5 mm long; stipules shortly adnate to the base of the petiole for up to 3 mm...***O. deflexa***
5b. Flowers remaining erect to spreading at late anthesis; pods erect or spreading; flowers usually larger, the banner 12.5–25 mm and keel 10–18 mm long (if smaller then the inflorescence 1-3-flowered and the leaves densely silky-pilose); stipules adnate to the base of the petiole for half their length or more, or for at least 4 mm...6

6a. Inflorescence 1–4-flowered, the flowering scape to 3 cm in length; leaves 0.5–7.5 cm long...7
6b. Inflorescence 6–35-flowered, the flowering scape 2–36 cm in length; leaves 5–26 cm long...9

7a. Flowers smaller, the banner 7–10 (12) mm long and the keel 8–9 mm in length; calyx 5–8 mm long, the tube portion 3–5 mm long; fruit lanceolate or lance-oblong, sessile...***O. parryi***
7b. Flowers larger, the banner 12–22 mm long and the keel 10–18 mm in length; calyx 7–13 mm long, the tube portion 5–10 mm long; fruit ovoid to ovoid-elliptic, stipitate (the stipe 0.5–3 mm long)...8

8a. Keel 12–18 mm long; calyx densely silky-pilose with white hairs mostly 1.5–2 mm long; fruiting calyx becoming inflated and enclosing the body and sometimes even the beak of the fruit; fruit 6–11 mm long...***O. multiceps***
8b. Keel 10–11 mm long; calyx pilose-hispidulous, with a mixture of shorter dark hairs and longer white hairs, the longer white hairs mostly 0.5–1 mm long; fruiting calyx not becoming inflated and enclosing the fruit; fruit 15–25 (30) mm long...***O. podocarpa***

9a. Most of the leaflets whorled with 3 or more per node on the rachis; leaflets, calyx, and scape densely pilose...***O. splendens***
9b. Leaflets opposite or alternate with no more than 2 per node; surfaces various but usually not as densely pilose...10

10a. Pubescence mostly of dolabriform hairs attached shortly above the base, composed of 2 arms with one arm considerably shorter than the other; fruit ovoid to lanceolate-acuminate...***O. lambertii***
10b. Pubescence entirely of basifixed hairs; fruit ovoid or semiorbicular...***O. besseyi* var. *obnapiformis***

***Oxytropis besseyi* (Rydb.) Blank. var. *obnapiformis* (C.L. Porter) S.L. Welsh**, BESSEY'S LOCOWEED. Erect to spreading, acaulescent perennials, 0.3–2.5 dm; *leaflets* 17–23, elliptic to oblong-lanceolate, 4–20 (25) mm long; *calyx* tube cylindric or campanulate, villous to pilose, or with loosely appressed hairs; *flowers* various shades of pink-purple or sometimes white, 16–25 mm long; *legumes* ovoid to semiorbicular, inflated, minutely hairy or villous. Rare in sandy soil, 5400–7000 ft. May–June. W.

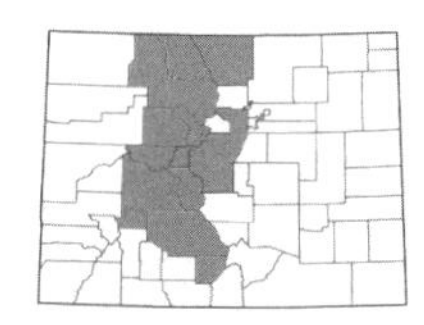

***Oxytropis campestris* (L.) DC.**, FIELD LOCOWEED. Erect to ascending perennials, 0.5–3 dm; *leaflets* 9-numerous, ovate to oblong-lanceolate; *calyx* tube 4–7 mm long, hairy; *flowers* whitish or ochroleucous to yellow, occasionally purple, (10) 14–18 (20) mm long; *legumes* lanceolate or ellipsoid, (5) 8–15 × 3.5–5 mm. Uncommon in meadows, along streams, and in the alpine, 8000–12,000 ft. June–Aug. E/W.

There are two varieties of *O. campestris* in Colorado:

1a. Leaflets usually more numerous, 17–33 in number; racemes 10–30-flowered; scapes 15–30 cm in length; stipules pilose or glabrate...**var. *spicata* Hook.**, YELLOW-FLOWERED LOCOWEED. [*O. campestris* var. *gracilis* (A. Nelson) Barneby; *O. monticola* A. Gray]. Uncommon in meadows, along streams, and in the alpine, 8000–12,000 ft.
1b. Leaflets usually fewer, 7–15 in number; racemes 3–15-flowered; scapes 5–10 (15) cm in length; stipules glabrate or sometimes ciliate on the margins...**var. *cusickii* (Greenm.) Barneby**, CUSICK'S LOCOWEED. [*O. cusickii* Greenm.]. Uncommon on grassy slopes and in meadows. 8500–9500 ft.

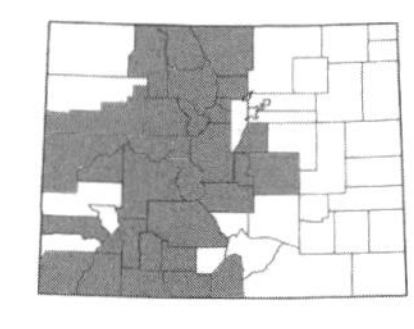

***Oxytropis deflexa* (Pall.) DC.**, NODDING LOCOWEED. Sprawling or ascending, shortly caulescent or subacaulescent perennials; *leaflets* 11-numerous, ovate to oblong-lanceolate, reduced in size distally, (1.5) 3–15 (25) mm long; *calyx* tube 2–4 mm long, pilose to black-puberulent; *flowers* purple to pale bluish, lavender, or whitish, 6–11 mm long; *legumes* ellipsoid to oblong, slightly obcompressed, 8–18 cm × 3–4.5 mm, hirsute with light or dark hairs. Locally common in meadows, open slopes, along streams, and in forest openings, 6000–14,000 ft. June–Aug. E/W.

There are two varieties of *O. deflexa* in Colorado. *Oxytropis deflexa* var. *foliolosa* is merely var. *deflexa* found at higher altitudes and falls within the variability of var. *deflexa* as a whole:

1a. Plants usually caulescent with 1–7 internodes; flowers pale purple to blue or whitish, sometimes just the banner bright purple...**var. *sericea* Torr. & A. Gray**, WHITE NODDING LOCOWEED. Locally common in mountain meadows and open slopes, along streams, and in forest openings in the central mountains, 6000–10,500 ft. June–Aug. E/W.

1b. Plants acaulescent; flowers bright purple, sometimes the claws of the banner paler...**var. *deflexa***, NODDING LOCOWEED. [*O. deflexa* var. *foliolosa* (Hook.) Barneby]. Locally common on open slopes and in forest openings, 8800–14,000 ft. June–Aug. W.

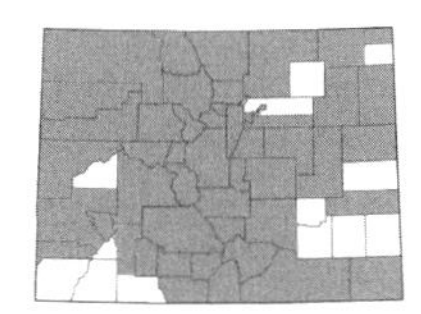

***Oxytropis lambertii* Pursh**, PURPLE LOCOWEED. Ascending or spreading, acaulescent perennials, (0.3) 0.5–2 dm, with mostly dolabriform hairs; *leaflets* (5) 7–15 (19), elliptic to linear-lanceolate, usually involute, 1–2.5 (3.5) cm long; *calyx* tube 5–9 mm long, strigose or villous; *flowers* various shades of pink-purple to violet or occasionally white, (12) 15–25 mm long; *legumes* ovoid to lanceolate-acuminate, obcompressed, 10–25 × 3–5 mm, silky hairy to strigose. Common from the plains to subalpine, often in rocky soil, 3500–11,000 ft. April–Aug. E/W. (Plate 45)

Oxytropis lambertii can hybridize with *O. sericea*, resulting in a plant with light purple flowers but usually with the basifixed hairs of *O. sericea*.

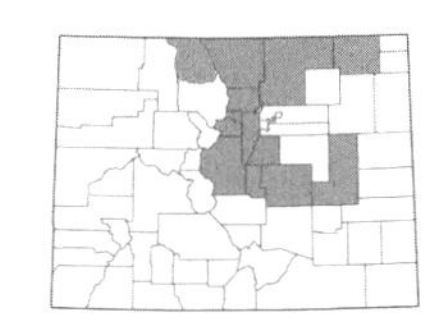

***Oxytropis multiceps* Nutt.**, NUTTALL'S OXYTROPE. Caespitose, acaulescent perennials; *leaflets* 5–9, elliptic to lanceolate or oblanceolate, usually involute or folded, (3) 5–15 mm long; *inflorescence* of subcapitate racemes with 2–4 erect flowers, ultimately humistrate; *calyx* tube 5–10 mm long, villous, pilose or with loosely appressed hairs; *flowers* pink-purple, 16–25 mm long; *legumes* ovoid-ellipsoid, 6–11 × 3–5 mm, villous. Locally common in rocky or sandy soil, 4400–11,000 ft. April–June. E. (Plate 45)

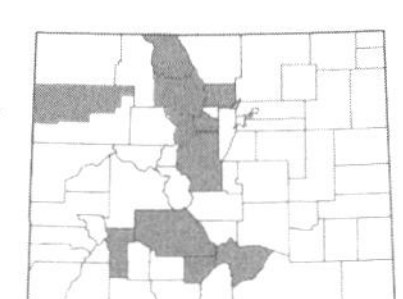

***Oxytropis parryi* A. Gray**, PARRY'S LOCOWEED. Caespitose, acaulescent perennials, 0.3–0.8 (1) dm; *leaflets* (7) 11–15, ovate to oblong, typically involute or folded, 3–10 (12) mm long; *inflorescence* of ascending-arcuate to reclinate racemes with 1–3 flowers; *calyx* tube 3–5 mm long; *flowers* purple or blue-purple, sometimes white, 7–10 (12) mm long; *legumes* lanceolate or lance-oblong, 15–20 × 5–7 mm, sericeous and villous with both black and white hairs. Uncommon on rocky, open slopes in spruce-fir forests and the alpine, 8500–12,500 ft. June–July. E/W.

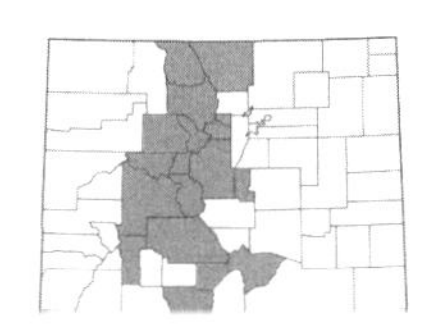

***Oxytropis podocarpa* A. Gray**, STALKPOD LOCOWEED. Erect to spreading, acaulescent perennials, 0.05–0.4 dm; *leaflets* (7) 9–13, elliptic to oblong, usually folded and recurved, 4–8 mm long; *inflorescence* of racemes with 1–3 erect to spreading flowers; *calyx* 5–7 mm long, with light to dark hairs; *flowers* purple or sometimes white, 13–16 (18) mm long; *legumes* ovoid, slightly obcompressed, 1.5–2.5 (3) cm, minutely hairy. Locally common in the alpine, 10,500–14,000 ft. June–Aug. E/W.

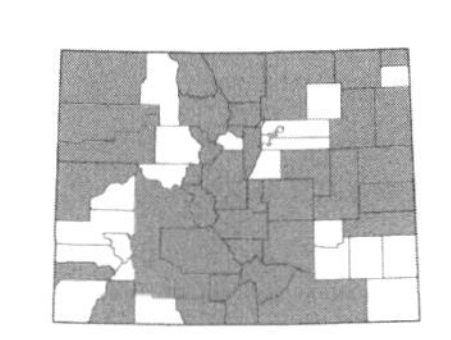

Oxytropis sericea* Nutt. var. *sericea, WHITE LOCOWEED. Acaulescent perennials, 1–3 (5) dm; *leaflets* (9) 11–19 (23), ovate to oblong-lanceolate, (0.7) 1–2.5 (4) cm long; *calyx* tube (5) 6–9 mm long, strigose to loosely sericeous with light to dark hairs; *flowers* white with a maculate keel, sometimes ochroleucous or pinkish, (15) 18–22 (25) mm long; *legumes* lanceolate or oblong, somewhat obcompressed, 1.2–2.5 cm × 4–8 mm. Common from the plains to mountains, in meadows and forest openings, and along roadsides, 3500–12,000 ft. (April) May–Aug. E/W. (Plate 45)

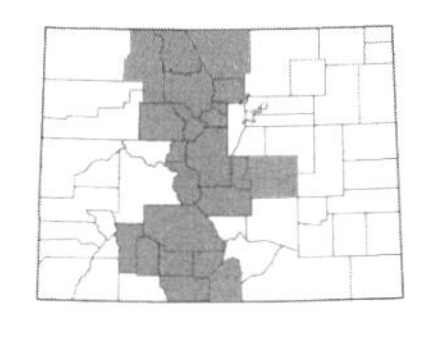

***Oxytropis splendens* Douglas ex Hook.**, SHOWY LOCOWEED. Erect, acaulescent perennials,1–4 dm; *leaflets* numerous, either paired or in whorls of 3–4, elliptic to oblong-lanceolate, reduced distally, (3) 6–20 cm long; *calyx* tube 5–6 mm long, sericeous-villous; *flowers* pink to bright red, drying violet, 12–16 mm long; *legumes* ellipsoid to lanceolate, obcompressed, 10–17 × 3–5 mm. Common in open meadows, forest openings, and on alpine slopes, 7600–13,000 ft. June–Aug. E/W. (Plate 45)

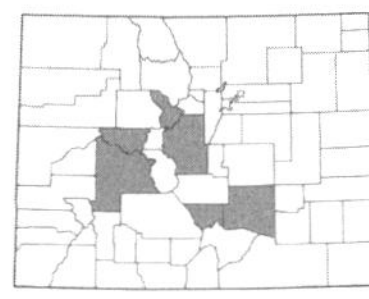

***Oxytropis viscida* Nutt.**, STICKY LOCOWEED. [*O. borealis* DC. var *viscida* (Nutt.) S.L. Welsh]. Erect to ascending, acaulescent perennials, 0.5–3 dm, with wart-like hairs; *leaflets* (21) 25–39, ovate to oblong-lanceolate, 2–15 mm long; *calyx* tube 4–7 mm long, with light or dark hairs; *flowers* purple, 10–16 mm long; *legumes* ovoid to lanceolate, obcompressed, 8–15 (20) × (3) 4–5 mm, light- to dark-hairy. Found in rocky alpine and subalpine meadows, 9000–14,000 ft. July–Aug. E/W.

PARRYELLA Torr. & A. Gray ex A. Gray – PARRYELLA

Shrubs, unarmed; *leaves* alternate, pinnately compound, leaflets gland-dotted; *flowers* actinomorphic, small, in terminal spicate racemes; *calyx* 5-lobed, 10-ribbed near the base; *petals* absent; *stamens* 10, distinct, attached to the base of the sepals; *legumes* indehiscent, 1-seeded.

***Parryella filifolia* Torr. & A. Gray**, COMMON DUNEBROOM. Shrubs, 10–15 dm; *leaflets* filiform, occasionally broader, folded or rolled, 0.3–1.5 cm long; *calyx* tube 2.5–3 mm long, glabrous, ciliate; *flowers* of 9–10 stamens and a style; *legumes* oval to ovoid, compressed, evidently glandular with large glands. Uncommon in sandy soil and on dunes, barely entering Colorado at Four Corners, 4300–4800 ft. June–Aug. W.

PEDIOMELUM Rydb. – INDIAN BREADROOT

Perennials, unarmed herbs; *leaves* alternate, palmately 3- to 7-foliate, gland-dotted; *flowers* zygomorphic, in axillary or spicate racemes; *calyx* 5-lobed, gibbous at the base on the upper side; *petals* 5, composed of a banner, wing, and keel, white, blue, or purple; *stamens* 10, diadelphous; *legumes* 1-seeded, usually included within the sepals.

1a. Plants acaulescent, from a taproot with an apically thickened globose portion 3–8 cm below the soil; inflorescence a dense raceme rarely reaching the leaves in height; calyx spreading-hairy, the lower tooth 3–4 mm wide and 3-veined...***P. hypogaeum***
1b. Plants caulescent or shortly caulescent with the lower portion of the stem hidden by long stipules, from rhizomes or taproots, sometimes with a fusiform-thickened portion; inflorescence a loose raceme or interrupted spike, or if a dense raceme then usually reaching the leaves in height; calyx appressed-hairy or spreading-hairy, the lower tooth 1-veined...2

2a. Leaflets linear, linear-lanceolate, or sometimes linear-oblanceolate, 0.1–0.8 cm wide; stems 3–10 dm tall, branched above the middle...3
2b. Leaflets elliptic, obovate, oblanceolate, or somewhat rhombic, 0.5–2.4 cm wide; stems 0.5–6 (8) dm tall, branched below or above the middle...4

3a. Inflorescence an interrupted spike, with flowers 2–7 per node, the flowers sessile or shortly pedicellate; upper 4 calyx teeth 3–4 mm long and the lower tooth 5–6 mm long...***P. digitatum***
3b. Inflorescence a loose raceme with 1–4 flowers per node, the flowers distinctly pedicellate; upper 3 calyx teeth 1.5–2 mm long, the lower 2 teeth nearly united and 2–2.5 mm long...***P. linearifolium***

4a. Inflorescence an interrupted spike with 2–5 well-separated whorls with 3–7 flowers each; plants densely silvery-sericeous to silky-villous throughout...***P. argophyllum***
4b. Inflorescence a dense spicate raceme; plants not densely silvery-sericeous throughout...5

5a. Stems with spreading hairs; plants from a deep taproot with a fusiform-thickened portion below the surface of the soil...***P. esculentum***
5b. Stems with appressed hairs; plants from rhizomes or a taproot with a fusiform-thickened portion...6

6a. Internodes hidden by long stipules (7–18 mm); flowers 14–21 mm long; keel with a dark purple spot at the tip...***P. megalanthum***
6b. Stipules to 7 mm long, not obscuring the internodes; flowers 7–13 mm long; keel without a dark purple spot at the tip...7

7a. Plants from a taproot with a fusiform-thickened portion below the surface of the soil, procumbent to somewhat ascending or rarely erect; calyx tube 4–7 mm long, the lower tooth 7–12 mm long...***P. cuspidatum***
7b. Plants from rhizomes and a branching caudex, usually erect; calyx tube 3–4 mm long, the lower tooth 5–7 mm long...***P. aromaticum***

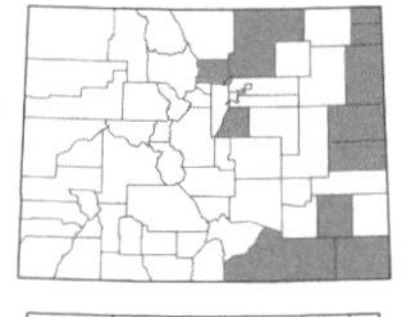

***Pediomelum argophyllum* (Pursh) J. Grimes**, SILVERLEAF INDIAN BREADROOT. [*Psoralea argophylla* Pursh]. Erect or ascending, caulescent perennials, 4–10 dm; *leaflets* 3, obovate to narrowly elliptic, 1–3 (4.5) cm long, thinly sericeous to glabrate above, sericeous below; *calyx* 5–8 mm long; *flowers* blue-violet but aging tawny, 7–11 mm long; *legumes* ovoid, the beak 3–4 mm long, eglandular. Found in sandy or rocky soil or in open prairie, 3500–5800 (6900) ft. June–Aug. E.

***Pediomelum aromaticum* (Payson) W.A. Weber**, AROMATIC INDIAN BREADROOT. [*Psoralea aromatica* Payson]. Ascending to decumbent, caulescent perennials, to 1.5 dm; *leaflets* 5–7, cuneate-obovate or rhombic, 0.5–2 cm long, glabrous above except along margin; *calyx* 7–9 (11) mm long, the tube 3–4 mm long; *flowers* blue-purple, (7) 8–12 mm long; *legumes* ovoid, the beak 5–6 mm long. Found on open rocky soil or clay outcroppings, 5000–5600 ft. May–June. W.

***Pediomelum cuspidatum* (Pursh) Rydb.**, LARGEBRACT INDIAN BREADROOT. [*Psoralea cuspidata* Pursh]. Erect to reclining perennials, 3–8 dm; *leaflets* (3) 5, obovate to elliptic-oblanceolate, 2–5 cm long, glabrate above; *calyx* 8–14 mm long, the tube 4–7 mm long; *flowers* blue-violet to violet or purple, 1.4–2 cm long; *legumes* ovoid, the beak ca. 2 mm long, distally hairy and glandular. Uncommon in rocky soil and open prairie, reported from Yuma Co., 3500–4000 ft. May–July. E.

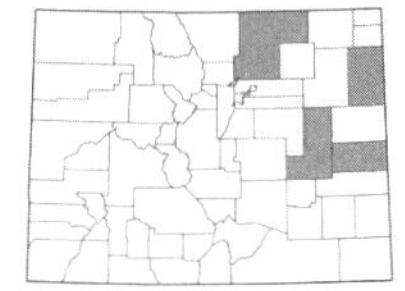

Pediomelum digitatum **(Nutt. ex Torr. & A. Gray) Isely,** PALMLEAF INDIAN BREADROOT. [*Psoralea digitata* Nutt. ex Torr. & A. Gray]. Erect perennials, 3–8 dm; *leaflets* 3 or 5, elliptic-oblong to oblong-oblanceolate, 1.5–4 cm × 2–6 mm, glabrate above except on the midvein; *calyx* 5–10 mm long, subappressed-villous, glandular; *flowers* purple to bluish, 9–10 mm long; *legumes* ovoid, the beak 2–4 mm long, strigose, sometimes glandular. Uncommon in sandy soil of the eastern plains, 3500–4800 ft. May–July. E.

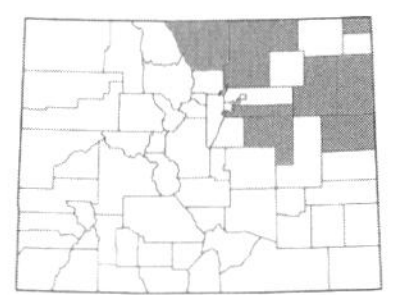

Pediomelum esculentum **(Pursh) Rydb.,** LARGE INDIAN BREADROOT. [*Psoralea esculenta* Pursh]. Erect, caulescent or subacaulescent perennials, 0.3–5 dm; *leaflets* 3 or 5, narrowly ovate to oblong-elliptic, 2–4 (5) cm long, glabrate above; *calyx* 1–1.5 mm long, hairy; *flowers* blue-purple but aging tawny, 1.4–1.8 (2) mm long; *legumes* ovoid, the beak 1–1.5 cm long, mostly glabrous. Uncommon in open prairie and sandy soil of the eastern plains, 3500–6000 ft. May–July. E. (Plate 45)

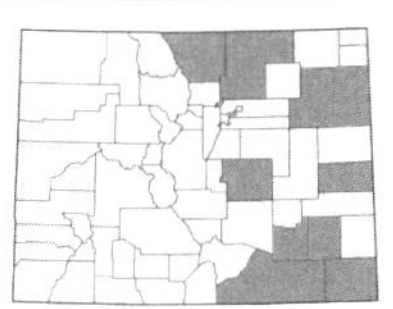

Pediomelum hypogaeum **(Nutt.) Rydb.,** LITTLE INDIAN BREADROOT. [*Psoralea hypogaea* Nutt. ex Torr. & A. Gray]. Acaulescent to subacaulescent perennials, 1–2 cm; *leaflets* (3) 5, rhombic, obovate or narrowly oblong, 2–6 cm long, obscurely glandular; *calyx* 8–15 mm long, appressed-pilose, eglandular; *flowers* lavender or dull lilac to purple, 11–18 mm long; *legumes* ovoid, the beak 8–18 mm long, glabrous or somewhat hairy. Uncommon in sandy or rocky soil of the eastern plains, 4200–5500 ft. May–July. E.

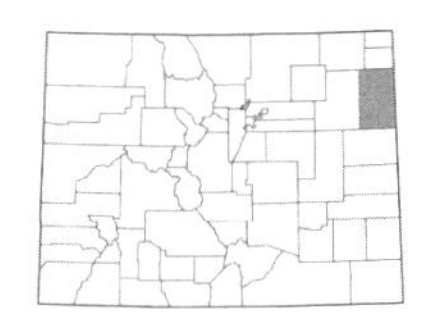

Pediomelum linearifolium **(Torr. & A. Gray) J. Grimes,** NARROWLEAF INDIAN BREADROOT. [*Psoralea linearifolia* Torr. & A. Gray]. Erect perennials, 5–10 (15) dm; *leaflets* linear, 2–6 cm × 2–4 mm; *calyx* 4.5–5.5 mm long; *flowers* blue-purple to pale blue, occasionally white, 9–11 mm long; *legumes* ovoid, the beak 3–3.5 mm long, glabrous, evidently glandular. Uncommon in rocky soil or sandstone outcroppings, sometimes in sandy soil, 3600–4000 ft. May–Aug. E.

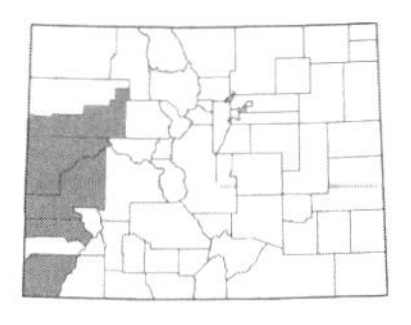

Pediomelum megalanthum **(Wooton & Standl.) Rydb.,** INTERMOUNTAIN INDIAN BREADROOT. [*Psoralea megalantha* Wooton & Standl.]. Spreading or decumbent, acaulescent to subacaulescent perennials, 1–3 (10) cm, spreading branches to 20 cm long; *leaflets* 5–6 (8), cuneate-obovate to broadly elliptic, 1–3 cm long, appressed villous, *calyx* 12–17 mm long, *flowers* purple to blue or bicolored blue and white, 14–21 mm long; *legumes* ovoid, the beak 5–8 mm long, apically hairy, eglandular. Locally common on rocky soil and sandstone outcroppings, often in pinyon-juniper, 4400–6400 ft. May–June. W. (Plate 45)

POMARIA Cav. – POMARIA

Perennial herbs; *leaves* alternate, bipinnately compound, the leaflets glandular below; *flowers* nearly actinomorphic, in axillary racemes; *calyx* 5-lobed; *petals* 5, the uppermost dissimilar, yellow, gland-dotted; *stamens* 10, distinct, shorter than petals; *legumes* lunate, widest above the middle, gland-dotted.

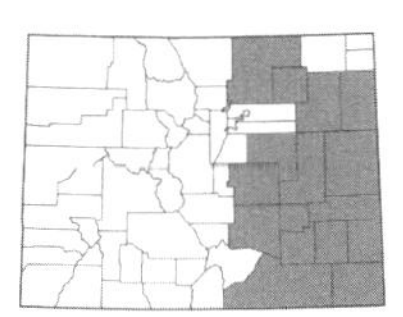

Pomaria jamesii **(Torr. & A. Gray) Walp.,** JAMES' HOLDBACK. [*Caesalpinia jamesii* (Torr. & A. Gray) Fisher; *Hoffmanseggia jamesii* Torr. & A. Gray]. Plants 0.5–3 (5) dm; *leaflets* ovate to elliptic-oblong, 3–5 (7) mm long, conspicuously punctate; *inflorescence* with 5–15 spreading or declined flowers; *calyx* 6–9 mm long; *flowers* yellow with a red-spotted banner, drying orange to orange-red, 6–9 mm long; *legumes* lunate to broadly half-ovate, (1) 2–3 cm × ca. 7 mm, punctate, pubescent with stellate hairs. Found in rocky or sandy soil, in open prairie, or along roadsides, 3500–5200 ft. May–Aug. E.

PROSOPIS L. – MESQUITE

Shrubs or small trees, armed; *leaves* alternate, bipinnately compound; *flowers* actinomorphic, in spicate racemes; *petals* 5, distinct or nearly so, yellow to ochroleucous; *stamens* 10, distinct, exserted; *legumes* indehiscent, spirally coiled or somewhat constricted between the seeds.

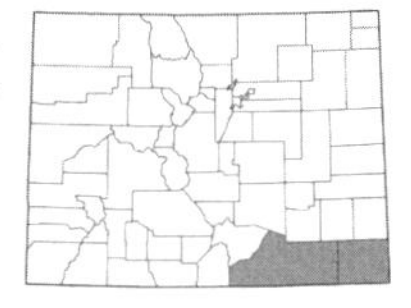

Prosopis glandulosa **Torr.,** HONEY MESQUITE. Shrubs or trees, 1–10 (20) m; *leaflets* oblong to narrowly oblong, sometimes obovate, 1–3.5 (6) cm × (1) 2–3 (4) mm, with evident venation; *inflorescence* of ascending to drooping ament-like spikes; *calyx* barely lobed; *flowers* cream-yellow to yellowish-green; *legumes* linear, subterete, 10–20 (30) cm long, ca. 8 mm diam. Uncommon on dry slopes, 4400–5200 ft. May–July. E.

PSORALIDIUM Rydb. – SCURFPEA

Perennials, unarmed, caulescent herbs; *leaves* palmately 3-foliate, gland-dotted; *flowers* zygomorphic, in axillary or terminal spicate racemes or racemes; *calyx* 5-lobed, the tube short; *petals* 5, composed of a banner, wing, and keel, bluish-purple to purple or sometimes white; *stamens* 10, diadelphous; *legumes* 1-seeded, indehiscent.

1a. Legumes globose, about as long as wide, 4–6 mm long; flowers white with a purple keel tip; leaflets narrowly linear to obovate...***P. lanceolatum***

1b. Legumes elliptic to ovoid, longer than wide, 7–9 mm long; flowers light blue to purple, rarely white with a purple keel tip; leaflets elliptic to oblanceolate, rarely linear...***P. tenuiflorum***

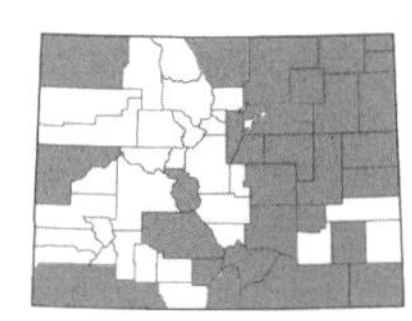

Psoralidium lanceolatum **(Pursh) Rydb.**, LEMON SCURFPEA. [*Psoralea lanceolata* Pursh]. Erect, often bushy perennials, 2–6 dm; *leaflets* obovate to filiform, 1.5–4 cm × 2–6 mm, yellow-green, evidently punctate; *inflorescence* of compact to elongate racemes with few-numerous flower clusters; *calyx* 2–2.5 mm long; *flowers* blue or white with a maculate keel, 4.5–8 mm long; *legumes* subglobose, 4–6 mm long, glabrous to hairy, evidently glandular. Found in sandy soil, 3400–8300 ft. May–July. E/W. (Plate 46)

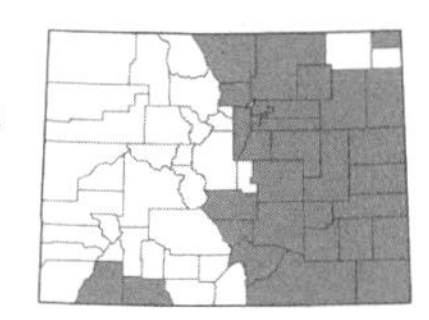

Psoralidium tenuiflorum **(Pursh) Rydb.**, SLIMFLOWER SCURFPEA. [*Psoralea tenuiflora* Pursh]. Ascending to erect perennials, 3–12 dm; *leaflets* obovate or elliptic to oblong-oblanceolate, 1.5–4 (5) cm long, punctate; *inflorescence* of numerous racemes with few–many flower clusters, 2–6 cm long; *flowers* lavender to violet, rarely white, 5–7 mm long; *legumes* elliptic to ovoid, laterally compressed, 7–9 mm long, brown, glabrous, glandular. Found in open prairies, fields, and along roadsides, 3800–6500 ft. June–Aug. E/W.

PSOROTHAMNUS Rydb. – DALEA

Usually armed shrubs; *leaves* alternate, odd-pinnately compound, gland-dotted; *flowers* zygomorphic, in racemes; *petals* 5, composed of a banner, wing, and keel; *stamens* 9 or 10, monadelphous; *legumes* 1–2-seeded, indehiscent.

Psorothamnus thompsoniae **(Vail) S.L. Welsh & N.D. Atwood var. *whitingii* (Kearney & Peebles) Barneby**, THOMPSON'S DALEA. [*Dalea whitingii* Kearney & Peebles]. Spreading shrubs, 0.5–1 (1.5) m; *leaflets* 7–15, narrowly oblanceolate, folded, appearing linear, 2–8 mm long; *calyx* 3.5–5 mm long, hairy; *flowers* pink-purple to violet, 6–8.5 cm long, glabrous; *legumes* obliquely ovoid or elliptic, glandular punctate. Grows in sandy soil of dunes and canyons, reported for Montezuma Co. but no specimens have been seen. May–Aug. W.

ROBINIA L. – LOCUST

Shrubs or trees, sometimes armed; *leaves* alternate, odd-pinnately compound; *flowers* zygomorphic, in axillary racemes; *calyx* 5-lobed, the teeth triangular; *petals* 5, composed of a banner, wing, and keel, white, pink, or pinkish; *stamens* 10, diadelphous; *legumes* flat, straight, often persistent through winter.

1a. Inflorescence axis minutely hairy, lacking glands; flowers white with a yellow patch on the banner; legume glabrous...***R. pseudoacacia***

1b. Inflorescence axis densely glandular or hispid; flowers purple or pink, rarely white; legume glandular-hispid...2

2a. Young twigs, inflorescence axis, calyx, and leaf axis densely hispid-setose with black hairs, older stems becoming less hispid with age...***R. hispida***

2b. Plants densely glandular on the inflorescence axis and calyx, the young twigs densely short-hairy but the stems becoming glabrous with age...***R. neomexicana***

Robinia hispida **L.**, BRISTLY LOCUST. Shrubs or small trees, 0.6–3 m; *leaflets* (7) 9–13, elliptic to lanceolate, hairy to glabrate below; *inflorescence* of axillary racemes with (3) 4–11 flowers; *calyx* lobes 3–7 mm long; *flowers* pink to rose-purple, occasionally white, drying lavender to blue, 20–25 (30) mm long; *legumes* rarely produced, but when present, oblong, laterally compressed, densely hispid with both glandular and eglandular hairs (2) 4–6 mm long. Cultivated plants, uncommon near towns, 5000–5500 ft. May–June. E. Introduced.

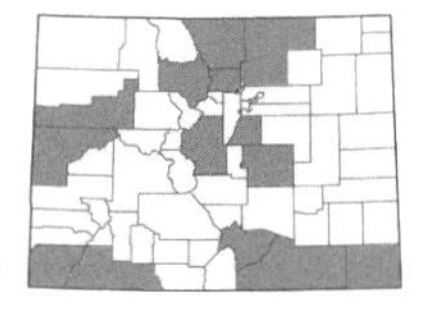

Robinia neomexicana **A. Gray**, NEW MEXICO LOCUST. Shrubs or small trees to 2 m; *leaflets* elliptic to lanceolate, appressed-sericeous above and below; *inflorescence* of axillary racemes with numerous, crowded flowers; *calyx* tube 6–7 mm long, glandular-hairy; *flowers* pink, 2–2.5 cm long; *legumes* oblong, laterally compressed, glabrous or glandular-hispid. Cultivated and escaping along roadsides, near towns, and old homesteads, often found along streams, scattered, 4500–9000 ft. May–July. E/W. Introduced northward but native in southern Colorado.

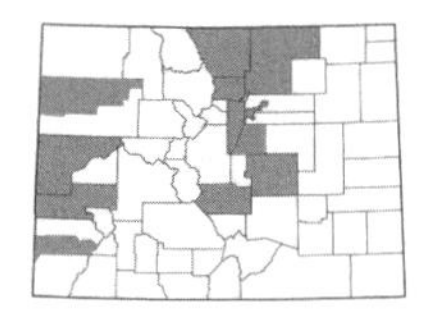

Robinia pseudoacacia **L.**, BLACK LOCUST. Trees, 12–18 m; *leaflets* 15–17 (21), elliptic to lanceolate, glabrous; *inflorescence* of pendant axillary racemes with 8-numerous flowers, to 1 dm; *calyx* tube 5–6 mm long, velvety-hairy; *flowers* white, sometimes pinkish, 1.5–2 cm long; *legumes* oblong, laterally compressed, glabrous. Cultivated and escaping around towns and old homesteads, scattered in the state, 4500–7000 ft. May–June. E/W. Introduced.

SECURIGERA DC. – CROWNVETCH

Perennial herbs, stems trailing to ascending; *leaves* alternate, odd-pinnately compound, with herbaceous, persistent stipules; *flowers* zygomorphic, in axillary umbels; *calyx* 5-lobed, bilabiate; *petals* 5, composed of a banner, wing, and keel, pink or white but drying purplish; *stamens* 10, diadelphous; *legumes* constricted between the seeds, 4-angled.

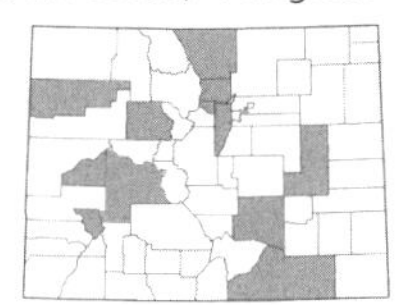

***Securigera varia* (L.) DC.**, CROWNVETCH. [*Coronilla varia* L.]. Sprawling or ascending perennials, 3–6 dm; *leaflets* 11–19, obovate to oblong, 0.8–2 cm long; *calyx* 2–2.5 mm long; *flowers* white and pink with purple marks, ca. 1 cm long; *legumes* with 3–10 segments, 2–5 cm long, longitudinally ridged. Found along roadsides and in disturbed areas, sometimes used in revegetation plantings, 5800–8000 ft. June–Sept. E/W. Introduced.

SOPHORA L. – NECKLACE-POD

Perennial herbs; *leaves* alternate, odd-pinnately compound; *flowers* zygomorphic, in terminal racemes; *petals* 5, composed of a banner, wing, and keel, the keel connivent in the distal ½, otherwise all petals distinct, white to ochroleucous; *stamens* 10, distinct or nearly so; *legumes* constricted between the seeds, stipitate.

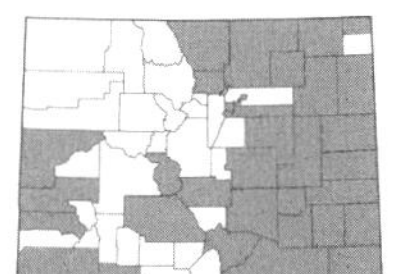

***Sophora nuttalliana* B.L. Turner**, SILKY SOPHORA. [*Vexibia nuttalliana* (B.L. Turner) W.A. Weber]. Plants 1–4 dm; *leaflets* (7) 11–23, ovate to oblanceolate, usually folded, 5–15 mm long; *calyx* 5–8 mm long; *flowers* 12–16 mm long; *legumes* oblong, laterally compressed, 5–7 cm × 5 mm, villous. Found in sandy soil, along roadsides, in open prairie, 3400–7500 ft. May–July. E/W. (Plate 46)

SPHAEROPHYSA DC. – SPHAEROPHYSA

Perennials, caulescent herbs; *leaves* alternate, odd-pinnately compound; *flowers* zygomorphic, in axillary racemes; *petals* 5, composed of a banner, wing, and keel; *stamens* 10, diadelphous; *legumes* stipitate, inflated with papery walls, globose.

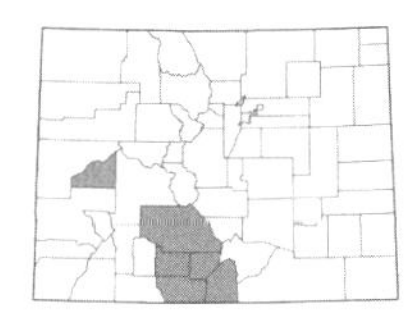

***Sphaerophysa salsula* (Pall.) DC.**, ALKALI SWAINSON-PEA. [*Swainsona salsula* (Pall.) Taubert]. Plants 4–10 (15) dm; *leaflets* 15–many, ovate-elliptic to oblong-elliptic, 0.6–2 cm long, glabrous above; *calyx* tube 4–4.5 mm long; *flowers* brick-red to orange-red, fading to lavender-brown, 12–14 mm long; *legumes* ovoid, 1.4–2.4 × 1–2 cm, often dark-mottled, glabrate. Uncommon along roadsides, ditches, and in fields, usually in saline soil, 4900–7800 ft. June–Sept. E/W. Introduced.

STROPHOSTYLES Elliott – WILD BEAN

Annuals, stems trailing; *leaves* simple and opposite below, pinnately trifoliate and alternate above; *flowers* zygomorphic; *calyx* 5-lobed, bilabiate; *petals* 5, composed of a banner, wing, and keel, the keel abruptly contracted and curved upward into a beak pointing back to the flower; *stamens* 10, diadelphous; *legumes* flattened, sessile.

***Strophostyles leiosperma* (Torr. & A. Gray) Piper**, SLICK-SEED FUZZYBEAN. Plants to 1 m; *leaflets* narrowly elliptic to oblong-lanceolate, 2–5.5 cm long, subappressed-villous below; *calyx* 2–3 (3.5) mm long, hairy; *flowers* pink or pale lavender, 5–8 mm long; *legumes* linear, laterally compressed, 2–4 cm × 5 mm, villous. Uncommon in sandy soil, 3500–4500 ft. May–Sept. E.

THERMOPSIS R. Br. – GOLDENBANNER

Perennial herbs; *leaves* alternate, ternately compound; *flowers* zygomorphic; *petals* 5, composed of a banner, wing, and keel, yellow; *stamens* 10, distinct; *legumes* flattened, sessile.

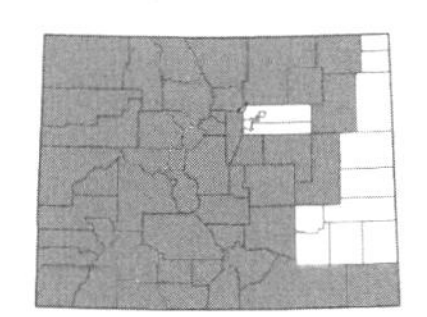

***Thermopsis rhombifolia* (Nutt. ex Pursh) Nutt. ex Richardson**, GOLDENBANNER. Plants 2–10 dm; *leaves* with large, foliaceous stipules 1–2.5 cm wide; *leaflets* broadly ovate to oblong-oblanceolate, 2–8 cm long; *inflorescence* a solitary, terminal raceme, 1–3 dm; *calyx* 8–11 mm long; *flowers* yellow, 18–22 (25) mm long; *legumes* oblong, laterally compressed, 4–6 cm × 5–8 mm, minutely hairy to glabrate. Common, 4300–10,600 ft. April–Aug. (Sept.). E/W. (Plate 46)

This is a polymorphic species with considerable variation and is often divided into three species (*T. divaricarpa*, *T. montana*, and *T. rhombifolia*). However, when one examines all three species together, considerable overlap in morphology is evident. At the extremes of this species complex, *T. rhombifolia* is generally smaller in stature with distinctly curved legumes and *T. montana* has strictly erect legumes and is generally taller with larger leaves. However, *T. divaricarpa* is intermediate between these two species and has spreading, somewhat curved fruit and is generally taller with larger leaves. Intermediates between all three taxa can be seen where their ranges overlap. Therefore, all species are here included within the *T. rhombifolia* complex:

1a. Fruit spreading-recurved, strongly curved, glabrous or pilose; plants 1–2.5 (3.5) dm; leaflets usually less than 3 cm in length...**var. *rhombifolia***, PRAIRIE GOLDENBANNER. Common in relatively dry, often barren places, known from the eastern plains to lower foothills, scattered in the San Luis Valley and known from Moffat Co. on the western slope, 4300–8900 ft. April–June (Sept.). E/W.

1b. Fruit erect or ascending, straight or slightly curved, pilose (sometimes sparsely so when mature); plants usually over 3 dm; leaflets usually 3 cm or more in length...2

2a. Fruit slightly curved and ascending...**var. *divaricarpa* (A. Nelson) Isely**, SPREADFRUIT GOLDENBANNER. Common in meadows, along streams, and in aspen forests, found throughout the central mountains and lower foothills, 5200–10,200 ft. May–Aug. E/W.
2b. Fruit straight and strictly erect...**var. *montana* (Nutt. ex Torr. & A. Gray) Isely**, MOUNTAIN GOLDENBANNER. Common in meadows, aspen forests, and along streams, found in the central mountains with its range extending into the western counties, 5500–10,600 ft. May–Aug. E/W.

TRIFOLIUM L. – CLOVER

Herbs, the stems caulescent or acaulescent; *leaves* alternate, palmately trifoliately or ternately compound; *flowers* zygomorphic, in terminal or axillary heads or racemes often subtended by a basal involucre; *petals* 5, composed of a banner, wing, and keel, pink, white, purple, or red; *stamens* 10, diadelphous; *legumes* often enclosed in the calyx.

1a. Heads sessile, subtended by stipules and a pair of leaves...2
1b. Heads pedunculate...3

2a. Leaflets (10) 20–35 mm long, the apex rounded with a very short mucronate tip; stipules large, 1–3 cm long, ovate-acuminate, scarious with several green nerves; heads solitary at the ends of each stem, with numerous (well over 12) flowers...***T. pratense***
2b. Leaflets 3.5–15 mm long, the apex acute or acuminate; stipules 0.5–1.4 cm long, without several green nerves; heads paired or rarely solitary at the end of each stem, 5–12-flowered...***T. andinum* var. *andinum***

3a. Flowers solitary or heads with 2–3 (rarely 4) flowers, reddish-purple to pinkish-purple; calyx glabrous, purplish; plants acaulescent, cushion- or mat-like from a branching caudex covered with scarious stipules...***T. nanum***
3b. Flowers 5 or more per head, or if fewer then the calyx hairy, white, pink, or purplish; plants various...4

4a. Calyx soon becoming inflated, papery, and membranous-reticulate in fruit, pilose on the dorsal side; banner 5–6 mm long; heads 1–1.6 cm in diam. in flower and becoming globose and 1.6–2.4 mm in diam. in fruit, subtended by an involucre of bracts (this inconspicuous in fruit); plants rooting at the nodes, without a whitish, inverted V-shaped blotch on the upper side of the leaves...***T. fragiferum***
4b. Calyx not becoming inflated, papery, and membranous-reticulate in fruit, usually glabrous or sometimes weakly pilose; banner usually over 6 mm long; heads not much larger in fruit than in flower, subtended by an involucre or not; plants not rooting at the nodes, or if rooting at the nodes then the calyx glabrous and usually at least some leaves with a whitish, inverted V-shaped blotch on the upper side...5

5a. Leaves glabrous above and below, or occasionally with just a few scattered hairs along the midvein below...6
5b. Leaves hairy below and glabrous or hairy above...11

6a. Plants creeping and rooting at the nodes; usually at least some leaves with a whitish, inverted V-shaped blotch on the upper side near the base; calyx with a whitish-green tube with a purple spot below each sinus, and green teeth; flowers whitish or pinkish...***T. repens***
6b. Plants erect or caespitose, not rooting at the nodes; leaves with or without a whitish, inverted V-shaped blotch on the upper side; calyx various; flowers various...7

7a. Flowers all reflexed or nearly so...8
7b. Flowers ascending or spreading, sometimes just the lower flowers of the head reflexed in age...9

8a. Caulescent plants 1–2 (4) dm; flowers numerous in a dense head; leaflets of upper leaves with acute tips...***T. kingii* var. *kingii***
8b. Acaulescent plants 0.5–1.5 dm; flowers 6–15 in a loose head; leaflets with rounded tips...***T. brandegeei***

9a. Heads not subtended by an involucre of bracts; calyx tube 1–1.8 mm in length and the longest teeth 1.2–2.5 mm; flowers white or pinkish, 5–10 mm in length...***T. hybridum***
9b. Heads subtended by an involucre, the bracts free or fused; calyx tube 1.5–4 mm long and the longest teeth 1.8–6 mm; flowers pink-purple, reddish-purple, or purple, 10–18 mm in length ...10

10a. Plants acaulescent or shortly caulescent with 1 elongate internode, 0.5–2 (2.5) dm; stipules entire on the margins; involucre bracts free or connate for about ⅓ their length, the margins entire and the apex bifid or toothed...***T. parryi***
10b. Plants caulescent, 1.2–4 dm; stipules toothed on the margins; involucre bracts fused for ⅓–½ their length, the margins and apex sharply dentate or lacerate...***T. wormskioldii* var. *wormskioldii***

11a. Leaflet margins entire; plants acaulescent...12
11b. Leaflet margins toothed their entire length or nearly so; plants caulescent or acaulescent...14

12a. Flowers light to dark violet-purple throughout, reflexed in age; leaflets (1) 2–4 (5) cm long...***T. attenuatum***
12b. Flowers bicolored, the banner whitish-lavender and the wing tips and keel purple, ascending in age; leaflets (0.6) 0.8–3 cm long...13

13a. Leaves hairy on both surfaces or glabrate on the upper surface of the leaves, but the plants green and not silvery in appearance...***T. dasyphyllum***
13b. Leaves closely white-hairy on both surfaces, the plants silky-silvery in appearance...***T. anemophilum***

14a. Flowers numerous, mostly 20–65 per head, the heads 15–35 mm wide; plants caulescent or rarely acaulescent, 0.5–3 (5) dm; stipules all green and herbaceous, 8–40 mm long; leaflets (2) 3–5 cm long; calyx teeth 2.9–6.5 mm long...***T. longipes***
14b. Flowers 4–15 per head, the heads 10–15 (20) mm wide; plants acaulescent or shortly caulescent, 0.4–0.8 (1) dm; stipules scarious or sometimes the upper also green and herbaceous, 6–23 mm long; leaflets (0.5) 0.8–1.5 (2) cm long; calyx teeth 1.8–3.5 mm long...***T. gymnocarpon***

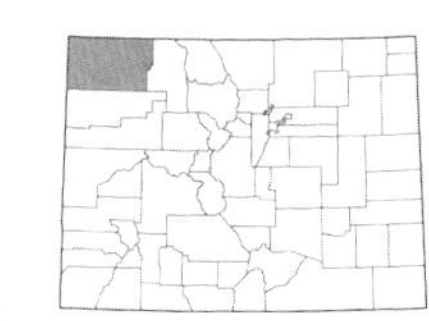

Trifolium andinum **Nutt. var. *andinum***, INTERMOUNTAIN CLOVER. Subacaulescent perennials, 0.5–4 cm; *leaflets* 3 (5) (7), oblanceolate, acute and mucronulate, 0.35–1.5 × 0.15–0.45 cm, the margins subentire or shortly toothed distally, silky-strigose to glabrescent; *heads* with 5–12 ascending flowers; *calyx* 7–10 mm long, villous to glabrate; *flowers* ochroleucous with a pink to purple-tinged banner, the banner 8.5–13 mm long; *legumes* ellipsoid, ca. 4.5 mm long, sparsely hairy. Locally common in sandy soil of dry, exposed places, usually in association with *Pinus edulis* and *Cercocarpus*, 5700–7300 ft. May–July. W.

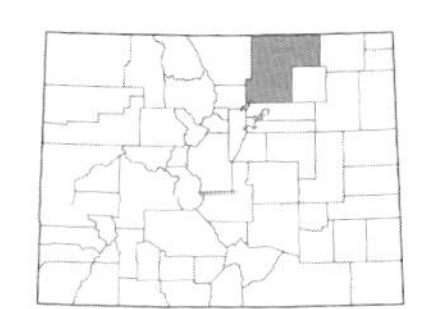

Trifolium anemophilum **Greene**, LARAMIE HILLS CLOVER. [*T. dasyphyllum* Torr & A. Gray var. *anemophilum* (Greene) J.S. Martin ex Isley]. Subacaulescent perennials, 1–5 cm; *leaflets* obovate to narrowly lanceolate, often folded, (0.6) 0.8–3 cm long, the margins subentire, silvery-strigose above and below; *heads* with 5–15 (20) ascending flowers; *calyx* 5–7 (10) mm long, hairy; *flowers* bicolored with purple wings and keel, the banner whitish-lavender, 10–16 mm long; *legumes* 4–6 mm long, hairy. Uncommon on rocky slopes and in rock crevices, 5400–6000 ft. May–June. E.

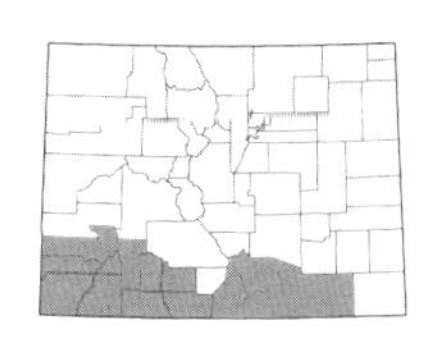

***Trifolium attenuatum*, Greene**, ROCKY MOUNTAIN CLOVER. Caespitose perennials, 1–2 dm; *leaflets* ovate-acute to linear-oblanceolate, (1) 2–4 (5) cm long, hairy; *heads* with 5–20 ascending-spreading to reflexed flowers; *calyx* tube 3–4.5 mm long, hairy; *flowers* light to dark violet-purple , 12–18 mm long; *legumes* obovate, 5–6 mm long, evidently hairy. Locally common in subalpine and alpine meadows of the southern mountains, 9500–14,000 ft. June–Aug. E/W.

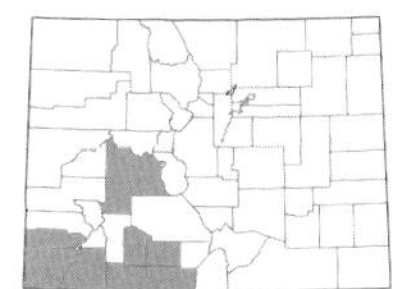

Trifolium brandegeei **S. Watson**, BRANDEGEE'S CLOVER. Acaulescent, caespitose perennials, 5–15 cm; *leaflets* obovate to elliptic-oblong, 1–2 (3) cm long, the margins entire to serrate, glabrous; *heads* with 5–10 (15) reflexed flowers; *calyx* 7–9 mm long, somewhat hairy; *flowers* purple to violet, ca. 1.5 cm long; *legumes* ca. 6 mm long, glabrate. Locally common in subalpine forests and alpine meadows, often in moist places, 9000–12,000 ft. June–Aug. E/W.

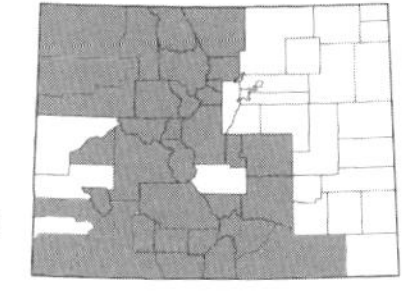

Trifolium dasyphyllum **Torr. & A. Gray**, ALPINE CLOVER. Subacaulescent, caespitose perennials, 1–5 cm; *leaflets* obovate to narrowly lanceolate, usually folded, (0.6) 0.8–3 cm long, the margins subentire, silvery-strigose above and below or glabrate above; *heads* with 5–15 (20) ascending flowers, 1.2–2.8 cm diam.; *calyx* 5–7 (10) mm long, hairy; *flowers* bicolored with purple wings and keel, the banner whitish-lavender; *legumes* 4–6 mm long, hairy. Common in subalpine and alpine meadows, largely replaced in the southern mountains by *T. attenuatum*, 9700–14,000 ft. June–Aug. E/W. (Plate 46)

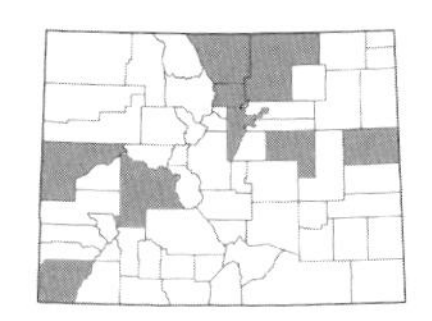

Trifolium fragiferum **L.**, STRAWBERRY CLOVER. Stoloniferous or tufted perennials, (0.5) 1–3 dm; *leaflets* obovate to elliptic, occasionally retuse, 5–20 mm long, evidently nerved; *heads* with numerous flowers, 10–24 mm diam.; *calyx* tube 1.8–2.3 mm long, pilose distally, glabrous ventrally, becoming inflated and papery; *flowers* pink, 5–6 mm long; *legumes* ca. 2 mm long. Locally common in moist places, lawns, and disturbed areas, 4100–5500 ft. May–Aug. E/W. Introduced.

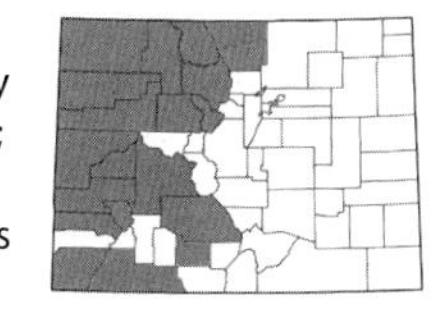

***Trifolium gymnocarpon* Nutt.**, HOLLYLEAF CLOVER. Subacaulescent, caespitose perennials, 4–8 (10) cm; *leaflets* elliptic to oblong, often folded, (5) 8–15 (20) mm long, the margins dentate, evidently nerved; *heads* with 4–15 ascending-spreading or partially reflexed flowers, 10–15 (20) mm diam.; *calyx* tube 2–2.5 (3) mm long, 10-nerved; *flowers* ochroleucous to flesh-colored, occasionally purple, (6) 8–10 (12) mm long; *legumes* ovoid, 4–5 × 2.5–4 mm, hairy. Found on dry, sandy slopes and in sagebrush meadows, 5500–8500 ft. April–June. E/W.

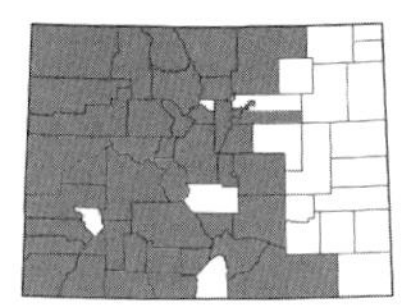

***Trifolium hybridum* L.**, ALSIKE CLOVER. Sprawling to erect, short-lived perennials, 1–6 dm; *leaflets* elliptic to obovate, 1–3 (4) cm long, the margins serrulate or entire; *heads* with numerous ascending to deflexed flowers, 1.5–2.5 (3) cm diam.; *calyx* tube 1–1.8 mm long; *flowers* white to pink or pale-reddish, 5–10 mm long; *legumes* 3–4 mm long. Common in meadows, along streams, and in disturbed places, 4500–10,500 ft. May–Oct. E/W. Introduced.

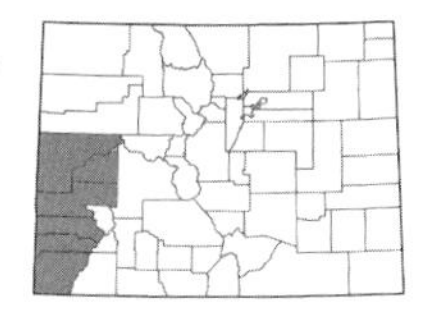

Trifolium kingii* S. Watson var. *kingii, KING'S CLOVER. Ascending, caulescent perennials, 1–2 (4) dm; *leaflets* slightly to strongly dimorphic, basal leaflets obovate to elliptic, 1–2 cm, cauline leaflets lanceolate, 3–4.5 cm long; *heads* with several-numerous reflexed flowers, 2–3 cm diam.; *calyx* often purple-mottled, 3.5–7 mm long, glabrous to villous; *flowers* light pink-purple to reddish, 12–17 mm long. Uncommon in moist forests and along streams, 8500–10,500 ft. June–Aug. W.

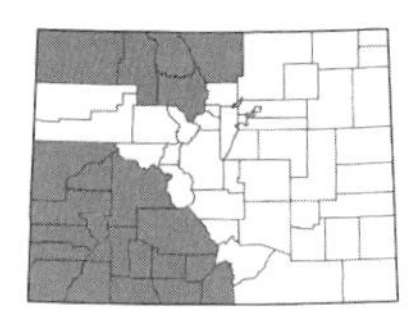

***Trifolium longipes* Nutt.**, LONGSTALK CLOVER. Decumbent or erect perennials, 0.5–3 (5) dm; *leaflets* obovate to elliptic-lanceolate, sometimes linear, (2) 3–5 cm long, the margins sharply serrate to subentire, glabrous above, glabrous to villous below; *heads* with ascending-divergent flowers, 1.5–3 (3.5) cm diam.; *calyx* tube 1–3 mm long, glabrate to villous; *flowers* whitish, ochroleucous, purple to lavender, or bicolored, 10–18 mm long. Found along streams, in meadows, and shaded forests, 7000–11,800 ft. June–Sept. E/W.

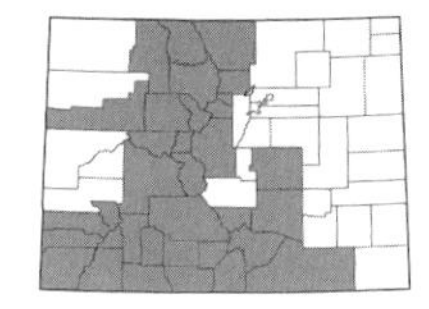

***Trifolium nanum* Torr.**, DWARF CLOVER. Subacaulescent perennials, to 8 cm; *leaflets* obovate to oblanceolate, 3–10 (14) mm long, the margins slightly toothed, evidently nerved; *heads* with (1) 2 (4) ascending flowers; *calyx* tube 3–4 mm long, evidently 10-nerved; *flowers* lavender to purple with pink wings and keel, sometimes white or ochroleucous, 14–18 (22) mm long. Common in the alpine, 10,500–14,200 ft. June–Aug. E/W. (Plate 46)

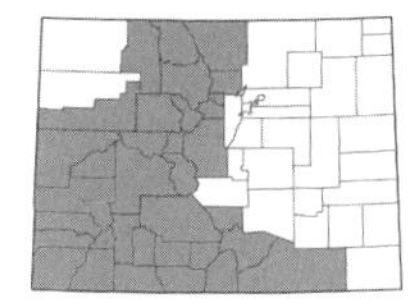

***Trifolium parryi* A. Gray**, PARRY'S CLOVER. Acaulescent to subacaulescent perennials, 0.5–2 (2.5) dm; *leaflets* elliptic to obovate, 1–2.5 (3.5) cm long, the margins toothed; *heads* with 15–20 ascending flowers, (10) 20–35 mm diam.; *calyx* tube 1.5–3 mm long, glabrous; *flowers* reddish-lavender or purple, drying purple; *legumes* oblong, 5–6 mm long. Common in subalpine and alpine meadows, 9500–14,000 ft. June–Aug. E/W.

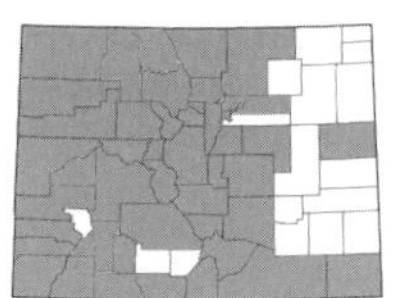

***Trifolium pratense* L.**, RED CLOVER. Ascending, short-lived perennials, 2 dm; *leaflets* ovate, elliptic or obovate, (1) 2–3.5 cm long, often with a central blotch; *heads* with numerous, ascending flowers, 2–3 cm diam.; *calyx* tube 2.5–3.5 mm long, 10-nerved, hairy; *flowers* red-purple, sometimes white, 13–16 mm long; *legumes* obconic or ovoid, 2–3 mm long. Common in moist places such as along streams, also found in meadows, fields and pastures, and disturbed places, 4100–10,500 ft. May–Oct. E/W. Introduced. (Plate 46)

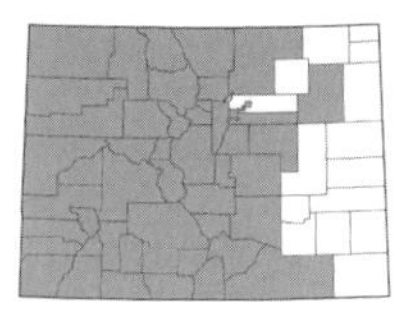

***Trifolium repens* L.**, WHITE CLOVER. Stoloniferous perennials; *leaflets* obovate, 0.5–2.5 cm long, the margins denticulate; *heads* with numerous, ascending-spreading to reflexed flowers, 1–2 (3) diam.; *calyx* tube 2–3 mm long; *flowers* white but aging pinkish, 7–10 (12) mm long; *legumes* broadly oblong, 3–5 mm long, constricted between seeds. Common in lawns and fields, along roadsides, and other disturbed places, occasionally in forests and meadows and appearing native, 4500–11,100 ft. May–Sept. E/W. Introduced.

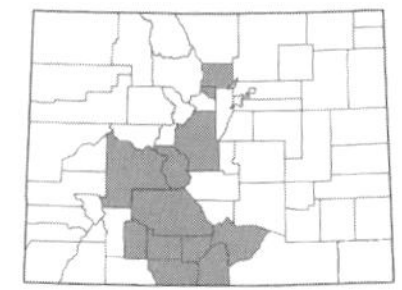

Trifolium wormskioldii* Lehm. var. *wormskioldii, COW'S CLOVER. [*T. fendleri* Greene]. Perennials, 1.2–4 dm; *leaflets* obovate to elliptic; *heads* with 10-numerous ascending-spreading flowers, 1.5–3 (3.5) cm diam.; *calyx* tube 3–4 mm long, membranous; *flowers* pink-purple and white-tipped, 12–15 (18) mm long; *legumes* ellipsoid to oblong, slightly obcompressed, 3–6 mm long. Found in moist meadows, along streams, and other wet places, 6500–8000 ft. June–Aug. E/W.

VICIA L. – VETCH

Annual or perennial herbs; *leaves* alternate, even-pinnately compound, the rachis terminating in a tendril; *flowers* zygomorphic; *petals* 5, composed of a banner, wing, and keel, the wings adnate to the keel, white, blue, pink, or purplish; *stamens* 10, diadelphous; *legumes* dehiscent, laterally compressed.

1a. Racemes densely flowered, with (8) 10–40 flowers per raceme; leaflets (10) 14–18, narrowly lanceolate or linear-oblong...***V. villosa***
1b. Racemes few-flowered, mostly with 1–9 flowers per raceme; leaflets 5–16, variously shaped...2

2a. Racemes subsessile, ascending, flowers solitary or paired, rarely up to 3 per raceme; stipules with a discolored nectary on the outer face; fruit sessile or nearly so...***V. sativa* ssp. *nigra***
2b. Racemes on long pedicels, loosely secund, 1–9-flowered; stipules lacking a nectary; fruit stipitate, the stipe about as long as the calyx tube...3

3a. Flowers small, 4.5–8 mm long, ochroleucous with a purplish tip or sometimes tinged with purple; stipules 2–4 mm long, linear-lanceolate, entire or toothed at the base; style thinly minutely hairy just below the stigma...***V. ludoviciana***
3b. Flowers larger, 15–22 (25) mm long, purple; stipules mostly over 4 mm long, triangular-acuminate to lanceolate, lobed or toothed at the base; style with a dense mass of hairs just below the stigma...***V. americana***

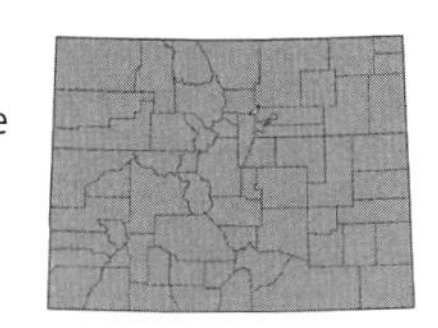

***Vicia americana* Muhl. ex Willd.**, American vetch. Scandent, sprawling or erect perennials, 2–10 dm; *leaflets* 8–16, broadly elliptic to linear, 1–3.5 cm long; *tendrils* prehensile or reduced to simple bristles; *inflorescence of* axillary racemes with 2–9 flowers; *calyx* 6–8 mm long; *flowers* various shades of blue-purple to red-purple, sometimes white, 1.5–2.2 (2.5) cm long; *legumes* oblong, 2–3.5 cm × 4–7 mm, glabrous or hairy. Common in meadows, along streams, in aspen forests, and on the shortgrass prairie, 3900–10,500 ft. April–Aug. E/W.

There are two varieties of *V. americana* in Colorado:
1a. Tendrils forked; racemes (4) 5–9-flowered...**var. *americana***, American vetch. Common in meadows, along streams, and in aspen forests, 4800–10,500 ft. May–Aug. E/W.
1b. Tendrils simple; racemes 2–5-flowered...**var. *minor* Hook.**, mat vetch. [*V. linearis* (Nutt.) Greene]. Common in meadows and on the shortgrass prairie of the eastern plains, 3900–8500 ft. April–June. E/W.

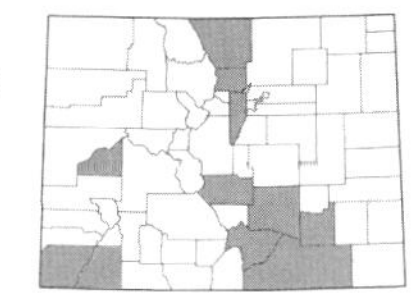

***Vicia ludoviciana* Nutt.**, Louisiana vetch. Sprawling, ascending or twining annuals, 1–6 dm; *leaflets* 6–16, elliptic to oblong-linear, (0.5) 1–2 (3) cm long; *tendrils* prehensile, or reduced to simple bristles; *inflorescence of* axillary racemes with 1–15 flowers, 2–10 cm long; *calyx* 2.8–3.5 mm long; *flowers* blue-purple to light lavender or pink, 4.5–8 mm long; *legumes* oblong, 1.5–3.5 cm × 4–7 (8) mm, glabrous. Uncommon in meadows and on dry hillsides, 4400–6500 ft. April–June. E/W.

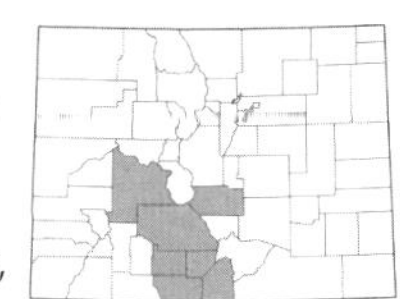

***Vicia sativa* L. ssp. *nigra* (L.) Ehrh.**, garden vetch. [*Vicia angustifolia* L.]. Erect, sprawling or climbing annuals, 1–5 dm; *leaflets* (6) 8–15, cuneate-obovate to oblong-lanceolate or linear, 1.5–3 (3.5) cm long; *tendrils* simple or branched; *inflorescence* axillary clusters with 2 flowers; *calyx* 7–12 (15) mm long; *flowers* pink-purple, lavender or whitish, 1–2.5 (3) cm long; *legumes* oblong, 2.5–6 cm × 3.5–8 mm, stigulose or minutely hairy to glabrous. Uncommon along roadsides and ditches, 6000–7700 ft. May–Aug. E/W. Introduced.

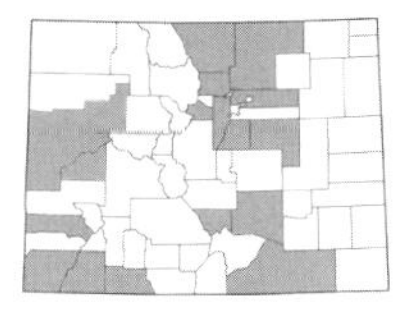

***Vicia villosa* Roth**, winter vetch. Scandent, sprawling or erect annuals, 5–10 dm; *leaflets* (10) 14–18, narrowly lanceolate to linear-oblong, 1–3 cm long; *tendrils* branched, vigorous; *inflorescence of* axillary racemes with (8) 10–40 reflexed flowers, 3–12 (15) cm long, secund; *calyx* 5–6 mm long; *flowers* violet or bicolored, occasionally white, (1) 1.2–1.6 (1.8) cm long; *legumes* elliptic to oblong, 1.5–4 cm × 6–10 mm. Uncommon along roadsides and sometimes planted in fields, 4800–7200 ft. June–Sept. E/W. Introduced.

FAGACEAE Dumort. – beech family

Monoecious, evergreen or deciduous trees or shrubs; *leaves* alternate, simple, petiolate, the margins entire to toothed or deeply pinnately lobed; *flowers* imperfect; *staminate flowers* subtended by a bract, with 4–6 sepals and 6–12 stamens, borne in catkins; *pistillate flowers* solitary or in threes, with 4–6 distinct or connate sepals; *pistil* 1; *ovary* inferior, 3-carpellate; *fruit* a nut.

QUERCUS L. – oak

Evergreen or deciduous trees or shrubs; *leaves* entire to toothed or lobed; *staminate flowers* with 2–8 connate sepals and 3–12 stamens surrounding a tuft of hairs; *pistillate flowers* with connate sepals, solitary, enclosed by a scaly involucre; *fruit* a nut (acorn), with the hardened scaly involucre covering the base. (Tucker, 1961, 1970, 1971; Welsh, 1986)

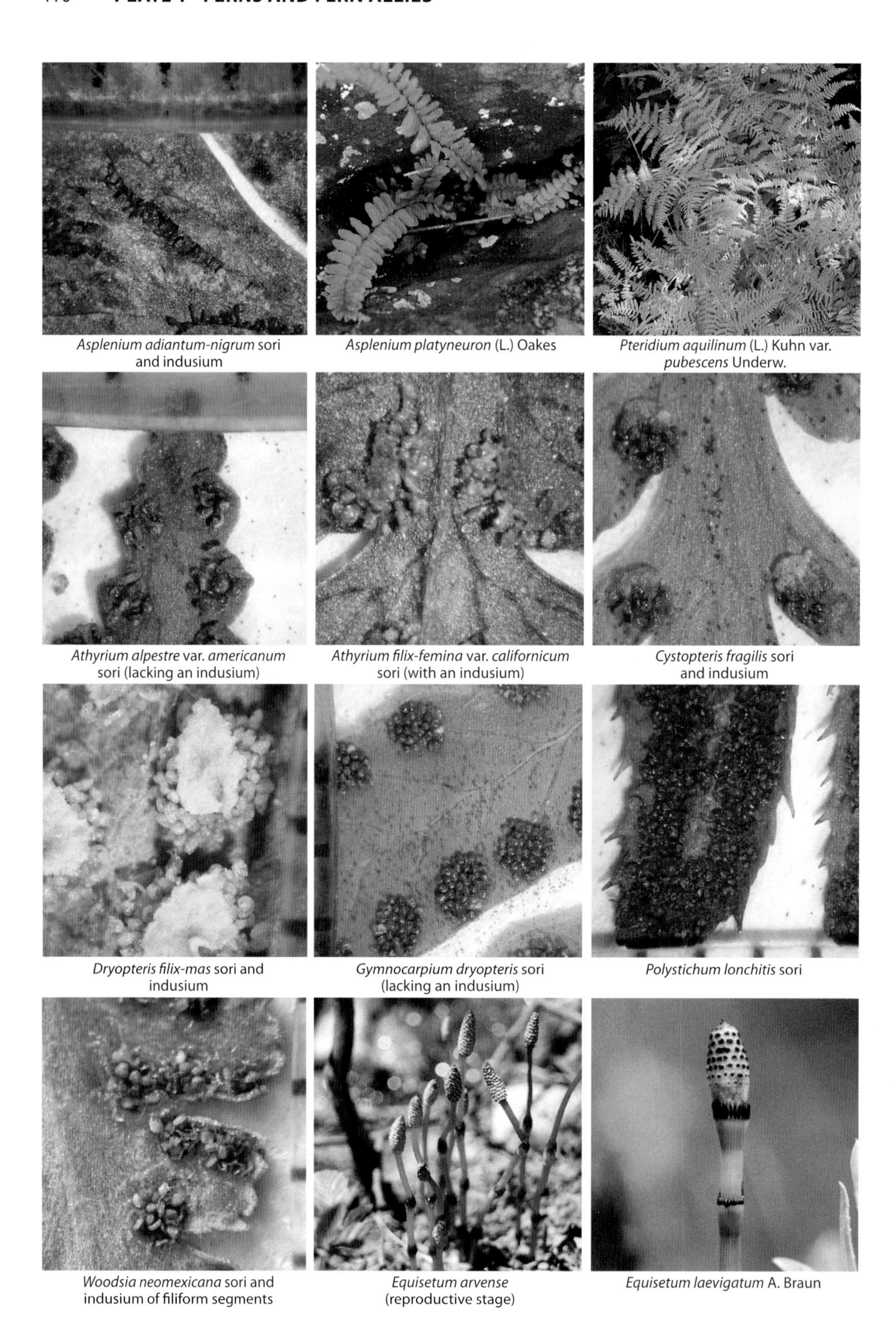

Asplenium adiantum-nigrum sori and indusium

Asplenium platyneuron (L.) Oakes

Pteridium aquilinum (L.) Kuhn var. *pubescens* Underw.

Athyrium alpestre var. *americanum* sori (lacking an indusium)

Athyrium filix-femina var. *californicum* sori (with an indusium)

Cystopteris fragilis sori and indusium

Dryopteris filix-mas sori and indusium

Gymnocarpium dryopteris sori (lacking an indusium)

Polystichum lonchitis sori

Woodsia neomexicana sori and indusium of filiform segments

Equisetum arvense (reproductive stage)

Equisetum laevigatum A. Braun

Lycopodium annotinum L.

Polypodium hesperium sori

Polypodium saximontanum Windham

Adiantum aleuticum sori and indusium

Argyrochosma fendleri (Kunze) Windham

Cheilanthes eatonii fronds (abaxial side)

Cheilanthes feei T. Moore

Cheilanthes fendleri fronds (with scales)

Cryptogramma acrostichoides R. Br.

Notholaena standleyi sori and abaxial frond surface

Pellaea glabella sori and false indusium

Pellaea truncata ultimate leaf segments

Botrychium ascendens W.H. Wagner

Botrychium campestre W.H. Wagner & Farrar ssp. *lineare* (W.H. Wagner) Farrar, ined.

Botrychium echo W.H. Wagner

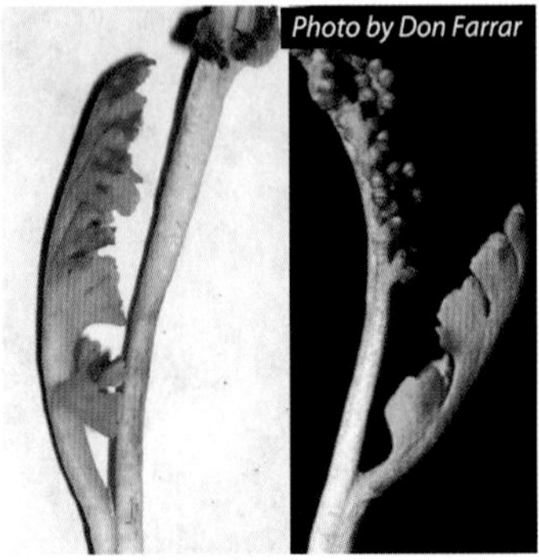

Botrychium furculatum Popovich & Farrar, ined.

Botrychium hesperium (Maxon & R.T. Clausen) W.H. Wagner & Lellinger

Botrychium lanceolatum (S.G. Gmel.) Angstr. ssp. *lanceolatum*

Botrychium lunaria (L.) Sw. var. *crenulatum* (W.H. Wagner) Stensvold, ined.

Botrychium minganense Vict.

Botrychium neolunaria Stensvold & Farrar, ined.

Botrychium pinnatum H. St. John

Botrypus virginianus (L.) Michx.

Sceptridium multifidum (S.G. Gmel.) M. Nishida ex Tagawa

Selaginella densa Rydb.

Juniperus scopulorum Sarg.

Ephedra viridis Coville var. *viridis* (pistillate cones)

Ephedra viridis Coville var. *viridis* (staminate cones)

Abies bifolia A. Murray

Picea pungens Engelm.

Pinus aristata Engelm.

Pinus contorta Douglas ex Loud. var. *latifolia* Engelm.

Pinus edulis Engelm.

Pinus flexilis E. James

Pinus ponderosa Douglas ex Laws. var. *scopulorum* Engelm.

Pseudotsuga menziesii (Mirb.) Franco var. *glauca* (Beissn.) Franco

Sambucus racemosa L.

Leucocrinum montanum Nutt. ex A. Gray

Yucca baccata Torr. var. *baccata*

Yucca glauca Nutt.

Allium acuminatum Hook.

Allium cernuum Roth

Amaranthus retroflexus L.

Froelichia gracilis (Hook.) Moq.

Rhus trilobata Nutt.

Toxicodendron rydbergii (Small ex Rydb.) Greene

Aletes acaulis (Torr.) J.M. Coult. & Rose

Angelica grayi J.M. Coult. & Rose

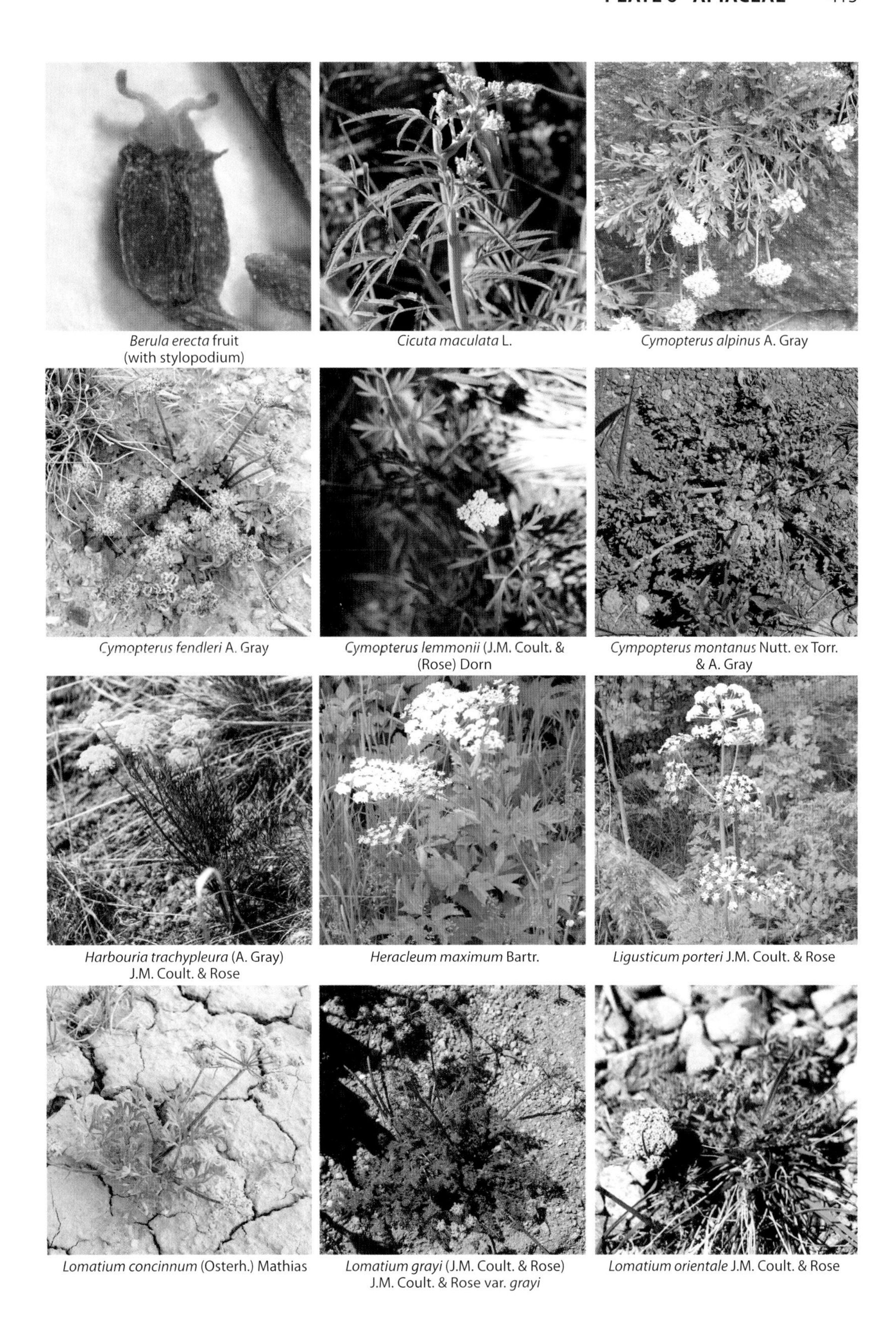

Berula erecta fruit (with stylopodium)

Cicuta maculata L.

Cymopterus alpinus A. Gray

Cymopterus fendleri A. Gray

Cymopterus lemmonii (J.M. Coult. & (Rose) Dorn

Cympopterus montanus Nutt. ex Torr. & A. Gray

Harbouria trachypleura (A. Gray) J.M. Coult. & Rose

Heracleum maximum Bartr.

Ligusticum porteri J.M. Coult. & Rose

Lomatium concinnum (Osterh.) Mathias

Lomatium grayi (J.M. Coult. & Rose) J.M. Coult. & Rose var. *grayi*

Lomatium orientale J.M. Coult. & Rose

Lomatium triternatum (Pursh) J.M. Coult. & Rose var. *platycarpum* (Torr.) Boivin

Musineon divaricatum (Pursh) Nutt. ex Torr. & A. Gray

Musineon tenuifolium Nutt.

Osmorhiza berteroi DC.

Osmorhiza depauperata fruit

Oxypolis fendleri (A. Gray) Heller

Podistera eastwoodiae (J.M. Coult. & Rose) Mathias & Constance

Sanicula marilandica L.

Apocynum cannabinum L.

Asclepias cryptoceras S. Watson ssp. *cryptoceras*

Asclepias engelmanniana Woodson

Asclepias incarnata L.

Asclepias latifolia (Torr.) Raf.

Asclepias speciosa Torr.

Asclepias tuberosa L. ssp. *interior* Woodson

Asclepias viridiflora Raf.

Achillea millefolium L.

Agoseris glauca (Pursh) Raf. var. *dasycephala* (Torr. & A. Gray) Jeps.

Ambrosia acanthicarpa head

Ambrosia trifida L. var. *trifida*

Anaphalis margaritacea (L.) Benth. & Hook.

Antennaria umbrinella Rydb.

Arnica cordifolia Hook.

Artemisia frigida Willd.

Artemisia ludoviciana Nutt. var. *ludoviciana*

Artemisia nova leaf with gland dots

Artemisia scopulorum head with dark-margined involucre bracts

Artemisia spinescens DC.

Bahia dissecta (A. Gray) Britt.

Balsamorhiza sagittata (Pursh) Nutt.

Bidens tenuisecta pappus

Brickellia eupatorioides involucre bracts

Carduus nutans L.

Centaurea diffusa involucre bracts

Chaenactis douglasii (Hook.) Hook. & Arn. var. *alpina* A. Gray

Chaetopappa ericoides (Torr.) G.L. Nesom

Chrysothamnus greenei head

Cichorium intybus L.

Cirsium arvense (L.) Scop.

Cirsium canescens Nutt.

Cirsium clavatum (M.E. Jones) Petr. var. *clavatum*

Cirsium eatonii (A. Gray) B.L. Rob.

Cirsium osterhoutii Rydb.

Cirsium scopulorum (Greene) Cockerell ex Daniels

Cirsium undulatum (Nutt.) Spreng.

Coreopsis tinctoria Nutt.

Crepis atribarba A. Heller

Dieteria bigelovii (A. Gray) D.R. Morgan & R.L. Hartm. var. *bigelovii*

Dyssodia papposa (Vent.) Hitchc.

Encelia nutans Eastw.

Ericameria nauseosa (Pallas ex Pursh) G.L. Nesom & Baird var. *nauseosa*

Erigeron caespitosus Nutt.

Erigeron compositus Pursh

Erigeron coulteri hairs on the involucre bracts

Erigeron eatonii A. Gray var. *eatonii*

Erigeron glacialis (Nutt.) A. Nelson

Erigeron grandiflorus Hook.

Erigeron leiomerus A. Gray

Erigeron melanocephalus (A. Nelson) A. Nelson

Erigeron pinnatisectus (A. Gray) A. Nelson

Erigeron pumilus Nutt.

Erigeron speciosus (Lindl.) DC.

Erigeron vetensis Rydb.

Gaillardia aristata Pursh

Gaillardia pulchella Foug.

Grindelia squarrosa (Pursh) Dunal var. *squarrosa*

Grindelia subalpina Greene

Gutierrezia sarothrae (Pursh) Britt. & Rusby

Helianthella quinquenervis (Hook.) A. Gray

Helianthus annuus L.

Helianthus pumilus Nutt.

Heliomeris multiflora Nutt. var. *multiflora*

Heterotheca villosa (Pursh) Shinners var. *villosa*

Hymenopappus filifolius Hook. var. *cinereus* (Rydb.) I.M. Johnst.

Hymenoxys grandiflora (Torr. & A. Gray ex A. Gray) Parker

Hymenoxys richardsonii (Hook.) Cockerell

Lactuca tatarica (L.) C.A. Mey. var. *pulchella* (Pursh) Breitung

Liatris punctata Hook.

Lygodesmia grandiflora (Nutt.) Torr. & A. Gray var. *grandiflora*

Lygodesmia juncea (Pursh) D. Don ex Hook.

Machaeranthera tanacetifolia (Kunth) Nees

Malacothrix torreya A. Gray

Melampodium leucanthum Torr. & A. Gray

Nothocalais cuspidata (Pursh) Greene

Packera cana (Hook.) W.A. Weber & Á. Löve

Packera crocata (Rydb.) W.A. Weber & Á. Löve

Packera fendleri (A. Gray) W.A. Weber & Á. Löve

Packera werneriifolia (A. Gray) W.A. Weber & Á. Löve

Parthenium alpinum (Nutt.) Torr. & A. Gray var. *alpinum*

Psilostrophe bakeri Greene

Ratibida columnifera (Nutt.) Wooton & Standl.

Rudbeckia hirta L. var. *pulcherrima* Farw.

Senecio amplectens A. Gray var. *holmii* (Greene) Harrington

Senecio atratus Greene

Senecio eremophilus Richardson var. *kingii* (Rydb.) Greenm.

Senecio fremontii Torr. & A. Gray var. *blitoides* (Greene) Cronquist

Senecio spartioides Torr. & A. Gray

Senecio taraxacoides (A. Gray) Greene

Senecio triangularis Hook.

Solidago canadensis L.

Solidago missouriensis Nutt.

Solidago rigida L. var. *humilis* Porter

Solidago simplex Kunth var. *simplex*

Stenotus armerioides Nutt. var. *armerioides*

Symphyotrichum ericoides involucre bracts

Symphyotrichum falcatum (Lindl.) G.L. Nesom

Symphyotrichum laeve (L.) Á. Löve & D. Löve var. *geyeri* (A. Gray) G.L. Nesom

Taraxacum officinale achene

Tetradymia canescens DC.

Tetraneuris acaulis (Pursh) Greene var. *caespitosa* A. Nelson

Thelesperma megapotamicum (Spreng.) Kuntze

Tonestus pygmaeus (Torr. & A. Gray) A. Nelson

Townsendia grandiflora Nutt.

Townsendia hookeri Beaman

Townsendia incana Nutt.

Townsendia rothrockii A. Gray ex Rothrock

Verbesina encelioides (Cav.) Benth. & Hook. f. ex A. Gray ssp. *exauriculata* (B.L. Rob. & Greenm.) J.R. Coleman

Wyethia amplexicaulis (Nutt.) Nutt.

Xanthisma spinulosum (Pursh) D.R. Morgan & R.L. Hartm.

Zinnia grandiflora Nutt.

Berberis repens Lendl.

Alnus incana (L.) Moench ssp. *tenuifolia* (Nutt.) Brietung

Corylus cornuta Marshall var. *cornuta*

Anchusa officinalis L.

Eritrichium nanum (Vill.) Schrad. ex Gaudin var. *elongatum* (Rydb.) Cronquist

Lithospermum incisum Lehm.

Mertensia ciliata (James ex Torr.) G. Don

Mertensia lanceolata (Pursh) DC. var. *lanceolata*

Onosmodium bejariense DC. ex A. DC. var. *occidentale* (Mack.) B.L. Turner

Oreocarya cana A. Nelson

Oreocarya osterhoutii Payson

Oreocarya thyrsiflora Greene

Cryptantha affinis nutlets, dorsal and ventral view

Cryptantha crassisepala nutlets

Cryptantha fendleri nutlets

Cryptantha kelseyana nutlets

Cryptantha minima nutlets, dorsal and ventral views

Cryptantha pterocarya nutlets

Cryptantha rudis nutlets, dorsal and ventral views

Cryptantha scoparia nutlet, ventral view

Hackelia floribunda nutlets

Lappula occidentalis var. *cupulata* nutlets

Lappula squarrosa nutlets

Lithospermum incisum nutlets

Oreocarya bakeri nutlet, ventral view

Oreocarya caespitosa nutlets

Oreocarya cana nutlets

Oreocarya celodioides nutlet, ventral view

Oreocarya elata nutlet

Oreocarya flava nutlet

Oreocarya rollinsii nutlet, ventral view

Oreocarya sericea nutlets

Oreocarya stricta nutlet, ventral view

Oreocarya suffruticosa nutlets

Oreocarya virgata (Porter) Greene

Oreocarya weberi nutlets

Alyssum alyssoides fruit

Alyssum desertorum fruit

Alyssum simplex Rudolphi

Barbarea orthoceras Ledeb.

Berteroa incana (L.) DC.

Boechera consanguinea lower stem hairs

Boechera gracilenta lower stem hairs

Boechera grahamii lower stem hairs

Boechera pendulina basal leaves

Boechera spatifolia (Rydb.) Windham & Al-Shehbaz

Cardamine cordifolia A. Gray

Chorispora tenella (Pall.) DC.

Draba aurea Vahl ex Hornem.

Draba exunguiculata (O.E. Schulz) C.L. Hitchc.

Draba incerta lower leaf surface

Draba oligosperma lower leaf surface

Draba reptans (Lam.) Fernald

Draba spectabilis lower leaf surface

Erysimum asperum leaf surface

Erysimum capitatum (Douglas ex Hook.) Greene

Lepidium alyssoides fruit

Lepidium appelianum fruit

Lepidium campestre fruit

Lepidium crenatum fruit

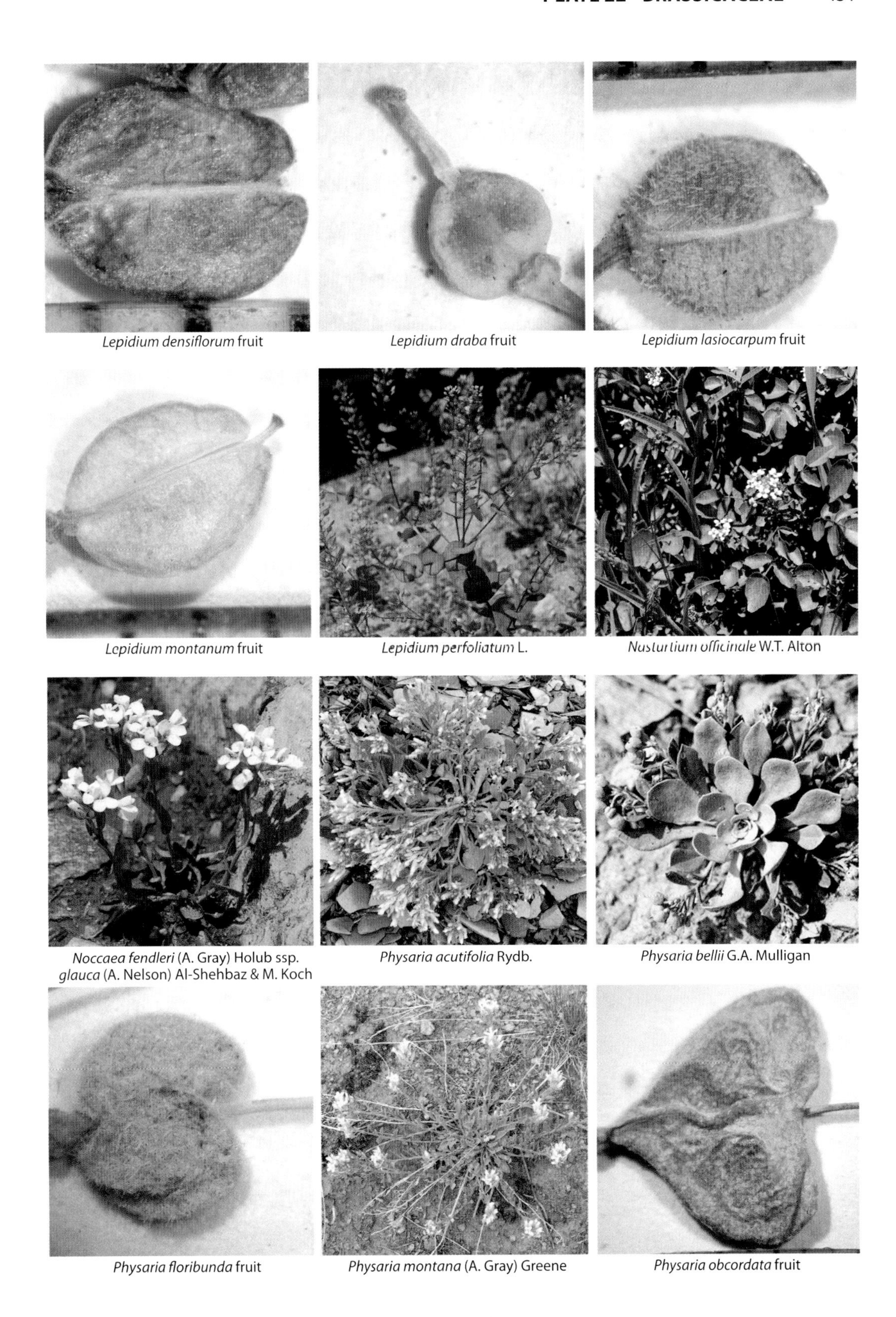

Lepidium densiflorum fruit

Lepidium draba fruit

Lepidium lasiocarpum fruit

Lepidium montanum fruit

Lepidium perfoliatum L.

Nasturtium officinale W.T. Alton

Noccaea fendleri (A. Gray) Holub ssp. *glauca* (A. Nelson) Al-Shehbaz & M. Koch

Physaria acutifolia Rydb.

Physaria bellii G.A. Mulligan

Physaria floribunda fruit

Physaria montana (A. Gray) Greene

Physaria obcordata fruit

Physaria parviflora fruit

Physaria reediana O'Kane & Al-Shehbaz

Physaria vitulifera Rydb.

Rorippa alpina (S. Watson) Rydb.

Smelowskia americana Rydb.

Stanelya pinnata (Pursh) Britton var. *pinnata*

Coryphantha vivipara (Nutt.) Britton & Rose

Cylindropuntia imbricata (Haw.) F.M. Kunth var. *imbricata*

Echinocereus triglochidiatus Engelm. var. *triglochidiatus*

Echinocereus viridiflorus Englem.

Opuntia polyacantha Haw. var. *polyacantha*

Pediocactus simpsonii (Engelm.) Britton & Rose

Campanula parryi A. Gray var. *parryi*

Campanula rapunculoides L.

Campanula rotundifolia L.

Humulus lupulus L. var. *neomexicanus* A. Nelson & Cockerell

Cleome serrulata Pursh

Polanisia dodecandra (L.) DC. ssp. *trachysperma* (Torr. & A. Gray) Iltis

Linnaea borealis L. var. *longiflora* Torr.

Lonicera involucrata (Richarson) Banks ex Spreng.

Lonicera tatarica L.

Symphoricarpos albus flower showing pubescence near the top of the corolla

Symphoricarpos occidentalis Hook.

Symphoricarpos rotundifolius flower showing pubescence near the base

Cerastium arvense L. ssp. *strictum* Gaudin

Dianthus armeria L.

Eremogone fendleri (A. Gray) Ikonn.

Eremogone hookeri (Nutt.) W.A. Weber

Minuartia obtusiloba (Rydb.) House

Paronychia jamesii Torr. & A. Gray

Paronychia pulvinata A. Gray

Pseudostellaria jamesiana (Torr.) W.A. Weber & R.L. Hartm.

Saponaria officinalis L.

Silene acaulis (L.) Jacq.

Silene drummondii Hook. ssp. *drummondii*

Silene menziesii Hook.

Atriplex argentea fruiting bract

Atriplex canescens fruiting bract

Atriplex confertifolia fruiting bract

Atriplex corrugata fruiting bract

Atriplex gardneri (Moq.) D. Dietr.

Atriplex graciliflora fruiting bracts

Atriplex heterosperma fruiting bracts

Atriplex obovata fruiting bract

Atriplex pleiantha fruiting bract (with exposed fruit)

Atriplex powellii fruiting bract (tuberculate)

Atriplex rosea fruiting bracts

Atriplex saccaria S. Watson

Atriplex subspicata
fruiting bract

Atriplex truncata
fruiting bracts

Atriplex wolfii
fruiting bracts

Axyris amaranthoides
fruit (lacking perianth)

Bassia hyssopifolia
flower

Corispermum americanum
fruit

Corispermum hyssopifolium
fruit

Corispermum navicula
fruit

Corispermum villosum
fruit

Cycloloma atriplicifolium
flower

Grayia spinosa (Hook.) Moq.

Halogeton glomeratus (M. Bieb.) C.A. Mey.

Kochia americana S. Watson

Krascheninnikovia lanata (Pursh) A. Meeuse & A. Smit

Micromonolepis pusilla fruit

Monolepis nuttalliana (Schult.) Greene

Salicornia rubra flowers

Salsola tragus fruit

Sarcobatus vermiculatus fruit

Sarcobatus vermiculatus (Hook.) Torr.

Suaeda calceoliformis flower

Suaeda nigra flower

Suckleya suckleyana fruit

Zuckia brandegeei fruit

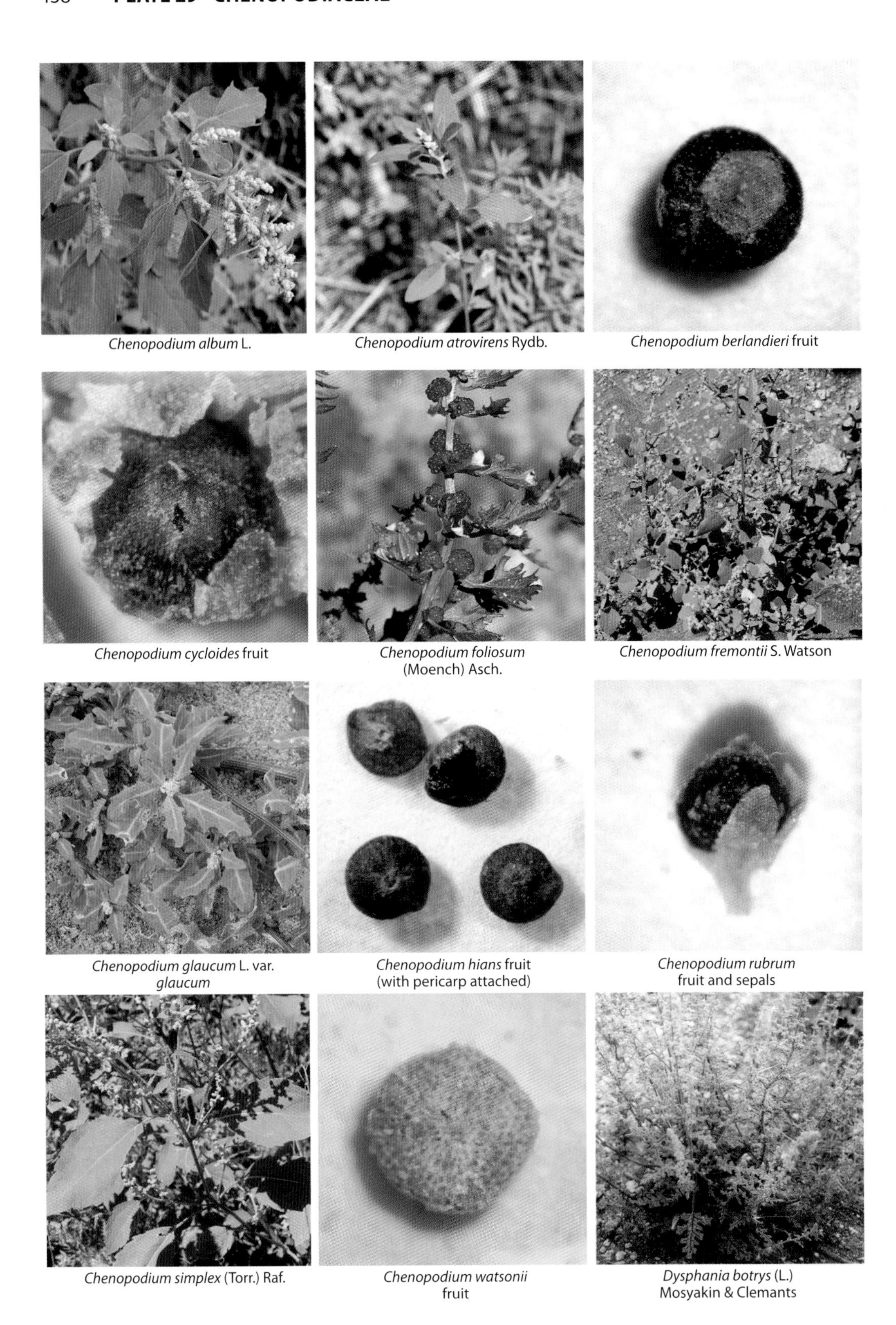

Chenopodium album L.

Chenopodium atrovirens Rydb.

Chenopodium berlandieri fruit

Chenopodium cycloides fruit

Chenopodium foliosum (Moench) Asch.

Chenopodium fremontii S. Watson

Chenopodium glaucum L. var. *glaucum*

Chenopodium hians fruit (with pericarp attached)

Chenopodium rubrum fruit and sepals

Chenopodium simplex (Torr.) Raf.

Chenopodium watsonii fruit

Dysphania botrys (L.) Mosyakin & Clemants

Hypericum perforatum L.

Tradescantia occidentalis (Britton) Smyth

Calystegia sepium (L.) R. Br.

Cuscuta cuspidata flower

Cuscuta indecora flowers

Cuscuta umbellata flowers

Evolvulus nuttallianus Schult.

Ipomoea leptophylla Torr.

Cornus canadensis L.

Rhodiola integrifolia Raf.

Sedum lanceolatum Torr.

Cucurbita foetidissima Kunth

Carex albonigra perigynia and pistillate scales

Carex aquatilis perigynia and pistillate scales

Carex arapahoensis perigynium and pistillate scale

Carex atherodes perigynia and pistillate scales

Carex athrostachya perigynium with pistillate scale removed

Carex aurea perigynium and pistillate scale

Carex bebbii perigynia and pistillate scales

Carex bella perigynia and pistillate scales

Carex brevior perigynia and pistillate scales

Carex brunnescens perigynium with pistillate scale removed

Carex buxbaumiii perigynium and pistillate scale

Carex canescens ssp. *canescens* perigynium with pistillate scale removed

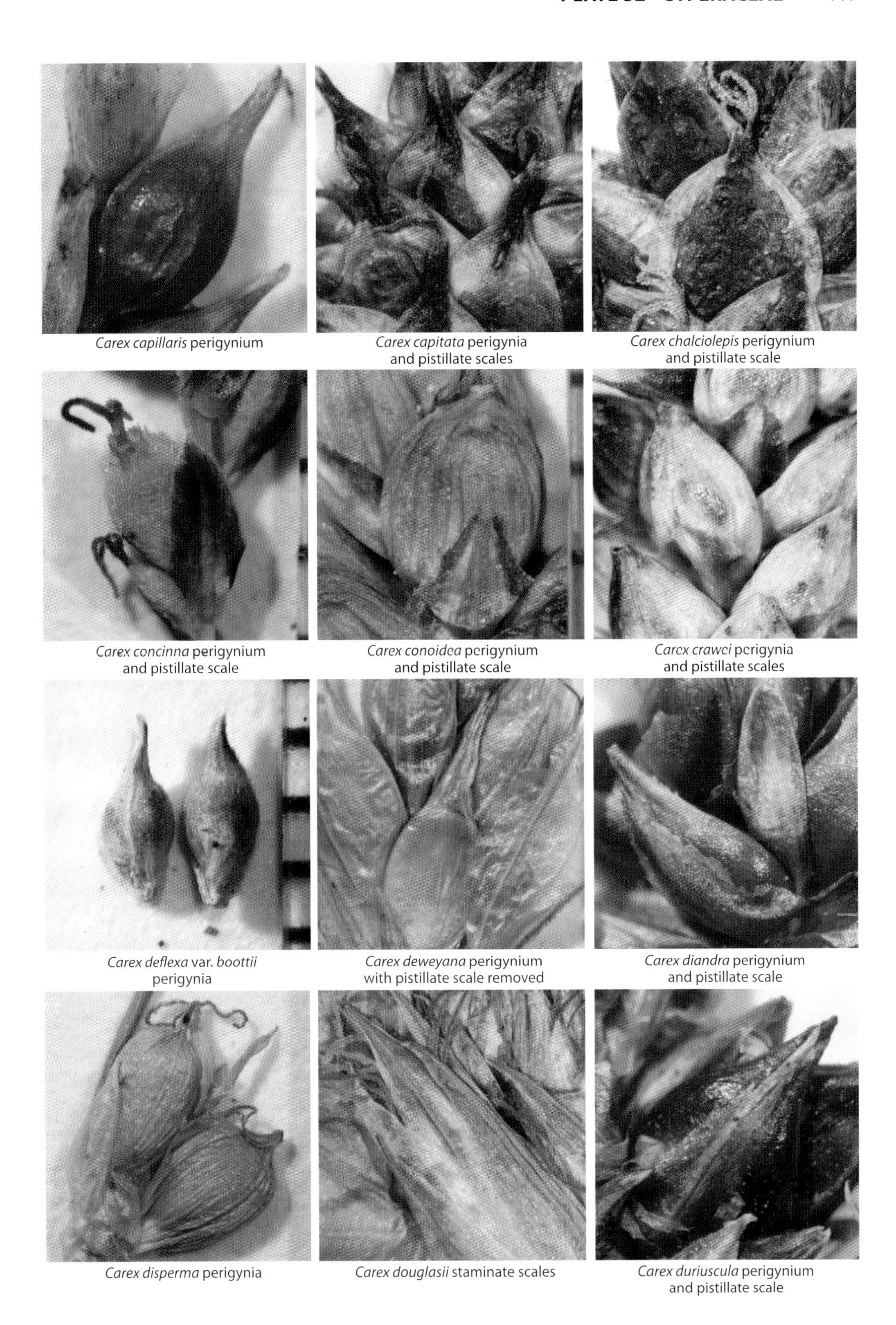

Carex capillaris perigynium

Carex capitata perigynia and pistillate scales

Carex chalciolepis perigynium and pistillate scale

Carex concinna perigynium and pistillate scale

Carex conoidea perigynium and pistillate scale

Carex crawei perigynia and pistillate scales

Carex deflexa var. *boottii* perigynia

Carex deweyana perigynium with pistillate scale removed

Carex diandra perigynium and pistillate scale

Carex disperma perigynia

Carex douglasii staminate scales

Carex duriuscula perigynium and pistillate scale

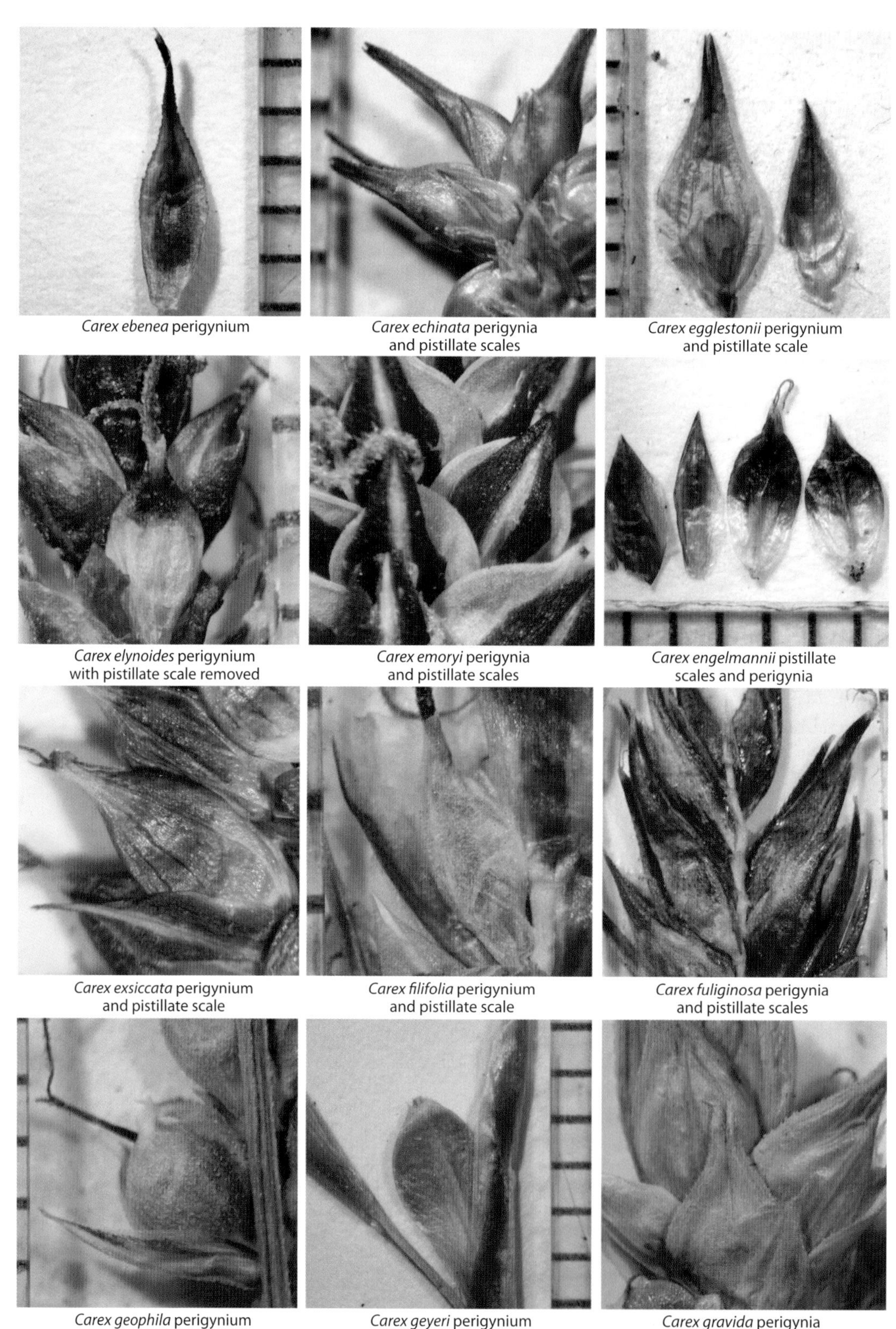

Carex ebenea perigynium

Carex echinata perigynia and pistillate scales

Carex egglestonii perigynium and pistillate scale

Carex elynoides perigynium with pistillate scale removed

Carex emoryi perigynia and pistillate scales

Carex engelmannii pistillate scales and perigynia

Carex exsiccata perigynium and pistillate scale

Carex filifolia perigynium and pistillate scale

Carex fuliginosa perigynia and pistillate scales

Carex geophila perigynium and pistillate scale

Carex geyeri perigynium

Carex gravida perigynia and pistillate scales

Carex gynocrates perigynia

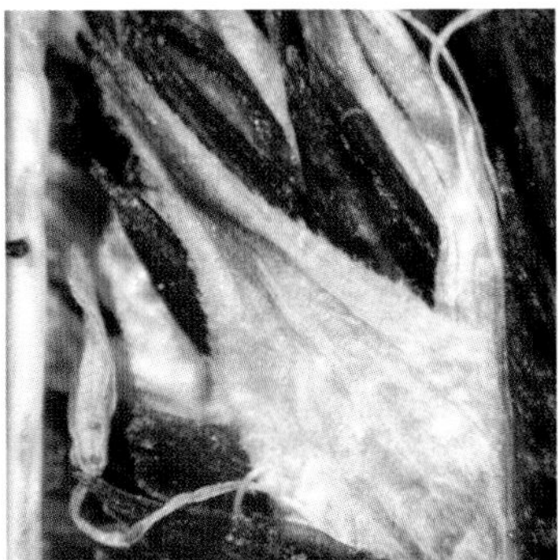
Carex haydeniana perigynium with pistillate scale removed

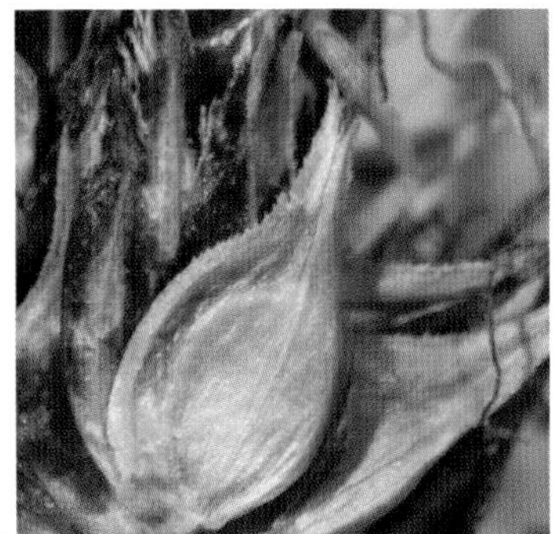
Carex hoodii perigynium with pistillate scale removed

Carex hystricina perigynia

Carex illota perigynium with pistillate scale offset

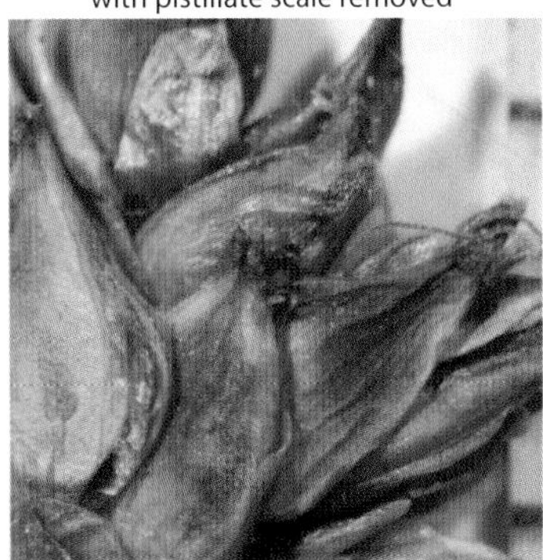
Carex incurviformis perigynia

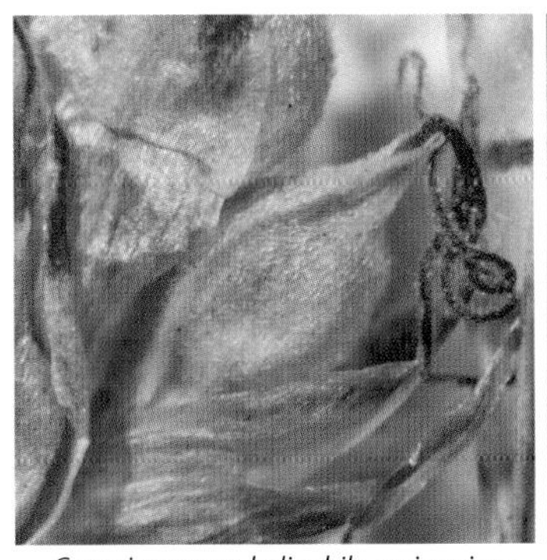
Carex inops ssp. *heliophila* perigynium and pistillate scale

Carex interior perigynia and pistillate scales

Carex jonesii perigynia

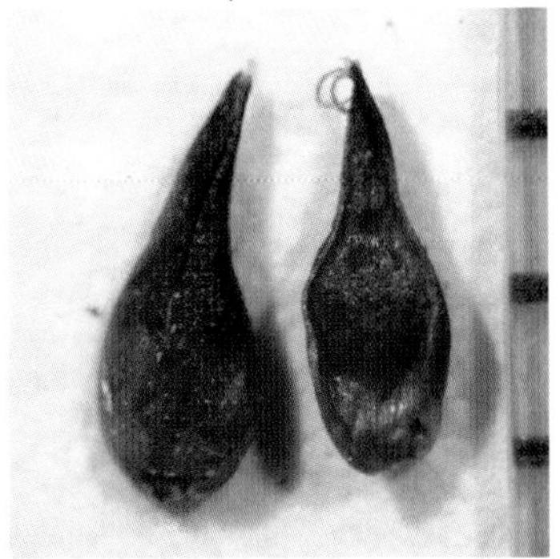
Carex lachenalii perigynia

Carex lenticularis perigynia and pistillate scales

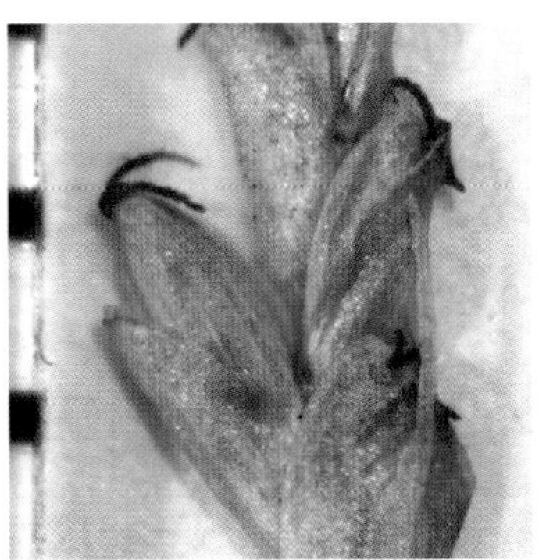
Carex leptalea perigynia and pistillate scales

Carex limosa perigynium and pistillate scale

Carex livida perigynium and pistillate scale

Carex magellanica ssp. *irrigua* perigynium and pistillate scale

Carex microglochin perigynia

Carex micropoda perigynium with pistillate scale removed

Carex microptera pistillate scale and perigynium

Carex nardina perigynia and pistillate scales

Carex nebrascensis perigynia and pistillate scales

Carex nelsonii perigynium with pistillate scale offset

Carex neurophora perigynia and pistillate scales

Carex nigricans perigynia

Carex nova perigynia and pistillate scales

Carex obtusata perigynia and pistillate scales

Carex occidentalis perigynia and pistillate scales

Carex oreocharis perigynia with pistillate scale removed

Carex pachystachya pistillate scale and perigynium

Carex parryana perigynium and pistillate scale

Carex paysonis perigynia and pistillate scales

Carex peckii perigynia and pistillate scales

Carex pellita perigynia and pistillate scales

Carex perglobosa perigynia with pistillate scale removed

Carex petasata perigynia and pistillate scales

Carex phaeocephala pistillate scale and perigynium

Carex pityophila perigynia and pistillate scales

Carex praeceptorum perigynia and pistillate scales

Carex praegracilis perigynium with pistillate scale removed

Carex praticola perigynia and pistillate scales

Carex raynoldsii perigynia and pistillate scales

Carex retrorsa perigynia

Carex rossii perigynia and pistillate scales

Carex rupestris perigynium

Carex sartwellii perigynium and pistillate scale

Carex saxatilis perigynium and pistillate scale

Carex saximontana perigynium

Carex scirpoidea perigynium with one pistillate scale removed

Carex scoparia pistillate scale and perigynium

Carex scopulorum perigynia and pistillate scales

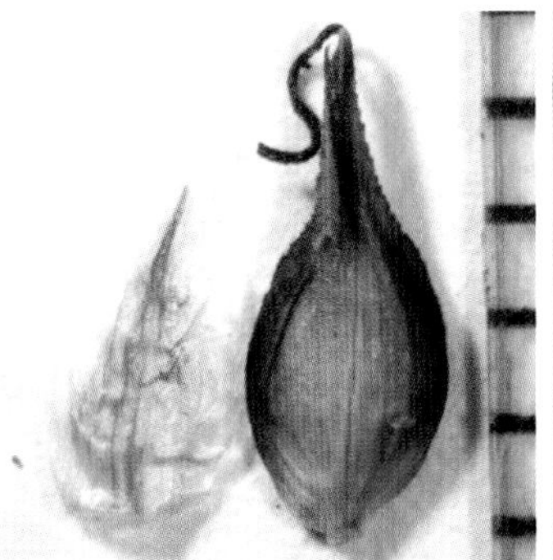
Carex siccata pistillate scale and perigynium

Carex simulata pistillate scale and perigynium

Carex sprengelii perigynium and pistillate scale

Carex stenoptila pistillate scale and perigynium

Carex stevenii perigynia and pistillate scales

Carex stipata perigynia and pistillate scales

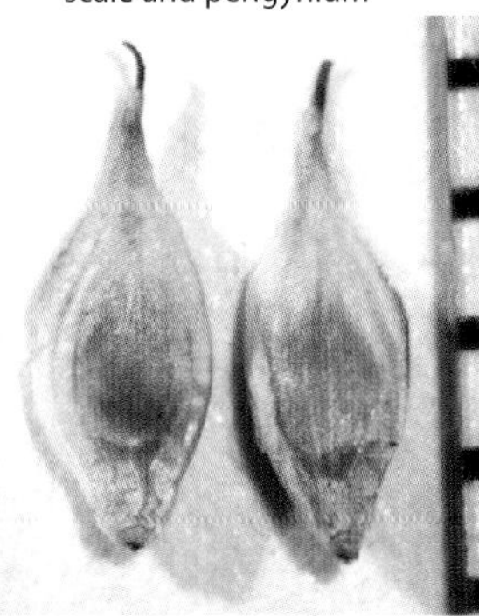
Carex tahoensis perigynia

Carex tenuiflora perigynia and pistillate scales

Carex torreyi perigynia and pistillate scales

Carex utriculata perigynia and pistillate scales

Carex vallicola perigynia and pistillate scales

Carex vernacula pistillate scale and perigynium

Carex vesicaria perigynium and pistillate scale

Carex viridula ssp. *viridula* perigynia and pistillate scales

Carex vulpinoidea perigynia and pistillate scales

Amphiscirpus nevadensis achene

Bolboschoenus maritimus achene

Cyperus acuminatus spikelet

Cyperus erythrorhizos spikelets

Cyperus esculentus L.

Cyperus fendlerianus spikelet

Cyperus lupulinus spikelets and rachilla

Cyperus odoratus spikelets and rachilla

Cyperus schweinitzii spikelets

Cyperus squarrosus spikelet

Eleocharis acicularis achene

Eleocharis bolanderi achene

Eleocharis compressa floral scales

Eleocharis engelmannii achene

Eleocharis obtusa achene

Eleocharis palustris achene

Eleocharis pauciflora achene

Eleocharis rostellata achene

Eleocharis angustifolium Honck.

Kobresia myosuroides perigynia (immature and mature)

Lipocarpha aristulata scale and achene

Rhynchospora alba achene

Schoenoplectus acutus (Muhl. ex J.M. Bigelow) Á Löve & D. Löve

Scirpus pallidus (Britton) Fernald

Shepherdia canadensis coppery-brown scaly-stellate hairs

Arctostaphylos uva-ursi (L.) Spreng.

Kalmia microphylla (Hook.) A. Heller

Moneses uniflora (L.) A. Gray

Orthilia secunda (L.) House

Pterospora andromedea Nutt.

Pyrola chlorantha Sw.

Vaccinium scoparium Leiberg ex Coville

Chamaesyce fendleri (Torr. & A. Gray) Small

Croton texensis (Kl.) Müll. Arg.

Euphorbia brachycera Engelm.

Euphorbia dentata Michx.

Euphorbia marginata Pursh

Tragia ramosa Torr.

Amorpha canescens Pursh

Astragalus argrestis Douglas ex G. Don

Astragalus alpinus L.

Astragalus asclepiadoides M.E. Jones

Astragalus bisulcatus (Hook.) A. Gray var. *bisulcatus*

Astragalus ceramicus Sheldon var. *filifolius* (A. Gray) F.J. Herm.

Astragalus convallarius Greene

Astragalus crassicarpus Nutt.

Astragalus drummondii Douglas ex Hook.

Astragalus flavus Nutt.

Astragalus flexuosus Douglas ex G. Don var. *flexuosus*

Astragalus laxmannii Jacq. var. *robustior* (Hook.) Barneby & S.L. Welsh

Astragalus lentiginosus Douglas ex Hook. var. *palans* (M.E. Jones) M.E. Jones

Astragalus linifolius Osterh.

Astragalus missouriensis Nutt. var. *missouriensis*

Astragalus parryi A. Gray

Astragalus pattersonii A. Gray

Astragalus pectinatus (Douglas ex Hook.) Douglas ex G. Don var. *pectinatus*

Astragalus sericoleucus A. Gray

Astragalus shortianus Nutt.

Astragalus spatulatus Sheldon

Astragalus tenellus Pursh

Astragalus thompsoniae S. Watson

Astragalus wingatanus S. Watson

Dalea candida Michx. ex Willd. var. *olgophylla* (Torr.) Shinners

Dalea jamesii (Torr.) Torr. & A. Gray

Dalea purpurea Vent.

Glycyrrhiza lepidota Pursh

Hedysarum boreale Nutt.

Lathyrus lanszwertii Kellogg var. *leucanthus* (Rydb.) Dorn

Lathyrus polymorphus Nutt. var. *incanus* (J.G. Sm. & Rydb.) C.L. Hitchc.

Lupinus argenteus Pursh var. *argenteus*

Lupinus plattensis S. Watson

Lupinus pusillus Pursh

Medicago sativa L.

Mimosa nuttallii (DC.) B.L. Turner

Oxytropis lambertii Pursh

Oxytropis multiceps Nutt.

Oxytropis sericea Nutt. var. *sericea*

Oxytropis splendens Douglas ex Hook.

Pediomelum esculentum (Pursh) Rydb.

Pediomelum megalanthum (Wooton & Standl.) Rydb.

Psoralidium lanceolatum (Pursh) Rydb.

Sophora nuttalliana B.L Turner

Thermopsis rhombifolia var. *divaricarpa* (A. Nelson) Isely

Trifolium dasyphyllum Torr. & A. Gray

Trifolium nanum Torr.

Trifolium pratense L.

Quercus gambelii Nutt.

Corydalis aurea Willd. ssp. *aurea*

Eustoma grandiflorum (Raf.) Shinners

Frasera speciosa Douglas ex Griseb.

Gentiana algida Pall.

Gentiana bigelovii A. Gray

Gentiana prostrata Haenke

Gentianella acuta (Michx.) Hiitonen

Gentianopsis detonsa (Rottb.) Ma var. *elegans* (A. Nelson) N.H. Holmgren

Geranium caespitosum James

Geranium richardsonii Fisch. & Trautv.

Ribes aureum Pursh var. *aureum*

Ribes cereum Douglas

Ribes inerme Rydb.

Ribes lacustre (Pers.) Poir.

Ribes montigenum McClatchie

Jamesia americana Torr. & A. Gray var. *americana*

Philadelphus microphyllus A. Gray

Najas guadalupensis ssp. *guadalupensis* fruit

Ellisia nyctelea (L.) L.

Hydrophyllum fendleri (A. Gray) A.A. Heller var. *fendleri*

Phacelia alba Rydb.

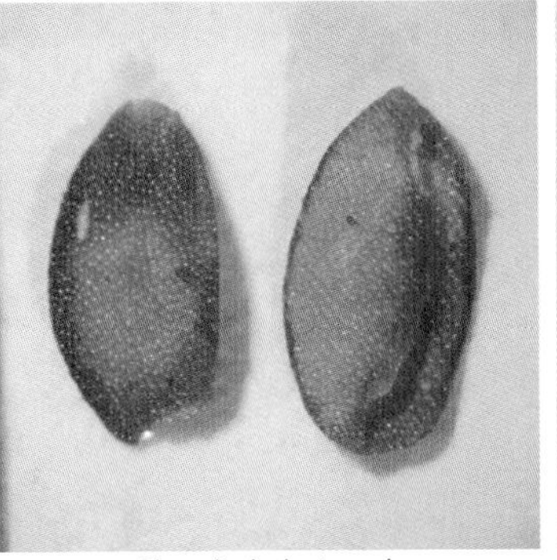

Phacelia bakeri seeds (dorsal and ventral views)

Phacelia crenulata var. *corrugata* seeds (dorsal and ventral views)

Phacelia formosula seeds (dorsal and ventral views)

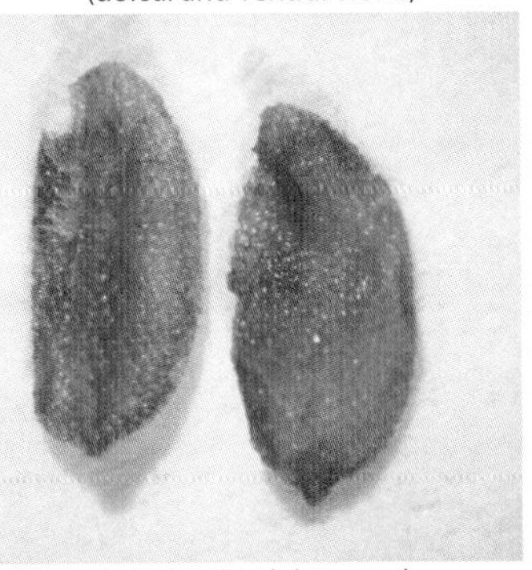

Phacelia glandulosa seeds (ventral and dorsal views)

Phacelia hastata Douglas ex Lehm.

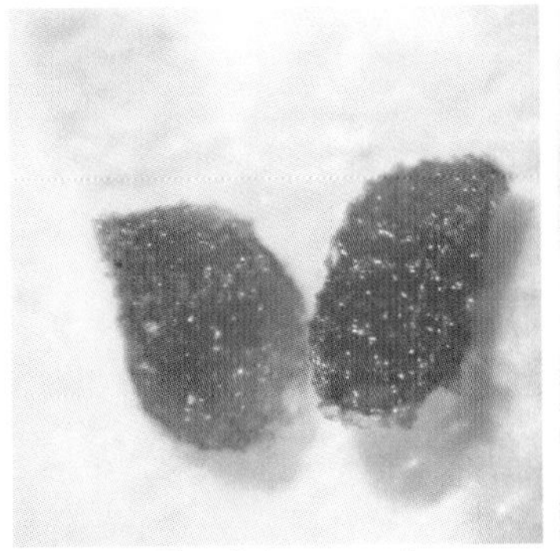

Phacelia incana seeds

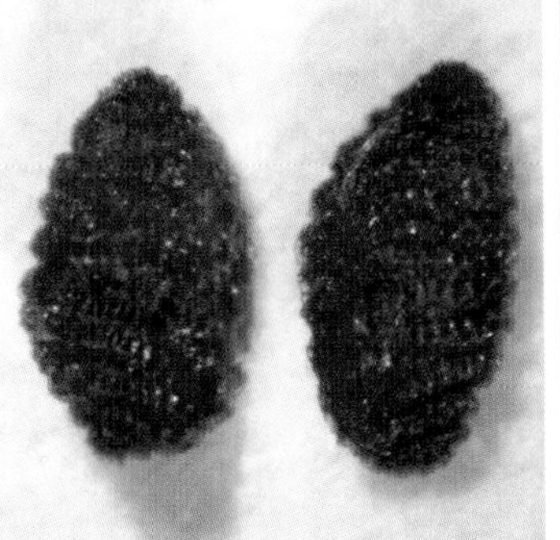

Phacelia scopulina var. *submutica* seeds

Phacelia sericea (Graham ex Hook.) A. Gray var. *sericea*

Iris missouriensis Nutt.

Sisyrinchium montanum Greene var. *montanum*

Juncus articus Willd. var. *balticus* (Willd.) Trautv.

Juncus brachycephalus seeds (tailed)

Juncus confusus tepals

Juncus drummondii bristle tip on leaf sheath

Juncus dudleyi leaf sheath auricles

Juncus interior flower and seeds (not tailed)

Juncus nodosus capsule

Luzula parviflora (Ehrh.) Desv.

Luzula spicata (L.) DC.

Triglochin palustris schizocarp

Dracocephalum parviflorum spine-tipped leaves

Hedeoma drummondii Benth.

Leonurus cardiaca L.

Mentha arvensis L.

Monarda fistulosa L. var. *menthifolia* (Graham) Fernald

Monarda pectinata Nutt.

Nepeta cataria L.

Prunella vulgaris L.

Salvia azurea Lam. var. *grandiflora* Benth.

Scutellaria brittonii Porter

Stachys palustris L. var. *pilosa* (Nutt.) Fernald

Teucrium laciniatum Torr.

Calochortus gunnisonii S. Watson

Calochortus nuttallii Torr. & A. Gray

Erythronium grandiflorum Pursh ssp. *grandiflorum*

Lilium philadelphicum L.

Lloydia serotina (L.) Salisb. ex Rchb.

Prosartes trachycarpa S. Watson

Streptopus amplexifolius (L.) DC.

Linum compactum A. Nelson

Linum lewisii Pursh

Linum puberulum (Engelm.) Heller

Mentzelia albicaulis (Douglas ex Hook.) Douglas ex Torr. & A. Gray

Mentzelia chrysantha seed

Mentzelia decapetala (Pursh ex Sims) Urb. & Gilg ex Gilg

Mentzelia dispersa seeds

Mentzelia laciniata seed

Mentzelia multicaulis seeds

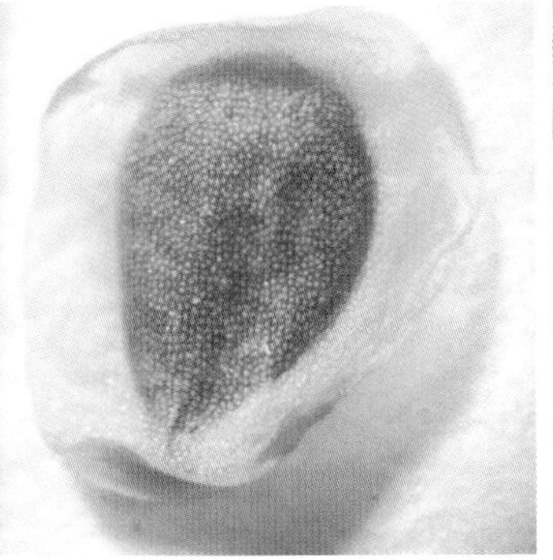

Mentzelia multiflora seed (rough in appearance)

Mentzelia nuda (Pursh) Torr. & A. Gray

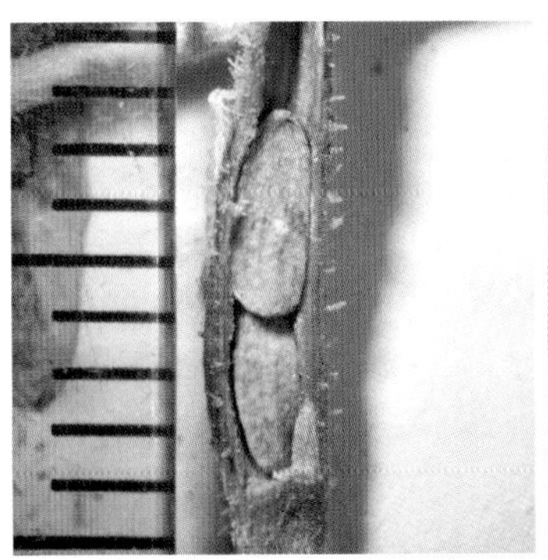

Mentzelia oligosperma capsule and exposed seeds

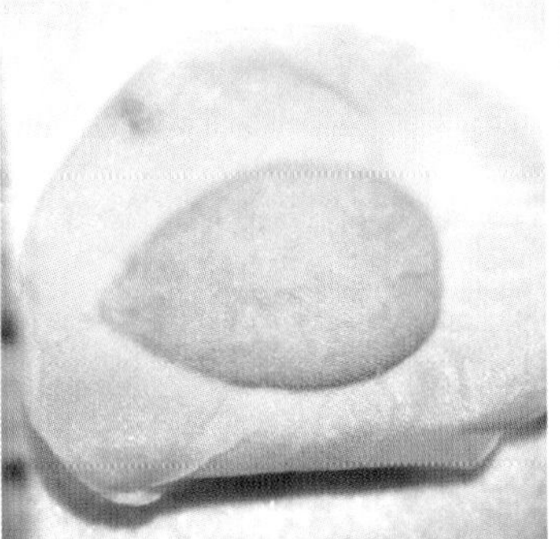

Mentzelia pterosperma seed

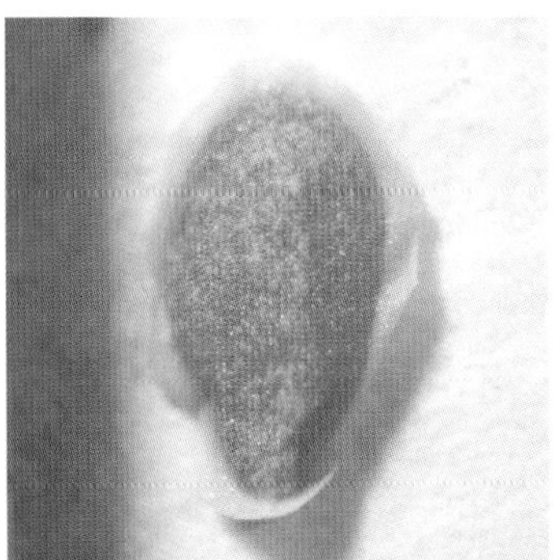

Mentzelia pumila seed

Mentzelia reverchonii seed (smooth in appearance)

Mentzelia speciosa Osterh.

Mentzelia thompsonii seeds

Hibiscus trionum L.

Malva neglecta Wallr.

Sidalcea neomexicana A. Gray

Sphaeralcea coccinea (Nutt.) Rydb.

Veratrum californicum T. Durand

Zigadenus elegans Pursh

Menyanthes trifoliata L.

Claytonia rubra (T.J. Howell) Tidestrom

Claytonia megarhiza (A. Gray) Parry ex S. Watson

Lewisia pygmaea (A. Gray) B.L. Rob. var. *pygmaea*

Lewisia rediviva Pursh var. *rediviva*

Lysimachia ciliata L.

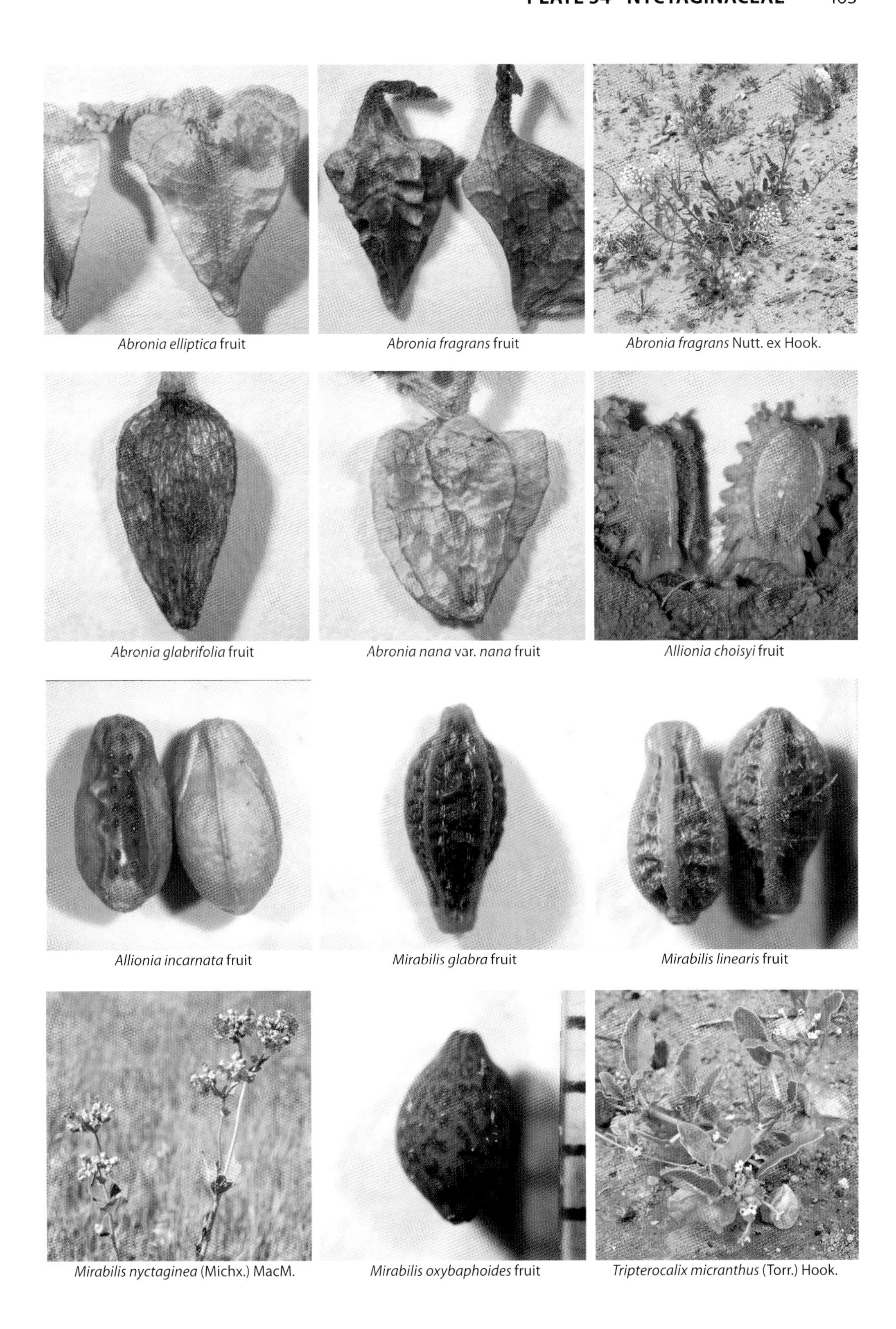

Abronia elliptica fruit

Abronia fragrans fruit

Abronia fragrans Nutt. ex Hook.

Abronia glabrifolia fruit

Abronia nana var. *nana* fruit

Allionia choisyi fruit

Allionia incarnata fruit

Mirabilis glabra fruit

Mirabilis linearis fruit

Mirabilis nyctaginea (Michx.) MacM.

Mirabilis oxybaphoides fruit

Tripterocalix micranthus (Torr.) Hook.

Chamerion angustifolium (L.) Holub.

Circaea alpina L.

Epilobium hornemannii Rchb. var. *lactiflorum* (Hausskn.) D. Löve

Oenothera albicaulis Pursh

Oenothera cespitosa Nutt. var. *macroglottis* (Rydb.) Cronquist

Oenothera coloradoensis Rydb.

Oenothera coronopifolia Torr. & A. Gray

Oenothera howardii (A. Nelson) W.A. Wagner

Oenothera lavandulifolia (Torr. & A. Gray) P.H. Raven

Oenothera serrulata Nutt. var. *serrulata*

Oenothera suffretescens (Ser.) W.L. Wagner & Hoch

Oenothera villosa Thunb. var. *strigosa* (Rydb.) Dorn

Calypso bulbosa (L.) Oakes var. *americana* (R. Br. ex Ait. f.) Luer

Corallorhiza maculta (Raf.) Raf.

Corallorhiza trifida Châtel.

Cypripedium fasciculatum Kellogg ex S. Watson

Cypripedium parviflorum Salisb. var. *pubescens* (Willd.) Knight

Photo by Al Schneider

Epipactis gigantea Douglas ex Hook.

Goodyera oblongifolia Raf.

Listera cordata (L.) R. Br. ex Ait. f. var. *nephrophylla* (Rydb.) Hultén

Piperia unalascensis (Spreng.) Rydb.

Platanthera dilatata (Pursh) Lindley ex L.C. Beck var. *albiflora* (Cham.) Ledeb.

Platanthera huronensis (Nutt.) Lindley

Spiranthes romanzoffiana Cham.

Castilleja chromosa A. Nelson

Castilleja occidentalis Torr.

Castilleja puberula Rydb.

Castilleja rhexifolia Rydb.

Castilleja sessiliflora Pursh

Orobanche fasciculata Nutt.

Pedicularis groenlandica Retz.

Pedicularis racemosa Douglas ex Benth. var. *alba* Pennell

Pedicularis sudetica Willd. ssp. *scopulorum* (A. Gray) Hultén

Oxalis violacea L.

Argemone polyanthemos (Fedde) G.B. Ownbey

Parnassia palustris L. var. *montanensis* (Fernald & Rydb. ex Rydb.) C.L. Hitchc.

Mimulus eastwoodiae Rydb.

Mimulus guttatus DC.

Besseya alpina (A. Gray) Rydb.

Callitriche heterophylla ssp. *heterophylla* fruit

Callitriche palustris fruit

Chionophila jamesii Benth.

Collinsia parviflora Lindl.

Linaria dalmatica (L.) Mill.

Penstemon albidus Nutt.

Penstemon angustifolius Nutt. ex Pursh var. *angustifolius*

Penstemon barbatus (Cav.) Roth var. *torreyi* (Benth.) A. Gray

Penstemon caespitosus Nutt. ex A. Gray

Penstemon eriantherus Pursh var. *eriantherus*

Penstemon glaber Pursh var. *alpinus* (Torr.) A. Gray

Penstemon hallii A. Gray

Penstemon procerus Douglas ex Graham var. *procerus*

Penstemon retrorsus Payson ex Pennell

Penstemon secundiflorus Benth.

Penstemon strictus Benth. ssp. *strictus*

Penstemon virens Pennell ex Rydb.

Penstemon whippleanus A. Gray (white-flowered form)

Plantago patagonica Jacq.

Veronica anagallis-aquatica L.

Veronica wormskjoldii Roem. & Schultes

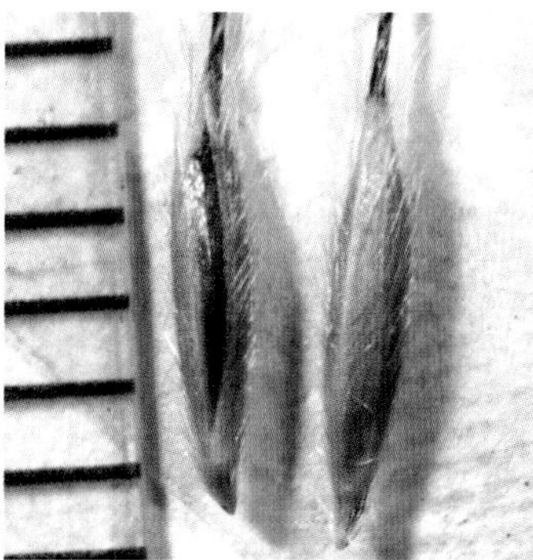
Achnatherum lettermanii florets

Agropyron cristatum (L.) Gaertn.

Agrostis stolonifera spikelet

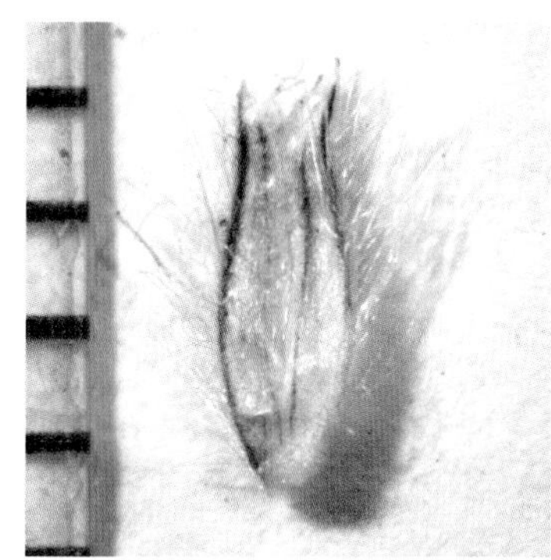
Alopecurus arundinaceus spikelet

Andropogon gerardii paired spikelets

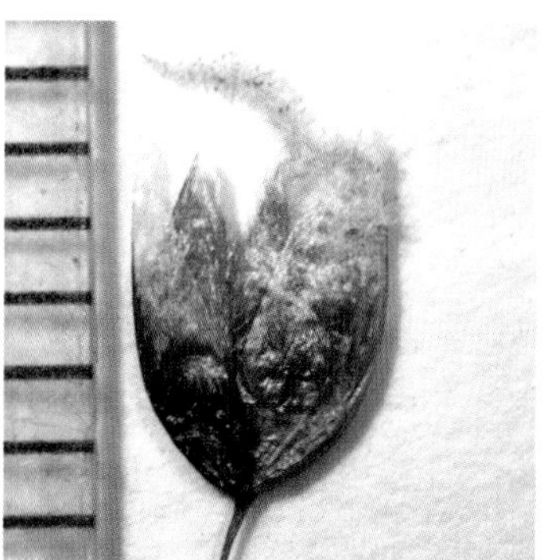
Anthoxanthum hirtum spikelet

Apera interrupta spikelet

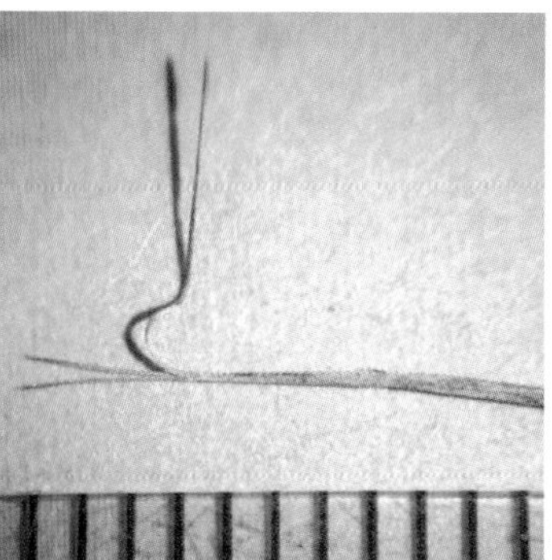
Aristida dichotoma lemma and awns

Arrhenatherum elatius spikelet

Avenula hookeri spikelet

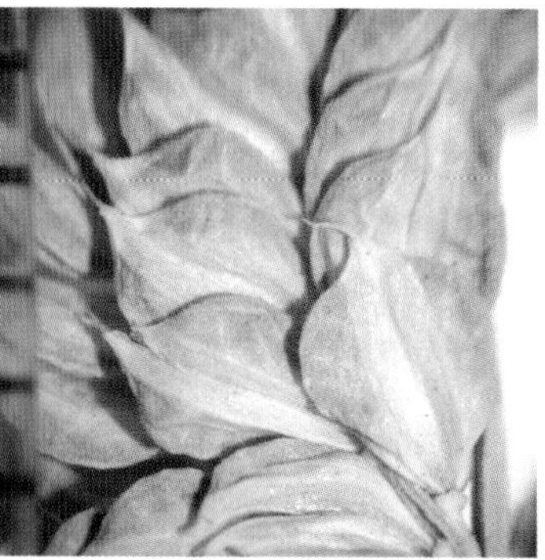
Beckmannia syzigachne spikelets

Bothriochloa barbinodis spikelet

Bouteloua gracilis (Kunth) Lag.

Bromus tectorum L.

Buchloë dactyloides staminate spikelets

Calamagrostis montanensis spikelet

Calamovilfa longifolia spikelet

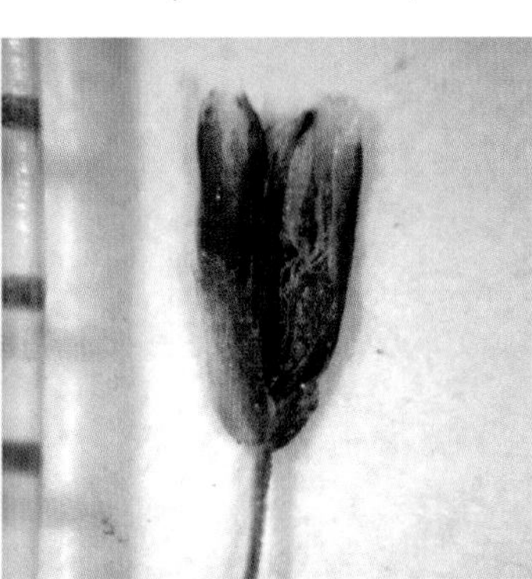
Catabrosa aquatica spikelet

Cenchrus longispinus (Hack.) Fernald

Chloris virgata Sw.

Cinna latifolia spikelet

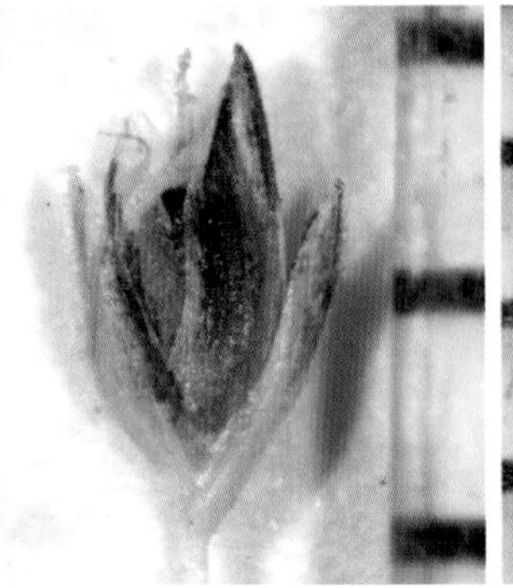
Crypsis alopecuroides spikelet

Cynodon dactylon var. *dactylon* spikelets

Dactylis glomerata spikelet

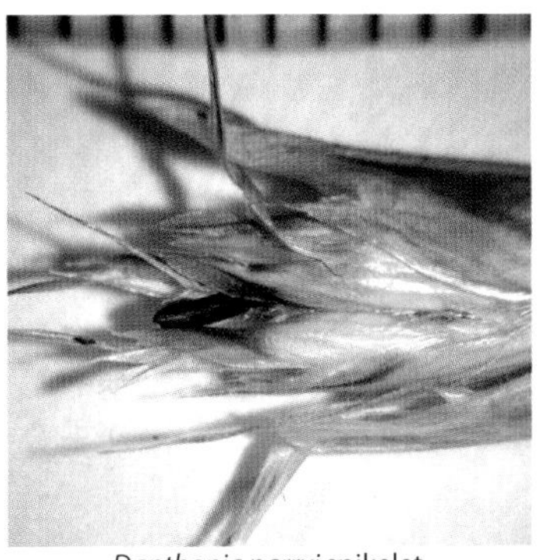
Danthonia parryi spikelet

Dasyochloa pulchella spikelet

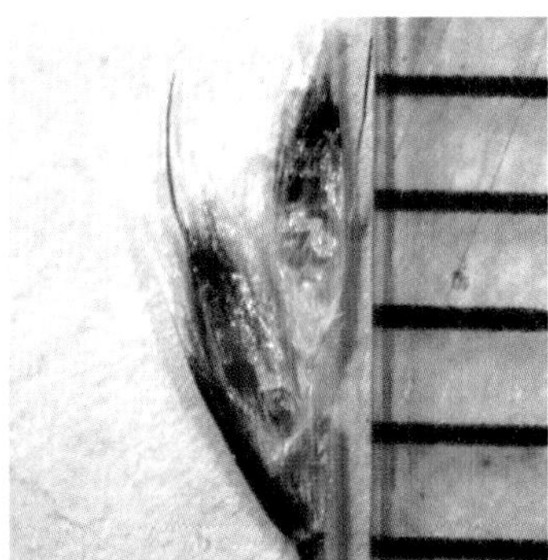
Dechampsia cespitosa spikelet with right glume removed

Dichanthelium oligosanthes var. *scribnerianum* spikelet

Digitaria sanguinalis spikelets

Distichlis stricta spikelets

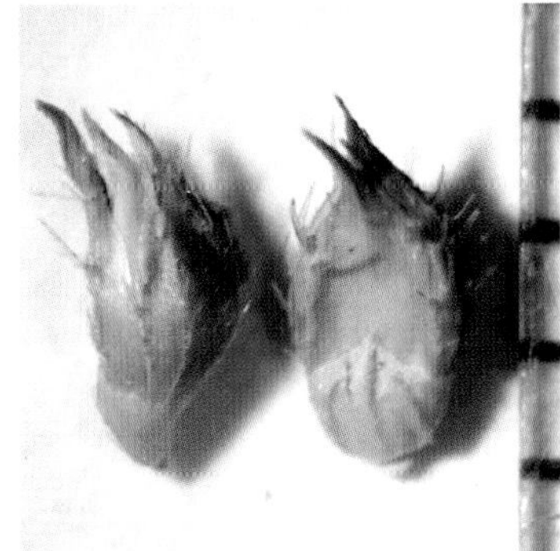
Echinochloa crus-galli spikelets

Elymus trachycaulus (Link) Gould

Enneapogon desvauxii spikelet

Eragrostis spectabilis spikelet

Eriochloa contracta spikelets

Erioneuron pilosum spikelet

Festuca idahoensis spikelet

Glyceria elata spikelet

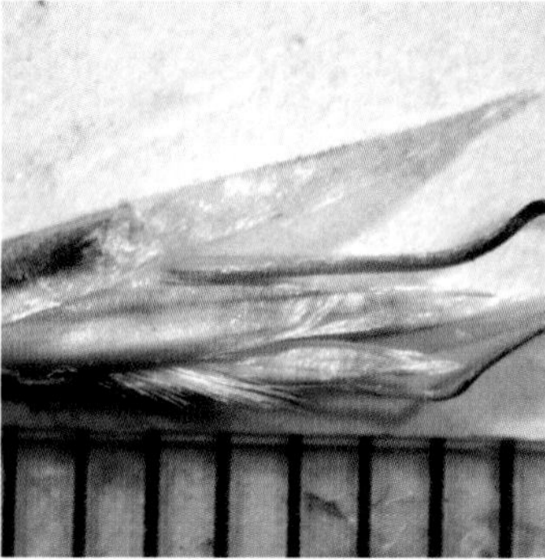
Helictotrichon mortonianum upper part of spikelet

Hilaria jamesii spikelet

Hopia obtusa spikelet

Hordeum jubatum L.

Koeleria macrantha (Ledeb.) Schult.

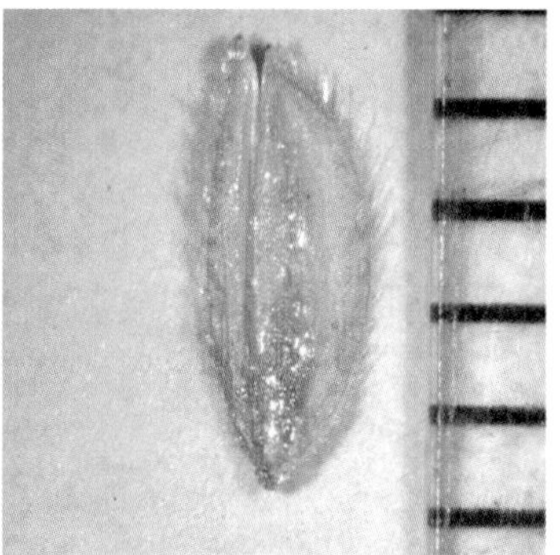
Leersia oryzoides spikelet

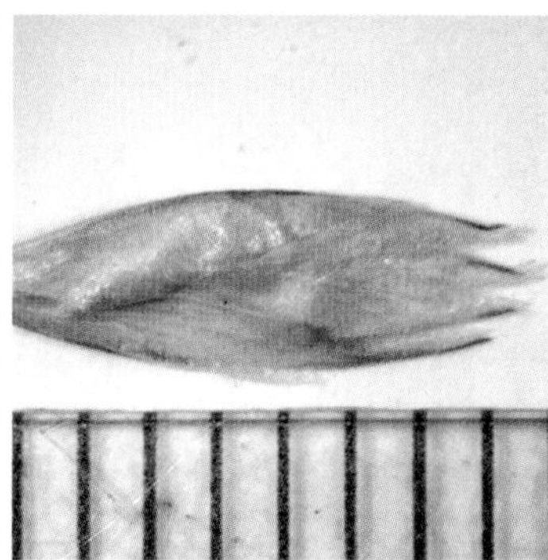
Leucopoa kingii spikelet

Lolium perenne spikelet (edgewise to the rachis)

Melica spectabilis lower half of the spikelet

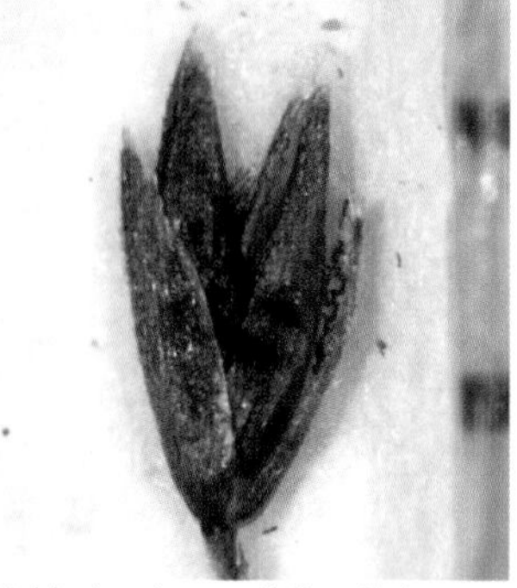
Muhlenbergia asperifolia spikelet

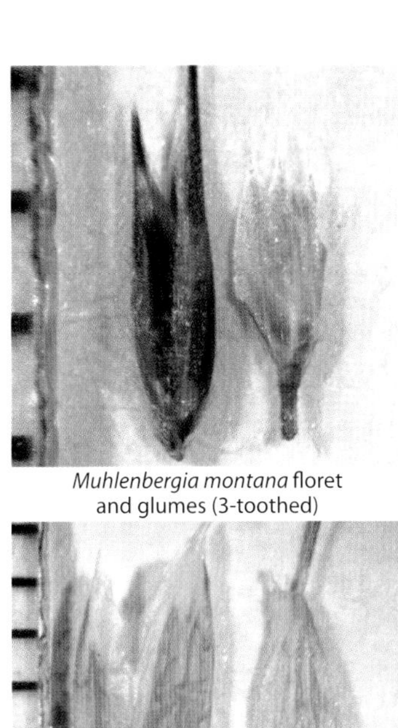
Muhlenbergia montana floret and glumes (3-toothed)

Muhlenbergia tricholepis spikelet

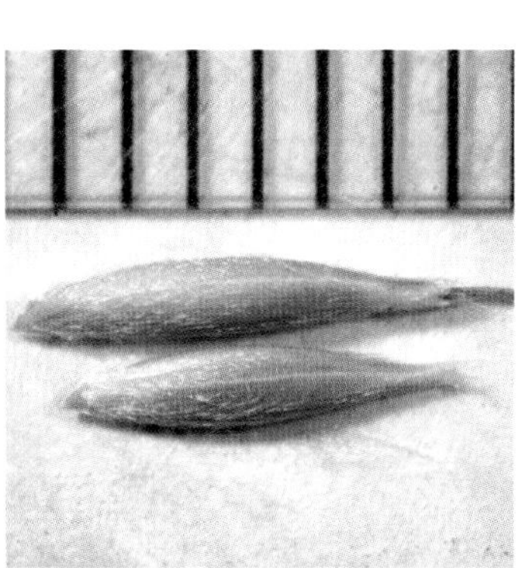
Nassella viridula florets

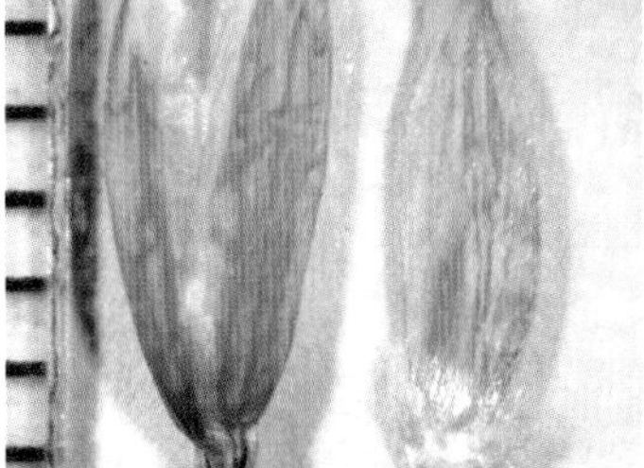
Oryzopsis asperifolia spikelet (glumes and floret separated)

Panicum capillare spikelet and *Panicum virgatum* spikelet

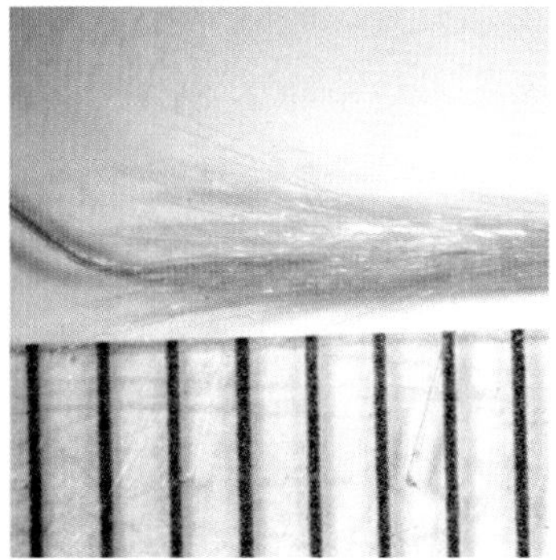
Pappostipa speciosa upper half of spikelet

Pascopyrum smithii spikelet

Paspalum setaceum spikelets

Phalaris arundinacea spikelet with the floret removed

Phippsia algida spikelets

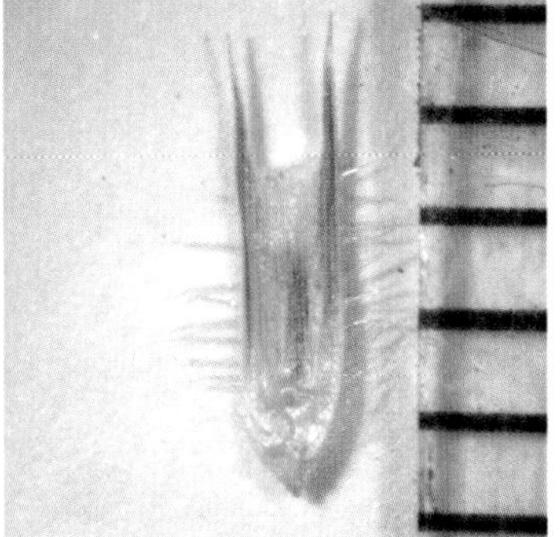
Phleum pratense spikelet

Phragmites australis spikelet

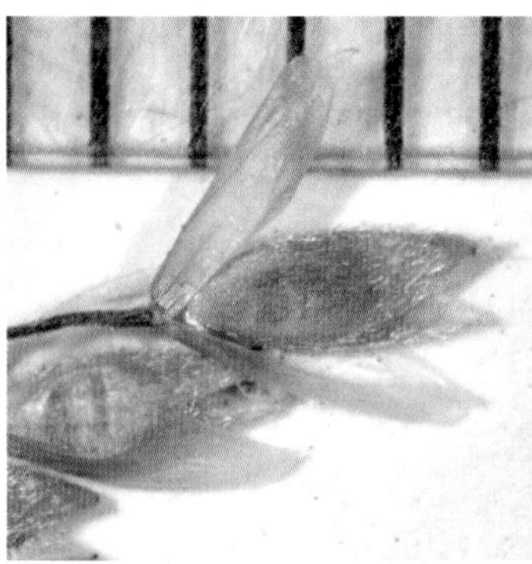
Piptatherum pungens spikelet

Poa pratensis spikelets (left spikelet with glumes removed)

Podagrostis humilis spikelet

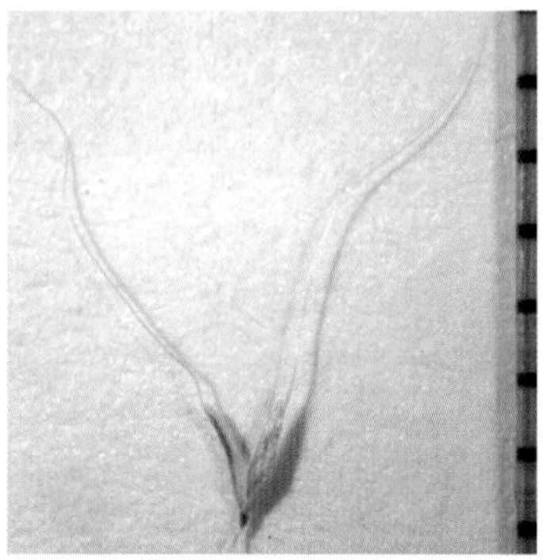
Polypogon monspeliensis spikelet

Psathyrostachys junceus spikelet

Pseudoroegneria spicata spikelet

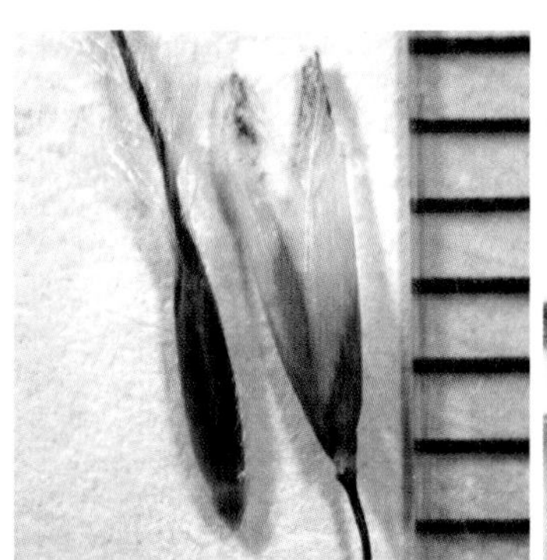
Ptilagrostis porteri floret and glumes

Puccinellia distans spikelet

Schedonorus arundinaceus spikelet

Schizachne purpurascens spikelet

Schizachyrium scoparium var. *scoparium* spikelet

Sclerochloa dura spikelets

Setaria viridis spikelets

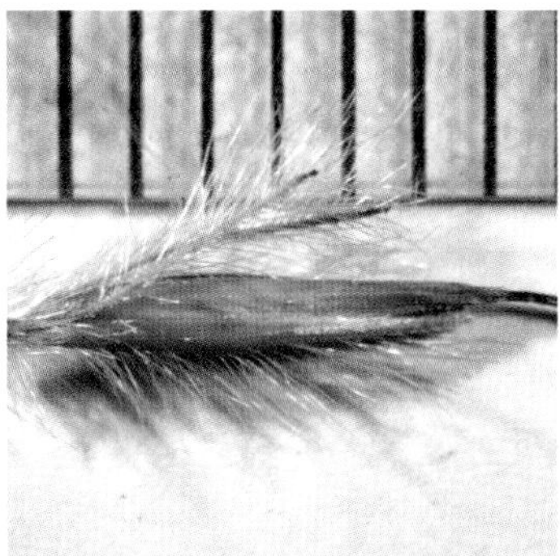
Sorghastrum nutans spikelet

Sorghum bicolor paired spikelets

Spartina pectinata Link

Sphenopholis obtusata spikelet

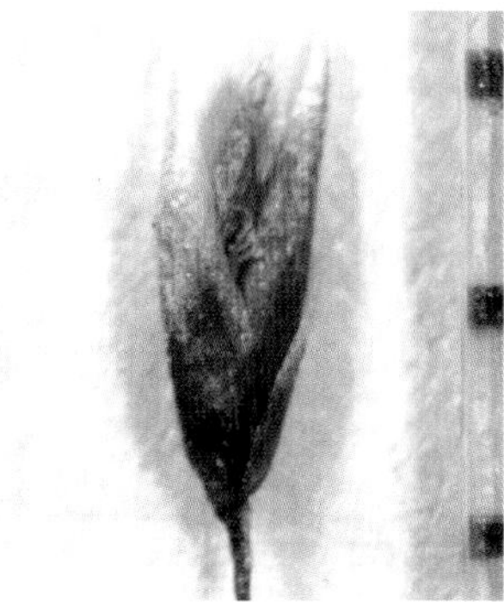
Sporobolus airoides spikelet

Torreyochloa pauciflora spikelet

Tridens muticus spikelet

Triplasis purpurea floret

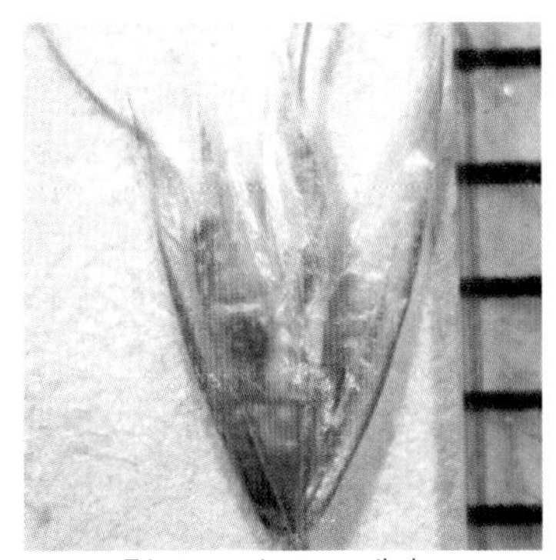
Trisetum spicatum spikelet

Vahlodea atropurpurea spikelet

Vulpia octoflora spikelet

Collomia linearis Nutt.

Giliastrum acerosum (A. Gray) Rydb.

Ipomopsis aggregata (Pursh) V.E. Grant

Ipomopsis spicata (Nutt.) V.E. Grant

Leptosiphon nuttallii (A. Gray) J.M. Porter & L.A. Johnson ssp. *nuttallii*

Linanthus pungens (Torr.) J.M. Porter & L.A. Johnson

Microsteris gracilis (Hook.) Greene

Phlox condensata (A. Gray) E.E. Nelson

Phlox longifolia Nutt.

Polemonium brandegeei (A. Gray) Greene

Polemonium pulcherrimum Hook. ssp. *delicatum* (Rydb.) Cronquist

Polemonium viscosum Nutt.

Polygala alba Nutt.

Bistorta bistortoides (Pursh) Small

Eriogonum brandegeei flowers and involucre

Eriogonum brevicaule Nutt. var. *brevicaule*

Eriogonum flavum Nutt. var. *flavum*

Eriogonum inflatum Torr. & Frém.

Eriogonum jamesii var. *jamesii* tepals

Eriogonum lachnogynum flowers and involucre

Eriogonum ovalifolium Nutt. var. *purpureum* (Nutt.) Durand

Eriogonum palmerianum flowers and involucre

Eriogonum umbellatum Torr. var. *majus* Hook.

Eriogonum umbellatum var. *umbellatum* flowers (with stipe-like base)

Fallopia convolvulus (L.) Á. Löve

Oxyria digynia (L.) Hill

Persicaria amphibia (L.) Delarbe

Persicaria maculosa A. Gray

Polygonum achoreum tepals

Polygonum kelloggii Greene var. *kelloggii*

Rumex acetosella L.

Rumex fueginus inner tepals and tubercle

Rumex occidentalis inner tepals (lacking a tubercle)

Rumex stenophyllus inner tepals and tubercle

Rumex triangulivalvis (Danser) Rech. f.

Rumex venosus Pursh

Portulaca oleracea L.

Potamogeton epihydrus Raf.

Androsace chamaejasme Wulfen ssp. *carinata* (Torr.) Hultén

Dodecatheon pulchellum (Raf.) Merr. var. *pulchellum*

Primula angustifolia Torr.

Primula parryi A. Gray

Aconitum columbianum Nutt.

Actaea rubra (Aiton) Willd.

Anemone narcissiflora L. var. *zephyra* (A. Nelson) B.E. Dutton & Keener

Anemone patens L. var. *multifida* Pritz.

Aquilegia coerulea James var. *coerulea*

Aquilegia elegantula Greene

Caltha leptosepala DC.

Clematis grosseserrata (Rydb.) Ackerfield

Clematis ligusticifolia Nutt.

Delphinium barbey (Huth) Huth

Ranunculus alismofolius Geyer ex Benth. var. *montanus* S. Watson

Ranunculus glaberrimus Hook. var. *ellpticus* (Greene) Greene

Ranunculus inamoenus Greene var. *inamoenus*

Ranunculus macauleyi A. Gray

Ranunculus macounii Britton

Ranunculus sceleratus L. var. *multifidus* Nutt.

Thalictrum fendleri Engelm. ex A. Gray

Trollius albiflorus (A. Gray) Rydb.

Ceanothus herbaceous Raf.

Amelanchier alnifolia (Nutt.) Nutt. ex M. Roem.

Cercocarpus montanus Raf.

Chamaerhodos erecta (L.) Bunge var. *parviflora* (Nutt.) C.L. Hitchc.

Crataegus succulenta Schrad.

Dryas octopetala L. var. *hookeriana* (Juz.) Breit.

Fragaria virginiana Mill.

Geum rossii (R. Br.) Ser. var. *turbinatum* (Rydb.) C.L. Hitchc.

Geum triflorum Pursh

Holodiscus dumosus (Nutt. ex Hook.) A. Heller

Physocarpus monogynus (Torr.) J.M. Coult.

Potentilla anserina L.

Potentilla fissa Nutt.

Potentilla fruticosa Pursh

Potentilla glaucophylla Lehm. var. *glaucophylla*

Potentilla nivea L.

Potentilla plattensis Nutt.

Prunus americana Marshall

Prunus virginiana L. var. *melanocarpa* (A. Nelson) Sarg.

Purshia tridentata (Pursh) DC.

Rosa blanda Aiton

Rubus deliciosus Torr.

Rubus idaeus L. var. *strigosus* (Michx.) Maxim.

Sibbaldia procumbens L.

Galium multiflorum Kellogg var. *coloradoense* (W. Wight) Cronquist

Galium triflorum Michx.

Hedyotis nigricans (Lam.) Fosberg

Ruppia cirrhosa fruit

Maianthemum racemosum (L.) Link ssp. *amplexicaule* (Nutt.) LaFrankie

Maianthemum stellatum (L.) Link

Ptelea angustifolia Benth.

Populus tremuloides Michx.

Salix bebbiana Sarg.

Salix exigua Nutt. ssp. *exigua*

Salix reticulata L. var. *nana* Anderss.

Comandra umbellata (L.) Nutt. ssp. *pallida* (A. DC.) Piehl

Acer glabrum Torr. var. *glabrum*

Anemopsis californica Hook. & Arn.

Heuchera bracteata (Torr.) Ser.

Micranthes rhomboidea (Greene) Small

Saxifraga bronchialis L. var. *austromontana* (Wiegand) Piper

Saxifraga hirculus L.

Saxifraga rivularis L.

Scrophularia lanceolata Pursh

Verbascum thapsus L.

Datura stramonium L.

Physalis longifolia Nutt.

Quincula lobata (Torr.) Raf.

Sparganium eurycarpum Engelm. ex A. Gray

Urtica dioica L. ssp. *gracilis* (Aiton) Selander

Valeriana occidentalis A. Heller

Glandularia bipinnatifida (Nutt.) Nutt.

Phyla cuneifolia (Torr.) Greene

Verbena stricta Vent.

Viola adunca Sm.

Viola canadensis L.

Viola nuttallii Pursh

Arceuthobium cyanocarpum (A. Nelson ex. Rydb.) A. Nelson

Vitis riparia Michx.

Zannichellia palustris fruit

Each of the following species freely hybridize where their ranges overlap, making it quite difficult to determine every specimen with confidence. *Quercus havardii* Rydb. does not occur in Colorado and is known from southeastern New Mexico, Texas, and southwestern Oklahoma. Specimens reported as *Q. havardii* from the west slope are *Q. gambelii* × *Q. turbinella*.

1a. Leaves mostly with entire margins, sometimes a few leaves with very shallowly lobed margins...***Q. grisea***
1b. Leaves mostly or all with lobed or sharply toothed margins...2

2a. Leaves lobed at least halfway to the midrib, the lobes rounded; lower leaf surface with stellate hairs with 5 or fewer arms...***Q. gambelii***
2b. Leaves 3–5-toothed or shallowly lobed, the teeth or lobes sharp-pointed or spinose-tipped; lower leaf surface with stellate hairs with 8 or more arms...***Q. turbinella***

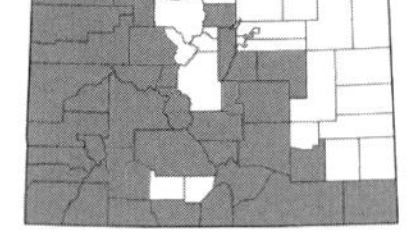

***Quercus gambelii* Nutt.**, GAMBEL OAK. [*Q. eastwoodiae* Rydb.; *Q. leptophylla* Rydb.; *Q. vreelandii* Rydb.]. Shrubs or small trees, 2–4 m, deciduous; *leaves* pinnately lobed at least halfway to the midrib, 2.5–17 cm long, the lobes rounded, lower leaf surface with stellate hairs with 5 or fewer arms; *staminate catkins* 3.5–5 cm long; *involucre cup* 3–10 mm long, 10–20 mm wide; *acorns* 8–20 mm long. Common on dry slopes, 4000–10,500 ft. May–July. E/W. (Plate 46)

Quercus gambelii and *Q. turbinella* (see below) hybridize where their ranges meet or overlap. These hybrids are called *Q.* ×*welshii* R.A. Denham and exhibit characteristics of both species. Often, hybrids will have leaves deeply lobed as in *Q. gambelii*, but the lobes are sharp-pointed as in *Q. turbinella*.

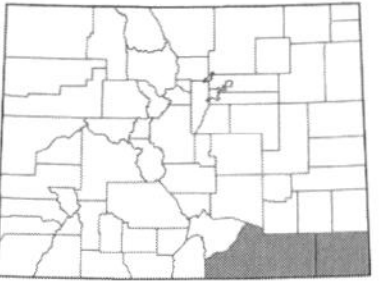

***Quercus grisea* Liebm.**, GRAY OAK. Shrubs or small trees; *leaves* oblong to elliptic or ovate, to 8 cm long, with mostly entire margins or sometimes a few leaves shallowly lobed, the apex acute or rarely rounded, sparsely stellate-hairy or rarely tomentose; *staminate catkins* to 7 cm long; *involucre cup* 4–10 mm long, 8–15 mm wide; *acorns* 12–18 mm long, 8–12 mm thick. Found on rocky slopes and canyon bottoms, 4500–5600 ft. May–July. E.

It is doubtful that our *Q. grisea* plants occur in Colorado in "pure form." *Quercus grisea* is usually found growing with numerous plants of hybrid origin (mostly *Q. grisea* × *Q. gambelii*) and varies considerably on the amount of lobing on the leaf margins. These hybrids are called *Q.* ×*undulata* Torr.

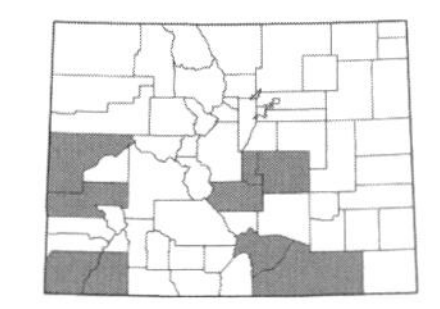

***Quercus turbinella* Greene**, SONORAN SCRUB OAK. [*Q. ajoensis* C.H. Muller]. Shrubs to 4 m, evergreen; *leaves* 3–5-toothed or shallowly lobed, 1.3–4 cm long, the teeth sharp-pointed or spinulose, lower leaf surface with stellate hairs with 8 or more arms; *staminate catkins* 1–3 cm long; *involucre cup* 6–8 mm long, 10–15 mm wide; *acorns* 12–25 mm long. Found on dry slopes, often with pinyon-juniper, 4500–7000 ft. May–July. E/W.

See discussion above under *Q. gambelii*.

FRANKENIACEAE Desv. – SEAHEATH FAMILY

Low shrubs; *leaves* simple, opposite or whorled, entire, exstipulate; *flowers* perfect, actinomorphic, solitary or in cymose clusters; *calyx* 4–7-lobed, persistent in fruit; *petals* 4–7, distinct, clawed; *stamens* 6, 2-whorled; *pistil* 1; *styles* 2–4; *ovary* superior, 2–4-carpellate; *fruit* a capsule enclosed by the persistent calyx.

FRANKENIA L. – SEAHEATH

With characteristics of the family.

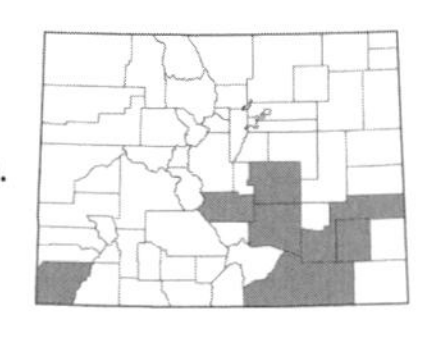

***Frankenia jamesii* Torr. ex A. Gray**, JAMES' SEAHEATH. Much-branched low shrubs to 3 dm; *leaves* opposite, fascicled, linear, 3–5 mm long, the margins strongly revolute; *calyx* tube 4–5 mm long, with lobes ca. 1 mm long; *petals* 5, white, about twice as long as the calyx tube; *capsules* linear, ca. 5 mm long. Found on limestone, gypsum, and shale slopes, 4000–6000 ft. May–July. E/W.

Closely resembling the genus *Linanthus* (Polemoniaceae), but the flowers are polypetalous.

FUMARIACEAE DC. – FUMITORY FAMILY

Herbs; *leaves* alternate, exstipulate; *flowers* perfect, zygomorphic; *sepals* 2, inconspicuous and bractlike; *petals* 4, the outer alike or dissimilar, one or both with a prominent spur or pouch at the base, the inner alike and apically connate over the stigmas and clawed; *stamens* 6, diadelphous with 3 per set; *pistil* 1; *style* 1; *ovary* superior, 2-carpellate; *fruit* a capsule. (Stern, 1997)

Dicentra uniflora Kellogg has been reported on the western slope, but no specimens have ever been found and its presence in Colorado would represent a significant disjunct from its range in Utah.

1a. Flowers yellow, white, or light pink, sometimes with a reddish-purple tip (and then the corolla 10–25 mm long); fruit an elongate, dehiscent capsule...***Corydalis***
1b. Flowers pink or white at the base and reddish-purple to maroon at the tips (drying purple), the corolla 5–6 mm long; fruit a globose, indehiscent capsule...***Fumaria***

CORYDALIS DC. – FUMEWORT

Annual, biennial, or perennial herbs; *leaves* simple or more often compound; *petals* distinct or somewhat connate at the base, the outer pair dissimilar with only 1 petal spurred or saccate at the base. (Ownbey, 1947)

1a. Flowers white to light pink, the spurred petal 16–25 mm long and the spur 9–16 mm long; ultimate leaf segments elliptic to oblong; plants mostly 10–15 dm tall...***C. caseana* ssp. *brandegeei***
1b. Flowers yellow, the spurred petal 13–16 mm long and the spur 4–5 mm long; ultimate leaf segments linear to oblong; plants 2–3.5 dm tall...***C. aurea***

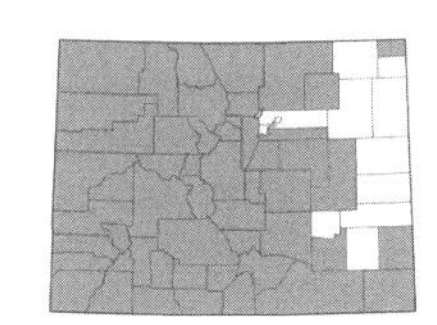

Corydalis aurea **Willd.**, GOLDEN CORYDALIS. Annuals, biennials, or short-lived perennials, 2–3.5 dm; *leaves* 1–4 times pinnately compound with linear to oblong leaflets; *sepals* 1–3 mm long; *petals* yellow, the spurred petal 13–16 mm long and the spur 4–5 mm long; *capsules* 1.5–3 cm long, torulose, usually curved-ascending. Common in rocky or sandy soil, forest clearings, and often seen in areas of previous burn, widespread across the state, 3600–11,000 ft. April–Aug. E/W. (Plate 46)

There are two subspecies of *C. aurea* in Colorado which are usually distinct but sometimes intergrading:

1a. Racemes generally not exceeding the leaves; capsules slender, pendant or spreading at maturity...**ssp. *aurea***
1b. Racemes generally exceeding the leaves; capsules stout, erect at maturity...**ssp. *occidentalis* (Engelm. ex A. Gray) G.B. Ownbey**. [*C. curvisiliqua* Engelm. ssp. *occidentalis* (Engelm. ex A. Gray) W.A. Weber]. Although the distribution of these two subspecies overlaps for the most part, ssp. *occidentalis* appears to be the only taxon present on the eastern plains. This taxon has been included in *C. curvisiliqua*, but *C. curvisiliqua* has rough seeds while *C. aurea* has smooth or obscurely muricate seeds. It is more appropriate to include this as a subspecies under *C. aurea*.

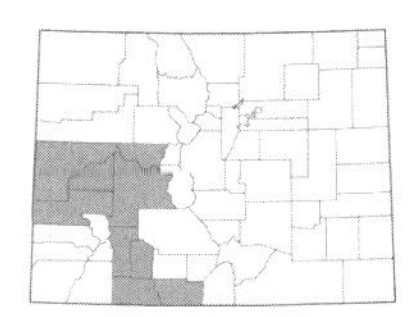

Corydalis caseana **A. Gray ssp. *brandegeei* (S. Watson) G.B. Ownbey**, FITWEED. Perennials, 10–15 dm; *leaves* 2–3 times pinnately compound with elliptic to oblong leaflets; *sepals* 2–4 mm long; *petals* white to light pink, the spurred petal 16–25 mm long with a spur 9–16 mm long; *capsules* 0.8–1.5 cm long, usually reflexed. Found in moist areas of forests and open meadows, often along streams and creeks, 8200–12,000 ft. June–Aug. E/W.

FUMARIA L. – FUMITORY

Annual herbs; *leaves* 2–3 times compound; *flowers* in terminal racemose inflorescences; *sepals* attached near the base; *petals* connivent, the outer inconspicuously crested with one spurred at the base.

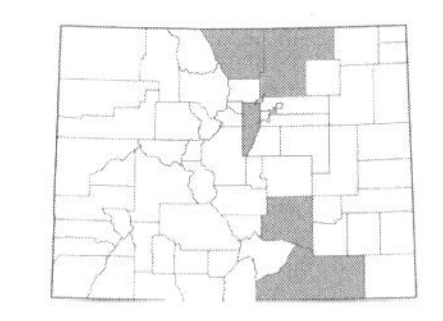

Fumaria vaillantii **Loisel.**, EARTHSMOKE. Plants 0.5–3 (5) dm, weak-stemmed; *leaflets* linear, mostly 1–2 mm wide; *sepals* 0.5–1 mm long; *petals* pink or white at the base and reddish-purple to maroon at the tips (drying purple), 5–6 mm long with a spur 1–1.5 mm long; *capsules* 1.7–2 mm diam. Uncommon along roadsides and in dry ditches, 4700–6000 ft. May–June. E. Introduced.

Our specimens were originally misidentified as *Fumaria officinalis* L., which has larger flowers (6–9.5 mm).

GENTIANACEAE Juss. – GENTIAN FAMILY

Herbs; *leaves* simple, usually opposite and decussate, sessile, exstipulate, the margins entire; *flowers* perfect, actinomorphic; *calyx* 4–5-lobed; *corolla* 4–5-lobed, usually contorted; *stamens* as many as the petals and opposite them, adnate to the petals; *pistil* 1; *ovary* superior, unilocular, 2-carpellate; *fruit* a septicidal capsule.

1a. Petals almost distinct, the corolla lobed nearly to the base, rotate or campanulate...2
1b. Corolla with a conspicuous connate tube and distinct lobes, funnelform to tubular or narrowly campanulate...5

2a. Corolla rotate, each lobe with 1 or 2 conspicuous fringed glands on the upper surface at the base...3
2b. Corolla campanulate or if rotate then without fringed glands at the base...4

3a. Flowers 5-merous, bluish-purple; stem leaves sometimes alternate...***Swertia***
3b. Flowers 4-merous, white or yellowish-green, sometimes dotted with purple or green; leaves all opposite or whorled...***Frasera***

4a. Corolla rotate with lobes 7–12 mm long, white or tinged with blue; leaves not glaucous, linear or the basal oblong-oblanceolate...***Lomatogonium***
4b. Corolla campanulate with lobes 35–50 mm long, blue-purple, pinkish, or whitish; leaves glaucous, elliptic to ovate...***Eustoma***

5a. Flowers salverform with pink to rose-pink lobes and a yellowish tube; anthers twisting at maturity; uncommon at low elevations (4500–6500 ft)...***Centaurium***
5b. Flowers campanulate, funnelform, or tubular, blue, purple, white, or yellowish; anthers not twisting at maturity; plants variously distributed...6

6a. Corolla lobes fringed on the margins...***Gentianopsis***
6b. Corolla lobes entire...7

7a. Corolla 8–50 mm long, with folded plaits in the sinuses, the lobes lacking a fringed scale or row of fringe; sinuses of the calyx with an inner membrane connecting the lobes...***Gentiana***
7b. Corolla 9–20 mm long, without folded plaits in the sinuses, the lobes with 1–2 fringed scales or a single row of fringe on the inside at the base, or the fringe sometimes absent in late blooming plants; sinuses of the calyx lacking an inner membrane...***Gentianella***

CENTAURIUM Hill. – CENTAURY

Annuals, glabrous herbs; *leaves* opposite; *flowers* in terminal cymes, 4–5-merous; *corolla* salverform, pink, rose,, or occasionally white; *stamens* inserted on the corolla throat, the anthers twisting at maturity.

1a. Corolla lobes (7) 8–12 mm long...***C. arizonicum***
1b. Corolla lobes 1–7 mm long...2

2a. Pedicels 1–6.5 cm long; calyx lobes 5–8 mm long...***C. exaltatum***
2b. Pedicels 0.3–0.5 cm long; calyx lobes 4–5 mm long...***C. pulchellum***

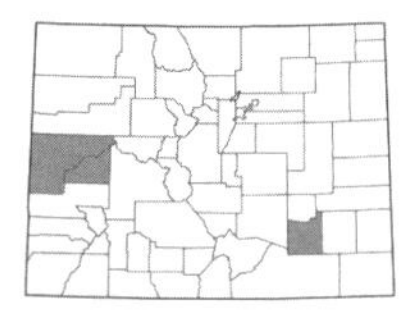

***Centaurium arizonicum* (A. Gray) A. Heller**, ARIZONA CENTAURY. [*C. calycosum* (Buckley) Fernald var. *arizonicum* (A. Gray) Tidestr.]. Plants 1.5–4 (6) dm; *leaves* oblanceolate to narrowly elliptic, 2–6.5 cm long; *pedicels* 0.5–3 cm; *calyx* 8–12 mm long, deeply divided into linear lobes; *corolla* tube and throat yellow, 9–15 mm long, lobes pink, (7) 8–12 mm long. Uncommon in moist places such as along streams and in wetlands, 4800–5500 ft. June–Sept. E/W.

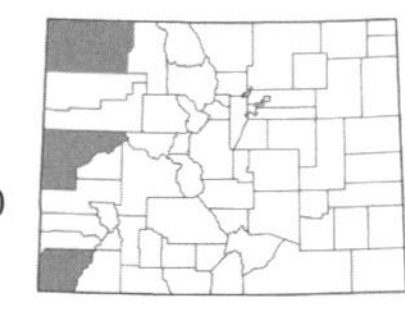

***Centaurium exaltatum* (Griseb.) W. Wight ex Piper**, DESERT CENTAURY. Plants 1–5 (6) dm; *leaves* linear to lanceolate, 1–3 (5) cm long; *pedicels* 1–6.5 cm; *calyx* 6–11 mm long, divided into linear lobes 5–8 mm long; *corolla* tube and throat greenish or yellowish, 6.5–14 mm long, lobes pink to rose, 3.5–7 mm. Uncommon in moist places along streams, in marshes, and wetlands, 6000–6500 ft. June–Sept. W.

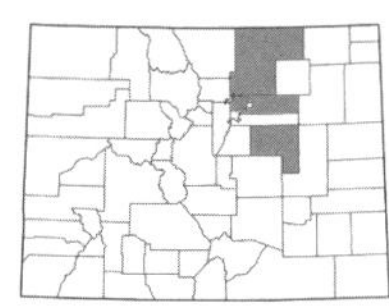

***Centaurium pulchellum* (Sw.) Druce**, BRANCHED CENTAURY. Plants 1–2 dm; *leaves* lanceolate to lance-ovate, 1–2 cm long; *pedicels* 0.3–0.5 cm; *calyx* 5–9 mm long, divided into linear lobes 4–5 mm long; *corolla* tube 8–13 mm long, the lobes pink, 1–4 mm long. Uncommon in moist places along creeks and streams, in often somewhat disturbed areas, 4500–5500 ft. June–Sept. E. Introduced.

EUSTOMA Salisb. – PRAIRIE GENTIAN

Annuals or short-lived perennials; *leaves* opposite; *flowers* in terminal cymes, 5-merous; *corolla* campanulate, deeply 5-parted, the lobes erose-dentate, much longer than the tube, blue-purple, pink, or rarely white.

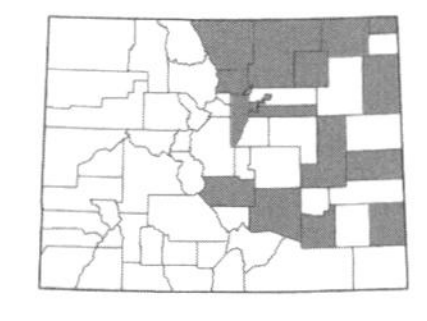

***Eustoma grandiflorum* (Raf.) Shinners**, SHOWY PRAIRIE GENTIAN. [*E. exaltatum* (L.) Salisb. ex G. Don ssp. *russellianum* (Hook.) Kartesz]. Plants 2.5–6 dm; *leaves* elliptic to ovate, 1.5–7.5 × 0.3–5 cm; *calyx* deeply divided into linear-lanceolate lobes, 1–2.3 cm long; *corolla* lobes 3.5–5 cm long; *capsules* to 2 cm. Uncommon in moist grasslands, often in alkaline soil, 3500–5500 ft. July–Aug. E. (Plate 46)

FRASERA Walter – GREEN GENTIAN

Biennials or perennials, glabrous to minutely hairy herbs; *leaves* opposite or whorled, mostly basal, entire; *flowers* 4-merous; *calyx* deeply cleft into lanceolate or linear lobes; *corolla* rotate or campanulate, white or yellowish-green to blue or purplish, the lobes with 1 or 2 depressed glands which are sometimes subtended by a corona of scales and surrounded by a fringed membrane. (Card, 1931)

1a. Leaves not white-margined, whorled; corolla lobes 10–30 mm long, each corolla lobe with 2 glands; inflorescence an elongated, compact or sometimes open, many-flowered thyrse...***F. speciosa***
1b. Leaves white-margined, opposite or whorled; corolla lobes 6–11 mm long, each corolla lobe with 1 gland; inflorescence a spreading panicle or cyme...2

2a. Leaves in whorls of 4; corolla gland lobed at the apex...***F. albomarginata***
2b. Leaves opposite; corolla gland lobed at the base...3

3a. Calyx 7–12 mm long, longer than the corolla; corolla white with purplish spots and an unfringed gland on each lobe; plants 1–2 dm tall...***F. coloradensis***
3b. Calyx 3–6 mm long, shorter than the corolla; corolla yellowish-green or white with greenish spots and a fringed gland on each lobe; plants 7–15 dm tall...***F. paniculata***

***Frasera albomarginata* S. Watson**, DESERT FRASERA. [*Swertia albomarginata* (S. Watson) Kuntze]. Plants 2–6 dm; *leaves* in whorls of 4, oblanceolate to linear above, mostly 4–11 cm long, white-margined; *inflorescence* a paniculate cyme; *calyx* 2–6 mm long; *corolla* lobes 7–11 mm long, white or greenish-white, with green spots and a solitary gland on each lobe. Uncommon on dry slopes, often with pinyon-juniper, 4700–7500 ft. May–July. W.

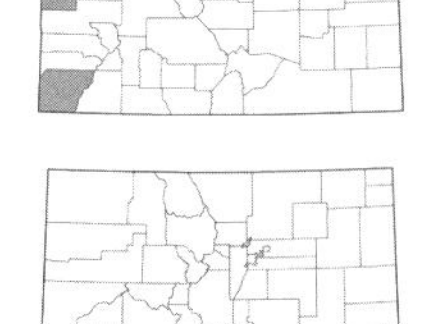

***Frasera coloradensis* (C.M. Rogers) D.M. Post**, COLORADO FRASERA. [*Swertia coloradensis* C.M. Rogers]. Plants 1–2 dm; *leaves* opposite, linear-oblanceolate to linear-lanceolate, 5–20 cm long, white-margined; *inflorescence* corymbose; *calyx* 7–12 mm long; *corolla* lobes 8–10 mm, white with purple spots, each lobe with a solitary gland. Locally common on rocky or sandstone outcrops, 4000–5500 ft. June–July. E. Endemic.

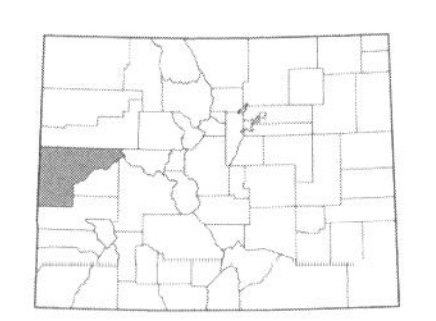

***Frasera paniculata* Torr.**, TUFTED FRASERA. [*Swertia utahensis* (M.E. Jones) H.H. St. John]. Plants 7–15 dm; *leaves* opposite, lanceolate to oblanceolate, reduced upward, mostly 7–20 cm long; *inflorescence* a pyramidal paniculate cyme; *calyx* 3–6 mm long; *corolla* lobes 6–10 mm long, white to yellowish-green, with green spots and a solitary, fringed gland on each lobe. Uncommon on open, sandy slopes and in dry washes, barely entering Colorado just east of the Utah state line in Mesa Co., 4500–4700 ft. June–July. W.

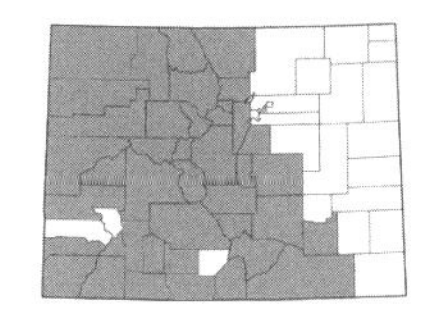

***Frasera speciosa* Douglas ex Griseb.**, ELKWEED. [*Swertia radiata* (Kellogg) Kuntze]. Plants 3–20 dm; *leaves* mostly whorled, elliptic to oblanceolate, reduced upwards, mostly 20–50 cm long, not white-margined; *inflorescence* an elongated thyrse of whorled cymes; *calyx* 10–22 mm long; *corolla* lobes 10–30 mm long, white with purple spots, with 2 fringed glands on each lobe. Common on open slopes and in forest openings from the lower foothills to alpine, 5900–12,500 ft. June–Sept. E/W. (Plate 46)

GENTIANA L. – GENTIAN

Annuals or perennials, glabrous or minutely hairy herbs; *leaves* opposite; *flowers* solitary or in axillary or terminal cymes, 4–5-merous; *calyx* with a well-developed tube, with a continuous membrane between the lobes; *corolla* plicate, tubular to funnelform or campanulate, the lobes shorter than the tube, blue, purple, white, or yellow. (Pringle, 1967)

1a. Corolla narrow, 2–5 mm wide, 8–22 mm long; flowers solitary and terminal, usually 4-merous; plants 2–12 cm tall, annuals or biennials...2
1b. Corolla 6–18 mm wide, 20–50 mm long; flowers solitary or usually 2 or more and clustered, 5-merous; plants 5–50 cm tall, perennials...3

2a. Leaves and sepals conspicuously white-margined; flowers whitish or greenish-purple to pale blue; capsule exserted from the corolla tube at maturity...***G. fremontii***
2b. Leaves and sepals not or only slightly white-margined; flowers deep blue or rarely white; capsule included in the corolla tube...***G. prostrata***

3a. Flowers whitish or pale yellowish with purple plaits and purple or green spots; leaves linear, in a loose basal rosette; plants 5–15 (20) cm tall, alpine...***G. algida***
3b. Flowers blue (rarely white), sometimes mottled or streaked with green; leaves linear, lanceolate, or ovate; plants 10–50 cm tall, variously distributed...4

4a. Flowers nearly closed at the apex, the lobes very short and nearly obsolete...***G. andrewsii***
4b. Flowers open at the apex with conspicuous lobes...5

5a. Calyx tube 10–18 mm long; leaf-like bracts subtending the flowers ovate to ovate-lanceolate and somewhat scarious, often hiding the calyx; corolla broadly tubular-funnelform or campanulate; anthers 3.5–5 mm long...***G. parryi***
5b. Calyx tube 4–10 mm long; leaf-like bracts subtending the flowers lanceolate or linear, not hiding the calyx; corolla tubular-funnelform, sometimes narrowly so; anthers 1.6–3 mm long...6

6a. Corolla lobes crenate-serrate along the upper edge just below the expanded portion; flowers light blue; leaf-like bracts subtending the flowers as long or longer than the flowers (the inflorescence crowded and congested at the top of the stem); leaves linear or linear-lanceolate or narrowly ovate...***G. bigelovii***
6b. Corolla lobes entire or inconspicuously serrulate along the upper edge; flowers purple to purplish-blue; leaf-like bracts subtending the flowers shorter than the flowers (the inflorescence usually elongate); leaves lanceolate to elliptic-ovate...***G. affinis***

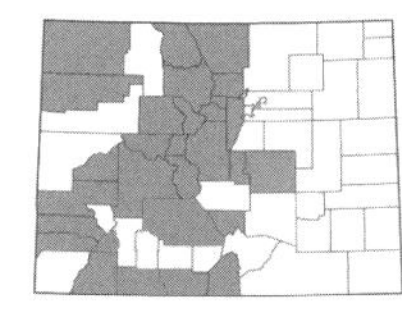

***Gentiana affinis* Griseb.**, Rocky Mountain gentian. [*Pneumonanthe affinis* (Griseb.) W.A. Weber]. Perennials, 1–5 dm; *leaves* lanceolate to elliptic-ovate, 1.5–4 cm × 6–12 (20) mm; *flowers* 5-merous; *calyx* tube 4–10 mm long, lobes 1.5–12 mm long; *corolla* 23–45 (48) mm long, tubular-funnelform, the lobes 3–5 (13) mm long, purple to purplish-blue and often streaked with green. Common in dry or more often moist meadows, 7500–10,500 ft. July–Sept. E/W.

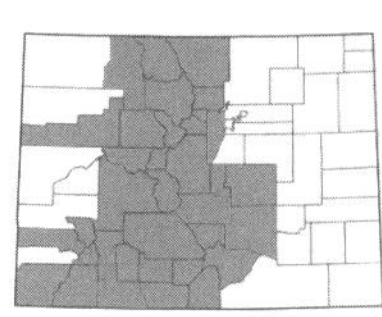

***Gentiana algida* Pall.**, arctic gentian. [*Gentianodes algida* (Pall.) Á. Löve & D. Löve]. Perennials, 0.5–1.5 (2) dm; *leaves* linear, in a basal rosette, 4–10 cm long; *flowers* 5-merous; *calyx* tube 12–20 mm long, lobes 7–12 mm long; *corolla* 35–50 mm long, tubular-funnelform, the lobes 2–5 mm long, whitish or pale yellowish with purple plaits and purple or green spots. Common in moist subalpine and alpine meadows and usually one of the last plants to bloom in the alpine, 10,000–14,000 ft. July–Sept. E/W. (Plate 46)

***Gentiana andrewsii* Griseb.**, closed bottle gentian. [*Pneumonanthe andrewsii* (Griseb.) W.A. Weber]. Perennials, 2–6 dm; *leaves* lanceolate to ovate, 5–15 cm long; *flowers* 5-merous; *calyx* tube 8–12 mm long, lobes 5–10 mm long; *corolla* 20–40 mm long, nearly closed at the apex, the lobes short, nearly obsolete, blue or rarely white. Uncommon in moist places, known only from an old collection near Boulder Creek (Boulder Co.), 5000–5200 ft. July–Sept. E.

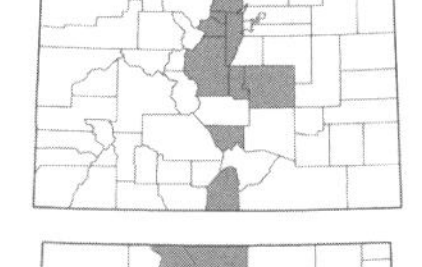

***Gentiana bigelovii* A. Gray**, Bigelow's gentian. [*Pneumonanthe bigelovii* (A. Gray) Greene]. Perennials, 1–5 dm; *leaves* linear to linear-lanceolate or narrowly ovate, to 4 cm long; *flowers* 5-merous; *corolla* 20–30 mm long, tubular-funnelform, light blue. Common on open hillsides, in meadows, and in forest openings, 5200–9500 ft. (July) Aug.–Oct. E. (Plate 46)

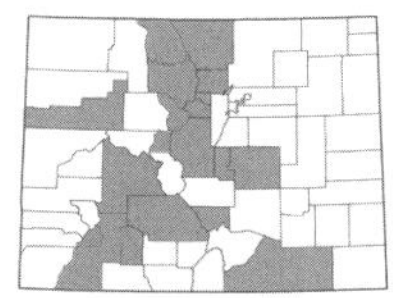

***Gentiana fremontii* Torr.**, moss gentian. [*Chondrophylla fremontii* (Torr.) A. Nelson; *G. aquatica* auct. non L.]. Annuals or biennials, 0.3–1 dm; *leaves* ovate below and lanceolate to oblong above, white-margined, 4–6 mm long; *flowers* usually 4-merous; *calyx* tube 5–7 mm long, the lobes ca. 2 mm long, white-margined; *corolla* 8–15 mm long, salverform, white to greenish-purple. Common in moist meadows, 7500–12,000 ft. May–June. E/W.

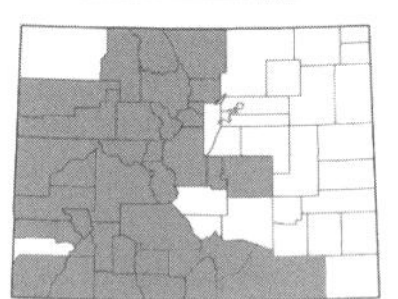

***Gentiana parryi* Engelm.**, Parry's gentian. [*Pneumonanthe parryi* (Engelm.) Greene]. Perennials, 1–4 (5) dm; *leaves* lanceolate to narrowly elliptic or ovate, 2–4 cm long; *flowers* 5-merous; *calyx* tube 10–18 mm long, lobes 1–8 mm long; *corolla* 33–50 mm long, broadly tubular-funnelform or campanulate, blue, sometimes streaked with green. Common in moist meadows, along streams, in meadows, and in forest openings, 7500–13,200 ft. July–Sept. E/W.

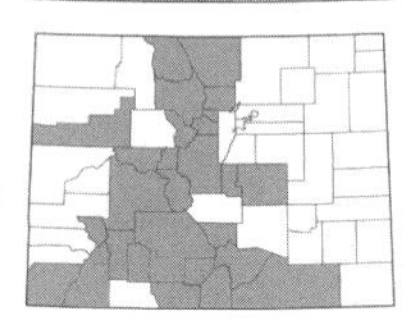

***Gentiana prostrata* Haenke**, pygmy gentian. [*Chondrophylla prostrata* (Haenke) J.P. Anderson; *G. nutans* Bunge]. Annuals or biennials, 0.2–1.2 dm; *leaves* orbicular to obovate, sometimes slightly white-margined; *flowers* usually 4-merous; *calyx* tube 4.5–7 (12) mm long; *corolla* 10–22 mm long, the lobes 1.5–2 mm long, deep blue or rarely white. Locally common in moist alpine, 10,500–14,000 ft. July–Sept. E/W. (Plate 47)

GENTIANELLA Moench – dwarf gentian

Annuals or biennials, glabrous herbs; *leaves* opposite, sessile; *flowers* in cymes, solitary, or solitary in leaf axils; *calyx* lobes nearly free or united at the base; *corolla* tubular or campanulate, with nectary glands located at the base of the tube, blue or purplish or rarely white.

1a. Plants mostly 2–5 cm tall, with short internodes hidden by leaves and flowers; corolla white or yellowish or rarely blue, 5.5–8.5 mm long, with lobes as long or longer than the tube...***G. tortuosa***
1b. Plants 4–50 cm tall, with elongated, evident internodes; corolla pinkish or purplish-blue or sometimes white, 7–16 mm long, with lobes shorter than the tube...2

2a. Pedicels longer than the subtending internodes; corolla with 2 fringed scales on the inside of each of the lobes; calyx lobes usually somewhat gibbous at the base; plants 4–20 cm tall...***G. tenella* ssp. *tenella***
2b. Most of the pedicels shorter than the subtending internodes; corolla with one fringed scale or a single row of fringe on the inside of each of the lobes or the fringe sometimes absent in late blooming plants; calyx lobes not gibbous at the base; plants 10–50 cm tall...3

3a. Calyx lobes all the same length or slightly unequal; corolla lobes 3–4.5 mm long, with a single row of fringe on the inside of each lobe or the fringe occasionally absent...***G. acuta***
3b. Calyx lobes conspicuously unequal, the outer lobes much larger and foliaceous, usually enclosing the inner 2 lobes; corolla lobes 4.5–7 mm long, with one fringed scale at the base of the inside of the lobe...***G. heterosepala***

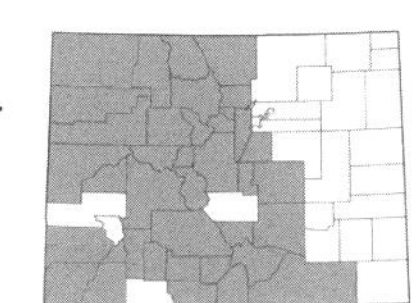

***Gentianella acuta* (Michx.) Hiitonen,** AUTUMN DWARF GENTIAN. [*Gentiana plebeia* Chamisso; *Gentianella amarella* (L.) Böerner ssp. *acuta* (Michx.) J.M. Gillett; *Gentianella strictiflora* (Rydb.) W.A. Weber]. Plants 0.8–5 dm; *leaves* elliptic to ovate or lanceolate above, 1–6 cm long; *calyx* 5–7 mm long, the lobes all the same length or nearly so; *corolla* 9–15 mm, the lobes 3–4.5 mm long, with a single row of fringe on the inside of each lobe (this sometimes absent), blue, purple, pinkish, or rarely white. Common along streams and in moist meadows, 6000–13,000 ft. July–Sept. E/W. (Plate 47)

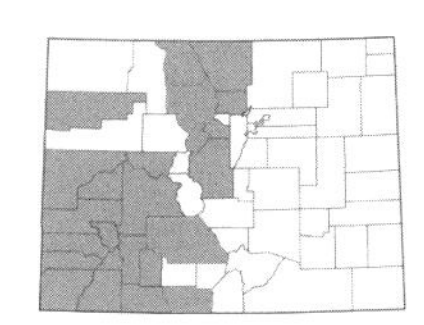

***Gentianella heterosepala* (Engelm.) Holub,** AUTUMN DWARF GENTIAN. [*Gentianella amarella* (L.) Böerner ssp. *heterosepala* (Engelm.) J.M. Gillett]. Plants 1–4.5 dm; *leaves* oblanceolate to elliptic, 2–6 cm long; *calyx* 8–15 mm long, the lobes conspicuously unequal; *corolla* 10–20 mm long, the lobes 4.5–7 mm long, with one fringed scale at the base of the inside of each lobe, blue, purple, or sometimes white. Common along streams and in moist meadows, 6000–13,000 ft. July–Sept. E/W.

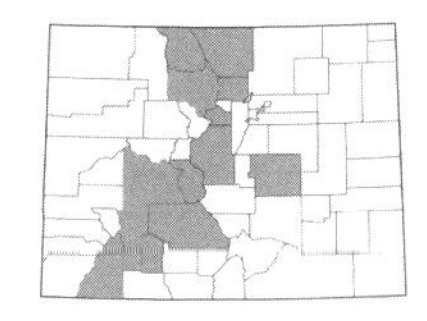

***Gentianella tenella* (Rottb.) Böerner ssp. *tenella*,** LAPLAND GENTIAN. [*Comastoma tenellum* (Rottb.) Toyok]. Plants 0.4–2 dm; *leaves* elliptic to obovate, lanceolate above, 4–10 mm long; *calyx* 4–10 mm long, the lobes conspicuously unequal (the outer foliaceous); *corolla* 7–16 mm long, the lobes 2.5–6 mm long, blue, purplish, or white. Locally common along streams and in moist subalpine and alpine meadows, 8500–13,000 ft. July–Sept. E/W.

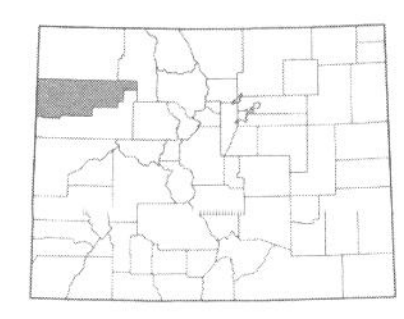

***Gentianella tortuosa* (M.E. Jones) J.M. Gillett,** CATHEDRAL BLUFF DWARF GENTIAN. Plants 0.2–0.5 (1) dm; *leaves* lanceolate, oblong, or narrowly elliptic, 0.5–3.5 cm long; *calyx* 4–8 mm long, the lobes subequal; *corolla* 5.5–8.5 mm long, the lobes as long or longer than the tube, each lobe with 2 fringed scales, white, yellowish, or blue. Rare on barren shale slopes of the Cathedral Bluffs, 8500–10,800 ft. July–Aug. W.

GENTIANOPSIS Ma – FRINGED GENTIAN

Annuals, biennials, or perennials; *leaves* opposite; *flowers* 4-merous; *calyx* usually 4-angled, in 2 dissimilar pairs; *corolla* tubular to funnelform or campanulate, the lobes fringed, not plicate, blue or rarely white.

1a. Flowers sessile or on short pedicels (less than 4 cm long); calyx lobes without a prominent purplish midvein; corolla tube 12–18 mm long...***G. barbellata***
1b. Flowers on long pedicels (1) 4–30 cm long; calyx lobes with a prominent purplish midvein; corolla tube 20–36 mm long...2

2a. Stem leaves linear-lanceolate or narrowly linear, 2–4 mm wide...***G. procera***
2b. Stem leaves oblanceolate to narrowly elliptic (linear above), usually wider than 4 mm...***G. detonsa* var. *elegans***

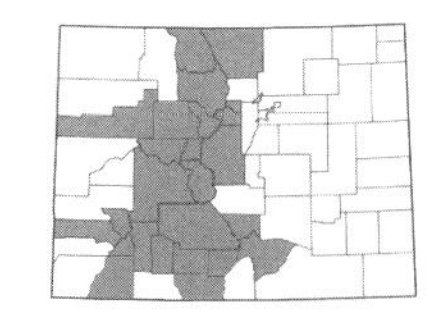

***Gentianopsis barbellata* (Engelm.) Iltis,** PERENNIAL FRINGED GENTIAN. [*Gentiana barbellata* Engelm.]. Perennials, 0.5–1.5 dm; *leaves* oblanceolate to narrowly ovate, 2–8 cm long; *flowers* sessile or on short pedicels to 4 cm long; *calyx* 11–25 mm long, the lobes without a purplish midvein; *corolla* 25–45 mm long with lobes 15–25 mm long, blue to purplish-blue. Found on moist slopes, meadows, alpine, and in aspen forests, 8600–13,000 ft. Aug.–Sept. E/W.

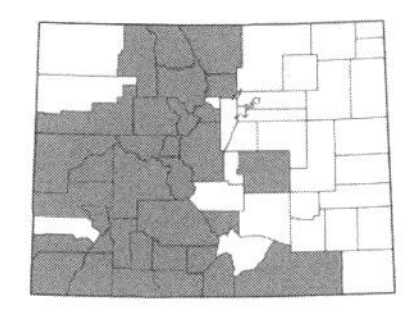

***Gentianopsis detonsa* (Rottb.) Ma var. *elegans* (A. Nelson) N. Holmgren,** ROCKY MOUNTAIN FRINGED GENTIAN. [*Gentiana thermalis* Kuntze]. Annuals or biennials, 0.5–5 (10) dm; *leaves* oblanceolate to narrowly elliptic, becoming linear above, 2–7 cm long; *flowers* on pedicels (1) 4–30 cm long; *calyx* tube 9–14 mm long, the lobes 8–20 mm long with a prominent purplish midvein; *corolla* 25–60 mm long with lobes 10–25 mm long, blue to purplish-blue. Common in moist meadows, bogs, and along streams, 7500–12,500 ft. July–Sept. E/W. (Plate 47)

***Gentianopsis procera* (Holm) Ma**, Great Plains fringed gentian. Annuals or biennials, 2–6 dm; *leaves* linear to linear-lanceolate, 1.5–7 cm × 2–4 mm; *flowers* on pedicels 12–20 cm long; *calyx* tube 10–20 mm long, lobes 15–22 mm long, dissimilar (the outer narrower), with a prominent purplish midvein; *corolla* 28–36 mm long, blue to purplish-blue. Rare along creeks, 6500–6800 ft. Aug.–Sept. E.

LOMATOGONIUM A. Br. – lomatogonium

Annuals, glabrous herbs; *leaves* opposite; *flowers* solitary or in terminal cymes, (4) 5-merous; *calyx* with nearly distinct lobes; *corolla* rotate, with 2 scaly appendages at the base of each lobes, blue to white.

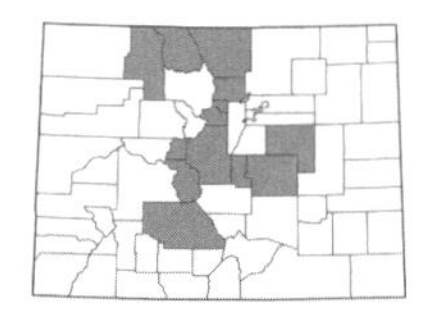

***Lomatogonium rotatum* (L.) Fries**, marsh felwort. [*Pleurogyne rotata* (L.) Griseb.]. Plants 0.5–5 dm; *leaves* oblanceolate to elliptic, becoming linear above, 0.5–2.5 cm long; *calyx* lobes 5–15 mm long; *corolla* lobes 7–12 mm long. Uncommon in moist meadows, along lake and stream margins, and in bogs, 8000–10,500 ft. Aug.–Sept. E/W.

SWERTIA L. – swertia

Annual or perennial, glabrous herbs; *leaves* alternate or opposite or nearly so; *flowers* (4) 5-merous; *calyx* deeply cleft; *corolla* rotate, deeply divided, each lobe with a pair of depressed, fringed nectary glands near the base, purple to bluish-purple or sometimes white.

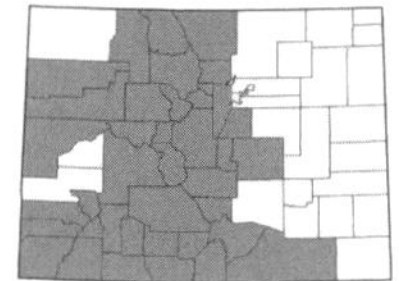

***Swertia perennis* L.**, felwort. Perennials, 1–6 dm; *leaves* narrowly elliptic to obovate, 4–25 cm long; *calyx* lobes 5–12 mm long; *corolla* 10–15 mm long. Common in moist places such as along streams and in wet meadows, 7700–14,000 ft. July–Aug. E/W.

GERANIACEAE Juss. – geranium family

Annual or perennial herbs (usually) or shrubs; *leaves* alternate (often the upper) or opposite (often the lower), simple or compound, petiolate, stipulate; *flowers* perfect, actinomorphic, solitary or in cymes; *sepals* 5, distinct; *petals* 5, distinct, clawed; *stamens* 5, 10, or 15, the filaments united at the base; *pistil* 1; *style* 1; *stigmas* 5; *ovary* superior, 5-carpellate; *fruit* a schizocarp, splitting and coiling along a central beak at maturity. (Hanks & Small, 1907)

1a. Leaves pinnately dissected; flowers small, the rose-purple petals 3–6 mm long…***Erodium***
1b. Leaves palmately lobed or divided; petals 3–25 mm long, white to pink-purple or rose-purple…***Geranium***

ERODIUM L'Her. – stork's bill

Annuals; *leaves* in basal rosettes and opposite and cauline, pinnately dissected; *stamens* 5, alternating with 5 staminodes; *fruit* with spirally twisted styles along a central beak at maturity.

***Erodium cicutarium* (L.) L'Her. ex Ait.**, filaria. Prostrate plants, 0.5–8+ dm long; *leaves* 1–12 cm long; *sepals* 3–6 mm; *petals* 5–7 mm, pink to light purple, with darker spots; *stylar column* 20–40 mm long. Common weed in disturbed areas, 3500–7500 ft. Mar.–Oct. E/W. Introduced.

GERANIUM L. – geranium; cranesbill

Annuals, bienniasl, or perennials; *leaves* alternate or opposite, mostly basal, palmately lobed or divided; *stamens* 10; *fruit* with long styles curved or coiling at maturity. (Aedo, 2000; 2001; Jones & Jones, 1943; Nebeker, 1974)

1a. Annuals or biennials from slender taproots; flowers smaller, the petals 3–9 mm long…2
1b. Perennials from a thick, usually branched caudex covered with old petiole bases and stipules; flowers larger, the petals (8) 10–25 mm long…4

2a. Sepals 2–3.7 (4) mm long, without a long awn tip but usually with a short callous tip; mature stylar column (measured from the receptacle to base of the beak) 9–11 mm long, without a beak; fertile stamens 5, alternating with 5 sterile ones; seeds smooth…***G. pusillum***
2b. Sepals 4.5–8 mm long, awn-tipped; mature stylar column 14–25 mm long, topped by a beak 1–5 mm long; stamens 10, all fertile; seeds finely reticulate…3

3a. Beak atop the stylar column 3–5 mm long; inflorescence open…***G. bicknellii***
3b. Beak atop the stylar column 1–2.5 mm long; inflorescence congested or open…***G. carolinianum***

4a. Petals white with purple veins, occasionally pinkish; nectaries usually hairy; pedicels glandular with purple-tipped glands…***G. richardsonii***
4b. Petals pink or purple with reddish veins, rarely white; nectaries glabrous but with a tuft of hairs at the apex; pedicels eglandular or if glandular then the glands yellow or white-tipped…5

5a. Basal leaves 1.7–7 (8) cm wide, the lobes obtuse or rounded and abruptly acuminate, 3-parted but not sharply toothed along the margins...***G. caespitosum***

5b. Basal leaves 5–15 cm wide, the lobes coarsely and sharply toothed along the margins...***G. viscosissimum* var. *incisum***

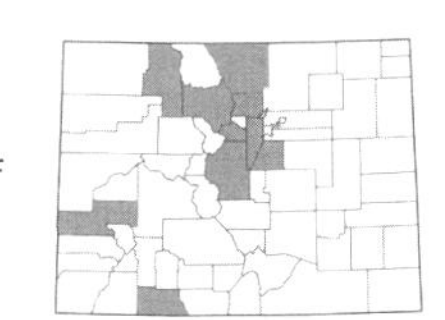

***Geranium bicknellii* Britton**, Bicknell's cranesbill. Annuals or biennials, 1–4 (6) dm; *stems* with spreading or retrorse hairs; *leaves* deeply palmately cleft, 2–6.5 cm wide; *pedicels* 1.5–3.5 cm long, densely glandular; *sepals* 4.5–8 mm long, with an awn tip 1–1.8 mm long, with a mixture of glandular and eglandular hairs; *petals* 6–9 mm long, pink to rose; *stylar column* 18–25 mm long with a beak 3–5 mm long. Uncommon in disturbed places and moist meadows, 5500–9000 ft. May–Aug. E/W. Possibly introduced.

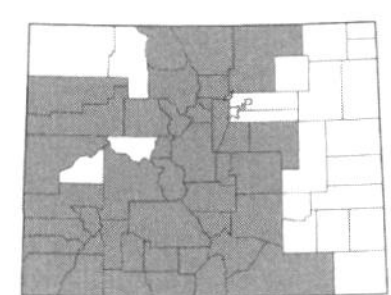

***Geranium caespitosum* James**, Rocky Mountain geranium. [*G. atropurpureum* Heller; *G. fremontii* Torr. ex A. Gray; *G. parryi* (Engelm.) Heller]. Perennials, 0.7–5 (8) dm; *stems* usually glandular above with yellow-tipped glands, sometimes eglandular throughout; *leaves* palmate 3-5-parted, 1.7–7 (8) cm wide; *pedicels* 1.3–4 (5) cm long; *sepals* 5.5–9 (11) mm long with an awn tip 0.2–1.2 (2) mm long; *petals* (8) 10–17 mm long, usually pink or purple; *stylar column* 22–30 (40) mm long with a beak 2.5–5.5 mm long. Common on dry, rocky slopes, in meadows, and in open forests, 5200–12,500 ft. May–Aug. E/W. (Plate 47)

Specimens that are eglandular or only with a few glandular hairs have typically been called *G. atropurpureum*. However, the forms of hairs present tend to be widely variable, and some plants without glandular hairs can be found directly next to plants with glandular hairs. There are no additional morphological characteristics that can be used to separate *G. atropurpureum* from *G. caespitosum*, and thus it is not recognized even at the infraspecific level.

***Geranium carolinianum* L.**, Carolina geranium. Annuals or biennials, 1.5–5 (7) dm; *stems* with retrorse hairs, sometimes glandular above; *leaves* palmately 5 (7)-parted, 3–7 cm wide; *pedicels* 3–7 (15) cm long, villous and glandular; *sepals* 4.5–7 mm long, with an awn tip 1–2 mm long; *petals* 4.5–7 mm long, pink to rose; *stylar column* 14–22 mm long with a beak 1–2.5 mm long. Uncommon weed in lawns and disturbed places, reported for Colorado from a 2006 collection in a lawn in Jefferson Co., 5000–6000 ft. May–Sept. E. Introduced.

***Geranium pusillum* L.**, small geranium. Annuals or biennials, 1–4 (6) dm; *stems* with retrorse hairs; *leaves* palmately 7–9-parted, 2–3.5 (7) cm wide; *pedicels* 0.6–1.2 cm long, glandular; *sepals* 2–3.7 (4) mm long with a short callous tip; *petals* 3–4 mm long, pink to purple; *stylar column* 9–11 mm long, lacking a beak. Uncommon in disturbed places, 4900–6500 ft. May–July. E. Introduced.

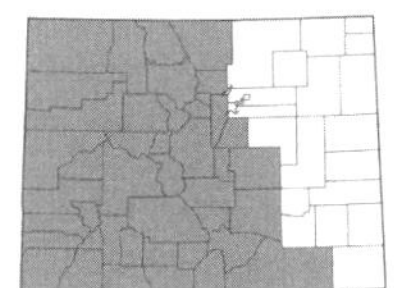

***Geranium richardsonii* Fisch. & Trautv.**, Richardson's geranium. Perennials, 3–7 (9) dm; *stems* glandular above, the glands purple-tipped; *leaves* palmately (3) 5–7-parted, (4) 6–12 (15) cm wide; *pedicels* 1–3.3 cm long, glandular with purple-tipped glands; *sepals* 6–10 mm long with an awn tip 1–2.5 mm long; *petals* 11–20 (25) mm long, white or light pinkish, with purple veins; *stylar column* 20–30 (35) mm long with a beak 1.7–4 mm long. Common in moist meadows, along streams, and in aspen forests, 6000–12,700 ft. May–Sept. E/W. (Plate 47)

Can form intermediates with *G. viscosissimum* and *G. caespitosum*.

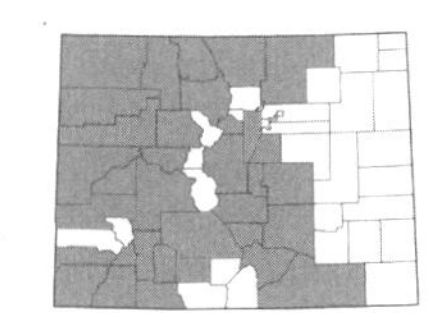

***Geranium viscosissimum* Fisch. & C.A. Mey. ex C.A. Mey. var. *incisum* (Torr. & A. Gray) N.H. Holmgren**, sticky purple geranium. [*G. nervosum* Rydb.]. Perennials, (2.5) 4–9 dm; *stems* glandular above, the glands yellow-tipped; *leaves* palmately 5 (7)-parted, 5–15 cm wide; *pedicels* 1.5–4 (5) cm long, glandular; *sepals* 7.5–10 mm long with an awn tip 1–3 mm long; *petals* 14–25 mm long, pink or purple; *stylar column* 28–40 mm long with a beak 4–7 mm long. Found in meadows, aspen forests, and dry slopes, 6500–10,000 ft. June–Aug. E/W.

GROSSULARIACEAE DC. – gooseberry family

Shrubs; *leaves* alternate, often fascicled, petiolate, simple, palmately veined; *flowers* perfect, epigynous; *sepals* (4) 5, often petaloid; *petals* (4) 5, small and shorter than the sepals, inserted near the top of the hypanthium and alternating with the sepals; *stamens* (4) 5, inserted at the mouth of the hypanthium; *pistil* 1; *styles* 2; *ovary* inferior, 2-carpellate, with parietal placentation; *fruit* a berry, usually with persistent perianth. (Berger, 1924)

Typical Grossulariaceae flower:

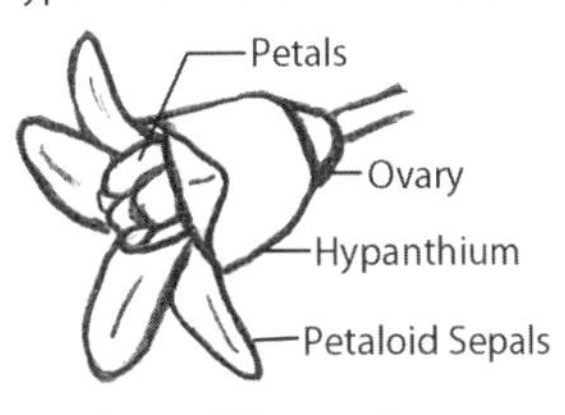

RIBES L. – currant; gooseberry

With characteristics of the family.

Note - In the following key, the hypanthium is measured from the top of the ovary to the base of the sepals.

1a. Stems unarmed, lacking spines or bristles; flowers subtended by bracts 0.9–10 (12) mm long…2
1b. Stems armed with spines, with at least 1 spine per node; flowers subtended by bracts 1–4 mm long…9

2a. Flowers with a bright yellow hypanthium and sepals, and reddish or yellowish petals; plants glabrous or nearly so; leaves deltate-ovate with a cuneate to somewhat truncate base, 3–5-lobed less than half their length; hypanthium with a narrowly cylindrical tube, mostly 5–10 mm long…***R. aureum***
2b. Flowers with a greenish-white, greenish-yellow, white, or pink hypanthium and sepals, and white to reddish-purple or yellowish-green petals; plants usually with hairs on the lower leaf surface (at least along the veins), often glandular-stipitate; leaves orbicular to reniform, cordate or truncate at the base, 3–7-lobed; hypanthium campanulate to cylindrical, 0.6–9.5 mm long…3

3a. Hypanthium narrowly cylindrical with a tube (6) 6.5–8.8 (9.5) mm long and 2–3.5 mm wide, pinkish or occasionally whitish; sepals mostly 2–3.2 mm long, pink or white; leaves mostly 1–2 cm long, shallowly lobed much less than half the length of the leaf, with crenate to crenate-dentate margins; fruit bright red or orange-red…***R. cereum***
3b. Hypanthium campanulate to tubular-campanulate or saucer-shaped, 0.6–7 mm long but usually wider, greenish-white, sometimes with a pink tinge; sepals 2–7 mm long, greenish-white or reddish-purple; leaves mostly 1.5–8 cm long, lobed much less than to almost half the length of the leaf, the margins crenate-dentate to coarsely serrate; fruit red, black, or bluish-black…4

4a. Ovaries and berries glabrous…5
4b. Ovaries and berries with glandular-stipitate hairs…7

5a. Leaves with large, sessile yellow gland-dots on the lower surface…***R. americanum***
5b. Leaves without large, sessile yellow gland-dots on the lower surface…6

6a. Racemes 8–20-flowered; hypanthium saucer-shaped, about 1 mm long; fruit red…***R. rubrum***
6b. Racemes 1–4-flowered; hypanthium tubular-campanulate, 2–3.5 mm long; fruit black or bluish-black…***R. inerme***

7a. Flowers large, the hypanthium (4.5) 5–7 mm long and sepals (3.5) 5–7 mm long; bracts subtending the flowers (5) 6–10 mm long; leaves sparsely to densely glandular-stipitate and finely tomentose, although sometimes nearly glabrous with only a few glandular hairs…***R. viscosissimum***
7b. Flowers smaller, the hypanthium 0.6–2 mm long and sepals 2–4 mm long; bracts subtending the flowers 0.9–5 (7) mm long; leaves glabrous to sparsely hairy, or with some glandular hairs below…8

8a. Racemes loosely 6–15-flowered; glandular-stipitate hairs on the ovary with purplish-red or red tips; flowers subtended by bracts mostly 0.9–3 mm long, less than half as long as the pedicels…***R. coloradense***
8b. Racemes densely 7–25-flowered; glandular-stipitate hairs on the ovary with yellowish or brownish tips; flowers subtended by bracts mostly 3–5.5 mm long, as long or longer than the pedicels…***R. wolfii***

9a. Hypanthium shallowly saucer-shaped or cup-shaped, 0.5–1.5 mm long, glabrous or glandular-hairy externally; ovary sparsely to densely glandular-stipitate; racemes 3–16-flowered; leaves cordate at the base…10
9b. Hypanthium a cylindric tube or narrowly campanulate, 2–6 mm long; ovary glabrous or occasionally glandular-hairy and then the hypanthium conspicuously hairy externally; racemes 1–4 (5)-flowered; leaves subcordate or truncate at the base…11

10a. Leaves with glandular and eglandular hairs on both surfaces; berry bright red…***R. montigenum***
10b. Leaves glabrous or sparsely hairy especially along the veins and margins, or sometimes glandular just on the veins below; berry red when young but purplish-black at maturity…***R. lacustre***

11a. Hypanthium conspicuously softly hairy externally and glabrous within, the cylindrical tube (2.3) 3–6 mm long; styles glabrous and connate nearly their full length; leaves small and orbicular, mostly 0.5–1.5 cm long and wide…***R. leptanthum***
11b. Hypanthium usually glabrous or sometimes sparsely hairy externally and hairy within, the tube cylindric or narrowly campanulate and 2–5.2 (6.5) mm long; styles pilose at least in the lower half, connate ⅓–¾ of their length; leaves usually larger, mostly 1.5–5 cm long…12

12a. Stamens shorter than or equal to the height of the petals; hypanthium tube cylindric, mostly 3.5–5.5 mm long; stems with numerous internodal bristles…***R. oxyacanthoides* ssp. *setosum***
12b. Stamens longer than the petals; hypanthium tube narrowly campanulate, 1.5–3.5 mm long; stems with or without internodal bristles…13

13a. Filaments of stamens usually glabrous, 2–3 mm long; sepals 2.5–4.5 (5) mm long; styles 5–7 mm long...***R. inerme***
13b. Filaments hairy, 6.5–8.5 mm long; sepals (5) 6–7.6 mm long; styles 8.5–15 mm long...***R. niveum***

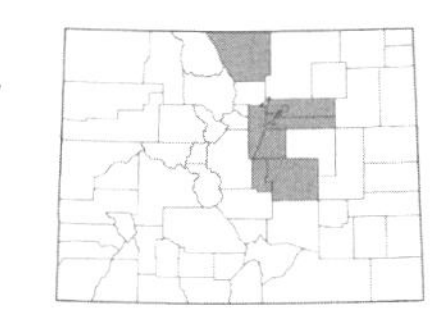

***Ribes americanum* Mill.**, AMERICAN BLACK CURRANT. Unarmed shrubs; *leaves* suborbicular, 3–5-lobed, 3–8 cm long, glabrate above and yellow gland-dotted below; *hypanthium* campanulate, 3–4.5 mm long, hairy, tawny to greenish-white; *sepals* 4–5 mm long, greenish-white; *petals* 2–3 mm long, white; *berries* 6–10 mm diam., black. Uncommon in shady places along streams and in moist meadows along the Front Range, 5500–7500 ft. May–July. E.

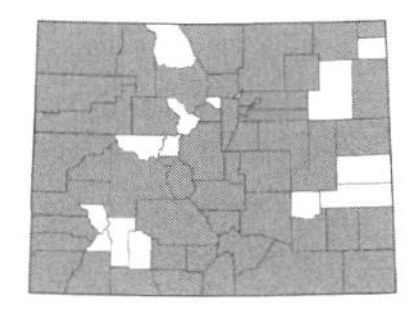

***Ribes aureum* Pursh**, GOLDEN CURRANT. Unarmed shrubs; *leaves* broadly deltate-ovate to obovate, 3 (5)-lobed, (1) 1.6–3.6 (5.7) × (0.8) 1.5–4.5 (7.8) cm, finely hairy to glabrous; *hypanthium* cylindrical, (4) 5–10 (13) mm long, glabrous, golden-yellow, occasionally aging orange or pinkish; *sepals* (4) 4.4–6.8 (7.8) mm long; *petals* (2) 2.3–3 (4) mm long, yellow to red or reddish-violet; *berries* 5.2–10 mm diam., red, orange, brown, black, or rarely yellow. Common on dry, often sandy slopes, along ditches and streams, and along roadsides, 3500–8500 ft. April–June. E/W. (Plate 47)

There are 2 varieties of *R. aureum* in Colorado, and var. *villosum* is often considered introduced:

1a. Hypanthium mostly less than 10 mm long, usually less than twice as long as the sepals; largest leaves usually 3-lobed with obtuse teeth...**var. *aureum***. Common on dry, often sandy slopes, and along ditches and streams, 3700–8500 ft. April–June. E/W.
1b. Hypanthium mostly 10 mm or more long, usually at least twice as long as the sepals; largest leaves usually 5-lobed with pointed teeth...**var. *villosum* DC**. [*R. odoratum* H. Wendl.]. Less common than var. *aureum*, usually escaped from cultivation or along roadsides on the eastern plains, 3500–6000 ft. April–June. E.

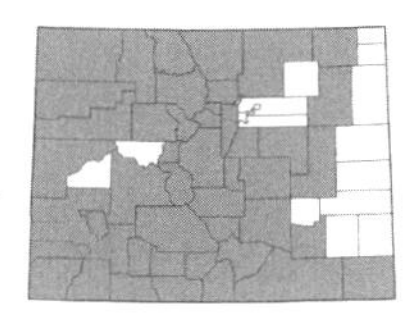

***Ribes cereum* Douglas**, WAX CURRANT. Unarmed shrubs; *leaves* reniform to orbicular, shallowly 3–5 (7)-lobed, (0.5) 1–2 (3.9) × (0.6) 1.2–3.1 (5) cm, sparsely to densely eglandular pubescent, glandular-stipitate hairy, or glabrate; *hypanthium* cylindrical, (6) 6.5–8.8 (9.5) mm long, glabrous within, densely hairy with scattered glandular hairs externally, white to pinkish-white or greenish-white with a pink tinge; *sepals* (1.4) 2–3.2 mm long, creamy pink to white; *petals* (1.2) 1.5–2.1 mm long, cream to white; *berries* 5–10 mm diam., bright red or orange-red. Common in rocky places and dry hillsides, sometimes in pinyon-juniper or sagebrush, 4300–12,000 ft. April–Aug. E/W. (Plate 47)

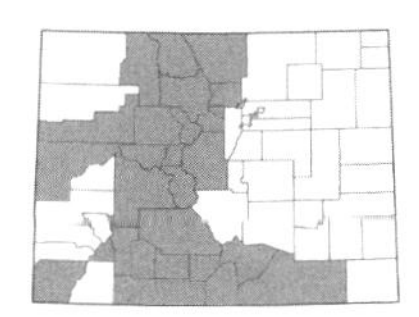

***Ribes coloradense* Coville**, COLORADO CURRANT. [*R. laxiflorum* Pursh]. Unarmed shrubs; *leaves* orbicular, 5 (7)-lobed, (2.5) 4–7 (9) × (3) 4–8 (11.8) cm, glabrate above, glabrous to sparsely eglandular and glandular-hairy below; *hypanthium* shallowly bowl- or saucer-shaped, 0.6–1.4 mm long, glabrous within, eglandular and glandular-stipitate hairy externally, greenish-white with a pinkish tinge to red or purple, lined with a reddish to brownish disk; *sepals* 2.4–3.8 (4) mm long, greenish-white, yellowish-green, or reddish-purple; *petals* 0.9–1.6 mm long, reddish to purplish with a greenish apex; *berries* (4) 5–10 (14) mm diam., purplish-black. Found on alpine ridges and in typically moist spruce-fir forests, 9500–12,500 ft. June–Aug. E/W.

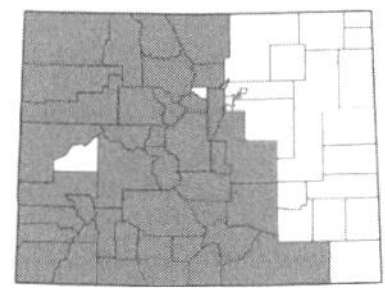

***Ribes inerme* Rydb.**, WHITESTEM GOOSEBERRY. Spiny shrubs; *spines* 0–3 spines per node, (1.5) 3–7 (12) mm; *leaves* suborbicular, 3–5-lobed, (0.9) 1.5–5 (7.5) × 1–5 (8.5) cm, glabrous to densely hairy; *hypanthium* narrowly campanulate to cup-shaped, 2–3.5 mm long, glabrous, occasionally densely pilose to tomentose within and sparsely hairy externally, greenish-yellow, yellowish-white, or white; *sepals* 2.5–4.5 (5) mm long, often reddish at the base; *petals* 1.2–2.3 mm long, white or pinkish; *berries* 6–11 (12) mm diam., reddish-purple to greenish-purple or black. Common along streams, moist roadsides, in meadows, and sometimes on dry slopes, 5500–10,500 ft. April–June. E/W. (Plate 47)

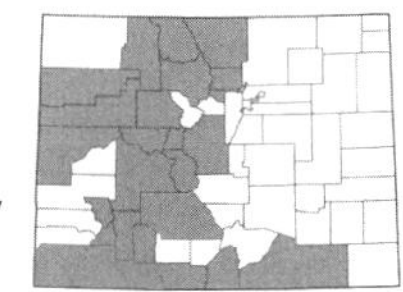

***Ribes lacustre* (Pers.) Poir.**, PRICKLY CURRANT. Spiny shrubs; *spines* 1–several per node, 3.8–6.5 (8) mm; *leaves* pentagonal, 3-5-lobed, (1) 2–5 (7.8) × (1.1) 2.1–5 (8.5) cm, glabrous or sparsely hairy; *hypanthium* shallowly saucer-shaped, (0.7) 1–1.2 (1.5) mm long, glabrous, lined with a pinkish to dark red disk; *sepals* 2.2–3.1 mm long, pale yellowish-green to reddish; *petals* 1–1.5 (1.7) mm long, reddish or pale yellowish-green apically and reddish basally; *berries* 5.5–8 (14) mm diam., red to black or dark purple. Locally common along streams and in meadows or occasionally found on open slopes, 7000–11,500 ft. June–Aug. E/W. (Plate 47)

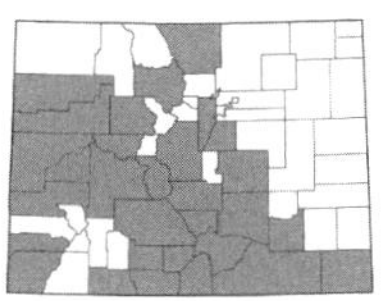

***Ribes leptanthum* A. Gray**, TRUMPET GOOSEBERRY. Spiny shrubs; *spines* 1–3 per node, to 19 mm; *leaves* orbicular or reniform-orbicular, (3) 5 (7)-lobed, 0.5–1.5 (2.7) × 0.5–1.5 (2.7) cm, glabrous, occasionally minutely hairy, rarely sparsely glandular-hairy; *hypanthium* cylindrical, (2.3) 3–6 mm long, glabrous within, softly hispid externally, greenish-white to white; *sepals* (2.5) 3.5–7 mm long; *petals* 2.5–4.4 mm long, white or pinkish; *berries* 5–10 mm diam., red to black. Found on dry, often sandy hillsides, sometimes with pinyon-juniper or pine, 4200–10,500 ft. April–June. E/W.

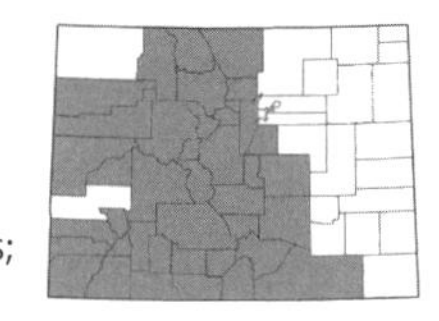

***Ribes montigenum* McClatchie**, ALPINE PRICKLY CURRANT. Spiny shrubs; *spines* (1) 3–5 per node, (1.5) 2.3–8 mm; *leaves* pentagonal, deeply 5-cleft, (0.5) 1–3.5 (4) × (0.5) 1–4 (5) cm, densely eglandular and glandular-hairy; *hypanthium* cup- to saucer-shaped, 0.5–1.5 mm long, glabrous within, eglandular and glandular-hairy externally, lined with a red, orange, or yellowish-pink disk; *sepals* (2) 2.4–4 mm long, yellowish-green to pink or red, occasionally with pale yellow scarious margins; *petals* 0.9–1.5 mm long, purplish-red, pink, rose, or purplish; *berries* 5–9.5 mm diam., bright red. Common on subalpine and alpine slopes and in spruce forests, 8500–13,200 ft. May–Aug. E/W. (Plate 47)

***Ribes niveum* Lindl.**, SNOW CURRANT. Spiny or occasionally unarmed shrubs; *spines* 1–3 (6) per node, sometimes lacking on older branches; *leaves* reniform to broadly ovate, 3–5-lobed, 0.3–3.5 (4.5) × (0.8) 1–4.2 (6.5) cm, glabrous or hairy; *hypanthium* narrowly campanulate, 1.5–3 mm long, glabrous or sparsely hairy, white, pale greenish, or ochroleucous; *sepals* (5) 6–7.6 mm long, white or cream, occasionally pinkish; *petals* 1.7–3.2 mm long, white, often with red veins; *berries* 5.5–12 mm diam., yellow-green to blue-black. Uncommon along streams and creeks near Canon City, 5000–6000 ft. April–July. E.

This species was collected by T.S. Brandegee in 1873 near Canon City. It was not rediscovered in Colorado until collections made by Tim Chumley (RM) in 1995. It is disjunct from the Pacific Northwest.

***Ribes oxyacanthoides* L. ssp. *setosum* (Lindl.) Sinnott**, INLAND GOOSEBERRY. Spiny shrubs; *spines* 1–7 per node, often with additional smaller spines per node, 2.2–9 (13) mm; *leaves* suborbicular to reniform, 3–5 (7)-lobed, (0.5) 1.5–4 (6) × (0.7) 1.7–4 (6.8) cm, pilose to villous or glabrate, often glandular; *hypanthium* cylindrical, (3) 3.5–5.5 (6.5) mm long, villous within, glabrous externally, greenish-white to white; *sepals* 3–4.2 (5.5) mm long; *petals* 1.8–2.6 (3) mm long, white or pinkish; *berries* (5) 6.7–13 mm diam., reddish, greenish-purple, or deep purplish-black. Not reported for Colorado but could occur in the Park and Rawah ranges (northcentral Colorado).

***Ribes rubrum* L.**, GARDEN RED CURRANT. [*R. sativum* Syme]. Unarmed shrubs; *leaves* cordate to suborbicular, usually 5-lobed, mostly 4–8 cm long, sparsely hairy below; *hypanthium* saucer-shaped, ca. 1 mm long; *petals* ca. 1 mm long, pinkish to cream; *berries* ca. 10 mm diam., red. Often cultivated and occasionally escaping into nearby areas, recently reported for the Roaring Fork Valley (Eagle Co.), April–June. W. Introduced.

***Ribes viscosissimum* Pursh**, STICKY CURRANT. Unarmed shrubs; *leaves* suborbicular to reniform, shallowly 3–5 (7)-lobed, (1) 2.6–6.5 (8.5) × (1.7) 3.6–8.5 (10.5) cm, sparsely to densely glandular-stipitate and finely tomentose to nearly glabrous; *hypanthium* bell-shaped to tubular-campanulate, (4.5) 5–7 mm long, glabrous within, sparsely to densely eglandular and glandular-stipitate externally, greenish-white to greenish-yellow; *sepals* (3.5) 5–7 mm long, whitish-green, occasionally pink- or purple-tinged; *petals* 2.5–3.5 (4) mm long, white or cream; *berries* (8) 10–15 mm diam., dark bluish-black. Uncommon on wooded slopes, 6500–9000 ft. June–July. W.

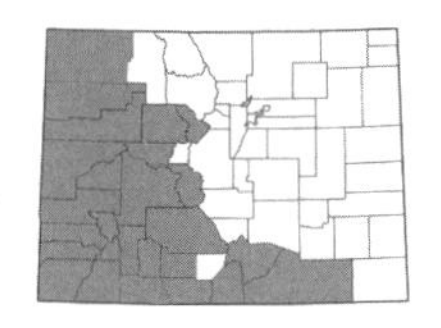

***Ribes wolfii* Rothr.**, WOLF'S CURRANT. [*R. mogollonicum* Greene]. Unarmed shrubs; *leaves* suborbicular, 3–5-lobed, 2–6 (7.5) × (2) 2.5–9.5 cm, glabrous or sparsely hairy; *hypanthium* cup-shaped to turbinate, (0.8) 1.2–1.5 (2.1) mm long, glabrous within, eglandular to scattered glandular-hairy externally, greenish-yellow to white or cream, occasionally drying pink; *sepals* (2) 3.3–4.1 mm long; *petals* 0.9–1.5 mm long, yellowish-green or pinkish; *berries* 3–6 (10) mm diam., black. Found along streams, on rocky slopes, and in aspen forests, 7000–11,000 ft. May–Aug. E/W.

HALORAGACEAE R. Br. – WATER MILFOIL FAMILY

Aquatic herbs; *leaves* usually whorled, sometimes alternate or opposite, simple and finely dissected or compound, exstipulate; *flowers* perfect or imperfect, actinomorphic, small, solitary or in spikes or racemes; *sepals* 4, distinct; *petals* 2–4 and distinct or absent; *stamens* 4 or 8; *pistil* 1; *styles* 2–4; *ovary* inferior, 2–4-carpellate; *fruit* a schizocarp splitting into 2–4 nutlets. (Aiken, 1981)

MYRIOPHYLLUM L. – WATER MILFOIL

Leaves whorled or alternate, finely pinnately dissected; *flowers* usually imperfect, in terminal spikes; *sepals* 4; *petals* 4 or absent; *stamens* 4 or 8; *styles* 4; *fruit* a schizocarp.

1a. Floral bracts entire to serrate or somewhat pectinate, shorter to about as long as the flowers; staminate flowers pinkish; pistillate flowers without sepals or these less than 0.5 mm long... ***M. sibiricum***

1b. Floral bracts strongly dissected into feather-like structures, much longer than the flowers; staminate flowers yellowish-green; pistillate flowers with sepals mostly 0.5–1 mm long... ***M. verticillatum***

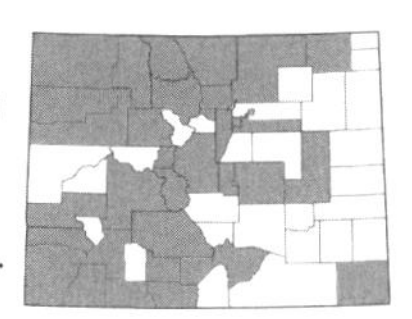

***Myriophyllum sibiricum* Komarov**, SHORTSPIKE WATER MILFOIL. Stems 3–8 dm long; *leaves* 1–3 cm long; *spikes* 3–10 cm long; *floral bracts* entire to serrate or somewhat pectinate, shorter to nearly as long as the flowers; *staminate flowers* pinkish; *pistillate flowers* to 0.5 mm or lacking sepals; *schizocarps* 2–3 mm long. Common in ponds, lakes, and still water, 4000–10,800 ft. May–Sept. E/W.

Myriophyllum spicatum L. has been reported for Colorado, but no specimens have been seen. It closely resembles *M. sibiricum* but differs in not producing turions. In addition, the leaves are more feathery with 12–20 segments on each side (vs. 5–12 segments in *M. sibiricum*).

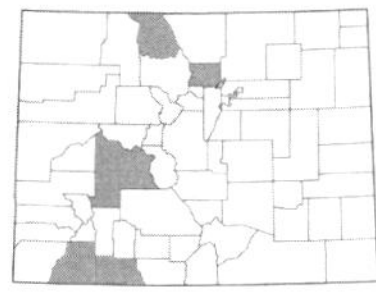

***Myriophyllum verticillatum* L.**, WHORLED WATER MILFOIL. Stems 3–9 dm long; *leaves* 1–4 cm long; *spikes* 4–12 cm long; *floral bracts* strongly dissected into feather-like structures, longer than the flowers; *staminate flowers* yellowish-green; *pistillate flowers* 0.5–1 mm long; schizocarps 2–3 mm long. Uncommon in ponds and lakes, 7800–10,000 ft. May–Aug. E/W.

HELIOTROPIACEAE Schrad. – HELIOTROPE FAMILY

Annual or perennial herbs or shrubs; *leaves* mostly entire, alternate; *flowers* in terminal helicoid spikes or racemes; *corolla* white to blue or purple; *calyx* shallowly to deeply cleft; *stamens* included; *style* terminal or lacking, the stigma disk-like and often surmounted by a small cone; *nutlets* 2 or 4, sometimes separating in pairs, smooth to roughened or hairy. (Ewen, 1942)

1a. Glabrous perennials from stout creeping roots, somewhat fleshy; flowers in an ebracteate coiled cyme; stems prostrate or loosely ascending; corolla without pointed tips at the well-defined lobes; corolla tube 2–4 mm long and scarcely surpassing the calyx, the corolla limb 3–15 mm wide; stigma sessile, expanded and as wide as the ovary; nutlets smooth...***Heliotropium***

1b. Pubescent annuals from a slender taproot, not fleshy; flowers in a bracteate cyme (not coiled); stems erect or the outer curved ascending; corolla with 5 small points marking the petal-tips, but otherwise hardly lobed; corolla tube 7–10 mm long and well-surpassing the calyx, the corolla limb 15–22 mm wide; stigma elevated on a slender style; nutlets loosely soft-hairy...***Euploca***

EUPLOCA Nutt. – EUPLOCA

Flowers in bracteate inflorescences or rarely solitary; *nutlets* 4.

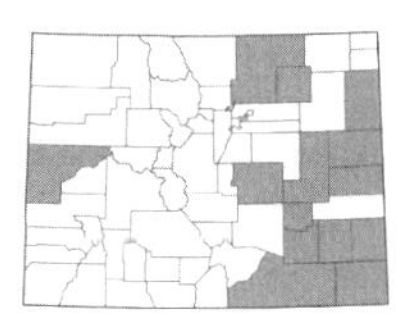

***Euploca convolvulacea* Nutt.**, BINDWEED HELIOTROPE. [*Heliotropium convolvulaceum* (Nutt.) A. Gray var. *convolvulaceum*]. Annuals to 3 dm; *leaves* lanceolate to ovate, 1–4 cm long, strigose; *inflorescence* a bracteate cyme; *calyx* ca. 5 mm long; *corolla* tube 7–10 mm long, the limb 15–22 mm wide, scarcely lobed; *stigma* elevated on a slender style; *nutlets* 4, ca. 3 mm long, soft hairy. Locally common in sandy soil, usually on sand dunes, 3500–5700 ft. June–Sept. E/W.

HELIOTROPIUM L. – HELIOTROPE

Flowers in ebracteate inflorescences; *nutlets* 2.

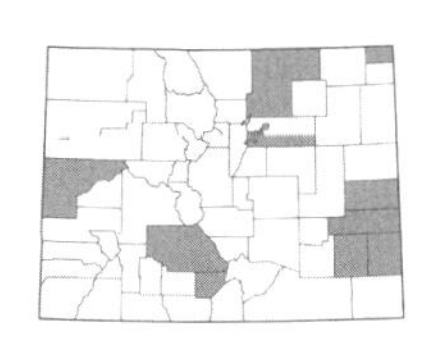

***Heliotropium curassavicum* L.**, SEASIDE HELIOTROPE. Perennials (annuals) from stout creeping roots, to 4 dm; *leaves* succulent, lanceolate to obovate, 2–6 cm long, glabrous; *inflorescence* an ebracteate cyme with rather crowded flowers; *calyx* 2–3 mm long; *corolla* tube 2–4 mm long, the limb 3–15 mm wide, with well-defined lobes; *stigma* sessile; *nutlets* 2–2.5 mm long, smooth or inconspicuously ribbed. Uncommon in saline soil along drying lake borders, 3500–7600 ft. June–Sept. E/W.

There are two varieties of *H. curassavicum* in Colorado. *Heliotropium curassavicum* var. *curassavicum* occurs in northern New Mexico and is separated by having small, narrow leaves and consistently white corollas to 3.5 mm wide:

1a. Corolla limb 5–15 mm wide, white or purplish-tinged or yellow on the throat; leaves tending to be broader, the larger ones 10–18 mm wide...**var. *obovatum* DC.** [*H. spatulatum* Rydb.].

1b. Corolla limb 3–6 mm wide, with a purple throat and eye; leaves narrower, the larger ones usually not over 10 mm wide...**var. *oculatum* (A.A. Heller) I.M. Johnst.**

HYDRANGEACEAE Dum. – HYDRANGEA FAMILY

Shrubs, subshrubs, herbs, or lianas; *leaves* alternate or more often opposite, petiolate, usually entire, exstipulate; *flowers* perfect, actinomorphic or the marginal ones sterile and zygomorphic, occasionally some flowers imperfect and the plants polygamo-dioecious; *sepals* 4–5 (10), distinct or united, imbricate or valvate; *petals* 4–5 (10), distinct, imbricate, contorted, or valvate; *stamens* 4, 8, or 10–200; *pistil* 1; *styles* (2) 3–5, distinct or partially connate below; *ovary* inferior or partly inferior, (2) 3–5-carpellate; *fruit* usually a septicidal or loculicidal capsule, or sometimes a berry. (Holmgren & Holmgren, 1997)

1a. Leaves toothed, pinnately veined, green above and densely grayish- or whitish-hairy below; petals 5...***Jamesia***
1b. Leaves entire, 3-veined from the base, green above and green and glabrous to densely white-hairy below; petals 4 or 5...2

2a. Flowers small and numerous in a compound, bracteolate cyme, with 5 petals, these 2–4 mm long; stamens 10; fruit a 3-valved capsule; at least some of the hairs on the leaves with pustulate, swollen bases...***Fendlerella***
2b. Flowers larger, solitary or in clusters of 2 or 3, with 4 petals, these 5–20 mm long; stamens 8 or 30–90; fruit a 4-valved capsule; hairs on the leaves without swollen bases or lacking...3

3a. Petals not clawed; stamens 30–90, the filaments terete and subulate...***Philadelphus***
3b. Petals clawed; stamens 8, the filaments petaloid, flat and broad...***Fendlera***

FENDLERA Engelm. & A. Gray – FENDLER-BUSH

Shrubs with reddish to gray bark; *leaves* opposite, deciduous, entire or nearly so; *sepals* 4; *petals* 4, white or sometimes tinged with red, clawed; *stamens* 8; *styles* 4; *ovary* inferior; *fruit* a septicidal capsule.

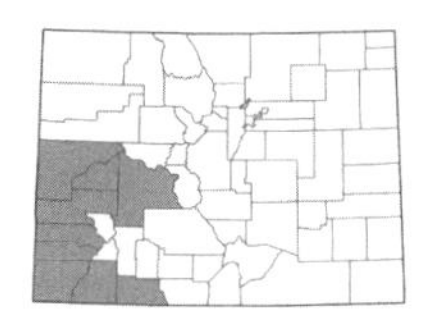

Fendlera rupicola **Engelm. & A. Gray**, CLIFF FENDLER-BUSH. [*F. wrightii* (Engelm. & A. Gray) Heller]. Shrubs, 1–2 m; *leaves* linear to narrowly elliptic, 1–3 cm × 2–7 mm; *inflorescence* solitary or in clusters of 2–3; *sepals* 3–5 mm long (to 8 mm long in fruit), tomentose above; *petals* white, 13–20 mm long, clawed; *stamens* with petaloid filaments; *capsules* 8–15 mm long. Found on dry hills, usually with oak, pinyon pine, and juniper communities, 4000–8300 ft. April–July. W.

FENDLERELLA Heller – FENDLER-BUSH

Shrubs with exfoliating bark; *leaves* opposite, deciduous, entire, small; *sepals* 5; *petals* 5, white to cream; *stamens* usually 10; *styles* 3 (4); *ovary* inferior; *fruit* a septicidal, 3-valved capsule.

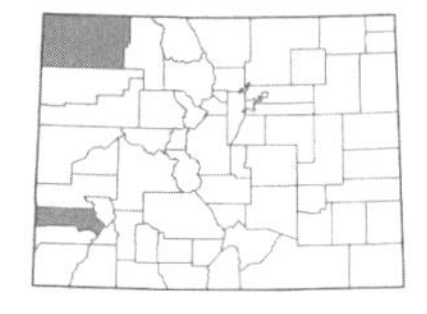

Fendlerella utahensis **(S. Watson) Heller**, UTAH FENDLER-BUSH; YERBA DESIERTO. Shrubs to 1 m; *leaves* linear to narrowly elliptic, 0.4–1.2 cm long, strigose, at least some hairs pustulate-based; *inflorescence* of compound, bracteolate cymes with numerous flowers; *sepals* 1–1.5 mm long; *petals* white, 2–4 mm long; *capsules* 3–4 mm long. Found in sandy soil with pinyon pine and juniper communities or on rocky cliffs and in rock crevices, 5100–8000 ft. June–Aug. W.

JAMESIA Torr. & A. Gray – CLIFFBUSH

Shrubs with exfoliating bark; *leaves* opposite, deciduous, simple with serrate or rarely entire margins; *sepals* 4 or 5; *petals* 4 or 5, white or pink; *stamens* 8 or 10; *styles* (2) 3–5; *ovary* partly inferior with parietal placentation; *fruit* a septicidal capsule beaked by long-persistent styles.

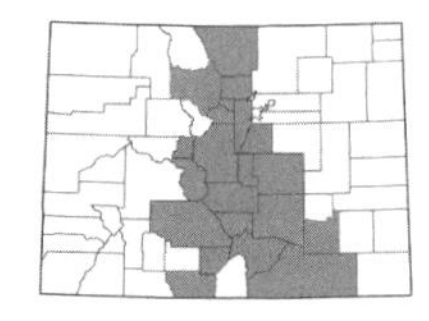

Jamesia americana **Torr. & A. Gray var. *americana***, FIVEPETAL CLIFFBUSH. Shrubs, 3–15 dm; *leaves* elliptic to ovate, 0.7–5 cm long, green above and gray- or white-tomentose below, the margins toothed; *inflorescence* few in cymes; *sepals* 3–4 mm long; *petals* white, 5–11 mm long, clawed; *capsules* 4–5 mm long. Found on rocky slopes, cliffs, and ledges, often in rock crevices, 5400–10,700 ft. May–July. E/W. (Plate 47)

PHILADELPHUS L. – MOCK ORANGE

Shrubs with exfoliating bark; *leaves* opposite, deciduous, simple with entire or toothed margins; *sepals* 4 (5), connate below; *petals* 4 (5), white; *stamens* 13–90; *styles* 4, rarely 3 or 5; *ovary* inferior or partly inferior; *fruit* a loculicidal capsule.

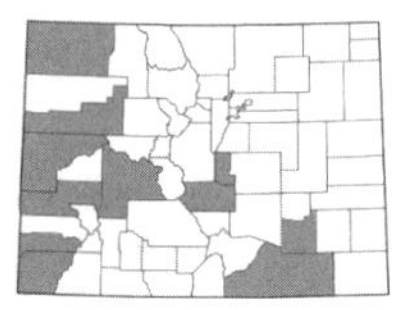

Philadelphus microphyllus **A. Gray**, LITTLELEAF MOCK ORANGE. [*P. argenteus* Rydb.; *P. occidentalis* A. Nelson]. Shrubs to 20 dm; *leaves* linear to ovate or narrowly elliptic, 0.4–3 cm long, glabrous to slightly hairy, the margins entire and slightly revolute; *sepals* 2–5 mm long; *petals* white, 8–15 mm long; *capsules* 6–8 mm long. Found on rocky cliffs and slopes, usually with oak, juniper, or pinyon pine communities, 5000–9500 ft. May–July. E/W. (Plate 47)

HYDROCHARITACEAE A.L. Juss. – FROGBIT FAMILY

Annual or perennial, monoecious or dioecious aquatic herbs; *leaves* basal, alternate, opposite, or whorled, sometimes stipulate; *flowers* imperfect or perfect, actinomorphic or rarely slightly zygomorphic; *sepals* mostly 3, distinct; *petals* 3 or absent, distinct; *stamens* 2–many, or absent; *ovary* inferior, 2–6-carpellate; *fruit* an achene or berrylike.

1a. Leaves opposite or whorled, often crowded at the branch tips, 5–30 mm long and 0.5–3 mm wide, entire to minutely toothed; inflorescence with a spathe, flowers pedunculate; fruit berrylike...***Elodea***

1b. Leaves mostly opposite, evenly spaced along the stem and not crowded toward the tip, only a few whorled, mostly less than 20 mm long and 0.5–1 mm wide, coarsely toothed or sometimes entire; flowers sessile; fruit fusiform, conspicuously areolate with areolae in longitudinal rows ...***Najas***

ELODEA Michx. – WATERWEED

Perennials, dioecious, aquatic herbs; *leaves* whorled or opposite, sessile; *flowers* projected to the surface of the water by an elongated floral tube base. (Cook & Unmi-König, 1985; St. John, 1962)

1a. Leaves mostly opposite at the upper nodes, to 30 mm long, the margins minutely serrate...***E. bifoliata***

1b. Leaves mostly in 3's or 4's at the upper nodes, to 15 mm long, the margins entire...2

2a. Middle and upper leaves 0.5–1.5 (2) mm wide, the leaves not particularly crowded at the branch tips; staminate flowers breaking from the pedicels at maturity...***E. nuttallii***

2b. Middle and upper leaves 1.5–3 mm wide, the leaves densely crowded at the branch tips; staminate flowers remaining attached at maturity...***E. canadensis***

***Elodea bifoliata* St. John**, TWOLEAF WATERWEED. [*E. longivaginata* H. St. John]. Leaves mostly opposite, to 30 mm long, 1–3 mm wide; *staminate flowers* with sepals 3.5–4 mm long and petals ca. 5 mm long; *pistillate flowers* with sepals 2.5–3 mm long; *capsules* 8–10 mm long; *seeds* 2.8–6 mm long, long-hairy. Found in ponds and ditches, 5000–8500 ft. June–Aug. E/W.

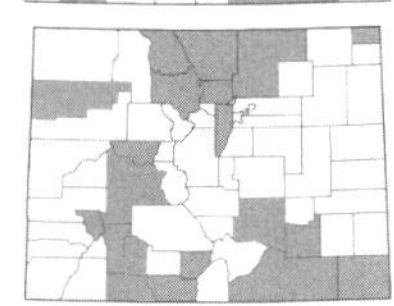

***Elodea canadensis* Michx.**, CANADIAN WATERWEED. Leaves mostly in 3's (at least above), to 15 mm long, 1.5–3 mm wide; *staminate flowers* 4–5 mm long; *pistillate flowers* with sepals 2–2.5 mm long; *capsules* 5–6 mm long; *seeds* ca. 4.5 mm long. Found in lakes and ponds, 4000–9500 ft. June–Aug. E/W.

***Elodea nuttallii* (Planch.) St. John**, WESTERN WATERWEED. Leaves in 3's or 4's above, to 10 (12) mm long, 0.5–1.5 (2) mm wide; *staminate flowers* ca. 2 mm long; *pistillate flowers* ca. 1 mm long; *capsules* ovoid; *seeds* hairy. Found in lakes, ponds, and creeks, 5000–8500 ft. June–Aug. E/W.

NAJAS L. – NAIAD; WATER-NYMPH

Annuals, dioecious aquatic herbs; *leaves* opposite or appearing whorled, sessile; *perianth* absent.

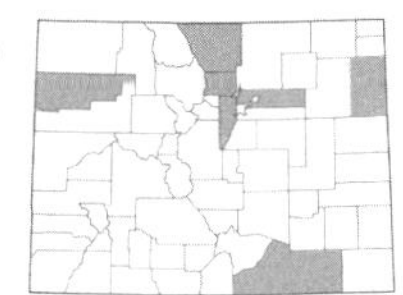

Najas guadalupensis* (Spreng.) Magnus ssp. *guadalupensis, SOUTHERN WATER-NYMPH. Leaves mostly opposite (sometimes appearing whorled), linear, to 20 mm long and 1 mm wide, toothed or sometimes entire; *flowers* sessile, axillary; *fruit* fusiform, the seeds conspicuously areolate with areolae in longitudinal rows. Found in slow-moving ditches, streams, and ponds, 4300–6000 ft. July–Sept. E/W. Introduced. (Plate 48)

HYDROPHYLLACEAE R. Br. ex Edwards – WATERLEAF FAMILY

Annual, biennial, or perennial herbs; *leaves* alternate or opposite, simple or pinnately compound or dissected; *flowers* perfect, actinomorphic; *sepals* 5, connate at the base or distinct; *corolla* 5-lobed; *stamens* 5, epipetalous, the filaments usually with basal appendages; *pistil* 1; *ovary* 2-carpellate, superior, with intrusive parietal placentation giving the ovary the appearance of having 2 locules although only 1 is present; *fruit* a capsule.

1a. Leaves entire...2

1b. Leaves with crenate margins or pinnately compound, lobed, or dissected...4

2a. Flowers solitary on long peduncles; leaves all basal; corolla rotate or saucer-shaped with widely spreading lobes, usually white with a yellowish, densely hairy center...***Hesperochiron***

2b. Flowers in cymes or if solitary then sessile or nearly so; leaves not all basal; corolla tubular to funnelform, variously colored, not densely hairy in the center...3

3a. Leaves narrowly oblanceolate or linear, to about 5 mm wide; flowers solitary...***Nama***

3b. Leaves ovate, elliptic, or lanceolate, over 5 mm wide; flowers more than one and arranged in a cyme...***Phacelia***

4a. Flowers solitary in or opposite the leaf axils; petals to 1 cm long, usually shorter than the calyx; stamens included...5

4b. Flowers in capitate or helicoid cymes, or sometimes in few-flowered cymes; petals various; stamens exserted or sometimes included...6

5a. Reflexed or spreading auricles (small lobes) present in between the calyx lobes; style cleft only at the tip...***Nemophila***
5b. Auricles lacking; style cleft to half its length...***Ellisia***

6a. Flowers in globose, capitate cymes; plants from fibrous roots...***Hydrophyllum***
6b. Flowers in elongate, helicoid cymes or few-flowered cymes; plants from a taproot...***Phacelia***

ELLISIA L. – ELLISIA

Annual herbs; *leaves* simple, pinnately dissected or divided, alternate above and opposite below; *flowers* solitary in the leaf axils; *corolla* white to purple; *stamens* included; *style* cleft to half its length; *capsules* globose.

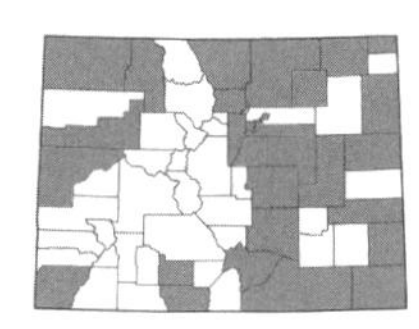

***Ellisia nyctelea* (L.) L.**, AUNT LUCY. Plants 0.5–4 dm, with diffusely branched, retrorsely hispid stems; *leaves* pinnately divided into lanceolate or linear-oblong segments; *calyx* 3–4 mm long at flowering, 5–7 mm long in fruit, hirsute near the margins; *corolla* campanulate; *capsules* 5–6 mm diam.; *seeds* ca. 4, globose. Found in moist, shaded places, meadows, open slopes, and along roadsides, sometimes in disturbed places, 3500–9000 ft. May–July. E/W. (Plate 48)

HESPEROCHIRON S. Watson – HESPEROCHIRON

Perennial herbs; *leaves* in a basal rosette, petiolate, entire to subentire; *flowers* solitary on long peduncles; *corolla* white, often tinged with purple, densely hairy within (ours); *stamens* included; *capsules* loculicidal.

***Hesperochiron pumilus* (Griseb.) Porter**, DWARF HESPEROCHIRON. Plants 2–12 cm; *leaves* linear to oblanceolate, 1.5–5 cm long; *corolla* 0.5–1.5 cm long; *seeds* numerous. Uncommon on moist, open slopes, known from one collection in Montezuma Co. made in 1914 [Milk Ranch on House Creek, *J. Ward Emerson 764* (RM)]. The site on which it was collected is perhaps near or even under what is now McPhee Reservoir, 5000–6000 ft. April–June. W.

HYDROPHYLLUM L. – WATERLEAF

Biennial or perennial herbs from fibrous roots and rhizomes; *leaves* alternate, mostly basal, petiolate, pinnately lobed to pinnately compound; *flowers* in capitate cymes; *calyx* divided nearly to the base; *corolla* white to purple; *stamens* exserted. (Beckmann, 1979)

1a. Leaf lobes sharply toothed along the margins; inflorescences subequal to or above the leaves...***H. fendleri* var. *fendleri***
1b. Leaf lobes with entire margins and usually with 1–2 deeply cleft lobes at the tips; inflorescences much shorter than the leaves...***H. capitatum* var. *capitatum***

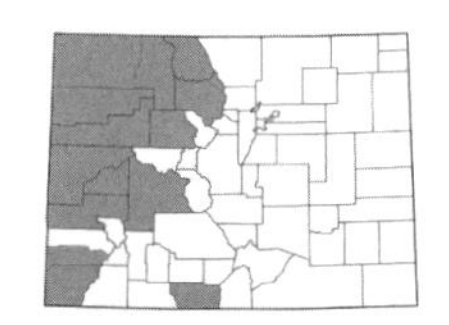

Hydrophyllum capitatum* Douglas ex Benth. var. *capitatum, WOOLLEN-BREECHES. Plants 1–5 dm; *leaves* 2.5–10 cm long, the lobes with entire margins and usually with 1–2 deeply cleft lobes at the tips; *peduncles* 1–5 cm long, shorter than the leaves; *corolla* 5–10 mm long; *seeds* ca. 2. Uncommon in forests and on open slopes, usually in moist places, 6200–10,500 ft. May–July. E/W.

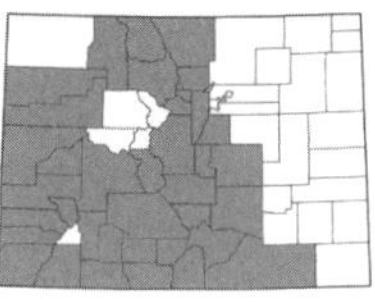

Hydrophyllum fendleri* (A. Gray) A.A. Heller var. *fendleri, FENDLER'S WATERLEAF. Plants 2–10 dm; *leaves* 6–30 cm long, the lobes sharply toothed along the margins; *peduncles* 3–20 cm long, usually longer than the leaves; *corolla* 6–10 mm long; *seeds* 1–3. Common in moist, often shady places, scattered across the state at 5500–11,500 ft. May–Aug. E/W. (Plate 48)

NAMA L. – FIDDLELEAF

Annuals (ours); *leaves* alternate, entire; *flowers* solitary or in small terminal cymes; *corolla* white to purple; *stamens* included. (Hitchcock, 1933)

1a. Plants low, prostrate or spreading, dichotomously branched from the base; hairs spreading-hispid, lacking shorter, fine retrorse hairs underneath; style undivided or nearly so...***N. densum* var. *parviflorum***
1b. Plants erect, simple or branched above; plants minutely glandular-hairy, or with spreading, hispid hairs and shorter, retrorse hairs underneath; style cleft to the base...2

2a. Plants minutely glandular-hairy; flowers pale lavender to light blue or white...***N. dichotomum***
2b. Plants with spreading-hispid hairs with shorter, retrorse hairs underneath; flowers reddish-violet...***N. retrorsum***

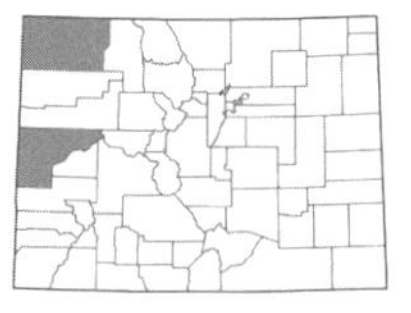

***Nama densum* Lemmon var. *parviflorum* (Greenm.) C.L. Hitchc.**, MATTED NAMA. Plants prostrate or spreading, dichotomously branched from the base, with spreading-hispid hairs; *leaves* oblanceolate, 4–30 × 1–5 mm; *corolla* white to lavender, 3–7 mm long, with a long tube and short lobes; *style* undivided or nearly so; *capsules* 2–4 mm long; *seeds* ca. 15, 0.5–1 mm long, reticulate and pitted. Found in dry, sandy soil, 5500–6500 ft. May–Aug. W.

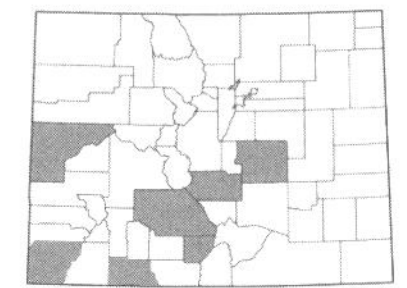

***Nama dichotomum* Choisy**, WISHBONE FIDDLELEAF. Plants erect, simple or branched above, 5–20 cm tall, minutely glandular-hairy; *leaves* linear to narrowly oblanceolate, 6–20 mm long; *corolla* white to light blue or pale lavender, 3–8 mm long with short lobes 1–2 mm; *style* cleft to the base; *capsules* 2–6 mm long; *seeds* 0.5–0.7 mm long, cross-ridged, pitted. Found in sandy soil and sandstone, 5300–7700 ft. June–Sept. E/W.

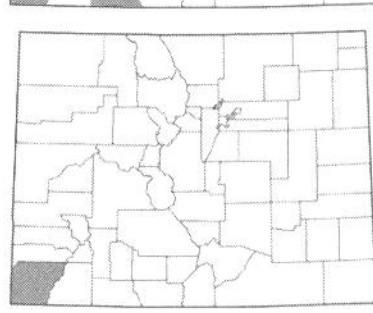

***Nama retrorsum* J.T. Howell**, BETATAKIN FIDDLELEAF. Plants erect, 10–30 cm tall, with spreading-hispid hairs and shorter, retrorse hairs underneath; *leaves* linear to narrowly oblanceolate, 15–50 × 2–5 mm; *corolla* reddish-violet, 4–7 mm long; *seeds* 0.6–0.9 mm long. Found on sandy soil and sandstone, 5000–5500 ft. May–Aug. W.

NEMOPHILA Nutt. – BABY BLUE EYES

Annual herbs from taproots; *leaves* opposite or alternate, pinnately dissected; *flowers* solitary or in a few-flowered terminal cyme; *corolla* white, bluish, or purple; *stamens* included.

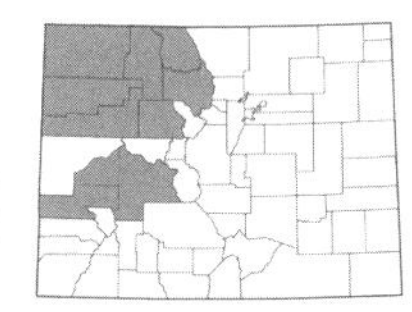

***Nemophila breviflora* A. Gray**, WOODLOVE. Plants weak-stemmed, stems 0.5–2 dm long; *leaves* pinnately divided into linear or lanceolate lobes ca. 3 mm long, sparsely hispid; *calyx* ca. 3 mm long, with reflexed or spreading auricles (small lobes) present in between the lobes; *corolla* 1.5–3 mm wide, shorter than the calyx; *style* cleft at the apex; *seed* ca. 1, globose, 2–4 mm diam. Found in meadows, moist forests, and along streams, 6500–9000 ft. May–July. E/W.

PHACELIA Juss. – PHACELIA

Annual, biennial, or perennial herbs; *leaves* usually alternate, entire to pinnately dissected; *flowers* in helicoid cymes or rarely solitary and terminal; *corolla* white, pink, blue, purple, or yellow; *stamens* included or exserted. (N.D. Atwood, 1975; Constance, 1949)

1a. Plants low, delicate annuals, usually branched from the base and lacking a definite central axis, and 2 dm or less in height; stamens included within the corolla; leaves mostly less than 2 cm long; plants of the western slope...2
1b. Perennials or if annuals then not low and delicate and usually with a main single stem, mostly 1–12 dm in height; stamens exserted from the corolla by at least 2 mm or rarely included and then the corolla lobes with toothed tips; leaves mostly longer than 2 cm; plants variously distributed...5

2a. Leaves all deeply pinnatifid into small, slender segments; flowers white; seeds cross-corrugated with horizontal ridges...***P. ivesiana***
2b. Leaves entire or some coarsely toothed or with a few lateral lobes; flowers white, yellow, or blue; seeds cross-corrugated or not...3

3a. Corolla (5) 6–11 mm long, evidently longer than the calyx, light or more often deep purple with a light yellow tube; plants glandular-hairy throughout...***P. demissa* var. *demissa***
3b. Corolla 2.5–4.5 mm long, about equaling the calyx in length, white or yellow or sometimes bluish; plants spreading-hairy, sometimes glandular-hairy as well...4

4a. Seeds pitted-reticulated but not cross-corrugated with horizontal ridges; stems with numerous gland-tipped hairs in addition to softly spreading hairs...***P. incana***
4b. Seeds pitted-reticulated as well as cross-corrugated with horizontal ridges; stems a few gland-tipped hairs or these lacking, in addition to softly spreading hairs...***P. scopulina***

5a. Leaves entire or sometimes a few of the lower with a pair of lateral lobes...6
5b. Leaves pinnatifid or with crenate or toothed margins...7

6a. Leaves usually all entire; plants from a taproot that is usually surmounted by a branching caudex; stems usually several...***P. hastata***
6b. Usually at least some of the lower leaves with a pair of lateral lobes at the base; plants from a taproot, without a branching caudex; usually with a single stem...***P. heterophylla***

7a. Perennials from a taproot and woody, often branching caudex; ovules and seeds 8–18; inflorescence a spicate thyrse composed of short helicoid cymes; flowers purple; stamens long-exserted about 2–3 times the length of the corolla...***P. sericea***
7b. Biennials or annuals from a taproot, lacking a woody caudex; ovules and seeds 4 or sometimes only 2 seeds maturing; inflorescence various; flowers blue, purple, or white; stamens included to long-exserted...8

8a. Stamens included; corolla 3.5–4.5 mm long, the lobes with sharply toothed apices...***P. denticulata***
8b. Stamens exserted by at least 2 mm from the corolla; corolla 4–10 mm long, the lobes entire to crenate or erose-fimbriate (fringed) at the tips but not sharply toothed...9

9a. Leaves shallowly crenate or deeply lobate with broad, rounded lobes that do not reach to the midrib...10
9b. Leaves pinnately or bipinnately divided with at least the lower lobes reaching to the midrib...12

10a. Flowers whitish to pale lavender, (4) 5–6 mm long; inflorescence narrow, the cymes short and crowded along the main axis and intermixed with well-developed leaves; filaments usually not blue or purplish...***P. constancei***
10b. Flowers blue-lavender to lavender, sometimes with a white base, 6–10 mm long; inflorescence branched, otherwise unlike the above; filaments usually blue or purplish...11

11a. Ventral surface of the seeds corrugated on the margins and on one side of the ridge; leaves with rounded lobes usually reaching more than halfway to the midrib...***P. crenulata* var. *corrugata***
11b. Ventral surface of the seeds lacking corrugations along the ventral ridge; leaves with rounded lobes not reaching more than about halfway to the midrib...***P. integrifolia* var. *integrifolia***

12a. Corolla white or pale lavender, 4–5 mm long, the lobes erose-fimbriate at the tips; leaves subbipinnatifid with the segments toothed or pinnatifid again into narrow segments...***P. alba***
12b. Corolla purple or blue, (4) 5–8 mm long, the lobes entire to crenate at the tips; leaves unlike the above...13

13a. Leaves glabrous or nearly so, rather thick and succulent; flowers bright blue to purplish-blue with a yellow or white tube...***P. splendens***
13b. Leaves evidently hairy or glandular-hairy, not succulent; flowers purple to dark blue, not bicolored...14

14a. Leaves lacking dark-tipped glandular hairs, densely hirsute-pubescent; seeds excavated on both sides of the ventral ridge...***P. deserta***
14b. Leaves with scattered to numerous dark-tipped glandular hairs, also usually with hirsute hairs as well; seeds excavated on both sides of the ventral ridge or not...15

15a. Seeds with a dorsal flap on one side of the ventral ridge but lacking excavations on both sides of the ventral ridge; style hairy on the lower ⅓; anthers greenish...***P. bakeri***
15b. Seeds excavated on both sides of the ventral ridge; style minutely hairy throughout or hairy on the lower ¼; anthers yellow or sometimes greenish...16

16a. Seeds dark brown; plants usually much-branched with a stout stem, usually biennials...***P. formosula***
16b. Seeds reddish-brown; plants usually few-branched with a slender stem, annuals or biennials...17

17a. Corolla 5–7 mm long; calyx 3–4 mm long; style hairy on the lower ¼ ...***P. glandulosa***
17b. Corolla ca. 4.5 mm long; calyx 4.2–5.5 mm long; style hairy throughout...***P. gina-glenneae***

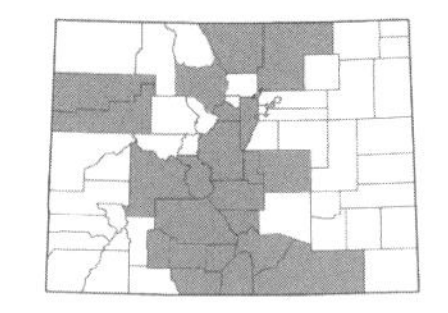

***Phacelia alba* Rydb.**, WHITE PHACELIA. Annuals, 1–7 dm; *stems* with numerous soft, short, spreading hairs and fewer longer, spreading hairs, glandular at the top and in the inflorescence; *leaves* pinnatifid; *calyx* 2–4 mm long; *corolla* 4–5 mm long, white to lavender or pale blue, the lobes erose-dentate; *seeds* (2) 4, 2.5–3.2 mm long, brown or black, alveolate, not corrugated, deeply excavated on each side of the conspicuous ventral ridge. Locally common on rocky and gravelly slopes, meadows, and along roadsides, 5600–10,200 ft. June–Aug. E/W. (Plate 48)

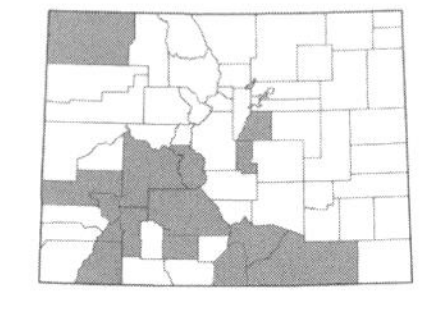

***Phacelia bakeri* (Brand) J.F. Macbr.**, BAKER'S PHACELIA. Annuals, 0.5–4.8 dm; *stems* pilose to somewhat hirsute with multicellular stipitate glands; *leaves* 2–8 cm long, pinnately divided, the lobes crenate to dentate; *calyx* 1–1.5 mm longer than the capsule; *corolla* 7–8 mm long, violet to dark blue, pubescent; *seeds* 4, 2.7–3 mm long, brown, pitted, lacking excavations on either side of the ventral ridge. Found on gravelly, usually dry slopes and along roadsides, 6500–12,600 ft. July–Sept. E/W. (Plate 48)

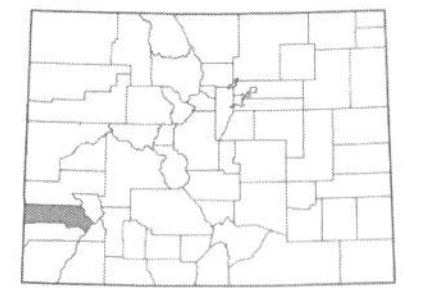

***Phacelia constancei* N.D. Atwood**, CONSTANCE'S PHACELIA. Biennials, 1–6 dm; *stems* with short, spreading, gland-tipped hairs and longer, spreading, eglandular hairs; *leaves* deeply lobate with broad, rounded lobes, or the margins simply crenate; *calyx* 3–4 mm long; *corolla* (4) 5–6 mm long, pale lavender or whitish; *seeds* 4, 2.5–4.2 mm long, brown to black, alveolate, the ventral ridge corrugated along one side. Found on gypsum outcroppings, currently known only from Big Gypsum Valley, 6400–6600 ft. July–Sept. W.

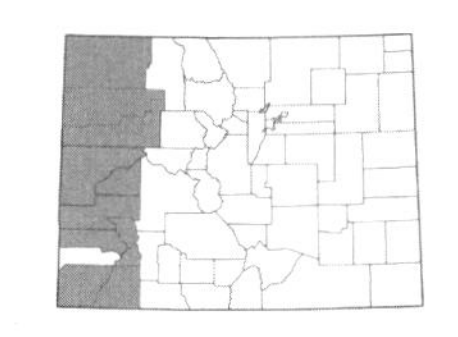

Phacelia crenulata **Torr. ex S. Watson var. *corrugata* (A. Nelson) Brand**, HELIOTROPE PHACELIA. Annuals, 0.5–4 (8) dm; *stems* densely glandular-stipitate, interspersed with short, spreading, eglandular hairs; *leaves* crenately pinnatifid with rounded, shallow lobes; *calyx* 4–5.5 mm long; *corolla* 6–10 mm long, blue-lavender to purple, sometimes proximally whitish; *seeds* (2) 4, (2.5) 2.7–4.5 mm long, dark brown with paler margins, alveolate-pitted, corrugated margins, deeply excavated on either side of the ventral ridge. Found on sandstone bluffs, gravelly hillsides, and clay slopes, 4600–6000 ft. April–June. W. (Plate 48)

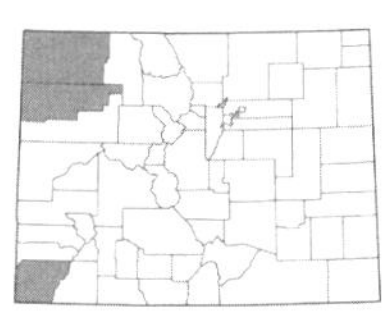

Phacelia demissa **A. Gray var. *demissa***, BRITTLE PHACELIA. Annuals to 1.5 dm; *stems* minutely hairy or villous with glandular hairs; *leaves* rounded to rounded-ovate with entire margins; *calyx* 2–5 mm long; *corolla* (5) 6–11 mm long, bright to pale lavender limb with a light yellow tube; *seeds* less than or equal to the number of ovules, rarely 4, 1–1.7 mm long, pitted-reticulate. Uncommon on clay hills and barren shale slopes, 4900–8800 ft. April–June. W.

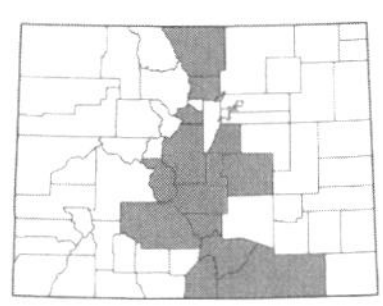

Phacelia denticulata **Osterh.**, ROCKY MOUNTAIN PHACELIA. Annuals, 0.5–5.4 dm; *stems* bristly and glandular-stipitate; *leaves* 1–7.5 cm long, pinnately cleft or divided; *calyx* 2.5 mm long; *corolla* 3.5–4.5 mm long, light blue, the lobes denticulate; *seeds* 4, alveolate, slightly excavated on either side of the curved ventral ridge. Uncommon in sandy or rocky soil, 5500–10,000 ft. June–July. E.

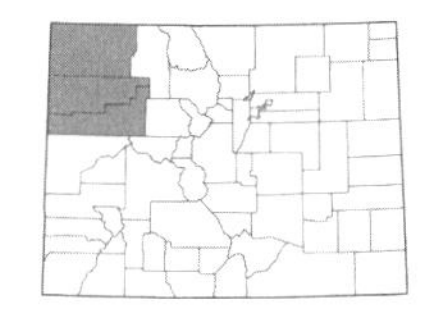

Phacelia deserta **A. Nelson**, DESERT PHACELIA. [*P. glandulosa* Nutt. var. *deserta* (A. Nelson) Brand]. Annuals or biennials, 0.6–3 dm; *stems* densely hirsute; *leaves* to 7 cm long, pinnatifid, densely hirsute; *calyx* 3–4 mm long; *corolla* 5–7 mm long, blue to purplish-blue, the lobes rounded and entire; *seeds* 4, 2.5–3 mm long, excavated on either side of the ventral ridge. Found on shale slopes, rocky or sandy hillsides, and along roadsides, 5900–8600 ft. June–Aug. W.

Phacelia formosula **Osterh.**, NORTHPARK PHACELIA. Biennials (rarely annuals), 1.5–2.2 dm; *stems* somewhat grayish, hirsute with glandular hairs; *leaves* pinnately divided; *calyx* 3.2–3.8 mm long; *corolla* ca. 6 mm long, violet, slightly glandular and pilose; *seeds* 4, 2.5–3 mm long, dark brown, pitted, excavated on either side of the ventral ridge, the margins rounded and smooth. Rare on shale slopes and sandstone outcroppings, 7900–8500 ft. June–Oct. E. Endemic. (Plate 48)

This species is listed as endangered under the Endangered Species Act.

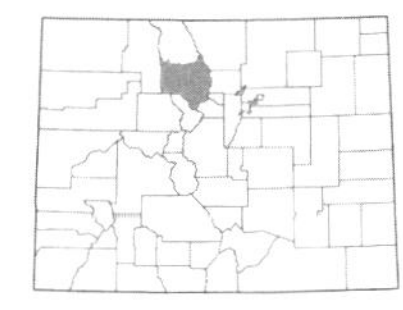

Phacelia gina-glenneae **N.D. Atwood & S.L. Welsh**, GINA'S PHACELIA. Annuals, 0.5–1.2 (1.6) dm; *stems* usually solitary or with 2–6 lateral branches from the base, glandular; *leaves* irregularly lobed to pinnatifid; *calyx* 4.2–5.5 mm long; *corolla* ca. 4.5 mm long; *seeds* 2.6–3.3 mm long, reddish-brown, pitted, excavated on either side of the ventral ridge, the margins entire. Rare on shale or clay slopes, 7400–7700 ft. June–Oct. W. Endemic.

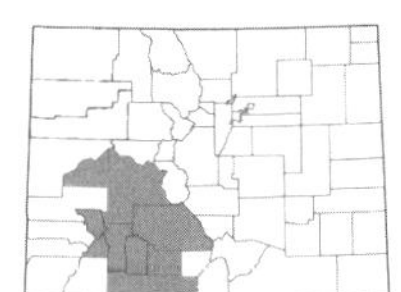

Phacelia glandulosa **Nutt.**, GLANDULAR PHACELIA. Annuals or biennials, 0.6–3.5 dm; *stems* densely hirsute with glandular hairs; *leaves* 1–7 cm long, pinnatifid; *calyx* 3–4 mm long; *corolla* 5–7 mm long, purple to bluish, the lobes crenate and pubescent; *seeds* 4, 2.4–3.3 mm long, reddish-brown, pitted, excavated on either side of the ventral ridge. Found on rocky slopes and along roadsides, 8500–9500 ft. June–Aug. W. (Plate 48)

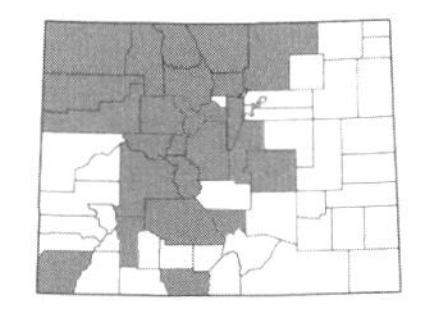

Phacelia hastata **Douglas ex Lehm.**, SILVERLEAF PHACELIA. Perennials to 5 dm; *stems* silvery with short, fine hairs, sometimes with ascending or appressed bristles; *leaves* entire or occasionally with a small pair of lateral lobes; *corolla* 4–7 mm long, dull whitish to lavender or dull purple; *seeds* 1–2 (4), 2–2.5 mm long. Common in meadows and on dry slopes, 5500–11,000 ft. June–Aug. E/W. (Plate 48)

This species and *P. heterophylla* belong to a large polyploid complex and might be better treated as a single species [*P. magellanica* (Lam.) Cov.] with many infraspecific taxa.

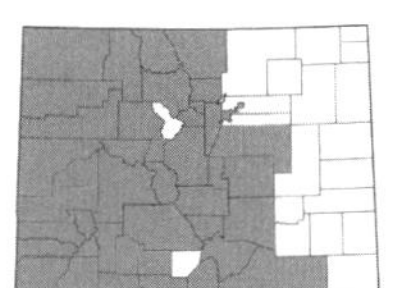

Phacelia heterophylla **Pursh**, WAND PHACELIA. Perennials, 2–12 dm; *stems* green or grayish, sometimes silvery, with long, spreading bristles and short, spreading, glandular hairs; *leaves* entire or with a pair of lateral lobes; *corolla* 3–6 mm long, dull whitish to ochroleucous or sometimes purplish; *seeds* 1–2 (4), 2–2.5 mm long. Common in meadows, on dry slopes, and along roadsides, 5200–12,000 ft. May–Aug. E/W.

Phacelia incana **Brand**, HOARY PHACELIA. Annuals to 1.5 dm; *stems* with soft, spreading, often glandular-stipitate hairs; *leaves* elliptic to ovate with entire margins; *calyx* 3.5–4.5 mm long; *corolla* 3.5–4.5 mm long, the tube white or yellowish, the limb white or bluish, hairy; *seeds* 16–24, 0.6–1 mm long, pitted-reticulate. Uncommon on shale slopes, 5500–6000 ft. May–June. W. (Plate 48)

Phacelia integrifolia* Torr. var. *integrifolia, GYPSUM PHACELIA. Annuals, 1–5 dm; *stems* densely glandular-stipitate or with short, spreading, sometimes glandular-stipitate hairs; *leaves* crenate with shallow lobes; *calyx* 3.5–4.5 mm long; *corolla* 6–8 mm long, dull blue-lavender; *seeds* 4, 3.2–4.4 mm long, dark brown to black with paler margins, alveolate, deeply excavated on either side of the curved ventral ridge, with distinct dorsal transverse ridges. Uncommon in sandy soil, barely entering Colorado near Four Corners, 5000 ft. April–July. W.

***Phacelia ivesiana* Torr.**, IVES' PHACELIA. Annuals to 2.5 dm; *stems* with spreading hairs, often finely glandular distally; *leaves* deeply pinnatifid, the lobes small, slender, few-toothed, or lobed; *calyx* 2.5–4 (4.5) mm long; *corolla* 2.5–4 (4.5) mm long, white, often with a yellowish tube or throat; *seeds* 1–1.5 mm long, alveolate-reticulate and cross-corrugated. Locally common in sandy or clay soil, often in dry washes, 4800–6500 ft. April–June. W.

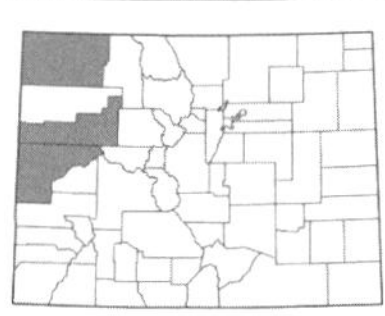

***Phacelia scopulina* (A. Nelson) J.T. Howell**, DEBEQUE PHACELIA. Annuals; *stems* with short, spreading hairs; *leaves* entire, coarsely crenate, pinnately lobed, or pinnatifid; *calyx* equal to or longer than the corolla; *corolla* 2.5–4.5 mm long, pale to deep yellow, hairy externally; *seeds* 8–25, ca. 1.2–2 mm long, pitted-reticulate and cross-corrugated. Found on barren clay slopes, 4700–6200 ft. April–June. W. (Plate 48)

There are two varieties of *P. scopulina* in Colorado:

1a. Capsules apiculate at the tip; corolla pale to deep yellow throughout...**var. *scopulina***. Known from Moffat Co.
1b. Capsules barely or not apiculate at the tip; corolla with a light yellow tube and whitish lobes...**var. *submutica* (J.T. Howell) R.R. Halse**. [*P. submutica* J.T. Howell]. Known from Garfield and Mesa cos. Endemic.

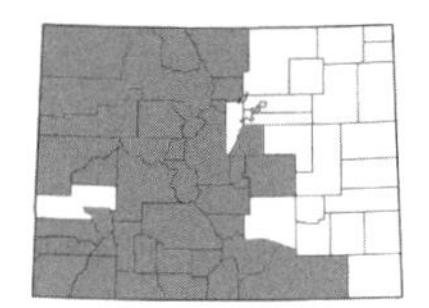

***Phacelia sericea* (Graham ex Hook.) A. Gray**, SILKY PHACELIA. Perennials, 1–4 dm; *stems* thinly strigose, loosely short-hairy, densely sericeous, or loosely woolly, with spreading hairs in the inflorescence; *leaves* pinnatifid with entire or cleft lobes; *corolla* 5–7 mm long, purple, hairy; *seeds* 8–18, 1–2 mm long, pitted-reticulate, with narrow ridges separating the longitudinal rows of alveolae. Common in meadows, on open slopes, and in alpine, 6500–13,500 ft. June–Aug. E/W. (Plate 48)

There are two varieties of *P. sericea* in Colorado:

1a. Plants usually densely hairy; leaf segments narrow; petioles lacking strongly ciliate hairs...**var. *sericea***. Common throughout the state at 6500–13,500 ft.
1b. Plants usually thinly hairy; leaf segments broader; petioles usually strongly ciliate...**var. *ciliosa* Rydb.** Known mostly from the northwestern counties, 6700–9800 ft.

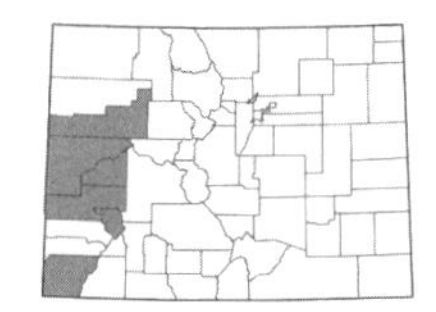

***Phacelia splendens* Eastw.**, PATCH PHACELIA. Annuals, 0.5–2.7 dm; *stems* minutely hairy, with scattered glandular-stipitate hairs; *leaves* 2–7.5 cm, pinnatifid; *calyx* 2.5–3 mm long; *corolla* 4–8 mm long, bright blue with a yellow tube, glabrous to sparsely hairy; *seeds* 4, 3–4 mm long, finely honeycomb-pitted, excavations on either side of the ventral ridge, the margins mostly revolute. Found on barren clay or shale slopes, 4500–6000 ft. April–June. W.

HYPOXIDACEAE R. Br. – STAR-GRASS FAMILY

Perennial herbs from corms or rhizomes; *leaves* basal; *flowers* actinomorphic, in few-flowered racemes; *tepals* 6, all alike and petaloid, yellow; *stamens* 6; *pistil* 1; *ovary* inferior, usually densely hairy; *fruit* a capsule.

HYPOXIS L. – STAR-GRASS

With characteristics of the family.

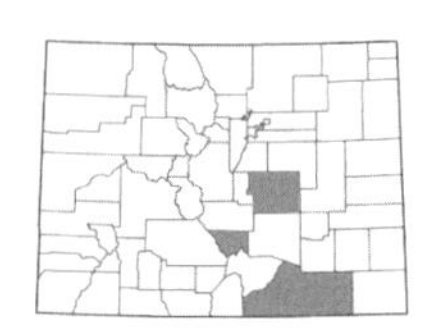

***Hypoxis hirsuta* (L.) Coville**, COMMON GOLDSTAR. Plants to 3 dm from a corm 8–20 mm wide; *leaves* 3–6, linear, 5–30 cm long; *tepals* 6–15 mm long, hairy below; *seeds* black. Rare in moist meadows and fens, 5000–7800 ft. April–July. E.

IRIDACEAE Juss. – IRIS FAMILY

Perennial herbs from rhizomes, corms, or bulbs; *leaves* alternate, usually distichous, entire, equitant; *flowers* perfect, actinomorphic (ours), usually enclosed in 2 spathes (bracts); *tepals* 6, in 2 whorls of 3 each; *stamens* 3, the filaments distinct or sometimes connate into a basal tube; *pistil* 1; *ovary* inferior, usually with 3 locules; *fruit* a loculicidal capsule.

1a. Outer tepals 4–8 cm long, in dissimilar whorls, the outer three spreading or reflexed and usually bearded in the middle with yellowish multicellular hairs, the inner three erect; style branches petaloid; plants from thick rhizomes...***Iris***
1b. Tepals to 3 cm long, all similar in appearance, all spreading; style branches not petaloid; plants from fibrous roots...***Sisyrinchium***

IRIS L. – IRIS
Perennial herbs from rhizomes; *leaves* mostly basal, linear to lanceolate; *tepals* in dissimilar whorls, the outer 3 sepal-like, often bearded with multicellular hairs, and spreading or reflexed, the inner 3 petal-like and erect; *stamens* distinct; *style branches* 3, petaloid, covering the stamens, divided at the tip into 2 lobes.
1a. Flowers purple, blue-purple, or rarely white...***I. missouriensis***
1b. Flowers yellow...***I. pseudacorus***

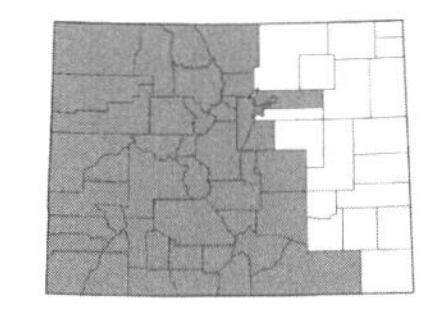

Iris missouriensis **Nutt.**, Rocky Mountain iris. Plants 2–6 dm; *leaves* 20–30 (45) cm × 4–8 mm; *flowers* blue-purple to lilac or rarely all white; *outer tepals* 4–6 cm long; *capsules* oblong, 3–5 cm long; *seeds* 4–5 mm long. Common in moist meadows, along streams, and in aspen forests, scattered across the state, 5200–11,000 ft. May–Aug. E/W. (Plate 49)

Iris pseudacorus **L.**, yellow iris. Plants to 10 dm; *flowers* yellow; *outer tepals* 4.5–8 cm long; *capsules* elliptic or cylindric, 5–8 cm long; *seeds* 6–7 mm long. Uncommon near streams along the Front Range where it has escaped from cultivation, 5000–5500 ft. May–July. E. Introduced.

SISYRINCHIUM L. – BLUE-EYED GRASS
Perennial herbs; *leaves* linear, grass-like; *spathes* equitant with scarious margins; *tepals* with similar whorls, blue-violet to white, lavender, or rose-pink; *stamens* distinct or basally connate into a tube; *styles* 3, not petaloid. (Cholewa & Henderson, 1984; Henderson, 1976)
1a. Outer and inner spathe bracts equal or nearly so in length, 11–25 mm long; stems usually branched near the top with a leaf-like bract subtending two–several pedunculate spathes...***S. demissum***
1b. Outer spathe bracts 14–75 mm long, evidently longer than the inner spathe bracts; stems simple and unbranched or branched with 1–2 nodes...2

2a. Outer spathe bracts less than twice as long as the inner spathe bracts...3
2b. Outer spathe bracts twice as long or more as the inner spathe bracts...4

3a. Stems 1–2 mm wide, not branched...***S. idahoense* var. *occidentale***
3b. Stems 2.3–5 mm wide, branched with 1–2 nodes (the first node usually longer than the leaves, and the uppermost node with 1–3 branches)...***S. angustifolium***

4a. Flowers pale blue, the outer tepals 7–10 mm long; inner spathe bracts with a narrow hyaline margin all the way to the tip...***S. pallidum***
4b. Flowers blue-violet, the outer tepals 8–15 mm long; inner spathe bracts with a narrow hyaline margin to within about 1 mm of the tip...***S. montanum* var. *montanum***

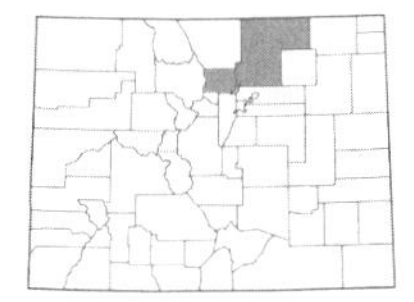

Sisyrinchium angustifolium **Mill.**, narrowleaf blue-eyed grass. Plants to 4.5 dm, branched with 1–2 nodes, the stems 2.3–5 mm wide; *outer spathe bracts* 18–40 mm long, 2–10 mm longer than the inner, basally connate 4–6 mm; *tepals* blue to violet or seldom white, with yellow bases, the outer tepals 7.5–12.5 mm long; *capsules* brown to black or purplish, 4–7 mm diam. Found in moist meadows and along creeks, 4500–6000 ft. June–Aug. E.

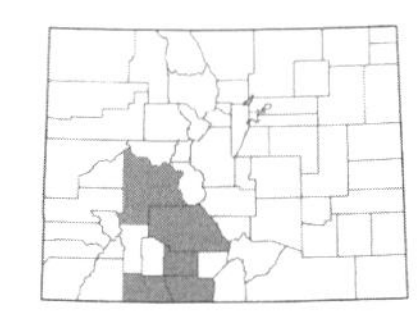

Sisyrinchium demissum **Greene**, stiff blue-eyed grass. Plants to 5 dm, usually branched near the top with a leaf-like bract subtending two–several pedunculate spathes; *outer spathe bract* 11–25 mm, 2.5 mm shorter to 5 mm longer than the inner, basally connate 3–8 mm; *tepals* light to dark blue with yellow bases, (6) 9–12 mm; *capsules* tan, 4–8 mm diam. Uncommon in moist meadows and along streams, sometimes found growing with *S. montanum*, 7500–7800 ft. June–Aug. E/W.

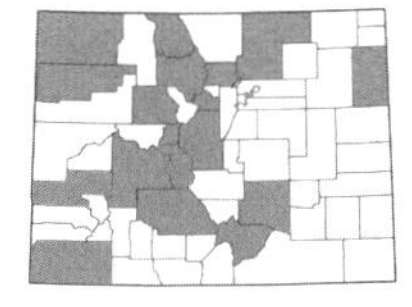

Sisyrinchium idahoense **E.P. Bicknell var. *occidentale*** **(E.P. Bicknell) Douglass M. Hend.**, Idaho blue-eyed grass. Plants to 4 dm, simple and unbranched, stems 1–2 mm wide; *outer spathe bracts* 14–30 mm long, 1.2–1.7 times longer than the inner, connate basally 2–5.5 mm; *tepals* blue-violet with yellow bases, (7) 9–15 mm long; *capsules* tan to brown, sometimes with purple splotches, 3–6 mm diam. Common in moist meadows and along streams, 3500–10,000 ft. June–Aug. E/W.

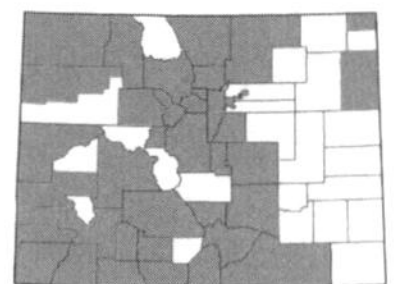

Sisyrinchium montanum **Greene var. *montanum*,** Rocky Mountain blue-eyed grass. Plants to 5 dm, simple and unbranched, the stems 1.5–3 mm wide; *outer spathe bracts* 35–75 mm long, ca. 1.8–3 times the length of the inner, connate basally 1.2–4.5 mm; *tepals* blue-violet with yellow bases, 8–15 mm long; *capsules* tan to brown, 4–7 mm diam. Common in meadows, marshes, and along streams, 4800–11,000 ft. May–Aug. E/W. (Plate 49)

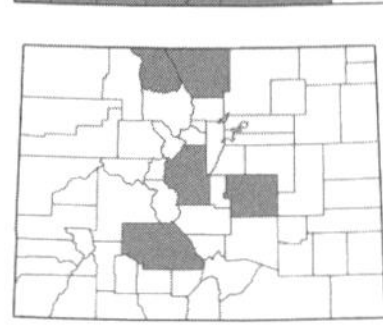

Sisyrinchium pallidum **Cholewa & Douglass M. Hend.,** pale blue-eyed grass. Plants to 3 dm, simple and unbranched, the stems 1–2 mm wide; *outer spathe bracts* (18) 28–40 (49) mm long, 1.8–2.5 times as long as the inner, connate basally 2–5 (6) mm; *tepals* pale blue with yellow bases, 7–10 mm long; *capsules* tan to brown, 3–5 mm diam. Locally common in moist meadows and fens and along streams, 7000–9500 ft. June–Aug. E.

It can be very difficult to tell *S. pallidum* from *S. montanum*, especially on herbarium specimens where the tepals have dried darker. This species may be best represented as a variety of *S. montanum*.

JUNCACEAE A.L. Juss. – rush family

Perennial or annual, grass-like herbs; *leaves* mostly basal, linear or sometimes much reduced or absent, sheathing; *flowers* small, usually perfect, actinomorphic, in headlike clusters or open and paniculate, subtended by one or more bracts; *tepals* 6, persistent, green, brown, or purplish-black, scale-like; *stamens* 3 or 6; *pistil* 1; *ovary* superior, with 1 or 3 locules; *fruit* a loculicidal capsule with 3–many seeds.

1a. Seeds and ovules numerous (more than 3); leaves glabrous; leaf sheaths usually open...***Juncus***
1b. Seeds and ovules 3; leaves usually with long, wavy hairs on the margins; leaf sheaths closed...***Luzula***

JUNCUS L. – rush

Stems round or flattened; *leaves* flat or terete, sometimes septate when terete, sometimes reduced to a bristle or absent, with open sheaths; *flowers* perfect, in headlike clusters or solitary; *fruit* a 1-locular or 3-locular capsule with many seeds. (Brooks & Clemants, 2000; Coville, 1896; Hermann, 1964, 1975)

1a. Tiny annuals 0.3–2.5 cm; inflorescence a solitary, terminal flower...***J. bryoides***
1b. Perennials (usually with rhizomes or densely caespitose) or if annuals then unlike the above, usually over 2.5 cm; inflorescence unlike the above...2

2a. Most of the leaf sheaths terminating in a bristle tip 2–10 mm long; inflorescence usually appearing lateral, the bract subtending the inflorescence erect, terete, and appearing to be a continuation of the stem; leaves absent, reduced to tan or brown leaf sheaths clustered at the base, or just the uppermost leaf sheaths with a well-developed blade, these not septate...3
2b. Leaf sheaths not terminating in a bristle tip; inflorescence usually appearing terminal, the bract subtending the inflorescence erect or spreading but usually not appearing to be a continuation of the stem, or if appearing lateral then the leaves septate; basal and/or stem leaves usually present...8

3a. Flowers 1–5 (7) per stem; bract subtending the inflorescence usually less than 5 cm long; seeds with a white tail at each end; plants densely tufted, lacking long-creeping rhizomes...4
3b. Flowers mostly 5 or more per stem; bract subtending the inflorescence usually longer than 5 cm; seeds not tailed; plants distinctly rhizomatous...6

4a. Uppermost leaf sheath with a bristle-tip about 1 cm long, or the bristle tip sometimes absent...***J. drummondii***
4b. Uppermost leaf sheath with a well-developed blade mostly over 4 cm long...5

5a. Tepals 4–5 mm long; capsules 3.5–5 mm long, oblong-ovoid...***J. hallii***
5b. Tepals 5.5–9 mm long; capsules 6–9 mm long, narrowly oblong...***J. parryi***

6a. Stamens 3; capsules 1.5–3.2 mm long; seeds 0.3–0.5 mm long; inflorescence with many flowers (30–100 per stem); plants 4–13 dm tall, lacking long-creeping rhizomes...***J. effusus***
6b. Stamens 6; capsules 2.5–5 mm long; seeds 0.5–0.8 mm long; inflorescence with 3–many flowers per stem; plants 0.2–10 dm tall, often with long-creeping rhizomes...7

7a. Bract subtending the inflorescence as long or longer than the stem; anthers 0.4–0.8 mm long, equal to or shorter than the filaments; capsule 2.5–3 mm long, shorter than the tepals...***J. filiformis***
7b. Bract subtending the inflorescence shorter than the stem; anthers 1.2–2.5 mm long, 3–5 times longer than the filaments; capsule 3.5–4.5 mm long, equal to or longer than the tepals...***J. arcticus***

8a. Inflorescence of a single solitary head, each head with 1–3 (5) flowers; seeds tailed; found above 9000 ft...9
8b. Inflorescence unlike the above; seeds tailed or not; plants variously distributed...11

9a. Capsules 6.5–8.5 mm long; tepals 5–6.6 mm long (to 10 mm long in fruit); usually at least 1 stem leaf present in addition to basal leaves; plants stoloniferous...***J. castaneus***
9b. Capsules 4–7 mm long; tepals 2.5–5 mm long; leaves all basal; plants caespitose, not stoloniferous...10

10a. Capsules shallowly notched at the apex; bract subtending the terminal head much longer than the inflorescence, foliaceous and erect...***J. biglumis***
10b. Capsules mucronate at the apex; bract subtending the terminal head usually equal to or shorter than the inflorescence, sometimes longer than the inflorescence, usually membranous and divergent...***J. triglumis***

11a. Leaves folded along the midrib with the margins fused toward the tips, such that the edge formed by the fused margins faces the stem (equitant, like the leaves of an *Iris*; look for a thin, white hyaline margin extending nearly the entire length of the inner leaf margin); flowers in panicles or racemes of 2–5 densely flowered heads, or rarely the heads solitary...***J. ensifolius***
11b. Leaves not equitant; inflorescence various...12

12a. Flowers each closely subtended by a pair of scarious bracteoles 1–2 mm long in addition to a bract subtending the pedicel; inflorescence usually open, the flowers borne singly on the branches of the inflorescence, not congested into ball-like heads; leaves never septate...13
12b. Flowers not subtended by a pair of scarious bracteoles; flowers often congested into ball-like or hemispheric heads; leaves often septate...22

13a. Auricles absent or rudimentary at the apex of the leaf sheath; inflorescence usually at least half of the total height of the plant; annuals...***J. bufonius***
13b. Auricles present at the apex of the leaf sheath (0.2–1.5 mm long), scarious or thick and yellowish; inflorescence less than half of the total height of the plant; perennials...14

14a. Tepals with an obtuse or hooded tip, brown or blackish, 1.7–3.5 mm long; plants with creeping rhizomes; capsules brown to dark brown...15
14b. Tepals lanceolate with a sharply acute to acuminate tip or bristle tip, green to tannish, 3–5.5 mm long; plants lacking creeping rhizomes; capsules tan to brown...16

15a. Anthers 0.6–1 mm long, 1–2 times the length of the filaments; capsules usually longer than the tepals; tepals 1.7–2.7 mm long...***J. compressus***
15b. Anthers 1.1–1.8 mm long, 2–4 times the length of the filaments; capsules usually shorter than or equal to the tepals; tepals 2.6–3.2 (3.8) mm long...***J. gerardii***

16a. Auricles at the top of the leaf sheath 2–5 mm long, with pointed tips; capsules with an acuminate tip...***J. tenuis***
16b. Auricles at the top of the leaf sheath usually 0.2–1.1 (2) mm long, the tips rounded or sometimes pointed; capsules various...17

17a. Auricles at the top of the leaf sheath yellowish, hard and leathery...***J. dudleyi***
17b. Auricles scarious or rarely leathery, whitish or purplish-tinged...18

18a. Seeds evidently tailed, the tails 0.2–0.5 mm long; uppermost bract usually much shorter than the inflorescence; leaves terete...***J. vaseyi***
18b. Seeds not tailed; uppermost bract usually longer than the inflorescence; leaves flat, channeled, or terete, sometimes becoming involute in dry conditions...19

19a. Tepals with a light brown or greenish center with dark brown strips on each side and scarious margins; capsules with a small, shallow notch at the summit...***J. confusus***
19b. Tepals greenish throughout with scarious margins; capsules rounded at the summit...20

20a. Capsule completely 3-locular; auricles at the top of the leaf sheath 0.4–1.1 (2) mm long; tepals (3.5) 4.5–5.7 mm long...***J. brachyphyllus***
20b. Capsule 1-locular or incompletely 3-locular with the placentae intruded halfway to the center; auricles 0.2–0.6 mm long; tepals 3–4.5 (5.5) mm long...21

21a. Capsules (2.5) 2.8–3.5 (4.5) mm long...***J. dichotomus***
21b. Capsules (3.3) 3.8–4.7 mm long...***J. interior***

22a. Leaves lacking auricles at the top of the sheath; capsules 6.5–8.5 mm long, narrowly oblong; inflorescence of 1–3 heads that are 2–12-flowered, subtended by a bract that is somewhat inflated at the base; tepals dark brown, 5–6.6 mm long in flower, to 10 mm long in fruit; leaves imperfectly septate, the septa not externally evident; plants stoloniferous...***J. castaneus***
22b. Leaves auriculate; capsules 1.8–5.7 mm long, variously shaped; plants otherwise unlike the above in all respects...23

23a. Leaves flattened above the basal sheath, not septate...24
23b. Leaves terete above the basal sheath, septate (with evident cross-partitions) at least distally if not throughout...25

24a. Tepals 5–6 mm long; capsules 3–5 mm long; stamens 6; plants with long-creeping rhizomes...***J. longistylis***
24b. Tepals 1.8–3.2 mm long; capsules 1.8–3 mm long; stamens 3; plants with short, often knotty rhizomes...***J. marginatus***

25a. Flowers tightly clustered in dense globose or spherical heads...26
25b. Flowers in panicles or if tightly clustered then in hemispherical heads (flat on the bottom)...30

26a. Tepals dark brown to black...27
26b. Tepals green to light brown...28

27a. Heads usually solitary with 12–60 flowers...***J. mertensianus***
27b. Heads usually 2–11 with 3–11 flowers per head...***J. nevadensis***

28a. Plants caespitose, not rhizomatous; capsule elliptic to narrowly ovoid, not narrowed at the top to a long beak; stamens mostly 3, rarely 6...***J. acuminatus***
28b. Plants rhizomatous or stoloniferous; capsule subulate, narrowing to a long beak; stamens 3 or 6...29

29a. Auricles 1–4 mm (usually over 2 mm), membranous; heads sessile and tightly clustered or on branches, 10–15 mm in diam., 25–100-flowered; outer tepals somewhat longer than the inner tepals, the outer ones 3.7–6 mm long, acuminate-tipped...***J. torreyi***
29b. Auricles 0.5–1.7 mm, membranous to cartilaginous; heads on branches, 3–10 (12) mm in diam., (2) 5–30 (50)-flowered; tepals all similar in size, 2.4–4.1 mm long, acute ...***J. nodosus***

30a. Bract subtending each flower brown, often scarious above but not throughout; stamens 6; seeds not tailed; plants mostly found above 6000 ft in elevation...31
30b. Bract subtending each flower white to yellowish-white and scarious throughout, or scarious with a narrow, greenish midrib; stamens 3 or 6; seeds tailed or not; plants found to 7000 ft in elevation...32

31a. Auricles 0.5–1.2 mm long; tepals 1.6–3 mm long; capsules usually evidently longer than the tepals; basal leaf sheaths usually reddish or purplish-red...***J. alpinoarticulatus***
31b. Auricles 1–3.2 mm long; tepals 2.4–6.2 mm long; capsules shorter than or only slightly longer than the tepals; basal leaf sheath tan to light brown...***J. nevadensis***

32a. Seeds not tailed; basal leaf sheaths usually reddish or purplish-red...33
32b. Seeds tailed; basal leaf sheaths usually tan to brown...34

33a. Stamens usually 3, rarely 6; capsules equal to or slightly exserted from the tepals; plants caespitose, not rhizomatous, not rooting at the lower nodes...***J. acuminatus***
33b. Stamens 6; capsules exserted about 1 mm beyond the tepals; plants loosely caespitose with short rhizomes, often rooting at the lower nodes...***J. articulatus***

34a. Tails on seeds 1/10–1/3 the length of the body; outer tepals usually rounded at the tip, with a broad, scarious margin...***J. brachycephalus***
34b. Tails on seeds about 1/3–1/2 the length of the body; outer tepals acute, with a narrow scarious margin...***J. brevicaudatus***

***Juncus acuminatus* Michx.**, TAPERTIP RUSH. Perennials, 1.4–10 dm; *leaves* 1–40 cm long, nearly terete, auricles 1–1.5 mm long; *inflorescence* of 5–50 hemispheric to spheric heads 3–10 mm diam.; *tepals* 2.5–4 mm long, light brown to greenish; *stamens* 3 (6); *capsules* elliptic to narrowly ovoid, 2.5–4 mm long; *seeds* 0.3–0.4 mm long, not tailed. Uncommon in swales and along lake shores, 5000–5500 ft. July–Sept. E.

Juncus alpinoarticulatus **Chaix**, NORTHERN GREEN RUSH. [*J. alpinus* Vill.]. Perennials, 0.5–5 dm; *leaves* 1.5–12 mm × 0.5–1.1 mm, terete, auricles 0.5–1.2 mm long; *inflorescence* of 5–25 obpyramidal heads, 2–6 mm diam.; *tepals* 1.6–3 mm long, greenish to straw; *stamens* 6; *capsules* oblong to oblong-ovoid, 2.3–3.5 mm long; *seeds* 0.5–0.7 mm long, not tailed. Occasional in moist seeps, along pond shores, and in moist meadows, 5800–9500 ft. July–Sept. E/W.

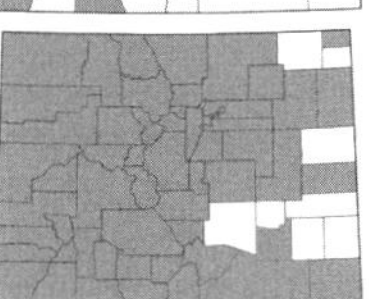

Juncus arcticus **Willd.**, ARCTIC RUSH. Perennials, 2–10 dm; *leaves* usually absent; *inflorescence* 3-many-flowered, lateral; *tepals* (2.5) 3.3–6 mm long, chestnut brown or lighter; *stamens* 6; *capsules* oblong to narrowly ovoid, 3.5–4.5 mm long; *seeds* 0.6–0.8 mm long. Common in moist meadows, along streams and lake shores, and in seepage areas, 3500–11,500 ft. May–Sept. E/W. (Plate 49)

There are two varieties of *J. arcticus* in Colorado:

1a. True leaves absent, only basal sheaths present...**var. *balticus* (Willd.) Trautv**. Widespread.

1b. Plants with 1 or 2 true leaves terminating the leaf sheaths...**var. *mexicanus* (Willd. ex Roem. & Schult.) Balslev**. Reported for southwestern Colorado, but no specimens have been seen.

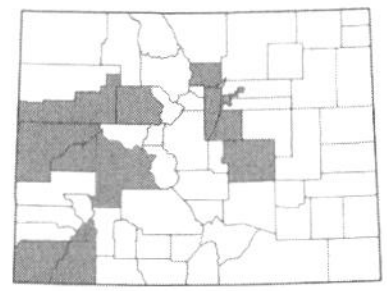

Juncus articulatus **L.**, JOINTLEAF RUSH. Perennials, 0.5–6 (10) dm; *leaves* 3.5–12 cm × 0.5–1.1 mm, terete, auricles 0.5–1 mm; *inflorescence* of 3–30 (50) obpyramidal to hemispheric heads, 6–8 mm diam.; *tepals* 1.8–3 mm long, green to straw or dark brown; *stamens* 6; *capsules* ellipsoid or ovoid, 2.8–4 mm long; *seeds* 0.5 mm long, not-tailed. Rather uncommon in moist places, especially along pond shores and in floodplains, 5000–6500 ft. June–Sept. E/W.

Juncus biglumis **L.**, TWO-FLOWERED RUSH. Perennials, 0.25–1.6 dm; *leaves* 2–7 cm × 0.5–1.5 mm, nearly terete, auricles 0.5 mm long or absent; *inflorescence* of 1–2 (4)-flowered heads; *tepals* 2.5–4 mm long, brown to blackish; *stamens* 6; *capsules* 4–5.5 × 1.7–2.3 mm; *seeds* 0.7–0.9 mm long, short-tailed. Uncommon in moist alpine and on rocky slopes, 11,000–14,100 ft. July–Sept. E/W.

Juncus brachycephalus **(Engelm.) Buchenau**, SMALLHEAD RUSH. Perennials, 2–7 dm; *leaves* 0.2–12 cm × 0.5–2 mm, terete to compressed, auricles 0.6–1.5 mm long; *inflorescence* of 5–80 heads, 2–5 mm diam.; *tepals* 1.8–2.8 mm long, green to light brown; *stamens* 3 or 6; *capsules* obconic, 2.4–3.8 mm long; *seeds* 0.8–1.2 mm, tailed. Rare in moist meadows and along streams, 6500–7000 ft. July–Sept. E. (Plate 49)

Juncus brachyphyllus **Wiegand**, TUFTED-STEM RUSH. Perennials to 8 dm; *leaves* 9–25 cm × 0.8–1.5 (2.4) mm, becoming involute in dry conditions, auricles 0.4–1.1 (2) mm long; *inflorescence* 10–150-flowered, 1.5–8 cm; *tepals* (3.5) 4.5–5.7 mm long, green to tannish; *stamens* 6; *capsules* ellipsoid to narrowly so, 2.6–4.7 × 1.3–2 mm; *seeds* 0.3–0.6 mm long, not tailed. Not reported for Colorado but occurs close to the border in New Mexico.

Juncus brevicaudatus **(Engelm.) Fernald**, NARROWPANICLE RUSH. [*J. tweedyi* Rydb.]. Perennials, 1.4–5.5 (7) dm; *leaves* 1.5–25 cm × 0.5–2.5 mm, terete, auricles 0.5–3 mm long; *inflorescence* of 2–35 ellipsoid to narrowly obconic heads, 2–9 mm diam.; *tepals* 2.3–3.2 mm long, green to light brown; *stamens* 3 (6); *capsules* narrowly ellipsoid to prismatic, 3.2–4.8 mm long; *seeds* 0.7–1.2 mm long, tailed. Rare along creeks and streams, known from El Paso Co., 6500–7000 ft. July–Sept. E.

Juncus bryoides **F.J. Herm.**, MOSS RUSH. Annuals, 0.03–0.25 dm; *leaves* to 0.9 cm long; *inflorescence* a terminal solitary flower; *tepals* 1.2–2.3 (2.8) × 0.4–0.6 mm, chestnut brown to black; *stamens* 3; *capsules* 1–1.9 mm × 0.5–1 mm; *seeds* 0.3–0.5 mm long. Rare on moist sandstone ledges, known from Mesa and Moffat cos., 7000–8500 ft. May–July. W.

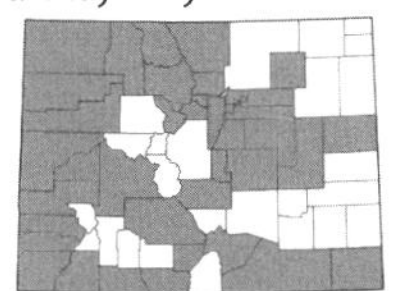

Juncus bufonius **L.**, TOAD RUSH. Annuals, 0.5–4 dm; *leaves* 3–13 cm × 0.3–1.1 mm, flat, auricles absent or rudimentary; *inflorescence* usually at least ½ height of plant; *tepals* 3.8–7 (8.5) mm long, greenish; *stamens* 3 to 6; *capsules* ellipsoid to narrowly so, 2.7–4 × 1–1.5 mm; *seeds* 0.26–0.49 mm long, not tailed. Common in moist areas and around drying pools, 4700–10,000 ft. June–Sept. E/W.

Juncus castaneus **J. Sm.**, CHESTNUT RUSH. Perennials, 1–4 dm; *leaves* to 20 cm, channeled, auricles absent; *inflorescence* of 1–3 (5) glomerules; *tepals* 4.5–6.6 mm long, brown or paler; *stamens* 6; *capsules* narrowly oblong, 6.5–8.5 × 1.8–2.3 mm; *seeds* 08–1.1 mm long, tailed, the tails 0.8–1.1 mm long. Found in bogs and moist alpine meadows, 9000–13,000 ft. June–Sept. E/W.

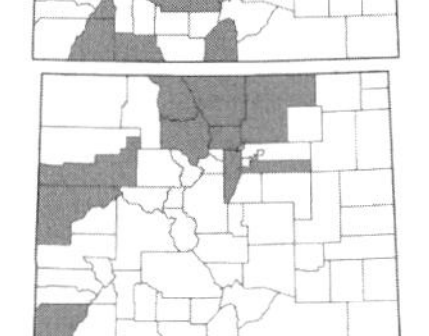

Juncus compressus **Jacq.**, ROUNDFRUIT RUSH. Perennials to 8 dm; *leaves* 5–35 cm × 0.8–2 mm, flat to slightly channeled, auricles 0.3–0.5 mm long; *inflorescence* 5–60-flowered, 1.5–8 cm; *tepals* 1.7–2.7 mm long, brownish; *stamens* 6; *capsules* widely ellipsoid to obovoid, 2.5–3.5 × 1.4–1.8 mm; *seeds* 0.3–0.6 mm long, not tailed. Found in wetlands, marshes, moist meadows, and along lake shores, 4500–6000 ft. June–Sept. E/W. Introduced.

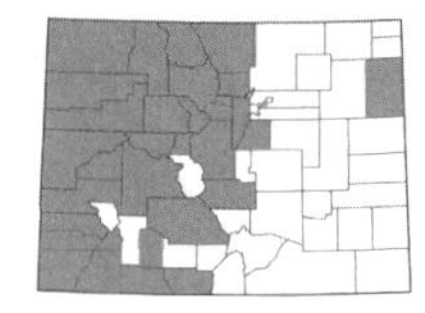

***Juncus confusus* Coville**, Colorado rush. Perennials, 1.5–4 dm; *leaves* 5–15 cm × 0.4–0.8 mm, channeled or nearly flat, auricles 0.3–0.7 mm long; *inflorescence* 3–25-flowered, 1–2.5 × 1–2 cm; *tepals* 3–4.5 mm long, green or straw-colored midstripe with brown side stripes; *stamens* 6; *capsules* equal or shorter than perianth; *seeds* ca. 0.5 mm long, not tailed. Common in moist meadows, grasslands, and along pond margins, 3800–10,700 ft. June–Sept. E/W. (Plate 49)

***Juncus dichotomus* Elliot**, forked rush. [*J. platyphyllus* (Wiegand) Fernald]. Perennials to 10 dm; *leaves* 10–25 (40) cm × 0.5–1.2 mm, channeled or flat, auricles 0.2–0.6 mm long; *inflorescence* terminal and (5) 10–100-flowered, (1) 2.5–13 cm long; *tepals* 3–4.5 (5.5) mm long, green; *stamens* 6; *capsules* ellipsoid to widely so, (2.5) 2.8–3.5 (4.5) × 1.6–2.2 mm; *seeds* 0.3–0.4 mm long, not tailed. Rare along lake shores, known from Boulder Co., 5000–6000 ft. July–Sept. E. Introduced.

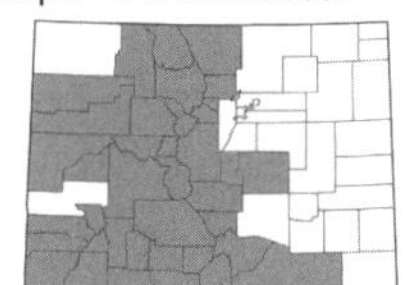

***Juncus drummondii* E. Mey.**, Drummond's rush. Perennials to 4 dm; *leaves* absent or rarely present, to 1 cm long; *inflorescence* 2–5-flowered; *tepals* (4) 5–8 mm long, brown to chestnut brown with green midstripe; *stamens* 6; *capsules* oblong, 4.5–7 (8) × 1.8–2.2 mm; *seeds* 0.5–0.6 mm long. Common in moist meadows, along lake shores and streams, in seepage areas, and in the alpine, (6000) 8500–13,500 ft. July–Sept. E/W. (Plate 49)

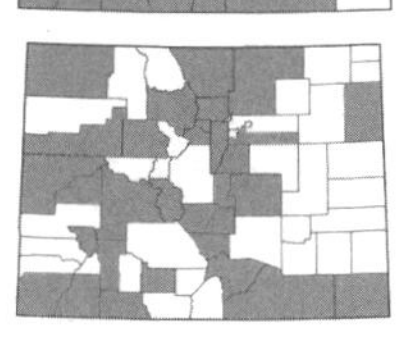

***Juncus dudleyi* Wiegand**, Dudley's rush. Perennials, 2–10 dm; *leaves* 5–30 cm × 0.5–1 mm, flat, auricles 0.2–0.4 mm; *inflorescence* with up to 80 flowers, 1.5–5 (9) cm long; *tepals* 4–5 mm long, greenish; *stamens* 6; *capsules* ellipsoid, 2.9–3.6 × 1.5–1.9 mm; *seeds* 0.4–0.7 mm long, not tailed. Common in marshes, wetlands, moist areas in grasslands, and along streams, 3800–9000 ft. June–Sept. E/W. (Plate 49)

***Juncus effusus* L.**, common rush. Perennials, 4–13 dm; *leaves* absent; *inflorescence* many-flowered and lateral; *tepals* 1.9–3.5 mm long, tan or darker, usually with a greenish midstripe; *stamens* 3; *capsules* broadly ellipsoid to oblong, 1.5–3.2 mm long; *seeds* 0.3–0.5 mm long. Uncommon in seepages, fens, and moist areas, currently known from Larimer, Boulder, and Fremont cos., 5000–8500 ft. June–Sept. E. Introduced.

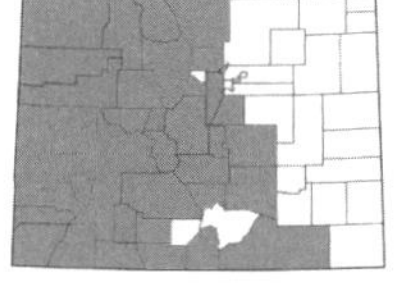

***Juncus ensifolius* Wikstr.**, swordleaf rush. [*J. saximontanus* A. Nelson]. Perennials, 2–6 dm; *leaves* 2–25 cm × 1.5–6 mm, auricles absent; *inflorescence* of 2–5 heads, 2–14 cm; *tepals* 2.2–3.5 mm long, green to brown or reddish-brown; *stamens* 3 or 6; *capsules* chestnut to dark brown, 2.4–4.3 mm; *seeds* 0.4–1 mm long, occasionally tailed. Common in moist meadows, along streams, and in marshy areas, 5500–9500 ft. June–Sept. E/W.

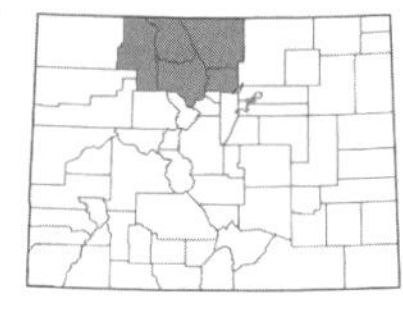

***Juncus filiformis* L.**, thread rush. Perennials, 0.2–3.5 dm; *leaves* absent; *inflorescence* 3–10-flowered, 1–2 cm; *tepals* 2.5–4.2 mm long, light brown or green; *stamens* 6; *capsules* nearly globose, 2.5–3 × 1.8–2.1 mm; *seeds* 0.5–0.6 mm long, not tailed. Found along lake margins and in moist meadows and fens, 7500–10,500 ft. Aug.–Sept. E/W.

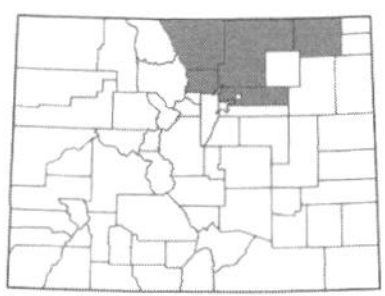

***Juncus gerardii* Loisel.**, saltmeadow rush. Perennials, 2–9 dm; *leaves* 10–40 cm × 0.4–0.7 mm, flat or somewhat channeled, auricles 0.4–0.6 (0.8) mm long; *inflorescence* 10–30 (80)-flowered, 2–16 cm; *tepals* 2.6–3.2 (3.8) mm long, dark brown or blackish; *stamens* 6; *capsules* widely ellipsoid, 2.2–3.5 × 1.3–1.9 mm; *seeds* 0.5–0.6 mm long, not tailed. Found in marshes and alkali flats, 4500–5000 ft. May–July. E. Introduced.

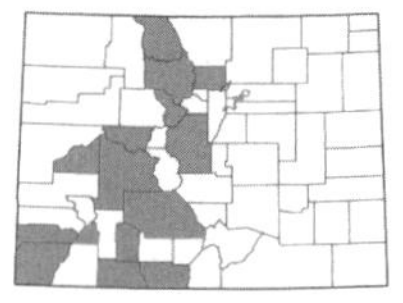

***Juncus hallii* Engelm.**, Hall's rush. Perennials to 4 dm; *leaves* 4–15 cm, auricles ca. 0.2 mm long; *inflorescence* 2–7-flowered; *tepals* 4–5 mm long, light brown with green midstripe; *stamens* 6; *capsules* oblong-ovoid, 3.5–5 × 1.5–1.9 mm; *seeds* ca. 0.5 mm long, tailed, the tails 0.3 mm long. Uncommon or locally abundant in moist meadows and fens, 8000–11,000 ft. June–Aug. E/W.

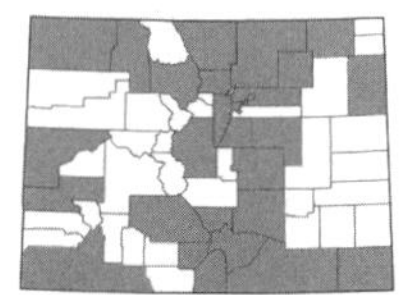

***Juncus interior* Wieg.**, inland rush. Perennials, 2–6 dm; *leaves* 5–15 cm × 0.5–1.1 mm, flat, auricles 0.2–0.6 mm long; *inflorescence* 1.5–7 cm; *tepals* 3.3–4.4 mm long, greenish; *stamens* 6; *capsules* (3.3) 3.8–4.7 mm long; *seeds* 0.4–0.7 mm long, not tailed. Common in moist places, meadows, and along creeks and pond margins, 3700–9000 ft. May–Aug. E/W. (Plate 49)

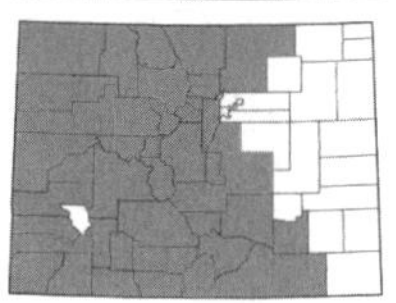

***Juncus longistylis* Torr.**, longstyle rush. Perennials, 2–6 dm; *leaves* 10–30 cm × 1.5–3 mm, flat, auricles 1–2.5 mm long; *inflorescence* of 1–4 (8) glomerules, 2–6 (10) cm long; *tepals* 5–6 mm long, brown with green midstripe; *stamens* 6; *capsules* obovoid, 3–5 mm long; *seeds* 0.4–0.6 mm long, not tailed. Common in moist meadows, marshes, and along streams, 5000–10,500 ft. June–Aug. E/W.

Juncus marginatus **Rostk.**, GRASSLEAF RUSH. Perennials, 3–13 dm; *leaves* 20–45 cm × 1.5–5 mm, flat, auricles 0.5–1.5 mm long; *inflorescence* of (2) 5–200 glomerules, 3–10 (15) cm; *tepals* 1.8–3.2 mm long, dark brownish, usually with green midstripe; *stamens* 3; *capsules* 1.8–3 mm long; *seeds* 0.4–0.7 mm long, not tailed. Uncommon along streams and in moist meadows, known from Boulder Co., 5000–5500 ft. June–Aug. E.

Juncus mertensianus **Bong.**, MERTENS' RUSH. Perennials, 0.5–4 dm; *leaves* 3–15 cm × 0.3–0.6 mm, terete, auricles 1–1.2 mm; *inflorescence* a terminal solitary head or rarely a cluster of 2 heads, 4.5–15 mm diam.; *tepals* 2.3–4.9 mm long, dark purplish-brown to black; *stamens* 6; *capsules* obovoid, 1.9–3.5 mm long; *seeds* 0.4–0.5 mm long, not tailed. Common in moist subalpine and alpine meadows, along streams, and near springs, 7500–13,000 ft. June–Sept. E/W.

Juncus nevadensis **S. Watson**, SIERRA RUSH. Perennials, 0.5–7 dm; *leaves* 1.5–31 cm × 0.5–2.2 mm, laterally flattened, auricles 1–3.2 mm long; *inflorescence* of 2–11 hemispheric to obpyramidal heads, 5–14 mm diam.; *tepals* 2.4–6.2 mm long, dark brown to white; *stamens* 6; *capsules* ellipsoid, 2.3–3.7 mm; *seeds* 0.4–0.5 mm long, not tailed. Uncommon or occasional along melting snowbanks and streams, near seepages, and in moist meadows, 7500–12,000 ft. June–Sept. E/W.

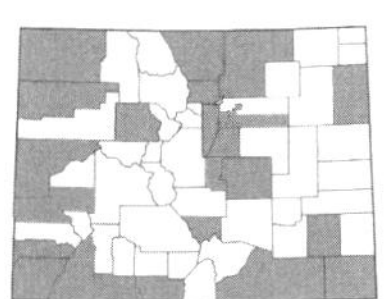

Juncus nodosus **L.**, KNOTTED RUSH. Perennials, 0.4–5.5 (7) dm; *leaves* 6–30 cm × 0.5–1.5 mm, terete, auricles 0.5–1.7 mm; *inflorescence* of 3–15 spheric heads, 3–10 (12) mm diam.; *tepals* 2.4–4.1 mm long, green to light brown; *stamens* 3 or 6; *capsules* lance-subulate, 3.2–5 mm long; *seeds* 0.4–0.5 mm long, not tailed. Common in moist meadows, fens, and along pond margins and streams, 3500–8500 ft. June–Sept. E/W. (Plate 49)

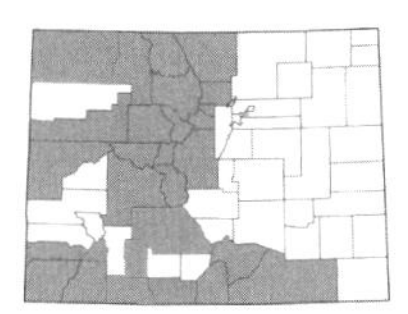

Juncus parryi **Engelm.**, PARRY'S RUSH. Perennials, 0.5–3 dm; *leaves* with absent blades, auricles 0.2–0.3 mm long; *inflorescence* 1–3-flowered; *tepals* 5.5–9 mm long, light brown with green midstripe; *stamens* 6; *capsules* narrowly oblong, 6–9 × 1.5–2 mm; *seeds* ca. 0.6 mm long, tailed, tails 0.4 mm long. Common in moist meadows, along streams, and in alpine, 9500–13,000 ft. July–Sept. E/W.

Juncus tenuis **Willd.**, POVERTY RUSH. Perennials, 1.5–5 dm; *leaves* 3–12 cm × 0.5–1 mm, flat, auricles 2–5 mm long; *inflorescence* 5–40 flowered, 1–5 cm or longer; *tepals* 3.3–4 mm long, greenish; *stamens* 6; *capsules* ellipsoid, 3.3–4.7 × 1.1–1.7 mm; *seeds* 0.5–0.7 mm long, not tailed. Uncommon in wetlands and moist places, 5000–6000 ft. June–Sept. E.

It is questionable whether *J. tenuis* is present in Colorado as all specimens of *J. tenuis* I have seen are *J. dudleyi*.

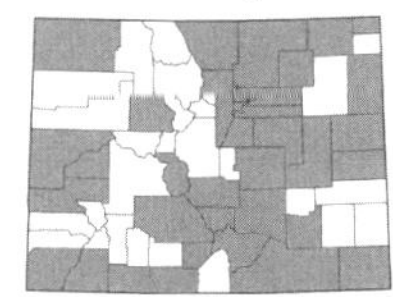

Juncus torreyi **Coville**, TORREY'S RUSH. Perennials, (3) 4–10 dm; *leaves* 13–30 cm × 1–5 mm, terete, auricles 1–4 mm; *inflorescence* of 1–23 globose heads, 10–15 mm diam.; *tepals* (3) 3.4–6 mm, green to straw or reddish; *stamens* 6; *capsules* 4.3–5.7 mm; *seeds* 0.4–0.5 mm, not tailed. Common in moist meadows and along streams, ditches, and pond margins, 3500–8500 ft. June–Sept. E/W.

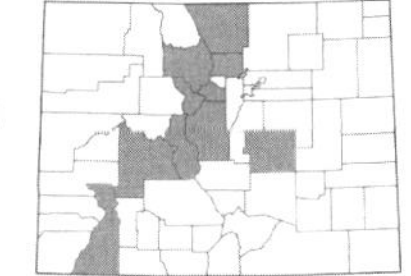

Juncus triglumis **L.**, THREE-HULLED RUSH. [*J. albescens* (Lang) Fernald]. Perennials, 0.3–3.5 dm; *leaves* 2–10 cm, deeply channeled, auricles slightly prolonged; *inflorescence* of solitary heads; *tepals* 3–5 mm long, pale brown or darker; *stamens* 6; *capsules* 3–7 mm long, the tips mucronate; *seeds* 0.5–1 mm long, tailed, tails 0.6–1 mm long. Common in bogs and moist alpine meadows, 10,000–13,500 ft. July–Sept. E/W.

There are two varieties of *J. triglumis* in Colorado:

1a. Capsules exserted from the perianth, 3.5–7 mm long; bract subtending the head shorter than the inflorescence...**var. *triglumis***

1b. Capsules included or barely exserted from the perianth, 3–5 mm long; bract subtending the head equal to or longer than the inflorescence...**var. *albescens* Lange**

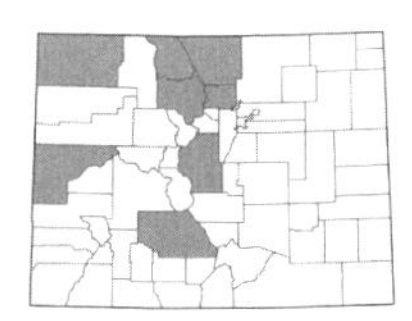

Juncus vaseyi **Engelm.**, VASEY'S RUSH. Perennials, 2–7 dm; *leaves* 10–30 × 0.5–1 mm, auricles 0.2–0.4 (0.6) mm long; *inflorescence* terminal, 5–15 (30)-flowered, 1–5 cm long; *tepals* 3.3–4.4 mm long, greenish to tan; *stamens* 6; *capsules* ellipsoid, (3.3) 3.8–4.7 × 1.1–1.7 mm; *seeds* 0.5–0.7 mm long, tailed, tails 0.2–0.5 mm long. Uncommon or rare in moist meadows, bordering ponds, and in seepage areas, 7500–10,000 ft. June–Sept. E/W.

LUZULA DC. – WOODRUSH

Perennials with round stems; *leaves* flat, never septate, the margins with long, multicellular hairs, with closed sheaths; *flowers* perfect, in headlike clusters or open and paniculate; *fruit* a globose capsule with 3 seeds, the base of which often have a tuft of hairs. (Hamet-Ahti, 1971)

1a. Flowers arranged in an open, arching panicle, usually in pairs at the ends of branches...***L. parviflora***

1b. Flowers densely crowded in a spike-like inflorescence or umbellate clusters...2

2a. Outer whorl of tepals with long bristle points; flowers in a dense, often interrupted spike-like inflorescence, this often nodding; bracts subtending flowers with ciliate-lacerate margins...***L. spicata***
2b. Outer whorl of tepals with acute tips, lacking long bristles; flowers in well-separated clusters, not nodding; bracts subtending flowers usually with entire or sometimes with ciliate-lacerate margins...3

3a. Clusters of flowers usually spherical; caruncle at the base of the seed absent...***L. subcapitata***
3b. Clusters of flowers usually cylindric; caruncle at the base of the seed 0.3–0.8 mm long...***L. comosa***

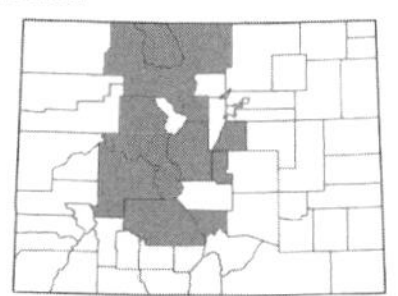

***Luzula comosa* E. Mey.**, PACIFIC WOODRUSH. Plants 1–4 dm; *leaves* 5–15 cm × 3–7 mm; *inflorescence* umbellate or of sessile cylindric clusters, 5–15 × 5–7 mm; *tepals* tan to brown with clear margins, 2–5 mm long; *capsules* to 2.5 mm long; *seeds* 1–1.5 mm long with a caruncle 0.3–0.8 mm long. Found along streams and creeks, meadows, and in moist forests, 7800–12,700 ft. July–Sept. E/W.

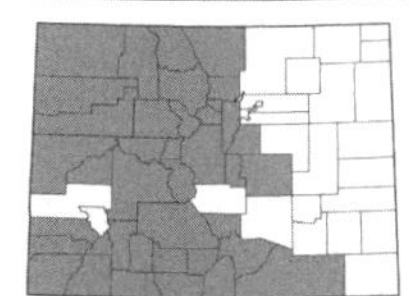

***Luzula parviflora* (Ehrh.) Desv.**, SMALL-FLOWERED WOODRUSH. Plants (2) 3–10 dm; *leaves* 12–17 cm × 5–10 mm; *inflorescence* paniculate, with lax, often arching branches; *tepals* brown, 1.5–2.5 mm long; *capsules* to 2.5 mm long; *seeds* 1–1.2 mm long, lacking a caruncle. Common along streams and lake margins and in moist forests, 6700–12,000 ft. June–Sept. E/W. (Plate 49)

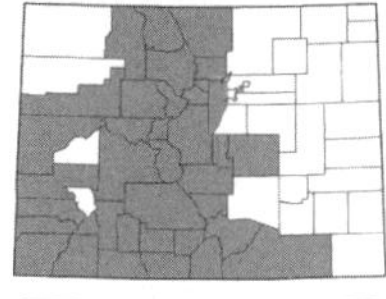

***Luzula spicata* (L.) DC.**, SPIKED WOODRUSH. Plants 0.3–3.5 dm; *leaves* 2–15 cm × 1–4 mm; *inflorescence* of dense, nodding, spike-like clusters; *tepals* brown with clear margins or pale throughout, the outer whorl with long bristle points; *capsules* slightly shorter than the tepals; *seeds* 1–1.2 mm long with a caruncle ca. 0.2 mm long. Common in alpine and moist meadows, 9500–14,200 ft. July–Sept. E/W. (Plate 49)

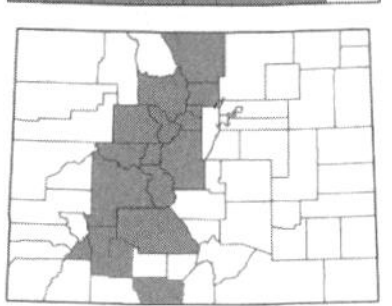

***Luzula subcapitata* (Rydb.) Harrington**, COLORADO WOODRUSH. Plants 0.8–4 dm; *leaves* ca. 5 cm long; *inflorescence* of 6–10 sessile spherical clusters; *tepals* brown with clear margins, 1.5–2 mm long; *capsules* about equaling the tepals; *seeds* ca. 1.3 mm long, lacking a caruncle. Found in moist meadows, bogs, fens, and along streams, 10,500–13,000 ft. July–Sept. E/W.

JUNCAGINACEAE Rich. – ARROWGRASS FAMILY

Monoecious, perennial or annual herbs from rhizomes; *leaves* mostly basal, sessile, 2-ranked, linear, sheathing; *flowers* perfect or imperfect, actinomorphic, in spikes or narrow spike-like racemes; *tepals* 6, rarely absent or 1; *stamens* 1, 4, or usually 6; *pistils* 1, 3, or 6; *ovary* 3–6-carpellate, superior, placentation basal; *fruit* nutlets or schizocarps.

TRIGLOCHIN L. – ARROWGRASS

Perennial herbs; *flowers* perfect, in narrow spike-like racemes; *tepals* 6 in 2 series of 3; *stamens* 4 or 6; *pistils* 6, united at first but separating when mature; *fruit* a schizocarp with 3 or 6 mericarps. (Loomin, 1976)

1a. Stigmas 6; schizocarps linear to more often almost globose, not narrowed at the base, 2–5 mm long; flowers usually densely grouped in the inflorescence; ligule unlobed or 2-lobed at the apex; fruiting receptacle without wings...***T. maritima***
1b. Stigmas 3; schizocarps linear, narrowed at the base, 5–8.3 mm long; flowers more spaced in the inflorescence, not densely grouped; ligule 2-lobed at the apex; fruiting receptacle with wings...***T. palustris***

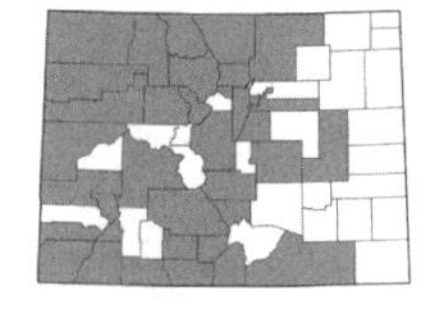

***Triglochin maritima* L.**, SEASIDE ARROWGRASS. [*T. concinna* Burtt-Davy]. Plants 0.7–11 dm; *leaves* 2–50 cm long, the ligule unlobed or 2-lobed at the apex; *pedicels* 1–4 mm long; *tepals* yellowish-green or greenish, 1–2.5 mm long; *stigmas* 6; *schizocarps* linear to globose, 2–5 mm long. Locally common in moist meadows, along lake shores, and in marshes and wet seeps, 5000–10,500 ft. May–Aug. E/W.

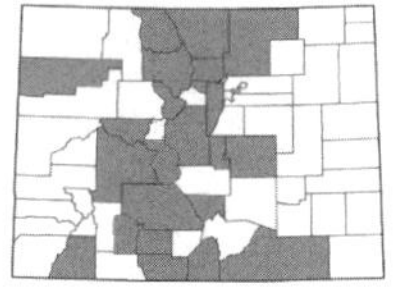

***Triglochin palustris* L.**, MARSH ARROWGRASS. Plants 0.8–7 dm; *leaves* 2–30 cm long, the ligule 2-lobed at the apex; *pedicels* 1–6 mm long; *tepals* yellowish-green or greenish, 1.5–2 mm long; *stigmas* 3; *schizocarps* linear, 5–8.3 mm long. Uncommon in moist meadows and marshes and along the borders of lakes, 5100–10,000 ft. June–Aug. E/W. (Plate 49)

KRAMERIACEAE Dum. – RATANY FAMILY

Shrubs or perennial herbs; *leaves* alternate, entire, usually simple; *flowers* perfect, zygomorphic, solitary in leaf axils; *sepals* (4) 5, distinct, unequal with the 3 outer larger than the 2 inner, petaloid, usually purplish; *petals* (4) 5, distinct, unequal, the upper 3 long-clawed, the lower 2 very different from the upper, broad and thick, often modified into glands, reddish; *stamens* (3) 4 (5); *pistil* 1; *ovary* superior, 2-carpellate,1 carpel developing; *fruit* achene-like, armed.

KRAMERIA L. – RATANY
With characteristics of the family.

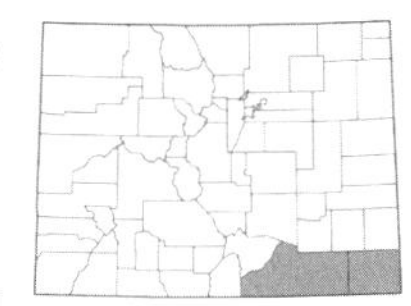

Krameria lanceolata **Torr.**, RATANY. Perennials, stems decumbent or scrambling, 2–5 (10) dm long; *leaves* linear to elliptic, 6–20 mm long, usually with a short spine at the tip, softly hairy; *sepals* 4 or 5, 8–10 mm long, hairy; *fruit* globose, 6–9 mm diam., tomentose, armed with sharp prickles. Uncommon in sandy or rocky soil, 4000–4900 ft. May–July. E.

Krameria erecta Willd. ex Schult. was erroneously reported for Colorado based on a specimen at RM (*Aven Nelson 1034*). This specimen was mistakenly labeled as collected in CO and actually occurs only in AZ, NM, and UT.

LAMIACEAE Lindl. – MINT FAMILY

Herbs (usually) or sometimes woody, the stem usually square; *leaves* opposite or whorled, simple or compound, exstipulate; *flowers* usually perfect; *calyx* usually 5-lobed, actinomorphic or zygomorphic, usually tubular; *corolla* 4–5-lobed, bilabiate, unequal but not bilabiate, or nearly actinomorphic (*Mentha*); *stamens* 2 or 4, epipetalous; *pistil* 1; *ovary* superior, 2-carpellate but appearing 4-carpellate due to the presence of false septa; *style* usually gynobasic; *fruit* of 4 nutlets.

1a. Calyx with an erect, prominent, shieldlike appendage 2–4 mm long on the upper side, inconspicuously nerved; corolla blue to purple or rarely white, the tube straight or more often sigmoidally curved, the upper lip with a rounded and helmet-like central lobe...***Scutellaria***
1b. Calyx without an appendage on the upper side, conspicuously nerved or not; corolla various...2

2a. Leaf margins entire...3
2b. Leaf margins toothed, cleft, or pinnatifid...8

3a. Shrubs; stems and leaves of the current season densely soft-canescent; sepals densely white-hairy, purple; flowers light blue-lavender...***Poliomintha***
3b. Annual or perennial herbs, or if shrubby then unlike the above; stems and leaves glabrous or hairy but not canescent; sepals glabrous or hairy, green or purplish-tinged; flowers various...4

4a. Flowers sessile in a dense head or spike and subtended by broad, leafy bracts, sometimes the heads not terminal but arranged in an interrupted spike and appearing axillary...5
4b. Flowers pedicellate (usually shortly so) in the axils of leaves or in a terminal raceme and not subtended by broad, leafy bracts...7

5a. Flowers in a spike (longer than wide), the bracts suborbicular with a long-acuminate tip...***Prunella***
5b. Flowers in a head (wider than long), the bracts elliptic or ovate with an acute, rounded, or spinose tip...6

6a. Corolla obscurely two-lipped, with 5 slender, linear to narrowly oblong lobes; stamens 4; leaves 1–3.5 cm long...***Monardella***
6b. Corolla conspicuously two-lipped; stamens 2; leaves 2–10 cm long...***Monarda***

7a. Stems ridged; calyx glabrous within, not gibbous at the base, the teeth without long, slender bristles at the tips...***Salvia***
7b. Stems not ridged; calyx with a ring of hairs within the throat, gibbous at the base, the teeth with long, slender bristles at the tips...***Hedeoma***

8a. Corolla appearing 1-lipped, the upper lip of the corolla deeply cleft and appearing to arise laterally from the margins of the 3-lobed lower lip; flowers white or rose-lavender; leaves narrowly elliptic or lance-ovate with a serrate margin, or deeply pinnatifid...***Teucrium***
8b. Upper lip of the corolla clearly discernable; flowers and leaves various...9

9a. Calyx with a sharp spine or bristle at the tip...10
9b. Calyx tips sometimes sharply acute to acuminate but lacking spines or bristles...13

10a. Leaves 3-cleft or palmately cleft or parted with serrate margins, 3–12 cm long; corolla 8–12 mm long, whitish to pink with dark purple spots, the upper lip hairy on the back; flowers in dense clusters in leaf axils...***Leonurus***
10b. Leaves merely toothed; plants otherwise unlike the above in all respects...11

11a. Flowers pale yellow or white; stem, leaves, and calyx densely white floccose-tomentose (the leaves becoming glabrate above in age); lower leaves ovate to elliptic and petiolate, the upper reduced and clasping; bracts 10–20 mm long, suborbicular with a long-acuminate tip...***Salvia*** (*aethiopis*)
11b. Flowers purple, pink, or rose; stem, leaves, and calyx not densely white floccose-tomentose; plants otherwise unlike the above in all respects...12

12a. Leaves sharply serrate, the teeth of at least the upper leaves ending with a long, slender bristle; stems lacking glandular hairs; corolla barely surpassing the calyx...***Dracocephalum***
12b. Leaves serrate but the teeth not ending in a long, slender bristle; stems usually with some glandular hairs also below the nodes and inflorescence; corolla well-exserted beyond the calyx...***Galeopsis***

13a. Inflorescences axillary in general appearance, the flower clusters arising from the leaf axils...14
13b. Inflorescences terminal in general appearance, the inflorescences spicate or racemose, or headlike...21

14a. Flowers and fruit pedicellate...15
14b. Flowers and fruit sessile...17

15a. Leaves with serrate margins, elliptic-ovate; stems erect...***Mentha*** (*arvensis*)
15b. Leaves with crenate margins, cordate-reniform or triangular-ovate with a rounded tip and cordate base; stems lax, sometimes creeping...16

16a. Leaves glabrous or sparsely hirsute, cordate-reniform; calyx green...***Glechoma***
16b. Leaves densely white-tomentose below, triangular-ovate with a rounded tip and cordate base; calyx bluish or purplish-tinged...***Nepeta***

17a. Stems, calyx, and lower leaves densely white-tomentose; calyx actinomorphic, 10-toothed with the teeth recurved-hooked at the tip; flowers white...***Marrubium***
17b. Stems, calyx, and lower leaves glabrous or variously hairy but not densely tomentose; calyx bilabiate or if actinomorphic then only 5-toothed; flowers various...18

18a. Corolla 2.5–3.5 mm long, regular or nearly so; stems ridged; flowers white; leaves pinnately cleft or sharply serrate; stamens 2...***Lycopus***
18b. Corolla 10–30 mm long, two-lipped and zygomorphic; stems usually not strongly ridged; flowers various; leaves crenate or serrate, but usually not sharply so, never pinnately cleft; stamens 2 or 4...19

19a. Calyx bilabiate with 3 teeth above and 2 teeth below, the teeth purplish-tinged; corolla glabrous; stems with spreading hairs...***Clinopodium***
19b. Calyx actinomorphic, green throughout; corolla pubescent at least on the upper lip; stems glabrous or appressed-hairy, or sometimes with spreading hairs above...20

20a. Stamens 2; leaves with serrate margins; stems with retrorse hairs; calyx 13–15-nerved; flowers white to pale pinkish or rarely purplish...***Monarda*** (*pectinata*)
20b. Stamens 4; leaves with mostly crenate margins; stems glabrous or nearly so or hairy but the hairs not retrose; calyx inconspicuously 5-nerved; flowers purple or pinkish-purple...***Lamium***

21a. Flowers in a head subtended by foliaceous bracts, dark lavender to rose-purple, pink-purple, or white, 12–35 mm long; stamens 2; calyx actinomorphic, green...***Monarda***
21b. Flowers in a spike or panicle, subtended by foliaceous bracts or not, variously colored, 6–35 mm long; stamens 2 or 4; calyx bilabiate or actinomorphic, green or purplish- to bluish-tinged...22

22a. Stems and leaves white- to grayish-canescent, especially on the lower surface of the leaves; flowers creamy-white and the lower lip usually with purple spots, or bluish to purplish; leaves triangular-ovate with crenate-serrate margins...***Nepeta***
22b. Stems and leaves glabrous or variously hairy but not canescent, if floccose-tomentose then the inflorescence in a terminal panicle subtended by foliaceous bracts; flowers various but if white then usually without purple spots; leaves various...23

23a. Corolla nearly actinomorphic, 4-lobed, 1.7–5 mm long, glabrous; flowers in a terminal spike not subtended by foliaceous bracts...***Mentha***
23b. Corolla distinctly bilabiate with the upper lip entire or emarginate and galeate and the lower lip 3-lobed, 6–30 mm long, usually hairy on the outside; flowers in a terminal panicle or spike, subtended by foliaceous bracts or not...24

24a. Calyx actinomorphic or nearly so...25
24b. Calyx bilabiate and zygomorphic...26

25a. Stems spreading-hairy and sometimes also with shorter glandular hairs; calyx green or sometimes purplish-tinged, with glandular hairs and longer nonglandular hairs; terminal spikes rather interrupted, subtended by foliaceous bracts...***Stachys***
25b. Stems glabrous or with short, retrorse hairs; calyx bluish to purplish or pinkish- to whitish-tinged, at least on the teeth, without glandular hairs; terminal spikes not interrupted, subtended by foliaceous bracts or not...***Agastache***

26a. Stamens 2; flowers in a terminal panicle or loose spike, subtended by foliaceous bracts or not; stems often ridged...***Salvia***
26b. Stamens 4; flowers in a dense terminal spike subtended by suborbicular bracts with long-acuminate tips; stems not ridged...***Prunella***

AGASTACHE Clayton ex Gronov. – GIANT HYSSOP

Perennial herbs; *leaves* simple; *flowers* borne in dense terminal spike-like inflorescences; *calyx* with 15 or more prominent veins, tubular or campanulate, the teeth and upper part of the tube often whitish and scarious or pink or blue; *corolla* pink to purple or white, bilabiate; *stamens* 4.

1a. Leaves white tomentulose below, green and glabrous above; flowers blue to violet; calyx teeth 1–1.5 mm long, blue to violet; corolla 7–10 mm long...***A. foeniculum***
1b. Leaves green above and below, scabrous or minutely hairy; flowers whitish or faintly pink; calyx teeth 1.5–5 mm long, rose, pink or green; corolla 10–15 mm long...2

2a. Stamens exserted 1–2 mm beyond the upper lip; stems finely and densely hairy; leaves scabrous hairy on both sides, more densely so below; petioles 0.5–1.5 cm long...***A. pallidiflora***
2b. Stamens exserted 4–7 mm beyond the upper lip; stems glabrous or hairy at the nodes or sparsely hairy throughout; leaves glabrous or nearly so above, glabrous or finely scabrous hairy below, mostly on the main veins; petioles 1–5 cm long...***A. urticifolia***

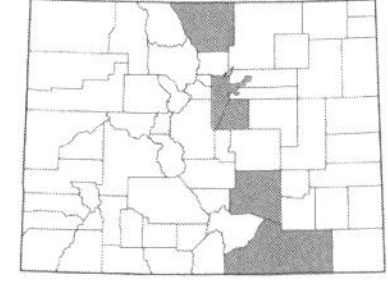

***Agastache foeniculum* (Pursh) Kuntze**, BLUE GIANT HYSSOP. Plants 6–12 dm; *leaves* ovate, 3–9 × 1.5–5.5 cm, white tomentulose below, green and glabrous above, serrate; *calyx* tube 4–7 mm long, with blue to violet teeth 1–1.5 mm long; *corolla* 7–10 mm long, blue to violet. Uncommon in forest openings, 6500–8000 ft. July–Sept. E.

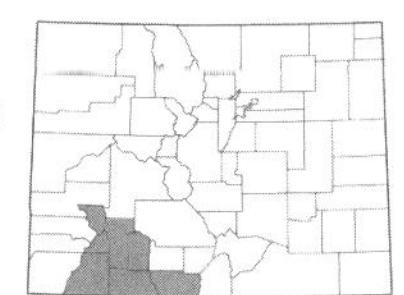

***Agastache pallidiflora* (A. Heller) Rydb.**, BILL WILLIAMS MOUNTAIN GIANT HYSSOP. Plants to 8+ dm; *leaves* deltoid-ovate, to 6 × 5.5 cm, green above and below, minutely hairy above, crenate-serrate; *calyx* tube 5–8 mm long, with pink to rose or rose-purple teeth 1–4 mm long; *corolla* 9–18 mm long, white to rose. Locally common on rocky outcroppings, along streams, and in meadows, 7600–10,800 ft. July–Sept. E/W.

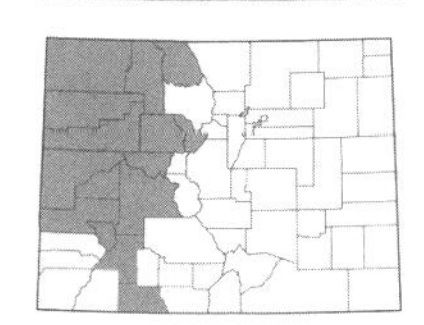

***Agastache urticifolia* (Benth.) Kuntze**, NETTLELEAF GIANT HYSSOP. Plants 5–20 dm; *leaves* ovate to deltoid-ovate, green above and below, glabrous or sparsely hairy below, crenate-serrate; *calyx* tube 4–7 mm long with green to pink teeth 2.5–5 mm long; *corolla* 8–15 mm long, white to rose or light purple. Common on rocky outcroppings and in meadows and forest openings, sometimes in moist places, 6700–10,900 ft. June–Sept. E/W.

CLINOPODIUM L. – CALAMINT

Perennial herbs; *leaves* simple; *flowers* borne in dense terminal cymules or an interrupted spike; *calyx* tubular, bilabiate; *corolla* white to purple or rose, bilabiate, the upper lip erect and flat or nearly so; *stamens* 4, didynamous.

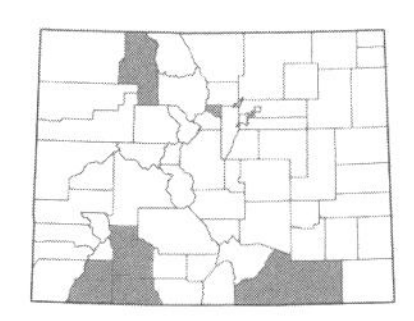

***Clinopodium vulgare* L.**, WILD BASIL. [*Satureja vulgaris* (L.) Fritsch]. Plants 1–5 dm; *leaves* ovate to deltoid-ovate,10–40 × 6–15 mm, entire to minutely toothed; *calyx* 8–9 mm long, villous, with subulate teeth; *corolla* 8–12 mm long. Uncommon in moist, disturbed places, 7500–8500 ft. July–Sept. E/W. Introduced.

DRACOCEPHALUM L. – DRAGONHEAD

Herbs; *leaves* petiolate, sharply serrate; *flowers* borne in dense clusters, with spinulose-pectinate bracts; *calyx* bilabiate, 15-nerved, 5-toothed, the upper tooth much larger; *corolla* blue or purple or sometimes white, bilabiate; *stamens* 4, included.

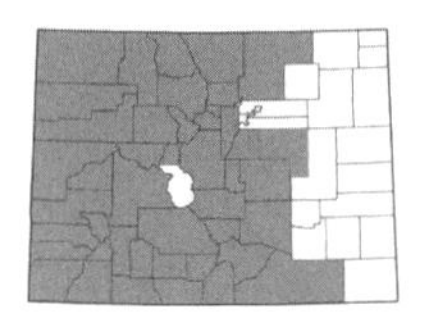

Dracocephalum parviflorum **Nutt.**, American dragonhead. [*Moldavica parviflora* (Nutt.) Britton]. Annuals or biennials, 3–8 dm; *leaves* lanceolate to elliptic, 3–10 × 0.7–3 (5) cm, glabrous above, glabrous to hairy below, sharply spinose-serrate; *calyx* 8–15 mm long with lobes 3–6 mm long; *corolla* just longer than the calyx; *nutlets* 2–3 mm long. Common along roadsides, in disturbed areas, often in moist places, 5500–10,800 ft. June–Aug. E/W. (Plate 50)

GALEOPSIS L. – hempnettle

Annual herbs; *leaves* petiolate, toothed; *flowers* sessile in dense terminal, bracteate interrupted spikes or clusters; *calyx* strongly 10-nerved, actinomorphic, 5-toothed; *corolla* purplish and white (ours), bilabiate; *stamens* 4, the upper pair longer.

1a. Stems swollen below the nodes...***G. bifida***
1b. Stems not swollen below the nodes...***G. ladanum***

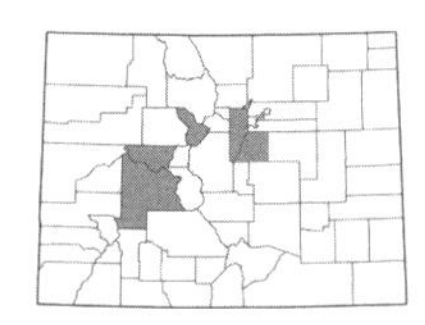

Galeopsis bifida **Boenn.**, splitlip hempnettle. Plants 4–8 dm; *stems* with recurved hairs, sometimes also with glandular hairs; *leaves* lanceolate to ovate, mostly 5–10 cm long, hirsute, crenate-serrate; *calyx* 7–15 mm long, with aristate lobes; *corolla* 13–17 mm long, hirsute, well-exserted from the calyx; *nutlets* ca. 3.5 mm long, smooth. Uncommon in moist areas, usually in disturbed areas, 7000–8500 ft. June–Aug. E/W. Introduced.

Galeopsis ladanum **L.**, red hempnettle. Plants 1–4 dm; *stems* with recurved and glandular hairs; *leaves* ovate, mostly 5–10 cm long, crenate-serrate; *calyx* with aristate lobes; *corolla* 15–28 mm long, well-exserted from the calyx. Uncommon in disturbed areas, recently found in Roxborough State Park (Douglas Co.), 6000–7000 ft. July–Oct. E. Introduced.

GLECHOMA L. – ground ivy

Creeping perennial herbs; *leaves* simple, crenate; *flowers* borne in small axillary clusters; *calyx* 5-toothed, the teeth unequal; *corolla* blue or purple, bilabiate; *stamens* 4, the upper pair longer.

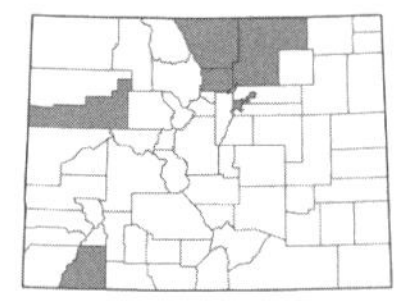

Glechoma hederacea **L.**, ground ivy. Stems to 4 dm long; *leaves* suborbicular with a cordate base, 1–2.5 (4) cm long, glabrous to sparsely hairy, punctate; *calyx* 4–7 mm long, with aristate teeth; *corolla* 7–17 mm long, hairy; *nutlets* ca. 2 mm long, smooth. Uncommon in disturbed, usually moist areas, 5000–6000 ft. May–July. E/W. Introduced.

HEDEOMA Pers. – false pennyroyal

Aromatic herbs; *leaves* simple, entire or obscurely toothed; *flowers* borne in the axils of leaves; *calyx* bilabiate, the lower 2 teeth longer than the upper 3, 13-nerved, gibbous at the base; *corolla* blue or purple, bilabiate; *stamens* 2. (Irving, 1980)

1a. Perennials; flowers present only on the upper ¾ of the stem; the upper and lower calyx teeth converging and closing the opening at maturity; bracteoles usually shorter, 1–2 mm long...***H. drummondii***
1b. Annuals; flowers present on almost the entire length of the stem; the upper and lower teeth not converging at maturity, leaving the mouth of the calyx open; bracteoles usually longer, (1.5) 2.5–6.5 mm long...***H. hispida***

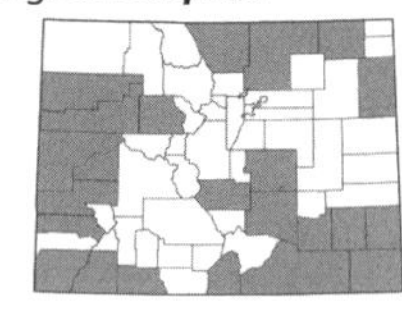

Hedeoma drummondii **Benth.**, Drummond's false pennyroyal. Perennials, 0.5–3 dm; *leaves* linear to narrowly elliptic, 5–20 × 1–3 (5) mm, punctate, glabrous above and hairy below, entire; *calyx* 5–7.5 mm long, hirsute, teeth subulate to triangular, 1–2 mm long; *corolla* 7–9 mm long; *nutlets* 1.3–1.6 mm long. Common, usually found in rocky soil, 3500–7800 ft. June–Aug. E/W. (Plate 50)

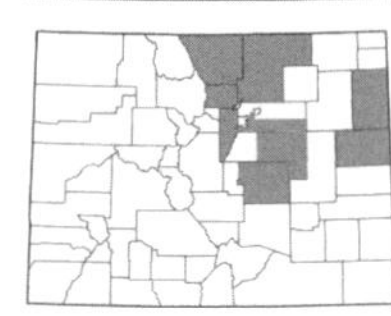

Hedeoma hispida **Pursh**, rough false pennyroyal. Annuals, 0.7–3.5 dm; *leaves* linear to narrowly elliptic, 7–20 × 1–3 mm, punctate, glabrous above and glabrous to hairy below, entire; *calyx* 4.5–6 mm long, hirsute, teeth triangular, 1–2 mm long; *corolla* 6–7 mm long; *nutlets* 1–1.3 mm long. Uncommon on the eastern plains and lower foothills, 3800–6000 ft. May–July. E.

LAMIUM L. – deadnettle

Annual or perennial herbs; *leaves* usually toothed; *flowers* whorled in axillary or terminal clusters; *calyx* actinomorphic, 5-toothed, the teeth all equal or the upper one larger; *corolla* purple, white, or rarely yellow, bilabiate; *stamens* 4.

1a. Rhizomatous perennials; at least some leaves with an irregular, broad whitish midstripe on the upper side; corolla 2–3 cm long with an upper lip 7–12 mm long...***L. maculatum***
1b. Annuals; leaves lacking a broad, whitish central stripe on the upper side; corolla 1–2 cm long with an upper lip 3–6 mm long...2

2a. Bracts and upper leaves sessile and clasping, rarely shortly petiolate, broad-based...***L. amplexicaule***
2b. Bracts and upper leaves petiolate...***L. purpureum***

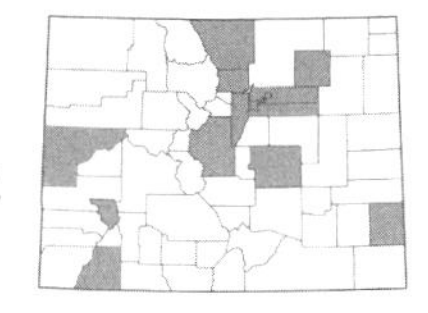

***Lamium amplexicaule* L.**, HENBIT. Annuals, 1–3.5 dm; *leaves* orbicular to ovate, 5–16 mm long, crenate, the upper leaves usually sessile and clasping; *calyx* 5–7 mm long, densely hairy; *corolla* pink-purple, 10–20 mm long, tube 10–15 mm long, upper lip 3–5 mm long, lower lip 1.5–2.5 mm long; *nutlets*1.5–2 mm long. Locally common in disturbed places, 3600–7800 ft. April–Sept. E/W. Introduced.

***Lamium maculatum* L.**, SPOTTED HENBIT. Perennials from rhizomes, 2–5 dm; *leaves* deltoid-ovate, 15–60 mm long, crenate, with a broad, whitish midstripe on the upper side; *calyx* 5–10 mm long; *corolla* pink-purple or rarely white, 20–30 mm long, upper lip 7–12 mm long. Uncommon but often cultivated, found in moist, disturbed places, known from Montrose Co., 5800–6000 ft. April–Aug. W. Introduced.

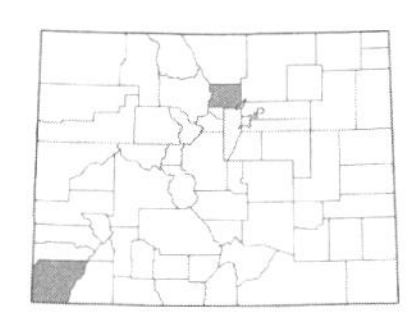

***Lamium purpureum* L.**, PURPLE DEADNETTLE. Annuals, 1–4 dm; *leaves* suborbicular to ovate, 5–35 mm long, hirsute, crenate, the upper leaves petiolate; *calyx* 5–8 mm long, sparsely hairy; *corolla* pink-purple, 10–20 mm long, the tube 7–12 mm long, upper lip 3–6 mm long, and lower lip 1.5–2.5 mm long; *nutlets* 1.8–2.5 mm long. Uncommon in disturbed places and gardens, 5000–6000 ft. April–June. E/W. Introduced.

LEONURUS L. – MOTHERWORT

Perennial herbs; *leaves* 3- to 5-cleft, toothed; *flowers* borne in dense leafy-bracteate, axillary clusters; *calyx* 5- or 10-nerved, 5-toothed, the teeth with spinulose tips; *corolla* pink or white, bilabiate, the tube barely exceeding the calyx; *stamens* 4.

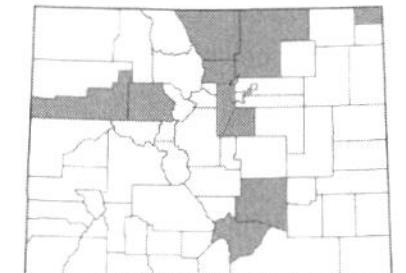

***Leonurus cardiaca* L.**, MOTHERWORT. Plants 5–20 dm, glabrous to sparsely hairy; *leaves* ovate to suborbicular, 3–12 cm long, the larger leaves 3–5-cleft with serrate margins; *calyx* 3.5–8 mm long, 5-nerved; *corolla* 8–12 mm long; *nutlets* 1.5–2.3 mm long. Locally common in moist, disturbed places, 4500–9000 ft. June–Aug. E/W. Introduced. (Plate 50)

LYCOPUS L. – WATER HOREHOUND; BUGLEWEED

Perennial herbs, not much aromatic; *leaves* simple, toothed or pinnatifid; *flowers* borne in dense sessile clusters in the axils of leaves; *calyx* actinomorphic, 4–5-toothed, campanulate to ovoid, 4–5-nerved; *corolla* usually white, 4–5-lobed; *stamens* 2.

1a. At least the lower leaves pinnatifid, the others irregularly sharply serrate; plants without tubers…***L. americanus***
1b. Leaf margins sharply but evenly serrate; plants usually with tubers (although difficult to collect)…2

2a. Calyx 2.5–4.5 mm long, much exceeding the nutlets, the teeth with a slender tip; lower leaves sessile…***L. asper***
2b. Calyx 1.3–1.6 mm long, equal to or shorter than the nutlets, the teeth without a slender tip; lower leaves shortly petiolate…***L. uniflorus***

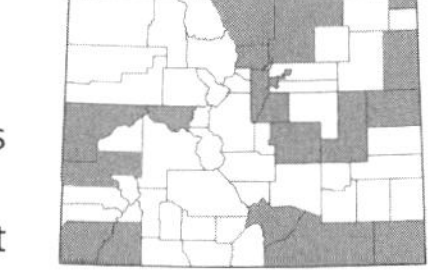

***Lycopus americanus* Muhl. ex W. Bartram**, AMERICAN WATER HOREHOUND. Plants 3–9 dm; *leaves* narrowly elliptic to ovate-lanceolate, 1.5–12 cm long, irregularly sharply serrate, the lower leaves pinnatifid; *calyx* 2–3 mm long; *corolla* mostly actinomorphic, 4-lobed, 2.5–3.5 mm long, just longer than the calyx; *nutlets* 1–1.5 mm long, with a narrow wing-like margin. Common in moist soil, sometimes in standing water, 3500–7500 ft. July–Sept. E/W.

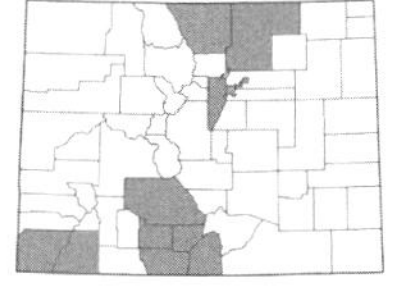

***Lycopus asper* Greene**, ROUGH BUGLEWEED. Plants 3–13 dm; *leaves* narrowly elliptic, 5–8 cm long, evenly serrate; *calyx* 2.5–4.5 mm long; *corolla* nearly actinomorphic, 4-lobed, 3.5–6 mm long, 1–2 mm longer than the calyx; *nutlets* 1.7–2.5 mm long, with a narrow wing-like margin. Locally common in moist places, 4800–7700 ft. July–Sept. E/W.

***Lycopus uniflorus* Michx.**, NORTHERN BUGLEWEED. Plants 3–6 dm; *leaves* narrowly elliptic to lanceolate-ovate, 2–10 cm long, evenly serrate; *calyx* 1.3–1.6 mm long; *corolla* nearly actinomorphic, 5-lobed, 2.5–3.5 mm long; *nutlets* 1–1.2 mm long with a narrow wing-like margin. Uncommon in moist places, 4600–6000 ft. July–Sept. E.

MARRUBIUM L. – HOREHOUND

Perennial herbs, usually densely woolly; *leaves* simple, toothed; *flowers* borne in dense axillary clusters; *calyx* 5- to 10-nerved, 5- or 10-toothed; *corolla* white or purplish or rarely yellow, bilabiate; *stamens* 4.

***Marrubium vulgare* L.**, COMMON HOREHOUND. Plants 3–7 dm; *leaves* ovate to orbicular, 1.5–5 × 1–4 cm, crenate-serrate; *calyx* tube 3–5 mm long, teeth alternately long and short with hooked apices; *corolla* 4–6 mm long, white; *nutlets* 1.8–2.3 mm long, smooth. Locally common in disturbed places, along roadsides, and in old pastures, 4200–9000 ft. May–Sept. E/W. Introduced.

MENTHA L. – MINT

Perennial herbs, very aromatic; *leaves* simple and toothed, punctate; *flowers* borne in dense axillary clusters or terminal spikes; *calyx* actinomorphic or obscurely bilabiate, 5-toothed, 10-nerved; *corolla* 4-lobed, bilabiate; *stamens* 4, exserted, equal.

1a. Flowers in dense axillary clusters at the nodes...***M. arvensis***
1b. Flowers clusters in a terminal spike...2

2a. Leaves sessile or shortly petiolate (to 3 mm long)...***M. spicata***
2b. Leaves conspicuously petiolate, the petioles mostly 4 mm or longer...***M. ×piperita***

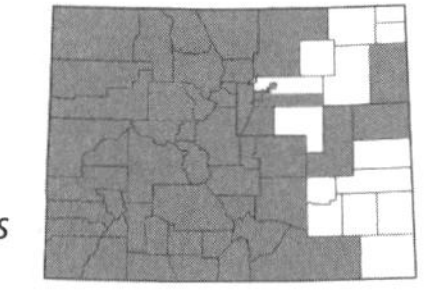

***Mentha arvensis* L.**, WILD MINT. Plants 3–5 (9) dm; *leaves* lanceolate to ovate, 2.5–12 cm long, serrate and ciliate; *flowers* in dense axillary clusters at the nodes; *calyx* 2.5–3.3 mm long; *corolla* 4.5–6.5 mm long, white to lavender, glabrous; *nutlets* 0.7–1.5 mm long, smooth. Common in moist places, especially along streams and ditches, 3500–9500 ft. June–Sept. E/W. (Plate 50)

Closely resembles *Lycopus* in appearance, but *Mentha* flowers have 4 stamens while *Lycopus* flowers have 2. In addition, the flowers of *Mentha* are on short pedicels and the middle cauline leaves are petiolate, while the flowers and leaves of *Lycopus* are sessile.

***Mentha ×piperita* L.**, PEPPERMINT. Plants 3–10 dm; *leaves* ovate, 1.5–7 cm long, serrate, with petioles mostly over 4 mm; *flowers* in terminal spikes; *calyx* 2–4 mm long; *corolla* 4–5 mm long, white to pink or pink-purple. Uncommon in moist places and disturbed areas, 5000–5800 ft. July–Sept. E. Introduced.

Apparently originating through hybridization between *M. spicata* L. and *M. aquatica* L.

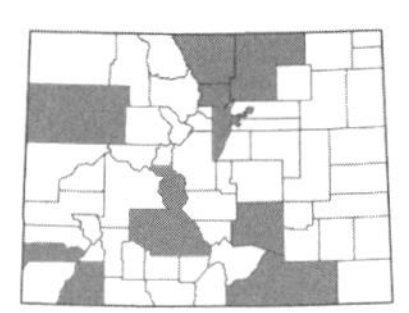

***Mentha spicata* L.**, SPEARMINT. Plants 3–7 dm; *leaves* lance-ovate to ovate, 3–9 cm long, serrate, with petioles 1–3 mm long or sessile; *calyx* 1–3 mm long; *corolla* 1.5–3 mm long, white to purple. Uncommon in moist places, especially along streams and ditches, 4800–6800 ft. July–Sept. E/W. Introduced.

MONARDA L. – BEEBALM

Annual or perennial herbs or shrubs; *leaves* simple, subentire or usually toothed; *flowers* in dense, headlike clusters; *calyx* 13- to 15-nerved, 5-toothed, the teeth nearly equal; *corolla* white, pink, or purple, strongly bilabiate; *stamens* 2. (McClintock & Epling, 1942)

1a. Heads solitary and terminal; stamens exserted from under the upper lip; flowers mostly 2–3.5 cm long, purple, pink, or rose-purple, rarely white, the upper lip erect, not helmet-like; calyx teeth short, about 1 mm long; perennials...***M. fistulosa* var. *menthifolia***
1b. Heads 2 or more in an interrupted spike; stamens included under the upper lip; flowers mostly 1.2–2 cm long, usually white, sometimes pale pinkish or rarely purple, the upper lip helmet-like; calyx teeth 2–4 mm long; annuals...***M. pectinata***

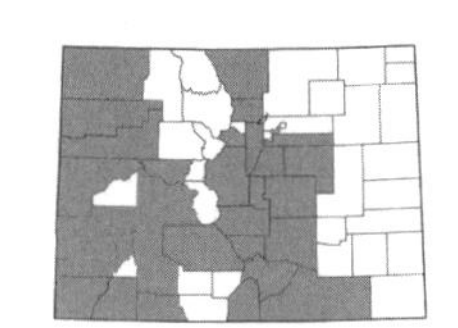

***Monarda fistulosa* L. var. *menthifolia* (Graham) Fernald**, WILD BERGAMOT. Perennials, 3–12 dm; *leaves* lanceolate to ovate, 3–10 cm long, serrate to nearly entire, with a petiole 2–20 mm long; *heads* solitary and terminal; *outer bracts* suborbicular to ovate, 1–2.5 cm long; *calyx* tube 5.5–11 mm long with teeth ca. 1 mm long; *corolla* purple, pink, or rose-purple, rarely white, 20–35 mm long, the tube 15–25 mm long and upper lip usually erect; *nutlets* 1.5–2 mm long. Common in meadows and along roadsides, often in moist places, 5400–8800 ft. June–Sept. E/W. (Plate 50)

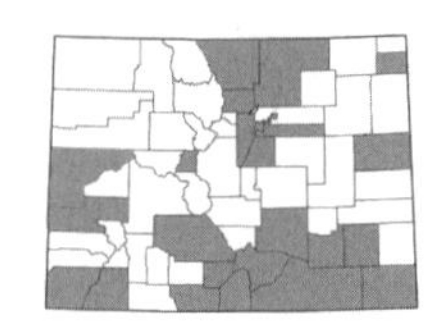

***Monarda pectinata* Nutt.**, PLAINS BEEBALM. Annuals, 2–5 dm; *leaves* lanceolate to ovate, 2–4 cm long, serrate to subentire, with a petiole 2–5 (15) mm long; *heads* 2 or more in an interrupted spike; *outer bracts* ovate to elliptic, 0.8–1.5 (2.5) cm long; *calyx* tube 6–8 mm long with teeth 2–4 mm long; *corolla* usually white, sometimes pale pink or rarely purple, 12–20 mm long, the tube 8–15 mm long and the upper lip helmet-like; *nutlets* 1.2–1.5 mm long. Common on open slopes and in meadows, 3500–8500 ft. May–Aug. E/W. (Plate 50)

MONARDELLA Benth. – MONARDELLA

Annual or perennial herbs; *leaves* simple, entire or toothed; *flowers* in a dense terminal head; *calyx* tubular, 5-toothed, 10- to 15-nerved; *corolla* pink-purple or sometimes white with purple spots, obscurely bilabiate; *stamens* 4. (Epling, 1925)

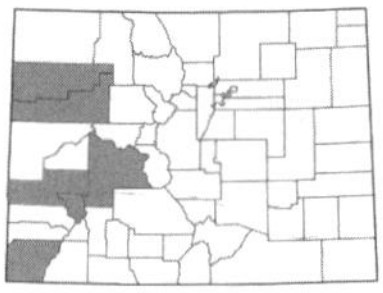

***Monardella odoratissima* Benth.**, STINKING HORSEMINT. [*Monardella glauca* Greene]. Perennials, 2–4 dm; *leaves* lanceolate to oblong, 2–3 cm long, minutely hairy; *bracts* orbicular to ovate, rose-purple or whitish; *calyx* 5–8 mm long, 13-nerved; *corolla* ca. 15 mm long. Uncommon on shale or rocky slopes and in sagebrush meadows, 5600–11,000 ft. July–Aug. W.

NEPETA L. – CATNIP

Perennial herbs, aromatic; *leaves* toothed, petiolate; *flowers* borne in a terminal interrupted spike; *calyx* 5-toothed, 15-nerved; *corolla* blue, white, or rarely yellow, bilabiate; *stamens* 4, the upper pair longer, barely exserted.

1a. Flowers white; calyx usually green…***N. cataria***
1b. Flowers blue or purple; calyx usually bluish-purple…2

2a. Calyx 9.5–11 mm long; leaves usually larger, 2–6 cm long; plants erect…***N. grandiflora***
2b. Calyx 8 mm long or less; leaves usually smaller, mostly 0.5–1.7 cm long; plants with decumbent or ascending stems…***N. racemosa***

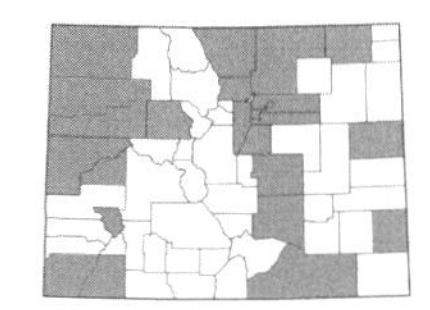

***Nepeta cataria* L.**, CATNIP. Plants 3–10 dm, erect; *leaves* deltoid-ovate, 2–7 (10) cm long, crenate to crenate-serrate, with a petiole 0.5–3 (5) cm long; *calyx* 5–7 mm long, green; *corolla* white with purple spots, 7–12 mm long; *nutlets* 1.3–2 mm long. Common in disturbed areas and along roadsides, often in moist places, 3500–8500 ft. July–Sept. E/W. Introduced. (Plate 50)

***Nepeta grandiflora* M. Bieb.**, GIANT CATMINT. Plants 4–8 dm, erect; *leaves* ovate, 2–6 (10) cm long, crenate, cordate; *calyx* 9.5–11 mm long, bluish-purple; *corolla* blue to bluish-purple, 14–18 mm long. Not yet reported for Colorado but present close to the border in southeastern Wyoming. June–Sept. Introduced.

***Nepeta racemosa* Lam.**, PERSIAN CATMINT. [*N. mussinii* Henckel]. Plants with decumbent or ascending stems; *leaves* ovate, mostly 0.5–1.7 cm long, crenate, cordate; *calyx* to 8 mm long, bluish-purple; *corolla* bluish-purple. Uncommon in disturbed areas, 7500–8500 ft. June–Sept. Introduced.

PHYSOSTEGIA Benth. – FALSE DRAGONHEAD

Perennial herbs; *leaves* toothed or sometimes entire, sessile; *flowers* borne in terminal spike-like racemes; *calyx* 10-nerved, 5-toothed, the teeth equal or unequal; *corolla* pink to purple or white, bilabiate; *stamens* 4, the uppermost pair shorter.

***Physostegia parviflora* Nutt. ex A. Gray**, WESTERN FLASE DRAGONHEAD. Plants 2–8 dm; *leaves* lanceolate to oblong, 4–12 cm long; *bracts* lanceolate to ovate; *calyx* 3–6 mm long, to 7 mm long in fruit, with triangular teeth ca. 1.5 mm long; *corolla* purple to pink-purple, 0–15 mm long, the tube 6–9 mm long; *nutlets* 2–3.3 mm long. Found along the shores of streams and lakes and in other moist places, not yet reported for Colorado but likely to be found in northern Larimer or Weld cos. July–Sept.

POLIOMINTHA A. Gray – ROSEMARY-MINT

Canescent shrubs; *leaves* entire, small; *flowers* borne in small axillary clusters in the upper leaf axils; *calyx* 5-toothed, the teeth nearly equal, 13- to 15-nerved; *corolla* pink, purple, or white, bilabiate; *stamens* 2.

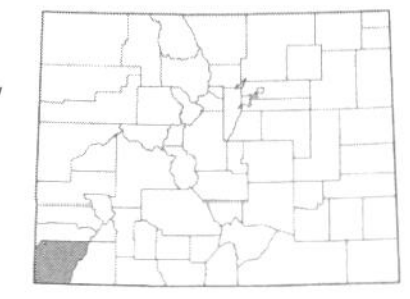

***Poliomintha incana* (Torr.) A. Gray**, PURPLE SAGE. Shrubs 3–10 dm; *leaves* linear to narrowly oblong, white-tomentose, 1–3 cm long, sessile; *calyx* 6–7 mm long with subulate teeth; *corolla* pink, purple, or white, with purple spots on the lower lip, 10–15 mm long. Uncommon in sandy, dry deserts and dunes, 4700–6000 ft. April–July. W.

PRUNELLA L. – SELFHEAL

Perennial herbs; *leaves* simple, entire or obscurely toothed (ours), petiolate; *flowers* borne in a dense, bracteate terminal spike; *calyx* 10-nerved, bilabiate; *corolla* pink, blue-violet, or white, bilabiate; *stamens* 4, the lower pair longer.

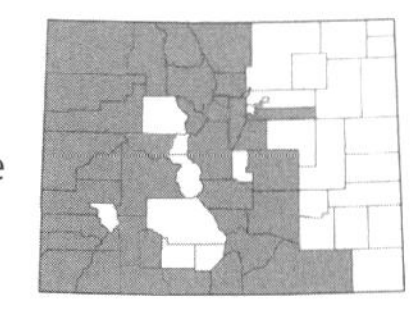

***Prunella vulgaris* L.**, HEAL-ALL. Plants prostrate to decumbent or erect, the stems to 10 dm long; *leaves* lanceolate to narrowly ovate or oblong, 2–9 cm long, entire to crenate-serrate, with a cuneate to attenuate base; *bracts* ovate to suborbicular, 5–15 mm long, abruptly short-acuminate at the apex; *calyx* 6–10 mm long; *corolla* 10–18 mm long; *nutlets* 1.7–2 mm long, carunculate. Common in moist places, 5300–10,500 ft. June–Aug. E/W. (Plate 50)

SALVIA L. – SAGE

Herbs or shrubs, usually aromatic; *leaves* entire or toothed; *flowers* in terminal bracteate racemes or interrupted spikes; *calyx* strongly 10- to 15-nerved, bilabiate; *corolla* strongly bilabiate; *stamens* 2.

1a. Flowers pale yellow or white; stem, leaves, and calyx densely floccose-tomentose (the leaves becoming glabrate above in age); lower leaves ovate to elliptic and petiolate, the upper reduced and clasping; bracts 1–2 cm long, suborbicular with a long-acuminate tip…***S. aethiopis***
1b. Flowers blue, purple, rose, or rarely white; stem, leaves, and calyx not densely floccose-tomentose; plants otherwise unlike the above in all respects…2

2a. Calyx glandular-hairy; stem glandular-hairy above; flowers large, the corolla 15–30 mm long…3
2b. Calyx glabrous or hairy but not glandular; stem glabrous or hairy but not glandular-hairy above; corolla 6–25 mm long…4

3a. Flowers violet-blue, rarely pink or white; bracts green, shorter than to nearly equaling the calyx in length…***S. pratensis***
3b. Flowers blue or rarely white; bracts usually dry and white or pinkish-tinged, twice as long as the calyx or more…***S. sclarea***

4a. Bracts broadly ovate, dark purple; calyx dark purple, the upper lip 3-toothed; leaves ovate to broadly elliptic with a cuneate base, glabrous to pilose above and paler and canescent below; flowers dark purple, blue, or rose…***S. ×sylvestris***
4b. Bracts linear-lanceolate or ovate-lanceolate, green or bluish; calyx green or bluish, the upper lip entire; leaves lanceolate, linear-lanceolate, or narrowly elliptic, with a tapering base, glabrous or minutely hairy above and glabrous or minutely hairy below; flowers dark to pale blue or rarely white…5

5a. Perennials; corolla 10–25 mm long, usually remaining after collecting; leaves lanceolate to linear-lanceolate…***S. azurea***
5b. Annuals; corolla 6–9 mm long, easily falling; leaves lanceolate to narrowly elliptic…***S. reflexa***

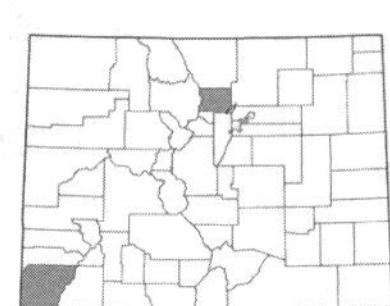

***Salvia aethiopis* L.**, Mediterranean sage. Perennials, 3–6 dm, tomentose; *leaves* ovate to elliptic, 5–30 cm long, irregularly toothed; *bracts* 1–2 cm long, suborbicular with a long-acuminate tip; *calyx* 8–15 mm long, with spine-tipped teeth; *corolla* 1–2 cm long, pale yellow or white; *nutlets* 2.5–3 mm long. Uncommon weed along roadsides and disturbed areas, 5000–6000 ft. May–July. E/W. Introduced.

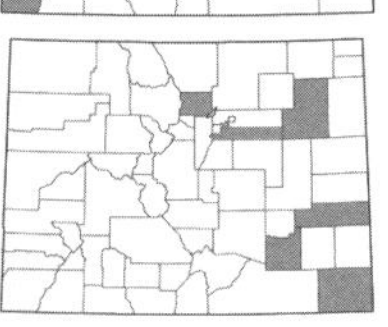

***Salvia azurea* Michx. ex Lam. var. *grandiflora* Benth.**, blue sage. Perennials, 5–15 dm; *leaves* lanceolate to linear-lanceolate, 3–7 cm long, serrate or minutely toothed; *bracts* 2–8 mm long; *calyx* 4–10 mm long, with acute, triangular teeth; *corolla* 1–2.5 cm long, blue or rarely white; *nutlets* 2–3 mm long. Uncommon along roadsides on the eastern plains and lower foothills, 4000–5200 ft. July–Oct. E. (Plate 50)

***Salvia pratensis* L.**, meadow clary. Perennials, 3–6 dm; *leaves* ovate to ovate-oblong, 7–12 cm long, crenate-serrate; *racemes* 10–20 cm long, interrupted; *bracts* shorter to nearly equal to the calyx; *calyx* 7–11 mm long, upper lip minutely 3-toothed and the lower lip with acute to aristate teeth; *corolla* 1.5–3 cm long, violet-blue, rarely pink or white. Uncommon weed in fields and gardens, known from the Denver/Boulder area, 5000–5500 ft. June–Sept. E. Introduced.

***Salvia reflexa* Hornem.**, Rocky Mountain sage. Annuals, 2–7 dm; *leaves* lanceolate to narrowly elliptic, 2–6 × 0.3–1.5 cm, irregularly toothed to subentire; *bracts* 2–5 mm; *calyx* 4–6 mm long in flower, to 8 mm long in fruit; *corolla* 6–9 mm long, blue or sometimes white; *nutlets* 2–3 mm long. Common in open, rocky places, 3500–8000 ft. June–Oct. E/W.

***Salvia sclarea* L.**, clary sage. Biennials, 5–15 dm; *leaves* ovate to ovate-oblong, 7–20 cm long, toothed, rugose; *bracts* 10–30 mm, broadly ovate, with an abruptly acuminate tip; *calyx* with teeth 1.5–3 mm long; *corolla* blue or rarely white, 1.5–3 cm long. Uncommon weed in gardens and could escape along roadsides, 5000–5500 ft. June–Aug. E. Introduced.

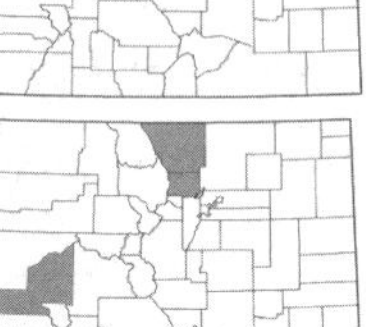

***Salvia ×sylvestris* L.**, sage. [*S. nemorosa* L.]. Perennials, 5–10 dm; *leaves* ovate to elliptic, 6–12 cm long, crenate-serrate; *bracts* broadly ovate, abruptly acuminate, 5–10 mm long; *calyx* 5–8 mm long, dark purple; *corolla* 0.9–1.5 cm long, dark purple, blue, or rose. Uncommon weed in fields and along roadsides, 5000–7000 ft. June–Sept. E/W. Introduced.

SCUTELLARIA L. – skullcap

Herbs or low shrubs; *leaves* simple; *flowers* 1–3 in leaf axils or sometimes in terminal bracteate racemes or spikes; *calyx* bilabiate, the lips entire, with an erect scutellum on the upper lip; *corolla* blue, purple, or white, bilabiate; *stamens* 4. (Epling, 1942)

1a. Leaf margins entire; leaves firm and slightly thickened, ovate to narrowly elliptic, 1–3.5 cm long…***S. brittonii***
1b. Leaf margins toothed; leaves thin, ovate to lance-ovate and truncate or obtuse at the base, (2) 3–11 cm long…2

2a. Flowers in axillary racemes; corolla 6–7 mm long…***S. lateriflora***
2b. Flowers 2 per node arising from the leaf axils; corolla 14–20 mm long…***S. galericulata***

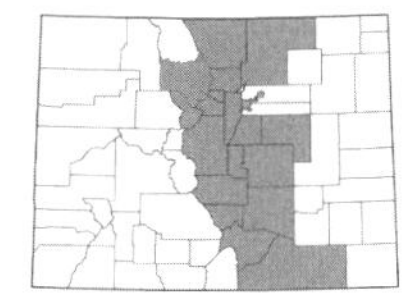

Scutellaria brittonii **Porter**, Britton's skullcap. Perennials from rhizomes, (6) 8–25 (33) cm; *leaves* lanceolate to narrowly elliptic, 1–3.5 × 0.5–1 cm, entire, petiole 5–10 mm long; *calyx* 4–6 mm long, to 6–8 mm long in fruit, with sessile glands and villous hairs, the scutellum 3–4 mm tall in fruit; *corolla* 20–30 mm long. Common in open meadows, forest openings, and prairies, 4800–9500 ft. May–July. E/W. (Plate 50)

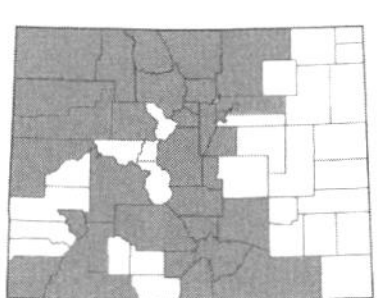

Scutellaria galericulata **L.**, marsh skullcap. Perennials, delicate, (10) 20–60 (100) cm; *leaves* lance-ovate, (2) 3–6 (9) × (0.5) 1–3 cm, shallowly toothed, petiole 0–4 mm long; *calyx* 3–4 mm long in flower and 4–6 mm long in fruit, with sessile glands and hairs, the scutellum ca. 3 mm tall in fruit; *corolla* 14–20 mm long. Locally abundant in moist places such as along pond shores, streams, and springs, 4800–9000 ft. June–Aug. E/W.

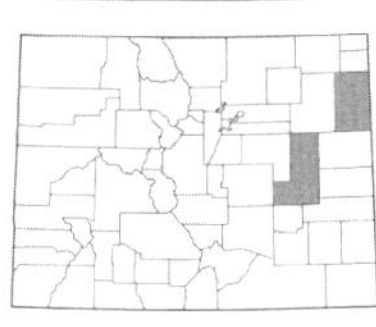

Scutellaria lateriflora **L.**, blue skullcap. Perennials, delicate, 10–60 (100) cm; *leaves* ovate, 3–11 × 1.5–5.5 cm, toothed, petiolate; *calyx* ca. 2 mm long in flower and 3 mm long in fruit, with coarse, antrorse hairs, the scutellum 2 mm tall in fruit; *corolla* 6–7 mm long. Uncommon along streams and near springs, 3500–5200 ft. July–Sept. E.

STACHYS L. – hedge-nettle

Perennial herbs from rhizomes (ours); *leaves* entire or toothed (ours), sessile or petiolate; *calyx* nearly actinomorphic, 5- to 10-nerved, 5-toothed, the teeth nearly equal and spinulose-tipped; *corolla* bilabiate; *stamens* 4.

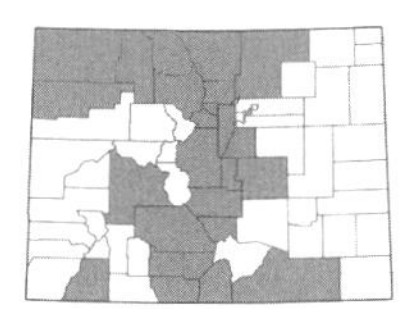

Stachys palustris **L. var. *pilosa*** **(Nutt.) Fernald**, hairy hedge-nettle; marsh betony. [*S. pilosa* Nutt. var. *pilosa*]. Plants 1.5–10 dm, spreading pilose and sometimes glandular, especially above; *leaves* lanceolate to ovate or elliptic, 3–10 cm long, densely pilose, crenate-serrate; *calyx* 5–10 mm long, pilose and usually glandular; *corolla* rose-purple, 8–15 mm long, the tube 7–10 mm long; *nutlets* 1.7–2.5 mm long. Common in moist places such as along streams, ditches, and lake shores, and in moist meadows, 4800–8500 ft. June–Aug. E/W. (Plate 50)

TEUCRIUM L. – germander

Herbs or rarely shrubs; *leaves* simple, the margins entire to deeply lobed; *flowers* in terminal bracteate spikes or racemes, or sometimes solitary in the axils of upper leaves; *calyx* 5-lobed with unequal teeth, usually actinomorphic, 10-nerved; *corolla* unequal but not bilabiate; *stamens* 4.

1a. Leaves deeply pinnatifid; flowers white, solitary in the axils of foliaceous bracts...***T. laciniatum***
1b. Leaves merely serrate or crenate; flowers rose, lavender, or purple, rarely white, in a terminal spike-like inflorescence...***T. canadense* var. *occidentale***

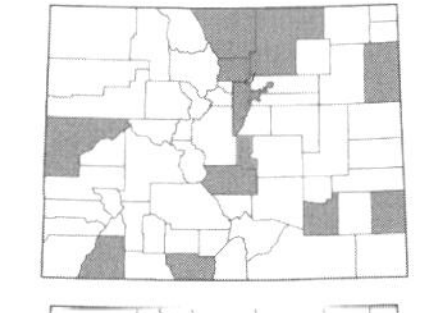

Teucrium canadense **L. var. *occidentale*** **(A. Gray) E.M. McClint. & Epling**, western germander. Perennials from rhizomes, 3–10 dm, with spreading hairs; *leaves* lanceolate to ovate, 3–15 × 1–5 cm, toothed, petioles 4–20 mm long; *calyx* 5–9 mm long, the tube 4–7 mm long; *corolla* 10–20 mm long, lavender, rarely white; *nutlets* 1.5–2.5 mm long, rugose. Found in moist places along streams, lakes, and ditches, 3500–5500 ft. June–Aug. E/W.

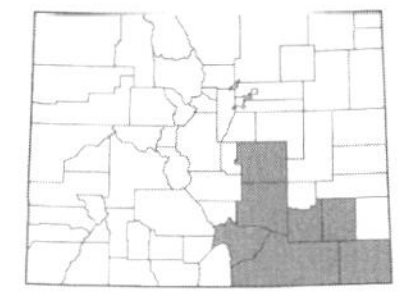

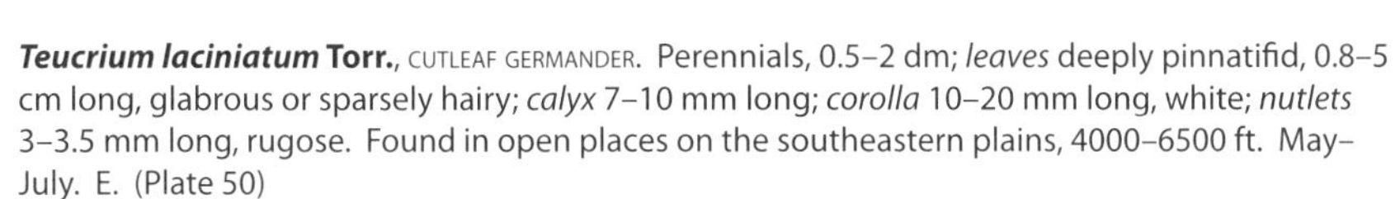

Teucrium laciniatum **Torr.**, cutleaf germander. Perennials, 0.5–2 dm; *leaves* deeply pinnatifid, 0.8–5 cm long, glabrous or sparsely hairy; *calyx* 7–10 mm long; *corolla* 10–20 mm long, white; *nutlets* 3–3.5 mm long, rugose. Found in open places on the southeastern plains, 4000–6500 ft. May–July. E. (Plate 50)

LENTIBULARIACEAE Rich. – bladderwort family

Perennials, often carnivorous herbs with or without roots, often aquatic, the traps in the form of submerged bladders; *leaves* alternate or basal, sometimes absent, exstipulate; *flowers* perfect, zygomorphic, solitary or in racemes; *sepals* 2; *petals* 5, bilabiate, saccately spurred; *stamens* 2 or 4 (if staminodes present), epipetalous; *pistil* 1; *stigma* 2-lobed; *ovary* superior, 2-carpellate, unilocular with free-central placentation; *fruit* a capsule. (Crow & Hellquist, 1985; Rossbach, 1939)

UTRICULARIA L. – bladderwort

Aquatic, carnivorous herbs, usually without roots; *leaves* submerged, finely dissected; *petals* yellow; *stamens* 2.

1a. Leaves generally large, 20–50 mm long, 2–3 times pinnately branched with a main rachis, the divisions terete; flowers 8–20, the corolla 10–18 mm long...***U. vulgaris***
1b. Leaves generally smaller, 2.5–30 mm long, 3 times palmately divided from the base (the divisions 1–3 times dichotomously branched), without a main rachis, the divisions flat; flowers 1–9, the corolla 6–16 mm long...2

2a. Margins of terminal leaf divisions entire; leaves alternating with bladder traps on the same branch; corolla spur 2.5–3 mm long...***U. minor***
2b. Margins of terminal leaf divisions minutely and sharply serrate or bristly; leaves and bladder traps on separate branches; corolla spur 3–5.5 mm long...3

3a. Spur about ½ the length of the lower corolla lip; corolla pale yellow; tips of leaf segments sharp and narrow; bristles of leaf margins on small teeth...***U. ochroleuca***
3b. Spur about as long as the lower corolla lip; corolla bright yellow; tips of leaf segments usually slightly rounded; bristles of leaf margins not on teeth...***U. intermedia***

***Utricularia intermedia* Hayne**, FLATLEAF BLADDERWORT. Leafy stems 1–5 dm long; *leaves* 3 times palmately divided from the base, with each division 1–3 times dichotomously branched, without a main rachis, 0.5–3 cm long, the ultimate segments with spinulose teeth on the margins, the tips slightly rounded; *calyx* lobes 2.5–3.5 mm long; *corolla* bright yellow, 8–12 mm long, the spur about as long as the lower lip. Uncommon in shallow fens, ponds, and bogs, 10,500–11,000 ft. July–Aug. W.

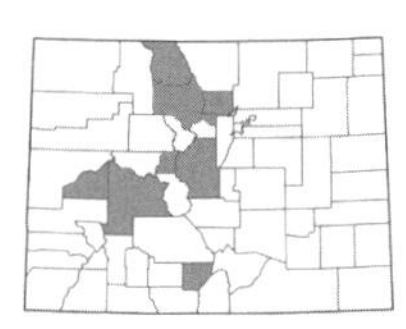

***Utricularia minor* L.**, LESSER BLADDERWORT. Leafy stems 1.5–8 dm long; *leaves* 0.25–1 cm long, 3 times palmately divided from the base (the divisions 1–3 times dichotomously branched), without a main rachis, the ultimate segments with entire margins, bladder traps alternating with leaves on the same branch; *calyx* lobes 0.5–2.5 mm long; *corolla* pale yellow, 5–8 mm long, the spur 2.5–3 mm long. Uncommon in shallow ponds, pools and fens, 7500–10,200 ft. July–Sept. E/W.

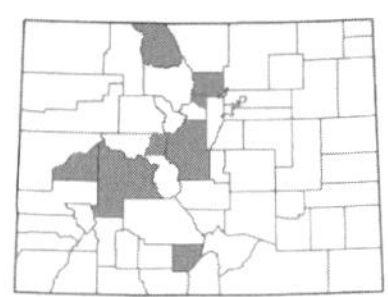

***Utricularia ochroleuca* R.W. Hartman**, YELLOWISH-WHITE BLADDERWORT. Leafy stems 1–5 dm long; *leaves* 3 times palmately divided from the base, with each division 1–3 times dichotomously branched, without a main rachis, 0.5–3 cm long, the ultimate segments with spinulose teeth on the margins, the tips sharp and narrow; *calyx* lobes 2.5–3.5 mm long; *corolla* pale yellow, 8–12 mm long, the spur about half the length of the lower lip. Uncommon in shallow pools, 9000–10,000 ft. July–Aug. E/W.

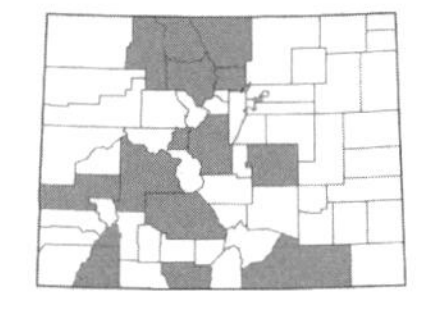

***Utricularia vulgaris* L.**, GREATER BLADDERWORT. [*U. macrorhiza* Le Conte]. Leafy stems 2–20 dm long; *leaves* 2–3 times pinnately divided with a main rachis, 2–5 cm long; *calyx* lobes 3–6 mm long; *corolla* bright yellow, 12–20 mm long, the spur ⅔ to nearly as long as the lower lip. Found in shallow ponds, lakes, marshes, fens, and slow-moving streams, 6500–10,000 ft. June–Aug. E/W.

LILIACEAE – LILY FAMILY

Perennial herbs from bulbs, corms, or rhizomes; *leaves* alternate, opposite, or whorled, simple, usually with parallel-venation, exstipulate; *flowers* usually perfect, actinomorphic or sometimes zygomorphic; *tepals* 6, in 2 series, sometimes the outer whorl greener and sepaloid, distinct; *stamens* 6, rarely 3 or 4; *pistil* 1; *ovary* superior to partly inferior, (2) 3 (4)-carpellate; *fruit* a capsule or fleshy berry.

Former members of this family are now separated in the families Alliaceae, Agavaceae, Asparagaceae, Hypoxidaceae, Melanthiaceae, Ruscaceae, and Themidaceae.

1a. Tepals in 2 distinct, dissimilar series (the inner tepals broad, the outer tepals narrow); inner tepals with a prominent, bearded or hairy gland at the base...***Calochortus***
1b. Tepals all alike and similar, not in dissimilar series; inner tepals lacking a prominent gland at the base...2

2a. Tepals orange, yellow, or purplish-brown with yellowish mottling on the inside...3
2b. Tepals white, greenish-white, or cream-colored...6

3a. Tepals large (4.5–9 cm long), orange or reddish-orange, yellowish at the base on the inside...4
3b. Tepals smaller, variously colored but unlike the above...5

4a. Tepals with maroon to blackish spots at the base on the inside, lacking a yellowish stripe in the center; leaves alternate with at least some leaves in whorls...***Lilium***
4b. Tepals lacking spots, with a central yellowish stripe running nearly the length of the tepal; leaves basal...***Hemerocallis***

5a. Leaves 2, basal and nearly opposite; tepals yellow, 2.5–4 cm long, showy; flowers nodding...***Erythronium***
5b. Stem leaves present, with some semiwhorled to whorled; tepals 0.8–2.5 cm long, yellow or purplish-brown with yellowish or white mottling on the inside...***Fritillaria***

6a. Leaves linear, mostly basal with the stem leaves reduced and grading into bracts; fruit a capsule; plants above 10,000 ft in elevation...***Lloydia***
6b. Leaves elliptic or ovate, stem leaves evidently present; fruit a reddish berry; variously distributed...7

7a. Flowers solitary or paired in the leaf axils, usually drying cream-colored; pedicels glabrous and sharply geniculate; stems glabrous or coarsely hairy below...***Streptopus***
7b. Flowers solitary or 2 and terminal, usually drying white; pedicels hairy, not sharply geniculate; stems densely short-hairy throughout...***Prosartes***

CALOCHORTUS Pursh – MARIPOSA LILY; SEGO LILY

Perennial herbs from bulbs; *leaves* alternate, linear; *flowers* actinomorphic; *tepals* 6, the outer three (sepals) narrower than the inner 3, the inner tepals (petals) with a gland near the base, usually hairy or bearded within; *fruit* a septicidal capsule. (Ownbey, 1940)

1a. Glands at the base of the inner tepals orangish and not depressed or surrounded by a membrane; inner tepals lacking a dark purplish band towards the base; stems weakly erect or twining, often decumbent and straggling on the ground; leaves often coiled...***C. flexuosus***
1b. Glands at the base of the inner tepals yellowish, depressed, and surrounded by a fringed membrane; inner tepals usually with a dark purplish band towards the base; stems erect; leaves not coiled...2

2a. Glands on the inner tepals circular, with a few scattered, simple hairs surrounding the gland, often bordered by a crescent-shaped reddish-purple botch...***C. nuttallii***
2b. Glands on the inner tepals elongate, oblong and arched, with numerous, distally branched hairs above, bordered by purple on one or both sides...***C. gunnisonii***

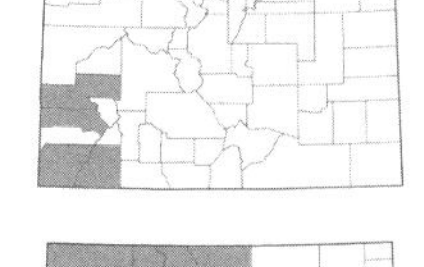

***Calochortus flexuosus* S. Watson**, WINDING MARIPOSA LILY. Plants weakly erect or twining, often decumbent, sprawling, stems 8–60 cm long; *leaves* 1-few, often coiled; *inner tepals* (petals) 2–4.5 cm long, purple or lavender; *glands* at the base of the inner tepals orangish and not depressed or surrounded by a membrane, densely hairy. Locally common on dry ridgetops and on shale or clay slopes, 5000–6000 ft. April–May. W.

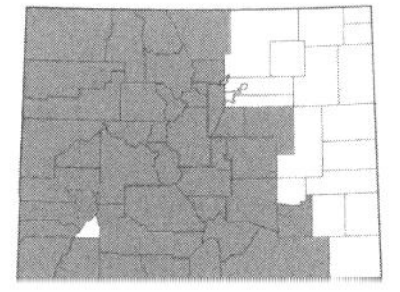

***Calochortus gunnisonii* S. Watson**, GUNNISON'S MARIPOSA LILY. Plants erect, 2–6 dm; *leaves* 2–4; *inner tepals* (petals) 2.5–4 cm long, truncate to rounded apically, purple to lavender; *gland* on the inner tepal elongate, oblong and arched, with numerous distally branched hairs above, bordered by purple on one or both sides. Common in meadows, aspen forests, grasslands, and on dry to moist slopes, 4700–11,000 ft. May–July. E/W. (Plate 51)

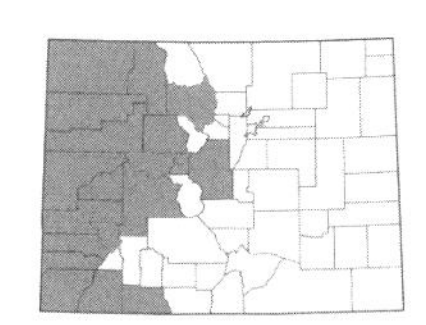

***Calochortus nuttallii* Torr. & A. Gray**, NUTTALL'S SEGO LILY. [*C. ciscoensis* S.L. Welsh & N.D. Atwood]. Plants erect, 0.8–5 dm; *leaves* usually 3; *inner tepals* (petals) 3.5–6.5 cm long, rounded or abruptly acuminate apically, white to lavender; *glands* at the base of the inner tepals circular, with a few simple hairs surrounding, in yellow often bordered by a crescent-shaped reddish-purple blotch. Common on dry slopes, ridgetops, and shale or clay hills, 4700–9200 ft. May–July. W. (Plate 51)

ERYTHRONIUM L. – FAWN-LILY; DOGTOOTH VIOLET

Perennial herbs from bulbs; *leaves* 2, basal, appearing opposite; *flowers* solitary or 2–5 in a terminal raceme, usually nodding; *tepals* all similar and petaloid, yellow (ours); *fruit* a loculicidal capsule. (Applegate, 1935)

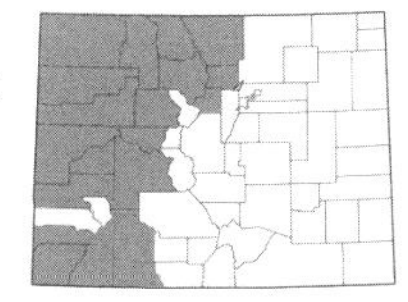

Erythronium grandiflorum* Pursh ssp. *grandiflorum, AVALANCHE LILY. Leaves 8–20 × 1–5 cm, elliptic or oblanceolate; *tepals* 2–3.5 cm long; *capsules* 2–5 (6) cm long. Common in moist meadows, along the edges of melting snowbanks, and in moist spruce forests, 7500–13,000 ft. May–July. E/W. (Plate 51)

FRITILLARIA L. – FRITILLARY

Perennial herbs from bulbs; *leaves* alternate or whorled; *flowers* usually nodding; *tepals* sometimes mottled; *fruit* a 6-angled or winged loculicidal capsule.

1a. Tepals purplish-brown, the inside mottled with yellow or white; styles 3...***F. atropurpurea***
1b. Tepals yellow to orange, not mottled but sometimes with brownish streaks, fading to red; style solitary...***F. pudica***

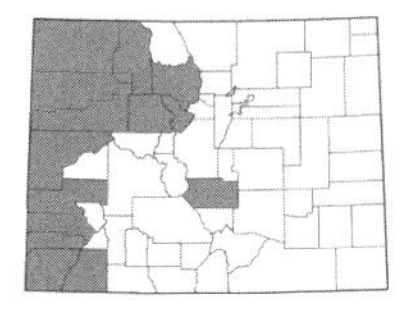

***Fritillaria atropurpurea* Nutt.**, CHECKER LILY. Plants 1–4.5 dm; *leaves* linear, 3–12 cm × 1.5–7 mm; *tepals* purplish-brown, the inside mottled with yellow or white; *styles* 3; *capsules* 10–20 mm long. Uncommon in forests, oak shrublands, and on grassy slopes, 6200–9800 ft. E/W.

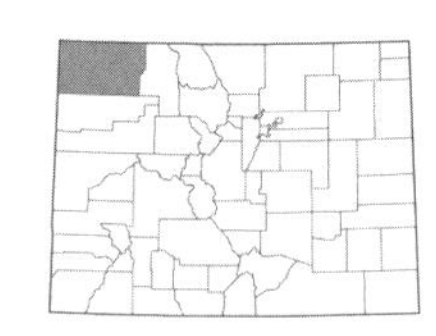

***Fritillaria pudica* (Pursh) Sprengel**, YELLOW BELL. Plants 0.5–2 dm; *leaves* usually oblanceolate, sometimes linear, 3–15 cm × 2–14 mm; *tepals* yellow to orange, sometimes with brownish streaks, fading red; *style* solitary; *capsules* 15–30 mm long. Uncommon in sagebrush and on sandy benches, 7500–7800 ft. May. W.

HEMEROCALLIS L. – DAYLILY

Perennial herbs; *leaves* basal, linear; *flowers* in a terminal cyme or solitary; *tepals* all alike and petaloid, variously colored (ours orange with a yellow base); *fruit* a loculicidal capsule.

***Hemerocallis fulva* (L.) L.**, ORANGE DAYLILY. Plants 8–15 dm; *leaves* mostly 20–50 cm long; *tepals* 5–9 cm long. Common ornamental that could potentially escape from cultivation and become established along roadsides or near old homesteads. No reported collections for the state, but to be expected. May–July. Introduced.

LILIUM L. – LILY

Perennial herbs from bulbs; *leaves* alternate or whorled, sessile; *flowers* large and showy, solitary or 2–many in a terminal raceme; *tepals* all alike and petaloid, variously colored (ours orange-red or reddish); *fruit* a loculicidal capsule.

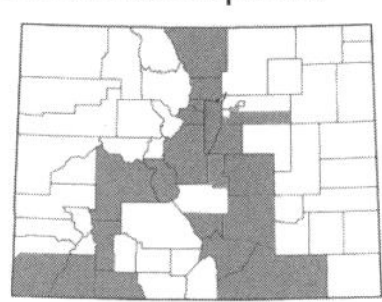

***Lilium philadelphicum* L.**, WOOD LILY. Plants 3–10 dm; *leaves* whorled above, 4–8 cm long; *tepals* to 5 cm long; *capsules* 3–4 cm long. Uncommon or locally common in moist, shaded forests and meadows, 6800–9800 ft. June–Aug. E/W. (Plate 51)

LLOYDIA Salisb. ex Rchb. – ALPLILY

Perennial herbs from short rhizomes simulating bulbs; *leaves* alternate, linear to lanceolate; *flowers* solitary or in a few-flowered terminal raceme; *tepals* all alike and petaloid, white to yellowish; *fruit* a loculicidal capsule.

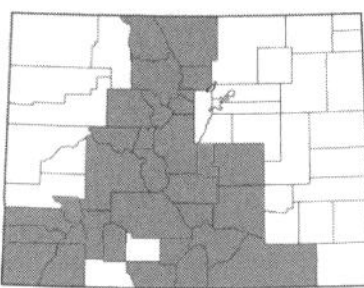

***Lloydia serotina* (L.) Salisb. ex Rchb.**, ALPINE LILY. Plants 5–20 cm; *leaves* 2–8 cm × 1–2 mm; *tepals* 8–13 mm long, white with purple veins, tinged purplish dorsally; *capsules* ca. 8 mm long. Common on alpine and in high meadows, 10,000–14,000 ft. June–Aug. E/W. (Plate 51)

PROSARTES D. Don – FAIRY BELLS

Perennial herbs from rhizomes; *leaves* alternate, oblanceolate to ovate; *flowers* solitary or 2–4 in a cyme, nodding on pedicels; *tepals* all alike and petaloid, white; *fruit* a dark blue to black or reddish berry.

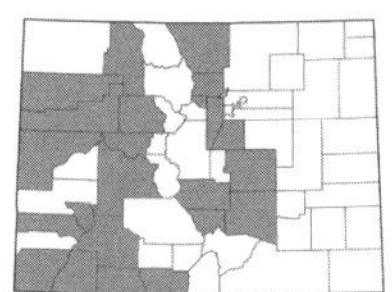

***Prosartes trachycarpa* S. Watson**, ROUGH-FRUITED FAIRY BELLS. [*Disporum trachycarpum* (S. Watson) Benth. & Hook.]. Plants 1–6 dm; *leaves* ovate, 3–10 × 1.5–5 cm, sessile; *tepals* 9–15 mm long; *berries* red. Uncommon in moist, shaded forests, often along streams, 5500–9500 ft. May–July. E/W. (Plate 51)

STREPTOPUS Michx. – TWISTED STALK

Perennial herbs from rhizomes; *leaves* alternate, with sessile and mostly clasping bases; *flowers* solitary in the leaf axils; *tepals* all alike and petaloid, white to greenish-yellow; *fruit* a fleshy, red or orangish berry.

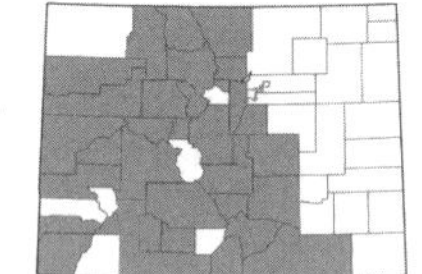

***Streptopus amplexifolius* (L.) DC.**, CLASPLEAF TWISTED STALK. Plants 4–10 dm; *leaves* ovate, 5–10 × 2–6 cm, clasping; *tepals* greenish-white, 7–15 mm long; *berries* red, ca. 10 mm diam. Common along streams and in moist forests and meadows, 6700–11,500 ft. June–Aug. E/W. (Plate 51)

LIMNANTHACEAE R. Br. – MEADOWFOAM FAMILY

Small, annual herbs; *leaves* alternate, petiolate; *flowers* perfect, actinomorphic, solitary, axillary on long pedicels; *sepals* 3–5, distinct or slightly connate at the base; *petals* 3–5, distinct; *stamens* 3 or 6–10; *pistil* 1 but the carpels almost separate and connected by a gynobasic style; *stigmas* 2–5; *ovary* superior to slightly inferior; *fruit* a schizocarp splitting into mericarps.

FLOERKEA Willd. – FLOERKEA

Leaves ternately to pinnately compound; *sepals* 3; *petals* 3; *stamens* 6; *ovary* 3-carpellate.

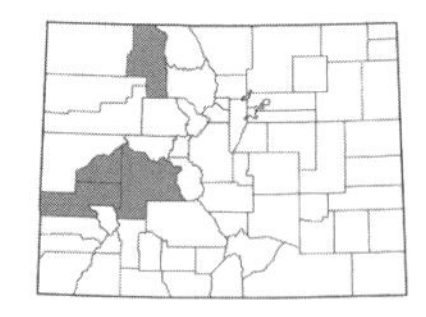

***Floerkea proserpinacoides* Willd.**, FALSE MERMAID-WEED. Plants delicate, 0.3–3 dm; *leaves* pinnately compound, the leaflets 3–15 mm long, narrowly elliptic to oblanceolate; *sepals* 2–6 mm long; *petals* 1–2 mm long. Uncommon in moist meadows, 6500–11,500 ft. May–July. W.

LINACEAE S.F. A. Gray – FLAX FAMILY

Herbs or shrubs; *leaves* alternate, simple and entire, sessile; *flowers* perfect, actinomorphic; *sepals* 5, distinct or sometimes basally connate; *petals* 5, distinct; *stamens* 5, connate by their filaments basally into a ring, sometimes with 5 staminodes as well; *pistil* 1; *ovary* superior, 2–5-carpellate; *fruit* a septicidal capsule, drupe, or nut.

LINUM L. – FLAX

Annual or perennial herbs; *petals* blue, yellow, or orange, easily falling; *ovary* 5-carpellate; *fruit* a septicidal capsule. (Rogers, 1968)

Two species have been excluded from the key below because they do not persist in Colorado. *Linum grandiflorum* Desf. has been found in revegetation project areas near Boulder but does not persist. It is easily identifiable by its large, red flowers. *Linum usitatissimum* L. is a cultivated plant known only from one collection where it was cultivated on the agricultural college farm at what is now Colorado State University. It has blue flowers and slender, elongate stigmas, and the inner sepals are toothed or ciliate at least above, unlike our other blue-flowered flaxes which have capitate stigmas and entire sepals.

1a. Flowers blue or white; sepals with entire margins; capsules 4–8 mm long, splitting into ten 1-seeded segments...2
1b. Flowers yellow or orange; sepals with glandular-toothed or ciliolate margins; capsule splitting into five 2-seeded segments, or if splitting into ten 1-seeded segments then the capsules 2.5–4 mm long...4

2a. Flowers whitish or light blue; petals 5–10 mm long; capsules 4–6 mm long; style 1–3 mm long, shorter than the stamens...***L. pratense***
2b. Flowers blue, sometimes with a whitish or yellowish base; petals mostly 12–23 mm long; capsules 6–8 mm long; style usually 3 mm or longer, usually equaling or exceeding the stamens...3

3a. Plants with multiple erect stems from the base that fan out above (vase-shaped)...***L. perenne***
3b. Plants with few to several stems which are usually ascending-spreading and seldom erect...***L. lewisii***

4a. Perennials; capsules splitting into 10 segments; leaves densely crowded towards the base; plants glabrous and glaucous; flowers bright yellow...***L. kingii***
4b. Annuals; capsules splitting into 5 segments; leaves usually not densely crowded at the base; plants hairy or glabrous; flowers various...5

5a. Plants finely and densely spreading-hairy throughout; flowers dull orange-yellow to salmon or peach with a deep red center...***L. puberulum***
5b. Plants glabrous or minutely hairy at the base of the stem only, or glandular-stipitate; flowers various...6

6a. Flowers orange or yellow-orange with a darker orange or reddish base, large with petals 12–18 mm long; leaves 1.5–2.5 mm wide, mostly linear-lanceolate, usually densely compacted on the stem; inflorescence dense and rather flat-topped; stipular glands present and conspicuous on the middle stem leaves, rarely absent...***L. berlandieri***
6b. Flowers yellow or orange-yellow without a reddish base, usually smaller, the petals 5–14 mm long; leaves 0.9–1.5 mm wide, linear, densely compacted on the stem or not, sometimes the plants "broom-like"; inflorescence usually not dense or flat-topped; stipular glands present or absent...7

7a. Plants short, to 2 (3) dm tall, densely branched with evident leaves...***L. compactum***
7b. Plants 1–4 dm tall (usually over 2 dm in height), plants usually "broom-like" with sparsely spaced leaves and wide branches...8

8a. Plants bushy, branched from the base with long, slender, stiffly spreading-ascending branches; style 4–7 mm long; stems glabrous below...***L. aristatum***
8b. Plants branched at or below the middle but the branches not grouped at the base; style 2.5–9 mm long; stems often minutely hairy below...9

9a. Petals mostly 12–15 mm long; style 6–9 mm long; plants usually branched from the middle of the stem or above...***L. rigidum***
9b. Petals 5–9 mm long; style 2.5–4 mm long; plants usually branched from below the middle of the stem...***L. australe***

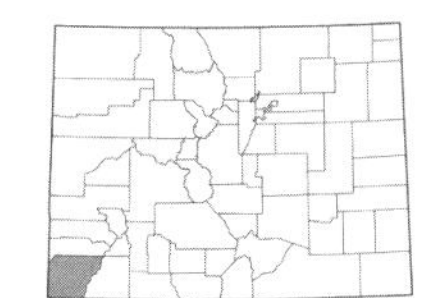

Linum aristatum **Engelm.**, BRISTLE FLAX. [*Mesynium aristatum* (Engelm.) W.A. Weber]. Annuals, 1–4 dm, glabrous; *leaves* linear, 0.7–1.5 cm × 1 mm; *sepals* 5–9 mm long, glandular-ciliate; *petals* yellow or orange-yellow, 8–12 mm long; *capsules* 3.5–4.5 mm long, splitting into 5 2-seeded segments. Found in open, dry soil, 4700–5000 ft. June–Aug. W.

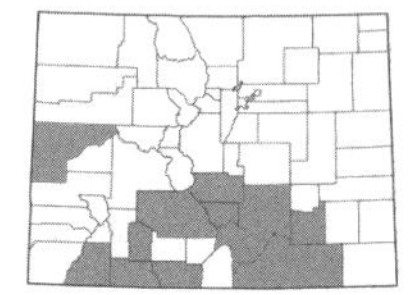

***Linum australe* Heller**, SOUTHERN FLAX. [*Mesynium australe* (Heller) W.A. Weber]. Annuals, 1–4 dm, minutely hairy on the lower stem; *leaves* linear, 1–2 cm × 1–1.5 mm; *sepals* 4.5–7 mm long, often glandular-ciliate; *petals* yellow or orange-yellow, 5–9 mm long; *capsules* 3.5–5 mm long, splitting into 5 2-seeded segments. Found in open places in ponderosa pine forests and on dry, open slopes, 6400–9000 ft. May–June. E/W.

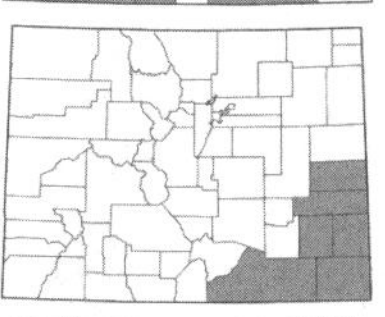

***Linum berlandieri* Hook.**, BERLANDIER'S YELLOW FLAX. Annuals, (0.5) 1–2.5 (4) dm, minutely hairy on the lower stem; *leaves* linear to narrowly lanceolate, mostly 3-nerved, 1.6–2.2 cm × 1.5–2.5 mm; *sepals* 7–9 mm long, glandular-toothed; *petals* yellow to orange with a darker orange or reddish base, 12–18 mm long; *capsules* 3.5–4.5 mm long, splitting into 5 2-seeded segments. Found in sandy soil of southeastern plains, 3500–5400 ft. May–June. E.

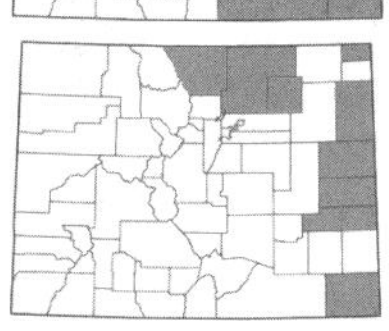

***Linum compactum* A. Nelson**, COMPACT FLAX. Annuals, (0.5) 1–2 (3) dm, glabrous throughout or minutely hairy near the base of the stem; *leaves* linear, (1) 1.5–2.5 cm × 0.8–1.5 mm; *sepals* (4) 5–6 (9) mm long, glandular-toothed; *petals* yellow, 6–11 mm long; *capsules* 3.5–4.5 mm long, splitting into 5 2-seeded segments. Found in open, dry places on the plains, 3500–5300 (6500) ft. May–Aug. E. (Plate 51)

***Linum kingii* S. Watson**, KING'S FLAX. [*Mesyniopsis kingii* (S. Watson) W.A. Weber]. Perennials, 0.5–2.5 dm, glabrous and glaucous; *leaves* linear to oblong-obovate, 0.2–1 cm × 1–1.5 mm; *sepals* 2.5–3.5 mm long, usually glandular-toothed; *petals* bright yellow, 6–12 mm long; *capsules* 2.5–4 mm long, splitting into 10 1-seeded segments. Uncommon on shale slopes and outcroppings, known from Eagle, Garfield, and Rio Blanco cos., 8400–8600 ft. May–Aug. W.

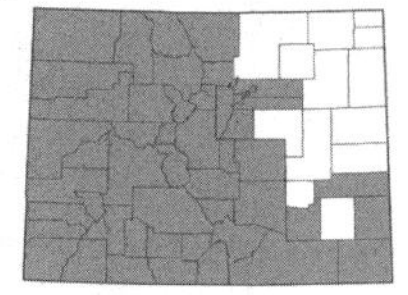

***Linum lewisii* Pursh**, LEWIS FLAX. [*Adenolinum lewisii* (Pursh) Á. Löve & D. Löve]. Perennials, 1.5–8 dm, glabrous; *leaves* linear, 0.5–3 cm × 0.5–2.5 (4.5) mm; *sepals* 4–6 mm long, margins entire; *petals* blue or whitish-blue, 10–23 mm long; *capsules* 6–8 mm long, splitting into 10 1-seeded segments. Common in open meadows and along roadsides, 3900–11,400 ft. May–Sept. E/W. (Plate 51)

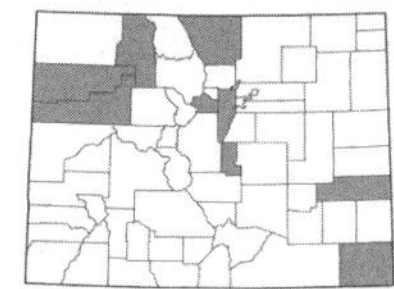

***Linum perenne* L.**, PERENNIAL FLAX. Perennials, 1.5–9 dm, glabrous; *leaves* linear, 1–3.5 cm × 1–2 mm; *sepals* 4–6.5 mm long, margins entire; *petals* blue, 12–25 mm long; *capsules* 6–9 mm long, splitting into 10 1-seeded segments. Found along roadsides, especially in revegetation areas, 4900–8600 ft. May–Sept. E/W. Introduced.

***Linum pratense* (J.B.S. Norton) Small**, MEADOW FLAX. [*Adenolinum pratense* (Norton) W.A. Weber]. Annuals, 0.5–4 dm, glabrous; *leaves* linear to narrowly lanceolate, 1–2 cm × 1–3 mm; *sepals* 3.5–4.5 mm long, margins entire; *petals* blue or pale blue, 5–10 mm long; *capsules* 4–6 mm long, splitting into 10 1-seeded segments. Uncommon in open meadows and along roadsides, known from Boulder, Jefferson, and Pueblo cos., 4000–5500 ft. May–June. E.

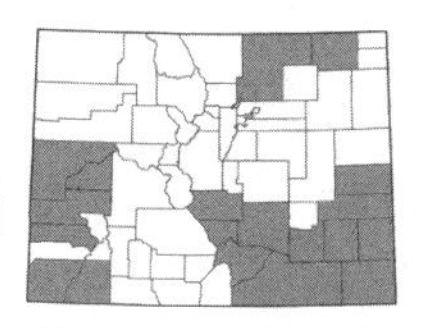

***Linum puberulum* (Engelm.) Heller**, PLAINS FLAX. [*Mesynium puberulum* (Engelm.) W.A. Weber]. Annuals, 1–2 (3) dm, densely spreading-hairy throughout; *leaves* linear, 0.5–1.5 cm × 1–2 mm; *sepals* 5–7 mm long, glandular-ciliate and often pubescent; *petals* dull orange-yellow to salmon or peach with a deep red center, 8–15 mm long; *capsules* 3.5–4 mm long, splitting into 5 2-seeded segments. Found in dry, sandy or gravelly soil, 3900–6500 ft. May–July. E/W. (Plate 51)

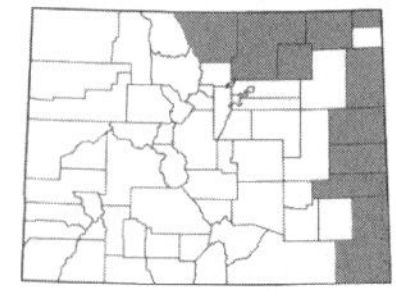

***Linum rigidum* Pursh**, STIFFSTEM FLAX. [*Mesynium rigidum* (Pursh) Á. Löve & D. Löve]. Annuals, (1.5) 2–4 (5.5) dm, glabrous throughout or minutely hairy near the base of the stem; *leaves* linear, 1.5–2.5 (3) cm × 0.7–1.5 mm; *sepals* (5) 6–8 mm long, glandular-toothed; *petals* yellow, (9) 12–15 (18) mm long; *capsules* 3.5–4.5 mm long, splitting into 5 2-seeded segments. Found in dry, open places on the eastern plains, 3500–5900 ft. May–Sept. E.

LOASACEAE Spreng. – STICKLEAF FAMILY

Herbs or shrubs; *leaves* alternate or opposite, simple, entire to sinuate, lobed, or dissected, exstipulate; *flowers* perfect, actinomorphic; *sepals* (4) 5, distinct, persistent in fruit; *petals* (4) 5 or 10 when petaloid staminodes are present, distinct, usually yellow, sometimes white, rarely orange or red; *stamens* 5 or 10–many, staminodes often present and sometimes petaloid; *pistil* 1; *ovary* inferior, unicarpellate or 2–5-carpellate; *fruit* a capsule.

MENTZELIA L. – BLAZINGSTAR

Annual, biennial, or perennial herbs, often with a woody base, glabrous or more often covered with large, multicellular hairs, the outer bark white and exfoliating; *leaves* alternate; *sepals* 5; *petals* 5 or 10 (if counting petaloid stamens as petals); *stamens* 10–many, often petaloid staminodes present, some cuspidate apically. (Darlington, 1934; Hill, 1976; Holmgren et al., 2005; Reveal, 2002; Schenk & Hufford, 2011; Thompson & Prigge, 1986; Thorne, 1986)

The genus *Mentzelia* poses great difficulties in writing a key. There are several species which can only be properly determined based on seed characteristics. It is also important to note the exact color of the petals on the label as they soon fade to brown.

1a. Seeds with narrow, linear, curved grooves, pendulous in the capsule, wingless; capsules curved at the base, small and narrow, mostly 0.7–1.2 cm long and 2–3 mm wide; leaves triangular-ovate, coarsely toothed or sometimes the basal leaves somewhat lobed; petals 5, 0.7–1 cm long; flowers orange or yellow-orange…***M. oligosperma***

1b. Seeds variously papillose, horizontal in the capsule and winged (this sometimes very thin) or pendulous in the capsule and unwinged; capsules erect, not curved at the base; leaves various but not triangular-ovate, entire or shallowly to deeply pinnately lobed; petals 5 or 10 (5 petals alternating with 5 petaloid stamens), 0.25–3 cm long; flowers yellow, cream, or white…2

2a. Petals 5, 1–7 mm long, glabrous; annuals or winter annuals; seeds angular, pendulous, not winged; capsule narrowly cylindric, 1–4 mm wide…3

2b. Petals 5 or 10, 9–80 mm long (or if shorter then hairy on the back); biennials or perennials, rarely annuals; seeds flattened and winged, horizontal in 1 or 2 rows; capsule usually thick-cylindric, bowl-shaped, or urn-shaped…5

3a. Leaves deeply pinnatifid into slender segments, sometimes a few upper ones entire or nearly so; with a dense basal rosette at anthesis; flowers subtended by linear or narrowly lanceolate to narrowly ovate bracts…***M. albicaulis***

3b. Leaves entire or shallowly toothed; usually without a dense basal rosette at anthesis; flowers subtended by oblanceolate or ovate to broadly ovate bracts…4

4a. Capsules 1–1.6 (2) mm wide; seeds appearing smooth at 10× with less pronounced papillae, in 1 row, triangular-prismatic, sharply angled with a groove along each angle; stamens numerous…***M. dispersa***

4b. Capsules 2–3.5 (4) mm wide; seeds appearing rough at 10× with pronounced papillae, in more than 1 row, irregularly angled; stamens few…***M. thompsonii***

5a. Plants of the eastern slope…6

5b. Plants of the western slope…15

6a. Petals large, 4–8 cm long; calyx lobes 15–40 mm long; capsules 30–50 mm long…7

6b. Petals smaller, 0.7–3 cm long; calyx lobes 5–15 mm long; capsules 5–30 mm long…8

7a. Bracts adnate with the capsule; seed with a narrow wing, ca. 0.2–0.3 mm wide; capsule 3–4.5 cm long, 13–18 mm wide; petals 4–8 cm long and 12–26 mm wide; leaves 5–20 cm long and 1–3.5 cm wide; calyx lobes 2–3.5 cm long; plants 4–10 dm tall…***M. decapetala***

7b. Bracts free from the capsule; seed with a wide wing, 0.6–1 mm wide; capsule (1.5) 2–3 cm long, 8–10 mm wide; petals 2–5 cm long and 3–10 mm wide; leaves 4–10 cm long and 1–2 cm wide; calyx lobes 1–2.5 cm long; plants to 5 dm tall…***M. nuda***

8a. Bracts subtending the flowers lobed or pinnatifid; flowers white, cream, or pale yellow…9

8b. Bracts subtending the flowers entire or nearly so; flowers white, cream, or pale to golden-yellow…10

9a. Flowers white to cream or rarely pale yellow; petals 2–5 cm long; seeds smoother in appearance at 10× with less pronounced papillae; plants of the eastern plains and foothills…***M. nuda***

9b. Flowers pale yellow; petals 1.5–3 cm long; seeds rough in appearance at 10× with numerous pronounced papillae; plants not of the eastern plains…***M. rusbyi***

10a. Plants branched from the base, the entire plant forming a rounded tuft; stem and branches white with the lower part of the stem strongly exfoliating, slender and flexuous; leaves narrowly linear-lanceolate, sinuate-pinnatifid with rather widely spaced, short lobes (mostly 1–3 mm long), and a narrow midrib (on most leaves 1–2 mm wide); capsule 13–20 mm long; petals 0.8–1.5 cm long, golden-yellow…***M. densa***

10b. Plants with solitary stems that are branched above or occasionally with a few branches at the base, not forming a rounded tuft; stem and branches white to yellowish and strongly exfoliating below or not, usually stouter; leaves with the midrib generally wider and the lobes usually longer; capsule 10–30 mm long; petals 1–3 cm long, pale to golden-yellow or white…11

11a. Seeds with a narrow wing (0.15–0.35 mm)...12
11b. Seeds with a wide wing (0.5–1 mm)...13

12a. Leaves mostly shallowly dentate or sometimes nearly entire; plants decumbent at the base with numerous old leaf bases...***M. chrysantha***
12b. Leaves mostly deeply pinnately lobed to near the midrib into narrow segments; plants erect or occasionally somewhat decumbent at the base, without numerous old leaf bases...***M. laciniata***

13a. Seeds rough in appearance at 10× with numerous pronounced papillae; flowers pale yellow, cream-colored, or occasionally white...***M. multiflora***
13b. Seeds smooth in appearance at 10× with less pronounced papillae; flowers golden-yellow...14

14a. Outer fertile stamens with a narrow filament; leaves (at least the upper) usually with a broad and more or less clasping base; plants generally tall, to 10 dm in height...***M. reverchonii***
14b. Outer fertile stamens with broad filaments, grading to inner stamens with narrow filaments; leaves usually without a broad and clasping base; plants generally shorter, 3–5 dm in height...***M. speciosa***

15a. Petals 5, rarely more, large (4–8 cm long), white or pale yellow outside and yellow inside except often lighter at the base; capsules 3–5 cm long...***M. laevicaulis***
15b. Petals 10 (5 petals alternating with 5 petaloid stamens) or sometimes 5, smaller (0.9–3 cm long), yellow, sometimes lighter at the base of the petal; capsules 0.5–2.8 cm long...16

16a. Petals hairy on the back, at least below the middle (this is easiest to see on unopened buds at the base of the petals)...17
16b. Petals glabrous on the back or sometimes with a tuft of hairs just at the apex...19

17a. Leaves mostly crenate or shallowly pinnately lobed less than halfway to the midrib, usually with clasping bases; petals 5, alternating with 5 petaloid stamens bearing functional anthers...***M. marginata***
17b. Leaves mostly pinnately lobed at least halfway to the midrib, or sometimes the upper linear and nearly entire, lacking clasping bases; petals 5 but appearing 10 (alternating with petaloid stamens lacking anthers)...18

18a. Plants columnar in appearance (with lateral branches short and not reaching the top of the plant); capsules 5–9 mm long...***M. paradoxensis***
18b. Plants bushy in appearance (with lateral branches mostly extending to near the top of the plant); capsules 6–16 mm long...***M. cronquistii***

19a. True perennials from woody branching caudices or deep rhizomes, freely branched with numerous stems...20
19b. Biennials from taproots, often with a single central main stem or branched above, lacking a branching, woody caudex...22

20a. Plants from rhizomes; seeds with a narrow wing 0.2–0.3 mm wide; leaves lanceolate to narrowly elliptic (3–15 mm wide), entire or shallowly sinuate-dentate; capsules 5–8 (10) mm wide...***M. rhizomata***
20b. Plants from a taproot and woody branching caudex; seeds essentially wingless; leaves linear (0.5–5 mm wide) or deeply pinnately divided into linear segments; capsules 3–5 mm wide...21

21a. Middle and upper leaves mostly entire, linear; petaloid stamens oblanceolate or narrowly spatulate, gradually tapering to the base...***M. multicaulis***
21b. Middle and upper leaves mostly deeply pinnatifid into linear segments; petaloid stamens obovate to rhomboidal, abruptly tapering at the base...***M. uintahensis***

22a. Most leaves deeply pinnatifid nearly to the midrib, with a narrow midrib mostly 1–3 mm wide...23
22b. Most leaves not deeply lobed nearly to the midrib, usually shallowly toothed to pinnatifid halfway to the midrib, or sometimes nearly entire, with a midrib mostly over 5 mm wide...25

23a. Seeds with a broad wing mostly 0.7–1 mm wide; plants mostly 5–10 dm; petals 9.5–16 mm long; capsules mostly 10–12 mm long (but reaching to 18 mm)...***M. procera***
23b. Seeds with a narrow wing mostly 0.1–0.35 mm wide; plants to 4 dm; petals 10–20 mm long; capsules 12–25 mm long...24

24a. Petals 10–13 mm long; bracts subtending the flowers usually pinnately lobed...***M. lagarosa***
24b. Petals 15–20 mm long; bracts subtending the flowers entire...***M. laciniata***

25a. Bracts subtending the flowers lobed or pinnatifid; plants 5–12 dm in height; petals 15–30 mm long…***M. rusbyi***
25b. Bracts subtending the flowers linear, entire; plants 1–3 dm; petals 10–20 mm long…26

26a. Leaves broadly oblanceolate to ovate, often rounded at the apex, very shallowly toothed; seeds 3.5–4.7 mm long, with a broad wing (0.7–1.3 mm wide); capsules bowl-shaped or rarely short-cylindric, 10–15 mm long and 7–10 mm wide…***M. pterosperma***
26b. Leaves linear to narrowly oblanceolate, shallowly pinnately lobed to nearly entire; seeds 2.3–3.5 mm long with a narrow wing 0.15–0.35 mm wide; capsules cylindric, 13–20 (25) mm long and 4–6 (7.6) mm wide…***M. pumila***

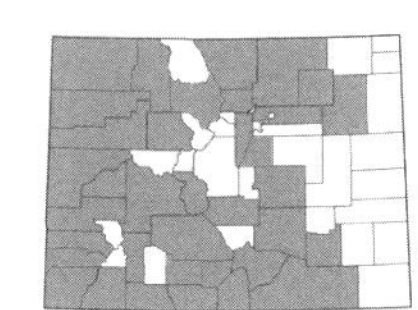

***Mentzelia albicaulis* (Douglas ex Hook.) Douglas ex Torr. & A. Gray**, WHITE-STEM BLAZINGSTAR. [*Acrolasia albicaulis* (Douglas ex Hook.) Rydb.; *M. montana* (Davidson) Davidson]. Annuals, 0.8–4 dm; *leaves* sessile, oblanceolate, lanceolate, or linear, 3–15 cm long, nearly entire to lobed or sinuate-pinnatifid; *petals* 5, 2–4 mm long, yellow; *capsules* narrowly cylindric, 8–22 × 2–3 mm; *seeds* irregularly angled, wingless. Common in a variety of soil types, often in dry or disturbed areas, 4500–7500 ft. May–July. E/W. (Plate 51)

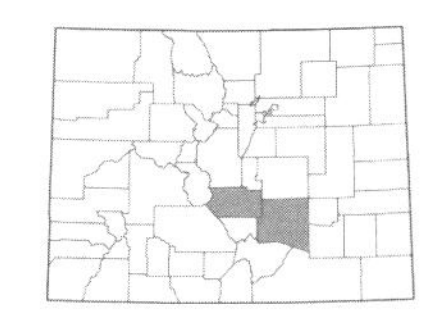

***Mentzelia chrysantha* Engelm. ex Brandeg.**, GOLDEN BLAZINGSTAR. [*Nuttalliia chrysantha* (Engelm. ex Brandeg.) Greene]. Biennials or short-lived perennials, 3–6 dm, usually decumbent at the base; *leaves* 5–15 cm long, lanceolate, nearly entire to shallowly dentate; *petals* 10, 15–20 mm long, yellow; *capsules* cylindric, 12–18 mm long; *seeds* with a narrow wing 0.15–0.3 mm wide, the surface with numerous papillae, giving the seed a rough appearance at 10×. Uncommon on limestone outcroppings between Canon City and Pueblo, 5100–5700 ft. July–Sept. E. Endemic. (Plate 51)

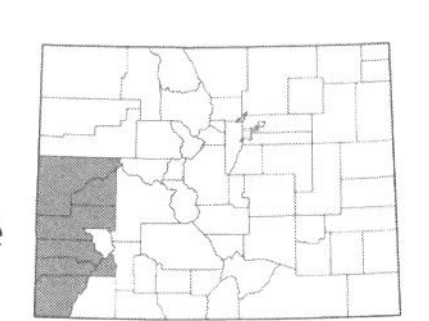

***Mentzelia cronquistii* H.J. Thomps. & Prigge**, CRONQUIST'S BLAZINGSTAR. [*M. marginata* (Osterh.) H.J. Thomps. & Prigge var. *cronquistii* (H.J. Thomps. & Prigge) N.H. Holmgren & P.K. Holmgren; *Nuttallia cronquistii* (H.J. Thomps. & Prigge) N.H. Holmgren & P.K. Holmgren]. Biennials or perennials, 1.5–4 (5) dm, usually branched at the base; *leaves* lanceolate, 2.5–8 cm long, lobed about halfway to the midrib, midrib 2–5 mm wide and lobes 2–10 mm long; *petals* 10, yellow, 7–15 mm long, hairy on the outer surface; *capsules* urceolate, 6–16 mm long; *seeds* 2.4–3 × 1.8–2 mm, with a narrow wing ca. 0.25 mm wide. Found on sandy or rocky slopes, 4600–7000 ft. June–Sept. W.

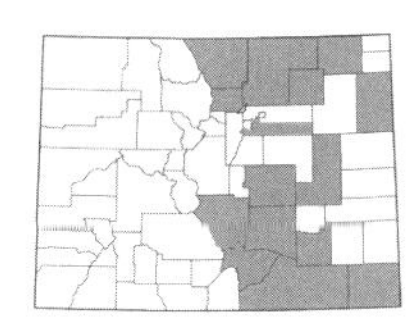

***Mentzelia decapetala* (Pursh ex Sims) Urb. & Gilg ex Gilg**, TEN-PETAL BLAZINGSTAR. [*Nuttallia decapetala* (Pursh ex Sims) Greene]. Biennials or perennials, 4–10 dm; *leaves* lanceolate, sinuate-pinnatifid, 5–20 cm × 1–3.5 cm; *petals* 10, 4–8 cm long, white; *capsules* cylindric, 30–45 × 13–18 mm; *seeds* 3 mm long, with a thin wing 0.2–0.3 mm wide. Common on dry slopes and along roadsides, 3500–7000 ft. July–Sept. E. (Plate 52)

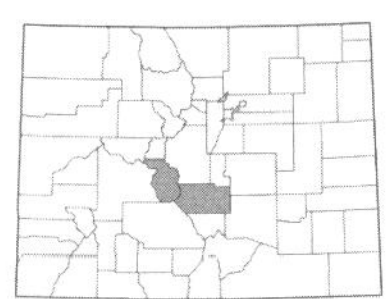

***Mentzelia densa* Greene**, ROYAL GORGE BLAZINGSTAR. [*Nuttallia densa* (Greene) Greene]. Perennials, much-branched from the base forming a hemispherical tuft, 2–5 dm; *leaves* narrowly linear-lanceolate, sinuate-pinnatifid into short, triangular segments 1–5 mm long; *petals* 10, 8–15 mm long, golden-yellow; *capsules* cylindric, 13–20 × 4–6 mm; *seeds* 2.5–3.5 mm long, with a wing ca. 0.5 mm wide. Uncommon in dry, rocky soil, known from the Arkansas River Canyon, 5800–7200 ft. July–Sept. E. Endemic.

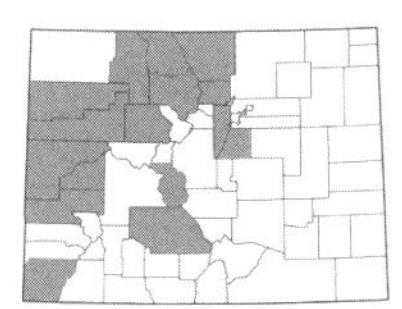

***Mentzelia dispersa* S. Watson**, NEVADA BLAZINGSTAR. [*Acrolasia dispersa* (S. Watson) Davidson]. Annuals, 1–4 dm, simple or branched; *leaves* oblanceolate to linear, 2–8 cm long, usually entire or sometimes lobed; *petals* 5, 1–3 (5) mm long, yellow; *capsules* narrowly cylindric, 8–20 × 1–1.6 (2) mm; *seeds* prismatic, with grooved angles. Rather uncommon, found on dry slopes, sometimes in disturbed areas, 4500–8300 ft. May–July. E/W. (Plate 52)

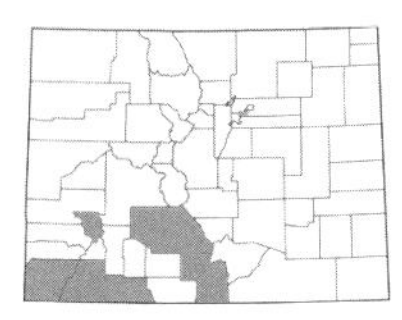

***Mentzelia laciniata* (Rydb.) J. Darl.**, CUTLEAF BLAZINGSTAR. [*Nuttallia laciniata* (Rydb.) Wooton & Standl.]. Biennials or perennials, 3–5 dm, branched above; *leaves* lanceolate, deeply pinnatifid into narrow segments; *petals* 10, 15–20 mm long, yellow; *capsules* urceolate, 15–20 × 6–8 mm; *seeds* rough in appearance at 10× with numerous pronounced papillae, with a narrow wing 0.15–0.35 mm wide. Uncommon in dry, open places, 5000–8000 ft. June–Aug. E/W. (Plate 52)

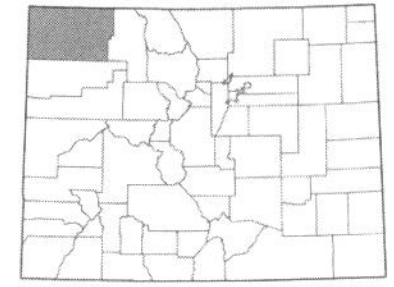

***Mentzelia laevicaulis* (Hook.) Torr. & A. Gray**, SMOOTH-STEM BLAZINGSTAR. Perennials, 3–10 dm, simple or branched above; *leaves* 2–20 cm long, lanceolate to oblanceolate, sinuate-pinnatifid with shallow lobes; *petals* 5, 4–8 cm long, light yellow; *capsules* cylindric, 30–50 × 7–13 mm; *seeds* 4–4.5 mm long, with a broad wing. Found in dry, open places, reported for Colorado from near Slater in Moffat Co., 7500–7800 ft. May–Sept. W.

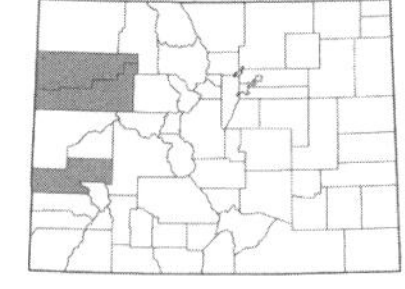

Mentzelia lagarosa **(K.H. Thorne) J.J. Schenk & L. Hufford**, DWARF MENTZELIA. [*M. pumila* Nutt. ex Torr. & A. Gray var. *lagarosa* Thorne]. Biennials or short-lived perennials, 1.5–3 dm; *leaves* deeply pinnatifid nearly to the midvein, 5–9 cm long; *petals* 10, 10–13 mm long, yellow; *capsules* thick-cylindric, 12–25 × 4.5–7 mm; *seeds* 3–3.5 mm long, with a wing ca. 0.5 mm wide. Uncommon on dry, sandy or rocky slopes, 5000–7000 ft. June–Sept. W.

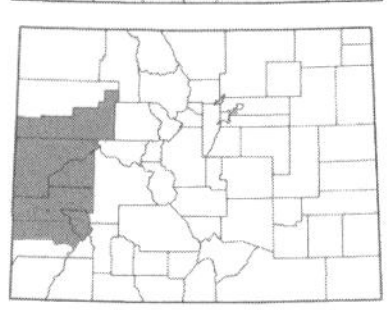

Mentzelia marginata **(Osterh.) H.J. Thomps. & Prigge**, COLORADO BLAZINGSTAR. [*Nuttallia marginata* Osterh.]. Perennials, 1–4 dm; *leaves* oblanceolate to obovate, shallow lobed or crenate-margined, 3–6 × 1–2.5 cm; *petals* 5, alternating with 5 petaloid stamens, 10–13 mm long, yellow, the outer ones hairy on the lower outer surface; *capsules* urceolate to subcylindric, 8–12 mm long; *seeds* 2.5–3 × 1.7–1.8 mm, with a narrow wing 0.15–0.3 mm wide. Uncommon on clay and shale slopes, 4600–6500 ft. May–Aug. W.

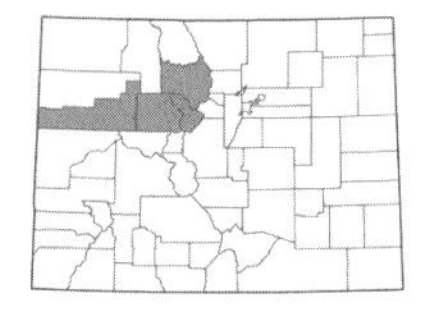

Mentzelia multicaulis **(Osterh.) A. Nelson ex J. Darl.**, MANYSTEM BLAZINGSTAR. [*Nuttallia multicaulis* Osterh.]. Perennial subshrubs from a woody, branching caudex, 2–4 dm, diffusely branched; *leaves* linear to narrowly lanceolate, entire or the lower often deeply pinnately lobed into narrow segments, 1.5–3 (5) cm long, mostly less than 5 mm wide; *petals* yellow, 7–20 mm long; *capsules* urceolate, (4) 6–9 × 3.3–5 mm; *seeds* 1.5–3.5 mm long, essentially wingless. Found on dry, shale or sandy slopes, 6000–8700 ft. June–Aug. W. (Plate 52)

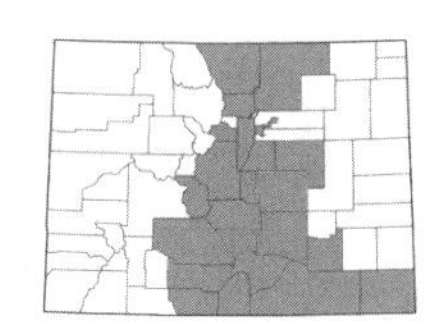

Mentzelia multiflora **(Nutt.) A. Gray**, ADONIS BLAZINGSTAR. [*Nuttallia multiflora* (Nutt.) Greene]. Perennials, 4–8 dm; *leaves* 2–12 cm long, lanceolate to oblanceolate, shallowly to deeply pinnatifid; *petals* 10, 9–15 (20) mm long, pale yellow, cream-colored, or occasionally white; *capsules* cylindric, 10–25 mm long; *seeds* 3–3.5 mm long, with a broad wing 0.7–1 mm wide, rough in appearance at 10× with numerous pronounced papillae. Common in the middle counties from the high plains to mountains, usually in sandy soil, 5000–9600 ft. May–Sept. E. (Plate 52)

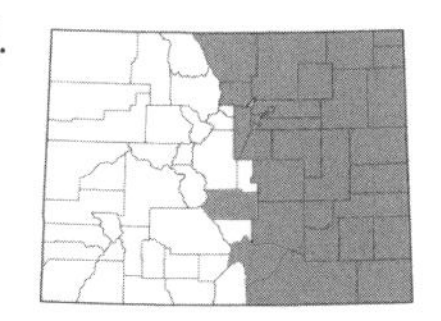

Mentzelia nuda **(Pursh) Torr. & A. Gray**, WHITE-FLOWERED BLAZINGSTAR. [*Nuttallia nuda* (Pursh) Greene]. Biennials or perennials, 1.5–5 dm; *leaves* 4–10 cm long, lanceolate to narrowly elliptic, shallowly pinnatifid; *petals* 10, 20–50 mm long, white to cream, subtended by pinnatifid bracts; *capsules* cylindric, (15) 20–30 × 8–10 mm; *seeds* 3–4.2 mm long, with a broad wing 0.6–1 mm wide, smoother in appearance at 10× with less pronounced papillae. Common on the eastern plains and foothills, 3500–6500 ft. July–Sept. E. (Plate 52)

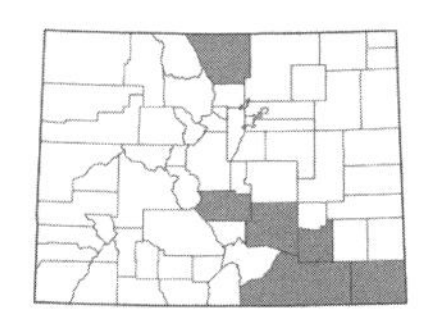

Mentzelia oligosperma **Nutt. ex Sims**, CHICKEN-THIEF. Perennials, 2–7 dm, much-branched; *leaves* 1–6 cm long, triangular-ovate, coarsely toothed or sometimes the basal leaves somewhat lobed; *petals* 5, 8–15 mm long, orange to yellow-orange; *capsules* 7–12 × 2–3 mm; *seeds* pendulous in the capsule, wingless, narrow, linear, with curved grooves. Found on rocky outcroppings and on rocky slopes and canyons, 4000–5700 ft. June–Aug. E. (Plate 52)

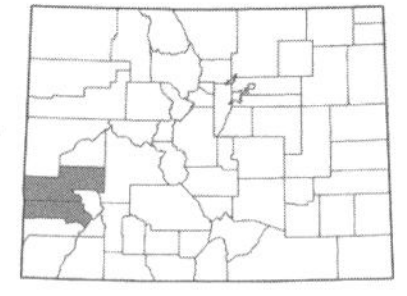

Mentzelia paradoxensis **J.J. Schenk & L. Hufford**, PARADOX VALLEY BLAZINGSTAR. Biennials or short-lived perennials, 4–9 dm; *leaves* 3.8–9 × 6–20 mm, lanceolate, deeply pinnatifid into narrow lobes, the lobes 2–8 (9) mm long; *petals* yellow, hairy on the outer surface, 8.3–15 (17) mm long; *capsules* urceolate, 5–9 × 3.7–6.5 mm; *seeds* 1.7–3 mm long, with a narrow wing 0.1–0.35 mm wide. Found in the Paradox and Gypsum valleys on dry, sandy or gypsum slopes, 5200–6500 ft. June–Sept. W. Endemic.

Mentzelia procera **(Wooton & Standl.) J.J. Schenk & L. Hufford**, TALL BLAZINGSTAR. Biennials or short-lived perennials, 5–10 dm; *leaves* deeply pinnatifid over halfway to the midrib, with 5–10 coarse, rounded teeth on each side, 5–9 cm long; *petals* yellow, 9.5–16 mm long; *capsules* cylindric, 10–12 (18) × 6–7 mm; *seeds* with a wide wing mostly 0.7–1 mm wide. Reported for the southern counties on dry slopes. June–Sept.

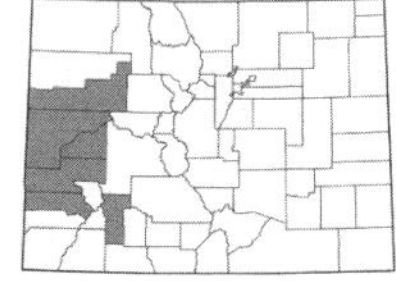

Mentzelia pterosperma **Eastw.**, WINGSEED BLAZINGSTAR. [*Nuttallia pterosperma* (Eastw.) Greene]. Biennials or perennials, 1–2 dm, divaricately branched from the base; *leaves* 3–7 cm long, broadly oblanceolate to ovate, shallowly pinnatifid to dentate; *petals* 10, 10–15 (20) mm long, golden-yellow; *capsules* cylindric, 10–15 × 7–10 mm; *seeds* 3.5–4.7 mm long, with a broad wing 0.7–1.3 mm wide, appearing smooth at 10×. Uncommon on dry or shale slopes, 4600–6500 ft. May–Aug. W. (Plate 52)

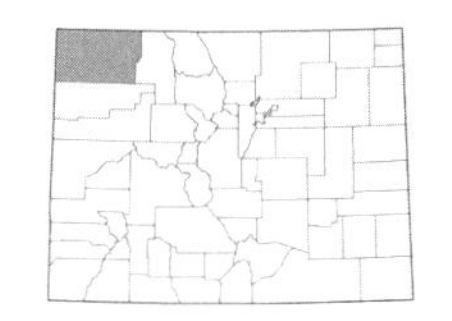

Mentzelia pumila **Nutt. ex Torr. & A. Gray**, Wyoming stickleaf. Biennials or short-lived perennials, 2–6 dm, mostly branched above; *leaves* lanceolate to oblong, 9–15 (20) cm long, shallowly pinnatifid to sinuate-dentate; *petals* 10, 9–15 mm long, yellow; *capsules* cylindric, 13–20 (25) × 4.5–6 (7.6) mm; *seeds* 2.3–3.5 mm long, with a narrow wing 0.15–0.35 mm wide, smooth or somewhat roughened in appearance at 10× with less pronounced papillae. Uncommon in dry, open places in sandy soil, barely entering Colorado in northwestern Moffat Co., 5500–7900 ft. May–Aug. W. (Plate 52)

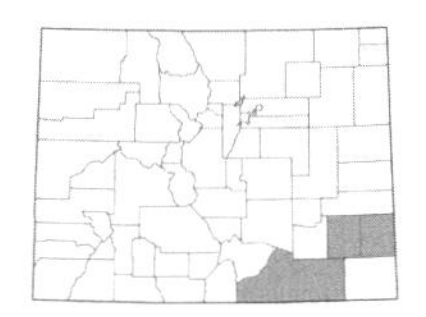

Mentzelia reverchonii **(Urb. & Gilg) H.J. Thomps. & Zavort.**, Reverchon's blazingstar. [*Nuttallia reverchonii* (Urb. & Gilg) W.A. Weber]. Perennials to 10 dm, branched above; *leaves* lanceolate to oblong, shallowly pinnatifid to dentate, 3–8 cm long; *petals* 10 or more, 10–30 mm long, golden-yellow; *capsules* cylindric, 15–30 mm long; *seeds* 3.5–5 mm long, with a broad wing ca. 1 mm wide. Uncommon on the shortgrass prairie, 3800–5500 ft. May–Aug. E. (Plate 52)

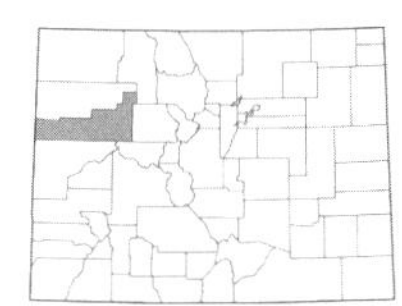

Mentzelia rhizomata **Reveal**, Roan Cliffs blazingstar. [Colorado specimens previously misassigned to *M. argillosa* J. Darl.]. Perennials from rhizomes, 1–2.5 dm, spreading, freely branched; *leaves* 1.5–4.5 cm long, lanceolate to obovate, entire or shallowly sinuate-dentate; *petals* 5, alternating with 5 petaloid stamens, 9–15 mm long, yellow; *capsules* urceolate, (5) 8–10 (13) × 5–8 (10) mm; *seeds* 2–3 mm long, with a narrow wing 0.2–0.3 mm wide, minutely papillate. Uncommon on talus and shale slopes of the Green River Formation, 5500–9100 ft. June–Aug. W. Endemic.

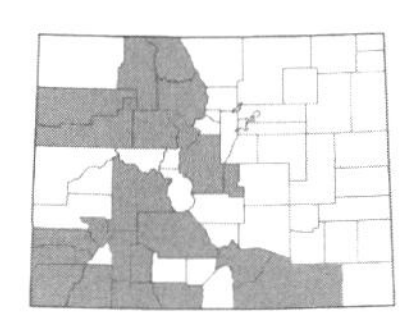

Mentzelia rusbyi **Wooton**, Rusby's blazingstar. [*Nuttallia rusbyi* (Wooton) Rydb.]. Biennials or short-lived perennials, 2–5 dm, branched above; *leaves* 5–25 cm long, oblanceolate to oblong, entire to shallowly sinuate-dentate; *petals* 5, 15–30 mm long, creamy yellow; *capsules* cylindric, (15) 20–30 × 7–10 mm; *seeds* 3–4 mm long, with a broad wing 0.8–1.5 mm wide. Found in dry, open places along roadsides, 6000–9500 ft. June–Sept. E/W.

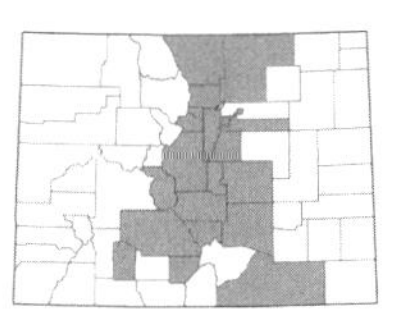

Mentzelia speciosa **Osterh.**, jeweled blazingstar. [*M. sinuata* (Rydb.) R.J. Hill; *Nuttallia sinuata* (Rydb.) Daniels; *N. speciosa* (Osterh.) Greene]. Biennials or perennials, 3–5 dm; *leaves* 8–15 cm long, linear to oblong-lanceolate, shallowly sinuate-dentate to dentate-pinnatifid; *petals* 10, 15–20 mm long, yellow; *capsules* cylindric, 15–28 × 5–9 mm; *seeds* 3–3.5 mm long, with a broad wing 0.7–1 mm wide, smooth in appearance at 10×. Common along the Front Range and high plains on rocky slopes or in sandy soil, 5000–9000 ft. June–Aug. E. (Plate 52)

Mentzelia speciosa and *M. sinuata* have traditionally been separated as distinct species. However, these two species are extremely difficult to separate based on leaf morphology. Specimens have been seen with a mixture of characteristics from each "species."

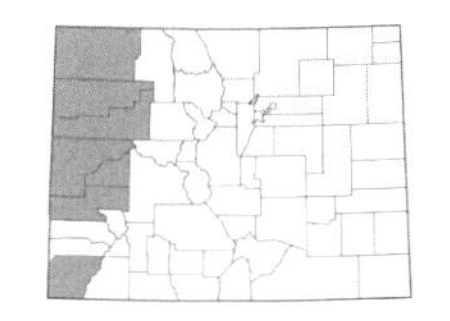

Mentzelia thompsonii **Glad**, Thompson's stickleaf. [*Acrolasia humilis* Osterh.]. Annuals, 1–2 dm, simple or branched above; *leaves* lanceolate to ovate, to 7 cm long, entire or with a few shallow lobes; *petals* 5, 1–4 mm long, yellow; *capsules* narrowly cylindric, 8–16 (20) × 2–3.5 (4) mm; *seeds* irregularly angled, 1.5–2 mm long. Rather uncommon, found on dry clay and shale slopes, 4900–8600 ft. May–June. W. (Plate 52)

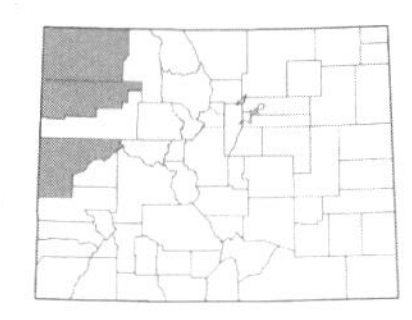

Mentzelia uintahensis **(N.H. Holmgren & P.K. Holmgren) J.J. Schenk & L. Hufford**, Uintah blazingstar. [*M. multicaulis* (Osterh.) J. Darl. var. *uintahensis* N.H. Holmgren & P.K. Holmgren]. Perennial subshrubs from a woody, branching caudex, 1.5–4 dm; *leaves* 1.5–3 (5) cm long, deeply pinnatifid into linear lobes with a narrow, linear rachis; *petals* 7–15 (17) mm long, yellow; *capsules* urceolate, (4) 6–9 × 3–5 mm; *seeds* 1.8–3 × 1.5–2 mm, with a narrow wing 0.1–0.3 mm wide. Found on shale and limestone slopes, 5900–7200 ft. June–Aug. W.

LYTHRACEAE Jaume St.-Hil. – loosestrife family

Herbs (usually); *leaves* usually opposite, simple and entire; *flowers* perfect, actinomorphic; *calyx* 4, 6, or 8-lobed, the lobes valvate, often an epicalyx of 2 bracteoles below; *petals* 4, 6, or 8, inserted on the inner surface of the hypanthium between the calyx lobes, plicate, usually red, purple, or orange; *stamens* usually twice the number of sepals or petals; *pistil* 1; *ovary* superior, 2–4-carpellate; *fruit* a capsule enclosed in the persistent calyx. (Graham, 1975)

1a. Perennials; petals purple to rose-purple; hypanthium cylindrical with conspicuous longitudinal striations...***Lythrum***
1b. Annuals; petals pale purple, white, or pink; hypanthium globose to campanulate, finely striate or smooth...2

2a. Hypanthium finely striate; upper leaves attenuate at the base, not clasping the stem; flowers solitary in axils...***Rotala***
2b. Hypanthium smooth; upper leaves auriculate and clasping the stem; 1–3 flowers per axil...***Ammannia***

AMMANNIA L. – TOOTHCUP

Annuals, glabrous herbs; *leaves* opposite and decussate, sessile; *flowers* in axillary cymes; *calyx* 4-lobed, alternating with small, hornlike appendages situated on each sinus of the hypanthium; *petals* 4; *stamens* 4 (8).

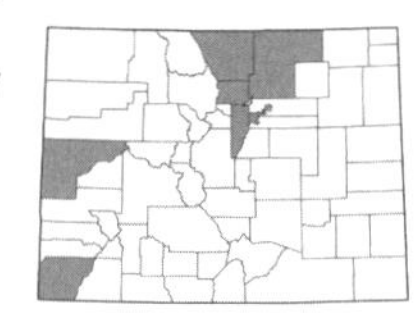

***Ammannia robusta* Heer & Regel**, GRAND REDSTEM. Plants to 10 dm; *leaves* linear-lanceolate, mostly 1.5–8 cm long, entire, with a clasping, auriculate base; *petals* lavender to rose, ca. 2.5 mm long; *capsules* 4–6 mm diam. Uncommon in moist places such as temporary pools or depressions and along the margins of ponds, 4500–5500 ft. July–Sept. E/W.

LYTHRUM L. – LOOSESTRIFE

Annual or perennial herbs; *leaves* opposite, alternate, or whorled; *flowers* solitary or in terminal spikes; *calyx* 5-8-lobed, alternating with appendages longer than the calyx lobes; *petals* 4 or 6, ours purple or rose-purple; *stamens* 4–14. (Shinners, 1953)

1a. Flowers solitary or paired in the axils; stamens 6...***L. alatum***
1b. Flowers 3 or more per axil above and solitary or paired below; stamens mostly 12 (6 long and 6 short, hidden in the flower)...***L. salicaria***

***Lythrum alatum* Pursh**, WINGED LOOSESTRIFE. Perennials to 12 dm; *leaves* lanceolate to oblong or ovate, 1.5–6 cm × 7–15 mm, sessile, mostly opposite; *flowers* solitary or paired in the axils; *petals* purple, 3–6 mm long; *stamens* 6. Uncommon in moist places such as along the margins of wetlands and wet meadows, 3500–5500 ft. July–Sept. E.

***Lythrum salicaria* L.**, PURPLE LOOSESTRIFE. Perennials to 12 dm; *leaves* lanceolate, 2–10 cm × 5–15 mm, opposite or whorled; *flowers* 3 or more per axil above and solitary or paired below; *petals* rose-purple, 5–10 mm long; *stamens* mostly 12. Uncommon in moist places such as along the margins of ponds, in irrigation ditches, and wetlands, an aggressive weed which should be eliminated if seen in an area, 5000–5500 ft. July–Sept. E/W. Introduced.

ROTALA L. – ROTALA

Annual herbs; *leaves* opposite and decussate; *flowers* solitary in axils, sessile; *calyx* 4-lobed, alternating with small appendages situated on each sinus of the hypanthium; *petals* 4, white or pink; *stamens* 4 (6).

***Rotala ramosior* (L.) Koehne**, LOWLAND ROTALA. Plants to 4 dm, glabrous; *leaves* linear to oblanceolate, 1–5 cm × 2–12 mm, attenuate at the base; *petals* just longer than the calyx lobes. Uncommon in moist places and easily overlooked, known from Boulder Co., 5200–5500 ft. June–Aug. E.

MALVACEAE Juss. – MALLOW FAMILY

Annual or perennial herbs or less often shrubs, often with mucilaginous sap, often with stellate or branched hairs; *leaves* alternate, simple, usually palmately lobed or veined, stipulate; *flowers* usually perfect, actinomorphic, sometimes subtended by an epicalyx; *calyx* 5-lobed, persistent; *petals* 5, distinct but attached at the base of the stamen column; *stamens* numerous, monadelphous; *pistil* 1; *ovary* superior, 2–many-carpellate; *fruit* a capsule or schizocarp.

1a. Plants with 2 different leaf shapes, the lower leaves reniform-orbicular and shallowly 5–7-lobed and the middle and upper leaves deeply 5–7-lobed with linear to lanceolate segments; flowers white, pale pink, or rose-purple...***Sidalcea***
1b. Plants with leaves all basically the same shape; flowers various...2

2a. Flowers orange to reddish-orange...3
2b. Flowers variously colored but not orange or reddish-orange...4

3a. Leaves with a cordate base; flowers lacking an epicalyx of involucel bractlets...***Abutilon*** (*incanum*)
3b. Leaves lacking a cordate base; flowers sometimes subtended by an epicalyx of involucel bractlets...***Sphaeralcea***

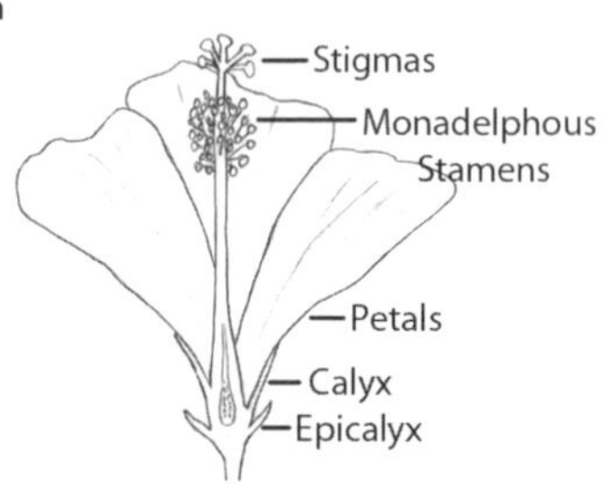

4a. Leaves deeply palmately lobed with the divisions well over halfway to the base of the leaf (sometimes nearly appearing palmately compound)...5
4b. Leaves unlike the above, if palmately lobed then the divisions not over halfway to the base of the leaf...6

5a. Involucel bractlets 5–13; calyx inflated or papery, with conspicuous purple veins; petals white to pale yellow with a dark reddish base; fruit a loculicidal capsule...***Hibiscus***
5b. Involucel bractlets 3; calyx not inflated or papery, lacking purple veins; petals pink to red-purple or rose-colored; fruit a schizocarp...***Callirhoe***

6a. Leaves distinctly palmately 3–7-lobed, the margins coarsely toothed; mericarps with long, stiff ascending hairs in addition to small stellate hairs on the dorsal side; perennials, 5–20 dm tall...***Iliamna***
6b. Leaves ovate to nearly orbicular, shallowly lobed, or if 3-lobed or palmately lobed then the margins entire or crisped; mericarps unlike the above; plants various...7

7a. Leaves triangular to hastate, usually 3-lobed with a longer middle lobe, hirsute; mericarps with conspicuous spreading hairs, with a long beak and dorsal spur; flowers pale blue to purple; annuals with mostly simple hairs...***Anoda***
7b. Leaves and mericarps unlike the above; flowers various...8

8a. Epicalyx of 6–9 involucel bractlets subtending the calyx, broadly triangular, united at the base and completely enclosing the flower bud at first; petals 3.5–6 cm long...***Alcea***
8b. Epicalyx absent or of 1–3 involucel bractlets and unlike the above; petals (0.3) 0.4–1.7 cm long...9

9a. Leaves with asymmetrical bases; plants often densely stellate-canescent, giving them a silvery appearance...***Malvella***
9b. Leaf bases symmetrical; plants with stellate hairs or not, but not silvery in appearance...10

10a. Ovules 2 or more per carpel, each mericarp of the fruit 2–6-seeded; epicalyx of involucel bractlets absent; flowers yellow, pink, or red...***Abutilon***
10b. Ovules 1 per carpel, each mericarp of the fruit 1-seeded; epicalyx of 1–3 involucel bractlets present and subtending the calyx; flowers white, pink, rose-purple, or light purple...***Malva***

ABUTILON Mill. – Indian mallow

Annual or perennial herbs with stellate or simple hairs; *leaves* entire or obscurely lobed; *flowers* solitary in the leaf axils or in leafy panicles, epicalyx lacking; *petals* yellow, pink, or red; *ovary* with 5–many carpels, these united in a ring around a central axis, rounded or beaked at the apex, dehiscent nearly to the base; *fruit* a schizocarp.

1a. Annuals, 2–20 dm; calyx tube 8–14 mm long, the tips acuminate or cuspidate; leaves velvety soft-pubescent, usually larger, mostly 5–17 cm long and to 20 cm wide with a long-acuminate tip; carpels 10–15, 10–17 mm long; flowers yellow...***A. theophrasti***
1b. Perennials to 6 dm; calyx tube 2–5 mm long, the tips acute to acuminate; leaves velvety soft or merely stellate-pubescent, usually smaller, mostly 2.5–5 (9) cm long and to 6 cm wide, the tips acute to shortly acuminate; carpels 5–9, 6–9 mm long; flowers pink, red, orange, or yellow...2

2a. Stems stellate-hairy, erect or nearly so; flowers orange or yellow with a reddish center, the petals 5–10 mm long...***A. incanum***
2b. Stems spreading pilose pubescent, slender and spreading or trailing, sometimes ascending; flowers pink or red, rarely orange or yellow, the petals 4–6 mm long...***A. parvulum***

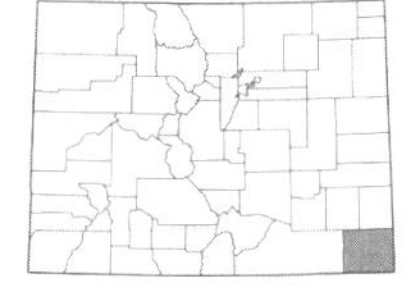

Abutilon incanum **(Link) Sweet**, pelotazo. Perennials to 6 dm, stellate-hairy; *leaves* ovate to deltoid, (1) 2.5–5 (9) cm wide, usually tomentose, toothed; *calyx* tube 2–5 mm long with lobes ca. 3 mm; *petals* orange or yellow with a reddish center, 5–10 mm long; *carpels* 5–9, 6–9 mm long. Uncommon in sandy canyons, known from a single collection in Baca Co., 4400–4500 ft. June–Sept. E.

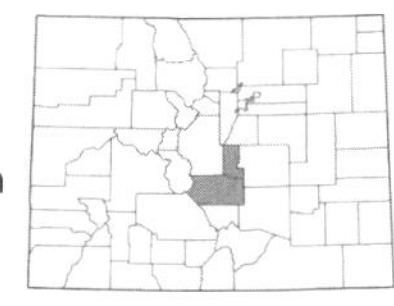

Abutilon parvulum **A. Gray**, dwarf Indian mallow. Perennials, spreading or trailing, sometimes ascending, pilose; *leaves* ovate, 0.5–5 cm long, toothed; *calyx* tube ca. 4 mm long with lobes ca. 2 mm; *petals* pink or red, rarely orange or yellow, 4–6 mm long; *carpels* 5, to 9 mm long. Uncommon on dry slopes and limestone hills, known from near Canon City, 5200–5500 ft. June–Sept. E.

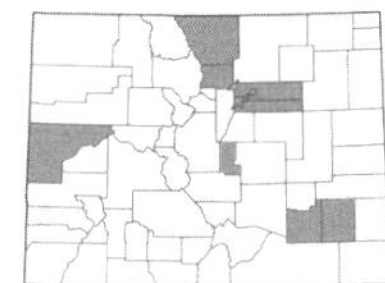

Abutilon theophrasti **Medic.**, velvet-leaf. Annuals, 2–20 dm, stellate-hairy; *leaves* ovate to orbicular, (3) 5–17 (25) cm long, acuminate; *calyx* tube 8–14 mm, the lobes ovate-acuminate; *petals* yellow, 6–15 mm long; *carpels* 10–15, 10–17 mm long. Uncommon weed found in gardens, disturbed places, in fields, and along roadsides, 3900–5300 ft. June–Oct. E/W. Introduced.

ALCEA L. – HOLLYHOCK

Biennial herbs, with stellate hairs; *leaves* lobed, petiolate; *flowers* solitary or in racemes, subtended by an epicalyx of 6–9 bractlets that are connate at the base; *petals* white, pink, purple, or red; *ovary* with numerous carpels, these united in a ring around a central axis, beakless; *fruit* a flattened, wheellike schizocarp.

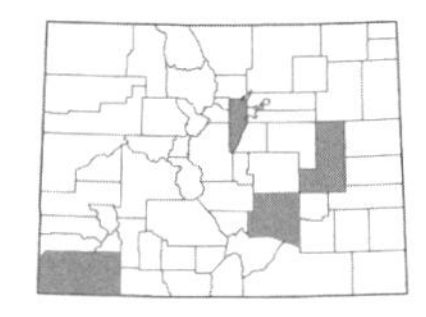

***Alcea rosea* L.**, HOLLYHOCK. [*Althaea rosea* (L.) Cav.]. Plants 1–3 m; *leaves* nearly orbicular, shallowly 5–7-lobed, 8–30 cm wide, toothed, cordate at the base; *petals* purple to rose or white, ca. 5 cm long. Garden plants occasionally escaping along roadsides and in disturbed areas, persisting around old buildings, 5000–6500 ft. May–Sept. E/W. Introduced.

ANODA Cav. – ANODA

Annual herbs with mostly simple hairs; *leaves* lobed with a truncate or cuneate base; *flowers* solitary in leaf axils, epicalyx lacking; *petals* blue to purple; *ovary* with 8–many carpels, these united in a ring around a central axis, conspicuously beaked with a dorsal spur; *fruit* a flattened, wheellike schizocarp.

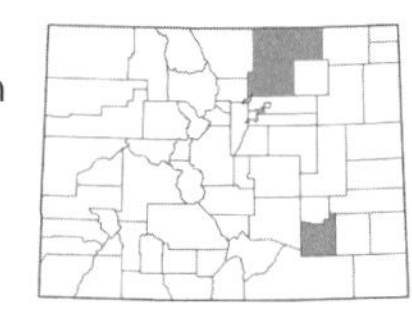

***Anoda cristata* (L.) Schltdl.**, CRESTED ANODA. Plants to 10 dm; *leaves* triangular or hastate, 4.5–10 cm long, irregularly toothed to entire, truncate or cuneate at the base; *petals* 1–2.5 cm long; *carpels* 8–20. Uncommon in disturbed areas, fields, and along roadsides, 4200–4500 ft. June–Oct. E. Introduced.

CALLIRHOE Nutt. – POPPY MALLOW

Perennial herbs, with simple or branched hairs or glabrous; *leaves* lobed or cleft; *flowers* subtended by an epicalyx of 3 bractlets or this absent; *petals* red, pink, white, or purple, erose; *ovary* with 10-numerous carpels, united in a ring around a central axis, indehiscent but separating at maturity, beakless; *fruit* a depressed schizocarp.

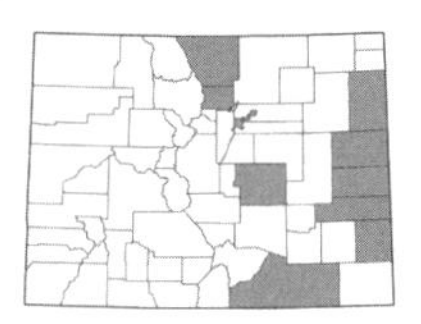

***Callirhoe involucrata* (Torr. & A. Gray) A. Gray**, WINECUPS; PURPLE POPPY MALLOW. Plants mostly sprawling or decumbent, the stems to 8 dm long, with simple and 4-rayed hairs; *leaves* orbicular, deeply 5–7-lobed, 1–6 cm long; *epicalyx* of 3 bractlets; *petals* pink to red-purple or rose, ca. 20–23 mm long; *carpels* 10–12. Uncommon along roadsides, in fields, and in open prairie, sometimes cultivated in gardens and escaping, 3400–6200 ft. May–Aug. E.

HIBISCUS L. – ROSE MALLOW

Annual or perennial herbs with simple or stellate hairs; *leaves* lobed or cleft; *flowers* solitary in leaf axils, subtended by an epicalyx of 5–13 distinct or united bractlets; *petals* showy, variously colored; *ovary* with 5 carpels, these permanently connate; *fruit* a loculicidal capsule.

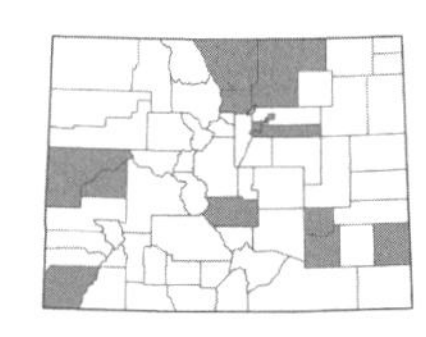

***Hibiscus trionum* L.**, FLOWER OF AN HOUR. Annuals, 1–6 dm; *leaves* orbicular to ovate, deeply 3–7-lobed, 2–6 cm long; *calyx* inflated or papery, with conspicuous purple veins; *petals* white to pale yellow with a dark reddish base, 1.5–2.5 cm long. Weed found in disturbed areas, along roadsides, in fields, and in gardens, 3500–5500 ft. Aug.–Sept. E/W. Introduced. (Plate 53)

ILIAMNA Greene – WILD HOLLYHOCK

Perennial herbs, minutely stellate pubescent; *leaves* lobed; *flowers* in interrupted spicate inflorescences, subtended by an epicalyx; *petals* white, pink, or rose; *ovary* with many carpels; *fruit* a loculicidal capsule with usually 3 seeds per carpel. (Wiggins, 1936)

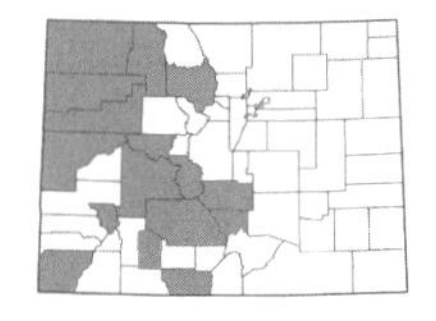

***Iliamna rivularis* (Douglas) Greene**, WILD HOLLYHOCK. [*I. crandallii* (Rydb.) Wiggins; *I. grandiflora* (Rydb.) Wiggins]. Plants 5–20 dm; *leaves* palmately 3–7-lobed, 2.5–15 cm × 2–16 cm, coarsely toothed; *calyx* lobes 3–5 mm long; *petals* rose to pink or rarely white, 16–35 mm long. Found in meadows, along streams and creeks, and on forest borders, 6000–9500 ft. June–Aug. E/W.

Iliamna crandallii and *I. grandiflora* are often treated as separate, distinct species from *I. rivularis*. However, they completely overlap morphologically and geographically in distribution.

MALVA L. – MALLOW

Annual, biennial, or perennial herbs, with simple, branched, or stellate hairs; *leaves* usually lobed; *flowers* subtended by an epicalyx of 3 persistent bractlets; *petals* white, pink, or rose-purple; *ovary* with 10–20 carpels, indehiscent but separating at maturity, beakless; *fruit* a flattened, wheellike schizocarp.

1a. Flowers rose-purple or red-purple with darker veins; involucel bractlets oblong to ovate; petals (2) 3–5 times the length of the calyx...***M. sylvestris***

1b. Flowers white, pinkish, or purplish; involucel bractlets usually linear to lanceolate, sometimes oblong; petals to 2 times the length of the calyx...2

2a. Leaves conspicuously 5-lobed, 5–20 cm wide, the margins very crisp or double-crenate; petioles usually with hairs in a line on the upper side; petals about 2 times the length of the calyx...***M. verticillata***
2b. Leaves shallowly or obscurely lobed, 1.5–10 cm wide, the margins not crisp; petioles glabrous or uniformly hairy; petals 2 times the length of the calyx or scarcely exceeding it...3

3a. Petals 2 times the length of the calyx; mature carpels smooth or only faintly reticulate...***M. neglecta***
3b. Petals scarcely longer than the calyx; mature carpels conspicuously reticulate...4

4a. Calyx lobes broadly ovate with a short mucronate tip...***M. parviflora***
4b. Calyx lobes narrowly triangular to triangular with an acute tip...***M. pusilla***

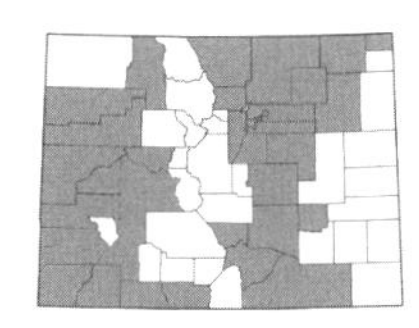

***Malva neglecta* Wallr.**, COMMON MALLOW. Annuals or biennials, prostrate or spreading, stems to 6 dm long; *leaves* orbicular, 0.6–3 cm long, sometimes shallowly 5–7-lobed, toothed; *calyx* 3–6 mm long; *petals* white, pink, or light purple, 6–12 mm long; *carpels* smooth or faintly reticulate, hairy. Common weed found in disturbed places, gardens, and fields, 3500–8500 ft. April–Oct. E/W. Introduced. (Plate 53)

***Malva parviflora* L.**, CHEESE-WEED MALLOW. Annuals, erect, 2–6 dm; *leaves* orbicular, to 6 cm long, shallowly 5–7-lobed, toothed; *calyx* lobes broadly ovate, mucronate at the tip; *petals* white to pink or light purple, (3) 4–6 mm long; *carpels* reticulate, glabrous or minutely hairy. Not reported for Colorado but could occur on the eastern plains in disturbed areas. April–Oct. Introduced.

***Malva pusilla* J. Sm.**, LOW MALLOW. [*M. rotundifolia* L.]. Annuals, prostrate or spreading to ascending; *leaves* orbicular, (1) 1.5–5 (6.5) cm long, obscurely lobed, toothed; *calyx* 3–4 (5) mm long, the lobes narrowly triangular to triangular, acute at the tip; *petals* white to light purple, scarcely exceeding the calyx; *carpels* reticulate, hairy. Known from one report in Denver but could occur on the eastern plains in waste areas or on disturbed ground, 5200 ft. May–Oct. E. Introduced.

***Malva sylvestris* L.**, HIGH MALLOW. Biennials, erect, 2–10 dm; *leaves* orbicular to reniform, shallowly 5–7-lobed, toothed; *calyx* 5–6 mm long, the lobes triangular; *petals* rose-purple or red-purple with darker veins, (2) 3–5 times the length of the calyx; *carpels* reticulate, glabrous or sparsely hairy. Garden plants, not reported for Colorado but could escape from cultivation. May–Oct. Introduced.

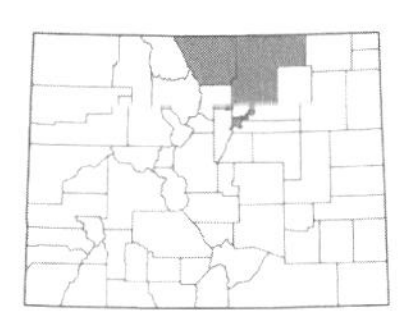

***Malva verticillata* L.**, CLUSTER MALLOW. [*M. crispa* L.]. Annuals, erect, to 20 dm; *leaves* orbicular to reniform, 5–20 cm wide, 5–7-lobed ca. ⅓ the blade length, the margins very crisp or double-crenate; *calyx* 3.5–5 mm long; *petals* white or purplish, about twice the length of the calyx; *carpels* smooth or weakly ridged, glabrous. Uncommon along roadsides where it has escaped from cultivation, 4800–5300 ft. July–Sept. E. Introduced.

MALVELLA Jaub. & Spach – LITTLE MALLOW

Perennial herbs with stellate-canescent or lepidote hairs; *leaves* crenate-serrate, with asymmetrical bases; *flowers* solitary in leaf axils, subtended by an epicalyx or not; *petals* white, ochroleucous, or fading to pink or purple, stellate-hairy; *ovary* with 6–many carpels, united in a single ring around a central axis and separating at maturity; *fruit* a schizocarp.

1a. Leaves mostly reniform to suborbicular, sometimes triangular-ovate; 1–3 linear involucel bractlets usually present; flowers white to cream or rose, often drying to brown or pink...***M. leprosa***
1b. Leaves mostly narrowly triangular or sometimes lanceolate to linear; involucel bractlets absent; flowers purple, yellow, or white with a purplish tinge...***M. sagittifolia***

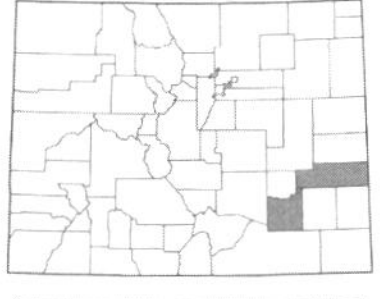

***Malvella leprosa* (Ortega) Krapov.**, ALKALI LITTLE MALLOW. Plants prostrate, the stems 5–45 cm long; *leaves* orbicular to reniform or triangular-ovate, to 4 cm long; *calyx* 4–7 mm long, with lobes 2–6 mm long; *petals* 10–16 mm long, white to cream or rose, often drying brown or pink; *carpels* 6–10. Uncommon in rocky places, usually in alkaline soil, 4000–4500 ft. April–Sept. E. Introduced.

***Malvella sagittifolia* (A. Gray) Fryxell**, ARROWLEAF LITTLE MALLOW. Plants prostrate, the stems to 40 cm long; *leaves* linear to lanceolate or narrowly triangular, 2–3 (5) cm long; *calyx* 3–8 mm long; *petals* purple, yellow, or white with purplish tinges, (9) 12–17 mm long; *carpels* 8–9. Uncommon in rocky places, known from a single collection near La Junta and probably not persisting, ca. 4100 ft. April–Sept. E. Introduced.

SIDALCEA A. Gray – CHECKER-BLOOM

Perennial herbs with usually stellate hairs and often somewhat hirsute; *leaves* lobed or cleft; *flowers* in terminal, bracteate racemes, epicalyx lacking; *petals* white, pinkish, purple, or rose-purple; *ovary* with 5–10 carpels; *fruit* a schizocarp. (Hitchcock, 1957; Roush, 1931)

1a. Flowers white to pale pink; anthers bluish-pink; middle and upper leaves palmately divided into lanceolate divisions (mostly 6–14 mm wide); stem retrorsely hispid below and finely stellate above; mericarps of the fruit with a short, erect beak at the apex…***S. candida***

1b. Flowers pink to rose-purple; anthers pale yellow or white; middle and upper leaves palmately divided into linear or narrowly lanceolate divisions (mostly 1–3 mm wide); stem sparsely hirsute with mostly simple or some forked hairs; mericarps of the fruit with a short, curved beak at the apex…***S. neomexicana***

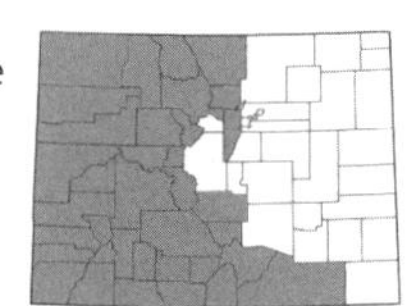

***Sidalcea candida* A. Gray**, WHITE CHECKER-BLOOM. Plants 4–10 dm, retrorsely hispid below and stellate above; *basal leaves* shallowly 5–7-lobed, crenate; *upper leaves* palmately divided into lanceolate divisions mostly 6–14 mm wide; *calyx* 7–10 mm long, stellate and glandular; *petals* white to pale pink, 12–20 mm long. Common along streams, in meadows, and in moist places, 6500–11,000 ft. June–Sept. E/W.

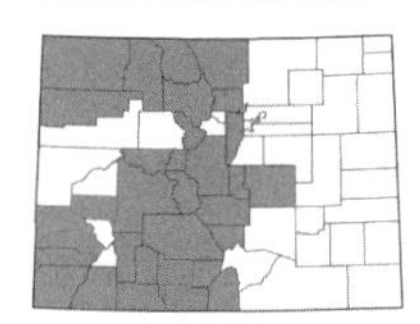

***Sidalcea neomexicana* A. Gray**, ROCKY MOUNTAIN CHECKER-BLOOM. Plants 2–10 dm, sparsely hirsute, sometimes with forked hairs; *basal leaves* shallowly 5–7-lobed or crenate; *upper leaves* palmately divided into linear or narrowly lanceolate divisions mostly 1–3 mm wide; *calyx* 5–10 mm long, with stellate and pustulate hairs; *petals* pink to rose-purple, 11–20 mm long. Common along streams, in marshes, meadows, and other moist places, 5500–9600 ft. June–Sept. E/W. (Plate 53)

SPHAERALCEA A. St.-Hil. – GLOBEMALLOW

Perennial herbs with stellate hairs; *leaves* lobed or cleft to toothed; *flowers* subtended by an epicalyx or not; *petals* orange to red or rarely pink; *ovary* with 10–many carpels, united in a ring around a central axis and often remaining attached to the axis after maturity by a thread; *fruit* a schizocarp.

1a. At least the middle and lower leaves very deeply palmately cleft or lobed to the base or nearly so …2

1b. Leaves all entire or shallowly lobed…3

2a. Leaves 3-foliate with linear, entire leaflets, the uppermost leaves simple and linear; inflorescence a terminal elongating raceme with one flower per node; plants densely stellate-hairy giving the plant a silvery appearance, the rays of the stellate hairs fused for most of their length; carpels 7–9…***S. leptophylla***

2b. Leaves 3-foliate with the lobes again deeply divided, giving the leaf a 5-foliate appearance, the ultimate segments oblanceolate to obovate; inflorescence a dense to somewhat elongating terminal raceme with lower flowers solitary in leaf axils; plants with stellate hairs, grayish-green, the rays of the stellate hairs free to the base; carpels 9–14…***S. coccinea***

3a. Leaves 2–3 times as long as wide, lanceolate or ovate-oblong; carpel apex ending in an abrupt, sharp point up to 1.2 mm long…4

3b. Leaves about the same length as width, orbicular to ovate or reniform; carpel apex obtuse or merely acute…5

4a. Leaves lanceolate, usually unlobed, thickish with prominent veins below; carpels usually prominently reticulate on the lower, indehiscent part; seeds usually glabrous…***S. angustifolia***

4b. Leaves ovate-oblong, usually shallowly 3-lobed with the terminal lobe much longer than the side lobes, thin or thickish with prominent veins below; carpels usually faintly and very finely reticulate on the lower, indehiscent part; seeds hairy…***S. fendleri***

5a. Hairs usually sparse on the leaves, giving the plant a bright green appearance; leaves relatively thin, the veins not prominent below; leaf margins coarsely and irregularly crenate or dentate; carpels nearly orbicular in outline, obtuse at the apex…***S. munroana***

5b. Hairs usually dense on the leaves, giving the plant a grayish-green appearance; leaves thickish with prominent veins below; leaf margins crenate-dentate and sometimes crisp; carpels broadly ovate in outline, acute or mucronate at the apex…***S. parvifolia***

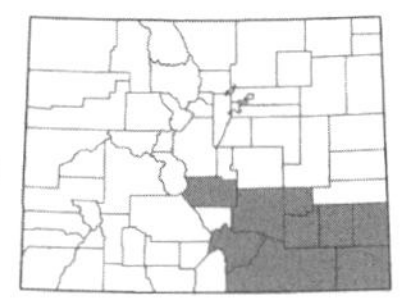

***Sphaeralcea angustifolia* (Cav.) G. Don**, COPPER GLOBEMALLOW. Plants 3–5 dm; *leaves* lanceolate, to 15 cm long, usually unlobed and subhastate at the base, with crenate margins; *calyx* 5–9 mm long; *petals* 6–20 mm long; *carpels* 10–15, usually prominently reticulate on the lower, indehiscent part. Found in sandy or rocky soil, often along roadsides, 4000–6500 ft. June–Sept. E.

***Sphaeralcea coccinea* (Nutt.) Rydb.**, SCARLET GLOBEMALLOW. Plants 0.5–5 dm; *leaves* 1–4 × 1.2–5 cm, deeply 3–5-lobed, the divisions toothed or lobed; *calyx* 3–10 mm long; *petals* 8–20 mm long; *carpels* 9–14, reticulate on the sides and back. Common in rocky or sandy soil, 3500–9000 ft. May–Sept. E/W. (Plate 53)

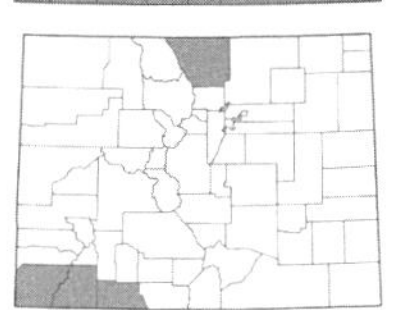

***Sphaeralcea fendleri* A. Gray**, FENDLER'S GLOBEMALLOW. Plants to 15 dm; *leaves* ovate-oblong, shallowly 3-lobed with the terminal lobe extended; *calyx* 4–6 mm long; *petals* 8–15 mm long; *carpels* 11–15, faintly reticulate. Found along dry riverbanks and in forest openings, native to the southwestern counties but also known from a collection near Ft. Collins (Larimer Co.) where it apparently escaped cultivation, 5000–7500 ft. May–Sept. E/W.

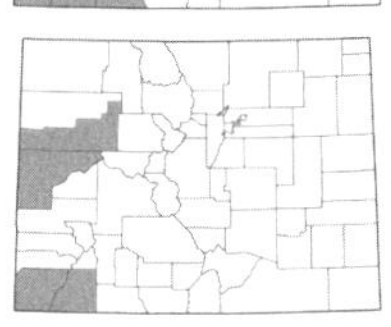

***Sphaeralcea leptophylla* (A. Gray) Rydb.**, SCALY GLOBEMALLOW. Plants 2–6 dm; *leaves* 1–3.5 cm long, deeply 3-lobed with each lobe further divided, giving a 5-foliate appearance; *calyx* 3–7 mm long; *petals* 8–12 mm long; *carpels* 7–9, reticulate or tuberculate on the back. Uncommon on sandy or rocky soil, usually on river benches, 4300–5000 ft. May–Sept. W.

***Sphaeralcea munroana* (Douglas) Spach**, MUNRO'S GLOBEMALLOW. Plants 1.5–8 dm; *leaves* orbicular to ovate, 1–6 × 0.8–6 cm, shallowly 3–5-lobed, sparsely hairy; *petals* 8–15 mm long; *carpels* 10–13, nearly orbicular in outline, reticulate on the sides. Uncommon in dry or gravelly soil, usually along roadsides, apparently introduced into Colorado at the Piceance Basin, known from Moffat, Rio Blanco, and Montezuma cos., 5500–8500 ft. May–Sept. W. Introduced.

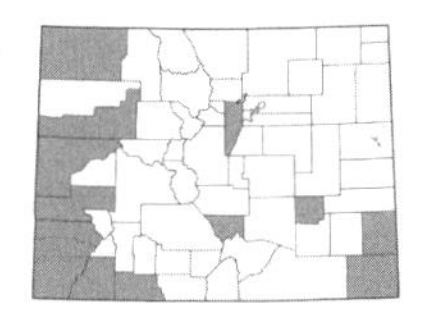

***Sphaeralcea parvifolia* A. Nelson**, SMALL-LEAF GLOBEMALLOW. Plants 2–7 dm; *leaves* orbicular to ovate or reniform, 1.5–5.5 × 1.2–5.4 cm, shallowly 3–5-lobed, usually densely hairy; *petals* 7–15 mm long; *carpels* 10–12, broadly ovate in outline, faintly reticulate on the sides. Found on rocky soil, in dry washes, and along roadsides, native to the western counties but occasionally found on the eastern slope where it has probably been introduced in revegetation projects, 4500–6500 ft. May–Sept. E/W.

MELANTHIACEAE – FALSE-HELLEBORE FAMILY

Perennial herbs from bulbs or rhizomes; *leaves* mostly basal, alternate, or whorled; *flowers* actinomorphic, perfect, in a raceme or panicle, or solitary; *tepals* 6; *stamens* 6; *ovary* superior or partly inferior, 3-carpellate; *fruit* a capsule.

1a. Tepals with 2 distinct, dissimilar series, the outer green and sepaloid; flowers solitary and terminal; plants with an elongated scape and 3 leaf-like, ovate, whorled bracts, the true leaves absent...***Trillium***
1b. Tepals all similar or nearly so, not in 2 distinct dissimilar series; flowers in terminal panicles; plants lacking an elongated scape, true leaves present...2

2a. Plants 1.5–2 m tall; leaves mostly elliptic to ovate, alternate and clasping on the stem; flowers numerous in a densely flowered terminal panicle 2–8 dm long...***Veratrum***
2b. Plants shorter; leaves linear, mostly basal; flowers in a terminal panicle shorter than 2 dm...***Zigadenus***

TRILLIUM L. – TRILLIUM

Perennial herbs from rhizomes, with an elongated scape and 3 leaf-like, ovate, whorled bracts; *flowers* solitary; *tepals* dissimilar, the outer three green and sepaloid and the inner 3 white and petaloid.

Trillium ovatum* Pursh ssp. *ovatum, PACIFIC TRILLIUM. Plants 2–4 dm; *leaves* ovate to elliptic, 6–12 cm long; *outer tepals* 1.5–4 cm long, linear-lanceolate; *inner tepals* 2–4 cm long. Uncommon in moist spruce-fir forests, 7200–11,000 ft. June–Aug. E/W.

Trillium ovatum in Colorado is disjunct from the Pacific Northwest.

VERATRUM L. – SKUNK CABBAGE; CORN-LILY

Leaves alternate, lanceolate to widely ovate, clasping; *flowers* in a terminal panicle or compound racemes; *tepals* whitish, greenish, or yellow, petaloid.

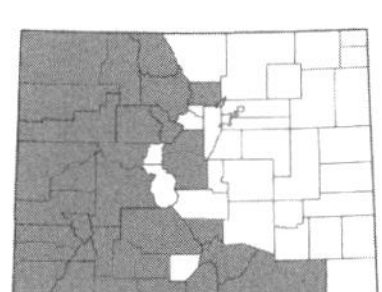

***Veratrum californicum* T. Durand**, CALIFORNIA FALSE HELLEBORE. [*V. tenuipetalum* A. Heller]. Plants 1.5–2 m; *leaves* elliptic to ovate, 10–30 × 6–20 cm; *tepals* white to greenish-yellow, 8–17 mm long; *capsules* 20–30 mm long, the seeds 10–15 mm long with a broad wing. Common in moist places along streams, in seepages, and meadows, 7500–11,000 ft. June–Aug. E/W. (Plate 53)

ZIGADENUS Michx. – DEATH CAMAS

Leaves alternate, mostly basal with the cauline leaves reduced and grading into bracts; *flowers* in a terminal raceme or panicle; *tepals* petaloid, greenish to white or yellowish-white.

1a. Pedicels recurved, flowers nodding; tepals 4–6 mm long, greenish with white margins or sometimes tinged with purple…***Z. virescens***

1b. Pedicels erect or ascending, flowers erect; tepals 2–12 mm long, cream to white or greenish…2

2a. Gland on the inside of the tepals obovate; flowers usually smaller, 5–10 mm in diam.; tepals 2–5 mm long; inflorescence a compact raceme or panicle; stamens equal to or longer than the tepals; ovary superior…***Z. paniculatus***

2b. Gland on the inside of the tepals deeply obcordate; flowers usually larger, 10–20 mm in diam.; tepals 5–12 mm long; inflorescence a loose raceme or panicle; stamens shorter than the tepals; ovary partly inferior…3

3a. Flowers 10–15 mm in diam., white; tepals 3–8 mm long; inflorescence usually paniculate with 1–4 lower branches, sometimes a raceme; leaves 25–40 cm long; plants of Moffat Co. below 6500 ft in elevation…***Z. vaginatus***

3b. Flowers 15–20 mm in diam., white, cream, or greenish; tepals 7–12 mm long; inflorescence usually a raceme, rarely paniculate with 1–4 lower branches; leaves 10–30 cm long; plants above 6500 ft in elevation, variously distributed but not found in Moffat Co…***Z. elegans***

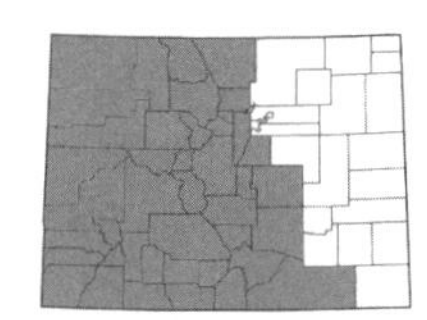

***Zigadenus elegans* Pursh**, MOUNTAIN DEATH CAMAS. [*Anticlea elegans* (Pursh) Rydb.]. Plants 2–8 dm; *leaves* 10–30 cm long; *inflorescence* loosely racemose or rarely paniculate with 1–4 lower branches; *flowers* 15–20 mm diam.; *tepals* cream to greenish, 7–12 × 4–5 mm; *gland* obcordate; *capsules* 10–20 mm. Common in meadows, dry to moist slopes, forests, and alpine, 6500–13,000 ft. June–Aug. E/W. (Plate 53)

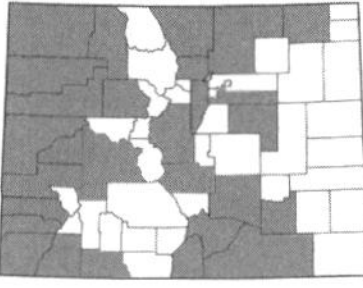

***Zigadenus paniculatus* (Nutt.) S. Watson**, FOOTHILL DEATH CAMAS. [*Toxicoscordion paniculatum* (Nutt.) Rydb.; *Z. venenosus* S. Watson]. Plants 2–7 dm; *leaves* 10–35 cm long; *inflorescence* a compact raceme or panicle; *flowers* 5–10 mm diam.; *tepals* cream to white, 2–5 × 1–4 mm; *gland* obovate; *capsules* 5–20 × 3–8 mm. Common on dry, grassy slopes, with sagebrush, and in forests, 4500–10,000 ft. April–July. E/W.

There are two varieties of *Z. paniculatus* in Colorado. These two varieties have traditionally been split out as separate species, but each overlap in morphological characters and distribution. *Zigadenus paniculatus* var. *venenosus* (S. Watson) Ackerfield occurs to the west of Colorado:

1a. Inflorescence a panicle with 2–4 lower branches, rarely a raceme; outer tepals usually not narrowed at the base to a short claw…**var. *paniculatus***. 5500–10,000 ft. April–June. W.

1b. Inflorescence usually a raceme or if paniculate then only with 1–2 lower branches; outer tepals usually narrowed at the base to a short claw…**var. *gramineus* (Rydb.) Ackerfield, comb. nov**. (see p. ix). [*Z. gramineus* Rydb.; *Z. venenosus* S. Watson var. *gramineus* (Rydb.) O.S. Walsh ex C.L. Hitchc.]. 4500–8500 ft. April–July. E/W.

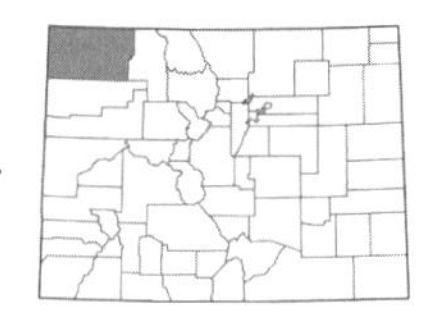

***Zigadenus vaginatus* (Rydb.) J.F. Macbr.**, ALCOVE DEATH CAMAS. [*Anticlea vaginata* Rydb.]. Plants 4–6 dm; *leaves* 25–40 cm long; *inflorescence* paniculate with 1–4 lower branches, occasionally a raceme; *flowers* 10–15 mm diam.; *tepals* white, 3–8 × 3–4 mm; *gland* obcordate; *capsules* 7–8 mm. Uncommon in hanging gardens in seeps and alcoves, 5200–6000 ft. July–Sept. W.

This species may simply represent a variety of *Z. elegans*, but it occurs in a much different kind of habitat.

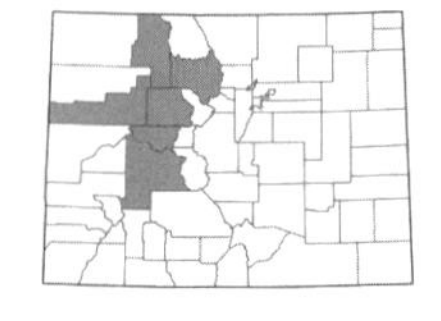

***Zigadenus virescens* (Kunth) J.F. Macbr.**, GREEN DEATH CAMAS. [*Anticlea virescens* (Kunth) Rydb.]. Plants 3.5–8.5 dm; *leaves* 10–30 cm long; *inflorescence* racemose to paniculate, with recurved pedicels 1.5–2 cm long; *flowers* 4–8 mm diam.; *tepals* greenish, sometimes tinged with purple, 4–6 × 1.5–3 mm; *gland* obcordate; *capsules* 10–20 mm. Uncommon in aspen and spruce forests, 8000–11,500 ft. July–Sept. E/W.

MENYANTHACEAE Dum. – BUCKBEAN FAMILY

Aquatic or semiaquatic herbs; *leaves* alternate, petiolate, simple or trifoliately compound, exstipulate; *flowers* perfect, actinomorphic; *calyx* 5-lobed; *corolla* 5-lobed; *stamens* 5, epipetalous, staminodes represented by fringed scales sometimes present in the corolla alternate with the stamens; *pistil* 1; *ovary* superior to half-inferior, 2-carpellate; *fruit* a capsule or a berry.

MENYANTHES L. – BUCKBEAN

Leaves all basal, emergent, trifoliately compound with a conspicuously sheathing base; *flowers* in a bracteate raceme; *corolla* densely hairy within; *ovary* partially inferior; *fruit* a capsule dehiscent by 2 valves.

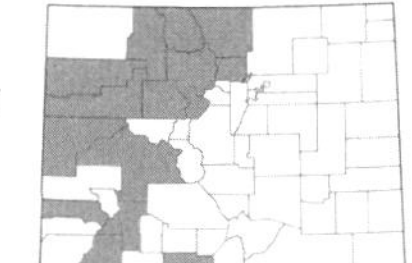

***Menyanthes trifoliata* L.**, BUCKBEAN. Plants to 3 dm; *leaflets* elliptic to obovate, 2–10 cm long; *calyx* 2–5 mm long; *corolla* lobes 5–8 mm long, white. Found in shallow water of ponds and lakes, slow-moving streams, marshes, and bogs, 8500–11,500 ft. June–Aug. E/W. (Plate 53)

MOLLUGINACEAE Raf. – CARPET-WEED FAMILY

Annual or perennial herbs; *leaves* simple, the margins entire; *flowers* usually perfect, actinomorphic; *sepals* 4–5, distinct or basally connate; *petals* absent; *stamens* 2–25; *ovary* superior, 1–5-carpellate; *fruit* a 3–5-valved loculicidal capsule.

MOLLUGO L. – CARPETWEED

Annual herbs; *leaves* whorled (ours); *flowers* in axillary, reduced umbellate cymes; *sepals* 5, distinct; *stamens* 3–5, alternate with the sepals; *styles* 3–5.

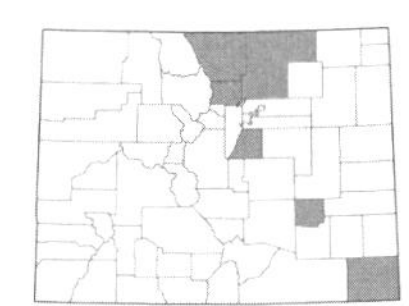

***Mollugo verticillata* L.**, GREEN CARPETWEED. Plants prostrate with dichotomously branched stems; *leaves* oblanceolate, 1–3 cm long; *sepals* to 2.5 mm long, white. Locally common in disturbed places and moist, sandy floodplains, 4200–5500 ft. June–Sept. E. Introduced.

This plant closely resembles a *Galium* at first sight, but the ovary is inferior in *Galium* and superior in *Mollugo*. *Mollugo* also has capsules with numerous small, red seeds.

MONTIACEAE Raf. – MINER'S LETTUCE FAMILY

Annual or perennial herbs, usually fleshy; *leaves* alternate, opposite, or basal, simple, entire; *flowers* perfect, actinomorphic; *sepals* usually 2 or 6–8 in *Lewisia*, distinct; *petals* 2–19, mostly 4–5, or rarely absent, distinct or basally connate; *stamens* 1–many, usually the same number as the petals, opposite the petals; *pistil* 1; *ovary* superior or half-inferior, 2–9-carpellate; *fruit* a circumscissle or valvate capsule. (Carolin, 1987; Packer, 2003)

1a. Plants with multiple pairs of stem leaves (3 or more); stems with slender stolons at the base; carpels 3...***Montia***
1b. Plants only with basal leaves or with basal leaves and 1–2 pairs of stem leaves; stems without stolons; carpels 3–8...2

2a. Plants with basal leaves (sometimes the basal leaves withering after flowering) and a single pair of stem leaves that are petiolate, sessile, or perfoliate; the leaves broader, ovate to oblanceolate or narrowly lanceolate...***Claytonia***
2b. Leaves all basal, clustered toward the base of the stem, or rarely with 2–3 narrowly linear stem leaves paired or whorled below the inflorescence; the leaves long and narrow, linear to narrowly linear-oblanceolate...3

3a. Leaves clustered near the base of the stem but not all basal; inflorescence extended well beyond the leaves on slender scapes (extending about 2–5 cm above the leaves); capsule longitudinally dehiscent from the apex...***Phemeranthus***
3b. Leaves all basal or paired or whorled below the inflorescence; inflorescence unlike the above; capsule circumscissile...***Lewisia***

CLAYTONIA L. – SPRING BEAUTY

Annual or perennial herbs; *leaves* generally fleshy, basal leaves few to many in rosettes, cauline leaves usually of a single opposite pair; *flowers* in terminal racemose or umbellate, bracteate inflorescences; *sepals* 2, persistent; *petals* 5; *stamens* 5, adnate to the base of the petals; *carpels* 3; *ovary* superior; *fruit* a 3-valved capsule. (Davis, 1996; Halleck & Wins, 1966; McNeill, 1972)

Care must be taken when collecting *Claytonia* as the underground tubers are often easily detached.

1a. Annuals from minute tubers; basal leaves rhombic to ovate, often with red or purplish pigmentation, stem leaves sessile or perfoliate; flowers generally smaller, 2–10 mm in diam....2
1b. Perennials from globose tubers or a woody caudex; basal leaves absent or linear to rhombic or oblanceolate, generally without reddish pigmentation except sometimes in *C. megarhiza* when the plants are more mature (and then plants of the alpine), stem leaves petiolate or sessile (never perfoliate); flowers 8–20 mm in diam....3

2a. Basal leaves in flattened to suberect rosettes, the red pigmentation very strong; stem leaves distinct and sessile, connate on just one side, or sometimes perfoliate...***C. rubra***
2b. Basal leaves in suberect to more often erect rosettes, the red pigmentation usually not as strong; stem leaves perfoliate...***C. perfoliata* ssp. *intermontana***

3a. Plants from a stout, elongate woody caudex; basal leaves numerous, oblanceolate or spatulate, narrowed at the base into a succulent sheath that surrounds the long petiole; stem leaves sessile and lanceolate to oblanceolate; flowers 12–20 mm in diam.; plants of the alpine...***C. megarhiza***

3b. Plants from globose tubers; basal leaves absent or few, linear to lanceolate and long-petiolate on a slender petiole; stem leaves ovate to narrowly lanceolate or linear, sessile or slightly petioled; flowers smaller, 8–14 mm in diam.; plants usually not of the alpine...4

4a. Stem leaves ovate to narrowly lanceolate, sessile; basal leaves usually absent at flowering; inflorescence 1-bracteate or rarely 2-bracteate...***C. lanceolata***

4b. Stem leaves linear and narrowed at the base into a short petiole; basal leaves usually present at flowering; inflorescence multibracteate with the bracts reduced to scales (this is only seen when the inflorescence is mature and elongate)...***C. rosea***

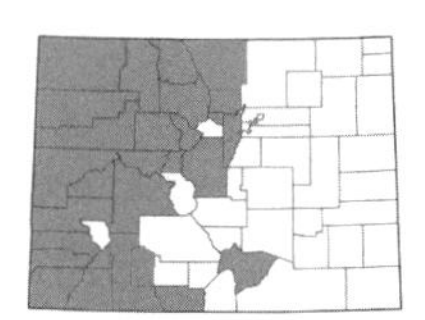

Claytonia lanceolata **Pall. ex Pursh**, WESTERN SPRING BEAUTY. Perennials, 1–10 cm, from globose tubers; *basal leaves* 1–6, often absent at flowering, linear to lanceolate, 5–40 × 0.2–1.6 cm; *cauline leaves* ovate to narrowly lanceolate, 1–6 × 0.5–2 cm, sessile; *inflorescence* 1 (2)-bracteate; *flowers* white, pink, rose, or magenta, 8–14 mm diam. Common in melting snowbanks in subalpine meadows, moist meadows, and montane forests, 7000–12,000 ft. May–July. E/W.

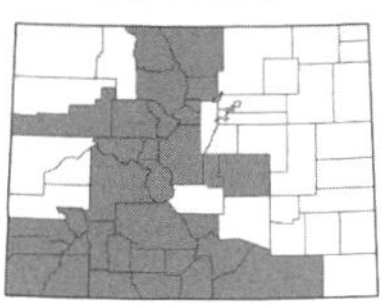

Claytonia megarhiza **(A. Gray) Parry ex S. Watson**, ALPINE SPRING BEAUTY. Perennials, 5–25 cm, from a stout, woody caudex; *basal leaves* numerous, oblanceolate to spatulate, 2–10 × 0.4–3 cm, dilated at the base into a succulent sheath; *cauline leaves* oblanceolate, 2–10 × 2–5 mm; *inflorescence* multibracteate; *flowers* 12–20 mm diam., white, pink, or rose. Common in rock crevices in the alpine, 11,000–14,000 ft. June–Aug. E/W. (Plate 53)

Claytonia perfoliata **Donn ex Willd. ssp. *intermontana* J.M. Mill. & K. Chambers**, MINER'S LETTUCE. Annuals, 5–50 cm, from minute tubers; *basal leaves* broadly rhombic to ovate, 20–30 cm long, in suberect to erect rosettes, sometimes with red or purplish pigmentation; *cauline leaves* perfoliate, 1–5 cm diam.; *inflorescence* 1-bracteate; *flowers* 3–10 mm diam., pink or white. Uncommon, recently reported for the state in southwestern Colorado, 6000–6500 ft. April–June. W.

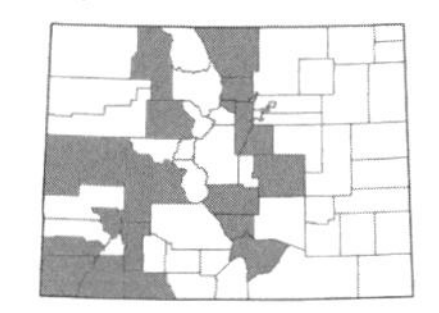

Claytonia rosea **Rydb.**, ROCKY MOUNTAIN SPRING BEAUTY. Perennials, 2–15 cm, from globose tubers; *basal leaves* linear to narrowly spatulate, or sometimes absent, 1–7 × 0.4–2 cm; *cauline leaves* linear, 2–5 cm long; *inflorescence* multibracteate or rarely 1-bracteate, reduced to scales; *flowers* 8–14 mm diam., pink or rose. Common in ponderosa pine forests in early spring, 5500–8500 ft. Mar.–May (June). E/W.

Claytonia rubra **(T.J. Howell) Tidestrom**, REDSTEM SPRING BEAUTY. Annuals, 1–10 cm, from minute tubers; *basal leaves* rhombic, 0.5–1.5 × 0.5–1 cm, in flattened to suberect rosettes, with strong red pigmentation; *cauline leaves* ovate, distinct or perfoliate; *inflorescence* 1-bracteate; *flowers* 2–5 mm diam., pink or white. Uncommon in moist, shady places, cool canyons, and ravines, 5500–6700 ft. April–Aug. E. (Plate 53)

LEWISIA Pursh – BITTERROOT

Perennial herbs; *leaves* often in basal rosettes or tufts, if cauline leaves present these alternate, opposite, or whorled, sessile or with a broad, clasping petiole; *flowers* solitary or in racemose, paniculate, or umbellate cymes; *sepals* 2 (9), persistent; *petals* (4) 5–10 (19); *stamens* 1–50; *carpels* 3–8; *ovary* superior; *fruit* a circumscissile capsule. (Davidson, 2000)

1a. Sepals (4) 6–9 and turning dry and scarious after flowering; flowers large and solitary, the 10–19 petals 15–35 mm long; plants from a large system of underground taproots, these thick, fleshy, and reddish-orange in cross-section...***L. rediviva* var. *rediviva***

1b. Sepals 2, remaining green and herbaceous after flowering (not turning dry and scarious); flowers smaller and in racemose cymes, panicles, or solitary, the 5–10 petals 4–15 (20) mm long; plants from shortly fusiform taproots or globose and cormlike roots...2

2a. Plants from globose, cormlike roots; basal leaves withering before flowering and stem leaves 2–3 (5) and paired or whorled below the inflorescence; flowers in paniculate cymes, these 3–25-flowered...***L. triphylla***

2b. Plants from shortly fusiform taproots; basal leaves present at flowering and stem leaves absent; flowers solitary or in racemose cymes, these 2–4 (7)-flowered...***L. pygmaea***

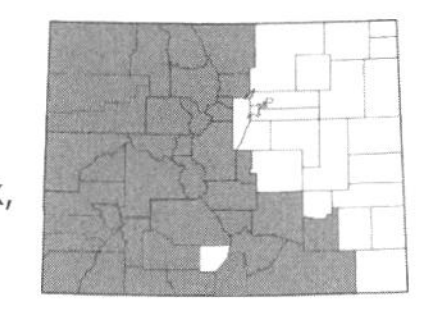

Lewisia pygmaea **(A. Gray) B.L. Rob.**, ALPINE LEWISIA. [*Oreobroma pygmaea* (A. Gray) T.J. Howell]. Plants 1–12 cm, from shortly fusiform taproots; *basal leaves* linear to linear-oblanceolate, 2–8 cm long, withering soon after flowering; *cauline leaves* absent; *inflorescence* solitary or a 2–4 (7)-flowered racemose cyme; *sepals* 2, 2–13 mm long; *petals* 5–10, 4–15 (20) mm long, white, pink, or magenta. Common in alpine meadows and forest openings, 7200–14,000 ft. May–Aug. E/W. (Plate 53)

Lewisia nevadensis is often treated as a separate species from the variable *L. pygmaea*. However I find that the morphological characteristics separating the two overlap considerably. Specimens of *L. pygmaea* sometimes have entire or merely dentate sepals that approach *L. nevadensis*. In addition, *L. pygmaea* specimens in Colorado are found as low as 7200 ft, and the two can hybridize. They are here treated as varieties of *L. pygmaea*:

1a. Sepals entire or with a few shallow, nonglandular teeth, 5–13 mm long; petals 10–15 (20) mm long, white or rarely pinkish...**var. *nevadensis* (A. Gray) Fosberg**, NEVADA LEWISIA. [*Oreobroma nevadensis* (A. Gray) T.J. Howell]. Locally common in moist meadows and forest openings or near springs, known from the southwestern counties and Moffat Co., 7500–8600 ft. May–June. W.

1b. Sepals toothed, the tips glandular or nonglandular, or rarely entire, 2–6 mm long; petals 4–10 mm long, pink, magenta, or white...**var. *pygmaea***, ALPINE LEWISIA. [*Oreobroma pygmaea* (A. Gray) T.J. Howell]. Common in alpine meadows and forest openings, 7200–14,000 ft. June–Aug. E/W.

Lewisia rediviva **Pursh var. *rediviva***, BITTERROOT. Plants 1–3 cm from thick, fleshy, reddish-orange taproots; *basal leaves* linear to clavate, withering soon after flowering, 2–5 cm long; *cauline leaves* absent; *inflorescence* solitary; *sepals* (4) 6–9, 10–25 mm long, scarious after flowering; *petals* 10–19, 15–35 mm long, rose to pink or white. Locally common in dry, gravelly soil of open places, 7300–9100 ft. June–July. E/W. (Plate 53)

Lewisia rediviva is at the southern limit of its range here, just extending into northern Colorado.

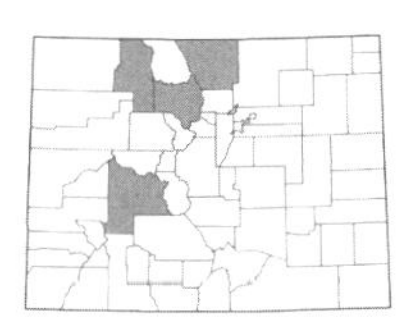

Lewisia triphylla **(S. Watson) B.L. Rob.**, THREELEAF LEWISIA. [*Erocallis triphylla* (S. Watson) Rydb.]. Plants 3–11 cm, from globose, cormlike roots; *basal leaves* solitary or absent; *cauline leaves* 2–3 (5), paired or whorled below the inflorescence, linear, 1–5 cm long; *inflorescence* of subumbellate to paniculate, 3–25-flowered cymes; *sepals* 2, 2–4 mm long; *petals* 5–9, 4–7 mm long, white or pinkish with darker veins. Uncommon in moist meadows, 8300–11,500 ft. June–Aug. E/W.

MONTIA L. – INDIAN LETTUCE

Annual, biennial, or perennial herbs; *leaves* usually opposite; *sepals* 2; *petals* 5 or sometimes absent, usually distinct; *stamens* (2) 3–5; *carpels* 3; *ovary* superior; *fruit* a 3-valved capsule. (Mill., 2003)

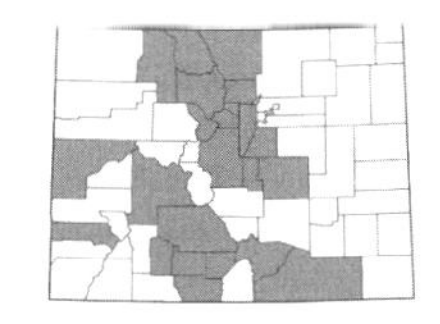

Montia chamissoi **(Ledeb. ex Spreng.) Greene**, WATER MINER'S-LETTUCE. [*Crunocallis chamissoi* (Ledeb. ex Spreng.) Rydb.]. Perennials, 5–25 cm, from slender rhizomes and stolons; *basal leaves* absent or reduced; *cauline leaves* oblanceolate to elliptic, 10–50 × 2–18 mm; *inflorescence* racemose with pedicels 8–30 mm long, nodding in bud; *sepals* 2–3 mm; *petals* white or seldom pinkish, 5–8 mm long; *capsules* 2–3 mm long. Common in moist places such as along streams and at the edges of lakes, often in shade, 6000–12,000 ft. June–Aug. E/W.

Montia linearis (Douglas ex Hook.) Greene has been reported for Colorado by Miller (2003), but this is probably an erroneous report as this would be out of range for this species.

PHEMERANTHUS Raf. – FAMEFLOWER

Perennial herbs from cormlike roots; *leaves* cauline, alternate; *flowers* in cymose or solitary and axillary inflorescences; *sepals* 2, usually deciduous; *petals* 5, soon withering, distinct or sometimes basally connate; *stamens* 4–many; *carpels* 3 (5); *ovary* superior; *fruit* a 3-valved capsule longitudinally dehiscent from the apex. (Carter & Murdy, 1985; Hershkovitz & Zimmer, 2000; McGregor, 1987)

1a. Petals to 7 mm long; stamens usually 5...***P. parviflorus***

1b. Petals 10–15 mm long; stamens numerous, 25–45...***P. calycinus***

Phemeranthus calycinus **(Engelm.) Kiger**, GREAT PLAINS FAMEFLOWER. [*Talinum calycinum* Engelm.]. Plants to 4 dm; *leaves* subterete, to 7 cm long; *sepals* persistent, ovate to suborbicualte, 4–6 mm long; *petals* 10–15 mm long, pink to reddish-purple; *stamens* numerous, 25–45. Reported for Weld Co. in the *Atlas of the Flora of the Great Plains* and Logan and Yuma cos. by Weber (2001), but no specimens exist at any herbarium in Colorado. Occurs in open, sandy places, and documented specimens should be sought after on the eastern plains. May–Sept. E.

Phemeranthus parviflorus **(Nutt.) Kiger**, SUNBRIGHT. [*Talinum parviflorum* Nutt.]. Plants to 2 dm; *leaves* terete, to 5 cm long; *sepals* deciduous or persistent, 2.5–4.5 mm long; *petals* 5.5–7 mm long, pink to purplish; *stamens* (4) 5 (6). Common (although easily overlooked) in sandy or rocky soil, 3500–8500 ft. May–Sept. E/W.

MORACEAE Link – MULBERRY FAMILY

Dioecious or monoecious trees, shrubs, herbs, or vines; *leaves* usually alternate, simple, stipulate; *flowers* imperfect, small, actinomorphic; *sepals* 2–6 (usually 4), distinct or partly connate near the base; *staminate flowers* with stamens equal in number to the sepals and opposite them; *pistillate flowers* with 1 pistil and 1–2 stigmas; *ovary* superior or inferior, 1–2-carpellate; *fruit* multiple, composed of individual achenes or drupelets enclosed by an enlarged common receptacle or calyx.

1a. Leaves entire; fruit a large, globose multiple with wrinkled rind, yellowish-green; stems usually thorny...***Maclura***

1b. Leaves serrate to lobed; fruit cylindric, resembling a blackberry, white to purple or nearly black; stems unarmed...***Morus***

MACLURA Nutt. – MACLURA

Dioecious trees; *leaves* alternate, entire; *staminate flowers* with 4 sepals and 4 stamens; *pistillate flowers* with 4 sepals; *ovary* unicarpellate; *fruit* a large, globose multiple, with a wrinkled rind comprised of old calyces.

Maclura pomifera **(Raf.) Schneid.**, OSAGE ORANGE. Trees to 15 (20) m, armed with stout thorns to 2.5 cm long; *leaves* elliptic to ovate, 4–10 × 2–5 cm, with an acuminate tip, entire; *flowers* in globose clusters; *multiple fruits* 6–15 cm diam., yellowish-green, with wrinkled rind. Sometimes escaping from cultivation, often grown as fence rows, with one population established near Grand Junction (Mesa Co.), 4500–5000 ft. May–June. W. Introduced.

MORUS L. – MULBERRY

Dioecious, deciduous trees or shrubs with milky sap; *leaves* alternate, serrate to dentate or lobed; *staminate flowers* with 4–5 sepals and 4 stamens; *pistillate flowers* with 4 sepals of 2 sizes; *ovary* superior, 2-carpellate; *fruit* composed of fleshy drupelets, resembling a blackberry.

Morus alba **L.**, WHITE MULBERRY. Trees or shrubs to 15 m, unarmed; *leaves* irregularly 2–5-lobed or sometimes unlobed, ovate to elliptic, 4–10 cm long, coarsely toothed; *staminate flowers* in catkins; *pistillate flowers* in axillary clusters; *multiple fruits* 0.5–2.5 cm long, cylindric, fleshy, resembling a blackberry (white, purple, or nearly black). Often planted in yards and escaping from cultivation, 4200–5700 ft. May–June. E/W. Introduced.

MYRSINACEAE R. Br. – MYRSINE FAMILY

Annual or perennial herbs, shrubs, trees, and woody vines; *leaves* alternate, simple, exstipulate; *flowers* perfect, actinomorphic; *calyx* (3) 5 (9)-lobed, usually persistent; *corolla* (3) 5 (9)-lobed or absent in *Glaux*; *stamens* (3) 5 (9), epipetalous (free in *Glaux*), opposite the corolla lobes, sometimes alternating with 5 staminodia; *pistil* 1; *ovary* superior (ours) or partly inferior, 3–5-carpellate; *fruit* a capsule.

1a. Petals absent, the sepals white or pinkish and petaloid, 3–5 mm long; flowers sessile in leaf axils; leaves tightly grouped on the stem, 0.3–2 cm long...***Glaux***

1b. Petals present, usually over 5 mm long; flowers sessile in leaf axils or not; leaves not tightly grouped on the stem, 0.2–15 cm long...2

2a. Annual herbs from taproots; flowers white, orangish, pinkish, or blue, solitary in the leaf axils; stems branched from the base...***Anagallis***

2b. Perennial herbs from rhizomes; flowers yellow (but fading white), in a raceme, paired, or solitary in the leaf axils; stems simple and erect (sometimes branched above but not at the base) or prostrate...***Lysimachia***

ANAGALLIS L. – POORMAN'S WEATHERGLASS; PIMPERNEL

Annuals, caulescent herbs; *leaves* opposite, alternate, or whorled; *flowers* solitary in leaf axils; *calyx* 4–5-lobed, deeply cleft; *corolla* 4–5-lobed, rotate and deeply-parted; *stamens* 4–5; *capsules* circumscissle, subglobose.

1a. Flowers orange, rose, blue, or sometimes white, with a pinkish center, 5-merous, pedicellate 10–30 mm, the petals conspicuously longer than the sepals; leaves opposite...***A. arvensis***

1b. Flowers white or pinkish, lacking a pinkish center, usually 4-merous, sessile or nearly so, the petals shorter than the sepals; leaves alternate or only the lower ones opposite...***A. minima***

Anagallis arvensis **L.**, SCARLET PIMPERNEL. Plants spreading or prostrate, the stems mostly 1–3 dm long, glabrous; *leaves* ovate to elliptic, 5–20 mm long, sessile and clasping; *calyx* 5-parted, the lobes 3–4 mm long; *corolla* 5-parted, the lobes orange, rose, blue, or sometimes white, with a pinkish center. Uncommon weedy plants found in disturbed areas and gardens, 5000–5500 ft. June–Aug. E. Introduced.

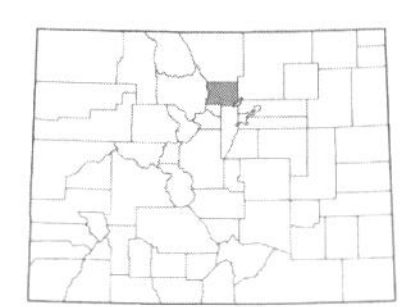

***Anagallis minima* (L.) Krause**, CHAFFWEED. [*Centunculus minimus* L.]. Plants ascending to spreading, the stems 0.3–1 dm long; *leaves* oblanceolate to ovate, 2–5 mm long; *calyx* 4-parted, the lobes ca. 3 mm long; *corolla* 4-parted, the lobes white or pinkish. Uncommon or overlooked in wet swales and along the edges of swamps, 5400–5700 ft. June–Aug. E.

GLAUX L. – SEA MILKWORT

Perennials, somewhat succulent, caulescent herbs; *leaves* opposite below and alternate above, entire, sessile; *flowers* solitary in leaf axils, sessile or nearly so; *calyx* 5-lobed, petaloid, white to pinkish; *corolla* absent; *stamens* 5; *capsules* ovoid to globose.

***Glaux maritima* L.**, SEA MILKWORT. Plants 3–20 cm, from slender rhizomes; *leaves* oblanceolate to elliptic, 0.3–2 cm long; *calyx* 3–4 mm long; *capsules* 2.5–3 mm long. Found in moist places, meadows, along streams, and in mudflats, 4700–9700 ft. May–July. E/W.

LYSIMACHIA – YELLOW LOOSESTRIFE

Perennials, caulescent herbs; *leaves* opposite or whorled, entire; *flowers* solitary in leaf axils, terminal, or in axillary racemes, yellow; *calyx* (3) 5 (9)-lobed, cleft nearly to the base; *corolla* (3) 5 (9)-lobed, cleft nearly to the base, densely glandular inside at the base; *stamens* (3) 5 (9); *capsules* subglobose to ovoid.

1a. Plants prostrate and rooting at the nodes; leaves orbicular to ovate with a rounded apex, glabrous; flowers solitary or paired in the leaf axils...***L. nummularia***
1b. Plants erect; leaves lanceolate to elliptic or ovate, acute to attenuate at the apex, glabrous or ciliate on the margins; flowers in a raceme, paired, or solitary in the leaf axils...2

2a. Flowers in an axillary raceme; corolla lobes linear, 3–5 (7) mm long; leaves sessile or nearly so, glabrous...***L. thyrsiflora***
2b. Flowers solitary or paired in the leaf axils; corolla lobes obovate, 5–12 mm long; leaves petiolate, the petiole coarsely ciliate and the margins of the leaves finely ciliate...***L. ciliata***

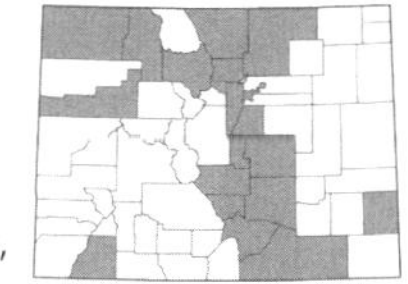

***Lysimachia ciliata* L.**, FRINGED LOOSESTRIFE. Plants 3–10 dm, erect, glabrous; *leaves* lanceolate to ovate, 3.5 15 × 1 6 cm, petiole 0.5 7 cm long and coarsely ciliate, margins ciliate; *flowers* solitary or paired in the leaf axils; *calyx* lobes (2) 4–9 mm long; *corolla* lobes obovate, 5–12 mm long. Locally common in moist places, floodplains, along rivers and streams, and in shady aspen groves, 5000–8800 ft. June–Aug. E/W. (Plate 53)

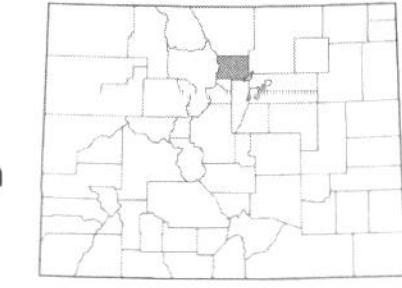

***Lysimachia nummularia* L.**, CREEPING JENNY. Plants prostrate, rooting at the nodes; *leaves* orbicular to ovate, glabrous, reddish glandular-punctate, 1–3 cm long, the petioles 1.5–4 mm long; *flowers* solitary from the middle axils; *calyx* lobes 6–9 mm long; *corolla* lobes 10–15 cm long. Uncommon along ditches and moist places, a commonly cultivated plant, 5000–5500 ft. May–July. E. Introduced.

***Lysimachia thyrsiflora* L.**, TUFTED LOOSESTRIFE. [*Naumburgia thyrsiflora* (L.) Rchb.]. Plants 2–8 dm, erect; *leaves* lanceolate to elliptic, 4–15 × 0.5–5 cm, glabrous or sparsely hairy along the midrib, sessile or nearly so; *flowers* in axillary racemes; *calyx* lobes 1.5–3 mm long; *corolla* lobes 3–5 (7) mm long, linear. Uncommon in moist places along rivers and streams and in marshes, 5000–7500 ft. May–Aug. E.

NITRARIACEAE Bercht & J. Presl – NITRARIA FAMILY

Shrubs or perennial herbs, sometimes spiny, usually succulent; *leaves* alternate, fleshy, simple, dissected or entire, stipulate; *flowers* perfect, actinomorphic; *sepals* 3–5; *petals* 3–5; *stamens* (10) 15; *ovary* superior, 2–6-carpellate; *fruit* a capsule.

PEGANUM L. – PEGANUM

Perennial herbs; *leaves* alternate, deeply irregularly pinnatifid; *petals* white to yellow; *ovary* (2) 3 (4)-carpellate; *fruit* a loculicidal capsule.

***Peganum harmala* L.**, AFRICAN RUE. Stems prostrate or ascending, to 9 dm long; *leaves* pinnatifid, the lobes 2–6 cm long; *sepals* entire or pinnatifid; *petals* to 1.7 cm long; *capsules* ca. 1.5 cm diam. Uncommon in disturbed places, 5000–5500 ft. May–Aug. E. Introduced.

NYCTAGINACEAE Juss. – FOUR-O'CLOCK FAMILY

Annual or perennial herbs or shrubs; *stems* usually swollen at the nodes; *leaves* usually opposite, exstipulate, simple, entire or sinuate; *flowers* perfect or imperfect, actinomorphic or zygomorphic, in bracteate clusters, the bracts distinct or connate and often forming an involucre; *calyx* usually 5-lobed, petaloid and colorful; *corolla* absent; *stamens* 1–18 (30); *pistil* 1; *ovary* superior; *fruit* an accessory fruit typically called an "anthocarp" consisting of an achene or utricle enclosed in the base of the calyx, the surface glabrous, hairy, or sticky glandular, the sides 5- or 10-ribbed or unribbed, the ribs varying from ridges to wings or rows or gland-tipped teeth. (Spellenberg, 2003; Standley, 1909)

1a. Stigmas linear; stamens included; flowers many per cluster, subtended by distinct bracts...2
1b. Stigmas capitate or hemispheric; stamens exserted; flowers 1–10 per cluster, subtended by bracts united for approximately 50% of their length, or only united at the base, rarely all distinct...3

2a. Fruit large, 1 cm or more in length and width, with conspicuous, translucent, prominently veined wings that extend above and below the body of the fruit...***Tripterocalyx***
2b. Fruit smaller, unwinged or with wings not extending above and below the body of the fruit...***Abronia***

3a. Flowers highly zygomorphic, in clusters of 3 simultaneously blooming flowers, each cluster subtended by a bract giving the inflorescence the appearance of a single flower; bracts 3, united at the base; fruit strongly compressed...***Allionia***
3b. Flowers actinomorphic, usually more than 3 per cluster, clearly distinguishable; bracts 4–5, usually united for 50% of their length or more, sometimes united only at the base, or rarely distinct; fruit round or angled...***Mirabilis***

ABRONIA Juss. – SAND-VERBENA

Perennial herbs; *flowers* perfect, in capitate clusters; *bracts* 5–10, distinct, persistent, not increasing in size in fruit, scarious; *calyx* 5-lobed, funnelform to salverform, actinomorphic; *stamens* 5–9, included; *styles* included; *fruit* leathery to hard and indurate, lobed or 2–5-winged, the wings not completely surrounding the body. (Ackerfield & Jennings, 2008; Galloway, 1975)

1a. Plants acaulescent, usually caespitose from a much-branched woody caudex; fruit extremely viscid and sticky glandular, with 5 broad wings that are flat and not dilated; leaves small, to 2.5 cm in length, very viscid glandular...***A. nana* var. *nana***
1b. Plants caulescent, spreading and procumbent or erect; fruit unwinged or winged, dilated or flat, glabrous to villous or glandular at the top, but not conspicuously viscid; leaves 1.5–12 cm in length, glabrous to glandular, but not densely so...2

2a. Fruit unwinged, dark brown to gray-brown with conspicuous small, appressed white lines but otherwise glabrous, and without a distinct reticulate vein pattern; flowers glabrous to slightly glandular on the upper tube and lobes; bracts obovate to ovate or elliptic and rounded at the tips...***A. glabrifolia***
2b. Fruit winged or wingless, white, greenish, or tan or brown, sometimes with appressed white lines but these not pronounced, usually villous and/or glandular especially at the top, with a conspicuous reticulate vein pattern throughout; flowers sparsely to densely glandular throughout; bracts ovate to obovate or elliptic with rounded or more often acuminate tips...3

3a. Fruit winged or unwinged, the wings not dilated distally but flattened, thick and indurate, the peripheral fruit often curved (S-shaped) in side view; plants spreading and procumbent to semierect; leaves ovate to triangular or sometimes narrow and lanceolate on the upper part of the stem; bracts ovate to linear-lanceolate, with acuminate tips...***A. fragrans***
3b. Fruit winged, the wings dilated (inflated) and expanded distally, thin and papery; plants erect or somewhat spreading; leaves ovate, elliptic, or sometimes orbicular; bracts ovate to obovate, with rounded or acuminate tips...***A. elliptica***

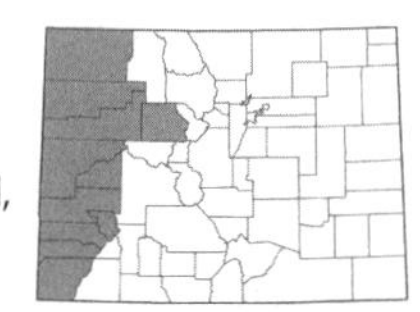

***Abronia elliptica* A. Nelson**, WESTERN SAND-VERBENA. Stems decumbent to erect, glandular or sometimes glabrous; *leaves* ovate to elliptic-oblong or orbicular, 1.5–6 × 0.5–3.5 cm; *bracts* ovate to obovate, 5–20 × 3–10 mm, scarious, glandular to short villous; *calyx* white with a rose to greenish tube, 10–20 mm long and 5–8 mm diam.; *fruit* 5–12 mm long, broadly turbinate, winged, the wings inflated, thin, and papery. Found on gypsum, clay, or sandy soil, 4500–6500 (8200) ft. April–June. W. (Plate 54)

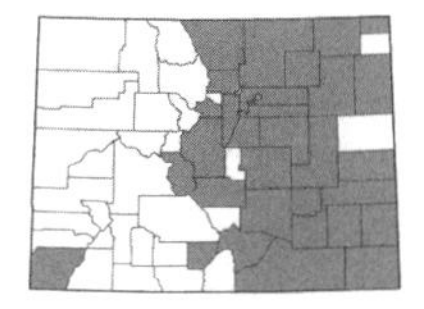

***Abronia fragrans* Nutt. ex Hook.**, FRAGRANT SAND-VERBENA. Stems decumbent to semierect, glandular or sometimes glabrous; *leaves* ovate to triangular or lanceolate, 3–12 × 1–8 cm; *bracts* linear-lanceolate to ovate, 7–25 × 2–12 mm, scarious, glandular to short villous; *calyx* white with a greenish to reddish tube, 10–25 mm long and 5–10 mm diam.; *fruit* 5–12 mm long, rhombic or S-shaped, winged or not, indurate. Found in sandy soil, 3500–8000 ft. May–Aug. (Sept.). E/W. (Plate 54)

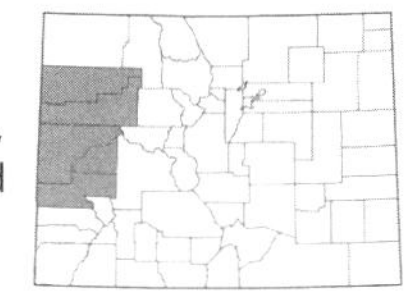

Abronia glabrifolia **Standl.**, CLAY SAND-VERBENA. [*A. argillosa* S.L. Welsh & Goodrich]. Stems erect to decumbent, glabrous; *leaves* elliptic to obovate, 1.5–5 × 0.3–2.5 cm; *bracts* obovate to ovate or elliptic, 7–15 × 6–15 mm, scarious, glandular; *calyx* white with a white to greenish or pinkish tube, 10–15 mm long and 5–7 mm diam.; *fruit* clavate or narrowly fusiform, unwinged, glabrous. Found on clay soil of the Mancos shale formation, 4500–6800 ft. May–June. W. (Plate 54)

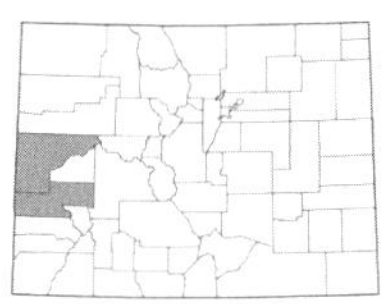

Abronia nana **S. Watson var. *nana***, DWARF SAND-VERBENA. Plants acaulescent, caespitose from a much-branched woody caudex; *leaves* elliptic-ovate to elliptic-lanceolate, 0.5–2.5 × 0.4–1.5 cm, viscid and glandular; *bracts* ovate to oblong-lanceolate, 4–9 × 2–7 mm, scarious, glandular; *calyx* white to pale pink, 8–30 mm long and 6–10 mm diam.; *fruit* obovate to obcordate, 6–10 × 5–7 mm, winged, viscid. Uncommon on gypsum outcrops in the Sinbad Valley, 5400–5800 ft. April–May. W. (Plate 54)

ALLIONIA L. – WINDMILLS

Herbs; *stems* procumbent; *flowers* perfect, in axillary clusters of 3 simultaneously blooming flowers, each cluster subtended by a bract giving the inflorescence the appearance of a single flower; *bracts* persistent, distinct, broadly ovate; *calyx* zygomorphic, 4–5-lobed; *stamens* 4–7; *fruit* 5-ribbed, the dorsal face with 2 ribs bearing glandular-tipped teeth. (Turner, 1994)

1a. Fruit shallowly convex adaxially (the side facing the bracts), with prominent, often gland-tipped, toothed wings along the margin and adaxial side...***A. choisyi***

1b. Fruit deeply convex adaxially (the side facing the bracts), the lateral wings curved and folded over the adaxial side, with 2 rows of prominent glands within the convex depression...***A. incarnata***

Allionia choisyi **Standl.**, TRAILING FOUR-O'CLOCK. Annuals or perennials, glabrous to viscid; *leaves* ovate to elliptic-ovate, 10–30 × 6–22 mm; *calyx* pink; *fruit* oblong, the adaxial face convex, keeled, with prominent, often gland-tipped, toothed wings along the margin and adaxial side. Known from a single collection made by George Osterh. at Rocky Ford (Otero Co.) in July of 1894, 4200–4400 ft. June–Aug. E. (Plate 54)

Whether or not this collection was actually made at Rocky Ford remains a mystery – Osterhout often left specimens untreated for years before being numbered and mounted, and he didn't number his specimens in the order of their collection (Williams, 1987). I have noticed several specimens of Osterhout's that couldn't possibly have been collected where the label says they were. In addition, in over 100 years, this species has not been recollected north of New Mexico.

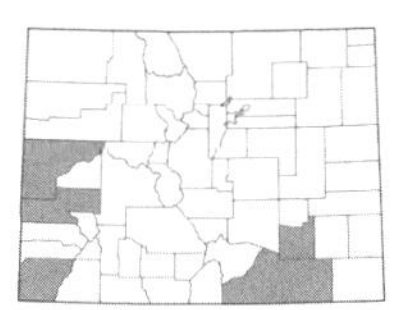

Allionia incarnata **L.**, PINK WINDMILLS. Annuals or perennials, glandular to viscid-villous; *leaves* ovate, 20–65 × 10–35 mm; *calyx* pink to magenta, 5–15 mm long; *fruit* oblong, deeply convex adaxially, the lateral wings curved and folded over the adaxial side, with 2 rows of prominent glands within the convex depression. Uncommon on dry slopes and canyonsides, 4000–5100 ft. May–Aug. E/W. (Plate 54)

MIRABILIS L. – FOUR-O'CLOCK

Perennial herbs; *stems* decumbent to erect; *flowers* perfect; *bracts* 5, connate or rarely distinct, usually ovate, persistent, herbaceous to scarious; *calyx* 5-lobed, salverform, funnelform or campanulate; *stamens* 3–6, exserted; *styles* exserted beyond the stamens; *fruit* smooth or 5-angled or 5-ribbed, glabrous or hairy, eglandular. (Pilz, 1978; Spellenberg, 1998; Spellenberg & Rodriguez, 2001; Turner, 1993)

1a. Flowers 6–9 per involucre, larger, the perianth 1.5–6 cm long; involucre bracts green and leaf-like, 11–35 mm long, not enlarging in fruit; fruit 5.5–11 cm long...2

1b. Flowers 1–3 per involucre, smaller, the perianth 0.5–1 cm long; involucre bracts tan, membranous with prominent reticulate-veins or less often green, 3–7 mm long in flower, enlarging in fruit to 15 mm; fruit 2.5–5.5 cm long...3

2a. Involucre bracts usually distinct or only connate up to 50% of their length; perianth 1.5–2 cm long, campanulate, white to magenta...***M. alipes***

2b. Involucre bracts connate for 50% of their length or more; perianth 2.5–6 cm long, funnelform, magenta...***M. multiflora***

3a. Plants decumbent or prostrate and spreading, often tangled in other vegetation; fruit nearly spherical, smooth and black-mottled or evenly black; involucre bracts green and not membranous, only slightly or not enlarged in fruit; leaves triangular with downward-pointing, rounded lobes and a cordate base...***M. oxybaphoides***

3b. Plants erect or ascending, only occasionally decumbent but not tangled in vegetation; fruit prominently ribbed; involucre bracts tan and membranous, enlarging in fruit; leaves various but unlike the above...4

4a. Hairs on the involucre bracts with dark purple or black cross-walls; stems pubescent below in 2 lines; leaves narrowly triangular to ovate, usually petiolate with the petiole 1–3 cm long...***M. melanotricha***
4b. Hairs on the involucre with pale cross-walls or rarely with dark purple cross-walls (and then the leaves linear to lanceolate); stems glabrous below or hirsute, only occasionally the hairs in lines; leaves various, sessile or petiolate...5

5a. Fruit glabrous; stems usually glabrous below but occasionally hairy; leaves linear to lanceolate, sessile or shortly petiolate...***M. glabra***
5b. Fruit hairy; stems glabrous to hairy; leaves various...6

6a. Petioles mostly over 1 cm long; at least some leaves broadly triangular, others ovate to ovate-lanceolate; stem and leaves glabrous or occasionally minutely hairy in 2 lines at the base of the stem...***M. nyctaginea***
6b. Leaves sessile or petioles to about 1 cm; leaves linear, lanceolate, ovate, or round; stem glabrous or hairy...7

7a. Lower leaves round, upper stem leaves broadly ovate or ovate-triangular, 3–6 cm wide; lower stems densely hirsute...***M. rotundifolia***
7b. Leaves all linear, lanceolate, lanceolate-oblong, or narrowly ovate, to 3 cm wide; lower stems glabrous, pilose, or densely hirsute...8

8a. Leaves linear to linear-lanceolate, mostly not more than 5 mm wide; lower stems usually glabrous, rarely hirsute...***M. linearis***
8b. Leaves lanceolate to narrowly ovate, the larger ones over 5 mm wide; lower stems hirsute or glabrous...9

9a. Lower stems glabrous; upper stem not hairy at the nodes...***M. lanceolata***
9b. Lower stems hirsute or pilose; upper stem densely hairy at the nodes...***M. hirsuta***

Mirabilis alipes **(S. Watson) Pilz,** WINGED FOUR O'CLOCK. [*Hermidium alipes* S. Watson]. Plants 2–4 dm; *leaves* ovate, 4.5–9 × 3.5–5 cm; *bracts* 6–9, distinct to 50% connate; *flowers* 6–9 per head; *calyx* usually pink-magenta, rarely white, 1.5–2 cm long; *fruit* elliptic, 5.5–7 mm long, glabrous, olive-green with 10 slender ribs. Uncommon on white shale slopes, 5500–5900 ft. May–June. W.

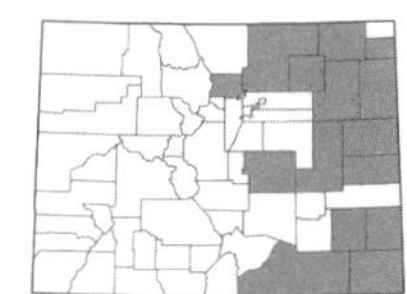

Mirabilis glabra **(S. Watson) Standl.,** SMOOTH FOUR O'CLOCK. [*M.* ×*carletonii* (Standl.) Standl.; *M. exaltata* (Standl.) Standl.; *Oxybaphus glaber* S. Watson]. Plants 5–20 dm; *stems* glabrous or hairy distally; *leaves* linear to ovate, 5–10 × 0.2–7.5 cm, glaucous; *bracts* 90–60% connate; *flowers* 1–3 per head; *calyx* white to pale pink, 0.6–1 cm long; *fruit* obovate, 4–5.5 mm long, usually glabrous, grayish to greenish-brown. Locally common in sandy soil of the eastern plains, 3500–5700 ft. June–Sept. E. (Plate 54)

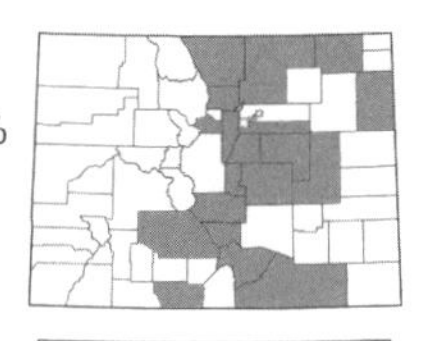

Mirabilis hirsuta **(Pursh) MacMill.,** HAIRY FOUR O'CLOCK. [*Oxybaphus hirsutus* (Pursh) Sweet]. Plants erect, to 12 dm; *stems* pilose; *leaves* lanceolate-oblong to ovate, ca. 8 × 3 cm; *bracts* ovate, 80–50% connate; *flowers* 1–3 per head; *calyx* rose-purple or rarely white; *fruit* obtuse, 5-ribbed, ca. 4 mm long, hispid. Common in open forests along the base of the Front Range and in sandy soil of the eastern plains, 3900–8200 ft. June–Sept. E.

Mirabilis lanceolata **(Rydb.) Standl.,** LANCELEAF FOUR O'CLOCK. [*Allionia sessilifolia* Osterh.]. Plants erect, to 12 dm; *stems* glabrous; *leaves* lanceolate to narrowly ovate; *bracts* 80–50% connate; *flowers* 1–3 per head; *calyx* rose-purple; *fruit* obovate, 3–5 mm long, hispid. A hybrid between *M. linearis* and *M. hirsuta*, common along the base of the northern Front Range where their ranges overlap, 5500–8000 ft. E.

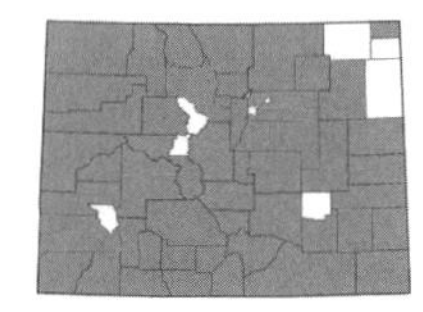

Mirabilis linearis **(Pursh) Heimerl,** NARROWLEAF FOUR O'CLOCK. Plants 1–13 dm; *stems* glabrous or rarely hirsute below; *leaves* linear to linear-lanceolate, 3–12 × 0.1–0.5 (1) cm; *bracts* 40–70% connate; *flowers* 3 per head; *calyx* white to pink-purple; *fruit* obovate to obovoid, 3–5.5 mm long, hairy. Common throughout the state, especially on the plains and in open places in mountain valleys, 3900–9500 ft. June–Sept. E/W. (Plate 54)

There are two varieties of *M. linearis* present in Colorado:

1a. Lower stem glabrous...**var. linearis**. [*Allionia divaricata* Rydb.; *M. decumbens* (Nutt.) Daniels; *Mirabilis diffusa* (Heller) Reed; *Oxybaphus bodinii* Holzinger; *Oxybaphus linearis* (Pursh) B.L. Robinson]. Common throughout the state, especially on the plains and in open places in mountain valleys, 3900–9500 ft. E/W.
1b. Lower stem hirsute...**var. *subhispida* (Heim.) Spellenberg**. [*Mirabilis gausapoides* Standl.]. Found on the southeastern plains (Baca Co.), 4000–5000 ft. E.

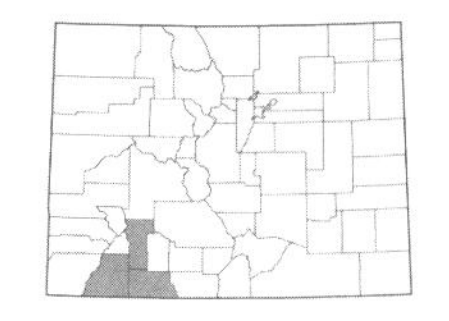

Mirabilis melanotricha **(Standl.) Spellenberg**, DARK-HAIRED FOUR O'CLOCK. [*M.* ×*decipiens* (Standl.) Standl.]. Plants erect; *stems* pubescent below in 2 lines, spreading glandular-hairy above; *leaves* triangular-ovate to ovate, 3–10 × 0.8–4 cm; *bracts* 40–50% connate, glandular-hairy, the cross-walls of the hairs dark purple or black; *flowers* 3 per head; *calyx* purple-pink; *fruit* obovoid, 3–4 mm long, spreading-pilose. Found in aspen woodlands and open slopes, uncommon in the southwestern counties, 6300–9100 ft. July–Sept. W.

This is the correct name for plants in Colorado that we have called *M. comata* Small.

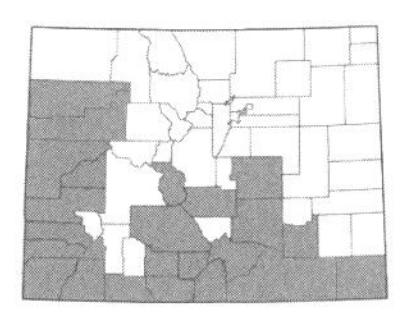

Mirabilis multiflora **(Torr.) A. Gray**, COLORADO FOUR O'CLOCK. [*Quamoclidion multiflorum* Torr.]. Plants 4–7 dm; *stems* glabrous or densely hairy; *leaves* ovate to suborbiculate, 5–10 × 4–8 cm; *bracts* 50–80% connate; *flowers* 6 per head; *calyx* pink-magenta, 2.5–6 cm long; *fruit* ovoid to globose, 6–11 mm long, glabrous or hairy. Common in open places, on shale outcrops, sometimes with pinyon-juniper, and often along roadsides, 4500–9000 ft. June–Sept. E/W.

There are 2 varieties of *M. multiflora* commonly recognized, although they are completely intergrading:

1a. Involucre bracts acute; fruit smooth to slightly tuberculate; inflorescence eglandular or only slightly glandular... **var. *multiflora***. E/W.

1b. Involucre bracts obtuse; fruit tuberculate; inflorescence conspicuously glandular... **var. *glandulosa*** **(Standl.) J.F. Macbr**. W.

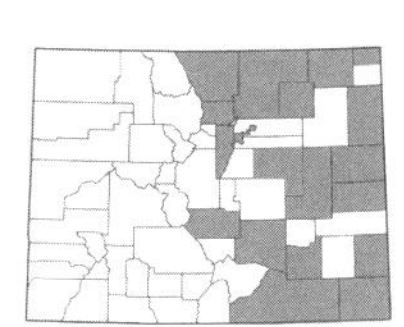

Mirabilis nyctaginea **(Michx.) MacMill.**, HEARTLEAF FOUR O'CLOCK. [*Oxybaphus nyctagineus* (Michx.) Sweet]. Plants 4–15 dm; *stems* usually glabrous or rarely spreading-hairy below; *leaves* triangular to ovate, 3–10 × 2–6.5 cm, petiolate; *bracts* 50–90% connate; *flowers* 2–5 per head; *calyx* pink to reddish-purple, rarely white; *fruit* obovate, 3–5 mm long, hairy. Common in sandy soil of open places, along roadsides, in fields, and often found in disturbed places, 3400–7000 ft. June–Sept. E. (Plate 54)

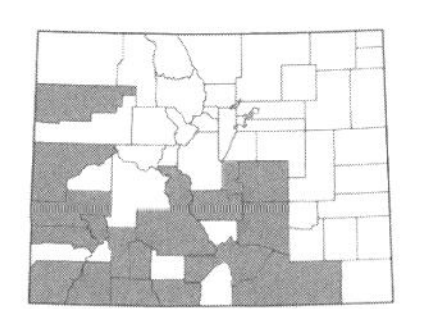

Mirabilis oxybaphoides **(A. Gray) A. Gray**, SMOOTH SPREADING FOUR O'CLOCK. Plants 2–12 dm, often tangled in other vegetation; *stems* hairy in lines or throughout, sometimes glandular; *leaves* ovate to deltate, 1.5–8 × 1–7.5 cm; *bracts* 70–50% connate; *flowers* 3 per head; *calyx* pale pink to purplish or rarely white; *fruit* obovoid to nearly spheric, 2.5–3.5 mm long, glabrous, dark-mottled. Uncommon, often found growing in shady places, usually with pinyon-juniper, 4500–8800 ft. Aug.–Oct. E/W. (Plate 54)

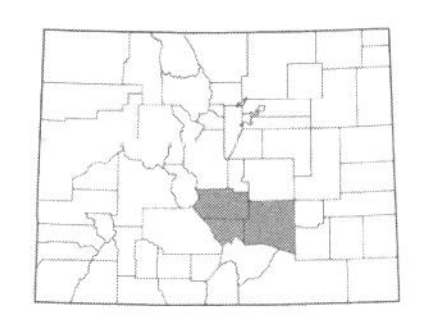

Mirabilis rotundifolia **(Greene) Standl.**, ROUNDLEAF FOUR O'CLOCK. [*Oxybaphus rotundifolius* (Greene) Standl.]. Plants 2–3 dm; *stems* spreading-hairy throughout; *leaves* ovate to triangular or nearly orbicular, 4–7 × 3–6 cm, thick; *bracts* 50–40% connate, spreading-hairy; *flowers* 3 per head; *calyx* purplish-pink; *fruit* obovoid, 4–5 mm long, hairy. Restricted to limestone plateaus or shale outcrops between Pueblo and Canon City in the Arkansas River Valley, 4800–5600 ft. June–Aug. E. Endemic.

TRIPTEROCALYX (Torr.) Hook. – SANDPUFFS

Annual herbs; *flowers* perfect, actinomorphic, in capitate clusters, the flowers maturing from one side of the inflorescence to the other; *bracts* 5–10, distinct; *calyx* 4–5-lobed; *stamens* 4–5, included; *styles* included; *fruit* fusiform, indurate, with 2–4 broad, thin and translucent, scarious, conspicuously reticulated wings completely surrounding the body. (Galloway, 1975)

1a. Perianth with inconspicuous lobes, the limb 3–5 mm in diam. and greenish to pink, the tube 6–18 mm long and greenish to pink... ***T. micranthus***

1b. Perianth with conspicuous lobes over 1.5 mm long, the limb 8–11 mm in diam. and white adaxially and often pink abaxially, the tube 12–25 mm long and pink to pinkish-red... ***T. wootonii***

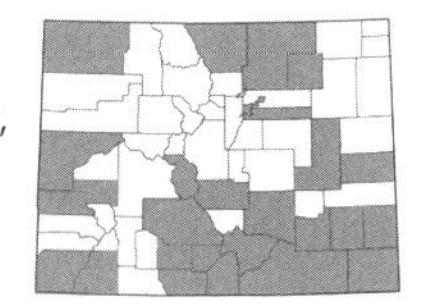

Tripterocalyx micranthus **(Torr.) Hook.**, SMALLFLOWER SAND-VERBENA. Stems glandular-hairy; *leaves* ovate to elliptic, 1–6 × 0.5–2.5 cm; *flowers* 5–15 per head; *bracts* lanceolate to ovate, 3–9 mm long, papery, glabrous to glandular-hairy; *calyx* pink with a greenish tube, 3–5 mm diam.; *fruit* spheric to oval, 1–2 cm long, with prominent wings. Found in sandy soil, 3500–8500 ft. May–Aug. E/W. (Plate 54)

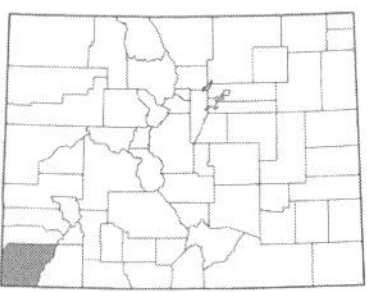

Tripterocalyx wootonii **Standl.**, WOOTON'S SANDPUFFS. [*T. carneus* (Greene) L.A. Galloway var. *wootonii* (Standl.) L.A. Galloway]. Stems glandular-hairy; *leaves* ovate to elliptic, 1.5–8 × 0.5–5 cm; *flowers* 10–25 per head; *bracts* linear-lanceolate to ovate, 5–14 mm long, papery, glabrous to glandular; *calyx* white with a pinkish tube, 8–11 mm diam.; *fruit* oval, 1.3–2.5 cm long, with prominent wings. Uncommon in sandy soil, 4500–5500 ft. W.

NYMPHAEACEAE Salisb. – WATERLILY FAMILY

Perennials, acaulescent aquatic herbs; *leaves* alternate, simple, floating, long-petiolate, stipulate or exstipulate; *flowers* perfect, actinomorphic, solitary, long pedicellate; *sepals* usually 4–9, distinct, green or petaloid; *petals* usually many, distinct, sometimes intergrading into stamens; *stamens* many; *pistil* 1; *stigma* disk-like; *ovary* superior or partly inferior, 3–35-carpellate; *fruit* a leathery berry or capsule. (Beal, 1956; Mill. & Standley, 1912)

1a. Sepals 5–12, the outer green and the inner yellow and petaloid; petals small and inconspicuous, stamenlike; leaves suborbicular to ovate...***Nuphar***
1b. Sepals 4, green; petals numerous, white, not stamenlike; leaves orbicular...***Nymphaea***

NUPHAR J. Sm. – POND-LILY; SPATTERDOCK

Sepals 5–12, green to yellow within, often reddish-tinged; *petals* 10–20, stamenlike; *fruit* a capsule.

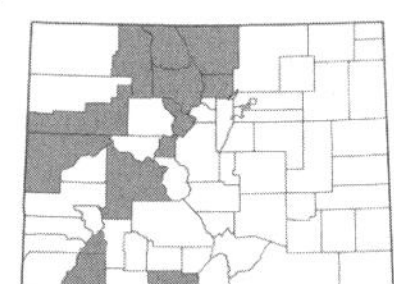

***Nuphar polysepala* Engelm.**, ROCKY MOUNTAIN POND-LILY. [*N. lutea* (L.) J. Sm. ssp. *polysepala* (Engelm.) E.O. Beal]. Leaves ovate to suborbicular, to 8–30 cm long; *sepals* 3–6 cm long; petals mostly 4–6 cm long, small and inconspicuous, stamenlike. Found in high altitude ponds and lakes, 8000–11,000 ft. July–Sept. E/W.

NYMPHAEA L. – WATERLILY

Sepals usually 4, green; *petals* 8–many, showy, white, pink, blue, or yellow; *fruit* a capsule.

Nymphaea odorata* Ait. ssp. *odorata, AMERICAN WHITE WATERLILY. Leaves orbicular, ca. 10–20 cm long; *sepals* to 8 cm long; *petals* ca. 8 cm long, numerous and white, not stamenlike. Uncommon in ponds, known from Otero Co., 4000–5200 ft. May–Oct. E. Introduced.

OLEACEAE Hoffmsg. & Link – OLIVE FAMILY

Trees and shrubs; *leaves* usually opposite, usually deciduous, petiolate, simple or compound, exstipulate; *flowers* usually imperfect, sometimes perfect, actinomorphic; *sepals* 4, 5–15, or absent, distinct or connate; *corolla* 4–6-lobed or absent; *stamens* usually 2 or rarely 4, distinct; *pistil* 1; *ovary* superior, 2-carpellate; *fruit* a samara, capsule, berry, or drupe.

1a. Herbs with a woody base; fruit a thin-walled circumscissle capsule; flowers yellow...***Menodora***
1b. Trees or shrubs; fruit a samara, berry, drupe, or loculicidal capsule; flowers purple or white, or the petals absent...2

2a. Fruit a samara; leaves odd-pinnately compound or simple and broadly ovate to suborbicular with a rounded apex; petals absent, flowers appearing before the leaves...***Fraxinus***
2b. Fruit a drupe, berry, or loculicidal capsule; leaves simple and unlike the above; flowers white or purple and appearing after the leaves, or petals absent and appearing before the leaves...3

3a. Leaves ovate to cordate with an acuminate apex, 3–12 cm long and 2–8 cm wide; flowers purple or sometimes white in large panicles mostly 10–20 cm long, appearing with the leaves, very sweet-smelling; fruit a loculicidal capsule...***Syringa***
3b. Leaves elliptic to ovate-lanceolate or oblanceolate with a rounded or shortly acute apex, 1.5–6 cm long and 0.5–2 cm wide; flowers white and in smaller panicles 3–6 cm long and appearing with the leaves, or petals absent and sessile or in cymes and appearing before the leaves; fruit a berry or drupe...4

4a. Petals white; flowers in panicles, appearing with the leaves; fruit a black berry; leaves entire...***Ligustrum***
4b. Petals absent; flowers sessile or in panicle-like cymes, appearing before the leaves; fruit a blue-black drupe; leaves entire to minutely toothed...***Forestiera***

FORESTIERA Poir. – SWAMP-PRIVET

Polygamo-dioecious or dioecious shrubs; *leaves* opposite, simple, serrate to entire; *flowers* imperfect or perfect, appearing before the leaves; *sepals* 4 or absent; *corolla* absent; *stamens* 2 or 4; *fruit* a drupe.

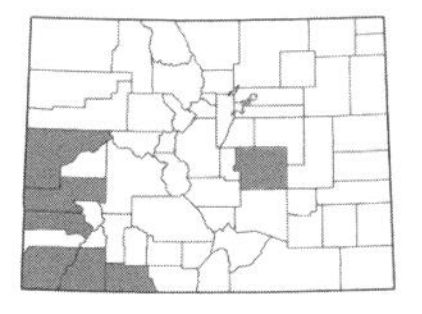

***Forestiera pubescens* Nutt.**, DESERT OLIVE; COYOTE BUSH. Shrubs to 2 m; *leaves* oblanceolate to elliptic, 1.3–4 cm long, sparsely to densely hairy, entire to minutely toothed; *flowers* usually imperfect; *staminate flowers* sessile, 4–8 in a dense cluster; *pistillate flowers* in panicle-like cymes, pedicellate; *drupes* blue-black, 5–7 mm long. Found in moist soil on canyon bottoms and floodplains along rivers, 4200–7300 ft. April–May. E/W.

FRAXINUS L. – ASH

Dioecious or polygamous trees; *leaves* opposite, deciduous, odd-pinnately compound or simple; *flowers* imperfect or perfect, appearing before or with the leaves; *sepals* 4 or absent; *corolla* absent; *stamens* 2; *fruit* a samara.

1a. Leaves simple or rarely 3-foliate; body of the samara flat...***F. anomala***
1b. Leaves odd-pinnately compound; body of the samara terete or subterete...2

2a. Leaf scars crescent-shaped, concave along the upper edge; leaflets whitened below...***F. americana***
2b. Leaf scars semicircular, truncate or only slightly concave along the upper edge; leaflets paler green below, not whitened...***F. pennsylvanica***

***Fraxinus americana* L.**, WHITE ASH. Trees to 30 m; *leaves* pinnately compound with 5–9 leaflets, leaflets 5–15 cm long, whitened below, serrate; *samaras* 3–5 cm long. Not persisting outside of cultivation but included in the key because of its close resemblance to *F. pennsylvanica* which can escape from cultivation.

***Fraxinus anomala* Torr. ex S. Watson**, SINGLELEAF ASH. Shrubs or trees to 4 m; *leaves* simple or rarely 3-foliate, ovate, 1.5–8 cm long, subentire to crenate-serrate; *samaras* 1.2–3 cm long. Found in dry, rocky places, often with pinyon-juniper, 4600–7000 ft. April–May. W.

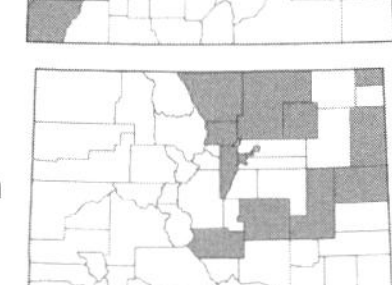

***Fraxinus pennsylvanica* Marshall**, GREEN ASH. Trees to 20 m; *leaves* pinnately compound with 5–7 (9) leaflets, leaflets 6–15 cm long, serrate to subentire; *samaras* 2–5 cm long. Cultivated and escaping along the Front Range, or native on floodplains of rivers or along the margins of lakes on the eastern plains, 3400–6000 ft. April–May. E.

LIGUSTRUM L. – PRIVET

Shrubs; *leaves* opposite, simple and entire; *flowers* perfect, white and showy, appearing after the leaves; *calyx* 4-lobed; *corolla* 4-lobed, funnelform; *stamens* 2; *fruit* a berry.

***Ligustrum vulgare* L.**, COMMON PRIVET. Shrubs to 3+ m; *leaves* oblong to ovate, 2–6 cm long, dark green; *corolla* lobes ca. 3 mm long; *berries* black. Cultivated and occasionally escaping in the Boulder area, 5500–6000 ft. May–June. E. Introduced.

MENODORA Bonpl. – MENODORA

Shrubs or herbs with woody bases; *leaves* alternate or the lowermost opposite, simple, short-petiolate; *flowers* perfect; *calyx* 5–15-lobed; *corolla* 5–6-lobed, yellow; *stamens* 2; *fruit* a circumscissile capsule.

***Menodora scabra* A. Gray**, ROUGH MENODORA. Subshrubs, 2–3.5 dm; *leaves* lanceolate to narrowly elliptic, 0.5–3 cm long, entire, glabrous or minutely hairy; *corolla* lobes 5–10 mm long. Found on gypsum soil and shale bluffs, known from the Arkansas River drainage, 4700–5700 ft. May–Sept. E.

SYRINGA L. – LILAC

Shrubs or small trees; *leaves* opposite, simple; *flowers* perfect; *calyx* 4-lobed, campanulate, persistent; *corolla* 4-lobed, tubular; *stamens* 2; *fruit* a loculicidal capsule.

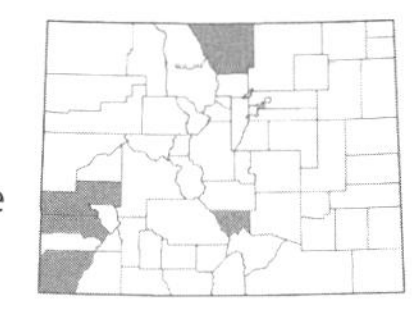

***Syringa vulgaris* L.**, COMMON LILAC. Shrubs to 6 m; *leaves* ovate to cordate with an acuminate tip, 3–12 × 2–8 cm, entire, glabrous; *flowers* in panicles 10–20 cm long; *corolla* white, pink, blue, rose-purple, or purple. Commonly cultivated and occasionally escaping along roadsides. A good persisting population can be found approximately 1 mile up the Hewlett Gulch trail in the Poudre Canyon alongside old cabins where it was presumably planted many years ago. 5000–7500 ft. May–June. E/W. Introduced.

ONAGRACEAE Juss. – EVENING PRIMROSE FAMILY

Herbs or shrubs, rarely trees; *leaves* alternate or opposite, occasionally whorled, simple, entire to toothed or pinnatifid; *flowers* perfect, actinomorphic or zygomorphic, with a hypanthium; *sepals* usually 4; *petals* 4 or rarely 2, yellow, white, pink, or purple, usually clawed; *stamens* usually 8; *pistil* 1; *ovary* inferior, usually 4-carpellate; *fruit* usually a capsule.

1a. Leaves all or mostly basal...2
1b. Stem leaves present or leaves not mostly basal...4

2a. Stigmas evidently 4-lobed; petals (1) 1.5–7.5 cm long, white, pink, or yellow...***Oenothera***
2b. Stigmas entire or shallowly 4-lobed; petals 0.15–1.5 cm long, yellow...3

3a. Annuals; flowers in a terminal raceme elevated above the leaves; petals yellow, usually with red dots at the base...***Chylismia***
3b. Perennials; flowers sessile in the leaf axils, the pedicels shorter than the leaves; petals yellow throughout...***Taraxia***

4a. Leaves all or mostly opposite (sometimes a few upper alternate), usually lanceolate to ovate with toothed margins...5
4b. Leaves mostly alternate (sometimes with a few opposite ones below), shape and margin various...6

5a. Fruit with hooked bristles, ovoid and 1.5–2.5 mm long; flowers 2-merous; seeds lacking a tuft of hairs at the upper end; petals white, to 2 mm long; leaves ovate...***Circaea***
5b. Fruit lacking hooked bristles, usually over 3 mm long; flowers 4-merous; seeds with a tuft of hairs at the terminal end; petals and leaves various...***Epilobium***

6a. Petals 1–6 mm long, not clawed; annuals; leaves linear to narrowly elliptic, entire or irregularly toothed...7
6b. Petals mostly over 6 mm long (or if shorter, then abruptly short-clawed at the base and flowers numerous in a terminal spike or raceme); perennials or if annuals then leaves usually unlike the above in all respects...11

7a. Seeds with a tuft of long white hairs (coma) at the terminal end; sepals erect at anthesis; petals pink or white (drying purple), 2–6 mm long...***Epilobium*** (*brachycarpum*)
7b. Seeds lacking a tuft of hairs at the terminal end, glabrous or hairy throughout; sepals reflexed or widely spreading at anthesis; petals various...8

8a. Petals yellow...9
8b. Petals white or pink..10

9a. Capsules 6–11 mm long, 4-angled; leaves usually in tufts towards the tips of the stem...***Holmgrenia***
9b. Capsules 15–30 mm long, rounded or somewhat 4-angled; leaves well-distributed along the stem, not crowded in tufts...***Camissonia***

10a. Leaves linear, to 5 mm wide; capsules straight; ovary 2-locular...***Gayophytum***
10b. Leaves mostly oblanceolate to narrowly elliptic, to 10 (15) mm wide; capsules usually contorted, occasionally straight; ovary 4-locular...***Eremothera***

11a. Seeds with a tuft of long white hairs (coma) at the terminal end; petals bright pink-purple (drying purple)...***Chamerion***
11b. Seeds lacking a tuft of hairs at the terminal end, glabrous or hairy throughout; petals various but usually not bright pink-purple...***Oenothera***

CAMISSONIA Link – SUNCUP
Annual herbs; *leaves* alternate, entire or weakly toothed; *flowers* actinomorphic; *petals* yellow. (Raven, 1969)

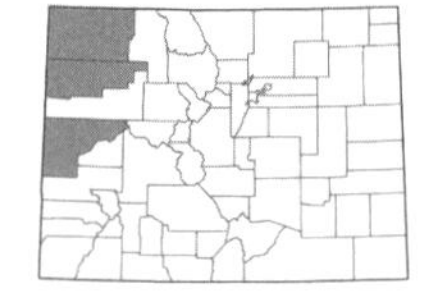

Camissonia parvula **(Nutt. ex Torr. & A. Gray) P.H. Raven**, LEWIS RIVER SUNCUP. Plants 1–18 cm; *leaves* linear, 5–30 × 0.4–1.2 mm, involute; *flowers* solitary in bract or leaf axils; *floral tube* 0.8–1.5 mm long; *sepals* 1–2.5 mm long; *petals* yellow, 2.5–4.5 mm long; *capsules* 1.5–3 cm long. Uncommon on sandy soil and dry, open slopes, 5000–6400 ft. April–June. W.

CHAMERION Raf. ex Holub – FIREWEED
Perennial herbs; *leaves* alternate or the lower opposite, entire or weakly toothed; *flowers* actinomorphic; *petals* rose-purple, pink, or rarely white, entire.
1a. Plants (5) 10–20 (30) dm; petals 1–1.5 (2) cm long; style 1–2 cm long, longer than the stamens; leaves (4) 5–10 times as long as wide; racemes elongate with numerous flowers (more than 15 flowers), lacking leafy bracts...***C. angustifolium***
1b. Plants mostly low, 1–4 (7) dm; petals 1.5–3 cm long; style short (to 1 cm), surpassed by the stamens; leaves mostly 2.5–4 (6) times as long as wide; racemes with leafy bracts and relatively few (usually less than 12) flowers, not much elongate...***C. latifolium***

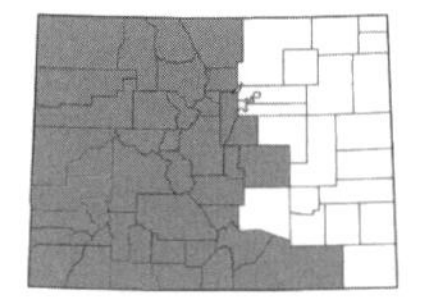

Chamerion angustifolium **(L.) Holub**, FIREWEED. [*C. danielsii* D. Löve; *Epilobium angustifolium* L.]. Plants (5) 10–20 (30) dm; *leaves* lanceolate to elliptic, 5–20 cm long, entire; *flowers* in terminal racemes; *sepals* 0.7–1.7 cm long; *petals* 1–1.5 (2) cm; *capsules* 8–10 cm long, hairy; *seeds* 1–1.5 mm long with a tuft of white or dingy hairs. Common in forests, meadows, and along streams, especially abundant in areas after fire, 5300–12,500 ft. July–Sept. E/W. (Plate 55)

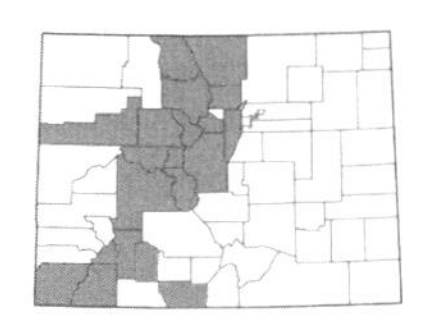

***Chamerion latifolium* (L.) Holub**, DWARF FIREWEED. [*C. subdentatum* (Rydb. Á. Löve & D. Löve; *Epilobium latifolium* L.]. Plants 1–4 (7) dm; *leaves* lanceolate to elliptic, 1.5–8 cm long, entire or minutely toothed; *flowers* in bracteates racemes; *sepals* 1–2 cm long; *petals* 1.5–3 cm long; *capsules* 3–10 cm long, usually purplish, hairy; *seeds* (1) 1.5–2 mm long with a tuft of dingy hair. Found along streams and creeks, often in gravelly soil, 7500–13,000 ft. July–Aug. E/W.

CHYLISMIA (Nutt. ex Torr. & A. Gray) Raimann – CHYLISMIA

Annual herbs; *leaves* mostly basal, alternate, simple or pinnately divided, with sharply toothed or entire margins; *flowers* actinomorphic; *petals* white, yellow, or lavender. (Raven, 1969)

1a. Basal leaves pinnatifid with a large, ovate terminal segment; lower stem with spreading, white hairs to 1.5 mm long; flowers yellow throughout...***C. walkeri***
1b. Basal leaves lanceolate, elliptic, or ovate, entire to toothed but not pinnatifid; lower stem glabrous or if hairy then the hairs mostly less than 0.5 mm long; flowers yellow and usually with red spots towards the base...2

2a. Petals mostly 5.5–15 mm long; stigma elevated above the anthers...***C. eastwoodiae***
2b. Petals mostly 2–5.5 mm long; stigma equal to or only slightly elevated above the anthers...***C. scapoidea* ssp. *scapoidea***

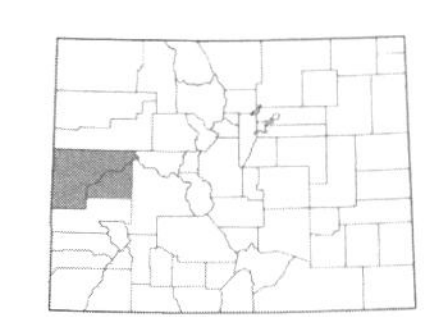

***Chylismia eastwoodiae* (Munz) W.L. Wagner & Hoch**, GRAND JUNCTION CHYLISMIA. [*Camissonia eastwoodiae* (Munz) P.H. Raven]. Plants 6–40 cm; *leaves* elliptic to ovate, 0.8–7 cm long, entire or irregularly toothed; *flowers* in terminal racemes; *floral tube* 3.5–9 mm long; *sepals* 4–9 mm long; *petals* yellow with red spots, 5.5–15 mm long; *capsules* 1.8–4 cm long, usually curved. Rare on Mancos shale and adobe clay hills, 4700–5000 ft. May–June. W.

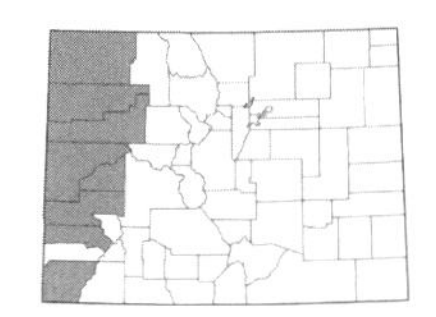

Chylismia scapoidea* (Torr. & A. Gray) Small ssp. *scapoidea, PAIUTE CHYLISMIA. [*Camissonia scapoidea* (Torr. & A. Gray) P.H. Raven]. Plants 3–30 (45) cm; *leaves* lanceolate to ovate or elliptic, 1–6 cm × 4–35 mm, entire to toothed; *flowers* in terminal racemes; *floral tube* 1–4.5 mm long; *sepals* 1.2–5 mm long; *petals* yellow, usually with red spots, 2–5.5 mm long; *capsules* 10–40 mm long. Found on Mancos shale, clay, gypsum, and dry hillsides, 4500–7000 ft. May–June. W.

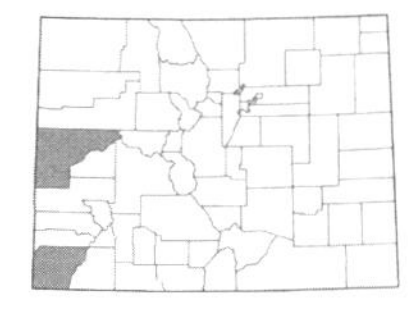

***Chylismia walkeri* (A. Nelson) W.L. Wagner & Hoch**, WALKER'S CHYLISMIA. [*Camissonia walkeri* (A. Nelson) P.H. Raven]. Plants to 7 dm; *leaves* oblanceolate to ovate, 2–22 cm long, usually pinnatifid with a larger, ovate terminal segment, often purple-dotted; *flowers* in bracteate panicles; *floral tube* 0.5–2 mm long; *sepals* 1.2–5 mm long, reflexed; *petals* 1.2–6 mm long, yellow, fading pink to purple; *capsules* 12–30 (40) mm long. Uncommon on shale, gypsum outcrops, rocky ledges, and dry slopes, 4700–7000 ft. May–June. W.

CIRCAEA L. – ENCHANTER'S NIGHTSHADE

Perennial herbs from rhizomes; *leaves* opposite and decussate, becoming alternate above, petiolate, simple, toothed; *flowers* zygomorphic; *petals* 2, white or pink; *fruit* indehiscent. (Boufford, 1983)

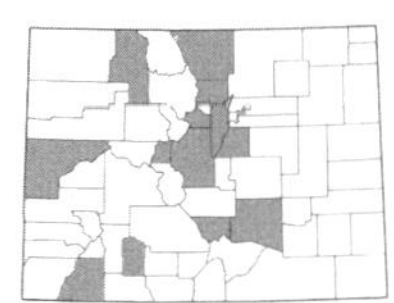

***Circaea alpina* L.**, SMALL ENCHANTER'S NIGHTSHADE. Plants 0.5–5.5 dm; *leaves* ovate to cordate, 1–9 × 0.8–6 cm, toothed to subentire; *flowers* in few-flowered racemes; *sepals* 1–2 mm long, white or pinkish or green at the tip; *petals* 1–2 mm long, white to pinkish, cleft to near the middle. Uncommon along streams and creeks, springs, and other shady, moist places, 5500–8800 ft. June–Aug. E/W. (Plate 55)

EPILOBIUM L. – WILLOW-HERB

Perennial or annual herbs; *leaves* mostly opposite or opposite below and alternate above, rarely all alternate; *flowers* actinomorphic; *petals* 4, rose-purple, pink, or white, notched at the apex. (Trelease, 1891)

Epilobium ciliatum, *E. halleanum*, *E. hornemannii*, *E. saximontanum*, and potentially other *Epilobium* species are known to hybridize, making certain identifications difficult. In addition, turions can easily become broken off when collecting.

1a. Leaves mostly or all alternate; annuals from taproots...***E. brachycarpum***
1b. Leaves mostly or all opposite; perennials...2

2a. Leaves and stems densely soft-hirsute; flowers pink-purple, large (about 2.5 cm wide), the petals notched at the tips; plants 5–20 dm; capsules 5–8 cm long...***E. hirsutum***
2b. Leaves and stems not densely hirsute; plants otherwise unlike the above in all respects...3

3a. Inflorescence and capsules (especially when young) densely silvery- or grayish-canescent with appressed hairs...***E. palustre***
3b. Inflorescence and capsules glabrous to glandular or if hairy then not silvery- or grayish-canescent...4

4a. Stems glabrous throughout (inflorescence can be minutely glandular though)...***E. oregonense***
4b. Stems hairy or glandular-hairy, at least above or in lines decurrent from the leaf bases...5

5a. Seeds longitudinally finely ridged (at 20×), these consisting of numerous confluent papillae; plants 0.5–20 dm; inflorescence often much-branched; capsules 3–10 cm long...***E. ciliatum***
5b. Seeds smooth or papillate (at 20×), the papillae sometimes in longitudinal rows but not forming distinct ridges; plants 0.5–5 dm; inflorescence not much-branched; capsules 1.5–6.5 cm long...6

6a. Pedicels stout, to 5 (9) mm long; turions present at the base of the plant, not at the ends of rhizomes; petals usually white or sometimes purple...***E. saximontanum***
6b. Pedicels slender, 5–55 mm long, or if shorter then the plants lacking turions and the petals pink or purple; plants lacking turions or if present then these at the ends of short, slender, rhizome-like shoots...7

7a. Capsules 1.7–3.5 (4) cm long; plants 0.5–2 dm tall, lacking turions, often curved at the base; petals purple or pink...8
7b. Capsules 4–10 cm long, or if as short as 2.5 cm long then the plants with turions; plants mostly 1–5 dm tall, usually erect; petals white, pink, or purple...9

8a. Pedicels 1–5.5 cm long in fruit; inflorescence usually nodding and glabrous or sparsely glandular; seeds 0.7–1.5 mm long...***E. anagallidifolium***
8b. Pedicels 0.2–1.8 cm long in fruit; inflorescence usually erect and glandular or hairy; seeds 1.3–2 mm long...***E. clavatum***

9a. Plants with turions at the ends of short, slender, rhizome-like shoots; floral tube 0.5–1 mm long...***E. halleanum***
9b. Plants lacking turions; floral tube 1–1.5 mm long...***E. hornemannii***

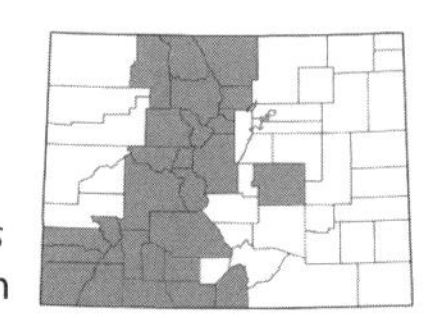

***Epilobium anagallidifolium* Lam.**, PIMPERNEL WILLOW-HERB. Perennials lacking turions, 0.5–2 dm; *stems* minutely hairy in lines decurrent from the leaf bases, often glandular above; *leaves* opposite, mostly 1–3 cm long, the margins entire or obscurely toothed; *pedicels* 1–5.5 cm long in fruit; *sepals* 2.5–5 mm long, often reddish; *petals* 3.5–6 (9) mm long; *capsules* 1.7–4 cm long; *seeds* 0.7–1.5 mm long, finely cellular-roughened or papillate at 20×. Found along streams and lakes, in moist meadows and other moist places, 8500–13,500 ft. July–Aug. E/W.

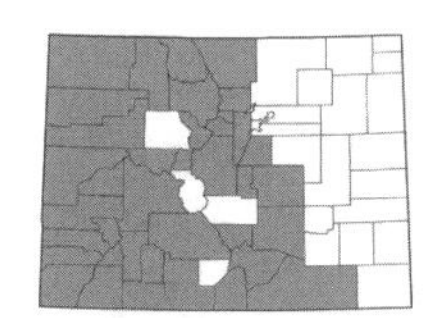

***Epilobium brachycarpum* C. Presl**, PANICLED WILLOW-HERB. Annuals, 2–10 (20) dm; *stems* glabrous below, usually glandular above; *leaves* alternate, linear to linear-lanceolate, 0.7–7 cm long, entire to toothed; *pedicels* 5–20 mm long; *sepals* 1–4 mm long; *petals* white or pink, 3–8 mm long; *capsules* 1.3–3 cm long; *seeds* 1.5–2.5 mm long. Found along creeks, on floodplains and mesas, and in meadows and forests, 5500–9700 ft. June–Aug. E/W.

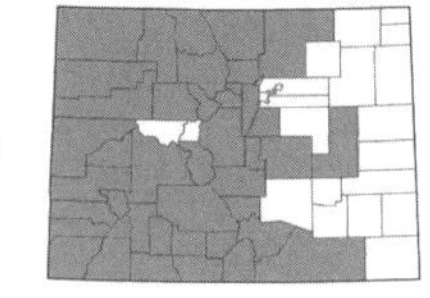

***Epilobium ciliatum* Raf.**, AMERICAN WILLOW-HERB. Perennials 0.5–20 dm, with or without turions; *stems* minutely hairy, usually in decurrent lines from leaf bases; *leaves* opposite, 3–12 cm long, serrulate; *pedicels* 2–20 mm long in fruit; *sepals* 2–6 mm long; *petals* white or pink, 2–5 mm long; *capsules* 3–10 cm long; *seeds* longitudinally finely ridges, consisting of numerous confluent papillae. June–Aug. E/W.

There are two varieties of *E. ciliatum* in Colorado:

1a. Inflorescence usually much-branched; turions absent; petals white to pink...**var. *ciliatum***. Found along streams and other moist places, on floodplains, and in meadows, 5000–11,500 ft.
1b. Inflorescence usually simple; large turions present at the base of the plant; petals rose-purple, pink, or rarely white...**var. *glandulosum* (Lehm.) Dorn**. Found in ravines, along streams, seepages, and in other moist places, 5200–10,500 ft.

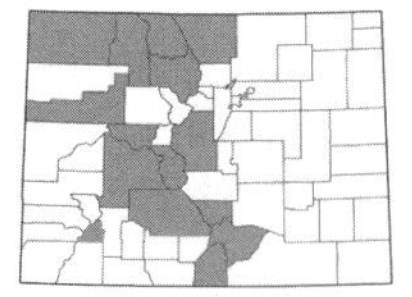

***Epilobium clavatum* Trel.**, TALUS WILLOW-HERB. Perennials, 0.5–2 dm, lacking turions; *stems* minutely hairy in lines decurrent from leaf bases, usually glandular above; *leaves* opposite, 1–3 cm long, entire or obscurely toothed; *pedicels* 0.2–1.8 cm long in fruit; *sepals* 2.5–5 mm long; *petals* pink, 3.5–6 (9) mm long; *capsules* 1.7–4 cm long; *seeds* 1.3–2 mm long, finely cellular-roughened or papillate at 20×. Found on talus and scree slopes, less often found along streams, 10,500–13,500 ft. July–Aug. E/W.

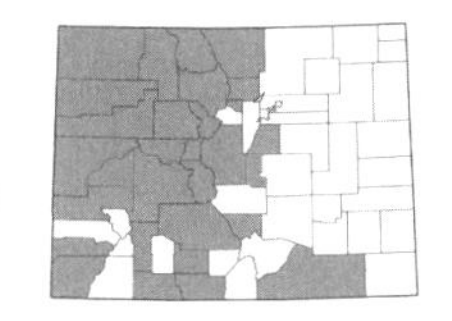

***Epilobium halleanum* Hausskn.**, GLANDULAR WILLOW-HERB. Perennials, 0.5–5 (7) dm, with turions; *stems* loosely hairy and often glandular above; *leaves* opposite, linear-oblong to lanceolate or narrowly elliptic, 1–4 cm × 2–11 mm, entire or toothed; *pedicels* 0.5–1.5 cm long in fruit; *sepals* 1.5–3 mm long; *petals* pink or lavender, rarely white, 3–5 mm long; *capsules* 2.5–5 cm long; *seeds* 1–1.5 mm long, very finely cellular-reticulate or minutely papillate. Common along streams, in bogs, meadows, and other moist places, 7000–12,000 ft. June–Aug. E/W.

***Epilobium hirsutum* L.**, HAIRY WILLOW-HERB. Perennials from rhizomes, 5–20 dm, lacking turions; *stems* densely hirsute above; *leaves* opposite, lanceolate to oblong, 5–12 × 1–3 cm, sharply toothed, densely soft-hirsute; *pedicels* to 10 mm long; *petals* pink-purple, 10–15 mm long, about 25 mm wide, notched at the tips; *capsules* 5–8 cm long. Weedy plants recently found in moist places, 5000–7000 ft. June–Sept. E. Introduced.

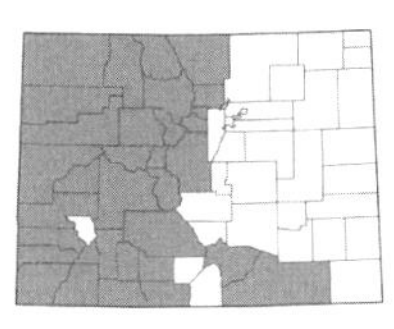

***Epilobium hornemannii* Rchb.**, HORNEMANN'S WILLOW-HERB. Perennials 1–5 dm, lacking turions; *stems* glandular above, hairs decurrent in lines from the leaf bases; *leaves* opposite, oblong-ovate to elliptic, 1.5–5.5 cm long, entire or remotely toothed; *pedicels* 0.5–4.5 cm long in fruit; *sepals* 1.5–4.5 mm long; *petals* pink, purple, or white, 3–9 mm long; *capsules* 2.5–6.5 cm long; *seeds* 1–1.5 mm long, minutely papillate. Common along streams, in meadows, and other moist places, 8000–12,000 ft. June–Aug. E/W. (Plate 55)

There are two varieties of *E. hornemannii* in Colorado, which can be difficult to separate from *E. clavatum*:

1a. Fruiting pedicels 0.5–2.5 cm long; petals pink or purple...**var. *hornemannii***
1b. Fruiting pedicels 2–4.5 cm long; petals white or sometimes pink...**var. *lactiflorum* (Hausskn.) D. Löve**

***Epilobium oregonense* Hausskn.**, OREGON WILLOW-HERB. Perennials, 1–3 dm, lacking turions; *stems* glabrous, often purplish above; *leaves* opposite, oblong-linear to narrowly ovate, 1–3 cm long, entire or minutely toothed; *pedicels* 1–3.5 cm long in fruit; *sepals* 1–2 mm long; *petals* pinkish or white, 4–7 mm long; *capsules* 2–5 cm long, often purplish; *seeds* ca. 1 mm long, smooth. Reported for Colorado, but no specimens have been seen.

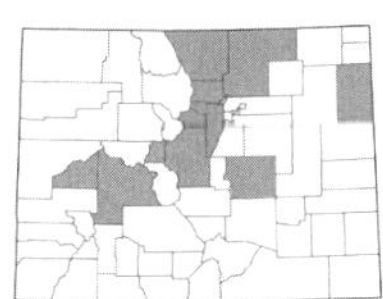

***Epilobium palustre* L.**, MARSH WILLOW-HERB. Perennials, 1–10 dm, with turions; stems sparsely minutely hairy, often in lines; *leaves* opposite or the upper sometimes alternate, linear to lanceolate, 1.5–7 cm × 1–15 mm, entire, often revolute; *pedicels* 0.5–2 cm long in fruit; *sepals* 2–4.5 mm long; *petals* white, pink, or light purple, 3.5–7 mm long, notched at the apex; *capsules* 3–8 cm long; *seeds* 1.5–2.5 mm, minutely papillate. July–Aug. E/W.

There are two varieties of *E. palustre* in Colorado:

1a. Leaves glabrous or sparsely hairy on the upper surface...**var. *palustre***. Uncommon in moist meadows, known from Larimer, Jackson, Routt, and Park cos., 8000–10,000 ft.
1b. Leaves more densely hairy on the upper surface...**var. *gracile* (Farw.) Dorn**. [*E. leptophyllum* Raf.]. Found along streams and lakes, seepage areas, and in moist meadows, 5400–10,000 ft.

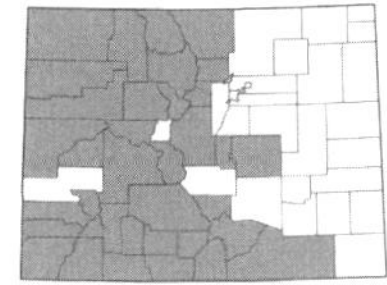

***Epilobium saximontanum* Hausskn.**, ROCKY MOUNTAIN WILLOW-HERB. Perennials, 1–5 dm, with turions; *stems* glandular above, glabrous below; *leaves* opposite, lanceolate to ovate, 1–6 cm × 5–25 mm, entire to minutely toothed; *pedicels* to 5 (9) mm long in fruit; *sepals* 1.5–3.5 mm long; *petals* white or sometimes purple, 2–5 mm long; *capsules* 2–6 cm long; *seeds* 1–1.5 mm long, minutely papillate. Common along streams and lake margins and in meadows, bogs, and other moist sites, 7000–12,000 ft. June–Aug. E/W.

EREMOTHERA – SMALL EVENING PRIMROSE

Annuals; *leaves* alternate, entire to weakly toothed; *flowers* actinomorphic, small; *petals* white or rarely red.

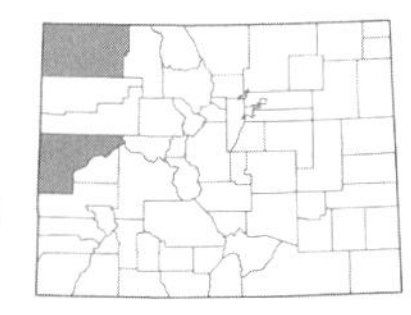

***Eremothera minor* (A. Nelson) W.L. Wagner & Hoch**, SMALL EVENING PRIMROSE. [*Camissonia minor* (A. Nelson) P.H. Raven]. Plants 2–30 cm, usually branched near the base; *leaves* linear to lanceolate, 0.5–5 × 0.2–1.5 cm, entire; *flowers* solitary in the upper axils; *floral tube* 0.5–2 mm; *sepals* 0.8–1.8 mm long; *petals* 0.8–1.3 mm long; *capsules* 8–25 mm long, twisted and contorted. Uncommon in dry, open places, often with sagebrush, 4500–7000 ft. May–June. W.

GAYOPHYTUM A. Juss. – GROUNDSMOKE

Annual herbs; *leaves* alternate or the lowest nearly opposite, entire; *flowers* actinomorphic; *petals* white, often fading pink or red, entire.

Gayophytum are very difficult to key out, as each species intergrades morphologically with other species.

1a. Capsules 3–6 (9) mm long, cylindric or clavate; pedicels nearly equal to or longer than the capsules; seeds usually in 2 rows per locule...***G. ramosissimum***
1b. Capsules 6–15 mm long, linear; pedicels equal to or shorter than the capsules; seeds usually in 1 row per locule...2

2a. Mature plants with branches below the middle or near the base of the main stem; pedicels 0.4–2 mm long...***G. racemosum***
2b. Mature plants branched throughout or with branches above the middle; pedicels 0.3–10 mm long...3

3a. Lowest flower in mature plants 1–3 (5) nodes above the base; stems usually with 2 or more nodes between lateral branches; pedicels 0.3–2 (5) mm long...***G. decipiens***
3b. Lowest flower in mature plants (3) 5–20 nodes above the base; stems usually branched at consecutive nodes or separated by only 1 node; pedicels 2–10 mm long...***G. diffusum* ssp. *parviflorum***

***Gayophytum decipiens* F.H. Lewis & Szweyk.**, DECEPTIVE GROUNDSMOKE. Plants 5–50 cm; *leaves* linear, 4–20 (30) × 1–3 (4) mm; *inflorescence* with the lowest flower in mature plants 1–3 (5) nodes above the base; *petals* 1–1.8 mm long; *capsules* 6–15 mm long. Found in moist places, forests, and on dry slopes, 7000–10,600 ft. June–Aug. E/W.

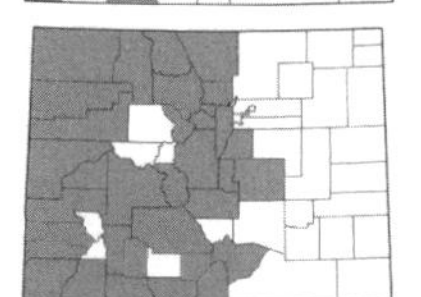

***Gayophytum diffusum* Torr. & A. Gray ssp. *parviflorum* F.H. Lewis & Szweyk.**, DIFFUSE GROUNDSMOKE. Plants 5–50 cm; *leaves* linear to oblong, 4–60 × 0.8–4 mm; *inflorescence* with the lowest flower in mature plants (3) 5–20 nodes above the base; *petals* 1.5–6 mm long; *capsules* 5–15 mm long. Found on dry or moist slopes, in forests, sagebrush, and grasslands, 5500–10,500 ft. June–Sept. E/W.

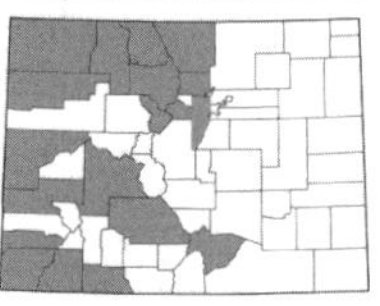

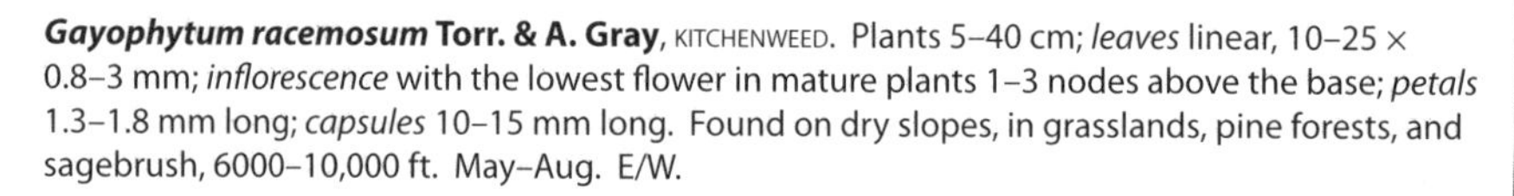

***Gayophytum racemosum* Torr. & A. Gray**, KITCHENWEED. Plants 5–40 cm; *leaves* linear, 10–25 × 0.8–3 mm; *inflorescence* with the lowest flower in mature plants 1–3 nodes above the base; *petals* 1.3–1.8 mm long; *capsules* 10–15 mm long. Found on dry slopes, in grasslands, pine forests, and sagebrush, 6000–10,000 ft. May–Aug. E/W.

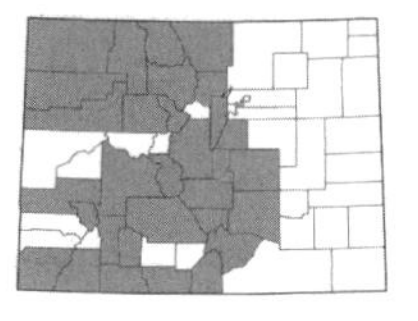

***Gayophytum ramosissimum* Torr. & A. Gray**, BRANCHING GROUNDSMOKE. Plants 5–50 cm; *leaves* linear, 10–40 × 1–5 mm; *inflorescence* with the lowest flower in mature plants 5–10 or more nodes above the base; *petals* 0.7–1.5 mm long; *capsules* 3–9 mm long. Found on dry slopes, in grasslands, pine forests, and sagebrush, 6300–10,500 ft. May–Sept. E/W.

HOLMGRENIA W.L. Wagner & Hoch – HOLMGREN'S EVENING PRIMROSE

Annual herbs, caulescent; *leaves* alternate, spirally arranged or subverticillate, densely tufted; *flowers* actinomorphic, axillary in leaves; *petals* yellow.

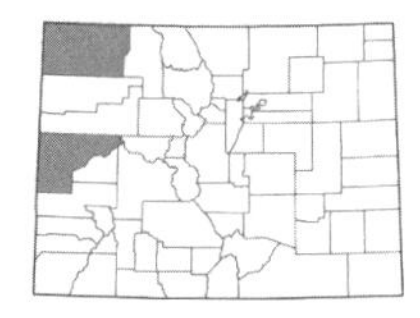

***Holmgrenia andina* (Nutt.) W.L. Wagner & Hoch**, BLACKFOOT RIVER EVENING PRIMROSE. [*Camissonia andina* (Nutt.) P.H. Raven]. Plants 1–8 cm; *leaves* linear, 6–30 × 1–3 mm, usually in tufts towards the tips of the stem, densely strigulose-puberulent, entire; *sepals* 1–2.5 mm long, reflexed; *petals* 1–3 mm long; *capsules* 6–11 mm long, 4-angled. Rare in open places, usually in sandy or clay soil, 5200–7000 ft. May–July. W.

OENOTHERA L. – EVENING PRIMROSE

Annual, biennial, or perennial herbs, caulescent or acaulescent; *leaves* alternate, usually with a basal rosette, mostly toothed to pinnatifid; *flowers* actinomorphic or slightly zygomorphic (former *Gaura* and *Stenosiphon*); *petals* white, yellow, purple, pink, or red. (Dietrich & Wagner, 1988; Wagner, 1983, 1984; Wagner & Mill, 1984; Wagner et al., 1985, 2013)
1a. Leaves all basal (plants acaulescent)...2
1b. Stems evidently leafy...5

2a. Flowers white (but often drying pink to rose-purple); capsules often with tuberculate ridges or rows of distinct tubercles...***O. cespitosa***
2b. Flowers yellow (but often drying bronze or purplish); capsules lacking tuberculate ridges or distinct tubercles...3

3a. Petals (1) 1.5–3 cm long; sepals 1–2 cm long...***O. flava***
3b. Petals (2.8) 3.5–8 cm long; sepals 1–7 (8) cm long...4

4a. Plants of the eastern slope; leaves 1–3 cm wide, entire or few-toothed; capsules with wings (2) 4–7 (11) mm wide...***O. howardii***
4b. Plants of the western slope; leaves 0.3–1.5 cm wide, irregularly toothed to somewhat pinnatifid; capsules with wings 1–2 mm wide...***O. acutissima***

5a. At least half the leaves deeply pinnatifid (with a narrow midrib less than 3 mm wide)...6
5b. Leaves entire to toothed or shallowly pinnatifid (with a midrib 5 mm or wider)...9

6a. Floral tubes densely bearded at the mouth with white hairs 1–2 mm long; perennials from deep rhizomes; petals 1–1.5 (2) cm long...***O. coronopifolia***
6b. Floral tubes glabrous; annuals or occasionally perennials from rhizomes; petals (1) 1.5–4 cm long...7

7a. Petals yellow (aging pink), (2) 2.5–4 cm long...***O. grandis***
7b. Petals white (aging pink), (1) 1.5–4 cm long...8

8a. Stems without an exfoliating outer layer; capsules with 2 rows of seeds in each locule; petals (1.5) 2–4 cm long...***O. albicaulis***
8b. Stems with a white exfoliating outer layer (sometimes inconspicuous on annual or young plants, look towards the base of the stem); capsules with 1 row of seeds per locule; petals (1) 1.5–2.5 (3) cm long...***O. pallida***

9a. Petals yellow (although sometimes drying or aging to orange, red, bronze, or purplish)...10
9b. Petals white, pink, or red...17

10a. Stigma peltate and discoid; flowers solitary in upper leaf axils...11
10b. Stigma 4-lobed; flowers in a leafy-bracteate spike...13

11a. Floral tubes 0.2–1.6 cm long; sepals with a keeled midrib; leaves mostly serrulate; petals 0.4–1.5 (2) cm long...***O. serrulata* var. *serrulata***
11b. Flora tubes 1.5–8 cm long; sepals lacking a keeled midrib; leaves mostly entire or occasionally serrate; petals 1–3.5 cm long...12

12a. Stems strigose and glandular, or glabrous; plants 0.5–5 dm tall...***O. hartwegii***
12b. Stems strigose to villous, lacking glandular hairs; plants 0.2–3 dm tall...***O. lavandulifolia***

13a. Petals 1–2 (2.5) cm long; sepals 0.9–2.5 cm long...14
13b. Petals (2.5) 3–6 cm long; sepals 2–6 cm long...16

14a. Petals rhombic obovate (diamond-shaped); capsules 1–1.6 (2) cm long...***O. rhombipetala***
14b. Petals obovate or obcordate; capsules 2–4.5 cm long...15

15a. Hairs usually with red, pustulate bases...***O. villosa***
15b. Hairs usually lacking red, pustulate bases...***O. biennis***

16a. Floral tubes 2–5.5 cm long...***O. elata* var. *hirsutissima***
16b. Floral tubes 6–13.5 cm long...***O. longissima***

17a. Petals pink or rarely white, with red spots or streaks throughout, 0.8–1.2 cm long; leaves 0.5–1.5 cm long; capsules 7–8 mm long; stems diffusely branched from the base...***O. canescens***
17b. Petals white or pink but lacking red spots or streaks, 0.3–4.5 cm long; leaves 1–15 cm long; capsules and stems various...18

18a. Capsules nut-like and indehiscent, 4-angled, 0.3–1 cm long; petals 0.3–1.5 cm long, abruptly narrowed to a slender claw 1–3 mm long; flowers in a spike, panicle, or spicate raceme...19
18b. Capsules loculicidal and splitting at the tips, 1.5–6 cm long; petals (1) 1.5–4.5 cm long, not narrowed to a slender claw; flowers solitary or solitary in upper leaf axils...24

19a. Stems glabrous with an exfoliating outer layer; petals white, 4–6 mm long; stigma shortly 4-lobed; capsule 3–4 mm long...***O. glaucifolia***
19b. Stems hairy and lacking an exfoliating outer layer; petals various; stigma rather deeply 4-lobed; capsule 4.5–18 mm long...20

20a. Petals 1.5–3 mm long; sepals 1.5–4 mm long; inflorescence branches with glandular hairs and longer hairs; anthers oval; plants (3) 5–20 dm tall...***O. curtifolia***
20b. Petals 3–15 mm long; sepals 5–15 mm long; inflorescence branches usually with the same type of hairs throughout or glabrous; anthers linear; plants 2–20 dm tall...21

21a. Leaves mostly less than 4 cm long; flowers pink to orange-red, drying deep red; inflorescence branches densely strigose, lacking glandular hairs; perennials...***O. suffrutescens***
21b. Larger leaves over 4 cm long; flowers white, aging pink or reddish; inflorescence branches usually with glandular hairs (rarely only strigose) or glabrous; annuals, biennials, or perennials...22

22a. Hypanthium (floral tube) 1.5–3 (5) mm long; inflorescence branches usually glabrous or rarely strigose; perennials...***O. cinerea***
22b. Hypanthium (5) 6–13 mm long; inflorescence branches with glandular hairs or strigose; annuals or biennials...23

23a. Inflorescence branches lacking glandular hairs...***O. coloradoensis***
23b. Inflorescence branches with glandular hairs...***O. dodgeniana***

24a. Stems lacking a white, exfoliating outer layer; capsules with prominent tubercles between the wings; leaves and flowers densely crowded on the stem...***O. harringtonii***
24b. Stems white or nearly so, the outer layer exfoliating (this sometimes inconspicuous on annual or young plants, look towards the base of the stem); capsules lacking tubercles; leaves and flowers not densely crowded on the stem...25

25a. Inflorescence and capsules glandular-hairy; stems glabrous...***O. nuttallii***
25b. Inflorescence glabrous or white-hairy but not glandular; stems hairy or glabrous...26

26a. Stems, sepals, floral tubes, and capsules conspicuously spreading-hairy with long hairs...***O. engelmannii***
26b. Stems, sepals, floral tubes, and capsules glabrous or appressed-hairy with a few long, spreading hairs...27

27a. Petals 3.8–4.5 cm long; capsules 4–8 cm long; leaves mostly 5–15 cm long; stems glabrous...***O. kleinii***
27b. Petals 1–4 cm long; capsules 1.5–5 cm long; leaves 1–8 cm long; stems usually hairy or sometimes glabrous...***O. pallida***

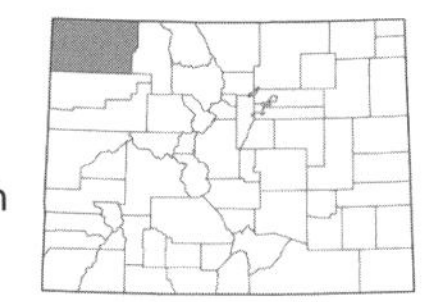

***Oenothera acutissima* W.L. Wagner**, FLAMING GORGE EVENING PRIMROSE. Perennials, acaulescent; *leaves* 7–15 cm × 0.3–1.5 cm, irregularly and coarsely toothed to somewhat pinnatifid; *flowers* actinomorphic; *sepals* 1–4 cm long, with a prominent red margin; *petals* yellow, becoming reddish-orange in age, 28–50 mm long; *stigma* 4-lobed; *capsules* loculicidal, 1.4–2.5 cm long, with wings 1–2 mm wide. Uncommon in seasonally moist places, 7200–8500 ft. May–June. W.

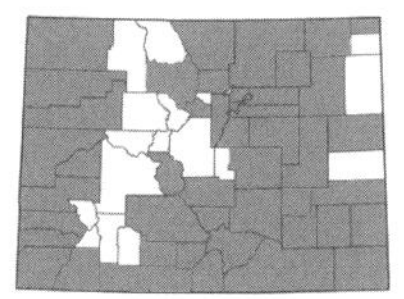

***Oenothera albicaulis* Pursh**, WHITEST EVENING PRIMROSE. Annuals, 0.5–3 dm, caulescent, finely minutely hairy throughout and usually with some longer, stiff hairs; *leaves* 1.5–10 cm × 3–25 mm, margins entire or with a few teeth, occasionally pinnatifid; *flowers* actinomorphic, axillary to cauline leaves; *floral tube* 1.5–3 cm long; *sepals* 15–30 mm long; *petals* white, aging pink, (15) 20–40 mm long; *stigma* 4-lobed; *capsules* loculicidal, 4-angled, 2–4 cm × 3–4 mm. Found on dry slopes, shortgrass prairie, often forming large masses on the plains, 3500–8500 ft. May–Aug. E/W. (Plate 55)

***Oenothera biennis* L.**, COMMON EVENING PRIMROSE. Biennials, 3–20 dm, caulescent, spreading-hairy and minutely strigose, glandular in the inflorescence; *leaves* 5–20 cm long, margins entire to dentate; *flowers* actinomorphic, in a terminal spike; *floral tube* (2) 2.5–4 cm long; *sepals* 1.2–2(2.8) cm long; *petals* yellow, aging orange, 1–2.5 (3) cm long; *stigma* 4-lobed; *capsules* loculicidal, 2–4 cm long. Reported for the southeastern plains in disturbed areas. June–Sept. Introduced.

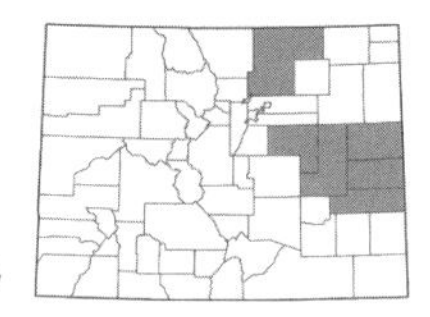

***Oenothera canescens* Torr. & Frém.**, SPOTTED EVENING PRIMROSE. Perennials, 1–2 dm, caulescent, densely strigose to canescent; *leaves* lanceolate, 0.5–1.5 cm × 2–7 mm, nearly entire to sinuate-dentate; *flowers* actinomorphic, axillary; *floral tube* 5–15 mm long; *sepals* 9–12 mm long; *petals* pinkish or rarely white, with red spots or streaks all over, 8–12 mm long; *stigma* 4-lobed; *capsules* loculicidal, 7–8 mm long, sharply 4-angled. Uncommon along drying shorelines and playa edges, 3900–5800 ft. June–Aug. E.

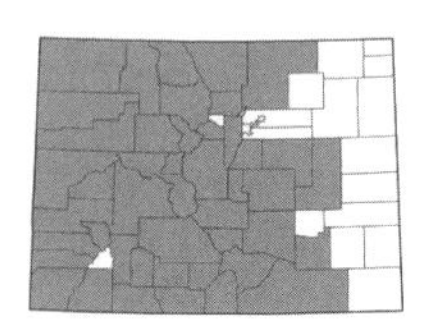

***Oenothera cespitosa* Nutt.**, TUFTED EVENING PRIMROSE. Perennials, acaulescent or with stems to 3 dm; *leaves* 3–30 cm × 5–40 mm, entire to pinnatifid; *flowers* actinomorphic, solitary in axils; *floral tube* 3–15 cm long; *sepals* (15) 20–50 (55) mm long; *petals* white, turning pink when drying, 20–50 (60) mm long; *stigma* 4-lobed; *capsules* loculicidal, 25–50 mm long, each valve with a tuberculate ridge. 4600–10,000 ft. May–Aug. E/W. (Plate 55)

There are 4 varieties of *O. cespitosa* in Colorado:

1a. Flowers smaller overall, the floral tube 3–7 (8.5) cm long and petals 2–4 (4.5) cm long...2
1b. Flowers larger overall, the floral tube 6–14 cm long and petals (2.5) 3–5 (6) cm long...3

2a. Plants usually lacking glandular hairs, often glabrous or with short hairs to 0.5 mm; petals usually aging dark rose-purple...**var. *cespitosa***. Found on dry slopes and in grasslands or sagebrush, 4600–10,000 ft. E/W.
2b. Plants with minute glandular hairs and hairs 0.5–1 mm long; petals usually remaining white or turning pink in age...**var. *navajoensis* (W.L. Wagner, Stockh., & W.M. Klein) Cronquist**. Found on shale slopes, clay hills, and in pinyon-juniper, mostly found in the southwestern counties, 5000–9500 ft. W.

3a. Capsule lacking distinct tuberculate ridges; leaves mostly toothed...**var. *macroglottis* (Rydb.) Cronquist**. Found on dry slopes and in oak, pinyon-juniper ,or spruce-fir forests, widespread, 5000–10,000 ft. E/W.
3b. Capsule with distinct tuberculate ridges; leaves usually mostly pinnatifid or deeply toothed...**var. *marginata* (Nutt. ex Hook. & Arn.) Munz**. Found on dry slopes, in sandy soil, canyons, and in sagebrush and pinyon-juniper, widespread, 5000–10,000 ft. E/W.

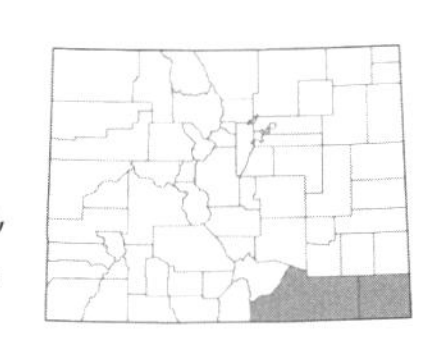

***Oenothera cinerea* (Wooton & Standl.) W.L. Wagner & Hoch**, WOOLLY BEEBLOSSOM. [*Gaura villosa* Torr.]. Perennials, 5–15 dm, sparsely to densely villous above, often glandular; *leaves* lanceolate to spatulate or ovate, 3–7 cm long, sinuate-dentate; *flowers* zygomorphic, in spicate, branched inflorescences 2–6 dm long; *floral tube* 1.5–3 (5) mm long; *sepals* 6–10 (15) mm long; *petals* white, fading pinkish, the blade 7–10 mm long; *stigma* 4-lobed; *capsules* nut-like, indehiscent, 4-angled, 10–18 mm long on a stipe 2–8 mm long. Found in sandy soil, 4200–5500 ft. June–Aug. E.

***Oenothera coloradoensis* (Rydb.) W.L. Wagner & Hoch,** COLORADO BEEBLOSSOM. [*Gaura coloradensis* Rydb.]. Biennials, 5–9 (10) dm, stems strigulose; *leaves* lanceolate to narrowly elliptic, 5–13 cm × 10–40 mm, subentire to minutely denticulate; *flowers* zygomorphic, in spicate or paniculately spicate inflorescences; *floral tube* 8–12 mm long, glandular; *sepals* 9–13 mm long; *petals* white, aging pink or reddish, 7–12 mm long; *stigma* 4-lobed; *capsules* nut-like, indehiscent, 4-angled, 6–8 mm long. Rare in moist meadows and along ditches, 5800–6500 ft. June–Sept. E. (Plate 55)

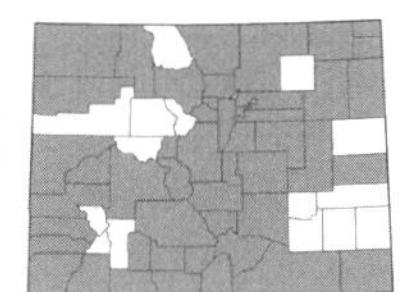

***Oenothera coronopifolia* Torr. & A. Gray**, CROWNLEAF EVENING PRIMROSE. Perennials from rhizomes, 1–6 dm, strigose throughout; *leaves* deeply pinnatifid, 2–7 cm × 8–15 mm; *flowers* actinomorphic, solitary in the upper axils; *floral tube* 1–3 cm long, densely hairy at the mouth with white hairs 1–2 mm long; *sepals* 10–20 mm long; *petals* white, aging pink, 10–15 (20) mm long; *stigma* 4-lobed; *capsules* loculicidal, 10–20 mm long, mostly round. Common on dry slopes, in sandy soil, and shortgrass prairies, 3500–9700 ft. May–Aug. E/W. (Plate 55)

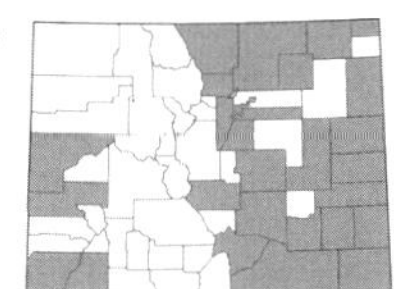

***Oenothera curtifolia* W.L. Wagner & Hoch**, VELVETWEED. [*Gaura mollis* James]. Annuals or biennials, to 2 m, glandular and spreading-hairy; *leaves* lanceolate to narrowly ovate or oblong, to 10 cm × 4 cm, remotely toothed; *flowers* zygomorphic, in an elongated spike; *floral tube* 1.5–5 mm long; *sepals* 1.5–4 mm long, reflexed at flowering; *petals* 1.5–3 mm long, pink; *stigma* 4-lobed; *capsules* nut-like, indehiscent, glabrous to short-hairy, 5–10 mm long, with a rounded rib on each of the 4 angles and a slender rib on each face. Found along streams, on sand hills, and in grasslands, fields, and disturbed areas, 3500–7500 ft. June–Sept. E/W.

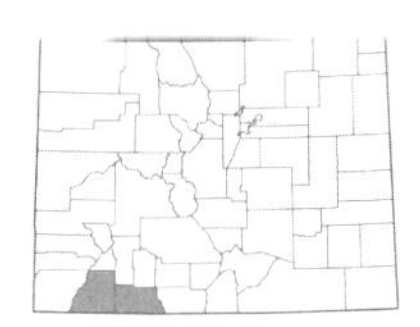

***Oenothera dodgeniana* Krakos & W.L. Wagner**, NEW MEXICO BUTTERFLY PLANT. [*Gaura neomexicana* Wooton]. Biennials, 5–12 dm, stems strigulose; *leaves* lanceolate to elliptic, 5–10 cm × 10–25 mm, subentire to minutely denticulate; *flowers* zygomorphic, in spicate or paniculately spicate inflorescences; *floral tube* 10–12 mm long, glandular; *sepals* 11–15 mm long; *petals* white, aging pink or reddish, 11–14 mm long; *stigma* 4-lobed; *capsules* nut-like, indehiscent, 4-angled, 6–8 mm long. Rare in moist meadows and seepage areas, 6500–7500 ft. June–Sept. W.

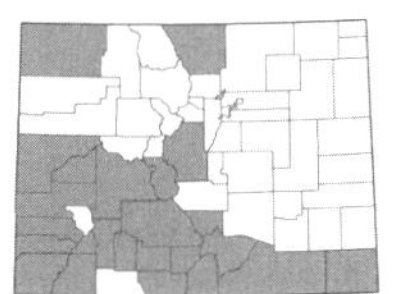

***Oenothera elata* Kunth var. *hirsutissima* (A. Gray ex S. Watson) W. Dietr.**, WESTERN EVENING PRIMROSE. Biennials or short-lived perennials, 2–25 dm, minutely hairy and with hairs 1–3 mm long that are sometimes pustulate; *leaves* lanceolate to narrowly ovate, 5–40 cm × 5–35 (60) mm, entire to irregularly toothed; *flowers* actinomorphic, in a bracteate spike; *floral tube* 2.5–5.5 cm long; *sepals* 2.5–5.5 cm long; *petals* yellow, fading orange or purplish, 3–5 (5.5) cm long; *stigma* 4-lobed; *capsules* loculicidal, hairy, 3–4 (6) cm long. Found in meadows, forests, and along roadsides and streams, 4000–10,000 ft. June–Aug. E/W.

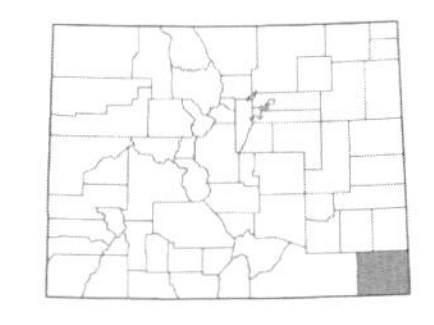

***Oenothera engelmannii* (Small) Munz**, ENGELMANN'S EVENING PRIMROSE. Annuals, 2–5 (8) dm, spreading-hairy; *leaves* lanceolate to oblong-lanceolate, 2–9 (12) cm × 10–20 (25) mm, entire to sinuate-dentate or the lower pinnatifid; *flowers* actinomorphic, solitary in upper axils; *floral tube* 20–30 mm long; *sepals* (10) 15–20 mm long; *petals* white, fading pink, 13–22 mm long; *stigma* 4-lobed; *capsules* loculicidal, with spreading hairs, 25–50 mm long. Uncommon in sandy soil, along roadsides, 4200–4500 ft. May–Aug. E.

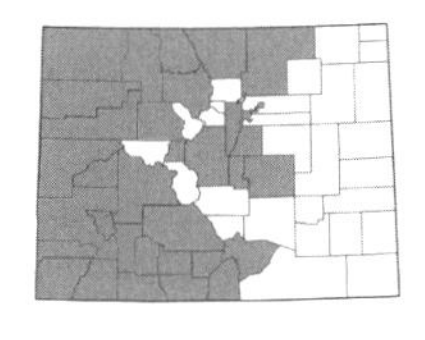

***Oenothera flava* (A. Nelson) Garrett**, YELLOW EVENING PRIMROSE. Perennials, acaulescent; *leaves* lanceolate, 5–30 cm long, entire to pinnatifid; *flowers* actinomorphic, solitary; *floral tube* 3–13 cm long; *sepals* 10–20 mm long; *petals* yellow, drying purple, (10) 15–30 mm long; *stigma* 4-lobed; *capsules* loculicidal, glabrous or minutely hairy, (10) 20–30 mm, winged distally. Found in moist places, meadows, and sagebrush, 5500–10,500 ft. June–Aug. E/W.

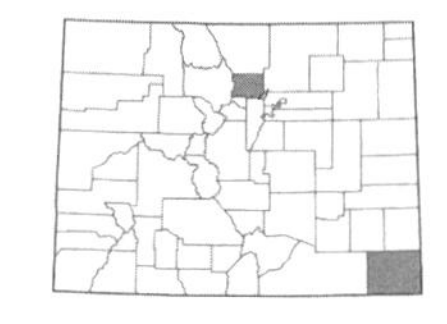

***Oenothera glaucifolia* W.L. Wagner & Hoch**, FALSE GAURA. [*Stenosiphon linifolius* (Nutt. ex James) Heynh.]. Biennials, 3–30 dm; *leaves* lanceolate to oblong, 3–10 cm × 4–18 mm, entire; *flowers* actinomorphic, in a spike or open paniculate spike, sparsely to densely glandular; *floral tube* 6–13 mm long, with short spreading hairs; *sepals* 4–6 mm long; *petals* white, 4–6 mm long, abruptly clawed; *stigma* shortly 4-lobed; *capsules* nut-like, indehiscent, hairy, 3–4 mm long, with prominent ribs on the angles. Found in sandy soil, native to southeastern Colorado but introduced in Boulder Co., 4500–5500 ft. July–Sept. E.

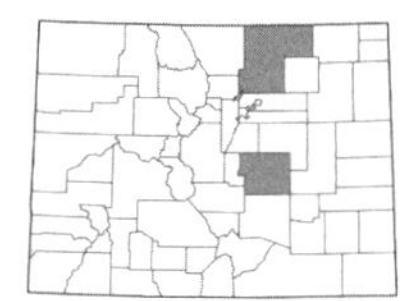

***Oenothera grandis* (Britton) Smyth**, SHOWY EVENING PRIMROSE. Annuals, 2–6 (8) dm; *leaves* oblanceolate to elliptic, 1.5–6 cm long, entire to sinuate-dentate, the lower sometimes pinnatifid; *flowers* actinomorphic, solitary in upper axils; *floral tube* 25–50 mm long, sparsely to densely villous; *sepals* (15) 20–30 mm long; *petals* yellow, fading pinkish, (20) 25–40 mm long; *stigma* 4-lobed; *capsules* loculicidal, with spreading hairs, (10) 20–40 mm long. Uncommon in grasslands and on dry slopes, 4000–7200 ft. May–Aug. E.

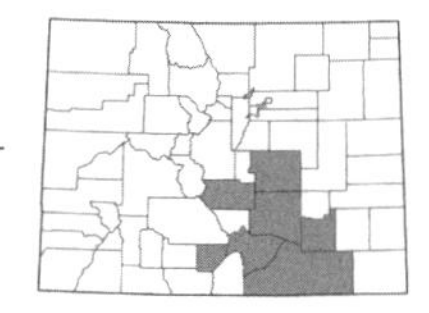

***Oenothera harringtonii* W.L. Wagner**, COLORADO SPRINGS EVENING PRIMROSE. Perennials, 1.5–4 dm, strigulose; *leaves* lanceolate to narrowly ovate, sinuate-dentate, to 30 cm long; *flowers* actinomorphic, solitary in upper axils; *floral tube* 30–60 mm long; *petals* white, fading pinkish, 20–25 mm long; *stigma* 4-lobed; *capsules* loculicidal, 20–35 mm long, winged and with prominent tubercle ridges between wings. Found in dry grasslands and in clay, rocky, or sandy soil, 4700–6100 ft. May–July. E. Endemic.

Closely resembling *O. cespitosa* except that *O. harringtonii* has a true stem 1.5–4 dm tall.

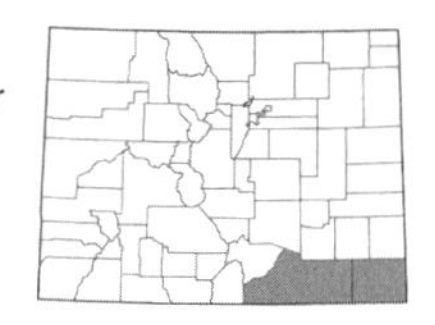

***Oenothera hartwegii* (Benth.) P.H. Raven**, HARTWEG'S SUNDROPS. [*Calylophus hartwegii* (Benth.) P.H. Raven]. Perennials, 0.5–5 dm, glabrous to hairy, usually glandular; *leaves* linear to oblanceolate or ovate, 0.3–5 cm long, entire or toothed; *flowers* actinomorphic, in axils of upper leaves; *floral tube* 1.5–6 cm long, hairy or glabrous; *sepals* 7–25 mm long; *petals* yellow, aging orange or pinkish, 10–35 mm long; *stigma* peltate, discoid; *capsules* loculicidal, 6–40 mm long. Found in grasslands, prairie, sandy soil, and on outcrops, 4000–5100 ft. May–Sept. E.

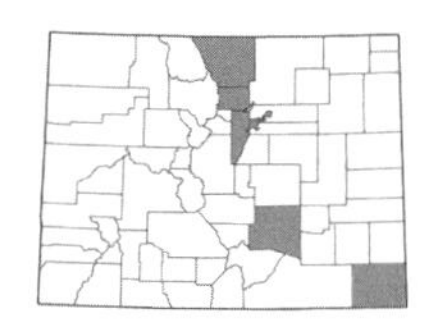

***Oenothera howardii* (A. Nelson) W.L. Wagner**, HOWARD'S EVENING PRIMROSE. Perennials, acaulescent; *leaves* lanceolate to oblanceolate, 5–23 cm × 10–30 mm, entire or few-toothed; *flowers* actinomorphic, solitary in upper axils; *floral tube* 4.3–13 cm long; *sepals* 30–70 (80) mm long; *petals* yellow, aging to orange or bronze, 30–75 mm long; *stigma* 4-lobed; *capsules* loculicidal, 20–50 (80) mm long, with broad wings along the sides. Found on dry hillsides, in sandy soil, and grasslands, 4700–6500 ft. May–July. E. (Plate 55)

***Oenothera kleinii* W.L. Wagner & S. Mill.**, WOLF CREEK EVENING PRIMROSE. Perennials, 1.5–3 dm; *leaves* lanceolate to elliptic, 5–15 cm long, toothed; *flowers* actinomorphic, solitary in upper axils; *floral tube* 30–37 mm long, villous; *sepals* 25–30 mm long, long-villous; *petals* white, fading pink, 38–45 mm long; *stigma* 4-lobed; *capsules* loculicidal, 40–80 mm long, nearly always curved upward. Known from near Wolf Creek Pass (Mineral Co.), 8700–8800 ft. W. Endemic.

This population of plants was destroyed in construction on Wolf Creek Pass. However, they may be specimens of *O. deltoides* Torr. & Frém. var. *deltoides*, which they are morphologically similar to, that were planted and do not naturally exist in Colorado.

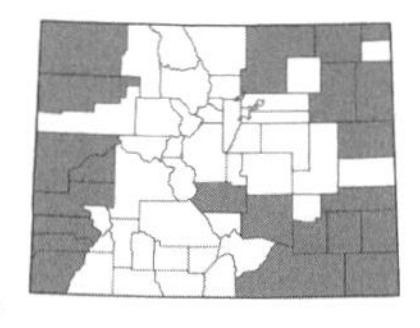

***Oenothera lavandulifolia* (Torr. & A. Gray) P.H. Raven**, LAVENDER-LEAF SUNDROPS. [*Calylophus lavandulifolius* (Torr. & A. Gray) P.H. Raven]. Perennials, 2–30 cm, with numerous stems; *leaves* linear to narrowly elliptic, 0.7–3.5 (5) cm × 1–6 mm, entire; *flowers* actinomorphic, solitary in the upper axils, appearing terminal; *floral tube* 2.5–8 cm long, glabrous to hairy or glandular; *sepals* 10–20 mm long; *petals* yellow, fading to pink or orange, 15–25 mm long; *stigma* peltate, discoid; *capsules* loculicidal, 10–20 mm long. Common on outcrops and in shortgrass prairie, pinyon pine, and other dry places, 3500–8700 ft. May–Aug. E/W. (Plate 55)

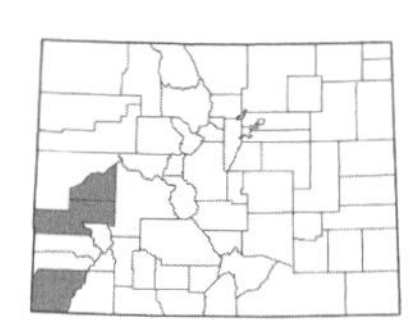

***Oenothera longissima* Rydb.**, LONGSTEM EVENING PRIMROSE. Biennials or short-lived perennials, 7–30 dm; *leaves* lanceolate to narrowly elliptic, 7–40 cm × 10–45 mm, mostly entire; *flowers* actinomorphic, sessile in terminal spikes; *floral tube* 6–13.5 cm long; *sepals* 25–55 mm long, reflexed; *petals* yellow, aging purplish or orange, 30–50 (65) mm long; *stigma* 4-lobed; *capsules* loculicidal, hairy, 30–55 mm long. Uncommon in moist places, 4800–5100 ft. July–Sept. W.

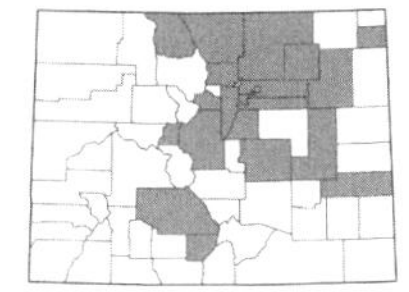

***Oenothera nuttallii* Sweet**, NUTTALL'S EVENING PRIMROSE. Perennials, 3–6 (10) dm; *leaves* linear to lanceolate, 2–6 (10) cm × 3–7 (10) mm, entire to denticulate; *flowers* actinomorphic, solitary in upper axils, the bracts glandular; *floral tube* 1.5–3 cm long; *sepals* 20–30 mm long, glandular, usually reddish; *petals* white, fading pink, 15–25 mm long; *stigma* 4-lobed; *capsules* loculicidal, glandular, 20–30 mm long. Found in sandy soil, grasslands, and meadows, 3500–8500 (9500) ft. June–Sept. E.

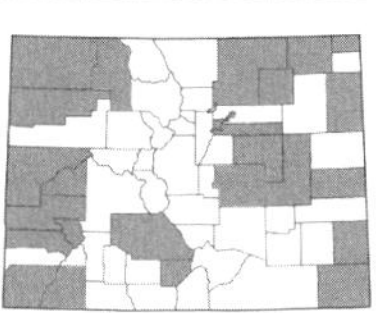

***Oenothera pallida* Lindl.**, PALE EVENING PRIMROSE. Annuals, biennials, or perennials, 1–5 dm; *leaves* lanceolate to ovate, to 8 cm long; *flowers* actinomorphic, solitary in upper axils; *floral tube* 1.5–4 cm long, glabrous to slightly hairy; *sepals* 10–30 mm long, glabrous to hairy; *petals* white, aging pink or reddish, 10–40 mm long; *stigma* 4-lobed; *capsules* loculicidal, 15–50 mm long, straight or curved. May–Sept. E/W.

There are 4 subspecies of *O. pallida* in Colorado. Some specimens will not key easily to subspecies, as considerable morphological overlap exists:

1a. Leaves mostly pinnatifid with a narrow midrib...2
1b. Leaves mostly entire or irregularly toothed to shallowly pinnatifid...3

2a. Perennials from creeping rhizomes; leaves 2–3.5 (5) cm long; stems usually pale...**ssp. *runcinata* (Engelm.) Munz & W. Klein**. Found in dry places, sandy soil, pinyon-juniper, 6000–9300 ft. E/W.
2b. Annuals or biennials from slender taproots, or short-lived perennials with creeping roots; leaves 3–6 cm long; stems usually reddish-tinged...**ssp. *trichocalyx* (Nutt.) Munz & W. Klein**. Found on dry slopes, along streams, and in pinyon-juniper, on the eastern slope in Alamosa and Saguache cos., 4700–9000 ft. E/W.

3a. Annuals, biennias, or short-lived perennials (occasionally with creeping roots); stems usually reddish tinged...**ssp. *trichocalyx* (Nutt.) Munz & W. Klein**. (See above for description).
3b. Perennials from creeping rhizomes; stems usually pale...4

4a. Leaves and stems densely hairy...**ssp. *latifolia* (Rydb.) Munz**. [*O. latifolia* (Rydb.) Munz]. Found in sandy soil and shortgrass prairie, 3500–6700 ft. E.
4b. Leaves and stems mostly glabrous or sparsely hairy...**ssp. *pallida***. Found in sandy soil and pinyon-juniper, 5000–6500 ft. W.

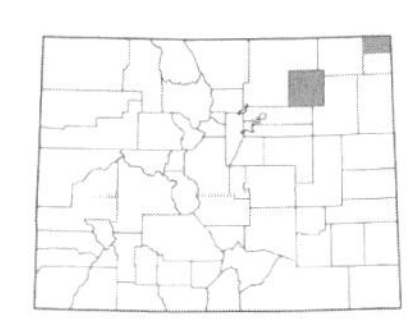

***Oenothera rhombipetala* Nutt. ex Torrey & A. Gray**, FOUR-POINT EVENING PRIMROSE. Biennials, 3–8 (12) dm; *leaves* narrowly oblanceolate to lance-ovate, 2–8 cm × 3–20 mm, remotely toothed to subentire; *flowers* actinomorphic, in a dense terminal spike 1–3 dm long; *floral tube* 2–4 cm long; *sepals* 1–2.2 cm long; *petals* yellow, rhombic-ovate, 12–25 mm long; *stigma* 4-lobed; *capsules* loculicidal, 10–16 (20) mm long. Uncommon in shortgrass prairie, 3800–4500 ft. June–Oct. E.

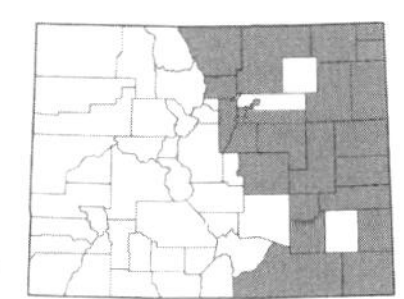

Oenothera serrulata* Nutt. var. *serrulata, YELLOW SUNDROPS. [*Calylophus serrulatus* (Nutt.) P.H. Raven]. Perennials, 0.5–8 dm; *leaves* narrowly lanceolate to oblanceolate, 1–10 cm × 0.1–12 mm, entire to serrulate; *flowers* actinomorphic, sessile in upper axils; *floral tube* 2–16 mm long; *sepals* 1.5–9 mm long, keeled; *petals* yellow, fading pinkish, 4–15 (20) mm long; *stigma* peltate, discoid; *capsules* 10–30 × 1–3 mm. Common in shortgrass prairie, grasslands, sandy soil, and on outcrops, 3500–7500 ft. May–Sept. E. (Plate 55)

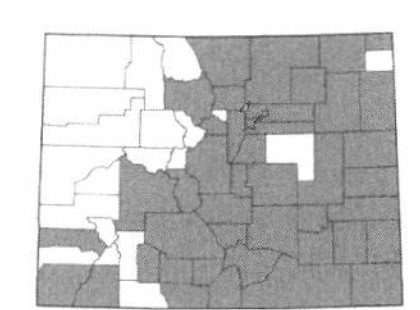

***Oenothera suffrutescens* (Ser.) W.L. Wagner & Hoch**, SCARLET BEEBLOSSOM. [*Gaura coccinea* Nutt. ex Pursh]. Perennials, 2–5 (10) dm; *leaves* lanceolate to narrowly oblong, 1–4 cm long, entire or few-toothed; *flowers* zygomorphic, in elongated spikes, often lax or nodding at the tip; *floral tube* 4–11 mm long; *sepals* 5–9 mm long, reflexed; *petals* pink to red, 3–7 mm long; *stigma* deeply 4-lobed; *capsules* nut-like, indehiscent, 1.5–2 mm long, strongly 4-angled. Common on dry slopes, outcrops, and in sandy soil, meadows, shortgrass prairie, and grasslands, 3500–9500 ft. May–Aug. E/W. (Plate 55)

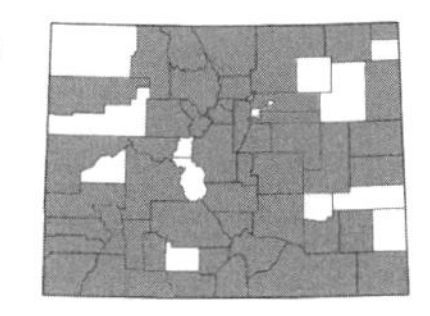

***Oenothera villosa* Thunb.**, HAIRY EVENING PRIMROSE. Biennials to short-lived perennials, 3–15 (20) dm; *leaves* lanceolate or narrowly oblong, 10–30 cm × 12–40 mm, entire to repand-dentate, densely strigose; *flowers* actinomorphic, in a bracteate spike; *floral tube* 23–45 mm long; *sepals* 9–18 mm long, with a slender terminal appendage 0.5–3 mm long; *petals* yellow, drying orange, 7–20 mm long; *stigma* 4-lobed; *capsules* loculicidal, hairy, 20–45 mm long. Found along streams, on dry slopes, sandy hillsides, often in disturbed areas, 3500–9500 ft. June–Sept. E/W. (Plate 55)

There are two varieties of *O. villosa* in Colorado:

1a. Leaves usually entire; sepals not densely hairy; internodes of the infructescence as long or longer than the capsules...**var. *strigosa* (Rydb.) Dorn**. 4800–9500 ft. E/W.

1b. Leaves usually toothed; sepals densely hairy; internodes of the infructescence shorter than the capsules...**var. *villosa***. 3500–8700 ft. E.

TARAXIA (Nutt. ex Torr. & A. Gray) Raimann – TARAXIA

Perennial herbs; *leaves* in a basal rosette, with subentire to deeply sinuate or pinnatifid margins; *flowers* actinomorphic; *petals* yellow (ours).

1a. Leaves pinnatifid throughout, sparsely to conspicuously hairy (sometimes glabrous with age)...***T. breviflora***
1b. Leaves entire or with a few lobes at the base of the blade, glabrous...***T. subacaulis***

***Taraxia breviflora* (Torr. & A. Gray) Nutt. ex Small**, FEWFLOWER TARAXIA. [*Camissonia breviflora* (Torr. & A. Gray) P.H. Raven]. Plants acaulescent; *leaves* 2–14 cm × 3–20 mm, pinnatifid; *flowers* sessile in leaf axils; *floral tube* 15–25 mm long; *sepals* 3–6.5 mm long, reflexed; *petals* 4–9 mm long; *capsules* 8–20 mm long. Uncommon in moist meadows and along the borders of ephemeral ponds, 7000–8000 ft. June–Aug. E.

***Taraxia subacaulis* (Pursh) Rydb.**, DIFFUSE-FLOWER TARAXIA. [*Camissonia subacaulis* (Pursh) P.H. Raven]. Plants acaulescent; *leaves* 2–22 cm × 3–35 (50) mm, entire to irregularly pinnately lobed; *flowers* sessile in leaf axils; *floral tube* 1.5–3 mm long; *sepals* 6–15 mm long, reflexed; *petals* 7–15 mm long; *capsules* 10–25 mm long. Uncommon in grasslands and moist meadows, 7500–8500 ft. May–July. W.

ORCHIDACEAE Juss. – ORCHID FAMILIY

Perennial herbs (ours), terrestrial (ours); *leaves* basal or cauline, usually alternate, simple and entire, exstipulate; *flowers* perfect or rarely imperfect, zygomorphic; *sepals*3; *petals* 3, the inner central petal quite distinct from the others and forming the lip; *stamens* 1–2 (3), more or less fused with the style and stigmas forming the column, the pollen grains aggregated in pollinia, pollen sacs 2; *stigmas* 1, 3-lobed with the apex of the central lobe forming the rostellum on the column; *ovary* inferior, 3-carpellate; *fruit* a capsule.

1a. Plants lacking chlorophyll (not green), saprophytic, yellowish-green, brown, or reddish in color; leaves absent, reduced to brown sheaths; from branching rhizomes resembling coral; lip petal often spotted or striped...***Corallorhiza***
1b. Plants green; leaves usually present although sometimes withered by flowering time; roots not resembling coral; lip petal various...2

Platanthera flower:

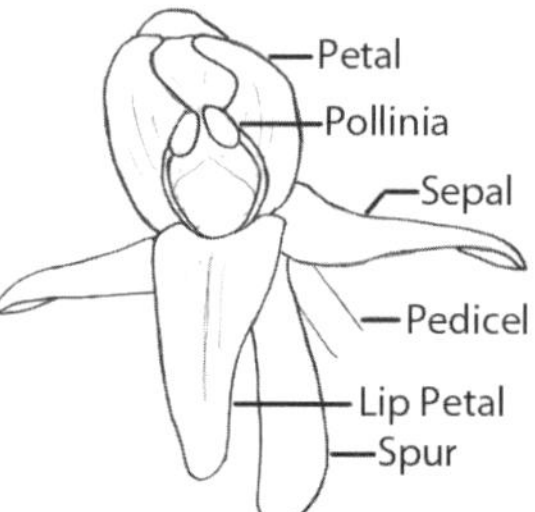

2a. Lip petal forming a slipper-shaped, saccate, inflated pouch 8–50 mm long, with a petaloid column or staminode hooded over the opening to the lip; flowers solitary or 2–4 in a short spike...3
2b. Lip petal usually flat or the margins somewhat upturned or saccate near the base but not forming a sac-like, inflated pouch, 2–20 mm long, the column unlike the above; flowers several in a spike or raceme...4

3a. Plants with a single elliptic to ovate leaf near the base; petals purplish to magenta, the lip petal purplish at the base and whitish toward the tip, with yellow hairs near the middle and with dark purplish stripes on the inner surface...***Calypso***
3b. Plants with 2–5 leaves; petals yellow or reddish-brown to purple, the lip lacking yellow hairs and otherwise unlike the above...***Cypripedium***

4a. Leaves 2, opposite near the middle of the stem, mostly elliptic to suborbiculate; flowers green to yellowish-green...***Listera***
4b. Leaves several and alternate along the stem, all basal, or plants with a single leaf at the base, variously shaped; flowers various...5

5a. Leaves in a basal rosette, with a white midrib and/or white reticulations, ovate to elliptic; dorsal petals connivent and forming a hood covering the lip petal, the flowers in a loose spike, not in spirally twisted rows...***Goodyera***
5b. Stem leaves present or plants with a single leaf at the base, if appearing to be in a basal rosette then the leaves linear-lanceolate, lacking a white midrib and/or white reticulations; if the dorsal petals connivent and forming a hood covering the lip then the flowers in a dense terminal spike with spirally twisted rows...6

6a. Flowers in a more or less dense terminal spike with spirally twisted rows, white; leaves linear to linear-lanceolate, mostly basal; petals sparsely hairy with short glandular hairs...***Spiranthes***
6b. Flowers not in a dense spike with spirally twisted rows, variously colored; leaves unlike the above in all respects; petals glabrous to sparsely glandular-hairy...7

7a. Petals pale pink to orange or light greenish, with prominent purplish or pinkish veins, 9–17 mm long, the outer whorl of tepals similar but more greenish; lip petal purplish to brown, 9–20 mm long and saccate at the base; bracts in the inflorescence large and leaf-like, 0.7–12 cm long...***Epipactis***
7b. Petals white, green, or yellowish-green, lacking prominent purplish veins, flowers smaller, 3–8 mm long; lip petal shorter than 9 mm, not saccate at the base; bracts in the inflorescence small or leaf-like, to 4 cm...8

8a. Lip petal lacking a spur at the base; plants with a single, ovate-elliptic leaf at the base, not tapering to the sheathing base...***Malaxis***
8b. Lip petal with a spur at the base; leaves 2–several along the stem or if plants with a single leaf at the base then the leaf narrowed and tapering to the sheathing base...9

9a. Lip petal 2–3-toothed at the tip; bracts in the inflorescence leaf-like, the lower ones at least twice as long as the flowers, 1.5–6 cm long...***Coeloglossum***
9b. Lip petal not 2–3-toothed at the tip; bracts in the inflorescence shorter than the flowers or if longer then the lower ones less than 2 cm long...10

10a. Sepals 1-nerved; plants with 2–4 leaves, these aggregated on the lower part of the stem and usually withered by flowering...***Piperia***
10b. Sepals at least 3-nerved; plants with several leaves along the stem or with a single basal leaf, not withered by flowering...***Platanthera***

CALYPSO Salisb. – FAIRY SLIPPER

Perennial herbs with chlorophyll; *leaves* solitary at the base; *flowers* solitary; *petals* pink, magenta, or rarely white, the lip slipper-shaped and deeply saccate with 3 rows or yellow hairs near the tip; *fruit* a capsule.

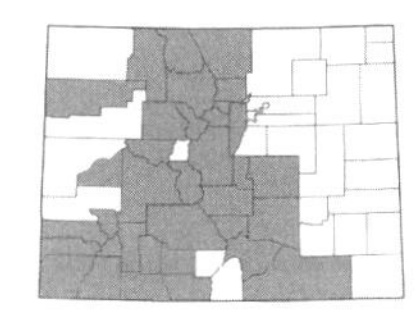

***Calypso bulbosa* (L.) Oakes var. *americana* (R. Br. ex Ait. f.) Luer**, FAIRY SLIPPER. Plants 0.4–2 dm; *leaves* elliptic to ovate, 2–7 cm long; *petals* 10–24 mm long, the lip 13–23 mm long; *capsules* 20–30 mm long. Found under pine/spruce trees in moist, shaded forests and along streams, 7000–10,500 ft. May–July. E/W. (Plate 56)

COELOGLOSSUM Hartm. – LONG-BRACTED ORCHID

Perennial herbs with chlorophyll; *leaves* cauline, alternate; *flowers* in a terminal, spike-like raceme with large bracts; *petals* greenish, with a saccate spur; *fruit* a capsule.

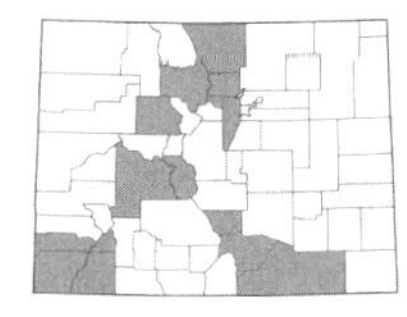

***Coeloglossum viride* (L.) Hartm.**, LONG-BRACTED FROG ORCHID. [*Dactylorhiza viridis* (L.) R.M. Bateman, A.M. Pridgeon, & M.W. Chase; *Habenaria viridis* (L.) R. Br.]. Plants 0.5–6 dm; *leaves* lanceolate to obovate, 2–18 cm long; *floral bracts* linear-lanceolate, 1.5–6 cm long (2–4 times as long as the flowers); *petals* 3–5 mm long, the lip narrowly oblong-spatulate, 2–3-toothed at the tip; *spur* short, saccate; *capsules* 7–15 mm long. Found in aspen forests, along streams, and in meadows, 6500–11,000 ft. June–Aug. E/W.

CORALLORHIZA Gagnebin – CORALROOT

Perennial herbs, saprophytic, the stems yellowish to brown or reddish or purplish, from branching rhizomes which resemble coral; *leaves* lacking; *flowers* in a terminal raceme; *petals* yellow to reddish or purple; *fruit* a capsule.

1a. Flowers prominently striped with 3–5 dark purplish or reddish veins, not spotted...***C. striata***
1b. Flowers spotted or not, but not prominently striped...2

2a. Lip petal 2.5–4 mm long, white and usually lacking purple spots but occasionally lightly purple spotted; plants usually yellowish-green...***C. trifida***
2b. Lip petal 4–9 mm long, white and usually purple or red spotted; plants usually brown to reddish but occasionally yellow...3

3a. Lip 3-lobed with 2 small lateral lobes; stems usually reddish-brown, sometimes yellowish-green...***C. maculata***
3b. Lip entire, lacking lateral lobes; stems usually bronze, sometimes yellowish...***C. wisteriana***

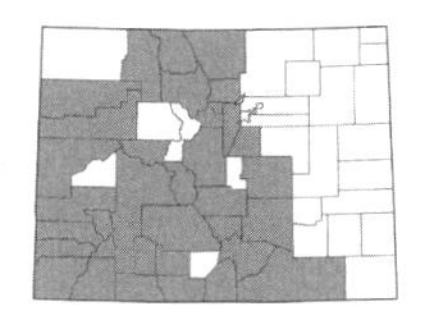

***Corallorhiza maculata* (Raf.) Raf.**, SPOTTED CORALROOT. Plants 2–6 dm, reddish to reddish-brown; *racemes* with 6–40 flowers, 10–65 cm long; *sepals* 4.5–15 mm long, 3-veined, brown, red, or yellowish; *lip petal* white with purple spots, 4–9 mm long, with 2 small lateral lobes near the apex; *capsules* 9–25 mm long. Common in moist or more often dry, shaded forests, 5500–11,000 ft. June–July. E/W. (Plate 56)

Corallorhiza striata **Lindl.**, STRIPED CORALROOT. Plants 1–6 dm, reddish-brown to reddish; *racemes* with 2–35 flowers, 10–65 cm long; *sepals* 3-5-veined, reddish-purple to yellowish; *lip petal* 3.5–16 mm long, white with 3–5 dark purplish or reddish veins; *capsules* 10–30 mm long. Uncommon in shaded forests, 6300–9000 ft. June–July. E/W.

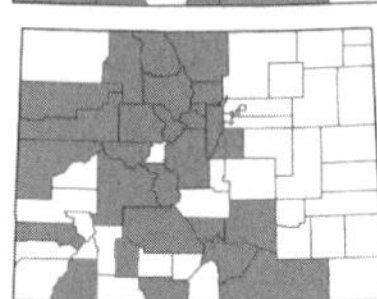

Corallorhiza trifida **Châtel.**, YELLOW CORALROOT. Plants 0.8–3 dm, usually yellowish-green; racemes with 3–18 flowers, 8–35 cm long; *sepals* 3.5–7 mm long, yellowish-green, sometimes spotted with purple; *lip petal* 2.5–4 mm long, white and usually lacking purple spots; *capsules* 4.5–15 mm long. Found in moist pine or spruce-fir forests and along streams, 7200–11,000 ft. June–July. E/W. (Plate 56)

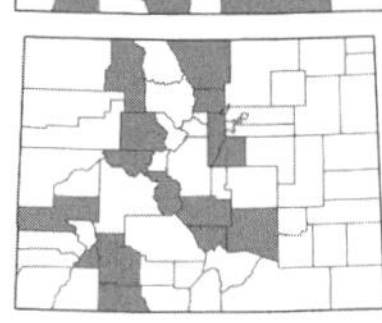

Corallorhiza wisteriana **Conrad**, SPRING CORALROOT. Plants to 4 dm, reddish to reddish-brown; *racemes* with 2–25 flowers, 10–55 cm long; *sepals* 4.5–10 mm long, 3-veined, purplish-brown to yellowish; *lip petal* white with purple spots, 4–7.5 mm long, entire; *capsules* 7–15 mm long. Uncommon in moist forests, 6000–9000 ft. May–July. E/W.

CYPRIPEDIUM L. – LADY'S SLIPPER

Perennial herbs with chlorophyll; *leaves* cauline, alternate or subopposite; *flowers* solitary and terminal or 2-several in a terminal raceme; *petals* with a pouch-like and very inflated lip; *fruit* a capsule.

1a. Flowers reddish-brown to purplish, 2–4 in a short spike, the lip mostly 8–15 mm long; leaves 2, subopposite...***C. fasciculatum***

1b. Flowers yellow, usually solitary or occasionally 2 per stem, the lip mostly 20–50 mm long; leaves 3–5, alternate...***C. parviflorum* var. *pubescens***

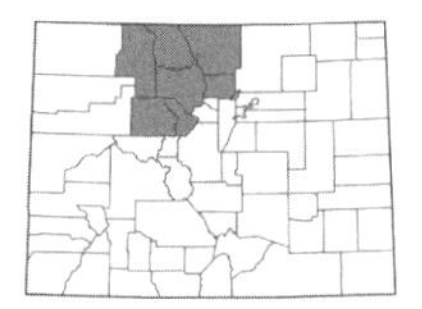

Cypripedium fasciculatum **Kellogg ex S. Watson**, PURPLE LADY'S SLIPPER. Plants 0.5–3.5 dm; *leaves* 2, subopposite, elliptic to ovate, 4–12 × 2.5–7.5 cm; *inflorescence* nodding in flower, with 2–4 flowers in a short spike; *sepals* 10–25 mm long, dull yellow, suffused with reddish-brown or purplish; *petals* similar in color to sepals, 10–23 mm long; *lip petal* 8–15 (25) mm long. Uncommon in shaded lodgepole pine or spruce-fir forests, 8000–10,500 ft. June–Aug. E/W. (Plate 56)

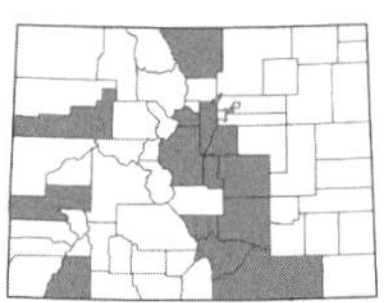

Cypripedium parviflorum **Salisb. var. *pubescens*** **(Willd.) Knight**, YELLOW LADY'S SLIPPER. [*C. calceolus* L. var. *pubescens* (Willd.) Correll]. Plants 7–70 dm; *leaves* 3–5, alternate; *inflorescence* usually solitary or occasionally 2 flowers per stem; *sepals* greenish or yellow, 11–80 mm long; *petals* similar in color to sepals, 2.5–10 cm long; *lip petal* yellow, mostly 20–50 mm long. Rare in aspen and ponderosa pine/Douglas-fir forests, 6000–9500 ft. June–July. E/W. (Plate 56)

EPIPACTIS Zinn – HELLEBORINE

Perennial herbs with chlorophyll; *leaves* cauline, alternate; *flowers* in a terminal raceme with foliaceous bracts; *petals* greenish-yellow, the lip saccate at the base; *fruit* a capsule.

1a. Lip petal 3-lobed, 14–20 mm long, the tip narrowly spatulate-oblanceolate or linear-oblanceolate; flowers larger, the lateral sepals 16–25 mm long...***E. gigantea***

1b. Lip petal not 3-lobed, 9–11 mm long, the tip broadly triangular-ovate; flowers smaller, the lateral sepals 10–13 mm long...***E. helleborine***

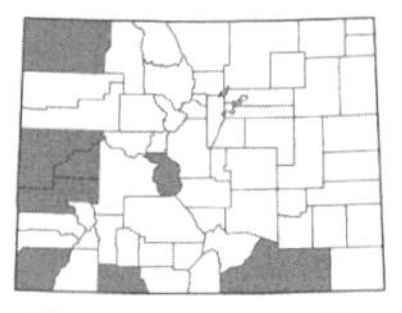

Epipactis gigantea **Douglas ex Hook.**, GIANT HELLEBORINE. Plants 3–14 dm; *leaves* lanceolate to ovate, 5–20 cm long; *floral bracts* 0.7–12 cm long; *lateral sepals* greenish to rose-colored, 16–25 mm long; *lip petal* 14–20 mm long, 3-lobed with a narrowly spatulate-oblanceolate or linear-oblanceolate tip; *capsules* 20–25 mm long. Uncommon in hanging gardens, marshes near hot springs, and on seepy slopes, 4800–8000 ft. June–July. E/W. (Plate 56)

Epipactis helleborine **L.**, BROADLEAF HELLEBORINE. Plants 2.5–8 (10) dm; *leaves* lanceolate to elliptic, 4–20 cm long; *floral bracts* 1–5 (7) cm long; *lateral sepals* greenish, usually tinged with purple, 10–13 mm long; *lip petal* 9–11 mm long, not 3-lobed; *capsules* 9–15 mm long. Uncommon in disturbed places, found as a weed in lawns, 5500–6000 ft. June–Aug. E. Introduced.

GOODYERA R. Br. – RATTLESNAKE PLANTAIN

Perennial herbs with chlorophyll; *leaves* basal; *flowers* sessile in a terminal spike; *petals* white to greenish, the dorsal ones connivent and forming a hood covering the lip; *fruit* a capsule.

1a. Leaves with a white midrib or sometimes also with white reticulations, 2.5–10 cm long; flowers larger, the hood 5–10 mm long and lip 5–8 mm long...***G. oblongifolia***

1b. Leaves lacking a white midrib but with white reticulations, 1–3.2 cm long; flowers smaller, the hood 3–5.5 mm long and lip 1.8–4.8 mm long...***G. repens***

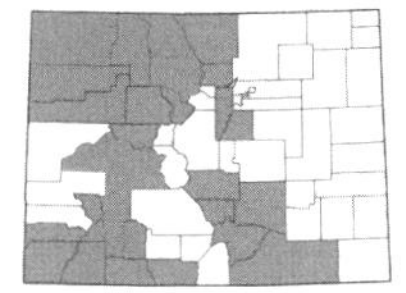

***Goodyera oblongifolia* Raf.**, WESTERN RATTLESNAKE PLANTAIN. Plants 2–3.5 dm; *leaves* elliptic to lanceolate, dark green with a white midrib or sometimes also with white reticulations, 2.5–10 cm long; *lateral sepals* 5.5–8 mm long; *petals* connivent, the hood 5–10 mm long, the lip 5–8 mm long. Locally common in shaded forests and along streams, 6000–10,700 ft. July–Sept. E/W. (Plate 56)

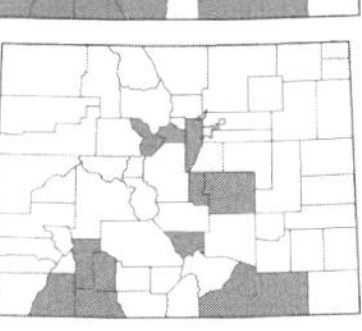

***Goodyera repens* (L.) R. Br. ex W.T. Aiton**, LESSER RATTLESNAKE PLANTAIN. Plants to 3 dm; *leaves* ovate to elliptic, green with white reticulations but lacking a white midrib, 1–3.2 cm long; *lateral sepals* 3–5 mm long; *petals* distinct, the hood 3–5.5 mm long and the lip 1.8–4.8 mm long. Uncommon in shaded forests and along streams, 8000–9500 ft. July–Aug. E/W.

LISTERA R. Br. – TWAYBLADE

Perennial herbs with chlorophyll; *leaves* 2, opposite or subopposite near the middle of the stem; *flowers* in a terminal raceme; *petals* green to yellowish-green, the lip notched or deeply lobed at the apex; *fruit* a capsule. (Wiegand, 1899)

1a. Leaves with a cordate or subcordate base; lip cleft to about the middle into 2 linear-lanceolate lobes, the margins glabrous...***L. cordata* var. *nephrophylla***
1b. Leaves rounded at the base, not cordate; lip notched or shortly cleft into 2 oblong or rounded lobes, the margins shortly ciliate...2

2a. Lip petal gradually narrowing to the base; leaves suborbiculate to broadly ovate or broadly elliptic, 1.5–6 cm wide...***L. convallarioides***
2b. Lip petal about the same width throughout; leaves lanceolate to elliptic, 0.7–3 cm wide...***L. borealis***

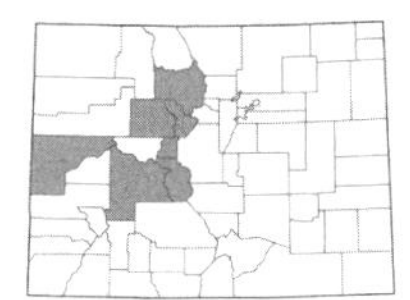

***Listera borealis* Morong**, NORTHERN TWAYBLADE. Plants to 25 cm; *leaves* elliptic to ovate, rounded at the base, 1.3–6 cm long, rounded at the tip; *floral bracts* 1–3 mm long; *petals* yellowish-green or green, 4.5–5 mm long, the lip about the same width throughout, cleft into 2 oblong lobes, margins shortly ciliate. Uncommon in moist, shady forests and in mossy seeps, 8700–10,800 ft. June–Aug. E/W.

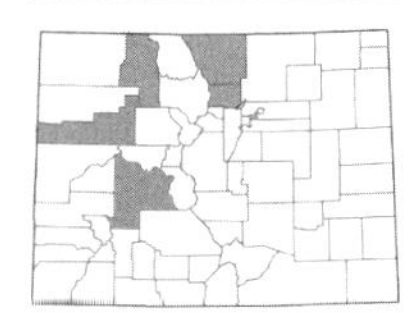

***Listera convallarioides* (Sw.) Nutt. ex Elliott**, BROAD-LEAVED TWAYBLADE. Plants to 35 cm; *leaves* ovate to elliptic, rounded at the base, 2–7 cm long; *floral bracts* 3–5 mm long; *petals* yellowish-green, 4–5 mm long, the lip gradually narrowing to the base, shallowly notched, the margins shortly ciliate. Uncommon in shady, moist forests and along streams, 6800–10,000 ft. June–Aug. E/W.

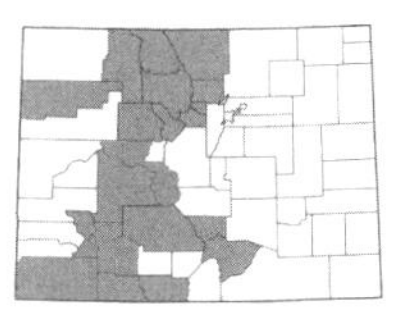

***Listera cordata* (L.) R. Br. ex Ait. f. var. *nephrophylla* (Rydb.) Hultén**, HEARTLEAF TWAYBLADE. Plants 5–30 cm; *leaves* ovate-cordate, 0.9–2 (4) cm long, with a mucronate tip; *floral bracts* 1–1.5 mm long; *petals* yellow-green to green, 1.5–2.5 mm long, the lip cleft to about the middle into 2 linear-lanceolate lobes, margins glabrous; *capsules* ca. 5 mm long. Locally common in moist, shady forests, and in mossy places along streams, 8500–11,500 ft. June–Aug. E/W. (Plate 56)

MALAXIS Sol. ex Sw. – ADDER'S MOUTH ORCHID

Perennial herbs with chlorophyll; *leaves* 1–3 near the base of the plant; *flowers* in a terminal spicate raceme with inconspicuous bracts; *petals* green or whitish-green and the lip 3-lobed with the central lobe longest with an acuminate tip (ours); *fruit* a capsule.

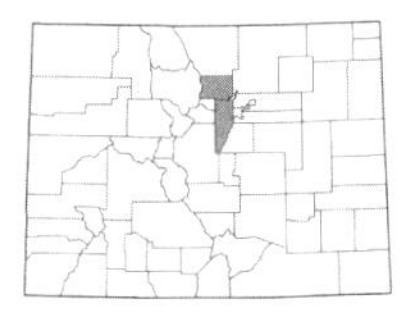

***Malaxis monophyllos* (L.) Sw. var. *brachypoda* (A. Gray) Morris & Eames**, WHITE ADDER'S MOUTH ORCHID. [*M. brachypoda* (A. Gray) Fernald]. Plants 0.3–3 dm; *leaves* usually solitary, ovate to elliptic, 1.5–10 cm long; *floral bracts* lanceolate, 1.5–2 mm long; *petals* green or whitish-green, strongly reflexed, 1.4–2.5 mm long. Rare along mossy, shaded streams, 7200–8000 ft. June–July. E.

PIPERIA Rydb. – REIN ORCHID

Perennial herbs with chlorophyll; *leaves* basal, often withered by flowering time; *flowers* in a spike-like raceme; *petals* white, green, or yellowish-green, the lip with a clavate to filiform spur at the base; *fruit* a capsule.

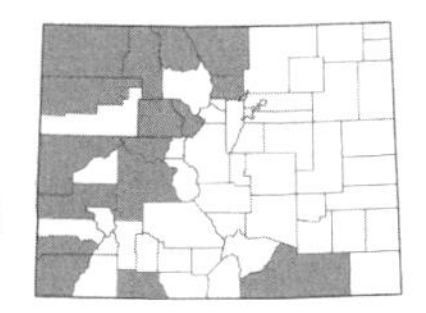

***Piperia unalascensis* (Spreng.) Rydb.**, SLENDER-SPIRE ORCHID. [*Habenaria unalascensis* (Spreng.) S. Watson; *Platanthera unalascensis* (Spreng.) Kurtz]. Plants 2–7 dm; *leaves* linear-lanceolate to oblanceolate, 5–20 cm long; *flowers* translucent green, fragrant; *lip petal* 2–5.5 mm long, with a small lobe on each side at the base; *spur* curved, 2–5 mm long. Uncommon in shaded forests and along streams, 6000–9800 ft. June–Aug. E/W. (Plate 56)

PLATANTHERA Rich. – FRINGED ORCHID

Perennial herbs with chlorophyll; *leaves* sometimes basal but usually cauline and alternate; *flowers* in a terminal spike; *petals* white to green or yellowish-green, the lip petal entire or lobed, with a saccate or cylindrical spur at the base; *fruit* a capsule. (Correll, 1943)

If one is collecting *Platanthera* it is best to note the color of the flower (since they often fade to brown later), the scent, and take a picture of the flower showing the lip and spur when fresh. This makes identifying them later much easier!

1a. Plants with a single basal, elliptic leaf (rarely with 2 leaves); lip petal linear, to 1.5 mm wide; spur equal to the lip in length or nearly so…***P. obtusata***
1b. Plants with several stem leaves; lip petal linear, linear-lanceolate to lanceolate, rhombic-lanceolate, or oblong, 1–5 mm wide; spur various…2

2a. Flowers white; lip petal broadened and dilated at the base…***P. dilatata***
2b. Flowers greenish-white, green, or yellowish-green; lip petal only slightly broadened at the base, usually not dilated…3

3a. Spur ½ to sometimes ¾ the length of the lip petal, short and saccate, scrotiform or strongly clavate, 2–3 mm long; flowers usually with lighter green sepals and darker green petals, the lip petal also sometimes bluish or marked with red; flowers with a very musky smell…***P. purpurascens***
3b. Spur ¾ or more as long or conspicuously longer than the lip petal, cylindric to slightly clavate, 2–17 mm long; flowers whitish-green or yellowish-green; flowers sweet-smelling or lacking an aroma…4

4a. Flowers whitish-green, the petals often whiter than the sepals; anther sacs vertical; lip petal slightly to conspicuously rounded-dilated at the base…***P. huronensis***
4b. Flowers green or yellowish-green; anther sacs vertical or horizontal in *P. aquilonis*; lip petal not rounded-dilated at the base…5

5a. Pollinia loose and free of anther sacs, trailing downward onto the stigma (flowers are self-pollinating); spur 2–5 mm long; lip petal rhombic-lanceolate to lanceolate, usually wider at the base than the tip…***P. aquilonis***
5b. Pollinia remaining enclosed in anther sacs, not trailing downward onto the stigma; spur 4.5–17 mm long; lip petal linear to linear-oblong, about the same width throughout…6

6a. Spur 12–17 mm long, much longer than the lip; lip petal 1–3 mm wide…***P. zothecina***
6b. Spur 4.5–7 mm long, equaling or only slightly longer than the lip; lip petal 0.8–1.6 mm wide…***P. tescamnis***

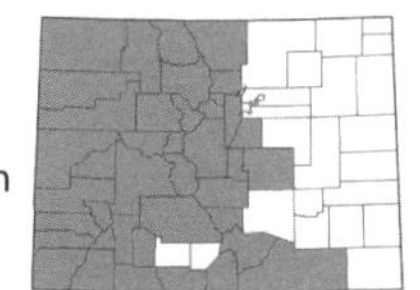

***Platanthera aquilonis* Sheviak**, NORTHERN GREEN ORCHID. Plants 0.5–6 dm; *leaves* narrowly lanceolate to oblong, 3–20 cm long; *petals* yellowish-green, the lip petal not rounded-dilated at the base, 2.5–6 mm long; *spur* usually clavate, 2–5 mm long; *pollinia* loose and free of anther sacs. Common in marshes, moist spruce-fir forests and meadows, and along streams, 6400–12,500 ft. May–Aug. E/W.

This is the correct name for the plant that has been called *P. hyperborea* (L.) Lindl. in Colorado.

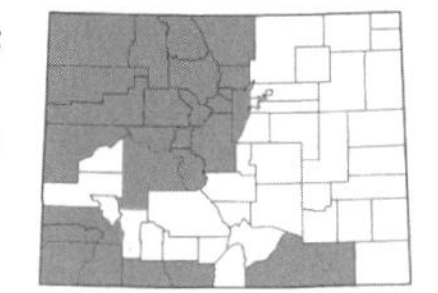

***Platanthera dilatata* (Pursh) Lindley ex L.C. Beck**, SCENTBOTTLE. [*Habenaria dilatata* (Pursh) Hook.; *Limnorchis dilatata* (Pursh) Rydb.]. Plants 1–15 dm; *leaves* linear-lanceolate to oblanceolate, 3.5–30 cm long; *petals* white, the lip petal 4–11 mm long, rounded-dilated at the base; *spur* clavate to cylindric, 4–12 mm long. Common in moist meadows and spruce-fir forests, along streams and creeks, and in marshes, 7500–12,500 ft. May–Aug. E/W. (Plate 56)

There are two varieties of *P. dilatata* in Colorado:

1a. Spur shorter than the lip…**var. *albiflora* (Cham.) Ledeb.**
1b. Spur equal to the lip or nearly so…**var. *dilatata***

***Platanthera huronensis* (Nutt.) Lindley**, HURON GREEN ORCHID. [*P. hyperborea* (L.) Lindl. var. *huronensis* (Nutt.) Luer]. Plants 1–12 dm; *leaves* linear-lanceolate to oblanceolate, 5–30 cm long; *petals* whitish-green, the lip 5–12 mm long; *spur* clavate to cylindric, 4–12 mm long; *pollinia* remaining enclosed in vertical anther sacs. Locally common in moist forests and meadows, marshes, and along creeks and streams, 6700–12,000 ft. June–Aug. E/W. (Plate 56)

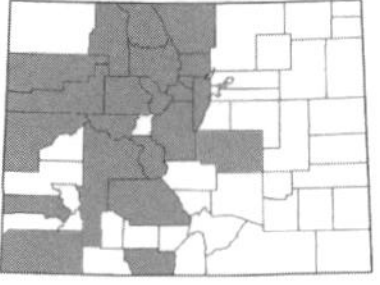

***Platanthera obtusata* (Banks ex Pursh) Lindl.**, BLUNT-LEAVED ORCHID. [*Habenaria obtusata* (Banks ex Pursh) Richardson; *Lysiella obtusata* (Banks ex Pursh) Rydb.]. Plants 0.5–3.5 dm; *leaves* usually single and basal, elliptic to oblanceolate, 3.5–15 cm long; *petals* greenish-white to yellowish-green, the lip 2.5–9 (10) mm long; *spur* 3–8 (10) mm long; *pollinia* remaining enclosed in anther sacs. Uncommon in moist spruce-fir forests and along creeks, 8000–11,500 ft. June–Aug. E/W.

Platanthera purpurascens **(Rydb.) Sheviak & W.F. Jennings**, PURPLE-PETAL BOG ORCHID. Plants 2.5–8 dm; *leaves* lanceolate to oblanceolate or ovate, 5.5–20 (28) cm long; *petals* green, the lip petal sometimes bluish or marked with red, 4–8 mm long; *spur* clavate to scrotiform, 2–3 mm long; *pollinia* remaining enclosed in anther sacs. Found in moist meadows and spruce-fir forests, along lakes and streams, in marshes, 7800–12,000 ft. June–Aug. E/W.

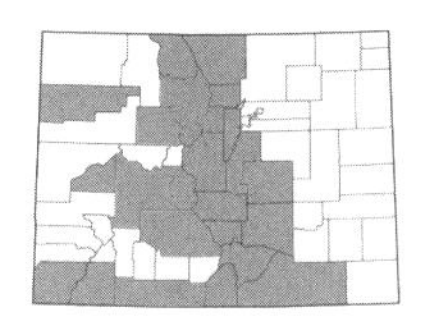

This is the correct name for the plant that has long been called *P. stricta* Lindl. or *P. saccata* (Greene) Hultén in Colorado.

Platanthera tescamnis **Sheviak & W.F. Jennings**, STREAM BOG ORCHID. Plants 2–15 dm; *leaves* linear-lanceolate to oblong or narrowly elliptic, 6.5–30 cm long; *petals* green or yellowish-green, the lip 0.8–1.6 mm wide; *spur* cylindric or slightly clavate, 4.5–7 mm long; *pollinia* remaining enclosed in the anther sacs. Uncommon in canyons, floodplains, meadows, and along streams, 5900–9700 ft. June–Aug. E/W. (See p. x for illustration)

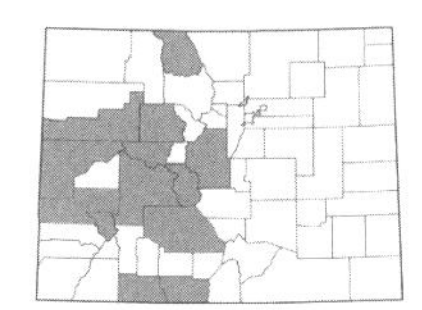

This is the correct name for the plant that has been called *P. sparsiflora* (S. Watson) Schltr. in Colorado.

Platanthera zothecina **(L.C. Higgins & S.L. Welsh) Kartesz & Gandhi**, ALCOVE BOG ORCHID. [*Habenaria zothecina* Higgins & S.L. Welsh; *Limnorchis zothecina* (Higgins & S.L. Welsh) W.A. Weber]. Plants 2.5–4 dm; *leaves* oblanceolate to elliptic, 7–17 cm long; *petals* green to yellowish-green, the lip petal not rounded-dilated at the base, 5–12 × 1–3 mm; *spur* cylindric, 12–17 mm long; *pollinia* remaining enclosed in anther sacs. Uncommon in hanging gardens and moist sandstone cliffs and overhangs, known from Dinosaur National Monument, 5300–6600 ft. June–Aug. W.

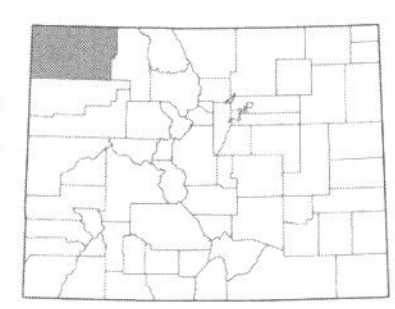

SPIRANTHES Rich. – LADY'S TRESSES

Perennial herbs; *leaves* basal or cauline and alternate; *flowers* in a dense terminal spike, usually in spirally twisted rows; *sepals* connivent with the 2 lateral petals and forming a hood enclosing the column and the majority of the lip (ours); *petals* white to yellowish; *fruit* a capsule. (Sheviak, 1984)

1a. Sepals distinct or united only at the base, not forming a hood above the lip, the outer lateral tepals spreading; flowers diverging at about a 90-degree angle from the rachis; inflorescence looser, the rachis usually visible...***S. diluvialis***

1b. Sepals and petals connivent, forming a hood above the lip, the outer lateral petals upcurved; flowers diverging at about a 45-degree angle or higher from the rachis; inflorescence tightly packed, the rachis usually not visible...***S. romanzoffiana***

Spiranthes diluvialis **Sheviak**, UTE LADY'S TRESSES. Plants 2–6 dm; *leaves* linear to lanceolate or oblanceolate, to 30 cm long; *spikes* loosely arranged, the flowers diverging at about a 90-degree angle from the rachis; *sepals* distinct or united at the base, 7.5–15 mm long; *petal* lip 7–12 mm long. Rare on floodplains, along streams, and in moist meadows and swales, 4500–7000 ft. July–Sept. E/W.

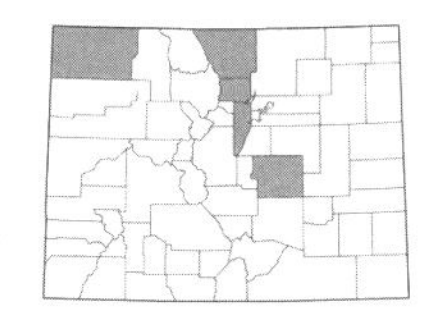

Listed as threatened under the Endangered Species Act.

Spiranthes romanzoffiana **Cham.**, HOODED LADY'S TRESSES. Plants 0.8–5 dm; *leaves* linear to oblanceolate or narrowly elliptic, 3–25 cm long; *spikes* tightly spiraled, the flowers diverging at about a 45-degree angle or higher from the rachis; *sepals* united at the base, 5.3–12 mm long; *petal* lip 4.8–10 mm long, with a broadly dilated tip. Locally common in moist meadows and forests, along streams and lake borders, and in bogs, 5500–11,000 ft. July–Sept. E/W. (Plate 56)

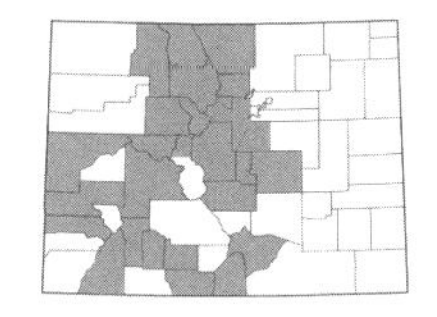

OROBANCHACEAE Vent. – BROOMRAPE FAMILY

Annual or perennial herbs, sometimes lacking chlorophyll, parasitic on the roots of other vascular plants, often turning black upon drying; *leaves* alternate, opposite, or basal, simple, exstipulate; *flowers* perfect or rarely imperfect, zygomorphic, strongly bilabiate, bracteate; *calyx* 2–5-lobed, the lobes subequal, persistent; *petals* 5-lobed, the tube often curved; *stamens* 4, epipetalous, didynamous; *pistil* 1; *ovary* superior, 2-carpellate, unilocular; *fruit* a capsule.

1a. Plants green, with chlorophyll...2

1b. Plants yellow to brownish, lacking chlorophyll...7

2a. Leaves opposite...3

2b. Leaves alternate; flowers unlike the above...4

3a. Flowers pink to purple or rarely white, with 2 yellow lines and reddish spots on the inside of the throat; leaves linear, entire; calyx 5-lobed, not inflated in fruit; flowers not subtended by bracts...***Agalinis***
3b. Flowers yellow, the upper lip of the corolla hooded and enclosing the anthers; leaves lance-triangular to ovate, crenate-serrate; calyx 4-lobed, becoming inflated in fruit; flowers subtended by sharply toothed bracts...***Rhinanthus***

4a. Leaves pinnatifid to bipinnatifid, the segments again toothed, or simple with crenate-serrate margins; anther sacs equal in size and similar in position; galea rounded at the top, sometimes prolonged into a beak; calyx 2- or 5-lobed...***Pedicularis***
4b. Leaves simple and entire or pinnately lobed with 3–7 linear, entire segments; anther sacs unequal in size or in different positions, the larger one attached by its middle and appearing terminal on the filament and the smaller one attached by its tip or absent; galea straight or rounded at the very tip, lacking a beak; calyx 2- or 4-lobed...5

5a. Perennial or rarely annuals; bracts and/or calyx more conspicuously colored than the corolla (appears that the plant has been dipped in paint), yellow, red, orange-red, pink, or purple; corolla green or yellowish-green and usually inconspicuous; galea usually longer than the lower lip of the corolla...***Castilleja***
5b. Annuals; bracts and calyx green; corolla yellow or purple and conspicuously colored; galea equal to or only slightly longer than the lower lip of the corolla...6

6a. Calyx cleft to the base or nearly so, forming 2 segments; stems paniculately branched from the base to near the middle...***Cordylanthus***
6b. Calyx 4-lobed; stems simple or branched above...***Orthocarpus***

7a. Calyx spathelike, deeply cleft below and the upper lip large and 4-toothed; plants yellow, the stems glabrous; flowers in a dense, bracteate spike; parasitic on *Quercus*, *Pinus*, and *Juniperus*...***Conopholis***
7b. Calyx regular, the lobes equal or nearly so; plants with some purplish tint, the stems viscid glandular-hairy; flowers solitary on long pedicels or in a dense spike-like inflorescence; parasitic on herbaceous plants (often Asteraceae species)...***Orobanche***

AGALINIS Raf. – FALSE FOXGLOVE
Annual herbs (ours), hemiparasitic; *leaves* opposite or nearly so, simple; *flowers* 5-merous, weakly zygomorphic; *calyx* 5-lobed; *stamens* 4, didynamous.

Agalinis tenuifolia **(Vahl) Raf. var. *parviflora* (Nutt.) Pennell,** SLENDERLEAF FALSE FOXGLOVE. Plants to 5 dm; *leaves* linear, entire, 3–7 cm × 1–2 mm; *calyx* 3.5–5.5 mm; *corolla* pink to purple or rarely white, with 2 yellow lines and reddish spots on the inside of the throat, 10–15 mm; *capsules* globose, 4–6 mm. Uncommon in moist places along creeks and the margins of ponds, ditches, and in wet meadows, 3500–6000 ft. Aug.–Oct. E.

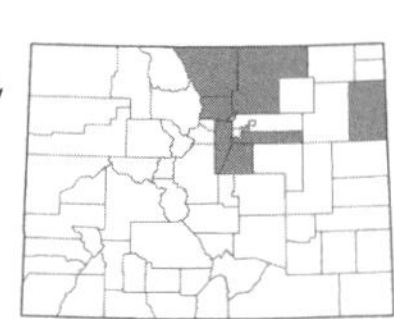

CASTILLEJA Mutis ex L. f. – INDIAN PAINTBRUSH
Perennial or occasionally annual herbs, hemiparasitic; *leaves* alternate, entire to pinnately dissected, sessile; *flowers* 4-merous, in a spike or spicate raceme, strongly bilabiate; *bracts* prominent, usually more conspicuously colored than the flowers, red, yellow, green, or purple; *calyx* cleft into 4 lobes, tubular; *corolla* usually inconspicuous, the upper lip hooded and beaklike, the lower lip with 3 short lobes; *stamens* 4, didynamous. (Heckard & Chuang, 1977)
1a. Annuals, densely glandular-hairy plants, usually with a single stem that is not decumbent at the base; bracts and calyx mostly green with just the upper ⅓–¼ of the bract red-tipped...***C. minor***
1b. Perennials, glabrous to hispid or tomentose, or rarely with some fine glandular hairs, the stems 1–several and often decumbent at the base; bracts and calyx unlike the above...2

2a. Bracts subtending the flowers entire or with a small pair of lateral lobes near the apex or above the middle, the middle lobe or entire lobe broadly rounded (more acute in hybrids); leaves all entire or sometimes the uppermost with a small pair of lateral lobes...3
2b. At least the lower bracts with 1–3 pairs of lateral lobes, usually cleft from the middle of the bract or below or if cleft from above the middle then the middle lobe acute and not rounded at the apex; leaves various...6

3a. Bracts orange-red; stems whitish-tomentose throughout; corolla (26) 29–50 mm long...***C. integra***
3b. Bracts yellow, white, purplish, rose, or crimson red; stems minutely hairy with small hairs but not whitish-tomentose; corolla 17–30 (36) mm long...4

4a. Bracts purplish, rose, or occasionally crimson red...***C. rhexiifolia***
4b. Bracts yellow or white, rarely streaked with red...5

5a. Plants 0.8–1.5 (2) dm; plants of the alpine, above 10,000 ft in elevation...***C. occidentalis***
5b. Plants (2) 2.5–5.5 (7) dm; plants of montane forests and meadows, occasionally above 11,000 ft in elevation...***C. sulphurea***

6a. Calyx more deeply cleft below than above, conspicuously colored, usually red but rarely yellow; corolla usually conspicuously projecting beyond the calyx, about 1.5–2 times the length of the calyx, the upper lip 10–24 mm long, the lower lip much reduced with short, incurved teeth; leaves linear to filiform, usually entire or sometimes a few of the upper with a pair of narrow lateral lobes...***C. linariifolia***
6b. Calyx cleft equally or nearly so below and above, or if cleft more deeply below than above then the upper lip of the corolla (galea) 5–8 mm long; corolla usually not 1.5–2 times the length of the calyx, if this long then the lower lip of the corolla prominent with flaring lobes; leaves various...7

7a. Lower lip of the corolla prominent, 5–6 mm long, with flaring lobes; corolla more colorful than the green bracts, yellow, greenish-yellow, purplish, or white, 35–55 mm long and conspicuously exserted beyond the bracts; calyx green or sometimes pinkish or purplish...***C. sessiliflora***
7b. Lower lip of the corolla much reduced with short, incurved teeth; corolla included or exserted beyond the bracts, usually green or sometimes yellowish, the bracts as or more colorful than the corolla and not entirely green...8

8a. Upper lip of the corolla (galea) 4–8 mm long; bracts yellow, rose-red, pink, purple, or sometimes orange-red, but never bright red...9
8b. Upper lip of the corolla (galea) (9) 10–20 mm long; bracts bright red, orange-red, or rarely yellow...12

9a. Stems, leaves, and inflorescence tomentose (drying grayish); bracts and sepals yellowish-green; inflorescence narrow and elongating; plants of the southern counties...***C. lineata***
9b. Stems and leaves glabrous, finely minutely hairy, or retrorsely to spreading-hispid, but not tomentose; sometimes the inflorescence whitish-tomentose; bracts and sepals yellow, pink, purple, rose-red, or sometimes orange-red; inflorescence short and broad or sometimes narrow and elongating; plants variously distributed...10

10a. Plants of sagebrush meadows, to 10,000 ft in elevation; stems retrorsely hispid; leaves entire below or with 1–2 pairs of lateral lobes; plants 1.5–5 dm; bracts yellow or sometimes orange-red...***C. flava***
10b. Plants of the alpine, above 10,000 ft in elevation; stems finely minutely hairy; leaves mostly entire, sometimes a few of the upper with 1–2 pairs of lateral lobes; plants 0.5–1.5 dm; bracts yellow, sometimes with reddish tips, or pink, purple, or rose-red...11

11a. Bracts rose-red, pink, or purple...***C. haydenii***
11b. Bracts yellow or sometimes the tips reddish...***C. puberula***

12a. Leaves mostly entire or a few of the upper with 1–2 pairs of lateral lobes; calyx more deeply cleft below than above; stems glabrous to pilose; plants found in forests, along streams, and in moist mountain meadows...***C. miniata***
12b. Leaves mostly with 1–2 pairs of lateral lobes; calyx cleft subequally above and below, or cleft deeper above than below; stems with spreading or reflexed hispid hairs; plants of sagebrush meadows or dry hillsides...13

13a. Stem leafy above the root crown; corolla usually included within the calyx tube or the galea only slightly exserted; plants 1.5–4 dm tall...***C. chromosa***
13b. Stem naked above the root crown; corolla usually conspicuously exserted from the calyx; plants 0.5–1.5 (2) dm tall...***C. scabrida* var. *scabrida***

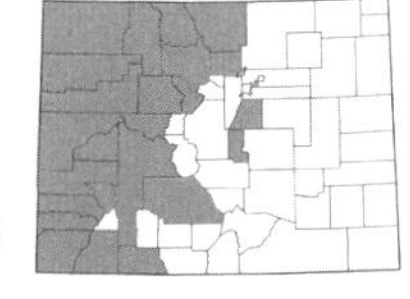

***Castilleja chromosa* A. Nelson**, RED DESERT PAINTBRUSH. Perennials, 1.5–4 dm; *leaves* linear to lanceolate, 2.5–7 cm long, with 1–2 (3) pairs of lateral lobes; *inflorescence* villous-hispid; *bracts* red to orange-red, occasionally yellowish, lanceolate with 1–3 pairs of rounded lateral lobes; *calyx* 15–27 mm long, the primary lobes more deeply cleft in back (6–12 mm) than in front (4–10 mm), the segments (1.5) 2–5 mm long, rounded; *corolla* 20–32 mm long, the galea 9–18 mm long (more or less half the corolla length), the lower lip much reduced with incurved teeth, the tube 10–13 mm long; *capsules* 9–15 (17) mm long. Common in sagebrush meadows, in pinyon-juniper woodlands, and on dry hillsides, 4800–9500 ft. May–July. E/W. (Plate 57)

Castilleja angustifolia (Nutt.) G. Don is not present in Colorado, although it is sometimes included as synonymous with *C. chromosa*. Specimens identified as this from Colorado are either *C. chromosa* or *C. scabrida*.

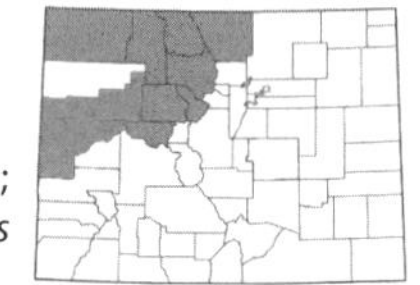

***Castilleja flava* S. Watson**, YELLOW INDIAN PAINTBRUSH. Perennials, 1.5–5 dm; *leaves* linear, 2.5–6.5 cm long, entire or with 1 (2) pairs of lateral lobes; *inflorescence* villous to tomentose; *bracts* usually yellow, sometimes orangish, with 1–2 pairs of lateral lobes; *calyx* 14–23 mm long, the primary lobes more deeply cleft in front (7–13 mm) than in back (4–12 mm), the segments 0.5–5 mm long; *corolla* 18–30 mm long, the galea 5–8 mm, the lower lip 1.5–3.5 mm, the tube 12–16 mm; *capsules* 8–15 mm long. Common in sagebrush and on dry hillsides, 6800–10,000 ft. June–Aug. E/W.

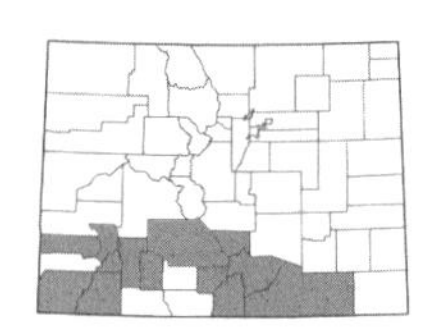

***Castilleja haydenii* (A. Gray) Cockerell**, HAYDEN'S INDIAN PAINTBRUSH. Perennials, 1–1.5 dm; *leaves* linear and entire or the upper with 1–2 pairs of spreading lobes, 2–8 cm long; *inflorescence* finely hairy; *bracts* red, rose-red, or purple, cleft; *calyx* ca. 20 mm long, about equally cleft above and below; *corolla* 20–25 mm long, the galea 7–8 mm long, lower lip 2–3 mm long. Locally common on alpine, 11,000–14,100 ft. July–Aug. E/W.

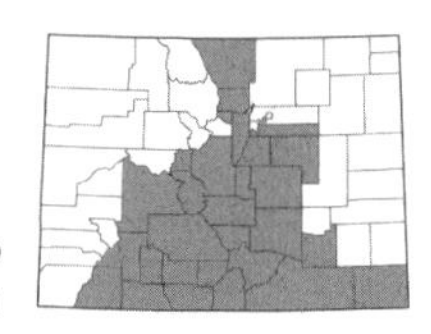

***Castilleja integra* A. Gray**, WHOLELEAF INDIAN PAINTBRUSH. Perennials, 1–3 (4.5) dm; *leaves* linear to narrowly lanceolate, 2–6 (8) cm long, entire, sometimes involute; *inflorescence* villous and finely glandular; *bracts* orange-red, entire or with a small pair of lateral lobes near the tip; *calyx* 23–35 (38) mm long, the primary lobes more deeply cleft in back (8–17 mm) than in front (6–16 mm), the segments 8–15 mm long; *corolla* (26) 29–50 mm long, the galea 10–18 mm long, the lower lip 1.3–2.8 mm long; *capsules* 12–16 mm long. Common on dry hillsides, in shortgrass prairie, and in grassy meadows, 5100–10,200 ft. April–July. E/W.

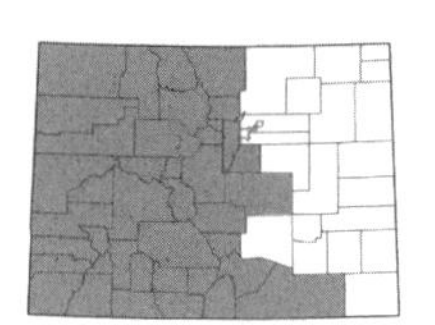

***Castilleja linariifolia* Benth.**, WYOMING PAINTBRUSH. Perennials, 3–8 (10) dm; *leaves* linear, 2–6 (8) cm long, entire or sometimes the upper with a pair of lateral lobes; *inflorescence* hispid or villous to glabrate; *bracts* red to reddish-orange, rarely yellowish, with a pair of lateral segments; *calyx* 18–30 mm long, often colored nearly throughout, the primary lobes unequally cleft (2–12 mm in back and 10–20 mm in front); *corolla* 24–44 mm long, the galea 10–24 mm long, the tube 11–23 mm long; *capsules* 9–13 mm long. Common in meadows, forests, on dry hillsides, or in sagebrush meadows, 4800–10,800 ft. June–Sept. E/W.

***Castilleja lineata* Greene**, NARROW INDIAN PAINTBRUSH. Perennials, 1–4 dm; *leaves* linear, 2–4 cm long, usually with 1–2 pairs of lateral lobes, tomentose; *inflorescence* tomentose, narrow and elongating; *bracts* yellowish-green, cleft at least 3 times; *calyx* 18–20 mm long, subequally cleft to the middle above and below, deeply cleft on the sides; *corolla* ca. 20 mm long, the galea 4–7 mm, shorter than the tube. Uncommon on dry hillsides and in grassy meadows, 6500–11,500 ft. July–Aug. E/W.

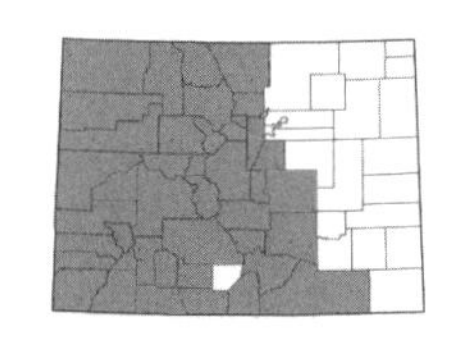

***Castilleja miniata* Douglas ex Hook.**, RED INDIAN PAINTBRUSH. Perennials, 2.5–7 dm; *leaves* linear to lanceolate, entire or rarely the uppermost cleft, 3–8 cm long; *inflorescence* villous, glandular-hairy; *bracts* red, red-orange, or rarely yellowish, with 1–2 pairs of lateral segments; *calyx* 20–30 mm long, the primary lobes more deeply cleft in front (9–17 mm) than in back (8–15 mm), the segments 2–8 mm long; *corolla* 25–45 mm long, the galea 12–20 mm, the tube 14–25 mm; *capsules* 10–13 mm long. Common in forests, along streams, and in moist mountain meadows, 6000–12,000 ft. June–Sept. E/W.

***Castilleja minor* (A. Gray) A. Gray**, LESSER INDIAN PAINTBRUSH. Annuals, 2–10 dm; *leaves* linear, entire, 3–8 (10) cm, densely glandular; *bracts* green with the upper part red; *calyx* 15–25 mm long, the primary lobes 6–15 mm long, subequally cleft; *corolla* 15–25 mm long, the galea 5–7 (10) mm long, the tube 10–16 (20) mm long; *capsules* 10–18 mm long. Uncommon in marshy places, seepages, and near springs, 4800–7000 ft. June–Aug. W.

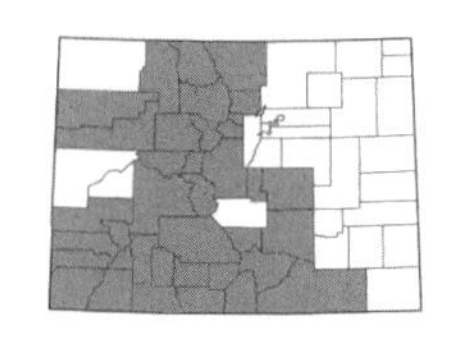

***Castilleja occidentalis* Torr.**, WESTERN INDIAN PAINTBRUSH. Perennials, 0.8–1.5 (2) dm; *leaves* linear, entire or sometimes the uppermost with a pair of lateral lobes, 2–4 (5.5) cm long; *inflorescence* viscid-villous; *bracts* yellow, rarely reddish, entire or with 1 pair of small lateral lobes near the tip; *calyx* (12) 15–20 mm long, the primary lobes more deeply cleft in front (6–11 mm) than in back (5–10 mm), the segments 1–3 (4.5) mm long; *corolla* 17–25 mm long, the galea 5–9 mm long, the tube 9–15 mm long; *capsules* 8–10 mm long. Common in alpine and mountain meadows, 10,000–14,000 ft. June–Sept. E/W. (Plate 57)

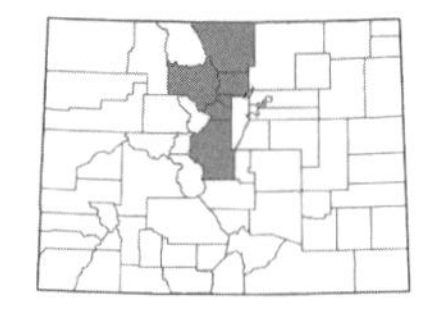

***Castilleja puberula* Rydb.**, SHORTFLOWER INDIAN PAINTBRUSH. Perennials, 0.8–1.5 dm; *leaves* linear and entire or the uppermost with a pair of lateral lobes, 2–3 cm long; *inflorescence* tomentose; *bracts* yellow or reddish, 3-lobed; *calyx* 10–15 mm long, cleft about halfway above and below or deeper below; *corolla* ca. 20 mm long, the galea 6–8 mm long, about half as long as the corolla tube. Locally common in the alpine, 10,200–13,000 ft. July–Aug. E/W. (Plate 57)

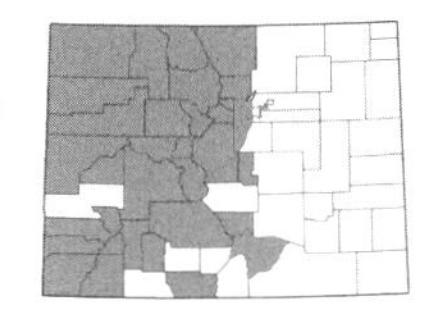

Castilleja rhexiifolia **Rydb.**, SPLITLEAF INDIAN PAINTBRUSH. Perennials, 2.5–6 (8) dm; *leaves* linear to lanceolate, entire, 3–7 cm long; *inflorescence* viscid-villous or rarely glabrate; *bracts* purplish, rose, or occasionally red, entire or with a pair of small lateral lobes near the tip; *calyx* 16–25 mm long, the primary lobes more deeply cleft in front (9–15 mm) than in back (8–11 mm); *corolla* 20–35 mm long, the galea 8–12 mm long, tube 13–24 mm long; *capsules* ca. 12 mm long. Common in forests, along streams, and in mountain meadows, often growing with *C. miniata* or *C. sulphurea*, 7500–13,000 ft. June–Aug. E/W. (Plate 57)

Castilleja rhexifolia often hybridizes or is found growing with *C. miniata*. They can be difficult if not impossible to separate. However, the bracts of *C. miniata* are usually crimson red while those of *C. rhexioflia* are rose or purple.

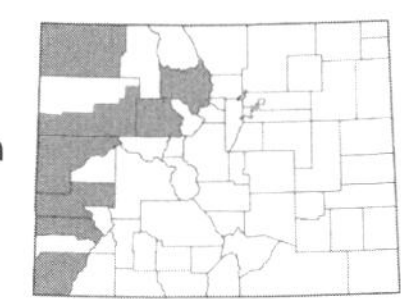

Castilleja scabrida **Eastw. var.** ***scabrida***, ROUGH INDIAN PAINTBRUSH. Perennials, 0.5–1.5 (2); *leaves* linear to lanceolate, entire or the upper with a pair of lateral lobes, 1.5–5 cm long; *inflorescence* villous; bracts red or reddish-orange; *calyx* 19–30 mm long, the primary lobes more deeply cleft in back (8–15 mm) than in front (6–12 mm); *corolla* 25–45 mm long, the galea 10–20 mm long, the tube 12–25 mm long; *capsules* 9–12 mm long. Found in sagebrush meadows and pinyon-juniper woodlands, rocky or sandy soil, and on rocky ledges, 4700–8000 ft. May–June. W.

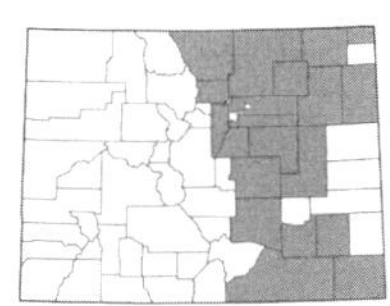

Castilleja sessiliflora **Pursh**, DOWNY PAINTEDCUP. Perennials, 1–3 dm; *leaves* linear, entire or the upper with a pair of narrow segments; *inflorescence* villous; *bracts* yellow, greenish-yellow, purplish, or white, with a pair of lateral segments; *calyx* 25–40 mm long, with long, linear segments 8–15 mm long; *corolla* 35–55 mm long, purplish to yellow or white, prominently exserted past the bracts, the galea 9–12 mm long. Found in dry grasslands, prairie, and on rocky outcroppings, 3500–7000 ft. May–June. E. (Plate 57)

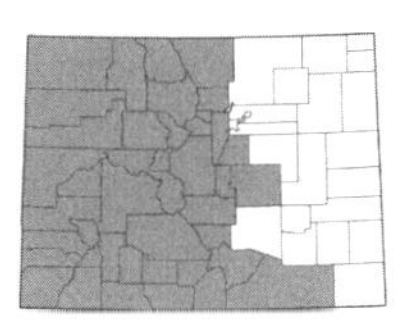

Castilleja sulphurea **Rydb.**, SULPHUR INDIAN PAINTBRUSH. Perennials, (2) 2.5–5.5 (7) dm; *leaves* lanceolate, entire, 2–6 (8) cm long; *inflorescence* villous and glandular; *bracts* pale yellow, entire or with a pair of small lateral lobes near the tip; *calyx* 13–25 (28) mm long, the primary lobes more deeply cleft in front (6–13 mm) than in back (5–11 mm); *corolla* 18–30 mm long, the galea 6–12 mm long, the tube 10–20 mm long; *capsules* 9–11 mm long. Common in forests, along streams, and in mountain meadows, often growing with *C. rhexifolia* or *C. miniata*, 6400–11,500 ft. May–Sept. E/W.

CONOPHOLIS Wallr. – CANCER ROOT

Plants yellowish, glabrous; *flowers* conspicuously bracted, in a dense spike-like inflorescence; *calyx* spathelike, deeply cleft on the lower side; *corolla* whitish, strongly bilabiate, the upper lip notched and arching and the lower lip 3-lobed and shorter.

Conopholis alpina **Liebm. var.** ***mexicana*** **(A. Gray ex S. Watson) Haynes**, MEXICAN CANCER ROOT. Uncommon, parasitic on *Quercus*, *Pinus*, and *Juniperus*, known from Las Animas Co. near the New Mexico border on Barela Mesa, 8500–8600 ft. May–June. E.

CORDYLANTHUS Nutt. ex Benth. – BIRD'S BEAK

Annual herbs; *leaves* alternate, deeply 3- to 5-parted or entire; *flowers* in a spike or sometimes solitary, with foliose bracts; *calyx* cleft to the base or nearly so; *corolla* bilabiate, yellow to purple; *stamens* 2 or 4.

1a. Outer bracts subtending each individual flower and floral bract; plants grayish-hairy; calyx 1–1.7 cm long…***C. ramosus***

1b. Outer bracts only subtending the spike, not each individual flower; plants glabrous to glandular or rarely minutely hairy with retrorse hairs; calyx 1.5–2.7 cm long…2

2a. Inner bract pinnately 3–7-divided; calyx cleft 1.2- 3 mm deep; plants glandular…***C. kingii***

2b. Inner bract entire or rarely pinnately 3–5-divided; calyx cleft 0.5–1 mm deep; plants glabrous to glandular…***C. wrightii***

Cordylanthus kingii **S. Watson**, KING'S BIRD'S BEAK. Plants 0.4–2.5 (4) dm, glandular; *leaves* 1.5–4 cm long, entire to pinnately lobed into filiform segments; *bracts* foliaceous, the inner bract pinnately 3–7-divided; *calyx* 1.5–2.7 cm long, cleft 1.2–3 mm deep; *corolla* violet to purple, 1.5–3 cm long. Found in dry shrubland communities. Not known from Colorado but occurs close to the Utah-Colorado border. June–Sept.

Cordylanthus ramosus **Nutt. ex Benth.**, BUSHY BIRD'S BEAK. Plants 0.8–3 (3.5) dm, grayish-hairy; *leaves* 1–4.5 cm long, entire to pinnatifid into filiform lobes; *bracts* foliaceous, 5–7-lobed, subtending each flower; *calyx* 1–1.7 cm long; *corolla* yellow or the lower lip purplish, 10–20 mm long. Found in sagebrush, rabbitbrush, and on sandy slopes, 5500–8500 ft. June–Aug. E/W.

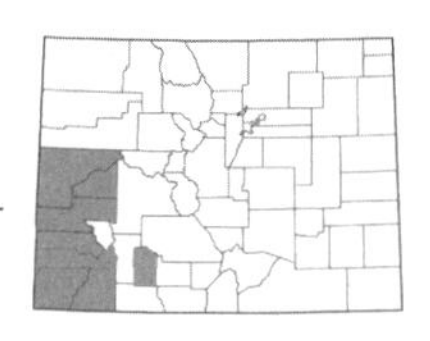

***Cordylanthus wrightii* A. Gray**, Wright's bird's beak. Plants 1.3–5 (12) dm, glabrous to glandular; *leaves* 1–3 cm long, entire or 3–5-lobed with filiform segments; *inflorescence* a spike or rarely solitary; *bracts* foliaceous, the inner bract entire or rarely pinnately 3–5-divided; *calyx* 16–20 mm long, cleft 0.5–1 mm deep; *corolla* yellow to purple, 15–27 mm long. Found in sagebrush, pinyon-juniper, and on sandy slopes, 4600–8000 ft. July–Sept. W.

OROBANCHE L. – broomrape

Perennials, more or less glandular-hairy throughout, the stem purplish to brown or yellowish-white; *leaves* scale-like; *flowers* solitary or in terminal spike-like inflorescences; *calyx* 5-lobed; *corolla* yellowish, reddish, or purplish, bilabiate, the tube slightly curved.

1a. Flowers 1–10 on pedicels 3–20 cm long; bracts absent below the calyx...2
1b. Flowers numerous in a dense spicate inflorescence, the lower ones sometimes on pedicels less than 3 cm long; 1–2 linear bracts subtending the calyx...3

2a. Flowers 1–3; stem leaves glabrous, 1–3 and overlapping at the top of the stem; calyx lobes longer than the calyx tube...***O. uniflora***
2b. Flowers 4–10; stem leaves glandular-hairy, usually 4 or more and distributed along the stem, not overlapping; calyx lobes shorter than the calyx tube...***O. fasciculata***

3a. Corolla lobes triangular-acute at the tips; corolla with glandular and nonglandular hairs; parasitic on annual members of the Asteraceae...***O. riparia***
3b. Corolla lobes rounded-obtuse at the tips; corolla with glandular hairs except on the inner lobe surface; parasitic on perennial members of the Asteraceae...4

4a. Corolla yellow to rose or pale purple, the tube conspicuously lighter than the lobes in color; style persistent in fruit until dehiscence; calyx 10–19 mm long, usually surpassing mature fruit in length...***O. multiflora***
4b. Corolla light to deep purple but rarely yellow, the upper lobes darker in color than the lower lobes; style deciduous on fruit before dehiscence; calyx 7–14 mm long, shorter than or equaling mature fruit in length...***O. ludoviciana***

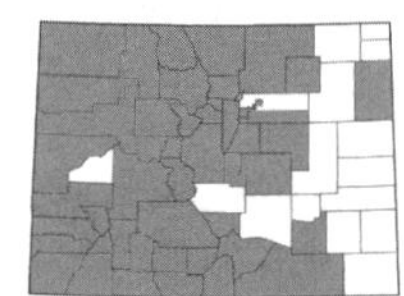

***Orobanche fasciculata* Nutt.**, clustered broomrape. [*Aphyllon fasciculatum* (Nutt.) Torr. & A. Gray]. Plants 7–25 cm; *inflorescence* of 4–10 flowers; *pedicels* 3–20 cm; *bracts* absent; *calyx* 8–11 mm long, with lobes 3–5 mm long, broadly triangular-acuminate; *corolla* 18–30 mm long, reddish with purple lines and yellow patches between the lobes of the lower lip, the lobes 2–6 mm long, broadly rounded; *capsules* 10–13 mm long. Common, usually parasitic on sagebrush, 3400–9500 ft. May–Aug. E/W. (Plate 57)

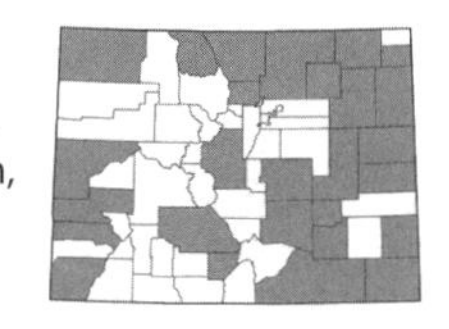

***Orobanche ludoviciana* Nutt.**, Louisiana broomrape. Plants 8–25 cm; *inflorescence* spicate, flowers numerous, sessile; *bracts* 1–2, linear, 4–8 mm long, subtending each flower; *calyx* 7–14 mm long, viscid; *corolla* 14–20 mm long, the lobes rounded-obtuse, light to deep purple or rarely yellowish, glandular; *style* deciduous in fruit. Found in sandy soil on the eastern plains and intermountain parks, parasitic on members of the Asteraceae (*Artemisia, Grindelia, Heterotheca*), 3500–8000 ft. April–Aug. E/W.

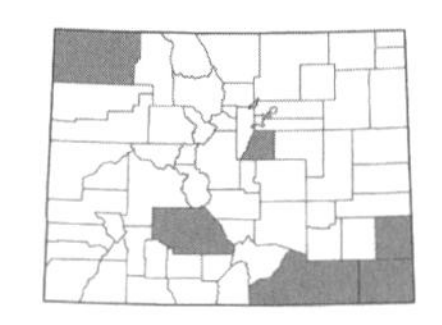

***Orobanche multiflora* Nutt.**, manyflower broomrape. Plants 5–30 cm; *inflorescence* spicate, flowers numerous, sessile; *bracts* 1–2, subtending each flower; *calyx* 10–19 mm long; *corolla* 13–30 mm long, with rounded-obtuse lobes 3–5 mm long, yellow to rose or pale purple, glandular; *style* persistent in fruit. Parasitic on members of the Asteraceae (*Artemisia, Heterotheca, Hymenopappus*), 4000–6500 ft. June–Sept. E/W.

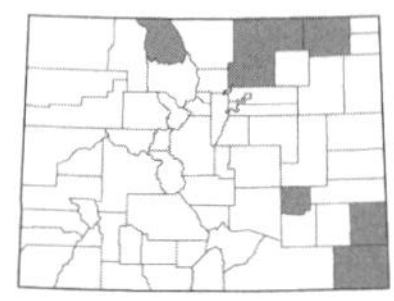

***Orobanche riparia* L.T. Collins**, riparian broomrape. Plants 5–25 cm; *inflorescence* racemose, with numerous flowers; *bracts* 1–2, subtending each flower; *calyx* 7–11 mm long; *corolla* 15–22 mm long, the lobes triangular-acute, 4–5 mm long, yellow to rose or purple, with glandular and eglandular hairs or glabrate. Found in sandy soil of stream banks and on sandbars, parasitic on annual Asteraceae (*Ambrosia trifida, A. artemisiifolia, Xanthium, Dicoria*), 3500–8500 ft. E.

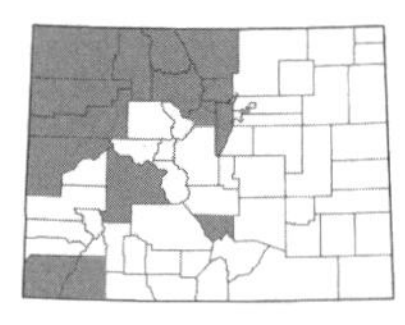

***Orobanche uniflora* L.**, naked broomrape. [*Aphyllon uniflorum* (L.) A. Gray]. Plants 5–15 (20) cm; *inflorescence* of 1–3 flowers; *pedicels* 3–15 cm long; *bracts* absent; *calyx* 6–12 mm long, with lobes 4–9 mm long; *corolla* 15–25 (35) mm long, with lobes 2–7 mm long, rounded and finely fringed, yellowish to purple; *capsules* 8–10 mm long. Uncommon in forests and open meadows, often in somewhat moist places, 5200–12,500 ft. May–July. E/W.

ORTHOCARPUS Nutt. – OWL'S CLOVER

Annual herbs, hemiparasitic; *leaves* alternate, sessile, entire to pinantifid; *flowers* in a bracteate spike; *calyx* 4-lobed; *corolla* bilabiate, yellow, white, or purple, the upper lip hooded and beaklike, the lower lip saccate; *stamens* 4, didynamous. (Chuang & Heckard, 1992)

1a. Flowers yellow, 9–12 (14) mm long; leaves mostly linear and entire with only the uppermost sometimes 3-cleft...***O. luteus***
1b. Flowers purple and white, 15–20 mm long; leaves mostly 3-cleft or sometimes only the lowermost linear and entire...***O. purpureoalbus***

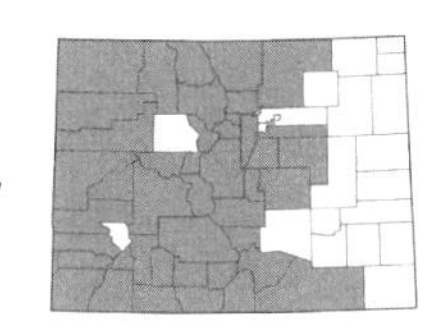

***Orthocarpus luteus* Nutt.**, YELLOW OWL'S CLOVER. Plants 1–4 dm; *leaves* 1–4 cm long, linear to narrowly lanceolate, usually entire; *bracts* 10–15 mm long, 3-5-lobed; *calyx* 5–8 mm long; *corolla* yellow, 9–12 (14) mm long, the galea 2.5–4 mm long; *capsules* 4–7 mm long. Found in meadows, dry grasslands, open woodlands, and in disturbed areas especially around old mines, often with sagebrush, 5500–10,800 ft. June–September. E/W.

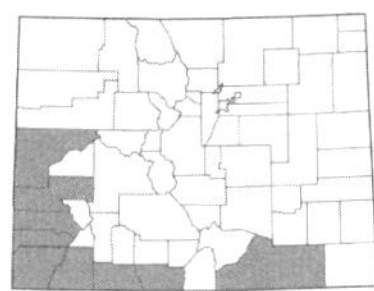

***Orthocarpus purpureoalbus* A. Gray ex S. Watson**, PURPLE-WHITE OWL'S CLOVER. Plants 1–4 dm; *leaves* 1.5–4 cm long, filiform to narrowly lanceolate, mostly 3-cleft or sometimes the lowermost entire; *bracts* 10–20 mm long, 3-lobed; *calyx* 6–10 mm long; *corolla* purple and white, 15–20 mm long; *capsules* 6–7 mm long. Found in ponderosa or pinyon-juniper woodlands and sagebrush meadows, 6500–9000 ft. June–September. E/W.

PEDICULARIS L. – LOUSEWORT

Perennial herbs; *leaves* alternate, opposite, or basal, toothed to bipinnatifid; *flowers* in a spike or spicate raceme; *calyx* 2-5-lobed, accrescent; *corolla* strongly bilabiate, yellow, white, purple, or red, the upper lip hooded, the lower lip 3-lobed; *stamens* 4, didynamous.

1a. Upper lip of the corolla (galea) with a prolonged, slender, upcurved beak 7–15 mm long (resembling an elephant's head with trunk); flowers 6–8 mm long (not including the beak), usually violet to purple but rarely white; calyx 3.5–5.5 mm long, with prominent veins, glabrous outside and white-ciliate inside...***P. groenlandica***
1b. Galea beakless or the beak smaller (the flower not resembling an elephant's head with trunk); flowers various...2

2a. Flowers rose, red, or purple throughout...3
2b. Flowers yellow, cream, or white, sometimes with a purple or reddish galea, or yellowish with reddish or purplish veins...5

3a. Leaves simple, not pinnately divided, linear to narrowly lanceolate, with serrate to crenate margins; stems with longitudinal lines of hairs...***P. crenulata***
3b. Leaves pinnatifid not quite to the midrib; stems glabrous...4

4a. Stems short, 0.4–0.7 (1) dm long, surpassed by the leaves; racemes usually surpassed by the leaves; calyx 15–22 mm long; flowers pale violet with dark tips on the galea and lobes of the lower lip, the corolla 30–42 mm long...***P. centranthera***
4b. Stems longer than the leaves; racemes longer than the leaves; calyx 8–10 mm long; flowers lacking dark tips on the galea and lobes of the lower lip, the corolla 15–20 mm long...***P. sudetica* ssp. *scopulorum***

5a. Leaves simple, not pinnately divided, linear to narrowly lanceolate, with serrate to crenate margins; galea prolonged into a slender, downcurved beak 5–7.5 mm long; flowers 9–15 mm long (not including the beak), in a relatively loose raceme...***P. racemosa* var. *alba***
5b. Leaves pinnatifid to bipinnatifid at least halfway to the midrib; galea beakless or if beaked then the beak short, straight, and about 2 mm long; flowers 16–36 mm long, in a rather dense spike or spicate raceme...6

6a. Bracts in the inflorescence pinnately 3–7-lobed; galea prolonged into a short beak 2 mm long; calyx with dark longitudinal lines; leaves mostly basal, pinnatifid to the midrib or nearly so...***P. parryi***
6b. Bracts in the inflorescence entire or serrate but not lobed; galea beakless; calyx usually lacking dark longitudinal lines; leaves various but if mostly basal then only pinnatifid to ⅔ the way to the midrib...7

7a. Calyx lobes 2; leaves pinnatifid to bipinnatifid with the lobes rather shallow, extending about halfway to ⅔ the way to the midrib; stems hairy; plants 1–3 (4) dm tall...***P. canadensis* var. *fluviatilis***
7b. Calyx lobes 5; leaves pinnatifid with the lobes extending to the midrib or nearly so, or bipinnatifid; stems glabrous; plants 3–12 dm tall...8

8a. Lower lip of the corolla 4–6 (7) mm long; calyx 7–10 mm long; flowers yellow or cream, without reddish or purplish veins...***P. bracteosa* var. *paysoniana***
8b. Lower lip of the corolla 7–12 mm long; calyx 10–12 (16) mm long; flowers yellow or cream and usually with darker, reddish or purplish veins, these usually drying brown when drying but remaining conspicuous...***P. procera***

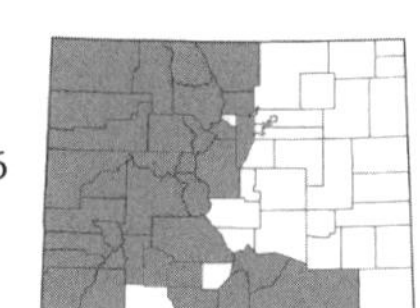

***Pedicularis bracteosa* Benth. var. *paysoniana* (Pennell) Cronquist**, Payson's lousewort. [*P. paysoniana* Pennell]. Plants 3–10 dm; *leaves* 7–12 (16) cm long, pinnatifid, the segments serrate to deeply incised; *inflorescence* a densely flowered spicate-raceme, 5–15 (45) cm long; *bracts* 8–16 mm long, simple, entire to toothed distally; *calyx* 7–10 mm long; *corolla* yellow or cream, 20–26 mm long, the galea 9–12 mm, the lower lip 4–6 (7) mm long; *capsules* 10–12 mm long. Common in moist spruce-fir forests, moist meadows, and along streams, 8000–12,500 ft. June–Aug. E/W.

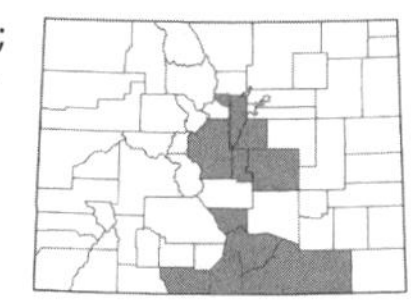

***Pedicularis canadensis* L. var. *fluviatilis* (A. Heller) Macbr.**, Canadian lousewort. Plants 1–3 (4) dm; *leaves* 4–15 cm long, pinnatifid to bipinnatifid, the lobes rather shallow, extending about halfway to ⅔ to the midrib, the ultimate segments crenate; *inflorescence* a dense spike 4–15 (30) cm long; *bracts* crenate to toothed; *calyx* 7–9 mm long; *corolla* pale yellow, sometimes with a purple galea or rarely all purple, 18–25 mm long, the galea 11–15 mm long and the lower lip 7–10 mm long; *capsules* 10–17 mm long. Found in moist meadows and marshes, moist forests, and along streams, 6400–10,500 ft. June–July. E.

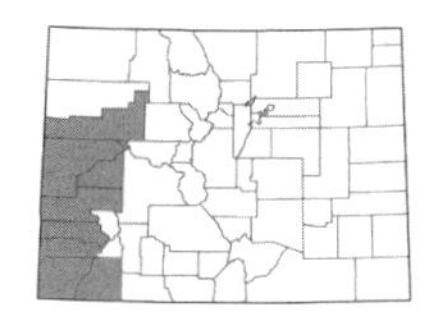

***Pedicularis centranthera* A. Gray**, dwarf lousewort. Plants 0.4–0.7 (1) dm; *leaves* 6–15 cm (including the petiole), the lowermost linear and entire, the upper pinnatifid, crenate-dentate, cartilaginous-margined; *inflorescence* a spicate-raceme, usually surpassed by the leaves; *calyx* 15–22 mm long; *corolla* violet, 30–42 mm long, the galea 10–18 mm long and the lower lip 9–15 mm long; *capsules* 10–13 mm long. Found on dry slopes, often in pinyon-juniper, 5500–8000 ft. April–May. W.

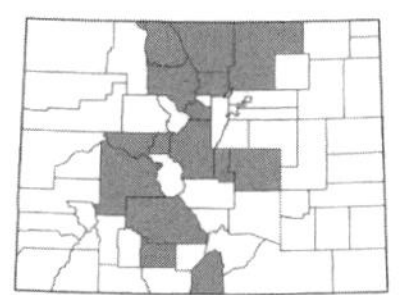

***Pedicularis crenulata* Benth.**, meadow lousewort. Plants 1.5–3.5 (5) dm; *leaves* 3–7 (9) cm long, linear to lanceolate, simple, serrate to crenate, often cartilaginous-margined; *inflorescence* a spicate-raceme, 2–10 cm long; *bracts* foliaceous; *calyx* 8–12 mm long; *corolla* rose, red, or purplish, 20–26 mm long, the galea 11–15 mm long and lower lip 7–12 mm long; *capsules* 10–20 mm long. Locally common in moist meadows and marshes, 7200–10,500 ft. June–Aug. E/W.

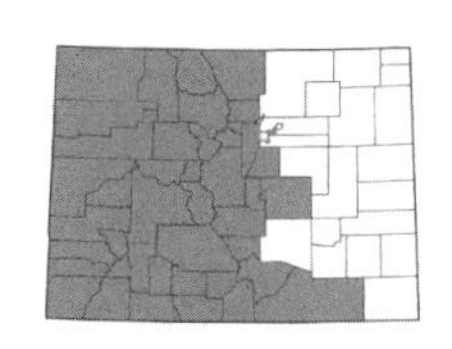

***Pedicularis groenlandica* Retz.**, elephant's head. Plants (1) 2–5 (7) dm; *leaves* 6–20 (25) cm long, pinnatifid, the segments toothed to crenate, sometimes cartilaginous-margined; *inflorescence* a dense spike 4–15 (25) cm long; *bracts* pinnately divided to entire above; *calyx* 3.5–5.5 mm long, prominently veined; *corolla* pink, purple, or rarely white, 6–8 mm long (not including beak), the galea 2–3.2 mm long, prolonged into an elongate, upcurved beak 7–15 mm long; *capsules* 7–9 mm long. Common in moist meadows, marshes, and along streams and creeks, 7500–13,000 ft. June–Aug. E/W. (Plate 57)

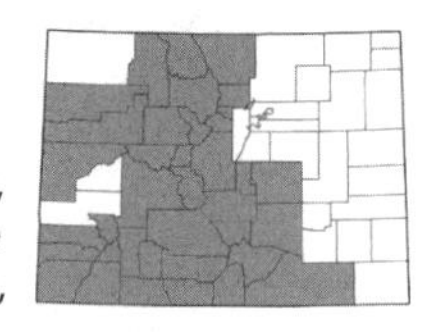

***Pedicularis parryi* A. Gray**, Parry's lousewort. Plants 1–4 (6) dm; *leaves* 3–15 (20) cm long, pinnatifid, the segments toothed and cartilaginous-margined; *inflorescence* a spicate raceme 5–25 cm long; *bracts* pinnately 3–7-lobed; *calyx* 7.5–10 (12) mm long; *corolla* pale yellow to white, 16–22 mm, the galea 6–10 (12) mm long, prolonged into a short, straight beak ca. 2 mm long, the lower lip 3–8 mm long; *capsules* 10–14 mm long. Common in the alpine and mountain meadows, often found in dry places, 8500–13,200 ft. July–Aug. E/W.

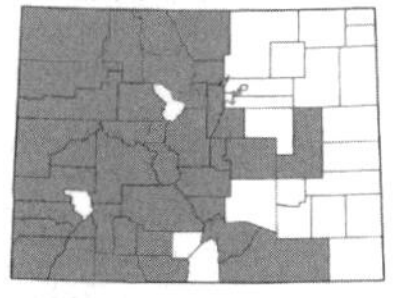

***Pedicularis procera* A. Gray**, giant lousewort. [*P. grayi* A. Nelson]. Plants 4–13 dm; *leaves* (10) 15–30 cm long, pinnatifid to bipinnatifid, the segments toothed; *inflorescence* a spicate raceme 10–35 (65) cm long; *bracts* linear, entire to toothed distally; *calyx* 10–12 (16) mm long; *corolla* pale yellow or cream with reddish or purplish veins, drying brown, 25–36 mm long, the galea 9–16 mm long and the lower lip 7–12 mm long; *capsules* 10–16 mm long. Common in moist forests and along streams, 6900–11,500 ft. June–Aug. E/W.

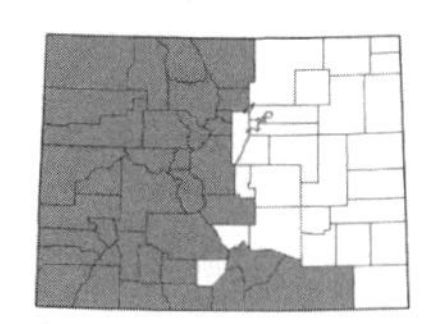

***Pedicularis racemosa* Douglas ex Benth. var. *alba* Pennell**, sickletop lousewort. Plants 1.5–5 dm; *leaves* (3) 4–8 (10) cm long, simple, lanceolate, serrate to crenate; *inflorescence* a loose raceme; *bracts* foliaceous below and reduced above; *calyx* 5–7.5 mm long; *corolla* pale yellow or white, 9–15 mm long, the galea 5–7.5 (9) mm long, prolonged into a downcurved beak 5–7.5 mm long; *capsules* 10–16 mm long. Common in spruce-fir forests and moist meadows, 7200–13,000 ft. June–Aug. E/W. (Plate 57)

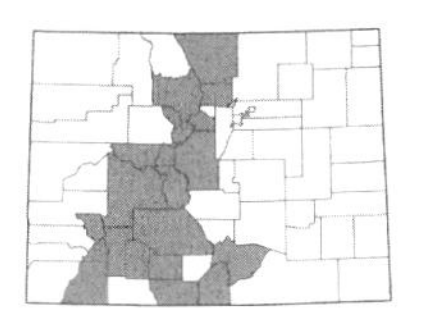

Pedicularis sudetica **Willd. ssp.** ***scopulorum*** **(A. Gray) Hultén**, SUDETIC LOUSEWORT. Plants 3–9 dm; *leaves* 3–8 cm long, pinnatifid nearly to the midrib, toothed; *inflorescence* spicate; *bracts* foliaceous below; *calyx* 8–10 mm long, arachnoid-hairy; *corolla* rose to purple, 15–20 mm long, the galea ca. 10 mm long. Found in bogs and marshes, moist meadows, and the alpine, 10,000–13,500 ft. June–Aug. E/W. (Plate 57)

RHINANTHUS L. – YELLOW RATTLE

Annual herbs; *leaves* opposite, sessile, toothed; *flowers* in a bracteate, spicate raceme; *calyx* 4-lobed, accrescent; *corolla* bilabiate, yellow, the upper lip hooded, the lower lip 3-lobed; *stamens* 4, didynamous.

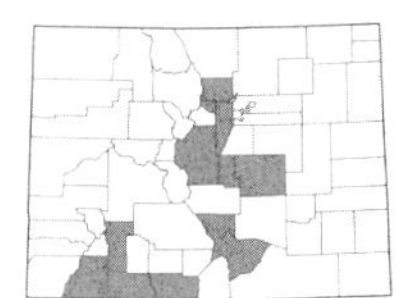

Rhinanthus minor **L. ssp.** ***minor***, LITTLE YELLOW RATTLE. Plants 1.5–8 dm; *leaves* ovate to lance-triangular, 2–6 cm × 4–15 mm, with crenate-serrate margins; *flowers* subtended by sharply toothed bracts; *calyx* 10–17 mm long; *corolla* 9–15 mm long. Found in moist meadows and fens, 7500–9200 ft. June–Aug. E/W.

OXALIDACEAE R. Br. – WOOD-SORREL FAMILY

Annual or perennial herbs; *leaves* alternate or basal, ternately compound, the leaflets obcordate and folding together at night, exstipulate; *flowers* perfect, actinomorphic; *sepals* 5, distinct; *petals* 5 or lacking in cleistogamous flowers, distinct; *stamens* 10, equal or unequal, the filaments basally connate; *pistil* 1; *styles* 5; *ovary* superior, 5-carpellate; *fruit* a loculicidal capsule.

OXALIS L. – WOOD-SORREL

With characteristics of the family. (Denton, 1973; Nesom, 2009, 2009; Wiegand, 1925)

1a. Flowers purple; leaves all basal; plants from scaly bulbs...***O. violacea***
1b. Flowers yellow; leaves present on the stem; plants from a taproot or rhizomes...2

2a. Seeds with brown transverse ridges; leaves and stem purplish-tinged; stipules conspicuous; plants trailing and usually rooting at the nodes...***O. corniculata***
2b. Seeds with white or grayish transverse ridges; leaves and stem green or purplish-tinged; stipules greatly reduced or inconspicuous; plants erect or trailing...3

3a. Capsules densely appressed-hairy...***O. dillenii***
3b. Capsules glabrous or with a few non-appressed, septate hairs...***O. stricta***

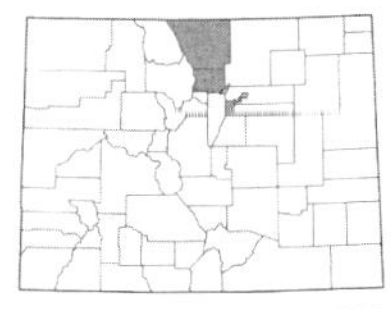

Oxalis corniculata **L.**, CREEPING YELLOW WOOD-SORREL. Plants to 5 dm, usually trailing and rooting at the nodes, with appressed or spreading hairs; *leaves* cauline, purplish-tinged; *stipules* conspicuous; *sepals* to 4.5 mm long; *petals* yellow, to 8 mm long; *capsules* 6–25 mm long; *seeds* with brown transverse ridges. Found as a weed in lawns and gardens, 4800–5200 ft. May–Oct. E. Introduced.

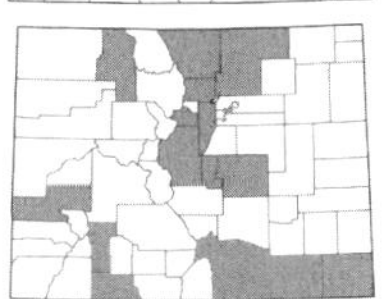

Oxalis dillenii **Jacq.**, SLENDER YELLOW WOOD-SORREL. Plants 1–2.5 dm, usually densely hairy; *leaves* cauline; *stipules* inconspicuous; *sepals* 3–7 mm long; *petals* yellow, 8–10 mm long; *capsules* 10–25 mm long, hairy; *seeds* with conspicuous white transverse ridges. Found in disturbed places, meadows, and open slopes, 4000–7600 ft. Mar.–Oct. E/W.

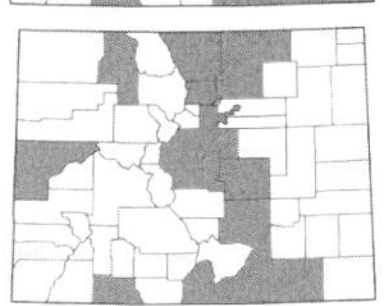

Oxalis stricta **L.**, YELLOW WOOD-SORREL. [*O. europaea* Jordan]. Plants to 5 dm, erect or trailing, glabrous to densely hairy; *leaves* cauline, sometimes purplish; *stipules* inconspicuous; *sepals* 3–5 mm long; *petals* yellow to orange-yellow, to 11 mm long; *capsules* 5–15 mm long, glabrous or with a few septate hairs; *seeds* with white transverse ridges. Uncommon or perhaps not often collected, found in gardens and disturbed places, 4500–5200 ft. April–Oct. E/W. Introduced.

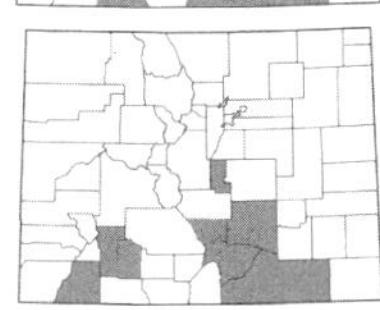

Oxalis violacea **L.**, VIOLET WOOD-SORREL. Acaulescent, rhizomatous, with scaly bulbs; *sepals* 4–6 mm long; *petals* purple or rarely white, 10–20 mm long; *capsules* 4–6 mm long; *seeds* reticulate. Uncommon in moist, shaded places, 7000–9000 ft. June–Aug. E/W. (Plate 57)

PAPAVERACEAE Juss. – POPPY FAMILY

Herbs, often with brightly colored sap; *leaves* alternate, opposite, or basal, entire to divided; *flowers* perfect, actinomorphic; *sepals* 2–3, distinct or connate, caducous; *petals* 4–12 or absent, distinct, white, yellow, red, purple, or other colors; *stamens* few to many, the filaments distinct; *pistil* 1; *stigmas* the same number as the carpels, usually sessile; *style* 1 or absent; *ovary* superior, 2–18 (22)-carpellate; *fruit* a poricidal or valvate capsule. (Kiger, 1997)

1a. Leaves, sepals, and capsules prickly; petals white...***Argemone***
1b. Leaves, sepals, and capsules not prickly, glabrous to hairy; petals red, yellow, orange, or rarely white...2

2a. Leaves glabrous and glaucous, deeply ternately dissected into narrowly linear divisions; sepals connate; petals orange...***Eschscholzia***
2b. Leaves glabrous and glaucous to hairy, pinnately dissected or lobed into broader divisions; sepals distinct; petals red, yellow, white, or orange...3

3a. Inflorescences umbelliform; capsules dehiscing from the base; flowers bright yellow...***Chelidonium***
3b. Inflorescences not umbelliform; capsules poricidal or dehiscing from the apex; flowers variously colored...4

4a. Stigmas velvety, 4–18, radiating on a sessile disc; fruit a poricidal capsule; plants with with white, orange, or red sap...***Papaver***
4b. Stigmas 2-lobed, not radiating on a sessile disc; fruit a 2-valved capsule; plants with yellow sap...***Glaucium***

ARGEMONE L. – PRICKLY-POPPY

Annual or perennial herbs, with yellow or orange sap; *leaves* alternate (cauline) and in basal rosettes, pinnately lobed, the divisions spinose-tipped, sessile; *sepals* 2 or 3, horn-tipped; *petals* 6, white, showy; *stamens* 20–many; *ovary* 3–7-carpellate; *fruit* an erect, 3–7 valved, prickly or rarely unarmed capsule. (Ownbey, 1958)

1a. Stems densely prickly, sometimes also hispid; leaves prickly and/or hispid on and between the veins; capsules densely prickly, obscuring the surface...2
1b. Stems sparingly prickly; leaves prickly only on the veins; capsules prickly but the surface clearly visible...3

2a. Largest prickles on the capsules simple; sepal horns 4–7 mm long; distal leaves not clasping...***A. hispida***
2b. Largest prickles on the capsules branched, with smaller prickles arising from their bases; sepal horns 8–15 mm long; distal leaves clasping...***A. squarrosa* ssp. *squarrosa***

3a. Stamens 150 or more; leaves somewhat succulent but not leathery; capsules 35–50 mm long (including the stigma and excluding prickles)...***A. polyanthemos***
3b. Stamens 100–120; leaves thick and leathery; capsules 25–35 mm long (including the stigma and excluding prickles)...***A. corymbosa* ssp. *arenicola***

***Argemone corymbosa* Greene ssp. *arenicola* G.B. Ownbey**, MOJAVE PRICKLY-POPPY. Plants 2–6 dm; *stems* sparingly prickly; *leaves* 3–16 cm long, prickly only on the veins, succulent and leathery; *flowers* (3) 4–6 (8) cm wide; *stamens* 100–120; *capsules* 25–35 mm long, prickly. Uncommon in sandy soil, 4500–4600 ft. May–June. W.

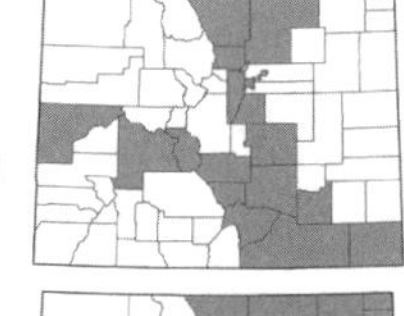

***Argemone hispida* A. Gray**, ROUGH PRICKLY-POPPY. Plants 2–6 dm; *stems* densely prickly and hispid; *leaves* 6–12 cm long, prickly and hispid between the veins, the distal leaves not clasping; *flowers* 4–10 cm wide; *capsules* 30–40 mm long, densely prickly. Common on dry, rocky slopes and along roadsides, 4900–7500 ft. May–July. E/W.

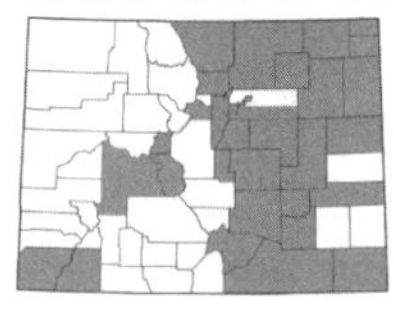

***Argemone polyanthemos* (Fedde) G.B. Ownbey**, CRESTED PRICKLY-POPPY. Plants 0.5–5 dm; *stems* sparingly prickly; *leaves* 7–20 (25) cm long, prickly only on the veins, glaucous and somewhat succulent; *flowers* (3) 5–10 cm wide; *stamens* 150+; *capsules* 35–50 mm long, prickly. Common in sandy soil of dry grasslands and open slopes, often found along roadsides, 3700–8000 ft. June–Aug. (Sept.). E/W.

Argemone squarrosa* Greene ssp. *squarrosa, HEDGEHOG PRICKLY-POPPY. Plants 3–12 dm; *stems* densely prickly and hispid; *leaves* 6–12 cm long, prickly and hispid between the veins, the distal leaves clasping; *flowers* 4–10 cm wide; *capsules* 30–40 mm long, densely prickly. Uncommon in sandy soil of the southeastern plains, 3800–4400 ft. June–July. E. (Plate 57)

CHELIDONIUM L. – CELANDINE

Biennials or perennials, caulescent herbs, with yellow to orange sap; *leaves* pinnately lobed; *flowers* of few-flowered axillary or terminal umbelliform inflorescences; *sepals* 2, distinct; *petals* 4, bright yellow (ours); *fruit* an erect, 2-valved capsule.

Chelidonium majus* L. var. *majus, CELANDINE. Plants 2–8 dm; *leaves* pinnatifid with 3–9 lobes, these orbicular to obovate, 2–8 cm wide, with crenate margins; *petals* 6–10 mm long; *capsules* 3–6 cm long. Planted in gardens and escaping to nearby yards in Boulder, 5500–6000 ft. May–July. E. Introduced.

ESCHSCHOLZIA Cham. – CALIFORNIA POPPY

Annual or perennial, caulescent or scapose herbs, with colorless, watery sap; *leaves* ternately divided, petiolate; *sepals* 2, connate, pushed off by the expanding petals; *petals* 4, orange or yellow; *fruit* a cylindric, 2-valved capsule.

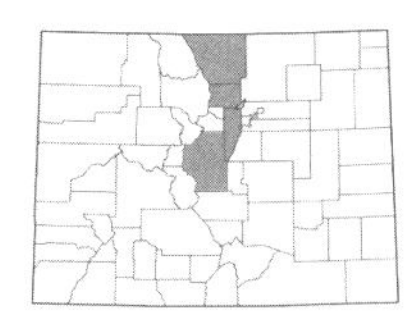

***Eschscholzia californica* Cham. ssp. *californica*,** CALIFORNIA POPPY. Annuals or winter annuals, 2–5 dm; *leaves* 2–8 cm long, ternately divided into narrow segments, glabrous; *petals* 2–7 cm long; *capsules* 3–10 cm long. A common garden plant sometimes escaping from cultivation or planted along roadsides, 5500–8500 ft. June–Aug. E. Introduced.

GLAUCIUM Mill. – HORNPOPPY

Annual or perennial, caulescent herbs, with yellow sap; *leaves* pinnately lobed, the basal leaves petiolate and the cauline sessile; *sepals* 2, distinct; *petals* 4; *fruit* a 2-valved capsule.

1a. Flowers red or orange (drying blackish) with a black spot at the base of the petals...***G. corniculatum***
1b. Flowers yellow or orange-yellow...***G. flavum***

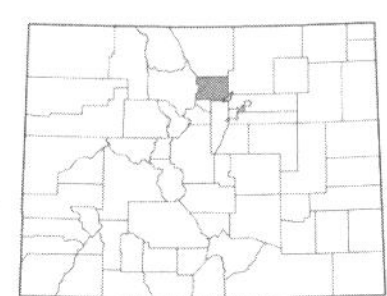

***Glaucium corniculatum* (L.) J.H. Rudolph**, BLACKSPOT HORNPOPPY. Annuals or biennials, 3–6 dm, densely hairy; *leaves* 4–20 cm long, the ultimate segments lobed or toothed, sessile above and petiolate below; *flowers* solitary; *petals* red or orange (drying blackish) with a black spot at the base, 2–3.5 cm long; *capsules* 15–20 cm long, hairy. Known from a single population near Boulder Reservoir, apparently not persisting but may escape from cultivation, 5200–5500 ft. April–May. E. Introduced.

***Glaucium flavum* Crantz**, YELLOW HORNPOPPY. Annuals or biennials, 3–9 dm, densely hairy; *leaves* sessile to clasping above and petiolate below; *flowers* 5–9 cm wide, solitary; *petals* yellow or orange-yellow; *capsules* 15–30 cm long, hairy. Known from a single collection made near downtown Boulder in 1965, not collected since and unlikely to occur outside of cultivation, 5200–5500 ft. April–June. E. Introduced.

PAPAVER L. – POPPY

Annual, biennial, or perennial, scapose or caulescent herbs, with white, orange, or red sap; simple to toothed or pinnately lobed, petiolate below and sessile or nearly so above; *flowers* solitary on long scapes or peduncles; *sepals* 2 (3), distinct; *petals* 4 (6), variously colored; *fruit* an erect poricidal capsule. (Kiger, 1975; 1985)

1a. Plants acaulescent; capsules hairy, ribbed; flowers orange, white, or yellow; filaments white or yellow...2
1b. Plants caulescent with at least a few stem leaves present; capsules glabrous, unribbed or ribbed; flowers orange, red, white, pink, or purple; filaments black, purple, or white...3

2a. Capsules with white hairs; leaves gray-green below and green above; plants generally taller, to 4 dm; flowers orange, white, or yellow, to 6 cm wide...***P. nudicaule* ssp. *americanum***
2b. Capsules with dark brown to black hairs; leaves green below and above; alpine plants only to 1.7 dm; flowers yellow, to 2 cm wide...***P. radicatum* ssp. *kluanensis***

3a. Leaves unlobed with shallowly to deeply dentate margins, with the distal leaves clasping the stem; filaments white; capsules stipitate...***P. somniferum***
3b. Leaves 1–2× pinnately lobed, distal leaves not clasping the stem; filaments black or dark purple; capsules sessile or substipitate...4

4a. Perennials; flowers large, 10 cm or more wide, the petals orange-red and usually with a pale or dark basal spot...***P. orientale***
4b. Annuals; flowers smaller, less than 10 cm wide, the petals orange, red, white, or pink, with or without a dark basal spot...5

5a. Peduncles spreading-hispid below and appressed-hispid above; capsules 2 times or more longer than wide; flowers orange to red, usually without a dark basal spot...***P. dubium***
5b. Peduncles spreading-hispid throughout; capsules less than 2 times longer than wide; flowers white, pink, orange, or red, usually with a dark basal spot...***P. rhoeas***

***Papaver dubium* L.**, BLINDEYES. Annuals, 2.5–6 dm, caulescent, hispid; *leaves* 1–2× pinnately lobed, 3–8 cm long, with 4–8 pairs of leaflets, the ultimate segments toothed or entire; *peduncles* 12–30 cm long, appressed-hispid above and spreading-hispid below; *petals* red, orange, or salmon, 1.5–2.5 cm long; *stamens* with black or dark purplish filaments; *capsules* 1–1.5 cm long, glabrous. Not reported for the state of Colorado yet but introduced in other states and could possibly escape from cultivation. Introduced.

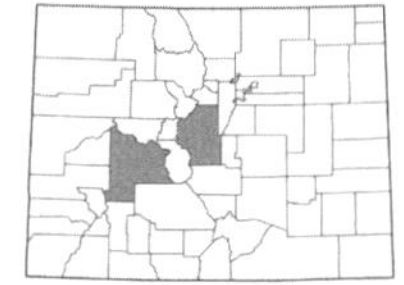

Papaver nudicaule* L. ssp. *americanum Rändal ex D.F. Murray, ICELANDIC POPPY. [*P. croceum* Ledeb.]. Annuals to 2 (4) dm, acaulescent; *leaves* 2–3× pinnately lobed, to 20 cm long, glabrous or with long bristles above; *flowers* to 6 cm wide; *petals* yellow or white; *stamens* with white or yellow filaments; *capsules* elliptic to subglobose, hairy. Found in old, high-elevation mining towns, 10,000–11,500 ft. June–Aug. E/W. Introduced.

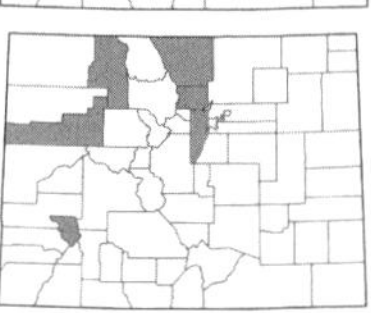

***Papaver orientale* L.**, ORIENTAL POPPY. Perennials, 6–10 dm, caulescent, hispid; *leaves* 1–2× pinnately lobed, 5–45 cm long, the lobes toothed; *flowers* 10+ cm wide; *petals* red to reddish-orange, with a pale or dark basal spot; *stamens* with dark purple or black filaments; *capsules* to 2.5 cm long, glaucous. Occasionally found persisting near old homesteads or along roadsides where it has escaped from cultivation, 5000–9000 ft. June–Aug. E/W. Introduced.

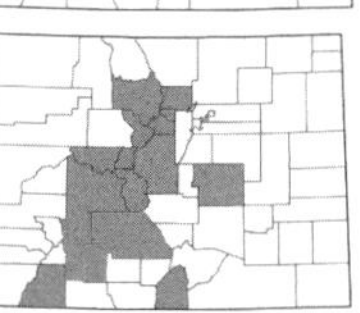

***Papaver radicatum* Rottb. ssp. *kluanensis* (D. Löve) D.F. Murray**, ALPINE POPPY. Perennials to 1.7 dm, hispid, acaulescent; *leaves* 1–2× pinnately lobed, to 7 cm long; *flowers* to 2 cm wide; *petals* yellow; *stamens* with white or yellow filaments; *capsules* to 1.2 cm long, with brown hairs. Rare in dry alpine meadows and scree slopes, distributed along the Continental Divide on the highest mountains,11,500–14,000 ft. June–Aug. E/W.

***Papaver rhoeas* L.**, CORN POPPY. Annuals, 3–7 dm, hispid, caulescent; *leaves* 1–2× pinnately lobed, 2–8 cm long; *peduncles* spreading-hispid; *flowers* 4–10 cm wide; *petals* white, pink, red, or orange, usually with a dark basal spot; *stamens* with black or dark purple filaments; *capsules* 1–1.5 cm long, glabrous. Not reported for Colorado yet but cultivated and could possibly escape. It is reported for several locations in Utah. Introduced.

***Papaver somniferum* L.**, OPIUM POPPY. Annuals, 2–10 dm, glaucous and glabrous, caulescent; *leaves* simple, unlobed with shallow to deeply dentate margins, 4–15 cm long, the upper leaves clasping the stem; *peduncles* sometimes sparsely hairy; *flowers* 6–10 cm wide; *petals* white to purple; *stamens* with white filaments; *capsules* stipitate, 2–5 cm long, glabrous. Reported as escaping from a few locations in Colorado, 5500–6500 ft. June–Aug. E. Introduced.

PARNASSIACEAE S.F. A. Gray – GRASS-OF-PARNASSUS FAMILY

Perennial herbs; *leaves* mostly basal, alternate, simple, long-petiolate, palmately veined, margins entire; *flowers* perfect, actinomorphic, solitary on long peduncles; *sepals* 5; *petals* 5, distinct, the margins entire or fringed; *stamens* 10, of 5 staminodia alternating with 5 functional stamens, the staminodia much branched with each branch tipped by a yellow ball; *pistil* 1; *ovary* superior, (3) 4-carpellate; *fruit* a capsule.

PARNASSIA – GRASS-OF-PARNASSUS

With characteristics of the family.

1a. Petals fringed on the lower half, 8–15 mm long, about twice as long as the sepals; leaves cordate or sometimes truncate at the base, mostly 1.5–5 cm wide; petioles mostly 3–15 cm long…***P. fimbriata***

1b. Petals entire, 3.5–12 mm long, shorter or longer than the sepals; leaves cuneate, truncate, or rounded at the base, mostly 0.5–2 cm wide; petioles 0.5–4 cm long…2

2a. Flowering stem bearing one leaf; petals 5–13-nerved, 6–16 mm long, usually longer than the sepals; staminodes divided into 6–20 filiform lobes…***P. palustris* var. *montanensis***

2b. Flower stem usually lacking a leaf (very rarely one will be present); petals 1–3-nerved, 3.5–6.5 mm long, as long as or shorter than the sepals; staminodes divided into 2–6 filiform lobes…***P. kotzebuei***

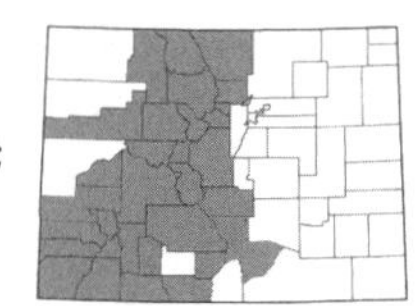

***Parnassia fimbriata* Koenig**, ROCKY MOUNTAIN GRASS-OF-PARNASSUS. Plants 1.5–4 dm; *leaves* reniform to orbicular, (1.2) 1.5–5 cm wide, cordate to truncate at the base; *petioles* mostly 3–15 cm long; *sepals* 4–7 mm long; *petals* 8–15 mm long (including the clawlike base), fringed on the lower half; *staminodes* scale-like, with a central, larger lobe and 7–9 short, rounded lateral lobes; *capsules* to 10 mm long. Locally common along streams, in marshes, moist alpine, and in wet meadows, 6500–13,500 ft. July–Sept. E/W.

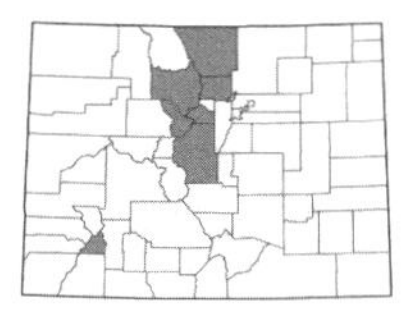

***Parnassia kotzebuei* Cham. ex Spreng.**, KOTZEBUE'S GRASS-OF-PARNASSUS. Plants 1–4 dm; *leaves* ovate to elliptic, 0.5–2 cm wide, cuneate to truncate at the base; *petioles* 0.5–2 cm long; *sepals* to 7 mm long; *petals* 3.5–6.5 mm long, as long or shorter than the sepals, entire, 1–3-nerved; *staminodes* divided into 2–6 filiform lobes; *capsules* to 10 mm long. Uncommon in moist alpine, on wet ledges, and along streams,10,500–12,000 ft. June–July. E/W.

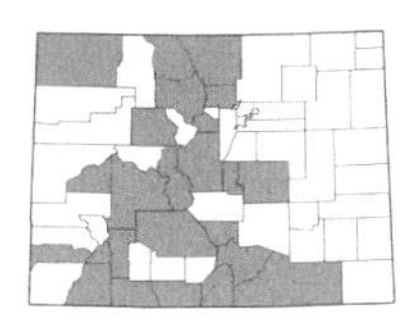

Parnassia palustris **L. var. *montanensis* (Fernald & Rydb. ex Rydb.) C.L. Hitchc.**, MOUNTAIN GRASS-OF-PARNASSUS. [*P. parviflora* DC.]. Plants 0.8–4.5 dm; *leaves* ovate to orbicular, 0.7–3 cm × 5–20 mm, cuneate or rounded at the base; *petioles* 0.7–4 cm long; *sepals* 3–10 mm long; *petals* 6–16 mm long, usually longer than the sepals, entire, 5–13-nerved; *staminodes* divided into 6–20 filiform lobes; *capsules* 8–12 mm long. Found along streams, lake margins, seepages, and in moist meadows, 5500–10,500 ft. June–Aug. E/W. (Plate 57)

PEDALIACEAE R. Br. – UNICORN-PLANT FAMILY

Mostly herbs (ours); *leaves* opposite, or the upper alternate, exstipulate; *flowers* perfect, bracteate, zygomorphic; calyx deeply 5-lobed; *corolla* 5-lobed; *stamens* 4 (paired) with the upper 5th represented by a small staminode, epipetalous, alternate with the corolla lobes; *pistil* 1; *ovary* superior, 2-carpellate; *fruit* a capsule or nut.

PROBOSCIDEA Schmid. – UNICORN-PLANT

Herbs; *leaves* opposite or the upper sometimes alternate; *corolla* somewhat bilabiate; *fruit* a drupaceous capsule with a long, incurved, hooked beak that exceeds the fruit body in length.

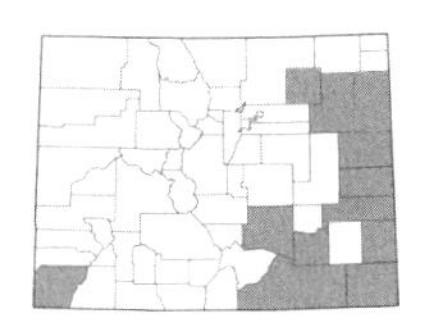

Proboscidea louisianica **(Mill.) Thellung**, RAM'S HORN. [*Martynia louisianica* Mill.]. Annuals, fetid with a strong odor, glandular, 1.5–6 (10) dm; *leaves* reniform to broadly ovate, 3–20 cm long, irregularly toothed to entire; *calyx* 10–22 mm long; *corolla* white to purplish and mottled with yellow or reddish-purple spots, the lower portion with yellow stripes, the lobes 1.5–2 cm long; *capsules* body to 10 cm long with incurved beaks. Found along roadsides, sometimes in moist places such as along ditches or river bottoms, often in disturbed places, 3400–5000 ft. June–Aug. E/W.

PHRYMACEAE – LOPSEED FAMILY

Herbs; *leaves* opposite or basal, simple, exstipulate; *flowers* perfect, zygomorphic, bilabiate; *calyx* 5-lobed; *corolla* 5-lobed; *stamens* 4, didynamous, epipetalous; *pistil* 1; *ovary* superior, 2-carpellate; *fruit* a capsule.

MIMULUS L. – MONKEY-FLOWER

Annual or perennial herbs; *flowers* in racemes; *calyx* tube plicate; *corolla* yellow to red, blue, or purple. (Grant, 1924)

1a. Petiole modified into a lenticular propagule 2–3 mm long; corolla often absent...***M. gemmiparus***
1b. Petiole not modified into a propagule; corolla usually present (sometimes dropping after flowering)...2

2a. Leaves pinnately veined, lanceolate, sessile with some auriculate-clasping; flowers blue to blue-purple, 20–30 mm long; plants 2–13 dm tall with 4-angled stems...***M. ringens***
2b. Leaves palmately veined or with a single vein, or if pinnately veined then obviously petiolate; flowers red, yellow, magenta, violet, or pink, 4–50 mm long; plants 0.1–8 dm tall with rounded stems...3

3a. Flowers bright red to orange-red, 30–40 mm long; leaves coarsely toothed; stoloniferous, glandular-hairy perennials...***M. eastwoodiae***
3b. Flowers yellow, magenta, pink, or violet, 4–50 mm long; leaves and plants various...4

4a. Flowers 33–50 (57) mm long, pink, magenta, or violet; rhizomatous perennials 3–8 dm tall; leaves mostly 3–7 (10) cm long, lanceolate to ovate...***M. lewisii***
4b. Flowers to 30 mm long, yellow or if magenta or violet then the plants annuals and 0.2–2.2 dm tall; leaves mostly 0.3–5.5 cm long, variously shaped...5

5a. Leaves pinnately veined, petiolate, ovate with dentate margins; plants slimy viscid-villous, smelling of musk, making the plants very unpleasant to handle; corolla 13–24 mm long, yellow with red stripes in the throat; rhizomatous perennials...***M. moschatus***
5b. Leaves palmately veined or with a single vein, or if pinnately veined and petiolate then the plant an annual and the corolla 7–14 mm long, variously shaped; plants not both slimy and musk-smelling; corolla yellow, often with red spots in the throat, or magenta to violet...6

6a. Upper calyx lobe conspicuously (2–3 times) longer than the lower lobe; calyx 5–16 mm long in flower, much-inflated in fruit, glabrous to shortly hairy; perennials or occasionally annuals, glabrous or glandular-hairy plants; corolla 10–30 mm long; pedicels 10–50 mm long...7
6b. Upper calyx lobe nearly the same length as the lower lobe; calyx 3–7 mm long in flower, not inflated in fruit, glandular-hairy at least on the ribs; glandular-hairy annuals; corolla 4–14 mm long; pedicels 2–20 mm long...9

7a. Corolla throat mostly open, not closed by the palate; calyx teeth of lateral lobes very short or nearly obsolete, mostly less than 1 mm long; calyx glabrous...***M. glabratus***
7b. Corolla throat closed by the palate; calyx teeth of lateral lobes mostly 1 mm or more in length; calyx glabrous or minutely hairy...8

8a. Plants usually with 5 or more flowers per stem; plants usually lacking stolons or rhizomes; corolla mostly 9–23 mm long; plants 0.5–5.5 (9) dm tall...***M. guttatus***
8b. Plants usually with 1- 3 or rarely 5 flowers per stem; plants with stolons or rhizomes; corolla 17–30 mm long; plants 0.7–2 dm tall...***M. tilingii***

9a. Leaves distinctly petiolate with a petiole 1–12 mm long, mostly 5–13 mm wide; corolla 7–14 mm long, yellow and often with red spots; calyx lobes ciliate...***M. floribundus***
9b. Leaves sessile, mostly 1–4 mm wide; corolla 4–8 (10) mm long, magenta to violet or yellow with red spots; calyx lobes ciliate or not...10

10a. Margins of the calyx teeth ciliate with white hairs; pedicels 7–20 mm long...***M. rubellus***
10b. Margins of the calyx teeth glandular-pubescent but not ciliate; pedicels 2–10 (13) mm long...11

11a. Flowers magenta to violet; pedicels straight in fruit...***M. breweri***
11b. Flowers yellow with red spots in the throat; pedicels curved or S-curved in fruit...***M. suksdorfii***

***Mimulus breweri* (Greene) Cov.**, Brewer's monkey-flower. Falsely reported from Colorado, specimens labeled this are misidentifications of *M. rubellus* or *M. suksdorfii*.

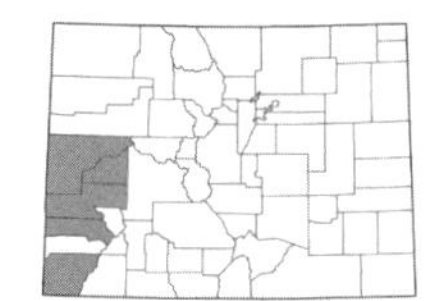

***Mimulus eastwoodiae* Rydb.**, Eastwood's monkey-flower. Perennials, stoloniferous, 0.5–3 dm, minutely glandular, occasionally viscid-villous; *leaves* obovate to oblanceolate, 1.7–4 cm × (5) 10–20 mm, sessile, with coarsely toothed margins; *pedicels* 1–3 (4) cm long; *calyx* 16–23 (27) mm long, the lobes subequal, 4–7 mm long; *corolla* 30–40 mm long, scarlet to orangish-red. Rare in moist cracks on overhanging walls of sandstone alcoves, 4700–5800 ft. July–Sept. W. (Plate 58)

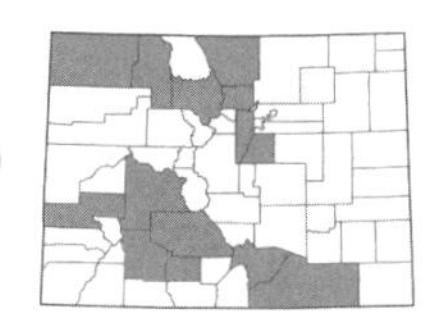

***Mimulus floribundus* Lindl.**, manyflowered monkey-flower. Annuals, 0.3–2.2 (4) dm, glandular-hairy, occasionally viscid and slimy; *leaves* ovate to lanceolate, (0.3) 0.8–2 (3) cm × (1) 5–13 mm, petiolate, with few-toothed margins; *pedicels* 0.5–1.4 (1.8) cm long; *calyx* 3.5–7 mm long, to 9 mm long in fruit, the lobes subequal, 0.8–1.6 (2) mm long; *corolla* 7–14 mm long, yellow, the palate pubescent and often with reddish spots. Found in moist places, cliff overhangs, along streams, and in moist rock crevices, 5500–9200 ft. June–Oct. E/W.

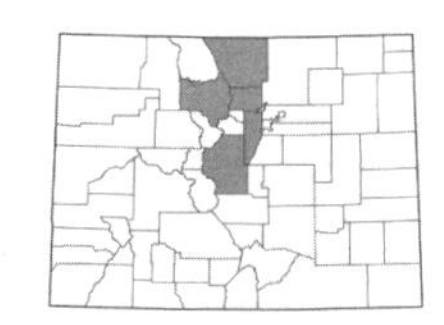

***Mimulus gemmiparus* W.A. Weber**, Rocky Mountain monkey-flower. Annuals, 0.1–1 dm, glabrous; *leaves* ovate, to 1 cm × 7 mm, petiolate with almost every petiole containing a lenticular propagule, the margins entire or remotely denticulate; *calyx* 3–4 mm long, the lobes subequal; *corolla* to 15 mm long, rarely flowering in the wild, yellow with red spots. Rare on shady ledges and overhangs, seepage areas, and rocky outcrops, 8400–11,000 ft. July–Sept. E/W. Endemic.

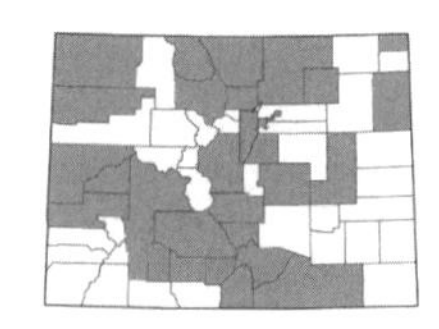

***Mimulus glabratus* Kunth**, smooth monkey-flower. Perennials, 1–5 dm, glabrous to sparsely viscid-hispid or glandular-hairy; *leaves* broadly ovate to orbicular, 1.3–2.8 cm × 15–30 mm, short-petiolate to sessile, with dentate, undulate, or entire margins; *pedicels* (1) 1.5–3 (6) cm long; *calyx* 5–11 (16) mm long, the lobes unequal, the lateral lobes very short and the posterior lobe more than 2 times as long as the others; *corolla* 10–20 mm long, yellow, the palate bearded and often with red-brown spots. Common along streams and in seepage areas and springs, 3500–8700 ft. May–Sept. E/W.

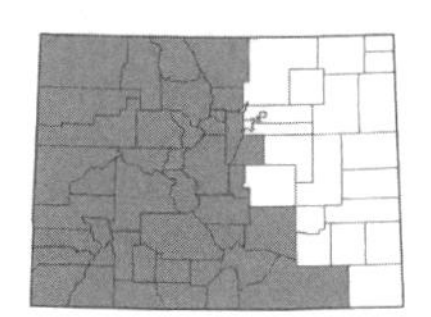

***Mimulus guttatus* DC.**, yellow monkey-flower. Annuals or perennials, 0.5–5.5 (9) dm, glabrous to hairy; *leaves* obovate, ovate, or orbicular, (0.5) 1.5–5.5 (10) cm × (5) 10–40 (85) mm, petiolate below and usually sessile above, with coarsely and irregularly toothed margins; *pedicels* 1–3.5 (5.5) cm long; *calyx* 6–16 mm long, to 20 mm long and inflated in fruit, the posterior lobe 2–3 times as long as the others; *corolla* 9–23 (30) mm long, yellow, the palate densely hairy with red spots. Common along streams, in seepage areas, and near springs, 5000–12,000 ft. (May) June–Sept. E/W. (Plate 58)

***Mimulus lewisii* Pursh**, LEWIS' MONKEY-FLOWER. Perennials, rhizomatous, 3–8 dm, glandular-hairy to viscid-villous; *leaves* ovate to lanceolate, 3–7 (10) cm × (10) 15–24 (32) mm, sessile, with denticulate margins or rarely entire; *pedicels* 3–7 (10) cm long; *calyx* 18–28 mm long, the lobes subequal, 3–7 mm long; *corolla* 33–50 (57) mm long, pink, magenta, or violet, the palate with 2 yellow-hairy ridges. Rare in moist meadows and along streams, 8400–11,000 ft. July–Sept. E/W.

***Mimulus moschatus* Douglas ex Lindl.**, MUSK FLOWER. Perennials, rhizomatous, 0.2–2 (3) dm, slimy viscid-villous with whitish hairs, musk-scented; *leaves* ovate, 1.2–5 cm × 5–25 mm, with dentate margins; *pedicels* 5–14 mm long; *calyx* 6–11 (13) mm long, the lobes slightly unequal, the posterior lobe slightly longer and broader than the others; *corolla* 13–24 mm long, yellow, the palate bearded with 2 red stripes in the throat. Uncommon in moist places, near springs and along streams, and along lake shores, 8000–10,000 ft. July–Aug. E/W.

***Mimulus ringens* L.**, ALLEGHENY MONKEY-FLOWER. Perennials, rhizomatous or stoloniferous, 2–13 dm, glabrous; *leaves* lanceolate to narrowly oblong, 2.5–8 (10) cm × 6–20 (35) mm, clasping or auriculate and sessile, with toothed margins; *pedicels* 2–3.5 (4.5) cm long; *calyx* 10–16 (20) mm long, the lobes more or less equal, 2–6 mm long; *corolla* 20–30 mm long, blue to blue-purple. Known from a few old collections made in Denver and has not been seen in the state since 1895, 5000–5500 ft. July–Aug. E.

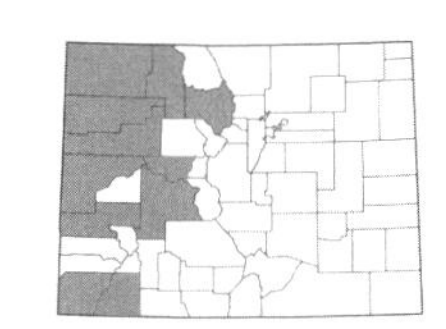

***Mimulus rubellus* A. Gray**, LITTLE REDSTEM MONKEY-FLOWER. Annuals, 0.1–2.2 dm, minutely glandular-hairy, occasionally viscid-villous; *leaves* narrowly lanceolate to linear, 0.3–1.5 (2.2) cm × 2–4 (7) mm, sessile, with entire to shallowly sinuate-denticulate margins; *pedicels* 7–18 (22) mm long; *calyx* 4–7 mm long, to 9 mm long in fruit, the lobes subequal, 0.5–1 mm long; *corolla* 6–8 (10) mm long, yellow with maroon spots, occasionally the limb pinkish or purplish. Uncommon in sagebrush or oak shrublands, rocky crevices, and along streams, 5500–8600 ft. May–June. W.

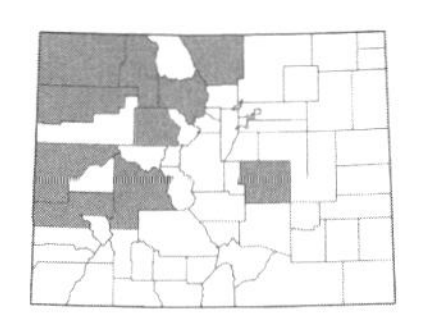

***Mimulus suksdorfii* A. Gray**, SUKSDORF'S MONKEY-FLOWER. Annuals, 0.1–1 dm, minutely glandular-hairy; *leaves* linear to lanceolate, 0.5–2 cm × 1–4 mm, sessile, with entire margins; *pedicels* 2–7 (10) mm long, sigmoid-curved in fruit; *calyx* 3–5 mm long, the lobes subequal, 0.2–0.7 mm long; *corolla* 4–6.5 mm long, yellow with red spots. Uncommon or overlooked in oak or sagebrush shrublands and on rocky outcrops, 7000–9000 ft. May–June. E/W.

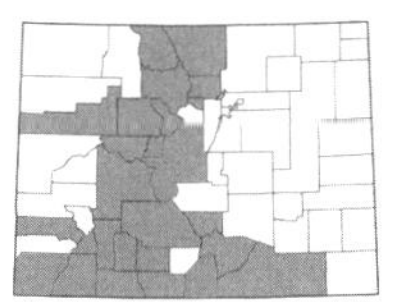

***Mimulus tilingii* Regel**, SUBALPINE MONKEY-FLOWER. Perennials, stoloniferous or rhizomatous, 0.7–2 dm, glabrous to minutely hairy; *leaves* ovate to suborbicular, 1–2.6 cm × 5–15 (25) mm, petiolate or sometimes sessile, with dentate, undulate, or entire margins; *pedicels* 1–4 (6) cm long; *calyx* 7–15 mm long, to 20 mm and inflated in fruit, the lobes unequal, the posterior lobe longer than the others, 3–7 mm long; *corolla* 17–30 mm long, yellow, the palate densely yellow-bearded with pale red spots. Common in the alpine and along streams, 9500–13,000 ft. June–Aug. E/W.

Mimulus tillingii can be difficult to separate from *M. guttatus* and might simply be an alpine form of *M. guttatus*.

PLANTAGINACEAE Juss. – PLANTAIN FAMILY

Annual or perennial herbs, shrubs or vines, sometimes aquatic; *leaves* alternate, opposite, or whorled, or on short stems and appearing basal, exstipulate; *flowers* perfect or imperfect, actinomorphic or zygomorphic; *calyx* 4–5-lobed, often persistent; *corolla* 4–5-lobed or absent; *stamens* 1, 2, 4 and didynamous, or with 4 fertile and 1 sterile staminode (*Penstemon*), epipetalous, alternate with the corolla lobes; *pistil* 1; *style* 1; *stigma* 2-lobed; *ovary* superior or sometimes inferior, 2-carpellate; *fruit* a capsule.

1a. Leaves whorled with 6 or more leaves per whorl, the emergent ones 1–3 cm long, submerged leaves often reduced to short scale-like projections; stems stout and thick, 0.7–2.5 cm wide; flowers clustered in the axils of emergent leaves; plants emergent aquatics…***Hippuris***

1b. Leaves not whorled with 6 or more leaves per whorl; plants otherwise unlike the above in all respects…2

2a. Plants true aquatics, often with dimorphic leaves, the submerged leaves usually linear-lanceolate and the floating leaves spatulate to oblong, or sometimes with all linear-lanceolate leaves; fruit of 4 flattened, winged mericarps sessile or nearly so in the leaf axils…***Callitriche***

2b. Plants not true aquatics, otherwise unlike the above in all respects…3

3a. Flowers distinctly spurred at the base of the corolla on the lower side, strongly two-lipped (bilabiate), purple, yellow, yellow and orange, or lavender and white…4

3b. Flowers not spurred but sometimes with a small pouch at the base on the upper side, two-lipped to nearly actinomorphic, variously colored…5

4a. Flowers solitary in the leaf axils; corolla 4.5–6 mm long, excluding the 1.7–2.8 mm long spur, pale lavender and white with a yellow throat; annuals, glandular-hairy throughout...***Chaenorrhinum***
4b. Flowers in a terminal raceme; corolla 7–24 mm long, excluding the 2–17 mm long spur, blue-purple, yellow, or yellow and orange; annuals or perennials, glabrous to sparsely pubescent or sometimes glandular-hairy just in the inflorescence...***Linaria***

5a. Leaves opposite (*careful*—in *Veronica* some species have leaf-like bracts subtending the flowers, giving the appearance of alternate leaves, but the lower, true leaves are opposite)...6
5b. Leaves alternate or in a basal rosette...12

6a. Flowers 4-merous and nearly actinomorphic, in axillary or terminal racemes; corolla blue, purple, white, or pink; stamens 2...***Veronica***
6b. Flowers 5-merous and weakly to strongly zygomorphic or bilabiate, or if 4-merous then the flowers strongly bilabiate; inflorescence various; corolla variously colored; stamens 2, 4, or 4 with 1 sterile staminode...7

7a. Flowers weakly zygomorphic, white with a yellow throat; leaves sessile, suborbicular or ovate, palmately veined, entire; inflorescence of 1–4 flowers in the upper leaf axils; plants prostrate and rooting at the lower nodes, of moist places or aquatic...***Bacopa***
7b. Flowers strongly zygomorphic or bilabiate, variously colored; leaves unlike the above in all respects; inflorescence and plants various...8

8a. Flowers with a small pouch at the base on the upper side, blue with the upper lobes white, mostly 4–6 mm long; leaves entire, rounded to oblanceolate, usually purplish on the lower side; annuals...***Collinsia***
8b. Flowers lacking a small pouch at the base, variously colored; leaves unlike the above; annuals or perennials...9

9a. Calyx mostly connate; flowers creamy-white or greenish-white, horizontally flattened at the apex; leaves mostly basal with the stem leaves reduced...***Chionophila***
9b. Calyx lobed to the base or nearly so; flowers not horizontally flattened at the apex; leaves various...10

10a. Stamens 4, fertile and with 1 sterile staminode, this often bearded with hairs; perennials; flowers in a terminal panicle, raceme, or cyme; corolla usually longer than 10 mm...***Penstemon***
10b. Stamens 2; annuals; flowers solitary in upper leaf axils; corolla 4–10 mm long...11

11a. Flowers yellow to white; plants glandular-hairy...***Gratiola***
11b. Flowers blue to purple; plants glabrous...***Lindernia***

12a. Corolla 4–6 cm long, constricted at the base, purple, pink, or sometimes white, with spots or mottling on the inside of the lower lip; plants mostly 6–12 dm tall...***Digitalis***
12b. Corolla smaller, variously colored but lacking spots or mottling on the inside of the lower lip; plants to 5 dm tall...13

13a. Leaves basal and cauline, but the stem leaves reduced and bract-like, with pinnate venation; corolla absent or two-lipped with the upper lip of the corolla flattened; stamens 2, conspicuously exserted...***Besseya***
13b. Leaves appearing all basal, narrowly linear with 1 nerve, or more often with 3 or more parallel veins; corolla actinomorphic, not two-lipped, petals 4; stamens 2 or 4, exserted or not...***Plantago***

BACOPA Aubl. – WATER HYSSOP
Aquatic perennial herbs; *leaves* opposite, entire, ovate to suborbicular (ours); *flowers* 5-merous, weakly zygomorphic; *calyx* nearly distinct; *corolla* white with a yellow throat (ours); *stamens* 4, didynamous.

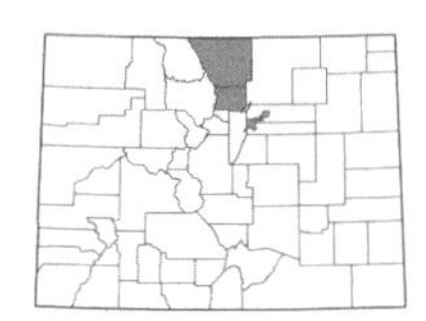

***Bacopa rotundifolia* (Michx.) Wettst.**, DISK WATER HYSSOP. Stems floating or prostrate, rooting at the lower nodes; *leaves* 12–30 (40) × (8) 12–20 mm, clasping; *calyx* lobes 3–4.5 mm long at flowering, to 6 mm long in fruit, the outer ovate and inner lanceolate; *corolla* 4.5–7 mm long; *capsules* subglobose, 3.5–5.5 mm long. Uncommon along the shores of drying ponds, 5000–5500 ft. Aug.–Sept. E.

BESSEYA Rydb. – KITTENTAILS
Perennials; *leaves* nearly all basal; *flowers* zygomorphic and bilabiate, in a dense spike; *calyx* deeply 2–4-lobed; *corolla* absent or present and bilabiate, purple, yellow, or white; *stamens* 2, exserted, the filaments purple in flowers lacking petals.

1a. Corolla lacking or greatly reduced and inconspicuous; calyx 2-lobed, positioned on the anterior side of the flower or capsule; filaments purple...***B. wyomingensis***
1b. Corolla present; calyx of 4 segments surrounding the flower or capsule; filaments purple or not...2

2a. Corolla blue-purple to purple; plants 0.5–2 dm tall, with 4–6 leaf-like bracts below the inflorescence; basal leaves 2–5 cm long...***B. alpina***
2b. Corolla white or yellow or sometimes pinkish or purplish-tinged; plants usually 2–4 dm tall, with 6 or more leaf-like bracts below the inflorescence; basal leaves 5–15 cm long...3

3a. Corolla pale yellow; filaments white; lateral sepals united for about ½ their length or more...***B. ritteriana***
3b. Corolla white or pinkish to purplish-tinged; filaments purplish; lateral sepals united for less than ⅓ their length...***B. plantaginea***

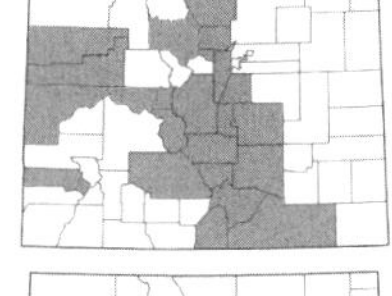

Besseya alpina **(A. Gray) Rydb.**, ALPINE KITTENTAILS. Plants 0.5–2 dm; *leaves* ovate to elliptic, 2–5 × 1.5–4 cm, the margins crenate; *calyx* 4-lobed, 4–6 mm long, densely villous; *corolla* blue-purple to purple, 5–8 mm long; *filaments* purple; *capsules* 3–5 mm long, orbicular. Found on rocky ridges and in the alpine, usually on scree slopes or among rocks, 9500–14,200 ft. June–Aug. E/W. (Plate 58)

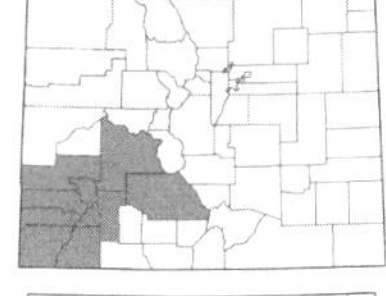

Besseya plantaginea **(James) Rydb.**, WHITE RIVER KITTENTAILS. Plants 1–4 dm; *leaves* ovate to elliptic or oblong, 5–15 cm long, with crenate margins; *calyx* 4-lobed, villous; *corolla* white or pinkish to purplish-tinged, 5–8 mm long; *filaments* rose to purple; capsules 3–6 mm long. Found in forest openings, dry meadows, and on grassy slopes, 5500–11,000 ft. May–Aug. E/W.

Besseya ritteriana **(Eastw.) Rydb.**, RITTER'S KITTENTAILS. Plants 2–4 dm; *leaves* elliptic to oblong, 5–15 cm long, with crenate margins; *calyx* 4-lobed, villous; *corolla* pale yellow; *filaments* white; *capsules* 3–6 mm long. Found on open slopes and in meadows and the alpine, 7000–12,500 ft. May–Aug. W. Endemic.

Besseya wyomingensis **(A. Nelson) Rydb.**, WYOMING KITTENTAILS. [*B. cinerea* (Raf.) Pennell]. Plants 1–3 dm; *leaves* ovate to oblong, 2.5–12 × 1.5–5 cm, the margins crenate to serrate; *calyx* 2-lobed, positioned on the anterior side of the flower or capsule, 3.5–7 mm long; *corolla* absent; *filaments* purple; *capsules* 3–6.5 mm long, villous. Found in openings in ponderosa pine forests and dry meadows, and on grassy hillsides, 6000–8600 ft. April–June. E.

CALLITRICHE L. – WATER-STARWORT

Aquatic annual or perennial herbs; *ovary* 4-loculed with false septae; *fruit* a schizocarp splitting into 4 mericarps at maturity, the mericarps flattened and usually winged. (Fassett, 1951)

1a. Leaves uniformly linear-lanceolate, all submerged, narrowed to a clasping base, the bases not connected by a ridge or wing; fruit 1–2.5 mm, more or less orbicular; flowers not subtended by bracts...***C. hermaphroditica***
1b. Leaves dimorphic with spatulate to oblong floating leaves and linear-lanceolate submerged leaves, rarely all uniformly linear and then the leaves more than 13 mm long and the bases connected by a ridge or wing; fruit 0.6–2 mm long, obovoid or orbicular; flowers subtended by bracts...2

2a. Fruit (0.8) 1–2 mm long, obovoid to oblong, longer than broad by about 0.2 mm, with thin wings or sharp edges at the apex; pits on fruit aligned in vertical rows...***C. palustris***
2b. Fruit 0.6–1.2 mm long, more or less orbicular, about as broad as long or not varying by more than 0.2 mm, not winged or sharp-edged at the apex; pits on fruit usually not aligned in vertical rows...***C. heterophylla* ssp. *heterophylla***

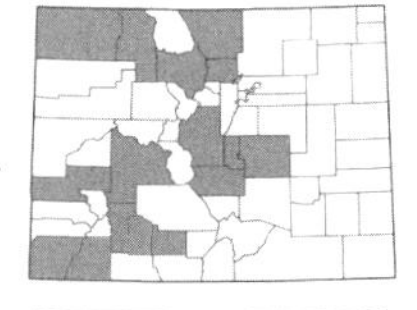

Callitriche hermaphroditica **L.**, NORTHERN WATER-STARWORT. Perennials; *leaves* uniformly linear-lanceolate, 5–13 × 0.5–1.5 mm, all submerged, the base clasping; *flowers* not subtended by bracts; *fruit* 1–2.5 mm long, more or less orbicular, obscurely pitted, the margins narrowly winged. Found in ditches and slow-moving streams, and along pond and lake margins, 6500–10,500 ft. June–Sept. E/W.

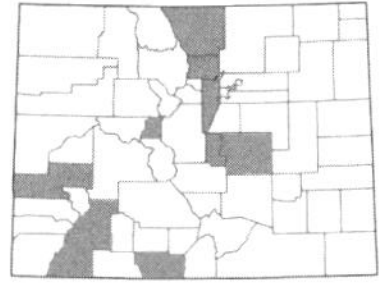

Callitriche heterophylla **Pursh ssp. *heterophylla***, TWO-HEADED WATER-STARWORT. Annuals or perennials; *leaves* dimorphic, the floating leaves spatulate to oblong, 6–15 mm long, and the submerged leaves linear-lanceolate; *flowers* subtended by bracts; *fruit* 0.6–1.2 mm long, orbicular, cordate, the margins unwinged or with a very narrow wing at the apex, pitted. Uncommon in slow-moving streams and lakes, 7500–9000 ft. June–Sept. E/W. (Plate 58)

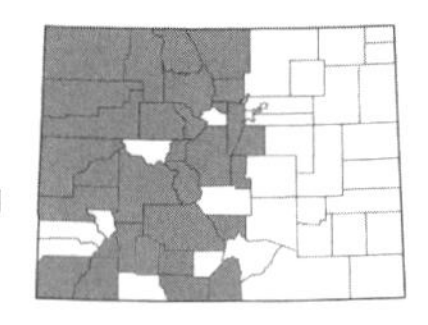

***Callitriche palustris* L.**, VERNAL WATER-STARWORT. [*C. verna* L.[. Perennials; *leaves* dimorphic, the floating leaves obovate to spatulate, 7–15 mm long, and the submerged leaves linear-lanceolate; *flowers* subtended by bracts; *fruit* (0.8) 1–2 mm long, obovoid to oblong, with thin wings or sharp edges at the apex, pitted with the reticulations aligned in vertical rows. Common in slow-moving streams, ditches, and along lake margins, 5000–12,000 ft. June–Sept. E/W. (Plate 58)

CHAENORHINUM (DC.) Rchb. – DWARF SNAPDRAGON

Annuals, glandular-hairy herbs; *leaves* alternate or the lowermost opposite, entire; *flowers* small, solitary in the leaf axils; *calyx* deeply 5-cleft; *corolla* strongly bilabiate, lavender and white with a yellow throat, spurred from the base of the tube ventrally; *stamens* 4, didynamous.

***Chaenorhinum minus* (L.) J. Lange**, DWARF SNAPDRAGON. Plants 1–3 dm; *leaves* linear to oblanceolate, 0.5–3 cm × 1–3 mm; *calyx* 2–3.5 mm at flowering, to 4.5 mm long in fruit; *corolla* 4.5–6 mm long, with a spur 1.7–3 mm long; *capsules* ca. 5 mm long, subglobose. Uncommon in alkaline meadows, known from a single location near Boulder but not seen since 1980, 5000–5500 ft. May–June. E. Introduced.

CHIONOPHILA Benth. – SNOWLOVER

Perennial herbs; *leaves* opposite and in a basal rosette, entire, lanceolate; *flowers* in a terminal raceme, bilabiate; *calyx* obscurely 5-lobed; *corolla* bilabiate, the upper lip erect and retuse, the lower lip bearded, creamy-white or greenish-white, horizontally flattened at the apex; *stamens* 5, 4 fertile and 1 sterile.

***Chionophila jamesii* Benth.**, ROCKY MOUNTAIN SNOWLOVER. Plants 0.5–1 (1.5) dm; *leaves* lanceolate to spatulate below, linear above; *calyx* 7–8 mm long; *corolla* 1–1.3 mm long. Locally common in the alpine, 11,000–14,000 ft. July–Aug. (Sept.). E/W. (Plate 58)

COLLINSIA Nutt. – BLUE-EYED MARY

Annual herbs; *leaves* opposite or partly whorled; *flowers* in a raceme or solitary in the axils of upper leaves; *calyx* 5-lobed; *corolla* bilabiate, pale blue or white; *stamens* 4, didynamous.

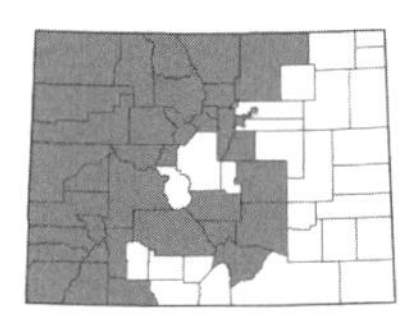

***Collinsia parviflora* Lindl.**, MAIDEN BLUE-EYED MARY. Plants 0.5–2 (3) dm, glabrous below and usually glandular in the inflorescence; *leaves* oblanceolate to oblong, rounded, 1.5–3 (5) cm long, with a short petiole; *pedicels* 4–10 mm long in flower, to 30 mm long in fruit; *calyx* 3.5–5.5 mm long; *corolla* with a white, erect upper lip and blue lower lip, 4–6 (7) mm long; capsules 4–5.5 mm long. Common in forests, meadows, and other open places, 5500–9700 ft. April–June. E/W. (Plate 58)

Often seen growing with *Microsteris gracilis*.

DIGITALIS L. – FOXGLOVE

Biennials or perennials, usually glandular and hairy herbs; *leaves* alternate, simple; *flowers* secund in a raceme; *calyx* 5-lobed; *corolla* more or less bilabiate, pink-purple with spots or mottling on the inside of the lower lip; *stamens* 4.

Digitalis purpurea* L. var. *purpurea, PURPLE FOXGLOVE. Biennials, 5–12 dm; *leaves* lanceolate to ovate or elliptic, 8–20 × 4–9 cm, tomentose below and green and sparsely hairy above, with crenate to serrate margins; *racemes* mostly 1–3 dm long; *corolla* 4–6 cm long. Uncommon in moist forests and along streams, usually found near old homesteads, 6500–7000 ft. June–Aug. E. Introduced.

GRATIOLA L. – HEDGEHYSSOP

Annuals; *leaves* opposite, sessile; *flowers* solitary or in pairs in the leaf axils, subtended by 2 bracts, pedicellate; *calyx* 5-lobed nearly to the base; *corolla* bilabiate, the upper lip shallowly 2-lobed, white to yellowish; *stamens* 2.

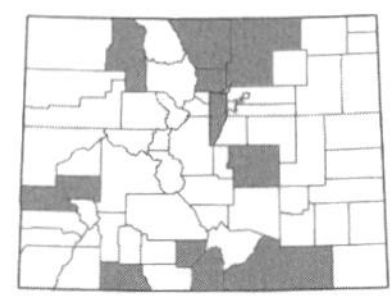

***Gratiola neglecta* Torr.**, CLAMMY HEDGEHYSSOP. Plants 0.5–2 dm; *leaves* narrowly ovate to elliptic, 1–5 cm long, the margins finely toothed above the middle; *pedicels* 8–20 mm long; *calyx* 3–5.5 mm long in flower, to 7 mm long in fruit; *corolla* 7–12 mm long, white with a yellowish tube. Uncommon in moist soil and along ditches and pond margins, 4500–8200 ft. June–Sept. E/W.

HIPPURIS L. – MARE'S TAIL

Emergent aquatic or amphibious perennial herbs; *leaves* whorled, simple and entire, linear; *flowers* inconspicuous in the upper leaf axils; *calyx* a minute 2–4-lobed rim at the top of the ovary; *corolla* absent; *stamen* 1; *ovary* inferior; *fruit* an achene.

***Hippuris vulgaris* L.**, COMMON MARE'S TAIL. Plants to 6 dm; *leaves* linear, sessile, 1–3 (4) cm long, in whorls of 6 or more per node; *achenes* 1.5–3 mm long. Found in ponds and lakes, emergent or sometimes completely submerged, 5500–10,500 ft. June–Aug. E/W.

LINARIA Mill. – TOADFLAX

Annual or perennial herbs; *leaves* alternate or the lowermost sometimes opposite, entire, sessile; *flowers* in a terminal raceme; *calyx* of 5 more or less distinct segments; *corolla* strongly bilabiate, spurred ventrally, yellow, blue, purple, or white, the throat nearly closed by a palate; *stamens* 4, didynamous.

1a. Flowers purple to blue-violet, the corolla 8–13 mm long excluding the spur and the upper lip mostly 3.5–5 mm long; slender annuals or winter annuals, 1–4 (5) dm; capsules 2.8–3.5 mm long...***L. canadensis* var. *texana***
1b. Flowers yellow or yellow and orange, the corolla 10–25 mm long excluding the spur and the upper lip 7–15 mm long; perennials, 3–8 dm; capsules 5–9 mm long...2

2a. Leaves ovate to lance-ovate, 10–20 (35) mm wide, not particularly crowded on the stem, sessile and clasping; flowers bright yellow, the corolla 14–25 mm long excluding the spur...***L. dalmatica***
2b. Leaves linear, 2–6 (8) mm wide, crowded on the stem, sessile but not clasping; flowers yellow with an orange palate, the corolla 10–15 (18) mm long excluding the spur...***L. vulgaris***

***Linaria canadensis* (L.) Dum.-Cours. var. *texana* (Scheele) Pennell**, BLUE TOADFLAX. [*Nuttallanthus texanus* (Scheele) D.A. Sutton]. Plants 1–4 (5) dm; *leaves* dimorphic, the basal leaves opposite, spatulate, 0.5–1 cm long, and the cauline leaves linear, 1–3 cm × 1–2 mm; *calyx* 2.8–3.5 mm long; *corolla* purple to blue-violet, 8–13 mm long with a spur 2–11 mm long; *capsules* 2.8–3.5 mm long. Uncommon in sandy soil or on grassy slopes, 5000–6000 ft. May–Aug. E.

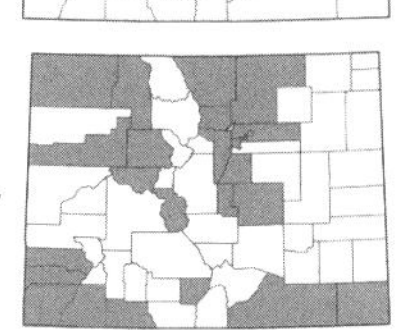

***Linaria dalmatica* (L.) Mill.**, DALMATIAN TOADFLAX. Plants 4–7 (10) dm; *leaves* lanceolate to ovate or elliptic, 2–5 cm × 10–20 (35) mm, clasping, sharply acute; *calyx* 5–9 mm long; *corolla* bright yellow, 14–25 mm long, with a spur 9–17 mm long; *capsules* 6–8 mm long. Common in meadows, along roadsides, and in other disturbed areas, 4500–9700 ft. May–Oct. E/W. Introduced. (Plate 58)

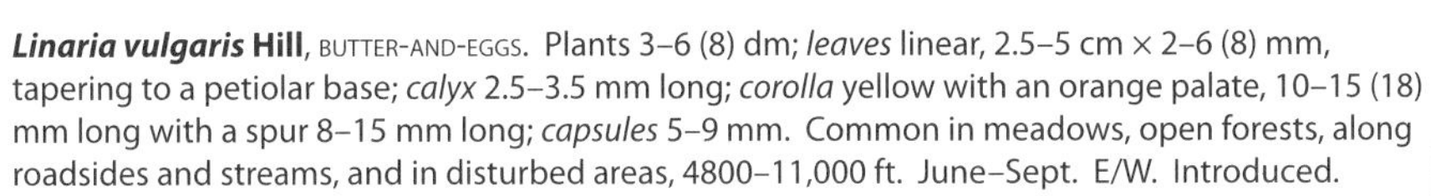

***Linaria vulgaris* Hill**, BUTTER-AND-EGGS. Plants 3–6 (8) dm; *leaves* linear, 2.5–5 cm × 2–6 (8) mm, tapering to a petiolar base; *calyx* 2.5–3.5 mm long; *corolla* yellow with an orange palate, 10–15 (18) mm long with a spur 8–15 mm long; *capsules* 5–9 mm. Common in meadows, open forests, along roadsides and streams, and in disturbed areas, 4800–11,000 ft. June–Sept. E/W. Introduced.

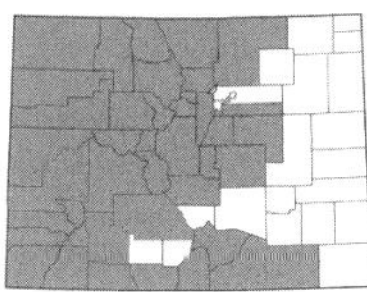

LINDERNIA All. – FALSE PIMPERNEL

Annual herbs; *leaves* opposite, entire or toothed; *flowers* solitary in the leaf axils; *calyx* of 5 distinct segments; *corolla* bilabiate, blue-violet; *stamens* 2 (ours).

***Lindernia dubia* (L.) Pennell var. *anagallidea* (Michx.) Copperr.**, YELLOWSEED FALSE PIMPERNEL. Plants 1–2 (3) dm; *leaves* lanceolate to narrowly ovate or elliptic, 10–20 (30) × 5–10 mm; *pedicels* 0.5–2.5 cm long; *calyx* 3–4.5 mm long in flower, to 5.5 mm long in fruit; *corolla* 4–10 mm long; *capsules* 3.5–6 mm. Uncommon in temporary pools and along pond margins, 5000–7500 ft. Aug.–Oct. E.

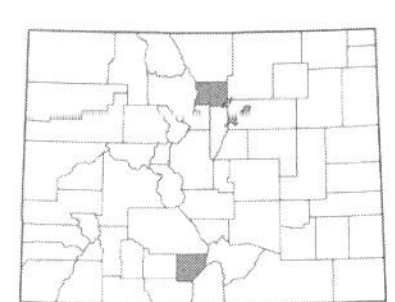

PENSTEMON Schmidel – BEARDTONGUE

Perennial herbs, or sometimes subshrubs; *leaves* opposite, entire to toothed; *calyx* of 5 distinct segments; *corolla* bilabiate, blue, purple, pink, white, red, or yellow; *stamens* 4, didynamous, with a fifth sterile staminode present.

1a. Flowers red or red-orange...2
1b. Flowers white, purple, pink, or blue...5

2a. Inflorescence and calyx glandular-hairy...***P. rostriflorus***
2b. Inflorescence and calyx glabrous to minutely hairy (sometimes the calyx obscurely glandular-hairy but the inflorescence not glandular)...3

3a. Lower lip of the corolla much shorter than the upper lip with strongly reflexed lobes; stem leaves linear; peduncles and pedicels usually elongate (as long or longer than the flowers in length)...***P. barbatus***
3b. Lower lip of the corolla subequal or equal to the upper lip, the lobes not strongly reflexed; stem leaves lanceolate to elliptic or ovate; peduncles and pedicels usually short (shorter than the flowers in length)...4

4a. Corolla 24–33 mm long; leaves not glaucous or thick and fleshy, usually with crisped margins; anther cells (1.4) 1.8–2.5 mm long, glabrous...***P. eatonii***
4b. Corolla 17–22 (25) long; leaves glaucous, thick and fleshy, lacking crisped margins; anther cells 0.6–1.1 mm long, minutely hairy...***P. utahensis***

5a. Plants essentially acaulescent, forming dense, caespitose mats; flowers usually shorter than the leaves, mostly 2–6 on the ends of very short branches (arising from basal leaf rosettes), blue to blue-purple...***P. yampaensis***
5b. Plants caulescent; plants otherwise unlike the above in all respects...6

6a. Middle stem and lower leaves sharply dentate, the upper leaves sessile and clasping, appearing nearly perfoliate; corolla 25–40 mm long, white or pale pink to lavender-pink, with prominent red guide lines, glandular-hairy externally or sometimes glabrous; staminode densely bearded at the tip with yellow hairs to 3 mm long...***P. palmeri***
6b. Plants unlike the above in all respects...7

7a. Inflorescence glandular-hairy on the axis, peduncles, pedicels, calyx, or corolla (the gland hairs with darker, elongate or rounded tips, sometimes interspersed with nonglandular hairs)...**Key 1**
7b. Inflorescence glabrous or pubescent but lacking glandular hairs on the axis, peduncles, pedicels, calyx, or corolla...8

8a. Anther sacs glabrous on the side opposite dehiscence, although often with minutely papillate-toothed sutures...**Key 2**
8b. Anther sacs short to long-hairy on the side opposite dehiscence, sometimes sparsely so, often also with minutely papillate-toothed sutures...**Key 3**

Key 1

(Inflorescence glandular-hairy on the axis, peduncles, pedicels, calyx, or corolla)

1a. Stems glabrous above and below (becoming glandular-hairy only in the inflorescence)...2
1b. Stems retrorsely or spreading-hairy at least below...7

2a. Calyx margins broadly scarious, erose to toothed...3
2b. Calyx margins not scarious or if scarious then narrowly so and the margins not erose or toothed...4

3a. Anther sacs with hirsute to villous, long (ca. 0.5–0.8 mm) hairs on the side opposite dehiscence; plants 1.5–5 dm tall, found below 9000 ft in elevation...***P. scariosus***
3b. Anther sacs glabrous with short hairs on the side opposite dehiscence; plants mostly 1–2 dm tall, found above 10,000 ft in elevation...***P. hallii***

4a. Plants glabrous except for the glandular-hairy external surface of the corolla; flowers deep pink; staminode glabrous or with short, papillate hairs at the tip...***P. utahensis***
4b. Plants glandular-hairy in the inflorescence, on the calyx, and on the external surface of the corolla; flowers blue, blue-purple, lavender, creamy yellow, or white; staminode with a tuft of hairs at the tip...5

5a. Calyx 7–10 (13) mm long; flowers white, creamy yellow, lavender, or black-purple...***P. whippleanus***
5b. Calyx 3–6 mm long; flowers dark blue or blue-purple...6

6a. Corolla 14–20 mm long; anther sacs short-hairy on the side opposite dehiscence; palate on the inside of the lower lip glabrous; plants 4–10 dm tall... ***P. mensarum***
6b. Corolla 10–15 (18) mm long; anther sacs minutely papillose over the entire surface; palate on the inside of the lower lip white-hairy; plants 1–4 dm tall...***P. virens***

7a. Corolla 35–55 mm long, white to pink or pale violet; leaves serrulate; capsule 13–18 mm long...***P. cobaea***
7b. Corolla usually shorter than 35 mm, if longer then the leaves entire and plants of the western slope, variously colored; capsule 3.5–17 mm long...8

8a. Flowers usually few in a short, crowded terminal thyrse subtended by ovate to lance-ovate leaves over 1 cm in length, lilac-purple; plants prostrate or decumbent with ultimately ascending stems, from creeping roots and forming mats, lacking a basal rosette of leaves...***P. harbourii***
8b. Flowers in an elongated thyrse or appearing solitary in the upper leaf axils, variously colored; plants erect or if prostrate and mat-forming then the leaves linear or spatulate to oblanceolate and mostly 1 cm or less in length (excluding the petiole), with or without a basal rosette of leaves...9

9a. Inflorescence conspicuously leafy, the leaves subtending flowers not reduced in size from the base to the top of the inflorescence, the flowers often appearing solitary in each axil (the thyrse not differentiated from the rest of the plant); plants lacking a basal tuft of leaves, subshrubs with decumbent, woody bases and herbaceous, ascending stems or the entire plant prostrate, usually mat-forming, or the plant erect from a woody caudex; leaves 0.5–3 cm long...10
9b. Leaves or bracts of the inflorescence obviously reduced in size from the base to the top, the flowers in terminal, well-differentiated thyrses; plants usually with a basal tuft of leaves but occasionally this lacking, plants of various habits; leaves various...14

10a. Staminode densely bearded with golden-yellow hairs at the apex and the rest of the staminode glabrous or sparsely bearded with yellow or whitish hairs; calyx margin conspicuously scarious, entire or somewhat erose; plants 1–3 (3.7) dm tall, from a stout woody, prostrate basal stem with erect branches; leaves linear or narrowly lanceolate...***P. linarioides* ssp. *coloradoensis***
10b. Staminode densely bearded with golden-yellow hairs for nearly the entire length; calyx margin sometimes conspicuously scarious; plants and leaves various...11

11a. Thyrse not densely leafy, usually with a clear distinction between the basal leaves and the inflorescence; plants usually grayish in appearance and covered with dense, retrorse hairs; stems erect...***P. retrorsus***
11b. Thyrse densely leafy, without a basal tuft of leaves; plants grayish in appearance or not, the leaves glabrous or with retrorse hairs; stems prostrate or erect...12

12a. Leaves glabrous or sparsely hairy above the petiole, sometimes densely short-hairy below but glabrous or nearly so above...***P. crandallii***
12b. Leaves densely short-hairy above and below...13

13a. Leaves linear to linear-oblanceolate, often involute; plants ascending-erect...***P. teucrioides***
13b. Leaves obovate to spatulate or oblanceolate; plants prostrate...***P. caespitosus***

14a. Anther sacs hirsute to villous on the side opposite dehiscence...15
14b. Anther sacs glabrous or papillate...16

15a. Corolla glabrous...***P. saxosorum***
15b. Corolla glandular-hairy...***P. gibbensii***

16a. Plants of the eastern slope, or present on the western slope only in Grand or Saguache counties...17
16b. Plants of the western slope except absent from Grand and Saguache counties...26

17a. Flowers white (sometimes just the tube light pink or blue), (12) 16–20 mm long; corolla palate densely glandular-hairy...***P. albidus***
17b. Flowers blue, violet, or pink, of various lengths, if white then the flowers less than 16 mm long (rare albino *P. virens*); corolla palate bearded with nonglandular or sometimes glandular hairs...18

18a. Lower leaves glabrous throughout or somewhat scabrous on the margins, the margins of at least some usually toothed...19
18b. Lower leaves minutely hairy throughout, or at least on the veins and margins, the margins entire or occasionally toothed...21

19a. Corolla 24–32 (35) mm long; calyx 8–12 mm long; staminode conspicuously exserted, with a dense tuft of tangled hairs at the apex followed by a short glabrous section and then densely hairy with backward pointing hairs; basal leaves similar to stem leaves, linear to lanceolate...***P. jamesii***
19b. Corolla 10–22 mm long; calyx 2–6 mm long; staminode included to barely exserted, densely hairy nearly the entire length with golden-yellow hairs; basal leaves usually different from the stem leaves, lanceolate to oblanceolate or ovate and the stem leaves linear to lanceolate...20

20a. Flowers pale lavender, the corolla 14–22 mm long; stems solitary or few, plants not forming mat-like clumps...***P. gracilis***
20b. Flowers blue to blue-violet, the corolla 10–15 (18) mm long; plants usually forming mat-like clumps...***P. virens***

21a. Staminode included, bearded with short hairs about 1 mm in length; corolla 14–25 mm long, blue to blue-violet, often white below, with 2 longitudinal grooves below and 2 corresponding raised ridges within; calyx lobes usually with prominent, scarious and somewhat erose margins about 0.5 mm wide, the scarious portion white or bluish...22
21b. Staminode barely to conspicuously exserted, bearded with long hairs mostly 1.5–5 mm in length; corolla (16) 18–35 mm long, pale lavender to violet or purplish-blue, lacking grooves below and ridges within; calyx lobes lacking scarious margins or with narrowly scarious margins just at the base...24

22a. Plants lacking a basal rosette; stem leaves well-developed, narrowly lanceolate; capsule 5–7 mm long; plants of the northern counties...***P. radicosus***
22b. Plants with a basal rosette; stem leaves usually reduced in size from the basal leaves, variously shaped; capsule 7–11 mm long; plants of the southern counties...23

23a. Stem leaves linear to linear-oblanceolate, mostly 2 mm wide; corolla pale blue or pale lavender, the palate densely yellow-bearded...***P. griffinii***
23b. Stem leaves lanceolate to lance-ovate, 4–10 (16) mm wide; corolla blue to blue-violet, the palate sparsely yellow-bearded...***P. degeneri***

24a. Stems and calyx with white villous-canescent, spreading hairs, especially above; staminode coiled at the tip, with long hairs to 5 mm in length; corolla 15–35 mm long; plants of Larimer and Weld counties...***P. eriantherus***
24b. Stems and calyx with short, retrorse hairs or glandular, but not villous-canescent; staminode coiled or not, with hairs to 3.5 mm in length; corolla 16–35 mm long; plants of the southern counties, absent from Larimer and Weld counties...25

25a. Staminode densely bearded nearly its entire length with golden-yellow hairs; corolla (16) 18–22 (24) mm long, the throat scarcely inflated and not broadly expanded, 4–9 mm wide (when pressed) at the widest point; basal leaves usually linear to narrowly lanceolate, (1) 2–5 (7) mm wide, the stem leaves usually entire or occasionally a few sharply toothed; capsule 6–10 mm long...***P. auriberbis***
25b. Staminode with a dense tuft of tangled hairs at the tip followed by a short glabrous section and then densely hairy with backward pointing hairs; corolla 24–35 mm long, the throat abruptly and broadly expanded from a narrow tubular base, 8–16 mm wide (when pressed) at the widest point; basal leaves mostly lanceolate, (2) 5–15 (30) mm wide, some stem leaves usually sharply toothed; capsule 14–17 mm long...***P. jamesii***

26a. Corolla 30–37 mm long, pale to deep lavender; staminode conspicuously exserted from the corolla, densely golden-yellow-bearded with short hairs; basal leaves broadly oval and stem leaves oblanceolate to narrowly elliptic...***P. grahamii***
26b. Corolla (7) 10–23 mm long, variously colored; staminode usually included or just reaching the opening of the corolla, sometimes conspicuously exserted and yellow-bearded; leaves various...27

27a. Plants grayish-cinereous throughout, from creeping woody rootstock, forming mats; basal leaves narrowly oblanceolate or spatulate, 0.8–1.5 cm long and 2–4 mm wide...***P. retrorsus***
27b. Plants green, not grayish-cinereous, from branched or thick, woody caudices, not forming mats; basal leaves absent or variously shaped, 1.5–12 cm long and 2–20 mm wide...28

28a. Plants lacking a basal rosette; corolla blue to blue-violet above and white below, with 2 longitudinal grooves below and 2 corresponding raised ridges within; plants of northern counties...***P. radicosus***
28b. Plants with a basal rosette; corolla not white below, with 2 grooves and corresponding ridges present or not; plants variously distributed...29

29a. Corolla with 2 longitudinal grooves below and 2 corresponding raised ridges within, the throat and limb narrow throughout, 2–4 mm wide at the widest point when pressed; calyx 3–5 (6) mm long; anther sacs dark purple or blackish on the back, 0.4–0.8 mm long; capsules 3.5–6 mm long; plants of Moffat County...***P. humilis* ssp. *humilis***
29b. Corolla lacking grooves below and ridges within, the throat expanded from the narrow limb, 3.5–7 mm wide at the widest point when pressed; calyx 5–10 mm long; anther sacs tan to brown or slightly purplish-tinged on the back, 0.6–1.4 mm long; capsules 6–10 mm long; plants absent from Moffat County...30

30a. Staminode sparsely to moderately bearded for half its length or less; anther sacs 1.1–1.4 mm long; lower leaves narrowly to broadly ovate or elliptic...***P. moffatii***
30b. Staminode densely bearded for over half its length; anther sacs 0.6–1.2 mm long; lower leaves lanceolate to narrowly ovate...***P. breviculus***

Key 2

(Anther sacs glabrous on the side opposite dehiscence, often with minutely papillate-toothed sutures)

1a. Leaves filiform, 0.5–1 mm wide; flowers white or pale pink to rose-pink...2
1b. Leaves various but not filiform, over 2 mm wide; flowers various but never white...3

2a. Corolla 11–15 mm long; calyx 4–6 mm long...***P. laricifolius* var. *exilifolius***
2b. Corolla (15) 20–25 mm long; calyx 2.5–4 mm long...***P. ambiguus***

3a. Corolla 35–48 mm long, pink to blue-lavender or pale blue; stem leaves usually orbicular and clasping...***P. grandiflorus***
3b. Corolla 6–26 mm long, variously colored; stem leaves various...4

4a. At least 2 of the 4 fertile stamens conspicuously well-exserted past the orifice of the corolla; anther sacs parallel and sagittate in appearance; leaves glaucous and thick…5
4b. Fertile stamens included to just reaching the orifice of the corolla, or sometimes 2 fertile stamens slightly exserted; anther sacs spreading and becoming opposite; leaves various…6

5a. Corolla 9–15 mm long; all 4 fertile anthers well-exserted; anther sacs 1.2–2 mm long; flowers in a crowded inflorescence…***P. cyathophorus***
5b. Corolla 18–24 mm long; only 2 of the 4 fertile anthers well-exserted; anther sacs 2–3 mm long; flowers in a rather loose inflorescence…***P. harringtonii***

6a. Leaves green, not blue-green and glaucous; staminode not widened at the tip or if widened then glabrous…7
6b. Leaves thick, firm or fleshy, and glaucous, often blue-green, occasionally scabrid-puberulent; staminode usually widened at the tip, bearded at the apex…10

7a. Inflorescence strongly one-sided; anther sacs about 1.5–2 mm long; corolla 17–26 mm long; stems glabrous; stem leaves linear-lanceolate to lanceolate…***P. virgatus* var. *asa-grayi***
7b. Inflorescence not one-sided, the verticillasters symmetrical on the rachis; anther sacs 0.3–1 mm long; corolla 6–16 (20) mm long; stems usually short-hairy below; stem leaves various…8

8a. Inflorescence of loosely flowered verticillasters with evident pedicels; stems glabrous or sparsely hairy below…***P. watsonii***
8b. Inflorescence of 1–7 very densely flowered verticillasters with obscured pedicels; stems usually sparsely hairy below…9

9a. Corolla 6–10 (11) mm long; flowers usually reflexed and pointing down; fertile stamens included or just reaching the orifice of the corolla; capsules 2–5 mm long…***P. procerus* var. *procerus***
9b. Corolla 10–16 (20) mm long; flowers usually horizontal; the 2 longer fertile stamens often exserted past the orifice or sometimes included; capsules 5–10 mm long…***P. rydbergii***

10a. Plants of the eastern slope or present on the western slope only in Saguache County…11
10b. Plants of the western slope except absent from Saguache County…15

11a. Inflorescence distinctly one-sided with the flowers all coming off of one side of the rachis…12
11b. Inflorescence not one-sided, the verticillasters symmetrical on the rachis…13

12a. Corolla blue; basal leaves rounded at the tips with an abrupt, mucronate point…***P. versicolor***
12b. Corolla pink to lavender or magenta; basal leaves tapering to an acute point at the tip…***P. secundiflorus***

13a. Stem leaves linear to linear-lanceolate; stem and leaves often scabrid-puberulent; corolla blue…***P. angustifolius***
13b. Stem leaves lanceolate to ovate or obovate; stem and leaves glabrous or rarely scabrid-puberulent; corolla various…14

14a. Inflorescence bracts with a short-acuminate to acute tip; corolla usually pale lavender or sometimes pale pink or very pale blue…***P. buckleyi***
14b. Inflorescence bracts tapering to a long-acuminate tip; corolla pink, lavender, or blue to deep blue…***P. angustifolius* var. *caudatus***

15a. Staminode broadly expanded and densely hairy at the apex with tangled hairs about 0.5–2 mm in length; plants of northwestern Colorado…16
15b. Staminode narrowly expanded at the apex, sparsely to densely hairy with straight hairs less than 1 mm long; plants variously distributed but not limited to northwestern Colorado…17

16a. Stem leaves spatulate, broadly lanceolate, ovate, or nearly orbicular, with a rounded, mucronate, or abruptly acute tip…***P. pachyphyllus***
16b. Stem leaves lanceolate to ovate with an acuminate, tapering tip…***P. osterhoutii***

17a. Corolla (10) 12–15 mm long; plants 0.8–1.8 (3) dm tall, often decumbent at the base; anther sacs 0.7–1 mm long; staminode slightly exserted from the corolla…***P. arenicola***
17b. Corolla 14–22 mm long; plants 1–5 (6) dm tall, usually not decumbent at the base; anther sacs (0.9) 1.1–1.5 mm long; staminode included or just reaching the orifice of the corolla…18

18a. Peduncles short (to 4 mm long), or obsolete and with the pedicels mostly 1–10 mm long; lower inflorescence bracts usually longer than the flowers and pedicels; basal and lower stem leaves linear to oblanceolate; upper stem leaves linear, linear-lanceolate, lanceolate, or lance-ovate; calyx lobes narrowly lanceolate; plants variously distributed... ***P. angustifolius***
18b. Both pedicels and peduncles elongated, the peduncles mostly 5–20 mm long; lower inflorescence bracts usually shorter than the flowers and pedicels, obviously reduced in size from the stem leaves; basal and lower cauline leaves ovate to obovate or oblong; upper stem leaves lanceolate to ovate; calyx lobes broadly lanceolate or ovate; plants of the southwestern counties...***P. lentus* var. *lentus***

Key 3

(Anther sacs short to long-hairy on the side opposite dehiscence)

1a. Plants mat-forming, from a slender, decurrent underground stem and caudex, the above-ground portion usually less than 4 cm; corolla white to light lavender; leaves obovate to elliptic, 1.2–2.5 (3.2) cm long, glabrous, glaucous, and rather succulent; plants restricted to Garfield County in distribution...***P. debilis***
1b. Plants not mat-forming or from slender, underground stems and caudices, plants over 8 cm; corolla blue, blue-violet, or occasionally pink; leaves various; plants variously distributed...2

2a. Leaves all narrowly linear, 1–1.5 mm wide, the margins involute or the leaves folded; corolla 12–15 mm long; anther sacs 1–1.2 mm long; plants restricted to Grand County in distribution...***P. penlandii***
2b. Leaves various but unlike the above; corolla 14–40 mm long; anther sacs 1–3 mm long; plants variously distributed...3

3a. Anther sacs sparsely to densely villous with long, tangled hairs usually longer than the width of the anther sac (usually 1 mm or more in length) sometimes nearly obscuring the surface of the anther sac; fertile stamens usually conspicuously exserted; corolla (18) 24–38 mm long...4
3b. Anther sacs sparsely to moderately hairy with straight hairs shorter than the width of the anther sac (usually 0.5 mm or less in length), these not tangled or obscuring the surface of the anther sac; fertile stamens included or just reaching the orifice of the corolla or sometimes exserted; corolla 12–40 mm long...5

4a. Lower peduncles elongate and spreading; flowers pale blue to lavender; inflorescence usually not densely crowded...***P. comarrhenus***
4b. Lower peduncles short and ascending; flowers blue to blue-violet or purple; inflorescence usually densely crowded with flowers...***P. strictus***

5a. Corolla 24–35 (40) mm long, hairy inside on the palate; anther sacs 1.8–3 mm long; inflorescence strongly one-sided; staminode often notched at the apex; calyx margins usually broadly scarious and conspicuously erose or toothed; plants of the eastern slope or present on the western slope only in Grand Co.... ***P. glaber***
5b. Corolla 14–25 mm long, glabrous within; anther sacs 1–1.7 mm long; inflorescence one-sided or not; staminode rounded at the apex; calyx margins scarious but usually entire, occasionally somewhat erose or toothed; plants of the western slope, including Grand County...6

6a. Stems densely short-hairy...***P. fremontii***
6b. Stems glabrous...7

7a. Calyx 3.5–4.5 mm long in flower; corolla 14–18 (20) mm long; most leaves usually with crisped or undulate margins; inflorescence glabrous throughout...***P. cyanocaulis***
7b. Calyx 5–7 (10) mm long in flower; corolla 18–24 mm long; leaves usually lacking crisped or undulate margins; inflorescence (pedicels and upper bracts) sparsely hairy to glandular...***P. saxosorum***

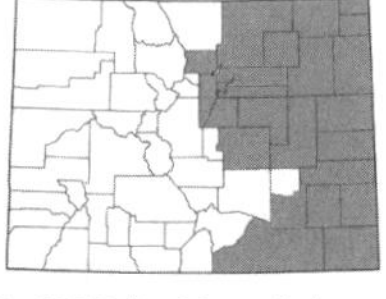

***Penstemon albidus* Nutt.**, WHITE PENSTEMON. Erect perennials, 1–5.5 dm; *stems* retrorsely hairy below and glandular-hairy near the inflorescence; *leaves* lanceolate to obovate, 2–9 (11) cm × (0.4) 0.7–2 cm, glabrous to scabrous, with entire to toothed margins; *thyrse* 4–25 (30) cm long, with 3–10 verticillasters, glandular-hairy; *calyx* 4–7 mm long, glandular-hairy; *corolla* (12) 16–20 mm long, white to faintly pink, glandular-hairy externally and internally, the palate rounded and densely glandular-hairy; *staminode* included, sparsely to moderately yellow-bearded; *anthers* glabrous, the anther sacs 0.7–0.9 mm long, black. Common on the shortgrass prairie and eastern plains, 3500–6800 ft. May–July. E. (Plate 58)

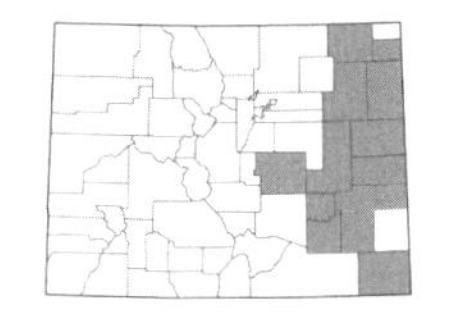

***Penstemon ambiguus* Torr.**, GILIA BEARDTONGUE. Spreading shrubs, 3–5 (8) dm; *stems* woody well above the base, glabrous to minutely hairy; *leaves* filiform, 1–3 (5) cm × 0.5–1 mm, glabrous to minutely hairy, with entire, involute margins; *thyrse* with 1 (2)-flowered cymes; *calyx* 2.5–4 mm long, glabrous; *corolla* (15) 20–25 mm long, white to rose-pink, glabrous externally, red-hairy internally at the orifice; *staminode* red, included, glabrous; *anthers* glabrous. Common on the eastern plains, 3500–5600 ft. June–Aug. E.

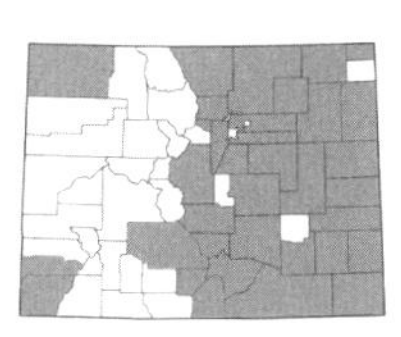

***Penstemon angustifolius* Nutt. ex Pursh**, BROADBEARD PENSTEMON. Erect, short-lived perennials, 1–4 (6) dm; *stems* glabrous and glaucous; *basal leaves* linear to oblanceolate, (4) 5–9 cm × 2–16 mm, glabrous and glaucous, with entire margins; *cauline leaves* linear to lanceolate or narrowly ovate, 3–10 cm × 3–20 (40) mm, glabrous and glaucous, with entire margins; *thyrse* with 6–16 verticillasters, glabrous; *calyx* (4) 5–7 mm long, to 8.5 mm long in fruit, glabrous; *corolla* 15–20 (22) mm long, blue to blue-purple or pink, glabrous externally, palate sometimes with a few scattered whitish hairs, *staminode* reaching orifice, bearded apically; *anthers* glabrous, sutures papillate-toothed. Found on sandy or rocky hillsides, 3500–7000 ft. May–June. E/W. (Plate 58)

There are two varieties of *P. angustifolius* present on the western slope. Varieties *vernalensis* and *venosus* have previously been separated from varieties *angustifolius* and *caudatus* respectively. However there is considerable morphological overlap between these varieties and they do not appear worthy of taxonomic rank:

1a. Stem leaves linear to narrowly lanceolate; flowers blue to blue-purple...**var. *angustifolius***. [var. *vernalensis* N. Holmgren]. Common in grasslands and on sandy or rocky hillsides, known from Moffat Co. on the western slope, 3500–7000 ft. May–June. E/W.

1b. Stem leaves lanceolate to ovate; flowers pink or lavender, sometimes bluish-tinged...**var. *caudatus* (Heller) Rydb.** [var. *venosus* (D.D. Keck) N. Holmgren]. Common on sandy slopes, known from Dolores and Montezuma cos. on the western slope, 3500–7000 ft. May–June. E/W.

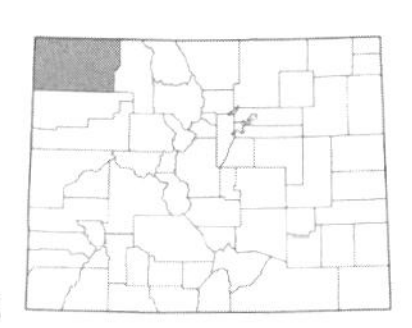

***Penstemon arenicola* A. Nelson**, SAND PENSTEMON. Ascending perennials, often decumbent at the base, 0.8–1.8 (3) dm; *stems* glabrous and glaucous; *leaves* narrowly elliptic to narrowly lanceolate, 2.5–5 (7) cm × 4–11 (15) mm, glabrous and glaucous, with entire margins; *thyrse* with 4–9 verticillasters, glabrous; *calyx* 4–6 (7.5) mm long, glabrous; *corolla* (10) 12–15 mm long, blue to bluish purple, with red-violet guide lines, glabrous externally, palate sparsely white bearded; *staminode* slightly exserted, densely golden-yellow-bearded at the apex; *anthers* glabrous, sutures papillate-toothed. Uncommon in sandy soil, 5500–6500 ft. May–June. W.

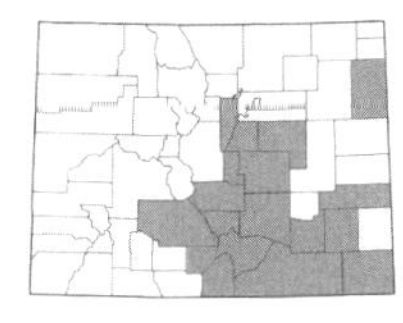

***Penstemon auriberbis* Pennell**, COLORADO BEARDTONGUE. Erect to assurgent perennials, 1–3 (3.5) dm; *stems* retrorsely short-hairy below, glandular-hairy near inflorescence; *leaves* linear to lanceolate, (1.5) 3–6 (10) cm × (1) 2–5 (7) mm, with entire or occasionally toothed margins; *thyrse* with 3–8 verticillasters, somewhat secund; *calyx* (6) 7–9 mm long, densely glandular-hairy; *corolla* (16) 18–22 (24) mm long, pale lilac to purplish-blue, glandular pubescent externally, pilose internally with yellow hairs on lower lip (beard); *staminode* barely to conspicuously exserted, bearded most of its length with stiff or twisted yellow-orange hairs to 2.5 mm long; *anthers* papillose along sutures. Found on dry hillsides and shale slopes and in grasslands, 4000–8600 ft. May–June. E.

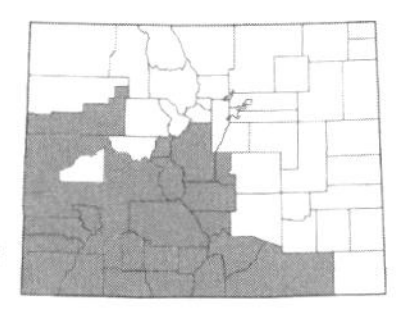

***Penstemon barbatus* (Cav.) Roth**, BEARDLIP PENSTEMON. Erect or ascending , short-lived perennials, 3–10 dm; *leaves* linear to lanceolate, 3–8 (14) cm × (6) 12–25 (30) mm wide, glabrous or short-hairy below, margins entire; *thyrse* with 6–10 verticillasters, glabrous, tending to be secund; *calyx* (3) 4–6 (9) mm long, glabrous to obscurely glandular-hairy; *corolla* 26–32 (36) mm long, red to orange-red, glabrous externally, palate yellow-bearded to glabrous; *staminode* included, glabrous; *anthers* pale yellow, glabrous or sparsely to moderately villous on sides, sutures papillate-toothed. Found on rocky slopes and in meadows and pinyon-juniper woodlands, 5200–10,000 ft. June–Aug. E/W. (Plate 58)

There are two varieties of *P. barbatus* in Colorado:

1a. Anthers hairy...**var. *trichander* A. Gray**. Found on rocky slopes and in pinyon-juniper woodlands in the southwestern counties, 5200–9800 ft. June–Aug. W.

1b. Anthers glabrous...**var. *torreyi* (Benth.) A. Gray**. Found on rocky or sandy slopes, in meadows, and along roadsides, known from the southwestern and southcentral counties, 6000–10,000 ft. June–Aug. E/W.

***Penstemon breviculus* (D.D. Keck) Nisbet & R. Jackson**, SHORTSTEM BEARDTONGUE. [*P. parviflorus* Pennell]. Erect or ascending perennials, 0.8–2 (3.5) dm; *stems* retrorsely hairy; *leaves* mostly basal, linear to narrowly lanceolate or oblanceolate, 3.5–7 cm × 3–12 mm, sometimes glabrate, rarely with few-toothed margins; *thyrse* glandular-hairy; *calyx* 5–7 mm long, glandular-hairy; *corolla* 10–15 (18) mm long, dark blue to purple, glandular-hairy externally, palate yellowish-white-bearded; *staminode* included, golden-yellow-bearded; *anthers* glabrous. Found in dry, sandy or clay soil and pinyon-juniper communities, 4900–6800 ft. May–June. W.

The type specimen of *P. parviflorus* was collected at the same locality as *P. breviculus* and is similar to *P. breviculus* morphologically except that it has leaves in whorls of 3 instead of opposite. *Penstemon parviflorus* has not been seen since the type collection was made and is most likely an odd form of *P. breviculus*. Plants that we have called *P. ophianthus* Pennell in Colorado are *P. breviculus*. True *P. ophianthus* has a very exserted staminode.

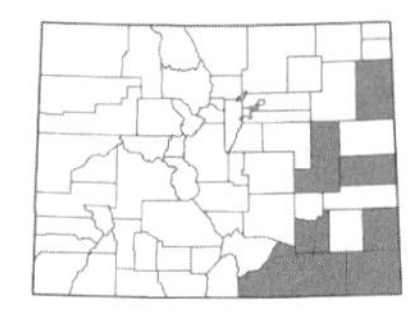

***Penstemon buckleyi* Pennell**, BUCKLEY'S PENTSEMON. Erect or ascending perennials, 1.5–5.5 (8.1) dm; *stems* glabrous and glaucous; *basal leaves* oblanceolate to spatulate, 1.9–11.6 (15) cm × 0.3–2.2 (3.1) cm, glabrous, thick and glaucous; *cauline leaves* lance-ovate to ovate, 2–9.5 cm × 1–3 cm, glabrous, thick and glaucous, with entire margins; *thyrse* with (2) 4–20 (35) verticillasters, glabrous; *calyx* 3.5–6 mm long, glabrous and glaucous; *corolla* (12) 14–20 mm long, pale pink or lavender to very pale blue, glabrous externally and internally; *staminode* included or reaching the orifice, densely to moderately golden-yellow-bearded for half its length; *anthers* minutely papillose. Uncommon on the shortgrass prairie of the southeastern plains, 3500–4800 ft. May–June. E.

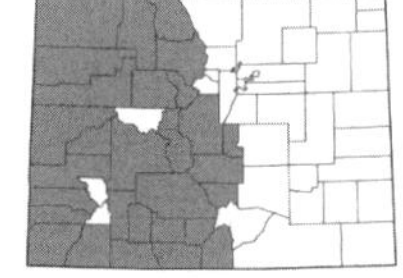

***Penstemon caespitosus* Nutt. ex A. Gray**, MAT PENSTEMON. Prostrate, mat-forming perennials, 0.2–0.8 (1) dm; *stems* rooting at nodes, puberulent with retrorsely curved hairs; *leaves* linear to obovate or spatulate, 0.4–1 (2) cm × 1–3 (4.5) mm, puberulent with retrorsely curved hairs; *thyrse* with subsessile, 1-flowered cymes, glandular-hairy; *calyx* 4–6.5 (8.5) mm long, retrorse-puberulent with interspersed viscid hairs; *corolla* 10–17 (21) mm long, blue to lavender externally, white with red lines internally, sparsely glandular-hairy externally, palate pale yellow-bearded; *staminode* reaching the orifice, golden-yellow-bearded; *anthers* glabrous, sutures papillate-toothed. Common on rocky, sandy, or shale slopes or on dry, open hillsides, 6000–9500 ft. May–Aug. E/W. (Plate 58)

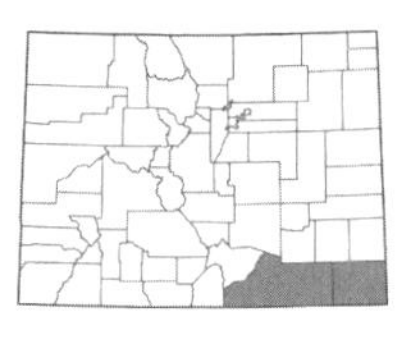

***Penstemon cobaea* Nutt.**, COBAEA BEARDTONGUE. Erect perennials, (1.5) 2.5–6.5 (10) dm; *stems* retrorsely puberulent below, glandular-hairy above; *basal leaves* lanceolate to spatulate, 3.5–18 (26.4) cm × 0.8–5.5 (7.6) cm, thick, margins subentire to sharply toothed; *cauline leaves* lanceolate to ovate, 3.5–12 (15) cm × 1–4.5 (5.4) cm, thick, with subentire to sharply toothed margins; *thyrse* with 3–6 (8) distinct to indistinct verticillasters; *calyx* (8) 10–16 mm long, densely glandular-hairy; *corolla* 35–55 mm long, white or pink to pale violet-purple, glandular-hairy; *staminode* included or slightly exserted, sparsely bearded nearly entire length with golden-yellow hairs; *anthers* light brown to brown, glabrous. Uncommon on the eastern plains, perhaps not persisting, 4200–4500 ft. April–June. E.

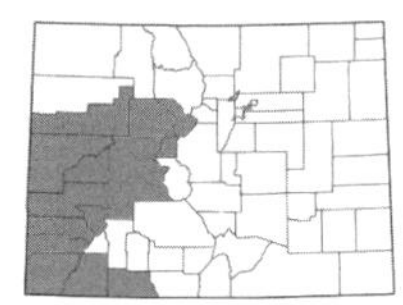

***Penstemon comarrhenus* A. Gray**, DUSTY BEARDTONGUE. Erect perennials, 4–8 (12) dm; *stems* glabrous to short-hairy below; *leaves* oblanceolate to linear, 4–8 (12) cm × 7–20 (30) mm, with entire margins; *thyrse* somewhat secund, glabrous; *calyx* (2.5) 3.5–6 (7) mm, glabrous; *corolla* 25–35 (38) mm long, pale blue to lavender with a pinkish-white throat, glabrous externally and internally; *staminode* white, included, glabrous to sparsely bearded on distal portion; *anthers* densely villous-woolly with white hairs, sutures papillate-toothed. Found on dry slopes, often with pinyon or sagebrush, sometimes found in moist places, 5000–9000 ft. June–Aug. W.

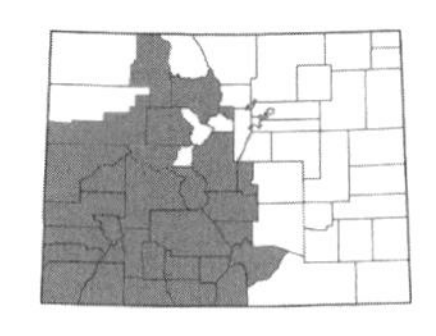

***Penstemon crandallii* A. Nelson**, CRANDALL'S BEARDTONGUE. Small, decumbent subshrubs, 0.7–1.7 dm; *stems* with minute, retrorse hairs; *leaves* narrowly oblanceolate to narrowly obovoid, 1–2.8 (3.5) cm × 1.5–3.4 (4.5) mm, with minute, retrorse hairs; *thyrse* secund; *calyx* 4.5–7 mm long, to 8.5 mm long in fruit, glandular-hairy; *corolla* 14–20 mm long, light to dark blue, glandular-hairy externally, palate pale yellow-bearded; *staminode* densely golden-yellow-bearded; *anthers* glabrous, sutures papillate-toothed. Common on rocky slopes and in open, dry places, 6000–12,000 ft. May–Aug. E/W.

There are 3 intergrading subspecies of *P. crandallii* in Colorado:

1a. Leaves all linear; sepals usually prominently scarious-margined and erose; stems ascending-erect...**ssp. *glabrescens* (Pennell) D.D. Keck**. [*P. ramaleyi* A. Nelson]. Found in rocky soil and open, dry places, known from the southwestern counties, 6000–10,500 ft. May–July. E/W.
1b. Leaves spatulate to oblanceolate; sepals usually narrowly or obscurely scarious-margined; stems ascending-erect or prostrate...2

2a. Stems ascending-erect; leaves usually narrowly oblanceolate to narrowly spatulate, with an ill-defined petiole...**ssp. *crandallii***. [ssp. *atratus* D.D. Keck]. Found on open, dry slopes and rocky hillsides, 6000–12,000 ft. June–Aug. E/W.
2b. Stems prostrate, often forming broad mats; leaves usually obovate to elliptic, with a clearly defined petiole...**ssp. *procumbens* (Greene) D.D. Keck**. Found on gravelly slopes and dry, flat places, known from Gunnison Co. northeast to Grand Co., 7000–12,000 ft. June–Aug. E/W.

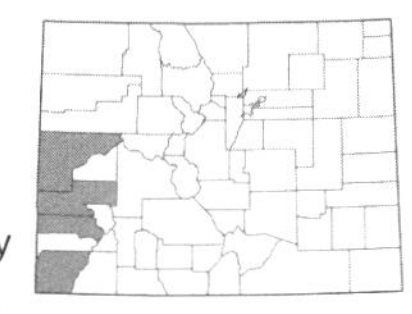

Penstemon cyanocaulis **Payson**, BLUESTEM BEARDTONGUE. Erect perennials, 2–6 dm; *stems* glabrous; *basal leaves* oblanceolate to spatulate, 5–10 cm × 10–15 (25) mm, glabrous, with entire, often crisped, margins; *cauline leaves* lanceolate to narrowly elliptic, 2–7 cm × 4–15 mm, glabrous, with entire, often crisped margins; *thyrse* with 6–11 verticillasters, glabrous; *calyx* 3.5–4.5 mm long, to 5.5 mm long in fruit, glabrous; *corolla* 14–18 (20) mm long, blue to blue-violet, glabrous externally and internally; *staminode* yellow-bearded at apex; *anthers* sparsely to moderately pilose on sides, sutures papillate-toothed. Found on dry slopes, often with pinyon-juniper, 4200–9000 ft. May–July. W.

Penstemon cyathophorus **Rydb.**, SAGEBRUSH BEARDTONGUE. Erect perennial, 2–6 dm; *stems* glabrous and glaucous; *basal leaves* oblanceolate or spatulate, 2–6 cm × 10–20 mm, glabrous and glaucous, with entire margins; *cauline leaves* ovate or orbicular, glabrous and glaucous; *thyrse* cylindrical and crowded; *calyx* 4–7 mm long, glabrous; *corolla* 9–15 mm long, violet-blue, glabrous externally and internally; *staminode* somewhat exserted, densely bearded near apex; *anthers* glabrous, well-exserted. Uncommon in rocky soil of sagebrush meadows, 7000–8500 ft. May–June. E/W.

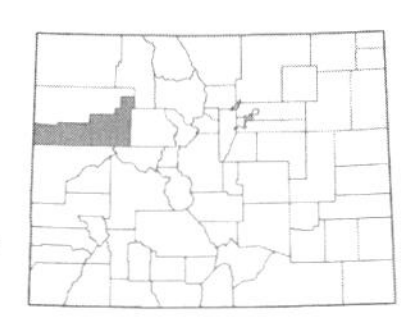

Penstemon debilis **O'Kane & J. Anderson**, PARACHUTE PENSTEMON. Mat-forming perennials; *stems* glandular-hairy distally, less so below; *leaves* entirely cauline, obovate, elliptic, or oblanceolate, 1.2–2.5 (3.2) cm × 4–12 mm, glabrous and strongly blue-glaucous; *thyrse* with 2–4 verticillasters, subsecund; *calyx* 7–11.5 mm long, glabrous; *corolla* (14) 17.5–20 mm long, lavender to nearly white, glabrous externally and internally; *staminode* barely included to scarcely exserted, sparsely golden-hairy on upper surface for most its length with tuft at apex; *anthers* bright magenta at anthesis, sparsely hispid. Uncommon on white shale talus slopes, 8000–9000 ft. June–July. W. Endemic.

Listed as threatened under the Endangered Species Act.

Penstemon degeneri **Crosswhite**, DEGENER'S BEARDTONGUE. Erect perennials, 2.5–4 dm; *stems* short-hairy; *leaves* lanceolate to lance-ovate, 2.5–6.5 (7) cm × 4–10 (16) mm, minutely hairy; *thyrse* with 2–10 flowers, sparingly glandular; *calyx* 4–7 mm long; *corolla* 14–19 mm long, blue, glandular but soon glabrescent; *staminode* included, pubescent for roughly half its length with golden hairs. Found in pinyon-juniper woodlands, meadows, and grasslands, 6000–9500 ft. June–July. E. Endemic.

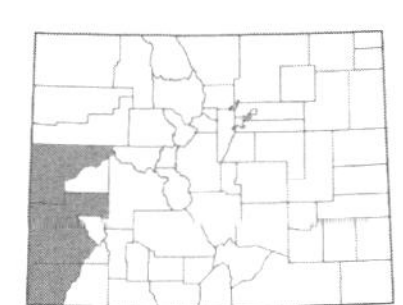

Penstemon eatonii **A. Gray**, FIRECRACKER PENSTEMON. Erect, short-lived perennials, 4–10 dm; *basal and lower cauline leaves* elliptic to broadly obovate, 5–10 (20) cm × (10) 16–28 (50) mm, glabrous or minutely hairy; *upper cauline leaves* broadly lanceolate to ovate, glabrous or minutely hairy; *thyrse* with 4–12 verticillasters, secund, glabrous to minutely hairy; *calyx* (3.5) 4–5 (6) mm long, glabrous; *corolla* 24–30 (33) mm long, red to orange-red, glabrous; *staminode* included, glabrous to sparsely bearded at apex; *anthers* minutely hairy on sides, sutures papillate-toothed. Found on dry, rocky slopes and mesa rims or sometimes with pinyon-juniper, 4700–7000 ft. May–June. W.

There are two varieties of *P. eatonii* in Colorado:

1a. Plants glabrous throughout...**var. *eatonii***. 5000–6000 ft. May–June. W.
1b. Plants minutely hairy...**var. *undosus* M.E. Jones**. 4700–7000 ft. May–June. W.

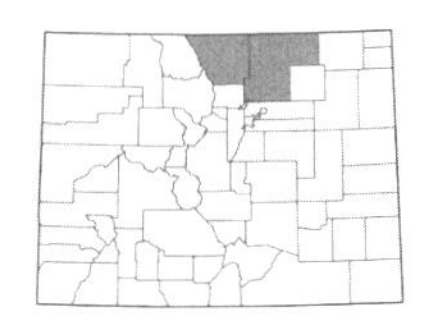

Penstemon eriantherus **Pursh**, FUZZYTONGUE PENSTEMON. Erect perennials, 0.8–2 (3.5) dm; *stems* puberulent with retrorse hairs, sometimes villous above; *basal and lower cauline leaves* oblanceolate to obovate, to 10 cm × 4–20 mm, puberulent with retrorse hairs, with entire to toothed margins; *upper cauline leaves* linear to lanceolate, 2.5–7 cm × 3–10 mm, cinereous-puberulent with retrorse hairs; *thyrse* with 2–5 densely flowered verticillasters, glandular; *calyx* 7–12 mm long, glandular; *corolla* 15–35 mm long, pale lavender to violet, glandular-hairy, palate yellow-bearded; *staminode* coiled apically, densely yellow-bearded; *anthers* glabrous. Found in grasslands, 6000–7200 ft. May–July. E. (Plate 59)

There are potentially two varieties of *P. eriantherus* in Colorado:

1a. Corolla (20) 25–35 mm long, the tube widest at the throat at the base of the corolla lobes...**var. *eriantherus***
1b. Corolla 15–25 (28) mm long, the tube constricted at the throat such that the tube is widest below the base of the corolla lobes...**var. *cleburnei* (M.E. Jones) Dorn**, CLEBURN'S PENSTEMON. Potentially present in northern Colorado.

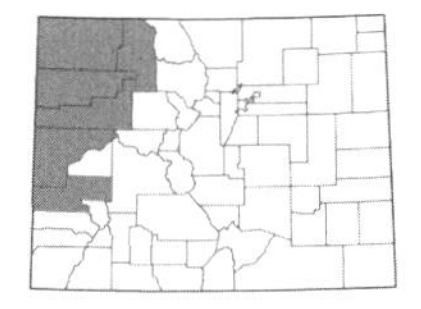

Penstemon fremontii **Torr. & A. Gray**, FREMONT'S PENSTEMON. Erect perennials, 0.8–2.5 (4) dm; *stems* densely cinereous-puberulent; *leaves* oblanceolate or lanceolate, 3–7.5 (10) cm × 5–18 (25) mm, cinereous-puberulent; *thyrse* densely short-hairy on bracts, axils, and sometimes on peduncles and pedicels; *calyx* (3.5) 4–6.5 mm long, glabrous to minutely hairy; *corolla* (14) 18–22 (25) mm long, blue to blue-purple, glabrous externally and internally; *staminode* included, yellow-bearded apically; *anthers* white-hispid on the sides, sutures papillate-toothed. Common on dry, sandy or shale slopes, often with sagebrush, 4900–8000 ft. May–July. W.

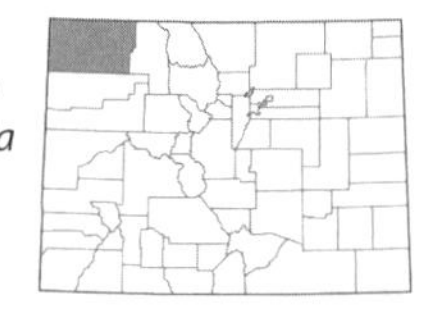

Penstemon gibbensii **Dorn**, GIBBENS' BEARDTONGUE. Erect perennials, 1–2 dm; *stems* hairy; *leaves* narrowly oblanceolate to linear, or appearing so due to involute margins, 3–5.5 (6) cm × 3–7 mm; *thyrse* with ca. 5 verticillasters, glandular-pubescent; *calyx* 6–8 mm long, white-puberulent; *corolla* 14–20 mm long, blue to blue-purple, hairy externally and internally; *staminode* sparingly villous apically with white, tortuous hairs; *anthers* villous with white hairs. Uncommon on white sandy loam and shale, 5500–7000 ft. June–July. W.

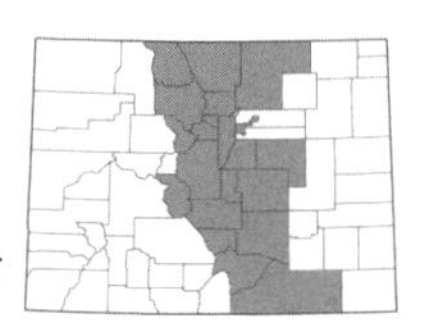

Penstemon glaber **Pursh**, SAWSEPAL PENSTEMON. Erect perennials, (1) 5–6.5 (8) dm; *stems* glabrous to minutely hairy; *basal leaves* lanceolate to obovate, 2–8 (15.5) cm × 0.5–2 (4.5) cm, glabrous or hairy, with entire margins; *cauline leaves* linear-lanceolate to lanceolate, 3–12 (15) cm × (0.7) 1.2–3.5 (4.8) cm, glabrous or hairy; *thyrse* with (5) 8–12 verticillasters, secund; *calyx* 2–7 (10) mm long, glabrous to minutely hairy; *corolla* 24–35 (40) mm long, bluish-purple or occasionally pink or deep blue, glabrous externally, palate glabrous to villous with white eglandular hairs; *staminode* included or slightly exserted, glabrous or terminal 2 mm sparsely bearded; *anthers* hirsute or rarely glabrous, papillate along sutures. Common in meadows and on open slopes, 5500–11,000 ft. June–Aug. E/W. (Plate 59)

There are two varieties of *P. glaber* in Colorado. *Penstemon glaber* var. *glaber* has long been reported for Colorado, but this was based on misidentifications of var. *alpinus*. *Penstemon glaber* var. *glaber* occurs in Wyoming and is similar to var. *alpinus* in having a rounded or obscurely notched staminode apex, but the calyx in var. *glaber* is much shorter (2–4 mm long) with a rounded or abruptly acuminate tip:

1a. Staminode distinctly notched at the tip; leaves ovate, usually with crisped or undulate margins upon drying; corolla 30–40 mm long...**var. *brandegeei* (Porter) Freeman**, BRANDEGEE'S PENSTEMON. [*P. brandegeei* (Porter) Porter ex Rydb.]. Found in mountain meadows and on open slopes, known from the southern Front Range, 5500–9800 ft. June–July. E.

1b. Staminode rounded or obscurely to slightly notched at the tip; leaves usually lanceolate or sometimes ovate, usually lacking crisped or undulate margins; corolla 24–35 mm long...**var. *alpinus* (Torr.) A. Gray**, ALPINE SAWSEPAL PENSTEMON. Common on open slopes and in meadows, known from along the Front Range south to Fremont Co., known on the west slope from Grand Co., 5500–11,000 ft. June–Aug. E/W.

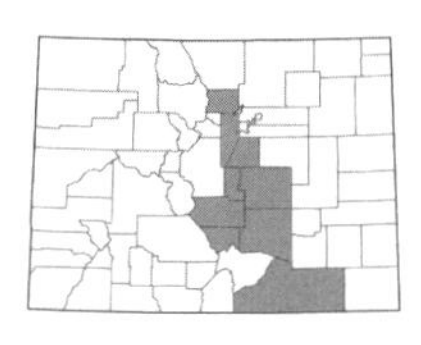

Penstemon gracilis **Nutt.**, LILAC PENSTEMON. Erect perennials, (1.5) 2–5 dm; *stems* with minute, retrorse hairs below and glandular-hairy near the inflorescence; *basal leaves* lanceolate to ovate or oblanceolate 2.5–7.5 cm × 0.4–1.5 cm, with subentire to toothed margins; *cauline leaves* linear to lanceolate, 2.5–8 (9) cm × (0.2) 0.4–1 (1.5) cm; *thyrse* with (2) 3–5 (7) verticillasters; *calyx* 4–6 mm long, glandular-hairy; *corolla* 14–22 mm long, pale lavender to mauve, glandular-hairy externally, palate bearded with whitish eglandular hairs; *staminode* reaching orifice or barely exserted, the terminal 7–9 mm densely bearded; *anthers* glabrous, sutures papillate. Uncommon in open forests and dry grasslands, 5800–9000 ft. May–July. E.

Penstemon grahamii **D.D. Keck**, UINTA BASIN BEARDTONGUE. Erect, short-lived perennials, 0.7–1.8 dm; *stems* retrorsely cinereous-puberulent; *leaves* mostly cauline, oblanceolate to elliptic, 2–4 cm × 20 mm, glabrous; *thyrse* glandular-hairy; *calyx* 7–9 mm long, to 14 mm long in fruit, glandular-hairy; *corolla* 30–37 mm long, pale to deep lavender with dark violet lines, glandular-hairy externally, palate sparsely white-bearded; *staminode* exserted, golden-yellow-bearded; *anthers* essentially glabrous. Rare on shale slopes, 5800–6000 ft. May–June. W.

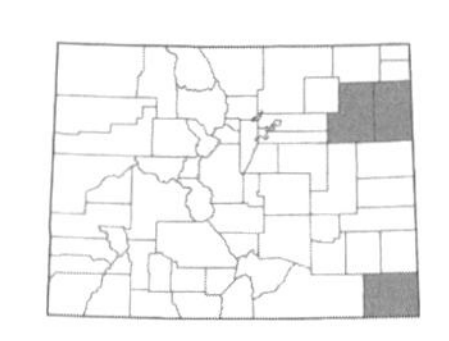

Penstemon grandiflorus **Nutt.**, LARGE BEARDTONGUE. Erect perennials, (4) 5–9.5 (12) dm; *stems* glabrous and glaucous; *basal leaves* spatulate or obovate, 3–16 cm × 6–50 mm, glabrous and glaucous, with entire margins; *cauline leaves* spatulate to orbicular,1.8–9 (11) cm × 15–50 mm, glabrous and glaucous; *thyrse* with 3–7 (9) verticillasters, interrupted; *calyx* 7–11 mm long, glabrous; *corolla* 35–48 mm long, pink to blue-lavender or pale blue, glabrous externally and internally; *staminode* included or reaching orifice, tip bearded; *anthers* minutely papillose on sutures. Uncommon on the eastern plains, 3500–4200 ft. May–July. E.

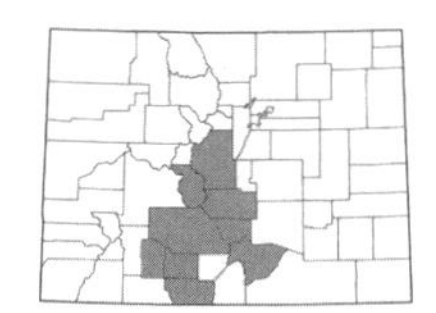

Penstemon griffinii **A. Nelson**, GRIFFIN'S BEARDTONGUE. Erect perennials, 2–5 dm; *stems* minutely hairy at least below, glandular-hairy in inflorescence; *basal leaves* mostly oblanceolate to elliptic, 2–6 cm × 5–10 mm, with entire margins; *cauline leaves* linear-oblanceolate to lance-linear; *thyrse* few flowered, glandular-hairy; *calyx* 4–6 mm long, sparsely glandular-hairy; *corolla* 15–25 mm long, pale blue or purplish, obscurely to moderately glandular-hairy externally, palate densely golden-bearded; *staminode* not exserted, densely bearded most of its length; *anthers* glabrous. Found on rocky hillsides, in open forests, grasslands, and meadows, 7400–10,000 ft. June–Aug. E/W.

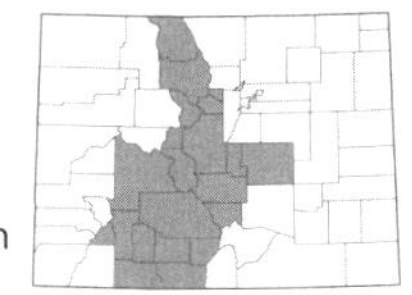

Penstemon hallii **A. Gray**, HALL'S BEARDTONGUE. Erect perennials, 1–2 dm; *stems* sparsely glandular-hairy above; *leaves* oblanceolate to linear-lanceolate, 2–5 cm × 3–8 mm; *thyrse* few-flowered to many-flowered, secund; *calyx* 5–9 mm long, glandular-hairy; *corolla* 18–24 mm long, bluish to blue-purple, minutely glandular to glabrous externally, glabrous internally; *staminode* often exserted, short-bearded its entire length, densely so at the apex; *anthers* minutely hairy. Common on alpine scree slopes and in meadows, 10,000–14,000 ft. June–Aug. E/W. Endemic. (Plate 59)

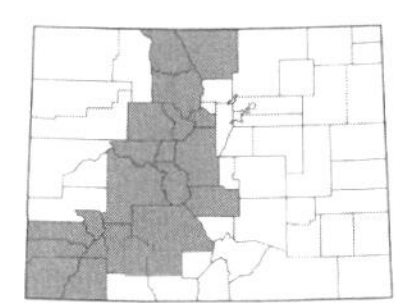

Penstemon harbourii **A. Gray**, HARBOUR'S BEARDTONGUE. Weakly erect perennial, 0.5–1.5 dm; *stems* minutely hairy below, viscid-villous above; *leaves* spatulate to oblanceolate below, ovate to lance-ovate above,1–2 cm × 5 mm; *thyrse* rather short (2–4 cm long) and leafy, mostly few-flowered and crowded, secund; *calyx* 6–9 mm long, canescently viscid; *corolla* 15–20 mm long, lilac-purple, glandular-hairy externally, densely bearded within on lower lip; *staminode* densely bearded nearly its entire length; *anthers* glabrous. Found on alpine scree or talus slopes, 10,500–13,800 ft. July–Aug. E/W. Endemic.

Penstemon harringtonii **Penland**, HARRINGTON'S PENSTEMON. Erect perennials, 3–7 dm; *stems* glabrous and glaucous; *basal leaves* spatulate to oblanceolate, 5–7 cm × 15–25 mm, glaucous; *cauline leaves* ovate, obovate, or elliptic, 2–5 cm × 10–20 mm, glaucous, with entire margins; *thyrse* with 5–10 verticillasters; *calyx* 5–9 mm long, glabrous; *corolla* 18–24 mm long, pale deep caerulean blue, often purple-tinged on tube, glabrous; *staminode* glabrous to middle then abruptly densely golden-yellow-bearded; *anthers* glabrous, 2 or the 4 fertile anthers well-exserted. Uncommon in sagebrush meadows and pinyon-juniper communities, 6800–9200 ft. June–July. W. Endemic.

Penstemon humilis **Nutt. ex A. Gray ssp.** ***humilis***, LOW BEARDTONGUE. Decumbent to erect perennial, 0.5–2.5 (3.5) dm; *stems* finely short-hairy; *basal leaves* ovate or obovate, (1.5) 2–5.5 (7.5) cm × (2) 4–18 (32) mm, glabrous to cinereous-pubescent, with entire margins; *cauline leaves* lanceolate, oblanceolate, or obovate, 1–3 (4.5) cm × 1–10 mm, glabrous to minutely hairy or cinereous-pubescent; *thyrse* with 3–9 verticillasters, glandular-pubescent; *calyx* 3–5 (6) mm long; *corolla* (7) 10–16 (18) mm long, blue, glandular-hairy externally, palate sparsely to densely bearded; *staminode* bearded distally; *anthers* dark purple on back, glabrous. Found in dry, open places, on rocky or shale slopes, often with sagebrush or pinyon-juniper, 5800–9000 ft. June–July. W.

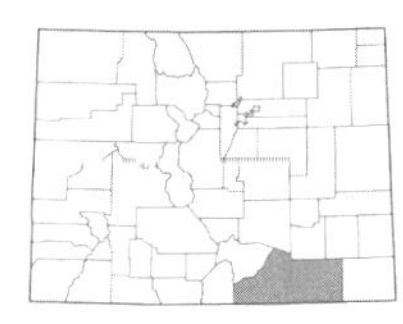

Penstemon jamesii **Benth.**, JAMES' BEARDTONGUE. Erect perennials, 1.4–5 (5.2) dm; *stems* glabrate to retrorsely puberulent below, glandular-hairy above; *leaves* linear to lanceolate, 2–10 (11) cm × (2) (2) 5–15 (30) mm, with entire to toothed margins; *thyrse* with 2–8 verticillasters, secund; *calyx* 8–12 mm long, glandular-hairy; *corolla* 24–32 (35) mm long, pinkish or pale lavender to violet-blue, glandular-hairy externally, usually sparsely glandular-hairy internally, palate moderately to densely pilose; *staminode* conspicuously exserted, terminal 10–14 mm bearded with hairs to 3.5 mm; *anthers* glabrous. Uncommon in dry grasslands and on shrubby hillsides, 5000–5500 ft. May–June. E.

Penstemon laricifolius **Hook. & Arn. var.** ***exilifolius*** **(A. Nelson) D.D. Keck**, LARCHLEAF BEARDTONGUE. Erect, cushion-forming perennial, 1–2 dm; *stems* minutely hairy to glabrous; *leaves* filiform, 1.5–2.5 cm long, glabrous, with involute margins; *thyrse* glabrous; *calyx* 4–6 mm long, glabrous; *corolla* 11–15 mm long, white, greenish-white, or sometimes light rose, glabrous externally, bearded internally on palate; *staminode* included or barely exserted, bearded strongly for ⅓–½ its length; *anthers* glabrous. Uncommon in grasslands and sagebrush meadows, 7000–9000 ft. July–Aug. E.

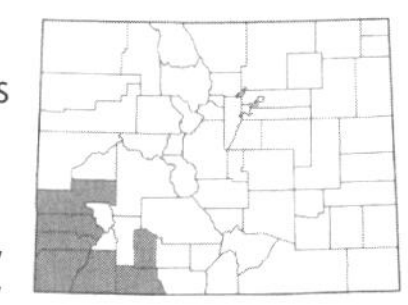

Penstemon lentus **Pennell var.** ***lentus***, HANDSOME BEARDTONGUE. Erect perennials, 2–4 (5) dm; *stems* glabrous and glaucous; *leaves* lanceolate to ovate or obovate, 2–6.5 (11) × 8–20 (30) mm, glabrous and glaucous, with entire margins; *thyrse* with 4–8 verticillasters, more or less secund, glabrous; *calyx* 5–6.7 (7.2) mm long, glabrous, becoming ribbed with age; *corolla* 17–20 (22) mm long, blue to bluish-violet, glabrous externally, palate sometimes white-bearded; *staminode* reaching orifice, shortly bearded distally; *anthers* glabrous, sutures papillate-toothed. Found on dry slopes, usually with pinyon-juniper, 5500–7800 ft. May–June. W.

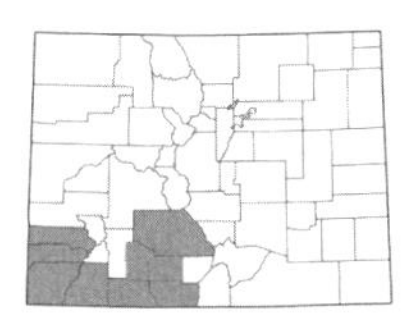

Penstemon linarioides **A. Gray ssp.** ***coloradoensis*** **(A. Nelson) D.D. Keck**, TOADFLAX PENSTEMON. Subshrubs, 1–3 (3.7) dm; *stems* minutely cinereous; *leaves* linear, 0.8–2 (3.7) cm × 0.8–2 (2.8) mm, sparsely short-hairy to glabrous; *thyrse* secund; *calyx* 5–7 mm long, to 8.5 mm long in fruit, glandular-hairy or sometimes nonglandular-hairy; *corolla* 13–17 mm long, blue to blue-violet, glandular-hairy externally, palate sparsely bearded, sometimes glabrous; *staminode* included to slightly exserted, with a tuft of yellow hairs at apex; *anthers* black on back and white along sutures, glabrous, sutures papillate-toothed. Common in sagebrush and pinyon-juniper, 5500–9500 ft. May–July. W.

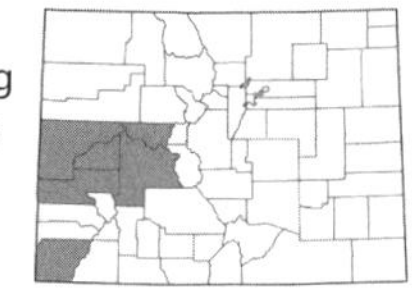

***Penstemon mensarum* Pennell**, TIGER BEARDTONGUE. Erect perennials, 4–10 dm; *stems* glabrous to glandular; *leaves* elliptic, oblong, or lance-elliptic basally, oblanceolate or oblong-lanceolate along stem, 6–20 cm × 10–15 mm, glabrous; *thyrse* glandular-hairy; *calyx* 3–5 mm long, glandular-hairy; *corolla* 14–20 mm long, dark blue to blue-purple, glandular-hairy externally, glabrous internally; *staminode* included, bearded most of its length; *anthers* sparsely short-hairy. Found in meadows, spruce-fir forests, and oak forests, 7400–10,200 ft. June–July. W. Endemic.

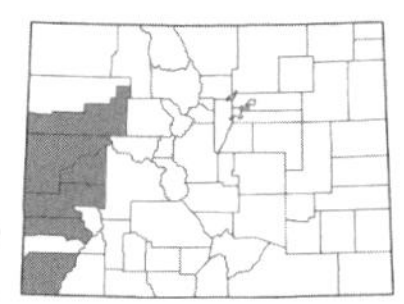

***Penstemon moffatii* Eastw.**, MOFFATT'S BEARDTONGUE. Erect perennials, 1–3 dm; *stems* retrorsely puberulent; *leaves* ovate or spatulate basally, linear to lanceolate or oblanceolate along stem, 1.5–4 (6) cm × 5–25 mm, retrorsely puberulent, sometimes glabrous, with entire or sinuate-toothed margins; *thyrse* glandular-hairy; *calyx* 6–8 mm long, glandular-hairy; *corolla* 15–20 (22) mm long, blue to blue-violet, rarely lavender, glandular-viscid externally, palate sparsely bearded; *staminode* included to barely excluded, moderately yellow-bearded; *anthers* glabrous with papillate-toothed sutures. Found on dry, open slopes, in sandy or clay soil, 4300–6800 ft. May–June. W.

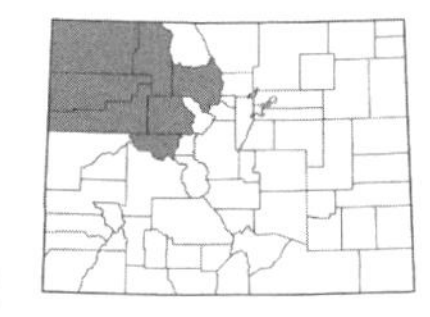

***Penstemon osterhoutii* Pennell**, OSTERHOUT'S PENSTEMON. Erect perennials, 3–7 (8) dm; *stems* glabrous and glaucous; *leaves* ovate, oblong-ovate, oblanceolate, or ovate-lanceolate, 3–12 cm × 10–40 mm, glabrous and glaucous; *thyrse* densely flowered; *calyx* 5–7 mm long, glabrous; *corolla* 14–20 (24) mm long, light purple to violet-blue, glabrous externally, sparsely bearded internally on palate; *staminode* barely exserted, densely short-bearded at apex; *anthers* glabrous. Uncommon on dry slopes, usually with sagebrush or pinyon-juniper, 5500–8200 ft. May–July. W.

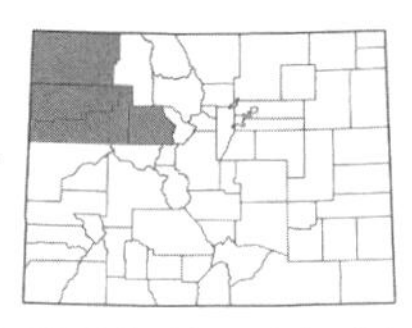

***Penstemon pachyphyllus* A. Gray ex Rydb.**, THICKLEAF BEARDTONGUE. [*P. mucronatus* N.H. Holgren]. Erect perennials, 1–5.7 dm; *stems* glabrous and glaucous; *basal leaves* spatulate to broadly lanceolate, 1.8–18 cm × 5–45 (53) mm, glabrous and glaucous; *cauline leaves* suborbiculate, ovate, or lanceolate, 1.5–8 cm × 7–32 mm, glabrous and glaucous; *thyrse* with (2) 4–10 verticillasters, glabrous; *calyx* 4–8 mm long, glabrous; *corolla* 15–23 mm long, blue, blue-violet, or purple, glabrous externally and internally, or with a few long hairs on the palate; *staminode* sparsely bearded; *anthers* glabrous. Found on sandy, rocky, or shale slopes, often with pinyon-juniper, 5500–7500 ft. May–July. W.

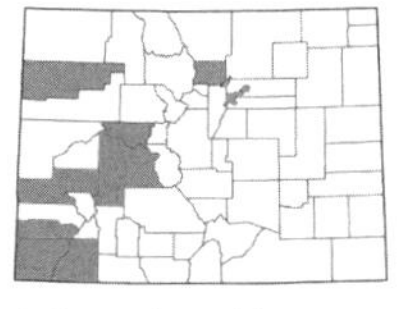

***Penstemon palmeri* A. Gray**, PALMER'S PENSTEMON. Erect perennials, 5–14 dm; *stems* glabrous and glaucous; *leaves* ovate, 60–100 (140) cm × 15–30 (60) mm, glabrous and glaucous, with dentate margins; *thyrse* with several verticillasters, glandular-hairy; *calyx* 4–6 (7.5) mm long, glandular-hairy or glabrous; *corolla* (25) 27–35 (40) mm long, white or pale pink to lavender-pink with prominent red guide-lines, glandular-hairy externally and internally, rarely glabrous, palate sparsely bearded; *staminode* exserted, base glandular-hairy, densely bearded with yellow hairs; *anthers* essentially glabrous. Uncommon in sandy soil, usually found along roadsides where it is introduced in revegetation projects, 6000–7700 ft. June–Aug. E/W. Introduced.

***Penstemon penlandii* W.A. Weber**, PENLAND'S BEARDTONGUE. Erect perennials, to 2.5 dm; *stems* retrorsely puberulent; *leaves* linear, involute, 10–15 mm wide, lower parts retrorsely puberulent; *thyrse* secund, glabrescent; *calyx* 4–5 mm long; *corolla* 12–15 mm long, blue-violet, glabrous externally and internally; *staminode* included, sparsely bearded ventrally on distal third with pale orange hairs; *anthers* moderately hairy with short, stiff hairs, sutures papillate-denticulate. Uncommon on barren seleniferous clay or shale slopes, 7500–7700 ft. June–July. W. Endemic.

Listed as endangered under the Endangered Species Act.

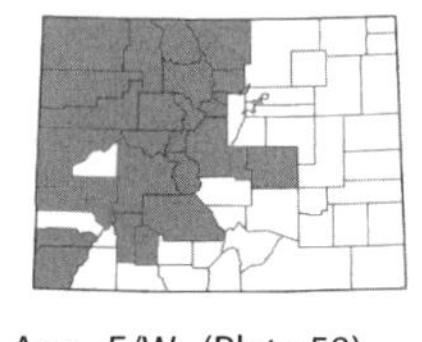

Penstemon procerus* Douglas ex Graham var. *procerus, PINCUSHION BEARDTONGUE. Ascending or decumbent perennials, 1.5–3.5 (4.5) dm; *stems* numerous and leafy, glabrous to minutely hairy or cinereous-puberulent; *leaves* oblanceolate to obovate or lanceolate, (0.5) 1–5 (8) cm × 2–10 (18) mm, glabrous to minutely hairy; *thyrse* with 1–3 verticillasters, glabrous to minutely hairy; *calyx* 3–5 mm long, glabrous to minutely hairy; *corolla* 6–10 (11) mm long, blue-purple, glabrous externally, palate white- to yellow- bearded; *staminode* glabrous or sparsely bearded; *anthers* black, glabrous. Common in forest openings, meadows, and along streams, 7000–12,000 ft. June–Aug. E/W. (Plate 59)

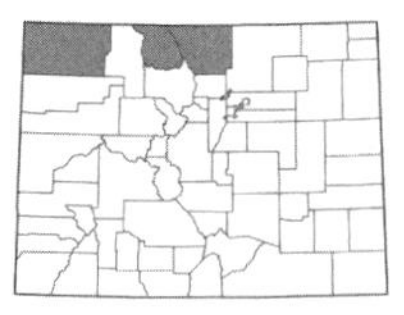

***Penstemon radicosus* A. Nelson**, MATROOT PENSTEMON. Erect perennials, 1.5–4 dm; *stems* retrorsely to spreading-puberulent; *leaves* mainly all cauline, lanceolate, (2) 3–6 cm × 3–10 (14) mm, retrorsely to spreading-puberulent, with entire, slightly revolute-creased margins; *thyrse* glandular-hairy; *calyx* 5–8 (10) mm long, glandular or viscid; *corolla* 17–23 mm long, dark violet to blue-violet, glandular-hairy externally, palate sparsely bearded; *staminode* included, sparsely pilose below with a dense apical tuft; *anthers* glabrous, sutures papillate-toothed. Uncommon in sagebrush and on dry, open shrubby slopes, 7900–9000 ft. June–July. E/W.

***Penstemon retrorsus* Payson ex Pennell**, Adobe Hills beardtongue. Erect perennials, 1–2 dm; *stems* cinereous; *basal leaves* oblanceolate or spatulate, 0.8–1.5 cm × 2–4 mm, cinereous, with entire margins; *cauline leaves* oblanceolate, 2.5–3.5 cm × 4 mm, cinereous; *thyrse* pubescent; *calyx* 3–5 mm long, densely cinereous-pubescent; *corolla* 15–20 mm long, blue-purple, glandular-hairy externally, palate moderately bearded; *staminode* bearded most of its length; *anthers* glabrous. Rare on Mancos shale and adobe slopes, 5000–6500 ft. May–June. W. Endemic. (Plate 59)

***Penstemon rostriflorus* Kellogg**, bridge penstemon. [*P. bridgesii* A. Gray]. Erect or ascending perennials, 3–10 dm; *stems* finely short-hairy below, sometimes glaucous; *leaves* linear to lanceolate or oblanceolate, 2–7 cm × 2–8 mm, glabrous; *thyrse* with 7–12 verticillasters, glandular-hairy; *calyx* (4) 4.5–6 mm long, to 8 mm long in fruit, glandular-hairy; *corolla* 22–33 mm long, scarlet-red or orange-red, sparingly glandular-hairy externally, glabrous internally; *staminode* included, glabrous; *anthers* glabrous, sutures finely ciliate. Found in dry, sandy or rocky soil and on sandstone ledges, 6000–7000 ft. July–Sept. W.

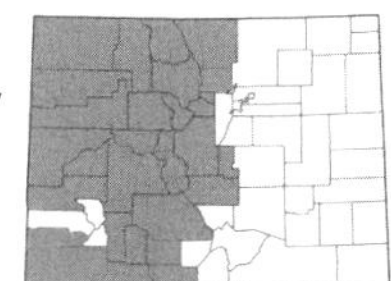

***Penstemon rydbergii* A. Nelson**, Rydberg's penstemon. Erect or decumbent perennial, 2–4 (6) dm; *stems* glabrous to minutely hairy; *leaves* oblanceolate or lanceolate, 2.5–7 (14) cm × 3–15 (22) mm, glabrous to minutely hairy; *thyrse* with 2–7 verticillasters, glabrous to minutely hairy; *calyx* 3–7 (8.5) mm long, glabrous to minutely hairy; *corolla* 10–16 (20) mm long, color, blue to blue-violet or purple, glabrous externally, palate white to yellow-bearded; *staminode* reaching the orifice, densely golden-yellow-bearded apically; *anthers* black, glabrous. Common in meadows, forest openings, and along streams, 7500–12,000 ft. June–Aug. E/W.

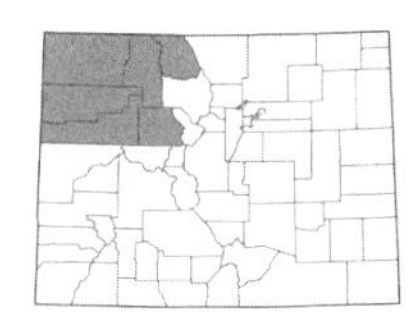

***Penstemon saxosorum* Pennell**, upland beardtongue. Erect perennials, 1–6 (8) dm; *stems* glabrous; *leaves* linear-oblanceolate to broadly oblanceolate, 0.4–1.5 cm × 3–14 mm, glabrous; *thyrse* glandular; *calyx* 5–7 (10) mm long, obscurely glandular; *corolla* 18–24 mm long, deep blue to purplish-blue, obscurely glandular externally near base of tube, glabrous internally; *staminode* bearded near the apex and sparsely hairy for roughly ½ its length; *anthers* sparsely hairy. Found on dry hillsides, in forest openings, and along roadsides, 6000–10,000 ft. June–July. W.

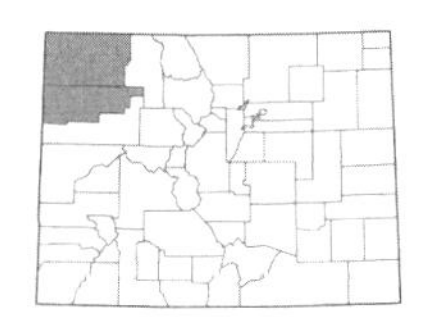

***Penstemon scariosus* Pennell**, White River beardtongue. Perennials, 1.5–5 dm; *stems* glabrous; *leaves* linear to narrowly oblanceolate, 4–17 cm × 3–15 (22) mm, glabrous; *thyrse* with 3–6 (8) verticillasters, secund, glandular-hairy or glabrous; *calyx* (3) 4–9 mm long , glandular-hairy, with broad, erose, scarious margins; *corolla* (18) 20–30 mm long, blue to lavender, usually glandular externally, palate glabrous; *staminode* reaching orifice, bearded at tip; *anthers* with long, tangled hairs. May–June. W.

There are two varieties of *P. scariosus* in Colorado, both considered rare:

1a. Corolla (18) 20–22 (24) mm long...**var. *albifluvis* (England) N.H. Holmgren**. [*P. albifluvis* England]. Rare on shale slopes in desert shrub and pinyon-juniper communities, known from Rio Blanco Co., 5000–7200 ft.

1b. Corolla 24–30 mm long...**var. *cyanomontanus* Neese**. Rare on slickrock and sandy slopes in pinyon-juniper and sagebrush communities, known from Moffat Co., 5800–8800 ft.

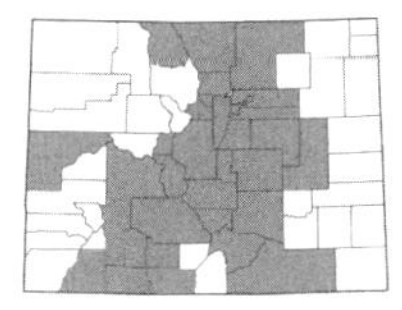

***Penstemon secundiflorus* Benth.**, sidebells penstemon. Perennials, 1.5–5 dm; *stems* glabrous and glaucous; *basal leaves* oblanceolate to spatulate, 2–8 (10.2) cm × 2–25 mm, glabrous and glaucous; *cauline leaves* lanceolate to ovate, (1.6) 2–7.8 cm × 3–24 mm, glabrous and glaucous, with entire margins; *thyrse* with (2) 3–12 verticillasters, secund; *calyx* 4–7 mm long, glabrous and glaucous; *corolla* 15–25 mm long, pink to lavender or pale to deep blue, glabrous externally, palate sparsely bearded; *staminode* densely bearded with pale yellow to golden-yellow hairs; *anthers* minutely papillose. Common in grasslands and meadows and on rocky hillsides, 4800–10,500 ft. May–July. E/W. (Plate 59)

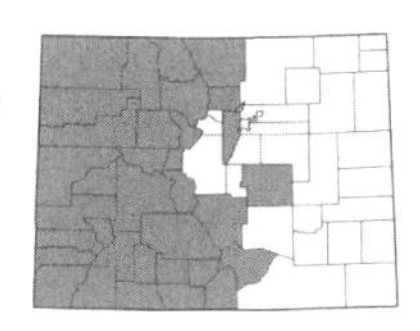

***Penstemon strictus* Benth.**, Rocky Mountain penstemon. Perennials, 3.5–7 dm; *stems* glabrous to retrorsely puberulent below; *leaves* narrowly oblanceolate to linear, often folded, 5–15 cm × 5–16 (20) mm; *thyrse* with 4–10 verticillasters, secund, glabrous; *calyx* (2.5) 3–5 mm long, glabrous; *corolla* (18) 24–32 mm long, blue to blue-violet or purple, glabrous; *staminode* included, sparsely bearded to glabrous; *anthers* lanate to villous on sides. Common on dry slopes and in meadows, sometimes found at lower elevations where it is cultivated or used in revegetation projects, 5500–11,000 ft. June–Aug. E/W. (Plate 59)

There are two varieties of *P. strictus*:

1a. Calyx tips long-acuminate or caudate...**ssp. *strictiformis* (Rydb.) D.D. Keck**, Mancos penstemon. [*P. strictiformis* Rydb.]. Found on dry slopes and in meadows, found in the southwestern counties, 6500–10,500 ft. June–Aug. W.

1b. Calyx tips obtuse and rounded to abruptly acuminate or mucronate...**ssp. *strictus***, Rocky Mountain penstemon. Common in meadows and on dry slopes, often with sagebrush, 5500–11,000 ft. June–Aug. E/W.

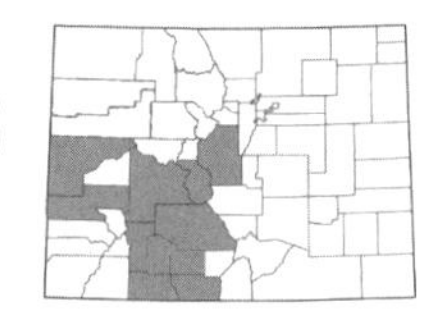

***Penstemon teucrioides* Greene**, GERMANDER BEARDTONGUE. Erect perennials, 0.4–1.2 dm; *stems* minutely hairy; *leaves* linear or narrowly oblanceolate, 0.8–1.5 cm × 1 mm, cinereous-puberulent, with involute margins; *thyrse* minutely hairy; *calyx* 3–5 mm long, viscid-puberulent; *corolla* 14–19 mm long, bluish or blue-purple, glandular-hairy externally, palate bearded; *staminode* bearded most of its length; *anthers* glabrous. Found on gravelly slopes and in open, dry places, 6000–10,000 ft. June–Aug. E/W.

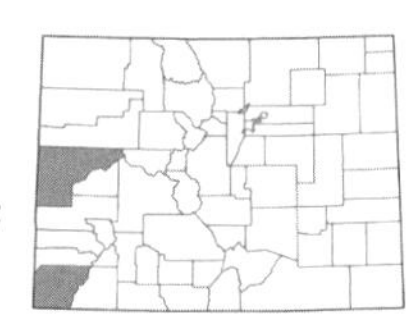

***Penstemon utahensis* Eastw.**, UTAH PENSTEMON. Erect perennials, 1.5–5 dm; *leaves* narrowly elliptic to lanceolate or oblanceolate, 3.5–8 (10) cm × 5–20 mm, glabrous and glaucous; *thyrse* with 5–15 verticillasters, often secund, glabrous; *calyx* 2.5–4 (5) mm long, glabrous, often anthocyanic; *corolla* 17–22 (25) mm long, red to crimson, sometimes deep pink or pink-purple, glandular-hairy; *staminode* included, glabrous or with short, papillate hairs at apex; *anthers* essentially glabrous. Uncommon on dry slopes with juniper or sagebrush, 4600–5600 ft. April–May. W.

***Penstemon versicolor* Pennell**, VARIABLE-COLOR BEARDTONGUE. Erect perennials, 2–3.5 dm; *stems* glabrous; *leaves* obovate, cordate-clasping, or ovate, 3–5 cm × 20–25 mm, glabrous and glaucous; *thyrse* with 6–12 verticillasters, glabrous; *calyx* 5–6 mm long, glabrous; *corolla* ca. 20 mm long, pink to blue, glabrous externally, lanate-hairy with few hairs or glabrous internally; *staminode* slightly exserted, bearded distally with 2 lines of short dense golden hairs; *anthers* glabrous. Uncommon on limestone outcroppings and shale or gypsum hillsides, 4500–6500 ft. May–June. E. Endemic.

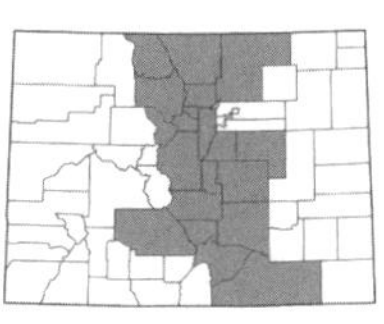

***Penstemon virens* Pennell ex Rydb.**, FRONT RANGE BEARDTONGUE. Erect perennials, 1–4 dm; *stems* spreading to retrorsely puberulent in lines below and glandular-hairy near inflorescence; *basal leaves* lanceolate to oblanceolate or spatulate, 2–10.2 cm × 4–15 mm, glabrous, with entire to toothed margins; *cauline leaves* lanceolate to lance-ovate, 1.8–5 (7) cm × 3–14 mm, glabrous with entire margins; *thyrse* with 3–6 (8) verticillasters; *calyx* 2–4.5 mm long, glandular-hairy; *corolla* 10–15 (18) mm long, pale to dark blue-tinged violet, glandular-hairy externally, palate moderately bearded; *staminode* included or reaching orifice, densely bearded at apex; *anthers* minutely papillose. Common on rocky slopes, in rock crevices, and in forest openings, 5200–11,000 ft. May–July. E/W. (Plate 59)

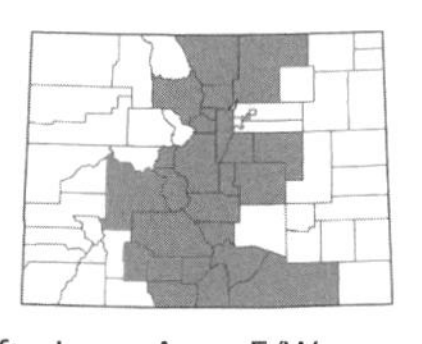

***Penstemon virgatus* A. Gray var. *asa-grayi* (Crosswhite) Dorn**, ONESIDE PENSTEMON. [*P. unilateralis* Rydb.]. Erect perennials, 2–6 (10) dm; *stems* glabrous; *basal leaves* narrowly to widely oblanceolate, 2–10 cm × 3–10 mm, glabrous; *cauline leaves* linear-lanceolate or ovate-lanceolate, glabrous with entire margins; *thyrse* secund, glabrous; *calyx* 3–5 mm long, glabrous; *corolla* 17–26 mm long, blue to blue-violet or pinkish, glabrous externally and glabrous to bearded with sparse, long hairs internally; *staminode* glabrous or with sparse short hairs near apex; *anthers* glabrous. Common in meadows and grasslands, on rocky hillsides, and pinyon-juniper forests, 4400–10,000 ft. June–Aug. E/W.

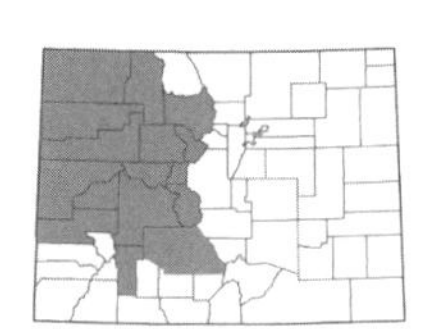

***Penstemon watsonii* A. Gray**, WATSON'S PENSTEMON. Erect perennials, 2.5–6 dm; *stems* sparsely hairy to glabrous; *leaves* oblanceolate to lanceolate, 3–7 (8) cm × 8–18 (35) mm, sparsely hairy to glabrous, nearly all cauline; *thyrse* with 2–6 (10) verticillasters, glabrous to minutely hairy; *calyx* 1.8–3.5 mm long; *corolla* 12–18 mm long, blue to violet, glabrous externally, palate lightly bearded; *staminode* reaching the orifice, with a dense tuft of yellow hairs at apex; *anthers* glabrous, sutures papillate-toothed. Common on dry slopes, often in sagebrush and pinyon-juniper, 6700–9100 ft. June–July. W.

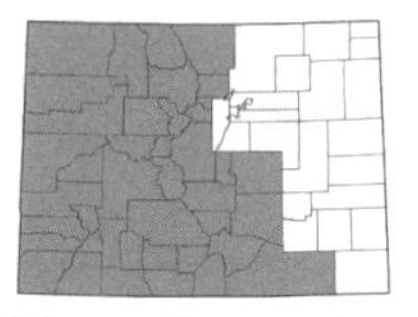

***Penstemon whippleanus* A. Gray**, WHIPPLE'S PENSTEMON. Erect perennials, 2–6 (10) dm; *stems* glabrous below, glandular-hairy above; *basal leaves* lanceolate to ovate, 4–9 (12) cm × 10–30 (75) mm, with entire or dentate margins; *cauline leaves* lanceolate to oblanceolate, 2.5–6 (8.5) cm × 3–15 (25) mm, with entire or dentate margins; *thyrse* with 2–5 (7) verticillasters, secund, glandular-hairy; *calyx* 7–10 (13) mm long; *corolla* 20–27 (30) mm long, creamy yellow, white, lavender, maroon, blue, black-purple, or violet, glandular-hairy externally, palate villous; *staminode* exserted to included, white, with a tuft of yellow hairs at apex; *anthers* blue, essentially glabrous. Common in mountain forests and meadows and on alpine, 8000–13,500 ft. June–Sept. E/W. (Plate 59)

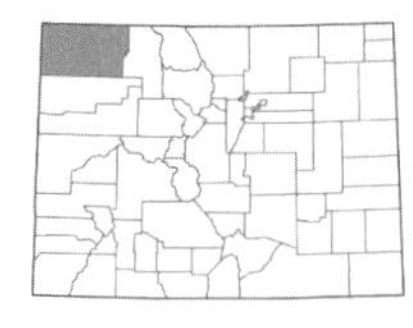

***Penstemon yampaensis* Penland**, YAMPA BEARDTONGUE. Mat-forming perennial, to 0.3 cm; *stems* spreading from branching rootstocks; *leaves* oblanceolate, 1.5–3 cm × 2–4 (5) mm, cinereous; *inflorescence* of 2–6 flowers on each of the very short ultimate branches, often exceeded by the foliage; *calyx* 5–9 mm long, densely viscid-pubescent; *corolla* 15–18 mm long, lilac or bluish-tinged, glandular-hairy externally, golden-bearded internally on throat; *staminode* exserted, bearded along the dorsal surface with golden hairs; *anthers* glabrous, sutures minutely denticulate. Uncommon on dry slopes, limestone outcrops, gypsum hills, or shale ridges, 5700–8300 ft. May–June. W.

PLANTAGO L. - PLANTAIN

Annual or perennial herbs; *leaves* alternate on short stems and appearing basal, simple, usually entire, parallel-veined; *flowers* actinomorphic or sometimes zygomorphic, in terminal spikes or heads; *calyx* 4-lobed, connate at the base; *corolla* white, 4-lobed, connate at the base, scarious; *stamens* 4 or 2, epipetalous, distinct. (Rahn, 1974)

1a. Plants white-woolly hairy throughout, the spike especially densely pubescent; bract subtending each flower linear and green; annuals; leaves linear to lanceolate, usually 3-nerved...***P. patagonica***

1b. Plants glabrous to hairy but not white-woolly; bracts ovate, often scarious-margined; perennials or if annuals then the leaves 1-nerved or obscurely nerved...2

2a. Annuals from slender taproots; leaves linear, 1-nerved or obscurely nerved...***P. elongata***

2b. Perennials; leaves lanceolate to elliptic or ovate, 3- to several-nerved...3

3a. Leaves broadly elliptic, broadly ovate, or cordate-ovate, 2.5–11 cm wide, the well-defined blade abruptly narrowed at the base to a well-defined petiole; seeds and ovules 6–30; plants glabrous or inconspicuously hairy at the crown (at the top of the caudex)...***P. major***

3b. Leaves lanceolate, narrowly elliptic, or elliptic but not cordate at the base, 1–5 cm wide, tapering at the base to an ill-defined petiole; seeds and ovules 2–4; plants glabrous or conspicuously woolly at the crown...4

4a. Sepals usually villous-ciliate along the strong, green midrib, the outer two sepals connate; inflorescence ovoid-conic (wider at the base) and very dense at first; corolla lobes 2–2.5 mm long; bracts acuminate at the tip; leaves usually lanceolate...***P. lanceolata***

4b. Sepals glabrous or merely shortly ciliate at the tip, the midrib not conspicuously strong and green, the outer two sepals distinct; inflorescence not ovoid-conic at first, about the same width throughout; corolla lobes 0.7–2 mm long; bracts acute or rounded at the tip; leaves lance-elliptic to elliptic...5

5a. Plants conspicuously and densely reddish-brown woolly at the crown (at the top of the caudex); spikes mostly 5–20 cm long...***P. eriopoda***

5b. Plants not or inconspicuously woolly at the crown; spikes mostly 2–7 cm long...***P. tweedyi***

Plantago elongata **Pursh**, PRAIRIE PLANTAIN. Annuals, 3–20 cm; *leaves* linear, succulent, 2–10 cm × 0.3–1.3 (3) mm, 1-nerved, glabrous to minutely hairy; *spikes* 1–10 cm long, less than 5 mm wide, with fleshy, broadly ovate, spurred-carinate bracts ca. 2 mm long; *corolla* lobes 0.5–1 mm long; *capsules* 1.5–3.5 mm long. Uncommon in moist, alkaline lowlands, 5000–6500 ft. May–July. E/W.

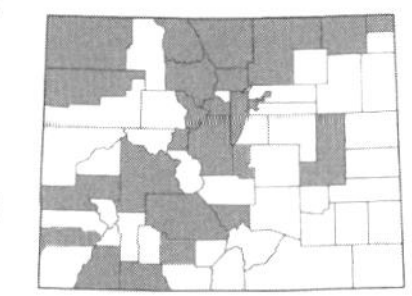

Plantago eriopoda **Torr.**, REDWOOL PLANTAIN. Perennials to 45 cm, usually conspicuously brown-woolly at the top of the caudex; *leaves* lance-elliptic, to 25 × 5.5 cm, 3–several-nerved, glabrous to villous; *spikes* mostly 5–20 cm long, with ovate, acuminate bracts; *corolla* lobes 1–1.5 mm long; *capsules* ca. 3 mm long. Found in moist, usually alkaline places, 5000–9500 ft. May–Aug. E/W.

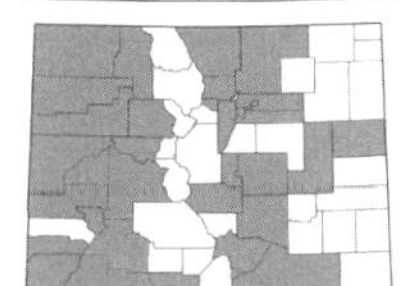

Plantago lanceolata **L.**, NARROWLEAF PLANTAIN. Perennials, 15–50 cm; *leaves* lance-elliptic, (5) 10–40 × 0.5–4 cm, 3–several-nerved, glabrous to villous; *spikes* ovoid-conic to cylindric, very dense, 1.5–8 cm long, with ovate, acuminate bracts; *corolla* lobes 2–2.5 mm long; *capsules* 2–4 mm long. Found in disturbed places and along roadsides, usually in more moist areas, 4000–9800 ft. May–Aug. E/W. Introduced.

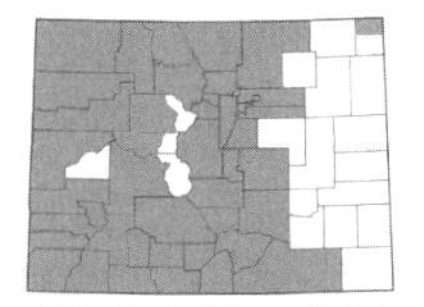

Plantago major **L.**, COMMON PLANTAIN. Perennials, mostly 10–50 cm; *leaves* cordate-ovate to broadly ovate or broadly elliptic, 4–18 × 2.5–11 cm, 3–several-nerved, glabrous or nearly so; *spikes* 5–30 cm long, less than 1 cm thick, with bracts mostly 2–4 mm long; *corolla* lobes ca. 1 mm long; *capsules* 3–4 mm long. Common weed in disturbed places and lawns and along roadsides, 3500–10,500 ft. June–Sept. E/W. Introduced.

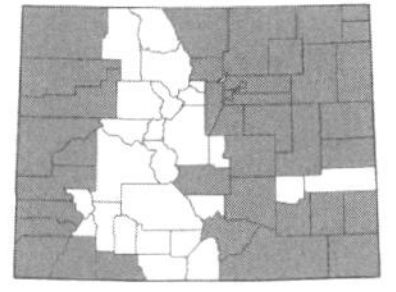

Plantago patagonica **Jacq.**, WOOLLY PLANTAIN. Annuals, 5–20 (30) cm; *leaves* linear to oblanceolate, 3–13 cm × 1–7 mm, densely woolly-villous; *spikes* mostly 1.5–10 cm long, less than 1 cm wide, densely woolly, with firm, scarious bracts ca. 5 mm long; *corolla* lobes 1.5–2 mm long; *capsules* ca. 3.5 mm long. Common in open, dry places and grassy meadows, 3400–8000 ft. May–July. E/W. (Plate 59)

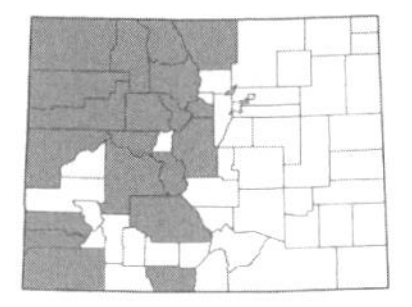

Plantago tweedyi **A. Gray**, TWEEDY'S PLANTAIN. Perennials; *leaves* elliptic, 8–22 × 1–4 cm, several-nerved, glabrous or with scattered hairs; *spikes* 2–7 (12) cm × 5–8 mm, with broad, 2 mm long bracts; *corolla* lobes 0.7–1.1 mm long; *capsules* narrowly ovoid. Usually found in nonalkaline moist places, 7500–11,500 ft. June–Aug. E/W.

VERONICA L. – SPEEDWELL

Annual or perennial herbs; *leaves* opposite; *flowers* in a terminal or axillary raceme or sometimes appearing solitary in the upper leaf axils (actually in a terminal raceme, but the flowers are subtended by foliaceous bracts); *calyx* of 4 nearly distinct lobes; *corolla* nearly actinomorphic, purple, blue, pink, or white, 4-lobed; *stamens* 2.

1a. Flowers in axillary racemes; plants glabrous, rhizomatous, aquatic perennials usually growing in water or along the margins of ponds and streams...2
1b. Flowers in a terminal raceme or appearing solitary in the upper leaf axils; plants glabrous or more often hairy or glandular, annuals or rhizomatous perennials, sometimes growing in water...4

2a. Leaves petiolate...***V. americana***
2b. Leaves sessile and clasping...3

3a. Capsule strongly 2-lobed with a conspicuous notch 0.4–1 mm deep; leaves linear to narrowly lanceolate, 4–20 times longer than wide...***V. scutellata***
3b. Capsule not evidently notched or the notch 0.1–0.3 mm deep; leaves lanceolate to elliptic or ovate, 1.5–5 times longer than wide...***V. anagallis-aquatica***

4a. Rhizomatous perennials; stems ultimately erect but usually decumbent or prostrate at the base...5
4b. Annuals from fibrous roots or short taproots, the stems erect to prostrate or weakly ascending, sometimes erect and decumbent at the base...6

5a. Stems finely puberulent with short, upturned hairs, or glabrous; style 2.2–3 mm long; flowers 4–8 mm across; capsule 2.8–3.7 mm long, not exceeding the calyx, with a conspicuous notch 0.3–0.8 mm deep; internodes usually elongate and visible...***V. serpyllifolia***
5b. Stems pilose with long, loosely spreading hairs; style 0.8–1.3 mm long; flowers 6–10 mm across; capsule (4.5) 5–6.5 mm long, exceeding the calyx, with a notch 0.1–0.3 mm deep; internodes usually short and covered by overlapping leaves...***V. wormskjoldii***

6a. Pedicels short, 0.5–1.5 (2) mm long; style 0.4–0.8 (1) mm long; notch of the capsule 0.2–0.8 mm deep; flowers subtended by oblong or narrowly lanceolate bracts, in an elongated raceme that often occupies more than half the stem...7
6b. Pedicels 4–20 mm long; style 0.8–3 mm long; notch of the capsule 0.7–3.5 mm deep; flowers subtended by lanceolate, ovate, or suborbicular bracts, the raceme occupying more than half the stem or not, or the flowers appearing solitary in the upper leaf axils (which are really foliose bracts)...8

7a. Flowers white; lower leaves narrowly oblong to oblanceolate, entire to irregularly toothed or crenate-serrate; capsule with a notch 0.2–0.5 mm deep; style 0.1–0.4 mm long...***V. peregrina* var. *xalapensis***
7b. Flowers blue to blue-violet; lower leaves ovate to broadly ovate, crenate-serrate; notch of the capsule 0.5–0.8 mm deep; style 0.4–0.8 (1) mm long...***V. arvensis***

8a. Flowers 8–12 mm across; style 2–3 mm long; petioles 15–30 mm long...***V. persica***
8b. Flowers 2–4 mm across; style 0.8–1.4 mm long; petioles 4–15 mm long...9

9a. Notch of the capsule 2.5–3.5 mm deep, reaching past the middle of the capsule; plants more or less erect; pedicels straight at maturity...***V. biloba***
9b. Notch of the capsule 1–1.6 mm deep, not reaching past the middle of the capsule; plants usually prostrate or sometimes weakly ascending; pedicels recurved at maturity...***V. polita***

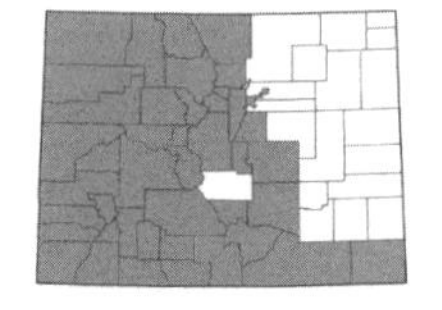

***Veronica americana* Schwien. ex Benth.**, AMERICAN SPEEDWELL. Rhizomatous, aquatic perennials, 0.5–3.5 (6) dm, glabrous; *leaves* petiolate, lanceolate to ovate, (0.5) 1.5–3 (5) cm × (3) 7–20 (30) mm, the margins crenate-serrate; *inflorescence* of axillary racemes; *calyx* 2.5–4.5 (5.5) mm long; *corolla* blue to violet, 5–10 mm wide; *capsules* 2.5–4 mm long, entire or scarcely notched; *style* 1.7–4 mm long. Common in shallow water, moist meadows, and along creeks and streams, 5000–11,000 ft. June–Sept. E/W.

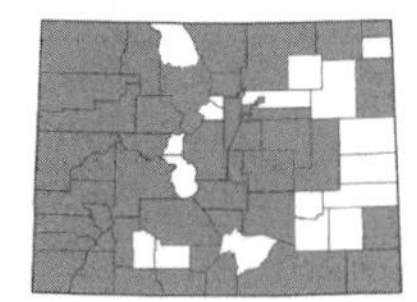

***Veronica anagallis-aquatica* L.**, WATER SPEEDWELL. [*V. catenata* Penn.; *V. salina* Schur.]. Rhizomatous, aquatic perennials, 1–6 (10) dm, glabrous; *leaves* sessile, elliptic to ovate, 2–7 (8) cm × 5–25 (40) mm, crenate-serrate; *inflorescence* of axillary racemes; *calyx* 3–5.5 mm long; *corolla* blue to violet, 5–10 mm wide; *capsules* 2.8–4 mm long, scarcely notched; *style* 1.5–3 mm long. Common in shallow water, streams, ditches, seepages, and other moist places, 3700–10,000 ft. May–Sept. E/W. (Plate 59)

***Veronica arvensis* L.**, CORN SPEEDWELL. Annuals, 0.5–2 (3) dm, somewhat villous-hirsute; *leaves* sessile or the lowermost petiolate, ovate, 0.5–2 cm × 3–13 mm, crenate-serrate; *inflorescence* a terminal raceme, glandular-hairy, the lower bracts foliaceous; *calyx* 3–5 mm long; *corolla* blue to violet, 2–3 mm wide; *capsules* 2.5–3.5 mm long, glandular-hairy on margins, with a notch 0.5–0.8 mm deep, strongly flattened; *style* 0.4–0.8 (1) mm long. Not known from Colorado but could be present, usually found as a weed in gardens and lawns. May–June. Introduced.

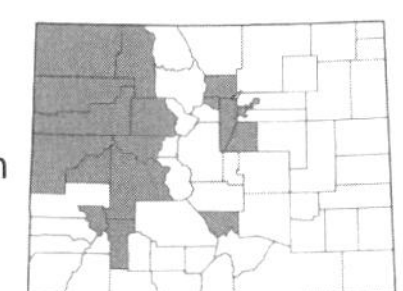

***Veronica biloba* L.**, BILOBED SPEEDWELL. [*Pocilla biloba* (L.) W.A. Weber]. Annuals, 0.7–3 dm, glandular-hairy; *leaves* short-petiolate, lanceolate to ovate, 0.8–3.5 cm × 3–12 (20) mm, serrate; *inflorescence* a terminal raceme, glandular; *calyx* 4–8 (12) mm long; *corolla* blue, 2–4 mm wide; *capsules* 3–5 mm long, ciliate with gland-tipped hairs, the notch 2.5–3.5 mm deep; *style* 0.8–1.4 mm long. Found in moist, disturbed places, and on floodplains, 5700–8500 ft. May–July. E/W. Introduced.

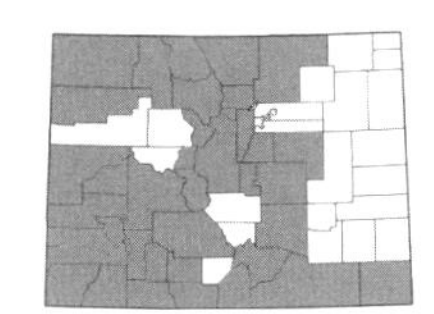

***Veronica peregrina* L. var. *xalapensis* (Kunth) Pennell**, PURSLANE SPEEDWELL. Annuals, 0.5–2 (3) dm, glandular-hairy; *leaves* sessile or the lowermost narrowed to a petiolar base, oblong to oblanceolate, 0.5–2.2 × 0.5–5 mm, entire to crenate-serrate; *inflorescence* a terminal raceme, glandular, with foliaceous bracts at the base; *calyx* 3–6 mm long; *corolla* whitish, 2–3 mm wide; *capsules* 3–4.5 mm long, with a broad notch 0.2–0.5 mm deep; *style* 0.1–0.4 mm long. Common along streams and creeks, in seepages and meadows, and other moist places, 4300–10,200 ft. May–Aug. E/W.

***Veronica persica* Poir.**, PERSIAN SPEEDWELL. [*Pocilla persica* (Poir.) Fourr.]. Annuals, 1–3 dm, pilose; *leaves* short-petiolate, 0.7–2.5 cm × 5–20 mm, ovate to suborbicular, crenate-dentate; *inflorescence* an elongate, terminal raceme, with foliaceous bracts at the lower nodes; *calyx* 4.5–8 mm long; *corolla* blue, 8–12 mm wide; *capsules* 3.5–4.5 mm long, with a notch 0.7–1.2 mm deep; *style* 2–3 mm long. Uncommon weed in yards and gardens, 4900–7000 ft. May–June. E/W. Introduced.

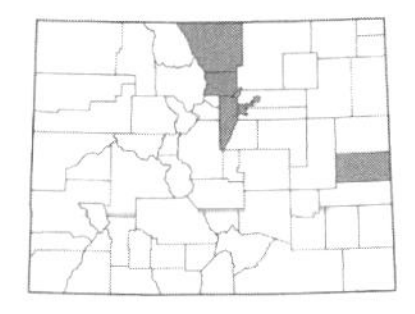

***Veronica polita* Fr.**, GRAY FIELD SPEEDWELL. [*Pocilla polita* (Fr.) Fourr.]. Annuals, 0.5–2.5 dm, prostrate or weakly ascending, hairy; *leaves* petiolate, ovate to orbicular, 4–20 × 2–12 mm, crenate-serrate; *inflorescence* a terminal raceme with foliaceous bracts; *calyx* 3.5–6.5 mm long; *corolla* blue, 2.5–3 mm wide; *capsules* 3–4 mm long, with a notch 1–1.6 mm deep, glandular-hairy; *style* 0.8–1.3 mm long. Uncommon weed in gardens and yards, 4100–5300 ft. May–Aug. E. Introduced.

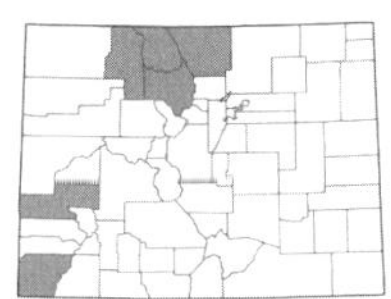

***Veronica scutellata* L.**, SKULLCAP SPEEDWELL. Rhizomatous, aquatic perennials, 1–4 dm, glabrous; *leaves* sessile, linear to lanceolate, 2–9 cm × 2–8 (15) mm, entire to remotely toothed; *inflorescence* of axillary racemes; *calyx* 2–3 mm long; *corolla* blue, pink, violet, or sometimes white, 4–5 mm wide; *capsules* 2.3–3.5 mm long, with a notch 0.4–1 mm deep. Uncommon in marshes and shallow water, 6700–9500 ft. June–Sept. E/W.

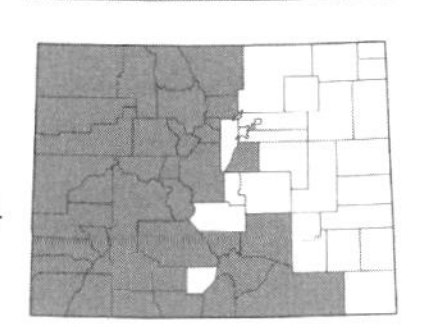

***Veronica serpyllifolia* L. var. *humifusa* (Dicks.) Syme**, THYME-LEAF SPEEDWELL. [*Veronicastrum serpyllifolium* (L.) Fourr. ssp. *humifusum* (Dicks.) W.A. Weber]. Rhizomatous perennials, 0.8–3 dm, glabrous to finely minutely hairy; *leaves* short-petiolate to subsessile, 1–2.5 cm × 8–15 mm, ovate to elliptic, entire to obscurely crenate; *inflorescence* a terminal raceme, glandular-hairy; *calyx* 2.5–4 mm long; *corolla* blue, 4–8 mm wide; *capsules* 2.8–4 mm long, with a notch 0.3–0.8 mm deep, sparsely glandular; *style* 2.2–3 mm long. Common in seepages, bogs, moist meadows, along creeks and streams, and in other moist places, 6000–11,500 ft. June–Aug. E/W.

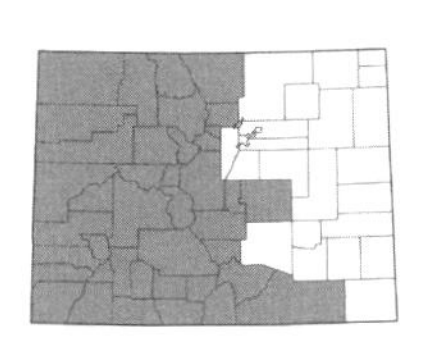

***Veronica wormskjoldii* Roem. & Schultes**, AMERICAN ALPINE SPEEDWELL. [*V. nutans* Bong.]. Rhizomatous perennials, 1–3 (4) dm, villous-hispid; *leaves* sessile, 2–4 cm × 8–18 mm, lanceolate to elliptic, entire to crenate; *inflorescence* a terminal raceme, glandular; *calyx* 3.5–5.5 mm long; *corolla* blue, 6–10 mm wide; *capsules* (4.5) 5–6.5 mm long, with a notch 0.1–0.3 mm deep, glandular; *style* 0.8–1.3 mm long. Common along creeks and streams, in moist meadows, seepages, and in the alpine, 8000–13,500 ft. July–Sept. E/W. (Plate 59)

PLUMBAGINACEAE Juss. – PLUMBAGO FAMILY

Herbs, or less often shrubs or lianas, often scapose; *leaves* alternate; *flowers* perfect, actinomorphic; *calyx* 5-lobed, often papery and plicate; *corolla* 5-lobed; *stamens* 5, distinct or epipetalous; *pistil* 1; *ovary* superior, 5-carpellate; *fruit* a capsule or a nut.

ARMERIA (DC.) Willd. – ARMERIA

Scapose herbs; *leaves* simple and entire; *flowers* in cymose heads; *corolla* connate at the base, scarious; *styles* distinct.

Armeria maritima **(Mill.) Willd. ssp. *sibirica* (Turcz. ex Boiss.) Nyman,** SEA-PINK. [*A. scabra* Pall. ssp. *sibirica* (Turcz. ex Boiss.) Hyl.]. Perennials, 3–20 cm; *leaves* 3–10 cm long, narrowly linear; *head* 10–20 mm wide; *corolla* pink. Rare in the alpine, usually near streams, 11,900–13,000 ft. E/W.

Resembles *Allium geyeri* which is from a bulb and has 3-merous flowers.

POACEAE Barnhart – GRASS FAMILY

Herbs; *leaves* alternate, linear, usually entire, the tubular sheath with usually free margins, nearly always ligulate; *flowers* usually perfect, aggregated in spikelets, with each spikelet having 1 or more florets (a flower subtended by two bracts, the outer lemma and inner palea), the spikelets subtended by 2 glumes and arranged in panicles, racemes, or spikes; *perianth* reduced to two scales (lodicules); *stamens* 2–3, 4, or 6; *pistil* 1; *ovary* superior, unicarpellate; *fruit* a caryopsis. (Barkworth et al., 2003; 2007; Harrington, 1946; Harrington & Durrell, 1944; Shaw, 2008; Wingate, 1994)

1a. Spikelets enclosed in burs…2
1b. Spikelets not enclosed in burs…3

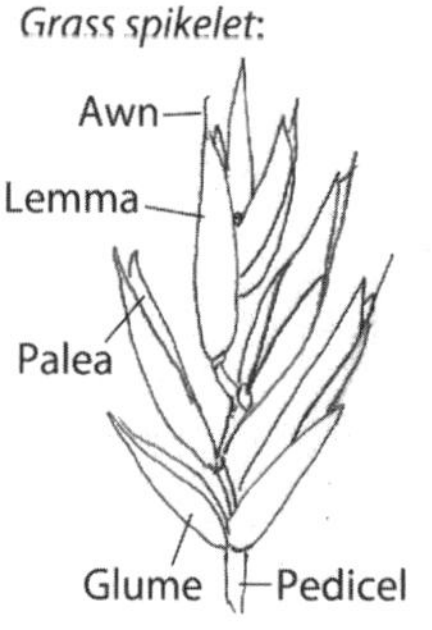

2a. Burs spiny; plants not strongly stoloniferous, monoecious…***Cenchrus***
2b. Burs not spiny; plants strongly stoloniferous, dioecious…***Buchloë***

3a. Inflorescence unisexual, the female spike with numerous spikelets in 8–24 rows (ears), with each spike surrounded by several leaf sheaths and husks, the male spikelets in panicles (tassels); blades flat, 2–12 cm wide, 3–9 dm long; corn…***Zea***
3b. Inflorescence unlike the above; not cultivated corn…4

4a. Spikelets in capitate clusters and subsessile in fascicles of leaves at each branch tip; plants stoloniferous…5
4b. Spikelets not in capitate clusters; plants stoloniferous or not…6

5a. Lemmas acuminate at the tip, with a tuft of hairs along the margins near the middle, otherwise glabrous, the spikelets not white-woolly…***Munroa***
5b. Lemmas deeply bifid at the tip, long-hairy, the spikelets appearing white-woolly overall…***Dasyochloa***

6a. Glumes absent; blades with strongly scabrous margins (can cause cuts); spikelets flat, 1-flowered; lemmas stiffly hispid-ciliate, awnless…***Leersia***
6b. Glumes (at least one) present; plants otherwise various…7

7a. Florets mostly forming bulblets with shiny, dark purple bases and exserted, linear green tips…***Poa*** (*bulbosa*)
7b. Florets not forming bulblets…8

8a. Plants with bulbous bases (resembling a small onion)…9
8b. Plants lacking bulbous bases…10

9a. Lemmas awnless; spikelets with 2 to several florets…***Melica***
9b. Lower lemma awned from near the middle; spikelets with 2 florets, the upper floret perfect and lower floret reduced and staminate…***Arrhenatherum***

10a. Tiny (mostly less than 3 cm tall), densely tufted, somewhat succulent, alpine grass; spikelets 1-flowered, awnless; glumes usually colorless, much smaller than the lemmas, the lowest glume minute or sometimes absent; leaf blades with boat-shaped tips…***Phippsia***
10b. Plants unlike the above in all respects…11

11a. Glumes equal, strongly ciliate along the margin, much longer than the lemmas, with horn-like awns; spikelets 1-flowered, flattened, tightly packed in a dense cylindric spike-like inflorescence…***Phleum***
11b. Plants unlike the above in all respects…12

12a. Lemmas strongly 9-nerved with 9 equal, plumose awns; spikelets with one fertile floret, arranged in a rather dense spike-like panicle…***Enneapogon***
12b. Lemmas not 9-nerved with 9 awns; spikelets various…13

13a. Glumes equal in size, broad and laterally compressed; spikelets suborbicular, one-flowered, tightly stacked together, crowded on short branches of a narrow, elongate panicle…***Beckmannia***
13b. Glumes unlike the above; spikelets various…14

14a. Inflorescence a spike-like panicle composed of 3–13 widely spaced, widely spreading to often curved, spicate unilateral branches; spikelets sessile, 3–5.5 mm long, 1-flowered, embedded and appressed to the slender rachis branches, awnless or with a short awn-tip; at maturity the spike-like panicle breaking off at the base and forming a "tumbleweed" like plants...***Muhlenbergia*** (*paniculata*)
14b. Plants unlike the above in all respects; if spikelets embedded and appressed to the rachis then the spikelets with 2–5 florets and the terminal spikelets with awns 2–8 cm long...15

15a. Spikelets arranged in a dense, cylindrical spike-like panicle, subtended by long bristles...16
15b. Spikelets unlike the above in all respects...17

16a. Bristles subtending the spikelets in 3 series – outer, inner, and with a central primary bristle 25–35 mm long; cultivated grasses occasionally persisting outside of gardens...***Pennisetum***
16b. Bristles subtending the spikelets in a single series; weedy or native grasses but not cultivated...***Setaria***

17a. Spikelets in pairs with one sessile fertile spikelet and one pediceled staminate or rudimentary spikelet, or sometimes the pediceled spikelet absent with just the pedicel remaining, or spikelets present in threes at the tips of branches; inflorescence usually with long hairs on the rachis and pedicel...**Key 1**
17b. Spikelets unlike the above; rachis and pedicel hairy or not...18

18a. Inflorescence branches bearing spikelets all on one side of the rachis...**Key 2**
18b. Spikelet arrangement not as above...19

19a. Spikelets truly sessile, arranged in 2-sided spikes or spike-like racemes...**Key 3**
19b. Spikelets pedicellate, arranged in panicles or racemes (sometimes the spikelets arranged in very dense spike-like, cylindrical panicles or racemes, but with short pedicels present upon dissection)...20

20a. Spikelets with one well-developed floret...**Key 4**
20b. Spikelets with two to several well-developed florets...**Key 5**

Key 1
(Spikelets in pairs)

1a. Spikelets arranged in one to several spike-like branches...2
1b. Spikelets not arranged in spike-like branches...4

2a. Spike-like branches solitary at the tips of long, slender peduncles...***Schizachyrium***
2b. Spike-like branches not solitary at the ends of long, slender peduncles...3

3a. Pedicels not grooved on both sides; pedicelled spikelets mostly over 5.5 mm long; spike-like branches more or less digitately arranged...***Andropogon***
3b. Pedicels grooved on both sides; pedicelled spikelets less than 5.5 mm long; spike-like branches digitately arranged or not...***Bothriochloa***

4a. Inflorescence densely hairy with tawny, long hairs; pedicelled spikelet absent, with just the pedicel remaining; leaf blades 1–4 mm wide...***Sorghastrum***
4b. Inflorescence not densely hairy with tawny, long hairs; pedicelled spikelet present; leaf blades 5–100 mm wide...***Sorghum***

Key 2
(Inflorescence branches bearing spikelets all on one side of the rachis)

1a. Inflorescence more or less digitate (in digitate whorls, or of a single terminal whorl of branches)...2
1b. Inflorescence not digitate...6

2a. Lemmas or glumes awned...3
2b. Lemmas or glumes not awned...4

3a. Glumes unawned; lemma awns 3–10 mm long, straight; lemma keels sparsely to densely hairy; leaf margins glabrous to shortly scabrous at the base...***Chloris***
3b. Upper glume awned (1–2.5 mm long); lemma awns 0.5–1 mm long, curved; lemma keels glabrous to scabrous; leaf margins scabrous to ciliate at the base...***Dactyloctenium***

4a. Spikelets 4–8 mm long, with 3–several florets...***Eleusine***
4b. Spikelets 2–3.5 mm long, with one fertile floret...5

5a. Ligules a fringe of hairs; plants mat-forming with stolons and rhizomes, perennials...***Cynodon***
5b. Ligules membranous; plants not mat-forming, lacking stolons and rhizomes, annuals...***Digitaria***

6a. Inflorescence densely hairy with silvery-white hairs 1.5–5 mm long...***Digitaria*** (*californica*)
6b. Inflorescence not densely silvery-white-hairy...7

7a. Spikelets or spikes pendulous and hanging to one side of the rachis...8
7b. Spikelets not as above...9

8a. Spikelets with 1 fertile floret and sometimes with 1 reduced floret above; fertile lemma with 3 awn tips...***Bouteloua*** (*curtipendula*)
8b. Spikelets with 2–5 fertile florets, the rachis prolonged above the fertile florets with several smaller sterile lemmas; lemmas awnless...***Melica*** (*porteri*)

9a. Inflorescence a spike-like panicle composed of 3–13 widely spaced, spreading to curved, spike-like unilateral branches; spikelets sessile, 3–5.5 mm long, embedded and appressed to the slender branches, awnless or with a short awn-tip; at maturity the panicle breaking off at the base and forming a "tumbleweed" like plants...***Muhlenbergia*** (*paniculata*)
9b. Plants unlike the above...10

10a. Plants dioecious, stoloniferous; staminate spikelets 2-flowered, 4–6 mm long, in 2 rows on each branch, disarticulating at the branch and falling as a single unit...***Buchloë***
10b. Plants monoecious, stoloniferous or not; spikelets unlike the above...11

11a. Spikelets arranged in a dense brush-like or eyebrow-like spike or spikes...12
11b. Spikelets not arranged as above...13

12a. Spikelets 1–4 per stem, spreading to ascending, with 1 or more reduced florets above the perfect one (these sometimes reduced to awns); rhizomes absent...***Bouteloua***
12b. Spikelets 4–30 per stem, strongly ascending, without reduced florets above the perfect one; plants strongly rhizomatous...***Spartina***

13a. Spikelets with two or more well-developed florets; lemmas not firm and cartilaginous, awned or not; ligule membranous...14
13b. Spikelets with one well-developed floret; lemmas awnless, firm and cartilaginous; ligule hairy, membranous, or absent...18

14a. Spikelets crowded in semi-orbicular to triangular clusters at the ends of stiff, wiry branches; glumes and lemmas coarsely ciliate on the keel, usually with a short awn-tip about 1 mm or less in length...***Dactylis***
14b. Spikelets in linear or oblong panicle branches, not crowded in dense clusters; glumes and lemmas unlike the above...15

15a. Lemmas with 5–9 prominent nerves, glabrous, blunt-tipped; plants small, 0.2–1.5 dm tall, usually prostrate or spreading...***Sclerochloa***
15b. Lemmas 3-nerved, the nerves usually hairy at least below, blunt or acute at the tips; plants usually larger, 0.5–15 dm tall, erect...16

16a. Leaf sheaths sparsely to densely hairy with papillose-based hairs; panicle branches filiform or narrowly linear...***Dinebra***
16b. Leaf sheaths glabrous to scabrous or sometimes with a pilose collar, but lacking papillose-based hairs; panicle branches not filiform or narrowly linear...17

17a. Ligules 1–2 mm long, truncate and erose; lemmas unawned or sometimes mucronate, sericeous but not silvery-hairy on the nerves...***Disakisperma***
17b. Ligules 2–8 mm long, attenuate and becoming lacerate at maturity; lemmas usually awned or sometimes acute, silvery-hairy on the nerves in the lower half...***Diplachne***

18a. Ligules absent; spikelets with stiff hairs, awned or with a sharp mucronate tip...***Echinochloa***
18b. Ligules present, hairy or membranous; spikelets glabrous or with soft hairs, otherwise various...19

19a. Spikelets with a small cuplike structure at the base, awned or with a sharp mucronate tip; leaf blades usually short-hairy on both sides...***Eriochloa***
19b. Spikelets lacking a cuplike structure, awnless, lacking a sharp mucronate tip; leaf blades various...20

20a. Lower glume present; rachis not flattened or broadly winged; plants with long stolons...***Hopia***
20b. Lower glume absent or inconspicuous; rachis often flattened and broadly winged; plants caespitose to rhizomatous...***Paspalum***

Key 3

(Spikelets sessile, arranged in 2-sided spikes or spike-like racemes)

1a. Spikelets subtended by 6–9 mm long white hairs, awned...***Hilaria***
1b. Spikelets not subtended by long white hairs, awned or not...2

2a. Spikelets embedded and appressed into the rachis flush with the nodes (resembling a cylinder); spikelets 9–12 mm long, with 2–5 florets, the terminal spikelet with awns 20–80 mm long, the other spikelets with shorter awns 2–9 mm long...***Aegilops***
2b. Spikelets not embedded and appressed into the rachis; plants otherwise unlike the above...3

3a. Spikelets arranged edgewise at the nodes of the rachis; first glume absent except in the terminal spikelet...***Lolium***
3b. Spikelets not arranged edgewise to the rachis; first glume usually present...4

4a. Inflorescence a spike-like raceme; lemmas stiffly ciliate along the upper margins...***Brachypodium***
4b. Inflorescence a terminal spike; lemmas various...5

5a. Spikelets widely divergent from the rachis at a wide angle (mostly about 35°-90°)...6
5b. Spikelets ascending, not widely divergent from the rachis at a wide angle...7

6a. Spikes ovate, 0.8–2.5 cm long; annuals...***Eremopyrum***
6b. Spikes rectangular to lanceolate, 3–10 cm long; perennials...***Agropyron*** (*cristatum*)

7a. Spikelets 3 per node, one-flowered, the central spikelet fertile and the two lateral spikelets reduced (often to awns) and on short pedicels...***Hordeum***
7b. Spikelets 1–7 per node, if with 3 spikelets per node then the spikelets all fertile and the lateral spikelets not reduced and pedicellate, or the spikelets with 2–several flowers...8

8a. Annuals; spikelets tightly packed into long, dense spikes 5–20 cm long, usually with long, ascending awns over 6 mm; cultivated crop occasionally escaping along roadsides...9
8b. Perennials; spikelets unlike the above in all respects; not a cultivated crop...11

9a. Spikelets 3 per node, one-flowered; lemmas with long awns mostly over 3 cm long, these tightly appressed and ascending, the lowest awns equal to or longer than the upper awns...***Hordeum*** (*vulgare*)
9b. Spikelets solitary at each node, with 2 to several flowers; lemmas various, if long-awned then unlike the above...10

10a. Glumes linear-subulate, 1-nerved; lemmas strongly ciliate along the upper keel...***Secale***
10b. Glumes ovate to lanceolate, 5–7-nerved, hard; lemmas sometimes slightly ciliate along the keel...***Triticum***

11a. Glumes and lemmas with conspicuous, long awns mostly over 6 mm long...***Elymus***
11b. Glumes and lemmas lacking awns or the awns shorter than 6 mm...12

12a. Spikelets 2–7 at most or all nodes...13
12b. Spikelets solitary at all or most nodes...17

13a. Glumes very narrow (about 1 mm wide or less) and gradually tapering to a sharp point, 1-nerved...14
13b. Glumes linear-lanceolate (about 2–4 mm wide), usually at least 3-nerved near the middle...15

14a. Spikelets 2 per node (or if 3 or more per node then the leaf blades 8–20 mm wide), 9–25 mm long, 2–7-flowered; base of the plants lacking numerous, old shredded leaf bases...***Leymus***
14b. Spikelets mostly 3 per node (look at multiple nodes, sometimes a few have 2 per node), 7–9 mm long, 1–2-flowered; leaf blades mostly 1.5–7 mm wide; base of the plants with numerous old, shredded leaf bases that become -fibrous...***Psathyrostachys***

15a. Spikelets 3–8 per node, arranged in a dense spike 15–35 cm long; glumes 12–25 mm long; lemmas softly hairy, at least toward the base...***Leymus*** (*racemosus*)
15b. Spikelets 2 per node, arranged in a spike 4–20 cm long; lemmas glabrous to scabrous...16

16a. Rhizomes present; glumes asymmetrical, slightly curving to one side toward the tip, tending to taper from below midlength to a pointed tip; spikelets solitary or sometimes 2 per node; leaf blades usually glaucous, bluish-green...***Pascopyrum***
16b. Rhizomes absent, plants caespitose or tufted; glumes symmetrical and straight, not curving to one side, tapering from midlength or higher to a pointed tip; spikelets 2 per node; leaf blades usually green...***Elymus***

17a. Glumes truncate or rounded at the tips, or blunt with a small mucro at the tip; one margin of the leaf sheath usually ciliate (at least on the middle or lower sheaths)...***Thinopyrum***
17b. Glumes gradually tapering to an acute or awned tip; one margin of the leaf sheath usually not ciliate...18

18a. Spikelets not imbricate, the tips of the lower spikelet barely reaching the base of the spikelet above or overlapping it by a small amount; lemma awns strongly divergent to arcuate at maturity...19
18b. Spikelets mostly all imbricate and closely overlapping except sometimes low on the spike, the tips of the lower spikelet reaching the middle of the spikelet above; lemmas awnless or awns various...20

19a. Glumes about ½ the length of the spikelet, acute to obtuse at the tips...***Pseudoroegneria***
19b. Glumes about ¼–⅓ the length of the spikelet, the tips acute or with a short awn 0.5–4 mm long...***Elymus*** (*albicans, arizonicus*)

20a. Rachis internodes short, 2.5–3.5 mm long; glumes lanceolate and 3-nerved, 3–5 mm long, somewhat twisted, with a short awn 1–3 mm long...***Agropyron***
20b. Rachis internodes more than 5 mm long; glumes various...21

21a. Glumes very narrow (about 1 mm wide or less) and gradually tapering to a sharp point, 1-nerved (subulate); spikelets occasionally paired at some nodes...***Leymus***
21b. Glumes linear-lanceolate (about 2–4 mm wide), 3–7-nerved near the middle; spikelets all solitary or sometimes 2 per node...22

22a. Glumes asymmetrical, slightly curving to one side toward the tip; leaf blades usually glaucous, bluish-green...***Pascopyrum***
22b. Glumes symmetrical and straight, not curving to one side; leaf blades usually green...***Elymus***

Key 4

(*Spikelets with one well-developed floret*)

1a. Tiny (mostly less than 3 cm tall), densely tufted, somewhat succulent, alpine grass; awnless; glumes usually colorless, much smaller than the lemmas, the lowest glume minute or sometimes absent...***Phippsia***
1b. Plants unlike the above in all respects...2

2a. Glumes equal, strongly ciliate along the margin, much longer than the lemmas, with horn-like awns; spikelets flattened, tightly packed in a dense cylindric spike-like inflorescence...***Phleum***
2b. Plants unlike the above in all respects...3

3a. Lemmas strongly 9-nerved with 9 equal, plumose awns; spikelets arranged in a dense spike-like panicle...***Enneapogon***
3b. Lemmas not 9-nerved with 9 awns; spikelets various...4

4a. Glumes equal in size, broad and laterally compressed; spikelets suborbicular, tightly stacked together, crowded on short branches of a narrow, elongate panicle...***Beckmannia***
4b. Glumes unlike the above; spikelets various...5

5a. Spikelets arranged in a dense, cylindrical spike-like panicle, subtended by long bristles...6
5b. Spikelets unlike the above in all respects...7

6a. Bristles subtending the spikelets in 3 series – with an outer, inner, and central primary bristle 25–35 mm long; cultivated grasses occasionally persisting outside of gardens...***Pennisetum***
6b. Bristles subtending the spikelets in a single series; weedy or native grasses but not cultivated...***Setaria***

7a. Lemmas 3-awned at the apex (the awns usually of nearly equal length, but the central one much longer in one species)...8
7b. Lemmas awnless or with a single awn...9

8a. Plants strongly stoloniferous, usually dioecious...***Scleropogon***
8b. Plants not stoloniferous, monoecious...***Aristida***

9a. Glumes densely hairy throughout, covered with silvery-white or purplish hairs 1.5–5 mm long; lemmas densely hairy on the nerves and margins, becoming dark brown when mature...***Digitaria*** (*californica*)
9b. Glumes and lemmas unlike the above...10

10a. Spikelets with one upper fertile terminal floret and one lower sterile floret consisting only of a single lemma (resembling a glume and usually equal to or slightly longer than the upper glume), usually with prominent nerves; lemma of fertile floret smooth and shiny, hard and indurate, with inrolled margins over the palea; disarticulation below the glumes...11
10b. Spikelets unlike the above...14

11a. Ligules absent; spikelets oval, with hispid to bristly hairs (these often pustulate at the base); inflorescence an open panicle with densely flowered branches...***Echinochloa***
11b. Ligules present (either a membrane or tuft of hairs, sometimes as short as 0.1 mm); spikelets and inflorescence various...12

12a. Spikelets with a small cuplike structure at the base, awned or with a sharp mucronate tip; leaf blades usually short-hairy on both sides...***Eriochloa***
12b. Spikelets lacking a cuplike structure at the base, variously tipped; leaf blades various...13

13a. Spikelets usually hairy, with mostly rounded tips; panicle 3–10 cm long, with relatively few spikelets...***Dichanthelium***
13b. Spikelets glabrous or merely scabrous on the nerves, with acute, acuminate, or awned tips; panicle 5–50 cm long, with numerous spikelets...***Panicum***

14a. Spikelets with one fertile, central floret with 1–2 bristle-like or narrowly lanceolate sterile florets below (these can be very difficult to see), awnless; glumes equal or nearly so, laterally flattened and keeled, the keel often with a thin, pale wing; leaves flat...***Phalaris***
14b. Spikelets lacking bristle-like or narrowly lanceolate sterile florets below, awned or not; glumes various, but never winged; leaves flat or involute...15

15a. Lemmas hard and indurate (much more so than the glumes), closely enclosing the palea and grain (usually with overlapping margins), without evident nerves, usually terminally awned (the awn sometimes over 2 cm long)...16
15b. Lemmas not hard and indurate or closely enclosing the palea, usually at least 1-nerved, awned or not (but the awn never over 2 cm long)...24

16a. Awns 9–25 cm long; glumes 15–60 mm long...***Hesperostipa***
16b. Awns less than 7 cm long (or occasionally fallen or absent); glumes 2.5–25 mm long...17

17a. Awns hairy at least below (with hairs 0.5–8 mm long)...18
17b. Awns not hairy, sometimes scabrous or with hairs less than 0.5 mm long...20

18a. Awns hairy their entire length, the hairs mostly less than 2 mm long; glumes 4–6 mm long...***Ptilagrostis***
18b. Awns hairy only the lower segment or lower ⅔, not the entire length, the hairs 0.5–8 mm long; glumes 9–25 mm long...19

19a. Awn hairs up to 8 mm long, located on the lower segment (about the lower ⅓) of the awn; glumes 15–25 mm long...***Pappostipa***
19b. Awn hairs 0.5–1.5 mm long, located on the lower ⅔ of the awn; glumes 9–15 mm long...***Achnatherum*** (*occidentale*)

20a. Lemma bases with a dense ring of hairs at the top of the callus; flowering stems widely spreading or sometimes prostrate, with only 2–3 leaf sheaths and reduced leaf blades...***Oryzopsis***
20b. Lemmas lacking a dense ring of hairs at the base; flowering stems usually unlike the above...21

21a. Lemmas densely covered with hairs 1.5–6 mm long...***Achnatherum***
21b. Lemmas glabrous or shortly hairy with hairs less than 1 mm long...22

22a. Glumes 2.5–5 mm long, with acute to rounded tips; lemma oblong to elliptic, often awnless or the awn early-deciduous, 1.8–6 mm long; callus rounded, not sharp...***Piptatherum***
22b. Glumes 5.5–15 mm long, the tips acuminate or shortly awned; lemma lanceolate, or rarely oblong-elliptic, awned, 2–11 mm long; callus pointed, blunt or often sharp...23

23a. Lemmas tightly closed (unable to pry apart, the margins strongly overlapping their entire length at maturity), glabrous to uniformly hairy, minutely papillate, the tips not lobed, tapering to a crown; palea not evident, ¼–½ the length of the lemma, glabrous...***Nassella***
23b. Lemma margins not strongly overlapping their entire length (able to pry the lemma apart, look for a line on the lemma), sparsely to densely uniformly hairy, not minutely papillate, the tips often 1–2-lobed; palea ⅓ to subequal to the lemma, usually hairy...***Achnatherum***

24a. Florets subtended by a tuft of long callus hairs, the hairs at least ⅓ as long as the lemmas...25
24b. Florets not subtended by a tuft of long callus hairs, or with a tuft of short callus hairs to 0.5 mm...27

25a. Lemmas or glumes awned from the tips...***Muhlenbergia*** (*andina, racemosa*)
25b. Lemmas awnless or awned from the middle or below; glumes awnless...26

26a. Lemmas awned, usually from the back or near the middle (this sometimes included in the spikelet and difficult to distinguish from callus hairs), or rarely the awn absent; glumes scabrous on the keel; plants rhizomatous or tufted, but the rhizomes lacking leaf-like scales, 1.5–15 dm; ligule membranous...***Calamagrostis***
26b. Lemmas awnless; glumes not scabrous on the keel; plants rhizomatous, the rhizomes covered with leaf-like scales, to 25 dm; ligule of hairs...***Calamovilfa***

27a. Lemmas and/or glumes awned, the awn 1 mm or longer...28
27b. Lemmas and glumes awnless, or the awn less than 1 mm long...38

28a. Lower glume with 2 awns, the upper glume awned...***Muhlenbergia*** (*alopecuroides, brevis, phleoides*)
28b. Lower glume with a single awn or awnless...29

29a. Glumes nearly equal, tipped with awns of nearly equal length...30
29b. Glume unlike the above...31

30a. Lemmas 0.5–1.5 mm long, glabrous; disarticulation below the glumes; plants tufted, lacking rhizomes...***Polypogon***
30b. Lemmas 1.9–4 mm long, long-hairy on the callus and sometimes also on the midnerve and margins; disarticulation above the glumes; plants rhizomatous with long, creeping scaly rhizomes...***Muhlenbergia***

31a. Lemmas awned from the back, near the middle or below...32
31b. Lemmas awned from the tip or above the middle...34

32a. Spikelets arranged in an open to contracted panicle, not dense, cylindric and spike-like; glumes scabrous on the keels but otherwise glabrous...***Agrostis***
32b. Spikelets arranged in a dense, cylindric spike-like panicle; glumes hairy on the keels and along the nerves, or uniformly hairy over the entire surface...33

33a. Lemmas glabrous; glumes equal; hairy on the keels and along the nerves, or occasionally hairy over the entire surface; spikelets with one fertile floret and no reduced florets...***Alopecurus***
33b. Lemmas with yellowish hairs; glumes conspicuously unequal, hairy throughout; spikelets with the middle floret perfect and two reduced florets on either side (reduced to sterile lemmas)...***Anthoxanthum***

34a. Awns arising from the tip of the lemma...35
34b. Awns arising from below the tip of the lemma...36

35a. Glumes equal to or longer than the lemma; disarticulation below the glumes; spikelets 2.5–4 mm long, strongly compressed; leaves flat, 4–15 mm wide...***Cinna***
35b. Glumes shorter than the lemma; disarticulation above the glumes; spikelets terete, not strongly compressed; leaves various...***Muhlenbergia***

36a. Lemma awns 6–15 mm long, about 2.5–5 times the length of the lemma, straight…***Apera***
36b. Lemma awns 1–5 mm long, equal to or shorter than the length of the lemma, straight or bent…37

37a. Paleas well-developed, 1–1.6 mm long; rachilla sometimes extended behind the palea as a bristle or up to 0.5 mm long, with a tuft of hairs at the tip…***Podagrostis***
37b. Paleas absent or minute and to 0.5 mm long; rachilla not extended…***Agrostis***

38a. Glumes of two distinctly different shapes, the lower glume narrow and 1-nerved with an acute tip, the upper glume broad and obovate, 3–5-nerved, with a rounded or broadly acute tip…***Sphenopholis***
38b. Glumes more or less similar in shape (equal or not in length)…39

39a. Low, mat-forming, prostrate annuals; spikelets 1.5–2.5 mm long, in a dense, cylindric, spike-like panicle 1.5–5 cm long, often purplish-black-tinged…***Crypsis***
39b. Plants unlike the above in all respects…40

40a. Ligules composed entirely or mostly of hairs, or shortly membranous and topped with long, ciliate hairs; apex of leaf sheaths with or without a tuft of white hairs…***Sporobolus***
40b. Ligules a membranous sheath, the top not long-hairy but often lacerate; apex of leaf sheath lacking a tuft of white hairs…41

41a. Glumes shorter than the lemma; lemma glabrous or often hairy, at least along the nerves…***Muhlenbergia***
41b. Glumes equal to or longer than the lemma; lemma glabrous or scabrous on the nerves, or just the callus sparsely hairy with hairs to 0.5 mm long…42

42a. Leaves mostly basal, involute or flat and narrow, mostly 0.5–2 mm wide; plants tufted, lacking rhizomes or stolons; ligule 0.5–3 mm long…43
42b. Stem leaves well-developed, flat, 2–10 mm wide; plants usually with rhizomes or stolons, occasionally tufted, sometimes decumbent at the base and rooting at the lower nodes; ligule 1–8 mm long…44

43a. Paleas well developed, 1–1.6 mm long, rachilla sometimes extended behind the palea as a bristle or up to 0.5 mm long, with a tuft of hairs at the tip…***Podagrostis***
43b. Paleas absent or minute and to 0.2 mm long; rachilla not extended…***Agrostis*** (*idahoensis*, *scabra*, *variabilis*)

44a. Disarticulation above the glumes; palea absent or to about ½ the length of the lemma; spikelets not strongly compressed…***Agrostis*** (*exarata*, *gigantea*, *stolonifera*)
44b. Disarticulation below the glumes; palea slightly shorter than the lemma; spikelets strongly compressed to subterete…45

45a. Spikelets 2.5–4 mm long; plants rhizomatous, 5–20 dm tall…***Cinna***
45b. Spikelets 1–2.2 mm long; plants stoloniferous or tufted, often decumbent basally and rooting at the lower nodes, 1–9 dm tall…***Polypogon*** (*viridis*)

Key 5

(*Spikelets with 2–several well-developed florets*)

1a. Rachillas long-hairy (hairs 6–10 mm long); leaves flat, 15–40 cm long and 2–4 cm wide; inflorescence large, 15–35 cm long; plants 2–6 m tall, "reedlike" with hollow internodes, found along ditches and rivers…***Phragmites***
1b. Rachillas various but not long-hairy; plants otherwise unlike the above in all respects…2

2a. Stems, leaves, and sheaths velvety-hairy; spikelets 2-flowered, the lower floret perfect and the upper floret staminate; glumes longer than the florets…***Holcus***
2b. Plants unlike the above in all respects…3

3a. Spikelets flattened, 9–20 mm long, with 5–20 florets, usually imperfect with staminate and pistillate flowers in separate inflorescences and often on separate plants; leaves with a tuft of long ciliate hairs on each side of the collar; ligule a short-ciliate membrane; leaf blades distichous; plants strongly rhizomatous, often found in alkaline swales…***Distichlis***
3b. Plants unlike the above in all respects…4

4a. Ligules composed mostly or entirely of a fringe of ciliate hairs…5
4b. Ligules membranous…12

5a. Lemma awns with the lower portion flattened and twisted; glumes about equal, longer than the florets; lemma margins usually hairy...***Danthonia***
5b. Lemmas awnless or if awned then the lower portion of the awn not flattened but sometimes twisted; glumes and lemmas various...6

6a. Lemmas glabrous or scabrous on the keel...7
6b. Lemmas hairy on the nerves or at the base...8

7a. Plants stoloniferous; florets all pistillate or staminate, more widely spaced, with at least 1 mm of rachilla showing between florets; inflorescence branches lacking a swollen base...***Scleropogon***
7b. Plants caespitose or sometimes stoloniferous; florets perfect, usually less than 1 mm of rachilla showing between florets; inflorescence branches with a small, swollen base...***Eragrostis***

8a. Glumes to 2 mm long; lemmas with a short awn to 2 mm long; leaf sheaths present the entire length of the stem and usually swollen at the base, the spikelets often hidden in the upper sheaths; annuals...***Triplasis***
8b. Glumes 3–10 mm long; lemmas unawned or awned to 4 mm; leaf sheaths unlike the above; perennials...9

9a. Panicle open and diffuse with wide-spreading branches; lemmas hairy just at the base...***Muhlenbergia***
9b. Panicle narrow or compact to capitate; lemmas hairy to at least the middle...10

10a. Panicle narrow and spike-like; lemmas unawned; leaves lacking thickened, white margins...***Tridens***
10b. Panicle a capitate raceme or compact panicle; lemmas with an awn 0.5–4 mm long; leaves usually with thickened, white margins...11

11a. Plants stoloniferous; lemmas densely hairy, deeply cleft with an awn 1.5–4 mm long...***Dasyochloa***
11b. Plants lacking stolons; lemmas densely hairy at the base and along the margins, acute with an awn 0.5–2.5 mm long...***Erioneuron***

12a. Spikelets densely crowded in 1-sided semi-orbicular to triangular clusters at the ends of stiff, wiry branches; glumes and lemmas coarsely ciliate on the keel, usually with a short awn-tip about 1 mm or less...***Dactylis***
12b. Spikelets not densely crowded in 1-sided semi-orbicular to triangular clusters; glumes and lemmas various...13

13a. Spikelets 1.8–3.5 cm long; glumes longer than the lowest floret and usually longer than the uppermost floret...***Avena***
13b. Spikelets less than 1.8 cm long, or if longer then the glumes shorter than the lowest floret...14

14a. Leaf auricles conspicuous and prominent (about 1 mm in length or longer, sometimes shrinking and breaking off when dried); plants to 1.5 (2) m tall...***Schedonorus***
14b. Leaf auricles absent or small and inconspicuous; plants various...15

15a. Lemmas awned, the awn over 0.5 mm long...**Key A**
15b. Lemmas awnless or tipped with a short mucro to 0.5 mm long...**Key B**

Key A

(*Spikelets with 2 or more florets; lemmas awned*)

1a. Lemmas awned from just below the apex, the awn to 2 mm long and equal to or slightly shorter than the lemma; leaves flat, 2–7 mm wide...***Trisetum*** (*wolfii*)
1b. Lemma awns unlike the above; leaves various...2

2a. At least some lemmas awned from the back (from near the middle or at the base of the lemma); awn usually twisted or bent; at least one glume equal to or longer than the lowest floret...3
2b. Lemmas awnless or awned from at or near the tip, or awned to ⅓ the length of the lemma (*careful*—sometimes awned from a bifid or toothed apex and appearing awned from the back at first glance); awn straight, twisted, or bent; glumes various...8

3a. Spikelets arranged in a dense, spike-like panicle, the middle perfect and lateral 2 reduced to sterile lemmas; lemmas densely yellow-hairy...***Anthoxanthum*** (*odoratum*)
3b. Spikelets in an open or narrow panicle, but not dense and spike-like, otherwise various but the lemmas not densely yellow-hairy...4

4a. Awns (of lower florets) 10–20 mm long; spikelets 7–15 mm long...5
4b. Awns 0.5–8 mm long; spikelets 2–7 mm long...7

5a. Spikelets mostly 3–6-flowered; lemmas 10–12 mm long; leaves flat, 2–4 mm wide...***Avenula***
5b. Spikelets mostly 2-flowered; lemmas 7–10 mm long; leaves involute or flat...6

6a. Leaves flat, 3–8 mm wide; plants 7–17 dm tall, found from 5000–10,000 ft in elevation; upper floret perfect, lower floret staminate...***Arrhenatherum***
6b. Leaves involute, plants 0.5–2 dm tall, found from 11,000–14,000 ft in elevation; mostly 1–2 mm wide; upper floret sometimes reduced and sterile, lower floret perfect...***Helictotrichon***

7a. Callus hairs short (about ⅕–¼ the length of the lemma) or absent; lemmas awned from near the base; leaves flat to involute, 0.3–4 mm wide...***Deschampsia***
7b. Callus hairs long (about ½ the length of the lemma); lemmas awned from near the middle; leaves flat, 1–8 mm wide...***Vahlodea***

8a. Leaf sheath closed to the top or nearly so...9
8b. Leaf sheath open more than ½ the length...10

9a. Callus long-hairy with hairs 1–2 mm long; leaf sheath glabrous...***Schizachne***
9b. Callus unlike the above; leaf sheath usually hairy or sometimes glabrous...***Bromus***

10a. Lemmas awned from a bifid apex (look closely, this can be minute)...11
10b. Lemmas awned from an entire apex...12

11a. Awns bent or twisted, 3–15 mm long; spikelets with 2–4 florets; lemmas 5-nerved, not silvery-hairy along the margins; uppermost floret with the rachilla hairy and prolonged...***Trisetum***
11b. Awns straight, to 3.5 mm long; spikelets with 5–9 florets; lemmas 3-nerved, silvery-hairy along the margins at least at the base; uppermost floret lacking a prolonged, hairy rachilla...***Diplachne***

12a. Annuals; inflorescence a narrow panicle, the spikelets often more or less situated on one side of the rachis...***Vulpia***
12b. Perennials; inflorescence various but the spikelets usually not situated on one side of the rachis...13

13a. Lemmas 5–7-nerved; spikelets 2–10-flowered...***Festuca***
13b. Lemmas 3-nerved; spikelets 2-flowered...***Muhlenbergia***

Key B

(*Spikelets with 2 or more florets; lemmas awnless*)

1a. Glumes very dissimilar, the upper glume broad and much wider than the lower, obovate and 3–5-nerved, the lower glume narrow, acute, and 1-nerved; spikelets 2–3-flowered; paleas colorless...***Sphenopholis***
1b. Glumes unlike the above; spikelets and paleas various...2

2a. Inflorescence a dense, spike-like panicle, the rachis densely short-hairy; paleas colorless, often as long as the lemmas; lemmas glabrous or scabrous just along the keel; leaves mostly basal, 1–3 mm wide...***Koeleria***
2b. Inflorescence not a dense, spike-like panicle, the rachis various; paleas with at least a green central nerve, shorter than to as long as the lemmas; lemmas various; leaves various...3

3a. Uppermost floret with the rachilla hairy and prolonged; upper glume about equal to the lowermost floret; spikelets with 2–3 florets...***Trisetum*** (*wolfii*)
3b. Uppermost floret without a prolonged rachilla, or the rachilla extended but glabrous; upper glume usually shorter than the lowermost floret; spikelets various...4

4a. Spikelets broadly ovate, golden-brown; glumes nearly equal in length, as long as the florets; plants sweet-smelling when dried...***Anthoxanthum*** (*hirtum*)
4b. Spikelets unlike the above...5

5a. Lemmas with 3 conspicuous nerves (with two lateral nerves and one central midnerve)...6
5b. Lemmas with 5 or more nerves, or sometimes the lateral nerves inconspicuous and just the central midnerve evident...8

6a. Leaf sheaths closed to the top or nearly so; inflorescence an open panicle with whorled branches; lemmas glabrous; glumes truncate; plants frequently rooting at the lower nodes...***Catabrosa***
6b. Leaf sheaths open to the base or nearly so; inflorescence branches not whorled; lemmas glabrous or hairy on the nerves; glumes acute; plants not rooting at the lower nodes...7

7a. Lemma tips truncate to notched; spikelets 4–12 mm long, 2–10-flowered; inflorescence an open panicle with narrow, primarily 1-sided branches...***Disakisperma***
7b. Lemma tips acute; spikelets 1.2–4 mm long, 2–3-flowered; inflorescence unlike the above...***Muhlenbergia***

8a. Leaf sheaths closed to the top or nearly so...9
8b. Leaf sheaths open more than ½ the length...12

9a. Plants with bulbous, swollen bases or the spikelets hanging on one side of the rachis; spikelets less than 2.6 mm wide and not strongly inflated...***Melica***
9b. Plants lacking bulbous, swollen bases; if the spikelets hanging on one side of the rachis, then these over 2.6 mm wide in side view and strongly inflated...10

10a. Lemmas 6–35 mm long, the tips usually not scarious, often awned...***Bromus***
10b. Lemmas 2–7 mm long, the tips usually scarious, awnless...11

11a. Lemmas glabrous, 7-nerved...***Glyceria***
11b. Lemmas cobwebby at the base or the keel and veins hairy below, 5-nerved...***Poa***

12a. Lower glume 3–5-nerved and upper glume 5–9-nerved; inflorescence a short panicle 1–4 cm long, the spikelets generally arranged on one side of the rachis, blunt-tipped; plants small, 0.2–1.5 dm tall, often prostrate or spreading...***Sclerochloa***
12b. Lower glume 1–3-nerved and upper glume 3-nerved; plants otherwise unlike the above in all respects...13

13a. Lemmas with nerves parallel and not converging at the tip; lower glumes 0.4–2.1 mm long and upper glumes 0.8–2.7 mm long...14
13b. Lemmas with nerves converging at the tip; lower glumes 2.5–10 mm long and upper glumes 3–9 mm...15

14a. Lemma nerves faint, not prominently raised; plants caespitose, lacking rhizomes...***Puccinellia***
14b. Lemma nerves conspicuous and prominently raised; plants rhizomatous...***Torreyochloa***

15a. Lemmas tips scarious, obtuse to broadly acute; lemma nerves usually conspicuously hairy at least below, or with a tuft of cobwebby hairs at the base...***Poa***
15b. Lemma tips usually not scarious, acute to acuminate; lemma nerves glabrous to scabrous but never hairy, lacking a tuft of cobwebby hairs at the base...16

16a. Plants incompletely dioecious, rhizomatous; inflorescence a contracted panicle with short, appressed branches; leaves flat...***Leucopoa***
16b. Plants monoecious, lacking rhizomes or with short rhizomes; inflorescence unlike the above; leaves involute or flat...***Festuca***

ACHNATHERUM P. Beauv. – NEEDLEGRASS

Perennials; *sheaths* open; *ligules* membranous, short-ciliate; *spikelets* in a terminal panicle, with one floret; *lemmas* long-awned, usually hairy; *paleas* ⅓ to as long as lemma, usually hairy; *disarticulation* above the glumes.

Achnatherum occidentale (Thurb.) Barkworth is reported for Colorado in *Flora of North America* based on misidentifications of *A. nelsonii*.

1a. Lemmas densely and evenly hairy throughout with white hairs 1.5–6 mm long...2
1b. Lemmas glabrous, or if hairy then the hairs in the middle of the lemma shorter than the hairs at the tip of the lemma...5

2a. Panicles open, the branches terminating in a pair of divaricately spreading spikelets; lemma awns readily deciduous (often not present at maturity)...***A. hymenoides***
2b. Panicles narrow, the branches appressed or ascending, not terminating in a pair of divaricately spreading spikelets; lemma awns readily deciduous or persistent...3

3a. Lemma awns 10–20 mm long, bent twice, persistent...***A. pinetorum***
3b. Lemma awns 4–18 mm long, straight or only weakly once bent, readily deciduous (often not present at maturity)...4

4a. Awns 4–7 (11) mm long, glabrous to scabrous; panicle 2.5–7 cm long; plants 1–3 dm tall...***A. webberi***
4b. Awsn 7–18 mm long, appressed short-hairy below; panicle 7–35 cm long; plants 3–10 dm tall...***A. ×bloomeri***

5a. Panicles open with flexuous, divergent, mostly paired branches, the longest branches 7–10 cm long...***A. richardsonii***
5b. Panicles narrow, contracted, with appressed to ascending branches shorter than 7 cm in length...6

6a. Awns 40–80 mm long, loosely twisted...***A. aridum***
6b. Awns less than 40 mm long, or if 40–45 mm long then bent twice with a straight terminal segment...7

7a. Leaf sheaths with a tuft of white hairs at the collar (hairs 0.5–2 mm long) and usually ciliate down the margins...8
7b. Leaf sheaths glabrous or scabrous at the collar, usually not ciliate down the margins...9

8a. Glumes 9–12 mm long; awns 20–30 mm long; panicle usually longer and dense, 15–30 cm long; lemma callus blunt; plants 10–20 dm tall...***A. robustum***
8b. Glumes 10–17 mm long; awns 10–20 mm long; panicle usually shorter and looser, 7–20 cm long; lemma callus sharp; plants 3–10 dm tall...***A. scribneri***

9a. Lemma hairs at midlength of the body 1.5–3.5 mm long, the hairs at the tips to 5 mm long; lemma callus sharp; plants 1–5 dm tall...***A. pinetorum***
9b. Lemma hairs 0.5–1 mm long at the midlength of the body, the hairs at the tips to 2 mm long; lemma callus sharp or blunt; plants 2–20 dm tall...10

10a. Glumes conspicuously unequal, the lower glumes 10–15 mm long and exceeding the upper by 1–4 mm; awn 10–20 mm long...***A. perplexum***
10b. Glumes subequal, the lower glumes 6–15 mm long and exceeding the upper by less than 1 mm; awn 12–45 mm long...11

11a. Awns 12–25 mm long; palea ⅔ to nearly as long as the lemma...***A. lettermanii***
11b. Awns 20–45 mm long; palea ⅓–⅔ as long as the lemma...***A. nelsonii*** (*Note—Achnatherum occidentale* (Thurb.) Barkworth will key to here and is reported for Colorado in *Flora of North America* based on misidentifications of *A. nelsonii*.)

***Achnatherum* ×*bloomeri* (Bol.) Barkworth**, HYBRID NEEDLEGRASS. [×*Stiporyzopsis bloomeri* (Bol.) B.L. Johnson]. Scattered on dry slopes, usually found growing with *A. hymenoides*, 4500–9000 ft. May–July. E/W.

This is a hybrid between *A. hymenoides* and some other species of *Achnatherum* (often *A. occidentale*). Numerous hybrids exist between *A. hymenoides* and other *Achnatherum* species. All of these hybrids are sterile. Sterile hybrids have anthers containing few, poorly formed pollen grains and do not dehisce.

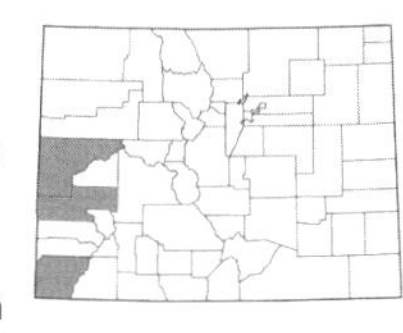

***Achnatherum aridum* (M.E. Jones) Barkworth**, MORMON NEEDLEGRASS. [*Stipa arida* M.E. Jones]. Plants 3–10 dm; *leaves* mostly involute, 10–20 cm long, with glabrous to sparsely hairy collars; *ligules* 0.2–1.5 mm, truncate with erose tips; *inflorescence* a narrow panicle 4–15 cm long; *spikelets* 7–12 mm long; *glumes* acuminate, the lower 8–15 mm long and 3-nerved, the upper 7–10 mm long and obscurely 5-nerved; *lemmas* 4–6 mm long, appressed-hairy on the margins and below the middle with hairs 0.2–0.5 mm long; *awns* 4–8 cm long; *paleas* 1–3.2 mm long, ½–¾ the length of the lemmas, hairy. Uncommon on pinyon-juniper and sagebrush slopes, 5000–6500 ft. June–Sept. W.

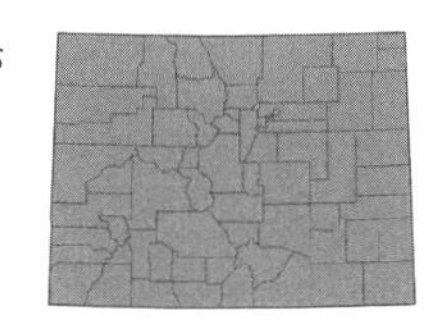

***Achnatherum hymenoides* (Roem. & Schult.) Barkworth**, INDIAN RICEGRASS. [*Oryzopsis hymenoides* (Roem. & Schult.) Ricker]. Plants 3–10 dm; *leaves* convolute, to 1 mm wide, collars glabrous or with tufts of hair on the sides; *ligules* 2–8 mm long, acuminate; *inflorescence* an open panicle with dichotomous, spreading branches; *spikelets* 4–8 mm long; *glumes* 5–9 mm long, 3-, 5-, or 7-nerved; *lemmas* 3–5 mm long, densely hairy with hairs 2.5–6 mm long; *awns* 0.6–1 cm long; *paleas* subequal to the lemmas, glabrous. Common on the plains, mesas, dry slopes, and in grasslands and pinyon-juniper woodlands, 3400–10,000 ft. May–July. E/W.

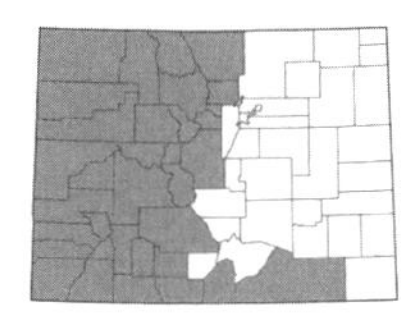

***Achnatherum lettermanii* (Vasey) Barkworth**, LETTERMAN'S NEEDLEGRASS. [*Stipa lettermanii* Vasey]. Plants 2–9 dm; *leaves* involute, to 2 mm wide, collars glabrous or sparsely hairy; *ligules* 0.2–2 mm long, truncate; *inflorescence* a narrow panicle 6–20 cm long; *spikelets* 5–10 mm long; *glumes* 6–10 mm long, often purple; *lemmas* 4–7 mm long, hairy with short hairs to 0.5 mm long, the hairs at the apex to 1.5 mm; *awns* 1.2–2.5 cm long, twisted below; *paleas* 3–4 mm long, ¾ to subequal to the lemmas, with apical hairs 0.5–1 mm long. Common in meadows, woodlands, and on dry slopes, 6000–12,500 ft. June–Aug. E/W. (Plate 60)

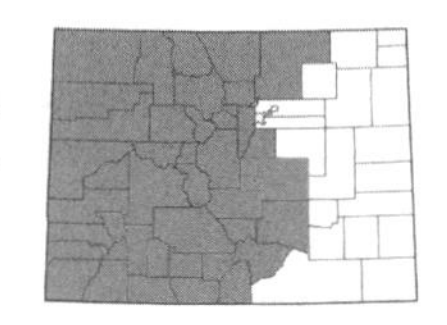

***Achnatherum nelsonii* (Scribn.) Barkworth**, NELSON'S NEEDLEGRASS. [*Stipa nelsonii* Scribn.]. Plants 3–20 dm; *leaves* usually involute, with glabrous or somewhat hairy collars; *ligules* 0.2–2 mm long, truncate; *inflorescence* a narrow panicle; *spikelets* 4–15 mm long; *glumes* 6–15 mm long, purplish, 3-nerved; *lemmas* 5–7 mm long, with hairs 0.5–1 mm and longer hairs at the apex; *awns* 2–4.5 cm long, twisted below; *paleas* 2–4 mm long, ⅓–⅔ the length of the lemmas, hairy. Common in meadows, grasslands, woodlands, on dry slopes, 6000–13,000 ft. June–Aug. E/W.

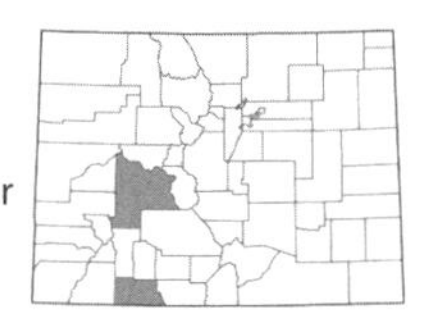

***Achnatherum perplexum* Hoge & Barkworth**, PERPLEXING NEEDLEGRASS. [*Stipa perplexa* (Hoge & Barkworth) J. Wipff & S.D. Jones]. Plants 3–10 dm; *leaves* 1–3 mm wide, collars glabrous; *ligules* 0.1–0.5 mm long on basal leaves and 0.2–3.5 mm long on upper leaves, acute or rounded; *inflorescence* a narrow panicle 10–25 cm long; *spikelets* 10–15 mm long; *glumes* unequal, the lower 10–15 mm long, 3–5-nerved; *lemmas* 5–11 mm long, with hairs ca. 1 mm long and longer hairs at the apex; *awns* 1–2 cm long, twisted below; *paleas* 2.8–5.6 mm long, ½–⅔ the length of the lemmas, hairy. Rare in pinyon-juniper woodlands, 6500–8500 ft. June–September. W.

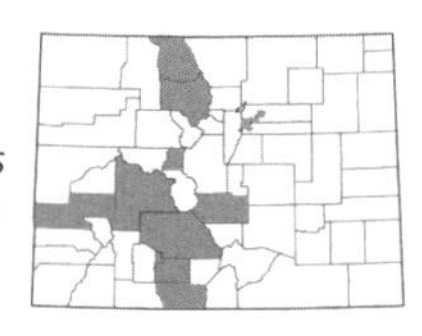

***Achnatherum pinetorum* (M.E. Jones) Barkworth**, PINE NEEDLEGRASS. [*Stipa pinetorum* M.E. Jones]. Plants 1–5 dm; *leaves* usually involute, 5–12 cm long, collars glabrous; *ligules* 0.1–1 mm long, rounded or truncate; *inflorescence* a narrow panicle 4–20 cm long; *spikelets* 7–12 mm long; *glumes* 7–12 mm long, often purplish, 1–3-nerved; *lemmas* 3–6 mm long, densely hairy with hairs 1.5–3.5 mm long, hairs at apex to 5 mm; *awns* 1–2 cm long, twisted below; *paleas* 2.5–4 mm long, ⅔ to subequal to the lemmas, hairy. Found on dry slopes, and in sagebrush shrublands and pinyon-juniper woodlands, 5500–10,000 ft. July–Sept. E/W.

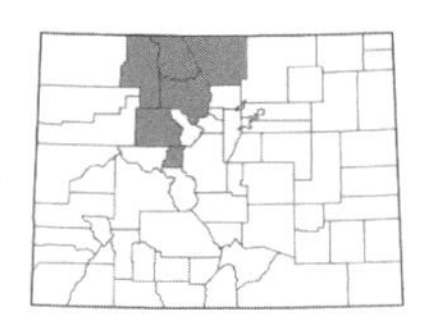

***Achnatherum richardsonii* (Link) Barkworth**, RICHARDSON NEEDLEGRASS. [*Stipa richardsonii* Link]. Plants 3–10 dm; *leaves* involute, collars glabrous; *ligules* 0.3–1 mm long, truncate; *inflorescence* an open panicle 7–25 cm long; *spikelets* 6–10 mm long; *glumes* unequal; *lemmas* ca. 5 mm long, with hairs 0.5 mm long; *awns* 1–3 cm long, twisted below; *paleas* 2.2–3.5 mm long, ½–⅗ the length of the lemmas, hairy. Uncommon in meadows and forests, 7500–10,000 ft. July–Sept. E/W.

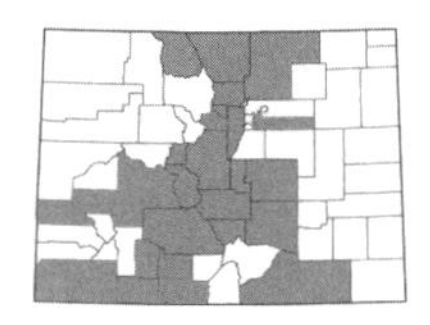

***Achnatherum robustum* (Vasey) Barkworth**, SLEEPYGRASS. [*Stipa robusta* (Vasey) Scribn.]. Plants 10–20 dm; *leaves* flat, to 10 mm wide, collars hairy; *inflorescence* a narrow panicle 15–30 cm long; *spikelets* 8–12 mm long; *glumes* 9–12 mm long, 3–5-nerved; *lemmas* 6–8 mm long, with hairs ca. 0.3–1 mm long; *awns* 2–3 cm long, twisted below; *paleas* 3.7–5.6 mm long, ⅔–¾ the length of the lemmas, hairy. Found in meadows, on dry slopes, and in canyons, 5400–10,000 ft. July–Sept. E/W.

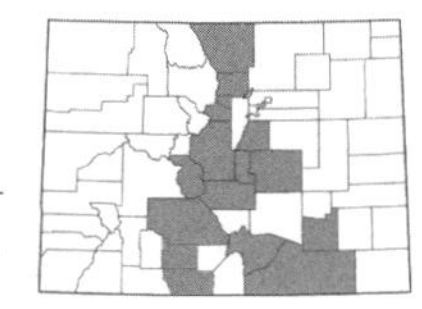

***Achnatherum scribneri* (Vasey) Barkworth**, SCRIBNER'S NEEDLEGRASS. [*Stipa scribneri* Vasey]. Plants 3–10 dm; *leaves* involute, collars glabrous; *inflorescence* a narrow panicle 7–20 cm long; *spikelets* 10–20 mm long; *glumes* unequal, the lower 10–17 mm long, 3-nerved; *lemmas* 6–10 mm long, with hairs ca. 1 mm long, to 3 mm long at the apex; *awns* 1–2 cm long, twisted below; *paleas* 2.5–3.5 mm, ⅓–½ the length of the lemmas, hairy. Found on rocky slopes and in dry forests, canyons, and pinyon-juniper woodlands, 5000–9000 ft. July–Sept. E/W.

***Achnatherum webberi* (Thurb.) Barkworth**, WEBBER'S NEEDLEGRASS. [*Stipa webberi* (Thurb.) B.L. Johnson]. Plants 1–3 dm, not rhizomatous; *leaves* involute, with a glabrous collar; *ligules* 0.2–2 mm long; *inflorescence* a narrow panicle 2.5–7 cm long, with appressed branches 1–2 cm long; *spikelets* 7–10 mm long; *glumes* 6–10 mm long, often purplish, 3–5-nerved; *lemmas* 4–6 mm, with long hairs 2.5–3.5 mm long; *awns* 0.4–0.7 (1.1) cm long, straight or twisted once, often deciduous; *paleas* 4–5.6 mm long, equal to or slightly longer than the lemmas. Reported for Mesa County.

AEGILOPS L. – GOATGRASS

Annuals; *sheaths* open; *ligules* membranous; *spikelets* solitary at nodes, in a cylindrical, bilateral spike with spikelets recessed into the rachis, with 2–8 florets; *glumes* awned; *lemmas* 5-nerved, awned from a bifid apex; *disarticulation* at the base of spikes or in rachis, the spikelets remaining attached to the rachis internode.

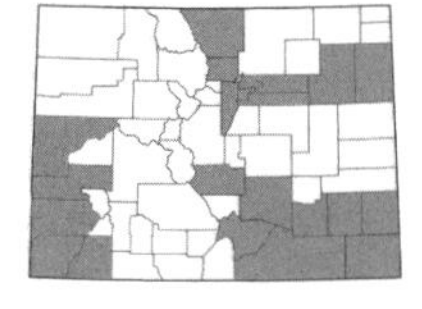

***Aegilops cylindrica* Host**, JOINTED GOATGRASS. [*Cylindropyrum cylindricum* (Host) Á. Löve]. Plants 3–7 dm; *leaves* flat, to 6 mm wide; *spikelets* 9–12 mm long; *glumes* 7–10 mm, 9–13-nerved, glumes of terminal spikelet with an awn 2–9 mm long; *lemmas of lower spikelets* 8–10 mm long, with an awn 0.1–0.5 cm long; *lemmas of upper spikelets* with long awns 2–8 cm long; *caryopsis* adhering to the lemmas and paleas. Found in disturbed areas, along roads, and in pastures, 3500–6700 ft. May–June. E/W. Introduced.

AGROPYRON Gaertn. – WHEATGRASS

Perennials; *sheaths* open; *ligules* membranous; *spikelets* solitary at nodes, in a bilateral spike, appressed or spreading at a right angle to the rachis, with 3–10 florets; *glumes* strongly keeled, with an acuminate or short awn tip; *lemmas* keeled, with an acuminate or short awn tip; *paleas* membranous, nearly as long as the lemma; *disarticulation* above the glumes.

1a. Spikelets widely diverging from the rachis at 30–95° angles; lemmas usually awned, the awns 1–6 mm long; spikes ovate to rectangular or lanceolate in outline…***A. cristatum***
1b. Spikelets diverging from the rachis at angles of less than 30°; lemmas unawned; spikes linear to narrowly lanceolate in outline…***A. fragile***

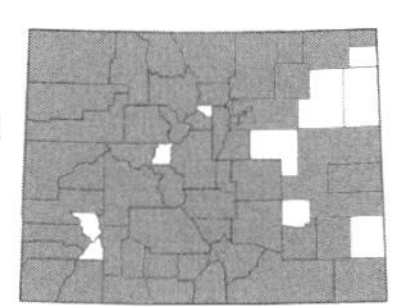

***Agropyron cristatum* (L.) Gaertn.**, CRESTED WHEATGRASS. [*A. desertorum* (Fisch.) Schult.]. Plants 3–10 dm; *leaves* flat, to 10 mm wide; *ligules* to 1.5 mm long; *spikelets* 7–15 mm long, spreading at a right angle to the rachis or nearly so; *glumes* 3–6 mm long, with a short awn tip; *lemmas* 5–9 mm long, 5-nerved, with a short awn tip to 6 mm long. Common in grasslands, meadows, woodlands, and along roadsides, 3500–9500 ft. May–Aug. E/W. Introduced. (Plate 60)

***Agropyron fragile* (Roth) P. Candargy**, SIBERIAN WHEATGRASS. Plants 3–10 dm; *leaves* flat, to 6 mm wide; *ligules* to 1 mm; *spikelets* 7–15 mm long, appressed or diverging from an angle up to 30 degrees from the rachis; *glumes* 3–5 mm long, with a short awn tip; *lemmas* 5–9 mm long, 5-nerved, unawned. Reported for Colorado, but no specimens have been seen. Introduced.

AGROSTIS L. – BENTGRASS

Annuals or perennials; *sheaths* open; *spikelets* in a panicle, with one floret; *glumes* 1- or 3-nerved, acute to acuminate, nearly equal; *lemmas* 3- or 5-nerved, acute to obtuse, glabrous or with a tuft of hairs at the base, usually awnless or awned from the middle or below; *paleas* membranous or absent; *disarticulation* above the glumes.

1a. Leaves mostly basal, involute or flat and narrow, mostly 0.5–2 mm wide; ligule 1–3 mm long; plants tufted and caespitose, lacking stolons or rhizomes…2
1b. Stem leaves well-developed, flat, 2–10 mm wide; ligule 2–8 mm long; plants usually with rhizomes (at least short ones) or stolons, or sometimes caespitose…5

2a. Lemma with a bent awn longer than the lemma…***A. mertensii***
2b. Lemma awnless or with a usually straight awn shorter than the lemma…3

3a. At least some branches of the panicle bearing spikelets to the base or nearly so…***A. variabilis***
3b. All panicle branches naked at the base…4

4a. Panicle branches forked below the middle; spikelets 1.5–2.5 mm long…***A. idahoensis***
4b. Panicle branches forked above the middle; spikelets 2.5–3.5 mm long…***A. scabra***

5a. Palea absent or inconspicuous (to 0.5 mm long); panicle narrow, compressed…***A. exarata***
5b. Palea about half the length of the lemma (0.7–1.5 mm long); panicle open or narrow at maturity…6

6a. Plants with strong creeping rhizomes, to 15 dm; panicle open at maturity…***A. gigantea***
6b. Plants lacking rhizomes but usually stoloniferous, spreading from a decumbent base and rooting at the lower nodes, mostly 2–10 dm; panicle narrower at maturity…***A. stolonifera***

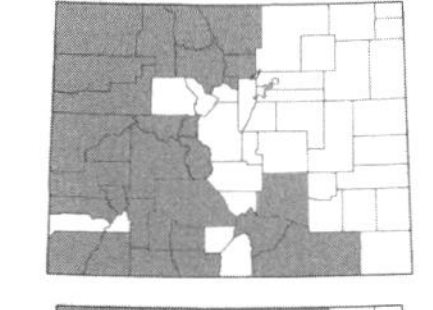

***Agrostis exarata* Trin.**, SPIKE BENTGRASS. Perennials, 2–12 dm, caespitose or sometimes with short rhizomes; *leaves* flat, to 10 mm wide; *ligules* 2–8 mm long; *inflorescence* a narrow panicle; *glumes* 2–4 mm long, the lower sometimes with a short awn tip; *lemmas* 1–3 mm long, with an awn 1–5 mm long from the middle or above; *paleas* absent or to 0.5 mm long. Common along streams and pond margins, near seeps, and in moist forests, 6000–12,000 ft. July–Sept. E/W.

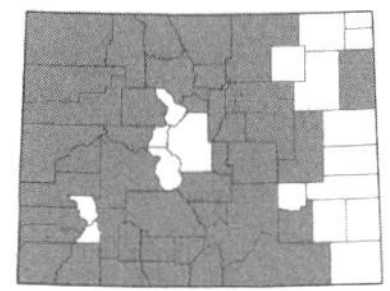

***Agrostis gigantea* Roth**, REDTOP BENT. Perennials to 15 dm, with creeping rhizomes; *leaves* flat, 3–8 mm wide; *ligules* 3–6 mm long; *inflorescence* an open panicle; *glumes* 2–2.5 mm long; *lemmas* 1.5–2 mm long; *paleas* 0.7–1.5 mm long. Common in grasslands, meadows, on floodplains, and along streams and rivers, 4500–9800 ft. July–Sept. E/W. Introduced.

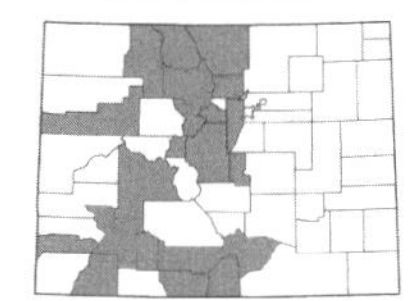

***Agrostis idahoensis* Nash**, IDAHO BENTGRASS. Perennials, 0.5–4 dm, caespitose; *leaves* involute to flat, 0.5–2 mm wide; *ligules* 1–3 mm long; *inflorescence* an open panicle; *glumes* 1.5–2.5 mm long, acute; *lemmas* 1–2 mm long, usually unawned; *paleas* absent or to 0.1 mm long. Found in moist meadows and alpine, 8500–12,000 ft. July–Sept. E/W.

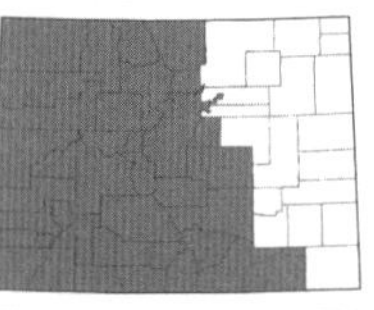

***Agrostis mertensii* Trin.**, NORTHERN BENTGRASS. Perennials, 2–3 dm, caespitose; *leaves* flat, 1–3 mm wide; *ligules* 1.5–2.5 mm long; *inflorescence* an open panicle; *glumes* 2–4 mm long, acute, purplish or reddish; *lemmas* 1.8–3.5 mm long, with a bent awn 3–4.5 mm long; *paleas* absent or to 0.1 mm long. Found in alpine and high meadows, 9000–13,500 ft. July–Sept. E/W.

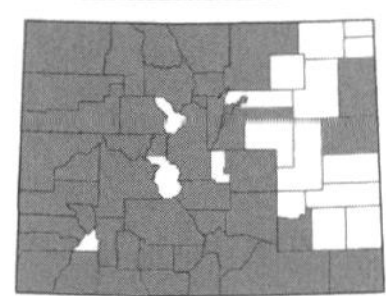

***Agrostis scabra* Willd.**, TICKLE GRASS. Perennials, 2–10 dm, caespitose; *leaves* usually involute, 1–2 mm wide; *ligules* truncate; *inflorescence* an open panicle; *glumes* 1.8–3.5 mm long, acuminate; *lemmas* 1.5–2 mm long, usually unawned; *paleas* absent or to 0.2 mm long. Common in meadows, forest floors, along lake shores, and in alpine, 5500–14,000 ft. July–Sept. E/W.

One of the first species to establish on burned-over sites, but is soon outcompeted and disappears.

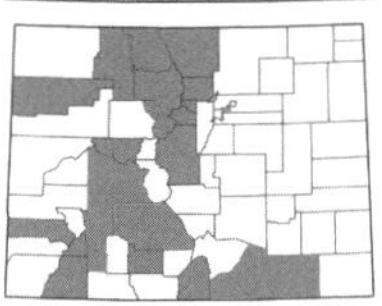

***Agrostis stolonifera* L.**, CREEPING BENTGRASS. Perennials to 10 dm, stoloniferous; *leaves* usually flat, to 10 mm wide; *ligules* 2–8 mm long; *inflorescence* narrow to spreading, 1–5 cm wide; *glumes* 1.5–3 mm long, purplish, acute; *lemmas* 1.4–2 mm long, usualy unawned; *paleas* 0.7–1.5 mm long. Found along streams and creeks, near springs, and on floodplains, 3500–10,000 ft. July–Sept. E/W. Introduced. (Plate 60)

***Agrostis variabilis* Rydb.**, MOUNTAIN BENTGRASS. Perennials, 1–3 dm, caespitose; *leaves* involute or rarely flat; *ligules* 1–3 mm long; *inflorescence* a narrow panicle 2–8 cm long; *glumes* 1.8–2.5 mm long, usually purplish; *lemmas* 1.5–2 mm long, usually unawned; *paleas* to 0.2 mm long. Found in moist subalpine meadows and alpine, 8500–13,000 ft. July–Sept. E/W.

ALOPECURUS L. – FOXTAIL

Perennials (ours); *sheaths* open; *ligules* membranous; *spikelets* in spike-like panicles, with one floret; *glumes* equal, awnless, ciliate on the nerves; *lemmas* awned from below the middle, about as long as the glumes; *paleas* usually absent; *disarticulation* below the glumes.

1a. Glumes woolly hairy throughout; panicle 1–4 cm long, about twice as long as wide...***A. magellenicus***
1b. Glumes hairy on the keel or nerves only; panicle 1.5–10 cm long, usually several times longer than wide...2

2a. Ligules truncate to rounded at the tips; glume tips obtuse to rounded; lemmas 3–6 mm long; spikelets 4–6 mm long...3
2b. Ligules attenuate or narrowed at the tips; glume tips acute; lemmas 1.5–3 mm long; spikelets 1.5–4 mm long...4

3a. Lemma tips truncate to obtuse; glume tips divergent; awns 1.5–7.5 mm long...***A. arundinaceus***
3b. Lemma tips acute; glume tips parallel to convergent; awns 5–8 mm long...***A. pratensis***

4a. Lemma awns straight, 0.7–3 mm long, often included in the spikelet...***A. aequalis***
4b. Lemma awns bent, 3–8 mm long, usually well-exserted about 2–3 mm from the spikelet...5

5a. Annuals; anthers about 0.5 mm long; leaves 0.9–3 mm wide...***A. carolinianus***
5b. Perennials; anthers about 1.5 mm long; leaves 2–6 mm wide...***A. geniculatus***

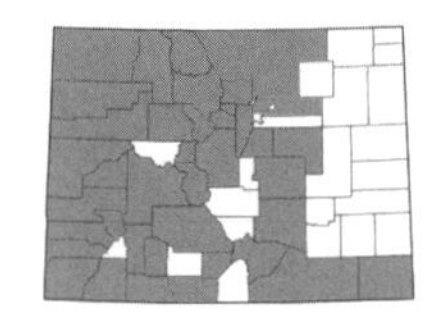

***Alopecurus aequalis* Sobol.**, SHORT-AWN FOXTAIL. Plants 2–6 dm; *leaves* flat, 1–5 mm wide; *ligules* 2–7 mm, acute; *inflorescence* 1–9 cm long; *glumes* 1.8–3 mm, hairy on the keels and nerves; *lemmas* 1.5–2.5 mm long, with a short awn 0.7–3 mm long (often included in the glumes), attached near the base; *paleas* absent. Common in moist places, along streams and ponds, and in wet meadows, 4500–11,000 ft. June–Sept. E/W.

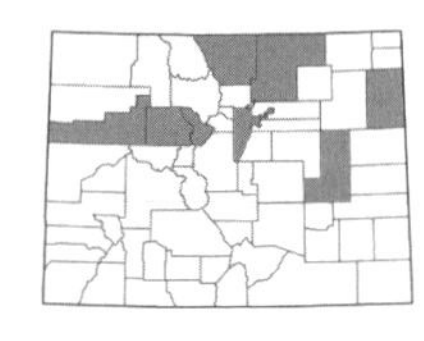

***Alopecurus arundinaceus* Poir.**, CREEPING MEADOW FOXTAIL. Plants 2–10 dm; *leaves* flat, 3–15 mm wide; *ligules* 1.5–5 mm, truncate; *inflorescence* 3–10 cm long; *glumes* 3.5–5 mm, long-hairy on the keel and sparsely hairy on the nerves; *lemmas* 3–4.5 mm long, with an awn 1.5–7.5 mm long, well-exserted from the glumes; *paleas* absent. Found along streams and lake margins, in disturbed areas, sometimes planted as a pasture grass, 4400–11,500 ft. May–July. E/W. Introduced. (Plate 60)

***Alopecurus carolinianus* Walter**, CAROLINA FOXTAIL. Plants 1–5 dm; *leaves* flat, 0.9–3 mm wide; *ligules* 3–4.5 mm; *inflorescence* 2–5 cm long; *glumes* 2–3 mm, hairy on the keel and nerves; *lemmas* 2–2.7 mm long, with an awn 3–5 mm long arising from slightly above the base, exserted from the glumes; *paleas* absent. Rare in moist places. Reported for Colorado, but no specimens have been seen.

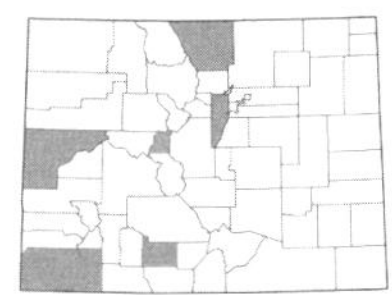

***Alopecurus geniculatus* L.**, MARSH FOXTAIL. Plants 2–6 dm; *leaves* flat, 2–6 mm wide; *ligules* 2–6 mm, acute; *inflorescence* 1.5–7 cm long; *glumes* 2–3.5 mm long, hairy on the keel and nerves; *lemmas* 2.5–3 mm long, with an awn 3–8 mm arising from near the base; *paleas* absent. Uncommon in moist meadows and along lake and pond margins, 5500–11,500 ft. June–Aug. E/W. Introduced.

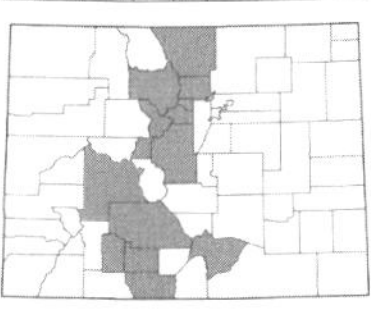

***Alopecurus magellenicus* Lam.**, ALPINE FOXTAIL. [*A. alpinus* J. Sm.]. Plants 1–8 dm; *leaves* flat, 2.5–7 mm wide; *ligules* 1.5–2 mm, truncate; *inflorescence* 1–4 cm long; *glumes* 3–5 mm, densely hairy; *lemmas* 2.5–4.5 mm long, with an awn 2–8 mm long arising from just below the middle of the lemma, exserted from the glumes; *paleas* absent. Found near seeps, along streams, and in wet meadows, 8500–13,000 ft. June–Aug. E/W.

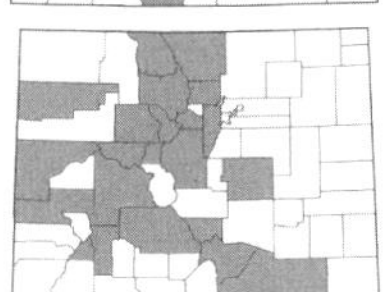

***Alopecurus pratensis* L.**, MEADOW FOXTAIL. Plants 3–11 dm; *leaves* flat, 2–10 mm wide; *ligules* 1.5–3 mm, rounded or truncate; *inflorescence* 3.5–9 cm long; *glumes* 4–6 mm, hairy on the keel and nerves; *lemmas* 4–6 mm long, with an awn 5–8 mm long arising from near the base and well-exserted from the glumes; *paleas* absent. Found in moist places and disturbed areas, sometimes planted as a pasture grass, 5500–11,000 ft. June–Aug. E/W. Introduced.

ANDROPOGON L. – BLUESTEM

Perennials; *sheaths* open; *ligules* membranous; *spikelets* paired, one sessile and fertile and one pedicelled and reduced or well-developed, in spike-like racemose panicles; *glumes* awnless, enclosing florets, firm; *lemmas* of fertile floret awned or awnless, membranous; *disarticulation* in the rachis.

1a. Spikelets with awns 8–25 mm long; ligules 0.5–2.5 mm long; rhizomes absent or sometimes present but short, the internodes usually less than 2 cm in length...***A. gerardii***

1b. Spikelets unawned or the awns less than 11 mm long; ligules (1) 2.5–4.5 mm long; rhizomes present, the internodes usually more than 2 cm in length...***A. hallii***

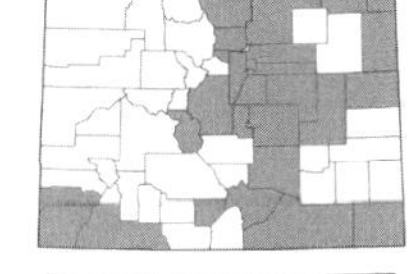

***Andropogon gerardii* Vitman**, BIG BLUESTEM. Plants 5–15 dm; *leaves* flat, 3–10 mm wide; *ligules* 0.5–2.5 mm long; *inflorescence* a panicle of 2–6 digitate branches, with hairy internodes; *awns* of fertile lemmas 8–25 mm long, twisted below. Common in prairies, grasslands, and meadows, 3400–6800 ft. July–Sept. E/W. (Plate 60)

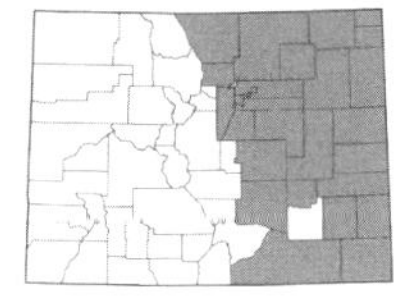

***Andropogon hallii* Hack.**, SAND BLUESTEM. Plants 5–20 dm; *leaves* flat, 3–10 mm wide; *ligules* (1) 2.5–4.5 mm long; *inflorescence* a panicle of 2–6 digitate branches, with hairy internodes; *awns* of fertile lemmas to 11 mm long. Common on sandhills and sand dunes of sandsage prairie, 3500–6500 ft. July–Sept. E.

ANTHOXANTHUM L. – SWEETGRASS

Perennials (ours); *sheaths* open; *ligules* membranous; *spikelets* with 3 florets per spikelet, the middle one perfect and the 2 lateral ones reduced to sterile lemmas; *glumes* unequal, the upper one 3-nerved and longer than the florets; *lemmas* of fertile floret broad, smooth, awnless, of sterile florets hairy, awned from the back and with a bifid apex; *paleas* present in fertile floret; *disarticulation* above the glumes.

1a. Panicle open and pyramidal in outline; lemmas unawned; glumes nearly equal in size; lowest 2 florets perfect and fertile...***A. hirtum***

1b. Panicle dense and congested, narrowly oblong in outline; lemmas awned; glumes conspicuously unequal in size, the lower glumes 3–4 mm long and the upper glumes 8–10 mm long; lowest 2 florets staminate and sterile...***A. odoratum***

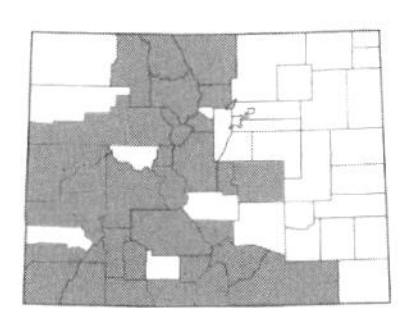

***Anthoxanthum hirtum* (Schrank.) Y. Schouten & Veldkamp**, NORTHERN SWEETGRASS. [*Hierochloe hirta* (Schrank) Borbás]. Plants 1.5–6 dm; *leaves* flat, 2–6 mm wide; *ligules* 2–4 mm long; *inflorescence* an open panicle; *glumes* subequal, glabrous, 3–6 mm long; *lemmas* of fertile florets 3–3.5 mm long, of sterile florets 3–5 mm long, hairy on the margins, unawned. Common along streams and in moist meadows, forest openings, and the alpine, 6500–13,000 ft. May–Aug. E/W. (Plate 60)

***Anthoxanthum odoratum* L.**, SWEET VERNALGRASS. [*Hierochloe odorata* (L.) P. Beauv.]. Plants 3–6 dm; *leaves* flat, 2–6 mm wide; *ligules* 1–5 mm long; *inflorescence* a narrow, spike-like panicle; *glumes* unequal, the lower 3–4 mm, half as long as the spikelet, the upper 8–10 mm, as long as the spikelet; *lemmas* of fertile florets ca. 2 mm long, of sterile florets 2–3 mm long, hairy, the first sterile lemma awned ⅓ from the apex with an awn 2–4 mm long, the second sterile lemma with a bent awn 4–9 mm from the base. No specimens currently reported for the state, old reports are based on a misidentification of the above species. Introduced.

APERA Adans. – SILKYBENT

Annuals; *sheaths* open; *ligules* membranous; *spikelets* in a panicle, the rachilla prolonged behind the floret as a naked bristle, with one floret; *glumes* acuminate; *lemmas* awned from below the tip; *disarticulation* above the glumes.

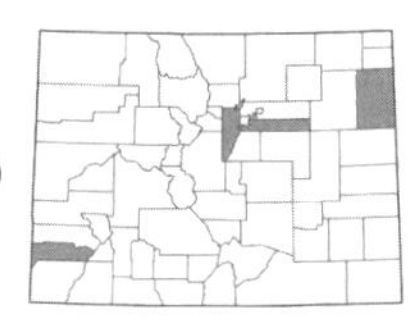

***Apera interrupta* (L.) P. Beauv.**, DENSE SILKYBENT. Plants 1–7 dm; *leaves* involute or flat, to 4 mm wide; *glumes* unequal, the lower 1.5–2.2 mm long and the upper 2–3 mm long; *lemmas* 1.5–2.5 mm long with an awn 6–16 mm long. Uncommon along streams and in riverbottoms, 3400–5200 ft. June–Aug. E/W. Introduced. (Plate 60)

ARISTIDA L. – THREE-AWN

Annuals or perennials; *sheaths* open; *ligules* partly or wholly composed of hairs; *spikelets* in an open or contracted panicle, with one floret; *glumes* 1-3-nerved; *lemmas* 3-nerved, with 3-parted awns; *disarticulation* above the glumes.

1a. Central lemma awn with 2–3 spiral coils at the base...2
1b. Central lemma awn straight or curved at the base but not spirally coiled...3

2a. Lateral lemma awns 5–10 mm long, curved and twisted at the base (but not coiled)...***A. basiramea***
2b. Lateral lemma awns 1–4 mm long, erect and straight...***A. dichotoma***

3a. Lower glume 3–7-nerved, 12–22 mm long; annuals; lemma awns 25–60 mm long...***A. oligantha***
3b. Lower glume 1–2-nerved, 4–15 mm long; annuals or perennials; lemma awns 10–140 mm long...4

4a. Annuals; central lemma awn 7–15 mm long; glumes unequal, the lower 4–8 mm long and the upper 6–11 mm long; inflorescence a narrow panicle...***A. adscensionis***
4b. Perennials; central lemma awn 10–140 mm long; glumes unequal to equal, 4–26 mm long; inflorescence various...5

5a. Lemmas topped by a 2–6 mm long column twisting 4 or more times before the juncture with the awns...6
5b. Lemmas not topped by a twisted column, or the column only 2–3 mm long with 1–2 twists...7

6a. Inflorescence a narrow, spike-like panicle with ascending branches; central lemma awn 20–35 mm long...***A. arizonica***
6b. Inflorescence an open, diffuse panicle with spreading branches; central lemma awn 10–20 mm long...***A. divaricata***

7a. Inflorescence with the pedicels flexuous and the central axis often curved; pedicels usually with axillary pulvini; central lemma awns 10–22 mm long...***A. havardii***
7b. Inflorescence with the pedicels stiff, or if flexuous then the central lemma awns usually over 20 mm long and the panicle branches lacking axillary pulvini...***A. purpurea***

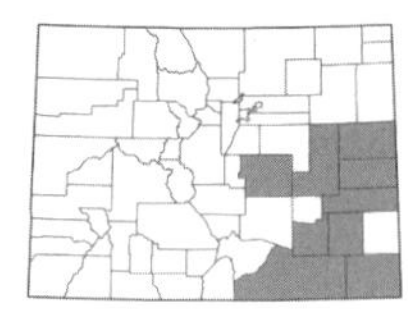

***Aristida adscensionis* L.**, SIXWEEKS THREE-AWN. Plants 1–5 dm; *leaves* involute or flat, to 2.5 mm wide; *ligules* 0.5–1 mm long; *inflorescence* a narrow panicle; *glumes* unequal, the lower 4–8 mm long, the upper 6–11 mm long; *lemmas* 6–9 mm long, scabrous along the keel, awned, the central awn 7–15 mm long, not coiled, the lateral awns slightly shorter. Found on the eastern plains, often in sandy soil, 3500–5600 ft. July–Sept. E.

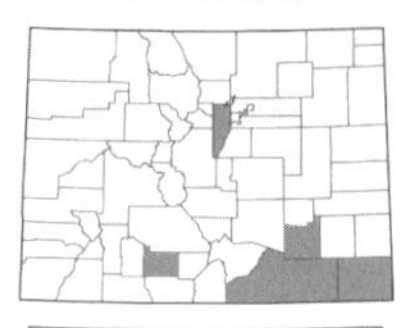

***Aristida arizonica* Vasey**, ARIZONA THREE-AWN. Plants 3–8 dm; *leaves* flat, 1–3 mm wide; *ligules* 0.2–0.5 mm long; *inflorescence* a narrow panicle; *glumes* equal or subequal, 10–15 mm long, acuminate or with an awn tip to 3 mm long; *lemmas* 12–18 mm long, glabrous, awned, the awn column 3–6 mm long and twisted, the central awn 20–35 mm long, lateral awns slightly shorter. Uncommon in grasslands and on dry slopes, often in sandy soil, 4500–6500 ft. July–Sept. E.

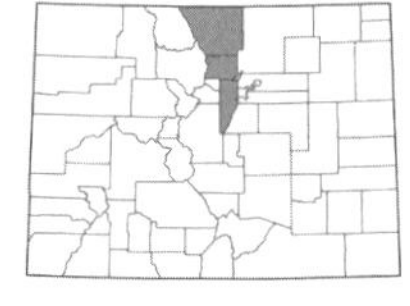

***Aristida basiramea* Engelm. ex Vasey**, FORKED THREE-AWN. Plants 2.5–4.5 dm; *leaves* involute to flat, 1–2 mm wide; *ligules* to 0.3 mm long; *inflorescence* a narrow panicle; *glumes* unequal, the lower 8–10 mm and the upper 12–14 mm long, awned with awns 1–2 mm long; *lemmas* 8–9 mm long, awned, the central awn 10–15 mm long, coiled at the base, the lateral awn 5–10 mm long. Rare on rocky or sandstone outcrops, 5200–6500 ft. July–Sept. E.

***Aristida dichotoma* Michx.**, CHURCHMOUSE THREE-AWN. Plants 1.5–6 dm; *leaves* involute to flat, 1–2 mm wide; *ligules* to 0.5 mm long; *inflorescence* a narrow panicle; *glumes* unequal, the lower 3–8 mm long and the upper 4–13 mm long, acuminate; *lemmas* 3–11 mm long, awned, the central awn 3–8 mm long, coiled, the lateral awns 1–4 mm long, erect. Rare on rocky or sandstone outcrops in open pine woodlands, 5200–6000 ft. July–Sept. E. (Plate 60)

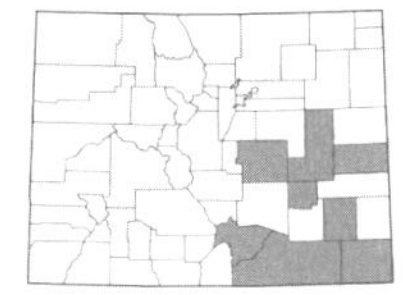

Aristida divaricata **Humb. & Bonpl. ex Willd.**, POVERTY THREE-AWN. Plants 2.5–7 dm; *leaves* involute or flat, 1–2 mm wide; *ligules* 0.5–1 mm long; *inflorescence* an open panicle; *glumes* 8–12 mm long, acuminate or short-awned; *lemmas* 8–13 mm long, awned, the awns 10–20 mm long, the lateral awns slightly shorter. Found on the eastern plains, on sand hills, and in grasslands, 3900–6500 ft. June–Sept. E.

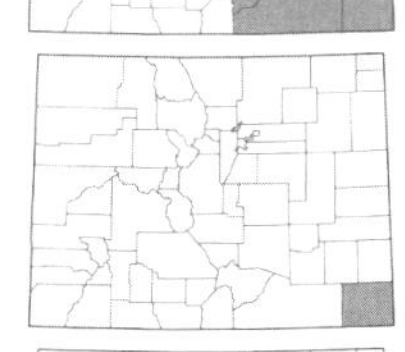

Aristida havardii **Vasey**, HAVARD'S THREE-AWN. Plants 1.5–4 dm; *leaves* involute or flat, 1–2 mm wide; *ligules* 0.5–1 mm long; *inflorescence* an open panicle; *glumes* 8–12 mm long, acuminate or short-awned; *lemmas* 8–13 mm long, awned, the central awn 10–22 mm long and the lateral awns slightly shorter. Rare on dry slopes and in sandy soil, 4000–4500 ft. June–Sept. E.

Aristida oligantha **Michx.**, PRAIRIE THREE-AWN. Plants 2–6 dm; *leaves* flat, 0.5–1.5 mm wide; *ligules* to 0.5 mm long; *inflorescence* a narrow panicle; *glumes* unequal or subequal, the lower 12–22 mm long with an awn 12–22 mm long, and the upper 11–20 mm long with an awn 4–10 mm long; *lemmas* 12–22 mm long, awned, the central awn 25–60 mm long, the lateral awns slightly shorter. Uncommon on dry slopes and disturbed areas, 5500–6000 ft. July–Sept. E.

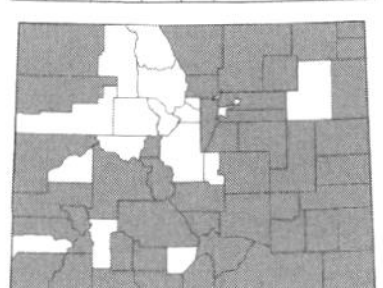

Aristida purpurea **Nutt.**, PURPLE THREE-AWN. Plants 1–10 dm; *leaves* involute or flat, 1–1.2 mm wide; *ligules* to 0.5 mm long; *inflorescence* an open panicle; *glumes* unequal, the lower 4–15 mm long and the upper 8–25 mm long, acuminate to shortly awned; *lemmas* 6–16 mm long, awned, the central awn 13–140 mm long, with lateral awns slightly shorter to ⅓ shorter. Our most common *Aristida*, found on dry slopes, in grasslands, and in open woodlands, 3500–9000 ft. May–Sept. E/W.

Aristida purpurea is a variable species with numerous varieties, 5 of which occur in Colorado:

1a. Central lemma awns over 40 mm long...2
1b. Central lemma awns less than 40 mm long...3

2a. Lemmas 12–16 mm long with apices 0.3–0.8 mm wide; central lemma awns 40–140 mm long; upper glumes 16–25 mm long...**var. *longiseta* (Steud.) Vasey**
2b. Lemmas 6–12 mm long with apices 0.1–0.3 mm wide; central lemma awns 20–60 mm long; upper glumes 7–16 mm long...**var. *purpurea***

3a. All or most panicle branches drooping, flexuous, or wavy...**var. *purpurea***
3b. All or most panicle branches stiff and straight, erect...4

4a. Plants 4.5–10 dm; panicle 14–30 cm long; leaves 10–25 cm long; leaves not primarily in a conspicuous short cluster at the base...**var. *wrightii* (Nash) Allred**
4b. Plants 1–4.5 dm; panicle 3–18 (20) cm long; leaves 4–15 cm long; leaves mostly in a short cluster at the base...5

5a. Lemmas with apices 0.1 mm wide, sometimes the upper portion twisted 1–2 times...**var. *nealleyi* (Vasey) Allred**
5b. Lemmas with apices 0.2–0.3 mm wide, the upper portion not twisted...**var. *fendleriana* (Steud.) Vasey**

ARRHENATHERUM P. Beauv. – TALL OATGRASS

Perennials; *sheaths* open; *ligules* membranous; *spikelets* in a panicle, with 2 florets, the lower floret staminate with a long, twisted awn near the middle, the upper floret perfect with a straight awn; *disarticulation* above the glumes.

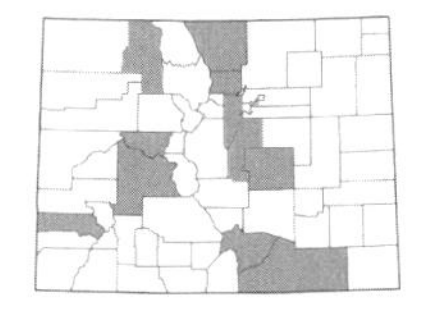

Arrhenatherum elatius **(L.) J. Presl & C. Presl**, TALL OATGRASS. Plants 7–17 dm; *leaves* flat, 3–8 mm wide; *ligules* 1–2 mm long, erose-ciliate; *spikelets* 7–9 mm long; *glumes* unequal, the lower 5–7 mm long, upper 8–9 mm long; *lemmas* 7–10 mm long, 7-nerved, the lower lemma awned from below the middle with a bent awn 10–20 mm long, the upper lemma awned just below the apex with a straight awn to 4 mm long. Found in meadows, shrublands, along streams, and in disturbed areas, 5000–10,000 ft. June–Aug. E/W. Introduced. (Plate 60)

AVENA L. – OATS

Annuals (ours); *sheaths* open; *ligules* membranous; *spikelets* in open, pendulous panicles, with 2–3 florets; *glumes* nearly equal, several-nerved, longer than the lower floret; *lemmas* 5–7-nerved, with a stout, often geniculate awn from below the bifid apex, or awn reduced or absent; *disarticulation* above the glumes and between florets.

1a. Lemma calluses long-hairy; spikelets with 2 bent awns (one on the lemmas of each floret)...***A. fatua***
1b. Lemma calluses glabrous; spikelets awnless or with a single straight or weakly twisted awn on the lower floret...***A. sativa***

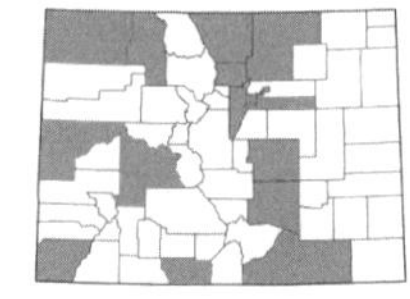

***Avena fatua* L.**, WILD OATS. Plants 3–10 dm; *leaves* flat, 3–10 mm wide; *ligules* 2–5.5 mm long; *spikelets* 1.8–3.2 cm long, reflexed; *glumes* 1.8–3.2 cm long, 9–11-nerved; *lemmas* 14–20 mm long, the callus long-hairy, with a bent awn 2.5–4.5 cm long arising from near the middle of the lemma. Found in disturbed areas, fields and pastures, and along roadsides, 4500–6000 ft. E/W. Introduced.

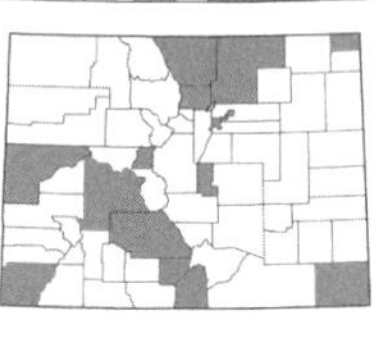

***Avena sativa* L.**, CULTIVATED OATS. Plants 4–10 dm; *leaves* flat, 3–20 mm wide; *ligules* 2–6 mm long; *spikelets* 2–3.5 cm long, reflexed; *glumes* 2–3.5 cm long, 9–11-nerved; *lemmas* 14–20 mm long, the callus glabrous, unawned or with a single straight or weakly twisted awn on the lower floret. Found in disturbed areas, fields, and along roadsides, 4500–8000 ft. E/W. Introduced.

AVENULA (Dumort.) Dumort. – OATGRASS

Perennials; *sheaths* open; *ligules* membranous; *spikelets* in a panicle, many branches with a single spikelet, with 2–7 florets, *glumes* 1–5-nerved, equaling or exceeding the florets; *lemmas* 5–7-nerved, with a bifid tip, awned from near the middle; *disarticulation* above the glumes and below the florets.

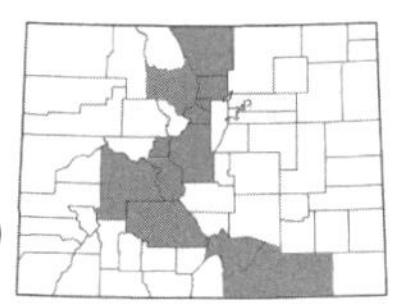

***Avenula hookeri* (Scribn.) Holub**, SPIKE OATGRASS. [*Helictotrichon hookeri* (Scribn.) Henrard]. Plants 0.5–5 dm; *leaves* flat, 2–4 mm wide; *ligules* 2–3 mm long, acute; *spikelets* 12–15 mm long; *glumes* 9–14 mm long, 3-5-nerved; *lemmas* 10–12 mm long, with an awn 10–18 mm arising from just above the middle. Found in meadows and the alpine, 10,500–12,500 ft. July–Sept. E/W. (Plate 60)

BECKMANNIA Host – SLOUGHGRASS

Annuals; *sheaths* open; *ligules* membranous; *spikelets* 1-flowered, in a panicle, with spikelets closely imbricate in 2 rows on one side of the panicle branches, flattened and suborbicular (stacked like coins); *glumes* equal, 3-nerved, keeled, apiculate; *lemmas* 5-nerved, acute or with a short awn-pointed; *disarticulation* below the glumes.

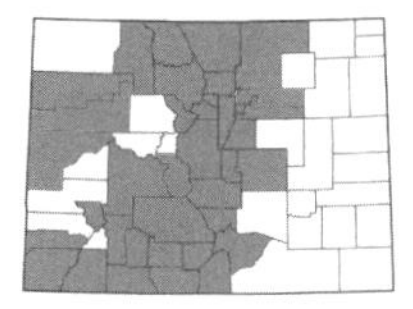

***Beckmannia syzigachne* (Steud.) Fernald**, AMERICAN SLOUGHGRASS. Plants 2–12 dm; *leaves* flat, 3–10 mm wide; *ligules* 5–9 mm long; *glumes* 2.5–3.5 mm long, laterally compressed; *lemmas* 2.5–3.5 mm long, acute or shortly awn-tipped, extending just slightly above the glumes. Common in moist places and wetlands, 5000–11,500 ft. June–Sept. E/W. (Plate 60)

BOTHRIOCHLOA Kuntze – BEARDGRASS

Perennials; *sheaths* open; *ligules* membranous; *inflorescence* with hairy pedicels; *spikelets* paired, in a terminal panicle of spicate branches, with one sessile, fertile, and awned spikelet and one pediceled, staminate or sterile spikelet; *lemmas* of fertile spikelet with a twisted awn; *disarticulation* in the rachis.

1a. Inflorescence branches purplish-red or reddish...2
1b. Inflorescence branches white to light tan...3

2a. Inflorescence branches racemose, not digitate...***B. bladhii***
2b. Inflorescence branches more or less digitate, clustered at the tip of the stem...***B. ischaemum***

3a. Leaf nodes densely bearded with spreading, long, silky hairs 3–7 mm long; fertile spikelet 5.5–8.5 mm long...***B. springfieldii***
3b. Leaf nodes glabrous to shortly hirsute or pilose with appressed or ascending hairs less than 3 mm long; fertile spikelet 2.5–7 mm long...4

4a. Fertile spikelet 2.5–4.5 mm long and sterile spikelet 1.5–2.5 mm long; awn of fertile spikelet 8–16 mm long; inflorescence branches racemose, not clustered at the tip of the stem...***B. laguroides***
4b. Fertile spikelet 4–7 mm long and sterile spikelet 3–4 mm long; awn of fertile spikelet 18–35 mm long; inflorescence branches more or less digitate, clustered at the tip of the stem...***B. barbinodis***

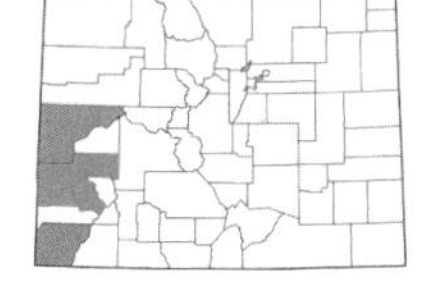

***Bothriochloa barbinodis* (Lag.) Herter**, CANE BLUESTEM. Plants 6–12 dm; *leaves* flat, 2–7 mm wide; *ligules* 1–2.5 mm long; *inflorescence* a terminal panicle of digitate branches, white; *spikelets* in pairs, the fertile one 4–7 mm long and the pediceled rudimentary one 3–4 mm long; *lemmas* of fertile spikelet with a bent awn 18–35 mm long. Uncommon on rocky slopes and in canyons, known from the southwestern counties, 4500–5500 ft. July–Sept. W. (Plate 60)

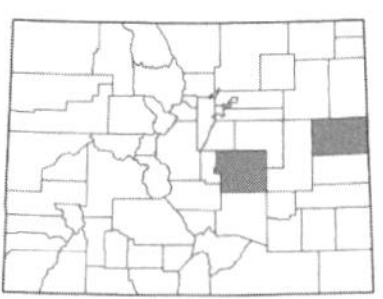

***Bothriochloa bladhii* (Retz.) S.T. Blake**, CAUCASIAN BLUESTEM. Plants 4–9 dm; *leaves* flat, 1–4 mm wide; *ligules* 0.5–1.5 mm long; *inflorescence* a terminal panicle of racemose branches, reddish at maturity; *spikelets* in pairs, the fertile one 3.5–4 mm long and the pediceled rudimentary one 2–4 mm long; *lemmas* of fertile spikelet with a bent awn 10–17 mm long. Uncommon along roadsides and in disturbed areas, 4000–6000 ft. July–Sept. E. Introduced.

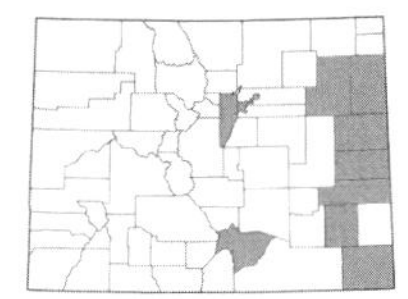

***Bothriochloa ischaemum* (L.) Keng**, YELLOW BLUESTEM. Plants 3–8 dm; *leaves* flat or folded, 2–5 mm wide; *ligules* 0.5–1.5 mm long; *inflorescence* a terminal panicle of more or less digitate branches, reddish-purple; *spikelets* in pairs, the fertile one 3–4.5 mm long, the pediceled rudimentary one 3–4.5 mm long; *lemmas* of fertile spikelets with a bent awn 9–17 mm long. Found in shortgrass prairie, along roadsides, and in disturbed areas, 3500–7500 ft. July–Sept. E. Introduced.

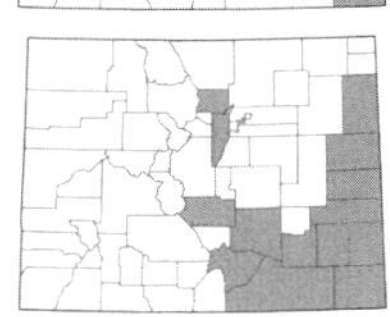

***Bothriochloa laguroides* (DC.) Herter**, SILVER BLUESTEM. Plants 3–12 dm; *leaves* flat or folded, 2–7 mm wide; *ligules* 1–3 mm long; *inflorescence* a terminal panicle of racemose branches, white to light tan; *spikelets* in pairs, the fertile one 2.5–4.5 mm long and the pediceled rudimentary one 1.5–2.5 mm long; *lemmas* of fertile spikelets with a bent awn 8–16 mm long. Found in shortgrass prairie, on rocky slopes, and along roadsides, 3500–6200 ft. June–Sept. E.

***Bothriochloa springfieldii* (Gould) Parodi**, SPRINGFIELD'S BEARDGRASS. Plants 3–8 dm; *leaves* flat or folded, 2–3 mm wide; *ligules* 1–3 mm long; *inflorescence* a terminal panicle of digitate racemes, white; *spikelets* in pairs, the fertile one 5.5–8.5 mm long, the pediceled rudimentary one 3.5–5.5 mm long; *lemmas* of fertile spikelets with a bent awn 15–30 mm long. Found in canyons and on grassy slopes, 4400–6000 ft. July–Sept. E.

BOUTELOUA Lag. – GRAMA

Annuals or perennials; *sheaths* open; *ligules* membranous or of hairs; *spikelets* sessile, with one fertile floret and 1-2 rudimentary florets above; *glumes* 1-veined, sometimes shortly awned; *lemmas* 3-nerved, often awned; *disarticulation* at the base of the branch rachis or above the glumes.

1a. Inflorescence of small (1-3 cm long), pendulous spikes attached to one side of the stem...***B. curtipendula***
1b. Inflorescence of dense, sessile spikelets in 2 rows along one side of a flattened rachis (resembling an eyelash)...2

2a. Stem internodes densely woolly-hairy...***B. eriopoda***
2b. Stem internodes glabrous...3

3a. Rachis extended beyond the terminal floret in the inflorescence 5–10 mm; leaf blades with long papillose-based hairs along the margins at the base...***B. hirsuta* var. *hirsuta***
3b. Rachis not extended beyond the terminal floret in the inflorescence; leaf blades usually lacking long papillose-based hairs along the margins at the base...4

4a. Glumes with papillose-based hairs along the nerves; perennials...***B. gracilis***
4b. Glumes lacking papillose-based hairs along the nerves; annuals...5

5a. Two or more spikes per stem; upper glumes 1.5–2.5 mm long...***B. barbata***
5b. One spike per stem; upper glumes 3.5–5.5 mm long...***B. simplex***

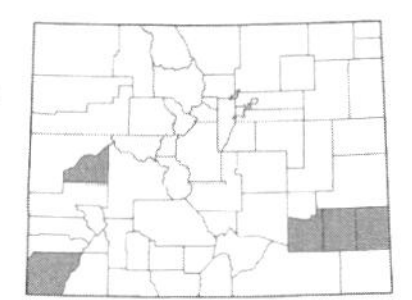

***Bouteloua barbata* Lag.**, SIXWEEKS GRAMA. [*Chondrosum barbatum* (Lag.) Clayton]. Annuals, 0.2–3 (6) dm; *leaves* involute to flat, 1–4 mm wide; *ligules* a fringe of hairs ca. 1 mm long; *inflorescence* of dense, sessile spikelets in 2 rows along one side of a flattened rachis; *glumes* unequal, the lower 1–2 mm and the upper 1.5–2.5 mm long, glabrous to strigose; *lemmas* of fertile spikelet1.5–2.5 mm, 3-awned, the central awn 0.5–3 mm long. Uncommon in disturbed areas and in sandy soil, 3500–6500 ft. July–Oct. E/W.

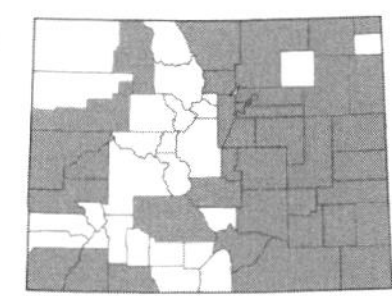

***Bouteloua curtipendula* (Michx.) Torr.**, SIDEOATS GRAMA. Perennials, 2–10 dm; *leaves* usually flat, 2–7 mm wide; *ligules* a fringe of hairs 0.2–1 mm long; *inflorescence* of small (1–3 cm long), pendulous spikes attached to one side of the stem; *glumes* unequal, the lower 2.5–6 mm and the upper 4.5–8 mm long, acute to awn-tipped; *lemmas* 3-nerved, the fertile lemma 3–7 mm long, 3-awned. Common on the plains, in the foothills, and in sagebrush and pinyon-juniper, 3500–7000 ft. July–Oct. E/W.

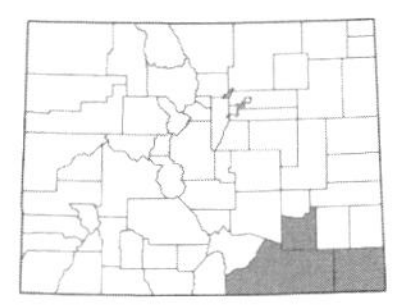

***Bouteloua eriopoda* (Torr.) Torr.**, BLACK GRAMA. [*Chondrosum eriopodum* Torr.]. Perennials, 2–6 dm; *leaves* flat to involute at the tips, 0.5–2 mm wide; *ligules* a ring of hairs ca. 1 mm long; *inflorescence* of dense, sessile spikelets in 2 rows along one side of a flattened rachis; *glumes* unequal, the lower 2–4.5 mm and the upper 6–9 mm long, glabrous to minutely scabrous; *lemmas* of fertile spikelets 4–7 mm long, 3-awned, the central awn 0.5–4 mm long. Uncommon on sandstone mesas, rocky slopes, and in canyons, 4000–5800 ft. July–Sept. E.

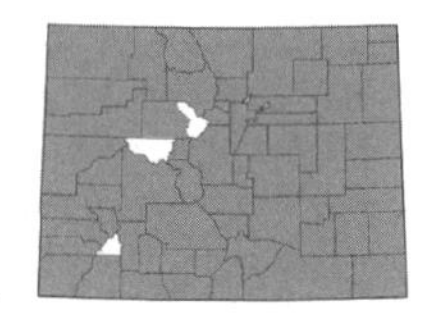

Bouteloua gracilis **(Kunth) Lag.**, BLUE GRAMA. [*Chondrosum gracile* Kunth]. Perennials, 2–7 dm; *leaves* flat or somewhat involute, 0.5–4 mm wide; *ligules* ciliate, 0.1–0.5 mm long; *inflorescence* of dense, sessile spikelets in 2 rows along one side of a flattened rachis; *glumes* unequal, the lower 1.5–3.5 mm and the upper 3.5–6 mm long, with papillose-based hairs on the nerve; *lemmas* of fertile spikelets 3.5–6 mm long, 3-awned, the central awn 0.5–1.5 mm long. Common on the plains, in the foothills, mountain meadows, dry grasslands, sagebrush, and pinyon-juniper, 3500–10,000 ft. July–Oct. E/W. (Plate 61)

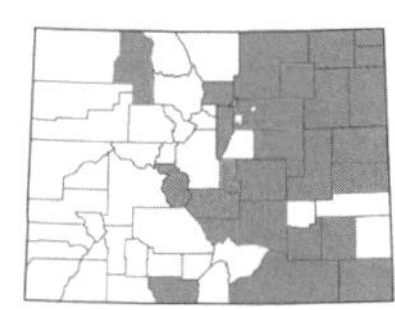

Bouteloua hirsuta **Lag. var. *hirsuta***, HAIRY GRAMA. [*Chondrosum hirsutum* (Lag.) Sw.]. Perennials, 1.5–7.5 dm; *leaves* involute or flat, 1–3 mm wide; *ligules* a fringe of hairs, 0.2–0.5 mm long; *inflorescence* of dense, sessile spikelets in 2 rows along one side of a flattened rachis; *glumes* unequal, the lower 1.5–3.5 mm and the upper 3–6 mm long, with papillose-based hairs along the keel; *lemmas* of fertile spikelets 2–4.5 mm long, 1–3-awned, awn 0.2–2.5 mm long. Common on the plains, on rocky slopes, in dry grasslands, and in pinyon-juniper, 3500–8500 ft. July–Oct. E/W.

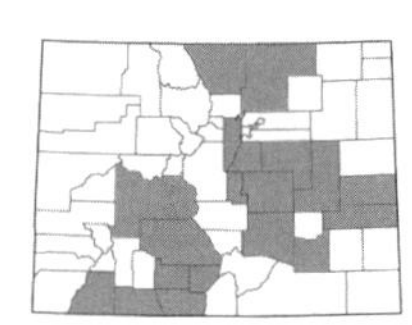

Bouteloua simplex **Lag.**, MATTED GRAMA. [*Chondrosum prostratum* (Lag.) Sweet]. Annuals, 0.1–3 dm; *leaves* flat, 0.5–1.5 mm wide; *ligules* of hairs with a membranous base, to 0.2 mm long; *inflorescence* of dense, sessile spikelets in 2 rows along one side of a flattened rachis; *glumes* unequal, the lower 2–3 mm and the upper 3.5–5.5 mm long; *lemmas* of fertile spikelets 2.5–3.5 mm long, pilose near the base, 3-awned, the central awn 1–2 mm long. Common in rabbitbrush, pinyon-juniper, and disturbed areas, 4500–8800 ft. July–Oct. E/W.

BRACHYPODIUM P. Beauv. – FALSE BROME

Annuals (ours); *sheaths* open; *ligules* membranous; *spikelets* in spicate racemes, usually solitary or sometimes 2–3 spikelets per node, with 15–24 florets; *glumes* 5–7-nerved, unequal; *lemmas* usually membranous, firm, 7-nerved, awned from an entire apex; *paleas* about as long as the lemma; *disarticulation* above the glumes.

Brachypodium distachyon **(L.) P. Beauv.**, PURPLE FALSE BROME. Plants 0.5–5 dm; *leaves* flat, 2–5 mm wide; *ligules* 1–2 mm long; *glumes* unequal, the lower 5–6 mm and the upper 7–8 mm long; *lemmas* 7–10 mm long, with a straight awn 8–20 mm long. Uncommon in dry, disturbed areas, 5000–5500 ft. July–Sept. E. Introduced.

BROMUS L. – BROME

Annuals, biennials, or perennials; *sheaths* connate for most their length, usually densely hairy; *ligules* membranous; *spikelets* usually in a terminal simple panicle, with 5–many florets; *glumes* subequal, usually acute, usually shorter than adjacent lemmas; *lemmas* 5–many-nerved, awned or not from an entire or bifid apex; *disarticulation* above the glumes.

Reports of *Bromus erectus* Huds. in Colorado were based on a misidentification of *B. carinatus.*

1a. Lemmas 6–8 mm wide (over 2.6 mm from the midnerve to the margin in side view), inflated and rounded over the midvein, glabrous or scabridulous; awns usually absent or sometimes to 1 mm; spikelets ovate, hanging on one side of the rachis; annuals...***B. briziformis***
1b. Lemmas 1–7 mm wide, not inflated; awns present or if absent then the plants perennial; spikelets variously shaped, sometimes hanging on one side of the rachis...2

2a. Perennials; awn of lemma arising from an entire or slightly emarginated apex, or lemma awnless...3
2b. Annuals; awn of lemma arising from a bifid apex, rarely awnless...13

3a. Plants rhizomatous...4
3b. Plants lacking rhizomes...5

4a. Lemmas glabrous or scabrous on the back; awns usually absent or sometimes to 3.5 mm long...***B. inermis***
4b. Lemmas usually hairy on the back, or hairy on the margins; awn usually present, to 7.5 mm long, sometimes absent...***B. pumpellianus***

5a. Upper glumes 5–9 (11)-veined; lower glumes 3–7 (9)-veined; spikelets strongly laterally compressed...6
5b. Upper glumes 3-veined; lower glumes 1–3-veined; spikelets terete to moderately laterally compressed...7

6a. Lemmas and sheath throats usually hairy...***B. carinatus***
6b. Lemmas and sheath throats glabrous...***B. polyanthus***

7a. Most of the lower glumes 3-veined (look at more than one spikelet)...8
7b. Most of the lower glumes 1-veined...9

8a. Glumes glabrous or rarely slightly hairy; lemmas glabrous or hairy; leaf blades often glaucous...***B. frondosus***
8b. Glumes and lemmas usually hairy; leaf blades not glaucous...***B. porteri***

9a. Lemmas dissimilar in pubescence, the upper lemmas in a floret more or less hairy throughout on the back and the lower lemmas glabrous across the back but with hairy margins...***B. richardsonii***
9b. Lemmas all similar in pubescence, glabrous or hairy throughout...10

10a. Lemmas glabrous, or with glabrous backs and hairy margins on the lower ½ or ¾; glumes glabrous...***B. ciliatus***
10b. Lemmas evenly hairy on the backs and margins; glumes glabrous or hairy...11

11a. Upper glume mucronate with a sharp point at the tip...***B. mucroglumis***
11b. Upper glume not mucronate at the tip...12

12a. Glumes usually glabrous (rarely hairy); awn 2–4 mm long...***B. lanatipes***
12b. Glumes usually hairy (rarely glabrous); awn (3) 4–8 mm long...***B. pubescens***

13a. Panicles in a compact terminal cluster (occasionally reduced to 1 or 2 spikelets), the branches shorter than the spikelets, erect...14
13b. Panicles open, the branches ascending, spreading, or often drooping or flexuous, at least some branches longer than the spikelets...15

14a. Panicle usually purplish or reddish-brown; awns 8–20 mm long; lower glumes usually 1-veined...***B. rubens***
14b. Panicle not purplish or reddish-brown (occasionally reduced to 1 or 2 spikelets); awns 5–9 mm long; lower glumes 3–5-veined...***B. hordeaceus* ssp. *hordeaceus***

15a. Lemmas linear to narrowly lanceolate, to 1.5 mm wide from the midnerve to the margin, 9–35 mm long; teeth of the bifid lemma apex 1–5 mm long; awns 10–65 mm long...16
15b. Lemmas elliptic to lanceolate, over 1.5 mm wide from the midnerve to the margin, 6.5–12 mm long; teeth of the bifid lemma apex shorter than 1 mm; awns 3–13 mm long...18

16a. Awns 30–65 mm long; lemmas 20–35 mm long; lower glumes 13–25 mm long...***B. diandrus***
16b. Awns 10–30 mm long; lemmas 9–20 mm long; lower glumes 4–14 mm long...17

17a. Lemmas 9–12 mm long; spikelets 10–24 mm long; awns 10–18 mm long...***B. tectorum***
17b. Lemmas 13–20 mm long; spikelets 20–35 mm long; awns 18–26 mm long...***B. sterilis***

18a. Leaf sheaths glabrous or sparsely hairy; florets spreading and exposing the rachilla joints in the spikelet; awn 3–10 mm long or rarely absent...***B. secalinus***
18b. Leaf sheaths densely pilose; florets not spreading; awn 3–30 mm long...19

19a. Awn 3–10 mm long, not flattened or twisted at the base, arising less than 1.5 mm below the lemma tips...20
19b. Awn 7–13 mm long, often flattened and twisted at the base, arising 1.5 mm or more below the lemma tips...21

20a. Lemmas 8–11.5 mm long; rachilla of some spikelets usually visible at maturity, with internodes 1.5–2 mm long...***B. commutatus***
20b. Lemmas 6.5–8 mm long; rachilla concealed at maturity, with internodes 1–1.5 mm long...***B. racemosus***

21a. Panicle branches usually with more than 1 spikelet, often drooping; hyaline margins of the lemma 0.3–0.6 mm wide...***B. japonicus***
21b. Panicle branches usually with 1 spikelet, not drooping; hyaline margins of the lemma 0.6–0.9 mm wide...***B. squarrosus***

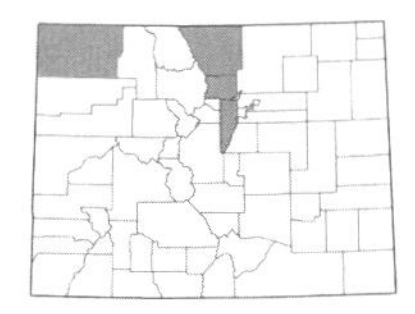

Bromus briziformis **Fisch. & C.A. Mey.**, RATTLESNAKE BROME. Annuals, 1–6 dm; *leaves* usually flat, 1–6 mm wide, with a hairy sheath; *ligules* hairy, 0.5–2 mm long; *spikelets* strongly compressed, 15–27 mm long, 7–15-flowered; *glumes* unequal, the lower 4.5–6 mm and the upper 5–9 mm long; *lemmas* 6.5–10 mm long, strongly inflated at maturity, unawned or shortly awn-tipped. Uncommon on dry slopes, in grasslands, open forests, and disturbed areas, 5000–7800 ft. June–Aug. E/W. Introduced.

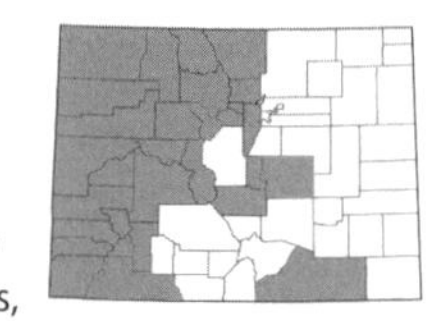

***Bromus carinatus* Hook. & Arn.**, California brome. [*Ceratochloa carinata* (Hook. & Arn.) Tutin]. Annuals or biennials, 5–10 dm; *leaves* usually flat, 2.5–6 mm wide, with a pilose sheath; *spikelets* strongly laterally compressed, 20–40 mm long, 4–7-flowered; *glumes* unequal, the lower 7–11 mm and the upper 9–13 mm long and 5–9 (11)-nerved; *lemmas* 11–15 mm long, usually uniformly hairy, sometimes scabrous, with an awn 4–8 mm long arising from an entire to slightly bifid apex. Common along roadsides and in disturbed areas, forests and meadows, and wetlands, 4500–11,500 ft. June–Aug. E/W.

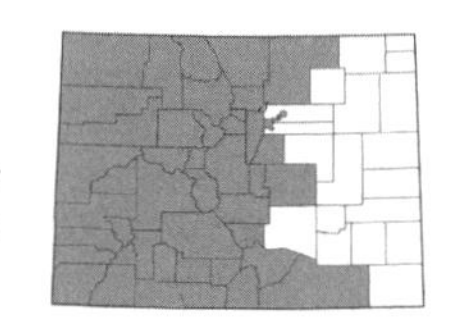

***Bromus ciliatus* L.**, fringed brome. [*Bromopsis ciliata* (L.) Holub.; *B. canadensis* Michx.]. Perennials, 4–12 dm; *leaves* flat, 4–10 mm wide, with a slightly pilose to glabrous sheath; *ligules* 0.3–2 mm, erose; *spikelets* 15–30 mm long, 4–9-flowered; *glumes* unequal, the lower 5–8 mm and the upper 7–8.5 mm long; *lemmas* 8–15 mm long, with hirsute margins, with an awn 3–5 mm long from an entire apex. Common in forests, meadows, and moist places, 4500–12,000 ft. July–Sept. E/W.

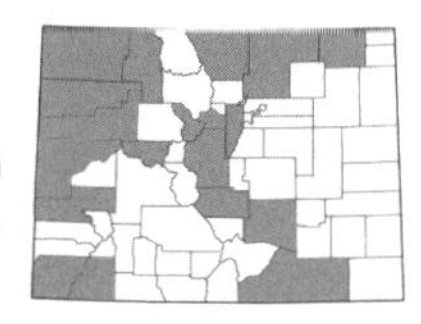

***Bromus commutatus* Schrad.**, meadow brome. Annuals, 3–10 dm; *leaves* flat, 1–6 mm wide, with a pilose sheath; *spikelets* slightly compressed, 10–25 mm long, 4–10-flowered; *glumes* unequal, the lower 5–6 mm and the upper 6–7 mm long; *lemmas* 8–11.5 mm long, with a straight awn 7–9 mm long arising from a bifid apex. Found in disturbed areas, forests, shrublands, and along roadsides, 3800–8500 ft. June–Aug. E/W. Introduced.

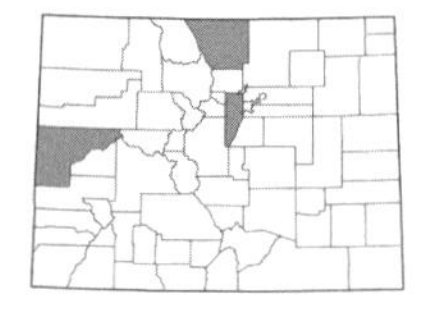

***Bromus diandrus* Roth**, great brome. [*Anisantha diandra* (Roth) Tutin ex Tzvelev]. Annuals, 2–9 dm; *leaves* flat, 3–8 mm wide, with a pilose sheath; *ligules* 2–6 mm; *spikelets* 30–50 mm long (not including awns), 4–8-flowered; *glumes* unequal, the lower 13–25 mm and the upper 20–35 mm long; *lemmas* 20–35 mm, with an awn 30–65 mm long arising from a bifid apex. Uncommon on dry slopes, in disturbed areas, and along roadsides, 5000–7600 ft. May–Aug. E/W. Introduced.

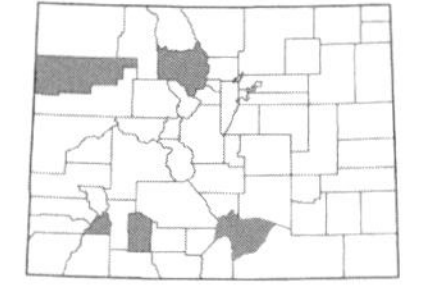

***Bromus frondosus* (Shear) Wooton & Standl.**, weeping brome. [*Bromopsis frondosa* (Shear) Holub.]. Perennials, 4–11 dm; *leaves* flat, 3–6 mm wide, with a glabrous sheath; *ligules* 1–3 mm; *spikelets* 15–30 mm long, 4–10-flowered, slightly compressed; *glumes* unequal, the lower 5–8 mm and the upper 6–9 mm long; *lemmas* 8–12 mm long, scabrous, with an awn 1.5–4 mm long arising from an entire apex. Found in forests and on open slopes, 6500–10,000 ft. June–Aug. E/W.

This species is often included within *B. ciliatus* L.

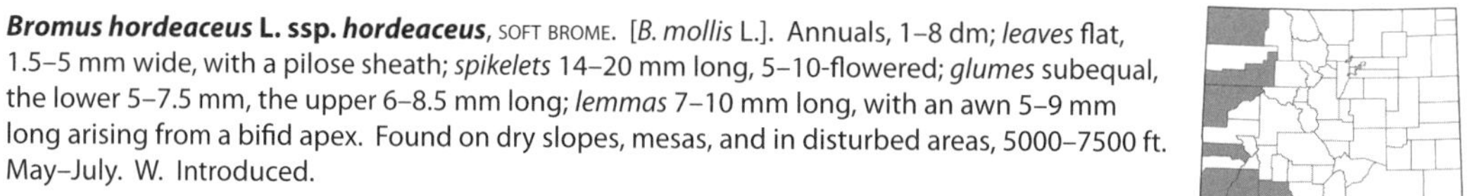

Bromus hordeaceus* L. ssp. *hordeaceus, soft brome. [*B. mollis* L.]. Annuals, 1–8 dm; *leaves* flat, 1.5–5 mm wide, with a pilose sheath; *spikelets* 14–20 mm long, 5–10-flowered; *glumes* subequal, the lower 5–7.5 mm, the upper 6–8.5 mm long; *lemmas* 7–10 mm long, with an awn 5–9 mm long arising from a bifid apex. Found on dry slopes, mesas, and in disturbed areas, 5000–7500 ft. May–July. W. Introduced.

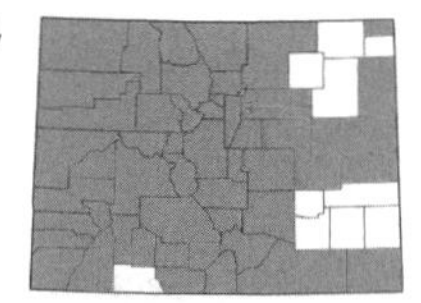

***Bromus inermis* Leyss.**, smooth brome. [*Bromopsis inermis* (Leyss.) Holub.]. Perennials, rhizomatous, 5–13 dm; *leaves* flat, 4–15 mm wide, with a "W" or "M" imprinted on the surface; *ligules* 1–2 mm long; *spikelets* 15–40 mm long, 5–13-flowered, slightly compressed; *glumes* unequal, the lower 5–8 mm and the upper 6–10 mm long; *lemmas* 9–13 mm long, unawned or with a short awn tip to 3.5 mm long, glabrous. Common in disturbed areas, meadows and grasslands, forests, and along roadsides, 4500–11,800 ft. June–Sept. E/W. Introduced.

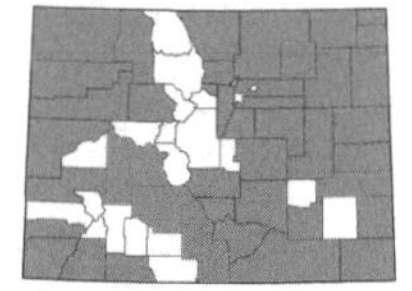

***Bromus japonicus* Thunb.**, Japanese brome. Annuals, 2–7 dm; *leaves* flat, 1.5–7 mm wide, with a pilose sheath; *ligules* 0.5–2 mm long, erose; *spikelets* 15–24 mm long, 5–12-flowered; *glumes* unequal, the lower 4–6 mm and the upper 5.5–8.5 mm long; *lemmas* 7–9 mm long, slightly compressed, with an awn 7–13 mm long arising from a bifid apex. Common on dry slopes, along roadsides, in shortgrass prairie, and in disturbed areas, 3500–8500 ft. May–July. E/W. Introduced.

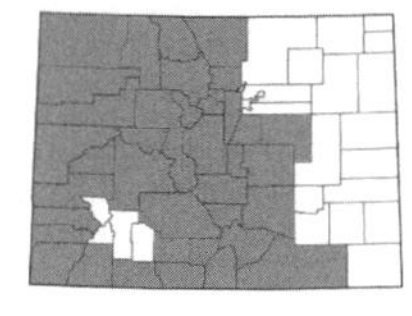

***Bromus lanatipes* (Shear) Rydb.**, woolly brome. [*Bromopsis lanatipes* (Shear) Holub.]. Perennials, 4–10 dm; *leaves* flat, 3–6 mm wide, with a pilose sheath; *ligules* 1–2 mm long; *spikelets* 10–30 mm long, 5–11-flowered; *glumes* unequal, the lower 5–8 mm and the upper 6–11 mm long; *lemmas* 8–11 mm long, hairy, with an awn 2–4 mm long. Found in forests, meadows, along moist drainages, and on mesas and dry slopes, 5000–11,200 ft. June–Aug. E/W.

***Bromus mucroglumis* Wagnon**, sharpglume brome. [*Bromopsis mucroglumis* (Wagnon) Holub]. Perennials, 5–10 dm; *leaves* flat, 4–11 mm wide, with a glabrous to hairy sheath; *ligules* 1–2 mm long; *spikelets* 20–30 mm long, 5–10-flowered; *glumes* unequal, usually hairy, the lower 6–8 mm and the upper 8–8.5 mm long; *lemmas* 10–12 mm long, with an awn 3–5 mm. Rare in forests and meadows, known from a single collection in Jackson Co., 8500–9000 ft. June–Aug. W.

Bromus polyanthus Scribn., COLORADO BROME. Perennials, 6–12 dm; *leaves* flat, 2–9 mm wide, with a glabrous to scabrous sheath; *ligules* (1) 2–2.5 mm; *spikelets* 20–35 mm long, 6–11-flowered, strongly laterally compressed; *glumes* unequal, the lower (5.5) 7–10 (11.5) mm, 3-nerved, and the upper (7.5) 9–12 mm long, 5–7-nerved, glabrous to scabrous; *lemmas* 12–15 mm long, with an awn 4–7 mm long arising from an acute apex. Common in forests and meadows and on open slopes, 7000–11,500 ft. June–Aug. E/W.

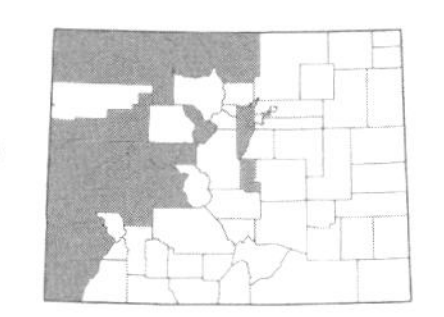

Bromus porteri (J.M. Coult.) Nash., NODDING BROME. [*Bromopsis porteri* (J.M. Coult.) Holub; this is the correct name for what we have referred to as *B. anomalus* Rupr. ex E. Fourn.]. Perennials, 4–9 dm; *leaves* flat, 2–5 mm wide, with a glabrous to hairy sheath; *ligules* 0.2–1 mm long; *spikelets* 15–30 mm long, 5–11-flowered; *glumes* unequal, usually hairy, the lower 5–8 mm and the upper 6–11 mm long; *lemmas* 7–12 mm long, hairy, with an awn 1–3.5 mm long. Found in forests, meadows, grasslands, shrublands, and on dry slopes, 5000–11,500 ft. July–Sept. E/W.

This species is often included within *B. lanatipes* (Shear) Rydb.

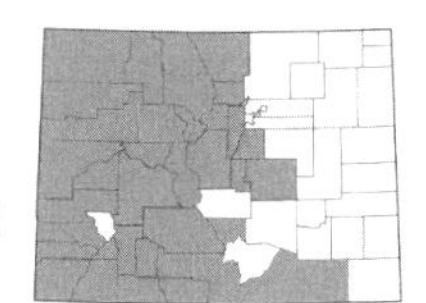

Bromus pubescens Muhl. ex Willd., HAIRY WOODLAND BROME. [*Bromopsis pubescens* (Muhl. ex Willd.) Holub.]. Perennials, 6–12 dm; *leaves* flat, 6–15 mm wide, with a pilose sheath; *ligules* 0.5–2 mm long; *spikelets* 15–30 mm long, 4–8-flowered; *glumes* unequal, usually hairy, the lower 4–8 mm and the upper 5–10 mm long; *lemmas* 8–11 mm long, usually hairy, with an awn 3–8 mm long. Found in forests, grasslands, and on dry slopes, 3400–9200 ft. June–Aug. E/W.

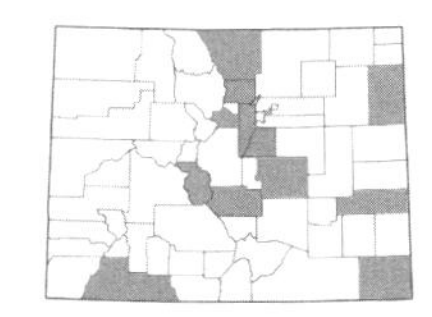

Bromus pumpellianus Scribn., PUMPELLY'S BROME. [*Bromopsis pumpelliana* (Scribn.) Holub.]. Perennials, rhizomatous, 5–12 dm; *leaves* flat, 4–15 mm wide, usually with a glabrous sheath; *ligules* to 3 mm long; *spikelets* 20–30 mm long, 7–11-flowered; *glumes* unequal, the lower 6–8 mm and the upper 7–9 mm long, glabrous or hairy; *lemmas* 9–14 mm long, hairy, with an awn 1–7.5 mm long, rarely unawned. Common on dry slopes and in grasslands, meadows, and forests, (5200) 7000–12,500 ft. July–Sept. E/W.

This is considered the native equivalent of *B. inermis*.

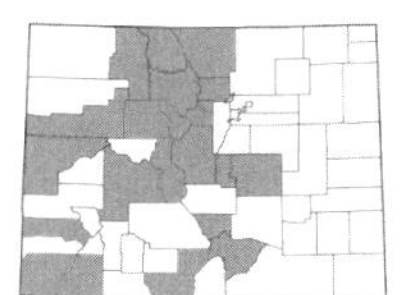

Bromus racemosus L., BALD BROME. Annuals to 10+ dm; *leaves* flat, 1–4 mm wide, the sheath with stiff, retrorse hairs; *ligules* 1–2 mm long, erose; *spikelets* 10–20 mm, 5–6-flowered; *glumes* subequal, the lower 4–6 mm and the upper 4–7 mm long; *lemmas* 6.5–8 mm long, rounded on the back, with an awn 5–9 mm long arising from a minutely bifid apex. Infrequent on dry slopes and in disturbed areas, 5000–9200 ft. June–Aug. E/W. Introduced.

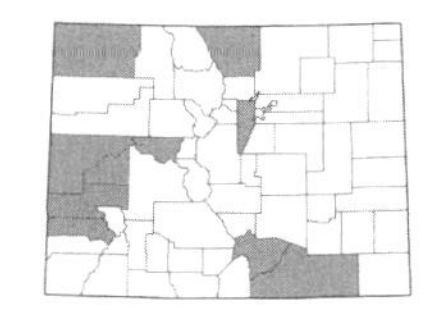

Bromus richardsonii Link, FRINGED BROME. [*Bromopsis richardsonii* (Link) Holub.; *Bromus ciliatus* L. var. *richardsonii* (Link) B. Boivin]. Perennials, 5–10 dm; *leaves* flat, 3–12 mm wide, with retrorsely hairy basal sheaths; *ligules* 0.4–2 mm long; *spikelets* 15–25 (40) mm long, usually 6–10-flowered; *glumes* unequal, glabrous or hairy, the lower 7.5–12.5 mm and the upper 9–11.3 mm long; *lemmas* 9–14 mm long, usually somewhat hairy, with an awn 3–5 mm long. Found in meadows and open forests, 6500–10,500 ft. June–Aug. E/W.

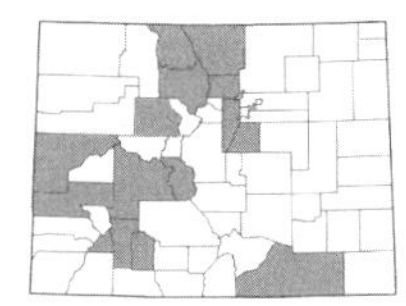

Bromus rubens L., RED BROME. [*Anisantha rubens* (L.) Nevski]. Annuals, 1–4 dm; *leaves* flat, to 15 cm long, with a pilose sheath; *ligules* 1–3 (4) mm long, hairy; *spikelets* 18–25 mm long, 4–8-flowered; *glumes* unequal, the lower 5–8 mm and the upper 8–12 mm long, 3–5-veined; *lemmas* 10–15 mm long, hairy, with an awn 8–20 mm long. Uncommon on dry slopes and in disturbed areas, known from a single collection in Montezuma Co., 6500–7000 ft. May–July. W. Introduced.

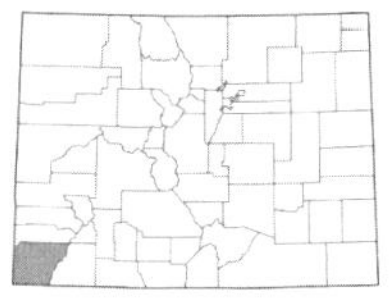

Bromus secalinus L., CHESS. Annuals, 3–10 dm; *leaves* flat, 2–9 mm wide, the sheath glabrous to loosely hairy; *ligules* 0.5–3 mm long, erose; *spikelets* 10–20 mm long, 4–10-flowered; *glumes* unequal, the lower 4–6 mm and the upper 5–7 mm long; *lemmas* 6–9 mm long, rounded on the back, with an awn 3–10 mm long arising from a bifid apex. Uncommon in fields, along roadsides, and in disturbed areas, 5000–7500 ft. June–Aug. E/W. Introduced.

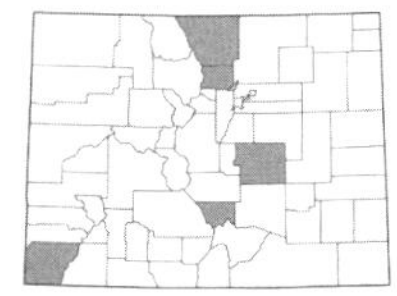

Bromus squarrosus L., CORN BROME. Annuals, 2–6 dm; *leaves* flat, 4–6 mm wide, with a pilose sheath; *ligules* 1–2.5 mm long, erose-ciliate; *spikelets* 15–70 mm long, 8–30-flowered; *glumes* unequal, the lower 4.5–7 mm and the upper 6–9 mm long; *lemmas* 8–11 mm long, rounded on the back, with an awn 8–10 mm long arising from a bifid apex. Uncommon along roadsides, on dry slopes, and in disturbed areas, 4500–6000 ft. June–Aug. E/W. Introduced.

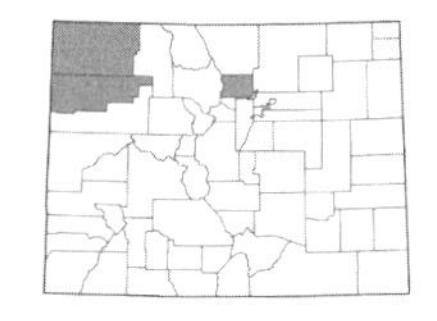

***Bromus sterilis* L.**, POVERTY BROME. [*Anisantha sterilis* (L.) Nevski]. Annuals, 2–10 dm; *leaves* flat, 2–5 mm wide, with a pilose sheath; *ligules* 1–2.5 mm long, erose; *spikelets* slightly compressed, 20–35 mm long, 3–9-flowered; *glumes* unequal, the lower 6–14 mm and the upper 10–20 mm long; *lemmas* 13–20 mm long, with an awn 18–26 mm long arising from a bifid apex. Uncommon along roadsides, in disturbed areas, and on dry slopes, 5000–7500 ft. June–Aug. E. Introduced.

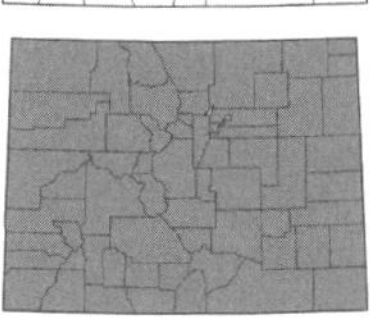

***Bromus tectorum* L.**, CHEATGRASS. [*Anisantha tectorum* (L.) Nevski]. Annuals, 0.5–10 dm; *leaves* flat or somewhat involute, 2–7 mm wide, with a pilose sheath; *ligules* 1–3.5 mm long, erose; *spikelets* slightly compressed, 10–24 mm long, 4–8-flowered; *glumes* unequal, the lower 4–8 mm and the upper 8–11 mm long; *lemmas* 9–12 mm long, with an awn 10–18 mm long arising from a bifid apex. Common in fields, grasslands, meadows, shrublands, forests, disturbed areas, and on dry slopes, 3800–10,500 ft. May–July. E/W. Introduced. (Plate 61)

BUCHLOË Engelm. – BUFFALOGRASS

Perennials, usually dioecious, stoloniferous; *sheaths* open; *ligules* membranous or of hairs; *pistillate spikelets* bur-like, with 1 floret; *staminate spikelets* in dense one-sided panicles, with 2 florets; *disarticulation* at the base of the panicle branches.

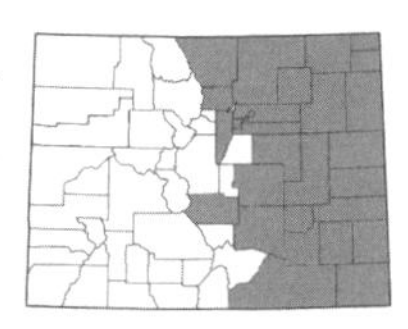

***Buchloë dactyloides* (Nutt.) Engelm.**, BUFFALOGRASS. [*Bouteloua dactyloides* (Nutt.) J.T. Columbus]. Plants 0.1–3 dm; *leaves* flat, 1–2.5 mm wide, often curling when dried; *ligules* 0.5–1 mm; *glumes* of staminate spikelets unequal, the lower 1–3 mm and the upper 2–5 mm long, of pistillate spikelets unequal, the lower rudimentary and the upper hard and enclosing the lemma, with 3 awn-tipped lobes; *lemmas* of staminate spikelets longer than the glumes. Common in the shortgrass prairie and foothills, sometimes grown as a lawn grass, 3400–5200 ft. May–July. E. (Plate 61)

CALAMAGROSTIS Adans. – REEDGRASS

Perennials from rhizomes; *sheaths* open; *ligules* membranous; *spikelets* in panicles, with one floret; *glumes* about equal, 1–3-nerved, acute; *lemmas* awned from the back, usually long-hairy on the callus; *paleas* about equaling the lemma; *disarticulation* above the glumes.

1a. Lemma awn straight, usually arising from near the middle of the lemma or above; callus hairs ½ to as long as or longer than the lemma; inflorescence an open or narrow panicle...2
1b. Lemma awn bent, arising from near the base of the lemma; callus hairs ⅓–½ as long as the lemma; inflorescence a narrow panicle...4

2a. Lemma awn arising from the upper two-fifths of the lemma (well above the middle), just reaching the tip of the lemma, or rarely the awn absent; spikelets 4.5–6 mm long; panicle spike-like and narrow...***C. scopulorum***
2b. Lemma awn arising from the middle of the lemma or below, longer or shorter than the lemma; spikelets 2–4.5 mm long; panicle open or narrow...3

3a. Callus hairs mostly as long or longer than the lemma; panicle open, usually over 2 cm wide...***C. canadensis***
3b. Callus hairs half to two-thirds as long as the lemma; panicle narrow, contracted to pyramidal in shape, rarely to 3 cm wide...***C. stricta***

4a. Lemmas 4–5.5 mm long; lemma awn 6–8 mm long, long-exserted and conspicuous in the inflorescence...***C. purpurascens***
4b. Lemmas 2–3.5 mm long; lemma awn 2–3.5 mm long, included or slightly exserted from the spikelet...5

5a. Leaves involute, 0.5–2 mm wide; at least some leaf sheaths hairy on the collar, the hairs longer than those present on the leaves; plants 1.5–4 dm tall...***C. montanensis***
5b. Leaves usually flat, or sometimes involute near the tip, 1.5–5 mm wide; leaf sheaths glabrous or with hairs similar to those present on the leaves; plants 4.5–11 dm tall...***C. rubescens***

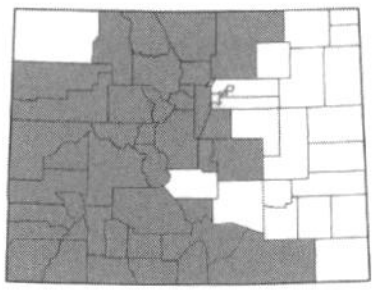

***Calamagrostis canadensis* (Michx.) P. Beauv.**, BLUEJOINT. Plants 6–15 dm; *leaves* flat, 2–8 mm wide; *ligules* 3–8 mm long; *glumes* 3–5 mm long, with a scabrous keel; *callus* hairs as long or longer than the lemmas; *lemmas* 3–4.5 mm long, with a delicate straight awn 1.2–2 mm long arising from the middle or below. Common in moist meadows, along streams, and in other moist areas, 5000–12,500 ft. July–Sept. E/W.

***Calamagrostis montanensis* Scribn.**, PLAINS REEDGRASS. Plants 1.5–4 dm; *leaves* involute, 0.5–2 mm wide; *ligules* 2–3 mm long; *glumes* 3.5–5 mm long, scabrous on the keel; *callus* hairs about half the length of the lemma; *lemmas* 3–3.5 mm long, with a geniculate awn 2–3 mm long arising from near the base. Rare on dry ridgetops and in the alpine, 8000–13,000 ft. July–Sept. E/W. (Plate 61)

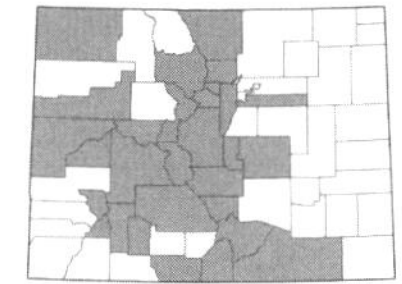

***Calamagrostis purpurascens* R. Br.**, PURPLE REEDGRASS. Plants 2–6 dm; *leaves* usually involute, 2–4 mm wide; *ligules* 2–5 mm long; *glumes* 5–7 mm long, scabrous on the keel; *callus* hairs 1–1.5 mm long; *lemmas* 4–5.5 mm long, with a geniculate awn 6–8 mm long arising from near the base. Common in forests, on rocky slopes, and on alpine ridges, 7500–13,000 ft. June–Aug. E/W.

***Calamagrostis rubescens* Buckley**, PINEGRASS. Plants 4.5–11 dm; *leaves* usually flat, 1.5–5 mm wide; *ligules* 2–6 mm long; *glumes* 3–5.5 mm long, with a scabrous keel; *callus* hairs ⅓ as long as the lemma; *lemmas* 2.5–3.5 mm long, with a geniculate awn 2.8–3.5 mm long arising from near the base. Rare in forests and on sagebrush slopes, known from Moffat, Rio Blanco, and Gunnison cos., 7000–9500 ft. July–Sept. W.

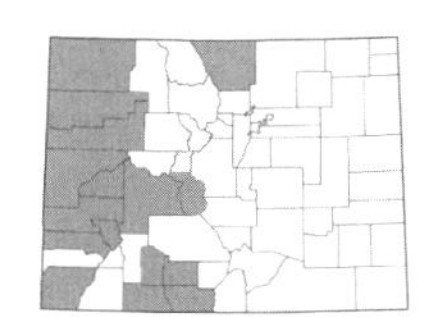

***Calamagrostis scopulorum* M.E. Jones**, DITCH REEDGRASS. Plants 5–9 dm; *leaves* flat, 3–7 mm wide; *ligules* 4–7 mm long; *glumes* 4.5–6 mm long, with a scabrous keel; *callus* hairs about half as long as the lemma; *lemmas* 3.5–5 mm long, with a straight awn to 2 mm long arising from above the middle. Found in moist places, wet meadows, often in rocky areas, 5500–10,500 ft. June–Sept. E/W.

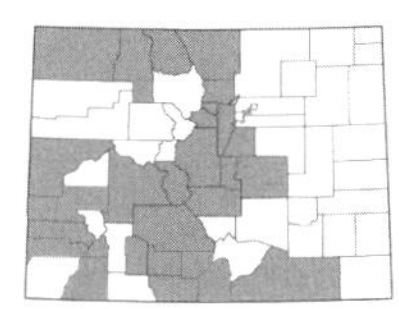

***Calamagrostis stricta* (Timm) Koeler**, SLIMSTEM REEDGRASS. Plants 2.5–9 dm; *leaves* involute or flat, 1–5 mm wide; *ligules* 0.5–6 mm long; *glumes* 2–4 mm long, scabrous on the keel; *callus* hairs half to as long as the lemma; *lemmas* with a straight awn arising from near the middle of the lemma. Common in moist meadows, grasslands, along lakes and creeks, and in other moist areas, 6800–11,500 ft. July–Sept. E/W.

CALAMOVILFA Hackel – SANDREED

Perennials from rhizomes; *sheaths* open; *ligules* a fringe of hairs; *spikelets* in open panicles, with one floret; *glumes* unequal, 1-nerved, acute; *lemmas* 1-nerved, firm, bearded on the callus; *paleas* greatly reduced; *disarticulation* above the glumes.

1a. Lemma and/or palea hairy...***C. gigantea***
1b. Lemma and palea glabrous...***C. longifolia***

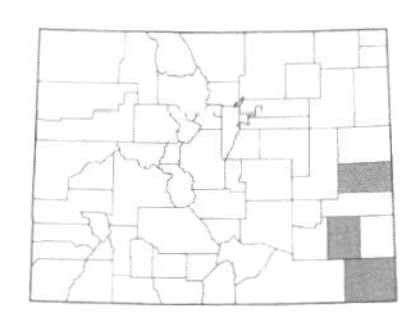

***Calamovilfa gigantea* Scribn. & Merr.**, GIANT SANDREED. Plants to 25 dm; *leaves* to 90 cm long; *ligules* 0.7–2 mm long; *glumes* unequal, the lower 4.5–10.5 mm and the upper 6.5–10 mm long, glabrous; *lemmas* 6–10 mm long, villous or rarely glabrous; *paleas* villous. Found on sand dunes of sandsage prairie, 3500–4500 ft. June–Aug. E.

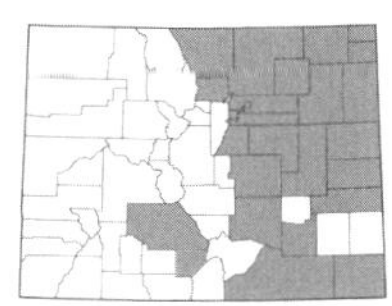

***Calamovilfa longifolia* (Hook.) Scribn.**, PRAIRIE SANDREED. Plants to 25 dm; *leaves* to 65 cm long; *ligules* 0.7–2.5 mm long; *glumes* unequal, the lower 3.5–6.5 mm and the upper 5–8.5 mm long, glabrous; *lemmas* 4.5–7 mm long, glabrous; *paleas* glabrous. Found in grasslands and on sand hills and dunes of sandsage prairie, 3500–7800 ft. July–Sept. E. (Plate 61)

CATABROSA P. Beauv. – BROOKGRASS

Aquatic perennials; *sheaths* closed at least half the length; *ligules* membranous; *spikelets* in open panicles, with two florets; *glumes* obtuse, unequal, 1-nerved or nerveless; *lemmas* with 3 parallel nerves, awnless, rounded on the back; *paleas* about as long as the lemma; *disarticulation* above the glumes.

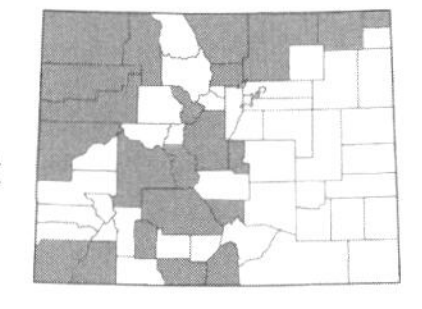

***Catabrosa aquatica* (L.) P. Beauv.**, BROOKGRASS. Plants 1–5 dm; *leaves* flat, 2–8 mm wide; *ligules* 2–6 mm long; *glumes* truncate, erose at the tips, unequal, the lower 0.5–1.5 mm and the upper 1.2–2.5 mm long; *lemmas* 2–3 mm long, glabrous, truncate. Uncommon along streams and moist places, sometimes in shallow water, 3700–10,500 ft. June–Sept. E/W. (Plate 61)

CENCHRUS L. – SANDBUR

Annuals (ours); *sheaths* compressed-keeled; *ligules* a fringe of hairs; *spikelets* enclosed in fascicles (spiny burs), these of bristles, flattened spines, or both fused together, several in each fascicle; *glumes* unequal.

1a. Inner bristles 0.5–1 (1.4) mm wide at the base; outer bristles usually round in cross-section; 45–75 bristles per fascicle...***C. longispinus***
1b. Inner bristles 1–3 mm wide at the base; outer bristles usually flattened; 20–40 bristles per fascicle...***C. spinifex***

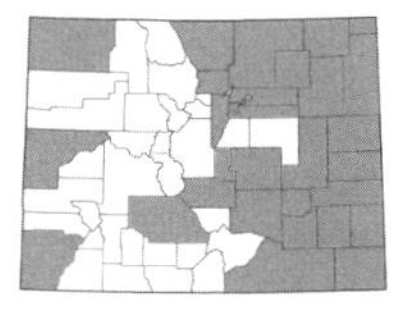

***Cenchrus longispinus* (Hack.) Fernald**, LONGSPINE SANDBUR. Plants decumbent or sometimes erect, the stems 1–5.5 dm long; *leaves* flat or folded, 3–6.5 mm wide; *ligules* 0.5–1.5 mm long; *spikelets* 6–8 mm long, the inner bristles 0.5–1 (1.4) mm wide at the base, the outer bristles usually round in cross-section, with 45–75 bristles per fascicle; *glumes* unequal, the lower 1.5–4 mm and the upper 4.5–6 mm long. Common in disturbed areas, fields, and along roadsides, 3500–6200 ft. July–Sept. E/W. (Plate 61)

***Cenchrus spinifex* Cav.**, COASTAL SANDBUR. Plants decumbent to erect, the stems 3–10 dm long; *leaves* flat or folded, 3–7 mm wide; *ligules* 0.5–1.5 mm long; *spikelets* 3.5–6 mm long, the inner bristles 1–3 mm wide at the base, the outer bristles usually flattened, with 20–40 bristles per fascicle; *glumes* unequal, the lower 1–3 mm and the upper 3–5 mm long. Uncommon in disturbed places, reported for Huerfano Co. July–Sept. E.

CHLORIS Sw. – WINDMILL GRASS

Annuals or perennials; *sheaths* open; *ligules* membranous; *spikelets* in digitate panicles of one-sided spicate branches, with a single perfect floret and 1–3 rudimentary florets above; *glumes* 1-nerved; *lemmas* awnless or awned from a minutely bifid apex, usually 3-nerved; *paleas* 2-nerved; *disarticulation* above the glumes.

1a. Inflorescence of 2–5 separate digitate whorls; keels of lemmas sparsely appressed-hairy with short hairs or glabrous; perennials...***C. verticillata***

1b. Inflorescence of a single terminal whorl of branches; keels of lemmas densely appressed-long-hairy, and usually with long hairs near the apex margin; annuals...***C. virgata***

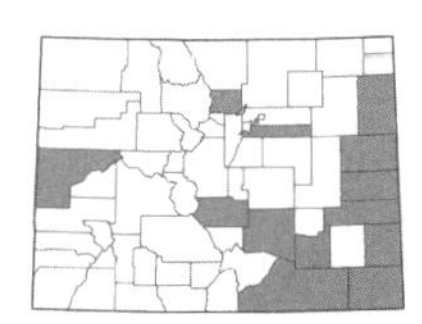

***Chloris verticillata* Nutt.**, TUMBLE WINDMILL GRASS. Perennials, 1.5–4 dm; *leaves* flat or folded, 2–3 mm wide; *ligules* 0.7–1.3 mm long; *glumes* unequal, the lower 2–3 mm and the upper 2.8–3.5 mm long, shortly awn-tipped; *lemmas* of perfect florets 2–3.5 mm long, the keel sparsely appresseed hairy or glabrous, with an awn 4–9 mm long. Found in gardens and lawns, vacant lots, and along roadsides, 3400–6000 ft. June–Aug. E/W.

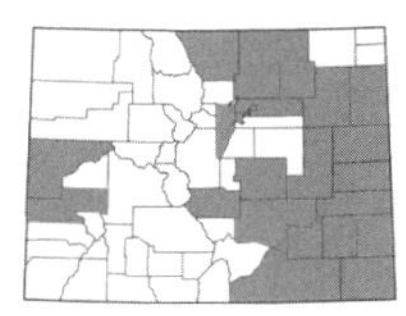

***Chloris virgata* Sw.**, FEATHER FINGERGRASS. Annuals, 1–10 dm; *leaves* flat or folded to 15 mm wide; *ligules* to 1 mm long; *glumes* unequal, the lower 1.5–2.5 mm and the upper 2.5–4.3 mm long, short-awned; *lemmas* of perfect florets 2.5–4.5 mm long, the keel densely appressed long-hairy, with an awn 2.5–15 long. Found in plantings, vacant lots and other disturbed areas, and along roadsides, 3400–6000 ft. June–Oct. E/W. (Plate 61)

CINNA L. – WOODREED

Perennials; *sheaths* open; *ligules* membranous; *spikelets* in panicles, with one floret; *glumes* about equal, 1–3-nerved, acute; *lemmas* 3–5-nerved; *paleas* slightly shorter than the lemma; *disarticulation* below the glumes.

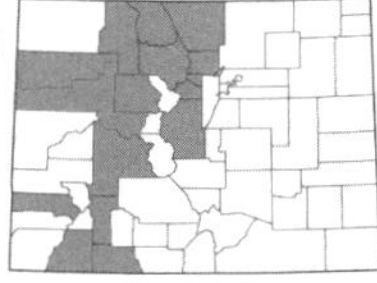

***Cinna latifolia* (Trevir. ex Goepp.) Griseb.**, DROOPING WOODREED. Plants 5–20 dm, rhizomatous; *leaves* flat, 4–15 mm wide; *ligules* 3–8 mm long; *glumes* 2.5–4 mm long, keel scabrous; *lemmas* slightly shorter than the glumes, unawned or with a short awn to 1.2 mm long arising just below the apex. Found along streams, ponds, and in moist woods, 7000–10,500 ft. July–Sept. E/W. (Plate 61)

CRYPSIS Aiton – PRICKLEGRASS

Annuals; *sheaths* open; *ligules* a ring of hairs; *spikelets* in short, dense panicles, with one floret; *glumes* strongly keeled; *lemmas* acute, awnless; *paleas* about equaling lemma; *disarticulation* above or below the glumes.

***Crypsis alopecuroides* (Piller& Mitterp.) Schrad.**, FOXTAIL PRICKLEGRASS. Plants prostrate to erect, stems 0.5–7.5 dm long; *leaves* involute or flat, 1–2.5 mm wide; *ligules* 0.2–1 mm long; *glumes* unequal, the lower 1.8–2.3 and the upper 1.4–2.5 mm long, keel ciliate; *lemmas* 1.7–2.8 mm long. Uncommon on drying mudflats and lake shores, 4000–5500 ft. Aug.–Oct. E. Introduced. (Plate 61)

CYNODON Rich. – BERMUDA GRASS

Perennials with rhizomes and stolons; *sheaths* closed for most of length; *ligules* membranous; *spikelets* in digitate panicles of one-sided spicate branches, with one fertile floret; *lemmas* awnless.

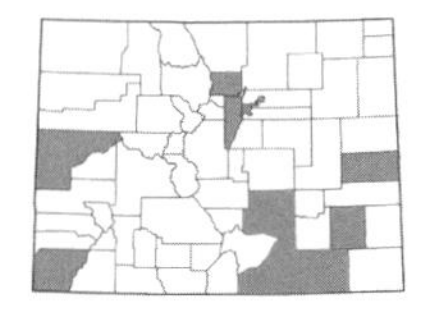

Cynodon dactylon* (L.) Pers. var. *dactylon, BAHAMA GRASS. Plants forming mats and rooting at lower nodes, *stems* 0.5–4 dm long; *leaves* flat to folded, 1–4 mm wide; *ligules* 0.2–0.5 mm long; *glumes* 1.4–2.5 mm long, shortly awned; *lemmas* 2–3 mm long. Found in lawns, disturbed areas, and along roadsides, 4000–5500 ft. July–Oct. E/W. Introduced. (Plate 61)

DACTYLIS L. – ORCHARD GRASS

Perennials; *sheaths* open; *ligules* membranous; *spikelets* in one-sided panicles, with 2–5 florets; *glumes* keeled, 1–3-nerved; *lemmas* keeled, 5-nerved; *paleas* well-developed; *disarticulation* above the glumes.

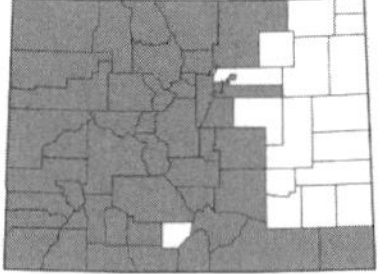

***Dactylis glomerata* L.**, ORCHARD GRASS. Plants 3–13 dm; *leaves* flat, 2–12 mm wide; *ligules* 2–6 mm long, lacerate; *glumes* 3–6 mm long, acute or shortly awn-tipped; *lemmas* 4–6.5 mm long, the keel coarsely ciliate, acute or shortly awn-tipped. Common along creeks, roadsides, in meadows, aspen woodlands, and disturbed areas, 4500–10,500 ft. June–Sept. E/W. Introduced. (Plate 61)

DACTYLOCTENIUM Willd. – CROWFOOT GRASS

Annuals; *sheaths* open; *ligules* ciliate; *spikelets* in digitately arranged, unilateral, spicate branches, in 2 rows, with 3–7 florets; *glumes* keeled, 1-nerved; *lemmas* 3-nerved, broad, short-awned tip; *paleas* about equaling the lemma; *disarticulation* between the glumes.

Dactyloctenium aegyptium **(L.) Willd.**, EGYPTIAN GRASS. Plants 1–3.5 (10) dm, usually rooting at the nodes; *leaves* flat, 2–10 mm wide; *ligules* 0.5–1.5 mm long; *spikelets* 3–4.5 mm long; *glumes* 1.5–2 mm long, the upper glume with an awn 1–2.5 mm long; *lemmas* 2.5–3.5 mm long, glabrous, with a short, curved awn tip 0.5–1 mm long. Uncommon in disturbed areas, known from near Fort Collins (Larimer Co.), 5000–5500 ft. June–Aug. E. Introduced.

DANTHONIA DC. – OATGRASS

Perennials; *sheaths* open; *ligules* a ring of hairs; *spikelets* in panicles or reduced to a single spikelet, with 2–several florets; *glumes* nearly equal, 1–5-nerved, longer than the florets; *lemmas* rounded, hairy on the back, with a twisted, geniculate awn flattened at the base; *paleas* well-developed; *disarticulation* above the glumes.

1a. Inflorescence usually consisting of a single spikelet (rarely with 2 spikelets); leaf sheaths usually densely hairy...***D. unispicata***
1b. Inflorescence consisting of more than one spikelet; leaf sheaths glabrous or hairy...2

2a. Inflorescence open with spreading or reflexed branches...***D. californica***
2b. Inflorescence narrow or contracted, with erect or ascending branches...3

3a. Glumes 16–23 mm long; lemmas 10–15 mm long, the apex deeply bifid with teeth 2.5–8 mm long; lemma awn 12–15 mm long...***D. parryi***
3b. Glumes 7–15 mm long; lemmas 3–6 mm long, the apex bifid with teeth 0.5–2.5 mm long; lemma awn 5–8 mm long...4

4a. Lemmas hairy only along the margins, glabrous on the back; spikelets usually purplish; glumes not chartaceous along the margins...***D. intermedia***
4b. Lemmas hairy along the margins and on the back; spikelets usually green; glumes chartaceous along the margins...***D. spicata***

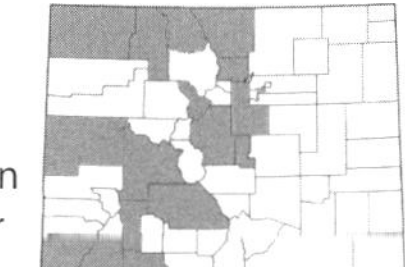

Danthonia californica **Bol.**, CALIFORNIA OATGRASS. Plants 3–13 dm; *leaves* flat to involute, 2–5 mm wide; *inflorescence* an open panicle; *spikelets* 14–25 mm long, 3–8-flowered, usually purple; *glumes* 14–18 mm long, usually glabrous; *lemmas* 5–10 mm long, hairy along the margins, with an awn 8–12 mm long arising from a deeply bifid apex. Uncommon in meadows and open aspen or spruce forests, 6500–10,500 ft. June–Sept. E/W.

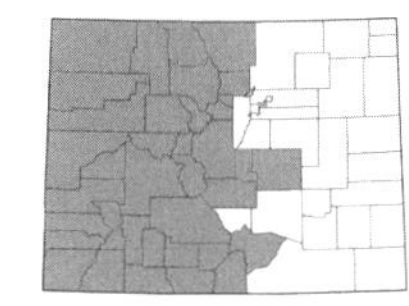

Danthonia intermedia **Vasey**, TIMBER OATGRASS. Plants 1–5 dm; *leaves* flat to involute, 1–3.5 mm wide; *inflorescence* a narrow panicle; *spikelets* 11–15 mm long, 3–6-flowered, usually purple; *glumes* 12–15 mm long; *lemmas* 3–6 mm long, hairy along the margins, with an awn 6.5–8 mm long arising from a bifid apex. Common in meadows, open forests, and alpine, 8500–14,000 ft. July–Sept. E/W.

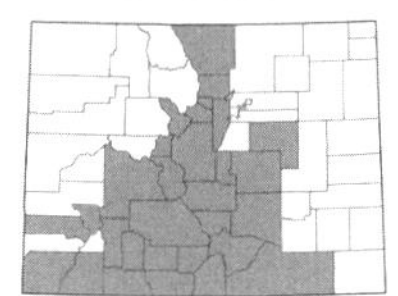

Danthonia parryi **Scribn.**, PARRY'S OATGRASS. Plants 3–8 dm; *leaves* flat to involute, 1–4 mm wide; *inflorescence* a narrow panicle; *spikelets* 16–25 mm long, 5–7-flowered; *glumes* 16–23 mm long; *lemmas* 10–15 mm long, pilose, with an awn 12–15 mm long arising from a deeply bifid apex. Common on dry slopes and in meadows, grasslands, and open forests, 6000–12,000 ft. June–Sept. E/W. (Plate 62)

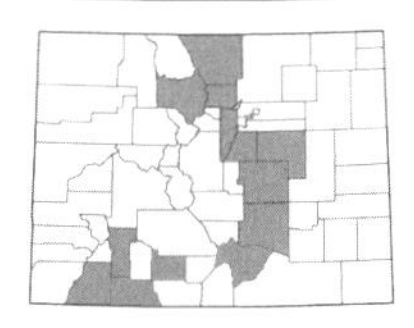

Danthonia spicata **(L.) P. Beauv. ex Roem. & Schult.**, POVERTY OATGRASS. Plants 1–10 dm; *leaves* flat to involute, 0.8–4 mm wide; *inflorescence* a narrow or contracted panicle; *spikelets* 7–15 mm long; *glumes* 10–12 mm long; *lemmas* 4–5 mm long, sparsely hairy, with an awn 5–8 mm long arising from a bifid apex. Found on mesas, dry slopes, and in grasslands and ponderosa pine forests, 5500–8500 ft. June–Sept. E/W.

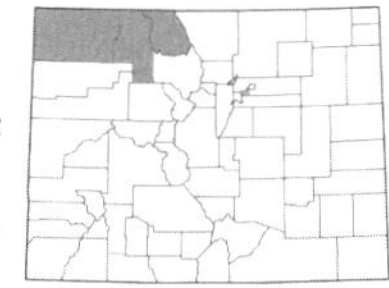

Danthonia unispicata **(Thurb.) Munro ex Vasey**, ONESPIKE OATGRASS. Plants 1.5–3 dm; *leaves* flat to involute, 1–3 mm wide; *inflorescence* usually consisting of a single spikelet (rarely with 2 spikelets); *spikelets* 12–26 mm long, usually purplish; *glumes* 14–20 mm long; *lemmas* 5.5–11 mm long, glabrous or rarely sparsely hairy, with an awn 5.5–13 mm long arising from a bifid apex. Found on sandstone ledges, 6500–9000 ft. June–Aug. E/W.

DASYOCHLOA Willd. ex Rydb. – WOOLLYGRASS

Perennials; *sheaths* open; *ligules* a ring of hairs; *spikelets* in small, compact panicles, with 5–15 florets; *glumes* 1-nerved, awn-tipped; *lemmas* deeply cleft, awned; *paleas* about half the length of the lemma, long-hairy below; *disarticulation* above the glumes.

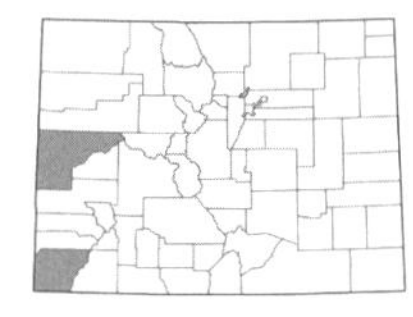

Dasyochloa pulchella **(Kunth) Willd. ex Rydb.**, LOW WOOLLYGRASS. [*Erioneuron pulchellum* (Kunth) Tateoka]. Plants stoloniferous, the stems 2–15 cm long; *leaves* involute, 1–6 cm long; *ligules* 3–5 mm; *spikelets* 5–10 mm long; *glumes* 6–9 mm long, glabrous; *lemmas* 2.5–5 mm, long-hairy, often purplish, with an awn 1.5–4 mm long. Uncommon in pinyon-juniper woodlands, desert shrublands, and dry areas, 4300–5500 ft. April–Aug. W. (Plate 62)

DESCHAMPSIA P. Beauv. – HAIRGRASS

Perennials (ours); *sheaths* open; *ligules* membranous; *spikelets* in a panicle, with two florets; *glumes* 1–3-nerved, equal to or longer than the lowermost floret; *lemmas* awned from the middle or near the base, 5–7 nerved, hairy on the callus, deeply cleft at the apex; *paleas* about as long as the lemma; *disarticulation* above the glumes.

1a. Lemmas with short hairs or minutely roughened, dull...***D. flexuosa***
1b. Lemmas glabrous, shiny...2

2a. Glumes mostly green with purple tips; panicles mostly narrow, 0.5–1.5 (2) cm wide, appearing greenish overall; anthers 0.3–0.5 (0.7) mm long...***D. elongata***
2b. Glumes purplish below or to more than ½ the surface, or with a purple midband, with white or yellowish tips; panicles 0.5–30 cm wide, usually appearing purple or bronze overall; anthers 1.2–3 mm long...3

3a. Spikelets strongly overlapping, often clustered at the tips of branches; glumes and lemmas purplish over more than ½ their length; panicles 0.5–2 (11) cm wide, usually dense, oblong-ovate to narrowly cylindrical...***D. brevifolia***
3b. Spikelets not strongly overlapping, not clustered at the tips of branches; glumes and lemmas usually purple over less than ½ their length; panicles 4–30 cm wide, usually open and pyramidal...***D. cespitosa***

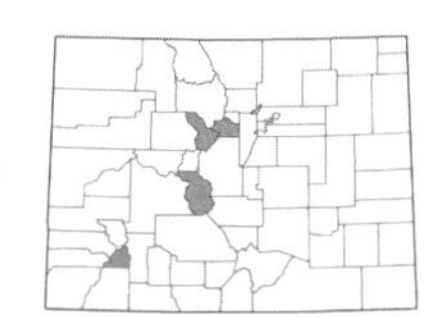

Deschampsia brevifolia **R. Br.**, BERING HAIRGRASS. Plants 1–6 dm; *leaves* usually folded, to 1 mm wide; *ligules* 1–5 mm long; *inflorescence* a dense, oblong-ovate to narrowly cylindrical panicle 0.5–2 (11) cm wide; *glumes* 2.5–5.6 mm long, purplish over ½ their length; *lemmas* 2–4 mm long, purplish over ½ their length, glabrous, with an awn 0.5–4 mm long arising from near the base to the middle. Found in the alpine, usually in moist places, 10,500–14,000 ft. July–Aug. E/W.

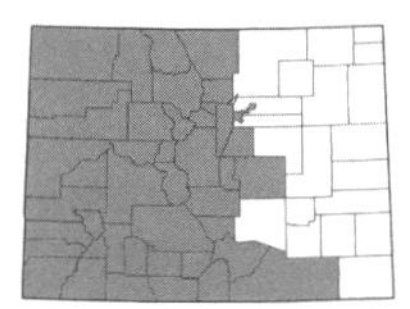

Deschampsia cespitosa **(L.) P. Beauv.**, TUFTED HAIRGRASS. Plants 1–15 dm; *leaves* flat or folded, 1–4 mm wide; *ligules* 2–13 mm long; *inflorescence* an open panicle 4–30 cm wide; *glumes* 2–7 mm long, green below with a purple midband and pale apex, glabrous; *lemmas* 2–5 mm long, glabrous, with an awn 1–8 mm long arising from near the base. Common in moist meadows, moist alpine, along lake margins, and in wetlands, 5000–14,000 ft. July–Sept. E/W. (Plate 62)

Deschampsia elongata **(Hook.) Munro**, SLENDER HAIRGRASS. Plants 1–12 dm; *leaves* usually involute, to 2 mm wide; *ligules* 2–9 mm long; *inflorescence* a narrow panicle; *glumes* 3–5.5 mm long, mostly green with purple tips; *lemmas* 2–4 mm long, glabrous, with an awn 2–6 mm long arising from near the middle. Uncommon in moist places, probably introduced in Colorado, reported from San Juan Co. July–Sept. W. Introduced.

Deschampsia flexuosa **(L.) Trin.**, WAVY HAIRGRASS. Plants 3–8 dm; *leaves* involute, to 0.5 mm wide; *ligules* 1.5–3.5 mm long; *inflorescence* a narrow to open panicle; *glumes* 2.7–5 mm long; *lemmas* 3.3–5 mm long, with short hairs or minutely roughened, with a bent awn 3.7–7 mm long arising from near the base. Uncommon on dry, rocky slopes, reported from Park Co. and probably introduced in Colorado, reported from Park Co. July–Sept. E. Introduced.

DICHANTHELIUM (Hitchc. & Chase) Gould – ROSETTE GRASS; PANICGRASS

Perennials; *sheaths* open; *ligules* a ring of hairs; *spikelets* in panicles, with one lower sterile floret and one upper perfect floret; *glumes* unequal, the lowermost often greatly reduced; *fertile lemmas* tightly inrolled over the palea, awnless; *disarticulation* below the glumes.

1a. Spikelets small, 1–2 mm long, the sterile and fertile lemmas to 1.7 mm long...***D. acuminatum***
1b. Spikelets usually larger, 2–4 mm long, the sterile and fertile lemmas 2–3.5 mm long...2

2a. Leaves mostly less than 10 times as long as wide, 4–15 mm wide, divaricately ascending; ligule a ring of hairs 1–3 mm long...***D. oligosanthes* var. *scribnerianum***
2b. Leaves mostly at least 10 times as long as wide, 2–5 mm wide, usually erect or slightly ascending; ligule a ring of hairs 0.3–1 mm long...3

3a. Basal leaves similar to stem leaves but shorter; spikelets conspicuously hairy...***D. wilcoxianum***
3b. Basal leaves similar to the stem leaves in both shape and size; spikelets glabrous to sparsely hairy...***D. linearifolium***

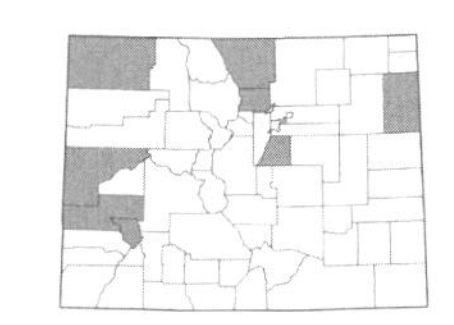

Dichanthelium acuminatum **(Sw.) Gould & C.A. Clark**, HAIRY PANICGRASS. [*Panicum lanuginosum* Elliott]. Plants 1.5–10 dm, decumbent to erect; *basal leaves* lanceolate to ovate, well-differentiated from the cauline; *cauline leaves* glabrous to hairy, 4–10 cm × 3–9 mm; *ligules* 1–5 mm long; *spikelets* 1–2 mm long, hairy; *glumes* unequal, the lower 0.3–1 mm long and the upper to 1.5 mm; *fertile lemmas* 1–1.7 mm long. Found on floodplains, in creek bottoms, moist grasslands and pastures, seepage areas, and near hot springs, 3500–7000 ft. June–Aug. E/W.

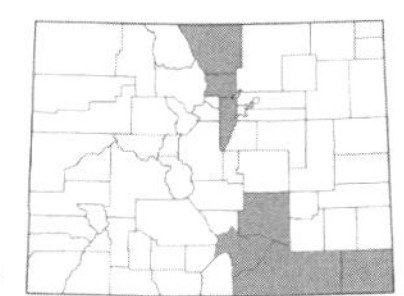

Dichanthelium linearifolium **(Scribn.) Gould**, SLIMLEAF PANICGRASS. Plants 1–5 dm, decumbent to erect; *basal leaves* similar to cauline; *cauline leaves* glabrous to densely hairy, 5–20 cm × 2–5 mm; *ligules* 0.3–1 mm long; *spikelets* 2–3.2 mm long, glabrous to sparsely hairy; *glumes* unequal, the lower 0.5–1 mm and the upper ca. 2 mm long; *fertile lemmas* 2.3–3.2 mm long. Found on dry, rocky hillsides, in canyons, ponderosa pine forests, and on mesa tops, 4000–6000 ft. June–Aug. E.

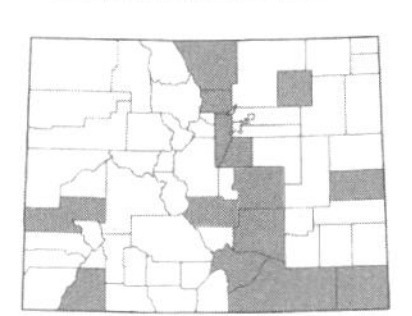

Dichanthelium oligosanthes **(Schult.) Gould var. *scribnerianum* (Nash) Gould**, FEW-FLOWERED PANICGRASS. Plants 2–8 dm, erect; *basal leaves* lanceolate to ovate, 2–6 cm long; *cauline leaves* glabrous to hairy below, 5–12 cm × 4–15 mm; *ligules* 1–3 mm long; *spikelets* 2.7–3.5 mm long, glabrous to sparsely hairy; *glumes* unequal, the lower 0.8–1.5 mm and the upper 2.2–3.3 mm; *fertile lemmas* 2.5–3 mm long. Found on dry, rocky slopes, canyon bottoms, sandstone walls, and near pools, 4000–6500 ft. June–Aug. E/W. (Plate 62)

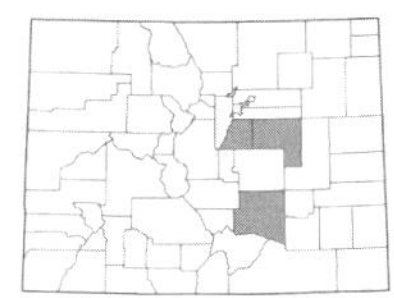

Dichanthelium wilcoxianum **(Vasey) Freckmann**, FALL ROSETTE GRASS. Plants 1.5–3.5 dm, erect; *basal leaves* similar to cauline, 2–4 cm long; *cauline leaves* hairy, stiffy erect, 4–8 cm × 2–5 mm; *ligules* 0.5–1 mm long; *spikelets* 2.4–3.5 mm long, densely hairy; *glumes* unequal, the lower 0.5–1.5 mm and the upper 2–3 mm long; *fertile lemmas* 2–3 mm long. Rare in ponderosa forests, 6000–7000 ft. May–July. E.

DIGITARIA Haller – CRABGRASS

Annuals or perennials; *sheaths* open; *ligules* membranous; *spikelets* in a panicle of one-sided racemes, with one lower sterile floret and one upper perfect floret; *glumes* unequal, the lowermost minute or absent; *fertile lemmas* awnless, firm and cartilaginous, not inrolled over the palea; *paleas* well-developed; *disarticulation* below the glumes.

1a. Glumes densely villous with long (1.5–5 mm) silvery-white to purple hairs; panicle with 4–10 alternating appressed to ascending spike-like primary branches...***D. californica***
1b. Glumes glabrous or appressed-hairy with short hairs; panicle with spreading subdigitate to racemose branches...2

2a. Spikelets in 3's; leaf sheaths glabrous to sparsely hairy; leaf blades glabrous except for a few hairs near the collar...***D. ischaemum***
2b. Spikelets paired; leaf sheaths sparsely to densely hairy; leaf blades usually hairy above and below, occasionally glabrous...***D. sanguinalis***

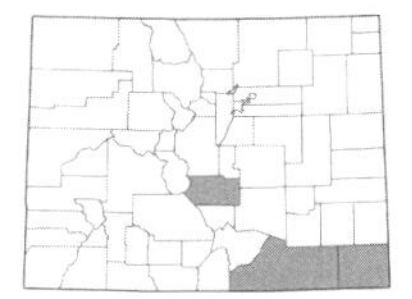

Digitaria californica **(Benth.) Henrard**, ARIZONA COTTONTOP. Perennials, 4–10 dm; *leaves* flat or folded, 2–18 cm × 2–5 mm, the sheaths sometimes with papillose-based hairs; *ligules* 1.5–6 mm long; *spikelets* 3–5.5 mm long; *glumes* unequal, the lower 0.4–0.6 mm long or absent, the upper 2–5 mm long and densely long-hairy; *fertile lemmas* 2.5–3.5 mm long. Uncommon on mesa tops and dry, rocky ridges, 4000–6000 ft. June–Aug. E.

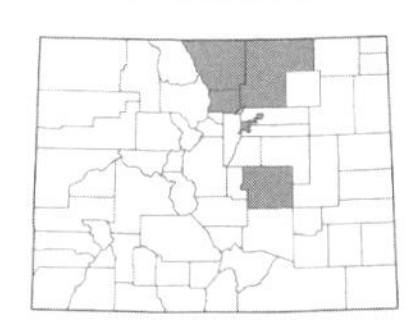

Digitaria ischaemum **(Schreb.) Muhl.**, SMOOTH CRABGRASS. Annuals, 2–6 dm; *leaves* flat, 2–9 cm × 3–5 mm, the sheaths glabrous to sparsely hairy; *ligules* 0.5–2.5 mm long; *spikelets* 1.7–2.5 mm long, in 3's; *glumes* unequal, the lower reduced to a rim or absent, the upper 1.7–2.5 mm long with appressed hairs; *fertile lemmas* 1.7–2.3 mm long. Uncommon weed in lawns and sidewalks, 4000–6500 ft. July–Sept. E. Introduced.

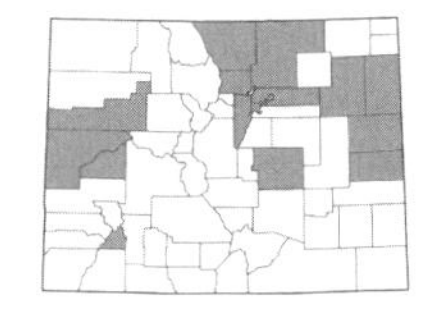

Digitaria sanguinalis **(L.) Scop.**, HAIRY CRABGRASS. Annuals, 2–7 dm; *leaves* flat, 2–11 cm × 3–8 mm, sheaths sparsely to densely hairy; *ligules* 0.5–2.5 mm long; *spikelets* 2.5–3.5 mm long, paired; *glumes* unequal, the lower 0.2–0.4 mm long and the upper 1–2 mm long with hairy margins; *fertile lemmas* 2.5–3.5 mm long. Found as a weed in lawns and gardens and in disturbed areas, 4000–6000 ft. E/W. (Plate 62)

DINEBRA Jacq. – SPRANGLETOP

Perennials or annuals; *sheaths* open; *ligules* membranous, sometimes ciliate; *spikelets* in panicles with one-sided spike-like branches, with 3–several florets; *glumes* 1-nerved, unawned; *lemmas* membranous, 3 (5)-nerved, unawned; *paleas* membranous, usually subequal to the lemmas; *disarticulation* below the florets.

***Dinebra panicea* (Retz.) P.M. Peterson & N. Snow,** MUCRONATE SPRANGLETOP. [*Leptochloa panicea* (Retz.) Ohwi]. Annuals, 1–15 dm; *leaves* flat, 6–25 cm × 2–21 mm, glabrous to sparsely hairy; *ligules* 0.5–3.2 mm long; *spikelets* 2–4 mm long, with 2–6 florets; *glumes* shorter to longer than the florets, 1.6–4 mm long; *lemmas* 1–1.7 mm long, glabrous to hairy. Reported for Montezuma and Montrose cos. June–Aug. W.

DIPLACHNE Beauv. – SPRANGLETOP

Perennials or annuals; *sheaths* open; *ligules* membranous, sometimes ciliate; *spikelets* in panicles with one-sided spike-like branches; *glumes* 1-nerved; *lemmas* chartaceous, 3(5)-nerved; *paleas* membranous, usually subequal to the lemmas; *disarticulation* below the florets.

***Diplachne fusca* (L.) P. Beauv. ex Roem. & Schult. ssp. *fasciculata* (Lam.) P.M. Peterson & N. Snow,** BEARDED SPRANGLETOP. [*Leptochloa fusca* (L.) Kunth ssp. *fascicularis* (Lam.) N. Snow]. Annuals, 0.5–11 dm; *leaves* flat or involute, 3–50 cm × 2–7 mm; *ligules* 2–8 mm long; *spikelets* 5–12 mm long, with 5–9 florets; *glumes* unequal, the lower 2–3 mm and the upper 2.5–5 mm long, sometimes with an awn 0.5–1.5 mm long; *lemmas* 3–5 mm long, with an awn to 3.5 mm long arising from a bifid apex. Found in moist alkaline soil, along pond margins, in standing water, and in disturbed areas, 3500–6200 ft. July–Oct. E/W.

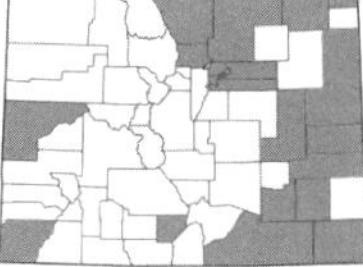

DISAKISPERMA Steud. – SPRANGLETOP

Perennials; *sheaths* open; *ligules* membranous, sometimes ciliate; *spikelets* in panicles with one-sided spike-like branches, with 3–several florets; *glumes* usually unequal, 1-nerved, unawned; *lemmas* membranous, unawned, 3 (5)-nerved; *paleas* membranous, usually subequal to the lemmas; *disarticulation* below the florets.

***Disakisperma dubia* (Kunth) P.M. Peterson & N. Snow,** GREEN SPRANGLETOP. [*Leptochloa dubia* (Kunth) Nees]. Plants 3–11 dm; *leaves* mostly flat, 8–35 cm × 2–10 mm; *ligules* 0.3–2 mm long; *spikelets* 4–12 mm long, with 2–10 florets; *glumes* unequal, the lower 2.3–5 mm and the upper 3.3–6 mm long, translucent with green nerves; *lemmas* 3–5 mm long, truncate, hairy on the nerves. Rare on rocky slopes, 4500–5000 ft. July–Sept. E.

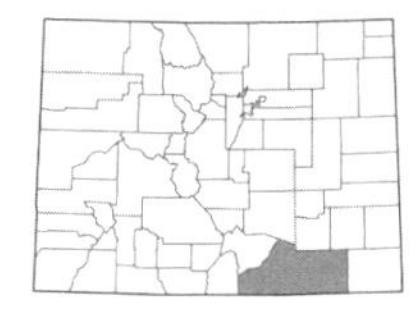

DISTICHLIS Raf. – SALTGRASS

Dioecious perennials; *sheaths* open; *ligules* a short, fringed membrane; *spikelets* strongly compressed, in panicles; *glumes* keeled, acute; *lemmas* awnless; *paleas* well-developed; *disarticulation* above the glumes.

***Distichlis stricta* (Torr.) Rydb.,** DESERT SALTGRASS. [*D. spicata* (L.) Greene]. Plants prostrate to erect, 1–6 dm, strongly rhizomatous; *leaves* flat to involute, distichous and stiffly spreading, 1–12 cm × 1–3 mm; *ligules* short-ciliate membranes, 0.2–0.5 mm long; *spikelets* 9–20 mm long, with 5–20 florets, usually imperfect with staminate and pistillate flowers in separate inflorescences and often on separate plants; *glumes* unequal, the lower 2–3 mm and the upper 3–4 mm long, 5–11-nerved; *lemmas* 3–6 mm long. Common in saline soil, along lake margins, and in shortgrass prairie, 3500–9000 ft. May–Aug. E/W. (Plate 62)

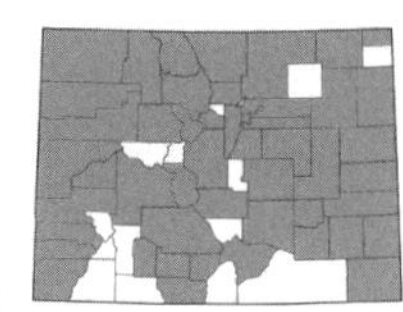

ECHINOCHLOA P. Beauv. – COCKSPUR GRASS

Annuals or perennials; *sheaths* open; *ligules* absent or of hairs; *spikelets* subsessile, in panicles of simple or compound spicate branches, with 2 (3) florets, the lower florets sterile or staminate and the upper florets fertile; *glumes* unequal, the lowermost minute or absent; *lemmas* awnless, firm and cartilaginous, those of fertile florets hard, smooth, and shiny with inrolled margins; *paleas* of fertile florets similar to the lemma in texture; *distarticulation* below the glumes.

1a. Spikelets not disarticulating at maturity (especially those near the bottom of panicle branches), purplish; lower lemmas unawned; upper lemmas wider and longer than the glumes, exposed at maturity...***E. frumentacea***
1b. Spikelets all disarticulating at maturity, purplish or greenish; lower lemmas awned or unawned; upper lemmas not or barely exceeding the upper glumes in length and width at maturity...2

2a. Sterile lemma with bristles, but these lacking a pustulate base...***E. crus-galli***
2b. Sterile lemma with pustulate-based bristles scattered over the surface...***E. muricata* var. *microstachya***

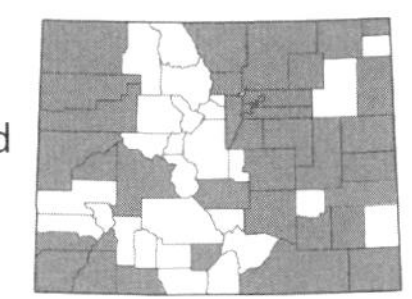

Echinochloa crus-galli **(L.) P. Beauv.**, BARNYARD GRASS. Annuals, 3–20 dm; *leaves* flat, 8–65 × 0.5–3.5 cm, usually glabrous; *ligules* absent; *spikelets* 2.5–4 mm long, usually greenish; *glumes* unequal, the lower 1.2–1.5 mm and unawned to short-awned, the upper 2.5–3.5 mm long, scabrous, awned or acuminate; *sterile lemmas* similar to upper glumes, 2.5–3.5 mm long, bristly, unawned or with an awn 0.5–50 mm long; *fertile lemmas* 2–3 mm long, short-awned. Common weed in disturbed areas, gardens, fields, and ditches, 3500–8000 ft. Aug.–Oct. E/W. Introduced. (Plate 62)

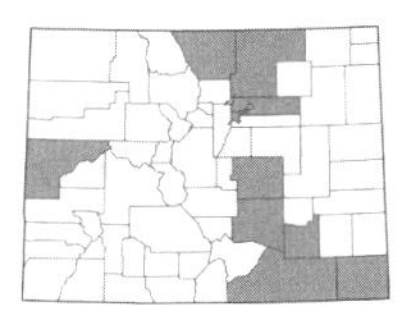

Echinochloa frumentacea **Link**, WHITE PANIC. Annuals, 7–15 dm; *leaves* flat, 8–35 × 0.3–2 cm, usually glabrous; *ligules* absent; *spikelets* 3–3.5 mm long, purplish; *glumes* unequal, the upper to 2.5 mm long, unawned; *sterile lemmas* unawned; *fertile lemmas* 2.5–3 mm long, unawned, hispid. Cultivated for grain and forage grass, occasionally escaping and persisting along roadsides and in fields, 4000–5500 ft. July–Sept. E/W. Introduced.

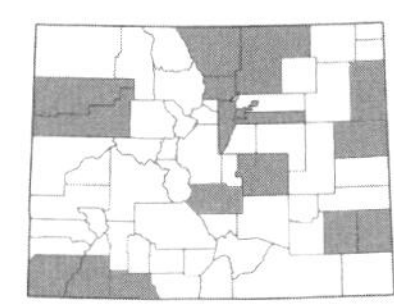

Echinochloa muricata **(P. Beauv.) Fernald var. *microstachya* Wiegand**, ROUGH BARNYARD GRASS. Annuals, 3–20 dm; *leaves* flat, 1–27 cm × 0.8–30 mm, usually glabrous; *ligules* absent; *spikelets* 2.5–4 mm long, purplish or streaked with purple, hispid with pustulate-based bristles; *glumes* unequal, the lower 1–1.5 mm and the upper 2.5–4 mm long, acuminate to awned to 2.5 mm; *sterile lemmas* awned to unawned; *fertile lemmas* 2–4 mm long, short-awned to 0.5 mm. Found in disturbed areas, ditches, and along roadsides, 3500–7500 ft. July–Sept. E/W. Introduced.

ELEUSINE Gaertn. – GOOSEGRASS

Annuals; *sheaths* open; *ligules* membranous; *spikelets* in one-sided, digitately arranged spikes, with 3-several florets; *glumes* firm, acute, the upper 3–7-nerved; *lemmas* awnless or mucronate, 3-nerved; *paleas* shorter than the lemma; *disarticulation* above the glumes.

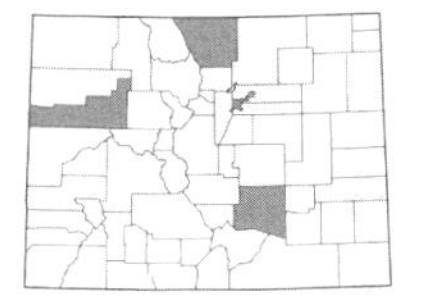

Eleusine indica **(L.) Gaertn.**, INDIAN GOOSEGRASS. Plants 3–9 dm; *leaves* mostly flat, 15–40 cm × 3–7 mm, with prominent white midveins, often with papillose-based hairs at the base; *ligules* 0.2–1.2 mm long; *spikelets* 4–8 mm long, with 5–8 florets; *glumes* unequal, the lower 1–2.3 mm and the upper 2–3 mm long; *lemmas* 2.5–4 mm long, glabrous. Uncommon weed in lawns and disturbed areas, 4000–5500 ft. Aug.–Oct. E/W. Introduced.

ELYMUS L. – WILDRYE

Perennials, sometimes with rhizomes; *sheaths* open; *ligules* membranous; *spikelets* in a 2-sided spike with one or more spikelets per node, with 2–7 florets; *glumes* lanceolate to subulate; *lemmas* awnless or mucronate, 3-nerved; *paleas* shorter than the lemma; *disarticulation* above the glumes.

Members of the genus *Elymus* readily hybridize, along with members of *Leymus* and *Hordeum*.

1a. At least some lemmas with awns over 5 mm long...2
1b. Lemmas awnless or with awns less than 5 mm long...12

2a. Glumes deeply 3–9 cleft into long awns of unequal length up to 70 mm long; lemma awn up to 100 mm long; rachis readily disarticulating at maturity; spikelets usually 2 or rarely 3–4 per node...***E. multisetus***
2b. Glumes entire or 2-cleft, but without 3–9 awns; plants otherwise various...3

3a. Culms usually decumbent-spreading or prostrate; plants 2–5 dm tall, found in the upper subalpine and alpine above 10,000 ft in elevation; spikelets usually purplish, solitary at each node or rarely with 2 per node; lemmas with divergent to recurved awns 15–25 mm long; rachis disarticulating at maturity...***E. scribneri***
3b. Plants variously distributed or if of high elevations then otherwise unlike the above in all respects...4

4a. Glumes subulate, 1-nerved, 35–85 mm long and tapering to a long awn (the entire glume resembling a long bristle); rachis readily disarticulating at maturity (the spikelets easily falling off in clusters with portions of the rachis)...***E. elymoides***
4b. Glumes linear-lanceolate to oblong-elliptic, 3–5-nerved, if awned then the awns to 30 mm long; rachis not disarticulating at maturity...5

5a. Spikelets 2–4 per node (or rarely a few nodes will have solitary spikelets)...6
5b. Spikelets solitary at each node (or rarely 2 at some nodes)...8

6a. Glumes curved outward at the base, the body longer than the adjacent lemma; awns straight; spike erect...***E. virginicus***
6b. Glumes more or less straight, the body shorter than the adjacent lemma; awns straight, flexuous, spreading, or divergent at maturity; the entire spike often nodding...7

7a. Glume awns 10–27 mm long; spikes thick and bristly, 3–7 cm wide; lemma awns 16–40 mm long...***E. canadensis***
7b. Glume awns 1–9 mm long; spikes narrow, 0.5–2 cm wide; lemma awns (5) 10–30 mm long...***E. glaucus* ssp. *glaucus***

8a. Awns straight or flexuous but not divergent or spreading at maturity...9
8b. Awns spreading, divergent, arcuate, or recurved at maturity...10

9a. Glumes 1.8–2.3 mm wide, with scarious or hyaline margins 0.2–0.3 mm wide; spikes sometimes somewhat one-sided; leaves usually green, flat or involute, 2–8 mm wide...***E. trachycaulus***
9b. Glumes 0.4–1.5 (2) mm wide, with scarious or hyaline margins 0.1–0.2 mm wide; spikes not one-sided; leaves usually glaucous, flat, 4–20 mm wide...***E. glaucus* ssp. *glaucus***

10a. Spikelets pretty much all overlapping, such that little to no rachis is visible in the spike; lemma awns 6–35 mm long...***E. bakeri***
10b. At least the lower spikelets not overlapping or overlapping by very little, such that the rachis is visible in the spike; lemma awns 5–25 mm long...11

11a. Plants usually rhizomatous; lemmas 7–10 mm long, usually with an awn 5–12 mm long...***E. albicans***
11b. Plants caespitose, lacking rhizomes; lemmas 8–15 mm long, with an awn 6–25 mm long...***E. arizonicus***

12a. Spikelets 2–4 per node (or sometimes solitary in the upper portion of the spike only); plants caespitose...13
12b. Spikelets mostly solitary per node, rarely a few with 2 per node; plants caespitose or rhizomatous...14

13a. Glumes awnless or with an awn to 5 mm long...***E. curvatus***
13b. Glumes with awns 4–20 mm long...***E. virginicus***

14a. Plants strongly rhizomatous; anthers (2.5) 3–7 mm long...15
14b. Plants usually caespitose or sometimes with short, thickened rhizomes; anthers 0.7–2.5 (3) mm long...16

15a. Leaves usually involute, mostly 1–3.5 mm wide, stiff; glumes ½–¾ the length of the adjacent lemmas, not keeled distally, the tips usually acute to mucronate...***E. lanceolatus***
15b. Leaves flat, 4–10 mm wide, thin and finely veined; glumes ¾ as long as to longer than the adjacent lemmas, keeled distally, the tips often with an awn 0.5–4 mm long...***E. repens***

16a. Glume margins equal, widest at the middle; glumes acute or if awned the awn to 11 mm long; leaves flat or folded, 2–8 mm wide; plants 3–15 dm tall, erect to ascending...***E. trachycaulus***
16b. Glume margins unequal, widest in the upper ⅓; glumes acute or if awned the awn to 2 mm long; leaves flat, 3–15 mm wide; plants 1.5–8 dm tall, often decumbent or abruptly bent...***E. violaceus***

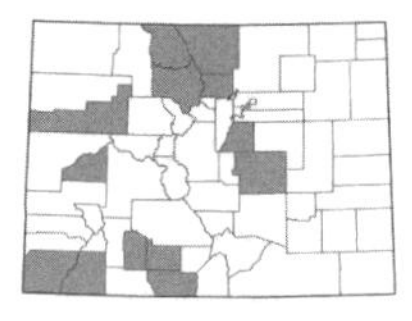

***Elymus albicans* (Scribn. & J.G. Sm.) Á. Löve**, Montana wheatgrass. Rhizomatous perennials, 4–10 dm; *leaves* usually involute, 1–3 mm wide; *ligules* to 0.5 mm long; *spikelets* usually 1 per node, 10–18 mm long, with 3–12 florets; *glumes* unequal or subequal, the lower 4–8 mm and the upper 5.5–8 mm long, mostly 3–5-nerved, glabrous to scabrous, usually terminating in an awn 0.5–3 mm long; *lemmas* 7–10 mm long, glabrous to hairy, usually terminating in a divergent awn 5–12 mm long. Found on dry slopes, in meadows, pinyon-juniper woodlands, spruce/fir forests, and sagebrush, 6500–12,000 ft. June–Aug. E/W.

This species is considered to be the result of hybridization between *Elymus lanceolatus* and *Pseudoroegneria spicata*.

***Elymus arizonicus* (Scribn. & J.G. Sm.) Gould**, Arizona wheatgrass. Perennials, lacking rhizomes, 4–10; *leaves* flat to folded, 2.5–6 mm wide; *ligules* 1–3 mm long; *spikelets* usually 1 per node, 14–25 mm long, with 4–6 florets; *glumes* unequal or subequal, the lower 5–9 mm and the upper 8–10 mm long, 3–5-nerved, acute or awned to 4 mm; *lemmas* 8–15 mm long, terminating in an awn 6–25 mm long. Rare in canyons and on moist, rocky slopes, reported from the southwestern counties but no specimens have been seen. June–Aug. W.

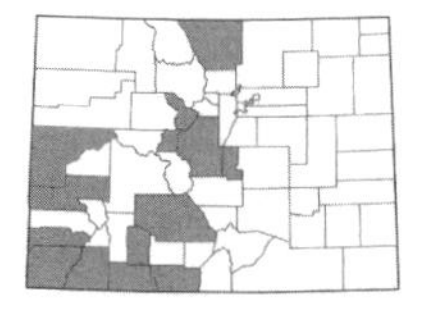

***Elymus bakeri* (E.E. Nelson) Á. Löve**, Baker's wheatgrass. Perennials, lacking rhizomes, 3–7 dm; *leaves* flat, 10–30 cm × 2–5 mm; *ligules* to 1 mm long; *spikelets* usually 1 per node, 10–20 mm long, with 3–6 florets; *glumes* unequal or subequal, the lower to 9 mm and the upper to 10 mm long, 3–7-nerved, with an awn 2–8 mm long; *lemmas* 9–12 mm long, with an awn 6–35 mm long. Uncommon in meadows, fir forests, and on sagebrush slopes, 8500–11,000 ft. July–Sept. E/W.

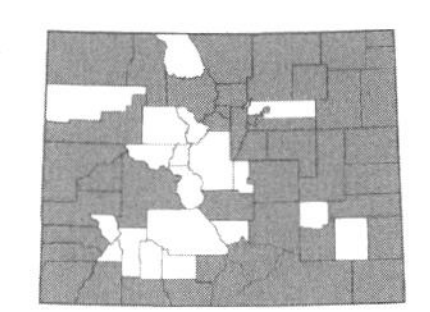

***Elymus canadensis* L.**, CANADA WILDRYE. Perennials, lacking rhizomes, 8–15 dm; *leaves* flat or folded, 5–40 cm × 7–20 mm; *ligules* 0.2–2 mm long; *spikelets* 2–4 per node, 12–15 mm long, with 2–5 florets; *glumes* subequal, 6–13 mm long, 3–5-nerved, with an awn 10–27 mm long; *lemmas* 8–15 mm long, scabrous, with a spreading or divergent awn 16–40 mm long. Common along roadsides, creeks, in shortgrass prairie, meadows, and on dry slopes, 3500–9000 ft. June–Sept. E/W.

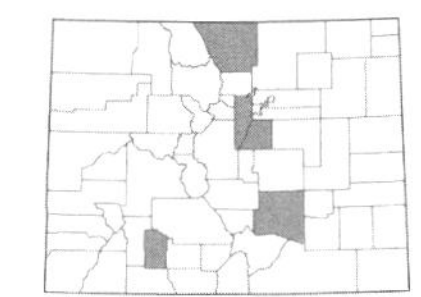

***Elymus curvatus* Piper**, AWNLESS WILDRYE. [*E. virginicus* L. var. *submuticus* Hook.]. Perennials, lacking rhizomes, 6–11 dm; *leaves* becoming involute, 15–20 cm × 5–15 mm; *ligules* to 1 mm long; *spikelets* usually 2 per node, 10–15 mm long, with 3–4 florets; *glumes* subequal to equal, 7–15 mm, 3–5-nerved, linear-lanceolate and curved outward, glabrous to scabrous, acute or with an awn to 5 mm long; *lemmas* 6–10 mm long, usually with an awn to 4 mm long, rarely to 10 mm. Uncommon in forests and moist meadows, 6000–10,000 ft. June–Aug. E/W.

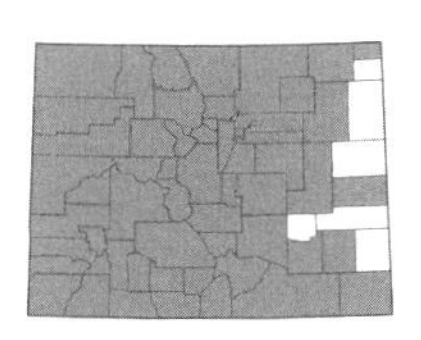

***Elymus elymoides* (Raf.) Swezey**, SQUIRRELTAIL; BOTTLEBRUSH. [*Sitanion hystrix* (Nutt.) J.G. Sm.]. Perennials, lacking rhizomes, 1–6 dm; *leaves* flat to involute, 5–20 cm × 1–5 mm; *ligules* 0.5–1 mm long; *spikelets* usually 2 or sometimes 3 per node, 10–15 mm long, with 2–6 florets; *glumes* equal, 3.5–8.5 cm, subulate, 1-nerved, with an awn 1–10 mm long; *lemmas* 8–10 mm long, the midnerve extending to an awn 5–15 mm long and the lateral nerves extending to bristles to 10 mm long. Common throughout the state on dry slopes, mesa tops, and in grasslands and meadows, 3500–11,000 ft. June–Aug. E/W.

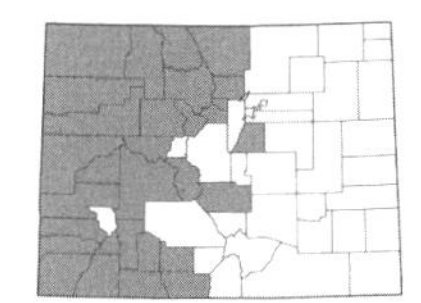

Elymus glaucus* Buckley ssp. *glaucus, BLUE WILDRYE. Perennials, sometimes with short rhizomes, 3–15 dm; *leaves* flat, 4–20 mm wide, glaucous; *ligules* 0.3–1 mm long; *spikelets* usually 2 per node, 8–25 mm long, with 2–6 florets; *glumes* subequal, 9–14 × 0.4–1.5 (2) mm, linear-lanceolate, 3–7-nerved, acute or with an awn to 9 mm; *lemmas* 8–15 mm long, with an awn (5) 10–30 mm long. Common in meadows and forests, and along creeks and lake margins, 7000–11,500 ft. June–Aug. E/W.

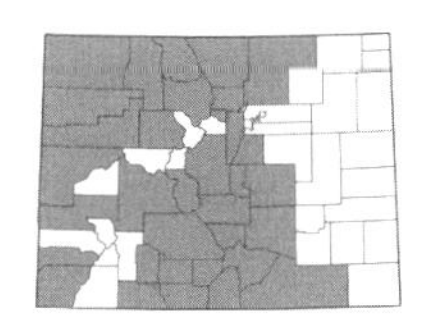

***Elymus lanceolatus* (Scribn. & J.G. Sm.) Gould**, THICKSPIKE WHEATGRASS. [*Agropyron dasystachyum* (Hook.) Scribn. & J.G. Sm.]. Rhizomatous perennials, 3–15 dm; *leaves* usually involute, 5–20 cm × 1–3.5 mm; *ligules* to 1 mm long; *spikelets* usually 1 per node, 11–18 mm long, with 3–12 florets; *glumes* subequal, ½–¾ the length of the adjacent lemmas, 3–5-nerved, acute; *lemmas* 7–10 mm long, glabrous to villous, occasionally with an awn tip to 2 mm long. Common on dry slopes, in grasslands, open pine woodlands, and along roadsides, 4500–10,500 ft. June–Aug. E/W.

***Elymus multisetus* M.E. Jones**, BIG SQUIRRELTAIL. [*Sitanion jubatum* J.G. Sm.]. Perennials, lacking rhizomes, 2–6 dm; *leaves* usually involute, 1.5–3.5 mm wide; *ligules* 0.2–0.7 mm; *spikelets* usually 2 (rarely 3–4) per node, 10–15 mm long, with 2–4 florets; *glumes* subequal, 10–100 mm long (including awns), deeply 3–9 cleft into long awns of unequal length up to 70 mm long; *lemmas* 8–10 mm long, the midnerve extending to an awn to 100 mm long with the 2 lateral nerves often extending to bristly awns to 20 mm long. Uncommon on dry slopes, 7000–8000 ft. June–Aug. W.

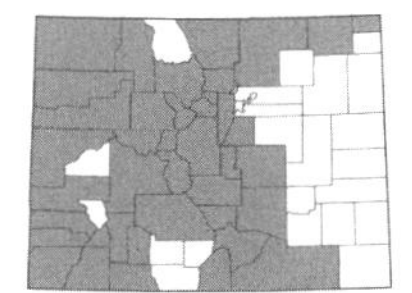

***Elymus repens* (L.) Gould**, QUACK GRASS. [*Elytrigia repens* (L.) Nevski]. Rhizomatous perennials, 5–10 dm; *leaves* flat, 10–30 cm × 4–10 mm; *ligules* to 1 mm long; *spikelets* solitary or sometimes 2 per node, to 27 mm long, with 3–7 florets; *glumes* subequal, 7–12 mm long, 5–7-nerved, awn-tipped or with an awn to 4 mm long; *lemmas* 7–12 mm long, acute or with an awn to 4 mm. Common in meadows, grasslands, forests, and riparian communities and along roadsides, 4000–10,000 ft. June–Aug. E/W. Introduced.

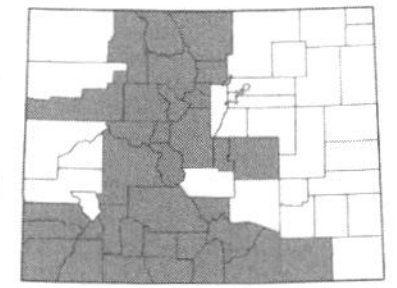

***Elymus scribneri* (Vasey) M.E. Jones**, SPREADING WHEATGRASS. Rhizomatous perennials, 2–5 dm; *leaves* flat to involute, 3–8 cm × 2–5 mm; *ligules* to 0.5 mm long; *spikelets* solitary or rarely 2 per node, 9–15 mm long, with 3–6 florets; *glumes* more or less equal, 4–9 mm long, 2–3-nerved, with an awn 8–30 mm long (sometimes the awn split into 2); *lemmas* 7–10 mm long, with an awn 15–25 mm long. Common in alpine and high subalpine meadows, 10,000–13,500 ft. June–Aug. E/W.

***Elymus trachycaulus* (Link) Gould**, SLENDER WHEATGRASS. Perennials, lacking rhizomes, 3–15 dm; *leaves* flat or folded, 5–25 cm × 2–8 mm; *ligules* 0.2–1 mm long; *spikelets* solitary at each node, 10–20 mm long, with 3–9 florets; *glumes* subequal, ¾ to as long as lemma, 6–15 × 1.8–2.3 mm, 5–7-nerved, acute or awned to 11 mm; *lemmas* 7.5–13 mm long, acute or awned to 40 mm. Common along roadsides, on dry slopes, and in meadows, grasslands, and forests, 4500–12,500 ft. June–Sept. E/W. (Plate 62)

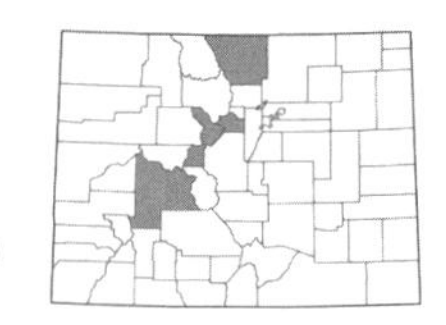

Elymus violaceus **(E.E. Nelson) Á. Löve,** ARCTIC WHEATGRASS. [*E. trachycaulus* ssp. *andinus* (Scribn. & Smith) Á. Löve & D. Löve]. Perennials, lacking rhizomes, 1.5–8 dm; *leaves* flat, 10–30 × 3–15 mm; *ligules* 0.5–1 mm long; *spikelets* solitary at each node, 11–20 mm long, with 4–5 florets; *glumes* subequal, 8–12 × 1.2–2 mm, ¾ to as long as lemmas, acute or with an awn to 2 mm long; *lemmas* 9–12 mm long, usually with an awn 0.5–3 mm long. Found in high subalpine and alpine meadows, 10,000–13,000 ft. June–Aug. E/W.

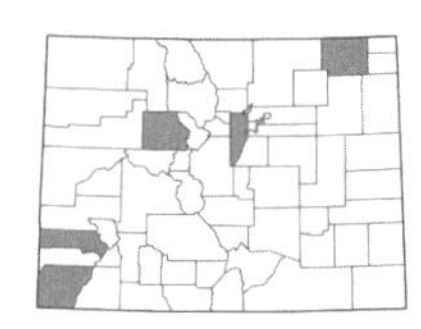

Elymus virginicus **L.,** VIRGINIA WILDRYE. Perennials, lacking rhizomes, 3–15 dm; *leaves* flat, 10–30 cm × 4–15 mm; *ligules* 0.2–1 mm long; *spikelets* usually 2 per node, 10–16 mm long, with 3–5 florets; *glumes* subequal, subulate to linear-lanceolate, 7–15 mm, curved outward, 2–5-nerved, with an awn 4–20 mm long; *lemmas* 6–10 mm long, with an awn 0.5–25 mm long. Uncommon in meadows and in grasslands, 4000–6500 ft. June–Aug. E/W.

ENNEAPOGON Desv. ex P. Beauv. – FEATHER PAPPUSGRASS

Perennials; *sheaths* open; *ligules* a ring of hairs; *glumes* with 3–many nerves; *lemmas* much shorter than the glumes; *disarticulation* above the glumes.

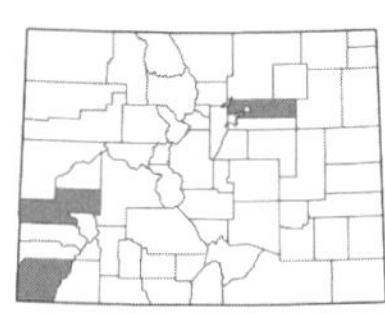

Enneapogon desvauxii **Desv. ex P. Beauv.,** NINEAWN PAPPUSGRASS. Plants 1–5 dm; *leaves* involute, 2–15 cm × 1–2 mm, more or less hairy; *ligules* 0.5–1 mm long; *spikelets* 5–7 mm long, with 3 florets, the lowest floret fertile and the upper 2 sterile, cleistogamous spikelets in lower sheaths; *glumes* unequal or subequal, the lower 3.5–5 mm and the upper 5–6 mm long, pilose, 3–7-nerved; *lemmas* 1.5–2.5 mm long, pilose, strongly 9-nerved with each nerve exserted into a plumose awn 2–5 mm long. Uncommon on mesa tops on the west slope and used in revegetation areas on the eastern plains, 4000–7500 ft. June–Sept. E/W. (Plate 62)

ERAGROSTIS Wolf – LOVEGRASS

Annuals or perennials; *sheaths* open; *ligules* a ring of hairs; *spikelets* in an open panicle, with 3–several florets, awnless; *lemmas* 3-nerved, acute or acuminate; *paleas* strongly 2-nerved and 2-keeled, as long as the lemmas.

1a. Plants stoloniferous and mat-forming, annuals; panicle 1–3.5 cm long...***E. hypnoides***
1b. Plants not stoloniferous or mat-forming, annuals or perennials; panicle (3) 4–80 cm long...2

2a. Pedicels with a yellowish or purplish band near or above the middle...***E. minor***
2b. Pedicels lacking a yellowish or purplish band...3

3a. Leaf margins and lemma keels with prominent glands (small yellowish bumps); spikelets 6–20 mm long with 10–40 florets...***E. cilianensis***
3b. Leaf margins and lemma keels lacking prominent glands; spikelets various...4

4a. Spikelets strongly flattened, ovate to oblong, with reddish-purple margins or entirely reddish-purple; lemmas 2–6 mm long with acuminate or attenuate tips; pedicels absent or 1 mm in length...***E. secundiflora* ssp. *oxylepis***
4b. Spikelets unlike the above in all respects; lemmas 1–3.5 mm long with acute or obtuse tips; usually at least some spikelets on pedicels over 1 mm in length...5

5a. Plants perennial, erect...6
5b. Plants annual, erect or decumbent...9

6a. Inflorescence with at least some pedicels under 1 mm long...7
6b. Inflorescence with all pedicels over 1.5 mm long...8

7a. Ligule 0.1–0.3 mm long; inflorescence branches viscid (look for dirt stuck to the branches)...***E. curtipedicellata***
7b. Ligule 0.6–1.3 mm long; inflorescence branches not viscid...***E. curvula***

8a. Inflorescence branches hairy at the junction with the main axis; leaf sheaths usually hairy; lemmas 1.3–2.5 mm long...***E. spectabilis***
8b. Inflorescence branches glabrous or with a few short hairs at the junction with the main axis; leaf sheaths usually glabrous or hairy only on the margins; lemmas 2.2–3.5 mm long...***E. trichodes***

9a. Lower glume 0.3–0.6 mm long, less than ½ the length of the first lemma and its tip reaching well below the midpoint of the first lemma...***E. pilosa***
9b. Lower glume 0.5–1.7 mm long, longer than or reaching the midpoint of the first lemma on most or all spikelets...10

10a. Leaf sheaths, underside of leaf blades at the base, and inflorescence branches with glandular pits...***E. lutescens***
10b. Leaves and sheaths lacking glandular pits, sometimes a few glandular depressions below the nodes in the inflorescence present...11

11a. Yellow dish-shaped glands present at the summit of at least some internodes in the inflorescence, these sometimes fused into a ring...***E. barrelieri***
11b. Yellow dish-shaped glands absent from the summit of the internodes in the inflorescence...12

12a. Spikelets 0.7–1.4 mm wide, with 5–11 (15) florets...***E. mexicana* ssp. *virescens***
12b. Spikelets 1.5–2.5 mm wide, with 6–22 florets...***E. pectinacea***

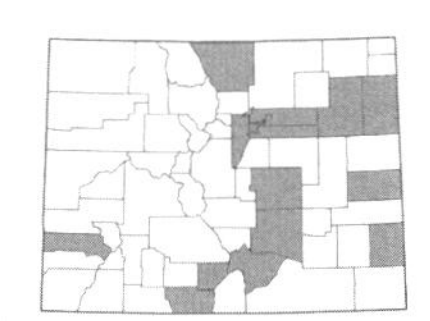

***Eragrostis barrelieri* Daveau**, Mediterranean lovegrass. Annuals, 0.5–6 dm; *leaves* involute or flat, 1.5–10 cm × 2–5 mm, lacking glands, the sheaths long-hairy at the apex; *ligules* 0.2–0.5 mm long; *inflorescence* branches with yellowish glandular rings or spots below the nodes; *spikelets* 4–11 mm long, with 6–20 florets, greenish to reddish-purple; *glumes* subequal, the lower 1–1.4 mm and the upper 1.4–1.8 mm long; *lemmas* 1.4–2 mm long, scabrous on the keel. Found in disturbed areas, sidewalk cracks, and along roadsides, 3400–7500 ft. Aug.–Nov. E/W. Introduced.

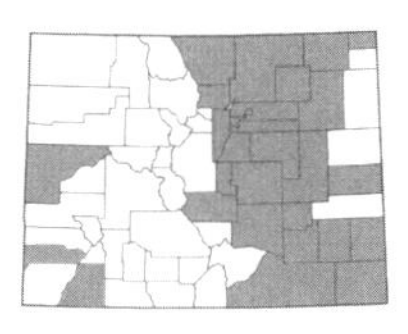

***Eragrostis cilianensis* (All.) Vignalo ex Janch.**, stinkgrass. Annuals, 1–4.5 dm; *leaves* flat to involute, 5–20 cm × 2–6 mm, the margins with prominent yellow glands, the sheaths with glands and long-hairy at the apex; *ligules* 0.4–0.8 mm long; *spikelets* 6–20 mm long, with 10–40 florets, greenish; *glumes* subequal, 1.2–2.6 mm long; *lemmas* 2–3 mm long, with glands on the keel. Common in shortgrass prairie and disturbed areas, 3400–8500 ft. July–Sept. E/W. Introduced.

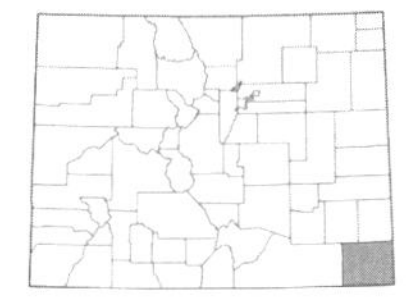

***Eragrostis curtipedicellata* Buckley**, gummy lovegrass. Perennials, 2–7 dm; *leaves* involute to flat, 5–18 cm × 2–5 mm, the sheaths viscid and hairy at the tip; *ligules* 0.1–0.3 mm long; *pedicels* 0.2–1 mm long; *spikelets* 3–6 mm long, with 4–12 florets; *glumes* subequal, 0.9–2 mm long, scabrous on the midnerve; *lemmas* 1.5–2.2 mm long. Uncommon in shortgrass prairie and sandy soil, 4000–4500 ft. July–Sept. E.

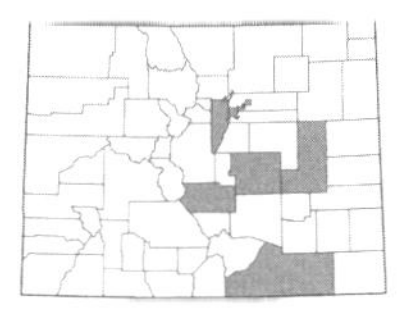

***Eragrostis curvula* (Schrad.) Nees**, weeping lovegrass. Perennials, 6–15 dm; *leaves* involute, 12–50 cm × 1–3 mm, the sheaths long-hairy at the apex; *ligules* 0.6–1.3 mm long; *pedicels* 0.5–5 mm long; *spikelets* 4–10 mm long, with 5–12 florets; *glumes* unequal, the lower 1.5–2 mm and the upper 2–3 mm long, scabrous on the midnerve; *lemmas* 1.8–3 mm long, scabrous on the midnerve. Found in shortgrass prairie and disturbed areas, 4500–6500 ft. July–Oct. E. Introduced.

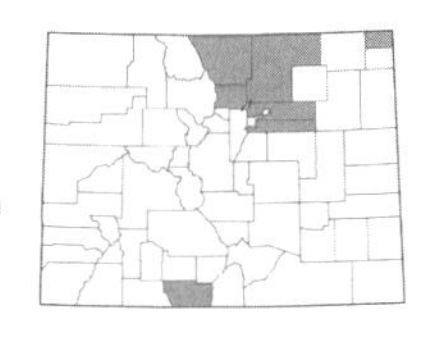

***Eragrostis hypnoides* (Lam.) Britton**, creeping lovegrass. Annuals, stoloniferous and mat-forming, prostrate; *leaves* involute to flat, 0.5–2.5 cm × 1–2 mm, appressed-hairy above, the sheath hairy; *ligules* 0.3–0.6 mm long; *inflorescence* a panicle 1–3.5 cm long; *spikelets* 4–13 mm long, with 7–35 florets; *glumes* unequal, the lower 0.4–0.7 mm and the upper 0.8–1.2 mm long; *lemmas* 1.5–2 mm long. Uncommon in sandy soil, along streambanks, and on drying pond margins, 4000–7500 ft. July–Oct. E.

***Eragrostis lutescens* Scribn.**, sixweeks lovegrass. Annuals, 0.5–2.5 dm; *leaves* involute to flat, 2–12 cm × 1–3 mm, the sheath with glandular pits and sparsely hairy at the apex; *ligules* 0.2–0.5 mm long; *inflorescence* with glandular pits on the branches; *spikelets* 3.5–7.5 mm long, with 6–11 florets; *glumes* subequal, 0.9–1.8 mm long; *lemmas* 1.5–2.2 mm long, scabrous on nerves. Uncommon along drying shores, known from the Boulder/Denver area, 5000–5500 ft. Aug.–Oct. E.

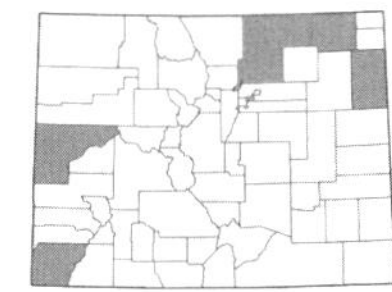

***Eragrostis mexicana* (Hornem.) Link ssp. *virescens* (J. Presl) S.D. Koch & Sánchez**, Mexican lovegrass. Annuals, 1–13 dm; *leaves* flat, 5–25 cm × 2–7 mm, the sheaths sometimes with glandular pits, long-hairy at the apex; *ligules* 0.2–0.5 mm long; *spikelets* 5–10 mm long, with 5–11 (15) florets, greenish-gray to purplish; *glumes* subequal, 0.7–2 mm long; *lemmas* 1.2–2.5 mm long. Uncommon in sandy soil and in disturbed areas, 3500–5500 ft. Aug.–Oct. E/W.

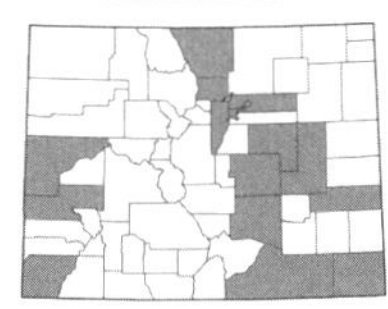

***Eragrostis minor* Host**, little lovegrass. Annuals, 1–5 dm; *leaves* flat or folded, 1.5–10 cm × 1–4 mm, usually with glands on the margins, the sheath long-hairy at the apex; *ligules* 0.2–0.5 mm long; *pedicels* with a yellowish or purplish band near or above the middle; *spikelets* 4–7 mm long, with 5–12 (20) florets; *glumes* subequal, 0.9–1.6 mm long; *lemmas* 1.4–1.8 mm long. Found in disturbed areas, sidewalk cracks, and along roadsides, 3500–6500 ft. July–Sept. E/W. Introduced.

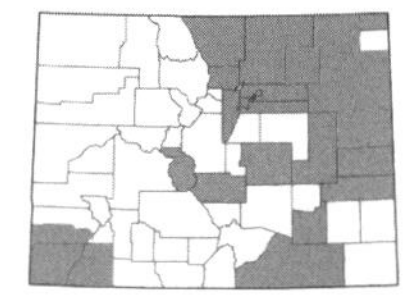

Eragrostis pectinacea **(Michx.) Nees,** TUFTED LOVEGRASS. Annuals, 1–8 dm; *leaves* involute to flat, 2–20 cm × 0.5–4 mm, the sheath long-hairy at the apex; *ligules* 0.2–0.5 mm long; *spikelets* 4.5–11 mm long, with 6–22 florets, yellowish-brown to reddish-purple; *glumes* subequal, the lower 0.5–1.5 mm and the upper 1–1.7 mm long; *lemmas* 1–2.2 mm long, scabrid on the nerves. Found on drying shores, in shortgrass prairie, and in disturbed areas, 3500–7800 ft. July–Oct. E/W.

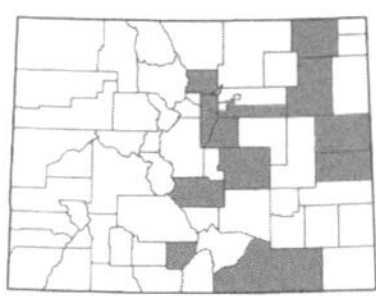

Eragrostis pilosa **(L.) P. Beauv.,** INDIAN LOVEGRASS. Annuals, 0.8–5 dm; *leaves* flat, 2–15 cm × 1–2 mm, the sheath usually glabrous and long-hairy at the apex; *ligules* 0.1–0.3 mm long; *spikelets* 3.5–6 mm long, with 3–17 florets; *glumes* unequal, the lower 0.3–0.6 mm and the upper 0.7–1.2 mm long; *lemmas* 1.2–1.8 mm long. Found in shortgrass prairie, disturbed areas, and along roadsides, 4000–7500 ft. July–Oct. E.

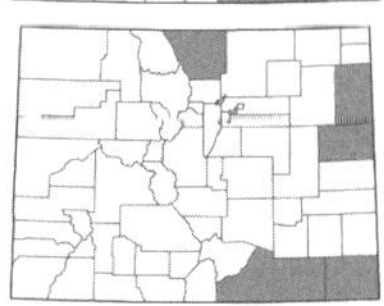

Eragrostis secundiflora **J. Presl ssp.** ***oxylepis*** **(Torr.) S.D. Koch,** RED LOVEGRASS. Perennials, 3–7.5 dm; *leaves* flat or involute near the tip, 5–25 cm × 1–5 mm, the sheath long-hairy at the apex when young; *ligules* 0.2–0.3 mm long; *spikelets* 6–20 mm long, with 8–40 florets, usually reddish-purple; *glumes* subequal, the lower 1.7–3 mm and the upper 2.2–4 mm long; *lemmas* 2–6 mm long, the keel scabrous. Found in sandy soil, sandsage prairie, and in canyons, 3500–5000 ft. July–Oct. E.

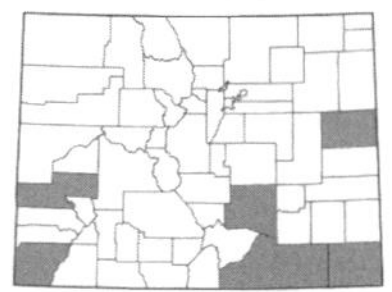

Eragrostis spectabilis **(Pursh) Steud.,** PURPLE LOVEGRASS. Perennials, 3–7 dm; *leaves* involute to flat, 10–30 cm × 3–8 mm, the sheath long-hairy at the apex; *ligules* 0.1–0.2 mm long; *spikelets* 3–7.5 mm long, with 6–12 florets, reddish-purple; *glumes* subequal or equal, 1.3–2.5 mm long; *lemmas* 1.3–2.5 mm long. Found on sandstone outcrops, rocky slopes, and in grasslands, 4000–6500 ft. Aug.–Oct. E/W. (Plate 62)

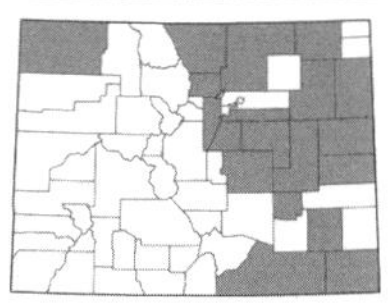

Eragrostis trichodes **(Nutt.) Wood,** SAND LOVEGRASS. Perennials, 3–15 dm; *leaves* involute to flat, 15–50 cm × 2–8 mm, the sheath long-hairy at the apex; *ligules* 0.3–0.5 mm long; *spikelets* 3–15 mm long, with 4–18 florets, purplish; *glumes* subequal, 1.8–4 mm long; *lemmas* 2.2–3.5 mm long. Found in sandy soil, shortgrass prairie, and along roadsides, 3400–7000 ft. Aug.–Oct. E/W.

EREMOPYRUM (Ledeb.) Jaub & Spach – FALSE WHEATGRASS

Annuals; *sheaths* open; *ligules* membranous; *spikelets* in distichous spikes with 1 spikelet per node; *glumes* awned; *lemmas* 5-veined; *paleas* shorter than the lemma; *disarticulation* in the rachises.

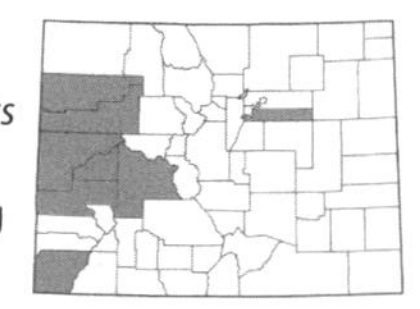

Eremopyrum triticeum **(Gaertn.) Nevski,** ANNUAL WHEATGRASS. [*Agropyron triticeum* Gaertn.]. Plants 1–4 dm; *leaves* flat or loosely involute, 4–8 cm × 1–4 mm, scabrous; *ligules* 0.2–1 mm long; *spikelets* 5–10 mm long, with 3–6 florets; *glumes* subequal, 5–7.5 mm long, 1-nerved, acute or with a small awn tip to 1.5 mm; *lemmas* 6–7.5 mm long, acuminate or with an awn 1–7 mm long. Found along roadsides, in disturbed areas, and on dry slopes, 4500–6700 ft. April–June. E/W. Introduced.

ERIOCHLOA Kunth – CUPGRASS

Annuals (ours); *sheaths* open; *ligules* a fringe of hairs with a membranous base; *spikelets* in one-sided panicle branches, with one lower sterile floret and one upper perfect floret; *glumes* unequal, the lowermost minute and fused to the upper part of the pedicel; *lemmas* awnless, firm and cartilaginous, villous, awned; *paleas* well-developed; *disarticulation* below the glumes.

Eriochloa contracta **Hitchc.,** PRAIRIE CUPGRASS. Plants 2–20 dm; *leaves* flat or folded, 6–12 cm × 2–8 mm, short-hairy; *ligules* 0.4–1 mm long; *spikelets* 3.5–4.5 mm; *glumes* unequal, the lower forming a cuplike structure, the upper 3–4.3 mm long, usually scabrous to hirsute, sometimes with an awn to 1 mm; *sterile lemmas* 3–4.3 mm long; *fertile lemmas* 2–2.5 mm long, with an awn 0.4–1 mm long. Uncommon in shortgrass prairie, 3800–5000 ft. May–July. E. (Plate 62)

ERIONEURON Nash – WOOLLYGRASS

Perennials; *sheaths* open; *ligules* of short hairs; *spikelets* in a short, capitate panicle or raceme, with 4–18 florets; *glumes* 1-nerved; *lemmas* 3-nerved; *paleas* shorter than lemmas, ciliate on the keels, long-hairy on the lower part between nerves; *disarticulation* above the glumes.

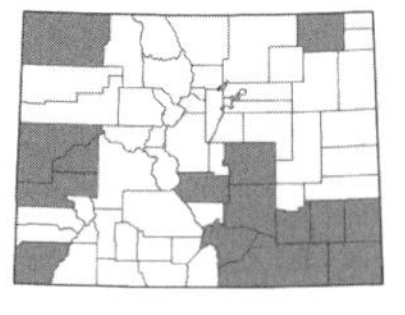

Erioneuron pilosum **(Buckley) Nash,** HAIRY WOOLLYGRASS. Plants 1–3 dm; *leaves* flat or folded, 1–6 cm × 1–2 mm, glabrous to sparsely hairy, with whitish margins; *ligules* 2–3.5 mm long; *spikelets* 6–12 mm long, with 6–12 florets, purplish; *glumes* subequal, 4–7 mm, acute to short-awned; *lemmas* 3–7 mm long, densely hairy on the nerves, with an awn 0.5–2.5 mm long. Found in shortgrass prairie, canyons, on rimrock, and sandstone, limestone, and gypsum outcroppings, 3800–7000 ft. May–July. E/W. (Plate 62)

FESTUCA L. – FESCUE

Perennials; *sheaths* open; *ligules* membranous; *spikelets* in an open or contracted panicle, with 3–several florets; *glumes* narrow, acute or acuminate, 1-3-nerved; *lemmas* usually 5–7-nerved, awned or awnless; *paleas* shorter than lemmas, free from caryopsis; *disarticulation* above the glumes.

1a. Panicle open and spreading, the branches divaricate and ciliate on the angles...***F. dasyclada***
1b. Panicle unlike the above in all respects...2

2a. Ligule 2–5 (9) mm long; lemma unawned or with a short mucro to 0.2 mm long...***F. thurberi***
2b. Ligule 0.1–1.5 mm long; lemma awned or unawned...3

3a. Leaves flat, 3–10 mm wide; panicle open with lax, spreading to reflexed branches...4
3b. Leaves usually folded or involute, rarely flat and then to 3 mm wide; panicle usually with erect, contracted branches, or sometimes open with spreading branches...5

4a. Lemmas unawned or the awns to 2 mm long...***F. sororia***
4b. Lemma awns (2.5) 5–15 mm long...***F. subulata***

5a. Glumes equal to or longer than the upper florets; spikelets rather closed (lemmas not much visible)...***F. hallii***
5b. Glumes conspicuously shorter than the upper florets; spikelets with the lemmas clearly visible...6

6a. Stems densely scabrous or densely hairy to shortly pilose below the inflorescence...7
6b. Stems smooth and glabrous, or sometimes sparsely scabrous near the inflorescence...8

7a. Plants 35–80 cm; ligule 0.5–1.5 mm long; panicle (4) 6–15 (20) cm long...***F. arizonica***
7b. Plants 5–25 (30) cm; ligule 0.1–0.3 mm long; panicle 1.5–4 (5) cm long...***F. baffinensis***

8a. Plants rhizomatous (sometimes the rhizomes short); sheaths of vegetative shoots closed for about ¾ of their length, shredding into fibers...9
8b. Plants lacking rhizomes; sheaths of vegetative shoots usually closed for less than ⅔ their length, sometimes up to ¾ of their length, sometimes shredding into fibers...10

9a. Lemmas 4–9.5 mm long; anthers 1.8–4.5 mm long; ovary tips glabrous...***F. rubra* ssp. *rubra***
9b. Lemmas 3–4.5 mm long; anthers 0.6–0.9 (1.4) mm long; ovary tips densely hairy...***F. earlei***

10a. Ovary tips densely hairy...***F. earlei***
10b. Ovary tips glabrous or sparsely and inconspicuously hairy...11

11a. Anthers 0.3–1.2 mm long; plants of the alpine, usually under 30 cm; lemmas 2–6 mm long; panicle 1–4 (5.5) cm long...12
11b. Anthers (0.8) 1.2–4.5 mm long; plants widespread, not exclusively of the alpine, often over 25 cm; lemmas (3) 3.5–8.5 mm long; panicle 3–16 cm long...13

12a. Lemma awns (0.8) 1–3.5 mm long; lemmas (3.5) 4–6 mm long; anthers 0.3–0.7 (1.1) mm long...***F. brachyphylla* ssp. *coloradensis***
12b. Lemma awns 0.5–1.5 (1.7) mm long; lemmas 2–3.5 (4) mm long; anthers (0.4) 0.6–1.2 mm long...***F. minutiflora***

13a. Lemma awn (1.5) 3–7 mm long, usually more than ½ as long as the body; lemmas 5–8.5 (10) mm...***F. idahoensis***
13b. Lemma awn 0.5–2.5 (3) mm long, usually less than ½ as long as the body; lemmas 3–6 (6.5) mm...14

14a. Anthers (0.8) 1.2–1.7 (2) mm long; plants widespread...***F. saximontana* var. *saximontana***
14b. Anther 2.2–3.5 mm long; plants scattered...15

15a. Leaves 0.4–0.8 mm in diam.; lemmas smooth or scabrous distally; ovary tips sparsely hairy...***F. calligera***
15b. Leaves (0.5) 0.8–1.2 mm in diam.; lemmas scabrous or hairy distally, especially on the margins; ovary tips glabrous...***F. trachyphylla***

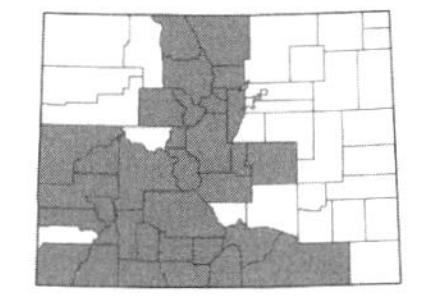

***Festuca arizonica* Vasey**, ARIZONA FESCUE. Plants 3.5–8 (10) dm, lacking rhizomes; *ligules* 0.5–1.5 (2) mm; *leaves* conduplicate, 0.3–0.8 mm diam.; *panicles* (4) 6–15 (20) cm, open or loosely contracted, with 1–2 branches per node; *spikelets* (6) 8–16 mm, with (3) 4–6 (8) florets; *lower glumes* (3) 3.3–5.5 mm; *upper glumes* 4.5–7 mm; *lemmas* 5.5–9 mm, glabrous, smooth or scabrous towards the tips, unawned or awned, awns 0.4–2 (3) mm long; *anthers* (2) 3–4.2 mm. Common on dry slopes, in meadows, often with pines or pinyon-juniper, 5900–10,500 ft. June–Aug. E/W.

***Festuca baffinensis* Polunin**, BAFFIN ISLAND FESCUE. Plants 0.5–2.5 (3) dm, lacking rhizomes; *ligules* 0.1–0.3 mm; *leaves* conduplicate, (0.4) 0.6–1 (1.2) mm diam.; *panicles* 1.5–4 (5) cm, contracted, with 1–2 branches per node; *spikelets* (4.5) 5–7.5 (8.5) mm, with (2) 3–5 (6) florets; *lower glumes* 2.2–3.7 (4) mm; *upper glumes* 3–5 mm; *lemmas* 3.5–6 mm, scabridulous apically, awned, awns (0.8) 1–3.5 mm; *anthers* 0.3–0.7 (1.1) mm. Found in the alpine, 11,500–14,000 ft. July–Aug. E/W.

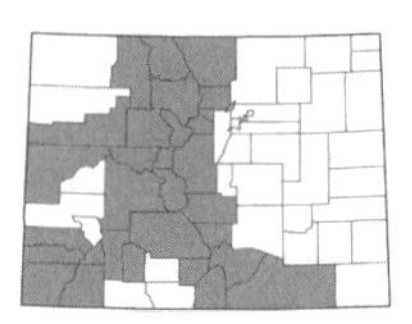

***Festuca brachyphylla* Schult. ex J.A. & J.H. Schultes ssp. *coloradensis* Frederiksen**, SHORTLEAF FESCUE. Plants (0.5) 0.8–3.5 (5.5) dm, lacking rhizomes; *ligules* 0.1–0.4 mm; *leaves* conduplicate, (0.3) 0.5–1.2 mm diam.; *panicles* 1.5–4 (5.5) cm, contracted, with 1–2 branches per node; *spikelets* 3.5–7 (8.5) mm, with 2–4 (6) florets; *lower glumes* (1.2) 1.8–3 (3.5) mm; *upper glumes* 2.4–4 (4.6) mm; *lemmas* (3.5) 4–6 mm, scabrous apically, awned, awns (0.8) 1–3.5 mm; *anthers* 0.3–0.7 (1.1) mm. Common in the alpine, meadows, and on ridges, 9500–14,000 ft. July–Sept. E/W.

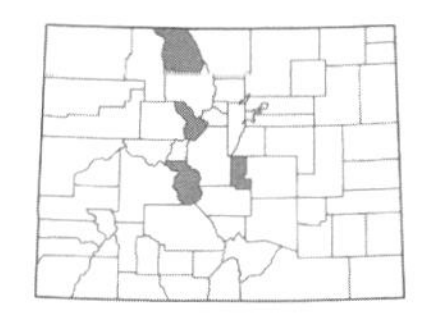

***Festuca calligera* (Piper) Rydb.**, CALLUSED FESCUE. [*F. ovina* L. ssp. *calligera* Piper]. Plants 1.5–6.5 dm, lacking rhizomes; *ligules* (0.2) 0.3–0.5 (1) mm; *leaves* conduplicate, 0.4–0.8 mm diam.; *panicles* 5–15 cm, loosely contracted, with 1–2 (3) branches per node; *spikelets* (6) 7–9 (11) mm, with (2) 4–6 florets; *lower glumes* 2.5–4 mm; *upper glumes* (2.8) 3–5 mm; *lemmas* (3.8) 4–6 mm, glabrous, smooth or scabrous distally, awned, awns 1–2.5 mm; *anthers* 2.2–3.5 mm. Found in meadows, grasslands, and forest openings, 7000–11,500 ft. July–Sept. E/W.

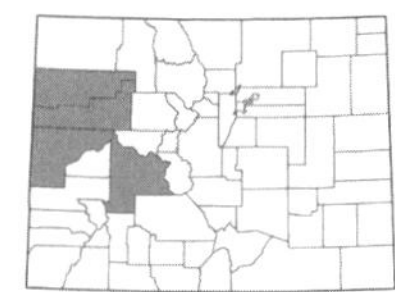

***Festuca dasyclada* Hack. ex Beal**, FALSE RICEGRASS. [*Argillochloa dasyclada* (Hack. ex Beal) W.A. Weber]. Plants 4–7 (10) dm, lacking rhizomes; *ligules* 0.3–0.5 mm; *leaves* conduplicate to flat, 1.5–3 mm diam.; *panicles* 8–15 cm, loosely contracted; *spikelets* 8–15 (18) mm, with (3) 4–6 (10) florets; *lower glumes* (2) 3.5–5.5 mm; *upper glumes* (4) 5.5–8 mm; *lemmas* (5.5) 8–11 mm, scabrous or puberulent, awned, awns 1–3.5 mm; *anthers* (3) 3.7–5.7 mm. Found on shale outcrops and dry slopes, often with sagebrush or pinyon-juniper, 6500–10,000 ft. June–Sept. W.

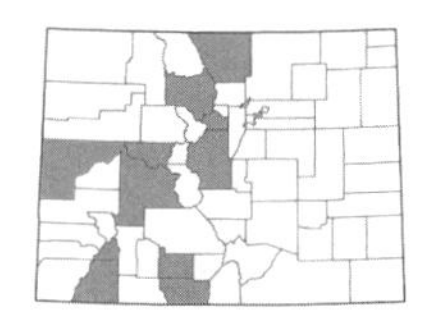

***Festuca earlei* Rydb.**, EARLE'S FESCUE. Plants (1.5) 2–4 (4.5) dm, often short-rhizomatous; *ligules* 0.1–0.5 (1) mm; *leaves* conduplicate to flat, 0.5–1 (1.5) mm diam. when conduplicate, to 3 mm wide when flat; *panicles* 3–5 (8) cm, contracted, with 1–3 branches per node; *spikelets* (4.5) 5–6.5 (7) mm, with 2–5 florets; *lower glumes* 1.5–3 mm; *upper glumes* 2.5–3.8 mm; *lemmas* 3–4.5 mm, glabrous or minutely hairy apically, awned, awns (0.3) 1–1.5 mm; *anthers* 0.6–0.9 (1.4) mm. Common in forests, grasslands, meadows, and the alpine, 6000–13,000 ft. June–Sept. E/W.

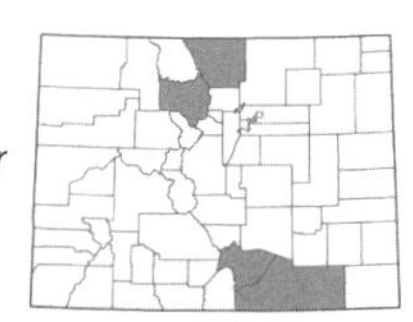

***Festuca hallii* (Vasey) Piper**, PLAINS ROUGH FESCUE. Plants (1.6) 2.5–6.5 (8.5) dm, usually short-rhizomatous; *ligules* 0.3–0.6 mm; *leaves* conduplicate, rarely flat, 0.5–1.2 mm diam.; *panicles* 6–16 cm, contracted, with 1–2 (3) branches per node; *spikelets* (6.5) 7–9.5 mm, with 2–3 (4) florets; *lower glumes* 5–8 (9.5) mm; *upper glumes* 6.2–8.5 (9.5) mm; *lemmas* 5.5–8 (9) mm, scabrous, unawned or awned, awns 0.5–1.3 mm; *anthers* 4–6 mm. Rare in open subalpine meadows and the alpine (where it is particularly subject to rodent disturbance), 10,500–12,000 ft. July–Sept. E/W.

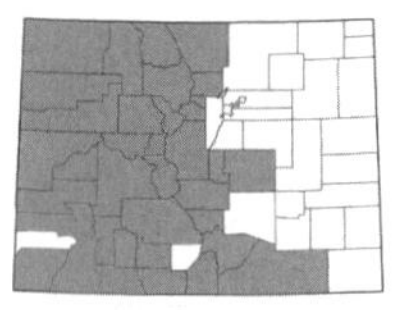

***Festuca idahoensis* Elmer**, IDAHO FESCUE. Plants 2.5–8.5 (10) dm, lacking rhizomes; *ligules* 0.2–0.6 mm; *leaves* conduplicate, (0.3) 0.5–0.9 (1.5) mm diam.; *panicles* (5) 7–15 (20) cm, loosely contracted or open, with 1–2 branches per node; *spikelets* (5.8) 7.5–13.5 (19) mm, with (2) 4–7 (9) florets; *lower glumes* 2.4–5 (6) mm; *upper glumes* 3–6 (8) mm; *lemmas* 5–8.5 (10) mm, scabrous apically, awned, awns (1.5) 3–7 mm; *anthers* 2.4–4.5 mm. Common in meadows, grasslands, and in forest openings, 5800–12,500 ft. July–Sept. E/W. (Plate 63)

Festuca idahoensis and *F. rubra* are difficult to separate. One trick is to look at the base; in *F. rubra* the base is decumbent and usually reddish-purple while *F. idahoensis* has an erect base that is usually tan.

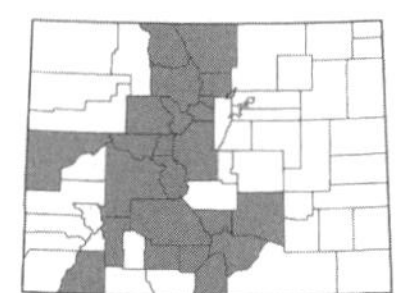

***Festuca minutiflora* Rydb.**, SMALLFLOWER FESCUE. Plants 0.4–3 dm, lacking rhizomes; *ligules* 0.1–0.3 mm; *leaves* conduplicate, 0.2–0.4 mm diam.; *panicles* 1–5 cm, contracted, with 1–2 branches per node; *spikelets* (2.5) 3–5 mm, with (1) 2–5 florets; *lower glumes* 1.3–2.5 mm; *upper glumes* 2–3.5 mm; *lemmas* 2–3.5 (4) mm, sparsely scabrous apically, awned, awns 0.5–1.5 (1.7) mm; *anthers* (0.4) 0.6–1.2 mm. Common in the alpine and on talus slopes, 11,000–14,000 ft. July–Sept. E/W.

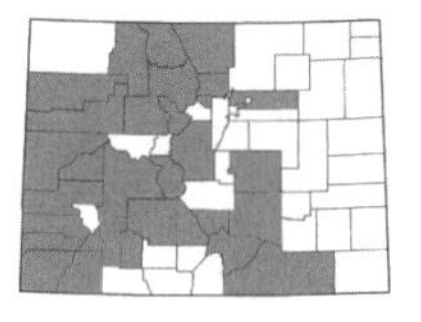

Festuca rubra* L. ssp. *rubra, RED FESCUE. Plants (0.8) 1–13 dm, rhizomatous; *ligules* 0.1–0.5 mm; *leaves* conduplicate or convolute, 0.3–2.5 mm diam.; *panicles* (2) 3.5–25 (30) cm, open or loosely contracted; *spikelets* (6) 7–17 mm, with 3–10 florets; *lower glumes* (1.5) 2–6 mm; *upper glumes* (3) 3.5–8.5 mm; *lemmas* 4–9.5 mm, smooth or scabrous apically, sometimes densely hairy throughout, awned, awns (0.1) 0.4–4.5 mm; *anthers* 1.8–4.5 mm. Common in forests, meadows, and in the alpine, often in moist places, 5000–13,000 ft. June–Sept. E/W. Introduced.

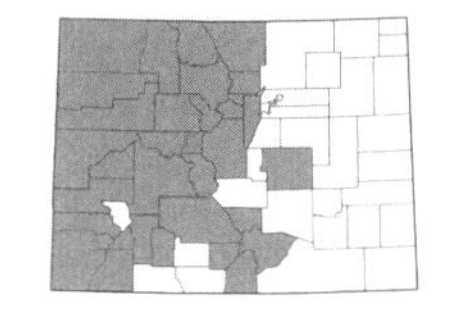

Festuca saximontana **Rydb. var.** ***saximontana***, ROCKY MOUNTAIN FESCUE. Plants (0.5) 0.8–5 (6) dm, lacking rhizomes; *ligules* 0.1–0.5 mm; *leaves* conduplicate, 0.5–1.2 mm diam.; *panicles* (2) 3–10 (13) cm, contracted, with 1–2 branches per node; *spikelets* (3) 4.5–8.8 (10) mm, with (2) 3–5 (7) florets; *lower glumes* 1.5–3.5 mm; *upper glumes* 2.5–4.8 mm; *lemmas* (3) 3.4–4 (5.6) mm, smooth or scabrous distally, awned, awns (0.4) 1–2 (2.5) mm; *anthers* (0.8) 1.2–1.7 (2) mm. Common in meadows, grasslands, and forest openings and on rocky slopes, 7000–12,500 ft. July–Sept. E/W.

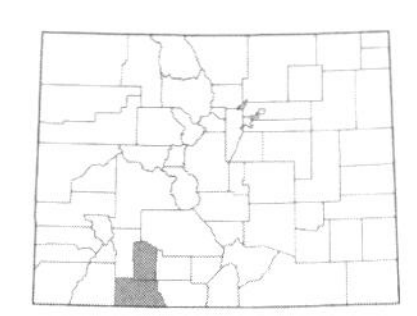

Festuca sororia **Piper**, RAVINE FESCUE. Plants 6–10 (15) dm, lacking rhizomes; *ligules* 0.3–1.5 mm; *leaves* flat, 3–6 (10) mm wide; *panicles* 10–20 (40) cm, open or somewhat contracted, with 1–2 (3) branches per node; *spikelets* 7–12 mm, with (2) 3–5 florets; *lower glumes* (1.5) 2.5–4 (4.5) mm; *upper glumes* (3) 4–6.5 mm; *lemmas* (5) 6–8 (9) mm, scabrous or puberulent, unawned or awned, awns to 2 mm; *anthers* 1.6–2.5 mm. Uncommon in spruce forests and along streams, 8000–9000 ft. July–Sept. W.

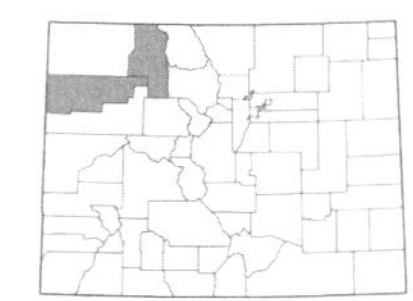

Festuca subulata **Trin.**, BEARDED FESCUE. Plants (3.5) 5–10 (12) dm, lacking rhizomes; *ligules* 0.2–0.6 (1) mm; *leaves* flat or loosely convolute, 3–10 mm; *panicles* 10–40 cm, open; *spikelets* 6–12 mm, with (2) 3–5 (6) florets; *lower glumes* (1.8) 2–3 (4) mm; *upper glumes* (2) 3–6 mm; *lemmas* 5–9 mm, glabrous, sometimes sparsely scabrous, awned, awns (2.5) 5–15 (20) mm; *anthers* 1.5–2.5 (3) mm. Uncommon along riverbanks and streams, 7400–9000 ft. July–Sept. W.

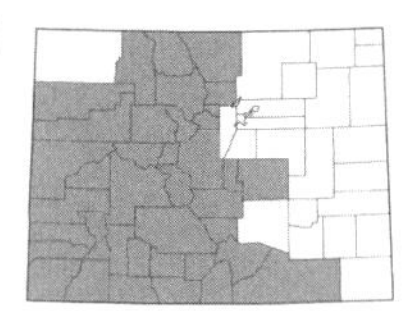

Festuca thurberi **Vasey**, THURBER'S FESCUE. Plants (4.5) 6–10 (12) dm, lacking rhizomes; *ligules* 2–5 (9) mm; *leaves* flat or conduplicate, 1.5–3 mm wide when flat, 0.8–1.8 mm diam. when conduplicate; *panicles* (7) 10–15 (17) cm, open, with 1–2 (3) branches per node; *spikelets* (8) 10–14 mm, with (3) 4–5 (6) florets; *lower glumes* (2) 3.5–5.5 mm; *upper glumes* (2.5) 4.5–6.5 (7) mm; *lemmas* 6–10 mm, scabrous or smooth, unawned, occasionally mucronate, mucros to 0.2 mm; *anthers* 3–4.5 mm. Common on dry slopes, in meadows, and in forest openings, 6500–12,500 ft. June–Aug. E/W.

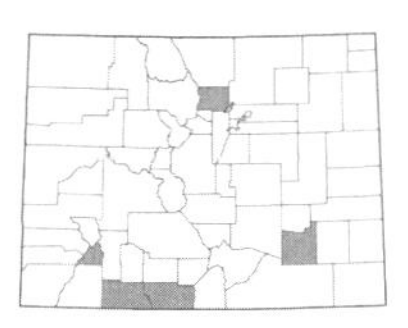

Festuca trachyphylla **(Hack.) Krajina**, HARD FESCUE. Plants (1.5) 2–7.5 (8) dm, lacking rhizomes; *ligules* 0.1–0.5 mm; *leaves* conduplicate, rarely flat, (0.5) 0.8–1.2 mm diam.; *panicles* (2.5) 3–13 (16) cm, contracted, with 1–2 branches per node; *spikelets* 5–9 (10.8) mm, with 3–7 (8) florets; *lower glumes* (1.8) 2–3.5 (4) mm; *upper glumes* 3–5 (5.5) mm; *lemmas* 3.8–5 (6.5) mm, smooth and glabrous proximally, scabrous or hairy distally, awned, awns 0.5–2.5 (3) mm; *anthers* (1.8) 2.3–3.4 mm. Found in meadows and forest openings, 6000–10,500 ft. July–Sept. E/W.

GLYCERIA R. Br. – MANNAGRASS

Perennials; *sheaths* closed; *ligules* membranous; *spikelets* in a large, open panicle, with 3–several florets; *glumes* 1-nerved, awnless, shorter than lemmas; *lemmas* 1-nerved, awnless, with 5 or more prominent parallel nerves; *paleas* large, often longer than the lemma; *disarticulation* above the glumes.

1a. Spikelets linear (or slightly laterally compressed at anthesis), 9–22 mm long, 8–12-flowered, usually green; panicle narrow, the branches appressed to erect...***G. borealis***
1b. Spikelets ovate to oblong, laterally compressed, 2–10 mm long, 3–8-flowered, often purplish at maturity; panicle open, the branches lax and spreading or even drooping at maturity...2

2a. Anthers 3; midnerve of the glume extending to the tip; spikelets 3.2–10 mm long; culms 10–15 dm tall...***G. grandis***
2b. Anthers 2; midnerve of the glume terminating below the tip; spikelets 2–6 mm long; culms 2–15 dm tall...3

3a. Leaf blades 6–15 mm wide; upper glume 1.5–2.5 mm long; culms 7.5–15 dm tall, spongy at the base, rooting freely at the lower nodes...***G. elata***
3b. Leaf blades 2–6 mm wide; upper glume 1–1.5 mm long; culms 2–10 dm tall, not or only slightly spongy at the base, not or sometimes rooting at the lower nodes...***G. striata***

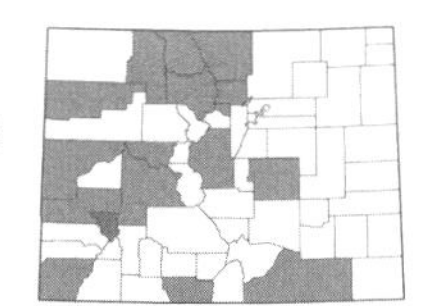

Glyceria borealis **(Nash) Batch.**, SMALL FLOATING MANNAGRASS. Plants 5–10 dm, rhizomatous; *leaves* flat or folded, 8–20 cm × 2–8 mm; *ligules* 4–12 mm long; *spikelets* linear, 9–22 mm long, with 8–12 florets; *glumes* unequal, the lower 2–2.5 mm and the upper 3–4 mm long; *lemmas* 3.5–4.5 mm long, 7-nerved, with scabrous nerves. Found in moist places, ditches, and along lake margins, sometimes in standing water, 5400–10,000 ft. July–Sept. E/W.

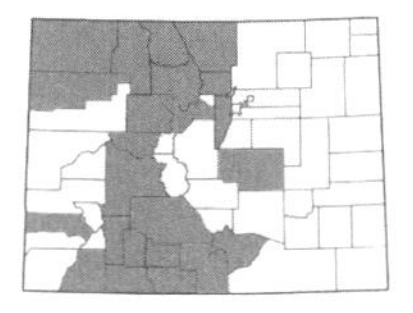

Glyceria elata **(Nash ex Rydb.) M.E. Jones**, TALL MANNAGRASS. Plants decumbent at the base, rooting at lower nodes, rhizomatous, 7.5–15 dm; *leaves* flat, 20–40 cm × 6–15 mm; *ligules* 3–4 mm long; *spikelets* ovate to oblong, 3–6 mm long, with 5–8 florets; *glumes* unequal, the lower 1–1.5 mm and the upper 1.5–2.5 mm long; *lemmas* 2–2.5 mm long, 7-nerved. Found along streams, lake margins, in marshes, and moist meadows, 6000–10,000 ft. June–Sept. E/W. (Plate 63)

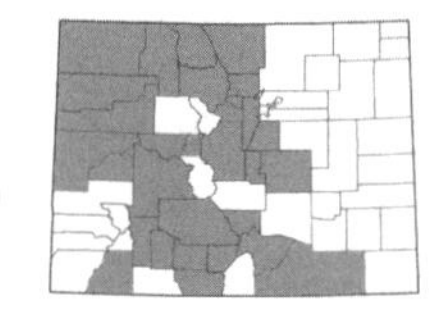

***Glyceria grandis* S. Watson**, American mannagrass. Plants decumbent at the base, rooting at the lower nodes, rhizomatous, 10–15 dm; *leaves* flat to folded, 25–40 cm × 6–12 mm; *ligules* 2–4 mm long; *spikelets* ovate to oblong, 3.2–10 mm, with 4–7 florets, purplish; *glumes* unequal, the lower 1–1.8 mm and the upper 1.5–2.5 mm long, the midnerve extending to the tip; *lemmas* 2–2.8 mm long, 7-nerved. Common in wet areas, along ditches and streams, and lake margins, 5500–9000 ft. June–Sept. E/W.

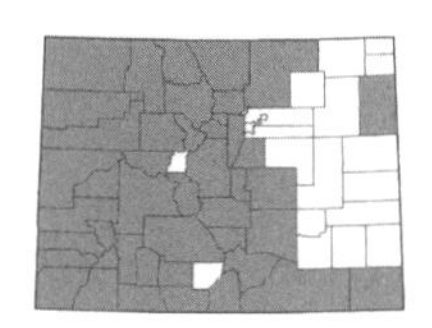

***Glyceria striata* (Lam.) Hitchc.**, fowl mannagrass. Plants erect or decumbent at the base, sometimes rooting at lower nodes, 2–10 dm, with short rhizomes; *leaves* flat, 5–30 cm × 2–6 mm; *ligules* 1–3 mm long; *spikelets* ovate to oblong, 2.5–4 mm, with 3–7 florets, purplish; *glumes* unequal, the lower 0.5–1.2 and the upper 1–1.5 mm long; *lemmas* 1.5–2.5 mm long, 7-nerved. Common along lake margins, streams, and in standing water, 4000–10,000 ft. June–Sept. E/W.

HELICTOTRICHON Besser ex Schult. & Schult. f. – alpine oatgrass

Perennials from rhizomes; *sheaths* open; *ligules* membranous; *spikelets* in narrow panicles or racemes, the rachis villous; *glumes* 1–3 (5)-nerved; *lemmas* 3–5-nerved, acute; *paleas* shorter than the lemma; *disarticulation* above the glumes.

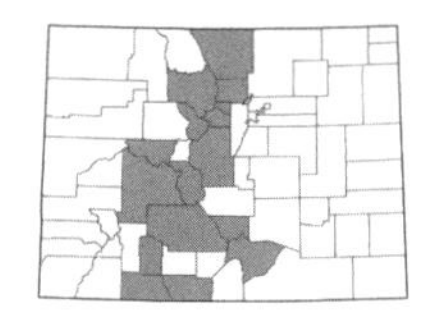

***Helictotrichon mortonianum* (Scribn.) Henrard**, Morton's alpine oatgrass. Plants 5–20 cm; *leaves* involute or convolute, 1–12 cm × 1–2 mm; *ligules* to 1 mm; *spikelets* mostly with 2 florets, the second often reduced and sterile; *glumes* subequal, the lower 8–10.5 mm and the upper 8.5–11.5 mm long; *calluses* with hairs 1–2 mm long; *lemmas* 7–10 mm long, with 4 bristly teeth at the tip, with an awn 1–1.5 mm long arising from near the middle. Found in the alpine, 11,000–14,000 ft. July–Aug. E/W. (Plate 63)

HESPEROSTIPA (Elias) Barkworth – needle and thread

Perennials; *sheaths* open; *ligules* membranous, often ciliate; *spikelets* in panicles, with one floret; *glumes* tapering from near the base to a hair-like apex; *lemmas* hard, smooth, the upper portion fused into a ciliate crown, awned, awns twice-geniculate; *paleas* equal to lemmas, hairy, with a hard apex; *disarticulation* above the glumes.

1a. Terminal segment of the awn with plumose hairs 1–3 mm long; leaf blades strongly involute, filiform, 0.5–1 mm wide...***H. neomexicana***

1b. Terminal segment of the awn glabrous or scabrous but lacking hairs 1–3 mm long; leaf blades flat to involute, 0.5–4.5 mm wide...2

2a. Lemmas 4–12 mm long, evenly appressed-hairy with white hairs to about 1 mm long...***H. comata***

2b. Lemmas 16–25 mm long, with beige to brown hairs on the lower portion and along the margins...***H. spartea***

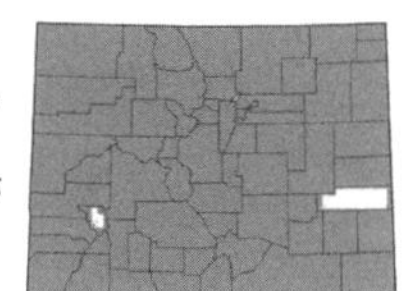

***Hesperostipa comata* (Trin. & Rupr.) Barkworth**, needle and thread. [*Stipa comata* Trin. & Rupr.]. Plants 1–10 dm; *leaves* flat or involute, 5–40 cm × 0.5–4 mm, scabrous above; *ligules* 1–7 mm long, acute; *glumes* subequal, the lower 18–30 mm and the upper 15–27 mm long; *calluses* 2–4 mm; *lemmas* indurate, 4–12 mm, evenly appressed-hairy, terminating in an awn 6.5–23 cm long; *paleas* subequal to the lemmas, hairy between the nerves. Common in shortgrass prairie, grasslands, sagebrush, and open forests and meadows, 3400–10,000 ft. May–July. E/W.

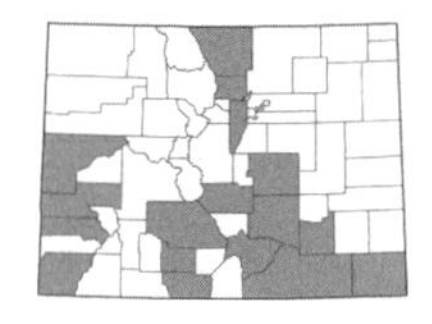

***Hesperostipa neomexicana* (Thurb.) Barkworth**, New Mexico feathergrass. [*Stipa neomexicana* (Thurb.) Scribn.]. Plants 4–10 dm; *leaves* involute, 10–30 cm × 0.5–1 mm; *ligules* 0.5–1 mm long; *glumes* subequal, 30–60 mm long; *calluses* sharp, 4–5 mm long; *lemmas* indurate, 15–18 mm long, evenly appressed-hairy, with an awn 12–22 cm long, the terminal segment with hairs 1–3 mm long; *paleas* subequal to lemmas. Uncommon on rocky ridgetops, open slopes, and in canyons and grasslands, 4500–8500 ft. May–July. E/W.

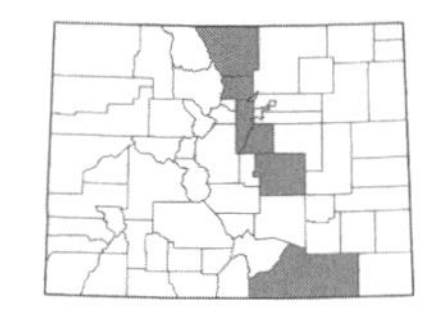

***Hesperostipa spartea* (Trin.) Barkworth**, porcupinegrass. [*Stipa spartea* Trin.]. Plants 4.5–9 dm; *leaves* flat to involute, 20–30 cm × 1.5–4.5 mm; *ligules* of the upper leaves 3–7.5 mm long, acute; *glumes* subequal, 22–45 mm long; *calluses* ca. 7 mm long, sharp; *lemmas* indurate, 16–25 mm long, hairy below, with an awn 9–19 cm long; *paleas* subequal to lemmas. Uncommon in meadows, grasslands, and pine forests, and on rocky hilltops, 5200–7500 ft. June–Aug. E.

HILARIA Kunth – curly-mesquite

Perennials; *sheaths* open; *ligules* membranous, ciliate; *spikelets* in 2-sided spike-like panicles, subtended by long hairs; *lateral spikelets* with 1–4 sterile or staminate florets; *central spikelets* sessile, with 1 pistillate or perfect floret; *glumes* deeply cleft into 2 or more lobes, awned; *lemmas* membranous; *disarticulation* at the base of branches, leaving a zigzag rachis.

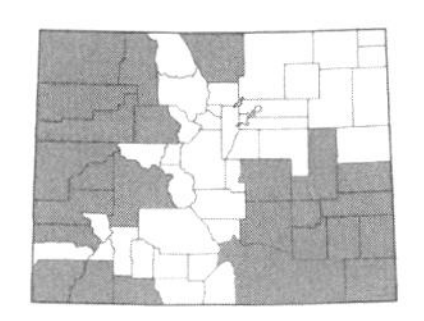

***Hilaria jamesii* (Torr.) Benth.**, GALLETA. [*Pleuraphis jamesii* Torr.]. Plants strongly rhizomatous or stoloniferous, 2–7 dm; *ligules* 1–5 mm long; *spikelets* 3 per node, the central fertile and the lateral 2 sterile, 6–9 mm long; *glumes* of fertile spikelet 4–5.5 mm long, with the nerves excurrent into awns 1.5–5 mm long, of sterile spikelets 5–7 mm long, the lower with an awn 3–5.5 mm long; *lemmas* of fertile spikelet 5.8–7.5 mm long, bifid, with an awn 1–2.5 mm long, of sterile spikelets 4.4–7 mm long. Common on dry slopes, in oak shrublands, sagebrush, and shortgrass prairie, 4000–7500 ft. May–Aug. E/W. (Plate 63)

HOLCUS L. – VELVETGRASS

Perennials; *sheaths* open; *ligules* membranous; *spikelets* in narrow panicles, with 2 florets; *glumes* strongly keeled, longer than the florets, 1–3-veined; *lemmas* firm, obscurely 3–5-veined; *disarticulation* below the glumes.

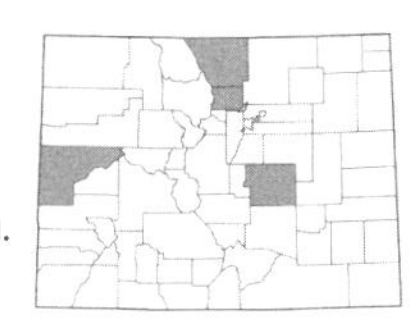

***Holcus lanatus* L.**, COMMON VELVETGRASS. Plants 3–10 dm; *leaves* flat, 2–20 cm × 3–10 mm, velvety-hairy; *ligules* 1–3 mm long; *spikelets* 3–6 mm long, the upper floret sterile and the lower fertile; *glumes* 3.5–5.5 mm long, 3-veined; *lemmas* 1.7–2.5 mm long, the upper with a hooked awn from below a cleft tip. Uncommon weed in disturbed areas and moist places, 5000–6000 ft. June–Aug. E/W. Introduced.

HOPIA Zuloaga & Morrone – VINE MESQUITE

Perennials, stoloniferous; *sheaths* open; *ligules* membranous; *spikelets* in panicles with one-sided spike-like branches, with one lower sterile floret and one perfect upper floret; *glumes* unequal, the lowermost usually absent; *upper lemmas* firm and cartilaginous; *paleas* well-developed, hard; *disarticulation* below the glumes.

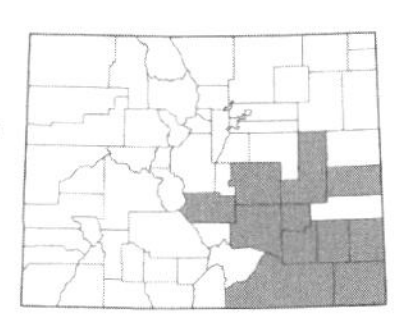

***Hopia obtusa* (Kunth) Zuloaga & Morrone**, VINE MESQUITE. [*Panicum obtusum* Kunth]. Plants decumbent to erect, sprawling or climbing, 2–20 dm long, with hairy nodes; *leaves* flat to involute, 3–26 cm × 2–7 mm; *ligules* 0.2–2 mm long; *spikelets* 2.8–4.5 mm long, the lower sterile and the upper fertile; *glumes* subequal, 2.8–4.4 mm long, glabrous; *sterile lemmas* 2.5–4 mm long; *fertile lemmas* indurate and shiny. Found in shortgrass prairie, canyons, and on mesas, 3400–6700 ft. June–Aug. E. (Plate 63)

HORDEUM L. – BARLEY; FOXTAIL

Perennials or annuals; *sheaths* open; *ligules* membranous; *spikelets* in dense 2-sided spikes with 3 spikelets per node, the central one perfect and fertile and the lateral 2 pedicelled and sterile or staminate; *glumes* awnlike; *lemmas* awned or unawned; *paleas* nearly equal to lemmas, keeled; *disarticulation* usually in the rachis.

1a. All three spikelets sessile at each node, fertile and plump; lemmas with long awns 60–160 mm long, the lowest awns equal to or longer than the upper awns; rachis generally not disarticulating at maturity...***H. vulgare***
1b. The lateral two spikelets at each node pedicelled, sterile and reduced, only the middle spikelet fertile; awns unlike the above; rachis usually readily disarticulating at maturity...2

2a. Glumes of lateral spikelets ciliate at the base; blades with auricles to 8 mm long, often completely surrounding the stem; lemma awns stout, 15–50 mm long...***H. murinum***
2b. Glumes of lateral spikelets scabrous or smooth but not ciliate at the base; blades lacking auricles or the auricles to 0.5 mm long; lemma awns various...3

3a. Glumes strongly spreading at maturity, 15–85 mm long; spikes 4–6 cm wide (nearly as wide as long)...***H. jubatum***
3b. Glumes straight, not or only slightly spreading at maturity, 8–20 mm long; spikes much longer than wide...4

4a. Glumes all similar and awnlike; perennials forming tufts...***H. brachyantherum***
4b. Glumes dissimilar, the outermost glumes awnlike throughout and the central glumes broadened above the base and then narrowed to slender awns; annuals, solitary or sometimes caespitose...***H. pusillum***

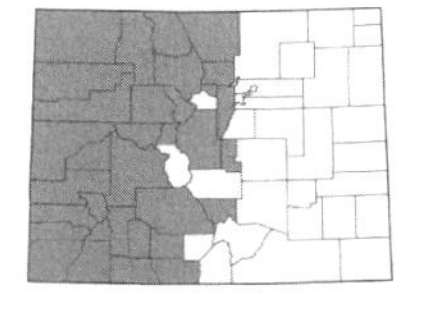

***Hordeum brachyantherum* Nevski**, MEADOW BARLEY. [*Critesion brachyantherum* (Nevski) Barkworth & D.R. Dewey]. Perennials, 3–7 dm; *leaves* flat, 4–12 cm × 2–5 mm; *ligules* 0.2–0.7 mm long; *spikelets* 3 per node, the central sessile and the lateral pedicellate; *glumes* awnlike, 7–20 mm long; *lemmas* of central spikelet 5–10 mm long, terminating in an awn 3.5–14 mm long. Common in grasslands, meadows, and riparian areas, 5500–11,000 ft. June–Aug. E/W.

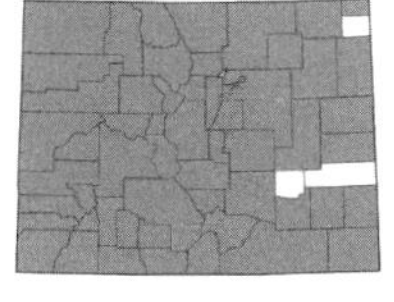

***Hordeum jubatum* L.**, FOXTAIL BARLEY. [*Critesion jubatum* (L.) Nevski]. Annuals or perennials, 2–8 dm; *leaves* flat or involute, 5–15 cm × 2–5 mm; *ligules* 0.2–1 mm long; *spikelets* 3 per node, the central sessile and the lateral pedicellate; *glumes* of central spikelet awnlike, 15–85 mm long; *lemmas* of central spikelet 4–8 mm long terminating in an awn 2–15 mm long. Common along roadsides, in grasslands, meadows, and in riparian areas, 3400–10,800 ft. June–Aug. E/W. (Plate 63)

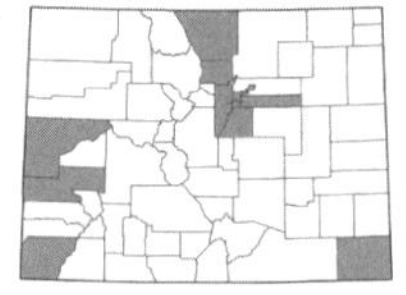

***Hordeum murinum* L.**, SMOOTH BARLEY. [*Critesion murinum* (L.) Á. Löve]. Annuals to 10 dm; *leaves* flat to involute, to 25 cm × 2–10 mm, auricles well-developed, to 8 mm; *ligules* 1–4 mm long; *spikelets* 3 per node, the central sessile and the lateral pedicellate; *glumes* 10–30 mm long, ciliate at the base; *lemmas* of central spikelet 8–14 mm long, terminating in an awn 15–50 mm long. Found along roadsides, in disturbed areas, and on dry slopes, 4500–6500 ft. May–July. E/W. Introduced.

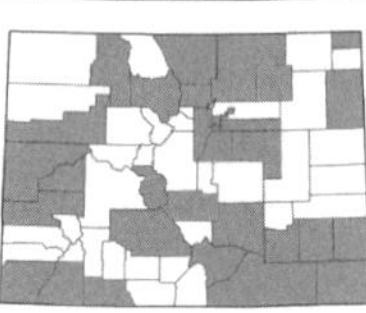

***Hordeum pusillum* Nutt.**, LITTLE BARLEY. [*Critesion pusillum* (Nutt.) Á. Löve]. Annuals, 1–6 dm; *leaves* flat, 1–12 cm × 2–5 mm, lacking auricles; *ligules* 0.2–0.8 mm long; *spikelets* 3 per node, the central sessile and the lateral pedicellate; *glumes* dissimilar, the lateral awnlike and the central broadened at the base and narrowed to a slender awn, 7–18 mm long with awns 7–15 mm long; *lemmas* of central spikelet 5–9 mm long, terminating in an awn 7–15 mm long. Found in disturbed areas, grasslands, and on dry slopes, 3400–6500 ft. May–July. E/W.

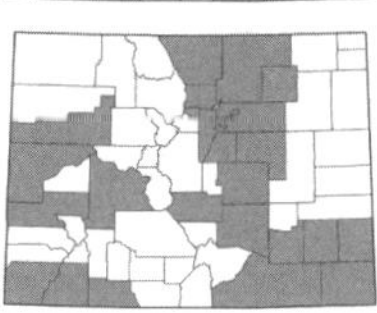

***Hordeum vulgare* L.**, BARLEY. Annuals, 6–12 dm; *leaves* flat, 10–30 cm × 5–18 mm, with well-developed auricles to 6 mm long; *ligules* 0.5–1.2 mm long; *spikelets* sessile; *glumes* linear, 6.5–20 mm long, terminating in awns 4–20 mm long; *lemmas* 6.5–12 mm long, terminating in a flattened awn 60–160 mm long with scabrous margins. Cultivated in fields and occasionally escaping along roadsides, 4000–8500 ft. June–Sept. E/W. Introduced.

KOELERIA Pers. – JUNEGRASS

Perennials; *sheaths* open; *ligules* membranous; *spikelets* in dense, spike-like panicles, with 2 florets; *lemmas* 5-nerved, membranous; *paleas* equal or subequal to the lemmas, hyaline; *disarticulation* above the glumes.

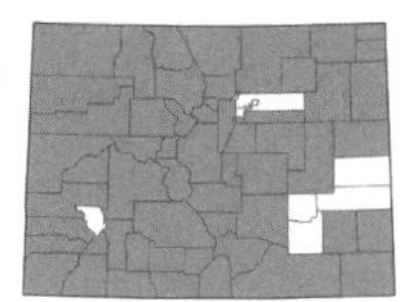

***Koeleria macrantha* (Ledeb.) Schult.**, JUNEGRASS. Plants 1.5–6.5 dm; *leaves* flat to involute, 5–25 cm × 1–3 mm, with hispid sheaths; *ligules* 0.5–1.5 mm long; *spikelets* 3–6 mm long; *glumes* subequal, the lower 2–4.5 mm and the upper 3–5 mm long; *lemmas* 3–5 mm long, acute or with a short awn tip; *paleas* colorless. Common in meadows, shortgrass prairie, woodlands, and sagebrush, 4000–12,000 ft. May–Aug. E/W. (Plate 63)

LEERSIA Sw. – CUTGRASS

Perennials; *sheaths* open; *ligules* membranous; *spikelets* in panicles, with one floret; *glumes* absent; *lemmas* 5-nerved, usually unawned, hard; *paleas* subequal to the lemmas, 3-nerved, unawned; *disarticulation* below the spikelets.

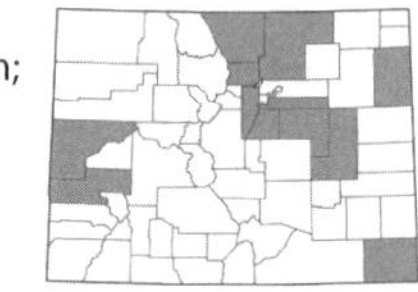

***Leersia oryzoides* (L.) Sw.**, RICE CUTGRASS. Plants 5–15 dm; *leaves* flat or folded, to 30 cm × 6–15 mm; *ligules* to 1 mm long, firm; *spikelets* flat; *lemmas* 4–5 mm long, hispid; *paleas* equal to or longer than the lemmas, with a stiffly ciliate keel. Found along pond margins, streams, and other wet areas, 4500–6000 ft. July–Sept. E/W. (Plate 63)

LEUCOPOA Griseb. – SPIKE FESCUE

Perennials; *sheaths* open; *ligules* membranous; *spikelets* in congested panicles, with 3–5 florets; *glumes* unawned; *lemmas* 5-nerved; *paleas* about equal to the lemmas, scabrous on the nerves; *disarticulation* above the glumes.

***Leucopoa kingii* (S. Watson) W.A. Weber**, SPIKE FESCUE. Plants 3–10 dm; *leaves* flat, 10–40 cm × 2–6 mm; *ligules* 1–3.5 mm long; *spikelets* 6–12 mm long, laterally compressed; *glumes* subequal, the lower 3–5.5 mm and the upper 4–6.5 mm long; *lemmas* 4.5–8 mm long, scabrous. Found in pine forests and meadows, 5500–10,000 ft. June–Aug. E/W. (Plate 63)

LEYMUS Hochst. – WILDRYE

Perennials, sometimes with rhizomes; *sheaths* open; *ligules* membranous; *spikelets* in a 2-sided spike usually with two or more spikelets per node, with 2–7 florets; *glumes* lanceolate to subulate; *lemmas* acute to short-awned; *paleas* shorter than the lemma; *disarticulation* above the glumes.

1a. Most or all spikelets at least 3 per node, with up to 8 per node…2
1b. Most or all spikelets solitary or paired at nodes…3

2a. Plants caespitose (bunchgrass) or occasionally with short, thick rhizomes; glumes subulate, 1-nerved, 0.5–2.5 mm wide…***L. cinereus***
2b. Plants strongly rhizomatous with long, creeping rhizomes; glumes linear-lanceolate, 3–5-nerved, 2–4 mm wide…***L. racemosus***

3a. Plants strongly rhizomatous with long, creeping rhizomes; glumes subulate to lanceolate, 1–3-nerved…***L. triticoides***
3b. Plants caespitose and densely tufted (bunchgrass) or occasionally with short, thick rhizomes; glumes subulate, 1-nerved…4

4a. Leaves flat to folded, with (9) 11–17 adaxial veins, not densely hairy just above the ligule; lemma awns (if present) 1.3–7 mm long...***L. ambiguus***
4b. Leaves strongly involute, with 5–9 adaxial veins, usually densely hairy just above the ligules; lemma awns (if present) 0.2–2.5 mm long...***L. salina* ssp. *salina***

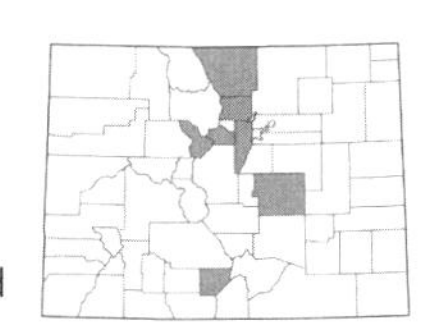

***Leymus ambiguus* (Vasey & Scribn.) D.R. Dewey**, Colorado wildrye. [*Elymus ambiguus* Vasey & Scribn.]. Caespitose perennials, sometimes with short rhizomes, 3–10 dm; *leaves* flat or folded, 10–30 cm × 2–6 mm, softly pilose; *ligules* 0.2–1.5 mm long; *spikelets* solitary at the base and tip of the rachis, otherwise 2 per node, 10–23 mm long, with 2–7 florets; *glumes* subulate, 6–15 mm long; *lemmas* 6–15 mm long, sometimes with an awn 1.3–7 mm long. Found on rocky slopes and in open pine woodlands, 5500–10,500 ft. June–Aug. E/W.

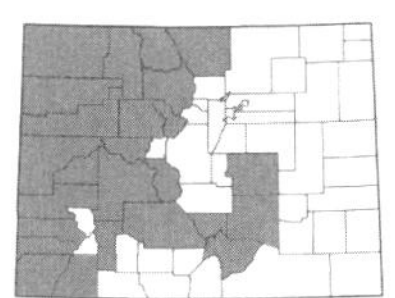

***Leymus cinereus* (Scribn. & Merr.) Á. Löve**, Great Basin wildrye. [*Elymus cinereus* Scribn. & Merr.]. Caespitose perennials, sometimes with short rhizomes, 7–20 dm; *leaves* flat, 15–40 cm × 8–20 mm, scabrous above; *ligules* 2–8 mm long; *spikelets* 2–7 per node, 9–25 mm long, with 3–7 florets; *glumes* subulate, 7–18 mm long; *lemmas* 6–12 mm long, awnless or with an awn 1–5 mm long. Common on rocky slopes, in meadows, canyons, and moist areas, 4600–10,000 ft. May–Aug. E/W.

***Leymus racemosus* (Lam.) Tzvelev**, Mammoth wildrye. [*Elymus giganteus* Vahl]. Strongly rhizomatous perennials, to 10 dm; *leaves* 6–15 mm wide, scabrous above; *ligules* to 2.5 mm long; *spikelets* 3–8 per node, 15–25 mm long, with 4–6 florets; *glumes* linear-lanceolate, 12–25 × 2–4 mm, 3–5-nerved; *lemmas* 10–20 mm long, hairy, awnless or with an awn to 2.5 mm. Uncommon on dry slopes, known from a collection near Craig (Moffat Co.), 6000–6500 ft. June–Aug. W. Introduced.

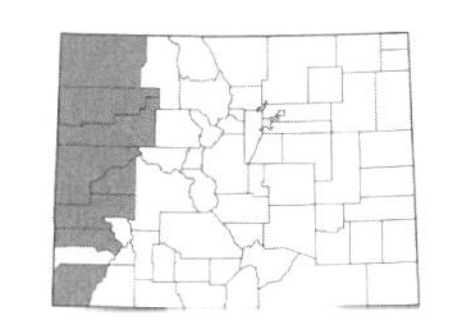

Leymus salina* (M.E. Jones) Á. Löve ssp. *salina, saline wildrye. [*Elymus salinus* M.E. Jones]. Caespitose perennials, sometimes with short rhizomes, 3–15 dm; *leaves* usually involute, 10–15 cm × 2–3 mm, glabrous to short-hairy; *ligules* 0.2–1 mm long; *spikelets* solitary per node or occasionally paired, 9–20 mm long, with 2–9 florets; *glumes* subulate, 4–12 mm long; *lemmas* 8–12 mm long, awn-tipped or awned to 2.5 mm. Common on rocky hillsides, cliffs, on shale slopes, and in pinyon juniper or sagebrush, 4900–8000 ft. May–June. W.

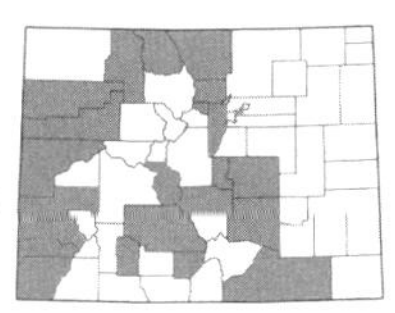

***Leymus triticoides* (Buckley) Pilg.**, creeping wildrye. [*Elymus simplex* Scribn. & T.A. Williams ex Scribn.]. Rhizomatous perennials, 5–12 dm; *leaves* flat or involute distally, 10–35 cm × 2–5 mm, scabrous to hairy; *ligules* 0.2–1 mm long; *spikelets* 1–2 (3) per node, 12–20 mm long, with 3–7 florets; *glumes* subulate to lanceolate, 6–16 mm long, 1–3-nerved; *lemmas* 5–16 mm long, awnless or with an awn to 3 mm long. Found in grasslands, canyons, pinyon-juniper, on clay flats, and at the edges of moist meadows, 6000–9500 ft. June–Aug. E/W.

LOLIUM L. – ryegrass

Perennials or annuals; *sheaths* open; *ligules* membranous; *spikelets* in two-sided spikes, positioned edgewise to the rachis; *glumes* usually one, 2 in the terminal spikelet, 3–9-nerved, unawned; *lemmas* 3–7-nerved, awned or unawned, chartaceous; *paleas* membranous, with ciliolate keels; *disarticulation* above the glumes.

1a. Lemmas awned, the awn 2–15 mm long; leaf blades convolute, 3–8 mm wide; spikelets with 10–22 florets...***L. multiflorum***
1b. Lemmas awnless or occasionally with an awn to 8 mm long; leaf blades flat, 1–6 mm wide; spikelets with 2–10 florets...***L. perenne***

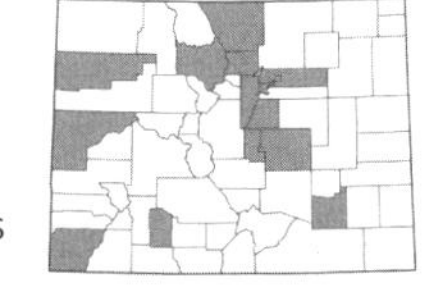

***Lolium multiflorum* Lam.**, Italian ryegrass. Annuals or short-lived perennials, 5–15 dm; *leaves* convolute, 10–30 cm × 3–8 mm, usually with auricles to 2 mm long; *ligules* 0.5–1.5 mm long; *spikelets* 9–25 mm long, with 10–22 florets; *glumes* present in terminal spikelet, 7–12 mm long; *lemmas* 3.5–9 mm, with an awn 2–15 mm long arising 0.2–0.7 mm below the tips. Weed in lawns and gardens and disturbed areas, 4500–9000 ft. June–Sept. E/W. Introduced.

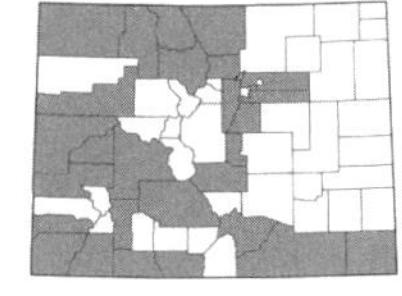

***Lolium perenne* L.**, perennial ryegrass. Perennials, 3–10 dm; *leaves* flat, 5–20 cm × 1–6 mm, usually with well-developed auricles; *ligules* 0.5–1.5 mm long; *spikelets* 9–15 mm long, with 2–10 florets; *glumes* 6–12 mm long; *lemmas* 4–9 mm long, acute or occasionally with an awn to 8 mm long. Weed of lawns, fields, and disturbed areas, 4500–9000 ft. June–Sept. E/W. Introduced. (Plate 63)

MELICA L. – oniongrass

Perennials; *sheaths* closed nearly to the top; *ligules* membranous, erose to lacerate; *spikelets* in panicles, with 2–7 florets; *glumes* 1–11-nerved, translucent; *lemmas* glabrous or hairy, 5–15-nerved, usually unawned; *paleas* usually with a ciliate keel, half as long to nearly equaling the lemmas; *disarticulation* below or above the glumes.

1a. Culms lacking bulbous swellings at the base; spikelets more or less secund and hanging to one side of the panicle, disarticulating below the glumes; pedicels sharply bent just below the spikelets…***M. porteri***
1b. Culms with bulbous swellings at the base; spikelets not secund, disarticulating above the glumes; pedicels straight or nearly so…2

2a. Lemma tips tapering to an acuminate point…***M. subulata* var. *subulata***
2b. Lemma tips rounded, obtuse to emarginate…3

3a. Bulb at the base of the culm attached to rhizome by a slender stem; glumes less than half the length of the spikelet…***M. spectabilis***
3b. Bulb at the base of the culm sessile on the rhizome; glumes more than half as long as the spikelet…***M. bulbosa***

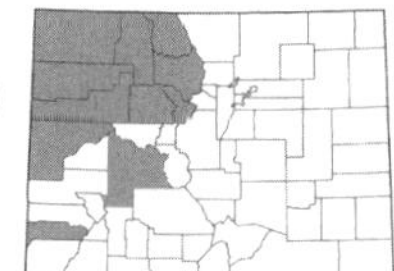

***Melica bulbosa* Geyer ex Porter & J.M. Coult.**, ONIONGRASS. [*Bromelica bulbosa* (Geyer ex Porter & J.M. Coult.) W.A. Weber]. Plants 3–10 dm, the base bulbous; *leaves* flat to involute, 10–30 cm × 2–5 mm; *ligules* 2–6 mm long; *spikelets* 6–25 mm long, with 2–9 florets, not secund; *glumes* unequal, the lower 5.5–10 mm and the upper 6–15 mm long; *lemmas* 6–12 mm long, with a purplish band near the tip. Found in forests, shrublands, meadows, and on open slopes, 4500–9700 ft. May–Aug. E/W.

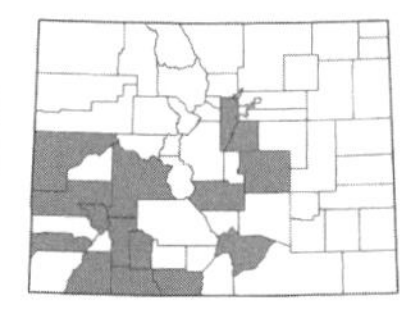

***Melica porteri* Scribn.**, PORTER'S MELIC. Plants 5.5–10 dm, lacking bulbous swellings at the base; *leaves* flat, 10–25 cm × 2–5 mm; *ligules* 2–7 mm long; *spikelets* 8–16 mm long, more or less secund and hanging to one side of the panicle, with 2–5 florets; *glumes* unequal, the lower 3.5–6 mm and the upper 5–8 mm long, scarious; *lemmas* 6–10 mm long. Locally common on rocky cliffs, along creeks, and in forests, 7000–10,000 ft. June–Aug. E/W.

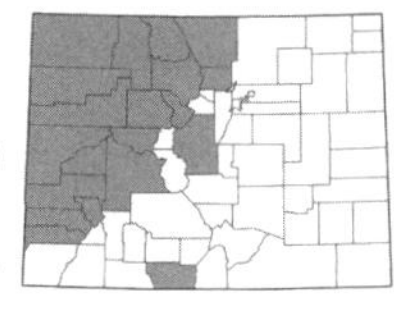

***Melica spectabilis* Scribn.**, PURPLE ONIONGRASS. [*Bromelica spectabilis* (Scribn.) W.A. Weber]. Plants 4–10 dm, with a bulbous base, the bulbs attached to rhizomes by a slender stem; *leaves* flat to involute, 5–18 cm × 2–4 mm; *ligules* 1–3 mm long; *spikelets* 7–20 mm long with 4–8 florets; *glumes* unequal or subequal, the lower 3.5–6.5 mm and the upper 5–7 mm long; *lemmas* 7–8.5 mm long. Found along streams, in meadows, and in aspen and spruce-fir forests, 6500–11,000 ft. June–Aug. E/W. (Plate 63)

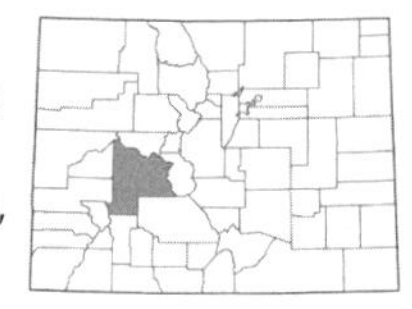

Melica subulata* (Griseb.) Scribn. var. *subulata, ALASKA ONIONGRASS. [*Bromelica subulata* (Griseb.) Farw.]. Plants 5–12 dm, with a bulbous base, rhizomatous; *leaves* flat, 2–10 mm wide; *ligules* 0.4–5 mm long; *spikelets* 10–28 mm long, with 2–5 florets; *glumes* unequal, the lower 4–8 mm and the upper 5.5–11 mm long; *lemmas* 5.5–18 mm long. Rare near seepages and in aspen-spruce forests, 8500–9000 ft. May–July. W.

MISCANTHUS Andersson – SILVERGRASS

Perennials; *sheaths* open; *ligules* membranous, ciliate; *spikelets* paired, in ovoid or corymbose panicles with spike-like branches, with sterile lower florets and perfect upper florets, with a hairy callus; *glumes* membranous to thick; *disarticulation* below the glumes.

***Miscanthus sinensis* Andersson**, CHINESE SILVERGRASS. Plants 6–20 dm; *leaves* flat, 20–70 cm × 6–20 mm; *panicles* 15–25 cm long; *spikelets* 3.5–7 mm long; *glumes* subequal, the lower ciliate on the margins; *calluses* with hairs 6–12 mm long, to twice the length of the spikelets; *lemmas* with a geniculate awn 6–12 mm long. Cultivated in gardens, occasionally persisting along roadsides in the Denver/Boulder area, 5000–6000 ft. Mar.–Aug. E. Introduced.

MUHLENBERGIA Schreb. – MUHLY

Perennials or annuals; *sheaths* open; *ligules* membranous or hyaline; *spikelets* in open to spike-like panicles, with 1 (2–3) florets; *glumes* usually 1-nerved, sometimes awned, lower glumes sometimes absent; *lemmas* glabrous to hairy, 3-nerved, awned or unawned; *paleas* shorter than to equal to the lemmas; *disarticulation* usually above the glumes.

1a. Inflorescence a spike-like panicle composed of 3–13 widely spaced, spreading to often curved, spike-like unilateral branches; spikelets sessile, 3–5.5 mm long, embedded and appressed to the slender rachis branches, awnless or with a short awn-tip; at maturity the panicle breaking off at the base and forming a "tumbleweed"-like plants…***M. paniculata***
1b. Plants unlike the above in all respects…2

2a. Upper glume 3-toothed at the apex, the teeth sometimes shortly awned…3
2b. Upper glume with an acute or obtuse apex, not 3-toothed, awned or not…4

3a. Lemma awn 6–25 mm long; spikelets 3–7 mm long; ligules 4–14 mm long…***M. montana***
3b. Lemma awn 1–5 mm long; spikelets 2.2–3.5 mm long; ligules 2–4 (5) mm long…***M. filiculmis***

4a. Panicle dense and spike-like, or narrow with ascending or closely appressed branches…5
4b. Panicle open with spreading branches, usually diffuse…19

5a. Lower glume with 2 unequal awns 1–3 mm long; upper glume with a single awn 2.5–4.5 mm long…6
5b. Glumes unlike the above…7

6a. Tip of the leaf ending in a fragile and easily broken awnlike tip mostly 4–7 mm long; ligules (2) 3–12 mm long, not deeply divided into 3 triangular lobes…***M. alopecuroides***
6b. Tip of the leaf acute or mucronate, or the midrib sometimes extended to 3 mm as a short bristle; ligules 1.5–3 mm long, usually deeply divided into 3 triangular lobes…***M. phleoides***

7a. Lemmas conspicuously hairy below and often on the margins (hairs to 3.5 mm long); rhizomatous perennials…8
7b. Lemmas glabrous or with short, appressed hairs below; plants various…12

8a. Calluses and lemma bases with dense long, silky hairs 2–3.5 mm long…***M. andina***
8b. Calluses and lemma bases lacking long, silky hairs or the hairs less than 1.2 mm long…9

9a. Leaves 0.2–1 mm wide, tightly involute; plants 1.2–3.6 dm; glumes unawned…***M. thurberi***
9b. Leaves 2–6 mm wide, flat; plants 3–12 dm; glumes awned or not…10

10a. Glumes equal to or shorter than the lemma, unawned, or sometimes awned to 2 mm…***M. mexicana***
10b. Glumes 1–2 times longer than the lemma, awned (the awns to 5 mm long)…11

11a. Stems smooth and polished, glabrous or slightly hairy just below the nodes, strongly keeled; ligules 0.6–1.7 mm long…***M. racemosa***
11b. Stems dull, usually sparsely hairy throughout (more densely so just below the nodes), rarely keeled; ligules 0.2–0.6 mm long…***M. glomerata***

12a. Lemmas with an awn tip 1.5–20 mm long…13
12b. Lemmas acute to mucronate, or with a short awn tip to 1 mm long…16

13a. Lemma awns 1.5–5 mm long; plants perennial but often appearing annual, not rhizomatous, often rooting at lower nodes; ligules 0.2–0.5 mm long, lacking lateral lobes…***M. schreberi***
13b. Lemma awns 6–25 mm long; ligules 1–3 (5) mm long, often with lateral lobes…14

14a. Plants perennial, sometimes rhizomatous, 30–70 cm; lower glume not bifid at the tip…***M. pauciflora***
14b. Plants annual, tufted, 3–20 cm; lower glume minutely to deeply bifid at the tip…15

15a. Glumes shorter than the lemma (to ⅔ the length of the lemma)…***M. brevis***
15b. Glumes equal to or longer than the lemma…***M. depauperata***

16a. Annuals, tufted, often rooting at lower nodes, 5–20 (35) cm tall…***M. filiformis***
16b. Perennials, caespitose or rhizomatous, 5–60 cm tall…17

17a. Plants rhizomatous, often mat-forming, 5–30 cm; lemmas glabrous; glumes ⅓–½ as long as the lemma…***M. richardsonis***
17b. Plants not rhizomatous, caespitose, not mat-forming, 15–60 cm; lemmas with short hairs or scabrous below; glumes ½–¾ as long as the lemma…18

18a. Glumes unawned or with a short mucro to 0.3 mm long; spike not dense; ligules 0.2–0.8 mm long…***M. cuspidata***
18b. Glumes abruptly narrowed to a short awn 0.5–1 mm long; spike dense; ligules 1–3 (5) mm long…***M. wrightii***

19a. Lemmas awned (awn 0.5–13 mm long, but usually over 1 mm)…20
19b. Lemmas unawned or with a short mucro to 0.3 mm…23

20a. Paleas 2-awned at the tip with awns to 1 mm; plants rhizomatous; pedicels 10–25 mm long…***M. pungens***
20b. Paleas not 2-awned at the tip; plants not rhizomatous; pedicels 1–15 (20) mm long…21

21a. Lemma awn 4–13 mm long; plants from a knotty base...***M. porteri***
21b. Lemma awn 0.5–4 mm long; plants not from a knotty base...22

22a. Leaf blades 4–16 cm long and 1–2.2 mm wide, flat or involute, not arcuate...***M. arenicola***
22b. Leaf blades 1–3 (5) cm long and 0.3–1 mm wide, tightly involute or folded, arcuate...***M. torreyi***

23a. Lemmas densely silky hairy on the veins and margins (the hairs to 1 mm long); paleas densely silky hairy between the 2 veins; glumes glabrous...***M. tricholepis***
23b. Lemmas and paleas unlike the above (sometimes hairy on the lower half of the margins and veins but not densely silky hairy nearly the entire length)...24

24a. Spikelets (3) 5–8 mm long; panicle 20–50 cm long; leaves 15–45 cm long...***M. ammophila***
24b. Spikelets 0.8–2.6 mm long; panicle 1–21 cm long; leaves 0.5–11 cm long...25

25a. Perennials, rhizomatous; lemmas 1.2–2.5 mm long...26
25b. Annuals; lemmas 0.8–1.5 mm long...27

26a. Leaf margins and midveins whitish and thickened; ligules with 1–2 mm long lateral lobes...***M. arenacea***
26b. Leaf margins and midveins greenish, not conspicuously thickened; ligules lacking lateral lobes...***M. asperifolia***

27a. Glumes minutely hairy at least near the tips; ligules 1–2.6 mm long; pedicels 2–7 mm long...***M. minutissima***
27b. Glumes glabrous or scabrous along the margins; ligules 0.2–0.5 mm long; pedicels 1–3 mm long...***M. ramulosa***

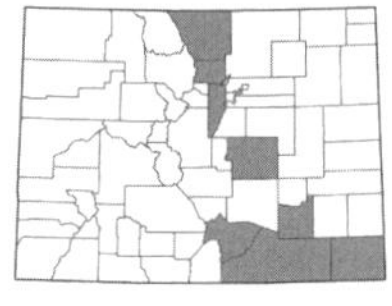

Muhlenbergia alopecuroides **(Griseb.) P.M. Peterson & Columbus**, BRISTLY WOLFSTAIL. [*Lycurus setosus* (Nutt.) C. Reeder]. Perennials, 3–5 (6) dm, caespitose; *ligules* (2) 3–10 (12) mm long, acuminate or sometimes shortly cleft; *leaves* 4–9 (13) cm × 1–2 mm, with prominent whitish midribs and scabrous margins, the midrib extending as an awnlike apex, (3) 4–7 (12) mm long; *panicles* 4–8 (10) cm × (5) 7–8 mm, dense and spike-like; *spikelets* 3–4 mm long; *glumes* unequal, lower glumes with 1–3 awns, awns 1–3.5 mm long, upper glumes awned, awns 2.5–4 (5)mm long, flexuous; *lemmas* 3–4 mm long, hairy, tapering to an awn, 1.5–2 (3) mm. Common on dry slopes, rocky hillsides, and in canyons, 4000–9000 ft. July–Sept. E.

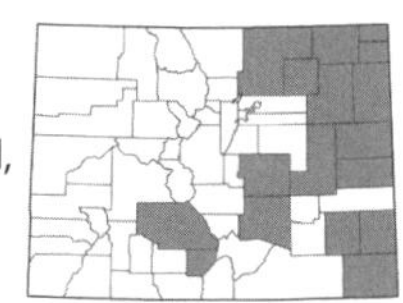

Muhlenbergia ammophila **P.M. Peterson**, BLOWOUT GRASS. [*Redfieldia flexuosa* (Thurb.) Vasey]. Perennials, 5–13 dm, rhizomatous; *ligules* to 1.5 mm, membranous, ciliate; *leaves* 15–45 cm × 2–8 mm, loosely involute; *panicles* 20–50 × 8–25 cm, open; *spikelets* (3) 5–8 × 3–5 mm; *glumes* unequal, lower glumes 3–4 mm, unawned, upper glumes 3.5–4.5 mm long, unawned; *lemmas* 4.5–6 mm long, with a tuft of hairs at the base, acute to awn-tipped, entire, or with 3 small teeth. Found in sandsage prairie, on sand dunes, and in sandy soil of floodplains, 3400–8300 ft. July–Sept. E.

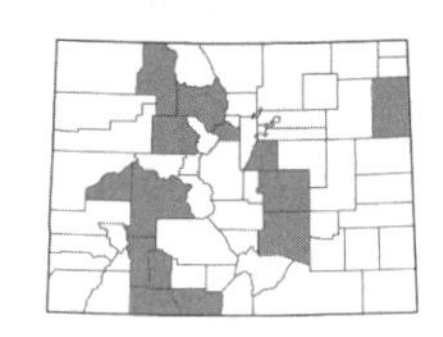

Muhlenbergia andina **(Nutt.) Hitchc.**, FOXTAIL MUHLY. Perennials, 2.5–8.5 dm, rhizomatous with scaly rhizomes; *ligules* 0.5–1.5 mm long, membranous, truncate, lacerate to ciliate; *leaves* 4–16 cm × 2–4 (5) mm, flat; *panicles* 2–15 × 0.5–2.8 cm, dense and spike-like; *spikelets* 2–4 mm long; *glumes* 2–4 mm long, acuminate to awn-tipped; *lemmas* 2–3.5 mm long, acuminate, awned, the awns 1–10 mm long, the calluses and bases with dense, long hairs 2–3.5 mm long. Found along streams and near moist seeps at the base of rocky outcrops, 5500–10,500 ft. July–Sept. E/W.

Muhlenbergia arenacea **(Buckley) Hitchc.**, EAR MUHLY. Perennials, 1–3 (4) dm; *ligules* 0.5–2 mm long, with lateral lobes, lobes 1–2 mm; *leaves* 0.7–4 (6) cm × 0.5–1.7 mm, flat or sometimes folded; *panicles* 5–15 × 4–14 cm, open with capillary branches; *spikelets* 1.5–2.6 mm long, sometimes with 2 florets; *glumes* 0.9–2 mm long, acute to acuminate or sometimes erose and mucronate, mucros to 0.2 mm; *lemmas* 1.5–2.5 mm long, acute to obtuse, occasionally shallowly bilobed, mucronate, sparsely appressed-hairy on the lower half of margins and nerves. Uncommon on shale slopes, alkaline flats, in moist depressions, and along the margins of ponds, 4400–6000 ft. June–Aug. E.

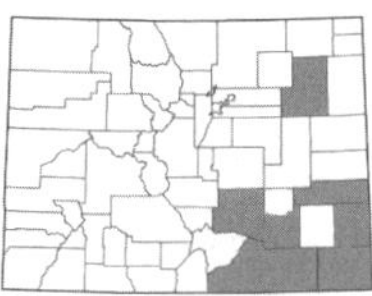

Muhlenbergia arenicola **Buckley**, SAND MUHLY. Perennials, (1.5) 2–6 (7) dm; *ligules* 2–9 mm long, acute, lacerate, often with lateral lobes; *leaves* 4–10 (16) cm × 1–2.2 mm, flat, folded, or involute, often glaucous; *panicles* 12–30 × 5–20 cm, open; *spikelets* 2.5–4.2 mm long; *glumes* 1.4–2.5 mm long, acute to acuminate, minutely erose, unawned to awned, awns to 1 mm; *lemmas* 2.5–4.2 mm long, appressed-hairy on the lower half, acuminate, awned, awns 0.5–4.2 mm long. Found in sandy soil of the shortgrass prairie, 3500–5200 ft. July–Sept. E.

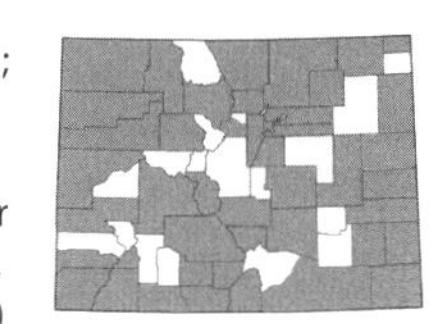

***Muhlenbergia asperifolia* (Nees & Meyen ex Trin.) Parodi**, SCRATCHGRASS. Perennials, 1–6 (10) dm; *ligules* 0.2–1 mm long, firm, truncate, ciliate; *leaves* 2–7 (11) cm × 1–2.8 (4) mm; *panicles* 6–21 × 4–16 cm, open, with capillary branches; *spikelets* 1.2–2.1 mm long, sometimes with 2 or 3 florets; *glumes* 0.6–1.7 mm, acute, unawned; *lemmas* 1.2–2.1 mm long, keel scabrous, acute, unawned or mucronate, mucros to 0.3 mm. Common in sandy soil, wet meadows, moist depressions, swales, near seeps and springs, and along ditches and streams, 3500–8000 ft. July–Sept. E/W. (Plate 63)

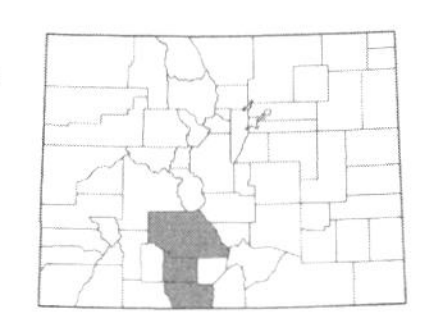

***Muhlenbergia brevis* C.O. Goodd.**, SHORT MUHLY. Annuals, 0.3–2 dm; *ligules* 1–3 mm, membranous, acute, lacerate, occasionally with lateral lobes; *leaves* 1–4.5 cm × 0.8–2 mm, flat to involute; *panicles* 3–11.5 × 0.8–1.8 cm, narrow, branches erect; *spikelets* 2.5–6 mm; *glumes* unequal, lower glumes 2–3.5 mm, bifid with 2 aristate teeth or awned, awns to 1.8 mm, upper glumes 2.4–4 mm long, acuminate, awned, awns to 2 mm; *lemmas* 3.5–6 mm long, acuminate, often bifid, awned, awns 10–20 mm, stiff. Uncommon on rocky slopes and outcrops, 7500–9000 ft. July–Sept. E.

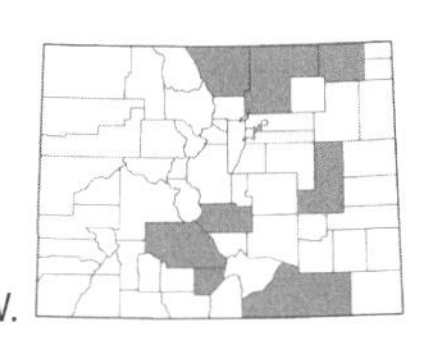

***Muhlenbergia cuspidata* (Torr. ex Hook.) Rydb.**, PLAINS MUHLY. Perennials, 2–6 dm, tufted; *ligules* 0.2–0.8 mm, truncate; *leaves* 2–22 cm × 0.5–2.7 (3.5) mm, flat to folded; *panicles* 4–14 × 0.1–0.8 cm, spike-like; *spikelets* 2.5–3.6 mm, dark green or plumbeous, sometimes with 2 florets; *glumes* 1.2–3 mm, acute to acuminate, occasionally mucronate, mucros to 0.3 mm; *lemmas* 2.5–3.6 mm, acuminate or occasionally mucronate, mucros to 0.6 mm. Found on dry slopes, sandstone and limestone outcrops, and near springs and along the edges of pools, 4000–9000 ft. July–Sept. E/W.

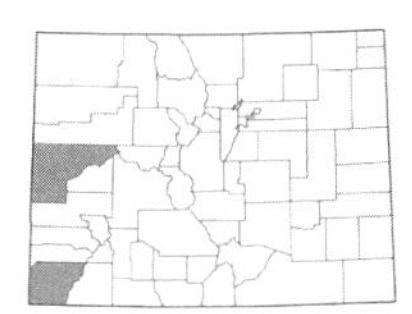

***Muhlenbergia depauperata* Scribn.**, SIXWEEKS MUHLY. Annuals, 0.3–1.5 dm; *ligules* 1.4–2.5 mm long, membranous, acute, with lateral lobes; *leaves* 1–3 cm × 0.6–1.5 mm, flat or involute; *panicles* 2.5–8.5 × 0.5–0.7 cm, spike-like; *spikelets* 2.5–5.1 mm; *glumes* unequal, lower glumes 2.3–4 mm, bifid, aristate or awned, awns to 1.3 mm, upper glumes 3–5.1 mm, acuminate, unawned; *lemmas* 2.5–4.5 mm long, acuminate, awned, awns 6–15 mm, stiff. Uncommon near moist depressions in sandstone outcrops, and near moist pools in sandstone depressions, 5500–6000 ft. July–Sept. W.

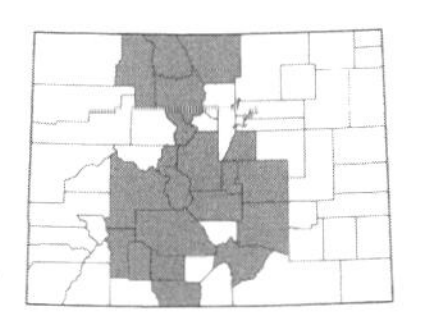

***Muhlenbergia filiculmis* Vasey**, SLIMSTEM MUHLY. Perennials, 0.5–3 (4) dm, caespitose; *ligules* 2–4 (5) mm long, membranous, acute; *leaves* 2–6 cm × 0.4–1.6 mm, tightly involute, filiform, stiff; *panicles* 1.5–7 × 0.4–2 cm, dense, contracted; *spikelets* 2.2–3.5 mm; *glumes* subequal, 0.8–2.5 mm, lower glumes awned, awns to 1.6 mm, upper glumes 3-toothed, the teeth occasionally awned, awns to 0.6 mm; *lemmas* 2.2–3.5 mm, acute to acuminate, awned, awns 1–5 mm long, straight or flexuous. Common on dry slopes, in pine forests, and in sagebrush, 6500–10,500 ft. July–Sept. E/W.

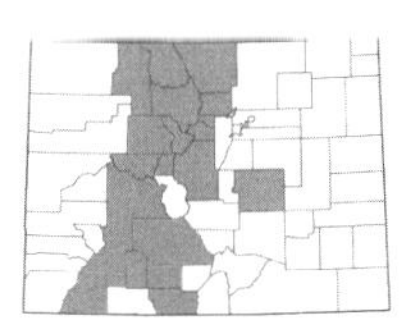

***Muhlenbergia filiformis* (Thurb. ex S. Watson) Rydb.**, PULL-UP MUHLY. Annuals, (0.3) 0.5–2 (3.5) dm, caespitose but often rooting at lower nodes; *ligules* 1–3.5 mm long, hyaline to membranous, rounded to acute; *leaves* 1–4 (6) cm × 0.6–1.6 mm, flat or involute; *panicles* 1.6–6 × 0.2–0.5 cm, spike-like; *spikelets* 1.5–3.2 mm long; *glumes* 0.6–1.7 mm long, rounded to subacute, unawned; *lemmas* (1.5) 1.8–2.5 (3.2) mm long, acute to acuminate, occasionally mucronate, mucros to 1 mm. Found in moist places, bogs, wet meadows, and along streams, 7000–12,000 ft. July–Sept. E/W.

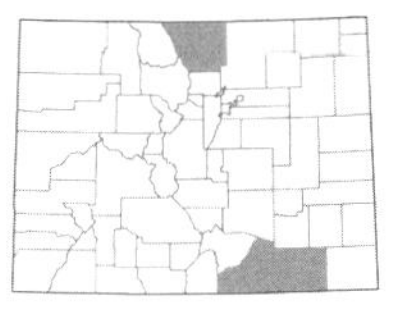

***Muhlenbergia glomerata* (Willd.) Trin.**, SPIKED MUHLY. Perennials, 3–12 dm, rhizomatous; *ligules* 0.2–0.6 mm, membranous, lacerate-ciliolate; *leaves* 2–15 cm × 2–6 mm, flat; *panicles* 1.5–12 × 0.3–1.8 cm, spike-like; *spikelets* 3–8 mm; *glumes* 3–8 mm long, acuminate, awned, awns to 5 mm; *lemmas* 1.9–3.1 mm long, hairy below with hairs to 1.2 mm, acuminate, unawned or awned, awns to 1 mm. Uncommon in moist meadows, along streams, and on rocky slopes, 4000–5000 ft. July–Sept. E.

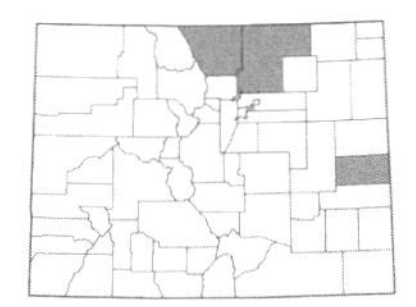

***Muhlenbergia mexicana* (L.) Trin.**, MEXICAN MUHLY. Perennials, 3–9 dm; *ligules* 0.4–1 mm, membranous, lacerate-ciliolate; *leaves* 2–20 × 2–6 cm, flat; *panicles* 2–21 × 0.3–3 cm, dense; *spikelets* 1.5–3.8 mm, often purple; *glumes* 1.5–3.7 mm, acuminate, unawned or awned, awns to 2 mm; *lemmas* 1.5–3.8 mm, hairy below with hairs to 0.7 mm long, acuminate, unawned or awned, awns to 10 mm. Uncommon on open slopes and in pine forests, 4500–6500 ft. July–Sept. E.

There are two varieties of *M. mexicana* in Colorado:

1a. Lemma awn 3–10 mm long...**var. *filiformis* (Torr.) Scribn.**
1b. Lemma awn less than 3 mm long...**var. *mexicana***

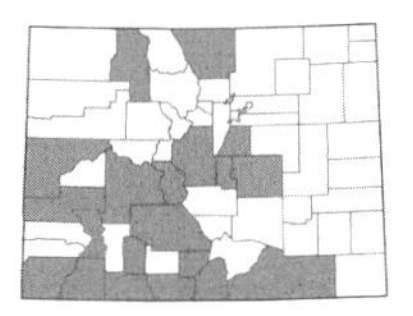

***Muhlenbergia minutissima* (Steud.) Swallen**, ANNUAL MUHLY. Annuals, 0.5–4 dm; *ligules* 1–2.6 mm long, occasionally with lateral lobes; *leaves* 0.5–4 (10) cm × 0.8–2 mm, flat or involute; *panicles* 5–16.2 (21) × 1.5–6.5 cm, open; *spikelets* 0.8–1.5 mm long; *glumes* 0.5–0.9 mm long, obtuse to acute, unawned; *lemmas* 0.8–1.5 mm, glabrous or veins and margins hairy, unawned. Common in sandy soil, pinyon-juniper, and in moist sandstone depressions, 5000–9000 ft. July–Sept. E/W.

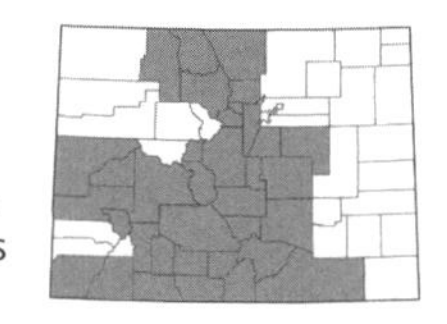

***Muhlenbergia montana* (Nutt.) Hitchc.**, MOUNTAIN MUHLY. Perennials, 1–8 dm, caespitose; *ligules* 4–14 (20) mm long, membranous, acute to acuminate; *leaves* 6–25 cm × 1–2.5 mm, flat to involute; *panicles* 4–25 × (1) 2–6 cm, contracted; *spikelets* 3–7 mm long; *glumes* subequal, (1) 1.5–3.2 (4) mm long, lower glumes occasionally awned, awns to 1 mm, upper glumes 3-toothed, occasionally awned, awns to 1.6 mm long; *lemmas* 3–4.5 (7) mm long, appressed-hairy on nerves below, awned, awns (2) 6–25 mm long, flexuous. Common on dry slopes, in pine forests, and in oak-sagebrush, 5500–10,500 ft. July–Sept. E/W. (Plate 64)

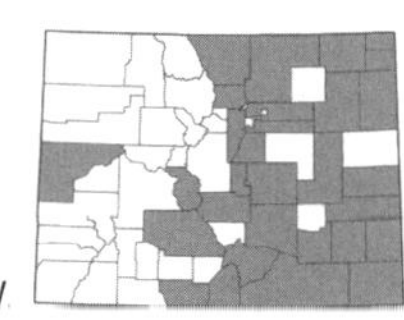

***Muhlenbergia paniculata* (Nutt.) P.M. Peterson**, TUMBLEGRASS. [*Schedonnardus paniculatus* (Nutt.) Trel.]. Perennials to 5 dm, caespitose; *ligules* 1–3.5 mm long, lanceolate; *leaves* 1–12 cm × 0.7–2 mm, usually folded; *panicles* widely spaced, of spike-like, strongly divergent branches; *spikelets* 3–5.5 mm long; *glumes* unequal, lower glumes 1.5–3 mm, upper glumes 1.5–5.5 mm long; *lemmas* 3–5 mm long, glabrous to scabrous, unawned or shortly awned. Common in shortgrass prairie, grasslands, and on dry hillsides, mesas, and rocky outcrops, 3400–8000 ft. June–Aug. E/W.

***Muhlenbergia pauciflora* Buckley**, NEW MEXICO MUHLY. Perennials, 3–7 dm, tufted or shortly rhizomatous; *ligules* 1–2.5 (5) mm long, membranous, obtuse, with lateral lobes; *leaves* (1) 5–12 (15) cm × 0.5–1.5 mm, flat to involute, often with dark brown necrotic spots; *panicles* (2) 5–15 × 0.5–2.8 cm, narrow; *spikelets* 3.5–5.5 mm long, sometimes with 2 florets; *glumes* 1.5–3.5 mm long, acuminate to acute, often erose, unawned or awned, awns to 2.2 mm long; *lemmas* 3–5.5 mm long, scabrous with short hairs on the callus, awned, awns (5) 7–25 mm long, flexuous. Uncommon in forests and on rocky slopes, 7000–7500 ft. July–Sept. W.

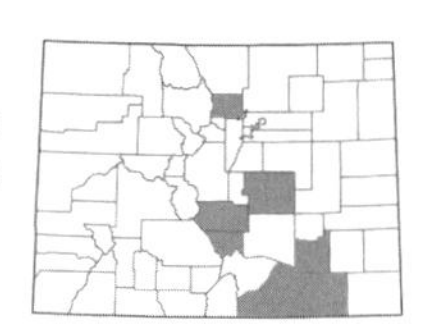

***Muhlenbergia phleoides* (Kunth) P.M. Peterson**, WOLFSTAIL. [*Lycurus phleoides* Kunth]. Perennials, 3–5 (6) dm, caespitose; *ligules* 1.5–3 mm long, with narrow triangular lobes 1.5–4 mm long; *leaves* flat or folded, 4–8 cm × 1–1.5 mm, the midrib sometimes extending up to 3 mm as a short bristle; *panicles* dense, spike-like, to 10 cm long; *spikelets* 3–4 mm long; *glumes* 1–2 mm long, with awns 1–4 (5) mm long; *lemmas* 3–4 mm long, hairy, with a terminal awn 1.5–2 (3) mm long. Found on dry slopes and rocky hillsides, 6000–9000 ft. July–Sept. E.

***Muhlenbergia porteri* Scribn. ex Beal**, BUSH MUHLY. Perennials, 2.5–10 dm, caespitose; *ligules* 1–2.5 (4) mm, truncate, with lateral lobes; *leaves* 2–8 cm × 0.5–2 mm, flat or folded; *panicles* 4–14 × 6–15 cm, open; *spikelets* 3–4.5 mm long; *glumes* 2–3 mm, acute to acuminate, occasionally mucronate, mucros to 0.4 mm; *lemmas* 3–4.5 mm, sparsely hairy below, acuminate, awned, awns 4–13 mm, straight. Uncommon on dry slopes, mesas, and in pinyon-juniper, 4500–6500 ft. July–Sept. E.

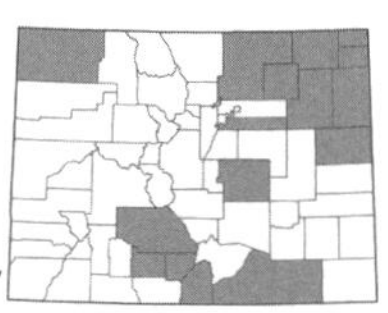

***Muhlenbergia pungens* Thurb. ex A. Gray**, SANDHILL MUHLY. Perennials, 1–7 dm, rhizomatous; *ligules* 0.2–1 mm, densely ciliate, obtuse, with lateral lobes; *leaves* 2–8 cm × 1–2.2 mm, flat to tightly involute, stiff, pungent; *panicles* (7) 8–16 (19) × (2) 4–14 cm, open; *spikelets* 2.6–4.5 mm; *glumes* 1.2–3 mm, purplish near the base, acuminate or acute, unawned or awned, awns to 1 mm; *lemmas* 2.6–4.5 mm long, purplish, scabrous on margins and above, acuminate, awned, awns 1–1.5 (2) mm, straight. Found in sandy soil, on sand hills, and in sandsage prairie, 3500–8500 ft. July–Sept. E/W.

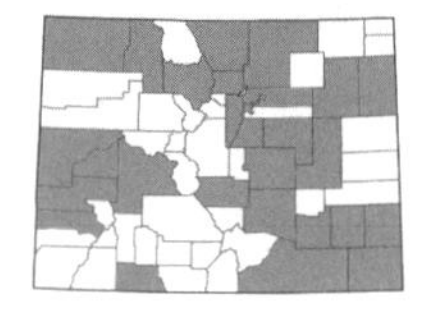

***Muhlenbergia racemosa* (Michx.) Britton, Sterns & Poggenb.**, MARSH MUHLY. Perennials, 3–11 dm, rhizomatous; *ligules* 0.6–1.5 (1.7) mm long, membranous, truncate, lacerate-ciliolate; *leaves* 2–17 cm × 2–5 mm, flat; *panicles* 0.8–16 × 0.3–1.8 cm, narrow, spike-like; *spikelets* 3–8 mm long; *glumes* 3–8 mm long, acuminate, awned, awns to 5 mm; *lemmas* 2.2–3.8 mm long, pilose below, acuminate, unawned or awned, awns to 1 mm. Common on rocky slopes, in forests, and along streams, 3500–9000 ft. July–Sept. E/W.

***Muhlenbergia ramulosa* (Kunth) Swallen**, GREEN MUHLY. [*M. wolfii* (Vasey) Rydb.]. Annuals, (0.3) 0.5–2.5 dm; *ligules* 0.2–0.5 mm, ciliate; *leaves* 0.5–3 cm × 0.8–1.2 mm, involute or flat; *panicles* (1) 2–9 × 0.6–2.7 cm, open; *spikelets* 0.8–1.3 mm; *glumes* 0.4–0.7 mm long, obtuse or subacute, unawned; *lemmas* 0.8–1.3 mm, usually glabrous, sometimes short-hairy on veins, acute, unawned. Uncommon in moist mossy mats and aspen-spruce forests, 8000–10,000 ft. July–Sept. E.

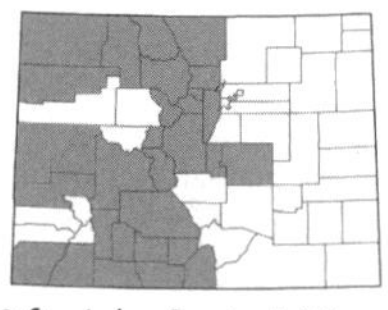

***Muhlenbergia richardsonis* (Trin.) Rydb.**, MAT MUHLY. Perennials, 0.5–3 dm; *ligules* 0.8–3 mm long, membranous, erose; *leaves* 0.4–6.5 cm × 0.5–4.2 mm, flat or involute; *panicles* 1–15 × 0.1–1.7 cm, narrow or spike-like; *spikelets* 1.7–3.1 mm long, sometimes with 2 florets; *glumes* 0.6–2 mm, acute, occasionally mucronate, mucros to 0.2 mm; *lemmas* 1.7–2.6 (3.1) mm long, glabrous with a scabrous keel, acute to acuminate, occasionally mucronate, mucros to 0.5 mm. Common on dry slopes, rocky ledges, in sagebrush and pinyon-juniper, and sometimes in moist places, 6000–10,000 ft. July–Sept. E/W.

***Muhlenbergia schreberi* J.F. Gmel.**, NIMBLEWILL. Perennials, 1–4.5 (7) dm, caespitose or stoloniferous; *ligules* 0.2–0.5 mm long, truncate, erose, ciliate; *leaves* (1) 3–10 cm × 1–4.5 mm, flat; *panicles* 3–15 × 1–1.6 cm, narrow; *spikelets* 1.8–2.8 mm long; *glumes* unequal, lower glumes lacking or rudimentary, rounded and often erose, unawned, upper glumes 0.1–0.3 mm long, unawned; *lemmas* 1.8–2.8 mm long, hairy below, acute to acuminate, awned, awns 1.5–5 mm long, straight. Uncommon weedy grass found in lawns near Boulder and Denver, 5000–5500 ft. July–Oct. E. Introduced.

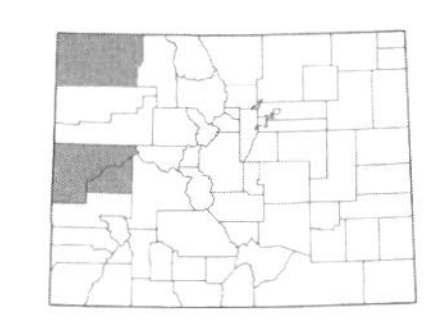

***Muhlenbergia thurberi* (Scribn.) Rydb.**, THURBER'S MUHLY. Perennials, 1.2–3.6 dm, rhizomatous; *ligules* 0.9–1.2 mm long, membranous, truncate to obtuse, erose; *leaves* 0.2–3.7 cm × 0.2–1 mm, tightly involute; *panicles* 0.7–5.5 × 0.2–0.7 cm, narrow and contracted; *spikelets* 2.6–4 mm long; *glumes* 1.6–3 mm long, acute, unawned; *lemmas* 2.6–4 mm long, pilose below and on margins, acuminate, unawned or awned, awns to 1 mm. Uncommon in pinyon-juniper and on rocky slopes, 5000–6500 ft. July–Sept. W.

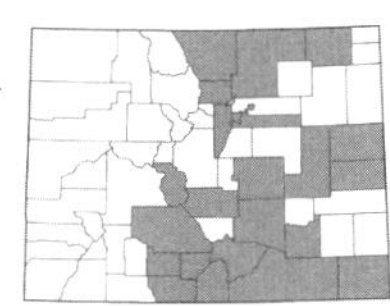

***Muhlenbergia torreyi* (Kunth) Hitchc. ex Bush**, RING MUHLY. Perennials, 1–4 (5) dm, caespitose; *ligules* 2–5 (7) mm long, acuminate, lacerate, sometimes with lateral lobes; *leaves* 1–3 (5) cm × 0.3–1 mm, tightly involute or folded; *panicles* 7–21 × 3–15 cm, open; *spikelets* 2–3.5 mm long; *glumes* 1.3–2.5 mm long, acute to acuminate, slightly erose, unawned or awned, awns to 1.1 mm; *lemmas* 2–3.2 (3.5) mm long, appressed-hairy below, acuminate, awned, awns 0.5–4 mm long. Common in sandy soil of the shortgrass prairie and in the San Luis Valley, 3500–8500 ft. June–Sept. E.

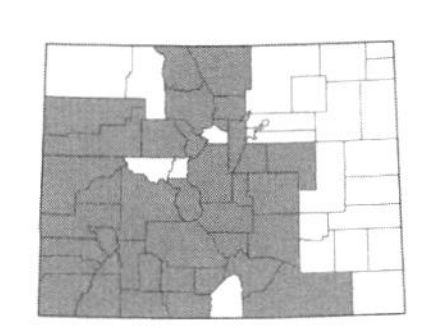

***Muhlenbergia tricholepis* (Torr.) P.M. Peterson**, PINE DROPSEED. [*Blepharoneuron tricholepis* (Torr.) Nash]. Perennials, 1–7 dm, caespitose; *ligules* (0.3) 0.7–2 (2.7) mm long, hyaline to membranous, entire; *leaves* 1–15 cm × 0.6–2.5 mm, flat to involute; *panicles* 3–25 × 1–10 cm, open; *spikelets* grayish-green; *glumes* (1.5) 1.8–2.6 (3) mm, unawned; *lemmas* (2) 2.3–3.5 (3.9) mm, densely thick hairy on the keel and nerves, acute to obtuse, sometimes mucronate. Common in grasslands, meadows, forest openings, alpine, and on dry slopes, 6400–12,000 ft. July–Sept. E/W. (Plate 64)

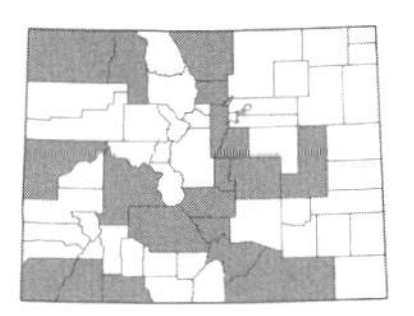

***Muhlenbergia wrightii* Vasey ex J.M. Coult.**, SPIKE MUHLY. Perennials, 1.5–6 dm, caespitose; *ligules* 1–3 (5) mm long, membranous, truncate; *leaves* 1.4–12 cm × 1–3 mm, flat to folded; *panicles* 5–16 × 0.2–1.2 cm, spike-like; *spikelets* 2–3 mm long, dark green or plumbeous; *glumes* 0.5–1.6 mm long, acute or obtuse, short-awned, awns 0.5–1 mm long; *lemmas* 2–3 mm long, appressed-hairy below, acute to acuminate, mucronate, mucros 0.3–1 mm long. Found on mesas, rocky slopes, and sandstone ledges and in ponderosa pine forests, 5000–9500 ft. July–Sept. E/W.

MUNROA Torr. – FALSE BUFFALOGRASS

Mat-forming annuals; *sheaths* with a tuft of hairs at the throat; *ligules* of hairs; *spikelets* in capitate-clustered panicles of spike-like branches, with 2–5 florets; *glumes* 1-nerved, keeled, unawned, the upper glume absent or reduced on terminal spikelet; *lemmas* awned; *disarticulation* above the glumes.

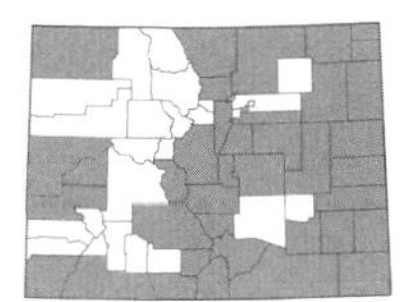

***Munroa squarrosa* (Nutt.) Torr.**, FALSE BUFFALOGRASS. Plants stoloniferous, decumbent; *leaves* usually involute, arising in pairs from each node, 1–5 cm × 1–2.5 mm; *ligules* ca. 0.5 mm; *spikelets* with lower florets pistillate or perfect and upper floret sterile and reduced; *glumes* 2.5–4.2 mm long; *lemmas* 3.5–5 mm long, with a tuft of hairs along the margins at the middle. Found in shortgrass prairie, pinyon-oak shrublands, and along sandy roadsides, 3400–8900 ft. June–Sept. E/W.

NASSELLA (Trin.) E. Desv. – NEEDLEGRASS

Perennials; *sheaths* open, a tuft of hairs usually present at junction of sheath and blade; *ligules* membranous; *spikelets* in panicles, with one floret; *glumes* longer than the floret, 3–5-nerved, acute to aristate, awned or not; *lemmas* hard, awned, the awn usually twice-geniculate, strongly convolute; *paleas* to half as long as the lemmas; *disarticulation* above the glumes.

1a. Awns 4–8 cm long; florets (1.5) 2.5–3 mm long; lemmas nearly glabrous, minutely tuberculate…***N. tenuissima***
1b. Awns 1.9–3.5 cm long; florets 3.4–5.5 mm long; lemmas sparsely hairy throughout…***N. viridula***

***Nassella tenuissima* (Trin.) Barkworth**, FINESTEM NEEDLEGRASS. Plants 3–10 dm; *leaves* involute, filiform, 7–60 cm × 0.5 mm; *ligules* 1–5 mm long; *glumes* 5–13 mm long, acuminate; *calluses* densely hairy; *lemmas* 2–3 mm long, glabrate to minutely papillose-roughened, with an awn 4–8 cm long. Rare on rocky slopes, perhaps escaped from cultivation, known from Larimer Co. June–Aug. E.

***Nassella viridula* (Trin.) Barkworth**, GREEN NEEDLEGRASS. [*Stipa viridula* Trin.]. Plants 3.5–12 dm; *leaves* flat to involute, 6–30 cm × 1–3 (6) mm; *ligules* 0.5–2 mm long; *glumes* 7–10 mm long, acuminate; *calluses* hairy; *lemmas* 4–7 mm, sparsely hairy, with an awn 1.9–3.5 cm long. Common in shortgrass prairie, on rocky slopes, and dry woods, 3500–8500 ft. June–Aug. E/W. (Plate 64)

ORYZOPSIS Michx. – RICEGRASS

Perennials; *sheaths* open, glabrous; *ligules* membranous; *spikelets* in panicles, with one floret; *glumes* 6–10-nerved, mucronate; *lemmas* 3–5-nerved, hard, the margins strongly overlapping, awned, with a dense tuft of hairs above the callus; *paleas* concealed by lemmas; *disarticulation* above the glumes.

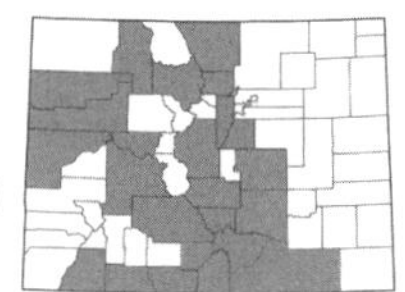

***Oryzopsis asperifolia* Michx.**, ROUGHLEAF RICEGRASS. Plants 2–7 dm; *leaves* flat or loosely involute, 30–90 cm × 4–9 mm; *ligules* 0.1–0.7 mm long; *glumes* subequal, 5–7.5 mm long; *calluses* with a dense collar of hairs; *lemmas* 5–7 mm long, hairy below, with an awn 7–15 mm long. Found along creeks, in forests, and in canyons, 5000–10,500 ft. May–Aug. E/W. (Plate 64)

PANICUM L. – PANICGRASS

Annuals or perennials; *sheaths* open; *ligules* a ring of hairs; *spikelets* in a panicle, with one lower sterile floret and one upper perfect floret; *glumes* unequal, the lowermost minute to nearly equaling the spikelets; *upper lemmas* awnless, firm and cartilaginous; *paleas* well-developed; *disarticulation* below the glumes.

1a. Leaf sheaths pubescent with papillose-based hairs (sometimes the hairs breaking off but leaving the papillose base), this easily visible with the naked eye…2
1b. Leaf sheaths glabrous or if pubescent then not densely hairy throughout or lacking papillose-based hairs…4

2a. Spikelets 4–6 mm long; sterile lemmas 4–4.8 mm long; panicles dense, often drooping at the tips when mature…***P. miliaceum***
2b. Spikelets 1.9–4 mm long; sterile lemmas 1.9–3 mm long; panicles open and diffuse…3

3a. Perennials; leaf sheaths sparsely hairy…***P. hallii* var. *hallii***
3b. Annuals; leaf sheaths densely hairy…***P. capillare***

4a. Glumes with long-acuminate tips, the lower glumes ¾ as long as the spikelets; sterile lemmas 3.3–8 mm long; plants rhizomatous…***P. virgatum***
4b. Glumes with rounded, acute, or short-acuminate tips, the lower glumes ¼–½ as long as the spikelets; sterile lemmas 1.8–3.2 mm long; plants tufted, not rhizomatous…5

5a. Lower glumes ¼–⅓ as long as the spikelets; annuals…***P. dichotomiflorum* var. *dichotomiflorum***
5b. Lower glumes about ½ as long as the spikelets; perennials…***P. hallii* var. *hallii***

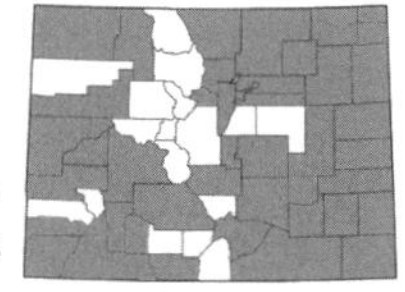

***Panicum capillare* L.**, WITCHGRASS. [*P. hillmanii* Chase]. Annuals, decumbent to erect, rooting at lower nodes, 2–13 dm; *leaves* flat or folded, 5–40 cm × 3–18 mm, the sheaths papillose-hirsute; *ligules* 0.5–1.5 mm long; *spikelets* 1.9–4 mm long, the lower floret sterile and the upper fertile; *glumes* unequal, the lower 1–2 mm and the upper 1.8–3 mm; *sterile lemmas* 1.9–3 mm long; *fertile lemmas* 1.5–2.3 mm long, indurate and shiny. Common in disturbed areas and along lake margins and roadsides, 3400–10,500 ft. June–Sept. E/W. (Plate 64)

Panicum dichotomiflorum* Michx. var. *dichotomiflorum, FALL PANICGRASS. Annuals, decumbent to erect, 5–20 dm; *leaves* flat, 10–65 cm × 3–25 mm, the sheaths glabrous to sparsely pilose; *ligules* 0.5–2 mm long; *spikelets* 2.2–3.8 mm long, the lower floret sterile and the upper fertile; *glumes* unequal, the lower ¼–⅓ as long as the spikelet, the upper 1.8–3.2 mm long; *sterile lemmas* longer than the upper floret; *fertile lemmas* 1.8–2.3 mm long, indurate and shiny. Uncommon in disturbed areas, parking lots, and along railroads, 5000–6500 ft. May–Sept. E. Introduced.

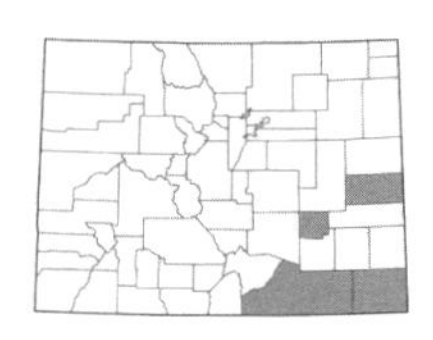

Panicum hallii* Vasey var. *hallii, HALL'S PANICGRASS. Perennials, decumbent to erect, 1.5–13 dm; *leaves* flat or folded, 5–40 cm × 3–18 mm, the sheath glabrous or papillose-hirsute; *ligules* 0.5–2 mm long; *spikelets* 2–4 mm long, the lower floret sterile and the upper fertile; *glumes* unequal, the lower 1–2.4 mm long, about ½ as long as the spikelet, the upper 1.8–3 mm long; *sterile lemmas* 1.9–3 mm long, similar to upper glumes; *fertile lemmas* 1.5–2.4 mm, indurate and shiny. Uncommon in canyons and grasslands of the eastern plains, 4000–5600 ft. May–Sept. E.

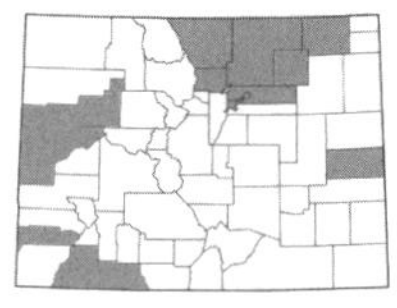

***Panicum miliaceum* L.**, BROOMCORN MILLET. Annuals, decumbent to erect, 2–12 dm; *leaves* flat, 15–40 cm × 7–25 mm, the sheath papillose-hirsute; *ligules* 1–3 mm long; *spikelets* 4–6 mm long, the lower floret sterile and the upper fertile; *glumes* unequal, the lower 2.8–3.6 mm long, ½–¾ the length of the spikelet, the upper 4–5 mm long; *sterile lemmas* 4–4.8 mm long; *fertile lemmas* 3–3.8 mm long, indurate and shiny. Cultivated and escaping along roadsides, often found in bird seed, 4000–5500 ft. June–Aug. E/W. Introduced.

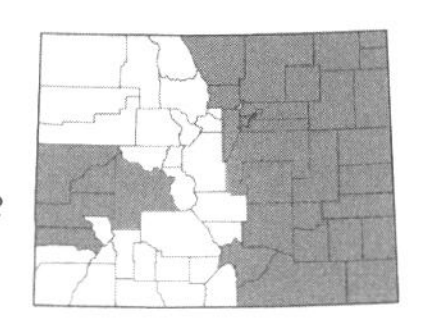

***Panicum virgatum* L.**, SWITCHGRASS. Perennials, rhizomatous, decumbent to erect, 4–30 dm; *leaves* flat, 10–60 cm × 2–15 mm, the sheath glabrous or pilose; *ligules* 2–6 mm long; *spikelets* 2.5–8 mm long, the lower floret sterile and the upper fertile; *glumes* unequal, the lower 1.8–3.2 mm long, ¾ to equal the length of the spikelet, the upper 3.3–8 mm long; *sterile lemmas* 3.3–8 mm long; *fertile lemmas* shiny and indurate. Common on the eastern plains, in sandy soil, and ponderosa pine forests, 3400–8000 ft. June–Sept. E/W. (Plate 64)

PAPPOSTIPA (Trin. & Rupr.) Romaschenko, P.M. Peterson, & Soreng – DESERT NEEDLEGRASS

Perennials; *sheaths* open; *ligules* membranous; *spikelets* in panicles, with one floret; *glumes* longer than the floret, 1–3 (5)-nerved; *lemmas* awned; *paleas* subequal to the lemmas, usually hairy; *disarticulation* above the glumes.

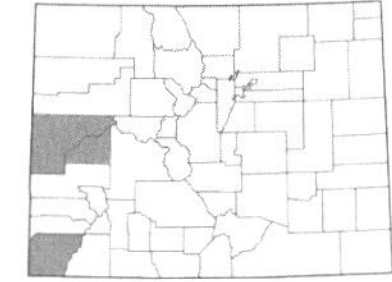

***Pappostipa speciosa* (Trin. & Rupr.) Romesch.**, DESERT NEEDLEGRASS. [*Achnatherum speciosum* (Trin. & Rupr.) Barkworth; *Jarava speciosa* (Trin. & Rupr.) Peñail; *Stipa speciosa* Trin. & Rupr.]. Plants 3–7 dm; *leaves* involute, firm, 10–40 cm long; *ligules* 0.5–2 mm long; *glumes* unequal, the lower 16–25 mm and the upper 13–20 mm long, long-acuminate; *lemmas* 7–10 mm, indurate, appressed-hairy below and glabrate above, terminating in a geniculate awn 3–5 cm long with a densely long-hairy lower segment. Uncommon on cliffs and in canyons, 5800–8500 ft. May–July. W. (Plate 64)

PASCOPYRUM Á. Löve – WHEATGRASS

Perennials, rhizomatous; *sheaths* open; *ligules* membranous; *spikelets* in a 2-sided spike with one spikelet per node or paired at lower nodes, with 2–12 florets; *glumes* narrowly lanceolate; *disarticulation* above the glumes.

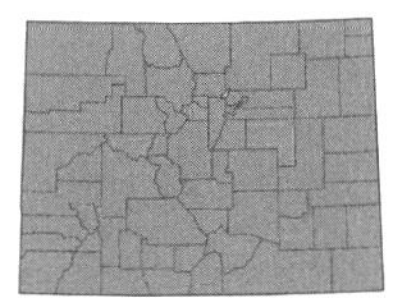

***Pascopyrum smithii* (Rydb.) Barkworth & D.R. Dewey**, WESTERN WHEATGRASS. [*Elymus smithii* (Rydb.) Gould]. Plants 2–10 dm; *leaves* flat to involute, 2–25 cm × 1–5 mm; *ligules* to 0.1 mm; *spikelets* 12–26 mm long; *glumes* 6–15 mm long, asymmetrical; *lemmas* 6–14 mm long, with an awn 0.5–5 mm long. Common along roadsides, in grasslands, and on dry slopes, 3500–9500 ft. June–Sept. E/W. (Plate 64)

PASPALUM L. – CROWNGRASS

Perennials or annuals; *sheaths* open; *ligules* membranous, *spikelets* in panicles with spike-like branches, these digitate or racemose, in 2 rows along one side of branches, with one lower sterile floret and one perfect upper floret; *glumes* unequal, the lowermost usually absent; *upper lemmas* firm and cartilaginous; *paleas* well-developed, hard; *disarticulation* below the glumes.

1a. Panicle of 40–75 unilateral spike-like branches with a widely winged rachis; lemma of the lower floret fluted or transversely wrinkled on either side of the midnerve...***P. racemosum***
1b. Panicle of 1–7 unilateral spike-like branches, the rachis not or only slightly winged; lemma of the lower floret smooth...2

2a. Spikelet margins ciliate with long pilose hairs...***P. dilatatum***
2b. Spikelet margins glabrous or with short hairs...3

3a. Spikelets elliptic to obovate, 2.8–3.6 mm long; panicle with 3–7 unilateral spike-like branches...***P. pubiflorum***
3b. Spikelets suborbicular to broadly obovate, 1.7–2.5 mm long; panicle with 1–3 unilateral spike-like branches...***P. setaceum***

***Paspalum dilatatum* Poir.**, DALLISGRASS. Perennials, 5–18 dm; *leaves* flat, to 35 cm × 2–16 mm, the sheaths glabrous to hairy; *ligules* 1.5–3.8 mm long; *panicles* of 2–7 unilateral spike-like branches; *spikelets* 2.3–4 mm long, the lower floret sterile and the upper fertile; *glumes* unequal, the lower absent and the upper 2.3–4 mm long, with long hairs along the margin; *sterile lemmas* 2.3–4 mm long, with long hairs along the margin; *fertile lemmas* indurate. Uncommon weed in lawns, known from Boulder (Boulder Co.), 5000–6000 ft. July–Sept. E. Introduced.

***Paspalum pubiflorum* Rupr.**, HAIRYSEED PASPALUM. Perennials, 3–13 dm; *leaves* flat, to 30 cm × 4–18 mm, the sheaths glabrous to hairy; *ligules* 1–3.3 mm long; *panicles* of 3–7 unilateral spike-like branches; *spikelets* 2.8–3.6 mm long, the lower floret sterile and the upper fertile; *glumes* unequal, the lower absent and the upper 2.3–4 mm long, glabrous or sparsely hairy; *sterile lemmas* 2.3–4 mm long, glabrous to sparsely hairy; *fertile lemmas* indurate. Uncommon weed in lawns, known from Boulder (Boulder Co.), 5000–6000 ft. June–Sept. E. Introduced.

***Paspalum racemosum* Lam.**, PERUVIAN PASPALUM. Annuals, 4–9 dm; *leaves* flat, 4–13 cm × 10–22 mm, sheaths glabrous; *ligules* 0.1–0.3 mm; *panicles* of 40–75 unilateral spike-like branches, rachis widely winged; *spikelets* 2.5–3 mm, the lower floret sterile and the upper fertile; *glumes* unequal, the lower absent, the upper 2.5–3 mm, the margins short-ciliate; *sterile lemmas* 2.5–3 mm, the margins short-ciliate, fluted or transversely wrinkled on either side of the midnerve; *fertile lemmas* indurate, 1.5–1.8 mm. Uncommon weed, known from the Denver area, 5000–6000 ft. July–Sept. E. Introduced.

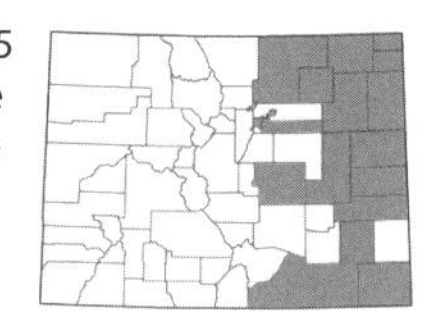

Paspalum setaceum **Michx.**, THIN PASPALUM. Perennials, prostrate to erect, 2–11 dm; *leaves* flat, 5–25 × 3–13 mm, the sheaths glabrous to hairy; *ligules* 0.2–0.5 mm; panicles of 1–3 unilateral spike-like branches; *spikelets* 1.7–2.5 mm long, sparsely hairy or sometimes glabrous, the lower floret sterile and the upper fertile; *glumes* unequal, the lower absent, the upper 1.7–2.5 mm long, glabrous to glandular-hairy; *sterile lemmas* 2.3–4 mm long, glabrous to glandular-hairy; *fertile lemmas* indurate. Found in shortgrass prairie and along sandy roadsides, 3500–4700 ft. June–Aug. E. (Plate 64)

PENNISETUM Rich. ex Pers. – FOUNTAINGRASS

Annuals or perennials; *sheaths* open; *ligules* a ring of hairs; *spikelets* in a spike-like panicle, with one lower sterile floret and one upper perfect floret, surrounded by an involucre of bristles; *glumes* unequal, the lowermost minute to nearly equaling the spikelets; *upper lemmas* awnless, firm and cartilaginous; *paleas* similar to lemmas; *disarticulation* below the glumes.

1a. Bristles conspicuously long-ciliate, the inner bristles 4–6 mm long; ligules 2–5 mm long; annuals...***P. glaucum***
1b. Bristles scabrous, not long-ciliate, the inner bristles 11–30 mm long; ligules 0.2–0.5 mm long; perennials...***P. alopecuroides***

Pennisetum alopecuroides **(L.) Spreng.**, FOXTAIL FOUNTAINGRASS. [*Panicum alopecuroides* L.]. Perennials, 3–10 dm; *leaves* flat or folded, 30–60 cm × 2–12 mm, the sheath glabrous or sometimes with ciliate margins; *ligules* 0.2–0.5 mm long; *spikelets* 5–8.5 mm long, glabrous, subtended by numerous scabrous bristles, the outer 1–15 mm and the inner 11–30 mm long; *glumes* unequal, the lower 0.2–1.4 mm and the upper 2–5 mm long (about ½ as long as the spikelet); *sterile lemmas* 5–8 mm long; *fertile lemmas* 5–7.5 mm long. A common ornamental grass occasionally persisting outside of cultivation. Introduced.

Pennisetum glaucum **(L.) R. Br.**, PEARL MILLET. [*Setaria glauca* (L.) P. Beauv.]. Annuals, 5–30 dm; *leaves* flat, 15–100 cm × 7–70 mm, the sheath glabrous or hairy, sometimes with ciliate margins; *ligules* 2–5 mm long; *spikelets* 3–7 mm long, subtended by numerous long-ciliate bristles, the outer 0.5–6 mm and the inner 4–6 mm long; *glumes* unequal, the lower absent or to 1.5 mm, the upper 0.5–3.5 mm long; *sterile lemmas* 1.5–6 mm long; *fertile lemmas* 4.3–7 mm long. Cultivated for grain and birdseed, occasionally persisting outside of cultivation. Introduced.

PHALARIS L. – CANARY GRASS

Perennials or annuals; *sheaths* open; *ligules* membranous; *spikelets* in spike-like panicles, with 1 well-developed floret and 2 sterile florets reduced to bristles; *glumes* equal, longer than the florets, keeled; *lemmas* coriaceous or hard, unawned; *paleas* similar to the lemmas; *disarticulation* above the glumes.

1a. Glumes broadly winged...2
1b. Glumes wingless or nearly so...3

2a. Glume wings irregularly toothed to crenate, rarely entire; sterile florets 1...***P. minor***
2b. Glume wings entire; sterile florets 2...***P. canariensis***

3a. Panicle evidently branched at least near the base; perennials...***P. arundinacea***
3b. Panicle dense and spike-like, not evidently branched; annuals...***P. caroliniana***

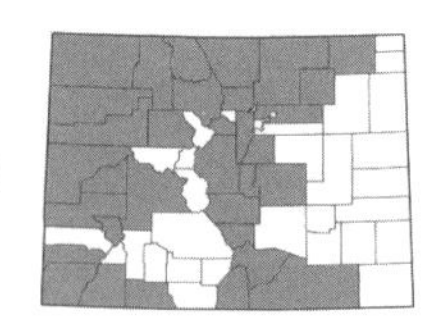

Phalaris arundinacea **L.**, REED CANARY GRASS. [*Phalaroides arundinacea* (L.) Rauschert]. Perennials, 5–20 dm; *leaves* flat, 10–30 cm × 6–15 mm; *ligules* 2–8 mm long; *panicles* compact but evidently branched near the base; *spikelets* 4–8 mm long, with a terminal fertile floret and 2 lateral reduced florets; *glumes* 4–6 mm long, wingless; *fertile lemmas* 3–4 mm long, somewhat hairy; *sterile lemmas* to 2 mm long, hairy, subulate. Common along streams, irrigation ditches, and the margins of ponds, 4500–10,000 ft. June–Aug. E/W. (Plate 64)

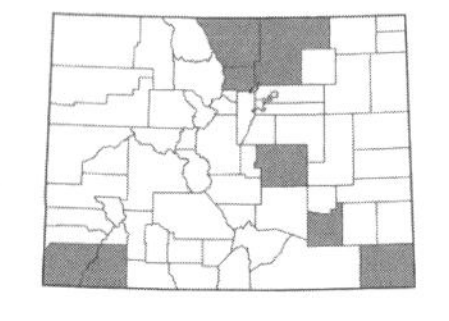

Phalaris canariensis **L.**, ANNUAL CANARYGRASS. Annuals, 3–6 dm; *leaves* flat, 4–20 cm × 4–10 mm; *ligules* 3–6 mm long; *panicles* dense and spike-like; *spikelets* 7–9 mm long, with a terminal fertile floret and 2 lateral sterile florets; *glumes* 7–9 mm long, the keel with a broad, entire wing; *fertile lemmas* 4–6 mm long, densely hairy; *sterile lemmas* linear, sparsely hairy, about half the height of the fertile lemmas. Found in commercial birdseed, occasionally found as a weed in gardens and lawns, 4000–7500 ft. June–Aug. E/W. Introduced.

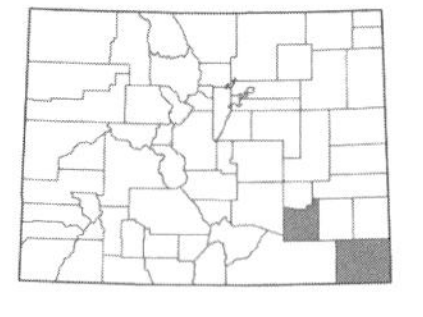

Phalaris caroliniana **Walter**, CAROLINA CANARYGRASS. Annuals, 2.5–10 (15) dm; *leaves* flat, 5–20 cm × 2–10 mm; *ligules* 2–6 mm long; *panicles* dense and spike-like; *spikelets* 4–6 mm long, with a terminal fertile floret and 2 lateral sterile florets; *glumes* 4–6 mm long, wingless; *fertile lemmas* 3–4 mm long, appressed-hairy; *sterile lemmas* 1.2–2 mm long, subulate, hairy. Uncommon in moist canyons, 4300–5000 ft. June–Aug. E.

***Phalaris minor* Retz.**, LITTLESEED CANARYGRASS. Annuals, 3–6 dm; *leaves* flat, 5–30 cm × 3–6 mm; *ligules* 2–5 mm long; *panicles* dense and spike-like; *spikelets* 4–6 mm long, with a perfect upper floret and sterile lower floret; *glumes* 4–6 mm long, the keel with a dentate to crenate wing (rarely entire); *fertile lemmas* to 3 mm long, sparsely hairy; *sterile lemmas* hairy towards the tip, about ⅓ the length of the fertile lemma. Uncommon weed, known from one record from Larimer Co., perhaps not persisting, 5000–5500 ft. June–Aug. E. Introduced.

PHIPPSIA (Trin.) R. Br. – ICEGRASS

Perennials; *sheaths* closed for ½ their length; *ligules* membranous; *spikelets* in panicles, with 1 floret, awnless; *glumes* nerveless, shorter than the florets, unawned; *lemmas* 1–3-nerved, unawned; *paleas* subequal to lemmas; *disarticulation* above the glumes.

***Phippsia algida* (Sol.) R. Br.**, ICEGRASS. Plants to 10 cm, somewhat thick and succulent; *leaves* folded with a prow-shaped tip, to 10 cm long; *ligules* 0.3–1.5 mm long, acute; *glumes* 1 or unequal, the lower short or absent, 0.05–0.3 mm long, the upper 0.3–0.6 mm long, truncate and erose at the tip; *lemmas* 1.3–1.8 mm long; *paleas* 1.1–1.3 mm long. Rare in moist alpine, found along melting snowbanks and streams, 11,500–14,000 ft. July–Sept. E/W. (Plate 64)

PHLEUM L. – TIMOTHY

Perennials or annuals; *sheaths* open; *ligules* membranous; *spikelets* in dense spike-like panicles, with 1 floret; *glumes* equal, longer than the florets, keeled, 3-nerved; *lemmas* 5–7-nerved, white, unawned; *paleas* subequal to the lemmas; *disarticulation* above the glumes.

1a. Spike-like panicles 2–6 cm long, usually 1.5–3 times as long as wide; lower internodes of culms not enlarged or bulbous; sheaths inflated...***P. alpinum***
1b. Spike-like panicles 2–15 cm long, usually 5–20 times as long as wide; lower internodes of culms often enlarged or bulbous; sheaths not inflated...***P. pratense***

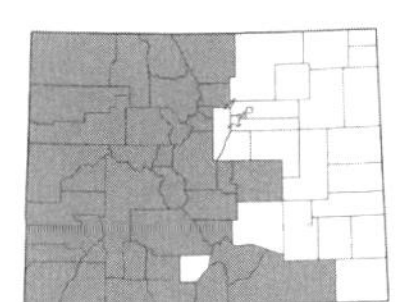

***Phleum alpinum* L.**, ALPINE TIMOTHY. [*P. commutatum* Gaudin]. Perennials, 2–6 dm; *leaves* flat, 2–10 cm × 2.5–8 mm, glabrous to scabrous, the sheaths inflated and separating from the culm; *ligules* 0.5–3 mm long, truncate; *panicles* 2–6 cm long, usually 1.5–3 times as long as wide; *glumes* 2.5–5 mm long, terminating in horn-like awns 1.5–3 mm long, pectinate-ciliate along the keel; *lemmas* 1.7–2.5 mm long, scabrous on the keel and sometimes nerves. Common in alpine and meadows, 8000–14,000 ft. June–Aug. E/W.

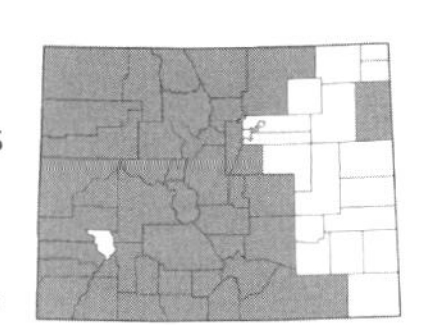

***Phleum pratense* L.**, TIMOTHY. Perennials, 4–10 dm; *leaves* flat, 5–20 cm × 4–8 mm, glabrous to scabrous, the sheaths not inflated; *ligules* 2–4 mm long; *panicles* 2–15 cm long, usually 5–20 times as long as wide; *glumes* 2.5–3.5 mm long, terminating in horn-like awns 1–2 mm long, pectinate-ciliate along the keel; *lemmas* 1.2–2 mm long, scabrous on the keel and nerves. Common in meadows, grasslands, and disturbed areas and along roadsides, 3800–11,000 ft. June–Sept. E/W. (Plate 64)

PHRAGMITES Adans. – REED

Perennials; *sheaths* open; *ligules* membranous, ciliate; *spikelets* in plumose panicles, with 2–10 florets, the rachilla with long, silky hairs; *glumes* shorter than the florets, 1–3-nerved, glabrous; *lemmas* 3-nerved, unawned.

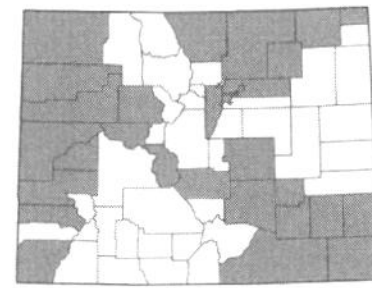

***Phragmites australis* (Cav.) Steud.**, COMMON REED. Plants 2–6 m; *leaves* flat, 15–40 × 2–4 cm, glabrous; *ligules* to 1 mm; *spikelets* 10–15 mm long, with 3–10 florets, the rachilla with hairs 6–10 mm long; *glumes* unequal, the lower 3–7 mm and the upper 5–10 mm long; *lemmas* linear-lanceolate, 8–15 mm long, glabrous, with a long-acuminate tip. Found along rivers, ditches, and lake shores, 3500–9000 ft. July–Sept. E/W. (Plate 64)

PIPTATHERUM P. Beauv. – RICEGRASS

Perennials; *sheaths* open, glabrous; *ligules* membranous; *spikelets* in panicles, with one floret; *glumes* 1–9-nerved, unawned; *lemmas* awned, 3–7-nerved; *paleas* as long or longer than the lemmas; *disarticulation* above the glumes.

1a. Lemma awns absent or 1–2 mm long...***P. pungens***
1b. Lemma awns 4–10 mm long...2

2a. Lemma awns strongly bent; lemma evenly short-hairy; spikelets 4–6 mm long; panicle 3–9 (11) cm long; ligules 1.5–4 mm long, minutely hairy and acute...***P. exiguum***
2b. Lemma awns straight; lemma usually glabrous, occasionally sparsely hairy; spikelets 3–4 mm long; panicle 10–15 cm long; ligules 0.2–2 mm long, truncate with a finely ciliate apex...***P. micranthum***

Piptatherum exiguum **(Thurb.) Dorn,** LITTLE RICEGRASS. [*Oryzopsis exigua* Thurb.]. Plants 1–5 dm; *leaves* involute, 5–10 cm × 0.5–2 mm, scabrous; *ligules* 1.5–4 mm long, acute, puberulent; *glumes* 4–5 mm long, glabrous to scabrous; *lemmas* 3–6 mm long, hard, appressed short-hairy, with a geniculate, early-deciduous awn 4–7 mm long. Uncommon in the alpine, on rocky ridge tops, and in spruce-fir forests, 8200–12,000 ft. June–Aug. E/W.

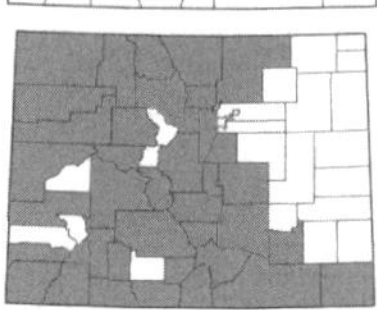

Piptatherum micranthum **(Trin. & Rupr.) Barkworth,** LITTLESEED RICEGRASS. [*Oryzopsis micrantha* (Trin. & Rupr.) Thurb.]. Plants 2–7 dm; *leaves* flat to loosely involute, 5–15 cm × 0.5–2 mm, glabrous to scabrous; *ligules* 0.2–2 mm long, truncate, finely ciliate; *glumes* 2.7–3.5 mm long, glabrous to scabrous; *lemmas* 1.8–3 mm long, glabrous to sparsely hairy, with a straight awn 5–10 mm long. Common in pinyon-juniper, canyons, and on rocky slopes, 4500–10,000 ft. June–Aug. E/W.

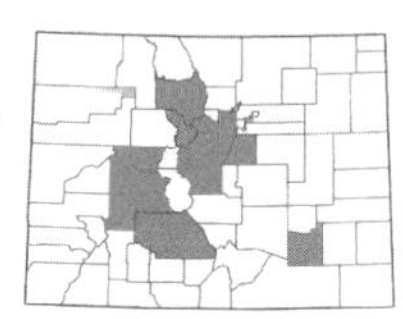

Piptatherum pungens **(Torr.) Dorn,** MOUNTAIN RICEGRASS. [*Oryzopsis pungens* (Torr.) Hitchc.]. Plants 1–9 dm; *leaves* flat to convolute, 18–45 cm × 0.5–1.8 mm, scabrous; *ligules* 0.5–2.5 mm long, acute to truncate; *glumes* 3.5–4.5 mm long, appressed-hairy; *lemmas* 2.5–5 mm long, evenly hairy, unawned or with a short awn to 2 mm long. Uncommon in pine forests, 7000–10,000 ft. June–Aug. E/W. (Plate 65)

POA L. – BLUEGRASS

Perennials or annuals; *sheaths* open to partially closed; *ligules* membranous; *spikelets* in panicles, with 2–several florets; *glumes* usually keeled, 1–3 (5)-nerved, unawned; *lemmas* usually keeled, 5-nerved, unawned; *paleas* subequal to the lemmas; *disarticulation* above the glumes.

In writing a key to the genus *Poa* one is met with several difficulties. First, it is strongly advised to feel the culm of your *Poa* in the field to verify if it is *P. compressa* (flattened stems) or not. This is very difficult to discern after your specimens have been pressed. Second, it is important to collect as much underground material as possible. The presence/absence of rhizomes is often used to discern between species.

1a. Florets mostly forming bulblets with shiny, dark purple bases and exserted, linear green tips; culms bulbous at the base...***P. bulbosa***
1b. Florets usually not forming bulblets; culms not bulbous at the base...2

2a. Panicle open (or loosely contracted), with ascending, spreading, or reflexed branches, often the lower branches long and capillary...**Key A**
2b. Panicle narrow and tightly contracted, usually with short, ascending or erect branches...**Key B**

Key A

(Panicle open with ascending, spreading, or reflexed branches)

1a. Lemmas lacking a tuft of cobwebby hairs at the base...2
1b. Lemmas with a tuft of cobwebby hairs at the base (this can be scant)...7

2a. Plants annual, densely tufted, lacking the remains of old culms...***P. annua***
2b. Plants perennial, densely tufted or not, sometimes rhizomatous, often with the remains of old culms...3

3a. Lower leaf sheaths retrorsely scabrous or softly puberulent; plants 3.5–8 dm tall, often dioecious with the florets mostly pistillate...***P. wheeleri***
3b. Lower leaf sheaths not retrorsely scabrous or softly puberulent; plants otherwise various...4

4a. Lemmas glabrous between the nerves; panicle usually congested, with erect or steeply ascending branches 0.5–4 (5) cm long...***P. cusickii***
4b. Lemmas sparsely to moderately short-hairy at least near the base between the nerves; panicle open or somewhat congested, with ascending to spreading branches 1–15 cm long...5

5a. Plants rhizomatous (although these sometimes short)...***P. arctica* ssp. *grayana***
5b. Plants lacking rhizomes...6

6a. Spikelets 6–10 mm, 3–3.5 times longer than wide; panicles 5–18 (25) cm long, usually with sparsely to densely scabrous branches...***P. stenantha* var. *stenantha***
6b. Spikelets 3.9–6.2 mm, ovate, 1.5–2.5 times longer than wide; panicles 2–6 (8) cm long, usually with smooth or sparsely scabrous branches...***P. alpina***

7a. Lemmas glabrous between the nerves...8
7b. Lemmas sparsely to densely hairy at least near the base between the nerves...13

8a. Lemma glabrous or only hairy along the keel…***P. trivialis***
8b. Lemma hairy on the keel and marginal nerves…9

9a. Lower panicle branches conspicuously reflexed at maturity; panicle branches usually single or in remote pairs, capillary…***P. reflexa***
9b. Lower panicle branches mostly spreading or ascending at maturity, rarely slightly reflexed; panicle branches various…10

10a. Plants rhizomatous; cobwebby hairs about ½ as long as the lemma; panicle branches usually in whorls of 4–5 (but ranging from 2–9)…***P. pratensis***
10b. Plants tufted or solitary, sometimes appearing rhizomatous because of a decumbent base rooting at the nodes, but not truly rhizomatous; cobwebby hairs various; panicle branches various…11

11a. Lower panicle branches mostly single or in pairs; leaves flat, lax; anthers 0.2–1.1 (1.7) mm long…***P. leptocoma***
11b. Lower panicle branches 2–9 per node; leaves flat, folded, or flat but becoming involute in dried specimens; anthers (1.1) 1.3–2.5 mm long…12

12a. Cobwebby hairs less than ½ the lemma length; ligule 0.5–1.5 (3) mm long; lower glumes 1.8–2.8 mm long; panicle 3–15 cm long…***P. interior***
12b. Cobwebby hairs mostly ⅔ the lemma length; ligule (1) 1.5–6 mm long; lower glumes mostly 1.5–1.7 mm long; panicle (9) 13–30 (41) cm long…***P. palustris***

13a. Leaf sheaths not compressed and keeled; leaves flat to involute; plants rhizomatous (although these sometimes short), of the subalpine and alpine; anthers 1.4–2.5 mm long…***P. arctica* ssp. *arctica***
13b. Leaf sheaths distinctly compressed and keeled, the keel often with wings up to 0.5 mm wide; leaves flat; plants lacking rhizomes or with short rhizomes, not exclusive to the subalpine and alpine; anthers vestigial (0.1–0.2 mm) or 0.3–3 mm long…14

14a. Plants lacking rhizomes; leaf sheaths closed for up to ½ their length, usually densely retrorsely scabrous; ligule 3–12 mm long, densely scabrous; anthers 0.3–1 mm long…***P. occidentalis***
14b. Plants shortly rhizomatous; leaf sheaths closed for ½–9/10 their length, glabrous or retrorsely scabrous; ligule 2–4.5 mm long, glabrous, scabrous, or softly hairy; anthers vestigial or (1.3) 2–3 mm long…***P. tracyi***

Key B
(Panicle narrow and contracted)

1a. Culms strongly flattened, 2-edged; plants extensively rhizomatous, usually with solitary shoots or sometimes loosely tufted…***P. compressa***
1b. Culms round or only slightly flattened; plants rhizomatous or not…2

2a. Plants annual, densely tufted, lacking the remains of old culms…***P. bigelovii***
2b. Plants perennial, densely tufted or not, often with the remains of old culms…3

3a. Glumes and lemmas not or obscurely keeled, rounded at the base; spikelets rounded, not compressed; lemmas lacking cobwebby hairs at the base; plants lacking rhizomes…***P. secunda***
3b. Glumes and lemmas distinctly keeled to the base; spikelets compressed; lemmas and plants various…4

4a. Lemmas with a conspicuous tuft of cobwebby hairs at the base ⅓–⅔ the lemma length…5
4b. Lemmas lacking a tuft of cobwebby hairs at the base, or sparsely webbed with the hairs to ¼ the lemma length…6

5a. Plants extensively rhizomatous, usually found below 11,000 ft in elevation…***P. pratensis***
5b. Plants lacking rhizomes, often found above 11,000 ft in elevation…***P. glauca* ssp. *glauca***

6a. Lemmas glabrous or scabrous, but not hairy on the keel and/or nerves…7
6b. Lemmas conspicuously hairy on the keel and/or nerves at least below…8

7a. Lemmas 2.5–3 mm long; panicle 1–3 cm long; plants 1–12 cm tall, alpine; anthers 0.2–0.8 mm…***P. lettermanii***
7b. Lemmas (3) 4–7 mm long; panicle 2–10 cm long; plants 10–60 (70) cm tall, found from the plains to the alpine; anthers vestigial (0.1–0.2 mm) or 2–3.5 mm…***P. cusickii***

8a. Plants rhizomatous (these sometimes short)…9
8b. Plants lacking rhizomes…11

9a. Lemmas usually with some tuft of cobwebby hairs at the base to ¼ the lemma length; culms 0.7–6 dm...***P. arctica* ssp. *aperta***
9b. Lemmas lacking a tuft of cobwebby hairs at the base; culms to 8 dm...10

10a. Leaves usually flat or folded, with the sheaths closed for ⅒–⅕ their length; florets usually perfect; anthers 1.3–2.2 mm long...***P. arida***
10b. Leaves usually mostly involute, with the sheaths closed for about ⅓ their length; florets mostly pistillate, the plants often dioecious; anthers usually vestigial or if present 2–3 mm long...***P. fendleriana***

11a. Plants mostly over 2 dm; florets mostly pistillate, the plants often dioecious; anthers usually vestigial or if present 2–3 mm long...***P. fendleriana***
11b. Plants 0.5–4 (8) dm; florets usually perfect; anthers 0.6–2.5 mm long...12

12a. Anthers 0.6–1.2 (1.8) mm long; plants usually not glaucous; lemmas often with a very short tuft of cobwebby hairs at the base...***P. abbreviata* ssp. *pattersonii***
12b. Anthers (1) 1.2–2.5 mm long; plants usually glaucous; lemmas lacking cobwebby hairs at the base...***P. glauca* ssp. *rupicola***

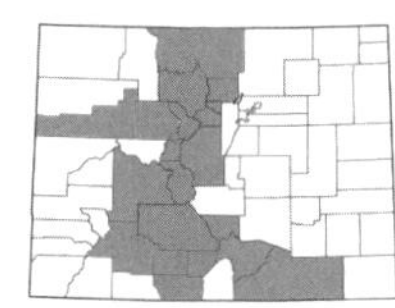

***Poa abbreviata* R. Br. ssp. *pattersonii* (Vasey) Á. Löve & D. Löve & B.M. Kapoor**, PATTERSON'S BLUEGRASS. [*P. pattersonii* Vasey]. Perennials, not rhizomatous, 0.5–1.5 (2) dm; *leaves* 0.8–1.5 (2) mm wide, involute; *ligules* 0.8–5.5 mm long; *panicles* 1.5–5 cm long, contracted and congested, the nodes with 1–3 branches, branches erect; *spikelets* 4–6.5 mm long, with 2–5 florets; *lemmas* 3–4.6 mm long, keeled, marginal veins usually villous, intercostal regions glabrous or short-hairy, base often shortly cobwebby; *anthers* 0.2–1.2 (1.8) mm long. Common in the alpine, 11,500–14,000 ft. July–Aug. E/W.

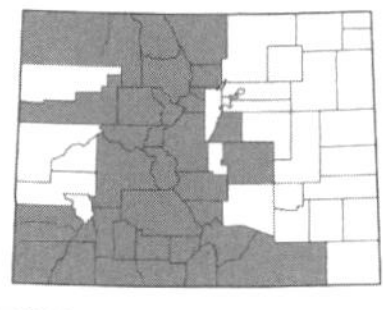

***Poa alpina* L.**, ALPINE BLUEGRASS. Perennials, not rhizomatous, 1–4 dm; *leaves* 2–4.5 mm wide, flat; *ligules* 1–4 (5) mm long; *panicles* 2–6 (8) cm long, open or loosely contracted, somewhat congested, the nodes with 1–2 branches, branches 1–3 (4) cm long, ascending to spreading; *spikelets* 3.9–6.2 mm long, plump, with 3–7 florets; *lemmas* 3–5 mm long, keeled, marginal veins villous, intercostal regions short-villous, not cobwebby at the base; *anthers* 1.3–2.3 mm long. Common in mountain meadows and the alpine, often in moist places, 8500–14,000 ft. June–Aug. E/W.

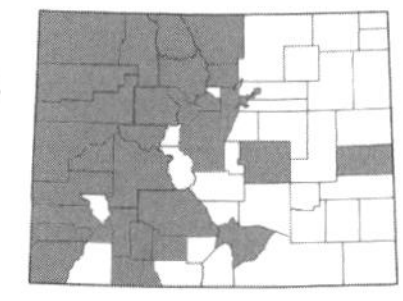

***Poa annua* L.**, ANNUAL BLUEGRASS. Annuals, not rhizomatous, occasionally stoloniferous, 0.2–2 (4.5) dm; *leaves* 1–3 (6) mm wide, flat or weakly folded; *ligules* 0.5–3 (5) mm; *panicles* 1–7 (10) cm, erect, the nodes with 1–2 (3) branches; *spikelets* 3–5 mm, with 2–6 florets; *lemmas* 2.5–4 mm, keeled, marginal and lateral veins minutely hairy to long-villous, rarely glabrous, intercostal regions glabrous, cobwebby at the base; *anthers* 0.6–1.1 mm. Found in lawns and gardens, disturbed areas, forests, and meadows, 4200–11,000 ft. June–Sept. E/W. Introduced.

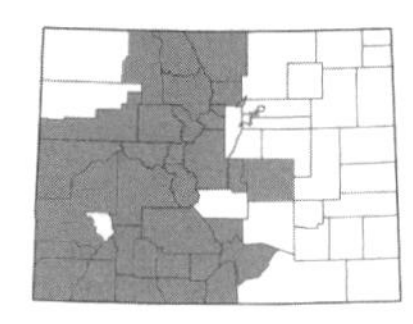

***Poa arctica* R. Br.**, ARCTIC BLUEGRASS. Perennials, rhizomatous, 0.7–6 dm; *leaves* 1–6 mm wide, flat or folded, somewhat involute; *ligules* (1) 2–7 mm long; *panicles* (2) 3.5–15 cm long, open and sparse, nodes with (1) 2–5 branches; *spikelets* (3.5) 4.5–8 mm long, with (2) 3–6 florets; *lemmas* 2.7–6 (7) mm long, keeled, marginal veins villous, intercostal regions hairy or glabrous, base cobwebby or not; *anthers* 1.4–2.5 mm long. Common in alpine and subalpine meadows and on scree slopes, 9500–14,000 ft. June–Aug. E/W.

There are three subspecies of *P. arctica* in Colorado:

1a. Panicle narrow and contracted, the branches erect, the longest branches of the lowest nodes ¼–½ the length of the panicles...**ssp. *aperta* (Scribn. & Merr.) Soreng**
1b. Panicle open, the branches ascending or widely spreading, the longest branches of the lowest panicle nodes ⅖–½ the length of the panicles...2

2a. Lemmas lacking cobwebby bases...**ssp. *grayana* (Vasey) Á. Löve, D. Löve & B.M. Kapoor**
2b. Lemmas with cobwebby bases...**ssp. *arctica***

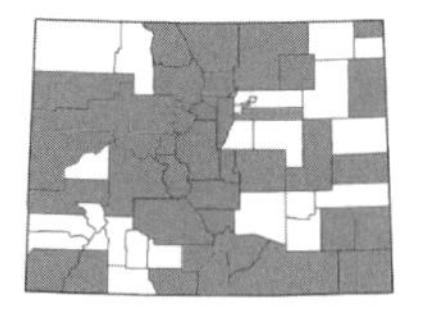

***Poa arida* Vasey**, PLAINS BLUEGRASS. [*P. glaucifolia* Scribn. & T.A. Williams]. Perennials, rhizomatous, 1.5–8 dm; *leaves* 1.5–5 mm wide, flat to folded; *ligules* (1) 1.5–4 (5) mm long; *panicles* (2.5) 4–12 (18) cm long, contracted and congested, nodes with 1–5 branches; *spikelets* 3.2–7 mm long, with 2–7 florets; *lemmas* 2.5–4.5 mm long, marginal veins villous, lateral veins glabrous to minutely hairy, intercostal regions glabrous, base cobwebby; *anthers* 1.3–2.2 mm long. Found on alkaline or salt flats, in moist meadows, and along pond margins, 3500–9500 ft. June–Aug. E/W.

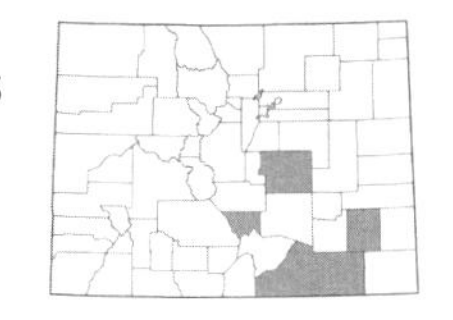

***Poa bigelovii* Vasey & Scribn.**, Bigelow's bluegrass. Annual to short-lived perennials, not rhizomatous, (0.2) 0.5–6 (7) dm; *leaves* 1.5–5 mm wide, flat; *ligules* 2–6 mm long; *panicles* (1) 5–15 cm long, contracted and congested, the nodes with 2–3 (5) branches; *spikelets* 4–7 mm long, with 3–7 florets; *lemmas* 2.6–4.2 mm long, keeled, marginal and often lateral veins short to long-villous, intercostal regions glabrous or softly minutely hairy, cobwebby at the base; *anthers* 0.2–1 mm long. Uncommon on dry slopes, 4000–6000 ft. May–July. E.

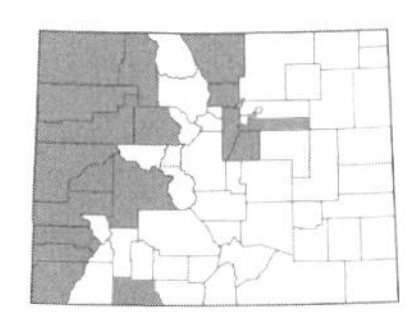

***Poa bulbosa* L.**, bulbous bluegrass. Perennials, not rhizomatous, 1.5–6 dm; *leaves* 1–2.5 mm wide, flat, soon withering; *ligules* 1–3 mm long; *panicles* 3–12 cm long, the nodes with 2–5 branches; *spikelets* 3–5 mm long, with 3–7 florets (mostly forming bulblets with shiny, dark purple bases and exserted, linear green tips); *lemmas* 3–4 mm long, keeled, marginal veins glabrous or short to long-villous, intercostal regions glabrous or softly puberulent; *anthers* 1.2–1.5 mm long. Found along roadsides and in disturbed areas, 4600–9000 ft. April–July. E/W. Introduced.

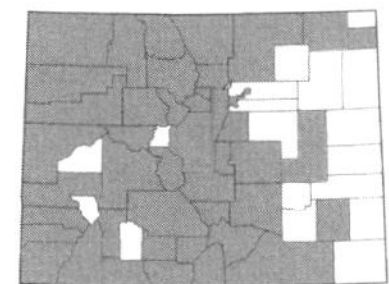

***Poa compressa* L.**, Canada bluegrass. Perennials, extensively rhizomatous, 1.5–6 dm, culms strongly flattened; *leaves* 1.5–4 mm wide, flat; *ligules* 1–3 mm long; *panicles* 2–10 cm long, sparse to congested, the nodes with 1–3 branches; *spikelets* (2.3) 3.5–7 mm long, with 3–7 florets; *lemmas* 2.3–3.5 mm long, keeled, marginal veins short-villous, intercostal regions glabrous, not cobwebby at the base; *anthers* 1.3–1.8 mm long. Found in grasslands, forests, meadows, and along streams, 5000–10,500 ft. May–Aug. E/W. Introduced.

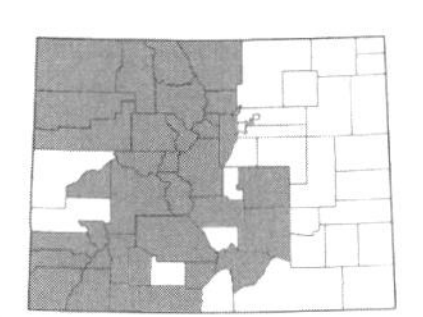

***Poa cusickii* Vasey**, Cusick's bluegrass. [*P. epilis* Scribn.]. Perennials, not or shortly rhizomatous, 1–6 (7) cm; *leaves* 0.5–3 mm wide, flat, folded, or involute; *ligules* 1–3 (6) mm long; *panicles* 2–10 (12) cm long, contracted to loosely contracted, congested, the nodes with 1–3 (5) branches; *spikelets* (3) 4–10 mm long, with 2–6 florets; *lemmas* (3) 4–7 mm long, keeled, marginal veins glabrous to proximally minutely hairy, intercostal regions glabrous, not cobwebby at the base; *anthers* 2–3.5 mm long, vestigial anthers 0.1–0.2 mm long. Common on dry slopes and in grasslands, meadows, and the alpine, 5800–13,500 ft. E/W.

There are two subspecies of *P. cusickii* in Colorado:

1a. Panicle branches smooth or slightly scabrous, stem leaves over 1.5 mm wide...**ssp. *epilis* (Scribn.) W.A. Weber**
1b. Panicle branches densely scabrous; stem leaves usually less than 1.5 mm wide...**ssp. *pallida* Soreng**

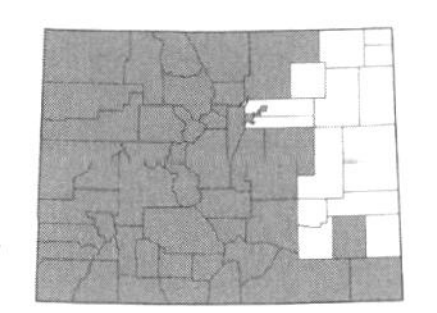

***Poa fendleriana* (Steud.) Vasey**, muttongrass. Perennials, shortly rhizomatous (these often inconspicuous), 1.5–7 dm; *leaves* (0.5) 1–3 (4) mm wide, involute; *ligules* 0.2–18 mm long; *panicles* 2–12 (30) cm long, contracted and congested, the nodes with 1–2 branches; *spikelets* (3) 4–8 (12) mm long, with 2–7 (13) florets; *lemmas* 3–6 mm long, keeled, marginal and lateral veins glabrous, short to long-villous, intercostal regions softly puberulent or glabrous, not cobwebby at the base; *anthers* 2–3 mm long, vestigial anthers 0.1–0.2 mm long. Common in grasslands, meadows, forests, the alpine, and on dry slopes, 4000–14,000 ft.

There are two subspecies of *P. fendleriana* in Colorado:

1a. Ligules of middle stem leaves 0.1–1.2 (1.5) mm long, not decurrent, usually scabrous...**ssp. *fendleriana***
1b. Ligules of middle stem leaves (1.5) 1.8–18 mm long, decurrent, usually smooth or sometimes sparsely scabrous...**ssp. *longiligula* (Scribn. & Williams) Soreng**

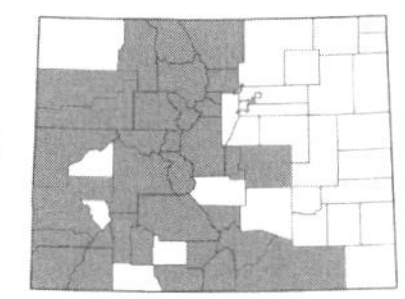

***Poa glauca* Vahl,** glaucous bluegrass. Perennials, not rhizomatous, 0.5–4 (8) dm; *leaves* 0.8–2.5 mm wide, flat or folded; *ligules* 1–4 (5) mm long; *panicles* 1–10 (20) cm long, contracted to somewhat open, sparse, the nodes with 2–3 (5) branches; *spikelets* 3–7 (9) mm long, with 2–5 florets; *lemmas* 2.5–4 mm long, keeled, marginal veins short-villous, lateral veins sparsely softly puberulent to short-villous, intercostal regions glabrous or minutely hairy, cobwebby at the base or not; *anthers* (1) 1.2–2.5 mm long.

There are two subspecies of *P. glauca* in Colorado:

1a. Lemmas with cobwebby hairs at the base...**ssp. *glauca*** Found in the alpine, 11,000–13,500 ft. July–Aug. E/W.
1b. Lemmas lacking cobwebby hairs at the base...**ssp. *rupicola* (Nash) W.A. Weber**. Common in mountain meadows and the alpine, 9500–14,000 ft. June–Aug. E/W.

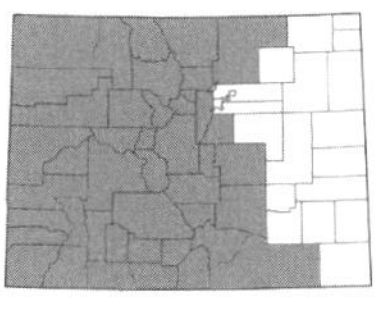

***Poa interior* Rydb.**, inland bluegrass. [*P. nemoralis* L. ssp. *interior* (Rydb.) W.A. Weber]. Perennials, not rhizomatous, 0.5–8 dm; *leaves* 0.8–3 mm wide, flat but aging involute; *ligules* 0.5–1.5 (3) mm long; *panicles* (1.5) 3–17 cm long, sparsely to moderately congested, the nodes with 2–5 branches; *spikelets* 3–6 mm long, with 2–3 (5) florets; *lemmas* 2.4–4 mm long, keeled, marginal veins short-villous, lateral veins glabrous to sparsely minutely hairy, intercostal regions glabrous, cobwebby at the base; *anthers* (1.1) 1.3–2.5 mm long. Common in forests, on rocky slopes, and in meadows, 5500–12,500 ft. June–Aug. E/W.

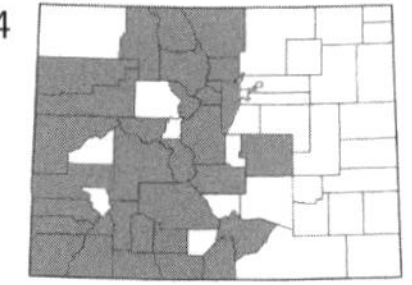

***Poa leptocoma* Trin.**, MARSH BLUEGRASS. Perennials, not to shortly rhizomatous, 1.5–10 dm; *leaves* 1–4 mm wide, flat; *ligules* 1.5–4 (6) mm long; *panicles* 5–15 cm long, open and sparse, the nodes with 1–3 (5) branches; *spikelets* 4–8 mm, with 2–5 florets; *lemmas* 3–4 mm long, keeled, marginal veins softly puberulent to long-villous, lateral veins and intercostal regions glabrous, cobwebby at the base; *anthers* 0.2–1.1 mm. Common in moist places and forests, 8500–13,500 ft. May–Aug. E/W.

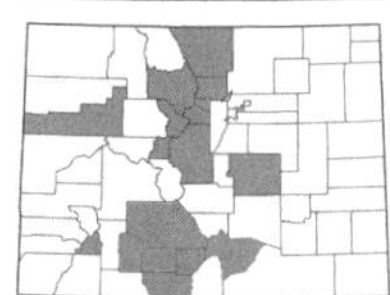

***Poa lettermanii* Vasey**, LETTERMAN'S BLUEGRASS. Perennials, not rhizomatous, 0.1–1.2 dm; *leaves* 0.5–2 mm wide, flat or folded; *ligules* 1–3 mm long; *panicles* 1–3 cm long, contracted; *spikelets* 3–4 mm long, with 2–3 florets; *lemmas* 2.5–3 mm, marginal veins glabrous or sparsely hairy, not cobwebby at the base; *anthers* 0.2–0.8 mm. Found in alpine meadows, 11,000–14,000 ft. July–Aug. E/W.

***Poa occidentalis* Vasey**, NEW MEXICO BLUEGRASS. Perennials, not rhizomatous, 2–11 dm; *leaves* (1.2) 1.5–6 (10) mm wide, flat; *ligules* 3–12 mm long; *panicles* (6) 12–40 cm long, open, the nodes with 2–7 branches; *spikelets* (3) 4–7 (8) mm long, with 3–7 florets; *lemmas* 2.6–4.2 mm, keeled, marginal veins short to long-villous, lateral veins and intercostal regions sparsely minutely hairy, cobwebby at the base; *anthers* 0.3–1 mm. Found in aspen forests, 8000–10,500 ft. June–Aug. E/W.

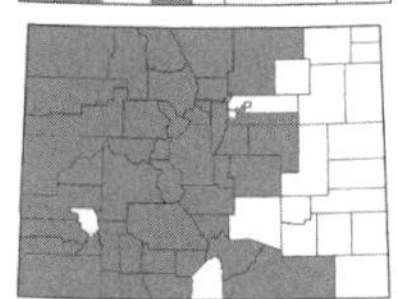

***Poa palustris* L.**, FOWL BLUEGRASS. Perennials, often stoloniferous, 2.5–12 dm; *leaves* 1.5–8 mm wide, flat; *ligules* (1) 1.5–6 mm long; *panicles* (9) 13–30 (41) cm long, open, the nodes with 2–9 branches; *spikelets* 3–5 mm long, with 2–5 florets; *lemmas* 2–3 mm long, keeled, marginal veins short-villous, intercostal regions glabrous, cobwebby at the base; *anthers* 1.3–1.8 mm long. Common in moist places, grasslands, and meadows, 4800–11,500 ft. June–Sept. E/W. (Plate 65)

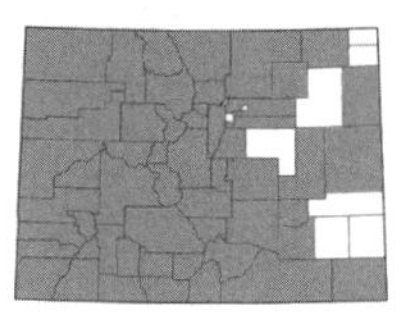

***Poa pratensis* L.**, KENTUCKY BLUEGRASS. [*P. agassizensis* B. Boivin & D. Löve]. Perennials, rhizomatous, 0.5–7 (10) dm; *leaves* 0.4–4.5 mm wide, flat, folded, or involute; *ligules* 0.9–2 (3.1) mm long; *panicles* 2–15 (20) cm long, loosely contracted to open, sparsely to moderately congested, the nodes with (1) 2–7 (9) branches; *spikelets* 3.5–6 (7) mm long, laterally compressed, with 2–5 florets; *lemmas* 2–4.3 (6) mm long, keeled, marginal veins long-villous, lateral veins glabrous, intercostal regions glabrous, cobwebby at the base; *anthers* 1.2–2 mm long. Common in lawns, moist places, grasslands, meadows, and forests, 3500–11,500 ft. May–Sept. E/W. Introduced.

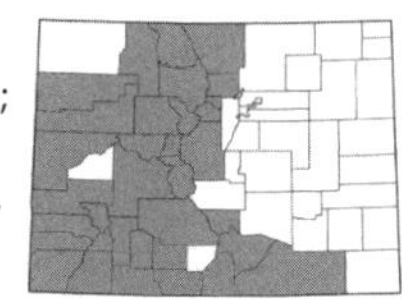

***Poa reflexa* Vasey & Scribn.**, NODDING BLUEGRASS. Perennials, not rhizomatous, 1–6 dm; *leaves* 1.5–4 mm wide, flat; *ligules* 1.5–3.5 mm long; *panicles* 4–15 cm long, open, the nodes with 1–2 branches; *spikelets* 4–6 mm long, with 3–5 florets; *lemmas* 2–3.5 mm long, keeled, marginal veins short to long-villous, lateral veins sparsely minutely hairy at least on one side, intercostal regions glabrous, cobwebby at the base; *anthers* 0.6–1 mm long. Common in meadows, subalpine forests, and the alpine, usually in dryer sites than *P. leptocoma*, 7500–13,500 ft. May–Aug. E/W.

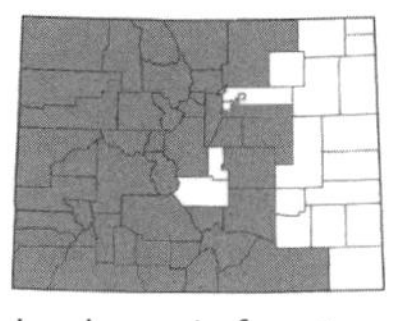

***Poa secunda* J. Presl**, SANDBERG BLUEGRASS. [*P. ampla* Merr.; *P. canbyi* (Scribn.) Howell.; *P. nevadensis* Vasey ex Scribn.; *P. sandbergii* Vasey; *P. scabrella* (Thurb.) Benth. ex Vasey]. Perennials, rarely rhizomatous, (1) 1.5–12 dm; *leaves* 0.4–3 (5) mm wide, flat, folded, or involute; *ligules* 0.5–6 (10) mm long; *panicles* 2–25 (30) cm long, contracted and moderately congested, the nodes with 1–3 branches; *spikelets* (4) 5–10 mm long, subterete to weakly laterally compressed, with (2) 3–5 (10) florets; *lemmas* 3.5–6 mm long, weakly keeled, marginal veins and intercostal regions glabrous to short-villous or softly puberulent, not cobwebby at the base; *anthers* 1.5–3 mm long. Common on dry slopes, in forests and meadows, and in the alpine, 5000–14,000 ft. May–Aug. E/W.

There are two subspecies of *P. secunda* in Colorado:

1a. Lemmas glabrous or scabrous...**ssp. *juncifolia* (Scribn.) Bettle**
1b. Lemmas minutely hairy towards the base...**ssp. *secunda***

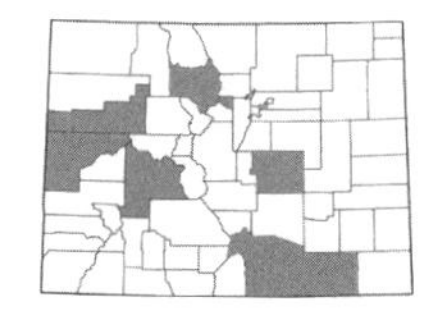

Poa stenantha* Trin. var. *stenantha, NORTHERN BLUEGRASS. Perennials, not rhizomatous, 2–6 (10) dm; *leaves* 1.5–4 (5) mm wide, flat or folded; *ligules* 2–5 mm; *panicles* 5–18 (25) cm, loosely contracted to open, sparse, the nodes with 2 (7) branches; *spikelets* 6–10 mm, laterally compressed, with 3–4 (7) florets; *lemmas* 4–6 mm, keeled, marginal and occasionally lateral veins short to long-villous, intercostal regions glabrous, sparsely minutely hairy, or hispidulous below, not cobwebby at the base; *anthers* 1.2–2 mm. Uncommon in forests and meadows, 8500–12,000 ft. June–Aug. E/W.

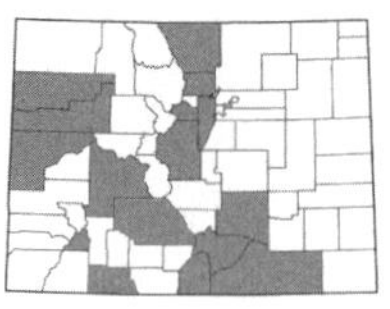

***Poa tracyi* Vasey**, TRACY'S BLUEGRASS. Perennials, shortly rhizomatous, 3–12.5 dm; *leaves* 1.5–5.5 mm wide, flat; *ligules* 2–4.5 mm; *panicles* (8) 13–29 cm, open, nodes with (1) 2–5 branches; *spikelets* 3–8 mm, laterally compressed, with 2–8 florets; *lemmas* 2.6–5 mm, marginal veins villous, intercostal regions sparsely hairy, base cobwebby; *anthers* (1.3) 2–3 mm, vestigial anthers 0.1–0.2 mm long. Found in conifer and aspen forests, meadows, and in moist areas, 5500–11,500 ft. May–Aug. E/W.

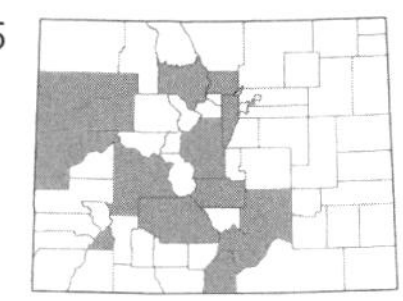

***Poa trivialis* L.**, ROUGH BLUEGRASS. Short-lived perennials, weakly stoloniferous, 2.5–12 dm; *leaves* 1–5 mm wide, flat; *ligules* 3–10 mm long; *panicles* 8–25 cm long, open, the nodes with 3–7 branches; *spikelets* 3–4.5 (5) mm long, laterally compressed, with 2–4 florets; *lemmas* 2.3–3.5 mm long, keeled, marginal veins and intercostal regions glabrous, cobwebby at the base; *anthers* 1.3–2 mm long. Uncommon in moist places, 3500–7000 (11,000) ft. May–July. E/W. Introduced.

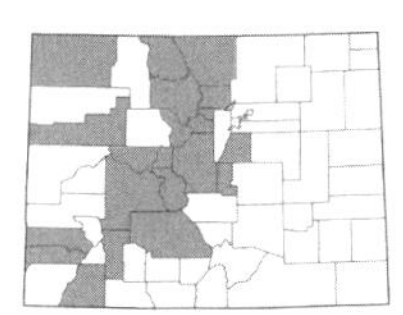

***Poa wheeleri* Vasey**, WHEELER'S BLUEGRASS. [*P. nervosa* (Hook.) Vasey]. Perennials, shortly rhizomatous, 3.5–8 dm; *leaves* 2–3.5 mm wide, flat or folded; *ligules* 0.5–2 mm long; *panicles* 5–12 (18) cm long, loosely contracted to open, the nodes with 2–5 branches; *spikelets* 5.5–10 mm long, laterally compressed, with 2–7 florets; *lemmas* 3–6 mm long, keeled, marginal veins glabrous or softly puberulent to short-villous, intercostal regions glabrous to hispidulous, not cobwebby at the base; *anthers* usually vestigial, 0.1–0.2 mm long, or up to 2 mm long. Common in forests, meadows, and the alpine, 6500–13,000 ft. June–Aug. E/W.

PODAGROSTIS (Briseb.) Scrib. & Merr. – BENTGRASS

Perennials; *sheaths* open; *ligules* membranous; *spikelets* in panicles, with one floret; *glumes* unawned; *lemmas* 5-nerved, awned or not; *disarticulation* above the glumes.

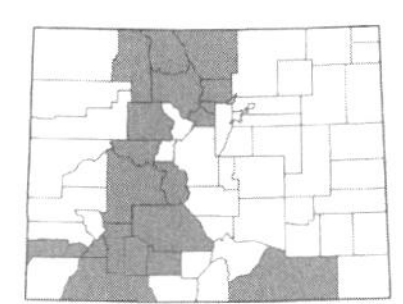

***Podagrostis humilis* (Vasey) Bjorkman**, ALPINE BENTGRASS. [*Agrostis humilis* Vasey; *P. thurberiana* (Hitchc.) Hultén]. Plants 0.5–6 dm; *leaves* flat or involute, 1.5–5 cm × 0.4–1.5 mm; *ligules* 0.2–4 mm long; *glumes* 1.6–2.3 mm long, usually purple, 1-nerved; *lemmas* 1.5–2.5 mm long, 5-nerved, unawned or with an awn to 1.3 mm long attached subapically or near the middle. Found in moist meadows, spruce-fir forests, and the alpine, 8600–12,000 ft. July–Sept. E/W. (Plate 65)

POLYPOGON Desf. – BEARDGRASS

Perennials or annuals; *sheaths* open; *ligules* membranous; *spikelets* tiny, in dense spike-like panicles, with one floret; *glumes* longer than the floret, awned; *lemmas* usually awned; *disarticulation* at the base of the stipes.

1a. Glumes with awns 4–10 mm long, the awns about 2–4 times the length of the glume; entire panicle dense and spike-like (the branches not discernible), soft and resembling a furry rabbit's foot; ligule 2.5–16 mm long; annuals...***P. monspeliensis***
1b. Glumes awnless or the awns 1–3.2 mm long; entire panicle not appearing soft and furry, in a compact and spike-like or interrupted panicle with evident branches; ligule 1–6 mm long; perennials...2

2a. Glumes and lemmas awned...***P. interruptus***
2b. Glumes and lemmas awnless...***P. viridis***

***Polypogon interruptus* Kunth**, DITCH BEARDGRASS. Perennials, 2–8 dm, usually rooting at lower nodes; *leaves* flat, 2.5–10 cm × 3–6 mm, scabrous; *ligules* 2–6 mm long; *inflorescence* a contracted to open panicle, 3–15 cm long; *glumes* 2–3 mm long, scabrous on the keel, with an awn 1.5–3.2 mm long; *lemmas* 0.8–1.5 mm long, glabrous, with an awn 1–3.2 mm long. Rare along streams, in marshy meadows, and other moist places, 4900–6000 ft. June–Aug. E/W.

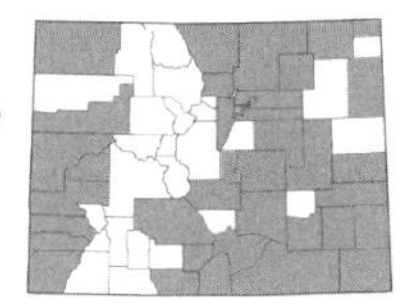

***Polypogon monspeliensis* (L.) Desf.**, ANNUAL RABBITSFOOT GRASS. Annuals, 0.5–6.5 dm, rooting at the lower nodes; *leaves* flat, 1–20 cm × 1–7 mm, glabrous to scabrous; *ligules* 2.5–16 mm; *inflorescence* a dense, spike-like panicle 1–17 cm long; *glumes* 1–2.7 mm, scabrous on the keel, with awns 4–10 mm long; *lemmas* 0.5–1.5 mm, unawned or with a short awn to 1 mm long. Common in moist places, ditches, and along pond margins, 3500–7600 ft. June–Sept. E/W. Introduced. (Plate 65)

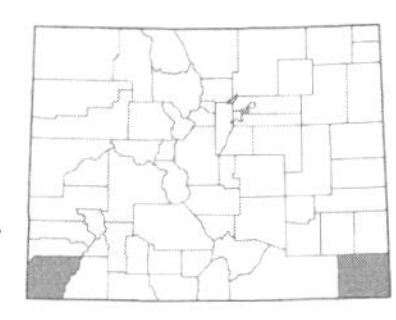

***Polypogon viridis* (Gouan) Briestr.**, WATER BEARDGRASS. Perennials, 1–9 dm, often rooting at the lower nodes; *leaves* flat, 2–13 cm × 1–6 mm; *ligules* 1–5 mm; *inflorescence* a compact to open panicle 2–10 cm long; *glumes* 1–2.2 mm long, scabrous on the keel, unawned; *lemmas* ca. 1 mm, unawned. Rare in moist places such as along streams, 4500–6000 ft. June–Aug. E/W. Introduced.

PSATHYROSTACHYS Nevski – WILDRYE

Perennials; *sheaths* closed; *ligules* membranous; *spikelets* in spikes, sessile, 2–3 per node, with 1–2 (3) florets; *glumes* subulate; *lemmas* 5–7-nerved; *disarticulation* at the rachis nodes above spikelets.

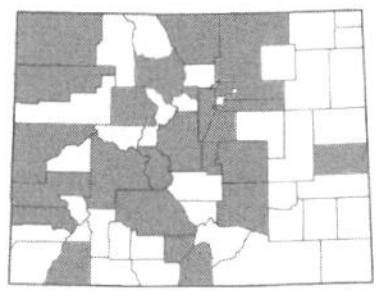

***Psathyrostachys juncea* (Fisch.) Nevski**, RUSSIAN WILDRYE. [*Elymus junceus* Fisch.]. Plants 3–12 dm; *leaves* flat to involute, 2.5–18 × 0.5–2 cm; *ligules* 0.2–0.3 mm long; *spikelets* 7–10 mm long; *glumes* 4–9 mm long, short-hairy or scabrous, acute or with a short awn-tip; *lemmas* 5–8 mm long, acute or with an awn 0.8–3.5 mm long. Found in disturbed areas, fields, and along roadsides, planted for revegetation use and as a pasture grass, 5000–9500 ft. May–Aug. E/W. Introduced. (Plate 65)

PSEUDOROEGNERIA (Nevski) Á. Löve – WHEATGRASS

Perennials; *sheaths* open; *ligules* membranous; *spikelets* in spikes, with one spikelet per node, with 4–9 florets; *glumes* 4–5-veined, unawned; *lemmas* awned or not; *disarticulation* above the glumes.

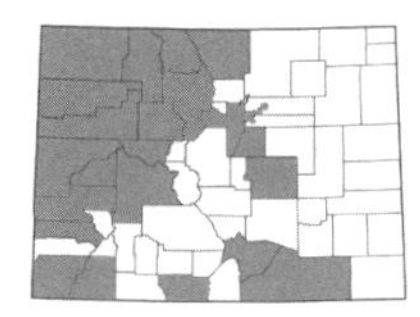

Pseudoroegneria spicata **(Pursh) Á. Löve**, BLUEBUNCH. [*Agropyron spicatum* Pursh]. Plants 3–10 dm; *leaves* flat, 5–25 cm × 2–4.5 mm; *ligules* 0.5–2 mm long; *spikelets* 8–22 mm long; *glumes* 4–13 mm long, about ½ the length of the spikelet; *lemmas* 7–15 mm long, glabrous, acute or with a strongly divergent awn 10–20 mm long. Common in grasslands, pinyon-juniper, sagebrush, and on rocky slopes and outcrops, 5200–10,000 ft. May–Aug. E/W. (Plate 65)

PTILAGROSTIS Griseb. – FALSE NEEDLEGRASS

Perennials; *sheaths* open, glabrous; *ligules* membranous; *spikelets* in open panicles, with one floret; *glumes* acute; *lemmas* awned from the apex; *paleas* shorter than to longer than the lemmas, hairy; *disarticulation* above the glumes.

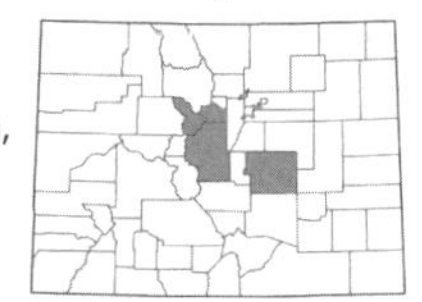

Ptilagrostis porteri **(Rydb.) W.A. Weber**, PORTER'S FALSE NEEDLEGRASS. [*Stipa porteri* Rydb.]. Plants 2–5 dm; *leaves* filiform, 2–12 cm × 0.3–0.5 mm; *ligules* 1.5–3 mm long; *glumes* 4.5–6 mm long, purplish, glabrous below and hairy above; *lemmas* 2.5–4 mm long, appressed-hairy, with an awn 5–25 mm long. Rare in moist meadows, fens, peat hummocks, and willow carrs, 9000–12,000 ft. July–Aug. E/W. (Plate 65)

PUCCINELLIA Parl. – ALKALIGRASS

Perennials or annuals; *sheaths* open; *ligules* membranous; *spikelets* in panicles, with 2–10 florets; *glumes* rounded or weakly keeled, with parallel nerves; *lemmas* with obtuse to truncate tips, 5-nerved, the nerves parallel; *paleas* subequal to the lemmas, scarious; *disarticulation* above the glumes.

1a. Lemmas densely hairy along the lower ½–¾ of the veins, the tips smooth or with a few scattered minute roughenings; annuals...***P. parishii***
1b. Lemmas glabrous or sparsely hairy on the lower ½ of the veins, the tips uniformly and densely roughened; perennials...2

2a. Lemmas 1.4–2 (2.2) mm long, the tips widely obtuse to truncate; lower panicle branches reflexed at maturity...***P. distans***
2b. Lemmas (2) 2.2–3.5 mm long, the tips acute to obtuse; lower panicle branches erect to diverging, sometimes reflexed at maturity...***P. nuttalliana***

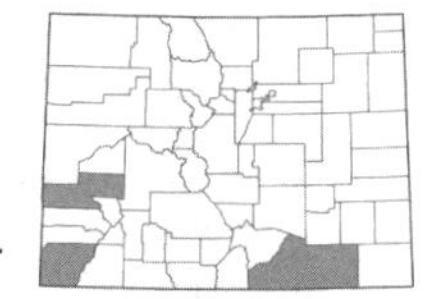

Puccinellia distans **(L.) Parl.**, WEEPING ALKALI GRASS. Perennials, 1–5 dm; *leaves* flat to involute, 2–10 cm × 1–3 mm; *ligules* 0.8–1.2 mm long; *spikelets* 3–7 mm long, with 2–7 florets; *glumes* unequal, the lower 0.5–1.5 mm and the upper 1–2 mm long; *lemmas* 1.4–2.2 mm long, glabrous, truncate or rounded. Uncommon along pond margins and in other moist places, 4500–8000 ft. May–Aug. E/W. Introduced. (Plate 65)

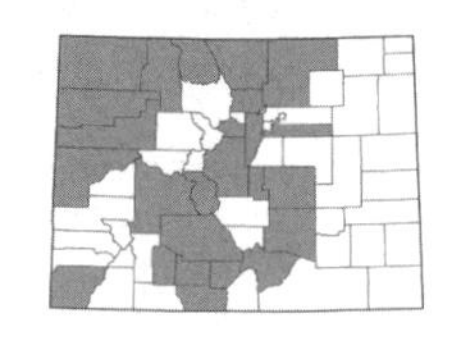

Puccinellia nuttalliana **(Schult.) Hitchc.**, NUTTALL'S ALKALI GRASS. [*P. airoides* (Nutt.) S. Watson & J.M. Coult.]. Perennials, 3–7 dm; *leaves* usually involute, 3–10 cm × 1–3 mm; *ligules* 1–3 mm long; *spikelets* 3–12 mm long, with 3–7 florets; *glumes* unequal, the lower 0.5–1.5 mm and the upper 1–2.5 mm long; *lemmas* (2) 2.2–3.5 mm long, glabrous or sparsely hairy on the lower half of the veins, the tips roughened. Common in moist meadows, along lake shores, ditches, and streams, usually in alkaline soil, 4500–9500 ft. June–Sept. E/W.

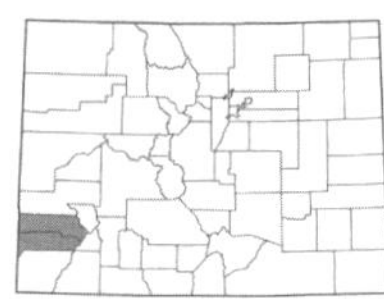

Puccinellia parishii **Hitchc.**, BOG ALKALIGRASS. Annuals, 0.3–2.5 dm; *leaves* flat to involute, 0.2–1.2 mm wide; *ligules* 1–2 mm long; *spikelets* 3.5–5 mm long, with 2–7 florets; *glumes* unequal, the lower 1–2 mm and the upper 1.8–2.2 mm long; *lemmas* 1.8–2.2 mm long, densely hairy along the lower ½ to ¾ of the veins, the tips smooth or with a few scattered minute roughenings. Rare in moist, alkaline places such as along the edges of ponds and drainages, 7500–8000 ft. June–Aug. W.

SCHEDONORUS P. Beauv. – FESCUE

Perennials; *sheaths* open; *ligules* membranous; *spikelets* in panicles, with 2–22 florets; *glumes* shorter than the lemmas, 3–9-veined, unawned; *lemmas* 3–7-veined, awned or not; *disarticulation* above the glumes.

1a. Auricles of at least some leaf collars ciliate; lemma awn (when present) to 4 mm long...***S. arundinaceus***
1b. Auricles all glabrous; lemma awn (when present) very short, to 0.2 mm long...***S. pratensis***

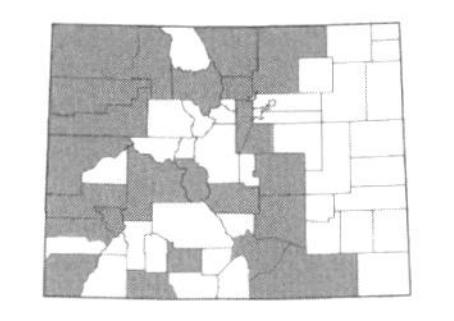

***Schedonorus arundinaceus* (Schreb.) Dumort.**, TALL FESCUE. [*Festuca arundinacea* Schreb.]. Plants 7–15 dm; *leaves* flat, 5–45 cm × 3–8 mm; *ligules* 0.2–0.6 mm long; *spikelets* 10–15 mm long, with 3–9 florets; *glumes* unequal, the lower 3–7 mm and the upper 5–7 mm long; *lemmas* 7–10 mm long, scabrous to hispid, unawned or with an awn to 4 mm long. Found in grasslands, meadows, shrublands, and along streams and pools, often used for forage and turf, 4000–10,500 ft. May–Aug. E/W. Introduced. (Plate 65)

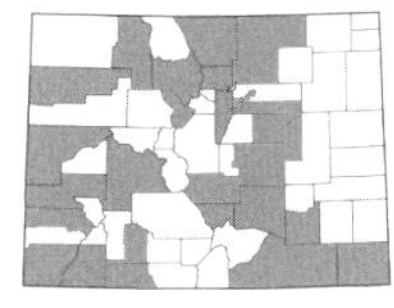

***Schedonorus pratensis* (Huds.) P. Beauv.**, MEADOW FESCUE. [*Festuca pratensis* Huds.; *Lolium pratense* (Huds.) Darbysh.]. Plants 6–12 dm; *leaves* flat, 5–20 cm × 2–5 mm; *ligules* 0.3–0.5 mm long; *spikelets* 10–15 mm long, with 4–10 florets; *glumes* unequal, the lower 2.5–4.5 mm and the upper 3.5–5 mm long; *lemmas* 5–8 mm long, unawned or with a short mucro to 0.2 mm. Found in grasslands, meadows, disturbed areas, and along roadsides and streams, 4800–9500 ft. May–Aug. E/W. Introduced.

SCHIZACHNE Hack. – FALSE MELIC

Perennials; *sheaths* closed to near the top; *ligules* membranous; *spikelets* in open panicles, with 3–6 florets; *glumes* chartaceous, the upper third hyaline; *lemmas* chartaceous, 7–9-veined, awned; *disarticulation* above the glumes.

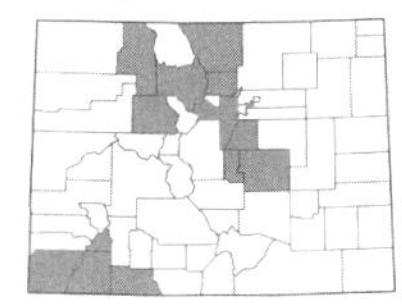

***Schizachne purpurascens* (Torr.) Swallen**, FALSE MELIC. Plants 5–8 dm; *leaves* flat or folded, 5–15 cm × 2–5 mm; *ligules* 0.5–1.5 mm long; *spikelets* 11–17 mm long, with a long-hairy callus; *glumes* unequal, the lower 4.2–6.2 mm and the upper 6–9 mm long; *lemmas* 8–10 mm long, with an awn 8–15 mm long arising from a bifid apex. Found in shady forests and along streams, 6500–11,000 ft. June–Sept. E/W. (Plate 65)

SCHIZACHYRIUM Nees – LITTLE BLUESTEM

Perennials (ours); *sheaths* open; *ligules* membranous; *spikelets* in spike-like racemes, with spikelets in sessile-pedicellate pairs; *disarticulation* below the spikelets.

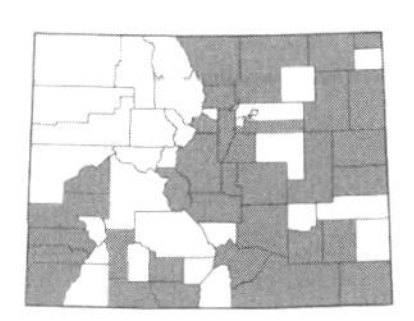

Schizachyrium scoparium* (Michx.) Nash var. *scoparium, LITTLE BLUESTEM. [*Andropogon scoparius* Michx.]. Plants 7–10 dm; *leaves* flat or folded, 7–30+ cm × 1.5–9 mm; *ligules* 0.5–3 mm long; *spikelets* paired, the sessile spikelet perfect, 6–8 mm long, the pedicellate spikelet sterile or staminate, 4–5 mm long; *glumes* of sessile spikelet 2.5–4 mm long, glabrous or scabrous; *lemmas* of sessile spikelet 4–6.5 mm long, terminating in a geniculate awn 2.5–17 mm long. Common in grasslands, prairie, on mesas, dry slopes, and in pinyon-juniper and canyons, 3400–9500 ft. July–Sept. E/W. (Plate 65)

SCLEROCHLOA P. Beauv. – HARDGRASS

Annuals; *sheaths* open or closed half their length; *ligules* membranous; *spikelets* in racemes, 1-sided, with 2–7 florets; *glumes* glabrous, with wide hyaline margins, obtuse, unawned; *lemmas* 7–9-veined, unawned; *paleas* shorter than to equal to the lemmas; *disarticulation* above the glumes.

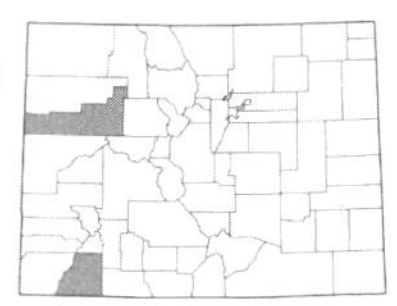

***Sclerochloa dura* (L.) P. Beauv.**, COMMON HARDGRASS. Plants prostrate, the culms 2–15 cm long; *leaves* flat or folded, 0.5–5 cm × 1–4 mm; *ligules* 0.7–2 mm long; *spikelets* 5–12 mm long, with 3–4 florets; *glumes* unequal, the lower 1.4–3 mm and the upper 2.5–5.5 mm long; *lemmas* 4.5–6 mm long, with prominent nerves. Uncommon in dry, shrubland communities, 5000–6000 ft. May–July. W. Introduced. (Plate 65)

SCLEROPOGON Phil. – BURROGRASS

Perennials; *ligules* of hairs; *spikelets* in spike-like racemes or panicles, with several florets, with staminate florets usually below the pistillate florets; *glumes* membranous, 1–3-veined; *lemmas* 3-veined, awned or not; *paleas* shorter than the lemmas; *disarticulation* above the glumes.

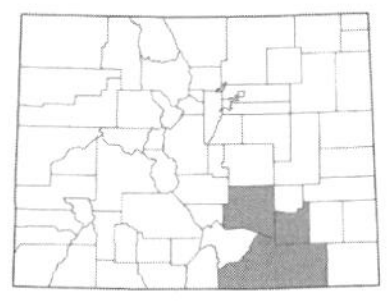

***Scleropogon brevifolius* Phil.**, BURROGRASS. Plants 1–2 dm; *leaves* flat or folded, 2–8 cm × 1–2 mm, firm; *ligules* to 1 mm long; *staminate spikelets* 20–30 mm long, with 5–10 florets, with glumes 3–10 mm long, lemmas unawned or awned to 3 mm; *pistillate spikelets* 20–30 mm long (excluding awns) with 1–4 florets, with glumes unequal and lemmas 3-awned. Uncommon in grasslands, 4000–5500 ft. June–Aug. E.

SECALE L. – RYE

Annuals (ours); *sheaths* open; *ligules* membranous; *spikelets* in a 2-sided spike, with 1 spikelet per node, with 2 florets; *glumes* keeled, linear; *disarticulation* below the florets or in the rachis.

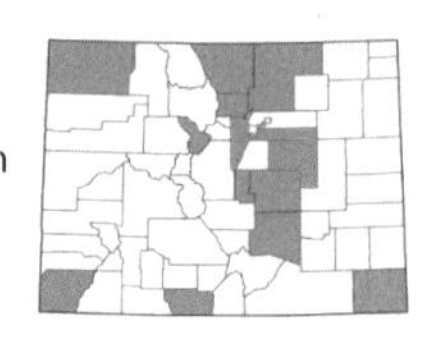

***Secale cereale* L.**, CULTIVATED RYE. Plants 5–12 dm; *leaves* flat to involute, 10–25 cm × 4–12 mm, usually glabrous; *ligules* 0.5–1.5 mm long; *spikelets* 10–18 mm long; *glumes* 8–18 mm long, with an awn 1–3 mm long; *lemmas* 10–19 mm long, with a ciliate keel, terminating in an awn 7–50 mm long. Found in disturbed areas, grasslands, and along roadsides and fields, 5000–8500 ft. May–July. E/W. Introduced.

SETARIA P. Beauv. – BRISTLEGRASS

Annuals or perennials; *sheaths* open; *ligules* membranous and ciliate or of hairs; *spikelets* in spike-like panicles, with one lower sterile floret and one perfect upper floret, subtended by bristles; *fertile lemmas* firm and cartilaginous; *paleas* well-developed, hard; *disarticulation* below the glumes.

1a. Bristles subtending the spikelets and branches retrorsely scabrid with the teeth pointing down, mostly shorter or equal to the spikelet in length...***S. verticillata***
1b. Bristles subtending the spikelets and branches antrorsely scabrid with the teeth pointing up, mostly longer than the spikelet...2

2a. Bristles purplish; inflorescence 12–30 mm wide (excluding the bristles); disarticulation between the upper and lower florets...***S. italica***
2b. Bristles yellowish or green, rarely purple; inflorescence usually less than 12 mm wide (excluding the bristles); disarticulation below the glumes...3

3a. Spikelets 3–3.5 mm long; upper glume about ½ the length of the spikelet...***S. pumila* ssp. *pumila***
3b. Spikelets 1.8–2.8 mm long; upper glume about ¾ the length of to as long as the spikelet...4

4a. Perennials; inflorescence often interrupted; leaves 2–5 mm wide...***S. leucopila***
4b. Annuals; inflorescence occasionally interrupted near the base, otherwise usually dense throughout; leaves 4–25 mm wide...***S. viridis* var. *viridis***

***Setaria italica* (L.) Beauv.**, FOXTAIL BRISTLEGRASS. Annuals, 1–10 dm; *leaves* flat, 10–40 × 1–3 cm, scabrous; *ligules* 1–2 mm long; *spikelets* 2.5–3 mm long, glabrous, subtended by 1–3 antrorsely scabrid, purplish bristles; *glumes* unequal, the lower 0.8–1.4 mm, the upper 2–2.6 mm long (about ¾ the length of the spikelet); *sterile lemmas* 2.2–2.8 mm long, 5–7-nerved; *fertile lemmas* 2.2–2.8 mm long. Uncommon weed in sidewalks, along roadsides, and sometimes cultivated as hay or pasture grass, 3500–6000 ft. July–Sept. E. Introduced.

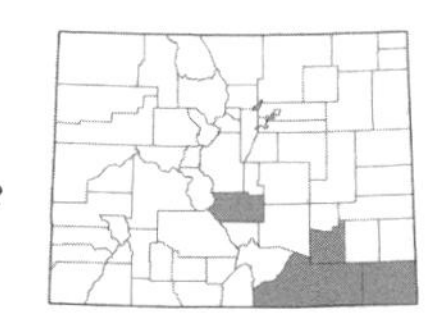

***Setaria leucopila* (Scribn. & Merr.) K. Schum.**, STREAMBED BRISTLEGRASS. Perennials, 2–10 dm; *leaves* flat to folded, 8–25 cm × 2–5 mm, scabrous; *ligules* 1–2.5 mm long; *spikelets* 2.2–2.8 mm long, glabrous, subtended by a solitary, antrorsely scabrid, green or yellowish bristle; *glumes* unequal, the lower 1.1–1.8 mm and the upper 1.7–2.5 mm long (about ¾ the length of the spikelet); *sterile lemmas* 2.2–3 mm long, 5-nerved; *fertile lemmas* 2.3–3.2 mm long. Found in canyons and on rocky slopes, 4000–5500 ft. July–Sept. E.

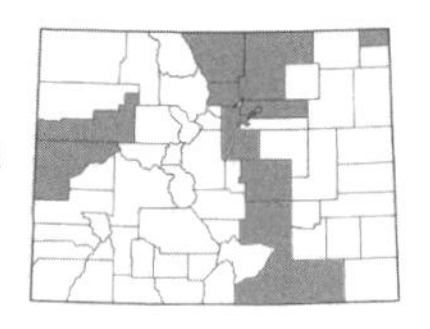

Setaria pumila* (Poir.) Roem. & Schult. ssp. *pumila, YELLOW BRISTLEGRASS. [*S. glauca* (L.) P. Beauv.; *S. lutescens* (Weigel) F.T. Hubbard]. Annuals, 0.3–1.5 dm; *leaves* usually flat, 5–20 cm × 4–10 mm, usually with papillose hairs near the throat; *spikelets* 3–3.5 mm long, glabrous, subtended by 4–12 antrorsely scabrid, greenish bristles; *glumes* unequal, the lower 1–1.7 mm and the upper 1.4–2.2 mm long (about ½ the length of the spikelet); *sterile lemmas* 2.2–3 mm long, 5-nerved; *fertile lemmas* 2.3–3.2 mm long. Found as a weed in lawns and gardens, fields, along roadsides, and in disturbed areas, 3500–6000 ft. June–Sept. E/W. Introduced.

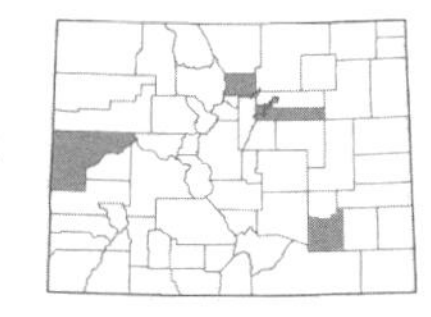

***Setaria verticillata* (L.) Beauv.**, HOOKED BRISTLEGRASS. Annuals, 3–10 dm; *leaves* flat, 8–20 cm × 5–15 mm, scabrous; *ligules* to 1 mm; *spikelets* 2.2–2.5 mm long, subtended by relatively short (4–7 mm long), retrorsely scabrid bristles; *glumes* unequal, the lower 0.7–1.2 mm and the upper 1.6–2 mm long (about as long as the length of the spikelet); *sterile lemmas* 1.9–2.5 mm long, 7-nerved; *fertile lemmas* 1.8–2.3 mm long. Uncommon in disturbed areas, gardens, and along roadsides, 5000–6500 ft. June–Sept. E/W. Introduced.

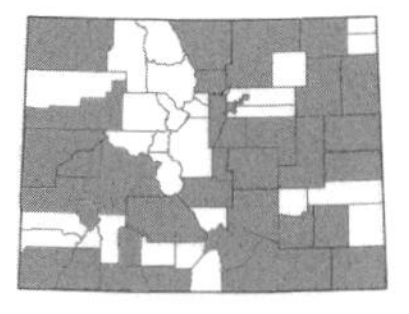

Setaria viridis* (L.) Beauv. var. *viridis, GREEN BRISTLEGRASS. Annuals, 2–25 dm; *leaves* flat, 8–20 cm × 4–25 mm, glabrous to scabrous; *ligules* 1–2 mm long; *spikelets* 1.8–2.2 mm long, glabrous, subtended by 1–3 antrorsely scabrid, greenish bristles; *glumes* unequal, the lower 1–1.2 mm and the upper 2–2.5 mm long (about as long as the spikelet); *sterile lemmas* 2–2.5 mm, 5–7-nerved; *fertile lemmas* 1.8–2.5 mm. long Common weed in lawns and gardens, sidewalks, fields, along roadsides, and in disturbed areas, 3400–9500 ft. June–Sept. E/W. Introduced. (Plate 66)

SORGHASTRUM Nash – INDIAN GRASS

Perennials; *sheaths* open; *ligules* membranous; *nodes* densely hairy; *spikelets* in panicles, with sessile-pedicellate spikelet pairs, subtending a hairy pedicel, awned; *disarticulation* below the spikelets.

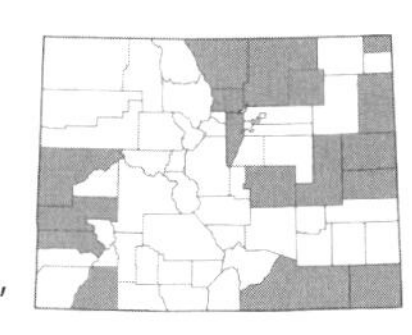

Sorghastrum nutans **(L.) Nash**, INDIAN GRASS. Plants 5–25 dm; *leaves* flat, 10–70 cm long, the sheaths terminating in prominent, erect auricles 2–7 mm long; *ligules* 1.5–6.5 mm long; *spikelets* paired, the pedicellate spikelet usually absent but leaving a hairy pedicel 3–6 mm long, sessile spikelet perfect and 5–9 mm long; *glumes* 5–8 mm long; *fertile lemmas* 4.5–6.5 mm long with a geniculate awn 12–18 mm long. Found in grasslands, prairie, canyons, and open pine woodlands, 3500–6800 ft. Aug.– Sept. E/W. (Plate 66)

SORGHUM Moench – SORGHUM

Perennials or annuals; *sheaths* open; *ligules* membranous, ciliate; *spikelets* in panicles, the primary branches whorled, with sessile-pedicellate spikelet pairs, sometimes awned; *disarticulation* below the spikelets.

1a. Annuals, lacking rhizomes; fertile lemma of sessile spikelet unawned or awned, the awn 5–8.5 mm long; panicle usually a dense terminal cluster...***S. bicolor***

1b. Perennials, with rhizomes; fertile lemma of sessile spikelet awned, the awn 9–12 mm long; panicle usually open, not in a dense terminal cluster...***S. halepense***

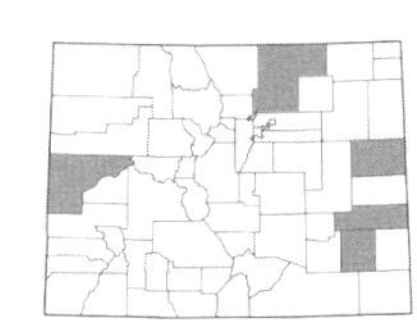

Sorghum bicolor **(L.) Moench**, GRAIN SORGHUM. Annuals, 10–30 dm; *leaves* flat, 5–100 × 0.5–10 cm; *ligules* 1–5.5 mm long; *panicle* a dense terminal cluster; *spikelets* hairy, 1 sessile and 2 pediceled; *sessile spikelet* perfect, 3–9 mm long, with one lower reduced floret and one upper perfect floret; *pedicellate spikelets* staminate or sterile, 3–6 mm long; *fertile lemmas* 3.9–4.3 mm long, unawned or with an awn 5–8.5 mm long. Cultivated in fields and occasionally escaping along roadsides, 3400–5000 ft. July–Sept. E/W. Introduced. (Plate 66)

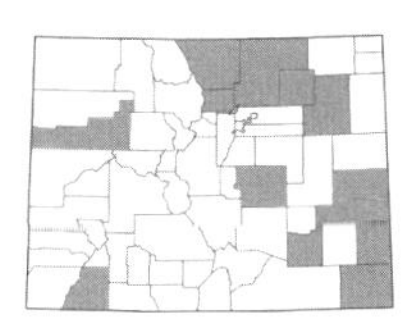

Sorghum halepense **(L.) Pers.**, JOHNSON GRASS. Perennials, 5–20 dm; *leaves* flat, 10–90 × 0.8–4 cm; *ligules* 1.5–6 mm long; *panicle* usually open; *spikelets* 1 sessile and 2 pediceled; *sessile spikelet* perfect, 3.8–6.5 mm long, with one lower reduced floret and one upper perfect floret; *pedicellate spikelets* staminate, 3.6–5.6 mm long; *fertile lemmas* with a geniculate awn 9–12 mm long. Found along roadsides and in ditches and other disturbed areas, 3400–6000 ft. July–Sept. E/W. Introduced.

SPARTINA Schreb. – CORDGRASS

Perennials; *sheaths* open; *ligules* of hairs; *spikelets* in panicles of one-sided spike-like branches, sessile, with 1 floret; *glumes* 1–6-veined, strongly keeled; *lemmas* 1–3-veined, shorter than the paleas; *disarticulation* below the glumes.

1a. Upper glume awnless or with a short awn to 1 mm long; plants 4–10 dm tall...***S. gracilis***

1b. Upper glume with a stout awn 3–10 mm long; plants to 25 dm tall...***S. pectinata***

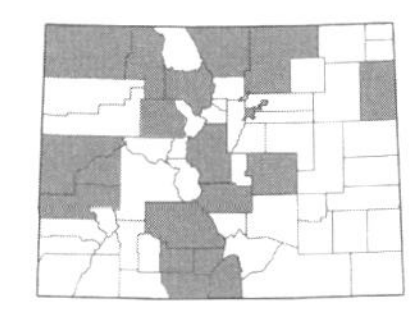

Spartina gracilis **Trin.**, ALKALI CORDGRASS. Plants 4–10 dm; *leaves* flat, 6–30 cm × 2.5–8 mm, with scabrous margins; *ligules* 0.5–1 mm long; *glumes* unequal, the lower 3–7 mm and the upper 6–10 mm long, awnless or with a short awn to 1 mm, with bristly keels; *lemmas* 6–7.5 mm long. Found along pond margins, creeks, in moist meadows, and on alkali flats, 3400–10,000 ft. June–Sept. E/W.

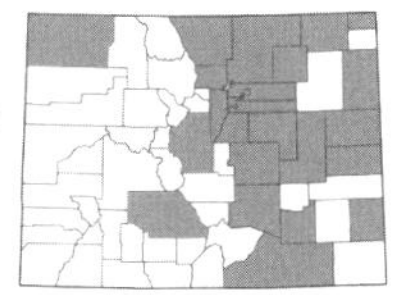

Spartina pectinata **Bosc ex Link**, PRAIRIE CORDGRASS. Plants to 25 dm; *leaves* mostly flat, 20–120 cm × 6–15 mm, with scabrous margins; *ligules* 2–4 mm long; *glumes* unequal, the lower 5–10 mm and the upper 10–20 mm long, with an awn 3–10 mm long, with bristly keels; *lemmas* 6–9 mm long. Found in moist meadows, ditches, on moist grassy slopes, and along creeks and pond margins, 3400–7000 ft. June–Sept. E/W. (Plate 66)

SPHENOPHOLIS Scribn. – WEDGEGRASS

Perennials; *sheaths* open; *ligules* membranous; *spikelets* in dense panicles, with 2–3 florets; *glumes* unlike in size and shape, the lower narrower than the upper glumes, unawned; *lemmas* unawned (ours); *disarticulation* below the glumes.

1a. Upper glumes shortly hooded; panicles often spike-like, the spikelets usually densely arranged...***S. obtusata***

1b. Upper glumes not hooded; panicles usually more open, the spikelets loosely arranged...***S. intermedia***

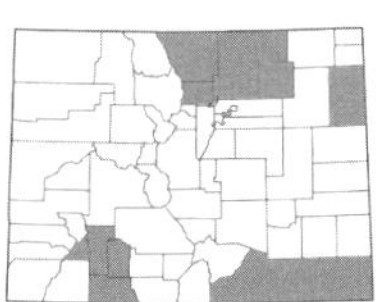

Sphenopholis intermedia **(Rydb.) Rydb.**, SLENDER WEDGEGRASS. Plants 3.5–12 dm; *leaves* flat, 8–15 cm × 2–5 mm; *ligules* 1.5–2.5 mm long; *inflorescence* a loose panicle; *glumes* 1.5–3 mm long; *lemmas* 2–3 mm long, glabrous or scabrous. Uncommon in moist meadows, near springs, in canyon bottoms, and on grassy slopes, 3500–9000 ft. June–Aug. E/W.

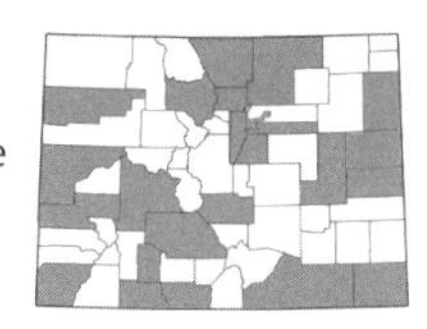

Sphenopholis obtusata **(Michx.) Scribn.**, PRAIRIE WEDGEGRASS. Plants 3–10 dm; *leaves* flat, 5–20 × 1.5–6 mm; *ligules* 1.5–3.5 mm long; *inflorescence* a dense, spike-like panicle; *glumes* 1.7–3 mm long; *lemmas* 2–3 mm long, glabrous or scabrous. Found along pond margins, creeks, in drainage ditches, canyon bottoms, and near springs, 3400–9000 ft. June–Aug. E/W. (Plate 66)

SPOROBOLUS R. Br. – DROPSEED

Perennials or annuals; *sheaths* open; *ligules* membranous or of hairs; *spikelets* in panicles, with one floret, unawned; *disarticulation* above the glumes.

1a. Panicle narrow, contracted, cylindrical, and often spike-like at maturity…2
1b. Panicle open and diffuse at maturity (the panicle sometimes enclosed in the sheath when young, but not dense and spike-like throughout), often pyramidal in shape…6

2a. Annuals, delicate; panicle 1–5 cm long, 0.2–0.5 cm wide; leaf sheaths usually inflated…3
2b. Perennials, often stout; panicle 5–75 cm long and 0.2–4 cm wide; leaf sheaths not particularly inflated…4

3a. Lemmas glabrous; spikelets 1.6–3 mm long…***S. neglectus***
3b. Lemmas hairy; spikelets 2.3–6 mm long…***S. vaginiflorus***

4a. Spikelets 4–6 mm long; leaf sheaths sparsely hairy at the tips…***S. compositus***
4b. Spikelets 1.5–3.5 mm long; leaf sheaths with conspicuous tufts of hair at the tips…5

5a. Plants 4–10 dm; stems 2–4 (5) mm thick near the base; anthers 0.3–0.5 mm long…***S. contractus***
5b. Plants 10–20 dm; stems (3) 4–10 mm thick near the base; anthers 0.6–1 mm long…***S. giganteus***

6a. Lower panicle nodes with 7–12 branches, appearing distinctly whorled; annuals or short-lived perennials…***S. pyramidatus***
6b. Lower panicle nodes with 1–3 branches (*careful*—if the branches are still in the sheath pull them apart to see if they originate at the same node), not appearing whorled; perennials…7

7a. Pedicels 6–25 mm long…***S. texanus***
7b. Pedicels 1–3 mm long…8

8a. Spikelets 3–6 mm long; lemma 3–4.5 mm long; panicle 5–22 cm long, 6–11 cm wide…***S. heterolepis***
8b. Spikelets 1.3–2.8 mm long; lemma 1.2–2.5 mm long; panicle 3–45 cm long, 1–25 cm wide…9

9a. Leaf sheath glabrous or sparsely hairy at the summit…***S. airoides***
9b. Leaf sheath with a conspicuous tuft of white hairs to 4 mm long at the summit…10

10a. Primary branches mostly reflexed, with white hairs at the base…***S. flexuosus***
10b. Primary branches appressed, spreading, or sometimes reflexed, glabrous or sometimes scabrous at the base, but lacking white hairs…11

11a. Plants with a hard, knotty base; panicle 3–10 cm long and 1–5 cm wide; leaves involute, 1–2 mm wide…***S. nealleyi***
11b. Plants lacking a hard, knotty base; panicle 10–40 cm long, 2–12 cm wide; leaves involute or flat, 2–6 mm wide…***S. cryptandrus***

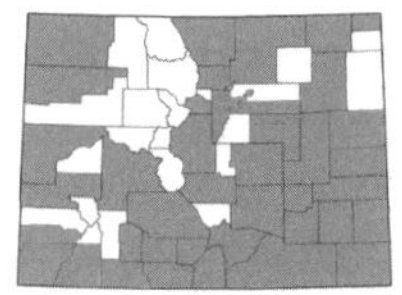

Sporobolus airoides **(Torr.) Torr.**, ALKALI SACATON. Perennials, 3–12 dm; *leaves* flat to involute, 10–50 cm × 2–5 mm, the sheath summit glabrous or sparsely hairy; *inflorescence* an open panicle; *glumes* unequal, the lower 0.5–1.8 mm and the upper 1–2.5 mm long; *lemmas* 1.2–2.5 mm long, glabrous, acute; *paleas* 1–2.5 mm long, equal to the lemma. Common in grasslands, shortgrass prairie, shrublands, and on alkaline flats, 3400–8000 ft. June–Sept. E/W. (Plate 66)

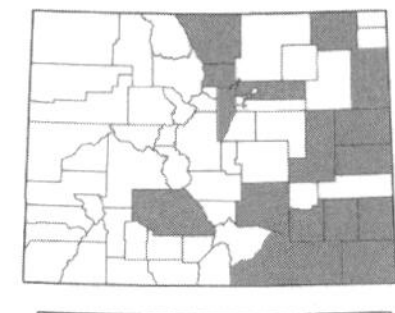

Sporobolus compositus **(Poir.) Merr.**, COMPOSITE DROPSEED. [*S. asper* (Michx.) Kunth]. Perennials, 3–13 dm; *leaves* flat to involute, 5–70 cm × 1.5–10 mm, the sheath sparsely hairy at the summit; *inflorescence* a narrow, spike-like panicle; *glumes* unequal, the lower 2–4 mm and the upper 2.5–5 mm long; *lemmas* 3–6 mm long, glabrous; *paleas* 2–3 mm long. Found on the eastern plains, on roadsides, and in grasslands, 3400–6000 ft. July–Sept. E.

Sporobolus contractus **Hitchc.**, SPIKE DROPSEED. Perennials, 4–10 dm; *leaves* flat to involute, 4–40 cm × 3–8 mm, the sheath with tufts of hair at the summit; *inflorescence* a narrow, spike-like panicle; *glumes* unequal, the lower 0.7–1.7 mm and the upper 2–3.2 mm long; *lemmas* 2–3.2 mm long, glabrous; *paleas* 1.8–3 mm long. Found on dry slopes, 5000–8000 ft. June–Sept. E/W.

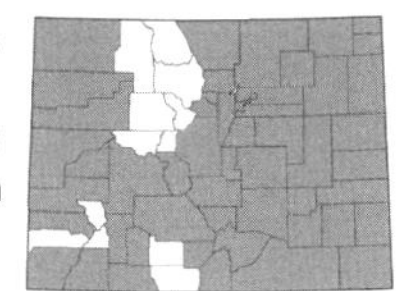

Sporobolus cryptandrus **(Torr.) A. Gray**, SAND DROPSEED. Perennials, 3–10 dm; *leaves* flat to involute, 5–25 cm × 2–6 mm, the sheath with conspicuous tufts of hair at the summit; *inflorescence* a contracted to open panicle; *glumes* unequal, the lower 0.5–1 mm and the upper 1.5–2.8 mm long; *lemmas* 1.4–2.5 mm long; *paleas* 1.2–2.5 mm long. Common on the eastern plains, dry hillsides, in grasslands, pinyon-juniper woodlands, and along roadsides, 3400–9000 ft. June–Sept. E/W.

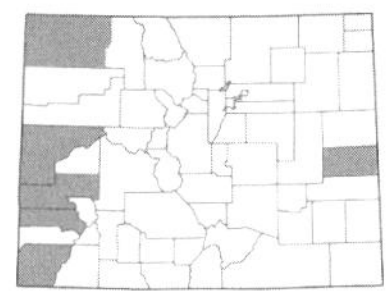

Sporobolus flexuosus **(Thurb. ex Vasey) Rydb.**, MESA DROPSEED. Perennials, 3–12 dm; *leaves* flat to involute, 5–25 cm × 2–4 mm, the sheath with conspicuous tufts of hair at the summit; *inflorescence* an open panicle; *glumes* unequal, the lower 1–1.5 mm and the upper 1.4–2.5 mm long; *lemmas* 1.4–2.5 mm long; *paleas* 1.4–2.5 mm long. Uncommon on dry slopes, in pinyon-juniper woodlands, and in canyons, 4000–6000 ft. June–Sept. E/W.

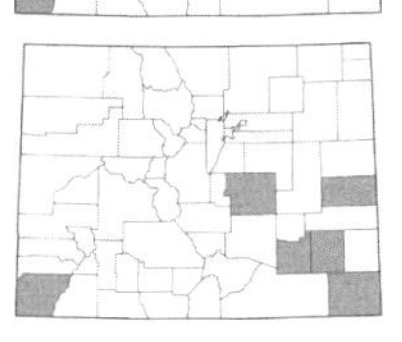

Sporobolus giganteus **Nash**, GIANT DROPSEED. Perennials, 10–20 dm; *leaves* flat, 10–50 cm × 4–10 mm, the sheath with conspicuous tufts of hair at the summit; *inflorescence* a narrow, spike-like panicle; *glumes* unequal, the lower 0.5–2 mm and the upper 2–3.5 mm long; *lemmas* 2.5–3.5 mm long; *paleas* 2.4–3.4 mm long. Uncommon on sand dunes, 3700–6000 ft. July–Sept. E/W.

Sporobolus heterolepis **(A. Gray) A. Gray**, PRAIRIE DROPSEED. Perennials, 3–8 dm; *leaves* flat to folded, 7–30 cm × 1.2–2.5 mm, the sheath glabrous to sparsely hairy; *inflorescence* an open panicle; *glumes* unequal, the lower 1.8–4.5 mm and the upper 2.5–6 mm long; *lemmas* 3–4.5 mm long; *paleas* 3–4.5 mm long. Found on mesas, in grasslands, and open ponderosa pine forests along the northern Front Range (Larimer, Boulder, Jefferson, Douglas, Elbert, Teller, and El Paso cos.), 5000–7500 ft. July–Sept. E.

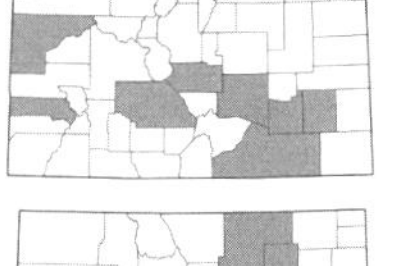

Sporobolus nealleyi **Vasey**, GYP DROPSEED. Perennials, 1–5 dm; *leaves* involute, 1.5–6 cm × 1–1.5 mm, the sheath with conspicuous tufts of hair at the summit; *inflorescence* an open panicle; *glumes* unequal, the lower 0.5–1 mm and the upper 1.3–2 mm long; *lemmas* 1.4–2 mm long; *paleas* 1.4–2 mm long. Uncommon on gypsum flats and on dry slopes, 4200–7500 ft. June–Sept. E/W.

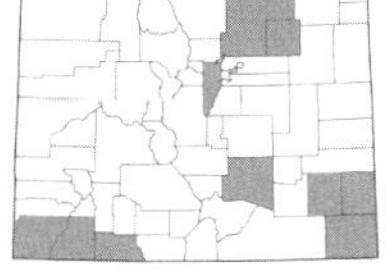

Sporobolus neglectus **Nash**, PUFF-SHEATH DROPSEED. Annuals, 1–5 dm; *leaves* flat to involute, 1–12 cm × 0.5–2 mm, the sheath with small tufts of hair at the summit; *inflorescence* a narrow panicle; *glumes* unequal, the lower 1.5–2.4 and the upper 1.7–2.7 mm long; *lemmas* 1.5–3 mm; *paleas* 1.5–3 mm long. Uncommon in moist depressions and shortgrass prairie, 3500–7500 ft. July–Sept. E/W.

Sporobolus pyramidatus **(Lam.) Hitchc.**, WHORLED DROPSEED. Annuals to short-lived perennials, 0.7–4 dm; *leaves* flat, 2–12 cm × 2–6 mm, the sheath with tufts of hair at the summit; *inflorescence* an open panicle at maturity; *glumes* unequal, the lower 0.3–0.7 mm and the upper 1.2–1.8 mm long; *lemmas* 1.2–1.7 mm; *paleas* 1–1.6 mm. Uncommon on dry slopes and along roadsides, reported from La Plata and Otero cos., 4000–6000 ft. July–Sept. E/W.

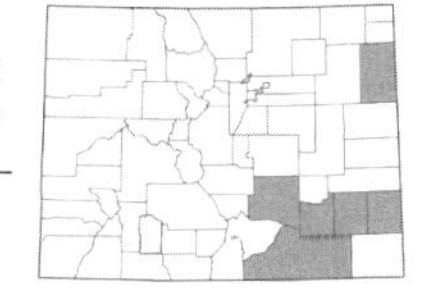

Sporobolus texanus **Vasey**, TEXAS DROPSEED. Perennials, 2–7 dm; *leaves* flat to involute, 2.5–13 cm × 1–4 mm, the sheath glabrous or with papillose-based hairs; *inflorescence* an open panicle; *glumes* unequal, the lower 0.5–1.7 mm and the upper 1.7–3 mm long; *lemmas* 1.8–3 mm long; *paleas* 1.7–3 mm long. Uncommon on the eastern plains and on alkaline flats, 3500–4700 ft. July–Sept. E.

Sporobolus vaginiflorus **(Torr. ex A. Gray) Alph. Wood**, POVERTY GRASS. Annuals, 1.5–7 dm; *leaves* flat to involute, 2–15 (25) cm × 0.6–2 mm, the sheath with small tufts of hair at the summit; *inflorescence* a narrow panicle; *glumes* unequal, the lower 2.2–4.7 and the upper 2.4–5 mm; *lemmas* (2) 3–5.4 mm long, strigose; *paleas* (2) 3–6 mm long, as long or longer than the lemma. Uncommon in disturbed areas and along roadsides, known from Jefferson Co., 7000–8000 ft. July–Sept. E. Introduced.

THINOPYRUM Á. Löve – WHEATGRASS

Perennials; *sheaths* open; *ligules* membranous; *spikelets* in spikes, with one or more spikelets per node, with several florets; *glumes* stiff, indurate, keeled, unawned; *lemmas* 5-veined, awned or not; *disarticulation* below the florets.

1a. Glumes with a small mucro at the tip, glabrous or hairy; lemmas sometimes awned; plants with rhizomes... ***T. intermedium***

1b. Glumes truncate, lacking a mucro at the tip, glabrous; lemmas unawned; plants caespitose, lacking rhizomes... ***T. ponticum***

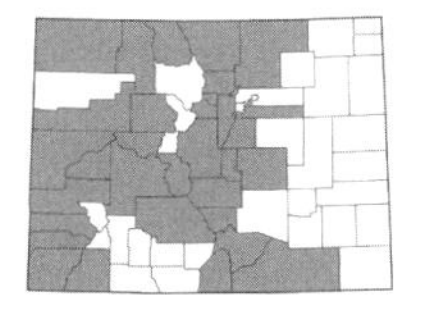

Thinopyrum intermedium **(Host) Barkworth & D.R. Dewey**, INTERMEDIATE WHEATGRASS. [*Elymus hispidus* (Opiz) Melderis]. Rhizomatous perennials, 5–11 dm; *leaves* flat to involute, 10–40 cm × 2–8 mm, stiff; *ligules* 0.1–1 mm long; *spikelets* 11–18 mm long, solitary at each node; *glumes* 4.5–8 mm long, glabrous to hairy; *lemmas* 7–10 mm long, sometimes with an awn to 5 mm long. Common along roadsides and in woodlands, 4500–10,500 ft. June–Aug. E/W. Introduced.

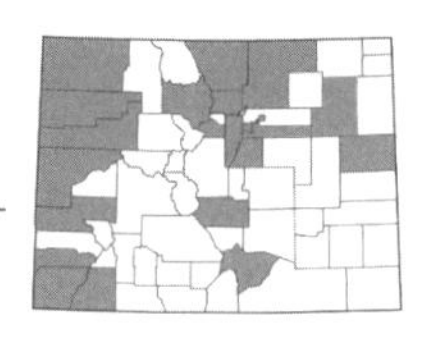

Thinopyrum ponticum **(Podp.) Z.-W. Liu & R.-C. Wang**, RUSH WHEATGRASS. [*Elymus elongatus* (Host) Runemark]. Caespitose perennials, 5–20 dm; *leaves* flat or involute, 10–40 cm × 2–6 mm; *ligules* 0.3–1.5 mm long; *spikelets* 13–30 mm long, solitary at each node; *glumes* 6.5–10 mm long, glabrous; *lemmas* 9–12 mm long, unawned. Found along roadsides and in disturbed areas, 4000–10,500 ft. June–Aug. E/W. Introduced.

TORREYOCHLOA G.L. Church – FALSE MANNAGRASS

Perennials; *sheaths* open; *ligules* membranous; *spikelets* in panicles, with 2–8 florets; *glumes* unequal, unawned, 1–3-veined; *lemmas* 7–9-veined; *disarticulation* above the glumes.

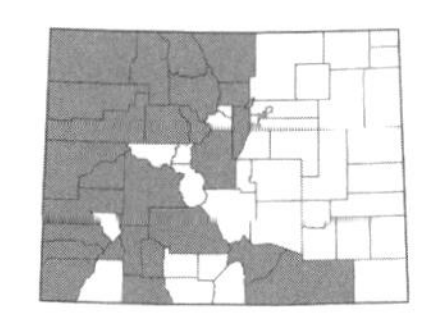

Torreyochloa pallida **(Torr.) G.L. Church var. *pauciflora* (J. Presl) J.I. Davis**, PALE FALSE MANNAGRASS. Plants 2–12 dm; *leaves* flat, 1.5–17 mm wide; *ligules* 2–9 mm long; *glumes* unequal, the lower 0.7–2 mm and the upper 1–3 mm long; *lemmas* 2–3.5 mm long, rounded or weakly keeled, truncate to acute, with parallel veins, unawned. Found along streams, pond margins, and in moist meadows and wetlands, 7000–11,500 ft. July–Sept. E/W. (Plate 66)

TRIDENS Roem. & Schult. – TRIDENS

Perennials; *sheaths* open; *ligules* membranous and ciliate or of hairs; *spikelets* in panicles, with 4–11 florets, usually with a pilose callus; *glumes* 1–3-veined, unawned; *lemmas* 3-veined, unawned; *disarticulation* above the glumes.

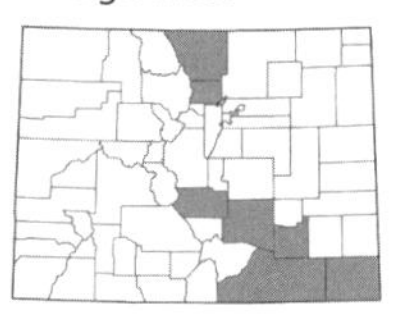

Tridens muticus **(Torr.) Nash**, ROUGH TRIDENS. Plants 4–8 dm; *leaves* flat to involute, 3–25 cm × 3–4 mm; *ligules* 0.5–1 mm long; *inflorescence* a narrow panicle; *glumes* unequal, the lower 3–8 mm and the upper 5–10 mm long, glabrous with a scabrous keel; *lemmas* 3.5–7 mm long, with hairy veins, with a rounded apex. Found on rocky or shale outcrops, talus slopes, and in grasslands, 4000–6000 ft. July–Sept. E. (Plate 66)

TRIPLASIS P. Beauv. – SANDGRASS

Annuals (ours); *sheaths* open; *ligules* of hairs; *spikelets* in small panicles, with cleistogamous flowers in upper sheaths, with 2–5 florets; *glumes* 1-veined, keeled; *lemmas* 3-veined, sometimes awned; *disarticulation* above the glumes.

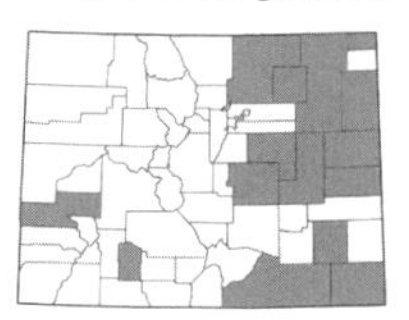

Triplasis purpurea **(Walter) Chapman**, PURPLE SANDGRASS. Plants 1.5–10 dm; *leaves* flat to involute, 1–8 cm × 1–5 mm; *ligules* to 1 mm long; *inflorescence* often included in the leaf sheaths; *glumes* to 2 mm long, glabrous or scabrous; *lemmas* 3–4 mm long, with a 2-lobed tip, the nerves densely hairy, with a short awn to 2 mm long. Found in sandy soil of floodplains, dunes, and sandsage prairie, 3500–5700 ft. July–Sept. E/W. (Plate 66)

TRISETUM Pers. – OATGRASS

Perennials or annuals; *sheaths* open; *ligules* membranous; *spikelets* in open or dense, spike-like panicles, with 2–5 florets; *glumes* with scabrous keels, acute and unawned; *lemmas* 3–7-veined, unawned or awned from above the middle; *disarticulation* above the glumes.

1a. Lemmas unawned, or the awns only to 2 mm and arising just below and rarely exceeding the tips...***T. wolfii***
1b. Lemmas awned, the awns usually 4–14 mm long, exceeding the lemmas tips, rarely 2–4 mm long...2

2a. Panicle dense, narrow and spike-like...3
2b. Panicle more or less open, not dense and spike-like...4

3a. Upper glumes about twice as wide as the lower glumes; annual, lacking sterile shoots...***T. interruptum***
3b. Upper glumes less than twice as wide as the lower glumes; perennial, with both fertile and sterile shoots...***T. spicatum***

4a. Upper glumes shorter than the lowest florets; lemma awns 7–15 mm long; spikelets 7–9 mm long...***T. canescens***
4b. Upper glumes as long as or longer than the lowest florets; lemma awns (3) 5–9 mm long; spikelets 4–8 mm long...***T. flavescens***

Trisetum canescens **Buckley**, TALL TRISETUM. Perennials, 4–15 dm; *leaves* flat, 1–3 dm × 4–10 mm; *ligules* 1.5–6 mm long; *inflorescence* a narrow, open panicle; *glumes* unequal, the lower 3–5 mm and the upper 5–9 mm long; *lemmas* 5–7 mm long, with an awn 7–15 mm long arising from the base of a deeply bifid apex. Rare along creeks and in moist forests, 8800–11,000 ft. July–Sept. E/W.

Trisetum flavescens **(L.) P. Beauv.**, YELLOW OATGRASS. Perennials, 5–10 dm; *leaves* flat or involute, 5–20 cm × 2–7 mm; *ligules* 0.5–2 mm long; *inflorescence* a narrow, open panicle; *glumes* unequal, the lower 2–5 mm and the upper 4–7 mm long; *lemmas* 3–6 mm long, with an awn 3–9 mm long arising from the base of a bifid apex. Uncommon along roadsides and in fields, introduced as a forage species, known from Larimer Co., 5000–6000 ft. July–Sept. E. Introduced.

Trisetum interruptum **Buckley**, PRAIRIE TRISETUM. Annuals, 1–5 dm; *leaves* flat, 3–12 cm × 1–6 mm; *ligules* 1–3 mm long; *inflorescence* a dense, narrow, spike-like panicle; *glumes* 4–5 mm, about as long as the lowest lemmas, the lower glume 0.5–1 mm wide and the upper glume about twice as wide as the lower; *lemmas* 3–4.5 mm long, with a geniculate awn 4–8 mm long arising from the base of a deeply bifid apex. Reported for Colorado from Hinsdale Co., perhaps not persisting. July–Sept. W.

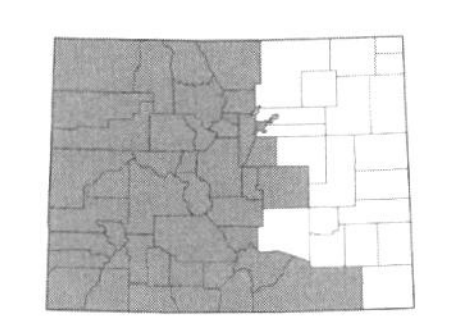

Trisetum spicatum **(L.) K. Richt.**, SPIKE TRISETUM. [*T. montanum* Vasey]. Perennials, 1–12 dm; *leaves* flat or involute, 10–20 (40) cm × 1–6 mm; *ligules* 0.5–4 mm long; *inflorescence* a dense, spike-like panicle; *glumes* unequal, the lower 3–5 mm and the upper 4–7 mm long, less than twice as wide as the lower; *lemmas* 3–6 mm long, with a geniculate awn 3–9 mm long arising from a deeply bifid apex. Common along streams, in forests, meadows, and the alpine, 7000–14,000 ft. July–Sept. E/W. (Plate 66)

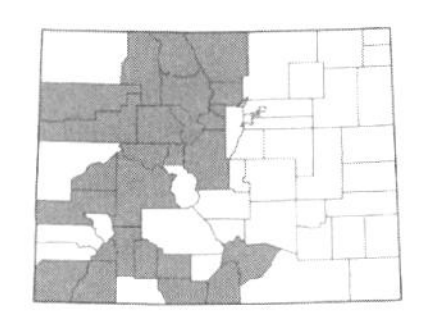

Trisetum wolfii **Vasey**, WOLF'S TRISETUM. [*Graphephorum wolfii* (Vasey) Vasey ex J.M. Coult.]. Perennials, 2–10 dm; *leaves* flat, 2–7 mm wide; *ligules* 1.5–6 mm; *inflorescence* a dense, narrow panicle; *glumes* 4–7 mm long, the upper as long as to slightly shorter than the lowermost floret; *lemmas* 4–7 mm, acute to slightly bidentate, unawned or with a short mucro to 2 mm. Found along creeks, lakes, in moist forests, willow carrs, and meadows, 7200–11,500 ft. July–Sept. E/W.

TRITICUM L. – WHEAT

Annuals; *sheaths* open; *ligules* membranous; *spikelets* in spikes, with 1 spikelet per node, with 2–9 florets; *glumes* hard, broad; *lemmas* hard or thick, keeled; *disarticulation* in the rachis.

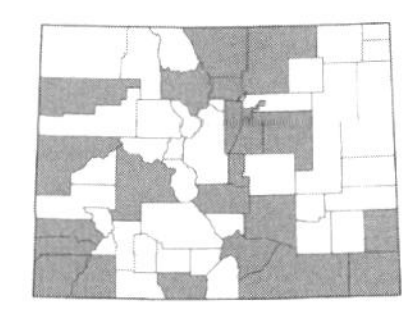

Triticum aestivum **L.**, WHEAT. Plants 5–15 dm; *leaves* flat, 2–15 mm wide; *ligules* 0.5–1.5 mm long; *spikes* 4–20 cm long; *glumes* 6–12 mm long, glabrous to hispid, with a short mucro or awned to 4 cm; *lemmas* 10–15 mm long, hard, acute or with an awn to 12 cm long. Cultivated and escaping along roadsides and field margins, sometimes in pinyon-juniper, 4000–7000 ft. June–Sept. E/W. Introduced.

VAHLODEA Fr. – HAIRGRASS

Perennials; *sheaths* open; *ligules* membranous; *spikelets* in panicles, with 2 florets; *glumes* 1–3-veined, equal to or longer than the florets; *lemmas* awned from near the middle; *disarticulation* above the glumes.

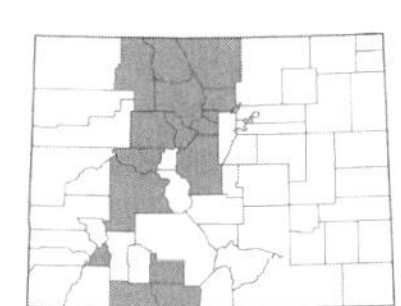

Vahlodea atropurpurea **(Wahlenb.) Fr. ex Hartman**, MOUNTAIN HAIRGRASS. Plants 1.5–8 dm; *leaves* flat, 1–30 cm × 1–9 mm; *ligules* 0.8–3.5 mm long; *spikelets* 4–7 mm long; *glumes* 4–5.5 mm long, the keel sometimes scabrous; *lemmas* 2–3 mm long, with long hairs at the base, with a twisted awn 2–4 mm long arising near the middle of the back. Found in moist high meadows, open spruce forests, and the alpine, 9500–13,500 ft. July–Sept. E/W. (Plate 66)

VULPIA C.C. Gmel. – FESCUE

Perennials or annuals; *sheaths* open; *ligules* membranous, ciliate; *spikelets* in narrow panicles, with 6–several florets; *glumes* lanceolate, awned or not; *lemmas* 3–5-veined, lanceolate, awned or not; *disarticulation* above the glumes.

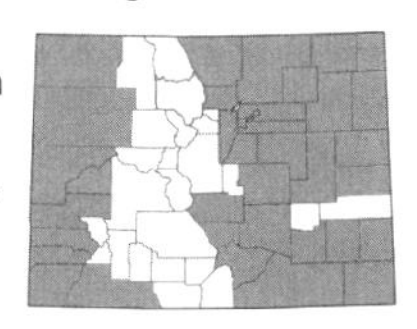

Vulpia octoflora **(Walter) Rydb.**, SIXWEEKS FESCUE. Annuals, 0.5–6 dm; *leaves* flat or involute, to 1 mm wide, 1–5 cm long; *ligules* 0.2–1 mm; *glumes* unequal, the lower 2–5 mm and the upper 2.5–7 mm long; *lemmas* 3–7 mm long, terminating in an awn 0.3–9 mm long. Common in shortgrass prairie, on dry slopes, or with pinyon-juniper or sagebrush, 3400–8000 ft. May–July. E/W. (Plate 66)

ZEA L. – CORN

Annuals (ours); *sheaths* open; *ligules* membranous; *spikelets* in spikes (ears), each spike surrounded by several leaf sheaths and a prophyll (husks), with numerous spikelets in 8–24 rows.

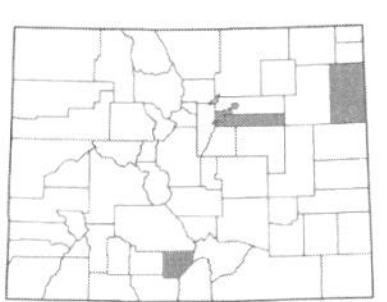

Zea mays **L.**, CORN. Plants 10–30 dm, the lower nodes with adventitious roots; *leaves* flat, distichous, 3–9 dm × 2–12 cm; *inflorescence* of two types, the staminate a terminal panicle of 1–many branches, the pistillate spike in leaf axils, with 4 or more rows of spikelets (ears). Cultivated and occasionally escaping along roadsides and field margins, 3400–5500 ft. June–Aug. E. Introduced.

POLEMONIACEAE Juss. – PHLOX FAMILY

Annual, biennial, or perennial herbs, or subshrubs; *leaves* alternate or opposite, simple or compound, entire to pinnatifid or palmatifid, exstipulate; *flowers* perfect, actinomorphic; *calyx* 5-lobed, either herbaceous throughout or with herbaceous costae separated by hyaline membranes; *corolla* 5-lobed, contorted; *stamens* 5, adnate to the petals, included or exserted; *pistil* 1; *ovary* superior, (2) 3-carpellate; *fruit* usually a loculicidal capsule.

1a. Plants with 2 persistent basal cotyledons and a whorl of bracts subtending a dense terminal head, the stem otherwise naked; small annuals 1–5 cm tall...***Gymnosteris***
1b. Plants with well-developed leaves, lacking persistent cotyledons; annuals, biennials, or perennials...2

2a. Leaves palmately 3–9-cleft to the base into linear or subulate segments (sometimes giving the appearance of whorled leaves) or fascicled in the axils of the main leaves...3
2b. Leaves entire, toothed, pinnately cleft to compound, or pinnatifid with 1–3 pairs of linear lateral segments...5

3a. Flowers in a dense headlike, white-tomentose terminal cluster, blue or white; annuals...***Eriastrum***
3b. Flowers either not in a headlike cluster or the cluster not densely white-tomentose, white to cream; annuals, perennials, or subshrubs...4

4a. Leaves firm, prickly, with a spinulose tip; cushion-forming or matted perennial herb or subshrub to shrub; flowers sessile and solitary in the axils or at the ends of branches...***Linanthus*** (*Leptodactylon*)
4b. Leaves soft, not sharp and prickly; perennial or annual herbs, at most loosely matted; flowers sessile or nearly so in a rather dense headlike cluster at the end of the stems or on slender pedicels and solitary in the leaf axils...***Leptosiphon***

5a. Leaves all simple and entire...6
5b. At least some leaves toothed, pinnatifid to bipinnatifid, or pinnately compound...12

6a. Perennial herbs, often cushion- or mat-forming; leaves opposite and usually densely crowded in cushion- or mat-forming plants...***Phlox***
6b. Annual or biennial herbs, or if perennials then the leaves alternate and in a basal rosette; plants not cushion- or mat-forming...7

7a. Flowers sessile or nearly so in a dense headlike cluster...8
7b. Flowers pedicellate, the inflorescence various but not a headlike cluster...9

8a. Leaves lanceolate to oblanceolate or elliptic, 3–13 mm wide, often with smaller fascicled leaves in the axils, the lowermost leaves usually opposite; calyx green or chartaceous and uniform throughout in texture, the lobes not spinulose-tipped...***Collomia***
8b. Leaves linear, 0.5–1.2 mm wide, lacking fascicles, alternate; calyx with herbaceous veins separated by hyaline sinuses, the lobes spinulose-tipped...***Ipomopsis***

9a. Leaves opposite below and alternate above; stamens unequally inserted on the corolla tube; plants villous with multicellular hairs and often glandular above...***Microsteris***
9b. Leaves mostly or all alternate, or in a basal rosette with alternate stem leaves; stamens more or less equally inserted on the corolla tube; plants glabrous or glandular-stipitate...10

10a. Biennials or perennials from stout taproots; plants glabrous; leaves thick and succulent, the basal leaves linear-spatulate and forming a rosette, the upper leaves linear and bract-like; flowers crowded in linear leaves from the middle of the stem upwards...***Aliciella*** (*sedifolia*)
10b. Annuals from slender taproots; plants glandular-stipitate to subglabrous; leaves not thick and succulent, linear to linear-oblong; flowers terminal and solitary or in a loose cyme...11

11a. Corollas 1.5–3 (3.5) mm long; filaments evidently attached well below the sinuses of the corolla tube; capsules 1.5–2 mm long...***Lathrocasis***
11b. Corollas 4–9 mm long; filaments attached just below the sinuses of the corolla; capsules 3–4 mm long...***Navarretia*** (*sinistra*)

12a. Leaves pinnately compound with broad, oblong, elliptic to ovate or suborbicular, more or less distinct leaflets, either the upper few leaflets confluent or the leaflets whorled on the rachis; calyx herbaceous throughout; perennial, glandular-hairy herbs...***Polemonium***
12b. Leaves toothed or pinnatifid to bipinnatifid into narrow segments; calyx with herbaceous veins separated by hyaline sinuses; annual or perennial herbs, variously hairy...13

13a. Flowers sessile or nearly so in a dense headlike cluster or spicate inflorescence...14
13b. Flowers not in a dense headlike cluster, sessile or usually at least some pedicellate...16

14a. Calyx lobes equal or nearly so in length; leaves pinnatifid, the segments sometimes shortly spinulose at the tips but not appearing prickly; annuals, biennials, or perennials...***Ipomopsis***
14b. Calyx lobes usually more or less evidently unequal in length; leaves pinnatifid or bipinnatifid with linear, spinulose segments, often giving the plant a prickly appearance; annuals...15

15a. Headlike clusters white-tomentose with interwoven hairs; flowers blue or white...***Eriastrum***
15b. Headlike clusters glabrous to villous but not white-tomentose; flowers yellow, white, pale lilac, or pale blue...***Navarretia***

16a. Corollas rotate, with spreading lobes and little to no tube; perennials, often woody at the base; stem leaves well-developed, pinnatifid with spinulose, linear segments, lacking a basal rosette or leaves; flowers blue-violet to purple...***Giliastrum***
16b. Corollas funnelform to salverform or campanulate, usually with a distinct tube and erect to spreading lobes; annuals, biennials, or perennials but not woody at the base; stem leaves often reduced, with or without a basal rosette of leaves, pinnatifid and sometimes shortly spinulose; flowers variously colored...17

17a. Annuals from slender taproots; flowers in a loose, open, diffusely branching cyme or solitary, small, the corolla 3–12 mm long; leaves mostly in a basal rosette, the stem leaves greatly reduced...18
17b. Biennials or perennials, or if annuals then unlike the above in all respects...19

18a. Plants with at least some fine, arachnoid, cottony hairs on the stems below...***Gilia***
18b. Plants lacking fine, arachnoid, cottony hairs on the stems below...***Aliciella***

19a. Calyx lobes spinulose-tipped; plants glabrous to glandular-hairy, usually also with nonglandular, white hairs in addition to glandular hairs (at least below on the stem); seeds becoming mucilaginous when wetted; corolla tube 5–45 mm long; flowers white, cream, blue, lavender, pink, or red...***Ipomopsis***
19b. Calyx lobes acute, lacking a spinulose tip or weakly spinulose tipped; plants usually glandular hairy throughout, lacking white, nonglandular hairs; seeds not becoming mucilaginous when wetted; corolla tube 4–15 mm long; flowers rose-purple, lavender, blue, or white...***Aliciella***

ALICIELLA Brand – ALICIELLA

Annual or perennial, glandular-hairy herbs; *leaves* alternate, entire to toothed or pinnatifid; *calyx* with a hyaline membrane on the sinuses; *corolla* salverform or funnelform. (Porter, 1998)

A report of *A. subnuda* (Torr. ex A. Gray) J.M. Porter, the red or crimson gilia, in Colorado is apparently based on a misidentification of *A. haydenii* ssp. *crandallii* and does not occur in the state.

1a. Annuals from slender taproots; stem leaves all or nearly all reduced to small, linear bracts; corolla 1.5–9 mm long, usually with a pinkish-lavender or white limb and yellow throat, the lobes mucronate or tridentate at the apex; calyx 1–3 mm long...2
1b. Biennials or perennials from stout taproots; stem leaves conspicuous, the lower ones similar to the basal leaves and gradually reduced up the stem, sometimes becoming linear and bract-like; corolla 4–22 mm long, blue, purple, white, or rose, the lobes entire at the apex; calyx 2.5–6 mm long...4

2a. Corolla lobes tridentate at the tips; corolla tube conspicuously longer than the calyx...***A. triodon***
2b. Corolla lobes entire or cuspidate at the tips but not tridentate; corolla tube longer or shorter than the calyx...3

3a. Corolla 4.5–9 mm long; corolla tube conspicuously longer than the calyx, (2) 3–6 mm long; terminal pedicel shorter than the calyx...***A. leptomeria***
3b. Corolla 1.5–4 mm long; corolla tube shorter or only slightly longer than the calyx, (0.8) 1.2–3 mm long; pedicels all longer than the calyx...***A. micromeria***

4a. Stamens included in the corolla; corolla rose-purple, sometimes drying dull blue, the tube 9–20 mm long; flowers mostly crowded at the tips of the branches...***A. haydenii***
4b. Stamens exserted from the corolla; corolla blue, purple, or white, the tube 4–10 mm long; inflorescence various...5

5a. Basal leaves glabrous, entire, linear-spatulate, thick and succulent; bracts linear, at least the lower ones longer than the flowers...***A. sedifolia***
5b. Basal leaves usually glandular-hairy (sometimes sparsely so), rarely glabrous, entire to pinnatifid, not thick or succulent; bracts unlike the above...6

6a. Flowers white or cream, rarely pale lavender, in a narrow, fairly dense cylindric thyrse; corolla lobes mostly erect...***A. stenothyrsa***
6b. Flowers blue to purple; inflorescence usually open, branching, not dense and narrow; corolla lobes erect or spreading...7

7a. Corolla lobes erect; basal leaves glaucous...***A. mcvickerae***
7b. Corolla lobes spreading at about a 90-degree angle from the tube; basal leaves not glaucous...8

8a. At least some basal leaves entire...***A. penstemonoides***
8b. Basal leaves pinnatifid...***A. pinnatifida***

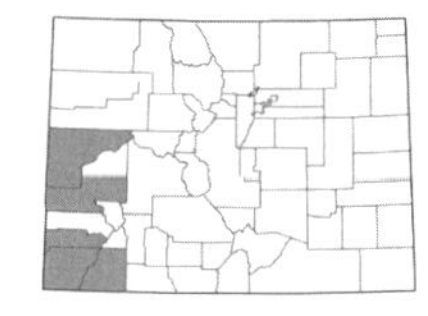

Aliciella haydenii **(A. Gray) J.M. Porter**, San Juan gilia. Biennials, 1–5 dm, sparsely glandular-stipitate throughout to subglabrous; *basal leaves* coarsely toothed to pinnately cleft or lobed, with a broad midstrip, 1.5–6 × 0.5–1.5 cm; *cauline leaves* abruptly reduced, mostly linear and entire, glandular; *calyx* 2.5–4.5 mm long, stipitate-glandular; *corolla* tube and throat 9–20 mm long, rose-purple to pink-lavender. Found on shale, rocky soil, barren outcroppings, sometimes with pinyon-juniper or sagebrush, 4500–8000 ft. April–Sept. W.

There are two subspecies of *A. haydenii* in Colorado:

1a. Corolla lobes 3.5–5.5 mm long, the corolla tube glabrous or sparsely glandular; flowers often drying dull blue...**ssp. *haydenii***, San Juan gilia. [*Gilia haydenii* A. Gray; *Gilia montezumae* Tidestr. & Dayton]. 4500–7500 ft.
1b. Corolla lobes 6–9 mm long, the corolla tube glandular; flowers drying pink...**ssp. *crandallii* (Rydb.) J.M. Porter**, [*Gilia crandallii* Rydb.; *G. montezumae* Tidestr. & Dayton]. Known from La Plata and Montezuma cos., 6500–8000 ft.

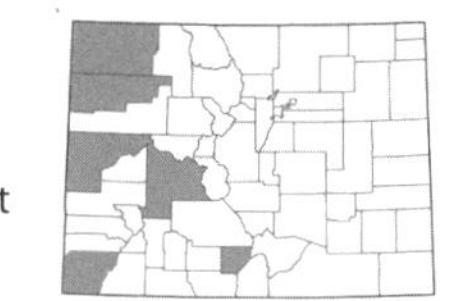

Aliciella leptomeria **(A. Gray) J.M. Porter**, sand gilia. [*Gilia leptomeria* A. Gray; *Gilia subacaulis* Rydb.]. Annuals to 3 dm, *stems* glandular-stipitate at least above; *basal leaves* shallowly pinnatilobed, bright green, usually glabrous, 6 (9) × 1.5 (2.5) cm; *cauline leaves* mostly reduced to bracts; *calyx* 2–3 mm long; *corolla* 4.5–9 mm long, the limb pink-violet to white and the throat usually yellow. Uncommon in dry, sandy and gravelly places, often with sagebrush or pinyon-juniper, 4700–6500 ft. April–June. E/W.

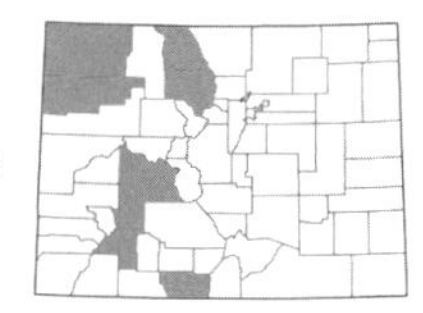

Aliciella mcvickerae **(M.E. Jones) J.M. Porter**, McVicker's gilia. [*Gilia calcarea* M.E. Jones]. Biennials, 1.5–7 dm; *basal leaves* deeply pinnatifid with narrowly elliptic lobes, mostly 1.5–8 cm × 7–30 mm, glaucous; *cauline leaves* reduced; *calyx* 2.5–4.5 mm long; *corolla* blue to blue-violet, the tube 4–9 mm and the lobes 3–5 mm long, erect. Uncommon in sandy soil, rock crevices, and on rocky outcroppings, 6500–7000 ft. June–Aug. (Sept.). W.

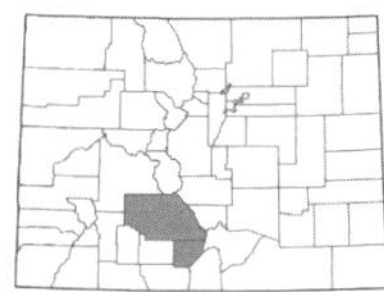

Aliciella micromeria **(A. Gray) J.M. Porter**, dainty gilia. [*Gilia micromeria* A. Gray]. Annuals to 1.5 dm, stems glandular-stipitate at least above; *basal leaves* shallowly pinnatilobed, bright green, usually glabrous, 5–9 × 1–2.5 cm; *cauline leaves* mostly reduced to bracts; *calyx* 1–2.5 mm long; *corolla* with a pink-violet or white limb and usually yellow throat, 1.5–4 mm long. Uncommon in alkaline soil or saline flats, or sandy places, often with sagebrush, 7000–8000 ft. April–June. E.

Aliciella micromeria might be better represented as a variety of *A. leptomeria*. The elevation and ranges of the above three species are based on very few specimens and may not be entirely accurate.

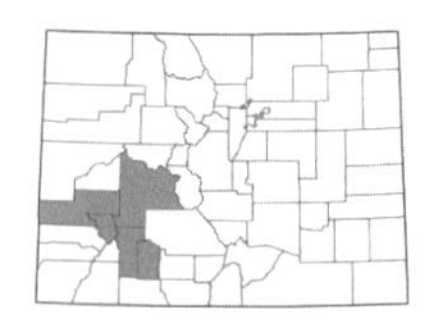

Aliciella penstemonoides **(M.E. Jones) J.M. Porter**, Black Canyon gilia. Biennials or perennials, 0.5–2 dm, sparsely glandular above; *basal leaves* linear to linear-lanceolate, entire or pinnately lobed, to 5 cm long; *cauline leaves* reduced, linear and entire or slightly lobed; *calyx* 2.5–4 mm long; *corolla* blue, lavender, or sometimes white, the tube ca. 10 mm long with ca. 5 mm long lobes. Found in rocky crevices of cliffs and rimrock, 6800–9000 ft. June–Aug. W. Endemic.

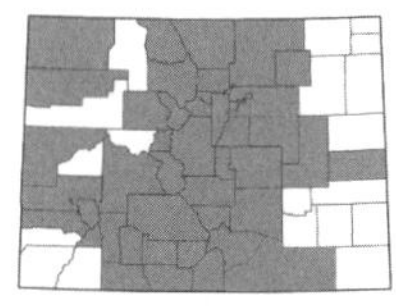

Aliciella pinnatifida **(Nutt. ex A. Gray) J.M. Porter**, sticky gilia. Biennials or sometimes annuals, 1.5–7 dm, glandular-hairy throughout; *basal leaves* deeply pinnatifid with narrow to elliptic lobes, mostly 1.5–8 cm × 7–30 mm; *cauline leaves* pinnatifid, reduced; *calyx* 2.5–4.5 mm long; *corolla* blue to blue-violet, the tube 4–9 mm and the lobes 3–5 mm long. Common on dry, sandy or gravelly slopes, alpine, in meadows, open forests, sagebrush, dry grasslands, and in cracks of rocky cliffs, 5000–14,000 ft. May–Aug. E/W.

Aliciella sedifolia **(Brandegee) J.M. Porter**, stonecrop gilia. [*Gilia sedifolia* Brandegee]. Biennials with glandular stems; *basal leaves* entire, linear-spatulate, sessile, succulent, ca. 1 cm long; *cauline leaves* reduced and bract-like; *calyx* ca. 4 mm long; *corolla* blue-violet, mostly 6–8 mm long with erect lobes. Uncommon on rocky, barren alpine slopes, 13,000–14,000 ft. July–Aug. W.

This species was previously known only from the type specimen and wasn't collected again for over 100 years. The type locality is most likely from San Juan Co. on Sheep Mt. The report of this species from Gunnison Co. is based on the type locality of Sheep Mountain (there are lots of Sheep Mountains in Colorado) but was incorrectly placed in the wrong county.

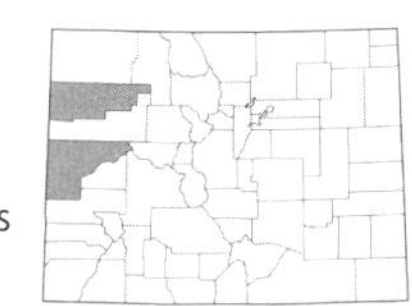

***Aliciella stenothyrsa* (A. Gray) J.M. Porter**, Uinta Basin gilia. [*Gilia stenothyrsa* A. Gray]. Biennials with glandular stems, mostly 1.5–6 dm; *basal leaves* deeply pinnatifid with linear segments, sometimes entire or few-toothed, mostly 1.5–5 cm × 3–13 mm (including lobes); *cauline leaves* reduced upwards; *calyx* 3.5–6 mm long; *corolla* white to pale purple, the tube 6–10 mm and lobes 3.5–5 mm long. Uncommon on dry, sandy or clay soil, sometimes in sagebrush/greasewood or pinyon-juniper, 4800–6000 ft. May–June. W.

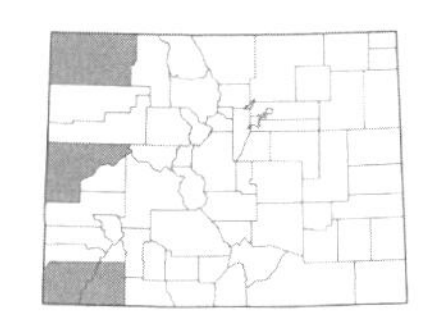

***Aliciella triodon* (Eastw.) Brand**, coyote gilia. [*Gilia triodon* Eastw.]. Annuals, 0.5–1.5 dm, glandular-hairy; *basal leaves* shallowly pinnatifid, 0.5–2 cm long; *cauline leaves* linear and entire; *calyx* 2–3 mm long; *corolla* white with a purplish tube, 5–7 mm long, the lobes trifid at the apex. Uncommon in dry, sandy places, 6500–7500 ft. April–June. W.

COLLOMIA Nutt. – trumpet

Annual or perennial herbs; *leaves* sessile, alternate or sometimes opposite below, simple, the margins entire to pinnatifid; *calyx* with a herbaceous tube throughout or hyaline just below the sinuses, in age the sinuses with conspicuous outwardly projecting folds; *corolla* lavender, pink, or white.

1a. Flowers 8–15 mm long, pink to purple or white...***C. linearis***
1b. Flowers 15–30 mm long, salmon or yellowish-white...***C. grandiflora***

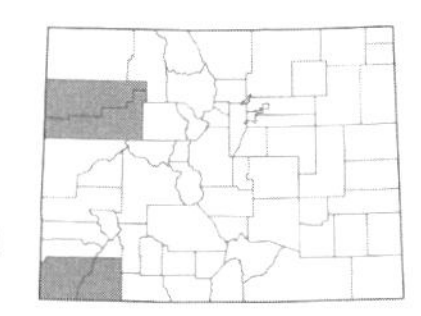

***Collomia grandiflora* Douglas ex Lindl.**, large collomia. Annuals, 1–5 dm; *leaves* lanceolate to narrowly elliptic, 1.5–6.5 cm × 3–13 mm, entire; *calyx* 7–10 mm long; *corolla* 15–30 mm long, salmon or yellowish-white; *capsules* ca. 5 mm long. Uncommon on rocky hillsides, often found in recently burned areas, 7000–8000 ft. May–July. W. (Plate 67)

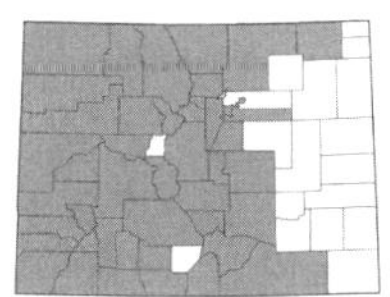

***Collomia linearis* Nutt.**, tiny trumpet. Annuals, 0.5–5 dm; *leaves* lanceolate to narrowly elliptic, 1–6 cm × 3–10 mm, entire; *calyx* 4–7 mm long; *corolla* 8–15 mm long, pink to purple or white; *capsules* ca. 5 mm long. Common in meadows, forests, and along streams, sometimes in disturbed places, widespread, 4500–11,000 ft. June–Aug. E/W.

ERIASTRUM Wooton & Standl. – woollystar

Annual herbs; *leaves* alternate or rarely some opposite, simple, linear and entire or pinnatifid; *calyx* with hyaline, membranous sinuses; *corolla* pink, purplish, blue, or white, the corolla funnelform.

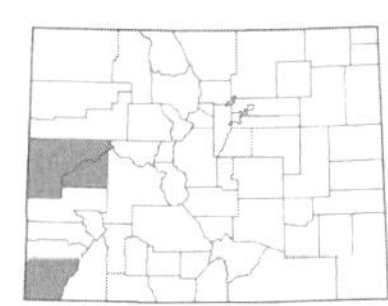

***Eriastrum diffusum* (A. Gray) H. Mason**, spreading eriastrum. Plants 3–15 cm tall, diffusely branched from the base; *leaves* linear or pinnatifid with 3–5 narrow lobes, 0.5–2.5 cm long; *inflorescence* a capitate, tomentose cyme; *calyx* 5–7 mm long, with spinulose-tipped lobes; *corolla* 7–9 mm long, blue to white; *capsules* 3–4 mm long. Uncommon on dry, sandy, shale or adobe slopes, 4900–6500 ft. April–June. W.

GILIA Ruiz & Pav. – gilia

Annual herbs, glandular- and often cobwebby-hairy herbs from a taproot; *leaves* alternate and mostly basal, usually pinnatifid or bipinnatifid; *calyx* with hyaline, membranous sinuses; *corolla* salverform or funnelform. (Grant & Grant, 1956)

1a. Corolla small, mostly 4–6.5 mm long, the tube and part of the throat included in the calyx; basal leaves usually few and not forming a dense tuft; stem leaves well-developed...***G. tweedyi***
1b. Corolla usually longer, 4.5–12 mm long, the tube usually longer than the calyx, the throat always exserted beyond the calyx; basal leaves usually well-developed, forming a rosette; stem leaves usually much reduced immediately above the rosette...2

2a. Stem leaves auriculate-clasping on the stem; rachis of leaves 2–6 mm wide at the broadest point; leaves pinnatifid or shallowly lobed, the lobes usually as long or only slightly longer than the rachis is wide; lower stems glabrous or with only a few cobwebby hairs...***G. sinuata***
2b. Stem leaves sessile but not auriculate-clasping; rachis of leaves mostly 1–2.5 mm wide at the broadest point; leaves pinnatifid or often bipinnatifid, the lobes longer than the rachis is wide; lower stems at least somewhat cobwebby-hairy, often densely so...3

3a. Corolla tube and throat extended approximately 3–5 mm beyond the calyx; capsules ovoid...***G. ophthalmoides***
3b. Corolla tube and throat extended approximately 1–2 mm beyond the calyx; capsules nearly globose...***G. clokeyi***

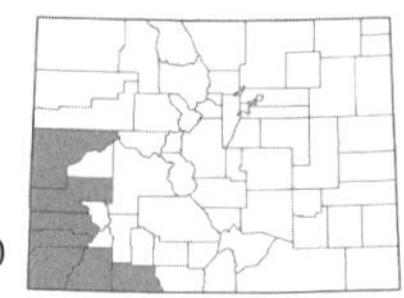

***Gilia clokeyi* H. Mason**, Clokey's gilia. Plants 8–17 (35) cm; *leaves* pinnatifid or bipinnatifid with a narrow rachis 1–2 mm wide, the lobes mostly 2–4 mm long; *inflorescence* diffuse with often divaricate branches; *calyx* 2.3–4.5 mm long, glabrous to gland-dotted or arachnoid; *corolla* violet to blue-violet or pink-violet with a yellow throat, 4.3–9.3 mm long with lobes 1–2.8 mm long. Common on dry slopes, rocky ridges, in pinyon-juniper woodlands, or with sagebrush, 4500–7000 ft. April–May. W.

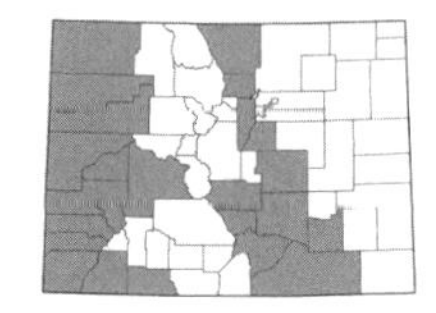

***Gilia ophthalmoides* Brand**, eyed gilia. [*G. inconspicua* (J. Sm.) Sweet]. Plants 15–30 cm; *leaves* bipinnatifid with a narrow rachis 1–2.5 (4) mm wide; *inflorescence* usually loose with 1–5 flowers borne above each bract; *calyx* 2.5–4.6 mm long, gland-dotted or arachnoid; *corolla* light violet to purple with a yellow throat, drying blue, 5.5–11.3 mm long, with lobes 1.5–2 mm. Common in dry, often sandy places, forest openings, prairie, sagebrush, and on rocky ridges, 4300–8500 ft. April–July. E/W.

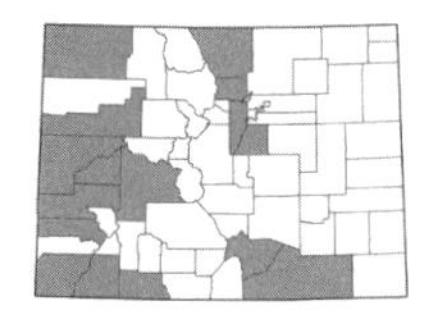

***Gilia sinuata* Douglas ex Benth.**, rosy gilia. Plants 10–30 cm; *leaves* pinnatifid or shallowly lobed, with a rachis 2–6 mm wide at the broadest point, auriculate-clasping; *inflorescence* diffuse with 1–3 flowers borne above each bract; *calyx* 2.5–4 mm long, glandular; *corolla* purple with a yellow throat, the lobes violet to pinkish or nearly white, 7–11.5 mm long with lobes 1.4–3.2 mm. Uncommon on dry, sandy soil, 4300–6500 ft. April–June. W.

***Gilia tweedyi* Rydb.**, Tweedy's gilia. Plants 5–20 cm; *leaves* pinnatifid into linear lobes with a narrow rachis 1–1.5 mm wide, or the uppermost reduced to small bracts; *inflorescence* loose with 1–2 flowers borne above each bract; *calyx* gland-dotted; *corolla* tube white, 4–6.5 mm long, included in the calyx, the throat yellow, and lobes pale violet. Uncommon in dry, sandy soil, 6500–7000 ft. April–June. W.

GILIASTRUM (Brand) Rydb. – giliastrum

Perennials, suffrutescent, glandular-hairy herbs from a woody base; *leaves* alternate, pinnatifid; *calyx* with herbaceous veins separated by hyaline membranes; *corolla* rotate, blue-violet to purple with a yellow throat.

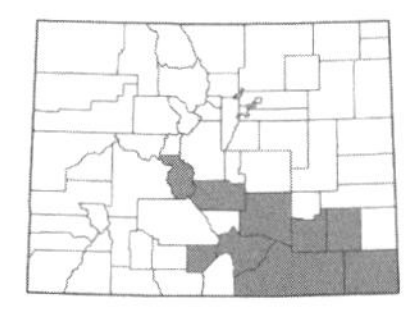

***Giliastrum acerosum* (A. Gray) Rydb.**, bluebowls. [*Gilia acerosa* (A. Gray) Britton; *Gilia rigidula* Benth.]. Plants 0.8–2.5 dm, diffusely branched from the base; *leaves* pinnatifid with 2–7 narrow segments 2–12 mm long; *inflorescence* solitary or a loose cyme; *calyx* tube 2–5 mm long; *corolla* tube 3–4 mm long, lobes 3–5 mm long; *capsules* 3–5 mm long. Uncommon on rocky hillsides, on sandstone outcrops, and in sandy soil of the shortgrass prairie, 3800–5000 (7300) ft. April–June. E. (Plate 67)

GYMNOSTERIS Greene – gymnosteris

Annual herbs; *leaves* represented by persistent connate-clasping cotyledons near the base of the stem and foliaceous bracts subtending the inflorescence; *flowers* few, in terminal heads, the 4–5 bracts scarious at the base and partly united as an involucre; *calyx* membranous basally, the lobes green or reddish, not folded in the sinuses; *corolla* white or yellowish.

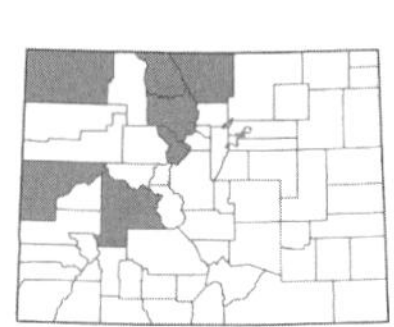

***Gymnosteris parvula* A. Heller,** smallflower gymnosteris. Plants 1–5 cm; *cotyledons* 1–3 mm long; *bracts* 4–13 × 1–5 mm, often purplish; *calyx* 3–5 mm long, the lobes bristle-tipped; *corolla* 5–7 mm long, persistent; *capsules* 3.5–5 mm long. Uncommon or perhaps just easily overlooked, found on gravelly soil, usually with sagebrush, 6500–9500 ft. May–July. E/W.

IPOMOPSIS Michx. – ipomopsis

Annual or biennial herbs or short-lived perennials; *leaves* alternate on the stem, in a rosette at the base, pinnatifid or palmatifid; *calyx* with herbaceous veins separated by hyaline membranes, the lobes equal to subequal; *corolla* white, pink, or red. (Grant, 1956)

1a. Flowers in a dense, terminal capitate cyme or head...2
1b. Flowers in a spicate or narrow thyrse or flat-topped, loose cyme...8

2a. Plants annual from a slender taproot...3
2b. Plants perennial, sometimes subshrubby at the base...5

3a. Leaves all basal except for 1–2 stem leaves, pinnatifid, 2–10 mm wide; corolla lobes mostly 1–2 mm long…***I. polycladon***
3b. Leaves mostly or all cauline, linear and entire or linear with 1–2 pairs of lateral linear segments, mostly 1–2 mm wide; corolla lobes 2–4 mm long…4

4a. Plants more or less densely white-tomentose on the stems and in the inflorescence; bracts and most leaves trifid or pinnately cleft with 5 narrow segments, sometimes a few linear and entire…***I. pumila***
4b. Plants glandular or inconspicuously white-tomentose; bracts and leaves linear and entire…***I. gunnisonii***

5a. Plants of the high alpine, at 12,000 ft or higher in elevation, with a strong, heavy scent; stems and calyx densely arachnoid-villous to white-tomentose, the hairs obscuring the sepals; corolla tube 5–7 mm long; flowers white to bluish when fresh but turning brown or blackish when pressed…***I. globularis***
5b. Plants below 11,000 ft in elevation, usually lacking a strong scent; stems and calyx glabrous to arachnoid-villous or tomentose but the tomentum usually not obscuring the sepals; corolla tube 3–7 mm long; flowers white, sometimes turning brown when pressed…6

6a. Filaments longer than the anthers, usually twice as long or more; anthers slightly but evidently elevated above the orifice of the corolla; corolla tube 3–5 (7) mm long, usually equal to or only slightly surpassing the calyx, rarely evidently surpassing it…***I. congesta***
6b. Filaments shorter than or equal to the anthers in length; anthers usually appearing sessile at the throat of the corolla and scarcely elevated above the orifice; corolla tube 5–9 mm long, equal to or more often evidently surpassing the calyx…7

7a. Stems white-tomentose to arachnoid-woolly; flowers white, usually drying brown; calyx green, 4-5.5 mm long; plants lacking a woody base…***I. spicata***
7b. Stems villous or finely glandular, not tomentose or arachnoid-woolly; flowers white, not drying brown; calyx often pinkish, 5.5-8.5 mm long; plants often woody at the base…***I. roseata***

8a. Flowers on evident pedicels mostly 1–4 cm long, solitary or paired in a flat- or round-topped, loosely aggregated cyme, white or light blue-lavender; plants usually diffusely branched from the base or near the middle, lacking a basal rosette…9
8b. Flowers sessile or nearly so, or on pedicels under 1 cm long, in an elongate, spicate or narrow thyrse, variously colored; plants with 1–several stems from a basal rosette, usually not branched throughout…10

9a. Corolla tube (8) 10–20 (25) mm long, the lobes 4–6 (7) mm long…***I. laxiflora***
9b. Corolla tube usually longer, (25) 30–40 (55) mm long, the lobes (7) 8–12 mm long…***I. longiflora***

10a. Corolla tube (10) 14–50 mm long and lobes 6–13 mm long; flowers white, white with pink dots, violet, pink, or red…11
10b. Corolla tube 4.5–10 mm long and lobes 2.5–6 mm long; flowers white, sometimes with purple spots, or pale blue to lavender…12

11a. Stamens deeply included within the corolla tube; corolla white with pink dots; plants strongly glandular…***I. ramosa***
11b. Stamens shortly exserted or if included within the corolla tube then usually inserted well above the middle of the tube; corolla red, white, or pink, sometimes white with pink dots but these usually restricted to the lobes and not on the corolla tube; plants glandular or not…***I. aggregata***

12a. Stamens included, the anthers appearing sessile at the throat of the corolla; stems white-tomentose to arachnoid-woolly; flowers white, usually drying brown and in a dense, spicate, bracteate cluster…***I. spicata***
12b. Stamens prominently exserted; stems minutely hairy to glandular but not white-tomentose; flowers white with purple spots or pale blue to lavender, not in a dense spicate cluster…13

13a. Flowers white or pale lavender and usually with purple or pink spots; leaves pinnatifid with 5 or more narrow segments; stems usually densely glandular-hairy and with nonglandular, white hairs towards the base of the stem…***I. polyantha***
13b. Flowers pale blue or lavender, usually with darker markings in the throat; leaves pinnatifid with 2–5 narrow segments or some entire and linear; stems usually with numerous nonglandular, white hairs throughout the stem in addition to a few short glandular hairs…***I. multiflora***

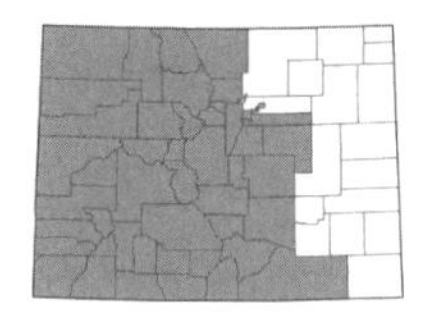

***Ipomopsis aggregata* (Pursh) V.E. Grant**, SKYROCKET; SCARLET GILIA. [*Gilia aggregata* (Pursh) Spreng.]. Biennials or short-lived perennials, usually 2–10 dm, the stems glandular-stipitate at least above or not; *leaves* pinnatifid or subpinnatifid with narrow segments, reduced upwards; *inflorescence* a narrow thyrse or paniculate; *corolla* white, pink, or red, the tube (1) 1.5–5 cm long, mostly 1.5–5 mm wide at the summit, the lobes spreading with acute to acuminate or attenuate tips. Common on open, gravelly slopes, in grassy or sagebrush meadows, forest openings, and along roadsides, 5100–10,500 ft. May–Aug. (Sept.). E/W. (Plate 67)

A highly variable complex composed of several intergrading varieties:

1a. Calyx lobes triangular, 2 mm long, the calyx united ⅕–¾ its length; anthers included; corolla tube 20–40 mm long; plants of the eastern slope…2

1b. Calyx lobes 3–5 mm long, the lobes usually distinctly longer than the tube; anthers included or exserted; corolla tube 10–45 mm long; plants of the eastern or western slope…3

2a. Corolla trumpet shaped with the tube flaring upward, red or pink, the tube 20–25 mm long…**ssp. *collina* (Greene) Wilken & Allred**, SCARLET GILIA. [*Gilia candida* Rydb. ssp. *collina* (Greene) Wherry]. Common in forest clearings, along roadsides, and in open, gravelly soil, found along the Front Range and east of the Continental Divide, 6000–9500 ft. May–Aug. E.

Similar to ssp. *aggregata* but with a usually longer tube and included stamens, and commonly intergrading with ssp. *candida* (especially in Jefferson and Park cos.).

2b. Corolla salverform with tube little expanded upward, white, pink, or rarely red, the tube 20–40 mm long…**ssp. *candida* (Rydb.) V.E. Grant & A.D. Grant**, WHITE SKYROCKET. [*Gilia candida* Rydb.]. Common on gravelly slopes and along roadsides, found along the Front Range and mountains east of the Continental Divide, 6000–10,000 ft. May–Aug. E.

3a. Corolla trumpet-shaped with the tube flared upward (3–5 mm wide at summit), the tube 15–22 mm long, red with yellow spots or rarely pink; anthers exserted or the highest anther slightly inserted…**ssp. *aggregata***, SCARLET SKYROCKET. [*I. aggregata* ssp. *formosissima* (Greene) Wherry]. Common in meadows, open forests, along roadsides, and in sagebrush meadows, mostly scattered on the west slope with a few collections from the east slope (Jackson Co.), 5100–10,500 ft. May–Aug. E/W.

Intergrades with ssp. *attenuata* in northern Colorado.

3b. Corolla salverform with the tube only slightly widened upward (1–3 mm wide at summit), the tube 10–45 mm long, white, violet, pink, or red; anthers well-included to exserted…4

4a. Corolla tube 10–25 mm long, white, pink, or red; anthers included or sometimes exserted…**ssp. *attenuata* (A. Gray) V.E. Grant & A.D. Grant**, SHORTTUBE SKYROCKET. [*Gilia attenuata* (A. Gray) A. Nelson; *I. aggregata* ssp. *weberi* V.E. Grant & Wilken]. Common in gravelly soil, along roadsides, and in meadows, found in the northern and northcentral counties, 6500–9700 ft. June–Aug. E/W.

4b. Corolla tube 24–45 mm long, white, violet, or pink; anthers mostly well-included… **ssp. *tenuituba* (Rydb.) Ackerfield, comb. nov.** (see p. ix), SLENDERTUBE SKYROCKET. [*Gilia tenuituba* Rydb.]. Common on open, gravelly slopes, in grassy or sagebrush meadows, and along roadsides, found in the northcentral and westcentral counties, 5500–10,500 ft. June–Aug. E/W.

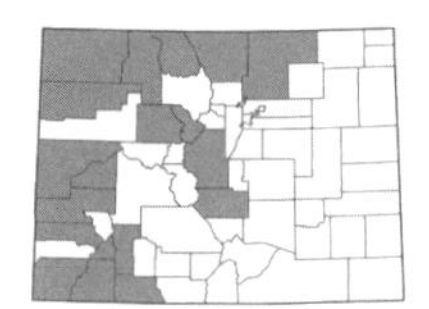

***Ipomopsis congesta* (Hook.) V.E. Grant**, BALL-HEAD IPOMOPSIS. [*Gilia congesta* Hook.]. Perennials, 0.5–5 dm, the stems white-tomentose to glabrate; *leaves* linear and entire to pinnatifid or trifid with narrow segments, 1–4 cm long; *inflorescence* of 1–several dense cymose heads; *calyx* mostly 3–5 mm long, the tips minutely spinulose; *corolla* white, the tube 3–5 (7) mm long, the lobes 1.5–3 mm long. Common on dry hillsides and open slopes, in meadows, often in sagebrush or pinyon-juniper, 4500–10,600 ft. May–Sept. E/W.

There are 3 subspecies of *I. congesta* in Colorado:

1a. Leaves mostly trifid to pinnatifid, sometimes with a few entire…**ssp. *congesta***. The most common and widespread variety, 4500–10,600 ft. May–Sept. E/W.

1b. Leaves mostly or all entire…2

2a. Leaves mostly green and glabrous; plants to 1.2 dm tall, usually with simple stems above the branching base…**ssp. *crebrifolia* (Nutt.) A.G. Day**. Uncommon on the western slope, currently known from La Plata and Montrose cos., 5500–6500 ft. May–Aug. W.

2b. Leaves sparsely to evidently arachnoid-pubescent or tomentose, rarely glabrous; plants 1–5 dm tall, usually with the stems branched above…**ssp. *frutescens* (Rydb.) A.G. Day**. Known from the western counties, 4500–10,600 ft. May–Aug. W.

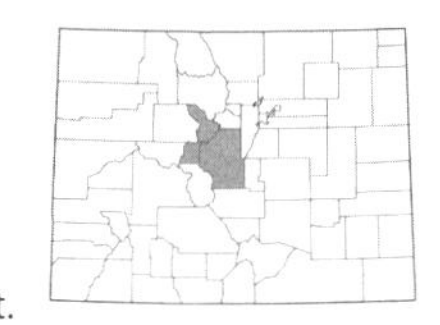

Ipomopsis globularis **(Brand) W.A. Weber**, Hoosier Pass ipomopsis. [*I. spicata* (Nutt.) V.E. Grant ssp. *capitata* (A. Gray) V.E. Grant]. Perennials, 0.6–1.2 dm, the stems densely arachnoid-tomentose; *leaves* mostly linear and entire, or rarely pinnatifid with 1–5 narrow lateral segments, densely arachnoid-villous to tomentose below; *inflorescence* capitate, globose, 2–2.5 cm wide; *calyx* tube 4–5 mm long with lobes 1.5–3 mm long; *corolla* white or cream but drying dark brown, the tube 5–7 mm and lobes 3–4 mm long. Uncommon on high alpine slopes and ridges, 12,000–14,000 ft. July–Aug. E/W. Endemic.

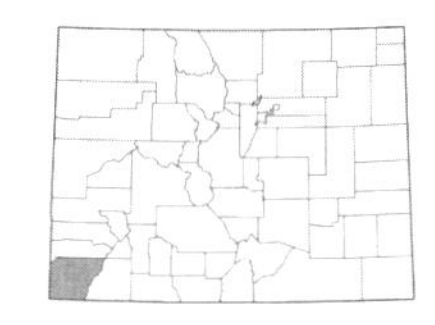

Ipomopsis gunnisonii **(Torr. & A. Gray) V.E. Grant**, sand dune ipomopsis. [*Gilia gunnisonii* Torr. & A. Gray]. Annuals, 6–25 cm, glabrous to minutely hairy or glandular-stipitate; *leaves* linear and entire, to 4 cm long, or occasionally some with 1–2 pairs of small lateral lobes; *inflorescence* a dense cymose cluster with leafy bracts; *calyx* 3.5–4.5 mm long; *corolla* white to pale lavender, 6–10 mm long. Uncommon in sandy soil and on dry slopes, sometimes with pinyon-juniper, 5000–6600 ft. May–Aug. W.

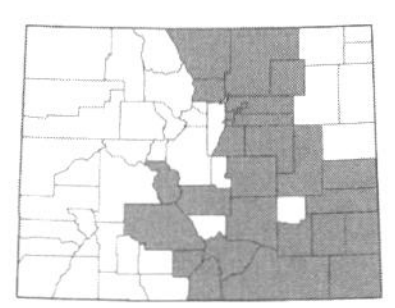

Ipomopsis laxiflora **(J.M. Coult.) V.E. Grant**, iron ipomopsis. [*Gilia laxiflora* (J.M. Coult.) Osterh.]. Annuals or biennials, 1–4 dm, glabrous to glandular-hairy; *leaves* pinnately divided into 3–5 filiform or linear segments 5–15 mm long; *inflorescence* solitary or a loosely aggregated cyme; *calyx* tube (2) 3–4 mm long with lobes 2–3 mm long; *corolla* white to light blue-violet, the tube (8) 10–20 (25) mm long and lobes 4–6 (7) mm long. Common on the shortgrass prairie, in sandy soil, and on sandstone outcroppings, 3600–7000 ft. May–Sept. E.

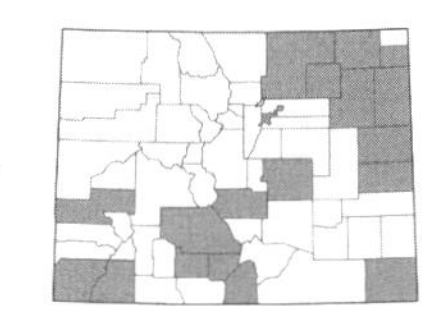

Ipomopsis longiflora **(Torr.) V.E. Grant**, long-flowered ipomopsis. [*Gilia longiflora* (Torr.) G. Don]. Annuals or biennials, 1–6 dm, glabrous or inconspicuously arachnoid-pubescent; *leaves* usually pinnately divided with 1–3 pairs of filiform to linear segments 0.5–2 cm long; *inflorescence* a flat-topped, loose cyme; *calyx* 5–8 mm long, glabrous to finely glandular; *corolla* light blue-violet to white, the tube (25) 30–40 (55) mm long and the lobes (7) 8–12 mm long. Found in sandy soil, often with sagebrush, 3800–8000 ft. June–Sept. E/W.

Ipomopsis multiflora **(Nutt.) V.E. Grant**, many-flowered ipomopsis. [*Gilia multiflora* Nutt.]. Perennials, 1.5–4 dm, with glandular and eglandular hairs; *leaves* pinnately divided into linear segments or sometimes linear and entire; *inflorescence* a narrow, elongated thyrse; *calyx* 4–5 mm long; *corolla* pale blue to lavender, usually with darker markings in the throat, the tube 5–10 mm long and lobes 3–5 mm long. Uncommon in sandy soil, 6500–8500 ft. June–July. E/W.

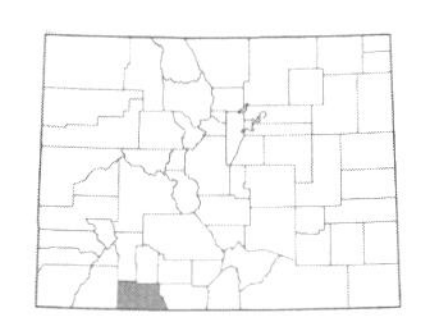

Ipomopsis polyantha **(Rydb.) V.E. Grant**, Pagosa sky rocket. [*Gilia polyantha* Rydb.]. Short-lived perennials, 1.5–4 dm, minutely hairy to villous or sometimes inconspicuously glandular; *leaves* pinnatifid into few linear segments 1–2.5 cm long; *inflorescence* an elongated thyrse; *calyx* 3.5–5 mm, glandular; *corolla* pale violet to white with purple spots, the tube 4.5–7 mm long and lobes mostly 3–6 mm long. Rare on Mancos shale slopes or dry hillsides with pinyon-juniper or oak, 6800–7300 ft. June–July. W. Endemic.

Listed as endangered under the Endangered Species Act.

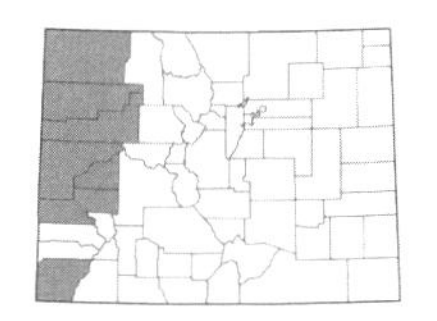

Ipomopsis polycladon **(Torr.) V.E. Grant**, spreading ipomopsis. [*Gilia polycladon* Torr.]. Annuals, 2.5–25 cm, glandular-stipitate and with flattened, multicellular hairs; *leaves* pinnatifid, mostly 1–3 cm long; *inflorescence* of dense, capitate cymose clusters; *calyx* 2–6 mm long; *corolla* white, the tube 3–6 mm long and lobes 1–2 mm long. Found in dry places, clay hills, shale slopes, and ridges, 4400–6100 ft. April–June. W.

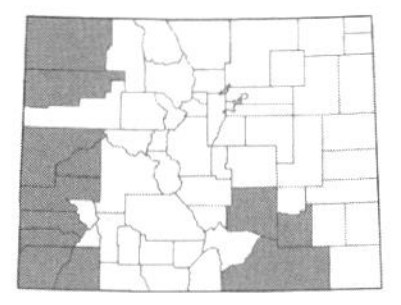

Ipomopsis pumila **(Nutt.) V.E. Grant**, dwarf ipomopsis. [*Gilia pumila* Nutt.]. Annuals to 2 dm, the stems woolly-villous especially above; *leaves* linear, most with 1–2 pairs of linear lateral segments to 1 cm long; *inflorescence* a dense, capitate cluster; *calyx* 3–5 mm long; *corolla* lavender to violet, the tube 5–9 mm long with lobes 2–4 mm long. Found on clay hills and flats and in sandy soil, often with sagebrush, pinyon-juniper, or *Atriplex*, 4500–8000 ft. April–June. E/W.

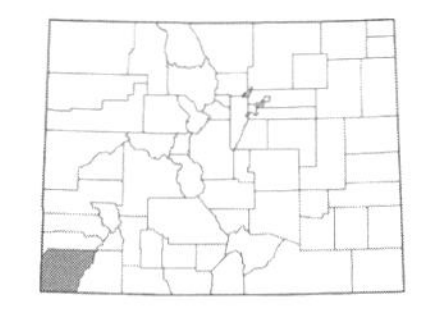

Ipomopsis ramosa **A. Schneider & Bregar**, coral ipomopsis. Biennials to perennials, 2.5–4.5 dm, villous and glandular above; *leaves* pinnatifid with linear lobes ca. 2.5 cm long; *inflorescence* a narrow thyrse or paniculate; *calyx* 6–7 mm long, densely glandular; *corolla* white with pink dots, 14–17 mm long. Uncommon in spruce, fir, or aspen forests, 8000–9500 ft. June–Aug. W. Endemic.

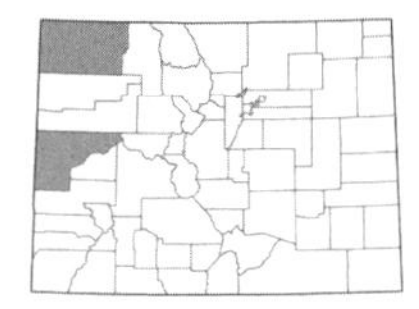

***Ipomopsis roseata* (Rydb.) V.E. Grant**, San Rafael ipomopsis. [*Gilia roseata* Rydb.]. Perennials, 1–3 dm, glabrate to villous or finely glandular; *leaves* pinnatifid with 1–3 pairs of narrow segments; *inflorescence* of dense cymose heads; *calyx* 5.5–8.5 mm long; *corolla* white, tube 7–9 mm long with lobes 2.5–4 mm long. Uncommon in sandy soil and on rocky slopes and sandstone cliffs, 4500–6000 ft. May–July. W.

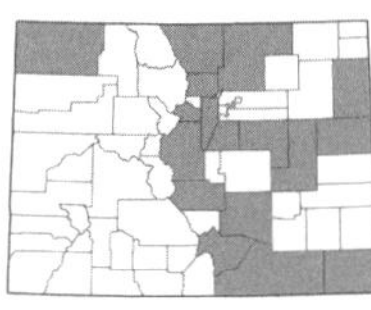

***Ipomopsis spicata* (Nutt.) V.E. Grant**, spiked ipomopsis. [*Gilia spicata* Nutt. var. *spicata*]. Perennials, 0.5–4 dm, the stems white-tomentose, glandular-stipitate in the inflorescence; *leaves* linear and entire or some pinnatifid with 1–several linear segments to 1 cm long; *inflorescence* dense, spiciform, and bracteate; *calyx* 4–5.5 mm long, glandular; *corolla* white to cream, the tube 5–7 mm long and lobes 2.5–4 mm long. Common in sandy or rocky soil, dry meadows, and on sandstone outcroppings, 5000–9000 ft. May–Aug. E/W. (Plate 67)

LATHROCASIS L.A. Johnson – lathrocasis

Annual herbs with conspicuous glandular hairs; *leaves* alternate and in a basal rosette, sometimes opposite below, entire; *calyx* with herbaceous veins separated by hyaline membranes; *corolla* funnelform, white, ochroleucous, or pale lavender.

***Lathrocasis tenerrima* (A. Gray) L.A. Johnson**, delicate gilia. [*Gilia tenerrima* A. Gray]. Plants 5–20 cm; *leaves* linear to lanceolate, 3–20 × 0.5–4 mm; *inflorescence* a diffuse, paniculate cyme; *calyx* 1–2 mm; *corolla* 1.5–3.5 mm; *capsules* subequal to the calyx. Uncommon on dry, often barren hills, usually with sagebrush, 6500–7500 ft. May–July. W.

LEPTOSIPHON Benth. – leptosiphon

Annual or perennial herbs; *leaves* opposite, usually palmatifid or trifid into linear segments; *calyx* subherbaceous or with herbaceous veins separated by hyaline membranes; *corolla* white, bluish, or purple, campanulate to salverform.

1a. Perennial with a woody base; calyx 6–10 mm long, mostly herbaceous with narrow hyaline sinuses; flowers sessile or nearly so in a rather dense headlike cluster at the end of the stems; corolla with a tube 6–10 mm long and spreading lobes 5–8 mm long...***L. nuttallii***
1b. Annual from a slender taproot; calyx 3–5 mm long, with conspicuously hyaline sinuses; flowers on slender pedicels, solitary in the leaf axils; corolla 2.5–6 mm long...***L. septentrionalis***

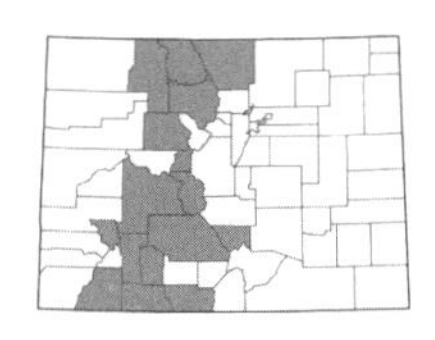

Leptosiphon nuttallii* (A. Gray) J.M. Porter & L.A. Johnson ssp. *nuttallii, Nuttall's leptosiphon. [*Linanthus nuttallii* (A. Gray) Greene ex Milliken; *Linanthastrum nuttallii* (A. Gray) Ewan]. Perennial subshrubs, forming clumps, 8–30 cm; *leaves* with 5–9 linear lobes, minutely spinulose at the tips; *inflorescence* a dense headlike cluster of sessile flowers; *calyx* 6–10 mm long, with narrow hyaline sinuses; *corolla* white, tube 6–10 mm long, lobes 5–8 mm long; *capsules* 4–5.5 mm long. Found on gravelly slopes, dry hillsides, in meadows and open forests, and in the alpine, 6500–11,500 ft. June–Aug. E/W. (Plate 67)

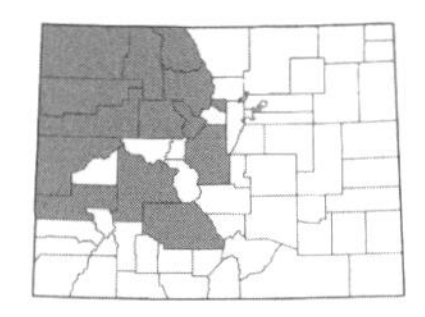

***Leptosiphon septentrionalis* (H. Mason) J.M. Porter & L.A. Johnson**, northern leptosiphon. [*Linanthus septentrionalis* H. Mason]. Annuals, 4–25 cm; *leaves* with linear segments mostly 3–20 mm long; *inflorescence* of solitary flowers arising from the leaf axils; *calyx* 3–5 mm long, with conspicuously hyaline sinuses; *corolla* white, 2.5–6 mm long; *capsules* shorter than the calyx. Found in dry meadows, open forests, and on dry hillsides, 6000–10,800 ft. May–July. E/W.

LINANTHUS Benth. – linanthus

Perennial herbs or subshrubs; *leaves* opposite below and alternate above, palmatifid or subpinnatifid with linear, pungent lobes; *calyx* broadly scarious below the sinuses, the tube ruptured by the maturing fruit; *corolla* funnelform, white or yellowish, or rarely purplish in the throat.

1a. Upper leaves alternate or subopposite; shrub or subshrub 1–5 dm tall, with more or less open, erect branches; flowers 5-merous...***L. pungens***
1b. Leaves opposite; cushion-forming or matted perennial herb or subshrub less than 1 dm; flowers usually 4- or 6-merous, less often 5-merous...2

2a. Corolla 15–28 mm long with lobes 7–15 mm long; flowers 6-merous or occasionally 5-merous; leaves 3–9-cleft, the segments 6–15 (20) mm long...***L. watsonii***
2b. Corolla 12–20 mm long with lobes 3–6 mm long; flowers 4-merous; leaves usually 3-cleft or rarely 5-cleft, the segments 2–9 mm long...***L. caespitosus***

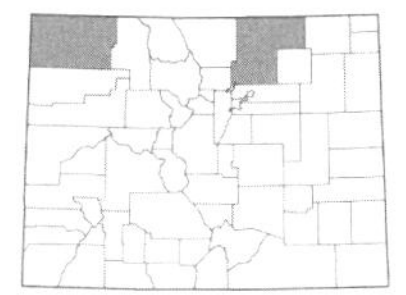

***Linanthus caespitosus* (Nutt.) J.M. Porter & L.A. Johnson**, CUSHION PRICKLY PHLOX. [*Leptodactylon caespitosum* Nutt.]. Subshrubs forming cushions or mats, mostly 1–5 dm wide; *leaves* opposite, 3–5-cleft into linear segments 2–9 mm long, spinulose; *flowers* 4-merous; *calyx* 6–8.5 mm long; *corolla* white or cream, 12–20 mm long with lobes 3–6 mm long. Locally common on barren sandstone outcroppings and in sandy soil, 4500–7700 ft. May–July. E/W.

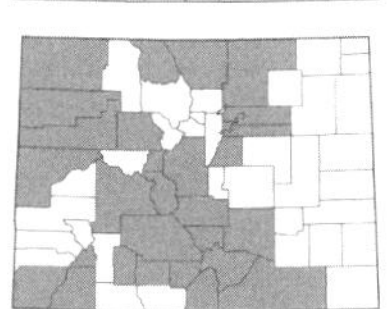

***Linanthus pungens* (Torr.) J.M. Porter & L.A. Johnson**, SHARP SLENDERLOBE. [*Leptodactylon pungens* (Torr.) Nutt.]. Shrubs or subshrubs 1–5 dm tall, with open, erect branches; *leaves* alternate or subopposite above, opposite below, 3–9-cleft into linear segments, spinulose; *flowers* 5-merous; *calyx* 7–12 mm long; *corolla* cream to yellowish, often tinged with purple below, 12–25 mm with lobes 5–15 mm long. Common on dry hillsides, in dry grasslands, and in open forests and dry meadows, 4000–9500 ft. May–Aug. E/W. (Plate 67)

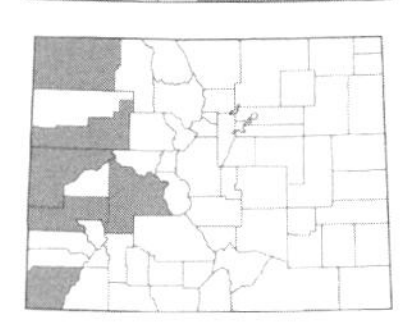

***Linanthus watsonii* (A. Gray) Wherry**, WATSON'S PRICKLY PHLOX. [*Leptodactylon watsonii* (A. Gray) Rydb.]. Subshrubs forming cushions; *leaves* opposite, 3–9-cleft into linear segments 6–15 (20) mm long, spinulose; *flowers* 6-merous; *calyx* 7–12 mm long; *corolla* white to cream, 15–28 mm long with lobes 7–15 mm long. Found in rocky crevices of cliffs, on shale slopes, and on rocky outcrops and ledges, 5600–8500 ft. May–July. W.

MICROSTERIS Greene – MICROSTERIS

Annual herbs; *leaves* opposite below and alternate above, entire; *flowers* solitary or in axillary pairs at the ends of branches; *calyx* with broad hyaline membranes between the herbaceous veins; *corolla* white to pink or purplish.

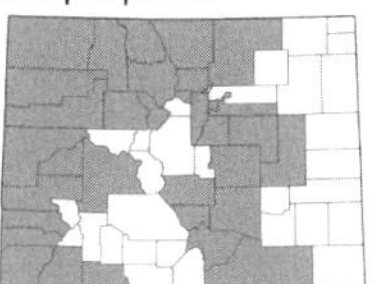

***Microsteris gracilis* (Hook.) Greene**, SLENDER PHLOX. [*Phlox gracilis* (Hook.) Greene]. Plants 1–20 cm, villous and often glandular above; *leaves* lanceolate to obovate or linear, 5–30 × 2–7 mm; *calyx* 4.5–8 mm long; *corolla* 6–12 mm long. Common in forests, meadows, shortgrass prairie, and on sagebrush slopes, 4900–9500 ft. April–June. E/W. (Plate 67)

NAVARRETIA Ruiz & Pav. – PINCUSHION PLANT

Annual herbs; *leaves* alternate or sometimes the lower opposite, irregularly bipinnatifid to pinnatifid, or sometimes entire, the segments linear and spine-tipped; *flowers* in dense, cymose heads; *calyx* with broad hyaline membranes between the herbaceous veins.

1a. Leaves simple, linear; inflorescence open, the flowers mostly solitary opposite a leaf...***N. sinistra***
1b. Leaves pinnately dissected, the lobes usually firm and prickly; inflorescence a dense, terminal cymose head...2

2a. Flowers yellow; stigmas 3...***N. breweri***
2b. Flowers white, pale lilac, or pale blue; stigmas usually 2, rarely 3...3

3a. Style 3.5–10 mm long; corolla longer than longest sepal lobes...***N. intertexta* ssp. *propinqua***
3b. Style 1.9–3 (3.4) mm long; corolla shorter than the longest sepal lobes...***N. saximontana***

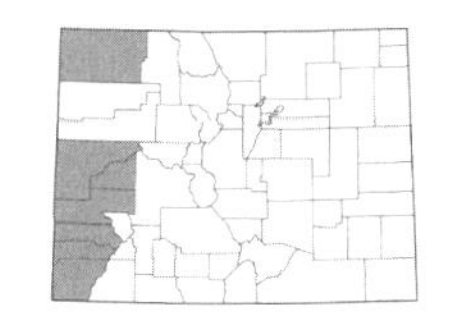

***Navarretia breweri* Greene**, YELLOW PINCUSHION PLANT. Plants 1–12 cm, glandular-hairy; *leaves* pinnately dissected, the 1–4 pairs of lateral lobes linear, firm and prickly; *bracts* 4–15 mm long, pinnatifid; *calyx* 6–10 mm long; *corolla* yellow, 5–7 mm long; *stigmas* 3. Found on dry slopes, in meadows, often with oak, 6500–9300 ft. June–July. W.

***Navarretia intertexta* Hook. ssp. *propinqua* (Suksd.) A.G. Day**, GREAT BASIN NAVARRETIA. Plants 3–15 cm, villous; *leaves* pinnately dissected into linear, firm and prickly lobes; *bracts* pinnatifid; *calyx* 4–7 mm long; *corolla* white, pale lilac, or pale blue, 4–9 mm long; *stigmas* 2. This may not occur in Colorado as most specimens that have long been identified as this are actually *N. saximontana*.

***Navarretia saximontana* S.C. Spencer**, PINCUSHION PLANT. Plants 3–10 cm, subglabrous to villous above; *leaves* pinnately dissected or bipinnatifid into linear, firm, and prickly lobes; *bracts* pinnatifid or bipinnatifid; *calyx* 4–7 mm long; *corolla* white to blue or pale lilac, 3–5 mm long; *stigmas* 2. Uncommon in open forests, on dry slopes, and along the margins of drying ponds, 5500–9000 ft. June–Aug. E/W.

***Navarretia sinistra* (M.E. Jones) L.A. Johnson**, MINIATURE GILIA. [*Gilia capillaris* Kellogg; *Gilia sinistra* M.E. Jones]. Plants 5–30 cm, glandular-stipitate; *leaves* simple, linear, 5–35 × 1–4 mm; *flowers* mostly solitary opposite a leaf; *calyx* 2–5 mm long; *corolla* lavender to bluish or white, 4–9 mm long. Uncommon on open slopes, in meadows, and along streams, 6700–8500 ft. June–July. W.

PHLOX L. – PHLOX

Perennial herbs; *leaves* opposite, simple, entire, sometimes needle-like with a spinose tip; *flowers* in terminal cymes or solitary; *calyx* with hyaline membranes between the herbaceous veins; *corolla* salverform, white, pink, purple, blue, or red.

1a. Flowers 3–12 in a cyme, on conspicuous, slender pedicels; stems with well-developed internodes, the longer ones well over 10 mm long; leaves mostly over 20 mm long; plants erect to tufted but not mat- or cushion-forming, usually over 10 cm tall…2
1b. Flowers 1–3 (5), sessile or short-pedicellate (on pedicels 3–10 mm long) at the ends of stems; stems with shorter internodes often hidden by the leaves, the longer ones to 10 mm long; leaves mostly 2–30 mm long; plants caespitose, often mat- or cushion-forming, usually to 10 cm above the ground…3

2a. Hyaline membranes between the calyx lobes carinate, forming a distinct raised ridge, bulging towards the base; plants glabrous or often glandular-hairy, especially in the inflorescence; flowers pink to white; not restricted to the southwestern counties…***P. longifolia***
2b. Hyaline membranes between the calyx lobes flat, not folded or forming a ridge, not bulging; plants pilose with multicellular hairs, not glandular; flowers bright purple or pink; known from the southwestern counties…***P. caryophylla***

3a. Leaves stiffly hirsute-ciliate along the margins, at least from the middle to the base; calyx usually glandular-hairy, occasionally glabrous; plants at or above 9000 ft in elevation…4
3b. Leaves glabrous, granular-scabrid, glandular, or arachnoid-pubescent with tangled hairs, but not stiffly ciliate or sometimes ciliate on the margins but also arachnoid-pubescent at the base or granular-scabrid throughout; calyx glabrous or villous- to arachnoid-tomentose; plants variously distributed, usually below 9500 ft in elevation (but up to 10,500 ft)…6

4a. Leaves more or less thick and succulent; style 3.5–7.5 mm long; plants of alkaline meadows and fens in South Park (Park Co.), about 9000 ft in elevation…***P. kelseyi***
4b. Leaves not succulent; style 1.5–5 mm long; plants of the subalpine and alpine above 10,000 ft in elevation…5

5a. Plants forming dense cushions with numerous erect, short, nearly parallel and densely compact shoots; corolla lobes 3–5 mm long and the tube 6–10 mm long; leaves less than 1.5 mm wide and 2–8 mm long; flowers usually white, sometimes pink or light blue to lavender…***P. condensata***
5b. Plants forming loose cushions with spreading caudices and ultimately erect branches; corolla lobes 5–7 mm long and the tube 9–13 mm long; leaves 1–2.5 mm wide and 5–15 mm long; flowers light blue or white…***P. pulvinata***

6a. Calyx 3–4.5 mm long; leaves lance-subulate to triangular, 2–4 (5) mm long, densely crowded, imbricate, and closely appressed to the stem; plants densely caespitose, forming low mounds or cushions, densely arachnoid-pubescent to woolly tomentose, the plants usually grayish; corolla lobes 3–5 mm long…***P. muscoides***
6b. Calyx 5–15 mm long; leaves linear to lance-subulate, 2–30 mm long, less crowded and spreading; plants forming loose to dense mats, sparsely arachnoid-pubescent in the leaf axils or on the calyx to glabrous or granular-scabrid, the plants green or sometimes grayish; corolla lobes 4–11 mm long…7

7a. Hyaline membranes between the calyx lobes carinate and forming a low, raised ridge towards the base (look at a few flowers, some ridges are not as conspicuous as others); plants of the western slope…***P. austromontana***
7b. Hyaline membranes between the calyx lobes flat, not carinate and forming a ridge; plants of the eastern or western slope…8

8a. Leaves mostly 4–10 mm long and 0.5–1 mm wide; internodes usually short and hidden by the leaves; leaf axils and calyx lobes arachnoid-woolly with tangled hairs; calyx 5–7.5 mm long, the hyaline membranes often obscured by the tangled hairs; style 2–5 mm long; flowers sessile or nearly so at the ends of stems…***P. hoodii***
8b. Leaves 10–30 mm long and 1–2 mm wide; internodes usually visible; leaf axils arachnoid-woolly or not; calyx 6–15 mm long, arachnoid-pubescent or not; style 5–9 mm long; flowers sessile to shortly pedicellate on pedicels 3–10 mm long…9

9a. Plants found below 5500 ft, known from the northeastern plains; calyx 6–11 mm long, the lobes arachnoid-pubescent with tangled hairs on the margins; flowers white…***P. andicola***
9b. Plants above 6000 ft, variously distributed but absent from the northeastern plains; calyx 10–15 mm long, glabrous or sparsely hairy but the hairs not tangled; flowers white, pink, or bluish…***P. multiflora***

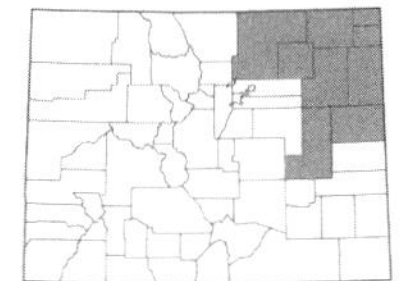

Phlox andicola **E.E. Nelson**, PLAINS PHLOX. Plants erect to decumbent, 4–10 (12) cm, minutely hairy to arachnoid, mat-forming; *leaves* subulate to linear, 10–25 (30) × 1–2 mm; *inflorescence* of 1–3 (5) flowers, subsessile or on pedicels to 2 (5) mm long; *calyx* 6–11 mm long; *corolla* white, the tube 6–13 (17) mm long and lobes 6–9 mm long; *style* 5–9 mm long. Common on sandhills and in sandy soil of the northeastern plains, 3500–5500 ft. May–July. E.

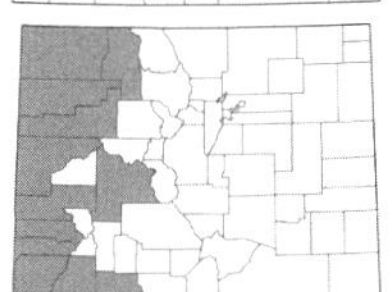

Phlox austromontana **Cov.**, DESERT PHLOX. Plants forming loose to dense mats, to 10 cm tall, subglabrous; *leaves* linear, sharply acute, (5) 8–20 × 0.5–1.5 mm; *inflorescence* solitary and sessile or nearly so; *calyx* intercostal membranes with a low, linear median keel; *corolla* white to bluish or pink, the tube 8–15 mm long and lobes 5–8 mm long; *style* 2–5 mm long. Found on dry ridgetops, shale slopes, and in sandy soil, 5000–9000 ft. May–July. W.

Phlox caryophylla **Wherry**, CLOVE PHLOX. Plants caespitose, pilose-pubescent; *leaves* linear to oblong, short-acuminate, 30–50 × 2–3 mm; *inflorescence* a cyme of 3–12 flowers on long pedicels to 2.5 cm; *calyx* 11–15 mm long; *corolla* bright purple or pink, the tube 14–18 mm long and lobes 7–10 mm long. Locally common on dry hillsides, 6500–8000 ft. May–July. W.

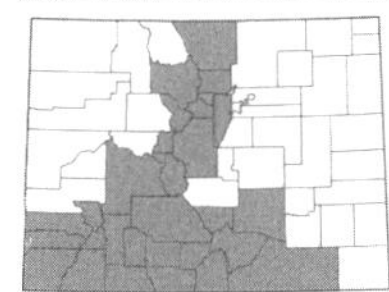

Phlox condensata **(A. Gray) E.E. Nelson**, ALPINE PHLOX. Plants densely pulvinate, forming cushion-mats, sparsely to densely glandular-hairy; *leaves* linear to oblong, with a sharp bristle tip, 2–8 × 0.5–1.5 mm; *inflorescence* solitary and sessile; *calyx* 4–8 mm long, glandular; *corolla* white to pink or light blue-lavender, the tube 6–10 mm long and lobes 3–5 mm long; *style* 1.5–3 mm long. Common in the alpine, 11,000–14,000 ft. June–Aug. E/W. (Plate 67)

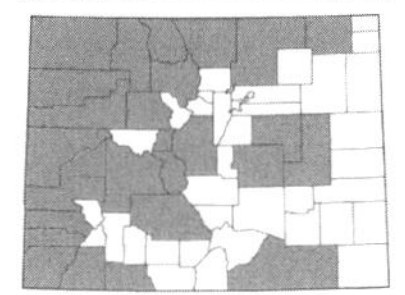

Phlox hoodii **Richardson**, CARPET PHLOX. Plants mat- or cushion-forming; *leaves* linear, (2.5) 4–10 (13) × 0.5–1 mm; *inflorescence* solitary and sessile; *calyx* 5-7.5 mm long, arachnoid-woolly; *corolla* lavender to pink-purple or white, the tube 4–15 mm long and lobes 4–8 mm long; *style* 2–5 mm long. Common on dry slopes or bluffs, ridgetops, in sagebrush meadows, shortgrass prairie, and sometimes with pinyon-juniper, 5200–9300 ft. April–June. E/W.

Phlox kelseyi **Britton**, MARSH PHLOX. Plants caespitose, to 10 cm; *leaves* lance-linear, succulent, to 25 mm long × 1–2.5 mm wide, ciliate at least toward the base; *inflorescence* solitary; *calyx* with flat intercostal membranes, glabrous to villous; *corolla* light blue to white, the tube 10–13 mm long and lobes 6–9 mm long; *style* 3.5–7.5 mm long. Uncommon in alkaline meadows and fens in South Park, 9000–9300 ft. May–July. E.

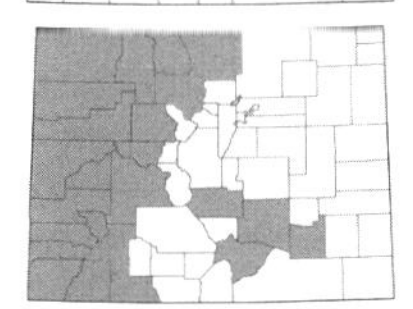

Phlox longifolia **Nutt.**, LONGLEAF PHLOX. Plants (0.5) 1–4 dm, caespitose; *leaves* linear to lance-linear or elliptic, 10–90 × 1–6 mm; *inflorescence* a cyme of 3–12 flowers on slender pedicels; *calyx* with intercostal membranes carinate, forming a distinct raised ridge, bulging towards the base; *corolla* white to pink, the tube 10–35 mm long (1–3 times as long as calyx) and lobes 7–15 mm long, entire to erose; *style* 6–25 mm long. Common in sagebrush meadows and on dry slopes, often with pinyon-juniper or oak, 4500–9000 ft. May–July. E/W. (Plate 67)

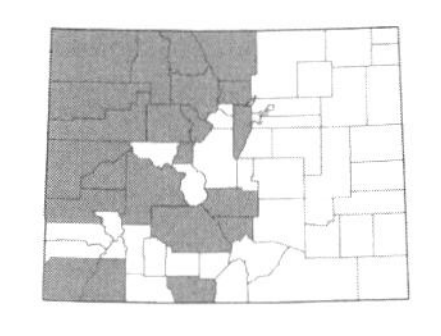

Phlox multiflora **A. Nelson**, MOUNTAIN PHLOX. Plants to 10 cm, mat-forming; *leaves* linear, 10–30 × 1–2 mm, glabrous to inconspicuously scabrous; *inflorescence* of 1–3 flowers at the ends of stems; *calyx* with flat intercostal membranes; *corolla* white or sometimes pink or bluish, the tube 10–15 mm long and lobes 6–11 mm long; *style* 5–8 mm long. Common in sagebrush meadows, open forests, and along streams, 6000–10,500 ft. April–July. E/W.

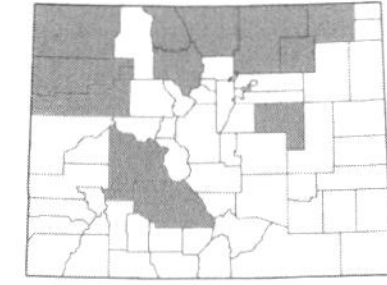

Phlox muscoides **Nutt.**, MOSS PHLOX. [*P. bryoides* Nutt.; *P. hoodii* Richardson ssp. *muscoides* Wherry]. Plants pulvinate, mat-forming; *leaves* lance-subulate to triangular, 2–4 (5) mm long, imbricate, usually white-margined; *inflorescence* solitary and sessile to short-pedicellate; *calyx* 3–4.5 mm long, arachnoid-woolly; *corolla* white to purple or bluish, the tube 5–10 mm long and lobes 3–5 mm; *style* 2–5 mm. Found on sandy, barren soil and dry bluffs, 5000–8700 ft. April–June. E/W.

Phlox opalensis Dorn may be present in Moffat Co. and is similar to *P. muscoides*. It differs in having longer corolla lobes (5.8–7 mm long), a longer calyx (6–6.5 mm long), and often longer leaves (2–10 mm long). *Phlox muscoides* can sometimes be difficult to separate from *P. hoodii*.

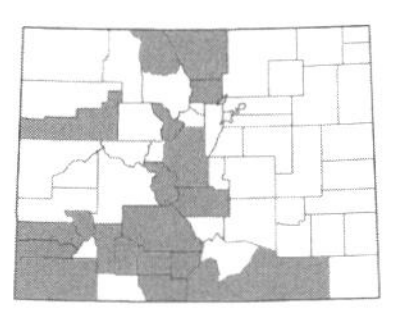

Phlox pulvinata **(Wherry) Cronquist**, CUSHION PHLOX. Plants mat-forming or pulvinate; *leaves* linear to oblong, 5–15 × 1–2.5 mm, ciliate at least toward the base; *inflorescence* usually solitary; *calyx* usually glandular or sometimes glabrous, the intercostal membranes flat; *corolla* white to light bluish, the tube 9–13 mm long and lobes 5–7 mm long; *style* 2–5 mm long. Found on alpine and rocky slopes, 10,000–13,000 ft. June–Aug. E/W.

POLEMONIUM L. – Jacob's ladder

Perennial herbs, usually glandular (at least in the inflorescence); *leaves* alternate, pinnatifid to pinnately compound; *flowers* in terminal cymes; *calyx* herbaceous; *corolla* funnelform to campanulate or tubular-funnelform. (Wherry, 1942)

1a. Leaflets opposite or subopposite, mostly 8–40 mm long, the upper 3–5 often confluent; flowers campanulate, as wide as long or wider than long; plants glandular-hairy in the inflorescence, sometimes throughout but usually not densely so...2

1b. At least some leaflets whorled, 1.5–10 mm long, the upper leaflets not confluent; flowers funnelform, tubular-funnelform, or tubular-campanulate, distinctly longer than wide; plants densely glandular-hairy throughout...4

2a. Plants 0.5–2.5 dm high; leaves mostly basal; flowers usually blue-violet with a yellow center, rarely white...***P. pulcherrimum***

2b. Plants usually over 3 dm high; leaves well-developed along the stem; flowers blue-violet or white, with a white or pale center, not yellow...3

3a. Inflorescence of 1–many small, compact, flat-topped cymes; stems usually several and clustered, from a branching caudex surmounting a taproot; leaflets hairy; leaves not much reduced above...***P. foliosissimum***

3b. Inflorescence an elongate, sometimes openly branched cyme; stems solitary from a short rhizome; leaflets usually glabrous; leaves usually markedly reduced above...***P. occidentale* ssp. *occidentale***

4a. Flowers white to ochroleucous or light yellow; inflorescence with some of the lower flowers removed from the terminal cluster; leaflets usually rather remotely placed on the rachis...***P. brandegeei***

4b. Flowers light to dark blue or blue-violet, rarely white; flowers all usually in a dense terminal cluster; leaflets usually densely crowded on the rachis, sometimes loose...5

5a. Corolla tubular to funnelform, mostly 5–8 mm wide at the widest point (on pressed specimens), deep purple to blue-violet; plants foul-smelling, pollinated by flies...***P. viscosum***

5b. Corolla tubular-campanulate, mostly 8–12 mm wide at the widest point (on pressed specimens), light purple to pale blue-violet; plants sweet-smelling, pollinated by ants...***P. confertum***

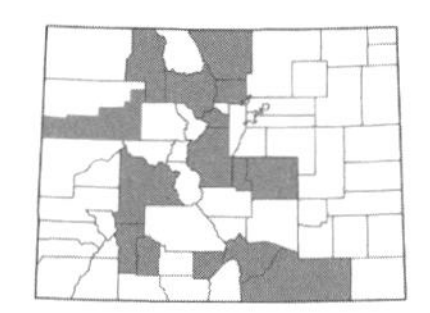

***Polemonium brandegeei* (A. Gray) Greene**, Brandegee's sky pilot. Plants to 3 dm; *leaflets* numerous, at least some whorled, 3–10 mm long; *inflorescence* a loose cyme, racemose or spicate, some of the flowers removed from the terminal cluster; *calyx* to 14 mm long; *corolla* white to light yellow, 20–35 (40) mm long, funnelform or tubular-funnelform. Found in rock crevices, on rocky cliffs and outcroppings, and on talus slopes, 6500–12,500 ft. June–Aug. E/W. (Plate 67)

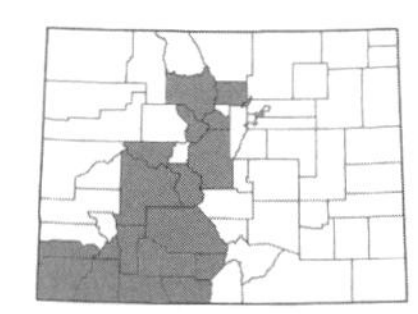

***Polemonium confertum* A. Gray**, Rocky Mountain sky pilot. [*P. grayanum* Rydb.]. Plants to 3 dm; *leaflets* numerous, whorled, 6 (9) mm long; *inflorescence* a dense compact terminal cyme; *calyx* 7–11 mm long; *corolla* light purple to pale blue-violet, mostly 8–12 mm wide, tubular-campanulate. Found in alpine, 10,500–14,000 ft. June–Aug. E/W. Endemic.

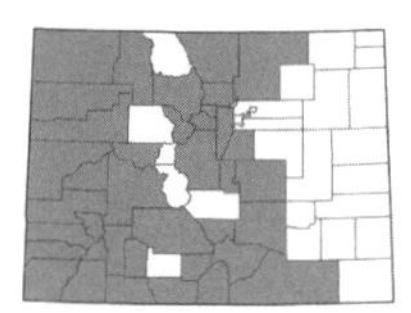

***Polemonium foliosissimum* A. Gray**, leafy Jacob's ladder. Plants (3) 5–12 dm; *leaflets* mostly 11–25, opposite or subopposite, mostly (8) 10–40 (45) mm long, the 3 or 5 upper ones confluent; *inflorescence* of 1-many small, compact cymes; *calyx* 5–8 mm long; *corolla* blue-violet, white, or light yellow, 10–16 mm long, campanulate. Common along streams, in meadows, and open forests, 6500–13,000 ft. June–Aug. E/W.

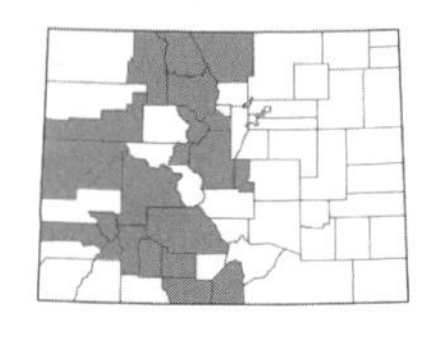

Polemonium occidentale* Greene ssp. *occidentale, western Jacob's ladder. [*P. caeruleum* L. ssp. *amygdalinum* (Wherry) Munz]. Plants (1.5) 4–10 dm; *leaflets* mostly 11–27, opposite or subopposite, (5) 10–40 mm long, the upper 3 often confluent; *inflorescence* in an elongate cyme; *calyx* 5–8 mm long; *corolla* blue or rarely white, 10–16 mm long, campanulate. Locally common in moist places such as along streams, in bogs, and moist meadows and forests, 7500–11,000 ft. June–Aug. E/W.

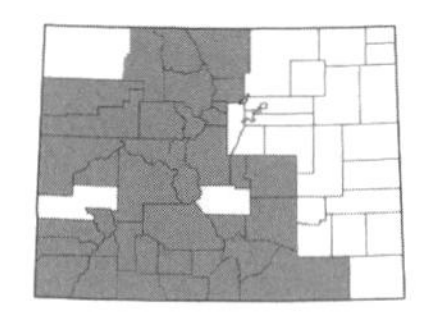

***Polemonium pulcherrimum* Hook. ssp. *delicatum* (Rydb.) Cronquist**, Jacob's ladder. Plants 0.5–2.5 dm; *leaflets* mostly 11–25, opposite or subopposite, (5) 10–45 mm long, the upper 3 sometimes confluent; *inflorescence* an open to congested cyme; *calyx* 4–7 mm long; *corolla* blue-violet with a yellow center, rarely white, 7–15 mm long, campanulate. Common in spruce and krummholz forests, subalpine meadows, and in rock crevices, 8500–13,300 ft. June–Aug. E/W. (Plate 67)

Usually quite malodorous and smells very much like a skunk!

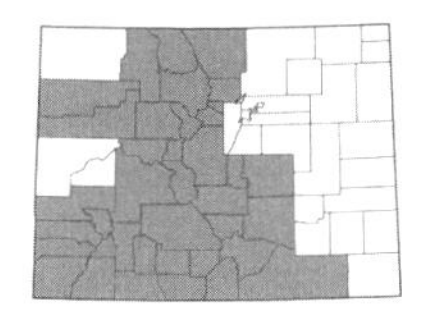

Polemonium viscosum **Nutt.**, SKY PILOT. Plants to 3 dm; *leaflets* numerous, whorled, 1.5–6 (9) mm long, mostly or all 2–5-cleft to the base; *inflorescence* a dense cyme with a terminal cluster of flowers; *calyx* 7–12 mm long; *corolla* deep purple to blue-violet, 15–30 mm long, funnelform to tubular, mostly 5–8 mm wide. Common in high meadows, alpine, and on talus or scree slopes, 9500–14,300 ft. June–Aug. E/W. (Plate 67)

POLYGALACEAE Juss. – MILKWORT FAMILY

Herbs, shrubs, or trees; *leaves* alternate, opposite, or whorled, simple and entire, sometimes highly reduced and the principal photosynthesizing function performed by the stems; *flowers* perfect, strongly zygomorphic; *sepals* 5, distinct or connate below; *petals* 3, consisting of two upper and a keeled, often fringed lower petal; *stamens* 8, in two whorls, derived by suppression of one member of each of the two whorls, epipetalous; *pistil* 1; *ovary* superior; *fruit* a loculicidal capsule, drupe, nut, or samara. (Wendt, 1979)

POLYGALA L. – MILKWORT

Herbs or shrubs; *flowers* in terminal or axillary racemes; *sepals* 5, the inner two large and petaloid; *petals* adnate to the filaments to form a tube; *stamens* with connate filaments; *ovary* 2-carpellate; *fruit* a capsule.

1a. Perennial herbs; flowers white, 2.2–4 mm long; leaves narrowly oblanceolate below and linear above, 1–2 mm wide; capsules 2.5–3 mm long...***P. alba***
1b. Shrubs or subshrubs with spinescent stems; flowers bright pink-purple with a yellow keel, 8–12 mm long; leaves oblanceolate, mostly 2–10 mm wide; capsules 5–8 mm long...***P. subspinosa***

Polygala alba **Nutt.**, WHITE MILKWORT. Perennials, 1–5 dm, glabrous; *leaves* linear above and narrowly oblanceolate below, 1–2 mm wide; *flowers* white; *keel petal* ca. 3 mm long; *capsules* 2.5–3 mm long. Found in rocky places and on sandstone outcroppings on the eastern plains, 3400–4900 ft. May–Aug. E. (Plate 68)

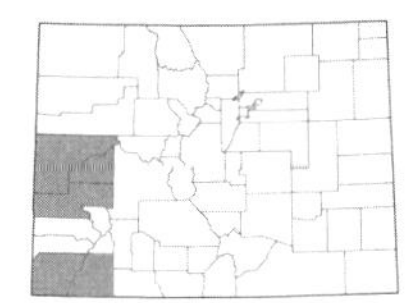

Polygala subspinosa **S. Watson**, SPINY MILKWORT. Spinescent shrubs or subshrubs, 0.5–2 dm tall, subglabrous to hirsute; *leaves* oblanceolate, 8–30 × 2–10 mm; *flowers* bright pink-purple with a yellow keel, 8–12 mm long; *capsules* 5–8 mm long. Uncommon on dry slopes, 5000–6000 ft. April–June. W.

POLYGONACEAE Juss. – BUCKWHEAT FAMILY

Herbs, shrubs, trees, or vines; *leaves* usually alternate, sometimes opposite or whorled, simple, the margins usually entire, stipulate with the stipule modified into an ocrea or exstipulate; *flowers* perfect or rarely imperfect, sometimes subtended by involucre bracts, actinomorphic; *tepals* 2–6, usually in 2 whorls, distinct or connate, sepaloid or petaloid, persistent in fruit; *stamens* usually 6–9; *pistil* 1; *ovary* superior, (2) 3 (4)-carpellate; *fruit* achenes, winged or wings lacking, sometimes lenticular.

1a. Leaves lacking a papery ocrea sheathing the stem at the nodes; tepals 6, petaloid; flowers enclosed in or subtended by involucre bracts...2
1b. Leaves with a papery ocrea sheathing the stem at the nodes; tepals 3–5 or if 6 then not petaloid; flowers not enclosed in or subtended by involucre bracts...3

2a. Involucre bracts united, the involucre tubular in shape; annuals or perennials...***Eriogonum***
2b. Involucre bracts distinct or nearly so, in 2 whorls of 3; annuals...***Stenogonum***

3a. Tepals 3, 1–1.8 mm long, greenish or tinged white or pink at the tips; alpine, reddish-tinged annuals usually less than 4 cm; leaves 1–6.5 mm long, spatulate-ovate...***Koenigia***
3b. Tepals 4–6; plants otherwise unlike the above...4

4a. Vines or sprawling, weak-stemmed annuals; leaves sagittate or cordate...5
4b. Annuals, biennials, or perennials, but not vines or sprawling annuals; leaves various...6

5a. Stems with numerous retrorse, yellow barbs...***Persicaria*** (*sagittata*)
5b. Stems lacking retrorse, yellow barbs...***Fallopia***

6a. Tepals 4, in 2 whorls of 2 with the outer 2 narrower than the inner, sepaloid; leaves mostly basal, orbiculate to reniform with a cordate base and rounded apex, 0.5–6 cm long; achenes winged...***Oxyria***
6b. Tepals 5–6, rarely 4 and then petaloid; leaves various but usually unlike the above; achenes unwinged or winged in one species (*Rheum*) and then the basal leaves 30–50 cm long...7

7a. At least some leaves hastate with basal lobes pointing outward; tepals 6, reddish, 0.5–1.5 mm long, the inner tepals not enlarging in fruit...***Rumex*** (*acetosella*)
7b. Leaves various but never hastate; tepals 5–6...8

8a. Achenes winged; leaves mostly basal, 30–50 cm long, palmately veined, with undulate margins; petioles red or pinkish-red; tepals 6, the inner tepals not enlarging in fruit...***Rheum***
8b. Achenes not winged (*careful*—sometimes the persistent tepals around the achene are winged); leaves various, if 30–50 cm long then pinnately veined; petioles usually not red; tepals 5 or if 6 then the inner tepals enlarging in fruit...9

9a. Tepals 6, the inner tepals persistent and usually enlarging in fruit, sometimes also bearing a tubercle in the center...***Rumex***
9b. Tepals 5, the inner tepals not enlarging in fruit...10

10a. Outer 3 tepals winged in fruit; flowers with a slender, stipelike base, whitish-green; flowering stems becoming horizontal and zigzag; inflorescence of racemes arranged in erect panicles from the axils of the upper leaves; leaves ovate with a truncate or cordate base...***Fallopia*** (*japonica*)
10b. Outer 3 tepals not winged or rarely keeled in fruit; flowers with or without a stipelike base, variously colored; flowering branches not becoming zigzag; plants otherwise unlike the above in all respects...11

11a. Tepals distinct; leaves sagittate, cordate, or triangular; achenes sharply 3-sided, exserted from the persistent perianth...***Fagopyrum***
11b. Tepals connate at least at the base; leaves various but not sagittate, cordate, or triangular; achenes lenticular or 3-sided, included or sometimes exserted from the perianth...12

12a. Flowers arranged in the axils of leaves or leaf-like bracts, or in loose terminal racemes...13
12b. Flowers in dense terminal or axillary spikes or racemes lacking leaf-like bracts...14

13a. Perianth and leaves glandular-punctate with numerous dots; ocrea brown...***Persicaria*** (*hydropiper, punctata*)
13b. Perianth and leaves not glandular-punctate; ocrea silvery or white...***Polygonum***

14a. Flowers in a single, terminal spike-like raceme, white or light pink; perennials...***Bistorta***
14b. Stems with more than one spike or spike-like raceme, if with a single terminal inflorescence then the flowers bright pink to rose; annuals or perennials...***Persicaria***

BISTORTA (L.) Scop. – BISTORT

Perennial herbs; *leaves* alternate, with an ocrea; *flowers* in a terminal, spike-like inflorescence; *tepals* 5, connate at the base, petaloid, white to pinkish, the outer larger than the inner; *stamens* 5–8; achenes not winged, triangular.
1a. Inflorescence short cylindric to ovoid, 8–25 mm wide; bulblets absent in the inflorescence...***B. bistortoides***
1b. Inflorescence narrowly elongate cylindric, 4–10 mm wide; pink, reddish-purple, or brown bulblets present below and sterile flowers above in the inflorescence...***B. vivipara***

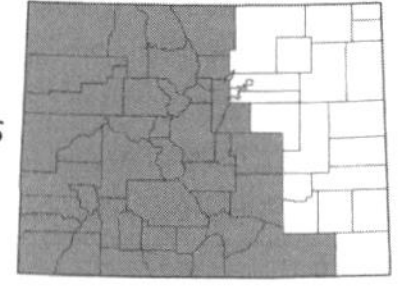

Bistorta bistortoides **(Pursh) Small**, AMERICAN BISTORT. [*Polygonum bistortoides* Pursh]. Plants (10) 20–75 cm; *leaves* elliptic to narrowly lanceolate or oblong-lanceolate, 3.5–22 × 0.8–5 cm, glabrous or hairy below; *inflorescence* short cylindric to ovoid, (1) 2–5 cm × 8–25 mm, bulblets absent; *tepals* white or pale pink, 4–5 mm long; *achenes* 3–4.2 mm, shiny. Common along streams, in moist meadows and marshes, aspen forests, and in alpine, 7000–14,000 ft. June–Sept. E/W. (Plate 68)

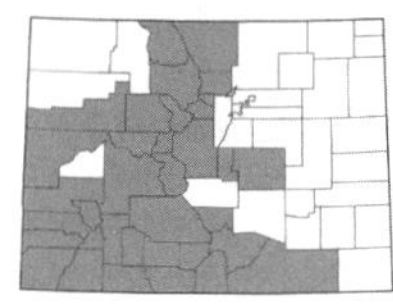

Bistorta vivipara **(L.) A. Gray**, ALPINE BISTORT. [*Polygonum viviparum* L.]. Plants (2) 8–35 (45) cm; *leaves* linear to lanceolate or narrowly ovate, 1–8 (10) × 0.5–2 cm, hairy below; *inflorescence* narrowly elongate-cylindric, (1.5) 2–9 cm × 4–10 mm, with pink, reddish-purple, or brown bulblets below and sterile flowers above; *tepals* 2–4 mm long; *achenes* rarely produced. Common in moist meadows, along streams, and in alpine, 8500–14,000 ft. June–Sept. E/W.

ERIOGONUM Michx. – WILD BUCKWHEAT

Annual or perennial herbs, subshrubs, or shrubs; *leaves* lacking an ocrea, alternate, opposite, or whorled; *flowers* 6–100 per tubular to campanulate involucre, in cymes or umbels; *tepals* 6, connate at the base, petaloid, sometimes with a stipelike base; *stamens* 9; *achenes* lenticular or triangular.
1a. Achenes 3-winged; plants (3) 5–25 dm tall, from a thick taproot; leaves all or mostly basal, linear-lanceolate to oblanceolate, hirsute, 5–20 cm long; flowers yellow or yellowish-green, or reddish in fruit...***E. alatum***
1b. Achenes wingless; plants of various heights, but usually not over 10 dm tall, from a slender taproot or branching caudex, often mat-forming; leaves various but unlike the above in all respects; flowers various...2

2a. Inflorescence not or barely exceeding the leaves; plants less than 3 cm tall, forming cushion-mats, known from Moffat County…3
2b. Inflorescence on evident floral stems exceeding the leaves; plants various, but usually more than 3 cm tall, variously distributed…4

3a. Tepals white to pink, 3–4 mm long; involucres with 7–10 spreading teeth…***E. tumulosum***
3b. Tepals yellow, 2–2.5 mm long; involucres with 4–5 erect teeth…***E. acaule***

4a. Annuals from taproots, not mat-forming…**Key 1**
4b. Perennials from woody, branching caudices, often mat-forming…5

5a. Stem leaves present (sometimes falling with age, but leaving behind the scale-like bases of the leaves on the stem) or the plants broom-like and lacking evident leaves; shrubs or subshrubs; flowers white to rose…**Key 2**
5b. Leaves all or mostly basal, or plants with basal leaves and a single whorl of leaves just below the inflorescence; usually perennial, often mat-forming herbs, sometimes shrubs or subshrubs; flowers various…6

6a. Inflorescence capitate or subcapitate, or tightly umbellate (flowers in ball-like clusters), or reduced to a single involucre…**Key 3**
6b. Inflorescence open with several to many exposed branches, or raceme- or spike-like…**Key 4**

Key 1
(Annuals)

1a. Stem leaves present and well-developed (sometimes falling with age, but leaving behind the scale-like bases of the leaves on the stem)…2
1b. Leaves all or mostly basal…3

2a. Plants spreading, 1–2 (3) dm tall, minutely hairy or short-hairy throughout; flowers yellow; leaves orbiculate or oblong-elliptic…***E. divaricatum***
2b. Plants erect, 1–10 dm tall, floccose or densely tomentose throughout; flowers white to cream or pink-tinged; leaves oblanceolate to oblong…***E. annuum***

3a. Flowers hairy, yellow and often with reddish midribs; stems usually distinctly inflated below the inflorescence, often glaucous…4
3b. Flowers glabrous, variously colored; stems the same width throughout, glaucous or not…5

4a. Flowering stems and inflorescence branches grayish and usually glaucous…***E. inflatum***
4b. Flowering stems and inflorescence branches yellowish-green or green, not glaucous…***E. fusiforme***

5a. Involucres sessile; stems glabrous, or floccose to tomentose…6
5b. Involucres pedicellate; stems glabrous or sparsely hairy…8

6a. Stems and inflorescence branches glabrous and glaucous; flowers yellow or reddish-yellow…***E. hookeri***
6b. Stems and inflorescence branches floccose to tomentose or scabrous; flowers white to pink or rose…7

7a. Involucres angled; leaves 5–15 (18) mm long, the margins usually not crisped and wavy…***E. palmerianum***
7b. Involucres smooth; leaves 10–30 (40) mm long, with crisped and wavy margins…***E. scabrellum***

8a. Leaves sparsely hairy below, green…***E. gordonii***
8b. Leaves densely white-tomentose or floccose below…9

9a. Flowers white to pinkish, turning red in age; peduncles reflexed in age; involucres 1.5–2 mm long…***E. cernuum***
9b. Flowers yellowish to red, turning rose in age peduncles erect; involucres 0.3–1 mm long…***E. wetherillii***

Key 2
(Stem leaves present or plants broom-like; shrubs or subshrubs)

1a. Leaves linear to linear-oblanceolate, with tightly revolute margins (such that the underside of the leaf is scarcely if at all visible)…2
1b. Leaves linear-lanceolate, elliptic, ovate, oblong, or obovate, revolute only on the margins if at all…5

2a. Inflorescence much-branched; plants usually yellowish-green; stems glabrous or thinly hairy…***E. leptophylum***
2b. Inflorescence of simple cymes or umbels, or not much-branched; plants greenish to grayish; stems usually floccose to tomentose…3

3a. Tepals dimorphic, the outer whorl broadly obovate and 2–2.5 mm wide and the inner whorl oblanceolate to spatulate and 1–1.5 mm wide; involucre (3) 3.5–5 mm long, glabrous; known from Montezuma Co.…***E. clavellatum***
3b. Tepals all similar; involucre 2–3.5 mm long, glabrous or floccose; variously distributed…4

4a. Subshrubs 0.5–1 (1.5) dm; leaves glabrous or nearly so above; tepals connate for ½ their length; known from Delta and Montrose counties…***E. pelinophilum***
4b. Subshrubs or shrubs (1) 2–15 dm; leaves usually whitish-floccose above; tepals connate in the lower ¼; widespread…***E. microthecum* var. *simpsonii***

5a. Leaves mostly 2.5–3.5 times longer than wide (at least the largest leaves 1 cm or more wide), lanceolate, oblanceolate, elliptic, ovate, or orbiculate, the margins often crenulate…***E. corymbosum***
5b. Leaves 4–9 times longer than wide (to 1 cm wide), linear-lanceolate, narrowly oblong, narrowly elliptic, or oblanceolate, the margins not crenulate…6

6a. Plants of the eastern slope…***E. effusum***
6b. Plants of the western slope…7

7a. Inflorescence a large, open cyme 10–50 cm wide; few leaves usually left on the shrubs at flowering time; flowers white or yellow…***E. leptocladon***
7b. Inflorescence a compact cyme, mostly 1–10 cm wide; numerous leaves usually left on the shrubs at flowering time; flowers white…***E. microthecum* var. *laxiflorum***

Key 3

(Inflorescence capitate or subcapitate, or tightly umbellate, or reduced to a single involucre)

1a. Inflorescence subtended by a whorl of leaf-like bracts; perianth usually narrowed at the base to a slender, stipe-like base about the same width as the pedicel (this sometimes obscure in *E. flavum*)…2
1b. Inflorescence usually not subtended by a whorl of leaf-like bracts; perianth directly attached to the pedicel, lacking a slender, stipelike base…6

2a. Perianth hairy externally…3
2b. Perianth glabrous externally…5

3a. Flowers white, cream, or pinkish…***E. jamesii* var. *jamesii***
3b. Flowers pale to bright yellow…4

4a. Involucres elevated on evident peduncles…***E. flavum* var. *flavum***
4b. Involucres sessile or nearly so…***E. arcuatum***

5a. Stems with a whorl of leaves in between the basal leaves and whorls of leaf-like bracts subtending the inflorescence; basal leaves linear to oblanceolate…***E. heracleoides* var. *heracleoides***
5b. Stems only with a whorl of leaf-like bracts subtending the inflorescence; basal leaves oblong-ovate, oblanceolate, or oval…***E. umbellatum***

6a. Perianth hairy externally…7
6b. Perianth glabrous externally…9

7a. Inflorescence branched; flowers yellow…***E. lachnogynum* var. *lachnogynum***
7b. Inflorescence not branched; flowers white, rose, or rarely yellow…8

8a. Achenes glabrous; leaves 1–4 cm long; plants 0.5–2 dm tall, of the eastern slope…***E. pauciflorum***
8b. Achenes hairy; leaves 0.2–0.8 (1.2) cm long; plants 0.3–0.5 (0.7) dm tall, of the western slope…***E. shockleyi***

9a. Leaves green and glabrous or thinly hairy above and white-tomentose below…10
9b. Leaves densely white-tomentose above and below (the same color above and below)…11

10a. Leaves oblanceolate or lanceolate (3–9 mm wide) with flat margins; peduncles absent; flowering stems glabrous…***E. coloradense***
10b. Leaves linear to linear-oblanceolate (1–3 mm wide) with revolute margins; peduncles 1–4 mm long; flowering stems often sparsely floccose or tomentose, sometimes glabrous…***E. exilifolium***

11a. Leaves linear-oblanceolate or narrowly elliptic, 1–3 mm wide, and 5–20 mm long; subshrubs…***E. bicolor***
11b. Leaves spatulate, elliptic, oblong, oval, or oblanceolate, 4–16 mm wide and 5–60 mm long; perennials, often forming mats; not restricted to Mesa Co.…12

12a. Plants erect to spreading but not forming mats; leaves oblanceolate or narrowly elliptic, 1.5–5 cm long; capitate cymes 0.1–1.5 cm wide, usually with branches…***E. brandegeei***
12b. Plants forming pulvinate or caespitose mats; leaves oblanceolate, oblong, or elliptic, 0.2–6 cm long; capitate cymes 1–3.5 cm wide, lacking branches…***E. ovalifolium***

Key 4

(Inflorescence open with several to many exposed branches, or raceme- or spike-like)

1a. Leaves short-hirsute at least below; stems usually distinctly inflated below the inflorescence, glaucous; flowers densely hirsute, yellow…***E. inflatum***
1b. Leaves densely white-tomentose, at least below; stems not inflated below the inflorescence, sometimes glaucous; flowers glabrous or tomentose, white, pink, cream or yellow…2

2a. Inflorescence raceme- or spike-like; leaves elliptic to ovate or nearly orbicular 1.5–6 (10) cm long, on petioles 3–10 (15) cm long; flowers white to pinkish, 2–5 mm long…***E. racemosum***
2b. Inflorescence unlike the above; leaves and flowers various…3

3a. Perianth densely hairy, yellow; stems tomentose to floccose; leaves densely white-tomentose above and below…***E. lachnogynum* var. *lachnogynum***
3b. Perianth glabrous, white, pink, or yellow; stems usually glabrous; leaves various…4

4a. Leaves white-tomentose above and below, elliptic, mostly 0.5–1.4 cm wide; flowers white or pink…5
4b. Leaves white-tomentose below, and green and glabrous or thinly hairy above, linear to lanceolate, oblanceolate, or narrowly elliptic, 0.1–0.7 cm wide; flowers white or yellow…6

5a. Involucres all on pedicels; leaves 0.3–3 cm long; tepals dimorphic, the outer whorl ovate to suborbiculate and the inner whorl narrowly oblong…***E. tenellum* var. *tenellum***
5b. Middle involucre sessile or nearly so in the inflorescence; leaves 1.5–4 cm long; tepals all similar in shape and size…***E. batemanii***

6a. Flowers white…***E. lonchophyllum***
6b. Flowers ochroleucous, pale yellow, or bright yellow…7

7a. Leaves linear to narrowly oblanceolate, tightly revolute such that the underside of the leaf is scarcely if at all visible; plants known from Mesa Co.…***E. contortum***
7b. Leaves linear, lanceolate, oblanceolate, or narrowly elliptic, flat or if with revolute margins then the underside of the leaf clearly visible (at least on most leaves); plants not restricted to Mesa Co.…8

8a. Leaves 1.5–2.5 cm long; flowers ochroleucous or pale yellow, rarely bright yellow; inflorescence narrowly cymose, the floral stems much-branched in the upper ⅔ (resembling a small *Ephedra*); plants from a robust, woody root system…***E. ephedroides***
8b. Leaves 3–10 cm long; flowers bright yellow (rarely cream); inflorescence an open cyme, the floral stems branched in the upper ½; plants from a rather slender root system…***E. brevicaule* var. *brevicaule***

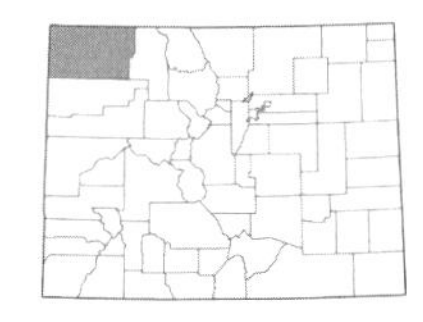

Eriogonum acaule **Nutt.**, SINGLE-STEM BUCKWHEAT. Perennials, forming cushion mats, 1–2 dm; *leaves* linear to oblanceolate, 0.3–0.8 × 0.1–0.2 cm, densely tomentose, margins revolute; *inflorescence* not or barely exceeding the leaves, bracts absent; *involucres* campanulate, 1.5–3 mm long, tomentose; *tepals* yellow, 2–2.5 mm long, white-tomentose; *achenes* 2–2.5 mm long, sparsely hairy. Found on sandy, barren slopes, often with sagebrush or saltbush, 5600–7000 ft. May–July. W.

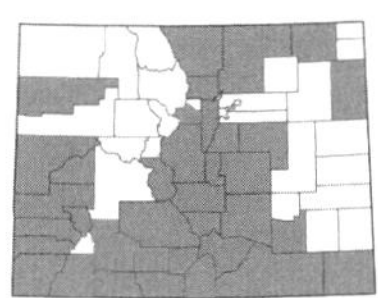

Eriogonum alatum **Torr.**, WINGED BUCKWHEAT. Perennials from taproots, (3) 5–25 dm; *leaves* narrowly oblanceolate to linear-lanceolate, 5–20 cm long, hirsute, mostly basal; *inflorescence* paniculate; *involucres* campanulate to obconic, 2–5 mm long, glabrous to pilose; *tepals* yellowish to yellowish-green, 1.5–3 mm long; *achenes* 5–9 mm long, 3-winged, glabrous. Common on open slopes, in grasslands, forests, and sagebrush, 3500–10,500 ft. June–Aug. E/W.

Eriogonum annuum Nutt., ANNUAL WILD BUCKWHEAT. Annuals, 1–10 dm, simple or few-branched; *leaves* oblanceolate to oblong, 2–5 cm long, the basal leaves often withering before flowering, tomentose; *inflorescence* an open cyme, with minute bracts; *involucres* campanulate, 2.5–3 mm long; *tepals* white to cream or sometimes pink-tinged, glabrous outside and hairy inside; *achenes* glabrous. Common on dry slopes, sand hills, and in grasslands and prairie, 3500–7500 ft. May–Oct. E.

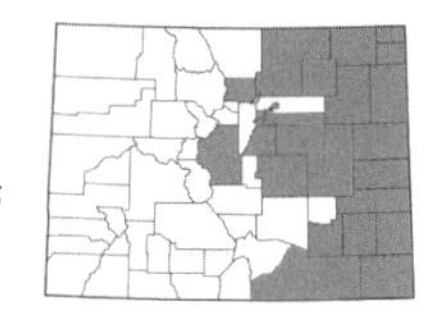

Eriogonum arcuatum Greene, BAKER'S BUCKWHEAT. Perennials, often mat-forming, 0.5–8 dm; *leaves* basal, oblanceolate to elliptic, 0.5–3 × 0.5–1.5 cm, tomentose below and floccose and greenish above; *inflorescence* capitate or umbellate, with 3–8 bracts; *involucres* campanulate, 3–7 mm long; *tepals* yellow, densely hairy, 1.5–5 mm long; *achenes* 4–5 mm long, glabrous with a sparsely hairy beak.

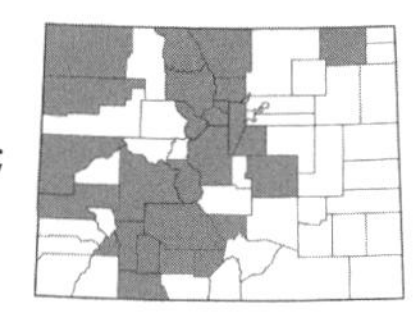

There are two varieties of *E. arcuatum* in Colorado:

1a. Inflorescences umbellate, branched 1–3 times; stems mostly 5–20 cm tall...**var. _arcuatum_**, BAKER'S BUCKWHEAT. [*E. jamesii* Benth. ssp. *bakeri* (Greene) S. Stokes]. Common on shortgrass prairie, dry slopes, rocky outcrops, sagebrush flats, and in grasslands and pinyon-juniper forests, 4000–9800 ft. June–Oct. E/W.

1b. Inflorescences capitate, unbranched, or reduced to a single involucre; stems 2–8 cm tall...**var. _xanthum_ (Small) Reveal**, IVY LEAGUE BUCKWHEAT. [*E. flavum* Nutt. var. *xanthum* (Small) S. Stokes]. Common in forests, sagebrush flats, and the alpine, 9500–13,500 ft. July–Sept. E/W. Endemic.

Eriogonum batemanii M.E. Jones, BATEMAN'S BUCKWHEAT. Perennials, 1–5 dm; *leaves* basal, elliptic, 1.5–4 × 0.6–1.2 cm, densely white-tomentose below, less so above; *inflorescence* cymose-paniculate with spreading-ascending branches; *involucres* campanulate to turbinate, 2–5 per cluster, 2–4 mm long, glabrous; *tepals* white, 1.5–3 mm long, glabrous; *achenes* 2.5–3 mm long, glabrous. Found on sagebrush flats, shale barrens, canyons, and dry slopes, 5400–7500 ft. June–Sept. W.

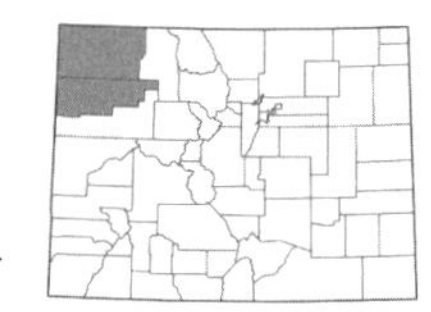

Eriogonum bicolor M.E. Jones, PRETTY BUCKWHEAT. Shrubs or subshrubs, 0.2–0.8 dm tall, with horizontal, spreading stems; *leaves* caulescent, linear-oblanceolate to narrowly elliptic, 0.5–2 × 0.1–0.3 cm, densely white-tomentose below and less so above, margins revolute; *inflorescence* umbellate to compact cymose; *involucres* 2–4 mm long, tomentose; *tepals* white, 2.5–4.5 mm long, glabrous; *achenes* 3–3.5 mm long, glabrous. Rare on barren, clay hillsides and shale slopes, 4500–5500 ft. April–June. W.

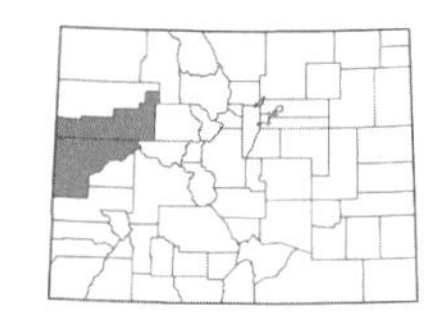

Eriogonum brandegeei Rydb., BRANDEGEE'S BUCKWHEAT. Perennials, 1–2.5 dm; *leaves* basal, oblanceolate to narrowly elliptic, 1.5–5 × 0.4–1.6 cm, densely white-tomentose below, less so and greenish above; *inflorescence* capitate or umbellate, with 3 scale-like or leaf-like bracts; *involucres* turbinate, 4–8 per cluster, 3.5–5 mm long, glabrous to floccose; *tepals* yellowish, 2.5–3.5 mm long, glabrous; *achenes* 3–3.5 mm long, glabrous. Found on dry slopes and barren clay hills, usually with pinyon-juniper, 5800–7600 ft. June–Aug. E. Endemic. (Plate 68)

Eriogonum brevicaule Nutt. var. _brevicaule_, SHORTSTEM BUCKWHEAT. Perennials, 0.5–4 dm; *leaves* basal, linear to oblanceolate, 3–10 × 0.1–0.7 cm, tomentose below and greenish and floccose above; *inflorescence* an open cyme, with 3 scale-like bracts; *involucres* turbinate to campanulate, 1.5–4 mm long, glabrous; *tepals* yellow or rarely cream, 2–3 mm long, glabrous; *achenes* 2–3 mm long, glabrous. Common on shale slopes and ridges, clay hillsides, and bluffs, 5200–9000 ft. June–Sept. E/W. (Plate 68)

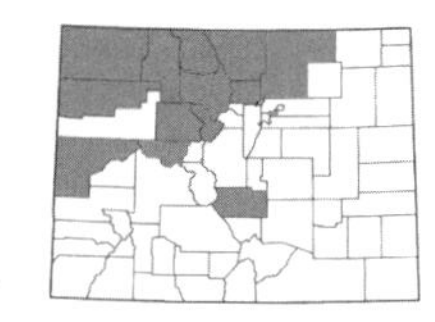

Eriogonum cernuum Nutt., NODDING BUCKWHEAT. Annuals, 0.5–6 dm; *leaves* mostly basal, orbiculate to broadly ovate, 1–2.5 × 1–2.5 cm, tomentose below and tomentose to floccose or glabrate above; *inflorescence* an open, diffuse cyme; *involucres* turbinate, 1.5–2 mm long, glabrous; *tepals* white to pink-tinged, turning reddish in age, 1–2 mm long, glabrous; *achenes* 1.5–2 mm long, glabrous. Common on dry hillsides, sandy slopes, open prairie, and in sagebrush, 4500–9300 ft. June–Sept. E/W.

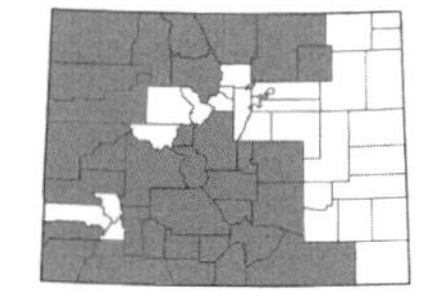

Eriogonum clavellatum Small, COMB WASH BUCKWHEAT. Subshrubs or shrubs, 0.7–2.5 dm; *leaves* cauline, oblanceolate, 1–2 × 0.08–0.2 cm, densely tomentose below, grayish or rarely glabrous above, the margins tightly revolute; *inflorescence* umbellate or a compact cyme, with 3 scale-like bracts; *involucres* turbinate to campanulate, (3) 3.5–5 mm long, glabrous; *tepals* white, 2.5–3.5 mm long, glabrous; *achenes* 3–3.5 mm long, glabrous. Rare on dry, clay or shale slopes, 5200–6500 ft. May–July. W.

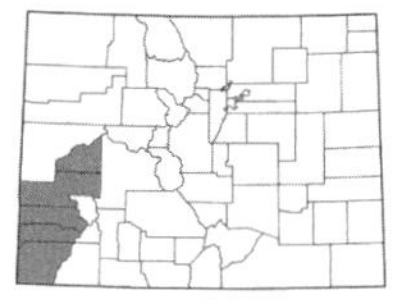

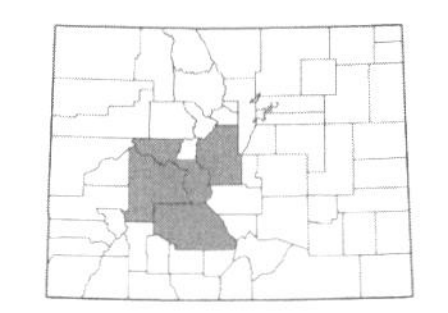

***Eriogonum coloradense* Small**, COLORADO BUCKWHEAT. Perennials, 0.3–0.6 dm; *leaves* basal, lanceolate to oblanceolate, 1–5 × 0.3–0.9 cm, densely tomentose below, thinly floccose or green and glabrous above; *inflorescence* capitate, with 3 small, leaf-like bracts; *involucres* turbinate to campanulate, 2–3.5 mm long, glabrous with tomentose teeth; *tepals* white to pinkish, 2.5–4 mm long, glabrous; *achenes* 2.5–3.5 mm long, glabrous. Found on dry flats and slopes, in grasslands, sagebrush meadows, and on talus slopes and alpine ridges, 8800–12,800 ft. July–Sept. E/W. Endemic.

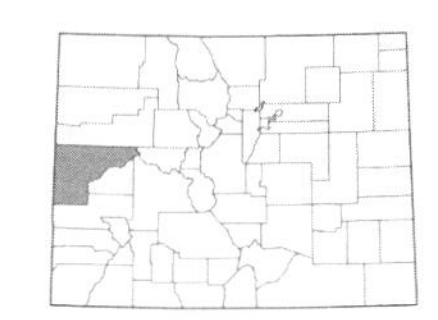

***Eriogonum contortum* Small**, GRAND BUCKWHEAT. Shrubs, 0.5–0.8 (2) dm; *leaves* cauline, linear to narrowly oblanceolate, 0.5–2.5 × 0.15–0.25 cm, densely tomentose below, greenish above, with revolute margins; *inflorescence* cymose, with 3 scale-like bracts; *involucres* turbinate, 1.5–2.5 mm long, glabrous; *tepals* yellow, 1.5–2.5 mm long, glabrous; *achenes* 2–2.5 mm long, glabrous. Uncommon on clay or shale slopes, usually with saltbush, 4500–5100 ft. May–Aug. W.

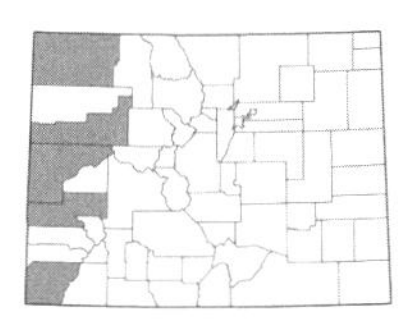

***Eriogonum corymbosum* Benth.**, CRISPLEAF BUCKWHEAT. Shrubs or subshrubs, 0.7–12 dm; *leaves* cauline, lanceolate, oblanceolate, elliptic, ovate, or orbiculate, 0.5–4.5 × 0.3–3.5 cm, densely tomentose on both surfaces or less so and green above, the margins sometimes crisped; *inflorescence* cymose, with 3 scale-like bracts; *involucres* turbinate, 1–3.5 mm long, glabrous or sparsely hairy; *tepals* white, cream, or pink, 1.5–3.5 mm long, glabrous or rarely sparsely hairy; *achenes* 2–3 mm long, glabrous. Found on dry hillsides, clay slopes, cliffs, often with sagebrush or juniper, 4500–8500 ft. July–Oct. W.

There are 3 varieties of *E. corymbosum* in Colorado:

1a. Leaves lanceolate to oblanceolate or elliptic, 0.5–1.5 cm wide...**var. *corymbosum***, CRISPLEAF BUCKWHEAT. Scattered in western Colorado. July–Oct.
1b. Leaves elliptic-oblong to oblong, ovate, or orbiculate, 1–3.5 cm wide...2

2a. Leaves broadly ovate to orbiculate...**var. *orbiculatum* (S. Stokes) Reveal & Brotherson**, ORBICULAR-LEAF BUCKWHEAT. Known from Mesa, Montezuma, San Juan, and San Miguel cos. Aug.–Oct.
2b. Leaves elliptic-oblong to oblong or rarely ovate...**var. *velutinum* Reveal**, VELVETY BUCKWHEAT. Known from Mesa, Montezuma, Montrose, and San Miguel cos. Aug.–Oct.

***Eriogonum divaricatum* Hook.**, DIVERGENT WILD BUCKWHEAT. Annuals, 1–2 (3) dm; *leaves* basal and cauline, orbiculate to oblong-elliptic, 1–2.5 × 1–2.5 cm, minutely hairy with crinkly hairs on both surfaces; *inflorescence* a diffuse cyme; *involucres* campanulate, 1–2 mm long, pilose, sessile; *tepals* yellowish or sometimes tinged with red, 1–2 mm long, hispid and glandular; *achenes* 1.5–2 mm long. Found on dry hillsides and clay or shale slopes, often with saltbush or sagebrush, 4500–6000 ft. May–Sept. W.

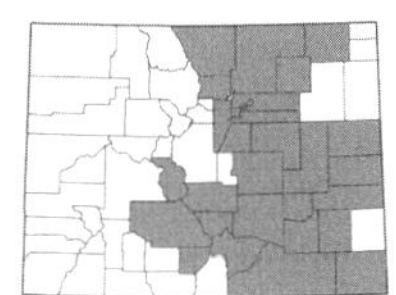

***Eriogonum effusum* Nutt.**, SPREADING BUCKWHEAT. Shrubs; *leaves* cauline, oblanceolate to oblong, 1–3 × 0.2–0.7 cm, densely tomentose below and floccose to glabrate and green above; *inflorescence* an open cyme; *involucres* turbinate, 1 per node, 2–2.5 mm long, tomentose; *tepals* yellow, 2–4 mm long, glabrous; *achenes* 2–2.5 mm long, glabrous. Common on dry slopes, sand hills, and in grasslands, shortgrass prairie, and sagebrush flats, 4000–8500 ft. July–Nov. E.

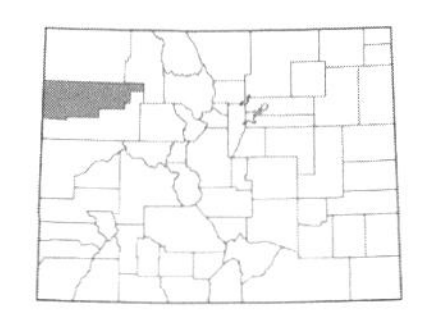

***Eriogonum ephedroides* Reveal**, EPHEDRA BUCKWHEAT. Perennials, 1–3 dm; *leaves* basal, lanceolate, 1.5–2.5 × 0.2–0.3 cm, densely tomentose below, glabrate and green above; *inflorescence* cymose; *involucres* turbinate, 2–2.5 mm long, glabrous; *tepals* pale yellow, rarely bright yellow, 2–2.5 mm long, glabrous; *achenes* 2–2.5 mm long, glabrous. Uncommon on white shale slopes, 5500–6500 ft. May–July. W.

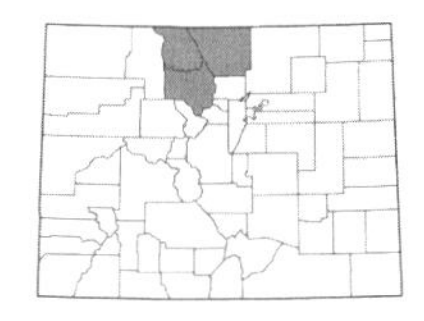

***Eriogonum exilifolium* Reveal**, DROPLEAF BUCKWHEAT. Perennials, matted, 0.3–1 × 1–2 dm; *leaves* basal, linear to narrowly oblanceolate, 2–6 × 0.1–0.3 cm, densely tomentose below and glabrate and green above, with revolute margins; *inflorescence* umbellate; *involucres* turbinate to campanulate, 2.5–4.5 mm long, glabrous and floccose between the teeth; *tepals* white, 2–3.5 mm long, glabrous; *achenes* 2.5–3.5 mm long, glabrous. Found on shale barrens, clay hills, and sandy slopes, sometimes with sagebrush, 7500–8800 ft. June–Aug. E/W.

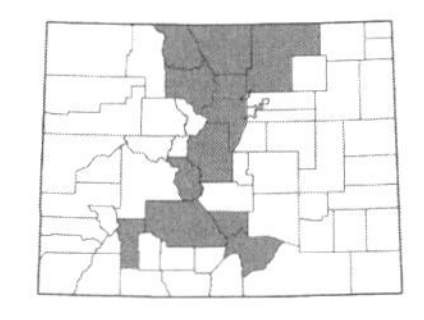

Eriogonum flavum* Nutt. var. *flavum, GOLDEN BUCKWHEAT. Perennials, matted, to 2 dm; *leaves* basal, elliptic, 1–5 × 0.3–1.5 cm, densely tomentose below and tomentose to floccose and green above; *inflorescence* umbellate; *involucres* turbinate to campanulate, 3–8 mm long; *tepals* bright yellow, 3–7 mm long, densely hairy; *achenes* 3–5 mm long, glabrous. Found on shortgrass prairie, rocky ridges, and in forests and sagebrush meadows, 4400–10,500 ft. June–Sept. E/W. (Plate 68)

***Eriogonum fusiforme* Small**, GRAND VALLEY DESERT TRUMPET. [*E. inflatum* Torr. & Frém. var. *fusiforme* (Small) Reveal]. Annuals, 0.5–4 dm, stems hollow and usually inflated below the inflorescence; *leaves* basal, orbicular, 0.5–3 × 0.5–2.5 cm, hirsute and green; *inflorescence* an open cyme; *involucres* turbinate, 1–1.5 mm long, glabrous; *tepals* yellow with green to reddish midribs, 1.3–1.6 mm long, densely hairy; *achenes* 1.3–2 mm long, glabrous. Found on shale or clay slopes, sometimes found in shrubs or pinyon-juniper, 4400–5500 ft. May–July. W.

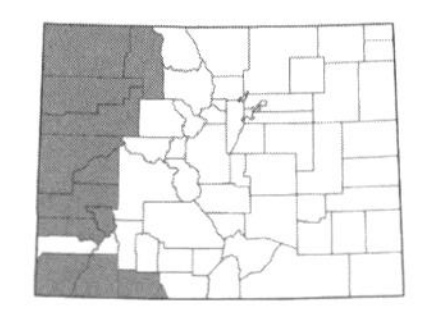

***Eriogonum gordonii* Benth.**, GORDON'S BUCKWHEAT. Annuals, 0.5–5 (7) dm; *leaves* basal, obovate to orbicular, 1–5 × 1–5 cm, sparsely hirsute and green; *inflorescence* an open cyme; *involucres* campanulate, 0.6–1.5 mm long, glabrous; *tepals* white with green to reddish midribs, rarely ochroleucous, turning pink in age, glabrous; *achenes* 2–2.5 mm long, glabrous. Common on dry slopes and clay flats, often with sagebrush or pinyon-juniper, 4500–7000 ft. May–Aug. W.

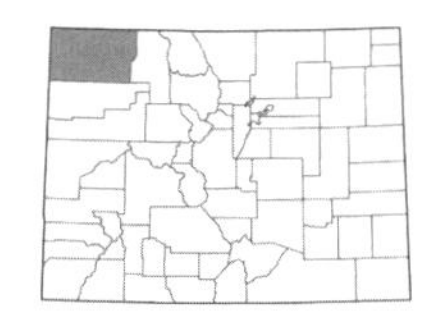

Eriogonum heracleoides* Nutt. var. *heracleoides, PARSNIP FLOWER BUCKWHEAT. Perennials, matted, stems 1–6 dm; *leaves* basal, linear to oblanceolate, 1.5–5 × 0.2–1.5 cm, densely white-tomentose below and green and floccose to glabrate above; *inflorescence* umbellate, with a whorl of leaf-like bracts ca. midlength along the stem; *involucres* turbinate to campanulate, 3–4.5 mm long, tomentose or rarely glabrous; *tepals* white to cream or ochroleucous, 4–8 mm long, glabrous; *achenes* 3–5 mm long, glabrous. Found in sandy or rocky soil, often with sagebrush, 5800–8800 ft. June–Aug. W.

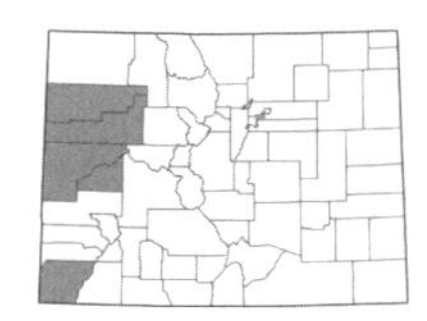

***Eriogonum hookeri* S. Watson**, HOOKER'S BUCKWHEAT. Annuals, 0.8–6 dm; *leaves* basal, orbicular to reniform, 2–5 × 2–6 cm, densely tomentose on both surfaces; *inflorescence* an open cyme; *involucres* campanulate to hemispheric, 1–2 mm long, glabrous; *tepals* yellow or reddish-yellow, 1.5–2 mm long, glabrous; *achenes* 2–2.5 mm long, glabrous. Uncommon on sandy or shale slopes, and in canyon bottoms, 4500–6500 ft. June–Aug. W.

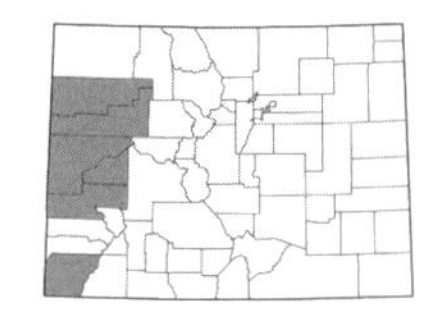

***Eriogonum inflatum* Torr. & Frém.**, DESERT TRUMPET. Annuals, 1–10 (15) dm, the stems usually hollow and inflated below the inflorescence; *leaves* basal, oblong to orbicular, 0.5–3 × 0.5–2.5 cm, hirsute and greenish, the margins usually undulate; *inflorescence* an open cyme; *involucres* turbinate, 1–1.5 mm long, glabrous; *tepals* yellow with green or reddish midribs, 2–4 mm long, densely hairy; *achenes* 2–2.5 mm long, glabrous. Found on dry hillsides, clay or shale slopes, and in dry washes, 3900–5500 ft. May–Aug. W. (Plate 68)

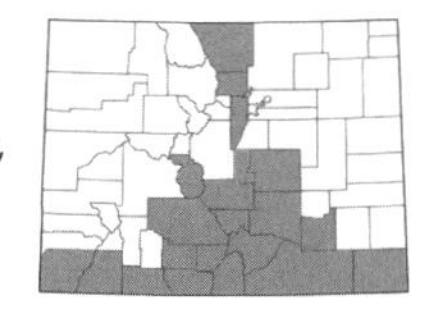

Eriogonum jamesii* Benth. var. *jamesii, JAMES' BUCKWHEAT. Perennials, mat-forming, 3–8 dm wide × 0.5–1.5 dm; *leaves* basal, narrowly elliptic, 1–3.5 × 0.3–1.2 cm, densely tomentose below and floccose to glabrate and greenish above; *inflorescence* compound-umbellate; *involucres* turbinate, 4–7 mm long, tomentose to floccose; *tepals* white to cream or pinkish, with a stipelike base, 3–8 mm long; *achenes* 4–5 mm long, glabrous. Found in sandy soil, on rocky or shale slopes, often with pinyon-juniper, 5000–10,000 ft. June–Oct. E/W. (Plate 68)

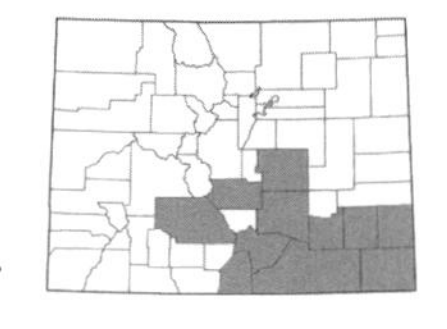

Eriogonum lachnogynum* Torr. var. *lachnogynum, WOOLLYCUP BUCKWHEAT. Perennials, matted, 1–2 (3.5) dm; *leaves* basal, lanceolate to oblanceolate or narrowly elliptic, 1–3 × 0.3–0.8 cm, densely tomentose on both surfaces; *inflorescence* umbellate; *involucres* broadly campanulate, (2) 3–4 mm long, tomentose; *tepals* yellow, densely hairy; *achenes* 3–4 mm long, hairy. Found on sandy or rocky soil, shale slopes, or gypsum flats, often with pinyon-juniper, 4000–8000 ft. May–Aug. E. (Plate 68)

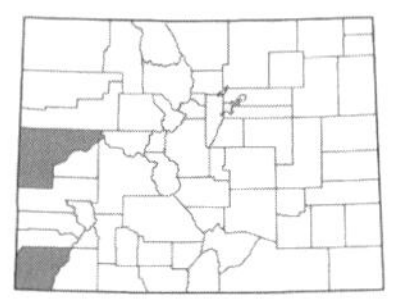

***Eriogonum leptocladon* Torr. & A. Gray**, SAND BUCKWHEAT. Shrubs, 2–10 dm; *leaves* cauline, linear-lanceolate to narrowly oblanceolate or oblong, 1.5–4 × 0.2–1 cm, densely tomentose below, floccose and greenish above; *inflorescence* an open cyme; *involucres* turbinate, 1.5–3 mm long, tomentose to glabrous; *tepals* white or yellow, 2–3.5 mm long, glabrous; *achenes* 2.5–3.5 mm long, glabrous. Found on dry hillsides and Mancos shale or sandy slopes, 4800–7000 ft. June–Oct. W.

There are 2 varieties of *E. leptocladon* in Colorado:

1a. Flowers yellow...**var. *leptocladon***. Known from Mesa Co. near the Utah border, 4800–5000 ft.
1b. Flowers white...**var. *ramosissimum* (Eastw.) Reveal**, Known from Montezuma Co., 6000–7000 ft.

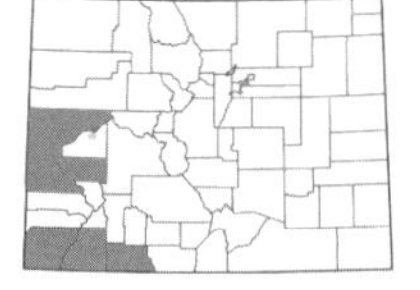

***Eriogonum leptophyllum* (Torr. & A. Gray) Wooton & Standl.**, SLENDERLEAF BUCKWHEAT. Shrubs or subshrubs, 1.5–6 dm; *leaves* cauline, linear to narrowly oblanceolate, 2–6 × 0.1–0.3 cm, densely tomentose below and glabrate and green above, the margins revolute; *inflorescence* a cyme; *involucres* narrowly turbinate, 2–4.5 mm long, glabrous; *tepals* white, 2.5–4 mm long, glabrous; *achenes* 3.5–4 mm long, glabrous. Uncommon on dry slopes, 4700–6500 ft. July–Sept. W.

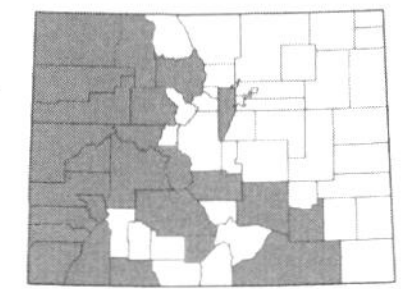

***Eriogonum lonchophyllum* Torr. & A. Gray**, SPEARLEAF BUCKWHEAT. Shrubs or subshrubs, 0.8–8 dm; *leaves* cauline, lanceolate to oblanceolate or narrowly elliptic, 1.5–7 × 0.2–2 cm, tomentose below, floccose to glabrate and green above; *inflorescence* usually an open cyme; *involucres* turbinate, 2.5–4 mm long, glabrous; *tepals* white, 2–4 mm long, glabrous; *achenes* 2–3 mm long, glabrous. Common on dry hillsides and shale and clay slopes, often with pinyon-juniper or sagebrush, 4800–9500 ft. May–Oct. E/W.

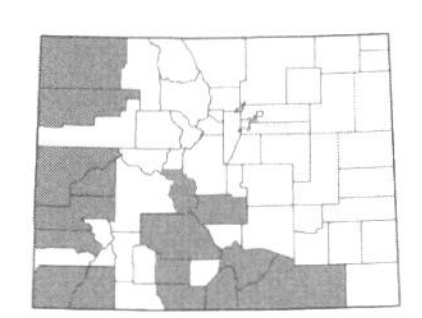

***Eriogonum microthecum* Nutt.**, SLENDER BUCKWHEAT. Shrubs or subshrubs, (1) 2–15 dm; *leaves* cauline, linear-lanceolate, oblong, obovate to elliptic, 0.3–3.5 × 0.1–1.2 cm, tomentose below and glabrate above, the margins sometimes revolute; *inflorescence* a cyme, often flat-topped; *involucres* turbinate, 1.5–4 mm long, tomentose to glabrous; *tepals* white, pink, or yellow, 1.5–4 mm long, glabrous; *achenes* 1.5–3 mm long, glabrous.

There are two varieties of *E. microthecum* in Colorado:

1a. Leaves linear-lanceolate, elliptic, ovate, oblong, or obovate, revolute only on the margins if at all...**var. *laxiflorum* Hook.**, SLENDER BUCKWHEAT. Found on dry, rocky slopes and sandy hillsides, often with sagebrush, known from the northwestern counties, 5800–9000 ft. June–Oct. W.

1b. Leaves linear to linear-oblanceolate, with tightly revolute margins (such that the underside of the leaf is scarcely if at all visible)...**var. *simpsonii* (Benth.) Reveal**, SIMPSON'S BUCKWHEAT. Found on dry slopes, often with pinyon-juniper, known from the western and southcentral counties, 5200–8500 ft. June–Oct. E/W. Intergrades with var. *laxiflorum* in the northwestern counties.

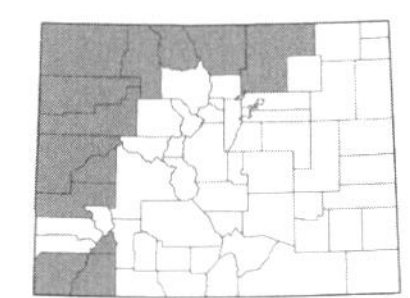

***Eriogonum ovalifolium* Nutt.**, CUSHION BUCKWHEAT. Perennials, mat-forming, 0.5–5 dm across; *leaves* basal, oblanceolate, oblong, or elliptic, 0.2–6 × 0.1–1.5 cm, tomentose to floccose; *inflorescence* capitate; *involucres* turbinate, (2) 3.5–5 (8) mm long, tomentose to floccose; *tepals* yellow, white, cream, or rose, 2.5–7 mm long, glabrous; *achenes* 2–3 mm long, glabrous. (Plate 68)

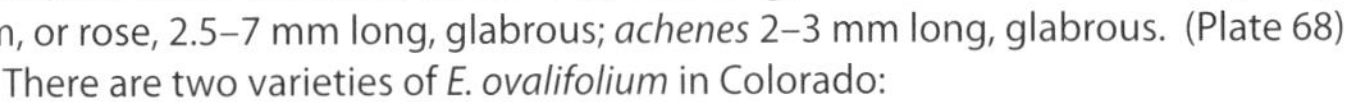

There are two varieties of *E. ovalifolium* in Colorado:

1a. Flowers yellow...**var. *ovalifolium***, CUSHION BUCKWHEAT. Found on dry slopes, in grasslands, sagebrush flats, and in pinyon-juniper, known from the northwestern counties, 4500–8500 ft. May–July. W.

1b. Flowers white, rose, or purple...**var. *purpureum* (Nutt.) Durand**, PURPLE CUSHION BUCKWHEAT. Common on dry slopes, sagebrush flats, and in washes, often with pinyon-juniper, 4500–10,200 ft. May–July. E/W.

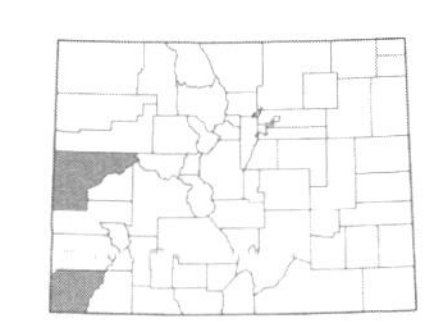

***Eriogonum palmerianum* Reveal**, PALMER'S BUCKWHEAT. Annuals, 0.6–3 dm; *leaves* basal, orbiculate to cordate, 0.5–1.5 (1.8) × 0.5–2 cm, densely tomentose below, tomentose to glabrate above; *inflorescence* an open cyme; *involucres* campanulate, 1.5–2 mm long, tomentose; *tepals* white to pink or ochroleucous, 1.5–2 mm long, glabrous; *achenes* 1.5–2 mm long. Uncommon on dry slopes, often with pinyon-juniper, 5000–6200 ft. May–June. W. (Plate 68)

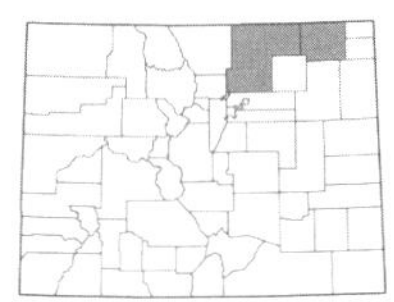

***Eriogonum pauciflorum* Pursh**, FEWFLOWER BUCKWHEAT. Perennials, forming loose mats, 0.5–2 dm; *leaves* basal, narrowly oblanceolate to oblanceolate or elliptic, 1–4 × 0.1–1 cm, tomentose; *inflorescence* capitate or umbellate; *involucres* turbinate, 3.5–5 mm long, tomentose; *tepals* white to rose, 2–2.5 mm long, hairy or rarely glabrous; *achenes* 2–2.5 mm long, glabrous. Uncommon on barren, rocky outcrops, sandstone, and bluffs, 4000–5400 ft. May–Sept. E.

***Eriogonum pelinophilum* Reveal**, CLAY-LOVING BUCKWHEAT. Subshrubs, 0.5–1 (1.5) dm; *leaves* cauline, oblanceolate, 0.5–1.5 × 0.08–0.3 cm, densely tomentose below, glabrate and green above, the margins revolute; *inflorescence* cymose; *involucres* turbinate, 2–3.5 mm long, glabrous to floccose; *tepals* creme, 2.5–3.5 mm long, glabrous; *achenes* 3–3.5 mm long, glabrous. Rare on Mancos shale slopes, 5500–6000 ft. May–July. W. Endemic.

Listed as endangered under the Endagered Species Act.

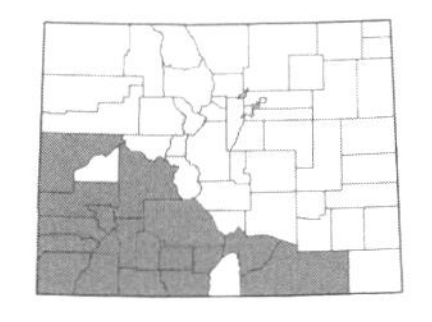

***Eriogonum racemosum* Nutt.**, REDROOT BUCKWHEAT. Perennials, 1.5–10 dm; *leaves* basal, ovate to elliptic or nearly orbicular, 1.5–6 (10) × 1–5 cm, thinly tomentose below, floccose to glabrous and green above; *inflorescence* racemose; *involucres* turbinate 2–5 mm long, tomentose; *tepals* white to pinkish, 2–5 mm long, glabrous; *achenes* 3–4 mm long, glabrous. Found on rocky slopes, dry hillsides, and sagebrush flats, often with pine or oaks, 6000–10,500 ft. June–Sept. E/W.

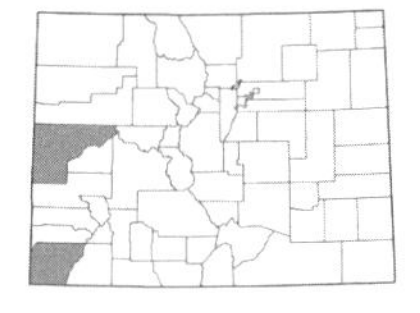

***Eriogonum scabrellum* Reveal**, WESTWATER BUCKWHEAT. Annuals, 1–3 (5) dm; *leaves* basal, orbicular, 1–3 (4) × 1–4 cm, densely tomentose below and floccose and green above, the margins crisped; *inflorescence* cymose, open, glandular and tomentose; *involucres* turbinate, 1.5–2.5 mm long, usually on decurved branchlets; *tepals* white, turning pink or red, 1–1.5 mm long, pustulose; *achenes* 2 mm long, glabrous. Uncommon on dry slopes, 4500–5000 ft. Sept.–Nov. W.

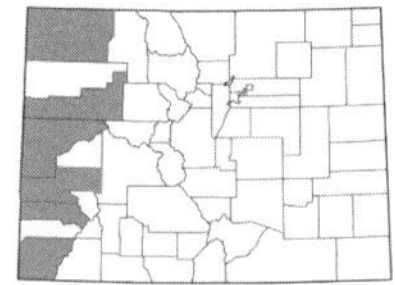

***Eriogonum shockleyi* S. Watson**, SHOCKLEY'S BUCKWHEAT. Perennials, mat-forming, 0.3–0.5 (0.7) dm; *leaves* basal, elliptic to oblanceolate, 0.2–0.8 (1.2) × 0.2–0.6 cm, tomentose; *inflorescence* capitate; *involucres* campanulate, 2–6 mm long, tomentose; *tepals* white to rose or yellow, 2.5–4 mm long, densely hairy; *achenes* 2.5–3 mm long, hairy. Found on gypsum hills, sandstone ledges, and shale slopes, often with sagebrush or juniper, 4600–5700 ft. May–Aug. W.

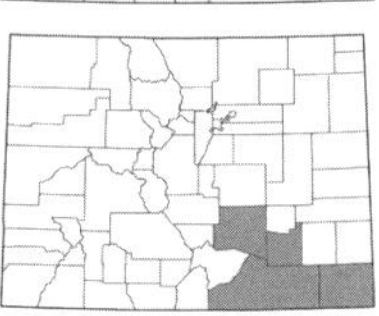

Eriogonum tenellum* Torr. var. *tenellum, TALL BUCKWHEAT. Perennials, 1–6 dm; *leaves* basal, elliptic to orbiculate, 0.3–3 × 0.2–3 cm, densely tomentose; *inflorescence* an open cyme; *involucres* turbinate, 2–4 mm long, glabrous; *tepals* white to pinkish, 1.5–3.5 mm long, glabrous; *achenes* 2–3 mm long, glabrous. Found on sandstone outcrops, mesas, rimrocks and cliffs, and in dry grasslands, 4000–6000 ft. May–Aug. E.

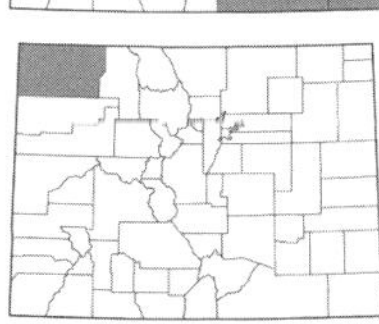

***Eriogonum tumulosum* (Barneby) Reveal**, WOODSIDE BUCKWHEAT. Perennials, mat-forming; *leaves* oblanceolate to elliptic, 0.3–0.5 × 0.07–0.1 cm; *inflorescence* capitate; *involucres* campanulate, 2–4 mm long, tomentose; *tepals* white to pink, 3–4 mm long, hairy; *achenes* 2–2.5 mm long, glabrous. Found in sandy soil, on barren slopes and gypsum hills, often with pinyon-juniper or sagebrush, 5400–6500 ft. May–July. W.

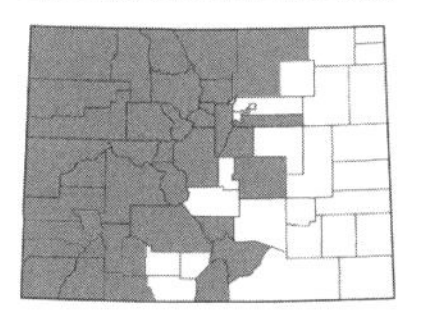

***Eriogonum umbellatum* Torr.**, SULPHUR FLOWER. Perennials or subshrubs; *leaves* basal, oblong-ovate to oblanceolate or oval, 0.3–4 × 0.1–2.5 cm, tomentose to glabrous; *inflorescence* usually umbellate; *involucres* turbinate to campanulate, 1–6 mm long, tomentose to glabrous; *tepals* yellow, white, or red, 2–12 mm long (including the stipelike base), glabrous; *achenes* 2–7 mm long, glabrous. (Plate 68)

There are several varieties of *E. umbellatum*, 6 of which occur in Colorado:

1a. Leaves glabrous above and below...2
1b. Leaves white-pubescent or floccose at least below...3

2a. Plants prostrate, 0.2–0.6 dm; leaves 0.4–1.1 cm long; inflorescence branches to 5 mm long...**var. *porteri* (Small) S. Stokes**, PORTER'S SULPHUR FLOWER. Found in sagebrush meadows, subalpine forests, and alpine, 8800–12,500 ft. July–Sept. E/W.
2b. Plants erect, 1–3.5 dm; leaves 1–2 cm long; inflorescence branches 3–20 mm long...**var. *aureum* (Gandoger) Reveal**, GOLDEN SULPHUR FLOWER. Common on dry slopes, in grasslands, sagebrush flats, pinyon-juniper or conifer forests, and alpine, 5500–11,500 ft. June–Sept. E/W.

3a. Flowers cream or pale yellow...**var. *majus* Hook.**, SUBALPINE SULPHUR FLOWER. [*E. subalpinum* Greene]. Common on dry slopes, in sagebrush, meadows, forests, or alpine, 6500–12,800 ft. June–Sept. E/W.
3b. Flowers bright yellow...4

4a. Inflorescences umbellate, not branched, with a single whorl of bracts below the pedicels...**var. *umbellatum***, COMMON SULPHUR FLOWER. Common on dry slopes, sagebrush flats, and in mountain meadows and forests, 5500–10,000 ft. June–Oct. E/W.
4b. Inflorescences branched, with a whorl of bracts below the pedicels and a whorl of bracts below the inflorescence branches...5

5a. Leaves thinly floccose below...**var. *subaridum* S. Stokes**, FERRIS'S SULPHUR FLOWER. Found on dry slopes, often with sagebrush, saltbush, or pinyon-juniper, known from the southwestern counties, 6000–8000 ft. June–Oct. W.
5b. Leaves densely gray-tomentose below...**var. *ramulosum* Reveal**, BUFFALO BILL'S SULPHUR FLOWER. Uncommon on dry slopes and sagebrush flats, found along the Front Range (Jefferson and Larimer cos.), 5200–9000 ft. June–Sept. E. Endemic.

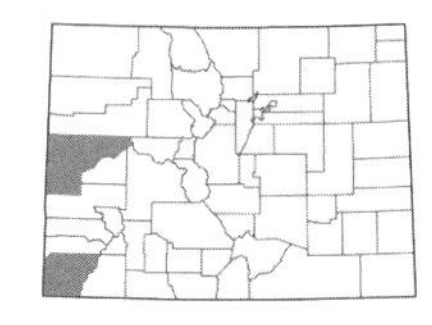

***Eriogonum wetherillii* Eastw.**, WETHERILL'S BUCKWHEAT. Annuals, 0.5–2.5 dm; *leaves* basal, oblong to orbiculular, 0.5–4 × 0.5–3 cm, tomentose below and glabrate and greenish above; *inflorescence* cymose; *involucres* turbinate, 0.3–1 mm long, glabrous; *tepals* yellow to red, becoming rose, 0.5–1.5 mm long, glabrous; *achenes* 0.6–1 mm long, glabrous. Found on dry slopes, often with pinyon-juniper or saltbush, 4500–5000 ft. April–Sept. W.

FAGOPYRUM Mill. – BUCKWHEAT

Annual herbs; *leaves* alternate, cauline, with an ocrea; *flowers* usually perfect, with a stipelike base; *tepals* 5, petaloid, white, pink, or green, the outer smaller than the inner; *stamens* 8, with white, pink, or red anthers; *achenes* triangular, not winged.

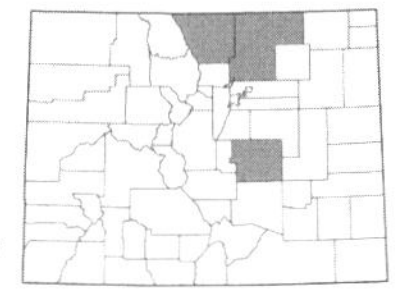

***Fagopyrum esculentum* Moench**, COMMON BUCKWHEAT. Uncommon in fields and waste places, or cultivated as a crop plant, 4500–7500 ft. July–Sept. E. Introduced.

FALLOPIA Adans. – FALSE BUCKWHEAT

Herbs or vines; *leaves* alternate, with an ocrea; *flowers* with a stipelike base; *tepals* 5, petaloid, green to white or pink, the outer 3 winged or keeled and larger than the inner 2, connate at least at the base; *stamens* 6–8; *achenes* triangular.

1a. Perennials from rhizomes; leaves ovate with a truncate or attenuate base, the lower surface minutely dotted and glaucous; stigmas fringed...***F. japonica* var. *japonica***

1b. Annual or perennial sprawling or climbing plants, lacking rhizomes; leaves sagittate, hastate, cordate, or ovate with a cordate base, the lower surface rarely minutely dotted, not glaucous; stigmas capitate or peltate...2

2a. Outer 3 tepals obscurely keeled; fruiting perianth angled but lacking wings or rarely with small wings 0.4–0.9 mm wide that are scarcely decurrent on the stipelike base; flowers arranged in a spike-like inflorescence...***F. convolvulus***

2b. Outer 3 tepals winged; fruiting perianth with conspicuous wings 1.5–4 mm wide that are decurrent on the stipelike base; flowers arranged in a raceme-like or paniclelike inflorescence...3

3a. Flowers arranged in a paniclelike inflorescence; wings of the fruiting perianth 1.8–4 mm wide at maturity...***F. baldschuanica***

3b. Flowers arranged in a raceme-like inflorescence; wings of the fruiting perianth 1.5–2.2 mm wide at maturity...***F. scandens***

***Fallopia baldschuanica* (Regel) Holub**, BUKHARA FLEECEFLOWER. [*Polygonum aubertii* Henry; *Polygonum baldschuanicum* Regel]. Perennial vines; *leaves* ovate to ovate-oblong, 3–10 × 1–5 cm, glabrous, rarely minutely dotted, cordate to sagittate at the base, the margins entire; *inflorescence* paniclelike, 3–15 cm; *tepals* greenish-white to pink, 5–8 mm long (including the stipelike base), the outer 3 winged; *stigmas* peltate; *achenes* 2–4 × 1.2–2.2 mm, dark brown to black and shiny. Escaping cultivation and becoming established along fences, 5200–5500 ft. June–Aug. E. Introduced.

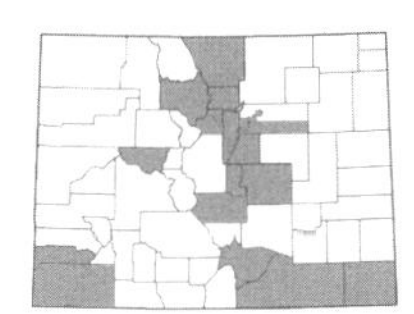

***Fallopia convolvulus* (L.) Á. Löve**, BLACK BINDWEED. [*Polygonum convolvulus* L.]. Annuals, sprawling, 0.5–1 m long; *leaves* cordate-hastate to sagittate, 2–6 (15) × 2–5 (10) cm, usually mealy below, rarely minutely dotted; *inflorescence* axillary, spike-like; *tepals* greenish-white, often pink-tinged, 3–5 mm long (including the stipelike base), the outer 3 obscurely keeled; *stigmas* capitate; *achenes* black, 4–6 × 1.8–2.3 mm, dull. Found in gardens, disturbed places, ravines, along streams, and in forest clearings, 4000–9000 ft. June–Sept. E/W. (Plate 69)

Fallopia japonica* (Houtt.) Ronse Decr. var. *japonica, JAPANESE KNOTWEED. [*Polygonum cuspidatum* Siebold & Zucc.]. Perennials, erect, 1.5–2 (3) m, profusely branched, rhizomatous; *leaves* ovate, 5–15 × 2–10 cm, glabrous to scabrous, minutely dotted below, cuspidate at the apex, the margins entire; *inflorescence* paniclelike, 4–12 cm long; *tepals* white to greenish or pink, 4–6 mm long (including the stipelike base); *stigmas* fringed; *achenes* 2.3–3.6 × 1.4–2 mm, smooth and shiny. Cultivated and sometimes escaping, reported for Colorado but no specimens have been seen. Introduced.

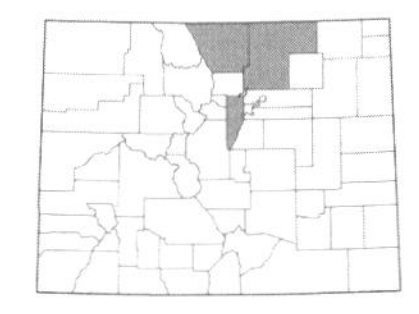

***Fallopia scandens* (L.) Holub**, CLIMBING FALSE BUCKWHEAT. [*Polygonum scandens* L.]. Annuals or perennials, sprawling, stems 1–5 m long; *leaves* cordate to hastate, 2–14 × 2–7 cm, glabrous to scabrid, rarely minutely dotted below; *inflorescence* axillary, raceme-like; *tepals* greenish-white to pinkish, 3.8–8 mm long (including the stipelike base), the outer 3 winged; *stigmas* capitate; *achenes* 2–6 × 1.4–3.5 mm, dark and shiny. Uncommon in gulches and in moist places, 5000–6000 ft. June–Sept. E.

KOENIGIA L. – KOENIGIA

Annual herbs; *leaves* cauline, with an ocrea; *flowers* without a stipelike base, in a terminal cyme or panicle; *tepals* 3, distinct, sepaloid, greenish or tinged white or pink at the tips, all equal in size; *stamens* 3; *achenes* not winged, lenticular.

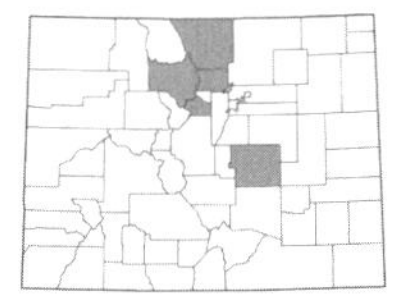

***Koenigia islandica* L.**, ISLAND KOENIGIA. Plants 0.5–8 (20) cm; *ocrea* 1–1.5 mm long; *leaves* spatulate-ovate to orbiculate, 1–6.5 × 1–5 mm, entire, glabrous; *tepals* 1–2 mm long; *achenes* 1–2 × 0.7–0.8 mm, brown to black. Uncommon and inconspicuous along streams and in moist alpine, 11,000–14,000 ft. July–Sept. E/W.

OXYRIA Hill – MOUNTAIN-SORREL

Perennial herbs; *leaves* reniform, with an ocrea; *flowers* perfect, with a stipelike base; *tepals* 4, in 2 whorls of 2 with the outer 2 narrower than the inner, sepaloid; *stamens* usually 6, the anthers white to red or purplish; *achenes* winged, lenticular.

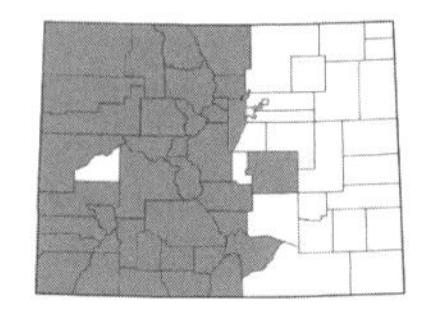

***Oxyria digyna* (L.) Hill**, ALPINE MOUNTAIN-SORREL. Plants 3–50 cm; *ocrea* 2.5–10 mm long; *leaves* basal, 0.5–6.5 cm; *inflorescence* terminal, paniculate or racemose, 2–20 cm long; *tepals* dimorphic, the outer 1.2–1.7 × 0.5–1 mm, the inner 1.4–2.5 × 0.7–1.7 mm; *achenes* 3–5 mm long, with reddish or pinkish, veiny wings. Common in rocky alpine and meadows, 8200–14,200 ft. June–Sept. E/W. (Plate 69)

PERSICARIA (L.) Mill. – SMARTWEED

Annual or perennial herbs; *leaves* alternate, cauline, with an ocrea; *flowers* lacking a stipelike base; *tepals* 4–5, connate at least at the base, petaloid, white, greenish, red, or purple, the outer larger; *stamens* 5–8; *achenes* triangular or lenticular.

1a. Stems with numerous retrorse yellow barbs; leaves narrowly sagittate; plants usually vine-like or sprawling, sometimes erect; flowers in a rounded, capitate inflorescence…***P. sagittata***
1b. Stems lacking retrorse yellow barbs; leaves lanceolate to ovate; plants not vinelike; inflorescence various…2

2a. Perennials, usually with rhizomes or stolons, usually aquatic or standing in water, or sometimes in moist places; inflorescence a single terminal raceme, rarely with 2 racemes; perianth bright pink to red, rarely light pink; leaves not glandular-punctate below, often hairy…***P. amphibia***
2b. Annuals or rarely perennials, terrestrial; inflorescence of more than one raceme, usually numerous, terminal and also usually axillary; perianth greenish-white, white, pink, rose, or bright pink; leaves usually glandular-punctate below, otherwise glabrous…3

3a. Perianth glandular-punctate with numerous golden-brown dots; flowers loosely arranged in a raceme; ocrea ciliate with bristles 1–11 mm long…4
3b. Perianth glabrous or rarely glandular but not punctate; flowers densely clustered in a raceme; ocrea ciliate or not…5

4a. Achenes dull and minutely roughened; annuals from fibrous roots…***P. hydropiper***
4b. Achenes shiny and smooth; perennials from rhizomes…***P. punctata***

5a. Ocrea with cilia or bristles 1–4 (5) mm long at the apex; leaves often with a prominent dark blotch on the upper side; peduncles glabrous…***P. maculosa***
5b. Ocrea entire or lacerate at the apex, lacking cilia or bristles, or the bristles sometimes a few to 1 mm long; leaves usually lacking a prominent dark blotch on the upper side; peduncles glandular-stipitate or glabrous…6

6a. Perianth segments 4 or rarely 5, the outer ones with the midvein divided at the top with the branches recurved (giving the nerve an anchor-shaped appearance); racemes nodding to drooping or arching; flowers greenish-white or pinkish…***P. lapathifolia***
6b. Perianth segments 5, without an anchor-shaped vein; racemes erect or rarely arching; flowers pink or rose, rarely greenish-white…7

7a. Stamens or styles exserted (flowers heterostylous); achenes with a central bump within a depression on one side…***P. bicornis***
7b. Stamens and styles included (flowers homostylous); achenes lacking a central bump on one side…***P. pensylvanica***

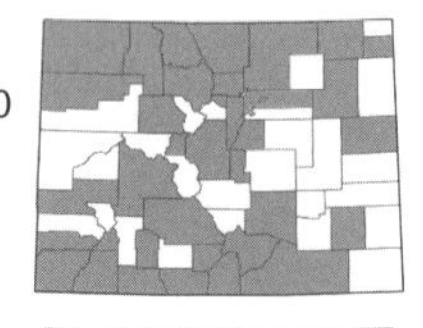

***Persicaria amphibia* (L.) Delarbe**, WATER SMARTWEED. [*Polygonum amphibium* L.]. Perennials, usually aquatic, to 30 dm; *ocrea* 5–50 mm long, glabrous or ciliate; *leaves* ovate-lanceolate to elliptic, 2–20 × 1–6 (8) cm, glabrous or strigose; *inflorescences* terminal racemes, 10–150 mm long; *tepals* bright pink to red, rarely light pink, 4–6 mm long; *achenes* dark brown, 2–3 mm long. Found in and along the margins of lakes, 3500–10,700 ft. June–Oct. E/W. (Plate 69)

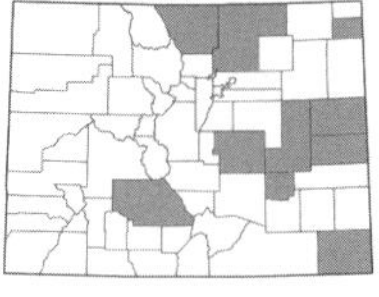

***Persicaria bicornis* (Raf.) Nieuwl.**, PINK SMARTWEED. [*Polygonum bicorne* Raf.]. Annuals, 2–20 dm; *ocrea* 6–20 mm long, ciliate or not, with bristles to 1 mm; *leaves* lanceolate, 2–15 (20) × 1–2.5 cm, glandular-punctate below; *inflorescences* terminal and axillary, 8–60 mm long, usually glandular-stipitate; *tepals* 5, pink, 3–5 mm long; *achenes* black, 2–3 mm long, with a central bump within a depression on one side. Common in moist places, 3500–6000 ft. July–Sept. E.

Persicaria hydropiper **(L.) Opiz**, WATER PEPPER. [*Polygonum hydropiper* L.]. Annuals, 2–10 dm; *ocrea* 8–15 mm long, ciliate with bristles 1–4 mm long; *leaves* lanceolate, (2) 4–10 (15) × 0.5–2.5 cm, glandular-punctate; *inflorescences* terminal and axillary, 30–180 mm long; *tepals* white or pinkish distally, glandular-punctate; *achenes* brownish-black, 2–3 mm long, dull and minutely roughened. Uncommon along creeks and in moist places, 4500–7600 ft. July–Sept. E/W. Introduced.

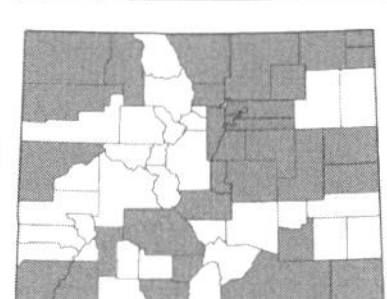

Persicaria lapathifolia **(L.) A. Gray**, PALE SMARTWEED. [*Polygonum lapathifolium* L.]. Annuals, 1–10 dm; *ocrea* 4–30 mm long, ciliate or not, with bristles to 1 mm long; *leaves* lanceolate, 4–15 (22) × 0.5–5 (6) cm, glandular-punctate below, glabrous or tomentose above, and sometimes with a dark blotch above; *inflorescences* terminal and axillary, 30–80 mm long, nodding to arching, sometimes glandular-stipitate; *tepals* usually 4, greenish-white or pinkish, 2.5–3 mm long, the outer ones with the midvein divided at the top with each branch recurved; *achenes* 1.5–3.5 mm long. Common in moist places, often in disturbed areas, 3500–10,000 ft. June–Sept. E/W. Introduced.

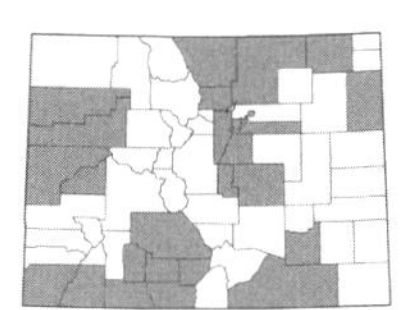

Persicaria maculosa **A. Gray**, LADY'S THUMB. [*Polygonum persicaria* L.]. Annuals, 0.5–10 (13) dm; *ocrea* 4–15 mm long, ciliate with bristles 1–4 (5) mm; *leaves* lanceolate to narrowly ovate, 5–15 × 1–3 (4) cm, often glandular-punctate below; *inflorescences* terminal and axillary, erect, 10–50 cm long; *tepals* 4–5, greenish-white, 2–3.5 mm long; *achenes* blackish, 2–3 mm long, shiny. Common in moist places, often in disturbed areas, 3500–9000 ft. July–Oct. E/W. Introduced. (Plate 69)

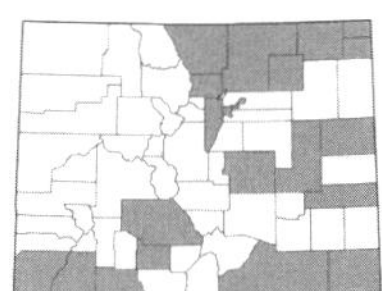

Persicaria pensylvanica **(L.) Small**, PENNSYLVANIA SMARTWEED. [*Polygonum pensylvanicum* L.]. Annuals, 1–20 dm; *ocrea* 5–20 mm long, ciliate or not with bristles to 0.5 mm long; *leaves* lanceolate, 4–20 × 1–5 cm, glabrous, glandular-punctate below, sometimes with a dark blotch on the upper side; *inflorescences* terminal and axillary, 5–50 cm long, usually glandular-stipitate; *tepals* 5, rose or pink, rarely greenish-white, 2.5–5 mm long; *achenes* brown or black, 2–3.5 mm long. Common in moist places, 4000–8000 ft. July–Sept. E/W.

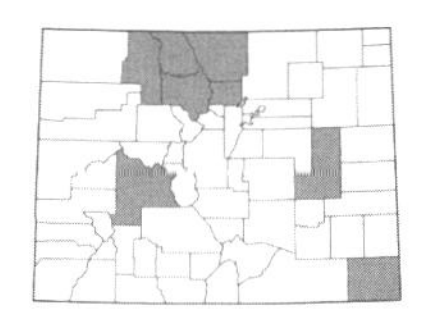

Persicaria punctata **(Elliot) Small**, DOTTED SMARTWEED. [*Polygonum punctatum* Elliot]. Perennials, 1.5–12 dm, rhizomatous; *ocrea* (4) 9–18 mm long, ciliate; *leaves* lanceolate to lanceolate-ovate, 4–10 (15) × 0.6–2.5 cm, glandular-punctate; *inflorescences* usually terminal and axillary, 5–20 cm long, glandular-punctate; *tepals* 5, greenish-white or rarely pink, glandular-punctate, 3–4 mm long; *achenes* brownish to black, 2–3.2 mm long, shiny and smooth. Uncommon in moist places, 4500–5500 ft. July–Sept. E/W. Introduced.

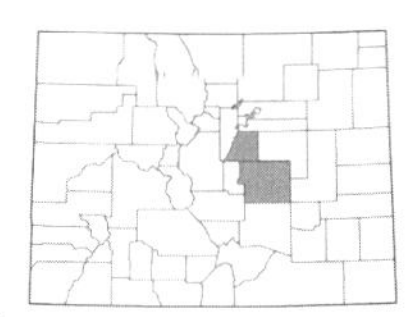

Persicaria sagittata **(L.) H. Gross**, ARROWLEAF TEARTHUMB. [*Polygonum sagittatum* L.; *Truellum sagittatum* (L.) Soják]. Annuals, vine-like or sprawling plants, sometimes erect, 3–20 dm, with numerous retrorse yellow barbs along the stem; *ocrea* 5–13 mm long, glabrous or ciliate apically; *leaves* narrowly sagittate, 2–8.5 × 1–3 cm, glabrous or hairy, usually with retrorse prickles along the midvein below; *inflorescence* capitate, 5–15 × 4–10 mm; *tepals* white or greenish-white, 3–5 mm long; *achenes* trigonous, 2.5–4 mm long, smooth or minutely punctate. Uncommon in moist, shady places, meadows, and along streams, 6000–7500 ft. June–Sept. E.

POLYGONUM L. – KNOTWEED

Annuals, perennials, subshrubs, or shrubs; *leaves* alternate or rarely opposite, cauline, with a 2-lobed ocrea; *flowers* perfect, the base not stipelike; *tepals* 5, connate at least at the base, sepaloid or petaloid, all the same or nearly so; *stamens* 3–8; *achenes* triangular.

1a. Stems obscurely ribbed or ribs lacking; leaves with a single midrib or parallel veins; anthers pink or purple…2
1b. Stems prominently 8–16-ribbed; leaves with pinnate venation; anthers yellowish…6

2a. Leaves ovate to elliptic or subrotund; ocrea 1–4 mm long; stem reddish-brown…***P. minimum***
2b. Leaves linear, oblanceolate to narrowly lanceolate, or narrowly oblong; ocrea 3–12 mm long; stem green to brownish…3

3a. Inflorescence a dense, terminal spike-like raceme, the bracts usually as long or longer than the leaves, crowded in overall appearance; tepals pink or white throughout; achenes yellow to greenish-brown or dark brown; plants 2–15 (20) cm tall, the stems simple or often divaricately branched…***P. kelloggii***
3b. Inflorescence loose, not crowded, the bracts usually much shorter than the leaves; tepals with a green center and white or pink margins; achenes black; plants 4–80 cm tall, the stems simple or branched from the base…4

4a. Pedicels of flowers erect or somewhat spreading, but not reflexed…***P. sawatchense* ssp. *sawatchense***
4b. Pedicels of flowers becoming reflexed in age (look at older flowers at the base of the inflorescence)…5

5a. Tepals 1.5–2.5 mm long; ocrea 3–5 mm long; achenes 1.2–2.5 mm long...***P. engelmannii***
5b. Tepals 3–5 mm long; ocrea 5–12 mm long; achenes 3–5 mm long...***P. douglasii***

6a. Tepals connate to above the middle, enlarged at the base and abruptly pinched and narrowed above; leaves elliptic to ovate or obovate, the same size throughout, often covered with a whitish, powdery mildew; achenes yellow-green to light brown or tan...***P. achoreum***
6b. Tepals connate below the middle or if connate to above the middle then not constricted above; leaves variously shaped, usually reduced above or sometimes the same size throughout, sometimes with a whitish, powdery mildew; achenes light to dark brown...7

7a. Stems strictly erect, branching above; tepals with yellow or greenish-yellow margins, rarely with whitish-green or pinkish-green margins...***P. ramosissimum***
7b. Stems prostrate or decumbent; tepals with pink, white, or red margins...8

8a. Achenes shiny and smooth; upper leaves abruptly reduced and shorter than or equaling the flowers in length (not overtopping the flowers); pedicels 1–2 mm long; leaves linear-lanceolate to lanceolate...***P. argyrocoleon***
8b. Achenes dull and usually roughened; upper leaves longer than the flowers (overtopping the flowers); pedicels 1.5–5 mm long; leaves lanceolate, elliptic, obovate, or spatulate...***P. aviculare***

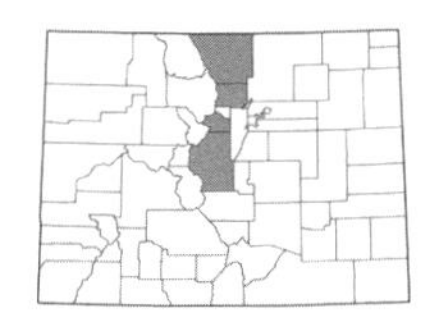

***Polygonum achoreum* Blake**, Blake's knotweed. Annuals, prostrate or decumbent, stems 5–7 dm; *ocrea* 5–12 mm long; *leaves* obovate to ovate or elliptic, 0.8–3.5 cm × 3–15 mm; *tepals* 2.5–4 mm long, connate to above the middle, enlarged at the base and abruptly pinched and narrowed above; *achenes* 2.5–3.5 mm long, yellow-green to light brown or tan. Uncommon weed in disturbed places, 5000–6000 ft. July–Sept. E. Introduced. (Plate 69)

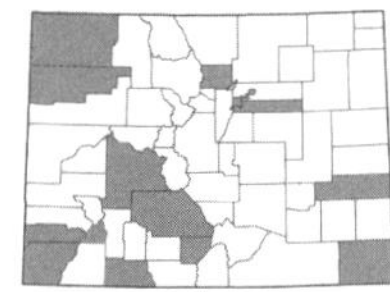

***Polygonum argyrocoleon* Steud. ex Kunze**, silversheath knotweed. Annuals, 1.5–10 dm; *stems* prominently 8–16-ribbed; *ocrea* 4–8 mm long, silvery distally; *leaves* linear-lanceolate to lanceolate, 1.5–5 cm × 2–8 mm; *tepals* 1.8–2.5 mm long, green or white, with pink, red, or white margins; *achenes* 1.3–2.5 mm long, shiny and smooth, light to dark brown. Uncommon weed in disturbed places, fields, along roadsides, and in drying pools, 4500–8000 ft. E/W. Introduced.

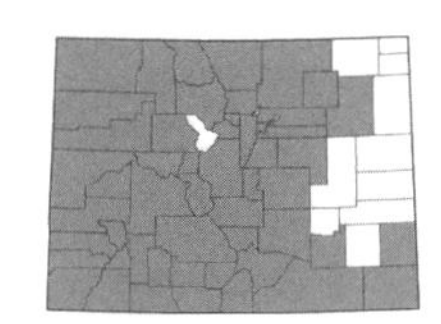

***Polygonum aviculare* L.**, prostrate knotweed. [*P. arenastrum* Jord. ex Boreau; *P. buxiforme* Small]. Annuals, 0.5–20 dm; *stems* prominently 8–16-ribbed; *ocrea* 3–15 mm long, silvery distally; *leaves* lanceolate, elliptic, obovate, or spatulate, 0.6–5 (6) cm × 0.5–22 mm; *tepals* 1.8–6 mm long, green with white, pink, or reddish margins; *achenes* 1–4.2 mm long, dull, usually conspicuously roughened, light to dark brown. Common weed in disturbed places, sidewalks, along roadsides, meadows, and near drying ponds, 3500–11,700 ft. May–Sept. E/W. Introduced.

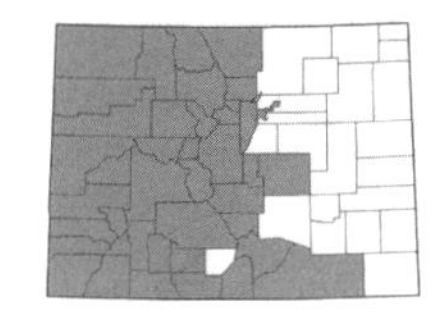

***Polygonum douglasii* Greene**, Douglas' knotweed. Annuals, 0.5–8 dm; *stems* obscurely ribbed; *ocrea* 5–12 mm long; *leaves* linear, narrowly oblong, or oblanceolate, 1.5–6 cm × 2–8 (12) mm, with revolute margins; *pedicels* reflexed; *tepals* 3–5 mm long, green with white or pink margins; *achenes* 3–5 mm long. Common on open slopes and in meadows, floodplains, and forests, often found in disturbed places, 5000–11,600 ft. June–Sept. E/W.

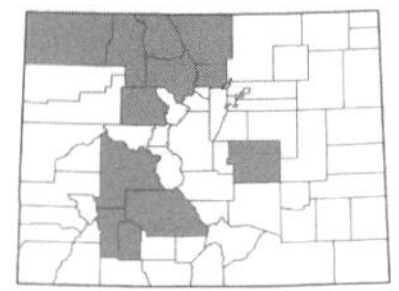

***Polygonum engelmannii* Greene**, Engelmann's knotweed. [*P. douglasii* Greene ssp. *engelmannii* (Greene) Kartesz & Gandhi]. Annuals, 0.4–3 dm; *stems* obscurely ribbed; *ocrea* 3–5 mm long; *leaves* linear-oblanceolate, 1–2.5 cm × 1–4 mm; *pedicels* reflexed; *tepals* 1.5–2.5 mm long, green with white margins; *achenes* 1.2–2.5 mm long. Found on open slopes, rocky hillsides, and in moist draws, often in disturbed places, 6000–10,500 ft. June–Aug. E/W.

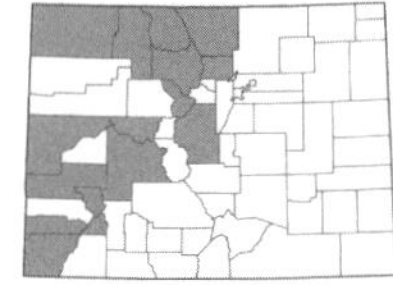

***Polygonum kelloggii* Greene**, Kellogg's knotweed. Annuals, 2–15 (20) cm; *stems* obscurely ribbed; *ocrea* 4–8 mm long, silvery distally; *leaves* narrowly linear, 1–4 cm × 1–2 mm; *tepals* 1.5–3 mm long, reddish; *achenes* 1.3–2.5 mm long, smooth or longitudinally ridged. Found in moist meadows, forests, vernal pools, and near seepages, 7000–10,700 ft. June–Aug. E/W. (Plate 69)

There are two varieties of *P. kelloggii* in Colorado:

1a. Bract margins green or if white and scarious then no more than a thin line to 0.2 mm wide; achenes shiny, smooth or obscurely reticulate with longitudinal ridges...**var. *kelloggii***. [*P. polygaloides* Meisn. ssp. *kelloggii* (Greene) J.C. Hickman]. Easily overlooked in moist meadows and forest openings and near drying pools and seepages, 7000–10,700 ft. June–Aug. E/W.
1b. Bracts with a white, scarious margin 0.25–0.5 mm wide; achenes dull, reticulate with longitudinal ridges...**var. *confertiflorum* (Nutt. ex Piper) Dorn**. Less common than var. *kelloggii*, found in meadows and vernal pools, 7000–10,500 ft. June–Aug. W.

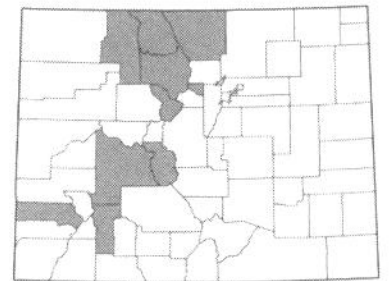

***Polygonum minimum* S. Watson**, BROADLEAF KNOTWEED. Annuals, erect or prostrate, 0.2–3 dm; *stems* obscurely ribbed; *ocrea* 1–4 mm long; *leaves* elliptic to ovate or subrotund, 0.6–3 cm × 3–8 mm; *tepals* 1.8–2.5 mm long, greenish with white or pink margins; *achenes* 1.8–2.3 mm long, shiny, smooth. Uncommon on rocky slopes and near moist drainages, 9000–11,500 ft. July–Sept. E/W.

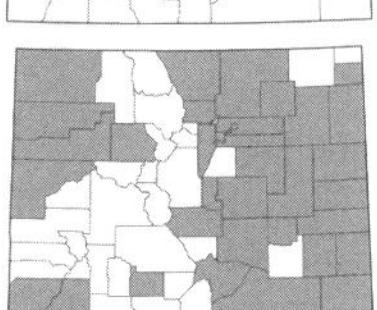

***Polygonum ramosissimum* Michx.**, BUSHY KNOTWEED. Annuals, 1–10 (20) dm; *stems* prominently 8–16-ribbed; *ocrea* 6–15 mm long; *leaves* linear, lanceolate, oblanceolate, or narrowly elliptic, 0.8–7 cm × 4–20 (35) mm; *tepals* 2–4 mm long, greenish-yellow, rarely pink or white; *achenes* 1.5–3.5 mm long, light to dark brown. Found in disturbed places, along roadsides, and in moist places, 3500–8000 ft. July–Sept. E/W.

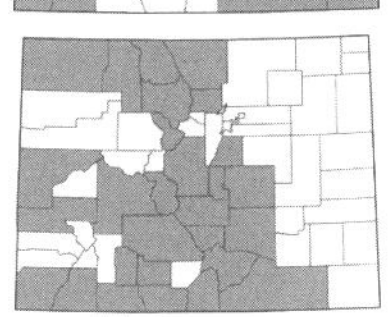

Polygonum sawatchense* Small ssp. *sawatchense, SAWATCH KNOTWEED. [*P. douglasii* Greene ssp. *johnstonii* (Munz) J.C. Hickman]. Annuals, 0.4–5 dm; *stems* obscurely ribbed; *ocrea* 4–10 mm long; *leaves* linear to oblong or oblanceolate, 0.8–4.5 cm × 2–10 mm, the margins flat or revolute; *pedicels* erect; *tepals* 2.5–4 mm long, green with white or pink margins; *achenes* 2.5–3.3 mm long, shiny, smooth. Found on open slopes and in dry meadows, 5500–11,500 ft. June–Aug. E/W.

RHEUM L. – RHUBARB

Perennial herbs; *leaves* with an ocrea; *flowers* with a stipelike base; *tepals* 6, sepaloid, the outer 3 tepals narrower than the inner 3, not enlarging in fruit; *stamens* usually 9; *achenes* winged, triangular.

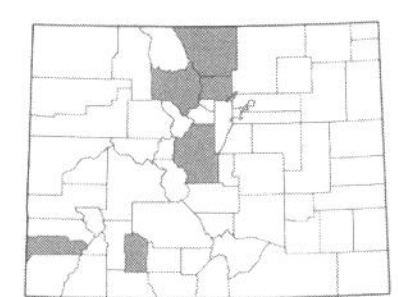

***Rheum rhabarbarum* L.**, RHUBARB. Plants 3–25 dm; *ocrea* 2–4 cm; *leaves* with pinkish or reddish-green petioles, 30–50 × 10–30 cm; *inflorescences* paniculate, with numerous flowers; *tepals* 2–4 mm, whitish-green; *achenes* 6–12 × 6–11 mm including wings. Cultivated and sometimes escaping to nearby valleys and forests, found in abandoned gardens or near old cabins, 5500–11,500 ft. June–July. E/W. Introduced.

RUMEX L. – DOCK

Herbs; *leaves* alternate, with an ocrea; *flowers* with a stipelike base; *tepals* usually 6, sepaloid, the outer 3 small and the inner 3 enlarging in fruit (and then called valves), the inner tepals sometimes with a central tubercle (resembling a grain); *stamens* 6; *achenes* unwinged, triangular.

1a. Plants dioecious, the flowers imperfect; tepals red or reddish-pink, mostly 0.5–1.5 mm long; inflorescence with slender branches...2
1b. Plants monecious, the flowers perfect or imperfect; tepals variously colored, 1–5 mm long; inflorescence branches slender or stout...3

2a. At least some of the leaves hastate with spreading lobes, the stem leaves well-developed; inner tepals (valves) not enlarged at maturity; stems several from the base...***R. acetosella***
2b. Leaves broadly lanceolate to elliptic, lacking hastate lobes, mostly basal; inner tepals (valves) distinctly enlarging at maturity; stems solitary or few from the base...***R. paucifolius***

3a. Inner tepals (valves) with 2–3 bristle-like teeth along the margins, with a prominent tubercle on each valve; inflorescence conspicuously leafy; flowers in dense golden or reddish-brown clusters, these in interrupted whorls; annuals; leaves lanceolate or linear-lanceolate...***R. fueginus***
3b. Inner tepals (valves) entire or with coarsely toothed margins, with or without a tubercle; inflorescence often leafy below but not throughout; flowers not in interrupted whorls, variously colored; perennials or biennials; leaves various...4

4a. Inner tepals (valves) conspicuously coarsely toothed along the margins, with a tubercle about ½ as long as the valve; leaves often undulate or crisped on the margins, sometimes flat...5
4b. Inner tepals (valves) with entire or very weakly erose margins, with or without a prominent tubercle; leaves with flat margins or crisped and undulate...6

5a. Leaves lanceolate to oblong-lanceolate, 2–7 cm wide, with a truncate or cuneate base; each inner tepal with a tubercle...***R. stenophyllus***
5b. Leaves oblong to ovate-oblong or broadly ovate, 10–15 cm wide, the largest leaves with a cordate base or occasionally the base rounded; only one inner tepal with a tubercle...***R. obtusifolius***

6a. Inner tepals 20–30 mm wide at maturity, reddish-pink, distinctly reticulate-veined, orbiculate; plants from creeping rhizomes, usually producing axillary shoots near the base...***R. venosus***
6b. Inner tepals 2–18 mm wide at maturity, variously colored, sometimes reticulate-veined, ovate to triangular or occasionally orbiculate; if plants from creeping rhizomes then not producing axillary shoots near the base...7

7a. Inner tepals with a tubercle; leaves lanceolate to elliptic-lanceolate, 1–6 cm wide...8
7b. Inner tepals lacking a tubercle; leaves various...10

8a. Leaves strongly undulate and crisped on the margins...***R. crispus***
8b. Leaves with flat or indistinctly crisped margins...9

9a. Inner tepals 2–3.5 mm long; leaves linear-lanceolate to lanceolate or rarely ovate-lanceolate; achenes 1.5–2.2 mm long...***R. triangulivalvis***
9b. Inner tepals 4.5–6 mm long; leaves lanceolate to ovate-lanceolate or elliptic-lanceolate; achenes 2.5–3.5 mm long...***R. altissimus***

10a. Leaves linear-lanceolate to lanceolate, 2–3 cm wide; inner tepals 2.5–3.5 mm long at maturity; plants lacking basal leaves...***R. utahensis***
10b. Leaves oblong, elliptic, oblong-lanceolate, ovate-lanceolate, or sometimes almost orbiculate; inner tepals 4–16 mm long at maturity; plants usually with a basal rosette of leaves...11

11a. Inner tepals 11–16 mm long and 9–18 mm wide at maturity; plants from clusters of deeply seated tuberous roots, usually found in dry places; ocrea persistent, whitish or silvery white at maturity...***R. hymenosepalus***
11b. Inner tepals 5–12 mm long and 4.5–10 mm wide at maturity; plants not from deeply seated tuberous roots, usually found in moist places; ocrea brownish or deciduous...12

12a. Plants from creeping rhizomes; inner tepals 5–7 mm long and 4.5–6 mm wide at maturity...***R. densiflorus***
12b. Plants from a vertical rootstock, lacking rhizomes; inner tepals 5–12 mm long and 5–10 mm wide at maturity...***R. occidentalis***

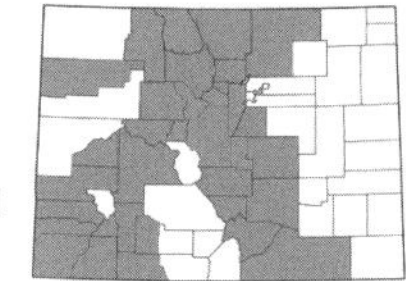

***Rumex acetosella* L.**, SHEEP SORREL. [*Acetosella vulgaris* (K. Koch) Fourr.]. Perennials, 1–4.5 dm; *leaves* lanceolate to ovate-lanceolate, 2–6 × 0.3–2 cm, hastate at the base; *inflorescences* terminal, paniculate; *inner tepals* not or only slightly enlarged, 1.2–2 × 0.5–1.3 mm, tubercles absent; *achenes* 1–1.5 × 0.6–1 mm, brown. Common in meadows and along roadsides, often in disturbed areas, 5500–12,000 ft. June–Aug. E/W. Introduced. (Plate 69)

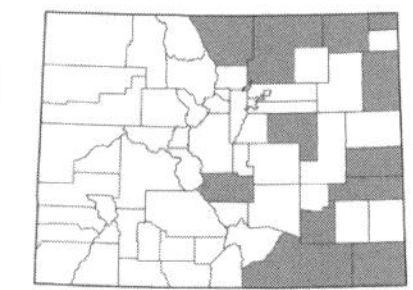

***Rumex altissimus* Wood**, PALE DOCK. Perennials, 5–10 (12) dm; *leaves* lanceolate to ovate-lanceolate or elliptic-lanceolate, 10–15 × 3–5.5 cm, margins flat and entire; *inflorescences* terminal and axillary, broadly paniculate; *inner tepals* enlarged, triangular, 4.5–6 × 3–5 mm, with tubercles; *achenes* 2.5–3.5 × 1.8–2.5 mm. Found along ditches and creeks, and near pond margins, 3500–5400 ft. May–Sept. E.

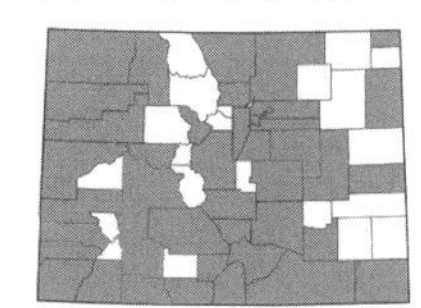

***Rumex crispus* L.**, CURLY DOCK. Perennials, 4–10 (15) dm; *leaves* lanceolate to elliptic-lanceolate, 15–35 × 2–6 cm, margins strongly crisped and undulate; *inflorescences* terminal, paniculate; *inner tepals* ovate-deltoid, 3.5–6 × 3–5 mm, tuberculate, the margins entire to weakly erose; *achenes* 2–3 × 1.5–2 mm. Common in disturbed places, fields, meadows, and along roadsides and ditches, 3500–9500 ft. May–Sept. E/W. Introduced.

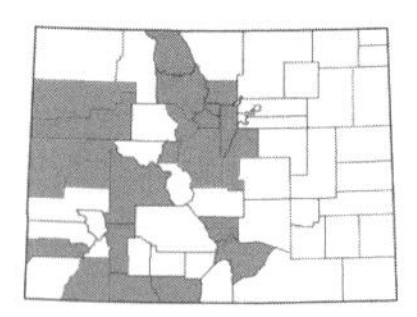

***Rumex densiflorus* Osterh.**, DENSE-FLOWERED DOCK. [*R. praecox* Rydb.; *R. pycnanthus* Rech. f.]. Perennials from creeping horizontal rhizomes, 5–10 dm; *leaves* oblong, 30–50 × 10–12 cm, the margins flat; *inflorescences* narrowly paniculate; *inner tepals* 5–7 × 4.5–6 mm, tubercles absent, with entire, erose, or minutely toothed margins; *achenes* 2.5–4.5 mm long. Found in moist meadows, swamps, and along streams, 5800–12,000 ft. June–Aug. E/W.

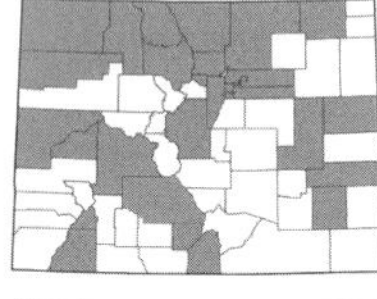

***Rumex fueginus* Phil.**, GOLDEN DOCK. [*R. maritimus* L. var. *fueginus* (Phil.) Dusen.]. Annuals, 1.5–7 dm; *leaves* linear-lanceolate to lanceolate, (3) 5–25 (30) × 1.5–3 (4) cm, the margins entire, usually crisped and undulate; *inflorescences* terminal, broadly paniculate; *inner tepals* triangular, 1.5–2.5 × 0.7–1 mm, prominently bristle-toothed; *achenes* 1–1.4 × 0.6–0.8 mm. Found along the shores of lakes and marshes, 3900–9000 ft. June–Aug. E/W. (Plate 69)

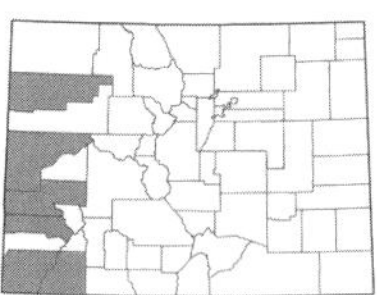

***Rumex hymenosepalus* Torr.**, CANAIGRE DOCK. Perennials from tuberous roots and short rhizomes, 2.5–10 dm; *leaves* oblong to obovate-lanceolate, 5–30 × 2–10 cm, the margins flat; *inflorescences* narrowly paniculate; *inner tepals* 11–16 × 9–18 mm, tubercles absent; *achenes* 4–7 mm long. Uncommon in clay or sandy soil and on dry hillsides, 4500–6000 ft. April–May (June). W.

***Rumex obtusifolius* L.**, BITTER DOCK. Perennials, 6–15 dm; *leaves* oblong to ovate, 20–40 × 10–15 cm, the margins entire, rarely somewhat crisped; *inflorescences* terminal, paniculate; *inner tepals* deltoid, 3–6 × 2–3.5 mm, the margins toothed, tuberculate; *achenes* 2–3 × 1.2–1.7 mm. Uncommon along roadsides, in meadows, and along creeks, 5000–9000 ft. June–Aug. E. Introduced.

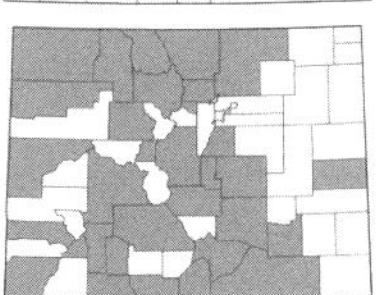

***Rumex occidentalis* S. Watson**, WESTERN DOCK. [*R. aquaticus* L. var. *fenestratus* (Greene) Dorn]. Perennials, 5–14 dm; *leaves* ovate-lanceolate to oblong-lanceolate, mostly 10–35 × 5–12 cm, the margins entire to undulate; *inflorescences* narrowly paniculate; *inner tepals* 5–12 × 5–10 mm, the margins entire to inconspicuously erose, tubercles absent; *achenes* 3–5 × 1.5–2.5 mm. Found in moist meadows, along pond margins, and in swampy areas, 4500–10,500 ft. June–Aug. E/W. (Plate 69)

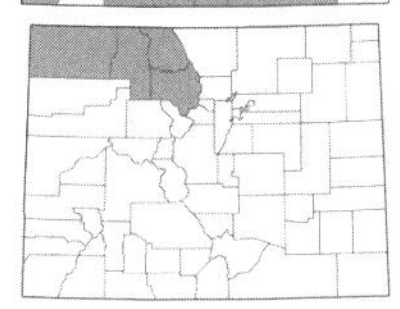

***Rumex paucifolius* Nutt.**, ALPINE SHEEP SORREL. [*Acetosella paucifolia* (Nutt.) Á. Löve]. Perennials, 1–5 (6) dm; *leaves* lanceolate to ovate-lanceolate, usually not hastate, 3–8 (10) × 1–3 (4) cm; *inflorescences* terminal, paniculate; *inner tepals* 2.8–4 × 2.7–3.5 mm, enlarging at maturity, tubercles absent; *achenes* 1.2–1.8 × 0.8–1 mm, brown. Uncommon in meadows and along streams, 8000–10,000 ft. June–Aug. E/W.

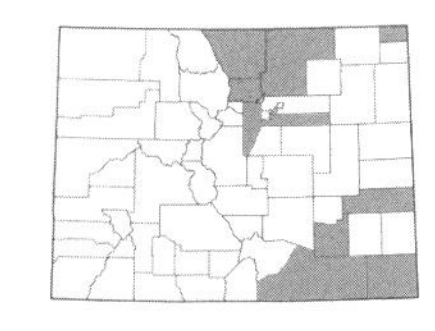

***Rumex stenophyllus* Ledeb.**, NARROWLEAF DOCK. Perennials, 4–8 (13) dm; *leaves* lanceolate to oblong-lanceolate, 15–30 × 2–7 cm, the margins entire or sparsely toothed; *inflorescences* terminal, narrowly paniculate; *inner tepals* ovate-deltoid, 3.5–5 × 3–5 mm, the margins toothed, tuberculate; *achenes* 2–3 × 1–1.5 mm. Uncommon along the shores of lakes and creeks and in marshes and ephemeral ponds, 3500–5200 ft. June–Sept. E. Introduced. (Plate 69)

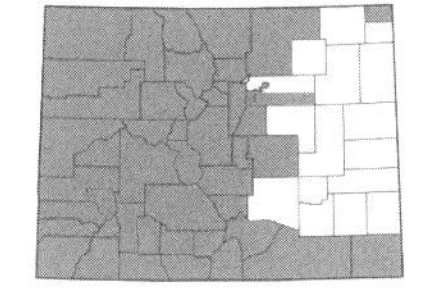

***Rumex triangulivalvis* (Danser) Rech. f.**, WILLOW DOCK. [*R. salicifolius* Weinm. ssp. *triangulivalvis* Danser]. Perennials, 3–10 dm; *leaves* linear-lanceolate, 6–17 × 1–4 (5) cm, the margins flat or weakly undulate to crisped; *inflorescences* paniculate; *inner tepals* enlarged, triangular, 2–3.5 × 2–3.5 mm, with a tubercle; *achenes* 1.5–2.2 mm long. Common along roadsides, in meadows, and along streams, 4500–11,500 ft. May–Sept. E/W. (Plate 69)

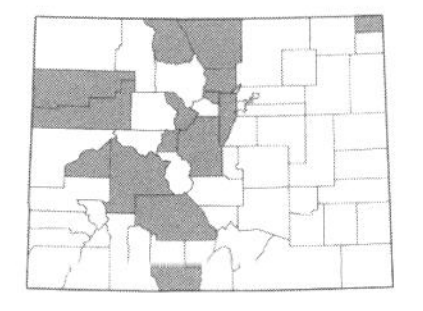

***Rumex utahensis* Rech. f.**, UTAH WILLOW DOCK. Perennials, 1.5–5 (6) dm; *leaves* linear-lanceolate to lanceolate, 6–15 × 2–3 cm, the margins flat; *inflorescences* broadly paniculate; *inner tepals* enlarged, deltoid, 2.5–3.5 × 2.5–3 mm, tubercles absent; *achenes* 1.8–2 mm long. Found in meadows, along roadsides, and along streams, 4000–11,500 ft. May–Sept. E/W. Similar in aspect to *R. triangulivalvis* except that the inner tepals lack a tubercle.

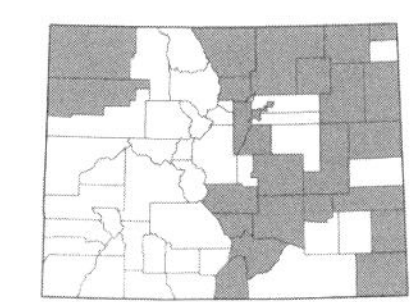

***Rumex venosus* Pursh**, WILD BEGONIA. Perennials, 1–4 dm, from creeping rhizomes; *leaves* ovate-elliptic to ovate-lanceolate, 2–15 × 1–6 cm, the margins flat or slightly undulate; *inflorescences* terminal and axillary, broadly paniculate; *inner tepals* reddish-pink, distinctly reticulate-veined, orbiculate, 13–20 × 20–30 mm, enlarged, tubercles absent; *achenes* 5–7 mm long, brown. Common in sandy soil, along roadsides, and in shortgrass prairie, 3400–7500 ft. April–June. E/W. (Plate 69)

STENOGONUM Nutt. – TWO-WHORL BUCKWHEAT

Annual herbs; *leaves* lacking an ocrea, basal or basal and cauline, alternate; *flowers* (6) 9–15 per involucre, in terminal cymes; *tepals* 6, petaloid, hairy on the back, yellow to reddish-yellow; *stamens* 9; *achenes* lacking wings, triangular.

1a. Leaves all basal; peduncles with a distinct, sharp curve forming a right angle; stems and peduncles glandular…***S. flexum***

1b. Basal and stem leaves present; peduncles straight or slightly curved; stems and peduncles glabrous, sometimes the inflorescence glandular just at the lower nodes…***S. salsuginosum***

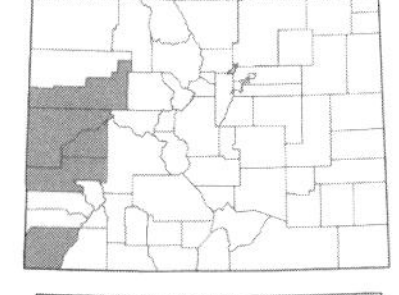

***Stenogonum flexum* (M.E. Jones) Reveal & J.T. Howell**, BENT TWO-WHORL BUCKWHEAT. [*Eriogonum flexum* M.E. Jones]. Plants 0.5–3 dm; *leaves* orbicular, 0.5–2 cm long, glabrous; *involucres* 2–3 mm high, glabrous; *tepals* 1.5–3.5 mm long; *achenes* 2–2.5 mm long. Found in dry washes and on barren, dry slopes, 4500–6600 ft. May–June. W.

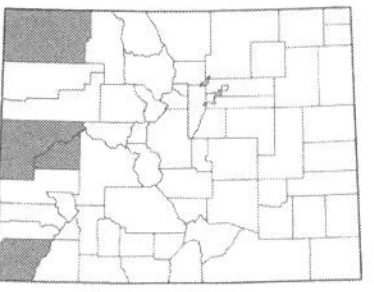

***Stenogonum salsuginosum* Nutt.**, SMOOTH TWO-WHORL BUCKWHEAT. [*Eriogonum salsuginosum* (Nutt.) Hook.]. Plants 0.5–2 dm; *leaves* lanceolate to spatulate, (1) 2–4 cm long, glabrous; *involucres* 2–8 mm high, glabrous; *tepals* 1.5–3 mm long; *achenes* 2–2.5 mm long. Found on dry, barren clay slopes, 4700–6800 ft. May–June. W.

PONTEDERIACEAE Kunth – PICKEREL-WEED FAMILY

Aquatic herbs; *stems* dimorphic; *leaves* dimorphic, stipulate, the sessile leaves linear, submerged and forming a basal rosette or alternate, the petiolate leaves cordate, reniform, or ovate, floating or emergent, sometimes lacking; *flowers* perfect, actinomorphic, with a bractlike spathe; *tepals* 6, yellow, blue, or white; *stamens* 3 or 6, epipetalous; *pistil* 1; *ovary* superior, 3-carpellate; *fruit* a capsule or utricle.

1a. Tepal limbs ovate, blue to purplish-blue, the upper tepal with a central yellow spot; petiolate leaves round to ovate with an obtuse apex, 3.5–10 cm wide, with a swollen or inflated petiole base; stamens 6, the upper 3 shorter than the lower 3 and upcurved; roots feathery...***Eichhornia***

1b. Tepal limbs linear to linear-lanceolate elliptic, yellow or purplish with the upper 3 tepals white at the base and the central tepal also with a yellow spot at the base; leaves linear to oblong or ovate with an acute to obtuse apex, 0.1–3.3 cm wide, lacking a swollen petiole base; stamens 3, equal or the lateral 2 shorter; roots not feathery...***Heteranthera***

EICHHORNIA Kunth – WATER HYACINTH

Petiolate leaves floating, cordate to ovate with an obtuse apex; *flowers* in a spike, spathes folded with an acuminate tip, open for a single day; *tepals* connate below for at least half their length, blue to pinkish, rarely white; *stamens* 6, with upcurved, purple, glandular-hairy filaments; *fruit* a capsule.

***Eichhornia crassipes* (Mart.) Solms**, COMMON WATER HYACINTH. Perennials to 3 dm, with feathery roots; *petiolate leaves* round to ovate, 3.5–10 cm wide, with a swollen or inflated petiole base; *perianth* 5–7 cm wide. Not persisting in our area but sometimes seen in lakes around populated areas. June–Aug. Introduced.

HETERANTHERA Ruiz & Pav. – MUD PLANTAIN

Petiolate leaves floating or emergent; *flowers* in a spike or solitary, spathes folded or clasping, with an acute apex; *tepals* connate below for at least half their length, yellow, blue, pinkish, or white; *stamens* 3, equal or the 2 lateral ones shorter, with yellow to purple, glabrous to glandular-hairy or hairy filaments; *fruit* a capsule.

1a. Flowers yellow; petiolate leaves lacking, leaves all sessile and linear to narrowly oblong; stamens all equal...***H. dubia***

1b. Flowers purplish; petiolate leaves present, oblong to ovate; stamens unequal, the lateral two shorter...***H. limosa***

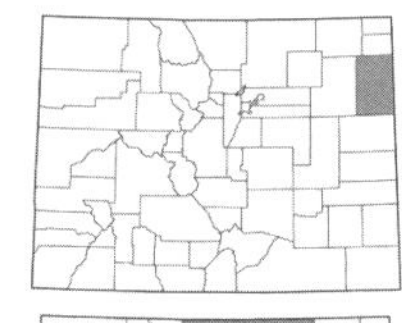

***Heteranthera dubia* (Jacq.) MacMill.**, GRASSLEAF MUD PLANTAIN. [*Zosterella dubia* (Jacq.) Small]. Annuals or short-lived perennials, grass-like, the stems rooting at the nodes; *leaves* narrowly oblong to linear, 3–12 × 0.2–0.7 cm; *perianth* yellow with linear lobes 6–15 mm long. Uncommon in lakes and reservoirs, 3500–4000 ft. June–Sept. E.

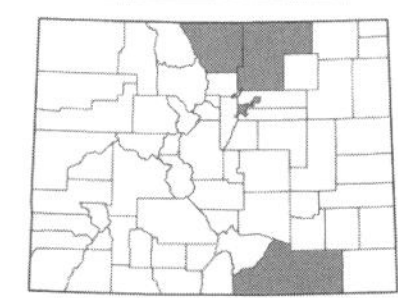

***Heteranthera limosa* (Sw.) Willd.**, BLUE MUDPLANTAIN. Annuals, not rooting at the nodes; *leaves* oblong to ovate, to 4 cm long and 3.3 cm wide; *perianth* purplish with linear-lanceolate lobes 10–40 mm long. Uncommon in ponds, 5000–5300 ft. June–Sept. E.

PORTULACACEAE Adans. - PURSLANE FAMILY

Annual or perennial herbs, usually fleshy; *leaves* alternate or opposite, simple, entire; *flowers* perfect, actinomorphic; *sepals* usually 2, distinct; *petals* mostly 4–5, distinct or basally connate; *stamens* 4–many, usually the same number as the petals, opposite the petals; *pistil* 1; *ovary* half-inferior, with free-central placentation, 2–9-carpellate; *fruit* a circumscissle capsule.

PORTULACA L. – PURSLANE

Usually annual herbs; *leaves* alternate or subopposite, the blade terete to flattened; *flowers* in terminal clusters or axillary on short branches; *stamens* 6–many; *ovary* 3–8-carpellate.

1a. Leaf blades terete, linear to linear-oblong; nodes and inflorescence with conspicuous tufts of hair; capsules 1.1–2 mm in diam....***P. halimoides***

1b. Leaf blades flat, spatulate to obovate; trichomes absent or inconspicuous at nodes and in the inflorescence; capsules 4–9 mm in diam....***P. oleracea***

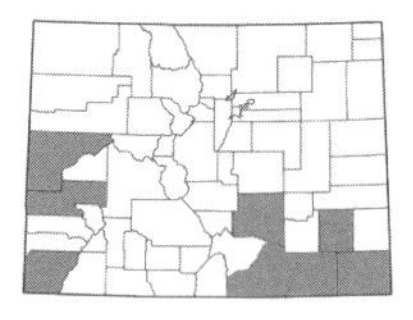

***Portulaca halimoides* L.**, SILKCOTTON PURSLANE. [*Portulaca parvula* A. Gray]. Annuals, prostrate to ascending, the stems mostly 5–15 cm long, nodes and leaf axils with tufts of hair 3–7 mm long; *leaves* terete, linear to linear-oblong, 4–13 mm long; *sepals* 2–2.5 mm long; *petals* yellow to orange, 2–2.5 mm long; *capsules* 1.1–2 mm diam. Uncommon in open places in sandy soil, 4500–5800 ft. June–Sept. E/W.

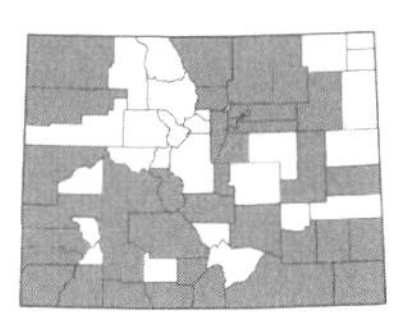

***Portulaca oleracea* L.**, COMMON PURSLANE; LITTLE HOGWEED. Annuals, prostrate, the stems 5–50 cm long, nodes and leaf axils glabrous; *leaves* flat, spatulate to obovate, 2.8–4.5 mm long; *sepals* 2.5–4.5 mm long; *petals* yellow, 3–4.5 mm long; *capsules* 4–9 mm diam. Common and widespread weed found in disturbed places, 3400–8500 ft. June–Sept. E/W. Introduced. (Plate 70)

POTAMOGETONACEAE Dumort. – PONDWEED FAMILY

Aquatic herbs; *leaves* alternate or opposite, simple, sessile or petiolate, stipulate, the stipules forming a tubular sheath around the stem; *flowers* perfect, actinomorphic, in terminal or axillary spikes; *tepals* 4 in 1 series; *stamens* 4; *pistils* 1 or 4, distinct; *ovary* superior; *fruit* a drupaceous achene. (Haynes & Hellquist, 2000; Matsumura & Harrington, 1955)

1a. Plants with broad, leathery, elliptic floating leaves, often with linear submerged leaves...***Potamogeton***
1b. Leaves unlike the above, if leaves floating then not differing from the submerged leaves in shape and size...2

2a. Submerged leaves wider, lanceolate to narrowly elliptic...***Potamogeton***
2b. Submerged leaves filiform and thread-like...3

3a. Stipular sheaths of submerged leaves free from the base of the leaf blade or adnate for less than ½ the length of the stipule; submerged leaves not channeled; peduncle stiff and often projected above the surface of the water (if long enough)...***Potamogeton***
3b. Stipular sheaths of submerged leaves adnate to the base of the leaf blade for ⅔ or more the length of the stipule (forming a sheath around the stem, often difficult to see); submerged leaves channeled; peduncle flexible, not projected above the surface of the water...***Stuckenia***

POTAMOGETON L. – PONDWEED

Perennial herbs with or without turions; *leaves* sugmerged or with some floating, often dimorphic with the sugmerged leaves translucent and linear, not channeled, and floating leaves broad and leathery, the stipular sheath of submerged leaves distinct, or if adnate then fused less than half the length of the stipule; *fruit* a beaked achene.

1a. Plants with one kind of leaf, if floating leaves are present then these are not markedly different than the submerged leaves...2
1b. Plants with 2 kinds of leaves, the floating leaves elliptic to oblong-elliptic, leathery, tapering to a distinct petiole, obviously different from the submerged leaves...9

2a. Leaves auriculate with small lobes projecting past the stem on either side, linear to lanceolate [3–4 (8) mm wide], and conspicuously 2-ranked; stipules conspicuous; inflorescence often branched...***P. robbinsii***
2b. Leaves not auriculate, linear to elliptic, not 2-ranked; stipules various; inflorescence not branched...3

3a. Leaves filiform to narrowly linear (less than 2 mm wide)...4
3b. Leaves wider, often ribbon-like...5

4a. Leaves with 2 small globose basal glands just below the point of attachment of the blade and the stem; achenes smooth; peduncles very slender or filiform, (0.4) 1–5 (8) cm long; spikes 0.6–1.5 cm long with 1–3 (5) whorls of 2–3 flowers...***P. pusillus***
4b. Leaves lacking basal glands; achenes with an undulate to dentate keel; peduncles rather stout, 0.3–1.5 (2) cm long; spikes 0.1–0.5 cm long, with 3–5 whorls of paired flowers...***P. foliosus***

5a. Leaf margins finely toothed; achene beak 2–3 mm long...***P. crispus***
5b. Leaf margins entire; achenes with shorter beaks...6

6a. Submerged leaves petiolate or if sessile then not clasping the stem; stem usually reddish-brown...***P. alpinus***
6b. Submerged leaves sessile and clasping the stem; stem usually green or whitish...7

7a. Stipules shredding and persisting as long fibers...***P. richardsonii***
7b. Stipules not shredding and persisting as long fibers...8

8a. Leaves (5) 10–30 (35) cm long, the apex hood-like and often splitting when pressed; stems usually zigzag; stipules large (2–8 cm long) and white, persisting, especially noticeable in the axils of upper leaves; fruit 4–5.7 mm long with an erect beak 0.3–1 mm long...***P. praelongus***
8b. Leaves 1–8 (10) cm long, the apex not hood-like or splitting when pressed; stems usually not zigzag; stipules inconspicuous, deciduous; fruit 1.5–3 mm long with an erect beak 0.4–0.6 mm long ...***P. perfoliatus***

9a. Submerged leaves filiform-linear (less than 2 mm wide)...10
9b. Submerged leaves broader, sometimes ribbon-like (over 2 mm wide)...11

10a. Floating leaves (1) 1.5–3 (4) cm long and 0.3–2 cm wide, with fewer than 20 nerves; fruit 1–1.8 mm long, with sharp to rounded points along the margins and keels...***P. diversifolius***
10b. Floating leaves (3) 5–11 cm long and (1.5) 2.5–6 cm wide, with 20–35 nerves; fruit 3–5 mm long, smooth...***P. natans***

11a. Upper submerged leaves folded along the midvein, falcate in outline, petiolate; stipules 3.5–11 cm long, those of floating leaves 8–20 cm long...***P. amplifolius***
11b. Submerged leaves not folded or falcate in outline, sessile or petiolate; stipules 0.5–8 cm long...12

12a. Submerged leaves on long petioles 2–13 cm long; stipules (3) 4–9 cm long...***P. nodosus***
12b. Submerged leaves sessile or tapering to a short petiole mostly less than 2 cm long; stipules 0.5–7 cm long...13

13a. Stems flattened; submerged leaves linear and ribbon-like with nearly parallel sides...***P. epihydrus***
13b. Stems terete or rarely flattened; submerged leaves elliptic to lanceolate, the sides not parallel...14

14a. Stipules 0.5–3 cm long; floating leaves (1.5) 2–5 (7) cm long; fruit 1.8–2.8 mm long...***P. gramineus***
14b. Stipules (2.5) 3–7 cm long; floating leaves (4) 6–12 cm long; fruit 2.5–4 mm long...***P. illinoensis***

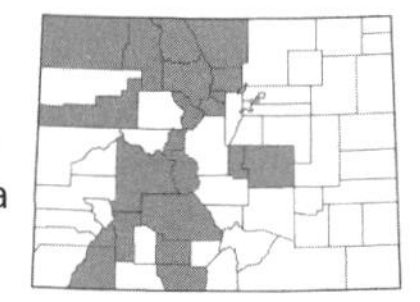

***Potamogeton alpinus* Balb.**, ALPINE PONDWEED. Stems mostly simple, usually reddish-brown; *leaves* floating and submerged, all linear-lanceolate to oblong-linear, 4–18 cm × 5–20 mm, with a short petiole to 3 cm long or sessile; *stipules* free from the leaf, persistent, inconspicuous, 1.2–2.5 (4) cm long; *peduncles* 3–15 cm long; *spikes* 1–3.5 cm long; *achenes* obovate, (2.5) 3–4 × 2–2.5 mm, with a recurved beak 0.5–1.3 mm long. Found in lakes and ponds, 6800–11,500 ft. June–Sept. E/W.

***Potamogeton amplifolius* Tuck.**, LARGELEAF PONDWEED. Stems 6–15 dm; *leaves* dimorphic, the floating leaves oblong-lanceolate to ovate, 7.5–18 × 3–4 cm, petiolate, the sugmerged leaves more or less recurved and falcate in outline, 12–20 cm × 30–50 mm; *stipules* 3.5–11 cm long, those of floating leaves 8–20 cm long; *peduncles* 4.5–22 cm long; *spikes* 3.5–6.5 cm long; *achenes* obovoid to obovate, 4–6.5 mm long, abaxially keeled, with a short, erect beak 0.5–0.8 mm long. Uncommon in lakes and ponds, 8000–9500 ft. June–Aug. E/W.

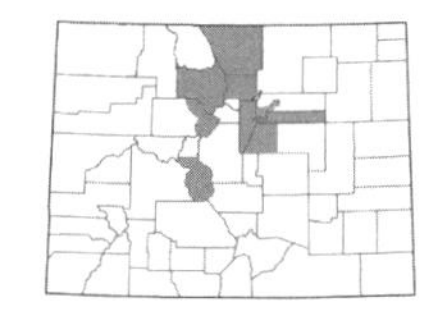

***Potamogeton crispus* L.**, CURLY PONDWEED. Stems much-branched, to 10 dm, flattened, reddish-brown; *leaves* all submerged, linear-lanceolate, 1.2–9 cm × 4–10 mm, sessile to clasping, with a rounded apex, the margins finely toothed; *stipules* to 0.5 cm long, bilobed at the apex; *peduncles* 2.5–4 cm long; *spikes* 1–1.5 cm long; *achenes* obovoid, 2.5–3 × 2–2.5 mm, with a recurved beak 2–3 mm long. Uncommon in ponds and lakes, 5500–7800 ft. May–Sept. E/W. Introduced.

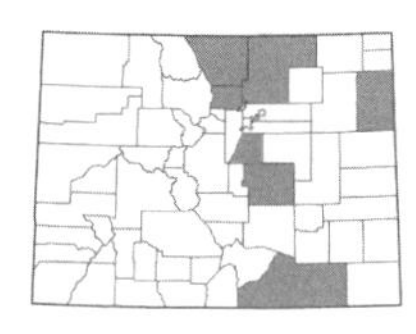

***Potamogeton diversifolius* Raf.**, WATERTHREAD PONDWEED. Stems many-branched; *leaves* dimorphic, floating leaves elliptic to oval, (1) 1.5–3 (4) × 0.3–2 cm, entire, petiolate, submerged leaves linear, 1–1.3 cm × 0.1–1 mm; *stipules* of submerged leaves mostly adnate to the leaf bases for about ½ their length, inconspicuous; *peduncles* dimorphic, the emergent 6–15 mm and the submerged 3–5 mm long; *spikes* dimorphic, the emergent 0.5–2 cm long, the submerged 0.2–0.3 cm long; *achenes* orbicular, 1–1.8 mm long, 3-keeled with sharp or rounded points along each keel, with a short beak to 0.1 mm long. Uncommon in ditches and ponds, 3500–7500 ft. June–Sept. E.

***Potamogeton epihydrus* Raf.**, RIBBONLEAF PONDWEED. Stems flattened, to 20 dm; *leaves* dimorphic, the floating leaves elliptic to oblong, 2–8 × 0.5–2.5 cm, petiolate, the submerged leaves linear, 10–13 cm × 2–4 mm, sessile; *stipules* free from the leaf, persistent, inconspicuous, 1–3 cm long; *peduncles* 1.5–5 (15) cm long; *spikes* 0.8–4 cm long; *achenes* obovoid, 3-keeled, with a short erect beak to 0.5 mm long. Uncommon in mountain ponds and lakes, 8500–11,500 ft. July–Sept. E/W. (Plate 70)

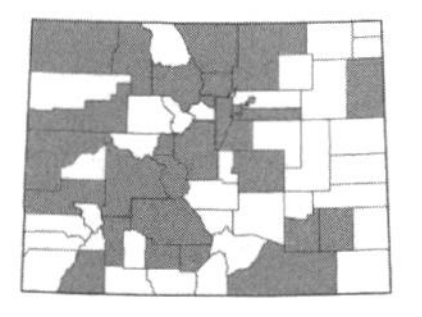

***Potamogeton foliosus* Raf.**, LEAFY PONDWEED. Stems to 10 dm; *leaves* submerged, filiform to linear, 1.3–8 cm × 0.3–2 mm, sessile, lacking basal glands; *stipules* at first forming fibrous sheaths around the stems but soon rupturing and deciduous, free from the leaf; *peduncles* 0.3–1.5 (2) cm long; *spikes* 0.1–0.5 cm long; *achenes* obovate, 1.5–2.7 × 1.5–2.2 mm, with an undulate to dentate keel and erect beak 0.2–0.6 mm long. Found in ditches, ponds, and streams, 3500–10,500 ft. June–Sept. E/W.

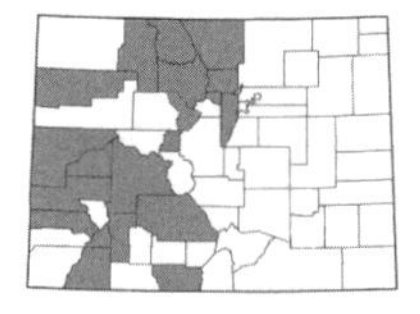

***Potamogeton gramineus* L.**, VARIABLE-LEAF PONDWEED. Stems to 15 dm; *leaves* dimorphic, floating leaves ovate to elliptic,, (1.5) 2–5 (7) × 1–2 cm, on petioles 2–10 cm long, and submerged leaves linear to narrowly ovate, 3–9 cm × 3–30 mm, sessile; *stipules* free from leaf bases, 0.5–3 cm long, persistent; *peduncles* 3–8 cm; *spikes* 1.5–3.5 cm long; *achenes* 1.8–2.8 × 1.8–2 mm, 3-keeled, with an erect beak 0.3–0.5 mm long. Common in lakes and ponds, 7500–10,000 ft. June–Sept. E/W.

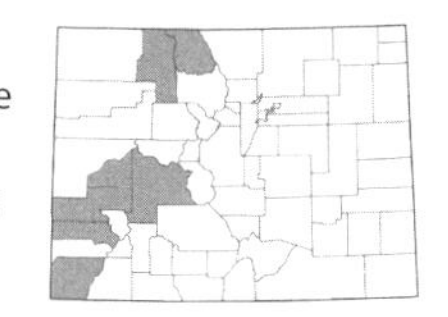

Potamogeton illinoensis **Morong**, ILLINOIS PONDWEED. Stems branching above; *leaves* dimorphic, the floating leaves ovate to elliptic, (4) 6–12 × 3–8 cm, thick and coriaceous, on short petioles, the submerged leaves elliptic to oblong-elliptic or lanceolate, 5–20 cm × 2–45 mm, sessile; *stipules* free, persistent, conspicuous, 1–8 cm long, brown; *peduncles* 4–30 cm long; *spikes* 2.5–7 cm long; *achenes* 2.5–4 × 2–3 mm, 3-keeled, with an erect to slightly recurved beak 0.5–0.8 mm long. Uncommon in ponds and lakes, 8000–9000 ft. June–Sept. E/W.

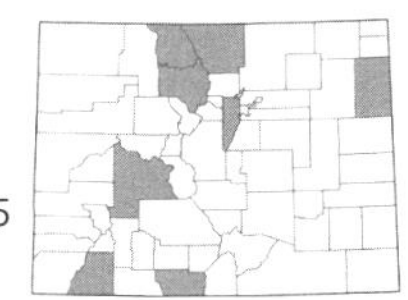

Potamogeton natans **L.**, FLOATING PONDWEED. Stems rarely branched, 6–20 dm; *leaves* dimorphic, the floating leaves ovate to oblong-ovate, thick and coriaceous, (3) 5–11 × (1.5) 2.5–6 cm, on long petioles, entire, the submerged leaves narrowly linear, 9–20 cm × 0.5–2 mm; *stipules* white, persistent, fibrous; *peduncles* terminal, 4.5–10 cm long; *spikes* 2.5–5 cm long; *achenes* obovoid, 3–5 mm long, smooth, with a short, erect to apically recurved beak 0.4–0.8 mm long. Found in lakes, ponds, and ditches, 3500–9000 ft. June–Sept. E/W.

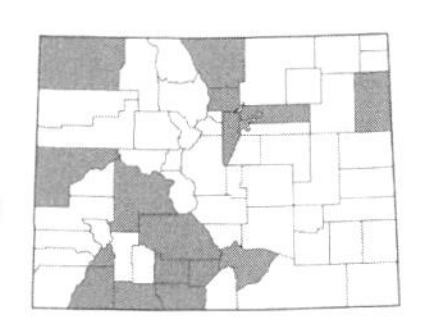

Potamogeton nodosus **Poir.**, LONGLEAF PONDWEED. Stems simple or little branched, to 10 dm; *leaves* dimorphic, the floating leaves elliptic to ovate, thick and coriaceous, 3–11 cm × 20–45 mm, on long petioles, the submerged leaves linear-lanceolate to lance-elliptic, 9–20 × 1–3.5 cm, on petioles 2–13 cm long; *stipules* free, conspicuous, (3) 4–9 cm long, light brown; *peduncles* 3–15 cm long; *spikes* 3–6 cm long; *achenes* obovoid to obovate, 3–4.5 × 2.5–3 mm, with a prominent keel and short, erect beak, reddish or reddish-brown at maturity. Found in lakes, ponds, and ditches, 3500–10,000 ft. June–Sept. E/W.

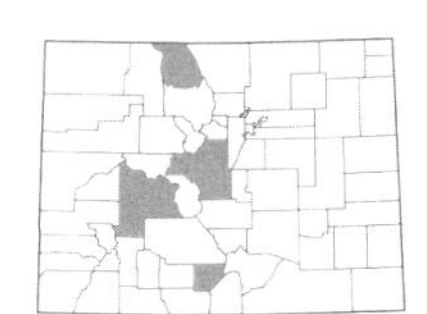

Potamogeton perfoliatus **L.**, CLASPINGLEAF PONDWEED. Stems branched above, to 25 dm; *leaves* all submerged, broadly lanceolate to ovate or orbiculate, 1–8 (10) × 0.7–4 cm, sessile and clasping; *stipules* inconspicuous, deciduous, 3.5–6.5 cm, free from the leaf; *peduncles* 1–7 cm long; *spikes* 0.5–5 cm; *achenes* obovoid, 1.5–3 × 1.3–2.2 mm, with a short, erect beak 0.4–0.6 mm long. Found in lakes and ponds, 6500–10,000 ft. June–Sept. E/W.

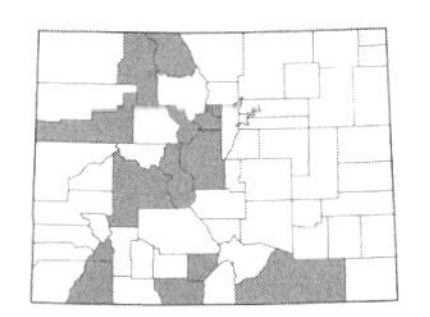

Potamogeton praelongus **Wulf.**, WHITESTEM PONDWEED. Stems often zigzag, green to white; *leaves* all submerged, ovate-oblong to oblong-lanceolate, (5) 10–30 (35) cm × 10–45 mm, entire, with a hood-like apex, sessile and clasping; *stipules* 2–8 cm long, white, persistent; *peduncles* 9.5–50 cm long; *spikes* 3–7.5 cm long with 6–12 whorls; *achenes* obovoid to obovate, rounded on the back, 4–5.7 × 2.6–4 mm, with an erect beak 0.3–1 mm long. Found in mountain lakes and ponds, 8000–11,500 ft. June–Sept. E/W.

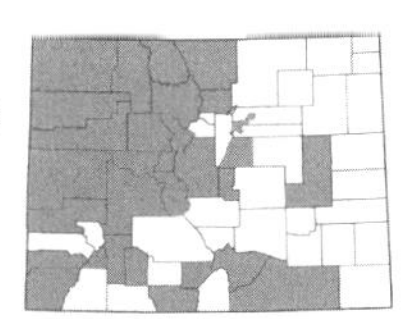

Potamogeton pusillus **L.**, SMALL PONDWEED. Stems freely branching, to 15 dm; *leaves* all submerged, linear, 1.5–6 cm × 0.2–2 mm, sessile, with 2 small globose basal glands on at least some nodes just below the point of attachment of the blade and stem; *stipules* tubular and united to above the middle, free from the leaf, 3–17 mm; *peduncles* (0.4) 1–5 (8) cm long; *spikes* 0.6–1.5 cm long with 1–3 (5) whorls of 2–3 flowers; *achenes* obliquely obovoid, 1.5–3 × 1.1–1.8 mm, with an erect beak 0.1–0.6 mm long. Found in ponds and shallow ditches, 5000–10,500 ft. June–Sept. E/W.

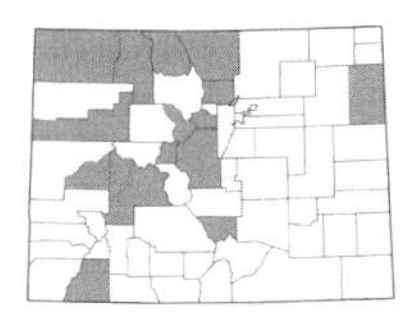

Potamogeton richardsonii **(A. Benn.) Rydb.**, RICHARDSON'S PONDWEED. Stems branched above, to 10 dm; *leaves* all submerged, lanceolate to ovate-lanceolate, 2–13 cm × 5–25 mm, sessile and clasping; *stipules* persistent, shredding and disintegrating into fibers; *peduncles* 1.5–15 cm long; *spikes* 1.3–4 cm long; *achenes* obovoid, 2.2–4.2 × 1.7–3 mm, rarely abaxially keeled, with a short, erect beak 0.4–0.7 mm long. Found in ponds and lakes, 3500–11,500 ft. June–Sept. E/W.

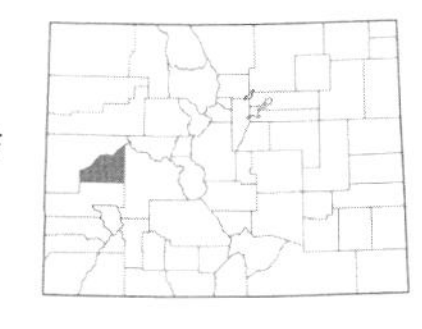

Potamogeton robbinsii **Oakes**, ROBBINS' PONDWEED. Stems to 10 dm; *leaves* all submerged, linear to lanceolate, 2–7 (12) cm × 3–4 (8) mm, with minutely spinulose to serrulate margins, auriculate at the base; *stipules* 0.5–2 cm long, persistent and conspicuous, fused with the lower part of the leaf forming a sheath; *peduncles* 3–5 (7) cm long; *spikes* 0.7–2 cm long, inflorescence often branched; *achenes* obliquely obovoid, 3–5 × 2–3 mm, with an erect beak recurved at the apex, 0.7–0.9 mm long. Recently discovered in Colorado in Delta Co. June–Sept. W.

STUCKENIA Börner – PONDWEED

Perennial herbs without turions; *leaves* all submerged, channeled, the stipular sheath adnate to the leaf base for ⅔ the length of the stipule or more; *fruit* beaked or not.

1a. Stipule sheath of lower leaves inflated and wider than the stem, 3–6 cm long; spikes with equally spaced whorls of flowers…***S. vaginata***

1b. Stipule sheath of lower leaves not inflated, 0.5–3 cm long; spikes with mostly unequally spaced whorls of flowers…2

2a. Stipule sheaths of lower leaves to 0.5–2 cm long, forming a conspicuous hyaline ligule 1–7 mm long; leaves usually blunt at the apex; achenes lacking a beak or the beak inconspicuous...***S. filiformis***
2b. Stipule sheaths of lower leaves (1) 2–3 cm long, forming a short hyaline ligule to 1 mm long; leaves with a sharp-pointed apex; achenes with a beak 0.4–1 mm long...***S. pectinata***

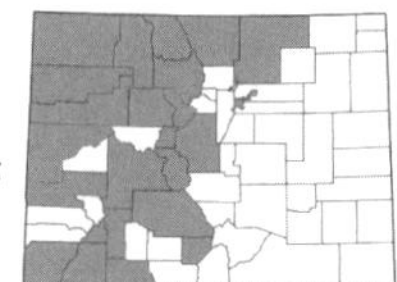

Stuckenia filiformis **(Pers.) Börner**, FINELEAF PONDWEED. Stems freely branching, especially near the base; *leaves* linear-filiform, 5–15 cm × 0.2–0.5 mm, blunt or obtuse at the apex, sessile and clasping; *stipules* united to the base of the leaves for 1 cm, forming ligules 1–7 mm long; *peduncles* 2–15 cm long; *spikes* 0.5–5.5 cm long, with 2–6 (9) whorls; *achenes* obovoid, 2–3 × 1.5–2.4 mm, with an inconspicuous beak or the beak absent. Found in streams and ponds, 5000–11,500 ft. June–Sept. E/W.

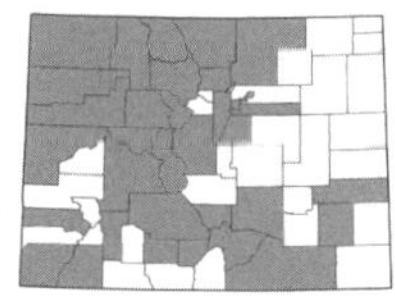

Stuckenia pectinata **(L.) Börner**, SAGO PONDWEED. Stems freely branching; *leaves* filiform to linear, 2–15 cm × 0.3–0.5 mm, with a sharp-pointed apex, sessile; *stipules* united to the base of the leaves for about ½ their length, forming ligules to 1 mm long; *peduncles* 4.5–11.5 cm long; *spikes* 1.5–2.2 cm long, with 3–5 whorls of flowers; *achenes* obliquely obovoid, 3.5–4.7 × 2.2–3 mm, with an erect beak 0.4–1 mm long. Found in streams and ponds, 3500–10,500 ft. June–Sept. E/W.

Stuckenia vaginata **(Turcz.) Holub**, SHEATHED PONDWEED. Stems freely branching; *leaves* linear, 5–10 cm × 0.5–1 mm, sessile; *stipules* with sheaths inflated and wider than the stem, 3–6 cm long; *peduncles* 3–15 cm long; *spikes* 1–8 cm long, with 3–12 whorls of equally spaced flowers; *achenes* 3–4 × 2–3 mm, obliquely obovate, with an inconspicuous beak or the beak absent. Found in lakes and ponds, 4800–10,800 ft. July–Sept. E/W.

PRIMULACEAE Vent. – PRIMROSE FAMILY

Herbs, usually scapose; *leaves* alternate, opposite, or whorled, simple, exstipulate; *flowers* perfect, actinomorphic, solitary or in various arrangements; *calyx* (4) 5-lobed, usually persistent; *corolla* (4) 5-lobed; *stamens* (4) 5, epipetalous, opposite the corolla lobes, sometimes alternating with 5 staminodia; *pistil* 1; *ovary* superior, 5-carpellate; *fruit* a capsule.
1a. Flowers small, the petals 0.5–5 mm long, white, entire at the tip; leaves 0.2–6 cm long...***Androsace***
1b. Flowers larger, the petals over 5 mm long (including both the tube and lobes), pink, purple, or lilac, with a yellow center; leaves 1.5–50 cm long...2

2a. Corolla lobes sharply reflexed, the tube inconspicuous; flowers nodding; stamens connivent around the style, protruding and erect from the corolla...***Dodecatheon***
2b. Corolla lobes spreading or erect, the tube conspicuous and equal to or longer than the calyx; flowers not nodding; stamens not connivent around the style and protruding from the corolla...***Primula***

ANDROSACE L. – ROCKJASMINE

Annual or perennial, scapose herbs; *leaves* in basal rosettes; *flowers* in terminal umbels subtended by bracts; *calyx* with 5 lobes nearly equal to the tube in length; *corolla* 5-lobed, shorter than the calyx, white to pink; *stamens* 5; *capsules* subglobose or globose, 5-valved.
1a. Mat-forming perennials; leaves ciliate on the margins and pilose below; flowers nearly sessile, on pedicels to 3 mm long; corolla lobes 1.5–3 mm long; plants with one peduncle per rosette...***A. chamaejasme* ssp. *carinata***
1b. Annuals or possibly biennials from a taproot; leaves glabrous or sparsely to densely minutely hairy; flowers on conspicuous pedicels 3–55 (90) mm long; corolla lobes 0.8–1.5 mm long; plants with one to several peduncles per rosette...2

2a. Pedicels with glandular hairs; leaves narrowed to a distinct petiole; sepals not keeled on the back...***A. filiformis***
2b. Pedicels glabrous or with small, minute or branched hairs; leaves tapering at the base, not clearly differentiated into a blade and petiole; sepals keeled on the back...3

3a. Bracts subtending the umbel elliptic to ovate...***A. occidentalis***
3b. Bracts subtending the umbel linear to lanceolate...***A. septentrionalis***

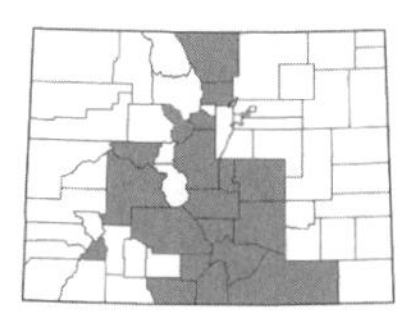

Androsace chamaejasme **Wulfen ssp. *carinata* (Torr.) Hultén**, BOREAL ROCKJASMINE. Mat-forming perennials, 1.2–5 cm; *leaves* 2–8 mm long, ciliate on the margins, pilose below; *pedicels* 1–3 mm long; *calyx* 2–3 mm long; *corolla* white to cream with a yellow center, 1.5–3 mm long; *capsules* 2–3 mm long. Locally common in the alpine, 11,200–14,000 ft. June–Aug. E/W. (Plate 70)

Androsace filiformis **Retz.**, SLENDER ROCKJASMINE. Annuals, 1–10 cm; *leaves* 6–20 (30) mm long, toothed, glabrous or sometimes glandular above; *pedicels* glandular, 10–40 mm long; *calyx* 1.5–2 mm long; *corolla* lobes 0.8–1.2 mm long; *capsules* longer than the calyx. Locally common in moist meadows and along streams, 7000–10,500 ft. June–Aug. E/W.

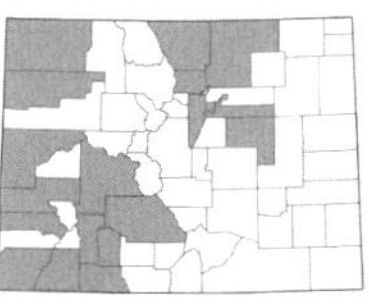

Androsace occidentalis **Pursh**, WESTERN ROCKJASMINE. Annuals, 1–10 cm; *leaves* 6–20 (30) mm long, entire to toothed, sparsely short-hairy with simple hairs; *pedicels* 3–25 (30) mm long; *bracts* elliptic to ovate; *calyx* 3.5–5 mm long, keeled below each lobe; *corolla* shorter than the calyx; *capsules* slightly shorter than the calyx. Locally common on dry, grassy or sagebrush slopes, in grasslands, or on pinyon-juniper slopes, 5000–9500 ft. April–May (June). E/W.

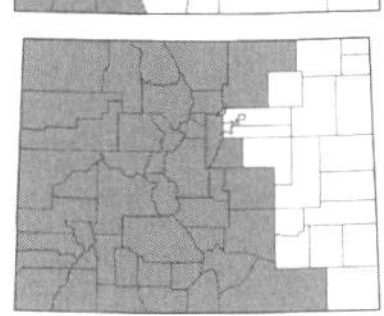

Androsace septentrionalis **L.**, PYGMYFLOWER ROCKJASMINE. Annuals, 1–20 (30) cm; *leaves* 5–60 mm long, toothed, sparsely to densely short-hairy with simple or forked hairs; *pedicels* (2) 3–60 (90) mm long; *bracts* linear to lanceolate; *calyx* 2.5–4 mm long, keeled below each lobe; *corolla* lobes 0.5–1 mm long; *capsules* 2–4 mm long. Common on open slopes, in meadows, along roadsides, in forests, and in the alpine, 5100–14,500 ft. May–July. E/W.

DODECATHEON L. – SHOOTING STAR

Perennials, scapose herbs; *leaves* in basal rosettes; *flowers* solitary or in terminal umbels; *calyx* (4) 5-lobed, reflexed in flower; *corolla* (4) 5-lobed, the lobes well exceeding the tube, reflexed; *stamens* (4) 5, connivent around the style, protruding.

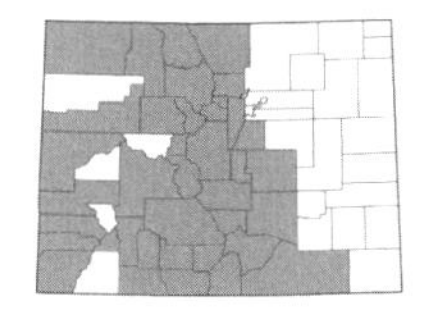

Dodecatheon pulchellum **(Raf.) Merr.**, DARKTHROAT SHOOTING STAR. Plants (0.5) 1–6.5 dm; *leaves* oblanceolate to elliptic, 3–35 cm long, entire; *umbels* 3–20-flowered; *calyx* tube 2–4 mm long with lobes 2.5–6 mm long; *corolla* 9–20 mm long; *capsules* 7–20 mm long, cylindric. Common in meadows, moist places, and aspen or spruce forests, 5500–12,500 ft. May–July. E/W. (Plate 70)

There are two varieties of *D. pulchellum* in Colorado:

1a. Leaves mostly less than 3 cm wide; inflorescence usually glabrous; plants usually above 6000 ft in elevation, not of hanging gardens...**var. *pulchellum***. Common in meadows, forests, and moist places, 5900–12,500 ft. May–July. E/W.

1b. Leaves usually over 3 cm wide; inflorescence minutely glandular; plants below 6000 ft in elevation, of hanging gardens...**var. *zionense* (Eastw.) S.L. Welsh**. Uncommon in hanging gardens, known from Moffat Co., 5500–5600 ft. May–July. W.

PRIMULA L. – PRIMROSE

Perennials, scapose herbs; *leaves* in basal rosettes, simple; *flowers* solitary or in terminal umbels; *corolla* tube equal to or longer than the calyx; *stamens* 5; *capsules* ovoid or cylindric.

1a. Flowers pink to pink-purple with a yellow center...2
1b. Flowers lilac with a yellow center...3

2a. Flowers solitary or sometimes paired; plants to 7 cm; leaves narrowly oblanceolate to linear-lanceolate, to 7 mm wide...***P. angustifolia***
2b. Flowers 3–20 per umbel; plants 7–60 cm; leaves broadly oblanceolate to elliptic, over 10 mm wide...***P. parryi***

3a. Lower surface of the leaves and sepals white-farinose; flowers on pedicels 2–8 mm long...***P. incana***
3b. Lower surface of the leaves and sepals not white-farinose; flowers on pedicels 3–23 mm long...***P. egaliksensis***

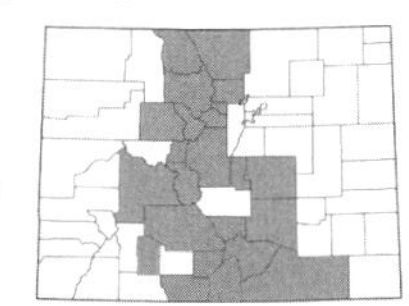

Primula angustifolia **Torr.**, ALPINE PRIMROSE. Plants 0.5–7 cm; *leaves* linear-lanceolate to narrowly oblanceolate, 1.5–5 × 0.2–0.7 cm, entire, glabrous to glandular, green above and below; *flowers* solitary or sometimes paired; *calyx* 5–8 mm long; *corolla* pink to pink-purple with a yellow center, the limb 10–20 mm wide. Common on rocky alpine slopes, 9500–14,500 ft. June–Aug. E/W. (Plate 70)

Primula egaliksensis **Wormsk. ex Hornem.**, GREENLAND PRIMROSE. Plants 0.5–5 dm; *leaves* spatulate to elliptic, 0.5–5 × 0.2–1.5 cm, entire or obscurely toothed, glabrous and green above and below; *umbels* (2) 3–9-flowered; *pedicels* 3–23 mm long; *calyx* 4–6 mm long, often glandular-ciliate on the teeth; *corolla* lilac with a yellow center, the limb 5–9 mm wide. Found in moist meadows and fens, known from Park Co., 9200–9800 ft. June–July. E.

Primula incana **M.E. Jones**, SILVERY PRIMROSE. Plants 0.5–5 dm; *leaves* elliptic to spatulate, 1.5–8 × 0.5–2 cm, toothed, white-farinose below and green above; *umbels* (2) 3–15-flowered; *pedicels* 2–8 mm long; *calyx* 5.5–10 mm long, white-farinose; *corolla* lilac with a yellow center, the limb 6–10 mm wide. Found in moist meadows and fens, 6500–10,000 ft. June–July. E/W.

***Primula parryi* A. Gray**, Parry's primrose. Plants 0.7–6 dm; *leaves* broadly oblanceolate to elliptic, 8–50 × 1–9 cm, entire, green and glandular above and below; *umbels* 3–20-flowered; *pedicels* 3–100 mm long; *calyx* 7–15 mm long, glandular; *corolla* bright pink with a yellow center, the limb 12–30 mm wide. Common along streams, in moist meadows, and in the alpine, 9200–14,000 ft. June–Aug. E/W. (Plate 70)

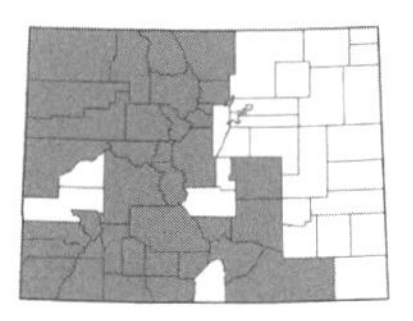

RANUNCULACEAE Juss. – BUTTERCUP FAMILIY

Herbs, shrubs, or vines; *leaves* usually alternate, simple, entire to variously dissected, usually sheathing at the base; *flowers* usually perfect, actinomorphic or zygomorphic; *sepals* 3–6 (20), distinct, often petaloid, sometimes spurred; *petals* 3–26 or absent, distinct; *stamens* several to many, distinct; *pistils* 1–many; *ovary* superior; *fruit* a follicle, achene, or berry.

1a. Leaves opposite, ternately to pinnately compound; vines or erect, perennial herbs...***Clematis***
1b. Leaves alternate or all basal, or basal and with a single series or stem leaves whorled below the inflorescence, simple to variously lobed or compound; annual or perennial herbs but not vines...2

2a. Delicate, annual herbs; leaves basal, linear or filiform; receptacle elongating in flower and fruit; sepal spurred at the base...***Myosurus***
2b. Perennial herbs or if annuals then unlike the above; leaves various; receptacle usually not elongating in flower and fruit; sepals spurred or not...3

3a. Flowers zygomorphic, white, blue, violet-blue, or violet; leaves palmately lobed; fruit a follicle...4
3b. Flowers actinomorphic, variously colored; leaves various; fruit a follicle, achene, or berry...5

4a. Upper sepal saccate and hood-like, not spurred; petals 2...***Aconitum***
4b. Upper sepal not hood-like, spurred at the base; petals 4...***Delphinium***

5a. Petals all spurred; leaves 1–3 times ternately compound, with crenate margins...***Aquilegia***
5b. Petals not spurred or absent (only sepals present); leaves various...6

6a. Leaves basal and with a single whorl of leaves on the stem below the inflorescence...***Anemone***
6b. Leaves unlike the above...7

7a. Single pistil present per flower; fruit a red or white berry; flowers in a terminal raceme; leaves pinnately compound...***Actaea***
7b. More than one pistil present per flower; fruit a follicle or achene, sometimes with a long, persistent style attached; inflorescence various; leaves various...8

8a. Both sepals and petals present; petals yellow, yellow and red, or blue, if white then the plant a true aquatic and submerged or floating in water...9
8b. Only sepals present, or petals also present but these small, linear, and staminode-like; sepals white, white-tinged with blue below, cream, or greenish-white; plants not true aquatics floating in water...10

9a. Fruit an achene; sepals 2.5–12 mm long, shorter or longer than the petals; petals 1–20 mm long or occasionally absent, yellow or occasionally white (and then the plant a true aquatic); leaves various but rarely ternately compound...***Ranunculus***
9b. Fruit a follicle; sepals 15–40 mm long, longer than the petals, yellow, yellow and tinged with red, or blue to blue-violet; petals 12–25 mm long, yellow or white; leaves 1–3 times ternately compound, with crenate margins...***Aquilegia*** (spurless forms)

10a. Flowers solitary or in 1–3-flowered cymes; sepals 10–20 mm long; leaves simple and unlobed, or palmately lobed or compound; fruit a follicle...11
10b. Flowers numerous in terminal corymbs, panicles, or racemes; sepals 1.5–6 mm long; leaves 2–3 times ternately compound or palmately lobed; fruit an achene...12

11a. Leaves all basal, simple and unlobed; sepals white and often bluish-tinged below...***Caltha***
11b. Leaves alternate on the stem, palmately lobed to compound; sepals white or cream...***Trollius***

12a. Leaves 2–3 times ternately compound; flowers perfect or imperfect...***Thalictrum***
12b. Leaves palmately lobed; flowers perfect...***Trautvetteria***

ACONITUM L. – MONKSHOOD

Perennial herbs; *leaves* alternate, palmately lobed; *flowers* perfect, zygomorphic, in terminal or axillary racemes or panicles; *sepals* 5, petaloid, the upper sepal saccate and hood-like; *petals* 2, hidden in the hooded sepal, with a coiled spur at the apex; *fruit* a follicle.

***Aconitum columbianum* Nutt.**, COLUMBIAN MONKSHOOD. Plants mostly 5–20 dm; *leaves* deeply 3–5 (7)-cleft, 5–25 cm long; *sepals* blue or sometimes white, the hood 11–30 mm high; *follicles* 10–20 mm long. Common in meadows and along streams in the mountains, 7200–12,500 ft. June–Aug. E/W. (Plate 70)

ACTAEA L. – BANEBERRY

Perennial herbs; *leaves* alternate, ternately or pinnately compound; *flowers* perfect, actinomorphic, in terminal or axillary racemes; *sepals* 3–5; *petals* 4–10, white to cream-colored; *fruit* a red or white berry.

***Actaea rubra* (Ait.) Willd.**, RED BANEBERRY. Plants 3–10 dm; *leaves* ternately compound, the leaflets ovate to elliptic, toothed or lobed, 2–10 cm long; *sepals* whitish-green, 2–4 mm long; *berries* 5–10 mm diam. Common in shady forests and along creeks and streams, 6800–12,000 ft. June–Aug. E/W. (Plate 70)

ANEMONE L. – WINDFLOWER

Perennial herbs; *leaves* basal and usually with a whorl of leaves on the stem below the inflorescence; *flowers* perfect, actinomorphic; *sepals* 4–20, petaloid; *petals* usually absent; *fruit* an achene with a persistent style.

1a. Sepals 20–40 mm long, usually lavender or blue-purple; persistent style of the achenes plumose, 20–40 mm long; stamens numerous, 150–200 per flower; plants usually softly villous throughout...***A. patens* var. *multifida***
1b. Sepals 5–20 (25) mm long, white, green, yellow, blue, purple, or red; persistent style of the achenes glabrous, 0.5–6 mm long; stamens 40–100 per flower; plants variously hairy or glabrous...2

2a. Basal leaves ternately compound, each leaflet shallowly lobed with crenate margins; stems with a single flower; plants 4–20 cm tall...***A. parviflora***
2b. Basal leaves unlike the above; stems often more than 1-flowered; plants usually over 20 cm tall...3

3a. Stem or involucre (leaves subtending the inflorescence) leaves petiolate or narrowed to a widened petiole at the base; achenes not winged, woolly to tomentose or villous...4
3b. Stem or involucre leaves sessile, not narrowed to a petiole; achenes winged, strigose to glabrous...5

4a. Receptacle in fruit subglobose, less than twice as long as wide; flowers red, purple, blue, yellow, or sometimes white; ultimate leaf divisions linear to lanceolate, usually less than 5 mm wide...***A. multifida***
4b. Receptacle in fruit long cylindric, 2 or more times as long as wide; flowers white; ultimate leaf divisions usually over 5 mm wide...***A. cylindrica***

5a. Stems and petioles thick (1.5–6 mm wide) with spreading hairs; stems with 2–4 flowers, arranged in an umbellate inflorescence; achenes and ovaries glabrous; style in fruit 0.8–1.5 mm long...***A. narcissiflora* var. *zephyra***
5b. Stems and petioles slender (1–2 mm wide) with ascending hairs; stems usually with a single flower or sometimes 2 flowers; achenes and ovaries strigose pubescent; style in fruit 2–6 mm long...***A. canadensis***

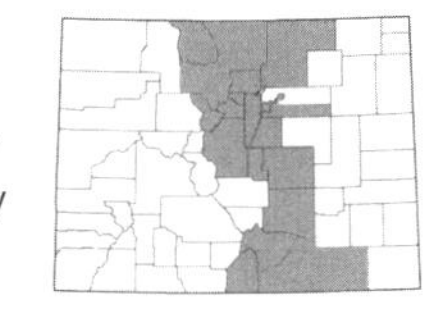

***Anemone canadensis* L.**, CANADIAN ANEMONE. Plants 1.5–8 dm; *stems* with spreading or ascending hairs; *basal leaves* petiolate, simple, deeply 3–5-cleft, 4–10 × 5–15 (20) cm, the segments lobed, with toothed margins; *involucre leaves* sessile, deeply 3-cleft, 3–10 cm long; *inflorescence* solitary or sometimes a cyme; *sepals* 5, white, (8) 10–20 (25) mm long; *achenes* winged, strigose or nearly glabrous, with a straight beak 2–6 mm long, not plumose. Common in meadows and along streams, 5000–9600 ft. May–Sept. E.

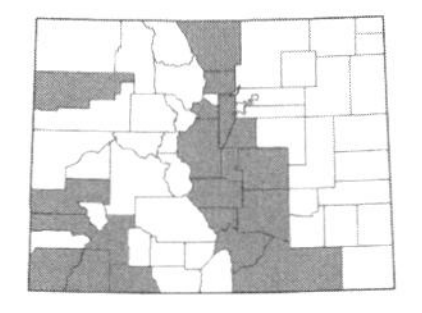

***Anemone cylindrica* A. Gray**, CANDLE ANEMONE. Plants (2) 3–8 dm; *stems* usually densely hairy; *basal leaves* petiolate, 3–7-lobed, the margins toothed to deeply incised; *involucre leaves* petiolate and similar to basal; *inflorescence* solitary or rarely a cyme; *sepals* 4–5 (6), white, 5–12 (15) mm long; *achenes* not winged, woolly, with a recurved beak 0.3–1 mm long, usually hidden by the achene hairs. Found in meadows, along streams, and in thickets, 5000–9000 ft. June–Aug. E/W.

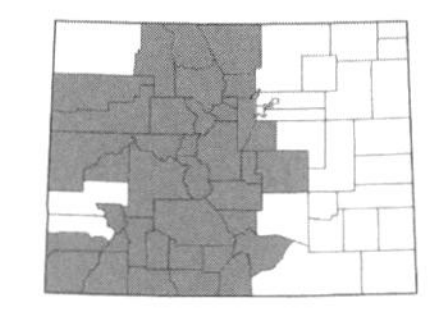

***Anemone multifida* Poir.**, CUT-LEAVED ANEMONE. Plants (2) 3–5 (6) dm; *stems* spreading-hairy; *basal leaves* petiolate, deeply 3–5-cleft into linear or lanceolate segments; *involucre leaves* petiolate, similar to basal; *inflorescence* of 2–7-flowered cymes; *sepals* 5 (9), red, purple, blue, yellow, or sometimes white; *achenes* not winged, woolly to villous, with a straight beak 1–6 mm long. Common in the alpine, in aspen, spruce-fir, or pine forests, and in meadows, 6500–14,000 ft. June–Aug. E/W.

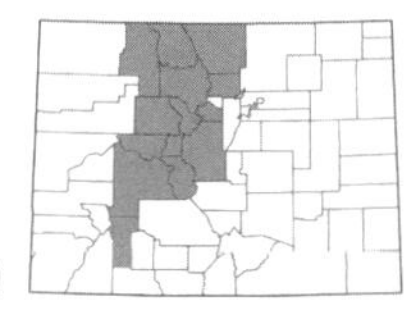

***Anemone narcissiflora* L. var. *zephyra* (A. Nelson) B.E. Dutton & Keener**, NARCISSUS ANEMONE. Plants 0.7–4 dm; *stems* glabrous to sparsely hairy or villous; *basal leaves* petiolate, ternately divided, the margins toothed to incised; *involucre leaves* sessile, otherwise similar to the basal leaves; *inflorescence* solitary or of 2–4-flowered umbels; *sepals* 5–9, white to yellowish, 8–20 mm long; *achenes* winged, with a curved to recurved beak 0.8–1.5 mm long. Common in moist meadows, spruce-fir forest openings, and the alpine, 10,000–14,000 ft. June–Aug. E/W. (Plate 70)

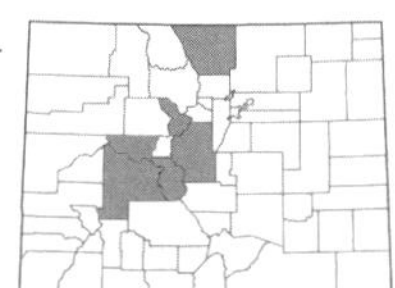

***Anemone parviflora* Michx.**, SMALL-FLOWERED ANEMONE. Plants 0.4–2 dm; *stems* glabrous to spreading-hairy; *basal leaves* petiolate, ternately compound, the leaflets shallowly lobed or cleft to near the middle, the margins crenate to broadly toothed on the distal ⅓; *involucre leaves* similar to basal; *inflorescence* solitary; *sepals* 4–7, white or tinged blue, 7–20 mm long; *achenes* not winged, woolly, with a straight beak 1–2.5 mm long. Uncommon in the alpine, 11,000–14,000 ft. June–July. E/W.

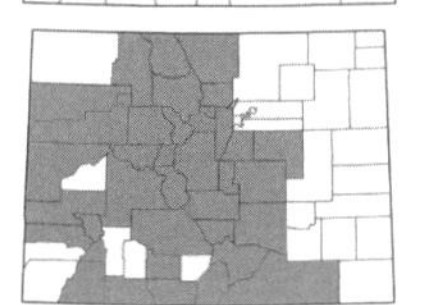

***Anemone patens* L. var. *multifida* Pritz.**, PASQUE FLOWER. [*Pulsatilla patens* (L.) Mill. ssp. *multifida* (Pritz.) Zamels]. Plants 0.5–5 dm; *stems* soft villous; *basal leaves* ternately compound, each leaflet dichotomously dissected into ultimate segments 2–4 mm wide; *involucre leaves* simple, clasping, with deeply laciniate margins; *inflorescence* solitary; *sepals* 5–8, lavender or blue-purple, 20–40 mm long; *achenes* not winged, villous, with a plumose beak 20–40 mm long. Common on open hillsides, in meadows, and in forests, 5400–13,000 ft. April–July. E/W. (Plate 70)

I have seen this blooming again in Oct. in warm years.

AQUILEGIA L. – COLUMBINE

Perennial herbs; *leaves* alternate, 1–3 times ternately compound; *flowers* perfect, actinomorphic; *sepals* 5, petaloid, white, blue, yellow, or red; *petals* 5, white, blue, yellow, or red, with a backward-pointing spur at the base; *fruit* a follicle.

1a. Leaflets viscid with sticky, glandular and pilose hairs (plants usually with much sand or dirt stuck to the leaves); lower stem, petioles, and petiolules glandular and viscid; sepals and spurs white or cream, or sometimes tinged with pink or blue; petals white or cream...***A. micrantha***
1b. Leaflets glabrous or occasionally pilose; lower stem, petioles, and petiolules glabrous or pilose or sometimes with glandular and viscid hairs; sepals and spurs white, blue, pink, yellow, or red; petals yellow, white, or cream...2

2a. Sepals and spurs yellow, red, or pink; petals yellow or cream...3
2b. Sepals and spurs white or blue; petals white or cream...5

3a. Sepals 20–35 mm long, yellow; spur 42–65 mm long, yellow; petals 13–23 mm long, yellow...***A. chrysantha***
3b. Sepals 7–18 mm long, pink to red below and yellow above or rarely all yellow; spur 14–27 mm long, pink to red; petals 8–16 mm long, yellow or cream...4

4a. Sepals reflexed or horizontally spreading, pink to red below and yellowish above or rarely all yellow; spurs gradually and evenly tapered from the base, pink to red; leaves usually blue-green and glaucous on both sides; beak of mature follicles 8–12 mm long...***A. barnebyi***
4b. Sepals erect or nearly so, red below and yellowish above; spurs abruptly narrowed near the middle, red; leaves green and not glaucous above; beak of mature follicles 13–15 mm long... ***A. elegantula***

5a. Spur 3–9 mm long, hooked at the end; sepals 9–18 mm long, blue; petals 7–10 mm long; follicles 7–10 mm long with a 3–5 mm long beak; peduncles glabrous...***A. saximontana***
5b. Spur 28–70 mm long, straight; sepals 25–50 mm long, white to blue; petals 13–28 mm long; follicles 20–30 mm long with an 8–12 mm long beak; peduncles more or less glandular-hairy...***A. coerulea***

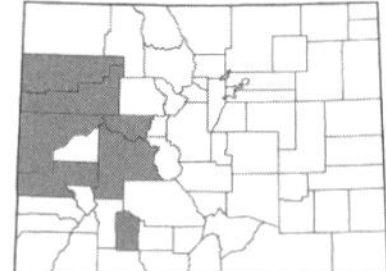

***Aquilegia barnebyi* Munz**, OIL SHALE COLUMBINE. Plants 3–8 dm; *leaves* glabrous and glaucous; *sepals* 10–18 × 5–7 mm, pink to red below and yellowish above, rarely all yellow, reflexed or horizontally spreading; *spurs* 14–27 mm long, pink to red, evenly tapered from the base; *petals* yellow or cream, 6–10 × 4–6 mm; *follicles* 18–25 mm long, the beak 8–12 mm long. Found on moist shale or limestone cliffs and along streams, 5800–9700 ft. May–July. W.

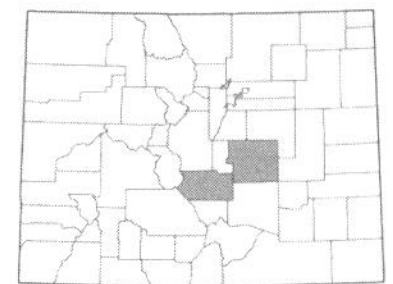

Aquilegia chrysantha **A. Gray**, GOLDEN COLUMBINE. Plants 3–12 dm; *leaves* glabrous or pilose distally; *sepals* 20–35 × 5–10 mm, yellow; *spurs* 42–65 mm long, yellow, parallel or divergent, evenly tapered from the base; *petals* yellow, 13–23 × 6–15 mm; *follicles* 18–30 mm long, the beak 10–18 mm long. Uncommon in moist gulches and ravines, often near waterfalls, sometimes introduced in garden plantings along the Front Range, 5400–7000 ft. June–July. E.

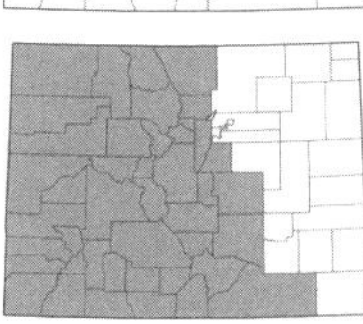

Aquilegia coerulea **James**, COLORADO BLUE COLUMBINE. Plants 1.5–8 dm; *leaves* glabrous or sometimes pilose; *sepals* 25–50 × 8–23 mm, white, blue, or sometimes pink; *spurs* 28–70 mm long, white, blue, or sometimes pink, parallel or divergent, evenly tapered from the base; *petals* white, 13–28 × 5–15 mm; *follicles* 20–30 mm long, the beak 8–12 mm long. Common in meadows, along streams, in forests, and in the alpine, 5500–13,500 ft. June–Aug. E/W. (Plate 70)

Aqilegia coerulea can hybridize with *A. elegantula*, resulting in a flower with the form of *A. coerulea* but with reddish sepals and spurs. It is also the state flower of Colorado. There are two varieties in Colorado:

1a. Sepals medium to deep blue...**var. *coerulea*** [var. *daileyae* Eastw.].
1b. Sepals white or pale blue...**var. *ochroleuca* Hook.**

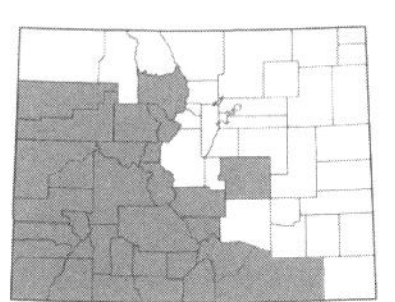

Aquilegia elegantula **Greene**, WESTERN RED COLUMBINE. Plants 1–6 dm; *leaves* glabrous or pilose; *sepals* 7–11 × 4–5 mm, red below and yellow above; *spurs* 16–25 mm long, red, abruptly narrowed near the middle; *petals* yellow, 6–8 × 3–4 mm; *follicles* 13–20 mm long, the beak 13–15 mm long. Common along streams and in moist, shady forests, 6500–11,800 ft. May–July. E/W. (Plate 70)

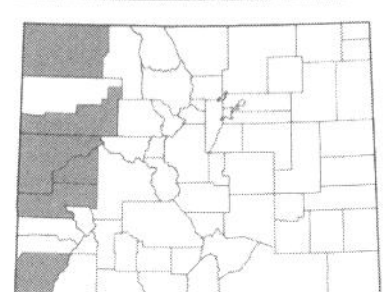

Aquilegia micrantha **Eastw.**, MANCOS COLUMBINE. Plants 3–6 dm; *leaves* viscid and pilose; *sepals* 8–20 × 3–6 mm, white, cream, pink, or blue; *spurs* 15–30 mm long (rarely absent), white or colored similar to the sepals, parallel or divergent, evenly tapered from the base; *petals* white or cream, 6–10 × 3–7 mm; *follicles* 10–20 mm long, the beak 8–10 mm long. Uncommon on moist cliffs and hanging gardens, 4200–7000 ft. May–July. W.

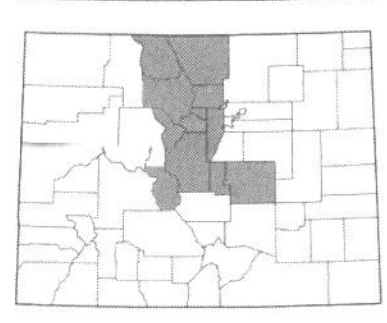

Aquilegia saximontana **Rydb.**, ROCKY MOUNTAIN BLUE COLUMBINE. Plants 0.5–2.5 dm; *leaves* glabrous; *sepals* 9–18 × 5–8 mm, blue; *spurs* 3–9 mm long, blue, hooked, evenly tapered from the base; *petals* yellow, 7–10 × 3–5 mm; *follicles* 7–10 mm long, the beak 3–5 mm long. Uncommon on rocky hillsides and alpine scree slopes, 9000–13,500 ft. July–Aug. E/W. Endemic.

CALTHA L. – MARSH MARIGOLD

Perennial herbs; *leaves* basal and cauline, alternate, simple; *flowers* perfect, actinomorphic, solitary or in 2-6-flowered cymes; *sepals* 5–12, petaloid; *petals* absent; *stamens* rarely petaloid; *fruit* a follicle.

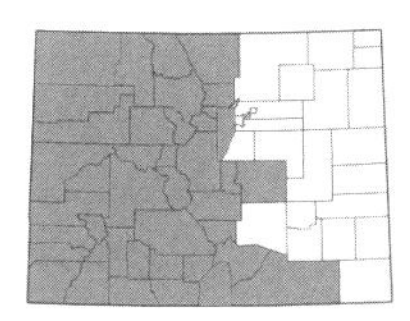

Caltha leptosepala **DC.**, MARSH MARIGOLD. Plants 0.5–5 dm; *leaves* oblong-ovate, 1.5–9 × 2–5 cm, crenate to nearly entire; *sepals* white, drying brown, tinged bluish below, 2–30 mm long; *follicles* 12–20 mm long. Common in moist meadows, along streams, in marshes, and near seepages, often found growing with *Trollius*, 8000–13,500 ft. June–Aug. E/W. (Plate 71)

CLEMATIS L. – CLEMATIS

Perennial herbs or vines; *leaves* opposite, ternately to pinnately compound; *flowers* perfect or imperfect, actinomorphic, solitary or in cymes; *sepals* 4–5, petaloid or greenish; *petals* absent; *fruit* achenes with persistent styles.

1a. Sepals yellow, 0.8–2 cm long...***C. orientalis***
1b. Sepals white, cream, or violet-blue, 0.6–6 cm long...2

2a. Flowers imperfect, numerous and crowded in axillary compound cymes; sepals 0.6–1.5 cm long, white to cream; clambering or climbing vines...***C. ligusticifolia***
2b. Flowers perfect, solitary, axillary or terminal; sepals 1.5–6 cm long, violet-blue, pale blue, or rarely white; vines or erect, herbaceous perennials...3

3a. Leaves once ternately compound, the margins entire or shallowly toothed; flattened, petaloid staminodes present between the stamens and sepals...***C. grosseserrata***
3b. Leaves 2–3 times ternately compound or 2–3 times pinnately compound, the margins conspicuously toothed or entire; petaloid staminodes present or not...4

4a. Leaves 2–3 times ternately compound, the leaflets with serrate margins; flattened, petaloid staminodes present between the stamens and sepals; trailing or climbing vines or sometimes plants tufted and not viny; sepals pale to medium violet-blue or rarely white...***C. columbiana***
4b. Leaves 2–3 times pinnately compound, the leaflets with entire margins; staminodes absent; erect or sprawling herbaceous perennials; sepals dark violet-blue or rarely white...5

5a. Leaflets linear to narrowly lanceolate, mostly 0.5–6 mm wide; flowers urn-shaped or narrowly cylindric...***C. hirsutissima***
5b. Leaflets narrowly to broadly ovate or elliptic, mostly 7–17 mm wide; flowers broadly cylindric...***C. scottii***

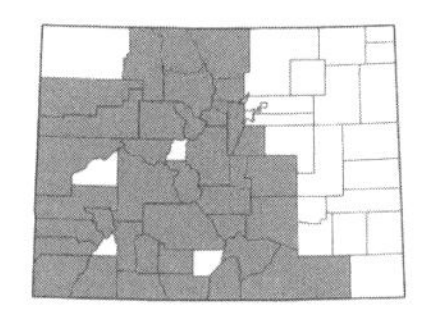

***Clematis columbiana* (Nutt.) Torr. & A. Gray**, ROCK CLEMATIS. [*Atragene columbiana* Nutt.; *Clematis pseudoalpina* (Kuntze) A. Nelson]. Plants viney; *leaves* 2–3 times ternately compound, the leaflets serrate; *inflorescence* terminal on short shoots; *sepals* 1.5–5 cm long, violet-blue or rarely white; *staminodes* present between the stamens and sepals; *achene* beak 2+ cm long. Common in forests, on rocky slopes, and in canyons, 6000–10,500 ft. May–June. E/W.

There are two varieties of *C. columbiana* in Colorado:
1a. Ultimate leaf segments mostly over 4 mm wide...**var. *columbiana***. Found in forests, on rocky slopes, and in canyons and gulches, widespread, 6000–10,000 ft. May–June. E/W.
1b. Ultimate leaf segments mostly less than 3 mm wide...**var. *tenuiloba* (A. Gray) Pringle**. Found in open forests, on rocky slopes, and in canyons, known from the central counties south of Denver, 7000–10,500 ft. May–June. E/W.

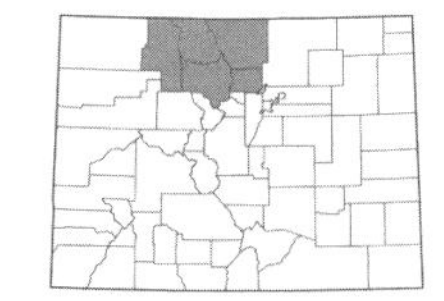

***Clematis grosseserrata* (Rydb.) Ackerfield, comb. nov.** (see p. ix), WESTERN BLUE VIRGIN'S-BOWER. [*Atragene grosseserrata* Rydb.; *C. occidentalis* (Hornem.) DC. var. *grosseserrata* (Rydb.) J. Pringle]. Plants viney; *leaves* once ternately compound, the segments entire or shallowly toothed; *inflorescence* terminal on short shoots; *sepals* 3–6 cm long, violet-blue to blue or rarely white; *staminodes* present between the stamens and sepals. Found in forests, canyons, and on rocky slopes, 6000–10,000 ft. May–July. E/W. (Plate 71)

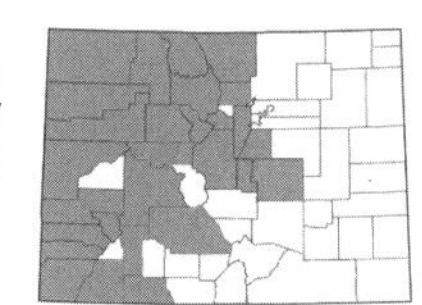

***Clematis hirsutissima* Pursh**, SUGAR BOWLS. [*Coriflora hirsutissima* (Pursh) W.A. Weber]. Plants erect, 1.5–6 dm; *leaves* 2–3 times pinnately compound, the leaflets linear to narrowly lanceolate, mostly 0.5–6 mm wide; *inflorescence* solitary, the flowers urn-shaped or narrowly cylindric; *sepals* 2.5–4.5 cm long, violet-blue or rarely white; *staminodes* absent; *achene* beak 4–9 cm long. Common on grassy slopes and in meadows and woodlands, 5000–10,000 ft. April–June. E/W.

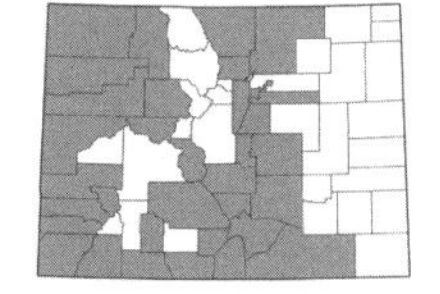

***Clematis ligusticifolia* Nutt.**, WESTERN WHITE VIRGIN'S-BOWER. Plants viney; *leaves* pinnately compound with mostly 5–7 leaflets or sometimes bipinnately compound, the leaflets entire to toothed; *inflorescence* of numerous flowers crowded in axillary compound cymes, the flowers imperfect; *sepals* 0.6–1.5 cm long, white to cream; *achene* beak 3–4 cm long. Common vine in thickets and woods, gulches, river bottoms, and along roadsides, 4500–8500 ft. June–Aug. E/W. (Plate 71)

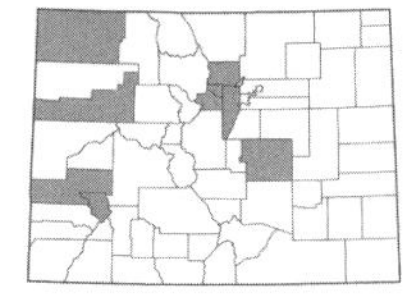

***Clematis orientalis* L.**, ORIENTAL VIRGIN'S-BOWER. [*Viticella orientalis* (L.) W.A. Weber]. Plants viney; *leaves* pinnately compound or sometimes ternately compound, the leaflets usually 2–3-lobed, otherwise entire to few-toothed; *inflorescence* solitary or axillary cymes; *sepals* 0.8–2 cm long, yellow; *achene* beak 2–5 cm long. Found as a weed on fences and along roadsides, 5900–9000 ft. July–Sept. E/W. Introduced.

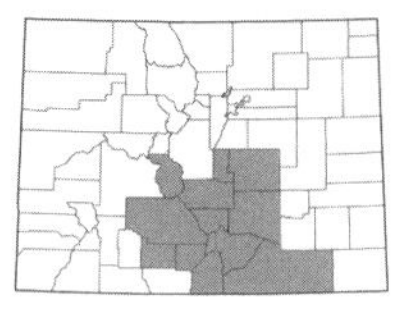

***Clematis scottii* Porter**, SCOTT'S CLEMATIS. [*C. hirsutissima* Pursh var. *scottii* (Porter) Erickson; *Coriflora scottii* (Porter) W.A. Weber]. Plants erect, 1.5–6 dm; *leaves* 2–3 times pinnately compound, the leaflets narrowly to broadly ovate or elliptic, mostly 7–17 mm wide; *inflorescence* solitary, the flowers broadly cylindric; *sepals* 2.5–4.5 cm long, violet-blue; *staminodes* absent; *achene* beak 4–9 cm long. Found on rocky slopes, in oak thickets, and in pine forests, 5500–10,000 ft. May–July. E.

DELPHINIUM L. – LARKSPUR

Annual or perennial herbs; *leaves* basal and/or cauline, alternate, palmately lobed; *flowers* perfect, zygomorphic, in terminal racemes or panicles; *sepals* 5, petaloid, the upper sepal spurred; *petals* 4, the upper petals hidden in the upper sepal and spurred, the lower petals clawed; *fruit* a follicle. (Ewan, 1945)

1a. Sepals white or cream-white, sometimes tinged with blue, with a distinctive green gland on upper part of the inside of the sepals; stems often glandular-hairy up to the inflorescence, sometimes the stems lacking glandular hairs altogether...***D. carolinianum* ssp. *virescens***
1b. Sepals blue, blue-purple, or purple, sometimes paler below but not white throughout; stems various but if glandular-hairy then also densely so in the inflorescence...2

2a. Plants from a small cluster of thickened, fleshy, tuberlike roots or a single, small, tuberlike root; stems blue-purple below; plants 10–50 cm; leaves few, the ultimate leaf divisions narrowly oblong to linear; follicles flaring outward from the base...***D. nuttallianum***
2b. Plants from stout or sometimes slender woody-fibrous roots; stems sometimes purplish or reddish below; plants 10–200 cm; leaves various; follicles erect or slightly divergent...3

3a. Inflorescence glandular-hairy...4
3b. Inflorescence glabrous to minutely hairy but not glandular-hairy...6

4a. Plants with short racemes (mostly 2–12 cm long) sometimes barely longer than the leaves; plants 0.5–3 (4) dm; found above 10,500 ft in elevation...***D. alpestre***
4b. Plants with an elongated raceme that much exceeds the leaves; plants 2.5–20 dm; found from 7400–12,700 ft in elevation...5

5a. Stems minutely hairy below the inflorescence; sepals 13–20 mm long, lanceolate to narrowly ovate and acuminate or acute at the tips; flowers dark violet-blue; racemes less than 3 times longer than wide (usually less than 10 cm long)...***D. barbeyi***
5b. Stems glabrous below the inflorescence; sepals 6–12 mm long, oblong to spatulate and blunt, rounded, or acute at the tips; flowers usually bicolored, dark violet-blue and paler white, rarely all dark violet-blue; racemes more than 3 times longer than wide (usually over 10 cm long)...***D. occidentale***

6a. Leaves all basal, or sometimes with a few very reduced stem leaves; stem and rachis of the inflorescence glabrous...***D. scaposum***
6b. Stem leaves present, sometimes also with a basal cluster; stem and rachis minutely hairy or the stem glabrous below the middle...7

7a. Leaves mostly basal, the stem leaves distinctly reduced in comparison to the basal leaves; stems densely minutely hairy throughout; plants found below 7500 ft in elevation...***D. geyeri***
7b. Basal leaves lacking (only stem leaves present), or with both stem and basal leaves but the stem leaves not greatly reduced in comparison to the basal leaves; stems minutely hairy or glabrous below the middle; plants usually found above 7000 ft in elevation...8

8a. Plants 5–30(-40) cm; racemes 2–8-flowered, short (mostly 2–12 cm long) and sometimes barely exceeding the leaves; follicles 7–12 mm long; seeds unwinged...***D. alpestre***
8b. Plants 50–200 cm; racemes (10) 15–90 (120)-flowered, elongate and much exceeding the leaves (usually over 15 cm long); follicles 10–18 mm long; seeds wing-margined (sometimes narrowly so)...***D. ramosum***

***Delphinium alpestre* Rydb.**, ALPINE LARKSPUR. Perennials, 0.5–3 (4) dm; *stems* minutely hairy to glandular above; *leaves* cauline, palmately divided, the segments again lobed, 2–3 cm wide; *sepals* 11–15 mm long, dark blue and brownish dorsally; *spurs* 7–9 mm long. Found in alpine and high meadows, 10,500–13,500 ft. July–Aug. E/W.

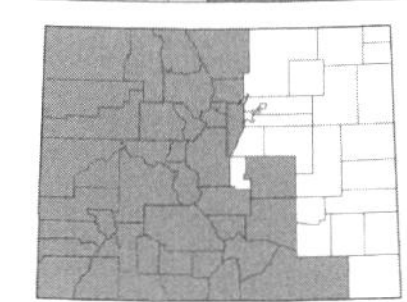

***Delphinium barbeyi* (Huth) Huth**, SUBALPINE LARKSPUR. Perennials, 5–20 dm; *stems* glandular-hirsute, densely so above; *leaves* cauline, 3-lobed, the segments cleft or coarsely toothed; *sepals* 13–20 mm long, dark purple to blue-purple; *spurs* ca. 10 mm; *follicles* 14–17 mm long. Common throughout the mountains in moist meadows and adjacent spruce-fir forests, along streams, and in moist alpine, 8000–12,700 ft. July–Sept. E/W. (Plate 71)

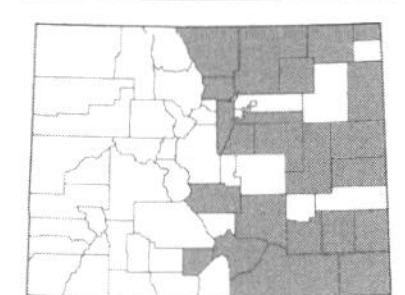

***Delphinium carolinianum* Walter ssp. *virescens* (Nutt.) R. Brooks**, PLAINS LARKSPUR. [*D. virescens* Nutt.]. Perennials, 3–8 dm; *stems* with short, curled hairs and often glandular above; *leaves* mostly basal, palmately divided, the segments again cleft into linear lobes; *sepals* 10–12 mm long, white or cream-white, sometimes tinged with blue; *spurs* 11–13 mm long; *follicles* 15–20 mm long. Common in dry grasslands and on open prairies, 4000–6800 ft. May–July. E.

***Delphinium geyeri* Greene**, GEYER'S LARKSPUR. Perennials, 2–7 dm; *stems* minutely hairy; *leaves* mostly basal, dissected with the ultimate segments linear to linear-filiform; *sepals* 10–15 mm long, blue-purple to blue; *spurs* 10–20 mm long; *follicles* 12–15 mm long. Common on dry, rocky or sandy hillsides and sagebrush slopes, 5000–7500 ft. June–July. E/W.

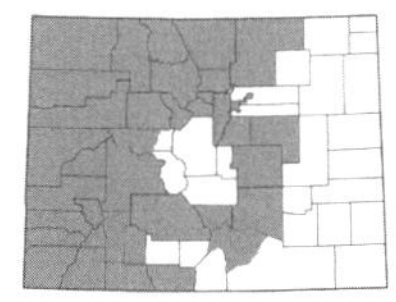

***Delphinium nuttallianum* Pritz.**, NUTTALL'S LARKSPUR. [*D. nelsonii* Greene]. Perennials, 1–5 dm; *stems* glabrous to minutely hairy; *leaves* cauline, 3–5 cm wide, deeply palmately cleft, each segment again cleft into linear or narrowly oblong lobes; *sepals* 11–17 mm long, blue-purple to blue; *spurs* 10–15 mm long; *follicles* 13–20 mm long. Common in meadows, open forests, oak thickets, and on sagebrush slopes, 5500–10,800 ft. April–July. E/W.

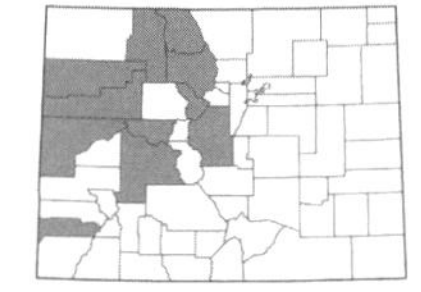

***Delphinium occidentale* (S. Watson) S. Watson**, WESTERN LARKSPUR. Perennials, 6–20 dm; *stems* glandular-hairy in the inflorescence, otherwise glabrous; *leaves* cauline, cleft into 3–7 divisions; *sepals* 6–12 mm long, blue-purple and paler white dorsally, rarely all dark violet-blue; *spurs* 9–12 mm long; *follicles* 9–12 mm long. Common in meadows, aspen and pine forests, and along streams, 7400–10,000 ft. July–Aug. E/W.

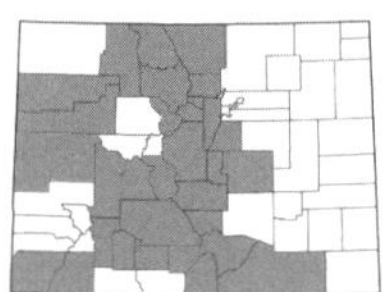

***Delphinium ramosum* Rydb.**, MOUNTAIN LARKSPUR. [*D. robustum* Rydb.]. Perennials, 2.5–20 dm; *stems* glabrous to thinly hairy; *leaves* cauline, cleft into 5–7 segments, the ultimate segments oblong to lanceolate and toothed; *sepals* 8–15 mm long, blue; *spurs* 9–15 mm long; *follicles* 10–18 mm long. Common in meadows, aspen forests, and along streams, 7000–11,000 ft. July–Sept. E/W.

Delphinium robustum supposedly has a glabrous stem and is 10–20 dm tall while *D. ramosum* has a minutely hairy stem and is 5–10 dm tall. A small form of *D. ramosum* (forma *sidalceoides* Ewan) is only about 2.5 dm tall.

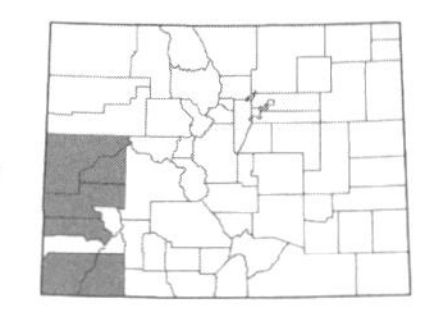

***Delphinium scaposum* Greene**, SCAPOSE LARKSPUR. Perennials, 2–5 dm, scapose or nearly so, glabrous; *leaves* 3-5-lobed, each segment lobed or toothed; *sepals* 10–15 mm long, blue and often with a bronze-colored spur; *follicles* 10–20 mm long. Found in desert scrub, pinyon-juniper communites, and on open, rocky or shale slopes, 4500–6500 ft. May–June. W.

MYOSURUS L. – MOUSETAIL

Annual herbs; *leaves* basal, linear to narrowly oblanceolate, the margins entire or toothed; *flowers* perfect, actinomorphic, solitary, the receptacle elongating in flower and fruit; *sepals* usually 5, herbaceous, spurred at the base; *petals* absent or 5–7, linear to narrowly spatulate, long-clawed, white; *fruit* an achene.

1a. Achenes with a cup-like border, the outer face of the achene orbiculate or sometimes square…***M. cupulatus***
1b. Achenes lacking a cup-like border, the outer face narrowly rhombic to elliptic or orbiculate…2

2a. Outer face of the achenes orbiculate to rhombic; petal claw 3 times as long as the blade; plants 0.5–2 cm; leaves 0.6–1.2 cm long…***M. nitidus***
2b. Outer face of the achenes narrowly rhombic to elliptic or oblong; petal claw 1–2 times as long as the blade; plants 1.5–16 cm; leaves 0.9–11 cm long…3

3a. Beak of the achenes spreading and divergent from the outer face of the achenes…***M. apetalus* var. *montanus***
3b. Beak of the achenes flat and parallel or nearly so to the outer face of the achenes…***M. minimus***

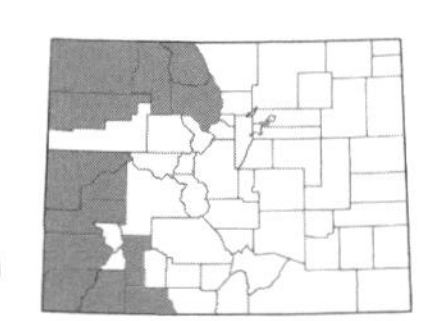

***Myosurus apetalus* Gay var. *montanus* (G.R. Campbell) Whittem.**, BRISTLY MOUSETAIL. Plants 1.5–6 (12) cm; *leaves* linear to narrowly oblanceolate, 0.9–4 cm long; *sepals* 1–3 mm long, greenish or purple-tinged, with a spur 1–3 mm long; *petals* 5 or absent, 1–1.5 mm long, white; *receptacle* mostly 0.4–2 cm long; *achenes* with a narrowly rhombic to elliptic outer face, with a divergent beak 0.5–1.4 mm long. Found along the margins of drying ponds, in dry ephemeral pools, and in moist meadows, 5400–10,500 ft. April–July. E/W.

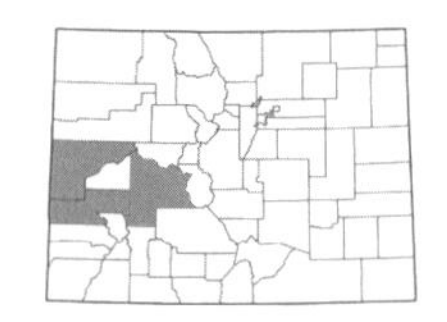

***Myosurus cupulatus* S. Watson**, ARIZONA MOUSETAIL. Plants 3–15 cm; *leaves* linear to narrowly oblanceolate, 1.8–10 cm long; *sepals* 1.5–3 mm long, white to greenish or brown, with a spur 1–2 mm long; *petals* 5, 2–2.5 mm long; *receptacle* 0.7–4 cm long; *achenes* with a cup-like border, the outer face orbiculate or sometimes square, with a weakly divergent beak 0.6–1.2 mm long. Uncommon on dry or saline hillsides, 5300–7500 ft. Mar.–May. W.

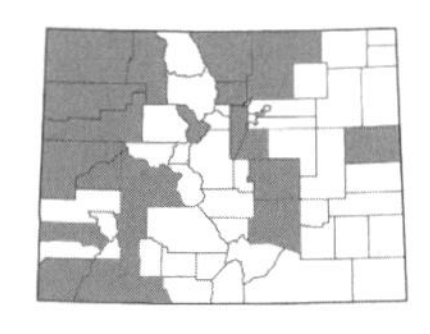

***Myosurus minimus* L.**, TINY MOUSETAIL. Plants 4–16 cm; *leaves* linear to narrowly oblanceolate, 2–11 cm long; *sepals* 2–3 mm long, white to greenish, with a spur 1–3 mm long; *petals* 5; *receptacle* greatly elongating in fruit; *achenes* with a narrowly rhombic to elliptic outer face, with a beak 0.05–0.4 mm long, parallel to the outer face of the achenes (appearing smooth). Found in alkaline meadows, along the margins of ponds, drying puddles, in moist fields and lawns, and near seepages and springs, 4000–9500 ft. April–June. E/W.

***Myosurus nitidus* Eastw.**, WESTERN MOUSETAIL. Plants 0.5–2 cm; *leaves* linear to narrowly oblanceolate, 0.6–1.2 cm long; *sepals* white or greenish, with a short spur 1–2 mm long; *petals* 5; receptacle elongating in fruit; *achenes* with an orbiculate to rhombic outer face, the beak 0.8–1 mm long, weakly divergent. Rare on sagebrush flats and pinyon-juniper slopes, 6500–7000 ft. Mar.–May. W.

RANUNCULUS L. – BUTTERCUP; CROWFOOT

Annual or perennial herbs, sometimes aquatic; *leaves* alternate or basal; *flowers* perfect, actinomorphic, solitary or in cymes; *sepals* 3–5, petaloid or herbaceous; *petals* 5–22 or absent, usually yellow; *fruit* an achene.

1a. Flowers white; plants true aquatics, submerged in water; leaves finely dissected into numerous filiform segments less than 1 mm wide; achenes cross-corrugated and hairy...***R. aquatilis* var. *diffusus***
1b. Flowers yellow; plants aquatic or terrestrial; leaves and achenes unlike the above...2

2a. Tomentose annuals; achenes 3-lobed, the middle lobe lanceolate with an acute tip and the lateral lobes much smaller and empty; leaves all basal, deeply 3-parted with the lateral lobes again dissected into linear segments; petals as long or shorter than the sepals...***R. testiculatus***
2b. Annuals or perennials but not tomentose; achenes not 3-lobed; leaves and petals various...3

3a. Lower surface of the sepals densely covered with brown hairs; leaves simple and entire with a 3-toothed apex; flowers usually solitary, the petals 10–20 mm long...***R. macauleyi***
3b. Lower surface of the sepals glabrous or if hairy then the hairs not brown; leaves unlike the above; flowers various...4

4a. Leaves all simple, the margins entire, crenate, or shallowly dentate...5
4b. At least some leaves compound, dissected, lobed, or divided at least to the middle...8

5a. Leaves ovate to cordate with crenate margins; achenes in a cylindrical cluster, the outer surface longitudinally striate with 3 or more nerves on each side; plants stoloniferous...***R. cymbalaria***
5b. Leaves variously shaped but with entire or shallowly dentate margins; achenes in a globose or subglobose cluster, the outer surface not longitudinally striate; plants stoloniferous or not...6

6a. Plants stoloniferous or creeping and rooting at the nodes; leaves linear to narrowly elliptic, 1–8 mm wide; sepals 2–5 mm long...***R. flammula* var. *ovalis***
6b. Plants not stoloniferous or rooting at the nodes; leaves lanceolate to elliptic or ovate, 5–35 mm wide; sepals 4.5–7.5 mm long...7

7a. Achenes glabrous, 15–45 in a subglobose cluster 6.5–8.5 mm wide; leaves lanceolate to elliptic...***R. alismifolius* var. *montanus***
7b. Achenes hairy, 50–180 in a globose cluster 8–11 mm wide; leaves orbicular to ovate, elliptic, or oblanceolate...***R. glaberrimus* var. *ellipticus***

8a. Plants of the alpine, 0.6–3.5 cm tall (sometimes to 10 cm in fruit); petals 1.2–4 mm long; leaves 0.4–1 cm long, 3-parted or 3-foliate with rounded lobes; achenes 1–1.2 mm long...***R. pygmaeus***
8b. Plants either not of the alpine or if so then unlike the above in all respects...9

9a. Most basal leaves compound with stalked leaflets (the leaves clearly compound and the leaflets stalked with petiolules)...10
9b. Basal leaves all simple, sometimes deeply cleft to near the base but not compound with distinct, stalked leaflets...14

10a. Stems and petioles glabrous; achenes longitudinally striate; petals 1–3 mm wide and 3–6 mm long, or rarely absent...***R. ranunculinus***
10b. Stems and petioles with long, spreading hairs or hairy at least on the stem below the flower, rarely glabrous and then the petals 7–11 mm long and the achenes strongly flattened and not longitudinally striate; petals 1–10 mm wide and 2–18 mm long...11

11a. Achenes strongly flattened with an abruptly recurved beak; ultimate leaf segments linear to linear-lanceolate, entire or with a small lobe on the margin...***R. acriformis***
11b. Achenes not strongly flattened, thick-lenticular or compressed-globose, with a straight or curved beak; ultimate leaf segments usually wider, with toothed or lobed margins...12

12a. Petals 6–18 mm long and 5–12 mm wide, about twice as long as the sepals, sometimes very numerous; achene beak curved...***R. repens***
12b. Petals 2–6 mm long and 1–5 mm wide, about equal to or only slightly longer than the sepals, never more than 5; achene beak straight or curved...13

13a. Petals 2–4 mm long and 1–2.5 mm wide, shorter than the sepals; achenes with a beak 0.5–0.8 mm long, in a cylindric cluster 5–7 mm wide; plants erect, not rooting at the nodes...***R. pensylvanicus***
13b. Petals 4–6 mm long and 3.5–5 mm wide, barely shorter or slightly longer than the sepals; achenes with a beak 0.8–1.5 mm long, in a subglobose to ovoid cluster 7–15 mm wide; plants prostrate to erect, sometimes rooting at the nodes...***R. macounii***

14a. Plants aquatic and submerged, or semiaquatic and creeping and rooting at the nodes...15
14b. Plants terrestrial or if semiaquatic then not creeping and rooting at the nodes...16

15a. Leaves 3–5 times ternately lobed, the lobes entire and rounded; receptacle glabrous; achenes with a beak 0.1–0.3 mm long; petals 2–4.5 mm long...***R. hyperboreus***
15b. Leaves 3–5 times palmately divided, the lobes again 3–5-lobed or forked, rarely dissected into linear segments; receptacle hairy; achenes with a beak 0.6–0.8 mm long; petals 4–8 mm long...***R. gmelinii***

16a. Basal leaves simple and orbicular to elliptic or oblanceolate with entire margins, or sometimes shallowly to deeply 3-lobed at the apex; stem leaves simple and similar to the basal leaves or deeply 3-parted with rounded lobes; achenes hairy, in a globose cluster 8–11 mm wide...***R. glaberrimus* var. *ellipticus***
16b. Basal leaves all divided or cleft to the middle or below, or if simple and oval and orbiculate then the margins crenate and the stem leaves deeply 3–7-parted; achenes various...17

17a. Basal leaves undivided or shallowly 3-lobed at the apex, oval to orbicular, with crenate margins (sometimes the innermost basal leaves 3–5-lobed below the middle); stem leaves deeply 3–7-parted into linear, lanceolate, or narrowly elliptic segments...18
17b. Basal and stem leaves unlike the above...20

18a. Petals 1.5–3.5 mm long; sepals glabrous below; achenes 1.4–1.6 mm long with a short beak 0.1–0.2 mm long...***R. abortivus***
18b. Petals (3) 5–15 mm long; sepals hairy below; achenes 1.6–3 mm long with a beak 0.2–1 mm long...19

19a. Basal leaves with cordate to broadly obtuse bases, usually densely hairy; petals 6–15 mm long or rarely absent; sepals 5–8 mm long; stems usually with long, spreading hairs, rarely glabrous...***R. cardiophyllus***
19b. Basal leaves with acute to rounded bases, glabrous or sometimes hairy; petals 5–8 mm long; sepals 2.5–5 mm long; stems usually glabrous or sometimes with long, spreading hairs...***R. inamoenus* var. *inamoenus***

20a. Basal and stem leaves 3-parted with each division again deeply dissected, the ultimate leaf segments filiform or narrowly linear (to 2 mm wide); petals 5–10, these 8–15 mm long; plants glabrous...***R. adoneus***
20b. Basal and stem leaves unlike the above, with wider ultimate leaf segments; petals 5, of various lengths; plants glabrous or hairy...21

21a. Achenes 0.8–1.3 mm long with a short beak to 0.1 mm long; style essentially absent; basal leaves 3-parted with the main lobes again lobed, the lobes rounded; stem leaves petiolate, similar to the basal leaves but reduced above to deeply, narrowly lobed leaves; bracts sessile, simple and entire; petals 0.5–5 mm long; stems 2–15 mm thick, hollow; plants glabrous, found growing in shallow water or in moist places...***R. sceleratus* var. *multifidus***
21b. Achenes 1–3.5 mm long with a beak 0.2–2.5 mm; style present; plants otherwise unlike the above...22

22a. Achenes strongly compressed and flattened, 2–3.5 mm long; stem leaves petiolate; nectary scale at the base of the petal free from the petal for ½ its length or more...23
22b. Achenes not strongly flattened, thick-lenticular or compressed-globose, 1–2.5 mm long; stem leaves sessile or subsessile, rarely with one stem leaf petiolate; nectary scale at the base of the petal adnate to the petal its entire length or sometimes free just at the tip...25

23a. Petals 1–4 (6) mm long, shorter than or barely longer than the sepals; achene beak 1.2–2.5 mm long, strongly curved...***R. uncinatus***
23b. Petals 7–11 mm long, conspicuously longer than the sepals; achene beak 0.2–1.2 mm long, curved or straight...24

24a. Achene beak 1–1.5 mm long, strongly curved; ultimate leaf segments linear to linear-lanceolate, entire or with a small lobe on the margin...***R. acriformis***
24b. Achene beak 0.2–1 mm long, straight or curved; ultimate leaf segments lanceolate to narrowly elliptic with toothed or lobulate margins...***R. acris***

25a. Sepals glabrous; basal leaves deeply divided into 3 oblanceolate segments, these sometimes again lobed; roots tuberous and thick, clustered; plants found from 6000–8000 ft in elevation...***R. jovis***
25b. Sepals hairy; basal leaves unlike the above; roots not tuberous and thick; plants usually found above 8000 ft in elevation...26

26a. Petals 3.5–6 mm long; stems abruptly curved near the middle such that the stem is horizontally spreading; plants 0.2–1 dm; leaves 0.5–1 (2) cm long...***R. gelidus***
26b. Petals 6–10 mm long; stems straight or curved above the middle but not horizontally spreading; plants 0.4–2.5 dm; leaves 0.8–5 cm long...27

27a. Pedicels glabrous; flowers usually solitary, occasionally to 3 per stem...***R. eschscholtzii* var. *eschscholtzii***
27b. Pedicels pilose, densely so directly under the flower; flowers solitary or 2–7 per stem...***R. pedatifidus* var. *affinis***

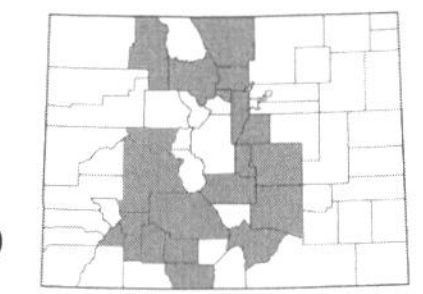

***Ranunculus abortivus* L.**, LITTLELEAF BUTTERCUP. Perennials, 1–6 dm; *stems* glabrous; *basal leaves* undivided or sometimes the innermost 3-parted, reniform to orbiculate, 1.4–4.2 × 2–5.2 cm, crenate; *cauline leaves* 3–5-cleft into linear-lanceolate, entire segments; *sepals* 2.5–4 mm long, glabrous; *petals* 5, yellow, 1.5–3.5 mm long; *achenes* 1.4–1.6 × 1–1.5 mm, with a curved beak 0.1–0.2 mm long, glabrous. Uncommon along streams, near seeps, and in shady forests, 6400–10,700 ft. May–July. E/W.

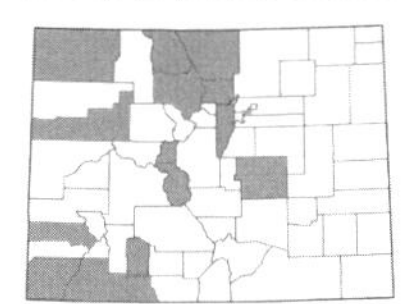

***Ranunculus acriformis* A. Gray**, SHARP BUTTERCUP. Perennials, 2–6 dm; *stems* spreading-hirsute; *basal leaves* 3-parted, each segment further deeply divided into linear-lanceolate segments, 3–5 × 1–5 cm; *cauline leaves* similar to basal leaves; *sepals* 4–5 mm long, pilose dorsally; *petals* 5, yellow, 7–11 mm long; *achenes* 3–3.5 mm long with abruptly curved beaks 1–1.5 mm long, glabrous. Found in moist meadows and along streams, 7500–9000 ft. June–Aug. E/W.

***Ranunculus acris* L.**, TALL BUTTERCUP. Perennials, 2–7 (9) dm; *stems* with spreading hairs; *basal leaves* 3-parted, the 3 main segments lobed or toothed; *cauline leaves* similar to basal; *sepals* 4–5.5 mm long, dorsally pilose; *petals* 5, yellow, 8–10 mm long; *achenes* 2.5–3 mm long, the beaks straight or curved, 0.2–1 mm long, glabrous. Uncommon in disturbed, moist areas, 5900–9800 ft. June–Aug. E/W. Introduced.

***Ranunculus adoneus* A. Gray**, ALPINE BUTTERCUP. Perennials, 0.5–3 dm; *stems* glabrous; *basal leaves* several times divided into linear segments, 1–5 × 1–5 cm; *cauline leaves* similar to basal; *sepals* 3.5–10 mm long, sparsely hairy; *petals* 5–10, yellow, 8–15 mm long; *achenes* 2.5–4 mm long with straight or curved beaks 1.4–2 mm long, glabrous. Common in the alpine, usually along the margins of melting snowbanks, 10,000–14,000 ft. June–Aug. E/W.

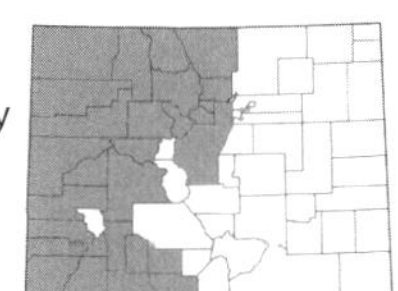

***Ranunculus alismifolius* Geyer ex Benth. var. *montanus* S. Watson**, WATERPLANTAIN BUTTERCUP. Perennials, 0.5–4.5 dm; *stems* glabrous; *basal leaves* simple, lanceolate to elliptic, entire or minutely toothed, 2–10 × 0.5–3.5 cm; *cauline leaves* lanceolate; *sepals* 4.5–7.5 mm, glabrous; *petals* 3–10, yellow, 4–14 mm; *achenes* 1.5–3.2 mm with a beak 0.4–1.2 mm long, glabrous. Common in moist meadows and along streams and melting snowbanks, 8500–13,000 ft. May–Aug. E/W. (Plate 71)

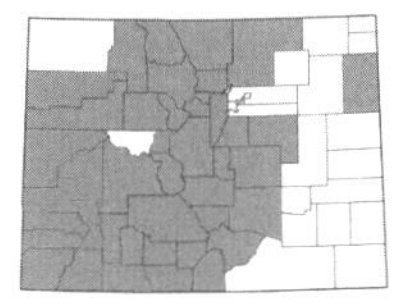

***Ranunculus aquatilis* L. var. *diffusus* With.**, WATER CROWFOOT. [*R. longirostris* Godron; *R. subrigidus* W.B. Drew; *R. trichophyllus* Chaix var. *hispidulus* (E. Drew) W. Drew]. Perennials, aquatic, 2–8 dm; *stems* glabrous; *basal leaves* absent; *cauline leaves* dissected into numerous filiform segments, 1–3 cm long; *sepals* 3–5 mm long, glabrous; *petals* 5, white; *achenes* 1.5–2 mm long with a beak 0.2–0.4 mm long or absent. Found in ponds, streams, and creeks, 3500–10,500 ft. June–Aug. E/W.

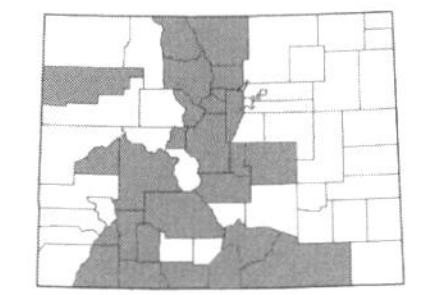

***Ranunculus cardiophyllus* Hook.**, SHOWY BUTTERCUP. Perennials, 1.3–5 dm; *stems* with spreading hairs, rarely glabrous; *basal leaves* simple, oval to orbicular, crenate or lobed, 2–3.5 (6) × 1.8–3.5 cm; *cauline leaves* 3–7-lobed into lanceolate segments; *sepals* 5–8 mm, dorsally villous; *petals* 5 or lacking, yellow, 6–15 mm, rarely absent; *achenes* 1.5–3 mm long with a beak 0.6–1.2 mm long, hairy. Common in meadows, along streams, and in the alpine, 5500–12,000 ft. May–Aug. E/W.

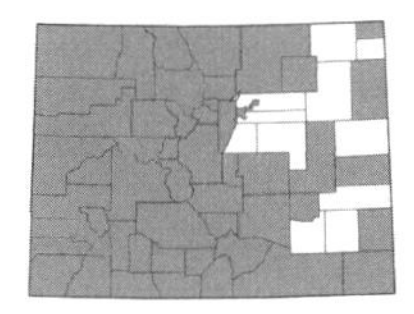

***Ranunculus cymbalaria* Pursh**, MARSH BUTTERCUP. [*Halerpestes cymbalaria* (Pursh) Greene]. Perennials, scapose, stoloniferous; *basal leaves* simple, ovate to cordate, crenate, 0.5–4 cm × 0.5–4 cm; *cauline leaves* absent; *sepals* 2.5–7.5 mm long, glabrous; *petals* 5+, yellow, 2.5–8 mm long; *achenes* 1–1.5 mm long with a straight or incurved beak ca. 0.3 mm long, glabrous. Common along the margins of streams, ponds, and lakes, in seepage or swampy areas, and in moist meadows, 3500–9700 ft. May–July. E/W.

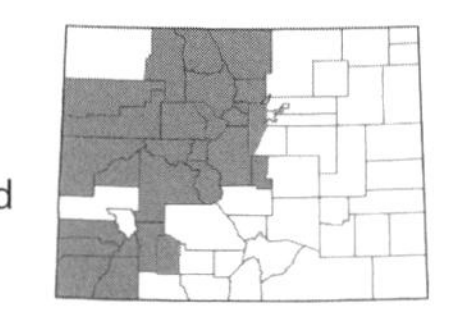

Ranunculus eschscholtzii* Schlect. var. *eschscholtzii, Eschscholtz's buttercup. Perennials, 0.4–2.5 dm; *basal leaves* divided into 3 main lobes, the linear to lanceolate segments lobed or toothed, 1–5 × 0.7–2.5 cm; *cauline leaves* similar to basal; *sepals* 4–6 mm long, dorsally pilose; *petals* 5, yellow, 6–10 mm long; *achenes* 1.3–1.8 mm long with a beak 0.7–1 mm, glabrous or hairy. Found along streams and in moist subalpine forests and the alpine, often found along the margins of melting snowdrifts, 8500–13,000 ft. June–Aug. E/W.

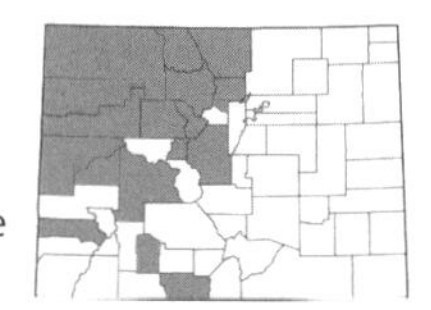

***Ranunculus flammula* L. var. *ovalis* (J.M. Bigelow) L.D. Benson**, creeping spearwort. [*R. reptans* L. var. *ovalis* Torr. & A. Gray]. Perennials, 0.1–1 dm, stoloniferous; *stems* glabrous; *leaves* linear to narrowly elliptic, entire, 0.5–7.5 × 0.1–0.8 cm; *sepals* 2–5 mm long, glabrous or hairy; *petals* 5, yellow, 2–8 mm long; *achenes* 1.2–2 mm long with a beak 0.1–0.7 mm, glabrous. Found along the margins of lakes and ponds and in marshy or seepage areas, 7500–11,000 ft. June–Aug. E/W.

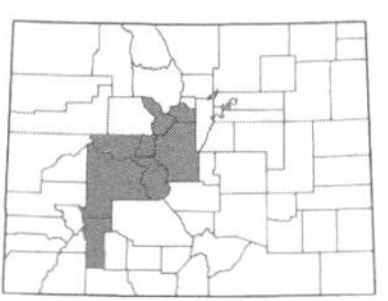

***Ranunculus gelidus* Kar. & Kir.**, tundra buttercup. [*R. karelinii* Czern.]. Perennials, 0.2–1 dm; *stems* glabrous or sparsely hairy; *basal leaves* 3-parted, the 3 main segments lobed or cleft, 0.5–1 (2) cm long; *sepals* 2.5–4.5 mm long, hairy; *petals* 5, yellow, 3.5–6 mm long; *achenes* 2–2.5 mm long with a beak 0.4–0.7 mm long, glabrous. Rare on fell fields and scree or talus slopes of the alpine, 12,500–14,000 ft. July. E/W.

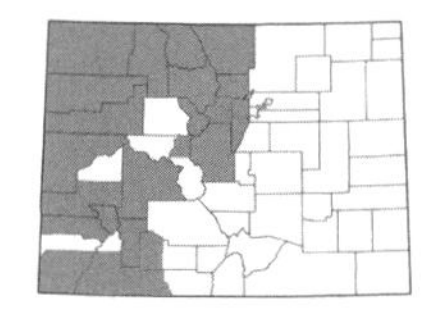

***Ranunculus glaberrimus* Hook. var. *ellipticus* (Greene) Greene**, sagebrush buttercup. Perennials, 0.5–2 dm; *stems* glabrous; *basal leaves* orbicular to ovate or elliptic to oblanceolate, entire or apically 3-lobed, 1–5 × 0.5–2 cm; *cauline leaves* simple or cleft into 2–3 linear-lanceolate lobes; *sepals* 4.5–7 mm long, hairy; *petals* 5, yellow, 6.5–12 mm long; *achenes* 1.3–3.5 mm long with a curved beak 0.5–1 mm long, hairy. Common in moist meadows, on rocky hillsides, and shrubby slopes, 5100–10,000 ft. April–June. E/W. (Plate 71)

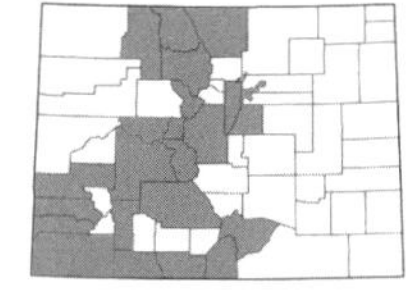

***Ranunculus gmelinii* DC.**, Gmelin's buttercup. Perennials, 2–5 dm, aquatic, rooting at the nodes; *stems* glabrous or hairy; *leaves* 3-parted, each segment again lobed or forked, sometimes dissected into linear segments; *sepals* 2.5–6 mm long, glabrous or hairy; *petals* 5, yellow, 4–8 mm long; *achenes* 1–1.5 mm long with a beak 0.6–0.8 mm long, glabrous. Found in ponds, along streams, and in ditches, 6000–10,500 ft. June–Aug. E/W.

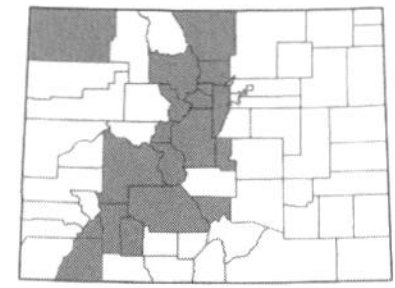

***Ranunculus hyperboreus* R. Br.**, floating buttercup. Perennials, 1–5 dm, aquatic, rooting at the nodes; *stems* glabrous; *leaves* with 3–5 rounded, entire lobes, 0.3–1.5 cm long; *sepals* 2–3 (4.5) mm long, glabrous; *petals* 5, yellow, 2–4.5 mm long; *achenes* 0.6–1.2 mm long with a beak 0.1–0.3 mm long, glabrous. Found in ponds and streams, (4900) 7000–12,000 ft. May–Sept. E/W.

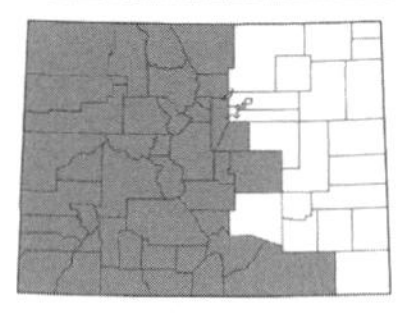

Ranunculus inamoenus* Greene var. *inamoenus, graceful buttercup. Perennials, 0.5–4.5 dm; *stems* glabrous or hairy; *basal leaves* reniform to orbiculate, crenate or 3-lobed, 1–5.5 × 1–4.5 cm; *cauline leaves* 3–5-cleft into linear-lanceolate, entire segments; *sepals* 2.5–5 mm long, dorsally pilose; *petals* 5, yellow, 5–8 mm long; *achenes* 1.5–2.5 mm long with a straight or curved beak 0.2–0.8 mm long. Common in meadows, spruce-fir forests, along streams, and occasionally in the alpine, 6000–12,500 ft. May–Aug. E/W. (Plate 71)

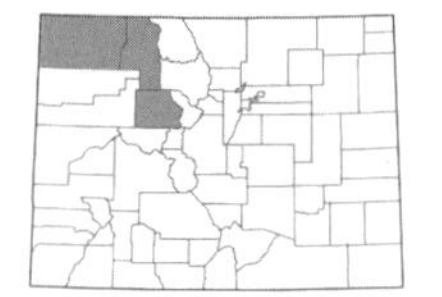

***Ranunculus jovis* A. Nelson**, Utah buttercup. Perennials, 0.2–1 dm; *stems* glabrous; *basal leaves* divided into 2–5 oblanceolate segments 0.8–4 cm long; *cauline leaves* usually 2–3-lobed; *sepals* 2–7.5 mm long, glabrous, reflexed; *petals* 5, yellow, 6–15 mm long; *achenes* 1–2.3 mm long with a beak 0.2–1 mm, glabrous or hairy. Uncommon on sagebrush or rocky slopes, 6000–8000 ft. April–May. W.

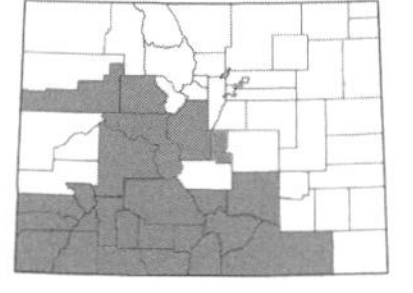

***Ranunculus macauleyi* A. Gray**, Rocky Mountain buttercup. Perennials, 0.5–1.5 dm; *stems* glabrous or pilose; *leaves* lanceolate to narrowly elliptic, undivided, the margins entire with a 3-toothed tip, 1.5–4.5 × 0.5–1 (2.5) cm; *sepals* 6–12 mm long, densely brown to black pilose dorsally; *petals* 5 (8), yellow, 10–20 mm long; *achenes* 1.5–1.7 mm long with a straight or recurved beak 0.5–1.5 (2.2) mm long, glabrous. Found in alpine, often along the edges of melting snowbanks, 10,500–13,500 ft. June–Aug. E/W. (Plate 71)

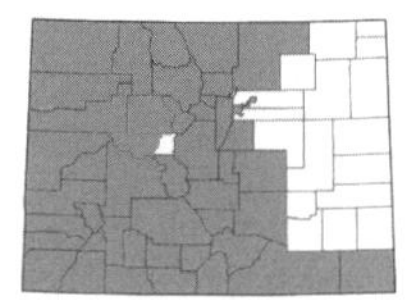

***Ranunculus macounii* Britton**, Macoun's buttercup. Perennials, 2–10 dm; *stems* with long, spreading hairs; *basal leaves* simple or pinnately 3–5-lobed, 3–9 cm long; *cauline leaves* similar to basal; *sepals* 3.5–8.5 mm long, hairy, reflexed; *petals* 5, yellow, 4–6 mm long; *achenes* 2–3 mm long with a beak 0.8–1.5 mm long, glabrous. Common in moist meadows and along streams, often in disturbed areas, sometimes emergent in shallow water, 5000–9800 ft. June–Aug. E/W. (Plate 71)

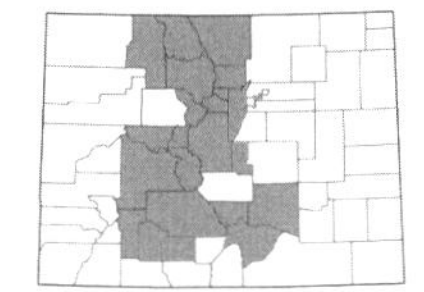

Ranunculus pedatifidus **Sm. var.** ***affinis*** **(R. Br.) L.D. Benson**, NORTHERN BUTTERCUP. Perennials, 0.5–2 dm; *stems* glabrous to minutely hairy; *basal leaves* divided into several linear to narrowly oblanceolate lobes, 0.8–1.2 (3) cm long; *sepals* 4–6 mm long, with spreading hairs; *petals* 5, yellow, 7–10 mm long; *achenes* ca. 2 mm long with a beak 0.6–1 mm long, glabrous. Found in meadows and alpine, 7600–13,000 ft. June–Aug. E/W.

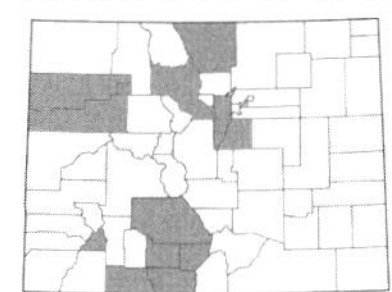

Ranunculus pensylvanicus **L. f.**, PENNSYLVANIA BUTTERCUP. Perennials, 2–10 dm; *stems* with spreading hairs; *leaves* ternately compound, 4–12 cm long, each segment coarsely toothed and usually cleft; *sepals* 3–5 mm long, with spreading hairs; *petals* 5, yellow, 2–4 mm long; *achenes* 1.8–3 mm long with a straight beak 0.5–0.8 mm long, glabrous. Uncommon in moist places, along streams, and in moist meadows, 5600–9500 ft. June–Aug. E/W.

Ranunculus pygmaeus **Wahlenb.**, PYGMY BUTTERCUP. Perennials, 0.6–3.5 cm; *stems* glabrous to sparsely hairy; *basal leaves* simple or compound, 3-parted or 3-foliate, the central lobe entire or 3-lobed and the lateral 2 lobes entire or 2–4-lobed, 0.3–1.2 cm long; *cauline leaves* absent or solitary; *sepals* 2–3.5 mm, sparsely hairy; *petals* 5, yellow, 1.5–4 mm; *achenes* 1–1.2 mm long with a beak 0.3–0.7 mm long, glabrous. Uncommon in the alpine, 11,000–13,500 ft. July–Aug. E/W.

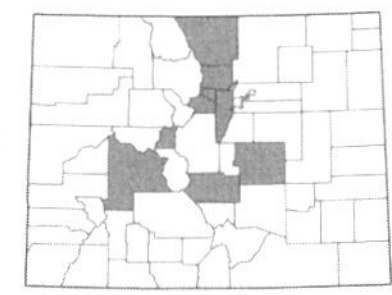

Ranunculus ranunculinus **(Nutt.) Rydb.**, TADPOLE BUTTERCUP. [*Cyrtorhyncha ranunculina* Nutt. ex Torr. & A. Gray]. Perennials, 0.8–3 dm; *stems* glabrous; *basal leaves* ternately compound, 3–5.5 × 4–6 cm, each segment pinnately to bipinnately lobed; *cauline leaves* similar to basal or reduced; *sepals* 2–4 mm, glabrous; *petals* 5, yellow, 3–6 mm; *achenes* 2.4–3 mm with a beak 0.6–1.2 mm. Found on rocky cliffs and outcroppings, open slopes, and in forests, 5200–9000 ft. April–June. E/W.

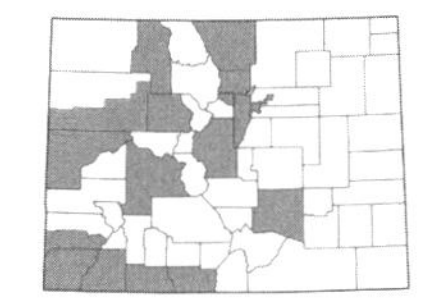

Ranunculus repens **L.**, CREEPING BUTTERCUP. Perennials, often rooting at the nodes, 1–8 dm; *stems* sparsely to densely spreading-hairy; *basal leaves* ternately compound, the leaflets lobed or toothed; *cauline leaves* similar to basal; *sepals* 5–8.5 mm long, spreading-hairy; *petals* 5+, yellow, 6-18 mm long; *achenes* 2.5–4 mm long with a beak 0.5–1.2 mm long. Uncommon in disturbed, moist areas, along streams, and in moist fields, 5000–8000 ft. May–July. E/W. Introduced.

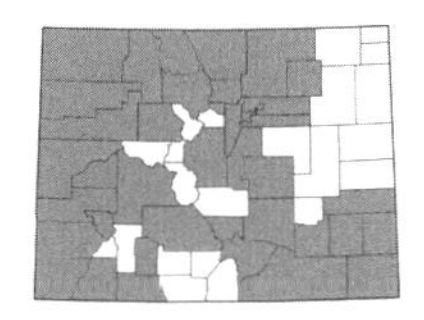

Ranunculus sceleratus **L. var.** ***multifidus*** **Nutt.**, BLISTER BUTTERCUP. Annuals, often growing in standing water, 0.5–7 dm; *stems* glabrous; *leaves* 3-lobed, the main segments toothed or lobed, 0.5–4.5 (6) × 0.5–6 (9.5) cm; *sepals* 2–5.5 mm long, spreading-hairy; *petals* 5, yellow, 0.5–5 mm; *achenes* 0.8–1.3 mm with a beak ca. 0.1 mm long, glabrous. Found in shallow water of streams and ponds, on floodplains, and in moist meadows, 3500–9300 ft. May–Aug. E/W. (Plate 71)

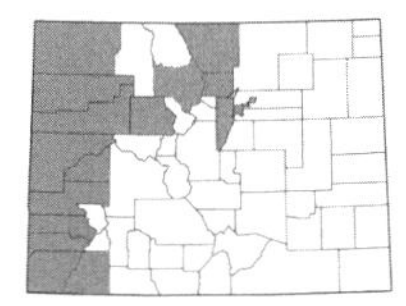

Ranunculus testiculatus **Crantz**, BUR BUTTERCUP. [*Ceratocephala testiculata* (Crantz) Roth]. Annuals, 1.5–10 cm, scapose; *leaves* deeply 3-cleft, the lateral segments lobed into linear segments, 0.3–4 cm long; *sepals* 2.5–8.5 mm long, tomentose; *petals* 2–5, yellow, 3.5–8 mm long; *achenes* 3-chambered, the lateral two chambers empty, with a prominent, straight beak 2.5–4 mm long, tomentose. Found in disturbed areas such as trailheads, parking lots, lawns, and forest clearings, and on dry slopes, 4500–8500 ft. April–May. E/W. Introduced.

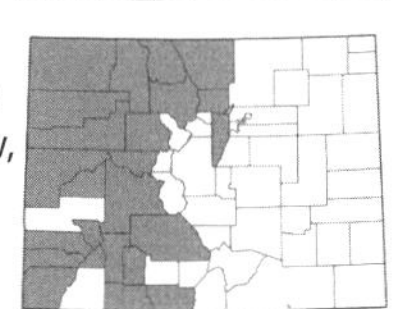

Ranunculus uncinatus **D. Don ex G. Don**, WOODLAND BUTTERCUP. Perennials; *leaves* deeply 3-cleft, the segments lobed, margins toothed, 1.8–5.5 × 2.8–8 cm; *sepals* 2–3.5 mm long, hairy; *petals* 5, yellow, 1–4 (6) mm; *achenes* 2–3 mm with a curved, hooked beak 1.2–2.5 mm long, glabrous or sparsely hairy. Found in moist meadows, marshes, and along streams, 7000–11,500 ft. June–Aug. E/W.

THALICTRUM L. – MEADOWRUE

Perennial herbs; *leaves* basal and cauline, alternate, 1–4 times ternately or pinnately compound, with entire to crenate margins; *flowers* perfect or imperfect, actinomorphic, in terminal racemes, panicles, or corymbs; *sepals* 4–10, green, greenish-yellow, or white; *petals* absent; *fruit* an achene.

1a. Plants scapose with leaves all basal; leaves twice-pinnately compound with the lowermost leaflets ternately lobed; flowers in a few-flowered raceme; plants 0.3–2 (3) dm; flowers perfect...***T. alpinum***

1b. Plants with stem leaves; leaves (2) 3–5 times ternately compound; flowers in a panicle with few to many flowers or sometimes in a raceme; plants 1.4–15 (20) dm; flowers perfect or imperfect...2

2a. Leaflets entire or shallowly 2–5-lobed with acute lobes, leathery or sometimes thin, with prominent veins on the abaxial side, usually hairy below, the margins often narrowly revolute; plants usually stout and 4–15 (20) dm tall...***T. dasycarpum***

2b. Leaflets 3 or more lobed with more or less rounded or mucronate lobes, thin or sometimes leathery, lacking prominent veins below, glabrous to glandular-hairy below, the margins not revolute; plants 1.4–10 (15) dm tall...3

3a. Flowers perfect; anthers 0.5–0.8 mm long with rounded tips; filaments whitish; flowers 1-few per panicle; sepals 5; achenes obliquely obovate-elliptic and strongly laterally compressed with a beak 1–1.5 mm long...***T. sparsiflorum***
3b. Flowers imperfect, plants dioecious; anthers 1.5–4 mm long, the tips apiculate with a slender point; filaments yellow or purplish; flowers many per panicle; sepals 4–6; achenes various or if shaped like the above then the beak 1.5–4 mm long...4

4a. Leaflets glabrous and glaucous, leathery, 5–8 mm long; carpels and achenes 4–5 or rarely 6; filaments brown, 2–3 mm long...***T. heliophilum***
4b. Leaflets glabrous or often glandular below, not glaucous, thin, (5) 10–20 mm long; carpels and achenes 5–17; filaments yellow or purplish, (1.8) 3–10 mm long...***T. fendleri***

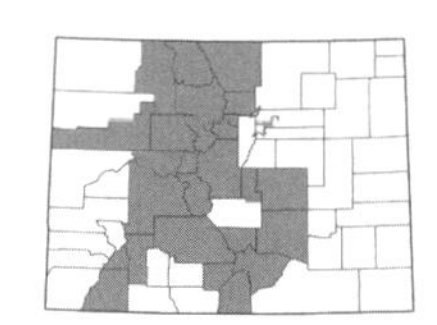

Thalictrum alpinum **L.**, ALPINE MEADOWRUE. Plants 0.3–2 (3) dm, scapose or nearly so; *leaves* twice-pinnately compound, the leaflets 2–10 mm long, cuneate-obovate to orbicular, 3-5-lobed at the tip, glabrous; *inflorescence* of racemes, few-flowered; *flowers* perfect; *sepals* 1–2.5 mm long; filaments purple; *achenes* 2–3.5 mm long, lance-obovoid. Found in moist alpine meadows, peat bogs, and along streams in spruce-fir forests, 7800–14,000 ft. June–Aug. E/W.

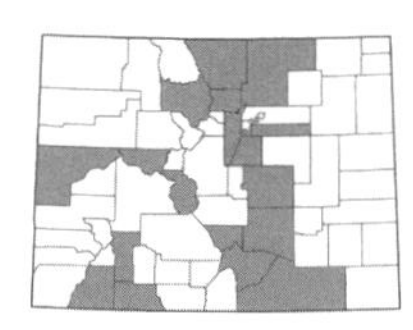

Thalictrum dasycarpum **Fisch. & Avé-Lall.**, PURPLE MEADOWRUE. Plants 4–15 (20) dm; *leaves* 3–5×-ternately compound, the leaflets ovate to cuneate-obovate, undivided or 2–5-lobed at the tip, 1.5–6 cm × 8–45 mm, usually hairy and/or papillose below; *inflorescence* of panicles; *flowers* usually imperfect; *sepals* 3–5 mm long; *filaments* white to purple; *achenes* 2–4.5 mm long, ovoid to fusiform, usually hairy and/or glandular. Found along irrigation ditches, gulches, and streams, often in disturbed areas, 4700–9700 ft. June–July. E/W.

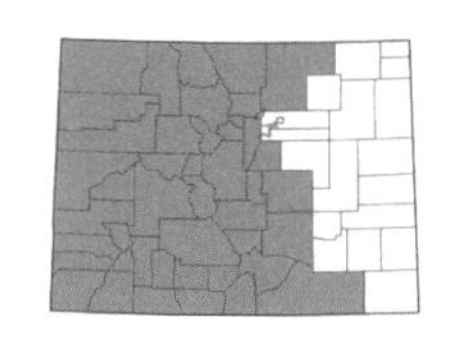

Thalictrum fendleri **Engelm. ex A. Gray**, FENDLER'S MEADOWRUE. Plants 3–6 (15) dm; *leaves* (2) 3–4×-ternately compound, the leaflets orbicular or cordate, 3-lobed at the tip, (0.5) 1–2 cm × (6) 8–12 (19) mm, often glandular below; *inflorescence* of panicles; *flowers* imperfect; *sepals* 1.5–5 mm long; *filaments* yellow or purplish, (1.8) 3-10 mm long; *achenes* 5–11 mm long with a beak 1.5–4 mm, oblanceolate to obovate-elliptic, glabrous or glandular. Common along streams, in meadows, and in aspen and spruce forests, 5500–12,700 ft. June–Aug. E/W. (Plate 71)

Thalictrum fendleri, *T. occidentale* A. Gray, and *T. venulosum* Trel. form a variable complex that would best be treated as a single species. *Thalictrum confine* (a species from the northeast) may also be better included within this complex.

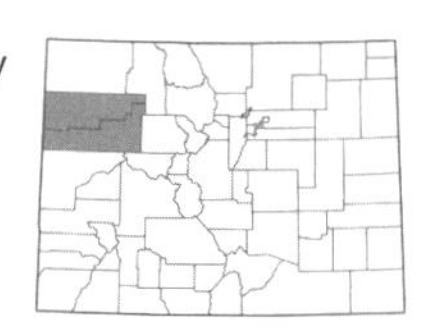

Thalictrum heliophilum **Wilken & DeMott**, SUN-LOVING MEADOWRUE. Plants 1.4–5 dm; *leaves* ternately compound, the leaflets obovate, 3-toothed at the tips, 5–8 × 4–5 mm, glabrous and glaucous; *inflorescence* of panicles; *flowers* imperfect; *sepals* 2–3 mm long; *filaments* brown, 2-3 mm long; *achenes* 4–5 mm long with a beak 1.5 mm long, strongly laterally compressed. Rare on barren, shale talus slopes of the Green River Formation, 6300–8800 ft. June–July. W. Endemic.

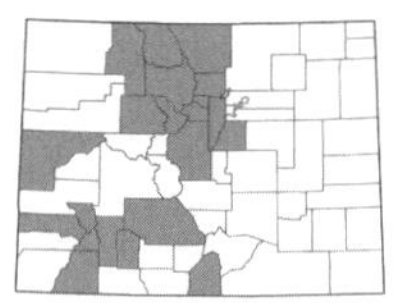

Thalictrum sparsiflorum **Turcz. ex Fisch. & C.A. Mey.**, FEW-FLOWERED MEADOWRUE. Plants (2) 3–12 dm; *leaves* (2) 3×-ternately compound, the leaflets obovate to orbicular, 3-lobed at the tips, 1–2 cm long, often glandular-hairy below; *inflorescence* axillary, with 1-few flowers; *flowers* perfect; *sepals* 2–4 mm long; filaments 3–5 mm long; *achenes* 4–6 mm long with a beak 1–1.5 mm, strongly laterally compressed, glabrous or glandular. Uncommon in moist meadows, along streams, and in moist spruce-fir forests, 7000–10,500 ft. July–Aug. E/W.

TRAUTVETTERIA Fischer & C.A. Mey. – FALSE BUGBANE

Perennial herbs from rhizomes; *leaves* basal and cauline, the cauline leaves alternate, palmately lobed; *flowers* perfect, actinomorphic; *sepals* 3–7, greenish-white, concave-cupped; *petals* absent; *fruit* a prominently veined, 4-angled achene with a curved to hooked, persistent style.

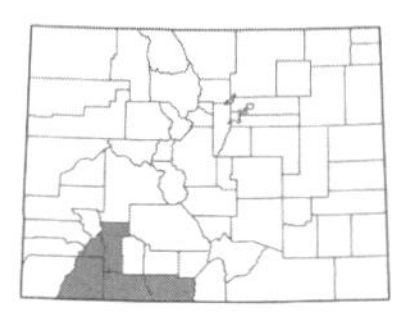

Trautvetteria caroliniensis **(Walter) Vail**, CAROLINA BUGBANE. Plants 5–15 dm; *leaves* 5–11-lobed, the segments serrate to lacerate, broadly cuneate; *inflorescence* of terminal, many-flowered corymbs; *sepals* 3–6 mm. Uncommon in moist spruce forests and along streams, 8000–12,000 ft. July–Aug. W.

TROLLIUS L. – GLOBEFLOWER

Perennial herbs; *leaves* alternate, deeply palmately divided; *flowers* perfect, actinomorphic; *sepals* 5–10 (30), petaloid; *petals* 5–25, yellow or orange, linear, resembling staminodia; *fruit* a follicle.

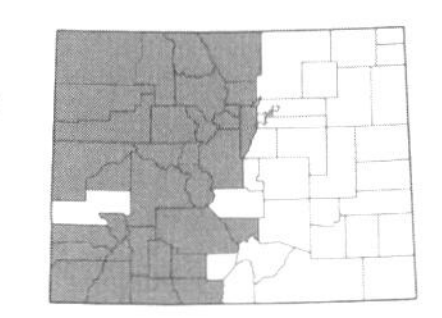

***Trollius albiflorus* (A. Gray) Rydb.**, WHITE GLOBEFLOWER. Plants 0.5–7 dm; *leaves* on a petiole 4–25 cm long; *sepals* 5–9, white to cream, 10–20 mm long; *petals* 2–5 mm long; *follicles* 8–15 mm long (including the beak). Common in moist meadows, along streams, and in marshes, often found growing with *Caltha leptosepala*, 7500–13,500 ft. June–Aug. E/W. (Plate 71)

RESEDACEAE S.F. A. Gray – MIGNONETTE FAMILY

Herbs; *leaves* alternate; *flowers* perfect, zygomorphic, in terminal racemes; *sepals* (4) 6 (8), distinct, persistent; *petals* (2) 6 (8), distinct, unequal but not bilabiate, the innermost largest, white or yellow, usually clawed with scale-like appendages; *stamens* 3–many, distinct; *pistil* 1; *ovary* superior, (2) 3–6-carpellate; *style* absent; *fruit* a capsule or berry.

RESEDA L. – MIGNONETTE

Herbs; *leaves* simple and entire or pinnately dissected; *flowers* numerous in elongate terminal racemes; *petals* usually 6, white or greenish-yellow; *ovary* usually 3-carpellate; *fruit* a capsule opening apically.

1a. Leaves pinnately dissected; flowers yellowish-green; sepals 6…***R. lutea***
1b. Leaves entire, lanceolate; flowers whitish-green; sepals 4…***R. luteola***

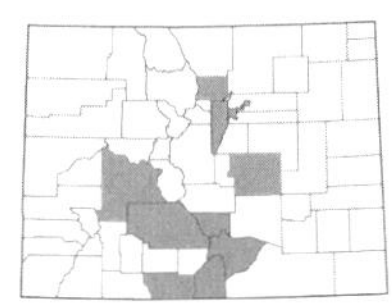

***Reseda lutea* L.**, YELLOW MIGNONETTE. Perennials, 2–8 dm; *leaves* deeply pinnately lobed, 2.5–8 × 1.5–5 cm, glabrous; *sepals* 6, 1–3 mm long; *petals* 6, greenish-yellow, 2–3 mm long; *stamens* 15–20. Uncommon along roadsides and in disturbed areas, 5000–8600 ft. May–Sept. E/W. Introduced.

***Reseda luteola* L.**, WELD; DYER'S-ROCKET. Biennials, 2–10 dm; *leaves* entire, lanceolate, glabrous; *sepals* 4, 1–3 mm long; *petals* 4, greenish-yellow to yellow, 2–4 mm long; *stamens* 20–25. Uncommon in disturbed areas, 5200–5500 ft. June–Aug. E. Introduced.

RHAMNACEAE Juss. – BUCKTHORN FAMILY

Shrubs (ours); *leaves* alternate or opposite, pinnately veined or with 3 veins arising from the base; *flowers* perfect or imperfect, actinomorphic, small; *sepals* (4) 5; *petals* (4) 5 or sometimes absent, distinct, usually more or less hooded or concave; *stamens* (4) 5, distinct, opposite the petals; *pistil* 1; *ovary* superior to inferior, 2–4 (5) carpellate; *fruit* usually drupaceous or sometimes a capsule or schizocarp.

1a. Fruit capsular; leaves 3-nerved from the base, sometimes also with additional lateral nerves above; petals long-clawed, ca. 2 mm long…***Ceanothus***
1b. Fruit drupaceous; leaves pinnately veined with 2–14 pairs of veins per side; petals short-clawed, small (ca. 1 mm long)…***Rhamnus***

CEANOTHUS L. – CEANOTHUS

Unarmed or thorny shrubs; *leaves* usually 3-nerved from the base; *flowers* perfect, white, blue, or purple; *sepals* 5, petaloid; *petals* 5, hooded; *stamens* 5; *ovary* 3-carpellate; *fruit* appearing drupaceous when immature, but capsular at maturity.

1a. Leaves entire or rarely glandular-serrate just near the apex…2
1b. Leaves glandular-serrate their entire length…3

2a. At least some branches thornlike; new growth grayish-green and hairy; leaves densely gray-tomentose below and green above, lanceolate-elliptic to oblong-ovate…***C. fendleri***
2b. Branches not thornlike, unarmed; new growth hairy, but not grayish-green; leaves green on both sides, lighter green and strigose-hairy below (at least along the main veins), elliptic to subrotund…***C. martinii***

3a. Inflorescence paniculate; leaves glabrous and glossy above, evergreen, elliptic or ovate-elliptic, 1.5–5 cm wide, strongly aromatic…***C. velutinus***
3b. Inflorescence corymbiform; leaves glabrous above but not glossy, deciduous, lanceolate to lanceolate-elliptic or oblong-elliptic, rarely over 2 cm wide, not strongly aromatic…***C. herbaceous***

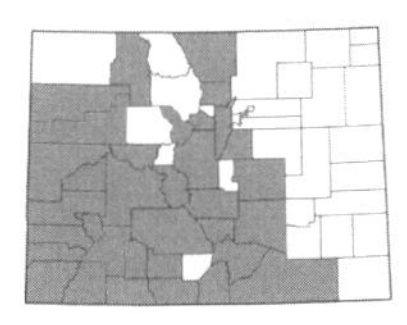

***Ceanothus fendleri* A. Gray**, FENDLER'S CEANOTHUS; BUCKBRUSH. Shrubs, 3–8 dm, some branches thornlike; *leaves* lanceolate-elliptic to oblong-ovate, 5–10 × 2.5–6 cm, entire or rarely irregularly glandular serrate near the tip; *inflorescence* an elongated panicle of corymbs; *sepals* ca. 1.3 mm long, white; *petals* ca. 1.5 mm long, white; *capsules* 4–5 mm long, black. Common on dry slopes, 5500–9500 ft. June–Aug. E/W.

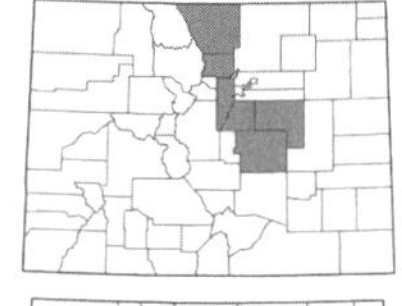

***Ceanothus herbaceous* Raf.**, New Jersey tea; redroot. [*C. ovatus* Desf.]. Shrubs, 2–10 dm, unarmed; *leaves* lanceolate to lanceolate-elliptic or oblong-elliptic, 2–6 × 2 cm, glandular-serrate; *inflorescence* of corymbose clusters; *sepals* 1–1.5 mm long; *petals* 2–2.5 mm long, white; *capsules* 3–4.5 mm long, brownish. Found on rocky slopes and in forest openings along the Front Range, 5000–8500 ft. May–July. E. (Plate 72)

***Ceanothus martinii* M.E. Jones**, Martin's ceanothus. Shrubs, 1.5–8 dm, rigidly branched but unarmed; *leaves* elliptic to subrotund, 1–2 (4) × 0.4–2 cm, entire or inconspicuously glandular-serrulate near the apex; *inflorescence* corymbose; *sepals* 0.5–1.1 mm long; *petals* 1.8–2.2 mm long, white; *capsules* ca. 3 mm long, brown. Uncommon on shale slopes, usually with oak, 7400–8500 ft. June–July. W.

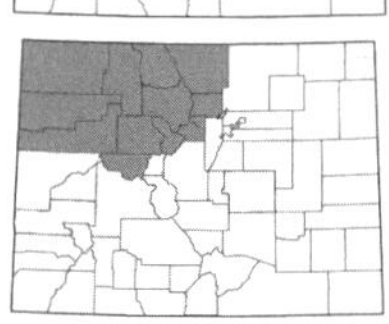

***Ceanothus velutinus* Douglas ex Hook.**, tobacco-brush; sticky laurel. Shrubs, 6–15 (20) dm, unarmed; *leaves* elliptic to ovate-elliptic, 3–8 × 1.5–5 cm, glossy above, evergreen, glandular-serrulate; *inflorescence* paniculate; *sepals* 1.5–2 mm long, white; *petals* 2–3 mm long, white; *capsules* 3–6 mm wide. Found on rocky slopes and in forest openings, sometimes in moist places, 6000–10,000 ft. June–Aug. E/W.

RHAMNUS L. – buckthorn

Unarmed or thorny shrubs or trees; *leaves* entire or more often serrate, pinnately veined; *flowers* perfect or imperfect, green or greenish-white; *sepals* 4–5; *petals* 4–5 or absent, not much clawed; *ovary* 2–4-carpellate, partly embedded in the nectary disk; *fruit* drupaceous with 2–4 stones.

1a. Leaf margins crenate with each tooth tipped with a dark-colored gland; branches usually ending in short thorns; leaf veins 2–4 per side; style 4-parted; fruit with 4 stones...***R. cathartica***
1b. Leaf margins entire or crenate to serrate, but without a dark-colored gland on each tooth; plants unarmed; leaf veins 7–14 per side; style 2–3-parted; fruit with 2–3 stones...2

2a. New growth and leaf petioles densely hairy; young leaves downy or hirsute-spreading-hairy below; flowers perfect...***R. betulifolia***
2b. Plants glabrous or nearly so; flowers perfect or imperfect...3

3a. Flowers 5-merous, perfect; style 3-lobed; fruit with 3 stones; leaves elliptic or obovate, 2–5 cm wide, the margins entire or faintly crenate...***R. frangula***
3b. Flowers 4-merous, imperfect; style 2-lobed; fruit with 2 stones; leaves lanceolate or narrowly elliptic, 1–3.5 cm wide, the margins crenate or serrate...***R. smithii***

***Rhamnus betulifolia* Greene**, birchleaf-buckthorn. Shrubs 1–2.5 m, unarmed; *leaves* alternate, broadly elliptic to elliptic-oblong, 4–11 (14) × 2.5–6.5 (7.5) cm, glabrous above and hirsute below, crenate; *flowers* perfect; *sepals* 5, 1.5–2.5 mm long; *petals* 5, ca. 1 mm long; *fruit* 7–10 mm diam., bluish-black. Not reported for Colorado but could occur near the Four Corners region in moist areas. May–June.

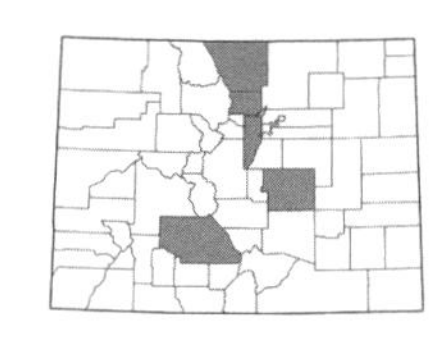

***Rhamnus cathartica* L.**, common buckthorn. Dioecious shrubs 2–6 m, the branches usually ending in short thorns; *leaves* opposite or some alternate, elliptic to elliptic-obovate, 2–6 × 1.5–5 cm, minutely hairy to glabrous, glandular-crenate; *flowers* imperfect; *sepals* 4, 2–3 mm long; *petals* 4, 1–1.3 mm long; *fruit* 5–6 mm diam., black. Escaping from cultivation into canyons and along streams, 5000–7700 ft. May–July. E. Introduced.

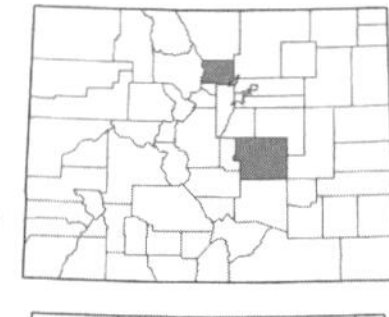

***Rhamnus frangula* L.**, glossy buckthorn. [*Frangula alnus* Mill.]. Shrubs to 3 m, unarmed; *leaves* alternate, elliptic to obovate, 3–7 × 2–5 cm, glabrous or nearly so, entire or crenulate; *flowers* perfect; *sepals* 5, 1.5–2 mm long; *petals* 5, ca. 1 mm long; *fruit* ca. 8 mm diam., black. Escaping from cultivation near Denver/Boulder and Colorado Springs, in usually moist places, 5300–6300 ft. May–July. E. Introduced.

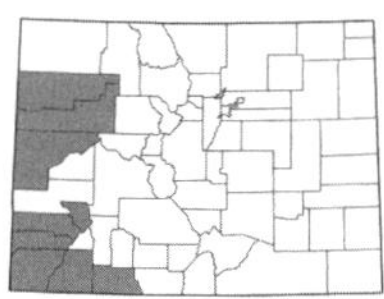

***Rhamnus smithii* Greene**, Smith's buckthorn. Dioecious shrubs 1–3 m, unarmed; *leaves* alternate, lanceolate to narrowly elliptic, 3–7 × 1–3.5 cm, glabrous, crenate to serrate; *flowers* imperfect; *sepals* 4, ca. 2 mm long; *petals* 4, ca. 1 mm long; *fruit* ca. 8 mm diam., black. Uncommon on shale slopes and dry hillsides, 6600–8300 ft. May–July. W.

<u>ROSACEAE</u> L. – ROSE FAMILY

Trees, shrubs, or herbs; *leaves* alternate or basal, rarely opposite, simple or compound, usually stipulate or sometimes exstipulate; *flowers* perfect or rarely imperfect, actinomorphic, with a hypanthium, perigynous or epigynous; *sepals* 3–10, usually 5, distinct, often appearing as lobes of the hypanthium; *petals* 3–10, usually 5, occasionally absent, distinct; *stamens* 5-numerous, distinct, inserted near the edge of the hypanthium; *pistils* 1–many; *ovary* inferior or superior, 2–5-carpellate; *fruit* a follicle, pome, achene, or drupe, sometimes an aggregate of achenes or drupelets. (Eriksson et al., 2003)

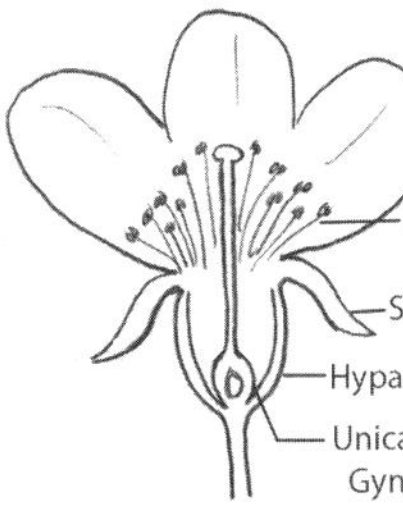

1a. Leaves simple, with entire or toothed margins, or shallowly lobed (less than halfway to the midrib)...2
1b. Leaves compound, or simple but deeply dissected or lobed over halfway to the midrib...18

2a. Leaves with entire margins...3
2b. Leaves with toothed or lobed margins...7

3a. Prostrate or densely caespitose shrubs forming dense cushions or mats; flowers white, in a dense cylindrical raceme, the petals 1.3–2.5 mm long...***Petrophytum***
3b. Erect shrubs not forming dense cushions of mats; flowers white, pink, red, or petals absent, in various inflorescences but not in a dense cylindrical raceme...4

Prunus flower:
Petals
Stamens
Sepals
Hypanthium
Unicarpellate Gynoecium

4a. Sepals 4, yellow; petals usually absent or if present then linear, yellow; fruit an achene but lacking a long, plumose style; branches often with spines; leaves linear to narrowly oblanceolate, in fascicles...***Coleogyne***
4b. Sepals 5, green; petals absent or present, white, pink, or red; fruit a pome or an achene with a long, plumose style; branches lacking spines; leaves various...5

5a. Leaves ovate to elliptic, the larger leaves 12–25 mm wide, shiny and glossy; pome black...***Cotoneaster***
5b. Leaves linear, narrowly lanceolate, or narrowly elliptic, 0.4–10 (12) mm wide, not shiny and glossy; fruit an achene with a long, plumose style, or a yellowish to reddish pome...6

6a. Petals absent; fruit an achene with a long, plumose style; leaf margins inrolled or revolute...***Cercocarpus***
6b. Petals present, white or pinkish; fruit a yellowish to reddish pome; leaf margins flat...***Peraphyllum***

7a. Low, mat-forming alpine plants with prostrate branches; leaves with revolute, crenate-serrate margins, white below and green and impressed above; flowers solitary, the 8–10 petals white or cream, 0.9–1.5 cm long...***Dryas***
7b. Erect shrubs or trees; leaves unlike the above; flowers various but unlike the above...8

8a. Leaves 3-lobed at the apex, wedge-shaped, white-tomentose below and green above, mostly 1–2.5 cm long; flowers yellow...***Purshia***
8b. Leaves unlike the above; flowers white, cream, or pink...9

9a. Leaves palmately veined and shallowly lobed...10
9b. Leaves pinnately veined, shallowly lobed or not...11

10a. Leaves with stellate hairs; fruit a follicle; pistils 1–5...***Physocarpus***
10b. Leaves lacking stellate hairs; fruit an aggregate of drupelets (e.g., raspberry); pistils numerous, more than 5...***Rubus***

11a. Petals absent; fruit an achene with a long, plumose style; leaves rhombic to obovate with serrate margins...***Cercocarpus***
11b. Petals present; fruit various or if an achene then lacking a long, plumose style; leaves various...12

12a. Branches armed with spines or thorns...13
12b. Branches lacking spines or thorns...14

13a. Ovary inferior; fruit a pome; leaves simply or doubly toothed, or lobed; petioles lacking a pair of glands; flowers (2) 5–10 in corymbose or racemose inflorescences...***Crataegus***
13b. Ovary superior; fruit a drupe (e.g., cherry, peach, or plum); leaves simply toothed; petioles often with a pair of glands at the summit; flowers solitary or in clusters of 2–5...***Prunus***

14a. Fruit an achene; pistils usually 5; flowers numerous in a diffuse, softly hairy panicle; petals 1.5–2.5 mm long, white or cream; leaves cuneate at the base, mostly 1–2.5 cm long, hairy below and also with sessile, glandular hairs (these can be difficult to see)...***Holodiscus***
14b. Fruit a follicle, pome, or drupe; plants otherwise unlike the above in all respects...15

15a. Leaves subsessile, narrowly oblanceolate or narrowly elliptic, mostly to 1 cm wide, 3–5 times as long as wide, shallowly serrate, the teeth usually with a dark tip, with a short point at the apex, usually in fascicles...***Peraphyllum***
15b. Leaves evidently petiolate, otherwise unlike the above...16

16a. Fruit a drupe (e.g., cherry, peach, or plum); ovary superior; style 1; petioles often with a pair of glands at the base of the leaf...***Prunus***
16b. Fruit a pome; ovary inferior; styles 2–5; petioles lacking a pair of glands at the base of the leaf...17

17a. Pome 4–7 cm thick (e.g., apple); petals 12–25 mm long; leaves densely hairy below, 5–10 cm long; small trees...***Malus***
17b. Pome 0.5–1 cm thick; petals 5–10 (12) mm long; leaves glabrous or hairy below, 1.5–6 cm long; shrubs 1–5 m tall...***Amelanchier***

18a. Leaves ternately compound (with 3 leaflets)...19
18b. Leaves pinnately or palmately compound with 5 or more leaflets, or simple but deeply divided...23

19a. Plants with stolons; fruit an aggregate of achenes on an enlarged, red, fleshy receptacle (a strawberry); flowers white...***Fragaria***
19b. Plants lacking stolons; fruit unlike the above; flowers white, yellow, pink, or red...20

20a. Calyx lobes not alternating with bracteoles; fruit an aggregate of drupelets (e.g., raspberry); stems often armed with prickles; flowers white, pink, or red...***Rubus***
20b. Calyx lobes alternating with bracteoles; fruit an achene or aggregate of achenes; stems lacking prickles; flowers yellow or white...21

21a. Stamens 5; pistils mostly 5–15; leaflets cuneate, 3–5-toothed at the apex; petals 1.5–3 mm long, usually shorter than the sepals; low, mat-forming plants of the subalpine and alpine...***Sibbaldia***
21b. Stamens more than 10; pistils numerous, usually more than 20; plants otherwise various...22

22a. Ovary and achene hairy; style terminal, persistent and elongating in fruit, sometimes hooked at the tip, or not hooked but plumose or glabrous...***Geum***
22b. Ovary and achene glabrous or nearly so; style terminal, lateral, or basally inserted, deciduous in fruit, not elongating in fruit, not hooked at the tip or plumose...***Potentilla***

23a. Woody shrubs or trees; fruit various...24
23b. Annuals, biennials, or perennial herbs; fruit an achene...30

24a. Stems armed with thorns and/or prickles...25
24b. Stems lacking thorns or prickles...26

25a. Fruit an achene enclosed within a fleshy hypanthium (rose hip); sepals conspicuously constricted above to a linear segment and then expanded at the tip; flowers pink, rose, or yellow...***Rosa***
25b. Fruit an aggregate of drupelets (e.g., raspberry); sepals unlike the above; flowers usually white or sometimes pinkish or rose...***Rubus***

26a. Fruit an achene enclosed within a fleshy hypanthium (rose hip); sepals conspicuously constricted above to a linear segment and then expanded at the tip; flowers pink, rose, or yellow...***Rosa***
26b. Fruit and sepals unlike the above; flowers white, yellow, or cream...27

27a. Leaves pinnately compound with 5–6 pairs of sharply serrate, elliptic leaflets 1.5–2.5 cm wide; fruit a reddish-orange pome; flowers numerous in a flat-topped corymb...***Sorbus***
27b. Leaves unlike the above; fruit an achene; flowers solitary...28

28a. Leaves densely brownish-yellow-tomentose below, pinnately lobed with linear divisions, the margins strongly revolute; fruit an achene with an elongate, plumose style; flowers white...***Fallugia***
28b. Leaves white or green below, various, the margins revolute or not; fruit an achene but lacking an elongate, plumose style; flowers yellow, cream, or white...29

29a. Leaves pinnately compound, usually with 5 leaflets, green below; petals yellow; pistils numerous...***Potentilla***
29b. Leaves simple and 3-lobed at the apex or pinnately lobed with narrow divisions, white below; petals white, cream, or yellow; pistils 1–12...***Purshia***

30a. Petals absent; sepals 4, often reddish; achenes enclosed in a 4-winged, dry hypanthium, with conspicuous ridges between the winged angles; flowers in a dense globose to ovoid spike; basal leaves pinnately compound with orbicular to ovate and crenately toothed leaflets...***Sanguisorba***
30b. Petals present, white, yellow, or pink; sepals 5; achenes and inflorescence unlike the above; leaves various...31

31a. Top of the hypanthium with ascending, hooked bristles; flowers yellow, numerous in an elongated raceme; leaves pinnately compound, the leaflets usually with sessile glands and nonglandular hairs; achene enclosed within a hardened hypanthium...***Agrimonia***
31b. Top of the hypanthium lacking hooked bristles; flowers yellow, white, or pinkish, variously arranged but not in an elongated raceme; leaves various; achenes not enclosed within a hardened hypanthium...32

32a. Flowers in a dense, globose cyme; leaves mostly basal, pinnately compound, the 10–25 pairs of leaflets palmately divided to near the base, the divisions with rounded tips; pistils usually 2, sometimes 1–6...***Ivesia***
32b. Flowers variously arranged but not in a dense, globose cyme; leaves unlike the above; pistils 5–many...33

33a. Leaves deeply ternately divided into small, linear lobes; stamens 5; pistils 5–10; plants conspicuously glandular-hairy throughout; petals white or pinkish, about as long as the sepals; stems reddish, much-branched above; base of the plant with a dense cluster of the previous year's leaves...***Chamaerhodos***
33b. Leaves truly compound or pinnately divided but the lobes unlike the above; stamens 5–many; pistils 5–many; plants otherwise various...34

34a. Flowers white, numerous in large panicles; pistils 5–15; achenes tightly packed and spirally twisting; leaves pinnately compound and usually with some tiny leaflets intermingled with the larger ones...***Filipendula***
34b. Flowers yellow or white, relatively few in open cymes; pistils numerous; achenes unlike the above; leaves various...35

35a. Ovary and achene hairy; style terminal, persistent and elongating in fruit, sometimes hooked at the tip, or not hooked but plumose or glabrous...***Geum***
35b. Ovary and achene glabrous or nearly so; style terminal, lateral, or basally inserted, deciduous in fruit, not elongating in fruit, not hooked at the tip or plumose...***Potentilla***

AGRIMONIA L. – AGRIMONY

Perennial herbs; *leaves* alternate, pinnately compound, stipulate with leaf-like stipules; *flowers* perfect, in a terminal spike-like raceme; *sepals* 5; *petals* 5, yellow; *stamens* 5–15; *pistils* 2; *ovary* superior; *fruit* a single achene enclosed within a hardened hypanthium.

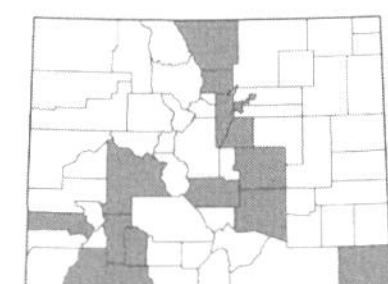

Agrimonia striata **Michx.**, ROADSIDE AGRIMONY. Plants 5–15 dm, hirsute and often glandular above; *leaflets* 5–13, serrate, glabrous to sparsely hairy above, gland-dotted and hairy below; *stipules* 1–2 cm long; *hypanthium* with ascending, hooked bristles at the top; *petals* yellow, (2) 3–4 mm long; *achenes* globose, 2.5–3.5 mm diam. Found along streams and ditches, in moist ravines, and in open woodlands, 5000–8000 ft. July–Sept. E/W.

AMELANCHIER Medik. – SERVICEBERRY

Shrubs or small trees, unarmed; *leaves* alternate, simple, with linear-attenuate stipules; *flowers* perfect, in short terminal racemes; *sepals* 5; *petals* 5, white or pinkish; *stamens* 10–20; *pistil* 1; *ovary* inferior, 5-carpellate; *fruit* a pome.

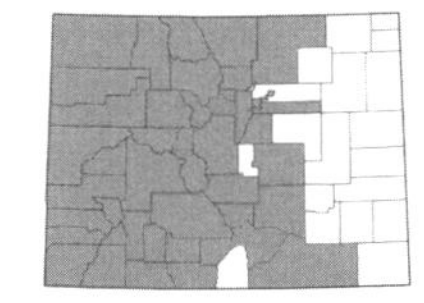

Amelanchier alnifolia **(Nutt.) Nutt. ex M. Roem.**, WESTERN SERVICEBERRY. Shrubs, 1–5 m; *leaves* elliptic to ovate or oblong, 1–6 cm long, serrate near the apex; *sepals* 1.5–3 mm long; *petals* 5–10 (12) mm long; *pomes* ca. 1–1.5 cm diam., dark purple at maturity. Common on dry hillsides, in canyons, aspen, pinyon-juniper, or with sagebrush, 4700–11,000 ft. May–July. E/W. (Plate 72)

Amelanchier alnifolia and *A. utahensis* overlap in morphology and distribution. Many specimens are difficul, if not impossible, to assign to one species or the other. Therefore, *A. utahensis* and *A. pumila* are included here as a varieties within the large *A. alnifolia* complex:

1a. Ovary tips glabrous (or sparsely hairy); leaves glabrous; shrubs 1–2(-4) m...**var. *pumila* (Torr. & A. Gray) C.K. Schneid.**, DWARF SERVICEBERRY. [*A. pumila* Nutt. ex Torr. & A. Gray].
1b. Ovary tips hairy (rarely glabrous); leaves sparsely to moderately hairy or sometimes glabrous; shrubs 0.5–8 m...2

2a. Leaves (2) 2.5–6 cm long, usually glabrous above but sometimes hairy; petals 7–18 mm long; styles usually 5, sometimes 4...**var. *alnifolia***, WESTERN SERVICEBERRY.
2b. Leaves 1–3 cm long, usually hairy above and below at maturity but occasionally glabrous above; petals often smaller, 5–9 (10) mm long; styles 2–4, rarely 5...**var. *utahensis* (Koehne) M.E. Jones**, UTAH SERVICEBERRY. [*A. utahensis* Koehne].

CERCOCARPUS H.B.K. – MOUNTAIN MAHOGANY

Shrubs or small trees; *leaves* alternate, usually clustered in fascicles, simple, stipulate; *flowers* perfect; *sepals* 5; *petals* absent; *stamens* 15–40; *pistil* 1; *ovary* superior, unicarpellate; *fruit* an achene with a persistent, elongate, and plumose style.

1a. Leaves toothed from the middle, ovate to rhombic, usually flat, deciduous...***C. montanus***
1b. Leaves with entire, revolute margins, linear to narrowly oblong, evergreen...2

2a. Leaves 0.4–1.5 (2) mm wide, nearly terete; mature, plumose style of achenes 3–4.5 cm long...***C. intricatus***
2b. Leaves (1.5) 2–10 mm wide; mature, plumose style of achenes 4–10 cm long...***C. ledifolius***

Cercocarpus intricatus **S. Watson**, DWARF MOUNTAIN MAHOGANY. Shrubs to 2 m; *leaves* linear to narrowly oblong, tightly revolute and nearly terete, 3–9 × 0.4–1.5 (2) mm, glabrous to villous, evergreen, with entire margins; *sepals* 0.6–1.2 mm; *stamens* 10–20; *style* of achenes 3–4.5 cm long. Found on rocky hillsides and sandstone rimrock, often with pinyon-juniper, 4900–8500 ft. May–June. W.

Cercocarpus ledifolius **Nutt. ex Torr. & A. Gray**, CURL-LEAF MOUNTIAN MAHOGANY. Shrubs to 5 m; *leaves* narrowly oblong, 10–42 × (1.5) 2–10 mm, glabrous to hairy, with revolute, entire margins, evergreen; *sepals* 1.2–2 mm; *stamens* 20–30; *style* of achenes 4–10 cm long. Found on rocky hillsides, canyon slopes, and on cliffs, often with pinyon-juniper, 6000–8600 ft. May–June. W.

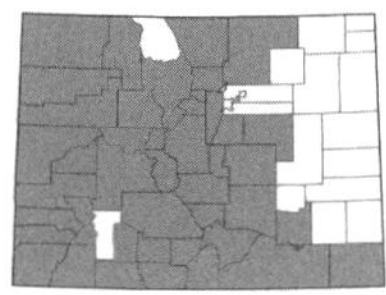

Cercocarpus montanus **Raf.**, BIRCHLEAF MOUNTAIN MAHOGANY. Shrubs or small trees to 5 m; *leaves* ovate to rhombic, 6–44 × 5–23 mm, usually flat, glabrous above and hairy below, the margins toothed above the middle, deciduous; *sepals* 1–1.7 mm long; *stamens* 25–40; *style* of achenes 3–10 cm long. Common and forming large stands on dry hillsides, often with pinyon-juniper, oaks, ponderosa pine, or Douglas-fir, 4000–9500 ft. May–June. E/W. (Plate 72)

CHAMAERHODOS Bunge – LITTLE ROSE

Biennial or perennial herbs; *leaves* primarily basal, simple but bi- or tri-ternately dissected into linear segments; *flowers* perfect, in branched cymes; *sepals* 5; *petals* 5, white or pinkish-tinged; *stamens* 5; *pistils* mostly 5–10; *ovary* superior; *fruit* an achene.

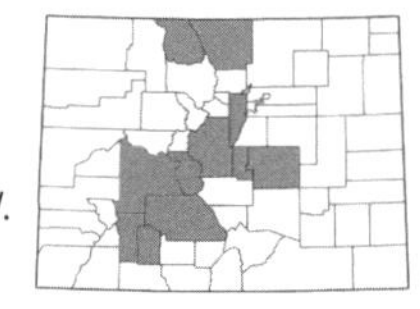

Chamaerhodos erecta **(L.) Bunge var. *parviflora* (Nutt.) C.L. Hitchc.**, LITTLE GROUND ROSE. Plants 1–3 dm, densely glandular; *leaves* deeply ternately divided into small, linear lobes; *sepals* 1.5–2 mm; *petals* subequal to the sepals; *achenes* ca. 1.5 mm long. Found on gravelly hillsides, in gravelly/sandy washes and creek bottoms, and sagebrush slopes, 6000–10,500 ft. June–Oct. E/W. (Plate 72)

COLEOGYNE Torr. – COLEOGYNE

Shrubs, spinescent; *leaves* alternate, clustered in fascicles, simple, stipulate; *flowers* perfect, solitary; *sepals* 4; *petals* usually absent; *stamens* 30–40; *pistil* 1; *fruit* an achene.

Coleogyne ramosissima **Torr.**, BLACKBRUSH. Shrubs mostly 3–12 dm; *leaves* narrowly oblanceolate, 3–12 mm long, usually mucronate at the tip, with dolabriform hairs, entire; *sepals* yellow inside, reddish-brown outside, 4.5–7 (8) mm long; *petals* (if present) 6–10 mm; *achenes* 5–8 mm long, glabrous. Found on dry, open hillsides and in canyon bottoms, 4300–5500 ft. April–June. W.

COTONEASTER Medik. – COTONEASTER

Shrubs, unarmed; *leaves* alternate, simple, entire, stipulate; *flowers* perfect, in terminal or axillary cymes; *sepals* 5; *petals* 5, white, pink, or red; *stamens* 10–20; *pistil* 1; *ovary* 2–5-carpellate, inferior; *fruit* a small pome, pink, red, orange, or black.

Cotoneaster lucidus **Schltdl.**, SHINY COTONEASTER. Shrubs 1–2 m; *leaves* elliptic-ovate, mostly 2–5 cm long, shiny and glabrous above; *sepals* villous dorsally; *petals* pale pink; *pomes* black, ca. 1 cm diam. Cultivated and sometimes escaping from gardens along the edges of woods west of Colorado Springs and Boulder, 6000–7000 ft. May–June. E. Introduced.

CRATAEGUS L. – HAWTHORN

Shrubs or small trees, usually armed with thorns; *leaves* alternate, simple, toothed or lobed; *flowers* perfect, in cymes; *sepals* 5; *petals* 5, white or rarely pinkish; *stamens* 5–25; *pistil* 1; *ovary* 5-carpellate, inferior; *fruit* a pome.

1a. Style 1; leaves with 1–3 lobes per side, the sinuses usually deep; pomes bright red, with a single seed; introduced...***C. monogyna***
1b. Styles 3–7; leaves various, sometimes unlobed; pomes usually not bright red, with more than 1 seed; native...2

2a. Mature fruit reddish-orange or purplish-red; thorns 2.5–5 cm long; leaves rhombic-ovate or ovate...***C. succulenta***
2b. Mature fruit purplish-black; thorns 0.5–4 cm long; leaves lanceolate or elliptic...3

3a. Petals 4.5–7.5 mm long; leaves mostly elliptic, with 4–5 veins per side...***C. rivularis***
3b. Petals 3–4 mm long; leaves lanceolate to narrowly elliptic, thick, with 6–9(-12) veins per side...***C. saligna***

Crataegus monogyna **Jacq.**, ONESEED HAWTHORN. Shrubs or small trees 4–10(-12) m tall, armed with thorns 0.7–2.5 cm long; *leaves* ovate to deltate, 1.2–5.5 cm long, with 1–3 lobes per side, the margins weakly or few-toothed; *sepals* 1.5–2.5 mm long; *petals* 4–8 mm long; *pomes* red, 6–11 mm diam. Cultivated shrub reported in Colorado. April–May. Introduced.

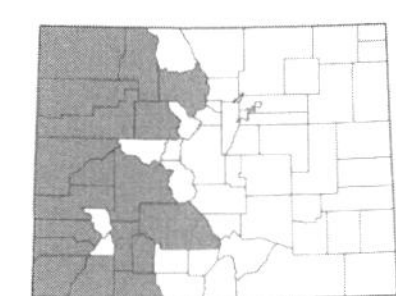

Crataegus rivularis **Nutt.**, RIVER HAWTHORN. Shrubs or small trees 3–5 m tall, armed with thorns 1.5–4 cm long; *leaves* elliptic or narrowly elliptic, 3–8 cm long, serrate or crenate-serrate; *sepals* 6–8 mm long; *petals* 4.5–7.5 mm long; *pomes* purplish-black, glaucous, ca. 1 cm diam. Common along streams, 5200–8500 ft. W.

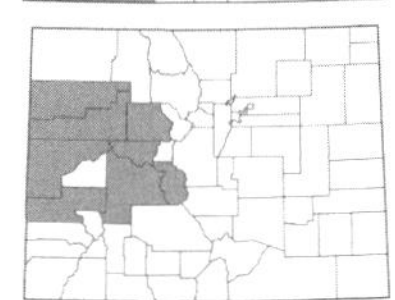

Crataegus saligna **Greene**, WILLOW HAWTHORN. Shrubs or small trees 3–5 m tall, armed with thorns mostly 0.5–3 cm long; *leaves* lanceolate, to narrowly elliptic, more or less coriaceous, 3–5 cm long, crenate-serrate; *sepals* 1.5–2 mm long; *petals* 3–4 mm long; *pomes* purplish-black, glaucous, ca. 8 mm diam. Common along streams and in canyon bottoms, mostly found in the Gunnison and upper Colorado River basins, 5500–8000 ft. E/W.

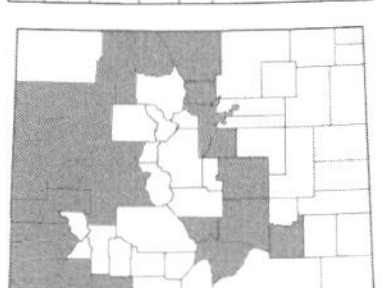

Crataegus succulenta **Schrad.**, ROCKY MOUNTAIN HAWTHORN. [*C. chrysocarpa* Ashe; *C. erythropoda* Ashe; *C. macrantha* Lodd. ex Loud.]. Shrubs or small trees 3–5 (8) m tall, armed with thorns 2.5–5 cm long; *leaves* rhombic-ovate or ovate, 3–6 × 2–5 cm, coarsely serrate; *sepals* 4–5 mm long; *petals* 6–8 mm long; *pomes* reddish-orange or purplish-red, ca. 1 cm diam. Common along streams and in gulches and canyons, 4500–8000 ft. May–June. E/W. (Plate 72)

DRYAS L. – MOUNTAIN AVENS

Undershrub, low and mat-forming with prostrate branches; *leaves* alternate, simple, entire to subpinnatifid, stipulate; *flowers* perfect, solitary; *sepals* (6) 8–10; *petals* (6) 8–10, white or sometimes yellow; *stamens* numerous; *pistils* numerous; *ovary* superior; *fruit* an achene with a persistent, plumose style.

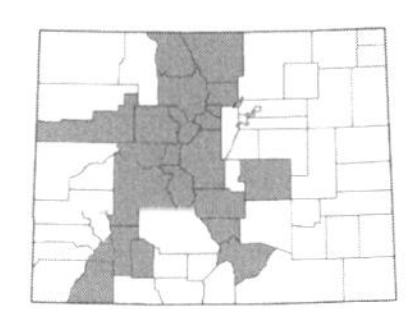

Dryas octopetala **L. var. *hookeriana* (Juz.) Breit.**, HOOKER'S MOUNTAIN AVENS. Scapes 1–11 cm; *leaves* lanceolate to oblong, mostly 10–40 × 3–10 mm, crenate, green and glabrous or sparsely hairy above, densely white-tomentose below; *petals* 9–15 mm long; *style* of achenes to 4 cm long. Found in the alpine, 10,600–14,000 ft. June–Aug. E/W. (Plate 72)

FALLUGIA Endl. – APACHE PLUME

Dioecious or polygamo-dioecious shrubs, unarmed; *leaves* alternate, often fascicled, simple, pinnatisect into narrow, linear lobes, stipulate; *flowers* perfect, solitary or of 2–3 flowers in leaf axils; *sepals* 5; *petals* 5, white; *stamens* numerous; *pistils* numerous; *fruit* an achene with a persistent, plumose style.

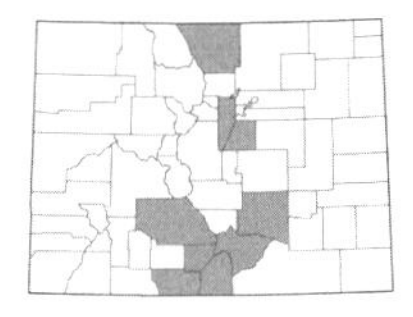

Fallugia paradoxa **(D. Don) Endl. ex Torr.**, APACHE PLUME. Shrubs to 2 m; *leaves* mostly 4–15 mm long, 3–5-lobed with scales; *sepals* 4–10 mm long; *petals* 11–15 mm long; *style* of achenes 2–4 cm long in fruit. Found in sagebrush, canyon bottoms, and on alluvial banks, native to the San Luis Valley but often cultivated as a landscape plant elsewhere, 4500–8000 ft. May–Sept. E.

FILIPENDULA Mill. – QUEEN OF THE MEADOW

Perennial herbs; *leaves* alternate; *flowers* perfect, in branched cymes; *sepals* 5; *petals* 5, white or pink; *stamens* 20–40; *pistils* 5–15; *ovary* superior; *fruit* an achene.

Filipendula ulmaria **(L.) Maxim. var. *denudata* (J. Presl & C. Presl) Maxim.**, QUEEN OF THE MEADOW. Plants 1–2 m; *leaves* pinnately compound with ca. 9 leaflets; *leaflets* ovate to elliptic, 2–8 cm long, doubly serrate, and usually with some tiny leaflets intermingled with the larger ones; *petals* 2–5 mm long; *achenes* 3–4 mm long, glabrous, spirally imbricate. Uncommon weed in moist places, 5000–5500 ft. June–Aug. E. Introduced.

FRAGARIA L. – STRAWBERRY

Perennial herbs with stolons; *leaves* mostly basal, ternately compound, stipulate; *flowers* usually perfect, in cymes; *sepals* 5; *petals* 5, white; *stamens* 20–40; *pistils* numerous; *ovary* superior; *fruit* an aggregate of achenes on an enlarged, fleshy, reddish receptacle.

1a. Terminal tooth of the leaflets smaller and shorter than the 2 adjacent teeth; flowers usually surpassed by the leaves...***F. virginiana***
1b. Terminal tooth of the leaflets about equal in size or extended past the 2 adjacent teeth; flowers usually equal to or surpassing the leaves...***F. vesca***

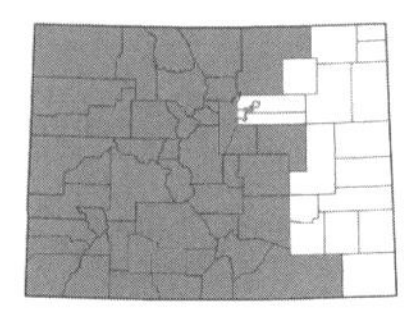

***Fragaria vesca* L.**, WOODLAND STRAWBERRY. [*F. americana* (Porter) Britton]. Leaflets 3, elliptic to obovate, the terminal one 1.3–6.5 × 1–4 cm, pilose, serrate with the terminal tooth equal to or longer than the 2 adjacent teeth; *flowers* in 3–15-flowered cymes, usually equal to or surpassing the leaves; *sepals* 3.5–7 mm long; *petals* 5–10 mm long; *fruit* to 1 cm long. Found along streams and in shady forests, 6400–10,500 ft. May–July. E/W.

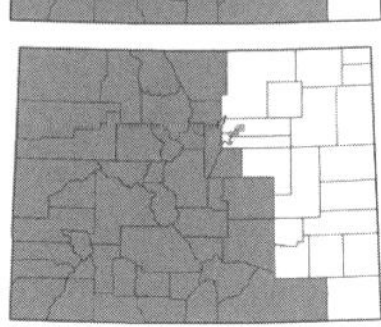

***Fragaria virginiana* Mill.**, MOUNTAIN STRAWBERRY. [*F. ovalis* (Lehm.) Rydb.]. Leaflets 3, elliptic to obovate, the terminal one 1–4.5 × 0.5–2.5 cm, pilose to glabrate, serrate with the terminal tooth shorter than the 2 adjacent teeth; *flowers* in 2–12-flowered cymes, usually surpassed by the leaves; *sepals* 3–6.5 mm long; *petals* 3.5–10 mm long; *fruit* to 1 cm long. Common along streams, in meadows, and in forests, 6000–12,700 ft. May–July. E/W. (Plate 72)

GEUM L. – AVENS

Perennial herbs; *leaves* alternate, sometimes opposite, pinnately compound, stipulate; *flowers* perfect; *sepals* 5; *petals* 5, yellow, white, or sometimes pinkish-tinged; *stamens* numerous; *pistils* numerous; *ovary* superior; *fruit* an aggregate of achenes.

1a. Flowering stem leaves opposite and in 2–4 pairs; flowers white to pinkish, the petals erect and converging at the tips; style strongly plumose, not jointed or hooked at the tip, 3–5 cm long at maturity; terminal leaflet of basal leaves 3-toothed or 3-cleft at the tip...***G. triflorum***
1b. Flowering stem leaves alternate; flowers yellow to pinkish with purple veins, the petals not erect and converging at the tips; style glabrous, glandular, or pilose below, usually less than 3 cm long at maturity; terminal leaflet of basal leaves various...2

2a. Plants nearly glabrous or with appressed hairs; style glabrous, not jointed or hooked; leaflets divided into narrow segments mostly less than 3 mm wide; plants 0.3–4 dm tall...***G. rossii* var. *turbinatum***
2b. Plants hirsute with spreading or retrorse hairs (at least on the petioles); style jointed at the middle and hooked at the tip; leaflets broadly lanceolate to ovate or nearly orbicular, the larger segments wider than 3 mm; plants 3–12 dm tall...3

3a. Sepals erect, purplish, 7–10 mm long; style plumose below; flowers nodding...***G. rivale***
3b. Sepals reflexed, green; style glabrous, sparsely hairy, or glandular; flowers erect...4

4a. Lower portion of the style glandular-hairy; terminal leaflet of basal leaves with a cordate or rounded base...***G. macrophyllum* var. *perincisum***
4b. Lower portion of the style glabrous or hairy but not glandular; terminal leaflet of basal leaves with a cuneate base...***G. aleppicum***

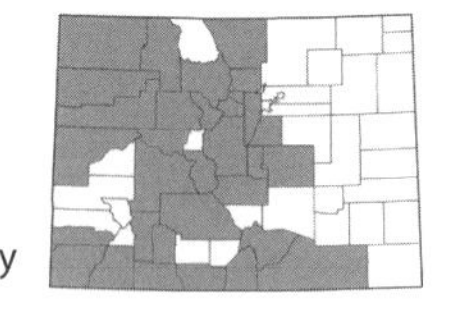

***Geum aleppicum* Jacq.**, YELLOW AVENS. Plants 4–8 (10) dm, spreading-hirsute on the stems and petioles; *leaves* alternate, the basal ones 8–20 cm long; *leaflets* broadly lanceolate to ovate, with toothed margins, the terminal leaflet with a cuneate base; *sepals* 4–8 mm long, green, reflexed; *petals* yellow, about equal to the sepals; *styles* jointed at the middle and hooked at the tip, glabrous or hairy on the lower half. Found along streams, in moist meadows, or sometimes in dry spruce-pine forests, 5500–10,000 ft. June–Aug. E/W.

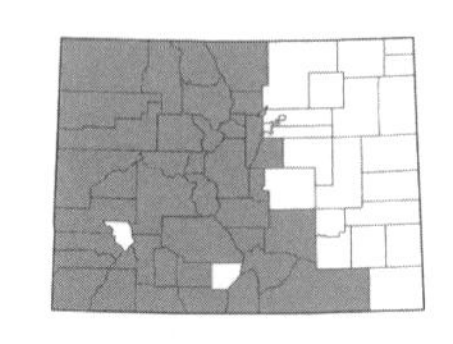

***Geum macrophyllum* Willd. var. *perincisum* (Rydb.) Raup**, LARGE-LEAVED AVENS. Plants 2–12 dm, spreading-hirsute on the stems and petioles; *leaves* alternate, the basal ones 4–30 cm long; *leaflets* ovate to nearly orbicular, the terminal leaflet 2.5–13 cm long, with a cordate or rounded base, with toothed margins; *sepals* 2.5–5 mm long, green, reflexed; *petals* yellow, 3–6 mm long; *styles* jointed at the middle and hooked at the tip, glandular-hairy on the lower half. Common along streams and in moist meadows, 5800–11,500 ft. June–Sept. E/W.

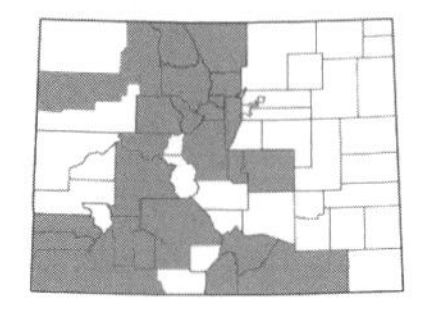

***Geum rivale* L.**, PURPLE AVENS. Plants to 4 dm; *leaves* alternate, the basal ones to 30 cm long; *leaflets* ovate to nearly orbicular, the terminal leaflet to 10 cm long, with toothed margins; *sepals* to 10 mm long, purplish, erect; *petals* yellow to pinkish, about as long as the sepals; *styles* jointed at the middle and hooked at the tip, plumose on the lower half. Found in moist meadows, willow thickets, and along streams, 7500–10,500 ft. June–Aug. E/W.

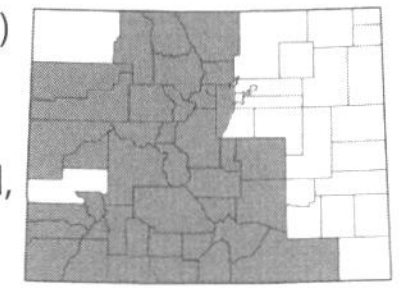

***Geum rossii* (R. Br.) Ser. var. *turbinatum* (Rydb.) C.L. Hitchc.**, ALPINE AVENS. [*Acomastylis rossii* (R. Br.) Greene ssp. *turbinata* (Rydb.) W.A. Weber]. Plants 0.3–4 dm; *leaves* alternate, basal leaves 2.2–12 cm long; *leaflets* linear to lanceolate, to 3 mm wide, glabrous or hairy on the veins below, ciliate, margins toothed; *sepals* 3–5 mm long; *petals* yellow, 6.5–8.5 mm long; *styles* not jointed or hooked, glabrous. Common in the alpine and subalpine, 8000–14,300 ft. June–Sept. E/W. (Plate 72)

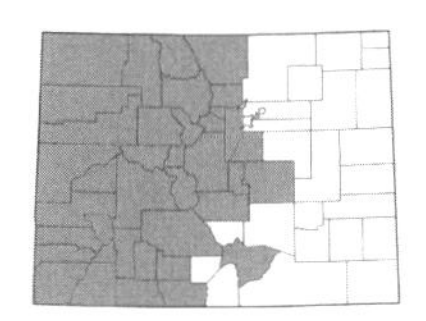

***Geum triflorum* Pursh**, OLD MAN'S WHISKERS; PRAIRIE SMOKE. [*Erythrocoma triflora* (Pursh) Greene]. Plants to 4 dm; *leaves* opposite, in 2–4 pairs, the basal ones 2–19 cm long, pinnatifid, softly hairy; *sepals* 7–10 mm long, reddish to pink or purple; *petals* white to pinkish, 7–11 mm long, erect and converging at the tips; *styles* not jointed or hooked, 3–5 cm long and strongly plumose at maturity. Common in meadows, forest openings, and along streams, sometimes found on sagebrush flats, 6400–12,500 ft. May–Aug. E/W. (Plate 72)

HOLODISCUS (K. Koch) Maxim. – OCEANSPRAY

Shrubs or small trees, unarmed; *leaves* alternate, simple, toothed or shallowly lobed, exstipulate; *flowers* perfect, in panicles or racemes; *sepals* 5; *petals* 5, white or pinkish; *stamens* 20; *pistils* usually 5; *ovary* superior; *fruit* an achene.

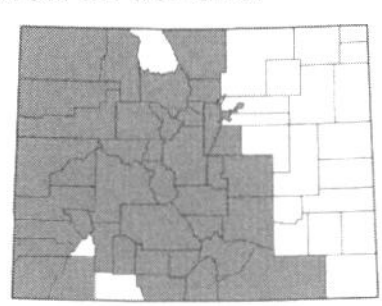

***Holodiscus dumosus* (Nutt. ex Hook.) A. Heller**, ROCKSPIREA. Shrubs mostly 5–15 dm, densely branched; *leaves* oblanceolate to obovate or elliptic, 0.5–3 × 0.2–2.5 cm, glabrous to hairy; *panicles* 3–15 cm long; *sepals* 1.3–2 mm long, villous, often pinkish; *petals* ca. 2 mm long; *achenes* villous. Common in rocky canyons and on cliffs, 5200–10,500 ft. June–Sept. E/W. (Plate 72)

IVESIA Torr. & A. Gray – MOUSETAIL

Perennial herbs; *leaves* alternate or mostly basal, pinnately divided into small lobes, stipulate; *flowers* perfect, in cymes; *sepals* 5; *petals* 5, yellow or white; *stamens* 5 or 15–35; *pistils* 2, sometimes 1-6; *ovary* superior; *fruit* an achene.

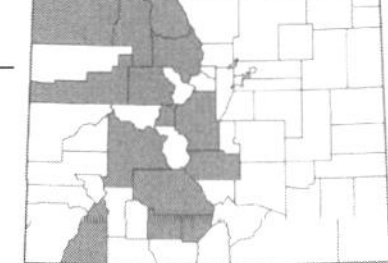

***Ivesia gordonii* (Hook.) Torr. & A. Gray**, ALPINE IVESIA. Plants 0.7–3 dm; *leaves* 1–25 cm long, with 10–25 leaflets 2–20 mm long, glandular or glabrous; *sepals* 2.5–5 mm long; *petals* about equal to the sepals; *achenes* ca. 2 mm. Found in the alpine and on dry slopes, 7500–12,800 ft. May–Sept. E/W.

MALUS Mill. – APPLE

Trees or shrubs, unarmed; *leaves* alternate, simple, stipulate; *flowers* perfect, in an umbellate or corymbose cluster; *sepals* 5; *petals* 5, white or pink; *stamens* few to numerous; *pistil* 1; *ovary* inferior; *fruit* a pome.

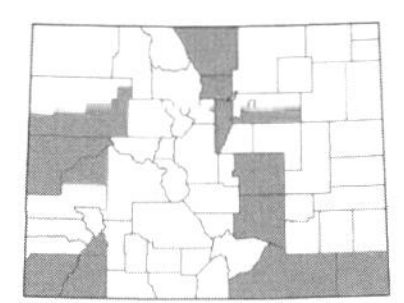

***Malus pumila* Mill.**, COMMON APPLE; CRABAPPLE. Trees to 10 m; *leaves* elliptic to ovate, 5–10 cm long, serrate, tomentose below and often above; *petals* 15–25 mm long; *pomes* red, yellow, or reddish-purple at maturity, mostly 3–12 cm diam. Cultivated and persisting near old homesteads or escaping along roadsides, 5000–8500 ft. May–June. E/W. Introduced.

PERAPHYLLUM Nutt. – PERAPHYLLUM

Unarmed shrubs; *leaves* alternate, simple, stipulate; *flowers* perfect, solitary or 2–3 in clusters; *sepals* 5; *petals* 5, white or pinkish; *stamens* 15–20; *pistils* 1–2; *ovary* inferior; *fruit* a pome.

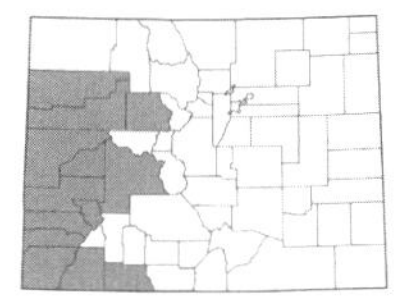

***Peraphyllum ramosissimum* Nutt.**, SQUAW APPLE. Shrubs 4–20 dm; *leaves* obovate to oblanceolate, 1–4 × 0.5–1 cm, entire or nearly so; *sepals* 3–4 mm long, entire to minutely toothed; *petals* 6.5–9 mm long; *pomes* yellow-orange, 8–20 mm diam. Found on dry slopes, usually with pinyon, juniper, or oak, 5500–9000 ft. May–Aug. W.

PETROPHYTUM (Nutt.) Rydb. – ROCK SPIRAEA

Prostrate or densely caespitose shrubs, unarmed; *leaves* evergreen, alternate, simple, entire, exstipulate; *flowers* perfect, in spike-like racemes; *sepals* 5; *petals* 5, white; *stamens* 20–40; *pistils* 3–7; *ovary* superior; *fruit* a follicle.

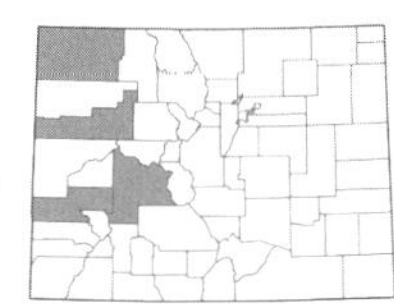

***Petrophytum caespitosum* (Nutt.) Rydb.**, TUFTED ROCKMAT. Mat-forming shrubs to 20 cm; *leaves* oblanceolate to obovate, 3–20 × 1.5–4.5 mm, pilose to nearly glabrous; *sepals* 1.2–2 mm long; *petals* 1.3–2.5 mm long; *follicles* 1.5–2 mm long. Found on rocky cliffs and in rock crevices, usually on limestone, 5800–9000 ft. July–Sept. W.

PHYSOCARPUS (Camb.) Raf. – NINEBARK

Unarmed shrubs, the bark exfoliating; *leaves* alternate, simple, palmately lobed, stipulate; *flowers* perfect, in an umbellate corymb; *sepals* 5; *petals* 5, white or cream; *stamens* 20–40; *pistils* 1–5; *ovary* superior; *fruit* a follicle.

1a. Pistil and style, or follicle, solitary; leaves 0.3–2 cm long and about as wide…***P. alternans***
1b. Pistils and styles 2–3; follicles 2–3; leaves 2–5.5 cm long…***P. monogynus***

Physocarpus alternans **(M.E. Jones) Howell,** DWARF NINEBARK. Shrubs 4–12 dm, the twigs stellate-hairy and often glandular; *leaves* ovate to orbicular-ovate, 0.3–2 cm long and nearly as wide, cordate at the base, doubly crenate on the margins; *inflorescence* 2–15-flowered; *pedicels* 2–10 mm long; *sepals* 1.3–3.5 mm long; *petals* 1.8–3.5 mm long; *follicles* solitary, 4–5 mm long, densely stellate-hairy. Found on cliffs and in rocky soil, 5500–8500 ft. May–July. W.

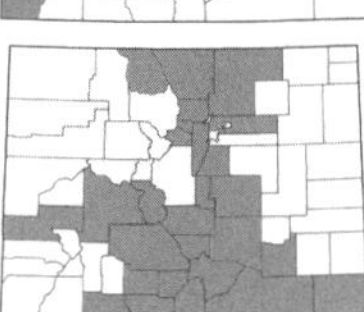

Physocarpus monogynus **(Torr.) J.M. Coult.,** MOUNTAIN NINEBARK. Shrubs 4–20 dm, the twigs glabrous to densely stellate-hairy; *leaves* ovate to orbicular-ovate, 2–5.5 cm long, doubly crenate on the margins; *inflorescence* 9–25-flowered; *pedicels* 3–10 mm long; *sepals* 2–3 mm long; *petals* 2–4 mm long; *follicles* 2–3, 3–4.5 mm long, densely stellate-hairy. Common along streams, on dry hillsides, and in canyons, 4800–10,500 ft. May–July. E/W. (Plate 72)

Physocarpus opulifolius (L.) Raf. does not occur in Colorado unless in cultivation. Its presence in Colorado was based on large-leaved *P. monogynous* specimens.

POTENTILLA L. – CINQUEFOIL

Annual, biennial, or perennial herbs, or rarely shrubs; *leaves* alternate or mostly basal, palmately or pinnately compound, stipulate; *flowers* usually perfect, in cymes or solitary, subtended by 5 bracts; *sepals* 5; *petals* 5, yellow, sometimes white or rarely reddish; *stamens* 5–many; *pistils* numerous or rarely 5–10; *ovary* superior; *fruit* an achene.

Potentillas can be difficult to key out due to rampant hybridization within the group. A maddening array of shapes and forms can result from hybridization events.

1a. Shrubs with reddish-brown, shredding bark; leaflets usually 5, margins entire; achenes densely hairy…***P. fruticosa***
1b. Herbs; leaflets various but usually lacking entire margins; achenes usually glabrous…2

2a. Petals reddish-purple; sepals reddish-purple; plants usually rooting at the nodes, prostrate or ascending, sometimes aquatic; stems and petioles reddish-brown, glandular and also with nonglandular hairs…***P. palustris***
2b. Petals yellow or sometimes white or cream; sepals usually green or sometimes anthocyanic with purplish-red tips; plants various but not aquatic; stems and petioles usually green or brownish, glandular or not…3

3a. Basal leaves pinnately compound with 5 or more leaflets (sometimes the stem leaves are reduced and simple or ternately compound), or subdigitate (with 3–5 terminal leaflets sharing a common basal point and a pair of additional leaflets separated by a short rachis)…4
3b. Basal leaves ternately compound or palmately compound with 5 or more leaflets (or sometimes both)…21

4a. Plants stoloniferous; flowers solitary; leaflets 7–25, green above and white-tomentose below…***P. anserina***
4b. Plants not stoloniferous; flowers solitary or more often in cymes; leaflets various…5

5a. Styles widest in the middle and tapering at both ends, laterally attached at or below the middle of the ovary; stems and petioles usually with multicellular, viscid hairs; leaflets nearly orbicular to elliptic or ovate, the terminal leaflet mostly 1.5–2.5 cm wide…6
5b. Styles filiform or widest at the base, variously attached; stems and petioles various; leaflets usually narrower…8

6a. Flowers in open cymes, each cyme usually with 4 or less flowers; leaflets 5–9…***P. glandulosa***
6b. Flowers in congested cymes, each cyme usually with 5 or more flowers; leaflets 7–13…7

7a. Plants 3–8 (10) dm; flowers white, cream, or pale yellow, the petals 4–8 mm long; leaflets (5) 7–9 (11)…***P. arguta***
7b. Plants 2–3 (4) dm; flowers bright yellow, the petals 7–12 (14) mm long; leaflets 9–13…***P. fissa***

8a. Plants densely glandular-puberulent and sparsely hirsute, 0.4–0.8 (1) dm; leaves subpinnately compound with 5 or rarely 7 leaflets, green above and below; known from Las Animas County…***P. subviscosa***
8b. Plants unlike the above in all respects; variously distributed…9

9a. Stipules deeply cleft or pinnatifid; leaflets with narrowly revolute margins (such that the leaflets appear whitish-gray below with a thin grayish-green line along the margin), deeply divided into linear segments…10
9b. Stipules entire or shallowly toothed; leaflets with flat margins (or nearly so), variously toothed or divided…11

10a. Lobes of the leaflets 2–5.5 (6) mm long; inflorescence conspicuously glandular…***P. pensylvanica***
10b. Lobes of the leaflets 6–14 mm long; glands absent or sparse in the inflorescence…***P. bipinnatifida***

11a. Leaflets shallowly toothed from the middle or more often 2–3-toothed only at the apex and entire for most of their length…***P. crinita***
11b. Leaflets toothed or lobed for most of their length…12

12a. Annuals or short-lived perennials from slender taproots; plants lacking a basal rosette of leaves (equally leafy throughout); style 0.5–0.6 mm long; petals equal to or more often shorter than the sepals...13
12b. Perennials from stout taproots or woody caudices; plants with a basal rosette of leaves, the stem leaves few and reduced; style 0.7–2.5 mm long; petals equal to or more often longer than the sepals...14

13a. Achenes with a large protuberance nearly as large as the achene on the ventral side; leaves usually pinnately compound with 5–9 leaflets, or sometimes a few of the upper leaves ternately compound; stamens 10, 15, or 20...***P. supina* ssp. *paradoxa***
13b. Achenes lacking a large protuberance on the ventral side; lower leaves usually pinnately compound with 5 leaflets and upper leaves ternately compound; stamens usually 10, rarely 15...***P. rivalis***

14a. Leaflets with decurrent bases, silky-villous throughout; plants large, 4–8 dm tall...***P. ambigens***
14b. Leaflets lacking decurrent bases, glabrous or variously hairy; plants 0.2–3.5 (5) dm tall...15

15a. Leaves pinnately compound with 5–23 leaflets, usually not distinctly bicolored...16
15b. Leaves subdigitate(with 3–5 terminal leaflets sharing a common basal point and a pair of lower leaflets separated by a short rachis), usually distinctly bicolored with white- or grayish-tomentose surfaces below and green to grayish surfaces above...19

16a. Leafets 5–13, densely white- or grayish-tomentose below...***P. hippiana***
16b. Leaflets 5–23, glabrate or silky-sericeous below but not densely tomentose...17

17a. Leaflets shallowly toothed or not deeply pinnatifid into linear segments, glabrous or sometimes with a few hairs below; stems erect or ascending; petioles from previous years usually remaining on the plant, such that the plant looks like it has numerous small "sticks" at the base...***P. rupincola***
17b. Leaflets deeply divided or pinnatifid at least halfway to the midrib into linear or lanceolate segments, glabrous, strigose, or silky-sericeous; stems usually decumbent; petioles unlike the above...18

18a. Pedicels recurved in fruit; leaflets 11–23, glabrous or appressed-strigose below...***P. plattensis***
18b. Pedicels straight or nearly so, spreading or erect in fruit; leaflets (5) 9–17, usually silky-sericeous below...***P. ovina***

19a. Leaflets mostly over 2.5 cm long...***P. hippiana* × *P. pulcherrima*.** A common hybrid between the two species.
19b. Leaflets 0.5–2.5 (3) cm long...20

20a. Styles 0.8–1.5 mm long; leaflets usually 3 at the tip in addition to a pair of leaflets separated by a short rachis...***P. saximontana***
20b. Styles 1.5–2 mm long; leaflets usually 5 at the tip in addition to 1 pair of leaflets separated by a short rachis (rarely with 3 leaflets at the tip)...***P. subjuga***

21a. Plants densely glandular-puberulent and sparsely hirsute throughout; leaflets 1.5–2.5 (5) cm long; plants 0.4–0.8 (1) dm tall, known from Las Animas Co...***P. subviscosa***
21b. Plants lacking glandular hairs, or if glandular hairs present then plants 2–8 dm tall with leaflets 2.5–5 dm long and not known from Las Animas Co...22

22a. Stems trailing and rooting at the nodes (often forming a ground cover); flowers solitary in the leaf axils...23
22b. Stems erect or decumbent-ascending, but not trailing and rooting at the nodes; flowers in cymes or solitary and terminal...24

23a. At least some flowers 4-merous; stems usually openly branched...***P. anglica***
23b. Flowers 5-merous; stems not branched... ***P. reptans***

24a. Leaflets densely white- or grayish-tomentose below, usually bicolored and the leaflets green or grayish-green above...25
24b. Leaflets glabrate or silky-sericeous below but not densely white- or grayish-tomentose, usually not bicolored and the leaflets the same color above and below or nearly so...31

25a. Stem leaves numerous, usually 5 or more; petals 2–4 mm long; achenes 0.5–1 mm long...***P. argentea***
25b. Stem leaves few, usually 2–3; petals 2–7 mm long; achenes 0.9–2.1 mm long...26

26a. Style over 1 mm long, filiform; basal leaves usually with 5 leaflets...27
26b. Style less than 1 mm in length, thickened at the base; basal leaves with 3 or 5 leaflets...30

27a. Leaflets toothed nearly their entire length, the larger ones with 6 or more teeth on each side, mostly (2) 2.5–6 cm long; plants 2–9 dm tall...28
27b. Leaflets toothed in the upper half, or toothed the entire length with mostly 3–5 (6) teeth on each side, mostly 1–2 cm long; plants 0.2–2.8 dm tall...29

28a. Glands in the inflorescence usually conspicuous, red-tipped...***P. pulcherrima***
28b. Glands in the inflorescence absent or inconspicuous and uncolored...***P. gracilis***

29a. Leaflets toothed the entire length or at least in the upper half...***P. concinna***
29b. Leaflets 2–3-toothed only at the tip or sometimes entire...***P. bicrenata***

30a. Glands in the inflorescence conspicuous and abundant; inflorescence 1–12 (20)-flowered; epicalyx bractlets 1.5–3 mm long; sepals 2.5–5 mm long...***P. hookeriana***
30b. Glands in the inflorescence usually absent; inflorescence 1–5 (7)-flowered; epicalyx bractlets (2) 4–7 mm long; sepals (2.5) 4–8 mm long...***P. nivea***

31a. Leaves all ternately compound; style 0.5–0.8 mm long; annuals, biennials, or short-lived perennials...32
31b. Leaves palmately compound with 5 or more leaflets, or sometimes just the upper stem leaves ternately compound with 3 leaflets; style 1.2–2.5 mm long; perennials...34

32a. Plants viscid-villous with long nonglandular hairs and shorter gland-tipped hairs, especially on the stem; stamens about 10...***P. biennis***
32b. Plants not viscid-villous, lacking glandular hairs; stamens 10–20...33

33a. Stems, especially at the base, with long, spreading, rather stiff, hirsute hairs; stems simple or branched above, erect; achenes brown, often becoming wrinkled at maturity; leaves all ternately compound...***P. norvegica***
33b. Stems with villous, often appressed hairs; stems usually much-branched, sometimes prostrate; achenes yellowish, usually smooth or lightly wrinkled at maturity; some leaves often pinnately compound with 5 leaflets...***P. rivalis***

34a. Stems usually with 5 or more well-developed stem leaves; stems, petioles, and calyx with spreading, hirsute hairs; stamens 25 or 30; mature achenes with a slightly winged margin, the surface obviously reticulate...***P. recta***
34b. Stems usually with 4 or fewer stem leaves, these usually conspicuously reduced in size from the basal leaves; stems, petioles, and calyx glabrous or with appressed, ascending hairs or sometimes with spreading hairs; stamens 20; mature achenes smooth or reticulate...35

35a. Leaflets blue-green and glaucous or less often greenish, 0.8–2.5 (4) cm long, usually toothed in the upper half only, with 1–3 (5) teeth per side; anthers 0.35–0.7 mm long...***P. glaucophylla* var. *glaucophylla***
35b. Leaflets green or grayish-green, (2) 2.5–5 (10) cm long, toothed most of their length, with (2) 5–10 teeth per side; anthers 0.7–1.4 mm long...***P. gracilis***

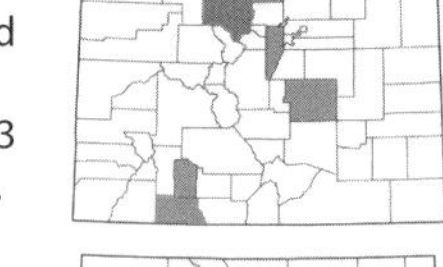

***Potentilla ambigens* Greene**, SILKYLEAF CINQUEFOIL. Perennials, 4–8 dm; *leaves* pinnately compound with 9–15 leaflets, the leaflets 3–7 cm long, with 6–18 teeth per side, decurrent on the rachis, silky-villous, green above and below; *sepals* 4–7 mm; *petals* yellow, 6–10 mm; *style* filiform, 1.8–3 mm; *achenes* 1.4–1.6 mm, smooth. Uncommon in grassy meadows, 6000–8600 ft. July–Aug. E.

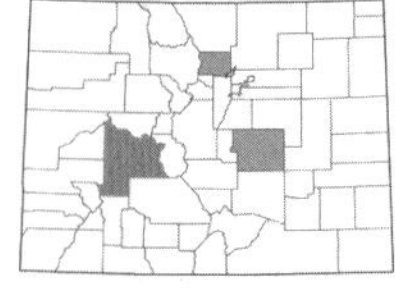

***Potentilla anglica* Laichard.**, ENGLISH CINQUEFOIL. Trailing perennials, stems openly branched; *leaves* palmately compound with (3)5 leaflets, the leaflets 1–2 cm long, sharply toothed apically, green above and below; *sepals* (3) 4–6 mm long; *petals* yellow, some flowers 4-merous, 6–9 mm long; *style* 0.9–1.5 mm long; *achenes* 1–1.8 mm long. Uncommon weed in lawns and disturbed areas, 5300–6800 ft. June–Aug. E/W. Introduced.

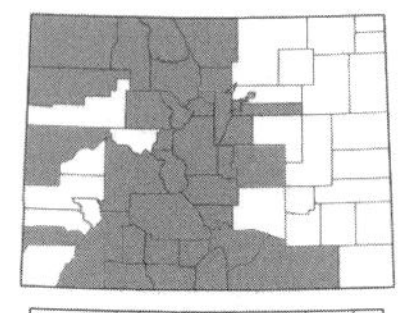

***Potentilla anserina* L.**, SILVERWEED. [*Argentina anserina* (L.) Rydb.]. Perennials, stoloniferous; *leaves* 3-25 cm long, pinnately compound with (9) 12–25 leaflets, these toothed or lobed, green above, white-tomentose below; *sepals* 3–5.5 mm long; *petals* yellow, 5.5–11 mm long; *style* 1.6–2.5 mm long; *achenes* ca. 2 mm long, glabrous. Common along pond and stream margins and in seepage or swampy areas, often in sandy soil, 5000–10,000 ft. June–Sept. E/W. (Plate 72)

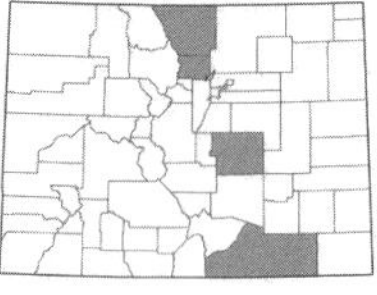

***Potentilla argentea* L.**, SILVER CINQUEFOIL. Perennials, 1–3 (6) dm; *leaves* palmately compound with 5 leaflets, leaflets 1.5–3.5 (5) cm long, lobed-serrate, with revolute margins, green and lightly tomentose above, densely gray-tomentose below; *sepals* 2–4.5 mm long; *petals* yellow, 2–4 mm long; *style* 0.6–0.9 mm long, thickened and somewhat warty at the base; *achenes* 0.5–1 mm long. Uncommon in meadows and along roadsides, 5000–8000 ft. June–Aug. E. Introduced.

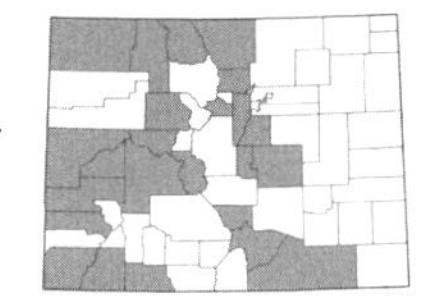

Potentilla arguta **Pursh**, TALL CINQUEFOIL. [*Drymocallis arguta* (Pursh) Rydb.]. Perennials, 4–8 (10) dm; *leaves* 10–25 cm long, pinnately compound with (5) 7–9 (11) leaflets, leaflets deeply serrate-dentate or double-serrate, brownish viscid-villous; *sepals* 4.7–10 mm long; *petals* white, cream, or pale yellow, 4–8 mm long; *style* 0.7–1.1 mm long, fusiform-thickened; *achenes* 0.8–1.5 mm long, glabrous. Uncommon in moist meadows and along streams, 4000–9000 ft. June–Aug. E/W.

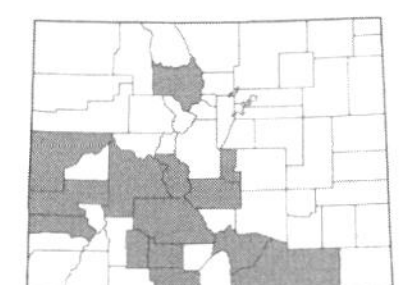

Potentilla bicrenata **Rydb.**, ELEGANT CINQUEFOIL. [*P. concinna* Richardson var. *bicrenata* (Rydb.) S.L. Welsh & B.C. Johnst.]. Perennials, 0.2–0.8 (2.8) dm; *leaves* palmately compound with 5–7 (9) leaflets, these 0.7–4 cm long, entire or 2–3-toothed at the tip, green above, white-tomentose below; *sepals* 2–5.5 mm long; *petals* yellow, 3.5–7 mm long; *style* 2–2.5 mm long, filiform; *achenes* ca. 2 mm. Found in pinyon-juniper, sagebrush, and meadows, 7400–11,000 ft. May–June. E/W.

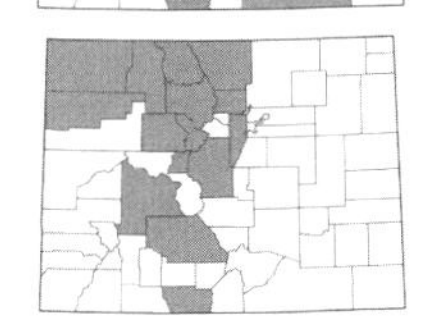

Potentilla biennis **Greene**, BIENNIAL CINQUEFOIL. Annuals or biennials, 1–7 (8.5) dm; *leaves* ternately compound, the leaflets (0.5) 1–3 (3.5) cm long, crenate-serrate to coarsely serrate, viscid-villous, green above and below; *sepals* (2) 3–5 mm long; *petals* yellow, 1.3–2.7 mm long; *style* 0.5–0.7 mm long, glandular-thickened below the middle; *achenes* pale yellowish, (0.5) 0.6–0.9 mm long, smooth. Found in moist meadows, floodplains, along pond shores and streams, and disturbed areas, 6500–8600 ft. June–Aug. E/W.

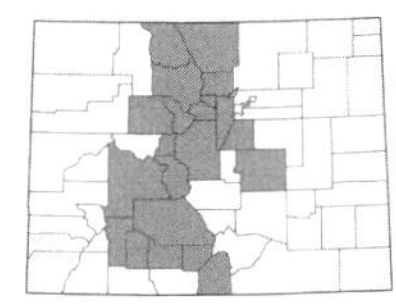

Potentilla bipinnatifida **Douglas ex Hook.**, TANSY CINQUEFOIL. Perennials, 1–5 dm; *leaves* (2) 3–6 (10) cm long, pinnately compound with 5–7 leaflets, these deeply pinnatifid with linear, rounded segments 6–15 mm long, green above and grayish-tomentulose below, margins revolute; *sepals* 3–6 mm; *petals* yellow, 3–5 mm; *style* 1–1.2 mm; *achenes* 1–1.2 mm, weakly wrinkled to smooth. Common in meadows, dry grasslands, and sagebrush flats, 7300–10,500 ft. June–Aug. E/W.

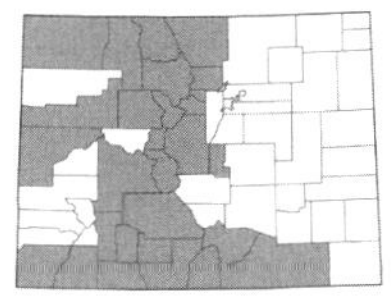

Potentilla concinna **Richardson**, EARLY CINQUEFOIL. [*P. proxima* Rydb.]. Perennials, 0.2–1.6 dm; *leaves* palmately compound with 5–7 leaflets, these 1–3 (7) cm long, toothed ¼–¾ to the midvein, green above and white-tomentose below; *sepals* 3.5–6 mm long; *petals* yellow, 4–9 mm long; *style* 1.5–2.5 mm long, filiform; *achenes* 1.5–2.1 mm long. Common on dry slopes and in meadows, sagebrush, and the alpine, 7500–13,000 ft. May–June. E/W.

Potentilla crinita **A. Gray**, BEARDED CINQUEFOIL. Perennials, 1.7–4 (5) dm; *leaves* 3-15 (20) cm long, pinnately compound with (9) 11–13 leaflets, these 1–2.5 (3) cm long, shallowly toothed apically or entire, strigose-sericeous, sometimes glabrous above; *sepals* (3) 4–6 (7) mm long; *petals* yellow, (3) 4.5–8 mm long; *style* 1.6–2.5 mm long, filiform; *achenes* 1.4–1.7 mm long, smooth to weakly wrinkled. Uncommon on shale outcrops and sagebrush flats, 6500–8000 ft. July–Aug. W.

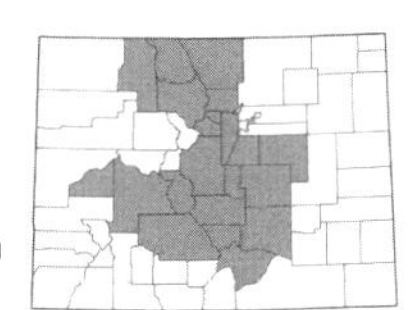

Potentilla fissa **Nutt.**, BIGFLOWER CINQUEFOIL. [*Drymocallis fissa* (Nutt.) Rydb.]. Perennials, 2–3 (4) dm; *leaves* 7-20 cm long, pinnately compound with 9–13 leaflets, these 1–4 cm long, with incised margins, hairy or glabrate above, short-hirsute and usually glandular below; *sepals* 5–7 (10) mm long; *petals* yellow, 7–12 (14) mm; *style* slender-fusiform, somewhat glandular-roughened near the middle; *achenes* brownish-yellow, 1.2–1.5 mm, slightly beaked, obscurely striate. Common on dry hillsides, rocky slopes, and in dry grasslands, 5500–10,000 ft. May–Aug. E/W. (Plate 73)

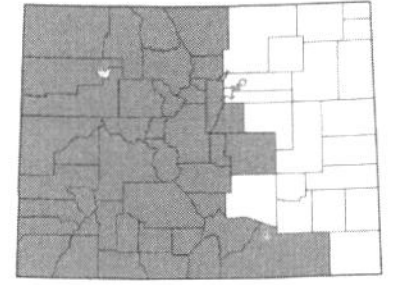

Potentilla fruticosa **Pursh**, SHRUBBY CINQUEFOIL. [*Dasiphora fruticosa* L. (Rydb.) ssp. *floribunda* (Pursh) Kartesz; *Pentaphylloides floribunda* (Pursh) Á. Löve]. Shrubs, (2) 4–7 (10) dm; *leaves* pinnately or ternately compound with (3) 5 (7) leaflets, the leaflets (0.5) 1–2 cm long, with entire or revolute margins, appressed- to spreading-hairy and grayish below, less hairy above; *sepals* 4–6 mm; *petals* yellow, (6) 8–10 (13) mm; *style* clavate; *achenes* light brown, 1.5–1.8 mm, densely white-hairy. Common in meadows, along streams, and in forests, 7500–13,500 ft. June–Sept. E/W. (Plate 73)

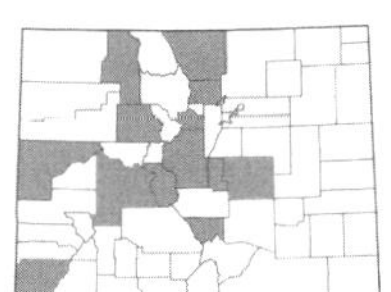

Potentilla glandulosa **Lindl.**, STICKY CINQUEFOIL. [*Drymocallis glandulosa* (Lindl.) Rydb.]. Perennials, 1–4 (5) dm; *leaves* 3–20 cm long, pinnately compound with 5–9 elliptic to nearly orbicular leaflets, these deeply serrate-dentate, glandular-hairy, and viscid-villous; *sepals* 3.7–7.5 (10) mm long; *petals* yellow, (3) 5–9.5 mm long; *style* 0.7–1 mm long, fusiform-thickened; *achenes* 0.8–0.9 mm long, glabrous. Uncommon on rocky slopes and in spruce-fir forests, 7500–9800 ft. July–Aug. W.

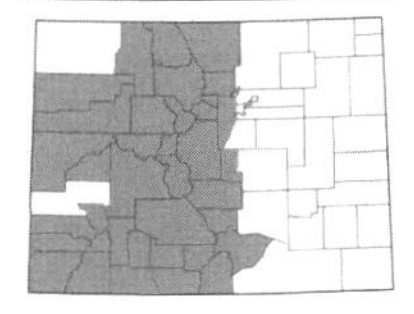

Potentilla glaucophylla **Lehm. var. *glaucophylla***, BLUELEAF CINQUEFOIL. [*P. diversifolia* Lehm. var. *glaucophylla* (Lehm.) Lehm.]. Perennials, 0.5–2.5 (4) dm; *leaves* palmately compound, leaflets 5–7, these 0.8–2.5 (4) cm long, toothed distally, sparsely to densely sericeous or glabrate, glaucous or greenish; *sepals* 3–5.5 (6.5) mm; *petals* yellow, 4.8–8.5 (10) mm; *style* 1.4–2.3 mm, filiform; *achenes* 1.2–1.6 mm. Common in meadows and alpine, 8700–14,000 ft. May–Aug. E/W. (Plate 73)

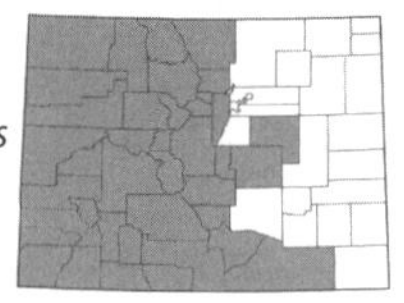

Potentilla gracilis **Douglas ex Hook.**, SLENDER CINQUEFOIL. Perennials, 2–9 dm, lacking glands or glands lacking red tips; *leaves* usually palmately compound with 5–9 leaflets, the leaflets (2) 2.5–5 (10) cm long, shallowly toothed to pinnatifid with linear segments, sparsely to densely hairy; *sepals* 4–6 (9) mm long; *petals* yellow, (4) 5–9 mm; *style* 1.6–2.5 mm long, filiform; *achenes* 1.1–1.4 (1.8) mm, smooth. Found in meadows, open forests, and the alpine, 5200–11,500 ft. June–Aug. E/W.

There are two varieties of *P. gracilis* in Colorado:

1a. Leaflets distinctly bicolored, white below, the margins incised about ¾ to the midvein...**var. *elmeri* (Rydb.) Jeps.**

1b. Leaflets usually about equally grayish-green above and below, the margins incised about ¼–½ to the midvein...**var. *fastigiata* (Nutt.) S. Watson**

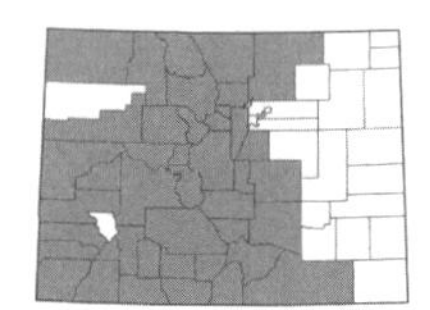

Potentilla hippiana **Lehm.**, WOOLLY CINQUEFOIL. Perennials, 0.6–3.5 (5) dm; *leaves* 5-15 (25) cm long, pinnately compound with 5–13 leaflets, the leaflets 1.2–4 cm long, toothed, grayish above, densely white-tomentose below; *sepals* 4–5.5 (6.5) mm long; *petals* yellow, 4–6.2 mm long, slightly longer than sepals; *style* 1.7–2.3 mm long, filiform; *achenes* 1.4–1.6 mm long, smooth. Common in meadows, grasslands, forests, and along streams, 6000–11,500 ft. June–Sept. E/W.

There are two varieties of *P. hippiana* in Colorado. *Potentilla hippiana* can also hybridize with *P. rupincola*:

1a. Bractlets the same color as the sepals or nearly so; leaflets usually greenish above and white below...**var. *hippiana***

1b. Bractlets markedly different in color than the sepals; leaflets often about equally greenish-gray above and below...**var. *effusa* (Douglas ex Lehm.) Dorn** [*P. effusa* Douglas ex Lehm.].

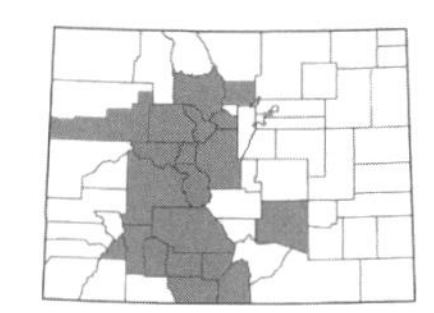

Potentilla hookeriana **Lehm.**, HOOKER'S CINQUEFOIL. [*P. modesta* Rydb.]. Mat- or cushion-forming perennials, (0.3) 0.5–3 (4) dm; *leaves* palmately or ternately compound, the leaflets 0.5-2.5 cm long, with margins incised ½–¾ to the midvein, green and sparsely hairy above, densely white-tomentose below; *sepals* 2.5–5 mm long; *petals* yellow, 3–6 (7) mm long; *style* 0.8–1.2 mm long, glandular-thickened at base; *achenes* 1–1.5 mm long, smooth. Common in rocky alpine and subalpine meadows, 10,500–14,000 ft. June–Aug. E/W. (Plate 73)

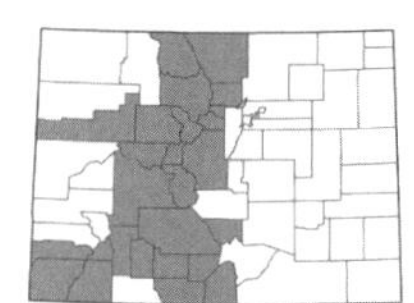

Potentilla nivea **L.**, SNOW CINQUEFOIL. [*P. subgorodkovii* Jurtzev]. Mat or cushion-forming perennials, 0.2–3 (4) dm; *leaves* ternately compound, leaflets 0.5-2 (4) cm long, margins revolute, incised ⅓–¾ to the midvein, green and sparsely hairy above, densely white-tomentose below; *sepals* (2.5) 4–8 mm; *petals* yellow, (3) 4–9 mm; *style* 0.7–1 mm, glandular-thickened at base; *achenes* 1.1–2 mm, smooth. Common in alpine and subalpine meadows, 9500–14,000 ft. June–Aug. E/W. (Plate 73)

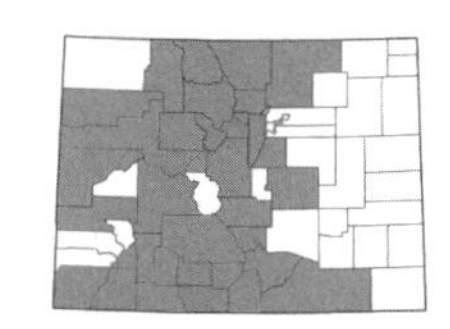

Potentilla norvegica **L.**, NORWEGIAN CINQUEFOIL. [*P. monspeliensis* L.]. Annuals, biennials, or short-lived perennials, 1–4 (9) dm; *leaves* ternately compound, the leaflets 1.5–4 cm long, shallowly lobed to toothed or coarsely serrate, hirsute, green above and below; *sepals* 5–6 (8.5) mm long; *petals* yellow, 2–5 mm long, shorter than sepals; *style* 0.7–0.8 mm long, glandular-thickened below the middle; *achenes* brown, 0.8–1.3 mm long. Common in moist meadows and along streams,4800–10,500 ft. May–Aug. E/W.

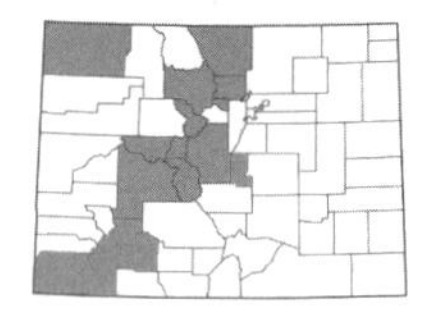

Potentilla ovina **J.M. Macoun**, SHEEP CINQUEFOIL. Mat-forming perennials, 0.5–1.5 (3.5) dm; *leaves* 2-10 (13) cm long, pinnately compound with (5) 9–17 leaflets, the leaflets (0.4) 0.7–2 (4) cm long, pinnatifid with linear segments, sericeous, strigose, or glabrate; *sepals* 3.5–5.5 mm long; *petals* yellow, 4–7 mm long; *style* 2–2.8 mm long; *achenes* 1.5–2 mm long, smooth. Found on rocky ridges and dry slopes, 7500–13,000 ft. June–Aug. E/W.

There are two varieties of *P. ovina* in Colorado:

1a. Leaflets grayish with sparse to numerous hairs, obovate; margins of distal leaflets incised ¾+ the distance to the midvein...**var. *ovina***

1b. Leaflets green and glabrate, narrowly cuneate-oblanc eolate to obovate; margins of distal leaflets incised mostly ¼–½ to the midvein...**var. *decurrens* (S. Watson) S.L. Welsh & B.C. Johnst.**

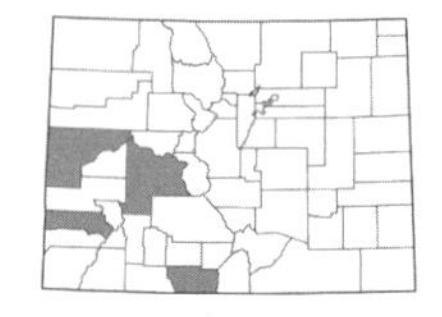

Potentilla palustris **(L.) Scop.**, PURPLE MARSHLOCKS. [*Comarum palustre* L.]. Perennials; *leaves* 2-15 cm long, pinnately compound with 5 (7) leaflets, the leaflets (2) 3–6 (8.5) cm long, sharply serrate, green above, paler and appressed-puberulent below; *sepals* 7–12 mm long; *petals* wine-purple, 2.5–6 mm long; *style* reddish, 1.3–2 mm long, filiform; *achenes* ca. 1.5 mm long, glabrous. Uncommon along pond margins and streams, 9000–11,500 ft. June–Aug. E/W.

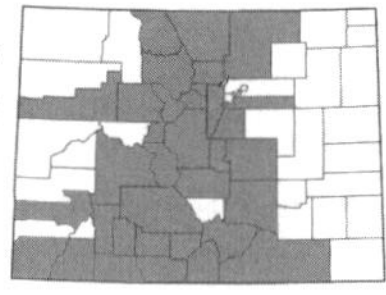

Potentilla pensylvanica **L.**, PENNSYLVANIA CINQUEFOIL. [*P. jepsonii* Ertter]. Perennials, 0.5–4 (6) dm; *leaves* 4-20 (25) cm long, pinnately compound with 4–10 (12) leaflets, leaflets 1–5 (6) cm long, cleft about halfway to midrib, green above and grayish-tomentulose below, margins revolute; *sepals* 2–7 mm; *petals* yellow, 2–6 mm; *style* 0.8–1.2 mm; *achenes* 1–1.2 mm, weakly wrinkled to smooth. Common in meadows, dry grasslands, sagebrush flats, and the alpine, 5300–13,000 ft. June–Aug. E/W.

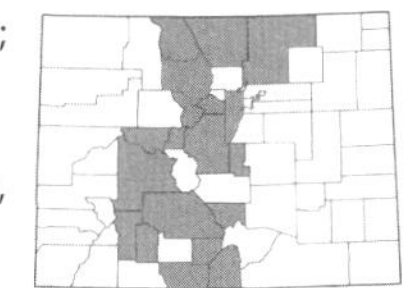

Potentilla plattensis **Nutt.**, PLATTE RIVER CINQUEFOIL. Perennials, 0.5–2.5 (4.5) dm, becoming prostrate; *leaves* 2-15 (20) cm long, pinnately compound with 11–23 leaflets, the leaflets 5–15 mm long, cleft to the midvein with linear segments, appressed-strigose or glabrous, green above and below; *sepals* 3–6 mm long; *petals* yellow, 3.5–7 mm; *style* 1.5–2.5 mm, filiform; *achenes* 1.3–2 mm, smooth. Found in moist meadows and along creeks, 6500–10,800 ft. June–Aug. E/W. (Plate 73)

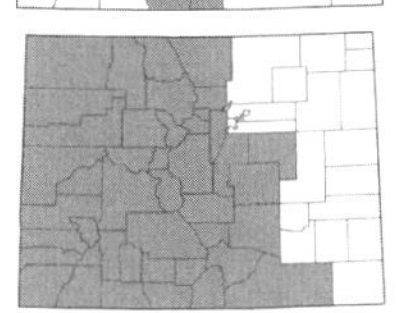

Potentilla pulcherrima **Lehm.**, BEAUTIFUL CINQUEFOIL. [*P. gracilis* var. *pulcherrima* (Lehm.) Fernald]. Perennials, 2–8 dm; *leaves* palmately compound with 5–9 leaflets, the leaflets 2.5–5 (7) cm long, shallowly toothed to pinnatifid with linear segments, green above and white-tomentose below; *sepals* 4–6 (9) mm long; *petals* yellow, 4–9 mm long; *style* 1.6–2.5 mm long, filiform; *achenes* 1.1–1.8 mm long, smooth. Common in meadows, along creeks, on open slopes, and in the alpine, 7000–13,000 ft. June–Aug. E/W.

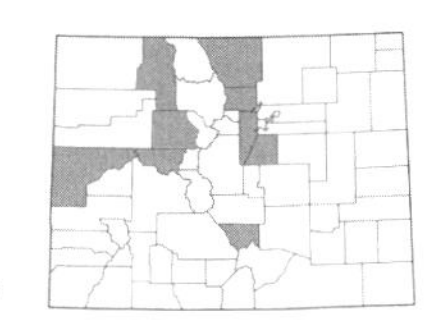

Potentilla recta **L.**, SULPHUR CINQUEFOIL. Perennials, (2) 3–5 (8) dm, hirsute; *leaves* palmately compound with 5–7 leaflets, the leaflets 3–8 cm long, serrate-lobate roughly halfway to midrib, glabrate or short-appressed-hairy above, sparsely to densely hirsute below; *sepals* 5–9 (12) mm long; *petals* yellow, 7–11 (13) mm long; *style* thickened and warty near base; *achenes* 1–1.8 mm long. Uncommon in meadows and on grassy slopes, 5200–8800 ft. May–Sept. E/W. Introduced.

Potentilla reptans **L.**, CREEPING CINQUEFOIL. Perennials, soon prostrate, rooting at the nodes, stems 1.5–10+ dm, not branched; *leaves* palmately compound with (3) 5 (7) leaflets, the leaflets (0.5) 2–4 (7) cm long, serrate nearly the entire length with 4–12 teeth per side, green, sparsely to moderatley hairy; *sepals* (3) 5–7 mm long; *petals* yellow, 7–10 (12) mm long; *style* 0.6–1.3 mm long; *achenes* 1.3–1.6 mm long. Uncommon weed found in floodplains and lawns, known from Boulder Co., 5000–5200 ft. April–Aug. E. Introduced.

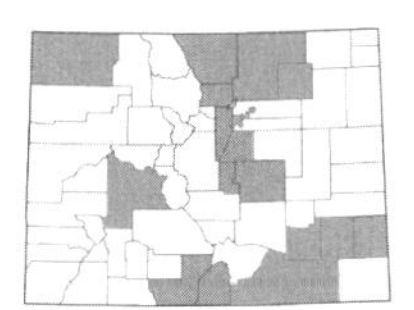

Potentilla rivalis **Nutt.**, BROOK CINQUEFOIL. Annuals or biennials, 1–4 (7) dm; *leaves* ternately compound above and mostly pinnately compound with 5 leaflets below, the leaflets 1–3 (4) cm long, coarsely toothed, villous, green above and below; *sepals* 3–5 mm long; *petals* yellow, 1.3–2.7 mm long; *style* 0.5–0.6 mm long, glandular-thickened below the middle; *achenes* yellowish, 0.7–0.9 mm long, lightly wrinkled or smooth. Found along streams, pond margins, and ephemeral pools, 5500–8000 ft. June–Oct. E/W.

Potentilla rupincola **Osterh.**, ROCK CINQUEFOIL. Perennials, 2–3 dm; *leaves* pinnately compound with 5–7 leaflets, the leaflets 1–6 cm long, sharply serrate, more or less glabrous, green above and below; *sepals* 3–4 mm long; *petals* yellow, 3–4 mm long; *style* filiform; *achenes* glabrous but covered in the wool of receptacle. Rare on rocky granite outcrops and crevices, 6800–10,000 ft. June–Aug. E. Endemic.

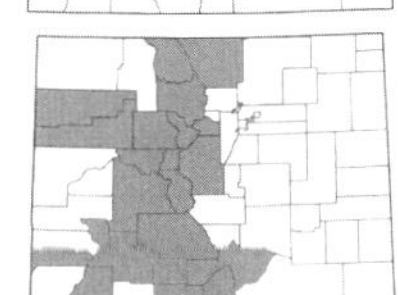

Potentilla saximontana **Rydb.**, ROCKY MOUNTAIN CINQUEFOIL. Perennials, 0.2–1 (1.5) dm; *leaves* pinnately to subdigitately compound with 5 leaflets, the leaflets 0.5–1.5 cm long, with 2–4 teeth on each margin, grayish-green to white below and green above; *sepals* 3.5–5 mm long; *petals* yellow, 4–6 (8) mm; *style* columnar-filiform, swollen at the base, 0.8–1.5 mm; *achenes* ca. 1.5 mm. Found in subalpine and alpine meadows and on rocky slopes, 10,500–14,200 ft. June–Aug. E/W.

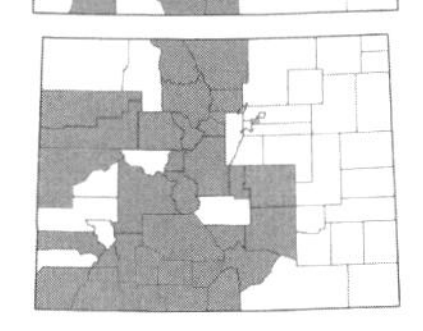

Potentilla subjuga **Rydb.**, COLORADO CINQUEFOIL. Perennials, (0.8) 1–2.5 (3.5) dm; *leaves* palmately compound or subdigitate with 5–7 leaflets (5 at the tip plus an additional pair separated on a short rachis), the leaflets 0.5–4 cm long, with deeply incised margins, silky-villous, green above, white-tomentose below; *sepals* 4–7 mm long; *petals* yellow, 4–8 mm long; *style* filiform; *achenes* 1.2–1.6 mm long. Common in meadows and the alpine, 9000–13,500 ft. June–Aug. E/W.

Potentilla subviscosa **Greene**, NAVAJO CINQUEFOIL. Perennials, 0.4–0.8 (1.1) dm; *leaves* palmately or subpinnately compound with 5 (7) leaflets, the leaflets 1.5–2.5 (5) cm long, prominently lobed, the lobes obovate to oblanceolate, glandular-puberulent and sparingly hirsute, green above and below; *sepals* 2.8–5 mm long; *petals* yellow, (3) 4–8 mm long; *style* 2–3 mm long, filiform; *achene* 1.5–1.6 mm long, striate. Uncommon on rocky slopes in pinyon-oak, 8000–8500 ft. May–July. E.

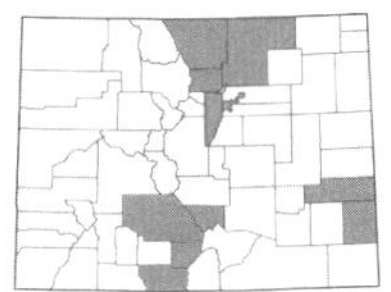

Potentilla supina **L. ssp. *paradoxa*** **(Nutt.) Soják**, BUSHY CINQUEFOIL. [*P. paradoxa* Nutt.]. Annuals or short-lived perennials, 1–4 (5) dm; *leaves* 4-10 cm long, pinnately compound with 5–9 leaflets, the leaflets 10–23 (27) mm long, crenate-toothed, villous-hirsute with spreading hairs, green above and below; *sepals* 2.8–5 (6) mm long; *petals* yellow, 2.5–3 mm long; *style* 0.5–0.6 mm long, glandular-thickened at base; *achenes* brown, 0.7–1.3 mm long, strongly rugose. Found along pond margins, in seepage areas, in floodplains, and along streams, 4500–7500 ft. June–Aug. E.

PRUNUS L. – CHERRY; PEACH; PLUM

Trees or shrubs; *leaves* alternate, simple, serrate or crenate-serrate, usually with a pair of glands on the petiole and at the base of the leaf; *flowers* perfect, in an umbel, corymb, raceme, or sometimes solitary; *sepals* 5; *petals* 5, white or pink; *stamens* 15–30; *pistil* 1; *ovary* superior, unicarpellate; *fruit* a drupe.

1a. Flowers in a raceme; fruit 0.8–1.5 cm long, dark purple, black, or deep red...2
1b. Flowers solitary or in clusters of 2–5 (8); fruit various...4

2a. Hypanthium hairy inside; petals 5–7.5 mm long; fruit black...***P. padus***
2b. Hypanthium glabrous or sparsely hairy inside; petals 2.5–4 mm long; fruit black, dark purple, or deep red...3

3a. Sepals soon deciduous, with a callous-tipped, fringed margin...***P. virginiana* var. *melanocarpa***
3b. Sepals persistent in fruit, usually lacking a fringed margin...***P. serotina* var. *virens***

4a. Leaf margins with gland-tipped teeth (each tooth with a dark tip surmounted with a clear, dew-drop like substance, look under 10× magnification to see)...5
4b. Leaf margins lacking gland-tipped teeth or simply with a cartilaginous, dark tip...8

5a. Sepals with glandular margins; petioles hairy, at least somewhat so above...6
5b. Sepals glabrous or with ciliate margins but not glandular; petioles glabrous...7

6a. Leaves densely hairy below (at least when young, sometimes losing hairs in age); pedicels densely hairy; petioles lacking glands...***P. gracilis***
6b. Leaves glabrous or with hairs below only along the veins; pedicels glabrous or sparsely hairy; petioles with 2 glands near the top or glands lacking...***P. rivularis***

7a. Leaves strongly trough-shaped or folded; branches usually armed with spines; fruit 2–3 cm long, red or yellowish; much-branched shrub 1–3 (5) m tall, usually forming large thickets...***P. angustifolia***
7b. Leaves flat; branches lacking spines; fruit 0.5–0.7 cm long, red; small tree or shrub 5–10 m tall...***P. pensylvanica***

8a. Petioles hairy; young branches densely and softly hairy; pedicels hairy; leaves densely hairy below and thinly hairy above (when young, becoming glabrous in age); fruit blue-purple, 2–3 cm long, glaucous...***P. domestica***
8b. Petioles glabrous or slightly hairy on the upper side; young branches glabrous or slightly hairy; pedicels glabrous or hairy; leaves glabrous or hairy below on the veins, glabrous above; fruit various but unlike the above...9

9a. Flowers solitary or rarely 2–3 per bud, pink, violet, red, or white; nonnative, cultivated trees or shrubs to 8 m tall, single plants not forming dense thickets...10
9b. Flowers in clusters of 2–5, white; native shrubs or trees, often forming dense thickets, 0.1–8 m tall...12

10a. Pedicels 5–12 mm long; ovary and fruit glabrous; fruit red to purple at maturity; leaves often purple...***P. cerasifera***
10b. Flowers and fruit sessile or on pedicels to 3 mm long; ovary and fruit velvety soft-hairy or tomentose; fruit yellow-orange or yellow at maturity; leaves green...11

11a. Leaves broadly ovate to nearly rotund; flowers white or pale pink; sepals glandular on the margins but otherwise glabrous; fruit 3–5 cm long...***P. armeniaca***
11b. Leaves lanceolate to elliptic-lanceolate or lanceolate-oblong; flowers pink to dark pink or red, seldom white; sepals hairy on the outer surface; fruit 5–8 cm long...***P. persica***

12a. Shrubs or small trees 3–8 m tall; sepals 3–4 mm long, hairy above, soon reflexed; petals 7–12 mm long; fruit 2–3 cm long; branchlets often with spines...***P. americana***
12b. Shrubs 0.1–0.4 (0.7) m tall; sepals 1.2–1.7 mm long, toothed and often glandular at the tip but otherwise glabrous, mostly erect; petals 6.5–8 mm long; fruit 1.3–1.5 cm long; branchlets lacking spines...***P. pumila* var. *besseyi***

***Prunus americana* Marshall**, WILD PLUM. Shrubs to 8 m, often spinescent; *leaves* obovate to lanceolate-ovate, 2.5–7 × 0.5–3 cm, with doubly serrate margins, green and glabrous above, slightly hairy below; *flowers* in clusters of 2–5; *pedicels* 7–20 mm long; *sepals* 3–4 mm, reflexed, hairy above; *petals* white, 7–12 mm long; *drupes* reddish-purple or yellowish, 2–3 cm diam. Common in canyons, on rocky slopes, and along streams, 3500–8700 ft. April–May. E. (Plate 73)

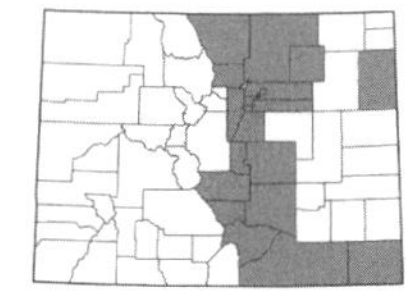

***Prunus angustifolia* Marshall**, SANDHILL PLUM. Shrubs with spines, forming thickets, 1–3 (5) m; *leaves* trough-shaped or folded, 2–7 × 1–2 cm, margins serrate with gland-tipped teeth, glabrous above, sparsely hairy below; *flowers* in clusters of 2–4; *pedicels* 3–6 (10) mm long; *sepals* 1–1.5 mm long; *petals* white, 3.5–6 mm long; *drupes* red to yellowish, 2–3 cm diam. Uncommon on sandy dunes and along roadsides, known from Baca Co., 3700–4500 ft. April–May. E.

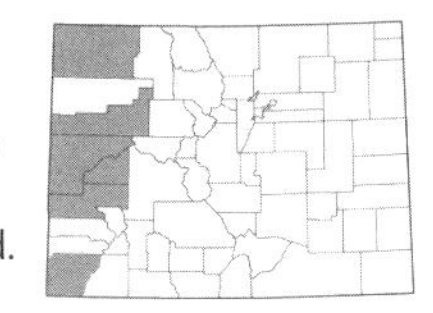

Prunus armeniaca **L.**, APRICOT. [*Armeniaca vulgaris* Lam.]. Small trees to 8 m, armed or unarmed; *leaves* broadly ovate to nearly rotund, 1.5–7 × 1–6 cm, margins serrate, glabrous or hairy along the veins below; *flowers* solitary, appearing before the leaves; *pedicels* to 3 mm long, hairy; *sepals* glabrous with glandular margins; *petals* white, 8–12 mm long; *drupes* yellow-orange, 3–5 cm diam., hairy. Escaping from cultivation along roadsides, 5500–6500 ft. April–May. W. Introduced.

Prunus cerasifera **Ehrh.**, CHERRY PLUM. Small trees to 6 m, unarmed; *leaves* ovate to elliptic, 2–6.5 × 1.5–4 cm, with serrate or doubly serrate margins, hairy along the veins below, often purple; *flowers* solitary or sometimes in clusters of 2–3; *pedicels* 5–12 mm long, glabrous; *petals* pink, violet, or white, 6–9 mm long; *drupes* red to purple. Commonly cultivated as an ornamental tree, occasionally escaping near urban areas, 5000–6500 ft. April–May. E. Introduced.

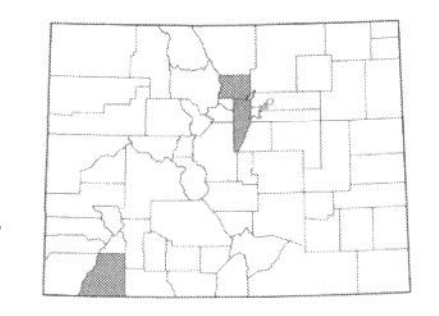

Prunus domestica **L.**, EUROPEAN PLUM. Small trees to 5 m, usually armed; *leaves* ovate to obovate, mostly 2–10 × 1–6 cm, with serrate margins, thinly hairy above and densely hairy below; *flowers* solitary or sometimes in clusters of 2–3; *pedicels* hairy; *petals* white, ca. 10 mm long; *drupes* blue-purple, 2–3 cm long, glaucous. Commonly cultivated and occasionally escaping along roadsides, 5500–6500 ft. April–May. E/W. Introduced.

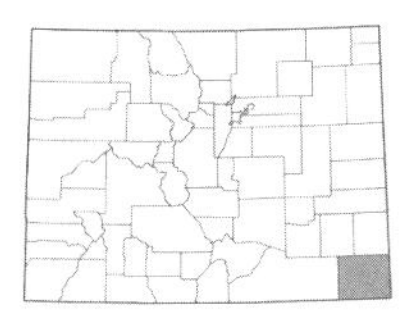

Prunus gracilis **Engelm. & A. Gray**, OKLAHOMA PLUM. Unarmed shrubs to 3 m; *leaves* elliptic to ovate, 2–5 × 1–2.5 cm, the margins serrate with gland-tipped teeth, sparsely hairy above and densely hairy below; *flowers* in clusters of 2–4 (8), appearing before or with young leaves; *pedicels* 6–12 mm, hairy; *sepals* 1.2–1.7 mm, slightly gland-toothed, hairy; *petals* white, 5–6.5 mm long; *drupes* yellow-red, 1.5–2 cm diam. Uncommon on sandy dunes, known from Baca Co. where it was found growing and potentially hybridizing with *P. angustifolia*, 3500–4000 ft. Mar.–April. E.

Prunus padus **L.**, EUROPEAN BIRD CHERRY. Unarmed trees to 8 m; *leaves* elliptic to obovate, 1.5–20 × 0.7–3.5 cm, the margins serrate, glabrous except sometimes hairy on the veins below; *flowers* in racemes mostly 7–12 cm long; *pedicels* 5–17 mm long, subtended by caducous bracts, glabrous; *hypanthium* hairy inside; *sepals* with fringed margins; *petals* white, 5–7.5 mm long; *drupes* black, ca. 1 cm diam. Commonly cultivated and occasionally escaping, known from Larimer Co., 5500–6500 ft. April–May. E. Introduced.

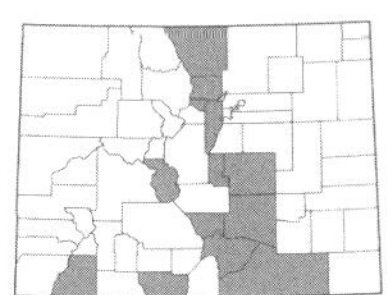

Prunus pensylvanica **L. f.**, PIN-CHERRY. [*Cerasus pensylvanica* (L. f.) Loisel.]. Unarmed small trees or shrubs, 5–10 m; *leaves* lanceolate to elliptic, 6–12 cm long, finely serrate with gland-tipped teeth, glabrous; *flowers* in clusters of 2–5, opening with the leaves; *pedicels* 10–15 mm long, glabrous; *sepals* glabrous; *petals* white, 5–7 mm long; *fruit* red, 0.5–0.7 cm diam. Found in forests, along creeks and streams, and in cool ravines, 5200–9500 ft. April–June. E/W.

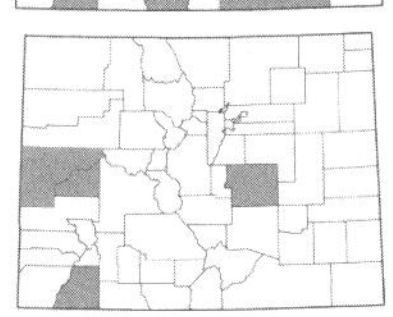

Prunus persica **(L.) Batsch.**, PEACH. Trees to 3 (4) m, unarmed; *leaves* lance-oblong to elliptic-lanceolate or lanceolate-oblong, mostly 3–15 × 0.5–6 cm, with serrate margins, glabrous; *flowers* usually solitary, appearing before the leaves; *sepals* hairy on the outer surface; *petals* pink to red or white, ca. 10 mm long; *drupes* yellow-orange, 5–8 cm diam, velvety soft-hairy. Widely cultivated and occasionally escaping, 5500–6500 ft. April–May. E/W. Introduced.

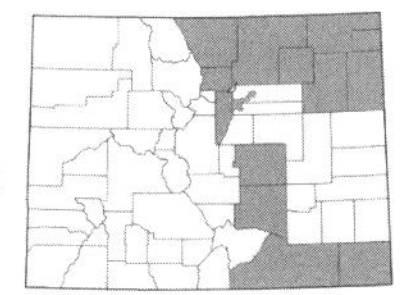

Prunus pumila **L. var. *besseyi* (L.H. Bailey) Gleason**, SAND CHERRY. [*Cerasus pumila* (L.) Michx. ssp. *besseyi* (L.H. Bailey) W.A. Weber]. Low shrubs 1–4 (7) dm, unarmed; *leaves* oblanceolate, obovate, or elliptic, 4–7 × 1–3 cm, with finely serrate to nearly entire margins, the upper surface glabrous, lower surface paler and glaucous; *flowers* in clusters of 2–4, opening with the leaves; *pedicels* 4–12 mm long, glabrous; *sepals* 1.2–1.7 mm long, toothed and often glandular at the tip but otherwise glabrous; *petals* white, 6.5–8 mm long; *drupes* dark purple, 1.3–1.5 cm diam. Found on rocky slopes and on sandstone cliffs and bluffs, 3500–7500 ft. April–May. E.

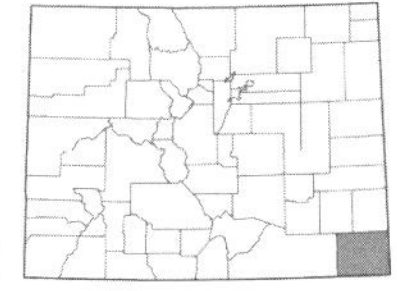

Prunus rivularis **Scheele**, CREEK PLUM. Shrubs 1–2 (3) m, forming dense thickets, unarmed; *leaves* trough-shaped or folded, rarely flat, 5–7 × 2–4 cm, serrate with gland-tipped teeth, glabrous above and hairy along the veins below; *flowers* in clusters of 2–4 (8), opening with the leaves; *pedicels* 6–15 mm long, glabrous to sparsely hairy; *sepals* 1.5–2 mm long, glandular, hairy near the base within; *petals* white, 5–6.5 mm long; *drupes* yellowish-red, 1.3–2 cm diam. Uncommon along creeks and streams, 3500–4500 ft. April–May. E.

Prunus serotina **Ehrh. var. *virens* (Wooton & Standl.) McVaugh**, BLACK CHERRY. [*Padus virens* Wooton & Standl.]. Trees 10–15 m, unarmed; *leaves* lanceolate to oblong or oblanceolate, 6–12 × 2–5 cm, serrate with callous-tipped teeth, glabrous; *flowers* in racemes (5) 8–15 cm long; *pedicels* 3–6 mm long, glabrous; *sepals* 0.6–1.2 mm long, sometimes glandular-erose, persistent; *petals* white, 2.5–4 mm long; *drupes* dark purple, 0.8–1.2 cm diam. Not reported for Colorado but could potentially occur in the southeastern counties (Baca, Las Animas). April–May.

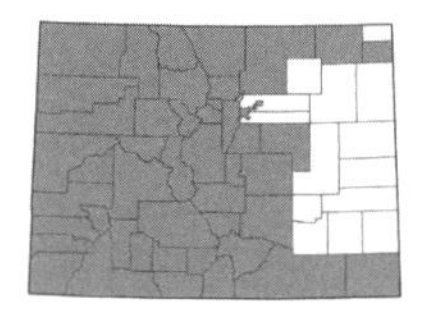

***Prunus virginiana* L. var. *melanocarpa* (A. Nelson) Sarg.**, CHOKECHERRY. [*Padus virginiana* (L.) Mill. ssp. *melanocarpa* (A. Nelson) W.A. Weber]. Shrubs or small trees 2–6 (10) dm, unarmed; *leaves* ovate to elliptic or obovate, 4–12 × 3–6 cm, serrate, glabrous or sometimes hairy on the veins below; *flowers* in racemes 4–10 (15) cm long; *pedicels* 3–6 mm long, hairy; *sepals* 1–1.5 mm long, glandular-erose, deciduous; *petals* white, 3–4 mm long; *drupes* red to dark purple or nearly black, ca. 1 cm diam. Common on rocky slopes, in canyons and gulches, open forests, and along streams, 4000–10,500 ft. April–June. E/W. (Plate 73)

PURSHIA DC. ex Poir. – BITTERBRUSH

Shrubs or small trees; *leaves* alternate, usually in fascicles, simple and 3-lobed at the apex or pinnately compound, white-tomentose below, the margins revolute; *flowers* perfect, solitary; *sepals* 5; *petals* 5, yellow, pink, white, or cream; *stamens* numerous; *pistils* 1–12; *ovary* superior; *fruit* an achene.

1a. Leaves pinnately 5–7-lobed, glandular-punctate above; pistils 4–10; styles of achenes 2–6 cm long, plumose; petals 7–14 mm long…***P. stansburiana***
1b. Leaves 3-lobed at the apex, hairy but not glandular-punctate above; pistils 1 or rarely 2; styles of achenes hard, ca. 0.5–1 cm long, not plumose; petals 4–7 (10) mm long…***P. tridentata***

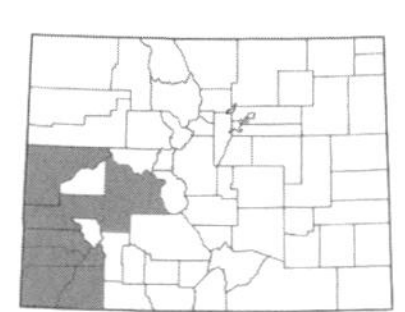

***Purshia stansburiana* (Torr.) Henr.**, STANSBURY'S CLIFFROSE. [*Cowania stansburiana* Torr.]. Shrubs or small trees to 4 m; *leaves* pinnately 5–7-lobed, 3–15 mm long, green and glandular-punctate above, white-tomentose below; *sepals* 4–6 mm long; *petals* white to cream or yellowish, 7–14 mm long; *styles* of achenes 2–6 cm long, plumose. Found on dry hillsides, mesa rims, and in canyons, often with pinyon-juniper or sagebrush, native to the western slope, introduced on the eastern slope, 4200–6600 ft. May–July. E/W. (with native distribution shown in map)

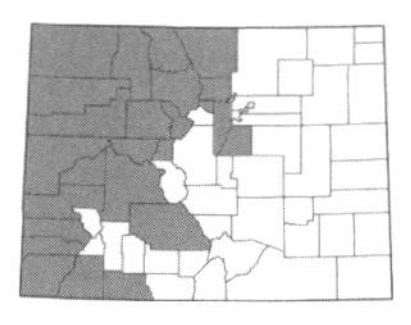

***Purshia tridentata* (Pursh) DC.**, ANTELOPE BITTERBRUSH. Shrubs to 2 m; *leaves* 3-lobed at the apex, 4–20 mm long, green and hairy above, white-tomentose below; *sepals* 4–6 mm long; *petals* white to cream or yellowish, 4–7 (10) mm; *styles* of achenes ca. 0.5–1 cm long, not plumose. Common on dry hillsides, often with sagebrush or pinyon-juniper, 4700–9500 ft. April–July. E/W. (Plate 73)

ROSA L. – ROSE

Shrubs or woody vines, armed with prickles or thorns; *leaves* alternate, odd-pinnately compound, stipulate; *flowers* perfect, solitary or in cymes; *sepals* 5; *petals* 5 (or more in cultivated forms), pink, rose, white, or yellow; *stamens* numerous; *pistils* numerous; *ovary* superior; *fruit* an achene enclosed within a fleshy hypanthium. (Joly & Bruneau, 2007)

Rosa nutkana Presl var. *hispida* Fernald either does not occur in Colorado or simply represents variation within the *Rosa blanda* complex.

1a. Flowers yellow, often double (with more than 5 petals)…***R. ×harisonii***
1b. Flowers pink, rose, or rarely white, usually with 5 petals…2

2a. Prickles present and usually numerous on new growth (these are long, slender, straight thorns without greatly enlarged, flattened bases)…3
2b. Prickles absent on new growth (sometimes densely present at the base of the stem but absent above), thorns usually stout, often sparsely spaced on the stem, with an enlarged, flattened base, and usually curved…4

3a. Leaflets mostly 5–7, rarely 9, the teeth on the margins often gland-tipped; flowers usually solitary or sometimes in clusters of 2–3…***R. acicularis* ssp. *sayi***
3b. Leaflets mostly 9–11, rarely as few as 7, the teeth on the margins rarely gland-tipped; flowers usually in clusters of 3 or more…***R. arkansana***

4a. Thorns smaller, straight or curved; sepals seldom with lateral lobes; native plants…***R. blanda***
4b. Thorns stout, the larger ones about 1 cm long, strongly curved or hooked; sepals with prominent lateral lobes; cultivated plants sometimes escaping…5

5a. Leaflets glandular-stipitate below; pedicels and hypanthium glandular-stipitate or glandular-hairy…***R. eglanteria***
5b. Leaflets glabrous or with a few deciduous glands on the main veins below; pedicels and hypanthium usually glabrous…***R. canina***

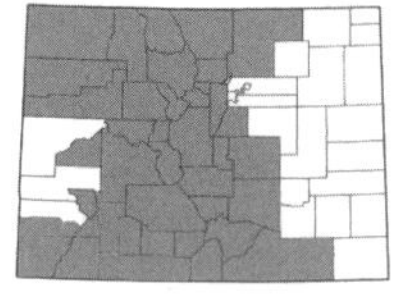

***Rosa acicularis* Lindl. ssp. *sayi* (Schwein.) W.H. Lewis**, PRICKLY ROSE. Stems densely bristly with straight, slender prickles; *leaflets* mostly 5–7 (9), 1.5–5 (8) cm long, serrate, the teeth often gland-tipped, glabrous above and glabrous to villous and often glandular below; *flowers* usually solitary or in clusters of 2–3; *sepals* 1.5–4 cm long, usually glandular; *petals* pink to rose, (1.5) 2–3.5 cm long. Common on dry hillsides, along streams, in aspen forests, and in meadows, 5200–11,500 ft. June–Aug. E/W.

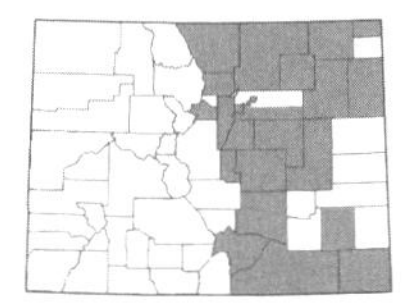

***Rosa arkansana* Porter**, PRAIRIE ROSE. Stems densely bristly with straight, slender, unequal prickles; *leaflets* (7) 9–11, 1–5 (6) cm long, serrate but the teeth rarely gland-tipped, glabrous to villous below; *flowers* in clusters of 3+; *sepals* 1.5–2 (3) cm long, glandular-stipitate; *petals* white to pink or rose, 1.5–3 cm long. Common on the shortgrass prairie and adjacent foothills, 3500–7500 ft. June–Aug. E.

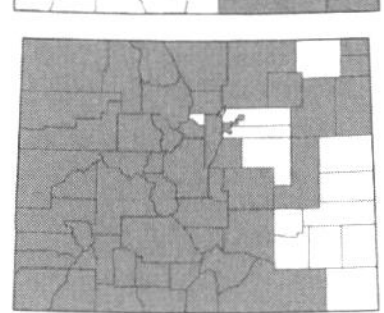

***Rosa blanda* Ait.**, SMOOTH ROSE. [*R. manca* Greene; *R. woodsii* Lindl.]. Stems with stout thorns with an enlarged, flattened base or absent, slender prickles absent on new growth; *leaflets* 5–9, 1.5–5 cm long, serrate with the teeth sometimes gland-tipped, glabrous to hairy below; *flowers* solitary or in clusters of 2–5; *sepals* 1–2.5 cm long, sometimes glandular-stipitate; *petals* pink to rose, 1.5–2.5 cm long. Common on dry hillsides, along streams and rivers, and in forests and meadows, 3500–11,700 ft. June–Aug. E/W. (Plate 73)

***Rosa canina* L.**, DOG ROSE. Stems with stout, flattened and strongly curved or hooked thorns, sometimes with slender, bristle-like prickles intermixed; *leaflets* 5 or 7, 1–2 cm long, serrate with gland-tipped teeth, glabrous or with a few glands on the veins below; *flowers* solitary or in few-flowered clusters; *sepals* 1–2 cm long, with prominent slender lateral lobes; *petals* pink, 1–2 cm long. Cultivated and escaping near Boulder, 5800–6500 ft. June–Aug. E. Introduced.

***Rosa eglanteria* L.**, SWEETBRIER ROSE. [*R. rubiginosa* L.]. Stems with stout, flattened and strongly curved or hooked thorns, sometimes with slender, bristle-like prickles intermixed; *leaflets* 5 or 7, 1–2 cm long, serrate with gland-tipped teeth, glandular-stipitate and hairy below, glabrous to sparsely hairy above; *flowers* solitary or in few-flowered clusters; *sepals* 1–2 cm long, with prominent slender lateral lobes; *petals* pink, 1–2 cm long. Often cultivated and occasionally escaping near Boulder, 5800–6500 ft. June–Aug. E. Introduced.

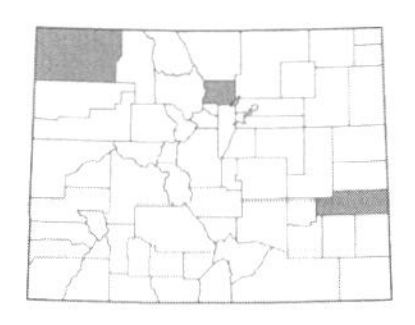

***Rosa ×harisonii* Rivers**, HARISON'S YELLOW ROSE. Stems with stout, flattened and strongly curved or hooked thorns, sometimes with slender, bristle-like prickles intermixed; *leaflets* 5–7 (9), 1–2.5 cm long, serrate with gland-tipped teeth, glandular-stipitate on the midrib below; *flowers* usually solitary; *sepals* entire; *petals* yellow, 2.5–3 cm long, usually double (with more than 5 petals). Cultivated and occasionally persisting near old homesteads and buildings, 4500–7000 ft. June–Aug. E/W. Introduced.

RUBUS L. – BLACKBERRY; RASPBERRY

Armed or unarmed shrubs, vines, or rarely herbs; *leaves* alternate, simple and lobed or more often pinnately or palmately compound, stipulate; *flowers* perfect, in a raceme, corymb, or sometimes solitary; *sepals* 5; *petals* 5, white, pink, or red; *stamens* numerous; *pistils* numerous; *ovary* superior; *fruit* an aggregate of drupelets.

1a. Leaves simple and palmately lobed; petals 8–30 mm long, white…2
1b. Leaves ternately compound with 3 leaflets, or rarely pinnately compound with 5 leaflets; petals 3–16 mm long or if longer then rose-colored…4

2a. Leaves 11–20 cm wide; sepal apices abruptly narrowed to a linear, bristle-like tip; flowers usually in clusters of 2–4 (7)…***R. parviflorus* var. *parviflorus***
2b. Leaves 2.5–6 cm wide; sepals acute or acuminate but not narrowed to a bristle-like tip; flowers usually solitary…3

3a. Leaves glabrous to slightly hairy below only on the veins, sparsely hairy and becoming glabrous above…***R. deliciosus***
3b. Leaves softly hairy above and below…***R. neomexicanus***

4a. Stems armed with prickles and/or thorns; leaves usually white- or grayish-tomentose below, green above…5
4b. Stems lacking prickles or thorns; leaves glabrous or nearly so…8

5a. Leaflets green below, deeply cleft on the margins; petals deeply trilobed at the tip; stems with stout, recurved, flattened prickles…***R. laciniatus***
5b. Leaflets white- or grayish-tomentose below, merely serrate on the margins; petals not trilobed at the tip; prickles various…6

6a. Plants glandular-hairy on the smaller stems and petioles as well as in the inflorescence; prickles slender, numerous, spreading; sepals armed with prickles; drupelets red at maturity…***R. idaeus* var. *strigosus***
6b. Plants hairy but lacking glandular hairs; prickles stout, flattened, recurved or sometimes spreading; sepals tomentose and lacking prickles; drupelets black at maturity…7

7a. Petals 9–16 mm long; flowers numerous in a panicle; fruit separating from the stem with the receptacle included...***R. armeniacus***
7b. Petals 3–5 mm long, shorter than the sepals; flowers 3–7 in a terminal cluster; fruit separating from the receptacle...***R. occidentalis***

8a. Petals white, 4–8 mm long; leaflets with acute or acuminate tips...***R. pubescens* var. *pubescens***
8b. Petals rose or pink, 10–20 mm long; leaflets with mostly rounded tips...***R. acaulis***

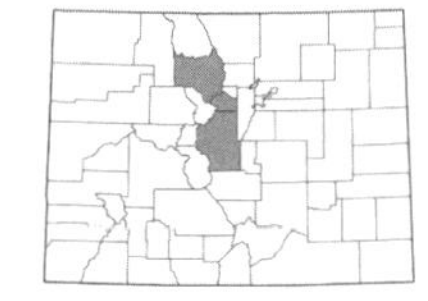

***Rubus acaulis* Michx.**, DWARF RASPBERRY. [*Cylactis arctica* (L.) W.A. Weber ssp. *acaulis* (Michx.) W.A. Weber]. Rhizomatous perennials, 2–15 cm tall, unarmed; *leaves* ternately compound; *leaflets* ovate to obovate, 1.5–3 (5) cm long, glabrous to sparsely appressed-hairy, the margins serrate; *inflorescence* usually solitary; *sepals* 7–10 mm long; *petals* rose or pink, 10–20 mm long; *drupelets* reddish, globular, ca. 1 cm wide. Uncommon in moist bogs and along creeks, 8500–10,000 ft. June–July. E/W.

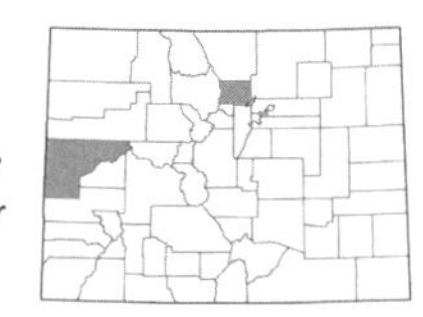

***Rubus armeniacus* Focke**, HIMALAYAN BLACKBERRY. [*R. discolor* Weihe & Nees]. Shrubs, often forming thickets, armed; *leaves* pinnately to palmately compound with 3–5 leaflets; *leaflets* ovate to elliptic, 3–12 cm long, densely tomentose below, green and glabrous above, serrate; *inflorescence* a panicle of numerous flowers; *sepals* 6–10 mm long; *petals* white, 9–16 mm long; *drupelets* red or black at maturity. Cultivated and occasionally escaping along roadsides and trails, 5200–8000 ft. June–Aug. E/W. Introduced.

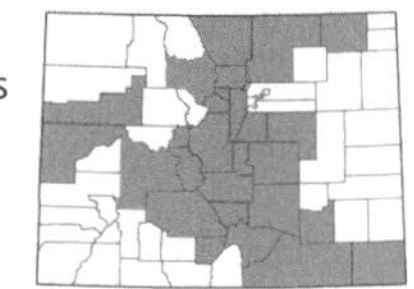

***Rubus deliciosus* Torr.**, DELICIOUS RASPBERRY. [*Oreobatus deliciosus* E. James]. Shrubs to 1.5 m, unarmed; *leaves* simple, 3–5-lobed, orbicular-reniform, 3–6 cm wide, glabrate above and glabrous or sparsely hairy on the veins below, with dentate margins; *inflorescence* of a single terminal flower on a short leafy branchlet; *sepals* 9–18 mm long; *petals* white, 10–30 mm long; *drupelets* 8–12 mm diam., dryish. Common in canyons, along streams, on rocky hillsides, and growing in cracks of boulders, 4400–11,000 ft. May–Aug. E/W. (Plate 73)

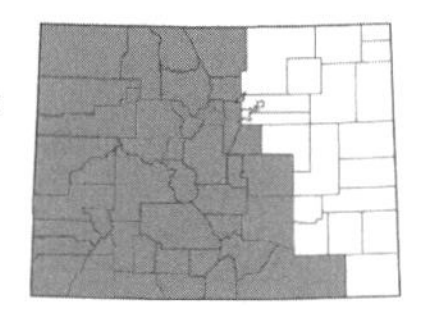

***Rubus idaeus* L. var. *strigosus* (Michx.) Maxim.**, RED RASPBERRY. Shrubs 2–15 (20) dm tall, armed with numerous, slender prickles; *leaves* pinnately or ternately compound with 3–5 leaflets; *leaflets* ovate to elliptic, 1.2–10 cm long, grayish-tomentose below and green and sparsely hairy above, serrate; *inflorescence* of few-flowered clusters or sometimes solitary; *sepals* 6–11 mm long; *petals* white, 3–7 mm long; *drupelets* red, edible. Common on dry hillsides, along creeks, and in forests, 6300–12,000 ft. June–Aug. E/W. (Plate 73)

***Rubus laciniatus* Willd.**, EVERGREEN BLACKBERRY. Shrubs, often trailing, with stems to 10 m long, armed with flattened, recurved prickles; *leaves* pinnately compound with 5 leaflets; *leaflets* green and glabrous to sparsely hairy above, green and hairy below, with deeply cleft or toothed margins; *inflorescence* of flat-topped racemes; *sepals* 8–15 mm long; *petals* white or pinkish, 9–15 mm long, deeply trilobed at the tip; *drupelets* 1–1.5 cm diam., edible. Uncommon in gulches and open forests, perhaps not persisting, known from near Boulder, 5500–6000 ft. May–July. E. Introduced.

***Rubus neomexicanus* A. Gray**, NEW MEXICO RASPBERRY. [*Oreobatus deliciosus* (E. James ex Torr.) Rydb. ssp. *neomexicanus* (A. Gray) W.A. Weber]. Shrubs to 1.5 m, unarmed; *leaves* simple, 3–5-lobed, 3–6 cm wide, softly hairy above and below, with dentate margins; *inflorescence* of 1–2 flowers in leaf axils or terminating short leafy branchlets; *sepals* 5–20 mm long; *petals* white, 8–30 mm long; *drupelets* red, 8–12 mm diam., dryish. Found in canyons and open valleys, 4400–5500 ft. May–Aug. E.

***Rubus occidentalis* L.**, BLACK RASPBERRY. Shrubs with arching and trailing stems, armed; *leaves* ternately to pinnately compound with 3–5 leaflets; *leaflets* ovate to elltiptic, 5–9 cm long, densely white-tomentose below, green and glabrous above, with serrate margins; *inflorescence* of 3–7 flowers in flat-topped clusters; *sepals* 6–8 mm long; petals white, 3–5 mm long; *drupelets* black at maturity, 1–1.5 cm diam. Cultivated and occasionally escaping, 5000–6000 ft. May–July. E. Introduced.

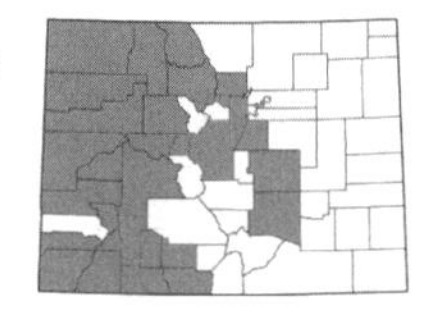

Rubus parviflorus* Nutt. var. *parviflorus, THIMBLEBERRY. [*Rubacer parviflorum* (Nutt.) Rydb.]. Shrubs to 1.5 (3) m tall, unarmed; *leaves* simple, 5-lobed, 6–15 (20) × 11–18 cm, glabrous to sparsely stipitate glandular, serrate; *inflorescence* a 2–4 (7)-flowered, flat-topped raceme; *sepals* 10–18 mm long, terminating in a linear, bristle-like tip; *petals* white, 8–20 mm long; *drupelets* red, hairy. Found in shady, cool gulches, canyons, and moist ravines, 6500–10,500 ft. June–Aug. E/W.

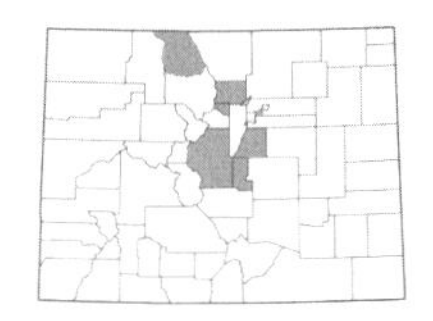

Rubus pubescens **Raf. var.** ***pubescens***, DWARF RED BLACKBERRY. [*Cylactis pubescens* (Raf.) W.A. Weber]. Shrubs with trailing, creeping stems, unarmed; *leaves* ternately to pinnately compound with 3–5 leaflets; *leaflets* obovate to ovate, 2–6 cm long, glabrous or rarely appressed-hairy, serrate; *inflorescence* of 1–2 (7) terminal flowers, the axis often glandular-stipitate; *sepals* 3–6 mm long; *petals* white, 4–8 mm long; *drupelets* red, 5–12 mm diam. Uncommon in shady, moist places, 7000–8800 ft. June–Aug. E.

SANGUISORBA L. – BURNET

Herbs; *leaves* alternate, pinnately compound; *flowers* in dense heads or spikes; *sepals* 4, petaloid, white, greenish, red, or purple; *petals* absent; *stamens* 2–12; *pistils* 1 or 2; *ovary* superior; *fruit* an achene enclosed in a 4-winged, dry hypanthium.

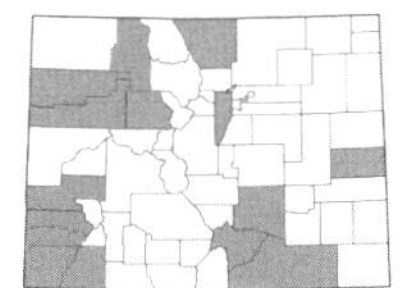

Sanguisorba minor **Scop.**, GARDEN BURNET. Perennials, 2–5 dm; *leaflets* 9–17, oval to obovate, 0.5–2 cm long, coarsely serrate; *spikes* subglobose, 0.8–4 cm long with ovate bracts; *flowers* imperfect, the lower staminate and the upper pistillate. Weedy plants found along roadsides and on open slopes, 4700–9500 ft. May–July. E/W. Introduced.

SIBBALDIA L. – SIBBALDIA

Low, mat-forming perennial herbs; *leaves* mostly basal, trifoliately compound; *flowers* perfect, in cymes or solitary, subtended by 5 bracts; *sepals* 5; *petals* 5, yellow; *stamens* 5; *pistils* 5–15; *ovary* superior; *fruit* an achene.

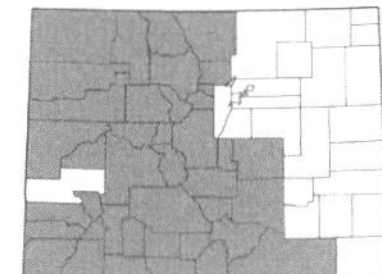

Sibbaldia procumbens **L.**, CREEPING SIBBALDIA. Plants 0.5–1.5 dm; *leaflets* oblanceolate to obovate, 1–3.5 cm long, usually 3-toothed at the apex, hairy; *sepals* 2.5–5 mm long; *petals* 1.5–3 mm long, shorter than the sepals; *achenes* ca. 1 mm long. Common in alpine and on rocky, subalpine slopes or spruce-fir forests, 9000–14,000 ft. June–Aug. E/W. (Plate 73)

SORBUS L. – MOUNTAIN ASH

Trees or shrubs; *leaves* alternate, odd-pinnately compound; *flowers* perfect, in a compound corymb; *sepals* 5; *petals* 5, white, cream, or yellow; *stamens* 15–20; *pistil* 1, 2–5-carpellate; *ovary* inferior or half-inferior; *fruit* a pome, orange to red.

1a. Bud scales hairy; leaflets mostly 13–17, rounded to acute at the tips, the margins serrate; cultivated trees 4–15 (20) m tall...***S. aucuparia***

1b. Bud scales glabrous but with ciliate margins; leaflets mostly 11–13, usually acuminate or sometimes acute at the tips, the margins sharply serrate; native shrubs or small trees 1–4 m tall...***S. scopulina* var. *scopulina***

Sorbus aucuparia **L.**, EUROPEAN MOUNTAIN ASH. Trees to 15 (20) m; *leaflets* mostly 13–17, oblong, (2) 3–6 cm long, rounded to acute at the apex, the margins serrate; *sepals* hairy; *petals* white, ca. 4–5 mm long; *pomes* red, ca. 1 cm diam. Commonly cultivated and occasionally escaping along the edges of woods and ravines, 6000–7000 ft. May–July. E/W. Introduced.

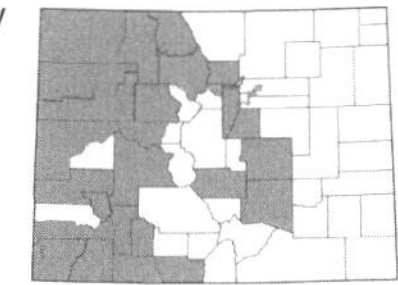

Sorbus scopulina **Greene var.** ***scopulina***, MOUNTAIN ASH. Shrubs or small trees 1–4 m; *leaflets* mostly 11–13, oblong to oblong-elliptic, (2) 3–7 cm long, acuminate or acute at the apex, the margins sharply serrate; *sepals* hairy; *petals* white, 4–6 mm long; *pomes* orange to red, ca. 1 cm diam. Found in cool ravines, along streams, and in forests and meadows, 6000–11,500 ft. June–Aug. E/W.

RUBIACEAE Juss. – MADDER FAMILY

Herbs, shrubs, or trees; *leaves* opposite or whorled, simple, entire or serrate, stipulate with leaf-like stipules; *flowers* perfect or sometimes imperfect, actinomorphic, in various cymose inflorescences or rarely solitary; *calyx* 4–8-lobed, persistent; *corolla* (3) 4 or 5, or 8–10-lobed; *stamens* 4 or 5, adnate to the corolla tube; *pistil* 1; *ovary* inferior (ours), 2 (5)-carpellate, with parietal or axile placentation; *fruit* a capsule, berry, drupe, or a schizocarp.

1a. Leaves whorled, or sometimes just the upper leaves merely opposite and the lower leaves whorled; corolla white, greenish-white, yellow, yellowish-green, or rarely purplish...***Galium***

1b. Leaves merely opposite; corolla light blue to purplish or white...***Hedyotis***

GALIUM L. – BEDSTRAW

Herbs or shrubs with square stems; *leaves* whorled; *flowers* perfect or sometimes imperfect; *calyx* minute or absent; *corolla* usually 4-lobed, the corolla rotate or nearly so; *stamens* 4; *ovary* 2-carpellate, 2-loculed; *fruit* a schizocarp. (Dempster & Ehrendorfer, 1965; Puff, 1976)

1a. Leaves mostly in whorls of 4, sometimes the uppermost merely opposite, the tips rounded or acute but usually not cuspidate…2
1b. Leaves mostly in whorls of 5 or more, rarely only in whorls of 4 on smaller branches, the tips usually cuspidate or with a short awn point…5

2a. Fruit glabrous; leaves 1-nerved, 0.5–1.2 cm long; stems lax and weak, often scrambling on other vegetation…***G. trifidum* var. *subbiflorum***
2b. Fruit hairy or bristly, rarely glabrous (and then the leaves strongly 3-nerved); stems erect or ascending; leaves usually larger, 1–5 cm long…3

3a. Slender annuals; uppermost leaves merely opposite, the lower leaves in whorls of 4 with one pair larger; flowers few and solitary in leaf axils; fruit covered with short, hooked hairs (0.2–0.4 mm long)…***G. bifolium***
3b. Perennials; flowers in terminal cymose panicles or numerous and solitary in leaf axils at the ends of branches; fruit hairy with long or short, straight or curled (but not hooked) hairs, rarely glabrous…4

4a. Fruit densely covered with long, straight or flexuous white bristles about 1.5–2.5 mm long; plants woody towards the base; flowers yellowish-green, imperfect; leaves 1-nerved or slightly 3-nerved…***G. multiflorum***
4b. Fruit covered with shorter, straight or curled hairs, rarely glabrous; plants not woody at the base; flowers white or cream-colored, perfect; at least the larger leaves strongly 3-nerved…***G. boreale***

5a. Flowers bright yellow, fading white, in dense terminal panicles; fruit glabrous or nearly so; leaves narrowly linear; stems erect, 3–9 dm tall, densely white-hairy at the nodes…***G. verum***
5b. Flowers white or greenish-white, in cymes or on axillary peduncles with divergent pedicels that are generally 3-flowered at the end, rarely terminal and then few-flowered; fruit covered with hooked bristles, rarely glabrous; leaves oblanceolate, narrowly elliptic, or nearly linear; stems erect or weak and more or less prostrate, often with a row of bristles at the nodes but not densely white-hairy…6

6a. Stems erect, retrorsely hispid just at the nodes, otherwise glabrous or nearly so; flowers 4–7 mm wide; leaves usually in whorls of 8, mostly oblanceolate, antrorsely scabrous-ciliate on the margins…***G. odoratum***
6b. Stems weak and tending to scramble on other vegetation, retrorsely hooked-scabrous pubescent on the angles, at least below; flowers 1–3 mm wide; leaves in whorls of 6 or 8, narrowly elliptic to somewhat oblanceolate, antrorsely or retrorsely scabrous-ciliate on the margins…7

7a. Annuals; leaves bristly-hispid above and strongly retrorsely scabrous-ciliate on the margins and midrib below; stems densely retrorsely scabrous on the angles the entire length…***G. aparine***
7b. Perennials; leaves scabrous-ciliate on the margins and midrib below, but otherwise glabrous; stems usually weakly retrorsely scabrous on the angles or scabrous mostly at the base…8

8a. Bristles on the fruit short, 0.15–0.3 mm long, the ovary at anthesis appearing smooth or minutely bristly; flowers generally more than three in cymose inflorescences at the ends of the main stem and branches; leaves often narrower, mostly 0.15–0.7 cm wide…***G. mexicanum* ssp. *asperrimum***
8b. Bristles on the fruit longer, 0.5–1 mm long, the ovary at anthesis conspicuously bristly; flowers generally in threes at the ends of axillary peduncles with divergent pedicels, sometimes more numerous and in cymose inflorescences; leaves usually wider, mostly 0.4–1.2 cm wide…***G. triflorum***

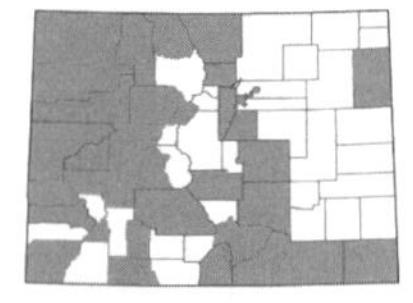

***Galium aparine* L.**, STICKYWILLY; CLEAVERS. [*G. spurium* L.]. Annuals, weak-stemmed and usually scrambling, the stems retrorosely scabrous; *leaves* (6) 8 per nodes, bristly-hispid above and retrorsely scabrous-ciliate on the margins and midrib below, 1–4 cm long; *corolla* greenish-white; *fruit* 1.5–5 mm long, covered with hooked bristles or rarely glabrous. Common in moist and shady places, sometimes in disturbed places, 3500–8000 ft. May–Aug. E/W. Introduced.

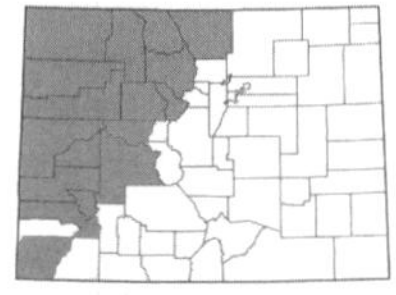

***Galium bifolium* S. Watson**, TWINLEAF BEDSTRAW. Annuals, mostly erect, 0.5–2 dm tall, the stems glabrous; *leaves* 4 per node or the upper opposite, glabrous, 1–2.5 cm long; *corolla* white, minute; *fruit* 2.5–3.5 mm long, covered with short, hooked hairs 0.2–0.4 mm long. Found in moist places, sometimes in drier areas, 6500–10,800 ft. May–Aug. E/W.

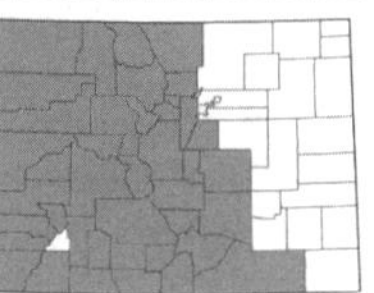

***Galium boreale* L.**, NORTHERN BEDSTRAW. [*G. septentrionale* Roem. & Schult.]. Perennials, erect from creeping rhizomes; *leaves* 4 per node, glabrous to scabrous, 1.5–4.5 cm long; *corolla* white; *fruit* ca. 2 mm long, covered with short, straight or curled hairs, or glabrous. Common in forests and meadows, and along streams, 5000–11,800 ft. May–Aug. E/W.

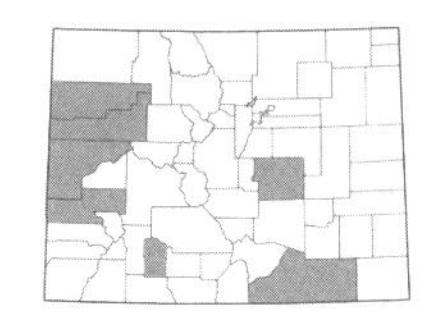

***Galium mexicanum* Kunth ssp. *asperrimum* (A. Gray) Dempster**, MEXICAN BEDSTRAW. Perennials, weak-stemmed and usually scrambling, the stems retrorsely scabrous; *leaves* usually 6 per node, retrorsely scabrous-ciliate on the margins and midrib below, rarely glabrous, 1.5–4 cm long; *corolla* white; *fruit* 1.5–2 mm long, covered with short, hooked bristles 0.15–0.3 mm long. Found in moist places and wooded slopes, 6900–8500 ft. July–Aug. E/W.

***Galium multiflorum* Kellogg**, SHRUBBY BEDSTRAW. Perennials, dioecious, erect from creeping rhizomes, the stems glabrous; *leaves* 4 per node, glabrous, the larger ones 1–2 cm long; *corolla* yellowish-green; *fruit* densely covered with bristles ca. 1.5–2.5 mm long. Common in dry, rocky places, 4700–8500 ft. May–July. W. (Plate 74)

There are two varieties of *G. multiflorum* in Colorado:

1a. Leaves mostly narrower, mainly 5–10 times longer than wide, mostly 1–2 mm wide and 10–20 mm long; plants glabrous or nearly so...**var. *coloradoense* (W. Wight) Cronquist**, COLORADO BEDSTRAW. [*G. coloradoense* W. Wight]. Common in dry, rocky places, 4700–8500 ft. May–July. W.

1b. Leaves mostly wider, 1.5–6 times longer than wide, mostly 1–5 mm wide and 3–15 mm long; plants often densely hispid or sometimes glabrous...**var. *multiflorum***, SHRUBBY BEDSTRAW. Uncommon in dry, rocky places, barely entering Colorado and currently known only from Moffat Co., 5200–5500 ft. May–July. W.

***Galium odoratum* (L.) Scop.**, SWEET WOODRUFF. Perennials, erect, 1.5–5 dm tall, the stems retrorsely hispid at nodes, otherwise glabrous; *leaves* (6) 8 (10) per node, antrorsely scabrous-ciliate on the margins, 1.5–5 cm long; *corolla* white; *fruit* ca. 3 mm long, covered with hooked bristles. A garden plant occasionally escaping from cultivation, doubtfully persisting, known from the Denver area, 5000–5500 ft. May–June. E. Introduced.

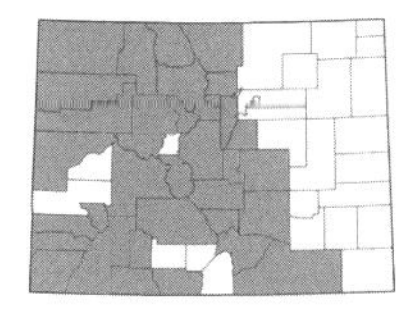

***Galium trifidum* L. var. *subbiflorum* Wiegand**, THREEPETAL BEDSTRAW. [*G. brandegeei* A. Gray; *G. columbianum* Rydb.]. Perennials, weak-stemmed and usually scrambling, the stems retrorse-scabrous on the angles or sometimes glabrous; *leaves* mostly 4 per node, retrorsely scabrous along the margins and midrib below, 0.5–1.5 cm long; *corolla* white; *fruit* glabrous. Common in moist, often shady, places, 5500–10,500 ft. June–Aug. E/W.

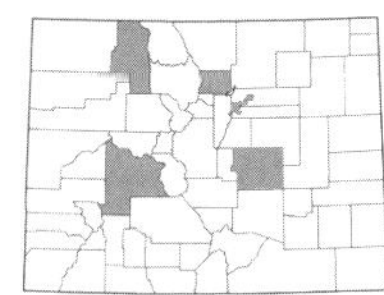

***Galium triflorum* Michx.**, FRAGRANT BEDSTRAW. Perennials, stems prostrate or scrambling, usually retrorsely scabrous; *leaves* mostly (5) 6 per node, antrorsely scabrous on the margins and midrib below, 1.5–4 cm long; *corolla* whitish; *fruit* 1.5–2.5 mm long, covered with hooked bristles 0.5–1 mm long. Common in moist, shady places, 5800–10,500 ft. May–Aug. E/W. (Plate 74)

***Galium verum* L.**, YELLOW SPRING BEDSTRAW. Perennials, erect, 3–9 dm tall, the stems usually densely hairy at the nodes; *leaves* (6) 8 (12) per node, often minutely hairy; *corolla* bright yellow, fading whitish; *fruit* 1–2 mm long, usually glabrous. Uncommon along roadsides and fencerows, 5500–8500 ft. May–Sept. E/W. Introduced.

HEDYOTIS L. – BLUETS

Herbs; *leaves* opposite, stipules short and adnate to the petioles or leaf bases; *calyx* 4-lobed; *corolla* 4-lobed, funnelform or salverform; *stamens* 4; *ovary* 2-carpellate, 2-loculed; *fruit* a capsule.

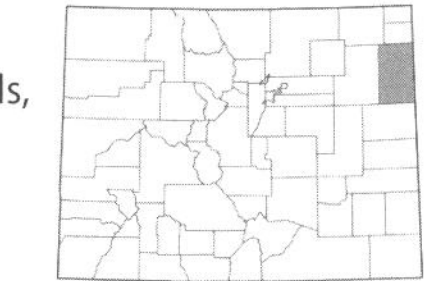

***Hedyotis nigricans* (Lam.) Fosberg**, NARROWLEAF BLUET. [*Stenaria nigricans* (Lam.) Terrell]. Perennials, 1–5 dm; *leaves* 1–3 (4) cm × 1–3 mm, usually revolute, glabrous or minutely scabrous; *corolla* white to light blue or purplish; *fruit* 2.5–4 mm long. Uncommon on limestone outcrops on the eastern plains, 3500–3600 ft. May–Aug. E. (Plate 74)

RUPPIACEAE Hutch. – DITCHGRASS FAMILY

Annuals, submerged aquatic herbs; *leaves* usually alternate, sessile, stipulate, the stipules forming a sheath adnate to the leaf base; *flowers* perfect, actinomorphic, the peduncles often spiraling and elongating; *perianth* absent; *stamens* 2; *pistils* 4–16, distinct; *stigmas* sessile, broad, and flat; *fruit* a drupaceous achene.

RUPPIA L. – DITCHGRASS

Leaves filiform; *flowers* with 2 sessile anthers and 4 pistils; *fruit* long-stipitate, beaked. (Fernald & Wiegand, 1914; Thorne, 1993)

***Ruppia cirrhosa* (Petagna) Grande**, SPIRAL DITCHGRASS. [*R. maritima* L. var. *occidentalis* (S. Watson) Graebn.]. Stems profusely branched, filiform; *leaves* 2–10 (15) cm × 0.5–1 mm, rounded or obtuse at the tips; *inflorescence* axillary; *fruit* 2–3 mm long. Found in ponds and small lakes, with occurrences scattered across the state, 3700–7500 ft. Aug.–Oct. E/W. (Plate 74)

RUSCACEAE – BUTCHER'S-BROOM FAMILY

Shrubs or perennial herbs; *leaves* basal or cauline and alternate; *flowers* actinomorphic, perfect or imperfect; *tepals* 4 or 6, white; *stamens* 6; *stigma* 3-lobed; *ovary* superior, 2–3-carpellate; *fruit* a berry or capsule.

1a. Perennial herbs; leaves cauline and alternate, distichous; fruit a red, fleshy berry...***Maianthemum***
1b. Shrubby plants; leaves numerous and narrowly linear, in dense basal rosettes; fruit a broadly 3-winged capsule...***Nolina***

MAIANTHEMUM F.H. Wigg. – MAYFLOWER

Perennial herbs from rhizomes; *leaves* alternate, distichous, *flowers* perfect, in terminal panicles or racemes; *fruit* a fleshy, red berry. (LaFrankie, 1986)

1a. Flowers in a 70-numerous-flowered terminal panicle; tepals 0.5–2 mm long...***M. racemosum* ssp. *amplexicaule***
1b. Flowers in a 6–15-flowered terminal raceme; tepals 4–7 mm long...***M. stellatum***

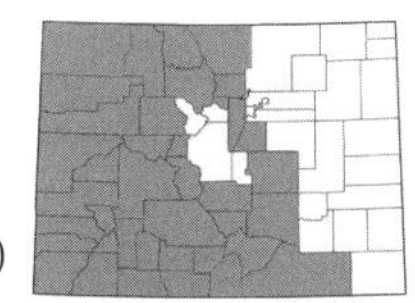

***Maianthemum racemosum* (L.) Link ssp. *amplexicaule* (Nutt.) LaFrankie**, LARGE FALSE SOLOMON'S SEAL. [*Smilacina amplexicaulis* Nutt.]. Plants 3–10 dm; *leaves* oblong-lanceolate, 7–20 cm long; *inflorescence* a numerous-flowered terminal panicle; *tepals* 0.5–2 mm long; *berries* 4–6 mm diam. Common along streams, in forests, and in shady places, 6000–10,500 ft. May–July. E/W. (Plate 74)

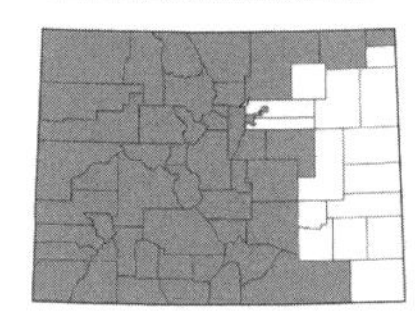

***Maianthemum stellatum* (L.) Link**, FALSE SOLOMON'S SEAL. [*Smilacina stellata* (L.) Desf.]. Plants 3–5 dm; *leaves* lanceolate, 5–15 cm long; *inflorescence* a 6–15-flowered terminal raceme; *tepals* 4–7 mm long; *berries* 7–10 mm diam. Common in forests, shady places, and along streams, 4500–12,500 ft. May–July. E/W. (Plate 74)

NOLINA Michx. – BEARGRASS

Shrubs; *leaves* numerous, narrowly linear with a greatly expanded base, margins entire or serrulate, the apex tapering to a short-pointed tip; *flowers* functionally imperfect, white to cream or greenish-white; *fruit* capsular, 3-loculed, 3-lobed, broadly 3-winged, often splitting irregularly.

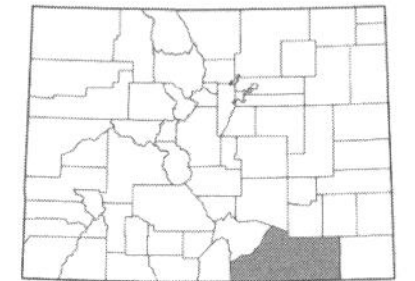

***Nolina texana* S. Watson**, TEXAS SACAHUISTE; BUNCHGRASS. Plants 3–6 dm; *leaves* 6–12 dm × 2–5 mm; *inflorescence* a dense compound panicle; *tepals* 3–3.5 mm long. Known only from Mesa de Maya (Las Animas Co.), ca. 6000 ft. April–June. E.

This is the northernmost locality for *Nolina texana*.

RUTACEAE Juss. – CITRUS FAMILY

Herbs, or mostly shrubs and trees; *leaves* alternate or opposite, simple or compound, with prominent to obscure oil glands, exstipulate or rarely stipulate; *flowers* perfect or imperfect, actinomorphic, in racemes or cymes; *sepals* 4 or 5, distinct or basally connate; *petals* 4 or 5, distinct or basally connate; *stamens* 4 or 5 or more, distinct; *pistil* 1 or 3–5; *ovary* superior (ours); *fruit* a follicle, capsule, or samara.

1a. Shrubs or small trees, strongly malodorous; leaves palmately 3-foliate; fruit a samara; flowers greenish-white...***Ptelea***
1b. Herbaceous plants, woody only at the base; leaves simple, linear, entire; fruit a capsule that is conspicuously gland-dotted (in the shape of inflated dutchman's breeches with legs protruding upwards); flowers yellow...***Thamnosma***

PTELEA L. – HOPTREE

Shrubs or small trees, without spines; *leaves* palmately 3-foliate, deciduous, alternate; *flowers* perfect or imperfect, in cymes; *sepals* 4 or 5; *petals* 4 or 5, greenish-white, distinct; *stamens* 4 or 5, alternate with the petals; *pistil* 1; *ovary* 2–3-carpellate; *fruit* a samara or capsule.

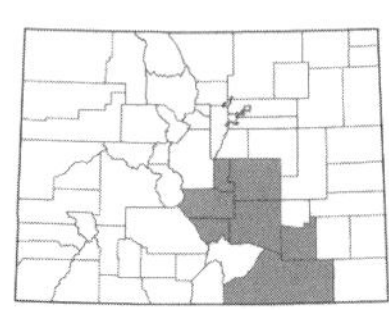

***Ptelea angustifolia* Benth.**, NARROW-LEAVED HOPTREE. [*P. baldwinii* Torr. & A. Gray; *P. trifoliata* L. ssp. *angustifolia* (Benth.) V.L. L.H. Bailey var. *angustifolia* (Benth.) M.E. Jones]. Shrubs or small trees, 1–3 m; *leaflets* 2–6 cm long, with prominent glands, margins serrate; *petals* greenish-white, 4–6 mm long; *fruit* a samara. Found in gulches and canyons, 4300–7700 ft. June–Aug. E. (Plate 74)

THAMNOSMA Torr. & Frém. – DESERTRUE; DUTCHMAN'S BREECHES

Shrubs or herbs; *leaves* simple, entire, alternate, strongly gland-dotted; *flowers* perfect, in short racemes or cymose clusters; *sepals* 4; *petals* 4, yellowish to purplish, with a well-developed disk; *stamens* 8, inserted on the disk; *pistil* 1; *ovary* 2-carpellate; *fruit* a capsule, conspicuously gland-dotted.

Thamnosma texana **(A. Gray) Torr.**, RUE OF THE MOUNTAINS; DUTCHMAN'S BREEECHES. Perennials or subshrubs, 0.5–3 dm; *leaves* linear, 0.5–1.5 cm long, with entire margins, strongly gland-dotted; *petals* yellow, 3–6 mm long; *fruit* a capsule, 3–7 mm long, strongly gland-dotted. Known only from 2 old collections from near Canon City (Fremont Co.) and likely extirpated in Colorado. 5300–5500 ft. Mar.–May. E.

SALICACEAE Mirb. – WILLOW FAMILY

Shrubs or trees, usually dioecious; *leaves* alternate, simple, with entire, toothed, sometimes with glandular margins, usually stipulate; *flowers* usually imperfect, arranged in catkins; *sepals* usually absent or sometimes present, or reduced to 1 or 2 nectaries; *petals* absent; *stamens* 1–60; *pistil* 1; *ovary* superior, 2–7-carpellate; *fruit* a capsule with comose seeds. (Argus, 2010)

1a. Buds 3–10-scaled; catkins usually pendulous; flowers subtended by a cup- or saucer-like disc; trees...***Populus***
1b. Buds 1-scaled; catkins usually erect or spreading; flowers subtended by a nectary or entire bract; trees or shrubs...***Salix***

POPULUS L. – COTTONWOOD

Trees; *sepals* reduced to a non-nectariferous disc; *stamens* 6–60; *ovary* 2–4-carpellate.

1a. Leaves palmately lobed or coarsely toothed, densely white-tomentose below; petioles densely white-tomentose...***P. alba***
1b. Leaves unlobed or usually not coarsely toothed, not densely white-tomentose below; petioles glabrous...2

2a. Leaves broadly triangular-ovate with an acuminate tip, truncate base, and crenate-serrate margins; bark deeply furrowed...***P. deltoides***
2b. Leaves unlike the above; bark smooth to furrowed...3

3a. Leaves nearly orbicular to cordate; petioles strongly laterally flattened just below the leaf blade; bark smooth, white to grayish-white...***P. tremuloides***
3b. Leaves lanceolate to ovate; petioles usually round or only slightly flattened below the leaf blade; bark furrowed, reddish-brown or light brown to dark gray...4

4a. Petiole usually less than ⅓ of the blade length (mostly 0.2–1 cm long); leaves usually lanceolate, sometimes narrowly ovate...***P. angustifolia***
4b. Petiole usually over ⅓ of the blade length (mostly 1–5 cm long); leaves ovate...5

5a. Leaf margins coarsely crenate; leaves about equally green above and below; leaf venation pinnate; petiole slightly flattened below the leaf blade...***P. ×acuminata***
5b. Leaf margins minutely and finely crenate-serrate or subentire; leaves dark green above and lighter green below; leaves with the three lowest distinct veins arising from a central point; petiole round or slightly flattened below the leaf blade...***P. balsamifera***

Populus ×acuminata **Rydb.**, LANCELEAF COTTONWOOD. Trees 10–25 m tall, bark furrowed; *leaves* lanceolate, (1) 5–10 (13) × (0.7) 3–7 cm, the margins coarsely crenate; *petioles* (1) 2–5 cm long, slightly flattened below the leaf blade; *catkins* 6.5–9 cm long in flower (to 16.5 cm in fruit); *capsules* 5–7 mm long. Found on floodplains and along creeks and streams, 4800–8500 ft. April–May. E/W.

Populus ×acuminata is a hybrid between *P. angustifolia* and *P. deltoides*.

Populus alba **L.**, WHITE POPLAR. Trees, 5–30 (45) m tall, bark light gray, furrowed on the lower trunk and white and smooth above; *leaves* ovate, (1) 3–5 (7) × (0.7) 2–05 (6) cm, densely white-tomentose below, the margins palmately lobed or coarsely toothed; *petioles* (0.5) 1–2.5 cm long, round, densely white-tomentose; *catkins* 2–9 cm long; *capsules* (2) 3–5 mm long. A cultivated tree sometimes escaping into nearby ditches and riparian areas or persisting near old homesteads, 4200–6500 ft. April–May. E/W. Introduced.

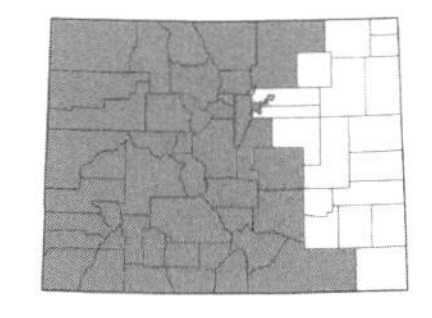

Populus angustifolia **James**, NARROWLEAF COTTONWOOD. Trees, 5–10 (20) m tall, bark furrowed below and smooth above; *leaves* lanceolate to narrowly ovate, (1.5) 4–10 (13.5) × 0.8–3 (4) cm, the margins crenate-serrate; *petioles* round, 0.2–1 (1.7) cm long; *catkins* 3–9 cm long; *capsules* 3–5 mm long. Common along streams and rivers and in floodplains, 5000–10,500 ft. April–May. E/W.

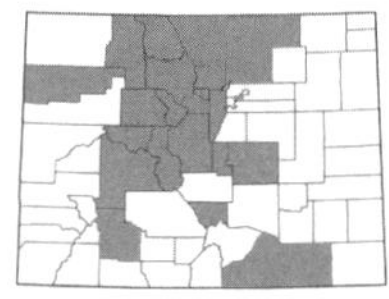

***Populus balsamifera* L.**, BALSAM POPLAR. Trees, 10–30 m tall, bark furrowed; *leaves* ovate, (2.5) 5–10 (14) × (0.7) 3–6 (10) cm, the margins crenate-serrate to subentire; *petioles* round or slightly flattened below the leaf blade, (0.2) 1.5–5 cm long; *catkins* 7.5–15 cm long; *capsules* (3) 5–8 mm long. Common along streams and rivers, seepage areas, and in aspen forests, 6500–11,800 ft. April–May. E/W.

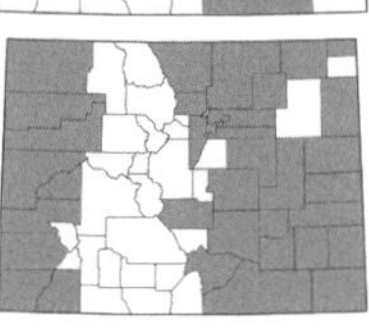

***Populus deltoides* W. Bartram ex Marshall**, COTTONWOOD. Trees, 20–40 m tall, bark deeply furrowed; *leaves* triangular-ovate, with acuminate tips and truncate bases, (1) 4–10 (14) cm long, the margins crenate-serrate; *petioles* flattened below the leaf blade, (1) 3–10 (13) cm long; *catkins* (0.7) 5–13 cm long; *capsules* (4) 8–15 mm long. Common along streams and creeks and on floodplains, 3500–7500 ft. Mar.–May. E/W.

There are two subspecies of *P. deltoides* in Colorado:

1a. Leaf tips long-acuminate; leaf bases usually with 2 round glands...**ssp. *monilifera* (Ait.) Eckenw.**, PLAINS COTTONWOOD. Common along creeks and streams and on floodplains, 3500–6000 ft. Mar.–May. E.

1b. Leaf tips short-acuminate; leaf bases lacking glands...**ssp. *wislizeni* (S. Watson) Eckenw.**, RIO GRANDE COTTONWOOD. [*P. fremontii* S. Watson]. Common along streams and on floodplains, 4200–7500 ft. Mar.–May. W.

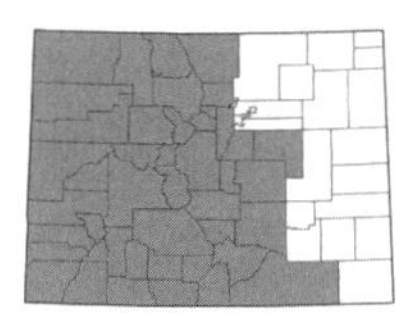

***Populus tremuloides* Michx.**, QUAKING ASPEN. Trees, 6–15 (30) m tall, bark mostly white or grayish-white and smooth; *leaves* nearly orbicular to cordate, (1) 2.5–5 (7) × (0.6) 2–6 (7) cm, the margins subentire to finely crenate-serrate; *petioles* strongly flattened below the leaf blade, (0.7) 1–6 cm long; *catkins* (2) 4–7 cm long in flower, to 11 cm long in fruit; *capsules* 2.5–5 (7) mm long. Common along streams and in the mountains where it forms large clonal stands, 6000–11,700 ft. Mar.–June. E/W. (Plate 74)

SALIX L. – WILLOW

Shrubs or trees; *pedicels* present or not; *sepals* reduced to an adaxial nectary; *stamens* 1, 2, or 3–10, the filaments distinct or connate; *ovary* 2-carpellate, sessile or stipitate. (Dorn, 1997)

Hybridization is common in the genus *Salix* and makes some specimens practically impossible to correctly identify.

1a. At least some leaves opposite (or nearly so)...***S. purpurea***
1b. Leaves all clearly alternate...2

2a. Leaves densely white-tomentose below, the upper surface dark green and shiny, with strongly revolute, entire margins; twigs of this year's growth densely woolly or tomentose; shrubs to 1.8 m tall; rare in fens...***S. candida***
2b. Leaves and twigs various but unlike the above; variously distributed...3

3a. Plants creeping shrubs 1–10 cm tall, usually with trailing or decumbent branches, strictly alpine (over 10,500 ft in elevation)...**Key 1**
3b. Plants erect (*S. glauca* sometimes creeping, but 20 cm or more tall), usually over 10 cm tall, the branches usually not trailing or decumbent, alpine or not...4

4a. Twigs pruinose (with a waxy, whitish coating that can be rubbed off) on the current or previous year's growth, this sometimes only present at the nodes behind the buds...**Key 2**
4b. Twigs not pruinose...5

5a. Leaves linear or linear-elliptic, mostly 1 cm wide or less, over 6 times longer than wide, the margins usually remotely toothed or occasionally entire, often grayish-green...**Key 3**
5b. Leaves various but usually less than 6 times as long as wide (or if over 6 times longer than wide, then lacking the other characteristics)...6

6a. Trees (often cultivated or introduced), usually with a single trunk, to 30 m tall; found to 7500 ft in elevation...**Key 4**
6b. Shrubs or small trees, native; variously distributed...7

7a. Leaves about equally green above and below, sometimes hairy below but not glaucous; capsules glabrous...**Key 5**
7b. Leaves conspicuously lighter below (glaucous or with silvery or white hairs) and green above; capsules glabrous or hairy...8

8a. Plants with fruiting pistillate catkins...**Key 6**
8b. Plants lacking fruiting pistillate catkins (with at least leaves)...**Key 7**

Key 1

(*Creeping shrubs, alpine*)

1a. Leaves entire or nearly so to the naked eye, but with white glands present on the margins at 10× magnification; stipules with a long, attenuate tip; capsules glabrous…***S. calcicola***

1b. Leaves various but unlike the above; stipules lacking a long, attenuate tip; capsules hairy…2

2a. Leaves with mostly rounded tips, usually thick and leathery, often with revolute margins, with conspicuous reticulate veins below; peduncles lacking leaves…***S. reticulata* var. *nana***

2b. Leaves with mostly acute tips, the margins flat, usually not leathery or reticulate-veined below; peduncles usually with at least one leaf…3

3a. Leaves narrowly elliptic or elliptic, 2–7 mm wide; pistillate catkins 1–2.3 (3) cm long…***S. cascadensis***

3b. Leaves elliptic, oblanceolate, or obovate, 7–21 mm wide; pistillate catkins 1.8–6 cm long…***S. petrophila***

Key 2

(*Twigs pruinose, waxy*)

1a. Capsules glabrous…2

1b. Capsules hairy…3

2a. Plants less than 1 m tall; leaf margins crenulate…***S. myrtillifolia***

2b. Plants over 2 m tall; leaf margins mostly entire, only a few sparsely toothed…***S. irrorata***

3a. Leaves usually with undulate or irregularly toothed margins (rarely entire), usually glabrous below; pistillate catkins flowering before the leaves emerge (the catkins not subtended by leaves and the emerging leafy branches lacking catkins)…***S. discolor***

3b. Leaves mostly entire, glabrous to hairy below; pistillate catkins various…4

4a. Capsules on a conspicuous stipe mostly 3–6 mm long…***S. bebbiana***

4b. Capsules on a stipe less than 3 mm long or nearly sessile…5

5a. Twigs usually strongly pruinose and grayish-blue throughout; leaves linear to narrowly elliptic, 3.5–9 times as long as wide; pistillate catkins 0.8–2 cm long (globose or subglobose)…***S. geyeriana***

5b. Twigs mostly weakly pruinose or only pruinose at the nodes; leaves various, 2–4.5 times as long as wide; pistillate catkins 0.6–10 cm long, elongate and cylindric or sometimes subglobose…6

6a. Lower leaf surfaces glabrous or sparsely hairy; upper leaf surfaces highly glossy…***S. planifolia***

6b. Lower leaf surfaces moderately to densely villous or long-silky; upper leaf surfaces slightly glossy or dull…7

7a. Pistillate catkins flowering before the leaves emerge (the catkins not subtended by leaves and the emerging leafy branches lacking catkins); floral bracts dark brown or black; lower leaf surfaces glaucous and usually obscured by dense, white hairs…***S. drummondiana***

7b. Pistillate catkins flowering with the leaves (the catkins on leafy branches); floral bracts tawny, brown, or greenish; lower leaf surfaces usually not obscured by dense, white hairs…8

8a. Pistillate catkins 0.6–2 cm long; petioles mostly 1–4 mm long; stipe 0–0.6 mm long; petioles mostly 1–4 mm long; leaves (1) 2–3 (4) cm long…***S. brachycarpa* var. *brachycarpa***

8b. Pistillate catkins 1.5–8 cm long; at least some petioles usually over 3 mm long; stipe 0.3–1.8 mm long; at least some petioles usually over 3 mm long; leaves 2.7–8.2 cm long…***S. glauca***

Key 3

(*Leaves linear, over 6 times longer than wide*)

1a. Leaves usually bright green, the older ones glabrous; floral bracts with rounded or blunt tips; absent from the eastern plains…***S. melanopsis***

1b. Leaves pale or grayish-green, the older ones usually hairy below, or if glabrous then of the eastern plains; floral bracts usually with acute tips…***S. exigua***

Key 4

(*Trees, often introduced*)

1a. Branches pendulous (weeping willow); capsules 2–3 mm long…***S. babylonica***

1b. Branches erect; capsules 3–7 mm long…2

2a. Tree crown spherical or nearly so...***S. matsudana***
2b. Tree crown not spherical...3

3a. Twigs of this year's growth glabrous or sparsely hairy; bud scale margins distinct and overlapping; stipes 1.2–3.2 mm long; native...4
3b. Twigs of this year's growth pilose to densely hairy; bud scale margins connate; stipes 0.2–0.8 mm long; introduced...5

4a. Lower leaf surfaces glaucous, paler than above; leaves 2.5–6 times as long as wide...***S. amygdaloides***
4b. Lower leaf surfaces not glaucous, the leaves about equally green on both sides; leaves mostly 5–12 times as long as wide...***S. goodingii***

5a. Older leaves silky hairy below...***S. alba***
5b. Older leaves glabrous or sparsely hairy below...***S. ×fragilis***

Key 5

(*Leaves about equally green above and below*)

1a. Leaves all entire, hairy above and below; pistillate catkins 0.8–3 cm long, globose or subglobose; plants less than 2 m tall...***S. wolfii* var. *wolfii***
1b. Leaves mostly toothed, glabrous or hairy above and below; pistillate catkins and plants various...2

2a. Leaf tips mostly acuminate; petioles with a pair or clusters of spherical glands where the petiole meets the blade; leaf margins strongly glandular-toothed; floral bracts yellow or green, deciduous in fruit...3
2b. Leaf tips acute; petioles lacking spherical glands; leaf margins glandular-toothed or not; floral bracts usually dark brown, persistent in fruit...5

3a. Bud scale margins distinct and overlapping; shrub or tree (3) 6–30 m tall; style 0.1–0.3 mm long...***S. goodingii***
3b. Bud scale margins connate; tall shrub or small tree 1–6 (10) m tall; style 0.3–1 mm long...4

4a. Leaf tips shortly acuminate (the acuminate portion 1 cm or less long); capsules 6–12 mm long, brown to dark brown...***S. serissima***
4b. Leaf tips long-acuminate (the acuminate portion 2 cm or more long); capsules mostly 4–7 mm long, greenish-brown...***S. lasiandra* var. *caudata***

5a. Decumbent shrubs 0.1–1 m tall; older leaves glabrous below; floral bracts 0.4–1.1 mm long...***S. myrtillifolia***
5b. Erect shrubs 0.3–6 m tall; older leaves usually hairy below; floral bracts 0.7–2.1 mm long...6

6a. Largest blades 2.5–5.2 times as long as wide, oblong to elliptic...***S. boothii***
6b. Largest blades 1.6–3 (3.6) times as long as wide, elliptic...***S. arizonica***

Key 6

(*Leaves paler below than above, with fruiting pistillate catkins*)

1a. Capsules glabrous...2
1b. Capsules hairy...6

2a. Floral bracts yellow, green, or pale, deciduous in fruit; leaves with gland-tipped, toothed margins, at least some tips acuminate; petioles usually with glands near the base of the blade...3
2b. Floral bracts dark brown or tawny, persistent in fruit; leaves with entire margins or sometimes with gland-tipped teeth, but the tips then not acuminate; petioles usually lacking glands...5

3a. Bud scale margins distinct and overlapping; mostly trees, 4–20 m tall...***S. amygdaloides***
3b. Bud scale margins connate; shrubs or small trees 1–10 m tall...4

4a. Leaf tips long-acuminate (mostly over 2 cm long); capsules mostly 4–7 mm long, greenish-brown...***S. lasiandra* var. *lasiandra***
4b. Leaf tips short-acuminate (mostly less than 1 cm long); capsules 6–12 mm long, brown... ***S. serissima***

5a. Styles mostly over 0.7 mm long; leaves evenly toothed, with rounded or somewhat acute tips, green or yellowish-green...***S. monticola***
5b. Styles 0.1–0.6 mm long; leaves entire or toothed, usually with sharply acute tips, darker and more bluish-green than *S. monticola*...***S. eriocephala***

6a. Pistillate catkins flowering with the leaves (the catkins on leafy branches); floral bracts tawny, brown, or greenish…7
6b. Pistillate catkins flowering before the leaves emerge (the catkins not subtended by leaves and the emerging leafy branches lacking catkins); floral bracts dark brown or black…10

7a. Stipes mostly 3–6 mm long; leaves narrowly oblong, oblanceolate, elliptic, or obovate (mostly 16–45 mm wide)…***S. bebbiana***
7b. Stipes less than 2 mm long or the capsules nearly sessile, if as long as 3 mm then the leaves lanceolate or narrowly elliptic (6–19 mm wide)…8

8a. Capsules short-hairy; leaves very narrowly elliptic or linear-lanceolate, 5–9 times as long as wide; shrubs 1–6 m tall; styles 0–0.5 mm long; branches red-brown…***S. petiolaris***
8b. Capsules densely white-tomentose; leaves variously shaped but less than 5 times as long as wide; shrubs less than 1.5 m tall; styles 0.3–1.6 mm long; branches various…9

9a. Pistillate catkins 0.6–2 cm long; petioles mostly 1–4 mm long; stipes 0–0.6 mm long…***S. brachycarpa*** **var.** ***brachycarpa***
9b. Pistillate catkins 1.5–8 cm long; at least some petioles usually over 3 mm long; stipes 0.3–1.8 mm long…***S. glauca***

10a. Stipes 0–0.8 mm long; pistillate catkins 1.5–6 cm long (to 7 cm in fruit); leaf margins usually entire or remotely toothed…***S. planifolia***
10b. Stipes 0.8–2.7 mm long; pistillate catkins 2–11 cm long (to 14 cm in fruit); leaf margins various…11

11a. Leaves mostly oblanceolate or obovate (rarely elliptic), widest above the middle, with entire or remotely toothed margins, usually hairy below and often with reddish hairs; fresh bark usually with a skunkish odor…***S. scouleriana***
11b. Leaves mostly elliptic (rarely oblanceolate or obovate), widest in the middle, usually with undulate or irregularly toothed margins (rarely entire), usually glabrous below; fresh bark lacking a skunkish odor…***S. discolor***

Key 7

(Leaves paler below than above, lacking fruiting pistillate catkins)

1a. Leaves regularly and distinctly toothed throughout…2
1b. Leaves entire, remotely or irregularly toothed, or only a few leaves with toothed margins…8

2a. Leaves with long-acuminate tips (the acuminate portion mostly over 2 cm long)…3
2b. Leaves with acute or short-acuminate tips (the acuminate portion under 1 cm long)…4

3a. Bud scale margins distinct and overlapping; mostly trees, 4–20 m tall; branches dull; leaves usually lacking prominent secondary veins above and below…***S. amygdaloides***
3b. Bud scale margins connate; shrubs or small trees 1–10 m tall; branches slightly to highly glossy; leaves with prominent secondary veins above and below…***S. lasiandra*** **var.** ***lasiandra***

4a. Leaves mostly elliptic and 4-11 (13.5) cm long, the teeth widely spaced along the leaf margins (usually 4 or fewer teeth per cm in the middle of the leaf); pistillate catkins flowering before leaves emerge (the catkins not subtended by leaves and the emerging leafy branches lacking catkins)…***S. discolor***
4b. Leaves unlike the above in all respects (either not elliptic, smaller, or with more teeth per cm); pistillate catkins flowering with the leaves (the catkins on leafy branches)…5

5a. Petioles with a pair of glands at the base of the leaf; leaf margins strongly gland-toothed; upper leaf surface highly glossy, glabrous…***S. serissima***
5b. Petioles lacking a pair of glands at the base of the leaf; leaf margins usually not strongly gland-toothed; upper leaf surface dull or slightly glossy, glabrous or hairy…6

6a. Leaves linear-lanceolate to very narrowly elliptic, 5–9 times as long as wide; capsules hairy; branches red-brown…***S. petiolaris***
6b. Leaves various but not linear-lanceolate, sometimes narrowly elliptic, 2–6 times as long as wide; capsules glabrous; branches various…7

7a. Leaves with rounded or somewhat acute tips, green or yellowish-green; styles mostly over 0.7 mm long…***S. monticola***
7b. Leaves usually with sharply acute tips, darker and more bluish-green than *S. monticola*; styles 0.1–0.6 mm long…***S. eriocephala***

8a. Leaves mostly obovate or oblanceolate (widest above the middle); fresh bark usually with a skunkish odor...***S. scouleriana***
8b. Leaves variously shaped but widest at the middle; fresh bark lacking a skunkish odor...9

9a. Upper leaf surfaces highly glossy; older branches usually red-brown and shiny; pistillate catkins flowering before the leaves emerge (the catkins not subtended by leaves and the emerging leafy branches lacking catkins); leaves mostly 2–6.5 cm long...***S. planifolia***
9b. Upper leaf surfaces dull or slightly glossy; older branches variously colored, usually dull; pistillate catkins flowering with the leaves (the catkins on leafy branches) except in *S. discolor* (and then the leaves mostly 4–11 cm long)...10

10a. Leaves mostly elliptic and 4–11 (13.5) cm long; pistillate catkins flowering before leaves emerge (the catkins not subtended by leaves and the emerging leafy branches lacking catkins); known from northern Larimer Co....***S. discolor***
10b. Leaves unlike the above in all respects (either not elliptic or smaller); pistillate catkins flowering with the leaves (the catkins on leafy branches); variously distributed...11

11a. Leaves linear-lanceolate to very narrowly elliptic, 5–9 times as long as wide; branches red-brown...***S. petiolaris***
11b. Leaves various but not linear-lanceolate, sometimes narrowly elliptic, 2–5.5 times as long as wide; branches various...12

12a. Shrubs less than 1.5 m tall, sometimes creeping; leaves usually hairy above and below...13
12b. Shrubs 2–7 m tall, never creeping; leaves various...14

13a. Petioles mostly 1–4 mm long; leaves mostly 2–3 (4) cm long; pistillate catkins 0.6–2 cm long; stipes 0–0.6 mm long...***S. brachycarpa* var. *brachycarpa***
13b. At least some petioles usually over 3 mm long; leaves mostly 3–8 cm long; pistillate catkins 1.5–8 cm long; stipes 0.3–1.8 mm long...***S. glauca***

14a. Branches hairy (sometimes sparsely so); capsules hairy, 5–9 mm long; stipes mostly 3–6 mm long...***S. bebbiana***
14b. Branches glabrous; capsules glabrous, 3–6 mm long; stipes 1–3.5 mm long...***S. eriocephala***

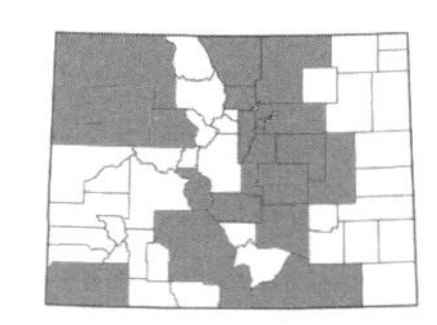

***Salix ×fragilis* L.**, CRACK WILLOW. Trees to 20 m; *stems* yellowish-brown to olive, glabrous in age; *leaves* lanceolate, acuminate, usually 70–130 × 15–30 mm, the margins serrate; *pistillate catkins* appearing with the leaves, 30–60 mm long, on leafy branchlets, bracts yellowish, hairy; *capsules* 4–5.5 mm long, glabrous, subsessile or on stipes to 1 mm. Cultivated, persisting near old homesteads, 4500–7000 ft. April–June. E/W. Introduced.

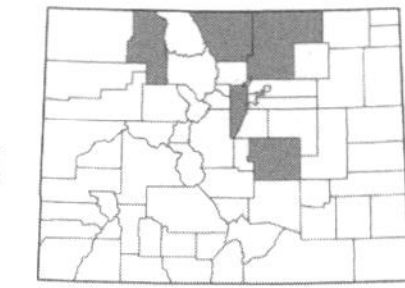

***Salix alba* L.**, WHITE WILLOW. Trees, 10–25 m; *stems* yellowish to gray or red-brown, glabrous or hairy; *leaves* narrowly oblong to narrowly elliptic, 63–115 × 10–20 mm, 4–7.3 times as long as wide, densely long-hairy to glabrescent below, sparsely long-hairy above, with toothed margins; *staminate catkins* 27–60 × 6–10 mm; *pistillate catkins* 31–51 × 4–8 mm, loosely flowered, the floral bracts 1.6–2.8 mm long; *capsules* 3.5–5 mm long, glabrous. Cultivated and rarely escaping, or persisting near old homesteads, 5000–7000 ft. April–June. E/W. Introduced.

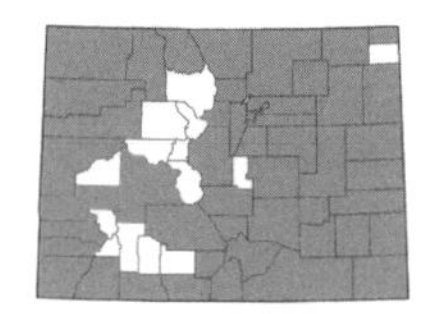

***Salix amygdaloides* Andersson**, PEACH-LEAF WILLOW. Trees, 4–20 m; *stems* yellow to gray-brown, glabrous; *leaves* elliptic or lanceolate, 55–130 × 24–37 mm, 2.5–6 times as long as wide, glaucous and glabrous below, glabrous or sparsely hairy along the midrib above, with serrulate margins; *staminate catkins* with 3–7 stamens, filaments basally hairy; *pistillate catkins* with deciduous floral bracts post-flowering; *capsules* 3–7 mm long, glabrous. Common along streams and rivers, pond margins, and in floodplains, 3500–8000 ft. April–June. E/W.

***Salix arizonica* Dorn**, ARIZONA WILLOW. Shrubs, 0.3–2.6 m; *stems* red-brown or yellow-brown, weakly glaucous or not, glabrous or pilose at nodes; *leaves* elliptic, 20–50 × 10–31 mm, 1.6–2–2.8 (3.6) times as long as wide, pilose or glabrous above and below, with flat, serrulate or entire margins; *staminate catkins* 7–17 × 6–10 mm, flowering as leaves emerge; *pistillate catkins* 12–38 × 6–12 mm, the floral bracts brown, black, or bicolor, 1–2 mm long; *capsules* 3.2–4.5 mm long, glabrous. Rare in moist meadows and along streams, 9500–10,500 ft. May–June. E.

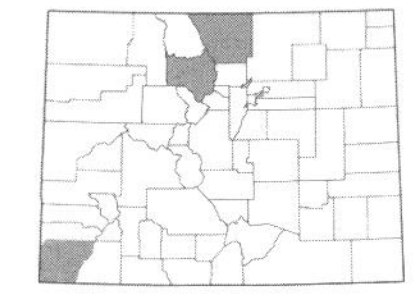

***Salix babylonica* L.**, WEEPING WILLOW. Trees; *stems* yellow-brown to red-brown; *leaves* lanceolate to narrowly elliptic, 90–160 × 5–20 mm, 5.5–10 times as long as wide, with serrulate or spinulose-serrulate margins; *staminate catkins* 13–35 mm; *pistillate catkins* 9–27 × 2.5–7 mm, flowering just before leaves emerge, the floral bracts 1.1–2 mm long; *capsules* 2–3 mm long. Cultivated tree rarely escaping, or persisting near old homesteads, 5000–6000 ft. May–June. E/W. Introduced.

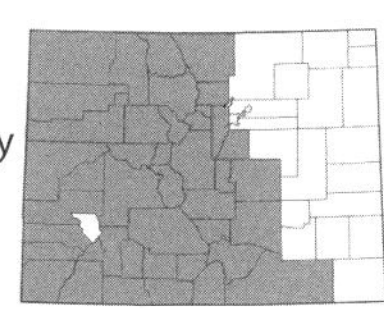

***Salix bebbiana* Sarg.**, BEBB WILLOW. Trees or shrubs; *stems* yellow-brown to dark red-brown, weakly glaucous or not, pilose to glabrescent; *leaves* narrowly oblong, elliptic, oblanceolate, or obovate, 20–44 (87) × 10–16 (45) mm, glaucous and moderately densely hairy to glabrescent below, densely hairy to glabrescent above, with entire, crenate, or irregularly toothed margins; *staminate catkins* 10–42 × 7–16 mm, flowering just before leaves emerge; *pistillate catkins* 16.5–85 × 9–32 mm, flowering as leaves emerge, the floral bracts tawny, 1.2–3.2 mm long; *capsules* 5–9 mm long, hairy. Common along creeks, streams, and in moist meadows and aspen groves, 5000–10,800 ft. April–June. E/W. (Plate 74)

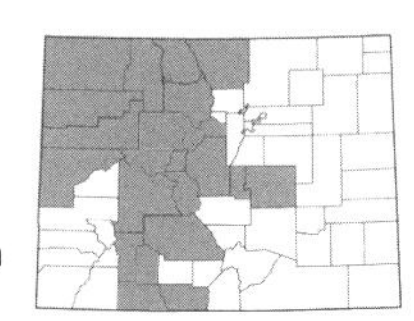

***Salix boothii* Dorn**, BOOTH'S WILLOW. Shrubs, 0.3–6 m; *stems* yellow-gray, yellow-brown, or red-brown, glabrous, pilose, or villous; *leaves* narrowly oblong or elliptic, 26–102 × 8–30 mm, 2.5–3-5.2 times as long as wide, glabrous to moderately densely pilose above and below, with flat or slightly revolute, entire or serrulate margins; *staminate catkins* 7–37 × 5–12 mm, flowering as or just before leaves emerge; *pistillate catkins* 12–62 × 7–17 mm, the floral bracts brown, 0.7–2.1 mm long; *capsules* 2.5–6 mm long, glabrous. Common in moist meadows, along creeks and streams, and in floodplains, 5000–10,500 ft. April–July. E/W.

This and *S. arizonica* are practically impossible to separate without chromosome counts.

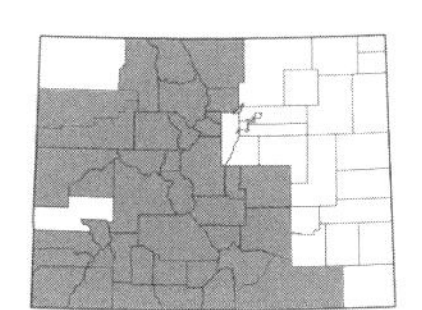

Salix brachycarpa* Nutt. var. *brachycarpa, SHORTFRUIT WILLOW. Shrubs, 0.2–1.5 m; *stems* erect or decumbent, gray-brown or red-brown, short-hairy or villous to glabrescent; *leaves* oblong, elliptic, narrowly oblanceolate, or obovate, (10) 20–30 (40) × 5–16 mm, (1.5) 2.8–3 (4) times as long as wide, moderately densely villous below, pilose, villous, or long-hairy to glabrescent above; *staminate catkins* 5.3–21 × 4–10 mm; *pistillate catkins* 6–20 × 4–15 mm, the floral bracts tawny, 1–3 mm long; *capsules* 3–6 mm long, hairy. Common along streams, on dry slopes, and in forests, moist meadows, and the alpine, 7000–13,500 ft. June–Aug. E/W.

Frequently hybridizes with *S. glauca* and completely overlaps with it in distribution.

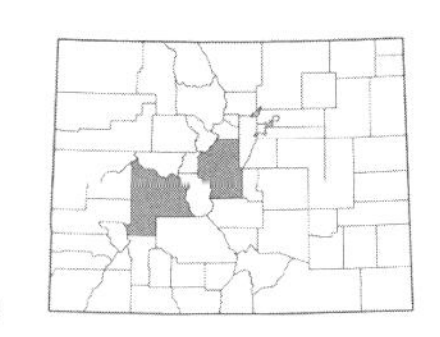

***Salix calcicola* Fernald & Wieg.**, LIME-LOVING WOOLLY WILLOW. Shrubs, 0.05–1.3 m; *stems* gnarled, yellow-brown, gray-brown, or red-brown, weakly glaucous or not, villous to glabrescent; *leaves* elliptic or suborbicular, 16–61 × 10–44 mm, 0.7–2.6 times as long as wide, glaucous and sparsely villous or pilose to glabrescent below, dull or slightly glossy above, with entire or serrulate margins; *staminate catkins* 18–45 × 13–21 mm, flowering before leaves emerge; *pistillate catkins* 32–75 (-100 in fruit) × 12–25 mm, the floral bracts brown or black, 1.2–3.2 mm long; *capsules* 4–8 mm long, glabrous. Rare on moist, calcareous soil, 11,200–12,500 ft. June–Aug. E/W.

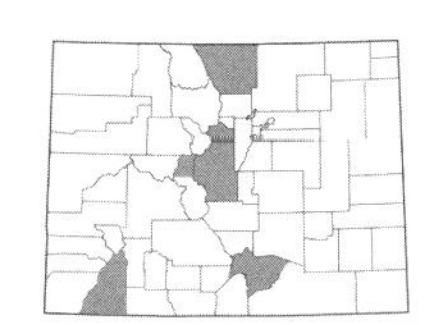

***Salix candida* Flügge ex Willd.**, SILVER WILLOW. Shrubs to 1.8 m; *stems* dark gray-brown to yellow-brown, woolly to glabrescent; *leaves* narrowly elliptic or oblanceolate, 47–103 × 5–20 mm, glaucous, very densely to sparsely tomentose-woolly above and below, with strongly to slightly revolute, entire or sinuate margins; *staminate catkins* 17–39 × 8–16 mm, flowering as leaves emerge; *pistillate catkins* 20–66 × 9–18 mm, the floral bracts brown, 1.2–1.8 mm long; *capsules* 4–6 mm long, hairy. Uncommon in rich fens and bogs, 8500–10,700 ft. April–June. E/W.

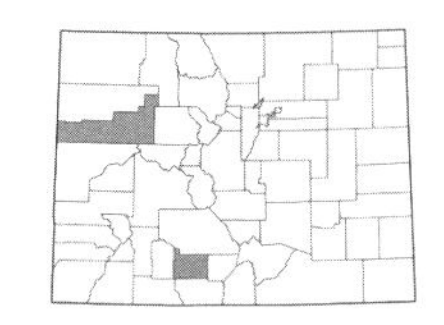

***Salix cascadensis* Cockerell**, CASCADE WILLOW. Shrubs, 0.03–0.1 m; *stems* erect or trailing, yellow-brown to gray-brown, glabrous; *leaves* elliptic, 9–26 × 2–7 mm, 2.4–4.3 times as long as wide, glabrous or pilose above and below, with flat, entire or ciliate margins; *staminate catkins* 12.5-26.5 × 5.5–9 mm; *pistillate catkins* 10–23 (-30 in fruit) × 5–8 mm, the floral bracts brown, 1.6–2.6 mm long; *capsules* 3.5–5 mm, hairy. Rare in the alpine, 11,000–13,000 ft. July–Aug. E/W.

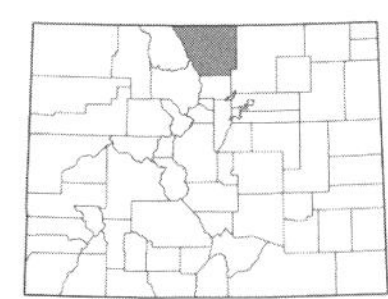

***Salix discolor* Muhl.**, PUSSY WILLOW. Shrubs, 2–4 (-8) m; *stems* dark red-brown or yellow-brown, not to strongly glaucous, villous to glabrescent; *leaves* mostly elliptic (rarely oblanceolate or obovate), 40–110 (135) × 12–33 mm, (2.3) 3–3.5 (4.5) times as long as wide, glaucous below, glabrous to pilose above and below, with crenate, irregularly toothed, sinuate, or entire margins; *staminate catkins* 23–52 × 12–22 mm, flowering before leaves emerge; *pistillate catkins* 25–108 (-135 in fruit) × 12–33 mm, the floral bracts brown, black, or bicolor, 1.4–2.5 mm long; *capsules* 6–11 mm, hairy. Uncommon along streams, 7000–8000 ft. April–May. E.

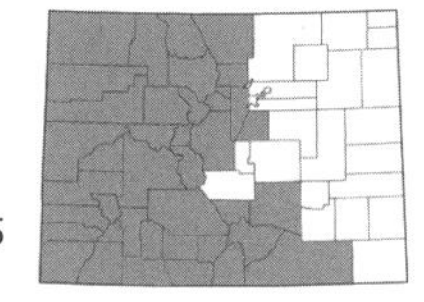

Salix drummondiana **Barratt ex Hook.**, DRUMMOND'S WILLOW. Shrubs, 1–5 m; *stems* yellow-brown or red-brown, strongly glaucous, glabrous; *leaves* elliptic or oblanceolate, 40–85 × 9–26 mm, 3–6.2 times as long as wide, glaucous and densely short- to long-hairy below, sparsely short-hairy to glabrescent above, with slightly revolute, entire, or shallowly crenate to sinuate margins; *staminate catkins* 19–40 × 8–20 mm, flowering before leaves emerge; *pistillate catkins* 22–87 (-105 in fruit) × 8–18 mm, densely flowered, the floral bracts brown or black, 1.2–2.8 mm long; *capsules* 2.5–6 mm long, hairy. Common along streams and pond edges in the mountains, 7000–11,000 ft. April–July. E/W.

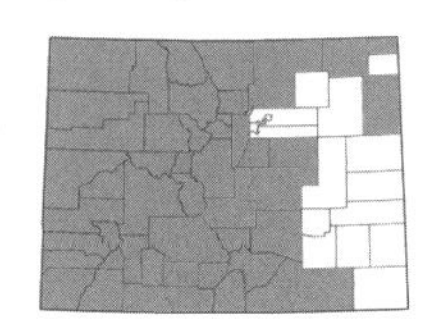

Salix eriocephala **Michx.**, STRAPLEAF WILLOW. Shrubs or trees, 1–8 m; *stems* yellow-brown to gray-brown or reddish, glabrous or sparsely to densely hairy; *leaves* lorate, oblong, narrowly elliptic, to lanceolate, 28–133 × 8–32 mm, glabrous or sparsely hairy above and below, the margins serrate to entire or crenulate; *staminate catkins* 15–44 × 8–14 mm; *pistillate catkins* 13.5–74 (-115 in fruit) × 7–18 mm, the floral bracts brown to tawny; *capsules* 3–6 mm long, glabrous. Common along streams and rivers, in floodplains, and in seepage areas, 4000–10,000 ft. April–June. E/W.

There are three varieties of *S. eriocephala* present in Colorado:

1a. Leaves mostly entire or indistinctly toothed... **var. *watsonii* (Bebb) Dorn**. [*S. lutea* Nutt.]. E/W.
1b. Leaves mostly distinctly toothed...2

2a. Leaves dull above; branches usually red-brown, sometimes yellow-brown and branchlets yellow-green or yellow-brown...**var. *ligulifolia* (C.R. Ball) Dorn**. [*S. ligulifolia* (C.R. Ball) C.R. Ball ex C.K. Schneid.]. E/W.
2b. Leaves glossy above; branches usually yellow, yellow-brown, or yellow-gray and branchlets usually red-brown...**var. *famelica* (C.R. Ball) Dorn**. [*S. famelica* (C.R. Ball) Argus]. Found on the northeastern plains. E.

Salix exigua **Nutt.**, COYOTE WILLOW. Shrubs or small trees, 0.5–5 (17) m; *stems* gray-brown, red-brown, or yellow-brown, villous to glabrescent; *leaves* linear or lorate, 30–143 × 2–14 mm, 6.5–28 (37.5) times as long as wide, glaucous below, densely long-hairy to glabrescent above and below, with slightly revolute, entire, or remotely spinulose-serrulate margins; *staminate catkins* 7–54 × 2–10 mm; *pistillate catkins* 14.5–70 × 3–12 mm, the floral bracts 1.2–2.6 mm long; *capsules* 4–8 mm long, glabrous. Common along rivers and in floodplains, 3400–9600 ft. April–July. E/W. (Plate 74)

There are two subspecies of *S. exigua* in Colorado:

1a. Older leaves hairy below...**ssp. *exigua***. 3400–9600 ft. E/W.
1b. Older leaves usually glabrous or nearly so below...**ssp. *interior* (Rowlee) Cronquist**. 3500–5800 ft. E.

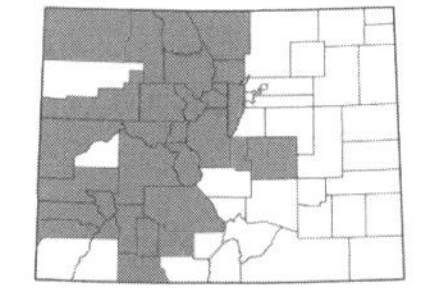

Salix geyeriana **Andersson**, GEYER WILLOW. Trees or shrubs, 0.6–5 m; *stems* yellow-green, gray-brown, red-brown, or violet, glaucous, glabrous or sparsely tomentose; *leaves* lorate, narrowly elliptic, or linear, 32–89 × 5.5–14 mm, 3.6–8.4 (11.3) times as long as wide, glaucous, glabrous to densely hairy above and below, with flat or slightly revolute, entire or distantly and shallowly serrulate margins; *staminate catkins* 11–18 × 6–11 mm, flowering before or as leaves emerge; *pistillate catkins* 8–21 × 7–17 mm, the floral bracts tawny or brown, 1.2–2.8 mm long; *capsules* (3) 4–6 mm long, hairy. Common in bogs, moist meadows, and along streams and ponds, 5800–10,000 ft. April–June. E/W.

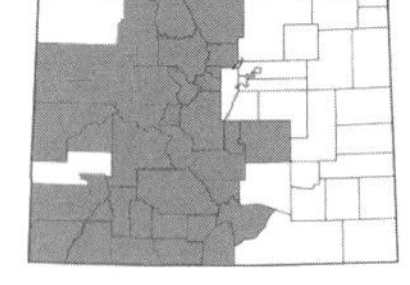

Salix glauca **L.**, GRAYLEAF WILLOW. Trees or shrubs, 0.2–6 m; *stems* erect or decumbent, brownish, yellow-brown, or red-brown, villous or pilose to glabrescent; *leaves* elliptic, oblanceolate, obovate, or narrowly oblong, 27–82 × 6–39 mm, 1.4–4.8 times as long as wide, with slightly revolute or flat, entire margins; *staminate catkins* 14–53 × 5–17 mm; *pistillate catkins* 15–80 × 7–21 mm, the floral bracts tawny, brown, bicolor, or greenish, 1–3.4 mm long; *capsules* 4.5–9 mm long, hairy. Common along streams, in forests, and the alpine, 7500–13,500 ft. May–July. E/W.

Frequently hybridizes with *S. brachycarpa*.

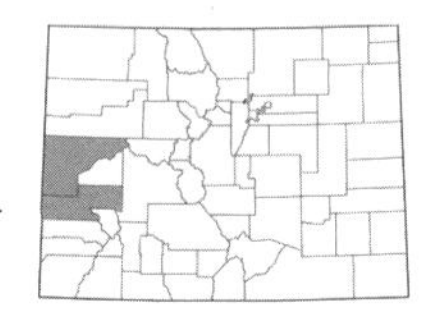

Salix goodingii **C.R. Ball**, BLACK WILLOW. Trees, (3) 6–30 m; *stems* yellow-brown to gray-brown, hairy to glabrescent; *leaves* narrowly elliptic, very broadly oblong, lorate, or linear, 67–130 × 9.5–16 mm, 5 -12 times as long as wide, glabrous or minutely hairy below, pilose to nearly glabrous above, with serrulate to serrate margins; *staminate catkins* 19–80 × 6–10 mm; *pistillate catkins* 23–82 × 6–15 mm, the floral bracts 1.4–2.4 mm long; *capsules* 6–7 mm long, glabrous. Uncommon along streams, in moist meadows, and in washes, 4500–6500 ft. April–June. W.

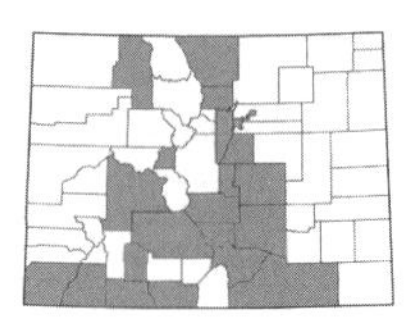

Salix irrorata **Andersson**, BLUESTEM WILLOW. Shrubs, 2–7 m; *stems* glaucous, glabrous; *leaves* lorate, narrowly oblong, elliptic, or oblanceolate, 47–115 × 8–22 mm, 3.5–7.7 times as long as wide, glaucous and glabrous to sparsely tomentose below, glabrous or hairy above, with flat to slightly revolute, entire and gland-dotted, serrulate or crenate margins; *staminate catkins* 15–34 × 8–22 mm; *pistillate catkins* 14–43 × 7–12 mm, floral bracts brown or black, 1.3–2.5 mm long; *capsules* 3.5–4 mm long, glabrous. Found along creeks and streams, 5500–9500 ft. April–May. E/W.

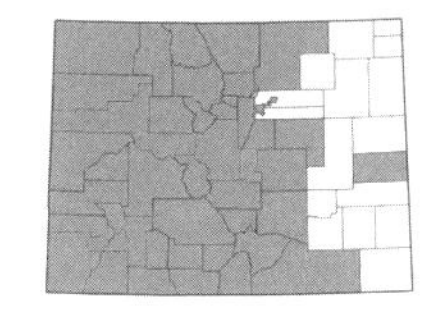

***Salix lasiandra* Benth.**, Pacific willow. Shrubs or trees, 1–9 (11) m; *stems* yellow-brown, gray-brown, or red-brown, slightly to highly glossy, glabrous or pilose to glabrescent; *leaves* narrowly oblong, narrowly elliptic, lanceolate, or oblanceolate, 53–170 × 9–31 mm, 3.1–9.8 times as long as wide, glaucous or not below, glabrescent to pilose above and below, with serrulate margins; *staminate catkins* 21–78 × 8–15 mm; *pistillate catkins* 18.5–103 × 6–17 mm, the floral bracts 1.7–4 mm long; *capsules* 4–11 mm long, glabrous.

There are two varieties of *S. lasiandra* in Colorado:

1a. Leaves about equally green above and below...**var. *caudata* (Nutt.) Sudw.**, greenleaf willow. [*S. lucida* Muhl. ssp. *caudata* (Nutt.) E. Murray]. Common along streams and in moist meadows, 4000–10,000 ft. May–June. E/W.

1b. Leaves paler below than above...**var. *lasiandra***, Pacific willow. [*S. lucida* Muhl. ssp. *lasiandra* (Benth.) E. Murray]. Found along rivers and in floodplains, 5400–9000 ft. May–June. E/W.

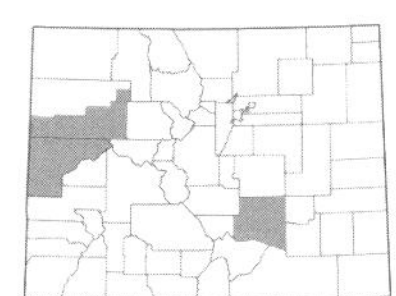

***Salix matsudana* Koidz.**, globe willow. Trees to 18 m, the crown spherical or nearly so; *stems* yellow to greenish or brown; *leaves* lanceolate, 5–10 × 1–1.5 cm, glaucous or slightly white below, green and shiny above, with glandular-serrulate margins, the apex long-acuminate; *staminate catkins* 15–25 (35) mm; *pistillate catkins* ca. 2 cm × 4 mm, flowering before leaves emerge; *capsules* ca. 3 mm long, glabrous. Cultivated tree grown in towns in the lower Colorado and Arkansas River valleys (Grand Junction, Pueblo), 4500–5000 ft. May–June. E/W. Introduced.

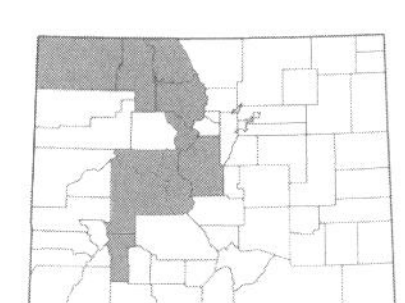

***Salix melanopsis* Nutt.**, dusky willow. Shrubs, 0.8–4 m; *stems* gray-brown or red-brown, glabrous or hairy; *leaves* lorate, narrowly oblong, narrowly elliptic, narrowly oblanceolate, or linear, 30–133 × 5–20, 3.4–8–15 times as long as wide, glaucous or not below, pilose or villous to glabrescent above and below, with spinulose-serrulate or entire margins; *staminate catkins* 18–48 × 5–13 mm; *pistillate catkins* 22–58 × 4–9 mm, the floral bracts 1.3–2.8 mm long; *capsules* 4–5 mm long, glabrous. Uncommon along streams and in moist meadows, 7000–10,000 ft. May–July. E/W.

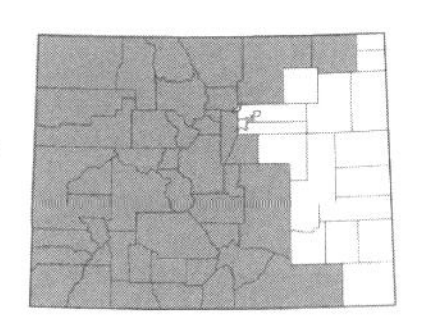

***Salix monticola* Bebb**, mountain willow. Trees or shrubs, 1.5–6 m; *stems* yellow-brown or red-brown, weakly glaucous or not, glabrous; *leaves* oblong, elliptic, lanceolate, oblanceolate, or obovate, 35–95 × 11–33 mm, 2–3.9 times as long as wide, glaucous and glabrous below, glabrous or pilose above, with slightly revolute or flat, serrulate, serrate, or sinuate margins; *staminate catkins* 14–39 × 9–17 mm; *pistillate catkins* 14–39 × 9–17 mm; *capsules* 4–7 mm long, glabrous. Common along streams, in floodplains, and in moist meadows, 5000–12,500 ft. April–July. E/W.

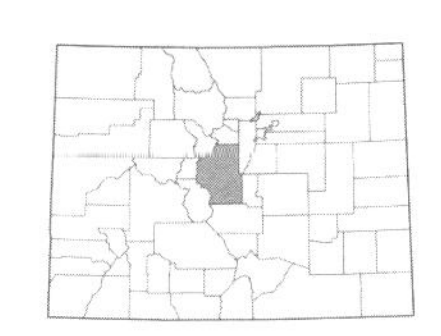

***Salix myrtillifolia* Andersson**, blueberry willow. Trees or shrubs, 0.1–0.6 (1) m; *stems* decumbent, gray-brown, red-brown, or yellow-brown, not to strongly glaucous, hairy; *leaves* elliptic or obovate, 17–74 × 8–30 mm, 1.2–4.5 times as long as wide, glabrous above and below, with serrulate, crenulate, or sinuate margins; *staminate catkins* 11.5–39 × 5–14 mm; *pistillate catkins* 16–46 (-50 in fruit) × 4–15 mm, the floral bracts brown, black, tawny, or bicolor, 0.4–1.1 mm long; *capsules* 4–6 mm long, glabrous. Rare in rich fens, 9000–10,000 ft. June–July. E.

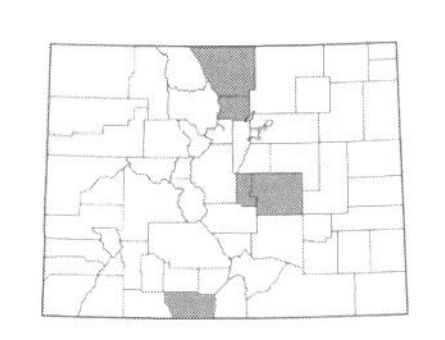

***Salix petiolaris* J. Sm.**, meadow willow. [*S. gracilis* Andersson]. Trees or shrubs, 1–6 m; *stems* red-brown or violet, not or weakly glaucous, minutely hairy; *leaves* lorate, very narrowly elliptic or linear-lanceolate, 38–110 × 6–19 mm, 5–9 times as long as wide, glaucous and densely long-silky to glabrescent below, glabrous or sparsely hairy above, with flat to slightly revolute, entire, serrate, serrulate, or spinulose-serrate margins; *staminate catkins* 12–29 × 6–17 mm; *pistillate catkins* 12–39 × 6–18 mm, the floral bracts brown, tawny, or bicolor, 1–2 mm long; *capsules* 5–9 mm long, hairy. Uncommon along streams and in moist meadows, 7000–8000 ft. April–June. E.

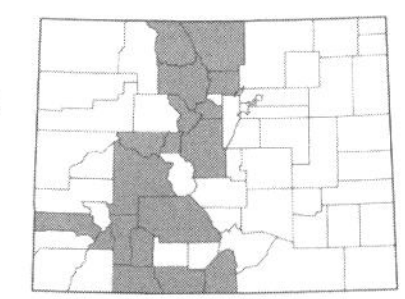

***Salix petrophila* Rydb.**, alpine willow. [*S. arctica* Pall. var. *petraea* (Andersson) Bebb]. Shrubs, 0.02–0.1 m; *stems* decumbent, trailing, glabrous; *leaves* elliptic, oblanceolate, or obovate, 19–44 × 7–21 mm, 1.5–4.6 times as long as wide, pilose to glabrescent above and below, margins entire; *staminate catkins* 18–32 × 6–13 mm; *pistillate catkins* 18–60 (-70 in fruit) × 6–15 mm, the floral bracts tawny, 0.5–3.6 mm long; *capsules* 3.6–5 mm long, hairy. Common in the alpine, 11,000–13,500 ft. July–Aug. E/W.

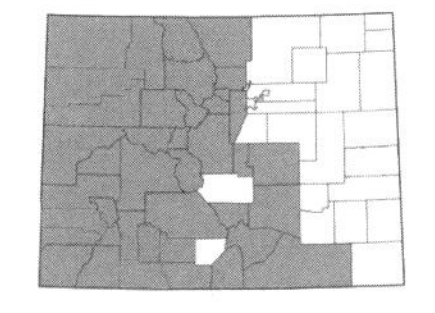

***Salix planifolia* Pursh**, planeleaf willow. Shrubs or trees, 0.1–9 m; *stems* yellow-brown, red-brown, or violet, not to strongly glaucous, glabrous or hairy; *leaves* narrowly oblong, elliptic, or oblanceolate, 20–65 × 5–23 mm, 1.7–4.7 times as long as wide, glaucous below, glabrous or sparsely hairy above and below, with entire, crenulate, or serrulate margins; *staminate catkins* 12–41 × 10–20 mm, flowering before leaves; *pistillate catkins* 15–67 (-70 in fruit) × 8–18 mm, the floral bracts dark brown or black, 1–3.2 mm long; *capsules* (2.5) 5.5–6 mm long, hairy. Common along streams and pond margins and in fens, bogs, moist meadows, and the alpine, 8000–13,500 ft. May–July. E/W.

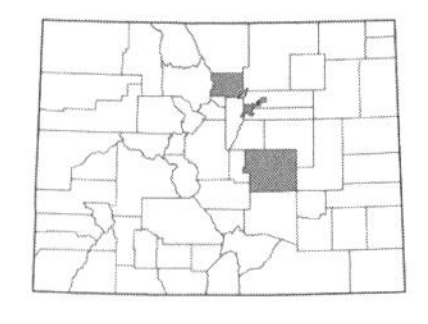

***Salix purpurea* L.**, BASKET WILLOW. Trees; *stems* yellow-brown to olive-brown, glabrous; *leaves* lorate, oblong, or oblanceolate, 35–77 × 5–20 mm, the margins strongly revolute, entire, or serrulate, glabrous or sparsely hairy abaxially; *staminate catkins* 25–33 × 6–10 mm; *pistillate catkins* 13–35 × 3–7 mm, flowering before leaves emerge, the floral bracts black, 0.8–1.6 mm long; *capsules* 2.5–5 mm long. Cultivated tree sometimes escaping along creeks, found near Colorado Springs, Boulder, and Denver, 5000–6000 ft. April–May. E. Introduced.

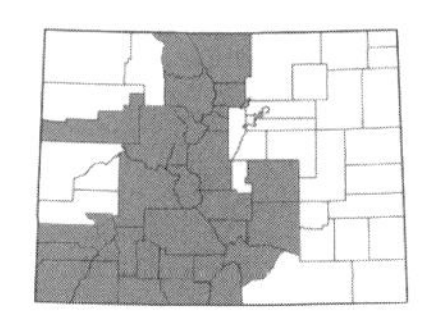

***Salix reticulata* L. var. *nana* Andersson**, SNOW WILLOW. [*S. nivalis* Hook.]. Subshrubs or shrubs, 0.03–0.15 m; *stems* trailing, yellow-brown or red-brown, glabrous; *leaves* oblong, broadly elliptic, suborbicular, or circular, (8) 12–66 × 8–50 mm, 1–1.5 times as long as wide, sparsely hairy to glabrescent below, glabrous or pilose above, with revolute, entire or crenulate margins and conspicuously reticulate veins; *staminate catkins* 11–54 × 4–9 mm; *pistillate catkins* 11–79 × 3–8 mm, the floral bracts tawny, 0.8–1.8 mm long; *capsules* 4.5–5 mm long, hairy. Common in the alpine, 10,500–13,500 ft. June–Aug. E/W. (Plate 74)

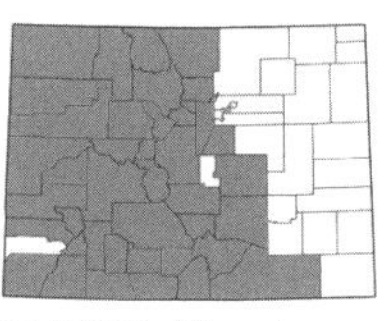

***Salix scouleriana* Barratt ex Hook.**, SCOULER'S WILLOW. Shrubs or trees, 1–10 (20) m; *stems* gray-brown, yellow-brown, or red-brown, glabrous or tomentose; *leaves* mostly oblanceolate or obovate (rarely elliptic), 29–100 × 9–37 mm, 1.7–3.9 times as long as wide, glaucous and sparsely to densely hairy below, pilose or moderately hairy above, with strongly to slightly revolute or flat, entire, remotely serrate, crenate, or sinuate margins; *staminate catkins* 18–40.5 × 8–22 mm; *pistillate catkins* 18–60 (-90 in fruit) × 10–22 mm, the floral bracts brown, black, or bicolor, 1.5–4.5 mm long; *capsules* 4.5–11 mm long, hairy. Common in forests, meadows, and along streams, 6500–11,500 ft. May–June. E/W.

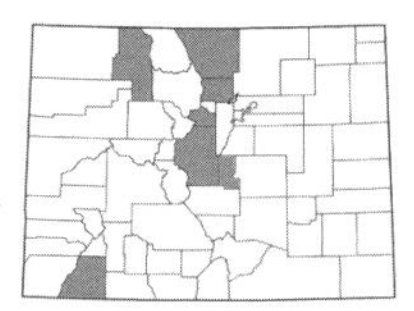

***Salix serissima* (L.H. Bailey) Fernald**, AUTUMN WILLOW. Shrubs, 1–5 m; *stems* yellow-brown, red-brown, or gray-brown, glabrous, slightly glossy or dull; *leaves* shallowly to deeply grooved above, 3–13 mm, slightly to highly glossy and glabrous above and below, with serrulate margins; *staminate catkins* 25–53 × 12–16 mm; *pistillate catkins* 17–42 (-65 in fruit) × 11–22 mm, the floral bracts 1.2–4 mm long; *capsules* 6–12 mm long, glabrous. Rather uncommon in fens and bogs and along streams, 7500–10,000 ft. June–July. E/W.

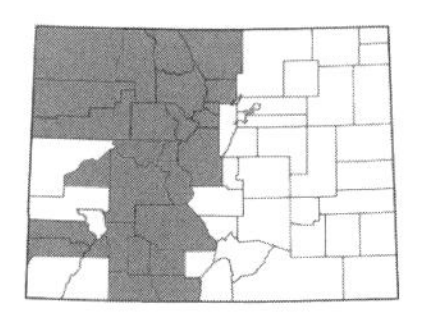

Salix wolfii* Bebb var. *wolfii, WOLF'S WILLOW. Shrubs, 0.1–2 m; *stems* red-brown, violet, yellow-gray, or yellow-brown, hairy to glabrescent; *leaves* narrowly oblong, elliptic, or oblanceolate, 26–56 × 8–16.5 mm, 2.5–3.7–5.6 times as long as wide, short-hairy or villous below, sparsely to densely hairy to villous above, with entire margins; *staminate catkins* 9.5–16 × 6–12 mm, flowering as leaves emerge; *pistillate catkins* 8.5–38 × 5–12 mm, the floral bracts brown, black, or bicolor, 0.8–2 mm long; *capsules* 3–5 mm long, glabrous. Common in moist meadows, along lakes and streams, and in seepage areas, 7000–11,500 ft. May–June. E/W.

SANTALACEAE R. Br. – SANDALWOOD FAMILY

Herbs (ours) partially parasitic on roots; *leaves* simple and entire, exstipulate; *flowers* perfect or imperfect, actinomorphic; *perianth* of 4–5 sepals, these distinct or connate and forming a fleshy cup or tube; *stamens* 4–5, distinct or epipetalous; *pistil* 1; *ovary* inferior or partly inferior, (2) 3 (5)-carpellate.

COMANDRA Nutt. – BASTARD TOADFLAX

Leaves alternate; *flowers* perfect; *calyx* (4) 5-lobed, petaloid, connate, white, whitish-green, or whitish-purple; *petals* absent; *stamens* (4) 5, opposite the sepals, adnate; *ovary* inferior; *fruit* a dry, subglobose drupe surmounted by persistent sepals.

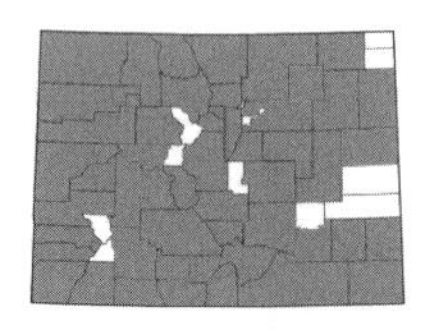

***Comandra umbellata* (L.) Nutt. ssp. *pallida* (A. DC.) Piehl**, PALE BASTARD TOADFLAX. Perennials from rhizomes, 0.7–5 dm; *leaves* linear to elliptic or narrowly ovate, 0.7–5 (6) cm long; *inflorescence* of terminal paniculate to corymbose cymes; *calyx* lobes 2–4.5 mm long; *fruit* 4–7 mm diam. Common from the plains to mountains, usually in dry places, 3500–11,000 ft. May–July. E/W. (Plate 74)

SAPINDACEAE Juss. – SOAPBERRY FAMILY

Shrubs, trees, lianas, or herbs; *leaves* alternate or opposite, petiolate, often gland-dotted, stipulate or exstipulate; *flowers* imperfect or rarely perfect; *sepals* 4–5, usually distinct; *petals* 4–5, usually united; *stamens* 8–10, in 2 whorls, distinct; *pistil* 1; *ovary* superior, (2) 3 (8)-carpellate; *fruit* a capsule, berry, drupe, nut, samara, or capsule.

1a. Leaves opposite, palmately lobed or ternately compound (occasionally with 5–7 leaflets); fruit a schizocarp of 2 winged, 1-seeded samaras...***Acer***
1b. Leaves alternate, pinnately compound; fruit a globose berry...***Sapindus***

ACER L. – MAPLE

Shrubs or trees; *leaves* opposite, simple, palmately lobed or ternately compound; *fruit* a schizocarp of 2 winged, 1-seeded samaras.

1a. Leaves ternately compound with 3 leaflets, or occasionally with 5 or 7 leaflets, the terminal leaflet with an evident petiolule; nectary disk absent...***A. negundo***
1b. Leaves simple and palmately lobed, or ternately compound with sessile leaflets; nectary disk present and well-developed...2

2a. Leaves glabrous, palmately 3–5-lobed or ternately compound, with numerous sharp teeth along the margins; sepals distinct...***A. glabrum* var. *glabrum***
2b. Leaves hairy below, usually 5-lobed and never ternately compound, with a few, blunt teeth; sepals irregularly connate...***A. grandidentatum* var. *grandidentatum***

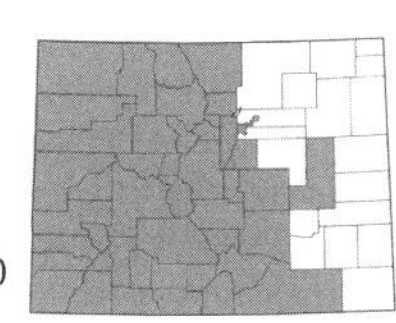

Acer glabrum* Torr. var. *glabrum, ROCKY MOUNTAIN MAPLE. [including *A. glabrum* Torr. var. *neomexicanum* (Greene) Kearney & Peebles]. Shrubs or small trees to 8 m; *leaves* palmately 3–5-lobed or ternately compound, mostly 2–8 cm wide, sharply toothed along the margins, glabrous; *sepals* 3–5 mm long, greenish, distinct; *samaras* glabrous, 2–3 cm long. Common and widespread across the state along streams, in gulches and ravines, and in dry forests, 5200–10,500 ft. April–June. E/W. (Plate 75)

Acer grandidentatum* Nutt. var. *grandidentatum, BIG-TOOTH MAPLE. Small trees, 4–8 m; *leaves* usually palmately 5-lobed, mostly 2.5–10 (13) cm wide, with a few, blunt teeth along the margins, hairy below; *sepals* 3–5 mm long, greenish, connate to near the middle or above; *samaras* hairy. Uncommon on sandstone cliffs, 7500–8000 ft. April–May. W.

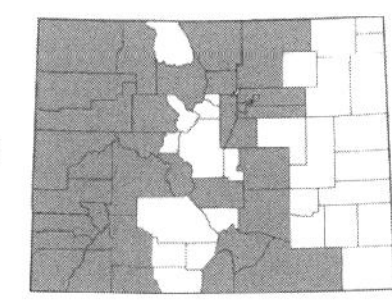

***Acer negundo* L.**, BOX ELDER. [*Negundo aceroides* (L.) Moench]. Trees to 12 m; *leaves* ternately compound, occasionally with 5 or 7 leaflets, the terminal leaflet on a petiolule, coarsely toothed, hairy or glabrous; *sepals* 1–2 mm long; *samaras* glabrous or hairy, 2.5–4 cm long. Common across the state along creeks and in canyon bottoms, 4800–7900 ft. April–May. E/W. Introduced.

There are two varieties of *A. negundo* in Colorado:

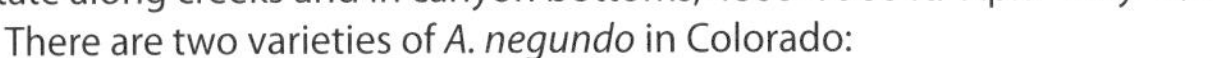

1a. Young branches glaucous...**var. *violaceum* (Kirchn.) Jaeg.**

1b. Young branches with soft, velutinous hairs...**var. *interius* (Britton) Sarg.**

SAPINDUS L. – SOAPBERRY

Shrubs or trees; *leaves* alternate, pinnately compound; *petals* white, with a pilose claw attached below the disk; *fruit* a globose berry.

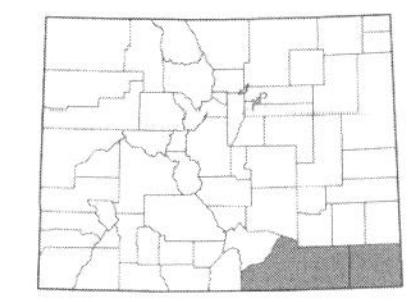

***Sapindus saponaria* L. var. *drummondii* (Hook. & Arn.) L.D. Benson**, WESTERN SOAPBERRY. Trees to 15 m; *leaflets* lanceolate to narrowly elliptic, to 9 cm long; *inflorescence* a terminal panicle; *berries* ca. 1.3 cm diam., yellowish. Found on river benches and in canyons, 4100–5500 ft. May–July. E.

SAURURACEAE Meyer – LIZARD-TAIL FAMILY

Perennial herbs; *leaves* simple, stipulate, margins entire; *flowers* perfect; *perianth* absent, each flower subtended by a petaloid bract; *stamens* (3) 6 (8); *pistil* 1; *ovary* superior or inferior, 3–5 (7)-carpellate; *fruit* a capsule or schizocarp.

ANEMOPSIS Hook. & Arn. – YERBA MANSA

Leaves mostly basal, long-petiolate; *flowers* in terminal congested, conic spikes; *stamens* mostly 6; *styles* 3 (4); *stigmas* 3 (4); *ovary* inferior, 3 (4)-carpellate; *fruit* a capsule.

***Anemopsis californica* Hook. & Arn.**, YERBA MANSA. Plants rhizomatous, to 5 dm; *leaves* oblong-elliptic, with rounded tips, to 30 cm long; *spikes* 1–3 cm long; *petaloid bracts* 4–8, white, 1–2 cm long. Uncommon in irrigation ditches and marshes, 3600–5300 ft. June–Aug. E. (Plate 75)

SAXIFRAGACEAE Juss. – SAXIFRAGE FAMILY

Annual or mostly perennial herbs; *leaves* usually basal or sometimes alternate or opposite, simple, entire to pinnately or palmately compound, the margins entire to crenate, serrate, or dentate; *flowers* usually perfect, usually actinomorphic, with a hypanthium; *sepals* (4) 5, distinct; *petals* 5 or absent, distinct, lobed or unlobed; *stamens* 10 or sometimes 5, rarely 4 or 8, distinct; *pistil* 1, sometimes appearing 2–3; *ovary* superior to inferior, usually 2-carpellate or rarely 3-carpellate; *fruit* a capsule.

1a. Leaves all basal or with very reduced, small scale-like stem leaves...2
1b. Leaves both basal and cauline, the stem leaves usually reduced distally but not scale-like...5

2a. Flowers in a narrow, loose simple raceme or spike...3
2b. Flowers in a panicle or thyrse, sometimes densely crowded into a dense, terminal conic or cylindric head-like thyrse...4

3a. Petals pinnately or ternately dissected or cleft into linear divisions, white or greenish-yellow, lacking a long claw; hypanthium broadly saucer-shaped; leaf margins glabrous or with glandular ciliate hairs...***Mitella***
3b. Petals entire, white, erect with a long claw exceeding the sepals; hypanthium narrowly campanulate to cylindric; leaf margins with stiffly ciliate, incurved hairs...***Conimitella***

4a. Stamens 10; leaves oblanceolate, triangular, or ovate to elliptic with cuneate bases, or orbiculate with a cordate to truncate base and regularly dentate margins (not double); plants from thick, fleshy rhizomes...***Micranthes***
4b. Stamens 5; leaves reniform to orbiculate or broadly ovate, with cordate or truncate bases, the margins doubly crenate to dentate; plants from a stout, often branched caudex or rhizome...***Heuchera***

5a. Petals deeply palmately lobed, white or pink; leaves deeply palmately lobed or compound...***Lithophragma***
5b. Petals entire, variously colored; leaves not deeply palmately lobed or compound...6

6a. Petals violet-purple to red-purple or bright pink, (5) 7–11 mm long; plants densely stipitate glandular-hairy...***Telesonix***
6b. Petals green, white, or yellow, sometimes with reddish-pink spots, if purple then the petals only 1.5–4.5 mm long; plants densely stipitate glandular or not...7

7a. Sepals 4, green or greenish-yellow; petals absent; stamens 4; plants lacking glandular-stipitate hairs; leaves reniform to orbicular with shallowly crenate margins...***Chrysosplenium***
7b. Sepals 5, green; petals present, green, white, yellow, or purple; stamens 5 or 10; plants with or without glandular-stipitate hairs; leaves unlike the above...8

8a. Flowers green or greenish-yellow, in a dense, one-sided thyrse; ovary unilocular with parietal placentation...***Heuchera*** (*bracteata*)
8b. Flowers white, yellow, rose, or purple, solitary or loosely arranged in an open cyme, thyrse, or panicle; ovary 2-locular with axile or marginal placentation...9

9a. Stamens 5; leaves palmately 9–12-lobed, 2–4 cm wide, 1–2 times dentate; plants of hanging gardens and wet cliffs on the western slope...***Sullivantia***
9b. Stamens 10; leaves unlobed and entire, apically lobed or toothed, or if palmately 3–7-lobed then 0.3–2.5 cm wide; plants variously distributed...***Saxifraga***

CHRYSOSPLENIUM L. – GOLDEN SAXIFRAGE

Rhizomatous or stoloniferous herbs; *leaves* cauline, opposite or alternate, *flowers* actinomorphic; *sepals* 4; *petals* absent; *stamens* usually 4 or 8; *pistil* 1; *ovary* ½–¾-inferior, unilocular.

Chrysosplenium tetrandrum **(Lund ex Malmgr.) Th. Fr.**, NORTHERN GOLDEN SAXIFRAGE. Perennials to 10 cm; *leaves* reniform to oval, mostly 5–12 mm wide, broadly crenate; *inflorescence* of terminal, few-flowered, leafy cymes; *flowers* ca. 3 mm wide. Uncommon or easily overlooked along mossy streambanks and inlets, 8000–12,800 ft. April–Aug. E/W.

CONIMITELLA Rydb. – CONIMITELLA

Herbs; *leaves* in a basal rosette; *flowers* actinomorphic, with a hypanthium; *sepals* 5; *petals* 5, white, entire, erect; *stamens* 5; *pistil* 1; *ovary* ½–¾-inferior.

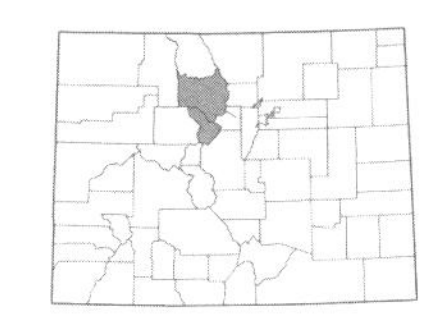

***Conimitella williamsii* (D.C. Eaton) Rydb.**, WILLIAMS' MITERWORT. Perennials, 2–5 (6) dm tall, finely glandular; *leaves* reniform, 1–4 cm wide, broadly crenate, glabrous with ciliate margins, on petioles 1–6 (10) cm long; *inflorescence* a loose raceme with 5–10 (12) flowers; *sepals* 4–6 mm long; *petals* 4–5 mm long, entire. Uncommon in Douglas-fir forests, apparently disjunct from northern Wyoming, 8000–9000 ft. June–July. W.

HEUCHERA L. – ALUMROOT

Perennial herbs; *leaves* in a basal rosette and cauline, the cauline leaves very reduced and scale-like; *flowers* actinomorphic or zygomorphic; *sepals* 5; *petals* 5 or sometimes absent or minute; *stamens* 5, opposite sepals; *pistil* 1, 2-carpellate; *ovary* ½-inferior.

1a. Petals finely dentate to fimbriate or moderately erose at the tips; hypanthium strongly zygomorphic and conspicuously longer on the adaxial side, 5–14 mm long, free 2–7 mm...***H. richardsonii***
1b. Petals entire; hypanthium weakly to slightly zygomorphic or actinomorphic, 2.5–6.5 mm long, free 0.4–2 mm...2

2a. Hypanthium and petals pink to rose; stamens exserted 0.8–3 mm...***H. rubescens***
2b. Hypanthium greenish, greenish-yellow, cream, or pinkish-yellow; petals green or white; stamens included or exserted 0.5–1 mm...3

3a. Hypanthium flat, saucer-shaped in flower; petals white or yellowish; inflorescence not one-sided; leaves with rounded teeth...***H. parvifolia***
3b. Hypanthium campanulate in flower; petals green, greenish-yellow, or white; inflorescence usually one-sided; leaves sharply to bluntly dentate...4

4a. Flowering stems leafless; petals white; styles and stamens included; leaves bluntly dentate; inflorescence usually with fewer flowers and loosely compact...***H. hallii***
4b. Flowering stems usually with 1–2 reduced, sometimes scale-like leaves; petals green or greenish-yellow; styles and stamens exserted 0.5–1 mm; leaves sharply dentate; inflorescence usually with numerous flowers and densely compact...***H. bracteata***

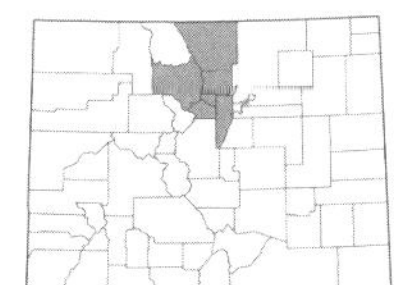

***Heuchera bracteata* (Torr.) Ser.**, ROCKY MOUNTAIN ALUMROOT. Plants 0.6–3 (3.8) dm; *leaves* reniform or broadly ovate, shallowly 5–7-lobed, 1.5–4 cm, with sharply dentate margins, short glandular-stipitate; *inflorescence* dense, secund; *hypanthium* narrowly campanulate, 3–5 mm long; *sepals* erect, 1–1.5 mm long; *petals* erect, green or greenish-yellow, ca. 2 mm long, entire. Common in rock crevices and on rocky cliffsides, 6000–12,300 ft. June–Aug. E/W. (Plate 75)

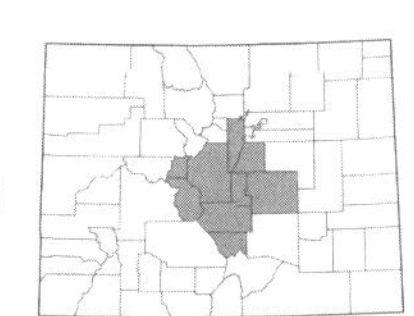

***Heuchera hallii* A. Gray**, FRONT RANGE ALUMROOT. Plants 1–3 dm; *leaves* rounded-reniform, deeply 5–7-lobed, 1–3.8 cm, with dentate margins, glabrous or short glandular-stipitate; *inflorescence* dense, occasionally secund; *hypanthium* broadly campanulate, 4–5.5 mm long; *sepals* erect, 1.2–2 mm long; *petals* spreading, white, 1.8–2.5 mm long, entire. Found in rock crevices and rocky cliffsides, 7500–12,000 ft. June–Aug. E. Endemic.

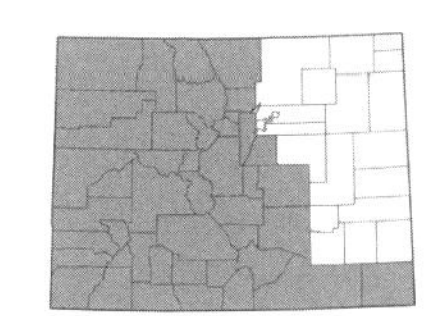

***Heuchera parvifolia* Nutt.**, COMMON ALUMROOT. Plants 0.4–7.1 dm; *leaves* orbiculate or reniform to broadly cordate, shallowly to deeply 5–7-lobed, 1–8 cm, with dentate margins, glabrate or short glandular-stipitate, occasionally long stipitate below; *inflorescence* dense at anthesis, spreading in fruit; *hypanthium* flat, saucer-shaped, 2.5–5 mm long; *sepals* reflexed, 0.5–1.5 mm long; *petals* reflexed, white or yellowish, 0.7–3 mm long, entire. Common on rocky outcroppings, among boulders, in meadows, and sometimes with sagebrush or pinyon-juniper, 5000–13,500 ft. May–Aug. E/W.

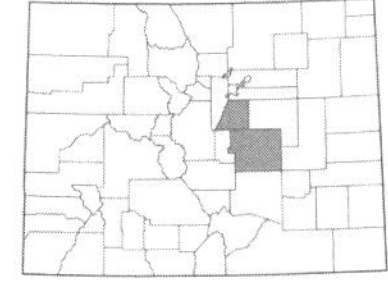

***Heuchera richardsonii* R. Br.**, RICHARDSON'S ALUMROOT. Plants (0.7) 2–9.5 dm; *leaves* broadly ovate or cordate, deeply 5–7-lobed, 2.5–10 cm, with dentate margins, glabrous or long glandular-stipitate above, long glandular-stipitate below; *inflorescence* dense to spreading; *hypanthium* campanulate, 5–14 mm long; *sepals* erect, 1.3–4.2 mm long; *petals* erect, green or greenish-white, rarely pink, 1.3–4 mm long, finely dentate, coarsely fimbriate, or moderately erose at the tips. Rare in pine forests and meadows, 6500–8000 ft. June–July. E.

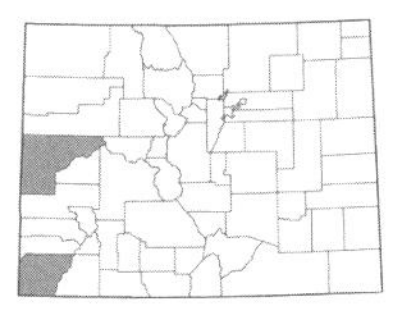

***Heuchera rubescens* Torr.**, RED ALUMROOT. Plants (0.6) 1–3.2 (5) dm; *leaves* suborbiculate or broadly ovate, shallowly 3–7-lobed, 0.6–4.5 cm, with dentate margins, glabrous or short to long glandular-stipitate; *inflorescence* dense and often secund, spreading in fruit; *hypanthium* narrowly campanulate, becoming urceolate, 3–6.5 mm long; *sepals* erect, 1–3.3 mm long; *petals* spreading, pink to rose, 2–3 (6) mm long, entire. Rare in rock crevices and on cliffsides, 7000–9000 ft. May–Aug. W.

LITHOPHRAGMA (Nutt.) Torr. & A. Gray – WOODLAND STAR

Rhizomatous perennial herbs with underground bulbils; *leaves* in a basal rosette and cauline, the cauline leaves usually alternate and palmately dissected; *flowers* actinomorphic; *sepals* 5; *petals* 5, white or pink, the limb palmately or pinnately 3–7-cleft or divided, sometimes entire; *stamens* 10; *ovary* superior to almost completely inferior.

1a. Reddish-purple bulbils usually present in the leaf axils and often replacing flowers; basal leaves usually glabrous below or sometimes sparsely hairy; stigma apex with numerous round, short bumps (papillae)...***L. glabrum***
1b. Bulbils usually absent from the leaf axils and not replacing flowers; basal leaves usually sparsely to densely hairy below; stigma glabrous at the apex with a subapical, narrow band of papillae below...2

2a. Hypanthium rounded at the base, campanulate or hemispheric; petals palmately 5–7-lobed; stem leaves deeply 3-lobed and appearing pinnatifid, ultimately with very narrow lobes and appearing markedly different than the basal leaves...***L. tenellum***
2b. Hypanthium acute at the base and gradually tapering to the pedicel, obconic-elongate; petals palmately 3-lobed; stem leaves palmately 3-lobed, similar to the basal leaves except with longer lobes...***L. parviflorum***

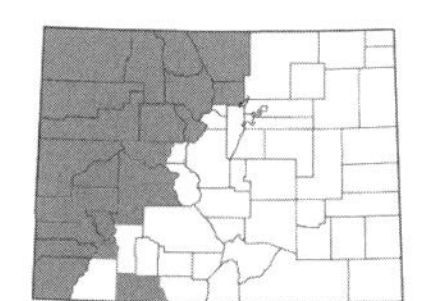

Lithophragma glabrum **Nutt.**, BULBOUS WOODLAND STAR. [*L. bulbifera* Rydb.]. Plants 0.8–3.5 dm; *leaves* round, trilobed or trifoliate, the ultimate segments1–4-lobed, glabrous or sparsely hairy; *inflorescence* 2–5 (7)-flowered racemes; *hypanthium* narrowly campanulate; *sepals* erect in bud, spreading in flower, triangular; *petals* widely spreading, pink or rarely white, 3.5–7 mm long, 3–5-lobed. Found in meadows, aspen groves, and sagebrush, 6500–10,000 ft. May–July. E/W.

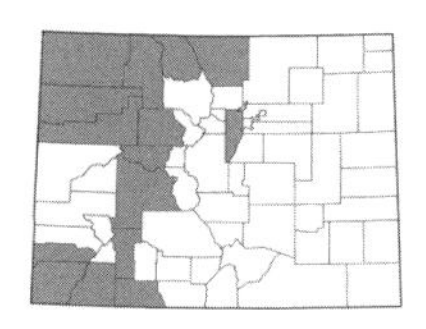

Lithophragma parviflorum **(Hook.) Nutt.**, SMALLFLOWER WOODLAND STAR. Plants 2–5 dm; *leaves* trilobed or trifoliate, the cauline leaves reduced, glabrous or sparsely to densely hairy; *inflorescence* 4–14-flowered racemes; *hypanthium* obconic-elongate at anthesis, elongate in fruit; *sepals* erect, triangular; *petals* widely spreading, white or pink, 7–16 mm long, 3-lobed. Uncommon in moist, rich soil, along streams, in meadows, or sometimes with sagebrush, 6500–9000 ft. May–June. E/W.

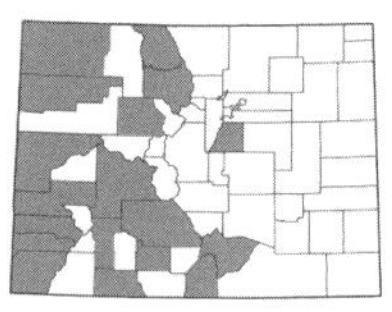

Lithophragma tenellum **Nutt.**, SLENDER WOODLAND STAR. Plants 0.8–3 dm; *leaves* trilobed, the cauline leaves reduced; *inflorescence* 3–12-flowered racemes; *hypanthium* hemispheric to campanulate; *petals* pink or sometimes white, 3–7 mm long, 5–7-lobed. Found in rich, often moist soil of meadows, forest openings, and sagebrush slopes, 6500–9500 ft. May–June. E/W.

MICRANTHES Haworth – SAXIFRAGE

Annual or perennial herbs; *leaves* basal; *flowers* actinomorphic; *sepals* 5; *petals* 5 or absent; *stamens* 10; *pistils* 2 (3), 2 (3)-carpellate; *ovary* superior or inferior. (Elvander, 1984)

1a. Inflorescences with most of the flowers replaced by bulbils, only 1–2 terminal flowers remaining; leaves oblanceolate with a cuneate base, 3–5-toothed at the apex...***M. foliolosa***
1b. Flowers not replaced by bulbils; leaves unlike the above...2

2a. Leaves round, regularly dentate and lacking ciliate hairs along the margins; petiole rounded, distinct from the leaves; flowering stem purple-gland-tipped above; flowers in an open, lax thyrse...***M. odontoloma***
2b. Leaves linear, oblanceolate, ovate, elliptic, or triangular, the margins ciliate and serrate to denticulate; petiole flattened, sometimes indistinct from the leaves; flowering stem yellow- or rarely purplish-gland-tipped above; flowers in open to crowded conic or cylindric thyrses or heads...3

3a. Flowers crowded into a dense, terminal conic or cylindric thyrse; leaves 1–6 cm long, gradually tapering to a distinct petiole...***M. rhomboidea***
3b. Flowers in a panicle; leaves 6–20 cm long, gradually tapering to an indistinct petiole...***M. oregana***

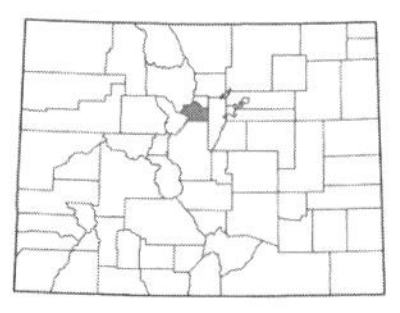

Micranthes foliolosa **(R. Br.) Gornall**, LEAFYSTEM SAXIFRAGE. [*Saxifraga foliolosa* R. Br.; *Spatularia foliolosa* (R. Br.) Small]. Perennials to 1 dm; *leaves* oblanceolate, 0.6–1.5 cm long, cuneate at the base, 3–5-toothed distally, eciliate or sparsely ciliate, glabrate; *inflorescence* 2 (5)-flowered, most flowers replaced by bulbils, glabrous or sparsely purple-tipped glandular-stipitate; *sepals* reflexed; *petals* white with 1–2 basal yellow spots, 3–8 mm long. Rare in moist, mossy alpine, known only from Mt. Evans and vicinity (Clear Creek Co.), 12,500–13,500 ft. July–Aug. E.

Disjunct from Alaska and Canada and found only on Mt. Evans in the lower 48 states.

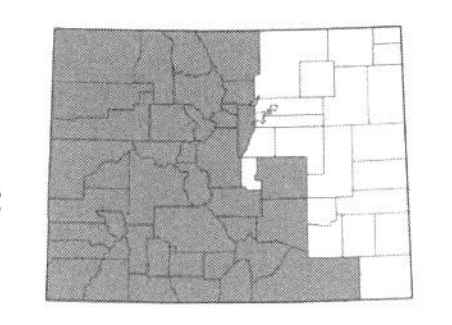

Micranthes odontoloma **(Piper) A. Heller**, BROOK SAXIFRAGE. [*Saxifraga odontoloma* Piper]. Perennials, (1.3) 2–7 dm, glandular-hairy above; *leaves* round, cordate to truncate at the base, the margins dentate and eciliate, glabrous above, sparsely brownish-hairy below; *inflorescence* 10–30+-flowered, glabrous below and purple-tipped subsessile-glandular above; *sepals* reflexed; *petals* white with 2 basal yellow spots, 3–4.5 mm long. Common in moist soil along streams and around lakes, 7500–13,000 ft. July–Sept. E/W.

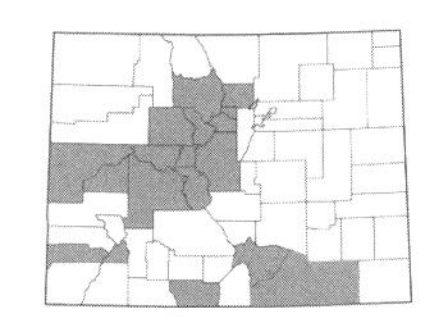

Micranthes oregana **(Howell) Small**, OREGON SAXIFRAGE. [*Saxifraga oregana* Howell]. Perennials, (1.5) 3–6 (10) dm, glandular-hairy above; *leaves* linear to oblanceolate, 6-20 cm long, cuneate at the base, the margins serrulate to denticulate and ciliate, glabrous to sparsely hairy; *inflorescence* (30) 50-flowered, hairy below, yellow- to purple-tipped glandular-stipitate above; *sepals* reflexed; *petals* white, not spotted, 2–5 mm long. Found along streams and in moist meadows, 9000–13,500 ft. June–Aug. E/W.

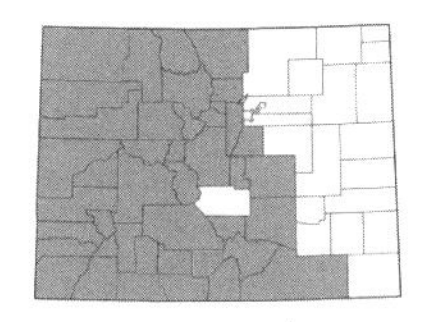

Micranthes rhomboidea **(Greene) Small**, DIAMONDLEAF SAXIFRAGE. [*Saxifraga rhomboidea* Greene]. Perennials, 0.25–3 dm, densely glandular above; *leaves* broadly ovate to triangular, 1–6 cm long, the margins coarsely serrate and ciliate, glabrous above, sparsely to moderately tangled, reddish-brown hairy below; *inflorescence* (5) 10–40-flowered, usually densely yellow-glandular-stipitate; *sepals* ascending; *petals* white, not spotted, 2–4 mm long. Common from the foothills to the alpine, 5000–14,000 ft. April–Sept. E/W. (Plate 75)

MITELLA L. – MITREWORT

Rhizomatous perennial herbs; *leaves* in a basal rosette; *flowers* actinomorphic, with a discoid hypanthium; *sepals* 5; *petals* 5, the limb usually pinnately or ternately cleft or dissected; *stamens* 5 or 10; *ovary* superior to nearly completely inferior. (Rosendahl, 1914)

1a. Petals greenish-yellow, pinnatifid into 5–11 narrow spreading divisions; stamens alternate with the sepals; flowers on evident pedicels 1–5 mm long; inflorescence not one-sided in flower; leaves with toothed or mucronate lobes and margins…***M. pentandra***
1b. Petals white, with 3 ternate divisions; stamens opposite the sepals; flowers sessile or on short pedicels to 1 mm long; inflorescences one-sided in flower; leaves with rounded, crenate to shortly toothed lobes and margins…***M. stauropetala* var. *stenopetala***

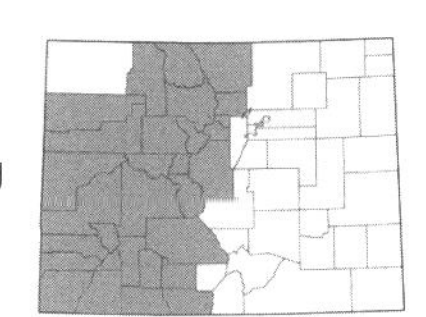

Mitella pentandra **Hook.**, FIVE-STAR MITREWORT. Plants (0.8) 1–4 dm, glandular above; *leaves* ovate-cordate, 2.5–7 cm long, doubly crenate-dentate, sparsely hirsute or glabrous; *inflorescence* 6–20 (25)-flowered; *hypanthium* strongly flattened, 2.4–3 mm diam.; *sepals* 0.5–0.6 mm long, spreading or reflexed; *petals* 1.5–3 mm long, pectinate-pinnatifid with 5–11 narrow spreading divisions, greenish-yellow. Common along streams, in shady forests, and in moist meadows, 8000–12,000 ft. June–Aug. E/W.

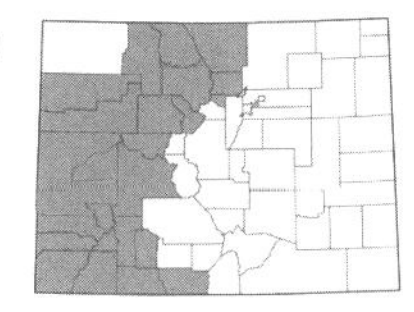

Mitella stauropetala **Piper var. *stenopetala*** **(Piper) Rosend.**, SIDE-FLOWERED MITREWORT. Plants 1–4.5 dm, glandular-hairy; *leaves* orbicular to ovate or reniform, shallowly 1-2-times crenate, hirsute or glabrous; *inflorescence* (7) 10–35-flowered, secund; *hypanthium* short-campanulate, 1.5–3 mm diam.; *sepals* 1–2.2 mm long, erect or recurved; *petals* 1.5–4 mm long, with 3 linear lobes, white. Common along streams, in moist meadows, and in forests, 7000–12,000 ft. June–Aug. E/W.

SAXIFRAGA L. – SAXIFRAGE

Perennial herbs; *leaves* in a basal rosette and cauline, the cauline leaves reduced distally, stipules absent; *flowers* actinomorphic or zygomorphic; *sepals* 5; *petals* 5 or absent; *stamens* 10; *pistil* 1, 2-carpellate; *ovary* superior to inferior. (Brouillet & Elvander, 2009)

1a. Leaves unlobed, the margins entire; flowers yellow, or white with purple or reddish spots…2
1b. Leaves lobed or toothed; flowers white (sometimes greenish- or yellowish-tinged at the tips) to purple, not spotted…5

2a. Flowers white with purple to reddish spots on the lower part of the petals and yellowish spots on the upper part of the petals; leaves with a white-spinulose tip and stiff, white-ciliate margins; plants mat-forming with trailing stems…***S. bronchialis* var. *austromontana***
2b. Flowers yellow or orange-yellow; leaves various; plants mat-forming or not but lacking trailing stems…3

3a. Plants with slender stolons; sepals erect, glandular-stipitate hairy; basal leaves stiffly ciliate…***S. flagellaris* ssp. *crandallii***
3b. Plants lacking stolons; sepals spreading to reflexed, glabrous to glandular-stipitate hairy; basal leaves not stiffly ciliate…4

4a. Flowering stem sparsely to densely reddish-brown-villous below the inflorescence; sepals glabrous or with reddish-brown-ciliate margins...***S. hirculus***
4b. Flowering stem sparsely purplish-glandular-stipitate below the inflorescence; sepals glandular-stipitate with glandular-ciliate margins...***S. chrysantha***

5a. Leaves oblanceolate to obovate or spatulate, 2–5-lobed or -toothed at the apex...6
5b. Leaves round, reniform, or orbiculate, 3–7-palmately-lobed...7

6a. Leaves shallowly toothed into ovate segments; plants solitary, or tufted with 2 stems but not strongly cushion-forming or caespitose...***S. adscendens***
6b. Leaves deeply lobed into linear segments; plants usually with more than 2 stems, strongly caespitose and densely or loosely cushion- or mat-forming...***S. cespitosa***

7a. Bulbils present in the leaf axils; petals 5–12 mm long; sepal margins glandular-ciliate...***S. cernua***
7b. Bulbils absent from the leaf axils; petals (1.5) 2–5 (6) mm long; sepal margins with or without glandular-ciliate hairs...***S. rivularis***

Saxifraga adscendens **L.**, WEDGELEAF SAXIFRAGE. [*Muscaria adscendens* (L.) Small]. Plants solitary or tufted, not stoloniferous, with a caudex; *leaves* oblanceolate to obovate, (2) 3 (5)-toothed or shallowly lobed, (2) 4–15 mm long, the margins entire, glabrate to glandular-stipitate; *inflorescence* a (2) 6–15 (40)-flowered thyrse, densely purple-tipped glandular-stipitate; *sepals* erect; *petals* white, not spotted, (3) 3–6 mm long. Uncommon along alpine streams, in alpine meadows, and on scree slopes, 10,500–13,500 ft. July–Aug. E/W.

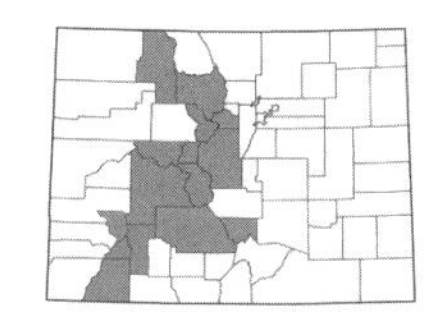

Saxifraga bronchialis **L. var. *austromontana* (Wiegand) Piper**, SPOTTED SAXIFRAGE. [*Ciliaria austromontana* (Wiegand) W.A. Weber]. Plants mat-forming, not stoloniferous; *leaves* linear or linear-lanceolate, unlobed, 3–15 mm long, the margins entire and often white- or glandular-ciliate, glabrous, occasionally sparsely glandular above; *inflorescence* a 2–15-flowered cyme or thyrse, sparsely short glandular-stipitate; *sepals* erect, purplish; *petals* yellowish-spotted proximally, purple- or red-spotted distally, 3–7 mm long. Common on rocky outcrops, in spruce-fir forests, and the alpine, 6600–13,500 ft. June–Sept. E/W. (Plate 75)

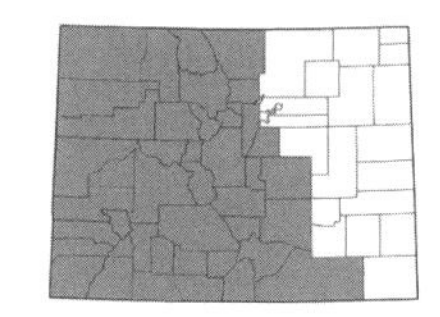

Saxifraga cernua **L.**, NODDING SAXIFRAGE. Plants solitary or in tufts, not stoloniferous; *leaves* round to reniform, 3–7-lobed, (3) 5–20 mm long, the margins entire and eciliate, or sometimes sparsely glandular-ciliate, glabrous or sparsely glandular-stipitate; *inflorescence* a 2–5-flowered paniculate or raceme-like thyrse, glandular-stipitate; *sepals* erect; *petals* white, not spotted, 5–12 mm long. Found in rocky alpine, on fell fields and talus slopes, and along alpine creeks, 10,500–14,000 ft. July–Aug. E/W.

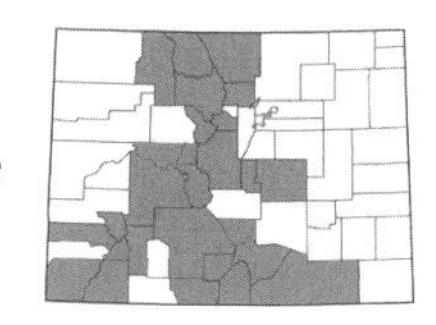

Saxifraga cespitosa **L.**, TUFTED ALPINE SAXIFRAGE. [*Muscaria delicatula* Small]. Plants densely cushion-to loosely mat-forming, not stoloniferous or rhizomatous; *leaves* elliptic to obovate or spatulate, 3 (5)-lobed apically, 4–16 (25) mm long, the margins entire and sparsely short glandular-ciliate, short glandular-stipitate; *inflorescence* a 2–5-flowered cyme, densely glandular-stipitate ; *sepals* erect to spreading; *petals* white, not spotted, greenish to yellowish-tinged proximally, (1.5) 3–7 (15) mm long. Found in rocky alpine, on fell fields, and talus and scree slopes, 11,000–14,000 ft. July–Aug. E/W.

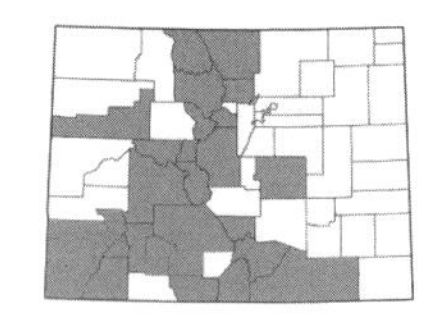

Saxifraga chrysantha **A. Gray**, GOLDEN SAXIFRAGE. [*Hirculus serpyllifolius* (Pursh) W.A. Weber ssp. *chrysantha* (A. Gray) W.A. Weber]. Plants mat-forming, slenderly rhizomatous, not stoloniferous; *leaves* linear-oblanceolate to narrowly spatulate, unlobed, (2) 3–12 mm long, the margins entire and eciliate, sometimes sparsely glandular-ciliate, glabrous or occasionally sparsely glandular-stipitate; *inflorescence* a 2–3-flowered cyme, sparsely glandular-stipitate; *sepals* strongly reflexed; *petals* golden-yellow, sometimes orange-tipped proximally, 4–8 mm long. Found in alpine meadows and on scree slopes, 11,500–14,200 ft. July–Aug. E/W.

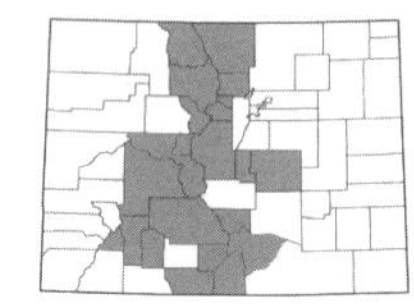

Saxifraga flagellaris **Willd. ssp. *crandallii* (Gandog.) Hultén**, STOLONIFEROUS SAXIFRAGE. [*Hirculus platysepalus* (Trautvetter) W.A. Weber ssp. *crandallii* (Gandog.) W.A. Weber]. Plants in solitary clumps, stoloniferous, slenderly rhizomatous; *leaves* oblong-lanceolate or elliptic to obovate, unlobed, 5–20 mm long, the margins entire and sparsely to densely glandular-ciliate, glabrous; *inflorescence* a 2–3 (5)-flowered lax cyme, densely purplish-glandular-stipitate; *sepals* erect; *petals* yellow, not spotted, 4–9 (10) mm long. Found in alpine meadows and on scree slopes, 11,500–14,000 ft. July–Aug. E/W.

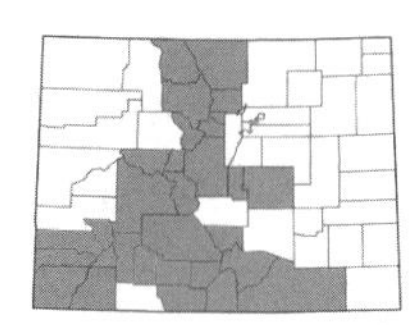

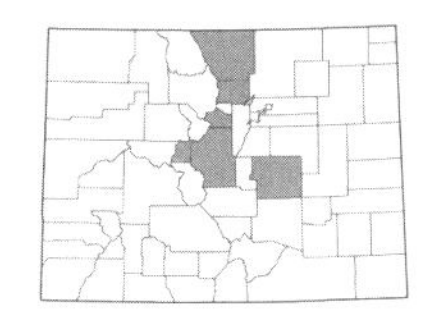

***Saxifraga hirculus* L.**, YELLOW MARSH SAXIFRAGE. [*Hirculus prorepens* (Fisher ex Sterb.) Á. Löve & D. Löve]. Plants loosely tufted, rhizomatous, sometimes shortly stoloniferous; *leaves* linear to spatulate, unlobed, (5) 10–30 mm long, the margins entire and eciliate, or sparsely reddish-brown-ciliate, glabrous or sparsely reddish-brown-villous; *inflorescence* a 2 (4)-flowered cyme, flowers initially nodding, sparsely to densely reddish-brown-villous; *sepals* ascending to spreading; *petals* yellow, often proximally orange-spotted, usually drying cream, 6–18 mm long. Uncommon in moist meadows, bogs, fens, and along creeks and lake shores, 9000–13,000 ft. July–Sept. E/W. (Plate 75)

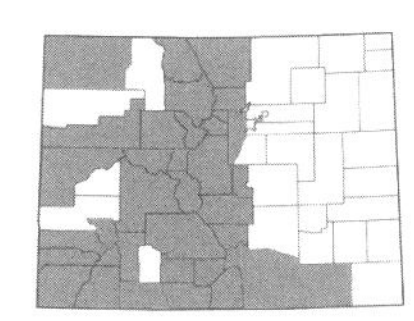

***Saxifraga rivularis* L.**, WEAK SAXIFRAGE. [*S. debilis* Engelm.; *S. hyperborea* R. Br.]. Plants solitary or in dense tufts, lacking stolons and rhizomes; *leaves* reniform to orbiculate, 3–7-lobed, 3–7 (10) mm long, the margins entire and eciliate or sparsely glandular-stipitate, glabrous or sparsely hairy; *inflorescence* a 2–5-flowered cyme, purple-glandular-stipitate or hairy and eglandular; *sepals* erect; *petals* white to purple, not spotted, (1.5) 2–5 (6) mm long. Found on rocky alpine, scree slopes, fell fields and talus slopes, and in shady, moist spruce-fir forests, 8300–14,000 ft. July–Aug. E/W. (Plate 75)

SULLIVANTIA Torr. & A. Gray ex A. Gray – COOLWORT

Perennials, rhizomatous herbs; *leaves* in a basal rosette and cauline; *flowers* actinomorphic; *sepals* 5; *petals* 5, white; *stamens* 5; *pistil* 1, 2-carpellate; *ovary* ½–⅘-inferior.

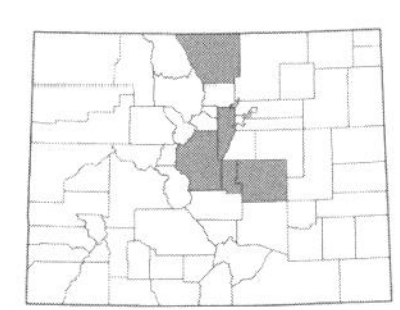

***Sullivantia hapemanii* (J.M. Coult. & Fish.) J.M. Coult. var. *purpusii* (Brand) Soltis**, HANGING GARDEN COOLWORT. Plants 2–3 dm; *leaves* reniform to orbiculate, 2–4 cm wide, 9–12-lobed, sharply toothed; *inflorescence* a paniculate cyme, glandular; *sepals* 3–4 mm long; *petals* 3–4 mm long. Uncommon in hanging gardens and on wet cliffs, 7000–10,000 ft. June–July. W. Endemic.

TELESONIX Raf. – BROOKFOAM

Perennial herbs; *leaves* in basal rosettes and cauline, the cauline leaves reduced; *flowers* actinomorphic; *sepals* 5; *petals* 5, entire; *stamens* 10; *pistil* 1, 2-carpellate; *ovary* ½-inferior, 2-locular.

***Telesonix jamesii* (Torr.) Raf.**, ALUMROOT BROOKFOAM. [*Boykinia jamesii* (Torr.) Engler; *Saxifraga jamesii* Torr.]. Plants 1–1.5 dm, glandular-hairy; *leaves* reniform to orbiculate, 2–3.5 cm wide, long-petiolate, with crenate margins; *inflorescence* a dense paniculate cyme; *sepals* 4–5 mm long; *petals* violet-purple, red-purple, or bright pink, (5) 7–11 mm long. Found on granite outcrops, 8700–12,800 ft. July–Sept. E.

SCROPHULARIACEAE Juss. – FIGWORT FAMILY

Herbs; *leaves* alternate or opposite, simple, exstipulate; *flowers* perfect, 5-merous, nearly actinomorphic or zygomorphic; *calyx* 5-lobed; *corolla* 5-lobed; *stamens* 4 with a staminode, or 5 and all fertile, epipetalous, alternate with the corolla lobes; *pistil* 1; *ovary* superior, 2-carpellate; *fruit* a capsule.

Former members of the Scrophulariaceae family have been split into the following families: Orobanchaceae, Phrymaceae, and Plantaginaceae.

1a. Aquatic or semi-aquatic, acaulescent herbs; leaves simple, entire, basal, on long petioles, usually oblong-elliptic; flowers solitary, found near the base of the plant and not surpassing the leaves; corolla white or pinkish...***Limosella***
1b. Plants unlike the above in all respects...2

2a. Leaves opposite, petiolate; corolla strongly zygomorphic and bilabiate, the upper lip 2-lobed and projected forward and the lower lip 3-lobed with erect lateral lobes and a deflexed middle lobe; stamens 4 with a 5th sterile staminode; stem 4-angled...***Scrophularia***
2b. Leaves alternate or in a basal rosette, sessile; corolla nearly actinomorphic to zygomorphic, but not bilabiate, otherwise unlike the above; stamens 5, all fertile; stem round...***Verbascum***

LIMOSELLA L. – MUDWORT

Perennials, aquatic or semi-aquatic herbs; *leaves* simple, entire, mostly basal, on long petioles; *flowers* solitary; *corolla* actinomorphic, 5-lobed, white or pinkish; *stamens* 4, subequal.

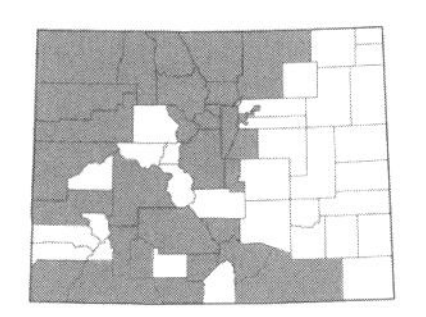

***Limosella aquatica* L.**, WATER MUDWORT. Stoloniferous herbs; *leaves* oblong-elliptic, 10–30 × 3–12 mm, on a petiole 3–10 (20) cm long, sometimes floating on the water's surface; *calyx* 2–2.5 mm long, with lobes 0.5–0.8 mm long; *corolla* slightly longer than the calyx; *capsules* ca. 3 mm long. Found in shallow water, temporary pools, along the shore of ponds and creeks, and on mud flats, 4500–10,500 ft. July–Sept. E/W.

SCROPHULARIA L. – FIGWORT

Perennial herbs with 4-angled stems; *leaves* opposite, petiolate, toothed (ours); *flowers* in a thyrsoid panicle; *corolla* strongly bilabiate, greenish-yellow to brownish or purple; *stamens* 4, slightly didynamous, with a 5th sterile staminode.

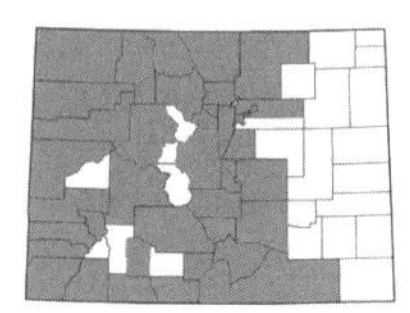

***Scrophularia lanceolata* Pursh**, LANCELEAF FIGWORT. Plants 8–15 dm, glandular-hairy; *leaves* lanceolate to narrowly ovate, with a truncate or cordate base, 8–13 × 3–5 (7) cm; *calyx* 2–4 mm long; *corolla* 8–12 mm long; *capsules* 5–10 mm long. Locally common in open places, forests, and along creeks, 4500–10,000 ft. May–Aug. E/W. (Plate 75)

VERBASCUM L. – MULLEIN

Biennial or perennial herbs; *leaves* alternate or in a basal rosette, simple, sessile, clasping or somewhat decurrent on the stem; *flowers* in a raceme or spike; *corolla* yellow or white, nearly actinomorphic; *stamens* 5, all fertile.

1a. Inflorescence a glandular-hairy raceme; leaves glabrous to minutely hairy; flowers usually white (rarely yellow) with purple filaments...***V. blattaria***

1b. Inflorescence woolly-tomentose, spicate panicle or panicle; leaves densely woolly-tomentose (at least below); flowers yellow or sometimes white, with white, yellow, or orange filaments...2

2a. Inflorescence a branched, pyramidal panicle; leaves white and tomentose below, green and glabrous or nearly so above...***V. lychnitis***

2b. Inflorescence a dense spike-like panicle; leaves tomentose above and below, although sometimes only thinly so above...3

3a. Leaves decurrent on the stem; filaments usually yellow...***V. thapsus***

3b. Leaves sessile or clasping but not decurrent on the stem; filaments usually orange...***V. phlomoides***

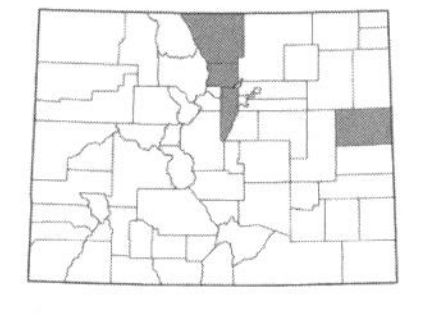

***Verbascum blattaria* L.**, MOTH MULLEIN. Biennials, 4–15 dm; *leaves* oblanceolate to narrowly ovate, 5–20 × 1–3 cm, glabrous to minutely hairy, the margins toothed to lobed or pinnatifid; *inflorescence* a simple raceme, glandular; *corolla* white or rarely yellow, 2.5–3 cm diam.; *stamens* with purple-hairy filaments. Found in disturbed places, along roads, in gardens, and along streams, 4000–6500 ft. June–Aug. E. Introduced.

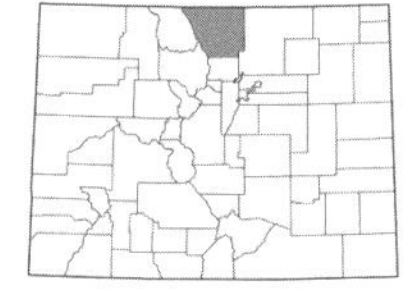

***Verbascum lychnitis* L.**, WHITE MULLEIN. Biennials, to 15 dm; *leaves* oblong-oblanceolate to elliptic, the margins entire to toothed, white-tomentose below, green and glabrous or nearly so above; *inflorescence* a branched, pyramidal panicle; *corolla* yellow or rarely white, 1–2 cm diam.; *stamens* with white or yellow filaments. Uncommon in disturbed places, 5000–5500 ft. June–Oct. E. Introduced.

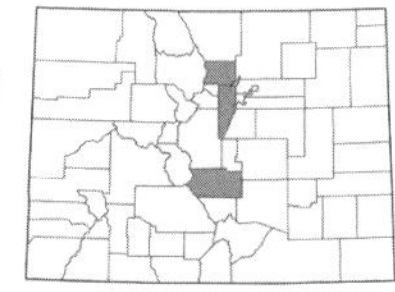

***Verbascum phlomoides* L.**, ORANGE MULLEIN. Biennials, 5–15 dm; *leaves* oblong to ovate, the margins toothed, to 4 dm long, tomentose on both sides but thinly so above; *inflorescence* a dense spike-like panicle; *corolla* yellow, 2.5–3.5 cm diam.; *stamens* usually with orange filaments. Uncommon along roadsides and in disturbed areas, 5500–8000 ft. June–Sept. E. Introduced.

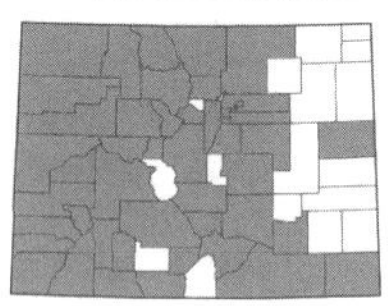

***Verbascum thapsus* L.**, WOOLLY MULLEIN. Biennials, 3–20 dm; *leaves* oblanceolate to obovate or ovate, the margins entire to shallowly toothed, to 5 dm long, tomentose, decurrent; *inflorescence* a dense spike-like panicle; *corolla* yellow, 1.2–2 (3.5) cm diam.; *stamens* usually with yellow filaments. Common in meadows, along roadsides, on open slopes, and in disturbed places, 4000–9500 ft. June–Oct. E/W. Introduced. (Plate 75)

SIMAROUBACEAE DC. – QUASSIA FAMILY

Trees or shrubs, malodorous; *leaves* alternate, usually compound, exstipulate; *flowers* usually imperfect, actinomorphic; *sepals* 3–5 (8), distinct or basally connate; *petals* 3–5 (8), distinct, imbricate; *stamens* (3) 10 (16), usually 2-whorled; *pistil* 1; *ovary* superior, 2–5 (8)-carpellate; *fruit* a schizocarp of samaras.

AILANTHUS Desf.– AILANTHUS

Trees or shrubs; *leaves* odd-pinnately compound; *flowers* in large terminal panicles; *stamens* 2 or 3 in perfect flowers or 10 in staminate ones; *ovary* 2–5-carpellate.

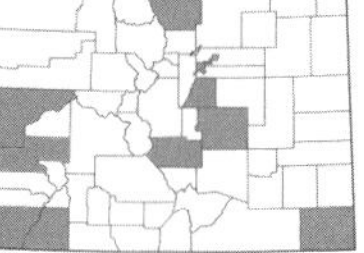

***Ailanthus altissima* (Mill.) Swingle**, TREE OF HEAVEN. Trees to 20 m; *leaves* 3–6 dm long; *leaflets* lanceolate to ovate, acuminate, with a conspicuous gland below near the tip, margins entire or crenate; *sepals* 5, 0.8–1.2 mm long; *petals* 5, 1.5–3 mm long. Very malodorous tree cultivated and sometimes escaping, 4500–7000 ft. June–July. E/W. Introduced.

SMILACACEAE Vent. – CATBRIER FAMILY

Perennial dioecious herbs, shrubs, or lianas often with tendrils; *leaves* alternate, evergreen, simple and entire; *flowers* imperfect, actinomorphic, 3-merous, small, yellowish-green; *sepals* 6, distinct, petaloid but inconspicuous; *petals* absent; *stamens* (3) 6; *pistil* 1; *ovary* superior, 3-carpellate or rarely unicarpellate; *fruit* a berry.

SMILAX L. – CATBRIER; GREENBRIER

Perennial herbs or lianas (ours); *stamens* 6; *ovary* 3-carpellate.

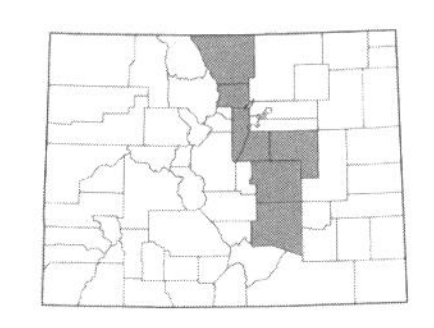

***Smilax lasioneura* Hook.**, BLUE RIDGE CARRIONFLOWER. [*Smilax herbacea* L. ssp. *lasioneuron* (Small) Rydb.]. Lianas with stems to 2 dm; *leaves* ovate, 7–12 × 3.5–9 cm, with a cuspidate tip; *flowers* in umbels, smelling of rotten flesh; *sepals* 3–6 mm long; *berries* blue, ca. 10 mm diam. Uncommon in the outer foothills along the Front Range, 5200–7500 ft. June–July. E.

SOLANACEAE Juss. – POTATO OR NIGHTSHADE FAMILY

Herbs, shrubs, trees, or lianas; *leaves* alternate or alternate below and becoming opposite above, simple to compound, exstipulate; *flowers* perfect, actinomorphic or slightly zygomorphic; *calyx* usually 5-lobed, usually persistent and sometimes accrescent or inflated in fruit; *corolla* 5-lobed; *stamens* 5, epipetalous, alternate with the corolla lobes; *pistil* 1; *ovary* superior, 2-carpellate or falsely 3- or 5-carpellate, with axile placentation; *fruit* a berry or capsule.

1a. Shrubs, sometimes sprawling, often spiny on older growth; leaves entire...***Lycium***
1b. Herbs, if somewhat woody below then lacking spines and at least some leaves with a pair of lobes at the base...2

2a. Fruit a circumscissile capsule surrounded by a persistent, urn-shaped calyx; flowers in a dense one-sided spike or raceme with leafy bracts, cream to greenish-yellow, with purple veins, somewhat zygomorphic...***Hyoscyamus***
2b. Fruit a berry or if a capsule then not circumscissile; flowers unlike the above in all respects...3

3a. Corolla narrowly long-tubular or funnelform; fruit a capsule...4
3b. Corolla campanulate or rotate; fruit a dry or fleshy berry...5

4a. Plants glabrous to hairy but lacking glandular hairs; corolla funnelform, 4–20 cm long, white or violet; capsules usually spiny or with prickles...***Datura***
4b. Plants with glandular hairs; corolla narrowly tubular, 2.5–4 cm long, white; capsules not spiny or prickly...***Nicotiana***

5a. Stems and leaves armed with spines or prickles...***Solanum***
5b. Stems and leaves lacking spines or prickles...6

6a. Calyx segments with small, downward-pointing lobes at the base; flowers blue to bluish-purple, with a white center; fruit a dry berry enclosed in a dry, inflated calyx...***Nicandra***
6b. Calyx segments lacking downward-pointing lobes at the base; flowers and fruit various...7

7a. Stamens connivent around the style, the anthers opening by terminal pores or rarely also with longitudinal slits; flowers usually 2 or more in a cymose or racemose inflorescence; calyx not becoming dry and inflated and completely enclosing the berry...***Solanum***
7b. Stamens not connivent around the style, the anthers opening by longitudinal slits; flowers solitary on axillary peduncles; calyx often becoming dry and inflated and completely enclosing the berry...8

8a. Calyx not becoming dry and bladdery-inflated and completely enclosing the berry; leaves mostly linear to lanceolate with pinnatifid margins; flowers white, yellow, or greenish-yellow...***Chamaesaracha***
8b. Calyx becoming dry, bladdery-inflated and papery, completely enclosing the berry; leaves unlike the above; flowers various...9

9a. Flowers yellow, nodding; plants with glandular or eglandular hairs, or glabrous; seeds glossy...***Physalis***
9b. Flowers purple or rarely white; plants (especially peduncles and calyx) with flat, spherical white hairs; seeds dull...***Quincula***

CHAMAESARACHA (A. Gray) Benth. – FIVE EYES

Perennial herbs; *leaves* alternate, simple, entire to pinnatifid; *flowers* actinomorphic, white, ochroleucous, or greenish-yellow; *fruit* a berry.

1a. Plants with eglandular and a few glandular hairs; corolla 10–15 mm wide...***C. coniodes***
1b. Plants sparsely hairy with stellate hairs; corolla 6–10 mm wide...***C. coronopus***

Chamaesaracha coniodes **(Moric. ex Dunal) Britton**, GRAY FIVE EYES. Plants 0.4–3 dm; *leaves* lanceolate, 20–60 × 5–20 mm, with simple hairs or sometimes with glandular hairs; *inflorescence* of 1–2 flowers in the leaf axils; *calyx* 3–4 mm long; *corolla* 10–15 mm diam.; *berries* 6–8 mm diam. Found in the shortgrass prairie and along roadsides, 3500–5800 ft. May–Sept. E.

Chamaesaracha coronopus **(Dunal) A. Gray**, GREEN FALSE NIGHTSHADE. Plants 1–3 dm; *leaves* linear to narrowly lanceolate, 15–65 × 1.5–10 mm, sparsely hairy with stellate hairs; *inflorescence* of 1–2 flowers in the leaf axils; *calyx* 2.5–4 mm long; *corolla* 6–10 mm diam.; *berries* 5–8 mm diam. Found on dry, open slopes, in shortgrass prairie, and along roadsides, 4500–7000 ft. May–Sept. E/W.

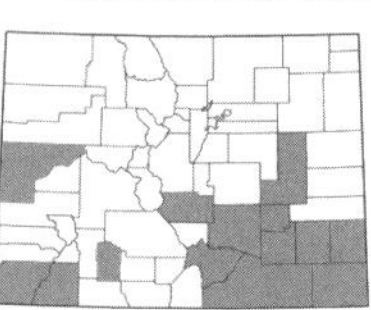

DATURA L. – JIMSONWEED

Herbs; *leaves* alternate, entire to toothed; *flowers* actinomorphic, solitary, white to violet; *fruit* a capsule.

1a. Calyx 6–13 cm long; corolla 8–25 cm long; capsules nodding; leaves softly cinereous-hairy, the margins shallowly sinuate-dentate; perennials from thick caudices…***D. wrightii***
1b. Calyx 2–5 cm long; corolla 4–10 cm long; capsules erect; leaves glabrous to hairy but not cinereous-hairy, the margins deeply cleft to pinnately lobed; annuals from thick taproots…2

2a. Capsules armed with large, unequal prickles, some to 20 mm long; sharp-pointed apex of the corolla lobes mostly 2 mm long…***D. quercifolia***
2b. Capsules armed with prickles 3–9 mm long and minutely hairy; sharp-pointed apex of the corolla lobes 3–8 mm long…***D. stramonium***

Datura quercifolia **H.B.K.**, CHINESE THORN-APPLE. Annuals to 15 dm; *leaves* ovate to elliptic, 6–20 × 4–12 cm, sinuate-dentate to pinnately lobed, moderately hairy; *calyx* 20–30 mm long; *corolla* violet to purple, 4–7 cm long, 1.5–2.5 cm diam.; *capsules* ovoid, 3–4 cm long, armed with flattened prickles to 2 cm long. Not known from Colorado but could occur in the southeastern counties. Aug.–Sept.

Datura stramonium **L.**, JIMSONWEED. Annuals to 12 dm; *leaves* lanceolate-ovate to ovate, 5–25 × 2.5–18 cm, sinuate-dentate, glabrous to slightly hairy; *calyx* 3.5–5 cm long; *corolla* white or sometimes purplish, 6–10 cm long, ca. 2.5 cm diam.; *capsules* ovoid, 3.5–5 cm long, smooth to sparsely hairy and prickly, the prickles ca. 3–9 mm long. Found in disturbed areas, along railroads, and in fields, 4800–6000 ft. June–Sept. E. Introduced. (Plate 75)

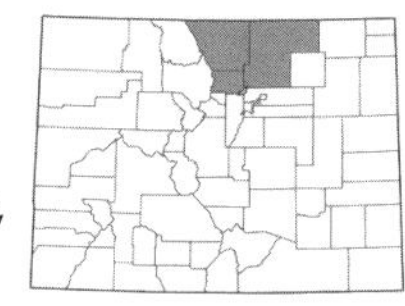

Datura wrightii **Regel**, INDIAN APPLE. Perennials from thick caudices, to 10 dm; *leaves* ovate, 5–25 cm long, entire to shallowly sinuate-dentate, softly cinereous-pubescent; *calyx* 6–13 cm long; *corolla* white to purplish, 8–25 cm long; *capsules* ovoid, armed with prickles 5–12 mm long. Found in fields and disturbed areas, sometimes planted, perhaps native to McElmo Canyon in Montezuma Co. but otherwise introduced, 3500–6000 ft. June–Sept. E/W. Introduced.

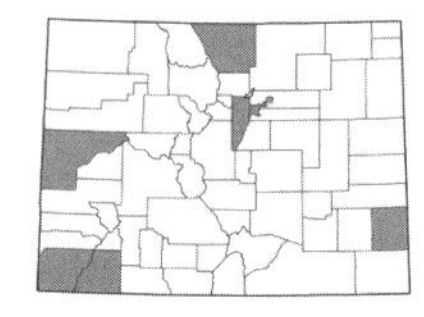

HYOSCYAMUS L. – HENBANE

Annual or biennial herbs, viscid-villous; *leaves* alternate or subopposite, simple; *flowers* slightly zygomorphic, cream to greenish-yellow, with purple veins; *fruit* a circumscissile capsule.

Hyoscyamus niger **L.**, FETID NIGHTSHADE; BLACK HENBANE. Plants 3–10 dm; *leaves* ovate to oblong, 5–20 × 2–10 cm, coarsely toothed to pinnatifid, the cauline leaves clasping; *inflorescence* a bracteate, one-sided raceme or spike; *calyx* 10–15 mm long in flower, accrescent and reticulate-veined in fruit; *corolla* 25–35 mm long; *capsules* enclosed in the accrescent calyx. Found along roadsides and in disturbed areas, 5000–9000 ft. June–Aug. E/W. Introduced.

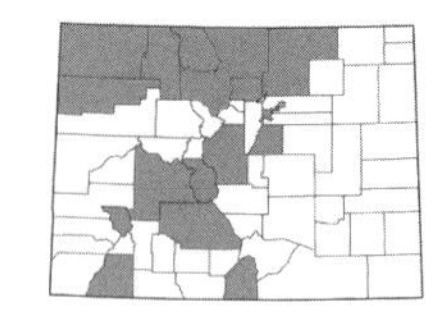

LYCIUM L. – WOLFBERRY

Shrubs, usually bearing spines; *leaves* alternate or fascicled, simple and entire; *flowers* actinomorphic, white, yellow, or greenish; *fruit* a dry or fleshy berry.

1a. Leaves densely fascicled on older branches, mostly spatulate or oblanceolate (widest above the middle); corolla greenish-white with pale to dark purplish veins, the tube 9–20 mm long…***L. pallidum***
1b. Leaves merely alternate, or sometimes in fascicles of 2–6 on older branches, elliptic or oblong (widest at the middle); corolla pink to purple, the tube 4–7 mm long…***L. barbarum***

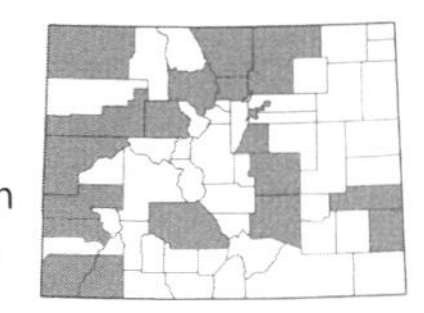

***Lycium barbarum* L.**, MATRIMONY VINE. [*L. halimifolium* Mill.]. Shrubs to 3 m; *leaves* solitary or in fascicles of 2–6, elliptic to oblong, 2–5 cm long, glabrous; *inflorescence* solitary or of clusters of 2–4 flowers; *calyx* 3–5 mm long; *corolla* violet to lavender or pinkish, funnelform, 8–12 mm long; *berries* red (drying blackish), 15–20 mm diam. Found along moist banks, in woody thickets, and in disturbed areas, sometimes cultivated and escaping along roadsides or fencerows, 4000–8500 ft. June–Oct. E/W. Introduced.

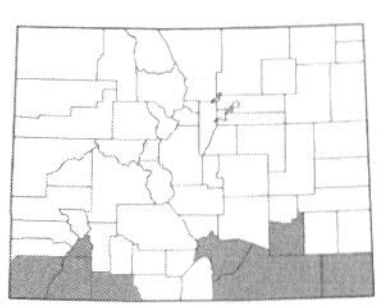

***Lycium pallidum* Miers**, PALE WOLFBERRY. Shrubs to 2 m; *leaves* mostly fascicled, oblanceolate, 1–4 cm long, glabrous; *inflorescence* solitary or sometimes flowers in pairs; *calyx* 5–9 mm long; *corolla* greenish-white or tinged with purple, 15–20 mm long; *berries* red (drying blackish), 8–12 mm diam. Found on dry slopes and in washes, 4500–7000 ft. April–June. E/W.

NICANDRA Adans. – NICANDRA

Annuals; *leaves* alternate, simple, toothed or shallowly lobed; *flowers* actinomorphic, blue; *fruit* a dry berry.

***Nicandra physalodes* (L.) Scop.**, APPLE OF PERU. Plants 2–10 dm, glabrous or sparsely hairy; *leaves* ovate, 5–25 × 3–12 cm, sinuate-dentate to shallowly lobed; *calyx* 10–20 mm, sharply angled, each segment with a small, downward-pointing lobe at the base; *corolla* blue to bluish-purple with a white center, broadly campanulate, 15–30 mm long; *berries* 10–20 mm diam., enclosed in the accrescent calyx. Cultivated, rarely escaping and probably not persisting, 4000–4500 ft. July–Sept. E. Introduced.

NICOTIANA L. – TOBACCO

Annual or perennial herbs or subshrubs; *leaves* alternate, simple; *flowers* actinomorphic; *fruit* a capsule.

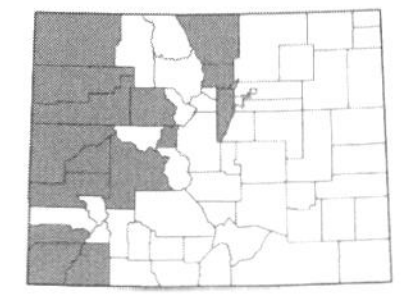

***Nicotiana attenuata* Torr. ex S. Watson**, COYOTE TOBACCO. Annuals, 3–16 dm, glandular to glabrate; *leaves* lance-ovate to ovate, 5–15 cm long; *calyx* 6–8 mm long; *corolla* white, 25–30 mm long, long-tubular; *capsules* 8–12 mm long. Found in dry, sandy soil, along roadsides, and on pinyon-juniper slopes, 5000–7500 ft. June–Sept. E/W.

PHYSALIS L. – GROUND CHERRY

Herbs; *leaves* alternate, simple, usually entire to coarsely toothed; *flowers* actinomorphic, usually solitary on nodding pedicels, yellow or rarely white; *calyx* becoming inflated, dry, and membranous, enclosing the fruit; *fruit* a berry. (Waterfall, 1958)

1a. Plants with glandular hairs on the leaves and stem, these sometimes mixed with nonglandular, multicellular hairs…2
1b. Plants with nonglandular hairs or glabrous…4

2a. Annuals from taproots; flowering calyx 3–5 mm long…***P. subulata* var. *neomexicana***
2b. Perennials from somewhat woody or deeply buried caudices or with rhizomes; flowering calyx 6–12 mm long…3

3a. Stems intermixed with glandular hairs and numerous long (1–3 mm), spreading, multicellular hairs; leaves mostly 5–10 cm long, the lower surface intermixed with glandular hairs and nonglandular hairs, densely so on the veins…***P. heterophylla***
3b. Stems with glandular hairs only or occasionally with only a few, sparse longer nonglandular hairs; leaves mostly 1.5–5 cm long, the lower surface with more or less evenly distributed glandular hairs…***P. hederifolia* var. *comata***

4a. Stem hairs mostly or all reflexed or decurved…5
4b. Stem hairs ascending or ascending-spreading, or the stem glabrous…6

5a. Hairs on the calyx not multicellular, forked or branched, mostly 0.1–0.2 mm long; stem hairs not multicellular, simple or a few forked or branched, mostly 0.1–0.2 mm long…***P. fendleri***
5b. Hairs on the calyx multicellular, mostly simple or with a few forked, mostly 0.7–1 mm long; stem hairs multicellular, flattened when pressed, simple or a few forked or branched, mostly 0.8–2 mm long…***P. virginiana***

6a. Calyx tube with 10 lines of short, ascending hairs or nearly glabrous; stems glabrous below and sparsely hairy above with ascending hairs…***P. longifolia***
6b. Calyx tube uniformly hairy over its entire surface; stems with ascending or ascending-spreading hairs throughout…***P. hispida***

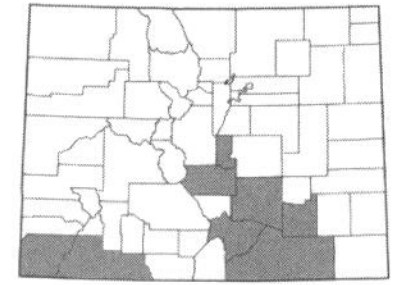

***Physalis fendleri* A. Gray**, Fendler's ground cherry. [*P. hederifolia* A. Gray var. *cordifolia* (A. Gray) Waterf.]. Perennials to 8 dm, with simple to forked or branched hairs; *leaves* ovate to ovate-lanceolate, to 5 cm long, with entire to sinuate-toothed margins; *calyx* 7–10 mm long at anthesis, to 30 mm long in fruit; *corolla* yellow, 1.2–1.5 cm wide; *berries* yellowish, 8–15 mm diam. Found in pinyon-juniper woodlands and on dry hillsides, 4200–7500 ft. May–Aug. E/W.

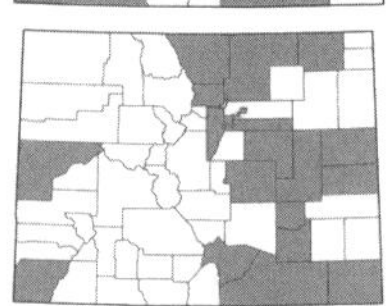

***Physalis hederifolia* A. Gray var. *comata* (Rydb.) Waterf.**, ivy-leaf ground cherry. Perennials to 7 dm, glandular; *leaves* ovate to rotund, mostly 1.5–5 cm long, entire to sinuate-toothed, the lower surface with short-stipitate-glandular hairs; *calyx* 7–10 mm in flowering, to 30 mm long in fruit; *corolla* yellow, 12–15 mm diam.; *berries* yellowish to brown, 8–10 mm diam. Found in grasslands, prairies, ravines, and pastures, 3900–6000 ft. June–Aug. E/W.

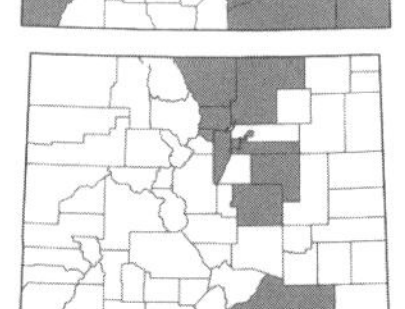

***Physalis heterophylla* Nees**, clammy ground cherry. Perennials, 1.5–5 (9) dm, usually glandular and with long, multicellular hairs; *leaves* ovate to rhombic, 5–10 cm long, sinuate dentate or entire; *calyx* 7–12 mm long at anthesis, 25–40 mm long in fruit; *corolla* yellow, 1–1.2 cm diam.; *berries* yellowish, 1–1.2 cm diam. Found in grasslands, open *Pinus ponderosa* woodlands, gulches, and disturbed areas, 4800–7500 ft. June–Aug. E.

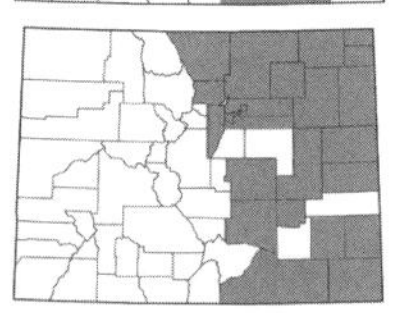

***Physalis hispida* (Waterf.) Cronquist**, prairie ground cherry. [*P. pumila* Nutt. ssp. *hispida* (Waterf.) Hinton]. Perennials to 4.5 dm, with multicellular hairs; *leaves* ovate to lanceolate or rhombic, 4–8 cm long, the margins entire or sinuate-dentate; *calyx* 10–15 mm long in flowers, 20–40 mm long in fruit; *corolla* yellowish; *berries* 10–15 mm diam. Common in grasslands, prairies, and along roadsides, 3500–6500 ft. May–Sept. E.

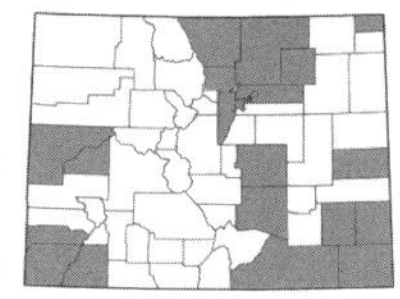

***Physalis longifolia* Nutt.**, longleaf ground cherry. Perennials, 2–8 dm, sparsely hairy with short, antrorse hairs; *leaves* lanceolate to elliptic-lanceolate, the margins entire to sinuate-dentate; *calyx* 4–12 mm, becoming 20–40 mm long in fruit; *corolla* yellow, with dark purplish-brown spots in the center, 10–15 mm long; *berries* 8–15 mm diam., greenish-yellow. Common in grasslands, prairies, disturbed areas, on open hillsides, and along roadsides, 3500–7000 ft. June–Sept. E/W. (Plate 75)

***Physalis subulata* Rydb. var. *neomexicana* (Rydb.) Waterf. ex Kartesz & Gandhi**, New Mexican ground cherry. Annuals, 1–6 dm, with short and long hairs, or sometimes glandular; *leaves* ovate to oblong-ovate or lanceolate-ovate, the margins toothed; *calyx* 3–5 mm long in flower, 20–30 mm long in fruit and sharply 5-angled; *corolla* yellow, with dark purplish-brown spots in the center; *berries* 6–10 mm diam. Not reported for Colorado but could occur in the southeastern counties. July–Sept.

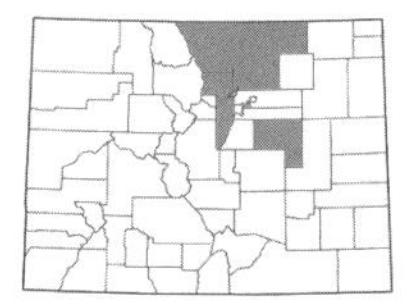

***Physalis virginiana* Mill.**, Virginia ground cherry. Perennials, 1–6 dm, with retrorse hairs on the stem; *leaves* narrowly lanceolate to ovate or elliptic, 2–5 cm long, the margins entire or sinuate-dentate; *calyx* 3–6 mm long in flower, hirsute, 25–40 mm long in fruit and 5-angled; *corolla* yellow with dark spots in the center, 10–20 mm long; *berries* 10–18 mm diam. Uncommon on dry hillsides and in grasslands, 4300–6000 ft. May–Aug. E.

Most specimens originally identified as *P. virginiana* actually belong to *P. longifolia*. *Physalis virginiana* appears to be restricted to the base of the northern Front Range in Colorado.

QUINCULA Raf. – Chinese lantern

Perennial herbs; *leaves* alternate, simple; *flowers* actinomorphic, usually in pairs from leaf axils; *calyx* becoming inflated, dry, and membranous, enclosing the fruit; *fruit* a berry.

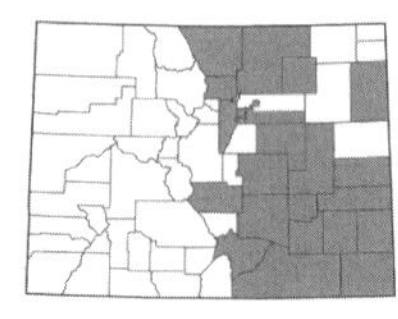

***Quincula lobata* (Torr.) Raf.**, purple ground cherry. [*Physalis lobata* Torr.]. Plants decumbent or spreading, 1–8 dm in diam.; *leaves* oblong to oblanceolate or spatulate, sometimes linear-lanceolate, mostly 4–10 cm long, the margins entire to pinnatifid, with spherical and white, flat hairs; *corolla* 1–2 (3) cm wide, purple or rarely white; *berries* greenish-yellow, 5–8 (10) mm diam. Common in fields, along roadsides, in grasslands, and on dry hillsides, 3500–7100 ft. May–July. E. (Plate 75)

SOLANUM L. – nightshade

Annual or perennial herbs, sometimes shrubby or tuber-bearing; *leaves* alternate, simple, entire to pinnatifid; *flowers* actinomorphic, white, yellow, blue, or purple; *calyx* sometimes enclosing the fruit but not inflated; *fruit* a berry. (Schilling, 1981)

1a. Stems and leaves armed with spines or prickles...2
1b. Stems and leaves lacking spines or prickles...5

2a. Calyx surface lacking spines; stamens all equal and similar in appearance; perennial herbs from deep, creeping rootstocks; leaves simple and shallowly few-lobed to sinuate or subentire...3
2b. Calyx surface covered with spines, closely invested or adherent to the berries giving them a spiny appearance; stamens unequal in size, one larger, darker, and spreading, the other 4 stamens similar and yellow; annual herbs from taproots; leaves pinnatifid to bipinnatifid...4

3a. Leaves ovate to elliptic-ovate; plants with stellate hairs but green in appearance, the hairs not concealing the leaf surface...***S. carolinense***
3b. Leaves narrowly lanceolate to oblong; plants silvery-canescent with a dense layer of stellate hairs concealing the leaf surface...***S. elaeagnifolium***

4a. Plants glandular-hairy; flowers violet...***S. heterodoxum***
4b. Plants with stellate hairs but lacking glandular hairs; flowers yellow...***S. rostratum***

5a. Plants usually somewhat woody at the base, climbing; flowers purple or bluish-purple with 5 pairs of yellow spots at the base of the corolla lobes; at least some leaves with 2 rounded lobes at the base; berries red...***S. dulcamara***
5b. Plants not climbing; flowers white or sometimes bluish-purple but lacking spots at the base of the corolla lobes; leaves pinnately compound or simple, usually lacking 2 rounded lobes at the base; berries purplish-black, green, or yellowish at maturity...6

6a. Leaves pinnately compound with 5–11 leaflets; perennials with slender rhizomes ending in small globose tubers; corolla lobes 8–16 mm long...***S. jamesii***
6b. Leaves simple or with a pair of basal lobes; annuals or if perennials then lacking tubers; corolla lobes 2–5 mm long...7

7a. Leaf margins deeply pinnatifid, the lobes extending more than halfway to the midrib...***S. triflorum***
7b. Leaf margins entire to sinuate-dentate but the lobes not extending more than halfway to the midrib...8

8a. Plants villous-hirsute throughout, sometimes also with glandular hairs; calyx enlarging in fruit; fruit greenish or yellowish at maturity...***S. physalifolium***
8b. Plants sparsely hairy, lacking glandular hairs; calyx not enlarging in fruit; fruit purplish-black at maturity...***S. nigrum***

***Solanum carolinense* L.**, CAROLINA HORSE-NETTLE. Perennials, mostly 3–10 dm, armed; *leaves* ovate to elliptic-pvate, (3) 5–15 × 3–10 cm, with sessile, stellate hairs, the margins shallowly few-lobed; *corolla* purple to white, the lobes 6–9 mm long; *berries* yellow, 1–2 cm diam. Uncommon weed in gardens and disturbed places, probably not persisting in Colorado, 4500–6000 ft. May–Sept. E/W. Introduced.

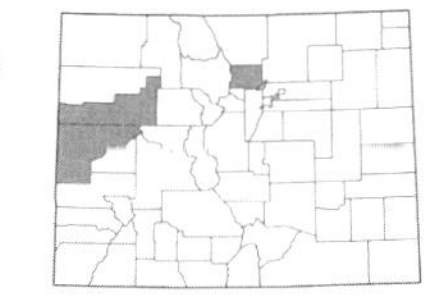

***Solanum dulcamara* L.**, CLIMBING NIGHTSHADE. Perennials, viney or scrambling, the stems to 3 m long, unarmed; *leaves* ovate, 7–10 (12) × 4–9 cm, glabrous to hairy with mostly simple or rarely stellate hairs, the margins entire to deeply 1–2-lobed at the base; *corolla* purple or bluish-purple with 5 pairs of yellow spots just below the base of each lobe, lobes 6–9 mm, reflexed; *berries* red, 8–12 mm. Found along streams, in moist ravines, and in thickets, 4500–7000 ft. May–Sept. E/W. Introduced.

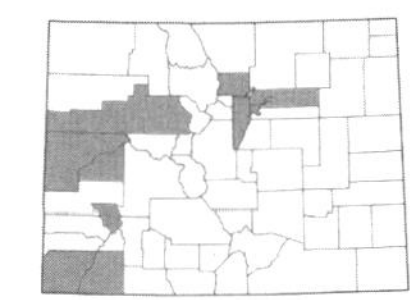

***Solanum elaeagnifolium* Cav.**, SILVERLEAF NIGHTSHADE. Perennials, 3–10 dm, armed; *leaves* narrowly lanceolate to oblong, 3–10 (15) cm long, silvery-canescent with numerous stellate hairs, the margins subentire to sinuate; *corolla* blue to lavender or rarely white, 2–3.5 cm wide; *berries* yellowish or reddish, black at maturity, 1–15 mm diam. Found in canyons, grasslands, along roadsides, in fields and pastures, and in disturbed areas, 3500–6500 ft. May–Sept. E/W. Introduced.

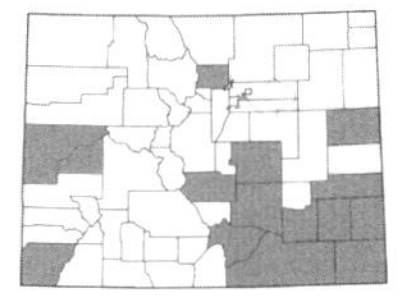

***Solanum heterodoxum* Dunal**, MELONLEAF NIGHTSHADE. Annuals to 7 dm, armed; *leaves* bipinnatifid with rounded lobes, glandular-hairy and with a few stellate hairs; *corolla* violet, 1.5–4 cm diam.; *berries* enclosed in a close-fitting and adherent, spiny calyx. Uncommon along roadsides, in fields, and in disturbed areas, 5000–6500 ft. June–Sept. E. Introduced.

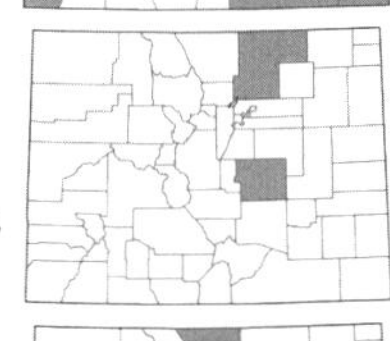

***Solanum jamesii* Torr.**, WILD POTATO. Perennials with slender rhizomes terminating in small, globose tubers, unarmed; *leaves* odd-pinnately compound, to 15 cm long, glabrous to sparsely hairy; *corolla* white, 1.2–3 cm diam.; *berries* ca. 1 cm diam. Found in canyons and moist thickets, often under pinyon-juniper trees, native in the southern counties and introduced in the northern counties, 5000–7500 ft. July–Sept. E/W.

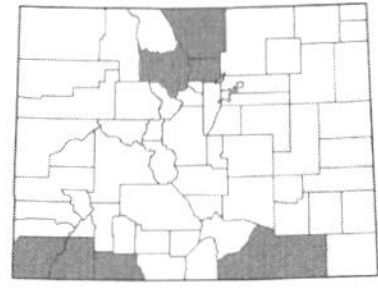

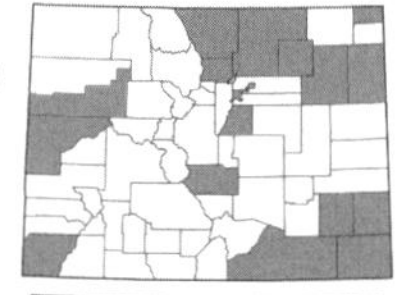

***Solanum nigrum* L.**, BLACK NIGHTSHADE. [*S. americanum* Mill.]. Annuals or short-lived perennials, to 10 dm, unarmed; *leaves* ovate to triangular-ovate or rhombic, 3–7 (10) cm long, the margins entire to sinuate-dentate, sparsely strigose; *corolla* white or sometimes bluish-purple, the lobes 3–7 mm long; *berries* purplish-black, 5–9 mm diam. Found in moist ravines, thickets, moist woodlands, and shortgrass prairie, 3400–5200 ft. July–Oct. E/W.

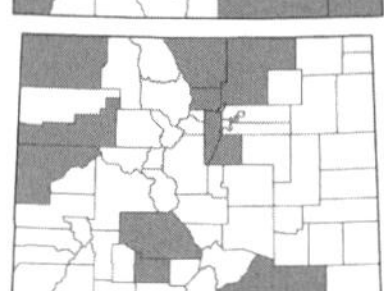

***Solanum physalifolium* Rusby**, HAIRY NIGHTSHADE. [*S. sarrachoides* Sendtner]. Annuals, 1–5 dm, unarmed; *leaves* ovate to deltoid, 2–8 cm long, villous-hirsute and often glandular, the margins subentire to sinuate-dentate; *corolla* white or greenish-white, 3–5 mm diam.; *berries* greenish or yellowish, 6–7 mm diam. Found as a weed in lawns and gardens, in fields and pastures, along roadsides, and other disturbed areas, 4300–8000 ft. July–Oct. E/W. Introduced.

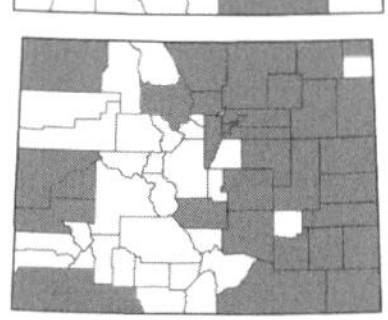

***Solanum rostratum* Dunal**, BUFFALO BUR. Annuals to 7 dm, armed; *leaves* pinnatifid to bipinnatifid, 4–15 cm long, covered with sessile or stalked stellate hairs; *corolla* yellow, 1.5–3 cm diam.; *berries* closely invested in a spiny calyx. Found in disturbed areas, fields and pastures, and along roadsides, 3500–8500 ft. June–Oct. E/W. Introduced.

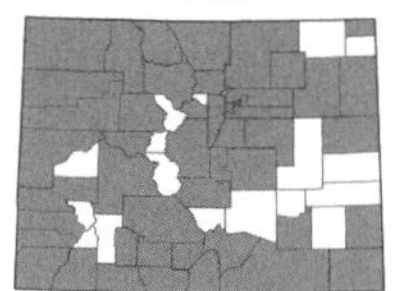

***Solanum triflorum* Nutt.**, CUTLEAF NIGHTSHADE. Annuals, 1–4 dm, unarmed; *leaves* pinnatifid with linear to lanceolate lobes, 2–5 cm long, hairy and often scurfy; *corolla* white, 7–9 mm diam. with lobes 2–3 mm long; *berries* greenish, 7–15 mm diam. Found along roadsides, in fields and pastures, prairie dog towns, grasslands, and in disturbed areas, native but with weedy tendencies, 3500–9500 ft. June–Sept. E/W.

TAMARICACEAE Link. – TAMARISK FAMILY

Shrubs or small trees; *leaves* minute or small, often scale-like, alternate, often evergreen, simple and entire; *flowers* usually perfect, actinomorphic, in spike-like racemes arranged in panicles; *sepals* 4–5, distinct; *petals* 4–5, distinct, white or pink; *stamens* 4–6 or 8–12; *pistil* 1; *ovary* superior, 3–5-carpellate; *fruit* a loculicidal capsule with comose seeds.

TAMARIX L. – TAMARISK

With characteristics of the family.

1a. Flowers 5-merous, appearing with or after the leaves; branches reddish-brown...***T. chinensis***
1b. Flowers 4-merous, appearing before the leaves; branches brown to deep purple...***T. parviflora***

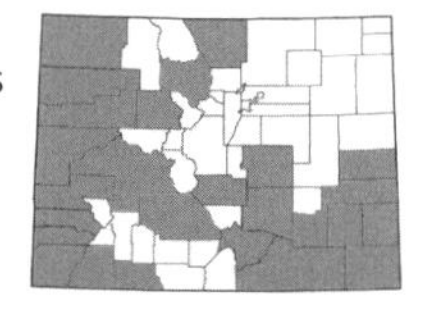

***Tamarix chinensis* Lour.**, SALT-CEDAR. [*T. ramosissima* Ledeb.; *T. pentandra* authors, not Pall.]. Shrubs or small trees to 8 m with reddish-brown branches; *leaves* 1.5–3 mm long; *flowers* 5-merous, appearing with or after the leaves; *sepals* 0.5–1.5 mm; *petals* 1–2 mm. Found along streams and lake and reservoir margins, 3500–7500 ft. May–Aug. E/W. Introduced.

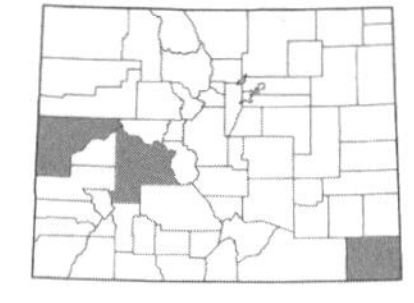

***Tamarix parviflora* DC.**, SMALL-FLOWERED TAMARISK. Shrubs or small trees to 5 m with brown to deep purple branches; *leaves* 2–3 mm long; *flowers* 4-merous, appearing before the leaves; *sepals* 1–1.5 mm; *petals* 2–2.3 mm. Escaping from cultivation where it is uncommon along streams and lake margins, 4500–6500 ft. May–July. E/W. Introduced.

THEMIDACEAE Salisb. – BRODIAEA FAMILY

Perennial herbs from fibrous-coated corms; *leaves* basal, linear to narrowly lanceolate; *flowers* actinomorphic, perfect, in umbels subtended by papery bracts; *tepals* 6, fused below into a tube; *stamens* 3 or 6; *ovary* superior, 3-carpellate; *fruit* a loculicidal capsule.

1a. Filaments partially united to form a tube, with erect, toothed appendages forming a crown between the anthers; tepals white to light purple with dark longitudinal stripes on each tepal, united into a tube 4–8 mm long with free lobes 5–12 (15) mm long and not in two series; leaves 1.5–2.5 mm wide; seeds flat...***Androstephium***
1b. Filaments distinct; tepals light to dark blue, united into a tube 9–12 (14) mm long with free lobes 10–14 (16) mm long and in two series with the inner series with strongly undulate margins; leaves 3–10 mm wide; seeds round...***Triteleia***

ANDROSTEPHIUM Torr. – FUNNEL LILY

Leaves linear, channeled; *flowers* white to purple; *stamens* with filaments fused into a nectar tube with erect, toothed appendages forming a crown between the anthers.

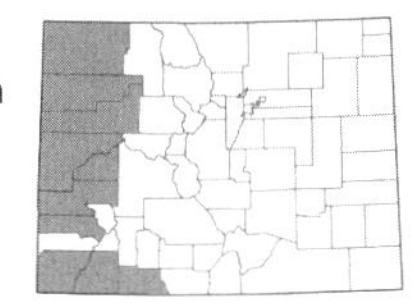

Androstephium breviflorum **S. Watson**, PINK FUNNEL LILY. Leaves 1–3 per scape, 1.5–2.5 mm wide; *flowers* in 3–8-flowered umbels; *tepals* white to light purple with dark longitudinal stripes on each tepal, united into a tube 4–8 mm long with free lobes 5–12 (15) mm long and not in two series; *capsules* 10–15 mm long. Found on dry, shale clay hills, 4500–6500 ft. April–June. W.

TRITELEIA Douglas ex. Lindl. – TRITELEIA

Leaves narrowly lanceolate, flattened and keeled; *flowers* blue to purple (ours), on jointed ascending pedicels; *stamens* with distinct filaments.

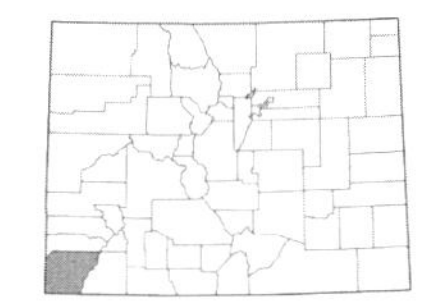

Triteleia grandiflora **Lindl. var.** ***grandiflora***, LARGEFLOWER TRITELEIA. Leaves 1–2 per scape, 3–10 mm wide; *flowers* in 5–16-flowered umbels; *tepals* light to dark blue, united into a tube 9–12 (14) mm long with free lobes 10–14 (16) mm long and in two series with the inner series with strongly undulate margins; *capsules* 5–10 mm long. Uncommon on open, dry slopes with pinyon-oak, recently found in Montezuma Co., 7500–8000 ft. May–July. W.

TYPHACEAE Juss. – CATTAIL FAMILY

Perennials, semiquatic or aquatic herbs from rhizomes; *leaves* 2-ranked, alternate, sheathing; *flowers* imperfect; *staminate flowers* with 1–8 stamens; *pistillate flowers* 1–2-carpellate; *ovary* superior; *fruit* an achene-like follicle or drupaceous achene.

1a. Flowers in terminal cylindric spikes; plants mostly 1–3 m tall...***Typha***
1b. Flowers in dense globose heads; plants usually less than 1 m tall...***Sparganium***

SPARGANIUM L. – BUR-REED

Flowers in dense globose heads, unisexual or staminate above the pistillate; *tepals* 3–6, distinct, white to green; *staminate flowers* with 2–8 distinct stamens; *fruit* a hardened, drupaceous achene. (Cook & Nicholls, 1986; 1986)

1a. Stigmas 2, rarely 1; fruit sessile, obpyramidal, the body 3–7-faceted, not constricted at the middle; tepals often with a dark subapical spot, the tips entire to subentire; staminate heads 8–40 or more; inflorescence rachis usually branched, the bracts not inflated near the base; leaves and inflorescences emergent and erect...***S. eurycarpum***
1b. Stigmas 1; fruit stipitate or short-stipitate, ellipsoid to fusiform, not faceted, constricted at the middle or not; tepals without a dark subapical spot, the tips erose; staminate heads 1–10; inflorescence rachis unbranched, the bracts often inflated near the base; leaves and inflorescences emergent and erect or floating and limp...2

2a. Staminate heads 1 or rarely 2 and appearing as one (contiguous), terminal, less than 1 cm in diam.; pistillate heads usually 1.2 cm or less in diam.; fruit beak (including stigma) 0.5–1.5 mm long, the mature fruit heads 1.5 mm or less in diam.; bract subtending the lowest head about equal to the length of the entire inflorescence...***S. natans***
2b. Staminate heads 2–10 (sometimes the heads are contiguous and appearing as one elongate head); pistillate heads 1–3.5 cm in diam.; fruit with a straight or curved beak (including stigma) 1.5–4.5 mm long, the mature fruit heads over 1.5 mm in diam.; bract subtending the lowest head about twice as long as the length of the entire inflorescence...3

3a. Staminate heads contiguous and appearing as one elongate head; leaves usually floating and limp, 2–5 (10) mm wide; fruit greenish, the beak (including stigma) 1.5–2 mm long...***S. angustifolium***
3b. At least some of the staminate heads not contiguous, distinctly separate on the rachis; leaves emergent or floating, 4–10 mm wide; fruit reddish to brown, the beak (including stigma) 2–4.5 mm long...***S. emersum***

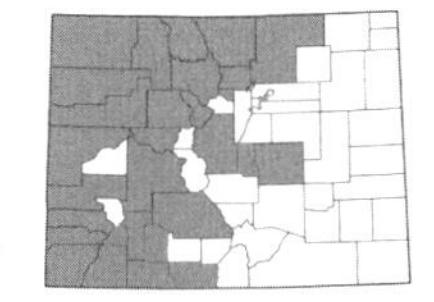

Sparganium angustifolium **Michx.**, NARROWLEAF BUR-REED. Plants to 2+ m; *leaves* 0.2–2 m × 2–5 (10) mm; *inflorescence* unbranched; *pistillate heads* 2–5, not contiguous, sessile or the lowermost peduncled, 1–3 cm diam. in fruit; *staminate heads* (1) 2–4, contiguous and appearing as one elongate head; *fruit* short-stipitate, ellipsoid to fusiform, constricted in the middle, 3–7 mm long, tapering to a straight beak 1.5–2 mm long. Common in shallow water of ponds and lakes, 6000–12,000 ft. June–Sept. E/W.

This species and *S. emersum* are difficult to distinguish from each other and would perhaps be better treated as a single species, *S. angustifolium*. They can also hybridize.

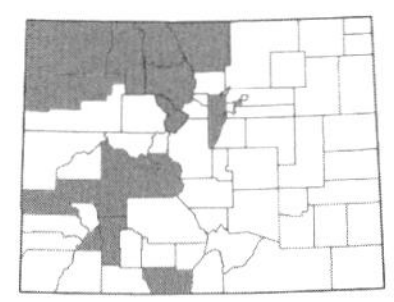

Sparganium emersum **Rehmann**, EMERGENT BUR-REED. [*S. multipedunculata* (Morong) Rydb.]. Plants to 2 m; *leaves* 0.8 -1 m × 4–10 mm; *inflorescence* unbranched; *pistillate heads* 1–6, contiguous or not, 1.6–3.5 cm diam. in fruit; *staminate heads* 3–7 (10), contiguous or not; *fruit* stipitate, fusiform, slightly constricted in the middle, 3–4 × 1.5–2 mm, tapering to a straight or curved beak 2–4.5 mm long. Found in shallow water of ponds and carrs, 3400–10,000 ft. June–Sept. E/W.

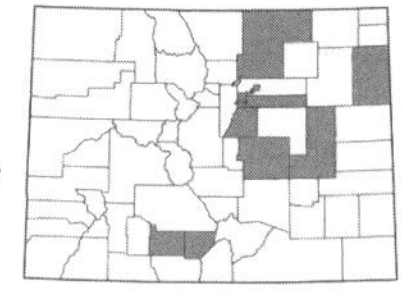

Sparganium eurycarpum **Engelm. ex A. Gray**, BROADFRUIT BUR-REED. Plants to 2.5 m; *leaves* 2–2.5 m × 6–20 mm; *inflorescence* usually branched; *pistillate heads* (1) 2–6 (8), not contiguous, 1.5–5 cm diam. in fruit; *staminate heads* 8–40+, the lowermost not contiguous; *fruit* sessile, obpyramidal, not constricted in the middle, 5–10 mm long, with a straight beak 2–4 mm long. Found in shallow water of ponds and wet meadows, 3400–7500 ft. June–Aug. E. (Plate 76)

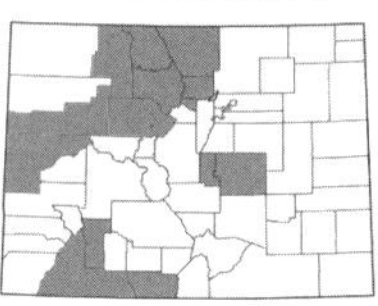

Sparganium natans **L.**, SMALL BUR-REED. [*S. minimum* (Hartman) Wallr.]. Plants to 0.6 m; *leaves* 0.04–0.6 m × 2–6 (10) mm; *inflorescence* unbranched; *pistillate heads* 1–3, not contiguous, 0.5–1.2 cm diam. in fruit; *staminate heads* 1, not contiguous with the uppermost pistillate head; *fruit* subsessile, ellipsoid to obovoid, not or barely constricted in the middle, 2–4 × 1–1.5 mm, with a curved beak 0.5–1.5 mm long. Uncommon in shallow water of high mountain ponds and fens, 7500–10,500 ft. July–Sept. E/W.

TYPHA L. – CATTAIL

Flowers in a terminal cylindric spike, the staminate flowers above the pistillate; *staminate flowers* with 1–5 stamens, the filaments connate, subtended by bristles; *pistillate flowers* stipitate, unicarpellate; *fruit* an achene-like follicle. (Smith & Kaul, 1986; Smith, 2000)

1a. Staminate and pistillate parts of the spike not separated by a naked segment of the axis, the spikes mostly 26–40 mm wide in fruit; pistillate bracteoles absent... ***T. latifolia***
1b. Staminate and pistillate parts of the spike separated by a naked segment of the axis, the spikes mostly 10–30 mm wide in fruit; pistillate bracteoles present (very small and inconspicuous in hybrids though)...2

2a. Staminate spikes (above) the same length as the pistillate spikes, pistillate spikes dark brown; mucilage glands absent from the leaves... ***T. angustifolia***
2b. Staminate spikes 1.4 times longer than the pistillate spikes, pistillate spikes light brown; mucilage glands present on the leaf blades (especially visible late in the season when the plant is dry)... ***T. domingensis***

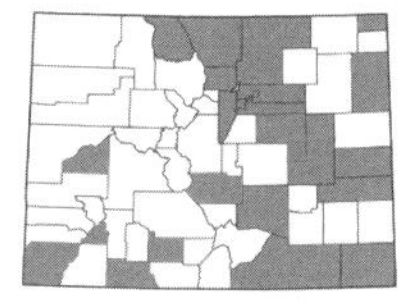

Typha angustifolia **L.**, NARROWLEAF CATTAIL. Plants 15–30 dm; *leaves* lacking mucilage glands, ca. 6–15 mm wide when fresh; *inflorescence* with staminate and pistillate parts separated by a naked segment of the axis; *pistillate spikes* 10–20 mm wide in fruit, dark brown; *staminate spikes* the same length as the pistillate spikes. Found in shallow water of small ponds, creeks, and streams, 3400–6000 ft. June–Aug. E/W.

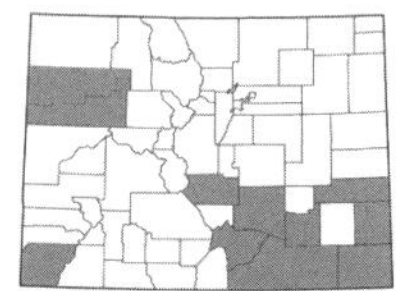

Typha domingensis **Pers.**, SOUTHERN CATTAIL. Plants 15–40 dm; *leaves* with mucilage glands, 6–18 mm wide when fresh; *inflorescence* with staminate and pistillate parts separated by a naked segment of the axis; *pistillate spikes* 20–30 mm wide in fruit, light brown; *staminate spikes* 1.4 times longer than the pistillate spikes. Found in shallow water of small ponds and slow-moving streams and creeks, 3400–6100 ft. June–Aug. E/W.

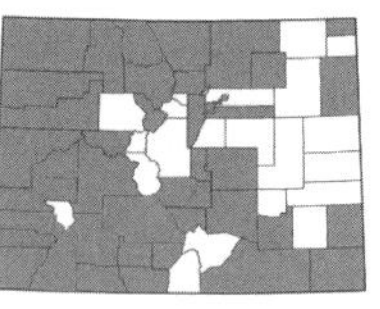

Typha latifolia **L.**, BROADLEAF CATTAIL. Plants 15–35 dm; *leaves* lacking mucilage glands, 10–30 mm wide when fresh; *inflorescence* with staminate and pistillate parts contiguous; *spikes* mostly 26–40 mm wide in fruit. Common in small ponds and slow-moving creeks and ditches, 3400–8600 ft. June–Sept. E/W.

Typha latifolia can hybridize with both *T. angustifolia* and *T. domingensis*. Specimens of these hybrids are intermediate between each parent and have a small gap in between the pistillate and staminate spikes. In hybrids, the bracteoles are more slender than the stigmas, while in both true *T. angustifolia* and *T. domingensis* the bracteoles are broader at the tips than the linear stigmas. Hybrids between *T. latifolia* and *T. angustifolia* lack mucilage glands on the leaf blades while hybrids of *T. latifolia* and *T. domingensis* have mucilage glands on the adaxial leaf surface.

ULMACEAE Mirb. – ELM FAMILY

Deciduous trees or shrubs; *leaves* alternate, simple, stipulate, the base oblique, the margins entire or toothed; *flowers* perfect or imperfect; *sepals* 4–9, distinct or connate; *petals* absent; *stamens* usually the same number as and opposite the calyx lobes; *pistil* 1; *styles* (1) 2; *ovary* superior, 2 (3)-carpellate; *fruit* a samara.

ULMUS L. – ELM

Leaves simply or more often doubly serrate, petiolate, with pinnate venation, each vein ending in a tooth; *flowers* perfect, in fascicles, cymes, or racemes, appearing before the leaves; *calyx* 3–9-lobed; *fruit* a samara, usually flattened with a membranous wing.

Ulmus americana L., American elm, is often planted as an ornamental tree. It differs from *U. pumila* in having larger leaves that are softly hairy below, samaras with ciliate margins, and in size (tall tree versus small tree to 15 m).

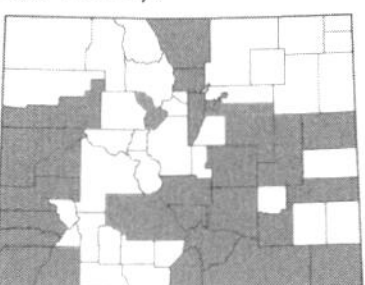

***Ulmus pumila* L.**, Siberian elm. Small trees to 15 m; *leaves* lanceolate to narrowly elliptic, (2.5) 4–7.5 × 1.5–3.5 cm, shortly acuminate, serrate; *calyx* 4–5-lobed; *samaras* 11–15 mm diam. with a notch at the apex. Found in disturbed places, along ditches, and often near old homesteads, 4200–8500 ft. Mar.–May. E/W. Introduced.

URTICACEAE A.L. Juss. – Nettle family

Herbs or small shrubs, often with stinging hairs; *leaves* alternate or opposite, simple, usually stipulate, petiolate; *flowers* perfect or imperfect; *staminate flowers* with 4–5 sepals and 4–5 stamens, white or green; *pistillate flowers* with 2–4 sepals, hypogynous, green or reddish; *perfect flowers* with 4 sepals and 4 stamens; *pistil* 1; *ovary* superior, unicarpellate; *fruit* an achene.

1a. Leaves alternate, entire; flowers in short, few-flowered clusters, subtended and exceeded by linear, green bracts; plants lacking stinging hairs...***Parietaria***

1b. Leaves opposite, serrate; flowers numerous in elongated racemes or panicles; plants with stinging hairs...***Urtica***

PARIETARIA L. – Pellitory

Annual herbs, lacking stinging hairs; *leaves* alternate, simple, entire; *flowers* imperfect, subtended and exceeded by linear, green bracts; *staminate flowers* with 4 sepals and 4 stamens; *pistillate flowers* with 4 sepals; *fruit* an ovoid achene enclosed in a persistent calyx.

***Parietaria pensylvanica* Muhl. ex Willd.**, Pennsylvania pellitory. Plants 0.5–5 dm; *leaves* lanceolate to narrowly ovate, 1.5–7 × 0.8–2 cm; *achenes* 1–1.2 mm long. Found in shady, often moist places, 3500–8500 ft. June–Aug. E/W.

URTICA L. – Nettle

Herbs with stinging hairs; *leaves* opposite, stipulate; *flowers* imperfect; *staminate flowers* with 4 distinct sepals and 4 stamens; *pistillate flowers* with 4 distinct sepals; *fruit* enclosed by a persistent calyx. (Fernald, 1926)

***Urtica dioica* L.**, Stinging nettle. Perennials, 5–25 dm; *leaves* lanceolate to ovate, 5–15 × 2–8 cm, serrate; *inflorescence* spike-like, raceme-like, or panicle-like; *achenes* 1–1.5 mm long. Found in moist places such as along rivers and lakes, often in disturbed places, 4700–11,500 ft. May–Sept. E/W. (Plate 76)

There are two varieties of *U. dioica* in Colorado:

1a. Staminate and pistillate flowers on different plants; stems hispid; lower leaf surfaces hispid...**ssp. *gracilis* (Aiton) Selander**. [var. *procera* (Muhl. ex Willd.) Wedd.]. Common and widespread across the state, 4700–11,500 ft. May–Sept. E/W.

1b. Staminate and pistillate flowers mostly on the same plant; stems softly hairy; lower leaf surfaces tomentose to strigose...**ssp. *holosericea* (Nutt.) Thorne**. [var. *occidentalis* S. Watson]. Barely entering Colorado in Moffat Co., 7000–7200 ft. June–Sept. W.

VALERIANACEAE Batsch. – Valerian family

Annual or perennial herbs; *leaves* opposite or in basal rosettes, petiolate, exstipulate; *flowers* perfect or imperfect, zygomorphic or nearly actinomorphic; *sepals* much reduced or absent; *corolla* 4–5-lobed; *stamens* 1–3, epipetalous, alternating with the corolla lobes; *pistil* 1; *ovary* inferior, usually 3-carpellate, only 1 of the 3 locules fertile; *fruit* achene-like or a samara.

VALERIANA L. – Valerian

Perennial herbs; *leaves* simple and entire or pinnatifid; *sepals* inrolled in flower but elongating to a plumose pappus in fruit; *corolla* funnelform to nearly salverform, basally saccate, white to pink; *stamens* 3; *ovary* 3-carpellate; *fruit* achene-like.

1a. Flowers in an open panicle; basal leaves gradually tapering to a petiolar base; stem leaves in 2–6 pairs, with narrow, linear lobes; plants from a stout taproot and branched caudex; stamens not well-exserted...***V. edulis***

1b. Flowers in a dense corymbose cyme, this becoming more open in fruit; basal leaves usually evidently petiolate; stem leaves in 1–4 pairs, with narrowly ovate or elliptic lobes; plants from a stout rhizome or caudex with numerous fibrous roots; stamens well-exserted...2

2a. Corolla mostly 10–15 mm long with a slender tube; plants 1–3 dm tall...***V. arizonica***
2b. Corolla generally smaller, 2–7 mm long; plants 1–10 dm tall...3

3a. Corolla mostly 4–7 mm long, usually somewhat hairy outside near the base, the lobes to half as long as the tube; flowers usually all perfect; stem usually with minutely spreading hairs...***V. acutiloba***
3b. Corolla mostly 2–4 mm long, glabrous outside, the lobes nearly or as long as the tube; flowers often chiefly all pistillate; stem glabrous or nearly so...***V. occidentalis***

Valeriana acutiloba **Rydb.**, SHARPLEAF VALERIAN. [*V. capitata* Pall. ssp. *acutiloba* (Rydb.) F.G. Mey.]. Plants 1–6 dm, the stems usually minutely spreading-hirtellous; *basal leaves* ovate to oblong or obovate-spatulate, to 8 cm long; *cauline leaves* in 1–3 pairs, mostly pinnatifid; *inflorescence* a corymb, 1.5–5 cm wide; *flowers* white or pinkish, 4–7 mm long, usually somewhat hairy outside; *plumose calyx segments* 10–17 mm long. Common in subalpine and alpine meadows and near streams, often in moist soil, 7500–13,000 ft. June–Aug. E/W.

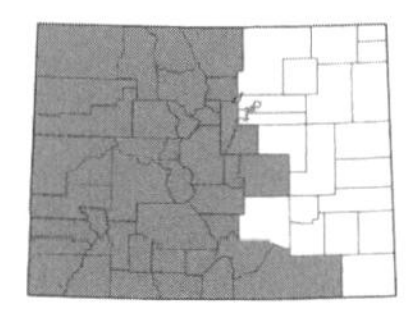

Valeriana arizonica **A. Gray**, ARIZONA VALERIAN. Plants 1–3 dm, the stems glabrous; *basal leaves* ovate to elliptic, undivided or sometimes with a pair of small lobes, 2–5 cm long; *cauline leaves* in 1–3 pairs, mostly pinnatifid; *inflorescence* a corymb, 2–4 cm wide; *flowers* pink, 10–15 mm long, glabrous. Found in moist soil, 6000–8600 ft. May–July. E.

Valeriana edulis **Nutt.**, TOBACCO ROOT. Plants 1–12 dm, the stems glabrous; *basal leaves* linear to obovate, firm, entire or with a few, narrow, forward-pointing lateral lobes, 7–40 cm long; *cauline leaves* in 2–6 pairs, pinnatifid with narrow segments; *inflorescence* a panicle; *flowers* white, 2.5–3.5 mm long, glabrous; *plumose calyx segments* 9–13 mm long. Common in meadows and forest openings, 6200–12,700 ft. June–Aug. E/W.

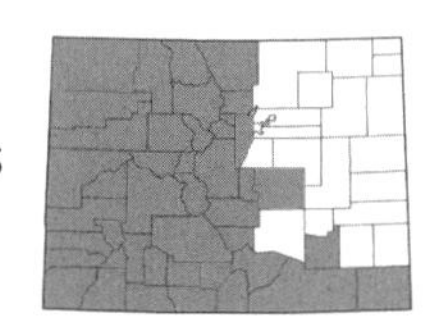

Valeriana occidentalis **A. Heller**, WESTERN VALERIAN. Plants 3–10 dm, glabrous; *basal leaves* undivided or with 1–2 pairs of small lateral lobes, to 11 cm long; *cauline leaves* in 2–4 pairs, pinnatifid or some undivided, the terminal lobe to 9 cm long; *inflorescence* a corymb, 1.5–3 cm wide; *flowers* white, 2–4 mm long, glabrous; *plumose calyx segments* 9–15 mm long. Common in meadows and open forests, sometimes along streams, usually in moist soil, 6400–11,000 ft. E/W. (Plate 76)

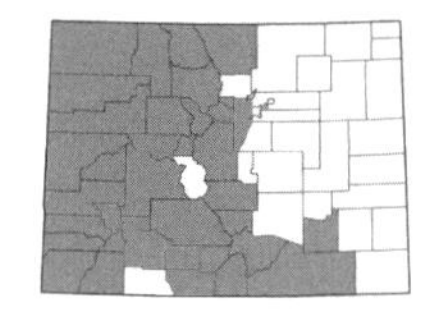

VERBENACEAE J. St.-Hil. – VERVAIN FAMILY

Herbs, shrubs, or small trees; *leaves* opposite or occasionally whorled, simple or compound; *flowers* usually perfect, usually zygomorphic, salverform or two-lipped; *calyx* 2–5-lobed; *corolla* (4) 5-lobed; *stamens* (2) 4 (5), didynamous when 4, epipetalous; *pistil* 1; *ovary* superior, 2 (4–5)-carpellate, often 2-carpellate but 4-lobed from the presence of a false septum; *fruit* a drupe or 2–4 nutlets.

1a. Flowers in dense, head-like, subglobose to cylindric spikes on long peduncles, the petals white to pink or pinkish-purple with a yellow center, strongly zygomorphic; plants procumbent or prostrate and often rooting at the nodes; fruit of 2 nutlets...***Phyla***
1b. Flowers in terminal corymbose or spicate inflorescences, blue or purple, mostly weakly zygomorphic; plants erect or procumbent but not rooting at the nodes; fruit of 4 nutlets...2

2a. Corolla tube 1.5 times longer than the calyx, the limb 7–15 mm wide; leaves deeply pinnatifid to bipinnatifid or 3-parted; nutlets with an evident cavity at the base...***Glandularia***
2b. Corolla tube shorter than or only slightly longer than the calyx, or if much longer then the limb 3–5 mm wide; leaves pinnately incised, 3-lobed, or the margins merely serrate; nutlets without a cavity at the base...***Verbena***

GLANDULARIA J.F. Gmel. – MOCK VERVAIN

Herbs; *leaves* opposite, simple; *flowers* weakly zygomorphic; *calyx* 5-lobed, persistent; *corolla* 5-lobed, salverform or funnelform, blue or purple; *stamens* 4, didynamous; *fruit* 4 nutlets with an evident cavity at the base.

Glandularia bipinnatifida **(Nutt.) Nutt.**, DAKOTA MOCK VERVAIN. [*Verbena ambrosifolia* Rydb.; *Verbena bipinnatifida* Nutt.]. Perennials, prostrate to ascending, 0.5–6 dm long; *leaves* 1–6 cm long, pinnatifid to bipinnatifid or 3-parted, hirsute; *calyx* 7–10 mm, glandular or eglandular; *corolla* 7–10 (15) mm wide; *nutlets* 2–3 mm long. Found in grasslands and sandy soil, 3800–7300 ft. May–Sept. E/W. (Plate 76)

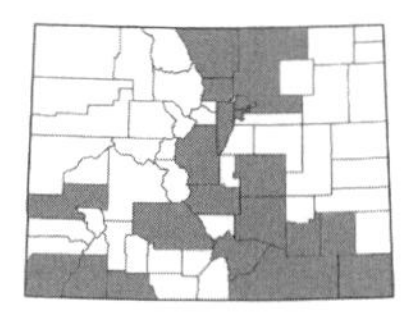

PHYLA Lour. – FOGFRUIT

Perennials, procumbent herbs; *leaves* opposite, toothed apically; *flowers* axillary, in a dense, headlike, globose to cylindric spike, zygomorphic; *calyx* 2–4-toothed and 2-lipped; *corolla* 5-lobed, white to pink or purplish; *fruit* 2 nutlets.

1a. Leaves linear-oblanceolate or cuneiform, 2–6 (8) mm wide, entire or with 1–4 teeth above the middle on each side, often fascicled, with one prominent midvein; mature flower spikes 8–12 mm in diam…***P. cuneifolia***

1b. Leaves lanceolate to lance-elliptic or ovate, 5–30 mm wide, toothed along the entire margin, not fascicled, with conspicuous secondary venation; mature flower spikes 5–7 mm in diam…***P. lanceolata***

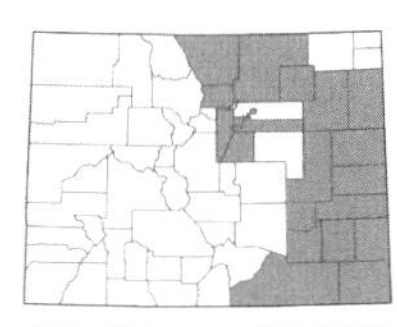

***Phyla cuneifolia* (Torr.) Greene**, WEDGE-LEAF FOGFRUIT. [*Lippia cuneifolia* (Torr.) Steud.]. Stems to 1 m long; *leaves* linear-oblanceolate or cuneiform, often fascicled, 1–5 cm × 2–6 (8) mm, entire or with 1–4 teeth above the middle on each side; *corolla* 2–4.5 mm wide; *nutlets* 1.7–2 mm long. Locally common in moist places along the margins of ponds and streams, sometimes in open prairie, 3500–6400 ft. June–Sept. E. (Plate 76)

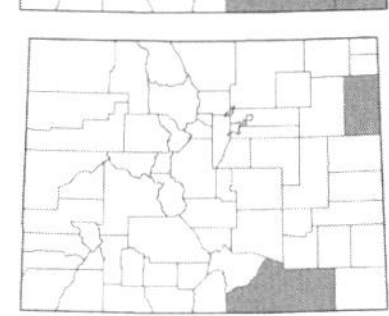

***Phyla lanceolata* (Michx.) Greene**, LANCE-LEAF FOGFRUIT. [*Lippia lanceolata* Michx.]. Stems 2–6 (10) dm; *leaves* lanceolate to lance-elliptic or ovate, 1.5–7.5 cm × 5–30 mm, toothed; *corolla* 2–2.5 mm long; *nutlets* 1–1.2 mm long. Uncommon along the margins of ponds and lakes, 4400–6000 ft. June–Sept. E.

VERBENA L. – VERVAIN

Herbs; *leaves* opposite, simple, toothed to lobed or pinnatifid; *flowers* weakly zygomorphic; *calyx* 5-toothed, persistent; *corolla* 5-lobed, salverform or funnelform, blue or purple; *stamens* 4, didynamous; *fruit* 4 nutlets.

1a. Bracts subtending the flowers 2–3 times longer than the calyx; plants prostrate, rarely erect, many-branched at the base; leaves pinnatifid or 3-lobed at the tip, the margins toothed…***V. bracteata***

1b. Bracts subtending the flowers equaling or only slightly longer than the calyx; plants erect or branching at the base; leaves elliptic to ovate with toothed margins, the lower occasionally incised-dentate, or rarely 3-lobed…2

2a. Leaves elliptic-ovate with a broadly obtuse apex, 1–4 cm long, mostly basal and reduced above, incised-dentate and often 3-lobed, with prominent white veins below, especially near the margins; plants decumbent or ascending, branched at the base…***V. plicata***

2b. Leaves lanceolate to ovate or elliptic with an acute or obtuse apex, 3–18 cm long, well-distributed along the stem, coarsely serrate to incised-serrate, without prominent white veins below near the margins; plants erect, simple to branched above…3

3a. Corolla tube about twice as long as the calyx, the limb 3–4.5 mm wide; leaves distinctly petiolate, not prominently veined below; fruiting spikes narrow, to 7 mm wide…***V. hastata***

3b. Corolla tube only slightly longer than the calyx, the limb 6–9 mm wide; leaves sessile or shortly petiolate, prominently veined below; fruiting spikes generally thicker, over 7 mm wide…4

4a. Calyx glandular and hirsute; leaves lanceolate to lance-elliptic, often somewhat triangular and widest at the base, with acute tips…***V. macdougalii***

4b. Calyx densely hirsute without glandular hairs; leaves ovate to elliptic, usually some of the lower leaves with rounded tips, the others acute…***V. stricta***

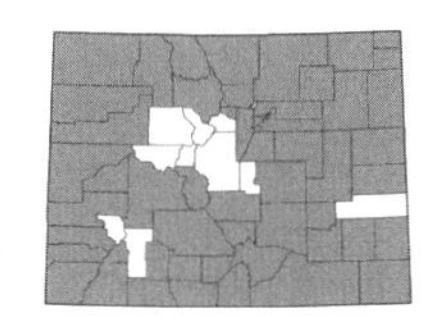

***Verbena bracteata* Lag. & Rodr.**, PROSTRATE VERVAIN. Annuals or short-lived perennials, prostrate or rarely erect, stems 1–5 (7) dm long; *leaves* lanceolate to ovate-lanceolate, 1–7 × 0.6–3 cm, pinnatifid or 3-lobed at the tip, toothed; *spikes* 2–20 cm long, hirsute; *bracts* 8–15 mm long, 2–3 times longer than the calyx; *calyx* 3–4 mm long; *corolla* 2–3 mm wide; *nutlets* 2–2.5 mm long. Common in disturbed places, along roadsides, in fields, and on the shortgrass prairie, 3400–8000 ft. May–Sept. E/W. Introduced.

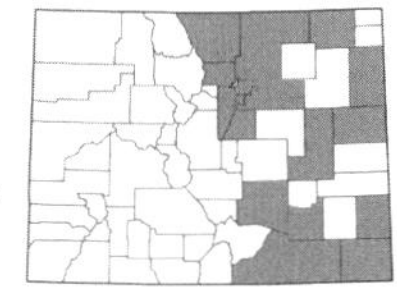

***Verbena hastata* L.**, SWAMP VERBENA. Perennials, 1–4 dm; *leaves* lanceolate to lance-ovate, 4–18 cm long, serrate to incised, often 3-lobed at the base, glabrous or minutely hairy above, conspicuously hairy below; *spikes* paniculate; *bracts* shorter than the calyx; *calyx* 2.5–3 mm long; *corolla* 3–4.5 mm wide; *nutlets* ca. 2 mm long. Found in moist places such as along the margins of ponds and lakes, streams, and ditches, 3400–5500 ft. July–Sept. E.

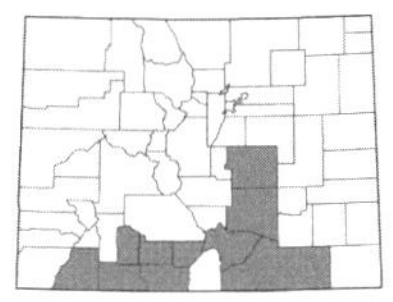

***Verbena macdougalii* Heller**, MACDOUGAL'S VERBENA. Perennials, 3–10 dm; *leaves* elliptic to ovate-lanceolate, 6–10 cm long, serrate-dentate, hirsute above and prominently veined below; *spikes* solitary or few; *bracts* 1–2 mm longer than the calyx; *calyx* 4–5 mm long; *corolla* ca. 6 mm wide. Found in dry, rocky places and along roadsides, 6400–9500 ft. June–Aug. E/W.

***Verbena plicata* Greene**, FANLEAF VERVAIN. Perennials, 1–4 dm; *leaves* elliptic-ovate, 1–4 cm long, incised-dentate to 3-lobed, broadly obtuse at the apex, sparsely hirsute above and hirsute below with prominent white veins; *bracts* 3.5–5 mm long, equal to or slightly longer than the calyx; *calyx* 3.5–4 mm long, glandular-hirsute; *corolla* 4–6 mm wide; *nutlets* 2–2.5 mm long. Uncommon in sandy prairie and canyon bottoms, 4200–6200 ft. May–Aug. E.

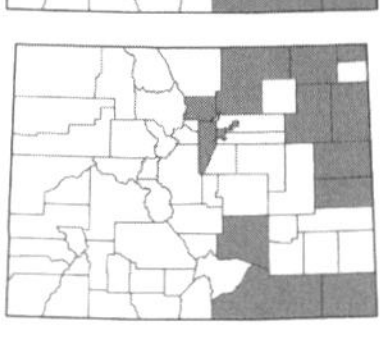

***Verbena stricta* Vent.**, HOARY VERBENA. Perennials, 2–25 dm; *leaves* elliptic, ovate, or suborbicular, 3–7 cm long, sharply toothed to incised-serrate, hirsute above, densely hirsute and prominently veined below; *bracts* about equal to the calyx; *calyx* (3) 4–5 mm long, densely hirsute; *corolla* 8–9 mm wide; *nutlets* 2–3 mm long. Found in open prairie and along roadsides, 3400–5700 ft. June–Aug. E. (Plate 76)

VIOLACEAE Batsch. – VIOLET FAMILY

Annual or perennial herbs; *leaves* usually alternate, simple, entire or dissected, stipulate; *flowers* perfect, zygomorphic, usually nodding, on axillary peduncles, later flowers often cleistogamous; *sepals* 5, distinct, persistent; *petals* 5, distinct, the lowest one spurred or gibbous at the base; *stamens* 5; *pistil* 1; *ovary* superior, 3-carpellate; *fruit* a loculicidal capsule.

1a. Sepals not auricled at the base; flowers 2.5–6 mm long, greenish-white or white with purplish tips; plants caulescent with alternate leaves above and opposite leaves below, often with smaller ones fascicled, the upper leaves mostly linear and 3 mm wide...***Hybanthus***
1b. Sepals auricled at the base; flowers 4–22 mm long, white to violet or bluish-purple; plants acaulescent or caulescent, lacking fascicled leaves; leaves lanceolate, cordate, or ovate to orbicular, over 3 mm wide...***Viola***

HYBANTHUS Jacq. – GREEN VIOLET

Perennial herbs; *flowers* solitary, axillary; *sepals* without auricles; *petals* subequal, the lowest one larger than the upper 4 and also gibbous at the base; *stamens* with united anthers.

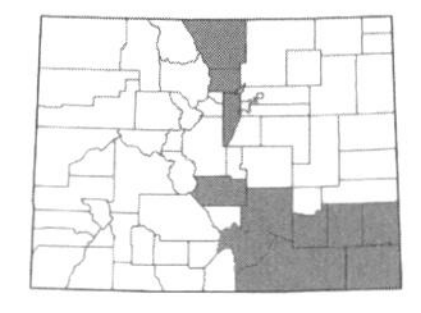

***Hybanthus verticillatus* (Ortega) Baill.**, BABY-SLIPPERS. Plants 1–3 dm from a much-branched subterranean caudex; *leaves* opposite or nearly so to mostly alternate above, sometimes fascicled, linear-lanceolate to lanceolate, 1–4 cm × 1–5 mm, entire to remotely toothed, ciliate, sessile; *sepals* 1.5–2.3 mm long; *petals* greenish-white to cream with purplish tips, the upper 4 petals 2–2.5 mm long, lower petal 2.5–5 mm long; *capsules* 4–6 mm long. Uncommon on dry hillsides, in grasslands, and on rocky outcrops, 5000–7000 ft. May–July. E.

VIOLA L. – VIOLET

Annual or perennial herbs; *leaves* alternate or sometimes the lower opposite, the blade entire to toothed or dissected; *sepals* with auricles at the base; *petals* unequal, the lowest one with a spur; *stamens* distinct but often cohering together by the broad filaments with conspicuous appendages. (Baker, 1935; 1940; 1957; Brainerd, 1921; McKinney, 1992; Russell, 1955; Spotts, 1939)

1a. Stipules leaf-like, pinnatifid or deeply toothed at the base; caulescent annuals or short-lived perennials...2
1b. Stipules unlike the above; acaulescent to caulescent perennials...3

2a. Flowers 1.5–3 cm long, usually tricolored, with two darker upper petals and three white or yellow lower petals and a yellow center; spur 3–6 mm long; plants 10–45 cm tall...***V. tricolor***
2b. Flowers 0.4–1 cm long, not tricolored, cream to pale violet with a yellow center; spur 1–2.5 mm long; plants 10–20 cm tall...***V. kitaibeliana***

3a. Leaves palmately divided into narrow lobes...4
3b. Leaves simple and not divided...5

4a. Flowers yellow (sometimes drying purplish), 6–12 mm long; plants of the western slope... ***V. sheltonii***
4b. Flowers violet, 10–20 mm long; plants of the eastern slope...***V. pedatifida***

5a. Flowers yellow, sometimes purple-tinged on the backs of the upper petals...6
5b. Flowers white to violet or bluish...8

6a. Plants conspicuously caulescent with a slender, elongated stem, usually with 2–3 stem leaves; leaves cordate at the base; flowers not purple-tinged on the backs of the upper petals...***V. biflora***
6b. Plants acaulescent or shortly caulescent; leaves cuneate to rounded or rarely subcordate at the base; flowers often purple-tinged on the backs of the upper petals...7

7a. Leaves mostly ovate to orbicular, with purplish, prominent veins below and crenate or rarely entire margins...***V. purpurea***
7b. Leaves mostly lanceolate to elliptic or oblong, lacking purplish veins, with irregularly toothed to entire margins...***V. nuttallii***

8a. Plants caulescent, the stem with 2 or more internodes; flowers white or occasionally pale violet, with purple veins...***V. canadensis***
8b. Plants acaulescent, or rarely caulescent but the flowers violet to bluish-purple...9

9a. Flowers white or occasionally pale lilac, with purple veins; slender stolons or rhizomes present; flowers 6–10 mm long; spur short, 1–2 mm long...***V. palustris***
9b. Flowers violet to deep blue; if stolons present or flowers white then the flowers larger, 12–22 mm long and the spur 2.5–4 mm long...10

10a. Plants with numerous leafy stolons rooting at the nodes; style with a hooked tip; flowers 12–22 mm long with a spur 2.5–4 mm long; plants hairy throughout...***V. odorata***
10b. Plants lacking leafy stolons; style not hooked at the tip; flowers 5–15 mm long, spur 1–7 mm long; plants glabrous or sometimes hairy...11

11a. Lower lateral petals not bearded; spur greatly enlarged and rounded at the base, 4–7 mm long; leaves deeply cordate at the base...***V. selkirkii***
11b. Lower lateral petals bearded; spur about the same width throughout or only slightly enlarged at the base, 1–7 mm long; leaves cordate to truncate or cuneate at the base...12

12a. Plants caulescent later in the season but acaulescent in early flowering; spur 4–7 mm long, often projected past the peduncle; stipules often toothed or spinulose-serrate on the margins; leaves truncate to cuneate or subcordate at the base...***V. adunca***
12b. Plants acaulescent; spur 1–5 mm long, not projected past the peduncle; stipules entire; leaves cordate at the base...***V. sororia***

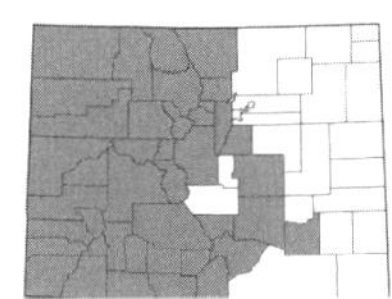

***Viola adunca* J. Sm.**, HOOK-SPURRED VIOLET. [*V. bellidifolia* Greene; *V. labradorica* Schrank]. Perennials, acaulescent or caulescent later in the season, 4–25 cm; *leaves* 1–4 cm long, round-ovate, truncate or cuneate at the base, with crenulate margins; *sepals* 5–7 mm long; *petals* violet, 7–12 mm long, with a spur 4–7 mm long; *capsules* 4–5 mm long. Common on open slopes, in meadows and open forests, and in the alpine, 6500–13,500 ft. May–Aug. E/W. (Plate 76)

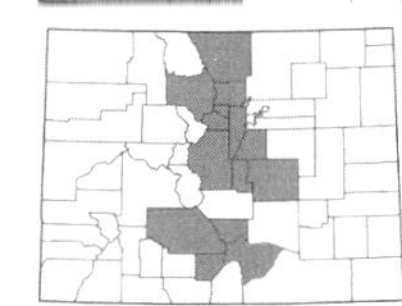

***Viola biflora* L.**, ARCTIC YELLOW VIOLET. Perennials, caulescent, the stems usually 2-leaved, 5–20 cm; *leaves* 1–2 cm long, round-reniform, cordate, with crenulate margins; *sepals* ca. 4 mm long; *petals* yellow-streaked with dark purplish-brown, 6–8 mm long, with a short-conical spur. Found in shady, moist forests, rocky crevices, and along shady streams, known mostly from the mountains along the Front Range, 6400–12,500 ft. May–July. E/W.

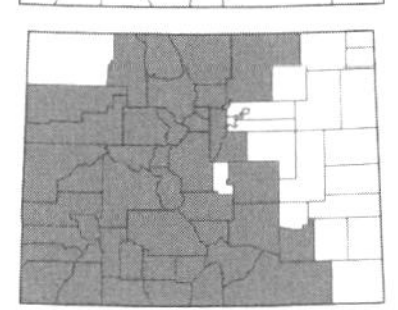

***Viola canadensis* L.**, CANADIAN WHITE VIOLET. [*V. rugulosa* Greene; *V. rydbergii* Greene; *V. scopulorum* (A. Gray) Greene]. Perennials, caulescent, 10–20 cm; *leaves* 1.5–6 cm long, cordate-ovate, crenate; *sepals* 4–7 mm long with white margins; *petals* white with purple lines, 6–14 mm long with a short spur; *capsules* 5–10 mm long. Common in moist, shady forests and along streams, 5000–10,500 ft. April–July. E/W. (Plate 76)

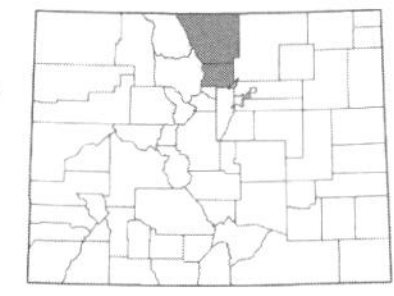

***Viola kitaibeliana* Schult.**, FIELD PANSY. [*V. bicolor* Pursh; *V. rafinesquii* Greene]. Annuals, caulescent, 10–20 cm; *leaves* 0.3–2 cm long, suborbicular to ovate, entire to crenate; *sepals* subulate, about half as long as the petals; *petals* cream, blue, or purplish, sometimes bicolored, 4–10 mm long, with a short spur 1-2.5 mm long. Uncommon on open hillsides and mesas, 5500–6000 ft. May–June. E. Introduced.

Viola arvensis Murray could be present in Colorado. It is very similar to *V. kitaibeliana* and differs in having larger flowers (1–1.5 cm long) that are all open and chasmogamous. *Viola kitaibeliana* has cleistogamous flowers present as well as chasmogamous flowers.

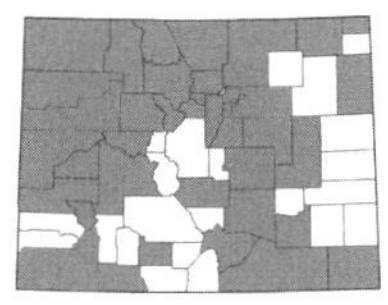

***Viola nuttallii* Pursh**, NUTTALL'S VIOLET. [*V. linguifolia* Nutt.; *V. praemorsa* Douglas ex Lindl.; *V. vallicola* A. Nelson]. Perennials, acaulescent or shortly caulescent, 5–25 cm; *leaves* linear-lanceolate to oblong-lanceolate, 2–7 cm long, entire to irregularly toothed; *sepals* acuminate, 5–10 mm long; *petals* yellow, 8–17 mm long, with a short spur; *capsules* 6–9 mm long. Common on open hills and in grasslands and meadows, widespread, 3500–11,500 ft. April–June. E/W. (Plate 76)

***Viola odorata* L.**, SWEET VIOLET. Perennials, acaulescent, 4–12 cm, from slender rhizomes and with stolons; *leaves* ovate-cordate to subreniform, 1.3–6 cm long, crenate; *petals* violet or rarely white, 12–22 mm long, with a spur 2.5–4 mm long. Uncommon along streams, known from Boulder Co., 6000–6500 ft. April–June. E. Introduced.

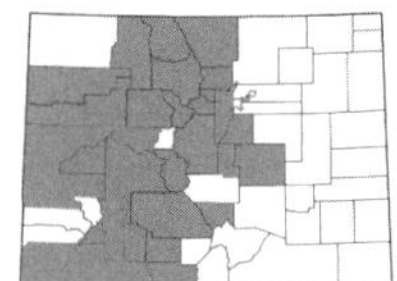

***Viola palustris* L.**, MARSH VIOLET. [*V. macloskeyi* Lloyd; *V. renifolia* A. Gray]. Perennials, acaulescent, 5–20 cm, with slender rhizomes and stolons; *leaves* ovate-cordate to reniform or orbicular, 2–4 cm long, crenulate; *sepals* 5–6 mm long, narrowly white-margined; *petals* violet to nearly white with purplish veins, 6–10 mm long, with a short spur 1–2 mm long. Found in moist moss along shady streams and in moist forests, 6000–12,600 ft. May–July. E/W.

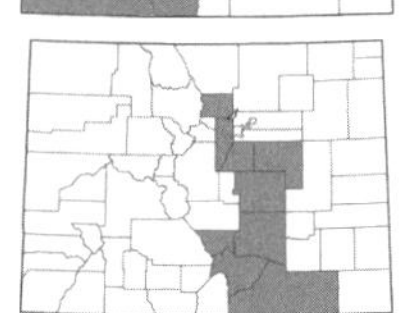

***Viola pedatifida* G. Don**, PRAIRIE VIOLET. Perennials, acaulescent, 7–20 cm; *leaves* ternately cleft into linear, entire to lobed segments, 1–7 cm long; *sepals* 6–10 mm long; *petals* violet, 10–20 mm long, with a short spur. Rare in dry grasslands and open ponderosa pine forests along the Front Range, 5800–8800 ft. April–June. E.

***Viola purpurea* Kellogg**, GOOSEFOOT VIOLET. [*V. utahensis* M.S. Baker & J.C. Clausen ex Baker; *V. venosa* (S. Watson) Rydb.]. Perennials, acaulescent or shortly caulescent, 4–15 cm; *leaves* ovate, 1–2 (3) cm long, the base obtuse to acute or rarely subcordate, margins crenate or sometimes entire; *sepals* white-margined; *petals* yellow and purplish-tinged dorsally, 6–10 mm long, with a short spur 1–2 mm long. Uncommon on rocky slopes, often with sagebrush, 7200–10,500 ft. May–July. W.

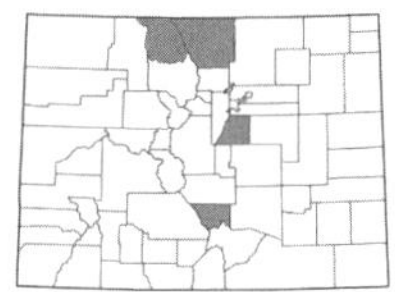

***Viola selkirkii* Pursh ex Goldie**, SELKIRK'S VIOLET. Perennials, acaulescent, 4–10 cm; *leaves* cordate-ovate, 1–4 cm long, with a deeply cordate base, crenate; *sepals* 4–6 mm long; *petals* pale violet, 5–10 mm long, the lower lateral petals not bearded, with a large spur 4–7 mm long, enlarged and rounded at the base. Rare in shady, moist forests, 8500–9100 ft. May–June. E.

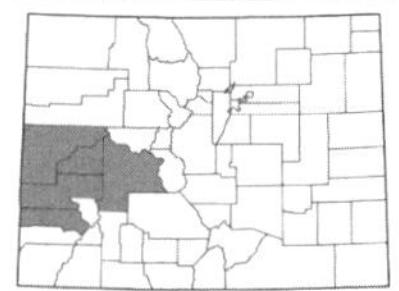

***Viola sheltonii* Torr.**, SHELTON'S VIOLET. [*V. biternata* Greene]. Perennials, caulescent or subacaulescent, 5–20 cm; *leaves* palmately 3-divided, the divisions palmately or pedately 3-parted and cleft, 1–4 cm long; *sepals* 4–8 mm long; *petals* yellow, 6–12 mm long, with a spur to ⅓ the length of the petals. Uncommon in aspen forests, under pinyon-juniper or oaks, or along streams, 6000–9000 ft. May–June. W.

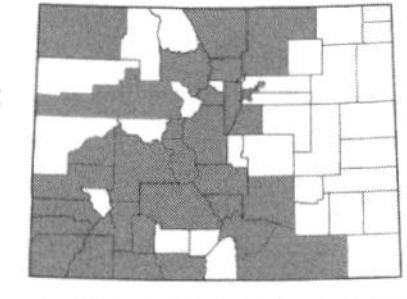

***Viola sororia* Willd.**, BOG VIOLET. [*V. affinis* Leconte; *V. nephrophylla* Greene; *V. papilionacea* Pursh]. Perennials, acaulescent, 4–30 cm; *leaves* cordate-ovate, 2–4 cm long, crenate; *sepals* 4–8 mm long; *petals* violet, 10–15 mm long with a short spur 1–5 mm long not projected past the peduncle. Common in bogs, along streams and creeks, in moist meadows, and on shady forest slopes, 4800–10,500 ft. May–Aug. E/W.

***Viola tricolor* L.**, JOHNNY JUMP-UP. Annuals, caulescent, 10–45 cm; *leaves* lanceolate to ovate, 0.8–3.5 cm long, obtuse to acute at the base, crenate; *sepals* lanceolate; *petals* usually tricolored, the lower white to yellow, the lower lateral white to yellow or lavender, and the upper dark violet, 15–30 mm long, with a spur 3–6 mm long. Commonly cultivated plants sometimes escaping, 5000–5500 ft. May–July. E. Introduced.

VISCACEAE Miq. – CHRISTMAS MISTLETOE FAMILY

Hemiparasitic perennials on the branches of trees, more or less green and photosynthetic, producing haustoria which penetrate the host and often induce gall-like growth on the host plants; *leaves* well-developed or reduced to scales, usually opposite, small, entire, evergreen, exstipulate; *flowers* imperfect, small, actinomorphic; *sepals* reduced to a mere rim; *petals* 3–6 (9), distinct and scale-like, often reduced to teeth, persistent; *pistil* 1; *ovary* inferior, 3–4-carpellate, unilocular; *fruit* a berry or a drupe.

1a. Plants parasitic on *Pinus* and *Pseudotsuga*; fruit bicolored with the proximal section darker and blue-glaucous; shoots usually 1 dm or less in length...***Arceuthobium***

1b. Plants parasitic on *Juniperus*; fruit white or pinkish, not bicolored; shoots 1–2.5 dm long...***Phoradendron***

ARCEUTHOBIUM M. Bieb. – DWARF MISTLETOE

Hemiparasitic on Pinaceae and Cupressaceae members, the stems decussately branched, plants dioecious; *leaves* reduced to small, connate scales; *staminate flowers* with 3–4 (7) petals and a sessile anther opposite each petal; *pistillate flowers* shallowly 2-lobed at the top; *fruit* a mucilaginous, bicolored berry. (Gill, 1935; Hawksworth & Wiens, 1972)

1a. Plants parasitic on *Pseudotsuga*, rarely on *Picea* or *Abies*; staminate buds subglobose; plants small, 0.5–3 (7) cm long overall; shoots closely scattered along the stem near the tip...***A. douglasii***
1b. Plants parasitic on *Pinus* species; staminate buds lenticular or if subglobose then the plants 4–10 cm long overall; shoots in individual clusters...2

2a. Branches whorled; staminate buds subglobose; the third internode relatively long, about 10 times as long as wide; plants parasitic on *Pinus contorta* and sometimes on *Pinus ponderosa*...***A. americanum***
2b. Branches flabellate (fan-like); staminate buds lenticular; the third internode usually under 10 times as long as wide; plants parasitic on various *Pinus* species, including *Pinus ponderosa*, rarely on *Pinus contorta*...3

3a. Plants parasitic on *Pinus ponderosa*, rarely on *Pinus contorta* or *Pinus edulis*; shoots usually orange; plants flowering early (May–June)...***A. vaginatum* ssp. *cryptopodum***
3b. Plants parasitic on *Pinus flexilis* or *P. edulis*; shoots yellowish-green, olive-green, or brown; plants flowering later (July–Sept.)...4

4a. Plants parasitic on *Pinus flexilis*; shoots yellowish-green, mostly 2–4 (5) cm long overall...***A. cyanocarpum***
4b. Plants parasitic on *Pinus edulis*; shoots olive-green or brown, mostly 5–10 cm long overall...***A. divaricatum***

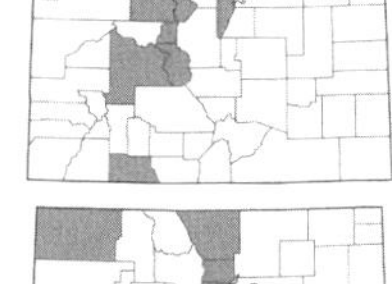

Arceuthobium americanum **Nutt. ex Engelm.**, LODGEPOLE-PINE DWARF MISTLETOE. Stems yellowish to olive-green with whorled branches; *staminate flowers* with lobes ca. 1 mm long, usually 3-merous; *pistillate flowers* 2-merous; *berries* 3.5–4 mm diam. Found throughout the range of lodgepole pine, sometimes found on ponderosa pine, 7000–9500 ft. April–June. E/W.

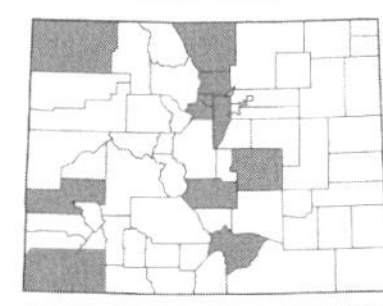

Arceuthobium cyanocarpum **(A. Nelson ex Rydb.) Abrams**, LIMBER PINE DWARF MISTLETOE. Stems yellowish-green with flabellate branches, the shoots mostly 2–4 (5) cm long overall; *staminate flowers* ca. 3 mm diam, 3-merous; *pistillate flowers* 3-merous; *berries* 3–5 mm diam. Found on limber pine, mostly distributed along the Front Range, 8500–10,000 ft. July–Sept. E/W. (Plate 76)

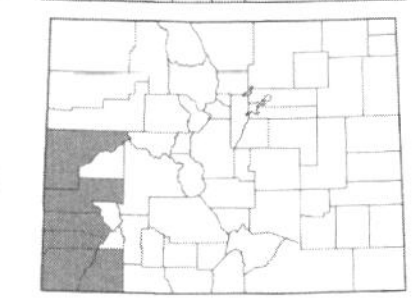

Arceuthobium divaricatum **Engelm.**, PINYON PINE DWARF MISTLETOE. Stems olive-green or brown with flabellate branches, the shoots mostly 5–10 cm long overall; *staminate flowers* ca. 2.5 mm long, 3-merous; *pistillate flowers* 3-merous; *berries* ca. 3.5 mm diam. Found on pinyon pine, 5500–8000 ft. Aug.–Sept. W (Not yet reported for the eastern slope but could possibly occur on *Pinus edulis* in the southcentral counties or in northern Larimer Co.).

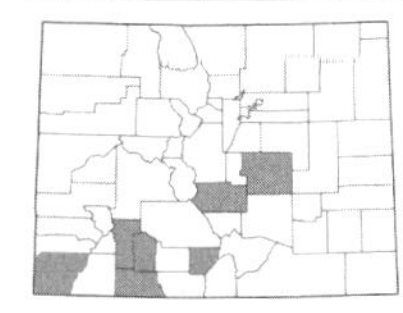

Arceuthobium douglasii **Engelm.**, DOUGLAS-FIR DWARF MISTLETOE. Stems olive-green, with flabellate branches and short stems 0.5–7 cm long; *staminate flowers* ca. 2 mm long, 3-merous; *pistillate flowers* ca. 1.5 mm long; *berries* 3–4 mm diam. Found on Douglas-fir, 6500–9000 ft. Mar.–May. E/W.

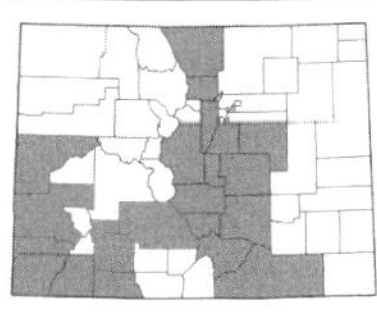

Arceuthobium vaginatum **(Willd.) J. Presl ssp. *cryptopodum* (Engelm.) Hawksw. & Wiens**, PINELAND DWARF MISTLETOE. Stems orange to reddish-brown with flabellate branches; *staminate flowers* 2.5–3 mm long; *berries* 4.5–5.5 mm diam. Found mostly on ponderosa pine, 5500–9300 ft. May–June. E/W.

PHORADENDRON Nutt. – MISTLETOE

Hemiparasitic on Cupressaceae members, stems decussately branched; *leaves* often well-developed; *staminate flowers* with 3 petals, and a sessile anther opposite each petal; *pistillate flowers* with 3 petals; *fruit* a drupe.

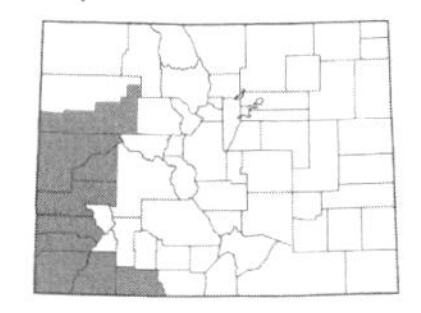

Phoradendron juniperinum **A. Gray**, JUNIPER MISTLETOE. Stems green or yellowish-green, with shoots 1–2.5 dm long; *pistillate flowers* 3–4-merous; *drupes* ca. 4 mm diam., white or pinkish. Parasitic on *Juniperus*, 5300–7500 ft. June–Sept. W.

VITACEAE Juss. – GRAPE FAMILY

Woody perennial vines, often with tendrils; *leaves* alternate, simple or compound; *flowers* perfect or imperfect, actinomorphic; *sepals* minute, often reduced to a collar; *petals* (3) 4–5 (7), distinct below or sometimes connate distally (*Vitis*); *stamens* as many and opposite the petals; *pistil* 1; *ovary* superior, 2-carpellate, with axile placentation; *fruit* a berry.

1a. Leaves palmately compound; stems with white pith; flowers mostly perfect, the petals distinct...***Parthenocissus***
1b. Leaves simple; stems with brown pith; flowers mostly imperfect, the petals connate at the apex and falling as a single unit without expanding...***Vitis***

PARTHENOCISSUS Planch. – VIRGINIA CREEPER

Woody vines with branched tendrils, with white pith; *leaves* palmately compound, the leaflets serrate; *flowers* mostly perfect, in cymose panicles; *sepals* minute, very slightly 5-toothed; *petals* 5; *stamens* 5.

1a. Inflorescence more or less dichotomously branched, with 10–60 flowers per cyme; tendrils lacking adhesive disks…***P. vitacea***

1b. Inflorescence with a central axis, with 25–200 flowers per cyme; tendrils terminating in adhesive disks…***P. quinquefolia***

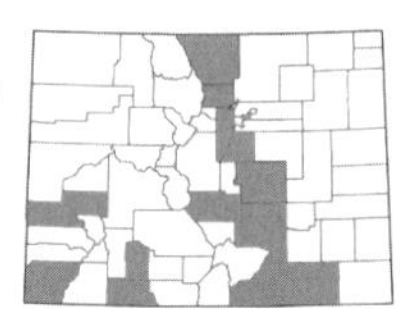

***Parthenocissus quinquefolia* (L.) Planch. & C. DC.**, VIRGINIA CREEPER. Vines with tendrils terminating in adhesive disks; *leaflets* oblanceolate, lanceolate, or ovate, 3–15 cm long; *inflorescence* with a central axis, with 25–200 flowers per cyme; *berries* 5–7 mm diam. Cultivated and probably not truly escaping but commonly grown on buildings and occasionally found on fencerows around towns, 5000–8000 ft. May–July. E/W. Introduced.

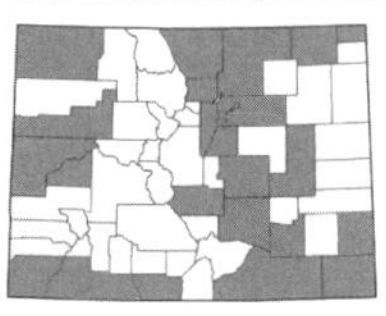

***Parthenocissus vitacea* (Knerr) Hitchc.**, WOODBINE; THICKET CREEPER. Vines with tendrils lacking adhesive disks; *leaflets* oblanceolate to obovate, 3–15 cm long; *inflorescence* more or less dichotomously branches, with 10–60 flowers per cyme; *berries* 8–10 mm diam. Found on rocky hillsides, in cool ravines, and along streams and creeks, 3500–7500 ft. May–July. E/W.

VITIS L. – GRAPE

Woody vines with usually bifurcate, often coiled tendrils, with brown pith; *leaves* simple, serrate; *flowers* mostly imperfect in cymose panicles; *sepals* minute; *petals* 5, cohering at the apex and falling without expanding.

1a. Upper leaf surface with cobwebby hairs, especially on young leaves, the lower leaf surface with cobwebby hairs at least along the veins…***V. acerifolia***

1b. Upper leaf surface glabrous, the lower leaf surface glabrous or hairy along the veins, or with cobwebby hairs just at the base…***V. riparia***

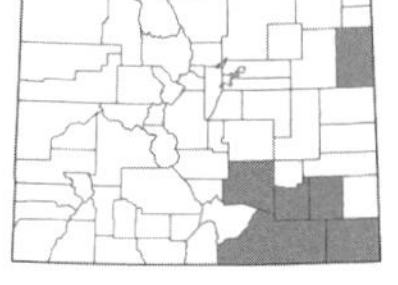

***Vitis acerifolia* Raf.**, BUSH GRAPE. [*V. longii* Prince]. Leaves suborbicular to deltate-ovate, 6–10 cm long, acute to acuminate, cobwebby pubescent above and below at least along the veins; *berries* 8–12 mm diam., black with a glaucous bloom. Found along streams and on rocky hillsides on the eastern plains, 3500–5000 ft. May–July. E.

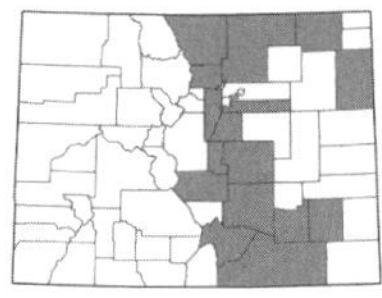

***Vitis riparia* Michx.**, RIVER BANK GRAPE. Leaves deltate-ovate to cordate-ovate, 7–15 cm long, acuminate, glabrous above, glabrous or pubescent along the veins below, sometimes cobwebby just at the base; *berries* 7–11 mm diam., purplish-black, with a glaucous bloom. Found along streams, in cool canyons, and on open hillsides, 3500–7000 ft. May–July. E. (Plate 76)

ZANNICHELLIACEAE Dumort. – HORNED PONDWEED FAMILY

Monoecious, annual aquatic herbs; *leaves* opposite or whorled, sessile, linear, stipulate, the stipules forming a sheath that is adnate to the leaf base; *flowers* imperfect, in axillary cymes; *perianth* absent; *staminate flowers* consisting of a single stamen; *pistillate flowers* with 1–9 distinct pistils; *ovary* superior, unicarpellate with a single ovule; *fruit* an achene.

ZANNICHELLIA L. – HORNED PONDWEED

With characteristics of the family.

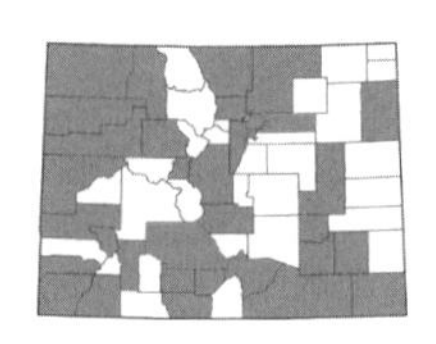

***Zannichellia palustris* L.**, HORNED PONDWEED. Stems 1–5 (8) dm long, much-branched; *leaves* opposite (sometimes appearing whorled where branches come off), to 10 cm long, entire, with a stipule sheath to 4 mm long; *achenes* oblong, 1.5–4 mm long, curved, flattened, beaked, and sharply ridged on the back. Found in slow-moving streams, ditches, and along pond margins, 3500–9800 ft. July–Sept. E/W. (Plate 76)

The leaves are typically bright green and stand out in the water.

ZYGOPHYLLACEAE R. Br. – CALTROP FAMILY

Trees, shrubs, or herbs (ours); *leaves* alternate or opposite, pinnately compound or pinnatifid, stipulate; *flowers* perfect, actinomorphic, solitary in leaf axils; *sepals* 4–5, distinct; *petals* 4–5, distinct, white, yellow, orange, or red; *stamens* 10–15; *pistil* 1; *styles* 1; *ovary* superior with axile placentation, 2–6-carpellate; *fruit* a capsule or schizocarp splitting into 5–10 mericarps.

1a. Erect, perennial herbs; leaves 2-foliate (with a single pair of leaflets), the leaflets mostly 1.5–4.5 cm long; fruit a capsule; flowers white and strongly marked with orange below...***Zygophyllum***
1b. Prostrate, annual herbs; leaves even-pinnately compound, the leaflets mostly 0.5–1.9 cm long; fruit a schizocarp splitting into mericarps; flowers yellow or orange...2

2a. Mericarps 5, hard with 2 horn-like spines; flowers yellow...***Tribulus***
2b. Mericarps 10, coarsely tuberculate but unarmed; flowers orange-yellow, drying white...***Kallstroemia***

KALLSTROEMIA Scrop. – CALTROP

Annual herbs; *leaves* opposite, even-pinnately compound; *petals* orange or yellow (drying white), rarely white; *ovary* 5-carpellate, with each carpel divided by a septum between the 2 ovules; *fruit* a schizocarp splitting into 10 coarsely tuberculate but unarmed mericarps.

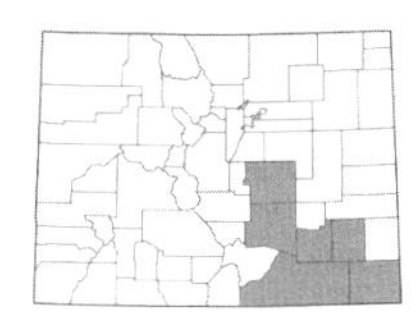

***Kallstroemia parviflora* J.B.S. Norton**, WARTY CALTROP. Plants prostrate, the stems to 1 m long, hirsute or becoming glabrous in age; *leaflets* elliptic to oblong or ovate, 8–19 × 3.5–9 mm, appressed-hairy, with uneven bases; *sepals* 4–7 mm long; *petals* orange-yellow, drying white, 5–11 mm long; *mericarps* 10, 3–4 mm high, coarsely tuberculate. Found along roadsides and in disturbed areas, usually in sandy soil, 4000–5300 ft. May–July. E.

TRIBULUS L. – PUNCTURE VINE; GOAT HEAD

Annual or perennial herbs; *leaves* opposite, even-pinnately compound; *petals* yellow or rarely white; *ovary* 5-carpellate; *fruit* a hard, spiny schizocarp splitting into 5 mericarps.

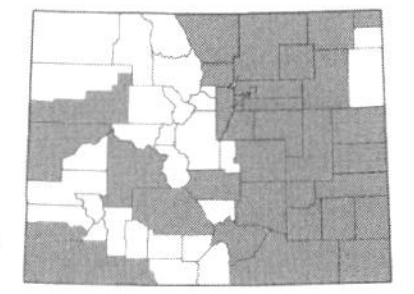

***Tribulus terrestris* L.**, PUNCTURE VINE. Annuals, prostrate, the stems to 1.5 m long, hirsute or becoming glabrous in age; *leaflets* elliptic to oblong or ovate, 4–11 × 1–4 mm, appressed-hairy, with uneven bases; *sepals* 2–3 mm long; *petals* yellow, 3–5 (6) mm long; *mericarps* 5, hard, with 2 horn-like spines. Found in disturbed places, fields, and along roadsides, 3500–7500 ft. June–Sept. E/W. Introduced.

ZYGOPHYLLUM L. – BEANCAPER

Perennial herbs; *leaves* opposite, 2-foliate; *petals* white and orange below; *ovary* 4–5-carpellate; *fruit* a capsule.

***Zygophyllum fabago* L.**, SYRIAN BEANCAPER. Plants erect, to 10 dm; *leaflets* oblong to ovate, 1.5–4.5 cm long, glabrous; *sepals* 6–7 mm long; *petals* white and strongly marked with orange below, about equal to the sepals in length; *capsules* oblong, 2.5–3.5 cm long. Known from a 1955 collection from near Fruita (Mesa Co.) where it was found along roadsides, 4500 ft. A note on the label states that neither 2, 4-D nor sodium chlorate could kill it. May–Aug. W. Introduced.

Lower Gunnison River Basin, Colorado

GLOSSARY

Abaxial. The side away from the axis.
Abortive. Imperfectly or not fully developed.
Acaulescent. Without a stem, or the stem short and all the leaves appear basal (i.e., as in a dandelion).
Accessory fruit. A fleshy fruit developing from the receptacle rather than the pistil.
Accrescent. Becoming enlarged with age.
Achene. A dry, 1-seeded, indehiscent fruit.
Acicular. Needle-shaped.
Acorn. A hard, dry indehiscent fruit of oaks, with a cup-like base and single, large, hard seed.
Actinomorphic. Radially symmetrical.
Acuminate. Gradually tapering to a sharp point, with concave sides along the tip.
Acute. Tapering to a pointed apex with more or less straight sides.
Adaxial. The side toward the axis.
Adnate. The fusion of unlike parts.
Adventitious. Structures or organs developing in unusual places, i.e., like roots developing on the stem.
Aggregate fruit. A cluster of small fleshy fruits starting from a number of separate carpels in a single flower (i.e., a raspberry or blackberry).
Alkaline. Salty.
Alpine. Areas above timberline in the mountains.
Alternate. One leaf, stem, or other structure per node on a stem.
Alveolate. Honeycombed.
Amphibious. Living in both water and on land.
Amplexicaul. Clasping the stem, usually as the bases of leaves.
Androecium. All of the stamens in a flower.
Androgynous. With both staminate and pistillate flowers, the pistillate flowers usually borne below the staminate.
Angiosperm. A plant producing flowers and bearing an ovary with ovules.
Annual. A plant which germinates from seed, flowers, sets seeds, and dies within the same year.
Annulus. A row of thick-walled cells along one side of a fern sporangium.
Anther. The pollen-bearing portion of the stamen.
Antheridium. The male reproductive structure in moss and fern gametophytes.
Anthesis. Flowering period for plants.
Anthocarp. A fruit with the lower portion of the perianth fused with the fruit pericarp.
Antrorse. Directed forward or upward.
Apetalous. Lacking petals.
Apex. The tip.
Apiculate. Ending abruptly in a small, pointed tip.
Appressed. Lying close and flattened to the surface of another organ.
Aquatic. Growing in water.
Arachnoid. With long, cobwebby, tangled hairs.
Archegonium. The female reproductive structure in fern gametophytes.
Arcuate. Curved into an arch.
Areole. Spine-bearing region of a cactus.
Aristate. With an awn or bristle at the tip.
Armed. Bearing thorns, spines, or prickles.
Ascending. Growing upward.
Asymmetric. Not divisible into equal halves.
Attenuate. Gradually tapering to a narrow tip or base.
Auricle. Small ear-shaped lobe at the base of a leaf.
Awl-shaped. Narrowly triangular, sharp pointed.
Awn. A slender, stiff bristle.
Axile placentation. Ovules attached to the central axis of an ovary with 2 or more locules.
Axillary. Arising in an axil.
Axis. The central supporting structure bearing organs.
Banner. Upper petal of the flower in legumes (Fabaceae).
Barbellate. With short, stiff hairs or barbs.
Basal placentation. Ovules positioned at the base of a single-chambered ovary.
Basifixed. Attached at the base.
Beak. Narrow or prolonged tip.
Bearded. Bearing tufts of long hairs.
Berry. Fleshy fruit developed from a single carpel with more than 1 seed. (i.e., a blueberry).
Bicarpellate. With 2 carpels.
Biennial. A plant which lives for 2 years, the first year growing vegetatively and the second year producing flowers and fruit.
Bifid. Deeply two-cleft or two-lobed.
Bifurcate. Two-forked or Y-shaped.
Bilabiate. A zygomorphic flower with two lips.
Bipinnatifid. Twice pinnately compound.
Biseriate. Arranged in 2 rows or series.
Bisexual. With both stamens and pistil.
Bladder. A thin-walled, inflated structure.
Blade. Exposed portion of a leaf.
Bloom. A whitish, waxy, powdery coating.
Bract. A reduced leaf at the base of a flower or inflorescence.
Bracteate. With bracts.
Bracteole. A small bract or secondary bract.
Bristle. A short, stiff hair.
Bud. An undeveloped shoot or unopened flower.
Bulb. Underground bud with thick, fleshy scales (i.e., an onion).
Bulbil. A small bulb arising from the base of a larger bulb.
Bulblet. A small bulb; a bulb-like structure occuring above ground, often in the axils of leaves.
Bundle scar. A scar left on a stem or branch by the vascular bundles when a leaf falls.
Bur. A structure armed with hooked spines.
Caducous. Early deciduous.
Caespitose. Growing in tufts.
Callus. Hardened area, i.e., the hardened downward extension of the base of the lemma in grasses.
Calyculate. With small bracts around the involucre or calyx.
Calyx. The outermost whorl of perianth; sepals.
Campanulate. Bell-shaped.
Canaliculate. Having longitudinal channels or grooves.
Canescent. Densely covered with short, fine gray or white hairs.
Capillary. Hair-like, thin and fine.
Capitate. In a head-like cluster.
Capsule. A dry, dehiscent fruit with more than one carpel.

Carinate. Keeled with at least one longitudinal ridge.
Carpel. A simple pistil; megasporophyll.
Carpophore. A thin prolongation of the receptacle between the carpels as a central axis; in Apiaceae a slender stalk supporting half of the fruit.
Cartilaginous. Tough and firm but flexible.
Caryopsis. A dry, one-seeded, indehiscent fruit with the seed coat fused to the pericarp; fruit type of the grass family (Poaceae).
Catkin. A dense spike or raceme of apetalous, usually unisexual flowers (seen in willows and birches).
Caudate. With a tail-like appendage.
Caudex. The woody base of a herbaceous perennial.
Caulescent. With a leafy stem.
Cauline. Pertaining to the stem, not basal.
Chaff. Thin, dry scales on the receptacle of the heads of Asteraceae.
Channeled. With one or more longitudinal grooves.
Chartaceous. Stiffly papery in texture, usually not green.
Chasmogamous. Flowers which open for pollination.
Chlorophyll. Green pigment of plants, associated with photosynthesis.
Ciliate. With a fringe of hairs on the margin.
Ciliolate. Slightly ciliate, the hairs minute.
Cinereous. Grayish-colored due to a covering of short hairs.
Circinate. Coiled from the tip downward.
Circumscissile. Dehiscing in a horizontal, circular line so that the top comes off like a lid.
Clasping. Partly or wholly surrounding the stem.
Clavate. Club-shaped, widest at the tip.
Claw. Narrowed base of some petals or sepals.
Cleft. Cut or split about half-way to the middle.
Cleistogamous. Flowers which self-fertilize without opening.
Coalescent. Fused together as a single unit.
Coherent. Sticking together of like parts.
Collar. The area on the outside of a grass leaf at the junction of the blade and sheath.
Column. A structure formed from the fusion of the united filaments and style in Orchidaceae.
Coma. A tuft of hairs.
Commissure. Surface along which two carpels join one another.
Comose. With a tuft of hairs.
Complete. A flower with all four floral whorls (sepals, petals, androecium, and gynoecium).
Compound leaf. A leaf separated into distinct leaflets.
Concave. Curved inward.
Conduplicate. Folded together lengthwise with the upper surface within.
Cone. An axis bearing a dense cluster of sporophylls.
Confluent. Running together; blending of one part into another.
Conic. Cone-shaped.
Connate. Fusion of like parts.
Connivent. Touching, but not actually fused.
Constricted. Narrowed or drawn together.
Convex. Rounded and curved outward on the surface.
Convolute. Rolled up longitudinally.
Cordate. Heart-shaped.
Coriaceous. Leathery.
Corm. A short, solid underground stem with thin papery leaves.
Corniculate. With small, horn-like projections.
Corolla. All of the petals in a flower.
Corona. Appendages between the petals and stamens in some flowers; a crown.
Corrugated. Wrinkled or folded with alternating ridges.
Corymb. A flat-topped inflorescence, with the lower pedicels longer than the upper.
Corymbiform. An inflorescence with the overall appearance of a true corymb.
Corymbose. With flowers arranged in corymbs.
Costa. A rib or prominent midvein.
Cotyledon. The first leaf of the embryo.
Crenate. With rounded teeth along the margin.
Crest. An elevated ridge on a surface.
Crisped. Curled, crinkled, or wavy.
Crown. The persistent base of a perennial herbaceous plant; pappus of Asteraceae flower which is short, scale-like, and usually continuous around the corolla.
Cruciform. Cross-shaped.
Cucullate. Hooded.
Culm. Stem of grasses, sedges, and rushes.
Cuneate. Wedge-shaped.
Cuspidate. Tapering to a short, sharp, abrupt point.
Cyathium. Inflorescence of some members of the Euphorbiaceae, consisting of a single pistil and several male flowers surrounded by a cup-like involucre.
Cyme. A somewhat flat-topped inflorescence in which the central or terminal flower opens first.
Deciduous. Falling off.
Decompound. More than once-compound.
Decumbent. Prostrate on the ground with the tip ascending.
Decurrent. Extending downward along a stem or branch from the point of insertion.
Decussate. Arranged along the stem in pairs, with each pair alternating at right angles to the pair below.
Deflexed. Abruptly bend downward.
Dehiscent. Opening at maturity.
Deltoid. Triangular.
Dendritic. Hairs with a branching pattern similar to that in a tree (with a central axis and shorter branches coming off both sides).
Dentate. Toothed along the margin, with the teeth directed outward.
Denticulate. Dentate with very small teeth.
Descending. Directed downward.
Determinate. An inflorescence in which the terminal flower blooms first.
Diadelphous. Stamens united into two sets by their filaments.
Dichotomous. Branched into two more or less equal divisions.
Didymous. Occurring in pairs.
Didynamous. With 4 stamens total, in two pairs of unequal length.
Digitate. Divided or lobed from a common point.
Dilated. Expanded or flattened.
Dimorphic. With two forms.
Dioecious. With staminate and pistillate flowers borne on different plants.

Disarticulate. Separating at maturity.
Disk flower. An actinomorphic flower in an Asteraceae head.
Dissected. Deeply divided into numerous narrow segments.
Distal. Toward the tip.
Distichous. Two-ranked.
Distinct. Separate.
Divaricate. Widely spreading apart.
Dolabriform. Hairs attached near the middle.
Dorsal. The back or outward surface away from the axis.
Dorsiventral. Flattened from top to bottom.
Drupe. Fleshy, indehiscent fruit with a stony pit.
Druplets. Small drupes.
Echinate. With small prickles or spines.
Elliptic. Broadest in the middle and narrower at the two equal ends.
Emarginate. Notched at the apex.
Emergent. Rising out of the water.
Entire. Not toothed, divided, or notched along the margin.
Epicalyx. An involucre of bracts resembling an outer calyx, as in some Malvaceae.
Epigynous. With an inferior ovary.
Epipetalous. With the stamens attached to the petals.
Equitant. Leaves in two ranks with each folded in half and subtending those to the inside.
Erose. With the margin irregularly toothed, fringed, or gnawed in appearance.
Evergreen. Having green leaves year-round.
Excurrent. Extending beyond the apex.
Exserted. Projecting beyond the surrounding parts; not included.
Falcate. Sickle-shaped.
Farinose. With a mealy, powdery covering.
Fascicle. Cluster.
Fertile. Capable of bearing seeds.
Fibrous roots. A root system in which all of the branches are approximately the same thickness.
Filament. Stalk of the stamen supporting the anther.
Filiform. Thread-like.
Fimbriate. Fringed.
Fistulose. Tubular.
Flabellate. Fan-shaped.
Fleshy. Thick and juicy; succulent.
Flexuous. Sinuous, curved.
Floccose. With tufts of long, soft, tangled hairs.
Floral tube. The elongated tubular portion of the perianth.
Floret. A grass flower consisting of the lemma, palea, and flower.
Foliaceous. Leaf-like.
Follicle. A dry, dehiscent fruit composed of a single carpel and splitting along a single side.
Forked. Divided into two essentially equal branches.
Fornix. A small crest or scale in the throat of a corolla.
Free-central placentation. With ovules borne on a central free-standing column in a unilocular ovary.
Fringed. With hairs or bristles along the margin.
Frond. A fern leaf.
Fruit. A ripened ovary with enclosed seeds.
Funnelform. Funnel-shaped.
Fusiform. Widest near the middle and tapering toward both ends.
Galea. The helmet-shaped or hood-like upper lip of a bilabiate corolla.
Geniculate. With abrupt bends.
Gibbous. Swollen or enlarged on one side.
Glabrate. Nearly glabrous.
Glabrous. Smooth; lacking hairs.
Gland. Organ which secretes sticky or oily substances.
Glandular. With glands.
Glaucous. Covered with a whitish or bluish waxy coating that rubs off.
Globose. Spherical.
Glochid. A barbed hair.
Glomerule. Dense cluster of flowers.
Glume. One of a pair of bracts at the base of a grass spikelet.
Glutinous. Sticky, gummy.
Grain. A seed-like structure, seen on the fruit of some *Rumex* species.
Gymnosperm. Plants producing seeds not borne in an ovary; usually cone-producing.
Gynaecandrous. An inflorescence with the pistillate flowers above the staminate flowers.
Gynandrous. With stamens adnate to the pistil.
Gynobase. An elongation or enlargement of the receptacle.
Gynophore. An elongated stalk subtending the pistil.
Gynostegium. Structure formed from the fusion of the anthers with the stigmatic region of the gynoecium, as in *Asclepias*.
Habit. The general appearance or mode of growth of a plant.
Hastate. Arrowhead-shaped, with the basal lobes turned outward.
Head. A dense cluster of sessile or nearly sessile flowers; inflorescence type of the Asteraceae.
Helicoid. Coiled like a spiral.
Herb. A plant with little or no above-ground woody portion.
Herbaceous. Not woody.
Heterosporous. Having two different kinds of spores, microspores and megaspores.
Hip. A berry-like structure formed from an enlarged hypanthium surrounding numerous achenes; fruit type of *Rosa*.
Hirsute. Hairy with coarse, stiff hairs.
Hispid. With rough, stiff, firm hairs.
Hood. A hollow, arched covering.
Horn. A tapering projection resembling a horn.
Humistrate. Lying on the ground.
Hyaline. Thin, membranous and transparent or nearly so.
Hypanthium. The fusion of the base of the sepals, petals, and stamens to form a cup-shaped structure.
Hypogynous. With a superior ovary and no hypanthium present.
Imbricate. Overlapping like shingles on a roof.
Imperfect. With either stamens or pistils; unisexual.
Incised. Cut deeply and sharply.
Included. Not projected beyond the surrounding parts.
Incomplete. Lacking one or more floral whorls.
Indehiscent. Not opening at maturity.

Indeterminate. An inflorescence in which the lower or outer flowers bloom first.
Indurate. Hardened.
Inferior ovary. Ovary that is attached below the sepals, petals, and stamens.
Inflated. Bladdery, swollen, or expanded.
Inflorescence. The flowering part of a plant.
Internode. The portion of the stem between nodes.
Introduced. Not native.
Involucel. A secondary involucre, as in the bracts of the secondary umbels in the Apiaceae.
Involucre. A whorl of bracts subtending a flower or cluster of flowers.
Involute. With inrolled margins.
Irregular. Bilaterally symmetrical.
Jointed. Having nodes or articulation points.
Keel. A conspicuous longitudinal ridge; two lower united petals of a Fabaceae flower.
Krummholz. Twisted or bent trees found in the subalpine.
Lacerate. Cleft or irregularly cut into narrow, pointed segments; torn in appearance.
Lactiferous. With milky sap.
Lamina. The expanded portion of a leaf or petal.
Lanate. Woolly.
Lanceolate. Lance-shaped, with the widest point below the middle.
Leaflet. A division of a compound leaf blade.
Leaf scar. The scar remaining on a branch after a leaf falls.
Legume. A dry, dehiscent fruit usually opening along two lines of dehiscence and derived from a single carpel; fruit type of the Fabaceae.
Lemma. The lower of the two bracts comprising a grass floret.
Leniticular. Lens-shaped.
Ligulate. Strap-shaped; with a ligule.
Ligule. The flattened part of a ray flower in the Asteraceae; the membranous appendage arising from the junction of the leaf and sheath in many grasses and sedges.
Limb. Expanded and spreading part of the corolla.
Linear. Long and narrow with more or less parallel margins.
Lobe. A rounded division or segment of a leaf or other plant part.
Locule. The chamber of an ovary.
Loculicidal. Dehiscing through the locules of a fruit.
Lodicule. Rudimentary scales at the base of the ovary in grass flowers.
Loment. A legume which is constricted between the seeds.
Lorate. Strap-shaped.
Lunate. Crescent-shaped.
Lyrate. Pinnatifid with the terminal lobe large and rounded, with the lower lobes smaller.
Maculate. Blotched or spotted.
Margin. The edge.
Marginal placentation. Ovules attached to the margins of a simple pistil.
Mealy. Powdery, dry, crumbly-looking.
Megasporangium. A spore-producing structure which bears megaspores.
Membranous. Thin, usually translucent.
Mericarp. A section of a schizocarp.
Microphyll. The narrow, small, single-veined leaf present in some lower vascular plants.
Microsporangium. A spore-producing structure which contains microspores.
Midrib. The central vein of a leaf or other organ.
Monadelphous. Stamens united by their filaments to form a tube around the style.
Moniliform. Cylindrical and constricted at regular intervals, giving the appearance of a string of beads.
Monoecious. With staminate and pistillate flowers on the same plant.
Mottled. With spots or blotches.
Mucilaginous. Moist and slimy.
Mucronate. Tipped with a short, sharp spine-point.
Muricate. With small, short, hard, sharp projections.
Needle. A slender, needle-shaped leaf found in the Pinaceae.
Net-veined. Reticulate.
Neutral. Lacking functional stamens or pistils.
Node. The point on a stem where leaves or branches are attached.
Nodule. A swelling or small knob.
Nut. A hard, dry, indehiscent fruit with a single seed.
Nutlet. A small nut.
Obcompressed. Compressed such that a structure is flattened dorsiventrally while other structures are flattened laterally.
Obcordate. Inversely cordate.
Oblanceolate. Inversely lanceolate; widest at the top.
Oblong. With nearly parallel sides and longer than wide.
Obovate. Inversely ovate; widest at the tip.
Obtuse. Rounded or blunt at the tip.
Ochroleucous. Yellowish-white.
Ocrea. A papery sheath surrounding the stem formed from the stipules, as in many members of the Polygonaceae.
Opposite. With two leaves or branches per node.
Orbicular. Nearly circular.
Oval. Broadly elliptic, widest in the middle.
Ovary. The part of the pistil containing the ovules.
Ovate. Egg-shaped; widest at the base.
Ovule. Immature seed.
Palate. Rounded projection on the inside lower lip of a 2-lipped corolla at the throat.
Palea. Uppermost bract in a grass floret.
Palmate. Divided or with veins arising from the same point (in a hand-like manner).
Panicle. A compound raceme with pedicellate flowers.
Papillate. With small, pimple-like projections.
Pappus. Modified calyx in the Asteraceae, consisting of a crown, bristles, scales, or awns at the top of the achene.
Parasitic. Attached to and obtaining nutrients from another plant.
Parietal placentation. With ovules borne on the wall of the ovary.
Pectinate. Pinnatifid with very narrow, close divisions (resembling a comb).
Pedicel. The stalk of an individual flower.
Peduncle. The stalk of the entire inflorescence.
Peltate. Attached with a stalk near the center rather than on the edge; shield-shaped.

Pendant. Hanging down.
Pepo. Fleshy, indehiscent fruit with a leathery or hard rind; fruit of the Cucurbitaceae family.
Perennial. A plant that lives for 2 or more years.
Perfect. A flower with both stamens and a pistil.
Perfoliate. A sessile leaf in which the base completely surrounds the stem.
Perianth. The calyx and corolla together.
Pericarp. The fruit wall.
Perigynium. The sheath enclosing the ovary and fruit in *Carex* and *Kobresia*.
Perigynous. A flower with a superior ovary and a hypanthium present.
Petal. One member of the corolla.
Petaloid. Resembling petals.
Petiole. Stalk of a leaf.
Petiolule. Stalk of a leaflet.
Phyllodia. Broad leaf-like petioles resembling narrow leaf blades.
Pilose. Covered with long, soft hairs.
Pinnate. Compound leaf with the leaflets arranged on both sides of the axis.
Pinnatifid. Divided or cleft halfway or more to the base or midrib in a pinnate manner.
Pistil. The organ of the flower consisting of the ovary, style, and stigma and containing the ovules.
Pistillate. A flower with pistils but lacking stamens.
Pith. Spongy center of a stem.
Placenta. The point of attachment for ovules within the ovary.
Placentation. Pattern of attachment of ovules within an ovary.
Plait. A flattened fold (like a fan).
Plano-convex. Flattened on one side and rounded on the other.
Plicate. Folded into plaits as in a fan.
Plumose. Feather-like.
Pod. A dry, dehiscent fruit with one carpel; legume.
Pollen. Dust-like grains borne in the anthers or in microsporangia of gymnosperms.
Pollinia. A mass of coherent pollen grains.
Polygamodioecious. Mostly dioecious but with a few bisexual flowers also present.
Pome. A fleshy fruit derived from an inferior ovary with more than one locule (e.g., apple).
Precocious. Developing early, appearing before the leaves.
Prickle. A small, sharp outgrowth of the epidermis.
Procumbent. Lying on the ground, creeping.
Prophyllum. Bracteole at the base of an individual flower (in *Juncus*).
Prostrate. Lying flat on the ground.
Pruinose. With a bluish-white bloom on the surface, a waxy, powdery secretion.
Pseudoscape. A false, naked scape between the roots and leaves.
Puberulent. Minutely hairy.
Pubescent. Hairy.
Pulvinate. Cushion-shaped.
Punctate. With colored or translucent dots or pits.
Pungent. Sharp-pointed.
Pustulate. With pimple-like elevations; hairs with an expanded or pimple-like base.
Pyriform. Pear-shaped.
Pyxis. A circumscissile capsule.
Raceme. Inflorescence with the stalked flowers all arising from the main axis individually.
Rachilla. Axis of a grass spikelet.
Rachis. The axis of an inflorescence or compound leaf.
Ray. Flower in the Asteraceae with a ligulate corolla, usually found along the margin of the head when disk flowers are also present.
Receptacle. Expanded part of the axis bearing the flower parts.
Recurved. Curved backward or downward.
Reflexed. Bent abruptly downward.
Regular. Actinomorphic; radially symmetrical.
Remote. Distantly spaced.
Reniform. Kidney-shaped.
Repand. Wavy.
Replum. The partition between two halves of a silique or silicle (Brassicaceae).
Resinous. Producing or containing resin.
Reticulate. Net-like.
Retrorse. Pointed downward or backward.
Retuse. With a rounded tip that has a shallow notch or indent.
Revolute. With the margins rolled toward the underside.
Rhizomatous. With rhizomes.
Rhizome. Underground, creeping stem.
Rhombic. Diamond-shaped.
Rib. A prominent nerve or vein.
Root. Underground portion of the plant; lacking nodes and internodes.
Rosette. A cluster of leaves radiating out in a circle, usually at the base of the plant.
Rostellum. In Orchidaceae, the extension from the upper edge of the stigma.
Rosulate. With a rosette.
Rotate. A corolla with a short tube and flattened limb.
Rotund. Round.
Rugose. Wrinkled.
Rugulose. Slightly wrinkled.
Runcinate. Sharply incised or serrate, with teeth pointing backward or toward the base.
Saccate. Sac-shaped.
Sagittate. Shaped like an arrowhead, with the basal lobes pointing downward.
Salverform. Corolla with an slender tube and abruptly flattened, expanded limb.
Samara. A dry, indehiscent, winged fruit.
Saprophytic. Living on dead organic matter.
Scaberulous. Slightly scabrous.
Scabrous. Rough to the touch.
Scale. A thin, dry, often membranous structure.
Scape. Flowering stem lacking leaves.
Scapose. With a scape.
Scarious. Thin, dry, membranous, and translucent.
Schizocarp. A dry, dehiscent fruit that splits into 2 or more indehiscent segments.
Scorpioid. Coiled at the tip in 2 rows along the outer side.
Scrotiform. Sac-like.
Scurfy. Covered with scale-like particles.
Scutellum. A shield-like protrusion on the calyx in Lamiaceae flowers.

Secund. Present on one side of the axis.
Sepal. One member of the calyx; outermost whorl of flower parts.
Sepaloid. Resembling a sepal.
Septate. Divided by partitions.
Septum. A partition or cross wall.
Seriate. Arranged in a series of rows.
Sericeous. Silky with soft, appressed hairs.
Serrate. With sharp teeth pointing forward.
Serrulate. Serrate with very small teeth.
Sessile. Attached directly, without a stalk.
Seta. A bristle.
Setaceous. Bristle-like.
Sheath. Portion of the grass leaf that surrounds the stem; tubular structure surrounding a plant part.
Sigmoid. Shaped like the letter S.
Silicle. A short silique.
Silique. Fruit of the Brassicaceae, elongate, with a membranous partition and ovules borne at its points of attachment to the fruit wall.
Simple. Not branched or compound.
Sinuate. Wavy-margined.
Sinus. Space or depression between 2 lobes.
Sorus. Cluster of sporangia.
Spathe. A leaf-like bract surrounding an inflorescence or smaller bract below the flower (e.g., *Iris*).
Spatulate. Spoon-shaped.
Spicate. Spike-like.
Spike. An unbranched inflorescence with sessile flowers.
Spikelet. In Poaceae, the portion consisting of the glumes and enclosed florets; a small or secondary spike.
Spinulose. Covered with small spines.
Sporangium. A body containing spores.
Spore. A reproductive body capable of giving rise to a new individual (ferns).
Sporocarp. An organ containing sporangia.
Sporophyll. A leaf that bears sporangia.
Spur. A tubular or sac-like projection of a petal or sepal.
Squamella. A tiny or secondary scale.
Stamen. The pollen-bearing organ of a flower.
Staminate. A flower with stamens but no pistil.
Staminode. A sterile stamen often lacking an anther, sometimes modified and petaloid.
Stellate. Star-shaped.
Stem. Main axis of the plant, contains nodes and internodes.
Sterile. Lacking reproductive parts.
Stigma. The tip of the pistil that receives the pollen.
Stipe. A stalk; petiole of a fern frond.
Stipitate. Borne on a stipe.
Stipule. Pair of appendages at the base of a petiole, varying from glandular to leaf-like.
Stolon. Above ground, horizontal stem that roots at the nodes and/or tip.
Stoloniferous. With stolons.
Stramineous. Straw-colored.
Striate. With fine, longitudinal lines or ridges.
Strigose. With sharp, stiff, straight, appressed hairs.
Style. The usually elongated portion of the pistil between the ovary and the stigma.
Stylopodium. A disk-like expansion at the base of the style in some Apiaceae flowers.
Subulate. Awl-shaped, narrow and gradually tapering to a sharp point.
Succulent. Thick and fleshy, juicy.
Suffrutescent. Somewhat woody.
Sulcate. Furrowed or grooved lengthwise.
Superior. Ovary with the perianth parts inserted below it.
Suture. Line of dehiscence.
Symmetrical. Regular in number and size of parts.
Sympetalous. With the petals united at least partially.
Talus. Rock slide.
Taproot. Primary root, larger than any branches of the root system.
Tawny. Dull yellowish-brown.
Tendril. A slender, twisting outgrowth of a leaf or stem, used by the plant for support.
Tepal. One member from a perianth with undifferentiated calyx and corolla, in two separate whorls.
Terete. Cylindrical, round in cross-section.
Ternate. In 3's.
Terrestrial. Growing on land.
Tetradynamous. With 4 long and 2 shorter stamens.
Thallus. A flat, leaf-like structure which is not differentiated into stem and leaves.
Throat. The opening of a corolla with united petals.
Thyrse. A compact, densely flowered panicle.
Tomentose. Densely pubescent with whitish, woolly, tangled hairs.
Tortuous. Twisted.
Torulose. Irregularly swollen and constricted at close intervals; knobby.
Translucent. Thin, transmitting light through.
Transverse. Horizontal; crosswise.
Tree. Woody plant with usually one or a few stout trunks.
Trichome. A hair or bristle.
Trifoliate. A compound leaf with 3 leaflets.
Trifurcate. With 3 branches.
Trigonous. 3-angled.
Truncate. Ending abruptly, as if cut off nearly straight across.
Tube. The basal portion of fused corollas.
Tuber. A thick, short, underground stem (e.g., potato).
Tubercle. A small, rounded bump; nipple-like structure on the stems of some cacti; enlarged base of the style in *Eleocharis*.
Turgid. Swollen by inner pressure.
Turion. A scaly, short shoot arising from a bud on a rhizome.
Ultimate segment. Segment or division that bears no smaller segments.
Umbel. Inflorescence in which pedicels or peduncles arise from a common point.
Umbo. Blunt protuberance at the ends of the scales of some pine cones.
Uncinate. Hooked at the tip.
Undulate. Slightly wavy.
Uniseriate. In a single series or row.
Unisexual. A flower with either pistils or stamens.
Urceolate. Urn-shaped.
Utricle. A bladder-like, indehiscent, 1-seeded fruit.
Valvate. Opening by valves; splitting open along regular vertical lines.

Valve. Inner perianth segments (e.g., *Rumex*) which enclose the achene.
Vein. Thread of vascular tissue in a leaf or flower.
Velutinous. Velvety.
Venation. Arrangement of veins.
Ventral. The lower or inner surface of a structure (pointing toward the axis).
Verticil. A whorl.
Verticillaster. A pair of cymes arising from the axils of opposite leaves or bracts and forming a false whorl.
Villous. With long, soft, unmatted hairs.
Virgate. Long, slender, and straight.
Viscid. Sticky (usually from glands).
Whorled. With 3 or more leaves or branches at a node.
Wing. A thin, membranous extension of an organ; the lateral petals of a Fabaceae flower.
Woolly. With soft, matted, interwoven hairs.
Zygomorphic. Irregular.

Parts of a Flower:

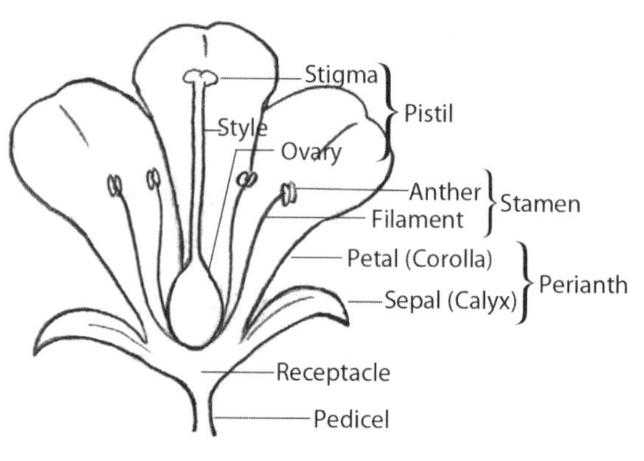

Floral Symmetry:

Actinomorphic

Zygomorphic

Placentation Types:

Parietal
1 locule
3 carpels

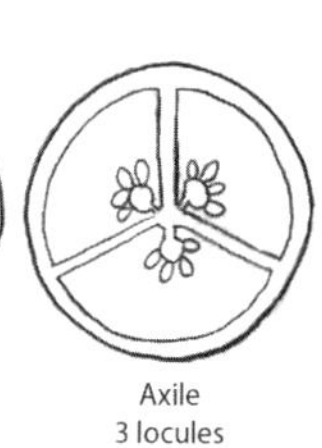
Axile
3 locules
3 carpels

Free-central
1 locule

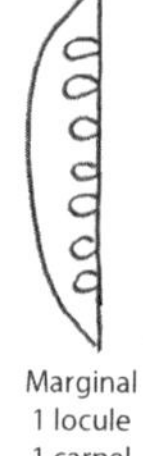
Marginal
1 locule
1 carpel

Ovary Position and Gynoecium Condition:

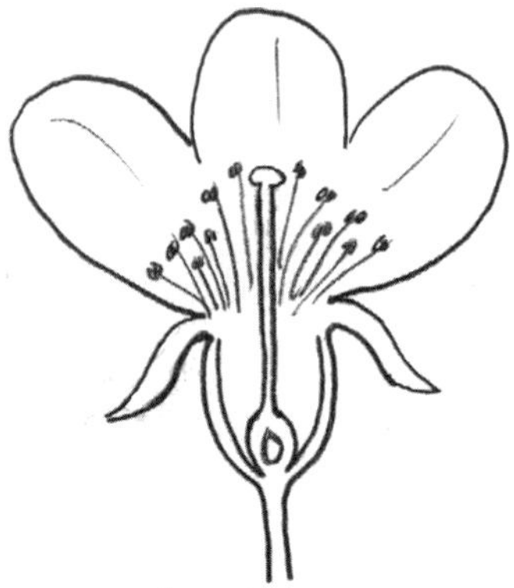
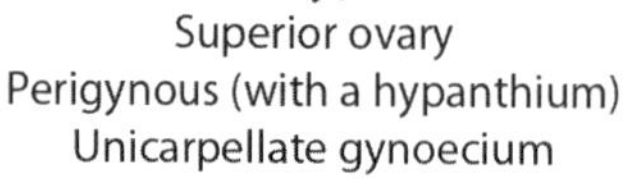
Superior ovary
Perigynous (with a hypanthium)
Unicarpellate gynoecium

Superior ovary
Hypogynous (lacking a hypanthium)
Polycarpous gynoecium

Inferior ovary
Epigynous
Syncarpous gynoecium

Inflorescence Types:

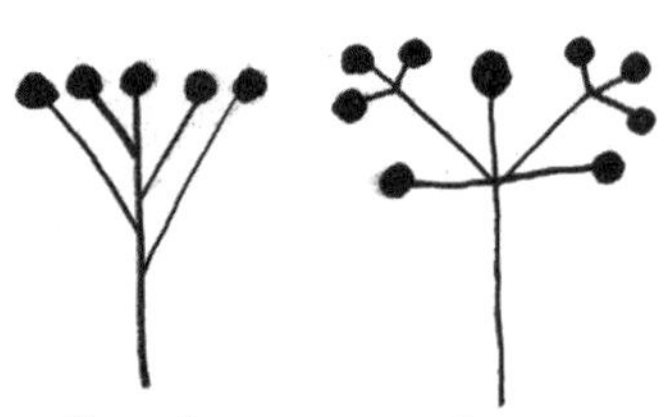
Corymb

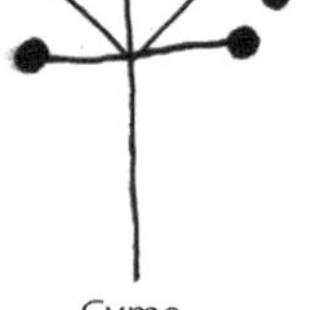
Cyme

Panicle

Raceme

Solitary

Spike

Umbel

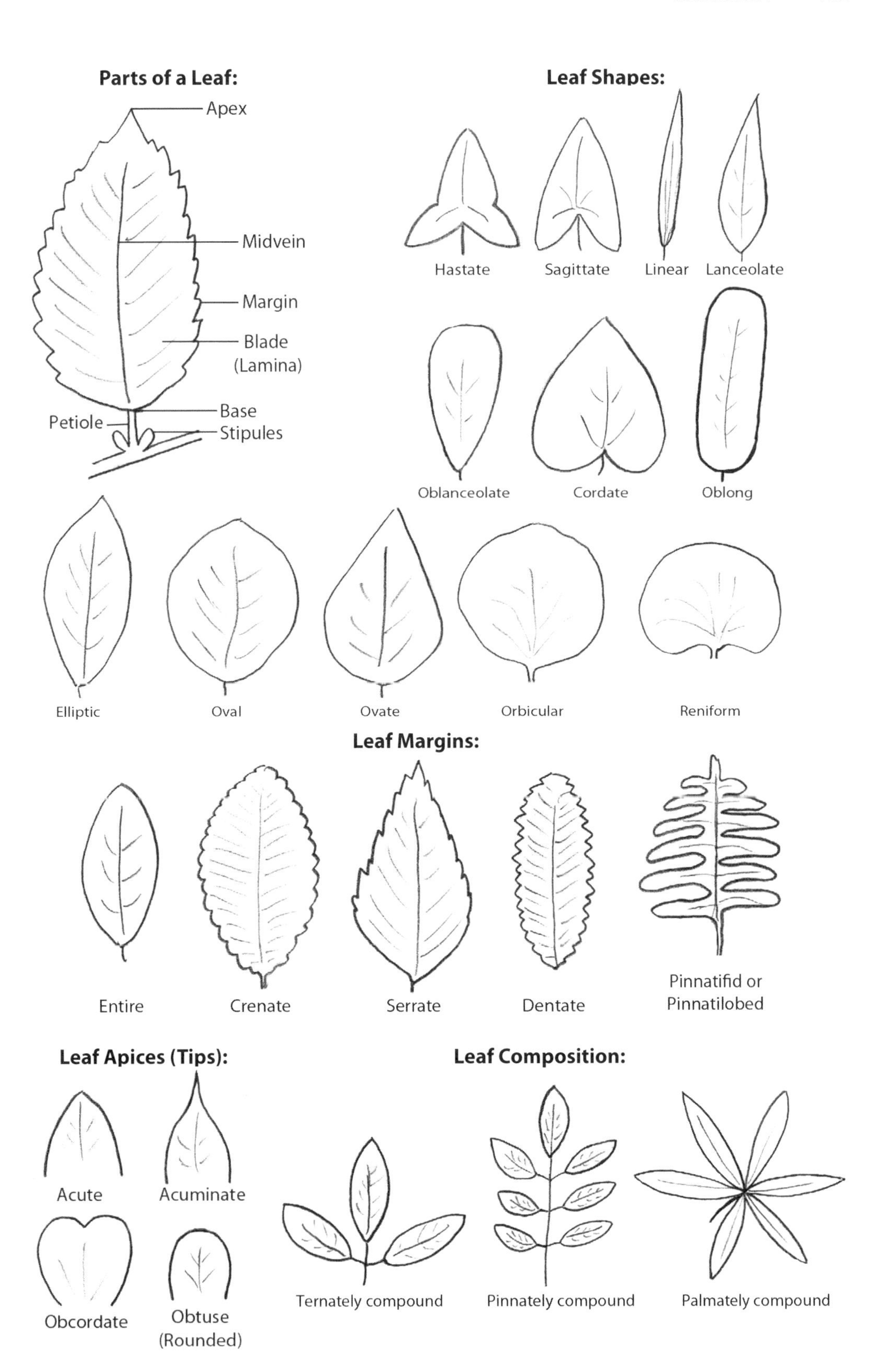
Parts of a Leaf:
Apex
Midvein
Margin
Blade
(Lamina)
Petiole
Base
Stipules
Leaf Shapes:
Hastate
Sagittate
Linear
Lanceolate
Oblanceolate
Cordate
Oblong
Elliptic
Oval
Ovate
Orbicular
Reniform
Leaf Margins:
Entire
Crenate
Serrate
Dentate
Pinnatifid or
Pinnatilobed
Leaf Apices (Tips):
Acute
Acuminate
Obcordate
Obtuse
(Rounded)
Leaf Composition:
Ternately compound
Pinnately compound
Palmately compound

REFERENCES

Ackerfield, J. & W.F. Jennings. 2008. The genus *Abronia* (Nyctaginaceae) in Colorado, with notes on *Abronia bolackii* in New Mexico. J. Bot. Res. Inst. Texas. 2(1):419–423.

Adams, J.E. 1940. A systematic study of the genus *Arctostaphylos* Adans. J. Elisha Mitchell Sci. Soc. 56:1–62.

Adams, R.P. 2014. Junipers of the world: The genus *Juniperus*. Trafford Publishing, Bloomington, Indiana, USA.

Aedo, C. 2000. The genus *Geranium* L. (Geraniaceae) in North America. I. Annual species. Anales Jard. Bot. Madrid 58(1):39–82.

Aedo, C. 2001. The genus *Geranium* L. (Geraniaceae) in North America. II. Perennial species. Anales Jard. Bot. Madrid 59(1):3–65.

Aellen, P. & T. Just. 1943. Key and synopsis of the American species of the genus *Chenopodium* L. Amer. Midl. Naturalist 30:47–76.

Ahrendt, L.W.A. 1961. *Berberis* and *Mahonia*: A taxonomic revision. Biol. J. Linn. Soc. 57(369):1–410.

Aiken, S.G. 1981. A conspectus of *Myriophyllum* (Haloragaceae) in North America. Brittonia 33:57–69.

Allen, G.A. 1985. The hybrid origin of *Aster ascendens* (Asteraceae). Amer. J. Bot. 72(2):268–277.

Allen, G.A., M.L. Dean, & K.L. Chambers. 1983. Hybridization studies in the *Aster occidentalis* (Asteraceae) polyploid complex of western North America. Brittonia 35(4):353–361.

Al-Shehbaz, I.A. 2010. Brassicaceae. In: Flora of North America Editorial Committee, eds. Flora of North America north of Mexico. Oxford University Press, New York, USA, and Oxford, U.K. 7:224–746.

Al-Shehbaz, I.A. & S.L. O'Kane. 2002. *Lesquerella* is united with *Physaria* (Brassicaceae). Novon 12:319–329.

Anderson, E.F. 2001. The Cactus family. Timber Press, Portland, Oregon, USA.

Anderson, L.C. 1978. New taxa in *Chrysothamnus*, section *Nauseosi* (Asteraceae). Phytologia 38(4):309–320.

Anderson, L.C. 1980. Identity of narrow-leaved *Chrysothamnus viscidiflorus* (Asteraceae). Great Basin Naturalist 40(2):117–120.

Anderson, L.C. *Chrysothamnus*-a key to the species. Unpublished manuscript.

Andresen, J.W. & R.J. Steinhoff. 1971. The taxonomy of *Pinus flexilis* and *P. strobiformis*. Phytologia 22(2):57–67.

Applegate, E.I. 1935. The genus *Erythronium*: A taxonomic and distributional study of the western North American species. Madroño 3(2):58–113.

Argus, G.W. 2010. *Salix*. In: Flora of North America Editorial Committee, eds. Flora of North America north of Mexico. Oxford University Press, New York, USA, and Oxford, U.K. 7:23–162.

Arp, G. 1973. Studies in the Colorado cacti V: The spineless hedgehog. Cact. Succ. J. (Los Angeles) 45:132–133.

Arp, G. 1973. Studies in the Colorado cactus IV: *Opuntia compressa* and *Opuntia macrorhiza*. Cact. Succ. J. (Los Angeles) 45:56–57.

Atwood, N.D. 1975. A revision of the *Phacelia* Crenulatae group (Hydrophyllaceae) for North America. Great Basin Naturalist 35(2):127–190.

Babcock, E.B. 1947. The genus *Crepis*. Univ. Calif. Publ. Bot. Vols. 21–22.

Baird, G.I. 2006. *Agoseris*. In: Flora of North America Editorial Committee, eds. Flora of North America north of Mexico. Oxford University Press, New York, USA, and Oxford, U.K. 19–21:323–335.

Baker, M.S. 1935. Studies in western violets—I. Madroño 3(2):51–56.

Baker, M.S. 1940. Studies in western violets—III. Madroño 5(7):218–231.

Baker, M.S. 1957. Studies in western violets—VIII. Brittonia 9:217–230.

Baldwin, B.G., D.H. Goldman, D.J. Keil, R. Patterson, T.J. Rosatti, & D.H. Wilken, eds. 2012. The Jepson manual: Vascular plants of California. Rev. 2, 2nd ed. University of California Press, Berkeley, California, USA.

Barkley, F.A. 1937. A monographic study of *Rhus*. Ann. Missouri Bot. Gard. 24:265–498.

Barkley, T.M. 1960. *Senecio*. North American Flora, Series II. 10:50–139.

Barkoudah, Y.I. 1962. A revision of *Gypsophila*, *Bolanthus*, *Ankyropetalum* and *Phyrana*. Wentia 9:1–203.

Barkworth, M.E., K.M. Capels, S. Long, L.K. Anderton, & M.B. Piep, eds. 2003. Magnoliophyta: Commelinidae (in part): Poaceae, part 2. Vol. 25. Flora of North America north of Mexico. Oxford University Press, New York, USA, and Oxford, U.K.

Barkworth, M.E., K.M. Capels, S. Long, & M.B. Piep, eds. 2007. Magnoliophyta: Commelinidae (in part): Poaceae, part 1. Vol. 24. Flora of North America north of Mexico. Oxford University Press, New York, USA, and Oxford, U.K.

Barneby, R.C. 1952. A revision of the North American species of *Oxytropis* DC. Proc. Calif. Acad. Sci. 27(7):177–312.

Barneby, R.C. 1964. Atlas of North American *Astragalus*. Mem. New York Bot. Gard. 13:1–1188.

Barneby, R.C. 1977. Dalea Imagines. Mem. New York Bot. Gard. 27:1–892.

Bassett, I.J. & C.W. Crompton. 1982. The genus *Chenopodium* in Canada. Canad. J. Bot. 60:586–610.

Bassett, I.J., C.W. Crompton, J. McNeill, and P.M. Taschereau. 1983. The genus *Atriplex* L. (Chenopodiaceae) in Canada. Agric. Canada Monogr. 31:1–72.

Bayer, R.J. 1982. A revised classification of *Antennaria* (Asteraceae: Inuleae) of eastern United States. Syst. Bot. 7:300–313.

Bayer, R.J. 1987. Morphometric analysis of western North American *Antennaria* (Asteraceae: Inuleae). I. Sexual species of Sections *Alpinae*, *Dioicae*, and *Plantaginifoliae*. Canad. J. Bot. 65:2389–2395.

Beal, E.O. 1956. Taxonomic revision of the genus *Nuphar* Sm. of North America and Europe. J. Elisha Mitchell Sci. Soc. 72:317–346.

Beaman, J.H. 1957. The systematics and evolution of *Townsendia* (Compositae). Contr. Gray Herb. 183:1–151.

Beaman, J.H. 1990. Revision of *Hieracium* (Asteraceae) in Mexico and Central America. Syst. Bot. Monogr. 29:1–77.

Beckmann, R.L., Jr. 1979. Biosystematics of the genus *Hydrophyllum* L. (Hydrophyllaceae). Amer. J. Bot. 66(9):1053–1061.

Beetle, A.A. 1940. Studies in the genus *Scirpus* L. I. Delimination of the subgenera *Euscirpus* and *Aphylloides*. Amer. J. Bot. 27(2):63–64.

Beetle, A.A. 1941. Studies in the genus *Scirpus* L. II. The section Baeothryon Ehrh. Amer. J. Bot. 28(6):469–476.

Beetle, A.A. 1941. Studies in the genus *Scirpus* L. III. The American species of the section *Lacustres* Clarke. Amer. J. Bot. 28(8):691–700.

Beetle, A.A. 1942. Studies in the genus *Scirpus* L. IV. The section Bolboschoenus Palla. Amer. J. Bot. 29(1):82–88.

Beetle, A.A. 1942. Studies in the genus *Scirpus* L. V. Notes on the section Actaeogeton Reich. Amer. J. Bot. 29(8):653–656.

Beetle, A.A. 1943. Studies in the genus *Scirpus* L. VI. The section*Schoenoplectus* Palla. Amer. J. Bot. 30(6):395–401.

Beetle, A.A. 1944. Studies in the genus *Scirpus* VII. Conspectus of sections represented in the Americas. Amer. J. Bot. 31(5):261–265.

Beetle, A.A. 1960. A study of sagebrush, the section *Tridentatae* of *Artemisia*. Bull. Wyoming Agric. Exp. Sta. B-779:1–68.

Beidleman, R.G. 1960. The Colorado blue spruce. The Green Thumb 17(3):81–84.

Benson, L. 1961. A revision and amplification of *Pediocactus* I. Cact. Succ. J. (Los Angeles) 33(2):49–54.

Benson, L. 1962. A revision and amplification of *Pediocactus* II and III. Cact. Succ. J. (Los Angeles) 34(1):17–19, 57–61.

Benson, L. 1962. A revision and amplification of *Pediocactus* IV. Cact. Succ. J. (Los Angeles) 34(6):163–168.

Berger, A. 1924. A taxonomic review of currants and gooseberries. Techn. Bull. New York State Agric. Exp. Sta. 109:3–118.

Biddulph, S.F. 1944. A revision of the genus *Gaillardia*. Res. Studies State Coll. Wash. 12(4):195–256.

Bierner, M.W. 1972. Taxonomy of *Helenium* sect. *Tetrodus* and a conspectus of North American *Helenium* (Compositae). Brittonia 24:331–355.

Bierner, M.W. 1994. Submersion of *Dugaldia* and *Plummera* in *Hymenoxys* (Asteraceae: Helenieae: Gaillardiinae). Sida 16(1):1–8.

Bierner, M.W. 2001. Taxonomy of *Hymenoxys* subgenus *Picradenia* and a conspectus of the subgenera of *Hymenoxys* (Asteraceae: Helenieae: Tetraneurinae). Lundellia 4:37–63.

Bierner, M.W. 2004. Taxonomy of *Hymenoxys* subgenus *Macdougalia* (Asteraceae: Helenieae: Tetraneurinae). Sida 21(2):657–663.
Bierner, M.W. & B.L. Turner. 2003. Taxonomy of *Tetraneuris* (Asteraceae: Helenieae: Tetraneurinae). Lundellia 6:44–96.
Bierner, M.W. & R.K. Jansen. 1998. Systematic implications of DNA restriction site variation in *Hymenoxys* and *Tetraneuris* (Asteraceae, Helenieae, Gaillardiinae). Lundellia 1:17–26.
Blasdell, R.F. 1963. A monographic study of the fern genus *Cystopteris*. Mem. Torrey Bot. Club 21(4):1–102.
Bock, J.H. & C.E. Bock. 1998. Tallgrass prairie: Remnants and relicts. Great Plains Res. 8:213–230.
Bogin, C. 1955. Revision of the genus *Sagittaria* (Alismataceae). Mem. New York Bot. Gard. 9(2):179–233.
Boissevain, C.H. & C. Davidson. 1940. Colorado cacti: An illustrated guide describing all of the native Colorado cacti. Abbey Garden Press, Pasadena, California, USA.
Boivin, B. 1956. *Stellaria* sectio Umbellatae schischkin (Caryophyllaceae). Svensk Bot. Tidskr. 50(1):113–114.
Boufford, D.E. 1983. The systematics and evolution of *Circaea* (Onagraceae). Ann. Missouri Bot. Gard. 69(4):804–994.
Brainerd, E. 1921. Violets of North America. Bull. Vermont Agri. Exp. Sta. 224.
Brashier, C.K. 1966. A revision of *Commelina* (Plum.) L. in the U.S.A. Bull. Torrey Bot. Club 93:1–19.
Brooks, R.E. & S.E. Clemants. 2000. *Juncus*. In: Flora of North America Editorial Committee, eds. Flora of North America north of Mexico. Oxford University Press, New York, USA, and Oxford, U.K. 22:211–255.
Brouillet, L. 2006. *Taraxacum*. In: Flora of North America Editorial Committee, eds. Flora of North America north of Mexico. Oxford University Press, New York, USA, and Oxford, U.K. 19:239–252.
Brouillet, L. 2014. *Potentilla*. In: Flora of North America Editorial Committee, eds. Flora of North America north of Mexico. Oxford University Press, New York, USA, and Oxford, U.K. 9:119–218.
Brouillet, L. & P.E. Elvander. 2009. *Saxifraga*. In: Flora of North America Editorial Committee, eds. Flora of North America north of Mexico. Oxford University Press, New York, USA, and Oxford, U.K. 8:132–146.
Brown, G.D. 1956. Taxonomy of American *Atriplex*. Amer. Midl. Naturalist 55:199–210.
Brown, G.K. & G.L. Nesom. 2006. *Oönopsis*. In: Flora of North America Editorial Committee, eds. Flora of North America north of Mexico. Oxford University Press, New York, USA, and Oxford, U.K. 20:410–412.
Camp, W.H. 1945. The North American blueberries with notes on other groups of Vacciniaceae. Brittonia 5:203–275.
Canne, J.M. 1977. A revision of the genus *Galinsoga* (Compositae: Heliantheae). Rhodora 79:319–389.
Card, H.H. 1931. A revision of the genus *Frasera*. Ann. Missouri Bot. Gard. 18:245–282.
Carolin, R.C. 1987. A review of the family Portulacaceae. Austral. J. Bot. 35:383–412.
Carter, J.L. 2006. Trees and shrubs of Colorado. Mimbres Publishing, Silver City, New Mexico, USA.
Carter, M.E.B. & W.H. Murdy. 1985. Systematics of *Talinum parviflorum* Nutt. and the origin of *T. teretifolium* Pursh (Portulacaceae). Rhodora 87:121–130.
Cassidy, J. 1888a. Insects native to Colorado as well as those injurious to useful vegetation and the means of controlling them. Bull. Colorado Agric. Exp. Sta. 6:1–20.
Cassidy, J. 1888b. Sugar beet and potato farming. Bull. Colorado Agric. Exp. Sta. 7:1–28.
Cassidy, J. & D. O'Brine. 1889. Some Colorado grasses. Bull. Colorado Agric. Exp. Sta. 12:1–138.
Chaudhri, M.N. 1968. A revision of the Paronychiinae. Meded. Bot. Mus. Herb. Rijks Univ. Utrecht 285:306–397.
Chen, Z., S.R. Manchester, & H. Sun. 1999. Phylogeny and evolution of the Betulaceae as inferred from DNA sequences, morphology, and paleobotany. Amer. J. Bot. 86:1168–1181.
Chinnappa, C.C. & J.K. Morton. 1991. Studies on the *Stellaria longipes* complex (Caryophyllaceae): Taxonomy. Rhodora 93:129–135.
Cholewa, A.F. & D.M. Henderson. 1984. Biosystematics of *Sisyrinchium* section *Bermudiana* (Iridaceae) of the Rocky Mountains. Brittonia 36(4):342–363.
Chuang, T.I. & L.R. Heckard. 1992. A taxonomic revision of *Orthocarpus* (Scrophulariaceae—Tribe Pediculareae). Syst. Bot. 17:560–582.
Clausen, R.T. 1938. A monograph of the Ophioglossaceae. Mem. Torrey Bot. Club. 19(2):1–177.
Clausen, R.T. 1975. *Sedum* of North America north of the Mexican Plateau. Cornell Univ. Press, Ithaca, New York, USA.
Colorado Natural Heritage Program (CNHP). 2005. Ecological system descriptions and viability guidelines for Colorado. Colorado Natural Heritage Program, Colorado State University, Fort Collins, Colorado, USA.
Constance, L. 1949. A revision of *Phacelia* subgenus *Cosmanthus* (Hydrophyllaceae). Contr. Gray Herb. 168:3–48.
Constance, L. & R.H. Shan. 1948. The genus *Osmorhiza* (Umbelliferae): A study in geographic affinities. Univ. Calif. Publ. Bot. 23(3):111–156.
Cook, C.D.K. & M.S. Nicholls. 1986. A monographic study of the genus *Sparganium* (Sparganiaceae). Part 1. Subgenus *Xanthosparganium* Holmberg. Bot. Helv. 96:213–267.
Cook, C.D.K. & M.S. Nicholls. 1986. A monographic study of the genus *Sparganium* (Sparganiaceae). Part 1. Subgenus *Sparganium* Holmberg. Bot. Helv. 97:1–44.
Cook, C.D.K. & K. Urmi-König. 1985. A revision of the genus *Elodea* (Hydrocharitaceae). Aquatic Bot. 21:111–156.
Core, E.L. 1941. The North American species of *Paronychia*. Amer. Midland Naturalist 26:369–397.
Correll, D.S. 1943. The genus *Habenaria* in western North America. Leafl. W. Bot. 3(11):233–247.
Coulter, J.M. & J.N. Rose. 1900. Monograph of the North American Umbelliferae. Contr. U.S. Natl. Herb. 7(1):1–256.
Coville, F.V. 1896. *Juncus confusus*, a new rush from the Rocky Mountain region. Proc. Biol. Soc. Wash. 10:127–130.
Crawford, D.J. 1975. Systematic relationships in the narrow-leaved species of *Chenopodium* of the western United States. Brittonia 27:279–288.
Crawford, D.J. 1977. A study of morphological variability in *Chenopodium incanum* (Chenopodiaceae) and the recognition of two new varieties. Brittonia 29:291–296.
Crawford, D.J. & H.D. Wilson. 1986. *Chenopodium*. In: Great Plains Flora Association, R.L. McGregor, & T.M. Barkley, eds. Flora of the Great Plains. Univ. Press of Kansas, Lawrence, Kansas, USA. Pp. 166–173.
Critchfield, W.B. & E.L. Little, Jr. 1966. Geographic distribution of the pines of the world. USDA Forest Service, Misc. Publ. 991.
Croat, T. 1972. *Solidago canadensis* complex of the Great Plains. Brittonia 24:317–326.
Croizat, L. 1945. *Euphorbia esula* in North America. Amer. Midland Naturalist 33(1):231–243.
Cronquist, A. 1943. Revision of the western North American species of *Aster* centering about *Aster foliaceus* Lindl. Amer. Midland Naturalist 29(2):429–468.
Cronquist, A. 1947. A revision of the North American species of *Erigeron*, north of Mexico. Brittonia 6:121–302.
Cronquist, A. 1994. *Cirsium*. In: A. Cronquist, A.H. Holmgren, N.H. Holmgren, J.L. Reveal, & P.K. Holmgren, eds. Intermountain Flora. New York Botanical Garden Press, Bronx, New York, USA. 5:388–414.
Cronquist, A. 1997. Apiaceae. In: A. Cronquist, N.H. Holmgren, & P.K. Holmgren, eds. Intermountain Flora. New York Botanical Garden Press, Bronx, New York, USA. 3(A):340–427.
Cronquist, A., A.H. Holmgren, N.H. Holmgren, J.L. Reveal, & P.K. Holmgren, eds. 1972. Intermountain Flora: Vascular Plants of the Intermountain West, U.S.A. Vol. 1. Geological and botanical history of the region, its plant geography and a glossary. The vascular cryptogams and the gymnosperms. Hafner Publishing, New York. New York. USA.

Cronquist, A., A.H. Holmgren, N.H. Holmgren, J.L. Reveal, & P.K. Holmgren, eds. 1977. Intermountain Flora: Vascular Plants of the Intermountain West, U.S.A. Vol. 6. The Monocotyledons. Columbia University Press, New York, New York, USA.

Cronquist, A., A.H. Holmgren, N.H. Holmgren, J.L. Reveal, & P.K. Holmgren, eds. 1984. Intermountain Flora: Vascular Plants of the Intermountain West, U.S.A. Vol. 4. Subclass Asteridae (except Asteraceae). New York Botanical Garden Press, Bronx, New York, USA.

Cronquist, A., A.H. Holmgren, N.H. Holmgren, J.L. Reveal, & P.K. Holmgren, eds. 1994. Intermountain Flora: Vascular Plants of the Intermountain West, U.S.A. Vol. 5. Asterales. New York Botanical Garden Press, Bronx, New York, USA.

Cronquist, A., N.H. Holmgren, & P.K. Holmgren, eds. 1997. Intermountain Flora: Vascular Plants of the Intermountain West, U.S.A. Vol. 3, Part A. Subclass Rosidae (except Fabales). New York Botanical Garden Press, Bronx, New York, USA.

Cronquist, A. & M. Ownbey. 1977. *Allium*. In: A. Cronquist, A.H. Holmgren, N.H. Holmgren, J.L. Reveal, & P.K. Holmgren, eds. Intermountain Flora. Columbia University Press, New York, New York, USA. 6:508–522.

Crow, G.E. 1978. A taxonomic revision of *Sagina* (Caryophyllaceae) in North America. Rhodora 80:1–91.

Crow, G.E. & C.B. Hellquist. 1985. Aquatic vascular plants of New England: Part 8. Lentibulariaceae. Bull. New Hampshire Agric. Exp. Sta. 528:1–22.

Culver, D.R. & J.M. Lemly. 2013. Field guide to Colorado's wetland plants. Colorado State University, Fort Collins, Colorado, USA.

Cutler, H.C. 1939. Monograph of the North American species of the genus *Ephedra*. Ann. Missouri Bot. Gard. 26:373–427.

Darlington, J. 1934. A monograph of the genus *Mentzelia*. Ann. Missouri Bot. Gard. 21(1):103–226.

Davidson, B.L. 2000. Lewisias. Timber Press, Portland, Oregon, USA.

Davis, R.J. 1966. The North American perennial species of *Claytonia*. Brittonia 18:285–303.

Dean, M.L. & K.L. Chambers. 1983. Chromosome numbers and evolutionary patterns in the *Aster occidentalis* (Asteraceae) polyploid complex of western North American. Brittonia 35(3):189–196.

Dempster, L. & F. Ehrendorfer. 1965. Evolution of the *Galium multiflorum* complex in western North America. II. Critical taxonomic revision. Brittonia 17:289–334.

Denton, M.F. 1973. A monograph of *Oxalis,* section *Ionoxalis* (Oxalidaceae) in North America. Publ. Mus. Michigan State Univ., Biol. Ser. 4(10):459–615.

Detling, L.E. 1939. A revision of the North American species of*Descurainia*. Amer. Midland Naturalist 22(3):481–520.

Detling, L.E. 1951. The cespitose lupines of Western North America. Amer. Midland Naturalist 45(2):474–499.

Dietrich, W. & W.L. Wagner. 1988. Systematics of *Oenothera* section *Oenothera* subsection *Raimannia* and subsection *Nutantigemma* (Onagraceae). Syst. Bot. Monogr. 24:1–91.

Donoghue, M.J., R.G. Olmstead, J.F. Smith, & J.D. Palmer. 1992. Phylogenetic relationships of Dipsacales based on *rbc*L sequences. Ann. Missouri Bot. Gard. 79:333–345.

Dorn, R.D. 1972. The nomenclature of *Isoetes echinospora* and *Isoetes muricata*. Amer. Fern J. 62:80–81.

Dorn, R.D. 1997. Rocky Mountain region willow identification field guide. USDA Forest Service, Fort Collins, Colorado, USA.

Dorn, R.D. 2001. Vascular plants of Wyoming, 3rd edition. Mountain West Publishing, Cheyenne, Wyoming, USA.

Dorn, R.D. & R.L. Hartman. 1988. Notes: Nomenclature of *Lomatium nuttallii*, *L. kingii*, and *L. megarrhizum* (Apiaceae). Madroño 35(1):70–71.

Dugal, J.R. 1966. A taxonomic study of western Canadian species in the genus *Betula*. Canad. J. Bot. 44:929–1007.

Duke, J.A. 1961. Preliminary revision of the genus *Drymaria*. Ann. Missouri Bot. Gard. 48:173–268.

Dunford, M.P. 1986. Chromosome relationships of diploid species of *Grindelia* (Compositae) from Colorado, New Mexico, and adjacent areas. Amer. J. Bot. 73(2):297–303.

Dunn, D.B. 1959. *Lupinus pusillus* and its relationship. Amer. Midland Naturalist 62(2):500–510.

Eastwood, A. 1893. A popular flora of Denver, Colorado. Zoe Publishing Co., San Francisco, California, USA.

Eastwood, A. 1903. New species of *Oreocarya*. Bull. Torrey Bot. Club 30:238–246.

Eastwood, A. 1934. A revision of *Arctostaphylos* with key and descriptions. Leafl. W. Bot. 1:105–128.

Ellison, W.L. 1964. A systematic study of the genus *Bahia* (Compositae). Rhodora 66:67–86, 177–215.

Elvander, P.E. 1984. The taxonomy of *Saxifraga* (Saxifragaceae) section *Borophila*, subsection *Integrifoliae* in western North America. Syst. Bot. Monogr. 3:1–44.

Ensign, M. 1942. A revision of the celastraceous genus *Forsellesia* (*Glossopetalon*). Amer. Midl Naturalist 27:501–511.

Epling, C.C. 1925. Monograph of the genus *Monardella*. Ann. Missouri Bot. Gard. 12:1–106.

Epling, C.C. 1942. The American species of *Scutellaria*. Univ. Calif. Publ. Bot. 20(1):1–146.

Eriksson, T., M.S. Hibbs, A.D. Yoder, C.F. Delwiche, & M.J. Donoghue. 2003. The phylogeny of Rosoideae (Rosaceae) based on sequences of the internal transcribed spacer (ITS) of nuclear ribosomal DNA and the *TRNL/F* region of chloroplast DNA. Int. J. Pl. Sci. 164:197–211.

Estes, J.R. 1969. Evidence for autoploid evolution in the *Artemisia ludoviciana* complex of the Pacific Northwest. Brittonia 21:29–43.

Ewan, J. 1942. A review of the North American weedy Heliotropes. Bull. S. Calif. Acad. Sci. 41(1):51–57.

Ewan, J. 1945. A synopsis of the North American species of *Delphinium*. Univ. Colorado Stud. Ser. D, Phys. Sci. 2:55–244.

Farrar, D. 2002. Systematics of western moonworts, *Botrychium* sub. *Botrychium*. Unpublished manuscript.

Farrar, D. & S. Popovich. 2002. *Botrychium*. In: W.A. Weber & R.C. Wittmann. 2012. Colorado flora: Eastern slope. University Press of Colorado, Boulder, Colorado, USA

Fassett, N.C. 1951. *Callitriche* in the New World. Rhodora 53:137–155, 161–182, 185–194, 209–222.

Fernald, M.L. 1926. *Urtica gracilis* and some related North American species. Rhodora 28:191–199.

Fernald, M.L. 1943. Notes on *Hieracium*. Rhodora 45:317–325.

Fernald, M.L. 1945. Some North American Corylaceae (Betulaceae). Rhodora 47:303–361.

Fernald, M.L. & K.M. Wiegand. 1914. The genus *Ruppia* in eastern North America. Rhodora 16:119–127.

Furlow, J.J. 1997. Betulaceae. In: Flora of North America Editorial Committee, eds. Flora of North America north of Mexico. Oxford University Press, New York, USA, and Oxford, U.K. 3:507–538.

Gaiser, L.O. 1946. The genus *Liatris*. Rhodora 48:165–183, 216–326, 331–382, 393–412.

Galloway, L.A. 1975. Systematics of North American desert species of *Abronia* and *Tripterocalyx*. Brittonia 27:328–347.

Gardner, R.C. 1974. Systematics of *Cirsium* (Compositae) in Wyoming. Madroño 22:239–265.

Gentry, J.L. 1974. Studies in the genus *Hackelia* (Boraginaceae) in the Western United States and Mexico. S. W. Naturalist 19(2):139–146.

Gentry, J.L. & R.L. Carr. 1976. A revision of the genus *Hackelia* (Boraginaceae) in North America, North of Mexico. Mem. New York Bot. Gard. 26(1):121–227.

Gill, L.S. 1935. *Arceuthobium* in the United States. Trans. Connecticut Acad. Arts Sci. 32:111–245.

Gottlieb, L.D. 1972. A proposal for classification of the annual species of *Stephanomeria* (Compositae). Madroño 21(7):463–481.

Grady, B.R. & S.L. O'Kane. 2007. New species and combinations in *Physaria* (Brassicaceae) from Western North America. Novon 17:182–192.

Graham, S.A. 1975. Taxonomy of the Lythraceae in the southeastern United States. Sida 6:80–103.

Grant, A. & V. Grant. 1956. Genetic and taxonomic studies in *Gilia*. VIII. The cobwebby gilias. Aliso 3:203–287.

Grant, A.L. 1924. A monograph of the genus *Mimulus*. Ann. Missouri Bot. Gard. 11:99–388.

Grant, V. 1956. A synopsis of *Ipomopsis*. Aliso 3:351–362.

Grant, V. & K.A. Grant. 1979. Systematics of the *Opuntia phaeacantha* group in Texas. Bot. Gaz. 140:199–207.
Gray, A. 1874. Contributions to the botany of North America. I. A synopsis of North American thistles. Proc. Amer. Acad. Arts Sci. 10:39–49.
Hall, H.M. 1928. The genus *Haplopappus*. A phylogenetic study in the Compositae. Section 6 Stenotopsis. Publ. Carnegie Inst. Washington 389(34):156–161.
Hall, H.M. & F.E. Clements. 1923. The phylogenetic method in taxonomy. The North American species of *Artemisia*, *Chrysothamnus*, and *Atriplex*. Publ. Carnegie Inst. Washington 326.
Halleck, D.K. & D. Wins. 1966. Taxonomic status of *Claytonia rosea* and *C. lanceolata* (Portulaceae). Ann. Missouri Bot. Gard. 53:205–212.
Haller, J.R. 1965. The role of two-needle fascicles in the adaptation and evolution of Ponderosa pine. Brittonia 17(4):354–382.
Hamet-Ahti, L. 1971. A synopsis of the species of *Luzula*, subgenus *Anthalaea* Griseb. (Juncaceae) indigenous in North America. Ann. Bot. Fenn. 8:368–381.
Hanks, L.T. & J.K. Small. 1907. Geraniaceae. In: North American Flora: Geraniales: Geraniaceae, Oxalidaceae, Linaceae, Erythrozylaceae. New York Botanical Garden, Bronx, New York, USA. 25(1):3–24.
Harms, V.L. 1965. Biosystematic studies in the *Heterotheca subaxillaris* complex (Compositae: Astereae). Trans. Kansas Acad. Sci. 68:244–257.
Harrington, H.D. 1946. Grasses of Colorado (including cultivated species). Colorado A&M College, Fort Collins, Colorado, USA. Pp. 1–167.
Harrington, H.D. 1950. Colorado ferns and fern allies. Colorado Agricultural Research Foundation, Fort Collins, Colorado, USA.
Harrington, H.D. 1954. Manual of the plants of Colorado. Sage Books, Denver, Colorado, USA.
Harrington, H.D. & L.W. Durrell. 1944. Key to some Colorado grasses in vegetative condition. Techn. Bull. Colorado Agric. Exp. Sta. 33:3–86.
Harris, J. & M. Harris. 2001. Plant identification terminology: An illustrated glossary. Spring Lake Publishing, Santaquin, Utah, USA.
Hartman, R.L. 1974. Rocky Mountain species of *Paronychia* (Caryophyllaceae): A morphological, cytological, and chemical study. Brittonia 26:256–263.
Hauke, R.L. 1993. Equisetaceae. In: Flora of North America Editorial Committee, eds. Flora of North America north of Mexico. Oxford University Press, New York, USA, and Oxford, U.K. 2:76–84.
Hawksworth, F.G. & D. Wiens. 1972. Biology and classification of dwarf mistletoes (*Arceuthobium*). USDA Forest Serv. Agric. Handb. 401, Washington, DC, USA.
Haynes, R.R. & C.B. Hellquist. 2000. Alismataceae. In: Flora of North America Editorial Committee, eds. Flora of North America north of Mexico. Oxford University Press, New York, USA, and Oxford, U.K. 22:7–25.
Haynes, R.R. & C.B. Hellquist. 2000. Potamogetonaceae. In: Flora of North America Editorial Committee, eds. Flora of North America north of Mexico. Oxford University Press, New York, USA, and Oxford, U.K. 22:47–70.
Heckard, L.R. & T.I. Chuang. 1977. Chromosome numbers, polyploidy, and hybridization in Castilleja (Scrophulariaceae) of the Great Basin and Rocky Mountains. Brittonia 29:159–172.
Heil, K., B. Armstrong, & D. Schleser. 1981. A review of the genus *Pediocactus*. Cact. Succ. J. (Los Angeles) 53:17–39.
Heil, K.D. & J.M. Porter. 1994. *Sclerocactus* (Cactaceae): A revision. Haseltonia 2:20–46.
Heiser, C.B., Jr. 1944. A monograph of *Psilostrophe*. Ann. Missouri Bot. Gard. 31:279–300.
Heiser, C.B., Jr. 1969. The North American sunflowers (*Helianthus*). Mem. Torrey Bot. Club. 22 (3):1–218.
Hellquist, C.B. & G.E. Crow. 1981. Aquatic vascular plants of New England: Part 3. Alismataceae. Bull. New Hampshire Agric. Exp. Sta. 518.
Henderson, D.M. 1976. A biosystematic study of Pacific Northwestern blue-eyed grasses (*Sisyrinchium*, Iridaceae). Brittonia 28:149–176.
Hermann, F.J. 1937. Extensions of range and a new species in *Carex*. Rhodora 39:491–495.
Hermann, F.J. 1964. The *Juncus mertensianus* complex in western North America. Leafl. W. Bot. 10(6):81–87.
Hermann, F.J. 1970. Manual of the carices of the Rocky Mountains and Colorado Basin. USDA Forest Serv. Agric. Handb. 374, Washington, DC, USA.
Hermann, F.J. 1975. Manual of the rushes (*Juncus* spp.) of the Rocky Mountains and Colorado Basin. Techn. Rep. USDA Forest Serv. RM-18:1–107.
Hershkovitz, M.A. & E.A. Zimmer. 2000. Ribosomal DNA evidence and disjunctions of western American Portulacaceae. Molec. Phylogen Evol. 15:419–439.
Hess, W.J. & R.L. Robbins. 2002. *Yucca*. In: Flora of North America Editorial Committee, eds. Flora of North America north of Mexico. Oxford University Press, New York, USA, and Oxford, U.K. 26:423–439.
Hickman, J.C., ed. 1993. The Jepson Manual: Higher plants of California. University of California Press, Berkeley, California, USA.
Higgins, L.C. 1971. A revision of *Cryptantha* subgenus *Oreocarya*. Brigham Young Univ. Sci. Bull. 13(4):1–63.
Hill, R.J. 1976. *Mentzelia sinuata* (Rydb.) R.J. Hill comb. nov. and *M. speciosa* Osterh. (Loasaceae), a species pair from the Rocky Mountain Front Range. Bull. Torr. Bot. Club 103:212–217.
Hitchcock, C.L. 1933. A taxonomic study of the genus *Nama* I. Amer. J. Bot. 20:415–430.
Hitchcock, C.L. 1933. A taxonomic study of the genus *Nama* II. Amer. J. Bot. 20:518–534.
Hitchcock, C.L. 1936. The genus *Lepidium* in the United States. Madroño 3:265–320.
Hitchcock, C.L. 1941. A revision of the Drabas of western North America. Univ. Washington Publ. Biol. 11:7–132.
Hitchcock, C.L. 1952. A revision of the North American species of *Lathyrus*. Univ. Washington Publ. Biol. 15:1–104.
Hitchcock, C.L. 1957. A study of the perennial species of *Sidalcea*. I: Taxonomy. Univ. Washington Publ. Biol 18:3–82.
Hitchcock, C.L. & B. Maguire. 1947. A revision of the North American species of *Silene*. Univ. Washington Publ. Biol 13:1–73.
Holmgren, A.H., L.M. Shultz, & T.K. Lowrey. 1976. *Sphaeromeria*, a genus closer to *Artemisia* than to *Tanacetum* (Asteraceae: Anthemideae). Brittonia 28:252–262.
Holmgren, N.H. & P.K. Holmgren. 1997. Hydrangeaceae. In: A. Cronquist, N.H. Holmgren, & P.K. Holmgren, eds. Intermountain Flora. New York Botanical Garden Press, Bronx, New York, USA. 3(A):5–12.
Holmgren, N.H. & P.K. Holmgren. 2002. New Mentzelias (Loasaceae) from the Intermountain Region of Western United States. Syst. Bot. 27:747–762.
Holmgren, N.H., P.K. Holmgren, & A. Cronquist. 2005. Intermountain Flora: Vascular Plants of the Intermountain West, U.S.A. Vol. 2, Part B. Subclass Dilleniidae. New York Botanical Garden Press, Bronx, New York, USA.
House, H.D. 1908. The North American species of the genus *Ipomoea*. Ann. New York Acad. Sci. 18(6):181–263.
Howell, J.T. 1959. Studies in *Cirsium*. Leafl. W. Bot. 9:9–15.
Hurd, E.G., N. Shaw, J. Mastrogiuseppe, L. Smithman, & S. Goodrich. 1998. Field guide to Intermountain sedges. Gen. Tech. Rep. RMRS-GTR-10. USDA Forest Serv., Rocky Mountain Res. Sta., Ogden, Utah, USA.
Iltis, H.H. 1958. Studies in the Capparidaceae IV. *Polanisia* Raf. Brittonia 10:33–58.
Irving, R.S. 1980. The systematics of *Hedeoma* (Labiatae). Sida 8(3):218–295.
Isely, D. 1998. Native and naturalized Leguminosae (Fabaceae) of the United States. Brigham Young Univ. Press, Provo, Utah, USA.
Johnson, W.M. 1964. Field key to the sedges of Wyoming. Bull. Wyoming Agric. Exp. Sta. 419:1–239.
Johnston, I.M. 1925. The North American species of *Cryptantha*. Contr. Gray Herb. 74:3–114.
Johnston, I.M. 1932. Studies in the Boraginaceae. IX. The *Allocarya* section of *Plagiobothrys* in the western United States. Contr. Arnold Arbor. 3:5–82.

Johnston, I.M. 1952. Studies in the Boraginaceae, XXIII. A survey of the genus *Lithospermum*. J. Arnold Arbor. 33:299–366.
Joly, S. & A. Bruneau. 2007. Delimiting species boundaries in *Rosa* Sect. *Cinnamomeae* (Rosaceae) in Eastern North America. Syst. Bot. 32(4):819–836.
Jones, A.G. 1975. The status of *Aster adsurgens* E.L. Greene. Taxon 24:357–360.
Jones, A.G. 1978. The taxonomy of *Aster* section *Multiflori* (Asteraceae)–II. Biosystematic investigations. Rhodora 80:453–490.
Jones, A.G. 1980. A classification of the new world species of *Aster* (Asteraceae). Brittonia 32(2):230–239.
Jones, G.N. 1940. A monograph of the genus *Symphoricarpos*. J. Arnold Arbor. 21:201–252.
Jones, G.N. & F.F. Jones. 1943. A revision of the perennial species of *Geranium* of the United States and Canada. Rhodora 45:5–26, 32–53.
Jones, S.B. & W.B. Faust. 1978. *Vernonia*. In: North American Flora: Compositae: Mutisieae, Senecioneae, Vernonieae. New York Botanical Garden, Bronx, New York, USA. Series II, Part 10:180–195.
Judd, W.S., R.W. Sanders, & M. J. Donoghue. 1994. Angiosperm family pairs: Preliminary cladistic analyses. Harvard Pap. Bot. 5:1–51.
Keck, D. 1946. A revision of the *Artemisia vulgaris* complex in North America. Proc. Calif. Acad. Sci. 17:421–468.
Keil, D.J. 2004. New taxa and new combinations in North American *Cirsium* (Asteraceae: Cardueae). Sida 21(1) 207–219.
Kelch, D.G. & B.G. Baldwin. 2003. Phylogeny and ecological radiation of New World thistles (*Cirsium*, Cardueae, Compositae) based on ITS and ETS rDNA sequence data. Molec. Ecol. 12:141–151.
Keller, A.C. 1942. *Acer glabrum* and its varieties. Amer. Midland Naturalist 27:491–500.
Kiger, R.W. 1975. *Papaver* in North America north of Mexico. Rhodora 77:410–422.
Kiger, R.W. 1985. Revised sectional nomenclature in *Papaver* L. Taxon 34:150–152.
Kiger, R.W. 1997. Papaveraceae. In: Flora of North America Editorial Committee, eds. Flora of North America north of Mexico. Oxford University Press, New York, USA, and Oxford, U.K. 3:300–339.
Kirschner, J. & J. Stepanek. 1987. Again on the sections in *Taraxacum* (Cichoriaceae). Taxon 36:608–617.
Kornkven, A.B., L.E. Watson, & J.R. Estes. 1998. Phylogenetic analysis of *Artemisia* section Tridentatae (Asteraceae) based on sequences from the internal transcribed spacers (ITS) of nuclear ribosomal DNA. Amer. J. Bot. 85(12):1787–1795.
LaFrankie, J.V., Jr. 1986. Transfer of the species of *Smilacina* to *Maianthemum* (Liliaceae). Taxon 35:584–589.
Landolt, E. 2000. Lemnaceae. In: Flora of North America Editorial Committee, eds. Flora of North America north of Mexico. Oxford University Press, New York, USA, and Oxford, U.K. 22:143–153.
Lane, M.A. 1985. Taxonomy of *Gutierrezia* (Compositae: Astereae) in North America. Syst. Bot. 10:(1):7–28.
Lane, M.A. & R.L. Hartman. 1996. Reclassification of North American *Haplopappus* (Compositae: Astereae) completed: *Rayjacksonia* gen. nov. Amer. J. Bot. 83(3):356–370.
Lellinger, D.B. 1985. A field manual of the ferns and fern allies of the United States and Canada. Smithsonian Inst. Press, Washington, DC, USA.
Loconte, H. & J.R. Estes. 1989. Phylogenetic systematics of Berberidaceae and Ranunculales (Magnoliidae). Syst. Bot. 14:565–579.
Loomin, J. 1976. Biological flora of the Canadian prairie provinces IV. *Triglochin* L., the genus. Canad. J. Pl. Sci. 56:725–732.
Lowry, P.P. II & A.G. Jones. 1979. Biosystematic investigations and taxonomy of *Osmorhiza* Rafinesque section *Osmorhiza* (Apiaceae) in North America. Amer. Midland Naturalist 101(1):21–27.
Lowry, P.P.II & A.G. Jones. 1984. Systematics of *Osmorhiza* Raf. (Apiaceae: Apioideae). Ann. Missouri Bot. Gard. 71(4):1128–1171.
Macbride, J.F. 1917. A revision of the North American species of *Amsinckia*. Contr. Gray Herb. 49:1–16.
Maguire, B. 1943. A monograph of the genus *Arnica*. Brittonia 4:386–510.
Maguire, B. 1947. Studies in the Caryophyllaceae. III. A synopsis of the North American species of *Arenaria* sect. *Eremogone* Fenzl. Bull. Torrey Bot. Club 74:38–56.
Mahler, W.F. & U.T. Waterfall. 1964. *Baccharis* (Compositae) in Oklahoma, Texas, and New Mexico. S. W. Naturalist 9:189–202.
Mathias, M.E. & L. Constance. 1944–45. Umbelliferae. In: North American Flora: Umbellales, Cornales. New York Botanical Garden, Bronx, New York, USA. 28B:43–297.
Matsumura, Y. & H.D. Harrington. 1955. The true aquatic vascular plants of Colorado. Tech. Bull. Colorado Agric. Exp. Sta. 57:1–130.
McClintock, E. & C. Epling. 1942. A review of the genus *Monarda* (Labiatae). Univ. Calif. Publ. Bot. 20(2):147–194.
McGregor, R.L. 1987. The status of *Talinum rugospermum* (Portulacaceae) in Kansas and Nebraska. Contr. Univ. Kansas Herb. 24:1–4.
McKinney, L.E. 1992. A taxonomic revision of the acaulescent blue violets (*Viola*) of North America. Sida, Bot. Misc. 7:1–66.
McNeal, D.W., Jr. & T.D. Jacobsen. 2002. *Allium*. In: Flora of North America Editorial Committee, eds. Flora of North America north of Mexico. Oxford University Press, New York, USA, and Oxford, U.K. 26:224–276.
McNeill, J. 1972. New taxa of *Claytonia* section *Claytonia* (Portulacaceae). Canad. J. Bot. 50:1895–1898.
McNeill, J. 1978. *Silene alba* and *S. dioica* in North America and the generic delimitation of *Lychnis*, *Melandrium* and *Silene* (Caryophyllaceae). Canad. J. Bot. 56:297–308.
McVaugh, R. 1936. Studies in the taxonomy and distribution of eastern North American species of *Lobelia*. Rhodora 38:241–298.
McVaugh, R. 1945. The genus *Triodanis* Rafinesque, and its relationships to *Specularia* and *Campanula*. Wrightia 1:13–52.
Meacham, C.A. 1980. Phylogeny of the Berberidaceae with an evaluation of classifications. Syst. Bot. 5:149–172.
Miller, G.S., Jr. & P.C. Standley. 1912. The North American species of *Nymphaea*. Contr. U.S. Natl. Herb. 16:63–108.
Miller, J.M. 2003. *Montia*. In: Flora of North America Editorial Committee, eds. Flora of North America north of Mexico. Oxford University Press, New York, USA, and Oxford, U.K. 4:485–488.
Mitton, J.B. & R. Andalora. 1981. Genetic and morphological relationships between blue spruce, *Picea pungens*, and Engelmann spruce, *Picea engelmannii*, in the Colorado Front Range. Canad. J. Bot. 59(11):2088–2094.
Mooring, J.S. 1980. A cytogeographic study of *Chaenactis douglasii* (Compositae: Helenieae). Amer. J. Bot. 67(9):1304–1319.
Moran, R.C. 1982. The *Asplenium trichomanes* complex in the United States and adjacent Canada. Amer. Fern J. 72:5–11.
Morgan, D.R. & R.L. Hartman. 2003. A synopsis of *Machaeranthera* (Asteraceae), with recognition of segregate genera. Sida 20:1387–1416.
Morton, J.K. & R.K. Rabeler. 1989. Biosystematic studies on the *Stellaria calycantha* (Caryophyllaceae) complex. Canad. J. Bot. 67:121–127.
Mosyakin, S.L. & S.E. Clements. 1996. New infrageneric taxa and combinations in *Chenopodium* L. (Chenopodiaceae). Novon 6:398–403.
Mulligan, G.A. 1975. *Draba crassifolia, D. albertina, D. nemorosa*, and *D. stenoloba* in Canada and Alaska. Canad. J. Bot. 53(8):745–751.
Mulligan, G.A. & I.J. Bassett. 1959. *Achillea millefolium* complex in Canada and portions of the United States. Canad. J. Bot. 37:73–79.
Murray, D.F. 1969. Taxonomy of *Carex* sect. *Atratae* (Cyperaceae) in the southern Rocky Mountains. Brittonia 21(1):55–76.
Mutel, C.F. & J.C. Emerick. 1992. From grassland to glacier: The natural history of Colorado and the surrounding region. Johnson Books, Boulder, Colorado, USA.
Nazaire, M. & L. Hufford. 2014. Phylogenetic systematics o e genus *Mertensia* (Boraginaceae). Syst. Bot. 39(1): 268-303.
Nebeker, G.T. 1974. Revision of the plant genus *Geranium* in Utah. Great Basin Naturalist 34:297–304.
Nesom, G.L. 1988. Synopsis of *Chaetopappa* (Compositae–Astereae) with a new species and the inclusion of *Leucelene*. Phytologia 64:448–456.
Nesom, G.L. 1990. Taxonomic summary of *Ericameria* (Asteraceae: Astereae), with the inclusion of *Haplopappus* sects. *Macronema* and *Asiris*. Phytologia 68(2):144–155.
Nesom, G.L. 1990. Taxonomy of *Laennecia*. Phytologia 68(3):205–228.

Nesom, G.L. 1990. Taxonomy of *Solidago petiolaris* (Astereae: Asteraceae) and related Mexican species. Phytologia 69(6):445–456.
Nesom, G.L. 1993. Taxonomic infrastructure of *Solidago* and *Oligoneuron* (Asteraceae: Astereae) and observations on their phylogenetic position. Phytologia 75(1):1–44.
Nesom, G.L. 1997. Taxonomic adjustments in North American *Aster* sensu latissimo (Asteraceae: Astereae). Phytologia 82(4):281–288.
Nesom, G.L. 2002. New combination in *Xylorhiza* (Asteraceae: Astereae). Sida 20(1):145–147.
Nesom, G.L. 2004. Taxonomic reevaluations in North American *Erigeron* (Asteraceae: Astereae). Sida 21(1):19–39.
Nesom, G.L. 2005. Infrageneric classification of *Liatris* (Asteraceae: Eupatorieae). Sida 21:1305–1321.
Nesom, G.L. 2009. Notes on *Oxalis* sect. *Corniculatae* (Oxalidaceae) in the southwestern United States. Phytologia 91(3):527–533.
Nesom, G.L. 2009. Again: Taxonomy of yellow-flowered caulescent *Oxalis* (Oxalidaceae) in eastern North America. J. Bot. Res. Inst. Texas 3(2):727–738.
Nesom, G.L. & G.I. Baird. 1993. Completion of *Ericameria* (Asteraceae: Astereae), diminution of *Chrysothamnus*. Phytologia 75(1):74–93.
Nesom, G.L. & G.I. Baird. 1995. Comments on "the *Chrysothamnus–Ericameria* connection". Phytologia 78(1):61–65.
Nesom, G.L. & D.R. Morgan. 1990. Reinstatement of *Tonestus* (Asteraceae: Astereae). Phytologia 68(3):174–180.
Nesom, G.L. & D.F. Murray. 2004. Notes on North American arctic and boreal species of *Erigeron* (Asteraceae: Astereae). Sida 21(1):41–57.
Norton, V.C. 1981. A century of botany at CSU highlights. Colorado State University, Fort Collins, Colorado, USA.
Olmstead, R.G., B. Bremer, K.M. Scott, & J.D. Palmer. 1993. A parsimony analysis of the Asteridae sensu lato based on *rbcL* sequences. Ann. Missouri Bot. Gard. 80:700–722.
Ownbey, G.B. 1947. Monograph of the North American species of *Corydalis*. Ann. Missouri Bot. Gard. 34:187–260.
Ownbey, G.B. 1958. Monograph of the genus *Argemone* for North America and the West Indies. Mem. Torrey Bot. Club (21)1:1–159.
Ownbey, M. 1940. A monograph of the genus *Calochortus*. Ann. Missouri Bot. Gard. 27:371–560. Ownbey, M. & W.A. Weber. 1943. Natural hybridization in the genus *Balsamorhiza*. Amer. J. Bot. 30:179–187.
Packer, J.G. 2003. Portulacaceae. In: Flora of North America Editorial Committee, eds. Flora of North America north of Mexico. Oxford University Press, New York, USA, and Oxford, U.K. 4:457–504.
Packer, J.G. & K.E. Denford. 1974. A contribution to the taxonomy of *Arctostaphylos uva-ursi*. Canad. J. Bot. 52:743–753.
Palmer, E.J. 1931. A conspectus of the genus *Amorpha*. J. Arnold Arbor. 12:157–197.
Parfitt, B.D. & A.C. Gibson. 2003. Cactaceae. In: Flora of North America Editorial Committee, eds. Flora of North America north of Mexico. Oxford University Press, New York, USA, and Oxford, U.K. 4:92–257.
Payne, W.W. 1963. The morphology of the inflorescence of ragweeds (*Ambrosia-Franseria*: Compositae). Amer. J. Bot. 50:872–880.
Payne, W.W. 1964. A re-evaluation of the genus *Ambrosia*. J. Arnold Arbor. 45(4):401–430.
Payson, E.B. 1917. The perennial scapose Drabas of North America. Amer. J. Bot. 4:253–267.
Payson, E.B. 1921. A monograph of the genus *Lesquerella*. Ann. Missouri Bot. Gard. 8:103–236.
Payson, E.B. 1922. Species of *Sisymbrium* native to America north of Mexico. Publ. Sci. Univ. Wyoming Bot. 1(1):1–27.
Payson, E.B. 1922. A monographic study of *Thelypodium* and its immediate allies. Ann. Missouri Bot. Gard. 9:233–324.
Payson, E.B. 1927. A monograph of the section *Oreocarya* of *Cryptantha*. Ann. Missouri Bot. Gard. 14:211–358.
Perry, L.M. 1937. Notes on *Silphium*. Rhodora 39:281–297.
Phillips, L.L. 1955. A revision of the perennial species of *Lupinus* of North America. Res. Stud. State Coll. Washington 23(3):161–201.
Pilz, G.E. 1978. Systematics of *Mirabilis* subgenus *Quamoclidion* (Nyctaginaceae). Madroño 25:113–132.
Pinkava, D.J. 1999. Cactaceae cactus family: Part 3. *Cylindropuntia*. J. Arizona-Nevada Acad. Sci. 32:32–47.
Porter, C.L. 1951. *Astragalus* and *Oxytropis* in Colorado. Publ. Univ. Wyoming 16(1):1–49.
Porter, C.L. 1964. The Cyperaceae of Wyoming. The Rocky Mountain Herbarium, University of Wyoming, Laramie, Wyoming, USA. Pp. 1–36.
Porter, J.M. 1998. *Aliciella*, a recircumscribed genus of Polemoniaceae. Aliso 17(1):23–46.
Preece, S.J., Jr. & B.L. Turner. 1953. A taxonomic study of the genus *Chamaechaenactis* Rydberg (Compositae). Madroño 12(3):97–103.
Price, R.A. 1980. *Draba streptobrachia* (Brassicaceae), a new species from Colorado. Brittonia 32(2):160–169.
Price, R.A. & R.C. Rollins. 1991. New taxa of *Draba* (Cruciferae) from California, Nevada, and Colorado. Harvard Pap. Bot. 3:71–77.
Prince, S.D. & R.N. Carter. 1977. Prickly lettuce (*Lactuca serriola* L.) in Britain. Watsonia 11:331–338.
Pringle, J.S. 1967. Taxonomy of *Gentiana*, section *Pneumonanthae*, in Eastern North America. Brittonia 19:1–32.
Puff, C. 1976. The *Galium trifidum* group (*Galium* sect. *Aparinoides*, Rubiaceae). Canad. J. Bot. 54:1911–1925.
Rabeler, R.K. & R.L. Hartman. 2005. Caryophyllaceae. In: Flora of North America Editorial Committee, eds. Flora of North America north of Mexico. Oxford University Press, New York, USA, and Oxford, U.K. 5:3–215.
Rahn, K. 1974. *Plantago* section *Virginica*: A taxonomic revision of a group of American plantains, using experimental, taximetric and classical methods. Dansk Bot. Arkiv 30(2):7–180.
Ramaley, F. 1939. Sand-hill vegetation of northeastern Colorado. Ecol. Monogr. 9(1):1–51.
Raven, P.H. 1962. The systematics of *Oenothera* subgenus *Chylismia*. Univ. Calif. Publ. Bot. 34(1):1–122.
Raven, P.H. 1969. A revision of the genus *Camissonia* (Onagraceae). Contr. U.S. Natl. Herb. 37(5):161–396.
Ray, P.M. & H.F. Chisaki. 1957. Studies on *Amsinckia*. I. A synopsis of the genus, with a study of heterostyly in it. Amer. J. Bot. 44:529–536.
Reveal, J. 1977. *Yucca*. In: A. Cronquist, A.H. Holmgren, N.H. Holmgren, J.L. Reveal, & P.K. Holmgren, eds. Intermountain Flora. Columbia University Press, New York, New York, USA. 6:527–536.
Reveal, J.L. 2002. *Mentzelia rhizomata* (Loasaceae: Mentzelioideae), a new species from Western Colorado. Syst. Bot. 27:763–767.
Reveal, J.L. & C.R. Broome. 2006. *Cryptantha gypsophila* (Boraginaceae: Boraginoideae), a new species from western Colorado. Brittonia 58(2):178–181.
Richards, A.J. 1985. Sectional nomenclature in *Taraxacum* (Asteraceae). Taxon 34:633–644.
Richards, E.L. 1968. A monograph of the genus *Ratibida*. Rhodora 70:348–393.
Robinson, B.L. 1917. A monograph of the genus *Brickellia*. Mem. Gray Herb. 1:1–151.
Rock, H.F.L. 1957. A revision of the vernal species of *Helenium* (Compositae). Rhodora 59:101–116, 129–158, 169–178, 203–216.
Rogers, C.M. 1968. Yellow-flowered species of *Linum* in Central America and Western North America. Brittonia 20:107–135.
Rollins, R.C. 1939. The Cruciferous genus *Physaria*. Rhodora 41:392–415.
Rollins, R.C. 1941. A monographic study of *Arabis* in western North America. Rhodora 43:289–481.
Rollins, R.C. 1950. The guayule rubber plant and its relatives. Contr. Gray Herb. 1972:1–73.
Rollins, R.C. 1981. Studies in the genus *Physaria* (Cruciferae). Brittonia 33(3):332–341.
Rollins, R.C. 1982. Studies on *Arabis* (Cruciferae) of western North America II. Contr. Gray Herb. 212:103–114.
Rollins, R.C. 1993. The Cruciferae of Continental North America: Systematics of the mustard family from the Arctic to Panama. Stanford Univ. Press, Stanford, California, USA.
Rosatti, T.J. 1987. Field and garden studies of *Arctostaphylos uva-ursi* (Ericaceae) in North America. Syst. Bot. 12:61–77.
Rosendahl, C.O. 1914. A revision of the genus *Mitella* with a discussion of geographical distribution and relationships. Bot. Jahrb. Syst. 50 (suppl.):375–397.
Rossbach, G.B. 1939. Aquatic utricularias. Rhodora 41:113–128.
Roush, E.M.F. 1931. A monograph of the genus *Sidalcea*. Ann. Missouri Bot. Gard. 18:117–244.

Russell, N.H. 1955. The taxonomy of the North American acaulescent white violets. Amer. Midl. Naturalist 54(2):481–494.
Rydberg, P.A. 1906. Flora of Colorado. Agricultural Experiment Station of the Colorado Agricultural College, Fort Collins, Colorado, USA.
Rydberg, P.A. 1922. *Cirsium.* In: Flora of the Rocky Mountains and adjacent plains. Hafner Publishing Company, New York, New York, USA. Pp. 1003–1014.
Sauer, J.D. 1955. Revision of the dioecious amaranths. Madroño 13:5–47.
Sauer, J.D. 1972. The dioecious amaranths: A new species name and major range extensions. Madroño 21(6):426–434.
Sauer, J.D. & R. Davidson. 1961. Preliminary reports on the flora of Wisconsin: No. 45. Amaranthaceae–amaranth family. Trans. Wisconsin Acad. Sci. 50:75–87.
Schenk, J.J. & L. Hufford. 2011. Phylogeny and taxonomy of *Mentzelia* section *Bartonia* (Loasaceae). Syst. Bot. 36(3):711–720.
Schenk, J.J. & L. Hufford. 2011. Taxonomic novelties from Western North America in *Mentzelia* section *Bartonia* (Loasaceae). Madroño 57(4):246–260.
Schilling, E.E. 1981. Systematics of *Solanum* sect. *Solanum* (Solanaceae) in North America. Syst. Bot. 6:172–185.
Schneider, A., P. Lyon, & G. Nesom. 2008. *Gutierrezia elegans* sp. nov. (Asteraceae: Astereae), a shale barren endemic of southwestern Colorado. J. Bot. Res. Inst. Texas 2:771–774.
Semple, J.C. 1996. A revision of *Heterotheca* sect. *Phyllotheca* (Nutt.) Harms (Compositae: Astereae): The prairie and montane goldenasters of North America. Univ. Waterloo Biol. Ser. 37:1–164.
Semple, J.C. & L. Brouillet. 1980. A synopsis of North American asters: The subgenera, sections, and subsections of *Aster* and *Lasallea*. Amer. J. Bot. 67(7):1010–1026.
Semple, J.C. & J.G. Chmielewski. 1987. Revision of the *Aster lanceolatus* complex, including *A. simplex* and *A. hesperius* (Compositae: Astereae): A multivariate morphometric study. Canad. J. Bot. 65:1047–1062.
Sharp, W.M. 1935. A critical study of certain epappose genera of the Heliantheae–Verbesiniae of the natural family Compositae. Ann. Missouri Bot. Gard. 22:51–152.
Shaw, R.B. 2008. Grasses of Colorado. University Press of Colorado, Boulder, Colorado, USA.
Sherff, E.E. 1920. North American species of *Taraxacum*. Bot. Gaz. 70:329–359.
Sherff, E.E. 1936. Revision of the genus *Coreopsis*. Publ. Field Mus. Nat. Hist., Bot. Ser. 11:279–475.
Sherff, E.E. 1937. The genus *Bidens*. Publ. Field Mus. Nat. Hist., Bot. Ser. 16:1–709.
Smith, E.B. 1976. A biosystematic study of *Coreopsis* in eastern United States and Canada. Sida 6:123–215.
Sheviak, C.J. 1984. *Spiranthes diluvialis* (Orchidaceae), a new species from the western United States. Brittonia 36(1):8–14.
Shinners, L.H. 1953. Synopsis of the United States species of *Lythrum* (Lythraceae). Field & Lab. 21(2):80–89.
Small, E. & A. Cronquist. 1976. A practical and natural taxonomy for *Cannabis*. Taxon 25(4):405–435.
Smith, E.C. & L.W. Durrell. 1944. Sedges and rushes of Colorado. Techn. Bull. Colorado Agric. Exp. Sta. 32:3–63.
Smith, J.G. 1894. North American species of *Sagittaria* and *Lophotocarpus*. Rep. (Annual) Missouri Bot. Gard. 6:1–38.
Smith, S.G. 2000. Typhaceae. In: Flora of North America Editorial Committee, eds. Flora of North America north of Mexico. Oxford University Press, New York, USA, and Oxford, U.K. 22:278–285.
Smith, S.G. & R.B. Kaul. 1986. Typhaceae. In: Great Plains Flora Association, R.L. McGregor, & T.M. Barkley, eds. Flora of the Great Plains. Univ. Press of Kansas, Lawrence, Kansas, USA. Pp. 1237–1239.
Solbrig, O.T. 1960. Cytotaxonomic and evollutionary studies in the North American species of *Gutierrezia* (Compositae). Contr. Gray Herb. 188:1–63.
Spackman, S., B. Jennings, J. Coles, C. Dawson, M. Minton, A. Kratz, & C. Spurrier. 1997. Colorado rare plant field guide. Prepared for the Bureau of Land Management, U.S. Forest Service, and the U.S. Fish and Wildlife Service by the Colorado Natural Heritage Program, USA.
Spellenberg, R. 1998. *Mirabilis melanotricha* (Nyctaginaceae), a new combination for a common four o'clock from southwestern North America. Phytologia 85:99–104.
Spellenberg, R.W. 2003. Nyctaginaceae. In: Flora of North America Editorial Committee, eds. Flora of North America north of Mexico. Oxford University Press, New York, USA, and Oxford, U.K. 4:14–74.
Spellenberg, R. & S.R. Rodriguez Tijerina. 2001. Geographic variation and taxonomy of North American species of *Mirabilis*, section *Oxybaphoides* (Nyctaginaceae). Sida 19:311–323.
Spotts, A.M. 1939. The violets of Colorado. Madroño 5(1):16–27.
Springfield, H.W. 1976. Characteristics and management of southwestern pinyon-juniper ranges: The status of our knowledge. U.S. Forest Service Research Paper RM-160.
St. John, H. 1962. Monograph of the genus *Elodea* (Hydrocharitaceae): Part I. The species found in the Great Plains, the Rocky Mountains, and the Pacific states and the provinces of North America. Res. Stud. State Coll. Wash. 30(2):19–44.
Standley, P.C. 1909. The Allioniaceae of the United States with notes on Mexican species. Contr. U.S. Natl. Herb. 12(8):303–389.
Stensvold, M.C. 2007. A taxonomic and phylogeographic study of the *Botrychium lunaria* complex. PhD dissertation, Iowa State University, Ames, Iowa, USA.
Stebbins, G.L. 1939. Notes on *Lactuca* in western North America. Madroño 5:123–126.
Stern, K.R. 1997. Fumariaceae. In: Flora of North America Editorial Committee, eds. Flora of North America north of Mexico. Oxford University Press, New York, USA, and Oxford, U.K. 3:340–357.
Steyermark, J.A. 1937. Studies in *Grindelia* II: A monograph of the North American species of the genus *Grindelia*. Ann. Missouri Bot. Gard. 21:433–608.
Stockwell, P. 1940. A revision of the genus *Chaenactis*. Contr. Dudley Herb. 3:89–168.
Stoutamire, W. 1977. Chromosome races of *Gaillardia pulchella* (Asteraceae). Brittonia 29:297–309.
Strother, J.L. 1974. Taxonomy of *Tetradymia* (Compositae: Senecioneae). Brittonia 26:177–202.
Strother, J.L. & M.A. Wetter. 2006. *Grindelia*. In: Flora of North America Editorial Committee, eds. Flora of North America north of Mexico. Oxford University Press, New York, USA, and Oxford, U.K. 20:424–436.
Stuessy, T.F., R.S. Irving, & W.L. Ellison. 1973. Hybridization and evolution in *Picradeniopsis* (Compositae). Brittonia 25:40–56.
Suh, Y. & B.B. Simpson. 1990. Phylogenetic analysis of chloroplast DNA in North American *Gutierrezia* and related genera (Asteraceae: Astereae). Syst. Bot. 15 (4):660–670.
Sundell, E. 1990. Notes on Arizona *Asclepias* (Asclepiadaceae) with a new combination. Phytologia 69(4):265–270.
Sundell, E. 1993. Asclepiadaceae Milkweed Family. J. Arizona-Nevada Acad. Sci. 27(2):169–187.
Svenson, H.K. 1932. Monographic studies in the genus *Eleocharis*—II. Rhodora 34:193–203, 215–227.
Svenson, H.K. 1934. Monographic studies in the genus *Eleocharis*—III. Rhodora 35:377–389.
Svenson, H.K. 1937. Monographic studies in the genus *Eleocharis*—IV. Rhodora 39:210–273.
Svenson, H.K. 1939. Monographic studies in the genus *Eleocharis*—V. Rhodora 41:1–110.
Svenson, H.K. 1947. The group of *Eleocharis palustris* in North America. Rhodora 49:61–67.
Taylor, C.E. & R.J. Taylor. 1983. New species, new combinations and notes on the goldenrods (*Euthamia* and *Solidago*—Asteraceae). Sida 10(2):176–183.

Taylor, C.E. & R.J. Taylor. 1984. *Solidago* (Asteraceae) in Oklahoma and Texas. Sida 10(3):223–251.

Taylor, N.P. 1985. The genus *Echinocereus*. Timber Press, Portland, Oregon, USA.

Taylor, W.C., N.T. Luebke, D.M. Britton, R.J. Hickey, & D.F. Brunton. 1993. Isoëtaceae. In: Flora of North America Editorial Committee, eds. Flora of North America north of Mexico. Oxford University Press, New York, USA, and Oxford, U.K. 2:64–75.

Thieret, J.W. 1993. Pinaceae. In: Flora of North America Editorial Committee, eds. Flora of North America north of Mexico. Oxford University Press, New York, USA, and Oxford, U.K. 2:352–434.

Thompson, H. & B. Prigge. 1986. New species and a new combination of *Mentzelia* section *Bartonia* (Loasaceae) from the Colorado Plateau. Great Basin Naturalist 46(3):549–554.

Thorne, K. 1986. New variety of *Mentzelia pumila* (Loasaceae) from Utah. Great Basin Naturalist 46(3):557–558.

Thorne, R.F. 1993. Potamogetonaceae. In: J.C. Hickman, ed. The Jepson Manual. Univ. California Press, Berkeley, California, USA. Pp. 1304–1310.

Tomb, A.S. 1980. Taxonomy of *Lygodesmia* (Asteraceae). Syst. Bot. Monogr. 1:1–51.

Trelease, W. 1891. The species of *Epilobium* occurring north of Mexico. Rep. (Annual) Missouri Bot. Gard. 2:69–117.

Trock, D.K. 2003. The genus *Packera* (Asteraceae: Senecioneae) in Colorado, U.S.A. Sida 20(3):1023–1041.

Tryl, R.J. 1975. Origin and distribution of polyploid *Achillea* (Compositae) in western North America. Brittonia 27:187–196.

Tryon, A.F. 1957. A revision of the fern genus *Pellaea* section *Pellaea*. Ann. Missouri Bot. Gard. 44:125–193.

Tryon, R.M., Jr. 1941. A revision of the genus *Pteridium*. Rhodora 43:1–67.

Tryon, R.M. & A.F. Tryon. 1982. Ferns and allied plants with special reference to tropical America. Springer-Verlag, New York, New York, USA.

Tucker, J.M. 1961. Studies in the *Quercus undulata* complex: I. A preliminary statement. Amer. J. Bot. 48:202–208.

Tucker, J.M. 1970. Studies in the *Quercus undulata* complex: IV. The contribution of *Quercus havardii*. Amer. J. Bot. 57(1):71–84.

Tucker, J.M. 1971. Studies in the *Quercus undulata* complex: V. The type of *Quercus undulata*. Amer. J. Bot. 58(4):329–341.

Turner, B.L. 1956. A cytotaxonomic study of the genus *Hymenopappus*. Rhodora 58:163–186, 208–242, 250–269, 295–308.

Turner, B.L. 1993. Texas species of *Mirabilis* (Nyctaginaceae). Phytologia 75(6):432–451.

Turner, B.L. & R.L. Hartman. 1976. Infraspecific categories of *Machaeranthera pinnatifida* (Compositae). Wrightia 5(8):308–315.

Turner, B.L. & M.I. Morris. 1976. Systematics of *Palafoxia* (Asteraceae: Helenieae). Rhodora 78:567–628.

Turner, B.L. & M. Whalen. 1975. Taxonomic study of *Gaillardia pulchella* (Asteraceae—Heliantheae). Wrightia 5:189–192.

Valdespino, I.A. 1993. Selaginellaceae. In: Flora of North America Editorial Committee, eds. Flora of North America north of Mexico. Oxford University Press, New York, USA, and Oxford, U.K. 2:38–63.

Vander Kloet, S.P. & T.A. Dickinson. 1999. The taxonomy of *Vaccinium* section *Myrtillus* (Ericaceae). Brittonia 51:231–254.

Wagenknecht, B.L. 1960. Revision of *Heterotheca*, section *Heterotheca* (Compositae). Rhodora 62:61–76, 97–107.

Wagner, W.H., Jr. & J.M. Beitel. 1993. Lycopodiaceae. In: Flora of North America Editorial Committee, eds. Flora of North America north of Mexico. Oxford University Press, New York, USA, and Oxford, U.K. 2:18–37.

Wagner, W.H., Jr. & F.S. Wagner. 1981. New species of moonworts, *Botrychium* subg. *Botrychium* (Ophioglossaceae), from North America. Amer. Fern J. 71(1):20–30.

Wagner, W.H., Jr. & F.S. Wagner. 1983. Two moonworts of the Rocky Mountains: *Botrychium hesperium* and a new species formerly confused with it. Amer. Fern J. 73(3):53–62.

Wagner, W.H., Jr. & F.S. Wagner. 1986. Three new species of moonworts (*Botrychium* subg. *Botrychium*) endemic in Western North America. Amer. Fern J. 76(2):33–47.

Wagner, W.L. 1983. New species and combinations in the genus *Oenothera* (Onagraceae). Ann. Missouri Bot. Gard. 70:194–196.

Wagner, W.L. 1984. Reconsideration of *Oenothera* subg. *Gauropsis* (Onagraceae). Ann. Missouri Bot. Gard. 71:1114–1127.

Wagner, W.L. & E.F. Aldon. 1978. Manual of the saltbushes (*Atriplex* spp.) in New Mexico. Gen. Techn. Rep. U.S. Forest Serv. RM-57:1–50.

Wagner, W.L., K.N. Krakos, & P.C. Hoch. 2013. Taxonomic changes in *Oenothera* sections *Gaura* and *Calylophus* (Onagraceae). PhytoKeys 28:61–72.

Wagner, W.L. & S.W. Mill. 1984. *Oenothera kleinii* (Onagraceae), a new species from southcentral Colorado. Syst. Bot. 9(1):50–52.

Wagner, W.L., R.E. Stockhouse, & W.M. Klein. 1985. The systematics and evolution of the *Oenothera caespitosa* species complex (Onagraceae). Monogr. Syst. Bot. Missouri Bot. Gard. 12:1–103.

Wahl, H.A. 1954. A preliminary study of the genus *Chenopodium* in North America. Bartonia 27:1–46.

Wallace, G.D. 1975. Studies of the Monotropoideae (Ericaceae): Taxonomy and distribution. Wasmann J. Biol. 33:1–88.

Ward, G.H. 1953. *Artemisia*, section *Seriphidium*, in North America, a cytotaxonomic study. Contr. Dudley Herb. 4(6):155–205.

Waterfall, U.T. 1958. A taxonomic study of the genus *Physalis* in North America north of Mexico. Rhodora 60:107–114; 128–142; 152–173.

Watson, T.J., Jr. 1977. The taxonomy of *Xylorhiza* (Asteraceae—Astereae). Brittonia 29(2):199–216.

Weber, W.A. 1946. A taxonomic and cytological study of the genus *Wyethia*, familiy Compositae, with notes on the related genus *Balsamorhiza*. Amer. Midland Naturalist 35(2):400–452.

Weber, W.A. 1950. A new species and subgenus of *Atriplex* from Southwestern Colorado. Madroño 10:187–191.

Weber, W.A. 1952. The genus *Helianthella* (Compositae). Amer. Midland Naturalist 48 (1):1–35.

Weber, W.A. & R.L. Hartman. 1979. *Pseudostellaria jamesiana*, comb. nov., a North American representative of a Eurasian genus. Phytologia 44:313–314.

Weber, W.A. & R.C. Wittmann. 2012. Colorado flora: Eastern slope, 4th edition. University Press of Colorado, Boulder, Colorado, USA.

Weber, W.A. & R.C. Wittmann. 2012. Colorado flora: Western slope, 4th edition. University Press of Colorado, Boulder, Colorado, USA.

Welsh, S.L. 1985. New species of *Astragalus* (Leguminosae) from Mesa County, Colorado. Great Basin Naturalist 45(1):31–33.

Welsh, S.L. 1986. *Quercus* (Fagaceae) in the Utah flora. Great Basin Naturalist 46(1):107–111.

Welsh, S.L., N.D. Atwood, S. Goodrich, & L.C. Higgins. 2003. A Utah flora, 3rd edition. Brigham Young University, Provo, Utah, USA.

Wendt, T.L. 1979. Notes on the genus *Polygala* in the United States and Mexico. J. Arnold Arbor. 60:504–514.

Wetmore, R.H. & A.L. Delisle. 1939. Studies in the genetics and cytology of two species in the genus *Aster* and their polymorphy in nature. Amer. J. Bot. 26:1–12.

Wheeler, L.C. 1941. *Euphorbia* subgenus *Chamaesyce* in Canada and the United States exclusive of Southern Florida. Rhodora 43:97–286.

Wherry, E.T. 1938. Colorado ferns. Amer. Fern J. 28:125–140.

Wherry, E.T. 1942. The genus *Polemonium* in North America. Amer. Midl. Naturalist 27:740–760.

Whittemore, A.T. 1997. *Berberis*. In: Flora of North America Editorial Committee, eds. Flora of North America north of Mexico. Oxford University Press, New York, USA, and Oxford, U.K. 3:276–286.

Wiegand, K.M. 1899. A revision of the genus *Listera*. Bull. Torrey Bot. Club 26:157–171.

Wiegand, K.M. 1925. *Oxalis corniculata* and its relatives in North America. Rhodora 27(319):113–140.

Wiggins, I.L. 1936. A resurrection and revision of the genus *Iliamna* Greene. Contr. Dudley Herb. 1(7):213–229.

Wiggins, I.L. 1940. Taxonomic notes on the genus *Dalea* Juss. and the related genera as represented in the Sonoran Desert. Contr. Dudley Herb. 3:41–64.

Williams, E.W. 1957. The genus *Malacothrix* (Compositae). Amer. Midland Naturalist 58(2):494–512.

Williams, F.N. 1896. A revision of the genus *Silene* Linn. Bot. J. Linn. Soc. 32:1–196.

Williams, L.O. 1937. A monograph of the genus *Mertensia* in North America. Ann. Missouri Bot. Gard. 24:17–159.
Williams, R.L. 1987. On the mountain top with Mr. Osterhout. Brittonia 39(2):149–158.
Wingate, J.L. 1994. Illustrated keys to the grasses of Colorado. Wingate Consulting, Denver, Colorado, USA.
Wolf, S.J. & K.E. Denford. 1984. *Arnica gracilis* (Compositae), a natural hybrid between *A. latifolia* and *A. cordifolia*. Syst. Bot. 9(1):12–16.
Wolf, S.J. & K.E. Denford. 1984. Taxonomy of *Arnica* (Compositae) subgenus *Austromontana*. Rhodora 86:239–309.
Woodson, R.E., Jr. 1941. The North American Asclepiadaceae: I. Perspective of the genera. Ann. Missouri Bot. Gard. 28:193–244.
Woodson, R.E., Jr. 1954. The North American species of *Asclepias* L. Ann. Missouri Bot. Gard. 41:1–211.
Wooton, E.O. & P.C. Standley. 1913. Descriptions of new plants preliminary to a report upon the flora of New Mexico. Contr. U.S. Natl. Herb. 16:114–115.
Yuncker, T.G. 1932. The genus *Cuscuta*. Mem. Torrey Bot. Club 18:111–331.
Zimmerman, A.D. 1985. Systematics of the genus *Coryphantha* (Cactaceae). Ph.D. dissertation. University of Texas, Austin, Texas, USA.

Independence Mountain, North Park, Colorado

INDEX

C

County map of Colorado (with continental divide in bold)

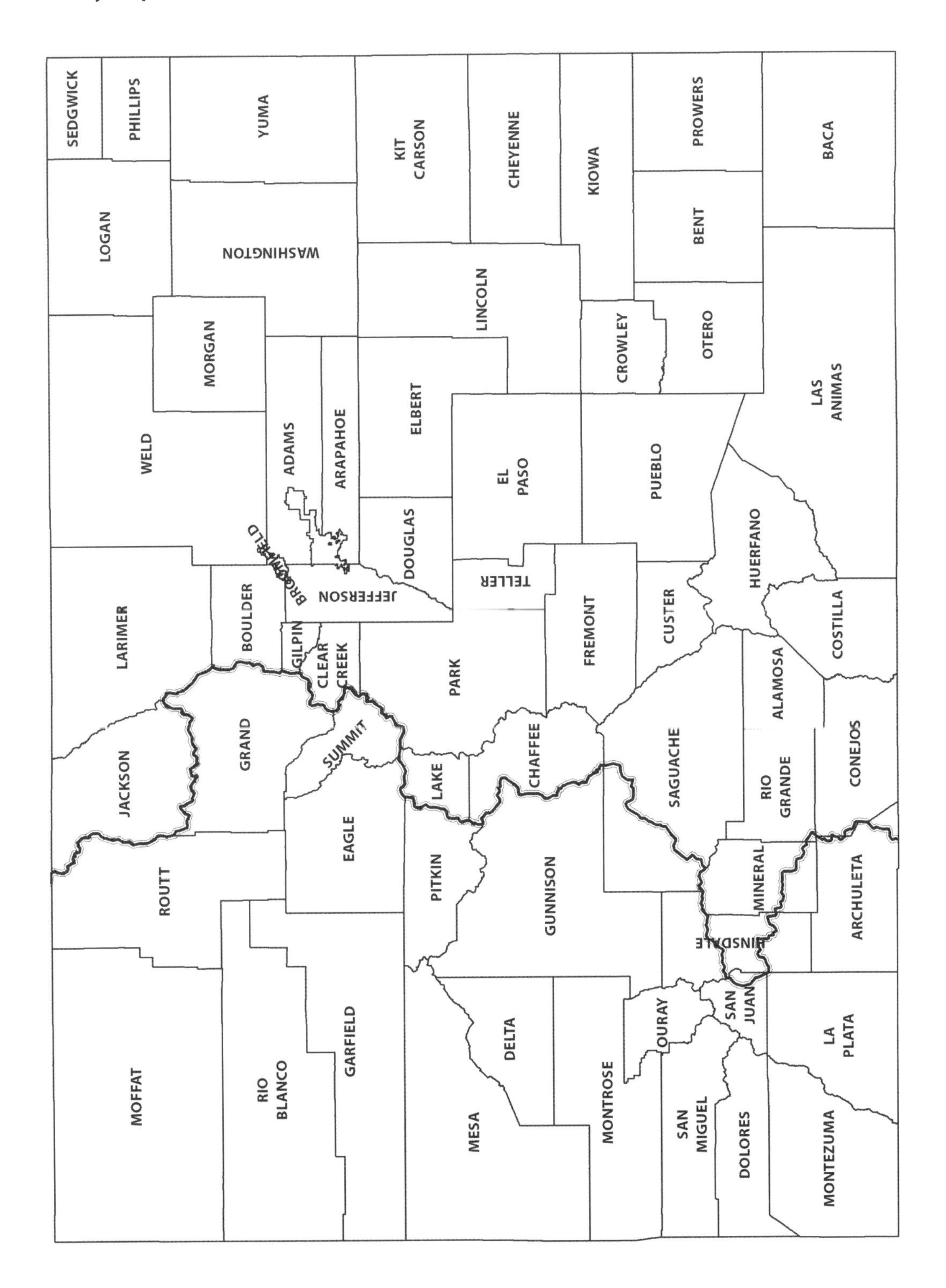

ABOUT THE AUTHOR

Jennifer Ackerfield has been studying the flora of Colorado for 20 years. During this time, she has traveled extensively across the state of Colorado documenting its rich floristic diversity. She is currently a curator at the Colorado State University Herbarium and also teaches Plant Identification at CSU where she is passionate about educating students on the wonderful world of botany. She received her Master's in Botany with a concentration in taxonomy and systematics in 2001 and is currently working on her PhD in Botany. Jennifer has served on the board for the Colorado Native Plant Society; participated in several botanical explorations and bioblitzes throughout Colorado; worked with the Colorado Natural Heritage Program, Mesa Verde National Park, the U.S. Forest Service, the Rocky Mountain Biological Research Station, and the Colorado BLM; and has written several articles on the flora of Colorado. She is also the mother of two wonderful twins. In her spare time she enjoys hiking, photographing wildflowers, rockhounding, painting, and botanical illustration.

NOTES